FLORA OF THE BRITISH ISLES

FLORA OF

THE BRITISH ISLES

A. R. CLAPHAM
University of Sheffield

T. G. TUTIN
University of Leicester

D. M. MOORE
University of Reading

THIRD EDITION

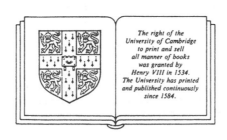

The right of the
University of Cambridge
to print and sell
all manner of books
was granted by
Henry VIII in 1534.
The University has printed
and published continuously
since 1584.

CAMBRIDGE UNIVERSITY PRESS
Cambridge
New York   Port Chester
Melbourne   Sydney

CAMBRIDGE UNIVERSITY PRESS
Cambridge, New York, Melbourne, Madrid, Cape Town, Singapore,
São Paulo, Delhi, Dubai, Tokyo

Cambridge University Press
The Edinburgh Building, Cambridge CB2 8RU, UK

Published in the United States of America by Cambridge University Press, New York

www.cambridge.org
Information on this title: www.cambridge.org/9780521389747

First published 1952
Reprinted 1957, 1958
Second edition 1962
Third edition 1987
First paperback edition (with corrections) 1989
Re-issued in this digitally printed version 2010

A catalogue record for this publication is available from the British Library

Library of Congress Cataloguing in Publication data
Clapham, A. R. (Arthur Roy), 1905–
Flora of the British Isles.
Includes index.
1. Botany—Great Britain.   I. Tutin, Thomas Gaskell,
1908–   .   II. Moore, D. M.   III. Title.
QK306.C57   1987   581.941   86–6070

ISBN 978-0-521-30985-1 Hardback
ISBN 978-0-521-38974-7 Paperback

Contents

To

HUMPHREY GILBERT-CARTER

*To whose stimulating teaching
and wide knowledge of plants we, his pupils,
owe so much*

Foreword

BY PROFESSOR A. G. TANSLEY

A new British flora has been a desideratum for the past half century and urgently needed during the last thirty years. What has been particularly required is a flora not primarily for specialists but a book of limited size easily usable by students and by everyone interested in our wild plants who is willing to learn the comparatively few technical terms necessary for the accurate description of species. The absence of such a flora has seriously hampered the teaching and learning of field botany. Time and again I have been asked by visiting foreign botanists to recommend a good modern British flora and have been ashamed to confess that no such thing existed. In this whole sphere the lack of an adequate handbook has indeed been something of a national scandal. Several attempts have been made to fill the gap but none has been carried through to success, largely because they were all too ambitious, aiming at a completeness and exhaustiveness unattainable except through many years of laborious effort and the collaboration of a large body of specialists.

It is often taken for granted by those who are unacquainted with the subject that the comparatively small British plant population is more or less completely known and has been fully and accurately described in the existing floras. It is not realized that modern work during the past half century, and increasingly since the end of the first world war, has revealed the existence of many distinct forms – species, subspecies and varieties – that had not previously been clearly recognized, or had not been recognized at all. Some of these were formerly described under the names of continental types which they resembled, but deeper knowledge and closer comparison have established that the British forms are in reality quite distinct. At the same time much new knowledge has been gained about many well-known species, especially about their genetics and ecology. This has been the result of the great revival of interest in field observation and work in the experimental garden among professional botanists and academic students of the subject. During the latter part of last century and the beginning of this, the study of British plants was very largely in the hands of enthusiastic amateurs to whom the subject owes a great deal, several of them having become the leading specialists in particular groups. With the rise of ecology and genetics, interest in British plants spread to the universities, and thus aroused the renewed attention to taxonomy among academically trained botanists which has been a marked feature of recent years.

At last it has been possible to stimulate three men, scarcely more than entering upon middle life, all with the modern training, all keenly interested in plants as they grow in the field, in ecology and genetics, to undertake the production of the much-needed flora as a matter of urgency. Though closely occupied with teaching, they have carried through the task in little more than three years and in a manner that seems to me excellently adapted to meet the need. A comparison of their book with any of the previous floras will make plain the distance that has been traversed since those were written.

Readers will find a good many unfamiliar specific names, but these changes were necessary if the rule of priority in nomenclature was to be followed. Personally, I should like to see the principle of *nomina conservanda* applied to specific epithets as well as to the names of genera, so that well-known names that have been in use for many years might be retained and cease to be subject to the risk of perpetual displacement as the result of literary research, often in obscure historical works. But botanists are not yet in agreement on this point, and the discovery of 'prior' names must, one supposes, come to an end some day. Meanwhile, the present generation of students has still to suffer in this respect, though the suffering may, we hope, be transient.

A. G. T.

GRANTCHESTER

January 1951

The reason for the addition of yet another flora to the long series which began in the seventeenth century is perhaps best explained by a brief historical survey.

Although many records of British plants are to be found in herbals, the first attempt at a true flora of these islands was John Ray's *Catalogus Plantarum Angliae et Insularum Adjacentium*, published in 1670. William Hudson's *Flora Anglica* (1762), and thus nearly a century later than Ray's *Catalogus*, was a worthy successor to that pioneer work and notable for the introduction of binomial nomenclature and the Linnean system of classfication into British floras. This was followed by Withering's *Botanical Arrangement of all the Vegetables naturally growing in Great Britain* (1776–92), the first of many floras written primarily for the amateur, and one which enjoyed considerable popularity.

Sowerby's *English Botany*, the first edition of which was published between 1790 and 1820, occupies a unique place. It presented for the first time a complete set of coloured illustrations of our plants, illustrations which are still unsurpassed in their delicacy of line and colouring. The third edition (1863–72), in which the text was completely rearranged, has inferior illustrations but is still a valuable work of reference more than 150 years after the first edition appeared.

The nineteenth century saw the production of three floras, all still in regular use, and a number of others which are now seldom seen. Bentham's famous *Handbook of the British Flora* (1858; revised by J. D. Hooker, 1886) was written as 'a before-breakfast relaxation' and was deliberately intended 'for the use of beginners and amateurs'. Its keys, a new feature in British floras, make it of value to anyone who desires to identify plants easily and with the minimum of previous knowledge but its treatment of species in many groups makes it of limited use to the ecologist, cytologist or serious taxonomist.

J. D. Hooker's *Student's Flora of the British Islands* (1870), familiar to many generations of botanists, has beautifully clear and concise descriptions but has not been revised since 1884. Babington's *Manual of British Botany* (1843) treats certain groups in greater detail than any other easily accessible work and was last revised as recently as 1922, but its scanty and frequently not very clear descriptions make it unsuitable for the average student of botany and particularly for the beginner.

During the past fifty years such advances have been made in all branches of botany that these floras are no longer adequate. The rise of ecology to a position of recognized importance has led to a demand not only for clear descriptions of species but for information of a kind not essential to identification, though of value to everyone interested in plants as living organisms.

There has also been among botanists a change of outlook so marked as to affect very seriously the usefulness of the existing books. When Babington and Hooker wrote their floras, 'systematic botany' was almost or (to Babington) quite synonymous with 'botany' and consequently these works are not primarily intended to permit the correct identification of plants but to teach the principles of classification and the technical characters of families and genera. Taxonomy is now only one branch, though an important and indeed a fundamental branch, of botany and many people who are not primarily taxonomists have need to identify plants correctly. Further, within the province of taxonomy itself there have been great changes. We now believe that the best way of learning general principles is by the recognition and study of individual species, so that from the point of view of the taxonomist also, a flora should provide a ready means of identifying plants. In the technique of description the value of measurements has been recognized and the general acceptance of the metric system has facilitated their use. To a systematic botanist a millimetre scale is now as essential a piece of equipment as a hand-lens. The experimental approach to taxonomic problems combined with the application of cytology and genetics provides a new method of attack. Though there are as yet only a few problems to which this method has been applied it has yielded valuable results and has greatly increased our understanding of certain species and their relationships.

There have also been changes in the flora itself, as well as in our knowledge of it, many of which will be apparent to every field botanist. A considerable number of introduced plants have become well established and some of them are now widespread. All those which persistently occur in natural or semi-natural communities must be regarded as integral parts of the flora of the country and so should be included in any account of it. Others, which only maintain themselves by repeated reintroduction, are of frequent occurrence on rubbish tips, near ports and in railway sidings. These, though in a different category from the naturalized plants and of less importance to the ecologist, are of interest to the systematist and should also be included in a British flora.

It is thus clear that there is a great need for a new flora of the British Isles and this need, at the suggestion of Professor A. G. Tansley, we have attempted to meet. Our aim has been to make accessible to students and amateurs a portion of the increased knowledge of our flora which has been gained since Hooker and Babington wrote. We have also included a considerably larger number of introduced plants, either because they are naturalized or because they are of frequent occurrence. Some information is also given about the time of flowering, fruiting and germination, the pollination and seed-dispersal mechanism as well as the life form and chromosome number.

It is necessary here to say something of the limitations of this book. In the first place it is intended primarily for students and amateur botanists who desire to gain an introduction to British plants and for botanists who are not taxonomic specialists but need to identify species without going into great detail in the so-called critical genera. It does not attempt to describe all named varieties or to give other details which a specialist might reasonably desire. Since it seemed desirable to complete the book as soon as possible, it has been written in the course of three and a half years in the intervals of teaching and other duties. Consequently, there has not been time to elucidate more than a few of the problems which have arisen and the work is, up to a point, a compilation of existing knowledge. In some groups we have had the benefit of expert help, but there remain a considerable number of families and genera where specialist knowledge was lacking; there is therefore some unevenness in the treatment of the different groups, and in a few (e.g. *Salicornia* and *Rhinanthus*) where existing knowledge is manifestly inadequate, the account given is necessarily unsatisfactory though, we hope, the best at present available.

The descriptions have, with few exceptions, been drawn up afresh from living material or herbarium specimens and the keys, wherever possible, make use of characters at least as easy to observe in the field as in the herbarium. For some of the larger families synopses of classification have been given, while in others descriptions of Tribes, etc., will be found in the text. It is hoped that the text-figures, drawn by Miss S. J. Roles, will prove of use in aiding identification.

The arrangement of families is in general similar to that adopted by Bentham and Hooker, though we have made a number of alterations to try to bring it more into line with modern ideas, and have always kept the doctrine of evolution in mind. Thus, instead of placing the Pteridophyta at the end, we start with them, as they are clearly the most primitive plants included in the book. It must be borne in mind, however, that no linear sequence of organisms, a sequence which must be used in a book, can be natural; often, particularly among the families of flowering plants, an arbitrary order has to be adopted. Within families and genera we have also followed the principle of starting with what appear to be the most primitive representatives in the British flora, though in some groups of which we have no intimate personal knowledge we have adopted the arrangement of a standard monograph.

In matters of nomenclature we have in general followed the *Check List of British Vascular Plants* issued by the British Ecological Society in 1946 and have also given synonyms in current use in British floras and in Druce's *British Plant List* (2nd edn, 1928).

In the spelling of certain specific epithets it has been customary to use an initial capital letter when the epithet concerned is derived from a personal name or is a noun, e.g. the name of another genus, or the pre-Linnean name for the plant. This custom is not made obligatory by the International Rules of Nomenclature but is mentioned in a recommendation attached to these Rules. The use of the initial capital has certain advantages; for instance it conveys some information about the origin of the name and explains the apparent lack of grammatical agreement between a generic name and a specific epithet which appears when written with a small initial letter to be adjectival (e.g. *Selinum Carvifolia*). We found upon inquiry, however, that many botanists in this country prefer, as a matter of convenience, to drop the initial capital. We have therefore adopted small initial letters for all specific epithets in the body of the book, but have indicated those which are commonly spelled with capitals.

English names have been given wherever possible for the benefit of agriculturists and others who prefer them for their special purposes. English names are frequently only of local use, and they give no reliable information of the relationship or otherwise of the plants, while frequently one name includes a number of distinct entities or is applied to different plants in various parts of the country. In addition only a small number of plants have English names which are in common use, though many others have names, often translations of the scientific names, which have been given to them mainly by the writers of nineteenth-century floras. We have tried to distinguish between the genuine English names and the invented ones by putting the latter in quotation marks. It cannot be emphasized too strongly that the scientific system of nomenclature has so many advantages over English names that it should be taught to university student and schoolchild alike.

Up to a point the limits of families and genera are a matter of personal opinion. For instance, Oxalidaceae and Balsaminaceae can be included in Geraniaceae or can be regarded as separate families. In such instances we have preferred to take the narrower view of family or generic limits when by doing so the groups obtained are more natural and are consequently easier to recognize. The genus *Antirrhinum* as established by Linnaeus was a large one from which the majority of species were soon removed and placed in the genus *Linaria*. This left *Antirrhinum* as a small homogeneous group in no way comparable with the vast assemblage of plants included in *Linaria*. There are only two reasonable courses open in such a case, either to keep the one large heterogeneous genus or to divide it up into a number of comparable and reasonably homogeneous groups. Wherever it seemed possible and convenient to do so, we have adopted the latter course.

As has already been pointed out, no attempt has been made to describe all the numerous named varieties of British plants, but when plants which are morphologically similar have been shown to differ cytologically or in geographical distribution or ecological preferences we have not hesitated to recognize them as subspecies.

In some genera we have placed two or more superficially similar species in an aggregate (agg.). This is simply a device for the convenience of those who do not wish to go into minute detail, and is of no taxonomic significance.

Hybrids between species have as far as possible been mentioned; descriptions have been given where the hybrid is common, usually owing to abundant vegetative reproduction (e.g. in *Mentha* and *Potamogeton*), where it is a highly distinct plant which has in the past been regarded as a species (e.g. × *Agropogon littoralis*), or where it is liable to lead to confusion between species (e.g. *Alopecurus* × *hybridus*). We have discarded as far as possible all names which appear to us to be ambiguous, either because there is doubt about what plant was originally intended by that name (e.g. *Orchis latifolia* L.), or because the name is currently applied to two or more distinct species (e.g. *Carex leporina* L.).

A volume of illustrations is in course of preparation but as it cannot be ready for some time yet, references to illustrations in easily accessible floras have been included wherever these drawings were sufficiently satisfactory to be a real aid to identification.

While we hope that this flora will prove useful, we are fully aware that it has many deficiencies and will doubtless be found to contain errors. As Bentham wrote nearly a hundred years ago 'the aptness of a botanical description, like the beauty of a work of imagination, will always vary with the style and genius of the author'. We should be most grateful if users of the book who detect any errors would inform us.

Acknowledgements

We should like to express our thanks to Professor Tansley for his constant encouragement, and to the many botanists who have given us the benefit of their expert advice. Among these we should specially like to name W. T. Stearn for much help with nomenclature and with the petaloid monocotyledons, and for reading the proofs; S. M. Walters for great assistance with *Alchemilla*, *Aphanes*, *Montia*, *Eleocharis*, etc.; and E. K. Horwood whose continuous help has enabled the work to be completed much more rapidly than would otherwise have been possible.

We are also greatly indebted to the following for help with special problems: A. H. G. Alston (Pteridophyta), Miss K. B. Blackburn, J. P. M. Brenan (*Chenopodium*), B. L. Burtt, Miss M. S. Campbell (*Salicornia*), J. L. Crosby (*Anagallis*), J. E. Dandy (Hydrocharitaceae, Najadaceae, Potamogetonaceae), J. S. L. Gilmour, C. E. Hubbard (Gramineae), Miss I. Manton, R. Melville (*Ulmus*, especially the key and the originals of Figs 48–9), E. Nelmes (*Carex*), C. D. Pigott, H. W. Pugsley (*Hieracium*), N. Y. Sandwith, H. K. Airy Shaw, W. A. Sledge, T. A. Sprague (classification), V. S. Summerhayes, G. Taylor, D. H. Valentine (*Primula, Viola*), A. E. Wade (Boraginaceae), W. C. R. Watson (*Rubus*), D. A. Webb, F. H. Whitehead, A. J. Wilmott (*Salicornia*, etc.), and many others who have assisted to a lesser extent in various ways. It should be added that these specialists cannot be held responsible for all the views expressed or implied in the accounts of those genera about which they have so freely given us their advice.

We should also like to express our indebtedness to the Director of the Royal Botanic Gardens, Kew, the Regius Keeper of the Royal Botanic Garden, Edinburgh, the Keeper of the Department of Botany at the British Museum, the Professors of Botany at the British Museum, the Professors of Botany at Oxford and Cambridge, and the Director of the Leicester City Museum for the loan of specimens and other help.

We are greatly indebted to P. W. Richards for the account of the Juncaceae.

November 1948

A. R. C.
T. G. T.
E. F. W.

Two of us wish at this point to acknowledge the special contribution of T. G. Tutin who, besides writing a substantial part of this flora, undertook in addition the arduous task of acting as general editor. It was he who collected and collated the various sections as they were completed, who strove to secure uniformity of treatment, who wrestled with text-figures, glossary and index, and who urged us on when we flagged. The work owes much to his patient and devoted labour.

November 1948

A. R. C.
E. F. W.

Preface to the second edition

Since the writing of the first edition of this flora was completed considerable progress has been made with the study of British plants. The most spectacular, though in some respects not the most important, addition to our knowledge has been the discovery of some plants not hitherto known to grow in these islands but occurring, to all appearances as native species, in a few localities. Among these, *Artemisia norvegica* and *Diapensia lapponica* deserve special mention not only for their phytogeographical interest but for the aesthetic pleasure that the field botanist will obtain from them.

Perhaps the greatest advance in the last few years has been brought about by the combination of intensive experimental, field and herbarium studies of variable species and 'critical groups'. This has resulted in a considerable clarification of the taxonomy of certain genera, for example *Dactylorchis* and *Polypodium* but, though much has been achieved, still more remains to be done.

Knowledge of the distribution of vascular plants in Britain has also been materially improved, largely as a result of the field work undertaken by the numerous helpers with the Botanical Society's Distribution Maps Scheme, which was made possible by the generosity of the Nature Conservancy and the Nuffield Foundation. The detailed results of this scheme are not yet fully available, but the major discoveries are incorporated here.

There have also been changes of various kinds in the flora itself. It is sad to record that *Schoenus ferrugineus* has apparently become extinct in Scotland, but to be set against this is the discovery of *Spiranthes romanzoffiana* in two new localities and the almost annual appearance of *Epipogium aphyllum*.

Changes among the alien plants have, of course, been even greater; some have become newly established while others have failed to maintain their footing. These changes are reflected, as accurately as our information permits, in this edition.

All this new work has resulted in the complete rewriting of many pages and innumerable minor alterations and additions. Several of the keys have also been considerably modified and, it is hoped, made easier to use and more certain to lead to correct identification.

The publication of Mr J. E. Dandy's *List of British Vascular Plants* marks another step towards nomenclatural stability, though changes of name for taxonomic reasons will inevitably continue to occur from time to time. We have in general followed this list, though differences of taxonomic opinion will be found here and there and, very occasionally, divergences of purely nomenclatural origin. We are, once again, greatly indebted to Mr Dandy for his help.

To those whose help was acknowledged in the first edition we should like to add the following: H. G. Baker (*Limonium*), P. W. Ball (*Cakile, Salicornia*), C. D. K. Cook (subgen. *Batrachium* and *Sparganium*), E. W. Davies (*Asparagus*), K. M. Goodway (*Galium*), R. A. Graham (*Mentha*), G. Halliday (*Minuartia*), J. Heslop-Harrison (*Dactylorchis*), I. H. McNaughton (*Papaver*), R. Melville (*Epilobium*), P. A. Padmore (*Ranunculus*), A. Pettet (*Viola*), C. D. Pigott (*Polemonium* and *Cirsium*), M. C. F. Proctor (*Helianthemum*), N. M. Pritchard (*Gentianella*), Peter Raven (Onagraceae), B. T. Styles (*Polygonum*), S. Walker (*Dryopteris*), D. P. Young (*Oxalis*).

In addition we should like to thank those correspondents, too numerous to mention individually, who pointed out errors, provided additional information or made suggestions for improvement.

November 1958

A. R. C.

T. G. T.

E. F. W.

Preface to the third edition

This third edition of the *Flora of the British Isles* comes more than thirty years after the original publication in 1952, the preface to which regarded the function of a national Flora as not merely to assist in the identification of native species but also to provide, preferably within the compass of a single volume, information of general use to those interested in their ecology, geographical distribution, evolutionary history, agricultural significance, etc. We still hold this view and have, therefore, continued to provide general descriptions which include life-form and chromosome numbers and some notes on phenology and mechanisms of pollination and seed-dispersal and also on variability, distribution within and outside the British Isles, preferred habitats and commonly associated species. There is still no attempt to include all named varieties or all known or putative hybrids, but descriptions are now given of most of the widely accepted subspecies. This limitation of taxonomic scope became less important with the publication of Dr C. A. Stace's excellent *Hybridization in the British Flora* (1975); we had also expected that it would have been even less so in view of the progress made towards the preparation of a *Critical Flora of the British Isles*, which intended a 'full taxonomic treatment of infra-specific variation'. Unfortunately, as we go to press, we learn that this project has now been abandoned.

The preface to the second edition (1962) noted a greatly increased knowledge of the native flora during the preceding ten years, this arising largely from detailed studies of certain 'critical' groups and from field-work undertaken for the Botanical Society's Distribution Maps Scheme. In both these directions there has been much further progress. The projected *Atlas of the British Flora* appeared in 1962 and its *Critical Supplement* in 1968, the latter containing provisional distribution maps of many members of critical genera, the treatment of *Hieracium* by P. D. Sell and C. West being of particular interest and value. These two volumes have assisted us greatly.

Of special significance for our work on a third edition has been the successive appearance of the five volumes of *Flora Europaea* (1964–1980). In the preface to the most recent edition of the *Excursion Flora* (1981) we suggested that all national and regional floras within the area covered by *Flora Europaea* should as soon as possible adopt its taxonomy and nomenclature, unless there seemed good grounds for doing otherwise, thus hastening the approach to a highly desirable uniformity. We stand by this view and have acted accordingly.

In providing the common names of plants we have followed *English Names of Wild Flowers* (Dony, Rob and Richards), published by the Botanical Society of the British Isles (1974). However, we have felt free to include other names which are well known, even if locally, or for which we have particular affection.

The recent statement from the Nature Conservancy Council entitled *Nature Conservation in Great Britain* (1984) has made it clear that the total area of natural and semi-natural vegetation in this country has been declining at a disturbingly rapid rate and that over 10% of our native flowering plants and ferns have been lost since 1930 from 20% or more of the 10 × 10-km grid-squares within which they then occurred. In view of this decline, and of the complete loss of at least seven native species since our *Flora* first appeared in 1952, we trust that all our readers will support the efforts of the Nature Conservancy Council and similarly motivated bodies to resist any further lowering of the level of nature conservation.

Finally, we wish once more to express our sincere thanks to all those friends and colleagues who have assisted us with information and advice. In addition to those mentioned in the prefaces to previous editions, many of whom have helped us again with this book, we should like to acknowledge our indebtedness to R. D. Meikle (Salicaceae), A. Newton (*Rubus*), F. J. Rumsey (*Orobanche*), A. O. Chater, Dr T. T. Elkington and Professor D. A. Webb, for general help, E. J. Clements for particular assistance with alien species, and Mrs L. M. Walters for preparing the index.

A. R. C.
D. M. M.
February 1985   T. G. T.

Dr E. F. Warburg, co-editor with us of the first and second editions and a close personal friend, sadly died in 1966. He is replaced by Professor D. M. Moore, of the Botany Department of Reading University, whom we are very happy to welcome as a collaborator.

A. R. C.
February 1985   T. G. T.

As in earlier editions of the *Flora*, his colleagues wish to acknowledge the central role played by Professor T. G. Tutin in coordinating our endeavours, collating the whole manuscript and urging us on when we flagged. We are grateful for his unstinted, kindly support.

A. R. C.
D. M. M.

Synopsis of classification

For signs and abbreviations see page xxix

PTERIDOPHYTA

Plants with an alternation of free-living generations. Sporophyte with vascular tissue, reproducing by spores which give rise to the small filamentous or thalloid gametophyte (prothallus) bearing archegonia and antheridia either on the same or on different prothalli.

PTEROPSIDA

Stems simple or dichotomously branched. Lvs small, spirally arranged; no lf-gap in stele. Sporangia borne singly in the axil of a lf (sporophyll) or on its upper-surface near the base.

LYCOPODIALES

Stems long with numerous small lvs; secondary thickening 0. Ligule 0. Homosporous. Spermatozoids biciliate.
1. Lycopodiaceae.

SELAGINELLALES

Stems long, with numerous small lvs; secondary thickening 0. Ligule present. Heterosporous. Spermatozoids biciliate.
2. Selaginellaceae.

ISOETALES

Stem short, tuberous, with secondary thickening. Lvs subulate. Ligule present. Heterosporous. Spermatozoids multiciliate.
3. Isoetaceae.

SPHENOPSIDA

Stem simple or with whorls of branches. Lvs small, in whorls. No lf-gap. Sporangia several on peltate sporangiophores borne in cones. Spermatozoids multiciliate.

EQUISETALES

Herbs. Homosporous.
4. Equisetaceae.

FILICOPSIDA

Lvs usually large, often compound, spirally arranged; lf-gap present. Sporangia often grouped in sori, borne on the underside of the lvs or on special lf-segments. Spermatozoids multiciliate.

FILICALES

Lvs large, flat, circinate in bud. Sporangia with wall of 1 layer of cells borne on the lf-surface (sometimes ±modified). Homosporous.
5. Ophioglossaceae.
6. Osmundaceae.
7. Adiantaceae.
8. Hymenophyllaceae.

9. Polypodiaceae.
10. Hypolepidaceae.
11. Thelypteridaceae.
12. Aspleniaceae.
13. Athyriaceae.
14. Aspidiaceae.
15. Blechnaceae.

MARSILEALES

Plants rooted. Lvs circinate in bud, not fern-like. Sporangia with wall of 1 layer of cells, borne in thick-walled sporocarps containing several sori. Heterosporous.
16. Marsileaceae.

SALVINIALES

Free-floating. Lvs small, not circinate. Sporangia with wall of 1 layer of cells borne in thin-walled sporocarps containing 1 sorus. Heterosporous.
17. Azollaceae.

GYMNOSPERMAE

Gametophyte not free-living. Woody plants with secondary thickening. Ovules not enclosed in an ovary. Female prothallus well developed, of numerous cells forming the food reserve of the seed. Xylem without vessels (except in Gnetales).

CONIFERAE

Stem usually freely branched. Lvs simple, usually small. Pollen-sacs borne on the under-surface of microsporophylls arranged in cones. Fertilization effected by means of a pollen-tube; male gametes not motile. Ovules usually on the surface of a scale.
18. Pinaceae.
19. Cupressaceae.
20. Taxaceae.

ANGIOSPERMAE

Ovules completely enclosed in an ovary which is usually crowned by a style and stigma. Microspores (pollen grains) adhering to the stigma and fertilization effected by means of a pollen-tube. Xylem containing vessels (except in some Winteraceae and Trochodendraceae).

DICOTYLEDONES

Embryo with 2 cotyledons (rarely one by reduction). Vascular bundles of the stem usually arranged in a single ring, cambium usually present. Lvs rarely parallel-veined. Fls typically 5–4-merous.

ARCHICHLAMYDEAE

Petals free from each other or 0, rarely united into a tube.

RANUNCULALES

Herbs, often with numerous vascular bundles and little or no cambium in the stems, less frequently woody. Lvs alternate (very rarely opposite), nearly always exstipulate. Fls hermaphrodite (rarely unisexual) hypogynous or rarely perigynous, actinomorphic (rarely zygomorphic). Perianth present (very rarely 0). Stamens numerous or less frequently definite in number, often spirally arranged. Ovary apocarpous; carpels often numerous and spirally arranged. Fr. various, but rarely fleshy. Seeds with copious endosperm and small embryo.

21. Ranunculaceae. 24. Nymphaeaceae.
22. Paeoniaceae. 25. Ceratophyllaceae.
23. Berberidaceae.

PAPAVERALES

Herbs or rarely ±woody. Lvs alternate (rarely opposite), exstipulate or with small stipules. Fls hermaphrodite, rarely unisexual, hypogynous, actinomorphic to zygomorphic. Petals and sepals usually in whorls of 2 or 4. Stamens numerous or 6 or 4, rarely 3 or 2. Ovary syncarpous (rarely apocarpous) 1–several-celled, often 2-celled and divided by a false septum; placentation parietal. Fruit dry. Seeds either with little or no endosperm and a large embryo or with abundant endosperm and a minute embryo.

26. Papaveraceae. 28. Cruciferae.
27. Fumariaceae. 29. Resedaceae.

VIOLALES

Herbs, shrubs or small trees. Lvs alternate (rarely opposite), stipulate. Fls hermaphrodite (rarely polygamous), hypogynous, actinomorphic to zygomorphic. Sepals 5, lowermost petal often larger and spurred. Stamens 5, ±connivent in a ring round the ovary. Ovary syncarpous, 1-celled, with 3–5 parietal placentae. Fr. a capsule or sometimes fleshy. Seeds with endosperm and a straight embryo.

30. Violaceae.

POLYGALALES

Herbs, shrubs and trees. Lvs alternate or rarely opposite; stipules 0 or small. Fls hermaphrodite, hypogynous to subperigynous, zygomorphic or rarely actinomorphic. Sepals 5, often unequal. Petals 1–5, free or sometimes some joined. Stamens up to 12, sometimes only 1 fertile, sometimes monadelphous. Anthers often opening by pores. Ovary syncarpous, 1–3-celled; placentation axile or apical. Fr. a capsule, drupe or samara. Seeds with a straight embryo; endosperm present or 0.

31. Polygalaceae.

CISTALES

Herbs or more often woody, juice often coloured. Lvs opposite or sometimes alternate; stipules usually present. Fls hermaphrodite, hypogynous, actinomorphic, usually large and showy. Sepals 3–5. Petals 5 (rarely fewer or 0). Stamens numerous, sometimes with their filaments joined in bundles. Ovary syncarpous, 1-celled or sometimes 3–5-celled; placentation parietal or sometimes axile or apical. Fr. a capsule, rarely fleshy. Seeds with a straight or, more often, curved or bent embryo; endosperm usually present and often abundant.

32. Hypericaceae. 33. Cistaceae.

TAMARICALES

Trees or shrubs, rarely herbs. Lvs opposite or alternate, small and scale-like or ericoid; stipules present or 0. Fls usually hermaphrodite, hypogynous, usually small, actinomorphic. Sepals 4–6. Petals 4–6. Stamens 5–10. Ovary syncarpous, 1-celled; placentation parietal or basal. Fr. a capsule. Seeds with a straight embryo; endosperm present or 0.

34. Tamaricaceae. 35. Frankeniaceae.

CARYOPHYLLALES

Herbs, rarely soft-wooded shrubs or trees. Lvs opposite or verticillate, or sometimes alternate; stipules present or 0. Fls hermaphrodite or occasionally unisexual, hypogynous to perigynous, actinomorphic or rarely slightly zygomorphic. Perianth in 2 whorls or 1 (rarely several) the outer sepaloid, the inner petaloid or sepaloid. Stamens usually definite. Ovary usually syncarpous, 1(–3–several)-celled; placentation axile to free-central or basal; ovules usually campylotropous. Seeds usually with endosperm and a curved embryo.

36. Elatinaceae. 40. Amaranthaceae.
37. Caryophyllaceae. 41. Chenopodiaceae.
38. Portulacaceae. 42. Phytolaccaceae.
39. Aizoaceae.

MALVALES

Trees, shrubs or herbs usually with stellate hairs. Lvs usually alternate, stipulate, often mucilaginous. Fls hermaphrodite or unisexual, hypogynous, actinomorphic. Calyx-lobes usually valvate. Petals present or, less frequently, 0. Stamens numerous, free or monadelphous, often some sterile or anthers 1-celled. Ovary syncarpous, 2 (or more)-celled; placentation axile. Fr. various. Seeds usually with endosperm; embryo straight or curved.

43. Tiliaceae. 44. Malvaceae.

GERANIALES

Herbs or small shrubs, rarely trees. Lvs usually alternate or basal, generally stipulate. Fls hermaphrodite, very rarely unisexual, hypogynous, actinomorphic to zygomorphic. Sepals imbricate or rarely valvate. Petals contorted or sometimes imbricate, often clawed, rarely some ±joined, rarely 0. Stamens as many to three times as many as petals, commonly twice as many. Ovary syncarpous, 3–5-celled; placentation axile. Fr. various, but very rarely fleshy. Seeds usually without endosperm; embryo straight.

45. Linaceae. 47. Oxalidaceae.
46. Geraniaceae. 48. Balsaminaceae.

RUTALES

Trees, shrubs or climbers, rarely herbs. Lvs usually alternate, often compound, and frequently gland-dotted. Fls hermaphrodite, rarely unisexual, hypogynous or weakly perigynous, usually actinomorphic. Sepals usually imbricate. Petals contorted or sometimes valvate, free or joined at base. Disk usually conspicuous. Stamens as many or twice as many as the petals. Ovary syncarpous, 1–5-celled; ovules usually 1–2 in each cell. Fr. various. Seeds with or without endosperm; embryo straight or curved.

    49. Simaroubaceae.

SAPINDALES

Trees or shrubs. Lvs usually pinnate and exstipulate. Fls polygamous or unisexual, hypogynous to slightly perigynous, sometimes zygomorphic, usually small. Sepals 4–5, imbricate. Petals usually 4–5, rarely 0. Stamens often twice as many as petals. Disk present. Ovary syncarpous (rarely apocarpous) with 1–2 ovules in each cell; placentation axile. Fr. various. Seeds usually without endosperm; embryo curved or variously bent.

    50. Aceraceae.   52. Hippocastanaceae.
    51. Staphyleaceae.

CELASTRALES

Trees or shrubs. Lvs simple, often entire; stipules small or 0. Fls hermaphrodite, rarely unisexual, hypogynous to perigynous, actinomorphic, usually small. Sepals usually imbricate. Petals often 4–5, rarely 0. Disk present or 0. Stamens often 4–5, opposite the sepals. Ovary syncarpous usually with 1–2 ovules in each cell; placentation axile or apical. Fr. various. Seeds usually with abundant endosperm; embryo straight.

    53. Aquifoliaceae.   55. Buxaceae.
    54. Celastraceae.

RHAMNALES

Trees, shrubs or woody climbers. Lvs usually stipulate. Fls similar to Celastrales but stamens 4–5, opposite the petals or alternating with the sepals in apetalous spp. Fr. usually a drupe or berry. Seeds with endosperm; embryo usually straight.

    56. Rhamnaceae.   57. Vitaceae.

LEGUMINOSAE

Trees, shrubs or herbs. Lvs often pinnate or bipinnate, sometimes trifoliolate or simple; stipules present or 0. Fls hermaphrodite, hypogynous to perigynous, actinomorphic to zygomorphic, often large and showy. Sepals often 5, often ±united into a tube. Petals usually 5, rarely 0, occasionally united. Stamens often 10, sometimes numerous, often monadelphous or diadelphous. Ovary of one carpel. Fr. a legume, often dehiscent. Seeds usually with little or no endosperm, rarely with abundant endosperm; embryo large.

    58. Leguminosae (Fabaceae).

ROSALES

Trees, shrubs or herbs. Lvs simple or compound. Fls hermaphrodite or rarely unisexual, perigynous to epigynous, actinomorphic (rarely ±zygomorphic). Sepals usually 4–5, free or united. Petals usually 4–5 (rarely 0), occasionally united. Stamens numerous to definite. Ovary apocarpous to syncarpous, with one or more ovules in each cell; placentation often axile. Fr. various. Seeds with or without endosperm.

    59. Rosaceae.   64. Hydrangeaceae.
    60. Platanaceae.   65. Escalloniaceae.
    61. Crassulaceae.   66. Grossulariaceae.
    62. Saxifragaceae.   67. Pittosporaceae.
    63. Parnassiaceae.

SARRACENIALES

Herbs. Lvs tubular or covered with viscid glands, adapted for trapping insects. Fls hermaphrodite, hypogynous to perigynous, actinomorphic. Sepals 4–5, ±united at base. Petals 5, rarely 0. Stamens 4–numerous. Ovary syncarpous; ovules usually numerous; placentation axile to parietal. Fr. a capsule. Seeds with endosperm; embryo straight.

    68. Droseraceae.   69. Sarraceniaceae.

MYRTALES

Trees, shrubs or herbs often with bicollateral vascular bundles. Lvs often opposite, usually exstipulate. Fls hermaphrodite or rarely unisexual, perigynous to epigynous often with a long receptacle, actinomorphic. Calyx-tube usually ±adnate to the ovary, lobes mostly 4–5, often valvate. Petals commonly 4–6 (rarely 0), sometimes united. Stamens 1–many, often 4 or 8. Ovary syncarpous, 1–many-celled; ovules numerous to 1; placentation axile or rarely parietal, apical or basal. Fr. various. Seeds with or without endosperm.

    70. Lythraceae.   74. Haloragidaceae.
    71. Thymelaeaceae.   75. Hippuridaceae.
    72. Elaeagnaceae.   76. Callitrichaceae.
    73. Onagraceae.

SANTALALES

Trees, shrubs or herbs often parasitic on other angiosperms or rarely on gymnosperms. Lvs usually opposite, sometimes scale-like, exstipulate. Fls hermaphrodite or unisexual, epigynous, actinomorphic. Calyx with valvate lobes or often reduced, sometimes to a ring. Petals present or 0, sometimes united into a tube. Stamens the same number as the calyx-lobes and opposite them or opposite the petals when present. Ovary 1-celled; ovules few, often imperfectly differentiated; placentation axile. Fr. a drupe or berry, less frequently a nut. Seeds with endosperm and a straight embryo.

    77. Loranthaceae.   78. Santalaceae.

UMBELLALES

Trees, shrubs or herbs. Lvs usually alternate, often much-divided; stipules present or 0. Fls hermaphrodite or unisexual, epigynous, actinomorphic to weakly zygomorphic, usually small and arranged in umbels or heads.

Calyx small, truncate or 4–10-toothed. Petals usually 4–5, rarely 0. Stamens usually the same number as the petals and alternate with them. Ovary usually 1–2-celled, sometimes many-celled; ovules solitary in each cell, pendent from the apex. Seeds usually with copious endosperm.

79. Cornaceae.
80. Araliaceae.
81. Umbelliferae (Apiaceae).

CUCURBITALES

Herbs, or sometimes small trees, often with bicollateral vascular bundles, and frequently climbing by tendrils. Lvs usually alternate, often large and deeply lobed or compound. Fls unisexual, epigynous, actinomorphic or rarely zygomorphic, often showy. Calyx variously lobed. Petals free or united into a tube. Stamens 1–numerous, free or variously united, sometimes epipetalous. Ovary 1–4-celled; ovules numerous, very rarely few; placentation parietal or axile. Fr. a capsule or berry. Seeds with little or no endosperm.

82. Cucurbitaceae.

ARISTOLOCHIALES

Woody climbers with broad medullary rays, or parasites, or epiphytes, rarely erect herbs. Lvs alternate, simple, exstipulate, sometimes 0. Fls hermaphrodite or unisexual, hypogynous to epigynous, actinomorphic or zygomorphic. Per.-segs in one whorl, usually petaloid. Stamens numerous to few. Ovary 1–6-celled; ovules numerous in each cell; placentation parietal or axile. Fr. a capsule or sometimes fleshy. Seeds with or without endosperm.

83. Aristolochiaceae.

EUPHORBIALES

Trees, shrubs or herbs. Lvs usually alternate and stipulate, simple or compound, sometimes reduced. Fls unisexual, hypogynous, actinomorphic. Sepals usually present. Petals usually 0. Stamens numerous to solitary, free or united. Ovary usually 3-celled; ovules 1–2 in each cell; placentation axile. Fr. a capsule or drupe. Seeds with copious endosperm.

84. Euphorbiaceae.

POLYGONALES

Herbs, shrubs or climbers, rarely trees. Lvs usually alternate, often with sheathing stipules. Fls hermaphrodite or unisexual, hypogynous, actinomorphic. Per.-segs 3–6, sepaloid or petaloid, free or united. Stamens usually 6–9. Ovary 1-celled with a solitary basal ovule. Fr. a trigonous or lenticular nut. Seeds with copious endosperm.

85. Polygonaceae.

URTICALES

Trees, shrubs or herbs. Lvs usually alternate and stipulate. Fls hermaphrodite or unisexual, hypogynous, actinomorphic. Per.-segs usually 4–5 ±united, sepaloid. Stamens usually the same number as and opposite to the per.-segs, erect or inflexed in bud. Ovary 1–2-celled; ovule solitary, erect or pendent. Fr. various. Seeds with or without endosperm.

86. Urticaceae.
87. Cannabaceae.
88. Ulmaceae.
89. Moraceae.

JUGLANDALES

Trees, often resinous and aromatic. Lvs alternate, pinnate, exstipulate. Fls unisexual, epigynous, actinomorphic. Perianth small and sepaloid or 0. Stamens 3–40. Ovary 1-celled; ovule solitary, erect. Fr. a drupe or rarely a nut. Seeds without endosperm.

90. Juglandaceae.

MYRICALES

Aromatic trees or shrubs. Lvs alternate, simple, exstipulate. Fls unisexual, arranged in dense bracteate spikes. Perianth 0. Stamens 2–many, free or connate. Ovary 1-celled; ovule solitary basal. Fr. a drupe. Seeds without endosperm.

91. Myricaceae.

FAGALES

Monoecious trees or shrubs. Lvs alternate, simple, stipulate. Fls epigynous or the female devoid of perianth, in catkins (rarely heads) or the female in cone-like spikes or few, often appearing before the lvs. Perianth very small or 0 in one sex; female fls often surrounded by an involucre of bracts. Stamens 2–many. Ovary 2–6-celled; ovules 1–2 in each cell, pendent. Fr. a nut, sometimes winged. Seeds without endosperm.

92. Betulaceae.
93. Corylaceae.
94. Fagaceae.

SALICALES

Dioecious trees or shrubs. Lvs alternate, simple, usually stipulate. Fls in catkins, often appearing before the lvs. Perianth 0, or very small. Stamens 2 or more. Ovary 1-celled; ovules numerous; placentation parietal. Fr. a capsule. Seeds without endosperm.

95. Salicaceae.

METACHLAMYDEAE

Petals united into a longer or shorter tube, very rarely free or 0.

ERICALES

Shrubs, rarely trees or herbs. Lvs simple, exstipulate, usually alternate. Fls hermaphrodite, rarely unisexual, hypogynous or epigynous, actinomorphic to zygomorphic. Calyx usually 4–6-lobed, sometimes of free sepals. Petals united, rarely free or 0. Stamens usually twice as many as the corolla-lobes, free, anthers often opening by pores. Ovary 3–several-celled; ovules 1–many in each cell; placentation usually axile. Fr. a capsule, berry or drupe. Seeds with abundant endosperm and small straight embryo.

96. Ericaceae.
97. Pyrolaceae.
98. Monotropaceae.
99. Empetraceae.
100. Diapensiaceae.

PLUMBAGINALES

Herbs, small shrubs, or sometimes climbers. Lvs alternate or basal, exstipulate. Fls hermaphrodite, hypogynous, actinomorphic. Calyx commonly 5-lobed, often strongly ribbed and membranous between the lobes. Petals 5, united into a longer or shorter tube, rarely free. Stamens 5, opposite corolla-lobes and ±adnate to the tube. Ovary 1-celled; ovule solitary; placentation basal; styles 5. Fr. dry, usually indehiscent. Seeds with or without endosperm.

101. Plumbaginaceae.

PRIMULALES

Herbs, shrubs or trees. Lvs variously arranged, but often basal, exstipulate. Fls hermaphrodite, hypogynous or very rarely perigynous, actinomorphic or very rarely zygomorphic. Calyx 4–9-lobed, persistent. Corolla 4–9-lobed, very rarely two-lipped or 0. Stamens as many as and opposite the corolla-lobes, adnate to the tube. Ovary 1-celled, very rarely adnate to the calyx; ovules 2–many on a free-central placenta; style 1. Fr. a capsule, variously dehiscent. Seeds with copious endosperm.

102. Primulaceae.

GENTIANALES

Trees, shrubs or herbs. Lvs simple, often opposite, usually exstipulate. Fls hermaphrodite (rarely unisexual), hypogynous, actinomorphic. Calyx tubular or rarely composed of separate sepals or 0. Corolla usually 4–5-lobed (rarely of free petals or 0), lobes contorted or valvate in bud. Stamens usually epipetalous, the same number as and alternate with the corolla-lobes, rarely fewer. Ovary mostly 1–2-celled, sometimes of 2 separate carpels; ovules numerous to 1 in each cell; placentation usually parietal or axile. Fr. various. Seeds with endosperm; embryo straight, often small.

103. Buddlejaceae.   106. Gentianaceae.
104. Oleaceae.   107. Menyanthaceae.
105. Apocynaceae.

SOLANALES

Herbs, less frequently trees, shrubs or woody climbers. Lvs often opposite, usually exstipulate. Fls usually hermaphrodite, hypogynous (rarely epigynous), actinomorphic to zygomorphic. Calyx usually 4–5-lobed, sometimes 2-lipped. Corolla 4–5-lobed, often 2-lipped. Stamens epipetalous, as many as or fewer than the corolla-lobes and alternate with them. Ovary usually 1–2-celled; ovules numerous to 1 in each cell; placentation usually axile, parietal or basal. Fr. various. Seeds with or without endosperm.

108. Polemoniaceae.   113. Orobanchaceae.
109. Boraginaceae.   114. Lentibulariaceae.
110. Convolvulaceae.   115. Acanthaceae.
111. Solanaceae.   116. Verbenaceae.
112. Scrophulariaceae.   117. Labiatae (Lamiaceae).

PLANTAGINALES

Herbs. Lvs simple, often sheathing at base. Fls usually hermaphrodite, hypogynous, actinomorphic. Calyx 4-lobed. Corolla 3–4-lobed, scarious. Stamens epipetalous, usually 4. Ovary 1–4-celled; ovules 1–several in each cell; placentation axile or basal. Fr. a capsule or nut. Seeds with endosperm.

118. Plantaginaceae.

CAMPANALES

Usually herbs. Lvs mostly alternate, simple, exstipulate. Fls hermaphrodite, rarely unisexual, epigynous (rarely hypogynous), actinomorphic to zygomorphic. Calyx usually 5-lobed. Corolla often 5-lobed, sometimes 2-lipped. Stamens as many as the corolla-lobes and alternate with them, free or inserted on the corolla-tube near its base; anthers often connivent and sometimes adhering in a tube. Ovary (1–)2–10-celled; ovules usually numerous; placentation axile. Fr. various. Seeds with endosperm.

119. Campanulaceae.

RUBIALES

Trees, shrubs or herbs. Lvs usually opposite, stipulate or not. Fls usually hermaphrodite, epigynous, actinomorphic to zygomorphic. Calyx often 4–5-lobed or reduced to a rim. Corolla sometimes 2-lipped. Stamens epipetalous, the same number as and alternate with the corolla-lobes, rarely fewer; anthers not connivent or cohering. Ovary (1–)2(or more)-celled; ovules numerous to 1 in each cell; placentation axile or apical, rarely basal. Fr. various. Seeds with or without endosperm.

120. Rubiaceae.   123. Valerianaceae.
121. Caprifoliaceae.   124. Dipsacaceae.
122. Adoxaceae.

ASTERALES

Herbs, shrubs, or rarely trees or woody climbers. Lvs exstipulate. Fls hermaphrodite or unisexual, epigynous, actinomorphic to zygomorphic, crowded in heads (rarely solitary) surrounded by 1 or more series of free or connate bracts. Calyx small, often with thread-like lobes (pappus). Corolla usually 4–5-lobed. Stamens epipetalous, 5(–4); anthers connate (rarely imperfectly so). Ovary 1-celled; ovule solitary; placentation basal. Fr. an achene. Seeds without endosperm.

125. Compositae (Asteraceae).

MONOCOTYLEDONES

Embryo with one cotyledon. Vascular bundles of the stem usually in several series or ±irregularly arranged, cambium usually 0. Lvs usually parallel-veined. Fls typically 3-merous.

ALISMATALES

Herbs living in water or wet places, sometimes marine. Fls actinomorphic, hermaphrodite or unisexual.

Perianth in two whorls, the outer usually sepaloid, the inner petaloid. Stamens 3, 6 or numerous. Ovary apocarpous and superior or syncarpous and inferior. Ovules 1–numerous, basal, parietal or scattered. Seeds without endosperm.

126. Alismataceae. 128. Hydrocharitaceae.
127. Butomaceae.

NAJADALES

Herbs living in water or wet places, sometimes marine. Lvs linear, with scales (*squamulae intravaginales*) in their axils. Fls hypogynous, hermaphrodite or unisexual. Perianth 0 or of one whorl, less often of 2 similar whorls. Stamens 1–6, rarely more. Ovary of few (often only one) free or ±connate carpels; ovules 1, rarely more; placentation usually basal or apical. Fr. usually dry. Seeds with little or no endosperm.

129. Scheuchzeriaceae. 133. Potamogetonaceae.
130. Juncaginaceae. 134. Ruppiaceae.
131. Aponogetonaceae. 135. Zannichelliaceae.
132. Zosteraceae. 136. Najadaceae.

ERIOCAULALES

Herbs with narrow lvs. Fls small, unisexual, arranged in heads. Perianth scarious or membranous, segments in 2 whorls, inner often united. Ovary superior, 3–2-celled. Ovules solitary, pendent. Seeds with endosperm.

137. Eriocaulaceae.

LILIALES

Herbs, often with corms, bulbs or rhizomes, rarely shrubs or small trees. Lvs mostly linear. Fls hermaphrodite or sometimes unisexual, hypogynous to epigynous, actinomorphic to zygomorphic. Perianth of two whorls, usually both petaloid, rarely both sepaloid, very rarely unlike. Stamens in one or 2 whorls, commonly 3 or 6. Ovary syncarpous, usually 3-celled; ovules 1–many in each cell; placentation axile or parietal. Seeds with endosperm.

138. Liliaceae. 141. Amaryllidaceae.
139. Pontederiaceae. 142. Iridaceae.
140. Juncaceae. 143. Dioscoreaceae.

ORCHIDALES

Herbs without bulbs but often with tubers, often epiphytes or saprophytes. Lvs simple, often rather thick. Fls mostly hermaphrodite, epigynous, zygomorphic. Perianth of two whorls, usually both petaloid, but sometimes the outer sepaloid. Stamens 2 or 1; pollen usually agglutinated into masses (*pollinia*). Ovary usually 1-celled often twisted through 180°; ovules numerous; placentation parietal. Fr. usually a capsule. Seeds minute, without endosperm and with undifferentiated embryo.

144. Orchidaceae.

ARALES

Herbs or occasionally woody climbers, rarely floating aquatics. Fls very small, hermaphrodite or unisexual, hypogynous, densely crowded on a spadix or rarely few together, infl. usually ±enclosed in a large bract (spathe). Perianth present and small, or 0. Ovary 1–many-celled; placentation various. Fr. usually a berry. Seeds with endosperm.

145. Araceae. 146. Lemnaceae.

TYPHALES

Rhizomatous marsh or aquatic herbs. Lvs linear, sheathing at base. Fls unisexual, hypogynous, small, densely crowded in spikes or heads. Perianth small, sepaloid, often of scales or threads. Stamens 2 or more. Ovary 1-celled; ovule solitary, pendent. Fr. dry. Seeds with endosperm.

147. Sparganiaceae. 148. Typhaceae.

CYPERALES

Mostly rhizomatous perennial herbs with solid stems Lvs usually linear and sheathing at base, sometimes reduced to sheaths. Fls hermaphrodite or unisexual, hypogynous, small, crowded in heads or spikes and each subtended by a bract. Perianth of scales, bristles or 0. Stamens usually 3; anthers basifixed. Ovary 1-celled; ovule solitary, erect. Fr. dry, indehiscent. Seeds with endosperm.

149. Cyperaceae.

POALES

Annual or more often perennial herbs, rarely woody; stems often hollow. Lvs usually linear and sheathing at base. Fls hermaphrodite or unisexual, hypogynous, small, distichously arranged, usually enclosed between 2 bracts. Perianth 0 or perhaps represented by minute scales. Stamens often 3; anthers versatile. Ovary 1-celled; ovule solitary, often adnate to the side of the carpel. Fr. a caryopsis, rarely a nut or berry. Seeds with endosperm.

150. Gramineae (Poaceae).

Artificial key to families

(For signs and abbreviations see page xxix)

1 Plant reproducing by spores; fls 0; always herbs.　2
　Plant reproducing by seeds; fls with stamens or carpels or both; often woody.　28

2 Stems jointed; lvs not green, forming a sheath at the nodes.　4. EQUISETACEAE
　Stems not jointed; lvs green, not connate into a sheath.　3

3 Plants free-floating on water, much-branched; lvs small imbricate.　17. AZOLLACEAE
　Plants rooted to the ground, terrestrial or aquatic.　4

4 Lvs not differentiated into lamina and petiole.　5
　Lvs with distinct lamina and petiole.　8

5 Lvs forming a basal rosette.　3. ISOETACEAE
　Lvs not forming a basal rosette.　6

6 Lvs filiform, with circinate vernation.　16. MARSILEACEAE
　Lvs lanceolate to ovate, vernation not circinate.　7

7 Stem robust; plant homosporous; lvs not ligulate.　1. LYCOPODIACEAE
　Stem slender; plant heterosporous; lvs ligulate.　2. SELAGINELLACEAE

8 Fertile lvs, or fertile parts of lvs, differing markedly from the sterile lvs or parts of lvs.　9
　Fertile lvs not markedly different from the sterile parts.　12

9 Lf looking like a stem with a fertile upper portion and a sterile lower portion, both of which may be simple or pinnate.　5. OPHIOGLOSSACEAE
　Lvs crowded at the end of a stout stock, the inner fertile sometimes with a few pairs of sterile pinnae at base, the outer sterile.　10

10 Lvs 1-pinnate; pinnae entire. 15. BLECHNACEAE
　Lvs 2- to 4-pinnate.　11

11 Fertile lvs with 2–3 pairs of sterile pinnae at base; growing in damp, ±peaty places.　6. OSMUNDACEAE
　Fertile lvs without sterile pinnae at base.　7. ADIANTACEAE

12 Lvs not more than 1 cell thick (except for midrib), translucent.　8. HYMENOPHYLLACEAE
　Lvs thicker, not translucent.　13

13 Lvs entire, or pinnatifid, or palmately lobed, or dichotomously forked 1–3 times.　14
　Lvs pinnately divided.　16

14 Lvs not pinnatifid.　12. ASPLENIACEAE
　Lvs pinnatifid.　15

15 Lvs covered with scales beneath.　12. ASPLENIACEAE
　Lvs not covered with scales beneath.　9. POLYPODIACEAE

16 Sori covered by the inflexed margin of the lf.　17
　Sori not covered by the inflexed margin of the lf.　18

17 Rhizome long, subterranean; pinnae not fan-shaped. Common.　10. HYPOLEPIDACEAE
　Rhizome short, erect; pinnae fan-shaped.　7. ADIANTACEAE

18 Indusium absent.　19
　Indusium present.　22

19 Pinnae entire.　9. POLYPODIACEAE
　Pinnae divided.　20

20 Lvs forming a crown.　13. ATHYRIACEAE
　Lvs solitary.　21

21 Lf divided into 3 nearly equal portions.　14. ASPIDIACEAE
　Lf pinnately divided.　11. THELYPTERIDACEAE

22 Indusium a ring of hair-like scales surrounding the base of the sorus. Small mountain plants; rare.　13. ATHYRIACEAE
　Indusium not as above.　23

23 Indusium hood-like, attached at lower side of sorus.　13. ATHYRIACEAE
　Indusium not hood-like.　24

24 Indusium peltate.　14. ASPIDIACEAE
　Indusium not peltate.　25

25 Sori orbicular.　26
　Sori ovate or linear.　27

26 Sori marginal; indusium lying along vein.　11. THELYPTERIDACEAE
　Sori not marginal; indusium lying across vein.　14. ASPIDIACEAE

27 Sori ovate; lower margin of indusium bent in the middle.　13. ATHYRIACEAE
　Sori linear or ovate; lower margin of indusium straight.　12. ASPLENIACEAE

28 Ovules naked, either on the upper surface of scales arranged in cones or solitary and terminal on a short scaly axillary shoot; pollen-sacs 2 or more on the lower surface of a flat sporophyll, or several pendent from the apex of a peltate sporophyll, the male sporophylls always in cones; monoecious or dioecious trees or shrubs with small needle-like or scale-like (but green) lvs; perianth 0.　CONIFERAE　29
　Ovules completely enclosed in a carpel; pollen-sacs 4 (or occasionally fewer) surrounding and adnate to a connective at the apex of a usually slender filament.　ANGIOSPERMAE　31

29 Lvs opposite or whorled; short shoots 0.　19. CUPRESSACEAE
　Lvs alternate or in clusters on short lateral shoots.　30

30 Ovules on the surface of scales arranged in cones; pollen-sacs two on the lower surface of a flat sporophyll; trunk usually single.   18. PINACEAE
Ovules solitary and terminal on short axillary shoots; pollen-sacs several on a peltate sporophyll; trunks usually several.   20. TAXACEAE

31 Herbs without chlorophyll, the lvs reduced to scales.   257 (J)
Green plants (if lfless at flowering time either trees or shrubs, or else herbs with only the fls showing above ground).   32

32 Plant free-floating on or below surface of water, not rooted in mud.   33
Land plants or aquatics rooted in mud.   35

33 Plant consisting of a discoid thallus (1–15 mm diam.), with or without roots from the lower surface; propagation mainly vegetative, so that several plants are often found joined together. 146. LEMNACEAE
Plants with obvious stems and lvs.   34

34 Plant with small bladders on lvs, or on apparently lfless stems; lvs divided into filiform segments.   114. LENTIBULARIACEAE
Plant without bladders; lvs sessile, in a rosette, or long-petiolate and orbicular.   128. HYDROCHARITACEAE

35 Small herb with lvs linear and all basal; fls solitary, unisexual, axillary, the male on long stalks, the female sessile (*Littorella*).   118. PLANTAGINACEAE
Not as above.   36

36 Perianth of 2 (rarely more) distinct whorls, differing markedly from each other in shape, size or colour.   37
Perianth 0, or of 1 whorl, or of 2 or more similar whorls, or segments numerous and spirally arranged.   41

37 Petals free (very rarely cohering at apex, free at base).   38
Petals united at least at the base.   40

38 Ovary superior.   39
Ovary inferior or partly so.   98 (C)

39 Carpels and styles free, or carpels slightly united at the extreme base.   43 (A)
Carpels or styles or both obviously united, or ovary of one carpel.   50 (B)

40 Ovary superior.   115 (D)
Ovary inferior.   148 (E)

41 Perianth corolla-like, at least the inner segments usually brightly coloured or white.   162 (F)
Perianth green and calyx-like, or scarious, or 0.   42

42 Trees or shrubs.   185 (G)
Herbs.   204 (H)

GROUP A

Petals free, ovary superior, carpels and styles free or nearly so.

43 Sepals and petals 3.   44
Sepals or petals more than 3.   45

44 Aquatic plants; fls conspicuous; at least the upper lvs broad, flat, stalked; carpels ±numerous.   126. ALISMATACEAE
Small land plants of mossy appearance; fls axillary, inconspicuous; lvs small, oblong, rather fleshy, sessile; carpels 3 (*Crassula*). 61. CRASSULACEAE

45 Stamens numerous.   46
Stamens twice as many as petals or fewer.   48

46 Herbs; stipules 0; fls hypogynous.   47
Herbs with stipules, or else shrubs; fls perigynous (sometimes only slightly so).   59. ROSACEAE

47 Fls c. 10 cm diam.   22. PAEONIACEAE
Fls much smaller.   21. RANUNCULACEAE

48 Lvs ternate, not fleshy; alpine plant (*Sibbaldia*).   59. ROSACEAE
Lvs simple.   49

49 Lvs ±succulent; carpels in 1 whorl.   61. CRASSULACEAE
Lvs not succulent; carpels spirally arranged on a slender elongated receptacle (*Myosurus*).   21. RANUNCULACEAE

GROUP B

Petals free, ovary superior, carpels or styles or both united, or ovary of one carpel.

50 Fls actinomorphic.   51
Fls zygomorphic.   90

51 Stamens more than twice as many as petals (always more than 6), or stamens and petals both numerous.   52
Stamens at most twice as many as petals (never more than 12); or petals 2, stamens 6.   60

52 Aquatic plants with large cordate floating lvs and floating fls; petals more than 10.   24. NYMPHAEACEAE
Plant not aquatic.   53

53 Stamens all united below into a tube; fls pink or purple; lvs usually palmately lobed.   44. MALVACEAE
Stamens free or in bundles; lvs never palmately lobed.   54

54 Lvs very succulent, 3-angled; fls 8–12 cm diam., with numerous narrow petals.   39. AIZOACEAE
Lvs not succulent; petals 5 or fewer.   55

55 Ovary surrounded by a cup-shaped hypanthium; ovule 1.   59. ROSACEAE
No cup-shaped hypanthium; ovules 2 or more.   56

56 Carpel 1; lvs 2-ternate, the lower lflets stalked.   21. RANUNCULACEAE
Carpels 2 or more; lvs not as above.   57

57 Trees; infl. with a conspicuous bract partly adnate to the infl.-stalk.   44. TILIACEAE
Herbs or low shrubs; bracts, if present, not adnate to the infl.-stalk.   58

58 Styles free; stamens united into bundles below.   32. HYPERICACEAE
Style 1 or 0; stigma simple; stamens free.   59

59 Sepals 2; petals 4; lvs toothed to pinnate.
                26. PAPAVERACEAE
   Sepals 5 (3 large, 2 small); petals 5; lvs entire.
                33. CISTACEAE

60 Trees or shrubs.   61
   Herbs.   68

61 Fls on the middle of lf-like cladodes; true lvs scale-
   like, colourless (*Ruscus*).   138. LILIACEAE
   Fls not on cladodes; lvs green.   62

62 Per.-segs in 2 or more whorls of 3; stamens 3 or 6. 63
   Per.-segs not in whorls of 3; stamens not 3 or 6.   64

63 Per.-segs in more than 2 whorls; stamens 6; lvs broad.
                23. BERBERIDACEAE
   Per.-segs in 2 whorls; stamens 3; lvs linear.
                99. EMPETRACEAE

64 Lvs small and scale-like; fls numerous in dense spikes.
                34. TAMARICACEAE
   Lvs not scale-like, not particularly small.   65

65 Lvs opposite.   66
   Lvs alternate.   67

66 Lvs palmately lobed.   50. ACERACEAE
   Lvs simple, not lobed.   54. CELASTRACEAE

67 Plant with rusty tomentum; fls cream; stamens more
   than 5 (*Ledum*).   96. ERICACEAE
   Plant not tomentose; fls greenish; stamens 4–5.
                56. RHAMNACEAE

68 Sepals 2, petals 5.   38. PORTULACACEAE
   Sepals more than 2; sepals and petals equal in
   number.   69

69 Lvs modified into pitchers, 10–20 cm; stigma very
   large, umbrella-like.   69. SARRACENIACEAE
   Lvs not modified into pitchers.   70

70 Sepals and petals normally 6; fls perigynous with a
   long tubular or bell-shaped hypanthium.
                70. LYTHRACEAE
   Sepals and petals normally fewer than 6; fls hypogynous, or if perigynous then with flat to cup-shaped
   hypanthium.   71

71 Lvs opposite or whorled.   72
   Lvs alternate or all basal.   79

72 Lvs compound or lobed.   46. GERANIACEAE
   Lvs entire.   73

73 Lvs in a single whorl of usually 4 on the stem; fl.
   solitary, terminal.   138. LILIACEAE
   Lvs opposite or in numerous whorls.   74

74 Stipules present.   75
   Stipules 0.   76

75 Stipules scarious; land plants.
                37. CARYOPHYLLACEAE
   Stipules not scarious; submerged aquatic plants.
                36. ELATINACEAE

76 Sepals free or united at the base; petals always white. 77
   Sepals united to above the middle; petals white, pink
   or purple.   78

77 Ovary 1-celled with free-central placentation; stamens usually twice as many as petals, if as many
   or fewer then lvs narrowly linear or plant ±hairy
   or sepals scarious-margined.
                37. CARYOPHYLLACEAE
   Ovary 4–5-celled with axile placentation; fertile stamens as many as petals; lvs obovate to oval; plant
   glabrous; sepals not scarious.   45. LINACEAE

78 Style long, simple (but stigmas free); placentation
   parietal; fls 5 mm diam., pink; stamens usually 6.
                35. FRANKENIACEAE
   Styles free; placentation free-central.
                37. CARYOPHYLLACEAE

79 Lvs 3-foliolate with obcordate or cuneiform and emarginate lflets.   47. OXALIDACEAE
   Lvs not 3-foliolate.   80

80 Sepals and petals 2–3; fls greenish or reddish, in many-
   fld terminal panicles.   85. POLYGONACEAE
   Sepals and petals 4–5.   81

81 Both floral whorls green and sepal-like (calyx and
   epicalyx); fls small, with conspicuous concave
   hypanthium; lvs palmate or palmately lobed
   (*Alchemilla* and *Aphanes*).   59. ROSACEAE
   Petals ±brightly coloured, never sepal-like.   82

82 Sepals and petals 4; stamens 6, rarely 4.
                28. CRUCIFERAE
   Sepals and petals 5; stamens 5 or 10.   83

83 Lvs covered with conspicuous red insectivorous glandular hairs.   68. DROSERACEAE
   Lvs not conspicuously glandular.   84

84 Style 1, stigma simple or shallowly lobed; anthers
   opening by pores.   97. PYROLACEAE
   Styles, or at least the stigmas, more than 1, free;
   anthers opening by slits.   85

85 Stigmas 5; petals blue, pink or purple, rarely white. 86
   Stigmas 2–4; petals white or yellow.   88

86 Lvs lobed or pinnate.   46. GERANIACEAE
   Lvs entire.   87

87 Calyx funnel-shaped or obconic, scarious; lvs all
   ±basal; fls in heads or panicles.
                101. PLUMBAGINACEAE
   Sepals free, not scarious or scarious only at the margins; stem lfy; fls in loose cymes.   45. LINACEAE

88 Fls with conspicuous glandular-fimbriate staminodes;
   lvs ovate, cordate, entire.
                63. PARNASSIACEAE
   Staminodes 0; lvs not as above.   89

89 Stamens 5; procumbent plant; lvs entire, linear-lanceolate; stipules scarious; fls very small (*Corrigiola*).   37. CARYOPHYLLACEAE
   Stamens 10; fls conspicuous; other characters not as
   above.   62. SAXIFRAGACEAE

90 Fls saccate or spurred at base.   91
   Fls not saccate or spurred.   93

91 Lvs much divided; corolla (apparently) laterally compressed; stamens 2, each with 3 branches bearing
   anthers, not connivent.   27. FUMARIACEAE
   Lvs simple; corolla not compressed; stamens 5, connivent round the style.   92

92 Sepals 5, ±equal, not spurred; petals 5, one spurred; stipules present; fls solitary, axillary; stem not translucent. 30. VIOLACEAE

Sepals 3, very unequal, one spurred; petals 3, not spurred; stipules 0; fls in few-fld infls; stem ±translucent. 48. BALSAMINACEAE

93 Stamens 8 or more all, or all but 1, united into a long tube; fls very zygomorphic, the petals ±erect. 94

Stamens free; fls less zygomorphic, petals spreading. 95

94 Fl. with upper sepal; anthers opening by pores; stigma tufted. 31. POLYGALACEAE

Fl. with upper petal; anthers opening by slits; stigma not tufted. 58. LEGUMINOSAE

95 Trees; lvs palmate. 52. HIPPOCASTANACEAE

Herbs; lvs not palmate. 96

96 Fls in cymes (often umbel-like); ovary 5-lobed with long beak. 46. GERANIACEAE

Fls in racemes; ovary not lobed or 2-lobed, rarely beaked. 97

97 Petals fimbriate or lobed; stamens more than 6. 29. RESEDACEAE

Petals entire or emarginate; stamens 6. 28. CRUCIFERAE

GROUP C
Petals free, ovary inferior or partly so.

98 Petals numerous. 99

Petals 5 or fewer. 100

99 Aquatic plants with floating fls and lvs. 24. NYMPHAEACEAE

Land plants with very succulent lvs. 39. AIZOACEAE

100 Petals and sepals 3. 101

Petals and sepals 2, 4 or 5. 104

101 Fls zygomorphic. 144. ORCHIDACEAE

Fls actinomorphic. 102

102 Both whorls of per.-segs petaloid. 103

Outer or both whorls of per.-segs sepaloid. 128. HYDROCHARITACEAE

103 Stamens 6. 141. AMARYLLIDACEAE

Stamens 3. 142. IRIDACEAE

104 Stamens numerous. 59. ROSACEAE

Stamens 10 or fewer. 105

105 Submerged aquatic with lvs pinnately divided into filiform segments; fls monoecious or polygamous, in terminal spikes projecting above water-surface. 74. HALORAGIDACEAE

Land plants, or, if aquatic, then fls hermaphrodite and in umbels. 106

106 Trees or shrubs. 107

Herbs. 111

107 Woody climber; fls in subglobose umbels, green. 80. ARALIACEAE

Not climbing; fls not in umbels. 108

108 Lvs palmately lobed; petals shorter than sepals. 66. GROSSULARIACEAE

Lvs simple not lobed. 109

109 Both perianth-whorls petaloid; hypanthium long and tubular (*Fuchsia*). 73. ONAGRACEAE

Outer perianth-whorl sepaloid. 110

110 Calyx-teeth very small; fls in corymbs; carpels 2, each with one ovule. 79. CORNACEAE

Calyx-teeth large; fls not in corymbs; ovules numerous in each carpel. 65. ESCALLONIACEAE

111 Both perianth-whorls green and sepaloid (calyx and epicalyx), or with an epicalyx as well as sepals and petals, or with a crown of long spines on the receptacle below the calyx; carpels 1 or 2, free from the receptacle and thus not truly inferior. 59. ROSACEAE

Inner perianth-whorl always petaloid, no epicalyx or crown of spines; ovary truly inferior. 112

112 Petals 5; styles normally 2, rarely 3. 113

Petals 4 or 2; style simple. 114

113 Fls in heads or umbels; stamens 5; ovules 1 in each carpel. 81. UMBELLIFERAE

Fls not in heads or umbels; stamens 10; ovules numerous. 62. SAXIFRAGACEAE

114 Fls deep purple, in umbels subtended by 4 conspicuous white petaloid involucral bracts. 79. CORNACEAE

Fls not in umbels; no petaloid involucral bracts. 73. ONAGRACEAE

GROUP D
Petals united, ovary superior.

115 Stamens more than 10; outer per.-segs longer than inner (*Consolida*). 21. RANUNCULACEAE

Stamens 10 or fewer. 116

116 Stamens united into a tube, or 9 united, 1 free. 117

Stamens all free. 118

117 Lvs simple; fl. with upper sepal; stamens 8. 31. POLYGALACEAE

Lvs 3-foliolate; fl. with upper petal; stamens 10. 58. LEGUMINOSAE

118 Stamens twice as many as corolla-lobes (i.e. 8–10). 119

Stamens as many as or fewer than corolla-lobes (i.e. 5 or fewer). 120

119 Shrubs or trees; lvs not peltate; carpels united. 96. ERICACEAE

Succulent herb; lvs peltate; carpels free (*Umbilicus*). 61. CRASSULACEAE

120 Sepals 2; fls actinomorphic. 121

Sepals more than 2 or fls zygomorphic (sometimes 2 conspicuous sepal-like bracts occur outside the calyx). 122

121 Petals 2; fls in heads; lvs linear, terete. 137. ERIOCAULACEAE

Petals 5; fls not in heads; lvs flat. 38. PORTULACACEAE

122 Ovary deeply 4-lobed with 1 ovule in each lobe. 123

Ovary not 4-lobed. 124

123 Lvs spirally arranged. 109. BORAGINACEAE

Lvs opposite. 117. LABIATAE

124 Trees or erect shrubs.   125
    Herbs or creeping or cushion-like dwarf shrubs.   128

125 Lvs opposite.   126
    Lvs alternate.   127

126 Stamens 2.   104. OLEACEAE
    Stamens 4.   103. BUDDLEJACEAE

127 Lvs usually spiny; fls actinomorphic; anthers opening
      by slits.   53. AQUIFOLIACEAE
    Lvs never spiny; fls zygomorphic; anthers opening by
      pores.   96. ERICACEAE

128 Stamens opposite the corolla-lobes.   129
    Stamens alternating with the corolla-lobes.   130

129 Style 1; stigma 1.   102. PRIMULACEAE
    Styles or stigmas more than 1.
                            101. PLUMBAGINACEAE

130 Lvs opposite.   131
    Lvs alternate or all basal.   136

131 Carpels 2, free; style expanded into a ring below the
      stigma; trailing evergreen plants.
                            105. APOCYNACEAE
    Carpels united; style not expanded into a ring below
      the stigma.   132

132 Cushion-like or creeping shrubs (high mountains).   133
    Herbs.   134

133 Creeping; lvs elliptical or oblong; fls pink (Loise-
      leuria).   96. ERICACEAE
    Cushion-like; lvs spathulate; fls white.
                            100. DIAPENSIACEAE

134 Flowers zygomorphic.
                            112. SCROPHULARIACEAE
    Flowers actinomorphic.   135

135 Land plants; lvs sessile.   106. GENTIANACEAE
    Aquatic plants with floating lvs on long petioles (Nym-
      phoides).   107. MENYANTHACEAE

136 Calyx- and corolla-lobes 4(–5); stamens 4 or 2.   137
    Calyx- and corolla-lobes and stamens 5.   143

137 Stamens 2; lvs and bracts not spine-toothed.   138
    Stamens 4.   139

138 Ovary 1-celled; corolla spurred; carnivorous bog or
      aquatic plants with lvs all basal or else divided into
      filiform segments. 114. LENTIBULARIACEAE
    Ovary 2-celled; corolla not spurred; lvs not as above.
                            112. SCROPHULARIACEAE

139 Lvs all basal.   140
    Lvs not all basal.   141

140 Corolla scarious; stamens exserted.
                            118. PLANTAGINACEAE
    Corolla not scarious; stamens included.
                            112. SCROPHULARIACEAE

141 Bracts spine-toothed; corolla 1-lipped.
                            115. ACANTHACEAE
    Bracts not spine-toothed; corolla weakly zygomor-
      phic or 2-lipped.   142

142 Ovules numerous.   112. SCROPHULARIACEAE
    Ovules 4.   116. VERBENACEAE

143 Ovary 3-celled; stigmas 3, or if only 1 then 3-lobed.   144
    Ovary 2-celled; stigmas 2 or 1, not 3-lobed.   145

144 Erect herb; lvs pinnate. 108. POLEMONIACEAE
    Cushion-like; lvs spathulate; fls white.
                            100. DIAPENSIACEAE

145 Ovules 4 or fewer; twining or prostrate herbs; lvs cor-
      date or hastate; corolla shallowly lobed.
                            110. CONVOLVULACEAE
    Ovules numerous; ±erect herbs or woody climbers;
      corolla-lobes conspicuous.   146

146 Aquatic or bog plants; lvs orbicular or ternate; corolla
      fringed.   107. MENYANTHACEAE
    Land plants; lvs neither orbicular nor all ternate (but
      some may be ternate in a woody climber); corolla
      not fringed.   147

147 Fls numerous, in terminal spikes or racemes (some-
      times aggregated into panicles); corolla-tube very
      short; stamens spreading (Verbascum).
                            112. SCROPHULARIACEAE
    Fls solitary or in cymes (sometimes scorpioid); cor-
      olla-tube long, or, if short, then anthers connivent.
                            111. SOLANACEAE

GROUP E

Petals united, ovary inferior.

148 Stamens 8–10, or 4–5 with filaments divided to the
      base.   149
    Stamens 5 or fewer, filaments not divided.   150

149 Herb; fls in heads, green; lvs ternate.
                            122. ADOXACEAE
    Low shrubs or prostrate creeping dwarf shrubs; fls
      pink or white, not in heads; lvs simple.
                            96. ERICACEAE

150 Fls in heads surrounded by an involucre; herbs (rarely
      slightly woody).   151
    Fls not in heads, or if in heads then with 2 bracts
      only and plant a woody climber.   154

151 Anthers coherent into a tube round the style.   152
    Anthers free.   153

152 Ovules numerous; calyx-lobes conspicuous, green; fls
      blue (Jasione).   119. CAMPANULACEAE
    Ovule 1; calyx represented by hairs or scales; fls rarely
      blue.   125. COMPOSITAE

153 Ovules numerous; corolla-lobes long and narrow,
      longer than tube.   119. CAMPANULACEAE
    Ovule 1; corolla-lobes shorter than tube.
                            124. DIPSACACEAE

154 Lvs in whorls; fls actinomorphic; petals 4.
                            120. RUBIACEAE
    Lvs not in whorls; fls zygomorphic, or if not then
      petals 5.   155

155 Fls zygomorphic.   156
    Fls actinomorphic.   158

156 Fls in corymbs.   123. VALERIANACEAE
    Fls in terminal racemes or spikes.   157

157 Anthers coherent into a tube round the style; pollen
      powdery.   119. CAMPANULACEAE
    Anthers 2, free; pollen cohering in pollinia.
                            144. ORCHIDACEAE

158 Herb, climbing by tendrils. 82. CUCURBITACEAE
     Herbs, shrubs or woody climbers; tendrils 0.   *159*

159 Lvs opposite.   *160*
     Lvs spirally arranged.   *161*

160 Stamens 4 or 5; usually shrubs or woody climbers;
     if herbs either prostrate and creeping or with lf-like
     stipules.   121. CAPRIFOLIACEAE
     Stamens 1–3; herbs, ±erect and without lf-like stip-
     ules.   123. VALERIANACEAE

161 Stamens opposite corolla-lobes; stigmas capitate; fls
     white (*Samolus*).   102. PRIMULACEAE
     Stamens alternating with corolla-lobes; stigmas 2–5;
     fls normally blue or purple.
          119. CAMPANULACEAE

GROUP F
Perianth entirely petaloid or in several series,
the inner petaloid.

162 Stamens numerous.   *163*
     Stamens 12 or fewer, or fls female.   *166*

163 Aquatic plants with floating lvs and fls.
          24. NYMPHAEACEAE
     Terrestrial plants.   *164*

164 Succulent prostrate plant with 3-angled lvs.
          39. AIZOACEAE
     Lvs not 3-angled.   *165*

165 Carpels free, rarely united and then per.-segs numer-
     ous.   21. RANUNCULACEAE
     Carpels united; petals usually 4; sepals 2, falling as
     fl. opens.   26. PAPAVERACEAE

166 Fls crimson, in ovoid heads without an involucre; lvs
     pinnate (*Sanguisorba*).   59. ROSACEAE
     Fls not in heads, or if so then with an involucre.   *167*

167 Ovary superior.   *168*
     Ovary inferior or fls male.   *176*

168 Perianth strongly zygomorphic, spurred or saccate at
     base; stamens 2, each with 3 anther-bearing
     branches; lvs much divided (sepals 2, but bract-like
     and soon falling).   27. FUMARIACEAE
     Perianth actinomorphic or slightly zygomorphic, and
     then neither spurred nor saccate.   *169*

169 Shrubs.   *170*
     Herbs.   *173*

170 Fls borne on the surface of lf-like cladodes; true lvs
     small and scale-like (*Ruscus*).   138. LILIACEAE
     Fls not on cladodes.   *171*

171 Per.-segs 4, continued below into a coloured hypan-
     thium.   71. THYMELAEACEAE
     Per.-segs 6 or more, free.   *172*

172 Low heath-like shrubs with inconspicuous axillary fls
     (if per.-segs 8, pink-purple, in 2 differing whorls,
     see *Calluna* in Ericaceae, p. 331).
          99. EMPETRACEAE
     Tall shrubs with yellow fls in racemes or panicles.
          23. BERBERIDACEAE

173 Per.-segs 5.   *174*
     Per.-segs 6, rarely 4.   *175*

174 Stigma 1, capitate; stipules 0 (*Glaux*).
          102. PRIMULACEAE
     Stigmas 2–3; stipules sheathing, scarious.
          85. POLYGONACEAE

175 Stamens 8(–9); ovules scattered over whole inner sur-
     face of carpels; aquatic plant.
          127. BUTOMACEAE
     Stamens 6, rarely 4; ovules on axile placentae; plants
     not aquatic.   138. LILIACEAE

176 Trees or shrubs; calyx present but very small and rim-
     like or with minute teeth. See *148* (Group E).
     Herbs.   *177*

177 Lvs in whorls of 4 or more.   120. RUBIACEAE
     Lvs not in whorls.   *178*

178 Fls in heads surrounded by a common involucre.   *179*
     Fls not in heads though sometimes shortly stalked
     in compact umbels.   *180*

179 Stamens free; fls hermaphrodite.
          124. DIPSACACEAE
     Anthers cohering in a tube round the style, or fls
     unisexual.   125. COMPOSITAE

180 Per.-segs 3, or perianth with a long tube swollen below
     and a unilateral entire limb; lvs ±orbicular, cor-
     date, entire.   83. ARISTOLOCHIACEAE
     Per.-segs 5 or 6; lvs not as above.   *181*

181 Per.-segs 5; fls small; ovules 1 or 2.   *182*
     Per.-segs 6; fls large, ovules numerous.   *184*

182 Fls in simple cymes; lvs spirally arranged, narrowly
     linear, small.   78. SANTALACEAE
     Fls in umbels or superposed whorls, or if in cymes
     then lvs opposite.   *183*

183 Stamens 5; per.-segs free; fls in umbels or superposed
     whorls; lvs spirally arranged.
          81. UMBELLIFERAE
     Stamens 1–3; per.-segs united; fls in cymes or pani-
     cles; lvs opposite.   123. VALERIANACEAE

184 Stamens 6.   141. AMARYLLIDACEAE
     Stamens 3.   142. IRIDACEAE

GROUP G
Trees or shrubs; perianth sepaloid or 0.

185 Parasitic on the branches of trees; lvs opposite, obo-
     vate or oblong, thick, leathery; stems green.
          77. LORANTHACEAE
     Not as above.   *186*

186 Root-climber; fls in umbels.   80. ARALIACEAE
     Not climbing; fls not in umbels.   *187*

187 Fls borne on the surface of flattened evergreen lf-like
     cladodes; true lvs colourless, scale-like (*Ruscus*).
          138. LILIACEAE
     Fls not on cladodes; lvs green.   *188*

188 Lvs opposite or subopposite.   *189*
     Lvs spirally arranged or in 2 ranks (alternate).   *193*

189 Lvs evergreen, thick, leathery, entire; styles 3.
          55. BUXACEAE
     Lvs deciduous; styles 4, 2 or 1.   *190*

190 Fls in catkins.   95. SALICACEAE
    Fls not in catkins.   191

191 Lvs pinnate; perianth 0; stamens 2 (*Fraxinus*).
                104. OLEACEAE
    Lvs simple; perianth present; stamens 4 or more.   192

192 Lvs palmately lobed.   50. ACERACEAE
    Lvs simple, not lobed.   56. RHAMNACEAE

193 Lvs evergreen, less than 10×2 mm, dense, oblong
    or linear, entire; shrubs to 1 m or less.   194
    Lvs relatively longer or broader, not particularly
    dense, usually deciduous and if evergreen then
    30 mm, or more.   195

194 Procumbent; stamens 3; stigmas 6–9; lvs leathery;
    moors, etc..   99. EMPETRACEAE
    Erect; stamens 5; stigmas 2; lvs fleshy; maritime
    (*Suaeda*).   41. CHENOPODIACEAE

195 Lvs pinnate (present at flowering time).
              90. JUGLANDACEAE
    Lvs simple (sometimes 0 at flowering time).   196

196 Fls, at least in the male, in catkins or in tassel-like
    heads on long pendent stalks.   197
    Fls not in catkins or stalked heads.   202

197 Dioecious; perianth 0; fls always solitary in the axil
    of each bract.   198
    Monoecious, though sexes usually in separate infls;
    perianth present at least in the fls of one or other
    sex.   200

198 Scales of catkins fimbriate or lobed at the tip; fls of
    both sexes with a cup-like disk; ovules numerous
    (*Populus*).   SALICACEAE
    Scales of catkins entire; disk 0.   199

199 Ovules numerous; lvs without resin glands, not aro-
    matic when crushed; fls of both sexes without brac-
    teoles but with nectaries at the base, placed above
    or below the fl.; stamens with long filaments (*Salix*).
              95. SALICACEAE
    Ovule 1; lvs dotted with resin glands, strongly aroma-
    tic when crushed; male fl. without nectaries or brac-
    teoles, female fl. with 2 lateral bracteoles; filaments
    short.   91. MYRICACEAE

200 Fls of both sexes with perianth; styles 3 or more; fr.
    large and nut-like, partly or completely enclosed
    in a hard cup or shell.   94. FAGACEAE
    Perianth present in one sex only; styles 2; fr. small,
    or large and nut-like; cup if present papery or lf-
    like.   201

201 Male fls 3 to each bract; perianth present; fr. small,
    in the axils of the accrescent bracts which persist
    till maturity and form cone-like structures.
              92. BETULACEAE
    Male fls solitary in the axil of each bract; perianth
    0; fr. not borne in cones, surrounded by a papery
    or lf-like cup formed from the bracts.
              93. CORYLACEAE

202 Lvs and twigs densely covered with silvery or brown
    peltate scales; dioecious; fls very small, male with
    2 free per.-segs; female with tubular perianth hav-
    ing 2 small lobes at its apex.
            72. ELAEAGNACEAE

Plant without peltate scales; fls hermaphrodite; per.-
    segs 4 or more.   204

203 Deciduous trees; fls in sessile clusters, appearing
    before the lvs; perianth ±bell-shaped, the stamens
    inserted at its base; styles 2.   88. ULMACEAE
    Evergreen shrub; fls in short-stalked racemes; per-
    ianth continued downwards into a long, cylindrical
    tube, the stamens inserted high on the tube; style
    1.   71. THYMELAEACEAE

GROUP H
Herbs, perianth sepaloid or 0.

204 Perianth 0 or represented by scales or bristles, minute
    in fl. but sometimes elongating in fr.; the fls in the
    axils of specialized chaffy bracts which are usually
    arranged along the rhachis of spikelets, sometimes
    themselves aggregated into compound infls; lvs
    always ±linear and grass-like, sheathing below.   205
    Perianth present, or if minute or absent then fls not
    arranged in spikelets nor the bracts chaffy; lvs var-
    ious.   206

205 Fls with bract above and below; lvs ±jointed at the
    junction with the sheath, commonly with a promi-
    nent projecting ligule; sheaths usually open; stems
    terete or flattened, usually with hollow internodes.
             150. GRAMINEAE
    Fls with a bract below only; lvs not jointed at the
    junction with the sheath; ligule, if present, not pro-
    jecting; sheaths usually closed; stem often 3-angled;
    internodes nearly always solid.
             149. CYPERACEAE

206 Aquatic plants; lvs submerged or floating; infl. some-
    times rising above the surface of the water.   207
    Land plants, or if aquatic then with stiffly erect stems
    and with lvs as well as fls rising above the surface
    of the water.   222

207 Lvs divided into numerous filiform segments.   208
    Lvs entire or toothed.   209

208 Lvs pinnately divided; fls in a terminal spike (bracts
    sometimes lf-like).   74. HALORAGIDACEAE
    Lvs dichotomously divided; fls solitary, axillary.
            25. CERATOPHYLLACEAE

209 Fls in a spike surrounded by a petaloid spathe (*Calla*).
             145. ARACEAE
    Without petaloid bracts or spathe.   210

210 Fls sessile or nearly so, arranged in heads.   211
    Fls in spikes or in the axils of the lvs.   213

211 Heads with many small fls, solitary at the ends of
    the lfless stalk.   137. ERIOCAULACEAE
    Heads few-fld and terminal, or lateral on lfy stems.   212

212 Fls unisexual, the male heads above, the female heads
    below.   147. SPARGANIACEAE
    Fls hermaphrodite.   140. JUNCACEAE

213 Fls in spikes.   216
    Fls axillary, solitary or in few-fld clusters.   214

214 Fls unisexual, arranged on one side of a flattened
spadix; perianth 0; marine.
132. ZOSTERACEAE
Fls hermaphrodite, arranged all round or on two sides
of a terete rhachis; fresh or brackish water but not
truly marine. 215

215 Per.-segs 4; carpels remaining sessile; usually fresh-
water. 133. POTAMOGETONACEAE
Perianth 0; fruiting carpels on long stalks; brackish
pools and ditches. 134. RUPPIACEAE

216 Female fls with very long filiform perianth-tube,
resembling a pedicel and raising them to the surface
of the water. 128. HYDROCHARITACEAE
Tube and pedicel short or 0. 217

217 Carpels 2–6, free; lvs narrowly linear; quite entire,
not whorled. 135. ZANNICHELLIACEAE
Carpels united or 1 only; lvs broader, or if narrowly
linear then finely toothed or whorled. 218

218 Perianth with 4–6 segments; stamens 4 or more. 219
Perianth 0, or entire, or with 2 segments; stamen 1. 220

219 Per.-segs 4; ovary inferior; lvs ovate (*Ludwigia*).
73. ONAGRACEAE
Per.-segs 6; ovary superior; lvs obovate.
70. LYTHRACEAE

220 Lvs in whorls of 8 or more; fls hermaphrodite; style
1. 75. HIPPURIDACEAE
Lvs opposite or in whorls of 3; fls unisexual; styles
2–3. 221

221 Lvs narrowly linear with sheathing base, finely (or
minutely) spiny-toothed, the apex acute; ovary terete, not lobed. 136. NAJADACEAE
Lvs (at least the upper) usually spathulate; if all linear,
then entire and with an emarginate apex; base not
sheathing; ovary flattened, 4-lobed.
76. CALLITRICHACEAE

222 Twining plants; fls unisexual. 223
Not climbing or, if climbing, fls hermaphrodite. 224

223 Lvs opposite, palmately lobed; per.-segs 5.
87. CANNABACEAE
Lvs spirally arranged, cordate, entire; per.-segs 6.
143. DIOSCOREACEAE

224 Lvs linear, ±grass-, rush- or iris-like; plants of wet
places. 225
Lvs not linear or, if so, small and not at all grass-like. 230

225 Fls unisexual, the male and female in separate infls
or in parts of the same infl. 226
Fls hermaphrodite. 227

226 Fls in globose heads, the male and female in separate
heads. 147. SPARGANIACEAE
Fls in dense cylindrical spikes, male above and female
below. 148. TYPHACEAE

227 Fls in dense spikes borne laterally on a flattened lf-like
stem (*Acorus*). 145. ARACEAE
Infl. not as above. 228

228 Carpels united only at extreme base; fls in racemes.
129. SCHEUCHZERIACEAE
Carpels ±completely united. 229

229 Fls in spikes; perianth herbaceous.
130. JUNCAGINACEAE
Fls not in spikes or racemes; perianth scarious.
140. JUNCACEAE

230 Lvs compound. 231
Lvs simple or 0. 234

231 Fls in heads. 232
Fls not in heads. 233

232 Lvs simply pinnate; style 1 (rarely 2), stamens 4 or
numerous. 59. ROSACEAE
Lvs ternate (sometimes 2 or 3 times); styles 3–5; stamens apparently 8–10 (4 or 5 with filaments divided
to base). 122. ADOXACEAE

233 Stamens numerous; no epicalyx.
21. RANUNCULACEAE
Stamens 4 or 5 (rarely 10); epicalyx present.
59. ROSACEAE

234 Infl. umbellate, consisting of several male fls (each
of 1 stamen) and one female fl. (appearing as a
stalked ovary) all surrounded by 4 or 5 crescent-
shaped or roundish glands; juice milky (*Euphor-
bia*). 84. EUPHORBIACEAE
Infl. not as above; juice not milky. 235

235 Infl. a dense spike with female fls below and male
fls above; lvs hastate (*Arum*). 145. ARACEAE
Infl. not as above; lvs not hastate. 236

236 Lvs 0; stem green and succulent, jointed; perianth
flush with the stem; salt-marsh plants (*Salicornia*).
41. CHENOPODIACEAE
Lvs obvious, green; stems not succulent. 237

237 Lvs spirally arranged or all basal (rarely the lower
opposite). 238
Lvs all opposite or whorled. 248

238 Stamens 12 or more. 239
Stamens 8 or fewer. 240

239 Per.-segs 5, with a whorl of honey-lvs within; lvs pal-
mately lobed (*Helleborus*). 21. RANUNCULACEAE
Per.-segs 3, without honey-lvs; lvs reniform, entire
(*Asarum*). 83. ARISTOLOCHIACEAE

240 Stipules ±scarious, united into a sheath.
85. POLYGONACEAE
Stipules free of 0. 241

241 Lvs large and rhubarb-like, all basal; fls in dense,
many-fld spikes from the base, much shorter than
the lvs (*Gunnera*). 74. HALORAGIDACEAE
Lvs not rhubarb-like; fls not in basal spikes. 242

242 Stamens twice as many as per.-segs; lvs reniform, cordate (*Chrysosplenium*). 62. SAXIFRAGACEAE
Stamens as many as per.-segs or fewer; lvs neither
reniform nor cordate. 243

243 Stipules lf-like; perianth of 4 segments with an epi-
calyx of 4 segments outside; lvs palmately lobed
(*Aphanes* and *Alchemilla*). 59. ROSACEAE
Stipules very small or 0; perianth without epicalyx. 244

244 Ovary inferior. 78. SANTALACEAE
Ovary superior. 245

245 Fls in simple ebracteate racemes (*Lepidium*).
                                    28. CRUCIFERAE
    Fls not in simple, ebracteate racemes.   246

246 Styles 2 or more, free or united below; stigmas simple;
        fls mostly 5-merous.   247
    Style 1; stigma feathery, tufted; fls 4-merous (*Parie-
        taria*).   86. URTICACEAE

247 Perianth herbaceous.   41. CHENOPODIACEAE
    Perianth scarious.   40. AMARANTHACEAE

248 Lvs toothed or lobed.   249
    Lvs entire.   252

249 Fls hermaphrodite; stems creeping or decumbent.   250
    Fls unisexual; aerial stems erect.   251

250 Ovary inferior, not lobed; styles 2; fls in dichotomous
        cymes(*Chrysosplenium*).
                                62. SAXIFRAGACEAE
    Ovary superior, 5-lobed, prolonged into a long beak
        bearing 5 stigmas; fls solitary or very few on long
        axillary peduncles (*Erodium*).
                                46. GERANIACEAE

251 Plant with stinging hairs; per.-segs 4 or 2; stamens
        4; style 1; stigmas feathery (*Urtica*).
                                86. URTICACEAE
    Plant without stinging hairs; per.-segs 3; stamens 9
        or more; styles 2, simple (*Mercurialis*).
                                84. EUPHORBIACEAE

252 Perianth 0 or obscurely 2-lobed or of 2–3 segments.   253
    Perianth of 4 or more segments.   255

253 Per.-segs 3; stamens 3 (*Koenigia*).
                                85. POLYGONACEAE
    Perianth 0 or of fewer than 3 segments; stamen 1
        (plants ±aquatic).   254

254 Lvs whorled; fls hermaphrodite; style 1.
                                75. HIPPURIDACEAE
    Lvs opposite; fls monoecious; styles 2.
                                76. CALLITRICHACEAE

255 Ovary inferior; style 1; per.-segs 4 (*Ludwigia*).
                                73. ONAGRACEAE
    Ovary superior.   256

256 Per.-segs 6 or 12, inserted on a bell-shaped hypan-
        thium; style 1; plant ±aquatic.
                                70. LYTHRACEAE
    Per.-segs 4 or 5, usually free (if on a bell-shaped
        hypanthium, then lvs linear); styles 2 or more, free;
        land-plants.   37. CARYOPHYLLACEAE

GROUP J
    Herbs without chlorophyll; lvs scale-like.

257 Fls zygomorphic.   258
    Fls actinomorphic.   259

258 Per.-segs free.   144. ORCHIDACEAE
    Per.-segs united into a tubular corolla.
                                113. OROBANCHACEAE

259 Erect saprophyte.   98. MONOTROPACEAE
    Twining parasites (*Cuscuta*).
                                CONVOLVULACEAE

Signs and abbreviations

agg.	aggregate, incl. 2 or more spp. which resemble each other closely.
C.	central.
c.	about (circa).
Ch.	Chamaephyte; see p. 655.
f.	forma, *filius*.
ff.	fragments (of chromosomes).
fl.	flower, flowering time; plural fls.
-fld	-flowered.
fr.	fruit, fruiting.
G.	Geophyte; see p. 655.
Germ.	time of germination.
H.	Hemicryptophyte; see p. 655.
Hel.	Helophyte; see p. 655.
Hyd.	Hydrophyte; see p. 655
incl.	including.
infl.	inflorescence, inflorescences.
lf	leaf; plural lvs.
lfless	leafless.
lflet	leaflet.
lfy	leafy.
M.	Microphanerophyte; see p. 655.
MM.	Mega- or Mesophanerophyte; see p. 655.
N.	Nanophanerophyte; see p. 655.
0	absent.
per.-seg.	perianth segment.
p.p.	*pro parte*.
Rep. B.E.C.	See Bibliography under *The Botanical Society and Exchange Club of the British Isles*.
sp.	species; plural spp.
subsp.	subspecies; plural subspp.

Th.	Therophyte; see p. 655.
var.	variety.
×	Preceding the name of a genus or sp. indicates a hybrid.
±	more or less.
*	Preceding the name of a sp. or a genus indicates that it is certainly introduced.
$2n$	The diploid chromosome number; when the number is followed by an asterisk it indicates that it refers to British material.
μm	1/1000 mm (micrometre).

Measurements without qualification (e.g. lvs 4–7 cm) refer to lengths; lvs 4–7 × 1–2 cm means lvs 4–7 cm long and 1–2 cm wide. Measurements or numbers enclosed in brackets (e.g. lvs 4–7(–10) cm) are exceptional ones outside the normal range. For distributions within Great Britain the counties have the boundaries and names in use prior to the local government reorganization of 1974. This allows ready incorporation of, and reference to, the Watsonian vice-counties long used for indicating the distributions of plants in the British Is. For further information see Dandy, J. E., *Watsonian vice-counties of Great Britain* (London: Ray Society, 1969). In the body of the book families are printed in bold capitals, genera and subgenera with a large initial capital followed by small capitals, and species in bold upper and lower case. Passages in italics indicate distinguishing features of the plant being described.

PTERIDOPHYTA

PTEROPSIDA
1. LYCOPODIACEAE

Herbs. Lvs small. Sporangia all alike, borne near the base of the upper surface of sporophylls, which vary from being like the foliage lvs and arranged among them to being strongly differentiated from the foliage lvs and arranged in terminal spikes. Sporangia unilocular, compressed, dehiscing by a split; spores numerous. Prothallus usually subterranean, saprophytic and with mycorrhiza; antheridia in the centre of the apical part of the prothallus, containing numerous biciliate spermatozoids; archegonia in a ring round the antheridia.

Six to eight or more genera. Temperate and tropical regions.

1 Stems ascending, divided dichotomously into branches of equal length; sporangia axillary.
　　　　　　　　　　　　　　3. HUPERZIA
　Stems creeping and rooted, with short lateral branches; sporangia in terminal spikes. 2
2 Lvs 4-ranked (opposite and decussate); branches dorsiventral. 4. DIPHASIASTRUM
　Lvs alternate or in whorls; branches radial. 3
3 Lvs subulate, curved upwards, sporophylls like lvs but rather wider and toothed at base.
　　　　　　　　　　　　2. LYCOPODIELLA
　Lvs flat, lanceolate, appressed or deflexed; sporophylls unlike lvs, ovate to broadly lanceolate, the margin scarious, toothed. 1. LYCOPODIUM

1. LYCOPODIUM L.

Stems creeping and rooted, with short lateral branches. Lvs arranged spirally or in whorls, flat, linear to lanceolate, appressed or deflexed. Sporophylls unlike lvs, ovate to broadly lanceolate, with scarious, toothed margin, forming terminal spikes. Prothallus without chlorophyll, disciform or tuberous, without appendages. Over 100 spp., cosmopolitan.

Spikes sessile; sporophylls acute; lvs acuminate.
　　　　　　　　　　　　　1. annotinum
Spikes pedunculate; sporophylls with long, white apical hair; lvs acute, with long, white, apical hair. 2. clavatum

1. L. annotinum L.　　　　Interrupted Clubmoss.
Stems 30–60 cm, *moderately branched* with many of the branches ascending. *Lvs* 4–6 mm, *spirally arranged*, denser on the branches than on the main stem, ±spreading, linear-lanceolate, acuminate, *ending in a short stiff point*, minutely serrulate or entire, dull green. Fertile branches 10–25 cm. Spikes 1.5–3 cm, *sessile*, solitary; *sporophylls ovate*, *acuminate*, *with wide scarious denticulate margin*. Spores ripe 6–8. $2n = $ c. 68*. Chh.

Native. Moors on mountains from 46 to 820 m, local; Cumberland, Mull, Argyll, Perth and Kincardine northwards, but not in the islands; extinct in Orkney and Caernarvon. Arctic and north temperate zone extending southwards in the mountains to the Pyrenees, N. Apennines and S. Carpathians; Caucasus, Himalaya, Oregon, Colorado and Maryland.

2. L. clavatum L.　　　　Stag's-Horn Clubmoss.
Stems 30–100 cm, much-branched, the branches all (except the fertile ones) procumbent. *Lvs* 3–5 mm, spirally arranged, dense, somewhat appressed or incurved, linear, acuminate, *with a long white apical hair*, minutely serrulate, rather bright green. Fertile branches 10–25 cm. Spikes 1–2(–3), 2–5 cm, *on long peduncles*; peduncles with distant, pale, appressed, linear-subulate, scale-like lvs; sporophylls ovate with *long, white apical hair*, with scarious denticulate margin. Spores ripe 6–9, $2n = 68$*. Chh.

Native. Heaths, moors and mountain grassland ascending to 840 m, common in mountain districts, rare and decreasing in lowland areas. Throughout the British Is., but absent or local in much of C. and E. England and C. and S. Ireland; N. and C. Europe, extending locally southwards to C. Spain and Portugal, C. Italy and Bulgaria.

2. LYCOPODIELLA Holub

Stems creeping and rooted, with short lateral branches. Lvs arranged spirally, subulate, curved upwards. Sporophylls like lvs but rather wider and toothed at base, forming terminal spikes. Prothallus with conical subterranean base and green leaf-like appendages at apex. About 50 spp.

1. L. inundata (L.) Holub　　　Marsh Clubmoss.
Lycopodium inundatum L.
Stems 5–20 cm, closely prostrate, sparingly branched. Lvs 4–6 mm, spirally inserted but secund towards the upper side of the stem, linear-subulate, acute, entire, green. Fertile branches 3–10 cm, their lvs spirally arranged, suberect. Spikes 1–3 cm, sessile, solitary; sporophylls similar to the foliage lvs but more spreading, toothed, and somewhat wider at the base. Spores ripe 6–9. $2n = 156$*. Chh.

Native. Wet heaths in the lowlands, local; from Cornwall to Surrey and E. Sussex, Buckingham, Norfolk, Lincoln, Pembroke, Merioneth, Cumberland, Scotland from the Clyde and Fife to Ross and Elgin; Mayo, Wicklow, Westmeath and Armagh. Most of Europe except the Mediterranean region and E. Russia; W. Caucasus; North America from Newfoundland and Alaska to Pennsylvania, Idaho and Oregon.

3. HUPERZIA Bernh.

Stems ascending, divided dichotomously into branches of equal length. Lvs imbricate, in many rows on stem. Sporangia axillary; sporophylls like foliage lvs. Prothallus without chlorophyll, subterranean, large and cylindrical. Over 100 spp., cosmopolitan.

1. H. selago (L.) Bernh. ex Schrank & Mart.

Fir Clubmoss.

Lycopodium selago L.

Stems 5–25 cm, erect from a decumbent base. Lvs 4–8 mm, suberect to spreading, linear- to ovate-lanceolate, acute, entire or very minutely serrulate, dull green, often bearing in their axils bud-like gemmae. Sporangia borne in the axils of many of the lvs, not forming a terminal spike but usually in fertile zones alternating with sterile ones on the stem. Spores ripe 6–8, but not functional, reproduction probably only by gemmae. $2n = 264^*$. Chh.

Native. Heaths, moors, mountain grassland, rock ledges and mountain tops, ascending to nearly 1310 m, usually in open habitats; common in the mountains, very rare and decreasing in lowland areas; throughout the British Is., but absent from most counties in S., E. and C. England. Most of Europe, but only on higher mountains in the south; Himalaya. North America, southern South America and adjacent islands.

4. DIPHASIASTRUM Holub

Stems long, creeping, dorsiventral; branches forked several times, forming caespitose tufts. Lvs in 4 rows, opposite and decussate, somewhat scale-like, those of the lateral rows keeled. Sporophylls unlike lvs, forming terminal spikes. Prothallus without chlorophyll, with conical subterranean base bearing a subglobose apical appendage. About 30 spp., North America, Europe, C. and E. Asia.

Finer branches cylindrical or slightly flattened; ventral lvs
  of vegetative branches 0.5 mm wide, lanceolate, petiolate.   **1. alpinum**
Finer branches distinctly flattened; ventral lvs of vegeta-

tive branches 1–1.25 mm wide, elliptic-lanceolate, sessile.   **2. × issleri**

1. D. alpinum (L.) Holub   Alpine Clubmoss.

Diphasium alpinum (L.) Rothm.; *Lycopodium alpinum* L.

Stems 15–50 cm, much-branched; branches mostly suberect, densely tufted. *Lvs* 2–4 mm, distant on the main stems, appressed and *strongly 4-ranked on the branches*, becoming ±equal, the ventral row well-developed, 0.5 mm wide, oblong-lanceolate, acute or acuminate, petiolate, often with a short wide hyaline point, concave, entire, *glaucous*. Fertile branches 4–7 cm. Spikes 1–2 cm, sessile, solitary; sporophylls lanceolate or ovate-lanceolate, acute or acuminate, gradually tapering, with narrow scarious denticulate margin. Spores ripe 6–8. $2n = c.\ 48^*$. Chh.

Native. Moors, mountain grassland and mountain tops, ascending to 1220 m, rather common; C. and N. Wales, Derby and from Westmorland and S. Northumberland northwards; N. and W. Ireland; Wicklow. N. and C. Europe, extending to the Pyrenees, N. Apennines, Bulgaria and Urals; Asia Minor, Caucasus, Altai, Japan, British Columbia and Quebec.

2. D. × issleri (Rouy) Holub

D. alpinum × D. complanatum (L.) Holub; *Lycopodium issleri* (Rouy) Lawalrée; *L. alpinum* var. *decipiens* Syme ex Druce; *L. complanatum* auct. brit.

Like *D. alpinum* but more robust and less caespitose; finer branches distinctly flattened, 2–2.5 mm wide; ventral lvs of vegetative branches 1–1.25 mm wide, ellipticlanceolate, sessile, green; sporophylls ovate, abruptly acuminate. Partly fertile.

Native. Heaths and moors, very rare. Scattered stations in N. Devon, Hampshire, Gloucester, Worcester and Caernarvon, but perhaps extinct, C. and S. Scotland; C. Europe, C. France, Ardennes.

Southern English material, described above, is ±intermediate between the presumed parents but many of the Scottish records seem to apply to atypical forms of *D. alpinum*.

2. SELAGINELLACEAE

Herbs with long, usually creeping stems producing lfless branches (rhizophores) which bear the roots. Lvs small with a minute ligule at the base, spirally arranged or 4-ranked and of two kinds. Sporangia of two kinds borne near the base of the upper-surface of sporophylls forming terminal spikes, usually with the megasporangia in the lower, the microsporangia in the upper part. Megaspores (1–)4(–42). Microspores numerous. Male prothallus contained in the microspore until maturity, with a vegetative cell and an antheridium containing numerous biciliate spermatozoids. Female prothallus many-celled, filling the megaspore and protruding from its split top; archegonia several, at the top of the prothallus. Fertili-

zation occasionally taking place before the shedding of the megaspore.

One genus.

1. SELAGINELLA Beauv.

The only genus. About 700 spp., cosmopolitan, but mainly tropical.

Lvs all similar, spirally arranged.   **1. selaginoides**
Lvs of 2 sizes in 4 rows.   **2. kraussiana**

1. S. selaginoides (L.) Link   Lesser Clubmoss.

S. spinosa Beauv.; *S. spinulosa* A. Braun

Stems 3–15 cm decumbent, slender, with short, vegetative and long, ascending, more robust fertile branches. Lvs 2–4 mm, spirally arranged, spreading or somewhat appressed, lanceolate, acute, spinulose-ciliate, *all alike*. Fertile branches 2–6 cm, suberect, their lvs larger than those of the stems and vegetative branches. Spikes sessile, solitary, 1–1.5 cm; sporophylls similar to the lvs but larger; megasporangia occupying the greater part of the spike; microsporangia few, in the upper part of spike, often 0. Spores ripe 6–8. $2n = 18^*$. Chh.

Native. Damp grassy or mossy ground, mainly on mountains, ascending to 1085 m, rather common; Merioneth to Yorkshire northwards; Clare, Tipperary and Wexford northwards. N. and C. Europe, Pyrenees; arctic and north temperate Asia and America, south to Caucasus, Colorado and New Hampshire.

*2. S. kraussiana (G. Kunze) A. Braun

Stems creeping, jointed at the nodes, flattened and dorsiventral. Lvs in 4 rows; those on the upper side of the stem c. 1 mm, unequal at base, with rounded auricle on outer margin, appressed; lateral lvs c. 2 mm, ovate-lanceolate, acute, spreading laterally, rounded at base. Spikes sessile, up to c. 2 cm, 4-sided; sporophylls ovate, cuspidate, keeled. $2n = 20$. Chh.

Introduced. Commonly grown as ground cover in greenhouses and conservatories; escaped and locally naturalized in Jersey, Cornwall, Dorset, Kent, Hertford, Berkshire, Cardigan, Caernarvon, Anglesey and oceanic sites in S. Ireland from Kerry to Dublin and in Mayo. Native of tropical and S. Africa and the Azores.

3. ISOETACEAE

Aquatic or terrestrial perennial heterosporous plants with short stout stems. Roots arising from the 2- or 3-lobed stem-base, slender, dichotomously branched. Stems with 1, rarely 2, rings of meristematic cells producing secondary tissue. Lvs crowded in a dense rosette, subulate or filiform, usually terete or subterete, often tubular and septate, sheathing at base. Ligule present. The first-produced lvs in any season bearing megasporangia, the next microsporangia, and the last sterile. Sporangia sessile, ±embedded in the lf-base below the ligule, usually covered by an indusium formed from the lf-base. Megasporangia traversed by strands of tissue. Outer layers of spore wall impregnated with silica. Spores on germination giving rise to prothalli. Male prothallus (from microspore) of 1 vegetative cell and an antheridium with a 4-celled wall surrounding 2 cells which give rise to 4 spermatozoids. The multiciliate spermatozoids liberated by the dehiscence of the spore and the breaking down of the antheridium wall. Female prothallus (from megaspore) many-celled, filling megaspore and bearing archegonia the necks of which protrude from the split top of the megaspore. The young plant developing without a resting stage from the fertilized archegonium.

Two genera and about 75 spp., distributed throughout the world.

1. ISOETES L.

The only genus, except for the monotypic *Stylites* from the Andes of Peru.

    1 Plant aquatic, never completely dormant and lfless; stem without persistent lf-bases; lvs 4–20 cm ×2–3 mm.   2
      Plant terrestrial, dormant and lfless in summer; stem ±covered by persistent lf-bases; lvs up to c. 3 cm, ×1 mm.   **3. histrix**
    2 Lvs very stiff; megaspores 530–700 µm, with short blunt tubercles.   **1. lacustris**
      Lvs rather flaccid; megaspores 440–550 µm, with long sharp spines.   **2. echinospora**

1. I. lacustris L.   Quillwort.

A submerged aquatic. Stem without persistent lf-bases, 2-lobed. *Lvs* 8–20(–45) cm × 3–5 mm, subulate, subterete, with 4 longitudinal septate tubes, *stiff*, dark green. Stomata 0. *Megaspores* 530–700 µm, yellowish, rarely white, *covered with short blunt tubercles*. Microspores smooth or finely furrowed. Spores ripe 5–7. $2n = c.$ 110*. Hyd.

Native. In lakes and tarns with water poor in dissolved salts, on substrata of stones with little silt, boulder clay, sand, or rarely thin peat, locally abundant. S. Devon; mountain districts of Wales; Shropshire, S.E. Yorks, Lake District; scattered throughout Scotland and Ireland north to Shetland. N. and C. Europe.

2. I. echinospora Durieu   Spring Quillwort.

Like *I. lacustris* but usually smaller. *Lvs* 4–12 cm ×2 mm, *rather flaccid*, pale green. *Megaspores* 440–550 µm, white or yellowish, *covered with long sharp fragile spines*. Spores ripe 5–7. $2n = 22$; c. 100*. Hyd.

Native. In lakes and tarns, usually on peaty substrata, local. E. Cornwall, S. Devon, Dorset, Glamorgan, Merioneth, Caernarvon, Cumberland; scattered throughout Scotland from Perth northward to Shetland and outer Hebrides; Ireland, mainly in the west. N. and C. Europe, southwards to N. Italy and Spain.

3. I. histrix Bory   Land Quillwort

Terrestrial. *Stem 3-lobed, covered with persistent* short blackish *lf-bases* each with 2 long points. Lvs 1–3 cm × 1 mm, ½-terete, dark green, shiny. Stomata present. *Megaspores* 400–560 µm, *ornamented with a regular net-like pattern*. Spores ripe 4–5. Period of vegetative growth 10–4. $2n = 20^*$. Hr.

Native. In peaty and sandy places, damp in winter but dry in summer, very local. Lizard district. W. Cornwall; Channel Is. Atlantic coast of Europe; Mediterranean region.

SPHENOPSIDA

4. EQUISETACEAE

Perennial herbs with creeping rhizome, bearing aerial stems at intervals. Stems all green and assimilating or the fertile stems without chlorophyll; all stems grooved, simple or branched from near the base, the branches resembling the stem, or with whorls of slender green branches from the nodes, with a central cavity surrounded by 2 rings of smaller cavities, the inner ring of 'carinal canals' alternating with the larger, outermost 'vallecular canals'. Lvs very small, usually not green, in whorls united into sheaths above the nodes, the sheaths ending in free teeth, usually of the same number as the grooves on the stem; sheaths of the branches much smaller, with fewer teeth. Spores all alike, overlaid by two spiral bands ('elaters') which show hygroscopic movement, numerous, in sporangia borne several together round the under surface of a peltate sporangiophore. Sporangiophores in whorls, closely aggregated together to form a spike ('cone') terminal on the main stem and occasionally on the branches also. Archegonia and antheridia borne on separate prothalli, the female being larger, or successively on the same prothallus. Prothalli with a cushion-like base having lobed green flat structures arising from the upper surface. Sex organs borne on the upper surface of the cushion. Spermatozoids multiciliate.

A single living genus.

1. Equisetum L.

About 23 spp., almost cosmopolitan, but absent from Australia, New Zealand, etc.

1	At least the green or whitish vegetative stems with whorls of green branches.	2
	No stems with whorls of branches.	9
2	Branches of vegetative stems branched. **6. sylvaticum**	
	Branches of vegetative stems not branched.	3
3	Fertile stems white, without whorls of branches.	4
	Fertile stems green, with whorls of branches.	6
4	Main internodes 10–30 mm in diam.; sheaths with long fine teeth c. 5 mm. **9. telmateia**	
	Main internodes up to 4 mm in diam., sheaths with triangular or fine teeth c. 2.5 mm.	5
5	Branch-internodes 1–2 mm in diam.,with (3)4 prominent narrow ridges, sheaths with triangular teeth. **5. arvense**	
	Branch-internodes 0.5–1 mm in diam., 3-angled; sheaths with lanceolate teeth. **7. pratense**	
6	Central cavity at least $\frac{4}{5}$ diam. of stem; sheath-teeth without ribs. **4. fluviatile**	
	Central cavity not more than $\frac{2}{3}$ diam. of stem; sheath-teeth with ribs.	7
7	Lowest internode of branches at least as long as adjacent stem-sheath; branches solid. **7. pratense**	
	Lowest internode of branches shorter than adjacent stem-sheath; branches hollow.	8
8	Central cavity more than $\frac{1}{2}$ diam. of stem; spikes apiculate, 6–12 mm (Lincoln). **2. ramosissimum**	

Central cavity not more than $\frac{1}{2}$ diam. of stem; spikes obtuse. 20–35 mm or more (widespread). **8. palustre**

9	Sheaths shallowly lobed, the teeth present only on juvenile stems or stems on some stunted plants of exposed places; internodes somewhat swollen. **1. hyemale**	
	Sheaths with persistent teeth; internodes not swollen.	10
10	Sheath-teeth with narrow dark centres and very wide, white, scarious margin. **3. variegatum**	
	Sheath-teeth without conspicuous, white margin.	11
11	Central cavity at least $\frac{4}{5}$ diam. of stem; sheath-teeth without ribs. **4. fluviatile**	
	Central cavity not more than $\frac{2}{3}$ diam. of stem; sheath-teeth with ribs.	12
12	Central cavity more than $\frac{1}{2}$ diam. of stem; spikes apiculate, 6–12 mm (Lincoln). **2. ramosissimum**	
	Central cavity not more than $\frac{1}{2}$ diam. of stem; spikes obtuse, 20–35 mm (widespread). **8. palustre**	

Subgenus 1. Hippochaete (Milde) Baker

Stomata sunk below level of other epidermal cells. Spikes apiculate. Stems all alike, hard, usually persisting through winter.

1. E. hyemale L.   Rough Horsetail

Stems (30–)70–100 cm, erect, 4–6 mm diam., glaucous-green, simple, persisting through the winter; internodes somewhat swollen; ridges rough, with 2 regular rows of conspicuous, angular tubercles; grooves 10–30, moderate; sheaths 3–9 mm, about as long as wide, soon whitish with a black band at top and bottom, appressed; teeth as many as the grooves, very quickly detached as a ring and carried up at tip of elongating shoot, leaving a crenulate upper edge to the sheath, sometimes persisting on some stems of depauperate plants; central cavity c. $\frac{2}{3}$ or more diam. of stem. Spike 8–15 mm. Spores ripe 7–8. $2n = 216^*$. Grh.

Native. Shady valley-sides and river banks, etc., ascending to 535 m. Hampshire, Glamorgan, Bedford and Norfolk northwards to Outer Hebrides, formerly much commoner in the south; E. Ireland from Wexford to Antrim and Fermanagh, Sligo, formerly in south and west. Most of Europe but rare in the Mediterranean region; Caucasus; N. and C. Asia; western North America southwards to California and New Mexico.

E. × moorei Newman (*E. hyemale × ramosissimum*; *E. occidentale* (Hy) Coste), differing from *E. hyemale* in its more slender, yellowish-green stems which die down in autumn, at least to near the base, non-swollen internodes, sheaths about twice as long as wide, and dark brown or black, usually persistent teeth, rarely producing small black spikes with abortive spores, occurs on low sandy and clayey banks near the sea between Wexford Harbour (Co. Wexford) and Ardmore Point (Co. Wicklow).

*2. E. ramosissimum Desf,

Stems 50–75 cm, *greyish-green*, usually dying in autumn, usually with axillary whorls of branches in lower half, sometimes whorls only partial or absent, rough with scattered tubercles; *grooves* 8–20, moderate; sheaths c. 8 mm, green, becoming brown with a black band at the bottom; teeth black with narrow white margins and a ±persistent hair-like apex; *central cavity* $\frac{1}{2}$–$\frac{2}{3}$ *diam. of stem. Branches hollow*; *lowest internode* c. $\frac{1}{3}$ *length of adjacent stem sheath.* Spike 6–12 mm. Spores ripe 5–8. $2n =$ c. 216*. Grh.

Introduced, probably with soil. Long grass by River Witham (S. Lincoln); first found 1947. C. and S. Europe, extending locally northwards to the Netherlands, Latvia and C. Russia; Asia; Africa; America.

E. × trachyodon A. Braun (*E. hyemale × E. variegatum*), Mackay's Horsetail, is generally intermediate between the parents. The dark green stems are 30–60(–90) cm, with non-swollen internodes; the sheaths are longer than wide, more so than in either parent, and the teeth are long, black, sometimes with narrow, white, scarious margin towards base, and mostly persist throughout the first season; the spikes are 4–5 mm, black or orange-tinged and contain abortive spores. This natural hybrid occurs on damp areas on calcareous coastal dunes and sandy river banks. It is local in Kincardine, Rhum, Skye and Harris, but more frequent in Ireland from Clare and Mayo to Donegal, Down and Antrim. It is recorded from many parts of N. and C. Europe, Greenland and North America.

3. E. variegatum Schleicher ex Weber & Mohr
Variegated Horsetail.

Stems 15–80 cm, decumbent or less often erect, up to 4 mm diam., green, *simple or rarely branched at the base, without axillary whorls*, persisting through the winter; grooves 4–10, moderate; ridges shallowly 1-grooved, rough, with 2 regular rows of minute tubercles; *sheaths* c. 2–4 mm, *green with a black band at the top, rather loose*; teeth as many as the grooves, with narrow dark centre and very wide, white, scarious margin, *triangular-ovate or triangular-lanceolate*, at first subulate, the tip falling and leaving an *obtuse* apex, 4-ribbed; central cavity c. $\frac{1}{3}$ diam. of stem. Spike 5–7 × c. 3–4 mm. Spores ripe 7–8. $2n = 216$*. Grh.

Native. Dunes, river-banks, wet ground on mountains, etc., ascending to 490 m in Kerry. Hampshire, Somerset, Berks, Wales; Cheshire and Yorks to Sutherland and Outer Hebrides, very local; widespread in C. Ireland, from Clare and Kilkenny to Donegal and Louth, Derry; var. *wilsonii* in Kerry only, usually in shallow water. N. and C. Europe, extending southwards to the Pyrenees, N. Apennines and S. Ural; N. Asia and North America.

A variable sp., which has been variously subdivided. Var. *variegatum*, the most frequent form in Britain, usually has stems 15–25 cm tall and c. 2 mm diam. Most Irish populations belong to var. *majus* Syme, in which the plants are more vigorous, with stems up to 40(–80) cm tall and 4 mm diam. Var. *wilsonii* Newman, described from Co. Kerry, was distinguished by stems,

up to 100 cm, with completely smooth internodes.

Subgenus 2. EQUISETUM.

Stomata not sunk below level of other epidermal cells. Spikes obtuse. Fertile and vegetative stems sometimes dissimilar, dying down in autumn.

4. E. fluviatile L. Water Horsetail.
E. limosum L.; *E. heleocharis* Ehrh.

Rhizome glabrous. *Stems* 50–150 cm, ±erect, 2–12 mm diam., green, *simple or with irregular whorls of branches in the middle, smooth*; *grooves* 10–30, *very fine*; sheaths 5–10 mm, green, tight; *teeth* as many as the grooves, subulate, c. 1 mm, black, with inconspicuous pale margin, *not ribbed*; *central cavity* $\frac{3}{4}$–$\frac{9}{10}$ *diam. of stem. Branches* ascending, slender, simple, usually 5-angled, hollow; lowest internode about as long as stem sheath or shorter; sheaths with 4–5 moderate, subulate, green or blackish teeth. Spike 1–2 cm. Spores ripe 6–7. $2n = 216$*. Hel.

Native. In shallow water at the edges of lakes, ponds and ditches, frequently dominant in these swamp communities; less often in marshes and fens, ascending to 915 m. Common throughout the British Is. Most of Europe; Caucasus; temperate Asia; North America from Labrador and Alaska to Virginia and Oregon.

E. × litorale Kühlew. ex Rupr. (*E. arvense × fluviatile*). Stems all alike, resembling those of *F. fluviatile* but more deeply grooved, and with twice the number of green bands, more branches and with loose sheaths. Teeth appressed, with minute black apex. Central hollow $\frac{1}{2}$–$\frac{2}{3}$ diam. of stem. Spike short; spores abortive. $2n = 216$. Scattered throughout British Is., especially in high rainfall districts, particularly common in Ireland.

E. × dycei C. N. Page (*E. fluviatile × palustre*) has been recorded from wet ditches in the Outer Hebrides (Harris, Lewis) and Perthshire. It is morphologically intermediate between the parents, but may be confused with *E. × littorale*, from which it can be distinguished by the lowest internode of the branches being about as long as the adjacent sheath, and the c. 9 shallow stem-furrows.

5. E. arvense L. Field Horsetail.

Rhizome pubescent, with ovoid tubers. Vegetative stems 20–80 cm, erect or decumbent, c. 3–5 mm diam., green, slightly rough; *grooves* 6–19, *deep*; sheaths 3–8 mm, green; teeth as many as grooves, subulate, acute, green below with blackish tips, 1-ribbed, often adhering together in pairs or threes; central hollow less than $\frac{1}{2}$ diam. of stem. *Branches* spreading, numerous, usually simple, solid, regular, (3–)4-*grooved*, the ridges prominent, narrow; *lowest internode longer than adjacent stem sheath*; sheaths pale, 4-toothed; *teeth triangular-lanceolate, acuminate*, somewhat spreading, pale. Fertile stem 10–25 cm, simple, brown; *sheaths* loose, pale brown, *with* 6–12 darker *teeth, few* (4–6), distant. Spikes 1–4 cm. Spores ripe 4. $2n =$ c. 216*. Grh.

Native. Fields, hedgebanks, waste places, dune-slacks, etc., ascending to over 900 m. Common throughout the British Is. Throughout Europe; C. China; North America, Greenland.

6. E. sylvaticum L.   Wood Horsetail.

Vegetative stems 10–50(80) cm, erect, 3–6 mm diam., green, rather rough; grooves 10–18; sheaths 10–15(–20) mm, green below; *teeth united into 3–6, broad, subacute*, brown *lobes*, each with 2–3 ribs; central cavity $\frac{1}{4}$–$\frac{1}{3}$ diam. of stem. *Branches drooping at ends*, numerous, *branched* except in small plants, regular, 3–4-grooved, solid; lowest internode longer than adjacent stem sheath; sheaths with 3–4 long subulate teeth. Fertile stem 10–40 cm, pale green, usually with short branches when the spores are ripe; sheaths loose, numerous, greenish below, brown above, with 3–6 broad brown teeth. Spike 1.5–2.5 cm. Spores ripe 4–5. $2n = 216^*$. Grh.

Native. Damp woods on acid soils, moors, etc., ascending to 915 m, throughout the British Is., common in Scotland, N. England, Wales and N. Ireland, becoming very local southwards and absent from several midland and southern counties and from the Channel Is. Most of Europe except S. Russia, but rare in the Mediterranean region; Caucasus; temperate Asia; North America from Newfoundland and Alaska to Virginia and Iowa; S. Greenland.

7. E. pratense Ehrh.   Shady Horsetail.

Vegetative stems (2–)10–30(–60) cm, erect, c. 1–2 mm diam., green, rough; *grooves* 8–20, *deep*; sheaths 3–8 mm; teeth as many as the grooves, brown with a blackish rib, subulate, acute; central cavity c. $\frac{1}{2}$ diam. of stem or rather more. *Branches* spreading or sometimes somewhat drooping, numerous, simple, regular, solid, 3(–4)-*grooved*; *lowest internode longer than adjacent stem sheath* (sometimes shorter in the lower part of stem); sheaths pale, 3(–4)-toothed; *teeth deltate, acute*. Fertile stem 10–25 cm, simple or with short branches when the spores are ripe, pale green; sheaths loose, numerous, yellowish-white with 10–20 pale teeth, the ribs dark. Spike 1.5–4 cm. Spores ripe 4. $2n = $ c. 216^*. Grh.

Native. Grassy stream banks, etc., ascending to over 900 m. In scattered localities from Yorks and Westmorland to Orkney, very local and mainly in the east; Down, Derry and Tyrone. N., C. and E. Europe; Caucasus; N. and C. Asia; North America from Nova Scotia and Alaska to New Jersey and Colorado.

8. E. palustre L.   Marsh Horsetail.

Rhizome glabrous. Stems 10–60 cm, erect or decum-bent, 1–3 mm diam., green, usually branched, often rather irregularly, occasionally simple, slightly rough; *grooves* 4–8, *deep*; sheaths 4–12 mm, green, loose; *teeth* 4–8, triangular-subulate, blackish with wide whitish scarious margin, 1-*ribbed*; *central cavity small, scarcely larger than the outer ones*. Branches spreading to sub-erect, often short, simple, 4–5-grooved, hollow; *lowest internode much shorter than the adjacent stem sheath*; sheaths with 4 short appressed black-tipped teeth. Spike 1–3.5 cm. Spores ripe 5–7. $2n = $ c. 216^*. Grh.

Native. Bogs, fens, marshes and wet heaths, woods and meadows, ascending to 915 m. Common throughout the British Is. Almost throughout Europe; Caucasus; temperate Asia; North America from Newfoundland and Alaska to Connecticut and Oregon.

E. × **rothmaleri** C. N. Page (*E. arvense* × *palustre*) which resembles a yellow-green *E. palustre* with more conspicuously angled branches, is known from marshy fields and ditches in the Isle of Skye.

9. E. telmateia Ehrh.   Great Horsetail.

E. maximum auct.

Rhizome pubescent, often with pyriform tubers. *Vegetative stems* 1–2 m, erect, c. 10–30 mm diam., *ivory white*, smooth; *grooves* 20–40, *fine*; sheaths 1.5–4 cm, ±appressed, greyish-green, blackish above; teeth c. 5 mm or more, as many as the grooves, blackish, subu-late, 2-ribbed; central cavity $\frac{1}{2}$–$\frac{1}{3}$ diam. of stem. *Branches* bright green, spreading, *numerous*, simple, *regular*, 4-grooved; lowest internode shorter than adjacent stem sheath; sheaths short, pale, 4-toothed. Fertile stem 15–25(–40) cm, simple, ivory white; *sheaths* loose, *numerous*, close together, pale brown *with* 20–30 dark *teeth*. Spike 4–8 cm. Spores ripe 4. $2n = 216^*$. Grh.

Native. Damp shady banks, etc., ascending to 365 m. Throughout England, Wales and Ireland, rather local; more local in Scotland and absent from much of the north, apart from the Inner and Outer Hebrides, Ross, Sutherland, Caithness and Aberdeen; Alderney. Europe except the extreme north and much of the USSR; Asia Minor and Caucasus; N. Africa; Azores, Madeira; W. America from British Columbia to California.

E. × **font-queri** Rothm. (*E. palustre* × *telmateia*), which resembles a rather slender form of *E. telmateia*, but has a long branchless part at the stem apex and the fertile stems resemble the vegetative stems of *E. telmateia*, is known as an extensive colony in the Isle of Skye and on a railway embankment in Worcester.

FILICOPSIDA

5. OPHIOGLOSSACEAE

Terrestrial (rarely epiphytic) herbs with short, usually erect, rhizome without scales; roots fleshy. Lvs one or more, stalked, not circinate in bud. Fertile lvs consisting of a sterile blade and one or more fertile spikes or a fertile panicle; (in our species the fertile spike or panicle appears as if terminal on a stem on which the sterile blade is borne laterally). Sporangia all alike, borne in 2 rows on the margins of the fertile spike or panicle-branches, sessile or nearly so, each derived from a group of cells; wall of several layers of cells; annulus 0; spores very numerous (1500–15 000). Prothallus usually subterranean, massive, without chlorophyll but with endotrophic mycorrhiza, bearing organs of both sexes, the antheridia sunk in the tissues.

Four genera, the two following the monotypic *Helminthostachys* Kaulf. from tropical Asia and Australia and *Rhizoglossum* Presl from S. Africa.

Sterile blade pinnate; fertile portion a panicle.
1. Botrychium
Sterile blade simple, entire; fertile portion a spike.
2. Ophioglossum

1. BOTRYCHIUM Swartz

Sterile blade pinnately lobed or *pinnate* (sometimes several times); veins free. *Sporangia subsessile, arranged in a panicle*, opening by a transverse slit.

About 40 spp., cosmopolitan, mainly north temperate regions.

1. B. lunaria (L.) Swartz   Moonwort.

Rhizome underground, ascending or creeping, very short, usually unbranched. Lvs (2–)5–15(–30) cm, solitary, rarely 2, erect, sheathed at the base by the brown remains of the previous year's lvs; sterile blade (1–)2–5(–12) cm, usually inserted about the middle of the lf, oblong in outline, pinnate; pinnae (2–)4–7(–9) pairs, fan-shaped, entire or shallowly and irregularly (rarely deeply) crenate, without midrib; fertile panicle (0.5–)1–5 cm (excluding stalk), overtopping sterile blade, 1–3 times branched. Spores ripe 6–8. $2n = 90^*$. Grh.

Native. Dry grassland and rock ledges throughout the British Is., ascending to 1020 m in Perth, rather local, more frequent in Scotland and N. England, becoming scarce southwards and in much of C. and S. Ireland. Almost throughout Europe, but rare in the Mediterranean region; Morocco (Atlas); Asia Minor, Himalaya, California and New York; Australia, Tasmania, New Zealand.

Reports of *B. matricariifolium* A. Braun ex Koch and *B. multifidum* (S. G. Gmelin) Rupr. in Scotland refer to aberrant forms of this sp.

2 OPHIOGLOSSUM L.

Sterile blade simple (rarely palmately lobed); veins reticulate. *Sporangia sunken, arranged in a simple spike* (or spikes), opening by a transverse slit.

About 30–50 spp., cosmopolitan.

1 Lvs single, rarely 2; blade more than 3.5 cm.
1. vulgatum

Lvs often 2–3 together; blade less than 3.5 cm.   2
2 Plant not more than 2 cm; sterile blade attenuate at base; spike 3–6 mm; sporangia c. 6–10.
3. lusitanicum
Plant usually 3–8 cm; sterile blade cuneate to rounded at base; spike 8–20 mm; sporangia 12–28(–40).
2. azoricum

1. O. vulgatum L.   Adder's-tongue.

Rhizome underground, erect, very short. Roots producing new plants from adventitious buds. *Lvs* 1(–2), 8–20(–45) cm, erect; *sterile blade* 4–15 cm, *ovate to ovate-lanceolate or oblong*, entire, obtuse or acute, sheathing the stalk of the fertile spike at the base, *with free vein endings inside the meshes of the network*; epidermal cells with sinuate walls; fertile spike 2–5(–7) cm (excluding stalk), overtopping sterile blade at maturity, with 10–40 sporangia on each side; apex sterile, acute. Spores with blunt tubercles. Spores ripe 5–8. $2n = 480–520^*$. Grh.

Native. Damp grassland, fens and dune-slacks. Throughout the British Is., but local or absent from much of mainland Scotland and C. and S. Ireland. Most of Europe, but rare in the Mediterranean region; Caucasus; Madeira; N. and W. Asia; N. Africa; North America.

2. O. azoricum C. Presl   Small Adder's-tongue.

O. vulgatum subsp. *ambiguum* (Cosson & Germ.) E. F. Warburg;
O. vulgatum subsp. *polyphyllum* auct., non A. Braun

Like *O. vulgatum* but smaller, usually 3–8 cm high; lvs (1)2–3 together; sterile blade 3–3.5 cm, lanceolate to ovate, usually acute, with free vein endings inside network, tapered at base or contracted so as to appear stalked. Spike 8–20 mm, with 6–14(–20) sporangia on each side. Spores moderately tuberculate. $2n = c. 720$. Grh.

Native. Dune-slacks and short coastal turf, very local; Scilly and Channel Is., Dorset, Lundy I., Pembroke, N. Wales, Cumberland, Northumberland; Caithness, Orkney and Shetland; Kerry, Galway, Mayo and Donegal. W. Europe, Azores.

3. O. lusitanicum L.

Like *O. vulgatum* but much smaller, not higher than 2 cm; lvs 1–3 together; sterile blade 1–3 cm., lanceolate to linear-lanceolate, obtuse, without free vein endings inside network, attenuate at base. Spike 3–6 mm, with 3–5 sporangia on each side. Spores smooth. $2n = 250^*–260^*$. Grh.

Native. Very local and rare in short turf on cliff-tops and rocky slopes in the Isles of Scilly and Guernsey. Mediterranean region and W. Europe.

6. OSMUNDACEAE

Rhizome large, erect, not scaly. Lvs pinnately divided, expanded at the base, the expansions covered with glandular hairs; veins free. Sporangia marginal or superficial, all alike and developing simultaneously; indusium 0; annulus consisting of a group of thick-walled cells near the apex; sporangia dehiscing by a slit running from the annulus across the apex. Spores rather numerous (up to 500), green. Prothallus green, cordate, fleshy.

Three genera and about 19 spp., cosmopolitan.

1. OSMUNDA L.

Sporangia marginal on reduced pinnules without chlorophyll or flat blade, the fertile pinnae usually occupying the top or middle portion of the lf or the whole lf. Outer lvs vegetative.

12 spp., absent from Australia.

1. O. regalis L.   Royal Fern.

Rhizome short, ascending or suberect, massive. Lvs 30–120(–400) cm, tufted, 2-pinnate, the outer vegetative, the inner with the lower pinnae vegetative, the upper fertile (often with a transition region with pinnae with some vegetative and some fertile pinnules); blade glabrous, ±lanceolate in outline, in the fertile lvs with c. 2–3 pairs of vegetative pinnae and 5–14 pairs of fertile

ones markedly decreasing in size upwards; petiole hairy when very young, soon glabrous; vegetative pinnae with 5–13 pairs of pinnules; rhachis narrowly winged; pinnules 2–6.5 cm, ±oblong, subobtuse, ±truncate at the base, often with a rounded lobe on the lower side at the base and occasionally shallowly crenately lobed on both sides, minutely and irregularly crenulate-serrulate, the veins prominent on both surfaces and repeatedly dichotomously branched, reaching the margin; fertile pinnules up to 3 cm, 2–4 mm wide, without blade, densely covered with clusters of brown sporangia. Spores ripe 6–8. $2n = 44^*$. H.

Native. Fens, bogs, wet heaths and woods, on peaty soil but in well-drained places, ascending to 365 m. Throughout the British Is., local, most frequent in W. Britain and, especially W. Ireland, rarer eastwards and much reduced by drainage, now almost extinct in most heavily populated areas owing to the depredations of collectors. Widespread in W. Europe, extending locally eastwards to E. Sweden, Poland and Turkey; Asia Minor, Transcaucasia; N. Africa; India, etc.; S. Africa, Madagascar, etc.; eastern North America from Newfoundland and Saskatchewan southwards; Central and South America to Uruguay.

7. ADIANTACEAE

Rhizome usually short, creeping or ascending, with hairs or scales. Lvs pinnately divided; veins free; petiole with 1 or 2 vascular strands. Sori on lower surface, without indusia but often protected by the reflexed lf-margin, not borne on a vein connecting the other vein endings. Spores tetrahedral.

About 3 genera and 220 spp. Cosmopolitan.

 1 Sori on lower surface of lf, not covered by the margin.
                                            2. ANOGRAMMA
   Sori on or near margin of lower lf-surface, covered by inrolled indusium-like margin or apparent margin.   2
 2 Lf-segments fan-shaped; recurved margin interrupted.   3. ADIANTUM
   Lf-segments ±oblong; recurved margin continuous along segment.   1. CRYPTOGRAMMA

1. CRYPTOGRAMMA R.Br.

Rhizome scaly. Fertile and vegetative lvs differing, 2- or more pinnate. *Sori borne on the apical part of the veins*, ±oblong, *protected by the reflexed, continuous lf-margin.*

Four spp. North temperate zone, temperate South America, S. Africa.

1. C. crispa (L.) Hooker   Parsley Fern.
Allosorus crispus (L.) Röhl.

Rhizome short, creeping or ascending, branched. Lvs densely tufted, bright green; outer vegetative; inner fer-

tile. Vegetative lvs 5–20 cm, 3-pinnatisect, naked except for a few brownish scales at the base of the petiole which is about as long as blade; blade triangular-ovate, pinnate; pinnae 3–7 on each side, usually 2-pinnatisect; segments c. 5–10 mm, obovate, obtuse, cuneate at base, pinnately lobed or toothed. Fertile lvs 10–25 cm (the blade about the same size as in the vegetative lvs, the petiole much longer); blade ovate, 3–4 pinnate; segments oblong-linear, stalked, appearing entire, the margins recurved, shallowly sinuately lobed, at first almost meeting and hiding the sori. Sori oblong, at first distinct but appearing to form a continuous band when mature. Spores ripe 6–8. $2n = 120^*$. H.

Native. Screes, etc., on acid soils on mountains from 90 to 1220 m; Devon, Somerset (artificial habitats, very rare); Wales; Yorks to Caithness, the Outer Hebrides and Orkney; locally abundant but absent from some areas; very rare in Ireland, (Galway, Down, Antrim). Mountains of Europe, at lower elevations in the extreme north, the Caucasus; W. Siberia (Ob region).

2. ANOGRAMMA Link

Small annuals but with perennial prothallus. *Rhizome very short with few scales. Fertile and sterile lvs somewhat different*, thin, 2–3 pinnate. *Sori linear, running along the length of the veins; lf-margin flat.*

Seven spp. scattered through the tropics and south

temperate zone, only the following reaching the north temperate zone.

1. A. leptophylla (L.) Link   Jersey Fern.

Gymnogramma leptophylla (L.) Desv.; *Grammitis leptophylla* (L.) Swartz

Rhizome very short, with a few narrow scales when young. Lvs few, slightly hairy when young, soon glabrous; outer vegetative, 7(–10) cm or less, pinnate; pinnae c. 1 cm long and almost as wide, deeply pinnatifid or almost palmatifid; segments lobed; inner lvs fertile, but not clearly marked off from the vegetative ones, 3–20 cm, ovate-oblong, 2–3-pinnate; pinnules or segments c. 5–10 mm, obovate, obtuse, cuneate at base, pinnately lobed or pinnatifid. Sori linear, along the ultimate veins, appearing confluent at maturity. Spores ripe 3–5. $2n = 58^*$. Th.

Native. Hedgebanks in the Channel Is (Guernsey and Jersey), rare. Mediterranean region and W. coast of Europe, Crimea; Macaronesia; Abyssinia, S. Africa, Madagascar; India; Australia, New Zealand; America from Mexico to Argentina.

3. ADIANTUM L.

Rhizome scaly. Lvs all alike, thin and translucent, usually with black glossy petiole and broad ±fan-shaped segments. *Sori* close to the ends of the veins *borne on the reflexed lf-margins.*

About 200 spp., tropical and warm temperate, only the following in Europe.

1. A. capillus-veneris L.   Maidenhair Fern.

Rhizome creeping, densely covered with narrow brown scales. Lvs close together, 6–40 cm, 2–3-pinnate; petiole (and rhachis) black and shining, as long as or shorter than blade; blade ±ovate in outline; pinnae and pinnules 5–30 mm, fan-shaped, often wider than long, crenately lobed in the upper part with wide rounded or truncate lobes which are recurved on the fertile pinnules; veins dichotomously branched, free; midrib 0. Sori borne close together along the veins of the recurved part of the lobes, 2–10 on each lobe. Spores ripe 5–9. $2n = 60^*$. H.

Native. Damp crevices of sea-cliffs and basic rocks, almost always near the sea, very local and rare. Cornwall, Devon, Glamorgan, Westmorland, Isle of Man; Cork, Clare, Galway, Donegal; Scilly and Channel Is.; also found rarely as an escape on walls, etc., elsewhere. Tropical and warm temperate zones of nearly the whole world; north in Europe to Caucasus, Crimea, S. Switzerland and W. France.

4. PTERIS L.

Lvs all alike; segments long, linear-lanceolate. Sori covered by scarious, deflexed lf-margin.

Two spp., natives of S. Europe, grown as house plants have spread to warm locations on walls in parts of S. and W. Britain: **P. cretica** L., with the blade ovate in outline, and up to 7 pairs of serrulate pinnae; **P. vittata** L., with the blade lanceolate in outline, and 10 or more pairs of entire pinnae, also seems to have been established on a slag-heap in the Forest of Dean (Gloucester).

Dicksonia antarctica R.Br., a 'tree-fern' from Australia, is cultivated in the mildest parts of S.W. Britain and W. Ireland and persists in abandoned gardens, as on Valencia Island (S. Kerry), Cornwall and the Isles of Scilly. The 2–3 pinnate lvs are borne on a 'stem' up to 2 m, high, the sori being borne on terminal pinnules of several lvs. Ripe spores sometimes produced.

8. HYMENOPHYLLACEAE

Rhizome usually creeping. Lvs thin and translucent, of 1 layer of cells without stomata, entire or divided, with the ultimate segments 1-veined. Sori marginal on the vein endings, often projecting from the lf; indusium ±cup-like, entire or 2-lipped or 2-valved, surrounding the base or whole of the sorus; sporangia all alike, shortly stalked, developed successively from apex to base; annulus oblique, without definite stomium, the sporangia opening laterally by a long slit. Spores 32–420. Prothallus green, either filamentous or strap-like.

Thirty-four genera and about 600 spp., mostly tropical, a few in moist parts of the temperate zones, absent from dry areas.

Indusium narrowly campanulate, not valved, the receptacle projecting from it as a long bristle; pinnae 1–2-pinnatisect; ultimate lobes short; lvs mostly over 10 cm; rhizome 2–4 mm diam.   **2. TRICHOMANES**
Indusium 2-valved, the receptacle not projecting; pinnae irregularly dichotomously divided, the ultimate lobes oblong; lvs rarely reaching 10 cm, usually much less; rhizome c. 1 mm diam.   **1. HYMENOPHYLLUM**

1. HYMENOPHYLLUM Sm.

Plants with habit of bryophytes, with which they often grow. Rhizomes filiform, less than 1 mm diam., smooth. Lvs ±procumbent, persistent after withering, deeply pinnatisect; segments asymmetrical, deeply lobed on side towards apex. Sori solitary, ±globose, near the base of distal segments of lvs. *Indusium of 2 ovate to suborbicular valves, united only at base.* Receptacle included. *Prothallus flat.*

About 25 spp., mostly temperate.

Valves of indusium orbicular, toothed; lf±flat.
                1. tunbrigense
Valves of indusium ovate, entire; pinnae bent back from the rhachis.   **2. wilsonii**

1. H. tunbrigense (L.) Sm.   Tunbridge Filmy-fern.

Rhizome creeping. *Lvs* 2.5–8(–12) cm, pinnate, the pinnae divided ±dichotomously but very irregularly into oblong segments, ±*flat*, persistent for some years; petiole occupying $\frac{1}{3}$–$\frac{1}{2}$ length, wiry, naked or with a few hairs; blade oblong or ovate-oblong in outline,

12–20 mm wide; rhachis winged; segments up to 3 mm, oblong, sharply and remotely serrulate, 1-veined, *the vein ceasing slightly below the apex*. Sori marginal on the tips of the segments, mostly near the rhachis of the lf; *indusium* c. 1 mm, flattened, *the valves ±orbicular, with a wide irregularly and sharply toothed mouth*. Spores ripe 6–7. $2n = 26^*$. Chh.

Native. Rocks, tree trunks, etc., in a moist atmosphere, ascending to 300 m. in Wales, local but often abundant where it occurs. Cornwall to Somerset; Sussex, Kent; S. and W. Wales; Lancashire to W. Inverness and Skye, east to Yorks, Peebles and Linlithgow; W. Ireland from Cork to Donegal, Waterford, Antrim. Locally in W. Europe southwards to Pyrenees, N. Italy, E. Germany; Macaronesia. The same or related spp. in North America, Australia, etc.

2. H. wilsonii Hooker   Wilson's Filmy-fern.

H. peltatum auct., vix Desv.; *H. unilaterale* auct.

Differs from *H. tunbrigense* as follows: Lvs usually narrower and appearing considerably so because the *pinnae are bent back from the rhachis*; pinnae usually with fewer and more unilateral segments; *vein reaching apex of segments*. Sorus somewhat projecting; *indusium ovoid*, not flattened, *the valves entire*. Spores ripe 6–7. $2n = 36^*$. Chh.

Native. In similar places to *H. tunbrigense*, requiring, in general, less sheltered conditions, and thus commoner, though in some places extending less far east, ascending to 1005 m in Kerry; Cornwall, Devon, Derby, Wales, Isle of Man; Lancashire and Yorks to Fair Is. (absent from N.E. Scotland and from several other eastern counties); throughout Ireland but absent from much of the centre and east. N.W. Europe; Azores.

2. TRICHOMANES L.

Like *Hymenophyllum*, but with stouter, hairy rhizomes; lvs erect, 1–2-pinnatisect; sori cylindrical; indusium tubular, slightly 2-lipped, enclosing basal part of sorus; receptacle exserted; prothallus filamentous.

About 25 spp., mostly temperate.

1. T. speciosum Willd.   Killarney Fern.

T. radicans auct.

Rhizome creeping, 2–4 mm diam., clothed with blackish hair-like scales. Lvs (7–)20–45 cm., ±irregularly 2–3-pinnatisect, persistent for some years; petiole occupying $\frac{1}{3}$–$\frac{1}{2}$ length, naked, winged above; blade ovate-triangular in outline, dark green, the rhachis winged; pinnae ±lanceolate in outline, the rhachis winged; pinnules or segments rather irregularly pinnatifid or pinnately lobed; lobes c. 1 mm or less, 1-veined, entire. Sori projecting from the margins of the upper pinnae; indusium 1–2 mm, narrowly campanulate; receptacle bristle-like, exserted. Spores ripe 7–9. $2n = 144^*$. Chh.

Native. Among shady rocks in places with a very humid atmosphere, ascending to 460 m in Kerry; very rare in Great Britain; Merioneth, probably extinct in Yorks, Westmorland, Cumberland, Arran and Argyll; more widespread in Ireland and formerly abundant in some places but now rare owing to the depredations of collectors; Kerry, Cork, Waterford, Kilkenny, Limerick, Tyrone, Antrim. W. French Pyrenees, W. Spain, Portugal (very rare in all); Macaronesia.

9. POLYPODIACEAE

Rhizome with thin-walled scales. Lvs in 2 ranks on upper side of rhizome; petioles with 1–3 main vascular strands. Sori orbicular to oblong, on lower surface of lf, near ends of veins if these are free; indusium 0; spores bilateral.

1. POLYPODIUM L.

Rhizome creeping, fleshy, with opaque scales. Lvs usually pinnatifid or 1-pinnate; veins usually regularly anastomosing with free endings inside the loops, sometimes free. Sori terminal on the veins, in 1(–3) rows on each side of the midrib.

About 75 spp., cosmopolitan, mainly tropical America, Asia and Polynesia.

1 Lvs ±oblong; sori orbicular when young, reddish-brown or bright orange when mature; annulus of sporangium dark reddish-brown.   **1. vulgare**
  Lvs oval to triangular-deltate; at least some sori oval when young, yellow to yellow-brown when mature; annulus of sporangium yellow to golden brown.   2
2 Sori containing long branched paraphyses; scales of rhizome lanceolate.   **3. australe**
  Sori without paraphyses; scales of rhizome abruptly narrowed above the wide base.   **2. interjectum**

1. P. vulgare L.   Polypody.

Rhizome creeping on or below the surface, rather stout, densely clothed when young with reddish-brown, ovate-lanceolate scales up to 6 mm. Lvs solitary, pinnatifid nearly to the rhachis or pinnate, persistent; petiole from about $\frac{1}{3}$ to nearly as long as blade, naked, ±erect; blade 10–25 cm, narrowly ovate to oblong or linear-lanceolate the lower $\frac{1}{3}$–$\frac{1}{2}$ with ±parallel sides, suberect, somewhat coriaceous, dull green; pinnae or segments c. 5–25 on each side, (10–)15–35(–45) mm, the lowest somewhat shorter, oblong to linear-lanceolate, obtuse, wide-based, entire to somewhat serrate; veins free. Sori on the ends of the lowest fork on the upper side of the main veins, about midway between the midrib and the margin, 1.5–3 mm, orbicular when young, reddish-brown or bright orange when mature. Annulus of sporangium dark reddish-brown, with 7–17 thick-walled cells, with 1 cell between its base and sporangium-stalk. $2n = 148^*$. Grh. or Ch.

Native. Woods, often on trees but also on the ground, rocks and walls, ascending to 855 m in Kerry; throughout the British Is., but absent from parts of C. and E. England, common in the wetter districts, less so in the drier ones. Most of Europe except the extreme north and south; Macaronesia; W. Siberia, Tibet, China, Japan; eastern North America from Newfoundland and Keewatin to Georgia and Missouri; S. Africa; Kerguelen Is.

2. P. interjectum Shivas

Similar to *P. vulgare* but scales on rhizome 3.5–11 mm, abruptly narrowed above wide base. Lf-blade (5–)15–30(–40) cm, oval; longest pinnae (20–)40–70(–90) mm, near the middle of the blade, subentire to shallowly serrate. Sori oval to suborbicular when young, yellow-brown when mature. Annulus of sporangium pale yellow to golden brown, with 4–13 thick-walled cells, with 2–3 cells between its base and sporangium-stalk. $2n = 222^*$. Grh. or Ch.

Native. Throughout the British Is. in similar habitats to *P. vulgare* but perhaps requiring higher humidity, often common on mature dune-systems. W. and W.C. Europe, extending locally eastwards to C. Russia.

P. × mantoniae Rothm. (*P. interjectum × vulgare*), is morphologically intermediate between the parents and very vigorous. Most spores are small, shrivelled and whitish. $2n = 185^*$. The hybrid is frequent throughout Britain but much less so in Ireland.

3. P. australe Fée

Similar to *P. vulgare* but scales on rhizome 5–16 mm, lanceolate. Lf-blade (5–)10–15 cm, triangular-dentate to broadly ovate; longest pinnae (35–)45–60(–75) mm, usually acute, usually serrate. Sori oval when young, bright yellow when mature, with long branched paraphyses. Annulus of sporangium yellow, with 4–19 thick-walled cells, with 3–4 cells between its base and the sporangium-stalk. $2n = 74^*$. Grh or Ch.

Native. On base-rich rocks usually on steep slopes and mostly below 150 m. S.W. England, Wales, N.W. England, W. Scotland northwards to Argyll, isolated occurrences further east in Sussex, Kent, Gloucester, Edinburgh and Perth; local throughout much of Ireland.

P. × fontqueri Rothm. (*P. australe × vulgare*), morphologically intermediate between the parents but with rather thin lvs often having twisted pinnae, has $2n = 111^*$ and all spores mis-shapen. It occurs locally on base-rich rocks, sometimes as an epiphyte in coastal areas, in E. Sussex, Monmouth, Anglesey, Yorks, Northumberland and Kirkcudbright.

P. × shivasiae Rothm. (*P. australe × interjectum*), which looks like a vigorous *P. interjectum* with light, yellow-green lvs up to 40–65 cm, has $2n = 148^*$ and is completely sterile. Occasionally found with the parents in Somerset, Pembroke, Anglesey, Kirkcudbright, Clare and Antrim.

10. HYPOLEPIDACEAE

Rhizome covered with hairs. Petioles with several vascular bundles below which fuse to form a single U-shaped strand above. Sori marginal, covered by deflexed lf-margin and an inner indusial flap. Spores tetrahedral.

1. PTERIDIUM Scop.

Rhizome creeping, subterranean. Lvs 3-pinnate; lower pinnae with basal nectaries; veins connected only by the marginal receptacle. Sori continuous round the margin of the lf-segment. Sporangia at first developing from apex to base of the sorus, later irregularly.

One sp.

1. P. aquilinum (L.) Kuhn   Bracken.

Pteris aquilina L.; *Eupteris aquilina* (L.) Newman
Rhizome underground, creeping for long distances, stout, hairy. Lvs (15–)30–180(–400) cm, (2–)3-pinnate, solitary, erect below with the blade bent towards the horizontal, pubescent and with numerous brown scales when young, dying in autumn; petiole to 2 m and c. 1 cm diam., about as long as blade, dark and tomentose at base, ½-cylindrical and soon glabrous above; blade ±deltate in outline; pinnae lanceolate or oblong; segments 5–15 mm, oblong, pectinately arranged, sessile, with a wide base, usually subacute, entire or the larger ones lobed at the base, rather thick, subglabrous above, pubescent below, at least on the main veins at maturity. Sori running all round the margins of the segments, both the recurved margin and the inner indusium membranous and ciliate. Spores ripe 7–8. $2n = 52, 104^*$. Grh.

Native. Woods, heaths, etc., mainly on light acid soils, rare on limestone and on wet peat; the commonest dominant in the field-layer of woods on acid soils. Dominant also over considerable areas formerly occupied by acid grassland or heather, spreading long distances vegetatively and favoured by the grazing of sheep or rabbits, neither of which animals eat it, and by fire; not tolerant of exposure, though ascending to 610 m in sheltered valleys in Scotland, nor of deep shade; common throughout the British Is. Cosmopolitan except for temperate South America and the Arctic, though just passing the Arctic circle in Norway; divisible into a number of geographical subspp.

11. THELYPTERIDACEAE

Rhizomes ascending or creeping, with hairs, or with relatively few papillate or hairy scales. Petioles with 2 vascular bundles below which fuse into a single U-shaped strand above; rhachis with a continuous groove above. Lvs pinnate, with pinnatifid pinnae, their lobes with wide bases; veins free. Sori up to 0.5 mm diam., submarginal; indusium ±reniform or 0; spores bilateral.

About 6 genera, cosmopolitan.

> *1* Rhizome ascending; lvs in a distinct crown; petiole c. ½ as long as blade; blade with dense, sessile, yellow glands beneath.   3. OREOPTERIS
> Rhizome creeping; lvs borne singly or in small tufts;

petiole about as long as blade or longer; blade
±eglandular. 2
2 Lvs bent at junction of rhachis and petiole; lowest
pair of pinnae turned downwards and forwards;
indusium absent. 2. PHEGOPTERIS
Lvs erect; pinnae all in one plane; indusium present.
1. THELYPTERIS

1. THELYPTERIS Schmidel

Rhizome long, creeping, subterranean, with few papillate scales. Lvs solitary, rarely in small tufts; petiole slender, about as long as to twice as long as blade; pinnae deeply pinnatifid, longest near middle of blade but only slightly shorter towards base, ±eglandular beneath. Sori ±midway between midrib and margin, but appearing closer to margin because this is recurved; indusium ±reniform, irregularly dentate.
About 500 spp., cosmopolitan.

1. T. palustris Schott   Marsh Fern.

Dryopteris thelypteris (L.) A. Gray; *Lastrea thelypteris* (L.) Bory; *Aspidium thelypteris* (L.) Swartz; *Nephrodium thelypteris* (L.) Strempel; *Thelypteris thelypteroides* Michx subsp. *glabra* Holub.

Rhizome with a few small scales, which soon disappear. Lvs erect, lanceolate in outline, pinnate, flat, thin, sparingly or not scaly, with a few white hairs on rhachis, midribs and margins, at least when young, the fertile up to 120 cm, the vegetative up to 80 cm; pinnae c. 5–9 cm, linear-lanceolate, all in one plane, with c. 10–20 segments on each side; segments 6–12 mm, smaller distally, oblong to triangular-lanceolate, wide at base, obtuse to acute, sinuate and occasionally with 1–2 small lobes at base; petiole as long as to twice as long as blade in fertile lvs, shorter in vegetative lvs, slender, brittle, blackish at base, sometimes with a few scales. Sori forming a row on each side of the segment; indusium small, thin, irregularly toothed. Spores ripe 7–8. $2n = 70^*$. Grh.

Native. Marshes and fens, often abundant in carr or alder wood. Cornwall and Kent northwards to Yorks and Kirkcudbright, but absent from much of the English Midlands and most of Wales except Caernarvon and Anglesey, also in Islay, Forfar and Easterness; in Ireland from Tyrone and Mayo to Wicklow and Clare, also in Kerry and Cork. Most of Europe except the extreme north and most Mediterranean islands; Caucasus; temperate Asia to Syria, Himalayas, S. China and Sakhalin; S. India; Algeria; tropical Africa; eastern North America. A var. in S. Africa and New Zealand.

2. PHEGOPTERIS (Presl) Fée emend. Ching

Rhizome long, creeping, subterranean, with brown, hairy, lanceolate scales when young. Lvs solitary, bent at junction of petiole and rhachis, pinnate; petiole as long as to twice as long as blade; pinnae pinnatifid almost to rhachis, the lowest pair reflexed backwards away from rest of blade, eglandular beneath. Sori close to somewhat recurved margin of segment; indusium 0.
Three spp., N. Temperate.

1. P. connectilis (Michx) Watt   Beech Fern.

Dryoptyeris phegopteris (L.) C.Chr.; *Polypodium phegopteris* L.; *Phegopteris polypodioides* Fée; *Thelypteris phegopteris* (L.) Slosson

Lvs 10–50 cm, pinnate, dying in autumn; petiole erect, slender, brittle, with a few scales at the base and sometimes a few small ones near the apex, the upper part usually clothed with reflexed white hairs; *blade triangular-ovate*, bent at nearly a right angle from the petiole, rather thin, light dull green, the rhachis scaly and ±hairy, *the pinnae ±hairy on both surfaces*, especially the midrib below; pinnae c. 10–20 on each side, *the lowest pair bent backwards away from the others*, 4–11 cm, lanceolate, *longer or slightly shorter than the one above it, the remainder decreasing rapidly in length*, all (except the lowest pair) *attached to the rhachis by a wide decurrent base*; segments c. 10–20 on each side of the lowest pinnae, longest near the middle of the pinna, c. 5–17 mm, oblong, wide at base, obtuse or subacute, usually entire or subentire but occasionally crenate or dentate. Sori c. 0.5 mm or less. Spores ripe 6–8. $2n = 90^*$. Apogamous. Grh.

Native. Damp woods and shady rocks, ascending to 1120 m, absent from limestone; Cornwall to Gloucester and Sussex; Wales and the border counties; Cheshire, Derby and Yorks northwards; rather common in the north, rare in S. England; Clare, Wicklow; Sligo, Cavan and Louth northwards. Europe southwards to the mountains of S. France, Corsica, N. Italy, the Balkan peninsula and the Caucasus; temperate Asia to Asia Minor, Himalaya and Japan; North America, south to Virginia and Oregon.

3. OREOPTERIS Holub

Rhizome short, stout, ascending, the apex covered with brown, papillate, ovate-lanceolate scales. Lvs in a distinct crown at apex of rhizome, pinnate; petiole c. ½ as long as blade; pinnae deeply pinnatifid, longest near middle of blade, all with numerous, sessile, golden yellow glands beneath. Sori close to margins of pinna-segments; indusium present or 0.
Three spp., this and 2 in E. Asia.

1. O. limbosperma (All.) Holub Lemon-scented Fern.

Dryopteris oreopteris (Ehrh.) Maxon; *Lastrea oreopteris* (Ehrh.) Bory; *Aspidium oreopteris* (Ehrh.) Sw.; *Nephrodium oreopteris* (Ehrh.) Desv.; *Lastrea montana* Newm.; *Thelypteris oreopteris* (Ehrh.) Slosson; *T. limbosperma* (All.) H. P. Fuchs

Lvs 30–100 cm, suberect not bent, dying in autumn; petiole rather stout, up to c. ¼ as long as blade, sparsely clothed with pale brown ovate or ovate-lanceolate scales which are most numerous below; *blade lanceolate or oblanceolate*, firm, yellowish-green, *smelling of lemon when crushed*, sparsely scaly on the rhachis below, the rhachis of the pinnae with short white hairs at least when young; *pinnae* c. 20–30 on each side, all in one plane and parallel, longest c. 5–12 cm linear-lanceolate, about the middle of the lf, *narrow at the base but scarcely* stalked, decreasing in length downwards, the *lowest distant*, *deltate*, c. 1 cm; segments c. 15–25 on each side of the longest pinnae, ±equal for some distance above the base or often the basal pair or the basal one on the upper side longer, c. 7–12 mm, oblong, wide at the base, obtuse or subacute, sinuate-crenate or subentire, the margins often recurved. Sori 0.5 mm or less; indusium small, thin, falling early, irregularly toothed, sometimes 0. Spores ripe 7–8. $2n = 68^*$. H.

Native. Woods, mountain pastures, screes, etc., especially characteristic of steep banks above streams, absent from limestone; ascending to 915 m; throughout Great Britain, common in the wetter districts, very local or absent in the S. English Midlands and East Anglia; Kerry, Wexford, Wicklow, Dublin, and from Clare to Down northwards. C. and N.W. Europe extending to N. Spain, Corsica and Serbia, east to S.W. Russia (Middle Dnieper region); Madeira; E. Siberia (Angara-Sayan region). A subsp. in Kamchatka, Japan and western North America.

12. ASPLENIACEAE

Rhizome erect to creeping, bearing firm scales with dark cell-walls. Petioles with 2 vascular bundles below which often fuse into a single X-shaped strand above. Sori borne along one or both sides of veins on underside of lf; indusium usually present; spores bilateral.

1. ASPLENIUM L.

Lvs in tufts, entire to 3-pinnate. Sori elliptical to linear, borne directly on a lateral vein; indusium a flap shaped like the sorus, usually opening towards middle of lf-segment.

About 650 spp., cosmopolitan.

1 Lvs densely covered beneath with brown, overlapping
   scales.   **10. ceterach**
   Lvs not or sparsely scaly beneath.   2
2 Lvs entire or slightly lobed.   **1. scolopendrium**
   Lvs pinnately lobed or divided, or irregularly forked. 3
3 Lvs irregularly dichotomous, forking into very narrow, linear segments.   **5. septentrionale**
   Lvs 1–3-pinnate, the segments not narrowly linear. 4
4 Lvs 1-pinnate, or 2-pinnate only near base. 5
   Lvs 2–3(4)-pinnate. 7
5 Rhachis reddish-black, brittle.   **8. trichomanes**
   Rhachis green. 6
6 Rhachis with conspicuous green wing.   **7. marinum**
   Rhachis not winged.   **9. viride**
7 Basal pair of pinnae shorter than those above it, frequently deflexed.   **4. billotii**
   Basal pair of pinnae longer than those above, not deflexed. 8
8 Petiole green, purple-brown at unwidened base; covered with small glands when young.
       6. ruta-muraria
   Petiole reddish-brown to almost black, widened at base, without glands. 9
9 Lvs, pinnae, pinnules and segments gradually tapering to long, slender apical points.   **3. onopteris**
   Lvs, pinnae, pinnules and segments not long-tapering at ends.   **2. adiantum-nigrum**

1. A. scolopendrium L.   Hart's-tongue Fern.

Scolopendrium vulgare Sm.; *Phyllitis scolopendrium* (L.) Newman.

Rhizome short, stout, ±erect, densely clothed with narrow brown scales. Lvs 10–60 cm, tufted, persistent, entire; petiole up to ½ as long as blade, usually much less, scaly; blade strap-shaped, cordate at base, tapered towards the usually obtuse apex, with scattered subulate scales at least when young; veins dichotomous, parallel, free. Sori linear, in close pairs on adjacent veins, usually occupying more than half the width of the lf; paired sori with conspicuous membranous indusia opening towards each other. Spores ripe 7–8. $2n = 72^*$. H.

Native. Rocky woods and hedgebanks, shady rocks and walls. Throughout the British Is.; common in the wetter districts, less so in the drier ones but local or absent in much of N. Scotland. S., W. and C. Europe; S. Sweden; Morocco (mountains); Macaronesia; Asia Minor to Caucasus and Iran; Japan.

2. A. adiantum-nigrum L.   Black Spleenwort.

A. cuneifolium auct. Brit., non Viv.

Rhizome short, creeping or decumbent, clothed when young with subulate, dark brown scales. Lvs 10–50 cm, tufted, persistent, somewhat coriaceous, 2–3-pinnate; petiole scaly and swollen at the extreme base, blackish, about as long as blade; blade triangular-ovate to triangular-lanceolate, naked, glossy green above; rhachis winged, blackish on the underside in the lower part; pinnae up to 15 on each side decreasing in size upwards, the lowest 2–6 cm, triangular-ovate or triangular-lanceolate, stalked; pinnules varying from lobed to pinnate; segments c. 4–12 mm, varying much in shape on different plants, ovate, obovate, elliptical or lanceolate, nearly always with a tendency to rhombic, with convex sides, acute or obtuse, cuneate at the narrow base, serrate. Sori 1–2 mm, linear-oblong or linear, occupying the greater part of the lateral veins, nearer the indistinct midrib than the margin of the segment, finally confluent and occupying the greater part of the segment; indusium whitish, entire or sinuate. Spores ripe 6–10. $2n = 144^*$. H.

Native. Rocky woods and hedgebanks and shady walls and rocks. Ascending to nearly 610 m. Throughout the British Is.; generally common but rather local in E. and C. England and C. Ireland. Europe from the Faeroes, Scandinavia, C. Germany and S. Russia (Middle Dnieper region and Crimea) southwards, absent from parts of the Mediterranean region; S.W. Asia to the Himalaya and Pamir-Altai; Macaronesia; N. Africa; mountains of tropical Africa, S. Africa, Réunion; Taiwan; Hawaii; Colorado.

Populations on serpentine rocks and screes in Scotland and W. Galway, with $2n = 144$, have been erroneously attributed to the C. and E. European **A. cuneifolium** Viv. They may merit subspecific status.

3. A. onopteris L.   Acute-leaved Spleenwort.

A. adiantum-nigrum subsp. *onopteris* (L.) Luerss.

Similar to *A. adiantum-nigrum*, but differs in having a more narrowly triangular, yellow-green lf-blade, with lvs, pinnae, pinnules and segments gradually tapering to long, slender apical points, in lacking the scattered, narrow, dark scales on the pinnae and midribs, and paler, smaller spores. $2n = 72^*$. H.

Native. Dry, lightly shaded earth banks and rockfaces, mainly of limestome. Cork, Kerry and Waterford; records from Down, Cornwall and Montgomery are uncertain. S. Europe, ?S.W. Poland.

A. × jacksonii (Alston) Lawalrée (*A. adiantum-nigrum × scolopendrium*; × *Asplenophyllitis jacksonii* Alston), a sterile hybrid with $2n = 108^*$, which resembles *A. adiantum-nigrum*, except that it has 1-pinnate, rather fleshy lvs, shorter, paler petioles and often paired sori, occurs locally in the Channel Is., Cornwall and N. Devon.

A. × ticinense D. E. Meyer (*A. adiantum-nigrum × onopteris*), a rare sterile hybrid with $2n = 108^*$, intermediate between the parents, has been recorded from 1 locality in Cork.

4. A. billotii F. W. Schultz   Lanceolate Spleenwort.

A. lanceolatum Hudson, p.p.; *A. obovatum* auct. angl., non Viv.

Rhizome short, erect to decumbent, with narrow, subulate, dark brown scales. Lvs 10–30 cm, tufted, persistent, 2-pinnate; petiole scaly at the extreme base, blackish, from about $\frac{1}{2}$ as long to nearly as long as the blade; blade lanceolate, with few scattered dark hairlike scales mainly on the rhachis, bright green; rhachis not or very narrowly winged, blackish on the underside in the lower part; pinnae up to 20 on each side, longest (c. 1–5 cm) about the middle of the blade but the lower ones little shorter, subsessile, the lowest pair usually markedly shorter than those above and deflexed; pinnules 4–10 mm, ovate or obovate, obtuse, cuneate at the base, dentate, the teeth mucronate, the lowest pinnule on the upper side of the pinna often larger than the others and lobed. Sori 1–2 mm, oblong, nearer the margin than the midrib of the pinnule, finally sometimes confluent; indusium whitish, entire. Spores ripe 6–9. $2n = 144^*$. H.

Native. Rocks, walls and hedgebanks, usually near the sea, very local; Cornwall, Devon, Dorset, Monmouth, W. Wales, extinct in Kent, Cumberland; Kintyre; Wexford, Clare, Wicklow, extinct in S.W. Ireland; Scilly and Channel Is. W. Europe, extending eastwards to Switzerland and C. Italy; Macaronesia.

A. × sarniense Sleep (*A. adiantum-nigrum × billotii*), a sterile hybrid with $2n = 144^*$, is similar to *A. adiantum-nigrum* but has a narrower lf, obtuse segments and oval pinnules on the middle pinnae. It is known from several sheltered roadside hedgebanks in Guernsey.

A. × microdon (T. Moore) Lovis & Vida (*A. billotii × scolopendrium*: × *Asplenophyllitis microdon* (T. Moore) Alston), a sterile hybrid with $2n = 108^*$, is morphologically intermediate between the parents. It is known from Guernsey, but may occur in S.W. England, W. Wales and S. Ireland if 19th-century records can be reconfirmed.

5. A. septentrionale (L.) Hoffm.   Forked Spleenwort.

Rhizome short, creeping, clothed with dark subulate scales when young. Lvs 4–15 cm, tufted, persistent, dark dull green throughout, except for the blackish base of the petiole, dichotomously divided, the forks often unequal; petiole usually several times as long as the blade (rarely only as long), with a few small hairs when young; blade 1–2(–3) times forked, the primary divisions stalked, glabrous; segments linear-cuneiform, 0.5–3 cm, very long-attenuate at the base, with a few long narrow teeth at the apex, or subentire without distinct midrib; veins dichotomous. Sori 5–20 mm, narrowly linear, covering almost the whole surface; indusium whitish, entire. Spores ripe 6–10. $2n = 144^*$. H.

Native. Crevices of siliceous rocks, sometimes on walls, rare; ascending to 915 m; Devon and Somerset (extinct), N. and C. Wales, Yorkshire, Cumberland, Roxburgh, Midlothian, E. Perth, S. Aberdeen, Westerness, ?W. Ross, Rhum; W. Galway. Throughout Europe except the extreme south; Caucasus; W. Asia from W. Siberia to the Himalaya and N. Syria; Morocco (Atlas and Rif); Japan; North America.

A. × contrei Callé, Lovis & Reichstein (*A. adiantum-nigrum × septentrionale*), morphologically intermediate between the parents, was recorded from 1 locality in Caernarvon a century ago, but could recur where its parents occur together.

6. A. ruta-muraria L.   Wall-Rue.

Rhizome short, creeping, clothed when young with dark subulate scales. Lvs 3–12(–15) cm, tufted, persistent, coriaceous, dark dull green throughout, except for the purple-brown base of the petiole, 2(–3)-pinnate; petiole 1–2 times as long as blade, glandular and with a few hair-like scales when young; blade triangular-ovate or triangular-lanceolate, naked; rhachis narrowly winged; pinnae 1–3 cm, 3–5 on each side, decreasing in size from the base upwards, stalked; pinnules rarely more than 5 and often only 3, even on the lowest pinnae, usually undivided but sometimes the lowest with 3–5 segments; segments 2–8 mm, varying considerably in shape on different plants, obovate-cuneiform to rhombic-lanceolate, obtuse, cuneate at the narrow base, crenate or dentate above the middle; veins dichotomous, without midrib. Sori c. 2 mm, linear nearer the base than the apex of the segment, finally confluent; indusium whitish, irregularly fimbriate. Spores ripe 6–10. $2n = 144^*$. H.

Native. Walls and mainly basic rocks, ascending to 610 m. Common throughout the British Is. Almost throughout Europe; N. and S. Asia to the Himalaya and Pacific; eastern North America from S. Ontario to Alabama and Missouri.

A. × **murbeckii** Dörfler (*A. ruta-muraria* × *septentrionale*), with triangular lvs composed of c. 3 narrowly wedge-shaped, stalked lobes, occurs rarely on rock-ledges in Cumberland.

7. A. marinum L.   Sea Spleenwort.

Rhizome short, ±erect, densely clothed with blackish subulate scales. Lvs 6–30(–100) cm, tufted, persistent, coriaceous, 1-pinnate; petiole scaly at extreme base, brown, $\frac{1}{3}$–$\frac{1}{2}$ as long as blade; blade lanceolate, naked, dark green; rhachis brown in the lower part, with conspicuous green wings; pinnae up to 20 on each side, longest (1–)1.5–4 cm about the middle of the blade, oblong or ovate-oblong, obtuse, markedly asymmetrical (more developed on the upper side where there is sometimes a small lobe) at the truncate or broad-cuneate narrow base, crenate-serrate, the apex of the frond often with a long lobed or pinnatifid point. Sori 3–5 mm, linear, on the upper fork of the secondary veins, about midway between the midrib and margin; indusium brownish, entire. Spores ripe 6–9. $2n = 72^*$. H.

Native. Crevices of sea-cliffs from Isle of Wight to Cornwall, thence to Shetland and southwards to N. Yorks, local, commonest along the west coast; all round the coast of Ireland and inland near Killarney (Kerry). W. Europe extending locally eastwards to S. Italy and Pantellaria; Morocco; Algeria; Macaronesia.

8. A. trichomanes L.   Maidenhair Spleenwort.

Rhizome short, creeping to ±erect, with dark narrow scales. Lvs 4–20(–40) cm, tufted, persistent, not coriaceus, simply pinnate; petiole not scaly, $\frac{1}{4}$ as long as blade or less, like the rhachis reddish-black, and with a narrow brownish wing; blade linear, deep green; pinnae 3–7(–10) mm, 15–40 on each side, ±equally long in the middle of the blade for some distance, oval or oblong, obtuse, somewhat asymmetrical at the truncate or cuneate base, crenate or crenate-dentate round the apex and on the upper margin, usually with a few short hairs below, finally falling from the rhachis. Sori 1–2 mm, oblong-linear, situated mainly on the upper branch of the veins, though sometimes continuing below the fork, about midway between the midrib and margin; indusium whitish, entire or nearly so. Spores ripe 5–10. H.

Native. Walls and crevices of mainly basic rocks, ascending to 870 m. Common throughout the British Is. Throughout Europe. N. and S. temperate zones and mountains of the tropics.

Subsp. **trichomanes**. Largest rhizome-scales up to 3.5 mm, with central reddish-brown stripe. Petiole slender, wiry, usually reddish-brown. Pinnae orbicular to rhombic, flat or concave. $2n = 72^*$.

Wales, Cumberland, Scotland northwards to Ross. Most of Europe.

Subsp. **quadrivalens** D. E. Meyer emend. Lovis. Largest rhizome-scales up to 5 mm, with central dark brown stripe. Petiole stout, dark brown. Pinnae usually oblong, or auriculate, convex with inrolled margins, rarely flat. $2n = 144^*$.

Thoughout the British Is. Most of Europe.

A. × **alternifolium** Wulfen (*A. septentrionale* × *trichomanes* subsp. *trichomanes*), a sterile hybrid ($2n = 108^*$), which can be distinguished from *A. septentrionale* by its alternate pinnae, pale green lvs and abortive spores, occurs in N. Wales and Cumberland.

9. A. viride Hudson   Green Spleenwort.

Differs from *A. trichomanes* as follows: lvs sometimes not persistent; petiole usually relatively longer, brownish or blackish near the base, green above; rhachis green, not winged; pinnae orbicular or ovate-orbicular, less unequal at the base, more deeply toothed all round, paler green, not falling from the rhachis, glabrous. Sori nearer the midrib than the margin of the pinna, situated mainly below the fork of the veins though sometimes extending along the upper fork. Spores ripe 6–9. $2n = 72^*$. H.

Native. Crevices of basic rocks in hilly districts, local, ascending to 960 m. Monmouth, Hereford, Stafford, Wales; Lancashire, Derby and Yorks northwards; Cork, Tipperary, Galway to Donegal. Most of Europe, mainly in the mountains; Caucasus; Morocco (Atlas); W. Asia east to Yenisei region of Siberia, south to the Himalaya; North America south to Vermont and Washington.

10. A. ceterach L.   Rustyback Fern.
Ceterach officinarum DC.

Rhizome short, ±erect, clothed with dark narrow scales. Lvs 3–20 cm, tufted, persistent, coriaceous, simply pinnate; petiole $\frac{1}{4}$ as long as blade or less, scaly; blade dull green above, below entirely covered by light brown, ovate, overlapping scales; pinnae up to 2 cm, ovate or oblong, rounded at apex, somewhat widened at the wide base, entire or crenate; veins anastomosing. Sori c. 2 mm, linear; indusium 0. Spores ripe 4–10. $2n = 144^*$. H.

Native. Crevices of limestone rocks and mortared walls, ascending to 440 m in Wales. Rather common in S. and W. England, Wales and Ireland; very local in E. England and absent from several counties; in Scotland very local and mainly in the south-west extending northwards to Kincardine, and Easterness. W., C. and S. Europe, northwards to S. Sweden, eastwards to the Crimea; east and north to the Himalaya, Tien-Shan, Caucasus; Madeira.

13. ATHYRIACEAE

Rhizome creeping to ±erect, with thin-walled, opaque scales. Lvs 1–3-pinnate, veins free; petioles grooved, with 2 vascular bundles which unite distally into a single strand. Sori on a side-branch from the vein, hooked, oblong or orbicular; indusium present or 0; spores bilateral.

1 Indusium rudimentary and falling early or 0.   2
　 Indusium distinct.   3
2 Lvs in a crown; rhizome ±erect, stout; sori oblong.
　　　　　　　　　　　　　　　　　　　　1. ATHYRIUM
　 Lvs solitary; rhizome creeping, slender; sori orbicular.   2. GYMNOCARPIUM
3 Indusium a ring of hairy scales around base of sorus.
　　　　　　　　　　　　　　　　　　　　4. WOODSIA
　 Indusium not a ring of scales.   4
4 Sori oblong; indusium flap-like, oblong (the lower reniform or hooked).   1. ATHYRIUM
　 Sori orbicular; indusium acuminate covering sorus like a hood but later becoming shrivelled.
　　　　　　　　　　　　　　　　　　　3. CYSTOPTERIS

1. ATHYRIUM Roth

Rhizome scales large and soft. Lvs 1–3-pinnate, usually rather limp, not hairy; veins free. *Sori borne on receptacles with a vascular strand branched off from the vein.* Indusium *reniform or horseshoe-shaped to J-shaped or linear, sometimes vestigial or absent.* Stele of rhachis U-shaped.

About 180 spp., cosmopolitan, mainly temperate E. Asia.

1 Lvs 8–20(–30) cm; petiole up to $\frac{1}{8}$ as long as blade, mostly bent sharply backwards at a distinct elbow; sori mainly on lower part of lf.   **3. flexile**
　 Lvs 20–120 cm; petiole c. $\frac{1}{4}$ or more as long as blade, suberect or arching outwards; sori well distributed over blade, or in the upper part.   2
2 Sori mostly oblong; indusium conspicuous at maturity.   **1. filix-femina**
　 Sori orbicular; indusium absent at maturity.
　　　　　　　　　　　　　　　　2. distentifolium

1. A. filix-femina (L.) Roth   Lady-fern.

Asplenium filix-femina (L.) Bernh.

Rhizome short, ±erect, stout, densely clothed with brown lanceolate scales. Lvs forming a crown at the apex of the rhizome, dying in autumn, usually spreading and often drooping at the ends but sometimes suberect, 20–100 (–150) cm, 2-pinnate, rarely 3-pinnate; petiole usually at least $\frac{1}{4}$ as long as blade, scaly, at least in the lower part, with brown lanceolate scales; blade thin and rather flaccid, light green, lanceolate; rhachis green or purplish-red, naked or with scattered hairs or scales, very rarely the midribs also with short whitish hairs; pinnae up to 30 on each side, the longest 3–25 cm, about the middle of the lf, the lowest considerably shorter, linear-lanceolate, tapered at the apex to an acute point; rhachis winged; pinnules regularly arranged, oblong or oblong-lanceolate, 3–20 mm, subobtuse to acute, sessile,

truncate at the narrow base, pinnately lobed or pinnatifid, the lobes often toothed. *Sori* c. 1 mm, forming a row down either side of the pinnule, nearer the midrib than the margin, ±*oblong*, the lower ones C- or J-shaped, the upper nearly straight; *indusium persistent*, covering the sorus till the spores are ripe, whitish, toothed; spores papillose. Very variable. Spores ripe 7–8. $2n = 80^*$. H.

Native. Damp woods and hedgebanks, shady rocks, screes, marshes, etc.; calcifuge, ascending to 1000 m in Caernarvon. Throughout the British Is.; common in most districts but rather local in E. England. Throughout the north temperate zone and south to the mountains of India and Java and in tropical America to Peru and Argentina.

2. A. distentifolium Tausch ex Opiz   Alpine Lady-fern.

A. alpestre Clairv.; *Polypodium alpestre* (Hoppe) Spenn.

Differs from *A. filix-femina*, from which it is only certainly distinguishable by the sorus, as follows: Petiole shorter, usually $\frac{1}{4}$ or less as long as blade; lobes of segments usually wider and blunter. *Sori* nearer the margin of the pinnule, *orbicular*, very small; *indusium rudimentary, falling long before the spores are ripe*; spores reticulate. Spores ripe 7–8. $2n = 80^*$. H.

Native. Screes and rocks, usually acid, on mountains from 365 to 1095 m. From Argyll, Stirling and Angus to Sutherland on the mainland and mainly in the east, local. N. Europe and Asia from Iceland to Kamchatka and in the high mountains of Europe south to the Pyrenees, Italy and Romania; Caucasus.

3. A. flexile (Newman) Druce

A. alpestre var. *flexile* (Newman) Milde

Differs from *A. distentifolium* in the very short petiole not more than $\frac{1}{8}$ as long as blade, usually bent sharply backwards at a distinct elbow; lvs generally smaller, 8–20(–30) cm, with sori borne only in lower part of the lf. Spores ripe 5–7. $2n = 80^*$. H.

Native. Local and rare in a few screes and corries at 1040–1140 m in the Highlands of Perth, Argyll and Westerness.

2. GYMNOCARPIUM Newman

Rhizomes slender, creeping; scales on younger part only, broadly ovate, brown, fringed with papillae. Leaves solitary, in 2 ranks, erect; petiole 1½–3 times as long as blade. Blade deltate; veins free. Sori orbicular or somewhat oblong, borne near margins of segments, often becoming confluent; indusium absent.

About 5 spp. N. temperate regions, Taiwan, Philippines and New Guinea.

Lvs bright or clear green, without glands; petiole shiny, dark.   **1. dryopteris**
Lvs dull green, glandular; petiole paler, dull, glandular.
　　　　　　　　　　　　　　　　2. robertianum

1. G. dryopteris (L.) Newman   Oak Fern.

Phegopteris dryopteris (L.) Fée; *Polypodium dryopteris* L.; *Dryopteris linnaeana* C.Chr.; *Thelypteris dryopteris* (L.) Slosson.

Rhizome long, creeping below the ground, with a few brown ovate scales when young. *Lvs 10–40 cm, 3-pinnate or 2-pinnate with deeply pinnatifid pinnae,* when young rolled up so as to resemble 3 small balls, dying in autumn; petiole erect, $1\frac{1}{2}$–3 times as long as blade, slender, brittle, shiny, dark brown, with a few scales near the base; *blade deltate,* bent at nearly a right angle to petiole, *brilliant yellowish-green when young,* becoming less bright *but remaining a clear green, thin, naked;* pinnae c. 5–10 on each side, the lowest pair much longer than the others, not deflexed, 3–15 cm, long-stalked, triangular-ovate; *lowest pinnule on lower side of lowest pinna much the longest,* 1–6 cm, about as long as the *3rd pinna from the base on either side;* upper pinnae not more than pinnate with lobed pinnules, sessile or shortly stalked; segments ±oblong, but not markedly parallel-sided, c. 5–15 mm, the lower ones rounded at the narrow base, entire, toothed or lobed; margins flat. Sori rather small (sometimes nearly 1 mm), near the margin of the segments; indusium 0. Spores ripe 7–8. $2n = 160^{*}$. G.h.

Native. Damp woods and shady rocks, ascending to 915 m; absent from limestone. Somerset; Berks and Kent and in a few localities to E. Norfolk and N. Lincoln; Wales and the border counties; Cheshire and Derby northwards; rather common in the north, rare in S. England; very rare in Ireland (Cavan, Antrim). Most of Europe, but rare in the south; N. Asia, Caucasus, W. Himalaya, China and Japan; North America south to Virginia and Oregon.

2. G. robertianum (Hoffm.) Newman Limestone Fern.

Phegopteris robertiana (Hoffm.) A.Br; *Polypodium robertianum* Hoffm.; *P. calcareum* Sm.; *Dryopteris robertiana* (Hoffm.) C.Chr.; *Thelypteris robertiana* (Hoffm.) Slosson.

Differs from *G. dryopteris* as follows: Lvs rather larger on an average (15–55 cm), when young each pinnule and pinna rolled up separately, the lf then rolled up as a whole; scales often ascending to nearly middle of petiole; *blade little bent relative to the petiole, dull green, rather firm, glandular on the rhachis and rhachises of the pinnae and on the surface beneath,* the glands short-stalked and appearing mealy; pinnae up to c. 15; *lowest pinnule on lower side of the lowest pinna the longest but less markedly so than in G. dryopteris and considerably smaller than the 3rd pinna from the base;* 2nd pair of pinnae often long-stalked; segments parallel-sided, the margins usually somewhat recurved. Spores ripe 7–8. $2n = $ c. 160^{*}. Grh.

Native. Limestone screes and rocks, ascending to 610 m, local; Somerset, Middlesex and Suffolk to Cumberland and Durham; Perth, Sutherland; E. Mayo.

Europe from Scandinavia and the Ladoga-Ilmen region of Russia to N. Spain, Corsica, Italy, Thessaly, Crimea and Caucasus; mountains of Asia from Afghanistan to the Far East of Russia; North America from Labrador to Alaska south to New Brunswick and Iowa.

3. CYSTOPTERIS Bernh.

Rhizome scales soft. Veins free. *Sori orbicular, borne on receptacles with a vascular strand branched off from the vein. Indusium attached at the base of the sorus, flap-like,* at first arching over the sorus like a hood, later becoming reflexed and exposing the sporangia.

About 18 spp., temperate and subtropical regions.

1 Rhizome far-creeping; lvs solitary; lowest pair of pinnae the longest.   **3. montana**
  Rhizome short, ±decumbent; lvs tufted; pinnae decreasing in length towards the base.   *2*
2 Spores rugose-verrucose; pinnules overlapping, crenate or shallowly crenately lobed. Kincardine.   **2. dickieana**
  Spores with numerous acute tubercles; pinnules not (or only a few of them) overlapping, dentate to deeply pinnatifid.   **1. fragilis**

1. C. fragilis (L.) Bernh.   Brittle Bladder-fern.

Rhizome short, rather stout, ±decumbent, clothed with thin brown lanceolate scales. *Lvs 5–30(–45) cm, tufted,* dying in autumn, suberect, 2(–3)-pinnate; petiole with a few scales at the base and usually a very few hair-like scales above, from $\frac{1}{3}$ as long to as long as the blade, dark brown at the base, paler brownish-green above, slender and brittle; blade thin, naked, lanceolate; pinnae up to 15 on each side, at least the lowest distant, the *longest 1–4 cm, about the middle of the lf* or rather below, ovate or lanceolate; *pinnules 4–10 mm not* (or only a few of them) *overlapping,* ovate, lanceolate or oblong, acute or obtuse, cuneate at the narrow base, ±decurrent, very variable in toothing from *dentate to deeply pinnatifid;* teeth obtuse or acute, rarely retuse or bidentate at the apex, but frequently toothed on the margins, the veins usually ending in the apices of the teeth. Sori in two rows on either side of the midrib of the pinnule, small; indusium whitish, ovate, acuminate, exceeding the sorus. *Spores with numerous narrow acute tubercles.* Very variable. Spores ripe 7–8. H. $2n = 168^{*}, 210^{*}, 252^{*}, 336$.

Native. Rocky woods, shady rocks and walls, especially on basic rocks, ascending to 1215 m; rather common in Scotland, N. England, Wales and N. and W. Ireland, becoming more local southwards and eastwards and absent from several counties in S. and E. England and S., C. and E. Ireland. Cosmopolitan, only on mountains in the tropics; the most widespread of all ferns, extending from 81° 47′ N in Greenland to Kerguelen Is. The plants with different chromosome numbers are possibly morphologically separable but the details have not yet been worked out.

2. C. dickieana Sim

Differs from *C. fragilis* as follows: Pinnae often all over-lapping; rhachis broadly winged; *pinnules overlapping, obtuse, crenate or shallowly crenately lobed*; lobes some-times indistinctly emarginate, the veins then ending in the notch; petiole c. $\frac{1}{4}$ as long as blade. *Spores rugose-verrucose.* Spores ripe 7–8. H. $2n = 168*$.

Native. Sea-caves in Kincardine, very rare. In a few isolated localities in arctic Europe, Asia and Siberia and on mountains in France, Germany, Switzerland, Spain, Sardinia and Sicily.

3. C. montana (Lam.) Desv.   Mountain Bladder-fern.

Rhizome long and creeping, blackish, rather slender (c. 2 mm diam.) with a few scattered ovate scales when young. *Lvs* 10–30(–45) cm, *solitary, distant*, dying in autumn, 3-pinnate; petiole with a few ovate-lanceolate scales mainly near the base, at least twice as long as the blade, dark brown at the base, pale brownish-green above; blade deltate, sparsely glandular below; pinnae up to 13 on each side, *the lowest much the longest*, 2.5–7 cm, triangular-ovate; segments c. 7 mm, ovate or oblong, obtuse, ±cuneate at the narrow base, pinnately lobed or pinnatifid, the lobes usually bidentate at the apex and often with 1 or 2 teeth on the margin. Sori small and widely separated; indusium whitish ovate-orbicular, acute or suboptuse, irregularly toothed. *Spores with low, rounded tubercles.* Spores ripe 7–8. $2n = 168$. Grh.?

Native. Damp, usually basic, rocks on mountains from 700 to 1100 m; now extinct in Westmorland; Argyll, Angus and Inverness; rare and very local. N. and C. Europe, extending southwards to the Pyrenees and C. Yugoslavia; Caucasus; Siberia (east to the Yenisei region); North America, south to Ontario and British Columbia.

4. WOODSIA R.Br.

Rhizome short, sparsely scaly, covered by persistent lf-bases. Lvs tufted with a joint in the petiole, 1–2-pinnate; veins free. Sori orbicular, borne on the lower surface of the lvs near the vein-endings; indusium divided into a fringe of hair-like scales surrounding the base of the sorus. Spores bilateral.

About 40 spp., Arctic and mountains of north temperate zone (to Himalaya) extending down the Andes to temperate South America, 1 in S. Africa.

Largest pinnae oblong or ovate-oblong, 1½–2 times as long as wide; rhachis and underside of pinnae with dense, long (2–3 mm) scales.   **1. ilvensis**
Largest pinnae triangular-ovate, 1–1½ times as long as wide; rhachis and underside of pinnae with sparse, short (c. 1 mm) scales, or the latter without scales.   **2. alpina**

1. W. ilvensis (L.) R.Br.   Oblong Woodsia.

Rhizome short, ±erect, sparsely scaly above. Lvs tufted, dying in autumn, suberect, 5–15 cm, dull green, simply pinnate or almost 2-pinnate; petiole clothed with brown lanceolate scales below and subulate scales above

and with flexuous hairs throughout, pale reddish-brown, jointed $\frac{1}{3}$ to $\frac{1}{2}$ from base, from $\frac{1}{2}$ as long as to as long as blade; blade oblong-lanceolate, the *rhachis and under surface* (at least the veins) *±densely clothed with long* (2–3 mm) *subulate pale brown scales* and flexuous hairs; pinnae c. 7–15 on each side, the longest about the middle of the lf, 7–17 mm (the lowest sometimes scarcely shorter, sometimes considerably so), *oblong or ovate-oblong*, 1½–2 times as long as wide obtuse or suboptuse, ±truncate at the narrow base, deeply pinnatifid into 7–13 oblong, obtuse, ±crenate lobes. Sori near the margin of the lobes; indusium surrounding the base of the sorus, divided nearly to the base into numerous irregular lobes which end in long jointed hair points which are longer than the rest of the indusium and arch over the sporangia. Spores ripe 7–8. $2n = 82*$. H.

Native. Rock crevices on mountains from 365 to 825 m, very rare; Merioneth, Caernarvon, Durham, Westmorland, Cumberland, Dumfries, Angus. Arctic and high mountains of north temperate zone, south to the S. Alps, S.W. Russia (Middle Dnieper region); Altai, Iowa and North Carolina.

2. W. alpina (Bolton) S. F. Gray   Alpine Woodsia.

W. hyperborea (Lilj.) R.Br.

Differs from *W. ilvensis* as follows: Usually smaller. Lvs 3–8(–15) cm; petiole with few scales, glabrescent, $\frac{1}{4}$–$\frac{2}{3}$ as long as blade; blade oblong-linear, the *rhachis sparsely clothed with short* (c. 1 mm) *subulate* (or a few of them lanceolate) *scales, the undersurface of the pinnae without or with very few* (up to c. 5) *scales*, the hairs also few; *pinnae* 5–12 mm, *triangular-ovate*, 1–1½ *times as long as wide*, obtuse, pinnately lobed or pinnatifid into 3–7 obovate or oblong-obtuse, ±crenate lobes. Spores ripe 7–8. $2n = $ c. 164*. H.

Native. Rock crevices on mountains from 580 to 915 m, very rare. Caernarvon, Perth, Angus, Argyll. Arctic and high mountains of north temperate zone, south to the Pyrenees, S. Alps, S. Russia (Black Sea region); Altai, W. Ontario and New York.

The following two members of the Athyriaceae are grown in gardens and appear to be locally naturalised.

***Matteuccia struthiopteris** (L.) Tod. Rhizome short, stout, erect, scaly, producing underground stolons and with an apical crown of lvs. Lvs dimorphic; vegetative lvs up to 1 m, 2-pinnatifid, dying down in autumn, the veins free; fertile lvs brown and much smaller (up to 60 cm), persistent. Sori globose, covered by recurved lf-margin, usually contiguous in longitudinal rows. Locally naturalised in N.E. Ireland. Native of N., E. and C. Europe.

***Onoclea sensibilis** L. Rhizome creeping, scaly. Lvs dimorphic; vegetative lvs up to 1 m, ovate-triangular, pinnate, with lobed pinnae, dying down in autumn, the veins anastomosing, the petiole longer than the blade; fertile lvs lanceolate, with narrow, brown pinnae. Sori globose, 1 on each pinna-lobe, the lobes recurved to form a globose structure enclosing several sori; indusium caducous. Locally naturalised in S.E. and N.W. England and in C.E. Scotland. Native of eastern North America and N. Asia.

14. ASPIDIACEAE

Rhizome short, erect or ascending, with densely opaque scales. Lvs in tufts at apex of rhizome, 1–4-pinnate; veins free; petiole with 5–7 vascular bundles. Sori orbicular; indusium peltate or reniform; spores bilateral.

Indusium peltate; teeth of lf-segments with apical bristle or spine.   1. POLYSTICHUM
Indusium reniform; teeth of lf-segments without spines.   2. DRYOPTERIS

1. POLYSTICHUM Roth

Rhizome stout, erect or ascending. Lvs 1–2-pinnate, often coriaceous; pinnae and pinnules asymmetrical, their teeth with apical bristle or spine. Indusium peltate, ±orbicular.

About 135 spp., cosmopolitan.

1 Lvs pinnate, narrowly oblong or linear-lanceolate; pinnae undivided.   **3. lonchitis**
Lvs 2-pinnate (at least below), lanceolate; pinnae divided.   2
2 Pinnae reducing in length towards base of lf; petiole $\frac{1}{4}$–$\frac{1}{5}$ or less as long as blade.   **2. aculeatum**
Pinnae not or scarcely reducing in length towards base of lf; petiole $\frac{1}{4}$ or more as long as blade. **1. setiferum**

1. P. setiferum (Forskål) Woynar   Soft Shield-fern.

P. angulare (Willd.) C. Presl; *Aspidium angulare* Kit. ex Willd.

Lvs 30–150 cm, persistent or subpersistent, *2-pinnate*; petiole c. $\frac{1}{4}$–$\frac{1}{3}$ as long as blade, clothed with ovate-lanceolate brown scales; *blade* deep green above, paler below, *rather flaccid and usually arching or drooping*, lanceolate, scaly on the rhachis, the lower surface with or without a few hair-like scales; pinnae c. 30–40 on each side, the longest c. 6–8 cm, about the middle of the lf, the lowest much shorter, linear-lanceolate, fully pinnate, straight or the upper curved towards the apex; *pinnules c. 12–20 on each side*, the basal one on the upper side usually the longest, 4–10 mm, ±ovate, curved, unequally cuneate at the shortly stalked base with the *margins meeting each other at about a right angle on most of the pinnules and at much more on one or more of the lowest ones of each pinna*, acute and spine-pointed at the apex, with a rounded or deltate lobe on the outer margin at the base reaching nearly to the midrib on the lowest pinnules, serrate with the teeth rounded on the outer edge but ending in a straight spine. Sori c. 0.5–1 mm, in a row down each side of the midrib of the pinnule and of its basal lobe; *vein on which the sorus is borne not or only shortly continued beyond it*. Spores ripe 7–8. $2n = 82^*$. H.

Native. Woods, hedgebanks, etc. Common in S.W. England becoming more local northwards and eastwards to Midlothian, Argyll, W. Ross, to Inverness and Moray; common throughout Ireland except parts of the centre. S.W. and C. Europe; Caucasus; the same or closely related spp. extending to many temperate regions of the world.

2. P. aculeatum (L.) Roth   Hard Shield-fern.

P. lobatum (Hudson) Chevall.; *Aspidium aculeatum* (L.) Swartz

Lvs 10–100 cm, persistent, *2-pinnate* (at least below); petiole up to $\frac{1}{5}$–$\frac{1}{4}$ as long as blade, clothed with ±ovate brown scales; *blade* dark green above and somewhat glossy, paler below, *rigid, somewhat coriaceous*, lanceolate or linear-lanceolate, scaly on the rhachis and with hair-like scales on the surface beneath; *pinnae* c. 25–50 on each side, the lowest much shorter, ovate-lanceolate to linear-lanceolate, *varying on different plants from fully pinnate to pinnately lobed with only a single free pinnule* and that sometimes only on a few of the lower pinnae, straight or curved towards the apex; *pinnules up to* c. 15 *on each side*, the basal one on the upper side of the pinnule the longest, 5–15 mm, rhombic-ovate to rhombic-lanceolate, somewhat curved, unequally cuneate at the narrow but scarcely stalked base *with the margins meeting each other at less than a right angle* (or sometimes at a right angle in the lowest pinnule of each pinna), acute and spine-pointed at the apex, often with a small deltate lobe on the outer margin at the base, serrate, with spine-pointed straight or somewhat incurved teeth. Sori c. 0.5–1 mm, in a row down each side of the midrib of the pinnule and sometimes also on its basal lobe; *vein on which the sorus is borne continuing well beyond the sorus*. Spores ripe 7–8. $2n = 164^*$. H.

Native. Woods, hedgebanks, etc., ascending to nearly 760 m; from Orkney southwards, rather common in the wetter districts, local in the drier ones; throughout Ireland but rather local, especially in the south. Most of Europe except the east and extreme north; Caucasus; N. Africa (mountains); S.W. Asia to India and the Tien Shan; China, Japan.

P. × bicknellii (Christ) Hahne (*P. aculeatum × setiferum*), which produces mostly abortive spores ($2n = 124^*$), differs from *P. aculeatum* in its longer petiole, longer lowest pinna-pair, more fully and deeply cut pinnae and often stalked lowest pinnules, occurs scattered throughout the British Is. in shaded lowland habitats.

3. P. lonchitis (L.) Roth   Holly Fern.

Aspidium lonchitis (L.) Swartz

Rhizome ascending. *Lvs* 10–30(–60) cm, persistent, *simply pinnate*; petiole $\frac{1}{6}$ as long as blade or less, clothed with ±ovate, reddish-brown scales; *blade deep green above, paler beneath, coriaceous*, linear-lanceolate or linear, with few or numerous scales on the rhachis and narrow hair-like ones on the lower surface; pinnae c. 20–40 on each side, the longest c. 1–3 cm, about the middle of the lf, decreasing in size downwards, the lowest 5 mm or less, but often pinnae of equal length throughout most of the blade, close together and often overlapping, ovate or ovate-lanceolate and somewhat curved towards the apex, acute, very unequal at the

narrow short-stalked base where the upper margin runs
±parallel with the rhachis and the lower diverges from
the rhachis at an angle of c. 45°, with a single ±deltate
lobe on the upper side at the base, serrate with straight
spine-pointed teeth, the teeth often crenulate or serrulate. Sori c. 0.5–1 mm, in a row down each side of the
midrib of the pinna (nearer the midrib than the margin)
and a double row down the basal lobe, usually confined
to the upper part of the lf; indusium irregularly toothed.
Spores ripe 6–8, $2n = 82^*$. H.

Native. Crevices of basic rocks on mountains from
60 m in Sutherland to 1030 m in Perth, local; Merioneth,
Caernarvon; Yorks and Westmorland to Northumberland and Dumfries; Dunbarton and Stirling to Orkney;
Kerry, Galway, Sligo, Leitrim, Donegal. Arctic and
mountains of north temperate zone, south to mountains
of S. Europe; Caucasus, Himalaya; N. California and
Nova Scotia.

P. × **lonchitiforme** (Halacsy) Becherer (*P. lonchitis* × *setiferum*), is a sterile diploid hybrid ($2n = 82^*$) and differs from
P. lonchitis in that the softer lvs have the pinnae cut at their
bases like *P. setiferum* and in having more finely spinose margins. One locality in Leitrim (Glenade).

P. × **illyricum** (Borbás) Hahne (*P. aculeatum* × *lonchitis*),
a largely sterile triploid ($2n = 123^*$), is morphologically intermediate between the parents and looks like a less-divided, narrower-lvd form of *P. aculeatum*. Leitrim (Glenade) and W
Sutherland (Inchnadamph).

*Cyrtomium falcatum (L. fil.) C. Presl. Sori like *Polystichum*
but lvs with anastomosing veins. lvs pinnate, pinnae 6–10 cm,
leathery, often with pointed lobe on one side at base. Much
grown as a pot plant and reported naturalized on warm walls
in S. and W. Britain. E. Asia, S. Africa, Madagascar, Hawaii.

2. DRYOPTERIS Adanson

Rhizome stout, erect or ascending, densely scaly, with
soft scales. Lvs forming a crown, deeply 2-pinnatified
to 3-pinnate (rarely 1-pinnate) with ±numerous scales
and without hairs; veins free; petiole with 4 or more
vascular bundles. Sori large, usually orbicular; indusium
reniform, nearer the midrib or about midway between
the midrib and margin of the segment.

About 150 spp., mainly north temperate, also tropical
and S. Africa, tropical Asia, tropical America.

 1 Lvs pinnate, with deeply pinnatifid pinnae, or 2-pin-
  nate; lobes or pinnules (except for basal pair of
  each pinna) at mostly slightly narrowed at base;
  longest pinnae near middle of lf.   2
  Lvs 3(–4)-pinnate, or 2-pinnate with deeply pinnati-
  fied pinnules; 1st-order pinnules distinctly narrowed or stalked at base; longest pinnae at or near
  base of lf.   5
 2 Lower pinnae markedly broadly based and triangular;
  fertile lvs longer and more erect than vegetative
  lvs; pinna-lobes and pinnules with mucronate marginal teeth; petiole sparsely scaly.   **6. cristata**
  Lower pinnae not markedly broadly based, triangular-lanceolate to lanceolate; fertile and vegetative
  lvs similar; petiole densely scaly (**filix-mas** group). 3

 3 Lf-blades thick, somewhat glossy and dark green
  above; petiole and rhachis with long, narrow scales
  having a dark base and, usually, a dark centre;
  lower margins of pinnules or pinna-segments parallel, with few teeth.   **3. affinis**
  Lf-blades thin, mid- or grey-green; petiole and rhachis
  with both narrow and wide pale-coloured scales;
  pinnules or pinna-segments tapering from base,
  usually with lobes or teeth on lower margins.   4
 4 Lf-blades grey-green; teeth of pinnules or pinna-segments obtuse, spreading in a fan-like arrangement
  at the apex; immature indusium with involute margin.   **1. oreades**
  Lf-blades mid-green; teeth of pinnules or pinna-segments acute, converging towards the apex; margin
  of immature indusium not involute.   **2. filix-mas**
 5 Lvs 2-pinnate, with deeply pinnatifid pinnules; basal
  pair of pinnules on lowest pair of pinnae subequal;
  pinnules or pinna-lobes with acute marginal teeth.
           4. submontana
  Lvs 2–3-pinnate; basal pair of pinnules on lowest pair
  of pinnae very unequal; pinnules or pinna-lobes
  with mucronate or aristate marginal teeth.   6
 6 Pinnae all markedly concave, giving lf a distinct crispate appearance; petiole, rhachis and midribs densely covered with sessile glands which also extend
  less densely to underside, and sometimes the upperside, of the blade.   **5. aemula**
  Pinnae not obviously concave; lvs sometimes with a
  few glandular hairs, otherwise eglandular.   7
 7 Pinnules with convex upper surface; scales at base
  of petiole with dark, often wide, central stripe; lvs
  blue-green.   **8. dilatata**
  Pinnules flat; scales at base of petiole concolorous,
  pale tan to reddish-brown; lvs yellowish-green.   8
 8 Petiole about as long as blade.   **7. carthusiana**
  Petiole up to c. ½ as long as blade.   **9. expansa**

(1–3). D. filix-mas group   Male-fern.
Rhizome ±erect. *Lvs* 15–150(–180) cm, *pinnate with
deeply pinnatifid pinnae or 2-pinnate*; blade naked
except for scales on the rhachis and midribs and sometimes some minute sessile glands on the under surface;
*pinnae 20–35 on each side, the longest 5–15 cm, linear-
lanceolate, about the middle of the lf, decreasing considerably in length downwards*; pinnules or segments c.
15–25 on each side of the longest pinnae, regularly
arranged, ±equal for some distance above the base
(except sometimes for a larger lowest pinna on the upper
side), oblong, 4–22 mm × 2–5 mm, all (except the lowest
pair of each pinna) *attached at the base by their whole
width and often somewhat widened* and decurrent,
obtuse or subtruncate, subentire or toothed (or rarely)
lobed, the *lobes not reaching half-way to the midrib and
entire or obscurely crenate or serrate*. Sori forming a
row down each side of the pinnule, c. 0.5–2 mm diam.;
indusium entire. Spores ripe 7–8. H.

Native. Woods, hedgebanks, rocks and screes.
Common throughout the British Is., ascending to 960 m
in Kerry. Europe; temperate Asia and in the moun-
tains to Java; Morocco; Madeira; Mascarene Is.,
Madagascar; America, south to Peru and Brazil;
Hawaii.

This group of species has been much misinterpreted in the past by British and European botanists. The following treatment is based on that provided by A. C. Jermy and C. R. Fraser-Jenkins in A. C. Jermy (1978) *Fern Atlas of the British Isles.*

1. D. oreades Fomin   Mountain Male-fern.

D. abbreviata auct.; *D. filix-mas* var. *abbreviata* Newman.

Rhizome much-branched. Lvs erect, dying down in autumn; petiole c. $\frac{1}{2}$ as long as blade, with dense pale scales; blade lanceolate, tapering to base, slightly coriaceous, pale grey-green. Pinnae usually concave, inclined to apex; pinnules obtuse or rounded at apex, the teeth obtuse and spreading in a fan-like arrangement at apex, the basal pair adnate or shortly stalked, longer than the pair above. Indusia 0.5–1 mm, strongly convex, thick, glandular, green when young, later grey-brown, scarcely shrinking at maturity. $2n = 82^*$.

Native. Open hillsides, rocky banks, screes and mountain ledges. S.E. and N. Wales, Shropshire; Westmorland, W. Durham and N. Northumberland northwards to Shetland; S. Kerry and Down. W. Europe from N.W. Germany and Spain eastwards to N. Italy, Caucasus and N.E. Turkey.

2. D. filix-mas (L.) Schott   Male-fern.

Lastrea filix-mas (L.) C. Presl; *Aspidium filix-mas* (L.) Swartz; *Nephrodium filix-mas* (L.) Strempel.

Rhizome little-branched. Lvs spreading, usually dying down during the winter; petiole c. $\frac{1}{3}$ as long as blade, with ±dense pale scales; blade ovate-lanceolate, truncate at base, ±herbaceous, mid-green. Pinnae flat, ±horizontal; pinnules acute, the teeth ±acute and converging towards the apex, the basal pair ±stalked, usually longer than the pair above. Indusia 0.5–2 mm, convex, thin, sometimes glandular, white or translucent when young, later brown, shrinking at maturity. $2n = 164^*$.

Native. Woodlands, hedgerows and open places amongst rocks. Throughout the British Is., most of Europe except the extreme north and some Mediterranean islands; eastwards to N. Iran, Himalayas and N.W. China; N.W. Africa; America southwards to N. Argentina.

D. × **mantoniae** Fraser-Jenkins & Corley (*D. filix-mas* × *oreades*), is sterile ($2n = 123^*$) and resembles *D. filix-mas*, but has a more branched rhizome, slightly more concave pinnae and pinna-segments and indusia intermediate between those of the parents. Recorded from scattered, mostly montane districts in N. Wales, the English Lake District, Lanark, Peebles, Mull and W. Sutherland.

3. D. affinis (Lowe) Fraser-Jenkins   Scaly Male-fern.

D. borreri (Newman) Newman ex von Tavel; *D. abbreviata* (DC.) Newman ex Manton; *D. pseudomas* (Wollaston) Holub & Pouzar; *Lastrea filix-mas* var. *paleacea* Moore.

Rhizome little-branched. Lvs ±erect, persisting through the winter; petiole up to c. $\frac{1}{4}$ as long as blade, with dense reddish-brown scales usually having a dark base; blade ovate to lanceolate, coriaceous, yellow-green when young, becoming blue-green. Pinnae flat, ±horizontal; pinnules with few or no teeth except at apex, the teeth acute, the basal segments not obvious. Indusia 1–2 mm, strongly convex, thick, often glandular, brown when young, later grey, scarcely shrinking at maturity.

Native. Woodland, hedgerows, open hillsides, especially on clay soils. Throughout the British Is. but confined to woodlands or shaded places on clay soils in lowland and S. and E. England. W., S.C. and S. Europe, eastwards to N. Turkey, Caucasus, Iran; N.W. Africa, Macaronesia.

A variable species, with many local races maintained because of its apomictic reproduction, within which the following three subspecies are recognized in the British Is.

Subsp. **affinis**: Petiole-scales dense, mostly narrow, brown usually ±dark-based; blade coriaceous, glossy. Pinnae symmetrical about their axes; pinna-lobes or pinnules often well separated, markedly rounded-truncate, usually with few teeth, unlobed or only shallowly lobed except for the basal auricles. Indusia splitting at maturity. $2n = 82^*$. Woodlands and rocky banks. Valleys and lower hills of mountainous parts of W. Britain.

Subsp. **borreri** (Newman) Fraser-Jenkins: Petiole-scales ±dense, both narrow and wide, dark-based or ±pale; blade slightly coriaceous, distinctly paler and less glossy than subsp. *affinis*. Basal pinnae often asymmetrical and developed on their proximal side; pinna-lobes or pinnules usually ±crowded; squarely truncate to ±acute, with more prominent, acute pinnule-teeth than subsp. *affinis*, usually with several rectangular lobes. Indusia shrivelling more than in subsp. *affinis* at maturity. $2n = 123^*$. Throughout much of the range of the sp. in the British Is.

Subsp. **stilluppensis** (Sabr.) Fraser-Jenkins (D: *affinis* subsp. *robusta* Oberholzer & von Tavel ex Fraser-Jenkins): Petiole-scales dense, lanceolate, reddish- or yellowish-brown, the darker base somewhat indistinct; blade ±coriaceous, somewhat grey-green, glandular on axes when young. Pinnae symmetrical about their axes; pinna-lobes and pinnules rounded-truncate, not lobed on the margins but with obtuse teeth at apex. Indusia shrinking slightly at maturity. $2n = 123^*$. Wales, N.W. England, W. Scotland, S.W. Ireland.

D. × **tavelii** Rothm. (*D. affinis* × *filix-mas*), which is morphologically intermediate between the parents, has large light green, glossy lvs, with the pinnules parallel-sided and acute, and the petiole-scales have dark bases. $2n = 164^*, 205^*$. The hybrids are apogamous and breed true. They occur wherever the parents grow together and are recorded in scattered localities from S. England, S.E. and N. Wales, northwards in the west to W. Ross; W. Galway.

4. D. submontana (Fraser-Jenkins & Jermy) Fraser-Jenkins   Rigid Buckler-fern.

D. rigida (Swartz) A. Gray; *D. villarii* (Bell) Woynar subsp. *submontana* Fraser-Jenkins & Jermy; *Aspidium rigidum* Swartz; *Lastrea rigida* (Swarz) C. Presl; *Nephrodium rigidium* (Swartz) Desv.

Rhizome decumbent or ascending. *Lvs* 20–60 cm, dying in autumn, erect or ±spreading, 2-*pinnate*, stiff and rather wiry; petiole from ⅓ to nearly as long as blade, densely scaly at base, sparsely so above, with pale-brown uniformly coloured, lanceolate scales; blade dull green, lanceolate or triangular-lanceolate, ±densely glandular on both surfaces with short-stalked glands, and with narrow scales on the rhachis; *pinnae* c. 15–25 on each side, *usually several basal pairs about equally long*, the lowest 4–7 cm, from somewhat longer to slightly shorter than the next, ±lanceolate; *pinnules* 8–12 × 3–5 mm, ±*equal in length for some distance above the base* (sometimes the lowest one on the upper side longer), oblong, several pairs on the lower pinnae rounded to broad-cuneate, obtuse, *those towards the base of each pinna stalked, regularly crenately lobed half-way to the midrib or nearly so, each lobe with 2–4 conspicuous acute teeth at the apex*. Sori 4–6 in a row down each side of the larger pinnules, crowded, c. 1 mm diam.; indusium glandular on the surface and margin. Spores ripe 7–8. $2n = 82, 164^*$. H.

Native. Clefts of limestone pavement, ascending to 475 m, very local; Caernarvon (?extinct), Denbigh, Derby, Lancashire, Yorks, Westmorland. Mountains of Europe and the Mediterranean region from the Jura, Alps and Caucasus to N. Africa.

5. D. aemula (Aiton) O. Kuntze

Hay-scented Buckler-fern.

Lastrea aemula (Aiton) Brackenr.; *Nephrodium aemulum* (Aiton) Baker; *Aspidium aemulum* (Aiton) Swartz

Rhizome erect or ascending. *Lvs* 15–60 cm, persistent, finally decaying from the apex downwards, 3-*pinnate* or almost 4-pinnate; *petiole with* few or numerous *narrow-lanceolate, lacerate, uniformly coloured, reddish-brown scales*, about as long as blade, *dark brown throughout*; *blade bright green, triangular-ovate or triangular-lanceolate*, scaly below on the rhachis and on the midribs of the segments below, *the petiole, rhachis and midribs densely covered with minute sessile glands*, which extend less densely to underside, and sometimes upperside, of blade; pinnae c. 15–20 on each side, *the lowest pair usually the longest*, 5–15 cm, triangular-ovate or triangular-lanceolate; *pinnules of the lowest pinna* c. 10–20 on each side, *the basal one on the lower side markedly longer than that on the upper* and usually longer than its neighbour (or about equal and the third shorter), c. 2–6 cm, triangular-ovate to lanceolate, *pinnate, narrow at the attachment* and often shortly stalked, *somewhat concave and thus giving the frond a distinctly crispate appearance*; segments lobed or almost pinnate, and toothed with narrow acuminate, scarcely or shortly mucronate *teeth of which some are ±straight, some somewhat incurved, and some somewhat curved outwards*. Sori c. 0.5–1 mm forming a row down each side of the midrib of the segment; *indusium irregularly toothed and fringed with* sessile glands. Spores ripe 7–9. $2n = 82^*$. H.

Native. Woods, hedgebanks, shady rocks, etc., ascending to 1005 m, local and mainly in the west; Cornwall to Somerset and Dorset; Sussex, Kent, Surrey; S.W. and N. Wales; N. Yorks, Cumberland, Northumberland; Isle of Man; Ayr; Kintyre and E. Inverness to Orkney; throughout Ireland but rare in the centre and parts of the east. N.W. and S.W. France, N. Spain, Madeira, Azores.

D. × pseudoabbreviata Jermy (*D. aemula × oreades*), which is morphologically intermediate between the parents, is sterile ($2n = 82$) and known only from the Isle of Mull.

6. D. cristata (L.) A. Gray   Crested Buckler-fern.

Lastrea cristata (L.) C. Presl; *Aspidium cristatum* (L.) Swartz; *Nephrodium cristatum* (L.) Michx.

Rhizome decumbent or shortly creeping. Lvs 30–100 cm, the outer vegetative, ±spreading, shorter than the suberect fertile inner ones, dying in autumn, *pinnate with deeply pinnatifid pinnae or* 2-*pinnate*; petiole ⅓ as long as blade, with pale-brown uniformly coloured ovate scales, numerous at the base, few above; blade dull green, lanceolate or oblanceolate, naked except for a few scales on the rhachis; *pinnae* c. 10–20 on each side, the longest 5–10 cm, triangular-ovate to lanceolate, about the middle of the frond, *decreasing in length downwards*; *pinnules or segments* c. 5–10 *on each side of the longest pinna*, ±regularly decreasing in size above the base, the lowest pair 10–25 × 7–10 mm, ±equal in length or that of the lower side longer, oblong or ovate-oblong, *attached at the base by their whole width* (rarely the basal pair somewhat narrowed), decurrent, obtuse, the larger *shallowly pinnately lobed*, all *serrate with acute mucronate teeth* which are often curved inwards. Sori forming a row down each side of the pinnule or more irregular on the larger pinnules, large; indusium entire or sinuate, without glands. Spores ripe 7–8. $2n = 164^*$. H.

Native. Wet heaths and marshes, very local, rare and much decreased; Kent, Surrey, Suffolk, Norfolk; extinct in Huntingdon, Nottingham, Cheshire and Yorks; Renfrew. Europe, southwards to S. France, N. Yugoslavia and S. C. Russia; W. Siberia, eastern North America from Newfoundland to Saskatchewan, southwards to Virginia and Idaho.

7. D. carthusiana (Villar) H. P. Fuchs

Narrow Buckler-fern.

Lastrea spinulosa C. Presl; *Nephrodium spinulosum* Strempel; *Aspidium spinulosum* Swartz; *Dryopteris spinulosa* Watt.

Rhizome decumbent or shortly creeping. *Lvs* 30–102(–150) cm, dying in autumn, the base of the petiole decaying first, 2-*pinnate with deeply pinnatifid pinnules or* 3-*pinnate*; *petiole* about as long as blade, *dark brown at the base, pale or green above, with pale brown, uniformly coloured, ovate or ovate-lanceolate, entire scales*, which are few above, more numerous (but not dense) at the base; *blade light or yellowish-green*, lanceolate or ovate-lanceolate, *naked* except for a few scales on the rhachis; *pinnae* c. 15–25 on each side, 5–10 cm, the 3 *or 4 basal pairs about equal in length* (the lowest slightly longer or slightly shorter than the next), triangular-ovate or triangular-lanceolate; *pinnules of the lowest pinna* c. 7–12, *the basal one* (or sometimes 2) *on the lower side markedly longer than that on the upper and than its neighbour*, c. 2–3 cm, ±lanceolate, *pinnatifid nearly to the midrib or pinnate, narrowed at the attachment, ±flat; segments ±oblong, serrate with incurved* acuminate, mucronate or aristate *teeth.* Sori c. 0.5–1 mm, forming a row down either side of the segment; *indusium* entire or sinuate, *without glands.* Spores ripe 7–9. 2n = 164*. H.

Native. Damp and wet woods, marshes and wet heaths, from Sutherland and the Hebrides southwards and scattered throughout Ireland, rather common but unrecorded for several counties and from the Channel Is. Most of Europe, but only on mountains in the south; Caucasus; Siberia (east to the Yenisei region); North America from Quebec to British Columbia, south to Virginia and Mississippi.

D. × **uliginosa** (A. Braun ex Doll.) Druce (*D. carthusiana* × *cristata*), which resembles a larger, narrower-lvd version of *D. carthusiana* with less-divided pinnae, and having abortive spores (2n = 164 *), is a rare and local hybrid confined to Norfolk.

D. × **brathaica** Fraser-Jenkins & Reichstein (*D. carthusiana* x *filix-mas*), which resembles *D.* × *uliginosa* but with more scaly lvs and ±erect rhizome, was recorded from woods by Windermere (Westmorland).

8. D. dilatata (Hoffman) A. Gray Broad Buckler-fern.

D. austriaca auct., non (Jacq.) Woynar; *Lastrea dilatata* (Hoffman) C. Presl.

Rhizome erect or ascending. *Lvs* (7–)30–150(–180) cm, dying in autumn or subpersistent, the base of the petiole decaying first, 3-pinnate; *petiole with ovate-lanceolate or lanceolate entire scales which are pale brown with a dark, often wide, central stripe*, usually dense at the base, less so above, from rather less than $\frac{1}{2}$ as long as blade to about as long, dark brown at base, pale or green above; *blade* usually *blue-green*, ±firm, triangular-ovate to lanceolate, sparsely or moderately scaly on the rhachis and sometimes glandular on the rhachis and lower surface; pinnae 6–20 cm, c. 15–25 on each side. 3 *or more of the basal pairs about equal in length* (the lowest rather longer to slightly shorter than the next), triangular-ovate or triangular-lanceolate; pinnules of the lowest pinna c. 10–20, *the basal one on the lower side markedly longer than that on the upper* and usually

longer than its neighbour, c. 2–5 cm, ±lanceolate, *pinnate, narrowed at the attachment, convex; segments* ±oblong, toothed or pinnately lobed, *with incurved* mucronate or aristate *teeth.* Sori c. 0.5–1 mm, in a row down either side of the segment; *indusium fringed with stalked glands* (sometimes indistinctly so) and usually irregularly toothed. Spores ripe 7–9. 2n = 164*. H.

Native. Woods, hedgebanks, wet heaths, shady rock ledges and crevices, etc., ascending to 1230 m. Common throughout the British Is. Europe and temperate Asia; in Europe southwards to Portugal, Bulgaria and C. Russia.

D. × **deweveri** (Jansen) Jansen & Wachter (*D. carthusiana* × *dilatata*; *Lastrea glandulosa* Newman) is rather common in Great Britain where the parents grow together. It is intermediate in all characters, having pale reddish-brown petiolar scales and abortive spores (2n = 164*).

9. D. expansa (C. Presl) Fraser-Jenkins & Jermy
Northern Buckler-fern.

D. assimilis S. Walker; *D. aristata* (Vill.) Druce; *Lastrea dilatata* var. *alpina* T. Moore.

Similar to *D. dilatata* but smaller, with lvs 7–60 cm; petiole up to c. $\frac{1}{2}$ as long as blade, the scales concolorous, pale tan to reddish-brown; blade yellowish-green, thin and membranous; pinnules flat, the basal on lower side of basal pinna at least twice as long as that on the upper side; sori c. 0.5–1.5 mm. 2n = 82*.

Native. Moist acidic woodland in lowlands of Scotland, N. England and Wales, and on mountain screes and rock-fissures up to c. 1000 m. Brecon to Caernarvon; Cumberland; Northumberland, Peebles and the Clyde Is. northwards to Shetland. N. Eurasia, extending southwards in the mountains to S.C. Europe, Corsica, N.E. Turkey, Caucasus; Greenland; North America from Alaska to California.

D. × **remota** (A. Braun) Druce (*D. affinis* × *expansa*), which resembles a narrower-lvd form of *D. expansa* with densely golden-scaly petioles, and has 2n = 123*, was recorded from Loch Lomond; Irish records (Kerry and Galway) need confirming, since *D. expansa* is not yet known from Ireland.

D. × **abrosiae** Fraser-Jenkins & Jermy (*D. dilatata* × *expansa*), which generally resembles a form of *D. dilatata* with somewhat more finely divided and yellow-green lvs, with abortive spores (2n = 123*), is known in scattered localities in the English Lake District, in W. Scotland from Kintyre to W. Ross, and in Moray.

D. × **sarvelae** Fraser-Jenkins & Jermy (*D. carthusiana* × *expansa*), with abortive spores (2n = 123*), occurs in one locality in Kintyre with its parents, between which it is morphologically intermediate.

15. BLECHNACEAE

Rhizome with firm scales having thickened cell-walls. Petiole with 2 vascular bundles. Sori free or confluent on a vein running parallel to the midrib; indusium present, flap-like. Spores bilateral.

Eight genera, cosmopolitan.

1. BLECHNUM L.

Rhizome short, suberect, densely scaly. Lvs pinnate, fertile and vegetative dissimilar; veins of vegetative lf free. Sori confluent into a continuous line on either side of the midrib, parallel to the pinna-margin; indusium linear, opening towards the midrib of the pinna.

About 220 spp., cosmopolitan, but only the following in north temperate regions.

1. B. spicant (L.) Roth   Hard Fern

Lomaria spicant (L.) Desv.

Rhizome robust, with brown lanceolate scales. Lvs numerous, tufted, the outer vegetative and spreading, the inner fertile and suberect; vegetative lvs 10–50 cm, pinnate, glabrous; petiole not more than $\frac{1}{3}$ as long as blade and often much shorter, scaly at base, dark brown; blade linear-elliptical; rhachis green; pinnae numerous, close together, pectinately arranged, somewhat coriaceous, entire or nearly so; longest pinnae about the middle of the blade, 1–2 cm × 3–5 mm, linear-oblong and somewhat curved towards apex of lf, acute or apiculate, widening at the base and there contiguous; lowest pinnae short, often wider than long, more distant. Fertile lvs 15–75 cm; rhachis blackish (except near apex); pinnae linear, the longest to 2.5 cm × 1–2 mm, distant, very suddenly widened near the base but not contiguous (except near the apex of the lf). Sporangia appearing to cover the entire under-surface of the pinna after the opening of the indusium, which is at first whitish, then brownish; margin of the lf outside indusium very narrow. Spores ripe 6–8. $2n = 68^*$. H.

Native. Woods, heaths, moors, mountain grassland and rocks ascending to 1190 m, calcifuge. Throughout the British Is., common in hilly districts, less so in lowland ones. Europe eastwards to S. Finland, Latvia, E. Carpathians and Turkey; Caucasus; Madeira, Azores, N. Morocco; Japan; western North America from Alaska to California.

Two alien species of *Blechnum*, grown as ornamentals, have escaped and are locally naturalized in warm, oceanic districts, though rarely producing fertile lvs.

B. chilense (Kaulf.) Mett., a native of S. Chile and the Falkland Is., is established in ditches and hedgerows in the Scilly Is., Guernsey and Cornwall. It is readily distinguished by the pinnae of the vegetative lvs which narrow at the base and are not contiguous.

B. penna-marina (Poiret) Kuhn has a creeping, wiry rhizome, from which arise clusters of vegetative lvs up to 17 × 1.7 cm, the pinnae reach 8.5 mm and have wide, usually contiguous bases. It is a native of cool temperate S. America, New Zealand, and many intervening sub-antarctic islands.

16. MARSILEACEAE

Aquatic or subaquatic perennial herbs with long, creeping, hairy rhizomes. Lvs alternate, 2-ranked, circinate in bud, subulate and entire or with 2 or 4, palmately arranged, wide, cuneiform lflts on a long petiole. Sporangia borne in globose or ovoid-oblong, hard, hairy 'sporocarps' at the base of the petiole. Each sporocarp contains 2 or more sori, each of which is surrounded by an indusium. Sporangia of two kinds, both occurring in the same sorus, the megasporangia below the microsporangia, developing in regular sequence from apex to base of the sorus. Megasporangia with 1 megaspore; microsporangia with numerous microspores; annulus 0 or rudimentary. Male prothallus of 2 vegetative cells and 2 antheridia each producing 16 multiciliate spermatozoids. Female prothallus with many vegetative cells and a single archegonium which is situated on a projection from the megaspore.

Three genera and about 70 spp., cosmopolitan (except Arctic).

1. PILULARIA L.

Lvs subulate, entire. Sporocarp solitary, divided by the indusia into 2 *or 4 compartments, each containing* 1 *sorus* and splitting separately longitudinally at maturity and releasing the sporangia in a mass of mucilage.

Six spp., N. and S. temperate regions, mountains in New World tropics.

1. P. globulifera L.   Pillwort

Rhizome to 50 cm, slender, creeping, often with short axillary branches. Lvs 3–8(–15) cm, subulate. Sporocarps c. 3 mm diam., globose, shortly (up to 1 mm) stalked, brown at maturity, hairy. Sori 4. Spores ripe 6–9. $2n = 26$. Hel. or Hyd.

Native. Edges of ponds and lakes, often submerged, on acid soils. From Easterness and the Outer Hebrides southwards, very local, absent from many counties including most of the English Midlands; Kerry, Galway, Mayo, Donegal and Antrim. W. Europe eastwards to S. Finland, Poland, Czechoslovakia and S. Italy; local everywhere and becoming more so eastwards.

17. AZOLLACEAE

Plant small, free-floating, with branched stems bearing roots and lvs. Lvs alternate, 2-ranked, imbricate, 2-lobed, the upper lobe floating, green and assimilating and with a hollow filled with mucilage and containing the threads of the blue-green alga *Anabaena*; lower lobe thin, submerged and bearing the sori in pairs on cylindrical receptacles, covered by a flange of the upper lobe of the lf. Sori surrounded by an indusium from the base and consisting of either numerous microsporangia developed successively from the apex to base of sorus or a single megasporangium; annulus 0; megaspore 1 in each sporangium, with 3 floats; microspores 64, grouped together into 'massulae', each massula being in some spp. furnished with projecting barbed hairs ('glochidia'). Massulae becoming fixed to the megaspore by the glochidia (when these are present), the spores germinating within them; male prothallus of a few vegetative cells and 8 spermatozoid mother cells. Female prothallus small, green.

One genus.

1. AZOLLA Lam.

The only genus. Six spp., tropical and subtropical.

*1. A. filiculoides Lam.   Water Fern.

Plant 1–10 cm diam., often growing in large masses, bluish-green, usually becoming red in autumn. Upper lf-lobes c. 2.5 × 0.9–1.4 mm, ovate, obtuse, covered above with unicellular hairs which make the surface non-wettable; margin wide, hyaline. Sori in each pair either both containing megasporangia or one of each kind of sporangium. Glochidia present, not divided into cells by transverse walls. Spores ripe 6–9. Hyd.

Introduced. Ditches, etc. Naturalized in many places, mainly in S. England, but apparently extending its range, now as far north as the Isle of Man and S.E. Yorks; Wicklow and Louth. Native of tropical America.

GYMNOSPERMAE

✕

CONIFERAE

18. PINACEAE

Monoecious trees, rarely shrubs; branches usually in regular whorls at the end of each year's growth. Buds scaly. Lvs spirally arranged, linear, entire or minutely serrulate, usually evergreen. Male and female fls in cones, formed of numerous scales, spirally arranged. Scales of male cone (microsporophylls) bearing 2 pollen sacs on the under-surface. Scales of female cones double, consisting of an ovuliferous scale bearing 2 inverted ovules on its upper surface, borne in the axil of a bract, the bract and scale free except at the extreme base. Pollen usually with 2 conspicuous bladderlike wings. Cones usually large and ±woody in fr.; seeds winged.

Ten genera and about 250 spp., northern hemisphere, mainly temperate.

1 Lvs all solitary on the main stems; short shoots 0.   2
  Assimilating lvs all or mainly on short lateral shoots.   5
2 Lvs without persistent woody decurrent prominent bases, the scars not or scarcely projecting after the lvs have fallen; bract large, often exserted in the fr. cone.   3
  Lvs with persistent woody decurrent bases, projecting as pegs after the lvs have fallen; bract small, never exserted in the fr. cone.   4
3 Cones erect; scales falling from the axis with the seeds; lf-scars quite flat; lvs rigid; buds ovoid, obtuse.   1. ABIES
  Cones pendent; scales and bracts persistent after the fall of the seeds; lf-scars slightly prominent; lvs rather soft; buds fusiform, acute.   2. PSEUDOTSUGA
4 Branches on the trunk all in whorls; lvs without a stalk above the persistent base, not 2-ranked on the upper side of the shoot; cones usually large.   3. PICEA
  Branches on the trunk not all whorled; lvs with a distinct stalk above the persistent base, spreading laterally and so appearing 2-ranked; cones small.   4. TSUGA
5 Short shoots prominent, with numerous lvs, continuing to grow for several years and producing fresh lvs each year; lvs deciduous (if evergreen, see *Cedrus*); lvs on the long shoots green and assimilating.   5. LARIX
  Short shoots much reduced, appearing as a cluster of 2(–5) lvs with a few scales at the base and finally falling as a whole; lvs evergreen; lvs on the long shoots brownish and scale-like.   6. PINUS

1. ABIES Miller

Evergreen trees of pyramidal habit, the branches in annual whorls. Bark of young trees smooth, often fissured on older ones. Twigs smooth (rarely grooved). Buds usually ovoid, obtuse or acute (fusiform only in *A. venusta* (Douglas) C. Koch). *Lvs spirally inserted on the stems* (short shoots 0) but often spreading into 2 lateral sets, ±flat, with 2 whitish waxy bands containing the stomata below and occasionally also above, contracted above the base, inserted directly on the stem and leaving a flat, ±circular scar on falling. *Ripe cones erect*; *scales* thin, densely imbricate, narrowed into a stalk below, *finally falling from the persistent axis*; *bracts long*, included or exserted. Seeds winged, ripening the first year. Cotyledons 4–10.

About 50 spp., north temperate zone. A number of species besides the following are sometimes planted in parks, etc.

*A. alba Miller (*A. pectinata* (Lam.) DC.)   Silver Fir.
Bark grey, smooth, becoming scaly on old trees. *Buds* small, ovoid, obtuse, *not resinous*. Twigs grey, pubescent. *Lvs* 1.2–3 cm, *arranged in two lateral sets, the lower spreading horizontally*, the *upper* much shorter, *pointing upwards* and outwards, notched at apex, *dark shining green above*. Cones 10–14 cm, cylindrical, greenish when young, brown when ripe; bracts exserted, reflexed.

Commonly planted for ornament and formerly for timber but now little used, though it thrives well in East Anglia. Native of the mountains of C. and S. Europe from S. Germany to the Pyrenees, Corsica, Apennines, Albania and Macedonia, and outlying forests in W. France and E. Poland.

*A. grandis (D. Don) Lindley   Giant Fir.
Bark becoming dark brown and scaly. *Buds* small, ovoid, obtuse, *resinous*. Twigs olive-green or brown, pubescent. *Lvs* 2–5 cm, *arranged in two lateral sets, all spreading horizontally*, the upper shorter than the lower, notched at apex, *dark shining green above*. Cones 5–10 cm, cylindrical, green; bracts hidden.
Planted for ornament and for timber. Native of western North America from British Columbia to California and Montana

*A. procera Rehder (*A. nobilis* (Douglas ex D. Don) Lindley, non A. Dietr.)   Noble Fir.
Bark reddish-brown and ridged on old trees. Buds small, ovoid, obtuse, resinous above. Twigs rusty-pubescent. Lvs 1.5–3.5 cm, the lower ones spreading laterally, *the upper ones appressed to the shoot for a short distance, then curving upwards* (so that the upper side of the shoot is hidden), entire or slightly notched at apex, *bluish-green and with stomata on both surfaces*. Cones 14–25 cm, oblong-cylindrical, purplish-brown at maturity. Bracts long-exserted, reflexed.

Commonly planted for ornament in Scotland and occasionally met with as small plantations. Native of western North America from Washington to N. California.

2. PSEUDOTSUGA Carr.

Similar to *Abies*. Bark of old trees thick, corky, furrowed. *Buds fusiform*, acute. Lf-scars slightly prominent after the fall of the lf. *Ripe cones pendent*; *scales* rigid, concave, *persistent*; *bracts 3-lobed*, *exserted*.

About 7 spp., western North America and E. Asia.

*1. P. menziesii (Mirbel) Franco   Douglas Fir.

P. douglasii (Lindley) Carr.; *P. taxifolia* Britton

Twigs yellowish, becoming brown, usually pubescent. Lvs 2–4 cm, ±spreading and arranged in two lateral sets, entire at apex, acute or obtuse, dark or bluish-green above. Cones 5–10 cm, ovoid. $2n = 26$. MM.

Introduced. Fairly frequently planted for timber and commonly for ornament and sometimes self-sown. Native of western North America from British Columbia to California, W. Texas and N. Mexico.

3. PICEA A. Dietr.

Evergreen trees of pyramidal habit, the branches in annual whorls. Bark scaly. *Twigs covered with lf-cushions which are separated by grooves and end in prominent peg-like projections on which the lvs are inserted and which remain after the lvs fall.* Buds ovoid or conical. *Lvs spirally inserted on the stems (short shoots 0), sessile on the peg-like bases,* 4-angled with stomata on all sides, or flat with stomata only beneath (but on the morphologically upper surface), *not divided into 2 lateral sets when seen from the upper side of the shoot;* resin canals 2. *Ripe cones pendent; scales persistent,* rather thin; *bracts minute.* Seeds winged, ripening the first year. Cotyledons 4–15.

About 50 spp., north temperate zone. Several are planted as ornamental trees.

Lvs 4-sided with inconspicuous stomatal lines on each side.
1. abies
Lvs flat, dark green above, with conspicuous glaucous lines beneath.   **2. sitchensis**

*1. P. abies (L.) Karsten   Norway Spruce.

P. excelsa (Lamb.) Link

Tree to 40 m. Bark brown. Buds ovoid, acute, not resinous. Twigs yellowish- or reddish-brown, glabrous or slightly pubescent. *Lvs 1–2 cm, lower ranks spreading laterally, upper ranks pointing forward,* 4-*angled and rhombic in section, with 2 or 3 rather inconspicuous glaucous stomatal lines on each side,* acute and with a short horny point. Cones 10–15 cm, cylindrical; scales thin, ±rhombic, irregularly toothed at apex. Fl. 5–6. Fr. 10. $2n = 24, 48$. MM.

Introduced. Commonly planted for timber but apparently rarely regenerating. Native of N. Europe from Scandinavia and N. Russia extending southwards to the S. Alps, Albania and Macedonia.

*2. P. sitchensis (Bong.) Carr.   Sitka Spruce.

P. menziesii (Douglas ex D. Don) Carr.

Tree to 40 m. Bark brown. Buds ovoid, acute, resinous. Twigs light brown, glabrous. *Lvs 1.5–2.5 cm, flat,* arranged as in *P. abies, dark green above, with 2 conspicuous stomatal bands beneath,* acute and with a pungent horny point. Cones 5–10 cm, oblong-cylindrical; scales thin, rhombic-oblong, irregularly toothed at apex. $2n = 24$. MM.

Introduced. Frequently planted for timber especially in the north and west, not reported as regenerating. Native of western North America from Alaska to California.

4. TSUGA (Antoine) Carr.

Evergreen trees of pyramidal habit, branching irregular, leading shoot and branches often pendent at tips. Bark furrowed, red-brown. *Twigs and lf-insertion as in Picea* but grooves often less conspicuous. Buds globose or ovoid, not resinous. *Lvs* spirally inserted on the stems (short shoots 0), *shortly stalked,* ±flat and usually with stomata only beneath, usually *spreading laterally into 2 sets;* resin canal 1. *Ripe cones pendent; scales persistent; bracts inconspicuous.* Seeds winged, ripening the first year. Cotyledons 3–6.

About 15 spp., temperate North America, Himalaya and E. Asia. Several spp. are sometimes planted for ornament.

*T. heterophylla (Rafin.) Sarg.   Western Hemlock.

T. albertiana (A. Murray) Sénécl.; *T. mertensiana* auct.

Tree to 70 m. Buds globose-ovoid, obtuse. Twigs light brown, densely pubescent for several years. Lvs 6–18 mm, spreading laterally into 2 sets, dark green, grooved above, with 2 broad, ill-defined glaucous bands beneath, obtuse, minutely serrulate. Cones 1.5–2.5 cm, scales obovate, concave, entire.

Sometimes planted for timber. Native of western North America from S. Alaska to Idaho and California.

5. LARIX Miller

Deciduous trees, ±pyramidal when young, branches horizontal. Bark scaly. *Twigs of two kinds,* long shoots with spirally arranged lvs and *short lateral spurs* increasing very slowly and producing a fresh tuft of numerous lvs at the apex each year. Lvs flat. Cones erect; scales persistent, suborbicular to oblong; bracts large, conspicuous in fl. and often brightly coloured. Seeds winged, ripening the first year. Cotyledons c. 6.

About 10 spp., cool north temperate regions.

Twigs yellowish, not pruinose; lvs bright green, without white bands beneath; ripe cone scales erect. **1. decidua**
Twigs reddish, pruinose; lvs glaucous, with 2 conspicuous white bands beneath; ripe cone scales curved back near the apex.   **2. kaempferi**

*1. L. decidua Miller   European Larch.

L. europaea DC.

Tree to 50 m. Bark greyish-brown, shed in small plates. *Young long shoots yellowish,* glabrous. *Lvs of short shoots 12–30 mm, bright green,* 30–40 on each shoot. Bracts bright pink (rarely cream) in fl. Cones 2–3.5 cm, ovoid; *scales erect,* 40–50, *suborbicular,* entire, pubescent outside; bracts almost concealed by the scale. Fl. 3–4. Fr. 9. $2n = 24, 48$. MM.

Introduced. Commonly planted for timber and ornament, frequently self-sown and naturalized in a number of places. Native of the Alps and Carpathians, planted elsewhere.

*2. L. kaempferi (Lamb.) Carr.   Japanese Larch.

L. leptolepis (Sieb. & Zucc.) Endl.

Tree to 30 m, of stiffer habit than *L. decidua*. Bark

reddish-brown, shed in flakes or strips usually rather larger than those of *L. decidua*. *Young long shoots reddish with a glaucous bloom*, slightly pubescent or glabrous. *Lvs* of short shoots 15–35 mm, *bluish-green with 2 whitish bands beneath*, broader than those of *L. decidua*, c. 40 on each shoot. Bracts cream tinged with pink in fl. Cones 1.5–3.5 cm, ovoid; *scales* numerous, *somewhat spreading and with the margin rolled back at the apex* so that the cone has a rather rosette-like appearance, truncate or slightly emarginate at apex, slightly pubescent outside; bracts mostly concealed. Fl. 3–4. $2n = 24$. MM.

Introduced. Rather frequently planted for timber, not known to be naturalized. Native of Japan.

CEDRUS Trew

Evergreen trees, pyramidal when young, becoming spreading and massive. Twigs and lf-arrangement as in *Larix*. Cones erect, taking 2 or 3 years to ripen, large; scales densely imbricate, falling from the persistent axis at maturity; bracts minute. Cotyledons 9–10.

Four closely allied spp., Mediterranean region and Himalaya; three of them are frequently planted and are occasionally met with in woods, etc., the following being the most common.

*C. libani A. Rich. (*C. libanensis* Mirbel)   Cedar of Lebanon.
Leading shoot of young trees usually somewhat curved, branches spreading, not drooping. Lvs mostly 2.5–3 cm. Cones 8–10 × 4–6 cm. $2n = 24$.

Native of S. Anatolia and Lebanon.

6. PINUS L.

Evergreen trees with regularly whorled branches, of pyramidal habit when young but often with spreading crowns at maturity. Bark rough, furrowed or scaly. Buds usually large. *Shoots of 2 kinds, ordinary long shoots bearing spirally arranged scale-lvs without chlorophyll and with woody decurrent bases, and short shoots borne in the axils of the scale-lvs and apparently consisting only of a definite number* (usually 2, 3 or 5) *of green needle-like lvs surrounded by sheathing scale-lvs at the base, the short shoot not growing further and finally falling off as a whole.* Male cones replacing short shoots at the base of a year's growth. Female cones replacing long shoots, taking 2(–3) years to ripen. Ripe cones with thin or thick and woody scales; in thick-scaled spp. the exposed part of the scale much thickened and provided with a prominent protuberance or scar ('umbo'); in other spp. the scale is flat with a small terminal umbo; bracts minute. Seeds usually winged. Cotyledons 4–15. Seedlings with green spirally arranged lvs.

70–100 spp., northern hemisphere, in the tropics mainly in the mountains. A number of spp. besides the following are planted for ornament.

1 Lvs glaucous, less than 10 cm; bark of upper part of
    trunk bright reddish-brown.   **1. sylvestris**
  Lvs dark green, mostly more than 10 cm; bark not
    bright reddish-brown.   2

2 Buds resinous, with appressed scales; lvs not 2 mm
    broad; bark of 3-year-old twigs conspicuously
    divided into plates; cone 5–8 cm.   **2. nigra**
  Buds not resinous, the tips of the scales strongly
    recurved; lvs very stout, more than 2 mm broad;
    bark of 3-year-old twigs not conspicuously divided
    into plates; cone 9–18 cm   **3. pinaster**

1. P. sylvestris L.   Scots Pine.
Tree to 30(–50) m; old trees with a flat crown. *Bark on the upper part of the trunk bright reddish-brown or orange*, shed in thin scales; on the lower part, dark brown, and fissured into irregular longitudinally elongated plates. Twigs greenish-brown when young, becoming greyish-brown in the 2nd year. *Buds* 6–12 mm, oblong-ovoid, reddish-brown, *resinous, with the upper scales free* (*but not reflexed*) *at the tips*. Lvs 2 on each short shoot, 3–8(–10) cm × 1–2 mm, stiff, usually twisted, *blue-green* from the continuous glaucous stomatal lines on the inner surface and the interrupted ones on the outer, finely serrulate; resin-canals marginal; sheath at first c. 8 mm, whitish, becoming grey and shortening. Cones 3–7 cm, ovoid-conical, symmetrical or somewhat asymmetrical, dull brown; scales oblong, the exposed portion flat or somewhat projecting; umbo small, with a small prickle or its remains. Seed c. 3–5 mm, with a wing about 3 times its length. Fl. 5–6. $2n = 24$. MM.

Native. The dominant tree of considerable areas in the Highlands from Perthshire to Ross, mainly in the east, ascending to 670 m; usually supposed to be introduced in England and Ireland but possibly persisting in small quantity since the Boreal period; now often forming woods on sandy soils in S.E. England, less frequent elsewhere but planted and naturalized in many places. N. and C. Europe, extending southwards in the mountains to S. Spain, N. Italy and Macedonia; temperate Asia. Several geographical variants exist, much of the variation following a clinal pattern.

Var. **scotica** (Willd.) Schott. Remaining pyramidal for a long period, then with a rounded crown. Lvs c. 4 cm. Cones c. 4 cm, symmetrical.

The native Scottish plant. English plants usually more quickly become flat-topped and have longer needles and cones, but are of too mixed origin to be referred to any particular variant.

*P. mugo Turra   Mountain Pine.
P. montana Miller
Usually a *low, spreading shrub* with irregular branches. Bark dark grey, scaly. *Buds* 6–12 mm, *very resinous*; scales appressed. *Lvs* 2 on each short shoot, 3–8 cm, rigid, *curved but not twisted, dark green*, finely serrulate; resin canals marginal; *sheath* at first 12–15 mm. Cones 2.5–5 cm, symmetrical or asymmetrical, yellowish or dark brown; scales with exposed portion flat or pyramidal; umbo small with a small prickle. $2n = 24$.

Sometimes planted for shelter in exposed situations. Native of the mountains of C. Europe and the Balkan peninsula.

*2. P. nigra Arnold

Tree to 50 m with short, spreading branches, usually remaining ±pyramidal. Bark of old trees thick, dark grey, deeply fissured. Twigs light brown, *the lf-bases persisting for several years and conspicuous as scale-like plates.* Buds 12–25 mm, oblong-ovoid or cylindrical, acuminate, light brown, *resinous; scales appressed. Lvs* 2 on each short shoot, 8–16 cm × 1–2 mm, stiff, twisted or not, *dark green,* the stomatal lines interrupted and inconspicuous, finely serrulate, resin canals deep in the tissues; sheath at first c. 12 mm, becoming shorter. *Cones* 5–8 cm, ovoid-conical, subsymmetrical, shining yellowish-brown; scales oblong, the exposed portion ±flat, transversely keeled; umbo rather small, usually with a small prickle. Fl. 5–6. $2n = 24$. MM.

Introduced. Native of S. Europe, Asia Minor and Morocco. A variable sp., divisible into several geographical subspp. of which the two following are commonly grown in this country.

Subsp. **nigra**   Austrian Pine.

P. laricio var. *nigricans* (Host) Parl.

Crown pyramidal. Lvs 8–16 cm, very rigid, nearly straight, dark dull green.

Commonly planted for making quick shelter but not suitable for forestry. S.E. Europe from Austria to C. Italy, Greece and Yugoslavia.

Subsp. **laricio** (Poiret) Maire   Corsican Pine.

P. nigra var. *poiretiana* (Antoine) Ascherson & Graebner; *P. laricio* Poiret

Crown narrowly ovoid. Lvs 18–16 cm, somewhat twisted, rather lighter in colour.

Fairly frequently planted for timber and also for ornament. Corsica, S. Italy and Sicily.

*3. P. pinaster Aiton   Maritime Pine.

Tree to 30 m, when mature usually bare of branches for the greater part of its height. Bark thick, dark reddish-brown, deeply fissured. Twigs light brown.

Buds 20–25 mm, cylindrical, acute, brown, *not resinous; the scales strongly reflexed at the tips. Lvs* 2 on each short shoot, 10–20 × 2–3 mm, very stout and rigid, curved, *dark green* and somewhat shining, the stomatal lines interrupted and rather inconspicuous, finely serrulate; resin canals deep in the tissues; sheath at first c. 2 cm. *Cones* often clustered, 8–18 cm, ovoid-conical, subsymmetrical, bright shining brown, often remaining closed for several years; scales oblong, the exposed portion ±pyramidal, transversely keeled; umbo prominent, with a prickle at first. Fl. 5–6. $2n = 24$. MM.

Introduced. Frequently planted on poor sandy soils, especially near the sea in S. England and completely naturalized near Bournemouth. S.W. Europe extending eastwards to Italy and Pantellaria; Algeria, Morocco.

*P. contorta Douglas ex Loudon

Tree to 25(–50) m. Bark dark red-brown, deeply fissured into plates. *Buds* c. 12 mm, *very resinous,* scales appressed. *Lvs* 2 on each short shoot, 3–8 cm, rigid, *twisted, dark green,* very obscurely serrulate; resin canals deep in the tissues; sheath at first 3–6 mm. Cones 2–5 cm, asymmetrical, light yellowish-brown; scales with exposed portion prominent on upperside of cone; umbo with a slender, often deciduous prickle. $2n = 24$.

Sometimes planted for timber. Native of western North America from Alaska to California and Colorado.

Of spp. with 3 needles on each short shoot **P. radiata** D. Don (*P. insignis* Douglas) with bright green soft lvs 10–15 cm, from S. California, is frequently grown in the south-west.

The two following spp. have 5 slender soft bluish-green lvs on each short shoot:

*P. strobus L. Twigs with minute tufts of hairs below the lf-insertions, not pruinose. Lvs 6–14 cm. Cones 8–20 cm, cylindrical, often curved; scales thin. $2n = 24$.

Formerly planted for timber, but unsuccessful in this country; old specimens are, however, met with in woods, etc. Native of eastern and central North America from Newfoundland and Manitoba to Georgia and Iowa.

*P. wallichiana A. B. Jackson (*P. excelsa* Wall., non Lam.). Twigs glabrous, pruinose. Lvs 10–20 cm. Cones 12–25 cm, cylindrical; scales thickened at apex. $2n = 24$.

Rather commonly planted for ornament in parks, etc. Native of the Himalaya.

TAXODIACEAE

Monoecious trees, rarely shrubs. Buds usually naked. Lvs spirally inserted, linear and needle-like or small and scale-like, usually evergreen. Fls of both sexes in cones formed of spirally arranged scales. Microsporophylls with 2–8 pollen sacs on the under-surface. Scales of female cone ±woody when ripe, bearing 2–12 erect or inverted ovules on their upper surface, the bract wholly or partially united with the scale. Pollen not winged.

Ten genera and about 16 spp., North America, E. Asia, Tasmania. Spp. of several of the genera are planted in gardens.

SEQUOIA Endl.

Evergreen trees of pyramidal habit with horizontal or slightly drooping branches. Bark very thick, spongy, fibrous. Twigs green. Winter buds small, scaly. Lvs of 2 kinds: those on leading and cone-bearing twigs spirally arranged, subulate, scale-like, keeled above, appressed or somewhat patent; those on lateral twigs distichous, linear to linear-oblong, flat, with 2 white bands of stomata beneath. Cones small (to 8 cm), short, pendent; scales 15–20, with flattened woody ±rhombic tops persistent; ovules 3–12 on each side. Seeds winged.

One sp.

*S. sempervirens (Lamb.) Endl.   Redwood.

Tree to 110 m in nature. Trunk ±cylindrical. Lvs of lateral twigs 6–18 mm, linear, spreading into 2 lateral rows, free, dark green above. Lvs of leading and cone-bearing twigs c. 6 mm. Cone 2–2.5 cm. $2n = 66$.

Less frequently planted for ornament than *Sequoiadendron* but has been planted for timber. Native of California.

SEQUOIADENDRON Buchholz

Similar to *Sequoia,* but winter buds naked; lvs of 1 kind, radially arranged, subulate, appressed or slightly patent at apex. Cone-scales 25–40. One sp.

*S. giganteum (Lindley) Buchholz   Wellingtonia, Big Tree.
Tree to 100 m in nature, with very wide, tapering trunk. Lvs

3–6 mm, lanceolate, densely imbricate, with flat bases adherent to the twig and completely covering it, tips free. Cone 5–8 cm, ovoid. $2n = 22$.

Commonly planted in parks and gardens, and occasionally among native vegetation. Native of California.

19. CUPRESSACEAE

Monoecious or dioecious evergreen trees or shrubs. Buds naked. Lvs opposite or in whorls of 3 or 4, usually small and scale-like, appressed to and completely hiding the stem, on young plants (and in some spp. throughout their life) needle-like, rarely the lvs of the sterile twigs spirally arranged. Fls in small cones formed of opposite or whorled scales. Microsporophylls in 4–8 whorls in each cone, each with 3–5 pollen sacs on the under-surface. Scales of female cone woody, rarely fleshy when ripe, bearing 2–many erect ovules on their upper sur-face, the bract united with the scale. Pollen not winged.

About 19 genera and about 130 spp.; north and south temperate regions and mountains of the tropics.

THUJA L.
Monoecious trees of narrow pyramidal habit. Bark scaly. Branchlets much divided near their tips into fine, ultimately deciduous, twigs arranged in one plane. Lvs scale-like, oppo-site, 4-ranked, those of the lateral ranks keeled or rounded at their back and partly concealing the flattened or grooved facial ones. Microsporophylls 3–6 pairs. Cones ±ovoid, com-posed of 3–6 pairs of imbricate scales of which only the 2 or 3 middle pairs are fertile; scales oblong, ±flat, thickened at apex; seeds 2–3 to each scale, usually winged.

Five spp., North America and E. Asia. Several are com-monly grown in gardens.

*T. plicata D. Don ex Lambert
                            Western Red Cedar, Giant Arbor-Vitae.
Tree to 60 m, the trunk buttressed at the base. Bark cinnamon-red, fissured into scaly ridges. Main axes of branchlet systems ±terete, their lvs up to 6 mm, ovate, long-acuminate, with an inconspicuous sunken resin gland. Ultimate twigs c. 2 mm across, their lvs up to c. 3 mm, scarcely acuminate, mostly eglandular, deep green on the upper side of the twig, with whitish marks on the lower, aromatic when bruised. Cones c. 12 mm; scales 5–6 pairs, 3 of them usually fertile, thin. Seeds winged. $2n = 22$.
Planted for timber. Native of western North America from Alaska to N. California and Montana.

CHAMAECYPARIS Spach
Breeding system, habit, branchlets and lvs as in *Thuja*. Cones globose, ripening the 1st year, up to 12 mm; scales 3–6 pairs, peltate, fitting together by their margins until maturity, with a central boss; seeds 1–5, winged. Cotyledons 2.

Seven spp., North America, Japan and Taiwan, most of which are ±frequently grown in gardens.

*C. lawsoniana (A. Murray) Parl.   Lawson's Cypress.
Cupressus lawsoniana A. Murray
Tree to 60 m. Bark red-brown, very thick and spongy, fissured into scaly ridges. Lvs of main axes of branchlet systems up to 6 mm, the lateral pair acuminate. Ultimate twigs 1–1.5 mm across; lvs up to 2 mm, with a sunken gland, deep green on the upper side of the twig, with indistinct whitish marks on

the lower. Cones c. 8 mm, reddish-brown, pruinose; scales 8; seeds 2–5 on each scale. $2n = 22$.
Very commonly planted in parks, gardens, etc., and some-times on a small scale for timber and shelter. Numerous forms differing in habit, colour of foliage, etc., are grown. Native of S.W. Oregon and N.W. California.

CUPRESSUS L.
Monoecious trees, rarely shrubs. Twigs in most spp. not arranged in one plane. Lvs scale-like, opposite, 4-ranked. Cones as in *Chamaecyparis* but larger and ripening the 2nd year; seeds usually 6–many, winged. Cotyledons 2–5.

About 15–20 spp., warm north temperate regions.

*C. macrocarpa Hartweg   Monterey Cypress.
Tree to 25 m, either pyramidal or becoming flat-topped. Branchlet systems not in 1 plane. Lvs all alike, 1–2 mm, deltate, obtuse, deep green, not glandular. Cones 2.5–3.5 cm; seeds c. 20 on each side.
Commonly planted near the sea in the south and west as an ornamental tree, windbreak or hedge-plant. Native of S. California.

1. JUNIPERUS L.
Dioecious trees or shrubs. Bark thin, usually shedding in longitudinal strips. Lvs either needle-like, subulate and spreading (and always so on young plants), in whorls of 3, or scale-like, appressed, opposite and resembling *Cupressus*. Female fls of 3–8 *scales* which are opposite or in whorls of 3, and *become fleshy and coalescent form-ing a berry-like fr*.

About 60 spp., northern hemisphere extending south to Mexico, W. Indies, mountains of E. Africa, Himalaya and Taiwan. Several spp. (including some with scale-like lvs) are sometimes planted in gardens.

1. J. communis L.   Juniper.
Shrub with the habit varying from procumbent to erect and narrow, rarely a small tree to 10 m. Bark reddish-brown, shredding. Lvs in whorls of 3, 5–19 mm, linear, sessile, jointed at the base, with a spiny point at the apex, spreading or ascending, entire, concave and with a broad white band above, green and keeled beneath. Male cones c. 8 mm, solitary, cylindrical, with 5–6 whorls of scales. Female cones c. 2 mm in fl., solitary. Fr. ripening the 2nd or 3rd year, c. 5–6 mm, globose or rather longer than broad, blue-black, pruinose, with 1–6 seeds. Fl. 5–6. Fr. 9–10. Germ. 3–4. $2n = 22$.
Native. Chalk downs, heaths, moorland, pine- and birch-woods, ascending to 975 m, often dominant in scrub on chalk, limestone and slate; W. Cornwall and from Kent and Dorset northwards to Hertford and Car-marthen, N. Wales, Lancashire and Yorks northwards to Orkney; in Ireland mainly in the west, Kerry, Cork,

from Clare and Tipperary northwards to Derry, Down and Antrim. Most of Europe, mainly on the mountains in the south; North America and N. Asia southwards to the mountains of N. California, Pennsylvania and N. Africa; Himalaya

A variable sp. probably divisible into several ecological and geographical subspp. The two following extremes are often regarded as distinct spp., but many intermediate forms occur.

Subsp. communis

Erect or spreading. Prickly to the touch. Lvs spreading almost at right angles to the stem, 8–19 × c. 1 mm, gradually tapering to a long point. Fr. globose. N. The lowland form.

Subsp. alpina (Sm.) Čelak

J. nana Willd.; *J. sibirica* Burgsdorf

Procumbent. Scarcely prickly to the touch. Lvs ascending or loosely appressed, 4–10 × c. 1.5 mm, more suddenly contracted to a shorter point. Fr. longer than broad. $2n = 22$. Chw.

Rocks and moors on mountains and lowland bogs in W. Ireland; N. Wales, N. England, Scotland, Ireland.

ARAUCARIACEAE

Dioecious evergreen trees with whorled branches. Lvs spirally arranged, needle-like or broad and flat. Fls of both sexes in cones of spirally arranged scales. Microsporophylls with 5–15 pendent pollen-sacs. Scales of female cone numerous, the bract large and woody, the scale ('ligule') fused to it; ovule 1, centrally placed, inverted.

Two genera and about 38 spp., Australasia, Malaysia and South America.

ARAUCARIA Juss.

Lvs needle-like or flat and lanceolate. Seeds adnate to the bract and falling with it.

About 18 spp., Australasia and South America.

*A. araucana (Molina) C. Koch   Monkey Puzzle.

A. imbricata Pavón

Tree to 50 m. Lvs 2.5–5 cm, lanceolate, spiny pointed, spreading all round the stem and persisting many years. Cones globose, 14–20 cm.

Commonly planted in parks and gardens and occasionally among native vegetation. Native of E. Chile and W. Argentina.

20. TAXACEAE

Usually dioecious evergreen trees or shrubs. Lvs linear, spirally inserted. Male fls in small cones (spikes or heads). Female fls solitary or in pairs in the lfs-axils, not forming cones; ovules erect, not borne on scales but wholly or partly surrounded when ripe by a fleshy aril.

Five genera and about 20 spp., north temperate zone and New Caledonia.

1. TAXUS L.

Trees or shrubs, the branches numerous and not regularly whorled. Buds small, with imbricate scales. Lvs ±spreading in 2 lateral ranks, linear, without resin canals. Male cones axillary, head-like, stalked, surrounded by scales at the base; *microsporophylls 6–14, peltate, each with 5–9 pollen sacs. Ovule solitary*, axillary, with scales at the base. *Seed surrounded by a red, fleshy, cup-like aril*, ripening the 1st year.

About 10 very closely related spp., north temperate zone.

1. T. baccata L.   Yew.

Tree to 20 m with massive trunk and usually rounded outline, capable of producing lfy branches from stools or old trunks (unlike most other gymnosperms). Bark reddish-brown, thin, scaly. Twigs green. Lvs 1–3 cm, shortly stalked, mucronate, dark green above, paler and yellowish beneath, midrib prominent on both sides, margins recurved. Seeds c. 6 mm, olive-brown, ellipsoid. Fl. 3–4. Fr. 8–9. $2n = 24$. M.

Native. Woods and scrub, mainly on limestone, sometimes forming pure woods in sheltered places on the chalk in S.E., and on limestone in N.W. England, tolerant of considerable shade; from Perth and Argyll southwards, rather local; in Ireland, mainly in the west, extending east to Cork, Offaly and Antrim; also Wicklow but possibly introduced. Europe northwards to 62° 45′ N in Norway, Estonia eastwards to White Russia and Turkey, only on mountains in the south; Caucasus; N. Persia, W. Himalaya; mountains of N. Africa.

ANGIOSPERMAE

DICOTYLEDONES: *ARCHICHLAMYDEAE*

21. RANUNCULACEAE

Annual to perennial herbs, occasionally small shrubs, and most spp. of *Clematis* are woody climbers. *Lvs* simple, often palmately lobed, to repeatedly and finely palmatisect or pinnatisect; *usually alternate and exstipulate* but opposite or whorled in *Clematis*, and stem-lvs in a single whorl, often close beneath the fl. or infl., in *Eranthis, Anemone, Pulsatilla* and *Hepatica*; in *Thalictrum, Ranunculus* subgen. *Batrachium, Caltha* and *Trollius* early-developing lobes of the primordial lf-base are regarded as somewhat atypical stipules. *Fls* solitary or in terminal infls; *usually actinomorphic and hermaphrodite* but zygomorphic in *Delphinium, Consolida* and *Aconitum*, and polygamous or ±dioecious in some spp. of *Clematis, Thalictrum*, etc. Inner members of the perianth bear nectaries in many genera, staminodes in *Pulsatilla* and carpels in *Caltha*, but there are no nectaries in *Anemone, Clematis, Adonis* or *Thalictrum*. Fls are visited by insects seeking nectar or the abundant pollen and most are insect-pollinated, though some *Thalictrum* spp. are at least partly wind-pollinated. *The perianth is either undifferentiated or consists of an outer set of green or petaloid sepals and an inner set of variously shaped nectar-secreting organs ('honey-lvs'), sometimes petal-like as in Ranunculus, or of typical petals not bearing nectaries, as in Adonis. Stamens numerous, hypogynous. Carpels 1–many, usually free* but connate basally in some *Helleborus* spp. and united to varying degrees in *Nigella*; ovules 1–numerous. Fr. various but usually of 1 or more follicles or a cluster of achenes; in *Nigella* the fr. is a group of partly united follicles or a capsule, and in *Actaea* a single berry. Seeds with a small embryo in oily endosperm. Germination usually epigeal. In some genera, including *Helleborus* and *Caltha*, all the floral parts arise in a spiral sequence on the conical receptacle, but in others the fl. is 'hemicyclic', sepals, stamens and carpels being arranged spirally but the petals forming a true whorl of members arising simultaneously. *Aquilegia* is unusual in that all parts, except only the sepals, are in 5-merous whorls. About 50 genera and 1900 spp.

Members of the Ranunculaceae are mostly acrid and often very poisonous plants; some, like *Aconitum* spp., contain alkaloids of medicinal importance.

Woody climbers of the genus *Clematis* are much grown in gardens, as are perennial herbaceous spp. of many genera as well as annual spp. of *Nigella, Consolida* and *Adonis*.

1 Woody climber with all lvs in opposite pairs.
  11. CLEMATIS
  Plants herbaceous or suffruticose, not climbing; lvs alternate (rarely ±opposite on creeping stems). *2*

2 Fl.-stem with a whorl of 3(–4) often much dissected lvs either close beneath the fl. and forming an involucre or separated from the fl. by a distinct length of stem. *3*
  Stem-lvs, if any, not in a whorl. *6*
3 Stem-lvs forming an involucre close beneath the fl. *4*
  Stem-lvs separated from the fl. by a distinct, if short, length of stem. *5*
4 Fls yellow; carpels c. 6, free; fr. a group of follicles.
  4. ERANTHIS
  Fls blue; carpels 5, united for most of their length; fr. a globose capsule. Garden plants, sometimes escaping. 5. NIGELLA
5 Stem-lvs petiolate; fls white or pinkish in native sp., yellow or blue in introduced spp.; stamens all fertile; styles not elongating in fr. 4. ANEMONE
  Stem-lvs sessile; fls campanulate, violet-blue; outer stamens sterile; styles becoming long and plumose in fr. 10. PULSATILLA
6 Fls with only 1 carpel. *7*
  Fls with 2 or more carpels. *8*
7 Fls small, whitish, not spurred; fr. a berry, eventually black; lvs commonly biternate. 8. ACTAEA
  Fls 2.5–4 cm diam., blue, spurred; fr. a follicle; lvs deeply palmatisect. 7. CONSOLIDA
8 Fls strongly zygomorphic, the posterior sepal a large erect helmet-shaped hood; plant 50–100 cm.
  6. ACONITUM
  Fls actinomorphic. *9*
9 Fls with 5 blue (sometimes white or reddish) sepals and 5 conspicuously spurred petals. 15. AQUILEGIA
  Fls not conspicuously spurred (but small green sepal-spurs, c. 2 mm, in *Myosurus*). *10*
10 Fls with per.-segs in a single undifferentiated series, with no petals or 'honey-lvs'. *11*
  Fls with 2 sets of per.-segs, either typical sepals and petals or the outer ±petaloid and conspicuous, the inner of smaller 'honey-lvs', often tubular. *14*
11 Per.-segs bright golden-yellow, 10–25 mm; fr. a group of 5–13 follicles. 1. CALTHA
  Per.-segs greenish, inconspicuous; fr. of numerous achenes. *12*
12 All lvs basal, linear; fls solitary; per.-segs with short spurs, c. 2 mm. 14. MYOSURUS
  Lvs neither all basal nor all linear; fls not solitary; per.-segs not spurred. *13*
13 Lvs 2–4-pinnate or biternate. 16. THALICTRUM
  All or most lvs palmately lobed or divided.,
  12. RANUNCULUS
14 Small plant with all lvs basal and linear; fls solitary, greenish-yellow, the sepals with appressed basal spurs, the petals tubular with short limb about equalling sepals; achenes numerous in a very long spike. 14. MYOSURUS
  Lvs not all basal and linear; fls not solitary. 15
15 Outer per.-segs larger and much more conspicuous than inner; fr. a group of follicles. *16*

Fls with green sepals and more conspicuous white or coloured petals; fr. a cluster of achenes. *17*

16 Fls globose; all per.-segs yellow, outer c. 10, incurved and broadly overlapping, inner narrow, shorter, not tubular.   2. TROLLIUS

Fls campanulate to spreading; outer per.-segs 5, greenish (but white or reddish in garden spp.), inner shorter, obliquely tubular.   3. HELLEBORUS

17 Petals yellow or white, with nectaries at base; lvs simple or palmately lobed or divided.
                                    12. RANUNCULUS

Petals red (or yellow), with no basal nectary; lvs bi- or tri-pinnate with linear segments.   13. ADONIS

Tribe 1. HELLEBOREAE. Fr. of one or more follicles, rarely a berry; chromosomes large, basic number 8.

1. CALTHA L.

Perennial herbs with stout creeping rhizomes and ±cordate lvs which in some spp. of the southern hemisphere have curiously prolonged basal lobes. Fls in a few-fld terminal cymose panicle, hermaphrodite, actinomorphic, with all parts spirally arranged. *Per.-segs 5 or more, petaloid, yellow* or *white*; stamens numerous; *carpels few, nectar-secreting*, with numerous ovules in 2 rows. *Fr. a group of follicles.* About 30 spp., north and south temperate regions.

1. C. palustris L.

Kingcup, Marsh Marigold, May Blobs.

A perennial glabrous herb with stout creeping rhizomes and erect ascending or prostrate aerial stems. Lvs chiefly basal, rounded, reniform or deltate, ±cordate at the base, long-stalked; upper lvs reniform-deltate, subsessile; all lvs ±crenate or toothed. Fls 16–50 mm diam. Per.-segs 5–8, 10–25 mm, bright golden-yellow above, often greenish beneath. Stamens 50–100. Carpels 5–13, erect or spreading; ovules numerous. Follicles 9–18 mm, dehiscing before they are dry. Seeds up to 2.5 mm. Fl. 3–7. Homogamous. Visited by a great variety of insects for pollen and for the nectar secreted from small depressions, one on each side of each carpel. $2n = 56^*$, rarely 64^* or aneuploid 52^*, 54^*, 55^*, 57^*, 58^* and 60^*; additional counts of $2n = 28, 32, 48, 53$ and 62 from European mainland.

Shows much variability in size of vegetative parts and fls and in habit, but much of it is environmentally determined. In particular it seems doubtful whether the small upland forms with decumbent stems and 1–few fls only 16–30 mm diam., often separated as *C. minor* Miller, merit specific or even subspecific rank, being connected by a range of intermediates with the larger and more erect lowland forms and becoming much larger when grown in lowland conditions. Plants rooting at nodes of prostrate or ascending stems, found in both lowland and upland habitats in the north and west of Great Britain and, especially by lakes, in Ireland, have been grouped as var. *radicans* (Forster fil.) Beck, the correct name for which is var. *flabellifolia* (Pursh) Torrey & Gray. Node-rooting has been shown to be genetically determined but it is largely independent of other morphological features.

Native. In marshes, fens, ditches, streamsides and wet woods, most luxuriant in partial shade; rare in very base-poor peat. Reaches 1100 m in Scotland. Common throughout the British Is. except Channel Is. The small plants with decumbent or procumbent stems, often rooting at the nodes, and few small fls with narrow per.-segs are found chiefly on mountains in the north and west. Europe, incl. Iceland and arctic Russia; temperate and arctic Asia; North America.

The flower of *Caltha palustris* is perhaps the most 'primitive' in the British flora.

2. TROLLIUS L.

Perennial herbs with ±erect woody stocks and alternate palmately-lobed lvs. Fls large, in 1–3-fld cymes, hermaphrodite, actinomorphic, with all parts spirally arranged. *Outer per.-segs* (sepals) *5–15, petaloid*, imbricate in bud, *yellow*; *inner per.-segs* (petals) *5–15, small, narrow, yellow, with a basal nectar-secreting depression*; stamens numerous; carpels numerous, sessile, free, each with numerous ovules in 2 rows. *Fr. a group of many-seeded follicles.*

About 25 spp. in temperate and arctic Europe, Asia and North America.

All species are acrid and poisonous.

1. T. europaeus L.   Globe Flower.

A perennial herb with a short erect woody stock, fibrous above, and an erect glabrous lfy usually simple shoot 10–60 cm. Basal lvs stalked, pentagonal in outline, palmately 3–5-lobed with the cuneate lobes ±deeply cut and toothed; stem-lvs ±sessile and usually 3-lobed; all lvs glabrous, dark green above, paler beneath. *Fls* 2.5–4 cm diam., terminal, solitary or rarely 2–3, *±globose. Sepals* c. 10 (5–15), pale or greenish-yellow, ±orbicular, very concave, *incurved and imbricate. Petals 5–15, yellow, equalling the stamens and hidden by the sepals*, ligulate, clawed, with a nectar-secreting cavity at the junction of blade and claw. Carpels numerous. Follicles c. 12 mm, keeled, transversely wrinkled, the persistent style forming a subulate beak c. 3 mm. Seeds blackish, shining, 1.5 mm. Fl. 6–8. Homogamous. Visited by various small insects. $2n = 16$. Hp.

Native. Locally common in wet pastures, scrub and woods in mountain districts northwards from S. Wales, Monmouth and Derby, and in N.W. Ireland; reaching 1130 m in Scotland. Throughout Europe to 71° N in Norway; Caucasus; arctic America.

T. caucasicus Steven, with *spreading yellow sepals* and *petals equalling the stamens*, and *T. asiaticus* L. (*T. giganteus* hort.), with *spreading orange sepals* and *petals longer than the stamens*, are also grown in gardens.

3. HELLEBORUS L.

Perennial herbs either with stout obliquely ascending rhizomes and aerial stems of short duration or with no rhizomes and overwintering stems which are somewhat woody below. *Lvs digitate or pedate with serrate segments*. Fls in cymes, hermaphrodite, actinomorphic, with all parts spirally arranged. *Outer per.-segs* (sepals) 5, *green or petaloid*, imbricate, *persistent*; *inner (nectaries)* 5–12(–20), *shorter, stipitate, obliquely tubular* and sometimes ±2-lipped, *green*; stamens numerous; carpels 3–10, free or connate at base; ovules numerous. *Fr. of 3–10 many-seeded follicles*; seeds initially with a fleshy white ridge ('elaiosome') along the raphe.

About 20 spp. in Europe and W. Asia, all calcicolous.

All the hellebores have a burning taste and are highly poisonous owing to the presence of the glycosides helleborin and helleborein. Both *H. viridis* and *H. foetidus* were formerly officinal as violent cathartics and emetics, but their use has long been discontinued.

No basal lvs; uppermost stem-lvs (bracts) simple, entire; fls numerous; perianth almost globose.   **1. foetidus**

Basal lvs usually 2; uppermost stem-lvs digitately lobed, serrate; fls 2–4; perianth spreading, almost flat.

2. viridis

1. H. foetidus L.   Bear's-foot, Stinking Hellbore.

A perennial *foetid* herb with a stout blackish ascending stock and a robust *overwintering* branched lfy stem 20–80 cm, glabrous below, glandular above. *No basal lvs. Lower stem-lvs evergreen*, pedate, long-stalked, with sheathing base; lf-segments 3–9, narrowly lanceolate, acute, serrate; middle stem-lvs with enlarged sheaths and reduced blades, transitional to the *uppermost* (bracts) which are *broadly ovate* and *entire* or with a small vestigial blade. *Fls* 1–3 cm *diam.*, *numerous*, drooping, in a corymb-like branched cyme. *Perianth campanulate or globose*, the erect concave broadly ovate sepals yellowish-green, usually bordered with reddish-purple. Nectaries 5–10, green, c. ½ as long as the stamens, short-stalked, curved, the outer lip slightly longer than the inner, both irregularly toothed. Stamens 30–55, equalling the perianth. Carpels 2–5, usually 3, slightly joined below. Follicles wrinkled, glandular, the persistent style forming a subulate beak ⅓ of the total length. Seeds black, smooth, with a fleshy white elaiosome. Fl. 3–4. Protogynous. Visited by early bees and other insects. Seeds dispersed by ants. Germ. spring. $2n = 32$. Chh.

Native. A local plant of woods and scrub on shallow calcareous soils and scree in S. and W. England and Wales, probably native northwards to Cumbria and N. Yorks but naturalized as a garden escape in N. England and E. Scotland to Aberdeen. W. and S. Europe from Belgium to Spain, S. Italy and Syria.

2. H. viridis L.   Bear's-foot, Green Hellebore.

A perennial herb with a short stout simple or branched ascending blackish stock and an erect stem 20–40 cm, glabrous or sparsely hairy above, *not overwintering*, slightly branched distally and lfless below the lowest branch at the time of flowering. *Basal lvs usually 2, arising with the flowers and dying before winter*, long-stalked, digitate-pedate, with 7–11 sessile segments, the central free, the lateral connected at their bases, all narrowly elliptical, acute or shortly acuminate, serrate, with

veins prominent on the paler lower surface. *Stem-lvs* (bracts) *smaller, sessile, ±digitate with narrow serrate segments. Fls usually 2–4, half-drooping, 3–5 cm diam.* Perianth spreading, almost flat, the sepals broadly elliptical, or ovate-acuminate, ±imbricate, yellowish-green. Nectaries 9–12, green, c. ⅔ as long as the stamens, shortly stalked, curved, the outer lip slightly longer than the inner, both irregularly toothed. Carpels 3, slightly joined at the base. Follicles more than ½ as broad as long, the persistent style forming a subulate beak ⅓ of the total length. Fl. 3–4. Protogynous. Visited by early bees. $2n = 32$. Hp.

The British plant described above is subsp. **occidentalis** (Reuter) Schiffner, with glabrous lvs, not hairy on the veins beneath, and with fls smaller than in the C. European forms.

Native. A rather local plant of moist calcareous woods and scrub chiefly in S. and W. England and Wales, and probably native as far north as Westmorland and N. Yorks, but more widely naturalized as a relic of cultivation and reaching Aberdeen. Subsp. *occidentalis* is native also in W. Germany, Belgium, France and Spain, where it replaces subsp. *viridis* of C. Europe from N.W. France to Switzerland and Hungary.

Forms of **H. niger** L. (C. and S. Europe, W. Asia), with pedate basal lvs, small entire bract-like stem-lvs and spreading white or pink-tinged sepals, bloom in winter and are grown in gardens as 'Christmas Roses'. The mostly later-flowering 'Lent Roses' consist of various spp. native in S.E. Europe and W. Asia and in particular of their numerous garden hybrids. They differ from *H. niger* in having toothed ±pedate bracts and are usually taller plants with fls of a wide range of colours from white or cream to deep purple. **H. lividus** Aiton subsp. **corsicus** (Willd.) Tutin (*H. corsicus* Willd.; *H. argutifolius* Hort.), with robust glabrous overwintering stems bearing pedate lvs with 3 closely spinescent-toothed glossy segments, and fls with spreading yellow-green sepals, is also grown in gardens.

4. ERANTHIS Salisb.

Perennial herbs with short tuberous rhizomes and stalked palmately divided basal lvs. *Stem-lvs 3, sessile, deeply palmate-lobed, forming an involucre-like whorl just beneath the solitary terminal fl.* Fls hermaphrodite, actinomorphic, with all parts spirally arranged. *Outer per.-segs 5–8, usually 6, yellow inner per.-segs (nectaries) stalked, tubular, 2-lipped, yellow*; stamens numerous; carpels c. 6, free, stalked, with many ovules in 1 row. *Fr. a group of many-seeded stalked follicles surrounded by the persistent stem-lvs.*

Eight spp. in S.E. Europe and Asia.

All spp. have a burning taste and are poisonous owing to the presence of an alkaloid.

*1. E. hyemalis (L.) Salisb.   Winter Aconite.

Helleborus hyemalis L.

A glabrous perennial herb with irregular tuberous rhizome and erect flowering stems 5–15 cm. Basal lvs arising after the fls, borne singly (or rarely 2) on dwarf vegetative shoots, long-stalked, orbicular in outline, palmately 3–5-lobed, the lobes further cut into contiguous segments; stem-lvs spreading horizontally. Fl. 20–30 mm diam. Sepals 10–15 mm, usually 6, yellow, narrowly ovate, enlarging during flowering. Nectaries 6 (5–9), shorter than the stamens, their outer lip considerably longer than the inner. Stamens c. 30. Carpels 3–11, usually 6, stalked (2 mm). Follicles brown, to 15 mm, each with numerous yellowish-brown seeds, 2.5 mm. Fl. 1–3. Homogamous. Fls very temperature-sensitive, opening above 10 °C. Visited by hive-bees and flies. Follicles dehisce in May, the lvs dying soon after. Germ. spring. $2n = 16$. Grh.

Introduced. A native of S. Europe, from S. France and C. Italy to Bulgaria, which has been naturalized in parks, plantations and woods in many parts of Great Britain northwards to Kincardine.

The closely allied **E. cilicica** Schott & Kotschy, of western Asia, is often grown in gardens for its larger fls.

5. NIGELLA L.

Annual herbs with alternate bi- or tri-pinnate lvs with linear or filiform segments. Fls usually solitary, terminal; hermaphrodite, actinomorphic. *Sepals 5, petaloid; nectaries 5, clawed,* opposite to and usually much smaller than the sepals; stamens numerous. *Fr. a group of partly joined follicles or a capsule.*

About 20 spp. in Europe and W. Asia, chiefly Mediterranean. None native in the British Is. Three blue-fld spp. which are sometimes found as casuals may be distinguished as follows:

1   Fls with an involucre of much-dissected lvs; carpels 5, united almost throughout their length; fr. a globose capsule. (Love-in-a-Mist.) *N. damascena L.
    Fls not surrounded by an involucre; fr. an angular capsule. 2
2   Carpels united only below the middle. *N. arvensis L.
    Carpels united almost to their summits.
                              *N. gallica Jordan
                    (*N. hispanica* auct., non L.)

N. damascena, and less frequently *N. gallica,* are grown in gardens.

6. ACONITUM L.

Perennial herbs with erect tuberous stocks and stout stems with alternate *palmately lobed or divided lvs. Fls in terminal racemes,* hermaphrodite, *zygomorphic,* with all parts spirally arranged. *Sepals 5 petaloid, the posterior sepal forming a large erect helmet-shaped hood; nectaries 2–8, the posterior pair included within the hood, long-clawed and with limbs prolonged into nectar-secreting spurs, the remainder very small or 0;* stamens numerous; carpels 3–5, sessile, free or slightly joined at the base, with numerous ovules. *Fr. a group of many-seeded follicles.*

About 300 spp. throughout the north temperate zone.

All species are highly poisonous and have often proved fatal owing to the presence of the powerful and deadly alkaloid aconitin and of other associated alkaloids. *A. napellus* L. has long been officinal as a narcotic and analgesic.

1. A. napellus L. subsp. napellus   Monkshood.
A. anglicum Stapf

A perennial herb often with paired blackish tuberous taproot-like stocks, up to 9 cm long by 3 cm diam. at the top, and erect lfy minutely downy flowering stems 50–100 cm, usually simple. *Lower stem-lvs up to 15 cm across, short-stalked, pentagonal in outline, palmately 3–5-partite, the middle segment wedge-shaped at the base then deeply laciniate with ±linear lobes,* the lateral segments similar but less divided (Fig. 1); upper lvs smaller,

Fig. 1. Leaf of *Aconitum napellus.* ×$\frac{1}{3}$.

±sessile; all *light green, soft,* and minutely hairy to glabrous. Infl. erect, moderately dense. *Fls mauve to blue-mauve,* minutely downy. Lower sepals strongly deflexed; *helmet 18–20 mm high, produced into a small point. Nectaries almost horizontal on the forward-curving tips of their erect stalks, their spurs forwardly directed with upturned capitate ends, their 2-lobed lips recurved.* Stamen filaments hairy above. Carpels 3, almost parallel, glabrous. Ripe follicles c. 2 cm. Seeds 4–5 mm, with a wide dorsal and narrower frontal wings. Fl. 5–6. Protandrous, the stamens erecting and dehiscing successively. Visited by long-tongued bumble-bees. $2n = 32^*$. Grh.

Native. Local in S.W. England and Wales, on shady stream banks.

A. napellus L., sens. lat., is a polymorphic aggregate of units which have been variously treated as varieties, subspp., or separate sp. In a detailed revision of the group, W. Seitz (1969) has recognized three spp. in all and, within *A. napellus* L. sens. str., nine subspp. The native British representative, which appears to be endemic, is referred to *A. napellus* subsp. *napellus.* The aggregate has a wide range in Europe and N.W. Asia eastwards to the Himalaya. Of the cultivated forms and hybrids which are grown in gardens and occasionally escape, *A. compactum* Reichenb., with ultimate lf-lobes linear to lanceolate, pedicels very short and appressed to the infl.-axis, and infl.-axis, pedicels and helmet all shortly appressed-hairy, is reported as established locally. Most cultivated forms have darker green and less narrowly divided lvs than subsp. *napellus* and have darker blue fls opening later in the season.

Besides *A. napellus,* **A. variegatum** L., with larger, often whitish fls in looser racemes, straight-clawed nectaries and seeds with prominent undulate ridges; and **A. vulparia** Reichenb. (*A. lycoctonum* auct., non L.), with pale yellow fls and a very tall conical-cylindrical helmet, are also seen in gardens.

7. Consolida (DC.) S. F. Gray

Annual herbs with alternate much-divided lvs and fls in simple or compound racemes. *Fls* hermaphrodite, *zygomorphic*, with all parts spirally arranged. *Outer per.-segs* (sepals) 5, *petaloid, only the upper (posterior) prolonged into a spur; inner per.-segs 2, coalescent into a single posterior petal with nectar-secreting spur contained within the sepal-spur;* stamens numerous in 5 curved rows; *carpel* 1. Fr. a single many-seeded follicle.

Several spp., chiefly in the Mediterranean region and W. Asia.

*1. C. ambigua (L.) P. W. Ball & Heywood Larkspur.
Delphinium ajacis auct.; *D. consolida* L., sec. Sm.; *D. gayanum* Wilmott

Annual finely pubescent herb with slender tap-root and erect stem 26–60(–100) cm, usually with a few ascending branches; upper part of stem ±glandular-hairy, the hairs with yellowish basal gland. Basal lvs long-stalked, deeply palmatisect into oblong segments; lower stem-lvs similar but with more numerous narrow-linear acute segments; upper stem-lvs ±sessile; all lf-segments finely pubescent. Infl.-branches 4–16-fld, ±*cylindrical; lower bracts much divided,* upper entire; *bracteoles on pedicels of lower fls rarely reaching beyond base of fl.* Fls 2.5–4 cm diam. Sepals 10–14(–20) mm, *usually deep blue,* sometimes paler blue, white or pink, ovate, clawed, the *posterior sepal with spur* 12–18 mm. Petal purplish-blue, 3-lobed, the middle lobe bifid and often marked with dark lines, the broader lateral lobes curving inwards. Stamens with broad filaments. *Follicle* 15–20 mm, *pubescent,* tapering into the persistent style, 2–3 mm. *Seeds* c. 2.5 mm, roundish, *nearly black,* with several ±*continuous* wavy membranous *transverse ridges.*

Introduced. Formerly established locally as a cornfield weed and still seen not infrequently as a casual and garden-escape, being the common larkspur of gardens. Mediterranean region; naturalized in many parts of Europe and in North America.

*C. orientalis (Gay) Schrödinger, native in S. and E. of the Iberian Peninsula and in S.E. Europe, resembles *C. ambigua* in its divided bracts and pubescent follicle but has bracteoles of lower fls reaching well beyond base of fl., sepal-spur not exceeding 12 mm and ripe follicle abruptly contracted into the very short style. Formerly a rare casual but not seen for some years. ***C. regalis** S. F. Gray (*Delphinium consolida* L.), native over much of continental Europe, differs in its ±entire bracts and glabrous follicle, the triangular outline of its infl.-branches and the broken transverse ridges of its seeds, like rows of scales. Like *C. orientalis* it was formerly an infrequent casual but has not been recorded in recent years.

Besides the annual 'Larkspurs', with only 2 inner per.-segs united into a single spurred nectary and with a single follicle, perennial herbs of the closely related genus **Delphinium** L. are also much grown in gardens. These differ in having 4 free inner per.-segs, the 2 upper independently spurred, the others unspurred, and 3 or more follicles; the 5 outer per.-segs (sepals), the posterior with a spur containing the two nectary-spurs, are as in *Consolida.* The commonly grown garden 'Delphiniums' are largely hybrid derivatives of the C. and E. European *D. elatum* L. with Asiatic spp. including *D. cheilanthum* Fischer.

8. Actaea L.

Perennial herbs with blackish rhizomes and 2–3-times ternate or pinnate lvs, the ultimate segments, often 3-lobed and irregularly toothed. *Fls small, in short racemes,* ±*actinomorphic,* usually hermaphrodite, with all parts spirally arranged. *Outer per.-segs* (sepals) 3–5, somewhat unequal, *petaloid, falling early; inner (petals)* 4–6(–10), *small,* spathulate, *sometimes 9; nectaries 0;* stamens numerous; *carpels* 1 with numerous ovules. *Fr. a single berry with several flattened seeds.*

The rhizome of *A. spicata* was formerly official as *Radix Christophorianae* and was used against skin diseases and asthma.

1. A. spicata L.   Baneberry, Herb Christopher.
A perennial foetid herb with stout blackish obliquely ascending rhizome and an erect glabrous flowering stem 30–65 cm. Basal lvs large, long-stalked, biternate or bipinnate, the secondary lflets ovate, acute, often 3-lobed, incise-serrate; stem-lvs 1–4, much smaller than the basal; all dark green above, paler beneath and ±hairy on the veins beneath. Raceme 25–30 cm, terminal or occasionally also axillary, dense-fld, elongating in fr., its axis and the fl.-stalks pubescent. Sepals 3–6, usually 4, whitish, blunt, concave, soon falling. *Petals* 4–6 or 0, *white, spathulate, clawed, shorter than the stamens,* not nectar-secreting. *Stamens white, clavate,* the filaments dilating upwards into the broad connective; anther-lobes small, distant, ±globose; carpel pyriform; stigma broad, sessile. *Berry* c. 1 cm, *ovoid, at first green, then blackish, shining.* Seeds c. 4 mm wide, semicircular, flattened. Fl. 5–6. Protogynous. Visited by small pollen-eating insects. $2n = 16$, c. 32. Grh.

Native. A local plant of ashwoods on limestone and of limestone pavements in Yorks, Lancashire and Westmorland; reaches 500 m in N. Yorks. Europe, northwards to Norway; temperate and arctic Asia to China.

Tribe 2. ANEMONEAE. Fr. a group of achenes, usually 1-seeded; chromosomes large, basic number 8.

9. Anemone L.

Perennial herbs with slender to tuberous creeping rhizomes or ±erect stocks, rarely spreading by root-buds. Basal lvs (rarely 0) borne either at intervals along a creeping rhizome or close together at base of fl.-stem; alternate, petiolate, usually palmately lobed or divided; *stem-lvs usually in a single whorl of* 3(–4), rarely with additional opposite pairs on stem-branches, sessile, and then sometimes partly united, or petiolate, closely resembling the basal lvs or differing ±markedly. Fl. usually solitary, terminal, sometimes in a few-fld umbel, hermaphrodite, actinomorphic, with all parts spirally arranged. *Per.-segs* 4–20, *petaloid, undifferentiated; nectaries* 0; *stamens* numerous, *all fertile;* carpels numerous, each with 1 functional ovule. *Fr. a cluster of achenes,* glabrous to densely woolly; *style neither elongating considerably nor becoming plumose in fr.*

Many spp. in temperate regions of both hemispheres and both Old and New Worlds, and on tropical mountains. Sharp-tasting plants, poisonous, and dangerous to grazing animals, owing to the presence of the lactone anemonin.

1 Per.-segs 5–9, oblong-elliptical to broadly ovate, rarely blue.   2
  Per.-segs 8–15, very narrow, blue (rarely white).
                    3. apennina
2 Per.-segs usually 6–7, white to pinkish or red-purple, rarely almost blue; common native woodland plant.
                    1. nemorosa
  Per.-segs 5(–8), yellow; non-native plant of gardens and private woodlands or orchards, occasionally becoming established.   **2. ranunculoides**

1. A. nemorosa L.   Wood Anemone.

Perennial herb with creeping brown rhizome from whose terminal bud develops the simple erect fl.-stem, 6–30 cm, glabrous or somewhat hairy. Basal lvs appearing only after fl., borne directly on the rhizome, often well away from fl.-stem, long-stalked, palmately 3-partite, the major divisions short-stalked and further cut or divided into coarsely toothed ±acute segments; stem-lvs 3, at c. ⅔ up from base of fl.-stem, stalked and resembling basal lvs but somewhat smaller; all lvs with sparsely hairy ciliate segments. Fls 2–4 cm diam., solitary. Per.-segs 5–12, usually 6 or 7, glabrous, oblong-elliptical, *white or pink-tinged*, occasionally reddish-purple, rarely almost blue. Stamens 60–70; anthers yellow. Carpels 10–30. Achenes c. 4 mm, downy, in a nodding globose cluster; beak c. 2 mm, curved. Fl. 3–5. Somewhat protandrous but self-incompatible. Fls yield no nectar and are visited for pollen by various bees, flies and beetles. Seed-germination in spring, hypogeal. $2n = 28–32, 37, 42, 45, 46$. Grh.

Native. A widespread gregarious herb of deciduous woodland and hedgerows, with lvs functional from early March until June. Found especially on good mull soils but occurring over the whole range from shallow rendzinas to podsols and tolerating a duration of waterlogging that excludes Bluebell and Dog's Mercury, its most serious competitors. Throughout Great Britain and the Inner Hebrides but very rare in the Outer Hebrides and not in Orkney or Shetlands; Ireland; Channel Is. Most of Europe but rare in Mediterranean and Arctic regions; W. Asia.

*2. A. ranunculoides L.   Yellow Wood Anemone.

Resembles *A. nemorosa* in its creeping brown rhizome and erect fl.-stem 7–30 cm but has basal lvs 0–1, rarely more, *stem-lvs with petioles not exceeding 5 mm*, and fls with 5(–8) *bright yellow oval per.-segs*, pubescent beneath. All lvs palmately and deeply 3(–5)-lobed, the lobes oblong-lanceolate, irregularly incise-toothed. Fl. 4. $2n = 32$.

Introduced. Sometimes planted in private woods, copses and orchards and also in gardens and maintaining itself or occasionally escaping and becoming established in a few widely scattered localities. Native in much of Europe but rare in the Mediterranean region; W. Asia.

*3. A. apennina L.   Blue Mountain Anemone.

Like *A. nemorosa* in habit but rhizome short, tuberous, blackish and basal lvs close to base of fl.-stem. All lvs petiolate, ternate, with triangular pinnatifid lobes, pubescent beneath; stem-lvs close to middle of fl.-stem which is 8–25 cm, erect. *Fls solitary, with 9–15 very narrow, blue*, rarely white, *per.-segs* which are *somewhat pubescent beneath*. Anthers pale yellow to white. Achenes in an erect cluster, shortly pubescent, broadly ovoid with short abruptly curved beak. Fl. 4. $2n = 16$. Grh.

Introduced. Like *A. ranunculoides* formerly often grown in private woods as well as gardens and occasionally establishing itself for a time. S. Europe eastwards from Corsica and Sicily.

Also much grown in gardens are the white-fld **A. sylvestris** L. of C. and E. Europe, superficially like *A. nemorosa* but with obliquely erect stock, lvs close round base of fl.-stem, larger fls (4–7 cm diam.) usually with 5 broadly ovate per.-segs, downy beneath, and an elongated head of densely woolly achenes; *A. blanda* Schott & Kotschy of S.E. Europe and Turkey, like *A. apennina* but fl. Jan.–March and with both lvs and the narrow blue, pink or white per.-segs glabrous beneath; the tall autumn-flowering Japanese Anemones of complex hybrid origin (*A.* 'Japonica', *A.* 'Hybrida'), and the brightly coloured 'florists' anemones' derived by hybridization and selection from S. European *A. coronaria* L., *A. pavonina* Lam. and the intermediate and probably hybrid *A. fulgens* Gay.

The related genus **Hepatica** Miller has 3- to 5-lobed overwintering basal lvs but the stem-lvs of *Anemone* are represented by a whorl of 3 ±entire green 'bracts' close beneath the perianth and simulating a calyx. **H. nobilis** Miller (*Anemone hepatica* L.), indigenous in much of continental Europe, has lvs usually bright purple beneath, with 3 broad entire lobes, rounded-triangular in outline and cordate at base. The fls are 15–25 mm diam. with 6–7(–10) elliptic-oblong per.-segs, usually blue but sometimes white, pink or purplish. It is often grown in gardens for its attractive lvs and fls.

10. PULSATILLA Miller

Perennial caespitose herbs with stout oblique to vertical often branching stocks. Basal lvs long-petiolate, 1–4 times pinnately or palmately divided, often silky-hairy when young, usually appearing at or just after flowering but sometimes over-wintering; *stem-lvs in a whorl of 3, usually sessile and ±united basally, smaller and more finely divided than the basal lvs and often held erect*, but in some spp. shortly petiolate and resembling the basal lvs. Fls solitary, terminal, often campanulate and erect, actinomorphic, usually hermaphrodite. Per.-segs usually 6, silky on the outer (lower) side; *outermost stamens usually replaced by nectar-secreting staminodes*; fertile stamens numerous; carpels numerous, each with 1 functional ovule. Fr. a cluster of *silky achenes with much elongated and plumose persistent styles*.

About 20 spp. in temperate Eurasia and North America.

Unpalatable and poisonous owing to the presence of anemonin, as in *Anemone*; some spp. officinal.

1. P. vulgaris Miller   Pasque Flower.
Anemone pulsatilla L.

Perennial herb with obliquely erect blackish usually

branching stock and erect silky-hairy fl.-stem 4–12 cm, lengthening to 15–30(–40) cm in fr. Basal lvs petiolate, bipinnate, the lflets further pinnatisect into many linear to linear-lanceolate segments; stem-lvs 3, sessile, united below, ±erect, deeply divided into linear segments; all lvs silky-hairy at least initially, the basal with pubescent petioles. Fl. campanulate, erect at first but drooping during fading. *Per.-segs* 6, c. 3 cm, elliptical, *blue-violet inside, paler and silky outside*. Staminodes and fertile stamens not more than half as long as per.-segs; anthers golden-yellow. The internode between stem-lvs and fl. elongates considerably after flowering. Achenes silky, with feathery style 3.5–5 cm. Fl. 4–5. Slightly protogynous and self-compatible. Visited chiefly by bees for pollen and nectar. $2n = 32$. H.

Native. Very local on dry grassy calcareous slopes on chalk from Berks to Cambs and on oolitic limestone from Gloucester to N. Lincoln; known in only 1 site on non-calcareous soil. Formerly more frequent and widespread and extending along the Magnesian limestone to near the N. Yorks–Durham border, but apparently lost from many sites through plough-up of old pasture. N.W. and C. Europe northwards to S. Sweden and eastwards to the Ukraine and W. Asia.

Various forms of *P. vulgaris*, some with maroon or creamy-white fls, are grown in gardens, as are **P. vernalis** (L.) Miller, with pinnate overwintering basal lvs and per.-segs white inside, violet or pink outside with silky yellow-brown hairs; and the more distinct **P. alpina** (L.) Delarbre, with shortly petiolate stem-lvs much like the basal lvs, and white or yellow fls 4–6 cm diam. which lack nectar-secreting staminodes.

11. Clematis L.

Woody climbers, less often non-climbing shrubs or perennial herbs, with exstipulate *lvs in opposite pairs, usually compound and ending in tendrils or with twining petiole and rhachis*, sometimes simple. Fls solitary, in simple cymes or in panicled infls; actinomorphic, usually hermaphrodite, with all parts spirally arranged. *Per.-segs (or sepals)* valvate, rarely imbricate, *in bud*, 4(–8), *petaloid; petals usually 0*, but petaloid staminodes with some intermediates replace the outer stamens in *C. alpina* (L.) Miller and other spp. of Sect. *Atragene* (L.) DC.; stamens numerous; carpels numerous, each with 1 functional ovule. Fr. a cluster of *achenes with persistent usually long and plumose styles*. Nectar, usually absent, is secreted by the stamen-filaments in Sect. *Atragene* and some other spp. and by special staminodal nectaries in a few tropical Asian spp.

200–300 spp., widely distributed especially in temperate regions of the northern hemisphere.

1. C. vitalba L.   Traveller's Joy, Old Man's Beard.
Perennial woody climber with stems up to 30 m. Lvs pinnate usually with 3 or 5 rather distant lflets; lflets 3–10 cm, narrowly ovate, acute to acuminate, rounded or subcordate at base, coarsely toothed or entire, glabrous or slightly pubescent. Fls c. 2 cm diam., in termi-

nal and axillary cymes, fragrant. Per.-segs greenish-white, pubescent above and more densely so beneath. Stamens whitish; anthers 1–2 mm. Achenes in large heads on the pubescent receptacle, scarcely compressed, with long whitish plumose styles. Fl. 7–8. Slightly protogynous. Nectar is secreted from the stamen-filaments and the fls are visited by various bees and flies for both pollen and nectar. $2n = 16*$. M.

Native. Hedgerows, thickets and wood-margins, chiefly on calcareous soil or rock-crevices, northwards to N. Wales and the Humber estuary; probably introduced in the scattered localities further north to C. Scotland as also in Ireland and Channel Is. Europe from S. Britain and Netherlands southwards; N. Africa; Caucasus.

The following, all ±woody climbers, sometimes escape from gardens and establish themselves locally.

*****C. flammula** L. (S. Europe and W. Asia) resembles *C. vitalba* but has bipinnate lvs and pure white fls which open Aug.–Oct. Anthers 3–4 mm, achenes strongly compressed and ±glabrous receptacle.

*****C. viticella** L. (S. Europe and S.W. Asia) has pinnate lvs with ±deeply lobed lflets which are sometimes ternate, and blue to reddish-purple fragrant fls 4 cm diam. which open in summer or early autumn.

*****C. montana** DC. (Himalaya, C. & W. China) with ternate lvs and ±ovate lflets and with a profusion of long-stalked axillary fl.-clusters, white or pink, produced in May, is a very popular garden plant.

The climbing Clematises most commonly grown in gardens, with large blue-purple to reddish fls, are chiefly hybrids of the Chinese *C. lanuginosa* Lindl. with *C. viticella* (*C.* × *jackmannii* Moore) or with the Chinese *C. patens* Morr. & Decne. The *viticella* hybrids are later-flowering than the *patens* hybrids and have usually only 4 per.-segs instead of 6–8. Non-climbing spp. often grown include *C. integrifolia* L., *C. heracleifolia* DC. and *C. recta* L.

12. Ranunculus L.

Annual to perennial herbs with alternate lvs, spirally arranged or distichous, rarely opposite, stipulate or exstipulate, often palmately lobed or divided, sometimes simple and ±entire. Fls solitary and terminal or in cymose infls, hermaphrodite, actinomorphic, with all parts spirally arranged or, more usually, with only the petals whorled. Sepals usually 5, but 3 in *R. ficaria*; *petals 5 or more, rarely fewer or 0, usually yellow or white, each with a nectar-secreting depression near its base*; stamens numerous; carpels numerous, each with 1 basally attached ascending ovule. *Fr. a head of achenes*.

About 400 spp., cosmopolitan, but chiefly in northern extratropical regions.

All species contain the glycoside ranunculin, converted into the lactone protoanemonin when the plant is crushed, as by grazing animals. They soon learn to avoid this acrid vesicant, but young stock may sometimes suffer severe blistering of the mouth and also intestinal disorders that may prove fatal, as may older animals able to graze only near ponds and ditches during very dry seasons. Protoanemonin dimerizes after a short time to the less toxic anemonin, so that even hay with large amounts of buttercup soon becomes harmless to stock.

1 Aquatic plants with white fls; achenes with distinct transverse ridges. 17
  Terrestrial or marsh plants with yellow fls, or petals 0; achenes lacking distinct transverse ridges. 2
2 Sepals 3; petals 7–12; lvs simple, usually cordate. **25. ficaria**
  Sepals 5; petals 5 or fewer, rarely 0; lvs various. 3
3 All or most lvs palmately lobed or divided. 4
  All lvs simple, subulate or narrowly linear to broadly ovate, entire or toothed. 14
4 Ripe achenes 6–8 mm, spiny on the faces; annual weeds of arable fields and disturbed ground. 5
  Ripe achenes not spiny though sometimes with small tubercles on the faces. 6
5 Spines long, on borders as well as faces of achenes; widespread cornfield weed with lemon-yellow fls. **6. arvensis**
  Spines short, on faces but not borders of achenes; introduced weed of bulb-fields in Scilly, casual elsewhere. **5. muricatus**
6 Ripe achenes with tubercles on the faces. 7
  Ripe achenes lacking tubercles though sometimes punctate or rugulose. 8
7 Tubercles few, just within the green border of the achene; fls 12–25 mm diam. Erect annual plant with strongly reflexed sepals. **4. sardous**
  Tubercles all over the red-brown faces of the achenes, each terminating in a minute hook; fls 3–6 mm diam. Decumbent or ascending annual with sepals eventually spreading or reflexed. **7. parviflorus**
  (*R. trilobus* and *R. marginatus* var. *trachycarpus*, p. 42, also have tubercles all over the faces of the achenes, but these are not hook-tipped.)
8 Lowest lvs glabrous or nearly so. 9
  Lowest lvs distinctly hairy (rare mountain variants of *R. acris* may have the earliest spring lvs glabrous). 10
9 Lowest lvs reniform in outline; fls 1.5–2.5 cm diam., but less when, as often, petals are reduced in number or 0; achenes in a globose cluster; woods and shady hedgerows. **9. auricomus**
  Lowest lvs angular in outline, shining; fls 0.5–1 cm diam.; achenes in an oblong-ovoid head; wet places and bare mud. **14. sceleratus**
10 Sepals strongly reflexed in fl. 11
  Sepals not strongly reflexed in fl. 12
11 Perennial; stem tuberous (corm-like) below-ground; achenes with narrow yellowish border and hooked beak **3. bulbosus**
  Annual; stem not or only slightly tuberous below-ground; achenes with conspicuous green border and almost straight beak. **4. sardous**
12 Stem-lvs several, at least the lower petiolate; achenes 2.5–3 mm, short-beaked, in a globose head. Common and widespread. 13
  Stem-lvs 1(–2), sessile; achenes c. 2 mm, long-beaked, in an elongated head. A rare perennial with root-tubers and slender hypogeal stolons, restricted to Jersey. **8. paludosus**
13 Plant with long epigeal stolons (runners); basal lvs with stalked terminal lobe; fl.-stalks furrowed. **2. repens**
  Plant without stolons; basal lvs with terminal lobe

sessile; fl.-stalks not furrowed. **1. acris**
14 Plant 60–90 cm; fls few, (2–)3–5 cm diam. **10. lingua**
  Plant usually not exceeding 60 cm; fls 0.5–2(–2.5) cm diam. 15
15 ±Erect annual with long-stalked broadly ovate often cordate basal lvs; fls 6–9 mm diam.; faces of ripe achenes covered with small tubercles. Very rare. **13. ophioglossifolius**
  Perennial herbs ranging from erect to creeping and basal lvs from subulate to broadly ovate; fls 5–20(–25) mm diam.; faces of ripe achenes smooth, lacking tubercles. 16
16 Plant with filiform stem rooting at every node and arching in the internodes; fls solitary, 5–10 mm diam.; achenes c. 1 mm. **12. reptans**
  Plant, if creeping, not rooting at every node; fls 1–several; achenes 1.5 mm or more. **11. flammula**
17 Plant with no finely dissected lvs. 18
  Finely dissected lvs present. 21
18 Lvs with 3–5 shallow lobes usually reaching much less than halfway to top of petiole; sepal-tips not clear blue; receptacle glabrous. 19
  Lvs with 3(–5) ±cuneate lobes usually reaching more than halfway to top of petiole; sepal-tips clear blue; receptacle pubescent. 20
19 Lf-lobes broadest at their base; petals about equalling sepals. **15. hederaceus**
  Lf-lobes narrowest at their base; petals 2–3 times as long as sepals. **16. omiophyllus**
20 Petals 6–10 mm, broadly obovate, contiguous in newly open fl.; ripe achenes 1–1.4 mm, winged; receptacle elongating in fr. Usually in brackish water. **18. baudotii**
  Petals less than 6 mm, narrowly obovate, non-contiguous; ripe achenes c. 1.5 mm, not winged; receptacle not elongating in fr. In fresh water. **17. tripartitus**
21 Floating lvs, with expanded blade circular to semi-circular in outline, present as well as finely dissected lvs. 22
  Floating lvs 0. 26
22 Petals rarely exceeding 5 mm, non-contiguous; sepal-tips clear blue; dissected lvs few, with very fine collapsing segments; floating lvs 3(–5)-lobed to more than halfway. **17. tripartitus**
  (Hybrids with *R. omiophyllus*, with rather larger fls, may rarely have dissected lvs with flattened linear non-collapsing segments. See 17, p. 46.)
  Petals usually exceeding 5 mm, contiguous in newly open fl.; dissected lvs usually numerous. 23
23 Dissected lvs yellow-green; floating lvs usually 3-lobed to more than halfway; sepal-tips clear blue; achenes glabrous even when immature, winged when ripe and dry; receptacle elongating in fr. Usually in brackish water. **18. baudotii**
  Dissected lvs not yellowish; floating lvs usually 5-lobed; sepal-tips not clear blue; achenes pubescent at least while immature, not winged when ripe; receptacle not elongating. 24
24 Petals not exceeding 10 mm, with circular nectaries; floating lvs with cuneate straight-sided usually dentate lobes; fr.-stalk usually less than 5 cm, shorter than petiole of opposed floating lf. **19. aquatilis**
  Petals usually exceeding 10 mm, with pyriform nectaries; floating lvs with convex-sided commonly

crenate lobes; fr.-stalk usually more than 5 cm, longer than petiole of opposed floating lf.   **25**

25 Mature dissected lvs shorter than stem-internodes, their segments usually rigid and divergent when removed from water.   **21. peltatus**

Mature dissected lvs longer than stem-internodes and up to 30 cm, their segments limp, subparallel, collapsing when removed from water. **22. penicillatus**

26 Lvs circular in outline, their rigid segments all lying in one plane.   **24. circinatus**

Lf-segments not all lying in one plane.   **27**

27 Petals rarely exceeding 5 mm, not contiguous in wholly open fl.; nectaries lunate (crescent-shaped).
    20. trichophyllus

Petals exceeding 5 mm, contiguous in wholly open fl.; nectaries variously shaped.   **28**

28 Mature lvs equalling or exceeding stem-internodes, their segments subparallel or somewhat divergent; nectaries long-ovate to pyriform.   **29**

Mature lvs shorter than stem-internodes, their segments divergent; nectaries various.   **30**

29 Lvs up to 50 cm, divided about 4 times into a few long firm segments; receptacle ±glabrous.
    23. fluitans

Lvs of variable length, each divided 7 or more times into more than 100 fine and usually flaccid segments; receptacle distinctly pubescent.
    22. penicillatus

30 Lvs yellow-green; nectaries lunate; achenes winged when dry, glabrous even when immature; receptacle elongating in fr. Usually in brackish water.
    18. baudotii

Lvs not yellowish; nectaries circular or pyriform; achenes not winged when dry, pubescent at least while immature; receptacle not elongating in fr. Usually in fresh water.   **31**

31 Ultimate lf-segments more than 100; nectaries pyriform.   **21. penicillatus**

Ultimate lf-segments fewer than 100; nectaries pyriform or circular (but fls rarely present if no floating lvs).   **32**

32 Petals usually exceeding 10 mm; nectaries pyriform; fr.-stalk usually exceeding 5 cm.   **21. peltatus**

Petals not more than 10 mm; nectaries circular; fr.-stalk usually less than 5 cm.   **19. aquatilis**

(Dissected lvs of these two spp. are indistinguishable. *R. peltatus* rarely flowers in the absence of floating lvs, though *R. aquatilis* does so occasionally.)

Subgenus RANUNCULUS

Annual or perennial usually terrestrial herbs. Lvs simple or palmately, less commonly pinnately, compound, very rarely divided into capillary segments. Sepals 5; petals 5, rarely fewer or 0, usually bright yellow but sometimes paler yellow or whitish, rarely reddish. Achenes neither transversely ridged nor striate.

Section *Ranunculus* (incl. *Echinella* DC.). Annual or perennial with roots often thick and fleshy but not tuberous. Nectary at base of petal covered by a flap or scale attached only at its base or laterally as well. Achenes distinctly beaked, at least moderately compressed, sometimes conspicuously bordered, the faces smooth or with tubercles, hooked hairs or spines. Receptacle

often elongating considerably in fr.

1. R. acris L. subsp. acris   Meadow Buttercup.

Perennial herb with overwintering lf-rosettes from the short oblique to erect premorse stock up to 5 cm, rarely longer and more rhizome-like; roots white, rather fleshy, little branched. *Stem* 15–60(–100) cm, branching above into the irregularly cymose infl., *with ±dense spreading or reflexed hairs below, appressed-hairy above*; rarely ±glabrous. Basal lvs (Fig. 2A) long-

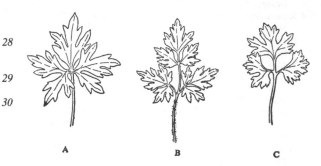

Fig. 2. Leaves of *Ranunculus*. A, *R. acris*; B, *R. repens*; C, *R. bulbosus*. ×½.

stalked, pentagonal to roundish in outline, palmately 3(–7)-lobed, the *primary lobes separate and ±sessile or shortly joined*, further cut or lobed, the degree of dissection increasing until shortly before fl-stems appear; lower stem-lvs similar but with shorter petioles; uppermost sessile, deeply cut into linear segments; all petioles and lf-segments ±hairy. Fls 16–25 mm diam., terminating infl.-branches which are appressed-hairy and *not furrowed*. Fl-buds globose. Sepals yellowish-green, hairy, appressed to the petals, *not reflexed*. Petals commonly 8–15 mm, broadly obovate, bright golden-yellow, occasionally paler or even white, glossy. *Receptacle glabrous*. *Achenes* 2.5–3.5 mm, ±obovate with short hooked beak, glabrous (Fig. 3A). Fl. 5–7(–10). Protogynous and self-

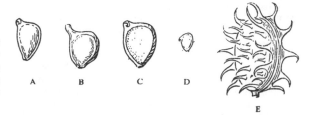

Fig. 3. Achenes of *Ranunculus*. A, *R. acris*; B, *R. repens*; C, *R. bulbosus*; D, *R. paludosus*; E, *R. repensis*. ×3.

incompatible. Plants may show the whole range from completely hermaphrodite to female (no viable pollen) or male (no functional ovules) and occasional plants produce neither viable pollen nor functional ovules. Visited by various insects, especially flies and small bees. $2n = 14^*$. Hp.

A very variable species with several subspp. based chiefly on the form of the underground stem, lvs and achenes. Of those now generally recognized, only subsp. *acris*, distinguished by combining a short premorse stock with lvs having the terminal lobe of the central primary segment usually at least $\frac{2}{3}$ of the total lf-length, is native to the British Is.

Native. Found commonly and often abundantly in damp meadows and pastures on calcareous and circum-neutral soils throughout the British Is. and occasional in mixed fen communities, on damp rock-ledges, in gullies and on mountain-top detritus, reaching 1200 m in Scotland. Rare in woodland and absent from areas very dry in summer. Crushed plants have a high content of the acrid vesicant protoanemonin and *R. acris* is avoided by most grazing animals, tending therefore to increase in heavily grazed communities. Subsp. *acris* occurs over much of Europe northwards to Iceland and the North Cape but extending southwards only to S. France, C. Italy and N. Greece; not in the Pyrenees or the Iberian Peninsula, where a form of the rhizomatous subsp. *friesianus* (Jordan) Rouy & Fouc. occurs. *R. acris* sens. lat. is widely distributed in temperate and arctic Eurasia and has become naturalized in North America, S. Africa and New Zealand.

2. R. repens L.   Creeping Buttercup.

Perennial herb overwintering as a short usually vertical stock bearing long stout roots and a rosette of lvs. In spring the apical and some of the upper axillary buds form fl.-stems, but some lower buds usually grow out later into *strong lfy epigeal stolons* (*runners*) which root at the nodes and commonly produce further fl.-stems in the current season. These die after flowering, as does the parent plant, the overwintering units being stolon-borne daughter-plants which have developed roots and lf-rosettes. Rosette-lvs and lower stem-lvs (Fig. 2B) petiolate, triangular-ovate in outline, divided into usually 3 primary segments, the central long-stalked, the lateral short-stalked or sessile and all further divided or lobed with the ultimate segments commonly 3-toothed; upper stem-lvs sessile with narrow ±entire segments; all lvs and petioles usually with spreading hairs, less commonly appressed. Fl.-stems 15–60 cm, erect or ascending, hairy. Fls 2–3 cm diam. at the ends of the hairy and furrowed branches of an irregularly cymose infl. *Fl.-buds ovoid. Sepals* hairy, *not reflexed.* Petals 5(–9 or more), 8–17 mm, ovate, suberect, glossy and more deeply golden-yellow than in **1** and **3**. *Receptacle pubescent. Achenes* in a globose head, each (2–)3 mm, with a ±curved beak up to 1 mm, obliquely obovate, bordered, *glabrous*, light to dark brown (Fig. 3B). Fl. 5–8. Protogynous; visited by various bees and flies; most plants highly self-incompatible. $2n = 32^*$. Hs.

Variation, especially in number and length of stolons, degree of dissection of lvs and hairiness, is considerable but too continuous for recognition of subspp.

Native. Common in damp meadows, pastures and woods on a wide range of soils but mainly on heavy wet soils of at least moderately high nutrient level and markedly less frequent in leached upland grasslands unless 'improved'; characteristic of furrow-bottoms in 'ridge and furrow' grassland; occurs also in marshes, open fen carr and dune-slacks. An efficient colonist of disturbed habitats such as ditch-sides, hedgerows, gravel-heaps, etc., and a troublesome weed of arable fields and gardens. Tolerant of trampling but not avoided by stock to the same extent as **1** and **3**. Reaches 1036 m on Snowdon. Throughout the British Is. Over almost the whole of Europe and extends in an interrupted broad band across Siberia to Kamchatka and Japan; introduced in eastern North America, Central and South America and New Zealand.

3. R. bulbosus L. subsp. **bulbosus**   Bulbous Buttercup.

Perennial herb overwintering as a *vertical corm-like tuberous stock*, somewhat flattened, 1–3 cm horizontal diam., with a rosette of lvs above and *±fleshy contractile roots up to* 3 mm *diam.* beneath. In late winter new lvs are added and in March or later the apical bud grows out into a lfy fl.-stem, sometimes with others from axillary buds. The aerial stems die back after fl. and the corm is dormant usually until Sept., when one lateral bud (rarely more) grows out, forms a lf-rosette and initiates a new corm. The old corm loses its food-reserves to the new one in Feb.–March and then dies and decays by the next fl. period. Basal and lower stem-lvs (Fig. 2C) petiolate, ovate in outline, *with usually 3 primary lflets, the central long-stalked*, the lateral short-stalked or sessile, all further divided or lobed; upper stem-lvs ±sessile, deeply cut into narrow often linear segments; all lvs usually hairy. Fl.-stems 15–40 cm, erect or ascending, ±hairy, often with spreading hairs below, appressed above. Fls 2–3 cm diam., at the ends of the hairy *deeply furrowed branches* of the irregularly cymose, often corymbose, infl. *Fl.-buds ovoid. Sepals* 5, pale yellowish, *strongly reflexed.* Petals 5, usually 10–15 mm, broadly obovate with cuneate base, bright glossy yellow, rarely paler or white. *Receptacle pubescent*, subglobose. Achenes 2–4 mm, obliquely obovate, their faces finely punctate, dark brown with paler border; beak short, somewhat hooked (Fig. 3C). Fl. (3–)5–6. ±Homogamous; highly self-incompatible. Freely visited by insects, especially bees and flies. $2n = 16^*$. Hs.

Very variable in size and shape of stem-tuber and number of fl.-stems arising from it, hairiness, etc. Subsp. *adscendens* (Brot.) Neves, with less strongly tuberized stock, rarely corm-like, and with thick tuberous roots 4–6 mm diam., is confined to the Mediterranean region though with intermediates further north.

Dry pastures, grassy slopes, fixed dunes, etc., especially on calcareous substrata, throughout the British Is. except Shetland; abundant in the south but becoming less common in the north; primarily lowland but reaching 580 m in Devon. Avoided by grazing animals but intolerant of trampling. Characteristic of ridge-tops in 'ridge-and-furrow' pasture (see **2** above). Throughout much of Europe northwards to c. 60°N but local in S.E. Europe; local also in N. Africa and W. Asia; introduced in North America and New Zealand.

4. R. sardous Crantz   Hairy Buttercup.

R. philonotis Ehrh.; incl. *R. hirsutus* Curtis & *R. parvulus* L.

An annual herb resembling **3** but with fibrous roots, with the corm-like stock at most very feebly developed and with stem-hairs spreading or reflexed, never appressed. Fl.-stems commonly several, 10–45 cm, erect or ascending, lfy, ±hairy. Basal lvs petiolate, ternate, with the central lflet longest-stalked, or ±deeply 3-lobed, the lflets or primary lobes further 3-lobed, the ultimate segments coarsely toothed; stem-lvs shorter-stalked, uppermost sessile with fewer and narrower segments; all ±hairy. *Fls* many, 12–25 mm diam., *terminating the hairy and furrowed infl.-branches. Receptacle pubescent. Sepals* 5, c. 8 mm, hairy, *strongly reflexed* in open fl. *Petals* 5 or more, 8–15 mm, obovate, *pale yellow. Achenes* 2.5–3 mm, almost orbicular, their brownish faces usually *with an incomplete ring of small blunt tubercles close to the conspicuous green border*, but tubercles sometimes 0; beak c. 0.5 mm, triangular, slightly curved (Fig. 4A). Fl. 6–10. Slightly protogynous; visited by flies and small bees. $2n = 16$. Th.

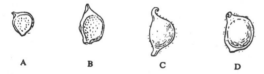

Fig. 4. Achenes of *Ranunculus*. A, *R. sardous*; B, *R. parviflorus*; C, D, *R. auricomus*. ×3.

Variable in size and hairiness and in size, number and distribution of tubercles on the achenes. *R. parvulus* L. is merely a dwarf form.

Native. A local and often casual weed of damp arable and waste land, especially near the coasts of S. and E. England, S. Wales and Cumbria, but also in a few localities further inland and further northwards to Argyll and E. Ross; Channel Is.; not in Ireland. In much of Europe northwards to S. Sweden; N. Africa; W. Asia.

R. marginatus D'Urv. var. **trachycarpus** Fischer & Meyer is an annual herb resembling **4** but almost glabrous, with golden-yellow petals exceeding the spreading sepals and numerous small achenes, 3–4 mm, with blunt wrinkled tubercles covering the faces and short beak c. 1 mm. Established as an arable weed in St Martin's, Scilly; a rare casual elsewhere. Native in Corsica, Sardinia, Sicily, S.E. Europe, W. Asia and N. Africa.
  The rare casual **R. trilobus** Desf. has petals about as long as the sepals and tubercles all over the faces of the achenes.

*5. R. muricatus L.
                    Scilly Buttercup, Prickly-fruited Buttercup.
Annual almost glabrous herb with fibrous roots and a

stout much-branched spreading or ascending stem, 10–50 cm. Basal lvs long-stalked, ±circular in outline with cordate base and 3(–5) shallow dentate lobes; successively higher lvs becoming shorter-stalked and narrower, grading from truncate to cuneate at base. Fls 10–15(–20) mm diam., with *petals slightly longer than the spreading or reflexed sepals*. Receptacle pubescent. *Achenes* 5–8 mm, usually 6–16, ovate, flattened, strongly keeled, with *spiny tubercles restricted to the brown faces*, none on the broad green margin; beak 2–3 mm, tapering and somewhat curving from the broad base. Fl. 3–6. $2n = 48, 64$. Th.

Introduced. Established in Scilly since 1923 or earlier, and now an abundant weed of bulbfields, and very locally in the Lizard area of Cornwall; casual elsewhere, but formerly established in Lancashire. Native in the Mediterranean region and in W. Asia westwards to India; widely naturalized.

6. R. arvensis L.   Corn Crowfoot.

An annual herb with fibrous roots and erect branching lfy stem, 15–60 cm, glabrous or pubescent. Lvs all petiolate, the lowest simple, broadly spathulate to obovate, toothed near the tip; the rest ±deeply 3-lobed or ternate to biternate, with narrow toothed or entire segments; uppermost short-stalked with few linear segments. *Fls* 5–12 mm diam., *terminating the hairy but non-furrowed branches* of an irregular cymose infl. *Sepals* pale yellowish-green, *spreading*, hairy. *Petals* c. 8 mm, exceeding the sepals, obovate-oblong, *shining lemon-yellow*. Receptacle with long hairs. *Achenes* 6–8 mm overall, few (4–8), reddish-brown, asymmetrically obovate, *spiny, the longest spines on the broad grooved border*; beak 3–4 mm, almost straight (Fig. 3E). Fl. 6–7. Protandrous to homogamous; sometimes gynomonoecious. Visited by small flies. $2n = 32$. Th.

Probably native. Long established as a cornfield weed, especially on calcareous soils, and sometimes colonizing disturbed ground of roadsides, etc. Common but decreasing in S. and E. England, local in Wales and S.W. England and now in only a few very scattered localities in N.W. England and Scotland, though once more frequent in S. Scotland; Orkney; no longer in Ireland or Channel Is. Throughout all C. and S. Europe; N. Africa; W. Asia eastwards to India.

7. R. parviflorus L.   Small-flowered Buttercup.

A pubescent annual herb with fibrous roots and numerous spreading to decumbent branching stems 10–40 cm. Basal lvs roundish-cordate, ±shallowly 3–5-lobed with obovate-cuneate toothed segments; upper lvs with fewer and narrower lobes, uppermost oblong; all petiolate, softly hairy, yellowish-green. Fls 3–6 mm diam., their *pedicels* lf-opposed or at forkings of the stem, hairy, *furrowed. Sepals* hairy, *reflexed* in the open fl. *Petals* 5 or fewer, 2–3 mm, *about equalling the sepals*, obovate-oblong, pale yellow. *Receptacle glabrous. Achenes* few, 2.5–3 mm, suborbicular, narrowly bordered, *with shortly-hooked tubercles all over the reddish-brown* faces; beak short, hooked at the tip (Fig. 4B). Fl. 5–6. $2n = 28$. Th.

Native. A local lowland plant, especially near the sea, of dry grassy banks, cliff-pastures, path-sides, etc., and a weed of arable land. Most frequent in S.W. England and S. Wales but scattered through S. England and extending northwards to N. Wales, Lincoln and S. Yorks; very rare in Ireland; Channel Is.; Iberian Peninsula and W. France, Mediterranean region; Macronesia.

Section *Ranunculastrum* DC. Perennial herbs with both tuberous and fibrous roots, often with filiform stolons. Nectary-scale attached only at its base, not laterally. Achenes compressed, keeled, beaked. Receptacle elongating in fr. and becoming ±cylindrical.

8. R. paludosus Poiret   Fan-leaved Buttercup.

R. flabellatus Desf.; *R. chaerophyllos* sensu Coste, non L.

A *perennial stoloniferous herb* with short erect stock bearing both fibrous roots and *a cluster of fusiform root-tubers*, 4–8 mm. Stolons hypogeal, very slender and with tiny scale-lvs. Stem 10–30(–50) cm, erect, simple or little branched, silky-hairy. Lvs chiefly basal, petiolate, the lowest shallowly 3-lobed, the remainder ternate with long-stalked central lflet or pinnatisect, the primary segments further divided into narrow dentate ultimate segments; stem-lvs 1–2, small, sessile. Fls commonly 1–4, 2.5–3 cm diam., with hairy non-furrowed pedicels. *Sepals 5, hairy, spreading.* Petals 5, exceeding the sepals, bright yellow, very glossy. *Receptacle glabrous, much elongating in fr.* Achenes c. 2 mm, suborbicular, ±glabrous, minutely punctate, *with long acute almost straight beak*, flattened below and *approaching half the overall length* (Fig. 3D). Fl. 5. $2n = 32$. G.-H.

Native. Found only in dry places near St Aubyns, Jersey. W. and S. Europe; N. Africa; W. Asia.

Section *Auricomus* Spach. Perennial fibrous-rooted herbs with long-stalked basal lvs, reniform to orbicular, entire or ±deeply palmate-lobed, and sessile stem-lvs. Pedicel not furrowed. Nectary-scale attached laterally as well as basally, so forming a pocket, but often very small or abortive. Achenes beaked, only slightly compressed, their faces neither tuberculate nor spiny; receptacle not or little elongating in fr.

9. R. auricomus L.   Wood Crowfoot, Goldilocks.

A perennial herb with short ascending premorse stock and many fibrous roots. Stems 10–40 cm, often numerous, ascending to erect, somewhat branched above, glabrous to sparsely pubescent. *Lvs chiefly basal*, variable in form; *outermost long-petiolate, reniform to suborbicular-cordate*, crenate to coarsely dentate, less commonly 3-lobed; the rest ±deeply 3–5-lobed, the lobes crenate or further lobed; *stem-lvs few, ±sessile, deeply divided into narrow usually entire segments*; all usually glabrous, at least beneath. Fls few, their pedicels pubescent, not furrowed. Sepals 5, spreading. *Petals 5 or fewer, or 0*, obovate, 5–10 mm, often unequal, golden-yellow; *nectary-scale very small or 0. Receptacle glabrous* to pubes-

cent, *with elongated projections to which the achenes are attached*. Achenes c. 4 mm overall, only slightly flattened, asymmetrically obovoid, very narrowly bordered, pubescent; beak shortish, ultimately recurved (Fig. 4C,D). Fl. 4–5. Homogamous or protogynous. Visited by various flies and small bees, but perhaps always apomictic. $2n = 16, 32^*, 40, 48$. G or Hs.

Variable, especially in shape and depth of lobing of basal lvs and number of petals, features shown to be inherited maternally, presumably because of apomixis. Large numbers of taxa have been named, many of them shown to be facultative or obligate apomicts, by continental European students of the *R. auricomus* group.

Native. A frequent herb of woodland on moist nutrient-rich and especially calcareous soils; also thickets, hedgebanks and, less commonly, shady pastures; locally on shaded limestone rock. Through much of Great Britain though local in Wales and N.W. Scotland but reaching N. Ebudes and W. Sutherland; Ireland; Channel Is.; not in Outer Hebrides, Orkney or Shetland. Widespread in Europe and N. Asia.

Section *Flammula* Webb & Berth. Annual or perennial herbs with fibrous roots; basal and stem-lvs simple, entire or toothed, not palmately lobed or divided; nectary-scale forming a pocket; achenes somewhat compressed, with distinct but very short beak, usually glabrous; fruiting receptacle much elongating.

10. R. lingua L.   Great Spearwort.

A perennial herb with stout stems (rhizomes and hypogeal stolons) creeping horizontally through the muddy substratum, ringed with fibrous roots at the nodes, finally erecting as aerial stems 50–120 cm, branching above, hollow, glabrous or with a few appressed hairs. Basal lvs up to 20×8 cm produced in autumn and often submerged, long-stalked, ovate to ovate-oblong, cordate, obtuse, disappearing before flowering; *stem-lvs* on non-flowering shoots like the basal, but those *on flowering shoots* up to 25×2.5 cm, *distichous, short-stalked or sessile, half-clasping, oblong-lanceolate*, acute to acuminate, entire or remotely denticulate. Infl. a few-fld cyme. *Fls 2–4(–5) cm diam.*, their *pedicels lf-opposed*, ±appressed-hairy, *not furrowed*. Sepals 5, c. 1 cm, strongly concave, ±hairy. Petals 5, c. 2 cm, broadly obovate, bright glossy yellow; nectary-scale rudimentary. Receptacle glabrous. Achenes in a globose head, each c. 2.5 mm, obovate, broadly membranous-bordered on the ventral, more narrowly on the dorsal margin, the faces glabrous, minutely pitted; beak short, broad, somewhat curved (Fig. 5A). Fl. 6–9. Protogynous. Visited by various flies. $2n = 128$. Hel.

Fig. 5. Achenes of *Ranunculus*. A, *R. lingua*; B, *R. flammula*; C, *R. reptans*; D, *R. ophioglossifolius*; E, *R. sceleratus*. ×3.

Native. A local plant of the reed-swamp zone of marshes and 'mixed fens' bordering streams, lakes and ponds where some silt is deposited; now decreasing through drainage. Scattered through England but sparsely in the south-west and in Wales and rare in most of Scotland though reaching E. Ross and N. Ebudes; Ireland; Channel Is.; not in Outer Hebrides, Orkney or Shetland. Most of Europe but very rare in the Mediterranean region; N. Asia.

11. R. flammula L.   Lesser Spearwort.

A perennial herb with short ±vertical stock and abundant fibrous roots; not stoloniferous. Aerial stems 8–50(–80) cm, erect, ascending or creeping, commonly rooting at least at the lower nodes, sometimes at irregular intervals throughout (f. *tenuifolius*), hollow, slightly branched, ±glabrous. *Basal lvs* up to 4(–5) × 2.5(–3) cm, stalked, *subulate or lanceolate to broadly ovate*, rounded to cordate at base; *stem-lvs* short-stalked, the uppermost sessile, *broadly lanceolate to linear-lanceolate*, ±acute, subentire to distinctly toothed, parallel-veined. *Fls* 8–20(–25) *mm diam.*, solitary or in a few-fld cyme, their *stalks slightly hairy, furrowed.* Sepals 5, greenish-yellow. Petals 5, obovate, pale yellow, glossy. Receptacle glabrous. Achenes 20–50(–60) in a globose head, each 1–2 mm long (excluding the short blunt beak), ovate, minutely pitted, obscurely bordered (Fig. 5B). Fl. 5–9. Protandrous. Visited by various flies and small bees. $2n = 32*$. Three subspp. may be distinguished.

Subsp. **flammula**: stem 8–50(–80) cm, erect to creeping; *basal lvs* usually 1–4 × 0.8–2.5 cm, *lanceolate to broadly ovate* (but narrowly elliptical in sterile submerged forms); achene about ⅓ longer than broad).

Subsp. **minimus** (A. Benn.) Padmore: stem 3–8(–14) cm, semi-prostrate but not rooting, with very short internodes; *basal lvs* short-stalked, *at least as broad as long*, *cordate at base*, thick and fleshy; fls 15 mm diam. or more; achenes about as broad as long.

Subsp. **scoticus** (E. S. Marshall) Clapham: stem erect, 20–60 cm; *lowest lvs subulate (with no expanded blade)*, caducous; later lvs sub-persistent but readily detachable, with short blunt linear-oblong blade; upper stem-lvs lanceolate, ±sessile, fls 1(–4), 10–15 mm diam.; achenes as in subsp. *flammula*.

Native. Common in wet places throughout the British Is. Europe, temperate Asia, Azores. Subsp. *minimus* in exposed places by the sea, often forming dense mats, in Caithness, Outer Hebrides, Orkney, Shetland and Co. Clare. Subsp. *scoticus* on lake-shores in Argyll and Inner Hebrides and perhaps in Co. Mayo.

12. R. reptans L.   Slender Creeping Spearwort.

A perennial stoloniferous herb resembling the slender creeping forms of *R. flammula* but with the *filiform stem* 5–20(–50) cm, *arching in the internodes and rooting at every node*. Lvs 0.5–2 cm, in small tufts at each rooting

point, long-stalked, *spathulate or narrowly elliptical. Fl.* 5–10 mm diam., *solitary* on the ascending tip of a stolon. Petals 5, narrowly obovate, pale yellow, glossy. *Achenes* 1–1.5 mm, ovate, glabrous, *with a curved ±terminal slender beak about* ¼ *as long as the rest of the achene* (Fig. 5c) (c. ⅛ in *R. flammula*). Fl. 6–8. $2n = 32$. H.

Plants referred to this sp. on the shores of Ullswater (Lake District) and Loch Leven (Kinross) do not show uniform agreement with the above description, based on continental material, but vary in size and in the extent of nodal rooting; some are indistinguishable from *R. flammula* f. *tenuifolius* Wallr. and these become erect in cultivation, unlike the more extreme forms. All plants show some pollen sterility and it is thought that the populations are of hybrid origin (*R. flammula* × *reptans*) with no remaining pure *reptans*, though some plants approach it closely.

Native or formerly native. A rare plant of lake margins in the Lake District and Scotland. N. and C. Europe.

13. R. ophioglossifolius Vill. Addderstongue Spearwort.

An *annual herb* with fibrous roots and erect or ascending branching stem 10–40 cm, often rooting at lower nodes, hollow, grooved, glabrous or somewhat hairy above. *Basal lvs long-stalked*, *ovate to suborbicular, cordate*, up to 20 × 12 mm, larger if floating; upper smaller, narrower and shorter-stalked upwards, the uppermost narrowly elliptical, ±sessile; all obscurely and distantly denticulate or entire. *Fls small*, 5–9 mm diam., some in many-fld axillary cymes, some solitary and lf-opposed, their pedicels appressed-pubescent, somewhat furrowed. Petals 5, narrowly ovate, pale yellow, exceeding the spreading ±glabrous sepals. Receptacle glabrous. *Achenes* c. 1.5 mm, compressed, narrowly bordered, *their faces with numerous very small tubercles*; beak very short and broad (Fig. 5D). Fl. 6–7. $2n = 16$. Th.

Native. A rare plant of marshes, now only in Gloucester; formerly also in Dorset and Jersey. Gotland, France, S. Europe; N. Africa; W. Asia.

Section *Hecatonia* (Lour.) DC. Annual or perennial small-flowered marsh or water plants. Nectary-scale often forked or forming a ring round the nectary. Achenes very small, little compressed, their faces smooth or somewhat rugose; beak usually very short.

14. R. sceleratus L.   Celery-leaved Crowfoot.

An *annual or overwintering herb* with fibrous roots and a stout erect stem 10–60 cm, hollow, branched above, ±glabrous, furrowed. *Basal lvs long-stalked, reniform or pentagonal in outline*, ±deeply 3-lobed, the lateral lobes often again 2–3-lobed, all crenate; stem-lvs short-stalked, more deeply divided into narrower segments, the *uppermost sessile with* 3 or fewer ±entire narrow segments; lower lvs glabrous, uppermost with a few scattered hairs; *all lvs shining*. Fls 5–10 mm diam., numerous, in branching cymes, their pedicels glabrous, furrowed. *Sepals* c. 4 mm, *reflexed*, hairy beneath. *Petals little longer than sepals*, ovate-oblong, pale yellow, the *open nectary surrounded by its scale*. Receptacle pubescent, elongating in fr. *Achenes very numerous* (70–100), c. 1 mm, ovoid, little compressed, glabrous, each face with a faintly rugose central area (Fig. 5E); the *head of ripe achenes oblong-ovoid*, 6–10 mm. Fl. 5–9. Protogynous. Visited by flies. $2n = 32$. Th. – Hel.

Native. Common in and by slow streams, ditches and shallow ponds of mineral-rich water over a muddy bottom. Through most of Great Britain but more local in the west and rare in N. Scotland; Ireland; Channel Is.; not in Shetland. Europe, Asia, N. Africa, North America.

Subgenus BATRACHIUM (DC.) A. Gray

Aquatic or semi-terrestrial annuals and perennials. Lvs stipulate, all with an expanded lamina ('floating lvs'), or all finely dissected ('submerged lvs'), or some of each type; lvs transitional between floating and submerged may also occur. Stipules membranous, adnate to the petiole over part of their length. Fls solitary, lf-opposed; protogynous and self-compatible with a tendency to cleistogamy. Sepals usually 5, ±caducous. Petals 5 or more, white, commonly with a yellow claw (rarely entirely pale yellow in non-British spp.); nectary-scale minute or 0. Ripe achenes not strongly compressed, their faces with regular transverse ridges c. 1 mm apart.

Our understanding of this taxonomically difficult group has been much increased by the considerable body of recent observational and experimental research, though problems remain, notably how to distinguish between certain species in the vegetative state. Much use has been made in the past of features of the dissected lvs such as the length and rigidity of their segments, but it is now clear that there is much variation, determined both genetically and environmentally, in such features. In particular, dissected terrestrial or aerial lvs commonly have shorter, stouter and more rigid segments than do submerged lvs, so that they become of little or no taxonomic value.

15. R. hederaceus L.   Ivy-leaved Crowfoot.

An overwintering or spring-germinating annual herb, sometimes perennial, with branched stem, 10–30 cm, creeping on mud or the upper part floating. *Lvs* 1–3 cm wide, *usually opposite*, petiolate, *reniform or suborbicular-cordate*, often with dark markings near the base, shallowly 3–5(–7)-lobed, the *lobes* rounded or bluntly triangular, entire or crenulate, *broadest at their base* (Fig. 6A,B); stipules shorter than wide, adnate over more than half their length. *No dissected lvs. Fls 4–8 mm diam.* Sepals not reflexed. Petals barely exceeding sepals, not contiguous; nectaries lunate. Stamens 5–10. *Receptacle glabrous*, rarely with a few hairs. Fr.-pedicel about equalling petiole of opposed lf, recurving so that fr. is buried. *Achenes* 1–1.5 mm, *glabrous, with short blunt lateral beak*. Fl. 6–9. $2n = 16^*$. Th. – Hyd.

Native. On wet mud and in shallow and often temporary water of small streams, ditches and ponds throughout the British Is., but somewhat local. W. Europe eastwards to C. Germany and S. Sweden.

16. R. omiophyllus Ten.

R. lenormandii F. W. Schultz

An overwintering or spring-germinating annual herb, sometimes perennial, growing on mud or in shallow water. Stem branched at the base, usually 5–25 cm but longer in robust floating plants, creeping or the upper part floating. *Lvs* usually 1–3 cm wide, opposite or alternate, petiolate, *roundish-reniform*, never with dark markings, 3–5(–7)-*lobed to less than* half-way, the *primary lobes rounded, narrowest at their base*, ±crenate or with a few secondary lobes (Fig. 6c); stipules shorter than wide, adnate to the petiole for half or less of their length. *No dissected lvs. Fls 8–15 mm diam. Sepals reflexed. Petals* 5 or more, *about twice as long as sepals*, not contiguous; nectaries lunate. Stamens 6–11. *Receptacle glabrous*. Fr.-pedicel equalling or falling short of petiole of opposed lf, recurving so that fr. is buried. *Achenes* 1–1.5 mm, *glabrous, with almost central* curved and pointed *beak*. Fl. (2–)4–10. $2n = 16, 32^*$. Th. – Hyd.

Native. Locally common in non-calcareous streams and muddy places in the south and west of Great Britain northwards to Argyll, but in only a few scattered localities in the east and absent from much of midland England; S. and S.E. Ireland. Restricted to Atlantic Europe from the British Is. southwards through W. France, N. Spain and Portugal to Sicily and S. Italy; Algeria.

17. R. tripartitus DC.

Incl. *R. lutarius* (Rével) Bouvet, pro parte

An annual to perennial herb with simple or branched stem, 5–60 cm, creeping on mud or in shallow water, usually with the upper part floating. '*Floating*' lvs opposite or alternate, petiolate, 5–20(–40) mm wide, reniform to suborbicular-cordate, *deeply* 3(–5)-*lobed* (*to more than half-way*), the lobes ±cuneate, distant, entire or with 2–4 rounded crenations (Fig. 6L); rarely 0. *Dissected lvs*, developed only on stems submerged in water, alternate, *few and restricted to lower nodes* except in deepish water, repeatedly trifid or forked, the *ultimate segments extremely slender and collapsing on removal from water*; stipules broadly rounded, adnate to the petiole over more than $\frac{2}{3}$ of their length. *Fls 3–10 mm diam. Sepals with clear blue tips, reflexed. Petals usually less than 5 mm, equalling or up to twice as long as sepals*, narrowly obovate, not contiguous; nectaries lunate. Stamens 5–10. *Receptacle hairy*, globose in fl. and fr. Fr.-pedicel about equalling petiole of opposed lf, recurved. *Achenes* 1.5–2 mm, *glabrous*, with small lateral or subterminal beak. Fl. 3–5. $2n = 48^*$. Th. – Hyd.

Native. Muddy ditches, small pools, cart-tracks, etc., where water stands only temporarily and then often lacking dissected lvs, or in more permanent water and then with dissected lvs and occasionally with floating lvs 0. Very local in coastal strips of S.W. England, S. Wales and S.W. Ireland; formerly also in S.E. England and Anglesey. Restricted to Atlantic Europe from N.W. Germany to Portugal through N.W. France and N. Spain.

Fig. 6. Leaves of *Ranunculus* subgenus *Batrachium*. *R. hederaceus*: A, robust floating form; B, small form on mud. *R. omiophyllus*: C. *R. aquatilis*: D, F, floating lvs; E, transitional lf. *R. peltatus*: G, I, floating lvs; H, var. *truncatus*. *R. baudotii*: J, floating lf; K, transitional lf. *R. tripartitus*: L, floating lf. All ×c. 1·5.

R. lutarius (Revel) Bouvet, described as differing from *R. tripartitus* in its larger and later fls and in the usual absence of dissected lvs, these when present having flattened non-collapsing segments, appears to consist partly of forms inseparable from *R. tripartitus* and partly of hybrids of *R. tripartitus* with *R. omiophyllus*, together with products of selfing and back-crossing. The populations in the New Forest, upon which the earliest British records were based, are almost certainly of this hybrid origin.

18. R. baudotii Godron   Maritime Water Crowfoot.

An annual or perennial herb with branching stem prostrate in terrestrial plants and ±erect in water. *Dissected lvs always* and floating lvs usually *present*, the latter on the upper part of the stem, sometimes 0 in deep water. All lvs alternate, petiolate, with rounded stipules adnate to the petiole for about half their length. *Floating lvs up to 3 cm wide, reniform to suborbicular in outline, deeply* 3(–5)-*lobed* (to $\frac{2}{3}$ or more), the *lobes distant, cuneate, usually crenate* (Fig. 6J); transitional lvs sometimes present (Fig. 6K); *dissected lvs usually yellowish-green, with fairly long spreading and non-collapsing segments. Fls* 12–18 mm *diam.*, with pedicel usually

opposite a floating lf. *Sepals with clear blue tips, reflexed or spreading*. Petals 6–12 mm, up to twice or more as long as sepals, broadly obovate, contiguous; nectaries lunate. Stamens 10–20. *Receptacle hairy, becoming ovoid in fr.* Fr.-pedicel longer than petiole of opposed lf, recurved. *Achenes numerous, small, glabrous*, with short lateral beak and, when mature and dry, with *thin membranous wings along dorsal and ventral edges*. Fl. (3–)5–9. $2n = 32^*$. Th. – Hyd.

Very variable in size and robustness, number of floating lvs, etc., but no clearly distinct infraspecific taxa seem recognizable.

Detached apical lengths of stem, with floating lvs only, may remain alive in water for some time and have been mistaken for *R. tripartitus*, from which they are readily distinguishable if fls are present.

Native. Locally in brackish ditches, ponds and slow-moving streams and on mud-flats in coastal areas throughout the British Is., and occasionally found where covered by sea water at high tide; said only rarely to root below water more than 30 cm deep. Coasts of most of Europe northwards to c. 65°N on the Baltic coasts of Sweden and Finland, but not in Norway or Iceland; N. Africa.

The next three species are all typical of early stages of aquatic succession in small streams and in newly dug or cleaned canals, drainage ditches, fish-pools, pits, etc., with still or slow-moving water not more than c. 1 m deep. They show similar patterns of environmentally and genetically determined variation in the length, fineness and rigidity of the segments of their dissected lvs (Cook, 1966) and no clear-cut comparative statements can be made about these features. *R. trichophyllus* never forms floating lvs and they are commonly absent in *R. aquatilis* and *R. peltatus* when growing terrestrially or in deep water. In these circumstances the three are morphologically indistinguishable unless fls are present, but even normally distinctive features of fl. and fr. may be so modified by low levels of light intensity, temperature or supply of mineral nutrients as to cause taxonomic problems. There is the additional complication that naturally occurring hybrids between *R. aquatilis* and *R. trichophyllus* have been recorded from this country and *R. aquatilis* and *R. peltatus* have been successfully crossed artificially (Cook, *Mitt. Bot. Staatssaml. München*, 1966, **6**, 47–237).

19. R. aquatilis L.

R. heterophyllus Weber; incl. *R. radians* Revel

An annual or perennial herb with branching stems decumbent in terrestrial plants but ±erect in water. *Dissected lvs always and floating lvs usually present*, the latter on the upper part of the stem and usually floating or emergent, sometimes 0 in plants in deep water. All lvs alternate, petiolate, with *±triangular stipules* adnate to the petiole for at least ¾ of their length. *Floating lvs* up to 3 cm wide, reniform to circular in outline, *rather deeply (3–)5(–7)-lobed* (but usually to less than ⅔), the *lobes broadly cuneate and usually straight-sided*, often dentate (Fig. 6D); transitional lvs rarely present. *Dissected lvs* 3–6(–8) cm, *shorter than the stem-internodes*, their segments not lying in one plane, spreading, collapsing or not on removal from water. *Fls* 12–18 *mm diam.*, their pedicels opposite either floating or dissected lvs. Sepals spreading, not blue-tipped. *Petals* 5–10 mm, *about twice as long as sepals*, broadly obovate, contiguous; *nectaries circular* (Fig. 7A). Stamens 13–20. *Receptacle hairy, globose in fr. Fr.-pedicel* 2–5 cm, rarely longer, *shorter than petiole of opposed floating lf. Achenes* 1.5–2 mm, *±distinctly hairy* (at least while immature), with short subterminal beak. Fl. 5–6. $2n = 48^*$. Th. – Hyd.

R. radians Revel comprises plants having submerged lvs with short rigid segments as well as transitional and almost circular floating lvs, and having achenes bristly at maturity, but no clear separation from *R. aquatilis* can be made.

Plants which have been named *R. drouetii* F. W. Schultz, lacking floating lvs and having dissected lvs with collapsing segments, are sometimes referable to *R. aquatilis* rather than to *R. trichophyllus* but fls are required for certain identification (see **20** below).

Native. In ponds, ditches and slow-moving streams throughout low-lying areas of the British Is. but absent

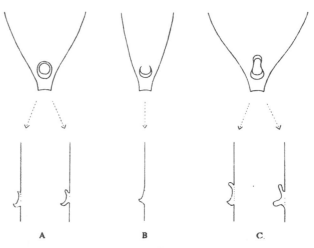

Fig. 7. Surface views and vertical sections of nectaries of *Ranunculus* subgenus *Batrachium*. A, 'circular' nectary of *R. aquatilis*; B, 'lunate' nectary of *R. trichophyllus*; C, '±pyriform' nectary of *R. peltatus*.

or very local in N.W. Scotland. In most of Europe except Faeroes, Iceland and Spitsbergen.

20. R. trichophyllus Chaix

R. paucistamineus Tausch, pro parte

An annual or perennial herb with branching stem, compact in terrestrial plants but spreading-erect in water. *Floating lvs* 0. *Dissected lvs* alternate, repeatedly trifid or forked to form a *globose or obconical cluster*, up to 4 cm diam., *of short spreading segments*, collapsing or not on removal from water; petiole varying from very short to c. 4 cm; stipules oblong to oblong-ovate, adnate to the petiole for ⅔ or more of their length, the free lobes rounded. *Fls* (5–)8–10(–15) *mm diam.* Sepals spreading, not blue-tipped. *Petals* 3.5–5.5 mm, not contiguous; *nectaries lunate* (Fig. 7B). Stamens 9–15. *Receptacle hairy*, globose in fr. *Fr.-pedicel less than* 5 cm, strongly recurved. *Achenes* 2–2.5 mm, ovate to obovate, *hairy at least while immature*, sometimes glabrous at maturity; beak very short, subterminal. Fl. 5–6. $2n = 32^*$. Th. – Hyd.

The British plant is placed in subsp. **trichophyllus** of temperate lowland habitats, robust and ±erect with fls opening normally. Subsp. **eradicatus** (Laest.) C. D. K. Cook is arctic–alpine, prostrate in the aquatic state, rooting at most nodes and with cleistogamous fls.

Variable, especially in rigidity of lf-segments and hairiness of mature achenes. Plants with the small fls of *R. trichophyllus*, collapsing lf-segments and little compressed obovoid achenes, glabrous at maturity, were formerly referred to *R. drouetii* F. W. Schultz, but no clear separation is possible (see also **19** above).

Native. In at least moderately nutrient-rich ponds, ditches and slow streams throughout the British Is. but most frequent in S.E. England and becoming very local in the west and north. Temperate regions of the northern hemisphere; S.E. Australia and Tasmania.

21. R. peltatus Schrank

Annual to perennial herb with branching stem, compact in the terrestrial state but spreading-erect in water. *Both floating and dissected lvs commonly but not invariably present*, all alternate; stipules oblong to triangular, adnate to the petiole for $\frac{3}{4}$ or more of their length. *Floating lvs* with petioles up to 7 cm and lamina up to 4 cm across, reniform to suborbicular in outline, (3–)5(–7)-lobed usually to less than $\frac{2}{3}$, the *lobes ±cuneate but usually convex-sided* with the outer margin crenate (Fig. 6G,H,I); floating lvs 0 in the terrestrial state and in deep water. Transitional lvs rarely present. *Dissected lvs shorter than stem-internodes*, with petioles up to 2.5 cm and *lamina a globose to obconical cluster of spreading and usually non-collapsing segments* which do not lie in one plane. *Fls 15–20(–30) mm diam.*, their pedicels usually opposite floating lvs. Sepals 3–6 mm, spreading, not blue-tipped. *Petals (8–)12–15(–18) mm, 5 or sometimes more, broadly obovate, contiguous; nectaries elongated, ±pyriform* (Fig. 7c). Stamens 15–30. *Receptacle hairy, globose in fr. Fr.-pedicel usually exceeding 5 cm, longer than petiole of opposed floating lf*, recurved. *Achenes 2–2.5 mm, moderately compressed, obovate, hairy or bristly at least when immature*, occasionally glabrous when mature; beak very short, subterminal. Fl. 5–8. $2n = 16, 32, 48^*$. Th. – Hyd.

Variable.

Native. Frequent in at least moderately eutrophic ponds, ditches, shallow streams and temporary pools throughout lowland Great Britain from Ross southwards; Ireland. In most of Europe but only in western USSR; N. Africa.

22. R. penicillatus (Dumort.) Bab.

R. pseudofluitans (Syme) Newbould ex Baker & Foggitt

Perennial plants of flowing water with branching stems 1–3(–6) m long, often rooting from internodes but prostrate at no season or only in summer. Dissected lvs always present but floating lvs frequently 0; all lvs; alternate; stipules ovate or oblicular, adnate to the petiole for $\frac{3}{4}$ or more of their length, the free lobes broadly rounded. *Dissected lvs* petiolate, variable in length but *usually at least equalling the mature stem-internodes* and often much longer; *ultimate segments commonly very numerous*, subparallel or ±spreading, *usually collapsing on removal from water* (but see below for var. *vertumnus*). Floating lvs with petiole 5–8 cm; lamina up to 4 cm across, reniform to suborbicular, occasionally with cuneate base, (3–)5-lobed to $\frac{2}{3}$ or less, the outer margin of the cuneate lobes variously entire, crenate or dentate. Transitional lvs, with lobes ending in filamentous appendages, sometimes present. *Fls usually 15–30 mm diam.*, like those of **21** or somewhat larger, borne opposite floating lvs when these are present. Sepals 3–7 mm,

spreading, not blue-tipped. Petals (5–)10–15(–20) mm, broadly obovate, contiguous; *nectaries ±pyriform*. Stamens usually 20–40. *Receptacle densely hairy*, globose in fr. *Fr.-pedicel 5–10 cm, usually exceeding petiole of opposed floating lf*, recurved. Achenes 2–2.5 mm, initially hairy, sometimes becoming glabrous at maturity; beak lateral to subterminal. Fl. 5–8. $2n = 32^*, 48^*$. Hyd.

Very variable, especially in length of stem, overall length of the dissected lvs and length and rigidity of their ultimate segments. British representatives fall into at least two groups (Cook, 1966). Var. *penicillatus* is long-stemmed, often bears floating lvs, and its dissected lvs exceed the mature stem-internodes, their ultimate segments being fine, ±parallel and collapsing on removal from water. Var. *calcareus* (R. W. Butcher) C. D. K. Cook, with stems very variable in length up to 5 m or so, has floating lvs 0 and dissected lvs about equalling the mature stem-internodes, their very numerous ultimate segments (up to 150) slightly divergent and not always collapsing on removal from water. A third group, var. *vertumnus* C. D. K. Cook, also lacking floating lvs, with dissected lvs shorter than the stem-internodes and with up to 200 widely divergent and non-collapsing ultimate segments, may not prove possible to separate clearly from var. *calcareus* (Holmes, *Nat. Conserv. Council CST Notes*, 1979, **14**, 1–31; *Watsonia*, 1980, **13**, 57–9.)

Native. Characteristically plants of fast-flowing streams, but var. *calcareus* is also found in water only intermittently swift-flowing and var. *vertumnus* is restricted to slow-flowing but clear rivers and canals. Var. *penicillatus* occurs in a number of rivers in W. England, Wales and Ireland; var. *calcareus* is common throughout England and local in Wales and S. Scotland but has not been recorded from Ireland, and var. *vertumnus* is so far known only from S. England, W. and C. Europe and in scattered localities eastwards to Lake Ladoga, but distribution still very imperfectly known.

Plants with floating lvs and fls are readily distinguishable from *R. fluitans*, which lacks floating lvs and has rather small fls with almost glabrous receptacles, but in the vegetative state confusion is possible. In general *R. fluitans* has longer stems and stem-internodes and longer but fewer lf-segments than *R. penicillatus*, the lvs rarely being forked more than 4 times (7–8 times in *R. penicillatus*), but these differences are not absolute. Confusion with *R. peltatus* is less likely since plants of *R. penicillatus* having short lvs with divergent and non-collapsing segments lack floating lvs, normally present in *R. peltatus*; the fls differ in no clear-cut way.

The combination of great variability with strong resemblances to *R. fluitans* on the one hand and *R. peltatus* on the other has led to the suggestion (Cook, 1966) that *R. penicillatus* might have arisen from hybrids between them: the sterile hybrid occurs naturally. A further suggestion (Holmes, 1980) is that *R. trichophyllus* may have been involved in the hybrid complex giving rise to vars *calcareus* and *vertumnus*, which lack floating lvs, but much further research is required.

23. R. fluitans Lam.

A perennial herb of flowing water which develops during spring and summer long branching submerged lfy stems up to 6 m, with internodes up to 35 cm, these dying back or becoming detached in autumn, the plant then overwintering as prostrate stems rooting at the nodes; also an annual terrestrial herb up to c. 6 cm. *Floating lvs* 0. *Dissected lvs* alternate, usually 8–30 cm and *equalling or exceeding the stem-internodes*, with petiole up to 20 cm; *lamina* up to 25 cm or more, initially trifid then forking 3–4 times into a *few very long firm sub-parallel segments forming dark-green obconical clusters*; stipules ovate to oblong, adnate to the petiole for most or all of their length. Lvs of terrestrial plants 15–45 mm with flattened rigid segments. *Fls* 15–25 *mm diam.* Sepals 4–6.5 mm, spreading, not blue-tipped. *Petals 5–10, usually* 7–16 mm, broadly obovate, contiguous; *nectaries ±pyriform.* Stamens 20–35. *Receptacle glabrous or nearly so,* globose in fr. Fr.-pedicel 4–10 cm, usually not exceeding the opposed lf. *Achenes* numerous (commonly 30–60), c. 2.5 mm, sparsely hairy while immature but *becoming glabrous or nearly so* when ripe; beak lateral. Fl. 6–8. $2n = 16*, 32*$. Hyd.

Terrestrial plants very rarely produce fls and then smaller than those of aquatic plants and almost invariably sterile.

Native. Most commonly in at least moderately fast-flowing non-calcareous streams (usually being replaced by *R. penicillatus* var. *calcareus* in limestone areas), and favouring a stable bottom; occasionally in very slow-flowing water of large drainage ditches. Local in Great Britain southwards from Lanark and Berwick but rare in Wales and S.W. England; very rare in Ireland. Decreasing, perhaps owing to river-pollution. W. and C. Europe eastwards to S. Sweden, Poland, Romania and Bulgaria.

Slender plants resembling *R. fluitans* but with shorter subsessile lvs and much smaller and invariably sterile fls have been named var. *bachii* (Wirtgen) Wirtgen. They may be hybrids between *R. fluitans* and *R. trichophyllus.*

24. R. circinatus Sibth.

R. divaricatus Schrank, nom. illegit.

A slender perennial herb overwintering by prostrate lfy stems which root at the nodes and forming during the summer simple ±erect stems of very variable length up to c. 3 m; said to behave occasionally as an annual. *Floating lvs* 0. *Dissected lvs* alternate, 0.5–2(–3) cm, *shorter than the stem-internodes*; *petiole short*, usually 2–5 mm and often enclosed by the stipules; *lamina circular in outline*, trifid and then repeatedly forked, *the short rigid divergent segments all lying in one plane*; stipules ovate, adnate to the petiole for ¾ or more of their length, the free lobes rounded or 0. *Fls* 10–20 *mm diam.*, sometimes borne below the water surface and then often cleistogamous. Sepals up to 6 mm, spreading, not blue-tipped. *Petals up to* 10 mm, obovate, contiguous; *nectaries lunate.* Stamens usually c. 20. *Receptacle hairy,* ovoid in fr. *Fr.-pedicel* 2–10 cm, *usually much exceeding the lvs.* Achenes up to 1.5 mm, hairy while immature but sometimes becoming glabrous; beak lateral. Fl. 5–8. $2n = 16*$. Hyd. (Th.).

Native. In mineral-rich water, commonly 1–3 m deep, of slow-flowing streams and canals and permanent ponds, gravel-pits, etc.; occasionally in mineral-poor water and then always sterile. Locally frequent in S. and E. England as far north as the Humber and Mersey estuaries, but very rare further northwards to Cumbria and Roxburgh; Ireland. In most of Europe except the Iberian Peninsula but extending only to c. 61° N in Sweden and Finland; central N. Asia.

Subgenus FICARIA (Hudson) L. Benson

Perennial terrestrial herbs with fusiform root-tubers and simple entire or crenate, rarely dentate, lvs. Sepals 3. Petals 7–12 or more, yellow. Nectary-scale small, forming a shallow pocket. Achenes little compressed, with minute beak. Cotyledon apparently single, emarginate or 2-lobed.

25. R. ficaria L.   Lesser Celandine, Pilewort.

Ficaria verna Hudson; *F. ranunculoides* Roth

Perennial mycorrhizal herb with both fibrous roots and numerous fusiform or clavate root-tubers, 10–25(–60) mm. Stems 5–30 cm, ascending, branching, often rooting at the decumbent base. Lvs all petiolate with sheathing bases, those of the lowest lvs very broad; *basal lvs* 1–4 cm long and wide, *long-stalked, cordate*, with the basal sinus wide or ±closed, bluntly angled or crenate, rarely dentate; stem-lvs similar but smaller and shorter-stalked, commonly in ±opposite pairs and often developing axillary tubercles ('bulbils') after fl.; all lvs somewhat fleshy, dark green, often with darker markings, glabrous. Fls 1.5–3(–5) cm diam., solitary at ends of main stem and branches. Sepals 3, rarely more, ovate, concave, spreading, falling early. *Petals* 8–12, rarely 0, ±narrowly ovate and up to twice as long as sepals, *bright shining yellow*, fading whitish. Receptacle pubescent. Achenes in a globose head, initially numerous but many often aborting; when ripe up to 4 mm, broadly obovoid, keeled, ±pubescent, minutely beaked. Fl. 3–5. Slightly protandrous, with both pollen and nectar accessible to short-tongued insects; visited by various bees, flies and small beetles. Some populations have male plants with large petaloid sepals, no petals and carpels abortive or 0. $2n = 16*$ (sometimes with 1–7 B chromosomes), 24*, 32*. Grt.

Variable, especially in the size and shape of lvs and in number and location of stem-lvs. There is marked variability in width of basal sinus, depth and number of crenations, colour, and presence or absence of dark markings. Fls vary in colour of sepals, length and degree of overlap of petals, pollen fertility and number of ripe achenes. Cytological studies have revealed the widespread occurrence in this country of both diploid and tetraploid plants, with $2n = 16$ and 32, respectively, and of occasional triploids. Much of the morphological variation seems independent of chromosome number, but certain features show a sufficiently close association to justify the recognition of the two following subspecies.

Subsp. **ficaria**: plant rather robust and compact; no aerial bulbils in axils of stem-lvs, which tend to be restricted to the lower third of the stem; fls 2–3 cm diam.; sepals green; pollen largely viable; carpels yielding a high proportion of well-developed achenes. Diploid ($2n = 16$).

Subsp. **bulbifer** (Marsden–Jones) Lawalrée: plant commonly lax and spreading; axillary bulbils developed after fl. in axils of stem-lvs, which are borne mostly in the upper third of the stem; fls 1.5–2.5 cm diam.; sepals green; a high proportion of the pollen non-viable; only a small percentage of carpels develop into ripe achenes. Tetraploid ($2n = 32$).

A third subspecies is claimed as almost certainly native in the Channel Is. and has been reported from a few other localities, but more work is needed.

Subsp. **ficariiformis** Rouy & Fouc.: plant very robust, lacking bulbils in axils of stem-lvs; fls 3–5 cm diam., with yellowish-white sepals; most carpels developing into ripe achenes.

Native. Locally abundant on nutrient-rich soils moist at least in spring and especially characteristic of old deciduous woodland and permanent pasture where the soil has a pH above c. 4.5 and is neither continually waterlogged nor dry for long periods. Lvs emerge in mid-January and fls in March–Apr., during the pre-vernal 'light phase' of native woodland. Common also in hedgerows and roadside verges and a successful colonist of suitably moist disturbed ground through its detachable root-tubers and the aerial bulbils of subsp. *bulbifer*, and so a locally troublesome garden-weed. Subsp. *ficaria* occurs throughout the British Is., reaching 730 m in Wales, but is local in Scotland and in C. Ireland. Subsp. *bulbifer* seems more easterly in distribution, being rare near the west coast of Gt. Britain and in Ireland, but more work is needed on both distributional and ecological distinctions between the two. Subsp. *ficaria* is restricted to W. Europe from S.W. Norway to the W. Mediterranean, but subsp. *bulbifer* is in much of Europe northwards to c. 60° N and southwards to Sicily but is absent from both eastern and western ends of the Mediterranean region.

13. ADONIS L.

Perennial or annual herbs with lfy stems and bi- or tripinnate lvs with linear segments. Fls usually solitary, terminal, hermaphrodite, actinomorphic. Sepals 5(–8), ±petaloid; *petals 3–20, yellow or red, not nectar-secreting*; stamens numerous; carpels numerous, each with 1 pendulous ovule. Fr. an elongated head of *short-beaked achenes*.

About 20 spp. in Europe and temperate Asia.

*1. A. annua L.   Pheasant's Eye.

A. autumnalis L.

An annual herb with slender tap-root and erect usually branched stem 10–40 cm, usually glabrous. Lvs tripinnate with linear acute segments. Fls *15–25 mm diam., terminal on stem and branches, erect. Sepals 5, ovate, spreading, green or purplish. Petals 5–8, suberect*, somewhat longer than the sepals, ovate, *bright scarlet with a dark basal spot*. Filaments dark violet. *Achenes*

Fig. 8. Achenes of A, *Adonis aestivalis* and B, *A. annua*. ×2.

3.5–5 mm in a ±lax head c. 18 mm, each *with a short straight beak and with no tooth at the base of the upper face* (Fig. 8B). Fl. 7. Homogamous. Visited by bees. $2n = 32$. Th.

Introduced. Naturalized as a cornfield weed in a few southern counties of England from Dorset to Bedford and Kent, and sometimes found as a casual elsewhere in S.E. and C. England but decreasing; no longer in Ireland. S. Europe and S.W. Asia.

The related **A. aestivalis** L. sometimes occurs as a casual and may be distinguished by its sepals being closely appressed to the ±spreading petals (in *A. annua* they do not touch the suberect petals), and by the sharp tooth at the base of the upper face of the achene (Fig. 8A). **A. vernalis** L., a perennial herb with 10–15 narrow yellow petals and rounded pubescent achenes with a strongly curved beak, is often grown in gardens.

14. MYOSURUS L.

Small annual herbs with linear lvs confined to a basal rosette. Fls solitary, terminal on lfless scapes, hermaphrodite, actinomorphic. Sepals 5 or more, with a small basal spur; *petals 5–7 or 0, tubular*, nectar-secreting; stamens few; carpels numerous, each with 1 pendent ovule. *Fr. of numerous small achenes in a much elongated spike*. A few spp. in both north and south temperate zones.

1. M. minimus L.   Mouse-tail.

A glabrous annual herb with a basal rosette of linear, entire, somewhat fleshy lvs and numerous erect lfless flowering stems 5–12.5 cm. Fls very small, pale greenish-yellow. Sepals 5, rarely 6–7, 3–4 mm, narrowly oblong, with their basal spurs appressed to the stem. Petals 5, greenish, tubular with a short strap-shaped limb about equalling the sepals. Stamens 5–10. Achenes 1–1.5 mm, numerous, brownish, keeled, shortly beaked, on a filiform receptacle 2.5–7 cm. Fl. 6–7. Homogamous or slightly protandrous. Visited by small flies for the nectar secreted in the petals, but usually self-pollinated. $2n = 28$. Th.

Probably native. Occurs locally in damp arable fields and open grassland, especially where water stands in winter; also in ditches, by sea-walls, etc. Decreasing, and now almost confined to south and south-midland England westwards to Hants, Wilts and Hereford, with a very few scattered outliers, but formerly reaching Cumbria and Northumberland; Channel Is. Widespread in C. and S. Europe, N. Africa and S.W. Asia. Naturalized in North America and Australia.

Tribe 3. THALICTREAE. Fr. of follicles or achenes; chromosomes small; basic number 7.

15. AQUILEGIA L.

Perennial herbs with erect woody stocks and alternate *bi- or triternately compound lvs. Fls* terminal, solitary, or in panicles; hermaphrodite, actinomorphic, *with all parts arranged in 5-merous whorls. Sepals 5, petaloid; petals 5, each with a long hollow backwardly directed nectar-secreting spur;* stamens numerous, the innermost staminodal; carpels 5(–10), sessile, free, each with many ovules. *Fr. a group of many-seeded follicles.*

About 100 spp., chiefly in the north temperate zone. Poisonous.

Secondary lflets stalked; spur of petal strongly curved into a knobbed hook; ripe follicles 15–20 mm.   **1. vulgaris**
Secondary lflets sessile; spur of petal almost straight; ripe follicles 12–15 mm.   **2. pyrenaica**

1. A. vulgaris L.   Columbine.

A perennial herb with a short stout erect often branched blackish stock and erect lfy flowering stems 40–100 cm, glabrous or softly hairy, branched above. *Basal lvs long-stalked, biternate,* with the *secondary lflets usually stalked,* irregularly 3-lobed, the lobes crenate; stem lvs smaller, short-stalked; the uppermost (bracts) ±sessile and narrowly 3-lobed; all lvs ±glabrous, somewhat glaucous above, pale and greener beneath, their stalks +hairy, broadening below into a sheathing base. Infl. irregularly cymose. Fls 3–5 cm diam., drooping, both sepals and petals usually blue but sometimes white or reddish. Sepals 15–30 mm, ovate, acute. Limb of petal oblong-truncate almost equalling the *spur,* which is *sharply and knobbed at its tip;* overall length of petals c. 30 mm. Fertile stamens c. 50, exceeding the petal-limbs; staminodes c. 10, white, blunt, crimped. Carpels 5(–10), shortly hairy. *Follicles* 15–25 mm, erect, beaked, dehiscing while green. Seeds 2–2.5 mm, black, shining. Fl. 5–6. Protandrous, the stamens rising and dehiscing from outside inwards. Visited by long-tongued humble-bees for pollen and nectar. $2n = 14$. Hp.

Native. A local plant of woods and wet places on calcareous soil or fen peat, and probably native throughout England, Wales and S. Scotland but naturalized further north to Caithness; Ireland; Channel Is.; reaches 915 m in Angus. S. and C. Europe and reaching S. Sweden; N. Africa; temperate Asia to China. Naturalized in North America.

*2. A. pyrenaica DC.   Pyrenean Columbine.

A. alpina auct. angl.

Stem 20 cm, slender, with small ternate or biternate basal lvs, the *secondary lflets sessile,* crenate or entire; stem-lvs much smaller. Fls 4.5–5 cm diam., 1–3, darker blue than in *A. vulgaris.* Petal-limb obovate-cuneate, rounded; *spur slender, very slightly incurved.* Stamens exceeding the petal-limb. *Follicles* 12–15 mm. Fl. 8–9. $2n = 14$. Hp.

Introduced. Sown or planted on rocky ledges in Caen-lochan Glen, Angus, in the late 19th century and persisting there. Native in the Pyrenees.

Several columbines besides *A. vulgaris* are grown in gardens. The 'long-spurred' types, with straight spurs much exceeding the petal limbs, are apparently derived by hybridization chiefly from the American *A. coerulea* James (pale blue), *A. chrysantha* A. Gray (yellow, often with red spurs) and *A. formosa* Fisch. (sepals and spurs red, petal-limbs yellow).

16. THALICTRUM L.

Perennial herbs usually with repeatedly pinnate stipulate lvs. Fls numerous, small, in terminal or axillary racemes or panicles; hermaphrodite, actinomorphic, hypogynous. No nectaries. Per.-segs 4–5, ±petaloid, spreading, readily falling; petals 0; stamens numerous with long erect or drooping filaments; carpels free, usually few, often stalked, with a sessile stigma. Fr. a group of sessile or stalked 1-seeded achenes.

About 150 spp. chiefly in the north temperate zone.

The genus is interesting in including some insect-pollinated fls, fragrant, with erect and conspicuous filaments (e.g. *T. aquilegifolium*); some wind-pollinated, non-fragrant, with drooping filaments (e.g. *T. alpinum*), and some which are intermediate and apparently both insect- and wind-pollinated (e.g. *T. flavum, T. minus*).

1 Fls in a simple raceme; plant usually not exceeding 15 cm.   **2. alpinum**
  Fls panicled; plant usually exceeding 15 cm.   2
2 Fls in dense clusters; stamens erect; achenes 1.5–2.5 mm; ultimate lflets of middle and upper lvs much longer than broad.   **1. flavum**
  Fls not densely clustered; stamens drooping; achenes 3–6 mm; ultimate lflets usually roundish, about as long as broad.   **3. minus**

1. T. flavum L.   Common Meadow Rue.

A perennial herb with creeping stoloniferous rhizome and erect robust furrowed usually simple stem 50–100 cm. Lower stem-lvs stalked, upper sessile, all bi- or tripinnate, with stipule-like structures at each branching; ultimate *lflets longer than broad,* obovate-cuneate, or oblong-lanceolate in the upper lvs, 3–4-lobed distally, dark green above, paler and almost glabrous beneath. *Panicle compact. Fls in dense clusters,* erect, fragrant, with 4 narrow whitish per.-segs. *Stamens ±erect, yellow; anthers not apiculate.* Achenes 1.5–2.5 mm, ovoid to elliptical, 6-ribbed, glabrous. Fl. 7–8. Homogamous or slightly protogynous. Pollen smooth. Visited by various bees and syrphids but probably wind- as well as insect-pollinated. $2n = 28, 84$. H.

There is variation in the shape of the panicle and the size and shape of the achenes, and several varieties have been described.

Native. Common in meadows and fens by streams; to 305 m in Derby. Great Britain northwards to Inverness. Europe and temperate Asia.

2. T. alpinum L.   Alpine Meadow Rue.

A perennial shortly stoloniferous herb with short slender rhizome and an erect slender wiry stem 8–15 cm, rarely taller. Lvs chiefly basal, stalked, *biternate,* the *ultimate lflets roundish,* shallowly and bluntly lobed or crenate, dark green above, whitish below. *Fls in a simple raceme*, at first drooping, then erect. Per.-segs 4, c. 3 mm, pale purplish, spreading. *Stamens 8–20, long and pendulous, with slender pale violet filaments and yellow anthers.* Carpels shortly stalked. *Fl.–stalks recurved in fr.* Achenes 2–3, 3–3.5 mm, stalked, narrowly oblong, curved, ribbed, with a short hooked beak. Fl. 6–7. Protogynous. Anemophilous. $2n = 14$. H.

Native. Rocky slopes and ledges especially on mountains but descending almost to sea-level in the north-west; to 1213 m in Scotland. Great Britain northwards from Westmorland, Yorks, and Durham; N. Wales; Inner and Outer Hebrides; Orkney; Shetland. Arctic and alpine Europe; Asia; North America.

3. T. minus L.   Lesser Meadow Rue.

Perennial caespitose to stoloniferous herb with horizontal or ascending rhizome and erect rigid often flexuous lfy stems 15–150 cm, ±terete or furrowed, green or pruinose, often glandular above. Lvs stipulate, 3 or 4 times pinnate, the ultimate *lflets* variable in size and shape but usually *about as broad as long*, 3–7-lobed or toothed, green or glaucous, glabrous or with stalked glands especially beneath. Fls in a loose spreading panicle, drooping at first but sooner or later erecting. Per.-segs 4, yellowish- or purplish-green. *Stamens* numerous, long and ±*pendulous*, with *apiculate anthers, Achenes* 3–6 mm, 3–15 per fl., *sessile, ±erect*, variable in shape from roundly and symmetrically ovoid to ±narrowly and asymmetrically ovoid-oblong with the ventral side gibbous and the dorsal side ±straight or gibbous only below, not or somewhat compressed, 8–10-ribbed, glabrous or with stalked glands. Fl. 6–8. Protogynous. Chiefly anemophilous but sometimes visited by insects; some races probably apomictic. $2n = 28, 42$. H.

A satisfactory treatment of this highly polymorphic group cannot be effected until it has been much more closely studied by modern taxonomic methods. It is not yet known to what extent the observed variation arises from genetic or from environmental differences, nor whether the group is cytologically uniform. Populations fall roughly into groups corresponding with the three main habitats; limestone rocks and grassland, dunes, and streamside or lakeside gravel and shingle. Within these groups plants show considerable inter- and intra-population variability in stature, mode of growth, leafiness of the stem-base, shape, colour and glandulosity of the lflets, size and shape of stigmas and achenes, etc. A statistical comparison of populations on the East Anglian chalk with those on the western limestones or the Pennine limestones would doubtless show different averages in respect of some of these characters but would show so much overlapping as to make specific or even subspecific distinction difficult to achieve or to justify. Unless better criteria of discrimination can be discovered it seems best merely to treat as subspecies the three habitat groups referred to, while recognizing that the distinctions are not clear-cut and are partly or even perhaps largely determined by the habitat differences.

Subsp. **minus** (including *T. babingtonii* Butcher, *T. montanum, T. collinum, T. saxatile, T. calcareum*, etc., auct. angl.). *Commonly subcaespitose* but often stoloniferous on soft or loose substrata. *Stem* (12–)25–50(–100) cm, branched, lfy to the base or with brown lfless sheaths below, terete or somewhat furrowed, pruinose or not, ±glabrous or with some stalked glands at least above. Lowest lvs c. 10–25 cm; lflets 4–15 mm across, rounded-cuneate to cordate at the base, green or glaucous, pruinose or not, ±glabrous or with stalked glands especially on the underside. *Panicle usually branching from above the middle of the stem.*

Native. Dry limestone slopes, limestone rocks, cliff-ledges, scree or shingle, chalk quarries and banks, etc.; to 850 m in Wales. Locally in suitable habitats throughout the British Is. except Shetland and Channel Is. Europe.

Subsp. **arenarium** (Butcher) Clapham (including *T. arenarium* Butcher = *T. dunense* auct., non Dumort). *Far-creeping by underground stolons. Stem commonly* 15–40 cm, usually lfy to the base, green or glaucous. Lowest lvs commonly 5–15 cm; lflets often only 4–8 mm across, rounded-cuneate to cordate at the base, green or glaucous, usually densely covered especially beneath with stalked glands. *Panicle usually branching from near or below the middle of the stem*, almost as broad as long.

Native. Open or closed dunes, especially in the north and west. Coasts of Great Britain, northwards from Devon and Cornwall in the west and from Suffolk in the east; Inner and Outer Hebrides; Orkney. N. and E. coasts of Ireland. N.W. Europe from France to Scandinavia.

Subsp. **majus** (Crantz) Rouy & Fouc. (including *T. umbrosum* Butcher, *T. expansum* Jordan, *T. flexuosum* p.p., *T. kochii, T. capillare*, etc., auct. angl.). *Caespitose to stoloniferous. Stems commonly* 50–120 cm, lfy to the base, often markedly striate or furrowed, ±glandular at least above. Lowest lvs commonly 20–40 cm; lflets commonly 10–30 mm across, truncate or cordate at the base, acutely or bluntly lobed, ±glabrous to densely covered with stalked glands, especially beneath. *Panicle often branching from about the middle of the stem*, about as broad as long.

This group is perhaps the most heterogeneous of the three, with the strongest tendency to differentiation between local populations.

Native. Damp places chiefly by streams and lakes and commonly in shade, local; to 610 m in Wales. Wales and W. and N. England northwards to Perth. Europe.

22. PAEONIACEAE

Large perennial herbs or shrubs with alternate exstipulate lvs. Fls large, terminal, usually solitary, hermaphrodite, actinomorphic, hypogynous. Calyx of 5 free sepals; corolla of 5–10(–13) large free petals; stamens numerous, often united below into a ring; carpels 2–5, free, with fleshy walls and mounted on a fleshy disk; ovules several, each with 2 integuments of which the outer projects beyond the inner. Fr. a group of 2–5 large follicles each with several seeds. Chromosomes large; basic number = 5.

One genus, *Paeonia*.

Formerly included in Ranunculaceae but differing markedly in anatomy, in the structure of the gynaecium and the peculiar outer integument of the ovules, as well as in the morphology and basic number of the chromosomes.

1. PAEONIA L.

Perennial herbs or shrubs with erect tuberous stocks and fleshy roots. Lvs large, ternately or pinnately divided, commonly biternate. Outer sepals often with a rudimentary lamina and inner sepals often grading into petals; petals red, purplish or white, rarely yellow; carpels with a short thick style and broad stigma. Fls visited by various insects chiefly for pollen, but some lick the fleshy disk at the base of the carpels. Follicles often hairy; seeds in 2 rows; at first red, then dark blue and shining.

About 33 spp. especially in C. and E. Asia, the Mediterranean region and S. Europe; 2 in western North America.

*1. P. mascula (L.) Miller   Peony.
P. corallina Retz.

A perennial herb with tuberous stock and fleshy roots, and an erect simple lfy glabrous shoot to 50 cm. Lvs biternate, the lflets ovate or elliptical, entire, dark green, glabrous and shining above, but glaucous and finely hairy below. Fl. solitary, terminal, c. 10 cm diam. Sepals 5, green, broadly ovate, imbricate, persistent, the outermost often transitional to the uppermost lvs. Petals 5–8, 4–5(–6) cm, broadly ovate, ±irregularly entire, deep purple-red, rarely whitish or yellowish. Stamens with crimson filaments and yellow anthers. Carpels usually 5, covered with white downy hairs; stigmas red, hooked or coiled. Follicles divergent, usually recurved. Seeds round, smooth, shining, at first red, then dark blue, finally black. Fl. 4–5. Protogynous. $2n = 20$. H.—G.

Introduced. Naturalized on Steep Holm (Severn Estuary). Native in S. Europe from Spain to Greece, Crete and the Near East, and extending northwards to N.C. France and Austria.

The cultivated 'tree peonies' are derived from the woody *P. suffruticosa* Andr. (*P. moutan* Sims), and the herbaceous types chiefly from *P. lactiflora* Pallas, both from N.E. Asia. *P. lactiflora* can be distinguished from *P. mascula* by its several-fld stem and usually ±glabrous carpels.

23. BERBERIDACEAE

Herbs or shrubs. Lvs simple or compound, alternate, usually exstipulate. Fls hermaphrodite, regular, hypogynous, solitary or in cymes, racemes or panicles, usually 3-, rarely 2-merous. Per.-segs free, in 2–7 whorls, the whorls often differing. Stamens usually in 2 (3 in *Achlys*) whorls opposite the inner per.-segs; anthers usually opening by valves, rarely by slits (*Podophyllum*, *Nandina*). Ovary of 1 carpel with 1–many anatropous ovules on a basal or lateral placenta; style short or 0. Fr. a berry or capsule; endosperm copious.

Four genera and about 575 spp., confined to the northern hemisphere (mainly temperate), except for *Berberis*.

A family with the incompletely differentiated perianth characteristic of many *Ranales*, distinguished from the other families by the single carpel.

Herbs; lvs 2-ternate; fr. a capsule.   1. EPIMEDIUM
Shrubs; fr. a berry.   2
Lvs simple; stem spiny.   2. BERBERIS
Lvs pinnate; stem unarmed.   3. MAHONIA

1. EPIMEDIUM L.

Perennial *herbs* with creeping rhizome. Lvs usually biternate, less often pinnate. *Fls 2-merous.* Perianth lvs in 6 whorls, the 2 outer whorls small, sepaloid; the two next petaloid; the two innermost petaloid and usually spurred, bearing nectar in the spur. *Ovules numerous. Fr. a capsule.*

About 21 spp. in the Mediterranean region and temperate E. Asia. Several spp. and many hybrids are cultivated.

*1. E. alpinum L.   Barren-wort.

Rhizome long. Lvs basal and cauline, usually biternate, rarely ternate; lflets to 13 cm, ovate, acute or acuminate, cordate at base, stalked, glabrous above, pubescent beneath at first, becoming subglabrous, remotely spinulose-serrate. Fl. stems 6–30 cm, with a single lf. Infl. a panicle, glandular, 8–26-fld; fls 9–13 mm across; pedicels 5–15 mm. Outer per.-segs 2.5–4 mm, greyish; middle 5–7 mm, dull red, ovate; inner c. 4 mm, yellow, slipper-shaped. Fr. c. 15 mm. Fl. 6–7. Pollinated by insects, protogynous. $2n = 12$*. Hs.

Introduced. Cultivated and sometimes escaping but apparently not thoroughly naturalized. Native from N. and C. Italy to Austria and Albania.

2. BERBERIS L.

Evergreen or deciduous *shrubs*. Wood yellow. *Shoots of 2 sorts, long shoots with the lvs represented by (usually tripartite) spines and short axillary shoots bearing clusters of simple lvs.* Fls in panicles, racemes, fascicles or solitary, 3-merous. Perianth lvs yellow or orange, usually in 5 whorls; the outermost small sometimes with 1 member absent); the 2 inner usually smaller than the intermediate, each bearing 2 nectaries near the base. Stamens springing inwards when touched at the base. *Ovules 1 or few, basal. Fr. a berry.*

About 450 spp., mainly in C. and E. Asia and South America, a few in Europe, the Mediterranean region, mts of tropical E. Africa and temperate North America. Many other spp. and hybrids are ±commonly cultivated.

The following are rarely naturalized:

*__B. buxifolia__ Lam. Lvs 1–2.5 cm, obovate, evergreen, entire. Fls 1–2, axillary, orange. Fr. subglobose, dark purple, pruinose. S. Chile and S. Argentina.

*__B. glaucocarpa__ Stapf. Lvs 3–5 cm, obovate-elliptical, somewhat shiny, subevergreen, with a few spiny teeth. Infl. a 15–25-fld. erect raceme, 2.5–4(5.5) cm. Fr. 9 mm, oblong, dark blue, very pruinose. Naturalized in Somerset. Himalaya.

*__1. B. vulgaris__ L.   Barberry.

Shrub 1–2.5 m; twigs grooved, yellowish. Spines (1–)3(–7)-partite, 1–2 cm. Lvs 2–4 cm, obovate or elliptical-obovate, usually obtuse, spinulose-serrate, reticulate; petiole to 1 cm, but usually much less. Infl. a pendent raceme 4–6 cm; pedicels 5–12 mm. Fls 6–8 mm diam., yellow. Fr. 8–12 mm, oblong, red. Fl. 5–6. Pollinated by various insects, possibly sometimes selfed, homogamous. Fr. 9–10. $2n = 28$. N.

Introduced. Hedges, etc., throughout Great Britain but everywhere very local and in small quantity; formerly much planted for its edible fr. and still often grown for ornament; possibly native in a few places in England; introduced in Ireland. Europe except the extreme north but rare in the Mediterranean region.

3. MAHONIA Nutt.

Evergreen *shrubs. Lvs all alike, pinnate.* Infl. a many-fld raceme or panicle. Fls and fr. as in *Berberis.*

About 70 spp. in E. Asia and North and Central America. Several others are sometimes grown in gardens.

*__1. M. aquifolium__ (Pursh) Nutt.   Oregon-grape.

Berberis aquifolium Pursh

Shrub 1–2 m. Lflets 5–9, ovate, 3.5–8 cm, stiff, coriaceous, dark green and glossy above, sinuately spiny-dentate. Fls in clustered terminal racemes, yellow. Fr. c. 8 mm, globose, blue-black, pruinose. Fl. 1–5. Pollinated by various insects, possibly sometimes selfed, homogamous. Fr. 9. $2n = 28$. N.

Introduced. Commonly planted for pheasant cover and naturalized in many places northwards to Aberdeen. Native of western North America from British Columbia to Oregon.

24. NYMPHAEACEAE

Water-lilies. Perennial water or marsh plants usually with stout creeping rhizomes and floating lvs; often also with submerged or with aerial lvs. Submerged lvs simple, thin, translucent; floating lvs peltate or cordate, usually long-stalked. Fls solitary, terminal, generally floating, hermaphrodite, actinomorphic, hypogynous to epigynous, the parts variously arranged. Perianth usually differentiated into 3–6 green sepals and 3 to many petals which may pass gradually into the usually numerous stamens with introrse anthers, and are either hypogynous or inserted at various heights on the wall of the ±inferior ovary. Carpels 8 or more, either free (but then sunk in the receptacle) or united and superior, or united into a many-celled ovary and inferior. Ovules 1–many, inserted all over the inner walls of the carpels. Fr. a group of achenes sunk in the receptacle or a spongy capsule dehiscing regularly or irregularly by the swelling of internal mucilage. Seeds often arillate, with or without endosperm, sometimes with perisperm as well as endosperm.

About 3 genera with c. 75 spp., cosmopolitan.

Sepals 4, green; petals white, the outermost longer than
   the sepals; lateral veins of lvs anastomosing at the margin.   1. NYMPHAEA
Sepals 5–6, yellowish; petals yellow, much shorter than
   the sepals; lateral veins of lvs forking near the margin.
                                                2. NUPHAR

1. NYMPHAEA L.

Rhizome stout. *Lvs stipulate,* ±orbicular, rarely peltate. Fls usually floating, often large and showy. *Sepals 4,* hypogynous, green; *petals numerous,* inserted at successively higher levels on the side of the ½-inferior ovary, *the outermost much longer than the sepals,* the innermost shorter and narrower and grading into the stamens; stamens numerous, the outer with broad petaloid filaments, all inserted towards the top of the ovary; ovary syncarpous, many-celled with numerous ovules in each cell; summit of the ovary concave but with a central boss from which the stigmatic surfaces radiate, and with marginal stylar processes. Fr. a spongy berry-like capsule ripening under water and splitting by the swelling of internal mucilage to release the seeds, which float because of the air-containing aril and seed-wall.

About 50 spp., cosmopolitan.

The cultivated water-lilies are chiefly varieties and hybrids of *N. alba* (white and red fls), *N. lotus* (white fls and toothed lvs), *N. rubra* (red fls and toothed lvs), *N. caerulea* (blue lfs, carpels free laterally, entire lvs), *N. capensis* (blue fls, carpels free laterally, sinuate lvs) and *N. mexicana* (yellow fls), etc.

1. N. alba L.   White Water-lily.

Castalia alba (L.) Wood

Lvs 10–30 cm, usually all floating, rarely some emergent, *±circular, entire,* with *basal sinus ⅓–½ of total length,* the *basal lobes* ±*contiguous*, sometimes overlapping; lf dark green above, paler and often reddish beneath; petioles (and peduncles) usually 0.5–2.5 m. *Fls* (5–)10–20 cm diam., *floating*. Sepals 3–5, usually 4, ovate to broadly elliptical, olive-coloured beneath, white above; *petals* (12–)20–25(–33), usually broadly elliptical, *white or pink-tinged*; stamens 45–125, innermost with ±linear yellow filaments; pollen-grains covered with short blunt tubercles; ovary subglobose, with 9–25 yellow stigmatic rays erecting at their ends in free triangular-ovate stylar processes which curve inwards at their tips. Nectaries 0; some exudation from stigmatic papillae. Fr. 18–40 mm diam., subglobose to obovoid, with stamen-scars usually almost to the top but small frs may have the upper part free. Seeds numerous, 2–3 mm, dark olive-green,smooth. Fl. 6–8. Homogamous or slightly protogynous, initially fragrant. Visited sparingly by bees, flies and beetles but probably often self-pollinated. $2n = $ c. 84*, c. 112*, 48, 64, 96, etc. Hyd.

Native. In lakes, ponds and slow-moving streams, canals and dykes, usually at water depths of 0.5–3 m and at pH 5—8; seems more tolerant of low nutrient levels than is *Nuphar lutea* and less of turbidity through wave action and water movement so that in wind-swept lakes it is commonly restricted to sheltered bays. Through most of the lowland British Is. but not in Orkney; reaches 456 m in Cumbria. Europe northwards to S. Scandinavia with a few scattered localities to 69° N, but not in Iceland or Faeroes.

There is considerable variation in size of lvs and fls, size and shape of fr., etc., and populations consisting entirely or largely of plants with lvs and fls well below average size are found in certain nutrient-poor lakes from Cumbria northwards and in the Hebrides, Shetland and W. Ireland. Such plants may remain below average size for several seasons when transferred to nutrient-rich conditions, so that genetic differences seem involved. Other isolated populations, too, may show unusual features but these are often attributable to the genetic make-up of original colonists or to subsequent non-adaptive 'genetic drift' and cannot be made the basis for infraspecific taxa. Subsp. *occidentalis* Ostenf. seems to have been based on restricted populations in W. Ireland and C. Scotland of very small plants with almost globose fr. less than 30 mm diam. having only 10–15 stigmatic rays and devoid of stamen-scars above. Similarly, small plants in lakes with larger populations of *N. alba* are often connected through a range of intermediates with more typical representatives of the species, so that recognition of a distinct subsp. is impossible, even though there may be an adaptive element in the smallness of size.

2. NUPHAR Sm.

Rhizome stout. *Lvs not stipulate*, all floating or some submerged, broadly elliptical or oblong, the lateral veins three times forked. *Fls yellow*, *globose*, *rising above the water*. Sepals 4–6, yellowish-green, hypogynous; *petals numerous*, *yellow*, *usually shorter than the sepals*, *spathulate*, all hypogynous, with a nectary on the lower side; stamens numerous, all hypogynous, with broad filaments; ovary superior, syncarpous, many-celled,

with many ovules in each cell; summit of ovary ±convex with a central depression and many radiating sessile stigmas. Insect-pollinated. *Fr. flask-shaped*, berry-like, ripening above water, splitting irregularly into parts each comprising the internal tissue and seeds of a single ovary-cell. Seeds without aril, with little endosperm and much perisperm.

About 25 spp. in temperate regions of the northern hemisphere.

1 Fls 4–6 cm diam.; stigma-rays 10–25; stigmatic disk
      circular, with entire margin.   **1. lutea**
    Fls 1.5–4 cm diam.; stigma-rays 7–14; margin of stig-
      matic disk wavy to deeply lobed.   2
2 Stigma-rays 7–12; stigmatic disk 6–8.5 mm diam.,
      ±stellate, its margin deeply lobed.   **2. pumila**
    Stigma-rays 9–14; stigmatic disk 7.5–11 mm diam.,
      circular with a wavy margin.   **3. ×spennerana**

1. N. lutea (L.) Sm.   Yellow Water-lily, Brandy-bottle.

Nymphaea lutea L.

Rhizomes 3–8 cm diam., branched. Floating lvs 12–40 × 8–30 cm, ovate-oblong, with a deep basal sinus, the basal lobes being about ⅓ the length of the lf, thick and leathery; submerged lvs shorter-stalked, broadly ovate or round, cordate, thin and translucent. *Fls 4–6 cm diam.*, rising out of the water, their stalks up to 2 m, rarely to 3 m. Sepals 2–3 cm, broadly obovate, persistent, bright yellow within. Petals broadly spathulate, one-third as long as the sepals. Stamens shorter than the ovary. Stigmatic disk wider than the top of the ovary (10–15 mm wide), with usually 15–20 *stigmatic rays which do not reach the entire margin* (Fig. 9A). Fr.

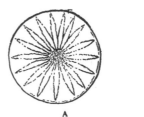

Fig. 9. Stigmatic disks of A, *Nuphar lutea* and B, *N. ×spennerana*. ×2·5.

3.5–6 cm, flask-shaped. Seeds c. 5 mm. Fl. 6–8. Homogamous or protogynous. Visited by small flies. Smelling of alcohol ('Brandy-bottle'). $2n = 34$*. Hyd.

Native. Mainly in lowland lakes, ponds and streams with water of pH 6.0 or higher throughout the British Is. but rare in N. Scotland northwards to Moray and W. Ross, very rare in Hebrides and not in Orkney or Shetland; reaches 510 m in Wales. Most of Europe to c. 67° N in Norway but not in Faeroes or Iceland; N. Asia; N. Africa.

2. N. pumila (Timm) DC.   Least Yellow Water-lily.

Rhizome 1–3 cm diam. Floating lvs 4–14 × 3.5–13 cm, broadly oval, with a deep basal sinus and lobes which are usually more divergent than in *N. lutea*; submerged lvs as in *N. lutea* but much smaller. *Fls* 1.5–3.5 cm *diam.* Sepals 4–5, roundish, yellow, persistent. Petals narrowly spathulate, one-third as long as the sepals. Stamens shorter than the ovary. Stigmatic disk 6–8.5 mm wide, little wider than the top of the ovary, with usually 8–10 *stigma-rays which reach the ends of the deep ±acute lobes of the margin.* Fr. pear-shaped, 2–4.5 cm. Seeds 2.5–4 mm. Fl. 7–8. Less fragrant than *N. lutea*. $2n = 34^*$. Hyd.

Native. In highland lakes in scattered localities of mainland N. and C. Scotland, also in one locality in Shropshire and formerly in Merioneth; reaches 523 m in Perth. N. and C. Europe southwards to C. France, S.W. Yugoslavia and S.E. Russia; N. Asia eastwards to Manchuria.

3. N. × spennerana Gaudin Hybrid Yellow Water-lily.

N. lutea var. *minor* Syme; *N. intermedia* Ledeb.

Intermediate between *N. lutea* and *N. pumila* and probably a hybrid between them, since the pollen is partly non-viable and fr. and seeds are produced sparingly. Lvs. intermediate in size. *Fls* 3–4 cm *diam.* Petals broadly obovate. *Stigmatic disk* 7.5–11 mm wide, *circular with a wavy margin* and with usually 10–14 *stigma-rays* not quite reaching the margin (Fig. 9ʙ). Fl. 6–8. $2n = 34^*$. Hyd.

Native. Found only in a few lakes in C. and S. Scotland and one in Northumberland. Some, though not all, of these are in the present area of overlap between *N. lutea* and *N. pumila*, but the locality in N.W. Argyll is outside the range of both. Recorded from C. and N. Europe northwards to c. 68.5° N in Lapland.

N. advena (Aiton) Aiton fil. (North America) is grown in gardens and has been found as an escape. Its lvs, 10–40 cm, are held erect, floating only in very deep water. Fls 6–10 cm diam., yellow and green, held above the water on erect peduncles. Plants reported as naturalized in Surrey have very red stamens and nectaries.

25. CERATOPHYLLACEAE

Submerged aquatic herbs. Lvs in whorls, divided, the segments linear or filiform. Fls unisexual, sessile, solitary in the whorls of lvs, the male and female fls at different nodes. Perianth herbaceous, of 8–12 narrow segments, often dentate or lacerate at apex. Male fls: stamens 10–20 on a flat torus; anthers subsessile, 2-locular, with the connective produced beyond the loculi, opening lengthwise, often coloured. Female fls: ovary sessile, unilocular; ovule solitary, pendent, anatropous. Fr. a nut. Seed pendent; endosperm 0.

One genus and 10 spp., cosmopolitan.

1. CERATOPHYLLUM L. Hornwort.

The only genus.

Lvs once to twice forked; fr. with 2 spines at base.
                                        1. demersum
Lvs 3–4 times forked; fr. without spines at base.
                                        2. submersum

1. C. demersum L.   Rigid Hornwort.

A dark green rather stiff densely lfy perennial. Stems 20–100 cm, slender. *Lvs* 1–2 cm, *once to twice forked*, the segments linear, rather closely but irregularly denticulate. *Fr.* c. 4 mm, ovoid, ±warty, somewhat shorter than or equalling the persistent style, *with two spines at base when ripe.* Fl. 7–9. $2n = 24$. Hyd.

Native. In ponds and ditches, local. Scattered throughout England, rare in Wales, Scotland and Ireland. Most of Europe, N. Africa, Asia, North America.

2. C. submersum L.   Soft Hornwort.

Similar to *C. demersum* but softer and brighter green. *Lvs* 3–4 *times forked*, the segments sparingly denticulate. *Fr.* warty, longer than the persistent style, *devoid of spines.* Fl. 7–9. $2n = 40$; 72. Hyd.

There has been much confusion between the two spp., but this plant appears to be the rarer.

Native. In ponds and ditches. Mainly in southern and eastern England. Most of Europe; temperate Asia, N. Africa.

26. PAPAVERACEAE

Annual to perennial herbs, rarely shrubs or small trees, with *exstipulate usually alternate and often lobed or deeply dissected lvs*; rarely some of lvs in opposite pairs or whorls of 3, and lvs occasionally ±entire. *Latex*, milky, yellow to reddish, or colourless, is *usually present throughout the plant* but only in the roots in *Glaucium* and *Eschscholzia*. Fls hermaphrodite, actinomorphic, hypogynous (slightly perigynous in *Eschscholzia*), with all parts in whorls. *Sepals* 2(–3), *commonly falling as the fl. opens*, usually free but cohering in a pointed hood over the fl. in *Eschscholzia*; *petals* 4 or 6 in 2 whorls, rarely more or 0, *crumpled in bud and often for a time after unfolding*, typically falling well before fr. ripens; *stamens usually* numerous, free; ovary superior, *syncarpous*, of 2 or more carpels *but 1-celled*, the numerous *ovules on projecting parietal placentae* (fl. apocarpous, with 6–16 carpels, in *Platystemon*). *Nectaries* 0. Fr. usually a capsule, often long and narrow, usually remaining 1-celled but 2-celled in *Glaucium* by meeting of the placentae. Seeds small, with minute embryo in oily endosperm; in *Chelidonium* and a few other genera the seeds have a fleshy oil-containing appendage (elaiosome) attractive to ants.

About 26 genera and 200 spp. cosmopolitan but

chiefly in temperate and subtropical regions of the northern hemisphere.

Many species are grown for their showy fls, including members of the native or adventive genera *Papaver*, *Meconopsis* and *Eschscholzia* and also the shrubby *Romneya coulteri* (Californian Tree-Poppy), with large solitary fragrant white fls; *Macleaya cordata* (Plume Poppy), a herb up to 2.5 m with large glaucous pinnate-lobed cordate lvs and long terminal panicles of apetalous fls, and the Prickly Poppies of the genus *Argemone*, with spine-toothed pinnatifid lvs and prickly capsules opening at the top by valves.

1 Latex milky (rarely yellowish after exposure to air); stigmas sessile on expanded disk at top of ovary; fr. a capsule opening by pores just beneath the stigmatic disk. **1. PAPAVER**
  Latex orange, yellow or colourless; stigmas usually not sessile; fr. opening by valves. 2
2 Fr. usually ovoid- to obovoid-oblong, opening from above by 3–6 rather short tooth-like valves. 3
  Fr. ±parallel-sided and more than 8 times as long as wide, opening by 2(–4) valves over all or most of its length. 4
3 Stem, lvs and capsule all spiny; latex yellow; petals pale yellow to orange. A rare casual.
    Argemone mexicana (p. 60).
  Plant not spiny; latex yellow; petals yellow (blue or red in cultivated spp.). **2. MECONOPSIS**
4 Petals deep violet-blue; fr. opening from above by 2–4, usually 3, valves. Now a very rare casual near ports. *Roemeria hybrida* (p. 60).
  Petals yellow to orange; fr. opening by 2 valves. 5
5 Sepals joined to form a hood which is displaced as the petals open; receptacle forming a distinct perigynous ledge round base of ovary; latex colourless. **5. ESCHSCHOLZIA**
  Sepals free; no perigynous receptacle-ledge; latex orange. 6
6 Fls 3.5–9 cm diam.; capsule 2-celled, opening from above; plant of maritime shingle. **3. GLAUCIUM**
  Fls 2–2.5 cm diam.; capsule 1-celled, opening from below; plant of banks, hedgerows, walls, etc.
    4. CHELIDONIUM

1. PAPAVER L.

Annual to perennial herbs usually with *white latex* and toothed or pinnately lobed or divided lvs, often hispid. Fls solitary, terminal or axillary, showy. Sepals 2, free, falling as the fl. opens; petals 2 + 2, fugacious, crumpled in bud; stamens very numerous, with anthers dehiscing outwards; *stigmas 4–20, sessile and over the placentae*; ovules very numerous. *Fr. a capsule opening by pore-like valves just beneath the persistent stigmatic disk.* Seeds small, without appendage.

About 100 spp. in the northern hemisphere with a few tropical spp. and one in Australia. Many are arctic or subarctic, and many are common weeds of cultivation.

1 Perennial; rosette-lvs lanceolate, ±coarsely toothed or at most shallowly pinnatifid, stiffly hairy, not glaucous; petals dull orange or apricot to brick-coloured. Rare aliens, locally established.
    7. atlanticum

Annual, sometimes overwintering; rosette-lvs usually deeply pinnatifid to pinnatisect or, if less deeply lobed, then ovate-oblong and glaucous; petals scarlet, crimson, orange-pink, pale lilac or white. 2
2 Plant not or only slightly glaucous; lvs once or twice deeply pinnatifid to pinnatisect; upper lvs not amplexicaul. 3
  Plant glaucous; lvs coarsely toothed or pinnately lobed, but not deeply pinnatifid or pinnatisect; upper lvs amplexicaul. **6. somniferum**
3 Capsule glabrous; stigmatic disk becoming flat at maturity. 4
  Capsule with stiff hairs or bristles; stigmatic disk remaining convex. 6
4 Capsule subglobose to broadly obovoid, less than twice as long as wide. **1. rhoeas**
  Capsule obovoid-oblong, at least twice as long as wide. 5
5 Latex white; anthers bluish to violet, usually below level of stigmatic disk; petals with their bases overlapping; seeds greyish- to bluish-black. **2. dubium**
  Latex yellow or soon turning yellow on exposure to air; anthers yellow or yellow-brown, usually at same level as stigmatic disk; petals with bases not overlapping; seeds chocolate-brown. **3. lecoqii**
6 Stigmatic rays 4–6(–7); capsule 1.5–2 cm, narrowly obovoid-oblong, rather sparsely covered, sometimes only above, with ascending bristles.
    4. argemone
  Stigmatic rays (4–)6–8; capsule 1–1.25 cm, subglobose or broadly obovoid, densely covered with spreading-ascending bristles. **5. hybridum**

1. P. rhoeas L. Field Poppy.

An annual or overwintering herb with slender tap-root and erect or ascending usually branched *stems* (5–)20–60(–90) cm, typically *with stiff spreading hairs throughout*. *Lvs* ovate to ovate-lanceolate in outline, *pinnatisect to pinnate*, the narrow acute and bristle-pointed *segments ±deeply toothed or pinnatifid to pinnatisect*; *terminal segment* usually *distinctly largest* (Fig. 10A,B); rosette- and lowest stem-lvs petiolate, other

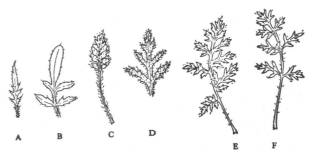

Fig. 10. Leaves of *Papaver*. A, B, *P. rhoeas*; C, D, *P. dubium*; E, *P. hybridum*; F, *P. argemone*. ×⅓.

stem-lvs sessile, diminishing upwards. *Fls* 5–10 cm diam., their *long axillary pedicels with spreading hairs*. Sepals bristly. *Petals* 2–6 cm, outer pair broader than inner but all *broader than long*, rounded above and broadly cuneate below, usually *scarlet, often with dark basal blotch*, rarely pink or white. *Stamens with slender* purplish-black *filaments* and dark violet *anthers held below level of stigmatic disk*; pollen brownish-green. *Capsule* 1–2 cm, *glabrous, subglobose to broadly obovoid, at most twice as long as wide; stigma-rays usually 8–14 on an initially convex disk with as many overlapping marginal lobes* within which the rays terminate. Seeds dark brown. Fl. 6–10. Anthers dehisce within unopened fl.-bud. Visited by numerous insects, especially bees; almost entirely self-sterile. $2n = 14^*$. Th.

Probably native; fossil seeds found in late Bronze Age settlements. A weed of arable fields, waste places, waysides, chalk and gravel pits, walls, etc., especially on medium to light calcareous soils. Throughout the British Is. and common in England and S.E. Scotland but in Wales mainly near coasts and rare and local in N.W. Scotland, Inner and Outer Hebrides, Orkney and Shetland; Ireland, especially in south-east; Channel Is. Most of Europe and certainly native in S. Europe; N. Africa; temperate Asia. Introduced in North America, Australia, New Zealand and other temperate countries.

P. rhoeas is very variable in shape and hairiness of the lvs, colour and blotching of the petals, shape of capsule, etc. The Shirley Poppies of gardens comprise a wide range of colour forms found occasionally in wild populations of the species. The fertile and otherwise typical plants which have appressed hairs on the fl.-stalks, variously named *P. strigosum* (Boenn.) Schur, *P.* × *strigosum* (Boenn.) Schur and *P. rhoeas* f. *strigosum* Boenn., seem similarly to fall within the natural range of variations of *P. rhoeas*, differing by only a single gene from plants with the more usual spreading hairs. Hybrids between *P. rhoeas* and *P. dubium* have been produced artificially and have been recognized in mixed wild populations. They are very variable, with either appressed or spreading hairs on the fl.-stalks, and are quite sterile: it seems very unlikely that *P. rhoeas* var. *strigosum* can be of this hybrid origin.

P. commutatum Fisch. & Mey. (E. Europe and W. Asia), found occasionally as a casual or garden escape, with appressed hairs on the fl.-stalks and narrow petals, at least as long as broad, which have a rounded or quadrangular blotch in the middle (not at the extreme base), does not seem to merit full specific distinction from *P. rhoeas*.

2. P. dubium L.   Long-headed Poppy.

An annual or rarely overwintering herb much like **1** but *stems with hairs spreading below, appressed above*. Root and *latex white*. Lvs as in **1** but seldom fully pinnate and with the segments shorter, broader and more abruptly acute and *with smaller terminal segments* (Fig. 10c,D). *Fls* 3–7 cm diam., their *pedicels appressed-hairy*. Sepals bristly. *Petals* 1.5–3 cm, *broadly obovate* with *outer pair slightly broader than long, always with overlapping bases; paler scarlet than in* **1** to orange-pink, and usually lacking dark basal blotch. *Stamens with slender* purplish-black *filaments* and *violet anthers held below level of stigmatic disk*; pollen lemon-yellow. *Capsule* 1.5–2 cm, *glabrous, ±narrowly obovoid-oblong, more than twice as long as wide; stigma-rays*

(4–)7–9(–12), each *ending a little short of the margin* of one of the shallow non-overlapping lobes of the initially strongly convex disk (later ±flat). Seeds bluish-black. Fl. 5–7. Anthers dehisce within unopened fl.-bud. Visited especially by bees and hover-flies; self-compatible but crossing favoured by position of anthers. $2n = 42^*$. Th.

Native or introduced. A weed of arable land, especially cereal crops, and waste places throughout lowland British Is., often with **1** but more common than **1** in the north and west. Favours light dry sandy or gravelly soils but also found on heavy wheat land. Reaches 450 m in Scotland. Through most of Europe to c. 65° N as a weed of cultivation, but only casual towards its northern limit.

3. P. lecoqii Lamotte   Babington's Poppy.

An annual herb closely resembling *P. dubium* but with *latex turning yellowish after exposure to the air; lvs sparsely hairy*, hardly glaucous, with deeper, narrower and more acute segments; *petals not overlapping at base*, deeper red than in *P. dubium; anthers yellow, equalling the stigmatic disk; seeds chocolate brown*. Fl. 6–7. Homogamous and self-compatible; more readily self-pollinated than *P. dubium* because of position of anthers. $2n = 28^*$. Th.

Native or introduced. An infrequent and decreasing weed of road-verges, quarries, headlands of arable fields, etc., but not found amongst cereals; often on calcareous soils. Probably ±confined to the southern and eastern counties of England, but detailed distribution unknown owing to confusion with *P. dubium*; local in Ireland. Europe.

4. P. argemone L.   Long Prickly-headed Poppy.

An annual herb with tap-root and erect or ascending usually branched stems 10–50 cm, stiffly appressed-hairy overall or hairs spreading only below. Lvs 1–2(–3)-pinnatisect with ultimate segments suddenly acuminate and bristle-pointed (Fig. 10F), rosette- and lowest stem-lvs petiolate, upper sessile; all stiffly hairy. Fls commonly 5–6.5 cm diam., their *pedicels appressed-hairy*. Sepals bristly. *Petals* 2–3 cm, obovate, *longer than broad and not contiguous in fully open fl., light scarlet*, usually with dark blotch at base. *Stamens with dark violet clavate filaments* and bluish anthers borne at or above level of stigmatic disk; pollen deep blue. *Capsule* 1.5–2 cm, *narrowly obovoid-oblong*, strongly ribbed and *with relatively few ascending bristles* between the ribs, often more numerous above, rarely 0; *stigmatic disk* remaining convex and *bearing* 4–6(–7) *stigmatic rays which reach to or beyond its margin*. Fl. 5–7. Anthers dehisce and self-pollination takes place within unopened fl.-bud. Frequent insect visitors, especially honey-bees; highly self-compatible. $2n = 42^*$. Th.

Probably introduced. A frequent weed of arable fields, waysides and waste places on sandy, gravelly and light loamy circumneutral soils, less commonly on calcareous soil. Throughout lowland England, Wales and

S. Scotland, thinning out towards the west and north but reaching Banff and E. Ross; Ireland; Channel Is. Probably native in S. Europe but a widespread weed of cultivation further north and reaching N. Scandinavia.

5. P. hybridum L. Round Prickly-headed Poppy.

P. hispidum Lam.

An annual or overwintering herb resembling **4** but *stems with spreading hairs at least below*, sometimes appressed-hairy or glabrescent above; lvs broader in outline, less stiffly hairy and with ultimate segments more gradually acute (Fig. 10E); *fls smaller*, only 2–5 cm diam., their *petals* 1–2(–2.5) cm, obovate, *dull crimson with dark purplish basal blotch; pollen pale powder-blue. Capsule* 1–1.25 cm, *broadly ellipsoid or obovoid to subglobose; not clearly ribbed; with numerous spreadingascending yellow bristles* in ±vertical bands; *stigmatic disk with (4–)6–8 broad stigma-rays reaching its sinuate margin*; seeds grey-brown. Fl. 5–8. Amongst our wild poppies this has the smallest and least conspicuous fls, visited by relatively few insects, but it is normally selfpollinated within the unopened fl.-buds and is highly self-compatible. $2n = 14^*$. Th.

Probably introduced. A rare and decreasing weed of arable land and waste places mainly on well-drained calcareous soils in S. and E. England with a few scattered localities, where it is mainly casual, westwards to S. Wales and Isle of Man and northwards to E. Ross; perhaps no longer in Ireland; Channel Is. Native in S. Europe and W. Asia but widely introduced elsewhere.

*6. P. somniferum L. Opium Poppy.

An annual herb with tap-root and simple or branched erect stem 30–100(–150) cm, glabrous or with a few spreading bristles. *Lvs undulate, ovate-oblong, ±pinnately lobed*, the lobes shallow, irregular, coarsely toothed; lowest lvs narrowed into a short stalk, upper sessile clasping the stem; all ±*glaucous*, glabrous or with stiff spreading hairs (subsp. *setigerum*). Fls to 18 cm diam., on glabrous or somewhat hairy stalks. *Petals white or pale lilac*, with or without a basal blotch. Filaments thickened above; anthers bluish. Capsule 3–9 cm, ovoid or obovoid to subglobose, glabrous; stigma-disk with deep non-contiguous marginal lobes; stigma-rays 5–12(–18), not quite reaching the ends of the lobes. Capsule dehiscent or not; seeds black or white. Fl. 7–8. $2n = 20, 22$. Th.

Very variable and treated variously by taxonomists. British plants have recently been assigned to the following subspp., both having lobes of the stigmatic disk furrowed, with toothed margin and truncate apex:

Subsp. **somniferum** (incl. subsp. *hortense* (Hussenot) Corb.). Up to 100(–150) cm, glabrous or nearly so, distinctly glaucous. Lvs coarsely toothed to shallowly and bluntly pinnatifid, subglabrous. Capsule up to 9 × 6 cm, ovoid to subglobose; stigmarays 8–12(–18); capsule sometimes indehiscent.

Subsp. **setigerum** (DC.) Corb. Up to 60 cm, ±stiffly hairy, barely glaucous. Lvs more deeply and acutely pinnatifid, the

lobes often bristle-pointed. Capsule 5–6 cm, subglobose to obovoid; stigma-rays 5–8.

Introduced. Subsp. *somniferum* is much grown in gardens and occasionally escapes. It is also a casual of cultivated and waste ground and rubbish-tips in scattered localities northwards to Caithness. Formerly grown in the East Anglian fens and elsewhere for medicinal purposes, and is still established locally as a relic of cultivation. Subsp. *setigerum*, the wild plant believed ancestral to subsp. *somniferum*, is a rare casual. The species as a whole is thought to be native in the west and central Mediterranean region but is now widespread as an introduced weed.

*7. P. atlanticum (Ball) Cosson

A perennial herb with a stout stock crowned by the remains of lvs of former years. Rosette-lvs 12–25 cm, oblong-lanceolate, gradually narrowed into a stalk-like base, coarsely and irregularly crenate-serrate or shallowly pinnatifid, covered on both sides with whitish bristle-like hairs; stem-lvs only on lower half of stem or 0, subsessile. Sepals hispid. *Petals* 2.5–4 cm, obovate, *dull orange* to reddish. Stigma-rays usually 6. *Capsule* to 2.5 cm, glabrous, narrowly clavate or perhaps sometimes approaching obovoid. Fl. 6–9. Hr.

Introduced. In gardens and churchyards, on waysides, waste ground and wall tops, etc., scattered localities in S. England and more sparsely northwards to S. Scotland, persisting for varying lengths of time. Native to Morocco.

P. lateritium C. Koch, a native of Armenia closely resembling **7** but said to differ in its more brightly coloured fls and broader capsule, has been reported from several localities but further investigation is necessary.

P. orientale L. (E. Mediterranean region, Caucasus and Near East), the Oriental Poppy of gardens, occasionally escapes. A robust hispid perennial with pinnatisect lvs and large fls with 4–6 petals, 5–8 cm, scarlet, usually with dark basal spot. In *P. nudicaule* L. (Siberia, C. Asia, and N. Pacific coasts), the Iceland Poppy of gardens, the lvs are confined to a basal rosette and the fls 3–8 cm diam., have yellow, orange or reddish petals. The closely related *P. radicatum* Rottb. (circumpolar and one of the 4 plant spp. found at 83° 24′ N on the N. coast of Greenland) has yellow, not white latex, smaller fls and a more rounded capsule. It is polymorphic with numerous subspp. in Scandinavia which seem to have arisen through isolation during the Ice Age.

2. MECONOPSIS Vig.

Annual or perennial herbs with yellow latex. Lvs entire, toothed or pinnately lobed. Fls variously arranged. Sepals usually 2, soon falling; petals usually 4, rarely to 10; stamens numerous; *capsule* subglobose to narrowcylindrical with a distinct style and 4–6 *stigmas* opposite the placentae, *dehiscing by valves which usually reach only a short distance below the top*. Seeds numerous, crested or not.

40 spp., one confined to W. Europe, the rest to southcentral temperate Asia, and especially the Himalaya.

Several of the Asiatic species are high alpines, and *M. horridula* was found at 5800 m on Everest. This species, known to horticulturalists as *M. racemosa* Maxim., is grown in gardens for its racemes of pale blue fls, but the more spectacular 'Blue Poppy' is *M. betonicifolia* Franchet (*M. baileyi* Prain) with sky-blue fls up to 7.5 cm diam., native at 3000–4000 m in S.E. Tibet, N.W. Yunnan and N. Burma.

1. M. cambrica (L.) Vig.   Welsh Poppy.
Papaver cambricum L.

A perennial herb with a branched tufted stock covered with the persistent lf-bases and erect branched leafy ±glabrous stems 30–60 cm. Basal lvs long-stalked, pinnately divided with pinnately lobed ovate acute segments, ±glabrous; upper lvs similar but shortly stalked. Fls 5–7.5 cm diam., arising singly in the axils of the upper lvs. Sepals hairy. Petals 4, yellow. Stamens with filiform filaments and yellow anthers. Capsule 2.5–3 cm, ovoid to ellipsoid, 4–6-ribbed, with 4–6 stigma-lobes, and splitting into 4–6 valves for about ¼ its length. Seeds pitted. Fl. 6–8. $2n = 22^*$. H.

Native. In damp, shady, rocky places in S.W. England and Wales, up to 610 m local in Ireland; widely introduced and established in N. England and in S. and C. Scotland, more sparsely in S. and C. England and N. Scotland; rare in Hebrides, Orkney and Shetland. Native only in N. Spain, W. France (to Massif Central) and British Is.

Roemeria hybrida (L.) DC., Violet Horned-poppy, is an annual herb 20–40 cm with yellow latex and three times pinnatisect lvs with almost linear segments. The solitary fls, 5–7.5 cm diam., have roundish deep violet-blue petals with dark basal blotch. The linear capsule, 5–7.5 cm, opens usually by 3 valves down to the base. Introduced and formerly established locally as a cornfield weed, but now a very rare casual. S. Europe.

Argemone mexicana L., Prickly Poppy, a robust annual herb with yellow latex, its erect spiny stem up to 1 m and its glaucous, white-veined, amplexicaul lvs pinnatifid, the segments with coarse prickly teeth. Fls 5–8 cm diam., solitary and ±sessile in lf-axils, their 4–6 petals 2–3 cm, obovate, bright yellow to orange. Capsule ellipsoid, prickly, opening by 4–6 short valves. A rare casual but already naturalized in several countries of W. and C. Europe. Native in S.W. USA and Central America.

3. GLAUCIUM Miller

Annual to perennial glaucous herbs with yellow latex. Lvs pinnately lobed or cut. Fls axillary, large. Sepals 2, caducous; petals 4, yellow to red; stamens numerous; stigma ±sessile, 2-lobed, over the placentae which meet in the centre of the ovary; ovules many. *Capsule linear, 2-celled, opening almost to the base by 2 valves, and leaving the seeds embedded in the septum.* Seeds pitted, without an appendage.

About 21 spp., chiefly Mediterranean.

Stem and capsule ±glabrous; fls yellow, 6–9 cm diam.
                                              1. flavum

Stem and capsule stiffly hairy; fls orange to scarlet, 3–5 cm diam.   **2. corniculatum**

1. G. flavum Crantz   Yellow Horned-poppy.
G. luteum Crantz; *Chelidonium glaucium* L.

A perennial or biennial herb with a deep stout tap-root and an erect branched stem 30–90 cm, glaucous, ±glabrous. Basal lvs roughly hairy, stalked, pinnately lobed or divided, sublyrate, the lobes pointing various ways and further lobed or coarsely toothed; upper *lvs* sessile, half-clasping, less deeply lobed or sinuate, rough; all *glaucous*. Fls 6–9 cm diam., their stalks short, glabrous. Sepals hairy. Petals roundish, *yellow*. Capsules 15–30 cm, glabrous but rough. Fl. 6–9. Protogynous. Visited by various flies and some small bees. $2n = 12$. Hs.

Native. A maritime plant, chiefly of shingle-banks, on the west coast of Great Britain from Arran and Kintyre southwards, the south coast from Cornwall to Kent, and the east coast from Kent northwards to the Wash, with an outlier in Berwick; formerly in a few other places northwards to Moray and Shetland; casual in some inland localities; local and decreasing in Ireland; Scilly; Channel Is. Coasts of W. and S. Europe northwards to c. 60° N in Norway and eastwards to the Black Sea; frequently naturalized inland in C. Europe.

*2. G. corniculatum (L.) J. H. Rudolph   Red Horned-poppy.
G. phoeniceum Crantz; *Chelidonium corniculatum* L.

An annual herb with an erect branched stem 25–30 cm, with stiff spreading hairs. Basal lvs stalked, sublyrate and pinnately and deeply lobed, the lobes distant, oblong, toothed; upper lvs sessile half-clasping, less deeply lobed, the lobes remotely toothed; all ±hairy. Fls 3–5 cm diam., their stalks very short. Sepals softly hairy. Petals roundish, *bright scarlet or orange-red* with a black spot at the base. Capsule 10–22 cm, slightly curved, hairy. Fl. 6–7. $2n = 12$. Th.

Introduced. Perhaps once established in Norfolk and at Portland but now a casual of waste places especially near ports. S. Europe and Mediterranean region.

4. CHELIDONIUM L.

Perennial herbs with short branching stocks and erect branched lfy stems. *Latex bright orange.* Lvs pinnately cut or lobed. Sepals 2, free, caducous; petals 4; stamens numerous; style very short with two spreading adnate stigma-lobes over the placentae. *Capsule linear, 1-celled, with no septum, opening from below by 2 valves which separate from the placentae;* seeds with a fleshy crested appendage on the raphe.

1 sp. in Europe and N. Asia.

1. C. majus L.   Greater Celandine.

A perennial herb with its branched woody stock covered with persistent lf bases and with erect branched lfy *brittle stems* 30–90 cm, slightly glaucous and sparsely hairy. Lvs almost pinnate with 5–7 ovate to oblong lflets; terminal lflet often 3-lobed, laterals usually with a stipule-like lobe on the lower side; all crenately-toothed and ±glabrous, somewhat glaucous beneath. Fls 2–2.5 cm diam., terminal. Sepals greenish-yellow, ±hairy. Petals bright yellow, broadly obovate. Stamens yellow; filaments thickened above. Capsule 3–5 cm. Seed black with a white appendage. Fl. 5–8. Homogamous. Visited by pollen-collecting flies and bees. Germ. spring. $2n = 12*$. Hs.

Native or introduced. A frequent plant of banks, hedgerows and walls, chiefly near habitations throughout lowland England and Wales and in scattered localities in Scotland northwards to Caithness; not in Outer Hebrides, Orkney or Shetland; frequent but local in Ireland; Channel Is. Most of Europe northwards to head of Baltic Sea in Sweden and Finland; N. Asia.

Formerly a herbalist's remedy for warts and eye troubles. The orange latex contains several alkaloids, including chelidonin and chelerythrin, and is poisonous.

Var. *laciniatum* (Miller) Koch (*C. laciniatum* Miller) is sometimes found as a casual. It has the lobes of the lvs deeply and narrowly pinnately cut, and petals with laciniate margins. Arose in a garden in Heidelberg in 1590.

5. ESCHSCHOLZIA Cham.

Annual or perennial glabrous and glaucous herbs with watery latex and very much divided lvs whose ultimate segments are ±linear. Fls solitary, terminal. *Receptacle forming a distinct ledge round the ovary. Sepals 2, coherent into a hood* which is pushed off when the petals open; petals 4; stamens numerous; style very short; stigma 4–6-lobed, spreading; ovules many. *Capsule linear, 1-celled, opening from below by two valves which separate from the placentae.*

About 15 spp., in western North America.

*1. E. californica Cham.   Californian Poppy.

An annual herb (in the British Is.) with a deep tap-root and erect or diffuse stems 20–60 cm. Lvs ternately dissected, the ultimate segments linear. Fls 5–7.5 cm diam., yellow to orange (ivory to scarlet in garden races), deeper-coloured at the base of the petals. Capsule 7–10 cm, ribbed. Fl. 7–9. $2n = 12$. Th.

Introduced. A frequent garden-escape and occasionally becoming established. Native in S.W. USA but widely naturalized in C. and W. Europe.

27. FUMARIACEAE

Herbs with usually brittle stems, sometimes climbing; juice watery. Lvs alternate, usually much divided. Fls in racemes, or spikes, rarely solitary, usually zygomorphic, hermaphrodite, hypogynous. Sepals 2, small caducous. Petals 4, in two dissimilar whorls, one or both of the outer whorl spurred or saccate, ±connivent, the inner narrower and often cohering. Stamens 2, tripartite, the central branch bearing a 2-celled anther, the lateral branches each bearing a 1-celled anther. Ovary 1-celled with 2 parietal placentae, each bearing 1–many anatropous ovules. Fr. a capsule or nutlet. Seeds with small embryo and copious endosperm.

Sixteen genera and about 450 spp. in the north temperate zone and S.E. Africa. Closely allied to Papaveraceae, differing mainly in the corolla, stamens and watery juice. The fls when zygomorphic are transversely so, but the axis is twisted so that one of the outer petals appears to be at the top of the fl.; it is called the 'upper petal' in the following account.

Fr. a many-seeded capsule; petals without dark tips.
                            1. CORYDALIS
Fr. a 1-seeded nutlet; inner petals with dark purple tips.
                            2. FUMARIA

DICENTRA Bernh.

Erect herbs with much-divided lvs. Fls racemose, actinomorphic; both the outer petals spurred or saccate. Fr. a capsule. Several spp. are grown in gardens and some have been found as escapes. The most common is ***D. spectabilis** (L.) Lemaire (Bleeding Heart). Roots fibrous. Glabrous, branched. Lf-lobes ovate or oblong. Fls pendent; petals free, outer pink, reflexed at apex, saccate; inner white, exserted. Rarely naturalized. Native of China, Korea and Manchuria.

1. CORYDALIS Vent.
(*Capnoides* Miller)

Glabrous, ±glaucous herbs. Lvs variously divided. Fls zygomorphic, in bracteolate racemes; only the upper petal spurred. *Ovules ±numerous. Fr. a 2-valved capsule.*

About 320 spp., north temperate, except for 1 sp. in the mountains of E. Africa; a few spp. are sometimes grown in gardens.

1 Stems simple or nearly so; stock tuberous; racemes terminal; fls purple; spur long (nearly as long as rest of corolla).   **1. solida**
   Stems usually much branched; tuberous stock 0; racemes lf-opposed; fl. yellow or cream; spur short.   2
2 Lvs ending in a branched tendril; fl. 5–6 mm, cream.
                         2. claviculata
   Lvs without tendrils; fl. 12–18 mm, yellow.   **3. lutea**

*1. C. solida (L.) Swartz
C. bulbosa auct.

Perennial herb 10–20 cm. *Tuber* globose, *solid.* Stem erect, usually simple with fleshy ovate scales below the lvs. Lvs 2-ternate, segments cuneate, lobed. *Infl.* solitary, *terminal;* bracts cuneate, incised. *Corolla* 15–22 mm, *dull purple, with a long nearly straight spur.* Fl. 4–5. Pollinated by long-tongued bees, homogamous, self-sterile. $2n = 16, 24*$. Grt.

Introduced. Sometimes grown in gardens, escaped and ±naturalized in a few places. Europe from Sweden and Finland to the Pyrenees, Italy, Serbia and Thrace; N. and W. Asia.

*C. bulbosa (L.) DC.
C. cava (L.) Schweigg. & Koerte

Resembling *C. solida.* Tuber hollow. Stem without scales at base. Bracts ovate, entire. Corolla 22–30 mm, spur curved at apex. $2n = 16$. A rare escape from gardens. Native of Europe, except the north and parts of the Mediterranean region.

2. C. claviculata (L.) DC.   Climbing Corydalis.

Annual, climbing, much-branched, delicate herb 20–80 cm; roots fibrous. Lvs pinnate, *rhachis ending in a branched tendril*; lflets distant, long-stalked, digitately divided into 3–5 segments; segments 5–12 mm, elliptical, mucronate, entire. *Infl.* c. 6-fld, *lf-opposed. Fls 5–6 mm*; *petals cream, the upper with a very short obtuse spur*; pedicels c. 1 mm. Fr. c. 6 mm, with 2–3 seeds. Fl. 6–9. Pollinated by bees, perhaps more often selfed. $2n = 32$. Th.

Native. Woods and shady rocks on acid soils over most of Great Britain from Sutherland southwards, ascending to 560 m in Aberdeen, rather local especially in the east; S.E. Ireland to Waterford and Dublin; E. Donegal. W. Europe extending eastwards to E. Denmark.

*3. C. lutea (L.) DC.   Yellow Corydalis.

Perennial herb 15–30 cm, much branched; root fibrous. *Lvs pinnate, ending in a lflet*; lflets distant, long stalked, ternately or pinnately divided into 3–5 segments; segments 8–20 mm, 2–3(–rarely more)-lobed, the smaller entire; lobes mucronate. *Infl.* 6–10-fld, *lf-opposed. Fls* 12–18 mm; *petals yellow, the upper with a short obtuse spur directed downwards*; pedicels to 10 mm in fr. Fl. 5–8. Pollinated by bees or selfed, self-fertile. $2n = 28$, 56? Hp.

Introduced. Commonly cultivated and naturalized on old walls in many places scattered over the British Is. Native in S. foothills of C. and E. Alps; naturalized in W. and C. Europe.

2. FUMARIA L.

Glabrous, ±glaucous annual herbs, often climbing by the petioles. Lvs irregularly 2–4-pinnatisect, all cauline. Fls zygomorphic, in lf-opposed bracteolate racemes; only the upper petal spurred. *Ovules 1, or 1 on each placenta. Fr. a nutlet*, with two apical pits when dry.

About 55 spp. mainly in Europe and the Mediterranean region but extending to Mongolia, N.W. India, the mountains of E. Africa, and the Cape Verde Is.

Although the spp. are well-defined, their determination is sometimes difficult. Ill-grown or shaded specimens often produce ±colourless, cleistogamous fls much smaller than normal and in shaded plants the pedicels may be straight in spp. where they are normally recurved. These states are ignored in the following account. The corolla provides important characters based on the colour and shape, best observed when fresh. The upper petal has a greenish keel and lateral, usually dark, wings which may be reflexed upwards so as to hide the keel or be ±spreading; towards the base of the petal the wings leave the margin and form a lateral ridge varying in position with the species and making the tube apparently dorsally or laterally compressed. The presence of dark tips on the lateral petals appears to be constant in the British spp., the dark colour often suffusing over other parts of the corolla after pollination. The shape of the lower petal is often very important; it is always provided with a keel and ±definite 'margins' but the outline may be ±spathulate or parallel-sided for most of its length. The direction of the margins in the latter may be ±erect, when the whole petal appears boat-shaped, or they may spread out laterally to form a rim. The fr. is sometimes joined to the expanded top of the pedicel by a distinct narrow fleshy neck which must be observed when fresh. The rugosity of the fr. in those spp. where it occurs can, however, only be seen when dry; the same is true of the apical pits.

Most spp. vary considerably. For a full account of the British spp. and varieties, see Pugsley, *J. Bot.* **50**, Suppl. 1 (1912).

1   Fls 9 mm or more; lower petal not spathulate (except *F.* × *painteri*); lf-lobes oblong, lanceolate or cuneiform.   **2**
   Fls 5–8(–9) mm; lower petal distinctly spathulate; lf-lobes lanceolate, linear-oblong or linear.   **8**
2   Lower petal with broad spreading margins; fr. c. 3 mm. (Cornwall).   **1. occidentalis**
   Lower petal with narrow margins; fr. 2–2.5(–2.75) mm.   **3**
3   Pedicels rigidly recurved in fr.; peduncles equalling or longer than raceme; fls numerous (c. 20 or more); fr. with a distinct fleshy neck when fresh.   **4**
   Pedicels rarely recurved in fr. and then usually flexuous; either peduncles shorter than raceme or fls few (c. 12); fr. with an indistinct fleshy neck.   **5**
4   Fls white; infl. rather dense; upper petal laterally compressed, wings not concealing keel.   **2. capreolata**
   Fls purple; infl. rather lax; upper petal not laterally compressed, wings concealing keel.   **3. purpurea**
5   Lower petal not spathulate.   **6**
   Lower petal subspathulate; very rare (see also *muralis* var. *cornubiensis*).   **×painteri**
6   Lower petal with spreading margins; fls numerous (c. 20).   **7**
   Lower petal with erect margins; fls usually few (c. 12); fr. smooth when dry.   **6. muralis**
7   Fr. rugose when dry; fls 9–11(–12) mm; upper petal laterally compressed, often without dark tip; sepals serrate.   **4. bastardii**
   Fr. smooth even when dry; fls 11–13 mm; upper petal not laterally compressed (always dark-tipped); sepals subentire   **5. martinii**
8   Sepals at least 2 × 1 mm; fls at least 6 mm.   **9**
   Sepals not more than 1.5 × 1 mm; fls 5–6 mm.   **10**
9   Bracts longer than pedicels; fr. rounded at apex; lf-segments channelled.   **7. densiflora**
   Bracts shorter than pedicels; fr. truncate or emarginate at apex; lf-segments flat.   **8. officinalis**
10   Lf-segments flat; bracts shorter than pedicels in fr.; fls pink; racemes shortly peduncled.   **9. vaillantii**
   Lf-segments channelled; bracts equalling pedicels in fr.; fls white or tinged with pink; racemes subsessile.   **10. parviflora**

Section 1. *Grandiflora* Pugsley. Lf-segments flat, relatively broad (from oval to lanceolate or cuneiform). Fls 9 mm, or more. Wings of upper petal ±reflexed upwards. Lower petal not spathulate.

1. F. occidentalis Pugsley

Very robust, often climbing. Lf-segments with oblong-lanceolate lobes. Infl. rather lax, 12–20-fld, about equalling its peduncle. Bracts lanceolate, acuminate, usually nearly equalling pedicels (rarely much shorter). Pedicels stout, straight and suberect to arcuate and slightly decurved, apex much dilated. Sepals 4–5.5 × 2–3.5 mm, ovate, acute or shortly acuminate, dentate towards the base. *Corolla* 12–14 mm, *white at first, becoming bright pink, wings and tips blackish-red, the wings at first white-edged; upper petal dorsally compressed,* subacute, wings concealing keel; *lower petal with broad spreading margins,* which are sometimes slightly dilated above. *Fr. fully* 3 mm., suborbicular; when fresh *subacute* with an indistinct fleshy neck; *when dry* with distinct lateral keel and short apical beak, *coarsely rugose.* Fl. 5–10. Th.

Native. Arable land and waste places in several localities in Cornwall. Endemic.

2. F. capreolata L.   White Ramping-fumitory.

F. pallidiflora Jordan

Robust, climbing, often to 1 m. Lf-segments with oblong or cuneiform lobes. *Infl. rather dense, many* (c. 20)-*fld, mostly shorter than its peduncle.* Bracts linear-lanceolate, acuminate, equalling or rather shorter than the fr. pedicels. *Pedicels stout, rigidly arcuate-recurved in fr.,* apex much dilated. *Sepals* 4–6 × 2.5–3 mm, ±oval, *acute or acuminate,* toothed at base, entire above. *Corolla* 10–12(–14) mm, creamy *white,* wings and tips blackish-red, sometimes becoming pink after pollination; *upper petal* acute, *strongly laterally compressed, wings not concealing keel; lower petal with erect narrow margins. Fr.* (2–)2.5 × 2 mm, subrectangular or suborbicular, *truncate* (rarely very obtuse but not truncate), *with a distinct fleshy neck, smooth or faintly rugulose when dry.* Fl. 5–9. 2*n* = 56. Th.

Native. Cultivated and waste ground and hedgebanks, scattered over the British Is. but local and absent from many districts, especially inland, commoner in the west than in the east. C. and W. Europe extending north to the Tirol, Denmark and Scandinavia (where it is only casual). Introduced in Florida and South America.

3. F. purpurea Pugsley   Purple Ramping-fumitory.

Differs from *F. capreolata* as follows: Usually shorter and more branched. Lf-lobes slightly narrower. *Infl.* c. 15–25-fld, *rather lax, about as long as its peduncle.* Lowest bracts sometimes lf-like. *Pedicels less recurved* (rarely divaricate). *Sepals* (4.5–)5–6.5 × 2–3 mm, oblong (rarely broadly oval), *obtuse or shortly acute. Corolla* 10–13 mm, pinkish-*purple,* wings and tips dark purple; *upper petal not laterally compressed, wings broader concealing keel.* Fr. c. 2.5 mm long, as broad or rather broader, subquadrate, truncate, faintly rugulose when dry; apical pits broader and shallower. Fl. 7–10. Th.

Native. Cultivated and waste ground and hedgebanks, scattered over the British Is. (not Channel Is.), but very local, especially so in E. England and in Ireland. Endemic. Of interest as an almost certainly endemic plant, entirely confined to artificial habitats.

4. F. bastardii Boreau   Tall Ramping-fumitory.

F. confusa Jordan

Rather robust, suberect or diffuse, not or scarcely climbing. Lf-segments with oblong lobes. *Infl.* rather lax, *usually 15–25-fld, longer than its peduncle.* Bracts linear-oblong, cuspidate, shorter than fr. pedicels. *Pedicels straight and suberect or ascending in fr.,* apex somewhat dilated. *Sepals* c. 3 × 1.5 mm, oval, acute, ±*serrate* nearly all round (the teeth directed forwards). *Corolla* 9–11(–12) mm, *pink,* tips of lateral petals blackish-red, *wings* blackish-red or more *frequently pink like the rest of the corolla; upper petal laterally compressed,* obtuse or acute; *lower petal with narrow spreading margins. Fr.* c. (2–)2.5 mm, *suborbicular,* subacute or obtuse, often scarcely narrowed below with the indistinct neck as broad as the apex of the pedicel, but sometimes narrowed; *rugose* with broad shallow apical pits *when dry.* Fl. 4–10. Th.

Native. Cultivated ground and waste places. Widespread in W. England, Wales and Ireland and common in many places; scattered over Scotland; very rare in E. England and absent from the Midlands; Channel Is. Mediterranean region (rare in the east) extending north in W. Europe to W. and C. France; Azores, Madeira, Canary Is.; introduced in Australia.

5. F. martinii Clavaud

F. paradoxa Pugsley

Robust, sometimes climbing. Lf-segments small with oblong or cuneiform lobes. *Infl.* lax, 15–20-fld, *longer than its peduncle.* Bracts linear-oblong, cuspidate, much shorter than fr. pedicels. *Pedicels usually arcuate-recurved in fl., ascending or spreading,* sometimes flexuous, *in fr.,* the apex somewhat dilated. *Sepals* 3–5 × 1.5–2.5 mm, oval, acute, *subentire* or with a few small teeth at the base. *Corolla* 11–13 mm, *pink,* tips and wings blackish-red; *upper petal broad but not compressed,* usually subacute; *lower petal with narrow spreading margins. Fr.* 2.5–2.75 × 2–2.5 mm, *oval,* acute (sometimes obtuse when dry), with a very indistinct neck; *smooth,* or rarely rugulose, *when dry,* with large shallow apical pits. Fl. 5–10. Th.

Native. Cultivated ground, very rare. W. Cornwall, Guernsey; has also occurred in Devon, Somerset and Surrey, but is perhaps impermanent in these. W. France, Spain and Portugal.

6. F. muralis Sonder ex Koch

Slender to robust, suberect to diffuse or climbing. Lf-segments with oblong, lanceolate or broadly cuneiform lobes. Infl. rather lax. *Pedicels slender, usually straight and ascending or spreading,* rarely flexuous and recurved, the apex somewhat dilated. *Sepals*

3–5 × 1.5–3 mm, *ovate*, rarely oval, *usually dentate towards the base*, entire above, rarely subentire (the teeth directed outwards). *Corolla 9–12 mm, pink*, tips and wings blackish-red; *upper petals dorsally compressed, spathulately dilated, wings concealing keel; lower petals with narrower erect margins*. *Fr.* 2–2.5 mm with an indistinct fleshy neck; *smooth or finely rugulose when dry*, with small apical pits. Fl. 5–10. Th.

A very variable species divisible into the three following subspp., of which the first two are themselves variable.

Subsp. **muralis**

Slender. Infl. c. 12-fld, about equalling peduncles. Bracts linear-lanceolate, acuminate, c. $\frac{2}{3}$ as long as fr. pedicels. *Sepals* 3–4 × 1.5–2.5 mm, *ovate*, ±acuminate, always toothed. Corolla 9–11 mm; upper petal not particularly broad, apiculate. *Fr.* 2–2.5 mm long, rather less broad, ovate-orbicular, *subacute or apiculate*, quite smooth when dry, the apical pits faint. $2n = 28$.

Native. Cultivated ground, etc. Rare and local in S. and W. England and Wales. W. Europe from Norway to Spain and Portugal, N. Germany (?native); Morocco; Algeria; Macaronesia where it is common and very variable. Introduced in St Helena, Ascension, S. Africa, Mauritius, S. India, Java, New Zealand, Bermuda, South America.

Var. *cornubiensis* Pugsley. Fls c. 9 mm, lilac-pink with broad wings; lower petal subspathulate with spreading margins (thus the plant may be confused with other spp.). Mevagissey, Cornwall.

Subsp. **boraei** (Jordan) Pugsley
Common Ramping-fumitory.

F. boraei Jordan

Robust to rather slender but always more robust than subsp. muralis. Infl. c. 12–15-fld, from shorter than to nearly equalling peduncles. Bracts linear-lanceolate, acuminate, usually a little shorter than the fr. pedicels but varying from much shorter to longer. *Sepals* 3–5 × 2–3 mm, *ovate*, acute or acuminate. Corolla 10–20 mm, upper petal broad, acute, acuminate or rarely obtuse. *Fr.* 2–2.5 mm long, less than to nearly as broad, *obovate* to orbicular-obovate, occasionally subquadrate, *obtuse* at least at maturity, smooth or finely rugulose when dry, apical pits usually distinct.

Native. Cultivated and waste ground, hedgebanks and old walls. Throughout Great Britain and the commonest member of this section, common in many places in the west, less so in the east; in Ireland less common and absent from the centre and west coast. W. Europe from Norway and W. Germany to C. Spain, Portugal and Sardinia.

Subsp. **neglecta** Pugsley

Robust, suberect and ascending. *Infl.* c. 20-*fld*, longer than peduncles. Bracts linear-oblong, cuspidate, about half as long as fr. pedicels, which are never recurved. *Sepals* c. 3 × 1.5–2 mm, *broadly ovate*, *subentire* or with

a few shallow teeth. Corolla c. 10 mm; upper petal not particularly broad, obtuse. *Fr.* c. 2 × 2 mm, orbicular-obovate, *almost truncate*, smooth or finely rugulose when dry, with distinct apical pits.

Native. Only known in a cultivated field near Penryn (W. Cornwall). Endemic.

F. × **painteri** Pugsley (*F. muralis* subsp. *boraei* × *F. officinalis*?)

Robust, climbing. Lf-segments of *F. muralis* subsp. *boraei* but rather narrower. Infl. rather lax, c. 20-fld, longer than peduncles. Bracts linear-lanceolate, acuminate, nearly equalling fr. pedicels. Fr. pedicels ascending, somewhat thickened at apex. *Sepals* 3–3.5 × 1.5 mm, *ovate-lanceolate*, acuminate, irregularly toothed below, entire above. Corolla 10–11 mm, pale pink, tips and wings blackish-red; upper petal dorsally compressed, wings spathulately dilated, concealing keel; *lower petal subspathulate with narrow spreading margins*. Fr. c. 2.5 mm, subquadrate but shortly apiculate with an obscure fleshy neck, faintly rugulose when dry, with broad shallow apical pits.

Only known from two places in Shropshire. This plant needs further study. It is the only fertile hybrid known in the genus and a sterile hybrid of the same parentage is known from Guernsey and S. England. It is possible that it will prove to be a distinct sp.

Section 2. *Fumaria* Pugsley. Lf-segments relatively narrow, lanceolate to linear. Fls 5–8(–9) mm. Wings of upper petal somewhat reflexed. Lower petal ±spathulate, margins spreading. Fr. with an obscure fleshy neck when fresh, ±rugose when dry, with shallow apical pits.

7. F. densiflora DC.   Dense-flowered Fumitory.

F. micrantha Lag.

Rather robust, usually suberect, rarely diffuse or climbing. *Lf-segments with channelled*, linear or linear-oblong *lobes*. Infl. very dense at first, becoming lax in fr., c. 20–25-fld, much longer than peduncles. *Bracts* linear-oblong, cuspidate, *longer than fr. pedicels* which are ascending and dilated above. *Sepals* (2–)2.5–3.5 × (1–)2–3 mm, *orbicular or broadly ovate* (rarely ovate), subentire or toothed at base, acute or mucronate. *Corolla 6–7 mm, pink*, tips and wings blackish-red; upper petal somewhat laterally compressed, obtuse or subacute, wings ascending. *Fr.* 2–2.5 mm, *subglobose*, very little compressed but somewhat keeled, *obtuse*. Fl. 6–10. $2n = 28$. Th.

Native. Arable land on dry soils in E. England and E. Scotland; very rare in the west and absent from Devon and Cornwall, S. and C. Wales and W. Scotland (except the south-west); only in the northern half of Ireland, very local. Mediterranean region extending eastwards to Iran and north (in the west only) to Germany and the Netherlands.

8. F. officinalis L.   Common Fumitory.

Rather robust, suberect, diffuse or climbing. *Lf-segments with flat*, lanceolate or linear-oblong *lobes*. Infl.

at first dense, becoming lax, 10–40 (usually more than 20)-fld, longer than peduncles. *Bracts* linear-lanceolate, acuminate, *shorter than the ascending fr. pedicels. Sepals* 2–3.5 × 1–1.5 mm, *ovate or ovate-lanceolate*, irregularly toothed at base, acuminate or cuspidate. *Corolla* 7–8(–9) mm, *pink*, tips and wings blackish-red; upper petal dorsally compressed, obtuse (rarely apiculate), wings concealing keel. *Fr.* 2–2.5 mm, *usually considerably broader* but rarely scarcely so, little compressed and obscurely keeled, *truncate or retuse* sometimes apiculate. Variable. Fl. 5–10. Pollinated by bees or probably more frequently selfed, homogamous, self-fertile. $2n = 28$. Th.

Native. Cultivated ground on the lighter soils. Throughout the British Is., common in the east, rather less so in the west and in Ireland. Europe (except Iceland, Facroe Is. and Svalbad); Mediterranean region, except Crete, east to Iran; Canary Is.; introduced in America.

9. F. vaillantii Loisel   Few-flowered Fumitory.

Slender, dwarf, much-branched, often very glaucous. *Lf-segments* distant, *with flat*, linear-oblong or lanceolate *lobes*. Infl. rather lax, c. 6–16-fld, longer than the *short peduncle. Bracts* linear-lanceolate, *acuminate, shorter than* the suberect or ascending *fr. pedicels. Sepals not more than* 1 × 0.3–0.5 mm, lanceolate, acuminate, ±laciniate-serrate. *Corolla* 5–6 mm, pale *pink* with tips of lateral petals blackish-red, the wings less markedly dark; upper petal dorsally compressed, emarginate but apiculate. *Fr.* c. 2 mm, suborbicular, compressed but *obscurely keeled*, obtuse. Fl. 6–9. Th.

Native. Arable land, usually on chalk. From Kent to Wilts, W. Gloucester and Norfolk, very local. Europe, except the extreme north; Mediterranean region (rare in N. Africa); temperate Asia to the Altai and Kashmir.

10. F. parviflora Lam.   Fine-leaved Fumitory.

Robust, suberect, diffuse or occasionally climbing, markedly glaucous. *Lf-segments with channelled*, linear or subulate *lobes. Infl.* dense in fl., lax in fr., often c. 20-fld, *subsessile. Bracts* linear-oblong, *cuspidate, about equalling* the suberect or ascending *fr. pedicels. Sepals* 1–1.5 × 0.6–0.8 mm, broadly ovate, acute, ±laciniate-dentate. *Corolla* 5–6 mm, *white or flushed with pink* with a blotch at the base of the wings and with the tips of the lateral petals blackish-red; upper petal dorsally compressed, truncate. *Fr.* c. 2 mm, suborbicular or ovate-orbicular, little compressed but *distinctly keeled*, obtuse or subacute and apiculate or beaked. Fl. 6–9. $2n = 28$. Th.

Native. Arable land, usually on chalk. From Kent to Dorset, Wilts, and W. Norfolk; E. Cornwall; Derby; S.E. Yorks; S. Aberdeen. Mediterranean region extending north to the Caucasus, Hungary, Germany and Belgium; Baluchistan, Arabia and the Sahara; Madeira, Canaries; introduced in Mexico and South America.

The following hybrids have occurred but very rarely; with the exception of *F. × painteri* (see above) all are nearly or quite sterile: *F. bastardii × muralis* subsp. *boraei, F. micrantha × officinalis, F. muralis* subsp. *boraei × officinalis* (see also *F. × painteri*), *F. officinalis × parviflora, F. officinalis × vaillantii.*

28. CRUCIFERAE (BRASSICACEAE)

Annual to perennial herbs, rarely woody plants, with spirally arranged exstipulate lvs and racemose usually ebracteate infl. Fls usually hermaphrodite, actinomorphic and hypogynous. Sepals 4, in 2 decussate pairs, the inner pair often with saccate bases in which nectar collects; petals 4, free, clawed, commonly white or yellow, placed diagonally and so alternating with the sepals; stamens usually 6, an outer transverse pair with short filaments and 2 inner pairs with long filaments, one pair on the anterior and one on the posterior side; stamens sometimes 4, by suppression of the outer pair, or fewer; filaments sometimes with tooth-like or wing-like appendages; ovary syncarpous with 1–many ovules on each of 2 parietal placentae, usually 2-celled, a false septum being formed by the meeting of outgrowths from the placentae; style single, rudimentary to long; stigma capitate, discoid, or ±2-lobed, the lobes standing over the placentae or sometimes with long decurrent lobes alternating with the placentae. Fr. usually a specialized capsule (called a *siliqua*, or, if less than 3 times as long as wide, a *silicula*) opening from below by 2 valves which leave the seeds attached to a framework consisting of the placentae and adjacent wall tissue (*replum*) and the false septum; sometimes 1 or more seeds develop also in an indehiscent beak at the base of the style, and the relative size of the dehiscent and indehiscent parts varies greatly, some Crucifers having their seeds confined to the beak which may then break transversely at maturity into 1-seeded joints (*lomentum*); sometimes again the valves are indehiscent and the fr. may break into 1-seeded halves or remain intact, when it may be 1–many-celled and often has spines or wings which facilitate dispersal. Seeds usually in 1 or 2 rows in each cell, non-endospermic, the relative positions of the radicle and cotyledons affording characters which have been used for taxonomic purposes. When the radicle is bent round so as to lie along the edges of the cotyledons, ⚲, it is said to be *accumbent*; when it lies on the face of one cotyledon, ⚲, *incumbent*; in the *Brassiceae* and a few other genera the radicle is incumbent but the cotyledons are folded longitudinally, ○». The fls usually secrete nectar from glands, variable in size, shape and position, which lie round the bases of the stamens and ovary. Visited by insects, but spp. with small fls are commonly self-pollinated.

About 3200 spp. in 375 genera, cosmopolitan but

chiefly in north temperate regions. Many are annual or ephemeral herbs of dry open habitats, or weeds of cultivation.

The Cruciferae are closely related to the Pavaveraceae and Capparidaceae and appear to be a recently evolved specialized offshoot from the Rhoeadalian stock.

Synopsis of Classification of Native and Introduced Genera*

(After O. E. SCHULZ)
See also artificial key (p. xx)

Tribe 1. BRASSICACEAE. Hairs simple or 0; stamen filaments very rarely appendaged (*Crambe*); median nectaries present, opposite the pair of long stamens; stigma capitate to 2-lobed, rarely with long decurrent lobes (*Eruca, Moricandia*); fr. very variable but commonly a siliqua with a distinct closed beak usually containing 1 or more seeds, rarely a silicula (*Succowia*); in many genera the fr. is divided into a small often sterile basal segment and an indehiscent seed-containing beak which may break transversely at maturity; rarely the basal segment is alone fertile but indehiscent, or the fr. is non-segmented and indehiscent (*Calepina*); cotyledons longitudinally folded or bent round the incumbent radicle.

I.   Fr. opening from below by 2 valves.
A. Fr. a ±beaked siliqua more than 1 cm and at least three times as long as wide.
1. Plants glaucous and quite glabrous with fully amplexicaul entire stem-lvs.
       Stigma deeply 2-lobed; fls violet. (Fig. 12D)
                                                         MORICANDIA
       Stigma shortly 2-lobed; fls cream or pale yellow.
                                                         12. CONRINGIA
2. Plants rarely both glaucous and quite glabrous, but if so with the stem lvs not more than $\frac{1}{3}$ clasping.
a. Stigma ±capitate or shortly 2-lobed; beak often seed-bearing.
a. Seeds in 1 row in each cell.
i. Valves each with a conspicuous midrib and much weaker lateral veins, and so ±1-veined.
       Siliquae usually with convex rounded valves; seeds spherical.   1. BRASSICA
       Siliquae with strongly keeled valves; seeds ovoid or ellipsoidal.   2. ERUCASTRUM
ii. Valves each with a midrib and only slightly weaker lateral veins and so 3–7-veined at least when young.
       Blade of petal contracted abruptly into a somewhat longer filiform claw; ovary with 14–54 ovules.
                                                         3. RHYNCHOSINAPIS
       Blade of petal narrowed gradually into a short claw; ovary with 4–17 ovules.
       Sepals spreading; seeds spherical.   4. SINAPIS
       Sepals suberect; seeds ovoid or oblong.
                                                         5. HIRSCHFELDIA
b. Seeds in 2 rows in each cell.   6. DIPLOTAXIS
b. Stigma deeply 2-lobed; beak flat, seedless. 7. ERUCA
B. Fr. an almost spherical silicula opening by 2 spiny valves; style long, persistent. (Fig. 12C). SUCCOWIA
C. Fr. less than 1 cm with 2 distinct ±equal segments of which only the lower opens by 2 valves.

* Several genera with species occurring rarely as casuals are included here, although they are not described in the text. Fruits of some are figured on pp. 70–1.

Upper segment usually 2-seeded. (Fig. 12A).
                                                         ERUCARIA
Upper segment seedless, broad and flat. (Fig. 12B).
                                                         CARRICHTERA
II. Fr. remaining intact or breaking transversely into joints, but not opening by valves.
A. Fr. of 2 or more distinct segments separated by transverse constrictions, the lowest segment often slender and seedless and then resembling a short stalk.
1. Fr. elongated, usually more than 3 times as long as wide.
       Lowest segment of fr. seedless, stalk-like; upper part either ±cylindrical and indehiscent or constricted between the seeds and then often breaking into 1-seeded joints.   8. RAPHANUS
       Fr. of 2 indehiscent segments, the lower with 1–3, the upper with 3 or more seeds. (Fig. 11D).
                                                         ENARTHROCARPUS
2. Fr. of 2 indehiscent segments, the upper less than 3 times as long as wide.
       Lower segment seedless, stalk-like; upper ±spherical, smooth-walled, non-septate, 1-seeded; large cabbage-like plants.   9. CRAMBE
       Lower segment 0–2-seeded, variable in diam.; upper segment ovoid or spherical, usually ribbed and rugose, septate but 1-seeded; plants not cabbage-like.   10. RAPISTRUM
       Lower segment small, top-shaped, ±flattened, 1-seeded; upper larger, 4-angled, narrowed upwards, 1-seeded, separating from the lower at maturity; succulent maritime plants. 11. CAKILE
B. Fr. not jointed, 1-seeded, indehiscent, ovoid with a short broadly rounded conical apex. (Fig. 11E).
                                                         CALEPINA

Tribe 2. LEPIDIEAE. Hairs usually simple or 0; stamen filaments often appendaged; median nectaries 0 or, if present, sometimes fused with the lateral; stigma capitate or shortly 2-lobed; fr. usually an angustiseptate silicula (i.e. laterally compressed, with narrow septum), but sometimes not or little compressed (*Subularia, Cochlearia* spp.) and sometimes breaking into 1-seeded halves or indehiscent; radicle incumbent or accumbent.

I.   Fr. a dehiscent silicula.
A. Filaments with tooth- or wing-like appendages.
       Lvs all in a basal rosette, toothed to pinnate.
                                                         19. TEESDALIA
       Lvs not confined to a basal rosette, small, narrow, entire.   AETHIONEMA
B. Filaments not appendaged.
1. Cells of fr. each with only 1 seed.
       Fr. of 2 ±circular halves, resembling a pair of spectacles; petals equal, yellow. (Fig. 12K). BISCUTELLA
       Fr. with winged valves; petals unequal, not yellow.
                                                         17. IBERIS
       Fr. with winged or strongly keeled valves; petals equal, whitish, or 0.   13. LEPIDIUM
2. Cells of fr. each with 2 or more seeds.
a. Valves of fr. turgid, not strongly compressed.
       Small submerged aquatic plants with subulate lvs.
                                                         23. SUBULARIA
       Maritime or alpine plants; lvs not subulate.
                                                         22. COCHLEARIA
b. Valves of fr. strongly compressed.
i. Valves winged.   18. THLASPI
ii. Valves not winged.

Fr. elliptical to broadly obovate; stem lvs not sagittate. 21. HORNUNGIA
Fr. triangular-obcordate, rarely elliptical; stem lvs sagittate at the base. 20. CAPSELLA

II. Fr. breaking into 1-seeded halves or indehiscent.
A. Infl. borne opposite the lvs. 14. CORONOPUS
B. Infl. terminal or axillary.
Fr. pendulous, flattened, winged; fls yellow. 16. ISATIS
Fr. ±turgid, broadly cordate or deltoid, not winged; fls white. 15. CARDARIA

Tribe 3. EUCLIDIEAE. Hairs simple, branched, glandular or 0; stamen filaments without appendages; median nectaries present or 0; stigma capitate or shortly 2-lobed, rarely with erect or spreading lobes; fr. indehiscent, neither beaked nor segmented, 1–4-seeded, hard-walled, often appendaged; radicle incumbent or accumbent.

Fr. small, spherical. NESLIA
Fr. ovoid, densely covered with branched hairs. (Fig. 13c). EUCLIDIUM
Fr. 1-celled and 1-seeded, ovoid-quadrangular with broad wavy wings along the angles. (Fig. 13A). BOREAVA
Fr. 1–4-celled with 1–4 seeds, irregularly ovoid, warty or with 4 irregularly toothed and crested wings. 24. BUNIAS
Fr. shortly and broadly compressed-clavate, 3-celled, the 2 upper side by side, empty. (Fig. 13B). MYAGRUM

Tribe 4. LUNARIEAE. Hairs simple or 0; stamen filaments without appendages, sometimes broadened; median nectaries present or 0; stigma capitate or 2-lobed; fr. a large latiseptate silicula (i.e. with broad septum); radicle accumbent.

Fr. with flat thin-walled translucent valves and a thin shining white septum; seeds flat. 25. LUNARIA

Tribe 5. ALYSSEAE. Hairs usually branched; stamen filaments sometimes appendaged; median nectaries 0; stigma usually shortly 2-lobed; fr. a small latiseptate silicula; epidermal cells of fr. septum with parallel walls; radicle accumbent.

Hairs stellate; petals yellow; filaments of at least some of the stamens appendaged; valves faintly net-veined. 26. ALYSSUM
Hairs bifid; petals white, entire; filaments without appendages; valves with a slender midrib. 27. LOBULARIA
Hairs stellate; petals white, deeply bifid; filaments toothed or winged; valves indistinctly veined. 28. BERTEROA

Tribe 6. DRABEAE. Hairs simple, or branched, or 0; stamen filaments sometimes appendaged; median nectaries present or 0; stigma capitate or shortly 2-lobed; fr. a latiseptate silicula; epidermal cells of fr. septum not with parallel walls; radicle accumbent.

I. Valves of fr. flat with the midrib vanishing above the middle.
Petals entire or slightly notched. 29. DRABA
Petals deeply bifid. 30. EROPHILA

II. Valves of the ±spherical fr. strongly convex with an indistinct network of veins. 31. ARMORACIA

Tribe 7. ARABIDEAE. Hairs simple, or branched, or 0; sepals somewhat spreading; stamen filaments without appendages; median nectaries present or 0; stigma capitate, ±2-lobed; fr. a siliqua; radicle accumbent.

I. Siliqua flat, opening suddenly, the valves coiling spirally from the base and flinging out the seeds; lvs often pinnately or palmately compound. 32. CARDAMINE

II. Siliqua neither opening suddenly nor the valves coiling spirally from below.
A. Siliquae with strongly compressed valves.
Septum of fr. thick and rigid; valves nodular (i.e. seeds forming protuberances on the faces of the valves); lvs commonly lyrate. 34. CARDAMINOPSIS
Septum thin; valves flat, not nodular; lvs not lyrate. 35. ARABIS
B. Siliquae with convex or keeled valves.
1. Valves with distinct midrib.
Siliqua bluntly 4-angled; plant grass-green. 33. BARBAREA
Siliqua ±cylindrical; plant glaucous. 35. ARABIS
2. Valves indistinctly veined.
Petals yellow. 36. RORIPPA
Petals white. 37. NASTURTIUM

Tribe 8. MATTHIOLEAE. Simple and branched hairs usually present; sepals erect; stamen filaments usually without appendages; median nectaries present or 0; stigma usually with long decurrent lobes; fr. a siliqua, rarely breaking transversely; radicle accumbent.

Fr. a short ellipsoidal siliqua with convex stellate-hairy valves; stigma capitate. AUBRIETA
Fr. a short (8–10 mm) curved 4-angled siliqua ending in 4 spreading horn-like points. TETRACME
Fr. a long (1–12 cm) compressed siliqua; stigma of 2 erect lobes each with a dorsal swelling or horn-like process which often enlarges in fr. 38. MATTHIOLA
Fr. long (3–4 cm), cylindrical, constricted between 2-seeded joints which separate and split vertically into 1-seeded nutlets at maturity. (Fig. 13G). CHORISPORA

Tribe 9. HESPERIDEAE. Hairs simple, or branched, or 0; sepals erect; stamen filaments sometimes broadened but not appendaged; median nectaries present or 0; stigma ±2-lobed or with long decurrent lobes; fr. a siliqua or breaking transversely; radicle incumbent (or sometimes accumbent in *Erysimum* and *Cheiranthus*).

I. Fr. a siliqua.
A. Median nectaries present, usually confluent with the lateral nectaries.
Stigma capitate or shortly 2-lobed. 40. ERYSIMUM
B. Median nectaries 0.
Stigma deeply 2-lobed, the lobes erect, joined below and forming a conical pointed beak; valves of fr. 3-veined. MALCOLMIA
Stigma deeply 2-lobed, the lobes ±erect and facing each other; valves of fr. 1-veined; plant ±glandular; radicle incumbent. 39. HESPERIS
Stigma often with 2 deep spreading lobes; valves of fr. 1-veined; plant with branched appressed hairs, not glandular; radicle ±accumbent. 41. CHEIRANTHUS
II. Fr. 4-angled, indehiscent or constricted and breaking transversely into 1-seeded joints. (Fig. 13J). GOLDBACHIA

Tribe 10. SISYMBRIEAE. Hairs simple, or branched, or 0; sepals spreading; stamen filaments without appendages; median nectaries present, confluent with the lateral; stigma capitate, often shortly 2-lobed; fr. a siliqua, rarely a silicula; radicle incumbent.

I. Fr. a siliqua.
  A. Plants glabrous or with simple hairs; valves of fr. ±distinctly 3-veined.
      Valves keeled; fls white.   42. ALLIARIA
      Valves convex; fls usually yellow.   43. SISYMBRIUM
  B. Lvs with branched hairs; valves of fr. 1-veined with faint lateral veins or an indistinct lateral network.
      Lvs entire or distantly toothed; fls white.
                                              44. ARABIDOPSIS
      Lvs finely pinnatisect; fls yellow. 46. DESCURAINIA

II. Fr. a latiseptate silicula.
    Silicula obovoid or pear-shaped.   45. CAMELINA

Key to Genera

1 A woodland plant with a white scaly rhizome, pinnate basal lvs, ternate or simple stem-lvs bearing brownish-violet axillary bulbils, and purple, pink or rarely white fls; fr. a flat veinless siliqua which rarely ripens in this country.   32. CARDAMINE
  Not as above.   2
2 A robust plant of gardens, waysides and waste places, with a very stout tap-root, large oblong waved lvs up to 60 cm, white fls in a large panicle, and obovoid fr. which does not ripen in this country.
                                                31. ARMORACIA
  Not as above.   3
3 A small submerged aquatic plant with subulate lvs, minute white fls and ±ovoid scarcely compressed siliculae.   23. SUBULARIA
  Not as above.   4
4 Infls borne opposite the pinnately cut lvs (as well as terminal and at forking of the stem); small ±prostrate plants with whitish fls.   14. CORONOPUS
  Infls not lf-opposed.   5
5 Fr. 8–20 mm, pendulous, flattened, winged; tall plants with yellow fls.   16. ISATIS
  Fr. not pendulous, flattened and winged.   6
6 Fr. at least 3 times as long as wide.   7
  Fr. less than 3 times as long as wide.   33
7 Stigma deeply 2-lobed or with long decurrent lobes, not ±capitate or discoid.   8
  Stigma ±capitate or discoid, at most shallowly 2-lobed.   11
8 Stigma-lobes spreading; fls yellow or red, without deep violet veins.   41. CHEIRANTHUS
  Stigma-lobes erect, facing one another, not spreading, or spreading only at their tips; fls purplish or white, or, if pale yellow, then with deep violet veins.   9
9 Stigma-lobes thickened or horned at the back; maritime plants.   38. MATTHIOLA
  Stigma-lobes not thickened at the back.   10
10 Biennial to perennial, with simple toothed lvs; petals violet or white, not veined with deep violet; fr. not beaked.   39. HESPERIS
  Annual or overwintering, with lyrate-pinnatifid lvs; petals pale yellow or white, with deep violet veins; fr. with a broad flat beak.   7. ERUCA
11 Fr. strongly flattened.   12
  Fr. quadrangular, cylindrical or slightly flattened.   14
12 Valves of fr. veinless, rolling suddenly into spirals on dehiscence; lvs pinnate; fls white, greenish or

mauve.   32. CARDAMINE
  Valves of fr. with a ±conspicuous central vein, not rolling into spirals on dehiscence; lvs simple, sometimes pinnatifid but not pinnate; fls white or very pale yellow.   13
13 Valves of fr. with a strong central vein and prominent seeds; basal lvs distinctly long-stalked, often lyrate-pinnatifid; a rare alpine plant. 34. CARDAMINOPSIS
  Valves of fr. with a weak central vein; seeds not prominent; basal lvs not pinnatifid, narrowed gradually into a short stalk.   35. ARABIS
14 Fls white or lilac.   15
  Fls yellow, apricot or cream.   19
15 Lvs simple.   16
  At least the basal lvs pinnate or pinnatifid.   17
16 Plant with both simple and branched hairs; basal lvs small, elliptical or spathulate; plant not smelling of garlic when crushed.   44. ARABIDOPSIS
  Plant glabrous or with simple hairs; basal lvs reniform or cordate, long-stalked; plant smelling of garlic when crushed.   42. ALLIARIA
17 Fls white or lilac and dark-veined, or white at first but later turning lilac; weeds of arable or waste land, with pinnatifid lvs.   18
  Fls white; plants of wet places, with pinnate lvs.
                                                37. NASTURTIUM
18 Fr. either distinctly constricted between the seeds or inflated and with a spongy wall, not flattened, indehiscent.   8. RAPHANUS
  Fr. a siliqua with somewhat flattened valves and seeds in 2 rows in each cell, dehiscent.   6. DIPLOTAXIS
19 Plant with branched or stellate hairs.   20
  Plant glabrous or with simple hairs only.   21
20 Lvs deeply and finely pinnatisect. 46. DESCURAINIA
  Lvs simple, entire or toothed.   40. ERYSIMUM
21 Fr. evidently with 2 valves (by which it will eventually open upwards from the bottom).   22
  Fr. not valved (and not opening from below upwards), cylindrical, constricted between the seeds and ultimately breaking into 1-seeded joints. 8. RAPHANUS
22 Valves of fr. veinless or with an indistinct central vein which vanishes above.   36. RORIPPA
  Valves of fr. with a distinct central vein.   23
23 Valves of fr. 1-veined, or with lateral veins very much weaker than the central so that the valves appear 1-veined.   24
  Valves of fr. with lateral veins only slightly weaker than the central vein and so appearing 3–7-veined, at least when young.   30
24 Seeds in 2 rows in each cell of the fr.   25
  Seeds in 1 row in each cell of the fr.   26
25 Upper stem-lvs sagittate at base, clasping the stem; fls cream or very pale yellow; fr. erect, appressed to the stem.   25. ARABIS
  Upper stem-lvs not clasping the stem; fls bright yellow; fr. not appressed to the stem.   6. DIPLOTAXIS
26 Fr. with convex rounded valves.   1. BRASSICA
  Central veins of valves prominent so that the valves are keeled and the fr. ±quadrangular in section.   27
27 Stem and lvs ±glabrous.   28
  Stem hairy, at least below.   29
28 Lvs all glaucous, entire; fls cream-coloured.
                                                12. CONRINGIA
  Lvs green, the basal lyrate-pinnatifid; fls yellow.
                                                33. BARBAREA

29 Stem-lvs deeply pinnatifid; fr. curving upwards but not stiffly erect nor appressed to the stem.
2. ERUCASTRUM
Stem-lvs shallowly lobed or ±entire; fr. held stiffly erect, appressed to the stem when ripe.
1. *Brassica nigra*
30 Fr. with a distinct beak, often containing 1 or more seeds, between the top end of the valves and the stigma. 31
Fr. not beaked and never with seeds beyond the ends of the valves, which open almost to the stigma.
43. SISYMBRIUM
31 Fr. short-stalked, erect, appressed to the stem; beak short, swollen; seeds ovoid. 5. HIRSCHFELDIA
Fr. not appressed to the stem; beak flat or conical, not swollen; seeds spherical. 32
32 Sepals spreading horizontally; weeds of arable land and waste places. 4. SINAPIS
Sepals erect; maritime plants, rarely casuals in waste places. 3. RHYNCHOSINAPIS
33 Fls yellow. 34
Fls not yellow, or petals 0. 40
34 Fr. of 2 distinct segments, the lower resembling a short stalk, the upper ovoid or almost spherical; weeds of arable land or waste places.
10. RAPISTRUM
Fr. not of 2 distinct segments. 35
35 Fr. almost spherical, tiny (1.5–3 mm diam.); casual.
36. *Rorippa austriaca*
Fr. not spherical and tiny. 36
36 Fr. distinctly flattened. 37
Fr. not or slightly flattened. 38
37 Tufted plant with lvs confined to basal rosettes; fl.-stems lfless; fr. elliptical. 29. *Draba aizoides*
Fl-stems lfy; fr. almost circular emarginate.
26. ALYSSUM
38 Lvs entire or nearly so; fr. pear-shaped.
45. CAMELINA
At least the basal lvs pinnatifid or distinctly toothed. 39
39 Fr. irregularly ovoid, either with crested wings or warty prominences. 24. BUNIAS
Fr. neither winged nor warty. 36. RORIPPA
40 Large glaucous cabbage-like maritime plant with much-branched infl. of greenish-white fls and fr. having 2 segments, the lower slender, stalk-like, the upper almost spherical (and indehiscent).
9. CRAMBE
Not as above. 41
41 Annual herb of sandy or shingly sea-shores with succulent lvs, at least the lower usually pinnatifid; fr. of 2 segments, the lower small, top-shaped, the upper larger, mitre-shaped. 11. CAKILE
Not as above. 42
42 Petals bifid almost to half-way, white. 43
Petals entire, or notched but not deeply bifid, or 0. 44
43 Stem lfless, 2–20 cm. 30. EROPHILA
Stem lfy, 20–60 cm. 28. BERTEROA
44 Two adjacent petals of each fl. conspicuously larger than the other two. 45
Petals of equal size, or 0. 46
45 Lvs scattered up the stem; smaller petals twice as long as sepals; fr. 4–5 mm, distinctly winged. 17. IBERIS
Lvs almost or quite confined to a basal rosette; smaller petals scarcely longer than sepals; fr. 3–4 mm, very narrowly winged. 19. TEESDALIA

46 Fr. rounded, not or slightly flattened.
22. COCHLEARIA
Fr. distinctly flattened. 47
47 Cells of fr. 1-seeded. 48
Cells of fr. each with 2 or more seeds. 50
48 Lvs all simple, narrow, entire, not clasping the stem; fr. latiseptate, i.e. with the septum across the widest diam. 27. LOBULARIA
At least some lvs pinnatifid or toothed or clasping the stem; fr. angustiseptate, i.e. with the septum across the narrowest diam. 49
49 Infl. a corymbose panicle of small white fls; fr. broadly cordate, deltate or ovate, tapering above into the persistent style, not winged, indehiscent.
15. CARDARIA
Infl. not corymbose; fr. not or hardly cordate below, often winged, dehiscent. 13. LEPIDIUM
50 Fr. 2–9 cm, with thin-walled translucent valves and silvery septum; petals purple (rarely white); lvs cordate. 25. LUNARIA
Fr. usually less than 2 cm, its valves not translucent. 51
51 Fr. winged. 18. THLASPI
Fr. not winged. 52
52 Fr. triangular-obcordate. 20. CAPSELLA
Fr. oval or elliptical. 53
53 Lvs deeply pinnatifid to pinnate; fr. angustiseptate, i.e. with the septum across the narrowest diam.
21. HORNUNGIA
Lvs simple, entire or toothed; fr. latiseptate, i.e. with the septum across the widest diam. 29. DRABA

1. BRASSICA L.

Annual or biennial, rarely perennial, herbs with a taproot and erect or ascending, usually branched stems, glabrous or with simple hairs, often pruinose. Infl. an ebracteate raceme. Sepals erect or somewhat spreading, the inner pair ±saccate at the base; petals clawed, usually yellow; stamens 6, without appendages; stigma capitate, or slightly 2-lobed. *Fr.* ±linear *with convex valves each with* 1 *prominent vein*; beak with 0–3 seeds; seeds almost *spherical*, unwinged, in 1 *row* in each cell. Cotyledons obcordate.

About 40 spp. especially in the Mediterranean region, usually calcicolous.

An important genus including many valuable vegetables and crop plants, and taxonomically very difficult because of the multitude of closely related races.

1 Upper stem-lvs stalked or narrowed into a stalk-like base. 2
Upper lvs rounded or deeply cordate at the base, often broadened and ±clasping; sometimes narrowed at the base, but then with convex margins. 6
2 Fr. stalked above the sepal scars; plant biennial or perennial. 3
Fr. not stalked above the sepal scars; plant annual. 4
3 Lower lvs densely covered with curved bristles; beak of fr. 0.5–2 mm. **4. elongata**
Lower lvs ±glabrous; beak 3–6 mm. **5. fruticulosa**
4 Fl.-stalks usually shorter than the sepals; fr. appressed to the stem; beak up to 4 mm, usually shorter.
8. nigra
Fl.-stalks longer than the sepals; fr. not appressed; beak 6–16 mm. 5

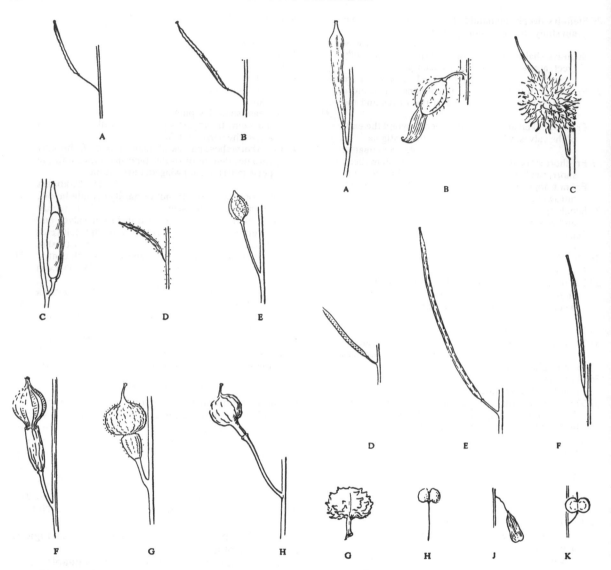

Fig. 11. Fruits of Cruciferae. A, *Erucastrum nasturtiifolium*; B, *E. gallicum*; C, *Eruca sativa*; D, *Enarthrocarpus lyratus*; E, *Calepina irregularis*; F, *Rapistrum perenne*; G, *R. rugosum*; H, *R. hispanicum*. A–D ×1; E ×2·5; F–H ×2.

Fig. 12. Fruits of Cruciferae. A, *Erucaria myagroides*; B, *Carrichtera annua*; C, *Succowia balearica*; D, *Moricandia arvensis*; E, *Conringia orientalis*; F, *C. austriaca*; G, *Coronopus squamatus*; H, *C. didymus*; J, *Isatis tinctoria*; K, *Biscutella laevigata*. A ×2·5; B, C, G, H ×2; K ×1; D–F, J ×½.

5 Lower lvs lyrate-pinnatifid, sparsely bristly; beak 4–10 mm, narrower than the stigma at its tip.
                                          6. juncea
  Lower lvs runcinate-pinnatifid, ciliate and densely bristly beneath; beak 10–16 mm, as wide as the stigma at its tip.   **7. tournefortii**
6 All lvs glabrous; middle and upper stem-lvs never more than ⅓-clasping; buds overtopping the open fls; all stamens erect.   **1. oleracea**
  Lowest lvs always ±bristly; middle and upper stem-lvs cordate, at least ½-clasping; filaments of outer stamens curved at the base.
7 All lvs glaucous; buds slightly overtopping the open fls; petals pale yellow or buff.   **2. napus**

Lowest lvs grass green; open fls overtopping the buds; petals bright yellow.   **3. rapa**

Section 1. *Brassica*. Sepals erect or half-spreading. Ovary with 9–45 ovules. Fr. 1.5–10 cm, usually with 1–2 seeds in the long beak.

1. B. oleracea L. subsp. **oleracea**   Wild Cabbage.

A biennial or perennial herb with a strong but not tuberous tap-root and a *thick* ±*decumbent* and irregularly curved *stem* which becomes *woody* and *covered with conspicuous lf scars below*. Basal lvs stalked, broad and

17

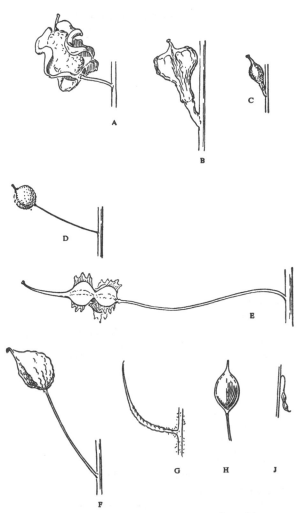

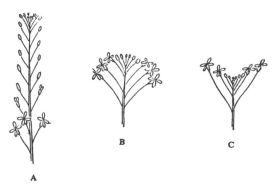

Fig. 14. Inflorescences of *Brassica*. A, *B. oleracea*; B, *B. napus*; C, *B. rapa*. $\times\frac{1}{2}$.

Fig. 13. Fruits of Cruciferae. A, *Boreava orientalis*; B, *Myagrum perfoliatum*; C, *Euclidium syriacum*; D, *Neslia paniculata*; E, *Bunias erucago*; F, *B. orientalis*; G, *Chorispora tenella*; H, *Camelina sativa*; J, *Goldbachia laevigata*. A, C, H ×2·5; B, D–F ×2; G. J $\times\frac{1}{2}$.

rounded, with sinuate margins, occasionally lyrate-pinnatifid with a few small basal lobes; upper lvs oblong, entire, sessile or up to $\frac{1}{3}$-clasping, *not broadened at the base*; all lvs glabrous and glaucous. Infl. lengthening so that *buds overtop the opened fls* (Fig. 14A). Sepals erect. Petals 12–25 mm, c. twice as long as the sepals, lemon yellow. *All stamens erect. Fr.* 5–10 cm, ±cylindrical, *with a short* tapering usually seedless *beak* 5–10 mm. Seeds 8–16 in each cell, dark grey-brown, 2–4 mm diam. Fl. 5–8. Homogamous. Visited by hive bees and other insects. $2n = 18$ (18, 36, 72 in cultivated races). Ch.

Doubtfully native on and below coastal cliffs in S. and S.W. England (recorded near Dover in 1551) and in Wales, but most southern coastal populations are close to habitations and seem recent and impermanent,

while those in inland and more northerly coastal localities are almost certainly derivative from cultivated races. The wild cabbage is regarded as native on maritime cliffs of the Atlantic and Mediterranean coasts of France and Spain and of the Mediterranean and Adriatic coasts of Italy.

Agreeing with the above description in their glaucous and glabrous lvs not broadened at the base, and in fls, frs and seeds, are the many and highly diversified cultivated races of cabbage, kale, cauliflower, broccoli, brussels sprouts, etc., which are treated taxonomically as varieties or subspecies of *B. oleracea*. Kohlrabi, with short stem forming an ovoid or almost globose edible tuber just above ground level, is sometimes treated as a distinct species (*B. caulo-rapa* (DC.) Pasquale). The branching kales closely resemble the 'wild' *B. oleracea* subsp. *oleracea*, as do populations originating from seed of cultivated plants, but virtually nothing is known of the history of cultivation. It is significant that all the cultivated races intercross more or less freely to yield fertile offspring but do not hybridize readily with *B. napus* or *B. rapa*.

*2. B. napus L.

Rape, Cole; and Swedish Turnip or Swede.

Incl. *B. napobrassica* (L.) Miller and *B. chinensis* L.

Annual to biennial herbs with a strong and often tuberous tap-root and erect ±branched stems up to 1 m with a few *not very conspicuous lf-scars* below. *Lvs all glaucous*, not very fleshy: lowest lvs stalked, *sparsely bristly*; middle and upper stem-lvs sessile, oblong-lanceolate with a broadened base which is more than half-clasping. Infl. lengthening so that *buds overtop the opened fls* (Fig. 14B). Sepals half-spreading; petals 11–14 mm, almost twice as long as the sepals, pale yellow or buff; *outer stamens curved outwards at the base*, shorter than the inner stamens. Fr. obliquely erect, 4.5–11 cm, distinctly flattened, with a beak about $\frac{1}{4}$ as long as the rest of the fr. and narrower than the stigma at its tip. Seeds blue-black with a whitish bloom, 1.5–2.5 mm diam. Fl. 5–8. Slightly protogynous. Visited by various bees. $2n = 38$ (38, 57, 76 in other cultivated races).

?Introduced. Of the various cultivated races the one hitherto most often found escaped or naturalized, commonly as an arable or wayside weed and on banks of streams or ditches, has been the biennial rape, cole or coleseed included in subsp. **oleifera** (DC.). It is grown as a fodder crop and has a non-tuberous root, lyrate-pinnatifid lowest lvs which usually drop before flowering begins, and almost entire middle and upper stem-lvs. The seeds are very rich in oil and both winter- and spring-sown races are now much cultivated as 'oil-seed rape' for the rape or colza oil expressed from them; rape-seed cake is made from the still oil-rich residue after extraction. Subsp. **rapifera** Metzger includes the Swedish turnip or swede, with a tuberized stem-base and tap-root which is yellow-fleshed and either violet, white or yellow outside, differing from the common turnip (3 below) in the small 'neck' formed by the tuberized base of the epicotyl. It is sometimes found as an escape or relic from cultivation.

There is good evidence that *B. napus* originated from crosses between *B. oleracea* and *B. rapa*.

3. B. rapa L.   Turnip, Navew.
Incl. *B. campestris* L.

Annual or biennial herbs with stout or tuberous tap-root and erect branched stems up to 1 m. *Basal lvs grass-green*, stalked, lyrate-pinnatifid, *bristly*; middle and *upper stem-lvs glaucous*, sessile, oblong-lanceolate, ±completely clasping the stem with the broadened deeply cordate base, ±glabrous. Infl. not lengthening so that *opened fls overtop the buds* (Fig. 14c). Sepals spreading; petals 6–10 mm, c. 1½ times as long as the sepals, bright yellow; *outer stamens curved outwards at the base*, much shorter than the inner stamens. Fr. obliquely erect, 4–6.5 cm, somewhat flattened, with a long tapering beak ¼–½ as long as the rest of the fr. and narrower than the stigma at its tip. Seeds blackish or reddish-brown, 1.5–2 mm diam. Fl. 5–8, slightly protogynous. Visited by various bees and hover-flies. $2n = 20$, 40. Th. – Ch.

Probably introduced. Common throughout Great Britain and Ireland.

Wild plants mostly belong to subsp. **sylvestris** (L.) Janchen, with a non-tuberous tap-root. They closely resemble the races grown on the Continent for their oil-containing seeds (colza, turnip-like rape), but have rather smaller grey or blackish, not red-brown seeds. There are two chief habitats, stream banks and arable or waste land; plants of the former being usually biennial with a well-defined basal rosette (var. *sylvestris* H. C. Watson ex Briggs), while those growing as weeds are often annual and lack the basal rosette (var. *briggsii* H. C. Watson ex Briggs). It is doubtful whether the two forms are genetically distinct. Similar plants are apparently wild throughout Europe, N. Africa, W. Asia and China, and have been introduced elsewhere. It is possible that their original home lies in N. and C. Europe.

The cultivated turnip is subsp. **rapa**, with a tuberous white-fleshed 'tap-root' lacking the epicotyledonary neck of *B. napus* subsp. *rapifera*. It is occasionally found as an escape from cultivation.

The following four casuals, *all with ±stalked non-clasping upper lvs*, also belong to section *Brassica*:

*4. B. elongata Ehrh.
Biennial to perennial, to 1 m high. *Lower lvs grass-green*, stalked, up to 20 cm, *elliptical in outline, sinuate or shallowly* pinnatifid, *densely covered with curved bristles*; *middle and upper lvs diminishing rapidly in size*, glaucous, narrowly oblong or lanceolate, narrowed into a long stalk-like base. Opened fls overtopping the buds. Sepals ±erect; petals pale yellow, 6–10 mm, twice as long as the sepals. Fr. obliquely erect, 1–2 cm, with a *stalk* 1–3 mm, *above the sepal scars* and a *tapering seedless beak* 0.5–2 mm. Seeds c. 6–8 in each cell, brown, 1–1.5 mm diam. Fl. 6–8. $2n = 22$. Ch.

Introduced. A casual in cornfields and near ports. S.E. Europe and the Near East. *B. elongata* resembles a *Diplotaxis* but has seeds in 1 row in each cell of its fr.

*5. B. fruticulosa Cyr.
Biennial to perennial, to 30 cm high, ±*woody at the base*. Lvs *stalked, lyrate-pinnatifid* with a large rounded terminal segment, *very sparsely bristly*; *upper lvs diminishing rapidly in size, the uppermost linear, ±entire*. Opened fls overtopping the buds. *Sepals erect*; petals pale yellow, later whitish, twice as long as the sepals. *Fr. almost erect, with a stalk* 1–3 mm, *above the sepal scars*, and with 0–2 seeds in the *ovate-lanceolate beak*, 3–6 mm. Seeds brown. Fl. 5–8. Ch.

Introduced. A rare casual, chiefly near ports. Mediterranean region.

*6. B. juncea (L.) Czern.
Sinapis juncea L.

Annual, to 1 m high, ±*glabrous*, with *numerous almost erect branches*. *Lower and middle lvs* stalked, up to 20 cm, usually lyrate-pinnatifid with a very *large ovate terminal segment* and only 1–2 small lateral segments on each side; upper lvs smaller, narrower, ±entire; all lvs ±glaucous. Open fls at about the same level as the buds. Sepals yellowish, half-spreading; petals pale yellow, 7–9 mm, almost twice as long as the sepals. *Fr. ascending*, 3–5 cm, *not stalked above the sepal scars, with a tapering seedless beak* 5–10 mm, *narrower than the stigma at its tip*. Seeds 8–12 in each cell, yellowish- or reddish-brown, 1.5 mm diam. Fl. 6–8. $2n = 36$. Th.

Introduced. A casual in cornfields and near ports. Asia. Cultivated, as Indian or Chinese Mustard, for the oil from its seeds.

*B. integrifolia (West) O. E. Schulz, very like *B. juncea* but basal lvs often obovate entire and fr. with a shorter (4–6 mm) broad flat beak, has been found as a casual in London.

*7. B. tournefortii Gouan
Annual. *Lower lvs runcinate-pinnatifid, densely bristly beneath, ciliate; upper lvs diminishing rapidly in size, the uppermost bract-like*. Sepals erect; *petals* pale yellow, later whitish, 5–8 mm, *only* 1.5 mm *wide*. Fr. 3–7 cm, *not stalked above the sepal scars*, with a *beak* 10–16 mm, as wide as the stigma at its tip. Fl. 6–8. $2n = $ Th.

Introduced. A rare casual, chiefly near ports. Mediterranean region and Asia.

Section 2. *Melanosinapis* (DC.) Boiss. Sepals half-spreading. Ovary usually with 5–11 ovules. Siliquae short, 0.5–3 cm, with a slender seedless beak.

8. B. nigra (L.) Koch   Black Mustard.
Sinapis nigra L.

An annual herb with a slender tap-root and an erect shoot up to 1 m high, bristly below, glabrous and glaucous above, with numerous ascending branches. *Lvs all stalked*; the lowest lyrate-pinnatifid with a large terminal lobe; grass-green, bristly, up to 16 cm; middle lvs sinuate, and uppermost lanceolate or narrowly elliptical, entire, glabrous, ±glaucous. Infl. subcorymbose. *Fl.-stalks shorter than the calyx.* Sepals half-spreading. Petals c. 8 mm, twice as long as the sepals, bright yellow. *Fr.* 12–20 mm, held *erect and appressed* to the stem on *stalks* 2–3 mm, ±quadrangular, with *strongly keeled valves* and with a slender *beak*, 1.5–3 mm, *always seedless.* Seeds 2–5 in each cell, dark red-brown. Chiefly visited by Diptera. $2n = 16$. Th.

Var. *bracteolata* (Fisch. & Mey.) Spach has the 1–6 lowest fls of each raceme subtended by linear bracts, the stem-lvs hastate or 5-lobed, the fl.-stalk equalling the calyx, fr. up to 2.5 cm, and seeds 1.5 mm diam.

Probably native. Apparently wild on cliffs by the sea, especially in S.W. England, and on stream banks throughout England and Wales, but common, probably as an escape, in waysides and waste places; only in S. Scotland and S. and E. Ireland. Widespread in C. and S. Europe and most regions with a temperate climate. Long cultivated for its seeds which yield the black mustard of commerce and also an oil used in medicine and soap-making. The original home of the species is unknown.

2. ERUCASTRUM C. Presl

Annual to perennial herbs with a tap-root and erect ascending stems, usually with numerous simple hairs. *Lower lvs lyrate-pinnatifid*, uppermost often linear. Infl. often with bracts, at least below. Sepals erect or spreading, the inner pair somewhat saccate; petals clawed, yellow; stamens 6, without appendages; stigma capitate or slightly 2-lobed. *Fr.* linear, ±*quadrangular, with keeled valves each with 1 prominent dorsal vein* and a lateral network; *beak* ±*conical*, with 0–3 seeds. *Seeds ovoid or ellipsoid*, in 1 row in each cell.

Sixteen spp. in C. and S. Europe, N. Africa and the Canary Is.

1. E. gallicum (Willd.) O. E. Schulz

Sisymbrium erucastrum sensu Poll.; *Erucastrum pollichii* Schimper & Spenner; *Diplotaxis bracteata* Godr.

An annual or biennial herb with erect stem 20–60 cm, hispid at least below with short white downwardly-directed hairs. Basal lvs lyrate with small lateral lobes; *stem-lvs deeply pinnatifid with rather distant oblong lobes, the basal pair not clasping the stem; all lvs ±hairy. Infl. usually bracteate below. Sepals almost erect; petals pale or whitish yellow*, 7–8 mm, twice as long as the sepals. *Fr.* (Fig. 11B) 2–4 cm, *curving upwards and continuing the direction of their pedicels, not stipitate above the sepal scars; beak* 3–4 mm, slender, *seedless.* Seeds red-brown, 1.3×0.7 mm. Fl. 5–9. $2n = 30$. Th. – Ch.

Introduced. A frequent casual in waste places and near ports, and occasionally establishing itself for a period. W. and C. Europe.

*E. nasturtiifolium (Poiret) O. E. Schulz differs in having the basal pinnae of the stem-lvs directed forwards and downwards, clasping the stem like auricles; infl. ebracteate or nearly so;

sepals spreading; petals 8–12 mm, bright yellow; and siliquae making an angle with their spreading stalks, shortly stipitate above the sepal-scars and with a usually 1-seeded beak. A rare casual. Native in S.W. Europe and northwards to N. France and S. Germany.

3. RHYNCHOSINAPIS Hayek

Annual to perennial herbs with long slender taproots or erect ±branching stocks, sometimes rhizomatous; stems glabrous or with simple hairs, often becoming woody below. Infl. ebracteate. *Sepals erect*, the inner pair distinctly saccate at base; petals long-clawed, bright to whitish yellow, often with darker veins; stamens 6, without appendages; stigma somewhat 2-lobed. Fr. a linear *siliqua with convex distinctly 3-veined valves and flattened sword-shaped beak with 0–6 seeds. Seeds globose, numerous, in 1 row in each cell.*

About 10 spp., chiefly in W. Europe eastwards to W. Germany and C. Italy, but with 1 sp. in N. Greece.

1 Plant ±glabrous, except for a few hairs on the sepals; stem ascending, with spreading branches from the basal rosette; few stem lvs. **1. monensis**
  Stem hairy, at least below; lvs hairy, at least beneath; stem erect, branched above; several stem-lvs. 2
2 Stem with scattered hairs; lvs hairy beneath, rarely also above; sepals equalling or exceeding the fl.-stalks; fr. 3–4 mm wide. **3. cheiranthos**
  Stem densely hairy; lvs hairy on both sides; sepals shorter than the fl.-stalks; fr. 2 mm wide.
  2. wrightii

1. R. monensis (L.) Dandy   Isle of Man Cabbage.

Sisymbrium monense L.; *Brassicella monensis* (L.) O. E. Schulz

A biennial herb with slender tap-root and ascending ±glabrous stems, 15–30 cm, branched at the base, the branches spreading-ascending. *Lvs glabrous* and glaucous, almost confined to a basal rosette, very deeply pinnatifid, the short oblong segments again lobed or coarsely toothed, the *lobes bristle-pointed; stem-lvs* 0–2, small, deeply pinnatifid. Infl. very lax. Fls about 18 mm diam. Sepals hairy at the tip, equalling or exceeding the fl.-stalks. Petals pale yellow, twice as long as the sepals. Ovary glabrous. Siliquae $4–7 \text{ cm} \times 2$ mm, spreading, on spreading stalks 6–10 mm; beak about $\frac{1}{3}$ of the overall length, with up to 5 seeds. Seeds 1.3–2 mm, dark brown. Fl. 6–8. $2n = 24^*$. Hs. (biennial).

Native and endemic to the British Is. Local on the W. coast of Great Britain from Merseyside to N. Ayrshire, on Isle of Man and on Arran; probably native in W. Glamorgan but probably or certainly introduced in Devon and Cornwall, elsewhere in S. Wales and in Anglesey.

2. R. wrightii (O. E. Schulz) Dandy   Lundy Cabbage.

Brassicella wrightii O. E. Schulz

A short-lived perennial herb with slender tap-root and erect stout *stems*, 20–90 cm, branched above, woody below, *densely covered, with simple deflexed hairs*, often purple early in the year. Lvs all stalked and hairy; basal lvs c. 12 cm, lyrate-pinnatifid with the lowest segments often runcinate, the segments irregularly toothed or lobed; *stem-lvs several*, the lower pinnatifid, the uppermost linear, entire, glaucous. Infl. lax. Fls up to 25 mm diam. *Sepals hairy, equalling or shorter than the fl.-stalks*. Petals yellow, twice as long as the sepals. *Ovary hairy*. Siliqua 6.5–8 mm, spreading or somewhat recurved on spreading stalks 10–12 mm; beak about $\frac{1}{3}$ of the overall length, with 1–3 seeds. Seeds 1.5 mm diam., purple-black. Fl. 6–8. Much visited by beetles (*Meligethes* spp.), but self-fertile. Siliquae dehisce very late in the season. $2n = 24^*$. Ch. – N.

Native and endemic to the British Is. On cliffs and slopes of the E. side of Lundy Island; known nowhere else.

*3. R. cheiranthos (Vill.) Dandy

Tall Wallflower Cabbage.

R. erucastrum Dandy, p.p.; *Brassica erucastrum* L. sec. O. E. Schulz; *B. cheiranthos* Vill.; *Brassicella erucastrum* auct.

A usually biennial herb with slender tap-root and 1 or more erect *stems* 30–90 cm, branched above, ±hispid below with scattered spreading or deflexed hairs, glabrous and somewhat glaucous above. Basal and lower stem-lvs about 10 cm, stalked, lyrate-pinnatifid, with 3–5 long, narrow, coarsely toothed or lobed segments on each side, hispid on the margins and on the veins beneath, rarely also above, grass-green or slightly glaucous; middle stem-lvs with fewer and narrower segments, uppermost lvs lanceolate, ±entire. Fls c. 18 mm diam. *Sepals equalling or exceeding the fl.-stalks*, usually hispid near the top. Petals pale yellow, almost twice as long as the sepals. *Ovary usually glabrous*. Silique 4–7 cm × 2 mm spreading or somewhat recurved on spreading stalks 6–10 mm; beak about $\frac{1}{3}$ of the overall length, 1–3-seeded. Seeds 1.3–2 mm, dark brown. Fl. 6–8. Homogamous. Visited by butterflies. $2n = 48$. Ch.

Introduced. Casual, often viatical, or locally established in a few chiefly coastal localities in S. and S.W. England, S. Wales and the Channel Is.; formerly more widely scattered northwards to East Lothian. W. Europe eastwards to Germany and C. Italy. Extremely variable.

4. Sinapis L.

Usually annual herbs with slender tap-root and stems bearing simple hairs; lvs all pinnatifid to pinnatisect or at least the uppermost ±simple. Infl. ebracteate. *Sepals spreading*, non-saccate at base; *petals bright yellow, clawed, the claw shorter than the limb*; stamens 6, without appendages; stigma somewhat 2-lobed. Fr. a *linear siliqua with convex distinctly 3–7-veined valves and a long beak* with 0–9 seeds. Seeds globose, in 1 row in each cell.

About 10 spp., chiefly in the Mediterranean region.

Uppermost lvs usually sessile, ovate to lanceolate, irregularly toothed or with a few shallow lobes; beak of fr.

conical, straight, rather more than half as long as the valves.   **1. arvensis**
All lvs petiolate, pinnately lobed or cut; beak of fr. strongly compressed, sabre-shaped, equalling or exceeding the valves.   **2. alba**

1. S. arvensis L.   Charlock, Wild Mustard.

Brassica arvensis (L.) Rabenh., non L.; *B. sinapistrum* Boiss.; *B. sinapis* Vis.; *B. kaber* (DC.) L. C. Wheeler

An annual herb with slender tap-root and erect simple or branched stem 30–80 cm, usually stiffly hairy at least at the base, but sometimes glabrous. Lvs up to 20 cm, all roughly hairy; *lower lvs stalked, lyrate*, with a large very coarsely toothed terminal lobe and usually with a few smaller lateral lobes; *upper lvs sessile*, usually simple, lanceolate, coarsely toothed. Petals 9–12 mm. Fr. 25–40 × 2.5–3 mm, spreading; *the conical straight beak rather more than $\frac{1}{2}$ as long as the valves*, often with 1 seed; valves glabrous or stiffly hairy but with glabrous beak in var. *orientalis* (L.) Koch & Ziz (*S. orientalis* L.), strongly 3(–5)-veined with 6–12 dark red-brown seeds in each cell. Fl. 5–7. Homogamous or protogynous. Freely visited by various flies and bees. $2n = 18$. Th.

Probably native. A weed of arable land, especially on calcareous and heavy soils. Throughout Great Britain and Ireland. Throughout Europe, N. Africa, S.W. Asia, Siberia. Introduced in North and South America, S. Africa, Australia and New Zealand. One of our most serious weeds, more troublesome in spring-sown than in autumn-sown crops. It may be controlled by spraying with acids, copper salts, or certain selective weed-killers.

*2. S. alba L.   White Mustard.

Brassica hirta Moench; *B. alba* (L.) Rabenh.

An annual herb with a pale slender tap-root and erect simple or branched stem 30–80 cm, glabrous or, more commonly, with stiff downwardly directed simple hairs. *Lvs* up to 15 cm, usually *stiffly hairy*, all stalked, *all lyrate-pinnatifid or pinnate with the terminal lobe larger than the laterals*. Petals 10–15 mm. Fr. 25–40 × 3–4 mm, spreading; the *strongly compressed sabre-like often curved beak equalling or exceeding the valves*, narrowing upwards from a broad base, often 1-seeded; *valves* usually *stiffly hairy, strongly 3-veined*, with 1–4 yellowish or pale brown seeds in each cell. Fl. 6–8. Homogamous. The vanilla-scented fls are visited by flies and bees. $2n = 24$. Th.

Introduced, or ? native: present in Roman times. Grown as a green fodder crop or green manure as well as for the 'mustard' derived from its ground seeds; and a weed of arable and waste land, especially on calcareous soils. Throughout Great Britain and Ireland. Probably native in the Mediterranean region and Near East, but in N.W. Europe (to 67° N in Norway) widespread. Introduced in Japan, North and South America and New Zealand.

The above description is of subsp. **alba**, with lvs lyrate-pinnatifid or -pinnate, fr. 3–4 mm wide and seeds yellow or pale brown. This is the widely cultivated variant and the common weed of arable land throughout the range of the species. Subsp. **dissecta** (Lag.) Bonnier (*S. dissecta* Lag.), with twice-pinnatifid non-lyrate lvs, fr. 3.5–6.5 mm wide and grey-brown seeds, is a weed of flax in the Mediterranean region and a rare casual in the British Is.

5. HIRSCHFELDIA Moench

Annual or overwintering herbs, very hairy below, with terminal ebracteate racemes. *Sepals almost erect, the inner pair slightly saccate; petals yellow or white; stamens 6, without appendages; ovary with 8–13 ovules; stigma capitate. Fr a siliqua with a short swollen beak containing 0–2 seeds; valves distinctly 3-veined when young, with a strong midrib and two weaker laterals, but obscurely veined when ripe. Seeds ovoid, in 1 row in each cell.*

Two spp., one in the Mediterranean region and one in Socotra.

*1. H. incana (L.) Lagrèze-Fossat   Hoary Mustard.

Sinapis incana L.; *Brassica incana* auct.; *B. adpressa* Boiss.

An annual or overwintering herb with a slender tap-root and a simple or branching *stem 30–100 cm, densely covered below with short downwardly directed stiff white hairs. Rosette- and lower stem-lvs with dense hairs,* stalked, deeply pinnately lobed or divided, with a large blunt broadly ovate terminal lobe, all lobes coarsely toothed; upper stem-lvs smaller, sessile, simple, narrowly lanceolate, glabrous or nearly so. Open fls overtopping the buds, their stalks shorter than the sepals; petals pale yellow, often with dark veins, 6–8 mm, twice as long as the sepals. *Fr. 8–15 × 1–1.5 mm, erect and appressed to the stem on short (2–3 mm) club-shaped stalks,* with a *short basally swollen 0–2-seeded beak* about half as long as the valves; valves thick-walled, obscurely 3-veined, hairy or glabrous. Seeds 3–6 in each cell, reddish-brown, about 1×0.7 mm. Fl. 6–9. Visited by bees. $2n = 14$. Th. Resembles *Brassica nigra* in habit but is greyer owing to the dense white hairs.

Introduced. Naturalized in London and parts of S. England, Dublin, and elsewhere. Native in the Mediterranean region and Near East, where it is a troublesome weed, and naturalized in France and in sandy places in Jersey and Alderney. A casual in the Netherlands, S. Germany, California, Australia, New Zealand, etc.

6. DIPLOTAXIS DC.

Annual to perennial herbs usually with pinnatifid lvs and terminal ebracteate racemes. Sepals somewhat spreading, the inner pair not or slightly saccate; petals clawed, yellow, white or lilac; stamens 6, without appendages; *stigma shortly 2-lobed. Fr. a long slender short-beaked siliqua with flattened 1-veined valves. Seeds numerous, ovoid, in two rows in each cell.*

27 spp., chiefly in C. Europe and the Mediterranean region but reaching India.

Stem usually hispid below; lvs not glaucous, almost confined to a basal rosette, all pinnatifid with triangular to oblong lobes or the outermost spathulate, ±sinuate-dentate; ripe fr. longer than its stalk, not stipitate above the sepal-scars.   **1. muralis**
Stem glabrous; lvs glaucous, not in a basal rosette, the lower deeply and narrowly pinnatifid; ripe fr. often shorter than its stalk, distinctly stipitate above the sepal-scars.   **2. tenuifolia**

*1. D. muralis (L.) DC.

Wall Rocket, Wall Mustard, Stinkweed.

Sisymbrium murale L.; *Brassica muralis* (L.) Hudson

An annual or biennial herb, occasionally perennial, with slender tap-root and erect or ascending *stems, 15–50 cm, branched from the base, usually with sparse stiff hairs below. Lvs up to 10 cm, elliptical- spathulate, toothed or with triangular lobes up to twice as long as broad, yellowish-green,* foetid when crushed, narrowed gradually into a long stalk; lobes variable in depth, the terminal longest, all entire or with a few distant teeth. Lvs at first confined to a basal rosette, but these dying and replaced by stem-lvs in perennial plants (f. *caulescens* Kit.). *Fls lemon-yellow,* c. 10 mm diam.; sepals half-spreading; *petals 5–8 mm, twice as long as the sepals. Fr.* (Fig. 15B) 30–40 × 2.4 mm, narrowed at both ends,

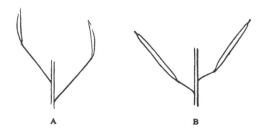

Fig. 15. Siliquae of A, *Diplotaxis tenuifolia* and B, *D. muralis*. ×$\frac{1}{2}$.

not stalked above the insertion-scars of the sepals, ascending to make an angle with the much shorter spreading stalk; beak slender, 2 mm, seedless; stigma hardly broader than style. Fr.-stalk much shorter than fr. Seeds yellow-brown, c. 1.2 × 0.6 mm. Fl. 6–9. Homogamous, fragrant. Visited by various flies and bees. $2n = 42$. Th.

Introduced. Naturalized especially in S. England on limestone rocks and walls and as a weed of arable and waste land. By railways in Ireland. S. and C. Europe.

2. D. tenuifolia (L.) DC.   Perennial Wall Rocket.

Sisymbrium tenuifolium L.; *Brassica tenuifolia* (L.) Fr.

A perennial herb with long stout tap-root and erect branching stem, 30–80 cm, entirely glabrous, glaucous. Rosette-lvs 0; *lower stem-lvs narrowed into a stalk-like base, deeply and narrowly pinnatifid with ±linear lobes never less than 3 times a s long as broad,* entire or with a few coarse teeth, the terminal lobe longer but little wider than the laterals; upper lvs less deeply divided or almost entire, linear-lanceolate; all glabrous, *glaucous* and foetid when crushed. Fls lemon-yellow; outer sepals half-spreading, inner erect; *petals 8–15 mm, twice as long as the sepals. Fr.* (Fig. 15A) 25–35 × 2 mm, *conspicuously stalked* (1–3 mm) *above the insertion-scars of the sepals; stigma much broader than style; fr.-stalk almost as long as fr*; beak 2–2.5 mm, slender, seedless. Seeds as in *D. muralis.* Fl. 5–9. Homogamous, fragrant. Visited by various insects. 2n = 22*. H. – Ch.

Doubtfully native. On old walls and in waste places in S. England. A casual further north and in Scotland and Ireland. Native in S. and C. Europe.

*D. erucoides (L.) DC., White Wall Rocket, an annual or over-wintering herb with pinnately lobed or cut lvs mostly in a basal rosette in the first season but replaced by stem-lvs in the second, *white violet-veined fls, later often wholly violet*, and *fr. much longer than its stalk* and with conical 0–1-seeded beak, was formerly a frequent and locally established weed of arable and waste land in S. England but has not been seen in recent years. Some other yellow-flowered species of *Diplotaxis* occur as rare casuals.

7. ERUCA Miller

Annual to perennial herbs usually with pinnatifid lvs and terminal ebracteate racemes. Sepals ±erect, the inner pair somewhat saccate at the base; *petals* long-clawed, usually *yellowish with violet veins*; stamens 6, without appendages; *stigma strongly 2-lobed*. Fr. a short *siliqua with strongly 1-veined valves* and a *broad flat seedless beak. Seeds* spherical or ovoid, *in 2 rows in each cell.*

Five spp. in the Mediterranean region.

*1. E. vesicaria (L.) Cav. subsp. sativa (Miller) Thell.

Brassica eruca L.; *E. sativa* Miller

An annual or overwintering *foetid* herb with slender tap-root and an erect usually stiff hairy stem, 10–60 cm, ±branched above. Lower lvs stalked, upper ±sessile; all lyrate-pinnatifid or rarely pinnate, with a large oblong or obovate terminal lobe and 2–5 narrow laterals on each side, all coarsely toothed or lobed, rarely entire. *Petals* 12–20 mm, *pale yellow or whitish, with deep violet veins*, twice as long as the sepals. *Fr.* (Fig. 11c) 12–25 × 3–5 mm, *erect and ±appressed to the stem on short erect stalks*, with a *sabre-shaped beak about half as long as the valves. Seeds* 1.5–2 mm, yellow-brown or reddish. Fl. 5–8. Homogamous. 2n = 22. Th.

Introduced. A frequent casual sometimes establishing itself for a period. Mediterranean region and E. Asia.

8. RAPHANUS L.

Annual to perennial herbs with tap-roots, erect stems usually hispid at least below, lyrate-pinnatifid lvs and terminal ebracteate racemes. Sepals erect, the lateral pair saccate at base; petals abruptly long-clawed; stamens 6, without appendages; stigma capitate, entire or somewhat 2-lobed. *Fr. a siliqua with slender seedless lower segment resembling a short stalk and an upper segment, circular in section, indehiscent but in some spp.*

constricted between seeds and commonly breaking transversely into 1-seeded joints*; beak narrow, seedless. Seeds ±globose.

Up to 10 species, chiefly in the Mediterranean region.

1 Tap-root tuberous; fr. 8–15 mm diam., not or barely constricted between seeds, not breaking into 1-seeded joints.   **3. sativus**
  Tap-root not tuberous; fr. 3–8 mm diam., markedly constricted between seeds, commonly breaking transversely at maturity into 1-seeded joints.   2
2 Lateral lf-lobes usually not contiguous; fr. with 3–8 seeds separated by shallow constrictions and breaking readily into 1-seeded joints; beak up to 4–5 times as long as top joint. A weed of cultivation
                                                    1. raphanistrum
  Lateral lf-lobes commonly contiguous; fr. with 1–5 seeds separated by deep constrictions but not breaking readily into 1-seeded joints; beak usually not more than twice as long as top joint. A plant of maritime drift-lines and cliffs.   **2. maritimus**

1. R. raphanistrum L.

Wild Radish, White Charlock, Runch.

An annual to biennial herb with a slender whitish tap-root and an erect simple or branched stem, 20–60 cm, rough with spreading or reflexed bristles, especially below, and somewhat glaucous above. *Lower lvs lyrate-pinnatifid with a large rounded terminal lobe and usually 1–4 pairs of much smaller distant laterals*; upper lvs smaller, oblong, pinnately lobed or toothed; all grass green and ±bristly. Petals 12–20 mm, twice as long as the sepals, yellow, lilac or white, usually dark-veined, but golden yellow and unveined in var. *aureus* Wilmott. *Fr. 3–6 mm diam. with long but not very deep constrictions between the 3–8 seeds, firm-walled, breaking readily into weakly ribbed 1-seeded joints; beak slender*, reaching 4–5 times the length of the uppermost joints (Fig. 16A). Seeds round-ovoid, 1.5–3 mm diam. Fl. 5–9.

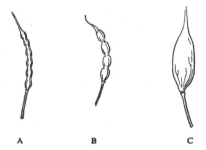

Fig. 16. Fruits of *Raphanus*. A, *R. raphanistrum*; B, *R. maritimus*; C, *R. sativus* ×$\frac{1}{2}$.

Homogamous. Visited especially by bees and flies. 2n = 18. Th.

Doubtfully native but present in prehistoric times. A common and troublesome weed, especially of non-calcareous soils. Throughout Great Britain and Ireland. Var. *aureus* in N. Scotland and Hebrides; Isle of Man.

Found throughout Europe, N. Africa, Australia, North and South America and Japan. The yellow-fld form is commoner in the northern and the white-fld in the southern part of the range.

2. R. maritimus Sm. Sea Radish.
A biennial or short-lived perennial with stout tap-root and erect simple or branched stem, 20–80 cm, bristly, especially above. *Lower lvs lyrate-pinnatifid or interruptedly pinnate with a large terminal lobe and usually 4–8 pairs of contiguous lateral lobes, often alternating in size, the basal lobes downwardly directed and very small;* upper lvs smaller; all *dark green.* Petals c. 20 mm, usually yellow (white in Channel Is.), less distinctly veined than in *R. raphanistrum. Fr. 5–8 mm diam.,* strongly ribbed with short but deep constrictions between the 1–5 *seeds,* fairly firm-walled, *not readily breaking into 1-seeded joints;* beak slender, *rarely more than twice as long as the top joint* (Fig. 16B). Fl. 6–8. The fr. are dispersed by sea water in which they will float for 7–10 days without loss of viability. $2n = 18$. Hs.
Native. A plant of the drift-line and cliffs on sandy and rocky shores from Argyll and Fife southwards, but very local on the E. coast of Great Britain; scattered round coast of Ireland but not in the N.W.; Isle of Man, Scilly (some white-fld) and Channel Is. (mainly white-fld); doubtfully in Hebrides and not in Orkney or Shetland. Native along the Atlantic coast of Europe from the Netherlands to N. Spain, and on the coasts of the Mediterranean and Black Seas.

*R. landra Moretti ex DC., which resembles *R. maritimus* but has non-contiguous and irregularly spaced lateral lobes to the basal lvs, petals not more than 15 mm, and a beak at least twice as long as the top joint, occurs as a casual near ports. It is close to *R. maritimus* and *R. sativus,* and the latter is perhaps derived from hybrids between the other two species.

*3. R. sativus L. Radish.
An annual or biennial bristly herb with a tuberous white or brightly coloured 'tap-root' and an erect simple or branched stem, 20–100 cm. *Petals white, lilac or violet,* rarely mixed purple and yellow, never pure yellow; usually with dark veins. *Fr. inflated, up to* 15 mm *diam., hardly or irregularly constricted between the 6–12 seeds and not breaking into 1-seeded joints; wall of fr.* ±spongy; beak long-conical (Fig. 16c). Seeds c. 3 mm diam. Fl. 6–8. Homogamous. Visited by bees, flies, etc. $2n = 18, 36$. Th.–Hs.
Introduced. A cultivated plant of doubtful origin and nowhere found wild but not infrequent as an escape from gardens in the British Is. and elsewhere.

9. CRAMBE L.
Annual to perennial cabbage-like herbs, often woody below, with swollen tap-roots and large entire or pinnately lobed lvs, and with white fls in large much-branched racemose infls. Sepals spreading, the inner hardly saccate; petals short-clawed; *stamens* 6, the inner

with toothed appendages; stigma capitate, sessile. *Fr. indehiscent, 2-jointed, the lower joint slender, stalk-like, seedless; the upper* ±spherical or ovoid, 1-seeded.
25 spp. in C. Europe, Mediterranean region, N. Tropical Africa, and W. Asia. Often very large herbs.

1. C. maritima L. Seakale.
A perennial herb with branched fleshy root-stock sprouting readily after burial under shingle. Stem erect, $40–60 \times 2–3$ cm, branching below, glabrous. *Lower lvs* up to 30 cm, *ovate, long-stalked,* glabrous, ±pinnately lobed, with irregular toothed and wavy margins; upper lvs narrow, the uppermost very narrow, entire and bract-like. Partial infl. corymbose. Fls 10–16 mm diam.; *petals* 6–9 mm, *white with green claws.* Fr. $12–14 \times 8$ mm, ascending to make an angle with the spreading stalk 20–25 mm. Fl. 6–8. Fr. dispersed by sea water in which it will float for many days without loss of viability. $2n = 30; 60^*$. Grh. or Hs.
Native. On coastal sands, shingle, rocks and cliffs and often on the drift-line. From Fife and Arran southwards and in Ireland; Channel Is., but in Scilly only from c. 1898; one locality in Outer Hebrides. Generally distributed along the Atlantic coast of Europe from Oslo Fjord to N. Spain, along the Baltic coast and round the Black Sea, but not round the Mediterranean. Cultivated for its asparagus-like shoots, blanched by darkening the sprouting stocks.

10. RAPISTRUM Crantz
Annual to perennial herbs with stiff bristly hairs, pinnately lobed or cut lvs, and fls in terminal ebracteate branching racemes. Sepals half-spreading, the inner pair slightly saccate; petals short-clawed; stamens 6, without appendages; stigma capitate, weakly 2-lobed. *Fr. of 2 joints;* the lower ±slender, with 0–2 seeds and therefore variable in diam. and length, separated by a constriction from the *larger upper joint which is usually 1-seeded and falls at maturity.* Seeds ovoid, that in the upper joint larger than any in the lower joint.
About three spp. in C. Europe, Mediterranean region and W. Asia, widely introduced as weeds of arable and waste land.

Upper joint of fr. strongly ribbed but not rugose, narrowing into a short broadly conical beak 0.5–1 mm; biennial or perennial. **1. perenne**
Upper joint of fr. commonly rugose, surmounted by a slender beak 1–5 mm; annual, rarely biennial.
2. rugosum

*1. R. perenne (L.) All.
A *biennial or perennial* herb with a deep stout tap-root and 1 or several branching stems, 30–80 cm, covered below with dense downwardly directed stiffish white hairs, but glabrous above. *Lower lvs* $10–15 \times 3–6$ cm, hairy, stalked, *with about 6 pairs of lateral lobes* smaller towards the base, and a terminal lobe somewhat larger than the adjacent laterals, all irregularly and coarsely toothed, the teeth horny-tipped; *upper lvs* glabrous,

short-stalked to sessile, less lobed or merely toothed. Infl. branched, much lengthening in fr. *Petals* twice as long as the sepals, *bright yellow* with darker veins. Fr.-stalk up to twice as long as the lower joint of the fr. Fr. (Fig. 11F) 7–10 mm; *upper joint* ovoid, strongly ribbed, *narrowing gradually into the short broadly conical beak*, 0.5–1 mm; lower joint narrower, usually 1-seeded. Fl. 6–8. Homogamous. G.

Introduced. Said to be established in one locality in the Breckland of East Anglia and perhaps a very rare casual elsewhere. Native of C. and S.E. Europe but locally introduced in W. Europe in dry sunny calcareous habitats. A steppe species.

*2. R. rugosum (L.) All.

An *annual* or rarely biennial herb with slender tap-root and simple or branched stem 15–60 cm, ±glaucous, stiffly hairy at least below. Basal and *lower lvs* stalked, pinnately lobed *with* a large terminal lobe and *about 3 pairs of much smaller laterals*, irregularly and coarsely toothed, *narrowed into a stalk-like base*; all dark green, stiffly hairy at least beneath. Infl. branched, much lengthening in fr. *Petals* twice as long as the sepals, *lemon-yellow* with darker veins. Fr.-stalk ±erect, 1–3 times as long as the lower joint of the fr. Fr. (Fig. 11G) 3–10 mm, hairy or becoming glabrous; lower joint cylindrical, seedless or with 1(–2) seed; *upper joint* 1-*seeded, ovoid, strongly rugose and ribbed, not more than twice as long as the slender style into which it narrows suddenly*. Fl. 5–9. Homogamous. ?Self-sterile. $2n = 16$. Th.

Introduced. A frequent casual and very locally established as a weed of arable and waste land. Variable, but there seem no clear lines of distinction between variants sometimes treated as separate species, sometimes as subspecies. The plants most commonly seen in the British Is. have the upper joint of the fr. ovoid and strongly rugose with the fr.-pedicel only 1–1½ times as long as the 1-seeded lower joint, and have been referred to subsp. *rugosum*. They are established in Kent and perhaps elsewhere and occur frequently as casuals on rubbish-tips etc. Also found as casuals are plants with the upper fr.-joint ±spherical and the fr.-pedicel 1½–3 times as long as the slender seedless lower joint (*R. orientale* (L.) Crantz, *R. rugosum* subsp. *orientale* (L.) Arcangeli). Less often seen are variants with the ±globose upper joint only slightly ribbed and rugose and with the fr.-pedicel 2–4 times as long as the seedless lower joint (*R. hispanicum* (L.) Crantz, *R. rugosum* subsp. *linnaeanum* Rouy & Fouc.). All variants are more or less restricted to the southern half of Great Britain, being absent from Ireland.

11. CAKILE Miller

Annual herbs with *glabrous succulent* simple or pinnately divided *lvs* and fls in terminal and axillary ebracteate racemes lengthening in fr. Sepals erect, the inner pairs somewhat saccate; *petals* clawed, *violet, pink or white*; stamens 6, without appendages; stigma small, capitate. *fr. indehiscent, of 2 unequal 1-seeded joints;*

the upper larger, ovoid, narrowing upwards; the lower smaller, top-shaped, ±flattened. At maturity the upper joint breaks off, the lower remaining attached to the plant.

15 spp.; on sandy shores throughout Europe, N. Africa, W. Asia, North America and Australia and one in C. Arabia.

1. C. maritima Scop.   Sea Rocket.

Bunias cakile L.

An annual herb with a very long slender tap-root and a prostrate or ascending branched stem, 15–45 cm. Lower lvs 3–6 cm, narrowed into a stalk-like base, ±entire, obovate or oblanceolate to deeply pinnate-lobed, the lobes oblong, distant, entire or distantly toothed; upper lvs less lobed or entire, sessile. Infl. dense, many-fld, terminating the main stem and branches. Petals 6–10 mm purple, lilac or white, twice as long as the sepals. *Fr.* 10–25 *mm overall* on short thick stalks 2–5 mm, upper joint up to twice as long as the lower, usually mitre-shaped with two broadly triangular basal teeth fitting over the convex top of the lower joint, which usually has two lateral projections just below the joint; lower joint sometimes seedless, and then small and stalk-like. Seeds yellow-brown, smooth. Fl. 6–8. Homogamous. Visited by various flies, bees and other insects. Fr. dispersed by floating in sea water without loss of viability.

The above description is of typical subsp. **maritima** of the W. and N.W. coasts of Europe from C. Portugal to C. Norway (other subspp. round the Baltic, Mediterranean and Black Seas). Plants from N. Scotland, Shetland and the Outer Hebrides commonly have ±entire lvs and the lower joint of the fr. lacking lateral teeth, but intermediates occur. Extreme forms have been incorrectly referred to *C. edentula* (Bigelow) Hooker, native in North America, Azores, Iceland, Faeroes, N. Norway and Arctic Russia.

Native. A plant of coastal drift-lines on sand and shingle all round the British Is.

12. CONRINGIA Adanson

Annual or overwintering herbs, *glabrous and glaucous*, with rounded lvs, entire and with transparent hoary margins. Basal lvs shortly stalked, middle and upper lvs sessile, clasping the stem with rounded basal lobes. Infl. a terminal ebracteate raceme. Sepals erect, the inner pair somewhat saccate; style short, stigma slightly 2-lobed. Fr. a long slender 4- *or* 8-*angled siliqua*; valves with strong midrib. Seeds dark brown in 1 row in each cell.

Six spp. in the Mediterranean region, C. Europe and C. Asia.

*1. C. orientalis (L.) Dumort.   Hare's-ear Cabbage.

Brassica orientalis L.; *Erysimum perfoliatum* Crantz

An annual or overwintering herb with stout tap-root and usually simple erect stem 10–50 cm. Basal lvs long-obovate blunt; *stem-lvs* 2.5–4 cm, obovate-oblong, blunt, with a *broadly clasping base*. Fls 10–12 mm diam.

Petals yellowish- or greenish-white. Siliquae (Fig. 12E) 6–12 cm × 2–2.5 mm, curving upwards on spreading stalks 6–12 mm; valves with prominent midrib so that the siliqua is *4-angled.* Seeds ovoid, dark brown, 2–2.5 mm. Fl. 5–7. Homogamous. Visited by hover-flies and Lepidoptera. $2n = 14$. Th.

Introduced. A frequent casual of arable and waste land and of cliffs by the sea, especially on calcareous and clayey soils, and occasionally established. Probably native in the E. Mediterranean region, but widely introduced in Europe and N. Africa.

13. LEPIDIUM L.

Annual to perennial herbs, sometimes woody below, often with simple hairs. Fls small, whitish, in dense terminal ebracteate racemes. Sepals non-saccate; petals sometimes shorter than the sepals or 0; stamens 2, 4 or 6, without appendages; ovary with 2 ovules, style short or 0, stigma capitate, sometimes slightly 2-lobed. Fr. an angustiseptate silicula, the valves strongly keeled to winged. Seeds usually 1 in each cell, hanging from the apex.

About 150 spp., cosmopolitan.

1 Middle and upper stem-lvs ±amplexicaul.   2
  Middle and upper stem-lvs not amplexicaul; fr. winged or not, the wing, if present, quite free from style.   4
2 Petals pale yellow; basal lvs bipinnatifid, upper stem-lvs broadly ovate, entire, with cordate base completely surrounding stem; fr.-wing very narrow. An infrequent casual.   **9. perfoliatum**
  Petals white; upper stem-lvs sagittate with narrow basal auricles which clasp but do not completely encircle stem; fr.-wing broad above and united with lower part of style.   3
3 Annual or biennial, usually with single erect stem branched only above; fr. covered with small scale-like vesicles; style not or barely exceeding apical notch; fr.-stalk no longer than fr.   **2. campestre**
  Perennial, usually with several ascending stems from the branched root-stock; fr. with few or no vesicles; style exceeding apical notch; fr.-stalk at least equalling fr.   **3. heterophyllum**
4 Annual or biennial; fr. ±distinctly winged and with apical notch; style at most equalling apical notch.   5
  Perennial; fr. not or very narrowly winged and with apical notch minute or 0.   8
5 Petals longer than sepals, at least before any fr. matures.   6
  Petals at most equalling sepals, often rudimentary or 0.   7
6 Uppermost stem-lvs often linear-oblanceolate, entire; stamens 6; fr. 5–6 mm, longer than the short stout fr.-stalk.   **1. sativum**
  Upper stem-lvs lanceolate to linear, usually toothed; stamens 2(–4); fr. 2.5–4 mm, shorter than the slender spreading-ascending fr.-stalk.   **5. virginicum**
7 Basal lvs once to twice pinnatisect; fr. 2–2.5 mm, ovate to broadly elliptical; plant foetid.   **4. ruderale**
  Basal lvs pinnatifid or deeply toothed; fr. 3–4 mm, broadly obovate to orbicular; plant almost scentless.   **6. densiflorum, s.l.**

8 Plant 40–130 cm with upper stem-lvs ovate-lanceolate to lanceolate; sepals broadly white-margined; petals up to twice as long as sepals; fr. 2 mm, suborbicular.   **7. latifolium**
  Plant 30–60 cm with upper stem-lvs linear; sepals narrowly white-margined only beyond half-way; petals up to 1½ times as long as sepals; fr. 2.5–4 mm, ovate-acute.   **8. graminifolium**

Section 1. *Cardamon* DC. Fr. ±broadly winged above; style free from wing, not exceeding apical notch; fr.-stalk stout, ascending, shorter than fr.; stamens 6; stem-lvs not amplexicaul; cotyledons deeply 3-lobed with central lobe longest.

*1. L. sativum L.   Garden Cress.

An annual pruinose herb with pale tap-root and a single erect stem 20–40 cm. Basal lvs soon falling, long-stalked, lyrate with toothed lobes; stem-lvs once or twice pinnate; *uppermost leaves sessile, not clasping stem,* linear, acute, entire. Petals white or reddish, twice as long as sepals. Stamens 6. Fr.-stalk little more than half as long as the fr., ascending; ripe siliculae 5–6 × 3–5 mm, broadly elliptical or nearly orbicular, narrowly winged above, with deep apical notch; style not projecting beyond the notch. Fl. 6–7. Protogynous. Fragrant and visited by small insects. $2n = 16$. Th.

Introduced. A common casual on tips, etc.; occasionally established. Cultivated as a salad plant all over the world. Wild forms seem native in Egypt and W. Asia.

Section 2. *Lepia* DC. Fr. broadly winged above, the wing united with the lower part of the style; fr.-stalk ±horizontal, shorter than or equalling fr.; stamens 6; middle and upper stem-lvs amplexicaul.

2. L. campestre (L.) R.Br.   Pepperwort.

Annual or biennial herb with pale slender tap-root and a single erect stem 20–60 cm, usually grey-green with dense short spreading hairs, branched above the middle, the branches curving upwards. Basal lvs entire or lyrate, falling before the fls open; lower stem-lvs narrowed into a short stalk; middle and upper stem-lvs narrowly triangular, sessile, clasping the stem with long narrow pointed basal lobes; all lvs softly hairy with small distant marginal teeth. Fls inconspicuous, 2–2.5 mm diam. Petals white, little longer than the sepals. Stamens 6, anthers yellow. Fr.-stalks hairy; *ripe siliculae 5 × 4 mm, densely covered with small white vesicles* becoming scale-like when dry; *style c. 0.5 mm, not or slightly projecting* beyond the apical notch of the fr. Seeds 2.5 mm (Fig. 17A). Fl. 5–8, slightly protogynous. $2n = 16$. Th.

Native. In dry pastures, on walls and banks, by waysides and in arable and waste land. Throughout Great Britain from Moray and Lanark southwards, rare in Scotland and Ireland. Throughout Europe and in Asia Minor and the Caucasus. Introduced in North America.

3. L. heterophyllum Bentham   Smith's Cress.

L. smithii Hooker; *L. hirtum* Sm., pro parte

A perennial herb with stout branching *woody stock* and *several ascending or decumbent stems,* 10–45 cm, usually

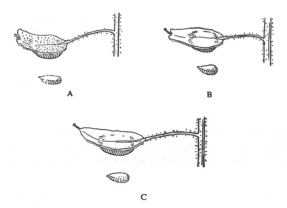

Fig. 17. Siliquae and seeds of *Lepidium*. A, *L. campestre*; B, C. *L. heterophyllum*

greyish with short spreading hairs, sometimes ±glabrous, and *often with ascending branches above.* Basal lvs oblanceolate to elliptical, entire or toothed, narrowed into a long petiole; *stem-lvs narrowly triangular, sagittate, clasping the stem with long narrow auricles,* sessile (or the lowest narrowed into a short petiole), usually rather distantly sinuate-toothed; all lvs shortly hairy to glabrous. Sepals c. 2 mm. Petals white, 1½ times as long as sepals. Stamens 6, with *violet anthers.* Fr.-stalks 3–6 mm, shortly hairy; fr. 4–7 mm, glabrous and with few or no vesicles, broadly winged above, the wing c. ⅓ of total length; apical notch broad and rather shallow, sometimes 0 (Fig. 17B,C); *style* free for c. 1 mm and *projecting beyond notch.* Seeds 2 mm.

Native. Cultivated ground, waysides and waste places on dry soils and dry banks by railways and elsewhere. Widely distributed throughout Great Britain and in much of Ireland except the north-west; Inner Hebrides; Channel Is., W. Europe eastwards to Denmark, but introduced in Belgium, Holland, Germany, Norway and Sweden, as also in North America and New Zealand.

The closely related **L. hirtum** Sm. (Mediterranean region), differing from **3** in being whitish-hairy and having yellow anthers, fr. hairy at least when young and fr.-wing up to half the total length of the fr., is a rare casual; as is **L. villarsii** Gren. & Godron, incl. *L. pratense* auct. (Alps and mountains of N.E. Spain), with long-stalked broadly ovate basal lvs, violet anthers and glabrous fr. with glabrous fr.-stalk. In the variant seen in the British Is. the fr. is distinctly notched and the wing is c. ¼ of the total length.

Section 3. *Dileptium* (Raf.) DC. Fr. either not winged, or, if narrowly winged above, with the style quite free; style not exceeding the apical notch; fr.-stalk ascending, equalling or longer than the fr.; stamens 2–4; middle and upper stem-lvs narrowed at the base, not amplexicaul.

4. L. ruderale L.   Narrow-leaved Pepperwort.

An annual or biennial ±*foetid* herb with pale slender tap-root and a single erect or ascending stem 10–30 cm, *almost glabrous or with sparse short spreading hairs.*

Stem branched above, the branches curving upwards. Basal lvs 5–7 cm, long-stalked, deeply pinnately divided, the narrow segments often again pinnately divided or lobed; lower stem-lvs pinnate with narrow entire segments; middle and upper *stem-lvs* sessile, simple, narrowly oblong up to 20 × 2 mm, blunt, *with entire margins.* Fls inconspicuous, greenish. *Petals usually* 0. Stamens 2, occasionally 4. Fr.-stalks hairy, about half as long again as the fr.; *siliculae* 2–2.5 × 1.5–2 mm, ovate or broadly elliptical, *deeply notched* above; valves sharply keeled and narrowly winged above; style very short, at the base of the notch. Fl. 5–7. Automatically self-pollinated. $2n = 32^*$. Th.-H.

Native. In waste places and on waysides, commonly near the sea. Locally frequent in England northwards to Humberside, Derby and S. Lancs, but especially in the south-east; very local in Wales and Isle of Man; very rare and doubtfully native in Scotland; Channel Is.; no longer in Ireland. Through most of Europe and in S.W. Asia; introduced in N. Africa, North America and Australia.

*5. L. virginicum L.

Annual or biennial herb with erect branching stem 30–70 cm, usually minutely pubescent. Basal lvs long-stalked, usually pinnately lobed or divided, sometimes only deeply incised; stem-lvs simple, lanceolate, incised or strongly toothed, the uppermost narrowly lanceolate to linear, often entire; all lvs ciliate, basal ±hairy. Fls in elongated spikes. *Petals* white, *up to twice as long as sepals.* Stamens 2, rarely 4. Fr. 2.5–4 mm, almost orbicular with broad but shallow apical notch; valves sharply keeled below, narrowly winged above; *style very short, at base of notch; fr.-stalk slender, spreading-ascending, longer than fr.* Fl. 5–7. Th–H.

Introduced. A rare casual. Native in North America, but introduced in Europe and widely naturalized.

*6. L. densiflorum Schrader, s.l. (incl. *L. neglectum* Thell. and *L. ramosissimum* Nelson)

Annual or biennial much-branched herbs resembling **5** but basal lvs simple, oblanceolate, deeply toothed to pinnatifid; petals filiform, shorter than sepals, or 0; fr. variable in shape from ovate-elliptical to broadly obovate or orbicular, about equalling the ascending fr.-stalk. Native in North America and a rare casual in the British Is. though widely naturalized in continental Europe. Very variable but with no clear lines of distinction between members of a closely related group of formerly recognized species. Intermediates also occur connecting it with *L. virginicum*, and there is some confusion with *L. ruderale*.

Section 4. *Lepidium*. Fr. not or very slightly winged, not or slightly emarginate; style short but usually longer than the notch; fr.-stalks slender equalling or exceeding the fr.; stamens 6; upper stem-lvs amplexicaul or not.

7. L. latifolium L.

Dittander, Broad-leaved Pepperwort.

Perennial herb with thick branched rootstock from which subterranean stolons arise; from each branch of the rootstock a single erect stem 50–130 cm, glabrous, much branched above. Basal lvs to 30 cm, long-stalked, simple and ovate with toothed margin or pinnately lobed with large terminal and 2 or more smaller lateral lobes, the lobes all rounded; lower stem-lvs like the basal but shorter stalked; middle and upper stem-lvs 5–10 × 1–2 cm, sessile, ovate or ovate-lanceolate, acute, entire or with distant teeth; uppermost lvs bract-like, white-margined near the apex. Infl. a large dense pyramidal panicle. Fls 2.5 mm diam. *Sepals broadly white-margined.* Petals white, up to twice as long as the sepals. Fr.-stalks very slender, twice as long as the fr.; *siliculae 2 × 2 mm, broadly elliptical or orbicular, with no or very slight apical notch;* valves somewhat keeled but not winged; style very short, with large rounded stigma. Fl. 6–7. 2n = 24. H.

Native. Coastal salt-marshes and wet sand and by streams and ditches near the sea. Local in Cleveland and from Norfolk and Wales southwards; no longer in Scotland or Ireland; Channel Is. Throughout Europe to c. 60°N, N. Africa, S.W. Asia. Formerly cultivated for use as a condiment ('Poor Man's Pepper').

*8. L. graminifolium L.

A perennial herb with several stiffly erect stems densely to sparsely hairy with short thick hairs, branched above. Basal lvs to 10 cm, long-stalked, often densely hairy, simple, lanceolate-spathulate, toothed or pinnately lobed; upper stem-lvs linear, entire. *Sepals narrowly white-margined* above. Petals white, half as long again as the sepals. Fr.-stalks ascending, equalling or slightly longer than the fr.; *siliculae ovate, 2.5–4 × 1.5–3 mm, with no apical notch;* valves keeled but not or scarcely winged; style short but projecting beyond the valves. Fl. 6–7. 2n = 16. H.

Introduced. A casual near docks, locally established in S. Wales. Mediterranean region and the Near East, naturalized in C. Europe.

*9. L. perfoliatum L.   Perfoliate Pepperwort.

An annual to biennial herb with single erect stem 15–40 cm, sparsely hairy, usually branched above. *Lvs strongly dimorphic:* basal to 10 cm, *long-petiolate, 2–3 times pinnatisect* with very narrow segments; lower stem-lvs smaller, less dissected and short-stalked to sessile; *middle and upper stem-lvs 1–1.5 cm, broadly ovate-acute, entire, sessile and encircling the stem with their large rounded and overlapping basal lobes.* Sepals c. 1 mm; *petals little longer, narrowly spathulate, pale yellow.* Fr. usually c. 4 × 3–4 mm, its valves keeled below and very narrowly winged above, the short style usually somewhat exceeding the apical notch; fr.-pedicel 4–5 mm, ascending, glabrous. Fl. 5–6. 2n = 16. Th.–H.

Introduced. A not infrequent casual near docks and in grass-seed mixtures, etc. Native in C., E. and S.E. Europe and in W. Asia.

Numerous other *Lepidium* spp. have been reported as casuals. Recent papers by T. B. Ryves (*Watsonia*, 1977, **11**, 367–72) and E. J. Clement (*B.S.B.I. News*, 1981, **29**, 8–10) will be found helpful.

14. CORONOPUS Haller

Annual to perennial herbs, glabrous or with simple hairs, with pinnately lobed or divided lvs. Fls inconspicuous in short ebracteate racemes, *mostly lf-opposed;* sepals ±spreading, non-saccate; petals whitish, often small or 0; fertile stamens 2–6, usually without appendages; ovary with 1 ovule in each loculus; stigma capitate or somewhat 2-lobed; style very short or 0. *Fr. an angustiseptate silicula constricted between the valves, indehiscent or splitting into 1-seeded halves;* valves subglobose, reticulate-pitted, rugose or verrucose.

About 12 species, mostly warm temperate or subtropical.

Fr. not notched above, longer than its stalk. **1. squamatus**
Fr. with an apical notch, shorter than its stalk. **2. didymus**

1. C. squamatus (Forskål) Ascherson
     Swine-cress, Wart-cress.

Cochlearia coronopus L.; *Senebiera coronopus* (L.) Poiret; *Coronopus procumbens* Gilib.; *C. ruellii* All.

An annual or biennial herb with slender tap-root and prostrate branched lfy stems 5–30 cm. Lvs stalked, deeply pinnatisect, the segments of the lower lvs obovate or oblanceolate, ±pinnatifid, with short lobes especially on the upper side, those of the upper lvs narrower, ±entire. Infl. sessile, one terminating the main stem the others opposite the lvs. Fls c. 2.5 mm diam. Petals white, longer than the sepals. *Fertile stamens usually 6. Fr. 2.5–3 × 4 mm, longer than its stalk, emarginate below but not above where it narrows abruptly into the short, pointed style;* valves rounded, reticulate-pitted or strongly and irregularly ridged, somewhat constricted at the septum, indehiscent (Fig. 12G). Seeds 2–2.5 mm. Fl. 6–9. Protogynous. Rarely visited by small flies, and automatically self-pollinated. 2n = 32*. Th.

Native. Waste ground, and especially trampled places such as gateways. Common in S. and E. England but becoming infrequent in the north though reaching Moray and Ross; Ireland; Channel Is. Throughout Europe; Mediterranean region; Canary Is. Introduced in North America, S. Africa and Australia.

*2. C. didymus (L.) Sm.   Lesser Swine-cress.

Lepidium didymum L.; *Senebiera didyma* (L.) Pers.

An annual or biennial foetid ±glabrous herb with slender tap-root and prostrate or ascending branched lfy stems 15–30 cm. Basal and lower stem-lvs stalked, very deeply pinnatisect, the 3–5 pairs of segments oblanceolate pinnatifid with slender acute lobes especially on the upper side; upper stem-lvs sessile, the segments narrower and ±entire. Main infl. terminal on the main stem, the others opposite the lvs or in the forks of branches. Fls 1–1.5 mm diam. Petals white, shorter than the sepals, or more usually 0. *Usually 2 fertile stamens,* rarely 4. *Fr. c. 1.5 × 2.5 mm, shorter than its stalk, emarginate above and below and constricted at the septum;* valves rounded, reticulate-pitted; style 0; the valves separating at maturity into 1-seeded achene-like nutlets (Fig. 12H). Seeds 1–2 mm. Fl. 7–9. automatically self-pollinated. 2n = 32*. Th.

Introduced. A weed of cultivated and waste ground.

Widespread, especially in S. and E. England, and extending northwards to Islay and Ayr. Ireland; Channel Is. Probably native only in South America but widely introduced.

15. CARDARIA Desv.

A perennial stoloniferous herb with deep tap-root and branched woody stock and several erect densely lfy stems with simple hairs. Infl. an ebracteate *corymbose panicle*. Sepals non-saccate; petals white, clawed; stamens 6, without appendages; ovary with 2(–4) ovules; style distinct; stigma capitate, broader than the style. *Fr.* broadly ovate to *cordate, indehiscent*, 2(–1)-seeded.

One species. Differs from *Lepidium* in the indehiscent fr.

*1. C. draba (L.) Desv.

Hoary Cress, Hoary Pepperwort.

Lepidium draba L.

A perennial herb with stout and deep primary tap-root, laterals from which turn downwards to become secondary tap-roots; *shoots arise from buds on any part of the root-system*. Flowering stems 30–90 cm, branching above, glabrous or with short appressed simple hairs. Basal lvs obovate, stalked, withering before fls open; *middle and upper stem-lvs ovate-oblong, somewhat narrowed towards the sessile base, then enlarging into amplexicaul auricles*; uppermost ovate-amplexicaul; all sinuate-toothed to ±entire, glabrous or sparsely hairy. Sepals 2–2.5 mm, erect. Petals white, twice as long as sepals. *Fr.* 3.5–4 mm, slightly wider than long, *broadly cordate* (emarginate below), tapering above into the persistent style; valves somewhat flattened laterally, reticulate when dry; fr. asymmetric when only 1 seed sets, *indehiscent* but often separating into 1-seeded parts. Seeds 1.5–2 mm. Fl. 5–6. Slightly protogynous. visited by small bees, beetles and other small insects but capable of self-pollination. $2n = 64*$. H–G.

Introduced. A weed of cultivated land and disturbed ground of roadsides, railway banks etc., spreading by root-buds as well as by seed. First seen in British Is. in early years of 19th century but becoming a serious field-weed only since about 1870. Now throughout Great Britain but most abundant south and east of a line from the Humber to the Severn estuary and in more scattered localities northwards to Moray Firth; local in Ireland; Orkney; Channel Is.

The above description is of subsp. **draba**, thought native in S. Europe but now very widely established elsewhere in Europe and in temperate climates all over the world. Subsp. **chalepensis** (L.) O. E. Schultz (S.W. Asia), differing in the ovate fr. with ±cuneate base, occurs only as a rare casual in the British Is.

16. ISATIS L.

Annual to perennial herbs with tall often glabrous and glaucous lfy shoots. Stem lvs sessile, sagittate, amplexicaul. Infl. a corymbose panicle. Sepals non-saccate; petals yellow, short-clawed; stamens 6, without appendages; ovary with 2 ovules, one aborting; style 0; stigma slightly 2-lobed. *Fr. indehiscent, flattened, broadly winged,* 1- (*rarely* 2-)*seeded*. Seeds large, unwinged.

About 45 spp. in C. Europe, the Mediterranean region, W. and C. Asia.

1. I. tinctoria L.   Woad.

A biennial or perennial herb with a stout tap-root and branched stock bearing several rosettes and flowering shoots; the latter lfy, branched above, 50–120 cm, softly hairy below, glabrous and glaucous above. Basal lvs lanceolate narrowed into a long stalk, ±sinuate-toothed, softly hairy; stem-lvs sessile, sagittate, amplexicaul, oblong-lanceolate, ±entire, the uppermost bract-like, with smaller basal lobes, all ±glabrous, glaucous. Infl. a much-branched corymbose panicle. Fls c. 4 mm diam. Petals yellow, up to twice as long as the sepals. *Fr.* (Fig. 12J) glabrous, purple-brown, 8–20 × 3–6 mm, *pendulous on deflexed stalks* 4–7 mm, slender below, clavate above; fr. oblong, usually broadest beyond the middle, truncate or rounded at the end, the 1-seeded cell surrounded by a broad thick wing. Seeds ellipsoid, c. 3 mm. Fl. 7–8. Homogamous. Visited by various small insects. $2n = 28$. H.

Status uncertain. Cultivated since prehistoric times for the blue pigments (woad) formed when partially dried lvs are crushed, made into a paste and exposed to air, and formerly widespread as a relic of this cultivation. Still established on limestone cliffs near Tewkesbury and in chalk-pits near Guildford, but elsewhere an extremely rare casual. Probably native in C. and S. Europe and W. Asia.

*I. aleppica Scop., with very long and narrow cuneate-based fr., has occurred as a casual.

17. IBERIS L.

Annual to perennial herbs with linear or spathulate lvs and simple hairs. Infl. ±corymbose. Sepals not saccate; petals white or pinkish, the 2 *towards the outside of the infl. much larger than the other* 2; stamens 6, without appendages; ovary with 2 ovules; stigma capitate, sometimes 2-lobed. Fr. an *angustiseptate silicula, the valves keeled and usually winged above*. Seeds large, flat, often winged, 1 in each cell.

About 30 spp., chiefly Mediterranean.

1. I. amara L.   Wild Candytuft.

An annual herb with slender tap-root and ±erect leafy stem, 10–30 cm, corymbosely branched, especially above, ±hairy below. Lvs scattered, the lower spathulate, narrowing into a stalk-like base, the upper oblanceolate-cuneate, sessile; all distantly pinnatifid or toothed, sometimes entire, ±ciliate but otherwise nearly glabrous. Fls 6–8 mm diam., white or mauve, in corymbs which elongate in fr. Outer petals 4 times and inner twice as long as the sepals. Siliculae 4–5 mm, suborbicular, with wings broadening upwards and ending in triangular lobes which leave a deep triangular apical notch; style about equalling the notch; valves convex, net-veined towards the margins. Seeds 2.5–3 mm, semiorbicular, slightly winged below, reddish-brown. Fl. 7–8. Homogamous. Visited by small bees. $2n = 14$; 16. Th.

Almost certainly native on dry calcareous hills and banks from Cambridge and N. Essex south-westwards to Oxford, Berks and N. Hants, where it is locally abundant, especially on the bare soil of rabbit-infested chalk slopes and in chalky fields. Perhaps formerly native at least in an area from S. Lincoln to Gloucester and W. Kent, but doubtfully so in the much wider area, extending into Scotland, from which it is still reported as a casual. W. and S. Europe; N. Africa. Introduced in S. Russia, Balkans, Japan, South America and New Zealand. Large-fld races are cultivated as *I. coronaria* hort. 'Rocket Candytuft' includes hybrids with *I. umbellata*.

Several spp. of *Iberis* occur as casuals or garden escapes. These include the annual or biennial *I. umbellata* L. with ±entire narrowly elliptical lvs, infl. not lengthening in fr., mauve fls, and fr. with the wings prolonged upwards into long triangular-acuminate lobes which equal or exceed the valves; and the perennial *I. sempervirens* L., Perennial Candytuft, woody below and evergreen, with entire linear to oblong-lanceolate lvs, infl. lengthening in fr., white fls and style exceeding the ±blunt apical lobes of the wing.

18. THLASPI L.

Annual to perennial, usually glabrous herbs with lfy stems and simple lvs, the stem lvs usually ±amplexicaul. Infl. ebracteate. Sepals non-saccate; petals short-clawed; *stamens* 6, *without appendages*, ovary with 2–16 ovules, stigma capitate, somewhat 2-lobed. Fr. an angustiseptate silicula, the *valves* keeled and *usually winged*, obscurely veined. *Seeds 1–5 in each cell.*

About 60 spp. in temperate Europe and Asia, North America and one in South America.

```
1 Fr. almost circular, 12–22 mm.           1. arvense
  Fr. obovate or obcordate, less than 8 mm.          2
2 Annuals; style less than 0.5 mm; included within the
  apical notch of the fr.                            3
  Biennial to perennial, with short branched woody
  stock; style usually more than 0.5 mm and equalling
  or exceeding the apical notch.        3. caerulescens
3 Stem 5–25 cm, glabrous, terete; stem-lvs glaucous,
  with blunt auricles; fr. broadly winged.
                                        2. perfoliatum
  Stem 20–60 cm, hairy below, grooved; stem-lvs with
  acute auricles; fr. narrowly winged.   4. alliaceum
```

1. T. arvense L.   Field Penny-cress.

An annual or overwintering glabrous herb with slender tap-root and erect lfy stem, 10–60 cm, simple or branched above, foetid when crushed. Basal lvs not in a compact rosette, oblanceolate or obovate, narrowed to a stalk-like base; stem-lvs oblong or lanceolate, sessile with sagittate amplexicaul base; all lvs glabrous, entire or distantly sinuate-toothed. Infl. greatly lengthening in fr. Fls 4–6 mm diam. Petals white, twice as long as the sepals. *Siliculae* 12–22 mm *diam., almost circular*, on upwardly curving stalks 5–15 mm; valves strongly compressed, with wings broadening upwards and leaving a *deep narrow apical notch which includes the very*

short style. Seeds 1.5–2 mm, brownish-black, 5–8 in each cell. Fl. 5–7. Homogamous. Visited by flies and small bees and automatically self-pollinated. $2n = 14^*$. Th.

Doubtfully native. A weed of arable land and waste places. Throughout the British Is., except the Outer Hebrides and Shetland, and sometimes a serious pest. Throughout Europe to 79°N, N. Africa, W. Asia. Siberia and Japan. Introduced in North America.

2. T. perfoliatum L.   Perfoliate Penny-cress.

An overwintering or annual herb with slender tap-root and one or more erect lfy glaucous usually simple stems, 5–25 cm. Basal lvs in a loose rosette, obovate, stalked; stem-lvs sessile, ovate-cordate, amplexicaul with contiguous rounded basal lobes; all lvs glabrous, entire or with very small distant teeth. Infl. greatly elongating in fr. Fls 2–2.5 mm diam. Petals white, narrow, c. twice as long as the sepals. *Siliculae* 4–6 × 3–5 mm, *broadly obcordate*, on slender spreading stalks 3–6 mm; wings of valves very narrow at the base, broadening above and leaving a *wide notch* between the broadly rounded apical shoulders; valves and wings conspicuously net-veined; *style very short at the base of the apical notch*. Seeds c. 1.5 mm, yellow-brown, $2n = $ c. 70. Th.

Native. Confined to limestone spoil in Oxford, Gloucester, Wilts and Worcester: casual elsewhere. Throughout Europe except for the far north, N. Africa, Near East. Introduced in North America.

3. T. caerulescens J. & C. Presl   Alpine Penny-cress.

T. alpestre auct., non L.

Biennial to perennial herbs with a short branched woody stock and erect lfy flowering stems to 40 cm. Basal lvs in a rosette, spathulate, contracted abruptly into the long stalk, ±entire; stem-lvs narrowly ovate-cordate, amplexicaul with usually subacute auricles, entire or sinuate-toothed. Petals white or lilac. Anthers often violet. *Siliculae* (Fig. 18) *obovate-obcordate*, 5–8 mm, with

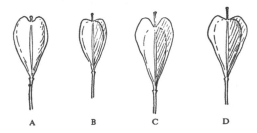

Fig. 18. Siliquae of *Thlaspi alpestre*. A, from Teesdale; B, from Matlock; C, from Mendips; D, from N. Wales. ×2.5.

valves winged from just above the base, the wings broadening upwards to apical lobes very variable in size and shape and causing variability in the depth, breadth and shape of the apical notch; *style usually equalling or exceeding the notch*. Seeds c. 1.5 mm, yellowish-brown, 4–6 in each cell. Fl. 4–8. Homogamous. Visited by various insects. $2n = 14$. Ch.

Native. A local subalpine or alpine plant of limestone and other basic rocks in various parts of Great Britain and in the Inner Hebrides. Reaches 914 m on Helvellyn. Mountains of C. and S. Europe, and S. Sweden. Related taxa in America.

A highly polymorphic sp. or aggregate of spp. the British representatives fall into 4 kinds often identified with named Continental types which, however, they do not match satisfactorily:

(a) A type ('*T. virens* Jordan, '*T. calaminare* Lej. & Court.') with a dense fruiting raceme equalling or falling short of the rest of the stem, siliculae (Fig. 18B) narrowly obovate, *shallowly notched or even truncate*, and with a *long style* (*to* 2 mm) *usually much exceeding the notch*. Fl. 4–7. Matlock and other localities in Derby, and in Inner Hebrides.

(b) A type ('*T. sylvestre* Jordan') with a much elongating fruiting raceme which exceeds the rest of the stem, and the siliculae (Fig. 18A) obcordate, the *apical lobes of the wing rounded* and the *style about equalling the notch*. Teesdale (Yorks), Northumberland, Scotland (?).

(c) A type ('*T. occitanicum* Jordan') with the fruiting raceme equalling or falling short of the rest of the stem, and the siliculae (Fig. 18D) obcordate, the *apical lobes of the wing divaricate*, rounded on the outside but ±straight on the inner side, and thus *forming a wide straight-sided shallow notch exceeded by the style*. Lake District, Yorks and N. Wales.

(d) A type with ±*densely tufted* rosettes, occasionally stoloniferous, and *lilac fls*, otherwise resembling (c). (Fig. 18c) Mendips.

*4. T. alliaceum L., an annual herb, 20–60 cm, *smelling of garlic*, has the stem hairy below, grooved, and its *stem-lvs have acute auricles*. The fls are white and the *narrowly obovate fr.* has a narrow wing and shallow apical notch. Introduced and very locally established as a weed of arable land in S. England. C. and S.E. Europe.

19. TEESDALIA R.Br.

Annual herbs, glabrous or with short simple hairs and with usually pinnatifid *lvs confined to a basal rosette* or with 1–3 stem-lvs on lateral shoots. Infl. ebracteate. Sepals half-spreading, non-saccate; petals white, the pair towards the outside sometimes larger than the inner pair; stamens 4 or 6, their *filaments with a white basal scale*; ovary with 4 ovules, style very short, stigma capitate. *Fr. an angustiseptate roundish-cordate silicula*; the valves thin-walled, *narrowly winged above*, concave on the upper and convex on the lower face; style very short or 0. Seeds 2 in each cell.

Two spp. in C. Europe and the Mediterranean region.

1. T. nudicaulis (L.) R.Br.   Shepherd's Cress.
Iberis nudicaulis L.

An annual ±glabrous or shortly pubescent herb with short tap-root and erect stem 8–45 cm, often with several ascending basal branches. *Lvs* 2–5 cm, mostly in a basal rosette, stalked, *narrowly lyrate-pinnatifid* with a few short *rounded lateral lobes* and a broader often 3-lobed terminal segment; stem-lvs 1–3 on the lateral stems (0 on the erect central stem), less lobed than the rosette lvs, the uppermost ±entire. Fls c. 2 mm diam. *Inner petals slightly longer than the sepals, outer twice as long.*

Ripe siliculae 3–4 mm, broadly elliptical-obovate, somewhat broader above the middle, narrowly winged, emarginate at the apex, style very short; fr.-stalks ±spreading, about as long as the fr. Seeds 1–1.2 mm, ovoid, pale brown. Fl. 4–6. Homogamous. $2n = 36^*$. Th.

Native. Locally common on sand and gravel. Throughout Great Britain northwards to Ross, but rare in Scotland; rare and local in N.E. Ireland; Europe to S. Norway and S. Sweden; N. Africa. Often associated with *Hypochoeris glabra*, *Ornithopus perpusillus*, *Rumex acetosella*, *Aira* spp., *Trifolium arvense*, etc.

20. CAPSELLA Medic.

Annual or biennial herbs with slender tap-root, and entire or pinnatifid basal lvs, and amplexicaul stem-lvs. Sepals non-saccate; petals usually white; stamens 6, without appendages; ovary with 12–24 ovules, style short, stigma capitate. *Fr. an angustiseptate silicula, obcordate* (rarely ovoid), the *valves keeled*, *net-veined*. Seeds unwinged, several in each cell.

Five (–10) spp, of which *C. bursa-pastoris* is a cosmopolitan weed and the others are restricted to the Mediterranean region, E. Europe and W. Asia.

1. C. bursa-pastoris (L.) Medic.   Shepherd's Purse.
Thlaspi bursa-pastoris L.

An annual to biennial herb, 3–40 cm, glabrous or with simple and branched hairs. Basal lvs in a rosette, oblanceolate in outline, narrowed into a stalk, varying from very deeply pinnatifid to quite entire; stem-lvs also very variable in shape and lobing but always clasping the stem with basal ±acute auricles. Fls c. 2.5 mm diam., white. Petals up to twice as long as the sepals. Siliculae 6–9 mm, triangular-obcordate, emarginate above, on spreading stalks 5–20 mm; style c. 0.5 mm. Seeds 0.8–1 mm, pale brown, up to 12 in each cell. Fl. 1–12. Homogamous. Stamens often reduced or abortive in cold weather. Visited by small insects and automatically self-pollinated. $2n = 32$. Th.

Native. Common everywhere on cultivated land, waysides and waste places. Throughout the British Is. Cosmopolitan.

Very variable with a strong tendency for distinctive local populations to arise because of self-pollination. Many of these have been named.

The Mediterranean *C. rubella* Reuter, with fl.-buds, and sometimes lf-lobes, bright red, reddish or red-margined, petals barely exceeding the sepals, and concave-sided siliculae with shallow apical notch, has been recorded occasionally as a casual and was recently observed in great abundance in meadows at Boxmoor, Herts.

21. HORNUNGIA Reichenb.

Annual herb with *pinnatisect lvs*. Infl. ebracteate. Sepals non-saccate; stamens 4 or 6, without appendages; ovary with 4 ovules, stigma small, sessile. *Fr. an oval unwinged angustiseptate silicula with 2 seeds in each cell.*

One species.

1. H. petraea (L.) Reichenb.   Rock Hutchinsia.

Lepidium petraeum L.; *Hutchinsia petraea* (L.) R.Br.

An overwintering annual herb with slender tap-root and simple or basally branching erect or ascending slender stems, 5–15 cm, glabrous or with small stellate hairs. Basal *lvs* in a rosette, stalked, *deeply pinnatisect with small elliptical segments*; stem-lvs numerous, similar but sessile. Fls c. 1.3 mm diam. Petals greenish-white, little longer than the sepals. Siliculae 2–4 mm, varying in shape from narrowly elliptical to oblong-obovate, not or hardly emarginate, strongly compressed, on horizontally spreading stalks 3–6 mm. Seeds ovoid, 0.6 mm, pale brown. Fl. 3–5. Homogamous. 2n = 12*. Th.

Native. A rare plant of limestone rock-ledges, scree and broken ground in Somerset, Wales, Derby and N.W. Yorks, and of calcareous sand-dunes in Wales; Jersey; not in Ireland. Europe northwards to S. Norway, S. Sweden and Estonia, Asia Minor, N. Africa.

22. COCHLEARIA L.

Annual to perennial, often maritime, herbs, glabrous or with unbranched hairs; lvs simple, often fleshy, the basal in a rosette. Sepals erect or somewhat spreading, not saccate; petals white or purplish, short-clawed; stamens 6, without appendages; ovary with 2–many ovules; stigma capitate. *Fr. a subglobose to narrowly ellipsoid silicula* usually with a fairly broad septum (but narrow in *C. anglica*); *fr.-valves ±rounded*, with distinct midrib and usually with a conspicuous network of lateral veins. Seeds verrucose or papillose, in 2 rows in each loculus.

About 24 spp. in Europe, Asia and North America, northwards to 82° 30′ N.

> *1* Basal lvs with cuneate blade narrowing into the long petiole; petals 5–7 mm; fr. compressed at right angles to the septum, which is at least 3 times as long as wide.   **1. anglica**
> Basal lvs petiolate with base of blade cordate, sinuate or truncate; petals usually not exceeding 5 mm; fr. little compressed, its septum rarely more than twice as long as wide.   2
> *2* Stem-lvs mostly petiolate, not amplexicaul, the lowest palmately 3–7-lobed; fls 3–5 mm diam.   **2. danica**
> At least the upper and commonly all but the lowest stem-lvs sessile and usually amplexicaul; fls more than 5 mm diam.   **3.–5. officinalis group**

1. C. anglica L.   Long-leaved Scurvy-grass.

A biennial or perennial maritime herb with slender tap-root and stiffish ±erect stems 8–40 cm. *Basal lvs with the base of the ovate to obovate blade tapering into a long petiole and ±cuneate*, never cordate, entire or with a few distant teeth, ±fleshy; *stem-lvs* ovate to elliptical, toothed or entire, the lowest commonly petiolate but the rest sessile and the *upper clasping the stem with basal auricles*. Fls 10–14 mm *diam*. Petals white or pale mauve, broadly oblong or ovate, narrowed abruptly into the short claw. *Fr.* 8–15 mm, ovoid-oblong but *distinctly flattened and constricted at the narrow septum which is 3–5 times as long as wide*; valves turgid, conspicuously

reticulate. Seeds reddish-brown, 2–2.5 mm, 5–6 in each cell. Fl. 4–7. 2n = 48*. Hs.

Native. Locally common on muddy shores and in estuaries all round England, Wales and Ireland; rare in Scotland and not in Orkney or Shetland; not in Channel Is. Atlantic, North Sea and Baltic coasts of N.W. Europe northwards to Finnmark and Öland.

2. C. danica L.   Early Scurvy-grass.

An overwintering annual herb with slender tap-root and ascending or spreading fl.-stems 5–20 cm, usually branched above. *Basal lvs* long-stalked, with suborbicular to rounded-triangular blades c. 1 cm wide, *cordate at base*, entire or with a few distant teeth or shallow lobes; *stem-lvs all petiolate* or only the uppermost sessile but not amplexicaul, *palmately 3–7-lobed*, the uppermost sometimes unlobed, oblong-lanceolate. Fls 4–5 mm diam. Petals white or pale mauve, with broadly oblong blade and short claw, hardly twice as long as the sepals. *Fr.* 3–6 mm, *broadly ovoid to ellipsoid*, ±narrowed at both ends; *septum broad*; valves rounded, finely net-veined when mature. Seeds reddish-brown, c. 1 mm. Fl. 1–6. The homogamous fls, scentless and with little or no nectar, are visited by a few small insects but are largely self-pollinated. 2n = 42*. Th.

Native. Locally common on sandy and rocky shores and on walls and banks near the sea all round the British Is.; also inland on railway ballast etc., chiefly in the S. Midlands of England. N. and W. Europe from Atlantic Spain and Portugal to S. Norway, and reaching c. 61° N on the Baltic coasts of Sweden and Finland.

Striking dwarf forms have been reported with basal lvs only 3 mm wide and slender stems 3–5 cm high, or 'like a moss', with stems less than 1 cm high and with vivid lilac fls.

C. officinalis group (3–5)

Biennial to perennial herbs with strong tap-root and one or more procumbent to ascending glabrous fl.-stems 5–60 cm. *Basal lvs* in a lax rosette, long-stalked, *orbicular to reniform with cordate to sinuate, less commonly truncate, base*, entire or nearly so; *stem-lvs* oblong to triangular-ovate, entire or with a few coarse teeth or small lobes, the lowest often petiolate (sometimes the middle also), the rest *sessile and usually amplexicaul*. Fls 5–10 mm diam. *Fr.* 3–7 × 2–6 mm., ovoid-globose to narrowly ellipsoid; *septum broad*.

A taxonomically difficult group comprising plants of very varied size and habit occupying two main habitats, coastal and subalpine to alpine. The former commonly have fleshy lvs and subglobose fr. which are somewhat narrowed above into the style-base, whereas the upland plants tend to have thinner lvs and fr. narrowed at both ends. There are by no means clear-cut distinctions and, moreover, the two habitat classes are heterogeneous both morphologically and cytologically. The following treatment is very provisional pending further investigation.

> *1* Plants of coastal salt-marshes, shingle, rocks, cliffs, banks, etc.; basal lvs usually fleshy; fr. usually subglobose, rounded below and somewhat narrowing above into the style-base, occasionally narrowing both above and below.   2

Plants of inland habitats, mostly at altitudes of 150–1200 m; basal lvs commonly non-fleshy; fr. usually narrowed both above and below but sometimes rounded below.   3

2 Basal lvs (0.5–)1–5 cm wide, cordate or sinuate at base; middle and upper stem-lvs sessile, most of them distinctly amplexicaul; petals 3–7 mm, usually white. Plants of varying size with decumbent to ascending stems up to 60 cm.

3. officinalis subsp. **officinalis**

Basal lvs usually 0.5–1.5 cm wide, truncate, rarely cordate, at base; middle and upper stem-lvs sessile but not truly amplexicaul; petals 2–3 mm, usually rose-lilac. Small plants forming compact tufts 5–10 cm diam. with ±prostrate fl.-stems.   **4. scotica**

3 Plants at altitudes of 170–750 m in N. England from S. Lancs and Derby northwards, with usually non-fleshy reniform basal lvs and ascending fl.-stems commonly 10–25 cm; fr. ovoid-ellipsoid to obovoid, ±narrowed above and below; diploid (2n = 12, with 0–2 accessory chromosomes).

3. officinalis subsp. **pyrenaica**

Plants of very variable size at altitudes up to 1200 m, mostly in mountains of Scotland, N. Wales and W. Ireland with thin or sometimes fleshy reniform to broadly cordate basal lvs and prostrate to ascending fl.-stems; fr. ellipsoid to broadly ovoid-ellipsoid or rarely subglobose, usually ±narrowed above and below; tetraploid (2n = 24 with 0–5 accessory chromosomes, or 26).   4

4 Fl.-stems decumbent to ascending up to 25 cm; basal lvs bright or pale green; stem-lvs 1–4-toothed on each side, only the lowest shortly petiolate; fr. broadly ellipsoid to ovoid-ellipsoid or subglobose; valves of ripe fr. net-veined.

3. officinalis subsp. **alpina**

Small plants with decumbent fl.-stems up to c. 10 cm; basal lvs dark green, shining; stem-lvs entire or with only 1 tooth on each side, the lower and middle often distinctly petiolate; fr. rather narrowly ellipsoid, up to 3 times as long as wide; valves of ripe fr. not net-veined.   **5. micacea**

3. C. officinalis L.   Common Scurvy-grass.

Basal lvs with base of blade cordate to sinuate; stem-lvs amplexicaul; fls usually white; fr. subglobose to ellipsoid, the valves distinctly net-veined at maturity. Fl. 5–8. The fragrant homogamous fls are visited chiefly by flies and beetles. 2n = 24* (with 0–5 B chromosomes) or 12*. Hs.

Widespread and abundant as a coastal plant but also occurring locally on hills and mountains inland, typical plants from the two habitats usually being morphologically distinguishable. There is nevertheless a considerable overlap in ranges of variation and some intergrading in near-coastal localities. It has been shown recently, moreover, that inland populations in N. England are diploid while morphologically similar inland populations elsewhere are tetraploid. It has therefore been thought appropriate, for the present, to treat the coastal and the two inland types as three subspecies of *C. officinalis* L. on the other hand, *C. scotica* and *C. micacea* (4. and 5. below) seem sufficiently distinct to justify separation at specific level, but there is certainly need for closer study of the whole complex.

Subsp. **officinalis**: fl.-stems up to 60 cm, decumbent to ascending; basal lvs typically fleshy; fr. usually subglobose, rounded below but somewhat narrowed above into the style-base, occasionally broadly ellipsoid and narrowed both above and below. 2n = 24 + 0–5 accessory or B chromosomes. Coastal salt-marshes, shingle, rocks, cliffs, banks, etc., all round the British Is. but local in S. England. Coasts of N.W. Europe eastwards to Poland.

Plants from Orkney recently referred to *C. islandica* Podeb. and described as having the lower and middle stem-lvs petiolate with cuneate-based blades and the fr. broadly ellipsoid to obovoid and narrowed at both ends, seem to be local variants of subsp. *officinalis* but need further study.

Subsp. **alpina** (Bab.) Hooker, emend.: fl.-stems usually not exceeding 25 cm, decumbent or ascending; basal lvs bright or pale green, thinnish but somewhat leathery or sometimes fleshy; fr. commonly broadly ellipsoid to ovoid-ellipsoid and narrowed at both ends, sometimes rounded below. 2n = 24 + 0–5 accessory chromosomes. At a range of altitudes up to 1200 m on mountains in Scotland, N. Wales and W. Ireland; on cliffs at Cheddar at only 75 m. Grades into subsp. *officinalis* especially at low levels and near the coast, and research is required into the phenotypic response of coastal tetraploids to changed environmental conditions. Probably also in Scandinavia and the Faeroes.

Subsp. **pyrenaica** (DC.) Rouy & Fouc.: distinguishable with certainty from subsp. *alpina* only by chromosome number but is more consistently characterized by non-fleshy lvs and fr. narrowed at both ends. 2n = 12 + 0–2 accessory chromosomes. Local at moderate altitudes (170–750 m) in N. England, from S. Lancashire and Derby northwards to Cumbria and Durham, on cliffs, banks, stream-sides and spoil-heaps of old lead-mines where the rooting-medium is alkaline or circumneutral. Widespread in mountains of C. Europe from the Pyrenees eastwards to the E. Carpathians and also at lower levels, often on zinc-rich substrata, in Belgium, the Netherlands and N.W. Germany.

4. C. scotica Druce   Scottish Scurvy-grass.

C. groenlandica auct. angl., non L.

Biennial to perennial herbs forming *small compact tufts* usually 5–10 cm diam. *with ±prostrate fl.-stems. Basal lvs* long-stalked, the blade varying in shape from reniform to triangular-ovate *with truncate base*, rarely cordate, *thick and fleshy*; stem-lvs sessile or the lowest short-stalked, elliptical to rhombic, entire or 1-toothed on each side, not truly amplexicaul. Fls 5–6 mm diam. *Petals 2–3 mm pale mauve*, with the *blade often almost square*, *abruptly contracted into the short claw*. Fr. c. 3 × 2 mm, broadly ovoid to broadly ellipsoid, narrowed at both ends; valves rounded, net-veined when perfectly ripe. Seeds reddish-brown, 1–2 mm. Fl. 6–8. 2n = 24* + 0–4 accessory chromosomes (an earlier reported count of 2n = 14 has not been confirmed). H.–Ch.

Native. Coasts of Great Britain northwards from Isle of Man and Berwick; Inner and Outer Hebrides, Orkney, Shetland. Apparently confined to the British Is.

Dwarf forms of *C. officinalis* subsp. *officinalis* on much exposed stretches of coast are sometimes confused with *C. scotica* but rarely show the same combination of truncate lvs, mauve fls and ellipsoid fr. Plants referred to *C. atlantica* Podeb., with sessile and amplexicaul stem-lvs, white spathulate petals, broadly ellipsoid to subglobose fr. and seeds c. 1 mm, so far known only from Outer Hebrides, Mull and Arran, may be mixtures of *C. scotica* and *C. officinalis* subsp. *officionalis* but should be further investigated.

5. C. micacea E. S. Marshall

Small biennial to perennial plants with decumbent fl.-stems up to c. 10 cm. *Basal lvs dark green, shining, thin but leathery, cordate to sinuate at base; stem-lvs* ovate-triangular to oblong or broadly elliptical, *entire or 1-toothed on each margin,* the lower and middle often distinctly petiolate, *the upper sessile but not truly amplexicaul.* Petals usually white. *Fr. rather narrowly ellipsoid,* up to 3 times as long as wide. Fl. 6–8. $2n = 26^*$. H.–Ch.

Native. Local on high mountains of C. and N.W. Scotland, probably always on calcareous substrata. Restricted to the British Is.

4. and 5. closely resemble *C. officinalis*, variants of which may in some circumstances be confused with one or other of them. It has therefore been suggested that they should be treated as subspecies of *C. officinalis* rather than as separate species, but they have special morphological as well as cytological and distributional features that have been thought to justify their specific status.

Hybrids between *C. officinalis* and *C. danica*, intermediate between the parents, and between *C. officinalis* and *C. anglica*, have been reported. The latter is probably *C. anglica* var. *horti* Syme, a large plant with basal lvs rounded, truncate or even somewhat cordate at base, and fr. intermediate in size and shape between those of *anglica* and *officinalis*.

23. SUBULARIA L.

Small annual or biennial *aquatic herbs* with very short stock and fibrous root system; *lvs* confined to a basal rosette, *subulate.* Infl. a few-fld ebracteate raceme, often submerged. Receptacle concave; sepals non-saccate; petals white, sometimes 0; stamens 6, without appendages; *ovary slightly sunk in the receptacle, its base surrounded by a fleshy ring;* 8–14 ovules; stigma sessile. *Fr. a latiseptate silicula with convex 1-veined valves.* Seeds 2–6 in each cell.

Two spp., one in Europe, N. Asia and North America, the other in the mountains of E. Africa.

1. S. aquatica L.   Awlwort.

A dwarf aquatic annual herb with lf-less fl.-stems 2–8 cm. *Lvs* 2–7 cm, numerous, *terete, subulate, entire, glabrous.* Fls 2–8, c. 2.5 mm diam, often submerged. Petals twice as long as the sepals. Siliculae 2–5 × 1.5–2.5 mm, oblong-elliptical, on ascending stalks

of about the same length; valves very convex, with a strong midrib. Seeds 2–6, c. 0.7 mm, in 2 rows in each cell. Fl. 6–8. Homogamous. Nectarless and rarely visited by insects; automatically self-pollinated and sometimes cleistogamous. Th.

Native. A local plant of base-poor lakes and pools in Wales, N. England and Scotland; Inner and Outer Hebrides, Shetland; N. and W. Ireland. Reaches 608 m in W. Ross. Often with *Isoetes lacustris, Littorella uniflora, Lobelia dortmanna,* etc. Mountains of W. Europe and in N. Europe from Ireland and Scandinavia to N. Russia; Siberia, Greenland, North America.

24. BUNIAS L.

Annual to perennial herbs with branched and simple hairs and some stout glandular hairs. Lvs ±pinnatifid, hairy. Infl. much elongating in fr. Sepals half-spreading, the inner pair not or slightly saccate; petals white or yellow; stamens 6, without appendages; ovary with 2 or 4 ovules; stigma capitate, slightly 2-lobed. *Fr. indehiscent,* irregularly ovoid, *warty or with wing-like crests,* 1–4-celled with 1–4 seeds.

Six spp. in the Mediterranean region and W. Asia, widely introduced.

Fr. with 4 irregularly crested wings and a long slender persistent style.   **1. erucago**
Fr. irregularly ovoid, warty, with a short, broad, asymmetrically-placed style.   **2. orientalis**

*1. B. erucago L.

An annual to biennial herb, 30–60 cm, roughly glandular-hairy, with the basal lvs runcinate-pinnatifid and the upper lvs oblong, entire or toothed, sessile, not amplexicaul. Fls yellow. *Fr.* (Fig. 13E) c. 11 mm, *quadrangular with irregularly toothed and crested wings on the angles,* and 4-celled with 1 seed in each cell; persistent style c. 5 mm; stalks ±horizontal. Fl. 5–8. Homogamous. Visited by various bees and automatically self-pollinated. $2n = 14$. Th.–H.

Introduced. A casual, scarcely established. Native in S. Europe but widespread as an adventive. The roots and shoots are eaten in Greece.

*2. B. orientalis L.

A biennial to perennial herb, 25–100 cm, sparsely glandular, with ±pinnatifid lvs, the lower stalked, upper ±sessile. Fls yellow. *Fr.* (Fig. 13F) 6–10 mm, *asymmetrically ovoid, unwinged, covered with warty prominences,* 1–2-celled with 1–2 seeds; style short, ±laterally placed. Fl. 5–8. Homogamous. Visited by various flies and small bees and automatically self-pollinated. $2n = 14, 42$. H.–G.

Introduced. Established in a number of places. Native in E. Europe and W. Asia, adventive in C. and W. Europe. Used as a salad and fodder plant.

25. LUNARIA L.

Annual to perennial herbs with simple hairs and large cordate lvs. Inner sepals deeply saccate; petals usually purple; stamens 6, the 4 inner with broad filaments; ovary with 4–6 ovules, style short, stigma ±deeply 2-lobed. *Fr. a large latiseptate silicula, with quite flat thin-walled translucent net-veined valves and a thin shining white septum.* Seeds large, strongly compressed, in 2 rows in each cell.

Three spp. in C. and S.E. Europe.

*1. L. annua L.   Honesty.
L. biennis Moench

A usually biennial herb (1–3 years) with a stiffly hairy stem, 30–100 cm and irregularly and coarsely toothed broadly cordate-acuminate lvs, rough with appressed hairs, the lower long-stalked, the *upper subsessile.* Fls c. 3 cm diam., in a loose raceme, scentless. Petals reddish-purple, rarely white. *Fr. broadly elliptical.* oblong, *or almost circular,* rounded at the ends, 30–45 × 20–25 mm. Fl.4–6. Homogamous. Visited for nectar by butterflies and long-tongued bees (working distance = 10 mm) and for pollen by smaller insects; automatically self-pollinated also. $2n = 28 + 2$ B. H.

Introduced. Much grown in gardens and often found as an escape. S.E. Europe. The infructescences with persistent silvery septa are used for winter decoration.

L. rediviva L., native in moist shady woods through much of continental Europe, is much less commonly grown and rare as a garden-escape. It differs from *L. annua* in the distinctly petiolate and fine-toothed upper lvs, the fragrant fls and the narrowly elliptical fr., 35–90 × 15–33 mm, with larger seeds.

26. ALYSSUM L.

Annual to perennial, often *mat-forming herbs with stems and simple lvs densely covered with usually stellate hairs.* Infl. ebracteate. Sepals non-saccate; *petals yellow,* clawed; *stamens 6, some or all with filament-appendages;* ovary with 1–16 ovules, style short, stigma small, somewhat 2-lobed. *Fr. a circular or oval latiseptate silicula;* the *valves flat, faintly net-veined.* Seeds compressed, winged.

About 100 spp., chiefly Mediterranean but some in C. Europe and C. Asia.

*1. A. alyssoides (L.) L.   Small Alison.
A. calycinum L.

A usually annual herb with slender tap-root and an erect or ascending stem, 7–25 cm, with several ascending branches near the base, the whole shoot grey-pubescent with stellate hairs. Lvs 6–18 mm, few, oblanceolate, narrowed below into a short stalk, entire, grey with stellate down. Fl. c. 3 mm diam. *Sepals erect, stellate-hairy, persistent in fr.* Petals pale yellow but becoming whitish, up to twice as long as the sepals, stellate-hairy on the outside. Outer stamens each with a pair of slender appendages from near the base of their filaments. *Siliculae* 3–4 mm, *almost circular,* emarginate above, on spreading stalks 2–5 mm, the valves bordered, stellate-hairy. Seeds 1–2 in each cell, reddish-brown, obovate,

1.2–1.5 mm, narrowly winged. Fl. 5–6. Slightly proto-gynous. Said to be devoid of nectar and little visited by insects. $2n = 32$. Th.–H.

Introduced. A rare and decreasing plant of grassy fields and arable land in scattered localities in S. and E. England northwards to Northumberland; 1 locality in Ireland. Europe, except the north-west (where it is introduced), and W. Asia. Introduced in New Zealand.

*A. saxatile L. (Golden Alyssum), a spring-flowering perennial with greyish oblanceolate lvs, 7–10 cm, and corymbose panicles of golden-yellow or lemon-yellow fls, is a native of C. Europe much cultivated in gardens and has been found as a casual.

27. LOBULARIA Desv.

Annual to perennial herbs wth *narrow entire lvs, the whole shoot covered with bipartite hairs.* Infl. ebracteate. Sepals spreading, non-saccate; *petals white, entire,* short-clawed; stamens 6, without appendages; ovary with 2–10 ovules, style short, stigma capitate. Fr. a circular or oval latiseptate silicula, the *valves ±flat with slender midrib.* Seeds compressed, narrowly winged.

Five spp. in the Mediterranean region and Macaronesia.

*1 L. maritima (L.) Desv.   Sweet Alison.
Clypeola maritima L.; *Alyssum maritimum* (L.) Lam.

An annual to perennial herb with slender tap-root and ascending stem 10–30 cm, branching freely near the base, the whole shoot greyish with dense bipartite appressed hairs. *Lvs 2–4 cm, scattered, linear-lanceolate,* subacute, narrowed below, subsessile. Fls c. 6 mm diam. *Sepals not persisting in fr.* Petals white, nearly twice as long as the sepals, with ±circular spreading limb. *Siliculae obovate,* c. 2.5 mm excluding the persistent style, on spreading stalks, the *valves slightly convex, 1-veined,* pubescent. Seeds pale reddish-brown, narrowly winged, 1 *in each cell.* Fl. 6–9. Homogamous. The fragrant nectar-bearing fls are much visited by small insects and are said to be self-sterile. $2n = 24$. Th.–H.

Introduced. An escape from cultivation naturalized in many scattered coastal and a few inland localities northwards to Moray Firth; rare in S. and W. Ireland. Native of the Mediterranean and Macaronesian regions but widely naturalized through cultivation. Much used as an edging plant.

28. BERTEROA DC.

Seven spp. from N. and C. Europe eastwards to C. Asia.

*1. B. incana (L.) DC.

Alyssum incanum L.; *Farsetia incana* (L.) R.Br.

An annual or overwintering herb with pale slender tap-root and erect branching stem, 20–60 cm, stellate-hairy, often reddish below. *Lvs lanceolate, narrowed below, subsessile, entire or distantly toothed, grey with stellate hairs.* Infl. ebracteate. Sepals non-saccate, hairy, not persistent. *Petals white, deeply bifid,* more than twice as long as the sepals. Stamens 6, the filaments of the outer pair toothed at the base, those of the 4 inner winged below. Ovary with 4–12 ovules; style long, stigma small, capitate. Fr. a broadly elliptical latiseptate silicula 7–10 mm, on a ±erect stalk 5–10 mm, the valves somewhat convex, stellate-hairy, indistinctly veined; persistent style 1.5–3 mm. Seeds brown, roundish, 1.5–2 mm, very narrowly winged. Fl. 6–9. Homogamous. Visited by hover-flies, etc. $2n = 16$. Th.

Introduced. Naturalized in waste places and a casual in cultivated ground in many scattered localities. Native in C., N. and E. Europe southwards to Italy and Greece, and W. Asia; naturalized in W. Europe.

29. DRABA L.

Annual to perennial, often densely tufted herbs or dwarf shrubs with stellate-hairy, rarely glabrous, shoots. Basal lvs in a rosette; stem-lvs sessile or 0; all lvs simple, entire, toothed or lobed. Infl. an ebracteate corymbose raceme lengthening in fr. Fls small, white or yellow; outer sepals ±saccate; *petals entire or slightly notched*; stamens 6 or 4, without appendages or rarely the filaments of the outer stamens toothed; ovary with 4–80 ovules; style distinct, stigma capitate or ±2-lobed. Fr. a latiseptate silicula, the *valves ±flat with a midrib conspicuous only in the lower half*. Seeds numerous, in 2 rows in each cell.

About 270 spp. of arctic and alpine plants of the northern hemisphere and in C. and S. America.

1 Lvs glabrous apart from marginal bristles; fls yellow.
    1. aizoides
  Lvs stellate-hairy; fls white.   2
2 Flowering stem lfless or occasionally with 1–2 small lvs.   **2. norvegica**
  Flowering stem lfy.   3
3 Stem-lvs lanceolate or narrowly ovate, not or hardly amplexicaul; fr. twisted.   **3. incana**
  Stem-lvs broadly ovate, amplexicaul; fr. straight.
    4. muralis

1. D. aizoides L.   Yellow Whitlow Grass.

A perennial tufted herb with slender tap-root and much-branched *glabrous* stems, 5–15 cm. Lvs confined to compact rosettes, rigid, linear, narrowed at each end, the midrib forming a keel below, terminating in a long white bristle and the margins fringed with similar bristles; dead lvs persisting. *Flowering stem* rigid, *lfless*. Infl. few-fld. Fls 8–9 mm diam. Sepals yellowish, usually non-saccate. *Petals bright yellow*, exceeding the sepals. Siliculae 6–12 × 2.5–4 mm, elliptical, compressed, narrowing above into the persistent style 2.5–4 mm, and held almost erect on ascending stalks 5–15 mm; valves usually glabrous, almost flat, with inconspicuous midrib and a faint lateral network. Seeds yellowish-brown. 1.5 mm, 6–12 in each cell. Fl. 3–5. Protogynous. Visited by flies and Lepidoptera. Ch.

Doubtfully native. Found only on limestone rocks and walls west of Port Eynon, Glamorgan. A calcicolous alpine plant of C. and S.E. Europe from the Pyrenees to the Balkans.

2. D. norvegica Gunn.   Rock Whitlow Grass.

D. hirta auct.; *D. rupestris* R.Br.

A perennial tufted herb with slender tap-root and short much-branched stellate-hairy stems bearing close basal rosettes of oblong-lanceolate lvs, hairy with mostly simple hairs, ciliate, usually entire. *Flowering stems* 2–5 cm, *usually lfless* but sometimes with 1–2 lvs, sessile, ovate-lanceolate, hairy and ciliate. Infl. few-fld. ±corymbose, lengthening in fr. Fls 4–5 mm diam. Petals white, slightly notched, nearly twice as long as the sepals. Stigma 2-lobed. *Siliculae* 5–6 × 2–2.5 mm, *straight*, elliptical, compressed, sparsely stellate-hairy, held obliquely on ascending stellate-hairy stalks c. 3 mm long; valves ±flat with inconspicuous midrib and lateral network; style very short. Seeds pale reddish-brown, 4–6 in each cell, Fl. 7–8. $2n = 48$. Ch.

Native. Found near the tops of a few Scottish mountains including Ben Lawers, Cairngorm and Ben Hope, and reaching c. 1210 m. The British plant belongs to a difficult arctic complex but is probably conspecific with plants in the Arctic and sub-Arctic of North America, Europe and Asia.

3. D. incana L.   Hoary Whitlow Grass.

A biennial to perennial herb with slender tap-root and short, prostrate, occasionally branched stock bearing the remains of dead rosettes. Basal lvs in a loose rosette, oblong-lanceolate, narrowing to a stalk, entire or distantly toothed, densely stellate-hairy, ciliate, dying before the fr. ripen. Flowering stems erect, 7–50 cm, simple or branched, stellate-hairy with numerous, ±erect, sessile, *narrowly elliptical or ovate stem-lvs*, *rounded at the base* or slightly amplexicaul, densely stellate-hairy, their margins ciliate and usually coarsely toothed, sometimes entire. Infl. dense, much elongating later. Fls 3–5 mm diam. Petals white, very slightly notched, twice as long as the sepals. Stigma ±entire. *Siliculae* 7–9 × 2–2.5 mm, variable in shape, elliptical to lanceolate, *twisted*, glabrous or stellate-hairy, held erect on ascending stalks 2–9 mm long; valves with conspicuous midrib; style not exceeding 1 mm. Seeds numerous, yellow-brown, 0.8–1 mm. Fl. 6–7. Homogamous. Self-pollinated. $2n = 32$. Ch.

Plants with stellate-hairy fr. have been named *D. confusa* Ehrh., but appear to be merely extreme types of a very variable species. Very small plants ('var. *nana* Lindbl.') are sometimes mistaken for *D. norvegica*.

Native. On rocky screes and cliffs and shelly dunes northwards from Caernarvon, Stafford and Derby to Sutherland and Shetland; Ireland. From sea-level to 1080 m on Ben Lawers. Mountains of C. Europe and C. Asia, Arctic and sub-Arctic Europe, Iceland, Greenland.

4. D. muralis L.   Wall Whitlow Grass

An annual or biennial herb with slender tap-root and an ascending simple or branched lfy stem 8–30 cm, densely stellate-hairy below, sparsely above. Basal lvs in a rosette, obovate or oblanceolate, narrowed into a stalk, entire or toothed; *stem-lvs* sessile, *broadly ovate with broad rounded semi-amplexicaul base* and sharply toothed margins; all lvs rough with stellate and a few simple hairs. Infl. many-fld, much elongating in fr. Fls 2.5–3 mm diam. Sepals hairy. Petals white, narrow, entire, almost twice as long as the sepals. *Siliculae* 3–6 × 1.5–2 mm, elliptical-oblong, *straight*, spreading on almost horizontal slender glabrous stalks 5–9 mm; valves glabrous, flat with distinct midrib; style almost 0. Seeds pale brown, c. 0.8 mm, 6–8 in each cell, Fl. 4–5. $2n = $ c. 32. Th.–H.

Native. A rare plant of limestone rocks and cliff-ledges in Dorset and Somerset, the Peak district, Cumbria and W. Yorks, where it seems truly native, but also on walls and roadside banks and verges in many scattered localities from Cornwall to Kent and northwards to Westerness, and in Ireland; occasionally a weed in nurseries and other gardens. Widely distributed in Europe from C. Scandinavia and Finland to S. Spain and Italy; N.W. Africa; W. Asia.

30. EROPHILA DC

Small annual or overwintering herbs with lvs confined to a basal rosette, producing 1–several scapes with ebracteate racemose infls. Sepals non-saccate; *petals white, deeply bifid*; stamens 6, without appendages; ovary with 10–60 ovules; style very short, stigma discoid, entire. *Fr. a latiseptate silicula*, the valves with a thin midrib vanishing above the middle. Seeds numerous, small, in 2 rows in each cell.

Perhaps 8 spp., some highly polymorphic, in Europe, W. Asia and N. Africa and 1 doubtful sp. in Peru.

Distinguished from annual scapigerous spp. of *Draba* by the deeply bifid petals. Taxonomically difficult because of the great phenotypic plasticity and the prevalence of inbreeding through early automatic self-pollination. The wide range in chromosome number provided a basis for the taxonomic treatment by Ö. Winge (*C.R. Trav. Lab. Carlsb.* (*Sér. Physiol.*, 1940, **23**, 41–73). The scheme adopted below follows the modification and updating of Winge's treatment by S. A. Filfilan & T. T. Elkington (*see New Phytol.*, **98**, 101).

1 Lvs densely pubescent, often greyish; petioles not exceeding half as long as lamina; scapes densely pubescent below and ±hairy up to lowest pedicel; seeds 0.3–0.5 mm.   **1. majuscula**
  Lvs at most moderately pubescent and always green; petioles at least half as long as lamina; scapes at most moderately pubescent below, sometimes ±glabrous; seeds 0.5 mm or more.   2
2 Petioles from c. half as long as lamina to about equally long; petals bifid for $\frac{1}{2}$–$\frac{3}{4}$ of their length; scapes always ±hairy below.   **2. verna**
  Petioles $1\frac{1}{2}$–$2\frac{1}{2}$ times length of lamina; petals bifid to at most half their length; scapes with sparse hairs below or glabrous throughout.   **3. glabrescens**

1. E. majuscula Jordan

E. simplex Winge

Lf-rosettes up to 3 cm diam. *Lvs* spathulate, *entire or with 1 tooth on each side*, usually densely pubescent with forked and stellate hairs and often greyish; petiole $\frac{1}{5}$–$\frac{1}{2}$ length of lamina. *Scapes* 1.5–9 cm, *densely pubescent below* with branched hairs and with at least scattered hairs above. Infl. with up to 15 fls; pedicels spreading in fr., glabrous or nearly so. Sepals with simple and forked hairs. *Petals* 1.5–2.5 mm, *bifid to half-way at most*. Fr. 2.5–6 mm, oblong to elliptic, 1.5–4 times as long as broad, with 15–60 or more seeds; *fresh seeds* 0.3–0.5 mm. $2n = 14^*$. Th.

Native. On rocks, walls, dry banks and waysides, also in open grassland and fallow arable land, usually on circumneutral or calcareous substrata; perhaps not on coastal dunes. Scattered throughout Great Britain northwards to Moray; local in Ireland. The least common of our three spp. Recorded also from Denmark and Netherlands.

2. E. verna (L.) Chevall.

Draba verna L.; *E. duplex* Winge; incl. *D. praecox* Steven; *E. spathulata* A. F. Láng; *E. inflata* H. C. Watson, etc.

Lf-rosettes up to 3.5(–7) cm diam. *Lvs* usually narrowly spathulate, acute, *entire or with 1–2(–3) teeth on each side*; petiole from half as long to about equalling lamina; both *lamina and petiole sparsely to moderately pubescent* with mainly branched hairs, *always green* (not grey). *Scapes* 1–10(–25) cm, *sparsely pubescent below* with simple and branched hairs but always glabrous, like the pedicels, above. Infl. with up to 24 fls. Sepals with scattered usually simple hairs. *Petals* 1.5–3.5 mm, *bifid usually for $\frac{1}{2}$–$\frac{3}{4}$ of their length*. Fr. 1.5–9 mm, oblanceolate to suborbicular, $1\frac{1}{2}$–3 times as long as broad, usually compressed but sometimes inflated (*E. inflata* H. C. Watson), with 15–50 seeds; *fresh seeds* 0.6–0.8 mm. $2n = 30^*$–46^*. Th.

Very variable, especially in size and shape of fr. The lectotype of *Draba verna* L. agrees with the above description and has narrowly elliptical fr. Plants with broadly elliptical to suborbicular fr., but otherwise similar and with $2n = 36^*$, are scattered throughout Great Britain but show some tendency to be characteristic of sand dunes in the north and west. Such plants have often been named *E. praecox* (Steven) DC. or *E. spathulata* Láng, but there seem no clear morphological discontinuities and provisionally the broad-fruited plants are referred to var. *praecox* (Steven) Filfilan & Elkington (since *Draba praecox* Steven antedates *E. spathulata* Láng), rather than to any higher taxonomic category.

Native. Throughout the British Is. in the same habitats as **1** but often also on coastal sand dunes; usually on circumneutral or calcareous substrata but occasionally on soils of acidity up to c. pH 5. Throughout Europe northwards to c. 66°N in Norway.

3. E. glabrescens Jordan

E. quadruplex Winge

Lf-rosettes 0.5–3 cm diam. *Lvs* usually narrowly spathulate acute, toothed or entire, *glabrescent or with scattered* branched and occasionally simple *hairs*, these often ±marginal only and lf appearing shiny; petiole

$1\frac{1}{2}$–$2\frac{1}{2}$ times as long as lamina. *Scapes* up to 9 cm, *with a few forked and simple hairs or ±glabrous below, upper parts and pedicels always glabrous. Petals 2–4 mm, bifid for at most half their length.* Fr 3–6 mm, elliptical to oblanceolate, compressed, 1.5–4 times as long as broad, with 20–60 seeds; fresh seeds 0.7–0.8 mm. $2n = 48$–56. Th.

Native. Generally distributed throughout Great Britain and in Inner and Outer Hebrides in same range of habitats as **1** and **2**, including coastal sand-dunes; local in Ireland. Recorded from Denmark and Netherlands.

31. ARMORACIA Gilib.

Tall glabrous perennial herbs with stout cylindrical sharp-tasting roots and lfy branching stems. Infl. a racemose panicle. Sepals non-saccate; *petals white*, short-clawed; stamens 6, without appendages; ovary with 18–50 ovules; style short, *stigma large, discoid*, shallowly 2-lobed. *Fr. an almost spherical or ellipsoid silicula, the valves strongly convex with an indistinct network of veins.* Seeds ovoid-compressed, unwinged, in 2 rows in each cell.

Three spp. in Europe and Siberia.

*1. A. rusticana P. Gaertner, B. Meyer & Scherb.

Horse-radish.

Cochlearia armoracia L.; *A. lapathifolia* Gilib.

A stout glabrous perennial herb with thick tap-root-like stocks, branched above and continuous downwards with the very long fleshy cylindrical roots. From the stocks arise subterranean stolons and erect lfy stems up to 125 cm, with numerous ±erect slender branches above. *Basal lvs* 30–50 cm, *ovate or ovate-oblong, long-stalked (to* 30 cm), *crenate-serrate*; stem-lvs short-stalked or sessile, elliptical or oblong-lanceolate, the lower often ±pinnatifid, the upper coarsely toothed or entire. *Infl. a much-branched corymbose panicle.* Fls 8–9 mm diam. Petals white, twice as long as the sepals. Silicula 4–6 mm, spherical to obovoid, on slender ascending stalks 8–20 mm, their convex valves ±unveined. Seeds 8–12 in 2 rows in each cell. Rarely ripens fr. in British Is. Fl. 5–6. Homogamous. Visited by various small insects. $2n = 32*$. H.

Introduced. Widely established as a relic of former cultivation and frequent in fields and waste places, on roadsides and stream-banks, etc., especially in and near villages, through most of England and Wales but thinning out in N. England and very local in Scotland northwards to Moray; rare in Ireland; not in Hebrides, Orkney or Shetland. Probably native in S.E. Europe and W. Asia but often grown in kitchen gardens for the condiment prepared from its roots and naturalized throughout Europe and in North America, New Zealand and many other temperate countries.

32. CARDAMINE L.

Annual to perennial herbs, glabrous or with simple hairs; perennials have horizontal to obliquely erect rhi-

zomes of varying length, often with epigela stolons; some spp. have tuberous roots. *Lvs usually pinnately or palmately compound* (often ternate), *sometimes simple*; basal lvs in a rosette or 0; stem-lvs few to many, alternate or the uppermost in an opposite pair or whorl of 3–5. Infl. an ebracteate raceme. Inner sepals slightly saccate; petals white, pink or purplish, rarely pale yellow; stamens 4–6, without appendages; stigma slightly 2-lobed. *Fr. a flattened siliqua which dehisces explosively, the inconspicuously veined valves coiling spirally from the base and thus flinging the seeds to some distance.* Seeds in 1 row in each cell.

Over 120 spp., cosmopolitan, the tropical spp. confined to mountains.

1 Petioles with stipule-like auricles clasping the stem; petals little longer than sepals or 0. **4. impatiens**
  Petioles lacking basal auricles; petals at least twice as long as sepals.   2
2 No basal lf-rosette; stem-lvs all pinnate or the upper ternate to simple; 'radical' lvs, borne singly on rhizome, sometimes present.   3
  Basal lvs in a ±distinct rosette.   4
3 All lvs pinnate; petals usually white, twice as long as sepals; anthers dark violet. **3. amara**
  Upper lvs ternate or simple with brownish-violet axillary bulbils; petals usually pale purple, 3 times as long as sepals; anthers yellow. **1. bulbifera**
4 Petals broad, spreading, usually lilac, rarely white, c. 3 times as long as sepals. **2. pratensis** *sens. lat.*
  Petals narrow, ±erect, white, c. twice as long as sepals, sometimes 0.   5
5 Annual with straight erect stem; basal lvs numerous in a compact rosette; stem-lvs 2–4, smaller than rosette-lvs; stamens 4(–6); young frs usually overtopping unopened fls. **6. hirsuta**
  Usually perennial with very flexuous erect or ascending stem; basal lvs few in a lax rosette; stem-lvs 4–10, often larger than rosette-lvs; stamens 6; young frs usually not overtopping unopened fls. **5. flexuosa**

Subgenus DENTARIA (L.) Bentham & Hooker

Rhizome hypogeal, usually with conspicuous scale-lvs; fl.-stem with cauline lvs but no basal rosette, though 'radical' lvs may arise singly from the rhizome; cotyledons distinctly stalked.

1. C. bulbifera (L.) Crantz

Coral-root, Coral-root Bitter-cress.

Dentaria bulbifera L.

A perennial herb with creeping branched *whitish rhizome bearing fleshy triangular scales* with laciniate margins and minute rudimentary pinnae at their tips. A single 'radical' lf, long-stalked and pinnate may be borne on the rhizome near the base of the fl.-stem. *Fl.-stem* 35–70 cm, erect or ascending, glabrous, usually unbranched, *lfless below.* Lowest stem-lvs pinnate with terminal and 2–3 pairs of smaller lateral lflets; middle lvs ternate, uppermost simple; all lvs short-stalked or the uppermost sessile, and all lflets lanceolate or narrowly oblong, acute, entire or ±irregularly gland-toothed, finely ciliate and often with sparse appressed hairs above. *At least the upper lvs with brownish-violet axillary bulbils.* Infl. a few-fld corymb. Fls 12–13 mm diam. Sepals violet-tipped. Petals lilac, pale pink or rarely white, 12–16 mm, c. 3 times as long as sepals. Fr. 20–35 × 2.5 mm, very rarely ripening in the British Is. Seeds 2.5 mm Fl. 4–6. The fls are rarely visited by insects and do not ordinarily set seed, reproduction being by bulbils. $2n = 96^*$. G.

Native. Very local in woods, usually on calcareous soils, in S.E. England and in Stafford; probably introduced in S.W. England and a few scattered localities further north. Widespread in Europe eastwards from France and Italy and northwards to 64° N in Norway, but rare in the Mediterranean region; characteristic of beechwoods on moist calcareous soils.

Subgenus CARDAMINE

Annuals or perennials with rhizomes of varying length, often creeping close to or above the soil surface and with scale-lvs few and distant or 0; fl.-stems commonly with a basal lf-rosette as well as cauline lvs; cotyledons very short-stalked or sessile.

2. C. pratensis L., sens.lat.

Cuckoo Flower; Lady's Smock.

Perennial, sometimes stoloniferous, herbs, usually with a short horizontal or ascending stock bearing numerous roots and giving rise at its apex to a ±well-defined lf-rosette from which a single fl.-stem develops, more rarely to 2 or more rosettes with a fl.-stem from each. Fl.-stem 15–60 cm, erect or ascending, usually simple but occasionally with basal branches from the axils of rosette-lvs, ±terete and, like the lvs, glabrous or nearly so. Rosette-lvs long-stalked, pinnate, with 1–7 pairs of ovate to orbicular often cordate-based lateral lflets and a larger ±reniform terminal lflet, all entire or with distant shallow lobes or broad teeth, the laterals short-stalked or sessile, the terminal usually distinctly stalked; stem-lvs (1–)2–6(–12), petiolate or the upper sessile, pinnae (sometimes pinnatifid), their lflets usually decreasing up the stem in number, width and stalk-length, those of the uppermost lvs often narrowly lanceolate to linear, short-stalked or sessile, the laterals usually ±entire, the terminal sometimes 3-toothed. Vegetative reproduction may take place by adventitious shoots on the upper side of rosette lvs, rarely of stem-lvs, which take root when the lvs fall. Alternatively rooting shoots may develop in the axils of stem-lvs, successful reproduction depending in both cases on a period of high atmospheric humidity and a moist substratum. Infl. of (1–)7–20(–35) fls, at first corymbose then lengthening. Sepals 2–6 mm, often with hyaline margins; *petals* 6–18 mm, obovate, *pink or purplish*, *sometimes white*; *anthers yellow*; style very short with broader capitate stigma. Fr. (10–)20–55 × 1–1.5 mm, held obliquely erect on fr.-pedicels of varying length. Fl. 4–7. Protogynous; often self-incompatible. Seeds may be thrown 1.5–2 m. $2n = 16^*–72^*(–108)$. Hs.

A very variable and taxonomically difficult polyploid complex with 8 as the basic chromosome number. Intercrossing can take place between plants differing in chromosome number and there are well-established aneuploid races: the two commonest British races have $2n = 30$ and 56, respectively. There is considerable phenotypic variability and it has proved difficult to distinguish component taxa with any high degree of certainty. Plants with $2n = 28–32$ are described by continental authors as being typically no more than 30 cm high, growing in only moderately damp grassland or in the shade of trees, having stem-lvs with sessile, narrow and ±entire lflets, small distinctly purplish fls and fr. only 20–40 mm long, and they have been referred to *C. pratensis sens.str.* Plants with $2n = 56–84$, commonly up to 50 cm or more high and growing in wetter grassland or reed-swamp, their stem-lvs with stalked, broader and sometimes markedly toothed lflets, larger whitish fls and fr. 30–55 mm long, seem referable to *C. palustris* (Wimmer & Grab.) Peterm. It has, however, been found in the Netherlands (Berg & Segal, 1966) that there are frequent transitional forms and individuals difficult to place and that it is therefore preferable to treat the two as subspecies of *C. pratensis sens.lat.* rather than as separate species. It has further been shown (Dale & Elkington, 1974) that in the British Is. the widespread and abundant race with $2n = 56$ (or 58) has itself a range of morphological variability and ecological tolerance spanning those recorded for the two taxa on the Continent. It is claimed that at least seven different taxa are provisionally recognizable in the British Is., including the above two and three which are rare and local and perhaps identical with named continental diploids ($2n = 16$). Much more investigation is clearly required before the appropriate taxonomic status and correct nomenclature of British components of the complex can be determined with any confidence.

Native. Members of the complex are common throughout the British Is. in damp meadows and pastures, by streams, in reed-swamps and in moist tree-shaded habitats, etc., and reaching c. 1000 m in Scotland. Plants with $2n = 30$ appear to favour drier habitats in S. England, those with $2n = 56–64$ being more widely distributed. Europe, N. Asia, North America.

3. C. amara L.   Larger Bitter-cress.

Herb perennating as a creeping stem (stolon) bearing small scale-leaves and fibrous roots, its tip then turning upwards to become the flexuous or ±erect fl.-stem, 10–60 cm, angled, usually almost glabrous, with stolons arising from the ascending base. Lower lvs not in a rosette, petiolate, pinnate, the 5–9 lflets ovate or orbicular, ±cordate, short-stalked, the terminal largest; upper lvs very shortly stalked, the 5–11 lflets ovate to lanceolate; all lflets angular in outline or ±crenate. Racemes 10–20-fld; fls c. 12 mm diam.; *petals white*, rarely purple, spreading, clawed, *twice as long as the sepals; anthers violet*; style long, slender. Siliqua 20–40 × 1–2 mm, narrowing upwards, straight, held obliquely on slender stalks 10–20 mm; seeds pale brown. Fl. 4–6. Homogamous. Visited by various small insects. $2n = 16$. Hs.

Native. Locally abundant in spring-line and other flushes, in fens and by streams, often tree-shaded and chiefly on eutrophic peat. Throughout much of Great Britain northwards to Aberdeen, but not in S.W. England, W. Wales or N.W. Scotland; N.E. Ireland.

Reaches 450 m in Scotland. Most of Europe northwards to c. 64° N in Scandinavia but with an isolated locality at 68° 30′ N near Abisko in Sweden; Asia Minor; Altai. A characteristic plant of certain alder-woods with moving ground-water and of subalpine spring-flushes.

4. C. impatiens L.   Narrow-leaved Bitter-cress.

A biennial or sometimes annual herb with a stout tap-root and an erect very lfy glabrous shoot, 25–60 cm, branched only above. Lvs pinnate with a winged rhachis; basal lvs in a rosette, dying in the second season, long-stalked, with 2–4 pairs of ±ovate, deeply toothed lflets and a larger ±lobed terminal lflet; stem-lvs ±sessile with 6–9 pairs of narrower toothed or entire lflets; all with *basal stipule-like auricles clasping the stem*, ciliate but otherwise glabrous. Infl. many-fld, inconspicuous. *Fls 6 mm diam.; petals white, erect, narrow, scarcely longer than the sepals, often* 0; stamens 6; *anthers greenish-yellow.* Siliquae 18–30 × c. 1 mm, at first ±erect, later wide-spreading, on stalks 3–5 mm; seeds unwinged, yellow-brown. Fl. 5–8. Little visited by insects. $2n = 16^*$. Th.–H.

Native. Very local in moist woods, especially of ash, on shaded stream banks and on moist limestone rocks and scree chiefly in the west of Great Britain from Dorset and Devon northwards to the Peak District, N.W. Yorks and Cumbria; very rare in S.E. England and in 1 locality in Angus; not in Ireland. N. and C. Europe from 64° N in Norway, S. Sweden and Estonia southwards to C. Spain, Italy and N. Balkans; C. Asia eastwards to Japan. In C. Europe reaching 1800 m.

5. C. flexuosa With.   Wood Bitter-cress.

C. sylvatica Link

Biennial to perennial, occasionally annual, herb with no persistent tap-root but with a short ±branched ascending stock and an erect or ascending *flexuous and very lfy fl.-stem* 10–50 cm, hairy especially below. *Basal lvs few in a lax rosette*, rather short-stalked, with 3–6 pairs of ovate to reniform lateral lflets and a larger terminal lflet; *stem-lvs* 4–10, short- stalked or sessile, with 5 or more pairs of *ovate-lanceolate lflets* becoming narrower up the stem, but *commonly longer than lflets of basal lvs*; all lflets gland-toothed or ±lobed and sparsely ciliate. Infl. at first corymbose. Sepals greenish-violet, white-margined. *Petals* 2.5–3 mm, equalling or up to *twice as long as sepals*, narrow, white. *Stamens usually 6. Young frs usually not overtopping the unopened fls.* Ripe fr. 12–25 × 1 mm, ±erect on slender upwardly curved stalks 6–13 mm. Seeds 1–1.2 mm, very narrowly winged. Fl. 4–9. $2n = 32^*$. (Th.–)H.

Native. Common in moist shady places, by streams, etc., throughout the British Is. except Shetland; reaches 1190 m on Ben Lawers. Most of Europe to c. 64° N in Scandinavia but only in W. Russia; C. Asia to China and Japan; probably introduced in North America.

6. C. hirsuta L.   Hairy Bitter-cress.

An annual herb with slender tap-root and erect simple or basally branched usually glabrous stem, 7–30 cm. *Basal lvs numerous in a compact rosette*, stalked, pinnate with 1–3(–5) pairs of obovate or orbicular lflets and a larger terminal lflet; *stem-lvs* 2–4, almost sessile, *with smaller and narrower lflets*; all lflets ±lobed or angled and (especially of basal lvs) sparsely hairy above and on the margins. Infl. at first corymbose. Sepals greenish-violet with narrow white margins. Petals white, narrow, up to twice as long as the sepals, sometimes 0. *Stamens usually 4. Young siliquae usually overtopping the unopened fls.* Ripe siliquae 18–25 × c. 1 mm, slightly beaded, ±erect on slender upwardly curved stalks 5–10 mm. Seeds 1 mm, very narrowly winged. Fl. 4–8. Homogamous. Rarely visited by small insects and automatically self-pollinated. $2n = 16^*$. Th.

Native. Common throughout the British Is. on bare ground, rocks, screes, walls, etc. Reaches 1160 m in Breadalbane. Through most of the northern hemisphere but not reaching the Arctic Circle.

Very locally established as garden-escapes are: *C. trifolia L., native in montane woods of C. Europe, especially on moist calcareous substrata, with creeping rhizome bearing distant scale-lvs and very long-stalked evergreen trifoliolate lvs, violet-tinged beneath, fl.-stems 20–30 cm with 0–2 stem-lvs, an infl. of a few large fls with white or pale pink petals c. 10 mm, and ripe fr. 20–25 mm, ascending on equally long pedicels; and *C. raphanifolia Pourret (*C. latifolia* Vahl, non Lej.), of spring-flushes and stream-sides in mountains of C. Europe and resembling *C. pratensis* but all lvs lyrate with very large orbicular-cordate terminal lflet, larger lilac fls and broader fr., 15–30 mm, lacking the thickened border of fr. of *C. pratensis*.

33. BARBAREA R.Br.

Biennial or perennial herbs with erect angular stems, glabrous or with sparse simple hairs. Basal lvs in a rosette, lyrate-pinnatifid; stem-lvs ±sessile, amplexicaul. Infl. dense. Inner sepals somewhat saccate; petals clawed, yellow; stamens 6, without appendages; ovary with 24–32 ovules; style longish, stigma slightly 2-lobed. *Fr. a bluntly 4-angled siliqua, the valves with strong midrib* and a lateral network. Seeds ellipsoid, unwinged, in 1 row in each cell.

About 12 spp. in Europe, the Mediterranean region, N. Asia and North America.

1 Upper stem-lvs ±simple, dentate to shallowly lobed, sometimes with 1–few longer pairs of narrow basal lobes.   2
   Upper stem-lvs pinnately lobed or cut throughout.   3
2 Upper stem-lvs obovate, dentate, usually with narrow lateral lobes near base; fl.-buds glabrous; petals 5–7 mm, almost twice as long as sepals; persistent style on ripe fr. 2–3 mm.   **1. vulgaris**
   Upper stem-lvs ovate, sinuate-toothed or -lobed; fl.-buds at least sparsely hairy; petals 3.5–6 mm, about half as long again as sepals; persistent style on ripe fr. 0.5–1.5 mm.   **2. stricta**
3 Petals up to twice as long as sepals; ripe fr. 10–30 mm.   **3. intermedia**
   Petals about three times as long as sepals; ripe fr. 30–70 mm.   **4. verna**

1. B. vulgaris R.Br.   Winter Cress, Yellow Rocket.
Erysimum barbarea L.

A biennial or perennial herb with stout yellowish tap-root and erect, branching, glabrous stem, 30–90 cm. *Rosette lvs stalked, lyrate-pinnatisect with a rounded often cordate terminal lobe* and 5–9 *oblong lateral lobes, the uppermost pair about equalling the width of the terminal lobe*; lower stem-lvs lyrate-pinnatisect with a few small lateral lobes; *uppermost ±simple, ovate*; all *deep-green, shining, glabrous, with coarsely toothed or sinuate margins. Infl. at first very dense, the width of the flowering part exceeding its length. Fl. 7–9 mm diam. Buds glabrous. Petals bright yellow, about twice as long as the sepals. *Siliquae* (Fig. 19A) 15–25 × 1.5–2 mm,

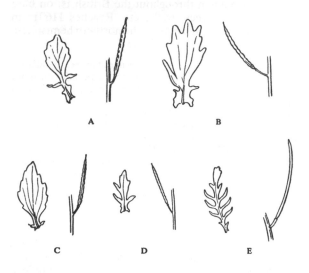

Fig. 19. Stem-leaves and siliquae of *Barbarea*. A, *B. vulgaris*; B, *B. vulgaris* var. *arcuata*; C, *B. stricta*; D, *B. intermedia*; E, *B. verna*. ×½.

standing ±*erect on their 4–5 mm long stalks*; *persistent style* 2–3 mm. Seeds yellowish-brown, ovoid, 1.3–1.5 mm. Fl. 5–8. Homogamous. Visited by hive-bees, beetles and flies. 2n = 16*. H.

Var. *arcuata* (Opiz) Fr. (*B. arcuata* (Opiz) Rchb.) differs from var. *vulgaris* in its *yellow-green* colour, the *cuneate base of the terminal* lobe of the rosette lvs, the laxer raceme whose length exceeds its breadth during flowering, the somewhat longer petals, and the *upwardly curving siliquae* (Fig. 19B), 5–8 *times the length of their spreading stalks* and the *persistent style only* c. 2 mm (2.5–3 mm in var. *vulgaris*).

Native. A plant of hedges, stream banks, wayside and other damp places. Throughout Great Britain and Ireland, common in the south but less so further north and absent from the Outer Hebrides and Shetland. Europe to 65° N in Finland, N. Africa, Asia. Introduced in North America, Australia and New Zealand.

2. B. stricta Andrz.   Small-flowered Yellow Rocket.

A biennial herb with stout yellow tap-root and erect simple or branched glabrous stem, 60–100 cm, the *branches stiffly ascending*. Rosette *lvs stalked, lyrate-pinnatifid with a large ovate-oblong non-cordate terminal lobe longer than the rest of the lf* and 1–3 *small lateral lobes, the uppermost pair falling short of the width of the terminal lobe*; lower stem-lvs pinnately lobed; *uppermost simple*, obovate, coarsely sinuate-toothed; all lvs yellow-green, shining, glabrous. Infl. at first dense. Fls 5–6 mm diam. Buds hairy. Petals bright yellow, about half as long again as the sepals. *Siliquae* (Fig. 19C) 20–30 × 1.5–2 mm, *held stiffly erect and appressed to the stem* on thick stalks 3–5 mm long; *persistent style* 0.5–1.5 mm. Seeds reddish-brown, ovoid, 1.5 mm. Fl. 5–8. Homogamous. 2n = 16*, 18*. Hs.

Native or introduced. Established locally on stream banks, waysides and waste places in the W. Midlands of England from Gloucester to Shropshire, in N. Wales (Flint) and in Middlesex; mostly casual in E. and C. England and northwards to S. Cumbria and N. Yorks; in Scotland only in Lanark; not in Ireland. C. and E. Europe eastwards from W. Germany, Austria and Italy and northwards to North Cape and Kola Peninsula; N. Asia.

*3. B. intermedia Boreau Intermediate Yellow Rocket.

B. sicula auct., vix C. Presl

A biennial herb with pale yellow tap-root and erect glabrous stem, 30–60 cm. *Rosette lvs* stalked, lyrate-pinnatisect *with 3–5 pairs of lateral lobes*, the uppermost pair exceeding the width of the ovate-cordate terminal lobe; stem-lvs all deeply pinnatifid with long narrow lateral lobes and a somewhat larger terminal lobe; all lvs ±glabrous, entire or sinuate-toothed. Infl. not very dense. Fl. 6 mm diam. *Petals* bright yellow, *up to twice as long as the sepals. Siliquae* (Fig. 19D) 10–30 × 2 mm. Fl. 5–8. Homogamous. 2n = 16*. Hs.

Introduced. A local plant of cultivated ground, stream banks and waysides throughout Great Britain northwards to Ross and Moray; Mull, Skye, Orkney; Ireland; Channel Is. Native in W. and S. Europe from S. Germany to N. Portugal, Italy and Yugoslavia; introduced elsewhere.

*4. B. verna (Miller) Ascherson
                    Early-flowering Yellow Rocket, Land-cress.
B. praecox (Sm.) R.Br.

A biennial herb with yellow tap-root and erect usually branching ±glabrous stem 20–70 cm. *Rosette lvs stalked, oblong in outline*, pinnatisect, *with 6–10 pairs of lateral lobes*, the uppermost pair at least equalling the width of the oblong-cordate terminal lobe; *stem-lvs all deeply pinnatifid* with 5–8 pairs of long narrow lobes. Infl. not dense. Fls 7–10 mm diam. *Petals* bright yellow, *about three times as long as the sepals. Siliquae* (Fig. 19E) 30–60 × 1.5–2 mm, *curving upwards on thick stalks* 4–8 mm long; *persistent style* 1–2 mm. Seeds dark red-brown, ovoid, to 2.5 mm. Fl. 5–7. Homogamous. $2n = 16^*$. Hs.

Introduced. Widespread but local on waste and cultivated ground and waysides in S. England and Wales and in a few scattered localities northwards to Angus and Perth; Ireland; Channel Is. Probably native in the W. Mediterranean region and Macaronesia but widely naturalized in W. and C. Europe, North America, S. Africa, Japan and New Zealand.

34. CARDAMINOPSIS (C. A. Meyer) Hayek

Biennial or perennial herbs with *simple or forked hairs* below and a *basal rosette of simple to pinnate often lyrate lvs*; lower and middle stem-lvs petiolate. Inner sepals somewhat saccate; petals clawed, white or pinkish; stamens 6, without appendages; ovary with 12–14 ovules; style short, stigma capitate. Fr. a *compressed unbeaked siliqua*, the *valves with a conspicuous midrib and prominent seeds*. Seeds compressed, not or slightly winged, in 1 row in each cell.

Ten spp. of arctic and alpine plants of the northern hemisphere. Differs from the closely related *Arabis* in the usually lyrate lvs and in the fr. whose septum is thicker and more rigid than in *Arabis*, so that the seeds cannot sink into it but instead bulge the valve-walls outwards.

1. C. petraea (L.) Hiit.   Northern Rock-cress.

Cardamine petraea L.; *Arabis hispida* L.; *A. petraea* (L.) Lam.

A perennial herb with a stout tap-root and a branching stock from which arise shoots with rosettes of basal lvs. Flowering stems erect or ascending, 10–26 cm, simple or branched, hispid below with spreading simple hairs, ±glabrous and pruinose above. *Basal lvs lyrate-pinnatifid* or oblong-ovate, *long-stalked*, glabrous or ±hispid with simple and stellate hairs; *stem-lvs spathulate or narrowly oblong*, entire or with a few distant teeth, *narrowed into a short stalk*, ±glabrous. Infl. few-fld, corymbose, elongating in fr. Fls c. 6 mm diam. Petals white or purplish, twice as long as the sepals. Siliquae $12–30 \times 1.5$ mm, curving upwards to continue the line of the slender stalks 6–10 mm long; *valves 3-veined* with conspicuous midrib and less distinct branching lateral veins. Seeds yellow, 1–1.5 mm, winged at the apex. Fl. 6–8. Homogamous. Probably self-pollinated. $2n = 16$. Ch.

Native. A local plant of alpine rocks in N. Wales, Scotland, the Inner and Outer Hebrides, and Shetland; Ireland: only on the Galtee Mountains (Tipperary) and the Glenade Mountains (Leitrim). Extending from near sea-level to over 1200 m in Scotland. An arctic-alpine, preferring calcareous rocks, on the mountains of C. Europe and in the Faeroes, Iceland, Scandinavia, Finland, Siberia and North America.

35. ARABIS L.

Annual to perennial herbs, commonly calcicolous, usually with simple, forked or stellate hairs or a combination of two or more of these. *Lvs simple*; basal lvs in a usually compact rosette; *stem-lvs 1–many, commonly ±sessile and often amplexicaul*, sometimes 0. Infl. dense in fl. but much elongating in fr.; basal 1–2 fls often bracteate. Sepals saccate or not; petals broadly spathulate to narrowly oblanceolate, white or variously coloured; stamens 6, without appendages; style very short, stigma capitate or emarginate. *Fr. an unbeaked siliqua* with *flat or slightly convex*, rarely strongly convex, *valves having a slender midrib*, sometimes indistinct or 0, and often also a network of lateral veins. *Seeds ovoid or compressed, often winged, in 1 row in each cell*, rarely in 2 rows.

About 120 species throughout the northern temperate and Arctic zones and in Africa.

1 Tall glaucous plant 50–120 çm; rosette-lvs stellate-hairy but the sagittate stem-lvs all or mostly glabrous; fls yellowish; fr. erect, with seeds in 2 rows in each cell.   **6. glabra**
  Plant not glaucous and not combining hairy rosette-lvs with glabrous and sagittate stem-lvs; seeds in 1 row.   2
2 Fls pale yellow; ripe fr. twisted to one side and curving downwards; a very rare plant of old walls. **1. turrita**
  Fls white or cream-coloured; ripe fr. erect or ascending.   3
3 Plants with non-flowering lf-rosettes and stems, often stoloniferous and mat-forming; stem-lvs strongly amplexicaul; petals 6–15 mm, white.   4
  Non-flowering lf-rosettes and stems 0; stem-lvs not or slightly amplexicaul; petals 4–6 mm, white or cream-coloured.   5
4 Lvs with 3–7 acute marginal teeth on each side; petals not exceeding 10 mm; seeds conspicuously winged; only in Cuillin Hills, Skye.   **2. alpina**
  Lvs with 2–5 obtuse marginal teeth on each side; petals 10–18 mm; seeds with wing inconspicuous or 0; a frequent garden-escape.   **3. caucasica**
5 Petals 4–5.5 mm, white; stem-lvs 3–many; seeds 0.9–1.3 mm.   **4. hirsuta**
  Petals 5–8 mm, cream-coloured; stem-lvs usually 1–3; seeds 1.5 mm; only on limestone near Bristol.   **5. stricta**

Subgenus ARABIS

Ripe fr. ±strongly compressed; seeds in 1 row in each cell.

*1. A. turrita L.   Tower Rock-cress, Tower-cress.

A biennial to perennial herb with slender tap-root and a horizontal or ascending stock from which arise non-flowering rosettes and 1 or more erect flowering stems, 20–70 cm, pubescent with simple and stellate hairs, often purple below. *Basal lvs in a rosette, broadly elliptical, sinuate-toothed, narrowing below into a long stalk; stem-lvs 3–5 cm, numerous, oblong, irregularly toothed, sessile, clasping the stem* with ±rounded basal lobes; *all lvs grey with short dense stellate hairs. Petals very pale*, twice as long as the sepals. Siliquae $8–12$ cm $\times 2–2.5$ mm, *all twisted to one side and curving downwards* on erect stalks 4–7 mm long; valves thick-walled, with many prominent veins but no distinct midrib. Seeds 2.5–3 mm, brown, with a membranous wing. Fl. 5–8. Homogamous. Probably self-pollinated. $2n = 16$. Hs.

Introduced. Naturalized on old walls at Cambridge and formerly at Oxford and Cleish Castle, Kinross. Native in C. and S. Europe, Asia Minor and Algeria.

2. **A. alpina** L.   Alpine Rock-cress.

A *perennial mat-forming herb* with slender horizontal branching stock giving rise to terminal and axillary lf-rosettes and then to extended shoots, *flowering and non-flowering*; prostrate *lfy stolons* from the axils of rosette-lvs develop further lf-rosettes and shoots of both kinds. *Fl-stems* 5–40 cm, ascending in fl. but sometimes decumbent in fr., their *basal lvs* in a ±compact rosette, *obovate-oblong, narrowing into a short broad petiole* (Fig. 20A); *stem-lvs* ovate-lanceolate, acute, *sessile, clasping*

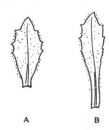

Fig. 20. Leaves of A, *Arabis alpina* and B, *A. caucasica.* $\times \frac{3}{5}$.

the stem with rounded basal lobes; *all lvs coarsely and irregularly dentate, with 5–7 acute teeth on each margin*; stems and lvs rather sparsely stellate-hairy and with a few forked and simple hairs. Non-flowering shoots are shorter and ±decumbent, the looser basal rosette merging into the stem-lvs. Infl.-axis and pedicels usually more densely hairy. Fls 6–10 mm diam. Sepals with white scarious margins. Petals white, 6–10 mm, twice as long as sepals, spreading. Siliquae 20–35(–60) × 1.5–2 mm, ascending on spreading fr.-pedicels 6–15 mm; valves flattened, with indistinct midrib and somewhat prominent seeds. Seeds 1–1.5 mm, brown, winged. Fl. 6–8. Homogamous. Visited by a few insects but probably for the most part self-pollinated. $2n = 16, 32$. Ch.

Native. found first in 1887 and still known only from two rock-ledges in the Cuillin Hills, Skye, at c. 840 m. A widespread arctic-alpine plant of the northern hemisphere, reaching 79° 35′ N in Spitsbergen, and favouring damp basic and circumneutral substrata.

*3. **A. caucasica** Willd.   Garden Arabis.

A. albida Stev. ex Jacq. fil.

Closely allied to *A. alpina* and resembling it in mode of growth, but with only 2–3 *teeth on each lf-margin* (Fig.20B); the *stem-lvs with pointed*, not rounded, *basal lobes*; and all the *lvs grey-green or whitish*, more densely hairy than in *A. alpina*. The fragrant fls are larger (c. 15 mm diam.) and the siliquae have a more prominent midrib. The seeds are at most very narrowly winged.

Fl. 3–5. Homogamous. Visited by small bees and other insects. Ch.

Introduced. A native of mountains of the Mediterranean region and Near East to Iran. Much cultivated on rock gardens and walls and occasionally naturalized.

4. **A. hirsuta** (L.) Scop.   Hairy Rock-cress.

Turritis hirsuta L., incl. *A. ciliata* R.Br., pro parte, *A. brownii* Jordan and *A. hibernica* Wilmott

A short-lived perennial, sometimes biennial, with slender ±erect woody stock and 1–several erect and usually unbranched fl.-stems, 8–60(–100) cm, rough with simple and bifid hairs below and with bifid and stellate hairs above, but sometimes ±glabrous below or throughout. *Basal lvs* (Fig. 21A) in a rosette, *oblong-*

Fig. 21. Leaves of *Arabis*. A, *A. hirsuta*; B, *A. brownii*; C, *A. stricta*. Leaves $\times \frac{2}{3}$.

elliptical to oblanceolate, narrowing into a stalk-like base; *stem-lvs* usually numerous, *erect, ovate-oblong to oblong, sessile and with rounded, truncate or slightly cordate base but not or barely amplexicaul*; all lvs usually with forked and simple hairs above and beneath, sometimes ±glabrous on the faces but with ciliate margins, and all usually distantly dentate or sinuate-dentate but sometimes entire. Sepals c. 2.5 mm, often violet with white margins. Petals 4–5.5 mm, white. *Siliquae* 15–35(–50) × 1–2 mm, *held erect and close to the infl.-axis* on ascending pedicels 3–8 mm, the *glabrous valves with distinct midrib* and somewhat prominent seeds. Seeds 0.9–1.3 mm, reddish-brown, *narrowly winged all round*. Fl. 6–8. Homogamous. Visited by small bees and other insects but probably for the most part self-pollinated. $2n = 32^*$. H.

Very variable, especially in density and distribution of hairs and in size and shape of lvs. Plants of calcareous sand-dunes along the west coast of Ireland from Donegal to Cork, while by no means uniform, show at least some of a set of characteristic differences from typical *A. hirsuta*. Their fl-stems are commonly only 8–25 cm high and ±glabrous, and their lvs are usually all conspicuously ciliate but otherwise glabrous or nearly so, and all entire or with very small and distant marginal teeth. Infl. and fls are much as in small plants of typical *A. hirsuta* but the ripe siliquae are somewhat shorter and broader and in a more compact cluster. These populations were formerly regarded as conspecific with a continental '*A. ciliata* R.Br.', but later as a separate and endemic species, *A. brownii* Jordan (*A. hibernica* Wilmott). Although plants of the two 'species' are normally self-pollinated they can readily be hybridized experimentally to give highly fertile progeny. Intermediates occur naturally in western Ireland and elsewhere in the

British Is., commonly in circumstances making it very unlikely that they can be of simple hybrid origin. It seems, rather, that *A. brownii* comprises genetic variants of *A. hirsuta* which are maintained by the combination of predominant self-pollination with topographical isolation and would be more appropriately treated as an infraspecific taxon, provisionally as a subspecies of *A. hirsuta*.

5 A. stricta Hudson   Bristol Rock-cress.

A. scabra All.

A perennial herb, initially with slender tap-root but later developing a ±branching woody stock bearing one or more compact overwintering lf-rosettes and erect simple or basally branched fl.-stems 3–20(–30) cm, rough below with simple and few stellate hairs, glabrous above. *Rosette-lvs dark shining green, oblanceolate, sinuate-lobed*, narrowing into a short petiole-like base; *stem-lvs usually 1–3, oblong*, less lobed than the rosette-lvs, *sessile with half-clasping base*; all lvs hispid and ciliate with simple and stellate hairs. Infl. of (1–)3–8(–12) fls, each 5–6 mm diam. Sepals with white scarious margins. *Petals cream-coloured*, 5–8 mm, twice as long as sepals. Siliquae (15–)25–50 mm, held ±erect on slender ascending pedicels 5–7 mm; valves of ripe fr. somewhat convex, with conspicuous midrib. Seeds 1.5 mm, dark brown, flattened and narrowly winged. Fl. 3–5. Slightly protogynous. Visited by a few small insects, but highly self-fertile. $2n = 16^*$. Ch.

Native. Known only from the Avon Gorge and Penpole Point, Shirehampton, both on the Carboniferous limestone near Bristol, but introduced in a few other localities in S.W. England. Intolerant of competition and of trampling and confined to ±unshaded south-facing rocky slopes where it roots in rock-crevices. Mountains of Spain, S. France and the French Jura.

Subgenus TURRITIS (L.)

Ripe fr. cylindrical or 4-angled, the convex valves with prominent midribs; seeds in 1–2 rows in each cell.

6. A. glabra (L.) Bernh.   Tower Mustard.

Turritis glabra L.; *Arabis perfoliata* Lam.

An overwintering to biennial *glaucous* herb with slender tap-root and a usually simple fl.-stem, 50–120 cm, sparsely hairy below with soft mostly simple hairs. *Basal lvs in a rosette*, mostly withering before fls appear, *oblanceolate to spathulate* in outline, narrowing into a short petiole-like base, *distantly denticulate to coarsely sinuate-toothed, downy* with stellate hairs; *stem-lvs 2–8 cm, ovate-lanceolate, ±acute, entire, glabrous* or nearly so, *sagittate at the sessile base and amplexicaul.* Fls c. 6 mm diam. *Petals yellowish-* or *greenish-white*, 4–6 mm, less than twice as long as the sepals. *Ripe siliquae* 30–70 × 1–1.5 mm, *held stiffly erect* on slender erect pedicels 6–10 mm, the convex valves with prominent midribs making them ±4-*angled*; beak 0. *Seeds in 2 rows in each cell*, brown, ovoid, unwinged but with a dark brown edge. Fl. 5–7. Homogamous. Rarely visited by insects and normally self-pollinated. $2n = 12, 16, 32$. H.

Native. A local and decreasing plant of dry banks, cliffs and stony places, roadsides and waste places, chiefly in England northwards to Cumbria and Northumberland, but rare in S.W. England and absent from Wales; probably only casual in Scotland; not in Ireland. Throughout Europe to 70° N in Norway and in N. Asia, North America and Africa. Introduced in Australia.

36. RORIPPA Scop.

Annual to perennial herbs, *glabrous or with simple hairs.* Lvs simple to pinnate, often pinnatifid. Infl. usually ebracteate. Inner sepals saccate at base; *petals yellow*; stamens 6, without appendages; *nectar-glands at base of median stamens* as well as of short transverse pair; ovary with many ovules; style short, stigma somewhat 2-lobed. *Fr. a silicula or short siliqua with valves convex and veinless or with very slender and often indistinct midrib* which may disappear beyond half-way; seeds in (1–)2 rows in each cell.

Many species, widespread in the northern hemisphere.

1 Fr. a slender siliqua 8–18 mm; petals twice as long as sepals; all lvs usually pinnate or deeply pinnatifid.   **1. sylvestris**
  Fr. rarely exceeding 9 mm, often much shorter; petals about equalling to twice as long as sepals; lvs various.   2
2 Middle and upper lvs oblong to lanceolate, ±toothed but usually not pinnatifid; fr. an ovoid-ellipsoid or globose silicula; petals exceeding sepals.   3
  All lvs pinnate or pinnatifid (or uppermost merely toothed); fr. shortly cylindrical to ellipsoid; petals about equalling sepals.   4
3 Lvs with conspicuous auricles; petals 3–4 mm, up to $1\frac{1}{2}$ times as long as sepals; fr. 1.5–3 mm, globose.   **5. austriaca**
  Lvs with auricles inconspicuous or 0; petals c. 5 mm, about twice as long as sepals; fr. 3–6 mm, ovoid or ellipsoid.   **4. amphibia**
4 Plant ±erect; sepals exceeding 1.6 mm; fr. at most twice as long as their pedicels; seeds distinctly papillose; widespread in British Is.   **2. palustris**
  Plant ±prostrate; sepals less than 1.6 mm; fr. 2–3 times as long as their pedicels; seeds very faintly papillose; a rare and local plant so far known from scattered western and northern localities, mostly near coasts.   **3. islandica**

1. R. sylvestris (L.) Besser   Creeping Yellow-cress.

Nasturtium sylvestre (L.) R.Br.

A perennial *stoloniferous* herb with horizontal or ascending slender stock and erect or ascending branched fl.-stems 20–50 cm, angled, ±glabrous. Lower lvs petiolate, pinnate to deeply pinnatifid with oblong or lanceolate segments, usually toothed or lobed; *upper lvs sessile, usually pinnatifid* with narrower segments but sometimes almost entire. Infl. a corymbose panicle with flexuous axis. Fls c. 5 mm diam. *Petals* 4–5 mm, *twice as long as sepals.* Fr. (Fig. 22A) *a slender ±cylindrical siliqua* 8–18 mm, *held obliquely or curving upwards on*

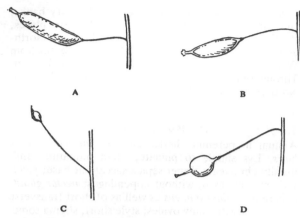

Fig. 22. Fruits of *Rorippa*. A, *R. sylvestris*; B, *R. palustris*; C, *R. amphibia*; D, *R. austriaca*. ×2.

horizontal to somewhat deflexed pedicels 5–12 mm. *Seeds* c. 0.7 mm, reddish-brown, *indistinctly 2-rowed*. Fl. 6–8. Homogamous. Visited by various small bees and flies; strongly self-incompatible. $2n = 48^*$. Hel.–H.

Native. Locally frequent in open vegetation by streams and lakes and in areas where water stands intermittently, as on riverside shingle; occasional on damp waysides and as a weed of cultivated ground. Throughout much of lowland England, Wales and S. Scotland but in only a few scattered localities further north and not in Orkney or Shetland; local in Ireland. Throughout most of Europe to c. 65° N but less frequent in the Mediterranean region.

2. R. palustris (L.) Besser

Common Marsh Yellow-cress.

R. islandica auct. mult., non Borbás; *Nasturtium palustre* (L.) DC., non Crantz

An annual or biennial *non-stoloniferous* herb with slender tap-root and ±erect fl.-stem 8–30(–60) cm, branched or not, angled, glabrous or sparsely hairy below, hollow. *Lower lvs* ±long-stalked, *deeply lyrate-pinnatifid* with narrow irregularly sinuate-toothed lateral segments and an ovate lobed terminal segment; *upper lvs similar but short-stalked or sessile and auricled*, the uppermost sometimes narrowly lanceolate and dentate, not pinnatifid. Fls c. 3 mm diam, in lax corymbs. *Sepals exceeding* 1.6 mm. *Petals equalling or slightly shorter than sepals*, pale yellow. Fr. (Fig. 22B) 4–7(–9) mm, oblong-ellipsoid or shortly cylindrical, turgid, narrowing abruptly into the short style, held obliquely erect or curving upwards on horizontal or slightly deflexed pedicels about equalling the fr. or a little shorter. Seeds c. 0.7 mm, their faces covered with distinct smoothly rounded papillae. Fl. 6–9. Homogamous. Visited by occasional flies and small bees but normally self-pollinated. $2n = 32^*$. Th.–Hel.

Native. In moist places, especially where water stands intermittently, throughout the lowlands of the British

Is. but very local in C. and N. Scotland and not in the Hebrides. Through most of Europe and in temperate zones of both hemispheres; reaches c. 1500 m in the Alps.

3. R. islandica (Oeder) Borbás, sens.str.

Northern Marsh Yellow-cress.

Sisymbrium pusillum Vill.; *Nasturtium palustre* (L.) DC., pro parte

Like **2** but *prostrate* or nearly so and with smaller fls, the *sepals* being *less than* 1.6 mm with petals about equally long. *Fr.* 2–3 *times as long as its pedicel*. Seeds only faintly papillose. $2n = 16^*$.

Native. Recognized only recently as meriting specific rank (Jonsell, *Symb. Bot. Uppsala*, 1968, **19**, 1–222) and so far known from a few localities in W. Ireland, Wales, Isle of Man, Outer Hebrides, Orkney, and in the Pentland Hills. Also in N. Norway, Iceland and Greenland and locally in the Alps and Pyrenees (at higher altitudes than *R. palustris*); a different subsp. in E. Russia and W. Siberia. Its diploid chromosome number and restricted distribution in C. Europe suggests that it may be a glacial relict, the tetraploid *R. palustris* having proved a more successful colonizer of ice-free lowland and montane habitats in post-glacial times.

4. R. amphibia (L.) Besser   Great Yellow-cress.

Sisymbrium amphibium L.; *Nasturtium amphibium* (L.) R.Br.

A perennial shortly *stoloniferous* herb, usually glabrous, with stout hollow fl.-stem 40–120 cm, ascending from a rooting base or ±erect. Lvs bright or yellowish-green, very variable in shape and size; *lower lvs broadly ovate or obovate to narrowly elliptic-oblong in outline*, narrowing below into the short petiole-like base, their *margins entire*, *sinuate or sharply dentate*, more rarely pinnatifid or even pectinate, especially if growing under water; *upper lvs smaller*, *narrowly elliptic-lanceolate*, more shortly petiolate to sessile, sometimes auricled and semi-amplexicaul, their *margins irregularly dentate to ±entire*. Fls c. 6 mm diam. in a terminal and several axillary corymbs. Sepals 2–3.5 mm. *Petals about twice as long as sepals, bright yellow*. Fr. (Fig. 22c) 3–6 × 1.5–3 mm, *ovoid-ellipsoid*, straight, with persistent style 1–2 mm; fr.-pedicel 6–17 mm, horizontal or somewhat deflexed. Seeds c. 1 mm. Fl. 6–8. Homogamous. Visited by various small bees but automatically self-pollinated. $2n = 16^*, 32$. Hel.

Native. Locally frequent by ponds, ditches and streams from Somerset and Kent northwards to Lancashire and N.E. Yorks; occasional and probably adventive further north; Ireland. Through most of Europe to c. 62° N, but rare in the Mediterranean region; Siberia; N. Africa. Local in marginal sedge-swamp by eutrophic streams, ditches and pools with very variable water level.

*5. R. austriaca (Crantz) Besser

Austrian Yellow-cress.

Nasturtium austriacum Crantz

A perennial stoloniferous herb with ascending to erect woody stock and erect ±glabrous fl.-stems 50–90 cm, sometimes hollow above, usually branching above. *Lvs diminishing upwards; lowest broadly lanceolate with broad petiole-like base; middle and upper narrower and with shorter base; all auricled and semi-amplexicaul, their margins sharply and irregularly serrate:* submerged lvs, if present, may be deeply and narrowly pinnatifid or pectinate. Fls 3–4 mm diam. Sepals yellow. *Petals 3–3.5 mm, little longer than sepals,* yellow. *Fr.* (Fig. 22D) 1.5–3 mm, *globose,* with persistent style equalling rest of fr.; fr.-pedicels 7–10(–15) mm, slender, spreading horizontally. Seeds c. 1 mm, finely verrucose, 6–12 in each cell. Fl. 6–8. Self-incompatible. $2n = 16$. H.

Introduced. An infrequent casual rarely naturalized, in eastern C. England from Berks, Surrey and Kent to S.W. Yorks; S. Wales; no longer in East Anglia. Native in Europe from the Elbe and Austria eastwards and in S.E. Asia.

The hybrid *R.* × *anceps* (Wahlenb.) Reichenb. (*R. amphibia* × *R. sylvestris*), intermediate between the parents, has been recorded from many localities. *R. amphibia* × *R. palustris*, with the habit of *R. amphibia* but intermediate in characters of lvs and fr., has been confirmed only in two localities.

37. NASTURTIUM R.Br.

Usually perennial herbs, glabrous or with a few simple hairs. *Lvs pinnate or pinnatisect.* Inner sepals saccate at base; *petals white or purplish;* stamens 6, without appendages; *nectar-glands only at base of the transverse pair of short stamens;* ovary with many ovules, short style and slightly 2-lobed stigma. Fr. a siliqua, its valves with slender median vein; seeds in 1–2 rows in each cell.

Differs from *Rorippa* in the white or pale purplish fls and the absence of nectar-glands at the base of the median stamens, and the lvs are more constantly pinnate or very deeply pinnatifid.

1 Fr. a well-formed siliqua with many good seeds. 2
  Fr. dwarfed and misshapen with no or very few good seeds. **3. microphyllum × officinale**
2 Lvs remaining green in autumn; anthers of all stamens dehiscing inwards (introrse); fr. 13–18 mm, broad and straight or slightly curving upwards; seeds usually in 2 rows in each cell, with c. 25 depressions in each face of the testa. **1. officinale**
  Lvs turning purple-brown in autumn; anthers of long stamens dehiscing outwards (extrorse); fr. 16–22 mm, narrow and usually curved; seeds usually ±in a single row in each cell, with c. 100 depressions in each face of the testa. **2. microphyllum**

1. N. officinale R.Br.

Green Water-cress, Summer Water-cress.

Sisymbrium nasturtium-aquaticum L.; *Rorippa nasturtium-aquaticum* (L.) Hayek

Perennial, rarely annual, herb of wet soil and shallow water overwintering by creeping or ascending rooted stems at the surface of the soil or water; fl.-stems 8–60(–200) cm, hollow, angular and glabrous, procumbent and rooting below then ascending or floating. Lvs pinnate, the lowest petiolate and with 1–3 lflets, the upper sessile, auricled, with 5–9 or more lflets; terminal lflet usually suborbicular to broadly cordate, laterals broadly elliptical to ovate; all entire or sinuate-toothed. *Lvs and stems remain green in autumn.* Fls c. 6 mm diam. Sepals c. 3 mm, erect. *Petals c.* 4 mm, white. *All stamens with introrse anthers.* Fr. 13–18 mm, *broad and straight or slightly upcurved* on horizontal or somewhat deflexed pedicels 8–12 mm, the valves with distinct slender midrib. *Seeds usually in 2* distinct rows in each cell, ovoid, somewhat compressed, *with c.* 25 *polygonal depressions in each face.* Fl. 5–10. Homogamous and self-compatible. Visited by various flies and small bees. $2n = 32*$. Hel. (or Th.).

Native. A common lowland plant at the edges of streams, ditches and springs and either just above or just below the level of the moving water; rarely in stagnant water and intolerant of heavy shade; usually in calcareous water. Throughout the Britsh Is. at altitudes below c. 300 m. Europe northwards to 61° N in Shetland, 58° N in Scandinavia and 55° N in Russia; N. Africa, W. Asia. Introduced in many parts of the world and a serious river weed in New Zealand. Cultivated on a large scale in S. England and in continental Europe.

2. N. microphyllum (Boenn.) Reichenb.

Rorippa microphylla (Boenn.) Hyl.; *Nasturtium uniseriatum* Howard & Manton

Scarcely distinguishable vegetatively from **1** except by the stomatal index of the lvs (ratio of number of stomata to total number of epidermal initials, expressed as a percentage). For the lower epidermis, this is c. 11% for *N. microphyllum* and c. 18% for *N. officinale. Lvs and stems turn purple-brown in autumn.* Fls as in **1** but larger. *Petals c.* 6 mm, narrowing below more abruptly than in **1**. *Long stamens with extrorse anthers.* Fr. 16–22 mm, *narrower and more distinctly upcurved* than in **1**, on pedicels 11–15 mm. *Seeds more or less in a single row* in each cell, *with c.* 100 *polygonal depressions in each face* of the testa. Fl. 6–10, starting about 2 weeks later than **1**. Self-compatible. $2n = 64*$. Hel.

An allotetraploid derivative of **1**, perhaps with a *Cardamine* sp.

Native. Common throughout the British Is. in similar habitats to **1**, the two spp. sometimes growing together. Not cultivated. W. Europe, but distribution not fully known, W. Asia.

3. N. microphyllum × officinale

Brown Water-cress, Winter Cress.

Rorippa ×*sterilis* Airy-Shaw

The triploid hybrid between **1** and **2** ($2n = 48*$) has been found wild and has also been raised by crossing the parent species. It resembles *N. microphyllum* in that the stem and lvs turn purple-brown in autumn but its stomatal index (lower epidermis) is .c 15% and its frs are dwarfed and deformed, with an average of less than

1 good seed per fr. The good seeds are intermediate in size and surface pattern between those of the parents. Pollen mostly abortive.

Native. Frequent through much of the British Is., sometimes in the absence of one or both parent species, and cultivated as brown or winter cress. France and perhaps C. Europe, but full distribution not yet known.

The two types of cultivated water-cress, green or summer cress (*N. officinale*) and brown or winter cress (*N. microphyllum × officinale*) are propagated vegetatively for the most part, though *N. officinale* may be grown from seed. 'Summer cress' is a misleading designation since it can be gathered throughout the year, but 'winter cress' is grown for gathering only in winter and early spring. An autotetraploid green cress has been produced artificially by colchicine treatment and a hybrid tetraploid by crossing this with *N. microphyllum*. Both may prove valuable commercially.

*Aubrieta deltoidea (L.) DC., native in the Aegean region and in Sicily, is much cultivated as 'Aubrietia' on walls and in rock-gardens and occasionally escapes. A perennial herb varying in habit from caespitose to spreading or mat-forming and with stem and lvs covered with stellate down; lvs spathulate to rhombic, usually with 1–3 teeth on each side. Fls few but large, the long-clawed petals 12–25 mm, usually rose-lilac to reddish-purple but sometimes pale pink. The outer stamens have a toothed appendage just beneath the anthers. Fr. a short ellipsoid siliqua 6–16(–22) mm.

38. MATTHIOLA R.Br.

Annual to perennial herbs or dwarf shrubs with entire, sinuate or pinnatifid lvs, the *stems and lvs grey with branched hairs*. Infl. an ebracteate raceme. Fls large; sepals erect, the inner pair saccate; petals long-clawed; stamens 6, without appendages but the filaments of the inner stamens branched; ovary with numerous ovules, *stigma of 2 erect lobes each with a dorsal swelling or horn-like process. Fr. a siliqua with hairy 1-veined valves*. Seeds compressed, in 1 row in each cell.

About 50 spp., chiefly in the Mediterranean region, some in C. Asia and S. Africa.

Stem woody below; all lvs entire; siliquae eglandular.
                                                    1. incana
Stem not woody; lower lvs sinuate or pinnatifid; siliquae glandular.   **2. sinuata**

1. M. incana (L.) R.Br.   Stock, Gilliflower.
Cheiranthus incanus L.

Annual to perennial with stout tap-root and *erect shoot* to 80 cm, stout, glandular-hairy, *woody and lfless below*, with several ascending branches from near the base. *Lvs* in rosettes on old branches and scattered on shoots of current season, *narrowly lanceolate*, lower stalked, upper ±sessile, *all entire or nearly so*, hoary. Fls 2.5–5.0 cm diam., in a loose raceme. Sepals hairy, with scarious margins. Petals c. 25 mm, purple, red or white. Siliquae 4.5–13 cm × 3–4 mm, held ±erect on ascending stalks 1–2.5 cm, the *valves* ±compressed, 1-veined, *downy but not glandular*; the persistent stigma-lobes with dorsal processes up to 3 mm. Seeds brown, 3 mm, compressed, broadly winged. Fl. 5–7. Homogamous.

The fragrant fls are visited by butterflies. $2n = 14$. Th.–Ch.

Possibly native. A rare plant of coastal cliffs, walls and sandy shores in Sussex and Isle of Wight; probably introduced elsewhere in S. England from Kent to Cornwall, in Scilly and Channel Is. and in Durham. Thought to be native on coasts of W. Europe and Mediterranean region and in Canary Is., but widely naturalized through cultivation. The 'Ten Weeks Stock' of gardens is an annual, and there are also double-fld varieties in cultivation.

2. M. sinuata (L.) R.Br.   Sea Stock.
Cheiranthus sinuatus L.

A biennial herb with stout tap-root and diffuse shoots, 20–60 cm, very lfy below, glandular-hairy. *Basal lvs* ±oblanceolate, narrowing below into a stalk, *sinuate, or pinnatifid with narrow oblong lobes*; stem-lvs narrowly elliptical, the uppermost linear-lanceolate and entire; all tomentose and *glandular*. Fls 2–2.5 cm diam., in a loose raceme. Sepals tomentose. Petals 20–25 mm, pale purple. Siliquae 7–12 cm × 3–5 mm held ±erect on ascending stalks c. 1 cm, the *valves* compressed, 1-veined, tomentose and roughly *glandular*; the persistent stigma-lobes with prominent tooth-like dorsal processes. Seeds 3–4 mm, brown, oval, much compressed, broadly winged. Fl. 6–8. Homogamous. Fragrant, especially in the evening. $2n = 14$. H.

Native. A very local plant of sea-cliffs and sand dunes in N. Devon and Mediterranean region.

Also found as a casual is **M. longipetala** (Vent.) DC. subsp. **bicornis** (Sibth. & Sm.) P. W. Ball (*M. bicornis* (Sibth. & Sm.) DC.), Night-Scented Stock, native in Greece and Asia Minor, whose siliquae, 8–30 cm, look like branches and have two long slender horn-like processes from the backs of the stigma-lobes.

Malcolmia maritima (L.) R.Br., often grown in gardens as Virginia Stock, is a frequent casual near gardens. It is native in the E. Mediterranean region and is a diffuse pubescent annual with narrow entire lvs, small violet, pink or white fls and siliquae c. 5 cm which are slender, flexuous, spreading or down-curved and constricted between the seeds.

39. HESPERIS L.

Tall biennial to perennial herbs with toothed or pinnatifid lvs; stem and lvs with simple, branched and glandular hairs. Infl. sometimes bracteate below. Inner sepals saccate; petals large, long-clawed; stamens 6, the filaments of the inner 4 broadened, ±winged; ovary with 4–32 ovules; style short; *stigma deeply 2-lobed, the lobes ±erect, facing each other, not appendaged at the back*. Fr. a linear, cylindrical or somewhat 4-angled siliqua, the valves with distinct midrib and ±distinct lateral veins, beaded. Seeds many in 1 row in each cell.

About 24 spp. chiefly in the Mediterranean region, some in C. Europe and C. Asia.

*1. H. matronalis L.   Dame's Violet.

A biennial or perennial herb with tap-root and branching ±woody stock and 1 or more erect lfy stems

40–90 cm, usually branched, ±hairy with short simple and stellate hairs. *Lvs* oblong-ovate to lanceolate, narrowing up the stem, all *short-stalked, finely toothed* and roughly hairy, the upper lvs usually with two glands at the base. Fls c. 18 mm diam., violet or white; their *stalks equalling or exceeding the sepals*. Siliquae 9 cm, curving upwards on spreading stalks 10–30 mm; valves ±glabrous, constricted between the seeds. Seeds brown, 3 mm. Fl. 5–7. Homogamous. The fls are fragrant, especially in the evening, and are visited by many kinds of insects including Lepidoptera. $2n = 24, 26, 28$. H.

Introduced. A garden-escape frequently establishing itself in hedgerows, shrubberies, plantations, waysides, etc., throughout lowland areas of the British Is. including Orkney and Shetland but only in Islay in the Western Is. Native in C. and S. Europe and in W. and C. Asia; a frequent escape from cultivation throughout most of Europe and in North America.

H. laciniata All., a casual from the Mediterranean region and the Iberian Peninsula, differs in the pinnatifid to sinuate-toothed lvs and the often yellowish fls with pedicels shorter than the sepals. The siliquae are glandular-hairy.

40. Erysimum L.

Annual to perennial herbs, sometimes woody below, with very lfy shoots bearing *appressed hairs at least some of which are stellate with 3–4 rays or once forked with the 2 arms aligned and attached near the midpoint (medifixed)*. *Lvs usually ±narrowly lanceolate*. Infl. ebracteate, often corymbose. Outer sepals often with a horny projection just beneath the tip, inner saccate or not at base; petals long-clawed, yellow to orange, rarely purple; stamens 6, without appendages; there are *nectarglands* round the bases of the short transverse stamens and also *outside the bases of the long median pairs*; ovary with many ovules, short style and capitate somewhat 2-lobed stigma. *Fr. a siliqua*, 4-angled or cylindrical, sometimes either laterally or dorsally compressed; valves usually strongly 1-veined, hairy or glabrous. *Seeds* ovoid, sometimes winged, *in 1 row* in each cell.

Over 100 spp., mainly in Europe, N. Africa, W. and C. Asia and North America.

Differs from *Cheiranthus* in the presence of median nectar-glands, the no more than slightly 2-lobed stigma and the 1-rowed seeds.

1. E. cheiranthoides L.   Treacle Mustard.

An annual or overwintering herb with short tap-root and one or more erect and usually branched lfy fl.-stems, 15–90 cm, dull green with scattered appressed 2(–3)-rayed medifixed hairs. Basal lvs in a rosette dying before the fls open, lanceolate-elliptical, acute, shortly petiolate, entire or irregularly sinuate-toothed; stem-lvs diminishing upwards, the lowest lanceolate, often shortly petiolate, sinuate-toothed, the upper narrowly oblong-lanceolate, sessile, denticulate; all sparsely pubescent with mostly 3-rayed hairs. Fls c. 6 mm diam., their slender ascending pedicels 4–8 mm, lengthening

in fr. Sepals 2–4 mm, the inner pair not or slightly saccate at base. *Petals* 3–6 mm, *bright yellow*, pubescent beneath. Siliquae 12–25(–50) × 1–1.5 mm, square in section, straight or somewhat upcurving and held almost erect on spreading-ascending fr.-pedicels up to 16 mm; valves sparsely hairy to ±glabrous. Seeds c. 1 mm, brown, numerous and often rather irregularly 1-rowed. Fl. 6–8. Homogamous. Visited by insects and automatically self-pollinated. $2n = 16$. Th.

Probably native. Locally frequent as a weed of cultivated ground and waste places at low altitudes, especially in S.E. England, but thinning out towards the west and north and very local in Wales, N. England and Scotland, though reaching E. Sutherland; C. Ireland; not in Hebrides, Orkney or Shetland. Throughout Europe to c. 70° N in Norway, N. Africa, N. Asia, North America.

E. linifolium (Pers.) Gay, native in Spain and Portugal, with purple or violet fls, is sometimes grown in gardens. Much more frequently grown is the orange-fld Siberian Wallflower, **Cheiranthus allionii** Hort., which is probably of intergeneric hybrid origin and is better named *E. allionii* Hort.

Several spp. of *Erysimum* have been recorded as casuals in the British Is., in particular the annual **E. repandum** L. with narrowly lanceolate lvs and short spreading fr.-pedicels almost as thick as the ripe siliqua.

41. Cheiranthus L.

Perennial herbaceous or suffruticose plants with the stems and *narrow entire lvs* covered with *appressed branched hairs*. Infl. an ebracteate raceme. Sepals erect, the inner pair saccate; petals long-clawed; stamens 6, without appendages; *nectaries only round the base of each of the 2 outer stamens*; ovary with 16–60 ovules, style short, *stigma often with 2 deep spreading lobes*. Fr. a ±flattened siliqua, the valves 1-veined, with conspicuous midrib and faint lateral network. Seeds compressed, in 1–2 rows in each cell.

About 10 spp. from Madeira and the Canary Is. to the Himalaya and in North America.

Very close to *Erysimum*, and best distinguished by the absence of median nectaries.

*1. C. cheiri L.   Wallflower.

A perennial herb with slender tap-root and an erect or ascending branched lfy stem, 20–60 cm, woody below, angled, and covered with forked appressed hairs. Basal lvs in a rosette, 5–10 cm, short-stalked; stem-lvs crowded, subsessile; all oblong-lanceolate, ±entire, with forked hairs especially beneath. Fls c. 2.5 cm diam. Petals bright orange-yellow, at least twice as long as the sepals. Siliquae 2.5–7 cm × 2–4 mm, ±erect on ascending stalks 5–15 mm, the valves conspicuously 1-veined, flattened, hairy. Seeds pale-brown, ±spherical, 3 mm, winged at the apex, 1-rowed or irregularly 2-rowed. Fls 4–6. Homogamous. The fragrant fls are visited freely by various bees and hover-flies. $2n = 14$. Ch.

Introduced. Well established on walls throughout lowland Great Britain, but not in Isle of Man, the Scottish Highlands, Hebrides, Orkney or Shetland. Widespread in Ireland. Channel Is. Probably native in the E. Mediterranean region, but widely naturalized through cultivation.

Besides *C. cheiri*, grown in many colour-varieties, **C. mutabilis** L'Hér. (Madeira, Tenerife), with petals at first whitish then passing through orange-brown to deep violet and fading to whitish again, is sometimes seen in gardens. (For Siberian Wallflower (*C. allionii* hort.) see *Erysimum*.)

42. ALLIARIA Scop.

Biennial or perennial herbs with slender tap-root and fl.-stems glabrous or with simple hairs; *lvs mostly reniform or cordate*, petiolate. Infl. bracteate or not. Sepals not saccate at base; *petals white*; stamens 6, without appendages; *nectar-glands* present round the bases of the short transverse stamens and also *outside the bases of the median pairs of long stamens*; ovary with short style and capitate slightly 2-lobed stigma. *Fr. an unbeaked siliqua* made ±4-angled by the prominent midribs of the 3-veined valves. *Seeds large* (3 mm), *in 1 row in each cell.*

About 5 spp. in Europe, N. Africa and Asia eastwards to Japan.

1. A. petiolata (Bieb.) Cavara & Grande
    Hedge Garlic, Garlic Mustard, Jack-by-the-Hedge.

A. officinalis Andrz. ex Bieb.; *Sisymbrium alliaria* (L.) Scop.

A biennial herb, often persisting from adventitious buds on roots, with tap-root smelling strongly of garlic and erect fl.-stem 20–120 cm, not or sparingly branched, sparsely hairy below but glabrous and pruinose above. *Basal lvs in a rosette, long-stalked, reniform*, crenate or distantly sinuate-toothed; stem-lvs short-stalked, triangular-ovate, acuminate, with cordate, sinuate or truncate base, deeply and irregularly sinuate-toothed; *all lvs thin, pale green, smelling of garlic when crushed.* Fls c. 6 mm diam. in racemes terminating main stem and branches, ebracteate except for 1–2 basal fls. Petals 5–7(–9) mm, about twice as long as sepals, obovate, abruptly narrowed into the claw. Fr. 30–70 × c. 2 mm, slightly torulose, straight or somewhat upcurved at the base and ascending on short stout spreading pedicels 4–8 mm; valves thinly hairy or glabrous, 3-veined, with prominent midrib and faint laterals. Seeds 3 × 1–1.5 mm, almost black. Homogamous. Visited by various small insects but automatically self-pollinated. $2n = 36^*$, c. 42. Hp.

Probably native, but earliest record of Roman date. A common plant of roadside hedgerows and wall-bases, wood margins and clearings, gardens, farmyards, waste places, etc., and locally frequent in some beechwoods on chalk escarpments; flourishing only where the soil has high levels of mineral nutrients, especially of phosphate. Throughout Europe to 68° N in Norway; N. Africa, W. and C. Asia.

43. SISYMBRIUM L.

Annual to perennial herbs, glabrous or with unbranched hairs and with simple to pinnatisect lvs. Infl. bracteate or not. Sepals not or slightly saccate at base; *petals usually bright or pale yellow*, rarely white, clawed; stamens 6, without appendages; *nectar-glands* round bases of transverse stamens and outside bases of median pairs, *joined in a complete ring*; ovary with distinct style and capitate or slightly 2-lobed stigma. Fr. a siliqua with beak short or 0; *valves convex with distinct midrib and usually 2 weaker laterals* or a faint lateral network. *Seeds numerous, small*, 0.5–2.5 mm, not winged, *in 1 row*, rarely 2, in each cell.

Many species, chiefly in temperate Europe, Asia and N. Africa, but there is much uncertainty about the native distribution of those occurring as widespread weeds.

1 All lvs simple, ±lanceolate, entire or dentate.
                                          7. strictissimum
   At least the lower lvs pinnately lobed or divided.   2
2 Siliquae 1–2 cm, held stiffly erect and appressed to
      the infl.-axis.   **1. officinale**
   Siliquae at least 2 cm, not appressed to infl.-axis.   3
3 Uppermost lvs sessile or nearly so, pinnately divided
      into linear or filiform segments.   **6. altissimum**
   Uppermost lvs petiolate or narrowed into a stalk-like
      base; simple or, if hastate or pinnatifid, then not
      with linear or filiform segments.   4
4 Ripe siliquae 5–10 cm, hairy at first but becoming gla-
      brous; uppermost lvs linear, entire.   **5. orientale**
   Ripe siliquae 2–5 cm, usually glabrous throughout.   5
5 Perennial; uppermost lvs linear-lanceolate, ±entire;
      tips of outer sepals horny; lateral veins of fr.-valves
      indistinct.   **4. volgense**
   Annual or overwintering; uppermost lvs hastate; tips of
      outer sepals not horny; fr.-valves distinctly 3-veined. 6
6 Petals pale yellow, little longer than sepals; young
      fr. overtopping open fls; ripe fr. 3–5 cm.   **2. irio**
   Petals bright yellow, c. twice as long as sepals; young
      fr. not overtopping open fls; ripe fr. 2–4 cm.
                                            3. loeselii

1. S. officinale (L.) Scop.   Hedge Mustard.

Erysimum officinale L.

An annual or overwintering herb with slender tap-root and stiffly erect stem, 30–90 cm, branched above, usually bristly with downwardly directed hairs. *Basal lvs in a rosette*, 5–8 cm, *deeply pinnatifid* with a round terminal lobe and 4–5 smaller lateral lobes on each side, all ±toothed; stem-lvs with a long hastate terminal lobe and 1–3 small oblong lateral lobes. *Infl. bractless*, at first corymbose but lengthening in fr. Fls 3 mm, diam., short stalked. Petals pale yellow, half as long again as the sepals. *Siliquae* 10–15 × 1 mm, held *stiffly erect* on short (2 mm) stalks and *±appressed to the axis*; their valves hairy (var. *officinale*) or glabrous (var. *leiocarpum* DC.), 3-veined. Seeds c. 1 mm, ovoid, orange-brown, c. 6 in 1 row in each cell. Fl. 6–7. Homogamous. Rarely visited by flies and small bees, and automatically self-pollinated. $2n = 14, 14 + 4$ ff. Th.

Native. In hedgebanks, by roadsides and in waste places and as a weed of arable land throughout the British Is. northwards to Orkney. Native throughout Europe to c. 67°N, and in N. Africa and the Near East. Naturalized in North and South America, S. Africa, Australia and New Zealand, Siberia and Greenland.

*2. S. irio L.   London Rocket.

An annual or overwintering herb with slender tap-root and erect usually branched stem 10–60 cm, glabrous or with short appressed hairs. Lower *lvs* not in a distinct rosette, stalked, deeply *pinnately lobed*, the terminal lobe larger than the 2–6 distant lateral lobes, all usually somewhat toothed; *upper lvs stalked* or narrowed into a stalk-like base, with hastate terminal lobe and a few smaller lateral lobes (Fig. 23A); uppermost lvs some-

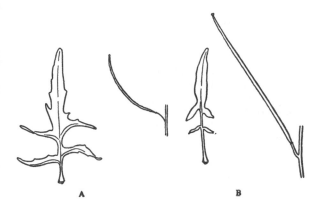

Fig. 23. Fruits and leaves of A, *Sisymbrium irio* and B, *S. orientale.* ×½.

times simply hastate. *Infl. ebracteate. Young fr. overtopping the open fls.* Fls 3–4 mm diam. Petals pale yellow, little longer than the sepals. *Siliquae* 3–5 cm × 1 mm, *glabrous*, narrowed at both ends, usually curved and ascending on slender stalks 7–10 mm (Fig. 23A); *valves 3-veined*, thin-walled and made nodular by the seeds which are visible through the translucent wall; *style not more than* 0.5 mm. Seeds ovoid, hardly 1 mm, yellowish-brown, c. 40, in a single row in each cell. Fl. 6–8. $2n = 14 + \text{ff}$. Th.

Probably not native. On roadsides, walls and waste places in a few scattered localities in England northwards to Cumbria and near Edinburgh; mainly casual but established in London and in Bedford. Formerly more widespread but now local and rare. Believed native in S. Europe but now occurring throughout Europe except the far north and in N. Africa, W. Asia and North America.

Called London Rocket from its abundance after the Great Fire of 1666; *S. orientale*, not *S. irio*, occurred on bombed sites in 1940–5.

*3. S. loeselii L. False London Rocket, Lösel's Rocket.

Like **2** but fl.-stems 20–100 cm or more and ±hispid; large terminal lobe of lvs triangular-hastate; *petals*

4–7 mm, *about twice as long as sepals*, *bright yellow*; *young siliquae not overtopping open fls*; ripe siliquae only 2–4 cm. Fl. 6–8. $2n = 14$. Th.

Introduced. A casual, often a grain-alien, widespread in the southern half of England and in Wales and frequent on tips near London; occasionally established. Native in C. and E. Europe and W. Asia but widely introduced.

*4. S. volgense Bieb. ex E. Fourn. Russian Rocket.

A perennial *rhizomatous patch-forming herb* with *fl.-stems* 30–75 cm, sparsely and minutely hairy at base, but *glabrous and glaucous* to pruinose elsewhere. Lower and middle lvs very variable, triangular to ovate in outline, irregularly toothed or lobed but usually pinnately lobed or cut towards the base and distinctly hastate; upper lvs linear-lanceolate to rhombic, long-cuneate, ±entire; the early-withering lowest lvs are sparsely and minutely hairy but all others *glabrous*; length of petiole diminishes upwards. Infl. ebracteate. Outer sepals with small horny swellings beneath tip. *Petals bright yellow*, 7–9 mm, *about twice as long as sepals*. *Ripe siliquae* 2.5–4 cm (but often shorter in British plants and with no viable seed), unbeaked, somewhat torulose, often slightly upcurved on spreading-ascending pedicels 4–8 mm; *valves 3-veined but with the lateral veins indistinct.* Seeds c. 1 mm. Fl. 6–8. $2n = 14$. Hs.

Introduced. A grain-alien occurring mainly near docks and flour-mills in a few localities in England and Wales northwards to S. Lancashire and Northumbria and tending to persist for many years. Native in S.E. Russia but widely introduced in C. and N. Europe.

*5. S. orientale L.   Eastern Rocket.

S. columnae Jacq.

An annual or overwintering herb with slender tap-root and erect branched stem 25–90 cm, ±hairy with short downwardly directed hairs. Basal lvs in a rosette but usually dying before flowering, long-stalked with large terminal lobe and c. 4 pairs of broadly triangular lateral lobes; *stem-lvs* with hastate terminal lobe and fewer and smaller lateral lobes (Fig. 23B); *uppermost* lvs hastate or simply *lanceolate*, entire; all hairy, grey-green, *stalked* or narrowing into a stalk-like base. Infl. ebracteate, much elongating in fr. Fls c. 7 mm diam. Petals pale yellow, twice as long as the sepals. *Siliquae* 4–10 cm × 1–2 mm, *at first hairy* but becoming ±glabrous, not narrowed at the lower end, ±straight, held obliquely erect on stalks 3–5 mm, which are as thick as the siliquae (Fig. 23B); valves 3-veined, thick-walled, neither beaded nor translucent. Seeds narrowly ovoid, c. 0.9 mm, yellowish-brown, c. 60 in a single row in each cell. Fl. 6–8. $2n = 14 + \text{ff}$. Th. Variable in hairiness and in the shape and size of the leaves and siliquae. Many varieties and 'forms' have been described, but their taxonomic status is not clear.

Introduced. A frequent casual on waste ground throughout England and Wales and in scattered localities northwards to Banff and Ross and establishing itself locally in waste places, on old walls, and on sea-cliffs in Scotland. Its frequent appearance on bombed sites during the war of 1939–45 led to its confusion with *S. irio* from which it differs in its simply lanceolate uppermost lvs and in its siliquae which are not narrowed at the lower end and not torulose and whose valves are not translucent. Native in S. and S.E. Europe, N. Africa and the Near East, but widely introduced elsewhere in Europe and in North America.

*6. S. altissimum L.   Tall Rocket.

S. pannonicum Jacq.; *S. sinapistrum* Crantz

An annual or overwintering herb with slender tap-root and erect branched stem, 20–100 cm, ±hairy below but usually glabrous and pruinose above. Basal lvs dying before flowering, stalked, roughly hairy, runcinate-pinnatifid with 6–8 pairs of narrowly triangular distantly toothed lobes; middle stem-lvs deeply pinnatisect with narrow distantly and irregularly toothed or hastate lobes; *uppermost sessile, glabrous, deeply pinnatisect with linear or filiform entire segments. Infl. ebracteate,* much elongating in fr. Fls c. 11 mm diam. Petals pale yellow, about twice as long as the spreading sepals. *Siliquae* 5–10 cm × 1–1.5 mm, ±glabrous, straight, and *held obliquely* erect on stalks 6–10 mm; valve thick-walled, with prominent midrib and much weaker lateral veins. Seeds narrowly ovoid, 0.8–1 mm, yellowish-brown, c. 60 in 1 row in each cell. Fl. 6–8. Homogamous. $2n = 14$. Th.

Introduced. A not infrequent casual of waste ground, rubbish-tips, etc., in many places in England and Wales and occasionally becoming established; in Scotland in a few scattered localities chiefly near the E. coast but extending northwards to Caithness; very local in Ireland. Native in S. and E. Europe and W. Asia, but widely introduced in Europe and North America.

*7. S. strictissimum L.

A *perennial* herb with a stout root-stock tasting like horse-radish and a stiffly erect, branched, very lfy stem reaching 1 m or more, ±pubescent below with short downwardly directed hairs. *Lvs* 3–8 × 1–3 cm, *simple, elliptical or ovate-lanceolate,* acuminate, entire or glandular-toothed, hairy beneath, shortly stalked; uppermost lvs lanceolate, sessile. Infl. of numerous racemose panicles terminating stem and branches. Fls 4–6 mm diam. Petals 5–10 mm, bright yellow, less than twice as long as the sepals. Siliquae 3–5 cm × 1–1.5 mm, reddish-brown, glabrous, curving upwards on stalks 4–6 mm; valves thick-walled, strongly 3(–5)-veined. Seeds narrowly ovoid, c. 2 mm, reddish-brown, 15–20 in 1 row in each cell. Fl. 6–8. Homogamous. H.

Introduced. A native of C. and E. Europe which has established itself in waste places in a few localities in Great Britain.

Microsisymbrium lasiophyllum (Hooker & Arn.) O. E. Schulz, with cream-coloured fls and slender deflexed siliquae; and **Lycocarpus fugax** (Lag.) O. E. Schulz (*Sisymbrium fugax*

Lag.) with the siliquae clavate and hooked distally, also occur as casuals.

44. ARABIDOPSIS (DC.) Heynh.

Slender annual to perennial herbs, glabrous or with *simple and branched hairs.* Infl. often bracteate. Inner sepals not or hardly saccate; petals white, lilac or yellow; stamens 6, without appendages; ovary with 20–72 ovules, style short, stigma entire or shortly 2-lobed. Fr. a *slender siliqua, the convex 1-veined valves having* ±*prominent midribs. Seeds ovoid, usually in 1 row in each cell.*

A few spp. chiefly in Europe, Asia and N. Africa, introduced elsewhere. The limits of this genus are still uncertain.

1. A. thaliana (L.) Heynh.
Thale Cress, Common Wall Cress.

Arabis thaliana L.; *Sisymbrium thalianum* (L.) Gay

An annual or occasionally biennial herb with slender tap-root and 1 or more erect stems, 5–50 cm, branched above in large plants, roughly hairy below with mostly simple hairs, ±glabrous above. Basal lvs in a rosette, elliptical or spathulate, stalked, grey-green with simple and branched hairs; stem-lvs sessile, narrowly oblong or lanceolate, narrowed to a non-clasping base, ±glabrous or with branched hairs beneath and on the margins; lower lvs distantly toothed, upper stem lvs entire. Fls 3 mm diam. Petals white, 2–4 mm, c. twice as long as sepals. Siliquae 10–18 × 0.8 mm, somewhat curved, held obliquely erect on slender spreading stalks 5–10 mm; valves glabrous, with conspicuous midribs and very faint lateral veins. Seeds yellowish-brown, 0.5 mm, 25–40 ±in 1 row in each cell. Fl. 4–5 and sometimes 9–10. Homogamous. Sometimes gynomonoecious. Visited by various small insects and automatically self-pollinated. $2n = 10$. Th.–H.

Native. Fairly common on walls and banks, in hedgerows and waste places, and on dry soils throughout the British Is. northwards to Orkney. Europe to 68° 30′ N in Sweden, Mediterranean region, E. Africa, N. and C. Asia to Japan. Introduced in North America, S. Africa and Australia.

45. CAMELINA Crantz

Annual or overwintering herbs with erect fl.-stems and *simple sagittate stem-lvs, sessile and* ±*amplexicaul.* Infl. ebracteate. Sepals erect, not saccate at base; *petals yellow,* often pale or almost white; stamens 6, without appendages; nectar-glands only at base of each short transverse stamen; ovary with distinct style and capitate stigma. *Fr. an obovate or pear-shaped latiseptate silicula* with persistent style; *valves convex, with midrib and lateral network of veins, their margins* ±*narrowly appressed and simulating a wing; fr. ending in a short beak round the base of the style.* Seeds numerous in (1–)2 rows in each cell.

Perhaps 10 spp. in Europe, the Mediterranean region and Asia, but introduced over a much wider area.

*1. C. sativa (L.) Crantz, *sens. lat.*   Gold-of-Pleasure

Myagrum sativum L.

Annual or overwintering herbs with slender yellow tap-root and erect lfy fl.-stems 25–80(–100) cm, branched above, ±glabrous or with simple and branched hairs. *Basal lvs oblong-lanceolate to spathulate*, entire, sinuate-toothed or somewhat pinnately lobed, *narrowed into the sessile base or petiolate; middle and upper lvs lanceolate to linear-oblong, sessile, semi-amplexicaul with short acute auricles*, entire or sinuate-toothed; all ±glabrous or with long simple and short stellate hairs. Petals up to 5 mm, c. 1½ times as long as sepals. Fr. usually pear-shaped or obovoid, sometimes almost globose, 3–10 mm, 2–4 times as long as the style; the convex valves usually becoming hard and woody at maturity. Seeds 0.7–2(–2.5) mm, numerous, usually in 2 rows in each cell. Fl. 5–7. Homogamous. Visited by bees and automatically self-pollinated. $2n = 40$ (28, 42). Th.

The above description covers several variants often regarded as separate species but not always clearly distinguishable and therefore better treated as subspecies of *C. sativa sens. lat.* One of these is a common casual and two others occur less frequently:

1 Stem and lvs ±glabrous or with sparse appressed mostly stellate hairs with few or no simple hairs; fr. 7–9 mm (excl. style), 3–4 or more times as long as style; spring-germinating annual.   subsp. **sativa**
  Stem and lvs ±densely covered with long simple and short stellate hairs; fr. 3–7 mm (excl. style), 2–3 times as long as style; autumn-germinating annuals.   2
2 Basal lvs oblong-lanceolate, short-stalked; fr. 4–6 mm, c. twice as long as style, the valves with broad wing-like margin, not strongly convex and with midrib not very prominent; seeds 0.7–0.8 mm.   subsp. **microcarpa**
  Basal lvs spathulate, long-stalked; fr. 4–7 mm, c. 3 times as long as style, the valves with narrow wing-like margin, strongly convex and with midrib forming a distinct keel; seeds 1.2–1.8 mm.   subsp. **pilosa**

Subsp. **sativa** (*C. sativa* (L.) Crantz, *sens. str.*)
Plant 30–90 cm, fresh green. Basal lvs oblong-lanceolate, entire or distantly denticulate, narrowed into a sessile base. Fr. pear-shaped, c. 1½ times as long as wide, with narrow wing-like margin; valves strongly convex, reticulate, with prominent midrib, woody and hard at maturity. Seeds 1–2 mm, brown.

Introduced. Formerly an occasional weed in fields of flax, cereals and lucerne but has become rare with improved cleaning of seed though still sometimes seen near docks, corn silos and flour-mills; now more often introduced in seed for cage-birds and a frequent casual in gardens and waste places and on rubbish tips. Through much of Great Britain and Ireland. Probably native in E. Europe and W. Asia.

Subsp. **pilosa** (DC.) Thell. (*C. sativa* var. *pilosa* DC.; *C. pilosa* (DC.) Vassilcz.)
Plant 30–60 cm, grey-green. Basal lvs spathulate, long-stalked. Fr. pear-shaped with narrow wing-like margin; valves strongly convex and ±keeled by the prominent midrib, woody and hard at maturity. Seeds 1.2–1.8 mm, brown.

Introduced. A rare casual, intermediate between subspp.

sativa and *microcarpa*, with the fr. of the former but the long simple hairs of the latter, and often difficult to distinguish.

Subsp. **microcarpa** (Andrz.) Thell. (*C. microcarpa* Andrz.)
Plant 30–60 cm, grey-green. Basal lvs oblong-lanceolate, short-stalked. Fr. commonly c. 5 mm, pear-shaped with broad wing-like margin; valves less strongly convex than in subsp. *sativa* and with the midrib less prominent; woody and hard at maturity. Seeds only 0.7–0.8 mm, dark red-brown.

Introduced. A rare casual, sometimes not readily distinguishable from subsp. *pilosa* or even subsp. *sativa*. Formerly a weed of autumn-sown crops, in contrast with *sativa*, which was characteristic of spring-sown crops.

Subsp. **alyssum** (Miller) Thell. (*C. alyssum* (Miller) Thell.; *C. dentata* Pers.; *C. foetida* (Schkuhr) Fries), with short simple and appressed stellate hairs and fr. 7–10 mm, often almost globose and with strongly convex but not hard-walled valves, formerly occurred as a casual but has not been seen for some years.

46. DESCURAINIA Webb & Berth.

Usually annual herbs with *finely pinnatisect lvs* and *branched or stellate* and often glandular as well as simple *hairs*. Infl. ebracteate. Fls small; sepals non-saccate; petals yellowish, usually not exceeding the sepals; stamens 6, without appendages, often exceeding the petals; ovary with 6–85 ovules; style very short, stigma capitate. *Fr. a short siliqua*, the somewhat *convex valves with a strong midrib* and a lateral network of smaller veins. Seeds small, in 1–2 rows in each cell.

About 50 spp. chiefly in North and South America with a few in Europe, Asia and Macaronesia.

Differs from *Sisymbrium* in the finely pinnatisect lvs and branched hairs.

1. D. sophia (L.) Webb ex Prantl   Flixweed.

Sisymbrium sophia L.

An annual or overwintering herb with slender tap-root and erect terete stem 30–80 cm, branched above, usually grey with stellate hairs below but ±glabrous above. Lvs greyish-green, ±stellate-hairy, bi- or tripinnatisect, the segments ±linear; uppermost lvs sometimes almost simply pinnate with linear segments. Fls 3 mm diam. Petals pale yellow, about as long as the sepals. Siliquae 15–25 × c. 1 mm, ±cylindrical, curving upwards and held almost erect, making an angle with their spreading very slender stalks c. 1 cm. Seeds narrowly ovoid, c. 1 mm, orange-brown, 10–15 in 1 row in each cell. Fl. 6–8. Said to be self-pollinated. $2n = 28$, (56). Th.

Probably native. Roadsides and waste places, especially in East Anglia but scattered throughout England and Wales and in E. Scotland northwards to Moray Firth, though nowhere common; very rare in Ireland; Channel Is. Decreasing. Apparently native throughout Europe to c. 65° N, in N. Africa and across Asia to China and Japan. Introduced in North and South America and N. Zealand. Several American spp. of *Descurainia* have been reported as casuals.

29. RESEDACEAE

Annual to perennial herbs, rarely woody, with alternate simple or pinnately divided lvs, often glandular-stipulate, and fls in bracteate racemes or spikes. Fls 4–7-merous, zygomorphic, hypogynous or perigynous, hermaphrodite or rarely unisexual. Calyx usually zygomorphic, of 2–8 persistent sepals; corolla of (1–)2–8 free petals, entire or laciniate, those at the posterior (top) side being largest; stamens 3–40, inserted on the zygomorphic disk, those at the anterior side largest and most crowded; carpels 2–6, superior, free or united below into a 1-celled ovary which often remains open at the top; ovules numerous, campylotropous, on 2–6 parietal placentae. Entomophilous; nectar secreted by a hypogynous disk. Fr. a capsule open at the top, rarely a berry or group of follicles; seeds numerous, with a curved embryo and no endosperm.

Six genera and about 70 spp., chiefly in the Mediterranean region but extending to Somalia and India and with a few spp. in the Canary Is., the Cape, and California and New Mexico.

1. RESEDA L.

Annual to perennial herbs with simple, pinnatifid or pinnate lvs with glandular stipules. Fls in spike-like racemes, hypogynous, usually hermaphrodite. Sepals 4–7. Petals 4–7, those at the back with larger and more deeply and repeatedly lobed limbs than those at the front. Stamens 7–40, ±free, crowded to the front, inserted on the fleshy nectar-secreting disk which is broadest at the back. Carpels 3–6, united below, each with an apical stigma-bearing lobe, the ovary being open between these lobes. Fr. a 1-celled capsule opening more widely by the spreading of the apical lobes, but not splitting.

About 60 spp. chiefly in the Mediterranean region and E. Africa.

1 Ovary and capsule usually with 4 apical lobes; petals
  white.   **3. alba**
  Ovary and capsule with 3 apical lobes; petals yellowish.   2
2 Lvs simple entire; sepals and petals usually 4; capsule
  c. 4 mm, subglobose.   **1. luteola**
  Lvs pinnately lobed or divided; sepals and petals
  usually 6; capsule 8–16 mm, ±oblong-ellipsoid.
                                    2. lutea

1. R. luteola L.   Dyer's Rocket, Weld.

A biennial glabrous herb with a deep tap-root, producing a rosette of lvs in the first season and a flowering stem in the second. *Stem* 50–150 cm, stiffly erect, ribbed, hollow, *simple* or with a few erect branches. Rosette lvs commonly 5–8 cm, narrowly oblanceolate, sessile; *stem-lvs* narrowly oblong, the lower narrowed into a stalk-like base, the upper sessile; all *with entire ±undulate margins.* Fls 4–5 mm diam., in long slender spike-like terminal racemes with or without shorter lateral racemes, yellowish-green, on ascending stalks hardly equalling the sepals. Sepals 4, not accrescent. Petals 4 (3–5) those at the back and sides with the limb divided

into 3 or more lobes, the front petal usually entire, linear; all with a small scale-like claw. Stamens 20–25, ±downwardly curved. Capsule c. 4 mm, ovoid to subglobose, divided nearly halfway into 3 acuminate lobes. Seeds 0.8–1 mm, brownish-black, smooth, shining. Fl. 6–8. Homogamous; visited by small bees and self-pollinated. $2n = 24$–28. Hs. (biennial).

Native. A locally common plant of roadsides and waste places, old quarries, gravel-pits and brickyards, dry banks and old walls and sometimes in woodland clearings, partly fixed sand dunes and cultivated ground; often but not exclusively on calcareous substrata. Throughout lowland England, Wales and S.E. Scotland and in scattered localities further north to Moray and Ross; Ireland; Channel Is. Most of Europe northwards to c. 66° N in Sweden and Finland; W. Asia; N. Africa; Canary Is. Introduced in North America. Formerly much cultivated for the yellow dye from its lvs.

2. R. lutea L.   Wild Mignonette.

A biennial to perennial glabrous herb with a deep tap-root, a woody stock and erect or ascending *diffusely branched stems*, 30–75 cm, ribbed, solid, ±rough with whitish tubercles, with ascending branches. Basal lvs commonly 2.5–8 cm, in a rosette withering early; stem *lvs numerous; all ±pinnatifid* with 1–3(–5) pairs of narrowly oblong, blunt, entire or pinnatifid lobes diverging rather widely from the central part of the lf which is narrowly elliptical to broadly oblanceolate in outline but constricted above the insertions of the lobes; the whole with ±undulate margins. Fls c. 6 mm diam., greenish-yellow, in short compact conical racemes; stalks erect, tubercled, exceeding the sepals. Sepals usually 6, linear, unequal. Petals usually 6, with ±rounded claws, the 2 upper with 3-fid, the 2 lateral with 2–3-fid, the 2 lower with entire linear limbs. Stamens 12–20, ±downwardly curved, inserted on the nectar-secreting disk which is broadest at the back of the fl. Capsule 8–16 mm, oblong, short-stalked, tubercled, opening by the further divergence of the 3 stigma-bearing lobes. Seeds 1.6–2 mm, black, smooth, shining. Fl. 6–8. ±Homogamous; visited by small bees and other insects, and self-pollinated. $2n = 48$. H.

Variable in lf-shape and in roughness of the stem and capsules.

Native. Locally frequent on dry calcareous, less commonly sandy, soils of roadsides, railway banks and sidings, arable land and waste places, and also in disturbed chalk and limestone grassland, e.g. near rabbit warrens, quarries or chalk-pits; occasional on partly fixed sand dunes. Throughout lowland England and Wales but only in scattered localities in Scotland, mainly near the E. coast northwards to Moray; very local in Ireland; Channel Is. Perhaps native only in W. Europe and the Mediterranean region but widely naturalized further north and east and reaching 66° N in Sweden and Finland; W. Asia; N. Africa. Introduced in North America.

*3. R. alba L.   Upright Mignonette.

Incl. *R. suffruticulosa* L.

An annual to perennial glabrous and somewhat glaucous herb with a deep tap-root and a branched ±woody stock from which arise several erect stems, 30–75 cm, ribbed, hollow. *Lvs deeply pinnatifid* with 5–8 pairs of narrow unequal decurrent lobes whose margins are entire and ±undulate. *Fls* c. 9 mm diam., *whitish*, in dense conical racemes, their ascending stalks about equalling the sepals. Sepals usually 5, linear, acute. Petals usually 5 with short roundish claws and 3-fid subequal limbs. Stamens 11–14 inserted on the funnel-shaped disk which is broadest at the back. Capsule c. 12 mm, oblong, contracted above, opening by the further divergence of the usually 4 short stigma-bearing lobes. Seeds c. 1 mm, reniform, brownish, dull and tubercled. Fl. 6–8. $2n = 20$.

(Th.–)H., sometimes biennial.

Introduced. Waste places. A frequent casual near ports and occasionally establishing itself, especially in S.W. England. Mediterranean region; W. Asia to Iran.

*R. phyteuma L., native in S. Europe, W. Asia and N. Africa, is a rare casual recognizable by its oblong-spathulate basal lvs, fls with 6 sepals much enlarging in fr. and 6 whitish deeply laciniate petals, and a drooping obovoid-cylindrical capsule c. 14 × 7 mm which is open above through the divergence of its 3 stigma-bearing lobes.

*R. odorata L., the very fragrant and much grown Mignonette, occasionally escapes from gardens. Native in Libya.

*Sesamoides pygmaea (Scheele) O. Kuntze (*Astrocarpus sesamoides* (L.) DC.), with 4–7 ±free carpels spreading widely in fr., each with a single seed, is recorded as a casual. Native in S.W. Europe.

30. VIOLACEAE

Trees, shrubs or herbs. Lvs usually stipulate, simple. Infl. of racemose type or fls solitary. Pedicels with 2 bracteoles. Fls 5-merous (except gynoecium) regular or zygomorphic, hermaphrodite, hypogynous, heterochlamydeous, corolla spurred when zygomorphic. Stamens 5, alternate with the petals, the 2 lower spurred in zygomorphic fls; anthers connivent round the ovary, introrse; filaments short; connective often elongated. Ovary 1-celled with (2–)3(–5) parietal placentae with 1–many anatropous ovules; style simple, often curved or thickened above; stigma very variously developed. Fr. a capsule or berry; endosperm fleshy, copious; embryo straight.

About 22 genera and 900 spp., almost cosmopolitan. A natural family not clearly related to any other.

1. VIOLA L.

Herbs, rarely undershrubs. Lvs alternate, stipulate (stalked in the British spp.). *Fls solitary*, rarely 2. *Sepals prolonged into appendages below their insertion. Corolla zygomorphic, the lower petal spurred.* Two lower stamens spurred; connective broad. Ovary with 3 placentae; ovules numerous; style thickened above, straight or curved. Fr. a 3-valved capsule; valves elastic.

About 500 spp., mainly temperate.

1 Style hooked or obliquely truncate; stipules entire to fimbriate, not lobed; lateral petals directed downwards.   *2*

　 Style expanded above into a globose head; stipules pinnatifid or palmatifid; lateral petals directed upwards.   *11*

2 Style hooked at apex.   *3*

　 Style straight, obliquely truncate at apex; lvs orbicular-reniform; plant with long slender creeping rhizome.   **9. palustris**

3 Sepals obtuse; plant acaulescent (i.e. lvs and fls all basal); petioles and capsules pubescent.   *4*

　 Sepals acute; plant normally caulescent though in small forms often appearing acaulescent; petioles and capsules glabrous (except *rupestris*).   *5*

4 Plant with long stolons; fls sweet-scented, dark violet or white; hairs on petioles deflexed.   **1. odorata**

　 Plant without stolons but with short rhizome; fls odourless, usually blue-violet; hairs on petioles spreading.   **2. hirta**

5 Petioles and capsules pubescent; lvs all or mostly obtuse.   **3. rupestris**

　 Petioles and capsules glabrous; lvs all or mostly acute or acuminate.   *6*

6 Main axis ending in a rosette of lvs, not growing out into a fl. stem; lvs ovate-orbicular; teeth of stipules usually filiform, flexuous, spreading.   *7*

　 Main axis without basal rosette, growing out into a fl. stem; lvs ovate to lanceolate; teeth of stipules triangular-subulate, ±straight, ascending.   *8*

7 Appendages of sepals 2–3 mm, accrescent in fr.; corolla blue-violet, spur paler, stout, furrowed or notched.   **4. riviniana**

　 Appendages of sepals not more than 1 mm, obsolete in fr.; corolla lilac, spur darker, slender, not furrowed or notched.   **5. reichenbachiana**

8 Corolla deep or bright blue; stipules rarely more than $\frac{1}{3}$ as long as petiole; lvs usually ovate.   **6. canina**

　 Corolla pale or nearly white; middle and upper stipules usually half as long as petiole or more; lvs lanceolate.   *9*

9 Lvs rounded or broad-cuneate at base, widest at about $\frac{1}{3}$ of the distance to the apex, rather thick. Heaths.   **7. lactea**

　 Lvs truncate or cordate at base, widest very near base. Fens.   *10*

10 Spur short, ±conical, not or scarcely longer than the appendages of the calyx; plant creeping below the ground; lvs thin.   **8. persicifolia**

　 Spur about twice as long as the appendage of calyx; plant not creeping; lvs thick.   **6. canina**

11 Perennial; stylar flap (see p. 110) large and conspicuous; spur usually twice as long as appendages of sepals or more; petals much longer than sepals.   *12*

　 Annual; stylar flap 0 to distinct; spur usually less than twice as long as appendages; petals shorter or longer than sepals.   *13*

12 Plant with long creeping rhizome; stipules ±divided

but terminal segment not large or crenate; corolla large (2–3.5 cm vertically) on long (5–9 cm) peduncles.　　　　　　　　　**10. lutea**

Rhizome 0 or short and scarcely creeping, plants ±tufted; stipules pinnately divided, the terminal segment usually crenate; corolla moderate (1.5–2.5 cm vertically) on often shorter peduncles.　　　　　　　　　**11. tricolor**

13 Corolla blue-violet or mainly so, moderate (1.5–2.5 cm vertically); petals longer than sepals; stylar flap distinct; mid-lobe of stipules usually lanceolate and entire, not lf-like.　　**11. tricolor**

Corolla cream or occasionally tinged blue in upper part (rarely pale blue-violet but then very small), usually smaller (0.5–2 cm vertically); petals usually shorter than or as long as sepals but sometimes longer; stylar flap 0 or indistinct; mid-lobe of stipules (at least the lower) usually ovate-lanceolate, crenate-serrate, lf-like.　　　　　　14

14 Corolla 8–20 mm vertically, ±flat; plant usually large and much branched.　　　　　　　**12. arvensis**

Corolla c. 5 mm vertically, concave; plant small, not or sparingly branched. (Scilly and Channel Is.)　　　　　　　　　**13. kitaibeliana**

Section 1. VIOLA　　　　　　　　Violet.

Stipules not lf-like. Lateral petals directed downwards. Style hooked or obliquely truncate at apex. Cleistogamous apetalous fertile fls produced in summer after the open (chasmogamous) ones with petals have ripened seed (fl. times refer to chasmogamous fls). The open fls of all spp. are pollinated by bees or, less often, by other insects (apparently never selfed in nature though self-fertile).

Subsection 1. *Viola*. Style hooked at apex, tapered, the hook about as long as the diam. of the style. Plants acaulescent, i.e. with lvs and fls all basal. Sepals obtuse. Capsules lying on the ground and spilling out their seeds.

1. V. odorata L.　　　　　　　Sweet Violet.

Perennial herb with short thick rhizome and *long procumbent stolons rooting at the ends*. Lf-blades 1.5–6 cm ovate-orbicular or the summer ones broadly ovate, deeply cordate at base, obtuse or the summer ones acute, crenate-serrate, dark green, sparingly hairy or subglabrous; *petioles* long, *with short deflexed hairs* or those of the spring lvs subglabrous; stipules ovate or ovate-lanceolate, glandular-fimbriate-toothed. Peduncles about equalling lvs; bracts about or above the middle. *Fls sweet-scented*. Sepals oblong, appendages spreading. *Corolla* c. 1.5 cm, *deep violet or* as commonly *white* with pale lilac or violet spur, rarely purple or pink, very rarely apricot; the base whitish; four upper petals obovate-oblong. Capsule globose, obscurely trigonous, pubescent. Fl. (9–)2–4. 2n = 20*. Hr.

Native. Hedgebanks, scrub and plantations (rarely natural woods), usually on calcareous soils. From Durham and Westmorland southwards and in C. Ireland, rather common; naturalized northwards to Morayshire, and sporadically throughout Ireland. Europe, except the extreme north and parts of the Mediterranean region; Asia Minor, Caucasus, Syria, Palestine; N. Africa; Macaronesia.

V. hirta × odorata (*V. × permixta* Jordan)

A rather frequent hybrid combining the characters of the parents in various ways. Fertility variable in different plants, usually partially sterile. Often very vigorous and floriferous.

2. V. hirta L.　　　　　　　Hairy Violet.

Differs from *V. odorata* as follows: *No stolons. Lvs narrower*, triangular-ovate or oblong-ovate, hairier, lighter green, small at fl., increasing greatly in size later; *petioles with* longer, more numerous, *spreading hairs*; stipules lanceolate. Peduncles often exceeding lvs; bracts usually below the middle. *Fls without scent*. Appendages of sepals appressed. *Corolla blue-violet* (paler and bluer) rarely white or white streaked with violet or pink. Capsule pubescent, rarely glabrous. Fl. 4–5. 2n = 30*. Hr.

Native. Calcareous pastures, scrub and open woods from Kincardine southwards, widespread and rather common on suitable soils, ascending to 595 m; rare in Ireland (Dublin, Leix, Limerick, Clare). Most of Europe; Caucasus, Turkestan, Siberia (to 95° E).

Plants with lvs scarcely longer than broad, the corolla less than 1 cm and with the spur scarcely longer than the calyx-appendages, have been recognised as a distinct taxon (*V. calcarea* (Bab.) Gregory, *V. hirta* subsp. *calcarea* (Bab.) E. F. Warb.). In view of the many intermediates and mixed populations, even varietal status seems dubious.

Subsection 2. *Rostratae* Kupffer. Style hooked at apex, the hook shorter than the diam. of the style. Plants caulescent, the fls solitary, axillary (dwarf specimens sometimes appear acaulescent). Sepals acute. Capsule erect, shooting out its seeds.

3. V. rupestris Schmidt　　　　Teesdale Violet.
V. arenaria DC.

Small perennial tufted herb, 2–4 cm in fl., to 10 cm in fr., *pubescent all over*; stock thick, clothed with the remains of dead lvs, not creeping; *with central nonflowering rosette*, fls on axillary branches. *Lf-blades* 5–10 mm, *reniform or ovate-orbicular*, *obtuse* (rarely a few shortly acuminate), *truncate or shallowly cordate at base*, crenulate; stipules ovate-lanceolate, fimbriate, much shorter than *pubescent petiole*. Sepals ovate-lanceolate, appendages subquadrate in fr. Corolla pale blue-violet, spur paler, lower petal without dark zone and with the veins fainter and shorter than in *V. riviniana*, petals obovate-oblong; spur thick, ±furrowed, at least twice as long as appendages. *Capsule* c. 6 mm, ovoid-oblong, trigonous, *pubescent*. Fl. 5. 2n = 20*. Hs.

Native. Open mossy sheep-grazed turf or bare ground on limestone in Upper Teesdale (Durham) and E. and S. Westmorland; 135–600 m. N. Europe and the mountains of C. Europe from 70° 20′ N in Norway to E.C. Spain, Corsica, Italian Alps and Macedonia; N. and

C. Asia to c. 145° E, south to the Himayala and Caucasus; North America from E. Quebec to Alaska, south to Maine and Oregon.

Glabrous forms (rather rarely) occur on the Continent but all the records from this country are errors.

4. V. riviniana Reichenb.   Common Dog-violet.

Perennial 2–20 cm in fl., to 40 cm in fr., *glabrous or somewhat pubescent*; stock rather slender, short, erect, not creeping, but shoots arise in some plants from adventitious buds on the roots; *with central, non-flowering rosette*, fls on axillary branches. *Lf-blades* 0.5–8 cm, *ovate-orbicular* (slightly longer than broad), shortly and often bluntly acuminate (or those of the rosette obtuse), *deeply cordate at base*, crenate; *stipules* much shorter than *glabrous petiole*, lanceolate, *fimbriate*, or the upper almost entire, the fimbriae long, spreading, filiform, often flexuous. Peduncle glabrous or hairy. Sepals 7–12 mm, lanceolate; *appendages large*, about $\frac{1}{4}$ to $\frac{1}{3}$ as long as sepals, often emarginate, the lowest subquadrate, *accrescent in fr.* Corolla (Fig. 24A) 14–22 mm,

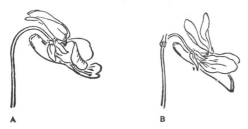

Fig. 24. Flowers of A, *Viola riviniana* and B, *V. reichenbachiana.* ×1.5.

usually *blue-violet* but variable in colour, *spur paler,* whitish or pale violet, or more rarely yellowish, lower petal usually with dark zone outside the whitish base which has numerous long dark veins; petals broadly obovate, usually overlapping (the lowest to 7 mm broad); *spur* c. 5 mm, *thick,* scarcely tapering, *furrowed or notched at apex. Capsule* 6–13 mm, trigonous, acute, *glabrous.* Very variable. Fl. 4–6 (rarely again 8–10). $2n = 40^*$. Hs.

Native. Woods, hedgebanks, heaths, pastures and mountain rocks, ascending to 1020 m, on all types of soil if not very wet. Common and often abundant throughout the British Is. except the south-east. Throughout Europe; Morocco (Atlas); Madeira.

Plants, with smaller lvs and fls, in more exposed habitats have been recognized as subsp. *minor* (Murb. ex Gregory) Valentine. The numerous intermediates suggest that such ecotypic variation does not merit formal recognition.

5. V. reichenbachiana Jordan ex Boreau
Early Dog-violet.

V. sylvestris Lam., *pro parte*

Scarcely distinguishable from *V. riviniana* except by the fls or fr. but often shorter (8–15 cm in fl., to 25 cm in

fr.), more slender and less hairy, the peduncles always glabrous; stipules narrower and the upper more fimbriate. *Appendages of sepals small, very short, entire, obsolete in fr. Corolla* (Fig. 24B) 12–15 mm, *lilac,* rarely pink or white, *spur dark purple,* lower petal with dark zone and with few short dark veins; petals narrower, not overlapping; *spur slender, laterally compressed,* tapering, *not furrowed or notched.* Fl. 3–5 (earlier than *V. riviniana*). $2n = 20^*$. Hs.

Native. Woods, hedgebanks, etc., usually on calcareous soils. Rather common in S., C. and E. England, more local in N. England, N., E. and S. Wales and in Ireland S. to Cork, very rare in S. Scotland. S., W. and C. Europe from S. Sweden southwards; Caucasus, Kashmir; Morocco (Atlas); Madeira, Canary Is. (a subsp.).

6. V. canina L.   Heath Dog-violet.

Perennial herb 2–30 cm (–40 cm in fr.), glabrous or sparingly and shortly pubescent; stems decumbent to erect, solitary to many together from a short creeping rhizome; *no central non-flowering rosette. Lf-blades* 0.5–8 cm, *ovate to ovate-lanceolate,* often ±triangular in outline, the upper sometimes markedly narrower than the lower, obtuse or subacute or a few bluntly acuminate, *truncate or widely and shallowly, rarely deeply, cordate at base,* often somewhat decurrent, shallowly crenate or crenate-serrate, bright or dark green, rather thick; *stipules* ±lanceolate, *subentire, distantly serrate-dentate or fimbriate-dentate, the teeth usually ascending, straighter, stouter, shorter and fewer than those of* V. riviniania, $\frac{1}{3}$ *the length of the glabrous petiole or less* (to $\frac{1}{2}$ as long at the middle of the stem, longer in the upper part, in some fen forms). Sepals ±lanceolate, appendages rather large. *Corolla* 7–18(–22) mm, *blue* with very little violet tint, *spur usually yellowish* (occasionally greenish-white), lower petal without a dark zone outside the whitish base, petals obovate (usually about twice as long as broad); spur thick, usually straight, obtuse, sometimes furrowed, usually about twice as long as the appendages of the calyx but sometimes little longer, ±cylindrical. Capsule c. 8–9 mm, glabrous, trigonous, blunt, often apiculate. Fl. 4–6 (later than *V. riviniana*). Hp.

A very variable sp., probably divisible into several subspp. The forms of the drier habitats are often ±constant and characteristic but the fen plants are extremely variable. Experimental work is much needed. The sp. is usually recognizable by its flower colour when once known (apart from other characters). The two following subspp. are generally recognized.

Subsp. canina

Upper margin of lf concave; stipules rarely long. Corolla 7–18 mm, deep or bright blue. Very variable. $2n = 40^*$. Heaths, dry grassland, dunes and fens, ascending to 430 m. Throughout the British Is. but rather local. Most of Europe except Greece and the Balearic, Crete and Sicilian Islands; Greenland; N.W. Asia Minor, C. Asia (rare).

Subsp. **montana** (L.) Hartman

Stems erect. Lvs ovate-lanceolate-triangular, often asymmetric, margins straight or convex; middle stipules half as long as petiole, upper often as long. Corolla 15–22 mm, pale blue, petals oblong; spur greenish, curved upwards. Only known from fens in Cambridge and Huntingdon. Much of Europe (not in W. France or Portugal); N. asia to Kamchatka and Manchuria.

7. **V. lactea** Sm.   Pale Dog-violet.

Perennial herb 4–20 cm, subglabrous; stems ascending, solitary or few; *no central non-flowering rosette*; foliage sometimes purplish tinged. *Lf-blades* (except the small ovate lowest ones which usually soon disappear) 1–4.5 cm, *ovate-lanceolate or lanceolate*, subacute, *rounded or broadly cuneate*, or occasionally subcordate, *at base*, broadest at about $\frac{1}{3}$ of the distance from base to apex, shallowly crenate-serrate, dark green, *rather thick*; stipules (except the lower) large, the middle ±lanceolate, c. $\frac{1}{2}$ as long as the rather short petioles, *the upper ovate-lanceolate, conspicuous, coarsely and irregularly fimbriate-serrate* or dentate, *as long or somewhat longer than the petioles*. Sepals lanceolate (rarely ovate-lanceolate); appendages rather large. *Corolla* 12–20 mm, *pale greyish-violet*; spur yellowish or greenish; *petals oblong, subacute*, c. 3–4 *times as long as broad*; *spur thick, obtuse, about twice as long as appendages of calyx*, cylindrical. Capsule glabrous, acuminate. Fl. 5–6 (later than *V. canina*). $2n = 58*$. Hp.

Native. On heaths; in scattered localities from Anglesey and Pembroke to Sussex and Essex, very local, commonest in S.W. England; Kerry, W. Cork, Clare and Connemara, Wexford; W. and C. France, N.W. Spain, N. Portugal.

8. **V. persicifolia** Schreber   Fen Violet.

V. stagnina Kit.

Perennial herb 10–25 cm, subglabrous, *creeping underground* and sending up stems at intervals from adventitious buds on the roots; *no central non-flowering rosette*. *Lf-blades* 2–4 cm, *triangular-lanceolate*, subacute, *truncate or subcordate at base*, often somewhat decurrent, shallowly crenate-serrate, light green, *rather thin*; stipules lanceolate, subentire or distantly serrate-dentate or fimbriate, the *middle ones usually* c. $\frac{1}{2}$ *as long as the petiole*, but variable, the upper often as long. Sepals lanceolate; appendages large. *Corolla* 10–15 mm, appearing almost circular in front view, *bluish-white or white*, spur greenish; *petals obovate-orbicular, scarcely longer than broad*; *spur obtuse, not or scarcely longer than the appendages of the calyx*, ±conical. Capsule c. 7 mm, glabrous, ovoid, acute. Fl. 5–6. $2n = 20*$. Hp.

Native. Fens and marshes in E. England from Oxford to S. Yorks; damp grassy hollows on the limestone in C. and W. Ireland from Fermanagh to Clare; very local. Most of Europe except the extreme north, the south-east and the Mediterranean region; C. Siberia to the Altai.

The following hybrids are often to be found where the parents occur together. They are often very vigorous

and in general sterile, though some may produce a small amount of seed: *V. canina* × *lactea*, *V. canina* × *riviniana*, *V. canina* × *persicifolia*, *V. lacta* × *riviniana*, *V. reichenbachiana* × *rivinica*, *V. riviniana* × *rupestris*.

Subsection 3. *Plagiostigma* (Godron) Kupffer. Style straight, obliquely truncate at apex. Plants acaulescent with very long creeping rhizome. Capsule erect, shooting out its seeds.

9. **V. palustris** L.   Marsh Violet.

Perennial herb with long slender creeping rhizome, emitting lvs at the nodes; *aerial stems* 0. Lvs usually 3–4, blades 1–4 cm, *orbicular-reniform*, usually very obtuse, sometimes some subacute or bluntly acuminate, cordate at base, obscurely crenate; stipules ovate, glandular-denticulate. *Sepals* oval, *obtuse. Corolla* 10–15 mm, lilac, rarely white, with darker veins; petals obovate; spur obtuse, longer than appendages. Capsule glabrous, subtrigonous. Fl. 4–7. Hr.

Native. Bogs, fens, marshes and wet heaths ascending to 1215 m; common almost throughout the British Is., but absent from several eastern and Midland counties and the Channel Is. Most of Europe but rare in the south and east; Morocco (mountains); Azores; Greenland; North America, south to the mountains of New England and Washington.

Subsp. **palustris**. Lvs all obtuse; petioles glabrous; bracts below middle of peduncle. Spur slightly longer than appendages. $2n = 48$. The common form.

Subsp. **juressii** (Link ex K. Wein) P. Coutinho

V. epipsila auct. angl.; *V. juressii* Link K. Wein

Summer lvs subacute or shortly and bluntly acuminate, usually with spreading hairs on the petioles; veins and teeth more prominent. Bracts often about the middle of the peduncle. Fls rather larger with longer spur. In scattered localities mostly in S. England and Wales but recorded as far north as Perth and Angus (distribution incompletely known); Ireland; usually growing with subsp. *palustris* and connected with it by intermediate forms. Portugal (the only subsp.), Spain, S. France and perhaps elsewhere.

Section 2. MELANIUM Ging.

Stipules lf-like, pinnatifid or palmatifid or at least deeply lobed. Lateral petals directed upwards; lowest very broad, cuneiform. Style geniculate near base, expanded at the apex into a subglobose head with a hollow at one side, with or without a flap below the hollow. Cleistogamous fls not produced.

A difficult group which cytogenetic investigations have helped to elucidate to a considerable extent.

10. **V. lutea** Hudson   Mountain Pansy

Perennial herb 7–20 cm with slender creeping rhizome sending up usually solitary, simple, slender, ±flexuous stems. Lowest lvs ovate, obtuse; lvs rapidly becoming narrower upwards; upper 1–2 cm, oblong-lanceolate, subacute, cuneate at base, crenate or crenate-serrate, sparsely clothed with short stiff hairs at least on the margins and veins below; stipules ±palmatipartite, *the middle lobe oblanceolate-linear, entire,* c. 2 mm broad. *Peduncles* (5–)5–9 cm, 1–2(–4) on each stem. Sepals triangular-lanceolate, acute; appendages toothed. *Corolla 2–3.5 cm vertically, flat, bright yellow, blue-violet or red-violet, or of these colours variously combined,* always yellow at the base of the lower petal; *spur 2–3 times as long as appendages.* 80–90% of pollen-grains with 4 pores, remainder with 3 or 5. Stylar flap large and conspicuous. Fl. (4–)5–8. Insect pollinated. $2n = 48^*$. Hp.

Native. Grassland and rock ledges in hilly districts, especially on base-rich but not strongly basic soils; ascending to c. 1050 m. From Monmouth, Shropshire, Nottingham and Yorks northwards and westwards (not Orkney or Shetland), widespread and locally common; ?W. Cork, Clare, Carlow, Wicklow, Kildare. W. and C. Europe, southwards to the Pyrenees.

*V. cornuta L., a native of the Pyrenees, resembles *V. lutea* but fls are always pale violet, the lower petal being white at the base, and the spur is c. 6 times as long as the calyx-appendages. Locally naturalized, as a garden-escape, in E. Scotland from Peebles to N. Aberdeen.

11. V. tricolor L.   Wild Pansy.

Annual or perennial herb, usually much branched, glabrous or pubescent, *rhizome 0 or short and scarcely creeping.* Lvs very variable, lowest ovate, obtuse, becoming slowly or rapidly narrower upwards; upper ovate, oblong-lanceolate, lanceolate or elliptical, obtuse to subacute, ±cuneate at base, crenate, crenate-serrate or crenate-dentate, glabrous or ±pubescent; stipules *variable but often palmately lobed, mid-lobe usually lanceolate and entire, not lf-like.* Peduncles 2–8 cm, several on each stem. Sepals triangular-lanceolate to linear-lanceolate, acute; appendages variable. *Corolla 1.5–2.5 cm vertically, longer than sepals, flat, blue-violet* (rarely pink) *or yellow or of combinations of these; spur longer than appendages* (up to twice as long). 80–90% of pollen-grains with 4 pores, remainder, with 3 or 5. *Stylar flap distinct.* Fl. 4–9. Pollinated mainly by long-tongued bees.

Subsp. tricolor

Incl. subsp. *saxatilis* (Schmidt) E. F. Warb., *V. lloydii* auct., *V. orcadensis* Drabble, *V. lejeunei* auct. p.p., *V. variata* auct. p.p., *V. cantiana* Drabble, *V. lepida* auct.

Annual, rarely perennial, with several long (15–45 cm) stems, but scarcely tufted; stems usually much branched. Fls blue-violet or mainly so, very rarely all yellow. Spur rather longer than appendages. $2n = 26^*$. Th. or Hp.

Native. Cultivated and waste ground and short grassland, mainly on acid and neutral soils, ascending to 445 m. Throughout the British Is., common in some districts, local in others. Most of Europe but rare in the south and only on mountains; Asia Minor, Caucasus, Siberia (to c. 103° E), Himalaya.

Subsp. curtisii (E. Forster) Syme

Subsp. *maritima* (Schweigger) Hyl.; *V. curtisii* E. Forster, incl. *V. pesneaui* Lloyd

Perennial 3–15 cm, usually branched from the base and ±tufted, the stock vertical, not or scarcely creeping, producing many slender lateral branches. Corolla less than 2 cm vertically, blue-violet, yellow or particoloured. Spur variable in length, often twice as long as appendages or rather more. $2n = 26^*$. Hp.

Native. Dunes and grassy places near the sea; S. Devon, Cornwall, whole west coast of Great Britain, east coast south to Northumberland; nearly the whole coast of Ireland; inland in the breckland of Norfolk and Suffolk and by a few lakes in N. Ireland. Shores of the Baltic, North Sea and English Channel.

*V. × wittrockiana Gams   Garden Pansy.

V. tricolor var. *hortensis* auct., ?DC.

A group of hybrids originating about 1830 probably from *V. tricolor* and *V. lutea* subsp. *sudetica* (Willd.) W. Becker. Sometimes escaping and hybridizing with *V. tricolor*. They can usually be distinguished by their large fls with overlapping petals.

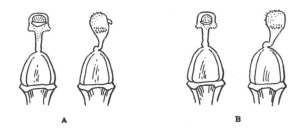

Fig. 25. Front and side views of the gynoecium of *Viola tricolor* (A) and *V. arvensis* (B), both ×6.

12. V. arvensis Murray   Field Pansy.

Incl. *V. agrestis* auct., *V. segetalis* auct., *V. obtusifolia* auct., *V. latifolia* Drabble, *V. ruralis* auct., *V. deseglisii* auct., *V. subtilis* auct., *V. anglica* Drabble, *V. arvatica* auct., *V. derelicta* auct., *V. contempta* auct., *V. monticola* auct., *V. vectensis* Williams, *V. variata* var. *sulphurea* Drabble, *V. lejeunei* auct. p.p.

Similar vegetatively to annual forms of *V. tricolor* subsp. *tricolor* but with the *mid-lobe of stipules usually ovate-lanceolate or ovate, crenate-serrate, lf-like. Corolla* 8–20 mm *vertically, usually small and not longer than calyx* but sometimes longer, *flat, cream* or with some blue-violet on the upper and lateral petals but never mainly blue, the spur sometimes violet; *spur about equalling appendages.* 85–95% of the pollen-grains with 5 pores, remainder with 4 or 6. *Stylar flap 0 or indistinct.* Very variable. Fl. 4–10. Pollinated by various insects, often selfed. $2n = 34^*$. Th.

Native. Cultivated and waste ground, mainly on basic and neutral soils. Common throughout the British Is.

Almost throughout Europe (except Iceland, Faeroes, Svalbard, Crete, Balearic Is.); Siberia (to c. 93° E), Turkestan, Caucasus, Asia Minor, Iran, Iraq; N. Africa; Madeira.

13. V. kitaibeliana Schultes   Dwarf Pansy.

Incl. *V. nana* (DC.) Godron

An erect annual, 3–10 cm, simple or little-branched, ±grey pubescent all over. Lowest lvs suborbicular, obtuse; upper to 2 cm, oblong-lanceolate, subacute, crenate-serrate; stipules, at least the lower, with large ovate-lanceolate, crenate-serrate lf-like mid-lobe and few lateral lobes. *Corolla* c. 5 mm, *vertically, about half the length of the sepals, somewhat concave, cream or pale blue-violet*; spur slightly longer than appendages. 70–85% of pollen-grains with 5 pores, remainder with 4 or 6. *Stylar flap* 0. Fl. 4–7. 2n = 48*. Th.

Native. Dunes in the Scilly Is. and Channel Is. S. and C. Europe, extending to W.C. France and E. Ukraine.

The hybrids *V. arvensis × tricolor* and *V. lutea × tricolor* are recorded but appear to be rare.

31. POLYGALACEAE

Herbs or small shrubs. Lvs alternate or rarely opposite, simple, exstipulate. Fls hermaphrodite, zygomorphic. Pedicels often jointed. Sepals 5, free, imbricate, the two inner larger than outer, often petaloid. Petals 3–5; two outer free or united with lower; two upper free or minute or 0. Stamens 8, monadelphous for more than half their length, rarely free, the tube split above and often adnate to petals; anthers usually opening by an apical pore. Ovary superior, usually 2-locular; style simple. Fr. a capsule, or sometimes fleshy. Seeds often hairy, with a conspicuous strophiole; endosperm usually present; embryo straight.

About 12 genera and 800 spp., throughout the world, except New Zealand, Polynesia and the Arctic.

1. POLYGALA L.   Milkwort.

Herbs or small shrubs. Fls bracteate and bibracteolate, in terminal or lateral racemes or spikes, rarely axillary. Inner sepals petaloid, much larger than outer. Petals 3, outer united with lower and adnate to staminal tube. Stamens 8; anthers 1–2-locular, opening by pores. Capsule 2-locular, compressed, obcordate and narrowly winged in our spp., splitting loculicidally at the edges. Seeds one in each loculus, ±hairy; strophiole 3-lobed.

About 500–600 spp. Distribution of the family.

The strophiole is sometimes used in distinguishing the spp.; its shape and size, however, seem to vary independently of other specific characters.

1 Lower lvs smaller than upper, not forming a rosette.   2
  Lower lvs larger than upper, forming a rosette.   3
2 At least the lower lvs opposite.   **2. serpyllifolia**
  All lvs alternate.   **1. vulgaris**
3 Fls 6–7 mm; rosette usually not at base of stem, the portion below the rosette being lfless; stem lvs usually less than 1 cm.   **3. calcarea**
  Fls not more than 5 mm; rosette at base of stem; stem lvs usually more than 1 cm.   **4. amarella**

1. P. vulgaris L.   Common Milkwort.

Incl. *P. oxyptera* Reichenb.

A perennial herb, 10–30 cm. Stems woody at base, erect or ascending, much-branched. *Lvs scattered, alternate*, lower 5–10 mm, narrowly obovate or spathulate, *upper longer* (up to c. 35 mm), lanceolate or linear-lanceolate, all ±acute. Infl. with 10–40 fls, lax. Fls 4–7(–8) mm, blue, pink, or white. Outer sepals c. 3 mm, green with coloured borders; inner sepals (Fig. 26A) c. 6 mm, ovate,

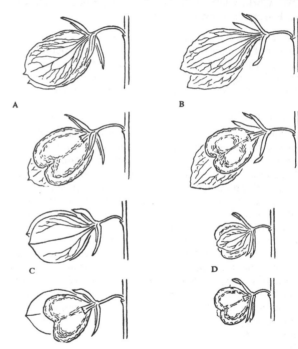

Fig. 26. Fruits of *Polygala*. A, *P. vulgaris*; B, *P. serpyllifolia*; C, *P. calcarea*; D, *P. amarella*. ×3.

apiculate, lateral veins much-branched, anastomosing. Capsule c. 5 mm. Seeds 2.5–3 mm. Fl. 5–9. 2n = 28, 32, 48, c. 56, 68, c. 70. Chh.

A variable plant.

Native. In grassland, on heaths, dunes, etc. Widely distributed throughout the British Is. Most of Europe, W. Asia, N. Africa.

2. P. serpyllifolia Hose   Heath Milkwort.

P. serpyllacea Weihe; *P. depressa* Wenderorth

Like *P. vulgaris* but usually smaller and more slender. Stems filiform, scarcely woody at base. Lvs 3–15 mm, the lower *opposite*, ovate, the upper larger, often opposite, elliptical-lanceolate. Infl. with 3–8 fls, usually short and rather dense. Fls 5–6 mm, commonly gentian blue or slate blue. Inner sepals (Fig. 26B) 4.5–5.5 mm, longer and slightly broader than capsule, veins much-branched and anastomosing. Capsule 4–5 mm. Seed c. 2.5 mm. Fl. 5–8. Chh.

Some Scottish variants, referred to this species, differ considerably from southern plants.

Native. On heaths and in grassy places, on lime-free soils. Widely distributed throughout the British Is. and the commonest species, except on calcareous soils. W. and C. Europe, Greenland.

3. P. calcarea F. W. Schultz   Chalk Milkwort.

A perennial herb, 5–20 cm. Stems somewhat woody at base. *Lower lvs* 5–20 mm, spathulate to obovate, obtuse, *crowded into an irregular rosette* from which the unbranched flowering stems arise; *rosette often not at base of stem, the portion below it being lfless; upper lvs smaller*, oblanceolate, ±acute. Infl. usually rather dense, with 6–20 fls. *Fls 6–7 mm, intense blue or sometimes bluish-white.* Inner sepals (Fig. 26C) c. 5 mm, longer and narrower than capsule, veins little-branched

and less anastomosing than in *P. vulgaris.* Capsule 4–5 mm. Seed c. 2.5 mm. Fl. 5–7. Ch.

Native. In calcareous grassland. S.E. England to S. Lincoln, local. W. Europe.

4. P. amarella Crantz

P. amara auct.; *P. uliginosa* Reichenb. Incl. *P. austriaca* Crantz

A perennial herb up to c. 10 cm. Stems erect or ascending, woody at base, rather stout, usually unbranched, sometimes very short so that infl. is condensed and subsessile. *Basal lvs* 5–20 mm, obovate, obtuse, *forming a rosette*; stem lvs somewhat smaller, narrowly obovate to lanceolate, acute. Main stem often ending in a short lfy shoot. Infl. lateral, many-fld. *Fls 2–5 mm, usually pink, purplish-pink, or blue.* Inner sepals (Fig. 26D) about as long and half as wide as capsule; veins sparingly branched. Capsule c. 4 mm. Seed 2 mm, oblong. Fl. 6–8. $2n = 34$. Ch.

Native. In chalk grassland and damp mountain pastures on limestone, rare. Kent, Surrey; N.W. and Mid-W. Yorks, Durham. Much of Europe except for most of the south.

32. HYPERICACEAE

Herbs or shrubs with resinous juice. Lvs opposite, simple, usually entire, often gland-dotted, exstipulate. Fls showy, usually terminal, solitary or in branched cymes, actinomorphic. Sepals usually 5, imbricate in bud. Petals usually 5, contorted in bud. Stamens numerous, often ±connate in bundles; anthers versatile. Ovary superior, 1-, 3- or 5-celled; ovules usually numerous, anatropous, on axile or parietal placentae. Fr. a septicidal capsule, rarely a berry or drupe.

1. HYPERICUM L.   St John's-wort.

Herbs or shrubs. Lvs sessile or nearly so. Fls yellow, usually without nectar. Sepals 5. Petals 5, generally very oblique. Ovary 1-celled with 3 or 5 parietal placentae, or 3- or 5-celled with axile placentae; styles 3 or 5, free or connate. Fr. a capsule, rarely a berry. Pollinated by a considerable variety of insects; eventually self-pollinated if crossing has not previously occurred.

About 400 spp. in temperate regions and on mountains in the tropics.

1 Plant ±shrubby; petals deciduous; stamens in 5 bundles united at base only.   2
  Stems not or slightly woody and then at base only; petals persistent; stamens in 3 bundles (very rarely 5).   5
2 Plant not rhizomatous; stems freely branched; stamens longer than or nearly equalling petals; styles 3.   3
  Plant rhizomatous; stems simple or nearly so; stamens distinctly shorter than petals; styles 5. **4. calycinum**

3 Stems quadrangular; lvs usually with strong goat-like smell when crushed; sepals deciduous; stamens exceeding petals.   **3. hircinum**
  Stems 2-edged or nearly terete; lvs ±aromatic; sepals persistent; stamens about equalling petals.   4
4 Stems with 2 raised lines; sepals longer than petals; styles shorter than stamens; fr. fleshy, red becoming black, indehiscent.   **1. androsaemum**
  Stems slightly 2-edged; sepals distinctly shorter than petals; styles longer than stamens; fr. reddish, fleshy, becoming dry, dehiscent.   **2. inodorum**
5 Plant pubescent or scabrid, at least on the lvs beneath.   6
  Plant glabrous; lvs smooth beneath.   8
6 Lvs glabrous above, puberulent or scabrid beneath; infl. dense, often subcapitate.   **13. montanum**
  Lvs pubescent on both sides; infl. lax.   7
7 Stems stiff, erect, ±branched; sepals with black glands; stamens free except at base (dry places).   **12. hirsutum**
  Stems soft, creeping and rooting below; sepals with red glands; stamens united in 3 bundles ⅓ way up (wet places).   **14. elodes**
8 Stems quadrangular.   9
  Stems terete or with 2 raised lines.   12
9 Petals shorter or little longer than sepals.   10
  Petals 2–3 times as long as sepals.   11
10 Stems up to 20 cm, very slender; plant without red or black glands.   **15. canadense**
  Stems usually 30–70 cm, stout; plant with black glands.   **8. tetrapterum**
11 Lvs abruptly narrowed at base, not amplexicaul, flat; sepals obtuse.   **6. maculatum**
  Lvs ½-amplexicaul, undulate; sepals acute.   **7. undulatum**

12 Lvs on main stems cordate-amplexicaul; petals red-
　　tinged.　　　　　　　　　　　　　　　**11. pulchrum**
　　Lvs not cordate-amplexicaul; petals without a red
　　tinge.　　　　　　　　　　　　　　　　　　　　　13
13 Stems slender, procumbent or ascending; sepals
　　unequal, the larger usually nearly as long as petals.
　　　　　　　　　　　　　　　　　　　9. humifusum
　　Stems usually stout and erect; sepals equal, usually
　　less than half as long as the petals.　　　　　　14
14 Sepals entire, not black-glandular-ciliate; lvs with
　　numerous large translucent glands.　　**5. perforatum**
　　Sepals black-glandular-ciliate; lvs usually without
　　large translucent glands.　　　　　　　　　　　15
15 Fls crowded in corymbose or shortly cylindrical heads;
　　lvs scabrid beneath.　　　　　　　　　　**13. montanum**
　　Fls in a lax panicle; lvs smooth beneath.
　　　　　　　　　　　　　　　　　　　10. linarifolium

Section 1. *Androsaemum* (Duh.) Godron. Shrubs, without black glands. Sepals free, unequal. Petals deciduous. Stamens connate in 5 bundles of 10–25, united
at base only, without scales alternating with the bundles.
Ovary incompletely 3-celled.

1. H. androsaemum L.　　　　　　　　　　　　Tutsan.

A glabrous, *slightly aromatic*, half-evergreen, shrub,
40–100 cm. *Stems with 2 raised lines.* (Lvs 2.5–)4–15 cm,
sessile, ovate, obtuse, with minute translucent glands.
Infl. few-fld. Fls c. 2 cm diam.; *sepals* 8–12 mm, ovate
obtuse, *very unequal, the larger about equalling the
petals*, persistent and deflexed in fr. Stamens about
equalling petals. *Styles shorter than ovary. Fr. fleshy*,
red, turning purple-black when ripe, *indehiscent.* Fl.
6–8. $2n = 40$. N.
　　Native. In damp woods and hedges, local. Scattered
throughout the British Is., northwards to W. Sutherland, rarer in the north and east. W. and locally in S.
Europe; Caucasus; Asia Minor; N.W. Africa.

*2. H. inodorum Miller　　　　　　　　Tall Tutsan.

H. elatum Aiton

Like *H. androsaemum* in general appearance but taller,
up to 200 cm. Stems slightly 2-edged. *Lvs 3–8 cm, ovate
to oblong-lanceolate, with a distinct aromatic smell when
crushed.* Fls 2–3 cm diam.; *sepals 5–8 mm, ovate, distinctly shorter than the petals,* persistent in fr. Stamens
longer than the petals. *Styles longer than ovary.* Fr.
reddish, eventually dehiscent. Fl. 7–8. $2n = 16$. N.
　　Introduced. Naturalized in hedges and thickets, particularly in S.W. England, Wales and W. Scotland.
Madeira and Canary Is.

*3. H. hircinum L.

Probably *H. androsaemum* × *hircinum*Stinking Tutsan.

Like *H. androsaemum* in general appearance. *Stems
quadrangular*, or 2-lined. Lvs 2.5–6 cm, ovate to lanceolate, acute, often *with a strong goat-like smell when
crushed.* Fls c. 3 cm diam.; *sepals* usually 3–6 mm, lanceolate, acute, shorter than petals, *deciduous in fr.* Sta-

mens longer than petals. Styles longer than ovary. Fr.
dehiscent. Fl. 5–8. $2n = 40$. N.
　　Introduced. Planted in woods in several localities.
Mediterranean region and western Asia; naturalized
elsewhere.

Section 2. *Eremanthe* (Spach) Endl. Like *Androsaemum*
but with fls usually solitary and ovary 5-celled.

*4. H. calycinum L.　　Rose-of-Sharon, Aaron's Beard.

An extensively creeping rhizomatous evergreen shrub
20–60 cm. Stems bluntly quadrangular. Lvs 5–8.5 cm,
ovate-oblong to elliptic, subsessile. *Fls usually solitary*,
7–8 cm diam.; sepals 10–20 mm, obovate, persistent. Fl.
7–9. $2n = 20^*$. Chw. or N.
　　Introduced. Naturalized in shrubberies, parks, and
by roads in many places. Turkey and S.E. Bulgaria.

Section 3. *Hypericum.* Perennial herbs usually with
axillary shoots. Black glands present. Sepals connate
at base, rarely free, generally subequal. Petals persistent. Stamens persistent, connate in 3 bundles at base
only, without scales alternating with the bundles. Ovary
completely 3-celled.

5. H. perforatum L.　　　　Perforate St John's-wort.

An erect glabrous rhizomatous perennial, 10–100 cm.
Stems woody at base, *with 2 raised lines. Lvs 8–30 mm,
sessile, ovate to linear, obtuse, sometimes mucronulate,
with abundant translucent glandular dots.* Fls c. 2 cm
diam.; sepals entire, lanceolate, glandular, acute to acuminate, much shorter than petals. Fl. 6–9. $2n = 32$
(mainly apomictic). Hp. Variable.
　　Native. In open woods, hedgebanks and grassland;
common, especially on calcareous soils. Throughout
most of the British Is., absent from a number of northern
counties. Europe (except the extreme north), temperate
Asia, N. Africa; Macaronesia. Widely naturalized in
temperate regions.

6. H. maculatum Crantz　　Imperforate St John's-wort.

H. quadrangulum auct., non L.; *H. dubium* Leers

An erect glabrous rhizomatous perennial, 15–100 cm.
*Stems quadrangular, not winged. Lvs 15–40 mm, sessile,
elliptic, obtuse, abruptly narrowed at base,* not clasping
stem, *glandular dots 0 or few.* Fls c. 2 cm diam.; petals
golden yellow, usually with numerous black dots or
streaks; *sepals* entire, *ovate, obtuse,* with few or no black
dots, $\frac{1}{3}-\frac{1}{4}$ length of petals. Fl. 6–8. Hp.
　　Native. In damp places at margins of woods and in
hedgebanks, local. Scattered throughout the British Is.
but mainly in Wales, the W. Midlands and S. Ireland.
Most of Europe, western Siberia.

Subsp. **maculatum**

Plant slender and rather sparsely branched. Lvs usually
with pellucid glands, venation densely reticulate. Infl.
branches making an angle of c. 30° with stem. Sepals
ovate, entire. Dark glands on margin of petals 0, on
surface in form of dots. $2n = 16$.

Apparently very local and mainly in W. Scotland. E., N. and C. Europe.

Subsp. obtusiusculum (Tourlet) Hayek

Plant stouter, branches numerous. Lvs often with pellucid glands, venation less dense. Infl. branches making an angle of c. 50° with stem. Sepals often lanceolate. Dark glands sometimes present on margin of petals, on surface mainly in form of lines. $2n = 32$.

Widely distributed.

H. × desetangsii Lamotte

H. maculatum subsp. *obtusiusculum* × *H. perforatum*

Stems with 2 very distinct and 2 faint raised lines. Lvs with a few translucent glands. Sepals narrowly oblong to ovate-lanceolate, apiculate and erose-denticulate at apex. $2n = 32$.

Native. Hedgebanks and open woods. Recorded from scattered localities, mainly in S.E. and N. England and S. Scotland. Recorded from Belgium, France, Switzerland, Austria and Hungary. Backcrossing occurs, giving a range of intermediates between the parents.

7. H. undulatum Schousb. ex Willd.

Wavy St John's-wort.

H. boeticum Boiss.

An erect glabrous stoloniferous perennial, 15–100 cm. *Stems quadrangular, narrowly winged.* Lvs 1–4 cm, sessile, oblong or ovate-oblong, obtuse, ½-amplexicaul, *with translucent glandular dots*; margins strongly undulate. Fls c. 2 cm diam.; *petals* rather narrow, one half *red-tinged beneath*; *sepals* entire, *ovate, acute*, glandular, c. ⅓ length of petals. Fl. 8–9. $2n = 32$. Hp.

Native. In small boggy patches or marshy ground beside streams on non-calcareous soils, very local. Cornwall, Devon, Pembroke, Cardigan, and Merioneth. W. Spain, Portugal, France, Madeira, Azores.

8. H. tetrapterum Fries   Square-stalked St John's-wort.

H. acutum Moench; *H. quadrangulum* L., nom. ambig.

An erect glabrous perennial, 10–100 cm. *Stems with slender stolons at base, quadrangular, angles winged.* Lvs 1–3.5 cm, sessile, ovate, obtuse, ½-amplexicaul, with small translucent glands. *Fls* c. 1 cm *diam.; petals pale yellow; sepals* entire, *lanceolate, acute,* ⅔ *length of petals.* Fl. 6–9. $2n = 16$. Hp.

Native. In damp meadows, grassy places beside rivers and ponds and in marshes. Throughout the British Is. except for most of the north of Scotland. W., C. and S. Europe, N. Africa, Caucasus.

9. H. humifusum L.   Trailing St John's-wort.

A glabrous *procumbent* or ascending perennial (?sometimes biennial), 5–30 cm. *Stems very slender*, rather woody at base, with 2 raised lines. Lvs 0.5–1(–1.5) cm, the upper oblong to lanceolate, the lower obovate to oblanceolate, usually *with translucent glands*. Fls c. 1 cm diam.; *sepals unequal*, oblong to lanceolate, obtuse or

acute, glandular, entire or toothed, teeth sometimes glandular at tip. Fl. 6–9. $2n = 16$. ?Chh.

Native. On heaths, dry moors and in open woods on non-calcareous soils. In suitable places throughout the British Is. W. and C. Europe; N. Africa, Macaronesia.

10. H. linarifolium Vahl   Flax-leaved St John's-wort.

An ascending glabrous perennial, 5–65 cm. Stems subterete, often reddish. *Lvs* 1–3.5 cm, *linear or linear-oblong*, obtuse, usually not *gland-dotted*. Fls c. 1 cm diam.; petals red-tinged, bordered with black glands; *sepals* lanceolate, acute or subobtuse, ⅓–⅓ length of petals, dotted with black glands and *fringed with slender black, stalked glands.* Fl. 6–7. Hp or Chh.

Native. On dry rocky slopes on acid soils, rare. E. Cornwall, S. Devon, Radnor and Caernarvon; Guernsey, Alderney. W. Europe to 4° E in France; Madeira.

Section 4. *Taeniocarpium* Jaub. & Spach. Perennial herbs. Black glands confined to margins of sepals and petals, and, rarely, the lf-apex. Stems terete or with 2 lines. Petals and stamens persistent. Stamens in 3 bundles, without scales alternating with the bundles. Styles 3.

11. H. pulchrum L.   Slender St John's-wort.

A glabrous erect of ascending perennial, 10–90 cm. Stems terete, often reddish. *Lvs* on main stem 0.5–2 cm, sessile *broadly ovate-cordate*, obtuse, ½-*amplexicaul, dotted with translucent glands*; lvs on axillary shoots smaller, oblong, shortly petiolate. Fls c. 1.5 cm diam.; petals red-tinged, with a row of black glands near margin; *sepals ovate*, obtuse, ⅓–⅓ length of petals, *margins with shortly stalked black glands.* Fl. 6–8. $2n = 18$. Hp. A very elegant plant.

Native. In dry woods and rough grassy places on non-calcareous soils, local. Widely distributed throughout the British Is. N.W. Europe, extending locally to S.W. Poland, Switzerland and C. Portugal.

12. H. hirsutum L.   Hairy St John's-wort.

An erect *pubescent* perennial, 40–100 cm. *Stems* terete, *little branched. Lvs* 2–5.5 cm, subsessile, ovate, obtuse, with sparse translucent glands but *without black marginal glands. Infl. lax.* Bracteoles with short-stalked black glands. Fls c. 1.5 cm diam.; petals pale yellow, sparsely glandular-ciliate at tips; sepals oblong-lanceolate, subacute, ½ length of petals, the margins with short-stalked black glands. Fl. 7–8. $2n = 18$. Hp.

Native. In woods and damp grassland, mostly on basic soils. Scattered throughout the British Is., particularly in the south and east; rare in Ireland. Most of Europe, except the extreme north and south; N. Africa; W. Asia; Siberia.

Section 5. *Adenosepalum* Spach. Perennial herbs, usually pubescent. Black glands present on lvs, sepals and anthers. Petals and stamens persistent. Stamens in 3 bundles, without scales alternating with the bundles. Styles 3.

13. H. montanum L.   Pale St John's-wort.

An erect slightly pubescent perennial, 20–80 cm. *Stems terete, rigid, glabrous, simple or nearly so, upper internodes longer than lvs. Lvs* 2.5–7 cm, sessile, ovate to lanceolate, obtuse, ½-amplexicaul, *puberulous and with scattered black and some translucent glands. Infl. dense, often subcapitate.* Bracts and bracteoles with glandular teeth. Fls 1–1.5 cm diam., fragrant; petals pale yellow, without black glands; sepals lanceolate, acute, ½–⅔ length of petals, strongly black glandular-ciliate. Fl. 6–8. $2n = 16$. Hp.

Native. In woods, scrub and hedgebanks on calcareous or gravelly soils, local. Scattered throughout England and Wales. W. and C. Europe, Caucasus, N. Africa.

Section 6. *Elodes* (Adanson) Koch. Perennial, usually tomentose herbs. Red glands present on sepals, bracts and bracteoles. Petals and stamens persistent. Petals with a 3-fid ligulate, nectar-secreting appendage. Stamens in 3 bundles, united for more than ½ their length and alternating with scales. Styles 3.

14. H. elodes L.   Marsh St John's-wort.

A decumbent stoloniferous usually *tomentose* perennial, 10–30 cm. *Stems soft,* terete, *rooting at nodes below. Lvs* 0.5–3 cm, sessile, *suborbicular to broadly ovate,* ½-amplexicaul. Infl. few-fld. Fls c. 1.5 cm diam.; *sepals* elliptical to ovate, erect, obtuse, *with fine, red or purplish glandular teeth.* Fl. 6–9. Hel. The hairs prevent the plant from being wetted when submerged.

Native. In bogs and wet places beside ponds and streams on acid soils. Scattered throughout the British Is. in suitable habitats, commoner in the west and absent from much of Scotland and the centre of Ireland. Lost through drainage in some localities. W. Europe from Scotland southwards and locally eastwards to Germany, and Italy; Azores.

Section 7. *Brathys* (Mutis ex L. fil.) Choisy. Glabrous annual or perennial herbs, with black glands. Petals and stamens persistent. Stamens 5 or in 5 bundles. Styles 3–5.

*15. H. canadense L.   Irish St John's-wort.

A *slender erect* glabrous annual (sometimes perennial), ±tinged with purplish-red, (3–)12–20 cm. *Stems quadrangular,* ±winged. *Lvs* up to 2 cm, elliptical to linear-oblanceolate, *mostly 3-veined,* semi-amplexicaul, with numerous translucent glands. Infl. 3–9(–30)-fld. Sepals 2–4 mm, ovate-lanceolate, with pale or reddish streaks, but without black glands. *Petals* 3–4 mm, deep golden yellow, usually with a crimson line on the back, *widely separated so that the fl. has a star-like appearance.* $2n = 16^*$.

Probably introduced; first recorded in 1954. In wet peaty places near Lough Mask. Native in eastern North America.

33. CISTACEAE

Shrubs or herbs. Lvs usually opposite, less frequently alternate, stipulate or not, simple, mostly with stellate indumentum. Fls hermaphrodite, actinomorphic, hypogynous, solitary or in raceme-like cymes. Sepals 5 or 3. Petals 5, rarely 3 or 0, usually quickly caducous (often lasting one day only). Stamens numerous. Ovary 1-celled or septate at base with 3 or 5 (rarely 10), often intrusive, parietal placentae; ovules few to numerous, usually orthotropous, rarely anatropous; style simple or 0, stigmas 3 to 5, free or united. Fr. a loculidal capsule. Seeds with ±curved embryo; endosperm present.

Eight genera and c. 200 spp., mainly Mediterranean region, a few in N. and C. Europe, C. Asia and North and South America.

Style 0; annual herb.   1. TUBERARIA
Style filiform; perennial undershrubs. 2. HELIANTHEMUM

CISTUS L.

Differs from *Helianthemum* in being taller shrubs (to 2 m) and in the 5- (rarely 10)-valved capsule. About 20 spp. in Mediterranean region and Canary Is. Most species and numerous hybrids are grown in gardens.

*C. incanus L. Shrub c. 1 m. Lvs 1–7 cm, stalked, ovate, elliptic or obovate, ±undulate. Sepals 5, ±equal. Petals large, purplish-pink. Style long, straight. Mediterranean region, rarely naturalized.

1. TUBERARIA (Dunal) Spach

Herbs with basal rosette. Infl. a raceme-like cyme. Sepals 5, unequal. Petals 5. *Style 0 or very short.* Placentae 3; ovules orthotropous; funicle stout, swollen in the middle; embryo curved but not plicate. About 10 spp., Mediterranean region, only the following more widespread.

1. T. guttata (L.) Fourr.   Spotted Rock-rose.

Helianthemum guttatum (L.) Miller

Erect or diffuse *annual* 6–30 cm, simple or branched, stellate-pubescent and pilose. Basal lvs forming a rosette, usually dead at fl.; lower cauline lvs 1.5–5 cm, elliptic-lanceolate, 3-veined, subacute, sessile, entire, opposite, exstipulate; upper shorter and narrower, alternate, with conspicuous linear stipules. Fls 5–12 in lax unilateral raceme-like cymes; pedicels slender; bracts conspicuous, oblong or ovate, or 0. Inner sepals ovate, pilose, accrescent; outer oblong, c. ½ as long as inner. Corolla 8–12 mm diam., pale yellow, often with a red spot at the base of the petals. Capsule ovoid. Fl. 5–8. $2n = 36^*$, 48. Th.

Native. Cliffs and exposed, open cliff-top moorland. Jersey, Alderney, Anglesey, Caernarvon, W. Cork, W. Galway, W. Mayo; very local and rare. S. Mediterranean region, extending north (in W. Europe only) to N.W. Germany; Canary Is.

The recognition of the Welsh and Irish plants as a distinct taxon, subsp. **breweri** (Planchon) E. F. Warb. (*Helianthemum breweri* Planchon), distinguished by a diffuse habit, usually exstipulate upper lvs and with bracts, cannot be maintained. The characters vary independently and ecotypic differentiation is suggested.

2. HELIANTHEMUM Miller

Shrubs or annual herbs. Infl. a raceme-like cyme. Sepals 5, the 2 outer smaller than the 3 inner. Petals 5. Placentae 3; ovules orthotropous; funicle stout, thickened at the apex; style usually filiform; stigmas large, capitate; embryo plicate. Capsule 3-valved. About 100 spp., mainly Mediterranean region, a few in N. and C. Europe, E. to C. Asia and the Cape Verde Is.

The garden 'rockroses', with fls of varied colours, are hybrids derived from *H. chamaecistus*, *H. apenninum* and allied spp.

1 Style nearly straight; stipules present; corolla normally large (2 cm diam. or more). 2
  Style markedly bent in the middle; stipules 0; corolla small (1–1.5 cm diam.). **3. canum**
2 Lvs green above; corolla yellow. **1. nummularium**
  Lvs grey-tomentose above; corolla white.
  2. apenninum

Subgenus 1. HELIANTHEMUM

Herbs or shrubs. Stipules present. Style filiform, straight or curved at base, then ascending, shorter than or equalling stamens. Embryo once plicate; cotyledons straight.

1. H. nummularium (L.) Miller   Common Rock-rose.

H. vulgare Gaertner; *H. chamaecistus* Miller

Dwarf shrub 5–30 cm, with ±numerous, procumbent or ascending, often rooting branches from a thick woody stock with vertical tap-root. *Lvs* 0.5–2 cm, oblong or the lower oval, *green* and glabrescent or somewhat pubescent *above*, densely white-tomentose beneath, obtuse or subacute, shortly stalked, margins entire, not or slightly revolute; stipules lanceolate or linear-lanceolate, c. twice as long as petiole. Fls 1–12 in lax unilateral raceme-like cymes; rhachis and pedicels tomentose; bracts lanceolate, shorter than pedicels. Inner sepals ovate, c. 6 mm, prominently veined, ±pubescent, outer subulate c. ⅓ as long. *Corolla* 2–2.5(–3) cm *diam.*, rarely considerably smaller, *bright yellow*, sometimes with a small orange spot at base, rarely cream, very rarely white or copper-coloured. Capsule ovoid, tomentose, about equalling the spreading calyx. Fl. 5–9. Generally self-incompatible and pollinated mainly by bees. $2n = 20*$. Chw.

Native. Basic grassland and scrub; ascending to 640 m. Common over most of Great Britain, but absent from Cornwall, Isle of Man, N.W. Scotland, Orkney, Shetland and one or two other counties; in Ireland known only from one locality in Donegal. Most of Europe except the extreme north; Asia Minor, Caucasus,

N. Iran.

H. apenninum × *nummularium* (*H.* × *sulphureum* Willd.) with pale yellow fls occurs with the parents at Purn Hill, Bleadon (Somerset).

2. H. apenninum (L.) Miller   White Rock-rose.

H. polifolium Miller

Differs from *H. nummularium* as follows: *Lvs* oblong-linear, with *grey indumentum, mostly of stellate hairs above*, more densely so beneath, margins often more revolute; stipules linear-subulate, the lower scarcely longer than the petiole. Sepals densely grey-tomentose. *Corolla white*. Fl. 5–7. Generally self-incompatible and pollinated mainly by bees. $2n = 20*$. Chw.

Native. Limestone rocks; Brean Down (Somerset) and near Torbay (Devon). S. and W. Europe northwards to W. Germany; Asia Minor; Morocco.

Subgenus 2. PLECTOLOBUM Willk.

Shrubs. Stipules 0. Style filiform with a strong sigmoid curve in the middle, longer than the stamens. Embryo biplicate, the cotyledons sharply reflexed in the middle.

3. H. canum (L.) Baumg.   Hoary Rock-rose.

Dwarf shrub 4–20 cm with ±numerous procumbent or ascending branches from a central stock. Lvs 2–12 mm, elliptical, obovate-elliptical or ovate, green or greyish above, densely grey or white-tomentose beneath, obtuse to acute, shortly stalked, margins entire, not or slightly revolute. Fls 1–6(–10) in rather lax, ±unilateral, raceme-like cymes; bracts ovate-lanceolate, small. Inner sepals oval, obtuse, prominently veined, ±pubescent; outer oblong-linear, less than half as long. *Corolla* 1–1.5 cm *diam.*, bright yellow. Capsule ovoid, about equalling the appressed sepals, pilose. Fl. 5–7. Visited by various bees, stamens irritable. $2n = 22*$. Chw.

Native. Rocky limestone pastures, very local; ascending to 535 m. Glamorgan, N. Wales; Yorks, Westmorland; Clare, W. Galway. C. and S. Europe, Öland; Asia Minor, Caucasus; Morocco, Algeria.

Subsp. **canum**

±Prostrate. Lvs with few or many long bristles above, sometimes stellate-pubescent as well (British plants are, in general, hairier than Irish); the longest 6–12 mm. Infl. c. 6–10 cm usually 2–6-fld. Sepals stellate-pubescent and pilose.

Distribution of the species, but absent from Upper Teesdale. Almost throughout the range of the species.

Subsp. **levigatum** M. C. F. Proctor

Closely prostrate (more so than in subsp. *canum*). Lvs small, the longest c. 6 mm, dark green and glabrous or nearly so above. Infl. c. 3–6 cm, mostly 1–3-fld. Sepals stellate-pubescent and slightly pilose.

Cronkley Fell (Upper Teesdale), Yorks only. Endemic.

34. TAMARICACEAE

Trees or shrubs. Lvs small, scale-like or needle-like, exstipulate, alternate. Fls regular, hermaphrodite, hypogynous, solitary or in slender spikes or racemes, 4- or 5-merous. Disk present. Petals imbricate, free. Stamens as many as petals and alternate with them or twice as many and diplostemonous, sometimes united below, anthers usually extrorse. Ovary unilocular with 2–5 basal or parietal placentae each with 2–many anatropous ovules; styles free or united below or stigmas sessile. Fr. a capsule. Seeds with long hairs, with or without endosperm; embryo straight.

Four genera and c. 120 spp., mainly Mediterranean region and C. Asia, extending to E. Asia, S. Africa and W. Europe.

1. TAMARIX L.

Deciduous, the smaller twigs falling with lvs. Fls in long catkin-like racemes, 5-merous. *Petals without ligules. Stamens free or nearly so*; anthers extrorse. *Styles short and thick. Seeds with a sessile tuft of hairs at apex*; *endosperm* 0.

About 54 spp., mainly E. Mediterranean extending to C. and E. Asia, S. Africa and W. Europe. A few others are sometimes grown.

1.*T. gallica L.   Tamarisk.
T. anglica Webb
Feathery shrub 1–3 m. *Lvs* of larger twigs 1–3 mm, triangular-lanceolate acute; those of ultimate twigs much smaller and densely imbricate, ovate-lanceolate, sessile, acute or acuminate; margins scarious. Infl. terminal on the current year's growth or the rhachis continuing as a lfy shoot, consisting of cylindrical dense-fld spike-like racemes, 1–3 cm, arranged in a panicle; pedicels c. 1 mm. Fls 1.5–2 mm, pink or white; petals caducous; styles 3; disk 5-lobed, the stamens inserted on the lobes which gradually taper to the insertion. Capsule ovoid, trigonous, acuminate. Fl. 7–9. M. or N.

Introduced. Often planted near the sea and naturalized in a number of places along the S. and E. coasts of England from Cornwall to Suffolk; Channel Is. S.W. Europe, extending to N.W. France.

T. africana Poiret, a native of S.W. Europe and N. Africa, which differs from *T. gallica* in having lvs 1.5–4 mm, fls subsessile, 2–3 mm, and at least some petals persistent, has been planted as a windbreak and is locally established in hedges near the sea in the Scilly Is.

35. FRANKENIACEAE

Herbs or small shrubs. Lvs opposite, exstipulate, often ericoid. Fls actinomorphic, hermaphrodite or rarely unisexual, solitary or cymose. Sepals 4–6, persistent, connate at base. Petals 4–6, clawed, with a scale-like appendage on the claw. Stamens usually 6; anthers 2-celled, opening lengthwise. Ovary superior, 1-celled, with 2–4 parietal placentae; ovules numerous. Capsule enclosed in the calyx, opening by valves. Seeds endospermous, embryo straight, axile.

1. FRANKENIA L.

Fls nearly always hermaphrodite. Petals and sepals 5 (rarely 4). Stamens in two whorls, the outer shorter; anthers emarginate at both ends. Ovary usually of 3 carpels with all the placentae bearing ovules; ripe capsule enclosed in the calyx.

About 80 spp. with the distribution of the family.

1. F. laevis L.   Sea-heath.
A *procumbent* slightly pubescent rather wiry perennial, dark green tinged with reddish-brown. Stems up to 15 cm, rarely more, puberulent, woody at base. *Lvs* 2–5 mm, *heath-like*, linear with revolute margins, glabrous above, puberulent beneath, ciliate at base, *mostly densely crowded on short lateral shoots*. Fls solitary or in small clusters, sessile, usually in the forks of the branches or terminating the short lateral shoots. Calyx somewhat fleshy; teeth erect, narrow, acute. *Petals 4–6 mm*, cuneate-obovate, obtuse and faintly crenate, *pink*. Stamens usually 6. Capsule trigonous-conical, concealed in calyx-tube and surrounded by the persistent flattened filaments. Fl. 7–8. Chh.

Native. At the landward margins of salt-marshes, usually on rather sandy or gravelly soils. Coast of S. and E. England, from the Isle of Wight to W. Norfolk, Channel Is., local. W. Europe from the English Channel southwards, eastwards to S.E. Italy; Macaronesia; N. Africa.

36. ELATINACEAE

Small herbs of wet places, or undershrubs. Lvs opposite or verticillate, simple, stipulate. Fls small, axillary, solitary or cymose. Sepals 3–5, free or connate at the base, imbricate. Petals 3–5, imbricate. Stamens as many or twice as many as petals; anthers versatile. Ovary superior, 3–5-celled; styles 3–5, free; ovules numerous; placentation axile. Fr. a septicidal capsule.

Two genera and about 40 spp., cosmopolitan.

1. ELATINE L.

Small, submerged, glabrous herbs. Sepals 3–4, membranous. Petals 3–4. Capsule ±globose, membranous.

About 15 spp., cosmopolitan.

Fl. pedicellate, 3-merous.   **1. hexandra**

Fl. sessile or subsessile, 4-merous　　**2. hydropiper**

1. E. hexandra (Lapierre) DC.
Six-stamened Waterwort.

A slender, decumbent annual or perennial, 2.5–20 cm. Stems rooting at nodes. Lvs opposite, spathulate, entire, *petiole shorter than blade. Fls pedicellate, 3-merous.* Petals pinkish-white. *Stamens 6.* Pedicels in fr. as long as or somewhat longer than fr. *Seeds straight or slightly curved.* $2n = 72$. Fl. 7–9. Hel. or Hyd.

Native. In ponds and on wet mud, very local. Scattered throughout the British Is. mainly in the south and west; north to the Outer Hebrides. W. and C. Europe extending to S. Sweden and N. Italy; Azores.

2. E. hydropiper L.　　Eight-stamened Waterwort.

Like *E. hexandra* in general appearance. Petiole usually as long as or longer than the blade. Fls sessile or subsessile, 4-merous. Petals pale red. Stamens 8. Seeds almost straight or unequally horseshoe-shaped. Fl. 7–8. $2n = c. 40$. Hyd.

Native. In ponds and small lakes, rare and local. W. Sussex, Surrey, Worcester, Anglesey, N.E. Ireland. N. and C. Europe extending southwards to C. Spain, N. Italy and S.E. Russia; W. Asia; N. Africa.

37. CARYOPHYLLACEAE

Annual to perennial herbs, sometimes suffruticose, rarely shrubs. *Lvs usually in opposite and decussate pairs*, rarely whorled, sometimes spirally arranged; *simple, entire*, usually narrow, commonly exstipulate but sometimes with scarious stipules. *Fls commonly in terminal bracteate dichasia*, sometimes in raceme-like cymes or dense cymose clusters, rarely solitary, terminal; usually hermaphrodite, hypogynous or sometimes perigynous, actinomorphic. Sepals 4–5, free or gamosepalous below; petals sometimes 0, usually 4–5, free, often bifid or emarginate, sometimes with coronal scales; stamens commonly twice as many or as many as the sepals, sometimes an intermediate number, occasionally 1–3; *ovary superior, syncarpous, 1-celled at least above*, with 1–many *campylotropous ovules on a basal or free-central placenta*; styles 2–5, free to the base or joined below (*Polycarpon*). Nectar is usually secreted by hypogynous glands and the fls are cross-pollinated by insects or self-pollinated. Fr. a capsule opening by as many or twice as many teeth or valves as styles, the valves rarely remaining joined above (*Paronychia* spp., *Illecebrum*); sometimes a berry (*Cucubalus*), or an indehiscent 1-seeded nutlet (*Corrigiola, Scleranthus*). Seeds usually with a curved embryo surrounding perisperm; embryo rarely straight (*Dianthus, Kohlrauschia, Illecebrum*, etc.).

About 1750 spp. in 70 genera, chiefly in temperate regions of the northern hemisphere.

A well-defined family with relationships to Portulacaceae, Phytolaccaceae, Amaranthaceae, Chenopodiaceae, etc., and perhaps to Polygonaceae and Plumbaginaceae.

Synopsis of Classification

Calyx gamosepalous:　　　　*Subfamily* SILENOIDEAE
　Styles 3–5.　　　　　　　　*Tribe* LYCHNIDEAE
　　Fr. a capsule, rarely indehiscent; plants not scrambling.
　　　Capsule dehiscing by as many teeth as styles.
　　　　Styles alternating with calyx-segments; no carpophore; petals entire; coronal scales 0; calyx with long herbaceous teeth.　　3. AGROSTEMMA
　　　　Styles opposite the calyx-segments; carpophore usually well developed; petals usually bifid and with coronal scales; calyx-teeth short.　　2. LYCHNIS
　　　Capsule dehiscing by twice as many teeth, or as many bifid teeth, as styles.　　1. SILENE
　　Fr. berry-like; plants scrambling.　　4. CUCUBALUS
　Styles 2.　　　　　　　　*Tribe* DIANTHEAE
　　Calyx-tube with white scarious seams between the teeth.
　　　Individual fls or heads of fls with an involucre of 2–5 or more membranous bracts; seeds peltate, convex above, concave below; embryo straight.
　　　　　　　　　　　　　　8. PETRORHAGIA
　　　Involucre 0; seeds reniform, convex on both sides; embryo curved.　　9. GYPSOPHILA
　　Calyx-tube lacking scarious seams between the teeth.
　　　Petals with coronal scales.　　7. SAPONARIA
　　　Coronal scales 0.
　　　　Epicalyx present; embryo straight.　　5. DIANTHUS
　　　　Epicalyx 0; embryo curved.　　6. VACCARIA
Calyx of free sepals.　　*Subfamily* ALSINOIDEAE
　Fr. a capsule opening by teeth or valves.
　　Styles free to the base.
　　　Stipules 0.　　　　　*Tribe* ALSINEAE
　　　　Petals bifid or deeply emarginate; sometimes 0.
　　　　　Styles 5, opposite the sepals, rarely 3, 4 or 6; capsule cylindrical.　　10. CERASTIUM
　　　　　Styles 5, opposite the petals; capsule ovoid, opening by 5 bifid teeth.　　11. MYOSOTON
　　　　　Styles 3; capsule ovoid, opening by 6 teeth.
　　　　　　　　　　　　　　12. STELLARIA
　　　　Petals irregularly toothed or jagged. 13. HOLOSTEUM
　　　　Petals entire or weakly emarginate; sometimes minute or 0.
　　　　　Styles as many as sepals.
　　　　　　Styles opposite the sepals; capsule opening by 8 short teeth.　　14. MOENCHIA
　　　　　　Styles alternating with the sepals; capsule opening by 4–5 valves.　　15. SAGINA
　　　　　Styles fewer than sepals, usually 3.
　　　　　　Capsule opening by as many teeth as styles.
　　　　　　　Capsule ovoid or cylindrical; lvs linear or lanceolate.　　16. MINUARTIA
　　　　　　　Capsule ±spherical; lvs ovate, fleshy; a maritime plant.　　17. HONKENYA
　　　　　　Capsule opening by twice as many teeth as styles.
　　　　　　　Seeds with an oily appendage (strophiole or elaiosome).　　18. MOEHRINGIA
　　　　　　　Seeds without an oily appendage.　19. ARENARIA
　　　Lvs with small scarious stipules.　*Tribe* SPERGULEAE
　　　　Styles 5.　　　　　　20. SPERGULA
　　　　Styles 3.
　　　　　Ovary 1-celled throughout.　　21. SPERGULARIA
　　　　　Ovary 3-celled below.　　(TELEPHIUM p. 148)

Styles joined below.   *Tribe* POLYCARPEAE
  Styles 3; sepals entire, keeled; petals 5.
                           22. POLYCARPON
Fr. indehiscent or opening by valves which remain joined
  above.
  Lvs with stipules.   *Tribe* PARONYCHIEAE
    Embryo curved.
      Styles 3.   23. CORRIGIOLA
      Styles 2.
        Sepals ±hooded, usually with dorsal points; fls in
          terminal or lateral clusters with scarious and often
          very conspicuous bracts.   (PARONYCHIA p. 149)
        Sepals not hooded, blunt; fls in lateral clusters with
          herbaceous bracts.   24. HERNIARIA
    Embryo straight.
      Sepals white, hooded, with dorsal points, thick and
      becoming hard in fr.; fls in axillary whorls.
                          25. ILLECEBRUM
  Lvs without stipules.   *Tribe* SCLERANTHEAE
    Perigynous; fr. 1-seeded.   26. SCLERANTHUS

Key to Genera

1 Lvs alternate or spirally arranged, not in opposite
    pairs or whorled.   2
   At least the lower lvs in opposite pairs or rarely
    whorled.   3
2 Annual; lvs oblong-lanceolate or strap-shaped; fr. tri-
    gonous, 1-seeded, indehiscent.   23. CORRIGIOLA
   Perennial; lvs ovate; fr. a capsule with many seeds.
                        (TELEPHIUM, p. 148)
3 Stipules 0.   4
   Stipules present.   28
4 Fls perigynous; petals 0; fr. 1-seeded, dry and indehis-
    cent, enclosed in the perigynous tube; ±prostrate
    herbs with subulate connate lvs. 26. SCLERANTHUS
   Fls hypogynous; fr. a capsule or berry with few to
    many seeds.   5
5 Calyx of joined sepals.   6
   Calyx of free sepals.   14
6 Styles 2.   7
   Styles 3–5 (or fls with stamens only).   11
7 Calyx-tube with whitish scarious seams alternating
    with the free teeth.   8
   Calyx-tube lacking scarious seams.   9
8 Individual fls or heads of fls with an epicalyx or invo-
    lucre respectively of 2 to many membranous scales.
                          8. PETRORHAGIA
   No epicalyx or involucre.   9. GYPSOPHILA
9 Base of calyx tightly enclosed by an epicalyx of 1–3
    pairs of bracteoles; petals with no scale-like appen-
    dages at the top of the claw (coronal scales).
                         5. DIANTHUS
   Epicalyx 0.   10
10 Calyx-tube winged: coronal scales 0.   6. VACCARIA
   Calyx-tube not winged: coronal scales present.
                         7. SAPONARIA
11 Fr. black, berry-like; plant scrambling.
                         4. CUCUBALUS
   Fr. a capsule; plant not scrambling.   12
12 Calyx with long narrow herbaceous teeth much longer
    than the petals.   3. AGROSTEMMA
   Calyx-teeth not exceeding the petals.   13
13 Styles 3, or (2–)5 in fls lacking stamens, or fls with
    stamens only; capsule opening by twice as many
    teeth, or as many bifid teeth, as styles.   1. SILENE

Styles 5; fls always with both stamens and styles; cap-
    sule opening by 5 teeth.   2. LYCHNIS
14 Petals present.   15
   Petals 0.   26
15 Petals ±deeply bifid.   16
   Petals entire, emarginate or irregularly toothed.   19
16 Styles 5.   17
   Styles 3, or varying from 3 to 6.   18
17 Lvs 2–5 cm, ovate-cordate; petals bifid almost to the
    base.   11. MYOSOTON
   Lvs rarely exceeding 2.5 cm, not cordate; petals bifid
    to less than half-way.   10. CERASTIUM
18 Styles 3–6; a decumbent alpine plant with small ellipti-
    cal lvs; rarely descending below 456 m; capsule
    oblong.   10. *Cerastium cerastoides*
   Styles 3; not exclusively alpine plants; capsule nar-
    rowly to broadly ovoid.   12. STELLARIA
19 Petals irregularly toothed or jagged. 13. HOLOSTEUM
   Petals entire or slightly emarginate.   20
20 As many styles as sepals (4–5).   21
   Fewer styles (2–3) than sepals.   22
21 Styles opposite the sepals; capsule opening by 8 short
    teeth; a small glaucous herb with strap-shaped or
    narrowly lanceolate lvs.   14. MOENCHIA
   Styles alternating with the sepals; capsule splitting to
    the base by 4–5 valves; small non-glaucous herbs
    with linear lvs.   15. SAGINA
22 Succulent maritime plants with broad lvs, greenish
    fls and ±spherical capsules.   17. HONKENYA
   Not as above.   23
23 Lvs linear.   24
   Lvs not linear.   25
24 Fls greenish; petals minute; nectaries 10, conspi-
    cuous; a densely caespitose alpine plant.
                   16. *Minuartia sedoides*
   Fls with white petals; nectaries not conspicuous.
                       16. MINUARTIA
25 Lvs ovate, 1–2.5 cm, 3-veined; seeds with a fleshy oily
    appendage (elaiosome); a slender woodland plant.
                   18. MOEHRINGIA
   Lvs not exceeding 1 cm; seeds without elaiosome.
                       19. ARENARIA
26 Lvs linear.   27
   Lvs not linear.   12. STELLARIA
27 Styles 3; nectaries 10, conspicuous; a densely caespi-
    tose alpine plant.   16. *Minuartia sedoides*
   Styles 4–5; nectaries not conspicuous.   15. SAGINA
28 Lower lvs usually in closely approximated pairs
    resembling whorls of 4, obovate; a small decumbent
    herb with tiny fls c. 3 mm diam.; styles joined
    below.   22. POLYCARPON
   At least the lower lvs in opposite pairs; styles free
    to the base.   29
29 Fls in terminal dichasia or monochasia; petals
    ±equalling the sepals or exceeding them.   30
   Fls aggregated in dense terminal, lateral or axillary
    clusters; petals minute or 0.   31
30 Styles 5; stipules free, often deciduous. 20. SPERGULA
   Styles 3; stipules united round node.
                   21. SPERGULARIA
31 Fls in terminal clusters made conspicuous by white
    scarious bracts.   (PARONYCHIA, p. 149)
   Fls not in terminal clusters; bracts not conspicuous.   32
32 Fls white, in axillary clusters or false whorls; sepals
    hooded with dorsal points and resembling follicles

of *Sedum*; upper lvs in equal pairs. 26. ILLECEBRUM
Fls greenish, in lateral clusters; sepals not hooded;
upper lvs commonly in unequal pairs.

25. HERNIARIA

1. SILENE L.

Annual to perennial herbs of very various habit, or rarely dwarf shrubs. Lvs opposite, exstipulate. Fls in cymose infls or solitary, hermaphrodite or unisexual, with the bracteole not forming an epicalyx. Sepals joined below, commissures not scarious, with 5 free teeth and either with 10 veins (i.e. alternating with the teeth as well as opposite them) or with 20, 30 or 60; petals long-clawed, usually bifid or emarginate and with or without coronal scales at the throat; stamens 10; *ovary 1-celled, at least above, usually 3–5-celled at base; styles 3, or 5 alternating with the sepals.* Fr. a *capsule opening by twice as many teeth, or as many bifid teeth, as there are styles,* usually with a stalk (carpophore) between its base and the base of the calyx. Seeds numerous, reniform.

About 300 spp. in Europe, extra-tropical Asia and Africa, and North and South America.

1 Styles 5; dioecious.   2
  Styles usually 3; fls usually hermaphrodite (but styles varying 2–5 and plants ±dioecious in 10. *otites*).   3
2 Stem eglandular; fls red; capsule-teeth revolute.   **1. dioica**
  Stem ±glandular above; fls white; capsule-teeth erect.   **2. latifolia**
  (Fls pink. *latifolia* × *dioica*.)
3 Fruiting calyx with 20–30 veins, strongly inflated at least below.   4
  Fruiting calyx with 10 veins, not strongly inflated.   5
4 Perennial herbs; fruiting calyx usually glabrous, bladdery, subglobose to broadly ellipsoid, conspicuously net-veined, teeth broadly triangular.   **4. vulgaris**
  Annual; fruiting calyx pubescent, swollen below but not bladdery, conical, strongly ribbed but not conspicuously net-veined, teeth narrowly triangular-acuminate or subulate.   **5. conica**
5 Perennial herbs with non-flowering shoots   6
  Annuals lacking non-flowering shoots.   9
6 Plant of mountains and maritime cliffs with linear lvs and forming moss-like cushions 2–10 cm high; fls solitary, pink or rarely white.   **8. acaulis**
  Plant with erect fl.-stems, not forming moss-like cushions.   7
7 Plant 10–50 cm; fls 3–4 mm diam., yellowish-green, in a long narrow panicle; calyx glabrous. On dry sandy soils in East Anglia.   **10. otites**
  Plant 20–60 cm or more; fls c. 18 mm diam. with petals whitish above, reddish or greenish beneath; calyx pubescent and often glandular.   8
8 Fls ±erect; carpophore about equalling capsule.   **12. italica**
  Fls horizontal or drooping; carpophore about ⅓ as long as capsule.   **11. nutans**
9 Lvs glabrous and glaucous, middle stem-lvs ovate-cordate and ±amplexicaul; fls pink in ±compact flat-topped cymes; calyx glabrous.   **9. armeria**

Lvs hairy, of varying shape but not amplexicaul; fls many in raceme-like monochasia or few in a terminal dichasium; calyx hairy and often viscid.   10
10 Infl. a few-fld dichasium; calyx 20–25 mm; petals inrolled during daylight hours.   **3. noctiflora**
  Fls in long raceme-like monochasia; calyx at most 15 mm; petals expanded during daylight hours.   11
11 Fl.-stems dichotomizing, the ±horizontal fls in monochasia arising in pairs at 1–2(–3) nodes; carpophore 1.5–4 mm, stout, glabrous.   **6. dichotoma**
  Fl.-stems not or rarely dichotomizing, the ±erect fls in monochasia arising singly at any one node; carpophore not exceeding 1 mm, pubescent.   **7. gallica**

Section 1. *Melandriformes* Boiss. Annual to perennial. Fls in dichasial cymes, unisexual (plants dioecious) or hermaphrodite. Calyx 10–20-veined, not bladdery in fr. Styles 3 or 5. Ovary usually 1-celled to the base. Carpophore short.

1. S. dioica (L.) Clairv.   Red Campion.

Lychnis dioica L., p.p.; *L. diurna* Sibth.; *Melandrium rubrum* (Weigel) Garcke; *M. dioicum* (L.) Cosson & Germ.

A biennial to perennial herb with a slender creeping stock producing numerous decumbent non-flowering shoots up to 20 cm and erect flowering stems 30–90 cm covered with soft spreading hairs and sometimes slightly viscid above. Basal lvs broadly elliptical with a long *winged stalk*; upper lvs ovate to oblong, short-stalked or ±sessile; all with blade 4–10 cm, acute or acuminate, hairy. *Fls* 18–25 mm diam., unisexual, *scentless*, in many-fld terminal dichasia, their stalks 0.5–1.5 cm. Calyx-tube 10–15 mm, hairy and slightly viscid; of male fl. cylindrical, faintly 10-veined; of female fl. ovoid and 20-veined, becoming rounded in fr.; *teeth triangular-acute*, 2–2.5 mm, broader in the female fl. *Petals* bright *pink*, rarely magenta or white, the limb broadly obovate, deeply bifid into narrow segments with 2 narrow acute scales at the base; claw auricled above. Styles 5. *Capsule* (Fig. 27A) *broadly ovoid, opening widely by*

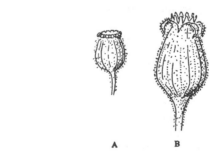

Fig. 27. Dehisced capsules of A, *Silene dioica* and B, *S. latifolia*. ×1.

10 *revolute teeth*; carpophore extremely short. Seeds black, densely and acutely tubercled. Fl. 5–6. Dioecious. Open by day and visited by long-tongued humble-bees and hover-flies. $2n = 24^*$, 48. Hp.–Chh.

Native. In deciduous woods on moist but well-drained mull and moder soils over a wide range of pH and appearing calcifuge in some areas though not in others; also in hedgerows and shaded gardens on fertile soil, on cliff-ledges and stabilized scree, especially of Carboniferous limestone, and in the grikes of limestone pavement. Requires high levels of mineral nutrients and is a characteristic plant of bird-cliffs in Scotland. Reaches 1060 m in Scotland. Locally abundant throughout the British Is. but rare in the dry sandy Breckland of East Anglia. Through most of Europe northwards to the Faeroes and Spitsbergen but absent from parts of the Mediterranean region; C. Asia eastwards to the Altai; N. Africa; Greenland. Introduced in North America.

There is much variation within and between populations, the latter being relatable both to geographic location and to ecological factors. Coastal populations in Shetland were at one time thought sufficiently distinct to justify treatment as a separate subspecies. Recent studies have shown, however, that *S. dioica*, in this country as in continental Europe, consists of 'local weakly differentiated ecogeographic races that intergrade at their margins' and cannot usefully be recognized taxonomically. (H. C. Prentice, *Watsonia*, 1980, **13**, 11–26).

2. S. latifolia Poiret subsp. **alba** (Miller) Greuter & Burdet   White Campion.

Lychnis alba Miller; *L. vespertina* Sibth.; *Melandrium album* (Miller) Garcke

A short-lived perennial or sometimes annual or biennial herb with a thick almost woody stock from which arise a few short non-flowering shoots and erect flowering shoots 30–100 cm, covered with soft spreading hairs and slightly glandular-viscid above. Lower lvs and those of barren shoots oblanceolate or elliptical, narrowed into long unwinged stalks; upper lvs lanceolate or elliptical, acuminate, sessile; all 3–10 cm, hairy. *Fls* 25–30 mm diam., unisexual, slightly *evening-scented* in a few-fld terminal dichasium. Calyx-tube 18–25 mm, downy and viscid; of the male fl. cylindrical, 10-veined; of the female narrowly ovoid, 20-veined; *teeth linear-lanceolate blunt*, c. 6.5 (5–8) mm in the female fls, c. 4.8 mm in male. *Petals white*, the limb obovate, deeply bifid into broad segments with 2-lobed coronal scales, the claw auricled above. Styles 5. *Capsule* (Fig. 27B) *ovoid-conical*, often breaking the calyx-tube, opening narrowly by 10 *suberect teeth*; carpophore very short. Seeds 1.3–1.5 mm diam., grey, bluntly tubercled. Fl. 5–9. Dioecious. Opening at night and pollinated by night-flying moths. $2n = 24^*$. Hp.

The above description is of the widespread subsp. **alba**. Subsp. *divaricata* (Reichenb.) Walters (*S. macrocarpa* (Boiss. & Reuter) E. H. L. Krause), a native in the Mediterranean region, differs in its long acuminate calyx-teeth and large capsule opening by teeth which are ultimately revolute. It is an occasional casual.

Probably introduced in Neolithic times. A weed of cultivated land, mainly on light calcareous or sandy and usually deep soils of circumneutral pH; frequent also in such open habitats as roadsides, field-borders, hedge-rows, grassy banks, quarries and waste places; occasionally on walls. Common throughout lowland Great Britain and northwards to Shetland; more local in Ireland; Channel Is. Reaches 424 m in Scotland. Europe northwards to c. 68°N in Scandinavia; W. Asia to Lake Baikal; N. Africa. Introduced in North America.

S. dioica and *S. latifolia* cross readily to give F₁ hybrids capable of intercrossing or of backcrossing with either parent species, there being no appreciable loss of fertility. The F_1 plants are intermediate between the parents in many observable features including petal colour, diameter of corolla, length of calyx, size and shape of capsule, length and degree of recurving of capsule-teeth and mean weight of seeds, which have, however, the large blunt tubercles of *S. latifolia*. Populations of hybrid origin are frequent where the parent species grow close together and are immediately recognizable by the presence of many pink-fld individuals.

3. S. noctiflora L.   Night-flowering Campion.

Melandrium noctiflorum (L.) Fr.

An annual herb 15–45 cm, with an erect simple or branched flowering stem, with soft spreading hairs, glandular and viscid especially above. Lower lvs obovate or ovate-lanceolate, narrowed to a stalk-like base, uppermost narrowly oblong-lanceolate, sessile, very acute; all 5–10 cm, with scattered hairs. Fls c. 18 mm diam., *hermaphrodite*, in a terminal few-fld dichasium. Calyx-tube c. 20 mm, woolly and viscid, cylindrical-ovoid, becoming swollen in fr., membranous and whitish between the 10 broad green veins, and with 5 slender ciliate teeth. *Petals yellowish beneath and rosy above, rolled inwards during the day*, spreading and scented at night, deeply bifid, with 2 coronal scales at the base of the limb and auricles at the top of the claw. *Styles* 3. Capsule ovoid-conical often bursting the calyx, opening by 6 ±recurved teeth; carpophore about ⅙ as long as the capsule. Seeds 1–1.2 mm across, tubercled. Fl. 7—9. Strongly protandrous. Probably visited by moths. $2n = 24^*$. Th.

Probably native. A local weed of arable fields on sandy soils or over chalk but also on heavier calcareous soils as over chalky boulder-clay; sometimes in waste places. Frequent in lowland S. and E. England but rare in W. England and Wales and in only a few scattered localities in mainland Scotland northwards to E. Ross; very rare in Ireland. Through much of Europe but perhaps introduced in Scandinavia; W. and C. Asia. Introduced in North America.

Section 2. *Inflatae* Boiss. Perennial. Fls in dichasia, often few-fld, hermaphrodite or unisexual. Petals not convolute-contorted in bud. Calyx bladdery in fr., faintly 20-veined. Styles 3. Ovary septate at base. Carpophore well-developed

4. S. vulgaris (Moench) Garcke, *sens. lat.*

Cucubalus behen L.; *S. cucubalus* Wibel; *S. inflata* Sm.

Perennial herbs with erect to almost prostrate fl.-stems, usually 10–80 cm, glabrous or hairy, often glaucous. Lvs variable in size and shape. Fls solitary or in few- to many-fld cymose infls, hermaphrodite or female by abortion of stamens, there being varying degrees of gynodioecism or gynomonoecism; male fls have been reported from C. Europe. Calyx pale green, often tinged and veined with red, *thin-walled and bladdery with 20 main veins* and a ±distinct connecting network of secondary veins, persistent round the ripe capsule. *Petals white, rarely pink or greenish, clawed, the blade deeply bifid* (rarely with additional small lobes) *with coronal scales or bosses at its base.* Styles 3(–5). The protandrous fls, open and fragrant at night and often open also by day, have nectaries at the base of the filaments and are visited especially by nocturnal moths and long-tongued bees. *Capsule ±broadly ovoid, opening by 6 teeth*; separated from the base of the calyx by a *carpophore* 2–4.5 mm. Seeds 1–2 mm, covered with tubercles of varying size and prominence or with flat plates ('armadillo'). $2n = 24, 48$.

A large complex of taxa hitherto variously treated as species, semi-species, subspecies or varieties. The only two native British representatives, Bladder Campion and Sea Campion, differ in a number of features, though with some overlapping in all of them. Naturally occurring hybrids are quite rare and, if a range of differential characters is considered, all populations can be assigned without difficulty to one or other of the two taxa. British botanists, while recognizing that they are very closely related, have therefore tended to regard them as distinct species. They are in fact highly interfertile and their maintenance as separate entities presumably arises from the effective isolation conferred by their ecogeographical differences. In Scandinavia, and elsewhere in continental Europe where the two coexist, they are somewhat less clearly distinct, and Bladder Campions in mountains of C. Europe, outside the range of the Sea Campion, are in many ways intermediate. All in all it seems best to follow *Flora Europaea* 1 (1964) and treat the major component taxa as subspecies of *S. vulgaris sens. lat.* rather than as separate species.

1 Plant erect or ascending to 50 cm or more; infl. of at least 5 and up to 80 fls; bracts ±scarious; capsule with narrowed neck 1.5–3 mm and erect or somewhat spreading teeth; carpophore 2–3 mm.   *2*

    Plant ±procumbent with slender ascending fl.-stems to 25 cm and infl. of 1–4 fls; bracts herbaceous; capsule with neck 3–4 mm and spreading or deflexed teeth; carpophore 3–4.5 mm. On coastal shingle or cliffs, rarely inland on cliffs and scree or by lakes and streams.   **Subsp. maritima**

2 Plant with branching underground stock but not stoloniferous; lvs elliptic-lanceolate to ovate; fls white; capsule up to 10 mm. Common plant of roadsides, banks, arable land, etc.   **Subsp. vulgaris**

    Plant with slender hypogeal stolons up to 30 cm; lvs narrowly lanceolate; fls pink or greenish; capsule 10–13 mm. Alien established on Plymouth Hoe.   **Subsp. macrocarpa**

Subsp. **vulgaris**   Bladder Campion.

S. vulgaris (Moench) Garcke, s. str.

A perennial herb with branching ±woody stock and *erect or ascending fl.-stems* 20–100 cm, simple or branched; *all above-ground parts normally dying down in late autumn.* Lvs very variable, commonly 3–10 × 0.7–2.5 cm, broadly lanceolate or ovate-lanceolate to narrowly elliptic or oblong-elliptical, rarely ovate or linear, acute to acuminate; lower lvs narrowed into a stalk-like base, upper sessile and often ±amplexicaul. *Stem and lvs usually glabrous* or with hairs only on lf-margins, sometimes ±pubescent. *Infl. a many-fld subcorymbose cyme*, often with 30 or more fls, rarely fewer than 5; *bracts usually becoming scarious.* Fls usually hermaphrodite but most populations include plants with female or both female and hermaphrodite fls, c. 18 mm diam., *female fls somewhat smaller, nodding and somewhat zygomorphic. Calyx* (Fig. 28B) broadly ovoid to

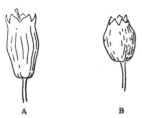

Fig. 28. Calyces of A, *Silene vulgaris* subsp. *maritima* and B, *S. vulgaris* subsp. *vulgaris.* ×1.

cylindrical, *the contracted mouth narrower than the base* and with 5 triangular teeth, glabrous. *Petals* white, *usually not overlapping, with pairs of small coronal bosses*, rarely small scales, at base of blade. *Capsule* up to 10 mm, broadly ovoid to subglobose, *with narrow neck 1.5–3 mm across and erect or slightly spreading teeth*; carpophore 2–3 mm. Seeds 1.2–1.5 mm, mostly tubercled, rarely with flat plates ('armadillo'), usually white when immature, sometimes purple. Fl. 6–8. $2n = 24^*$. Hp.

Probably native (identified in Late Glacial of Ireland). Throughout the British Is. except Outer Hebrides and Shetland and common in arable fields, rough pasture and hedgerows, on banks by roads, railways and canals, on walls and in quarries, gravel-pits, etc., on a wide range of substrata, in most of England, Wales and S. Scotland but only in scattered localities in N. Scotland, though reaching Caithness and Sutherland; Ireland, but local in the north; reaches 330 m in Yorks. Through most of Europe to c. 70° N; N. Africa; temperate Asia.

Subsp. **maritima** (With.) Á. & D. Löve   Sea Campion.

S. maritima With.

A perennial herb resembling subsp. *vulgaris* but with a more spreading or sometimes cushion-like habit because of the *numerous ±prostrate shoots*, some ascending as fl.-stems up to 25 cm; all glabrous and glaucous; *some prostrate shoots remain green throughout winter.* Lvs commonly 0.5–3 × 0.2–1 cm, *lanceolate to linear-lanceolate*, usually acute; margin narrowly membranous, often ciliate, but lvs otherwise glabrous. *Fls solitary or in 2–4(–7)-fld cymes; bracts thick, ±herbaceous. Fls* 20–25 mm diam., *usually hermaphrodite,* sometimes female, *erect and actinomorphic. Calyx* (Fig. 28A) as in subsp. *vulgaris* but *with mouth wider than base. Petals usually white, overlapping, with pairs of well-developed coronal scales at base of blade. Capsule broadly ovoid, usually with wide mouth and reflexed teeth*; carpophore 3–4.5 mm. *Seeds* c. 1.6 mm, *more commonly with flat plates than tubercles* but proportions varying widely from one population to another; immature seeds purplish, rarely white. Fl. 4–10. $2n = 24^*$. Chh.

Native. Locally abundant on coastal shingle or sandy gravel and on cliffs all round the British Is.; rarely inland on upland cliff-ledges, stony ground by lakes and streams and old mine-workings; reaches 970 m in Scotland. Chiefly coastal in W. Europe from Spain to the Kola Peninsula; Azores. Related subspp. in Iceland and on coasts of N. Spain and W. France. Other subspp. usually on mountains of C. and S. Europe resemble subsp. *maritima* in having few-fld infls and wide-mouthed capsules with reflexed teeth but are closer to subsp. *vulgaris* in other features.

*Subsp. **macrocarpa** Turrill

S. angustifolia (Ten.) Guss., non Poiret; *S. linearis* auct., non Sweet

Like subsp. *vulgaris* but with *hypogeal stolons* up to 30 cm and with the well-developed *lvs linear-lanceolate to narrowly lanceolate*. The zygomorphic fls are in infls of 3–25 and have *dull red or greenish-yellow petals with very small coronal bosses*. The capsule is broadly ovoid with contracted mouth, like that of subsp. *vulgaris* but larger (10–13 mm), with erect or somewhat spreading teeth. Seeds 1.6 mm, tubercled, white when immature.

Introduced, perhaps from Cyprus, and established on Plymouth Hoe for more than 40 years. May not be clearly distinguishable from subsp. *commutata* (Guss.) Hayek of S. Europe, though this typically has broader lvs.

Section 3. *Conomorpha* Otth. Annuals. Fls in dichasia. Petals convolute-contorted in bud. Fr.-calyx conical, swollen below but not bladdery, 20–30(–60)-veined, with long acuminate teeth. Styles 3. Ovary septate at base. Carpophore very short.

5. S. conica L.   Sand Catchfly.

An annual herb with slender tap-root and erect simple or branched grey-pubescent stem 3–20(–35) cm, usually glandular-viscid at least above. *Lvs oblong- to linear-lanceolate*, acute, uppermost long-acuminate; all pubescent. Fls solitary or in few-fld dichasia, 4–5 mm diam., short-stalked, erect, actinomorphic, hermaphrodite. *Calyx* 8–15 mm, *at first cylindrical, later ovoid-conical*, swollen below, 30-*veined, densely glandular-hairy, with long subulate teeth. Petals rose-coloured*, rarely white, *long-clawed*, the blade 3–5 mm, bifid, with pairs of coronal scales at its base, the claw auricled above. *Capsule* 7–12 mm, somewhat shorter than calyx, *ovoid-conical*, opening by 6 ±erect teeth; carpophore less than 1 mm.

Seeds pale grey, c. 0.8 mm across, tubercled. Fl. 5–7. Protandrous. $2n = 24^*$. Th.

Native. A very local plant of sand dunes and sandy pastures, dry sandy and chalky fields, roadsides, waste places, etc., coastal and inland in East Anglia and on the coast of Kent and Sussex; Channel Is. Decreasing in most areas and now only a casual elsewhere apart from three localities in S. England and one in Scotland (Moray) where it still persists. Europe from Spain and France eastwards to the Ukraine; doubtfully native in Denmark and S. Sweden.

S. conoidea L. differs from *S. conica* in its broadly lanceolate lvs, less densely pubescent and often subglabrous, entire or slightly emarginate petals and longer capsule, 12–18 mm, long-attenuate above. It is a rare casual, reports of its being established in Jersey having arisen from a misidentification.

Section 4. *Dichotomae* (Rohrb.) Chowdhuri. Annuals. Fls almost sessile in paired raceme-like cymes. Petals convolute-contorted in bud. Fruiting calyx 10-veined. Styles 3. Ovary septate at base. Carpophore well developed.

***6. S. dichotoma** Ehrh.   Forked Catchfly.

An annual herb with a stiffly erect sparsely hairy branching stem, 20–60 cm. Lower lvs spathulate, stalked, middle 3–5 cm, ovate-lanceolate, upper lanceolate acute; all 3–5-veined, hairy. Infl. of *long terminal cymes dichasial below (with fls in the forks) then raceme-like and one-sided*. Fls 15–18 mm long and wide, ±horizontal, short-stalked, erecting in fr. Calyx 11–15 mm, cylindrical, 10-veined, hairy, the veins stiffly ciliate. Petals whitish, deeply bifid, usually with short coronal scales. Styles 3. Capsule ovoid, opening by 6 acute teeth; carpophore $\frac{1}{5}$ to $\frac{1}{3}$ as long as the capsule. Seeds 1.3 mm, reniform, furrowed along the back. Fl. 7–8. Protandrous. $2n = 24$. Th.

Introduced. A frequent casual. E. and S.E. Europe and W. Asia.

Section 5. *Silene*. Annuals. Fls almost sessile in simple raceme-like cymes. Petals convolute-contorted in bud. Fruiting calyx 10-veined. Styles 3. Ovary septate at base.

7. S. gallica L., *sens. lat.*   Small-flowered Catchfly.

Incl. *S. anglica* L. and *S. quinquevulnera* L.

Annual herbs with slender tap-root and erect or ascending fl.-stem 15–45 cm, simple or branched, pubescent or with long white hairs, viscid above. Lower lvs oblong-spathulate, narrowed into a stalk-like base; *middle lvs* 2.5–5 cm, *lanceolate, sessile*; uppermost narrowly oblong-lanceolate to linear, sessile; all ±hairy. *Infl. of simple raceme-like often secund monochasia with lf-like bracts and single short-stalked fls at successive nodes* of the sympodial axis. *Calyx* 7–11 mm, *cylindrical at first but becoming ovoid in fr., with long acuminate teeth and 10 viscid-hairy veins*. Petals clawed, with entire or emarginate limb up to 6 mm, whitish or pale rose, sometimes with dark red basal blotch; coronal scales distinct, acute. Stamen-filaments hairy. *Capsule ovoid, about equalling calyx*; fr.-pedicel erect, spreading or deflexed; *carpophore very short*. Seeds reniform, 0.8 mm across, dark brown with deeply concave faces. Fl. 6–10. $2n = 24$. Th.

British representatives of this highly polymorphic complex have been placed in three 'series' (Lousley, *J. Botany, London*, 1936, **74**, 194–201), convenient groupings of local or more widespread races corresponding roughly with the three species recognized by Linnaeus but freely interfertile and insufficiently distinct to warrant higher than varietal rank. Of these the most widespread is var. *anglica* (L.) Mert. & Koch, with branched spreading stems, *fls c.* 8 mm *diam.*, *petals dingy white to yellowish or pale pinkish-purple*, and *lower fr.-pedicels exceeding calyx, commonly spreading or deflexed*. Var. *quinquevulnera* (L.) Boiss. is like var. *anglica* but may have larger fls and *each petal-limb has a deep crimson basal blotch*. Var. *sylvestris* (Schott) Ascherson & Graebner (?var. *gallica*) has the stem simple or with a *few erect branches, fls* 12–13 mm *diam.*, *petals pale pink or rose, almost entire*, and *fr.-pedicels shorter than calyx, erect*.

Probably native but casual in many localities and decreasing, now almost confined in England and Wales to the area southwards from S. Lancashire and the Humber and a rare casual further north and in Scotland; Channel Is.; in a few scattered localities in Ireland. The widespread race is var. *anglica*, native or casual in sandy and gravelly arable fields and in waste places, chiefly with whitish fls in S. and S.E. England but with pale purplish fls in S.W. England. Var. *quinquevulnera*, formerly grown in gardens for its ornamental fls, is now a rare casual except in the Channel Is. Var. *sylvestris* is thought native in the Channel Is. and is established in Scilly but is a rare casual elsewhere. *S. gallica, sens. lat.*, appears native in much of S. and C. Europe northwards to Denmark, Poland and C. Russia.

Section 6. *Nanosilene* Otth. Caespitose perennials. Fls solitary. Petals convolute-contorted in bud. Fruiting calyx faintly 10-veined. Styles 3. Ovary septate at base.

8. S. acaulis (L.) Jacq. Moss Campion.

A perennial, densely tufted and much branching, ±glabrous, somewhat woody below, forming *bright green moss-like cushions* 2–10 cm high. *Lvs* 6–12 mm, *in dense rosettes, linear-acute*, channelled above, stiffly ciliate especially towards the base. Hermaphrodite, male and female fls on different plants. Fls 9–12 mm diam., solitary, erect, their stalks very short at first then lengthening. Calyx 7–9 mm, ±reddish, campanulate, faintly 10-veined, glabrous, its 5 blunt teeth scarious-margined. *Petals deep rose* or whitish with bifid or emarginate blade, small coronal scales and claw without auricles. Capsule up to twice as long as the calyx, subcylindrical, opening by 6 erect teeth; carpophore downy, shorter than the capsule. Seeds pale yellowish, reniform. Fl. 7–8. Protandrous. Visited by various Lepidoptera and other insects. $2n = 24$. Chc.

Native. On cliffs, ledges, scree and summit detritus of mountains in N. Wales, the Lake District and Scotland and also on cliffs, rocks and stabilized sand-dunes down to sea-level in the Hebrides, Orkney and Shetland. Behaves as a ±strict calcicole in the southern part of its range but becomes increasingly abundant on poor acid rocks and soils from the Cairngorms north-westwards; very rare in W. and N. Ireland. Reaches 1280 m

in Scotland. An arctic-alpine of arctic Europe, E. Siberia and North America, reaching 83° N in Greenland, and of the mountains of W. and C. Europe and North America.

Section 7. *Compactae* Boiss. Annuals. Fls in compact subcorymbose cymes. Petals convolute-contorted in bud. Fruiting calyx membranous, 10-veined. Styles 3 or 5. Ovary septate at base.

*9. S. armeria L. Sweet-William Catchfly.

An annual herb with slender tap-root and erect hollow stems, 10–60 cm, simple or branched above, slightly viscid above, glaucous. *Lvs* 2–5 cm, sessile, the lowest rather crowded, spathulate, the upper *ovate-lanceolate to -cordate*, ±amplexicaul, acute; all *glabrous and glaucous*. Fls short-stalked, erect, c. 15 mm diam. and 18 mm long, in rather crowded subcorymbose dichasia. Calyx 12–15 mm, reddish, 10-veined, cylindrical-clavate with short blunt teeth. Petals deep rose, emarginate, with 2 narrow erect scales at the base of the lamina, the claw not auricled. Styles 3. Capsule cylindric-ovoid just enclosed by the calyx, opening by 6 revolute teeth; carpophore almost equalling the capsule. Seeds small, blackish, furrowed down the back. Fl. 6–7. Protandrous. Nectar accessible only to Lepidoptera. $2n = 24$. Th.

Introduced. A garden escape in a few places in S. England and occasionally establishing itself. S. and C. Europe, but widely introduced by cultivation in gardens.

Section 8. *Otites* Otth. Perennials. Fls in opposite pairs of cymes simulating whorls. Petals convolute-contorted in bud. Fruiting calyx faintly 10-veined. Styles 3 (2–5). Ovary septate at base.

10. S. otites (L.) Wibel Spanish Catchfly.

Cucubalus otites L.

A perennial herb with a thick branching woody stock from which arise non-flowering rosettes and erect simple flowering stems, 20–90 cm, shortly hairy below and viscid up to about the middle. Basal lvs 1.5–8 cm, narrowly spathulate, narrowing to a stalk; lower stem lvs similar but smaller; upper stem lvs in distant pairs, linear-lanceolate, sessile; all with dense short hairs. *Infl. of numerous fls in a long narrow panicle, interrupted below, the opposite pairs of short cymes simulating false whorls*. ±Dioecious, but some male plants with a few hermaphrodite fls. *Fls* 3–4 mm *diam.* and 4–5 mm long, erect, their glabrous stalks not much exceeding the calyx. Calyx glabrous, narrowly campanulate, faintly 10-veined, with 5 short blunt teeth. *Petals pale yellowish-green*, narrow, entire, with no coronal scales, the claw without auricles. Female fls have no stamens and male fls a usually vestigial ovary. Stamens and the 3 (3–5) styles exserted. Capsules ovoid, longer than and rupturing the calyx, opening by 6 (4–10) short teeth; carpophore almost 0. Seeds reniform, furrowed along the back, finely rugose, 0.7 mm across. Fl. 6. Hermaphrodite fls protandrous. Evening-scented and nectar-secreting and visited in some districts by Lepidoptera, but perhaps also wind-pollinated (male fls said not to secrete nectar in C. Germany). $2n = 24$. Hs.

Native. Confined to the Breckland heaths of Norfolk, W. Suffolk and Cambridge, but occasionally casual elsewhere. S. and C. Europe and W. Asia. A 'steppe' species.

Section 9. *Siphonomorpha* Otth. Perennials. Fls drooping, usually in a panicle with paired dichasia. Petals convolute-contorted in bud. Fruiting calyx 10-veined, ±cylindrical. Styles 3. Ovary septate at base.

11. S. nutans L.   Nottingham Catchfly.

A perennial herb with a slender branched woody stock from which arise short non-flowering shoots and erect flowering shoots, 25–80 cm, downy below and viscid above. Basal lvs spathulate narrowing to a long stalk, c. 7.5 cm; upper lvs narrowly lanceolate, subsessile, acute; all softly hairy and ciliate. Infl. a lax subsecund panicle with opposite branches ending in 3–7-fld dichasial cymes. Fls hermaphrodite or unisexual, 18 mm diam., c. 12 mm long, drooping, their stalks short, viscid. Calyx cylindrical-clavate, glandular-pubescent, with 10 purplish veins and 5 acute white-margined teeth. *Petals* white or pink-tinged (rarely red or yellow), bifid, *with narrow ±inrolled lobes* and 2 *acute basal scales*, the *claws not auricled*. Styles 3. Capsule ovoid, somewhat exceeding the calyx and opening by 6 spreading or reflexed teeth; *carpophore downy, about* ⅓ *as long as the capsule*. Seeds 1–2.2 mm across, reniform, furrowed along the back, tubercled. Fl. 5–7. Fls heterostylous, nectarless, protandrous, 'opening and fragrant for 3 nights, 5 stamens ripening on each of the two first nights, the styles protruding on the third'. Visited by Lepidoptera and humble-bees. $2n = 24*$. Hs.

Extremely variable. British representatives are all assigned to subsp. **nutans**, described above. They fall into two distinct varieties: var. *salmoniana* Hepper, with narrow sparsely hairy lvs and ripe capsules 11–14 mm (carpophore 3–4 mm); and var. *smithiana* Moss, with usually broader, more pubescent and less acute leaves and ripe capsules 8–10 mm (carpophore 2–3 mm). Individual populations, especially of var. *smithiana*, show further intravarietal differences. Var. *salmoniana* in some respects approaches subsp. *dubia*, native only in the Carpathians.

Native. A local plant of dry slopes, rocks, cliff-ledges, shingle, walls and field-borders. Var. *salmoniana* in W. Kent, Sussex, S. Hants, Isle of Wight and Surrey; var. *smithiana* in E. Kent, Dorset, Devon, N. Wales, Peak District, Nott, W. Yorks, Fife, Angus, Kincardine. The form in Jersey is close to var. *salmoniana*. *S. nutans* is also introduced in several localities. Europe, Canary Is., N. Africa, Caucasus, N. Asia to Japan.

Section 10. *Paniculatae* Boiss. Perennials. Like *Siphonomorpha* but fls erect.

12. S. italica (L.) Pers.   Italian Catchfly.

Cucubalus italicus L.

A perennial herb with slender branched woody stock from which arise long non-flowering shoots and erect flowering shoots 25–70 cm, softly hairy and viscid.

Lower lvs 3–5 cm, lanceolate-spathulate narrowed to a long stalk; uppermost linear-lanceolate sessile; all pubescent. Infl. a pyramidal panicle, the opposite, long, ascending, viscid branches ending in c. 3-fld dichasia. Fls c. 18 mm diam. and 20 mm long ±erect, their stalks viscid. Calyx narrowly clavate, glandular-pubescent, 10-veined with 5 blunt ovate teeth. *Petals* yellowish-white above, often reddish-green beneath, deeply bifid *with two small bosses at the base* of the blade, the *claw auricled* above. Styles 3. Capsule ovoid, about equalling the calyx, opening by 6 spreading teeth; *carpophore at least equalling the capsule*. Fl. 6–7. Fls heterostylous, opening only in the evening, and then fragrant. $2n = 24$. H.

Almost certainly introduced. A rare and local plant of arable fields, quarry-sides, waste places, etc., in a very few localities in S.E. England, S. Wales and Midlothian; usually casual but persistent in N. Kent. Mediterranean region and Near East.

A few other *Silene* spp. have been recorded as casuals, but their frequency varies greatly from one decade to the next. Amongst these are **S. cserei** Baumg., a perennial like *S. vulgaris* subsp. *vulgaris* but with broadly ovate stem-lvs and calyx with 10 long and 10 short veins; and **S. cretica** L. and **S. muscipula** L., annuals, glabrous at least above and viscid above, and both pink-fld, the former with glabrous, latter with pubescent carpophores.

Grown in gardens and sometimes escaping are **S. pendula** L. (Mediterranean region), annual, with glandular-pubescent fl.-stems decumbent below then ascending to 15–40 cm, ovate stem-lvs, lax raceme-like monochasia of fls which become pendent, usually pinkish petals with acute coronal scales, an ovoid-conical capsule 9–12 mm and carpophore 3–6 mm; **S. coeli-rosa** (L.) Godron (S.W. Europe), also annual but glabrous and with linear-lanceolate lvs, fls in very lax dichasia, calyx subclavate and deeply furrowed between veins, rose-pink petals with long linear coronal scales, styles 5, capsule opening by 5 bifid teeth and carpophore 7–12 mm, more than ⅔ as long as the capsule; and **S. schafta** S. G. Gmelin, Moss Campion (Caucasus), a pubescent tufted rock-garden perennial with rosettes of narrow spathulate lvs, decumbent fl.-stems to 15 cm and rose-pink fls borne singly or in pairs.

2. LYCHNIS L.

Annual to perennial herbs with opposite lvs and cymose infls which are sometimes spike-like or head-like. Fls hermaphrodite, 5-merous. Epicalyx 0. *Sepals joined below* into a calyx-tube and with 5 free teeth above; *petals* red or white, long-clawed, *with coronal scales* at the base of the limb; stamens 10; ovary 1-celled throughout or 5-celled only at the base; *styles usually 5. Capsule opening* loculicidally, usually *by 5 teeth*; carpophore present but sometimes rudimentary; seeds tubercled or rugose, often very small.

About 15 spp. in the northern hemisphere.

1 Stems usually less than 15 cm; fls in a compact head-like panicle.   **2. alpina**
   Stems usually more than 30 cm; fls in a spike-like panicle or long-stalked dichasia.   *2*
2 Stems very viscid beneath each node; fls in an interrupted spike-like panicle; petals slightly notched.
                                          3. viscaria

Stems not viscid beneath the nodes; fls in long-stalked dichasia; petals deeply 4-cleft with narrow segments. **1. flos-cuculi**

1. L. flos-cuculi L. Ragged Robin.

A perennial herb with slender branching stock from which arise decumbent non-flowering shoots to 15 cm and erect flowering shoots 30–75 cm, rough above and with a few downwardly directed hairs. Lvs 2–10 cm, glabrous, somewhat rough; lower lvs and those of barren shoots oblanceolate, acute, narrowing to a stalk-like base; upper stem lvs narrower, oblong-lanceolate subsessile. Fls 3–4 cm diam. on stalks 1–1.5 cm, in terminal and lateral long-stalked dichasia. Calyx-tube c. 6 mm, reddish, sub-membranous, strongly 10-veined, with 5 ovate-acuminate teeth 3 mm. *Petals* rose-red, rarely white, *deeply 4-cleft* with narrow spreading segments and 2 bifid subulate coronal scales at the base of the limb. Capsule broadly ovoid, enclosed by the calyx, 1-celled throughout, opening by 5 short acute revolute teeth; *carpophore extremely short.* Seeds 0.5–0.7 mm across, brown, tubercled. Fl. 5–6. Protandrous. Visited by butterflies and long-tongued bees and flies. $2n = 24$. Hp.

Native. A common plant of damp meadows, marshes, fens and wet woods throughout the British Is. Reaches 610 m in Scotland. Europe to Norway and Iceland, Siberia, Caucasus. Introduced in North America.

2. L. alpina L. Red Alpine Catchfly.
Viscaria alpina (L.) G. Don
A perennial caespitose herb with tap-root and woody stock whose short branches end in non-flowering rosettes or erect lfy *glabrous flowering stems* 5–15(–20) cm. Lvs 1.5–5 cm, sessile, oblong-lanceolate or linear, acute, in dense basal rosettes with 1–6 pairs on the flowering stems; all glabrous or slightly ciliate at the base. *Fls 6–12 mm diam., crowded into almost head-like corymbose panicles,* their stalks very short. Bracts ovate-acuminate, rose-coloured. Calyx narrowly campanulate, its tube faintly veined, glabrous, the 5 teeth short, broad and rounded with purplish scarious margins. *Petals* rose-coloured, long-clawed, the limb obovate *deeply bifid* with 2 short *tubercle-like coronal scales* at its base. Capsule ovoid, 5-celled at base, opening by 5 recurved teeth, on a *carpophore less than half its length.* Seeds 0.3–0.4 mm, dark brown, bluntly tubercled. Fl. 6–7. Protandrous. Larger fls hermaphrodite, smaller usually with abortive stamens. Visited by butterflies and automatically self-pollinated. $2n = 24$. Chh.

Native. A very rare alpine plant of the Lake District and Angus, reaching 870 m in Angus. A subarctic-alpine species found in the Alps and Pyrenees and in subarctic Europe, W. Asia and North America, reaching 73° 10′ N in E. Greenland. Frequently on serpentine and other substrata with unusually high content of copper, zinc, nickel or other heavy metals.

3. L. viscaria L. Red German Catchfly.
Viscaria vulgaris Bernh.
A perennial tufted herb with a woody stock whose short

erect branches end in very short densely lfy non-flowering shoots or erect lfy *flowering shoots*, 30–60 cm. *Stems* dark green or purplish, glabrous, *very viscid beneath each node.* Lvs of barren shoots 5–12.5 cm, narrowly elliptic-lanceolate, acute or acuminate narrowing below into a long stalk-like base; upper lvs oblong-lanceolate; all with woolly margins at the base, otherwise glabrous. *Infl.* an interrupted spike-like panicle of axillary cymes which simulate whorls. Bracts broadly lanceolate, acuminate. Fls 18–20 mm, diam., their stalks extremely short. Calyx c. 12 mm, cylindrical-clavate, membranous, purplish, 10-ribbed, with short ovate teeth. *Petals* purple-red, claw long, auricled above, limb obovate, *slightly notched, with 2 conspicuous coronal scales* 3 mm *long* at its base. Capsule ovoid, 5-celled at base, bursting the calyx, opening by 5 ±spreading teeth, on a *carpophore almost equalling it in length.* Seeds 0.5 mm across, acutely tubercled. Fl. 6–8. Protandrous. Visited by butterflies and long-tongued bumble-bees. $2n = 24$. Chh.

Native. A very rare and local plant of cliffs, dry rocks and rock débris, chiefly on igneous rocks, in a few places in Wales and in Scotland from Roxburgh and Edinburgh northwards to Perth and Angus. Reaches 427 m in Scotland. Europe northwards to S. Scandinavia and Finland; W. Asia. Said to be calcifuge on the Continent, and characteristic of open sandy habitats.

Several spp. and hybrids are grown in gardens for their attractive fls, amongst them **L. chalcedonica** L., perennial with ovate lvs and bright scarlet fls in dense terminal heads; **L. coronaria** (L.) Desr., a white-tomentose perennial, 30–100 cm, with ovate-oblong lvs and long-stalked crimson (or white) fls 2.5 cm or more in diam. in a lax few-fld infl.; **L. flos-jovis** (L.) Desr., also white-tomentose but with lanceolate lvs and 4–10 crimson or scarlet fls in compact head-like dichasia, and **L. × haegeana** Lemaire (*L. fulgens* Fischer × *L. coronata* Thunb.) with fls c. 5 cm diam. in various shades of red. *L. viscaria* and *L. alpina* may also be seen in gardens.

3. AGROSTEMMA L.

Annual herbs with tall erect stems and narrow opposite lvs. Fls conspicuous, solitary or in few-fld dichasia. *Epicalyx* 0; sepals joined below into a 10-ribbed tube and with *five narrow spreading teeth much longer than the petals; coronal scales* 0; stamens 10; styles 5, alternating with the sepals; ovary 1-celled to the base. Capsule opening by 5 teeth, carpophore 0; seeds numerous, black.

Three spp. in Europe and W. Asia.

*1. A. githago L. Corn Cockle.
Lychnis githago (L.) Scop.; *Githago segetum* Link
An annual herb with a strong tap-root and an erect simple or sparingly branched flowering stem 30–100 cm, covered with appressed white hairs. Lvs 5–12.5 cm, linear-lanceolate, acute, with appressed hairs. Fls 3–5 cm diam., usually solitary at the ends of main stem and branches, their stalks long, hairy. *Calyx-tube* cylindrical-ovoid, coriaceous, woolly, 10-ribbed, *with long*

spreading linear acute lf-like teeth 3–5 cm long. *Petals pale reddish-purple*, long-clawed, the limb shorter than the sepals, obovate, *slightly notched, with no coronal scales*. Capsule ovoid, exceeding the calyx-tube, opening by 5 ±erect teeth. Seeds 3–3.5 mm across, black, tubercled. Fl. 6–8. Protandrous to homogamous. Visited by butterflies and automatically self-pollinated. $2n = 24, 48$. Th.

Probably introduced, but identified in deposits of Roman date. A weed of arable crops, especially cereals, and of field-margins and waysides; now decreasing rapidly because of improved seed-cleaning. Through much of lowland England and Wales northwards to S. Lancashire and E. Yorks and in a few scattered localities further north to Aberdeen and Orkney; Ireland; Channel Is. Probably native in the Mediterranean region and perhaps derived from *A. gracilis* Boiss., but now naturalized as a weed in most temperate regions of the world. The seeds are said to be poisonous and also to affect deleteriously the physical properties of wheat flour.

4. CUCUBALUS L.

Lvs opposite. Fls hermaphrodite, in lax dichasia terminating the main stem and branches. *Epicalyx* 0; *sepals joined* below, with 5 teeth; *petals* 5, long-clawed, *with coronal scales*; stamens 10; ovary 3-celled below; styles 3. *Fr. berry-like*, indehiscent, becoming dry when mature; seeds reniform, black.

One sp.

*1. C. baccifer L.   Berry Catchfly.

A large perennial herb 60–100 cm, with a branched creeping stock and brittle diffusely branched, scrambling flowering shoots, pubescent with short curved hairs. Lvs ovate, acuminate, narrowed into a short stalk, sparsely hairy, entire or somewhat sinuate. Fls c. 18 mm diam., shortly stalked, ±drooping. Calyx-tube 8–15 mm, widely campanulate, indistinctly veined and rough with reflexed points, unequally 5-toothed above, the teeth almost twice as long as the tube, blunt. Petals greenish-white, their spreading distant limbs narrowly spathulate and deeply bifid, with 2 boss-like scales at the base, their claws cylindrical, broadening at the junction with the limb. *Fr.* 6–8 mm diam., *black, globular*, on a short carpophore, not enclosed by the widely open ±rotate calyx with revolute teeth. Seeds 1.5 mm across, white at first, turning yellow then black. Fl. 7–9. Protandrous. Fr. taken by birds. $2n = 24$. Hp.

Possibly native. Recorded as a rare British species since 1570, and in the period 1837–55 often seen on the banks of a ditch in the Isle of Dogs, Middlesex, a locality long since destroyed. Now known only from two partly wooded areas in the brecklands of S.W. Norfolk. S. and C. Europe northwards to the Netherlands and to c. 58° N in C. Russia, chiefly in fen-woodland and streamside communities.

5. DIANTHUS L.

Herbs, usually perennial, or small shrubs, often with ±linear glaucous lvs. Fls solitary, in loose few-fld cymes or in compact involucrate heads, mostly hermaphrodite but some spp., have plants with smaller male-sterile fls. Base of calyx tightly enclosed by an *epicalyx* of 1–3 pairs of usually mucronate or awned scales; sepals joined below into a cylindrical tube which is neither strongly ribbed nor has scarious commissures alternating with the 5 usually short calyx-teeth; petals 5, pink, red or sometimes white, long-clawed; *coronal scales* 0; stamens 10; *styles* 2, *Capsule* 1-*celled, opening by* 4 *teeth*; carpophore often present; seeds compressed, concave on one side.

Perhaps 300 spp. in Europe, Asia eastwards to Japan, and Africa. Many spp. are grown in gardens as 'carnations' and 'pinks'.

1 Fls in ±head-like cymose clusters surrounded by an
   involucre of bracts.   2
  Fls solitary or 2–5 in a lax cyme.   3
2 Involucral bracts and epicalyx scales hairy.
                                                   1. armeria
  Involucre and epicalyx glabrous.
                               D. barbatus, etc. (p. 130)
3 Petals deeply cut into long narrow segments up to
   one third or more of the length of the limb.   4
  Petals entire, crenate or toothed, not deeply cut.   5
4 Lvs 10–15 × 1.5–3 mm, ±blunt, rigid; fls 1–3; petal-
   limb cut to c. $\frac{1}{3}$ of its length into narrow lobes.
   Grassy sand dunes in Jersey.   **4. gallicus**
  Lvs to 50 mm or more × c. 1mm, ±acute, not very
   rigid; fls usually solitary (but 1–5 in cultivars); petal-
   limb cut to c. $\frac{1}{2}$ its length into narrow lobes. Naturalized on old walls; a garden escape.   **2. plumarius**
5 Stems rough with short hairs; lvs mostly 1–2 cm long;
   fls 15–20 mm diam., scentless; epicalyx-teeth long-
   awned.   **6. deltoides**
  Stems glabrous; lvs mostly more than 2 cm long; fls
   more than 20 mm diam., fragrant; epicalyx-teeth
   shortly mucronate.   6
6 Fl.-stems 5–25 cm; lvs mostly 2–5 cm long, rough-
   edged; fls 20–30 mm diam.; petal-limb bearded at
   base and with teeth not more than $\frac{1}{6}$ of its length.
   Confined to limestone cliffs at Cheddar.
                     5. gratianopolitanus
  Fl.-stems 20–80 cm; lvs mostly 5–15 cm long, smooth-
   edged (or rough only near base); fls 25–50 mm
   diam., or more in some cultivars; petal-limb not
   bearded, with teeth up to $\frac{1}{3}$ of its length. Occasionally naturalized on old walls.   **3. caryophyllus**

1. D. armeria L.   Deptford Pink.

An annual or overwintering, rarely perennial, herb with slender tap-root and rigidly erect fl.-stem 30–60 cm, simple or branched above and sometimes also below, not glaucous, shortly hairy at least above. Basal lvs linear-oblanceolate, in a rosette; stem lvs linear-lanceolate acute, keeled, 3–5 cm × 1–3 mm, obliquely ascending; all with short hairs. Fls 8–13 mm diam., ±sessile in terminal and lateral 2–10-fld short-stalked cymose clusters.

Involucral bracts erect, lfy, hairy, equalling the fl.-clusters. *Epicalyx* scales 2, lanceolate-subulate, *hairy*, ribbed, equalling the *calyx-tube* which is 13–20 mm, cylindrical, narrowing upwards, *woolly*, strongly ribbed, with 5 lanceolate-acute teeth. Petals bright rose-red with pale dots; limb 4–5 mm, narrowly ovate, shallowly and irregularly toothed, not contiguous; claw white. Capsule ±cylindrical, equalling the calyx. Seeds 1.5 mm across. Fl. 7–8. Protandrous. stamens of one or both whorls abort in some fls. Infrequently visited by butterflies and automatically self-pollinated. $2n = 30$. Th.

Native. A rare and decreasing local plant of disturbed and open habitats such as hedgerows, banks by roads and railways, gravel-pits and banks, waysides and the edges of sandy fields; more rarely in open pasture on light soil or on drained fen-peat. All recent records are for lowland England and Wales south of E. Suffolk, Huntingdon and N. Dyfed, apart from scattered localities, northwards to Angus, where it clearly is casual or introduced. Most of Europe northwards to c. 60°N in Sweden and Finland; Caucasia and Armenia. Introduced in North America.

*2. D. plumarius L.   Common Pink.

A perennial tufted herb 15–30 cm, with a woody stock, procumbent rooting non-flowering shoots and erect or ascending 4-angled flowering shoots, glabrous and glaucous, usually branched above. *Lvs* to 5 cm, ascending, linear-subulate, very acute, *with rough margins*. Fls 1–5 in a lax cyme, 25–35 mm diam., *strongly fragrant*. Epicalyx ¼ to ⅓ as long as the calyx, of 4–6 broadly ovate, shortly cuspidate, herbaceous, scarious-margined scales. Calyx-tube $25–30 \times 3–4$ mm, narrowing upwards, glabrous, violet-coloured, with narrow ciliate teeth. *Petals* pale pink or white, the *limb digitately cut almost to the middle* into slender lobes, the entire part obovate, *hairy at the base*; claw up to 25 mm. Capsule cylindrical, slightly longer than the calyx. Seeds flat, ±orbicular. Fl. 6–8. Protandrous. $2n = 30, 60, 90$. Ch.

Introduced. Naturalized on old walls in many parts of the country. S.E. Europe southwards from N. Italy, Styria and Bohemia. It and its hybrids are much cultivated for their fragrant fls.

*3. D. caryophyllus L.   Clove Pink, Carnation.

A perennial tufted herb with the habit of *D. plumarius* but with shoots 20–50 cm and linear-lanceolate *lvs with their edges smooth* or rough only near the base. *Fls 1–5* in a lax cyme, 35–40 mm diam., *strongly fragrant*. Epicalyx ¼ as long as the calyx, of 4–6 broadly ovate, abruptly mucronate, membranous scales with herbaceous tips. Calyx-tube 25–30 mm, its teeth not ciliate. *Petals* rose-pink, the limb *obovate, crenate or dentate*, the teeth not exceeding ⅓ of its length. Capsule cylindrical-ovoid, longer than the calyx. Seeds pear-shaped, nearly flat. Fl. 7–8. Protandrous. Visited by butterflies and hawk-moths, and said to be self-sterile. $2n = 30, 90$. Ch.

Introduced. Occasionally naturalized on old walls. S.

Europe and N. Africa. The cultivated carnations are derived from *D. caryophyllus* and its hybrids with *D. plumarius*, etc.

*4. D. gallicus Pers.   Western Pink.

A perennial tufted herb with erect flowering stems 15–25 cm, glaucous, *downy below with minute hairs*. Lvs linear, short, stiff, ±blunt, 3-veined. Fls 1–3, fragrant. Epicalyx ¼ as long as the calyx, of 4 ovate-oblong abruptly mucronate scales. Calyx cylindrical, striate. Petals pink, the *limb cut to ⅓ of its length into narrow lobes*, the entire part ±orbicular. Capsule cylindrical. Fl. 6–8. $2n = 60, 90$. Ch.

Probably introduced. known only on dry grassy dunes in Jersey. Atlantic coast of Europe from Brittany to Spain and Portugal.

5. D. gratianopolitanus Vill.   Cheddar Pink.

D. caesius Sm.; *D. glaucus* sensu Hudson, non L.; *D. caespitosus* Poiret

A perennial *densely tufted* herb with a woody stock, *long procumbent non-flowering shoots* and ascending flowering shoots. 10–20 cm, *glabrous and glaucous*. Lvs 2–6 cm, linear, bluntish, rough at the edges. Fls usually solitary, c. 25 mm diam., fragrant. Epicalyx hardly ¼ as long as the calyx, its 4–6 scales herbaceous, roundish, abruptly mucronate or blunt. Calyx-tube 16–20 mm, striate glabrous, usually violet in colour, teeth broadly triangular. Petals pale to deep rose, the *limb ±obovate with irregular teeth reaching not more than ⅙ of its length*, ±hairy at the base; claw c. 15 mm. Capsule cylindrical. Seeds ovate. Fl. 6–7. Protandrous. Strongly clove-scented and visited by butterflies and day-flying hawkmoths. $2n = 90$. Ch.

Native. A rare and local plant confined to Carboniferous limestone cliffs at Cheddar Gorge (N. Somerset); occasionally naturalized elsewhere. W. and C. Europe from France to Moravia and Hungary.

6. D. deltoides L.   Maiden Pink.

A perennial *loosely tufted* green or glaucous herb with a creeping branched slender stock, *short procumbent non-flowering shoots* and decumbent then ±erect flowering shoots, 15–45 cm, *rough with short hairs*. Lower lvs and those of barren shoots 10–16 mm, narrowly oblanceolate, blunt; upper lvs 10–25 mm, linear-lanceolate, acute; all *roughly hairy on the margins* and on the underside of the midrib. Fls c. 18 mm diam., scentless, solitary or rarely 2–3 terminating the main stem and branches. *Epicalyx about ½ as long as the calyx*, of 2–4 broadly ovate, long-cuspidate scales, herbaceous with scarious margins. Calyx 12–17 mm, cylindrical, glabrous, green or reddish; teeth lanceolate, acute. Petals rose or white with pale spots and a dark basal band, the limb shortly and irregularly toothed. Capsule equalling or somewhat exceeding the calyx. Seeds 2–2.5 mm obovate, black. Fl. 6–9. Protandrous. Visited by butterflies and moths. $2n = 30, 60$, Ch.

Native. A local and decreasing lowland plant in grass-land of dry fields, banks, hilly pastures, waysides, etc., chiefly on shallow soil over limestone, chalk or basic igneous rocks, very locally on maritime cliffs or stabilized sand dunes; also on spoil-heaps of quarries, chalk-pits and old lead-workings and occasionally a casual or garden-escape. Throughout Great Britain northwards to E. Perth and Kincardine; probably introduced in Moray and in S.E. Ireland. Most of Europe; W. Asia. Introduced in North America.

Amongst the several other spp. often grown in gardens and sometimes escaping are **D. barbatus** L., Sweet William, a perennial with broadly elliptical sessile rough-edged lvs, fls almost sessile in dense cymose heads with involucre of ±linear spreading bracts, epicalyx of 4 long-awned scales at least equalling the calyx, and dark red or pink petals, often variously spotted or barred; **D. carthusianorum** L., closely related but with linear-acute long-sheathing lvs, involucral bracts ±obovate, abruptly awned, coriaceous, and epicalyx scales similar and only about half as long as the dark purplish calyx; **D. chinensis** L., Rainbow Pink, and especially the annual var. *heddewigii* Regel, with strikingly patterned fls 5–8 cm across, the bearded petals irregularly toothed or cut; and **D. superbus** L., with 4–12 fragrant fls, 5 or more cm across, in a cymose panicle, often in pairs, their rose or white petals with the bearded limb cut more than half-way into narrow segments.

6. VACCARIA Medicus

Differs from *Saponaria* in the winged calyx-tube and absence of coronal scales.

Three spp. in the Mediterranean region.

*1. V. hispanica (Miller) Rauschert (*V. pyramidata* Medicus; *Saponaria vaccaria* L.; *S. segetalis* Necker)

An annual herb with a slender tap-root and glabrous branched flowering stems 30–60 cm. Basal lvs oblong-lanceolate, upper ovate-lanceolate cordate, all sessile, acute, glabrous and glaucous. Fls in loose dichasia. Epicalyx 0. *Calyx-tube* inflated, glabrous, *with 5 sharp angles or wings* and 5 triangular teeth. Petals pale rose-coloured, the limb cuneate, rounded or somewhat emarginate, toothed; *coronal scales* 0. Stamens 10. Styles 2. Capsule globular, 4-celled below, opening by 4 teeth. Seeds 2 mm across, black. Fl. 6. ±Homogamous. Visited by butterflies and automatically self-pollinated. $2n = 30$. Th.

Introduced. A not infrequent casual from bird-seed and a cornfield weed, occasionally becoming established. Europe northwards to Denmark and Sweden; Asia. Introduced in North America, Australia and New Zealand.

7. SAPONARIA L.

Annual or perennial herbs with opposite lvs. Fls in loose or condensed dichasia. *Epicalyx* 0; *sepals joined* below into a green *tube without scarious seams* and with 5 teeth above; petals 5, with the limb narrowing abruptly into the long claw; coronal scales present; stamens 10; *styles* 2(–3); ovary 1-celled. Capsule opening by 4(–6) teeth; carpophore short; seeds reniform.

About 20 spp., especially in the Mediterranean region.

1. S. officinalis L.   Soapwort, Bouncing Bett.

A perennial herb with a stout branched creeping rhizome from which arise long stolons and erect or ascending ±glabrous, flowering shoots 30–90 cm, simple or branched above. Lvs 5–10 cm, broadly ovate to elliptical, acute, 3(–5)-veined, ±glabrous. Fls c. 2.5 cm diam., in compact terminal corymbs on the main stem and branches. Calyx-tube 18–20 mm, cylindrical, often reddish, with 5 short triangular teeth. Petals pink or flesh-coloured, the claw exceeding the calyx-tube, the limb obovate, entire or slightly emarginate, not contiguous, with 2 small blunt coronal scales at the base. Capsule oblong-ovoid, equalling the calyx-tube, opening by 4(–5) ±unequal teeth, but often failing to ripen; carpophore short. Seeds 1.8 mm across, blackish. Fl. 7–9. Protandrous. Visited chiefly by day- and night-flying hawkmoths. $2n = 28$. Hp.

Native or introduced. A fairly common plant of hedgebanks and waysides near villages, probably as an escape from cultivation, northwards to Aberdeen; and perhaps native in Devon and Cornwall and in N. Wales, where it grows by streams. Europe northwards to Scandinavia; Asia. Introduced in North America. Naturally a plant of streamsides and damp alluvial woods.

S. ocymoides L., a perennial mat-forming plant with ±prostrate pubescent shoots bearing obovate or spathulate lvs and small bright pink fls, is much grown on rock-gardens and walls and occasionally escapes. Native in the mountains of C. and S.W. Europe.

8. PETRORHAGIA (Ser. ex DC.) Link

Annual or perennial herbs with opposite ±*linear lvs* and usually hermaphrodite fls either in lax or fasciculate cymose panicles, the individual fls with or without epicalyx, or in heads with a basal involucre of bracts. *Sepals* 5, *joined below into a 5- or 13-veined calyx-tube, whitish scarious commissures alternating with the 5 free teeth* into which the veins run; petals clawed or not; *coronal scales* 0; stamens 10; *styles* 2. Capsule opening by 4 teeth; *seeds compressed, scutate; embryo straight.*

About 25 spp. in Europe, N. Africa, W. and C. Asia, Canary Is., Madeira.

Differs from *Dianthus* and *Saponaria* in the scarious commissures of the calyx-tube and from *Gypsophila* in the scutate seeds and straight embryo (reniform and curved respectively in *Gypsophila*).

1 Plant annual; fls in ovoid heads almost completely enclosed in a loose involucre of very broad pale brown membranous bracts.   **2**
  Plant perennial; fls not in heads, each with an epicalyx of 4 (2–5) whitish mucronate bracts.   **3. saxifraga**
2 Lf-sheaths up to twice as long as wide; middle of stem often tomentose; seeds tuberculate.   **1. nanteuilii**
  Lf-sheaths about as long as wide; stem glabrous or roughly pubescent; seeds reticulate or rugulose.
                               2. prolifera

1. P. nanteuilii (Burnat) P. W. Ball & Heywood
                          Childing or Proliferous Pink.

Dianthus prolifer, Tunica prolifera, Kohlrauschia proli-fera, auct. eur. occident., *pro parte*; *Dianthus nanteuilii* Burnat; *Tunica nanteuilii* Gürke

An *annual* herb with slender tap-root and one or more erect wiry fl.-stems 10–50 cm, simple or branched above, glabrous to somewhat roughly pubescent but *often tomentose near the middle*, slightly glaucous. Lvs 1–2 cm, linear-lanceolate, acute, 3-veined, with scabrid margins; *lf-sheaths up to twice as long as wide. Ovoid heads of up to 11 fls are enclosed in a loose involucre* the outermost bracts of which are mucronate and only half as long as the *very broadly elliptical blunt inner bracts*, each almost completely encircling a fl. and equalling its calyx; *all bracts brown, shining and of parchment-like texture.* Fls 6–9 mm diam., opening one at a time and only then visible above the involucre. Calyx-tube 10–13 mm, faintly 15-veined, glabrous, reddish, with 5 blunt membranous teeth. Petals pale purplish-red with obovate emarginate limb 2–3 mm, abruptly narrowed into the long slender claw 10–13 mm. Capsule ovoid-ellipsoid, shorter than the calyx. *Seeds* 1.5 mm across, blackish, *tuberculate*. Fl. 6–8. Homogamous. Visited sparingly by butterflies and other insects. $2n = 60*$. Th.

Native. A very rare plant of sandy and gravelly places, now thought confined as a native to a few coastal localities in W. Sussex and E. Hants and in the Channel Is. (Jersey); probably introduced in W. Norfolk and Kent. W. Europe northwards to Great Britain and eastwards to Corsica and Sardinia; Morocco; Canary Is., Madeira.

Recognized as distinct from *P. prolifera* (2 below) by Burnat in 1892 but frequently confused with it and still regarded by some authors as insufficiently distinct to merit separate specific rank despite the constant differences in seed-coat features and chromosome number and the strongly Atlantic distribution. Probably an allotetraploid derived from *P. prolifera* × *P. velutina* (Guss.) P. W. Ball & Heywood.

*2. P. prolifera (L.) P. W. Ball & Heywood

Dianthus prolifer L.; *Tunica prolifera* (L.) Scop.; *Kohlrauschia prolifera* (L.) Kunth

Very closely resembling **1** but the middle of the stem never tomentose, lf-sheaths usually only about as long as wide, and seeds reticulate or rugulose, not tuberculate. $2n = 30$.

Introduced. A rare alien in a few scattered localities chiefly in S. and E. England. C. Europe from E. Spain to the Ukraine and northwards to S. Sweden, and of mountains in S. Europe; also in W. Asia. Naturalized in North America.

*3. P. saxifraga (L.) Link

Dianthus saxifragus L.; *Tunica saxifraga* (L.) Scop.; *Kohlrauschia saxifraga* (L.) Dandy

A *perennial mat-forming herb* with numerous decumbent or ascending much-branched shoots 10–35 cm, glabrous or roughly pubescent. Lvs up to 1 cm, linear, acute, rough-edged. *Fls solitary at ends of branches*, each *with epicalyx* of 4(2–5) whitish mucronate bracts. Calyx 3–6 mm, narrowly campanulate with blunt scarious-margined teeth. *Petals white or pale pink* with obcordate limb 3–4 mm *narrowing gradually into the* claw. Capsule ovoid, barely as long as calyx, with blunt oblong teeth; seeds blackish, tuberculate. Fl. 6–9. $2n = 60$. Chh.

Introduced. Native in C. and S. Europe and W. Asia; established near Tenby, Pembroke.

9. GYPSOPHILA L.

Annual to perennial herbs or dwarf shrubs, with opposite lvs. Fls hermaphrodite in cymose infls. Epicalyx 0; *sepals joined below into a campanulate tube with pale scarious seams between the main veins* and with 5 teeth above; petals with the *limb narrowing gradually into the claw*; coronal scales 0; stamens 10; styles 2, rarely 3. Capsule opening by 4 (6) teeth; seeds reniform.

About 10 spp, chiefly in S.E. Europe and W. Asia.

No sp. of *Gypsophila* is native in the British Is. The annual **G. muralis** L. (Europe) with somewhat glaucous shoots 4–18 cm high, shortly hairy below, ±linear lvs to 2 cm, and pink fls 5 mm diam. on long stalks; and the annual **G. porrigens** (L.) Boiss. (Near East) with diffusely branched procumbent stems covered with long dense spreading hairs, tiny fls and very large seeds, occur as casuals. The perennial **G. paniculata** L. (Maiden's Breath; C. Europe and W. Asia), 60–90 cm high with lanceolate lvs and numerous white fls 4–5 mm diam. in crowded corymbose panicles, is much grown in gardens and sometimes escapes, as does the annual **G. elegans** Bieb. (Caucasus), 30–45 cm high, with lanceolate, acute lvs and white or pink fls.

10. CERASTIUM L.

Annual to perennial herbs or dwarf shrubs, commonly hairy, with opposite *sessile lvs*. Fls 5- or 4-merous, usually in cymose infls, sometimes solitary. Sepals free; petals white, emarginate or bifid up to half-way, sometimes 0; stamens usually 10 or 8, sometimes 5 or fewer; nectaries at base of outer stamens; ovary 1-celled; *styles usually 5 or 4, opposite the sepals, sometimes 3*, rarely 6. *Capsule almost cylindrical* or distinctly tapering upwards, *longer than calyx, often curved, opening by twice as many short teeth as styles.* Seeds numerous, spherical or reniform, usually tuberculate, sometimes rugose.

Perhaps c. 100 spp, chiefly in north temperate and arctic regions of the Old World but some in North America, South America and on mountains of tropical Africa and Papua New Guinea.

1 Styles usually 3 (varying 4–6); capsule-teeth usually
    6. A rare plant of high mountains in Scotland.
                                    1. cerastoides
  Styles usually 5 or 4; capsule-teeth 10 or 8.   *2*
2 Petals 10 mm or more, 2–3 times as long as calyx;
    perennials with procumbent to ascending non-
    flowering shoots.   *3*
  Petals at most c. 10 mm, usually much less and shorter
    than or about equalling calyx (but c. $1\frac{1}{2}$ times calyx
    in the very rare Scottish *C. fontanum* subsp. *scoti-
    cum*); annuals apart from *C. fontanum*.   *6*
3 Stem and lvs densely whitish-tomentose. Garden-
    escapes.   **3. tomentosum** (and *C. biebersteinii*)
  Stem and lvs not densely tomentose but sometimes
    woolly with long soft ±crisped white hairs
    ('lanate'), sometimes shortly pubescent or almost
    glabrous.   *4*
4 Whole plant or at least the young lvs with a distal
    brush of entangled white hairs; sepals oblong-
    lanceolate. Plant of high mountains in N. Wales,

Lake District and Scotland.   **4. alpinum**
Plant pubescent or ±glabrous, not lanate (some *alpinum* hybrids may be included here); sepals broadly lanceolate to ovate-lanceolate.   5
5 Lvs 1.5–3 mm wide, the lower commonly with axillary lf-clusters (i.e. very short lateral shoots); bracts ciliate and with scarious margins and tips. Mainly lowland.   **2. arvense**
Lvs 4–5 mm wide, the lower usually lacking axillary lf-clusters; bracts wholly herbaceous. Mainly on high mountains in N. Wales, Scotland and Inner Hebrides but near sea-level in Shetland.
                                          5. arcticum
6 Usually perennial with short decumbent non-flowering shoots and decumbent to ascending fl.-shoots to 50 cm, ±hairy but very rarely glandular; capsule (6–)9–12 mm.   **6. fontanum**
Annuals lacking non-flowering shoots, often glandular-hairy; capsule rarely reaching 10 mm, usually distinctly less.   7
7 Fls in compact clusters remaining compact in fr.; fr.-pedicel at most equalling calyx; sepals with long white hairs extending beyond the tip. **7. glomeratum**
Fls in spreading cymes; fr.-pedicel usually much exceeding calyx.   8
8 Plant shaggy with spreading silvery hairs, not glandular; petals c. half as long as sepals; stamens 10.
                                       8. brachypetalum
Plant not shaggy but viscid with glandular hairs; petals more than half as long as sepals; stamens 4 or 5.   9
9 Bracts entirely herbaceous; fls usually 4-merous, sometimes 5-merous; fr.-pedicels usually erect throughout.   **9. diffusum**
At least the upper bracts with scarious tips or margins; fls 5-merous; fr.-pedicels at first recurved or sharply deflexed.   10
10 Bracts with upper half scarious; petals c. $\frac{2}{3}$ as long as sepals, slightly notched; fr.-pedicels at first sharply deflexed from the base.
                                     11. semidecandrum
Upper bracts with narrow scarious margin; petals about equalling sepals, up to $\frac{1}{4}$ bifid; fr.-pedicels at first curving downwards but not sharply deflexed; plant erect, reddish at least below.   **10. pumilum**

1. C. cerastoides (L.) Britton
                       Starwort Mouse-ear Chickweed.
Stellaria cerastoides L.; *Cerastium trigynum* Vill.
A perennial herb with much branched creeping ±woody stems from which arise prostrate non-flowering shoots, 5–15 cm, and decumbent or ascending flowering shoots 5–10 cm; all *shoots* rooting, *glabrous except for a line of small hairs down each internode*. Lvs 6–12 mm, pale green, elliptical-oblong or linear-lanceolate, blunt, usually curving to one side. Fls 9–12 mm diam., 1–3, with slender glandular pedicels to 8 cm. Bracts herbaceous. Sepals 4–5 mm, narrowly lanceolate, 1-veined, with a narrow scarious margin. Petals white, deeply bifid, almost twice as long as the sepals. Stamens 10, *Styles usually* 3, sometimes 4–6. Capsule oblong, straight, up to twice as long as the calyx, opening by 6(–10) spreading valves. Seeds brown, 0.5 mm across. Fl. 7–8. ±Homogamous. Visited by flies and automati-

cally self-pollinated. $2n = 38*$ Chh.
Native. Very local in spring-head communities close to the latest snow-beds on high mountains of C. Scotland northwards to Ross, always above 850 m and usually in beds of *Pohlia albicans* var. *glacialis* with *Deschampsia caespitosa* and *Saxifraga stellaris*; said also to grow on high grassy slopes or on fine scree below them. Arctic and subarctic Europe, eastern North America and W. Asia and mountains of C. Europe and Asia. Reaches 73° 15′ N in E. Greenland.

2. C. arvense L.   Field Mouse-ear Chickweed.
A perennial herb with a branched creeping stock from which arise long prostrate rooting non-flowering shoots up to 30 cm, and ascending flowering shoots 4–30 cm high, prostrate below; all ±hairy and glandular. *Lower lvs commonly with axillary lf-clusters* (i.e. non-elongated axillary shoots). Lvs 5–20 mm, linear-lanceolate or narrowly oblong, hardly narrowed to the base, usually soft and downy but not woolly, ±glandular. Fls 12–20 mm diam., in lax dichasia, their stalks glandular-hairy. Sepals 5–8 mm, oblong-lanceolate subacute, glandular-hairy, with membranous margins and tips. Petals white, obovate, bifid, about twice as long as the sepals. Stamens 10. Styles 5. Capsule cylindrical, slightly exceeding the calyx. Fr.-stalks erect but curved just beneath the calyx. Seeds 0.8–1 mm across, tubercled. Fl. 4–8. Protandrous. Visited by various insects, chiefly flies and small bees. $2n = 36, 72*$. Chh.
Variable. Var. *latifolium* Fenzl, with lvs up to 25 × 6 mm, has been collected in W. Norfolk.
Native. On dry banks and waysides and in grassland especially on calcareous or slightly acid sandy soils. Through most of Great Britain to Sutherland, but with a distinct eastern tendency; Orkney; very local in Ireland. Europe, N. Africa, W. and temperate Asia, North America.

*3. C. tomentosum L.
                       Dusty Miller, Snow-in-Summer.
A perennial mat-forming herb resembling luxuriant *C. arvense* but with *white-tomentose stems and lvs*. Stems prostrate below, rooting freely; flowering shoots ascending, 15–30(–45) cm. *Lvs 10–30 × 2–5 mm, linear-lanceolate to lanceolate with slightly revolute margins*; lower lvs commonly with axillary lf-clusters. Infl. a lax dichasial cyme, the tomentose bracts with scarious margins. Fls 12–17 mm diam., with pedicels more than twice as long as sepals. Sepals 5–7 mm, lanceolate, tomentose, with broad scarious margins. Petals narrow, bifid, c. twice as long as sepals. Capsule-teeth usually spreading and with revolute margins. Stamens 10, Styles 5. Fl. 5–8. Protandrous or homogamous. $2n = 38, 72, c. 108$.
Introduced. Much grown in rock-gardens and on walls and often escaping. Native in Italy (S. Apennines and Sicily).

*C. biebersteinii DC., native in the Crimea, closely resembles *C. tomentosum* and is also much grown in gardens. It differs in having flowering stems rarely exceeding 30 cm, larger lvs

(20–50 × 3–8 mm) which are rather more densely white-tomentose, somewhat larger fls up to 25 mm diam. with sepals 6–10 mm, and capsule-teeth usually ±erect and with flat margins. Some cultivated material may well be of hybrid origin.

4. C. alpinum L.   Alpine Mouse-ear Chickweed.

Incl. *C. lanatum* Lam.

A perennial mat-forming herb with numerous short procumbent non-flowering shoots up to 6 cm and decumbent then ascending flowering shoots up to 15 cm; *hypogeal stolons* 0. Lvs commonly c. 10 × 5 mm but up to 18 × 7 mm; smallest on non-flowering and at base of flowering shoots, these obovate to elliptic-oblanceolate, others oval or elliptic. *Infl.-axes, pedicels and lvs ±densely 'lanate'*, i.e. covered with long curled and entangled shining white hairs, especially conspicuous where they project beyond the tips of young shoots; glandular hairs sometimes present on pedicels and elsewhere. *Bracts and bracteoles with scarious margins.* Fls 1–5, up to 25 mm diam., their pedicels extending 1–4 cm beyond the bracteoles. *Calyx square-based; sepals oblong-lanceolate*, 7–10 mm, with narrow scarious margins and often with violet tips. *Petals c. twice as long as sepals*, not very deeply bifid. *Capsule* (Fig. 29A)

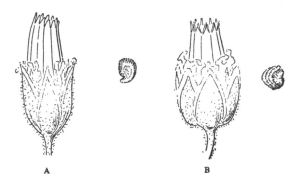

Fig. 29. Capsule and seeds of A, *Cerastium alpinum* and B, *C. arcticum*. Capsules ×2.5; seeds ×5.

8–14 mm, *half or more as long again as sepals, its upper part narrow and somewhat curved.* Seeds 1–1.4 mm, finely and acutely tuberculate. Fl. 6–8. Protandrous. $2n = 54, 72^*, 108$. Ch.

The above description is of subsp. **lanatum** (Lam.) Ascherson & Graebner, to which all British plants seem referable except for some with few or no lanate hairs and somewhat larger fls. These usually have hypogeal stolons and are probably hybrids with *C. arcticum* or their derivatives. Hybrids with *C. fontanum* have also been recorded.

Native. An alpine plant of outcrops and ledges of limestone and other calcareous rocks and of well-drained soil on stable ground over such rocks. Very rare on high mountains in N. Wales, the Lake District and S. Scotland; local in C. and N. Scotland northwards to Sutherland, especially on mica-schists; reaches 1213 m on Ben Lawers. Northern Europe from British Is., Iceland and Jan Mayen eastwards to Novaya Zemlya, and

mountains in S.W., C. and S.E. Europe from S. Spain and S. France to the Balkan peninsula; also in eastern North America and Greenland.

5. C. arcticum Lange   Arctic Mouse-ear Chickweed.

C. latifolium auct., non L.; *C. edmondstonii* auct., non (H. C. Watson) Murb. & Ostenf. Incl. *C. nigrescens* Edmondston ex H. C. Watson.

A perennial *low-growing caespitose herb with long slender hypogeal stolons*, flowering shoots 2–15 cm and numerous non-flowering shoots up to 7 cm ascending from the creeping basal stems. Lvs very variable in size and shape, commonly narrower than in *C. alpinum* but sometimes obovate to suborbicular at least on non-flowering shoots. Stem and lvs usually covered with short stiff 3–4-celled hairs with a variable proportion of still shorter gland-tipped hairs (all glandular in subsp. *edmondstonii*); *lvs ciliate but with faces often almost glabrous*. Plants with some longer hairs like those of *C. alpinum* are probably of hybrid origin. *Bracts and bracteoles wholly herbaceous, without scarious margins.* Fls up to 30 mm diam., their pedicels extending 1–3 cm beyond the bracteoles, usually with spreading yellowish hairs and often glandular. *Calyx semiglobose, with rounded base; sepals 6–9 mm, broadly lanceolate to ovate-lanceolate*, hairy and sometimes glandular, *with broad scarious margins.* Petals 2 or more times longer than sepals, shallowly bifid. Stamens 10. Styles 5. *Capsule* (Fig. 29B) up to twice as long as calyx, its *upper half broad and almost straight.* Seeds 1–1.5 mm, rugose. Fl. 6–8. Protandrous. $2n = 108^*$. Ch.

Native. Very local, mainly on mountains, in N. Wales, Scotland, Inner Hebrides and Shetland. Pure *C. arcticum* with no evidence of introgression from related spp. (see below) appears to be restricted to N. W. Europe from British Is. to N.W. Finland, Faeroes, Iceland and Spitsbergen, Greenland and arctic N.E. Canada.

Variable. Two subspp. have been recognized in the British Is:

Subsp. **arcticum**: plant *loosely to ±densely caespitose*, stem and lvs *with short stout non-glandular hairs* and usually some shorter gland-tipped hairs as well; *lvs up to 15 mm, oblanceolate to broadly elliptical or obovate, pale or yellowish-green*, ciliate but *with faces only sparsely pubescent to glabrous*; pedicels not or somewhat glandular; *sepals ±acute.* Mainly on wet rock-ledges and scree on high mountains in N. Wales, C. and N. Scotland and Inner Hebrides (Skye), reaching c. 1070 m in Scotland.

Subsp. **edmondstonii** (H. C. Watson) A. & D. Löve (*C. nigrescens* Edmondston ex H. C. Watson): *plant densely and compactly caespitose* and rarely exceeding 5 cm high, *stem and lvs densely and exclusively clothed with short stout gland-tipped hairs; lvs 4–7 mm broadly ovate-elliptical to suborbicular, dark and purple-tinged*; pedicels densely glandular; *sepals blunt.* Confined to serpentine debris at c. 15 m in Unst (Shetland), where it is endemic.

C. alpinum and C. arcticum with about 6 other spp. constitute the circumpolar 'C. alpinum complex' of E. Hultén (*Svensk Bot. Tidskr.*, 1956, **50**, 411–95) whose members form hybrids and hybrid-swarms and cause taxonomic difficulties especially in areas where two or more coexist. Hultén regarded as characteristic of *C. alpinum sens. str.* that the lvs bear '*alpinum* hairs', i.e. long woolly hairs that are multicellular, thin-walled and air-filled and therefore shining white. The calyx is square-based, the bracts and bracteoles have scarious margins and there are no hypogeal stolons. Typical *arcticum* is distinguished by its almost glabrous or only sparsely hairy lvs and lack of *alpinum* hairs, its wholly herbaceous bracts and bracteoles, round-based calyx and long hypogeal stolons. Plants like *arcticum* except only for a few *alpinum* hairs Hultén named *C. arcticum* var. *alpinopilosum*, at the same time suggesting that they carry introgressive genes from *alpinum*. Such plants have been recorded from Scotland; and also others, usually combining *alpinum* hairs and scarious-margined bracts and bracteoles with a round-based calyx and hypogeal runners, generally recognized as of hybrid origin.

6. C. fontanum Baumg.

Common Mouse-ear Chickweed.

C. vulgatum L., nom. ambig.; *C. caespitosum* Gilib., nom. illegit.; *C. triviale* Link, nom. illegit.; incl. *C. holosteoides* Fries

Usually a *short-lived perennial* herb, rarely annual, with slender creeping stock bearing decumbent non-flowering shoots up to 15 cm and decumbent then ascending flowering shoots up to 45(–60) cm; shoots pubescent to almost glabrous, glandular or not. *Lvs* ±sessile. 10–25(–40) × 2–10 mm, those on non-flowering shoots oblanceolate, blunt, narrowed below into a stalk-like base, on flowering shoots lanceolate or narrowly elliptical to ovate-oblong, subacute; *commonly dark grey-green*, hairy or ±glabrous on one or both faces. Infl. a dichasial cyme becoming lax in fr.; at least the upper bracts usually with scarious margins. *Sepals* 3–9 mm, *ovate-lanceolate*, blunt, pubescent or ±glabrous, *with scarious margins*. Petals rather deeply bifid, *equalling or somewhat exceeding sepals*, rarely up to twice as long. Stamens 10, rarely 5. Styles 5. *Ripe capsule* 7–18 mm long, *narrowly cylindrical and usually somewhat curved*. Seeds 0.4–1.2 mm, brown, tuberculate. Fl. 4–9. Protandrous and visited chiefly by flies but often self-pollinated. $2n = 144*$ (134–152). Ch.(–Th.).

Native. A very common plant of mesotrophic grass-lands, shingle, sand dunes, waysides, waste places and cultivated ground throughout the British Is.; reaching c. 1200 m in Scotland. Cosmopolitan.

A very variable complex within which several spp. and subspp. have at times been recognized but without general agreement as to their number or grade. This polymorphism presumably arises from its very wide geographical and ecological range and its frequent autogamy. British material has recently been deemed to comprise two subspp.

Subsp. **triviale** (Link) Jalas: the common lowland plant here and in continental Europe, with *flowering shoots up to* 50 cm; stems and lvs usually pubescent with or without glandular hairs; *lvs* commonly 10–25 × 3–10 mm, *dark grey-green*; sepals commonly 4–7 mm, pubescent; *petals shorter to slightly longer than sepals*; *ripe fr. usually up to* 12 mm and *seeds* 0.4–0.9 mm *with tubercles* 15–40 µm *wide and high*. Very variable and tending to form local races of which var. *holosteoides* (Fries) Jalas, an almost glabrous plant of river-sides and other wet places in N. England with stems pubescent only down two opposite lines in each internode, lvs ±glabrous above or on both faces though sometimes with ciliate margins, sepals 5–8 mm and often glabrous, and ripe fr. up to c. 15 mm, is accorded subspecific rank by some continental authors (*C. fontanum* subsp. *glabrescens* (G. W. F. Meyer) Salman). A rare annual variant with only 5 stamens (*C. triviale* var. *pentandrum* Syme), 'principally on the sea-coast', may also merit subspecific status.

Subsp. **scoticum** Jalas & Sell: non-flowering shoots short, rarely rooting; *flowering shoots ascending*, 3–12 cm, *pubescent*, but often almost glabrous at the base and sometimes also down two lines in each internode; *glandular hairs* 0; *stem-lvs* sessile, (6–)7–8(–10) *mm*, lanceolate and 3 or more times longer than wide, blunt to subacute, sparsely hairy or subglabrous, ciliate on the margins and with median veins prominent beneath. *Bracts herbaceous or the upper with scarious margins towards the apex*. Infl. commonly 3-fld with slender pedicels up to twice as long as calyx; sepals 4.5–6.5 mm, lanceolate, acute, distinctly keeled near the base, sparsely hairy, margins scarious; *petals c.* 1.4(–1.7) *times as long as calyx*; *capsule* 6–10 mm, *straight or slightly curved*; *seeds* 0.8–1 mm *with large tubercles c.* 50 µm *high and* 125 µm *long*. Mountains of the Clova area. Plants from coastal rocks at Strathy Point, W. Sutherland, are similar but with still larger tubercles on the seeds.

7. C. glomeratum Thuill.

Sticky Mouse-ear Chickweed.

C. viscosum auct.

An annual or overwintering herb with *pale yellowish-green lvs* and erect or ascending flowering shoots, 5–45 cm, hairy, *glandular* at least above, rarely eglandular. Lvs 0.5–2.5 cm; basal lvs oblanceolate or obovate narrowed below; stem lvs broadly ovate or elliptical-ovate; all ±sessile, apiculate, covered with long white hairs. *Fls aggregated into compact cymose clusters*; pedicels shorter than sepals. Bracts herbaceous, hairy. *Sepals* 4–5 mm, lanceolate, very acute, with a narrow scarious margin, usually *with* glandular hairs as well as *long white hairs which extend to and project beyond the tip*. Petals white, about equalling sepals, c. ¾ bifid; rarely 0. Stamens 10. Styles 5, Capsule 6–10 mm, narrowly cylindrical, curving upwards out of the line of its stalk which it exceeds in length. Seeds 0.4–0.5 mm, pale brown, finely tubercled. Fl. 4–9. Homogamous. Little visited by insects; automatically self-pollinated and sometimes cleistogamous. $2n = 72*$. Th.

Native. A common weed of arable land and waste places, on walls and banks and on open sand dunes.

Throughout the British Is. Cosmopolitan.

*8. C. brachypetalum Pers. subsp. **brachypetalum**
Grey Mouse-ear Chickweed.

An annual herb with simple or basally branched stems,
5.5–30 cm, erect, *shaggy* with spreading-ascending simple hairs, eglandular purple-tinged below. Lvs 4–14 mm,
those at the base ±crowded, lanceolate-acute, narrowed
into a rather broad stalk; stem lvs lanceolate or elliptic,
acute, sessile; all thinly covered with long hairs. Infl.
a lax dichasium of up to 30 fls on erect or ascending
shaggy stalks, 6–15 mm. *Bracts entirely herbaceous*,
thinly hairy. Fls 5-merous. *Sepals* 4–5.5 mm, lanceolate,
concave, herbaceous and *thinly hairy to the tip*, with
or without scarious margins. *Petals about ¾ as long as
the sepals*, bifid for ⅓ their length, sparsely ciliate below.
Stamens 10, with a few long hairs near the base of the
filament. Styles 5. Capsule 6–8 mm, slightly exceeding
the sepals, broad, cylindrical, slightly curved near the
apex, its stalk bent near the upper end. Seeds 0.5 mm
across, pale brown, acutely tubercled. Fl. 5. Probably
self-pollinated. $2n = 90^*$. Th.

Probably introduced. Found in 1947 on the bank of
a railway-cutting between Sharnbrook and Irchester,
Bedford. C. Europe, from Spain, N. Italy and Romania
to Denmark. The range of the species is wider, from
the Mediterranean to S. Scandinavia, and in N. Africa
and the Caucasus.

The wholly herbaceous bracts distinguish it from *C. semidecandrum* and *C. pumilum*, the ciliate bases of petals and filaments
from *C. atrovirens* and the lax infl. from *C. glomeratum*. From
all it differs in the long spreading silvery hairs which give it
a greyish appearance.

9. C. diffusum Pers.
Sea Mouse-ear, Dark-green Mouse-ear Chickweed.

C. tetrandrum Curtis; *C. atrovirens* Bab.

An annual herb with a slender tap-root and diffusely
branched decumbent or ascending flowering shoots
7.5–30 cm, dull dark green in colour, densely covered
and viscid with short glandular hairs. Lvs 5–10 mm; basal
oblanceolate; stem lvs ovate or oblong-ovate, acute; all
covered with short glandular and non-glandular hairs.
Fls 3–6 mm diam., in few-fld dichasia, their stalks short,
slender, glandular. *Bracts and bracteoles wholly herbaceous. Fls usually 4-merous*, sometimes 5-merous.
Sepals 4–7 mm, lanceolate, acute or acuminate, with
narrowly scarious white margins and tips, glandular-
hairy, but with hairs usually not projecting beyond the
glabrous tip. Petals white, c. ¾ length of the sepals, *about
⅓ bifid*, with several branching veins. Stamens 4(–5).
Styles 4(–5). Capsule 5–7.5(–10) mm, narrowly cylindrical, often not much exceeding the calyx, nearly straight.
Fr. stalks much longer than the capsule, straight, ultimately erect. Seeds 0.5–0.7 mm, reddish-brown, bluntly
tubercled. Fl. 5–7. Homogamous. Rarely visited by
insects and automatically self-pollinated. $2n = 72^*$. Th.

Native. Locally common near the coast all round the
British Is., especially on sand dunes and shingle but

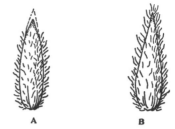

Fig. 30. Sepals of A, *Cerastium diffusum* and
B, *C. glomeratum*. ×5.

also on rocks, cliffs and walls; occasionally inland on
limestone rocks and in dry heath or grassland and, more
frequently, on railway tracks and ballast. Reaches
c. 290 m in Wilts. *C. diffusum sens. lat.* is native in much
of Europe northwards to Norway and S. Sweden and
eastwards to the Ukraine; subsp. *diffusum* mainly in
W. and S. Europe.

Very variable. In *Flora Europaea* **1** (1964), 3 subspp. are recognized, one restricted to Sicily. The above description is of
subsp. **diffusum**, native in W. and C. Europe, with which most
British material clearly conforms. Tall slender plants from
Sutherland and Shetland have been referred to subsp. *tetrandrum* (Lange) P. D. Sell & Whitehead (*C. tetrandrum* (Lange)
Murb.), which they resemble in habit in the long sepals
(7–9 mm) and capsules (7–10 mm). They differ, however, in
having petals relatively narrower than in continental material
and especially in having smaller pollen grains and larger and
darker seeds. Petal-length and capsule-size vary considerably
and independently over the whole distribution-range of *C. diffusum* in this country and there are at present inadequate
grounds for recognizing more than one subsp.

10. C. pumilum Curtis Curtis's Mouse-ear Chickweed.

Incl. *C. glutinosum* Fries

An annual or overwintering herb with slender tap-root
and erect or ascending stems, 2–12 cm, branching near
the base, covered with glandular and non-glandular
hairs, *often reddish below*. Lvs 0.5–1.5 cm; the basal
oblanceolate with a long stalk-like base; stem lvs elliptical or ovate-oblong; all hairy. Fls 6–7 mm diam., in few-
fld dichasia, their stalks short, glandular; lower bracts
usually lf-like: *upper bracts small with narrow scarious
margins and tip*. Fls 5-merous. Sepals 4–5 mm, oblong-
lanceolate, acute, glandular-hairy with fairly broad scarious margins; hairs usually not projecting beyond the
glabrous tip. *Petals* white or purple-tinged, about equalling the sepals, *about ¼ bifid*, with branching veins. Stamens 5. Styles 5. Capsule 6–8 mm, narrowly cylindrical,
slight curved upwards. *Fr. stalks* longer than the capsule,
at first curved downwards then nearly erect but with
a slight curve just beneath the capsule. Seeds
0.5–0.6 mm, dark brown, finely tubercled. Fl. 4–5.
Homogamous. Automatically self-pollinated. $2n = 90^*$.
Th.

The above description is of subsp. **pumilum** , with only 5 stamens and none but the upper bracts scarious-margined, to
which all British plants seem referable.

Native. Rare and local in open calcareous grassland, in chalk and limestone quarries and on railway tracks and banks southwards from the Oolite of Leicester (incl. Rutland) and S. Lincoln to Devon, Dorset, Hants and Isle of Wight; locally common in Gloucester and adjacent counties. *C. pumilum sens. lat.* is native in most of Europe southwards from c. 60° N in Sweden and Finland.

11. C. semidecandrum L. Little Mouse-ear Chickweed.

An annual or overwintering herb with slender tap-root and erect or ascending stems, 1–20 cm, branching at the base, covered with short dense glandular and non-glandular hairs. Lvs 0.5–1.8 cm; the basal narrowly oblanceolate with a long stalk-like base; stem lvs ovate to elliptical-oblong; all with short white hairs. Fls 5–7 mm diam., in ±spreading dichasia, their stalks glandular. Lowest bracts with a broad scarious margin; *upper bracts largely scarious with a small green central portion*. Sepals 3–5 mm, narrowly lanceolate, acute, with broad scarious margins, glandular-hairy. *Petals* white. *c. ⅔ length of the sepals*, narrow, *slightly notched*, with *unbranched veins*. Stamens 5. Styles 5. Capsule 4.5–7 mm, cylindrical, very slightly curved. *Fr. pedicels* longer than the capsule, *at first sharply deflexed from the base but not curved*, later almost erect. Seeds 0.4–0.5 mm, pale yellowish-brown, finely tubercled. Fl. 4–5. Homogamous. Sparingly visited by small insects, and automatically self-pollinated. $2n = 36*$. Th. Variable in habit and stature.

The above description is of subsp. **semidecandrum**, characterized by a dense covering of glandular and non-glandular hairs. The whole plant is glabrous in subsp. *macilentum* (Aspergren) Möschl, of S. Sweden.

Native. Frequent in dry open habitats especially on calcareous or sandy soils (including dunes) through most of Great Britain; mainly coastal in Ireland; not in Shetland. Europe southwards from c. 63° N in Norway; N. Africa; W. Asia.

11. MYOSOTON Moench

An annual to perennial herb with opposite ovate-cordate lvs and a lfy cymose infl. Fls 5-merous. Sepals free; *petals* white, *bifid*; stamens 10; ovary 1-celled; *styles 5, alternating with the sepals. Fr. an ovoid capsule opening by 5 bifid teeth*. Seeds numerous, small.

One(–2) spp.

Differs from *Stellaria* in the 5 styles and from *Cerastium* in their alternation with the sepals and in the ovoid capsule; from both in the 5 bifid teeth by which the capsule opens.

1. M. aquaticum (L.) Moench   Water Chickweed.

Cerastium aquaticum L.; *Stellaria aquatica* (L.) Scop.; *Malachium aquaticum* (L.) Fries

A usually perennial herb with prostrate overwintering stem-bases from which lfy flowering and non-flowering shoots arise. Flowering shoots 20–100 cm, decumbent or ascending, weak and fragile and often trailing over other plants, ±glabrous below, glandular-hairy above.

Lvs 2–5(–8) cm, thin, ovate-acuminate with a cordate base and often a wavy margin; lower lvs with a short ciliate stalk, upper ±sessile. Infl. a spreading dichasial cyme with lf-like bracts so that the fls appear solitary at forks of the stem. Fls 12–15 mm diam., their stalks glandular. Sepals narrowly ovate, blunt, with broad membranous margins. Petals up to half as long again as the sepals, bifid almost to the base with diverging lobes. Capsule longer than the sepals, drooping, its stalk at first reflexed then ±horizontal. Seeds brown, 0.4 mm across, densely covered with barbed papillae. Fl. 7–8. Protandrous. Visited by flies and small bees, etc. $2n = 28$. H.–Ch.

Native. In marshes and fens, streamsides, ditches and damp woods at low altitudes through England and Wales northwards to Durham. Europe (except Arctic) and temperate Asia.

12. STELLARIA L.

Annual to perennial herbs, usually slender and fragile and often glabrous. Lvs opposite, simple, entire. Fls in dichasial cymes, rarely solitary, usually 5-merous. Sepals free; *petals* white, usually *very deeply bifid*, sometimes 0; stamens 10 (8) or fewer; nectaries present; ovary 1-celled; *styles 3. Fr. a ±rounded capsule opening by 6 valves*. Seeds numerous, roundish-reniform.

About 85 spp., cosmopolitan.

1 Lvs ovate or cordate, at least the lower stalked; stem terete.   2

   Lvs narrow, linear to oblong, sessile; stem quadrangular.   5

2 Petals about twice as long as the sepals; capsule narrowly ovoid, up to twice as long as the sepals.   **1. nemorum**

   Petals not or little exceeding the sepals, or 0; capsule ovoid, longer than the sepals.   3

3 Sepals 5–6.5 mm; stamens 10; seeds not less than 1.3 mm diam.   **4. neglecta**

   Sepals 2–5 mm; stamens 1–7; seeds not more than 1.3 mm diam.   4

4 Sepals 4.5–5 mm; petals usually present; stamens 3–7 with red-violet anthers; seeds 0.9–1.3 mm diam.   **2. media**

   Sepals 2–3.5 mm; petals 0; stamens 2(1–5) with grey-violet anthers; seeds 0.6–0.8 mm diam.   **3. pallida**

5 Petals deeply bifid with divergent lobes, distinctly shorter than the sepals; calyx with a funnel-shaped base.   **8. alsine**

   Petals equalling or exceeding the sepals; calyx with a rounded base.   6

6 Bracts wholly herbaceous; fls 15–30 mm diam.; petals notched to less than half-way; capsule globular.   **5. holostea**

   Bracts with scarious margins; fls 5–18 mm diam.; petals bifid at least half-way and usually to near the base; capsule oblong-ovoid.   7

7 Bracts ciliate; plant not glaucous; fls 5–12 mm diam.   **7. graminea**

   Bracts not ciliate; plant usually glaucous; fls 12–18 mm diam.   **6. palustris**

1. S. nemorum L. Wood Stitchwort, Wood Chickweed.

A *perennial stoloniferous* herb with weak decumbent to ±erect pale green lfy shoots, 15–60(–80) cm, terete or obscurely angled, ±hairy all round, especially below the distinctly swollen nodes, and somewhat glandular, sometimes almost glabrous. Most *stolons hypogeal* for up to c. 15 cm, white or reddish with small or scale-like lvs, then emerging as decumbent or ascending non-flowering lfy shoots up to 60(–80) cm. *Most lvs ovate or ovate-oblong, acuminate,* commonly 4–8 × 3–4 cm, with *petioles up to 4 cm or more,* but *upper lvs of flowering shoots ovate-elliptical with rounded to cuneate base, short-stalked or ±sessile;* all thin, glabrous or sparsely hairy and often somewhat ciliate. Fls 10–18 mm diam. in lax few-fld terminal dichasia, their pedicels 1–2 cm, slender, glandular-pubescent. Sepals c. 6 mm, broadly lanceolate, blunt, with rather narrow scarious margins. *Petals c. twice as long as sepals, very deeply bifid* with narrow divergent lobes. Stamens 10, rarely 0. Styles 3. Capsule narrowly ovoid, up to twice as long as sepals, opening almost to the base by 6 valves. Fr.-pedicels up to 4 cm, spreading or reflexed. Seeds 1–1.3 mm across, orange-brown, papillate. Fl. 5–6. Protandrous. Visited by flies and beetles. $2n = 26^*$. H.

Two subspp. have been recognized in the British Is.: subsp. **nemorum**, commonly up to 60 cm and reaching 80 cm, with bracts decreasing in size at each branching of the infl., the 3rd pair usually at least 1 cm long, and seeds with marginal rows of ±hemispherical tubercles; and subsp. **glochidisperma** Murb., only 15–30(–50) cm, with bracts decreasing abruptly in size after the first pair, the 3rd pair scale-like and usually not more than 1.5 mm long, and with seeds having long cylindrical papillae 0.15 mm long round the margin. The 2 subspp. are fully interfertile and putative hybrids have been reported from Wales.

Native. Locally in damp woods and by streams in the west and north of Great Britain. Subsp. *nemorum* from N. Devon, Derby, Nottingham and N. Lincoln northwards to Moray and Banff, reaching c. 900 m in Scotland; subsp. *glochidisperma* very local and only in Wales. *S. nemorum* sens. lat. in most of Europe northwards to the North Cape and eastwards to the Caucasus, chiefly in montane and subalpine woods on moist and mineral-rich mull soils; subsp. *glochidisperma* from the British Is. to C. Spain and eastwards to S. Sweden, Austria and Yugoslavia.

S. media group (2–4). Usually annual or overwintering herbs with terete stems, at least the lower lvs petiolate and very deeply bifid petals; in all British representatives the stem has a single vertical line of hairs, rarely 2 lines, down each internode.

2. S. media (L.) Vill. Chickweed.
Alsine media L.

An annual or overwintering herb with a slender tap-root and diffusely branched decumbent or ascending lfy *stems* 5–40 cm, *with a single line of hairs down each internode.*

Lower lvs 3–20 mm, ovate-acute, *long-stalked; upper ovate or broadly elliptical, usually larger (up to 25 mm), acute or shortly acuminate, ±sessile;* all ±glabrous or ciliate at the base. Fls numerous, in terminal dichasia, their pedicels usually with a line of hairs. *Sepals* 3–5 mm, ovate-lanceolate with narrow scarious margins, usually glandular. Petals from ¾ as long as to ±equalling sepals, very rarely minute or 0. Stamens 3–8 with red-violet anthers. Styles 3. Capsule ovoid-oblong, somewhat longer than the calyx. Fr. stalks downwardly curved, often wavy. *Seeds reddish-brown, 0.9–1.3 mm across, usually with rounded or flat-topped tubercles.* Fl. 1–12. ±Homogamous. Visited by numerous flies and small bees, etc., and automatically self-pollinated. $2n = 42$. Th.

A very polymorphic species, varying in size, habit, hairiness, length of petals, number of stamens, size and surface-detail of seeds, etc.

The above description is of subsp. **media**, the only one of the 3 European subspp. recognized in *Flora Europaea* 1 (1964) which is native in the British Is. It is distinct in having a single line of hairs, rarely 2, down each stem-internode, the other subspp. having at least the upper internodes hairy all round.

Native. A common weed of cultivated ground, roadsides and waste places; locally on bare patches in nitrogen-rich pastures with abundant *Lolium perenne* which are periodically flooded with fresh or brackish water, and conspicuous in the upper parts of some coastal salt-marshes; abundant throughout the British Is. Cosmopolitan.

3. S. pallida (Dumort) Piré Lesser Chickweed.
S. apetala auct.; incl. *S. boraeana* Jordan

An annual much-branched prostrate herb resembling *S. media* but with more slender often filiform stems and smaller pale green ovate *lvs* which are usually less than 7 mm and *all or most* of which are *short-stalked.* Stem up to 40 cm, usually with a single line of hairs along each internode. Fls smaller than in *S. media*, never opening widely, their stalks hairy or almost glabrous. *Sepals* 2–3.5 mm, lanceolate, hairy or glabrous. *Petals 0 or minute.* Stamens 1–3, commonly 2; anthers grey-violet. Styles 3. Capsule exceeding the calyx. Fr. stalks short, usually not reflexed. *Seeds pale yellowish-brown,* rarely dark-brown, 0.6–0.8 mm *across, with small blunt tubercles.* Fl. 3–5. Automatically self-pollinated and often cleistogamous. $2n = 22$. Th. Readily distinguished from apetalous *S. media* by the small sepals and small pale seeds.

Native. A locally common plant of dunes and an inland weed of waste places and cultivated ground on light sandy soils. Chiefly in S. and E. England but reaching Kintyre and N. Aberdeen; very local in E. Ireland; Channel Is. C. and E. Europe. A characteristic species of woods of *Pinus sylvestris* on light glacial sands.

4. S. neglecta Weihe Greater Chickweed.
S. umbrosa Opiz & Rupr.; incl. *S. elisabethae* F. W. Schultz
An overwintering annual (to perennial) herb like *S.*

media but larger in all its parts, with weak branching stems 25–90 cm, procumbent below then ascending, each internode with a single line of hairs. *Lower lvs ovate, subcordate, acute or shortly acuminate, long-stalked*, with the blade 1–2.5 cm and petiole up to twice as long; *upper lvs* to 5 cm, ovate-oblong, acuminate, *with short flattened petioles or subsessile*; all glabrous. Fls c. 10 mm diam. in lax dichasia, their long slender pedicels glabrous or pubescent. *Sepals* 5–6.5 mm, lanceolate, glabrous or pubescent. *Petals* deeply bifid, *equalling or slightly exceeding the sepals*, rarely 0. *Stamens usually* 10. Styles 3. Capsule exceeding calyx. Fr.-pedicels to 30 mm, at first straight, ±spreading, finally deflexed. *Seeds* (1.1–)1.3–1.6 mm, *usually dark reddish-brown* with slender conical tubercles. Fl. 4–7. Protandrous. $2n = 22^*$. Th.(–H).

Native. A local plant of hedgerows, wood-margins, streamsides and shady places. Found throughout England and Wales but commonest in the west; very local in Scotland northwards to W. Ross and Aberdeen; very rare in S.W. Ireland. Most of Europe northwards to S. Sweden and eastwards to the Ukraine.

5. S. holostea L.

Satin Flower, Adder's Meat, Greater Stitchwort.

A perennial herb with a slender creeping stock and weak brittle ascending stems, some short and non-flowering, others 15–60 cm, flowering. Stems with 4 rough angles, more slender below, glabrous or hairy above. *Lvs* 4–8 cm, slightly glaucous, rigid, *lanceolate-acuminate, tapering from a wide base to a long fine point, very rough at the margins* and on the underside of the midrib. Fls 15–30 mm diam., long-stalked, in loose terminal dichasia, their *bracts lf-like*. Sepals 6–9 mm, ovate-lanceolate, inconspicuously 3-veined, with a narrow membranous margin. Petals 8–12 mm, white, bifid to about half-way, occasionally laciniate, rarely 0. Stamens 10, sometimes fewer by degeneration. Styles 3. Capsule subglobose, about equalling the calyx. Seeds 1.5–2 mm across, reddish-brown, papillose. Fl. 4–6. \Protandrous to homogamous. Visited by many flies, small bees and beetles. $2n = 26$. Chh.

Native. A common plant, especially of woods and hedgerows throughout the British Is. except Orkney, Shetland and Outer Hebrides. Reaches 640 m in Scotland. Europe northwards to 63° N in Finland but rare in Mediterranean region; N. Africa, the Near East. A woodland herb on a wide range of mull soils.

6. S. palustris Retz.   Marsh Stitchwort.

Incl. *S. dilleniana* Moench, non Leers, and *S. glauca* With.

A perennial herb with slender creeping stock and glabrous *smoothly 4-angled* ±erect flowering and non-flowering *shoots*. Flowering shoots 20–60 cm, weak and brittle. *Lvs* 1.5–3.5 cm, usually glaucous, sessile, *linear-lanceolate*, ascending, glabrous and *with smooth edges*. *Fls* 12–18 mm *diam*. in a terminal few-fld cyme, or solitary, axillary, their stalks 3–10 cm. *Bracts* narrowly lan-

ceolate *with broad membranous non-ciliate margins and a narrow central green strip*. Sepals 5–8 mm, lanceolate-acute, distinctly 3-veined, with broad white membranous margins. Petals white, equalling or up to twice as long as the sepals, bifid almost to the base. Stamens 10. Styles 3. Capsule ovoid-oblong, about equalling the calyx. Seeds 1.2–1.4 mm across, dark reddish-brown, with rounded tubercles. Fl. 5–7. Protandrous. Visited by flies. $2n = $ c. 130. Hel.

Native. A local plant of marshes and base-rich fens southwards from Perth. Ireland, chiefly in central districts and absent from the south-west. C. and N. Europe to c. 69° N in Scandinavia; temperate Asia. Associated with *Carex* spp. *Phalaris arundinacea*, *Galium palustre*, *Lythrum salicaria*, etc., in the sedge-dominated communities which follow the *Scirpus–Phragmites* reed-swamp.

7. S. graminea L.   Lesser Stitchwort.

A perennial herb with slender creeping stock from which flowering shoots and numerous non-flowering shoots arise. Flowering *shoots* slender and brittle, diffusely branched, decumbent or ascending, *smoothly 4-angled*, glabrous, 20–90 cm. *Lvs* 1.5–4 cm, not glaucous, *linear-lanceolate* or narrowly elliptical, acute, *smooth at the margins but often ciliate near the base*. *Fls* 5–12 mm *diam.*, rarely larger, in a loose spreading terminal cyme, their stalks 1–3 cm, 4-angled, glabrous. *Bracts wholly scarious, margins ciliate*. Sepals 3–7 mm, lanceolate, acute, distinctly 3-veined, with membranous margins. Petals white, bifid more than half-way, equalling or exceeding the sepals, being largest in hermaphrodite and smallest in male-sterile fls. Stamens 10, all, some or 0 of which may be fully developed and fertile. Styles 3. Capsule ovoid-oblong, longer than the calyx. Seeds c. 1 mm across, reddish-brown, rugulose. Fl. 5–8. Protandrous. Hermaphrodite and partially or completely male-sterile plants are found. Visited chiefly by flies. $2n = 26$. H.

Native. A common plant of woods, heaths and grassland especially on light soils. Throughout most of the British Is., Europe and Asia.

8. S. alsine Grimm   Bog Stitchwort.

S. uliginosa Murray

A perennial herb with slender creeping stock and numerous decumbent and ascending smoothly 4-angled glabrous shoots 10–40 cm. Lvs 5–10(–20) mm, sessile, elliptical or oblanceolate, acute, slightly ciliate at the base, otherwise glabrous. Fls c. 6 mm diam. in terminal few-fld cymes. Bracts scarious with a green central stripe. *Calyx funnel-shaped at the base*; sepals 2.5–3.5 mm, lanceolate, acute, 3-veined. *Petals white, shorter than the sepals, bifid almost to the base, with widely divergent lobes*. Stamens 10. *Capsule* equalling the calyx, ovoid, narrowed below and *with a short carpophore*. Fr. stalks 0.5–3 cm, at first reflexed, then erect. Seeds 0.3–0.4 mm across, pale reddish-brown, with small tubercles. Fl. 5–6. Protandrous or homogamous. Sparingly visited by flies. $2n = 24$. Hel.

Native. A frequent plant of streamsides, flushes, wet tracks and woodland-rides, etc., throughout the British Is., reaching 1003 m in Scotland. Often with *Juncus bufonius*, *J. articulatus*, *Montia*, *Callitriche stagnalis*. Europe, temperate Asia, North America.

13. HOLOSTEUM L.

Small annual herbs, glaucous and usually glabrous below but viscid with glandular hairs above. Lvs opposite, extipulate. *Fls in umbel-like terminal cymes*, 5-merous. Sepals free; *petals* white, *irregularly toothed or jagged*; stamens 10 or fewer; nectaries present; ovary 1-celled; styles 3, rarely 4 or 5. *Fr. a cylindrical capsule opening by usually 6 revolute teeth. Seeds peltate*, papillose.

Six spp., in Europe, N. Africa and temperate Asia to the Himalaya.

1. H. umbellatum L.   Jagged Chickweed.

An annual herb with a simple or basally branched erect or ascending stem 3–20 cm, glaucous, glabrous below, viscid above. Basal lvs 10–25 mm, oblanceolate, narrowed into a short broad stalk; stem-lvs (2–3 pairs) oblong or elliptical, sessile; all acute, glabrous or glandular-ciliate. Petals white or pale pink, longer than the sepals. Stamens 3(–5). Capsule cylindrical, narrowing above, twice as long as the sepals, its stalk at first deflexed then re-erecting. Seeds reddish-brown, peltate, furrowed on one side and broadly keeled on the other, 0.5 mm across. Fl. 4–5. ±Homogamous. Visited by small bees and flies and automatically self-pollinated. $2n = 20$. Th.

Doubtfully native. A very rare plant of walls, roofs and sandy soils. Formerly in Surrey, Norfolk and Suffolk, where it may still persist. Europe northwards to S. Sweden, N. Africa, W. Asia. Commonly with other small annuals such as *Erophila verna*, *Cerastium semidecandrum*, *Veronica* spp., and in C. Europe usually in man-made habitats.

14. MOENCHIA Ehrh.

Annual herbs, *glabrous and glaucous*, with opposite narrow entire lvs. Fls solitary or in few-fld spreading cymes, 4- or 5-merous. Sepals with broad membranous margins; *petals* white, *entire*; stamens 8 (10) or 4(–5); nectaries present; ovary 1-celled; *styles 4 or 5, opposite the sepals*. Fr. an oblong straight *capsule opening by* as many or twice as many *short blunt teeth* as styles. Seeds numerous, reniform, papillose.

About 5 spp. in Europe, N. Africa and the Near East.

1. M. erecta (L.) Gaertn., Mey. & Scherb.
                Upright Chickweed.
Sagina erecta L.; *Cerastium quaternellum* Fenzl

An annual herb with erect main stem 3–12 cm, and usually with a few ascending or decumbent basal branches. Basal lvs 6–20 mm, strap-shaped, short-stalked; upper lvs shorter, narrowly lanceolate, sessile, ascending; all rigid, acute, glabrous and glaucous. Fls

about 8 mm diam., usually 5-merous 1–3 on long stalks. *Sepals* lanceolate, acute, *with broad white margins.* Petals shorter than the sepals. Stamens usually 4. *Styles* 4, short and recurved. *Capsule* a little longer than the sepals, *opening by 8 teeth.* Seeds reddish-brown, papillose, c. 0.5 mm across. Fl. 5–6. Protogynous. Visited by flies. Automatic self-pollination takes place in dull weather in fls which do not open. $2n = 36^*$. Th.

Native. A local plant of gravelly pastures, maritime cliffs and dunes throughout England and Wales to Northumberland and Cumberland. C. and S. Europe, northwards to the Netherlands and Brandenburg.

15. SAGINA L.   Pearlwort.

Small annual or perennial herbs, often tufted, with slender prostrate or ascending flowering shoots and subulate exstipulate opposite and connate lvs. Fls ±spherical in bud, 4–5-merous, in dichasial cymes or solitary and terminal. Sepals free; petals white, entire, often minute, sometimes 0; stamens as many or twice as many as the sepals; ovary 1-celled; *styles 4–5, alternating with the sepals*; ovules many. Fr. a *capsule splitting to the base* into 4–5 valves. Seeds very small.

About 20 spp., chiefly in the north temperate zone.

1   Annual plants lacking vegetative shoots at time of
    flowering; fls usually 4-merous; petals minute or
    0.   **2**
    Perennial plants with vegetative shoots at time of
    flowering; fls commonly 5-merous with petals at
    least equalling sepals (but in **3** fls usually 4-merous
    and petals minute or 0).   **3**
2   Lvs distinctly awned; seeds averaging less than
    0.4 mm across.   **1. apetala**
    Lvs blunt or very shortly mucronate, not distinctly
    awned; seeds averaging more than 0.4 mm across;
    usually maritime.   **2. maritima**
3   Plants densely tufted with crowded and rigid strongly
    recurved lvs. Problematic Scottish alpine, retained
    in cultivation.   **4. boydii**
    Lvs not strongly recurved.   **4**
4   Stem-lvs at 2 uppermost nodes less than 2.5 mm and
    with short dense axillary lf-clusters; petals almost
    twice as long as sepals.   **9. nodosa**
    Stem-lvs at 2 uppermost nodes not distinctly shorter
    than those at lower nodes and lacking dense axillary
    lf-clusters; petals at most about equalling sepals,
    sometimes 0.   **5**
5   Lvs with terminal awn at least $\frac{3}{4}$ as long as maximum
    lf-width; fls 5-merous with 10 stamens.   **8. subulata**
    Lvs blunt or with awn less than $\frac{3}{4}$ of maximum lf-width;
    fls 4–5-merous with 10 or fewer stamens.   **6**
6   Plants forming small dense tufts up to 5 cm diam.;
    sepals with purple margins; stamens 8 or 10, some-
    times fewer; fr.-pedicels not reflexing. Rare Scot-
    tish alpine.   **7. intermedia**
    Plants forming lax tufts or mats; sepals with white
    margins; fr.-pedicels reflexing at some stage.   **7**
7   Fls usually 4-merous; petals usually minute or 0; sta-
    mens 4(5); ripe capsule less than 3 mm.
                               3. procumbens
    Fls usually 5-merous; petals about equalling sepals;
    stamens 10, rarely 5; ripe capsule at least 3 mm.   **8**

8 Plants with decumbent then ascending non-rooting stems; sepals more than 2.8 mm; ripe capsule more than 3.5 mm, setting seeds freely.   **5. saginoides**

Plants with ±prostrate rooting stems; sepals less than 2.8 mm; capsule less than 3.5 mm with few or no good seeds.   **6. ×normaniana**

1. S. apetala Ard.   Common Annual Pearlwort.

Incl. *S. ciliata* Fries, *S. erecta* Hornem. and *S. filicaulis* Jordan

Annual herbs with slender tap-root, *main stem* 2–10(–20) cm ±*erect and with terminal fl.*, and several decumbent to ascending, rarely prostrate, lateral stems from the axils of a *usually transient rosette-like central lf-cluster*. *Lvs linear-subulate*, flattened above, *tapering into an awned or long-mucronate tip*, sometimes ±ciliate towards the base. *Fls usually 4-merous, solitary*; pedicels ±erect, filiform, often glandular-hairy at least above. Sepals ovate to ovate-oblong; *petals minute and often falling early*, rarely 0; *stamens usually 4. Ripe capsule equalling or somewhat exceeding the spreading or appressed sepals.* Seeds ±reniform, averaging less than 0.4 mm across, tubercled. Fl. 5–8. Slightly protandrous and visited by small crawling insects; usually self-pollinated. $2n = 12^*$. Th.

An extremely variable and largely autogamous complex within which a number of varieties, subspp. and spp. have from time to time been deemed distinguishable. *Flora Europaea* **1** (1964) tentatively recognizes 2 widespread subspp. both native in the British Is.

Subsp. **apetala** (*S. ciliata* Fries); at least the outer *sepals subacute, all appressed to the ripe capsule or slightly spreading; capsule on main stem of well-grown plants* c. 1¼ *times as long as sepals; seeds averaging more than* 0.3 mm *across.*

Subsp. **erecta** (Hornem.) F. Hermann (*S. apetala* auct.): *sepals hooded and ±blunt at tip, spreading horizontally in ripe fr.; capsule on main stem of well-grown plants more then* 1¼ *times as long as sepals; seeds averaging less than* 0.3 mm *across.*

Native. Both subspp. are frequent on bare ground in dry heaths, grassland and waysides and on gravel paths, walls and wall-bases, in quarries, railway-sidings, etc., subsp. *erecta* throughout the British Is. except Shetland but subsp. *apetala* less frequent in general and absent from much of W. Scotland and the Hebrides and very local in Ireland.

A variant of subsp. *apetala*, smaller in all its parts and with stems, lvs, pedicels and sepals all glandular-hairy (*S. ciliata* var. *minor* Rouy & Fouc., *S. filicaulis* auct. angl., non Jordan) is found locally in S. and W. England and may merit subspecific rank. '*S. reuteri* Boiss.' of British authors appears to be merely a dwarf growth-form of subsp. *apetala* growing in dry places in S.W. England.

S. apetala sens. lat. occurs throughout Europe northwards to Scotland, Denmark, southernmost Sweden and Estonia; W. Asia; Canary Is.

2. S. maritima G. Don fil.   Sea Pearlwort.

An *annual* herb with slender tap-root with or without a central rosette of lvs, the *main stem flowering* and it and the numerous laterals varying from prostrate to ±erect and sometimes densely tufted; usually glabrous. *Lvs* linear-lanceolate, fleshy, blunt or apiculate but *not awned*, usually glabrous, rarely ciliate, shorter than the internodes. Pedicels remaining erect, glabrous. Fls 4-merous. *Sepals hooded, blunt*, glabrous often with a purplish margin, *half-spreading in fr.* Petals white, minute or 0. Capsule (Fig. 31c) about equalling the sepals. Seeds

A   B   C   D

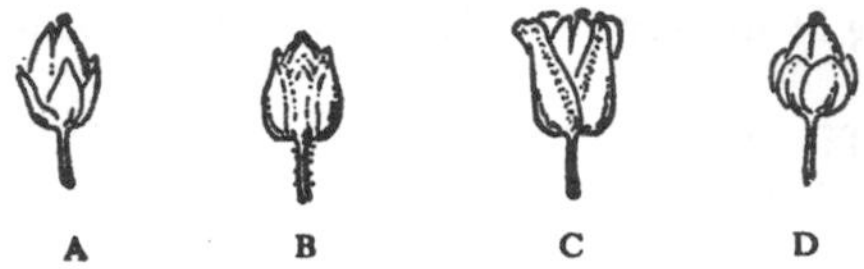

Fig. 31. Unripe capsules of *Sagina*. A, *S. apetala*; B, *S. ciliata*; C, *S. maritima*; D, *S. procumbens*. ×2.5.

averaging more than 0.4 mm across, papillose. Fl. 5–9. Automatically self-pollinated. $2n = 28^*$; 22, 24. Th.

Native. A local plant of coastal dune-slacks, cliffs and rocks all round the British Is. and occasional on Scottish mountains to 1300 m. Atlantic, Channel and North Sea coasts of Europe from S. Portugal to 64° N in Norway but not in Iceland or Faeroes, and local on coasts of Baltic to c. 60° N in Sweden, Finland and Estonia; local in the Mediterranean region eastwards to Turkey and on Black Sea coast of Bulgaria; N. Africa, W. Asia.

A very variable species. *S. debilis* Jordan and *S. densa* Jordan are names given to types with few weak slender decumbent branches and numerous densely tufted branches respectively, both described as having lvs more acute than in *S. maritima*.

3. S. procumbens L.   Procumbent Pearlwort.

A *perennial* tufted herb *with a dense central rosette* of lvs, the *main stem never* elongating or *flowering*, the *laterals* up to 20 cm, *prostrate and rooting at the base*, then ascending. Lvs 5–12 mm, not exceeding the internodes, linear-subulate narrowing abruptly to a short awn, glabrous or ciliate. Fl.-stalks glabrous, slender, erect during flowering, then recurved at the tip, and finally re-erecting in ripe fr. Fls 4-, occasionally 5-merous. *Sepals* ovate, *hooded, blunt, spreading in ripe fr.* Petals white, minute or 0. Stamens 4. Capsule (Fig. 31D) longer than the sepals, opening with 4 blunt valves. Seeds 0.3–0.5 mm, brown, strongly papillose. Fl. 5–9. Homogamous. Automatically self-pollinated. $2n = 22$. H.

Native. Throughout the British Is. where a bare substratum or the soil of open vegetation is moist for prolonged periods. Locally common in garden-beds and arable fields, especially where the soil is nutrient-rich and somewhat compacted, by streams and on banks and walls; resistant to treading and often on paths, roadside verges and lawns and in trampled agricultural and re-creational grassland; occasional on wet rocks and reaching 1150 m in Scotland. Through most of Europe; Asia, North America, Greenland. Widely introduced in the southern hemisphere.

4. S. boydii F. B. White   Boyd's Pearlwort.

A dwarf *perennial densely tufted* glabrous herb with erect stems and *crowded* imbricate *rigid* short curved *lvs* which are somewhat fleshy and *strongly recurved*. Fl.-stalks short, slightly curved. Fls 4–5-merous. Sepals ovate, blunt, never opening widely. Petals 0 or very minute. Stamens 5–10. Stigmas often rudimentary. Capsule enclosed by the *sepals* which are *closely appressed* to it, never ripening seeds. Fl. 5–7. $2n = 22^*$. Ch.

Probably native. Presumed to have been collected near Braemar, Aberdeen. Not found since 1878 but retained in cultivation with its distinctive characters unchanged.

5. S. saginoides (L.) Karst.   Alpine Pearlwort.

S. linnei Presl

A *perennial* tufted herb with a slender woody branching stock and numerous decumbent then ascending non-rooting glabrous lfy *shoots* 2–7 cm *high* arising from *basal* rosettes of linear *lvs up to* 2 cm. Stem-lvs 0.5–1 cm, linear; all lvs glabrous, mucronate or awned. Fls usually 5-merous, c. 4 mm diam., usually solitary *on slender stalks* 1–2.5 cm. Sepals ovate, rounded above, glabrous, with narrow scarious margins. Petals white, broad, rounded, sometimes slightly emarginate contracted below into a short claw, equalling or somewhat shorter than the sepals. Stamens 10. Styles 5. *Capsule* 3.5–4 mm, ovoid, exceeding the sepals, and almost twice as long when the valves have straightened after dehiscence; *sepals appressed to the ripe capsule*. Fr.-pedicel at first recurved just beneath the capsule and ±prostrate, later becoming straight and erect. Seeds 0.3 mm across, with rows of fine tubercles. Fl. 6–8. Homogamous. Visited by small flies, etc., and automatically self-pollinated; remaining closed in dull weather. $2n = 22^*$. Ch.

Native. A very rare arctic-alpine plant of rock-ledges and fine scree on mountains in Scotland from Perth and Angus to Sutherland; Skye. Reaches 1215 m on Ben Lawers. A subarctic circumboreal plant of Europe, Asia and North America, reaching 71°N in N. Norway; and an alpine plant of the mountains of C. Europe, Asia and Mexico.

6. S. ×normaniana Lagerh.

S. scotica (Druce) Druce; *S. saginoides* subsp. *scotica* (Druce) Clapham. Probably *S. procumbens × saginoides*

A perennial loosely tufted or mat-forming herb differing from *S. saginoides* only in the *longer more slender* ±*prostrate rooting stems* 2.5–10(–15) cm, the longer lvs of the basal rosettes (0.5–3 cm), the shorter sepals (2–2.5 mm), shorter capsules (3.3–5 mm), and longer fr.-stalks, 1.5–3(–4) cm, and the *sepals somewhat spreading in ripe fr*. *Rarely sets seed* and the great majority of the capsules remain undeveloped. Fl. 7–10. Chh.

Native. Occurs in the same mountain regions as *S. saginoides*, growing with it in wet subalpine mat-pasture or on water-splashed rocks. Does not reach as high as *S. saginoides* (to 1067 m on Ben Lawers) and flowers later. The small differences in habit, lf-length and time of flowering are retained in cultivation. May be the hybrid *S. procumbens × saginoides* as suggested by its morphological features and many abortive capsules, but a higher proportion of well-developed capsules and good seed is set in cultivation.

7. S. intermedia Fenzl   Lesser Alpine Pearlwort.

S. nivalis auct., ?(Lindblad) Fries; *S. caespitosa* auct. angl., non (J. Vahl) Lange

A dwarf perennial herb with a much-branched woody stock and numerous densely tufted erect or ascending glabrous lfy shoots, forming a *small cushion* 1.5–5 cm *across and* 1–3 cm *high*. Basal lf-rosette only in first season; stem-lvs 3–6 mm linear, acute, usually shortly mucronate, glabrous. Fls 3–4 mm diam., 4–5-merous, borne singly *on short stalks* 2–5 mm. Sepals 1.5–2 mm, *glabrous*, ovate, rounded at the apex, with narrow scarious often violet-coloured margins, strongly concave, *appressed to the ripe capsule*. Petals white, somewhat shorter than the sepals, *narrowly elliptical*. Stamens 8–10. Styles 4–5. Capsule 2.5–3 mm, about half as long again as the sepals, greenish- or whitish-yellow, dull. Fr-pedicels usually ±erect. Seeds 0.5–0.6 mm across, yellowish-brown, tubercled. Fl. 6–8. Homogamous. Rarely visited and automatically self-pollinated; remaining closed on dull days. $2n = 88$. Chh.

Native. A very rare alpine plant of Ben Lawers and neighbouring Scottish mountains, reaching 1130 m. Circumpolar, reaching 82° 29' N on the N. coast of Greenland; Norway, Sweden, Faeroes, Finland, Iceland, Alps (very rare).

8. S. subulata (Swarz) C. Presl   Awl-leaved Pearlwort.

A perennial mat-forming herb with a branched creeping stock and rosettes of linear lvs 0.5–1.5 cm, from which arise numerous decumbent then erect or ascending ±*glandular-hairy shoots* 2–7.5(–12.5) cm high; the main stem short and non-flowering. Stem-*lvs* diminishing upwards, 0.3–1.2cm, linear, *narrowing to a long awn* somewhat channelled above and keeled below, ±*ciliate* and with scattered hairs elsewhere. Fls 5-merous, usually solitary on glandular-hairy stalks 2–4 cm. *Sepals* 2–2.5 mm, oblong-oblanceolate, hooded and blunt at the apex, with narrow scarious margins, *glandular-hairy*, *appressed to the ripe capsule*. Petals white, ovate blunt, short-clawed, *equalling or somewhat shorter than the sepals*. Stamens 10. Styles 5. Capsule c. 3 mm, somewhat longer than the sepals. Fr.-stalks at first curved just beneath the capsule, then erect. Seeds 0.3–0.4 mm, yellowish-brown, tubercled. Fl. 6–8. Homogamous. Rarely visited and automatically self-pollinated. $2n = 22^*$; 18. Chh.

Native. A local and decreasing plant of dry grassy and heathy places over sand or gravel and of rocks and cliffs, both coastal and inland. Widespread in Scotland, the Hebrides and the northern isles but chiefly coastal elsewhere and absent from most of England except the south and south-west; local and coastal in Ireland; Channel Is. Reaches 820 m in Scotland. W. and C. Europe from Iceland to N. Spain and Portugal and eastwards to c. 64° N in Norway, S. Sweden, S.E. Poland and S. Yugoslavia; Corsica.

9. S. nodosa (L.) Fenzl   Knotted Pearlwort.

Spergula nodosa L.

A perennial tufted herb with an occasionally branched stock and numerous procumbent or ascending stems 5–15(–35) cm high, glabrous or glandular-hairy above, arising from basal lf-rosettes. *Main stem short and non-flowering.* Rosette-lvs 5–20 mm; stem-lvs diminishing upwards from 10–15 mm to 1–2 mm, with lateral lf-clusters in the axils of the upper lvs giving the characteristic 'knotted' appearance; all lvs linear-subulate, abruptly and shortly mucronate, glabrous or somewhat glandular at the base. *Fls* 0.5–1 cm *diam.*, 1–3 at the ends of the stems and upper branches, 5-merous; their stalks glabrous or glandular, 3–10 mm. *Sepals* 2–4 mm, ovate, concave, blunt, glabrous or glandular, *appressed to the ripe capsule. Petals* white, narrowly ovate, 2–3 *times as long as the sepals.* Stamens 10. Styles 5. Capsule 4 mm, ovoid, exceeding the sepals. Fr.-stalks erect, straight. Seeds 0.4 mm, dark brown, tubercled. Fl. 7–9. Protandrous. Large fls are hermaphrodite, but smaller fls may lack fertile stamens. Sparingly visited by insects and automatically self-pollinated. $2n = 56^*$; 20–24, 44. H.–Ch.

Var. *moniliformis* (G. F. W. Meyer) Lange has the stems ±procumbent and the upper axillary buds with their lf-fascicles become detached as bulbils which propagate the plant vegetatively.

Native. Occasional to locally frequent in open vegetation on damp ±calcareous sand, including dune-slacks, on mineral-rich mud or peat of flushes, seepage-lines and valley-fens, by streams, lakes and canals, on ledges of coastal cliffs and less commonly in moist open woods; an infrequent weed of cereal crops. Throughout the British Is. and reaching 640 m on Cross Fell. N. and C. Europe from Iceland to N. Spain and Portugal and north-eastwards to northernmost Norway, the Kola Peninsula and C. Russia; not in the Mediterranean region.

16. MINUARTIA L.

Mostly perennial tufted or mat-forming herbs, but some annuals or biennials, with usually *linear to narrowly lanceolate, often subulate,* opposite and exstipulate *lvs. Fls usually 5-merous* in few-fld dichasia or solitary, terminal. Sepals free; petals usually white, entire or slightly emarginate, sometimes minute or 0; stamens 10, rarely fewer, the outer with paired nectaries at their base; styles 3(–5). *Capsule* ovoid to oblong, *opening by as many wide blunt teeth as styles.* Seeds 1 to many, usually reniform, tuberculate to verrucose-aculeate.

Up to 100 spp. in temperate and arctic regions of the northern hemisphere.

1 Slender annual herb of walls and dry places; shoots erect or ascending, all flowering.   **5. hybrida**
  Perennial herbs forming tufts or cushions with some decumbent or ascending non-flowering shoots.   *2*
2 Fls greenish, short-stalked to ±sessile, usually unisexual; sepals blunt, hooded at the tip; petals minute or 0; nectar-glands 10, linear, conspicuous. Compact alpine cushion-plant.   **6. sedoides**
  Fls white, distinctly stalked, hermaphrodite; sepals ±acute; petals not minute.   *3*
3 Lvs obscurely 1-veined or apparently veinless; pedicels glabrous, non-glandular. Only in calcareous flushes in upper Teesdale.   **3. stricta**
  Lvs distinctly 3-veined, especially when dry; pedicels glandular.   *4*
4 Lvs subulate, recurved and ±secund; sepals 5(–7)-veined. Very rare densely caespitose plant of mountains in S.W. Ireland.   **4. recurva**
  Lvs linear, somewhat recurved but not secund; sepals 3-veined.   *5*
5 Lvs 6–15 mm; petals equalling or somewhat exceeding sepals.   **1. verna**
  Lvs 4–8 mm; petals usually c. $\frac{2}{3}$ as long as sepals. A rare plant of mountains in C. Scotland.   **2. rubella**

1. M. verna (L.) Hiern   Vernal Sandwort.

Arenaria verna L.; *Alsine verna* (L.) Wahlenb.

A loosely caespitose herb with stoutish tap-root and a branching almost woody stock from which tufts of ascending flowering and non-flowering shoots arise. Flowering shoots 5–15 cm, glabrous or glandular-hairy. *Lvs* 6–15 mm, linear-subulate, ±acute or apiculate, *strongly 3-veined*, rather rigid, often curved. Fls 8–12 mm diam., in a few-fld cyme; *pedicels slender,* ±*glandular-hairy.* Sepals ovate-lanceolate, acute, strongly 3-veined, scarious-margined. *Petals* white, obovate, short-clawed, a little *longer than the sepals. Anthers red.* Seeds 0.6–0.9 mm, reniform, papillose. Fl. 5–9. Protandrous. $2n = 24^*$, 48 (78). Chh.

Variable, especially in ciliation and flower size. The above description is of the widespread subsp. **verna.** Plants from the Lizard (Cornwall) are smaller, more glaucous and ciliate, with the lower lvs closely appressed to the stem, but many of the distinctive features disappear in cultivation.

Native. A local plant of base-rich rocks, screes and pastures, and often abundant on the rocky debris of old lead workings. Mainly in N. Wales, the Peak District, the N. Pennines and the Lake District, with outliers in Cornwall (the Lizard), Somerset and a few scattered localities in Scotland from Berwick northwards to Banff; in Ireland only in Clare and the north-east. Through much of Europe from N. Spain to Greece and northwards to the British Is. and N.E. Russia but not in Scandinavia or the Baltic region; N. Africa, Caucasus, Siberia.

2. M. rubella (Wahlenb.) Hiern   Alpine Sandwort.

Alsine rubella Wahlenb.; *Arenaria rubella* (Wahlenb.) Sm.; *A. hirta* Wormsk.

A *perennial* herb forming compact tufts 2–8 cm across, with leathery tap-root and woody branched stock from which arise decumbent non-flowering and ascending flowering stems 2–6(–9) cm high. Lvs 4–8 mm, linear, glabrous or sparsely glandular, strongly 3-veined when dry, crowded below but distant on the glandular flowering stems. Bracts glandular, ovate. Fls 5–9 mm diam., in 1–2(–3)-fld cymes; *pedicels* 4–15 mm, *glandular*. Sepals 3.5–5 mm, broadly lanceolate, acute, strongly 3-veined when dry, glandular at least below. *Petals* white, ovate-oblong, short-clawed, $\frac{2}{3}$ *as long as sepals. Anthers red.* Styles 3–4(–5). Capsule ovoid, 3–4-valved, equalling or just exceeding sepals. Seeds 0.55–0.7 mm, brown, reniform, with low rounded papillae, Fl. 6–8. Homogamous. $2n = 24^*, 26$. Chh.

Native. A rare plant of base-rich rock ledges and detritus near the tops of a few Scottish mountains in E. and Mid Perth; probably extinct in E and W. Sutherland and Shetland. Circumpolar, in arctic and subarctic Europe southwards to 60° N in Norway and to 56° 30′ N in Scotland; Siberia, Greenland and North America south to Colorado.

3. M. stricta (Swarz) Hiern — Bog Sandwort.

Spergula stricta Swarz; *Alsine stricta* (Swarz) Wahlenb.; *Arenaria uliginosa* Schleicher ex DC.

A perennial glabrous herb forming small loose tufts, with slender tap-root and short somewhat woody stock from which arise short ascending non-flowering and slender erect flowering stems 5–10 cm. *Lvs* 6–12 mm, *filiform*, often curved, *obscurely* 1-*veined or apparently veinless*, rather crowded below but few and distant on the flowering stems. Fls 5–8 mm diam. in 1–3(4)-fld cymes; *pedicels* 15–50 mm, *filiform, glabrous*. Sepals 2.5–4 mm, ovate, acute, veinless when fresh, 3-veined when dry. *Petals* white, elliptical-oblong, narrowed below, *about equalling sepals. Anthers white.* Styles 3. *Capsule* ovoid, equalling or slightly exceeding sepals, *divided nearly to the base by the 3 spreading blunt valves.* Seeds 0.75–0.85 mm, reddish-brown, finely reticulate, almost smooth. Fl. 6–7. Homogamous. $2n = 22^*, 26$. Chh.

Native. Confined to calcareous flushes on Widdybank Fell, Upper Teesdale (Durham), mainly round the base of moss hummocks at 450–500 m. Arctic and sub-Arctic Europe, including Iceland and Spitsbergen, southwards to 60° N in Norway, and formerly also in the Jura and S.W. Germany but now lost from most or all of its localities; arctic North America to Alaska and Greenland; Siberia. Interesting as an 'arctic-prealpine' species not found in the Alps proper.

4. M. hybrida (Vill.) Schischkin — Fine-leaved Sandwort.

Arenaria tenuifolia L.; *M. tenuifolia* (L.) Hiern, non Nees ex Mart.

An *annual* herb with slender tap-root and slender erect or ascending stems 3–12 cm, branching especially below, usually glandular-pubescent at least above, sometimes glabrous. Lvs up to 12 mm, crowded below but distant above, linear-subulate, acute, 3(–5)-veined near the enlarged base, often recurved. Fls c. 6 mm diam. in lax much-branched subcorymbose dichasia, their pedicels 5–20 mm, glandular or glabrous. Sepals 3–4 mm, linear to ovate-lanceolate but commonly lanceolate-acuminate, 3-veined and with narrow scarious margins, usually glandular. *Petals* white, oblong, *slightly to distinctly shorter than sepals.* Stamens 10, rarely 5, with *yellow anthers.* Styles 3. Capsule oblong, 1–1½ times as long as sepals. Seeds 0.4–0.6 mm, red-brown, minutely tuberculate. Fl. 5–6. $2n = 46, 70$. Th.

Very variable. The above is a description of the native and widespread subsp. **hybrida**.

Native. In dry and usually calcareous rocky or stony places and sandy arable fields and on road verges, walls, railway banks, railway sidings, etc. Local in E. and S. England from Durham and W. Yorks southwards to W. Sussex and Devon and very rare in S.E. Wales; local in C. Ireland; Jersey. Europe northwards to 54° N in the British Is. and eastwards to S. Ukraine; N. Africa; W. Asia.

5. M. recurva (All.) Schinz & Thell.

Curved-leaved Sandwort.

Arenaria recurva All.; *Alsine recurva* (All.) Wahlenb.

A *caespitose perennial* with branching ±woody and often blackish stock and crowded ascending lfy stems, the fl-stems commonly 2.5–12 cm, non-fl. much shorter, all sparsely glandular-pubescent at least above. *Lvs* 4–8 mm, *subulate*, 3-*veined, mostly falcate-secund*, curving downwards and to one side. Cymes 1–3(–5)-fld with glandular-pubescent pedicels 1½ times to twice as long as the 3–5-veined white-margined bracts and up to 3 times as long as the sepals. *Sepals* 3–6 mm, ovate-lanceolate, acuminate, scarious-margined and *indistinctly* 3(–5)-*veined. Petals* white, *equalling or somewhat longer than the sepals*, elliptical, blunt, narrowed abruptly into a very short claw. Capsule narrowly ovoid, slightly exceeding sepals. Seeds 1.2 mm, reniform, brown, weakly tuberculate. Fl. 6–8. $2n = 30$. Chh.

Native. In small cracks in Old Red Sandstone at 550–610 m in three localities in Ireland, two in W. Cork and one in Kerry. Mountains of S. and S.C. Europe from Portugal and N. Spain through S. France, Switzerland, Austria and Italy (incl. Sicily) and eastwards to Bulgaria; W. Asia. A subalpine and alpine calcifuge on rock and rock-debris and in short turf with *Carex curvula* or *Kobresia myosuroides*, often in open vegetation on wind-exposed ridges with no permanent snow cover; at 1700–3165 m in the Swiss Alps.

6. M. sedoides (L.) Hiern — Mossy Cyphel.

Cherleria sedoides L.; *Arenaria sedoides* (L.) F. J. Hanb.

A perennial herb with very long tap-root and a branching woody stock from which arise *densely lfy shoots*, some ±prostrate and non-flowering, others short, erect, flowering, the whole *forming a yellow-green mossy cushion* 8–25 cm diam. and 4–8 cm high. Lvs 5–15 mm, somewhat fleshy, crowded, imbricating, linear-subulate, blunt, keeled beneath and channelled above, the margins horny and ciliate, otherwise glabrous. Fls

4–5 mm diam. Sepals 5, ovate, blunt, 3-veined, with narrow membranous margins. *Petals usually* 0, but sometimes present in male fls and then minute and subulate. Stamens 10, shorter than the sepals, sometimes fewer, and 0 in female fls. Styles very short. Capsule up to twice as long as the sepals, but often abortive in fls with fertile stamens, opening half-way by 3 valves. Seeds few, c. 0.5 mm, nearly smooth. Fl. 6–8. Protandrous to homogamous. Visited for nectar by small flies. $2n = 48^*$; 26. Chh.

Native. Locally common on rocks and rocky debris on mountains in Scotland from Perth and Argyll northwards to Sutherland and in Skye, Rhum and Eigg; reaching 1190 m on Ben Lawers but locally on gravel or shingle at sea-level; ±indifferent to pH of substratum; usually in moss- or grass-heath and often on very wind-exposed slopes, ridges and cols with little winter snow cover. Pyrenees, Alps and Carpathians. One of our few alpine plants not found also in the Arctic.

M. rubra (Scop.) McNeill (*Alsine fasciculata* auct., non L.; *Arenaria fastigiata* Sm.), an erect annual or biennial herb with linear lvs and fls in dense axillary and terminal clusters, is said to have been collected by G. Don in Clova and Fife. Several dried specimens exist but the plant has never been found in the British Is. since Don's day. It is a native of sub-Mediterranean Europe northwards to Burgundy, the Rhine Palatinate and Czechoslovakia, and of N. Africa, growing in dry usually calcareous stony or sandy ground and on sunny hills.

17. HONKENYA Ehrh.

A perennial *maritime* ±dioecious herb with opposite *fleshy ovate glabrous lvs* and greenish 5-merous fls. Sepals free; *petals entire*; stamens 10, abortive in female fls; nectaries present; ovary 1–celled, abortive in male plants; *styles usually* 3. *Fr. a globular capsule opening by 3 teeth*. Seeds few, large, pyriform.

One sp.

Differs from *Minuartia* in being succulent and in the shape of the capsule and the large seeds.

1. H. peploides (L.) Ehrh.   Sea Sandwort.
Arenaria peploides L.

A succulent stoloniferous herb with lfy flowering and non-flowering shoots, decumbent below, then erect, arising terminally from long pale slender *stolons creeping in sand or shingle*. Flowering shoots 5–25 cm, with *sessile, very fleshy, ovate-acute lvs*, 6–18 mm, having translucent wavy margins and downwardly pointing tips. Fls 6–10 mm diam., solitary in the axils of the upper lvs and in terminal 1–6-fld cymes. Sepals 3–5 mm, ovate, blunt, 1-veined, equalling pedicel. *Petals greenish-white*, obovate, equalling the sepals in male fls but shorter in female. Stamens alternately longer and shorter, each with a yellow, oblong nectary-gland at its base. Styles 3, sometimes 4 or 5. Capsule up to 8 mm diam., longer than the sepals, usually with 6 or fewer chestnut-coloured pear-shaped seeds 3–4 mm across. Fl. 5–8. Protandrous. Rarely visited by flies; hermaphrodite fls are automatically self-pollinated when they close in dull

weather. $2n = 68$ (?48, 64, 66). H.

Native. Common all round the British Is. on mobile sand and sandy shingle. Forms miniature dunes on its own or more commonly is associated with *Elymus farctus* on fore-dunes and persists to the 'yellow dune' stage. Tolerant of short periods of immersion in salt water. Coasts of the temperate and arctic regions of Eurasia and North America to 78° 20′ N in W. Greenland and Spitsbergen.

18. MOEHRINGIA L.

Annual to perennial, usually slender, herbs with opposite exstipulate lvs and 4–5-merous fls either solitary axillary or in terminal cymes. Sepals free; *petals* white, ±*entire*; stamens 8 or 10; nectaries present; ovary 1-celled; styles usually 2 or 3, rarely 4 or 5. *Fr. a rounded capsule opening by 4 or 6 recurved or revolute teeth. Seeds* dark, reniform, *with a variously-shaped oily appendage (elaiosome) in the sinus*.

About 20 spp. in temperate and arctic Eurasia and North America.

Differs from *Arenaria* chiefly in the appendaged seeds.

1. M. trinervia (L.) Clairv.   Three-nerved Sandwort.
Arenaria trinervia L.

A usually annual herb with weak slender diffusely branching shoots 10–40 cm, prostrate or ascending, pubescent. *Lvs* 6–25 mm, *ovate, acute, 3(–5)-veined*, ciliate; lower stalked, upper subsessile. Fls c. 6 mm diam., solitary in the forks of the stem and axils of the upper lvs, usually 5-merous; their stalks long, slender, pubescent. Sepals lanceolate-acuminate, 3-veined with the central vein hairy and the margins membranous and ciliate. Petals white, entire, $\frac{1}{2}$–$\frac{2}{3}$ as long as sepals. Stamens 10. Styles 3(–4). *Capsule almost globular*, shorter than the sepals, opening by 6(–8) revolute teeth. *Seeds* c. 1 mm across, blackish, shining, almost smooth with a *small laciniate appendage*. Fl. 5–6. ±Homogamous. Visited by small flies and beetles and automatically self-pollinated. $2n = 24$. Th.

Native. A woodland herb of well-drained nitrate-rich mull soils. Throughout Great Britain and Ireland; Islay, Jura and Raasay, but not in Outer Hebrides, Orkney or Shetland. Europe, W. Asia, Siberia.

19. ARENARIA L.

Annual to perennial herbs with slender branched ascending shoots and small usually *ovate or lanceolate*, rarely linear, opposite exstipulate *lvs*. Fls solitary terminal or in dichasial often few-fld cymes, usually 5-merous. Sepals free; *petals* white or pink, *entire* or slightly emarginate; stamens 10, rarely 8; ovary 1-celled; *styles* 3, rarely 4 or 5. *Capsule opening by twice as many teeth, or as many bifid teeth, as styles*. Seeds numerous, reniform.

About 160 spp., cosmopolitan but chiefly in temperate and Arctic regions of the northern hemisphere.

1 Perennial mat-forming herb with slender procumbent shoots bearing small broadly ovate to suborbicular lvs, 2–4 mm; fls up to 6 mm diam. with petals about twice as long as sepals. Garden-escape naturalized in several widely scattered localities.   **4. balearica**

Not as above. Annual to perennial herbs of varying habit with lvs 3–8 mm, broadly ovate or narrower; fls usually in 2–many-fld cymes, sometimes solitary; petals shorter than sepals or up to twice as long.   *2*

2 Lvs usually less than 3 times as long as wide, acute or acuminate; fls 3–6 mm diam., in dichasial cymes; petals shorter than sepals.   **1. serpyllifolia**

Lvs usually at least 3 times as long as wide, blunt; fls 9–16 mm diam., solitary or in few-fld cymes; petals at least equalling sepals, usually longer.   *3*

3 Lvs usually ciliate to the tip and distinctly veined, especially when dry; sepals strongly ciliate; petals 5–7.5 mm. Only in the Ben Bulben range, Co. Sligo, Ireland.   **2. ciliata**

Lvs glabrous or ciliate only in the basal third, obscurely veined; sepals glabrous or with a few marginal hairs near base; petals 4.5–5 mm. In a very few localities in Great Britain and one in Co. Clare, Ireland; Rhum; Shetland.   **3. norvegica**

1. A. serpyllifolia L.   Thyme-leaved Sandwort.

Incl. *A. leptoclados* (Reichenb.) Guss.

Slender annual herbs, rarely overwintering or perennial, often much-branched and bushy in habit with decumbent or ascending grey-green shoots 3–30 cm, rough with short stiff hairs, usually eglandular. Lvs 2.5–8 mm, broadly ovate to ovate-lanceolate, acuminate, subsessile or the upper shortly petiolate, ±roughly hairy and ciliate. Fls 3–8 mm diam. in many-fld cymes, 5 (rarely 4)-merous. Petals white, shorter than the 3–5-veined sepals. Stamens 10. Styles 3. Capsule conical to flask-shaped. Fl. 6–8. Homogamous. Nectar-secreting and occasionally visited by small insects but normally self-pollinated. Usually Th.

Variable and often regarded as comprising 2 or more spp. but these are not clearly distinct and are better treated as subspp. (as by Perring & Sell, *Watsonia*, 1967, **6**, 294; *Critical Supplement to the Atlas of the British Flora*, 1968, Nelson). The following key will usually serve for identifying British representatives:

1 Ripe capsule more than 3 mm long, ±swollen at base, its curving wall firm, fracturing if pressed. Pedicels stout, c. 0.5 mm diam., straight. Ripe seeds reniform, at least 0.5 × 0.4 mm.   *2*

Ripe capsule less than 3 mm long, not swollen at base and therefore straight-sided, its wall not firm and possible to indent without fracturing. Pedicels longer than sepals, slender, c. 0.3 mm diam., often upturning towards the tip. Ripe seeds 0.4 × 0.4 mm.   Subsp. **leptoclados**

2 Infl. diffuse, with pedicels longer than sepals. Capsule distinctly swollen at base and therefore flask-shaped, 1.5–2 mm wide. Ripe seeds 0.5 × 0.4 mm. Common throughout Britain Is. Subsp. **serpyllifolia**

Infl. usually very dense, with pedicels shorter than sepals. Capsule only slightly swollen at base but at least 2 mm wide. Ripe seeds at least

0.6 × 0.4 mm. Local and restricted to coastal habitats.   Subsp. **macrocarpa**

Subsp. **serpyllifolia** has fls commonly 5–8 mm diam. with ovate-lanceolate to ovate sepals, capsules more than 3 mm long, clearly flask-shaped and firm-walled, and ripe seeds 0.5 × 0.4 mm. $2n = 40$. The most widely distributed of the native subspp. and common on bare ground and roadsides, in arable fields, gardens and waste places and on walls and cliffs throughout the British Is. Abundant on bare soil round rabbit burrows, especially on chalk downs. Europe except the Arctic, temperate Asia; N. Africa; naturalized in North America, Australia and New Zealand.

Subsp. **macrocarpa** (Lloyd) Perring & Sell is usually distinguishable from subsp. *serpyllifolia* by its very compact infl. but more reliable features are the wider capsules and distinctly larger seeds. In Great Britain it is a local plant of coastal sands and walls, etc., southwards from N. Wales and N.W. Norfolk with outliers on the coasts of Fife, Angus and N. Uist; in Ireland known only from Co. Wexford; Guernsey. Coasts of W. Europe.

Subsp. **leptoclados** (Reichenb.) Nyman is typically more slender and diffuse than the two other subspp., with smaller fls only 3–5 mm diam. and narrower lanceolate sepals. The straight-sided conical and thin-walled capsule is less than 3 mm long and the ripe seeds are smaller than in the other subspp. It is cytologically distinct in having $2n = 20$. A less common plant than subsp. *serpyllifolia* but in similar habitats and often associated with it, though showing greater preference for calcareous soils. Throughout England and Wales but rare in N. England and Scotland though extending northwards to Sutherland; in several localities in E. Ireland but rare further west; Isle of Man; Channel Is. Mainly in W., C. and S. Europe; W. Asia.

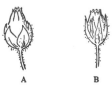

Fig. 32. Capsules of A, *Arenaria serpyllifolia* subsp. *serpyllifolia* and B, subsp. *leptoclados*. ×2.5.

2. A. ciliata L. subsp. ciliata   Irish Sandwort.

A low *perennial* herb with *prostrate hairy non-flowering shoots*, somewhat woody at the base, and ascending flowering shoots to 5 cm high. *Lvs* 4–6 mm, the *lateral veins usually visible in dry material. Fls* 12–16 mm *diam.*; cymes 1(–2)-fld. *Sepals* ovate-lanceolate with 3 prominent veins, *strongly ciliate on the margin and back.* Petals 5–7.5 mm, white, oblong, clawed, half as long again as sepals. Conspicuous *orange nectaries* at the base of the

5 outer stamens. Capsule ovoid, equalling or somewhat exceeding sepals, opening by 6 teeth. Seeds 0.95 mm, black, with low tubercles. Fl. 6–7; fr. 7–8. $2n = 40*$. Chh.

Native. Confined to limestone cliffs on the north side of the Ben Bulben range, Co. Sligo, Ireland, at 360–600 m. Ostenfeld & O. C. Dahl recognized the Irish plant as an endemic subsp. *hibernica*, distinguishable from their subsp. *pseudofrigida* of arctic Europe, but *Flora Europaea* 1 (1964) groups both with *A. tenella* Kit. (Alps and Carpathians) in subsp. *ciliata*. *A. ciliata* is very polymorphic, both morphologically and cytologically, and its poses many taxonomic problems.

3. A. norvegica Gunn.

Annual to perennial herbs 3–6 cm. *Lvs obovate, glabrous or ciliate only in the lower half, obscurely veined. Shoots and pedicels almost glabrous.* Cymes 1–2(–4)-fld. Sepals ovate-lanceolate, acute, 3-veined but the laterals often obscure, *glabrous or with a few cilia at the basal margin.* Petals 4–5.5 mm, white, ovate-oblong, clawed, exceeding sepals. *Nectaries green.* Seeds 0.8–1.0 mm, black, with low broad tubercles. $2n = 80*$.

Subsp. **norvegica**   Norwegian Sandwort.

Perennial, with slender tap-root and short basal internodes; stem branching to the base forming a compact tuft; flowering shoots to 6 cm high. Lvs 3–4.5(–6) mm, somewhat succulent, dark green. *Fls 9–10 mm diam.* Sepals glabrous or rarely with a few cilia at the basal margin. Petals 4–4.5 mm, broadly ovate, slightly exceeding sepals. Styles 3–5. Capsule-teeth 6–10. Fl. 6–7; fr. 7–8. Chh.

Native. A rare and local plant of base-rich screes and river shingle in Argyll, Westerness, Rhum, W. Sutherland and Shetland; the Burren, Co. Clare, Ireland. Norway, Sweden, Iceland.

Subsp. **anglica** Halliday   English Sandwort.

Winter annual or biennial with few non-flowering but many short erect flowering shoots to 5 cm high; laxer in habit and lighter green than subsp. *norvegica.* Lvs 4.5–5(–6) mm, narrowly obovate. *Fls 11–12 mm diam. Outer sepals slightly ciliate at base*, rarely quite glabrous. Petals 5–5.5 mm, oblong. Styles 3. Capsule-teeth 6. Fl. 5–10. Th.–Chh.

Native. An endemic subsp. confined to limestone depressions and tracks in upper Ribblesdale, Yorks. Quite distinct from *A. gothica* Fries ($2n = 100$), of Sweden and the Jura, in which it was formerly included.

*4. A. balearica L.   Balearic Pearlwort.

A perennial herb with slender stock and diffusely branching prostrate filiform ±hairy shoots, ascending only at the tips of flowering branches. *Lvs small, ovate or suborbicular, blunt, stalked, the stalks almost equalling the blade.* Fls 5-merous, solitary, terminal, on filiform stalks 6–10 times as long as the sepals. Sepals ovate, subacute. Petals white, obovate, longer than the

sepals, 5 or sometimes fewer. Capsule ovoid, drooping, hardly exceeding the sepals. Fl. 6–8. Chh.

Introduced. Often grown on rock-gardens, paved garden-paths, etc., and naturalized in many scattered localities. Balearic Is., Corsica, Sardinia, Tyrrhenian Is.

A. montana L., native in W. Europe southwards from N.W. France, is grown in gardens and occasionally escapes. It is a perennial grey-green herb with long prostrate non-flowering shoots and decumbent or ascending flowering shoots 10–30 cm, all shortly pubescent but eglandular. Lvs 1–2(–4) cm, *oblong-lanceolate to linear*, acuminate, puberulent. Fls 2 cm diam. or more with white petals at least twice as long as the sepals.

20. Spergula L.

Annual herbs, the linear blunt opposite lvs with *small deciduous free stipules* and *conspicuous axillary clusters of lvs.* Fls in loose terminal dichasia, 5-merous, hermaphrodite, their stalks after flowering at first deflexed then re-erected. Sepals 5, free; *petals* 5, *white, entire*; stamens 5–10; ovary 1-celled with 5 short styles. Fr. an ovoid capsule splitting deeply into 5 valves; seeds black, sharply keeled or winged.

About 5 spp. in the north temperate zone.

Stem weak, ascending; lvs furrowed beneath; seeds
  sharply keeled or with a very narrow wing.   **1. arvensis**
Stems usually stiffly erect from a decumbent base; lvs not
  furrowed beneath; seeds with a broad wing somewhat
  narrower than the diam. of the seed.   **2. morisonii**

1. S. arvensis L.   Corn Spurrey.

An annual herb with ascending geniculate stems 5–60 cm, branching close to the base; internodes weak, 3–12 cm. *Lvs appearing whorled*, 1–3(–6) cm, grass-green to grey-green, linear, fleshy, blunt, convex above and *channelled beneath*. Stem and lvs slightly to strongly viscid with glandular hairs. Fls 4–7 mm diam., in forked or umbel-like cymose panicles; stalks 1–2.5 cm; bracts small, ±scarious. Sepals 3–5 mm, ovate, obtuse, faintly veined, ±glandular, with a narrow scarious margin. Petals white, obovate, slightly longer than the sepals. Stamens (4–)5–10. Capsule ovoid-conical, up to twice as long as the sepals. Fr. stalks at first strongly deflexed from the base, then re-erecting. *Seeds 1–1.5(–2) mm across, blackish, papillose or not, with very narrow wing,* very rarely unwinged. Fl. 6–8. Homogamous and normally self-pollinated but visited occasionally by syrphids and some other insects. $2n = 18*$. Th.

Very variable. Plants with seeds bearing pale club-shaped papillae (*S. vulgaris* Boenn.) and those with minutely tubercled seeds lacking papillae (*S. sativa* Boenn.) have been regarded as specifically or at least varietally distinct, it being claimed that papillosity is associated with a lower level of hairiness and viscidity than in plants with tubercled but non-papillose seeds which tend to be densely hairy and therefore grey-green rather than grass-green in colour. In the British Is. the proportions of densely hairy plants and of plants with non-papillose seeds both increase towards the west and north, but there is no association between the two features in experimental crosses (New, *Ann. Bot., London,* N.S., 1959, **23**, 23–33; *Watsonia,* 1978, **12**, 137–43). There is also variation in seed-weight. Native British plants are referable to Zinger's *Spontaneae* with seeds weighing 0.3–0.4 mg, but the larger seeds of his *Cultae*, grown as fodder plants and found here as rare casuals, reach 0.7–1 mg. The *Linicolae* occurs as weeds of flax in continental Europe and have still larger seeds weighing 1.5–2 mg.

Native. A locally abundant and often troublesome calcifuge weed of arable fields and gardens and also found on roadsides and in other open habitats; reaches c. 450 m in Shropshire. Throughout the British Is.; Channel Is. (often as a dwarf variant, var. *nana* Linton), in short turf of cliff slopes and fixed dunes). Almost cosmopolitan.

2. S. morisonii Boreau

S. vernalis auct.; *S. pentandra* auct., non L.

An annual herb with a much-branched stem 10–20 cm, erect from a decumbent base, ±glandular above. *Lvs* 1–2 cm, linear-subulate, dark green, apparently in dense whorls, slightly fleshy, *not furrowed beneath*, ±glandular, often ±deflexed; stipules short. Infl. usually of many fls on long slender stalks, spreading in fr. Sepals ovate, shortly acuminate, ±glabrous, with scarious margins. Petals white, elliptical, subacute, contiguous, ±equalling the sepals. Stamens 5 or 10. Styles 5. Capsule slightly exceeding the calyx, opening by 5 valves. *Seeds* 1–1.6 mm across (including the wing), nearly circular, blackish-brown, *flattened*, *smooth* except for minute whitish marginal papillae, *with a scarious striate whitish or brownish wing*, whose breadth falls short of the diam. of the seed. Fl. 4–6. Probably self-pollinated. $2n = 18$. Th.

Probably introduced. Found in 1943 in Sussex, on sandy cultivated ground on the heathland between Crowborough and Tunbridge Wells. C. Europe from Spain, N. Italy and Hungary northwards to S. Scandinavia, Germany and Poland.

The closely related **S. pentandra** L. has laxer whorls of longer and less deflexed lvs, narrower non-contiguous petals which exceed the sepals, and black non-papillose seeds with a broader wing which equals the diameter of the actual seed. *S. arvensis* is larger and coarser with longer, furrowed lvs and subglobose seeds with the wing narrow or lacking.

21. SPERGULARIA (Pers.) J. & C. Presl

Annual to perennial often decumbent herbs whose opposite linear lvs have *pale scarious stipules* connate round the node. Fls 5-merous, in cymose infls. Sepals free; *petals* white or pink, *entire*, sometimes 0; stamens 5–10; ovary 1-celled; *styles* 3. Capsule opening by 3 valves; seeds spherical or pyriform.

About 20 spp., cosmopolitan, mostly halophytic.

1 Perennial salt-marsh herb with thick woody stock; sepals 4–6 mm; stamens 10, rarely 7–9; capsule 7–9 mm; seeds 0.7–1 mm (excl. wing), almost all broadly winged (?very rarely all unwinged).
        4. marginata
  Annual to perennial herbs, maritime or not; sepals at most 4.5 mm; capsule not exceeding 7 mm; seeds

not exceeding 0.7 mm, all unwinged or a few winged at base of capsule.   2

2 Perennial herb of maritime cliffs, rocks and walls, often glandular-hairy throughout; sepals 4–4.5 mm, equalled or slightly exceeded by the deep pink petals; stamens 10; seeds 0.5–0.7 mm unwinged.
        3. rupicola
  Annual to biennial herbs with prostrate or decumbent shoots, sometimes glandular at least above; sepals 2.5–4 mm, equalling or exceeding petals; seeds unwinged or a few broadly winged at base of capsule.   3

3 Annual salt-marsh herb with lvs acute but not or very shortly mucronate; stamens 1–5(–8); capsule 4–5(–6) mm, usually exceeding sepals; seeds 0.6–0.7 mm, unwinged or a few broadly winged.
        5. marina
  Annual to biennial herbs of coastal or inland habitats, not in salt-marshes; lvs distinctly awned; capsule 2–4 mm, about equalling sepals; seeds 0.3–0.6 mm, unwinged.   4

4 Common non-halophytic calcifuge of open sandy or gravelly habitats with conspicuous silvery stipules; pedicels longer than sepals; petals uniformly pink; stamens (5–)10.   **1. rubra**
  Rare and local coastal plant with stipules not silvery; pedicels usually shorter than sepals; petals pink with white base or entirely white; stamens usually 2–5.   **2. bocconii**

1. S. rubra (L.) J. & C. Presl   Sand-spurrey.

Arenaria rubra L.; *A. campestris* auct.; *Buda rubra* (L.) Dumort.

An annual or biennial herb with a slender tap-root and several decumbent branched ±hairy stems 5–25 cm, usually glandular above. *Lvs* 4–25 mm, linear, *awned*, not fleshy; *interpetiolar parts of stipular sheath lanceolate, acuminate*, usually torn at the tip, *silvery* and conspicuous. Fls 3–5 mm diam., in terminal rather few-fld cymes, their *stalks longer than the sepals*. Sepals 3–4 mm, ovate-lanceolate, usually glandular, with a broad scarious margin. Petals rose-coloured, paler at the base, ovate, shorter than the sepals. Stamens usually fewer than 10. Capsule about equalling the sepals. Fr.-stalks spreading or reflexed then becoming ±erect. Seeds 0.45–0.55 mm across, brownish, tubercled, unwinged, with a raised rim. Fl. 5–9. ±Homogamous. Visited by flies and automatically self-pollinated; the fls sometimes fail to open. $2n = 36$. Th.

Native. A common calcifuge plant of open sandy or gravelly habitats. Throughout most of Great Britain; Skye; very local in Ireland, and absent from the Outer Hebrides, Orkney and Shetland. Europe, N. Africa, Asia, North America. Introduced in Australia.

2. S. bocconii (Scheele) Ascherson & Graebner
        Boccone's Sand-spurrey.

S. campestris auct., non Ascherson; *S. atheniensis* Ascherson

An annual or biennial herb resembling *S. rubra* but with stems and lvs very glandular; lvs more shortly awned;

stipules broadly triangular, not silvery; *fls c.* 2 mm *diam.*, ±*numerous*, *the infl. branches lengthening to resemble* 1-*sided racemes*; *fl.-stalks usually shorter than the sepals.* Sepals 2–3.5 mm. Petals pale rose or whitish, shorter than the sepals. Seeds 0.35–0.45 mm across, pale greyish-brown, tubercled, unwinged. Fl. 5–9. Th.

Probably native. A rare and local plant of dry sandy and rocky coastal areas in Cornwall and ?Scilly Is.; introduced in Devon, Suffolk and Glamorgan. France, S.W. Europe and Mediterranean region.

3.   S. rupicola Lebel ex Le Jolis   Cliff Sand-spurrey.
S. rupestris Lebel, non Cambess.

A *perennial* herb with a stoutish branched ±woody stock and numerous decumbent shoots 5–15 cm. *Stems* ±*terete*, often dark purple, *densely glandular-hairy*. Lvs 5–15 × 1.5–2 mm, linear-acute, fleshy, ±flattened, with a horny tip prolonged into a short mucro, sparsely glandular-hairy especially near the base; *stipules* ovate-triangular, acuminate, somewhat *silvery*; the lvs of lateral shoots form axillary fascicles at each node. Infl. a terminal cyme becoming monochasial above, with up to 20 fls, each 8–10 mm diam. on a glandular-hairy stalk 7–8 mm. Sepals 4–4.5 mm, lanceolate, blunt, glandular-hairy, with white scarious margins. Petals equalling or slightly exceeding the sepals, deep pink. Stamens 10, with yellow anthers and filaments broadened below. *Capsule* 5–7 mm, slightly exceeding the erected sepals but shorter than its *stalk which reflexes strongly after flowering then gradually erects during ripening.* Seeds c. 0.6 mm, pyriform-triangular, black, minutely tubercled, with thickened borders along two sides. Fl. 6–9. Chh.

Native. Maritime cliffs, rocks and walls, local, and chiefly in the south and west. Coasts of Great Britain from Hants to Cornwall and up the W. coast to Ross; on the E. coast only in Norfolk, Mid and East Lothian and Aberdeen; Isle of Man, Inner and Outer Hebrides; coasts of Ireland. S.W. Europe from S. Italy and Spain to the N. coast of France.

4.   S. marginata (DC.) Kittel   Greater Sea-spurrey.
S. media (L.) C. Presl; *S. dillenii* Lebel

A *perennial* herb with a ±stout branching root-stock and many decumbent or geniculate-ascending, stout, flattened *shoots* to 30 cm or more, usually *glabrous* except in the inflorescence. Lvs 1–2.5 cm × 1–2 mm, linear, fleshy, horny-tipped, blunt to acute, often shortly mucronate, flat above and rounded beneath, usually glabrous; stipules broadly triangular, not silvery. Infl. lax, dichasial, its bracts becoming very small. Fls 7.5–12 mm diam. Sepals 4–6 mm, blunt, glabrous or hairy. *Petals equalling or somewhat exceeding the sepals*, ovate, blunt, whitish or pink above. Stamens (7–)10. Capsule 7–9 mm, much exceeding the persistent erect sepals and about equalling or exceeding its stalk. *Seeds* 0.7–1 mm (excl. wing), pale yellowish-brown, smooth, almost *all with a broad scarious wing*. Fl. 6–9. Protandrous and

visited by small flies; sometimes gynodioecious. $2n = 18$. G.–Chh.

Native. A component of coastal (but not inland) salt-marsh communities from the *Spartina* and *Salicornia* to the *Juncus maritimus* zone all round the British Is. but not in Scilly or the Channel Is.; also locally inland on the sides of major roads treated with rock-salt in winter. Coasts and inland salt-areas of the temperate zones of both hemispheres.

5.   S. marina (L.) Griseb.   Lesser Sea-spurrey.
S. salina J & C. Presl.

An annual herb with a slender tap-root and many prostrate or decumbent slender flattened shoots to 20 cm or more, subglabrous or ±glandular-hairy above. Lvs 1–2.3 cm × 1–2 mm, linear, very variable in stoutness and fleshiness, horny-tipped, ±acute, not or very shortly mucronate, flat above and rounded beneath, glabrous to glandular-hairy; stipules broadly triangular-acuminate, not silvery. Infl. dichasial at first but becoming ±monochasial, its bracts resembling the lvs or much smaller. Fls 6–8 mm diam. Sepals 3.5–4 mm, lanceolate, blunt, often pink-tinged, glabrous or glandular-hairy. *Petals* 2.5–3 mm, *shorter than sepals*, ovate, blunt, commonly pink or deep rose with white base. Stamens usually 4–8 or fewer. Capsule 4–5(–6) mm, somewhat exceeding the persistent sepals which are usually somewhat spreading at maturity, but distinctly shorter than its glabrous or glandular stalk. *Seeds* 0.8 mm (unbordered) to 1.5 mm (bordered) across, brownish, smooth, slightly rugose or strongly tubercled, *all unbordered or a small proportion of basal seeds with a broad scarious border* as in *S. media*. Fl. 6–8. Occasionally visited by small insects but usually self-pollinated. $2n = 36$. Th.

Native. In suitable habitats all round the coasts of the British Is., including Scilly and the Channel Is. A widespread but inconstant and often inconspicuous component of coastal salt-marsh communities from the *Salicornia* to the *Juncus maritimus* zone, sometimes mistaken for *S. marginata*; more prominent, and usually associated with *Puccinellia distans*, where very high salinity is developed in drying pans and depressions both in coastal marshes and in inland saline areas of Cheshire, Worcester, etc. It also occurs inland on the sides of major roads treated with rock-salt in winter and as a rare casual in waste places. Coasts and inland saline areas of the temperate zone of the northern hemisphere.

S. marina, though very variable, is typically a much less robust plant than *S. marginata* with more slender lvs and often with some glandular hairs, especially above. Its fls are conspicuously smaller and deeper pink than those of *S. marginata*, its capsules considerably shorter, its sepals less closely appressed to the ripe capsule, apart from the differences in the seeds. It extends further from the coast into brackish marshes than does *S. marginata*, and is sometimes found inland. Intermediates are sometimes encountered where the 2 spp. occur together, and these may be hybrids, but further investigation is required.

Telephium imperati L., a perennial dwarf shrub with procumbent to ascending glabrous shoots 15–40 cm, native on calcareous rocks and walls in S.W. Europe, is a rare casual in the British Is. It resembles the usually annual *Corrigiola litoralis* (p. 149) in having alternate glaucous stipulate lvs but these are obovate, the white fls are larger with sepals and petals 4–7 mm, and the fr. is a trigonous capsule with 15–20 seeds.

22. POLYCARPON L.

Small herbs with forking stems and obovate or oblong *opposite or whorled lvs with scarious stipules*. Fls small, in terminal dichasia, with scarious bracts; hermaphrodite, hypogynous, 5-merous. *Sepals keeled and hooded*; *petals narrow, shorter than the sepals*; stamens 3–5; ovary 1-celled; *styles trifid with 3 stigmas. Fr. a capsule with numerous seeds*, opening by 3 valves.

About 36 spp. in warm and temperate regions throughout the world.

1. P. tetraphyllum (L.) L. Four-leaved All-seed.
Mollugo tetraphylla L.

An annual herb with a slender tap-root and a much-branched slender erect or ascending shoot 5–15 cm, glabrous but with rough angles. *Lvs 8–13 mm, obovate*, narrowed into a stalk-like base, blunt, in opposite pairs, but the *lower pairs approximated so as to simulate whorls of 4*; stipules very small, narrowly triangular-acuminate, scarious. Infl. a much-branched dichasium, the fls 2–3 mm diam., short-stalked. Sepals with broad white scarious margins. Petals white, narrowly oblong, emarginate, soon falling. Stamens 3–5. Capsule ovoid, about equalling the calyx. Seeds brownish, finely papillose. Fl. 6–7. The homogamous fls are automatically self-pollinated and are often cleistogamous. Th.

Native. A very rare and decreasing plant of sandy habitats, roadsides and waste places near the coast in Cornwall, S. Devon and Dorset; common on sand dunes, roadside walls and in bulb-fields in Scilly and abundant in the Channel Is.; also casual in a few localities northwards to S. Scotland. Mediterranean region and W. Europe, but introduced elsewhere in Europe and widely in Asia, Africa, Australia and South America.

23. CORRIGIOLA L.

Annual to perennial glabrous herbs with decumbent shoots and *alternate obovate to linear glaucous stipulate lvs*. Fls in axillary clusters often further aggregated at the ends of the stems and branches; hermaphrodite, slightly perigynous, 5-merous. Sepals with white margins; petals equalling or exceeding the sepals; stamens 5; ovary 1-celled with 1 basal long-stalked ovule; stigmas 3, subsessile. Fr. an indehiscent ±trigonous 1-seeded nutlet enclosed in the calyx.

About 10 spp. in Europe, Africa, W. Asia and America.

1. C. litoralis L. Strapwort.

An annual, rarely biennial, *glaucous* herb with slender tap-root and several slender decumbent branching shoots 5–25 cm, often reddish. *Lvs 0.5–3 cm, linear-*

oblanceolate, blunt, entire, hardly fleshy, narrowed gradually into a stalk-like base; stipules small, scarious, whitish, half-sagittate, acuminate, denticulate. Fls very small, in crowded terminal and axillary bracteate cymes. Sepals c. 1 mm, blunt, green or red in the centre with broad white margins. Petals white or red-tipped, almost equalling the sepals. Anthers violet. Ovary surrounded at its base by the perigynous zone; styles small, sessile. Fr. 1–1.5 mm, obscurely trigonous. Seed papillose. Fl. 7–8. Homogamous. Automatically self-pollinated and often cleistogamous. $2n = 18^*$; 16. Th.

Native. Very local. On sandy and gravelly margins of 'leys' at Slapton Ley (S. Devon) and near Helston (Cornwall); Channel Is. Also adventive on railway tracks, etc. S.W. Europe northwards to Denmark; N. and E. Africa; W. Asia.

C. telephiifolia Pourret, native in the W. Mediterranean region and like **1** but perennial with narrowly obovate ±fleshy basal lvs, ebracteate infl. and fr. 1.5–2.5 mm, has been recorded as a casual on waste ground and railway sidings in Cornwall, Devon and Gloucester.

Paronychia polygonifolia (Vill.) DC., a spreading procumbent perennial native in mountains of S. Europe, W. Asia and N. Africa, is a not infrequent casual recognizable by the large silvery scarious stipules of its opposite lanceolate lvs and the similarly conspicuous silvery scarious bracts of its axillary fl.-clusters. The fls are apetalous with 5 stamens and 2 styles or stigmas, and the fr. is an achene. Other spp. also occur as rare casuals.

24. HERNIARIA L.

Annual or perennial mat-forming herbs with opposite stipulate lvs, the upper lvs often alternate through abortion of one member of a pair. *Fls in dense axillary clusters*, hermaphrodite or unisexual, *perigynous*, 5-merous; bracteoles scarious. Sepals 5, on the the edge of the bowl-shaped perigynous zone; petals 5, subulate, shorter than the sepals; ovary free at the base of the perigynous zone which closely surrounds its lower half, 1-celled with 1 basal long-stalked ovule; styles 2-branched with 2 stigmas. Fr. an indehiscent nutlet with 1 shining black seed.

About 15 spp. in Europe, N. and S. Africa and Asia.

1 Calyx ±glabrous; plants green. 2
  Calyx with spreading hairs; plant grey or whitish with hairs. 3
2 Stipules ovate, often greenish; fl-clusters ±confluent on short lateral spike-like branches; fr. acute, considerably exceeding the calyx. **1. glabra**
  Stipules broadly ovate-acuminate, white; fl.-clusters roundish, distinct; fr. blunt, little exceeding the calyx. **2. ciliolata**
3 Plant prostrate; lvs lanceolate; lower lvs opposite; sepals hair-pointed. **3. hirsuta**
  Branches ascending; lvs ovate-oblong, mostly alternate; sepals not hair-pointed. **4. cinerea**

1. H. glabra L. Glabrous Rupture-wort.

An annual or biennial, rarely perennial herb with tap-root and numerous prostrate *shoots 6–15(–30) cm*, rarely slightly woody at the base, *glabrous or slightly hairy all round*, green, with short alternate branchlets. Lvs 3–7 mm, elliptical-obovate, ±acute, narrowing to the base, glabrous or ciliate; stipules small, ovate, shortly fringed, often greenish; upper lvs alternate. *Fls* c. 2 mm diam. almost sessile, up to 10 in axillary clusters *on short lateral branches, the clusters ±confluent into oblong spikes. Sepals* c. 6 mm, *ovate, blunt, ±glabrous.* Petals minute, white. Stigmas slightly divergent. *Fr.* (Fig. 33B) *acute, distinctly exceeding the sepals.* Seeds

Fig. 33. Fruits of *Herniaria*. A, *H. ciliolata*; B, *H. glabra*; C, *H. cinerea*. ×5.

c. 6 mm, lenticular, shining, red then black. Fl. 7. Homogamous. Visited by various tiny insects and automatically self-pollinated. $2n = 18^*$. Th.–H.

Native. A very local and decreasing plant of dry sandy places in Cambridge, Suffolk, Norfolk and S. Lincoln and a rare casual elsewhere. Europe northwards to S. Scandinavia; N. Africa; Asia.

2. H. ciliolata Melderis   Ciliate Rupture-wort.

H. glabra var. *scabrescens* Roemer; var. *setulosa* Beck; *H. ciliata* Bab., non Clairv.

A perennial evergreen dwarf shrub with stout erect branched woody stock and numerous spreading prostrate *shoots* 5–20(–40) cm, woody and rooting at the base, usually *hairy only on the upper side*, each with short alternate ascending branches. *Lvs* 2–6(–10) mm, broadly ovate to ovate-elliptical, usually *ciliate*, the upper lvs often alternate though less generally than in *H. glabra*; stipules ovate-acuminate, silvery-white, fringed, conspicuous. *Fls* c. 2 mm diam., almost sessile in *distinct roundish axillary clusters* on the lateral branchlets. *Sepals* ±ciliate, *bristle-tipped*, but otherwise glabrous. Petals minute. Stigmas strongly divergent. Fr. (Fig. 33A) obtuse, hardly exceeding the sepals. Seeds c. 0.7 mm, lenticular, black, shining. Fl. 7–8. Pollination as in *H. glabra*. $2n = 72^*$. Ch.

Var. *angustifolia* (Pugsley) Melderis has narrowly elliptical lvs and is hairy all round the stem.

Native. A very rare plant of maritime sands and rocks; var. *ciliolata* known only from Lizard Point (Cornwall) and the Channel Is., var. *angustifolia* in Jersey. Coasts of W. Europe from Spain and Portugal to N. Germany.

*3. H. hirsuta L.   Hairy Rupture-wort.

An annual to perennial herb resembling *H. glabra* but with the wholly prostrate *shoots covered with dense straight spreading hairs*. Lvs ±lanceolate, the lower opposite. Fls in distinct roundish axillary clusters. *Calyx burr-like* with straight spreading hairs, and each *sepal ending in a long bristle*. Fl. 7–8. Pollination as in *H. glabra*. $2n = 36^*$. Th.–Ch.

Introduced. Naturalized on sandy ground at Christchurch, Hants, casual elsewhere. C. and S. Europe, Asia, Africa.

*4. H. cinerea DC., very like *H. hirsuta* but *ashy-grey* with most of the ovate-oblong lvs alternate and the hairy *sepals not ending in a conspicuously long bristle*, is naturalized in waste places at Burton-on-Trent, Stafford. S. Europe, N. Africa, W. Asia. Intermediates between **3** and **4** are reported.

25. ILLECEBRUM L.

A small herb with opposite stipulate lvs and *white fls in axillary cymose clusters.* Fls hermaphrodite or unisexual, 5-merous, with 2 ±sessile stigmas and a 1-celled ovary with 1 ovule. *Fr.* 1-*seeded, dehiscent.*

One species.

1. I. verticillatum L.   Illecebrum.

An annual glabrous herb with a slender tap-root and many slender spreading decumbent branches 5–20 cm, rooting at the basal nodes, often reddish. Lvs 2–6 mm, obovate, blunt, entire; stipules ovate, scarious. Fls 4–5 mm diam., 4–6 in each axillary cluster, the 2 clusters at a node forming a shining white whorl. Bracteoles scarious, silvery. Sepals 5, 2–2.5 mm, shining white thick and spongy, hooded, with a fine awn on the dorsal side just behind the incurved tip. Petals 5, white, filamentous, shorter than the sepals. Stamens 5, opposite the sepals, short, with roundish anthers. Styles very short, with 2 stigmas. Fr. a 1-seeded capsule enclosed by the persistent erect sepals and opening below by 5(–10) valves which remain cohering above. Seeds 0.8–1 mm, brown. Fl. 7–9. Homogamous. Automatically self-pollinated and sometimes cleistogamous. Th.

Native. A very local and decreasing plant of moist sandy places in Cornwall and Hants; casual in a few widely scattered localities. W. and C. Europe from Spain to Denmark, C. Germany, Bohemia and N. Italy; Canary Is.

The fruiting sepals resemble the carpels of *Sedum*.

26. SCLERANTHUS L.

Annual to perennial herbs with diffusely branched stems and *opposite subulate slightly connate exstipulate lvs.* Fls very small, green or whitish, in dense terminal and axillary cymes; hermaphrodite, perigynous, (4–)5-merous. Sepals usually 5, inserted on the rim of the urceolate perigynous zone; *petals* 0; stamens 1–10; ovary with its apex hardly reaching the insertion-level of the sepals, 1-celled with 1(–2) basal long-stalked ovules; styles and stigmas 2. *Fr. an indehiscent* 1-*seeded nutlet* enclosed by the hardened wall of the perigynous zone and the persistent sepals, which are shed with it; seed lenticular, smooth.

Perhaps as many as 150 spp. in Europe, Asia, Africa and Australia.

Annual or biennial; sepals acute with narrow scarious margins, suberect in fr.   **1. annuus**

Perennial, woody below; sepals blunt with broad white margins, incurved in fr. **2. perennis**

1. S. annuus L. Annual Knawel.

An *annual or biennial* herb with slender tap-root and one or more branched decumbent or ascending glabrous or shortly hairy stems 2.5–25 cm. Lvs 5–15(–20) mm, subulate, acute, usually ciliate, the members of a pair slightly connate by their narrow scarious margins. Fls c. 4 mm, subsessile, solitary in the forks of the stem and in terminal and axillary clusters. Bracts usually longer than the fls. *Sepals triangular, ±acute, glabrous, narrowly scarious-margined.* Stamens 10 or fewer, much shorter than the sepals. *Sepals suberect* or slightly incurved *in fr.*, the ±glabrous perigynous *tube below them becoming ±deeply 10-furrowed* (Fig. 34A). Fl. 6–8.

Fig. 34. Fruits of A, *Scleranthus annuus* and B, *S. perennis*. ×2.5.

Homogamous. Secretes a little nectar and is visited by a few insects: automatically self-pollinated. $2n = 22, 44$. Th.

Very variable in height, mode and extent of branching, size and orientation of leaves, width of scarious margin of sepals, depth of furrowing of perigynous tube, etc. It seems very doubtful whether distinct overwintering (var. *hibernus* Reichenb.) or biennial (var. *biennis* Fries) plants merit taxonomic recognition.

S. polycarpos L., a slender erect or decumbent plant, annual or biennial, with the short sepals (2 mm) not or scarcely white-bordered and ±connivent in fr., has been reported, but it has not yet been convincingly demonstrated that the British plants are not variants of *S. annuus*.

Native. In dry sandy and gravelly places and in cultivated and waste ground on sandy soil. Throughout Great Britain, but not in the Outer Hebrides, Orkney or Shetland; local in Ireland; Channel Is. Europe, N. Africa, Asia. Introduced in North America. Calcifuge.

2. S. perennis L. Perennial Knawel.

A *perennial* herb closely resembling *S. annuus* but somewhat *woody below* and usually more robust and more glaucous, and becoming reddish. *Lvs glaucous*, ±ciliate at the base, *often curved to one side of the stem* and with axillary lf-clusters. Bracts shorter than the fls. *Sepals oblong obtuse with broad white margins, incurved over the ripe fr.*, their tips in contact; *perigynous tube hairy* with 10 *shallow furrows* in fr. (Fig. 34B). Stamens 10. Fl. 6–8. Homogamous. The more conspicuous fls which yield more nectar than those of *S. annuus* are visited by many small flies and automatically self-pollinated. $2n = 44^*$. Ch.

Native. A rare plant of arable fields and other open ground and of short semi-closed vegetation on dry non-calcareous sand in W. Norfolk and W. Suffolk, and on dolerite rocks near Radnor. The Radnor plant, with ascending to erect stems up to 22 cm and distinctly ciliate lvs 5–9 mm, is referred to the widespread subsp. **perennis,** but the East Anglian plant appears to be restricted to the British Is. and has been named subsp. **prostratus** P. D. Sell. It has ±procumbent stems, glabrous or only slightly ciliate lvs 3–5(–7) mm, and fr. usually 2–3 mm and thus shorter than in subsp. *perennis* (3.5–4.5 mm). Most of Europe northwards to c. 66° N in Sweden and Finland; W. Asia.

38. PORTULACACEAE

Annual to perennial herbs, sometimes suffruticose, usually glabrous, often ±fleshy. Lvs alternate or opposite and decussate and then sometimes connate, simple, entire; stipules scarious, bristle-like or 0. Fls hermaphrodite, actinomorphic. Sepals usually 2, free or united below, anterior-posterior; petals 2–6(–18), free or united below, usually fugacious, sometimes minute or 0; stamens (1–)3–20 or rarely more, opposite the petals when equalling them in number, hypogynous or half-epigynous; ovary superior or half-inferior (*Portulaca*), usually 1-celled, with 1–many campylotropous ovules on a basal placenta; style simple or with 2 or more branches. Fr. a capsule opening by valves or transversely (*Portulaca*); seeds 1–many, with a curved embryo surrounding mealy perisperm.

About 200 spp. in 17 genera, chiefly in temperate and subtropical regions and especialy in W. America.

Distinguished from Caryophyllaceae by the calyx of only 2 sepals.

1 Fls yellow, sessile; sepals deciduous; stamens 6–15; capsule opening transversely; seed numerous.
　　　　　　　　　　　　　　　　3. PORTULACA
　Fls white or pink, stalked; sepals persistent; stamens 3–5; capsule opening by valves; seeds usually 3.　2
2 Small aquatic or subaquatic herbs with several pairs of stem-lvs.　1. MONTIA
　Herbs with 1 pair of stem-lvs, the members of a pair free or broadly connate.　3
3 At least the later-formed basal lvs with broadly elliptical or ovate blades.　1. MONTIA
　All basal lvs with linear or linear-oblanceolate blades.
　　　　　　　　　　　　　　　　2. CLAYTONIA

1. MONTIA L.

Annual to perennial glabrous herbs, *fibrous-rooted, rhizomatous or stoloniferous, never with corms*; stem-lvs in 1 or more opposite and decussate pairs; stipules 0. Fls in terminal infls which may appear lateral through overtopping by a branch. Sepals persistent. Petals (3–)5, sometimes unequal, free or united into a short tube. Stamens 3–5, adhering to the base of the petals. Ovary superior; styles 3-cleft. *Fr. a 3-valved capsule* with 1–3(–6) seeds.

About 40 spp., chiefly in America.

1 Stem-lvs many, narrowed into a short stalk-like base; fls 2–3 mm diam.; stamens 3.　　　　**1. fontana**
Basal lvs long-stalked; stem-lvs 2, sessile or broadly connate; fls more than 4 mm diam.; stamens 5.　　　*2*
2 Stem-lvs broadly connate; fls 5–8 mm, diam.; petals white, entire or slightly notched.　　**2. perfoliata**
Stem-lvs sessile but not broadly connate; fls 15–20 mm diam.; petals pink or white, bifid.　　**3. sibirica**

1. M. fontana L.　　　　　　　　　　Blinks.

Annual to perennial herb with branching shoots 2–50 cm, short and erect in land forms, weaker, decumbent and rooting below in aquatic forms (which may have non-flowering shoots and behave as biennials or perennials); sometimes floating. *Stem-lvs in few to many pairs*, 2–20 × 1.5–6 mm, narrowly spathulate to obovate, *narrowed below into a stalk-like base*, the members of a pair free or somewhat connate for a short distance. Fls 2–3 mm diam., inconspicuous, in terminal cymes which may be displaced to an apparently lateral position in vigorously growing shoots. *Petals 5*, white, *unequal, joined into a short tube which is cleft to the base in front. Stamens usually* 3, opposite the 3 smaller petals and adhering to their bases. Capsule 1.5–2 mm, exceeding the calyx, globose. Seeds 3, dark brown or black. Fl. 5–10. Homogamous; cleistogamous in dull weather; little visited by insects. $2n = 18$. Th.–Hel.–Hyd.

Native. Streamsides, springs, flushes, wet places among rocks, moist pastures, etc., especially on noncalcareous substrata; also in arable fields in S.W. England; to c. 1080 m in Scotland. Common in one or more of its forms throughout the British Is. Probably in temperate regions throughout the world.

M. fontana has often been divided into 2 or more spp. or subspp. on the basis of differences chiefly in vegetative and seed-coat characters. The following is based closely on the treatment by S. M. Walters (*Watsonia*, 1953, **3**, 1–6), now generally accepted as the most satisfactory. All 4 subspp. have been recorded from Ireland but their detailed distribution there is not yet known.

Subsp. **fontana** (*M. lamprosperma* Cham.; *M. rivularis* auct., ?C. C. Gmelin)

Plant variable in habit but usually loosely tufted and often submerged, green or yellow-green. Ripe *seed* 1.1–1.35 mm, *smooth and shining* as seen under a lens; under higher magnification the individual cells of the coat easily seen as a reticulate pattern, irregular in size and shape, those near the keel elongated and in rows; seed-coat thin and brittle. In trickles of water or very wet places on acid soil or rock. Common in Scotland and N. England southwards to Cheshire and E. Yorks, and in N. and W. Wales; Inner and Outer Hebrides, Orkney and Shetland. Circumpolar Arctic and north

temperate zones; also in the southern hemisphere into the sub-Antarctic.

Subsp. **chondrosperma** (Fenzl) S. M. Walters (*M. verna* auct.; *M. minor* auct.)

Plant usually annual, tufted, yellowish-green, with short erect branches terminating in cymes. Rarely submerged and then looser and longer-lived. Ripe *seed* 1.0–1.2 mm, *dull and entirely covered with rather coarse tubercles*; individual cells more or less hexagonal. On light acid soils, usually sandy or gravelly, with a high water-table at least in spring. The common, and in some areas the only, subsp. in southern England; rarer in N. England and Scotland but reaching Caithness and Orkney; Scilly; Channel Is. C. and S. Europe; North and South America; Australia, New Zeland.

Subsp. **amporitana** Sennen (*M. lusitanica* Sampaio; *M. limosa* Decker; *M. fontana* subsp. *intermedia* (Beeby) Walters)

Plant usually loose in habit, often more or less aquatic with long bright-green trailing branches bearing apparently axillary cymes only. Ripe *seed* 0.85–1.1 mm, *finely tuberculate at the edge and rather shiny* under the lens; under higher magnification (2–)3–4(5) rows of cells on each side of the keel are seen to bear small but relatively narrow and high tubercles; cells rather elongated. In trickles of water or very wet places on acid soil or rock. Common in W. England and Wales but extending to the south and east coast and northwards to Bute and Perth. W., C. and S. Europe; W. North America; Australia, New Zealand.

Subsp. **variabilis** S. M. Walters (*M. rivularis* auct. mult., ?C. C. Gmelin)

Plant usually loose in habit, like subsp. *amporitana*. Ripe *seed* 0.9–1.1 mm, *more or less smooth* under the lens *but not as shining as in subsp. fontana*; under higher magnification showing variable development of small usually very low tubercles along the keel. In similar habitats to subsp. *amporitana*. Locally common in N. England and Wales but in scattered localities southwards to Surrey and the south-west and northwards to the Outer Hebrides, Sutherland, Orkney and Fair Isle; in some areas the commonest or only subsp. W. and C. Europe; W. North America. Fertile intermediates between these subspp. have been reported (see maps in the *Critical Supplement to the Atlas of the British Flora*, Ed. F. H. Perring, Bot. Soc. Br. Is., 1968).

*2. M. perfoliata (Willd.) Howell (*Claytonia perfoliata* Donn ex Willd.)

Annual glabrous herb with erect or ascending flowering stems 10–30 cm. Basal lvs very long-stalked, their blades 1–2.5 cm, elliptical to ovate-rhombic, entire, rather fleshy, faintly veined; *stem-lvs 2, opposite, broadly connate* to form a concave suborbicular involucre beneath the terminal raceme-like infl. which often has 1 or more separate fls at its base. Fls 5–8 mm diam., on stalks about twice as long as the broadly ovate sepals. *Petals* 5, 2–3 mm, somewhat exceeding the sepals, entire or slightly notched, white, ±equal, joined into a short complete basal tube. Stamens 5, opposite the petals. Capsule shorter than the sepals, subglobose. Seeds c. 2 mm, 1(–3) per capsule, black, shining. Fl. 5–7. Visited by small insects. $2n = 18*$. Th.

Introduced. Cultivated, disturbed and waste ground, especially on light sandy soils. Scattered throughout Great Britain, and locally abundant, northwards to Aberdeen and Sutherland; Channel Is.; very local in Ireland. Pacific North America from Alaska to Mexico; Cuba.

*3. M. sibirica (L.) Howell (*Claytonia sibirica* L.; *C. alsinoides* Sims)

Annual to perennial glabrous herb with erect or ascending flowering stems 15–40 cm. Basal lvs very long-stalked, their blades 1–3 cm, ovate-acuminate, entire, rather fleshy, distinctly veined; *stem-lvs 2, opposite, sessile but not connate*. Infl. a terminal bracteate raceme-like cyme. Fls 15–20 mm diam., on stalks 2–3 cm. Sepals 4 mm, broadly ovate, blunt. *Petals* 5, 8–10 mm, clawed, white or pink *with dark veins, deeply notched or bifid*, ±equal, joined into a very short complete basal tube. Stamens 5, opposite the petals. Capsule about as long as the sepals, ovoid. Seeds c. 2.5 mm. Fl. 4–7. Protandrous and nectar-secreting; visited by flies and other insects. Th.–H.

Introduced. Damp woods, shaded streamsides, etc., especially on acid sandy soils. Scattered throughout much of Great Britain from Cornwall to Kent and northwards to Sutherland and Caithness but most frequent in Devon and from Stafford and Derby northwards to Aberdeen and very local elsewhere in S. and C. England and in Wales; absent from East Anglia and Lincoln; Jersey; Inner Hebrides, Orkney; very rare and casual in Ireland. E. Siberia; Pacific North America from Alaska to S. California.

2. CLAYTONIA L.

Perennial glabrous herbs with basal lvs and flowering stems from *deep-seated globose tubers* (*corms*). Flowering stems with 1 pair of opposite lvs and a terminal raceme or umbel of showy fls. Sepals ovate, persistent. Petals 5, free, equal. Stamens 5, opposite the petals and adherent to their clawed bases. Ovary superior, 1-celled; style 3-cleft. Fr. a capsule opening by 3 valves; seeds 3–6. About 160 spp. in America, Asia, Australia and New Zealand.

*C. virginica L., with 2–40 flowering stems from a corm 1–5 cm diam., linear to linear-oblanceolate basal lvs 4–15 cm, and racemes of fls with rose-coloured dark-veined petals which are usually longer than the blunt sepals, is grown in gardens and is established locally. Eastern North America from Quebec to Texas.

Spp. of *Calandrinia*, Rock Purslane, from western North and South America, are grown as rock-garden or edging plants for their attractive reddish or white fls. The capsule is dehiscent by 3 valves but differs from that of *Claytonia* in having numerous seeds. The annual *C. ciliata* DC. has been found as a casual in the Channel Is. and elsewhere.

3. PORTULACA L.

Fleshy low-growing herbs with spirally arranged or ±opposite lvs, the uppermost forming a kind of involucre below the fls; stipules scarious, or reduced to small bristles. Fls yellow or red, solitary terminal or in cymose infl. Sepals 2, the anterior larger and overlapping the posterior; petals 4–6, free or united basally, deliquescent after flowering; *stamens 4–many*, their filaments often hairy below. Style ±deeply divided into 3–8 branches. Ovary ½-inferior. *Capsule* 1-celled with a thin membranous wall, *opening by a transverse lid; seeds numerous*, on a free-central placenta which often has 3–8 branches.

About 20 spp. chiefly in tropical and subtropical America but with some in the Old World and some cosmopolitan.

*1. P. oleracea L.

An annual glabrous fleshy herb with prostrate or ascending branching stems 10–30 cm. Lvs 1–2 cm, spirally arranged or subopposite, the uppermost crowded beneath the fls, obovate-oblong with a cuneate sessile base, blunt, fleshy, shining; stipules often reduced to bristles. Fls 8–12 mm diam., 1(–3), terminal or at the forkings of the stem. Sepals with blunt hooded tips, falling in fr. *Petals* 4–6, ±free, *yellow*, soon falling. Stamens 6–15. Style with 3–6 branches. Ovary half-inferior. Capsule 3–7 mm. Seeds 0.7 mm, brownish-black, bluntly tubercled, shining. Fl. 6–9. Homogamous; no nectar; visited by some small insects but opening only on sunny mornings and probably often self-pollinated. $2n = 54$. Th.

Like *Portulaca* in having the capsule transversely dehiscent (though closer to its base) but differing in the 2 or more *free sepals* and *more numerous* (4–18) pink or white *petals*, are the fleshy-lvd perennial spp. of *Lewisia*, native in western North America and often grown on rock-gardens and walls.

39. AIZOACEAE

Herbs or shrubs, often fleshy. Lvs usually opposite. Fls usually hermaphrodite, actinomorphic. Calyx-tube free or adnate to ovary; lobes (1–)5–8, herbaceous and often fleshy. Petals numerous, inserted in the calyx-tube in 1 or more series, sometimes 0. Ovary usually inferior.

About 130 genera and 1200 spp., mainly in S. Africa.

Lvs 8–10 cm; stigmas 8–20.　　　　1. CARPOBROTUS
Lvs up to 2.5 cm; stigmas 5 or fewer.　　2. DISPHYMA

1. CARPOBROTUS N.E.Br.

Perennials. Stems trailing, branched. Lvs distinct, opposite, equal, 3-angled. Fls showy; stigma 8–20; ovules borne on *placentae on the outer wall or floor of the cell*. *Fr. indehiscent, fleshy, edible*.

About 24 spp., in S. Africa, Australia and temperate South America.

Lvs glaucous, broadest at or above the middle; stamens purple.　　　　　　　　.2. **acinaciformis**
Lvs not glaucous, not broadening above the base; stamens yellow.　　　　　　　　1. **edulis**

*1. C. edulis (L.) N.E.Br.　　　　Hottentot Fig.

Mesembryanthemum edule L.

A perennial herb. *Stems trailing*, woody, angled. *Lvs*

8–10 cm, not glaucous, narrow, upwardly curved, *triangular in section* and serrulate on the keel, *fleshy*, opposite and connate at base. Peduncle c. 3 cm, swollen upwards. *Fls* c. 9 cm across, solitary, *magenta* or (less frequently) yellow. Calyx-tube adnate to ovary; lobes 5, unequal, ±lf-like. Petals numerous, linear. *Stamens numerous, yellow.* Fr. fleshy and edible. Fl. 5–7. $2n = 18$. Ch.

Introduced. Naturalized and locally abundant on cliffs and banks by the sea. Cornwall, Devon, Dublin. Native of S. Africa but naturalized in many of the warmer temperate regions.

*2. C. acinaciformis (L.) L. Bolus

Like **1** but lvs glaucous, broadest at or above the middle, abruptly contracted to the acute apex; fls purple; stamens purple.

Naturalized on dunes in N. Devon and perhaps elsewhere. Native in S. Africa.

2. DISPHYMA N.E.Br.

Lvs fleshy, flat above, rounded or somewhat keeled beneath. Stigmas 5 or fewer; fr. a capsule, opening by 5 valves.

*1. D. crassifolium (L.) L. Bolus

A perennial with procumbent stems c. 20 cm, woody and rooting below. Lvs up to 2.5 cm, flat above, rounded or weakly keeled beneath. Fls c. 2 cm diam., pinkish-purple. Fr. woody; seeds not embedded in mucilage.

Naturalized on cliffs in W. Cornwall and Scilly. Native in S. Africa.

Several other members of this family are found as garden-escapes, particularly in W. Cornwall and Scilly, but it is not at present clear if any of these will persist.

40. AMARANTHACEAE

Annual or rarely perennial herbs. Lvs opposite or alternate, entire, exstipulate. Fls hermaphrodite or unisexual. Perianth of 3–5 segments, free or connate at base, *dry and scarious, often brightly coloured.* Stamens 3–5, opposite the per.-segs. Ovary unilocular; ovules 1–several, basal. Fr. dry, membranous, indehiscent or circumscissile.

About 65 genera and 850 spp., mainly in tropical and warm temperate regions.

1. AMARANTHUS L.

Infl. cymose, the small cymes forming short dense axillary spikes, or the upper spikes making a large, dense, terminal, lfless infl. Fls bracteolate, mostly unisexual. Per.-segs 3 or 5. Stamens free, usually as many as the per.-segs. Stigmas 2–3. Fr. 1-seeded, splitting transversely or indehiscent.

About 60 spp. in tropical and temperate regions.

More than 20 spp. have been found in the British Is., but none has become established. The following 4 spp. seem to occur most frequently. For a comprehensive key see J. P. M. Brenan, *Watsonia*, **4**, 261–280 (1961).

1 Fr. circumscissile.		2
Fr. indehiscent.	**2. bouchonii**	
2 Per.-segs 3.	**4. albus**	
Per.-segs (4–)5.		3
3 Per.-segs of female fls lanceolate to ovate-lanceolate, acute; stems not densely lanate.	**1. hybridus**	
Per.-segs of female fls widened towards the apex, ±spathulate, obtuse to truncate; stems densely and shortly lanate.	**3. retroflexus**	

*1. A. hybridus L.   Green Amaranth.

A glabrous or sparsely pubescent annual 20–100 cm. Lvs rhombic-ovate. Infl. elongate-spicate, with long branches. Bracteoles (2–)4–6 mm, ovate, with a long mucro, about twice as long as the perianth. Per.-segs (4–)5, narrowly ovate, usually acute. Fr. circumscissile. Fl. 7–9. $2n = 32$. Th.

Introduced. A casual on rubbish tips, etc. Native of tropical and subtropical America.

*2. A. bouchonii Thell.

Like **1** but bracteoles linear-lanceolate, with a long spinous apex. Per.-segs elliptic-lanceolate to linear. Fr. indehiscent. Fl. 7–9. Th.

Introduced. A casual on rubbish tips, etc. Origin unknown.

*3. A. retroflexus L.   Common Amaranth.

A rather stout somewhat pubescent grey-green annual 15–100 cm. *Stem* erect or with spreading branches, rough and *shortly lanate*. Lvs up to 15 cm, ovate or ovate-oblong, obtuse, mucronulate; petiole long, rough. *Infl.* much-branched, forming dense, stout spikes, *the upper part nearly or quite lfless.* Bracteoles 3–6 mm, equalling or exceeding the per.-segs, stiff, acuminate or aristate. *Fls usually 5-merous.* Per.-segs of female fls 2–3 mm, spathulate. Fr. c. 1.5 mm, suborbicular, compressed. Fl. 7–9. $2n = 32, 34$. Th.

Introduced. A casual of cultivated land and waste places, rare and impermanent. North America.

*4. A. albus L.

Differs from *A. retroflexus* as follows: nearly or quite glabrous. Lvs obovate, oblong or spathulate, mucronate. Infl. of small axillary spikes, lfy to top. Fls 3-merous. Per.-segs of female fls 3, narrowly elliptical.

Introduced. A rare and impermanent casual. North America.

41. CHENOPODIACEAE

Annual or perennial herbs or shrubs, rarely arborescent, frequently ±succulent or with bladder-like hairs which give the plant a 'mealy' appearance. Lvs usually alternate, simple, exstipulate. Fls often bracteolate, small and greenish, hermaphrodite or unisexual, usually actinomorphic. Perianth 3–5-lobed, rarely 0 in female fls, persistent, often accrescent in fr. Stamens usually the same number as the per.-segs, usually free; anthers 2-celled, opening lengthwise. Ovary superior or half-inferior, 1-celled; stigmas 2–3, rarely 1; ovule solitary, basal. Fr. an achene or occasionally with circumscissile dehiscence. Perisperm present or 0. Embryo curved round the outside of the perisperm.

About 100 genera and 1400 spp. Cosmopolitan, usually halophytic and mainly in arid regions.

1 Lvs flattened, neither subterete (or ½-terete) and succulent, nor plant apparently lfless but with succulent stems.　2

　Lvs subterete (or ½-terete) and succulent or plant apparently lfless but stems succulent.　5

2 Fls mostly hermaphrodite; fr. ±surrounded by 2–5 persistent per.-segs.　3

　Fls all unisexual; fr. enclosed between 2 ±vertical appressed bracteoles.　4

3 Lower lvs usually toothed or lobed, if entire triangular and ±hastate or cordate at base or else not more than 5 cm; per.-segs neither conspicuously thickened at base in fr. nor adhering in groups of 2–4.　　　1. CHENOPODIUM

　Lower lvs entire, ovate, often ±cuneate at base, some at least more than 5 cm; per.-segs conspicuously thickened at base in fr. and adhering in groups of 2–4.　　　2. BETA

4 Lvs toothed or if entire not elliptic; bracteoles not united above the middle; annual herbs.　3. ATRIPLEX

　Lvs entire, elliptic or nearly so; bracteoles united above the middle; small shrub or annual herb with long-pedicellate fr. (the latter probably extinct).　　　4. HALIMIONE

5 Lvs alternate, spreading.　7

　Lvs opposite, appressed to stem and fused along the margins.　6

6 Annual; stems not rooting at the nodes; fls solitary or arranged in a triangle, the middle fl. the largest and not completely separating the laterals.　　　7. SALICORNIA

　Perennial; stems often rooting at the nodes; fls ±equal in height, arranged in a row, the middle fl. completely separating the laterals.　8. SARCOCORNIA

7 Lvs acute or obtuse, not spinescent at apex.　　　5. SUAEDA

　Lvs spinescent at apex.　6. SALSOLA

1. CHENOPODIUM L.

Herbs usually ±mealy and very variable. Stem usually grooved or angled and often striped with white, red or green. Lvs lobed or toothed, less frequently entire. Fls hermaphrodite and female, in small cymes (glomerules) arranged in a ±branched infl. Bracteoles absent. Perianth herbaceous, of 2–5 segments joined at the base,

or sometimes to half way up or nearly to apex. Stamens (0–)2–5. Pericarp thin and membranous. Stigmas 2(–5). Seeds usually horizontal, often vertical in terminal fls, rarely all vertical; testa variously sculptured. Generally in open communities on disturbed ground, rubbish tips, or by the sea.

About 150 spp., mainly in temperate regions.

The markings on the testa of the seeds provide valuable specific characters. These can be seen with the low power of a microscope when the pericarp has been removed. The removal of the pericarp can often be effected by rubbing the seed between the finger and thumb ('pericarp easily removable'), but in some spp. it is necessary to boil the seed before the pericarp can be got off by this means, or else to scrape it off with a needle ('pericarp persistent').

1 Perennial; lvs triangular-hastate; stigmas long, exserted; testa: Fig. 35A.　　1. bonus-henricus

　Annual; lvs rarely hastate; stigmas short.　2

2 Larger stem-lvs cordate or truncate at base, coarsely sinuate-dentate; testa with large, deep, nearly circular pits (Fig. 35L).　　10. hybridum

　Lvs never cordate, ±cuneate at base.　3

3 Infl. axis and perianth glabrous (rarely perianth ±mealy in C. urbicum and then not completely enclosing fr.)　4

　Infl. axis and perianth mealy at least when young; fr. usually entirely or almost entirely enclosed by perianth.　9

4 Lvs entire or at most with a single obscure tooth on each side, green on both sides (or purple); stems 4-angled; seeds black; testa: Fig. 35B.　　2. polyspermum

　Lvs, except the uppermost, not entire (very rarely entire or nearly so in C. rubrum and C. botryodes but then seeds red-brown); stems ridged but not 4-angled.　5

5 All fls with 5 per.-segs and 5 stamens; seeds 1.2–1.5 mm diam., black; testa: Fig. 35K.　9. urbicum

　All fls except the terminal ones with 2–4 per.-segs and 2–3 stamens; seeds red-brown, 0.75–1.1 mm diam.　6

6 Lvs mealy-glaucous beneath, green above; testa: Fig. 35N.　　13. glaucum

　Lvs green on both sides.　7

7 Fruiting perianth fleshy, turning scarlet; fls in sessile heads forming a spike, lfless at top; testa: Fig. 35O.　　14. capitatum

　Fruiting perianth not fleshy, not turning scarlet; infl. branched; testa: Fig. 35M.　8

8 Per.-segs of lateral fls usually free to middle or below, not or only feebly ridged on back; lvs usually much toothed (not uncommon in various habitats).　　11. rubrum

　Per.-segs of lateral fls connate almost to apex, forming a sack closely investing the fr., ±distinctly ridged or keeled on back (at least when young); lvs entire or ±deltate and only slightly toothed; (rare, in salt-marshes).　12. botryodes

9 Lvs entire or almost so; plant stinking of bad fish.　　3. vulvaria

　Lvs toothed or lobed.　10

10 Lvs toothed, but not distinctly 3-lobed.*   *11*
   At least some of the lvs distinctly 3-lobed.   *14*
11 Infl. lfy almost to top, its branches short, numerous,
   divaricate; pericarp strongly adherent; seeds dull,
   with sharp, rather prominent keel; testa densely
   covered with small pits (Fig. 35J).   **8. murale**
   Infl. usually lfless in upper part, branches usually long
   and themselves little branched; seeds shining,
   obtuse or subacute at margin, but keel not prominent; testa not densely pitted.   *12*
12 Lvs usually markedly longer than broad, often longer
   than 3 cm, if nearly entire then scarcely mealy.
                                        (**album** group)   *13*
   Larger stem lvs often as long as or broader than long,
   up to c. 3 cm, much toothed to nearly entire, but
   some at least ±3-lobed, usually grey-green and very
   glaucous-mealy when young (as is the infl.); stems
   never red; testa: Fig. 35G.   **6. opulifolium**
13 Plant usually deep green (though often ±masked by
   grey meal); stems often reddish; lvs variable,
   usually ovate-lanceolate, toothed or entire; testa
   with shallow, spaced radial furrows (Fig. 35D) or
   a ±quadrate reticulum (Fig. 35E).   **4. album**
   Plant usually rather bright glaucescent green; stems
   not red; larger stem lvs always ovate-rhombic with
   sharp ascending teeth; testa with more numerous,
   closer and deeper furrows than in *C. album* (Fig.
   35F).   **5. suecicum**
14 At least the larger stem lvs as broad as or broader
   than long, lateral lobes short; plant usually grey-
   green; stems never red; testa: Fig. 35G.
                                               6. opulifolium
   Lvs usually markedly longer than broad.   *15*
15 Seeds c. 1.15 mm diam.; testa with narrow, radially
   elongate pits (Fig. 35H); mid-lobe of lvs elongate,
   ±parallel-sided, often obtuse; glomerules small.
                                                7. ficifolium
   Seeds 1.2–1.85 mm diam.; mid-lobe of lvs very rarely
   (in var. of *C. album*) ±parallel-sided but then testa
   not pitted.   *13*

Section 1. *Agathophytum* (Moq.) Ascherson. Perennial. Per.-segs and stamens 4–5. Per.-segs not or scarcely keeled on back. Stigmas 2–3, long. Seeds vertical except in terminal fls.

*1. C. bonus-henricus L.

All-good, Mercury, Good King Henry.

An erect perennial 30–50(–80) cm. *Lvs* up to 10 cm, mealy when young, *broadly hastate*, obtuse or acute, margins sinuous, entire. Infl. mainly terminal, narrowly pyramidal and tapering, lfless except at base. Seeds 1.5–2.2 mm diam., red-brown, not enclosed by the perianth. Pericarp persistent. Testa irregularly roughened (Fig. 35A). Fl. 5–7. $2n = 36$. Hs.

Introduced. In nitrogen-rich habitats, rich pastures, farmyards, roadsides, etc., long-established and well naturalized, though usually near buildings. Throughout England and Wales, northwards to S. and E. Scotland, rather local; rare in N. and W. Scotland and local in

* From this point onwards in the key the student is recommended, at least until he is familiar with the appearance of the spp. (which is commonly plastic), to check all his determinations by examining the testa of the ripe seed, whose markings appear to be constant.

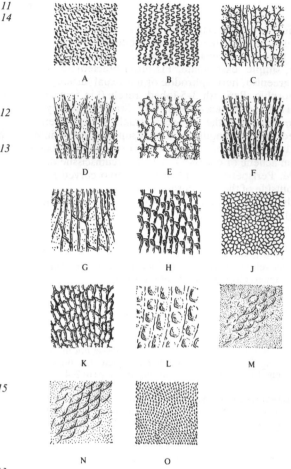

Fig. 35. Sculpturing of the testa of *Chenopodium* seeds. A, *C. bonus-henricus*; B, *C. polyspermum*; C, *C. vulvaria*; D, E, *C. album*; F, *C. suecicum*; G, *C. opulifolium*; H, *C. ficifolium*; J, *C. murale*; K, *C. urbicum*; L, *C. hybridum*; M, *C. rubrum*; N, *C. glaucum*; O, *C. capitatum*. All drawn with illumination from the left. ×c. 40.

Ireland. Most of Europe; W. Asia; introduced in North America.

Section 2. *Chenopodium*. Annual. Per.-segs and stamens 5. Per.-segs often prominently keeled on the back. Stigmas short.

2. C. polyspermum L.

Many-seeded Goosefoot, All-seed.

An erect or decumbent glabrous annual up to 100 cm. *Stems usually 4-angled. Lvs* up to 8 cm, *ovate or elliptical, entire* or occasionally with a single tooth-like angle on one or both sides towards the base, obtuse or subacute, thin. Infl. elongate, lax, mostly of axillary, divaricately-branched cymes shorter than their subtending lvs; glomerules small and indistinct. Per.-segs rounded on back. Seeds 0.8–1.25 mm diam., horizontal, not enclosed by the perianth. Pericarp easily detached.

Testa with close radially elongate pits with sinuous margins (Fig. 35B). Fl. 7–10. 2n = 18. Th.

Native. In waste places and cultivated ground. England, mainly south of a line from the Humber to the Severn, locally abundant; probably introduced in the north and in Ireland. Most of Europe; Asia; introduced in North America.

3. C. vulvaria L.   Stinking Goosefoot.

An ascending mealy annual 5–65 cm, *smelling strongly of bad fish*. Lvs 1–2.5 cm, *rhombic or ovate, entire*, or with a single tooth-like angle on one or both sides towards the base, *acute, very mealy beneath*. Infl. 0.5–2.5 cm, terminal and axillary, lfless, usually subtended by a full-sized lf. Per.-segs rounded on back. Seeds 1–1.5 mm diam., enclosed by perianth. Pericarp rather persistent. *Testa with faint radial furrows and irregular thickenings between them* (Fig. 35C). Fl. 7–9. 2n = 18. Th.

This plant contains trimethylamine, which gives it the nauseous odour of stale salt fish.

Native. At the landward margins of salt-marshes and shingle beaches, inland in waste places. Mainly S. and E. England, often only a casual; rather rare. Most of Europe; N. Africa; S.W. Asia; introduced in North America.

C. album group (spp. 4–5).

An erect ±mealy annual up to 150 cm. Lamina rhombic to ovate, ovate-lanceolate or lanceolate, ±toothed or occasionally entire. Infl. ±branched, usually dense. Per.-segs with a raised ridge or keel running down the back. Seeds 1.2–1.6(–1.85) mm diam., enclosed by perianth. Pericarp easily detached. Testa never closely pitted. 2n = 18; 54.

4. C. album L.   Fat Hen.

Incl. *C. reticulatum* Aellen

Plant usually ±deep green, ±mealy, usually with rather short strict branches. *Stem often reddish*. Lvs most often ovate-lanceolate and toothed but varying from rhombic to lanceolate. Glomerules usually densely crowded, usually spicate but not infrequently cymose. Seeds (1.25–)1.5–1.85 mm diam. *Testa very faintly radially striate, otherwise nearly smooth* (Fig. 35D), sometimes with raised lines forming a quadrate reticulum (Fig. 35E). Fl. 7–10. 2n = 54*.

Very common and variable.

Native. In waste places and cultivated land. By far the commonest sp. throughout the British Is. Europe; N. and S. Africa; Asia; Macaronesia. Introduced in North America and Australia.

*5. C. suecicum J. Murr

C. viride auct., non L.

Plant usually a rather bright glaucescent green, nearly or quite glabrous when mature, somewhat mealy when young and on the infl. *Stem without red coloration*; branches usually rather long, slender and ±spreading.

Larger stem lvs always ovate-rhombic with few to several sharp ascending teeth, sometimes somewhat 3-lobed; upper lvs oval to linear, ±toothed to entire. Glomerules usually rather small, in rather lax cymes. Seeds 1.5–1.7 mm diam. *Testa marked with more numerous and deeper furrows than in C. album* (Fig. 35F). 2n = 18*.

Introduced. Rather rare and most frequently on rubbish tips and in waste places. From N. Somerset and Essex to Selkirk; Down; very thinly scattered. N., C. and E. Europe and Asia.

*6. C. opulifolium Schrader ex Koch & Ziz
                                          Grey Goosefoot.

A variable, erect or decumbent mealy annual 30–150 cm. Lvs (0.7–)2–3 cm, ±*strongly glaucous-mealy beneath*, especially when young, rhombic, cuneate and subentire below, *often 3-lobed, almost as wide as long*; *lateral lobes short*; middle lobe triangular to shortly half oval, obtuse or subacute, subentire to coarsely and irregularly toothed, teeth often shortly mucronate. Infl. usually very mealy, branched; glomerules in dense interrupted spikes. *Per.-segs keeled on back.* Seed 1.1–1.5 mm diam. Pericarp somewhat persistent. *Testa with radial furrows and finely and irregularly papillose* (Fig. 35G). Fl. 8–10. 2n = 54*. Th.

Introduced. A not infrequent alien in waste places, chiefly in the south of England. Most of Europe; N. Africa; E. tropical and S. Africa; Asia Minor; C. Asia.

7. C. ficifolium Sm.   Fig-leaved Goosefoot.

C. serotinum auct., non L.

An erect or decumbent mealy annual 30–90 cm. *Blades of lower lvs up to c. 8 cm, 3-lobed; lateral lobes* near the base, *short, oblong or triangular, usually with 1 tooth on the lower margin; middle lobe oblong*, coarsely toothed or subentire; *lobes and teeth obtuse* or subacute; upper lvs slightly lobed or subentire. Infl. slender, much-branched; axillary branches longer than subtending lvs; *glomerules rather small* and distant. Per.-segs with a ridge-like keel running down the back. Seed 1–1.15 mm diam. Pericarp easily removable. *Testa with narrow radially elongate pits* (Fig. 35H). Fl. 7–9. 2n = 18*. Th.

Native. On waste ground and arable land, particularly round manure heaps. S. and E. England, north to Lincoln, local; rare and casual in Wales, Ireland and northern England. Most of Europe; Asia east to the Altai; N. Africa.

8. C. murale L.   Nettle-leaved Goosefoot.

An erect slightly mealy annual up to 90 cm. *Lvs* 1.5–6(–9) cm, *usually rhombic, acute, entire below, coarsely and irregularly toothed above*, rarely nearly entire; *teeth acute and ±incurved*. Infl. of axillary and terminal, *cymosely and divaricately branched panicles* up to c. 5 cm; glomerules rather densely crowded. *Per.-segs bluntly keeled on back*. Seed 0.9–1.3 mm diam., horizontal. *Pericarp very persistent. Testa with dense,* *minute, not radially elongate pits* (Fig. 35J). Fl. 7–10. $2n = 18^*$. Th.

Native. On dunes and in waste places, chiefly on light soils. S. England, introduced elsewhere in the British Is., and often only casual. Europe; N. and S. Africa; S.W. and S. Asia; introduced in America and Australia.

9. C. urbicum L.   Upright Goosefoot.

An erect annual 15–100 cm, *glabrous or very nearly so.* Lvs up to 14 cm, lower triangular, truncate to cuneate at base, usually toothed, teeth often long and hooked, acute or subobtuse. Infl. branched, branches crowded, short, erect; axillary branches mostly shorter than the subtending lvs. Glomerules small, distant. *Per.-segs not keeled on back. Seed* 1.2–1.5 mm diam., horizontal, *black, not completely enclosed by the perianth.* Pericarp persistent. *Testa marked with shallow grooves forming a slightly elongate reticulum enclosing convex areas* (Fig. 35K). Fl. 8–9. Th.

?Native. On waste ground and arable land especially round manure heaps. In lowland districts of England, local; rare and casual in Wales, Scotland and Ireland. Most of Europe; S.W. and C. Asia; introduced in North (and probably South) America.

10. C. hybridum L.

                    Maple-leaved Goosefoot, Sowbane.

An erect, scarcely mealy annual 10–100 cm. *Lvs* up to 18 cm (usually smaller), *broadly triangular to cordate-ovate*, acute or acuminate, *with few very large teeth. Infl. lax, cymose with divaricate branches*, nearly lfless. *Per.-segs not keeled on back.* Seed 1.7–2 mm diam., horizontal, black, not enclosed by the perianth. Pericarp usually easily removable, sometimes very persistent. *Testa with very large deep thick-walled pits* (Fig. 35L). Fl. 8–10. $2n = 18$.

?Native. In waste places and cultivated ground, rare. Mainly south-east of a line from the Humber to the Severn; casual elsewhere. Most of Europe; S.W. Asia; C. Asia; N. Africa.

Section 3. Pseudoblitum (Gren.) Ascherson. Annual. Perianth of terminal fls with 5, of lateral fls with 2–4 segments, not or scarcely ridged on back, or rarely ±distinctly keeled. Stamens equalling the per.-segs in number. Stigmas short. Seeds all vertical or those of the terminal fls horizontal, red-brown (at least in our spp.).

11. C. rubrum L.   Red Goosefoot.

A prostrate, ascending or erect *nearly or quite glabrous, usually reddish* annual up to 90 cm. Lvs very variable in size and shape, commonly 2–5 cm, ovate to rhombic or broadly triangular, coarsely and irregularly toothed, sinuate or subentire. Infl. variable, often dense, simple to much-branched, lfy or lfless; glomerules usually crowded. *Per.-segs of lateral fls free to middle or almost to base, not or scarcely keeled. Seed* of lateral fls 0.7–1.1 mm diam., *vertical*, red-brown. Pericarp easily

removable. *Testa with shallow non-radial grooves forming a reticulum especially near the hilum, sometimes nearly smooth, but always with numerous minute pits* (Fig. 35M). Fl. (5–)7–9. $2n = 36$. Th.

Native. In waste places, cultivated ground, and often near the sea; frequently abundant on rubbish tips and in farmyards. Often abundant in S. and E. England; Wales, Scotland and Ireland, rare and often only as a casual. Most of Europe, S.W. Asia, C. Asia, North America.

12. C. botryodes Sm.

Incl. *C. crassifolium* Hornem.; ?*C. chenopodioides* (L.) Aellen

A glabrous or nearly glabrous annual 5–30 cm. Stems erect or, more often, with spreading branches from the base. Lvs broadly triangular, entire or slightly toothed, usually obtuse, rather thick. Glomerules usually in distinct, rather distant groups. *Perianth of lateral fls saccate, closely investing the fr., the segments connate almost to the apex, each one marked by a distinct ridge or keel in the upper part* (at least when young). Otherwise very similar to forms of *C. rubrum* and not distinguishable by the seeds. Fl. 7–9. Th.

Native. On the muddy margins of salt-marsh ditches and creeks by the sea, very local. South and east coasts of England from Hants to S. Lincoln. C. and S. Europe, but absent from the C. and E. Mediterranean.

13. C. glaucum L.

                    Oak-leaved Goosefoot, Glaucous Goosefoot.

A prostrate, ascending or erect annual 5–50 cm. *Lvs* 1–5 cm, lanceolate to narrowly rhombic, obtuse or subacute, *sinuate to repand-dentate, mealy-glaucous beneath, green above.* Infl. of copious, axillary and terminal, little-branched, spiciform partial infl. up to c. 3 cm. Seed very similar to that of *C. rubrum* but differing slightly in the sculpturing of the testa (Fig. 35N). Fl. 6–9. $2n = 18$. Th.

?Native. On rich waste ground, rarely on seashores. Scattered localities northwards to Angus, but often only a casual. Most of Europe, Asia, America.

Section 4. Morocarpus Ascherson. Similar to *Pseudoblitum* but the perianth becoming succulent and bacciform in fr.

*14. C. capitatum (L.) Ascherson   Strawberry Blite.

An erect nearly or quite glabrous annual 10–60 cm. *Lvs* long-petiolate, *narrowly triangular, acuminate*, toothed or entire, *usually with 2 narrow spreading lateral lobes near the base. Infl. of dense sessile rather large subglobose heads of fls, becoming scarlet at maturity*, lfless in the upper part. Seed 1–1.3 mm diam., red-brown, oval not circular. Pericarp very persistent. *Testa densely punctate-pitted* (Fig. 35O). Fl. 7–8. $2n = 18$. Th.

Introduced. Naturalized in fields in Fermanagh, and as a rare casual on rubbish tips elsewhere. Origin unknown but naturalized in scattered localities in Europe; also in North America.

2. BETA L. Beet.

Herbs. Lvs almost entire. Fls hermaphrodite, in small cymes arranged in branched spike-like infl. *Perianth of 5 segments, becoming thicker especially towards the base as the fr. ripens.* Stamens 5. *Ovary half-inferior.* Fr. 1-seeded, adhering in groups (glomerules) by the swollen perianth bases.

About 6 spp. Europe and Asia.

1. B. vulgaris L.

A glabrous or slightly hairy annual, biennial or perennial of very varied habit, up to 200 cm. Root stout, conspicuously swollen or not. Stems decumbent, ascending or erect, ±branched and lfy. Lvs very varied in size, shape and colour, often dark green or reddish and rather shiny, frequently forming a basal rosette. Infl. varied, usually large and ±branched; partial infl. sessile, subtended by a small narrow lf, each of 1–several, frequently 2–4, fls. Per.-segs green. Stigmas usually 2. Fl. 7–9. Wind pollinated. $2n = 18$. Th. (biennial) or Hs.

Subsp. **vulgaris**. Root conspicuously swollen at junction with stem; stems ascending or erect; lvs up to c. 20 cm, often ovate and cordate at base, margins commonly wavy (except in spinach beet) and tissue puckered between the veins; partial infl. usually of 3–4 fls. Petiole, lamina and infl. axis usually green. Sugar-beet, beet-root, spinach-beet, chard, mangold. Wild plants of the sugar-beet type occur sporadically with subsp. *maritima*. All the variants appear to be interfertile and to cross freely when opportunity occurs.

Subsp. **maritima** (L.) Thell. (*B. maritima* L.), Sea Beet. Root usually not conspicuously swollen at junction with stem; stems commonly decumbent; lvs usually up to c. 10 cm, rhombic, thick and leathery, glossy or matt; partial infl. usually of 2–3 fls. Petiole, lamina and infl. axis usually coloured red. The common subsp., except in cultivation.

Native. On seashores. Coasts of the British Is., north to E. Inverness and N. Hebrides. Europe to c. 59° N in W. Sweden; Azores; N. Africa; S.W. Asia to the East Indies.

3. ATRIPLEX L. Orache.

Herbs or small shrubs, often mealy. Stems frequently striped white and green or red and green. *Lvs* toothed or lobed, sometimes entire and then linear to triangular but *not elliptical.* Infl. like that of *Chenopodium. Fls unisexual*; male fls with a perianth of (3–)5 segments; female fls without perianth but enclosed by two persistent bracteoles. *Bracteoles not connate above the middle, not obdeltate, entire or else with the lateral lobes smaller than the middle one.* Stigmas 2. Seeds vertical (except in *A. hortensis*), usually of 2 sizes.

About 200 spp., cosmopolitan.

Self-fertilization is frequent and numerous genetically distinct forms are thus perpetuated. The polymorphism of the genus is said to be further increased by the not infrequent occurrence of hybrids (see G. Turesson, *Hereditas*, **3**, 238–60, 1922).

1 Bracteoles herbaceous, connate at base only. 2
　Bracteoles thickened below, connate for some distance above the base. 6
2 Lower lvs linear to linear-oblong. **1. littoralis**
　Lower lvs narrowly rhombic to triangular or hastate. 3
3 Lower lvs ±truncate at base. **3. prostrata**
　Lower lvs cuneate at base. 4
4 Some or all bracteoles with stalks 5–20 mm.
　　　　　　　　　　　　　　　　　　　5. longipes
　Bracteoles sessile or with stalks not more than 3 mm. 5
5 Plant usually more than 10 cm; bracteoles much longer than seed. **2. patula**
　Plant usually 3–10 cm; bracteoles little longer than seed. **6. praecox**
6 Plant greenish, mealy; bracteoles in fr. not hardened below. **4. glabriuscula**
　Plant silvery; bracteoles in fr. hardened below.
　　　　　　　　　　　　　　　　　　　7. laciniata

Section 1. *Teutliopsis* Dumort. Stems striped white or red and green. Bracteoles connate less than half-way up and not hardened in fr., except in *A. glabriuscula* where they are connate half-way up and become slightly hardened.

1. A. littoralis L. Grass-leaved Orache, Shore Orache.

A ±mealy annual up to 100 cm. Root deep and spreading. Stems usually stout, much-branched ±erect. *Lvs linear to linear-oblong*, entire or dentate, the lower shortly petiolate, the upper sessile. Infl. up to 20 cm, spike-like, lfless except at base. *Bracteoles in fr. rhombic-ovate, strongly muricate*, sometimes smooth towards the top. Fl. 7–8. Fr. 8–9. $2n = 18$. Th.

Native. Near the sea. Around the coasts of the British Is., locally abundant, but not on rocky or sandy shores. C., W. and S. Europe from c. 69° N in Norway southwards; W. and C. Asia.

2. A. patula L. Iron-root, Common Orache.

Annual, up to 150 cm, ±mealy, much-branched, prostrate to erect, usually green. Lvs entire or toothed, the lower rhombic-lanceolate, with a projecting lobe on each side above the cuneate base, the upper narrow, entire. Petiole 1–10 mm. Bracteoles in fr. 2–5(–10) mm, broadly rhombic, the lateral angles entire or toothed, smooth or slightly warty on the back. Fl. 8–10. $2n = 36$. Th.

Native. Cultivated and waste ground and in open habitats near the sea, where it is generally less common than *A. prostrata*. Throughout the British Is. Europe from c. 71° N southwards; Azores, N. Africa, W. Asia; naturalized in North America.

3. A. prostrata Boucher ex DC. Hastate Orache.

A. hastata auct., non L.

Like 2 but lvs glabrous or mealy beneath only, the lower triangular-hastate, truncate or nearly so at base; petiole usually 10 mm or more; bracteoles ovate, with a subcuneate to subcordate base. Seeds mostly small. Fl. 7–9.

$2n = 18$. Th.

Native. Cultivated and waste ground and near the sea on sand, shingle and mud above high-tide mark, common; mainly coastal in the north and west. Europe from c. 71° N southwards; N. Africa, Asia, Macaronesia.

4. A. glabriuscula Edmondston　Babington's Orache.

Seldom more than 20 cm, often reddish. Lower lvs triangular to hastate, truncate, or nearly so at base. Bracteoles up to 8 mm, rhombic, thickened at base, connate for some distance above the base, sessile; seeds mostly large. Fl. 7–9. $2n = 18$. Th.

Native. On sandy or gravelly shores at or somewhat above high-tide mark. Around the coasts of the British Is. Coasts of W. and N.W. Europe and North America.

5. A. longipes Drejer

Stems usually 15–60 cm. Lower lvs lanceolate to rhombic. Bracteoles up to 20 mm, much longer than the seeds, not thickened, connate at the base only, at least some with stalks 5–20 mm. Fl. 7–9. $2n = 18$. Th.

Native. Very local but widely distributed in N.W. Europe. Distribution imperfectly known.

6. A. praecox Hülphers

Stems usually 3–10 cm. Lower lvs lanceolate, cuneate at base. Bracteoles 3–6 mm, little longer than the seeds, not thickened, connate at the base only, sessile or shortly stalked. Fl. 7–9. $2n = 18$. Th.

Native. Just above high-tide mark in N.W. Scotland. Coasts of N.W. Europe and the Baltic.

Hybrids between **4, 5** and **6** are frequent.

Section 2. *Obionopsis* Lange. Stems whitish or pale brown, occasionally with red patches. Bracteoles united to the middle, hardened in the lower half.

7. A. laciniata L.　Frosted Orache.

A. sabulosa Rouy; *A. arenaria* Woods, non Nutt.

A mealy *white or almost silvery* decumbent annual up to 30 cm or rarely more. Stems much-branched, yellowish or reddish. Lvs usually 1.5–2 cm, rhombic to ovate, sinuate-dentate, obtuse, rather thick and very mealy on both surfaces. Infl. axillary, much shorter than the lvs. *Bracteoles* in fr. 6–7 mm long and usually rather broader, lateral angles rounded and sometimes toothed, mealy, *hardened in the lower half.* Fl. 8–9. Fr. 9–10. $2n = 18$. Th.

Native. On sandy and gravelly shores at about high-tide mark. Scattered round the shores of Great Britain, local; south and east coasts of Ireland. W. and N.W. Europe from Norway to Spain. North America.

*A. hortensis L., Garden Orache, with dimorphic fls, about ¼ without bracteoles but with perianth, the remainder with free, ovate bracteoles but no perianth, is cultivated and sometimes escapes.

*A. halimus L., a mealy shrub with ovate-rhombic lvs, is planted near the sea in southern England and occasionally escapes and becomes naturalized.

4. HALIMIONE Aellen

Like *Atriplex* but lvs entire, elliptical or nearly so; bracteoles obdeltate in fr., connate nearly to the top, 3-lobed with lateral lobes usually larger than middle one.

Woody perennial; fr. sessile.　　　　　**1. portulacoides**
Annual; fr. long-pedicellate (probably extinct).
　　　　　　　　　　　　　　　　　　2. pedunculata

1. H. portulacoides (L.) Aellen　Sea Purslane.

Atriplex portulacoides L.; *Obione portulacoides* (L.) Moq.

A very mealy small *shrub* up to 80 or rarely 150 cm. Rhizome short, creeping. Stems decumbent, branches ascending, terete below, angled above. *Lower lvs opposite*, shortly (5–10 mm) petiolate, elliptical; upper linear, entire, obtuse or apiculate. Infl. of terminal and axillary compound spikes; partial infl. dense. *Fr. sessile.* Bracteoles in fr. 3–5 mm long and rather broader, obdeltate, united ⅔ of the way from the base, usually 3-lobed. Fl. 7–9. Fr. 9–10. $2n = 36$. Chw. or N.

Native. In salt-marshes, especially fringing channels and pools, ordinarily flooded at high tide. England north to N. Northumberland and Westmorland, locally abundant; Scotland: Wigtown, Ayr; Ireland: S. and E. coasts from W. Cork to Down. Europe from Denmark southward; N. Africa; S.W. Asia; introduced in North America.

2. H. pedunculata (L.) Aellen.

Atriplex pedunculata L.; *Obione pedunculata* (L.) Moq.

A silvery-mealy erect *annual* up to 50 cm. *Lvs all alternate*, sessile or shortly petiolate, elliptic to oblong, entire; apex rounded, apiculate. Partial infl. lax. *Fr. pedicellate*, pedicels up to 12 mm when mature. Bracteoles in fr. obdeltate, united almost to the top, 3-lobed, middle lobe very small, lateral lobes long and spreading. Fl. 8–9. Fr. 9–10. $2n = 18$. Th.

Native. In the drier parts of salt-marshes with *Puccinellia maritima.* Formerly recorded from Kent, Suffolk, Norfolk, Cambridge and Lincoln; probably now extinct. Occasionally also in other places introduced with ballast. N.W. and N. Europe from N. France to Estonia; Black Sea; saline places inland; Turkestan; Siberia.

5. SUAEDA Forskål ex Scop.

Herbs or small shrubs growing in saline places. Lvs fleshy, terete or ½-terete, small. Fls hermaphrodite and female, small, axillary; bracteoles 2, minute. Per.-segs 5, not keeled, small and ±succulent. Stamens 5. Stigmas 2–3(–5); achenes with a thin membranous pericarp. Seed horizontal or vertical; embryo in a flat spiral; endosperm present or 0.

About 110 spp., cosmopolitan.

Annual herb; lvs acute or subacute and narrowed at base; stigmas 2; seed horizontal.　　　　　　　**1. maritima**
Perennial, shrubby; lvs rounded at tip and base; stigmas 3; seed usually vertical.　　　　　　　**2. vera**

1. S. maritima (L.) Dumort   Annual Seablite.

A prostrate to erect ±glaucous or red-tinged glabrous *annual* 7–30(–60) cm. *Lvs* 3–25 mm × 1–2(–4) mm, $\frac{1}{2}$-terete, *acute or subacute*. Fls 1–3 together in small axillary cymes. *Stigmas* 2. *Seed* 1.1–2 mm diam., *biconvex*, nearly circular in outline but with a small curved beak, black, shining, *with fine reticulate sculpturing, horizontal*. Fl. 7–10. Fr. 8–11. Germ. spring. $2n = 36$. Th.

The following three varieties, differing considerably in appearance and fr. can be recognized. They all fall within subsp. *maritima*.

Var. maritima. Plant large, branches spreading. Seed not more than 1.5 mm diam. Fl. 8–10.

Var. *macrocarpa* (Desv.) Moq. Plant decumbent or prostrate, rarely erect and then small. Lvs up to 10 mm. Seed c. 2 mm diam. Fl. 7–8.

Var. *flexilis* (Focke) Rouy. Plant usually erect, often unbranched and never branched from the base; branches short, erect. Lvs 10–25 mm. Seed 1.1–1.4 mm diam. Fl. 8–10.

Native. In salt-marshes and on seashores, usually below high-water mark spring tides, common. Around the coasts of the British Is., north to the Outer Hebrides and Shetland; ascends the Severn estuary to W. Gloucester. Coasts of Europe (except the Arctic); inland in saline areas of C. and E. Europe; Asia; North America.

2. S. vera J. F. Gmelin   Shrubby Seablite.

S. fruticosa auct., non Forskål

A small much-branched glabrous shrub 40–120 cm. Stems suberect or ascending, very lfy; subterranean stems rooting freely. *Lvs* 5–18 × 1 mm, almost terete, *rounded or shortly apiculate at apex*, glaucous, evergreen. Fls 1–3 together in small axillary cymes. *Stigmas* 3. *Seed* 1.7–1.8 mm, *ovoid*, beaked near the hilum, *smooth*, black, shining, usually *vertical*. Fl. 7–10. Fr. 9–11. Germ. spring. $2n = 36$. N.

Native. On shingle banks and other well-drained substrata by the sea but above high-water mark spring tides, local. Dorset, E. and W. Kent, N. and S. Essex, E. Suffolk, E. and W. Norfolk, Lincoln; extinct in Isle of Wight, S. Hants and Channel Is. Coasts of S. and W. Europe from England southwards, inland in Spain; Madeira, Canary Is., St Helena, Angola, Somalia, S.W. Asia; inland in S. Russia, Transcaucasia, Afghanistan and India.

6. SALSOLA L.

Herbs. *Lvs sessile, succulent.* Fls small, sessile; per.-segs 5, usually developing a transverse dorsal wing in fr. Stamens 5. Ovary subglobose, style elongate, stigmas 2. Seed usually horizontal.

About 150 spp., cosmopolitan in saline habitats.

1. S. kali L.   Prickly Saltwort.

A decumbent or prostrate, seldom erect, prickly annual up to 60 cm. Stems with pale green or reddish stripes, usually much-branched. Lvs 10–40 mm, subulate, sessile, succulent, subterete, narrowed into a little spine at the apex. Fls usually solitary in the axil of a lf, each with 2 lf-like bracteoles. Per.-segs becoming tough in fr. and thickened transversely about the middle, the thickening forming either a ridge or a horizontal wing of variable size. Fr. c. 3.5 mm, turbinate, enclosed in the persistent perianth. The stems and lvs are either asperous (var. *hirsuta* Hornem.) or nearly glabrous (var. *glabra* Detharding). Fl. 7–9. Fr. 8–10. $2n = 36$. Th.

Native. On sandy shores. Around the coasts of the British Is., except Shetland. Europe to 63° 30′ N, Azores, N. Africa, Asia, North America.

Subsp. **kali**, with the per.-segs becoming stiff and ±spinose in fr., is the native plant in Britain. Subsp. **ruthenica** (Iljin) Soó (*S. pestifer* A. Nelson, *S. tragus* auct., non L.) occurs occasionally as a casual on waste ground. It has per.-segs that remain soft in fr.

7. SALICORNIA L. Glasswort, Marsh Samphire.

Annuals inhabiting salt-marshes. Stems usually much-branched. *Lvs succulent*, translucent and glabrous, *opposite, the pairs connate along their margins and enveloping the stem* forming the 'segments', tips usually free. *Infl.* of terminal, ±branched spikes with *axillary cymes* of 3, rarely 1, fls. Perianth indistinctly 3–4-lobed and ±immersed in the bracts. Bracteoles 0. *Stamens* 1–2. Seed covered with short curved or hooked hairs; radicle incumbent; endosperm 0.

Probably about 50 spp., cosmopolitan in saline districts. Many of the spp. are critical and need much further work for their elucidation. Herbarium specimens dried in the ordinary way are almost valueless.

Phenotypic variation is great and may render the identification of single specimens difficult, if not impossible. For instance, *S. ramosissima*, though typically much-branched, bushy and erect, becomes prostrate on hard stony mud and very much dwarfed and often quite unbranched when growing in crowded pure stands or in competition with other plants. Crowding reduces the degree of branching and this is often accompanied by an increase in the length of the terminal spike. Damage to or removal of the terminal spike often increases the abundance and size of the lateral branches and a similar effect is produced by the plant falling over. The typical colour of most species is not fully developed until the seed is nearly ripe and then only in the absence of shade. Until some familiarity with the genus has been obtained attempts at identification should be confined to well-grown specimens with undamaged main stems collected in Sept. or Oct. Dried specimens, however carefully prepared, lose much of their character and are difficult to interpret. Preservation in a fluid, such as equal parts by volume of alcohol, glycerin and water or sea water with 2% chrome alum and 2% formaldehyde, is more satisfactory, particularly if accompanied by colour photographs or detailed notes of colours.

Much work remains to be done on the taxonomy and distribution of the species.

1 Cymes regularly 1-fld; terminal spikes up to c. 6 mm.
                                                    4. pusilla
   Cymes normally 3-fld; terminal spikes usually more
   than 6 mm.   2
2 Central fl. of cyme usually much larger than the visible
   part of the lateral fls; terminal spikes usually with
   up to 10 fertile segments; stamens normally 1.   3
   Central fl. of cyme little larger than visible part of
   the lateral fls; terminal spike usually with (6–)12
   or more fertile segments; stamens normally 2.
3 Plant dark shining green, often becoming dark purplish-red; free part of lf with a conspicuous scarious
   border c. 0.2 mm wide; normally much branched
   and bushy, sometimes simple but then dark purplish-red.   **1. ramosissima**
   Plant green, becoming yellowish or sometimes pink;
   free part of lf with a narrow scarious border not
   more than 0.1 mm wide; moderately to freely
   branched.   4
4 Plant dark green, fertile spikes pink when mature;
   bushy with lower primary branches up to as long
   as the main stem, tertiary branches usually present.
                                                 2. europaea
   Plant dull glaucous green, fertile spikes dull yellow,
   occasionally purplish round the fls when mature;
   lower primary branches $\frac{1}{4}(-\frac{1}{2})$ as long as the main
   stem, tertiary branches 0, secondary branches pre-
   sent or 0.   **3. obscura**
5 Plant pale green or yellow with, at most, a slight purplish tinge when mature; fertile segments usually
   more than 3 mm.   6
   Plant brownish-purple to brownish-orange when
   mature; fertile segments rarely exceeding 3 mm.
                                                  5. nitens
6 Plant bushy and much-branched with numerous primary and usually secondary and some tertiary
   branches; lowest primary branches generally $\frac{1}{2}$-as
   long as main stem.   7
   Plant not bushy; tertiary branches usually 0; lowest
   primary branches $\frac{1}{4}(-\frac{1}{2})$ as long as main stem.   8
7 Sterile segments becoming dull yellow and finally
   brown; terminal spike distinctly tapering; fertile
   segments usually more than 12.   **7. dolichostachya**
   Sterile segments becoming bright yellow; terminal
   spike almost cylindrical; fertile segments usually
   less than 12.   **8. lutescens**
8 Sterile segments becoming dull yellow and finally
   brown; terminal spike with up to 20 segments,
   usually ±tapering.   **6. fragilis**
   Sterile segments becoming bright yellow; terminal
   spike usually with less than 12 fertile segments,
   almost cylindrical   **8. lutescens**

1. S. ramosissima Woods

S. gracillima (Towns.) Moss, *S. prostrata* auct., *S.
appressa* Dumort., *S. smithiana* Moss

An erect or prostrate annual 3–40 cm, typically abundantly branched and bushy with the lowest branches
c. as long as the main stem, but when crowded reduced
to a single stem. *Plant dark green*, usually *becoming
deep purplish-red. Segments with a conspicuous broad
scarious border* c. 0.2 mm wide. Terminal spike
(5–)10–30(–40) mm, ±tapering, with (1–)4–9(–12) fertile segments. Cymes 3-fld. Central fl. rounded-rhombic
to almost circular, much larger than the visible part of

the lateral fls. Stamens 1, exserted or not. Seeds
1.2–1.5(–1.7) mm. Fl. 8–9. $2n = 18^*$. Th.

Native. In the upper part of salt-marshes on bare,
rather firm mud and behind sea walls. Common and
widespread in suitable habitats in S. and E. England
and S. Wales. N.W. Europe. A somewhat similar sp.
occurs in S. and E. Ireland.

2. S. europaea L.   Glasswort.

S. herbacea (L.) L., ?*S. stricta* Dumort.

An erect annual (10–)15–30(–35) cm, *fairly richly
branched*, with the *lowest branches up to c. as long as
the main stem. Plant* dark green *becoming yellow-green
and ultimately flushed with pink or red.* Segments with
an inconspicuous scarious border not more than 0.1 mm
wide. Terminal spike (10–)15–50 mm, slightly tapering,
obtuse, with (3–)5–9(–12) fertile segments. Cymes 3-fld.
Central fl. rounded above, cuneate below, much larger
than the visible part of the lateral fls. Stamens usually
1, exserted or not. Seeds 1.2–1.8(–2.0) mm. Fl. 8.
$2n = 18^*$. Th.

Native. On open sandy mud in salt-marshes, local.
Coasts of the British Is.; W. Europe.

3. S. obscura P. W. Ball & Tutin

An erect annual 10–40(–45) cm, *typically with primary
branches only, the lowest branches up to c.* $\frac{1}{2}$ *as long as
the main stem. Plant dull slightly glaucous green with
a matt surface, becoming dull yellow,* rarely purple
round the pores of the fls. Segments with an inconspicuous scarious border not more than 0.1 mm wide. Terminal spike 10–40(–45) mm, nearly cylindrical, obtuse,
with (3–)5–9(–14) fertile segments. Cymes and fls rather
similar to those of *S. europaea.* Stamen 1, rarely
exserted. Seeds 1.1–1.7 mm. Fl. 8–9. $2n = 18^*$. Th.

Native. On bare damp mud, in pans and at the sides
of channels in salt-marshes. In suitable habitats on the
E., S. and W. coasts from Lincoln to Cheshire. Distribution elsewhere unknown.

Perhaps a variant of **2**.

4. S. pusilla Woods

S. disarticulata Moss

An erect or rarely prostrate annual up to 25 cm, usually
*much-branched and bushy. Plant yellowish-green,
becoming brownish- or pinkish-yellow,* often with bright
pink tips to the branches. *Terminal spike up to* c. 6 mm,
with 2–4 fertile segments. *Cymes* 1-fld. Fls almost circular. *Fertile segments all disarticulating* shortly before the
seeds are ripe. Stamen 1. Fl. 8–9. $2n = 18^*$. Th.

Native. In the drier parts of salt-marshes, chiefly along
the drift-line. Dorset to Kent, Essex, Norfolk, N. Lincoln; Glamorgan, Carmarthen; W. Cork, Waterford.
N.W. France.

5. S. nitens P. W. Ball & Tutin

An erect annual 5–25 cm, *typically with primary
branches only,* the lowest branches usually less than $\frac{1}{4}$
as long as the main stem. *Plant smooth, shining and*

somewhat translucent, green or yellowish-green, becoming clear light brownish-purple to brownish-orange. Sterile segments conspicuously swollen near the top. Terminal spike 12–40 mm, cylindrical, obtuse, with 4–9 fertile segments. Cymes 3-fld. Central fl. semicircular above, cuneate below, little larger than the triangular visible part of the lateral fls. Stamens usually 2, exserted. Seeds 1.5–1.7 mm. Fl. 9. Th.

Native. On bare damp mud and in pans in the upper parts of salt-marshes. S. and E. England, apparently common in suitable habitats from E. Suffolk to Hants. Distribution elsewhere unknown.

6. S. fragilis P. W. Ball & Tutin

An erect annual (10–)15–30(–40) cm, *usually with primary branches only, the lowest branches normally less than ¼ as long as the main stem. Plant dull green becoming dull yellowish-green. Terminal spike* (20–)30–80(–100) mm, distinctly tapering, with (6–)8–16(–20) fertile segments. Cymes 3-fld. Central fl. semicircular to triangular above, cuneate below, the visible part of the lateral fls triangular and almost as long as the central fl. Stamens usually 2, exserted. Seeds 1.5–2.0 mm. Fl. 8–9. $2n = 36^*$. Th.

Native. On soft mud in the lower levels of salt-marshes and, particularly, on the sides of channels below the fringe of *Halimione*; often forming pure stands. E. Suffolk to Kent; W. Ireland.

7. S. dolichostachya Moss

An erect annual 10–40(–45) cm, *abundantly branched and bushy, the lowest branches usually about as long as the main stem. Plant dark green, becoming paler green or dull yellow and finally brownish,* the fertile spikes occasionally with a slight purplish flush. *Terminal spike* (25–)50–120(–200) mm, *distinctly tapering* but often obtuse, with (7–)12–25(–32) fertile segments. Cymes and fls similar to *S. fragilis*. Stamens usually 2, exserted. Seeds 1.5–2.3 mm. Fl. 7–8. $2n = 36^*$. Th.

Native. Usually on rather firm mud and on muddy sand at the lowest levels in salt-marshes, occasionally on the sides of narrow channels at the middle levels. Coasts of Great Britain from Lancs to Devon, Kent and E. Ross; S. and E. coasts of Ireland. N.W. Europe. Characterized by the bushy growth, dull coloration and very long tapering spikes but varying in habit from straggling to strongly fastigiate.

8. S. lutescens P. W. Ball & Tutin

An erect annual (10–)15–30(–40) cm, typically with the habit of *S. dolichostachya* from which it differs in the *bright green to yellow-green* colour, *soon becoming bright yellow*, the shorter *terminal spikes (usually 25–60 mm with 8–12 fertile segments)* and in having the *spikes cylindrical* and obtuse. Fl. 7–8. Th.

Native. On firm and comparatively dry mud or muddy sand. Coasts of England from S. Lincoln to S. Hants; Glamorgan; ?elsewhere.

8. SARCOCORNIA A. J. Scott

Like *Salicornia* but perennial, with the stems often rooting at the nodes; fls ±equal in height, arranged in a row, with the middle fl. completely separating the laterals.

About 15 spp., cosmopolitan in temperate regions.

1. S. perennis (Miller) A. J. Scott

Salicornia perennis Miller, *S. radicans* Sm.; incl. *S. lignosa* Woods

A somewhat woody perennial often forming tussocks up to c. 1 m diam. Stems up to c. 30 cm, ascending or decumbent; segments dark green becoming red or brownish, the basal keeled. Terminal segments 10–20 mm, cylindrical, blunt; segments c. 8. Cymes 3-fld. Stamens 2. Stigma bifid. Seeds with a thin membranous testa covered with curved or hooked hairs. Fl. 8–9. Fr. 10. $2n = 18$. Ch.

Native. Gravelly foreshores and salt-marshes. S. and E. coasts of England from S. Devon to Lincoln; Durham; Merioneth; one locality in S.E. Ireland. S.W. coasts of Europe from c. 53° N in England; N. and S. Africa; North America.

42. PHYTOLACCACEAE

Trees, shrubs, herbs or woody climbers. Lvs alternate, entire; stipules 0 or very small. Fls usually regular, usually hypogynous. Perianth usually in 1 whorl; segs 4–5 free, or united at base; stamens 4–∞; carpels 1–10, free or united; ovule 1 in each carpel, campylotropous. Fr. various; embryo large, curved, surrounding the perisperm.

About 12 genera and 100 spp., mainly tropical.

1. PHYTOLACCA L.

Usually herbs. Fls regular, in lf-opposed racemes. Per.-segs 5. Stamens 5–30. Carpels 5–16, free or joined at base. Fr. berry-like.

*P. esculenta Van Houtte   Pokeweed.

Coarse glabrous perennial up to 3 m. Lvs 10–30 cm, ovate to oblong-lanceolate, stalked. Infl. 10–20 cm. Fls c. 6 mm, stamens and carpels usually 10. Fr. dark purple. Cultivated and occasionally naturalized. Native of eastern North America.

43. TILIACEAE

Trees, shrubs or rarely herbs. Lvs spirally arranged or distichous and alternate, rarely opposite; stipules usually small and caducous, often functioning as bud scales; sometimes 0. Infl. usually cymose. Fls hermaphrodite, actinomorphic, hypogynous. *Sepals* 5(–3), free or united, usually *valvate in bud*; petals as many as the sepals, free, rarely 0; stamens 10 or usually more, their *filaments free or united only at the base; anthers with 4 pollen sacs*; ovary superior 2–10-celled, each cell with 1–many anatropous ovules on axile placentae; style 1, with as many radiating stigma-lobes as ovary-cells, or the stigmas sessile. Fr. a capsule, a drupe or a nut with 1–5 seeds, rarely a berry or separating into drupelets. Seeds endospermic.

About 300 spp. in 35 genera, widely distributed in temperate to tropical regions.

Jute is made from the pericycle fibres of *Corchorus capsularis* and other spp., native in tropical Asia. *Sparmannia africana* is often grown as an ornamental plant in conservatories, etc. Distinguishable from Malvaceae by the absence of a stamen-tube or more certainly by the stamens with 4 pollen sacs.

1. TILIA L.   Lime, Linden.

Deciduous trees with sympodial growth owing to the abortion of the terminal bud. Winter buds large, blunt. Lvs distichous, alternate, usually cordate or truncate at the base and with a slender stalk; stipules (bud-scales) caducous. Fls yellowish or whitish, fragrant, in cymose infl. whose stalk is adnate for about half-way to a large oblong ±membranous bracteole. Sepals 5, free; petals 5, free; stamens many, free or in bundles opposite the petals, the filaments often forked distally; epipetalous staminodes sometimes present; ovary 5-celled, each cell with 2 ovules; style slender, with a 5-lobed stigma. Entomophilous, many insects visiting the fls for their copious nectar. *Fr. ovoid, indehiscent and nutlike*, 1-celled, with usually 1–3 seeds. *The infructescence is shed as a whole, with the adnate bracteole acting as a wing.* Cotyledons broad, lobed.

About 30 spp. in the temperate regions of the northern hemisphere.

Lime timber was formerly in great demand for woodcarving.

1 Lvs pubescent beneath and often also above; fls 2–5, usually 3; fr. strongly 3–5 ribbed.   **1. platyphyllos**
Lvs glabrous beneath except for tufts in the axils of the veins; fls 4–10; fr. slightly 3–5-ribbed or ribs 0.   *2*

2 Lvs 6–10 cm, broadly ovate-acuminate, bright green beneath, with the tertiary veins prominent; petioles 3–5 cm; cymes pendulous; fr. slightly ribbed.
                                                        3. × europaea
Lvs 3–6 cm, suborbicular, abruptly acuminate, often broader than long, ±glaucous beneath with the tertiary veins not prominent; petioles 1.5–3 cm; cymes obliquely erect; fr. not or barely ribbed. **2. cordata**

1. T. platyphyllos Scop.   Large-leaved Lime.
T. grandifolia Ehrh. ex Hoffm.

A large tree reaching 30 m, with spreading branches and a ±smooth dark bark. Young *twigs pubescent*, rarely glabrous. *Lvs* 6–12 cm, broadly ovate, abruptly acuminate, obliquely cordate at the base, dark dull green and glabrescent above, pale green and *pubescent beneath with simple hairs all over the surface* as well as in whitish tufts in the axils of the veins; *tertiary veins prominent beneath*; margin sharp-toothed; stalk 1.5–5 cm, pubescent. Cymes usually 3-fld (2–5) *pendulous*, the adnate bracteole 5–12 cm, pubescent on the midrib below. Fls yellowish-white. Stamens exceeding the petals. *Fr.* 8–10 mm, subglobose to pyriform, apiculate, densely pubescent, *strongly 3–5-ribbed*, woody when mature. Fl. late June, before *T. cordata* and *T. × europaea*. Freely visited by bees. $2n = 82$. MM.

Probably native. In woods on good calcareous or base-rich soils and on limestone cliffs, but perhaps often planted originally since its pollen-grains have rarely been found in peat from this country. Naturalized (or possibly native) in the Wye Valley and its neighbourhood (Gloucester, Hereford, Monmouth, Worcester), in Brecon and Radnor and on the Magnesian Limestone in S. Yorks, and established in old plantations northwards to Perth. C. and S.E. Europe from N. Spain, S. Italy and Greece northwards to c. 51°N in France, Belgium, Germany and Poland; Crimea; Asia Minor; Caucasus.

2. T. cordata Miller   Small-leaved Lime.
T. parvifolia Ehrh. ex Hoffm.

A large tree reaching 25 m with spreading branches and a smooth bark. Young *twigs* usually downy at first, then becoming *glabrous*. *Lvs* 3–6 cm, suborbicular, abruptly acuminate, cordate at the base, dark ±shining green and glabrous above, somewhat glaucous and *glabrous* beneath *except for tufts of rusty hairs* in the axils of the veins; *tertiary veins not prominent*; margin sharp-toothed; stalk 1.5–3 cm, glabrous. *Cymes 4–10 fld, obliquely erect or spreading*, the adnate bracteole 3.5–8 cm, glabrous below. Fls yellowish-white. Stamens about equalling the petals. *Fr.* c. 6 mm, globose, apiculate, *thin-shelled, ribs obscure or 0*. Fl. early July. Visited by bees. $2n = 82$. MM.

Native. In woods on a wide range of fertile soils but especially over limestone; commonly on wooded limestone cliffs. Scattered throughout England and Wales northwards to the Lake District and Yorks, but planted northwards to Perth. C. and E. Europe from N. Spain, N. Italy, N. Balkans and S. Russia northwards to c. 63°N in Norway, Sweden and Finland, eastwards across C. Russia to c. 75°E in Siberia; Crimea; Caucasus.

Distinguishable without much difficulty from *T. platyphyllos* and their hybrid by the smaller, relatively broader, tougher lvs, ±glaucous beneath, with relatively longer stalks, as well as by characters of the infl. and fr.

3. T. × europaea L.　　　　　　Common Lime.

T. cordata × *T. platyphyllos*; *T. vulgaris* Hayne; *T. intermedia* DC.

A large tree reaching 25 m, with arching lower branches and the trunk often covered with irregular bosses. Young *twigs* usually *glabrous*. *Lvs* 6–10 cm broadly ovate, shortly acuminate, obliquely cordate or ±truncate at the base, dark green and glabrous above, light green and *glabrous* or nearly so beneath *except for tufts of whitish hairs* in the axils of the veins; *tertiary veins*

±*prominent* beneath; margin sharp-toothed; stalk 3–5 cm, ±glabrous. *Cymes* 4–10-fld, *pendulous*. Fls yellowish-white. Stamens equalling or somewhat exceeding the petals. *Fr.* c. 8 mm, subglobose to broadly ovoid, apiculate, pubescent, *slightly ribbed*, woody when mature, producing some viable seed. Fl. early July. Freely visited by bees. $2n = 82$. MM.

Introduced or very doubtfully native. Widely planted over a long period, especially in copses, parks, gardens, roadsides, etc. Throughout the British Is. except the far north. Europe.

A variable tree, reproducing by seed despite its hybrid origin. Much planted despite the unsightly bosses on its trunk and the frequent infestation of its lvs by aphids which cause a copious rain of honey-dew.

44. MALVACEAE

Herbs, shrubs or trees, usually with mucilage canals and stellate pubescence. Lvs spirally arranged, usually palmately veined and commonly palmately lobed; stipules free, usually small, often caducous. Fls in racemes or racemose panicles, or solitary, axillary, usually hermaphrodite, actinomorphic, hypogynous. Sepals 5(–3), free or united, valvate in bud; often with an 'epicalyx' of 3 to several segs resembling an outer set of sepals; petals 5, free, convolute, commonly adherent to the base of the staminal tube; stamens numerous (–5), the filaments united below into a staminal tube which divides above into branchlets each bearing a single 1-celled anther-lobe; ovary superior, with 2–many cells each with 1–many amphitropous ovules on axile placentae; styles and stigmas as many or rarely twice as many as cells. Entomophilous. Fr. a capsule or schizocarpic by separation of 1-seeded nutlets, rarely fleshy. Seeds with very little endosperm and a curved embryo; cotyledons lf-like.

About 900 spp. in 40 genera, widely distributed in temperate to tropical regions.

Native British Malvaceae all belong to subfamily Malveae, with schizocarpic fr. Members of the Malveae much grown in gardens include spp. of *Sidalcea* and *Malvastrum* with indehiscent 1-seeded schizocarps, and of *Sphaeralcea* with dehiscent 2–3-seeded schizocarps. A few half-hardy spp. of **Abutilon**, with dehiscent 3–9-seeded schizocarps, are grown in mild districts, and **A. theophrasti** Medicus, an annual weed with small yellow fls, widespread in the Mediterranean region and elsewhere, occurs as a fairly frequent casual, mostly from imported wool. Subfamily Hibisceae, with fr. a capsule, includes the very large and mainly tropical genus *Hibiscus*. *H. syriacus* L. is a hardy shrub grown here for its large attractive fls, and the annual *H. trionum* L., with hispid palmatisect lvs and violet-centred white corolla, is occasional and as a casual from wool and bird-seed. Cotton is spun from hairs on the seed-coats of spp. of *Gossypium*, also in the Hibisceae.

Malvaceae are distinguished from Tiliaceae by the united filaments and 1-celled anthers of their stamens.

1 Epicalyx of 3 segs.		2
Epicalyx of 6–9 segs.	3. ALTHAEA	
2 Epicalyx-segs free to the base.	1. MALVA	
Epicalyx-segs united below.	2. LAVATERA	

1. MALVA L.

Annual to perennial subglabrous herbs with palmately lobed or divided very mucilaginous lvs. *Epicalyx of 3 segs free to the base*, arising close beneath the calyx. Calyx of 5 sepals, united below. Petals 5, free, cuneate or obovate, emarginate or deeply notched, purple, rose or white. Fr. of numerous unbeaked 1-seeded nutlets arranged in a flat whorl round the short conical apex of the receptacle.

Thirty spp. in north temperate regions.

1 Stem-lvs deeply palmatisect with slender segments.
　　　　　　　　　　　　　　　　1. moschata
　　Stem-lvs ±palmately lobed or roundish-crenate.　　2
2 Perennial; petals 15–30 mm, 3–4 times as long as the sepals.　　　　　　　　　　**2. sylvestris**
　　Annual to biennial; petals usually 4–13 mm (rarely to 20 mm), up to twice as long as the sepals (rarely three times as long).　　　　　　　3
3 Epicalyx-segments ovate-lanceolate; lvs not or slightly cordate at base.　　　　**3. nicaeensis**
　　Epicalyx-segments linear-lanceolate; lvs distinctly cordate at base.　　　　　　4
4 Petals 8–13(–20) mm, 2–3 times as long as the sepals; nutlets pubescent but neither reticulate nor ridged on their dorsal face.　　　　**4. neglecta**
　　Petals 3–9 mm, usually about as long as the sepals and never more than twice as long; nutlets distinctly reticulate or at least transversely ridged on their dorsal face.　　　　5

5 Fls 1–1.5 cm diam., in dense axillary clusters, their stalks, even in fr., not more than twice as long as the sepals; nutlets transversely ridged or obscurely reticulate; robust erect plants, sometimes with crisped lvs.   **7. verticillata**
   Fls c. 0.5 cm diam., not in dense clusters, their stalks in fr. many times longer than the sepals; nutlets distinctly reticulate.   6
6 Calyx herbaceous, little enlarged in fr.; nutlets with acute but neither winged nor toothed dorsal angles.
                                              5. pusilla
   Calyx almost scarious, much enlarged in fr.; nutlets with distinctly winged and toothed dorsal angles.
                                              6. parviflora

1. M. moschata L.   Musk Mallow.

A *perennial* herb with a branching stock producing several erect lfy shoots 30–80 cm, terete, often purple-spotted, with sparse simple hairs. Basal lvs 5–8 cm diam., reniform in outline, long-stalked, with 3 contiguous crenate lobes; *stem-lvs* successively shorter-stalked and more deeply divided, the 3–7 primary divisions *deeply pinnatifid into ±linear ultimate segments*; allsubglabrous; stipules small, lanceolate. Fls 3–6 cm diam., usually solitary in the axils of the upper lvs and in an irregularly racemose terminal cluster. Epicalyx-segments linear-lanceolate, narrowed at both ends, half as long as the calyx. Calyx-lobes ovate-deltate, erect and enlarging in fr. Petals rose-pink, rarely white, obovate-cuneate, truncate and deeply emarginate distally, c. three times as long as the calyx. Nutlets blackish when ripe, not rugose, hispid on the back, the back and sides not separated by an angle. Fl. 7–8. Visited by bees and other insects. $2n = 42$. H.

A very variable plant, especially in the degree of cutting of the lvs.

Native. Grassy places, pastures, hedgebanks, etc.; not uncommon on the more fertile soils. Throughout Great Britain northwards to Sutherland and the Inner Hebrides. Europe. N. Africa.

*M. alcea L., with stellate pubescence, ovate epicalyx-segments and glabrous nutlets, resembles *M. moschata* otherwise and sometimes occurs as a casual. Europe, especially central and southern. Both spp. are grown in gardens.

2. M. sylvestris L.   Common Mallow.

A *perennial* herb with an erect ascending or decumbent stem 45–90 cm, with sparse spreading hairs. Basal lvs 5–10 cm diam., roundish with very shallow crenate lobes, cordate at base, somewhat folded, long-stalked; stem-lvs with 5–7 rather deep crenate lobes, sparsely hairy, ciliate. *Fls* 2.5–4 cm diam., stalked, in axillary clusters, the whole forming an irregular raceme. Epicalyx-segments oblong-lanceolate, $\frac{2}{3}$ as long as the calyx. Calyx-lobes ovate-deltate, connivent and not enlarging in fr. *Petals* rose-purple with darker stripes, obovate-cuneate, deeply emarginate, 2–4 *times as long as calyx*. Nutlets brownish-green when ripe, the reticu-

late rugose usually glabrous back separated by a sharp angle from the transversely wrinkled sides. Fl. 6–9. Visited by various insects, especially bees. $2n = 42$. H.

Var. *lasiocarpa* Druce has the nutlets hairy.

Native. Roadsides, waste places, etc.; common in the south, less so in the north. Throughout the British Is., though local and coastal in N. Scotland. Not in Outer Hebrides, Orkney or Shetland. Throughout Europe.

*3. M. nicaeensis All.

An *annual* herb with ±ascending *hispid* stems 20–50 cm, and long-stalked palmately 5–7 lobed *lvs* which are *not or scarcely cordate* at the base; lobes of upper lvs ±acute. Fls in axillary clusters of 2–6, rarely solitary. *Epicalyx-segments ovate-oblong to broadly lanceolate*, about equalling the triangular-ovate sepals which enlarge little in fr. Petals 10–12 mm, narrowly cuneate, up to twice as long as the sepals, pale bluish-lilac. *Nutlets* glabrous or puberulent, *strongly reticulate*, the dorsal angles sharp but not toothed. Fl. 6–9. Th.

Introduced. A not infrequent casual, occasionally establishing itself. Mediterranean region and eastwards to Baluchistan, but introduced in C. and W. Europe.

4. M. neglecta Wallr.   Dwarf Mallow.

M. rotundifolia auct., non L.; *M. vulgaris* Fries, non S. F. Gray

An *annual* or longer-lived herb with stems 15–60 cm, decumbent or with the central stem ascending, *densely clothed in stellate down*. Lvs 4–7 cm diam., long-stalked, roundish-reniform, deeply cordate, with 5–7 shallow acutely crenate lobes, ±pubescent; stipules ovate. *Fls* 1.8–2.5 cm diam., in irregular racemes. *Epicalyx-segments linear-lanceolate*, c. $\frac{1}{2}$–$\frac{2}{3}$ as long as the calyx. *Calyx-lobes* broadly ovate-deltate with stellate pubescence, slightly connivent with reflexed tips, *not enlarging in fr. Petals* whitish with lilac veins or pale lilac, obovate-cuneate *with bearded claws*, deeply emarginate, c. *twice* (or up to three times) *as long as the calyx*. Nutlets brownish-green, in a disk 6–7 mm diam., with the pubescent but non-reticulate back and sides of each nutlet separated by a blunt angle, and with a straight line of contact between neighbours. Fl. 6–9. Little visited by insects and automatically self-pollinated. $2n = 42$. Th.–H.

Native. Waste places, roadsides, drift-lines, etc.; frequent in the south, less so in the north. Throughout the British Is. except Hebrides, Orkney and Shetland. Europe, Asia, N. Africa; introduced in North America.

The following annual spp., often mistaken for *M. neglecta*, occur as casuals or locally naturalized aliens:

*5. M. pusilla Sm.

M. rotundifolia L., nom. ambig.; *M. borealis* Wallr.

An annual herb resembling *M. neglecta* but the *fls only* c. 0.5 *cm diam.*, epicalyx segments equalling the deltate ciliate calyx-lobes; *petals pale pinkish, barely exceeding the calyx,* claws glabrous; and nutlets in a disk 7–9 mm diam., each *nutlet reticulate,* ±glabrous or more usually pubescent (var. *lasiocarpa* Salmon), with *sharp but not winged or toothed dorsal angles. Calyx hardly enlarged in fr., herbaceous,* slightly connivent but with reflexed tips. Fl. 6–9. Self-pollinated. $2n = 42, 76$. Th.

Introduced. Waste places, foreshores, etc.; local northwards to Aberdeen. N. and C. Europe.

*6. M. parviflora L.

An annual herb with pale lilac fls 4–5 mm diam., and glabrous petal-claws, closely resembling *M. pusilla* but with broadly ovate calyx-lobes, spreading *calyx, scarious and greatly enlarging in fr.,* and glabrous or pubescent reticulate *nutlets with toothed and winged dorsal angles,* so that there is a wavy line of contact between neighbours.

Introduced. A casual, occasionally becoming established. Mediterranean and W. Asia, widely introduced.

*7. M. verticillata L.

Like **6** but with *erect stems* to 80 cm or more, *glandular-hairy to almost glabrous.* Lower lvs long-stalked, shallowly 5–7-lobed, cordate at base, green and sparsely hairy above, bluish-green and more densely hairy beneath. *Fls* 1–1.5 cm diam., *short-stalked in dense axillary clusters.* Epicalyx-segments linear. Calyx-lobes ovate, acute. Petals pale rose, up to twice as long as the calyx. *Calyx enlarging in fr. Nutlets* glabrous, obscurely reticulate, *with transverse ridges extending all or part of the way to the median longitudinal ridge down the dorsal face;* dorsal angles squarish but neither winged nor toothed. Fl. 7–9. $2n = $ c. 84.

Var. *crispa* L., with crisped lvs and whitish petals hardly exceeding the calyx, has been grown as a salad plant.

Introduced. Casual and occasionally becoming established. E. Asia, but very widely naturalized owing to its use as a vegetable and as a medicinal plant.

2. LAVATERA L.

Annual to perennial herbs or woody plants usually covered with stellate pubescence. Closely resembling spp. of *Malva* but commonly larger and with the 3 *epicalyx segments united at the base,* forming an involucre-like lobed cup below the calyx.

About 25 spp., chiefly in the Mediterranean region but also in the Canary Is., Australia and the islands off S. and Baja California.

Plant 60–300 cm, erect, woody below; epicalyx enlarging in fr.; nutlets transversely wrinkled with acute raised dorsal angles.   **1. arborea**
Plant 50–150 cm, erect or ascending, herbaceous, hispid; epicalyx not enlarging in fr.; nutlets not transversely wrinkled, with blunt dorsal angles.   **2. cretica**

1. L. arborea L.   Tree Mallow.

A suffruticose biennial, almost tree-like, with stout erect stems 60–300 cm, reaching 2.5 cm diam., *woody* below,

thinly and softly stellate-pubescent above. Lvs to 20 cm diam., roundish-cordate, stalked, with 5–7 broadly triangular ±acutely crenate lobes, softly velvety with stellate pubescence, ±folded like a fan. Fls 3–4 cm diam. in terminal simple or compound racemes. *Epicalyx* with broadly ovate blunt segments *exceeding the calyx,* joined almost half-way, *much enlarging in fr.* Calyx-lobes ovate acute. Petals broadly obovate-cuneate, overlapping, 2–3 times as long as the calyx, pale rose-purple with broad deep purple veins confluent below. *Nutlets* yellowish, *transversely wrinkled* and with acute *raised dorsal angles,* glabrous or pubescent. Fl. 7–9. Protandrous. $2n = 36, 42$. H. (biennial).

Native. Maritime rocks or waste ground near the sea, to 150 m. S. and W. coasts of Great Britain from Dorset to Cornwall and northwards to Cumberland; Isle of Man; Ailsa Craig; probably introduced elsewhere on the S. coast and on the E. coast northwards to Bass Rock and E. Ross; S. Ireland and Antrim, casual elsewhere. Coasts of Mediterranean region and Atlantic coast of W. Europe from Spain northwards to N. France and British Is.; Canary Is.

2. L. cretica L.

L. sylvestris Brot.

An annual or biennial herb with erect or ascending (rarely prostrate) stellate-pubescent *herbaceous* stems 50–150 cm. Lower lvs roundish-cordate; upper ±truncate below, with 5 triangular-acute toothed lobes. Fls in irregular racemose panicles. *Epicalyx*-segments broadly ovate, spreading, *somewhat shorter than the calyx, not enlarging in fr.* Calyx-lobes ovate, abruptly acuminate. Petals obovate-cuneate, not contiguous, deeply emarginate, 2–3 times as long as the calyx, lilac. *Nutlets* yellowish, *almost smooth,* with *blunt dorsal angles,* glabrous or pubescent. Fl. 6–7. $2n = 40-44$; c. 112. Th.–H. (biennial).

Native. Waysides and waste places near the sea. W. Cornwall, Scilly Is. and Jersey; casual elsewhere. Mediterranean region and Near East; W. Europe northwards to Brittany and W. Cornwall.

L. trimestris L., an erect bushy annual up to 1 m or more, native in the Mediterranean region and Portugal, is much grown in gardens for its abundant bright pink fls up to 10 cm diam. It is found occasionally as a casual, perhaps from birdseed.

3. ALTHAEA L.

Annual to perennial herbs resembling *Malva* and *Lavatera* but with an *epicalyx of 6–9 segments joined below into a cup-like involucre.*

About 15 spp. in temperate Europe, Asia and N. Africa.

1 Annual; hispid with swollen-based bristles.   **2. hirsuta**
  Biennial or perennial; not hispid.   *2*
2 Softly pubescent; lvs with shallow acute lobes.
                1. officinalis
  Sparsely hairy; lvs with blunt lobes.   **3. rosea**

1. A. officinalis L.   Marsh Mallow.

A *perennial* herb with a thick fleshy pale stock tapering downwards into a tap-root, and erect stellate-pubescent velvety stems 60–120 cm, simple or slightly branched. Lower lvs 3–8 cm across, roundish, stalked; upper *lvs* narrower and ±ovate; all slightly 3–5-lobed, the upper more deeply and acutely so, irregularly toothed, *velvety*, folded like a fan. Fls 2.5–4(–5) cm diam., 1–3 in each upper lf-axil to form an irregularly racemose infl.; their *stalks shorter than the lvs*. *Epicalyx-segments* 5–7 mm, 8–9, narrowly triangular. *Sepals* 8–10 mm, *ovate-acuminate*, velvety like the lvs and epicalyx. Petals 2–3 times as long as the sepals, obovate-cuneate, truncate and somewhat emarginate, pale pink. *Nutlets* brownish-green, *pubescent*, with the rounded *smooth* back separated from each face by a marginal ring. Calyx curved over the fr. Fl. 8–9. Protandrous; visited by various bees and automatically self-pollinated. $2n = 42$. H.

Native. Upper margins of coastal salt and brackish marshes and on ditch sides and banks near the sea. Locally frequent on S. coast of England from Dorset eastwards and on the E. coast northwards to the Wash; also in Somerset and Gloucester, S. Wales and S. Lancs; S.W. Ireland. Probably introduced in other coastal and inland stations northwards to S. Scotland. Through much of Europe southwards of England, Denmark and C. Russia; N. Africa, W. Asia. Introduced in North America.

The roots yield abundant mucilage used in confectionery and medicinally as a demulcent and emollient. Also grown in gardens.

2. A. hirsuta L.   Hispid Mallow.

An *annual or biennial* herb with many ascending slender *hispid* stems 8–60 cm. Lower lvs 2–4 cm across, long-stalked, reniform, ±5-lobed, the lobes blunt to subacute, crenate; upper lvs short-stalked, deeply 3–5-lobed or palmately divided, the crenate-toothed segments becoming narrower and more acute towards the top of the stem. Fls 2.5 cm diam., solitary, axillary and in a terminal racemose cluster, their *stalks* long, slender, *exceeding the lvs*. Epicalyx-segments 7–10 mm, narrowly triangular, hispid, enlarging in fr. *Sepals* 12–15 mm, narrowly *lanceolate-acuminate*, hispid, like the stems and epicalyx, with swollen-based bristles. Petals exceeding the sepals, obovate-truncate, pale rosy-purple, becoming bluish. *Nutlets* dark brownish-green, *glabrous*, with the rounded slightly keeled back and the lateral faces transversely ridged; *fruiting calyx erect*. Fl. 6–7. Th.–H. (biennial).

Doubtfully native. A lowland weed of cultivated ground and dry places up to 245 m; rare and local, N. Somerset and W. Kent; Jersey. Casual elsewhere, northwards to Humberside and Cumberland. S. and S.E. Europe; W. Asia.

*3. A. rosea (L.) Cav.   Hollyhock.

A tall biennial to perennial herb with stout erect hairy stems to 3 m. Lvs to 30 cm across, long-stalked, roundish-cordate, 5–7-lobed or angled, crenate, roughly hairy. Fls 6–7 cm diam., ±sessile in axillary clusters forming a long irregular spike-like infl. Petals variable in colour, often white, yellow or red, sometimes very dark. Fl. 7–9. H. (sometimes biennial).

Introduced. Native in China. Forms of *A. rosea*, of **A. ficifolia** (L.) Cav. (W. Asia), with deeply 5–9-lobed lvs, and of hybrids between them (*A.* × *cultorum* Bergmans) are much grown in gardens and may escape.

---

45. LINACEAE

Herbs, shrubs or trees. Lvs usually alternate, rarely opposite, simple, entire or weakly toothed, with or without stipules. Infl. a cyme or pseudo-raceme, rarely fls solitary. Fls hermaphrodite, often heterostylous, heterochlamydeous, actinomorphic, hypogynous, 5-(rarely 4-)merous. Sepals usually imbricate, persistent. Petals usually contorted, often shed early. Stamens equal in number to and alternating with petals, often with small staminodes in addition, less often twice as many; filaments ±connate at base; anthers introrse. Ovary (2–)3–5-celled; cells often nearly halved by a false septum; ovules (1–)2 in each cell, axile, collateral (when 2), anatropous, micropyle directed upwards, raphe ventral; styles free, less often ±united. Fr. a loculicidal capsule or dehiscing also down the false septum (and so with twice as many lines of dehiscence as carpels), rarely a drupe. Endosperm usually scanty, sometimes copious or 0; embryo straight or slightly curved.

About 12 genera and 290 spp., tropical and temperate.

Most like the *Oxalidaceae* but can be distinguished from them by the simple lvs and dehiscence of the capsule and, in our spp., by the fewer stamens.

Fls 5-merous; petals longer than calyx; sepals entire at
   apex.   1. LINUM
Fls 4-merous; petals not longer than calyx; sepals toothed
   or lobed at apex; plant small and inconspicuous.
                                    2. RADIOLA

1. LINUM L.

Herbs, rarely undershrubs. *Lvs sessile, usually narrow, parallel-veined or with only the midrib prominent. Fls cymose, 5-merous. Sepals imbricate, entire. Petals clawed. Stamens 5, united into a tube at the base, with tooth-like staminodes between. Ovary 5-celled, ovules 2 in each cell, separated by false septa. Fr. dehiscing by 10 valves. Seeds flat.*

About 230 spp., subtropical and temperate, mainly northern hemisphere.

A number of spp. including the annual *L. grandiflorum* Desf. with crimson fls and others with yellow fls are grown in gardens.

1 Petals blue, 8 mm or more; lvs linear, alternate.   2
    Petals white, 6 mm or less; lvs oblong or obovate,
      opposite.   **3. catharticum**
2 Sepals all acuminate, more than half as long as capsule
    in fr., inner glandular-ciliate.   **1. bienne**
    Inner sepals very obtuse, outer obtuse to acuminate,
      all entire, less than half as long as capsule in fr.;
      inner entire.   **2. perenne**

1. L. bienne Miller   Pale Flax.

L. angustifolium Hudson

Glabrous and subglaucous, biennial or perennial (rarely annual), 30–60 cm with several stems from the base. Stems ascending or suberect, rather rigid, often flexuous, ±branched. *Lvs alternate*, numerous, *linear*, acute, 1–2.5 cm and mostly 3-veined on the main stem, entire. Fls in a lax cyme; bracts like the lvs but smaller; pedicels slender, c. 1–2 cm. *Sepals all ovate, acuminate* or the inner apiculate; *inner with a scarious glandular-ciliate border*, outer entire; *more than ½ as long as the capsule in fr. Petals 8–12 mm, pale blue*, obovate, with a short claw. *Stigmas club-shaped.* Capsule c. 6 mm, globose-conical. Seeds c. 3 mm, ovate, shining, not beaked. Fl. 5–9. $2n = 30$. Th. or Hp.

Native. Dry grassland, especially near the sea. From Isle of Man, N. Wales and Nottingham southwards, local and with a western tendency, commonest in the south and west; east and south Ireland from Meath to Kerry. South Europe extending further north only in the west to W. and C. France; Madeira, Canary Is.

L. usitatissimum L.   Flax.

Differs from *L. bienne* as follows: More robust, always annual. Stem usually solitary. Lvs and fls larger (petals c. 1.5 cm). Inner sepals ciliate or not. Capsule larger, 1 cm or more, seeds with a short obtuse beak. $2n = 30, 32$. Th.

Cultivated for the production of fibre for linen or of oil (linseed) and sometimes persisting for a year or two or occurring as a casual but scarcely naturalized. Origin unknown, possibly derived from *L. bienne*.

2. L. perenne L.   Perennial Flax.

L. anglicum Miller

Glabrous glaucous perennial 30–60 cm with ±numerous stems. Stems ascending, rather rigid, often curved, ±branched or nearly simple. *Lvs alternate*, very numerous, 1–2 cm on the main stems, linear, acute, 1-veined, entire. Fls in a lax cyme; bracts like the lvs but smaller; pedicels slender, c. 1–2 cm. Outer sepals oval or elliptical, obtuse, acute or apiculate; *inner broader oval or obovate, rounded at apex*, with an *entire* scarious margin; *all less than ½ as long as capsule in fr. Petals 15–20 mm, sky blue*, obovate with a short claw. *Stigmas*

capitate. Capsule c. 7 mm, globose or oval. Seeds 4–5 mm, oblong-ovate, matt, obscurely beaked. Fl. 6–7. Hp.

Native. Calcareous grassland; E. England from N. Essex to Durham, extending west to Cambridge, Leicester, Westmorland and Kirkudbrightshire, very local. E. and C. France southwards to C. Spain, northwards to N. Germany, and eastwards to the Urals.

British plants belong to the endemic subsp. **anglicum** (Miller) Ockendon.

3. L. catharticum L.   Fairy Flax.

Slender glabrous annual 5–25 cm. Stems usually solitary, sometimes several, simple (except in infl.), erect, wiry. *Lvs* 5–12 mm, *opposite*, distant, *oblong or obovate*, subobtuse, 1-veined, entire. Fls many, in a lax often dichasial cyme; bracts partly alternate, upper small, linear, lower passing into the lvs; pedicels slender, c. 5–10 mm. Sepals ovate-lanceolate, acuminate. *Petals* 4–6 mm, *white*, narrowly obovate. Stigmas capitate. Capsule c. 3 mm, globose. Fl. 6–9. Pollinated by various insects or selfed, homogamous. $2n = 16$, c. 57. Th.

Native. Grassland, heaths, moors, rock-ledges and dunes, ascending to 855 m, especially common and characteristic of calcareous grassland but by no means confined to basic soils. Common throughout the British Is. Europe northwards to 69° N in Fennoscandia, mainly on mountains in the south; Caucasus; Asia Minor, Iran.

2. RADIOLA Hill

Small annual herb, differing from *Linum* as follows; *fls 4-merous; sepals (2–)3(–4)-lobed or toothed at the apex; seeds ovoid.*

One sp.

1. R. linoides Roth   Allseed.

R. millegrana Sm.

Very small delicate annual 1.5–8 cm. Stems filiform, sometimes simple but more frequently several times dichotomously branched so that the habit is bushy. Lvs opposite, 3 mm or less, ovate-elliptical or elliptical, acute to subobtuse. Fls very numerous in a dichasial cyme; pedicels short; bracts not differentiated. Sepals scarcely 1 mm. Petals about equalling sepals, white. Capsule globose, scarcely 1 mm. Fl. 7–8. Probably almost always self-pollinated. $2n = 18$. Th.

Native. Damp bare sandy or peaty ground on grassland and heaths. Spread over the whole of Great Britain but local everywhere and absent from a number of counties particularly in the Midlands and in S. Scotland; in Ireland frequent near the west coast and in Wexford. Most of Europe but not the north-east and extreme north; N. Africa; Madeira; Tenerife; temperate Asia; mountains of tropical Africa.

46. GERANIACEAE

Herbs or small shrubs (very rarely tree-like). Lvs usually alternate, lobed or compound, lflets not joined to rhachis, usually stipulate. Infl. cymose or an umbel, sometimes fls solitary. Fls hermaphrodite (very rarely plant dioecious), heterochlamydeous, actinomorphic or slightly zygomorphic, 5- (rarely 4- or 8-)merous, hypogynous. Sepals persistent, usually imbricate. Petals usually imbricate, occasionally 0. Stamens usually obdiplostemonous, rarely in 3 whorls, sometimes some sterile. Filaments usually ±connate at the base. Ovary superior, of 5 united carpels, each with a long beak ending in the free lingulate stigma; ovules axile, usually 2 in each cell, superposed, rarely solitary, anatropous, micropyle directed upwards, raphe ventral. Fr. of five 1-seeded mericarps, usually opening septicidally from base to apex. Endosperm scanty or 0; embryo usually curved, rarely straight.

Five genera and about 750 spp., mainly temperate and subtropical. The garden scarlet and ivy-leaved 'Geraniums' belong to the genus *Pelargonium* L'Hérit., differing from *Geranium* in the zygomorphic fls with a spur adnate to the pedicel.

Lvs palmate or palmately lobed; beak of mericarp rolling upwards in dehiscence and releasing the seeds.
1. GERANIUM
Lvs pinnate or pinnately lobed; beak of mericarp twisting spirally in dehiscence and remaining attached to seeds.
2. ERODIUM

1. GERANIUM L.

Herbs. Lvs palmate or palmately lobed. Infl. cymose, ultimate peduncles usually 2-fld. *Fls actinomorphic or nearly so, without spur. Stamens 10, all fertile (except G. pusillum). Beak of mericarp usually rolling upwards at dehiscence remaining attached by its apex, releasing the seeds*, beak sometimes curling away at the apex also but never coiling spirally.

About 400 spp. mainly temperate. Some are grown in gardens and about 15 have been found as escapes or casuals. The relative lengths of the lobes and entire portion of the lvs in the following account are measured down the main axis of the lobes.

1 Perennial; corolla large (petals usually 10 mm or more). 2
  Annual; corolla small (petals less than 10 mm, sometimes to 12 mm but then with a claw as long as or longer than limb). 11
2 Petals with claw at least ½ as long as limb; stamens curving downwards; rhizome very stout, creeping; petals pink. **8. macrorrhizum**
  Petals without or with short claw; stamens spreading radially. 3
3 Lvs reniform or orbicular, the lobes ±cuneiform; petals 7–10 mm, deeply emarginate; root vertical, fusiform; rhizome inconspicuous. **9. pyrenaicum**
  Lf-lobes not cuneiform; petals 10 mm or more; rhizome conspicuous, oblique or horizontal. 4
4 Fls solitary; petals purplish-crimson, rarely pink or white; lobes of lvs narrow throughout their length. **7. sanguineum**
  Fls mostly in pairs; petals not purplish-crimson; lobes of lvs broader about the middle. 5
5 Petals blackish-purple, apiculate at apex, spreading or slightly reflexed so that the fl. is ±flat; sepals mucronate (the point less than 0.5 mm). **6. phaeum**
  Petals not blackish-purple, rounded or emarginate at apex, less spreading, fl. ±cup-shaped; sepals aristate (the point c. 1 mm). 6

6 Petals rounded at apex (sometimes emarginate in *G. endressii*), veins not darker than rest of petal; lvs 5–7-lobed. 7
  Petals emarginate, veins darker than rest of petal; lvs 3–5-lobed. 9
7 Fls pink; rhizome horizontal, creeping; plant almost eglandular. **3. endressii**
  Fls violet or blue; rhizome oblique, not creeping; plant glandular above. 8
8 Pedicels recurved after fl.; lowest segments of the lf-lobes c. 4–5 times as long as entire portion. **1. pratense**
  Pedicels erect after fl.; lowest segments of lf-lobes 1–2 times as long as entire portion. **2. sylvaticum**
9 Fls pink with darker veins. **endressii × versicolor**
  Fls white or pale lilac with lilac veins. 10
10 Stems with spreading hairs; lobes of lvs lobed or deeply dentate. **4. versicolor**
  Stems nearly glabrous; lobes of lvs shallowly crenate-dentate. **5. nodosum**
11 Sepals ±spreading; lvs dull- or grey-green. 12
  Sepals erect, somewhat connivent near the apex; lvs bright green. 17
12 Lvs divided nearly to base (at least ⅚) into linear lobes; lobes (except sometimes those of the lowest lvs) pinnatifid with linear segments; sepals aristate (the point 1–2 mm). 13
  Lvs divided to ¾ or less, lobes widened above, trifid or 3-lobed; sepals mucronate (the point less than 0.5 mm). 14
13 Pedicels 2 cm or more; carpels usually glabrous. **10. columbinum**
  Pedicels not more than 1.5 cm; carpels pubescent. **11. dissectum**
14 Petals entire; seeds pitted. **12. rotundifolium**
  Petals emarginate; seeds smooth. 15
15 Perennial; petals 7–10 mm. **9. pyrenaicum**
  Annual; petals not more than 7 mm. 16
16 Petals 3–7 mm mericarps glabrous; stamens all with anthers. **13. molle**
  Petals 2–4 mm; mericarps pubescent; 5 of the stamens without anthers. **14. pusillum**
17 Lvs 5–7-lobed to halfway or less. **15. lucidum**
  Lvs ternate or palmate. 18
18 Petals (8–)9–12 mm; anthers orange to purple. **16. robertianum**
  Petals 6–9 mm; anthers yellow. **17. purpureum**

1. G. pratense L.   Meadow Crane's-bill.

Perennial herb 30–80 cm; *rhizome stout, oblique. Stems* erect or ascending with short deflexed hairs below, densely pilose and *glandular-hairy above* (like the fl.-stalks and calyx). *Basal lvs* (Fig. 36A) polygonal in outline, blade 7–15 cm diam., appressed hairy on both sides, long-petiolate, *deeply 5–7-lobed* (lobes c. 5 times as long as entire portion) the 2 basal lobes often contiguous; lobes 7–10 mm across at base, ovate-rhombic in outline, pinnately lobed; secondary lobes with the lower margin usually concave, *the basal ones 4–5 times as long as the entire portion, themselves lobed*; ultimate lobes ±oblong, acute, apiculate; cauline lvs smaller, decreasing upwards, on shorter petioles, the uppermost 3-lobed, subsessile. Fls in pairs on axillary peduncles.

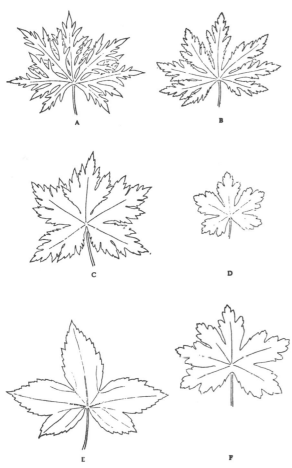

erect, with short deflexed hairs below, pilose and *glandular-hairy above* (like the fl. stalks and calyx). *Basal lvs* (Fig. 36B) polygonal in outline, blade 7–12 cm diam., hairy on both sides, long-petiolate, *deeply (5–)7-lobed* (lobes c. 5 times as long as entire portion) the 2 basal lobes separated; lobes 8–15 mm across at base, obovate-rhombic in outline, pinnately lobed; secondary lobes ±triangular in outline (but with the lower edge much longer than the upper), the lower edge mostly straight, *the basal ones 1–2 times as long as the entire portion*, deeply and coarsely dentate but scarcely lobed; cauline lvs smaller, decreasing upwards, on shorter petioles, the uppermost subsessile. Fls in pairs on axillary peduncles, more numerous than in *G. pratense* and forming a lax cyme. *Pedicels erect after fl.* Sepals ovate-lanceolate, aristate, spreading. *Petals* 12–18 mm, *blue-violet* (usually smaller and more reddish than in *G. pratense*), obovate, *apex rounded*; claw very short; corolla ±cup-shaped. Filaments of stamens somewhat widened below into a lanceolate base. Mericarps glandular-hairy, smooth; beak c. 2.5 cm. Seeds minutely reticulate. Fl. 6–7. Pollinated by various insects, protandrous; homogamous and unisexual fls also recorded. $2n = 28$. Hs.

Native. Meadows, hedgebanks, damp woods and mountain rock-ledges, ascending to c. 1060 m; from Caithness to C. Yorks; Worcester and Gloucester; introduced elsewhere; Antrim. Most of Europe, but only on mountains in the south and absent from many islands; Caucasus, Siberia to the Yenisei region; naturalized in North America.

*3. G. endressii Gay   French Crane's-bill.

Perennial herb 30–80 cm; *rhizome slender, horizontal, long.* Stems ±erect with ±spreading hairs, *not glandular* (but the pedicels and calyx somewhat glandular). Blades of basal lvs (Fig. 36c) 5–8 cm diam., polygonal in outline, hairy on both sides, long petiolate, deeply 5-lobed (lobes c. 5 times as long as entire portion); lobes 10–15 mm across at base, broadly ovate-rhombic in outline, pinnately lobed; secondary lobes with the lower margins straight or slightly concave, the basal ones 1–2 times as long as the entire portion, deeply dentate or lobed; cauline lvs smaller, decreasing upwards, the uppermost 3-lobed, subsessile. Fls in pairs on long peduncles. Pedicels erect after fl. Sepals ovate-lanceolate, aristate, spreading. *Petals* c. 16 mm, *deep pink, the veins not darker*, obovate, *apex rounded* or somewhat emarginate, tapered at base into a short claw; corolla ±cup-shaped. Mericarps pubescent, somewhat glandular, smooth; beak c. 2 cm. Fl. 6–7. $2n = 28$. Hs.

Introduced. Grown in gardens, sometimes escaping and ±naturalized by roadsides, etc., in a number of places throughout Great Britain and in Down. Native of the W. Pyrenees.

G. endressii × versicolor.

Petals pale pink with darker veins, emarginate. Cultivated in gardens, sometimes escaping and locally naturalized in Britain and Ireland.

Fig. 36. Leaves of *Geranium*. A, *G. pratense*; B, *G. sylvaticum*; c, *G. endressii*; D, *G. versicolor*; E, *G. nodosum*; F, *G. phaeum*. ×⅓.

Pedicels reflexed after fl., becoming erect in fr. Sepals ovate, aristate, spreading. *Petals* 15–18 mm, *bright violet-blue*, obovate, *apex rounded*; claw very short; corolla ±cup-shaped. Filaments of stamens much widened below into a broadly ovate base. Mericarps glandular-hairy, smooth; beak c. 2.5 cm. Seeds minutely reticulate. Fl. 6–9. Pollinated mainly by Hymenoptera, protandrous. $2n = 28^*$. Hs.

Native. Meadows and roadsides, ascending to 535 m, widespread but rather local in England (introduced in Devon and Cornwall), Wales and S. and C. Scotland; introduced in N. Scotland (to Orkney); Antrim (native); an escape elsewhere in Ireland. Europe from Scandinavia (64° 12′ N) and Finland southwards but rare in the Mediterranean region; N. and C. Asia to Ussuri region, Japan and the W. Himalaya; naturalized in North America.

2. G. sylvaticum L.   Wood Crane's-bill.

Perennial herb 30–80 cm; *rhizome stout, oblique.* Stems

*4. G. versicolor L.   Pencilled Crane's-bill.
G. striatum L.

Perennial herb 30–60 cm; rhizome slender, horizontal. *Stems* ±erect, *with spreading hairs*, not glandular. Blades of basal lvs (Fig. 36D) 4–8 cm diam., polygonal in outline, with scattered short hairs on both sides, deeply (3–)5-lobed (lobes 3–4 times as long as entire portion), basal sinus wide; *lobes* 10–15 mm across at base, obovate in outline, *pinnately lobed*; secondary lobes with the lower margins ±straight, the basal ones less than half as long as the entire portion, coarsely dentate; cauline lvs often 5-lobed with the 2 basal lobes much shorter than the other 3. Fls in pairs on long peduncles. Pedicels erect after fl., with small sessile glands. Sepals lanceolate-aristate, sparingly hairy, spreading. *Petals* 15–18 mm, *white or pale lilac with violet veins*, obovate-cuneiform, *deeply emarginate*, tapered at the base into a short claw; corolla ±cup-shaped. Mericarps sparingly pubescent, smooth; beak 2–3 cm. Fl. 5–9. Visited by bees. $2n = 28$. Hs.

Introduced. Grown in gardens and naturalized on hedgebanks, etc., in a number of places, mainly in S. England and S. Wales (common in Cornwall), Anglesey, Isle of Man; E. and S. Ireland. Native of C. and S. Italy, Sicily and the southern part of the Balkan peninsula.

*5. G. nodosum L.   Knotted Crane's-bill.

Perennial herb 20–50 cm. Differs from *G. versicolor* as follows: *stems nearly glabrous, or with short reflexed and appressed hairs*, swollen at the nodes. Lvs (Fig. 36E) more often 3-lobed, *the lobes crenate-dentate*, not lobed. Petals lilac with violet veins. Mericarps with a transverse ridge at the apex. Fl. 5–9. $2n = 28$. Hs.

Introduced. Sometimes found as a garden escape and ±naturalized but apparently rare. It has been much confused with *G. endressii*. Native of the mountains from C. France to the Pyrenees, C. Italy and C. Yugoslavia.

*6. G. phaeum L.   Dusky Crane's-bill.

Perennial herb 30–60 cm; rhizome short, stout, oblique. Stems erect, with spreading hairs, glandular above (like the fl.-stalks). Blades of basal lvs (Fig. 36F) 7–12 cm, polygonal in outline, with numerous long hairs above, shortly pubescent beneath with a few long hairs mainly on the veins, deeply and often irregularly 5–7-lobed (deepest lobes usually c. twice as long as entire portion), often with a blackish blotch at the sinuses, basal sinus wide; lobes 10–20 mm across at base, usually ±oblong in outline, irregularly lobed towards the apex; secondary lobes dentate; cauline lvs soon smaller, the uppermost sessile, very small. Fls in pairs on long axillary peduncles, ±numerous and forming a terminal infl. Pedicels erect or ascending after fl. *Sepals* oblong-lanceolate, *mucronate*, spreading. *Petals* c. 1 cm, *blackish-purple*, obovate-orbicular, *apiculate at apex*, claw very short or obsolete, *spreading at right angles so that the corolla is flat* or slightly reflexed. Mericarps hairy, with

several strong transverse ridges near the apex; beak c. 1.5 cm. Fl. 5–6. Pollinated by bees, protandrous. $2n = 28$. Hs.

Introduced. Grown in gardens and naturalized on hedgebanks, etc. throughout Great Britain north to Inverness, and in N. Ireland and Mayo. Native of the mountains of C. and S. Europe; naturalized elsewhere.

7. G. sanguineum L.   Bloody Crane's-bill.

Perennial herb 10–40 cm, often of rather bushy habit; rhizome stout, horizontal, creeping. Stems procumbent to erect, often branched from the base, and geniculate at the nodes, with numerous long, spreading or slightly deflexed hairs. Basal lvs 0. Lf-blades (Fig. 37A) 2–6 cm

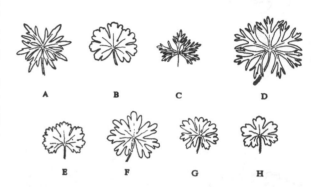

Fig. 37. Leaves of *Geranium*. A, *G. sanguineum*; B, *G. pyrenaicum*; C, *G. columbinum*; D, *G. dissectum*; E, *G. rotundifolium*; F, *G. molle*; G, *G. pusillum*; H, *G. lucidum*. ×$\frac{1}{3}$.

diam., orbicular or somewhat polygonal in outline, with stiff appressed white hairs on both sides, petiolate, deeply 5–7-lobed (the lobes c. 4–7 times as long as the entire portion); *lobes narrow*, not widened in the middle, 1.5–6 mm across at the base, mostly trifid but some bifid; secondary lobes entire or themselves bifid or trifid; *ultimate lobes oblong or linear* or sometimes somewhat broadened towards the apex. Fls solitary or rarely a few in pairs, on long peduncles. Sepals oval or elliptical aristate, spreading. *Petals* 12–18 mm, *bright purplish-crimson*, rarely pink or white, emarginate, with very short claw. Mericarps sparingly hairy, smooth; beak c. 3 cm. Seeds minutely dotted. Fl. 7–8. Pollinated by various insects, protandrous but capable of self-pollination. $2n = 84^*$. Hp.

Native. Grassland, woods and among rocks on basic soils and on fixed dunes, etc., ascending to 365 m. From Caithness and the Inner Hebrides southwards to England and Wales west of a line from N. Lincoln to Devon, Cambridge; C.W. and C.E. Ireland; introduced elsewhere. Europe from S. Scandinavia to C. Spain and Portugal, Sicily and Greece, east to the Urals; Caucasus.

The form of maritime sands (var. *prostratum* (Cav.) Pers.) differs from the inland form in its procumbent (not ascending or erect) stems and its often broader lf-lobes; it occasionally occurs with pink fls (var. *lancas-trense* (With.) Druce).

*8. G. macrorrhizum L.

Perennial 10–30 cm. Rhizome very thick, horizontal. Basal lvs numerous; petioles long. Lf-blades 5–10 cm, orbicular in outline, shortly pubescent, deeply 5–7-lobed; lobes crenate-dentate or lobulate. Fls in pairs on rather short peduncles, forming a corymbose cyme. Pedicels erect after fl. Sepals ovate or oval, aristate, reddish, ±erect. *Petals* c. 15 mm (including claw) *pink*; limb obovate-orbicular, rounded at apex, as long as claw. Mericarps glabrous, ±smooth. Fl. 6–8. $2n = 46$. Hs.

Introduced. Escaped from gardens and established on village walls in a number of places in S. Devon. Native of the S. Alps, Apennines, S. and E. Carpathians and the Balkan peninsula.

9. G. pyrenaicum Burm. fil.   Hedgerow Crane's-bill.

G. perenne Hudson

Perennial herb 25–60 cm, with short, inconspicuous rhizome. Stem erect, with numerous short glandular and numerous or few long spreading glandular hairs. Blades of basal lvs (Fig. 37B) 5–8 cm, orbicular in outline, hairy on both sides, 5–9-lobed, the lobes 1–1½ times as long as the entire portion; lobes obovate-cuneiform in outline, irregularly 3-lobed at apex; margins of lower ⅔ quite entire; secondary lobes ¼–⅓ as long as the entire portion, themselves usually 3-lobed or -toothed at apex (additional lateral lobes sometimes occur); upper cauline lvs often 3-lobed. Fls in pairs, forming a loose cyme. Pedicels deflexed after fl., curving upwards near the apex so that the fr. is erect. Sepals ovate-oblong, mucronate, spreading. *Petals* 7–10 mm, *purple*, less often purplish-white, obovate-cuneiform, *deeply emarginate*. Mericarps *pubescent, smooth*; beak c. 1 cm. Seeds smooth. Fl. 6–8. Pollinated by various insects, protandrous but capable of self-pollination. $2n = 26, 28^*$. Hs.

Native. Hedgebanks, waste places and field margins. Common in S. and E. England, rarer westwards and northwards, extending north to E. Inverness and the Clyde Is.; scattered over Ireland but absent from the north-west and south-west. S. and W. Europe northwards to Sweden; Asia Minor; N. Africa; naturalized in North America.

10. G. columbinum L.   Long-stalked Crane's-bill.

Annual herb 10–60 cm, usually branched. Branches ascending or erect, with reflexed appressed hairs or subglabrous. Lower *lf-blades* (Fig. 37C) 2–5 cm diam., ±polygonal in outline, appressed hairy on both sides, on long petioles, *divided nearly or quite to the base* into 5–7 lobes, dull or grey-green; lobes 1–2 mm across at base, pinnately lobed; secondary lobes c. 5 times as long as entire portion, linear or oblong-linear, usually entire but occasionally lobed; upper lvs smaller and sometimes less lobed, on shorter petioles. Fls in pairs. *Peduncles* 2–12 cm; *pedicels* 2–6 cm, spreading and often curving upwards after fl. Sepals ovate, *aristate* (awn c. 2 mm), *with appressed hairs* mainly on the veins, *eglandular*, spreading. Petals 7–9 mm, purplish-pink,

obovate-cuneiform, rounded or truncate and often apiculate or with a few small teeth at the apex with short claw. *Mericarps glabrous or sparsely hairy*, smooth; beak 1.5–2 cm. Seeds minutely pitted. Fl. 6–7. $2n = 18^*$. Th.

Native. Open habitats in dry grassland and scrub, mainly on basic soils; throughout England and Wales but rather local, especially in the Midlands and north; in Scotland local and mainly in the S.E., extending north to Inverness; widespread in Ireland but local and mainly in the south. Europe, except the extreme north; Algeria, Tunisia; naturalized in North America.

11. G. dissectum L.   Cut-leaved Crane's-bill.

Annual herb 10–60 cm, usually branched and often of straggling habit. Branches ascending, densely clothed with reflexed but not appressed hairs. Lvs dull or grey-green. Lower lf-blades (Fig. 37D) 2–7 cm diam., orbicular or reniform in outline, ±hairy especially on the veins beneath, on long petioles, deeply divided into (5–)7 lobes; *lobes* 5–10 *times as long as entire portion*, 2–4 mm across at the base, pinnately lobed (or those of the lowest lvs trifid); secondary lobes of the trifid lobes often oblong and only twice as long as entire portion but of the pinnate ones often linear and 7 times as long, often again lobed; upper lvs smaller, on shorter petioles, with narrower lobes so that they are similar to the corresponding lvs of *G. columbinum*. Fls in pairs. *Peduncles* 0.5–2(–3) cm; *pedicels* 0.5–1.5 cm, spreading or ascending after fl. *Sepals* ovate-lanceolate, *aristate* (awn 1–2 mm), *pilose and glandular-hairy* (like the fl. stalks), spreading. Petals c. 5 mm, reddish-pink, obovate, emarginate at apex, with short claw. *Mericarps pubescent*, smooth; beak 7–12 mm. Seeds pitted. Fl. 5–8. Homogamous and protogynous, probably usually selfed, insect visitors few. $2n = 22^*$. Th.

Native. Cultivated and waste ground, grassland, hedgebanks, etc., ascending to 380 m. Common throughout the British Is. but local in N.E. Scotland. Europe except the extreme north and much of European Russia); S.W. Asia to Iran and Pamir-Alai region; N. Africa, Macaronesia; naturalized in North and South America.

12. G. rotundifolium L.   Round-leaved Crane's-bill.

Annual herb 10–40 cm, usually much-branched from the base. Branches erect or ascending, pubescent and glandular-hairy. Lvs dull or grey-green. Lower lf-blades (Fig. 37E) 3–6 cm diam., reniform in outline, pubescent on both sides, on long petioles, 5–7-*lobed to half-way or less*; lobes c. 1 cm across at base, cuneiform, contiguous, 3-lobed at apex; secondary lobes short, obtuse, often 2- or 3-toothed or shortly lobed; upper lvs smaller, often rather more deeply and acutely lobed on shorter petioles. Fls numerous, in pairs; peduncles 0.5–3 cm; pedicels 0.5–1.5 cm, spreading or somewhat deflexed after fl., turning upwards at apex. *Sepals* ovate-oblong, mucronate (mucro not 0.5 mm), pilose and glandular-hairy, spreading. *Petals* 5–7 mm, pink, obovate-cuneiform, *rounded or slightly retuse at apex*, with short claw. Mericarps pubescent, smooth; beak 10–15 mm. *Seeds pitted*. Fl. 6–7. Homogamous, probably usually selfed, insect visitors few. $2n = 26$. Th.

Native. Hedgebanks, wall-tops, etc. From Cornwall and Kent to Glamorgan, Northampton and Suffolk, local; casual further north; S. Cork to S. Wexford. Most of Europe except the north; S.W. and C. Asia to the W. Himalaya and the Balkhash region; N. Africa; Macaronesia; naturalized in North and South America.

13. G. molle L.   Dove's-foot Crane's-bill.

Annual herb 10–40 cm, branched from the base, ±densely clothed with long soft white hairs, glandular above. Branches decumbent or ascending. Lvs dull or grey-green. Lower lf-blades (Fig. 37F) 1–5 cm diam., orbicular or reniform in outline, on long petioles, rather irregularly 5–9-lobed; *lobes from rather longer than to twice as long as the entire portion*, cuneiform, contiguous or with narrow sinuses, 2–5 mm across at base, 3-lobed at apex; secondary lobes short, obtuse or subacute, often somewhat widened above, mostly entire; upper lvs smaller, on shorter petioles, more deeply and acutely lobed, with wider sinuses, the lobes often entire. Fls numerous, in pairs; peduncles 0.5–3 cm; pedicels 0.5–1.5 cm, ±ascending and often curved upwards after fl. *Sepals* ovate, *mucronulate* (or some without mucro), densely clothed with long white hairs and with some shorter glandular ones, spreading. *Petals* 3–6(–7) mm, bright rosy-purple, or sometimes whitish, *deeply emarginate*, with short claw. *Mericarps glabrous*, wrinkled or rarely (var. *aequale* Bab.) smooth, keeled; beak 5–8 mm. Seeds smooth. Fl. 4–9. Visited by various insects, ±homogamous, probably often selfed. $2n = 26^*$. Th.

Native. Dry grassland, dunes, waste places and cultivated ground, ascending to 535 m; common throughout the British Is. Var. *aequale* rare. Europe, except the extreme north; S.W. Asia to the Himalaya; N. Africa; Macaronesia. Naturalized or escaped in North and South America; New Zealand and Japan.

14. G. pusillum L.   Small-flowered Crane's-bill.

Annual herb 10–40 cm branched from the base, shortly and softly pubescent, glandular above. Branches decumbent or ascending. Lvs dull or grey-green. Lower lf-blades (Fig. 37G) 1–4 cm diam., ±orbicular in outline, on long petioles, 7–9-lobed; *lobes 2–3 times as long as entire portion*, somewhat widened above, not contiguous, 2–5 mm across at base, trifid towards apex; secondary lobes oblong, mostly undivided, much longer than the entire portion; upper lvs smaller, on shorter petioles but divided like the lower. Fls rather numerous, in pairs; peduncles 0.5–1(–3) cm; pedicels 0.5–1.5 cm, spreading or ascending and often curved upwards after fl. *Sepals* ovate, *mucronulate*, clothed with long white hairs (but less densely than in *G. molle*) and shorter glands, spreading. Petals 2–4 mm, pale dingy lilac, deeply emarginate with short claw. Filaments of 5 of

the stamens without anthers. *Mericarps pubescent*, smooth; beak c. 5–6 mm. *Seeds smooth*. Fl. 6–9. Protogynous, probably usually selfed, insect visitors few. $2n = 26$.

Native. Cultivated and waste ground and open habitats in dry grassland. Widespread and rather common in England and E. Wales; E. Scotland north to Caithness, rare in the west; in a few scattered places in coastal E., S. and S.W. Ireland. Most of Europe, except the extreme north; Asia Minor to the Caucasus, Himalaya and Israel. Naturalized in North America.

15. G. lucidum L.   Shining Crane's-bill.

Lfy annual 10–40 cm, usually branched from the base, bright green and shining, often reddish tinged, fragile. Stems erect or ascending, subglabrous. Lvs bright green. Lower lf-blades (Fig. 37H) 2–6 cm diam., orbicular in outline, with few short scattered hairs or subglabrous, on long petioles, *palmately 5-lobed to* $\frac{1}{2}$ *to* $\frac{3}{4}$; lobes 5–12 mm across at base, cuneiform, 3-lobed at apex; secondary lobes shorter than entire portion, 2–3-lobed or toothed at apex, teeth short, broad, obtuse, mucronate; upper lvs smaller on shorter petioles. Pedicels spreading or ascending after fl., often curving upwards near apex. Sepals oblong-ovate, aristate, subglabrous, erect and somewhat connivent near apex, forming an angled calyx with transverse ridges towards the margins. Fls in pairs. Petals 8–9 mm (including claw), pink; limb 3–4 mm, obovate, rounded at apex. Mericarps 5-ridged at puberulent apex, the ridges continuing down the back but less well-marked, the rest of the surface reticulate, glabrous; beak c. 1 cm, not remaining attached at the apex. Seeds smooth. Fl. 5–8. $2n = 20^*$. Th.

Native. Shady rocks, walls and hedgebanks, calcicole, ascending to 760 m. From Ross and Cromarty southwards; locally common but uncommon in many districts, rare in N. Scotland; widespread in Ireland but rare in the north; Jersey. Europe, except the north-east; S.W. Asia to the Himalaya; N. Africa; Madeira.

16. G. robertianum L.   Herb-Robert.

Lfy annual or biennial 10–50 cm, usually branched from the base, bright or dark green, usually ±reddish tinged, fragile, with a strong disagreeable smell. Stems decumbent or ascending, ±clothed with dense hairs below, often nearly glabrous above, but very variable in indumentum. *Lvs palmate*, bright green, lower mostly with 5 lflets, polygonal in outline, with scattered appressed hairs on both sides, on long petioles; lflets 1.5–6.5 cm, ±ovate in outline, deeply pinnatisect, the segments pinnately lobed; upper lvs mostly ternate, smaller, on shorter petioles. Pedicels ascending after fl., mostly straight. Fls in pairs. Sepals oblong-ovate, aristate, ±pilose and glandular, erect and somewhat connivent near apex, without ridges; calyx not angled. *Petals* (8–)9–12 mm (including claw), bright pink or occasionally white; limb 4–6 mm, obovate-cuneiform, rounded at apex. *Anthers orange or purple*. Mericarps reticulately ridged (but often with fewer longitudinal than transverse ridges), clothed with short stiff hairs, or glabrous; beak 1–2 cm. Seeds smooth. Fl. 5–9. Visited by various insects, protandrous, but self-pollination possible. $2n = 64*$ (all 3 subspp.).

Native. Woods, hedgebanks, among rocks and on shingle, etc. Common throughout the British Is., ascending to 700 m. Europe, except the extreme north; temperate Asia to Japan and the Himalaya; N. Africa (rare and only in the mountains); Macaronesia; naturalized in North and South America.

Subsp. **robertianum**. Usually biennial with a rather dense mound-like rosette of lvs in the first year. Fl. stems ascending. Stems and petioles dull deep red. Lf-segments broad. Petals twice as long as sepals or more. Fr. ±hairy. The common inland form.

Subsp. **maritimum** (Bab.) H. G. Baker. Usually biennial with an open procumbent rosette of lvs in the first year. Fl. stems procumbent or arcuately ascending. Lf-segments narrow, petioles shorter. Fls usually rather smaller. Fr. usually glabrous. Shingle beaches, less frequently cliffs or walls near the sea probably all round the British Is. W. Europe, Madeira.

Subsp. **celticum** Ostenf. Usually annual. Reddish only on the nodes and petiole-bases. Petioles rather short. Fls pale. Fr. hairy, large. Sunny limestone rocks. Glamorgan, Brecon, Carmarthen; Clare, ?Galway. Endemic?

17. G. purpureum Vill.   Little-Robin.

Differs from *G. robertianum* as follows: often less reddish. Lf-segments often narrow. Pedicels and calyx more densely pilose. *Petals smaller*, 6–9 mm, purplish-pink. *Anthers yellow. Mericarps glabrous, typically with much more pronounced and more numerous ridges.* Fl. 5–9. $2n = 32*$.

Native. In coastal regions from Cardigan and W. Sussex westwards.

The recognition of plants with procumbent stems, ascending at the tips, as subsp. *forsteri* (Wilmott) H. G. Baker, seems to be inappropriate. Indeed, such plants grade into *G. robertianum* and support the opinion that *G. purpureum* is a subsp. of that species.

2. ERODIUM L'Hérit.

Herbs. Lvs usually pinnate or pinnately lobed. Infl. an umbel, sometimes reduced. Fls actinomorphic or slightly zygomorphic, *without spur. Stamens 10, those opposite the petals sterile and without anthers. Beak of mericarp twisting spirally at maturity; mericarps indehiscent,* separating from the base upwards, *the beaks remaining attached to them as awns.*

About 90 spp., mainly Mediterranean but extending over Europe and to C. Asia, a very few in S. Africa, North America and Australia. Several have been found as casuals.

1 Lvs simple, lobed or pinnatifid.   **1. maritimum**
    Lvs pinnate.   2
2 Most lflets divided less than half-way to the midrib; mericarps with glandular apical pits. **2. moschatum**
    Most lflets divided more than half-way to the midrib; mericarps with eglandular apical pits.
                                              3. cicutarium

1. E. maritimum (L.) L'Hérit.   Sea Stork's-bill.

Small annual or biennial. Stems decumbent, to 30 cm but usually much less and often almost 0, usually several from the base with rather stiff white hairs. *Lf-blades 5–15 mm, ovate, pinnately incised-lobed to about half-way,* the lobes coarsely toothed or lobulate, with white appressed hairs on both sides, on long petioles. Fls 1–2 on axillary peduncles about equalling the lf. Sepals c. 4 mm, oblong, mucronate, with white hairs. Petals pink, not longer than the calyx, often 0. Mericarps hairy, the apical pit usually eglandular, surrounded by a furrow; beak 1 cm or less in fr. Fl. 5–9. Th.

Native. Fixed dunes and open habitats in short dry grassland, mainly near the sea. S. and W. coasts from Cornwall to Essex and Wigtown, local; inland in Durham, Worcester and probably introduced in Norfolk and Stafford; S. and E. coasts of Ireland from Down to Cork, local; Channel Is. W. coast of France from the Somme to the Vendée, N. Spain, eastwards along the Mediterranean to Sicily.

2. E. moschatum (L.) L'Hérit.   Musk Stork's-bill.

Annual to 60 cm, usually robust, but sometimes dwarf and almost stemless, branching from base, covered ±all over with long white hairs and stalked glands, *smelling of musk*. Stems decumbent or ascending. *Lvs 5–15 cm, pinnate; lflets ovate, incised-dentate, the lower teeth rarely reaching half-way to the midrib; stipules scarious, whitish, broad at the base, obtuse.* Peduncles exceeding lvs, with 2–8 fls in a terminal umbel; bracts resembling the stipules. Sepals 4–7 mm, accrescent, ±oblong, mucronate. Petals rosy-purple, rather longer than the calyx. *Filaments of the fertile stamens with a tooth each side at the base;* pollen yellow. Mericarps densely glandular clothed with long hairs, the pit surrounded by a furrow, beak 2.5–4 cm in fr. Fl. 5–7. $2n = 20$. Th.

Native. Waste places, etc., mainly near the sea. Cornwall to Suffolk and S. Lancs, local; introduced in inland England and Wales; round most of the Irish coast from Down to Kerry; Channel Is. S. and W. Europe, south to Abyssinia; Macaronesia. Escaped in N. Europe, S. Africa, North and South America, Australia and New Zealand.

3. E. cicutarium (L.) L'Hérit.

Annual to 60 cm, very variable in habit, at first stemless, later usually branching from the base and often robust, with few to many long, white hairs and eglandular to densely glandular, *not smelling of musk*, often somewhat foetid. *Lvs 2–20 cm, pinnate; lflets ±ovate, most divided more than half-way to midrib; stipules scarious, whitish, sometimes broad at base, acute or acuminate.*

Peduncles equalling or exceeding lvs, with 1–9 fls in a terminal umbel; bracts resembling the stipules. Sepals 3–7 mm, accrescent. *Filaments of fertile stamens ±dilated below, but not toothed*; pollen orange, rarely yellow. Mericarps ±hairy, 5–7 mm, the pit eglandular. Fl. 6–9. Pollinated by various insects or selfed; protandrous or weakly protogynous, apparently depending on genetic and habitat factors. Th.

Native. Dunes, dry grassland and arable fields and waste places, mainly on sandy soils, ascending to 365 m; common near the sea all round the British Is.; inland, widespread in Great Britain but rather local, in Ireland from Limerick to Kilkenny and Kildare. Throughout Europe to c. 70°N in Norway, but perhaps introduced in much of the centre, north and east; temperate Asia to N.W. Himalaya and Kamchatka; N. Africa to Abyssinia; Macaronesia; naturalized in North and South America.

Subsp. **cicutarium**   Common Stork's-bill.

E. pimpinellifolium (With.) Sibth., *E. triviale* Jordan, *E. ballii* Jordan.

Plant usually robust, eglandular or somewhat glandular.

Fls ±zygomorphic. Petals purplish-pink, rarely white, the two upper often with a blackish basal spot. Mericarp with a large apical pit with a furrow at its base, the beak 22–40 mm. $2n = 20^*$, 40. The common form.

Subsp. **bipinnatum** (Willd.) Tourlet   Sticky Stork's-bill.

E. glutinosum Dumort.

Plant usually rather slender, densely glandular. Fls ±actinomorphic. Petals pale pink or white, without basal spot. Mericarp with a small apical pit without or with indistinct furrow at its base, the beak 22–28 mm. $2n = 20$, 40^*. Sandy places near the sea, mainly on the W. coast.

Hybrids, with characters intermediate between these two subspp., are found with the parents on coastal dunes between Glamorgan and S. Lancs. They have been called **E. cicutarium** subsp. **dunense** Andreas.

*E. botrys** (Cav.) Bertol.

Plant robust. Lvs pinnatifid. Fls 1–4, rather large (petals 10–15 mm), dull bluish purple. Fr. with 2 furrows round the pit; beak 70–110 mm. $2n = 40$.

The most frequent of the casual spp. Native of the Mediterranean region.

47. OXALIDACEAE

Herbs often with fleshy stock, rarely small shrubs, very rarely trees. Lvs alternate, pinnate or palmate (sometimes 1-foliate), often showing sleep movements, with or without stipules; lflets entire, jointed to petiole or rhachis. Fls solitary or in cymes or cymose umbels, hermaphrodite, often heterostylous, heterochlamydeous, actinomorphic, 5-merous, hypogynous. Sepals usually imbricate. Petals contorted. Stamens 10, weakly obdiplostemonous, rarely 15, connate at base. Ovary (3–)5-celled; ovules 1–many, axile, anatropous, micropyle directed upwards, raphe ventral; styles (3–)5; stigmas capitate. Fr. a loculicidal capsule, rarely a berry; seeds often with elastic testa; endosperm copious, fleshy; embryo straight.

Three genera and nearly 900 spp., tropical and temperate.

Allied to Geraniaceae, differing in the lvs with separate lflets, the absence of a beak to the ovary, the separate styles, often numerous ovules, dehiscence of the capsule and presence of endosperm.

1. Oxalis L.

Lvs palmate; lflets 1–many (3 in our spp. except *O. tetraphylla*). *Stamens 10. Carpels 5, completely united. Fr. a capsule.*

About 800 spp., temperate and tropical, especially S. Africa, South America and Mexico. Some are grown in gardens. Homogamous, some foreign spp. heterostylous.

1 Fls white, solitary in the axils of lvs borne on a slender creeping rhizome. Native plant of woods, etc.

                              1. acetosella

Fls rarely white, in umbels or cymes or, if solitary, stems lfy, aerial and rhizome 0. Introduced plants of cultivated ground, walls, etc.   2

2 Fls yellow.   3
  Fls pink, lilac or red, rarely white.   8

3 Stem and leaves succulent; 3 sepals cordate.
                          6. megalorrhiza
  Not succulent; sepals not cordate.   4

4 Lvs all basal; bulbils present at base of plant.
                          11. pes-caprae
  Stems lfy; bulbils 0.   5

5 Stem creeping and rooting; lvs alternate; stipules oblong.   6
  Stem ±erect, not rooting; lvs whorled or clustered; stipules 0 or inconspicuous.   7

6 Lflets 5–18 mm; capsule 10–25 mm   **2. corniculata**
  Lflets 2–5 mm; capsule 5–7 mm   **3. exilis**

7 Peduncles not reflexed after fl.; stipules 0.
                          4. europaea
  Peduncles reflexed after fl.; stipules narrow-oblong, adnate to petiole.   **5. stricta**

8 Stem lfy, aerial; fls solitary.   **12. incarnata**
  Aerial stem 0 or a stout woody rhizome with lvs and fls from apex; fls cymose or umbellate.   9

9 Lflets 3.   10
  Lflets 4.   **10. tetraphylla**

10 Lvs and infl. from apex of stout woody rhizome.
                          7. articulata
  Stem 0; lvs and infls from bulbs or bulbils.   11

11 Lflets broadest near apex, lobes meeting each other at more than a right angle, nearly glabrous, dots 0.   **9. latifolia**
  Lflets broadest at about the middle, lobes meeting each other at a narrow angle, hairy, with dark dots near margin beneath.   **8. corymbosa**

1. O. acetosella L.   Wood-sorrel.

Perennial herb with *slender creeping rhizome* clothed with swollen, fleshy, scale-like petiole bases, *bearing the lvs and fls on long stalks.* Lvs ternate; petioles 5–15 cm; lflets 1–2 cm cuneiform-obcordate, broader than long, entire except at apex, ciliate and with scattered appressed hairs, bright yellow-green. *Fls solitary*; stalks about equalling petioles, with 2 small bracts in the middle. Sepals oblong-lanceolate, subobtuse. *Petals* 10–16 mm, *white* veined with lilac, rarely lilac or purple, obovate. Capsule 3–4 mm, ovoid-globose, 5-angled, glabrous. Fl. 4–5(–8). Insect visitors apparently few; cleistogamous fls occur abundantly in summer and produce most of the seed. $2n = 22$. Hr. or Chh.

Native. Woods, hedgebanks, shady rocks, etc., sometimes epiphytic, frequently growing in humus and very tolerant of shade, avoiding very heavy and wet soils; ascending to 1215 m; a characteristic dominant of some phases of plateau beechwoods; sometimes also dominant in sessile oakwoods. Common and often abundant throughout the British Is. Most of Europe but rarer in the south; N. and C. Asia to Sakhalin and Japan.

*2. O. corniculata L.   Procumbent Yellow Sorrel.

Annual or perennial herb 5–15 cm; rhizome and stolons 0. *Stems* many, weak, *procumbent*, sometimes ascending at the ends, rooting at the nodes below, ±densely clothed with long spreading hairs, lfy. *Lvs alternate*, oblong, ternate; petioles 1–8 cm; *stipules* c. 2 mm, oblong adnate to petiole; *lflets* 5–18 mm cuneiform-obcordate, broader than long, *bilobed at apex to* $\frac{1}{4}$ *or* $\frac{1}{3}$ with a narrow sinus, otherwise entire, ±pilose on the margins and undersurface, the lateral sometimes smaller than the terminal. *Fls 1–6 in umbels on axillary peduncles*, equalling or exceeding lvs. *Bracts linear-lanceolate*, mostly 2 mm or more. *Pedicels* c. 1 cm, clothed with appressed hairs or subglabrous, *±deflexed after fl.* Sepals ±lanceolate, acute. *Petals* 8–10 mm, *yellow*, narrowly cuneiform. Capsule 12–15 mm, cylindrical, pubescent. Fl. 6–9. Probably usually self-pollinated. $2n = 24^*$. Th. or Chh.

Introduced. Waste places and as a garden weed; first recorded c. 1585; rather common in S.W. England, northwards to S. Yorks, S. Lancs and Anglesey, occasional in S.W. and E.C. Scotland and N.E. Ireland. Tropical and warmer temperate regions of the whole world (country of origin not definitely known).

*3. O. exilis A. Cunn.

O. corniculata var. *microphylla* Hooker fil.

Like *O. corniculata* but much smaller in all its parts, forming a low compact mat. Petioles 6–13 mm, lflets 2–5 mm. Peduncles 1-fld. Capsules short, 5–7 mm abruptly narrowed at top.

A garden weed. Locally naturalized in S. Britain,, Scilly Is. and Channel Is. Native of New Zealand.

*4. O. europaea Jordan   Upright Yellow Sorrel.

O. stricta auct. non L.

Perennial herb 5–40 cm, with slender underground stolons, but probably often annual. *Stems* solitary, *erect*, simple or sparingly branched, glabrous or hairy, lfy. *Lvs whorled or clustered*, ternate; petioles 1–12 cm; *stipules* 0; *lflets* 8–20 mm, cuneiform-obcordate, broader than long, *bilobed at apex to c.* $\frac{1}{6}$, with a broad or narrow sinus, otherwise entire, glabrous or appressed hairy. *Fls 2–5 in short cymes on axillary peduncles,* equalling or exceeding lvs. *Bracts subulate,* c. 1 mm. *Pedicels* c. 1 cm, glabrous or hairy, *erect to spreading after fl.* Sepals lanceolate, acute. *Petals* 4–10 mm, *yellow*, narrowly cuneiform. Capsule 8–12 mm, cylindrical, glabrous or sparsely pubescent. Fl. 6–9. Probably usually self-pollinated. $2n = 24$. Hp or Th.

Introduced. Waste places and as a garden, rarely arable, weed. Recorded from a number of counties, mainly in S. and C. England, locally northwards to Scotland and in Wales, E.C. and W.C. Ireland. Most of Europe except the extreme north and south; N. China, Korea (but not native in the Old World); North America from Nova Scotia to Colorado, Texas and Florida; escaped in Tropical Africa and New Zealand.

*5. O. stricta L.

O. dillenii Jacq.

Similar to *O. europaea* but without stolons. *Stipules narrow-oblong*, adnate to petiole. Fls (1–)2–3(–4) *in umbels. Pedicels* often longer, *deflexed in fr.* Capsules 15–25 mm. Fl. 7–10. Th. or Chh.

Introduced. Arable fields near Pulborough, W. Sussex and in Sark. Native of eastern North America.

*O. valdiviensis Barnéoud, a glabrous annual with a short unbranched stem and yellow fls in long forked cymes, occurs as a casual or as a weed in gardens where it has been grown. $2n = 18$. Native of Chile.

*6. O. megalorrhiza Jacq. (*O. carnosa* auct.)

Succulent glabrous herb to 20 cm. Stems 1–2 cm diam., fleshy, with lvs and fls at apices. Lflets up to 2 cm, broad, obcordate, fleshy, covered below with crystalline papillae. Fls 1–3(–5), in an umbel. *Three outer sepals cordate. Petals* c. 15 mm, bright *yellow*. Capsule 7 mm oblong. Fl. 5–9. Chh.

Introduced. Walls and banks in Scilly Is. Native of Chile.

*7. O. articulata Savigny

O. floribunda Lehm.

Tufted perennial herb with *short thick fleshy rhizomes* up to 2 cm diam. with lvs and fls at apices. Lflets 1–4 cm, obcordate, ±pubescent, with raised orange spots, especially near the margins beneath. Fls many, in an umbel. *Petals* 10–15 mm, *deep bright pink*, rarely pale or white. Capsule 10 mm. Fl. 5–10. Hr.

Introduced. Much grown in gardens and sometimes naturalized on waste ground, roadsides etc., occasionally among native vegetation, mainly in S.W. England, Scilly Is. Native of E. temperate South America.

*8. O. corymbosa DC.

Perennial herb with lvs and fls from a *bulb developing later into a mass of sessile bulbils.* Petioles 5–15 cm; lflets 2.5–5 cm wide, orbicular-obcordate, *lobes rounded with a deep narrow indentation, sparsely hairy, with small raised reddish spots near the margin beneath.* Fls in short cymes on peduncles up to 30 cm. *Petals* 15–20 mm, purplish-*pink*. Fr. unknown. Fl 7–9. Gb.

Introduced. Naturalized in gardens in England, especially near London. Native of South America.

*9. O. latifolia Kunth.

Habit of *O. corymbosa* but *bulbils on stolons up to* 2 cm. Petioles 10–30 cm; *lflets* 2–4.5 cm, wider, usually much broader than long, very broadly triangular *with a wide shallow indentation*, often with straight sides meeting at an obtuse angle, ciliate but otherwise *glabrous, raised spots* 0. Fls in umbels. *Petals* 8–13 mm, *pink*. Fr. unknown. Fl. 5–9. Gb.

Introduced. A horticultural weed, mainly in S.W. England and Jersey. Native of tropical South America.

*10. O. tetraphylla Cav.

Similar to *O. latifolia* but stolons longer; *lflets* 4, thinly pubescent; *petals* c. 20 mm, *bright rose-red*. Fl. 5–9. Gb.

Introduced. Naturalized in fields in Jersey. S. Mexico.

*11. O. pes-caprae L.   Bermuda Buttercup.
O. cernua Thunb.

Perennial herb with a *bulb producing* an underground stem bearing bulbils below and at ground level and *lvs and infls at ground level*. Petioles up to 20 cm; lflets up to 2 cm obcordate, sparsely hairy below. Fls in umbels on peduncles 10–30 cm. *Petals* 20–25 mm, *bright lemon-yellow*. Fr. not produced in Britain. Fl 3–6. $2n = 35$. Gb.

Introduced. A weed in bulb-fields in Scilly Is., occasionally in S. Devon and Channel Is. Native of S. Africa, naturalized in many of the warmer parts of the world.

*12. O. incarnata L.

Perennial, nearly glabrous herb 10–20 cm with a bulb producing an *aerial branching stem bearing bulbils in the lf-axils*. Lvs opposite; petiole 2–6 cm, lflets 5–15 mm obcordate. *Fls solitary*; peduncles 3–7 cm. *Petals* 12–20 mm, *pale lilac* with darker veins. Fr. not produced in Britain. Fl. 5–7. Gb.

Introduced. Naturalized on walls and hedgebanks in S.W. England and the Channel Is., occasionally escaped elsewhere. Native of S. Africa.

48. BALSAMINACEAE

Herbs with translucent stems. Lvs alternate, opposite or in whorls of 3, simple; stipules 0 or represented by glands. Fls solitary or in racemes, hermaphrodite, heterochlamydeous, strongly zygomorphic, hypogynous. Sepals 5 or 3, often petaloid; the lowest (the posterior sepal is at the bottom of the fl. because of the twisting of the pedicel) large and spurred; the lateral small. Petals 5, the upper large, the 4 lower usually united in pairs on each side of the fl. Stamens 5, alternate with the petals; filaments broad, short, connate above; anthers connate round the ovary. Ovary 5-celled; ovules axile, numerous (rarely 2–3) in 1 row in each cell, anatropous, the raphe in the reversed position as compared with allied families. Fr. a loculicidal capsule with the valves dehiscing elastically and coiling, very rarely a berry. Endosperm 0; embryo straight.

About 4 genera and 600 spp., tropical to temperate.

1. IMPATIENS L.

Sepals usually 3. Lower petals usually united in pairs, so petals apparently 3. *Ovules numerous. Fr. an elastically opening capsule.*

Over 500 spp., mainly tropical Asia and Africa, a very few in the north temperate zone and S. Africa.

```
1 Lvs alternate; fls yellow or orange.                    2
  Lvs opposite or whorled; fls purplish-pink, rarely
    white.                                   4. glandulifera
2 Fls large, 2 cm or more; lvs with 15 or fewer teeth
    on each side.                                          3
  Fls small, 1.5 cm or less, pale yellow; lvs with 20 or
    more teeth on each side.               3. parviflora
3 Fls bright yellow, the spur gradually contracted and
    curved through c. 90°; teeth of lvs c. 2–3 mm deep,
    usually 10–15 on each side.            1. noli-tangere
  Fls orange, the spur suddenly contracted and bent
    through 180°; teeth of lvs 1–2 mm deep, rarely more
    than 10 on each side.                     2. capensis
```

1. I. noli-tangere L.   Touch-me-not Balsam.
Erect glabrous annual 20–60 cm; stem swollen at nodes.

Lf-blades 5–12 cm, alternate, ovate-oblong, subobtuse or subacute, cuneate at base, petiolate, coarsely crenate-serrate *with* c. 10–15 *teeth on each side* and a few ciliations at the base; *teeth* very obtuse, mucronate, c. 2–3 mm *deep*. *Fls* in few-fld axillary racemes usually consisting of both normal and cleistogamous ones, c. 3.5 cm, bright yellow with small brown spots. *Lower sepal* conical below, *tapered into a slender spur gradually curving* downwards or upwards through c. 90° sometimes recurved, c. 2.5 cm (including spur). Lateral lobes of corolla clawed, bilobed. Fl. 7–9. Pollinated by bees (normal fls). $2n = 20$. Th.

Native. By streams and in wet ground in woods, very local, probably native in the Lake District and N. Wales; recorded from many other counties in England, C. Wales and S. Scotland but introduced and mostly only casual. Most of Europe but absent from the extreme north and parts of the south; temperate Asia to the Pacific.

*2. I. capensis Meerburgh   Orange Balsam.
I. biflora Walter: *I. fulva* Nutt.

Differs from *I. noli-tangere* as follows: Lvs smaller, 3–8 cm, ovate-oblong or elliptical-oblong; *teeth fewer, usually* 10 *or less*, rather less obtuse, *shallower*, 1–2 mm *deep*; margins often somewhat concave at base. *Fls* often all cleistogamous, the normal ones 2–3 cm overall, *orange*, strongly blotched or spotted with reddish-brown inside. *Lower sepal* rather *suddenly contracted into a spur which is suddenly bent or curved* upwards or downwards through 180°, c. 2 cm (including spur). Fl. 6–8. $2n = 20$. Th.

Introduced. Completely naturalized on river banks, etc., especially the Thames and its tributaries, extending to Leicester, Stafford, Glamorgan, S. Yorks, Dorset and N. Devon; probably still spreading. Native of eastern North America from Newfoundland to Saskatchewan, Florida and Nebraska.

*3. I. parviflora DC Small Balsam.

Erect glabrous annual herb 30–100 cm; stem simple or branched. *Lf-blades* 5–15 cm, alternate, ovate, acuminate, tapered at the base into a winged petiole, serrate or crenate-serrate *with numerous* (20 or more) *teeth* on each side and a few ciliations at the base; teeth directed forwards, mostly acute. *Fls* small, 4–10 in axillary racemes which often become elongated in fr. 5–15 mm *overall, pale yellow*, not spotted. Lower sepal ±conical; spur straight or nearly so, very variable in length. Fl. 7–11. Visited mainly by hover-flies but probably also selfed. $2n = 24, 26$. Th.

Introduced. Completely naturalized in woods and waste shady places, local; mainly in S. and E. England but found also in N. and W. England and Wales; rare in Scotland N. to Inverness; probably still increasing. Native of C. Asia; naturalized in N. and C. Europe.

*4. I. glandulifera Royle Indian Balsam.

I. roylei Walp.

Erect glabrous robust annual herb 1–2 m; stem reddish, stout. *Lvs opposite or in whorls of* 3, blades 6–15 cm, lanceolate or elliptical, acuminate, rounded or cuneate at base, stalked, sharply serrate with numerous teeth on each side. *Fls* large (2.5–4 cm), *purplish-pink*, rarely white, 5–10 together in racemes on long peduncles in the axils of all the lvs of a few of the upper whorls, which are much smaller than the lower ones. Lower sepal very broad, obtuse, with a short tail-like spur. Fl. 7–10. Pollinated by humble-bees; cleistogamous fls not known to occur. $2n = 18, 20$. Th.

Introduced. Completely naturalized on river banks and in waste places; locally common in C. England and Wales, less common in E. England, Scotland and Ireland but still increasing. Native of the Himalaya.

49. SIMAROUBACEAE

Trees or shrubs, usually with bitter bark. Lvs usually alternate and pinnate. Fls regular, in panicles or spikes, 3–7-merous. Stamens usually obdiplostemonous. Ovary superior, surrounded by prominent disk. Carpels 2–5 often free below. Endosperm 0 or scanty.

About 20 genera and 120 spp., mainly tropical.

*Ailanthus altissima (Miller) Swingle, Tree of Heaven. Tree to 20 m; bark with pale stripes. Lvs 45–60 cm, pinnate; lflets 7–12 cm, ovate-lanceolate with 2–4 coarse teeth near base, ±glabrous. Infl. a large terminal panicle. Fls 7–8 mm diam., greenish. Fr. a group of usually 3 reddish-brown samaras, 3–4 cm. Fl. 6–7.

Much planted in gardens and parks, rarely naturalized. Native of China.

50. ACERACEAE

Trees or shrubs apparently andromonoecious (but functionally monoecious), androdioecious or dioecious. Lvs opposite, simple (often palmately lobed) or pinnate, exstipulate. Infl. of racemose type. Fls regular, perigynous or hypogynous; disk usually present. Sepals 4 or 5. Petals equal in number to sepals or 0. Stamens 4–10, usually 8 in two whorls even in fls with a 5-merous perianth, inserted on or inside the disk. Ovary 2-celled and 2-lobed, compressed with the septum along the short axis; ovules orthotropous or anatropous, 2 in each cell, axile; styles 2, free or united below. Fr. a schizocarp splitting into 2 samaras. Endosperm 0; embryo with flat, folded or rolled cotyledons and long radicle.

Three genera and about 200 spp.

1. ACER L.

Samaras winged on one side only.

About 200 spp. in the north temperate zone and tropical mountains, a few in S.E. Asia. Besides the following, a number of others are ±frequently planted. Two of the commonest are **A. palmatum** Thunb. (Japanese Maple), a shrub with 5–11-lobed lvs with the lobes acuminate and doubly serrate, and **A. negundo** L. (Box Elder), from North America, with pinnate lvs. Both are much grown in small gardens, the former often as a purple-lvd form, the latter as a variegated-lvd one.

The female fls in our spp. have apparently well-developed anthers but they do not open. The four following spp. (and the two mentioned above) are each members of different sections.

1 Twigs glabrous; lvs large (5–16 cm) glabrous except in the axils of the veins beneath, lobes acute or acuminate; wings of fr. spreading at an acute angle. 2
  Twigs pubescent; lvs small (mostly 4–7 cm), pubescent beneath, lobes obtuse; wings of fr. spreading horizontally. **3. campestre**
2 Fls in pendent panicles; lf-lobes irregularly crenate-serrate, juice not milky. **1. pseudoplatanus**
  Fls in erect corymbs; lf-lobes sinuate-dentate with few large acuminate teeth; juice milky. **2. platanoides**

*1. A. pseudoplatanus L. Sycamore.

Large monoecious, deciduous tree to 30 m, with broad spreading crown. Bark grey, smooth for a long time, finally scaling. Buds rather large (to 12 × 8 mm), ovoid, acute; scales green with blackish margins, pubescent near the tip. *Twigs* soon light brown, *glabrous*. *Lf-blades* 7–16 cm and about as broad, 5-lobed to about half-way, cordate at base, dark green and glabrous above (slightly pubescent on veins at first), *glaucescent* and soon *glabrous except in the axils of the veins beneath; lobes* ±ovate, *acute, coarsely and irregularly crenate-serrate*, often lobulate; *petiole* 10–20 cm, often red, *without milky juice*. Fls 60–100 *in a narrow pendent panicle* 5–20 cm, terminal on a short lfy branch, appearing with or after the lvs. Fls c. 6 mm diam., yellowish-green; sepals and petals 5; *stamens 8, hypogynous, inserted within the disk*; ovary tomentose. *Fr. glabrous, wings spreading at an acute angle* or incurved at apices; each samara 3.5–5 cm. Fl. 4–6, pollinated mainly by bees. Germ. early spring. $2n = 52$. MM.

Introduced in the fifteenth or sixteenth century. Now common in woods, hedges, plantations, etc., throughout the British Is., preferring deep, moist, well-drained rich soils, but growing on all but very poor soils, very tolerant of exposure and salt spray, ascending to 490 m; often planted but completely and commonly naturalized. Native of the mountains of C. and S. Europe from Belgium and N. Poland to C. Portugal, Sicily, C. Greece; Asia Minor, Caucasus; often naturalized in the lowlands of C. and N. Europe.

*2. A. platanoides L.   Norway Maple.

Deciduous tree to 30 m with broad spreading crown, monoecious. Bark deep grey, with numerous short, shallow fissures, not scaling. Buds to c. 10 mm, ovoid; scales greenish at base, reddish-brown above. *Twigs* brownish, *glabrous. Lf-blades* 5–15 cm and about as broad, 5(–7)-lobed to about ⅓, cordate at base, *bright green and somewhat shining on both surfaces, glabrous except in the axils of the veins beneath; lobes* triangular or parallel-sided below, *acuminate, sinuate-dentate with few large acuminate teeth; petiole* 5–20 cm, often red, *with milky juice. Fls many in a broad erect corymbose panicle,* terminal on a short lfy branch, opening before the lvs have expanded. Fls c. 8 mm diam., bright greenish-yellow; sepals and petals 5; *stamens* 8, perigynous, *inserted about the middle of the disk;* ovary glabrous. *Wings of fr. spreading at an acute angle,* divergent; each samara 3.5–5 cm. Fl. 4–5 (c. 3 weeks earlier than *A. pseudoplatanus*), pollinated mainly by bees. Fr. 9–10. $2n = 26$. MM.

Introduced. In similar situations to *A. pseudoplatanus* and liking similar conditions but much less commonly planted and so less often met with though readily becoming naturalized. Most of Europe except the extreme north and west, and the islands; only on mountains in the south; Asia Minor, Caucasus, N. Iran.

3. A. campestre L.   Field Maple.

Small deciduous tree 9–15(–25) m with a ±ovoid crown, or a shrub, monoecious. Bark light grey with shallow fissures, finally flaking off. Buds c. 5 mm, brown. *Twigs* brown, *pubescent,* often developing corky wings at 4 or 5 years. *Lf-blades* 4–7(–10) cm, as broad or rather broader, (3–)5-lobed to about half-way, cordate at base, dull green above, pubescent at first, finally subglabrous, paler *beneath* and persistently *pubescent* especially on the veins; lobes ovate or obovate, *obtuse* (or on some lvs subacute), *entire or with a very few broad shallow rounded teeth or shallowly trilobed;* petiole 2–8 cm, pubescent, *juice milky.* Fls rather few (c. 10–20) *in an erect corymbose panicle,* terminal on a short lfy branch, appearing with the lvs. Fls c. 6 mm diam., pale green; sepals and petals 5; *stamens* 8, perigynous, *inserted on the disk near the inner edge. Fr.* pubescent or sometimes glabrous; *wings spreading horizontally, collinear;* each samara 2–4 cm. Fl. 5–6; visited by small insects. Fr. 9–10. Germ. late spring. $2n = 26$. MM.

Native. Woods, hedges and old scrub, mainly on basic soils and only abundant on these, frequently coppiced. From Westmorland and Durham southwards; common in S., E. and C. England, becoming rarer westwards and northwards; rare in Scotland and probably introduced; rare and not native in Ireland. Most of Europe from S. Sweden and C. Russia southwards, but rare in the Mediterranean region; Asia Minor and Caucasus to N. Iran and Turkestan; Algeria (very rare).

*A. saccharinum L. (*A. dasycarpum* Ehrh.)   Silver Maple.

Tree. Lf-blades 8–14 cm, deeply 5-lobed, green above, silvery beneath; lobes deeply and doubly serrate. Fls before the lvs in lfless fascicles from axillary buds; petals 0.

Occasionally planted in parks, etc. as an isolated tree among native vegetation, but not naturalized. Native of eastern North America.

51. STAPHYLEACEAE

Trees or shrubs. Lvs opposite or alternate, pinnate or ternate, with stipules and stipels. Fls in panicles, regular, usually hermaphrodite; disk present. Sepals and petals 5, imbricate. Stamens 5, alternating with petals, inserted on the edge of the disk, anthers introrse. Ovary 2–3-celled with 1–many axile anatropous ovules; styles free or united below; stigmas capitate. Fr. a capsule. Embryo large, straight; endosperm scanty.

Five genera and about 60 spp., north temperate zone and south to the E. Indies and Peru.

1. STAPHYLEA L.

Deciduous shrubs. *Lvs opposite.* Infl. terminal. *Ovules numer-* ous. *Fr. an inflated, membranous capsule. Seeds without aril.*

Ten spp., north temperate.

*1. S. pinnata L.   Bladder-nut.

Shrub to 5 m. Lvs with 5–7 lflets 5–10 cm, ovate-oblong, acuminate, serrulate, glabrous. Panicles 5–12 cm, pendent. Fls c. 1 cm; petals and sepals whitish, of about equal length, erect. Fr. 2.5–3 cm, subglobose, much inflated, 2–3-lobed. Fl. 5–6. Visited by Diptera. $2n = 26$. N or M.

Introduced. Sometimes planted in shrubberies, etc., and perhaps ±naturalized in a few places. Native of C. and parts of S. Europe.

52. HIPPOCASTANACEAE

Andromonoecious trees or shrubs. Lvs opposite, palmate, exstipulate. Fls in large terminal erect panicles, composed of scorpioid cymes, zygomorphic; disk present. Sepals 5; petals 5 or 4, clawed. Stamens 5–8, hypogynous; anthers introrse. Ovary superior, 3-celled; ovules 2 in each cell, axile, anatropous; style simple, long; stigma simple. Fr. a large leathery, usually 1-seeded loculicidal capsule, opening by 3 valves. Seed large, nut-like, shining; hilum large; embryo curved; endosperm 0.

Two genera, the other consisting of 2 spp. from Mexico to northern South America.

1. AESCULUS L.

Deciduous trees. Buds large. *Sepals united into a tubular or campanulate calyx.*

About 13 spp., E. Asia, S.E. Europe and North America. Several spp. and hybrids are sometimes planted.

*1. A. hippocastanum L.   Horse-chestnut.

Large deciduous tree to 25 m with broad crown, andromonoecious. Bark dark greyish-brown, finally cast in scales. Buds to 3.5 cm, ovoid, deep red-brown, very sticky. Twigs pale grey or brown, glabrous. Lvs palmate with 5–7 lflets; lflets 8–12 cm, obovate-lanceolate, acuminate, long cuneate at base, irregularly crenate-serrate, dark green and glabrous above, somewhat woolly tomentose beneath when young, often glabrous at maturity; petiole long. Panicle 20–30 cm. Fls c. 2 cm across. Petals 4, white with basal spots which are at first yellow then pink. Stamens long, arched downwards; pollen red. Fr. c. 6 cm, subglobose, prickly; seeds 1 or 2. Fl. 5–6. Fr. 8–9. Pollinated by humble bees. Infl. protandrous, hermaphrodite fls protogynous. $2n = 40$. MM.

Introduced. Commonly planted for ornament and often self-sown, throughout the British Is. but less common in the west and in N. Ireland. Native of Albania, Greece, Bulgaria and Yugoslavia.

*A. carnea Hayne

Usually smaller tree than *A. hippocastanum*. Lflets usually 5, darker and firmer. Panicles 12–20 cm. Fls pink or red. Fr. 3–4 cm with few prickles. $2n = 80$.

Often planted. Of garden origin, by hybridization between *A. hippocastanum* and *A. pavia* L. (from eastern North America) and chromosome doubling, so that it now behaves as a species.

53. AQUIFOLIACEAE

Trees or shrubs. Lvs alternate, simple, exstipulate. Fls in few-fld axillary cymes (rarely solitary), regular, hermaphrodite or unisexual, 4–5-merous, hypogynous, disk 0. Sepals and petals imbricate; sepals small; petals free or united at base. Stamens equalling in number and alternate with corolla-lobes (rarely numerous), free or united with the extreme base of the petals. Ovary (2–)4(–many)-celled, with 1–2 apical ovules in each cell; style very short or 0. Fr. a drupe with 3 or more stones. Embryo small, straight; endosperm copious, fleshy.

Two genera and over 400 spp., cosmopolitan (except the Arctic).

1. ILEX L.

Plants dioecious rarely polygamous. *Petals united below. Stamens the same number as the petals.* Carpels 4–6(–8).

About 400 spp., cosmopolitan (except the Arctic). Some spp. are occasionally cultivated.

1. I. aquifolium L.   Holly.

Evergreen small tree or shrub 3–10 m., more in cultivation; crown cylindrical or conical. Bark grey, smooth for a long time, eventually finely fissured. Buds ovoid, c. 2–3 mm. Twigs green, glabrous or puberulent when young. Lvs 3–10 cm, thick and coriaceous, ovate, elliptical or oblong; margin undulate, sinuate-dentate with large triangular spine-pointed teeth or on old trees largely entire, dark green and glossy above, paler beneath, glabrous, with a narrow cartilaginous border; short-stalked. Cymes cluster-like, axillary on the old wood. Fls c. 6 mm diam., white, 4-merous. Fr. 7–12 mm, scarlet, globose. Fl. 5–8. Visited by honey-bees. Fr. 9–3. $2n = 40^*$. MM. or M.

Native. Woods, scrub, hedges and among rocks, on all but wet soils, ascending to 550 m., sometimes dominant in the lower tree- or shrub-layer of woods, tolerating a considerable amount of shade; throughout the British Is. except Caithness, Orkney and Shetland, common. Often planted, as are numerous varieties with variegated lvs, etc. S. and W. Europe extending north-eastwards to N. Germany and Austria.

54. CELASTRACEAE

Shrubs, trees or woody climbers. Lvs alternate or opposite, simple; stipules small or 0. Fls usually in cymes, small, regular, hermaphrodite or unisexual, 4–5-merous with a well-marked fleshy disk, ±perigynous. Sepals and petals imbricate (very rarely valvate), inserted on or below the margin of the disk (petals rarely 0). Stamens equalling the corolla-lobes in number and alternate with them (very rarely 10). Ovary 2–5-celled with

(1–)2(–many) axile anatropous ovules in each cell; style very short; stigma ±capitate, often lobed. Fr. a loculicidal capsule, samara, drupe or berry; seeds often surrounded by an aril; embryo large, straight; endosperm fleshy.

About 55 genera and 850 spp., cosmopolitan (except the Arctic).

1. EUONYMUS L.

Trees or shrubs, rarely climbing by rootlets. Lvs opposite (rarely alternate). Fls 4–5-merous; *disk wide*, fleshy. Stamens short, inserted on disk. *Carpels isomerous with other parts. Ovules 1–2 in each cell. Fr.* a fleshy, often brightly coloured, loculicidal *capsule. Seed completely enclosed in a fleshy aril.*

About 176 spp., Europe, N. Africa (1 sp.), Madagascar (1 sp.), Asia, North and Central America, Australia (1 sp.). Some are grown in gardens, especially *E. japonicus* L. fil. from Japan, with glossy evergreen lvs, much planted near the sea, often as a variegated form.

1. E. europaeus L.   Spindle tree.

Much-branched, deciduous, rather stiff, glabrous shrub or rarely small tree 2–6 m. Bark grey, smooth. Buds 2–4 mm, ovoid, greenish, ±4-angled. Twigs green, 4-angled. Lvs opposite, 3–13 cm, ovate-lanceolate to oblong-lanceolate or elliptical, acute or acuminate, crenate-serrulate, cuneate at base, often turning reddish in autumn; petiole 6–12 mm. Fls 8–10 mm diam., usually 4-merous, hermaphrodite or polygamous, 3–10 together in axillary dichotomous pedunculate cymes. Petals greenish, ±oblong, widely separated. Fr. 4-lobed, deep pink, 10–15 mm across, exposing the bright orange aril after opening. Fl. 5–6. Pollinated by small insects. Fr. 9–10. $2n = 64$. M.

Native. Woods and scrub, mostly on calcareous soil, ascending to 365 m. Throughout England, Wales and Ireland, rather common; extending north in E. Scotland to the Forth. Most of Europe except the extreme north and much of the Mediterranean region; Caucasus; W. Asia.

55. BUXACEAE

Monoecious evergreen shrubs or small trees, rarely herbs. Lvs simple, exstipulate. Fls regular, small, without disk, in spikes, racemes or clusters. Calyx usually of 4 sepals, sometimes 0 or sepals more in female fl. Petals 0. Stamens 4, opposite the sepals, or more numerous. Ovary superior, (2–)3(–4)-celled with 1–2 pendent anatropous ovules in each cell; raphe dorsal; styles free. Fr. a loculicidal capsule or drupe. Seeds with fleshy endosperm and straight embryo.

Four genera and about 100 spp., scattered over tropical and temperate regions. A family of obscure relationships, very variously placed by different authors.

1. BUXUS L.

Shrubs or small trees with *opposite entire* coriaceous *lvs.* Fls in axillary clusters consisting of a terminal female fl. and several male fls in the axils of bracteoles. Male fl. with 4 sepals, 4 *stamens*, and rudimentary ovary. Filaments thick. Female fl. without clearly defined perianth but subtended by spirally arranged bracteoles; ovary 3-celled; styles short, thick, *ovules 2 in each cell.* Fr. a capsule, the valves 2-horned. Seeds black shining, with small caruncle.

About 70 spp. in W. Europe, Mediterranean region, temperate E. Asia, Socotra, Madagascar, West Indies, North and Central America.

1. B. sempervirens L.   Box.

Evergreen shrub or small tree 2–5(–10) m. Twigs pubescent, somewhat 4-angled. Lvs 1–2.5 cm, oblong or elliptical, shortly stalked, obtuse to emarginate. Fls whitish-green; stamens exserted. Fr. c. 8 mm., ovoid, 3-horned. Fl. 4–5. Pollinated by bees and flies. Fr. 9. $2n = 28$. M.

Native. In beech woods and scrub on chalk and oolitic limestone in S. Essex, Kent, Surrey, Berks, Bucks and Gloucester, locally abundant; elsewhere commonly planted and sometimes naturalized. S.W. and W.C. Europe; N. Africa (in the mountains).

56. RHAMNACEAE

Trees or shrubs with simple, usually stipulate lvs. Infl. cymose. Fls small, green, yellow or blue, sometimes unisexual. Calyx tubular, 4–5-lobed, lobes valvate in bud. Petals 4–5, sometimes 0, small, inserted at mouth of calyx-tube and often hooded. Stamens 4–5, opposite the petals and often ±enclosed by them; anthers versatile. Ovary 2–4-locular, free or sunk in disk; ovules solitary, basal, erect, anatropous. Fr. often fleshy.

About 58 genera and 900 spp., cosmopolitan.

For an account of the ecology of the British spp. see H. Godwin, *J. Ecol.,* **31**, 66–92. (1943).

Thorny; buds with scales; lvs serrate; blaze of bark orange.
         1. RHAMNUS
Unarmed; buds naked; lvs entire; blaze of bark lemon-
yellow.   2. FRANGULA

1. RHAMNUS L.

Trees or shrubs. Buds with scales. Lvs alternate or subopposite. Fls inconspicuous, greenish, usually 4-merous, polygamous or dioecious. Calyx-tube urceolate, adnate to base of ovary. Styles 3–4. Fr. a drupe with 2–4 pyrenes. Germination epigeal. About 100 spp. Cosmopolitan.

1. R. catharticus L.   Buckthorn.

A rather *thorny* deciduous bush or small tree commonly 4–6 m, but sometimes up to nearly 10 m. *Branches* opposite, *spreading almost at right angles to main stem, many of the laterals forming short lfy spurs or ending in thorns.* Young twigs grey or brown, glabrous to densely pubescent; *bark of old branches fissured and scaling; blaze orange. Buds with dark scales. Lvs* 3–6 cm, petiolate, *ovate to nearly elliptical*, obtuse, sometimes ±cuspidate, serrate, somewhat pubescent to glabrous, *dull green turning yellow or brownish in autumn; large lateral veins 2–3 pairs* curving upward and running ±parallel nearly to tip of lf. *Fls* c.4 mm diam., on slender pedicels, solitary or in axillary fascicles *on the previous year's wood of the short shoots. Calyx* greenish; *lobes* 4, lanceolate. *Petals* 4, small. Fr. 6–8 mm diam., changing from green to black on ripening, 3–4-seeded. Fl. 5–6. Pollinated by various insects. Fr. 9–10. Germ. autumn and spring. $2n = 24$. N. or M. Bark and berries purgative.

Native. On fen peat, in scrub, hedges, and ash and oak woods on calcareous soils. England, except the S.W. and N.E.; frequent in many midland, southern and eastern counties; Scotland: locally introduced; Ireland: frequent in the central plain, rare in the south and north. Most of Europe to 61° 45′ N in Sweden and eastwards to the River Ob; N. Africa, on high mountains in Morocco and Algeria.

*R. alaternus L., an evergreen shrub with elliptical to ovate-lanceolate lvs 2–5 cm, dark green and shiny above, paler but glabrous beneath, and very small clusted fls, sometimes escapes from cultivation. Mediterranean region.

2. FRANGULA Miller

Like *Rhamnus* but buds naked; fls usually 5-merous and hermaphrodite; style 1; germination hypogeal. About 50 spp.

1. F. alnus Miller Alder Buckthorn, Black Dogwood. *Rhamnus frangula* L.

An *unarmed* deciduous shrub or small tree commonly 4–5 m. *Branches* sub-opposite, *ascending at an acute angle to the main stem, without marked distinction into long and short shoots.* Young twigs green becoming grey-brown, appressed-puberulent; *bark of old branches smooth*, except in very old trees; *blaze lemon-yellow. Buds without scales*, densely covered with brownish hairs. *Lvs* 2–7 cm, petiolate, *obovate*, bluntly apiculate, *entire*, undulate, with a caducous brownish tomentum particularly beneath, *shiny green turning clear yellow and red in autumn; large lateral veins about 7 pairs. Fls* c. 3 mm diam., on rather stout pedicels, in axillary fascicles *on the young wood. Calyx* greenish; *lobes* 5, ovate. *Petals* 5, small. Fr. 6–10 mm diam., changing from green to red and then violet-black on ripening, 2(–3)-seeded. Fl. 5–6(–9). Pollinated by various insects, especially bees. Fr. 8–11. Germ. spring. $2n = 20*, 26$. N. or M. Purgative. Wood yields a high-grade charcoal.

Native. In scrub on fen peat and around the margins of raised bogs and valley bogs, on moist heaths and commons, in limestone scrub and as undergrowth in open woods; usually on damp and ±peaty soils. Generally distributed throughout England and Wales, northwards to S.W. Scotland; local, though often abundant in suitable habitats; Ireland: rare, and absent from the south-eastern and north-western counties. Europe north to c. 67° in Sweden and 66° 50′ in Russian Lappland, eastwards to the Urals and Siberia; N. Africa, in Morocco and Algeria; rare in the Mediterranean region.

57. VITIDACEAE

Like Rhamnaceae but usually climbers with tendrils. Infl. lf-opposed. Fr. a berry.

About 12 genera and 700 spp., mostly in warm temperate or tropical regions.

*Vitis vinifera L., the Grape Vine, a woody climber with palmately lobed lvs and tendrils which twine round any convenient support, is ±naturalized on the banks of the Thames at Kew.

Spp. of **Parthenocissus** (Virginia Creeper, 'Ampelopsis') are commonly cultivated.

*P. tricuspidata (Siebold & Zucc.) Planchon, which has the tendrils ending in adhesive disks and the lvs varied in shape but usually simple, is the commonest. 'Hundreds of miles of suburban architecture are happily hidden in summer by its foliage, which turns red in autumn' (Gilbert-Carter, *British Trees and Shrubs*). This sp. is also ±naturalized by the Thames at Kew.

*P. quinquefolia (L.) Planchon is also grown fairly frequently and sometimes escapes. It has all the lvs 5-foliolate and the young branches terete.

58. LEGUMINOSAE (FABACEAE)

Herbs, shrubs or trees. Lvs simple or (usually) compound, often 3-foliolate or pinnate, sometimes ending in a tendril. Fls papilionate, with a large, often erect, adaxial petal (standard), 2 lateral petals (wings) and 2 lower petals usually ±connate by their lower margins (keel); standard outside and enclosing the other petals in bud. Sepals usually 5, ±connate into a tube. Stamens 10, either all connate (monadelphous) or 1 free and 9 connate (diadelphous), rarely all free. Fr. usually dehiscent. Seeds often large; endosperm 0 or very scanty.

A large family (of 600 genera and 12000 spp.), represented in Britain only by subfam. Papilionoideae (Lotoideae), to which the above description applies. The fls are usually insect-pollinated. The subfamilies, Mimosoideae and Caesalpinioideae, are mainly tropical in distribution and contain few herbs.

1 Trees or shrubs, sometimes small and prostrate.   *2*
  Herbs, at most slightly woody at base.   *10*
2 Trees with pendent racemes.   *3*

Shrubs; fls in erect racemes or umbels or solitary in lf-axils.   4

3 Fls white; lvs pinnate; bark of trunk deeply fissured.   7. ROBINIA

Fls yellow; lvs palmate, bark of trunk ±smooth.   1. LABURNUM

4 Fls solitary in lf-axils.   5

Fls in erect racemes or umbels.   8

5 Spines branched.   5. ULEX

Spines simple or 0.   6

6 Twigs angled; style spirally coiled; valves of legume twisted after dehiscence.   2. CYTISUS

Twigs terete; style not spirally coiled; valves of legume not twisted after dehiscence.   7

7 Calyx split down the upper side.   4. SPARTIUM

Calyx 2-lipped, the upper lip with 2 teeth, the lower with 3.   3. GENISTA

8 Fls in racemes.   9

Fls in umbels.   24. CORONILLA

9 Lvs digitate; legume hard, not inflated.   6. LUPINUS

Lvs pinnate; legume papery, strongly inflated.   9. COLUTEA

10 Lvs consisting of a tendril and a pair of lf-like stipules.   13. LATHYRUS

Lvs not as above.   11

11 Lvs linear, grass-like.   13. LATHYRUS

Lvs compound.   12

12 Lvs pinnate.   13

Lvs with 3 or rarely 5 lflets, or digitate with up to 16 lflets.   27

13 Lvs paripinnate, often with a terminal tendril.   14

Lvs imparipinnate.   18

14 Stem winged.   13. LATHYRUS

Stem not winged.   15

15 Lflets with strong parallel veins.   13. LATHYRUS

Lflets pinnately veined.   16

16 Calyx-teeth lf-like; stipules up to 10 cm.   14. PISUM

Calyx-teeth not lf-like; stipules not more than 2 cm.   17

17 Style pubescent all round or on the lower side or glabrous.   12. VICIA

Style pubescent on the upper side only.   13. LATHYRUS

18 Fls in terminal heads closely subtended by palmatisect bracts; calyx inflated, woolly.   22. ANTHYLLIS

Fls in pedunculate axillary infls, not closely subtended by bracts; calyx not inflated.   19

19 Legume breaking transversely into 1-seeded segments when mature.   20

Legume not breaking transversely when mature.   22

20 Segments of fr. horseshoe-shaped.   25. HIPPOCREPIS

Segments of fr. ±straight.   21

21 Annual; infl. with 1–8 fls; corolla 3–8 mm.   23. ORNITHOPUS

Perennial; infl. with 10–20 fls; corolla 10–15 mm.   24. CORONILLA

22 Fls in umbels.   20. LOTUS

Fls in racemes.   23

23 Stipules scarious; fls pink or red; legume suborbicular, indehiscent.   25. ONOBRYCHIS

Stipules green; fls purple, white or yellowish; legume oblong, dehiscent.   24

24 Keel mucronate.   11. OXYTROPIS

Keel obtuse.   25

25 Stems not more than 35 cm; plant hairy.   10. ASTRAGALUS

Stems 60 cm or more; plant glabrous   26

26 Stems erect; infl. at least as long as subtending lf.   8. GALEGA

Stems procumbent; infl. much shorter than subtending lf.   10. ASTRAGALUS

27 Lvs digitate, with 6–16 lflets.   6. LUPINUS

Lvs with 3 or 5 lflets.   28

28 Principal lateral veins of lflets anastomosing and not reaching the margin, or veins obscure.   29

Principal lateral veins of lflets conspicuous, reaching the margin and ending in a tooth.   31

29 Lvs with 5 lflets; fls not more than 18 mm.   30

Lvs with 3 lflets; fls 25–30 mm, pale yellow.   21. TETRAGONOLOBUS

30 Lflets linear; standard and wings white.   19. DORYCNIUM

Lflets lanceolate to obovate; standard and wings yellow.   20 LOTUS

31 Plant glandular-hairy, at least above.   32

Plant without glandular hairs.   33

32 Fls pink; legume straight.   15. ONONIS

Fls yellow; legume spirally coiled.   17. MEDICAGO

33 Infl. a raceme, elongating after flowering.   16. MELILOTUS

Infl. a head, not elongating after flowering.   34

34 Filaments of at least 5 stamens dilated at the apex; corolla usually persistent in fr.   18. TRIFOLIUM

Filaments not dilated at the apex; corolla deciduous.   17. MEDICAGO

Genisteae (Adanson) Bentham. Shrubs, small trees or herbs. Lvs digitately 3–many-foliolate, unifoliolate or simple. Fls in terminal racemes, or heads, or solitary. Stamens monadelphous; anthers alternately long, basifixed and short, dorsifixed. Fr. usually linear or oblong, dehiscent, or subglobose and ±indehiscent. Seedlings epigeal.

1. LABURNUM Fabr.

Small unarmed trees with smooth bark. Lvs 3-foliolate. Racemes simple, *lfless, pendent.* Fls yellow. Calyx 2-lipped, shortly toothed. Stamens monadelphous; anthers alternately long and basifixed and short and versatile. *Pod subterete, bulging over the seeds;* seeds several, poisonous, dispersed by explosive dehiscence.

Two species in C. and S. Europe.

*1. L. anagyroides Medicus Laburnum, Golden Rain.

Cytisus laburnum L.; *L. vulgare* J. Presl

A small unarmed tree up to 7 m. Bark dark brownish-green. Twigs appressed-pubescent. Petioles 5–8 cm; lflets 4–8 cm, elliptical or elliptical-ovate, obtuse and mucronulate, appressed-pubescent beneath. Racemes 10–30 cm; pedicels appressed-pubescent. Fls c. 20 mm. Pods 3–5 cm, appressed-pubescent, the upper suture thickened. Fl. 5–6. Pollination mainly by humble bees. The pods remain on the trees throughout the winter. Germ. spring. $2n = 48$. M.

Introduced. Planted and ±naturalized in waste, bushy places. Scattered localities in Great Britain and C. Ireland. C. and S. Europe.

2. CYTISUS L. (*Sarothamnus* Wimmer)

Unarmed shrubs with small readily deciduous lvs and green stems. Lvs 1–3-foliolate. Fls axillary, yellow.

Calyx herbaceous, 2-lipped, minutely toothed, lower lip 3-, upper 2-toothed. Stamens as in *Genista*. *Style long, spirally coiled*; stigma capitate. Pod 2-valved, *valves coiled after dehiscence*; seeds several, dispersed by explosive dehiscence.

About 30 spp., mostly in Spain and Portugal.

1. C. scoparius (L.) Link   Broom.

Sarothamnus scoparius (L.) Wimmer ex Koch

A much-branched erect shrub 60–200 cm. Twigs glabrous, green, 5-angled. Lvs usually 3-foliolate distinctly petiolate; lflets 6–20 mm narrowly elliptical to obovate, acute, glabrous or with appressed hairs. Stipules 0. Fls c. 20 mm, yellow. Pedicels up to 10 mm, slender, glabrous; bracteoles minute. Calyx ¼ length of corolla, glabrous. Standard 16–18 mm. Pod 2.5–4 cm, black, with brown or white hairs on the margins. Fl. 5–6. Pollination by large bees. $2n = 48^*$. N.

Native. On heaths and waste ground and in woods, usually on sandy soils, strongly calcifuge. Generally distributed throughout the British Is. except for Orkney and Shetland. Europe, northwards to S. Sweden, eastwards to C. Ukraine. Broom tops (*Scoparii Cacumina* of the Pharmacopoeia) are used as a diuretic.

Subsp. **maritimus** (Rouy) Heywood with prostrate stems and densely silky lvs and young twigs, occurs on cliffs in W. Cornwall, Pembroke, Lundy, W. Cork and the Channel Is. It maintains its characters in cultivation and breeds true. $2n = 48^* (24^*)$.

3. Genista L.

Small shrubs, unarmed or *sometimes with simple spines*. Lvs 1-*foliolate* stipules minute or 0. Fls yellow, solitary in the axils of the lvs, bracteolate, bracteoles minute. Calyx shortly 2-lipped; *upper lip deeply bifid*, lower shortly 3-toothed. Wings oblong, deflexed after flowering; keel-petals separating and not resilient after deflection. Stamens monadelphous; anthers alternately long and basifixed and short and versatile. Style curved; stigma oblique. Pod 2-valved and dehiscent; seeds several, dispersed by explosive dehiscence.

About 75 spp. mainly in the Mediterranean region.

1 Spiny (rarely unarmed); lvs glabrous; legume inflated.   **2. anglica**
  Unarmed; legume not inflated.   2
2 Lvs oblong-lanceolate, ciliate; fls on long branches; standard glabrous.   **1. tinctoria**
  Lvs ovate, densely appressed-sericeous beneath; fls on short lateral shoots; standard hairy.   **3. pilosa**

1. G. tinctoria L.   Dyer's Greenweed.

An unarmed erect or ascending shrub 10–200 cm. Stems slender, branched, brown; *young twigs* green, sparsely hairy, *striate*. Lvs up to 30 mm, subsessile, *oblong-lanceolate, acute or mucronate, margins ciliate*, glabrescent when old. *Stipules* 1–2 mm, *thin, subulate*. Fls axillary, towards the ends of the main branches, without nectar; *standard* 8–15 mm, equalling keel. Pedicels 2–3 mm, almost glabrous; bracteoles minute, basal. Calyx and

corolla glabrous, the former deciduous when the fr. is ripe. *Pod* 25–30 mm, glabrous, *flat, tapering and obtuse at both ends*. Fl. 7–9. Pollination by diverse pollen-collecting insects. $2n = 48$. N.

Native. In rough pastures. Throughout England and Wales, and north to Wigtown and Berwick; absent from Ireland. Most of Europe from Estonia southwards (absent from Portugal) and eastward to the Urals, Caucasus and Asia Minor.

Very variable.

2. G. anglica L.   Petty Whin, Needle Furze.

A spiny, rarely unarmed, *erect or ascending shrub* 10–50(–100) cm. Stems slender, brown; *young twigs terete*, glabrous or pubescent; spines 1–2 cm, axillary, spreading or recurved, rarely branched, lfy when young. Lvs 2–8 mm, ovate, *acute or apiculate, glabrous and glaucous*; those on the spines linear-lanceolate. Fls glabrous, axillary towards the ends of the main branches; standard 6–8 mm, shorter than keel. Pedicels c. 2 mm, sparsely hairy, bibracteolate. Calyx-lobes fringed; corolla glabrous. *Pod* 12–15(–20) × 5 mm, *glabrous, obliquely narrowed and acute at both ends*. Fl. 5–6. Pollinated by bees. $2n = 42^*$, 48. N.

Native. On dry heaths and moors. Scattered throughout Great Britain north to E. Ross; absent from Ireland. W. Europe eastwards to S. Sweden, N. Germany and S.W. Italy; N.W. Africa.

3. G. pilosa L.   Hairy Greenweed.

An unarmed prostrate shrub 10–40(–150) cm. Stems rather stout, much-branched and tortuous, greyish when young, brown when old; *young twigs* pubescent, *grooved*. Lvs 3–5 mm, subsessile, ovate, *obtuse, with appressed hairs beneath, glabrous above. Stipules* 0.5 mm, *thick, ovate, obtuse*. Fls axillary, on dwarf lateral branches. Pedicels c. 5 mm, densely hairy. *Calyx and corolla pubescent.* Standard 8–10 mm. *Pods* 14–18(–28) mm, *pubescent, not inflated but bulging over the seeds, rounded at the base, acute at the apex.* Fl. 5–6. Pollinated by honey-bees. $2n = 24$. Chw. or N.

Native. On cliffs and dry sandy and gravelly heaths on poor soils. W. Cornwall, E. Sussex, W. Suffolk, Pembroke and Merioneth; formerly in W. Kent; rare and very local. W. and C. Europe from S. Sweden to C. Italy and Macedonia.

4. Spartium L.

Unarmed shrub. Lvs 1-foliolate. Fls yellow, in lax lfless terminal racemes. Calyx spathe-like, split down the upper side, with 5 short teeth. Stamens monadelphous. Legume linear-oblong, dehiscent, with several seeds.

*1. S. junceum L.   Spanish Broom.

Shrub up to c. 3 m, with cylindrical, striate, flexible, rush-like stems. Lvs 10–30 mm, few, oblong-linear to lanceolate, glabrous above, appressed-sericeous beneath, caducous. Fls fragrant; pedicels in the axils of small caducous bracts, with 2 bracteoles at the apex.

Corolla 20–25 mm. Legume flat, sericeous, becoming glabrous. Fl. 6. N or M.

Introduced. Naturalized locally in S. England. Mediterranean region and S.W. Europe.

5. ULEX L.

Densely spinous shrubs; spines green and branched. Lvs 3-foliolate on young plants, spinous or reduced to scales on mature plants. Stipules 0. Fls yellow, axillary, shortly pedicellate; bracteoles 2, at apex of pedicel. Calyx membranous, yellow, bipartite; lower lip minutely 3-toothed, *upper minutely 2-toothed*. Corolla persistent. Stamens as in *Genista*. *Style somewhat curved*; stigma capitate. Legume 2-valved and dehiscent; seeds 1–6, dispersed by explosive dehiscence. About 15 spp., in W. Europe and N.W. Africa. Pollination by insects.

1 Spines deeply furrowed; bracteoles at least 2 mm wide (fl. winter and spring). **1. europaeus**
 Spines faintly furrowed or striate; bracteoles less than 1 mm wide (fl. late summer and autumn). 2.
2 Spines rigid; calyx usually at least 10 mm; corolla deep yellow. **2. gallii**
 Spines weak; calyx 6–9 mm; corolla clear yellow. **3. minor**

1. U. europaeus L. Furze, Gorse, Whin.

A densely spinous rather glaucous shrub 60–200 cm or sometimes more. Main branches erect or ascending, with rather sparse, blackish hairs; *spines* 1.5–2.5 cm, rigid, *deeply furrowed*. Fls 11–20 mm. Pedicels 3–5 mm, densely velvety; *Bracteoles 2–4 mm wide, much wider than the pedicels*. Calyx ⅔ *length of corolla, with spreading hairs*, teeth connivent. Wings rather longer than keel. Legume 11–20 mm, bursting in summer, black, with grey or brown hairs. Fl. 3–6 and sporadically during a mild winter. Pollinated especially by humble-bees. Fr. 7. 2*n* = 96. N.

Represented in Britain by subsp. **europaeus**.

Native. In rough grassy places on light, usually rather acid soils. Generally distributed throughout the British Is., northwards to Shetland. W. Europe from the Netherlands southwards and eastwards to Italy. Introduced in many temperate countries and frequently naturalized. Occasionally cut to the ground by severe frosts.

2. U. gallii Planchon Western Gorse.

A densely spiny, dark green shrub 10–200 cm. Main branches usually ascending, with abundant brown hairs; spines 1–2.5 cm, faintly furrowed or striate. Fls 10–13 mm. Pedicels 3–5 mm, appressed-hairy; *bracteoles c.* 0.75 mm usually wider than the pedicels. Calyx slightly shorter than corolla, appressed-hairy. *Wings when straightened usually longer than keel*. Legume c. 10 mm, bursting in spring. Fl. 7–9. Fr. 4–5. 2*n* = 80. N.

Native. On heaths and siliceous hill grasslands, strongly calcifuge. In suitable habitats throughout England and Wales, mainly in the west; Scotland north to Sutherland; throughout Ireland but commonest in the south and east. From Scotland to N.W. Spain.

3. U. minor Roth Dwarf Gorse.
U. nanus T. F. Forster

A densely spinous shrub 5–100 cm. Main branches usually procumbent, with abundant, brown hairs; spines c. 1 cm, faintly furrowed or striate. Fls 8–10 mm. Pedicels 3–5 mm, hairs appressed; *bracteoles c.* 0.5 mm *wide*, usually narrower than the pedicels. Calyx 6–9.5 mm; hairs appressed, rather sparse. *Wings about as long as keel, straight*. Legume c. 8 mm, persistent for nearly a year. Fl. 7–9. 2*n* = 32. N.

Native. On heaths. S.E. England, East Anglia, Nottingham, Cumberland, W. France, N.W. Spain, W. Portugal and S.W. Spain. Introduced in the Azores and Brazil.

6. LUPINUS L.

Usually herbs. *Lvs digitate*; stipules adnate to base of petiole. *Fls in terminal racemes, showy*. Calyx deeply 2-lipped; stamens monadelphous; style curved, stigma capitate. *Legume flattened* and often constricted between the seeds.

About 200 spp., America and Mediterranean region.

1 Shrub; lflets 7–11; fls yellow or white, rarely bluish-tinged. **3. arboreus**
 Herb; fls blue, pink or white. 2
2 Lflets 6–8; fls blue and white. **1. nootkatensis**
 Lflets 10–17; fls blue, pink or white. **2. polyphyllus**

*1. L. nootkatensis Donn ex Sims Nootka Lupin.

A stout lfy pubescent *perennial*. *Lflets* 2.5–5 cm, 6–8, *cuneate-elliptical*, mucronate. *Petioles somewhat longer than lflets*; stipules subulate to linear-acuminate. Infl. up to c. 10 cm. *Fls* 12–16 mm, ±whorled, *blue and white*; bracts caducous, exceeding the buds. *Pods c.* 5 cm, brown and silky. Fl. 5–7. 2*n* = 48. Hs.

Introduced. Naturalized on river shingle in N.E. Scotland, Orkney and N.W. Ireland. Native of N.E. Asia and western North America.

*2. L. polyphyllus Lindley Common Lupin.

Like **1** but lvs with 10–17 lflets, the petioles about twice as long as the oblanceolate lflets; infl. 15–60 cm; fls 12–14 mm, blue, pink or white. Fl 5–7. 2*n* = 48. Hs.

Introduced. Locally common on railway banks and similar habitats. Native of western North America from California to British Columbia.

*3. L. arboreus Sims Tree Lupin.

An upright *shrub* up to 2.5 m. *Lflets c.* 3 cm, 5–12, *oblanceolate*, mucronulate, glabrous above, silky beneath. *Petioles about as long as lflets*. Racemes up to 25 cm. *Fls* 14–17 mm, *yellow* or white, rarely bluish tinged. Pod 4–7 cm, pubescent, 8–12-seeded. Fl. 6–9. 2*n* = 40. N or M.

Introduced. Naturalized on dunes and in waste places north to Wigtown and Fife; S.E. Ireland. Native of California.

Robinieae (Bentham) Hutch. Trees. Lvs usually pulvinate; stipules often becoming spinose. Fls in pendent racemes. Stamens diadelphous, almost uniform. Pod 2-valved, several-seeded, laterally compressed. Seedlings epigeal.

7. ROBINIA L.

Deciduous trees or shrubs with scaleless buds hidden by the petiole base. Lvs imparipinnate. *Stipules often spiny and persistent. Racemes pendent.* Stamens diadelphous. *Legume compressed.* The fls have nectar and are pollinated by bees.

Twenty spp. In North America.

*1. R. pseudoacacia L.   Acacia.

A deciduous tree with coarsely fissured bark, up to 20 m. Stipules often eventually spiny. Racemes pendent. Fls fragrant, white. Legume strongly compressed, glabrous, persistent. Fl. 6. $2n = 20$. M.

Introduced. Commonly cultivated and sometimes planted in thickets. Central and eastern North America; naturalized throughout much of North America, Europe, N. Africa, Asia and New Zealand.

Galegeae (Bronn) Torrey & Gray. Herbs or shrubs. Lvs usually epulvinate; stipules often connate round the stem. Fls in racemes. Stamens usually diadelphous; anthers usually uniform. Pod 2-valved, several-seeded. Seedlings epigeal.

8. GALEGA L.

Erect perennial *herbs.* Lvs imparipinnate. *Fls* blue, lilac or white, *in erect racemes*; petals clawed; calyx-teeth subequal. Stamens monadelphous. Legume linear, terete. The fls are without nectar.

Six spp. in S. Europe, W. Asia and E. Africa.

*1. G. officinalis L.   Goat's-rue, French Lilac.

A stout erect glabrous perennial 60–150 cm. Lflets oblong or ovate-oblong, obtuse or acute, mucronate. Peduncles usually equalling or exceeding lvs, many-fld; infl. racemose. Fls 12–15 mm, lilac or white bracts subulate, persistent; calyx gibbous at base on upper side, teeth setaceous, shorter than tube. Legume 2–3 cm, linear, terete. Fl. 6–7. $2n = 16$. Hp.

Introduced. Naturalized in waste places. E.C. and S. Europe; W. Asia, N.W. Africa.

9. COLUTEA L.

Deciduous shrubs. Lvs imparipinnate; lflets entire. *Fls* yellow or brownish-red *in small erect racemes. Legume membranous, much inflated.* The fls have nectar and are cross-pollinated by humble-bees.

About 28 spp. in Europe and Asia. Poisonous.

*1. C. arborescens L.   Bladder Senna.

A small shrub up to 4 m. Lflets silky beneath. Racemes 2–8-fld. Fls yellow, standard with red markings. Stamens diadelphous. Legume c. twice as long as broad, greenish, closed at apex. Fl. 5–7. $2n = 16$. N. or M.

Introduced. Naturalized on railway banks and in waste places particularly in the neighbourhood of London. Mediterranean region.

10. ASTRAGALUS L.

Herbs. Lvs imparipinnate, stipulate; lflets entire. Fls in axillary racemes, bracteoles small. Calyx tubular with 5 subequal teeth. *Keel obtuse.* Stamens diadelphous; anthers all similar. Legume 2-valved, *often longitudinally 2-celled, septum developed from suture next the keel*; seeds 2–many. Fls with nectar, usually cross-pollinated by bees.

About 2000 spp., cosmopolitan except for Australia.

1 Plant glabrous or nearly so; fls creamy with a greenish-grey tinge.   **3. glycyphyllos**
Plant pubescent; fls blue or purplish.   2
2 Stipules connate below; fls erect; peduncle usually much longer than subtending lf.   **1. danicus**
Stipules free to base; fls spreading or deflexed; peduncles shorter to slightly longer than subtending lf.   **2. alpinus**

1. A. danicus Retz.   Purple Milk-vetch.

A. hypoglottis auct., non L.

A slender ascending perennial 5–35 cm. Stock slender, branched. Stems and lvs with sparse soft white hairs. Lvs 3–7 cm; lflets 5–12 mm, 6–13 pairs; stipules connate below; peduncle pubescent with soft white hairs below mixed with black above. *Fls* 15 mm, *blue-purple, erect*; bracteoles oblong to triangular. Calyx with appressed black and white hairs. *Legume* 7–10 mm, *covered with crispate white hairs*, up to 7-seeded. Fl. 5–7. $2n = 16$. Hp.

Native. In short turf on calcareous soils and dunes. Locally abundant in suitable habitats from Hants to Sutherland, but mainly in the east; Isle of Man; Ireland: only Aran Islands. Denmark and S. Sweden eastwards to the Baikal region; S.W. Alps.

2. A. alpinus L.   Alpine Milk-vetch.

Like *A. danicus* but more slender. Stipules free to base. Fls c. 10 mm, pale blue tipped with purple, spreading or deflexed; peduncle with sparse, ±appressed hairs. Legume 10–15 mm, with short appressed brown or blackish hairs. few-seeded. Fl 7. $2n = $ c. 56. Hp.

Native. In grassy places on mountains from 715 to 793 m, rare. E. Perth, Angus and S. Aberdeen. Arctic Europe and on mountains south to the Pyrenees and Alps, Carpathians; Caucasus; temperate Asia; Greenland, North America.

3. A. glycyphyllos L.   Wild Liquorice, Milk-vetch.

A stout *glabrous* prostrate or ascending perennial 30–100 cm. Stock short, stout. Lvs 10–20 cm; lflets 15–40 mm, oblong-elliptical obtuse or mucronate; stipules 15–20 mm, free. Infl. 2–5 cm; peduncles much shorter than subtending lvs. *Fls* 11–15 mm, *creamy-white with a greenish-grey tinge*, spreading; *bracteoles subulate*. Legume 30–40 mm, slightly curved, acuminate, many-seeded, 2-celled with a longitudinal septum. Fl. 7–8. $2n = 16$. Hp.

Native. In rough grassy and bushy places, local. Scattered throughout Great Britain north to Kincardineshire. Europe to c. 65°N, rare in the Mediterranean region, eastwards to the Altai, Caucasus and Asia Minor.

11. Oxytropis DC.

Herbs similar to *Astragalus* in habit, etc., but differing in *the mucronate keel*. The fls have nectar and are pollinated by bees. About 300 spp. in north temperate regions.

Legume with both dorsal and ventral septum; fls pale purple.   **1. halleri**
Legume with ventral septum only; fls yellow tinged with purple.   **2. campestris**

1. O. halleri Bunge   Purple Oxytropis.

O. sericea (DC.) Simonkai, non Nutt.; *O. uralensis* auct., non (L.) DC.

A soft hairy perennial with a stout rootstock and very short branches. Lvs up to c. 10 cm; lflets 5–8 mm, 10–14 pairs, elliptical, subacute; stipules lanceolate, persistent and clothing the stock. Infl. c. 3 cm, 5–15-fld; peduncle stout, erect. *Fls* 15–20 mm, *blue to purple*, the keel tipped with dark purple. Legume 15–20 mm, pubescent, many-seeded. Fl. 6–7. $2n = 32$. Hr.

Native. In dry rocky pastures up to 610 m, local. Argyll, E. Perth, E. Ross, W. Sutherland and Caithness. Pyrenees, Alps from France to Austria, Carpathians.

Represented in Britain by subsp. **halleri**.

2. O. campestris (L.) DC.   Yellow Oxytropis.

Like *O. halleri* but larger; lvs up to 15 cm; lflets 10–20 mm, linear-oblong or oblong-lanceolate; fls yellow tinged with purple; legume up to 18 mm. Fl. 6–7. $2n = 48$. Hr.

Native. On rocky ledges up to 610 m very rare. E. Perth and Angus. N. Europe and mountains of C. and S. Europe; North America: Rocky Mountains, Labrador, Maine; absent from Greenland.

Represented in Britain by subsp. **campestris**.

Vicieae (Adanson) DC. Perennial or annual herbs, usually climbing or scrambling. Stems with cortical vascular bundles in the internodes, usually winged primary stem of limited growth. Lvs epulvinate, almost always paripinnate, with a terminal tendril or mucro. Fls usually in secund axillary racemes. Stamens diadelphous; anthers uniform, versatile. Fr. 2-valved, 2–many-seeded, ±linear, laterally compressed. Seedlings hypogeal.

12. Vicia L.   Vetch, Tare.

Annual or perennial herbs, often climbing. Lvs pinnate without a terminal lflet; tendrils usually present, simple or branched. Lflets opposite or alternate in 2–many pairs. Fls solitary or in axillary clusters or racemes; keel obtuse. Calyx-teeth subequal or the lower longer. Upper stamen ±free; filaments not dilated and anthers all similar. *Style* cylindrical or flattened, *glabrous or* *equally downy all round or bearded on the lower side.* Legume compressed, 2-valved, dehiscent, several-seeded. The fls have nectar and are usually cross-pollinated by bees.

About 140 spp. in north temperate regions and South America.

1 Infl. sessile or with a peduncle shorter than the fls.   *2*
　Infl. with a peduncle longer than the fls.   *6*
2 Corolla pale yellow, sometimes with a slight purple tinge.   **8. lutea**
　Corolla bluish or purplish.   *3*
3 Mouth of calyx-tube oblique.   **7. sepium**
　Mouth of calyx-tube not oblique.   *4*
4 Fls 6–9 mm; seeds tuberculate.   **10. lathyroides**
　Fls 9–26 mm; seeds smooth.   *5*
5 Lflets 1–3 pairs; wings and keel whitish. **11. bithynica**
　Lflets of upper lvs 3–8 pairs; wings and keel purple.   **9. sativa**
6 Calyx-teeth equal, all at least as long as the tube.   **1. hirsuta**
　Calyx-teeth unequal, at least the upper shorter than the tube.   *7*
7 Fls not more than 9 mm.   *8*
　Fls at least 9 mm.   *10*
8 Infl. with 10–30 bluish-purple fls.   **4. cracca**
　Infl. with 1–5 pale purple fls.   *9*
9 Infl. about equalling the subtending lf, with 1–2 fls.   **2. tetrasperma**
　Infl. longer than the subtending lf, with (1–)2–5 fls.   **3. tenuissima**
10 All lvs without tendrils; plant erect.   **5. orobus**
　Lvs with tendrils; plant climbing or scrambling.   *11*
11 Stipules entire.   **4. cracca**
　Stipules toothed.   *12*
12 Lvs with 2–3 pairs of lflets; infl. with 1–3 fls.   **11. bithynica**
　Lvs with 5–12 pairs of lflets; infl. with 5–20 fls.   **6. sylvatica**

Section 1. *Ervum* (L.) S. F. Gray. Annuals. Lflets usually more than 4 pairs. Peduncles long, 1–8-fld. Fls not exceeding 9 mm. Calyx-tube not gibbous at base. Style pubescent all round or sometimes nearly glabrous.

1. V. hirsuta (L.) S. F. Gray   Hairy Tare.

A slender trailing nearly glabrous annual 20–30(–70) cm. Lflets 5–20 mm, 4–8(–10) pairs, usually alternate, linear-oblong or rarely ovate-oblong, truncate to emarginate and often mucronulate; tendrils usually branched; stipules often 4-lobed. Racemes 1–8-fld; peduncles 1–3 cm, slender. Fls 4–5 mm, dirty white or purplish; *calyx-teeth subequal*, subulate, *somewhat exceeding tube. Legume* 6–11 mm, oblong, *pubescent, usually 2-seeded.* Fl 5–8. $2n = 14$. Th.

Native. In grassy places. Throughout the British Is., northwards to Shetland, but local in Scotland and Ireland. Almost all Europe; W. Asia; N. Africa. A weed of cultivation throughout much of the world.

2. V. tetrasperma (L.) Schreber   Smooth Tare.

V. gemella Crantz

A slender ±glabrous annual (15–)30–60 cm. Lflets

5–20 mm, (3–)4–6 pairs, usually alternate, linear-oblong, obtuse, mucronate, or rarely linear, acute (var. *tenuissima* Druce); tendrils usually simple; stipules half-arrow-shaped. Racemes 1–2-fld; fruiting peduncles 2–4 cm, about equalling the lvs. *Fls 4–8 mm, pale blue; calyx-teeth unequal, triangular, shorter than tube. Legume (9–)12–16 mm, oblong, shortly stipitate, glabrous, 3–5-seeded; hilum oblong.* Fl. 5–8. 2n = 14. Th.

Native. In grassy places. Throughout England and Wales, rare in the west and in southern Scotland, naturalized in a few places in Ireland. Europe north to 62°; W. Asia; Japan; N. Africa; Madeira, Canary Is.

3. V. tenuissima (Bieb.) Schinz & Thell. Slender Tare.

V. gracilis Loisel., non Banks & Solander

Like *V. tetrasperma* but lflets up to 25 mm, 2–5 pairs, usually linear-acuminate; racemes 1–5-fld; fruiting peduncles up to 8 cm, longer than lvs; fls 6–9 mm; legume 5–8-seeded; hilum almost circular. Fl. 6–8. 2n = 14. Th.

Native. In grassy places. From Devon and W. Kent to Warwick and S. Yorks., local. S. and W. Europe, Macaronesia; introduced in C. Europe.

Section 2. *Cracca* S. F. Gray. Perennials. Lflets 6–12 pairs. Peduncles long, 6–40-fld. Fls more than 10 mm. Calyx-tube gibbous at base on upper side. Style equally pubescent all round.

4. V. cracca L. Tufted Vetch.

A somewhat pubescent scrambling perennial 60–200 cm. Lflets 5–30 mm, 6–12(–15) pairs, lowest almost at base of petiole, oblong-lanceolate to linear-lanceolate, acute or mucronate; tendrils branched; *stipules half-arrow-shaped, entire.* Racemes 2–10 cm, 10–30-fld, rather dense; peduncles 2–10 cm, stout. Fls 8–12 mm, blue-purple, shortly pedicellate and drooping; calyx-teeth very unequal, upper minute; tube short. Legume 10–20 mm, ovate, obliquely truncate, with stipe shorter than the calyx-tube, glabrous, 4–8-seeded. Fl. 6–8. 2n = 14, 27, 28 (30). Hp.

Native. In grassy and bushy places. Generally distributed and common throughout the British Is. Almost all Europe; Asia to Sakhalin and Japan; Greenland; introduced in North America.

The 3 following resemble *V. cracca* in general appearance:

***V. tenuifolia** Roth, Fine-leaved Vetch. Usually more hairy. Racemes rather lax. Fls larger (12–15 mm), pale blue or violet with whitish wings. Legume more gradually narrowed at the base, with a stipe equalling the calyx-tube. A casual on disturbed ground, and locally naturalized. Europe from Sweden southwards, W. Asia, N.W. Africa.

***V. villosa** Roth (incl. *V. dasycarpa* Ten.), Fodder Vetch. Racemes longer than the subtending lf, calyx strongly gibbous, claw of standard twice as long as limb and stipe of legume longer than the calyx-tube. Variable in hairiness, size of lflets, number of fls in the raceme and pubescence of pod. Several spp. have been described, based on these characters and differing somewhat in distribution, but intermediates are of frequent

occurrence. Not infrequent in cultivated land and waste places. C. and S. Europe, W. Asia, N. Africa.

***V. benghalensis** L. (*V. atropurpurea* Desf.) differs from the preceding in the shaggy stipules and calyx and in having racemes shorter than their subtending lvs. Field margins, waste places, etc. S. Europe, N.W. Africa.

5. V. orobus DC. Wood Bitter Vetch.

Orobus sylvaticus L.

An erect branched ±pubescent perennial 30–60 cm. Stem stout, lfless or with reduced lvs at base. Lflets 10–20 mm, 6–15 pairs, elliptical-mucronate; *tendrils 0; stipules half-arrow-shaped, slightly toothed.* Racemes 1–3 cm, 6–20-fld, secund, rather dense; peduncles about as long as lvs. *Fls 12–15 mm, white tinged with purple,* distinctly pedicellate and drooping; calyx-teeth unequal. Legume 20–30 mm, oblong-lanceolate, acute at both ends, glabrous, few-seeded. Fl. 6–9. 2n = 12. Hp.

Native. In rocky and wooded places. Scattered throughout the W. and N. of the British Is. but very local; rare in Ireland. W. Europe from N. Spain to Norway.

6. V. sylvatica L. Wood Vetch.

A glabrous perennial 60–200 cm. Lflets 5–20 mm, 6–9(–12) pairs, oblong-elliptical, mucronate; tendrils much branched; *stipules lanceolate, semi-circular with many setaceous teeth at base.* Racemes 1–7 cm, up to 20-fld, secund, rather lax; peduncles stout, up to c. 10 cm. *Fls 15–20 mm, white with blue or purple veins,* distinctly pedicellate and drooping; *calyx-teeth setaceous,* shorter than the tube, *the upper about ½ length of lower. Legume* c. 30 mm, oblong-lanceolate, acuminate at both ends, glabrous, few-seeded, black when ripe. Fl. 6–8. 2n = 14. Hp.

Var. *condensata* Druce is a small densely lfy few-fld variant found on shingle banks.

Native. In rocky bushy places, woods and shingle and cliffs by the sea. Scattered throughout the British Is., local N., C. and E. Europe, southwards to Italy, east to Siberia (Baikal region).

Section 3. *Vicia.* Annuals or rarely perennials. Lflets 3–9, rarely 2 pairs. Peduncles very short or almost 0, 1–6-fld. Fls 10–25 mm, rarely less. Calyx-tube usually gibbous at base on upper side. Style bearded on the lower side.

7. V. sepium L. Bush Vetch.

A climbing or trailing, nearly glabrous perennial 30–100 cm. Lflets 10–30 mm, 5–9 pairs, ovate to elliptic, acute, obtuse or truncate, mucronate; tendrils branched; stipules half-arrow-shaped, sometimes toothed, with a dark spot. Racemes 1–2 cm, 2–6-fld, subsessile. Fls 12–15 mm, pale bluish-purple, shortly pedicellate; *calyx-teeth unequal, shorter than tube. Legume* 20–35 mm, oblong-lanceolate, beaked, *glabrous,* 6–10-seeded, black when ripe. Fl. 5–8. 2n = 14. Hp.

Native. In grassy places, hedges and thickets. Through out the British Is., common. Almost all Europe, but rare and local in the Mediterranean region; temperate Asia and Kashmir; Greenland.

8. V. lutea L. Yellow Vetch.

A tufted nearly glabrous to villous annual 10–45(–60) cm. Lflets 10–25 mm, 3–10 pairs, linear-oblong to elliptical, obtuse and mucronate, rarely acute; tendrils simple or branched; *stipules small, triangular, the lower with a basal lobe. Fls* 1–3 together, 20–35 mm, *pale* dirty *yellow*, shortly pedicellate; *standard glabrous*; *calyx-teeth setaceous, unequal, the lower longer than the tube. Legume* 20–40 mm, oblong-lanceolate, beaked, *pubescent*, 4–8-seeded. Fl. 6–8. 2n = 14. Th.

Native. On cliffs and shingle by the sea. Scattered round the coasts of Great Britain north to Wigtown and Kincardine, local. S. and W. Europe; Caucasus; N. Africa; Macaronesia.

The British plant is subsp. **lutea.**

The 2 following resemble *V. lutea* in general appearance:

*V. hybrida L. has emarginate lflets, the fls solitary, yellowish, the standard pubescent on the back and the legume with spreading hairs. A casual. S. Europe.

*V. pannonica Crantz, with entire stipules, fls usually 2–4 together (often brownish-violet) and the legume with appressed hairs, is found occasionally and is established in Kent. C. and S. Europe to Asia Minor.

9. V. sativa L. Common Vetch.

A tufted trailing or climbing sparsely hairy annual 15–120 cm. Lflets 10–20 mm, 3–4 pairs, linear to obo-vate, acute, obtuse or retuse and mucronate; tendrils simple or branched; *stipules* half-arrow-shaped, toothed or entire, *often with a dark blotch.* Fls solitary or in pairs (rarely 4), 10–30 mm, purple; *calyx-teeth subequal, as long as or shorter than the tube.* Legume 25–70 mm, linear-oblong, beaked, sparsely hairy to glabrous, 4–12-seeded, breaking the calyx when ripe. Fl. 5–9. 2n = 12. Th.

In hedges and grassy places, common and widely distributed. Throughout Europe to 69°N in Russia; temperate Asia; North Africa; introduced in North America.

Very variable. Represented in Britain by the two following subspp.:

(a) Subsp. **sativa: Legumes constricted between the seeds. Cultivated for forage and often escaping.

(b) Subsp. **nigra** (L.) Ehrh. (*V. angustifolia* L., *V. sativa* subsp. *angustifolia* (L.) Gaudin): Legumes not constricted between the seeds. The native plant.

10. V. lathyroides L. Spring Vetch.

A slender ±pubescent spreading annual 5–20 cm. *Lflets* 4–14 mm, 2–4 *pairs*, linear-oblong or obovate, obtuse or emarginate and mucronulate; tendrils small and unbranched or 0; stipules small, half-arrow-shaped, entire, unspotted. Fls solitary, 5–7 mm, lilac; *calyx-teeth*

equal, nearly as long as tube. Legume 15–30 mm, taper-ing at both ends, *glabrous*, 8–12-seeded, not breaking the calyx when ripe. Fl. 5–6. 2n = 10, 12. Th.

Native. In dry grassy places, particularly on sandy soils. Scattered throughout Great Britain, mainly in the east; N. and E. coasts of Ireland. Widely distributed throughout Europe north to S.W. Finland, eastwards to the Caucasus and Asia Minor; N. Africa.

Section 4. *Faba* (Miller) S. F. Gray. Perennials or annuals. Lflets 1–3 pairs. Fls more than 10 mm, solitary or in few-fld racemes. Calyx equal at base. Style bearded below the stigma.

11. V. bithynica (L.) L. Bithynian Vetch.

A trailing or climbing nearly glabrous tufted annual 30–60 cm. Stem angled, ±winged. Lflets 20–50 mm, 1–3 pairs, elliptic or ovate, acute or obtuse and mucronate, or sometimes linear-acuminate; tendrils branched; *stipules large* (1 cm *or more*) *ovate-acuminate, toothed.* Fls solitary or in pairs, c. 20 mm; standard purple, wings and keel white; peduncle very short or up to c. 5 cm, stout; *pedicels as long as calyx-tube; calyx-teeth longer than tube.* Legume 30–40 mm, abruptly beaked, 4–7-seeded. Fl. 5–6. 2n = 14. Hp.

Native. On bushy cliffs and in hedges. From Cornwall and Brecon to Essex, mainly near the coast, Nottingham and N.E. Yorks; Wigtown, local. S. and W. Europe northwards to Scotland; Caucasus; Algeria.

*V. faba L. Broad bean, Horse bean, an erect annual without tendrils, is cultivated and sometimes persists for a few years at the margins of fields. Stems square; fls white with a purplish-black blotch, in axillary clusters; legume 10–20 cm with few large (2–3 cm) seeds. Cultivated in the Mediterranean region since prehistoric times.

*V. narbonensis L. An erect annual with dark purple fls, ten-drils on the upper lvs only and a legume 5–6 cm occurs as a casual and persists in some localities. It is rather similar to, but smaller in all its parts than, *V. faba.* S. Europe, N. Africa, W. Asia.

13. LATHYRUS L.

Like *Vicia* but usually with fewer lflets and winged or angled stems. *Style* flattened and *bearded on its upper side only.* The fls have nectar and are usually cross-pollinated by bees.

About 150 spp. in north temperate regions and the mountains of tropical Africa and South America.

1 Lflets 0.	*2*
Lflets present.	*3*
2 Stipules lf-like; fls yellow.	**1. aphaca**
Stipules minute; phyllodes grass-like; fls crimson.	
	2. nissolia
3 Tendrils present on most lvs.	*4*
Tendrils 0, the lvs ending in short point.	*9*
4 Lflets 1 pair.	*5*
Lflets 2 or more pairs.	*8*
5 Stem flattened and distinctly winged.	*6*
Stem ±square and sharply angled.	*7*
6 Fls (3–)5–7(–12) together; calyx-teeth shorter than tube; legume glabrous.	**6. sylvestris**

Fls 1–2(–3) together; calyx-teeth as long as tube; legume densely silky.   **3. hirsutus**

7 Lflets lanceolate; fls yellow; legume compressed.
                                               4. pratensis

Lflets obovate; fls crimson; legume nearly cylindrical.
                                               5. tuberosus

8 Stem winged; stipules lanceolate; legume compressed.   **7. palustris**

Stem angled; stipules broadly triangular; legume turgid.   **8. japonicus**

9 Stem winged; stipules lanceolate.   **9. montanus**

Stem angled; stipules linear-lanceolate.   **10. niger**

Section 1. *Aphaca* (Miller) Dumort. Annual. Lvs with tendrils but devoid of lflets except in young seedlings. Calyx not gibbous at base.

*1. L. aphaca L.   Yellow Vetchling.

A glabrous scrambling annual up to 100 cm. Lflets 0 in mature plants; tendrils simple; *stipules 6–50 × 4–40 mm, ovate-hastate*, acute or obtuse, entire. Peduncles equalling or exceeding stipules, 1-fld. *Fls 10–12 mm, yellow*, usually erect; *calyx-teeth linear-lanceolate, equal, 2–3 times as long as the tube. Legume 20–30 × 5–8 mm, somewhat falcate, 6–8-seeded*; seeds smooth. Fl. 6–8. Germ. autumn. A large proportion of the young plants is killed by severe frost. 2*n* = 14. Th.

Introduced. In dry places on sand, gravel or chalk. Devon and Kent to Oxford and Cambridge; very local. Europe; Mediterranean region; Caucasus, temperate Asia, N. Africa.

Section 2. *Nissolia* (Miller) Dumort. Annual. Lvs reduced to grass-like phyllodes; tendrils 0. Calyx ±gibbous at base.

2. L. nissolia L.   Grass Vetchling.

An erect or ascending nearly glabrous annual 30–90 cm. Lflets 0; *tendrils 0; stipules minute; phyllodes up to c. 15 cm, grass-like.* Peduncles shorter than or equalling phyllodes, 1–2-fld. *Fls 8–18 mm, crimson*, usually erect; *calyx-teeth triangular, much shorter than the tube. Legume 30–60 × 2–3 mm, straight, 15–20-seeded*; seeds rough. Fl. 5–7. Germ. autumn. 2*n* = 14. Th.

Native. In grassy and bushy places. South of a line from the Humber to the Severn; very local. W., C. and S. Europe, east to the Caucasus and Syria; N. Africa; introduced in Belgium, the Netherlands and North America.

Section 3. *Lathyrus.* Perennial or rarely annual. Lvs always with lflets and tendrils. Calyx-tube gibbous at base.

*3. L. hirsutus L.   Hairy Vetchling.

An almost glabrous scrambling annual 30–120 cm. Stem winged. Lflets 1 pair, 15–80 mm, linear-oblong, mucronate; tendrils branched; stipules 10–18 mm, subulate, half-arrow-shaped. Peduncles exceeding lvs, 1–2-fld. *Fls 7–15 mm, with crimson standard and pale blue wings; calyx-teeth* subulate, about as *long as tube. Legume*

30–50 × 6–8 mm, *tubercled and densely silky*, 5–10-seeded; seeds rough. Fl. 5–7. 2*n* = 14.

Introduced. In fields and waste places. Rare and usually casual; north to Edinburgh. C. and S. Europe, N. Africa, temperate Asia.

4. L. pratensis L.   Meadow Vetchling.

A scrambling finely pubescent perennial 30–120 cm. Stem angled. *Lflets* 1 pair, 10–30 mm, *lanceolate*, acute; veins parallel, scarcely anastomosing; tendrils simple or branched; stipules 10–30 mm, lf-like, sagittate. Peduncles exceeding lvs, stout, (2–)5–12-fld. *Fls* 10–18 mm, *yellow*; lower calyx-teeth triangular-subulate, shorter to longer than the tube. *Legume* 25–35 mm, glabrous or finely pubescent, *compressed*, 5–10-seeded, seeds smooth. Fl. 5–8. 2*n* = 7, 8, 9, 12, 14, 16, 21, 28*, 42. Hp.

Native. In hedges and grassy places. Throughout the British Is., common. Almost all Europe; Siberia to the Arctic Circle, south to the Himalaya; N. Africa, Ethiopia; introduced in North America.

*5 L. tuberosus L.   Tuberous Pea.

A glabrous or subglabrous scrambling perennial 30–120 cm. Roots bearing small *tubers.* Stem angled. *Lflets* 1 pair, 15–45 mm *obovate*, obtuse or subacute, mucronate; tendrils simple or branched; *stipules* up to 20 mm, *narrowly lanceolate*, half-arrow shaped. Peduncles exceeding lvs, 2–7-fld. *Fls* 12–20 mm, *crimson*; calyx-teeth triangular, equalling tube. *Legume* 20–40 mm, glabrous, *nearly cylindrical.* Fl. 7. 2*n* = 14. Hp.

Introduced. Naturalized in cornfields and hedges since about 1800 at Fyfield, Essex. Rare and scattered, usually as a casual, north to S. Scotland. Most of Europe and W. Asia but absent from the islands.

6. L. sylvestris L.   Narrow-leaved Everlasting-pea.

A glabrous often glaucous scrambling perennial 100–200 cm. *Stem broadly winged. Lflets* 1 pair, 50–150 mm, *ensiform*; tendrils usually large and branched; *stipules* up to 30 mm, less than ½ as wide as stem, *narrowly ensiform with a spreading basal lobe.* Peduncles 10–20 cm, 3–12-fld. Fls 13–20 mm, rose-pink; *calyx-teeth* triangular, *shorter than tube.* Legume 40–70 mm, glabrous, compressed, narrowly winged along upper side, 8–14-seeded. Fl. 6–8. 2*n* = 14. Hp.

Native. In thickets and woods; sometimes naturalized in hedges near dwellings. Scattered throughout Great Britain northwards to Angus, local. Most of Europe, east to the Caucasus; N.W. Africa.

*L. latifolius L., Broad-leaved Everlasting-pea, closely related to *L. sylvestris* but with ovate lflets, stipules more than ½ as wide as stem, fls 20–30 mm, and calyx-teeth equalling or exceeding tube, is naturalized in some localities. 2*n* = 14. S. Europe.

*L. odoratus L., Sweet Pea, is a widely cultivated plant with fls c. 30 mm. Native in S. Italy and Sicily.

7. L. palustris L.   Marsh Pea.

A glabrous scrambling perennial 60–120 cm. *Stem winged. Lflets 35–70 mm, 2–5 pairs, narrowly lanceolate,* acute or obtuse and mucronate; tendrils usually branched; *stipules* 10–20 mm, *lanceolate,* half-arrow-shaped. *Peduncles usually longer than lvs, 2–6-fld.* Fls 12–20 mm, pale purplish-blue; lower calyx-teeth subulate, almost equalling tube. *Legume 25–60 mm, compressed,* glabrous, 3–12-seeded. Fl. 5–7. 2n = 42. Hp.

Native. In fens and damp grassy and bushy places. E. England from Cambridge to S.E. Yorks.; N. Somerset; C. Ireland, very local. Europe, rare in the Mediterranean region; arctic Russia and Siberia east to Sakhalin and Japan; North America.

Represented in Britain by subsp. **palustris**.

8. L. japonicus Willd.   Sea Pea.

incl. *L. maritimus* Bigelow

A creeping and ascending glabrous glaucous perennial 30–90 cm. *Stem angled. Lflets 20–40 mm, 3–4 pairs, elliptical,* obtuse; tendrils simple or branched, sometimes 0; *stipules* 10–25 mm, *broadly triangular,* half-hastate. *Peduncles usually shorter than lvs,* 5–12-fld. Fls 14–18 mm, purple to blue; lower calyx-teeth narrowly triangular, about equalling tube. *Legume* 30–50 mm, glabrous, *turgid,* 4–8-seeded. Fl. 6–8. 2n = 14. Hp.

Native. On shingle beaches. Coasts from Cornwall to Suffolk; Glamorgan; Kerry; very local. Circumpolar; China, Japan and North America.

Subsp. **maritimus** (L.) P. W. Ball, described above, occurs in W. Europe, the Baltic region and subarctic Russia.

Section 4. *Orobus* (L.) Godron. Perennial. Lf ending in a short point, without tendrils. Calyx gibbous at base.

9. L. montanus Bernh.   Bitter Vetch.

L. macrorrhizus Wimmer; *Orobus tuberosus* L.

An erect glabrous perennial 15–50 cm. *Rhizome creeping and tuberous. Stem winged.* Lflets 10–50 mm, 2–4 pairs, narrowly lanceolate to elliptic, acute or obtuse and mucronate; *stipules* 5–25 mm, *lanceolate,* half-arrow-shaped, *variable but usually somewhat toothed below. Peduncles 2–6-fld, glabrous. Fls 10–16 mm, crimson becoming green or blue; calyx-teeth unequal, lower about equalling tube.* Legume 25–45 mm, subcylindrical, glabrous, 4–10-seeded. Var. *tenuifolius* (Roth) Garcke (*Orobus tenuifolius* Roth) has linear acuminate lflets and narrow stipules Fl. 4–7. 2n = 14. Hp.

Native. In woods, thickets and hedgebanks in hilly country. Scattered throughout the British Is., commoner in the west and north and absent from East Anglia. S., W. and C. Europe, extending to the Baltic region and White Russia; N. Africa.

10. L. niger (L.) Bernh.   Black Pea.

Orobus niger L.

An erect nearly glabrous perennial 30–60 cm. *Rhizome*

short. Stem angled. Lflets 10–30 mm, 3–6 pairs, lanceolate or elliptic, acute or obtuse and mucronate; *stipules* 4–10 mm, *linear-lanceolate* to elliptical, half-arrow-shaped, *entire. Peduncles* 2–10-fld, *pubescent with appressed crisped hairs. Fls* 10–15 mm, *purple becoming blue; calyx pubescent with appressed crisped hairs, teeth unequal, lower longer than tube.* Legume 35–60 mm, turgid, rugose, 6–10-seeded. Fl. 6–7. 2n = 14. Hp.

Native or introduced. Rocky woods in mountain valleys up to 360 m. Formerly perhaps native in a few localities in Scotland but now extinct or very rare; naturalized in E. Sussex. Most of Europe but absent from much of the north-east and many islands; east to the Caucasus; N.W. Africa, very rare.

14. Pisum L.

Like *Lathyrus* but stems terete; *wings adnate to keel; calyx-teeth ±lf-like;* alternate filaments and style dilated at the top with recurved margins. The fls have nectar but are very rarely visited by bees and are completely self-fertile.

Two spp. in the Mediterranean region and W. Asia.

*1. P. sativum L.   Garden Pea, Field Pea.

A climbing glabrous glaucous annual 30–200 cm. Lflets 1–3 pairs, often with whitish markings; tendrils branched; stipules up to 10 cm, lf-like. Peduncle 1–3-fld. Fls 15–35 mm, white or (var. *arvense* (L.) Poiret) coloured. Legume subterete, several-seeded. 2n = 14. Th.

Introduced. In waste places as an occasional escape from cultivation. Cultivated from ancient times, native in S. Europe and N. Africa.

Trifolieae (Bronn) Bentham. Annual or perennial herbs, rarely small shrubs. Lvs usually pinnately or digitately trifoliolate. Fls in capitate or spicate axillary racemes. Stamens diadelphous or monadelphous, the filaments sometimes dilated at the apex; anthers uniform or dimorphic. Fr. straight, falcate, spirally coiled or ovoid, dehiscent or indehiscent. Seeds 1–many. Seedlings epigeal.

15. Ononis L.

Herbs or small shrubs. Lvs 3-foliolate, sometimes reduced to the terminal lflets, the veins ending in teeth. Stipules adnate to the petiole. *Fls* axillary, *pink* (in our spp.), without nectar. Standard broad, the wings oblong, the keel pointed. *Stamens monadelphous*; 5 or all the *filaments dilated above.* Style curved; stigma terminal. Legume 1-celled, 2-valved; seeds 1–many. Pollination by bees. About 75 spp.

1 Stout woody perennials; fls 10–20 mm; legume erect.   *2*
  Slender annual; fls 5–10 mm; legume deflexed.
                                                    3. reclinata
2 Stem usually rooting, hairy all round; lflets obtuse
  or emarginate.   **1. repens**
  Stem not rooting, with 2 lines of hairs; lflets acute.
                                                    2. spinosa

1. O. repens L.   Common Restharrow.

O. arvensis auct., non L.

A *procumbent* or rarely ascending perennial 30–60 cm. *Rhizomatous. Stems usually unarmed, rooting at the base, uniformly hairy,* woody and much-branched, sometimes with usually soft spines above. Petioles 3–5 mm, pubescent; stipules 3–5 mm, clasping the stem, serrate and ±glandular-pubescent. Lvs up to 20 mm; *lflets* obovate, serrate, pubescent and ±glandular, *obtuse or emarginate.* Fls 15–20 mm. Pedicels 1–5 mm. Legume 1–2-seeded, *shorter than the enlarged calyx.* Fl. 6–9. $2n = 30, 60$. Hp or Chh.

Native. In rough grassy places. Scattered throughout the British Is., except Orkney and Shetland, common in calcareous districts. W. and C. Europe.

2. O. spinosa L.   Spiny Restharrow.

O. campestris Koch & Ziz

An *erect or ascending* perennial 10–80 cm. Like *O. repens* but *not rhizomatous. Stems usually spiny, not rooting, with 2 lines of hairs. Lflets acute. Legume exceeding calyx.* Fl. 6–9. $2n = 30$. Hp. or Chh.

Native. In rough grassy places. Scattered throughout England and Wales, north to Midlothian. Most of Europe, except the extreme north and high mountain regions; Asia.

Intermediates (probably hybrids) between *O. repens* and *O. spinosa* occur.

3. O. reclinata L.   Small Restharrow.

An erect or ascending viscid *annual* 2–8 cm. *Stems slender,* unarmed, pubescent. Petioles slender, short; stipules 1–4 mm, pubescent, serrate, clasping the stem. Lvs 3–5 mm; lflets narrowly obovate, cuneate, serrate at apex, densely pubescent and glandular. *Fls 5–7(–10) mm. Pedicels strongly deflexed after flowering. Calyx as long as the fl. or longer.* Legume with up to 20 seeds, glandular, pubescent, about equalling the calyx. Fl. 6–7. $2n = 64, 60$. Th.

Native. By the sea in sandy soil. S. Devon; Glamorgan, Pembroke; Alderney and Guernsey, Wigtown. Mediterranean region eastwards to Iran, Syria and Israel; south to Arabia and Ethiopia, N. Africa, Canaries.

16. MELILOTUS MILLER

Annual, biennial or short-lived perennial herbs. Lvs 3-foliolate, veins ending in teeth. Stipules adnate to the petiole. Fls in racemes, small, yellow or white. Like *Medicago*, but *racemes usually longer and laxer*; stamens diadelphous, the filaments not dilated, *and the legume short, straight, thick* and usually indehiscent, *never spiny.* Pollination mainly by bees. Many spp. smell strongly of coumarin (new-mown hay), especially when drying. About 20 spp.

1 Young legume pubescent.	**1. altissima**	
Young legume glabrous.		*2*
2 Fls white.	**3. alba**	
Fls yellow.		*3*
3 Fls 2–3 mm; legume 2–3 mm.	**4. indica**	
Fls 4–7 mm; legume 3–5 mm.	**2. officinalis**	

1. M. altissima Thuill.   Tall Melilot.

M. officinalis auct., non (L.) Pallas

An erect branched biennial or short-lived perennial 60–150 cm. Lflets 15–20(–30) mm, oblong or obovate, those of the upper lvs nearly parallel-sided, serrate. Stipules subulate, dentate at base. Racemes 20–50 mm, compact, lengthening in fr. Fls 5–7 mm, yellow; *wings, standard and keel all equal. Legume* 5–6 mm, *pubescent,* reticulate, compressed ovoid, *black when ripe*; style long, persistent. Fl. 6–8. Fr 8–10. $2n = 16$. Hp.

?Introduced. Naturalized in waste places and woods. Generally distributed in England but rare in Wales and Scotland; Ireland, naturalized in the east. Throughout most of Europe.

*2. M. officinalis (L.) Pallas   Ribbed Melilot.

M. arvensis Wallr.

A decumbent or erect branched biennial similar in appearance to *M. altissima.* Lflets of upper lvs oblong-elliptical narrowed at both ends. Racemes rather lax and slender. Fls 4–7 mm, yellow; *wings and standard equal but longer than the keel. Legume* 3–5 mm, glabrous, *transversely rugose,* ovoid, slightly compressed, brown when ripe; style usually deciduous. Fl. 7–9. $2n = 16$. Hs.

Introduced. Naturalized in the British Is., mostly south-east of a line from the Humber to the Severn in fields and waste places. Most of Europe, though often only as a weed of cultivation, eastwards to W. China.

*3. M. alba Medicus   White Melilot.

An erect branched annual or biennial 30–150 cm. Stipules setaceous, entire. Racemes rather lax and slender. *Fls 4–5 mm, white; wings and keel nearly equal, somewhat shorter than standard. Legume* 3–5 mm, glabrous, reticulate, ovoid, compressed, mucronate, *greyish-brown when ripe*; style usually deciduous. Fl. 7–8. $2n = 16, 24$. Th. or H.

Introduced. Naturalized in fields and waste places, mainly in England. Throughout most of Europe, but doubtfully native; N. Africa; Asia eastwards to Tibet, introduced in many places; introduced in America and Australia.

*4. M. indica (L.) All.   Small Melilot.

M. parviflora Desf.

An erect branched annual, similar in general appearance to, but smaller in all its parts than, *M. altissima.* Racemes slender, usually lax, sometimes dense in fr. *Fls 2–3 mm, pale yellow; wings and keel equal, shorter than standard. Legume* 1.5–3 mm, glabrous, strongly reticulate, subglobose, *olive green when ripe*; style usually persistent. Fl. 6–10. $2n = 16$. Th.

Introduced. Naturalized in fields and waste places, in scattered localities in England. Mediterranean region and S.W. Europe to India; introduced and naturalized in most of the rest of the world.

17. MEDICAGO L.

Herbs. Lvs trifoliolate, the veins ending in teeth. Stipules adnate to the petiole. Fls in racemes, yellow or purple. Calyx-teeth 5, nearly equal. *Petals caducous*; keel obtuse, shorter than the wings. Upper stamen free; filaments not dilated; anthers all equal. Style glabrous; stigma subcapitate. *Legume spirally curved or coiled, rarely falcate, often spiny*, longer than the calyx, *usually indehiscent*, 1–many-seeded.

About 100 spp. in Europe, W. Asia, N. Africa.

1 Fls (5–)7–11 mm, yellow or purple.　　**1. sativa**
　Fls 2–5 mm, yellow.　　2
2 Racemes 10 or more-fld; legume unarmed, 1-seeded, black when ripe.　　**2. lupulina**
　Racemes 1–5-fld; legume spiny or tubercled, several-seeded, brown when ripe.　　3
3 Fr. densely pubescent.　　**3. minima**
　Fr. ±glabrous.　　4
4 Lflets not blotched; stipules laciniate; legume flat.　　**4. polymorpha**
　Lflets usually blotched; stipules toothed; legume subglobose.　　**5. arabica**

1. M. sativa L.　　Lucerne, Alfalfa.

A deep-rooted erect or ascending perennial up to 90 cm. Lflets up to 30 mm, obovate, long-cuneate, toothed in the upper third. Stipules linear-lanceolate, ±toothed. Pedicels short, stout, erect in fr. Fl. 4–11 mm, yellow, blue or violet. Legume nearly straight, falcate or a spiral of 1½–3 turns. Fl. 6–7. Pollinated mainly by bees. Fr. 8–9. Germ. autumn. Hp.

Native in East Anglia, naturalized in much of England and locally in Wales, Scotland and Ireland. Most of Europe, temperate Asia and N. Africa; introduced in North America.

Represented in Britain by the following subspp.

*Subsp. **sativa**: Corolla 7–11 mm, blue to violet; legume in a spiral of 1½–3 turns. 2n = 32. Cultivated as forage and widely naturalized throughout the range of the species. Origin unknown.

Subsp. **falcata** (L.) Arcangeli (*M. falcata* L.): Corolla 5–8 mm, yellow; legume almost straight to falcate. 2n = 16, 32. Native in grassy places on gravelly soils in East Anglia. Introduced elsewhere, mainly in England. Throughout most of the range of the species.

Fertile hybrids between the 2 subspp. (*M.* × *varia* Martyn) occur frequently and have fls which are yellow, purple, or yellow changing to purple through a series of dark greens and black.

2. M. lupulina L.　　Black Medick.

A procumbent or ascending, usually downy annual or short-lived perennial 5–60 cm. Lflets 3–20 mm, obovate, apiculate, finely serrate in the upper half. Stipules lanceolate, acuminate, half-cordate at base, ±toothed. Racemes 10–50-fld, compact; peduncles exceeding the petioles. Fls 2–3 mm, bright yellow. *Legume* 1.5–3 mm diam., *reniform*, reticulate, *coiled in almost 1 complete turn*, 1-seeded, *black when ripe*. Fl. 4–8. Mainly self-pollinated. Fr. 5–9. Germ. autumn or spring. 2n = 16, 32. Th. or Hp.

Native. Generally distributed and common in grassy places and roadsides north to Orkney, but local in Scotland, ascends to 360 m in Derby. Europe, N. Africa, temperate Asia, Macaronesia.

3. M. minima (L.) Bartal.　　Bur Medick.

An erect or procumbent villous annual 5–20 cm. Lflets 3–6 mm, narrowly or sometimes broadly obovate, often emarginate, apiculate, serrate at the apex. *Stipules* lanceolate to ovate-lanceolate, acuminate, *entire* or shallowly dentate. Racemes 1–6-fld; peduncles equalling or somewhat exceeding the petioles. Fls 4–4.5 mm, bright yellow. *Legume* 3–5 mm diam., *subglobose*, sparsely villous and ±glandular in a spiral of 3–5 turns, faintly reticulate, with a double row of hooked spines. Fl. 5–7. Fr. 6–8. Germ. autumn. 2n = 16. Th.

Native. In sandy fields and on heaths. Kent to Norfolk, Channel Is.; casual elsewhere. Most of Europe, except the north, eastwards to western Asia; N. Africa, Macaronesia.

4. M. polymorpha L.　　Toothed Medick.

M. hispida Gaertner; incl. *M. denticulata* Willd.

A procumbent nearly glabrous annual 5–60 cm. Lflets up to 25 mm, obovate to obcordate, serrate towards the top. *Stipules ovate-lanceolate*, acuminate, laciniate. Racemes 1–5-fld; peduncles about equalling petioles. Fls 3–4.5 mm. *Legume* 4–8 mm diam., *flat*, in a lax spiral of 1½–6 turns, *strongly reticulate*, with a double row of longer or shorter hooked or curved spines. Fl. 5–8. Mainly self-pollinated. Fr. 6–9. Germ. spring or autumn. 2n = 14. Th. Very variable.

Native. In sandy or gravelly ground near the sea in E. and S. England from Norfolk to Cornwall; Channel Is. Introduced and naturalized elsewhere. C. and S. Europe, N. Africa, Asia.

5. M. arabica (L.) Hudson　　Spotted Medick.

M. maculata Sibth.

A procumbent nearly glabrous annual up to 50 cm. *Lflets* up to 25 mm, obovate or obcordate, serrate towards the top, *usually blotched. Stipules half-cordate, toothed*, acuminate. Racemes; 1–4(–6)-fld; *peduncles shorter than petioles*. Fls 5–7 mm. *Legume* 4–6 mm diam., *subglobose, faintly reticulate*, in a lax spiral of 3–7 turns, with a double row of curved or hooked spines. Fl. 4–8. Pollinated mainly by bees. Fr. 5–9. Germ. spring. 2n = 16. Th.

Native. In grassy places and waste ground, especially on gravelly or sandy soils in S.E. England; Ireland, naturalized in a few localities. S. Europe northwards to Britain and the Netherlands; N. Africa; introduced elsewhere.

*M. arborea L., a sericeous shrub, up to 400 cm, is established near Clevedon, N. Somerset.

18. TRIFOLIUM L. Clover, Trefoil.

Annual or perennial, usually low-growing herbs. Lvs trifoliolate, the veins ending in teeth; stipules adnate to petiole. *Fls sessile or shortly pedicellate in usually dense racemose heads.* Calyx-teeth 5, subequal or unequal. *Petals usually persistent,* wings longer than keel, the claws of both adnate to the staminal tube. Upper stamen free, *all or 5 of the filaments dilated at the top.* Legume small, 1–4(–10)-seeded, indehiscent, opening by 2 valves or by the top falling off, ±*enclosed in the calyx and often covered by the persistent standard.* The fls have nectar and are usually cross-pollinated by bees.

About 250 spp. in temperate and subtropical regions but mainly N. temperate.

1 Fls yellow or yellowish-white.   *2*
   Fls white, pink or purplish.   *6*
2 Fls 15–20 mm, yellowish-white.   **20. ochroleucon**
   Fls not more than 7 mm, yellow.   *3*
3 Terminal lflet of upper lvs with a distinct stalk, longer
     than that of the lateral lflets.   *4*
   Terminal lflet of upper lvs subsessile, like the lateral
     lflets.   *5*
4 Standard nearly smooth after flowering. **11. dubium**
   Standard deeply furrowed after flowering.
                       10. campestre
5 Fls 2–3(–4) mm; pedicels in fr. 1–1½ times as long as
     upper limb of calyx-tube.   **12. micranthum**
   Fls 6–7 mm; pedicels in fr. distinctly shorter than
     upper limb of calyx-tube.   **9. aureum**
6 Infl. with 2–6 fls.   *7*
   Infl. with at least 7 fls.   *8*
7 Plant almost glabrous; peduncles up to 8 mm; legume
     6–8 mm, with 5–8 seeds.   **1. ornithopodioides**
   Plant hairy; peduncles 10 mm or more; legume
     2.5 mm, with 1 seed.   **22. subterraneum**
8 Peduncles recurved after flowering, burying the fruit-
     ing head; some fls sterile.   **22. subterraneum**
   Peduncles not recurved after flowering; all fls fertile.   *9*
9 Stems creeping and rooting at the nodes.   *10*
   Stems not creeping and rooting at the nodes.   *12*
10 Calyx not 2-lipped, not inflated in fr.; lateral veins
     of lflets not thickened towards the margin.   *11*
   Calyx ±2-lipped, much inflated and reticulate-veined
     in fr.; lateral veins of lflets thickened towards the
     margin.   **8. fragiferum**
11 Upper calyx-teeth narrowly lanceolate; lflets with
     translucent lateral veins and light or dark markings.
                       3. repens
   Upper calyx-teeth ovate-lanceolate or triangular;
     lflets with opaque lateral veins and usually
     unmarked.   **4. occidentale**
12 Heads all terminal.   *13*
   Some heads axillary.   *17*
13 Heads subtended by a pair of ±opposite lvs.   *14*
   Heads subtended by one lf.   **17 incarnatum**
14 Corolla up to 15 mm, much longer than the calyx   *15*
   Corolla up to 7 mm, little longer than the calyx.
                     21. squamosum
15 Fls yellowish-white; lflets without a whitish spot.
                     20. ochroleucon
   Fls pink or red, sometimes pale; lflets usually with
     a whitish spot.   *16*

16 Free part of stipules triangular, with a setaceous apex;
     calyx-tube hairy.   **18. pratense**
   Free part of stipules subulate; calyx-tube glabrous,
     at least below.   **19. medium**
17 Heads distinctly pedunculate.   *18*
   Heads sessile.   *20*
18 Heads softly downy, cylindrical.   **14. arvense**
   Heads not downy, ovoid or globose.   *19*
19 Stipules oblong, entire with triangular acuminate
     apex; corolla 8–10 mm.   **5. hybridum**
   Stipules ovate, glandular-denticulate; corolla
     4–5 mm.   **2. strictum**
20 Lateral veins of lflets strongly thickened and recurved
     near the margin.   **16. scabrum**
   Lateral veins of lflets not thickened and ±straight
     near the margin.   *21*
21 Lflets glabrous.   *22*
   Lflets hairy, at least on the veins beneath.   *23*
22 Calyx-teeth broad, abruptly narrowed to a short fine
     point; corolla distinctly longer than calyx.
                     6. glomeratum
   Calyx-teeth narrow, gradually tapering; corolla
     shorter than calyx, whitish.   **7. suffocatum**
23 Lflets hairy above.   **13. striatum**
   Lflets glabrous above.   **15. bocconei**

Subgenus 1. FALCATULA (Brot.) D. E. Coombe.

Fls bracteolate. Calyx without a ring of hairs or thickening in the throat. Legume exserted, curved, 5–9-seeded, not concealed by the corolla.

Section 1. *Falcatula.* Infl. few fld. Fls subsessile. Calyx-teeth subequal. Petals white or pink.

1. T. ornithopodioides L.   Fenugreek.

Trigonella ornithopodioides (L.) DC.; *T. purpurascens* Lam.; *Falcatula falsotrifolium* Brot.; *F. ornithopodioides* (L.) Brot. ex Bab.

A slender almost glabrous annual or short-lived perennial 2–20 cm. Petioles up to 40 mm; stipules lanceolate, acuminate. Lflets 4–10 mm, obovate to obcordate, cuneate at base, sharply serrate. *Fls 6–8 mm, white and pink. Heads* 1–3(–5)-*fld,* axillary; peduncles shorter than petioles. Calyx-tube narrowly conical; teeth narrowly triangular, acuminate, longer than the tube. *Legume* 6–8 mm, oblong, *somewhat curved,* slightly pubescent, 5–8-*seeded.* Fl. 5–9. $2n = 18*$. Th.

Native. In sandy and gravelly places, chiefly near the coast. Local in suitable habitats north to the Mersey and the Wash; S. and E. coasts of Ireland. S. and W. Europe from the Netherlands southwards and eastwards to Italy; N.W. Africa.

Subgenus 2. LOTOIDEA Pers.

Fls subtended by small bracts or red glandular hairs. Calyx without a ring of hairs or callosity in the throat. Legume included within the calyx or shortly exserted; seeds (1–)2–4(–10).

Section 2. *Paramesus* (C. Presl) Godron. Annual. Infl. of 1–4 closely superposed whorls, each subtended by a small involucre of ±connate bracts. Legume with a swollen indurated wall, indehiscent, exceeding the calyx-tube; seeds (1–)2.

2. T. strictum L.   Upright Clover.

An erect or ascending rather stiff glabrous annual 3–15(–25) cm. *Lflets* 5–15(–25) mm, *narrowly elliptical*, sharply toothed, the veins ending in stalked glands. Petioles 5–10(–40) mm, slender; *stipules broadly ovate or rhombic*, glandular-denticulate, *whitish*. Heads pseudo-terminal and axillary, 7–10 mm, ovoid; peduncles stout, 10–20(–60) mm. Fls purplish. *Calyx-tube strongly ribbed and angled*, about ½ as long as the subulate spinescent teeth. Corolla 5–6 mm, the standard slightly longer than calyx. Legume 2.5 mm, beaked, 1–2-seeded. Fl. 5–7. Th.

Native. In grassy places near the Lizard, Cornwall; Radnor; Channel Is. W. and S. Europe, N. Africa.

Section 3. *Lotoidea*. Annual or perennial. Fls umbellate, numerous, pedicellate, subtended by small lanceolate membranous bracts. Seeds (1–)2–4(–5).

3. T. repens L.   White Clover, Dutch Clover.

A *creeping* ±glabrous *perennial* up to 50 cm, *rooting at the nodes*. *Lflets* 10–30 mm, obovate or obcordate, *usually with a whitish or dark angled band towards the base*, lateral veins thin, translucent and nearly straight (cf. *T. fragiferum* and *T. occidentale*). Petioles erect, up to 14 cm or sometimes more; *stipules* ovate to oblong *with a short subulate point*. Heads 15–20(–35) mm, with 40–80 fls, all axillary, globose. Peduncles up to 30 cm, usually shorter. Fls 7–10 mm, white or pink, rarely purple (var. *rubescens* Ser.), fragrant; pedicels up to 3(–6) mm. *Calyx-tube* campanulate, *white with green veins*; teeth narrowly lanceolate c. half as long as tube. Standard folded over the legume. Legume 4–5 mm, oblong, (1–)3–4-seeded. Fl. 6–9. 2n = 32. H. or Chh.

Native. In grassy places, particularly common on clayey soils. Throughout the British Is. Europe to 71° N, ascending to 2750 m; N. and W. Asia; N. Africa; introduced in S. Africa, Macaronesia, North and South America and E. Asia.

4. T. occidentale D. E. Coombe.

Like *T. repens* but lflets 6–10 mm, thicker, almost orbicular, obtuse or emarginate, without light or dark markings; lateral veins not translucent; stipules vinous-red; heads with 20(–40) fls; fls unscented; upper calyx-teeth ovate-lanceolate or triangular, often with 1–2 teeth on the upper margin. Fl. 4–6. 2n = 16.

Native. On dunes and in dry grassy places near the sea. Coasts of S.W. England, Wicklow and Dublin, Scilly and Channel Is. N.W. France to N.W. Spain.

*5. T. hybridum L.   Alsike Clover.

An *erect or decumbent* nearly glabrous *perennial* 20–40(–90) cm. Lflets 10–20(–35) mm, obovate or obcordate. Petioles up to c. 10 cm; *stipules* ovate to oblong *with long acuminate tips*. Heads pseudo-terminal and axillary, up to c. 25 mm, globose; peduncles up to 15 cm. Fls purple or white at first, becoming pink; pedicels up to twice as long as calyx-tube. *Calyx-tube* campanulate, *whitish, feebly ribbed; teeth subulate*, greenish, subequal, *longer than the tube*. Standard (5–)7–10 mm, c. twice as long as calyx, folded over the legume. Legume 3–4 mm, oblong, (1–)2–4-seeded. Fl. 6–9. 2n = 16. Hs.

Introduced. Naturalized by roadsides etc. throughout the British Is., but local in the west and north. Apparently formerly rather local in Europe but cultivated for forage since the eighteenth century and now widespread in temperate regions.

Subsp. **hybridum**: Usually erect, 30 cm or more. Lflets mostly more than 20 mm with c. 20 pairs of lateral veins. Fls white and pink, sometimes deep pink in stunted plants. Commonly cultivated and often naturalized.

Subsp. **elegans** (Savi) Ascherson & Graebner: Usually decumbent, 15–30 cm. Lflets mostly less than 20 mm with c. 40 pairs of lateral veins. Fls bright pink. Rather uncommon.

6. T. glomeratum L.   Clustered Clover.

A procumbent or ascending glabrous annual 5–25 cm. Lflets 5–10(–20) mm, obovate-cuneate, mucronate, sharply serrate. Petioles up to c. 20 mm; *stipules ovate with long points*. Heads 8–12 mm wide, terminal and axillary, globose, sessile, usually remote. *Fls purplish*, sessile. *Calyx-tube* glabrous, broadly campanulate, with 10(–12) *strong* veins, *whitish; teeth triangular, spinescent*, green, a little shorter than the tube. Standard 4–5 mm, a little longer than calyx. Legume c. 2 mm, oblong, (1–)2-seeded. Fl. 6–8. 2n = 16. Th.

Native. In grassy places on sandy and gravelly soils, rare and mainly near the sea. Cornwall and Kent to Norfolk; Channel Is., S. and W. Europe northwards to 52° 30′ in England; Caucasus; N. Africa.

7. T. suffocatum L.   Suffocated Clover.

A prostrate glabrous annual 1–5 cm. *Lflets* 3–8 mm, *obovate–cuneate*, sharply serrate and sometimes emarginate. Petioles 10–20(–60) mm, slender; *stipules ovate, acuminate*, whitish when old. *Heads* terminal and axillary, *sessile, densely crowded on the short stems and often confluent*. Fls whitish, shortly pedicellate. *Calyx-tube* nearly cylindrical *weakly veined; teeth subulate*, about as long as tube. *Standard* c. 2 mm, *shorter than calyx*. Legume 1.5–2 mm, ovoid, 2-seeded. Fl. 4–8. Th.

Native. In grassy places on sandy and gravelly soils, rare. Coasts from Cornwall to S.E. Yorks; inland in East Anglia. S. and W. Europe northwards to c. 53° 30′ N; Caucasus; N. Africa; Macaronesia.

Section 4. *Vesicastrum* Ser. Annual or perennial. Fls umbellate, numerous, subsessile, subtended by small free or connate bracts. Calyx inflated in fr., the upper lip usually tomentose, scarious and reticulately veined. Seeds 1–2.

8. T. fragiferum L.   Strawberry Clover.

A creeping perennial, lfless in winter, *rooting at the nodes*, up to c. 30 cm. Lflets (3–)8–20 mm, ovate or obcordate with fine forward-directed teeth, lateral veins thickened markedly and curved backwards towards the lf margin (cf. *T. repens*). Petioles up to c. 10 cm, often shorter, erect; *stipules* ovate-oblong, *long-acuminate*. Heads 10–22 mm wide in fr., axillary; *peduncles longer than petioles*. Fls pale pink; *bracts* 3–4 mm. Upper lip of calyx strongly inflated, reticulate, downward-curved, tomentose. Standard 6–7 mm, longer than calyx. Legume 3 mm, oblong, (1–)2-seeded. Fl. 7–9. 2n = 16. H.

Native. In grassy places mainly on heavy clay and often rather saline soils. Scattered throughout Great Britain, mainly S.E. of a line from the Humber to the Severn, locally common; and local in Ireland, chiefly on the coast. Europe, to about 60° 30′ N; temperate Asia; N. Africa; Madeira and Canary Is.

*T. resupinatum L., Reversed Clover, a slender annual with narrow obovate lvs, *pink fls twisted so that the standard is below*, and calyx becoming somewhat inflated in fr. Occurs occasionally as a casual, chiefly near docks. 2n = 14, 16. Probably native in Asia.

*T. tomentosum L., a slender annual with obovate lflets, pink fls c. 2.5 mm, peduncles shorter than petioles and densely tomentose fruiting calyces occurs as a casual. Mediterranean region, W. Asia.

Section 5. *Chronosemium* Ser. Annual or biennial. Fls umbellate, pedicellate, subtended by a few short red glandular hairs. Calyx 5-veined. Corolla becoming dark and scarious in fr. Legume stipitate; seeds 1–2.

*9. T. aureum Pollich

T. agrarium L., nom. ambig.

Like *T. campestre* in general appearance, but usually larger. Stipules linear-oblong, acuminate. Fls 6–7 mm, subsessile. Legume oblong, c. twice as long as style, usually 2-seeded. Fl 7–8. Germ. spring and autumn. 2n = 14. Th.

Introduced. Naturalized in fields and waste places, rather uncommon. Scattered throughout the British Is., but casual in many places. Most of Europe, except the extreme north, most of the west and the Mediterranean region; Asia Minor.

10. T. campestre Schreber   Hop Trefoil.

T. procumbens L., nom. ambig.

A rather stout erect or ascending ±hairy annual up to 35(–50) cm. Lflets commonly 8–10 mm, obovate, sometimes obcordate, cuneate at base, the terminal petiolulate; petiole up to 10 mm; *stipules ½-ovate*, the apex triangular, acute. Heads axillary, 20–30-fld, (7–)10–15 mm; peduncles exceeding petioles. Fls 4–5 mm, yellow, turning rather light brown; *standard broad, not folded, sulcate, much exceeding legume*. Pedicels rather stout, half the length of the calyx-tube. *Legume* 2–2.5 mm, ovoid, usually 1-seeded, 3–6 *times*

as long as the style. Seeds yellow, rather shiny. Fl. 6–9. Germ. spring and autumn. 2n = 14. Th.

Native. In grassy places and roadsides. Generally distributed throughout the British Is., except the north. Europe, except the extreme north and east; W. Asia; N. Africa; Macaronesia; introduced in North America.

11. T. dubium Sibth. Suckling Clover, Lesser Trefoil.

T. minus Sm.

A slender procumbent or ascending ±hairy annual up to 25(–50) cm. Lflets up to 11 mm, obcordate or obovate, cuneate at base, the terminal petiolulate; stipules broadly ovate, acuminate. *Heads* axillary, 3–15-fld, usually 8–9 mm diam.; peduncles exceeding petioles. Fls 3–3.5 mm, yellow, turning dark brown; *standard narrow, folded over legume*, scarcely sulcate; *pedicels rather stout, shorter than calyx-tube*. Legume 2.5–3 mm, ovoid, usually 1-seeded; seeds light brown. Fl. 5–10. Germ. spring. 2n = 28, 32. Th.

Native. In dry grassy places. Throughout the British Is., rarer in N. and N.W. Scotland. Most of Europe except the extreme north; N. Africa; Macaronesia.

12. T. micranthum Viv.   Slender Trefoil.

T. filiforme L., nom. ambig.

A slender procumbent or ascending sparsely hairy annual 2–10(–20) cm. Lflets up to 5(–8) mm, obcordate to obovate, the terminal subsessile; stipules oblong to ovate. *Heads* axillary, 1–6-fld, c. 4 mm diam.; peduncles filiform, equalling or exceeding lvs. Fls 2–4 mm, yellow; *standard deeply notched*, folded over the legume, scarcely sulcate, *pedicels slender, about as long as calyx-tube*. Legume 2–2.5 mm, ovoid, usually 2-seeded; seeds dull brown, smooth. Fl. 6–7. Germ. spring. 2n = 16. Th.

Native. In rather open grassy places on sandy and gravelly soils. England, Wales and Ireland, rarer in the north; Scotland: Wigtown. W. and S. Europe; N. Africa; Caucasus.

Subgenus 3. TRIFOLIUM.

Fls ebracteate. Calyx usually ±closed by a ring of hairs or a callosity in the throat. Legume nearly always included within the calyx; seeds 1–2.

Section 6. *Trifolium*. Heads usually elongate, rarely globose. Fls usually sessile, all fertile.

13. T. striatum L.   Knotted Clover.

A softly hairy procumbent to erect annual 5–30(–50) cm. Lflets 5–15 mm, obovate, emarginate to acute or apiculate, the *lateral veins nearly straight and thin*. Petioles of lower lvs up to 30 mm; upper lvs often subsessile; stipules membranous, ovate or triangular with subulate points. *Heads* pseudo-terminal and axillary, up to 15 mm, sessile, ovoid, *±enfolded in the dilated stipules of the subtending lvs when young, becoming obtusely conical-ovoid in fr. Fls pink.* Calyx-tube becoming inflated and readily abscissing in fr., ribbed, throat slightly thickened, hairy; teeth somewhat shorter than tube, subulate, spinescent and suberect in fr. Corolla 4–5 mm, persistent. Legume 2–5 mm, membranous, indehiscent; style short, slender, asymmetrical. Seed 2 mm, ovoid. Fl. 5–7. $2n = 14$. Th.

Native. In rather open habitats on well-drained soils. Scattered throughout the British Is., north to Kincardine; very local in Ireland and only on the east coast. S., W. and C. Europe northwards to S. Sweden; Caucasus; N.W. Africa; Macaronesia.

14. T. arvense L.   Hare's-foot Clover.

A softly hairy erect or ascending annual or biennial (5–)10–20(–40) cm, usually with spreading branches. Lflets 5–15(–25) mm, narrowly obovate-oblong. Petioles 2–10 mm; stipules ovate with long setaceous points. *Heads* terminal and axillary, up to 25 mm, *cylindrical and softly downy*; *peduncles equalling or exceeding the lvs*, elongating in fr. Fls white or pink. Calyx-tube campanulate, feebly ribbed, the throat glabrous and unthickened; teeth 1–3(–5) times as long as the tube, setaceous, long-ciliate. Corolla c. 4 mm, c. $\frac{1}{2}$ as long as calyx, persistent. Legume c. 1.3 mm, membranous, indehiscent ovoid; seed c. 0.8 mm. Fl. 6–9. $2n = 14$. Th.

Native. In sandy fields, pastures and on dunes, somewhat calcifuge. Scattered throughout the British Is., north to E. Ross and locally common; Ireland, chiefly on the E. coast. Europe, except the extreme north; N. and W. Asia; N. Africa; Macaronesia; naturalized in North America.

15. T. bocconei Savi   Twin-flowered Clover.

A ±erect somewhat pubescent annual 5–10(–25) cm. *Lflets* 5–15 mm, obovate, *glabrous above, sparsely appressed-pubescent beneath*; *lateral veins thin and ±straight*. Petioles c. 5 mm; *stipules ovate-oblong with ciliate subulate points*. Heads terminal and axillary, 9–15 mm, sessile, ovoid. Fls pinkish. *Calyx-tube* cylindrical, coriaceous, pubescent, *strongly ribbed*, the throat thickened, slightly hairy; teeth triangular, spinescent and erect in fr. Corolla 4–5 mm, rather longer than calyx, persistent. Legume c. 1.8 mm, membranous, indehiscent, obovate-oblong, asymmetrical; style slender, nearly central. Seed similar in size and shape to pod. Fl. 5–6. Th.

Native. In grassy places near the Lizard, Cornwall, and in Jersey. S.W. Europe north to S.W. England; N.W. Africa.

16. T. scabrum L.   Rough Clover.

An erect or decumbent ±pubescent annual up to c. 20 cm. *Lflets* 5–10 mm, obovate, apiculate, *pubescent on both surfaces*, *lateral veins backward curving and thickened towards the margins*. Petioles up to c. 10 mm; *stipules* rather rigid, *ovate or oblong, cuspidate*. *Heads* mostly axillary, 5–12 mm, sessile, *ovoid*. Fls whitish. Calyx-tube campanulate, coriaceous, ribbed, throat thickened and somewhat hairy; teeth triangular, erect in fl., rigid and recurved in fr. Corolla about as long as calyx, persistent. Legume c. 1.7 mm, membranous,

indehiscent, obovate-oblong, asymmetrical; style slender, nearly central. Seed c. 1.5 mm, similar in shape to pod. Fl. 5–7. $2n = 10$. Th.

Native. In dry places on shallow and sandy soils. Scattered throughout Great Britain, north to Angus, mainly coastal; E. coast of Ireland. S. and W. Europe, Mediterranean region; N. Africa; W. Asia; Macaronesia.

*T. stellatum L., Starry Clover, a pubescent annual with ovate subacute stipules, terminal pedunculate heads, 2–4 cm in fr., and densely villous calyx-tube with ciliate triangular-subulate teeth 3 times as long as the tube persisted near Shoreham, Sussex for many years. S. Europe, W. Asia, N. Africa.

17. T. incarnatum L.   Crimson Clover.

A pubescent *annual* up to 50 cm. Lflets 5–35 mm, broadly obovate or obcordate. Petioles up to c. 8 cm; *stipules often dentate, ovate, obtuse. Heads* 10–40(–70) mm, terminal, ovoid or cylindrical, subtended by a single lf; peduncles long. Fls crimson, pink or pale cream, sessile. Calyx-tube nearly cylindrical, villous, ribbed; throat somewhat hairy but scarcely thickened; teeth setaceous $1\frac{1}{2}$–2 times as long as tube, spreading in fr. Corolla 10–12 mm, exceeding calyx, deciduous. Legume 2.5 mm, membranous in the lower part, thickened above, often rupturing at the junction of the two parts. Seed 2 mm. Fl. 5–9. $2n = 14$. Th.

*Subsp. incarnatum: Usually 30 cm or more, erect. Lflets obovate; hairs on stems and petioles usually spreading. Fls crimson. Introduced. Cultivated and ±naturalized, especially in the south. Mediterranean region.

Subsp. **molinerii** (Balbis ex Hornem.) Syme: Usually under 20 cm, decumbent or suberect. Lflets often obcordate; hairs on stems and petioles usually appressed. Fls pink or pale cream. Fl. 5–6. Native. Grassy places near the Lizard, Cornwall, and in Jersey. W. and S. Europe.

18. T. pratense L.   Red Clover.

An erect or decumbent ±pubescent perennial 5–100 cm. *Lflets* 10–30(–50) mm, elliptical to obovate, obtuse, hairy beneath, glabrescent above, *often with a whitish crescentic spot towards the base*. Petioles up to c. 20 cm; stipules ovate to oblong, *free portion triangular with a setaceous point, usually applied to the petiole. Heads* terminal, 20–40 mm, *globose* becoming ovoid, sessile and subtended by a pair of short-petiolate lvs. Fls pink-purple, sometimes whitish, sessile. Calyx-tube narrow-campanulate, ribbed; throat hairy, with longitudinal folds; teeth unequal, the longest c. twice as long as tube, all ciliate. Corolla 12–15 mm, exceeding calyx, persistent. Legume 2.2 mm, obovoid, membranous in the lower part, thickened above, often rupturing at the junction of the two parts. Seed 1.7 mm, ovoid. Very variable. The cultivated variant var. *sativum* Sturm, has the lflets usually entire, the stems fistular, and is larger than var. *pratense*, which has toothed lflets and usually solid stems. Fl. 5–9. $2n = 14$*. Hp.

Native. Generally distributed in grassy places throughout the British Is.

Var. *sativum* cultivated for hay and often naturalized. Most of Europe; W. Asia to the Altai; Baikal and Kashmir; N. Africa. Introduced in North and South America and New Zealand.

19. T. medium L.   Zigzag Clover.

A straggling ascending ±appressed-pubescent perennial with flexuous stems up to c. 50 cm and extensive slender rhizomes. *Lflets* 20–60 mm, *ovate to elliptical*, obtuse or acute, ciliate, scarcely toothed, often with a faint whitish spot. Petioles up to c. 8 cm, often less; *stipules oblong or linear-oblong* (the upper ovate), *free part subulate, spreading. Heads* terminal, 25–35 mm, *subglobose*, shortly pedunculate and subtended by a pair of lvs. Fls reddish-purple. Calyx-tube ±globose, 10-veined, hairy but not thickened in the throat; teeth subulate, ciliate, spreading in fr., the lower longest. Corolla 20 mm, 2–3 times as long as calyx, tardily deciduous. Legume c. 2 mm, membranous, dehiscing longitudinally, obovoid, truncate, asymmetrical. Seed c. 1.7 mm. Fl. 6–9. $2n = $ c. 70, 78–80, c. 84. Hp.

Native. In grassy places. Widely distributed but rather local, commoner in the north but not recorded from the Isle of Man and Shetland; very local in Ireland. Europe, except the extreme north and south; Caucasus; W. Siberia; naturalized in North America.

20. T. ochroleucon Hudson   Sulphur Clover.

An erect or ascending ±pubescent perennial up to c. 50 cm. Lflets 15–30 mm, elliptical to lanceolate, sometimes emarginate. Petioles up to 10 cm; *stipules* oblong or ovate-oblong *with a linear-lanceolate, herbaceous apex. Heads* terminal, 20–40 mm, globose becoming ovoid, subsessile and subtended by a pair of nearly sessile lvs. *Fls whitish-yellow*, sessile. Calyx-tube campanulate, strongly ribbed, the throat hairy, becoming somewhat thickened in fr.; lowest calyx-tooth usually 2–3 times as long as the others, all spreading in fr. Corolla 15–20 mm, longer than calyx, eventually deciduous. Legume c. 2.5 mm, obovoid, with a ±central stout style-base and slender style, membranous in the lower part, thickened above, often rupturing at the junction of the two parts. Seed c. 1.8 mm, obovoid. Fl. 6–7. $2n = 16$. Hp.

Native. In grassy places mainly on boulder clay in E. England from Essex to Lincoln and west to Northampton, probably introduced elsewhere. W.C. and E. Europe; Caucasus; N.W. Africa.

21. T. squamosum L.   Sea Clover.

T. maritimum Hudson

An erect or ascending ±pubescent annual up to c. 40 cm. Lflets 10–20 mm, usually narrowly obovate, often apiculate. Petioles up to c. 10 cm; *stipules oblong, the free herbaceous part longer than the rest*, spreading. Heads terminal, 10–20 mm, ovoid, shortly pedunculate and subtended by a pair of lvs. Fls pink. *Calyx-tube campa-nulate, strongly ribbed and coriaceous in fr.*; throat minutely hairy becoming strongly thickened in fr.; *teeth triangular*, subequal, *green, spreading in fr.* Corolla 5–7 mm, caducous. Legume 2.5 mm, obovoid, membranous in the lower part, thickened above, often rupturing at the junction of the two parts. Seed c. 2 mm, obovoid. Fl. 6–7. $2n = 16$. Th.

Native. In turf near the sea or tidal estuaries, local. Coasts of S. England from Dorset and Kent to Glamorgan and Lincoln. W. and S. Europe; Macaronesia; N. Africa.

Section 7. *Trichocephalum* Koch. Heads capitate. Fls sessile, the outer fertile, the inner consisting of calyces only.

22. T. subterraneum L.   Subterranean Clover.

A hairy prostrate annual 3–20(–30) cm. Lflets 5–12(–20) mm, broadly obcordate. Petioles 1–5 (–10) cm; stipules ½-ovate, acute. *Heads* axillary, *few-fld*, globose and appressed to or buried in the soil in fr. Fertile fls cream-coloured, subsessile, sometimes cleistogamous, reflexed after fl. Calyx-tube narrow-cylindrical becoming ±globose in fr.; teeth setaceous, ciliate, curved, subequal, c. ¼ length of tube. Corolla 8–12 mm, exceeding calyx, caducous. *Sterile fls consisting of slender, rigid, palmately-lobed calyces* which enlarge after fl. Legume 2.5 mm, ±globose with an asymmetrical keel near the top, style ±central, small. Seed 2 mm, ovoid. Fl. 5–6. $2n = 16$. Th.

Native. In sandy and gravelly pastures and on cliff tops, local. Cornwall and Kent to Anglesey and Lincoln; Wicklow. S. and W. Europe; Caucasus; N. Africa; Macaronesia; widely cultivated for forage in warm temperate regions.

Loteae DC. (incl. Coronilleae (Adanson) Boiss.). Herbs or small shrubs. Lvs epulvinate; stipules membranous or herbaceous, usually small. Fls usually in pedunculate heads or umbels. Stamens usually diadelphous; anthers uniform, all or alternate filaments dilated upwards. Fr. 2-valved, several-seeded, dehiscent, 1-seeded, indehiscent, or breaking transversely into 1-seeded, indehiscent portions. Seedlings epigeal.

19. DORYCNIUM Miller

Differs from *Lotus* in the small white fls with blackish keels crowded in dense heads, in the short rounded pod whose valves do not twist after dehiscence, and in the small number of seeds in the pod. Four spp. in Europe, S.W. Asia and S. Africa.

*D. pentaphyllum Scop. subsp. **gracile** (Jordan) Rouy

D. suffruticosum Vill., *D. gracile* Jordan

A greyish-pubescent perennial 30–60 cm, usually with appressed hairs. Stems herbaceous, ascending or erect. Lvs 5-foliolate, with minute stipules; lflets linear, entire. Heads with 12–25 fls. Pedicels about as long as calyx-tube. Calyx-teeth as long as tube, acuminate. Standard apiculate, keel bluish-black, curved, mucronate. Legume obtuse but mucronate. Fl. 6–7. $2n = 14$. Hp.

Introduced. Naturalized in a few places in S.E. England. France, Spain.

20. LOTUS L.

Annual or perennial herbs sometimes woody at the base
Lvs 5-foliolate, margins entire; stipules brown, minute.
Fls in axillary, pedunculate, *cymose heads*, yellow or
reddish (in our spp.), protandrous; bracts 3-foliate.
Calyx 5-toothed. Keel incurved, beaked. Upper stamen
free; alternate filaments dilated at the top. Style
attenuate at the top. *Legume elongate*, 2-valved, *many-
seeded, septate between the seeds*. The fls have nectar
and are usually cross-pollinated by bees.

About 100 spp. in temperate Europe, Asia, N. and
S. Africa, North America and Australia.

1 Perennial, with a woody stock; fls usually more than
   10 mm.   2
   Annual; fls not more than 10 mm.   4
2 Lflets of upper lvs linear or linear-lanceolate, at least
   (3–)4 times as long as broad.   **2. tenuis**
   Lflets of upper lvs lanceolate to obovate, usually not
   more than 3 times as long as broad.   3
3 Stem hollow; upper 2 calyx-teeth separated by an
   acute sinus.   **3. uliginosus**
   Stem solid, sometimes with a small hollow near the
   base; upper 2 calyx-teeth separated by an obtuse
   sinus.   **1. corniculatus**
4 Keel obtusely angled on the lower edge; legume not
   more than 3 times as long as calyx; seeds 8–12.
                                             4. subbiflorus
   Keel with a right angle on the lower edge; legume
   at least 3 times as long as calyx; seeds more than
   12.   **5. angustissimus**

1. L. corniculatus L.
Common Birdsfoot-trefoil, Bacon and Eggs.

A decumbent almost glabrous or rarely hairy perennial
5–35 cm. Stock stout, scarcely stoloniferous, stem solid
or nearly so. Lflets 3–8 mm, lanceolate to broadly ovate;
petioles short. Heads (1–)2–6(–8)-fld; peduncles up to
c. 8 cm, stout. *Fls 10–16 mm*, yellow, often streaked or
tipped with red; pedicels 1–2.5 mm. Calyx-teeth triangular, erect in bud, the two upper with an obtuse sinus.
Legume up to 30 mm. Fl. 6–9. $2n = 24^*$. Hp. Variable.

Native. In pastures and grassy places. Generally distributed throughout the British Is. Almost throughout
Europe; Asia; N. and E. Africa; in the tropics only
on mountains.

2. L. tenuis Waldst. & Kit. ex Willd.
Narrow-leaved Birdsfoot-trefoil.

Like *L. corniculatus* but stems slender, often taller (up
to 90 cm), and more branched; lflets linear or linear-
lanceolate, acuminate, or rarely narrowly obovate;
heads rarely more than 4-fld; peduncles slender; fls
6–10 mm; calyx-teeth narrower, often subulate. Stems
much more wiry than in *L. corniculatus*. Fl. 6–8.
$2n = 12^*$. Hp.

Native. In dry grassy places. Mainly south of a line
from the Humber to the Severn; Cumberland; introduced in Scotland and Ireland. Most of Europe, except
the extreme north and north-east; W. Asia; N. Africa.

3. L. uliginosus Schkuhr   Greater Birdsfoot-trefoil.
L. pedunculatus auct., non Cav.; *L. major* auct., non
Scop.

An erect or ascending glabrous or pubescent perennial
15–60(–100) cm. *Stock slender, producing numerous stolons. Stem hollow*. Lflets usually 15–20 mm, obovate,
often obliquely so, obtuse or mucronate, lower pair
ovate; petioles up to 10 mm. Heads (1–)5–12(–15)-fld;
peduncles up to c. 15 cm, rather slender. Fls 10–18 mm.
*Calyx-teeth spreading in bud, the 2 upper with an acute
sinus*. Otherwise much the same as *L. corniculatus*. Fl.
6–8. $2n = 12^*$. Hp.

Native. In damp grassy places. Throughout the British
Is., except C. Ireland and the extreme north, but less
common than *L. corniculatus*. W., C. and S. Europe,
north to 60° N. in Fennoscandia and east to c. 25° E.
in Ukraine; N. Africa; Canary Is.

4. L. subbiflorus Lag.   Hairy Birdsfoot-trefoil.
L. hispidus Desf. ex DC.

A villous much-branched annual 3–30(–90) cm. Lflets
up to 20 mm, narrowly obovate to lanceolate, often obliquely so. Heads (1–)2–4-fld; *peduncles exceeding the lvs*.
Fls 5–10 mm, yellow. Calyx-teeth subulate, longer than
the tube. Keel obtusely angled on its lower edge.
Legume 6–12×1.5–2 mm, $1\frac{1}{2}$–3 times length of calyx;
seeds 8–12. Fl. 7–8. $2n = 24^*$. Th.

Native. In dry grassy places near the sea. England,
from Cornwall to Hants; Pembroke; Channel Is.; W.
Ireland. W. Europe from 52° N. southwards and east
to Sicily; N. Africa.

5. L. angustissimus L.   Slender Birdsfoot-trefoil.

Like *L. subbiflorus* but heads 1–2-fld; peduncles usually
shorter than lvs; keel with a right angle on its lower
edge; and legume 20–40×1–1.5 mm, 4–7 times length
of calyx; seeds more than 12. Fl. 7–8. $2n = 12^*$. Th.

Native. In dry grassy places near the sea. Cornwall,
Devon, E. Kent and Channel Is. S. Europe, W. Asia,
N. Africa, Macaronesia.

21. TETRAGONOLOBUS Scop.

Like *Lotus* but with 3-foliolate lvs, large green stipules,
fls 1–2, legume 4-winged and tardily dehiscent and style
thickened at the top. About 6 spp., mainly Mediterranean region.

*1. T. maritimus (L.) Roth   Dragon's-teeth.
Lotus siliquosus L.; *T. siliquosus* (L.) Roth

Slightly pubescent perennial 10–25(–40) cm. Lvs shortly
petiolate; lflets up to 30 mm, the middle lflet rhombic-
ovate, the lateral ones ovate, asymmetrical. Peduncles
much exceeding the subtending lf, usually 1-fld with a
sessile 3-foliolate bract. Pedicel c. 1 mm. Fls 25–30 mm,
pale yellow with faint brown veins in the standard.
Legume 30–60 mm, oblong, acute, 4-winged, many-
seeded, ultimately dehiscing. Fl. 5–8. $2n = 14$. Hp.

Introduced. Very local in rough calcareous grassland from Somerset and Glamorgan to Sussex and Essex; Lincoln; well naturalized. Europe from S. Sweden southwards, rare in the Mediterranean region; Caucasus; N. Africa.

22. ANTHYLLIS L.

Herbs or shrubs with *imparipinnate lvs,* the lower sometimes reduced to terminal lflet. *Fls in capitate cymes,* yellow or red and *surrounded by an involucre (in our sp.).* Calyx *inflated, mouth oblique,* shortly 5-toothed. Petals with long claws, the 4 lower adnate to the staminal tube; keel incurved, gibbous at the sides. Stamens united or the upper free for up to $\frac{1}{2}$ its length; anthers uniform. *Legume enclosed by the calyx,* 1–many-*seeded.* Pollinated mainly by humble-bees.

About 25 spp. in Europe, N. Africa and W. Asia.

1. A. vulneraria L.   Kidney-vetch, Ladies' Fingers.

An erect or decumbent pubescent herb up to 60 cm, usually perennial. Lvs up to c. 14 cm, pinnate or the lower sometimes reduced to the terminal lflet. Lflets of lower lvs ovate or elliptic, acute or obtuse, alternate, all equal or the terminal one much the largest. Petioles short or 0. Infl. a capitate cyme; heads in pairs, rarely solitary, up to 4 cm across, sessile or subsessile within a lfy involucre, each pair pedunculate. Fls 12–15 mm, yellow or red. Calyx ±woolly, contracted at the mouth; teeth unequal. Petals exceeding calyx. Legume c. 3 mm, semiorbicular, compressed, glabrous, reticulate, 1–2-seeded. Fl. 6–9. $2n = 12^*$. Hs.

Native. In dry places on shallow soils. Scattered throughout the British Is., particularly on calcareous soils and near the sea. Throughout Europe, east to the Caucasus; N. Africa.

Very variable. The following 4 subspp. occur in Britain and intermediates between them are found where their ranges overlap:

1 Calyx 2–4(–5) mm wide, the lateral teeth small, appressed to the upper teeth; lobes of bracts narrowly deltate, acute; upper cauline lvs equifoliolate.   2

Calyx (4.5–)5–7 mm wide, the lateral teeth obvious; lobes of bracts parallel-sided, obtuse; upper cauline lvs inequifoliolate.   3

2 All hairs on stems spreading; lvs confined to the lower part of the stem.   **(b)** subsp. **corbierei**

At least some of the hairs on the stem appressed; lvs evenly distributed along the stem.

(a) subsp. **vulneraria**

3 Calyx appressed-sericeous.   **(c)** subsp. **carpatica**

Calyx with spreading hairs.   **(d)** subsp. **lapponica**

(a) Subsp. **vulneraria**: Stems 5–55 cm, sericeous. Lowest lvs inequifoliolate with 5–7 lflets; uppermost lvs equifoliolate with 9–15 lflets. Calyx usually with a red apex; corolla usually yellow, sometimes red (var. *coccinea* L.). Most of the British Is. N. Europe from Ireland to Finland and Latvia.

(b) Subsp. **corbierei** (Salmon & Travis) Cullen: Stems 20–25 cm, ascending, hirsute throughout with spreading

hairs. Lvs fleshy, the lowest with 1–3 lflets; uppermost lvs equifoliolate with 9–11 lflets. Calyx without a red apex; corolla yellow. Anglesey and Cornwall; Channel Is. Probably N. & W. coasts of France.

(c) Subsp. **carpatica** (Pant.) Nyman: Stems 10–30 cm, sparsely sericeous. All lvs inequifoliolate with 1–7(–9) lflets. Calyx usually without a red apex; corolla usually pale yellow, rarely red. N.W. and C. Europe.

(d) Subsp. **lapponica** (Hyl.) Jalas: Stems 15–40 cm, sericeous. All lvs inequifoliolate with 1–9 lflets. Calyx sometimes red at apex; corolla yellow. N. Britain to N. Russia.

23. ORNITHOPUS L.

Slender herbs. Lvs imparipinnate, minutely stipulate. Fls small, in axillary pedunculate heads. Calyx with 5 equal teeth. Keel obtuse, sometimes very short. Upper stamen free; alternate filaments dilated towards the top, anthers all similar. *Legume curved,* ±*constricted between the seeds* and breaking up into 1-seeded portions when ripe.

Six spp. in Europe, North and tropical Africa, W. Asia and South America.

Plant pubescent; fls white; bracts pinnate; legume strongly jointed.   **1. perpusillus**

Plant nearly glabrous; fls yellow; bracts 0; legume slightly jointed.   **2. pinnatus**

1. O. perpusillus L.   Bird's-foot.

A *finely pubescent* slender spreading prostrate annual 2–45 cm. Lvs 15–30 mm; lflets up to 4 mm, 4–7(–13) pairs, elliptic to linear-oblong, *lowest pair often at the base of the rhachis, distant from the others and recurved.* Heads 3–8-fld, *subtended by a sessile pinnate bract*; peduncles filiform, longer or shorter than lvs. *Fls* 3–5 mm, *white veined with red*; pedicels very short, stout. *Legume* 10–20 mm, curved and *strongly constricted between the seeds.* Fl. 5–8. Self-pollinated. $2n = 14^*$. Th.

Native. In dry sandy and gravelly places. Rather local but widely distributed in Great Britain, rare in most of Scotland; S. and E. Ireland. W. and W.C. Europe eastwards to Poland and S. Sweden; Macaronesia.

2. O. pinnatus (Miller) Druce   Orange Bird's-foot.

O. ebracteatus Brot.; *Arthrolobium ebracteatum* (Brot.) Desv.

A *nearly glabrous* slender ascending annual 5–15 cm. Lvs 10–25 mm, lflets up to 5 mm, 2–7 pairs, linear-lanceolate to narrowly obovate, *the lowest pair always distant from the base of the petiole. Heads* 1–2-fld, rarely more, *ebracteate*; peduncles filiform, as long as the lvs. *Fls* 6–8 mm, *yellow*, veined with red; pedicels very short, slender. *Legume* 20–35 mm, slender, curved and *scarcely constricted between the seeds.* Fl. 4–8. Th.

Native. In short open turf or disturbed ground on sandy soils. Scilly Is. and Channel Is. Atlantic coast from France to Spain, Mediterranean region; N.W. Africa; Macaronesia.

24. CORONILLA L.

Herbs or shrubs. Fls yellow, purple or white, in axillary umbels; *calyx* campanulate, the *teeth almost equal*. Legume terete or 4-angled, breaking up into 1-seeded joints. Pollinated by bees.

About 20 spp. in the Mediterranean region, W. Asia and Madeira.

Fls white, purple or pink; legume 4-angled.   **1. varia**
Fls yellow; legume with 2 obtuse angles.   **2. valentina**

*1. C. varia L.   Crown Vetch.

A straggling glabrous perennial 20–120 cm. Lflets 7–12 pairs oblong or elliptical, mucronate. *Peduncles 10–20-fld; infl. capitate*. Fls 10–15 mm, *white, purple or pink*; calyx-tube campanulate, the teeth very short. Legume 20–60 mm, slender, 4-angled. Fl. $2n = 24$. Hp.

Introduced. Well naturalized in a number of localities scattered throughout Great Britain; rare in Ireland. Native of C. and S. Europe and W. Asia.

*2. C. valentina L.

A shrub up to 100 cm. Lflets 2–3 pairs, obovate, emarginate, glaucous. Peduncles 4–12-fld. *Fls 7–12 mm, yellow*. Legume 10–50 mm, with 2 obtuse angles. Fl. 6. $2n = 24$. N.

Introduced. Locally naturalized in S. Devon. Mediterranean region, S. Portugal, N. Africa.

The above description applies to subsp. **glauca** (L.) Batt., to which the naturalized plant belongs.

25. HIPPOCREPIS L.

Perennial herbs. Lvs imparipinnate, the lflets entire; stipules small free. Fls yellow, capitate; calyx tubular-campanulate, with 5 subequal teeth; petals with long claws. Upper stamen free, alternate filaments dilated towards the top, anthers uniform. *Legume* several-seeded, breaking up into *3–6 horseshoe-shaped segments*. Pollinated mainly by bees.

Twenty spp., mostly in the Mediterranean region.

1. H. comosa L.   Horseshoe Vetch.

An almost glabrous diffuse perennial 10–40 cm. Stock

woody, branched. Lvs 3–5 cm; lflets 5–8 mm, usually 3–8 pairs, obovate to oblong; stipules lanceolate, spreading. Heads 5–8(–12)-fld; *peduncles exceeding lvs*, slender, curved. Fls 6–10(–14) mm, *yellow*, shortly pedicelled and bracteate. Legume 15–30 mm, compressed, minutely papillose, the segments horseshoe-shaped; style curved, persistent. Fl. 5–7. $2n = 28^*$. Chh.

Native. In dry calcareous pasture and on cliffs, local. In suitable habitats north to N.E. Yorks and Westmorland. W., C. and S. Europe, northwards to N.C. Germany.

Hedysareae DC. Herbs. Lvs epulvinate; stipules usually scarious. Fls in axillary racemes. Stamens diadelphous; anthers uniform. Fr. 1-seeded, indehiscent or breaking transversely into 1-seeded, indehiscent portions. Seedlings epigeal.

26. ONOBRYCHIS Miller

Herbs or shrubs. Lvs imparipinnate; lflets numerous, entire; *stipules scarious. Infl. an* axillary *raceme. Wings short*, keel obliquely truncate, equalling or exceeding standard. Upper stamen free, *filaments not dilated*, anthers all similar. *Legume indehiscent, not jointed*, often *spiny or tuberculate*, 1–3-seeded. The fls have nectar and are usually cross-pollinated by bees.

About 130 spp. in temperate Europe and Asia.

*1. O. viciifolia Scop.   Sainfoin.

O. sativa Lam.

An erect ±pubescent perennial 10–60 cm. Lflets 10–35 mm, 6–14 pairs, ovate to linear-oblong, mucronate, shortly petiolulate; stipules ovate-acuminate, scarious. Peduncle stout, exceeding lvs, racemes with c. 50 fls, dense. Fls 10–14 mm, bright pink or red, the wings half as long as calyx; calyx-teeth subulate, 2–3 times as long as the short, often woolly tube. Legume 5–8 mm, strongly reticulate, pubescent, tuberculate on the lower margin, 1-seeded. Fl. 6–8. $2n = 28$. Hp.

Introduced. Often a relic of cultivation, but possibly native in chalk and limestone grassland in S. and E. England, north to Yorks; apparently not even naturalized elsewhere. Perhaps native in C. Europe; cultivated and often naturalized elsewhere.

59. ROSACEAE

Trees, shrubs or herbs. Lvs nearly always alternate and stipulate. Fls regular, usually hermaphrodite, perigynous or epigynous. Epicalyx sometimes present outside the sepals. Sepals usually 5, usually imbricate. Petals equalling in number and alternate with sepals, imbricate, rarely convolute, sometimes 0. Stamens usually 2, 3 or 4 times as many as the sepals, rarely indefinite or 1–5 only. Carpels 1–many, free or sometimes united to each other and to the receptacle (very rarely united and free from the receptacle); ovules usually 2, sometimes 1 or more, anatropous; styles free, very rarely

united. Fr. of one or more achenes, drupes or follicles or a pome (very rarely a capsule), the receptacle sometimes becoming coloured and fleshy; endosperm 0, rarely scanty.

About 100 genera and over 2000 spp., cosmopolitan, especially temperate.

A very diverse but natural family usually recognizable by its perigynous or epigynous fls with ±numerous stamens. Closest to the Saxifragaceae (q.v. for distinctions). Liable to be confused with the Ranunculaceae by beginners but apart from the perigynous fls, the presence of stipules will separate any member likely to be taken for a member of the Ranunculaceae.

1 Trees or upright shrubs.   2
  Herbs or sometimes slightly woody and prostrate.   12
2 Fls bright yellow; epicalyx present; lvs pinnate; low
  unarmed shrub.   4. POTENTILLA
  Fls white (occasionally tinged yellow), or pink; epicalyx 0.   3
3 Receptacle strongly convex; lvs compound.   3. RUBUS
  Receptacle concave.   4
4 Receptacle slightly concave; carpels 5; fls pink, very
  numerous in a dense panicle.   1. SPIRAEA
  Receptacle strongly concave (carpels 1 or numerous),
  or ovary truly inferior (carpels 1–5).   5
5 Carpels and styles numerous (the latter occasionally
  connate into a column); prickly shrubs with pinnate
  lvs.   15. ROSA
  Carpels and styles 1–5; unarmed or thorny shrubs or
  trees (if lvs pinnate, then unarmed).   6
6 Carpel 1, free from receptacle; lvs simple, not lobed.
                                            16. PRUNUS
  Carpels 1–5 (if 1 then lvs lobed), united to the receptacle so that the ovary is inferior (at least partly).   7
7 Fls in compound corymbs; false septum 0.   8
  Fls solitary or 2–3 or in short simple umbel-like corymbs; false septum 0.   9
  Fls in racemes; carpels with false septum.
                                       20. AMELANCHIER
8 Thorny; lvs lobed; carpel-wall hard in fr.
                                       18. CRATAEGUS
  Unarmed; lvs entire; carpel-wall hard in fr.
                                       17. COTONEASTER
  Unarmed; lvs toothed, lobed or pinnate; carpel-wall
  cartilaginous in fr.   21. SORBUS
9 Fls less than 1 cm diam.; fr. c. 1 cm diam., red.
                                       17. COTONEASTER
  Fls c. 3 cm diam. or more; fr. larger, brown or green
  (sometimes tinged red).   10
10 Fls solitary; carpel-wall hard in fr.; sepals long, often
   lf-like.   19. MESPILUS
   Fls several; carpel-wall papery in fr.; sepals short,
   not lf-like.   11
11 Styles free; anthers purple; fr. gritty.   22. PYRUS
   Styles connate below; anthers yellow; fr. not gritty.
                                            23. MALUS
12 Carpels (4–)5 or more on the surface of a convex
   to weakly concave receptacle.   13
   Carpels 1–2(–4) enclosed in the strongly concave
   receptacle.   19
13 Epicalyx 0.   14
   Epicalyx present   16
14 Lvs pinnate; fls small in many-fld panicles.
                                       2. FILIPENDULA
   Lvs not pinnate; fls not in many-fld panicles.   15
15 Petals c. 8; plant prostrate; fr. a group of achenes
   with long feathery awns.   8. DRYAS
   Petals (4–)5; plant not prostrate; fr. a group of drupelets.   3. RUBUS
16 Style terminal, persistent on the fr. as a long, jointed
   awn; basal lvs pinnate with the terminal lflet much
   larger than any of the lateral ones.   7. GEUM
   Style ±lateral, not persistent; if lvs pinnate, terminal
   lflet not or scarcely larger than lateral ones (though
   these are sometimes unequal).   17
17 Stamens and carpels few (10 or less); petals inconspi-

cuous, lvs ternate, lflets tridentate; alpine plant.
                                            5. SIBBALDIA
   Stamens and carpels usually numerous; petals usually
   conspicuous; if lvs ternate, lflets with more than
   3 teeth.   18
18 Receptacle fleshy in fr.; fls white; lvs ternate; receptacle glabrous.   6. FRAGARIA
   Receptacle dry in fr.; if fls white and lvs ternate,
   receptacle hairy.   4. POTENTILLA
19 Petals present; fls in racemes or few-fld cymes.   20
   Petals 0; fls in heads or many-fld cymes, rarely spikes.   21
20 Epicalyx present; receptacle without spines.
                                            10. AREMONIA
   Epicalyx 0; receptacle with a crown of spines.
                                            9. AGRIMONIA
21 Fls in cymes (sometimes dense and head-like, then
   lf-opposed); epicalyx present.   23
   Fls in terminal heads or spikes; epicalyx 0.   22
22 Plant erect; receptacle not spiny in fr.
                                            13. SANGUISORBA
   Plant prostrate; fr. receptacle with conspicuous
   spines.   14. ACAENA
23 Perennial; fls in terminal cymes; stamens 4.
                                            11. ALCHEMILLA
   Annual; fls in dense lf-opposed clusters; stamen 1.
                                            12. APHANES

Subfamily 1. SPIRAEOIDEAE. Unarmed shrubs, rarely herbs. Stipules usually 0. Fls usually small and numerous in compound infl. Receptacle flat or slightly concave. Carpels 1–12, usually 5, in a single whorl; ovules several (rarely 2) in each. Fr. a group (rarely 1) of follicles, rarely a group of achenes or a capsule.

Tribe 1. SPIRAEAE. Fr. follicular; seeds not winged.

Species of *Neillia* D. Don, *Stephanandra* Siebold & Zucc., *Sibiruea* Maxim., *Aruncus* L. and *Sorbaria* (Ser.) A.Br. are sometimes grown. The two other tribes of this subfamily are represented in gardens by *Exochorda* Lindley (Tribe *Quilla-jeae*) with capsular fr. and winged seeds, and *Holodiscus* (C. Koch) Maxim. (Tribe *Holodisceae*) with the fr. of 5 achenes (stipules 0).

1. SPIRAEA L.

Deciduous shrubs. Lvs alternate, simple, exstipulate. Infl. a simple or compound corymb or a panicle. Receptacle somewhat concave. Sepals and petals 5. Stamens 15–many. *Carpels 5, free, alternate with the sepals. Fr. membranous, dehiscent along the inner edge;* seeds several, oblong, very small; endosperm 0. Nectar secreted by a ring inside the stamens.

About 100 spp., north temperate. Many spp. (and some hybrids) are grown in gardens and a few have been found as escapes.

*1. S. salicifolia L.   Bridewort.

Shrub 1–2 *m* with numerous erect suckering stems. Branches strict, yellowish-brown, somewhat angled, puberulent when young. Buds small, ovoid. Lvs 3–7 cm, oblanceolate-oblong or elliptical-oblong, cuneate at base, acute or subobtuse, sharply and sometimes doubly serrate, glabrous. Infl. a narrowly conical or cylindrical many-fld terminal panicle, 2–10 cm, dense or lax below, the fls always dense on the branches; rhachis, etc., pubescent. Sepals triangular-ovate, puberulent, erect in fr. Corolla pink, c. 8 mm diam. Stamens c. twice as long as petals. Carpels subglabrous, erect. Fl. 6–9. Visited by various insects. $2n = 36$. N.

Introduced. Frequently planted in hedges, etc., and ±naturalized in woods, etc. Throughout Great Britain but more frequent in Wales, N.W. England and S. and E. Scotland; S.W. and N. Ireland, occasional elsewhere. Native of C. and E.C. Europe and temperate Asia from Czechoslovakia and Austria eastwards; naturalized in other parts of Europe.

*S. douglasii Hooker, a native of western North America, is often cultivated in gardens and has been found as a ±naturalized escape. It is more robust than *S. salicifolia*, with lvs white-tomentose beneath and sepals deflexed in fr.

S. × billiardii Hérincq. (*S. douglasii × salicifolia*), which is grown in gardens and sometimes found as an escape (e.g. Surrey, Lincoln and Aberdeen), is intermediate between the parents in leaf-indumentum and is highly fertile.

PHYSOCARPUS (Cambess.) Maxim.
Deciduous shrubs. Lvs lobed; stipules caducous. Fls white or pinkish, in corymbs. Frs 1–5, dehiscing along both sutures.

*P. opulifolius (L.) Maxim. Lvs 2–7(–10) cm, 5-lobed. Carpels 4–5. Fr. glabrous. Grown in gardens and occasionally naturalized. Eastern North America.

Subfamily 2. ROSOIDEAE. Herbs or shrubs. Stipules present. Receptacle from very convex to very concave. Carpels usually numerous and spirally arranged, but sometimes few or 1, free from each other and from the receptacle but sometimes enclosed in it; ovules 1 or 2 in each carpel. Fr. of 1 or more achenes or a group of drupes (drupelets) adhering to each other; seeds always 1.

Tribe 2. ULMARIEAE. Receptacle flat or slightly concave. Filaments narrowed at base. Carpels 5–15, in one whorl; ovules 2, pendent. Fr. a group of achenes.

2. FILIPENDULA Miller
Perennial herbs with short rhizome. *Lvs alternate, pinnate with small pinnae between the large ones.* Infl. a cymose panicle. Sepals and petals 5–8. Stamens 20–40. Homogamous pollen-fls without nectar; both British spp. are visited by various insects but automatic self-pollination occurs if visits fail.

About 10 spp., north temperate zone.

Basal rosette-lvs with 8 or more pairs of the larger lflets; lflets less than 1.5 cm; carpels straight, erect. **1. vulgaris**
Basal rosette-lvs with 5 or fewer pairs of the larger lflets; lflets 2 cm or more; carpels twisted together after fl.
2. ulmaria

1. F. vulgaris Moench   Dropwort.
F. hexapetala Gilib.; *Spiraea filipendula* L.; *Ulmaria filipendula* (L.) Hill.
Perennial herb (7–)15–80 cm, nearly glabrous. *Roots bearing ovoid tubers.* Lvs of basal rosette ±numerous, 2–25 cm (including the short petiole), *with 8–20 pairs of main lflets*; lflets 5–15 mm, ±oblong in outline, pinnately lobed, with acute, sparingly dentate lobes, green on both sides; terminal lflets trifid, resembling 3 fused lflets; smaller lflets 1–3 mm; *stem-lvs very few*, the upper simple, lobed, very small, the lower usually with 2 or 3 pairs of lflets, sometimes a few at the extreme base resembling lflets of basal rosette. Fl.-stems usually simple. Fls ±numerous in an irregular, cymose panicle much wider than high. Sepals usually 6, triangular-ovate, spreading, then deflexed. Petals usually 6, 5–9 mm, obovate-spathulate, cream-white, tinged reddish-purple outside. Stamens about as long. *Carpels 6–12, erect, pubescent*, c. 4 mm in fr. Fl. 5–8. $2n = 14$, 16*. Hs.

Native. Calcareous grassland, ascending to 365 m in Yorks, widespread in England but rather local though often abundant, rare or absent in much of the southwest; N. Wales, Pembroke; very local in E. Scotland northwards to Angus; Ireland: Clare and Galway only; Jersey. Most of Europe northwards to c. 64° N in Norway; Caucasus; N. Africa (mountains); Asia Minor, Siberia (to the Angara-Sayan region).

2. F. ulmaria (L.) Maxim.   Meadowsweet.
Spiraea ulmaria L.; *Ulmaria pentapetala* Gilib.
Perennial herb, 60–120 cm. *Roots not tuberous.* Lvs of basal rosette ±numerous, mostly 30–60 cm (including the rather long petiole), *with 2–5 pairs of main lflets*; lflets 2–8 cm, ovate, acute, sharply doubly serrate, dark green and glabrous above, usually white-tomentose beneath but frequently green and pubescent or glabrous; terminal lflet 3- (or obscurely 5-)lobed to about $\frac{2}{3}$, resembling 3 fused lflets; smaller lflets 1–4 mm; lower stem-lvs resembling those of basal rosette but shorter and with fewer lflets, uppermost simple or with only small lflets in addition to the terminal one. *Fl.-stems lfy*, simple or branched above, glabrous or nearly so. Fls very numerous in an irregular cymose panicle, usually rather higher than wide, the fls dense on the ultimate branches. Sepals usually 5, triangular-ovate, deflexed, pubescent (like the fl.-stalks). Petals usually 5 (sometimes 6), 2–5 mm, obovate, clawed, cream-white. Stamens c. twice as long as petals. *Carpels 6–10, erect in fl. but soon becoming twisted together spirally*, c. 2 mm in fr., *glabrous*. Fl. 6–9. $2n = 14$, 16. Hs.

Native. Swamps, marshes, fens, wet woods and meadows, wet rock ledges and by rivers, etc. (absent from acid peat); ascending to nearly 915 m. Common throughout the British Is., sometimes locally dominant in fens and wet woods. Throughout Europe, except some of the islands and much of the Mediterranean region; Caucasus; temperate Asia to Asia Minor and Mongolia; an escape in eastern North America.

Tribe KERRIEAE. Receptacle weakly concave. Stamens numerous; filaments with broad base. Carpels 4–6, in one whorl. Fr. a group of achenes, often with hard coat.

Kerria japonica (L.) DC. Lvs alternate. No epicalyx. Petals 5, yellow. C. and W. China.

Rhodotypos scandens (Thunb.) Makino. Lvs opposite. Epicalyx present. Petals 4, white. Japan, C. China.

Both are monotypic genera, commonly grown in gardens, *Kerria* most commonly as a double-fld form.

Tribe 3. POTENTILLEAE. Receptacle convex to slightly concave. Stamens and carpels numerous (except *Sibbaldia* where both are few). Filaments with broad base (filiform in some *Rubus*). Fr. a group of achenes or of drupelets.

3. RUBUS L.

Herbs or shrubs; if the latter, usually with biennial stems dying after fl. Epicalyx 0. Sepals 5, imbricate. Petals 5(–8). Stamens numerous. Carpels ±numerous, spirally arranged; *ovules 2 in each carpel. Fr. of ±numerous 1-seeded drupelets, aggregated together* into a compound fr. Cotyledons elliptical, ciliate.

Spp. variously estimated at from 250 upwards. Cosmopolitan, especially north temperate zone.

1 Lvs simple, palmately lobed; dioecious herb with creeping rhizome and solitary terminal fls.
              1. chamaemorus
  Lvs pinnate or palmately compound; fls hermaphrodite, usually in infl.   **2**
2 Stems annual, herbaceous; fl. branches arising from ground level; stipules attached to the stem.   **3**
  Stems usually biennial, woody, bearing axillary fl. branches in the 2nd year; stipules attached only to petiole.   **4**
3 Plant with long above-ground stolons; fls 2–8, white; mature fr. scarlet.   **2. saxatilis**
  Plant without above-ground stolons; fls 1–3, bright pink; mature fr. dark purple.   **3. arcticus**
4 Lvs pinnate, with 3–7 lflets, white beneath; fr. bright red, separating from receptacle when ripe.
              4. idaeus
  Lvs palmate, with 3–5 lflets (rarely 7 and then the 4 lower from the same point), or all ternate; fr. black or blue-black (rarely deep red), not separating from receptacle when ripe.   **5**
5 Fr. very pruinose, with 2–5 or 14–20 drupelets; stems very pruinose, terete, weak; lvs ternate; stipules lanceolate.   **5. caesius**
  Fr. not or slightly pruinose; drupelets usually more than 20; stipules linear or linear-lanceolate (if the latter lvs not all ternate).   **6. fruticosus** *sens. lat.*

Subgenus 1. CHAMAEMORUS (Hill) Focke

Dioecious. Stems annual, herbaceous. Lvs simple, lobed. Stipules attached to the stem, slightly adherent to the petiole above. Fl. solitary, terminal. Filaments flattened, tapering. Stone smooth.

1. R. chamaemorus L.   Cloudberry.

Unarmed somewhat pubescent herb with creeping rhizome. Fl. stems 5–20 cm, annual, erect. *Lvs* few; blade 1.5–8 cm, *simple*, ±orbicular, palmately 5–7-lobed, deeply cordate at base, somewhat rugose; lobes dentate, those of the male plant reaching c. ⅓ of the way to the base, those of the female plant shallower; petiole 1–7 cm; stipules ovate, scarious. Fl. solitary, terminal, (4–)5-merous. Sepals ovate, acuminate. Petals 8–12 mm, obovate, white, much longer than sepals. Fr. orange when ripe, red earlier; drupelets few, large. Fl. 6–8. Visited by humble-bees and flies. $2n = 56^*$.

Native. Mountain moors and blanket bogs, locally abundant; ascending to 1160 m. N. Wales; Lancs, Derby and Yorks to Sutherland but absent from the Isle of Man, S.W. Scotland, Outer and Inner Hebrides, etc.; Ireland: only known from Tyrone, where it is very rare. N. Europe, extending southwards to N.W. Czechoslovakia; N. Asia (east to Kamchatka and Sakhalin); North America.

Subgenus 2. CYLACTIS (Rafin.) Focke

Stems annual, herbaceous. Lvs ternate, rarely simple. Stipules attached to the stem, slightly adhering to the petiole above. Fls cymose or solitary, terminal and axillary, hermaphrodite. Filaments flattened below. Receptacle flat. Stone smooth or reticulate.

2. R. saxatilis L.   Stone Bramble.

A herb with long overground stolons. Fl. stems annual, ±erect, 8–40 cm; non-fl. stems procumbent, much longer, producing axillary branches which root at the tips; both kinds pubescent, with weak prickles or unarmed. Lvs ternate; *terminal lflet* 2.5–8 cm, *usually stalked*, ovate to obovate, acute or subacute, broadly cuneate at base, irregularly and often doubly dentate, green and subglabrous above, paler and pubescent beneath; lateral lflets subsessile, often with a shallow rounded lobe on the lower margin; petiole 2–7 cm, armed like the stem; stipules ovate, green. Fls 2–8 in a compact terminal cyme, sometimes also with axillary cymes; peduncle up to 2.5 cm. Sepals triangular-ovate, acute or acuminate, shortly pubescent. *Petals* 3–5 mm, oblanceolate, *white*, shorter or scarcely longer than the sepals. Fr. scarlet, translucent; drupelets 2–6, large, glabrous, separate. Fl. 6–8. Visited by bees and flies, self-pollination possible. $2n = 28^*$.

Native. Stony woods and shady rocks, especially basic, in hilly districts, rather local; Cornwall to Somerset; Wales and the Border Counties; N. Lincoln, Derby and Lancs northwards to Shetland; widespread in Ireland from Kerry, Leix and Wicklow northwards, but perhaps extinct in much of the centre and east. Europe from Iceland and arctic Russia to the Pyrenees, mountains of Italy, N. Greece and the Caucasus; N. Asia Minor, temperate Asia to the Himalaya and Japan; Greenland.

3. R. arcticus L.   Arctic Bramble.

Similar to *R. saxatilis*, but creeping underground and without overground stolons. Lvs smaller; terminal lflet 1–5 cm, subsessile. Fls solitary. Petals 7–10 mm, bright pink. Fr. dark purple.

Appears formerly to have occurred in the Scottish Highlands but has not been seen for many years. Arctic and subarctic Europe, Asia, and North and South America.

Subgenus ANOPLOBATUS Focke

Stems biennial, woody, unarmed; bark peeling. Lvs simple, palmately lobed. Stipules attached to the petiole. Fls hermaphrodite. Receptacle flat.

The 2 following spp. are grown for their fls and sometimes become ±naturalized:

*R. parviflorvus Nutt. (*R. nutkanus* Moc. ex Ser.)

Stems erect, up to 2 m. Lvs 6–20 cm broad, 3–5-lobed. Infl. 3–10-fld. Fls 3–6 cm diam., white. Native of western North America.

*R. odoratus L.

Stems erect, up to 3 m. Lvs 10–30 cm broad, 5-lobed. Infl. many-fld. Fls 3–5 cm diam., purple. Native of eastern North America.

Subgenus 3. IDAEOBATUS Focke

Stems biennial, woody, normally prickly. Lvs pinnate, rarely palmate or simple. Stipules attached to the petiole. Fls hermaphrodite. Filaments subulate. Fr. red or yellow, separating from the dry conical receptacle when ripe. Stone rugose.

4. R. idaeus L.   Raspberry.

Suckering by adventitious buds from the roots. Stems 100–160 cm, erect, terete, somewhat pruinose, armed with slender, straight, subulate prickles, rarely unarmed. Lvs pinnate, with 3–5(–7) lflets; lflets 5–12 cm, the terminal one larger than the others, ovate or ovate-lanceolate, acuminate (rarely (var. *anomalus* Arrh.) orbicular, obtuse), rounded or subcordate at base, irregularly dentate with mucronate teeth, green and somewhat pubescent above, densely white-tomentose beneath; petiole 2–7 cm cylindrical; stipules filiform. Fls 1–10 in dense axillary and terminal racemes, forming a compound infl. on a short lateral branch with ternate lvs. Sepals ovate-lanceolate-triangular, long acuminate. Petals oblanceolate or oblong, white, erect, about as long as sepals. Fr. red, rarely pale yellow, opaque; drupelets numerous, pubescent. Fl. 6–8. Pollinated by various insects and selfed. $2n = 14^*$.

Native. Woods and heaths, especially in hilly districts; ascending to 825 m. Common almost throughout the British Is. Most of Europe, but only on mountains in the south; Caucasus; N. Asia Minor, N. and C. Asia east to the Yenisei region; subspp. or allied spp. in E. Asia and North America.

*R. phoenicolasius Maxim.   Japanese Wineberry.

Stems densely covered with red gland-tipped bristles. Lflets 3(–5), ovate, green and slightly pilose above, white-tomentose beneath. Petals pinkish, shorter than sepals, appressed to stamens. Fr. orange-red, translucent, edible.

Sometimes grown in gardens and occasionally naturalized. Native of Japan, Korea and N. China.

*R. spectabilis Pursh

Stems with slender prickles below. Lflets 3, subglabrous. Fls solitary, bright rose-purple, c. 2.5 cm diam. Fr. large, orange, edible.

Grown in gardens for its fls and sometimes naturalized. Native of western North America.

*R. loganobaccus L. H. Bailey   Loganberry.

Stems robust, long, arching. Lflets 5, pinnate (as in *R. idaeus*), large. Fr. purplish-red, large, coming away with the receptacle (as in subgenus *Rubus*). $2n = 42$.

Commonly grown for its fr. and rarely naturalized. Originated in 1881 in a Californian garden by hybridization between *R. idaeus* subsp. *strigosus* (Michx.) Focke (an American subsp.) and *R. vitifolius* Cham. & Schlecht. (a Californian sp. of subgenus *Rubus*); a hexaploid, now behaving as a sp.

Subgenus 4. RUBUS

Stems biennial (sometimes lasting longer or flowering the first year), woody, prickly. Lvs ternate or palmate (at least in their primary division). Stipules attached to the petiole. Infl. usually compound. Fls hermaphrodite. Filaments filiform. Fr. black (occasionally deep red or blue with bloom), coming away with the fleshy conical receptacle when ripe. Pollinated by various insects. (Description applying to European spp. only).

5. R. caesius L.   Dewberry.

Stem procumbent, rooting, *weak*, *terete*, glabrous, very *pruinose*; bark finally exfoliating; stalked glands and acicles few or 0; *prickles very* weak, subulate, straight to falcate, *scattered*. Lflets 3, sparsely pilose above, green and subglabrous or closely pubescent beneath, irregularly, coarsely and doubly dentate or shallowly lobed; terminal variable in shape but ±ovate or rhombic, acute or slightly acuminate, mostly emarginate at base, sometimes trilobed; *stipules lanceolate*, those of the lower lvs wider, those of upper lvs narrower. Infl. short, lax, few-fld, shortly pubescent, glandular or nearly eglandular and with weak prickles; peduncles ascending; pedicels slender. Sepals grey-green, white-margined, cuspidate with long points, clasping the fr. Petals elliptical to suborbicular, white or pinkish. Stamens about as long as styles; filaments white; pollen perfect. Styles greenish. Fr. densely pruinose and so appearing bluish, partially separating from receptacle; lateral frs with few drupelets, terminal fr. with many more. Very variable. Fl. 6–9. $2n = 28$.

Native. Dry grassland and scrub mainly on basic soils, also common in fen carr; widespread and common in C. and E. England more local in the south-west and Wales, local in E. Scotland northwards to E. Sutherland; widespread but rather local in Ireland. Europe from Scandinavia (62° N) and N. Russia to Spain, Portugal, Sicily and Greece; W. Siberia and Russian C. Asia to Asia Minor.

R. × idaeoides Ruthe (*R. caesius × idaeus*) resembles *R. caesius* in habit and stem characters, while the lvs are like those of the other parent but usually have 3 or 5 lflets, the infl. is short and corymbose. This hybrid is rare from Hants and Essex to Derby, Yorks and Durham.

6. R. fruticosus sens. lat.   Blackberry, Bramble.

Bark not exfoliating. Lvs rarely all ternate. Stipules filiform, linear or linear-lanceolate. Fr. glabrous when ripe, black (rarely deep red), not or slightly pruinose adhering to the receptacle and coming away with it when ripe. Fl. 5–9. Pollinated by various insects.

Native. Woods, scrub, hedges and heaths. Common throughout the British Is. Europe, Mediterranean region, Macaronesia.

The spp. included in this aggregate are very difficult to determine and the number of British spp. has been very variously estimated. Bentham (*Handbook of the British Flora*, Ashford, Kent: L. Reeve & Co. Ltd., 1980) includes them all under one sp., *R. fruticosus* L.; W. C. R. Watson (*Handbook of the Rubi of Great Britain and Ireland*, Cambridge University Press, 1958) gives 386 spp., whereas the most recent estimate by A. Newton (*Watsonia*, **13**, 35–40, 1980) suggests that there are about 290 spp., excluding the myriad highly localized entities. Though pollination is necessary for development, the vast majority of forms are apomictic (*R. ulmifolius* is the only known exception in the British Is.) and thus constant, but can on occasion reproduce sexually and give rise to hybrids (and presumably also at times to new apomicts). In the following account the sections and series used by Watson are given, and a few of the common species mentioned. It should be noted that the 'sections' are not comparable with the sections in genera with completely or mainly sexual reproduction but are at a much lower taxonomic level.

It must also be remembered that the spp. appear to form a ±continuous network of forms from the *Suberecti* and *Rubus* on the one hand to the *Glandulosi* on the other and that it is often a matter of doubt to which section a particular sp. should be referred, although the central forms in each section are usually distinct enough. The following are some of the more important characters used in the group:

Habit. The majority of spp. have stems ±arching over and rooting at the tips but the height of the arch varies much, in some spp. reaching c. 1 m, in others being negligible so that the stems lie along the ground unless supported. The *Suberecti* differ from the other sections in the suberect, not rooting, stems.

Angularity of stem varies from terete to sharply angled in different species and in proceeding from the base to the tip of the stem.

Armature and indumentum of first-year stem (descriptions of armature always refer to this and not to the branches). Apart from eglandular hairs and pruinosity, the organs clothing the stem may be grouped as follows:

Prickles. The largest type of armature. In the earlier sections they are confined to the angles of the stem, uniform in size and the only type present. In the later sections they are scattered round the stem, unequal, and pass gradually into the other types. Numerous intermediate stages occur. The size, shape, direction and abundance of the prickles are often most important characters.

Pricklets. Small prickles, often with stout bases, scattered round the stem.

Acicles. Very slender prickles or very stiff bristles, not stout-based, sometimes gland-tipped.

Stalked glands.

The last three types may be all ±equal in length to each other and among themselves and sharply marked off from the prickles (as in the *Radulae*) or very unequal and passing into each other and into the prickles (as e.g. in the *Hystrices*).

The quantity of stalked glands, acicles and pricklets present on the stem varies much from the lower part to the upper part of the same internode as well as from the lower part to the middle part of the same stem and also between a young plant and an older one of the same sp. Prickles always become more curved towards the tip of the stem. The quality of the armature is more constant.

The above characters of habit and armature are those mainly used for delimiting the sections.

Leaves. Number of lflets per lf. Nature and colour of the indumentum especially of lower surface. Shape of lflets especially of the terminal one. Depth and type of toothing. Length of stalks of lflets and especially whether the basal lflets are stalked or sessile. The lvs used are normally those about the middle of the first-year stems (not those of the branches).

Infl. The shape and branching of the infl., its armature and indumentum and leafiness.

Sepals. Armature and indumentum which varies considerably so that the sepals range from being green and sparingly hairy with grey-tomentose margins to uniformly grey- or white-tomentose. The direction of the sepals at the time the fr. swells is a character of considerable importance and it is this that is intended in the following descriptions unless otherwise mentioned.

Petals. Shape, colour (when first expanding) and clothing.

Stamens. Length relative to the styles. Colour (when first expanding) of filaments. Whether the anthers are glabrous or pilose (the latter occurs only in a few spp. and unless otherwise mentioned the anthers are glabrous).

Styles. Colour, especially at the base.

Specimens of brambles should consist of (1) pieces of stem from the middle of the first-year stem with lvs attached; (2) infl., preferably at two stages so that there is at least one lf without an axillary peduncle below the infl.; (3) separate petals. Care should be taken to collect them from the same bush. Note should be made, when collecting, of the habit of the plant, of the number of lflets per lf (many spp. have a variable number of lflets and this does not always show on the specimen), and of the colour of the various parts of the fl. in the opening bud and of the stem, etc. The direction of the sepals and the relative length of the stamens and styles are also not always easy to see when dry. The ideal procedure is to determine at least the section of the plant in the field and to prepare the specimen bearing in mind the characters used for determination.

In view of what has been said above it will be recognized that great caution is necessary in using the key given here.

While many of the bramble spp. are widespread (some of the *Suberecti*, for example, extend from Britain to Russia), others are very restricted in range, although they may be abundant where they occur. In any one locality the spp. are usually easily separable but in another neighbouring locality, an entirely different selection of spp. may occur. Indeed, Newton (*Watsonia*, **13**, 35–40, 1980) has recognized 6 major *Rubus* floristic regions and 8 more local area endemic complexes in Great Britain. He also notes that the number of species decreases markedly from south to north in the British Is., with former areas of oak woodland being richest in species. High, wet or calcareous areas have few or no brambles.

For a full account of the plants of this genus, Watson's book should be consulted, together with additions and amendments by E. S. Edees and A. Newton (*Watsonia*, **12–13**, 1978–80).

In the following key and descriptions 'stems' refers throughout to the first-year stem and 'lvs' to well-developed lvs from about the middle of such stems. If a character is qualified by 'usually' in the description of a section it means that the great majority of spp. in the section possess it.

Key to the Sections

1 Stalked glands absent from stem and infl.   2
   Stalked (or subsessile) glands present at least on infl.   4
2 Stem suberect or high-arching, not rooting, glabrous or nearly so; lvs green or somewhat greyish beneath; some lvs often with 7 lflets; infl. simple or only slightly branched; sepals green or greyish-green outside; fl. early.   **1. Suberecti**
   Stem arching and rooting; lflets never 7, white-felted beneath; infl. branched, fls large and usually brightly coloured; sepals grey or greyish-white, always deflexed; fl. late.   **4. Rubus**
   Not as either of above.   3
3 Stem usually low; prickles scattered; basal lflets usually sessile or subsessile; infl. usually corymbose with comparatively few large long-stalked fls; drupelets often few, large, partly defective (common on calcareous soils).   **2. Corylifolii**
   Stems with the prickles confined to the angles; basal lflets usually stalked; fls usually numerous; drupelets numerous (rare on calcareous soils).   **3. Sylvatici**

4 Basal lflets usually sessile or subsessile, often overlapping the others; stipules usually linear-lanceolate; infl. usually corymbose with comparatively few large long-stalked fls; drupelets often large, partly defective; stem usually low, terete or obscurely angled, prickles scattered.   **2. Corylifolii**
   Basal lflets nearly always stalked; stipules usually linear; other characters not combined as above.   5
5 Stem prickles equal or somewhat unequal, seated on or near angles of stem, much larger than and clearly marked off from the pricklets, acicles and stalked glands.   6
   Stem prickles very unequal, scattered all round the stem, usually connected with the pricklets, acicles and stalked glands by intermediates.   **7. Glandulosi**
6 Stalked glands, acicles and pricklets 0 or very few on the stem, not more numerous than the prickles.   7
   Stalked glands, acicles and pricklets (or some of them) ±numerous on the stem (at least about as many as pricklets).   **6. Appendiculati**
7 Stems slender, obscurely angled, ±hairy, with 0 or few stalked glands, pricklets and acicles; fls small (less than 2 cm diam.); sepals never deflexed; stamens shorter than styles.   **5. Sprengeliani**
   Stems mostly more robust, hairy or not, stalked glands, pricklets and acicles 0 or very few; fls more than 2 cm diam.; stamens usually longer than styles.   **3. Sylvatici**
   Stalked glands, acicles and pricklets few but some of them always present; stem and lvs usually very hairy; fls usually more than 2 cm diam.   **6. Appendiculati** ser. *Vestiti*

Section 1. **Suberecti** P. J. Mueller. Often spreading by suckers. *Stems suberect*, arching only near the apex, *not rooting*, angled, glabrous or more rarely slightly pilose at first or with sessile glands; *prickles confined to the angles* (except in *R. scissus* W. C. R. Watson), *all equal*; *pricklets, acicles and stalked glands 0*. Lflets (3–)5(–7) but always 5 on most lvs, green on both sides, or occasionally somewhat greyish beneath, the basal pair usually subsessile or shortly stalked. Infl. usually nearly simple, subracemose; stalked glands 0. *Sepals green or greyish-green* with conspicuous white-tomentose margins, spreading or deflexed after fl. Petals usually white or pale pink. $2n = 21, 28, 42$. Fl. 5–7.

Usually on heaths or in woods, especially high forest, usually on very acid soils.

A natural group of comparatively few spp., clearly marked off from all other groups except the *Silvatici*. They are believed to have originated by hybridization with *R. idaeus*, but some spp. have a wide distribution and they are clearly of ancient origin.

The 3 following spp. are common over most of the British Is.

R. nessensis W. Hall

Prickles short (to 1.5 mm), straight, purplish-black. Lflets large, 3–5(–7), green on both sides. Petals glabrous, white. Stamens white, longer than greenish styles. Fr. dark red.

R. scissus W. C. R. Watson

Similar to *R. nessensis* but with scattered prickles, the stamens shorter than the styles and pubescent petals.

R. plicatus Weihe & Nees

Prickles slender but stronger than in the 2 preceding spp., falcate, yellow and red. Terminal lflet wide, plicate. Petals hairy. Stamens about equalling styles. Fr. black.

Section 2. **Corylifolii** Lindley (*Triviales* P. J. Mueller). Stems low-arching or procumbent, rooting at end, terete or slightly angled; *prickles scattered round the stem, usually straight and spreading*; stalked glands, acicles and pricklets variable (from 0 to numerous). Lflets 3–5, *the basal ones sessile or subsessile*, usually wide, green or grey beneath; *stipules usually lanceolate or linear-lanceolate* (wider than in the other sections) but filiform in some spp. Infl. various but mostly either short, or narrow with the branches few-fld, stalked glands few or many, mostly unequal. Sepals grey, usually spreading or erect. Petals usually large, often suborbicular. *Drupelets usually large and few, partly abortive, not, or sometimes later, slightly pruinose.* $2n = 28, 35, 42$. Fl. 5–9 (the fl. period of the individual spp. is also long).

Mainly hedges and scrub, frequent on basic soils.

This section (19 spp.) is believed to have originated comparatively recently by hybridization between *R. caesius* and various members of the other sections.

Very local hybrids involving *R. caesius* or members of this section occur fairly frequently and only those with a fair distribution have been given names. Several spp. are widespread and common. The commonest may be distinguished as follows:

1 Prickles equal or nearly so.   2
  Prickles numerous, unequal; glands numerous.   4
2 Stalked glands 0 or very few.   3
  Stalked glands rather numerous; lvs large, soft; fls very large (to 5 cm diam.), pink; stamens pink, anthers hairy; styles red. Fl. begins early May.
      R. nemorosus Hayne & Willd.
      (*R. balfourianus* Bloxam ex Bab.)
3 Petals pink; prickles short; basal lflet stalked, terminal not lobed.   **R. conjungens** (Bab.) Rogers
  Petals white; prickles deep purple; terminal lflet often lobed.   **R. sublustris** Lees
4 Terminal lflet obovate.   **R. tuberculatus** Bab.
      (*R. myriacanthus* Focke)
  Terminal lflet ovate (south-east).
      R. britannicus Rogers

Section 3. **Sylvatici** P. J. Mueller. *Stem arching*, often low or scrambling, *rooting at the end*, angled (occasionally only weakly so), *subglabrous or hairy*; *prickles confined to the angles, usually without pricklets, acicles or stalked glands*, rarely with a few. Lflets 5 on most lvs in the great majority of spp., green or greyish beneath, rarely greyish-white, the basal pair stalked except in a very few spp. Infl. varying in shape, from little- to much-branched and from eglandular to rather glandular with stalked glands. Sepals usually uniformly grey-tomentose, sometimes green or greyish-green with

whiter margins. Stamens longer than the styles in most spp. $2n = 21, 28, 42$. Fl. 7–8.

Woods, scrub, heaths, hedges, etc.

A large and varied group. Variously subdivided by different authors. Watson divides it into 2 subsections and 7 series, the differences between which are ill-defined and difficult. It may be noted that all eglandular brambles which are clearly not *Suberecti*, *Corylifolii* or *Rubus* (*Discolores*) belong here.

The following are common spp.:

Subsection A. *Virescentes* Genev. Lvs green beneath, hairy, rarely the upper lvs of stem and panicle greyish felted.

R. platyacanthus P. J. Mueller & Lefèvre
(*R. carpinifolius* Weihe & Nees)

Eglandular. Stem tall. Prickles strong, numerous, yellow or reddish. Lflets 5(–7), unequally serrate, hairy beneath with longer hairs along the midrib and main veins. Panicle little-branched above, with numerous straight prickles. Fls rather large, faintly pink in bud. Sepals spreading.

R. nemoralis P. J. Mueller

Stem furrowed, becoming deep purple. Prickles strong, curved. Lflets 5(–7), serrate, pubescent to grey-felted beneath, the terminal suborbicular. Panicle wide, lfy, interrupted, with scattered glands. Sepals loosely deflexed, then patent, with small spines and long lf-like tips. Petals and filaments pink. Carpels pilose.

*R. laciniatus Willd.

Lflets irregularly and deeply pinnatifid (and so unlike all other spp.). Origin unknown, not known as a wild plant but fairly frequently as a bird-sown escape.

R. lindleianus Lees

Very prickly. Stem very tall, shining, prickles large. Lflets greyish-felted beneath, irregularly toothed. Panicle very large, branches spreading at right angles, densely hairy with a few glands. Fls white, very late (7). Sepals deflexed.

R. macrophyllus Weihe & Nees

Very robust. Stem hairy, prickles short, subulate. Terminal lflet very large, ovate, acuminate with long point, unequally serrate. Infl. little branched, densely hairy, with a few glands and weak prickles. Fls usually pink. Sepals deflexed or spreading. Receptacle hairy.

R. pyramidalis Kaltenb.

Stem low-arching, sometimes with a few glands and acicles. Prickles many, straight. Lflets softly hairy beneath, pectinately hairy on the veins, irregularly serrate. Infl. dense, pyramidal, with straight narrow-based prickles. Sepals loosely deflexed or spreading. Petals pink; filaments white. Receptacle hairy.

Subsection B. *Discoloroides* Genev. Lvs, at least the upper, greyish-white-felted beneath.

R. polyanthemus Lindeb.

Stem angled, somewhat pilose, yellow or red, slightly glandular. Lflets 3–6(–7) grey-felted beneath, simply serrate; terminal usually obovate, cuspidate. Panicle long, many-fld, with many prickles, glandular. Sepals deflexed, glandular and bristly. Petals pink. Carpels pilose.

R. cardiophyllus P. J. Mueller & Lefèvre

Eglandular. Stem red, glabrescent. Prickles rather few, straight. Lflets 5(–7) sharply and rather deeply serrate; terminal long-stalked, ovate-orbicular, cordate at base. Panicle rather short and wide. Sepals deflexed. Petals hairy, abruptly clawed, white or pinkish. Receptacle hairy.

Section 4. Rubus (*Discolores* P. J. Mueller). Eglandular and without acicles or pricklets. Stems arching, often scrambling, often red or purple, angled, rooting at ends. Prickles equal, confined to angles. *Lflets* 3–5, sub-coriaceous, *white-felted beneath*, the basal pair stalked; stipules linear. Infl. compound and well-developed. Sepals grey-tomentose, deflexed. Petals large, often bright pink. Stamens longer than or equalling styles. Fl. 7–8. $2n = 14, 21, 28, 35$.

A small group in Britain, only likely to be confused with some of the *Silvatici*.

The following is very common and the only known diploid sexual sp. of bramble in Britain. It is the only common sp. in this section.

R. ulmifolius Schott

Stem arching, often climbing or forming dense tangled bushes, furrowed, very *pruinose*, with a very fine often scarcely visible stellate pubescence; prickles strong, spreading to recurved. *Lflets* 3–5, small, dark green, somewhat rugose and glabrescent above, *densely and closely white-tomentose beneath*. Infl. cylindrical, lfy only at base; branches spreading; prickles strong, falcate. Petals crumpled, suborbicular, clawed, usually bright purplish-pink. *Stamens* pink, purplish or rarely white, *about equalling styles*. Styles usually pink or purple, sometimes greenish. Very variable. Fl. late. $2n = 14^*$; sexually reproduced, largely (?entirely) self-sterile.

The commonest sp. of bramble in general (except in Scotland) and unlike most sp. growing on chalk and heavy clay soils. Frequently hybridizing with other spp. and forming fertile hybrids, especially with *R. caesius*, and the *Triviales*.

Netherlands and S.W. Germany to Spain and Portugal, Italy and Dalmatia; N.W. Africa; Macaronesia.

Section 5. Sprengeliani (Focke) W. C. R. Watson. *Stems* arching, *slender*, procumbent or climbing, rooting at the ends, obscurely angled, subglabrous or hairy; prickles confined to the angles, often somewhat unequal, a few pricklets and stalked glands often present. Lflets 3 or 5, thin, green or more rarely greyish or greyish-white beneath, the basal pair stalked. Infl. compound, usually short and lax, sparsely glandular. *Sepals* uniformly grey or greyish-green with white margins, *spreading*, *or clasping the fr. Petals small*, or if larger, crumpled. *Stamens shorter than or equalling styles*. $2n = 21, 28$. Fl. 7–8.

Woods, scrub, heaths, etc.

A small natural group, closest to some *Silvatici* and *Suberecti*, but differing in the combination of characters given above.

No sp. is common but *R. sprengelii* Weihe is widespread in England, though rare in Wales, Scotland and Ireland. It has slender prickles, 3–5 lflets, the terminal elliptic-obovate, a short wide infl., and narrow crumpled bright rose-pink petals. The other spp. are all very local or rare.

Section 6. Appendiculati (Genev.) Sudre. Stems arching to procumbent, rooting at ends. Prickles varying from equal and confined to the angles to somewhat scattered. Pricklets, acicles and stalked glands always present but varying from few (comparable in number to prickles) to numerous but not passing into the prickles. Sepals grey or greenish-grey.

A very large and varied section, formerly grouped in several sections which were reduced by Watson to series. As the range of variations is great, these series are considered separately below.

Series 1. Vestiti Focke. *Stems* low-arching or sometimes prostrate, rooting at the end, angled (occasionally nearly terete), *usually markedly hairy*, *with prickles confined to the angles, and pricklets, acicles and stalked glands present but few* (±comparable in number to the prickles). Lflets 5 on most lvs in most spp., the basal pair stalked. *Infl.* ±glandular with stalked glands and aciculate, *mostly conspicuously pilose* with the hairs longer than the glands. Sepals usually grey-tomentose, usually deflexed but sometimes spreading. Stamens usually longer than styles. Fl. 7–8.

Woods, scrub, hedges, heaths, etc.

This series links the *Sylvatici* with the other series in this section, being less glandular than the latter but more so than the former. The plants are often, but not always, very hairy. The following is a very common and usually easily recognized sp.

R. vestitus Weihe & Nees

Stems densely hairy with short and long hairs. Lflets 3—5, feeling thick and soft, densely greenish-tomentose beneath and pectinately pilose on the veins, the terminal suborbicular, shallowly and evenly serrate-dentate with wide teeth, shortly cuspidate. Infl. long, with long slender declining prickles. Fls large. Sepals deflexed. Petals suborbicular, either like the filaments deep rose-pink or more commonly (var. *albiflorus* Boulay) pinkish and the filaments white. Common in England and Wales, scattered throughout Ireland, apparently absent in Scotland, tolerant, unlike most spp., of some lime and of clay.

Series 2. Mucronati W. C. R. Watson (Section *Rotundifolia* W. C. R. Watson). Stem low- (rarely high-)arching, rooting at end, angled or subterete, glabrous or hairy, with prickles confined to angles, *with pricklets, acicles and stalked glands present*, unequal, but very variable in number, from very few (and usually few on the basal part of the stem) to rather numerous in the upper part of the stem. Lflets 3–5, grey or green beneath with *fine or rather fine*, ±*even*, *mucronate teeth; terminal orbicular or broadly ovate or obovate, cuspidate or abruptly acuminate.* Infl. ±glandular and aciculate. Sepals greenish-grey with white margins or wholly grey, spreading or deflexed. Stamens usually longer than styles. Fl. 7–8.

Mainly woods, scrub and hedges.

A small group separated by Watson, partly characterized by variability in armature. Different members have been referred by various authors to *Sprengeliani*, *Vestiti* and *Apiculati*. The rounded terminal lflet is usually characteristic.

No sp. is more than locally common and several are rare. There are c. 3 Irish endemic spp.

Series 3. *Dispares* W. C. R. Watson. Stem arching, often angled, usually hairy, the prickles are usually numerous and *subsessile glands, short acicles and small pricklets are present and irregularly distributed on stem and panicle* usually being sparse on lower part of stem; longer glands, acicles and pricklets present or 0. Lflets (3–)5, green or grey- or white-felted beneath, often coarsely serrate. Stamens longer than styles.

A small group, first separated by Watson in the *Handbook*. It appears to be chiefly characterized by the presence of subsessile glands. No sp. is common and most are rare.

Series 4. *Radulae* Focke. Stems arching, usually rather robust, rooting at end, angled, with *prickles confined to angles; acicles and stalked glands numerous, short and ±equal,* sharply differentiated from the prickles, making the stem rough to the touch between the prickles; pricklets mostly few, equalling the acicles and glands. Lflets usually 5, grey or whitish beneath in most spp., but green in some, the basal pair stalked. Infl. usually well-developed, armed much like the stem but the arms more unequal, the glands mostly shorter than the hair. Sepals grey or greenish-grey. Stamens usually longer than styles. Fl. 7–8.

Mainly hedges, scrub and edges of woods.

A rather well-defined group; the clothing of the stem is characteristic in the commoner spp.

Several spp. are widespread and locally common. The 3 commonest can be distinguished as follows:

1 Stalked glands, acicles and pricklets all short and nearly equal; lvs finely but unequally serrate-dentate.   **R. radula** Weihe ex Boenn.
Stalked glands, acicles and pricklets more unequal; lvs coarsely and jaggedly compound-serrate.   2
2 Robust; stem hairy; infl. wide, petals wide, pink, entire; carpels glabrous.   **R. echinatus** Lindley
         (*R. discerptus* P. J. Mueller)
Slender; stem glabrous; infl. narrow; petals narrow, emarginate, pinkish; carpels pilose.
        R. echinatoides (Rogers) Dallmann

Series 5. *Apiculati* Focke. *Stem usually low-arching or procumbent and not very strong,* rooting at end, angled or subterete, the *prickles subequal to somewhat unequal and ±confined to the angles but not strictly so; pricklets, acicles and stalked glands* (or some of them) *numerous and ±unequal but not passing into the prickles.* Lflets 3–5, mostly green but often grey beneath, the basal pair stalked (occasionally subsessile). Infl. often wide with a truncate top, clothed with numerous stalked glands and a varying number of prickles. Many spp. have white petals and red styles. Sepals grey or greenish-grey, rarely with long points. Fl. 7–8.

Mainly woods and scrub.

A large group distinguished from series 1–4 by the much greater development of glands. Few spp. extend to Scotland and Ireland but a number are ±frequent in S.E. England. Two common, widespread and usually easily recognized spp. are:

R. flexuosus P. J. Mueller & Lefèvre (*R. foliosus* Weihe & Nees)
Stem deep purple with numerous purple glands, acicles and pricklets. Lflets 3(–5), thick, grey beneath, the terminal ±elliptical, gradually acuminate. Infl. frequently narrow with a zig-zag rhachis, but sometimes wide and pyramidal. Sepals long-pointed. Petals usually pink, narrow, fimbriate; filaments usually white; styles usually red or pinkish.

R. rufescens P. J. Mueller & Lefèvre
Lflets 3(–5), rather large, yellowish-green on both sides, densely and softly hairy beneath with glands on the midrib. Panicle pyramidal with long pedicels. Sepals erect and clasping. Petals clear pink, narrow; filaments white; styles red. C. and S. England.

Series 6. *Grandifolii* Focke. Stem arching, usually more robust than in *Apiculati*, rooting at end, ±angled, *the prickles more unequal and often less strictly confined to the angles than in the preceding sections; pricklets, acicles and stalked glands numerous and very unequal but not passing into the prickles.* Lflets 3–5, often rather sharply and irregularly serrate, the basal stalked. *Infl. well-developed, often pyramidal with racemose top* with numerous unequal stalked glands, and usually many prickles. *Sepals tapered at the apex into long points.* Fl. 7–8.

A group ±intermediate in armature between the *Radulae* and the *Hystrices* and not very clearly marked off from them or from the *Apiculati*, but generally with characteristic foliage and panicle.

No species is very common and some are rare.

Section 7. **Glandulosi** P. J. Mueller. Stem low-arching to procumbent, rooting at end. Stalked glands, acicles and pricklets numerous, passing into each other and into the prickles which are distributed all round the stem. Lflets 3–5, the basal pair stalked. Infl. densely glandular with unequal stalked glands, the longer ones usually exceeding the diam. of the pedicel. Sepals grey or greenish-grey.

Mainly in woods and scrub.

A group (medium-sized) fairly easily distinguishable by the armature of the stem. Contains 2 series formerly regarded as distinct sections.

Series 1. *Hystrices* Focke. Relatively robust. Stem usually low-arching, angled; largest prickles strong and broad-based. Infl. well-developed, the middle branches cymose. Fls usually large, often pink. Petals relatively wide. Stamens long.

Several spp. are widespread and frequent and the following is very common and widespread.

R. dasyphyllus (Rogers) E. S. Marshall

Stem red, hairy, with long glands, acicles and slender prickles. Lflets 3–5, thick, glabrous above, softly hairy beneath, unequally serrate. Infl. long, lax, narrow. Sepals spreading, then deflexed. Petals deep pink, hairy; filaments pink, styles salmon. Carpels hairy.

Series 2. *Euglandulosi* W. C. R. Watson. Slender. Stem usually procumbent, terete; prickles slender and weak. Infl. usually with middle branches racemose. Petals small, nearly always white. Stamens often short; styles often red.

Almost confined to woods.

None of the spp. is common. The series is not known north of Yorks in Great Britain and is rare in Ireland.

4. POTENTILLA L.

Perennial, rarely annual or biennial herbs, or small shrubs. Lvs palmate, pinnate or ternate. Fls solitary or in cymes, usually 5-, sometimes 4-merous, usually hermaphrodite. *Epicalyx present*; segments the same number as the sepals. *Stamens* 10–30 (often 20). *Carpels numerous*, indefinite (4–)10–80; *ovule pendent*; *style* jointed at its insertion, ±*lateral* but varying in position from nearly basal to nearly apical, *withered or not persistent in fr. Fr. a group of achenes*, inserted in the hemispherical or conical, *dry or spongy, receptacle*. Usually homogamous, nectar secreted by a ring on the receptacle.

About 500 spp., northern hemisphere (mainly temperate), a very few extending into the southern hemisphere, to Peru and New Guinea. Several spp. and hybrids are cultivated.

1	Lvs pinnate.	2
	Lvs palmate or ternate.	6
2	Shrub; lflets entire.	**1. fruticosa**
	Herbs, sometimes woody at base; lflets toothed.	3
3	Calyx and corolla purple; woody at base; bogs.	**2. palustris**
	Calyx green, corolla white or yellow; not woody; plants of relatively dry ground.	4
4	Fls in terminal cymes white or yellow.	5
	Fls solitary, yellow.	**5. anserina**
5	Fls white; petals longer than sepals.	**4. rupestris**
	Fls yellow; sepals at least as long as petals.	**8. norvegica**
6	Fls white; carpels hairy.	**3. sterilis**
	Fls yellow; carpels glabrous.	7
7	At least some fls with 4 sepals and 4 petals.	8
	Sepals and petals 5.	9

8	Sepals and petals usually 4; carpels 4–8(–20); most lvs with 3 lflets (but with stipules resembling lflets); stems decumbent to erect, not rooting at nodes.	**12. erecta**
	Sepals and petals 4–5; carpels 20–50; some lvs with 4–5 lflets; stems decumbent or prostrate, rooting at nodes in late summer.	**13. anglica**
9	Fls solitary, axillary; fl.-stems procumbent, rooting at nodes; most lvs with 5 lflets.	**14. reptans**
	Fls in terminal cymes; fl.-stems usually not rooting at nodes.	10
10	Lvs densely white-tomentose beneath.	**6. argentea**
	Lvs green on both sides.	11
11	Basal lvs ternate, rarely a few palmate with 4–5 lflets.	12
	Basal lvs mostly digitate with 5 or more lflets.	13
12	Petals longer than sepals; epicalyx-segments not longer than sepals in fr.	**11. crantzii**
	Sepals at least as long as petals; epicalyx-segments longer than sepals in fr.	**8. norvegica**
13	Fl.-stems terminal; rosettes of lvs absent or few, lateral.	14
	Fl.-stems lateral; stock with terminal rosettes of lvs, the plant often with numerous rosettes of lvs.	15
14	Petals 4–5 mm; sepals and epicalyx-segments accrescent, 15–20 mm in fr.	**9. intermedia**
	Petals 5–14 mm; sepals and epicalyx-segments not markedly accrescent.	**7. recta**
15	Stems from the woody stock long, usually rooting and forming mats; stipules of basal lvs with free part linear to linear-triangular; infl. scarcely raised above lvs.	**10. tabernaemontani**
	Stems from the woody stock short, not or scarcely rooting and never forming mats; stipules of basal lvs with free part lanceolate to ovate; infl. raised well above lvs.	**11. crantzii**

Subgenus 1. TRICHOTHALAMUS (Lehm.) Reichenb.

Shrubs. Lvs pinnate. Petals yellow or white. Carpels densely pilose. Style clavate (gradually widening upwards and contracted below the stigma), about as long as mature carpels, sub-basal. Receptacle hairy, not spongy.

1. P. fruticosa L.   Shrubby Cinquefoil.

Dioecious, deciduous shrub c. 1 m, ±pilose, much-branched; branches erect or ascending; bark peeling off in about the 3rd year. Lvs numerous, with (3–)5–(7) lflets; *lflets* 1–2 cm, oblong-lanceolate, acute, elliptical, *entire*, the margins revolute; stipules scarious, sheathing, entire, persisting for 2 or 3 years; petiole 5–10 mm. Fls few in a terminal cyme, or solitary, 5-merous, unisexual but with the sterile organs of both sexes conspicuous. Epicalyx-segments oblanceolate-linear, green, about equalling the triangular-ovate, yellowish sepals. Petals 8–12 mm, yellow, orbicular. Anthers oblong. Fl. 6–7. Visited by various insects. $2n = 28$. N.

Native. Damp rocky ground nearly always on basic rocks, ascending to 650 m in Ennerdale, very local; Upper Teesdale; Lake District (very rare); Clare, Galway, Mayo. S. Sweden, Estonia, Latvia, Urals, Caucasus, Maritime Alps, Pyrenees; N. and C. Asia to Armenia, Himalaya and Japan; North America from Labrador to Alaska south to New Jersey and (in the mountains) to California and New Mexico; Greenland.

In the Pyrenees plants are diploid ($2n = 14$) and bear hermaphrodite fls.

Subgenus 2. COMARUM (L.) Syme

Plants with a persistent woody base and annual herbaceous stems. Fl.-stems terminal. Lvs pinnate. Petals purple. Carpels glabrous or hairy. Style filiform, longer than mature carpels, lateral. Receptacle hairy, spongy.

2. P. palustris (L.) Scop.   Marsh Cinquefoil.

Comarum palustre L.; *P. comarum* Nestler

Rhizome woody, long-creeping. Stems 15–45 cm, ascending, dying back nearly to the base in winter. Lower lvs with 5 or 7 lflets; lflets 3–6 cm, oblong, sharply and coarsely serrate, subglaucous beneath, nearly glabrous to (rarely) densely villous beneath; petiole longer than blade; stipules scarious, adnate to petiole, long; upper lvs smaller and on shorter petioles, passing into the ternate bracts, their stipules green, short. Fls in a lax terminal cyme, 5-merous; fl.-stalks glandular-pubescent; upper bracts small (c. 1 cm). *Sepals* 1–1.5 cm, ovate, acuminate, *purplish*, accrescent; epicalyx-segments much smaller than sepals, linear. *Petals* ovate-lanceolate, acuminate, *deep purple*, shorter than sepals. Stamens, carpels and styles deep purple. Anthers ovate. Carpels glabrous. Fl. 5–7. Protandrous, visited by various insects, self-pollination apparently not possible. $2n = 28, 35, 42, 64$. Hel.

Native. Fens, marshes, where it is occasionally locally dominant, bogs, wet heaths, and moors throughout the British Is., ascending to 915 m in Perth; common except in S. and C. England. Europe from C. Spain, N. Italy, and S. Bulgaria northwards; Caucasus; temperate Asia to Armenia, Turkestan and Japan; North America south to New Jersey and N. California; Greenland.

Subgenus 3. FRAGARIASTRUM (Heister ex Fabr.) Reichenb.

Perennial herbs. Fl.-stems lateral, axillary from a terminal rosette. Lvs palmate or ternate. Petals white or pink. Carpels ±hairy. Style filiform, longer than mature carpels, subterminal. Receptacle hairy, not spongy.

3. P. sterilis (L.) Garcke   Barren Strawberry.

P. fragariastrum Pers.

Perennial herb 5–15 cm, softly pilose. Stock thick, oblique, somewhat woody, ending in a rosette of lvs, frequently emitting prostrate stolons. Fl.-stems axillary, very slender, decumbent, 1–3-fld. Rosette *lvs ternate*, with long petiole; lflets 0.5–2.5 cm, orbicular or broadly obovate, ±truncate at apex, crenate-dentate with 5–7 teeth on each side, the terminal tooth much smaller

than its neighbours, bluish-green and pilose above, paler and more densely pilose beneath, somewhat silky at least when young, shortly stalked; stipules ovate-lanceolate, scarious, yellowish, partly adnate to petiole; cauline lvs 1–2, smaller, otherwise similar to rosette-lvs. Fls 10–15 mm diam., 5-merous. Sepals triangular-ovate, acute; epicalyx-segments lanceolate, shorter than sepals. *Petals white*, obcordate, widely separated, about equalling or rather longer than sepals. Filaments glabrous. Carpels hairy only near the tip, the style scarcely longer. Fl. 2–5. Visited by various insects, self-pollination finally possible. $2n = 28$. Hs.

Native. Scrub, wood margins, open woods, etc., usually on rather dry soils, ascending to over 610 m. Throughout the British Is. except Shetland, common except in N. Scotland. W., C. and S. Europe, eastwards to Poland and Macedonia and northwards to S. Sweden; Asia Minor?; Algeria?

Distinguishable in fl. from *Fragaria vesca* by the blue-green lvs with spreading hairs beneath, small terminal tooth, distant petals, hairy receptacle and the different stolons.

Subgenus 4. POTENTILLA

Perennial herbs. Fl.-stems terminal. Lvs pinnate or palmate. Petals usually yellow. Carpels glabrous. Style fusiform, about equalling or longer than mature carpels, sub-basal. Receptacle hairy, not spongy.

4. P. rupestris L.   Rock Cinquefoil.

Pubescent perennial herb 20–50 cm. Stock thick, somewhat woody, branched. Fl.-stems terminal, erect, glandular-pubescent above. *Lvs of basal rosette 7–15 cm, pinnate* with 2–4 distant pairs of lflets, sometimes with smaller ones in addition, petiolate; lflets 2–6 cm, decreasing in size from apex to base of lf, ±ovate, obtuse, doubly dentate, the lateral asymmetrical at the base, green and pubescent on both sides; stipules ovate, adnate, subscarious. Cauline lvs few, the upper ternate, passing into the small simple upper bracts; stipules herbaceous. Fls several in a lax dichotomous cyme, 5-merous. Sepals ovate-oblong, mucronate; epicalyx-segments linear-lanceolate, much smaller than sepals. *Petals white*, 8–14 mm, obovate, somewhat longer than calyx. Fl. 5–6. Insect visits few, probably usually self-pollinated. $2n = 14$. Hs.

Native. Basic rocks in Montgomery and Radnor, extremely rare. W. and C. Europe and Balkan peninsula extending to S. Sweden, N. Italy and White Russia; Morocco (Atlas, very rare); subspp. in Asia Minor, Caucasus, W. and C. Asia (to c. 110° E) and in the Rocky Mountains from Canada to Colorado.

5. P. anserina L.   Silverweed.

Perennial ±silky herb with short, thick, simple or branched stock, ending in a rosette of lvs and emitting long (to 80 cm) creeping rooting and flowering stolons.

Lvs of basal rosette 5–25 cm, *pinnate*, with 7–12 pairs of main lflets, alternating with smaller ones; *lflets* 1–6 cm, the lower smaller, ovate or oblong, deeply and regularly serrate with narrow teeth, *silvery silky* on both sides or beneath only, rarely green and sparingly hairy or glabrous on both sides; smaller lflets 2–5 mm, 2–5-fid; stipules brownish, scarious, adnate, entire; lvs of stolons smaller, the upper much reduced or 0, the stipules herbaceous, multifid, connate at the base and shortly sheathing. *Fls 5-merous, solitary, axillary*, on long stalks. Sepals ovate; epicalyx-segments triangular-lanceolate, often toothed, about as long as sepals. *Petals yellow*, c. 1 cm, obovate. Style shorter than carpel. Fl. 6–8. Visited by various insects, self-incompatible. $2n = 28$, rarely 42 (low pollen fertility, no seed set). Hr.

Native. Waste places, roadsides, damp pastures, dunes, etc., ascending to 410 m; common throughout the British Is. Most of Europe except the extreme northeast and much of the south; N. and C. Asia to the Caucasus, N. Iran, W. Himalaya, Manchuria and Japan; Lebanon; Greenland; North America south to New Jersey and N. California; South America; Australia (Victoria), Tasmania, New Zealand.

6. P. argentea L.   Hoary Cinquefoil.

Perennial herb 15–50 cm with short thick branched stock. *Fl. stems terminal, decumbent or ascending,* ±tomentose. *Basal lvs palmate with 5 lflets; lflets* 1–3 cm, obovate-cuneiform, pinnately lobed with 2–5 lobes on each side, the lobes ±oblong, entire or more rarely with 1 or 2 small teeth, varying from green and glabrous to appressed pilose and greyish above, *densely whitetomentose beneath*, the margins narrowly recurved; petiole longer than blade; stipules brownish, scarious, adnate; cauline lvs similar but the upper sometimes ternate, the segments narrower, oblanceolate-cuneiform, the petioles short; upper lvs subsessile, the stipules herbaceous, ±ovate, acuminate, entire or with a few small teeth. Fls 10–15 mm diam., 5-merous, ±numerous in a tomentose dichotomous cyme; upper bracts small, entire, passing into lvs below. Calyx tomentose; sepals ovate; epicalyx-segments oblong-lanceolate, about as long as sepals. Petals yellow, obovate, slightly longer than the calyx. Carpels minutely rugulose; style conical at base, usually papillose, rather shorter than mature carpel; stigma dilated. Fl. 6–9. $2n = 14, 28, 35, 42, 56$. Some strains apomictic but pollination necessary for development, other strains amphimictic. Morphological differences exist between plants with different chromosome numbers and it is likely that several spp. or subspp. may ultimately be recognized. Hs.

Native. Dry sandy grassland from Moray and Cumberland southwards, local, especially so in the west, absent from Ireland and most of Wales; Jersey. Europe from Scandinavia to C. Spain, C. Italy and Greece; W. and C. Asia to Asia Minor, Turkestan and L. Baikal; North America from Nova Scotia to District of Columbia, Kansas and N. Dakota.

*7. P. recta L.   Sulphur Cinquefoil.

Perennial herb 30–70 cm with short thick branched stock. *Fl.-stems terminal, stiffly erect*, simple except in the infl., pubescent with short stiff hairs and sparingly villous with long flexuous white hairs. Basal and lower cauline *lvs palmate with 5–7 lflets*; lflets large (the middle one 5–10 cm, the basal shorter), oblong or oblanceolate, regularly serrate-dentate with 7–17 teeth on each side, green and appressed hairy on both sides, the veins prominent beneath; petiole very long; stipules long-adnate; lvs becoming smaller and petioles shorter upwards, the uppermost ternate, sessile; stipules of upper lvs green, shortly adnate, ovate-lanceolate, entire or cut. Infl. a many-fld dichotomous cyme, glandular-pubescent and villous; bracts simple, rather small. Fls 5-merous, 20–25 mm diam. Calyx glandular and ±densely villous; sepals ovate-lanceolate, acuminate; epicalyx-segments lanceolate, acute, about as long as sepals. Petals yellow, often sulphur-coloured, obovate-orbicular, deeply emarginate, as long as to much longer than the calyx. Carpels rugose; style rather thick, more so at the base, shorter than mature carpel; stigma slightly dilated. A very variable sp. on the Continent. Fl. 6–7. $2n = 28$, 42. Hs.

Introduced. A garden escape or casual sometimes becoming ±naturalized in waste or grassy places from Cornwall and Sussex northwards to Lancs and Yorks. C., E. and S. Europe; N. Africa (mountains); W. and C. Asia to c. 100° E, south to Asia Minor and N. Iran; casual in eastern North America.

*8. P. norvegica L.   Ternate-leaved Cinquefoil.

Annual, biennial or short-lived perennial herb 20–50 cm, with simple vertical root which is often rather thick in the biennial and perennial forms, ±hirsute and occasionally with scattered sessile glands. Stems 1 or several, terminal, erect or ascending, robust, often branched. *Lvs all ternate*; lflets 1–7 cm, obovate, elliptical or oblong, coarsely serrate or serrate-dentate, green on both sides, the lower on long petioles, the upper subsessile; stipules of lower lvs long-adnate, of upper large, ovate, entire or dentate. Infl. a dichotomous cyme, often much branched; bracts lf-like, the lower ternate, the upper simple; pedicels short or some of them longer (–2 cm). Fls 5-merous. Calyx 7–8 mm diam. in fl., accrescent in fr. to 15–20 mm, covered with long hairs; sepals ovate, acute; epicalyx-segments ±oblong, subobtuse, about as long as sepals in fl., becoming much longer in fr. Petals yellow, obovate, small, not longer than the calyx. Carpels rugulose; style with a very thick conical base, as long or shorter than mature carpels; stigma dilated. Fl. 6–9. $2n = 70$. Th. or Hs.

Introduced. Naturalized on waste ground in a number of places from Moray southwards, mostly in S.E. England; in Ireland apparently only in Down. N., C. and E. Europe; N. Asia to Kamchatka; naturalized in W. and S. Europe; a subsp. in North America from Newfoundland to S. Carolina, California and Alaska.

*9. P. intermedia L.

Biennial or perennial herb 20–50 cm finally developing a thick root. *Stems terminal*, robust, ascending from an arcuate base, dichotomously branched from low down, softly pubescent. *Lower lvs mostly palmate with 5 lflets*, a few ternate; lflets 1–4 cm, obovate or obovate-oblong, serrate-dentate or incised-serrate, green and softly pilose on both sides or more densely pilose and greyish beneath; petiole long; stipules small, shortly adnate; upper cauline lvs ternate, subsessile, the lflets oblong-lanceolate, the stipules ovate, mostly incised-dentate on the outer margin, rarely entire. *Infl. a lax many-fld corymbose dichotomous cyme*, softly pubescent; upper bracts simple, often trifid, the lower ternate; pedicels 0.5–2 cm. Fls 5-merous, c. 10 mm diam. Calyx pubescent and usually villous, not or scarcely accrescent; sepals ovate, acute; epicalyx-segments about as long as sepals, subobtuse or subacute. Petals yellow, about as long as calyx or rather shorter, obovate, slightly emarginate. Receptacle pilose. Carpels rugulose; style thickened and papillose at base, about as long as mature carpel. Fl. 6–9. $2n = 28$. Hs.

Introduced. A rather frequent casual, sometimes becoming ±naturalized. Native of N. and C. Russia; naturalized elsewhere in N. and C. Europe and in eastern North America.

*P. thuringiaca Bernh. ex Link, with a thick stock, 7–9 lflets and short style, is sometimes naturalized. C. Europe, S.W. Alps, N. and C. Italy.

10. P. tabernaemontani Ascherson   Spring Cinquefoil.

P. verna auct., non L.

Perennial herb 5–20 cm. *Stock thick, much branched, emitting prostrate usually rooting branches and forming mats. Fl.-stems axillary* from the lvs of a terminal rosette, decumbent, branched from below the middle, pubescent and villous with ascending hairs. *Basal lvs palmate with 5 lflets*; lflets 0.5–2 cm, obovate-cuneiform, dentate or deeply serrate with 2–9 teeth on each side, the terminal tooth markedly smaller than its neighbours, green and ±appressed hairy on both sides or subglabrous above; petiole long; stipules long-adnate, *the free part long and linear to linear-lanceolate*; cauline lvs few, the lower like those of the rosette, the upper ternate, subsessile; stipules ovate, ovate-lanceolate or oblong, acute to subobtuse, usually entire. Infl. a lax few-fld cyme, not or little raised above the lvs, pubescent and villous; pedicels slender, 1–2 cm; bracts simple. *Fls* 10–15 mm *diam.*, 5-merous. Calyx pilose; sepals ovate, acute; epicalyx-segments oblong, obtuse, shorter than sepals. Petals yellow, obovate, deeply emarginate, longer than the calyx. Receptacle pilose. Carpels rugose, not keeled; style clavate, shorter than mature carpel; stigma dilated. Fl. 4–6. $2n = 28, 42, 49, 63, 84$. Apomictic but pollination necessary for seed development. Hs.

Native. Dry basic grassland and rocky outcrops on sunny slopes from Banff and Roxburgh to Suffolk, Hants and Somerset, very local. N., W. and C. Europe, extending to the Baltic region, White Russia and Bulgaria.

11. P. crantzii (Crantz) G. Beck ex Fritsch

Alpine Cinquefoil.

P. alpestris Haenke.; *P. maculata* auct.; *P. salisburgensis* Haller fil.; *P. verna* L., nom. ambig.

Perennial herb 5–25 cm. *Stock thick, branched, emitting short branches not or scarcely rooting and never forming mats. Fl.-stems axillary* from the lvs of a terminal rosette, arching and ascending, often flexuous, branched only in the infl. above the middle of the stem, pubescent, the hairs crisped or spreading. *Basal lvs palmate with 5 lflets*; lflets 1–2 cm, obovate-cuneiform, dentate with 2–5 teeth on each side, the terminal tooth scarcely smaller than its neighbours, green on both sides, subglabrous or hairy above, hairy with ±spreading hairs especially on the veins beneath; petiole long; stipules long-adnate, *the free part ovate, often obtuse*; cauline lvs few, the upper ternate, subsessile, the stipules ovate, acute to subobtuse, usually entire. Infl. a lax few-fld cyme (or 1-fl.), raised well above the lvs, pubescent like the stem; pedicels long (1.5–3.5 cm) and slender; bracts simple. *Fls* 15–25 mm *diam.*, 5-merous. Calyx pilose; sepals triangular-ovate, acute; epicalyx-segments oblong or oblong-lanceolate, acute or obtuse, nearly as long as sepals. Petals yellow, often with an orange spot at base, obovate, deeply emarginate, longer than calyx. Receptacle pilose. Carpels rugulose or nearly smooth; style clavate, shorter than mature carpel; stigma dilated. Fl. 6–7. $2n = 42$ (49). Apomictic but pollination necessary for seed development. Hs.

Native. Mountain rock ledges and crevices and occasionally grassland, usually on basic soils from 80 to 1025 m, very local. N. Wales; Yorks and Lancs to Selkirk and Kirkcudbright; Argyll, Stirling and Angus to Skye and Sutherland. N. Europe, mountains of C. and S. Europe; N. Asia and North America but absent between c. 150° E and 130° W; Caucasus; Asia Minor, Iran.

Intermediates between *P. crantzii* and *P. tabernaemontani* in habit, flower-size and stipule-shape are known from Yorks, Northumberland and Aberdeen. Since the parents are pseudogamous apomicts, such plants apparently arise when an unreduced egg-cell is fertilized by normal (reduced) pollen.

12. P. erecta (L.) Räuschel   Tormentil.

P. tormentilla Stokes

Perennial herb; stock very thick (1–3 cm), woody, vertical to nearly horizontal, with reddish flesh, bearing a terminal *rosette of lvs which often withers and disappears before fl. time. Fl.-stems* (5–)10–30(–50) cm, several, axillary, slender, flexuous, decumbent to suberect, *never rooting, dichotomously branched above*, rather silkily laxly appressed-pilose. *Lvs all ternate* or rarely a few of the rosette-lvs palmate with 4 or 5 lflets; rosette-lvs on long petioles; lflets 5–10 mm, broadly obovate-cuneiform, coarsely dentate near the truncate apex, with 3–4 teeth on each side; stipules long-adnate, the free part lanceolate, entire; *cauline lvs sessile or subsessile*, the lflets 1–2 cm, narrowly obovate to oblanceolate-oblong, incised-serrate above the middle, with (2–)3–5(–6) teeth on each side, narrowly cuneate at base, green and glabrous or sparingly pilose above, appressed silky pilose on the margins and veins beneath, the stipules large, palmately lobed, appearing like extra lflets. *Fls many in terminal cymes*; *pedicels long and slender*, appressed pilose; bracts lf-like, the upper simple, the lower passing into the cauline lvs. *Fls 7–11(–15) mm diam., all 4-merous* (sometimes a few 3-, 5- or 6-merous). Calyx laxly appressed-pilose; sepals ovate-lanceolate, acute to subobtuse; epicalyx-segments linear-oblong, shorter to longer than sepals. Petals yellow, cuneiform, obovate or orbicular, emarginate, usually rather longer than calyx. Stamens 14–20, usually 16. Receptacle pilose. Carpels 4–8(–20), ovoid, rugose, obscurely keeled; style somewhat dilated at the base or at the apex or nearly equally slender throughout, about equalling mature carpel. Fl. 6–9. $2n = 28^*$. Hs.

Var. *sciaphila* (Zimm.) Druce is a small, slender decumbent form with the *cauline lvs shortly* (1–4 mm) *petiolate* and their *stipules entire* or bifid. It is liable to be confused with *P. anglica* but differs in the lvs all ternate and smaller fls. It deserves further study.

Native. Grassland, heaths, bogs, fens, mountain tops and sometimes open woods; absent or rare on heavy and strongly calcareous soils, very common on light acid ones; ascending to nearly 1067 m. Throughout the British Is. Almost throughout Europe, but rare in the Mediterranean region; W. Asia from W. Siberia to N. Asia Minor; Algeria, Morocco; Azores.

13. P. anglica Laicharding   Trailing Tormentil.
P. procumbens Sibth.

Perennial herb; stock rather thick (3–10 mm), branched, bearing a *persistent terminal rosette* of lvs. *Fl.-stems* 15–80 cm, axillary, slender, at first ascending, *soon decumbent, usually rooting at the nodes in late summer and autumn and producing new plants, usually dichotomously branched*, rarely simple, ±pilose. *Rosette lvs* on long petioles, *partly* (c. 50%) *palmate with 5 lflets, the remainder ternate or with 4 lflets*; lflets 0.5–2 cm, obovate-cuneiform, rounded or truncate at apex, coarsely dentate with 4–6 teeth on each side; stipules shortly adnate, the free part ovate or lanceolate, entire. *Cauline lvs shortly* (1–2 cm on the lower lvs, less on the upper) *petioled, mostly ternate* but some of the lower ones palmate with 4–5 lflets; lflets obovate or oblong-obovate (wider than in *P. erecta*) incised-serrate or serrate-dentate in the upper part, with 3–5(–6) teeth on each side, entire towards the cuneate base, subglabrous on both sides or ±pilose beneath; *stipules entire* or a few bifid or trifid. *Fls solitary on long slender pedicels, the upper*

often forming a few-fld cyme, the upper bracts smaller than the lvs but rarely simple. *Fls 4-merous* (c. 75%) *and 5-merous mixed*, (10–)14–18 mm *diam*. Sepals ovate-lanceolate, acute; epicalyx-segments linear-oblong or lanceolate, about as long as sepals. Petals yellow, obovate, emarginate, usually nearly twice as long as calyx. Stamens 15–20. Receptacle pilose. Carpels 20–50, oblong-ovoid, rugose; style slightly dilated at apex, about equalling mature carpel. Fl. 6–9. $2n = 56^*$. Hr. or Hs.

Native. Woods, especially their edges, heaths, hedge-banks, etc., on similar soils to *P. erecta* but less tolerant of extreme acid conditions and not recorded above 390 m, widespread but rather local, becoming very rare in N. Scotland and absent north of Perth and Argyll. W. and C. Europe, extending to S.W. Finland and the extreme west of the USSR; Madeira, Azores.

P. × suberecta Zimm. (*P. anglica × erecta*), intermediate between the parents in such characters as lflet and petal numbers, petiole length and fl. diam., has up to c. 40% fertile pollen and usually forms a few achenes. It is widespread in the British Is. wherever the parents occur together.

14. P. reptans L.   Creeping Cinquefoil.
Perennial herb; stock rather thick (2–6 mm), branched, bearing a persistent terminal rosette of lvs. *Fl.-stems* 30–100 cm or more, axillary, stoloniform, *prostrate, quickly rooting at the nodes and producing new plants, simple*, pubescent or subglabrous. *Rosette-lvs* on long petioles, *mostly palmate with 5(–7) lflets*, only a few ternate or with 4 lflets; lflets 0.5–3 cm, obovate or oblong-obovate, dentate in the upper part or all round, with 6–10 teeth on each side, subglabrous or sparingly pilose on both sides; stipules long-adnate, the free part lanceolate, entire. *Cauline lvs scarcely differing* but with rather shorter petioles, mostly with 5 lflets, a few of the uppermost often with 3–4 lflets; *stipules* herbaceous, usually *entire*, rarely dentate. *Fls solitary* on long slender pedicels, never forming a cyme. *Fls 5-merous*, 17–25 mm *diam*. Calyx often accrescent; sepals variable in shape, usually acute; epicalyx-segments usually obtuse, varying in length. Petals yellow, obovate, emarginate, to twice as long as calyx. Stamens c. 20. Receptacle pilose. Carpels 60–120 oblong-ovoid, rugose; style slightly dilated at apex, about as long as or rather shorter than mature carpel. Fl. 6–9. $2n = 28$. Hr.

Native. Hedgebanks, waste places and sometimes grassland mainly on basic and neutral soils, ascending to 410 m. Common throughout most of the British Is., rare in C. and N. Scotland, absent from Orkney and Shetland. Europe, except the extreme north; N. Africa; W. Siberia, Turkestan, Iran, Himalaya; introduced in North and South America.

P. × mixta Nolte ex Reichenb. (*P. erecta × reptans*; *P. anglica × reptans*) is widely distributed in the British Is. It is like *P. reptans* but has 3- and 4-nate, as well as

5-nate, lvs, 4- and 5-merous fls and a low (up to 10%) pollen fertility. Experimental hybridizations show that either or both of the parentages given above are possible.

5. SIBBALDIA L.

Small perennial herbs. *Lvs palmate or ternate. Fls in cymes, 5-merous, hermaphrodite, occasionally unisexual. Epicalyx present. Stamens 5, rarely 4 or 10. Carpels 5–12,* otherwise as in *Potentilla.* Nectar secreted by the disk.

About 20 spp., all except the following in the higher mountains of Asia.

1. S. procumbens L.   Sibbaldia.

Potentilla sibbaldii Haller fil.; *P. procumbens* (L.) Clairv., non Sibth.

Compact perennial tufted herb 1–3(–10) cm, clothed with stiff appressed hairs. Stock woody, branched, each branch ending in a rosette of lvs. Fl.-stems axillary, often shorter than lvs. Lvs ternate, stalked; lflets 0.5–2 cm, obovate-cuneiform, tridentate at the truncate apex, the central tooth markedly smaller than the lateral, bluish-green, often purple-tinged beneath; stipules adnate, the free part ovate-lanceolate, entire, often purplish. Fls few (2–5), in a rather dense cyme, c. 7 cm diam. Sepals oblong-lanceolate, acute, often purple-tinged; epicalyx-segments linear-lanceolate, shorter than sepals. Petals yellow or yellowish-green, small, narrow, inconspicuous, or 0. Carpels rugulose, style basal. Fl. 7–8. Visited by flies, ants, etc., homogamous, self-pollination apparently difficult. $2n = 14$. Hr.

Native. Mountain tops and grassland, rock crevices, etc., from 470 m to over 1350 m. Westmorland; Kirkcudbright, Peebles; Dunbarton and Stirling to Sutherland on the mainland; Shetland, Skye. N. Europe, mountains of C. and S. Europe; Greenland; arctic Asia and North America (in the mountains to New Hampshire and California).

6. FRAGARIA L.

Perennial stoloniferous herbs. Lvs ternate, in a basal rosette. Fls in few-fld cymes on axillary scapes, 5-merous. *Epicalyx present. Stamens numerous (c. 20). Receptacle glabrous. Carpels as in Potentilla, glabrous. Fr. a group of achenes on the surface of a much enlarged, fleshy, juicy, brightly coloured receptacle.* Protogynous, nectar secreted by a ring on the receptacle.

About 15 spp., north temperate and subtropical and in South America.

> 1 Lflets hairy above, the terminal one cuneate at base; fls 12–20(–25) mm diam.; fr. 2 cm or less, with projecting achenes.   2
> Lflets usually glabrous above, the terminal one rounded at base; fls 20–35 mm diam.; fr. c. 3 cm, with the achenes sunk in the flesh.   **3. × ananassa**
> 2 Pedicels (at least the upper) with appressed hairs; lateral lflets sessile or nearly so; fr. with achenes all over.   **1. vesca**

Pedicels with spreading hairs; lateral lflets stalked; fr. without achenes at base.   **2. moschata**

1. F. vesca L.   Wild Strawberry.

Perennial herb 5–30 cm; stock rather thick and woody, producing very long arching runners, rooting at the nodes and forming fresh plants. *Lflets 1–6 cm, ovate, obovate or oblong, coarsely serrate-dentate, the terminal tooth not or scarcely shorter (though sometimes narrower) than its neighbours, bright green above and somewhat pilose,* pale and glaucous beneath and clothed with silky appressed hairs, the lateral sessile or subsessile, the terminal one sessile or shortly stalked, cuneate at base; petiole long, clothed with spreading white hairs; stipules scarious, often purplish. Fl.-stems ±erect, not much exceeding lvs, clothed with spreading hairs; cauline lvs 0 but the lowest bracts usually lf-like; upper bracts ovate-lanceolate, entire; *pedicels, at least the upper, clothed with ±appressed hairs. Fls 12–18 mm diam.,* hermaphrodite. *Calyx spreading or deflexed in fr.;* sepals ovate, acuminate; epicalyx-segments ±lanceolate, acute, about as long as sepals. Petals white, obovate, obtuse, nearly contiguous to overlapping. Fr. receptacle 1–2 cm, ovoid or subglobose, red (rarely white), *covered all over with achenes which project from the surface.* Fl. 4–7. Visited by various insects. $2n = 14$. Hr.

Native. Woods and scrub on base-rich soils and on basic grassland, sometimes becoming locally dominant in woods on calcareous soils; ascending to 730 m. Almost throughout Europe (except Crete, Balearic Is., Spitzbergen and the Faroes); Asia to L. Baikal; Madeira, Azores; apparently native also in eastern North America; introduced in other parts of the world. The Alpine Strawberry, still sometimes grown for its fr., is a form of this sp.

*2. F. moschata Duchesne   Hautbois Strawberry.

F. elatior Ehrh.

Differs from *F. vesca* as follows: more robust, 10–40 cm high. Stolons few or 0. Fl.-stems much exceeding lvs. *Lflets all shortly stalked. Pedicels clothed with spreading or somewhat deflexed hairs. Fls 15–25 mm diam.,* usually unisexual. *Fr. receptacle without achenes at the base* and more contracted below, purplish-red or partly greenish, of a musky flavour. Fl. 4–7. Visited by various insects. $2n = 42$. Hr.

Introduced. Formerly grown for its fruit but now seldom seen. Reported as naturalized in many places in England and a few in Wales and Scotland but many of the records probably refer to *F. × ananassa.* Native of C. Europe from N. France, C. Germany and N. Russia to N. Italy, Turkey and C. Russia (Volga-Don region).

*3. F. × ananassa Duchesne
(*F. chiloensis × virginiana*).   Garden Strawberry.

F. chiloensis auct. angl.; *F. grandiflora* Ehrh., non Crantz.

Habit of *F. vesca* but much larger. *Lflets* 5–8 cm, orbicular or ovate, coarsely serrate-dentate, somewhat *bluish-green and glabrous above*, pale beneath and somewhat appressed hairy, *all stalked, the terminal one rounded at the base*; petiole long, sparsely hairy. Fl.-stems ascending in fl., usually decumbent in fr., not or scarcely longer than the lvs, sparingly hairy, the hairs or some of them appressed; bracts not lf-like. *Fls* relatively numerous, 20–35 mm *diam.*, hermaphrodite, often more than 5-merous. *Calyx appressed to the fr. Fr. receptacle* c. 3 cm *covered all over with achenes which do not project from the surface.* Fl. 5–7. $2n = 56$. Hr.

Introduced. The commonly cultivated strawberry originated in France as a hybrid between the two octoploid spp., *F. virginiana* from eastern North America and *F. chiloensis* (L.) Duchesne from Chile. It frequently escapes and often occurs ±naturalized on railway banks, etc.

F. virginiana Duchesne. Forms of this (e.g. 'Little Scarlet') are still sometimes grown for preserves. The fr. is much smaller than in *F. × ananassa* and with pink flesh and the fr. calyx spreading. It has not been reported as escaped.

*Duchesnea indica (Andrews) Focke.

Resembling *Fragaria*. Spreading by runners. Lvs ternate. Lflets rhombic or elliptical, ±hairy. Fls solitary, axillary, on long stalks, bright yellow. Epicalyx segs in fr. c. 5 × 5 mm, toothed at apex. Fr. receptacle red, but dry and insipid.

Grown in gardens and rarely an escape. ?Native of mountains of Asia from Afghanistan to Indonesia and Japan.

7. Geum L.

Perennial herbs. Lvs unequally pinnate. *Fls* solitary or cymose, 5-*merous*. *Epicalyx present.* Stamens numerous (20 or more). Receptacle flat. Carpels numerous (20 or more); *ovule* 1, *erect*; *style terminal, enlarged and* (all or the lower part) *persistent as an awn on the fr.* Fr. a group of achenes on a dry receptacle. Protogynous fls; nectar secreted by receptacle.

About 40 spp., temperate regions of both hemispheres (a few arctic). Cultivated forms or hybrids of *G. coccineum* Sibth. & Sm., often with double red fls, are much grown in gardens. The awn on our spp. is hooked and the fr. is dispersed by animals.

Fls erect; petals 5–9 mm, yellow, spreading, not emarginate or clawed.   **1. urbanum**
Fls nodding; petals 10–15 mm, reddish, erect, emarginate, clawed.   **2. rivale**

1. G. urbanum L.   Herb Bennet, Wood Avens.

Perennial herb 20–60 cm, ±pubescent; rhizome short, thick. Stems ±erect. Rosette-lvs pinnate, with 2–3 pairs of unequal lateral lflets 5–10 mm long and large suborbicular lobed terminal lflet 5–8 cm, or with the upper pair of laterals also large; lflets all crenate or dentate. Cauline lvs large, the lower like the rosette-lvs or ternate or deeply trilobed, the upper usually simple; *stipules large* (1–3 cm), lf-like with large irregular triangular teeth or lobes. *Fls erect*, few, on long stalks in very

open cymes. *Calyx green*; sepals triangular-lanceolate; epicalyx-segments oblong-linear, c. ⅓ as long as sepals. *Petals* 5–9 mm, *yellow, spreading, obovate or oblong, not clawed or emarginate*, about as long as sepals. *Carpels hirsute, remaining as a sessile head in fr.*; awn purplish 5–10 mm, jointed near the apex, the lower part hooked and persistent, glabrous throughout. Fl. 6–8. Visited by various insects but visitors few, self-pollination probably usual. $2n = 42$. Hs.

Native. Woods, scrub, hedgebanks and shady places on good damp soils; ascending to 520 m. Common throughout the British Is., but absent from Shetland. Most of Europe, except the extreme north, Azores, Balearic Is. and Crete; W. Asia from W. Siberia to the Himalaya and Syria; N. Africa.

2. G. rivale L.   Water Avens.

Perennial herb 20–60 cm, pubescent; rhizome short, thick. Rosette-lvs pinnate, with 3–6 pairs of unequal lateral lflets 2–20 mm long and large suborbicular lobed terminal lflet 2–5 cm; lflets all ±dentate. Cauline lvs few, small, simple or with a few very small lateral lflets, often trilobed; *stipules small* (c. 5 mm) green but scarcely lf-like, dentate or entire. *Fls nodding*, few, in a narrow cyme. *Calyx purple*; sepals triangular-lanceolate; epicalyx-segments linear, c. ⅓ as long as sepals. *Petals* 1–1.5 cm, *dull orange-pink, erect, ±spathulate with a long claw, retuse or emarginate*, about as long as sepals. *Carpels* hirsute, *head becoming stalked in fr.*; awn jointed rather above the middle, the lower part glabrous, hooked and persistent, the upper joint hairy. Fl. 5–9. Visited by various insects, especially humble-bees; self-pollination possible if insect visits fail. $2n = 42$. Hs.

Native. Marshes, streamsides, wet rock-ledges, and damp woods on base-rich soil, most frequently in shade; ascending to 975 m. Widespread and rather common in Scotland, N. England and Wales (absent only from the Outer Hebrides and the Isle of Man), local in S. England and very rare in the south-east; widespread in Ireland but rather local. Most of Europe except the Mediterranean region; Caucasus; Asia Minor, Siberia (to the Yenisei region); North America from Newfoundland to British Columbia, Colorado and New Jersey.

G. × intermedium Ehrh. (*G. rivale × urbanum*) is usually to be found when the parents grow together, though preferring *rivale* habitats. It is highly fertile and forms hybrid swarms which frequently link the parents by a range of intermediates.

*G. macrophyllum Willd.

Resembling *G. urbanum* but more robust. Rosette-lvs with 5–7 pairs of lateral lflets, the terminal lflet often more rounded. Stipules of stem-lvs smaller and less toothed or entire. Fls usually rather larger. Carpels more numerous, bristly at apex; lower part of style minutely glandular.

Cultivated in gardens and rarely naturalized. Native of North America and N.E. Asia.

8. DRYAS L.

Evergreen prostrate dwarf shrubs. Lvs simple. *Fls* solitary, axillary, hermaphrodite or unisexual, 7–10- (*often 8-)merous. Epicalyx* 0. Stamens c. 20. Receptacle slightly concave. Carpels numerous; *ovule* 1, *erect*; *style terminal, persistent on the fr.* and covered with long white feathery hairs. Fr. a group of achenes on a dry receptacle. Nectar secreted by a ring below the stamens.

Two spp., Arctic and mountains of north temperate region.

1. D. octopetala L.   Mountain Avens.

Much branched, tortuous, creeping dwarf shrub. Lvs numerous, blade 0.5–2 cm, oblong or ovate-oblong, obtuse, rounded or truncate at base, stalked, deeply crenate or crenate-dentate, dark green and glabrous above with impressed veins, densely white-tomentose beneath; stipules scarious, brownish, adnate. Fl.-stalk 2–8 cm, erect, tomentose, with blackish glandular hairs above. Fls 2.5–4 cm diam. Sepals oblong, tomentose and with blackish glandular hairs. Petals oblong, white. Fl. 6–7. Homogamous or nearly so, visited by various insects, self-pollination possible. $2n = 18^*$. Chw.

Native. Ledges and crevices on mountains of basic rocks, local; ascending to 975 m, descending to sea-level in Sutherland and Clare. Caernarvon (very rare), W. Yorks, Westmorland, Cumberland; Argyll and Perth to ?Orkney; N. and W. Ireland from Derry to Clare. Arctic and subarctic Europe, Asia, America; high mountains of Europe south to N. Spain, C. Italy and S. Bulgaria; Rocky Mountains, south to Colorado.

Tribe 4. SANGUISORBEAE. Receptacle deeply concave, enclosing the carpels and (usually) becoming dry and hard in fr. Carpels 1–4, often 2. Filaments narrowed at base. Fr. of 1–2 (rarely more) achenes enclosed in the persistent receptacle and shed with it.

9. AGRIMONIA L.

Perennial, rhizomatous herbs. Lvs unequally pinnate. Fls in terminal spike-like racemes, 5-merous, hermaphrodite. *Epicalyx* 0. *Petals* 5. Stamens 10–20. *Receptacle covered above with small spines*, becoming hard in fr. Carpels 2; *styles terminal*; ovule 1, pendent.

About 15 spp., in north temperate regions and South America.

Lvs not (or slightly) glandular beneath; fr. receptacle obconical, deeply grooved throughout, the basal spines spreading laterally.   **1. eupatoria**
Lvs with numerous sessile glands beneath; fr. receptacle campanulate, without grooves at the base, the basal spines deflexed.   **2. procera**

1. A. eupatoria L.   Agrimony.

Erect perennial herb 30–60 cm, ±villous, not or sparingly glandular. Stems usually simple, often reddish, densely lfy below, sparsely so above. Lower *lvs* pinnate, with 3–6 pairs of main lflets which become larger upwards, and 2–3 pairs of smaller lflets between each pair of main ones; largest lflets 2–6 cm, ±elliptical,

deeply and coarsely serrate or serrate-dentate, usually densely villous and often greyish beneath, *not glandular*; upper lvs smaller with fewer lflets; stipules lf-like. Fls numerous, c. 5–8 mm diam., yellow. *Fr. receptacle obconical, deeply grooved almost throughout its length*, covered above with hooked spines, the *lowest spines ascending or spreading horizontally*.

Var. *sepium* Bréb. is a more robust plant resembling *A. procera* in habit, in being distinctly aromatic and somewhat glandular. The glands on the lower surface of the lf are, however, absent or very sparse and the spines of the receptacle are those of *A. eupatoria*.

Fl. 6–8. Visited by Diptera and Hymenoptera, self-pollination frequent. $2n = 28^*$. Hs.

Native. Hedgebanks, roadsides, edges of fields, etc.; ascending to 490 m. Common throughout the British Is. except N. Scotland where it is rare (absent from Orkney and Shetland). Almost throughout Europe, except Iceland, Crete, the Faeroes and Spitzbergen; Asia Minor, Iran; N. Africa; Macaronesia.

2. A. procera Wallr.   Fragrant Agrimony.
A. odorata auct., non Miller

Differs from *A. eupatoria* as follows: more robust than most forms of that sp. (often 1 m high), sweet-smelling. Stems often more lfy. *Lvs* larger; lflets larger and relatively narrower, with narrower more acute teeth, green and usually less hairy beneath, and always *with numerous small sessile shining glands* (which also occur on other parts of the plant). Fls larger (to 1 cm diam.). *Fr. receptacle campanulate, more shallowly grooved, the grooves ceasing well above the base*, or almost without grooves; *lowest spines deflexed* (the others erect or ascending). Fl. 6–8. $2n = 56^*$. Hs.

Native. In similar places to *A. eupatoria* but usually absent from calcareous soils. From Argyll and Inverness southwards, most common in S. England; widespread but local in Ireland and rare in the Central Plain; Channel Is. Europe from S. Scandinavia and Finland to N. Portugal, Spain, N. Italy, Macedonia and S.W. Russia (but absent from many areas).

A. × wirtgenii Ascherson & Graebner (*A. eupatoria × procera*), which is intermediate between the parents in indumentum, petal length and the toothing of the lflet-margins, was recorded in the 1940s in S. Northumberland. No fruits are formed.

10. AREMONIA DC.

Differs from *Agrimonia* as follows: fls in few-fld cymes, each fl. surrounded by an involucre. *Epicalyx present. Receptacle without spines.* Stamens 5–10.

One sp.

*1. A. agrimonoides (L.) DC.

Agrimonia agrimonoides L.

Perennial herb 20–40 cm; stem ascending. Lvs pinnate, with 2–4 pairs of main lflets, the lower smaller than the upper, with smaller ones between; lflets serrate-dentate. Fls 7–10 mm diam., yellow. Involucre with

6–10 entire sepal-like lobes, concealing the receptacle. Fl. 6–7. Hs.

Introduced. Naturalized in woods in C. Scotland. Native of S. and C. Europe, extending from Sicily and S.W. Germany eastwards, and northwards to Czechoslovakia; N. Asia Minor.

11. ALCHEMILLA L.

Perennial herbs. Lvs palmate or palmately lobed. Fls ±numerous, in cymes, 4-merous, small, green. *Epicalyx present. Petals* 0. *Stamens* 4(–5), *alternate with sepals, inserted on the outer margin of the disk; anthers introrse.* Receptacle not spiny, small and dry in fr. Carpel 1; *style basal;* stigma capitate; ovule 1, erect.

About 250 spp., north temperate zone and the mountains of tropical Africa.

Nectar secreted by a ring on the receptacle, and fls sometimes visited by various insects, but most of the European spp. are apomictic.

1 Lvs silvery-silky beneath, palmately divided to more than halfway.	2
   Lvs green on both sides, palmately divided to halfway or less.	(*A. vulgaris sens. lat.*)	3
2 Lf-segments 5–7, 3–6 mm wide, free to the base or almost so, shiny green above.	**1. alpina**
   Lf-segments 7(–9), 6–15 mm wide, usually joined towards their bases, dull green above.	**2. conjuncta**
3 Stem and petioles with appressed or subappressed hairs.	4
   Stem and petioles with spreading hairs.	6
4 Stem and petioles with dense silky hairs; lvs undulate, hairy above; fl.-clusters dense.	**11. glomerulans**
   Stem and petioles subglabrous, hairy only below; lvs glabrous or almost so; fl.-clusters lax.	5
5 Lvs ±reniform, with open basal sinus; lf-lobes with few, wide, unequal, not connivent teeth, the teeth extending to the base of the lobes.	**12. glabra**
   Lvs orbicular, with ±closed basal sinus; lf-lobes with numerous, narrow, equal, connivent teeth, the sinuses between them with a basal, toothless, V-shaped prolongation.	**13. wichurae**
6 Receptacle ±densely hairy.	7
   Receptacle glabrous or sparsely hairy.	12
7 All pedicels ±densely hairy.	8
   At least some pedicels glabrous or subglabrous.	9
8 Lvs ±orbicular; lf-lobes with 4–6 teeth; petioles green at base.	**3. glaucescens**
   Lvs ±reniform; lf-lobes with 6–9 teeth; petioles reddish at base.	**9. filicaulis**
9 Dwarf plant up to 15 cm.	10
   Medium-sized plant up to 50 cm.	11
10 Sinuses between lf-lobes with basal, toothless, V-shaped prolongation; base of petiole not reddish.	**10. minima**
   Lf-lobes toothed to base of sinus; base of petiole reddish.	**9. filicaulis**
11 Lvs ±orbicular, with basal sinus closed or nearly so, densely and evenly hairy above; lobes with regular teeth and short basal toothless prolongation of sinuses between them.	**4. monticola**
   Lvs ±reniform, with wide basal sinus, unevenly hairy above; lobes with irregular teeth extending to their base.	**9. filicaulis**

12 Epicalyx-segments at least as long as sepals; mature achene much longer than receptacle.	**14. mollis**
   Epicalyx-segments usually shorter than the sepals; mature achene as long as receptacle.	13
13 Dwarf plant up to 15 cm.	14
   Medium-sized plant up to 50 cm.	15
14 Sinuses between lf-lobes with basal, toothless, V-shaped prolongation; base of petiole not reddish.	**10. minima**
   Lf-lobes toothed to base of sinus; base of petiole reddish.	**9. filicaulis**
15 Lvs glabrous or subglabrous above; fls up to 3 mm diam.	**8. xanthochlora**
   Lvs hairy above, at least in the folds; fls usually more than 3 mm diam.	16
16 Hairs on stems and petioles all patent or erecto-patent, none deflexed.	18
   At least some hairs on stems and petioles deflexed.	17
17 Lvs rather sparsely hairy above, often only on folds; fls 2.5–3.5 mm diam.	**6. subcrenata**
   Lvs densely and evenly hairy above; fls 1.5–2.5 mm diam.	**5. tytthantha**
18 All lvs densely and evenly hairy on both surfaces.	**4. monticola**
   Some lvs sparsely or unevenly hairy above.	19
19 Lf-lobes almost triangular; base of petiole not reddish; receptacle glabrous.	**7. acutiloba**
   Lf-lobes ±rounded; base of petiole reddish; receptacle usually somewhat hairy.	**9. filicaulis**

1. A. alpina L.	Alpine Lady's-mantle.

A. alpina subsp. *glomerata* (Tausch) Camus

Perennial herb 10–20 cm; stock somewhat woody, rather thick, branched, shortly creeping; lvs mostly basal; stems ascending, with few lvs. *Rosette-lvs with blades* (Fig. 38A) 2.5–3.5 cm diam., orbicular or reniform in

A	B	C

Fig. 38. Leaves of *Alchemilla*. A, *A. alpina*; B, *A. conjuncta*; C, *A. glaucescens*. ×½.

outline, *palmately divided almost or quite to the base into 5–7 oblanceolate-oblong segments*, green and glabrous above, *densely silvery-silky beneath; segments* 1–2 cm × 3–6 mm, sharply serrate at the extreme apex; petiole long; stipules brown, scarious. Cauline lvs few, small, sometimes with only 3 segments, on short petioles. Stems, fl.stalks, receptacle and calyx appressed silky. Fls c. 3 mm long and 3 mm diam., in rather dense clusters forming a terminal cyme. Fl. 6–8. $2n = $ c. 120. Hs.

Native. Mountain grassland, where it is sometimes locally dominant, rock crevices, screes and mountain tops; ascending to over 1215 m in the Cairngorms, descending nearly to sea-level in Skye. Yorks, Westmorland, Cumberland; Stirling and Arran northwards, widespread and locally abundant; Kerry, Wicklow, very rare. N., W. and W.C. Europe, southwards to S. Spain, the Apennines and E. Alps, only on mountains, except in the extreme north; Greenland.

*2. A. conjuncta Bab.

A. argentea (Trevelyan) D. Don, non Lam.

Similar to *A. alpina* but often more robust, reaching 40 cm. *Rosette-lvs* with *blades* (Fig. 38B) 4–8 cm diam., palmately divided to c. $\frac{2}{3}$–$\frac{4}{5}$ (the sinuses varying in depth on the same lf) *into 7–9 obovate or oblong segments; segments 6–15 mm wide*, the serrations less sharp and extending further down the lf than in *A. alpina*. Fls larger (c. 4 mm diam.), the clusters often larger. Fl. 6–7. Hs.

Widely cultivated in gardens and naturalized among rocks by streams in Glen Clova (Angus) and Glen Sannox (Arran), very rare; sometimes occurring as an escape elsewhere. Native of the S.W. Alps and Jura.

(3–13). A. vulgaris *sens. lat.* Lady's-mantle.

Perennial herb 5–45 cm, variously hairy to nearly glabrous; stock very thick and woody; lvs mostly in a basal rosette; stems ascending or decumbent with few lvs. *Rosette-lvs* with *blades* 1–15 cm diam., orbicular or reniform in outline, *palmately lobed to $\frac{1}{2}$ or less, green on both sides*; lobes (5–)7–11, wide, serrate, the teeth with a tuft of hairs even on otherwise glabrous lvs; petiole long; stipules scarious, brownish, sometimes purplish tinged. Cauline lvs few, smaller, on short petioles. Fls 3–4 mm diam., in a compound terminal cyme, made up of dense or lax small cymes. Fl. 6–9. Hs.

Native. Damp grassland, open woods, rock-ledges, etc., especially on basic and neutral soils; ascending to nearly 1215 m. Almost throughout the British Is., rather common in the north and west, becoming rare in S.E. England and absent from the Channel Is. Europe, N. and W. Asia, Greenland, eastern North America.

The following spp. are in general quite clear-cut. The most useful characters are derived from the nature and distribution of the hairs, the shape of the lvs and their lobing and toothing, and to a lesser extent the infl. and fls, and the habit. In the following descriptions 'lvs', if not qualified, refers to well-developed summer rosette-lvs. It should be noted that when hairs are said to be 'spreading' they all stick out from the stem at 45° or more, if 'appressed' most of the hairs are ±closely appressed to the stem though occasional individual hairs may stick out.

3. A. glaucescens Wallr.

A. hybrida auct., non (L.) L.; *A. minor* auct.; *A. anglica* Rothm.; *A. pubescens* auct., non Lam.

Plant small. *Stems, pedicels, receptacle and petioles densely clothed with spreading silky hairs. Lvs* (Fig. 38c) *hairy on both surfaces*, orbicular in outline, *the basal sinus closed or nearly so*; lobes (5–)7–9, wide (c. twice as wide as long), rounded, often overlapping; *teeth 9–11 on each of the middle lobes, almost straight, subobtuse.* Fls c. 3 mm, in dense clusters, almost silvery outside. $2n = 103$–110.

Native. Limestone grassland in Yorks (Ingleborough area); W. Ross, W. Sutherland; Leitrim; an escape in one or two other places. C. and N.E. Europe, extending locally to C. France, N. Italy, Bulgaria and Crimea. Quebec (introduced).

4. A. monticola Opiz

A. pastoralis Buser

Plant medium-sized. *Stem, petioles, and usually peduncles, with spreading hairs; pedicels glabrous;* receptacles usually hairy. *Lvs* (Fig. 39A) orbicular in outline, *densely*

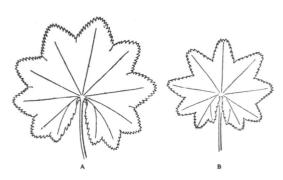

Fig. 39. Leaves of *Alchemilla*. A, *A. monticola*; B, *A. acutiloba*. ×$\frac{1}{2}$.

hairy on both surfaces, the basal sinus closed; lobes 9–11, the two basal often small, wider than long, ±rounded, the teeth 15–19 on each of the middle lobes, acute, somewhat curved towards the apex. Fls 2–3 mm diam., clusters dense. $2n = 101$; c. 103–9.

Native. Grassland in Upper Teesdale and Weardale; a casual introduction in Surrey and Bucks. Most of Europe, except the islands, rare or local in the west, and only on mountains in the south, east to W. Siberia.

A. gracilis Opiz is like *A. monticola* but the hairs are erecto-patent or even sub-appressed, the receptacles are always glabrous and the lvs are more nearly reniform. A native of N.E. and C. Europe, recently recorded from several places in Northumberland; recorded also (single specimen) from Upper Teesdale, but not re-found there.

*5. A. tytthantha Juz.

A. multiflora Buser ex Rothm.

Plant up to 50 cm. *Lvs* suborbicular, rather shallowly lobed, with a narrow basal sinus, *densely and evenly hairy above* and beneath; lobes c. 9, rounded to almost triangular, with 6–7 small, acute, subequal teeth; petioles and stems with ±deflexed hairs. *Fls* 1.5–2.5 mm diam.

Introduced. Locally naturalized in S. Scotland (Selkirk, Berwick and Stirling). Native of the Crimea.

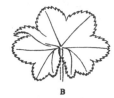

Fig. 40. Leaves of *Alchemilla*. A, *A. filicaulis*; B, *A. subcrenata*; C, *A. minima*. ×$\frac{1}{2}$.

Fig. 41. Leaves of *Alchemilla*. A, *A. xanthochlora*; B, *A. glomerulans*; C, *A. glabra*. ×$\frac{1}{2}$.

6. A. subcrenata Buser

Plant up to 50 cm. *Lvs* (Fig. 40B) suborbicular, usually very undulate, *rather sparsely hairy above, often only on the folds*, more densely and evenly hairy beneath; lobes 7–9, rather wide and long, with wide, coarse and unequal teeth, extending to the base. *Petioles and lower internodes* with spreading hairs and usually *with some hairs directed slightly downwards*. Upper part of stem, pedicels and receptacle glabrous. Fls 2.5–3 mm diam. $2n = 96, 104–110$.

Native. Grassland in Upper Teesdale and Weardale. N. and C. Europe, extending southwards to the S.W. Alps and S.W. Bulgaria; W. Siberia (Ob region).

7. A. acutiloba Opiz

A. acutangula Buser

Plant rather large, up to 65 cm. *Lower part of stem and petioles with spreading hairs*, the upper part of stem, *peduncles, pedicels and receptacle glabrous*. Lvs (Fig. 39B) ±reniform in outline, very variable in hairiness, the early lvs ±hairy on both sides, the late summer lvs often glabrous; *basal sinus open; lobes* 7–11, somewhat wider than long, *straight-sided*, ±truncate at apex, the *teeth* (13–)15–19(–21) on each of the larger lobes, acute, almost triangular, *straight* or nearly so, *those at the middle of each margin of the lobe much larger than the top and bottom ones*, the apical tooth markedly smaller than its neighbours. Fls c. 3–4 mm diam., the clusters lax. $2n = c. 100, 105–9$.

Native. Grassland in Upper Teesdale, Weardale and mid-Durham. N., E. and C. Europe, extending to the S.W. Alps and Macedonia.

8. A. xanthochlora Rothm.

A. pratensis auct., vix Opiz

Plant usually robust. *Stems and petioles densely clothed with spreading hairs*; peduncles, pedicels and receptacle glabrous or very sparingly hairy. Lvs (Fig. 41A) reniform in outline, *glabrous above* (rarely with a very few hairs on the folds), hairy on the veins and sometimes thinly on the surface beneath, the basal sinus wide (45° or more); *lobes* usually 9 (the two lowest small) *rounded*, rather wide, not overlapping; *teeth* 13–15 on each of the middle lobes, acute, *somewhat curved towards apex of lobe*, all ±equal; stipules brown. Fls 2.5–3 mm diam., numerous, in dense clusters. $2n = c. 105$.

Native. Grassland, etc., at low altitudes. Throughout much of the British Is. from Sutherland southwards, except for most of S.E. and E. England and S. Ireland. W. and C. Europe extending to S. Sweden, Latvia and C. Greece; naturalized in eastern North America.

9. A. filicaulis Buser

Plant 5–40 cm. Lower internodes, and petioles of all but the earliest lvs, with ±dense, spreading hairs; upper internodes and infl. usually distinctly less hairy and even glabrous. Stipules and base of petioles normally purplish-red. Lvs (Fig. 40A) reniform, with wide basal sinus, variably hairy; lobes 7(–9), with convex sides, 3 times wider than long to 1$\frac{1}{2}$ times as long as wide, the middle lobes with c. 11 acute, ±equal teeth somewhat curved towards the apex. Fls 3–4 mm diam., in dense clusters; receptacle usually with some spreading hairs.

Native. Grassland from sea-level to 915 m. N., N.W. and N.C. Europe and on the principal mountains of W. and C. Europe from the Pyrenees to the Sudeten Mountains; Greenland, Newfoundland, Labrador.

Subsp. **filicaulis**

A. salmoniana Jaquet

Upper internodes, infl.-branches and pedicels glabrous or very sparsely hairy. Lvs very often hairy only on the folds above and on the veins beneath. $2n = c. 96, 103–110, c. 150$.

Local in mountain grassland and on rock-ledges up to 915 m. Brecon, N. Wales, Derby, Lancs, Yorks and Durham, northwards to Shetland; Sligo (Ben Bulben). Iceland, Fennoscandia and N. Russia, locally on the mountains of W. and C. Europe.

Subsp. **vestita** (Buser) M. E. Bradshaw

A. minor auct., *A. pseudominor* Wilmott, *A. vestita* (Buser) Raunk.

Upper internodes, infl.-branches and pedicels markedly hairy. Lvs ±uniformly hairy on both surfaces. $2n = 102–110$.

Grassland throughout the British Is., except the Outer Hebrides, Orkney and Shetland, ascending to 840 m. N.W. Europe, extending to arctic Norway, Finland, E. Austria and S. France.

10. A. minima S. M. Walters

Plant very small, rarely exceeding 5 cm. *Stems, pedicels, receptacle and petioles rather sparsely covered with* *spreading hairs. Lf-blades* (Fig. 40c) 1–2.5 × 1.5–3 cm, reniform in outline, *hairy on the folds and edges above and veins beneath*, and sometimes sparsely on the rest of the surfaces, the *basal sinus wide*; lobes usually 5 (sometimes with small basal ones in addition), wider than long, rounded above, not overlapping, *the sinuses between them ending in a deep toothless incision*; teeth (7–)9–11 on each of the middle lobes, *subacute, somewhat unequal-sided but scarcely curved towards the apex of the lobe*; *stipules brownish. Pedicels usually with very few hairs*, occasionally glabrous; receptacles more hairy, not at all silvery. Fls c. 2 mm diam., in small, rather dense clusters. $2n = $ c. 103–9*.

Native. Limestone grassland in the Craven district (Yorks). Endemic.

11. A. glomerulans Buser

Robust or medium-sized. *Stem and petioles densely clothed with appressed* rather silky *hairs*; pedicels, receptacle and often peduncles glabrous. Lvs (Fig. 41B) reniform in outline, usually hairy all over above, *less hairy beneath* and with the hairs sometimes confined to the veins, the spring lvs often less hairy especially above, the basal sinus wide; lobes usually 9, usually wide and rounded (often c. twice as wide as long), often overlapping and folding when dried; teeth 13–15 on each of the middle lobes, wide but subacute, somewhat curved, those on either side of the apical tooth the largest, the lowest considerably smaller. Fls c. 4 mm diam., in dense clusters; infl. short. $2n = $ c. 64, 96, c. 101–9.

Native. Damp rock-ledges, usually acid, in the mountains, usually from 600–915 m. From Perth to Sutherland; Teesdale. N. Europe, extending southwards in the mountains to the Pyrenees and C. Alps; Greenland (to 70° 15′ N), Labrador.

12. A. glabra Neygenfind

A. alpestris auct.; *A. obtusa* auct. (incl. ed. 1)

Robust, up to 60 cm. *Stems sparsely* and closely *appressed hairy on the lowest 1–2 internodes, or almost glabrous*; *upper part of plant glabrous*; petioles appressed hairy or glabrous. Lvs (Fig. 41c) reniform in outline, glabrous except for the distal portion of the veins beneath, the *basal sinus wide* (usually 60° or more and often nearly 180°); lobes 7–9 rather wide (but rarely twice as broad as long), rounded or straight-sided, not or rarely overlapping, the *sinus between them wide and not suddenly narrowed at the base, toothed throughout*; teeth 11–17 on each of the middle lobes, ±wide, somewhat curved, *those in the middle of the lobe larger and wider than the upper and lower ones, the apical tooth conspicuously narrower and usually shorter than its neighbours*. Fls 3–4 mm, in lax clusters or scarcely clustered (except when the first fls expand). $2n = 96$, c. 100, c. 102–110.

Native. Grassland, open woods and rock ledges, ascending to nearly 1215 m. Almost throughout the British range of *A. vulgaris sens. lat.*, the commonest sp.

on mountains but the rarest of the three widespread spp. in C. and S. England; local or absent from much of C. and S. Ireland. N. and C. Europe, extending southwards to the Pyrenees and N. Balkan peninsula; N. Urals; N.E. America (?native).

British plants referred to *A. obtusa* Buser have been proved on cultivation to be forms of this sp.

13. A. wichurae (Buser) Stefánsson

A. acutidens auct., non Buser; *A. acutidens* var. *alpestriformis* C. E. Salmon

Rather small. *Stems appressed hairy on the two lowest internodes* (hairier than *A. glabra*), *glabrous or with a few hairs above*; *infl. glabrous*; petioles appressed hairy. Lvs (Fig. 42) orbicular in outline, glabrous except on

Fig. 42. Leaf of *Alchemilla wichurae*. ×$\frac{1}{2}$.

the veins beneath, which are usually hairy throughout their length but sometimes only in the distal portion; *basal sinus closed or nearly so*; lobes 7–9, usually wide and rounded (often twice as long as wide), the *sinus between them with a narrow V-shaped (or almost closed) entire prolongation at the base*; *teeth* 17–19 on each of the middle lobes, rather narrow, *strongly curved towards the apex, all ±uniform in size, the terminal tooth about equalling its neighbours* (sometimes somewhat shorter). Fls 3–4 mm, in lax clusters or scarcely clustered.

Native. Grassland and rock-ledges on mountains, apparently confined to basic soils, ascending to 915 m, very local. Yorks, Westmorland, Perth to Sutherland. N. Europe, southwards to the Sudeten mountains; E. Greenland (very rare).

Plants of *A. acutidens sens. lat.* not referable to *A. wichurae* occur in Scotland, particularly on the Ben Lawers range, and require further study.

*14. A. mollis (Buser) Rothm.

Up to 80 cm, robust, usually with dense, spreading hairs throughout, except on pedicels. Lvs up to 13 × 15 cm, shallowly lobed up to c. $\frac{1}{4}$ radius; lobes 9–11, semicircular, with 7–9 wide, ovate teeth. Fls 3.5–5 mm diam.; *epicalyx-segments at least as long as sepals. Mature achene much longer than receptacle*.

Introduced. Commonly grown in gardens, escaped and apparently naturalized in a few places (Devon, Surrey, Gloucester, Edinburgh and Linlithgow). Native of E. Carpathians.

12. APHANES L.

Differs from *Alchemilla* as follows: *Annual. Fls in dense lf-opposed cluster-like cymes. Stamen* 1(–2), *opposite a sepal, inserted on inner margin of disk; anthers extrorse.*

About 20 spp., cosmopolitan.

Lobes of stipules surrounding infl. triangular-ovate; fr. 2.2–2.6 mm, including the divergent sepals, bottle-shaped.   **1. arvensis**

Lobes of stipules surrounding infl. oblong; fr. 1.4–1.8 mm, including the convergent sepals, ovoid.   **2. microcarpa**

(1–2). A. arvensis *sens. lat.*

Small, rather inconspicuous annual 2–20 cm, pilose, usually much-branched from the base, pale green; branches decumbent or ascending. Lvs 2–10 mm, fan-shaped, shortly stalked, trisect, the segments divided at the apex into 3–5 oblong lobes; stipules fused into a lf-like cup, with 5–7 obtuse lobes. Fls in sessile clusters, which are half-enclosed by the stipular cup, very small (less than 2 mm); segments of epicalyx much smaller than the sepals, difficult to see. Fl. 4–10. Th.

Native. Arable land and bare places in grassland, mainly on dry soils; ascending to 520 m. Common throughout the British Is.

1. A. arvensis L.   Parsley-piert.

Alchemilla arvensis (L.) Scop.

Usually relatively robust, greyish-green. *Lobes of stipules surrounding infl. triangular-ovate, little longer than wide, c. half as long as the entire portion. Fr. (including sepals)* 2.2–2.6 mm, *the sepals divergent so that the whole appears bottle-shaped.* $2n = 48$. Facultatively apomictic.

Throughout the British Is. on both acid and basic soils, though rare in N. Scotland and absent from Shetland. S., W. and C. Europe, extending to S. Sweden, Latvia and N.E. Poland; Asia Minor, Iran; introduced in North America.

2. A. microcarpa (Boiss. & Reuter) Rothm.
    Slender Parsley-piert.

Usually more slender, not greyish. *Lobes of stipules surrounding infl. oblong, c. twice as long as wide, nearly as long as the entire portion. Fr. (including sepals)* 1.4–1.8 mm, *the sepals convergent so that the whole appears ovoid.* $2n = 16$; amphimictic.

Throughout the British Is. on acid sandy soils, though less common than *A. arvensis*. Europe northwards to S. Sweden and eastwards to N.E. Poland, the Carpathians and the Adriatic; local in the Balkan peninsula; Morocco, Algeria; Macaronesia; North America.

13. SANGUISORBA L.

Erect perennial herbs. Lvs pinnate. *Fls* in dense terminal spikes or heads, hermaphrodite or unisexual, with 2 or 3 bracteoles below each fl. *Epicalyx* 0. Sepals 4. Petals 0. *Stamens* 4 or numerous. *Carpels* 1–2(–3); ovule 1, pendulous; style terminal, stigma simple or feathery. *Receptacle* becoming corky in fr., enclosing the achene,

without spines. Nectar secreted by a ring round the style or not produced.

About 27 spp., in north temperate regions.

Fls all hermaphrodite; sepals dull red; stamens usually 4; carpel 1, with simple, disk-like, shortly papillate stigma.   **1. officinalis**

Upper fls of each head female, the middle hermaphrodite, the lower male or hermaphrodite; sepals green; stamens numerous; carpels 2(–3); stigma feathery, with long papillae.   **2. minor**

1. S. officinalis L.   Great Burnet.

Poterium officinale (L.) A. Gray

Glabrous perennial 30–100 cm; stock thick. Stems erect, branched above. Rosette and lower cauline lvs pinnate, with 3–7 pairs of lflets; lflets increasing in size upwards, the larger 2–4 cm, ovate or oblong-ovate, stalked, mostly cordate at base, obtuse, crenate-dentate or serrate-dentate; stipules of rosette-lvs scarious, adnate, of cauline lf-like, dentate; upper cauline lvs small and few. *Fls hermaphrodite*, in oblong heads 1–2 cm. *Sepals dull red. Stamens usually* 4, *equalling or somewhat longer than calyx. Carpel* 1, *with simple, disk-like, shortly papillate stigma.* Receptacle in fr. 4-winged, smooth between the wings. Fl. 6–9. Visited by Diptera and Lepidoptera. $2n = 28$. Hs.

Native. Damp grassland, ascending to 460 m. From Hants, Surrey and Suffolk westwards and northwards to Ayr and Berwick, locally common; W. Mayo, Down, Antrim. Most of Europe, except most of the Mediterranean region and parts of the north; temperate Asia to N. Iran, China and Japan; escaped in North America.

*S. canadensis L. with the white fls in cylindrical spikes and stamens exserted is sometimes grown in gardens and has been found as an escape. Native of eastern North America.

2. S. minor Scop.

Subglaucous perennial herb 15–80 cm, smelling of cucumber when crushed, glabrous except for some long flexuous hairs in the lower part; stock somewhat woody. Stems erect, simple or branched above. Rosette-lvs pinnate, with 4–12 pairs of lflets; lflets increasing in size upwards, the larger 0.5–2 cm, orbicular or shortly oval, shortly stalked, rounded at base, ±truncate at apex, deeply serrate or serrate-dentate with the terminal tooth smaller than the lateral; stipules scarious, adnate. Cauline lvs few (sometimes 0), the lower like the rosette-lvs, a few of the upper often with narrower (oblong) lflets, passing into the small bracts; stipules lf-like, small. Fls in globose heads 7–12 mm diam., the lower fls male or hermaphrodite, the middle hermaphrodite, the upper female. *Sepals green. Stamens numerous. Carpels* 2(–3); *stigma feathery, with long papillae. Fr. receptacle* ovoid, 4-angled.

Subsp. minor   Salad Burnet.

Poterium dictyocarpum Spach; *P. sanguisorba* L.

Stems 15–40(–60) cm. Cauline lvs few or 0, the upper small. Fr. receptacle c. 4 mm, with entire ridges down the angles, the surface between them with fine, raised, entire reticulations, not pitted (Fig. 43A). Fl. 5–8. $2n = 28^*$. Hs.

A   B

Fig. 43. Fruits of *Sanguisorba minor*. A, subsp. *minor*; B, subsp. *muricata*. ×2.5.

Native. Calcareous grassland, occasionally becoming locally dominant, less frequently neutral grassland; ascending to 505 m in Yorks. Widespread and common in England and Wales (not Isle of Man); very local in Scotland, extending north to Dunbarton and Angus (a rare introduction farther north); widespread in S.E. and C. Ireland to mid Cork, E. Galway and Louth; Donegal. C., W. and S. Europe, extending to S. Sweden and C. Russia; Armenia, N. Iran; Morocco (Atlas, subspp.); naturalized in North and South America.

*Subsp. **muricata** Briq.   Fodder Burnet.

Poterium polygamum Waldst. & Kit.

Stems 30–80 cm. Cauline lvs several, some with narrow lflets. Fr. receptacle c. 6 mm, with sinuate wings down the angles, the surface between them strongly rugose, with coarse, raised, toothed ridges or reticulations (Fig. 43B). Fl. 6–8. $2n = 28, 56$. Hs.

Introduced. Formerly grown for fodder and completely naturalized on field borders, etc., in many places in S. England and Wales and a few in N. England and S. Scotland, north to Lanark and Angus; Jersey. S. Europe, extending to Russian C. Asia; naturalized in C. and parts of N. Europe.

14. ACAENA L.

Herbs or dwarf shrubs. Lvs usually pinnate. Fls in terminal spikes or heads, hermaphrodite. *Epicalyx* 0. Sepals 3–4. *Petals* 0. Carpels 2; stigmas feathery. *Receptacle dry in fr., with 4 spines, 1 below each sepal, or more numerous scattered spines.*

About 100 spp., south temperate, extending north to Hawaii and California. A few spp. are occasionally grown in gardens.

Distal lflets with (12–)17–23 teeth; fr. cupule with 2 spines.
        2. ovalifolia
Distal lflets with 8–12(–13) teeth; fr. cupule with 4 spines.
        1. novae-zelandiae

*1. A. novae-zelandiae T. Kirk Pirri-pirri-bur.

A. anserinifolia auct. brit., non (Forster & Forster fil.) Druce

Prostrate creeping much-branched dwarf-shrub, emitting short, ascending or erect lfy pilose stems 2–15 cm high. Lvs pinnate with 7–9 lflets, bright glossy green

above lflets increasing in size upwards, the larger 5–10 mm, oblong, sessile, cuneate or rounded at base, obtuse, deeply crenate-serrate, *the distal with 8–12(–13) teeth*, glabrous above, appressed pilose beneath; stipules adnate and scarious with a free lf-like apex. Fls in globose heads which are solitary on erect scapes 4–7 cm long; heads 5–10 mm diam., greenish, the spines of the receptacle not evident at fl. time. Heads larger in fr., *the fr. cupule with 4 spines*, c. 10 mm, which are barbed at the tip. Fl. 6–7. $2n = 42$. Chw.

Introduced. Probably originally imported with wool and now locally naturalized on roadsides, river banks, wood-margins, etc., from Devon to Kent northwards in England to Yorks and Northumberland, and in Roxburgh and Berwick. Native of New Zealand and montane Australia.

*A. anserinifolia (Forster & Forster fil.) Druce, a native of New Zealand and Australia, which differs from *A. novae-zelandiae* in the lvs having distal lflets light matt green above with brownish edges and veins, is a casual or locally naturalized in Yorks, Northumberland and Perth.

*2. A. ovalifolia Ruiz & Pavón

Prostrate stems with long, white hairs. Lvs pinnate, with (7–)9 lflets, bright green and sometimes glossy above; lflets increasing in size upwards, the larger (10–)15–30 mm, elliptical or oblong, *with (12–)17–23 teeth*, glabrous above, subglabrous beneath; stipules adnate, the lf-like free part 2–4-lobed. Fls in heads solitary on erect scapes (2–)6–12 cm long; fl. heads 8–10 mm diam.; fr. heads 18–30 mm diam. (incl. spines), *the fr. cupule with 2 spines*, 8–10 mm, which are barbed at the tip. $2n = 42$. Chw.

Introduced. Casual or locally naturalized on roadsides, walls and in woods in Cornwall, Devon, Haddington, Inverness, Argyll and Dumbarton in Great Britain, and Kerry and Carlow in Ireland. Native of South America from Tierra del Fuego and the Falkland Is. northwards along the Andes to Peru.

Two spp. which may be distinguished from the above by the lvs having distal lflets at most $1\frac{1}{2}$ times as long as wide (rather than at least $1\frac{1}{2}$ times as long as wide) are cultivated in gardens and may locally occur as ±naturalized escapes. **A. inermis** Hooker fil., a native of New Zealand, which has lvs glaucous above, with the distal lflets as wide as long, 2(–4?) white stamens and spineless fr. capsules, has been found locally in the Inner Hebrides (Mull, Raasay). **A. magellanica** (Lam.) Vahl, a native of cool temperate South America, the Falkland Is. and the circum-Antarctic islands, which has lvs glaucous or not above, with at least some distal lflets longer than wide, 4 dark red stamens and 4-spined fr. cupules, has been reported from Bedford, Selkirk and Edinburgh.

Tribe 5. ROSEAE. Receptacle deeply concave, often almost closed at the mouth, becoming coloured and fleshy in fr. Carpels numerous. Fr. of numerous achenes enclosed in the fleshy receptacle ('hip').

15. ROSA L.

Shrubs, sometimes trailing or scrambling, usually deciduous. Lvs pinnate (very rarely simple); stipules (usually) adnate. Stems usually prickly. Fls terminal, solitary or in corymbs, hermaphrodite, (4–)5-merous. Stamens numerous. Styles protruding through the orifice of a disk. Ovule 1, pendent. Fls homogamous, mostly without nectar, visited by various insects for pollen, self-pollination possible if insect visits fail. Fr. of numerous achenes enclosed in the coloured, fleshy receptacle ('hip').

About 250 spp. (according to some authors many more), north temperate and subtropical regions. Our garden roses are hybrids of complex origin, derived from a number of spp., and have been developed over a very long period. A number of foreign spp. are also fairly often grown.

For a full account of the British forms see A. H. Wolley-Dod, *J. Bot.*, **68–69**, Suppl. (1930–31).

In sections 1–3 (spp. **1–3**) below, reproduction is sexual with normal behaviour of the chromosomes at meiosis. Members of section 4 (Caninae), on the other hand, show a very distinctive meiotic mechanism which leads to a complex pattern of variation, as will be described later (p. 227).

1	Styles exserted beyond at least lower stamens, united into a column in fl.	2
	Styles not united into a column, though sometimes exserted.	3
2	Stems weak, scrambling; stylar column slender, as long as stamens, persisting in fr., on a flat disk. **1. arvensis**	
	Stems strong, arching, 1–4 m; stylar column stout, shorter than stamens, the styles separating in fr., on a conical disk. **4. stylosa**	
3	Fls solitary, without bracteoles; lflets (3–)4–5 pairs; stems densely prickly and bristly, not tomentose. **2. pimpinellifolia**	
	Fls solitary or in corymbs, with bracteoles resembling stipules; lflets 2–3(–4) pairs; stems without, or with few, bristles, if with numerous bristles, then tomentose.	4
4	Stems prickly and bristly, tomentose; fls 6–8 cm diam.; sepals entire; fr. 2–2.5 cm. **3. rugosa**	
	Stems with few or no bristles, not tomentose; fls not more than 5 cm diam.; outer sepals usually with pinnate lobes; fr. usually not more than 2 cm.	5
5	Lflets glabrous, sometimes with a few reddish, stalked, non-scented glands along petiolule and midrib.	6
	Lflets tomentose or hairy beneath, at least on midrib.	7
6	Styles glabrous, hispid or sometimes woolly, forming a lax group; orifice of disk about $\frac{1}{5}$ its width. **5. canina**	
	Styles forming a dense, woolly, hemispherical mass almost covering the disk; orifice of disk about $\frac{1}{3}$ of its width. **7. afzeliana**	
7	Lflets with prominent, viscid, brownish or translucent glands beneath, giving a sharp, fruity scent when rubbed.	8
	Lflets eglandular or with ±inconspicuous glands which may give an aromatic scent when rubbed.	11
8	Lflets cuneate at base; pedicels glabrous.	9
	Lflets rounded at base; pedicels glandular-hispid.	10
9	Stems erect; styles hairy; sepals persisting until fr. ripe, spreading to erect. **15. elliptica**	
	Stems arching; styles glabrous or sparsely hairy; sepals soon falling, usually deflexed. **16. agrestis**	
10	Stems erect; prickles unequal; styles hispid; sepals erect or spreading, persistent at least until fr. reddens. **13. rubiginosa**	
	Stems arching; prickles uniform; styles usually glabrous; sepals deflexed, usually falling before fr. reddens. **14. micrantha**	
11	Lflets hairy or tomentose, eglandular (rarely a few scentless glands on midrib), serrate or doubly serrate.	12
	Lflets usually doubly serrate, the teeth glandular, usually with some glands beneath at least on midrib.	13
12	Styles glabrous or hispid, rarely woolly, sometimes slightly exserted on a low conical disk; orifice of disk about $\frac{1}{5}$ its width; sepals deflexed, soon falling. **6. dumetorum**	
	Styles woolly, in dense hemispherical mass almost covering disk; orifice of disk about $\frac{1}{3}$ its width; sepals spreading to erect after fl., falling late. **8. coriifolia**	
13	Prickles strongly hooked; pedicels glabrous; fr. glabrous. **9. obtusifolia**	
	Prickles usually straight or slightly curved; pedicels usually glandular-hispid; fr. glandular-hispid.	14
14	Sepals ±spreading or deflexed after fl.; orifice of disk about $\frac{1}{5}$ its width. **10. tomentosa**	
	Sepals erect and persistent after fl.; orifice of disk about $\frac{1}{3}$ its width or more.	15
15	Sepals ±simple, ±fused to fr. and persisting until it falls or decays; orifice of disk up to $\frac{1}{2}$ its width; prickles straight; main branches straight. **12. mollis**	
	Sepals usually with small lateral pinnae, persisting until fr. ripens but finally falling; orifice of disk about $\frac{1}{3}$ its width; at least some prickles inclined or slightly curved; main branches arching. **11. sherardii**	

Section 1. *Synstylae* DC. Stems scrambling, climbing or creeping; prickles hooked. Fls in bracteate corymbs; two outer sepals usually with pinnately arranged projecting lobes, all caducous. *Styles united into a column, reaching at least to the lower stamens.*

1. R. arvensis Hudson　　　　　Field Rose.

Deciduous shrub, glabrous or nearly so, with *weak, scrambling*, subglaucous, often purple-tinted stems, either decumbent and forming low bushes 50–100 cm high, or climbing over other shrubs, rarely more erect and reaching 2 m. Prickles hooked, all ±equal. Lflets 2–3 pairs, 1–3.5 cm, ovate or ovate-elliptical simply, rarely doubly serrate, glabrous on both sides, or pubescent on the veins (rarely all over) beneath, rather thin; petiole usually with some stalked glands; stipules narrow, the auricles straight. Fls 1–6, white, 3–5 cm diam.; pedicels 2–4 cm, with stalked glands, rarely smooth; buds short. Sepals short (not 1 cm), ovate, acuminate, the tip not expanded, often purplish, caducous; lobes very few. Styles glabrous, equalling the stamens, *the column slender*. Fr. small, red, smooth, globose or sometimes ovoid or oblong. Fl. 6–7. $2n = 14^*$. N.

Native. Woods, hedgebanks and scrub; ascending to

380 m. Rather common in S. England and Wales, less common from Yorks and Lancs to Cumberland and Durham; probably introduced in Scotland northwards to Aberdeen; throughout Ireland but rather local, C., W. and S. Europe.

*R. sempervirens L.

Lvs evergreen, coriaceous; lflets 2 pairs; stipules very narrow, with divergent auricles. Our plant (var. *melvinii* Towndr.) is characterized by its narrowly elliptical lflets and glabrous styles (they are usually hairy in this sp.) $2n = 14$. Naturalized in Worcester. Native of the Mediterranean region.

*R. multiflora Thunb., with pectinate stipules and many-fld infl., is rarely naturalized. $2n = 14$. Japan, Korea.

Section 2. *Pimpinellifoliae* DC. Usually low shrubs with ±erect stems, and ±numerous straight prickles and stiff bristles. *Lflets 3 or more pairs.* Fls solitary; *bracteoles 0 or small and scale-like.* Sepals entire, erect and persistent on the fr. Styles free, short.

2. R. pimpinellifolia L.   Burnet Rose.

R. spinosissima L., *pro parte*

Low, erect, deciduous shrub 10–40(–100) cm, *spreading by suckers and forming large patches*; *prickles numerous, straight, mixed with numerous stiff bristles and passing into them* (rarely stems almost unarmed). *Lflets* (3–)4–5 *pairs*, small, 0.5–1.5(–2) cm, suborbicular or ovate, obtuse, simply serrate, rarely doubly glandular-serrate, glabrous on both sides or sparingly pubescent beneath, rarely somewhat glandular; stipules narrow, the auricles expanded and divergent. Fls cream-white, rarely pink, 2–4 cm diam.; pedicels 1.5–2.5 cm, glandular-hispid or smooth. Styles woolly. Fr. 1–1.5 cm, subglobose, *purplish-black*. Fl. 5–7. $2n = 28^*$. N.

Native. Dunes, sandy heaths, limestone pavement, etc., especially near the sea, but ascending to 520 m. From Caithness and the Outer Hebrides southwards and throughout Ireland, but rather local, and absent from a number of counties. Europe except the north-east, most of Fennoscandia, the extreme south-west and many of the islands; Caucasus; temperate Asia east to Manchuria and N.W. China, south to Asia Minor.

Hybrids between *R. pimpinellifolia* and the members of the section *Caninae* occur occasionally. They are usually recognizable by the presence of bristles and by the habit and lf-shape showing the influence of *R. pimpinellifolia*. The other parent is frequently impossible to determine exactly. The main groups are as follows:

R. pimpinellifolia × subsection *Caninae* (*R.* × *hibernica* Templeton etc.). Occasional in S. England, more frequent further north and in N. Ireland.

R. pimpinellifolia × subsection *Villosae* (*R.* × *involuta* Sm. etc.). The most frequent group from Sussex northwards to the Inner Hebrides and Sutherland.

R. pimpinellifolia × subsection *Rubiginosae*. Rare, from Kent to Angus.

Section 3. *Cassiorhodon* Dumort. (*Cinnamomae* Crépin). Erect deciduous shrubs; prickles straight; bristles often present. Lflets 2–5 pairs. Fls in bracteolate corymbs. Sepals usually entire, erect and usually persistent after fl. Styles free, short. Carpels on sides and bottom of receptacle.

*3. R. rugosa Thunb.   Japanese Rose.

Deciduous shrub 1–2 m. *Stems densely prickly and bristly, tomentose.* Lflets 2–5 cm, 2–4 pairs, ±elliptical, dark green and rugose above, pubescent beneath. Upper stipules broad. *Fls 6–8 cm diam.*, 1 or few, bright purplish-pink to white; pedicels bristly. *Fr.* 2–2.5 cm, red, crowned by the erect sepals. Fl. 6–7. $2n = 14$. N.

Introduced. Frequently grown in gardens and often as a stock for other roses, sometimes occurring as an escape; naturalized in several places. Native of N. China and Japan; naturalized in parts of N., W. and C. Europe.

R. canina × *rugosa* (*R.* × *praegeri* W.-Dod) occurs spontaneously in Antrim.

Section *Carolinae* Crépin. Differs from *Cassiorhodon* in: Sepals spreading and caducous after fl. Carpels confined to bottom of receptacle.

*R. virginiana Miller (*R. lucida* Ehrh.)

Shrub to 2 m. Prickles straight or curved, slender, bristles confined to young shoots. Lflets 2–4 pairs, 2–6 cm, ±elliptical, coarsely serrate, dark green and shining above, glabrous or pubescent on the veins beneath. Fls bright pink, 5–6 cm diam.; pedicels and receptacle glandular-hispid. Sepals lobed. Fr. red, 1–1.5 cm diam. Sometimes grown in gardens and occurring as an escape. $2n = 28$. Native of eastern North America.

Section 4. *Caninae* DC. Deciduous shrubs with erect or arching stems, and ±numerous, usually hooked, but sometimes straight prickles; bristles usually 0. Lflets 2–3(–4) pairs. Fls in bracteolate corymbs. Outer sepals with pinnately arranged projecting lobes. Styles free or fused into a very short column.

The preceding sections contain normally reproducing, diploid or tetraploid species. The plants pertaining to this section are all 'unbalanced' polyploids, mostly pentaploids ($2n = 35$). In these, 2 of the 5 sets of 7 chromosomes pair at meiosis, while the other 3 sets remain unpaired. During pollen-formation the unpaired sets are lost, so that each pollen grain contains only 1 set of 7 chromosomes. On the female side, the 3 unpaired sets, together with one of the paired sets of chromosomes, are incorporated into the embryo-sac (which thus has 28 chromosomes). On fertilization the zygote, and consequently the resultant plant, is pentaploid (35 chromosomes) once again. Unbalanced hexaploids, with 42 chromosomes (*R. stylosa, R. agrestis, R. micrantha, R. sherardii*) have 4 unpaired sets of chromosomes at meiosis, giving rise to pentaploid embryo-sacs, while the octoploid races of *R. villosa* have 4 paired and 4 unpaired sets of chromosomes so that the embryo-sac is hexaploid and the pollen diploid (14 chromosomes). This method of reproduction results in progeny inheriting a preponderance of maternal characters, and this probably explains suggestions that the unbalanced polyploids are apomictic, though most workers consider that this is not so in most British spp.

Although many early authorities, including Wolley-Dod, failed to recognize the extent of hybridization in the *Caninae*, most modern works appreciate that hybrids are very common, in some areas being more abundant than the spp. These are not treated below; an authoritative account of British rose hybrids is provided by R. Melville in C. A. Stace (1975) (pp. 212–27). It should be noted that hybrids vary from being completely sterile to rather highly fertile and, because they are largely matroclinal, reciprocal hybrids can differ quite significantly.

Subsection 1. *Stylosae*. (Crépin) Crépin. Prickles stout, hooked. Lflets glabrous or pubescent, not tomentose or glandular. Pedicels usually glandular. *Styles at first fused into a stout column, becoming free in fr.* Stigmas in a narrow head.

4. R. stylosa Desv.

R. systyla Bast.

Shrub 1–4 m; stems *strong, arching*; prickles hooked, some with very stout bases. Lflets 1.5–5 cm, 2–3 pairs, narrowly (rarely broadly) ovate to lanceolate, usually acuminate, simply (rarely doubly) serrate, usually pubescent beneath, at least on the veins, rarely also above, or glabrous on both sides, eglandular; *stipules and bracts rather narrow*. Fls 3–5 cm diam., 1–8 or more, white or pale pink; pedicels long, usually glandular-hispid, rarely smooth. Sepals reflexed after fl., falling before fr is ripe. *Stylar column* glabrous, *shorter than the stamens*; *stigmas in an ovoid head*; *disk conical, prominent.* Fr. c.1–1.5 cm, ovoid, rarely globose, red smooth. Fl. 6–7. $2n = 35^*$, 42. N. or M.

Native. Hedges, etc.; from Denbigh, Worcester, Leicester, and Suffolk (?Norfolk) southwards and from Limerick, Offaly and Wicklow southwards, local; Channel Is. Local from France and W. Germany to S. Spain and Bulgaria.

Subsection 2. *Caninae* Crépin (*R. canina sens. lat.*). Dog Rose. *Prickles curved or hooked*, equal, usually stout. *Lflets glabrous or pubescent, eglandular or with a few glands on the main veins beneath*, rarely (only in *R. obtusifolia*) with numerous glands, the glands not scented. Pedicels glabrous or less frequently glandular. Styles free.

5. R. canina L.   Dog Rose.

Shrub 1–3 m; stems arching. *Prickles strongly curved or hooked*. *Lflets* 1.5–4 cm, 2–3 pairs, ovate, obovate or elliptical, acute or acuminate, simply to doubly (and then often glandular-) serrate, *glabrous* on both sides, *eglandular*, or rarely with a few, *reddish, stalked, non-scented* glands along petiolule and midrib; upper stipules and bracts wide. Fls 1–4 or more, pink or white; *pedicels* (0.5–)1–2 cm, smooth or less often glandular-hispid. *Sepals falling before the fr. ripens and usually before it reddens, deflexed* or rarely spreading after fl.; *lobes*

narrow, entire or nearly so. Petals c. 2–2.5 cm. *Styles glabrous*, hispid or sometimes woolly, *forming a lax group*; *orifice of disk about* $\frac{1}{5}$ *its width*. Fr. usually c. 1.5–2 cm, globose, ovoid or ellipsoid, scarlet, very rarely glandular-hispid. Fl. 6–7. $2n = 35^*$. N. or M.

Native. Woods, hedges, scrub, etc., ascending to 550 m. Common in England, Wales and Ireland (in general much the commonest sp.), becoming rare in Scotland though reported from as far north as Shetland; Channel Is. Europe northwards to c. 62° N; N. Africa; S.W. Asia (south to Palestine, north to Russian C. Asia); Madeira; naturalized in North America.

6. R. dumetorum Thuill.

Differs from *R. canina* as follows: Lflets broadly elliptical to suborbicular, subacute, simply rounded-serrate, eglandular, *hairy or tomentose beneath, and usually above*; petioles without prickles. Pedicels smooth. Fr. subglobose. $2n = 35$. N. or M.

Native. Woods, hedges, scrub, etc. Probably locally frequent and widespread in England and Wales, becoming rarer in Scotland northwards to Moray; widespread and locally frequent in S. Ireland and probably common in many parts of the island. Throughout Europe but rarer in the north and north-west.

Often included in *R. canina* as a hairy-lvd variant, but most modern British and Irish rhodologists consider it a distinct sp.

7. R. afzeliana Fries

R. dumalis auct.; *R. glauca* Vill. ex. Loisel., non Pourret; *R. caesia* Sm.

Differs from *R. canina* as follows: Prickles usually smaller. Upper stipules and bracts usually wider, often tinged red. *Pedicels short*, 0.5–1(–1.5) cm, often concealed by the bracteoles. *Sepals persistent at least till the fr. reddens*, usually becoming ±erect after fl., but sometimes spreading or deflexed. *Styles villous*, not at all exserted, *forming a dense, woolly, hemispherical mass almost concealing the disk*; orifice of disk about $\frac{1}{3}$ of its width. Fl. 6–7. Fr. earlier than *R. canina* $2n = 35^*$. N. or M.

Native. Woods, hedges, scrub, etc.; from Sussex and Gloucester northwards, very rare in S. England, becoming commoner in the north and largely replacing *R. canina* in the Scottish Highlands; Ireland (recorded from a few places in the north and west, but incompletely known). Most of Europe northwards to Iceland (very rare) and Fennoscandia, but rare in the south-west; W. Asia (mountains).

8. R. coriifolia Fries

Differs from *R. canina* as follows: *Lflets oblong, simply serrate, with fine, narrow teeth, hairy, at least on midrib beneath*; petioles villous, with fine spines. *Sepals spreading to erect after fl., falling late. Styles woolly, in dense hemispherical mass almost covering disk; orifice of disk about* $\frac{1}{3}$ *its width.* Fr. subglobose. $2n = 35$. N. or M.

Native. Woods, hedges, scrub, etc. Apparently rare in S. England but present from Yorks and Durham northwards to Perth and Aberdeen; distribution in Ireland still unclear. Most of Europe eastwards to Estonia and C. Ukraine.

9. R. obtusifolia Desv.

R. tomentella Léman

Differs from *R. canina* as follows: *Prickles usually more strongly hooked* with stout bases. *Lflets* 1.5–3.5 cm, ovate, usually more rounded in outline and less acute or even obtuse, doubly glandular serrate or, less frequently, simply serrate, *always pubescent* at least on the veins beneath and usually so on both sides, *usually glandular on the main veins beneath* but often eglandular, rarely with glands on the whole lower surface. Fls usually white; pedicels 0.5–1.5 cm, glabrous. *Sepals short; lobes wide, dentate or lobed.* Petals c. 1.5 cm. Styles pilose, rarely subglabrous. Fr.1–1.8 cm, globose or ovoid, red, glabrous. Fl. 6–7. $2n = 35^*$. N. or M.

Native. Hedges, scrub, etc.; from Northumberland and Cheshire southwards, local; Ireland (distribution uncertain). C., S. and N.W. Europe.

Subsection 3. *Villosae* (DC.) Crépin (*R. villosa sens.lat.*) Downy Rose. *Prickles straight or curved, usually equal, rather slender. Lflets doubly glandular-serrate, usually densely tomentose, frequently glandular beneath, the glands often small, always scentless or little scented.* Pedicels nearly always, and fr. often, glandular-bristly. Styles free.

10. R. tomentosa Sm.   Downy Rose.

Shrub 1–2 m; *stems arching; young stems and lvs pale green; prickles usually straight* or *slightly curved, stouter* than in the other spp. of this subsection. Lflets 2–3 pairs, 1.5–4 cm, ovate, ovate-lanceolate or elliptical, rarely obovate, acuminate or acute, usually densely pubescent or tomentose on both sides but sometimes only sparsely so or with the pubescence confined to the midrib, but then harsh to the touch, usually glandular all over beneath but sometimes eglandular; upper stipules not very wide, flat, their *auricles* short, triangular, *straight or diverging.* Fls c. 4 cm diam, 1–4 or more, pink or white; *pedicels relatively long* (1–2 cm), glandular-hispid. *Sepals falling before the fr. ripens* but frequently persisting until it reddens, ±spreading or deflexed *after fl.*, relatively long, *markedly constricted at the attachment*, with projecting pinnately-arranged lobes. *Styles pilose or glabrous*, rarely villous. Orifice of disk about $\frac{1}{3}$ its width. Fr. c. 1–2 cm, ovoid, less frequently globose or ellipsoid, red, usually glandular-hispid but often smooth. Fl. 6–7. $2n = 35^*$. N.

Native. Woods, hedges, scrub, etc.; widespread and rather common in England, Wales and Ireland, becoming rare in Scotland but extending at least to Ross. Most of Europe except the extreme north; Caucasus; Asia Minor.

11. R. sherardii Davies

R. omissa Déségl.

Intermediate in most respects between *R. tomentosa* and *R. mollis*. Habit intermediate, lower and more compact than in *R. tomentosa*; *main branches arching; lvs and young stems somewhat glaucous.* Prickles and lvs as *R. tomentosa* but lflets usually more markedly doubly serrate. Fls often several, frequently deep pink; *pedicels* short (up to 1.5(–1.7) cm but some of them, at least, usually less than 1 cm). *Sepals usually shorter* than in either of the other spp., *slightly constricted at the attachment*, rounded on the back, lobed as in *R. tomentosa, erect or ascending after fl., persistent until the fr. is ripe but finally falling. Styles villous*, rarely pilose. *Orifice of disk about $\frac{1}{3}$ its width.* Fr. most frequently globose, obovoid or pyriform. less often ovoid. Fl. 6–7. $2n = 28^*$, 35^*, 42^*. N.

Native. Woods, hedges, scrub, etc.; throughout Great Britain, rather common in Scotland, becoming very rare in S. England; Ireland (only recorded from a few places in the north and west but incompletely known). N., W. and C. Europe extending eastwards to S.W. Finland and southwards to Bulgaria.

12. R. mollis Sm.

R. villosa L.; *R. mollissima* Willd.

Erect shrub 40–100(–200) cm, somewhat suckering; *main branches straight*; young stems and lvs glaucous; *prickles quite straight, slender, subulate, their bases little thickened.* Lflets (1.5–)2–4 cm, 2–3(–4) pairs, oblong, oval or elliptical, more rounded than in the other spp. of the subsection, subobtuse, tomentose on both sides, usually glandular beneath, but sometimes eglandular, markedly doubly glandular-serrate; upper stipules very wide, their *auricles falcate and incurved.* Fls 3–5 cm diam., 1–3, usually deep pink or almost red; *pedicels short*, 0.5–1.0(–1.5) cm, glandular-hispid. *Sepals* 1.5–2.5 cm, *erect and persistent on the fr. till it falls or decays*, not constricted at the attachment, usually less lobed than in the other spp. of the subsection (lobes 1–3 pairs) and sometimes entire. *Styles villous*; stigmas in a wide flat head. *Orifice of disk up to $\frac{1}{2}$ its width.* Fr. 1–2 cm, red, usually globose, usually glandular-hispid, sometimes eglandular-hispid or smooth. Fl. 6–7. $2n = 28^*$. N.

Native. Woods, hedges, scrub, etc., ascending to over 610 m; rather common in Scotland, extending south to Glamorgan, Hereford and Derby, occasionally naturalized farther south; scattered over Ireland but apparently local and mainly in the north. Europe from 69° 10′ N. in Fennoscandia to the mountains of S. Europe; Caucasus; Asia Minor, Iran.

Subsection 4. *Rubiginosae* Crépin (*R. rubiginosa sens. lat.*). Sweet-briar. Prickles usually hooked, rarely arched or straight, equal or unequal sometimes mixed with a few stout bristles mainly on the fl. branches. *Lflets* always doubly glandular-serrate, glabrous or somewhat pubescent, never tomentose, ±*thickly clothed below with sweet-scented viscid brownish glands.* Pedicels glandular-hispid or less frequently glabrous. Styles free.

13. R. rubiginosa L.   Sweet Briar.

R. eglanteria L., nom. ambig.

Shrub 1–2 m; *stems erect. Prickles* hooked, *usually unequal*, usually mixed with scattered stout bristles on the fl. branches below the infl. and sporadically elsewhere. *Lflets* 2–3(–4) pairs, 1–2(–2.5) cm, suborbicular to ovate-elliptical, rarely ovate, *rounded at the base*, pubescent on the veins beneath and slightly so above, very glandular beneath. Fls 1–3, bright pink; *pedicels* rather short, c. 1 cm, *glandular-hispid. Sepals usually persistent at least until the fr. reddens, erect or spreading after fl.* Petals c. 1.5 cm. *Styles hispid,* short. Fr. subglobose or ovoid, c. 1–1.5 cm, scarlet, smooth or glandular-hispid at the base, rarely bristly all over. Fl. 6–7. Differs from most spp. in having nectar secreted by the edge of the receptacle, but is apparently mainly visited for pollen. $2n = 35^*$. N.

Native. Scrub, rarely woods and hedges, mainly on calcareous soils; an early colonizer of chalk grassland; widespread in England and Wales and locally common, rarer in Scotland but reported to extend to Caithness and the Outer Hebrides; widespread in Ireland but rather local; Channel Is. Frequently planted and an introduction in many of its localities. Most of Europe northwards to 61°N; Caucasus; W. Asia (to N.W. India); naturalized in North America.

14. R. micrantha Borrer ex Sm.

Shrub 1–2 m; *stems arching. Prickles* hooked, ±*equal*; stout bristles rarely present. *Lflets* 1.5–3(–3.5) cm, 2–3 pairs, ovate, obovate or elliptical, *rounded at base*, ±pubescent beneath, usually glabrous or nearly so above, glandular beneath. Fls 1–4, pink; *pedicels* rather long, 1–2 cm, *glandular-hispid*, very rarely smooth. *Sepals usually falling before fr. reddens, deflexed.* Petals 1–1.5 cm. *Styles usually glabrous*, very rarely thinly pilose, somewhat exserted. Fr. c. 1–1.5 cm, ovoid, scarlet, smooth or glandular-hispid. Fl. 6–7. $2n = 35, 42$. N.

Native. Woods, scrub and hedges mainly on calcareous soils but more frequent off them than *R. rubiginosa*: widespread and rather common in England and Wales; rare in Scotland and reported from only a few counties, north to Ross; in Ireland apparently confined to Kerry and Cork, though many records may refer to hybrids; Channel Is. W., S. and C. Europe, extending to N. Ukraine.

15. R. elliptica Tausch

R. inodora auct.

Shrub 1–2 m; *stems erect.* Prickles hooked. *Lflets* 2–3 pairs, obovate, *cuneate at base*, pubescent on both sides, glandular beneath. Fls solitary, white or pale pink; *pedicels smooth*, 6–12 mm. *Sepals persisting until the fr. is ripe, spreading to erect after fl. Styles hairy*, short. Fr. subglobose, red, smooth. Fl. 6–7. $2n = 35$. N.

Native. Very rare; Somerset, Huntingdon, Warwick; Lough Derg. Mountains of W. and C. Europe; extending south-eastwards to Albania and W. Ukraine, local everywhere.

16. R. agrestis Savi

R. sepium Thuill., non Lam.

Shrub 1–2 m; *stems arching.* Prickles hooked, ±equal. *Lflets* 1–3(–5) cm, 2–3 pairs, narrowly obovate, oblong or elliptical, *cuneate at base*, glabrous or pubescent, glandular beneath. Fls 1–3 or more, white or pale pink; *pedicels* 1–2 cm, *smooth*. Sepals soon falling, deflexed, rarely suberect after fl. Petals c. 1.5 cm. *Styles glabrous or slightly hairy*, somewhat exserted. Fr. 1–1.5 cm, subglobose, ovoid or ellipsoid, red, smooth. Fl. 6–7. $2n = 35$. N.

Native. Scrub, etc., mainly on calcareous soils; from Berwick southwards, local and rare; C. Ireland from E. Mayo and Westmeath to N. Tipperary and Kilkenny. Most of Europe, rare in the north and east; N. Africa (mountains).

Subfamily 3. PRUNOIDEAE. Trees or shrubs. Lvs simple; stipules present. Receptacle flat or concave. Carpel 1, rarely 2 or 5, free from the receptacle; ovules 2 in each carpel, pendent. Fr. a drupe.

16. PRUNUS L.

Fls hermaphrodite. *Petals and sepals* 5. Stamens usually 20. *Carpel* 1; style terminal. Fr. with stony endocarp, 1-seeded. Nectar secreted by the receptacle.

About 450 spp. mainly north temperate, a few extending into tropical Asia and the Andes. A number of spp. are ±commonly planted.

Measurements of lvs and petioles are when mature; they are smaller at fl.

1 Fls in clusters or umbels or solitary.	2
Fls in racemes.	6
2 Fls 1–3 from axillary buds; bud-scales caducous, not forming involucre around infl.; pedicel shorter to slightly longer than ripe fr.	3
Fls 2–6 in sessile umbels; infl. surrounded at base by the persistent bud scales, forming an involucre; pedicel at least twice as long as ripe fr.	5
3 Twigs and lvs dull; plant usually ±hairy on twigs, lvs or pedicels; lvs nearly always widest above the middle.	4
Twigs and upper surface of lvs somewhat glossy; plant glabrous except for petioles and lower part of midrib; lvs widest above or below the middle.	**3. cerasifera**
4 Very thorny shrub; bark blackish; fls usually before the lvs, mostly solitary, rarely 2; petals 5–8 mm; fr. 10–15 mm, ±erect.	**1. spinosa**
Unarmed or somewhat thorny shrubs or small trees; bark brown; fls with the lvs, 1–3; petals 7 mm or more; fr. usually 2 cm or more, pendent.	**2. domestica**
5 Lvs light dull green, somewhat pubescent beneath; most of the infls without lf-like scales; petals obovate; receptacle constricted at apex; usually a tree.	**4. avium**

Lvs dark green, rather glossy, soon nearly glabrous beneath; all the infls with the inner scales lf-like; petals orbicular; receptacle not constricted at apex; usually a shrub.   **5. cerasus**

6 Lvs deciduous, rather thin, closely and sharply serrate; peduncles with 1 or 2 lvs near base. **6. padus**
Lvs evergreen, thick and coriaceous, distantly serrate or subentire; peduncles lfless. **7. laurocerasus**

Subgenus 1. PRUNUS.

Axillary buds solitary; terminal bud 0. Lvs rolled or folded in bud. Fls solitary or in a few-fld axillary cluster, appearing with or before the lvs. Ovary with a groove down one side. Fr. usually with a whitish bloom (pruinose). Our spp. belong to Section *Prunus* with rolled lvs, 1–3 stalked fls and glabrous fr. *P. armeniaca* L. (Section *Armeniaca* (Lam.) Koch, with lvs rolled and fr. pubescent), the apricot, with broadly ovate lvs rounded or subcordate at base, subsessile fls and orange-yellow fr., is occasionally grown for its fr.

1. P. spinosa L.   Blackthorn, Sloe.

Rigid, deciduous much-branched shrub 1–4 m, often suckering and forming dense thickets. *Twigs shortly pubescent when young*, the pubescence usually persisting for c. 1 year, *becoming blackish or dark brown or grey in the 1st winter* (sometimes remaining green in shade), *dull, with numerous short lateral shoots which become thorns.* Buds obovate, hairy. Lf-blades 2–4 cm, oblong-obovate, elliptical-obovate or oblanceolate, cuneate at base, acute or obtuse, crenate-serrate, ±pubescent, at least on the midrib beneath, dull above; petiole 2–10 mm. *Fls usually appearing before the lvs, solitary* (–2); *pedicels glabrous*, rarely sparingly pubescent, shorter to slightly longer than ripe fr. *Petals* 5–8 mm, pure white, oblong-obovate. *Fr.* 10–15 mm, *globose*, blue-black, strongly pruinose, *erect*, very astringent, the flesh greenish, adherent to the stone; stone nearly globose, 7.5–10 × 6–8 mm, little flattened, nearly smooth or slightly pitted. Fl. 3–5. pollinated by various insects, protogynous. $2n = 32$. N. or M.

Native. Scrub, woods and hedges on a great variety of soils (not acid peat), not tolerant of dense shade; ascending to 415 m in Yorks. From Sutherland and Caithness southwards and throughout Ireland. Europe, except the north-east and extreme north; eastwards to Iran; S.W. Siberia (Upper Tobol region).

Var. *macrocarpa* Wallr. comprises rather less thorny more pubescent shrubs with larger lvs (–5 cm) and fls, and larger 12–16(–20) mm fr., the stone more flattened and the fls often appearing with the lvs. It occurs sporadically with more typical plants.

*2. P. domestica L.   Wild Plum.

Deciduous shrub or small tree 2–6(–12) m, often suckering. *Twigs* usually ±pubescent when young, sometimes glabrous, but amount and persistence of pubescence very variable, *becoming grey or brown in the* 1st winter, rarely remaining green, *dull; thorns* 0 *or few.* Lf-blades 4–10 cm, obovate or elliptical, cuneate at base, usually

acute, crenate-serrate, ±pubescent on both sides when young, becoming glabrous and dull above; petiole 5–20(–25) mm. *Fls appearing with the lvs*, 1–3 *together;* pedicels 0.5–2 cm, pubescent or glabrous, shorter to slightly longer than ripe fr. *Petals* 7–12(–15) mm, white, obovate. *Fr.* globose, ovoid or oblong, 2–4(–8 in cultivated forms) cm, blue-black, purple, red, green or yellow, *pendent;* flesh greenish; stone flattened, somewhat pitted. Fl. 4–5. Pollinated by various insects, protogynous or homogamous, some vars self-sterile, others self-fertile. $2n = 48$ (both subspp.). M.

There is evidence to show that this sp. originated in the past from *P. spinosa* and *P. cerasifera* by hybridization and chromosome doubling.

Introduced. Hedges, etc. Widespread but usually in small quantity in England, Wales, Ireland and S. Scotland; sporadic from Perth and Fife northwards to Caithness; Channel Is. Probably always ultimately descending from cultivated plants. Doubtfully wild anywhere, but the triploid hybrid from which it probably arose occurs in the Caucasus. The following subspp. are rather ill-defined and intermediates occur. They are completely interfertile .

Subsp. **insititia** (L.) C. K. Schneider   Bullace.

P. insititia L.

Usually a shrub, often somewhat thorny. *Twigs densely and conspicuously pubescent*, the pubescence persisting for 1 year or more. Lvs usually ±hairy above, more densely so beneath. Pedicels conspicuously pubescent. Petals pure white. Fr. (1.5–)2–3(–4) cm, globose or shortly ovoid, usually blue-black and pruinose or purple. Stone bluntly angled, the flesh adherent to it.

More thoroughly naturalized and probably longer established than the other subspp. and often found more remote from houses, often considered native. The bullace does not appear to be much cultivated at the present day, but the damson is usually regarded as a cultivated form, as is the greengage, which is not thorny, has finely hairy lvs and a usually green fr. 3–5 cm.

Subsp. **domestica**   Plum.

Usually a small tree, not thorny. Twigs sparingly hairy in the first year, or glabrous. Lvs large; petiole long. Pedicels sparingly pubescent or glabrous. Petals tinged greenish (especially in bud). Fr. 4–8 cm, oblong-ovoid, of varied colours. Stone much flattened, sharply angled, usually free from the flesh.

Widespread and rather frequent but mostly near houses. Very commonly grown for its fr.

P. ×italica Borkh. em. Kárpáti (*P. domestica* subsp. *domestica* × *subsp. insititia*, *P. domestica* subsp. *italica* (Borkh.) Hegi) is intermediate in the pubescence of the twigs, pedicels and sepals, in the shape of the fr. and stone, and in the adherence of the flesh to the stone. Hybrids are common in S. England, where they probably escaped after having arisen in cultivation.

P. ×fruticans Weihe (*P. domestica* × *spinosa*) comprises a series of plants variously intermediate between the parents in habit, thorniness, pubescence, size of fls, lvs and fr., and in stone shape, which are found throughout southern Britain northwards to at least Yorks. The plants are apparently fertile, but also spread rather vigorously by suckers.

*3. P. cerasifera Ehrh.   Cherry Plum.

P. divaricata Ledeb.

Deciduous shrub or small tree to 8 m, rarely thorny. *Twigs glabrous from the first, usually green* (often reddish on the exposed side) *in the* 2nd *year, rather glossy. Lf-blades* 3–7 cm, ovate-elliptical or obovate, rounded or broadly cuneate at base, acute, unequally crenate-serrate with rather deep obtuse teeth, glabrous and *somewhat glossy above*, usually pubescent along the lower part of the midrib beneath, at least when young; petiole 5–10 mm. *Fls appearing with the lvs, usually solitary* (–2); pedicels 5–15 mm, glabrous, shorter to slightly longer than ripe fr. *Petals* 7–11 mm, white, ovate. Fr. 2–2.5 cm, globose, yellow or reddish. Stone little flattened, ±orbicular, quite smooth on the faces. Fl. 3–4, earlier than *P. spinosa* and *P. domestica*. Fr. only produced in favourable years. $2n = 16$. M.

Introduced. Frequently planted in hedges in many parts of England, often far from houses and apparently naturalized from Hants and Sussex to Norfolk and Derby; also in Scilly Is., Berwick and Edinburgh. Native of Russian C. Asia and Iran through the Caucasus to the Balkan peninsula. A form with pink fls and purple lvs (var. *atropurpurea* Jaeger = var. *pissardii* (Carr.) L. H. Bailey) is very commonly grown in small gardens.

Subgenus 2. AMYGDALUS (L.) Focke

Axillary buds 3, the lateral ones fl-buds; terminal bud present. Lvs folded in bud. Fls 1–3 in axillary clusters. Fr. usually tomentose. Stone often deeply pitted. The two following are frequently cultivated.

*P. dulcis (Miller) D. A. Webb (*P. amygdalus* Batsch)
Almond.
Lvs 7–12 cm, ±lanceolate, widest rather below the middle, serrulate. Fls pink, 3–5 cm diam., appearing before the lvs. Fr. 3–6 cm, ellipsoid, compressed, green and dry when ripe, finally splitting. Native of C. and S.W. Asia and N. Africa.

*P. persica (L.) Batsch   Peach.
Lvs 8–15 cm, oblanceolate-elliptical to narrowly oblong, widest about or rather above the middle, serrate. Fls pink, 2.5–3.5 cm diam., appearing before the lvs. Fr. 5–7 cm, subglobose, reddish on the sunny side, juicy, not splitting. The nectarine is a glabrous-fruited form (var. *nectarina* (Aiton) Maxim.). Besides the forms cultivated for fr., forms with semi-double fls, ranging from white to red, are frequently grown. Native country probably China, or derived in cultivation from the Chinese *P. davidiana* (Carr.) Franchet.

Subgenus 3. CERASUS (Miller) Focke

Lvs folded in bud. Terminal bud present. Fls solitary or in umbels or short racemes. Style (and sometimes ovary) grooved. Fr. usually glabrous, not pruinose.

Our spp. belong to section *Cerasus* with solitary buds, teeth of lvs obtuse, bud scales persisting round the infl. as an involucre, fls in umbels and sepals deflexed.

Of the other spp. **P. serrulata** Lindley (Section *Pseudocerasus* Koehne, sepals upright, teeth of lvs acute), Japanese Cherry, is commonly planted, usually in pink double-fld forms. Other spp. are less frequently grown.

4. P. avium (L.) L.   Gean, Wild Cherry.

Deciduous *tree* 5–25 m, of rather open habit, suckering rather freely. Bark smooth, reddish-brown, peeling off in thin strips. Branches spreading or ascending. *Lf-blades* 6–15 cm, obovate-elliptical, acuminate, crenate-serrate, *light dull green* and glabrous above, sparingly but persistently appressed-*pubescent beneath*, rather thin, drooping; petiole *to* 5 cm, with 2 large glands near the apex. Fls 2–6, cup-shaped, in usually sessile umbels; *bud-scales* greenish, persisting below the infl., those of most of the infls *not lf-like*, but some infls with 1 or 2 inner ones with small lf-like blades; pedicels 2–5 cm. *Receptacle constricted at apex.* Sepals oblong, deflexed, entire. *Petals* 8–15 mm, white, *obovate*, gradually narrowed to the base, emarginate or retuse. Fr. c. 1 cm, subglobose, bright or dark red, glabrous, sweet or bitter. Fl. 4–5. Pollinated by various insects, homogamous, all vars completely self-sterile. $2n = 16$. MM. or M.

Native. Woods and hedges on the better soils, rather common in England, Wales and Ireland, becoming rare in northernmost Scotland but extending to Caithness and Sutherland. Most of Europe, except the extreme north and east, but rare as a native in the Mediterranean region; N. Africa (mountains); W. Asia. The garden Sweet Cherries are referable to this sp. but sometimes have an extra chromosome ($2n = 17$), probably because of past hybridization with *P. cerasus*.

*5. P. cerasus L.   Dwarf Cherry.

Differs from *P. avium* as follows: *Usually a shrub*, rarely a small tree to c. 7 m, of rather bushy habit, suckering very freely. Branches often somewhat drooping. *Lf-blades* 5–8 cm, relatively wider, *dark green* and somewhat shining *above, soon glabrous* or nearly so *beneath*, firm, spreading; petiole 1–3 cm, the glands smaller or 0. Fls flat; umbels usually shorter-stalked; *inner bud scales* of all the infls *with lf-like blades*; pedicels 1–4 cm. *Receptacle not or scarcely constricted.* Sepals wider, sometimes crenate. *Petals* ±orbicular, *rounded at the base*, entire, emarginate or regularly notched. Fr. bright red, acid. Fl. 4–5. Pollinated mainly by bees, protogynous (or homogamous?). $2n = 32$. M.

Introduced. Hedges, etc.; widespread but local in Britain northwards to Lincoln and Westmorland, introduced further north; throughout Ireland, but local. Native of S.W. Asia. The garden Sour and Morello Cherries belong to this sp., and all our wild plants are, in all probability, descended from them.

Subgenus 4. PADUS (Miller) Focke

Lvs folded in bud. Terminal bud present. Fls in long racemes. Style not grooved. Fr. glabrous, not pruinose.

Section *Padus*. Lvs deciduous. Peduncle lfy. Sepals deciduous.

6. P. padus L.   Bird Cherry.

Deciduous tree 3–15 m. Bark brown, strong smelling, peeling. Twigs brown or grey. Lf-blades 5–10 cm, elliptical or obovate, acuminate, rounded or cordate at base, closely and sharply serrate, glabrous except sometimes for tufts of hairs in the axils of the primary veins beneath; petiole 1–2 cm, with a gland on each side of the apex. Fls 10–40 in long, lax, ascending to drooping, glabrous racemes, 7–15 cm; peduncle with 1 or 2 lvs. Receptacle hairy within. Sepals short, obtuse, ascending, gland-fringed, deciduous. Petals 4–6(–10) mm, white, irregularly toothed. Fr. 6–8 mm, ovoid, black, astringent. Fl. 5. Pollinated by various insects, mainly flies, protogynous. $2n = 32$. M. or MM.

Native. Woods, etc. ascending to 610 m, widespread and rather common in Britain from Sutherland and Caithness southwards to York, Stafford and Monmouth, and in Suffolk, Norfolk and Cambridge; sporadic and local throughout Ireland; widely planted elsewhere. Most of Europe except the Mediterranean region, Balkan peninsula and S.E. Russia; Caucasus; Asia from W. Siberia to the Himalaya and N.E. Asia Minor; Morocco (Atlas, very rare).

Section *Calycopadus* Koehne. Lvs deciduous. Peduncle lfy. Calyx persistent.

*P. serotina Ehrh.

Deciduous tree to 22 m. Lf-blades 5–12 cm, oblong-ovate or oblong-lanceolate, cuneate or rounded at base, serrulate with appressed teeth, shining above. Receptacle glabrous within. Calyx persistent in fr. Sometimes planted among native vegetation. Native of eastern North America.

Subgenus 5. LAUROCERASUS (Duh.) Rehder.

Lvs folded in bud, evergreen. Peduncle lfless. Fls in long raceme. Calyx deciduous. Fr. glabrous.

*7. P. laurocerasus L.   Cherry Laurel.

Evergreen glabrous shrub or small tree 2–6 m. Lvs 5–18 cm, oblong-obovate, dark green and very glossy above, acuminate, cuneate or rounded at base, distantly serrate or subentire, thick and coriaceous; petiole 5–10 mm, green. Fls numerous, in ±erect racemes, 5–12 cm; peduncle lfless. Petals white, c. 4 mm. Fr. c. 8 mm, ovoid, purple-black. Fl. 4–6. $2n = $ c. 176. M.

Introduced. Commonly planted in gardens, woods, etc., frequently self-sown and ±naturalized in Great Britain northwards to Dunbarton and Kincardine, but absent from much of C. and N. England, Scotland and E. Wales; scattered in Ireland northwards to Galway and Louth. Native of E. part of Balkan peninsula and S.W. Asia.

*P. lusitanica L.   Portugal Laurel.

Evergreen shrub. Lvs 6–12 cm, oblong-ovate, dark green, serrate; petioles 15–25 mm, deep red. Less frequently planted than *P. laurocerasus* and rarely naturalized. $2n = 64$. Native of the

Iberian peninsula, S.W. France, the Azores, Madeira and the Canaries.

Subfamily 4. MALOIDEAE. Trees or shrubs with long and short shoots, the latter sometimes modified as thorns. Lvs simple or pinnate; stipules present, usually small and caducous. Receptacle deeply concave. Sepals 5, imbricate. Petals 5. Carpels 1–5, united to the receptacle at least in the lower half so that the ovary is inferior or ½-inferior. Ovules (1–)2(–20) in each carpel, axile or basal. Fr. a pome or drupe, usually crowned by the persistent calyx.

A very distinct subfamily. Besides the morphological characters, the basic chromosome number (17) is constant and is not found elsewhere in the Rosaceae.

17. COTONEASTER Medicus

Shrubs, rarely small trees, *not thorny. Lvs entire*, shortly petiolate. Fls solitary or few or in corymbs. Sepals short, ±triangular. Petals imbricate in bud. Stamens c. 20. *Carpels 2–5, free on the inner side, the wall stony in fr.; ovules 2 in each carpel*, similar; styles free. Fr. with mealy flesh, red, purple or black, with 2–5 stones. Nectar secreted by the inner wall of the receptacle.

About 50 spp., north temperate Old World. A number are grown in gardens.

1 Petals spreading, white; lvs 8 mm or less, widest near
  the apex.   **5. microphyllus**
  Petals ±erect, ±tinged with red or pink; lvs 10 mm
  or more, widest near the middle.   2
2 Shrub not more than 0.7 m, usually procumbent; fls
  solitary or in pairs.   **3. horizontalis**
  Erect shrub usually at least 1 m; fls in clusters of 2–5
  or more, rarely solitary.   3
3 Lvs green and sparsely hairy or appressed-pubescent
  beneath.   **2. simonsii**
  Lvs grey-tomentose beneath.   4
4 Fls in cymes of 2–3(–4); lvs 1.5–4 cm, ovate to suborbicular.   **1. integerrimus**
  Fls in corymbs of 10–30; lvs 3.75–8.75 cm, oblong to
  lanceolate-ovate.   **4. bullatus**

Section 1. *Cotoneaster*. Petals erect, usually pink.

1. C. integerrimus Medicus   Wild Cotoneaster.
C. vulgaris Lindley

Bushy deciduous shrub 15–100(–200) cm. *Twigs tomentose when young, soon glabrous. Lvs 1.5–4 cm, ovate to suborbicular, rounded at base, obtuse or acute*, usually mucronate, green and glabrous above, persistently *grey-tomentose beneath*; petioles 2–4 mm. Fls 1–4 in short cymes. Petals pink, c. 3 mm. Styles 2(–5). Fr. c. 6 mm, subglobose, red. Fl. 4–6. Visited by various insects, especially wasps, protogynous or homogamous, self-pollination possible. Fr. 8. $2n = 68$. N.

Native. Limestone ledges on Great Orme's Head (Caernarvon) in very small quantity. Rocky places in much of Europe, but absent from most of the USSR, the extreme north, and much of the Mediterranean region; Caucasus; N. Asia Minor, N. Iran.

*2. C. simonsii Baker   Himalayan Cotoneaster.

Deciduous or semi-evergreen shrub 1–4 m; stems sub-erect. *Twigs strigose-pubescent, the pubescence persisting for 2 or 3 years. Lvs* 1–3 cm., ovate or broadly elliptical, broadly cuneate at base, acute, deep green and glabrous or slightly pubescent above, paler *green and sparsely strigose beneath*, somewhat coriaceous; petioles 2–4 mm. Fls 2–4 in short cymes. Sepals pubescent. Petals pink. Fr. c. 8 mm, obovoid, scarlet. Fl. 5–7. Fr. 10. N. or M.

Introduced. Commonly planted and naturalized in many places in S. England northwards to Hereford and Norfolk, Wales, sporadically in N. England and along the Scottish coasts, and in S. Ireland (Kerry, Cork, Wexford). Native of the Khasia Hills (E. India).

*3. C. horizontalis Decaisne   Wall Cotoneaster.

Deciduous or semi-evergreen *shrub up to* 0.7 m, *usually procumbent*; stems spreading, with branches flattened horizontally in herringbone pattern. Lvs up to 1.2 cm, suborbicular or broadly elliptical, otherwise as for *C. simonsii. Fls solitary or in pairs*, subsessile. Petals reddish or whitish. Fr. 5–6 mm, subglobose, bright red. Fl. 6–7. Fr. 9–10. N.

Introduced. Very commonly grown in gardens and a casual escape or locally naturalized. Native of W. China.

*4. C. bullatus Bois

Deciduous shrub up to 3 m; stems suberect. Twigs pubescent. Lvs 3.75–8.75 cm, *oblong to lanceolate-ovate*, cuneate at base, acute, bullate between veins, dark green above, *grey-hairy beneath. Fls* 10–30 *in corymbs*. Petals pink. Fr. c. 8 mm, globose, red. Fl. 6–7.

Introduced. Railway lines, dune-slacks, roadsides and rocky areas (incl. walls). Grown in gardens and sometimes escaping. Locally naturalized in at least Warwick, Lancs, Kintyre, Argyll, Down, Antrim and Louth. Native of W. China.

*C. divaricatus Rehder & E. H. Wilson

Deciduous shrub to 2 m; branches spreading, some at ground level; twigs rigid, alternate, brown, pubescent in 1st and 2nd years, later glabrous. Lvs 1–2 cm., ±ovate, acuminate, glossy green above, paler and with scattered white hairs beneath. Fls small, pink, in groups of 3 (often 1–2 in shade-plants). Fr. red. Fl. 5.

Native of China (introduced 1904). Cultivated in gardens and locally ±naturalized in W. Kent.

Section 2. *Chaenopetalum* Koehne. Petals spreading, white.

*5. C. microphyllus Wallich ex Lindley
                              Small-leaved Cotoneaster.

Low evergreen shrub to 1 m, with rigid spreading or drooping branches. Twigs strigose-pubescent. *Lvs* 5–8 mm, *obovate-cuneiform*, cuneate at base, *obtuse to retuse*, dark green and glabrous above, glaucous and appressed hairy beneath, coriaceous; petiole 1–2 mm. Fls 1(–3), c. 1 cm diam. Anthers purple. Styles 2. Fr.

c. 6 mm, globose, crimson. Fl. 5–6. Fr. 9–10. $2n = 68$. N.

Introduced. Commonly grown in gardens (introduced 1824) and now naturalized in many places, especially on limestone near the sea. Scattered throughout S. England, Wales and W. Ireland, sporadic elsewhere and absent from much of C. Ireland, the English Midlands, and S. and inland Scotland northwards to Caithness and the Outer Hebrides. Native of the Himalaya.

*C. frigidus Wallich ex Lindley

Deciduous tall shrub or small tree to 6 m. Lvs ±oblong, 6–12 cm, dull green, glabrous or nearly so when mature. Fls numerous, in corymbs. Fr. red.

Native of the Himalaya. Frequently planted and has occurred as an escape.

C. affinis Lindley, *C. lacteus* W. W. Sm. and *C. multiflorus* Bunge have been reported as garden escapes and may spread by bird-dispersal.

Pyracantha M. J. Roemer

Differs from *Cotoneaster* as follows: Thorny. Lvs usually toothed. Fls always in many-fld corymbs. Carpels always 5. Petals white, spreading. Several spp. are grown in gardens, the commonest being ***P. coccinea** M. J. Roemer. Lvs 3–4 cm, lanceolate to oblanceolate, acute, glabrous or slightly pubescent beneath. Infl. pubescent. Fls c. 8 mm diam. Fr. 5–6 mm, subglobose, scarlet.

Sometimes escaping, rarely ±naturalized. Native of S. Europe, westwards to N.E. Spain; eastwards to W. Asia.

18. Crataegus L.

Deciduous trees or shrubs, usually *with thorns. Lvs lobed or serrate.* Fls in corymbs (rarely solitary). Sepals short, ±triangular. Stamens 5–25. *Carpels* 1–5, free at the apex, *united* at least *at the base on the inner side, the wall stony in fr.*; *ovules* 2, *the upper sterile*; styles free. Fr. usually with mealy flesh, red, yellow or black. Nectar secreted by a ring in the receptacle; fls strong-smelling, visited by various Diptera, Hymenoptera and Coleoptera, protogynous.

About 200 spp. in the north temperate zone, c. 90 in the Old World; a large number of highly critical North American spp. have been described, but many are believed to be hybrids. They have mostly larger, less-divided lvs and larger fls than our spp. and are sometimes confused with *Sorbus* spp. A number are cultivated and some have been found as escapes. The pink and double-fld forms so commonly grown are forms of our native spp.

Lobes of lvs of short shoots wider than long, usually
     rounded, the sinuses not reaching half-way to midrib;
     styles mostly 2 (often 1 or 3 in some fls).   **1. laevigata**
Lobes of lvs of short shoots longer than wide, ±triangular,
     the sinuses reaching more than half-way to midrib; style
     1 (sometimes 2 in some fls).   **2. monogyna**

1. C. laevigata (Poiret) DC.   Midland Hawthorn.
C. oxyacanthoides Thuill.

Thorny much-branched shrub or small tree 2–10 m. Twigs glabrous or with a few long hairs when young.

Lf-blades of short shoots 1.5–5 cm, obovate in outline, 3–5-lobed, glabrous except for scattered hairs on the main veins on both sides when young, *without axillary hair-tufts*; *lobes shallow (rarely reaching half-way to the midrib), usually rounded in outline, wider than long, serrate* except in the sinuses; main veins straight or curving upwards; lvs of the long shoots usually more deeply lobed, with conspicuous lf-like stipules. Fls rarely more than 10. Receptacle and pedicels glabrous (?rarely woolly). Sepals deltate, about as wide as long. Petals 5–8 mm, white. Anthers pink or purple. *Styles mostly 2* (often 1 or 3 in some fls). Fr. (6–)8–10(–13) mm, deep red; stones mostly 2. Fl. 5–6, about a week earlier than *C. monogyma*. Self-incompatible. $2n = 34$. M.

Native. Woods, less frequently scrub or hedges, much less common than *C. monogyna* though locally more frequent in woods and more tolerant of deep shade; mostly on clay or loam. In England local and mainly in the east, extending from Kent, Norfolk and Lincoln westwards to Hants, Somerset, Gloucester and Shropshire, also in Cheshire and Westmorland; in Wales only in Monmouth and Glamorgan; extinct in Ireland (?except Macroon, Co. Cork), but hybrids suggest formerly occurred with oak forests. N.W., N.C. and C. Europe, from C. Sweden and Latvia to the W. Pyrenees and N. Italy.

2. C. monogyna Jacq.   Hawthorn.

Habit of *C. laevigata*. Twigs glabrous. *Lf-blades* of short shoots 1.5–3.5 cm, ovate or obovate in outline, 3–7-lobed, glabrous except for *patches of hairs in the axils of the lower veins beneath* or with occasional scattered hairs; *lobes deep (usually reaching more than half-way to the midrib), tapered to an acute or subobtuse apex, longer than wide, entire or sparingly serrate near their apices*; main veins curving downwards; lvs of the long shoots more deeply lobed, with conspicuous lf-like stipules. Fls often up to 16, sometimes more. Receptacle and pedicels glabrous to woolly. Sepals triangular, 1–2 times as long as wide. Petals 4–6 mm, white. Anthers pink or purple. *Style 1* (or 2 in a few fls). Fr. (6–)8–10(–13) mm, deep red; calyx appressed or spreading; stone 1, rarely 2. Fl. 5–6. Self-incompatible. $2n = 34$. M.

Native. Scrub, woods and hedges, ascending to 550 m; the commonest scrub-dominant on most types of soil, rare only on wet peat and poor acid sands; the shrub most commonly planted for hedges. Throughout the British Is., very common in England, rare in northernmost Scotland. Europe except the northern and eastern margins; extending to Afghanistan.

British material belongs to subsp. **nordica** Franco.

C. ×media Bechst. (*C. laevigata × monogyna, C. ×intermixta* Beck), is intermediate between the parents in all characters, of which lf-shape, particularly the degree of lobing, is the most useful. However, there is a complete range of forms linking the parents and all seem to be fertile. The hybrid occurs wherever *C. laevigata* has occurred, northwards to Derby, and in Ireland where there were formerly oak forests.

*C. orientalis Pallas ex Bieb.

Lvs pinnatifid, with narrow lobes, grey-tomentose beneath. Styles 4–5. Fr. 1.5–2 cm diam, orange-red. Self-sown in one or two places. Native of S.E. Europe and S.W. Asia.

19. MESPILUS L.

Deciduous trees or shrubs, sometimes thorny. Lvs entire or finely serrulate above. *Fls solitary*, large. *Sepals large*, lf-like. Stamens 30–40. *Carpels 5, almost completely united, the walls stony in fr.*; ovules 2, the upper sterile; styles free. Fr. brown, crowned by the persistent calyx, at first hard, finally soft (bletting). Nectar secreted by a ring in the receptacle.

One sp.

*1. M. germanica L.   Medlar.

Pyrus germanica (L.) Hooker fil.

Thorny shrub 2–3 m (in cultivation a thornless tree to 6 m). Twigs densely pubescent when young, glabrous and blackish in the second year. Lvs 5–12 cm, lanceolate or oblanceolate, cuneate at base, acute, entire or serrulate near the apex, appressed-pubescent on both sides or subglabrous above; petiole c. 2 mm. Pedicels 5 mm or less, densely pubescent like the receptacle and calyx. Sepals 10–16 mm, linear-lanceolate to linear-triangular, somewhat lf-like. Petals c. 12 mm, suborbicular. Anthers red. Fr. 2–3 cm, subglobose, crowned by the calyx. Fl. 5–6. $2n = 34$. M.

Introduced. Grown for its fr. and naturalized (recorded 1597) in hedges in S. England, northwards to Middlesex, Oxford and Gloucester, and in Stafford and S. Yorks, rare; Channel Is. Native of S.E. Europe (extending to Sardinia and Sicily) and S.W. Asia (to Turkestan); cultivated and naturalized in C. and W. Europe.

20. AMELANCHIER Medicus

Deciduous shrubs or small trees, not thorny. Lvs simple, serrate. *Fls in racemes*, rarely solitary. Sepals ±triangular. Petals narrow. Stamens 10–20. Carpels (2–)5, free from the receptacle above and so ½-inferior, the *walls cartilaginous in fr.*; ovules 2 in each carpel, *separated by a false septum*; styles united below or free. Fr. bluish- or purplish-black, usually sweet and juicy.

About 25 spp., north temperate mainly North America; others, in addition to the following, are sometimes grown.

*1. A. ×lamarckii Schroeder   Juneberry.

A. laevis auct., non Wieg.; *A. grandiflora* Rehder; *A. canadensis* auct., non (L.) Medicus; *A. confusa* Hyl.

Shrub or small tree to 12 m. Lf-blades 3–7 cm, ovate-elliptical or ovate-oblong, acute, shortly acuminate, rounded or subcordate at base, finely and sharply serrate, glabrous or tomentose beneath when very young, purplish when unfolding. Infl. slender, spreading or drooping, glabrous or nearly so, many-fld. Petals 10–22 mm, linear-oblong, white. Top of ovary glabrous. Fr. globose, blackish-purple. Fl. 4–5. Fr. 6–7. M.

Introduced. Grown in gardens and sometimes becoming naturalized on light acids soils; common over several sq. miles in the Hurtwood (Surrey).

The British plant appears to be a complex hybrid, which probably arose in cultivation, combining features of *A. laevis* Wieg., from north-eastern North America, and *A. canadensis* (L.) Medicus and *A. arborea* (Michx. fil.). Fernald, from the eastern USA.

21. SORBUS L.

Deciduous trees or shrubs, not thorny. Lvs pinnate or simple and lobed or toothed. *Fls in compound corymbs.* Sepals ±triangular. Petals white (pink in some foreign spp.). Stamens 15–25. *Carpels 2–5,* united at least to the middle, inferior or ½-inferior, *the walls cartilaginous in fr.*; ovules 2, without false septum; styles free or joined below. Fr. variously coloured, with 1 or 2 seeds in each cell.

About 100 spp., north temperate regions. Several foreign spp. are sometimes planted.

The British spp. of this genus consist of three widespread diploids. *S. aria, S. aucuparia* and *S. torminalis* and a number of polyploids, some of which (perhaps all) are certainly apomictic. There are three main groups of polyploids: (A) resembling *S. aria* (*S. aria sens.lat*), (B) intermediate between (A) and *S. aucuparia* (*S. intermedia sens. lat.*) and (C) those intermediate between (A) and *S. torminalis* (*S. latifolia sens. lat.*); *S. pseudofennica* is intermediate between *S. aucuparia* and *S. arranensis* (*S. intermedia sens. lat.*). The groups (B) and (C) probably originated as hybrids of *S. aria* or other species of group (A) with *S. aucuparia* and *S. torminalis* respectively. In contrast to *S. aria* which is very variable, the polyploid spp. are ±constant.

The most useful characters are:

Lvs. Depth and character of toothing and lobing of the basal lvs of short shoots. The depth of the lobes in the following account is measured perpendicular to the midrib. The main types of toothing are (*a*) ±symmetrical teeth projecting perpendicular to the lf margin (e.g. *S. porrigentiformis*), and (*b*) with the outer margin longer than the inner and somewhat curved so that the tooth is directed more towards the lf-apex (e.g. *S. aria*); there can be much variation between the teeth even on one lf, and this character is best observed at a point about ⅓ the distance from apex to base.

Colour of upper and lower surfaces of lvs, and the density of the hairs beneath.

Fls. These have been little studied. The colour of the anthers is often constant in a sp., and the petal size may be another useful character. Other characters will probably also be available when the fls have been studied further.

Fr. Colour: usually a constant and important character. Size and shape of fr. when ripe. Number, size and distribution of the lenticels on the fr. None of these characters preserves well in herbarium specimens, however.

The following key is intended for use primarily on fruiting specimens which are the best for beginning the study of the genus.

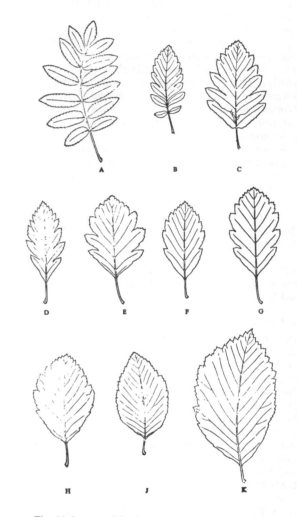

Fig. 44. Leaves of *Sorbus*. A, *S. aucuparia*; B, *S. aria* × *aucuparia*; C, *S. pseudofennica*; D, *S. arranensis*; E, *S. leyana*; F, *S. minima*; G, *S. intermedia*; H, *S. anglica*; J, *S. aria*; K, *S. leptophylla*. ×⅓.

1 Lvs pinnate with 4 or more pairs of lflets, the terminal lflet ±equalling the lateral ones (Fig. 44A).
                                  1. aucuparia
  Lvs simple or with up to 3 pairs of free lflets at the base and with the terminal part several times as large.   2
2 Lvs or many of them with at least 1 free lflet.   3
  Lvs without free lflets.   4
3 Veins of lvs 7–9(–10) pairs; lvs (Fig. 44c) 5.5–8.5 cm, mostly 1.5–1.7 times as long as wide; fr. longer than wide; Arran.   **2. pseudofennica**

Veins of lvs 10–12 pairs; lvs (Fig. 44B) 7–11 cm, 1.6–2.3 times as long as wide; fr. subglobose; occurring occasionally as a single tree.   **aria × aucuparia**

4 Lvs (Fig. 44B) green on both sides, subglabrous beneath except when very young, deeply lobed; lobes acuminate; fr. brown.   **20. torminalis**

   Lvs persistently grey- or white-tomentose beneath.   5

5 Fr. red.   6

   Fr. orange or brown; lvs grey-tomentose beneath, with ±triangular lobes or at least doubly serrate with the primary teeth acuminate and very prominent.   **(latifolia** sens. lat.)   19

6 Lvs grey beneath, lobed, with the deepest lobes extending at least $\frac{1}{6}$ of the way to the midrib.   **(intermedia** sens. lat.)   7

   Lvs white beneath, not lobed or shallowly crenately lobed with the lobes not reaching more than $\frac{1}{6}$ of the way to the midrib (and then lvs very white beneath).   **(aria** sens. lat.)   11

7 Lobes of the lvs extending nearly half-way to the midrib or more.   8

   Lobes of the lvs not extending to more than $\frac{1}{3}$ of the way to the midrib (sometimes to half-way on a few lvs).   9

8 Lvs (Fig. 44D) ±elliptical, acute, with 7–8(–9) pairs of veins; fr. longer than wide; petals c. 4 mm; Arran.   **3. arranensis**

   Lvs (Fig. 44E) ±oblong, obtuse or subacute, with (8–)9–10 pairs of veins; fr. subglobose; petals c. 5 mm; Brecon.   **4. leyana**

9 Lvs (fig. 44F) ±elliptical, 6–8 cm, mostly 1.8–2.2 times as long as wide; fr. 6–8 mm, subglobose; petals c. 4 mm; Brecon.   **5. minima**

   Lvs 7–12 cm, mostly not more than 1.8 times as long as wide; fr 7–15 mm; petals c. 6 mm.   10

10 Lvs (Fig. 44G) ±elliptical, many of them rounded at the base, yellowish-grey-tomentose beneath; fr. much longer than wide; anthers cream.   **6. intermedia**

   Lvs (fig. 44H) ±obovate, nearly all cuneate at the base, whitish-grey-tomentose beneath; fr. subglobose or wider than long; anthers pink or tinged pink.   **7. anglica**

11 Fr. longer than wide.   12

   Fr. subglobose or wider than long.   15

12 Lvs (Fig. 45F) obovate, tapering gradually to a cuneate base from about the middle of the lf and entire in the basal $\frac{1}{4}$ or more, rounded above; veins 8–9(–10) pairs; S. Somerset and N. Devon.   **16. vexans**

   Lvs not tapering at the base for a long distance (or if so also tapering above) and entire at the base for $\frac{1}{3}$ or less; veins rarely less than 10 pairs (except in S. wilmottiana).   13

13 Lvs with the terminal tooth and those terminating the main veins markedly prominent (except in S. wilmottiana), greenish-white-tomentose beneath; fr. with the lenticels mainly towards the base.   14

   Lvs (Fig. 44J) with the terminal tooth and those terminating the main veins not or scarcely prominent, pure white-tomentose beneath; fr. with scattered lenticels.   **8. aria**

14 Lvs (Fig. 44K) yellow or dark green above, mostly 1.5–1.7 times as long as wide; teeth somewhat curved on the outer margin, the basal margins of

the lf not arched inwards; veins c. 11 pairs; Brecon.   **9. leptophylla**

   Lvs bright green above, mostly 1.6–2.0 times as long as wide; teeth tending to be curved on the outer margin, the basal margins of lf not arched inwards; veins usually 8–9 pairs; Avon Gorge.   **10. wilmottiana**

   Lvs (Fig. 45A) bright green above, mostly 1.1–1.4 times as long as wide; teeth ±symmetrical, the basal margins of lf frequently arched inwards; veins usually 10–11 pairs; Wye Valley and Avon Gorge.   **11. eminens**

15 Lvs obovate or oblanceolate, evenly tapered to a cuneate base for at least $\frac{1}{3}$ and usually $\frac{1}{2}$ their length and entire in the lower $\frac{1}{4}$, bright or dark green above.   16

   Lvs variously shaped (sometimes obovate), with base rounded or cuneate, entire for $\frac{1}{5}$ or less, dull or yellow-green above.   18

16 Lvs (Fig. 45C) bright green above, rather thinly greenish-white-tomentose beneath, 6–9.5 cm long, 1.3–1.7 times as long as wide; the teeth ±symmetrical, those terminating the main veins markedly prominent; fr. 12 mm or less, crimson with few large lenticels towards the base; S.W. England and Wales.   **13. porrigentiformis**

   Lvs dark green above, rather thickly greyish- or pure white-tomentose beneath, mostly more than 8 cm long and 1.5 times as long as wide or more; the teeth terminating the main veins, not or scarcely prominent; fr. 12 mm or more, with at least a moderate number of lenticels.   17

17 Teeth of lvs ±symmetrical and directed outwards, the veins mostly 8–10 pairs (Fig 45D); fr. with a moderate number of large lenticels towards the base and scattered smaller ones; Lancashire and Westmorland.   **14. lancastriensis**

   Teeth of lvs mostly somewhat curved on the outer margin and directed towards the apex of the lf, the veins mostly 7–9 pairs (Fig. 45E); fr. with numerous scattered moderate and small lenticels (if lenticels few see S. vexans).   **15. rupicola**

18 Teeth of lvs ±symmetrical and directed outwards, (Fig. 45B); lvs greenish-white-tomentose beneath; fr. with rather few lenticels mainly towards the base; Ireland.   **12. hibernica**

   Teeth of lvs mostly curved on outer margin and directed towards apex of lf (Fig. 45J); lvs very white beneath; fr. usually with few or many scattered lenticels.   **8. aria**

19 Lvs (Fig. 45J) ±obovate; anthers pink; fr. longer than wide, bright orange; Avon Gorge. **17. bristoliensis**

   Lvs not (or only a few of them) broadest above the middle; anthers cream; fr. dull orange to brown.   20

20 Lvs (Fig. 45H) rhombic-elliptical, c. 1.6–1.9 times as long as wide, mostly cuneate at base, whitish-grey beneath; fr. subglobose, with numerous large lenticels; N. Devon and Somerset.   **18. subcuneata**

   Lvs (Fig. 45G) ±ovate, 1.3–1.6 times as long as wide, rounded at base, greenish-grey beneath; fr. subglobose, with numerous large lenticels; Devon, E. Cornwall, S.E. Ireland (a closely allied plant sometimes naturalized elsewhere).   **19. devoniensis**

   Not as above; lower margin of lf often arched slightly inwards near base (Fig. 46A); fr. in most plants

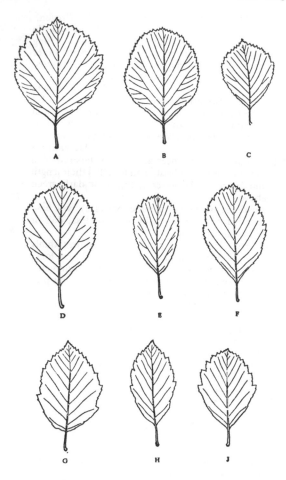

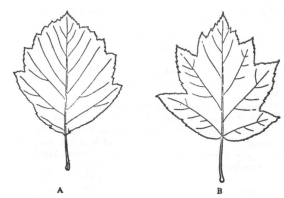

Fig. 46. Leaves of *Sorbus*. A, *S. aria* × *torminalis*;
B, *S. torminalis*. ×⅓.

Fig. 45. Leaves of *Sorbus*. A, *S. eminens*; B, *S. hibernica*;
C, *S. porrigentiformis*; D, *S. lancastriensis*; E, *S. rupicola*;
F, *S. vexans*; G, *S. devoniensis*; H, *S. subcuneata*;
J, *S. bristoliensis*. ×⅓.

longer than wide, with few small lenticels; Wye Valley.

aria × torminalis

1. S. aucuparia L.   Rowan, Mountain Ash.
Pyrus aucuparia (L.) Ehrh.

Slender tree to 15(–20) m with narrow crown and ±ascending branches. Bark greyish, smooth. Twigs pubescent when young, then glabrous and greyish-brown. Buds 10–14 mm, ovoid or ovoid-conical, dark brown, somewhat pubescent. *Lvs* (Fig. 44A) 10–25 cm, *pinnate; lflets* (4–)6–7(–9) *pairs, the terminal lflet ±equalling the lateral* (never larger), 3–6 cm, oblong, acute or subacute, ±rounded at the often somewhat unequal base, serrate, sometimes doubly so, dark green and glabrous above, subglaucous beneath and pubescent at first, especially on the midrib, usually becoming subglabrous; petiole 2–4 mm. Infl. dense, many-fld, woolly-pubescent in fl. Petals c. 3.5 mm, ovate. Anthers

cream. Styles 3–4. Fr. 6–9 mm, subglobose, scarlet, with a few inconspicuous lenticels. Fl. 5–6, Fr. 9. $2n = 34$. MM. or M.

Native. Woods, scrub, mountain rocks, ascending to 975 m (higher than any other tree), mainly on the lighter soils, rare or absent on clays and soft limestones, common in the north and west, rare and perhaps not native in some eastern and central English counties and less common in C. Ireland. Most of Europe except the extreme north and some Mediterranean islands; Caucasus; Morocco (high mountains); N. Asia Minor.

*S. domestica L.   Service-tree.

Resembling *S. aucuparia* in its lvs but a larger tree. Branches spreading. Bark rough, scaling. Fls 16–18 mm diam. Styles 5. Fr. c. 25 mm, apple- or pear-shaped, green or brown with numerous large lenticels and stone cells in the flesh. $2n = 34$.

A native of S. Europe, N. Africa and Asia Minor, which is locally planted for its fr. and for ornament.

S. aria × aucuparia = S. × thuringiaca (Ilse) Fritsch

Lvs (Fig. 44B) 7–11 cm, ±oblong in outline, 1.6–2.3 *times as long as wide, mostly with 1–3 pairs of free lflets at the base* but on some trees many of the lvs are lobed nearly to the base but without free lflets, upper part with ±oblong lobes, serrate, obtuse, dull green and glabrous above, greenish-grey-tomentose beneath; *veins* (including free lflets) 10–12 *pairs*; petiole 1.5–3 cm. Infl. woolly-pubescent. Petals 4–6 mm. Anthers cream, more rarely pink. Styles 2–3. *Fr.* 11 mm, *ovoid, brownish-red*, with few inconspicuous lenticels. Occurring very rarely as a single tree with the parents from Somerset to Banff. Sometimes also planted. A fertile hybrid showing marked segregation from seed but F_2 plants have not been found wild in this country. $2n = 34^*$.

S. aucuparia × intermedia = S. × pinnatifida auct., is intermediate between the parents in having (0–)1–2(–3) free, obtuse pinnae, which are greyish-green-tomentose beneath; the fr. is 9–10 mm, globose and red. It differs from *S. × thuringiaca* in having more ovate, narrower lvs, distant basal pinnae, smaller fr. and no pollen. It occurs occasionally with the parents in S. and W. Britain northwards to Merioneth.

2. S. pseudofennica E. F. Warburg
Bastard Mountain Ash.

S. fennica auct. angl., *S. arranensis* × *S. aucuparia*
Small slender tree. *Lvs* (Fig. 44c) 5.5–8.5 cm, oblong or ovate-oblong in outline, (1.3)1.5–1.7(–2.2) *times as long as wide, mostly with* 1(–2) *pairs of free lflets at the base* but some of the lvs without free lflets, lobed above, serrate with sharp rather variously directed teeth most marked at the apex of the lobes, obtuse or acute at the wide apex, at maturity dark yellowish-green and glabrous above, unevenly and rather sparsely grey-tomentose beneath; *veins* 7–9(–10) *pairs*; petiole 8–20 mm. Petals c. 4 mm. Anthers cream. *Fr.* 8–10 mm, *longer than wide,* ovoid, red, with few inconspicuous lenticels. Fl. 5–6. Fr. 9. M.

Native. Steep granite stream-bank. Glen Catacol (Arran). Endemic.

*S. hybrida L.
S. fennica Fries

More robust than *S. pseudofennica.* Lvs larger, 7.5–10.5 cm, deep bluish-green above, more densely tomentose and with a whiter tomentum beneath. Fls much larger, petals c. 6 mm. Anthers pink (?). Fr. globose, red. $2n = 68.$

Occasionally planted but much less frequently than *S.* × *thuringiaca* with which it is often confused. Native of Scandinavia and Finland.

(3–7). S. intermedia sens. lat.

Lvs lobed in various ways to at least $\frac{1}{5}$ of the way to the midrib, grey-tomentose beneath; veins 7–10(–11) pairs. Styles 2. Fr. red.

3. S. arranensis Hedl.

Pyrus arranensis (Hedl.) Druce
Small slender tree. *Lvs* (Fig. 44d) 6–9 cm, *elliptical or rhombic-elliptical,* 1.5–2.1(–2.6) *times as long as wide, acute,* cuneate at base, *lobed* with oblong or oblong-lanceolate acute lobes, *the deepest lobes extending* $(\frac{1}{3}-)\frac{1}{2}-\frac{3}{4}$ *of the way to the midrib,* and occasionally nearly to the base, sharply serrate, the teeth mainly on the upper part of the lower margin of the lobe and directed towards its apex and mostly with straight outer margin; at maturity dark yellowish-green, glabrous and somewhat glossy above, usually rather evenly grey-tomentose beneath, when young with a few scattered hairs above and rather dense whitish-grey tomentum beneath; *veins* 7–8(–9) *pairs*; petiole 9–18 mm. Infl. small and rather narrow, pubescent at first; receptacle densely tomentose. Sepals deltate. Petals c. 4 mm. Anthers cream or pink. *Fr.* 8–10 mm, *ovoid,* with few inconspicuous lenticels. Fl. 5–6. Fr. 9. $2n = 51*$. M.

Native. Steep granite stream banks. Glen Easan Biorach and Glen Catacol (Arran). Endemic.

Perhaps this and the next 2 spp. have resulted from the cross *S. aucuparia* × *rupicola.*

4. S. leyana Wilmott
Shrub c. 2 m. *Lvs* (Fig. 44e) 7–9 cm, *oblong, ovate-oblong or elliptic-oblong,* 1.3–1.7(–1.9) *times as long as wide, obtuse or subacute,* cuneate at base, lobed with oblong or oblong-lanceolate acute or obtuse lobes, *the deepest lobes extending at least to nearly half-way to the midrib,* except on occasional lvs, and sometimes nearly to the base, serrate, the teeth mainly on the upper part of the lower margin of the lobe, directed rather irregularly and mostly with curved outer margin, at maturity dull dark yellowish-green and glabrous above, rather unevenly grey-tomentose beneath; *veins* (8–)9–10 *pairs*; petiole 10–20 mm. Petals c. 5 mm. Anthers pinkish. *Fr.* c. 10 mm, *subglobose,* with few small lenticels. Fl. 5–6. Fr. 9. N.

Native. On carboniferous limestone crags. Near Dan-y-Graig (Brecon). Endemic.

5. S. minima (A. Ley) Hedl.
Pyrus minima A. Ley
Shrub to c. 3 m of more slender habit than allied spp. *Lvs* (Fig. 44f) 6–8 cm, *elliptical or oblong-elliptical* (1.4–)1.8–2.2 *times as long as wide,* acute or sub-acute, cuneate or somewhat rounded at base, shallowly lobed with obtuse or acute lobes, *the deepest lobes extending* $\frac{1}{3}-\frac{1}{4}(-\frac{1}{2})$ *of the way to the midrib,* serrate, the teeth mostly curved on the outer edge and directed towards the apex of the lobe; at maturity dull green and subglabrous above, ±evenly and rather sparsely grey- (not whitish- or yellowish-) tomentose beneath, when young sparsely pubescent above; veins (7–)8–9(–10) pairs; petiole 12–17 mm. Infl. small, rather narrow and round-topped, sparsely tomentose, densely so on the receptacle. Sepals triangular-lanceolate. *Petals c.* 4 mm. Anthers cream. *Fr. small,* 6–8 mm, *subglobose,* with few small lenticels. Fl. 5–6. Fr. 9. $2n = 51.$ M. or N.

Native. Carboniferous limestone crags. Near Crickhowell (Brecon). Endemic.

*6. S. intermedia (Ehrh.) Pers.   Swedish Whitebeam.
Pyrus intermedia Ehrh.; *Sorbus scandica* (L.) Fries; *S. suecica* (L.) Krokies

Tree to 10 m, with spreading branches and rather broad crown. Twigs stout. *Lvs* (Fig. 44g) 7–12 cm, *elliptical or oblong-elliptical,* (1.5–)1.6–1.8(–1.9) *times as long as wide,* obtuse or acute, *rounded or broadly cuneate at base (some lvs always rounded),* lobed with wide acute ascending lobes, *the deepest lobes extending* $\frac{1}{4}-\frac{1}{3}(-\frac{1}{2})$ *of the way to the midrib,* serrate, the teeth ascending and nearly all with curved outer margin; dark yellowish-green and glabrous above (very sparingly pubescent when young), rather *sparsely yellowish-grey-tomentose beneath,* very densely so when young; *veins* 7–9 *pairs*; petiole 10–15 mm. Lvs of long (especially sucker) shoots and sometimes also those of short shoots when shaded sometimes lobed nearly to the base or even with a free lflet. Infl. rather large densely tomentose in fl. Sepals triangular-lanceolate. *Petals c.* 6 mm. *Anthers cream.*

Fr. 12–15 mm, *oblong*, much longer than wide, *with few scattered small lenticels.* Fls. 5. Fr. 9. 2*n* = 68. MM. or M.

Introduced. Rather commonly planted, sometimes bird-sown and freely naturalized in a few places from Somerset and Hants to Orkney, and in Ireland widely naturalized near Galway and south of Dublin. Native of S. Sweden (to c. 61° N), Bornholm, Baltic states and N.E. Germany.

7. S. anglica Hedl.

S. mougeotii Soy.-Will. & Godr. var. *anglica* (Hedl.) C. E. Salmon

Shrub c. 1–2 m. Twigs rather stout. *Lvs* (Fig. 44H) 7–11 cm, *obovate or rhombic-obovate*, 1.3–1.7(–1.8) *times as long as wide*, obtuse or acute, *nearly all cuneate at base*, lobed, with acute ascending lobes, the *deepest lobes extending c.* $\frac{1}{6}$–$\frac{1}{4}$ *of the way to the midrib*, serrate, the teeth projecting and acuminate or with straight outer margin; dark green and glabrous above (very sparingly pubescent when young), rather sparsely *whitish-grey-tomentose beneath; veins* (7–)9–11 *pairs*; petiole 12–20 mm. Infl. small to rather large, sparsely tomentose in fl.; receptacle densely tomentose. Sepals triangular-lanceolate. *Petals c.* 6 mm. *Anthers pink or tinged pink*, at least in bud. *Fr.* 7–12 mm, *subglobose, with few to rather many small*, or *small and moderate, lenticels mainly near the base* (the fr. varies considerably in size, colour and number of lenticels in different localities but is constant in each). Fl. 5 (rather earlier than most spp.). Fr. 9. 2*n* = 68*. M. or N.

Native. Crags and rocky woods, nearly always Carboniferous limestone, very local. Devon, N. Somerset (Cheddar), Avon Gorge, Wye Valley, Brecon, Montgomery, Shropshire, Denbigh; Kerry (Killarney). Endemic; closely allied spp. occur in Norway and in the Alps, Pyrenees, Carpathians and Balkan peninsula.

(8–16). S. aria *sens. lat.*   Whitebeam

Lvs various toothed, rarely lobed and then with rounded lobes extending not more than $\frac{1}{5}$ the way to the midrib, white-tomentose beneath; veins 7–14 pairs. Styles 2. Fr. red.

8. S. aria (L.) Crantz   Common Whitebeam.

Pyrus aria (L.) Ehrh.

Tree to 15(–20) m with wide dense crown, or a large shrub. Bark dark grey, shallowly fissured. Twigs tomentose when young with rather numerous small lenticels, becoming chestnut-brown, later dark grey. Buds up to 2 cm, ovoid, greenish, glabrous or tomentose, the scales ciliate. *Lvs* (Fig. 44J) very variable in size and shape, 5–12 cm, *ovate, elliptical or ovate*, very rarely obovate, commonly 1.1–1.6 times as long as wide but up to 2.4 times and on some trees constantly more than twice as long as wide, obtuse or acute, rounded or cuneate at base, *the sides mostly gradually curved and the upper and lower portions of the lf rarely triangular*, doubly crenate-serrate or shallowly lobed, then the lobes wider

than long and not reaching more than $\frac{1}{6}$ of the way to the midrib, *the apical tooth scarcely projecting beyond its neighbours* and those terminating the main veins broader based than in allied spp. and not projecting markedly beyond the longest of the others, the teeth all acute, broad based, mostly somewhat curved on the outer margin which is longer than the upper, thus directed upwards; *lf entire for the basal* $\frac{1}{3}$ *or less; dull yellow-green above*, subglabrous or tomentose when young, usually glabrescent but sometimes remaining thinly tomentose, *densely and evenly pure white-tomentose beneath; veins* (9–)10–14(–15) *pairs*, usually impressed above at maturity; petiole 7–20 mm. Infl. large and broad, tomentose in fl., remaining so or glabrescent in fr. Sepals deltate. Petals c. 6 mm, ovate. Anthers cream or, less frequently, pink. *Fr.* 8–15 mm, *usually longer than wide* but sometimes subglobose, with usually numerous but sometimes few *scattered small lenticels* sometimes with a few moderate ones in addition but *not markedly more numerous towards the base*, often somewhat woolly at base and apex. Fl. 5–6. Fr. 9. 2*n* = 34*. MM. or M. A very variable sp., contrasting markedly in this respect with the other British spp. and, unlike them, varying considerably in the same locality. It is thus sometimes difficult to separate them except by considering the characters as a whole. Any plant of the group not agreeing with the description of one of the other spp. is probably to be referred here.

Native. Woods and scrub usually on chalk or limestone, and on these soils common within its range, more local on sandstone hills. Kent and Hertford (?native in Norfolk and Suffolk) to Dorset, Monmouth and Worcester; doubtfully native in Ireland, common around Galway, rare elsewhere; elsewhere rather frequently planted and sometimes becoming naturalized. From Belgium and Spain eastwards to the Carpathians; southern limit uncertain owing to confusion with allied spp.; ?N. Africa (mountains) (perhaps not this sp.).

9. S. leptophylla E. F. Warburg

Near *S. aria.* Shrub. *Lvs* (Fig. 44K) (8–)9–12(–14) cm, *usually obovate*, (1.3–)1.5–1.7(–1.9 or –2.5 on sterile shoots) *times as long as wide*, acute, cuneate at base, *usually straight-sided for about* $\frac{1}{3}$ *of lf at base and apex so that the apical and basal portions of the lf are ±triangular*, doubly crenate-serrate, the teeth increasing in size upwards, the toothing sharper and coarser than is usual in *S. aria*, especially near the apex, and the secondary teeth fewer, the *apical tooth projecting markedly (c.* 3 mm) *beyond its neighbours*, those terminating the main veins also prominent, larger and longer than those of *S. aria; lf entire for basal* $\frac{1}{3}$ *or less, yellow- or dark-green*, loosely and sparsely tomentose when young, glabrescent above, *more sparsely tomentose than in S. aria and greenish-white beneath*, thin in texture; *veins* (9–)11(–13) *pairs*, not markedly impressed above. Sepals triangular-lanceolate. Petals c. 6 mm. Anthers cream or slightly pink-tinged. *Fr.* c. 20 mm, *longer than wide, with rather few scattered moderate and small* lenticels, *the former mostly in the lower half*, patchily woolly at base and apex. Fl. 5. Fr. 9. $2n = 68^*$. M.

Native, Shady limestone crags. In 2 localities in Brecon; probably also in Montgomery. Endemic.

10. S. wilmottiana E. F. Warburg

Shrub or small tree. Lvs 7–12 cm, *elliptical or obovate-elliptical*, 1.6–2.0(–2.7) *times as long as wide*, acute, cuneate at base, doubly serrate with rather sharp teeth, the teeth mostly directed somewhat upwards, the *apical tooth* and those terminating the main veins *scarcely prominent*; *lf entire for basal $\frac{1}{3}$ or less*; *lvs bright green and somewhat glossy above, evenly greenish-white-tomentose beneath*; veins 8–9(–10) pairs, not markedly impressed; petiole 1–2 cm. *Fr. 10–13 mm, longer than wide, with rather few large lenticels mostly towards the base.* Fr. 9. M.

Native. Rocky limestone woodland and scrub. Avon Gorge. Endemic.

11. S. eminens E. F. Warburg

Shrub or small tree. *Lvs* (Fig. 45A) of fertile shoots (5.5–)7–9(–12) cm, *ovate-orbicular, obovate-orbicular or subrhombic*, 1.1–1.4(–1.7) *times as long as wide*, obtuse or subobtuse, mostly broadly cuneate at base, *often with one or both margins arched inwards, the margins becoming rounded where the teeth begin at c. $\frac{1}{3}$ of the distance or less, the upper part of lf rounded in outline* or becoming broadly triangular near the apex, doubly serrate with rather shallow primary teeth, the *teeth terminating the main veins markedly longer than the others, acute or subacuminate, ±symmetrical and directed outwards*; *lf bright green above*, with some tomentum at fl., soon glabrescent and glossy, evenly *greenish-white-tomentose beneath*; lvs of non-flowering short shoots often smaller and narrower, up to 1.8 times as long as wide and more obovate in outline; *veins* (9–)10–11(–12) *pairs*, not markedly impressed above at maturity; petiole 10–20 mm. Infl. c. 15–20-fld, laxly and patchily woolly in fl.; receptacle densely woolly, glabrescent. Petals c. 5 mm. Anthers pink. *Fr. c. 20 mm, slightly longer than wide, with a moderate number of large and small lenticels mostly towards the base*, slightly woolly at the apex. Fl. 5–6. Fr. 9. $2n = 68^*$. M. or N.

Native. Woods on carboniferous limestone. Wye Valley and Avon Gorge. Endemic.

12. S. hibernica E. F. Warburg

Small tree of much more slender and open habit than *S. aria*. Lvs (Fig. 45B) (7–)8–10(–11) cm, *ovate to obovate*, (1.1–)1.2–1.5(–1.8) *times as long as wide, rounded at apex*, broadly cuneate or more rarely rounded at the base, *the basal cuneate portion short, the margins becoming curved where the teeth begin at c. $\frac{1}{3}$ of the distance to the apex*, doubly serrate, *the teeth triangular, acute*, ±symmetrical, directed outwards, becoming longer and narrower towards the apex, those terminating the main veins larger and longer than the others at least towards the apex of the lf; *lf sparsely tomentose above* when

young, glabrescent, *at maturity dull green*, densely tomentose and *whitish-green beneath*; *veins* (8–)9–11(–12) *pairs*; petiole 10–20 mm. Petals c. 4–5 mm. Anthers tinged pinkish. *Fr.* c. 15 mm, *wider than long, with rather few moderate and a few small lenticels especially near the base.* Fl. 5. Fr. 9. M.

Native. Rocky open woods, hedges and scrub. Widespread in Ireland but commonest in the centre and rare in the north. Endemic.

13. S. porrigentiformis E. F. Warburg

S. porrigens Hedl., *pro parte*

Shrub, more rarely a small tree to c. 5 m. Lvs (Fig. 45c) (5–)6–9.5 cm, *obovate*, 1.3–1.7 *times as long as wide, rounded in the upper part to an obtuse or shortly acuminate apex, tapering uniformly to a cuneate base from near the middle of the lf* (occasionally more rounded on flowering shoots), doubly and rather finely serrate, *the teeth terminating the main veins markedly longer than the others*, acute or acuminate, *±symmetrical and directed outwards, entire or nearly so for at least* ($\frac{1}{4}$–)$\frac{1}{3}$ *of the distance from the base, the lf bright green above*, slightly tomentose at fl., soon glabrescent and glossy, evenly and *rather sparsely greenish-white-tomentose beneath*; *veins* (7–)8–10(–11) *pairs* not markedly impressed above at maturity; petiole 8–20 mm. Infl. ±woolly at fl., glabrescent. Sepals triangular-lanceolate. Petals c. 5–6 mm. Anthers pink or tinged pink. *Fr.* 8–12 mm, *subglobose or broader than long, with rather few large lenticels towards the base of the fr.* Fl. 5–6. Fr. 9. $2n = 68^*$. N. or M.

Native. Crags and rocky woods on limestone, usually Carboniferous. S. Devon (Babbacombe), N. Somerset (Mendips), Monmouth, Glamorgan, Brecon, Carmarthen and Anglesey. Endemic.

14. S. lancastriensis E. F. Warburg

Shrub. *Lvs* (Fig. 45D) (6.5–)8–11(–12.5) cm, *obovate*, (1.4–)1.5–1.8(–2.0) *times as long as wide, rounded in the upper part to a usually obtuse apex, tapering to a cuneate base from about $\frac{1}{3}$ of the distance above the base* on many of the lvs but some lvs usually more rounded (though always cuneate at the attachment), unequally serrate, *the teeth terminating the main veins scarcely prominent* and the toothing coarser than in *S. porrigentiformis*, acute or sub-acuminate, *±symmetrical and directed outwards*; *lf entire or nearly so in the basal $\frac{1}{4}$, dark green* and glabrous *above* at maturity, usually evenly and *rather densely greyish-white-tomentose beneath*, the tomentum occasionally becoming patchy at fr.; *veins* (7–)8–10 *pairs*; petiole 10–20 mm. Infl. woolly at fl., particularly on the receptacle, usually with some persistent wool particularly at the base and apex of the fr. Sepals triangular-ovate. *Fr.* 12–15 mm, *subglobose or wider than long, with a moderate number of large lenticels towards the base of the fr. and scattered smaller ones.* Fl. 5–6. Fr. 9. N. or M.

Native. Carboniferous limestone rocks. Near Morecombe Bay (Lancs and Westmorland). Endemic.

15. S. rupicola (Syme) Hedl.

Pyrus rupicola (Syme) Bab.; *S. salicifolia* (Hartm.) Hedl.

Shrub c. 2 m, rarely a small tree, of rather stiff habit. *Lvs* (Fig. 45E) (6–)8–14.5 cm, *obovate or oblanceolate*, (1.6–)1.8–2.1(–2.4) *times as long as wide* (shade lvs often much larger and wider than sun lvs), *rounded in the upper part to a usually obtuse apex, mostly tapering uniformly to a cuneate base from the middle of the lf or above, coarsely* and unequally *serrate, the teeth terminating the main veins not prominent; teeth* acute or subacuminate, *mostly somewhat curved on the outer margin and directed towards the apex of the lf*, in basal ⅓ of lf much smaller than elsewhere and sometimes obsolete; *lf dark green above*, slightly tomentose at first, soon glabrescent, *rather densely white-tomentose beneath*, the tomentum often becoming patchy at fr.; *veins* (6–)7–9(–10) *pairs*, not markedly impressed above at maturity; petiole 8–20 mm, usually reddish at fr. Infl. woolly in fl., especially on the receptacle, glabrescent. Sepals triangular-ovate or triangular-lanceolate. Petals c. 7 mm. Anthers cream or slightly pinkish-tinged. *Fr.* 12–15 mm, *wider than long*, when ripening green on one side reddish on the other (like many apples), *with numerous scattered moderate and small lenticels*. Fl. 5–6 (rather later than most spp.). Fr. 9. 2 *n* = 68*. N. or M.

Native. On crags and among rocks, usually limestone and nearly always basic, very local, ascending to 460 m. S. Devon, Wales, Pennines from Stafford and Derby to Durham and Cumberland. Scotland from Roxburgh to Sutherland; N. and W. Ireland. Norway, S. Sweden and Estonia.

S. aria × rupicola, intermediate between the parents, is recorded from the Avon Gorge (N. Somerset) and the Wye Valley (W. Gloucester). It seems to be very close to *S. vexans* and *S. lancastriensis*, which may have the same ancestry.

16. S. vexans E. F. Warburg

Small tree. *Lvs* (Fig. 45F) (7–)8–10(–11) cm, *obovate*, (1.4–)1.5–1.9(–2.0) times as long as wide, rounded in the upper part to an obtuse apex, *tapering uniformly to a cuneate base from about the middle of the lf*, coarsely and unequally, sometimes doubly serrate, the teeth terminating the main veins not or scarcely prominent; *teeth* acute, *mostly somewhat curved on the outer margin and directed towards the apex of the lf; lf entire or nearly so in the basal ¼–⅓, yellow-green above* and glabrous at maturity, *white-tomentose beneath; veins* 8–9(–10) pairs, not markedly impressed above at maturity; petiole 8–15 mm. Infl. somewhat woolly in fl., the receptacle densely so, with some tomentum remaining on the pedicels in fr. Sepals deltate. Petals c. 6 mm. Anthers cream. *Fr.* 12–15 mm, *longer than wide, with few moderate lenticels towards the base of the fr. and a few small ones above*. Fl. 5. Fr. 9. M.

Native. Rocky woods near the coast. Between Culbone (Somerset) and Lynmouth (Devon). Endemic.

(17–19). S. latifolia *sens. lat.* Broad-leaved Whitebeam.

Lvs lobed, with ±triangular, acute or acuminate lobes, or doubly serrate with the teeth terminating the main veins very prominent, straight, acute or acuminate, grey-tomentose beneath; veins 7–9(–10) pairs. Styles 2, sometimes joined below. Fr. orange or orange-brown, often becoming brown when fully ripe. (Description including *S. aria × torminalis*.)

17. S. bristoliensis Wilmott

Tree. *Lvs* (Fig. 45J) (6–)7–9(–10.5) cm, *obovate, oblong-obovate or rhombic-obovate*, 1.4–1.7(2.0) times as long as wide, subacute, broadly cuneate or slightly rounded at base, lobed mainly above the middle with short ascending broadly triangular-acute lobes, the deepest lobes extending c.⅛–⅙ of the way to the midrib, finely serrate, the teeth mainly curved on the outer margin and directed upwards; at maturity bright yellowish-green and somewhat glossy above, subglabrous except when very young, densely and evenly grey- (not whitish-) tomentose beneath; veins (7–)8–9(–10) pairs; petiole 12–20 mm. Infl. rather small, rather sparsely woolly in fl., except the densely woolly receptacle, glabrescent. Sepals triangular-lanceolate. Petals c. 6 mm. *Anthers pink. Fr.* 9–11 mm, intermingled with considerably smaller fr., *longer than wide, with rather many moderate and small lenticels mainly towards the base* of the fr. Fl. 5–6. Fr. 9. 2*n* = 51* (apomictic). M.

Native. Rocky woods and scrub on Carboniferous limestone. In the Avon Gorge (N. Somerset, W. Gloucester). Endemic.

18. S. subcuneata Wilmott

Small rather graceful tree. *Lvs* (Fig. 45H) 7–10.5 cm, *rhombic-elliptical* (1.5–)1.6–1.9(–2.5) *times as long as wide*, acute, *cuneate or somewhat rounded at the base but always tapered*, lobed in the upper ⅔ with ascending broadly triangular acute or subacuminate lobes, the deepest lobes extending (⅐–)⅙–¼(–⅓) of the way to the midrib, sharply serrate, the teeth with mainly straight outer margin but directed upwards, those terminating the main veins projecting markedly; at maturity bright green above (deep green in shade), subglabrous except when very young, densely and evenly *whitish-grey*-tomentose *beneath*, the tomentum rather creamy when young; veins (6–)8–9(–10) pairs; petiole 12–25 mm. Infl. large but narrow, rather sparsely woolly in fl., except the densely woolly receptacle, glabrescent. Sepals triangular-ovate. Petals c. 6 mm. *Anthers cream. Fr.* 10–13 mm, *subglobose*, few on each infl. or if more numerous then smaller, *brownish-orange becoming brown, with numerous lenticels which are large towards the base of the fr. becoming smaller upwards*. Fl. 5–6. Fr. 9. M.

Native. Open sessile oakwoods. Minehead (S. Somerset) to Watersmeet (N. Devon). Endemic.

19. S. devoniensis E. F. Warburg   French Hales.

Tree with dense crown. *Lvs* (Fig. 45G) 7–11(–12) cm, *ovate or oblong-ovate*, occasionally a few of them obovate, 1.3–1.6(–1.8) *times as long as wide*, acute or sub-acuminate, *rounded at base*, shallowly lobed (sometimes ¼ of the way to the midrib but usually much less) with very broadly triangular acute or acuminate lobes, the lowest often spreading, or lf scarcely lobed but doubly serrate, with very prominent straight acuminate teeth terminating the main veins in the upper part, the other teeth much smaller, straight-sided; lf at maturity deep green above, subglabrous almost from the first, rather unevenly and sparsely *greenish-grey*-tomentose *beneath*; veins 7–10 pairs; petiole 10–30 mm. Infl. large, rather sparsely woolly in fl., except the densely woolly receptacle, glabrescent. Sepals triangular-lanceolate. Petals c. 7 mm. *Anthers cream.* Styles 2, united below. *Fr.* 10–15 mm, *subglobose*, *brownish-orange becoming brown, with numerous lenticels which are very large towards the base of the fr.*, becoming smaller upwards. Fl. 5–6. Fr. 9. MM. or M.

Native. Woods, etc. Widespread in Devon where it is the commonest member of the genus; E. Cornwall; S.E. Ireland (Kilkenny, Wexford, Carlow and Waterford). Endemic.

An allied form differing in the lvs scarcely ever lobed, with broader teeth terminating the main veins, the surface raised between the veins at maturity and the veins (8–)9–10(–11) pairs, is rather frequently planted and sometimes becomes naturalized. Its name and origin are uncertain at present. Other forms of this group are occasionally found bird-sown and ±naturalized.

S. aria × torminalis = S. × vagensis Wilmott

A variable hybrid. Tree. *Lvs* (Fig. 46A) 6–12 cm, ovate to elliptical or rhombic-elliptical, (1.1–)1.2–1.7(–2.1) times as long as wide, acute or subacute, rounded or cuneate at the base, *on many plants with the margins arched slightly inwards towards the base*, variously lobed but usually shallowly so in distal ½, with ±triangular acute or acuminate lobes, from ⅐ to over ¼ of the way to the midrib, finely serrate with the teeth usually small and somewhat appressed; usually yellow-green above, closely yellowish- or greenish-grey-tomentose beneath, in some plants partially denuded at maturity; veins (6–)7–10(–12) pairs; petiole (15–)20–40 mm. *Fr.* 7–12 mm, *longer than wide or subglobose, brownish-orange to brown, usually with few small lenticels* but sometimes with a moderate number of large ones. $2n = 34*$.

Not infrequent in woods on limestone in the Wye Valley (W. Gloucester, Monmouth, Hereford) with the parents; not known elsewhere in Britain (the parents rarely occur together).

20. S. torminalis (L.) Crantz   Wild Service Tree.
Pyrus torminalis (L.) Ehrh.

Tree to 25 m with wide crown and spreading branches. Bark dark grey, shallowly fissured. Twigs woolly-pubescent when young, soon glabrous and dark brown. Buds greenish, glabrous. *Lvs* (Fig. 46B) 7–10 cm, ±ovate in outline, up to 1.3 times as long as wide but often wider than long, acuminate, rounded or cordate (rarely broadly cuneate) at base, *deeply lobed with acuminate lobes*, *the lowest pair of lobes much deeper than the others* (usually half-way to the midrib), and *spreading* often at right angles to the midrib, the others ascending, finely and often doubly serrate, the teeth small; *lf green on both sides*, somewhat pubescent beneath when young mainly near the base of the midrib, subglabrous at maturity; veins 4–6 pairs; petiole 15–40 mm. Infl. moderate, sparsely woolly in fl., glabrescent, the receptacle tomentose, Sepals deltate. Petals c. 6 mm. Anthers cream. Styles 2, united to about half-way. Fr. 12–16 mm, considerably longer than wide, *brown*, with numerous large but not very conspicuous lenticels. Fl. 5–6. Fr. 9. $2n = 34$. MM.

Native. Woods, usually on clay, sometimes on limestone, local, and usually with only a few trees in any locality. Westmorland and from Flint, S. Yorks and Lincoln southwards. S., W. and C. Europe extending to E. Denmark; Caucasus, Asia Minor, N. Syria; Algeria.

22. PYRUS L.

Deciduous trees or shrubs, sometimes thorny. Lvs simple. *Fls in short, umbel-like, simple corymbs.* Sepals ±triangular. Stamens 20–30. *Carpels 2–5, completely united with each other and with the receptacle, the walls cartilaginous in fr.*; ovules 2, without false septum; *styles free.* Fr. brownish or greenish, not deeply indented at base; *flesh with numerous stone-cells.* Nectar secreted by receptacle.

About 30 spp., north temperate Old World.

Petals 10–15 mm; fr. 2.5 cm or more, the calyx persistent; infl.-rhachis short (rarely more than 1 cm). **1. pyraster**
Petals 8–10 mm; fr. 12–18 mm, the calyx finally caducous; infl.-rhachis 1 cm or more; Devon and Cornwall.
                                                  2. cordata

1. P. pyraster Burgsd.   Wild Pear.
P. communis auct., non L.

Tree of broadly pyramidal outline, or a large shrub 5–15 m, sometimes thorny, with spreading or ascending branches. Bark rather deeply fissured, scaly. Twigs and buds glabrous or slightly pubescent, yellowish-brown; short shoots numerous. Lf-blades 2.5–6 cm, orbicular-ovate, ovate or elliptical, rounded or subcordate at base, cuspidate, acuminate or acute, crenate-serrulate or subentire, woolly-tomentose when young, especially beneath, finally glabrous or sparsely tomentose beneath; petiole nearly as long as blade. *Infl.* a short *umbel-like* corymb, with the *rhachis* usually obvious but *rarely more than* 1 cm; fls c. 5–9; pedicels 1.5–3 cm, woolly-tomentose to subglabrous. Sepals c. 5 mm. *Petals* 10–15 mm, white, obovate or orbicular, shortly clawed. anthers purple. *Fr.* 2–4 cm (much larger in cultivated forms), pyriform or globose, rounded or tapered at base, brownish, with numerous lenticels, *the calyx persistent.* Fl. 4–5, earlier than *Malus sylvestris.* Pollinated by various Diptera and Hymenoptera, protogynous, self-pollination finally possible, some varieties self-sterile, others self-fertile. $2n = 34, 51$, rarely 68 (all from garden vars). MM. or M.

?Introduced. Hedges, etc. Widespread in England, less so in Wales, but usually as isolated trees; rarely reported in Scotland (Edinburgh, Linlithgow, Angus) and Ireland (Cork, Tipperary, Antrim), even as an escape; Channel Is. The cultivated pears belong to this sp. S., W. and C. Europe and S. half of USSR; N. Asia Minor, N. Iran, Russian C. Asia.

2. P. cordata Desv.

Differs from *P. pyraster* as follows: Thorny shrub 3–4 m. Buds glabrous. Lf-blades 1–4 cm, ovate or rounded or occasionally subcordate at base, conspicuously crenate-serrulate, nearly glabrous from the first; petiole often longer than blade. *Infl. more slender, not or scarcely umbel-like* and with the *rhachis more elongate* (1–3 cm). *Petals* 8–10 mm, obovate. *Fr.* 10–18 mm, globose or obovoid, not tapered into the pedicel; *calyx finally caducous*. Fl. 4–5. M.

?Native. Hedges near Plymouth (in Devon and Cornwall) and there very rare (plants from the Wye Valley reported as this sp. are referable to *P. pyraster*). Western margin of Europe from C. Portugal to W. France; a plant said to be the same occurs in Iran.

23. MALUS Miller

Deciduous trees or shrubs, occasionally thorny. Lvs simple, toothed to lobed. *Fls in short umbel-like corymbs.* Sepals ±triangular. Stamens 15–50. Carpels 3–5, completely united with each other and with the receptacle, *the walls cartilaginous in fr.*; ovules 2, without false septum; *styles united below*. Fr. green to red, usually deeply indented at base, *flesh without stone-cells* (or in a few foreign spp. with some). Nectar secreted by receptacle.

About 35 spp., north temperate regions. Several spp. and a number of hybrids are grown for ornament.

Mature lvs glabrous on both surfaces.	**1. sylvestris**
Mature lvs tomentose, at least beneath.	**2. domestica**

1. M. sylvestris Miller　　　　　Crab Apple.

Pyrus malus L.

Small tree with dense round crown, or shrub 2–10 m, usually thorny; twigs and buds glabrous or laxly hairy when young. Bark grey-brown, irregularly fissured and scaly. Twigs reddish-brown with numerous short shoots. *Lf-blades* 3–4 cm, ovate or oval, broadly cuneate or rounded at base, acuminate or cuspidate, crenate-serrate sparsely hairy on veins when young, *soon glabrous*; petiole c. half as long. Infl. almost an umbel, fls c. 4–7; pedicels 1–3 cm, glabrous. Sepals c. 3–7 mm, tomentose within, glabrous or subglabrous outside. Petals 1.3–2 cm, obovate, white, usually ±suffused with pink.

Anthers yellow. Receptacle glabrous or subglabrous. Fr. 2.5 cm. diam., subglobose with a depression at each end, yellowish-green often speckled or flushed with red, sour; calyx persistent. Fl. 5. Pollinated by various Diptera and Hymenoptera, protogynous; most vars partially self-sterile, self-pollination sometimes possible. $2n = 34$. M.

Native. Woods (especially of oak), hedges and scrub; ascending to 380 m. Scattered throughout England (except the south-west) Wales and E., W. and S.C. Ireland; north to Ross but generally uncommon in Scotland; Channel Is. Most of Europe, northwards to Fennoscandia, but rather local on mountains in the south; S.W. Asia.

In many parts of the British Is. the commonest Crab Apples shows signs of introgressive hybridization with *M. domestica*.

*2. M. domestica Borkh.　　　　　Apple.

M. sylvestris subsp. *mitis* (Wallr.) Mansf.

Small to medium tree differing from *M. sylvestris* as follows: Rarely thorny. Twigs tomentose at first, soon glabrous; buds hairy. *Lvs* obtuse to acuminate, *±persistently tomentose, at least beneath*. Pedicels, receptacle and outside of calyx tomentose. Fr. usually more than 5 cm diam., often sweet. Fl. 4–5. $2n = 51$. M.

Introduced. Widely cultivated for its fr., often escaping and becoming naturalized. Apparently hybridizes frequently with *M. sylvestris*.

Apparently of hybrid origin, with *M. sylvestris*, and the east European *M. dasyphylla* Borkh. and *M. praecox* (Pallas) Borkh., together with some Asiatic species, as parents. A very large number of cultivars are grown.

The two following frequently cultivated genera differ from all the preceding by the numerous ovules in each cell.

CHAENOMELES Lindley. Styles united below; sepals deciduous; lvs toothed.

*C. speciosa (Sweet) Nakai　　　　Japonica.

C. lagenaria (Makino) Koidz.; *Cydonia japonica* auct., non (Thunb.) Pers.

Twigs smooth. Lf-blades 3–8 cm, ovate, sharply serrate. Fls few in clusters, 3.5–5 cm diam., scarlet or, less often, pink or white. Fr. 3–7 cm, globose or ovoid, green becoming yellow. Native of China.

CYDONIA Miller. Styles free; sepals persistent; lvs entire.

*C. oblonga Miller　　　　　Quince.

Lf-blades 5–10 cm, ovate or oblong, villous beneath. Fls solitary, 4–5 cm diam., white or pink. Fr. large, pyriform, villous, green becoming yellow when ripe. Cultivated for its fr., also used as a stock for pears; it has occurred as an escape. Native of S.W. and C. Asia. The only sp.

60. PLATANACEAE

Monoecious trees with bark coming off in large flakes, giving a mottled appearance. Buds within base of petiole. Lvs alternate, palmately lobed, stipulate. Fls in unisexual infls consisting of 1–7 short-stalked dense globose heads arranged in a raceme.

Perianth inconspicuous. Female fl. hypogynous; calyx a cup or of free sepals; petals present or 0; staminodes 3–4; carpels 5–9, free, in 2–3 whorls, the styles long slender, the stigma along the inner face; ovules 1–2. Male fl. with calyx cupular or 0; petals present; stamens 3–5, alternating with petals; rudimentary; rudimentary carpels sometimes occur in some fls. Fr. a group of nutlets; endosperm scanty. Wind-pollinated.

One genus with about 10 spp., S.E. Europe, S.W. Asia, North America.

*Platanus hybrida Brot. (*P. acerifolia* (Aiton) Willd.)
London Plane.

Tree to 35 m. Lvs 12–25 cm, 3–5-lobed, ±glabrous, truncate or cordate at base; lobes ±triangular, entire or somewhat toothed, extending c. $\frac{1}{3}$ of way to lf-base; petiole 3–10 cm. Female heads (1–)2(–3) per infl.; calyx cupular; petals present strap-shaped. Male fls: sepals present; petals fleshy, 3-lobed. Fr.-heads c. 2.5 cm diam, with numerous hairs surrounding the frs.

Originated c. 1700, much planted, especially in towns, and occasionally self-sown and ±naturalized. Perhaps a hybrid (*P. occidentalis* L. × *P. orientalis* L.) or a cultivar of *P. orientalis*.

61. CRASSULACEAE

Herbs or small shrubs, usually succulent. Lvs exstipulate, usually simple and entire. Infl. usually cymose. Fls actinomorphic, hermaphrodite or rarely unisexual, 3–32-merous, usually 5-merous, hypogynous or slightly perigynous. Petals free or united. Stamens as many as petals and alternating with them or twice as many; anthers introrse. Carpels as many as petals, free or united at the base, each with nectariferous scale at the base; ovules usually numerous (rarely 1 or few), anatropous. Fr. a bunch of follicles or a capsule; carpels dehiscing along upper edge; seeds very small; embryo straight; endosperm scanty or 0.

About 35 genera and 1500 spp., cosmopolitan, mainly in dry warm temperate regions. Spp. of several genera are grown under glass.

A very natural family usually easily recognized by the succulent lvs and free or nearly free carpels equal in number to the petals.

1 Basal and lower stem-lvs orbicular, peltate; petals united into a cylindrical tube for more than half their length. 4. UMBILICUS
  No lvs peltate; petals free or united only at base. 2
2 Lvs opposite, connate; fls inconspicuous, solitary in lf-axils; stamens as many as petals. 3. CRASSULA
  Lvs alternate, not connate, or crowded into rosettes; fls conspicuous, in cymes or corymbs; stamens twice as many as petals. 3
3 Plants with ±elongated lfy stems; fls 4–5-merous. 1. SEDUM
  Plants without lfy stems, the lvs crowded in a basal rosette; fls usually 13-merous. 2. SEMPERVIVUM

1. SEDUM L.

Herbs or small shrubs of varied habit. Lvs usually alternate but sometimes opposite or whorled, succulent. Infl. cymose; fls 3–10-(*usually* 5-)*merous*, usually hermaphrodite but occasionally unisexual. Sepals shortly united at base, somewhat fleshy. *Petals free* (shortly united at base in some foreign spp.). *Stamens twice as many as petals* (rarely as many in a few foreign spp.), free or adnate to petals below. Nectar-scales small, entire or slightly toothed. Carpels free or shortly united at base; ovules many (rarely few or 1). Fr. a group of follicles.

About 600 spp., north temperate, 1 in Peru and Bolivia. Many are cultivated and several have occurred as escapes.

1 Lvs at least 8 mm broad, flat, usually toothed. 2
  Lvs less than 8 mm broad, terete or flat above, entire. 4
2 Stems erect, several from a central stock. 3

Stems numerous, creeping and rooting; no central stock. 3. **spurium**
3 Stock very thick and fleshy, scaly; fls 4-merous, unisexual, greenish. 1. **rosea**
  Stock not fleshy, short, bearing carrot-like tubers, not scaly; fls 5-merous, hermaphrodite, reddish-purple. 2. **telephium**
4 Infl. and usually lvs hairy, with some glandular hairs. 5
  Infl. and lvs glabrous. 6
5 Lvs opposite; petals white with pink streaks. 4. **dasyphyllum**
  Lvs alternate; petals pink. 11. **villosum**
6 Petals white. 7
  Petals yellow. 8
7 Lvs 3–5 mm, glaucous; infl. with 2(–3) main branches. 5. **anglicum**
  Lvs 6–12 mm, green; infl. with several branches. 6. **album**
8 Lvs 7 mm, ovoid or cylindrical, obtuse. 9
  Lvs 8 mm or more, linear, terete or flat above, acute or apiculate. 10
9 Lvs ovoid, imbricate. 7. **acre**
  Lvs linear-cylindrical, spreading. 8. **sexangulare**
10 Lvs flat above, forming a congested cluster at the apex of the sterile shoots. 9. **forsteranum**
  Lvs terete; sterile shoots equally lfy throughout. 10. **reflexum**

Section 1. *Rhodiola* (L.) S. F. Gray. Perennial herbs. Stock fleshy, erect, crowned with lvs which are often reduced to scales in the axils of which the annual lfy shoots are borne. Lvs alternate, flat. Fls 4- or 5-merous, unisexual or hermaphrodite.

1. S. rosea (L.) Scop.   Roseroot, Midsummer-men.
S. rhodiola DC.; *Rhodiola rosea* L.

Dioecious, glabrous, glaucous perennial 15–30 cm. *Stock thick, fleshy*, branched, projecting above ground level, marked with the scars of old stems, with a crown of brownish chaffy scales at apex; roots not tuberous. *Stems erect*, simple, usually several from each branch of the stock. Lvs 1–4 cm, alternate, numerous, dense, obovate or obovate-oblong, acute or sometimes obtuse, rounded at base, sessile, dentate near apex or subentire decreasing in size downwards, the lowest brown and scale-like. Infl. terminal, compact. *Fls 4-merous, greenish-yellow*. Male fl. c. 6 mm diam.; sepals linear; petals linear c. $1\frac{1}{2}$ times as long as sepals, both often tinged purple on back; stamens rather longer than petals, the filaments yellow, the anthers purple; scales emarginate, conspicuous; abortive carpels erect, shorter than petals.

Female fl. with petals equalling sepals; stamens 0; scales as in male; carpels 4–6 mm, erect, greenish; style short, spreading, reflexed in fr. Fl. 5–8. Visited by small flies. $2n = 22, 36$ (North America). Hp.

Native. Crevices of mountain rocks, ascending to 1175 m, and of sea cliffs in W. Scotland and Ireland; common in such habitats; N. and S. Wales and from Yorks, Lancs and the Isle of Man northwards; all the Irish mountains (thus absent from the centre). Arctic Europe, Asia and North America and in the mountains of these continents south to the Pyrenees, C. Italy, Bulgaria, Japan and New Mexico.

Section 2. *Telephium* S. F. Gray. Perennial herbs. Stock short, without lvs or scales, bearing thick tuberous roots. Stems annual from buds near the apex of the stock. Lvs flat. Fls 5-merous, hermaphrodite.

2. S. telephium L.   Orpine, Livelong.

Glabrous subglaucous perennial 20–60 cm, often tinged with red. Stock short, stout; *roots tuberous, carrot-like. Stems erect*, simple, clustered. Lvs 2–8 cm, alternate, numerous, ovate-oblong to oblanceolate-oblong, sessile or shortly stalked, rounded or cuneate at base, obtuse to subacute, irregularly dentate except at base. Infl. terminal, composed of compact subglobose terminal and axillary cymes. Fls 9–12 mm diam. Sepals green, lanceolate, acute. *Petals* spreading, *reddish-purple* or lilac, lanceolate, acute. Scales yellow, ligulate, emarginate. Carpels erect, purple; styles very short. Fl. 7–9., Visited by Diptera and Hymenoptera; protandrous. Hp.

Native. Woods and hedgebanks, ascending to 460 m; spread over the greater part of the British Is. but rather local and becoming very rare in N. Scotland northwards to Argyll and Caithness; introduced in Ireland; certainly native in many places but often an escape. Almost throughout Europe; temperate Asia, North America. The habitats and distributions (either as native or introductions) of the two following subspp. in this country are not well known.

Subsp. **telephium**

S. purpurascens Koch; *S. purpureum* auct., ?(L.) Schultes

Lvs sessile; upper ovate-oblong, rounded at base; lower obovate-oblong, cuneate at base. Carpels grooved on back. $2n = 36$.

Chiefly in C. and E. Europe and Asia (to Japan).

Subsp. **fabaria** (Koch) Kirschleger

S. fabaria Koch

Lvs shortly stalked or subsessile, all cuneate at base, obovate-elliptical to oblanceolate-oblong. Carpels not grooved. Chiefly W. and C. Europe.

Section 3. *Sedum*. Perennial evergreen herbs. Stock 0. Sterile stems numerous, creeping and rooting; fl.-stems mostly ascending, annual. Fls hermaphrodite, 5-merous.

*3. S. spurium Bieb.

S. stoloniferum auct.

Creeping perennial forming large mats with numerous lfy ascending puberulent branches; fl.-branches c. 15 cm, the sterile ones shorter. Lvs 2–3 cm, opposite, *flat, obovate-cuneiform*, obtuse, *crenate-serrate* above the middle, ciliate, otherwise glabrous, tapered into a short petiole. Infl. a flat dense terminal corymb. Fls 9–12 mm long, sessile or subsessile. *Petals pink*, less frequently white or reddish-purple, *suberect*. Carpels erect. Fl. 7–8. $2n = 28$. Chh.

Introduced. Commonly planted and frequently escaping, probably increasing. Native of the Caucasus and Transcaucasia.

*4. S. dasyphyllum L.   Thick-leaved Stonecrop.

Small tufted perennial 2–5 cm, glaucous, often tinged pink. Sterile branches short, ascending, with imbricate lvs; fl.-stems taller, with distant lvs. *Lvs opposite* (or some alternate on fl.-stems), 3–5 mm, *ovoid or obovoid*, slightly flattened above, entire, sessile, ±*glandular-pubescent*, rarely glabrous, readily caducous and then often rooting and forming fresh plants. Infl. 2–4-fld. Fls c. 6 mm diam. *Petals white*, tinged pink on back, oblanceolate, spreading. Carpels erect. Fl. 6–7. Visited by small Diptera and Hymenoptera; protandrous. $2n = 28, 42, 56$. Chh.

Introduced. Occasionally naturalized on old walls, mainly in S. England, N. Wales. Yorks and Durham; on limestone rocks in Cork, Galway and King's Co. S. Europe northwards to C. France and S. Germany; N. Africa.

5. S. anglicum Hudson   English Stonecrop.

Glabrous glaucous evergreen perennial often tinged red, 2–5 cm, with numerous slender creeping and rooting stems forming mats, and ascending barren and fertile branches. *Lvs* 3–5 mm, *alternate*, ovoid or suborbicular, terete, entire, spreading, clasping the stem and with a small spur at the base. *Infl.* with 2(–3) *main* branches each with 3–6 fls, and with a fl. at the fork. Fls c. 12 mm diam. Pedicels stout, c. 1 mm. Sepals ovate, obtuse, free. *Petals white*, tinged pink on back, lanceolate, spreading. Scales red. Carpels ascending in fl., erect in fr., pale at first, becoming red after fl. Fl. 6–8(–9). Visited by various insects. $2n = 48, 144$. Chh.

Native. Rocks, less frequently dry grassland, ascending to 1070 m; also dunes and shingle, absent from strongly basic soils; common in the west. From the Hebrides southwards extending east to Moray, Argyll, Northumberland, Derby and Gloucester; also in several places near the south and east coasts; throughout Ireland, but rare in the centre. W. Europe extending to S.W. Sweden; Morocco; the southern forms differ from ours (subsp. *anglicum*).

6. S. album L.   White Stonecrop.

Glabrous bright *green*, evergreen perennial, sometimes tinged red, 7–15 cm, with numerous creeping stems forming large mats; fl-branches ascending. *Lvs* 6–12 mm, *alternate*, ovoid-globose to cylindrical, terete or somewhat flattened above, obtuse, entire, sessile but not clasping or spurred. *Infl. with several main terminal and axillary branches* forming a much branched, *many-fld*, flat-topped cyme 2–5 cm across. Pedicels slender, 1–4 mm. Fls 6–9 mm diam. *Petals white*, sometimes tinged pink on back, (2–)3–4 mm, ovate or lanceolate, spreading. Scales yellow. Carpels erect, pale, becoming tinged with red after fl. Fl. 6–8. Visited by various insects, protandrous. $2n = 32, 64$. Chh.

?Introduced. Walls and rocks, possibly native in the Malverns, Mendips and Devon, widely naturalized elsewhere northwards to Sutherland and Caithness, and throughout Ireland. Europe, except for parts of the north and east; N. and W. Asia; N. Africa.

Plants with oblong-ovate to subglobose lvs and petals 2–3 mm have been called subsp. *micranthum* (Bast.) Syme but the validity of such infraspecific taxa is doubtful.

*S. lydium Boiss.

Small, bright green. Lvs c. 6 mm, terete, linear. Infl. compact, flat. Petals white. Commonly cultivated and reported as naturalized. Native of Asia Minor.

7. S. acre L.   Wall-pepper, Biting Stonecrop.

Glabrous green evergreen perennial 2–10 cm, with numerous creeping stems forming mats, and ascending or erect sterile and fl.-branches; *taste hot and acrid*. *Lvs* 3–5 mm, alternate, *ascending*, *imbricate* (except sometimes on the fl.-stems), *ovoid-trigonous*, *obtuse*, entire, sessile, spurred at base. Infl. with 2 or 3 main branches, each with 2–4 fls at the fork. Fls c. 12 mm diam. Sepals lanceolate, obtuse. *Petals bright yellow*, lanceolate, acute, spreading. Scales whitish. Carpels ascending in fl., spreading in fr. Fl. 6–7. Pollinated by Diptera and Hymenoptera; protandrous, perhaps sometimes selfed. $2n = 16$, c. 48. Chh.

Native. Dry grassland, dunes, shingle and walls, especially on basic soils; ascending to 460 m. Throughout the British Is. except Shetland, common. Almost throughout Europe; N. and W. Asia; N. Africa; naturalized in North America.

*8. S. sexangulare L.   Tasteless Stonecrop.

S. mite Gilib.; *S. boloniense* Loisel

Differs from *S. acre* as follows: *Without biting taste. Lvs* 3–6 mm, *linear-cylindrical*, terete, *spreading*, *dense*, *but not imbricate*, obtuse. Fls c. 9 mm diam. Scales yellow. Carpels erect in fl., spreading in fr. Fl. 7–8. Chh.

Introduced. Naturalized on old walls in a few places in England and Wales. Native of C. Europe extending locally to Finland, C. Greece and C. France.

9. S. forsteranum Sm.   Rock Stonecrop.

S. rupestre L., *pro parte*

Glabrous, glaucous or green, evergreen perennial 15–30 cm, with numerous creeping stems forming mats, and short ascending sterile branches and long fl. ones. *Sterile shoots* clothed with withered lvs below, *with lvs crowded above* into dense, terminal rosettes. *Lvs* 8–20 mm, alternate, ascending, *linear* or oblanceolate-linear, *flat on upper surface*, *apiculate*, entire, sessile, spurred at base; on the fl. shoots more distant and broader, ascending. Infl. a many-fld umbel-like cyme with about 5 main branches, drooping in bud, becoming erect and flat-topped or convex in fl., concave in fr. Fls occasionally 6–8-merous, c. 12 mm diam. Sepals triangular, nearly free. *Petals bright yellow*, oblong-linear, obtuse, spreading. Scales yellow. Carpels erect in fl. and fr. Fl. 6–7. Chh.

Native. Rocks and screes, very local, ascending to 440 m; Devon, Dorset, Somerset, Gloucester, Shropshire, Wales; often planted and sometimes naturalized elsewhere northwards to Edinburgh and in N., E. and S. Ireland. W. Europe; Morocco.

*10. S. reflexum L.   Reflexed Stonecrop.

Differs from *S. forsteranum* as follows: Usually more robust. *Sterile shoots equally lfy for a considerable distance* so that the tips are not rosette-like, dead lvs not persistent. *Lvs terete* or nearly so, sometimes reflexed on the fl.-stems. Fls rather larger. Petals linear-lanceolate, acute, bright or pale yellow. Fl. 6–8. Visited by various insects; weakly protandrous. $2n = 34, 68$, c. 112. Chh.

Introduced. Commonly cultivated; naturalized on old walls and rocks in Great Britain northwards to Stirling and Kincardine; scattered localities in Ireland. C. Europe extending to Finland, Sicily, and W. France. W. Ukraine.

11. S. villosum L.   Hairy Stonecrop.

Small, usually reddish, not glaucous perennial 5–15 cm, perhaps sometimes biennial, *glandular-pubescent*. Stem erect, often branched at the base, the branches both sterile and flowering; sterile ones short. *Lvs* 6–12 mm, *alternate*, linear-oblong, flat above, obtuse, entire, sessile, not spurred. Infl. lax, rather few-fld, with terminal and axillary branches; pedicels long, slender. Fls c. 6 mm diam. Sepals lanceolate, obtuse. *Petals pink*, ovate, apiculate, spreading. Scales yellowish. Carpels erect. Fl. 6–7. Chh.

Native. Streamsides and wet stony ground, etc., in the mountains; ascending to 1100 m. From Yorks and Lancs to Perth and Banff, widespread within its range but rather local. W. and C. Europe extending to W. Finland, White Russia and N. Italy; Morocco, Algeria (rare); Greenland.

*S. dendroideum Sessé & Moç. ex DC.: small shrub with lvs 2–5 cm, oblanceolate, thick; infl. a large panicle; petals 4–8 mm, yellow; naturalized in Jersey. Mexico.

2. SEMPERVIVUM L.

Herbs. *Lvs thick and fleshy, alternate, mostly densely crowded into a basal rosette*, which often persists for several years and reproduces by stolons, finally flowering and dying. Fl.-stems erect, bearing a branched infl.

of scorpioid cymes often forming a panicle. *Fls 6–18-merous. Petals connate for a short distance at the base. Stamens twice as many as petals*, epipetalous. Scales toothed to fimbriate. Carpels free; ovules many. Fr. a group of follicles.

About 25 spp., Europe, mainly in the mountains, Atlas and Caucasus. Some are grown on rock gardens.

*1. S. tectorum L.   Houseleek,
Welcome home husband, however drunk you be.
Rosettes 5–14 cm diam. Lvs 2.5–6 cm, obovate-lanceolate, mucronate, dark or glaucous green tinged dull red above, ciliate, otherwise glabrous. Fl.-stems 30–60 cm, glandular-pubescent above. Infl. a many-fld panicle of scorpioid cymes. Fls 2–3-cm diam., 8–18- (mostly c. 12-)merous. Petals dull pink or purple, lanceolate. Inner whorl of stamens mostly sterile, often replaced by carpels. Fl. 6–7. Visited by various insects; protandrous. $2n = 72*$. Chh.

Introduced. Has been planted for centuries on old walls, roofs, etc., but scarcely naturalized. The form (with partly aborted stamens), described above, to which the Linnaean type belongs, extends over Europe but is probably nowhere wild. Other subspp. or allied spp. occur over the whole range of the genus.

3. CRASSULA L.

Herbs or shrubs of varying habit. *Lvs opposite*, usually entire. Fls in cymes or, more rarely, solitary and axillary, usually small, (3–)5(–9)-merous. Sepals free or nearly so. *Petals free or connate at base only. Stamens as many as petals*, alternate with them. Scales usually small. Carpels free or joined at base; ovules 1–many; styles short, apical; *stigmas apical*, small. Fr. a group of follicles.

About 300 spp., cosmopolitan but the great majority in S. Africa.

  1 Lvs 1–2 mm, ovate-oblong, concave; fls usually 3-merous; petals shorter than sepals.   **1. tillaea**
    Lvs 3–15 mm, linear to linear-ovate, flat; fls 4-merous; petals longer than sepals.   *2*
  2 Annual, up to 5 cm; lvs 3–6 mm; fls subsessile.
                      2. aquatica
    Perennial, up to 30 cm; lvs 4–15(–20) mm; pedicels 2–8 mm.   **3. helmsii**

1. C. tillaea L.-Garland   Mossy Stonecrop.
Tillaea muscosa L.

Small, often tufted, glabrous, usually reddish annual 1–5 cm. Stems decumbent or ascending, with short axillary branches very densely clothed above by the lvs and fls. *Lvs 1–2 mm, ovate-oblong*, thick, concave, connate in pairs, subacute. Fls solitary, sessile in the axils of nearly all the lvs and about the same length, 3(–4)-merous. Sepals ovate, mucronate. Petals smaller, lanceolate, whitish, acute. Scales 0. Follicles 2-seeded. Fl. 6–7. Self-pollinated. Th.

Native. Bare sandy or gravelly ground, very local; N. Somerset, Dorset; apparently extinct in S. Devon and S. Wilts; Hants, Norfolk, Suffolk, Nottingham, Channel Is. S. and W. Europe extending locally to N.E. Germany; Macaronesia.

2. C. aquatica (L.) Schönl.
Tillaea aquatica L.

Small slender, glabrous annual 2–5 cm. Stems decumbent, simple or branched from the base, with few axillary branches, *the lvs and fls distant. Lvs 3–6 mm linear*, somewhat fleshy, connate, subacute. Fls 1–2 mm, solitary, subsessile in the lf-axils, 4-merous. sepals ±triangular, obtuse. Petals larger, ovate, whitish, acute. Follicles with numerous seeds. Fl. 6–7. Probably self-pollinated. $2n = 42$. Th.

?Native. Muddy margins of a pool near Adel Dam, Yorks, where it was first found in 1921 but apparently extinct by 1945; found on wet mud in Westerness in 1969 and apparently naturalized. N. and C. Europe; N. Asia; North America.

*3. C. helmsii (T. Kirk) Cockayne
C. recurva (Hooker fil.) Ostenf., non N.E. Br.

Succulent, glabrous perennial herb. Stems simple or branched, decumbent, rooting, up to 5 cm on wet mud, usually erect and up to 15(–30) cm in water. *Lvs 4–15(–20) mm, linear- to ovate-lanceolate*, connate, acute. Fls 1.5–2 mm, solitary, 4-merous; *pedicels 2–8 mm*. Sepals oblong, subacute. Petals larger, ovate, white or pinkish, obtuse. Follicles 2–5-seeded. Fl. 8–9. Probably self-pollinated.

Introduced. Sold for aerating garden ponds since 1927. First found as an established alien in 1956 and now known around and in ponds from Wilts and Hants to Surrey and Essex northwards to Worcester, Yorks and Moray, probably expanding. Native of Australia and New Zealand.

*C. radicans Haw., a native of S. Africa, is locally established on dry stony areas in Jersey. It is a ±spreading small shrub up to c. 30 cm, with the lvs up to 1 cm, obovate, minutely pubescent towards their base, and clustered towards the ends of the branches. The white, oblong petals, each with an apical appendage, form a tube-like corolla; infl. a cyme.

*C. decumbens Thunb., a native of S. Africa, was observed as a bulbfield weed in the Scilly Is. in 1959 and seems to be spreading along tracks, etc. It is a slender, glabrous, fleshy annual 5–12 cm, with lanceolate, subacute, slightly connate lvs, and numerous, axillary, 4-merous fls with pale pink petals shorter than the sepals.

4. UMBILICUS DC.

Perennial herbs with *tuberous stock. Lvs alternate*, stalked, *peltate or cordate*, fleshy. *Infl. a narrow terminal spike or raceme. Fls 5-merous. Petals connate for most of their length* into a cylindrical or nearly campanulate tube. *Stamens 10 (rarely 5).* Scales small. Carpels free; styles short; ovules numerous. Fr. a group of follicles; seeds very small.

About 18 spp. in Mediterranean region extending to Iran, Abyssinia, Macaronesia and W. Europe.

1. U. rupestris (Salisb.) Dandy   Pennywort, Navelwort.
U. pendulinus DC.; *Cotyledon umbilicus-veneris* L., *pro parte*

Glabrous perennial herb 10–40 cm. Stock tuberous, subglobose. Stem usually solitary, simple or occasionally branched in the infl. Lvs mostly basal, blades 1.5–7 cm diam., orbicular, peltate, depressed at the junction with the long petiole, crenate with very broad teeth, or sinuate; stem lvs few, becoming smaller upwards, with shorter stalks, the uppermost often not peltate. Infl. long, usually occupying more than half the length of the stem, many-fld, sometimes branched below; pedicels 2–4 mm; upper bracts linear, entire, somewhat exceeding pedicels, lower often lf-like; fls pendent; sepals lanceolate, acute, less than half as long as corolla. Corolla 8–10 mm, whitish-green, tubular; lobes ovate, short. Stamens inserted on corolla-tube, included. Fl. 6–8(–9). Protandrous, but probably usually self-pollinated, only thrips known as visitors. Hs.

Native. Crevices of rocks and walls, especially acid; ascending to 550 m. Rather common in W. England and Wales, rarer eastwards, extending to Kent, Essex, Huntingdon, Lincoln and W. Yorks; S.W. Scotland from Kirkcudbright to Argyll and the mid Inner Hebrides; throughout Ireland but local in the centre. S. and W. Europe; Madeira, Azores.

62. SAXIFRAGACEAE

Herbs. Lvs usually alternate, exstipulate or with small sheathing stipules. Fls hermaphrodite, actinomorphic or more rarely somewhat zygomorphic, usually 5-, sometimes 4-merous, hypogynous to epigynous, often perigynous in varying degrees. Petals sometimes 0. Stamens equalling and alternate with the petals or twice as many and obdiplostemonous. Carpels 2, very rarely 3–5, free to united, very frequently united at the base, free above and gradually tapering to the stigmas, free from or adnate to the hypanthium; ovules numerous, placentation various; styles usually free. Fr. a capsule; seeds numerous with small straight embryo and copious endosperm. Nectar secreted by ovary.

About 30 genera and 580 spp., cosmopolitan but mainly temperate, some on tropical mountains. A family often easily recognized by its perigynous fls and partly fused carpels. Distinguished from the *Rosaceae* by the capsular fr., the almost constantly 2 carpels and the usually fewer stamens. It is often regarded as including Parnassiaceae, Hydrangeaceae, Escalloniaceae and Grossulariaceae, q.v. for the distinctions.

Spp. and hybrids of *Bergenia*, *Astilbe*, *Heuchera*, etc., are commonly grown in gardens.

Fls 5-merous; petals present.   1. SAXIFRAGA
Fls 4-merous; petals absent.   2. CHRYSOSPLENIUM

1. SAXIFRAGA L.

Usually perennial herbs. Lvs often thick, simple. *Bracteoles present.* Fls 5-merous (very rarely 4-merous in foreign spp.). Hypanthium flat or cup-shaped and adnate to the ovary. Sepals imbricate. *Petals present, narrow at the base. Stamens* 10 (very rarely 8). *Carpels* (usually) 2, ±*united below*; *placentation axile*; styles free, at first erect then spreading. Capsule opening along the inner edges of the carpels. Nectar secreted at the base of the ovary or by an epigynous disk. Usually protandrous.

About 370 spp. north temperate, Arctic and in the Andes. Many spp. and hybrids are grown in rock gardens, etc.

1 Lvs alternate; fls white or yellow.   2
  Lvs opposite; fls purple.   **15. oppositifolia**
2 Fls yellow.   3
  Fls white, sometimes spotted with yellow or purple.   4
3 Lower lvs with petiole as long as or longer than blade; fl.-stems reddish-hairy above; fls 1(–3); sepals reflexed in fr.; ovary superior.   **3. hirculus**
  Lvs sessile; fl.-stems without red hairs, fls usually several; ovary partly inferior.   **14. aizoides**
4 Sepals reflexed at or soon after fl.; ovary superior; fl.-stems naked.   5
  Sepals erect or spreading; ovary partly inferior.   8
5 Lvs subsessile, with distant teeth; stock not woody, short; filaments subulate; capsule inflated.
                                                 2. stellaris
  Lvs stalked, with contiguous teeth; stock somewhat woody; filaments clavate; capsule not inflated.   6
6 Lvs cuneate at base, glabrous; petiole 2 mm broad or more; petals spotted with red.   7
  Lvs cordate at base, hairy on both sides; petiole scarcely 1 mm broad; petals often without red spots.
                                                   6. hirsuta
7 Teeth of lvs obtuse, the terminal tooth shorter than its neighbours; petiole densely ciliate throughout, usually shorter than blade; styles ascending in fr.
                                                 4. umbrosa
  Teeth of lvs acute, the terminal tooth as long or longer than its neighbours; petiole sparsely ciliate at base only, longer than blade; styles divaricate in fr.
                                               5. spathularis
8 Fl.-stem naked; lvs ±spathulate, coarsely and closely crenate; plant purple-tinged.   **1. nivalis**
  Fl.-stem lfy; lvs not as above; plant not purple-tinged.   9
9 Plant without numerous procumbent or rosette-like lfy barren shoots.   10
  Plant with numerous procumbent or rosette-like lfy barren shoots, of mossy habit.   13
10 Annual; lvs cuneate at base.   **7. tridactylites**
   Perennial; basal lvs cordate at base.   11
11 Ovary c. $\frac{3}{4}$-inferior; basal lvs crenate-lobulate with 7 or more teeth; fls 2–12; lowland.   **8. granulata**
   Ovary less than $\frac{1}{2}$-inferior; basal lvs palmately 3–5(–7)-lobed; fls 1–3; high montane.   12
12 Plant with numerous red bulbils in the axils of the bracts; petals 8 mm or more; fl.-stems erect.
                                                   9. cernua
   Bracts without bulbils; petals 5 mm or less; fl.-stems ascending.   **10. rivularis**
13 Lobes of lvs ±oblong, the largest always more than 1 mm broad, obtuse or subacute, sometimes mucronate, never aristate; sterile shoots erect or ascending, without axillary bulbils.   14

Lobes of lvs ±linear, rarely more than 1 mm broad, acute or acuminate, aristate; sterile shoots procumbent, usually with axillary fusiform lfy bulbils.
                                    13. hypnoides
14 Plant very compact; lobes of lvs very obtuse.
                                    11. cespitosa
Plant usually rather lax; lobes of lvs subacute.
                                    12. rosacea

Section 1. *Micranthes* (Haw.) D. Don. Perennial herbs with basal rosette and few or *no stem-lvs*. Lvs alternate, toothed, without pits or lime glands. Fls usually in compact racemes or panicles. Hypanthium flat to cup-shaped. Sepals often reflexed after fl. *Petals ±white.* Ovary free or up to ⅓-inferior. *Capsule ±inflated, dehiscing to the middle or below.* Seeds cylindrical.

1. S. nivalis L.   Alpine Saxifrage.
Perennial herb 3–15 cm, with short thick stock and 1 or few basal rosettes of lvs. *Lf-blades* 1–2 cm, orbicular- or obovate-spathulate, *coarsely* and closely *crenate-serrate*, obtuse, green above and glabrous at maturity, *±purple below* and *glandular-pubescent near the margins*, rather thick; petiole broad about as long as blade. Fl.-stems naked, glandular-pubescent, often purple. Fls 3–12 in a dense head-like panicle; bracts linear; pedicels very short. *Sepals* triangular-ovate, often purplish, *erect or spreading. Petals* c. 3 mm, *greenish-white, not spotted,* reddish or purplish on the back, suborbicular. Anthers dull orange. *Ovary about ⅓-inferior;* styles divergent; carpels often purple or reddish, in fr. 3–4 mm, ovoid. Fl. 7–8. Visited by flies, slightly protandrous to slightly protogynous. $2n = 60$. Hr.

Native. Wet rocks on mountains from 365 to 1310 m, very local and usually rare; Caernarvon, Lake District, Dumfries, Scottish Highlands from Dunbarton and Stirling to Sutherland, Skye; in Ireland only in Sligo. Arctic and sub-Arctic Europe, Asia and eastern North America, Greenland; Sudeten Mountains.

2. S. stellaris L.   Starry Saxifrage.
Perennial herb with *short stock* and 1 to several basal rosettes of lvs often becoming elongated into lfy shoots, *not purple-tinged. Lvs* 0.5–3 cm, obovate-cuneiform, obovate-spathulate, or oblanceolate, acute or subacute, *remotely serrate or dentate, with scattered long hairs* especially above, rather thick, *scarcely petiolate.* Fl-stems naked, with scattered long hairs below and glandular hairs above. Infl. an open panicle up to 12- or more-fld, the branches cymose, but often simple and few-fld; bracts linear or linear-lanceolate; pedicels 4–10 mm, slender. *Sepals* ovate- or oblong-lanceolate, *reflexed.* Petals 4–5 mm, ovate-lanceolate, pure white with two yellow spots near the base. Anthers orange-red; filaments subulate. *Ovary superior,* free; styles somewhat divergent; fruiting carpels ovoid, c. 6 mm. Fl. 6–8. Visited mainly by flies; protandrous to protogynous; pseudoviviparous plants occur in W. Galway. $2n = 28$. Hr. or Chh.

Native. Common on mountains by streams, in

springs, on wet rock ledges and wet stony ground, ascending to the top (1343 m) of Ben Nevis. W. Wales from Cardigan to Caernarvon; Westmorland, Cumberland and Roxburgh to Caithness and the Outer Hebrides; most of the Irish mountains. N. Europe from Iceland to arctic Russia and in most major mountain-ranges elsewhere in Europe; Greenland, Labrador.

Section 2. *Hirculus* (Haw.) Tausch. Perennial herbs. Stems usually lfy. Lvs alternate, entire or nearly so, without pits or lime glands. Fls solitary or in panicles. Hypanthium flat or nearly so. Sepals spreading or reflexed. *Petals yellow,* often with 2 projections near the base. Ovary *free* or nearly so. Capsule not inflated, dehiscing only near the apex. Seeds fusiform.

3. S. hirculus L.   Marsh Saxifrage.
Perennial herb 10–20 cm, with short prostrate or ascending sterile shoots and erect lfy fl-stems. Lower lf-blades 1–3 cm lanceolate, oblanceolate, or oblong, obtuse, entire, narrowed below into a sheathing *petiole as long or longer than blade;* upper lvs shorter and narrower. *Fl.-stems reddish-hairy above.* Fls 1(–3), terminal. Sepals oblong, ciliate with reddish hairs, spreading in fl., reflexed in fr. Petals 10–13 mm, narrowly oblong or obovate, yellow, often with orange spots near the bituberculate base. *Ovary superior,* free; styles divergent. Fr. c. 1 cm, oblong-ovoid. Fl. 8. Visited mainly by flies; protandrous. $2n = 32$. Hp.

Native. Wet grassy ground on moors, very local and rare; ascending to over 610 m; N. Yorks, Durham, Westmorland and Cumberland; Lanark, Aberdeen, Linlithgow, Banff; Ireland in Mayo and Antrim; apparently extinct elsewhere in the British Is. N., E. and C. Europe, N. Asia and America, Caucasus, Himalaya.

Section 3. *Gymnopera* D. Don. Perennial herbs with a rather woody stock and short branches bearing rosettes of lvs. Lvs alternate, toothed, without pits or lime glands; *stem lvs 0.* Fls in panicles. Hypanthium flat. *Sepals reflexed* after fl. *Petals white.* Filaments clavate. *Ovate superior,* free. Capsule not inflated, dehiscing only near the apex. Seeds ovoid or ellipsoid.

***4. S. umbrosa** L.
Perennial herb 8–40 cm. *Lf-blades* 1–6 cm, spreading at maturity, *obovate-oblong, cuneate at base,* subtruncate at apex, crenate or crenate-serrate with 4–10 *obtuse teeth* on each side, *the terminal tooth shorter* and broader *than the lateral, glabrous,* dark green, thick, with a broad cartilaginous border; *petiole usually shorter than blade, flat, broad* (2–4 mm), *densely long-ciliate.* Scape reddish, glandular-pubescent especially above. Sepals oblong, obtuse. *Petals* 3–4 mm, elliptical, white with a yellow spot at the base and a few small *crimson spots* above. Fr. 5–6 mm, oblong-ovoid, reddish; styles ascending. Fl. 6–7. Visited by Diptera and Hymenoptera. $2n = 28$. Hr.

Introduced. Heseldon and Linn Gills, Yorks, where it has been known since 1792. Native of W. and C. Pyrenees.

*S. ×urbium D. A. Webb   London Pride.

S. spathularis× umbrosa; *S. umbrosa* L. var. *crenoserrata* Bab.

Differs from *S. umbrosa* in the larger, less spreading, subacute lvs with up to 12, more angular teeth on each side. Scape taller, less glandular. Petals large with more numerous crimson spots. Fr. very rarely produced. Commonly grown in gardens, sometimes escaping and becoming naturalized. Of unknown origin; not known wild.

5. S. spathularis Brot.   St. Patrick's-cabbage.

S. umbrosa auct. brit.; *S. umbrosa* var. *serratifolia* D. Don

Differs from *S. umbrosa* as follows: Lvs ascending at maturity, suborbicular or spathulate to obovate-cuneate, obtuse or acute, dentate with 4–7 *acute triangular teeth* on each side, *the terminal tooth as long as or longer than the lateral; petiole much longer than blade, sparsely ciliate towards the base*. Scape glandular-hairy throughout. Petals 4–5 mm, white with 1–3 yellow spots at base and numerous crimson spots above. Fr. oblong, greenish; *styles divaricate*. Fl. 6–8. Hr.

Native. Among rocks, mainly acid, in damp or shady places in mountain districts, very local but often abundant, ascending to over 1040 m; Kerry, Cork, Waterford, S. Tipperary, Galway, Mayo, Donegal, Wicklow. N.W. Spain and mountains of N. Portugal.

S. hirsuta × **spathularis** = *S.* × *polita* (Haw.) Link

S. hirsuta auct., non L.

Fertile and very variable, combining the characters of the parents in various ways. Common with the parents in Kerry and W. Cork and often commoner than *S. hirsuta*; found also in Galway and Mayo where *S. hirsuta* is unknown.

6. S. hirsuta L.   Kidney Saxifrage.

S. geum L., 1762, non 1753; incl. *S. lactiflora* Pugsley

Perennial herb 12–30 cm. *Lf-blades* 1–4 cm, ascending, *orbicular or reniform, cordate at base*, obtuse, crenate or crenate-dentate with 6–13 obtuse or apiculate teeth on each side, *with long hairs on both surfaces*, bright green, not very thick, with a narrow cartilaginous border; *petiole 2–4 times as long as blade, not flattened, slender* (scarcely 1 mm broad), hairy. Scape glandular-pubescent. Sepals oblong, obtuse. Petals c. 4 mm, elliptical, white with a yellow basal blotch, with or without crimson markings. Fr. 4–5 mm, oblong-ovoid, green; styles divergent. Fl. 5–7. 2n = 28. Hr.

Native. Shady rocks, woods and by mountain streams, in Kerry and W. Cork, locally common, ascending to 915 m. Cultivated in Great Britain and sometimes naturalized in Scotland and N. England, but this form has larger lvs with more truncate crenations than the Irish plant. Pyrenees and N. Spain.

Section *Cymbalaria* Griseb. Annual or biennial herbs; stems short, usually very lfy and much branched. Lvs rather thick, usually reniform, coarsely toothed, without pits or lime glands. Fls small, on long pedicels. Sepals often reflexed. Petals yellow. Ovary free or up to $\frac{1}{3}$ inferior. Capsule globose or oblong, dehiscing only near the apex. Seeds globose.

*S. cymbalaria L.

Annual or biennial to 30 cm. Lvs 5–25 mm, reniform, crenate or dentate with 5–13 teeth. Sepals ±spreading. Petals c. 5–6 mm, yellow, 3–4 times as long as sepals. Ovary superior. 2n = 18.

Occasionally ±naturalized as a garden weed. Native of the Carpathians, Asia Minor, Caucasus, Algeria.

Section 4. *Saxifraga*. Annual to perennial herbs. Stems lfy. Lvs alternate, entire to 3–5(–11)-lobed or -fid, without pits or lime glands. Fls solitary and terminal, or in simple or branched monochasial cymes. Hypanthium cup-shaped or campanulate. Sepals spreading or erect. *Petals* usually *white*. Ovary from about $\frac{1}{3}$ to almost completely inferior. Seeds oblong or fusiform.

7. S. tridactylites L.   Rue-leaved Saxifrage.

Erect annual herb 2–15 cm, simple or with axillary branches, ±glandular hairy all over. *Basal lvs* ±congested, ±*spathulate*, with a broad petiole, the blade 1 cm or less, orbicular or obovate, palmately 3–5-lobed or -fid, or the smaller lvs entire; upper smaller, passing into entire bracts. Fls in simple or branched monochasial cymes or solitary and terminal. Pedicels slender, twice as long as fl. or more. Hypanthium campanulate. Sepals ovate. Petals 2–3 mm, white, obovate-cuneiform. Fr. subglobose. Fl. 4–6. Probably usually self-pollinated. 2n = 22. Th.

Native. Dry open habitats in sandy grassland, walls, limestone rocks etc., mainly on basic soils, rather local, usually lowland but ascending to 730 m. Throughout much of England and Wales; E. Scotland north to Caithness; Inner and Outer Hebrides; almost throughout Ireland but local in the north-east and south-west; Channel Is. Europe, except the extreme north and C. and E. Russia, eastwards to W. Iran.

8. S. granulata L.   Meadow Saxifrage.

Perennial herb 10–50 cm, *overwintering by bulbils produced in the axils of the basal lvs*. Basal lvs in a rosette, sometimes dead at fl.; *blade* 0.5–3 cm, reniform, cordate at base, obtuse, *crenate-lobulate* with 7 or more teeth, with scattered flexuous long white hairs and glandular-ciliate; petiole several times as long as blade; cauline lvs few, decreasing in size upwards, more acutely toothed, cuneate at base, the petioles short. Bulbils brown, ovoid, c. 5 mm. Fl.-stems solitary, simple, hairy below, glandular especially above. Infl. a lax, open, terminal cyme; *fls* 2–12; bracts small, linear; pedicels 4–20 mm, densely glandular. Hypanthium densely glandular. Sepals ovate. *Petals* 10–17 mm, white, obovate. *Ovary* c. $\frac{3}{4}$-*inferior*; styles long, ascending; stigmas conspicuous. Fr. 6–8 mm, ovoid or subglobose. Fl. 4–6. Visited by various insects; protandrous. 2n = 46, 48*, 49* to 60 (very variable). Hs.

Native. Basic and neutral grassland ascending to 460 m, rather local and with an eastern tendency; usually on well-drained soils. From Moray and Renfrew southwards (not Pembroke or Cornwall), probably introduced in Cornwall and in Devon; in Ireland native in Dublin and probably Wicklow, introduced elsewhere. N., C. and W. Europe, eastwards to N.W. Russia, N.W. Ukraine, W. Hungary and Sicily; Morocco.

9. S. cernua L.   Drooping Saxifrage.

Perennial herb 3–15 cm, overwintering by bulbils produced in the axils of the lvs. Basal lvs in a rosette, the *blade* 5–18 mm, reniform, shallowly cordate at base, *palmately* 3–5-lobed, with ovate, acute lobes, subglabrous; petiole 2–3 times as long, slightly dilated below, not sheathing or toothed; cauline lvs on shorter petioles, passing into the sessile, oblong or linear, entire, bulbiliferous bracts. Basal bulbils c. 5 mm, solitary, narrowly ovoid. *Fl. stems solitary, erect. Fls solitary and terminal or more frequently 0, being wholly or partially replaced by numerous small red bulbils* several together in the axils of the bracts. Sepals ovate, spreading. *Petals* 8–13 mm, white, obovate-oblong. *Ovary less than ½-inferior.* Fr. unknown in the wild. Fl. 7. Visited by flies, but apparently never fruiting. $2n = 50, 60, 64,$ c. 66. Hs.

Native. Shaded places among basic rocks on mountains, from 915 to nearly 1215 m, very rare. Perth, Argyll, Inverness. Arctic and sub-Arctic Europe, Asia and eastern North America; Greenland; Alps, Carpathians; mountains of C. Asia, Himalaya, Japan; Rocky Mountains.

10. S. rivularis L.   Highland Saxifrage.

Perennial herb 2–8 cm, sometimes producing rooting stolons. Basal lvs scarcely forming a rosette; *blade* 5–20 mm, reniform, shallowly cordate at base, *palmately* 3–7-*lobed* with ovate, obtuse or subacute lobes, subglabrous; petiole 3–10 times as long, dilated at base and sheathing, with a tooth on each side of the expanded part at the junction with the petiole; cauline lvs sometimes entire, petioles shorter. Basal bulbils small, concealed by lvs or 0. *Fl.-stems usually several, ascending*, not or scarcely longer than basal lvs, *not bulbiliferous. Fls* 1–3; pedicels glandular-pubescent above, variable but mostly long; bracts ovate or oblong, entire. Sepals ovate, ascending, subobtuse. *Petals* 3–5 mm, *white*, obovate or obovate-oblong. *Ovary less than ½-inferior.* Fr. ovoid or ovoid-oblong; styles spreading. Fl. 7–8. Visited by flies, slightly protogynous. $2n = 26, 52, 56.$ Hs.

Native. Wet, shady rocks on mountains from 915 to nearly 1215 m; very rare and local. Perth and Aberdeen, N.E. to Ross. Arctic and sub-Arctic Europe, Asia and eastern North America (including Greenland); Cascade, Rocky and White Mountains.

11. S. cespitosa L.   Tufted Saxifrage.
S. groenlandica L.
Compact perennial 4–10 cm, *forming dense cushions.*

Sterile shoots short, erect, densely lfy, thickly clothed with dead lvs below the living ones, which are in a compact rosette. Axillary bulbils 0. *Lvs* to c. 1 cm, 3–5-lobed, narrowed at base into a short petiole, densely and uniformly *clothed with short glandular hairs*, the lobes of 3-fid lvs spreading at about 50° or less; *lobes* oblong, *obtuse*, more than 1 mm broad; upper cauline lvs entire. Stem, pedicels and calyx shortly glandular. Fls 1–5; pedicels mostly shorter than calyx. Sepals oblong-ovate, obtuse. *Petals* c. 5 mm, obovate *whitish*. Fl. 5–7. Visited by flies; usually protogynous. $2n = 80.$ Chc.

Native. High mountain rocks from 610 to 1065 m, very local and rare. Caernarvon, Aberdeen, Inverness, Ross. Arctic and sub-Arctic Europe, Asia and America.

12. S. rosacea Moench   Irish Saxifrage.
S. decipiens Ehrh. ex Pers.; including *S. hirta* Donn ex Sm., non Haw., *S. affinis* D. Don, *S. incurvifolia* D. Don, *S. sternbergii* Willd., *S. drucei* E. S. Marshall
Perennial herb 4–20 cm. More robust and less compact than *S. cespitosa* and forming laxer tufts, but very variable. *Sterile shoots* short or long (to 10 cm.), *erect or ascending* at fl., densely or laxly lfy. Axillary bulbils 0. *Lf-blades* 1–1.5 cm, 3–7-lobed, narrowed at base into a petiole about as long as blade, very variable in indumentum but often with long flexuous eglandular hairs; with *few (or 0) rather long glandular hairs*; lobes of 3-fid lvs spreading at c. 50° or less; *lobes* oblong or oblong-lanceolate, *subacute*, sometimes mucronate, or those of the lowest lvs obtuse, the largest more than 1 mm broad; upper cauline lvs entire. Stem, pedicels, and calyx densely or sparingly glandular. Fls 1–8; pedicels short or long. Sepals triangular-ovate, subobtuse or acute. *Petals* 6–8 mm, obovate, *pure white*. A very variable plant. Fl. 6–8. Protandrous. Chc. or Chh.

Native. Damp rock ledges and screes on mountains; limestone pavement; ascending to 1065 m. Caernarvon (extinct); Kerry, Tipperary, Limerick, Clare, Galway, Mayo, very local but often abundant. N.W. and C. Europe.

Subsp **rosacea**: Plant compact to rather lax; lvs without or with very few glandular hairs. $2n = $ c. 46*, c. 64*.
Distribution as given above.

Subsp. **hartii** (D. A. Webb) D. A. Webb (*S. hartii* D. A. Webb): Plant usually rather compact; lvs with numerous glandular hairs. $2n = $ c. 48*.
Native. Sea-cliffs. N.W. Ireland (Arranmore Island, Donegal). Endemic.

13. S. hypnoides L.   Dovedale Moss, Mossy Saxifrage.
Incl. *S. platypetala* Sm.; *S. sponhemica* auct. brit., non C. C. Gmelin
Perennial 5–20 cm, forming mats with flowering rosettes and *numerous axillary procumbent or decumbent sterile shoots* (up to 15 cm), *usually bearing axillary fusiform lfy bulbils.* Lf-blades of flowering rosettes to c. 1 cm,

3–5(–9)-lobed, contracted at base into a petiole longer than the blade, sparsely ciliate on the petiole with long flexuous eglandular hairs, otherwise subglabrous, the lobes of 3-fid lvs spreading at c. 75°; *lobes* linear or linear-oblong, *acute or acuminate, aristate*, usually c. 1(–1.5) mm broad. Lvs of barren shoots all, or at least those in the middle of the shoot, entire, rarely all 3-lobed, elliptic-linear, acute, aristate. Upper cauline lvs entire. Stem glabrous or sparingly glandular, pedicels and calyx more glandular. Fls 1–5, pedicels mostly slender, longer than fl. Sepals triangular or triangular-ovate, acute or acuminate, aristate. Petals 4–10 mm, oblong or obovate, pure white. Fl. 5–7. Protandrous. $2n = 48^*$, c. 64^*. Chh.

Native. Rock-ledges, screes and stony grassland in hilly districts, rarely on dunes, more frequent on basic than acid substrates, ascending to 1215 m, locally common; Somerset, Wales and the border counties, Derby and Lancs to Caithness; local in Ireland from Clare, Waterford and Wicklow northwards, mainly in the mountains. N.W. Europe, south-eastwards to the Vosges.

S. hypnoides × *rosacea* has been recorded from Tipperary (Galtee mountains) and Clare (Burren), the only places where the parents occur together.

Section 5. *Xanthizoon* Griseb. Perennial herb with numerous lfy stems. Lvs alternate, entire, with a pit at the apex, but not secreting lime. Hypanthium cup-shaped. *Sepals spreading. Petals yellow* (rarely red). *Ovary* $\frac{1}{4}$- *to* $\frac{1}{3}$-inferior. Seeds fusiform.

14. S. aizoides L. Yellow Saxifrage.

Perennial herb 5–20 cm with numerous decumbent sterile, and ascending fl-stems. *Lvs* 1–2 cm, numerous, dense on the sterile stems, less so towards the top of the fl. ones, oblong-linear, acute, *sessile*, usually remotely ciliate with stiff hairs, otherwise glabrous, rather thick. Fl.-stems simple or branched above, pubescent. *Fls* 1–10, *in a lax terminal cyme*; bracts like the lvs but rather smaller. Sepals triangular-ovate, obtuse, spreading. Petals yellow, often spotted with red, distant, 4–7 mm, obovate, oblanceolate, or oblong. Styles divergent. Fr. c. 7 mm, ovoid. Fl. 6–8(–9). Pollinated mainly by flies; protandrous. $2n = 26$. Chh.

Native. Streamsides and wet stony ground on mountains; ascending to 1175 m, locally common. Yorks, Durham, Westmorland and Cumberland, Ayr and Perth to Orkney (not S.E. Scotland or Outer Hebrides); N. and N.W. Ireland from Sligo to Donegal, Antrim. Arctic and sub-Arctic Europe, Greenland, eastern North America, W. Asia; high mountains of N. and C. Europe to Pyrenees, C. Italy, Albania and S. Carpathians; N. Rocky Mountains, Vermont.

Section 6. *Porphyrion* Tausch. Perennial herbs with numerous lfy stems. *Lvs opposite*, small, entire, with 1 or more pits near the apex, *secreting lime*. Hypanthium cup-shaped. Sepals erect or spreading. *Petals purple* or pink, rarely white. Ovary about $\frac{1}{2}$-inferior.

15. S. oppositifolia L. Purple Saxifrage.

Perennial herb with numerous stems with long prostrate and short erect branches forming lax mats or flat cushions. Lvs 2–6 mm, very dense, in 4 rows, suborbicular to obovate-lanceolate, concave, the apex thickened and flattened, with a single lime-secreting pit, dark blue-green, setulose-ciliate, otherwise glabrous, sessile. Fls solitary, terminal on stems 1–2 cm, which are less densely lfy than the sterile ones. Sepals ovate, suberect. Petals 6–10 mm, rosy-purple, ovate or ovate-oblong. Fr. 3–6 mm, ovoid; styles spreading. Fl. 3–5 (sometimes a few fls again in 7). Visited mainly by Lepidoptera, probably often self-pollinated; protandrous to protogynous. $2n = 26^*$, 52. Chh.

Native. Damp rocks and stony ground on mountains, ascending to nearly 1215 m, local but often abundant. Brecon, N. Wales, Yorks, Lake District, Dumfries, Lanark; Stirling northwards; N.W. Ireland from Galway to Londonderry. Arctic and sub-Arctic Europe, Asia and America; high mountains of Europe to Sierra Nevada, C. Italy and Bulgaria (the more southern forms belong to distinct subspp.); mountains of C. Asia to Kashmir (a subsp.); Rocky Mountains, Vermont.

2. CHRYSOSPLENIUM L.

Herbs. Lvs simple, stalked. Fls in terminal, lfy, corymbose cymes, epigynous. Hypanthium cup-shaped, adnate to the ovary. *Sepals 4–5. Petals 0.* Stamens 8–10. Carpels 2, united; *placentae parietal*; styles free, short. Capsule opening along the inner edges of the carpels. Nectar secreted by a disk round the styles.

About 55 spp., north temperate regions and temperate South America, mainly in E. Asia.

Lvs opposite; basal lvs truncate or broadly cuneate at base.
        1. oppositifolium
Cauline lvs alternate (usually only 1); basal lvs cordate
  at base.   **2. alternifolium**

1. C. oppositifolium L.

Opposite-leaved Golden-saxifrage.

Perennial herb 5–15 cm, with numerous *decumbent lfy rooting stems* forming large patches; fl.-stems ascending. *Lvs opposite*; blades of lower 1–2 cm, *orbicular, entire or shallowly sinuate-dentate*, very obtuse, *truncate or broadly cuneate at base*, often shortly decurrent, with scattered appressed hairs above; *petiole as long as or shorter than blade*; lvs of the fl.-stems smaller, on shorter petioles, usually glabrous, 1–3 pairs. Bracts similar but smaller and decreasing upwards, bright greenish-yellow. Infl. lax below, dense above. Calyx 3–4 mm diam., coloured like the bracts; sepals mostly 4, usually 5 in some fls. Fl. 4–7. Visited by various small insects; protogynous, self-pollination possible. $2n = 42$. Chh.

Native. Streamsides, springs, wet rocks, wet ground in woods, usually in shade, ascending to 1035 m; throughout the British Is. but unrecorded from Cambridge, Huntingdon and Shetland; common except in S.E., E. and C. England, and C. Ireland. W. and parts of E.

Europe extending eastwards to W. Poland and C. Czechoslovakia.

2. C. alternifolium L.

Alternate-leaved Golden-saxifrage.

Similar to *C. oppositifolium* but rather more robust and without creeping lfy stems but spreading by *lfless stolons*. Basal lvs 2–4; *blades* 1–2.5 cm, reniform, *crenate, cordate at base*, not decurrent; *petiole several times as long as blade*; lvs on the fl.-stems usually only 1, smaller, truncate or shallowly cordate at base, the petiole shorter. Bracts like those of *C. oppositifolium* but rather larger and more deeply toothed. Calyx 5–6 mm diam. Fl. 4–7. Pollinated by small insects; ±homogamous. $2n = 48$. Hs.

Native. In similar places to *C. oppositifolium*, but local, ascending to 990 m; Aberdeen to Elgin southwards but absent from the extreme west (Wigtown, W. Wales and Cornwall), from Ireland, and from all the smaller islands. Most of Europe but absent from the extreme north and west and most of the Mediterranean region; N. and C. Asia to Caucasus and Himalaya.

The two following monotypic genera of rhizomatous herbs with shallowly lobed broadly ovate-cordate lvs, racemose fls with a large perigynous zone and parietal placentation are grown in gardens and occasionally naturalized. Natives of western North America.

*Tolmiea menziesii (Pursh) Torrey & Gray Pick-a-back-plant.
Hypanthium ±cylindrical, greenish, tinged with purple. Sepals 5, unequal. Petals 4(–5), filiform, brown. Reproducing by bulbils on the lvs. Locally naturalized in S. Scotland and N.W. England.

*Tellima grandiflora (Pursh) Douglas ex Lindley
Hypanthium broadly campanulate or sometimes narrowed at mouth, glandular. Fls regular, 5-merous. Petals lanceolate in outline, laciniate, pale green becoming dark purple. Bulbils 0. Naturalized in woods and on shady walls in N. Devon and locally elsewhere in S. England and Ireland.

63. PARNASSIACEAE

Perennial herbs. Lvs simple, alternate, exstipulate, the basal numerous, those on the fl.-stems few. Fl. solitary, terminal on a stem from the axil of a basal lf. Fls actinomorphic, hermaphrodite, hypogynous or perigynous, 5-merous. A ring of 5 large usually fringed staminodes, each with nectaries on the upper surface, occurs inside and opposite the petals. Fertile stamens 5, alternating with the staminodes. Carpels (3–)4(–5), united into a superior or ½-inferior ovary, septate below, free above; placentae axile below, parietal above, with numerous anatropous ovules; style short, thick or 0; stigmas usually 4. Fr. a loculicidal capsule.

One genus.

An isolated family probably most closely allied to the Saxifragaceae, in which it is often included, but showing also resemblances to the Droseraceae and Hypericaceae. The staminodes are characteristic.

PARNASSIA L.

About 50 spp., north temperate regions.

1. P. palustris L.

Grass-of-Parnassus.
Glabrous perennial herb 10–30 cm, with short vertical stock, often somewhat tufted. Basal lvs ±numerous; blades 1–5 cm, ovate, cordate at base, sub-acute, entire; petiole longer than the blade. Fl.-stem erect, straight, with 1 sessile deeply cordate lf near the base. Petals 7–12 mm, broadly elliptical or oblong, white, with conspicuous veins. Staminodes c. ⅓ as long as petals, ±spathulate at the apex with 7–15 long setaceous processes tipped with yellowish glands. Ovary ovoid, superior. Fl. 7–10. Pollinated by various insects, protandrous. $2n = 18$. Hs.

Var. *condensata* Travis & Wheldon has very numerous rather leathery basal lvs on shorter petioles, shorter (2.5–15 cm) fl.-stems and rather larger fls.

Native. Marshes, fens, lake shores and wet moors or grassland; ascending to 790 m. Widespread but rather local. Wilts and Berks to Essex and Suffolk northwards (absent from the Isle of Man), apparently extinct in S. England and S. Wales and in the E. and S. Midlands; Anglesey, Caernarvon and Flint; Ireland (absent from Cork, Kerry, Wexford, Waterford, Sligo, Leitrim, Tyrone and Armagh). Most of Europe, but rare in the south; Morocco (Atlas); temperate Asia.

64. HYDRANGEACEAE

Shrubs or trees. Lvs opposite (very rarely alternate) simple, exstipulate. Fls hermaphrodite or some sterile, actinomorphic, 4–5(–10)-merous, perigynous or epigynous. Stamens twice as many as petals and obdiplostemonous, or more numerous. Carpels (2–)3–5(–10), united; styles free or united; ovules numerous (very rarely only 1), anatropous; placentation axile or parietal. Fr. a capsule, rarely a berry; seeds usually numerous; endosperm copious.

About 10 genera and 115 spp., mainly north temperate, a few in tropical Asia (to New Guinea) and Central and South America. Spp. and hybrids of *Hydrangea* and *Deutzia* are commonly grown in gardens. Differs from the Saxifragaceae in the woody habit, opposite lvs, tendency to more numerous carpels and stamens and to a more complete fusion of the ovary.

1. PHILADELPHUS L.

Fls 4–(6)-merous, all fertile. Petals contorted in bud. Stamens numerous. Ovary inferior, (3–)4(–5)-celled with numerous ovules on axile placentae.

About 75 spp. in E. Asia and North America, 1 in Europe. A number of spp. and hybrids are commonly cultivated.

*P. coronarius L.   Syringa, Mock Orange.

Deciduous shrub to 3 m, with brown peeling bark. Buds hidden in the base of the petiole. Lvs 4–8 cm, ovate, remotely toothed, glabrous except for tufts of hairs in the axils beneath, and sometimes pubescent on the veins. Fls 5–7, in terminal racemes, white, 2.5–3.5 cm diam., very fragrant. Calyx glabrous. Styles united to about half-way, shorter than stamens. Fl. 6. $2n = 26$.

Commonly planted, sometimes among native vegetation and possibly sometimes naturalized. Native of S.E. Europe and Caucasus.

*Hydrangea sargentiana Rehder, a shrub up to 3 m, with ovate lvs 10–25 cm pubescent above and with dense stiff hairs beneath, and pinkish-white fls in flattish corymbs 15–22 cm in diam., is ±naturalized in S.W. Ireland (Valencia Island, S. Kerry).

65. ESCALLONIACEAE

Shrubs or trees. Lvs alternate, rarely opposite or whorled, often with glandular teeth, exstipulate or rarely with small stipules. Fls hermaphrodite, rarely unisexual, actinomorphic (4–)5(–9)-merous, perigynous or epigynous, mostly with a well-marked disk. Stamens as many as corolla-lobes. Carpels 2–6, united, very rarely free, superior to inferior; ovules numerous (rarely 1 or 2). Fr. a capsule or berry; seeds usually numerous; embryo straight; endosperm copious.

Seven genera and 150 spp., southern hemisphere (rare in continental Africa) and S.E. Asia, a very few in Mexico and eastern North America. Differs from Hydrangeaceae in the alternate, often glandular lvs and few stamens.

1. ESCALLONIA Mutis ex L. fil.

Shrubs. Lvs alternate, usually evergreen, glandular-toothed. Fls in racemes or panicles, 5-merous. Petals with erect claw and spreading limb, pink or white. Stamens inserted under the edge of the disk. Ovary inferior, of 2–3 carpels, the placentation parietal; style 1; ovules numerous. Fr. a septicidal capsule.

About 60 spp., South America; some spp. and especially hybrids are frequently grown in gardens.

*1. E. rubra (Ruiz & Pavón) Pers.

E. macrantha Hooker & Arn.

Evergreen shrub 2–4 m. Twigs pubescent and glandular, sticky. Lvs 2.5–7.5 cm, broadly elliptical to obovate, doubly serrate, dark green, shining and glabrous above, dotted with resinous glands beneath, sessile. Infl. 5–10 cm. Fls c. 15 mm long and across, bright pinkish-red. Calyx campanulate, the lobes narrow, acute, glandular. Fl. 6–9. $2n = 24$.

Commonly planted as a hedge near the sea in S.W. England and Ireland and sometimes self-sown; apparently naturalized in Cornwall, Glamorgan, Cardigan, Merioneth, Scilly and Channel Is., and in Ireland from Wexford and Cork to Galway.

The above description refers to var. *macrantha* (Hooker & Arn.) Reiche, introduced from the island of Chiloe, Chile about 1846. Var. *rubra*, from S. Chile and adjacent Argentina, is less dense, with long, flexuous branches, obovate-lanceolate lvs, which are glandular-ciliate towards the base, and a glabrous calyx. It is widely cultivated as a garden ornamental but is not known to be naturalized.

66. GROSSULARIACEAE

Deciduous shrubs. Lvs alternate, simple, often palmately lobed, stalked, exstipulate or with small stipules adnate to the petiole, plicate or convolute in bud. Fls in racemes, hermaphrodite or unisexual, actinomorphic, epigynous, with the conspicuous and often coloured hypanthium prolonged beyond the ovary, 4–5-merous. Sepals coloured like the hypanthium. Petals usually shorter than sepals. Stamens equal in number to and alternating with petals. Ovary inferior, of 2 carpels with parietal placentae; styles 2, connate below; stigmas entire; ovules numerous. Fr. a berry, the calyx persistent at the apex; endosperm copious, fleshy; embryo rather small.

Two genera and 150 spp.

Nearest *Escalloniaceae*, usually recognizable by the conspicuous hypanthium and small petals.

1. RIBES L.

About 150 spp., in the north temperate zone and on mountains in Central and South America. Some spp.,

in addition to the following, are cultivated. Nectar secreted by an epigynous disk.

1 Unarmed; fls 5 or more in racemes.   2
  Branches with spines at the nodes; fls in axillary clusters of 1–2(–3).   **5. uva-crispa**
2 Fls hermaphrodite, usually in spreading or drooping racemes; bracts less than half as long as pedicel.   3
  Plant dioecious; racemes erect in fl. and fr.; bracts exceeding pedicels.   **4. alpinum**
3 Lvs and ovary glandular, not strong smelling; fr. red (rarely white).   4
  Lvs and ovary with sessile glands, strong smelling; fr. black.   **3. nigrum**
4 Hypanthium nearly flat, with raised rim round the style; anther-lobes widely separated; lvs cordate at base with narrow sinus.   **1. rubrum**
  Hypanthium basin-shaped, without raised rim, anther-lobes contiguous on inner side; lvs truncate at base, or cordate with wide sinus.   **2. spicatum**

Subgenus 1. RIBES.

Unarmed shrubs. Fls in many-fld racemes.

1. R. rubrum L. Red Currant.

R. sylvestre (Lam.) Mert & Koch; *R. vulgare* Lam.; *R. domesticum* Jancz.

Shrub 1–2 m. Lf-blades to 6 cm, rather broader than long, 3–5-lobed, pubescent at least when young, *not strong-smelling or glandular*, deeply *cordate at base with narrow sinus*, glabrous at maturity or slightly pubescent beneath; lobes triangular-ovate, dentate, with broadly ovate apiculate teeth; petiole usually with stalked glands, somewhat expanded and ciliate with long, often gland-tipped, hairs. Infl. pendent or spreading and curving downwards, with 6–20 fls; bracts ±ovate, somewhat curved, *less than half as long as pedicel*; rhachis and pedicels glabrous or slightly pubescent. Fls hermaphrodite, c. 5 mm diam., pale green or slightly tinged with purple, not pubescent or glandular; *saucer-shaped*, obscurely pentagonal with a *raised pentagonal rim round the style*; sepals inserted on lobes of receptacle, broader than long, contracted at base; petals cuneiform, very small; stamens shorter than petals, slightly inclined inwards, inserted on lobes of hypanthium, with a broad connective so that *anther-lobes* are *widely separated*. Fr. 6–10 mm diam., *red*, rarely brownish-white, globose; young fr. not contracted above. Fl. 4–5. Visited by Hymenoptera, homogamous. $2n = 16$. N.

?Native. Woods and hedges, ascending to 460 m. From Caithness southwards (not recorded from the Scottish islands), widespread (except in N. Scotland); locally frequent and perhaps native by streams in woods, etc., and in fen carr in many parts of England and Wales; probably introduced in Scotland; certainly introduced in Ireland and naturalized in Kerry, Cork, Waterford, Wexford and Antrim. Native in France, Belgium, Netherlands, Germany and Italy, widely cultivated and naturalized elsewhere in W. Europe. The majority of cultivated red and white currants belong to this sp.

2. R. spicatum Robson Downy Currant.

R. petraeum auct. (incl. Sm.), non Wulfen; *R. pubescens* (Hartm.) Hedl.

Like *R. rubrum* but differs as follows: *lf-blades* to 10 cm, *truncate at base to shallowly cordate with wide sinus*, usually persistently pubescent beneath but sometimes nearly glabrous. Infl. erect, ascending or spreading in fl., sometimes somewhat arched, but not pendent, except in fr.; rhachis and pedicels ±pubescent and with minute whitish glands. Fls c. 7 mm diam.; *hypanthium cup-shaped*, *without raised rim*; sepals ±orbicular, contracted at base; stamens erect, inserted on margin of hypanthium, with a *very narrow connective* so that anther-lobes are contiguous on inner side. Young fr. contracted above. $2n = 16$. N.

Native. Woods on limestone, ascending to 425 m. Lancs, Yorks, Cumberland and Durham, Inner Hebrides and Lanark to Moray, local; an occasional escape elsewhere, but rarely grown for fruit. N. Europe and Asia from Scandinavia to Manchuria, south to Poland and Romania.

R. spicatum × rubrum.

Some cultivated red currants belong here and it might occur as an escape.

Hybrids of *R. rubrum* with foreign spp. are also occasionally grown for fr.

3. R. nigrum L. Black Currant.

Shrub 1–2 m. Lf-blades 3–10 cm, rather broader than long, 3–5-lobed, cordate at base, glabrous above, pubescent on the veins beneath, at least when young, *with scattered sessile brownish glands*, especially beneath, *strong smelling*; lobes deltate or ovate-deltate, acute, irregularly and coarsely serrate-dentate; petiole pubescent, below conspicuously expanded and with sessile-brownish glands. *Racemes lax*, pendent, 5–10-fld; rhachis and pedicels pubescent, often glandular; *bracts* ovate, straight, *less than half as long as pedicel*. Fls hermaphrodite, c. 8 mm, dull purplish-green; *ovary with sessile glands*; *hypanthium* broadly campanulate, *pubescent outside*; sepals oblong, pubescent, recurved at apex; petals whitish, ovate, c. half as long as sepals; stamens about equalling petals. *Fr. black*, globose, 12–15 mm diam. Fl. 4–5. Fr. 6–7. Visited by bees, etc., but visits few; often self-pollinated; homogamous. $2n = 16$. N.

?Native. Woods and hedges from Caithness and Outer Hebrides southwards, widespread (except in N. Scotland); commonly cultivated for its fr. and often an escape but probably native by streams in woods, etc., and in fen carr, in many places; scattered localities but not native in Ireland. Most of Europe except the Mediterranean region; native certainly in C. and E. Europe; N. and C. Asia to the Himalaya.

*R. sanguineum Pursh Flowering Currant.

Lvs obtusely lobed, doubly serrate, pubescent beneath. Fls bright pink, 10–15 mm; hypanthium tubular. Fr. bluish-black, very pruinose. Commonly planted and sometimes naturalized. Native of western North America.

4. R. alpinum L. Mountain Currant.

Dioecious shrub 1–3 m. Lf-blades 3–5 cm, as broad as long, deeply 3(–5)-lobed, truncate to subcordate at base, with scattered hairs; lobes dentate, with acute teeth; petiole scarcely expanded below, with stalked and subsessile glands. *Fls in erect racemes*, the male 20–30-fld, the female 8–15-fld; rhachis and pedicels densely glandular with subsessile glands, otherwise glabrous; *bracts* lanceolate or linear, acute, *longer than pedicels*. Fls yellowish-green, 4–6 mm diam., the female smaller than the male; hypanthium nearly flat in male fl., cup-shaped in female; sepals ovate; petals very small, cuneiform; stamens short; style present in male fl. and abortive anthers in female. Fr. red, globose, 6–8 mm diam., insipid. Fl. 4–5. Pollinated by Diptera and Hymenoptera. Fr. 7. $2n = 16$. N.

Native. Cliffs and rocky woods on limestone; ascending to 380 m. Denbigh, Derby, Cheshire and Yorks; very local; an escape elsewhere but not now often cultivated. N. and C. Europe, extending southwards to the mountains of N. Spain, C. Italy and Bulgaria; Morocco (Atlas).

Subgenus 2. GROSSULARIA (Miller) A. Richard

Twigs with spines at the nodes. Fls solitary or in 2–4-fld axillary clusters.

5. R. uva-crispa L.   Gooseberry.

R. grossularia L.

Much-branched shrub c. 1 m. Spines 1–3 at each node, 5–15 mm. Twigs glabrous, pubescent, or bristly, with many short axillary shoots. Lf-blades 2–5 cm, as broad as or rather broader than long, 3–5-lobed, broadly cuneate to subcordate at base, glabrous or pubescent; lobes ovate-rhombic, obtuse, deeply and obtusely dentate in the upper part. Fls 1–2(–3) on each peduncle, pendent; rhachis pubescent and glandular; bracts small,

ovate. Fls c. 1 cm, hermaphrodite, greenish, purple tinged; ovary glandular-bristly or pubescent, rarely glabrous; hypanthium and calyx hairy on both surfaces; hypanthium broadly campanulate; sepals oblong, reflexed; petals small, whitish; stamens twice as long as petals. Fr. globose or ovoid, 10–20 mm (more in some cultivated forms), green, yellowish-green or reddish-purple, bristly, pubescent or smooth. Fl. 3–5. Visited mainly by Hymenoptera, protandrous. $2n = 16$. N.

Native. Woods and hedges from Sutherland southwards, widespread (rare in N.W. Scotland and not recorded from the Scottish Is.); commonly cultivated for its fr. and often an escape but probably native by streams in woods, etc., in many places; not native in Ireland where it is locally naturalized except in the extreme south and parts of the centre. Native in S., C. and W. Europe; naturalized elsewhere; Morocco, Algeria (mountains, rare); Caucasus.

67. PITTOSPORACEAE

Trees, shrubs or woody climbers with resin in the cortex. Lvs alternate. Fls usually hermaphrodite, regular, hypogynous, 5-merous. Sepals and petals imbricate. Petals clawed. Stamens 5, alternate with petals. Carpels 2(–5); placentation parietal or rarely axile; ovules usually with one integument, numerous; style simple. Fr. a capsule or berry; endosperm copious.

Nine genera and 200 spp., warmer parts of Old World, especially Australia, where 8 genera are endemic.

1. PITTOSPORUM Gaertner

Trees or shrubs. Ovules numerous on parietal placentas. Fr.

a leathery capsule. Seeds not winged.

About 150 spp. with the range of the family.

*1. P. crassifolium Putterl.

Evergreen shrub or small tree to 10 m, or erect habit. Lvs 4–8 cm, narrowly obovate, obtuse, coriaceous, dark green and shining above, white- or buff-tomentose beneath; margins recurved. Fls 1–6 in terminal umbel-like cymes, blackish-purple. Fr. 1.5–3 cm, woody, white-tomentose, (2–)3(–4)-valved.

Planted in the warmer parts of the country and naturalized in the Scilly Is. New Zealand (N. Island).

68. DROSERACEAE

Perennial glandular carnivorous plants. Lvs often circinate in bud. Fls actinomorphic, hermaphrodite, often in unbranched circinate cymes, 4–8-merous. Sepals and petals imbricate in bud, the sepals connate at base, the petals free. Stamens 4–20, often 5; anthers 2-celled, extrorse. Ovary superior, 1-celled, placentation parietal or sub-basal; styles 3–5, rarely connate below. Fr. a loculicidal capsule with numerous small seeds.

Four genera with about 105 spp., cosmopolitan in acid sandy, stony and boggy places.

1. DROSERA L.

Slender carnivorous herbs. Lvs densely glandular and fringed with long glandular hairs ('tentacles'), in a rosette or alternate; stipules usually present. Fls in circinate cymes, rarely solitary, 4–6- or 8-merous. Stamens as many as the petals. Ovary 1-celled; styles 3–5, free or connate below. Fr. a many-seeded capsule; seeds minute, testa often inflated, forming a wing.

About 100 spp. in temperate and tropical regions, specially abundant in S. Africa and Australia.

The British spp. are usually cleistogamous. For accounts of the carnivorous habits of the genus see Darwin, *Insectivorous Plants*, John Murray (1875) and Lloyd, *The Carnivorous Plants* (1954, 2nd edn).

1 Lvs not appreciably longer than broad, narrowed abruptly at base.   **1. rotundifolia**
Lvs distinctly longer than broad, gradually narrowed at base.   2
2 Lvs narrowly obovate to linear-oblong; scape in fl. up to twice as long as lvs, apparently arising from the centre of the rosette.   **2. anglica**
Lvs obovate; scape in fl. little longer than lvs, arising laterally below the rosette.   **3. intermedia**

1. D. rotundifolia L.   Round-leaved Sundew.

A slender reddish scapigerous perennial 4–8 cm. Stem short, usually unbranched or rarely (in floating forms) sending out slender stolons. *Lvs* long-petiolate, *spreading horizontally* and forming a rosette, *blade orbicular*, 4–8 mm diam.; *petioles* 15–30 mm, *hairy*. *Scape* erect, simple or rarely branched above, 2–4 *times as long as the lvs*, appearing to arise from the centre of the rosette. Fls white, shortly pedicellate, usually 6-merous, in 2 rows; petals 5 mm. Capsule smooth, equalling or exceeding the sepals; seeds reticulate, winged. Fl. 6–8. $2n = 20$. Hr. or Hel.

Native. In bogs and wet peaty places on heaths and moors, often amongst *Sphagnum*, occasionally forming a floating fringe to small ponds. Throughout the British Is. wherever suitable habitats exist. Europe, except for some of the islands and the extreme south; temperate Asia; Japan; Greenland; North America.

2. D. anglica Hudson   Great Sundew.

A reddish scapigerous perennial 10–18 cm. Stems short, unbranched. Lvs ±erect, *blade up to* c. 30 mm, *linear-oblanceolate*, gradually attenuate into the glabrous petiole, 50–100 mm. *Scape erect, up to twice as long as lvs*, appearing to arise from the centre of the rosette. Fls similar to those of *D. rotundifolia*, 5–8-merous; petals 6 mm. Capsule obovoid, smooth, exceeding the sepals; seeds reticulate, winged. Fl. 7–8. 2n = 40. Hr. or Hel.

Native. Usually amongst *Sphagnum* in the wetter parts of bogs. Mainly in N.W. Scotland and Ireland but scattered throughout the wetter parts of the British Is. N. and C. Europe; N. Asia; Japan; North America.

D. × obovata Mert. & Koch, a hybrid between *D. rotundifolia* and *anglica*, occurs occasionally. It resembles *D. intermedia* in general appearance but has a

straight scape 2–3 times as long as the lvs, arising from the centre of the rosette. It is sterile.

3. D. intermedia Hayne   Long-leaved Sundew.
D. longifolia L., pro parte

A reddish scapigerous perennial 2–3 cm. Stem short, slender, branched. Lvs ±erect, *blade* c. 7 mm, *obovate*, gradually attenuate into a glabrous petiole 2.5–4 cm. *Scape curved or decumbent at base, then erect, little longer than the lvs*, arising laterally from below the terminal rosette. Fls 5–8-merous, similar to those of *D. rotundifolia*; petals 4–5 mm. Capsule pyriform, grooved, equalling or slightly exceeding the sepals; seeds granulate, not winged. Fl. 6–8. 2n = 20. Hr.

Native. In damp peaty places on heaths and moors, but often in rather drier habitats than the two preceding spp., though occasionally in wet *Sphagnum* bogs especially in Ireland. Mainly in the west of the British Is., rather local but often abundant in suitable habitats. N.W. and C. Europe; North America; Asia Minor.

69. SARRACENIACEAE

Insectivorous herbs with basal lvs forming tubular 'pitchers' with a small blade at the top. Fls hermaphrodite, solitary, actinomorphic, nodding. Sepals 4–5, free, imbricate, persistent, often coloured. Petals 5 or 0, free, imbricate. Stamens numerous, free. Ovary superior, 3–5-celled; style simple, often much enlarged and peltate at apex. Ovules numerous, axile, anatropous. Capsule loculicidal. Seeds small, endosperm fleshy.

Three genera and about 17 spp. in Atlantic North America, California and the Pakaraima mountains.

1. SARRACENIA L.

Fls solitary, closely subtended by 3 bracts. Sepals 5, spreading, coloured. Petals 5, connivent. Ovary 5-celled;

style greatly dilated and umbrella-like above.

Ten spp. in Atlantic North America, some widely cultivated and not infrequently naturalized. Hybrids are of frequent occurrence.

*1. S. purpurea L.   Pitcherplant.

A stout low-growing perennial herb. Pitchers 10–20 cm. Scape 15–60 cm. Fl. c. 5 cm diam. Sepals dark purple outside, pale green within; petals purple on both surfaces. Umbrella-like top of style c. 3 cm diam. Fl. 6. Hel.

Introduced. Abundantly naturalized in bogs in C. Ireland, where it was planted in 1906. Roscommon and Westmeath. Native of North America; naturalized in Switzerland.

70. LYTHRACEAE

Herbs, shrubs, or, less frequently, trees. Lvs usually opposite or in whorls; stipules 0 or minute. Fls actinomorphic, hermaphrodite, perigynous. Hypanthium tubular or cup-shaped, the teeth valvate, often with appendages alternating with them. Petals present or 0, inserted at or near the top of the hypanthium, crumpled in bud. Stamens (2–)4 or 8(–12), inserted below the petals; filaments usually inflexed in bud. Ovary superior, usually 2–4-locular; style simple; ovules numerous, placentation axile. Fr. usually a capsule, opening by a transverse slit, or by valves, or irregularly. Seeds small; endosperm 0; embryo straight.

Twenty-five genera and 550 spp., throughout the world except for cold regions.

1. LYTHRUM L.

Annual or perennial herbs. Stems 4-angled, at least when young. Lvs entire, usually opposite. Fls axillary,

solitary or in small cymes, purple or pink. Hypanthium straight, teeth 4–6. Petals 4–6, rarely 0. Stamens 2–12, the filaments sometimes unequal in length. Ovary sessile, 2–4-celled; style filiform.

About 35 spp., cosmopolitan.

1 Perennial; stems 60–120 cm; fls in whorl-like cymes forming a long terminal spike; stamens exserted.
                                                    1. salicaria
  Usually annual; stems not more than 25 cm; fls solitary, axillary; stamens included.   2
2 Lvs obovate-spathulate, all opposite; hypanthial tube hemispherical to cup-shaped.   **3. portula**
  Lower lvs oblong-ovate, opposite, the upper linear, alternate; hypanthial tube obconical in fl., cylindrical in fr.   **2. hyssopifolia**

1. L. salicaria L.   Purple Loosestrife.

A ±pubescent *perennial* 60–120 cm. Stem erect, simple or with relatively short, slender branches. Lvs 4–7 cm, sessile, lanceolate to ovate, acute, shortly pubescent to subglabrous, ±cordate at base, the lower opposite or in whorls of 3, the upper sometimes alternate. Infl. up to 30 cm or sometimes more, spike-like, usually dense and many-fld. *Fls 10–15 mm, arranged in whorl-like cymes* in the axils of bracts. Bracteoles caducous. Hypanthium pubescent, the tube 4.5 mm, cylindrical, ribbed; teeth 2–3 mm. *Petals 8–10 mm*, obovate, *purple*. Stamens 12. Capsule 3–4 mm, oblong-ovoid, enclosed in hypanthium. Fl. 6–8. $2n = 60^*$. Hel. or Hp.

Three forms of fls are found on different plants: 1. Style short, stamens long and medium. 2. Style medium, stamens long and short. 3. Style long, stamens short and medium. Three sizes of pollen grain are found, one in each of the three types of stamen.

Native. In reed-swamp at the margins of lakes and slow-moving rivers, and in fens and marshes, often forming large stands. England, Wales and Ireland, locally abundant, in Scotland mainly in the west and absent from much of the north. Almost throughout Europe, Asia, N. Africa, naturalized in North America.

2. L. hyssopifolia L.   Grass Poly.

A glabrous *annual* 10–25 cm, sometimes perennating. Stem erect or decumbent, usually much-branched. Lvs 1–2.5 cm, the lower oblong-ovate, ±obtuse, opposite, the upper linear or linear-lanceolate, subacute, alternate, all sessile and rounded or narrowed at base. *Fls* c. 5 mm, *solitary* in the axils of the lvs, subsessile, not heterostylous. Bracteoles persistent. Hypanthium 4–6 mm, narrowly funnel-shaped, the teeth c. 0.75 mm. *Petals 2–3 mm, pale pinkish-lilac*. Stamens 6, included. Capsule c. 4 mm, subcylindrical, enclosed in the hypanthium. Fl. 6–7. $2n = 20$. Th.

Native. On bare ground or among open vegetation in hollows where water lies in winter, very local and uncertain in appearance from year to year. Recorded from a number of counties, undoubtedly a casual in most, but native in Cambridge and the Channel Is. C. and S. Europe, temperate Asia, N. Africa; probably introduced in America and Australia.

3. L. portula (L.) D. A. Webb   Water Purslane.

Peplis portula L.

A glabrous *creeping* annual 4–25 cm. *Stems* 4-angled, branched, *rooting freely at nodes*. Lvs 1(–2) cm, *obovate-spathulate*, opposite, rather thick; stipules minute, gland-like. Fls c. 1 mm, subsessile, solitary in the axils of almost all the lvs. Hypanthium hemispherical to cup-shaped; teeth 6, triangular, about as long as tube; appendages longer or shorter, setaceous. Petals 6 (5 or 0), fugacious. Stamens 6 or 12. Capsule c. 1.5 mm, subglobose. Fl. 6–10. $2n = 10$. Th.

Native. At muddy margins of pools or puddles, in open communities, or, more often, on bare ground, locally common but always absent from calcareous soils. Scattered throughout the British Is. from Orkney southwards. Europe except the extreme north; Azores; W. Siberia.

71. THYMELAEACEAE

Trees or shrubs, rarely herbs. Lvs usually alternate, entire, sessile or shortly stalked, exstipulate. Infl. a raceme, spike, head, cluster, or fls solitary. Fls regular, usually hermaphrodite, 4–5-merous, with an elongated, often coloured hypanthium. Calyx often coloured like the hypanthium. Petals scale-like or 0. Stamens usually twice as many as sepals, in 2 whorls, sometimes as many, rarely only 2. Ovary superior, 1-celled with a single apical pendent ovule, rarely 2-celled; style simple or 0; stigma usually capitate. Fr. a nutlet, drupe or berry, very rarely a capsule. Seed with a straight embryo; endosperm present or 0.

About 50 genera and 500 spp., cosmopolitan (except the Arctic). Usually easily recognized by the elongated, often coloured, hypanthium and single ovule. Placed by Bentham & Hooker among the apetalous families, but its true affinity seems to be with the Lythraceae which has a similar hypanthium.

1. DAPHNE L.

Shrubs. Lvs usually alternate, shortly stalked Fls terminal or axillary in short racemes or clusters, hermaphrodite. *Hypanthium* brightly coloured or green, *cylindrical or campanulate*, usually caducous, *without or with an inconspicuous scale-like or ring-like outgrowth at the base of the ovary*. Sepals 4, well-developed, coloured like the hypanthium. *Petals 0. Stamens 8, inserted near* the top of the hypanthium; filaments very short. Ovary 1-celled; *stigma large, capitate, sessile or nearly so*. Fr. a drupe; endosperm sparse.

About 70 spp. in Europe, N. Africa, Asia and Australia. Several spp. are cultivated.

Fls pinkish-purple, appearing before the lvs, in 2–4-fld clusters in the axils of the fallen lvs of the previous year; lvs thin, light green; fr. red.   **1. mezereum**

Fls yellowish-green, in short racemes in the axils of the persistent lvs; lvs coriaceous, dark green; fr. black.   **2. laureola**

1. D. mezereum L.   Mezereon.

Deciduous shrub 50–100 cm, with few ±erect branches. *Lvs 3–10 cm, oblanceolate, acute, light green, not coriaceous*, glabrous. *Fls 8–12 mm diam., appearing before the lvs, 2–4 in a subsessile cluster in the axil of a fallen lf of the previous year, pinkish-purple* rarely white, pubescent outside, fragrant. Sepals about as long as tube of the hypanthium. *Fr. 8–12 mm, ovoid, bright red*. Fl. 2–4. Pollinated by Lepidoptera and long-tongued bees, probably often selfed. Fr. 8–9. $2n = 18$. N.

Native. Woods on calcareous soils. Yorks, Lancs, Derby, and from Gloucester and Bucks to Dorset and E. Sussex, very local and rare; extinct or introduced elsewhere in England; a rare alien in Wales, Scotland and N. Ireland. Most of Europe except the extreme west, south and north; temperate Asia (east to the Altai Mountains).

2. D. laureola L. Spurge-laurel.

Evergreen shrub 40–100 cm, with erect little-branched stems, the lvs often all near the top of the plant. *Lvs* 5–12 cm, obovate-lanceolate or oblanceolate, acute or subacute, *dark glossy green*, *coriaceous*, glabrous. *Fls* 8–12 mm, *in short axillary 5–10-fld racemes*, *yellowish-green*, glabrous outside; peduncle c. 1.5 cm; bracteoles obovate-lanceolate, obtuse, caducous; pedicels 1–3 mm.

Sepals $\frac{1}{3}$–$\frac{1}{2}$ as long as tube of the hypanthium. Anthers yellow. *Fr.* c. 12 mm, ovoid, *black*. Fl. 2–4. Pollinated by Lepidoptera and humble bees. $2n = 18$. N.

Native. Woods, mainly on calcareous soils, widespread and rather common in England, extending north to Cumberland and Durham but seldom abundant; N. Wales, Pembroke, Glamorgan; Channel Is.; alien in Scotland and E. Ireland. W., S.C. and S. Europe northwards to Hungary; Asia Minor; N. Africa (rare); Azores.

D. laureola × *mezereum* is reported from Sussex and Yorks.

72. ELAEAGNACEAE

Trees or shrubs, densely covered with peltate or stellate silvery-brown scale-like hairs. Lvs entire, exstipulate. Fls solitary or in few-fld clusters or racemes, actinomorphic, hermaphrodite or variously unisexual, regular, with a well-developed hypanthium. Sepals 2 or 4, valvate. Petals 0. Stamens equal in number to sepals and alternate with them, or twice as many. Ovary superior, 1-celled with a single basal anatropous ovule; style long; stigma simple. Fr. drupe-like, consisting of the dry true fr. surrounded by the lower half of the hypanthium which becomes fleshy; endosperm 0 or scanty; embryo straight.

Three genera and about 50 spp., north temperate and subtropical regions, extending to tropical E. Asia.

Species of *Elaeagnus* L., with 4 sepals and stamens, hermaphrodite or polygamous fls and broad lvs, are often cultivated.

Resembling Thymelaeaceae but easily distinguished by the characteristic indumentum and basal ovule.

1. HIPPOPHAË L.

Thorny deciduous shrubs or trees, dioecious. *Lvs alternate*, narrow. Fls unisexual borne on the wood of the previous year; female in short axillary racemes, the axis often subsequently developing into a thorn; male in short spikes, the axis usually deciduous. Female fl. with conspicuous elongated hypanthium, bearing 2 minute sepals at the apex; disk 0. Male fl. with short hypanthium and *two* large *sepals*; disk small. Stamens 4. Style filiform; stigma cylindrical.

Three spp., the others in C. Asia.

1. H. rhamnoides L. Sea-buckthorn.

Much-branched shrub 1–3 m, suckering freely. Lvs 1–8 cm, linear-lanceolate, subsessile, covered with silvery scales on both sides or becoming subglabrous and dull green above. Fls before the lvs, greenish, very small. Fr. 6–8 mm, subglobose or ovoid, orange. Fl. 3–4. Wind-pollinated. Fr. 9, often persisting through the winter. $2n = 24$. M. or N.

Native. Fixed dunes and occasionally on sea cliffs. From S. Northumberland to Sussex, local but sometimes dominant; introduced widely on other coasts but mainly in E. Ireland. Coasts of Atlantic and Baltic from Norway (67° 56' N) to N. Spain, C. Italy and Bulgaria; river shingles in C. Europe in the Rhone Valley, Alps, etc.; Black Sea coast; temperate Asia to Kamchatka, Japan and the N.W. Himalaya.

73. ONAGRACEAE

Annual or more usually perennial herbs, often marsh or water plants; sometimes, as in *Fuchsia*, shrubs or small trees. Lvs alternate, opposite and decussate, or whorled; simple, usually exstipulate, but with small caducous stipules in *Fuchsia* and *Circaea*. In *Epilobium*, *Oenothera*, *Clarkia*, etc., the epigeal cotyledons of some species develop large blades, resembling those of the true lvs, by intercalary growth proximal to the original cotyledonary blades. Fls solitary axillary or in bracteate racemose infls (ebracteate in *Circaea alpina*); usually hermaphrodite and actinomorphic, but polygamous in some sections of *Fuschia* and ±zygomorphic in *Chamaenerion*, *Zauschneria*, etc.; epigynous and usually also perigynous, with an often brightly coloured hypanthial tube ('calyx-tube') of varying length, shape and persistence, rarely 0 as in *Ludwigia*; *commonly 4-merous*, sometimes 2-merous or 5-merous. Calyx of 2, 4 or 5 free *valvate* sepals; petals as many as sepals, free, rarely 0; stamens most commonly twice as many as sepals in two alternating whorls, sometimes as many (*Circaea*, *Ludwigia palustris*), rarely only 1 fertile and 1 staminode (*Lopezia*); ovary inferior, (1–)2-, 4- or 5-celled with 1–many anatropous ovules in each cell; style usually single with an entire capitate to discoid or ±distinctly 2-lobed stigma. Nectaries usually present round base of style or near the bottom of the hypanthial tube, and fls visited by insects, often by night-flying moths, or, in *Fuchsia* spp., by small birds and then di- or tristylic; some *Fuchsia* spp. are wind-pollinated. Fr. usually a loculicidal capsule but a berry in *Fuchsia* and indehiscent in *Circaea* and *Gaura*; seeds non-endospermic, with a chalazal plume of hairs in *Epilobium*, *Chamaenerion* and *Zauschneria*, a wing in North American *Hauya*, and outgrowths of the raphe in *Clarkia*, etc.

Commonly cultivated in gardens are spp. of *Fuchsia* (q.v.); *Zauschneria californica* C. Presl., suffruticose, with large scarlet *Fuchsia*-like fls; the 'evening primroses' of the genus *Oenothera*; annual spp. and hybrids of *Clarkia*, large-fld and with non-plumed seeds, the former with long-clawed petals, the latter with petal-claws short or 0; and spp. of *Gaura*, tall herbs with spikes or racemes of usually white fls and indehiscent fr. lacking the barbed bristles of *Circaea*.

About 20 genera and 500 spp. in the temperate and subtropical zones and especially in the New World.

1 Shrubs, usually with pendent fls; sepals spreading, red; petals purple, pink or white, rarely 0; fr. a berry. **5. FUCHSIA**
Herbs; fr. a capsule or dry and indehiscent. 2
2 Petals 0 (4–6, often minute, in non-native spp.); hypanthial tube 0. Native sp. An aquatic herb resembling *Lythrum portula*, with solitary axillary fls only 3 mm diam. **1. LUDWIGIA**
Petals present; hypanthial tube usually evident but sometimes very short or virtually 0. 3
3 Petals 2, white or rarely pinkish; stamens 2; fr. indehiscent, bristly. **6. CIRCAEA**
Petals 4; stamens 8; fr. usually a capsule. 4
4 Petals usually yellow; seeds not plumed; hypanthial tube exceeding 15 mm (but only 4–8 mm in the rare pink-fld casual *O. rosea*). **4. OENOTHERA**
Petals usually purplish, rose or white; seeds with a chalazal plume of hairs; hypanthium short or virtually absent. 5
5 All lvs alternate; fls held horizontally and somewhat zygomorphic through deflexing of stamens and style; hypanthial tube virtually 0. **3. CHAMAENERION**
At least the lowest lvs in opposite pairs; fls ±erect and actinomorphic; hypanthial tube very short.
2. EPILOBIUM

1. LUDWIGIA L.

Chiefly perennial aquatic herbs, often creeping or free-floating. Lvs opposite or alternate, simple, entire or toothed. Fls commonly solitary, axillary, often forming a terminal lfy infl.; actinomorphic and 4(–6)-merous; not at all perigynous, hypanthial tube 0. Petals often minute and green or purplish, or 0; sometimes larger and brightly coloured. Stamens usually 4, opposite the sepals, but 8–12 in two whorls in some non-British spp. Ovary 4(–6)-celled, with short style and entire stigma. Fr. variously shaped, many-seeded, dehiscing rather irregularly by valves or apical pores; seeds not plumed, non-endospermic.

About 20 spp. in temperate and warmer regions, especially of the New World. Only 1 sp. in the British Is.

1. L. palustris (L.) Elliott

Isnardia palustris L.

A perennial aquatic herb with slender glabrous reddish stems 5–30(–60) cm, prostrate and rooting below, ascending or floating above. Lvs opposite and decussate, 1–3(–5) × 0.5–2 cm, ovate to broadly elliptical, narrowed abruptly into a short petiole, acute or shortly acuminate, entire, glabrous, shining; lvs of wholly submerged shoots elliptical to oblanceolate. Fls 3 mm, solitary and sessile in lf-axils, with 2 bracteoles at base of ovary; 4-merous. Sepals broadly ovate, abruptly acuminate or cuspidate, green often with red margins, persistent in fr. Petals 0. Stamens 4, opposite the sepals but shorter. Capsule 2–5 mm, oblong-obovoid, truncate above, with 4 blunt green angles between the yellowish faces and crowned by the horizontally spreading sepals. Seeds less than 1 mm. Fl. 6. Probably self-pollinated. Hel.–Hyd.

Native. Shallow pools in acid fen, very local. Now known only in the New Forest (Hants) and Jersey, but formerly in Sussex. W., C. and S. Europe, N. Africa, W. Asia; somewhat different forms in North, Central and South America.

2. EPILOBIUM L.

Herbs, rarely suffruticose, with at least the lower lvs in opposite pairs or in whorls of 3; upper lvs or bracts commonly alternate. Fls usually ±erect, solitary axillary or in terminal bracteate racemes or spikes; actinomorphic, 4-merous, perigynous with a very short hypanthial tube, a break across which causes the withering floral parts to fall. Sepals free. Petals purple, rose or white, equal, usually 2-lobed. Stamens 4 + 4, the antepetalous longer. Ovary 4-celled, with numerous ovules; stigma 4-lobed or entire, capitate to clavate. Fr. a long capsule dehiscing loculicidally by 4 valves; seeds with a chalazal plume of long hairs and thus readily wind-carried.

About 200 spp. in temperate and Arctic/Alpine regions of both hemispheres.

1 Stem quite prostrate or only branch-tips ascending, slender, rooting at nodes; lvs commonly 2.5–10 mm, broadly ovate to orbicular; fls solitary, axillary. **12. brunnescens group**
Stem usually ±erect, sometimes ascending from a decumbent base, never wholly prostrate; lvs at least 10 mm; fls in terminal and axillary racemes, rarely solitary and then terminal. 2
2 Stem with spreading hairs, mostly non-glandular but some shorter glandular hairs on upper part of stem; stigma 4-lobed. 3
Stem glabrous or with ±appressed non-glandular hairs; spreading hairs, if present, glandular only; stigma either 4-lobed or entire. 4
3 Lvs usually 6–12 cm, somewhat amplexicaul and slightly decurrent; petals at least 1 cm, deep rose. **1. hirsutum**
Lvs usually 4–8 cm, neither amplexicaul nor decurrent; petals less than 1 cm, pale rose. **2. parviflorum**
4 Stigma 4-lobed. 5
Stigma entire, capitate to clavate. 6
5 All or most lvs in opposite pairs, ovate-lanceolate, the whole margin toothed, including the rounded base; petioles less than 3 mm. **3. montanum**
Only the lower lvs in opposite pairs; lvs elliptical-lanceolate, toothed apart from the entire cuneate base; petioles 3–10 mm. **4. lanceolatum**
6 Lvs with petioles usually 4–15 mm. **5. roseum**
Lvs sessile or with petioles usually not exceeding 3 mm. 7

7 Stem with no distinct ridges or raised lines decurrent from the lvs, but sometimes with 2 rows of curled hairs; lvs narrowly elliptical-lanceolate; very slender stolons, finally ending in bulbils, produced in summer.   **9. palustre**

Stem with 2 or 4 ridges or raised lines decurrent from the lvs; lvs usually lanceolate to ovate, but if narrowly lanceolate then not markedly narrowed at base.   8

8 Stems erect or ascending, usually 20–80 cm; top of infl. and unopened fls erect or slightly drooping.   9

Small plants of upland flushes and rill-sides with decumbent or ascending stems only 5–20(–30) cm; top of infl. and unopened fls markedly drooping.   11

9 Upper part of stem with numerous slender spreading glandular hairs as well as ±appressed curled hairs.
                                                    6. ciliatum

Spreading glandular hairs 0 or only on the hypanthial tube.   10

10 Plant wholly without spreading glandular hairs; capsule 7–10 cm; almost sessile lf-rosettes formed at base of stem in autumn.   **7. tetragonum**

Plant with a few spreading glandular hairs on the hypanthial tube; capsule 4–6 cm; elongating epigeal stolons formed near base of stem in summer.
                                                    8. obscurum

11 Stem 1–2 mm diam.; lvs 1–2.5 cm, yellowish-green, ±narrowly ovate-oblong, obtuse; epigeal stolons with green lvs produced in later summer by elongation of basal rosettes; petals 3–4.5 mm.
                                            10. anagallidifolium

Stem 2–3 mm diam.; lvs 1.5–4 cm, bluish-green, ovate-lanceolate to ovate, acute; stolons mostly hypogeal with fleshy yellowish scale-like lvs; petals 7–11 mm.   **11. alsinifolium**

1. E. hirsutum L.

Great Hairy Willow-herb, Codlins and Cream.

A tall perennial herb producing in summer *white fleshy underground stolons*. Stem 80–150 cm, erect, almost terete, branching above, ±densely glandular-pubescent and *with numerous spreading hairs*. Lvs 6–12 × 1.5–2.5 cm, smaller above and on branches, mostly opposite, oblong-lanceolate, acute, *sessile, semi-amplexicaul and slightly decurrent*, ±hairy on both sides and especially on the veins, the margins ciliate and with distant, unequal, slender incurved teeth. Bracts alternate, resembling the upper lvs. *Fls* 15–23 mm diam., *erect in bud*, in ±corymbose racemes terminating the main stem and branches. Sepals 7–9 mm, hooded and shortly apiculate. Petals 12–16 mm, deep purplish-rose, broadly obovate, distinctly but not deeply notched. Outer stamens twice as long as inner. *Stigma of 4 revolute lobes exceeding the longest stamens*. Capsule 5–8 cm, downy. Seeds c. 1 mm, oblong-obovoid, acute at the base, brownish-red, densely and acutely tubercled. Fl. 7–8. Protandrous and visited chiefly by bees and hover-flies; ±homogamous forms with stigmas about equalling the stamens have been reported. $2n = 36$. H. Variable in hairiness from densely villous to subglabrous.

Native. Stream banks, marshes, drier parts of fens, etc., to 380 m in Derby. Common throughout most of

Great Britain and reaching Caithness, but absent in the extreme north-west; not in Hebrides, Orkney or Shetland. Europe northwards to S. Sweden; temperate Asia; N., E. and S. Africa. Introduced in North America.

2. E. parviflorum Schreber
Small-flowered Hairy Willow-herb.

A fairly tall perennial herb producing in autumn basal lf-rosettes, often reddish, which are at first ±sessile but later terminate *short above-ground lfy stolons*. Stem 30–60(–90) cm, erect, ±terete, glandular-pubescent above and ±covered *with soft short spreading hairs* throughout; branching above. Lvs commonly 3–7 × 1–1.5 cm, but variable in size and usually smaller above and on branches; the lower ones opposite but those above the middle usually alternate; oblong-lanceolate, acute, rounded but *neither amplexicaul nor decurrent* at the sessile base, softly hairy on both sides, the margins shortly ciliate and ±denticulate with short and distant horny teeth. Bracts alternate, like the lvs but smaller. *Fls* 6–9 mm diam., *erect in bud*, in ±corymbose terminal racemes. Sepals 4–6 mm, acute. Petals 6–9 mm, pale purplish-rose, obovate, deeply notched. *Stigma of 4 non-revolute spreading lobes, about equalling the stamens*. Capsule 3.5–6.5 cm, subglabrous to downy. Seeds 0.9–1 mm, narrowly obovoid, rounded at the base, brownish-red, densely and acutely tubercled. Fl. 7–8. Homogamous. Visited sparingly by hive bees, etc., and often self-pollinated. $2n = 36$. H.–Ch. Very variable in hairiness. The subglabrous form has been named *E. rivulare* Wahl. but does not deserve specific rank. Sometimes the lvs are almost all alternate.

Native. Stream banks, marshes and fens, to 380 m in Derby. Common; throughout the British Is. except Shetland. Europe northwards to S. Sweden; N. Africa; W. Asia to India.

3. E. montanum L.   Broad-leaved Willow-herb.

A perennial herb producing in late autumn short stolons which may be underground with fleshy pink and white scales, or above ground, very short and terminating in subsessile lf-rosettes. Stem (5–)20–60 cm, erect, slender, terete, often reddish, subglabrous, or with ±sparse short curved hairs. Lvs commonly 4–7 × 1.5–3 cm, mostly opposite or occasionally in whorls of 3, ovate-lanceolate to ovate, acute, rounded at the *short-stalked* base, the stalks to 6 mm, and with narrow slightly connate wings, sharply and irregularly toothed, subglabrous but usually hairy on the margins and veins. Bracts alternate, like the lvs but smaller. *Fls* 6–9 mm diam., in terminal lfy racemes, ±*drooping in young bud*. Sepals 5–6.5 mm, ±acute, often reddish. Petals 8–10 mm, pale rose, longer than broad, deeply notched. *Stigma of 4 short non-revolute lobes*, exceeded by the longer stamens. Capsule 4–8 cm, downy with short curved hairs. Seeds 1–1.2 mm, reddish-brown, narrowly obovoid, blunt at the base, densely but shortly tubercled. Fl. 6–8. Homogamous. Sparingly visited by insects and commonly self-pollinated. $2n = 36*$. H.–Ch.

Native. In woods on the more base-rich soils, hedgerows, walls, rocks, and as a weed in gardens; to c. 800 m in Wales. Common throughout the British Is. Europe to Norway and Finland; W. Asia, Siberia and Japan.

4. E. lanceolatum Sebastiani & Mauri
Spear-leaved Willow-herb.

A fairly tall perennial herb producing in late autumn short above-ground stolons terminating in spreading lf-rosettes which appear subsessile. Stem 20–60(–90) cm, erect, simple or slightly branched, with 4 hardly raised lines; subglabrous below and downy with short crisped hairs above. *Lvs* commonly 2–5 × 0.8–1.6 mm, only the lower usually opposite, elliptical to elliptical-lanceolate, blunt, *cuneate* at the base, narrowing gradually into a *stalk* 4–8 mm; hairy on the veins and margins, and with small ±equal rather distant marginal teeth except at the entire base. Bracts alternate, like the lvs but smaller. Fls 6–7 mm diam., in a terminal raceme drooping in bud. Sepals c. 4 mm, lanceolate, ±acute, often reddish. Petals 6–8 mm, pale pink becoming deeper, shortly 2-lobed. *Stigma of 4 short spreading lobes.* Capsule 5–7 cm, downy. Seeds c. 1 mm, reddish-brown, narrowly oblong-obovoid, finely tubercled. Fl. 7–9. Probably self-pollinated. $2n = 36^*$. H.–Ch.

Native. Roadsides, railway banks, walls, dry waste places, etc. S. and S.W. England from Cornwall, Devon, Dorset, Somerset, Gloucester and Monmouth to Sussex, Surrey, and Kent; Glamorgan. W. and S. Europe from France and Belgium to the Balkans; N. Africa; Caucasus and the Near East.

5. E. roseum Schreber   Small-flowered Willow-herb.

A rather slender perennial herb producing in late autumn lax subsessile lf-rosettes from very short stolons. Stem 25–60(–80) cm, erect, fragile, ±branched, usually with two distinct and two indistinct raised lines; glabrous below but with whitish crisped hairs and numerous spreading *glandular hairs* above. Lvs c. 3–8 × 1.5–3 cm, the lower opposite or all alternate, ovate-elliptical to lanceolate-elliptical, narrowed both to the acute apex and to the cuneate base and ±*long stalk* 3–20 mm; *glabrous* or hairy only on the veins, the margin finely and sharply toothed. Fls 4–6 mm diam., their buds cuspidate, drooping. Sepals 3–3.5 mm, lanceolate, acute. Petals 4–5 mm, at first white then streaked with rose-pink, shortly 2-lobed. *Stigma entire*, about equalling the style. Capsule 4–7 cm, downy with crisped and glandular hairs. Seeds c. 1 mm, oblong-obovoid, ±blunt at the base, finely tubercled. Fl. 7–8. $2n = 36$. H.–Ch.

Native. Damp places, woods and copses, railway banks and cultivated ground; lowland. Throughout lowland Great Britain northwards to Perth and Angus. N.E. Ireland. Europe northwards to Norway and S. Sweden; Asia Minor.

Distinguished from other British spp. by the long-stalked lvs, from *E. montanum* and *E. lanceolatum* also by the entire stigma, from *E. ciliatum* by the cuneate base of the lvs and the paler fls, and from the next 3 spp. by the crisped and glandular hairs on the upper part of the stem.

*6. E. ciliatum Rafin.
E. adenocaulon Hausskn.

A tall perennial *non-stoloniferous* herb, usually reddish, with numerous ±erect branches above, abundant small fls, and sessile or subsessile basal rosettes, formed in late summer, having at first small fleshy rounded lvs but later developing normal lvs. Stem (30–)60–90(–150) cm, strictly erect, with 4 raised lines except at the base where there are only 2; ±glabrous below, ±densely clothed with *crisped hairs* and with *short spreading glandular hairs* above. Lvs 3–10 × 1.8–3 cm, all but the uppermost opposite, oblong-lanceolate, gradually tapering to an acute apex, suddenly narrowed at the rounded to subcordate base into a *short-stalk* (1.5–3 mm); ±glabrous, with numerous small irregular forwardly directed teeth. Fls 4–6 mm diam., erect in bud, numerous, in racemes terminating the main stem and branches. Petals 3–6 mm, pale pink edged with purplish-rose, divided almost half-way into 2 ±parallel lobes. *Stigma entire, less than half the length of the style.* Capsule 4–6.5 cm, spreading, covered with crisped and glandular hairs or becoming ±glabrous when ripe. Seed c. 1 mm, pale reddish-brown, acute at the base, when ripe with a usually paler translucent beak below the plume of hairs; densely tubercled. Fl. 6–8. Probably self-pollinated. $2n = 36^*$. H.–Ch.

Introduced. First record 1891. Damp woods, copses, stream-sides, railway banks, gardens, waste places; spreading rapidly, especially in S.E. England. From Hants to Kent and northwards to Wilts, Gloucester, Hereford and Worcester, and to Essex, Bedford, Huntingdon and Derby. Reported also from Coll (Inner Hebrides) and S. Uist (Outer Hebrides). Almost throughout Ireland. Well established in Scandinavia, Poland, and the Baltic States. Native in North America.

Very close to the North American *E. glandulosum* Lehm., but a much less robust plant. Distinguishable from *E. roseum* by the shorter-stalked lvs with a rounded base, and the deeper-coloured smaller fls; and from *E. tetragonum*, and *E. obscurum* by the numerous glandular hairs (few or 0 in those spp.), the crisped (not ±appressed) non-glandular hairs, and the stigma much shorter than the style instead of ±equalling it as in those spp. The pellucid beak of the seed is found elsewhere amongst British spp. only in *E. palustre*, *E. anagallidifolium* and *E. alsinifolium*.

7. E. tetragonum L.   Square-stemmed Willow-Herb.
E. adnatum Griseb.

A fairly stout narrow-leaved perennial herb producing in late autumn several lax lf-rosettes on very short stolons. *Stem* 25–60(–80) cm, erect, *firm*, tough, branched, with 4(–2) conspicuously raised lines or wings; ±glabrous below but downy above with silky ±appressed whitish hairs. *Lvs* 2–7.5 × 0.3–1 cm, the lower and usually the middle lvs opposite, the rest alternate, *strap-*

shaped to narrowly oblong-lanceolate, blunt at the apex, narrowed to the ±*sessile* base and decurrent into the raised lines of the stem; *shining* and greasy-looking above, glabrous or slightly hairy on the veins, the margin strongly and irregularly denticulate. Bracts alternate, like the lvs but smaller. Fls 6–8 mm diam.; buds erect, acute. Sepals c. 4 mm, lanceolate, acute; '*calyx-tube*' *without glandular hairs*. Petals 5–7 mm, pale lilac, shallowly notched. *Stigma entire, equalling the style.* Capsule 7–9(–11) cm, downy with short crisped hairs. Seeds 1–1.2 mm, reddish, densely but rather bluntly tubercled. Fl. 7–8. Homogamous, the anthers dehiscing before the fls open. Self-pollinated. H.–Ch.

The above description is of subsp. **tetragonum**, the widespread form. A much less frequent form, perhaps confined to the southern half of Great Britain, is often treated as a separate sp. (*E. lamyi* F. Schultz) but seems to merit no more than subspecific rank as subsp. **lamyi** (F. Schultz) Léveillé. It differs from subsp. *tetragonum* in having at least the *upper lvs narrowed into a short stalk*. The lower lvs are somewhat shorter (2–5 cm), less acute and less strongly toothed than in subsp. *tetragonum* and the fls are often somewhat larger (10–12 mm diam.), but these differences are by no means constant.

Native. Damp woodland clearings and hedgebanks, stream and ditch-sides, etc., and cultivated ground; lowland. Great Britain northwards to Inverness and Argyll, locally common in the south but rare in the north; casual and rare in Ireland. Europe northwards to S. Sweden; W. Asia.

Subsp. *lamyi* in S. England from Cornwall and Kent northwards to Brecon, Hereford, Worcester and Leicester, eastwards to Bedford and Huntingdon; Outer Hebrides. Europe northwards to S. Sweden; Asia Minor; Madeira.

E. obscurum is distinguished by the presence of glandular hairs on the 'calyx-tube', by its shorter capsules and by the elongated stolons which it produces in summer.

8. E. obscurum Schreber

A tall perennial herb producing in late summer slender ±*elongated stolons* above or below ground, bearing ±distant pairs of small lvs which do not form distinct rosettes. Stem 30–60(–80) cm, erect from a curving base, with 4(–2) distinctly raised lines; glabrous below but somewhat downy above with ±appressed hairs. *Lvs* commonly 3–7 × 0.8–1.7 cm, the lower opposite, the upper usually alternate, lanceolate to ovate-lanceolate or the lower oblong-lanceolate, tapering to a usually blunt apex, rounded at the *sessile* base and suddenly contracted to become *decurrent* into the raised lines of the stem; *dull* above, glabrous or hairy only on the margins and veins, the margins with a few distant irregular small teeth. Bracts alternate, like the lvs but smaller. Fls 7–9 mm diam.; buds erect, acute. Sepals 3–4 mm, lanceolate, acute; '*calyx-tube*' *with glandular hairs*. Petals 5–6 mm, deep rose, shortly 2-lobed. Stigma entire, equalling the style. Capsule 4–6 cm, downy with short crisped hairs. Seeds 1 mm, reddish, densely and

acutely tubercled. Fl. 7–8. Homogamous and self-pollinated. H.–Ch.

Native. Marshes, stream and ditch banks, moist woods; to 780 m in Ireland. Locally common throughout the British Is. except Orkney and Shetland. Europe northwards to C. Norway; N. Africa; Madeira; Caucasus.

Recognizable by the glandular 'calyx-tube', the elongated summer stolons and the short capsules.

9. E. palustre L.   Marsh Willow-herb.

A perennial narrow-leaved plant producing in summer *filiform hypogeal stolons* bearing distant pairs of yellowish scale-lvs and terminating in autumn in a *bulbil-like bud* with fleshy scales. Stem 15–60 cm, erect from a curved base, simple or branched, *terete* and without raised lines but often with 2 rows of crisped hairs; subglabrous or downy with short crisped hairs. *Lvs* 2–7 × 0.4–1(–1.5) cm, mostly opposite, lanceolate to linear-lanceolate, narrowed to the blunt apex, *cuneate* at the base, ±sessile or the uppermost very shortly stalked; sometimes limp and somewhat drooping, subglabrous or with crisped hairs on the margins and veins, entire or very obscurely denticulate. Bracts alternate, like the lvs but smaller. Fls 4–6 mm diam., held almost horizontally; buds blunt, initially erect, soon drooping so that the top of the raceme hangs over to one side. Sepals c. 4 mm, lanceolate, acute. Petals 5–7 mm, pale rose or lilac, rarely white, shortly notched. Stigma entire, shorter than the style. Capsule 5–8 cm, downy with short crisped hairs. Seeds 1.6–1.8 mm, ±fusiform but narrowed from above the middle to an acute base, rounded above and with a short beak formed by the projecting inner integument; pale reddish-brown, finely and acutely tubercled. Fl. 7–8. $2n = 36^*$. H.

Native. In marshes and acid fens, ditches, etc.; calcifuge; reaching 762 m in Ireland. Locally common throughout the British Is. Europe northwards to Iceland and Lapland; Asia; North America and Greenland.

10. E. anagallidifolium Lam.   Alpine Willow-herb.

E. alpinum L., et auct. plur.

A perennial alpine herb with a stem 4–10(–20) cm, slender (1–2 mm diam.), ascending from a decumbent base, ±glabrous except for 2 lines of hairs down the 2 faint ridges; producing in summer numerous ±prostrate slender *epigeal stolons* at first forming small subsessile rosettes but soon elongating and then with *distant pairs of small green lvs*. Lvs usually 1–2 cm, often yellowish-green, mostly opposite, lanceolate or elliptical-lanceolate, gradually narrowing into a short stalk-like base, entire or distantly and faintly sinuate-toothed, ±glabrous. Fls c. 4–5 mm diam., 1–3, the top of the raceme drooping in fl. and young fr. Sepals 2.5–4 mm, reddish. Petals 3.5–4.5 mm, rose-red. Ripe capsule 2.5–4 cm, reddish, on an erect stalk 2.5–5 cm. Seeds c. 1 mm, obovoid, not beaked, acute below, very faintly tuberculate. Fl. 7–8. Probably self-pollinated. $2n = 36$. H.

Native. By streams and springs on mountains, commonly with *Montia fontana* and *Philonotis fontana*; from c. 155 to 1220 m in Scotland. Great Britain northwards from N.W. Yorks and Durham; Inner and Outer Hebrides. N. Europe and mountains of C. Europe; Asia; Greenland; North America.

11. E. alsinifolium Vill.   Chickweed Willow-herb.

E. alpinum L., p.p.

A perennial alpine herb with a stem 5–20(–30) cm, rather slender (2–3 mm diam.), ascending from a decumbent base, ±glabrous except for 2 rows of hairs down the 2 faint ridges and a few crisped hairs above; producing in summer slender *yellowish hypogeal stolons* with *distant pairs of yellowish scale-lvs*. Lvs usually 1.5–4 cm, somewhat bluish-green and shining above, mostly opposite, ovate to ovate-lanceolate, rounded at the base and narrowed into a short stalk, distantly sinuate-toothed, ±glabrous. Fls 8–9 mm diam., usually 2–5, the top of the raceme drooping in fl. and young fr. Sepals c. 4–6 mm. Petals 7–9 mm, bluish-red. Ripe capsule 3–5 cm, subglabrous, on an erect stalk 2–3 cm. Seeds c. 1 mm, narrowly obovoid with an apical beak upon which the plume of hairs is borne, minutely tuberculate. Fl. 7–8. Probably self-pollinated. $2n = 36^*$. H.

Native. By streams and springs on mountains; from 122 m on Eigg to 1100 m in Perth. Caernarvon, and northwards from Durham, N.W. Yorks and Cumberland; Inner and Outer Hebrides; Shetlands. Ireland (one locality in Leitrim). N. and C. Europe.

12. E. brunnescens group

*12. E. brunnescens (Cockayne) Raven & Engelhorn.

E. pedunculare auct., non A. Cunn.; *E. nerteroides* auct., non A. Cunn.

A prostrate mat-forming perennial herb with slender often reddish creeping stems to 20 cm, rooting at the nodes and much branched, the branches often with ascending tips; stems with evident lines of short stiff appressed hairs. Lvs 2.5–10 mm, broadly ovate to suborbicular, entire or obscurely sinuate-toothed, green or somewhat bronzed and smooth above, often ±flushed with purple and with obvious midrib but faint lateral veins beneath, glabrous; petioles 0.5–3 mm. Fls solitary, axillary but usually in only one lf-axil of a pair; pedicels erect or briefly drooping, 0.5–4 cm, lengthening up to 7.5 cm in fr., glabrous. Sepals 2–3 mm, often reddish. Petals 2.5–4 mm, white or pale pink. Ripe capsule 2.5–4.5 cm, glabrous. Seeds 0.7 mm, numerous, papillose. Fl. 5–10. Self-pollinated. $2n = 36^*$. H.

Introduced from New Zealand. Locally frequent, especially in upland areas, on moist stony ground such as rocky beds and banks of streams, stony hedgebanks, damp walls, quarries, gravel pits, etc., over a wide range of pH from moderately acid to strongly alkaline and from sea-level to over 900 m. Throughout much of Great Britain from Cornwall to Kent and northwards to Caithness, but absent from lowland E. and C. England; Ireland; very rare in Inner and Outer Hebrides and

Orkney. First recorded as a garden weed in 1904, and a few years later as an established alien which spread rapidly and is still spreading.

The true *E. nerteroides* A. Cunn. differs in its almost glabrous stems and broadly elliptical to ovate lvs with petioles 1–6 mm.

*E. pedunculare A. Cunn. (*E. linnaeoides* auct.), also native in New Zealand, was recorded in 1938 as a garden weed both in Cheshire and Co. Dublin, and in 1953 was found by a road in Galway and later nearby in Mayo. It may be distinguished from *E. brunnescens* by its larger orbicular lvs, 4–8(–15) mm diam., with margins closely and sharply denticulate and the underside bright purple with prominent lateral veins, and by the longer fruiting pedicels, 5–10 cm.

*E. komarovianum Léveillé (*E. inornatum* Melville), yet another introduction from New Zealand, is a rather smaller plant than *E. brunnescens* and has ±elliptical lvs, bronzed and rugose above but usually green beneath with the lateral veins not showing. Its seeds are smooth. Established in a few localities in S. Devon, Derby and Lanark, and in N. Ireland.

The frequent occurrence in the wild of plants intermediate between two species and the results of experimental hybridization have made it clear that the British species of *Epilobium*, including the introduced *E. ciliatum*, can intercross freely. Hybrids with *E. brunnescens* are also known to be possible but have not yet been found here, though the cross *E. brunnescens* × *E. ciliatum* is known as a wild plant in New Zealand. It is of interest, however, that our *Epilobia* never cross successfully with *Chamaenerion angustifolium*.

Artificial hybridization has yielded healthy first-generation progeny from most of the many crosses attempted and has for the most part confirmed the status of the suspected hybrids found as wild plants. These are, to varying extents, intermediate between the known or supposed parents and are rarely fully fertile, commonly yielding only a few viable seeds. All our species have the same chromosome number, $2n = 36$, and meiosis in the hybrids is normal, the observed sterility resulting from early abortion of developing embryos. Some wild hybrids, however, do seem fairly fertile and may backcross with the parent species and even give rise locally to hybrid swarms. That hybrids are nevertheless far from abundant, and that our species seem in no danger of losing their individuality, is doubtless the result of several factors including the reduced fertility of most hybrids, differences in the ecological requirements of parent species, and perhaps especially the prevalence of self-pollination in the genus.

Valuable morphological features for recognizing hybrids are the stigmas, hybrids between species with 4-lobed and entire stigmas showing obscure or irregular lobing; the presence of long spreading eglandular stem-hairs, often with their distal ends turned up or down (*E. hirsutum* or *E. parviflorum* as one parent); presence of spreading glandular stem-hairs (*E. roseum* or *E. ciliatum* as a parent); mode of toothing of lvs, useful, for example, for hybrids between *E. hirsutum* and *E. parviflorum*, fl-size also being informative here. Features of leaf-shape, petiole-length and perennating structures must be relied on in identifying hybrids between *E. tetragonum*, *E. obscurum* and *E. palustre*.

3. CHAMAENERION Adanson

Perennial herbs resembling *Epilobium* but with *lvs all spirally arranged*, ±*entire*, and *fls held horizontally*.

Hypanthial tube very short; petals spreading, the upper 2 broader than the lower 2, so that the corolla is somewhat *zygomorphic*; stamens and style exserted and ultimately bending downwards.

About 8 spp. in temperate and arctic regions of the northern hemisphere.

1. C. angustifolium (L.) Scop.

Rosebay Willow-herb, Fireweed.

Epilobium angustifolium L.

A tall perennial herb with long horizontally spreading roots which give rise to erect lfy stems 30–120 cm, sub-terete, glabrous below, ±pubescent above. Lvs 5–15 cm, spirally arranged, numerous, ±ascending, narrowly oblong-lanceolate or oblong-elliptical, narrowed at each end, entire or with small and distant horny teeth, often ±waved at the margins, glaucous beneath with conspicuous veins, lateral veins numerous, joining into a continuous wavy intra-marginal vein. Fls 2–3 cm diam., numerous, in a long rather dense spike-like bracteate raceme; stalks 1–1.5 cm, ascending, pubescent. Sepals 8–12 mm, acuminate, dark purple. Petals 10–16 mm, obovate, clawed, entire or slightly notched, the upper pair broader than the lower, rose-purple. Stigmas finally exceeding the anthers, at first spreading then recurved or revolute. Capsule 2.5–8 cm, ±4-angled, pubescent, at first spreading, then ±erect. Seeds c. 2 mm, almost smooth, with a white plume. Fl. 7–9. Protandrous; visited by various insects for nectar secreted by the epigynous disk. $2n = 36^*$. H.

Variable. 'Epilobium angustifolium var. *macrocarpum* Syme' was distinguished from 'var. *brachycarpum* Syme' in fr.-length. The latter was probably a sterile cultivar, thus accounting for its short capsule. Efforts to relate these differences to supposed wild and naturalized forms in Britain have failed. Although certain British populations, especially in mountainous districts, are native the status of the common lowland plants is in doubt. There are, however, no constant morphological differences between these and the undoubtedly native populations.

Native. Rocky places, scree slopes, wood-margins and wood-clearings, disturbed ground, gardens, bombed sites, etc.; to 975 m in Scotland. Throughout the British Is., but commonest in the south. A century ago the species was a local plant though scattered throughout the country, especially in rocky places and on scree. Its phenomenal spread in the last few decades may be related to the increasing areas of cleared woodland and waste land. Europe, Asia, North America. Often cultivated as an ornamental garden plant.

C. angustifolium does not hybridize successfully with any sp. of *Epilobium*

*C. dodonaei (Vill.) Schur (C. rosmarinifolium Moench), with linear, indistinctly veined lvs, shorter lfy raceme, and unclawed petals, is sometimes grown in gardens and has been reported as a casual.

4. OENOTHERA L. Evening Primrose.

Annual to perennial herbs, rarely suffruticose, with *alternate exstipulate lvs* and solitary fls, or sometimes in pairs, in the axils of the upper lvs to form a large lfy spike; fls 4-merous. Sepals at first ascending, later strongly reflexed and often falling early; petals broad and overlapping, commonly yellow but sometimes turning reddish and sometimes rose-coloured or white from the start; stamens 4 + 4, equal or the inner shorter than the outer; ovary 4-celled; stigma entire or ±4-lobed. Fls usually open and become fragrant in the evening and are visited by moths, but some are diurnal and some cleistogamous; most are self-compatible. Fr. a many-seeded loculicidal capsule, ±cylindrical or clavate; seeds not plumed.

Of the c. 100 spp., largely American, many are genetically and cytologically normal, the 14 chromosomes forming 7 pairs at meiosis. Several, however, are permanently heterozygous for segmental interchanges affecting all the chromosomes and form 14-membered rings at meiosis with alternate members passing to the same pole. Zygotes from the fusion of similar gametes fail to survive, so that such spp. are true-breeding although highly heterozygous, but the occasional cross-over may give rise to genetically very different combinations or 'mutations' and to taxonomic problems. Between these extremes are other spp. forming chromosome pairs as well as rings of 4 or more at meiosis.

The following key is of species well established in the British Is. Some frequently encountered casuals are mentioned briefly at the end of accounts of the established species they most closely resemble. The very rare casual *O. rosea* L'Hér. ex Aiton, placed in Subgen. HARTMANNIA (Spach) Munz, is readily recognized by having petals only 5–10 mm long, pink from the start, and a clavate capsule up to 1 cm long.

1 Capsule 6–8 mm diam. below, ±cylindrical but usually narrowing upwards; seeds sharply angled; petals usually remaining yellow throughout. (Subgen. OENOTHERA)   2

Capsule 2–4 mm diam. below, often enlarged in upper half; seeds not sharply angled; petals initially yellow but reddish later or on drying. (Subgen. RAIMANNIA (Rose) Munz)   **1. stricta**

2 Hairs with red bulbous bases even on green parts of stem; numerous red spots on infl.-axis and ovaries; sepals entirely red or red-striped.   3

No hairs with red bulbous bases on green parts of stem and no red spots on infl.-axis and ovaries; sepals entirely green.

   2. biennis (see also *O. cambrica* var. *impunctata*, p.268)

3 Petals 30–50 mm; style long, with stigma-lobes spreading well above anthers; sepals normally red-striped.   **4. erythrosepala**

Petals rarely exceeding 30 mm; style shorter, with stigma-lobes spreading between anthers; sepals red-striped or entirely green.   4

4 Sepals green; petals as broad as long or slightly narrower; oldest frs without glandular hairs.

   5. cambrica

Sepals usually red-striped; petals broader than long;

all frs densely glandular with some stiff hairs.

                                              3. fallax

Subgenus RAIMANNIA (Rose) Munz

Usually annual or biennial. Fls nocturnal; petals initially yellow but becoming reddish later and on drying; hypanthium cylindrical. Capsule usually enlarged in upper half, sessile; seeds pendent, subcylindric or narrowly obovoid, not sharply angled.

1. O. stricta Ledeb. ex Link

                          Fragrant Evening Primrose.

O. odorata auct. europ., non Jacq.

Annual or biennial. *Stem* to 60(–150) cm, erect, slender, simple or sparingly branched, reddish, *shortly pubescent below and glandular-pubescent above* and with *some long spreading hairs but none with red bulbous bases.* Lvs 3–10(–18) cm, diminishing up the stem; *basal lvs linear-oblanceolate*, narrowed into a stalk-like base; stem-lvs narrowly lanceolate, ±sessile; *all with the margins ciliate*, distantly but sharply toothed *and sometimes undulate.* Infl. erect throughout; bracts ovate-lanceolate, reddish-margined, shorter than capsule; *fls fragrant.* Sepals becoming reddish, their terminal 1–3 mm free and erect or divergent in bud. *Petals* 15–40 mm, *at first yellow with basal red spot*, later reddening throughout. Stigma-lobes about equalling anthers. *Capsule* 2–3(–5) cm, c. 3 mm diam. at base but *conspicuously enlarged in upper half*, with both silky-appressed and spreading hairs. Seeds c. 1.5 mm, smooth, brown. Fl. 6–9. $2n = 14$. Th.–Hs.

Introduced. Locally established on sand dunes and in waste places in scattered coastal localities in S. England and Wales northwards to E. Norfolk and S. Gwynedd, more rarely inland; also near Selkirk, S. Scotland; Channel Is. Native in Chile.

Subgenus OENOTHERA.

Annual or biennial. Fls nocturnal; petals remaining yellow. Capsule 2–4 cm, ±cylindrical or fusiform, narrowing upwards from a largest diam. of 6–8 mm; seeds horizontal, sharply angled. Species and hybrids established in the British Is. have sepals touching laterally in bud, with only their extreme apices arching apart.

2. O. biennis L.

Usually biennial. Stem erect, 20–100(–150) cm, robust, simple or branched, reddish below, ±pubescent, with *some hairs on green parts of stem having green, but not red, bulbous bases* (bases red on red patches of stem). Basal lvs in a rosette, commonly 7–20 cm, obovate-lanceolate to oblong-elliptical, obtuse, narrowing into a long petiole; stem-lvs lanceolate to elliptical, ±acute, shortly petiolate or subsessile; *all lvs hairy with* ciliate entire or distantly denticulate margins and *reddish midribs* (except in shade). Infl. erect throughout; bracts ovate-lanceolate; fls somewhat fragrant. Hypanthial tube 25–40 mm. *Sepals green. Petals* 15–30 mm, obcordate, *broader than long. Stigma-lobes about level with*

anthers and spreading between them. Capsule 20–35 mm, sessile. *Infl.-axis, ovary, hypanthial tube, sepals and capsule all strongly glandular*; capsule with simple or bulbous-based stiff hairs. Fl. 6–9. $2n = 14$, forming rings of 6 and 8 at meiosis. Hs (biennial).

Introduced, probably from continental Europe. Dunes, roadsides, railway banks, waste places; established in many localities but apparently decreasing as a casual. In Great Britain northwards to N. Lancs, S.W. Yorks and N. Lincoln, with local outliers northwards to Perth. Europe northwards to 65° N in Finland; temperate Asia. *O. biennis* L. *sens. str.* (*O. biennis* subsp. *biennis* of Munz, 1965) appears not to be native in North America.

3. O. fallax Renner emend. Rostański

(*O. erythrosepala* (female) × *O. biennis* (male)). *O. cantabrigiana* B. M. Davis

Stem to 100 cm or more, simple or often branching near the base, green or reddish, *with numerous long hairs having red bulbous bases* and shorter simple hairs. Lvs elliptical to ovate-lanceolate with white or reddish midrib and the lower with crinkled margins. Infl.-axis green but reddened above, with both glandular and red-based hairs. *Sepals red-striped*, with stiff and glandular hairs; sepal-tips 2–4 mm. *Petals* 20–30 mm, obcordate, *broader than long. Stigma-lobes spreading between anthers.* Capsule 20–30 mm, red-striped when young, becoming green, *glandular with some hairs having red bulbous bases.* Fl. 6–9. $2n = 14$.

Resembles *O. erythrosepals* in having hairs with red bulbous bases and red-striped sepals but has fls almost as small as those of *O. biennis* and the stigma-lobes spread between the anthers. It is constant in its characters, thus behaving like a distinct species in contrast with the great variability of the reciprocal hybrid with *erythrosepala* as male parent, which forms extensive hybrid swarms linking the parent spp.

Native, being found wild as a spontaneous hybrid, though also escaping from cultivation. Sand dunes and waste places in a few mainly coastal localities in England and Wales northwards to Lancs and Cumbria; also in E. Ross; Channel Is. Known in continental Europe from Poland, E. and W. Germany and Czechoslovakia.

4. O. erythrosepala Borbás

O. lamarckiana auct., non Ser.; *O. grandiflora* auct., non L'Hér.

Usually biennial. *Stem* erect, 30–180 cm, robust, simple or sparingly branched, pubescent and *with numerous long stiff hairs having red bulbous bases.* Basal lvs in a rosette, 7–15 cm, broadly elliptical-lanceolate to ovate-lanceolate, petiolate, with *margins distantly toothed and often strongly crinkled*; stem-lvs ovate-lanceolate, subsessile; *midrib usually white*, sometimes reddish. Infl. erect throughout; bracts ovate-lanceolate. Hypanthial tube 30–40(–50) mm. *Sepals usually red-striped in bud*, sometimes entirely red, strongly glandular-pubescent. *Petals* 30–50 mm, *obcordate, broader than long*; sometimes much smaller in autumn. Style long, with *stigma-lobes spreading well above anthers.* Capsule 20–35 mm, green or red-striped when young. *Infl.-axis, ovaries and capsules covered with both glandular and non-glandular hairs, some of the latter with red bulbous bases.* Fl. 6–9. 2*n* = 14, forming a ring of 12 and 1 pair at meiosis. Hs (biennial).

Introduced. Sand dunes, roadsides, railway tracks, waste places, etc.; the commonest British sp., naturalized locally and apparently spreading. England and Wales northwards to Lancs, W. Yorks and Durham; Isle of Man; very local in Scotland northwards to E. Ross; N. Ireland; Channel Is. Introduced from North America in mid-19th century and now locally common in W. and C. Europe.

*5. O. cambrica Rostański

O. ammophila auct. angl., non Focke; *O. parviflora* auct. angl., non L.

Usually biennial. *Stem* erect, to 100(–150) cm, branched, with lower branches sometimes prostrate; green or red-streaked below, *with numerous stiff hairs having red bulbous bases.* Lower lvs elliptical-lanceolate, middle and upper lanceolate; all ±pubescent *with reddish midribs and* ±entire or slightly denticulate *flat margins.* Infl.-axis erect throughout, green at tip; infl. dense and blunt-tipped. *Sepals green. Petals* 15–30 mm, obcordate, *as broad as long or slightly narrower.* Style short, with *stigma-lobes spreading between anthers. Capsule* 25–40 mm, widest near middle, *green, with hairs having red bulbous bases;* only the upper (younger) with glandular hairs. Hypanthial tube and sepals covered with glandular and stiff non-glandular hairs. Fl. 6–9. 2*n* = 14. (Th.–)Hs.

Introduced, probably from Canada in 18th century, but origin uncertain. Dunes and waste places, especially in S. Wales but scattered further north in Wales, on coasts of S. England from Cornwall to Hants and in a few other coastal and inland localities northwards to Lancs, N.E. Yorks and near Edinburgh; Channel Is. Not known in continental Europe.

O. cambrica var. *impunctata* Rostański, very rare in S. Wales and near Oxford, lacks hairs with red bulbous bases but may be distinguished from *O. biennis* by its petals no broader than long and its older capsules without glandular hairs.

The hybrid *O. cambrica* × *O. erythrosepala* has been reported from a few localities in S. Wales.

The rare casual **O. perangusta** Gates resembles *O. erythrosepala, fallax* and *cambrica* in having stiff hairs with red bulbous bases on stems, infl.-axis and capsules and the first two in the red-striped sepals, but differs from all in the smaller petals up to only 20 mm long and broad, in the capsules hairy only along the angles and in the cylindrical shape of the red bases of stem-hairs. Its lvs are ±narrowly lanceolate. The related **O. rubricaulis** Klebahn, to which the nom. ambig. *O. muricata* L. seems originally to have referred, has sepals always green, the enlarged bases of stem-hairs funnel-shaped and broader elliptical to elliptical-lanceolate lvs. It is known here only as a rare casual, but is frequent in C. Europe. A few other spp. have also been recorded as casuals.

5. FUCHSIA L.

Shrubs or small trees, sometimes dioecious, with usually opposite or whorled glabrous simple stalked lvs with deciduous stipules. *Fls* 1 to several in lf-axils, usually stalked, *pendent*; actinomorphic, epigynous and with a perigynous (hypanthial) tube, hermaphrodite or unisexual. *Sepals* 4, *spreading, often red*; *petals* 4, free, *erect,* or 0; *stamens* usually 8, *exserted*; stigma capitate or 4-lobed. Nectar-secreting and pollinated by insects or small birds. *Fr. a 4-celled berry* with numerous seeds.

About 100 spp. in tropical and subtropical America and a few in New Zealand; 1 in Tahiti. Many spp. are cultivated but most are frost-tender.

*1. F. magellanica Lam.   Hedge Fuchsia.

Incl. *F. macrostema* Ruiz & Pavón; *F.* 'Riccartoni'; *F. gracilis* Lindl., etc.

A small shrub to 3 m or so. Lvs in opposite pairs or whorls of 3, 2–5.5 cm, ovate-oblong, acuminate, ±denticulate, often purplish, slightly pubescent on margin and underside of midrib; petiole 3–15 mm. Fls solitary in upper lf-axils, pendent, their pedicels 2–5 cm; hypanthial tube 5–9 mm, red. Sepals 12–20(–25) mm, red. Petals 6–20 mm, obovate, violet. Stamens much exserted, reddish. Berry 1–2 cm, black, 4-angled. Fl. 6–9. Protandrous. 2*n* = 44. N. or M.

Introduced and much grown in gardens and, especially in the west, as a hedge plant; frequently escaping and now well established in much of Ireland and in scattered localities in S.W. England, Wales, Isle of Man, S. and W. Scotland; Hebrides and Orkney; Channel Is. Native in Chile and Argentina. Planted as a hedge plant in the Azores.

Variable, some forms probably of hybrid origin. '*F. gracilis* Lindl.', with longer hypanthial tube and narrower sepals, is established in S.W. Ireland. Its origin and taxonomic status are uncertain.

6. CIRCAEA L.

Perennial rhizomatous or stoloniferous herbs with stalked ovate lvs in opposite pairs and terminal racemes of small white or pinkish fls; bracts short, setaceous, or 0. Hypanthial tube not exceeding 1.2 mm, slender and ±completely fused round the base of the style; sepals 2, caducous; petals 2, back and front, 2-lobed or notched; stamens 2, lateral; ovary inferior, 1- or 2-celled, with 1 ovule in each cell; style 1, with an entire, emarginate or distinctly 2-lobed stigma. Fr. indehiscent, obovoid, 1–2-seeded, ±densely covered with bristles which are commonly hook-tipped.

About 10 spp. in north temperate regions of Europe, Asia and North America.

1 Open fls in a terminal cluster; pedicels and sepals glabrous; petals less than 1.5 mm; no nectar-secreting ring round style; fr. 1-celled; petioles glabrous.
                                                          3. alpina
  Open fls spaced on the elongating infl.-axis; pedicels and sepals at least sparsely glandular-pubescent; petals 1.8–4 mm; nectar-secreting ring visible round

base of style; fr. with 2 equal or unequal cells; petioles hairy at least above. **2**

2 Nectar-secreting ring dark and prominent just above top of hypanthium; ovary equally 2-celled, persisting and ripening seed; lvs not or slightly cordate, sinuate-toothed or distantly denticulate, gradually acuminate; petiole hairy all round. **1. lutetiana**

Nectar-secreting ring low and inconspicuous; ovary unequally 2-celled, falling early without ripening good seed; lvs shallowly cordate, dentate, abruptly acuminate; petiole hairy above, subglabrous beneath. **2. ×intermedia**

1. C. lutetiana L. Common Enchanter's Nightshade.

Overwintering rhizomes non-tuberous; stolons from lower axils of aerial stem *usually* 0. Stem commonly 20–60 cm, erect or ascending, swollen at nodes, ±densely pubescent with some glandular hairs. *Lvs* usually 4–10 cm, ovate, *gradually acuminate, truncate or slightly cordate at base*, sinuate-toothed or distantly denticulate, thin, dull above but paler and shining beneath, subglabrous or with margins and underside of veins hairy; *petioles furrowed above, hairy all round.* Infl. usually ebracteate; infl.-axis elongating before petals drop, so that *open fls are well spaced*; pedicels densely glandular-pubescent, reflexed in fr. *Sepals pale green, glandular-pubescent.* Petals 2–4(–5) mm, truncate to rounded at base, deeply notched. Stamen filaments 2.5–5.5 mm. Stigma usually deeply 2-lobed. Fr. 3–4 mm diam., obovoid, equally 2-celled, densely covered with stiff hook-tipped white bristles; *most frs persisting and ripening seed.* Fl. 6–8. Nectar is secreted from a *prominent dark green ring round the style* just above the hypanthium, and fls are visited by small Diptera. $2n = 22$. Grh.

Native. Woods and shady places on moist nutrient-rich soils, often stony at surface; reaching 300–400 m; common. Throughout England, Wales and S. Scotland but very local in N. Scotland; Ireland; Inner Hebrides. Most of Europe except the north-east and eastwards to C. Asia; N. Africa.

2. C. × intermedia Ehrh.
Hybrid Enchanter's Nightshade.

(C. alpina × C. lutetiana)

Intermediate between the parents and often confused with *C. alpina*. Overwintering rhizome more slender than in **1**; *stolons produced from lower axils* of aerial stem. Stem 10–45 cm, ±erect, rather sparsely glandular-pubescent. *Lvs* usually 3–8 cm, ovate, *abruptly acuminate, shallowly cordate at base*, dentate, subglabrous; *petioles furrowed and hairy above, subglabrous beneath.*

Infl. commonly with very small setaceous bracts; its axis elongating before petals drop so that *open fls are well spaced*; pedicels sparsely glandular-pubescent, somewhat reflexed in fr. *Sepals whitish, sparsely glandular-pubescent.* Petals 2–4 mm, rounded to cuneate at base, deeply notched. Stamen filaments 2–5 mm long. Stigma ±deeply 2-lobed. *Fr.* up to 2 mm, obovoid, unequally 2-celled, *falling without ripening seed.* Fl. 7–8. Some nectar is secreted from an *inconspicuous ring round style*, but the pollen is sterile and no good seed is set. $2n = 22$. Grh.

Native. Shaded rocky places and upland woods. Not uncommon in W. and N. Great Britain from Gloucester and S. Wales eastwards to Stafford, Derby, W. Yorks and Berwick and northwards to Sutherland; Inner Hebrides; very local in Ireland. N.W. and C. Europe eastwards to C. Russia.

The hybrid origin of *C. intermedia*, inferred from its intermediate features and its normal failure to produce good pollen or seed, is strongly supported by success in raising plants referable to it from seed resulting from pollination of *C. alpina* with pollen from *C. lutetiana*. That it occurs in many localities outside the present range of *C. alpina* is explicable by the latter having lost ground during the post-glacial climatic amelioration.

3. C. alpina L. Alpine Enchanter's Nightshade.

Overwintering only as fusiform tubers formed *at the tips of short-lived slender stolons*, some from lower axils of the aerial stem. Stem 5–30 cm, erect or ascending, glabrous at least below. *Lvs* usually 2–6(–8) cm, ovate, subacuminate to acute, *cordate at base*, deeply dentate, thin and ±translucent, somewhat shining above, glabrous; *petioles flat above, glabrous.* Infl. commonly with small setaceous bracts; its axis elongating only after petals have dropped, so that *open fls are in a terminal cluster*; pedicels glabrous, little reflexed in fr. *Sepals white, glabrous.* Petals up to 1.5 mm, cuneate at base, often only shallowly notched. Stamen filaments 1–1.5 mm. Stigma usually ±entire or shallowly 2-lobed, rarely deeply lobed. Fr. 2 mm diam., narrowly obovoid, 1-celled, covered with soft bristles which are sparser, shorter and less consistently hook-tipped than in **1**; *fr. freely ripening seed* though sometimes falling prematurely. Fl. 7–8. *No nectar-secreting ring visible*, and probably largely self-pollinated: late in the season fls often produce fr. cleistogamously. $2n = 22$. ?G.

Native. Shaded rocky places and upland woods, very local. Only in S. and N. Wales, the Lake District and N.W. Yorks, Arran, and 1 locality in Westerness. N. Europe and some mountains in C. Europe.

74. HALORAGIDACEAE

Herbaceous plants, usually aquatic or subaquatic, often very large. Lvs spirally arranged, opposite and decussate or whorled, very variable in size and shape, exstipulate, sometimes with intravaginal scales (*Gunnera*). Fls inconspicuous, solitary axillary, in axillary dichasia or in terminal spikes, racemes or panicles; hermaphrodite or unisexual (monoecious or polygamous), actinomorphic, epigynous. Sepals 4, 2 or 0, free, small; petals free, as many as the sepals, but often much larger, commonly caducous, sometimes 0; stamens 4 + 4, 4, or 2, free, filaments short; ovary inferior, usually 4-celled with 1 pendulous anatropous ovule in each cell, but 1-celled with 1 ovule in *Gunnera*; styles 1–4, often very short, stigmas 1–4, feathery or coarsely papillose. Apparently anemophilous. Fr. a nut or drupe, sometimes separating into 1-seeded nutlets (*Myriophyllum*). Seed with much endosperm.

About 170 spp. in 7 genera, cosmopolitan but with a concentration in the southern hemisphere.

The relationships of the family are obscure, though the basic floral diagram, as seen in *Haloragis*, is that of the Onagraceae.

Water plants, submerged except for the infl.; lvs whorled, pinnate, with capillary segments. 1. MYRIOPHYLLUM
Large rhubarb-like marsh plants with enormous ±peltate exclusively basal lvs. 2. GUNNERA

1. MYRIOPHYLLUM L.

Perennial *aquatic* herbs, free-floating or with rhizomes in the substratum and the lfy shoots submerged apart from the infl. *Lvs in whorls of* 3–6, exstipulate, *pinnately divided into unbranched capillary segments*; aerial lvs and bracts sometimes simple, toothed or entire. Fls in lfy or bracteate terminal spikes, sessile in axillary whorls, more rarely alternate; hermaphrodite, polygamous or monoecious, rarely dioecious, the upper fls commonly male, the lower female. Calyx inconspicuous, of 4 small lobes in the male fl., minute in the female fl.; corolla of 4 boat-shaped caducous petals in the male fl., minute or 0 in the female fl.; stamens usually 8, sometimes 4 or 6; ovary 4-celled; style very short or 0; stigmas 4, subsessile, oblong, recurved, persistent. Anemophilous. Fr. separating into 1-seeded nutlets, usually 4, sometimes fewer by abortion.

About 45 spp., cosmopolitan.

1 All lvs and bracts pinnate to pectinate; lvs 4–6 in a whorl, often longer than internodes. 2
　 At least the upper bracts simple, entire or serrate; lvs 3–6 in a whorl, usually about equalling internodes. 3
2 Lvs commonly in whorls of 5, each lf with 24–35 segs; emergent lvs sparsely glandular; a few hermaphrodite fls usually present between male and female fls; fr. smooth. **1. verticillatum**
　 Lvs in whorls of 4–6, each lf with 8–30 segs; emergent lvs densely glandular; plants dioecious (only female in Europe); fr. finely tuberculate. A rare alien in S. England. **2. aquaticum**
3 Lvs commonly 4 in a whorl, all pinnatisect with up to 38 segs; stamens 8; bracts shorter than fls. 4
　 Lvs 4–6 in a whorl, submerged pinnate with 5–12 segs, emergent commonly, like the bracts, simple and lanceolate to linear, entire or serrate; stamens 4; bracts, longer than fls. Known only from a canal in S.W. Yorks. **5. heterophyllum**
4 Lvs (3–)4(–5) in a whorl; lf-segs 13–38; fl.-spike usually exceeding 4 cm; all fls in whorls. **3. spicatum**
　 Lvs (3–)4 in a whorl; lf-segs 6–18; fl.-spike not more than 3 cm; upper fls in opposite pairs or alternate. **4. alterniflorum**

1. M. verticillatum L. Whorled Water-milfoil.

Rhizome elongated, creeping in the muddy substratum. Lfy shoots 50–300 cm, branched. *Lvs* 2.5–4.5 cm, *usually 5 in a whorl*, rarely 4–6, commonly exceeding the internodes, simply pinnatisect with 25–35 rather distant segments. Spike 7–25 cm, emergent. *Fls* usually in whorls of 5 *in the axils of shortly pinnate or pectinate bracts* of very variable length, from little shorter than the lvs to little longer than the fls, but never entire and never shorter than the fls even at the tip of the spike. A few hermaphrodite fls usually present between the male and female fls. Petals of female fls 0; of male fls 4, c. 2.5 mm, greenish-yellow, rarely reddish, caducous. Stamens 8. Fr. c. 2 mm, subglobose, 4-lobed, at length separating into 4 nutlets. Fl. 7–8. Perennation and vegetative reproduction by clavate turions, 0.5–5 cm, with closely appressed lvs. $2n = 28$. Hyd.

Native. Ponds, lakes and slow streams of lowland districts, especially in base-rich water; not common. England and Wales; Ireland. Europe northwards to the Arctic Circle; N. Africa; Asia; North and South America.

*2. M. aquaticum (Velloso) Verdcourt

Parrot's Feathers.

M. brasiliense Camb.; *M. proserpinacoides* Gillies ex Hooker & Arnott

Rhizomatous water-plant with lfy shoots to 200 cm, often woody at base. *Lvs* usually exceeding internodes, 4–6 in a whorl, *pinnatisect* with 8–30 segs, stiffish; *emergent lvs* pale blue-green, *densely covered with minute hemispherical glands*. *Bracts resembling lvs. Plants dioecious*, with solitary axillary fls, *only the tiny* (1.5 mm) *white apetalous female fls so far seen in the British Is.*; male fls with petals 5 mm and 8 stamens. Ripe fr. 1.8 mm, ovoid, finely tuberculate.

Introduced. Grown in aquaria and garden pools and occasionally found wild in streams or ponds in S. England, where it has been known to persist for at least 7 years. Native in South America. Naturalized in S.W. France and casual elsewhere in Europe.

3. M. spicatum L. Spiked Water-milfoil.

A rhizomatous water-plant with branching lfy shoots 50–250 cm, naked below through decay of lvs. *Lvs* 1.5–3 cm, *usually 4 in a whorl*, rarely 3 or 5, about equalling internodes, simply pinnatisect *with 13–35 segments*. Spike 5–15 cm, emergent, erect throughout even in bud. Fls usually in whorls of 4 in the axils of *bracts all but the lowest* of which are *entire and shorter than the fls*, the lowest usually pectinate and somewhat larger than the fls. About 4 basal whorls are of female fls with 4 very small petals, then 1 whorl of hermaphrodite fls, the upper whorls being of male fls with larger dull red caducous petals, c. 3 mm. Stamens 8. Fr. subglobose, 4-lobed. Turions 0. Fl. 6–7. $2n = 28, 36$. Hyd.

Native. Lakes, ponds, ditches, etc., to 475 m in the Lake District; locally common, especially in calcareous water. Throughout most of the British Is. except Shetland; Channel Is. Europe, Asia, N. Africa, North America.

4. M. alterniflorum DC.

Alternate-flowered Water-milfoil.

A rhizomatous water-plant with slender branching lfy shoots 20–120 cm, not exceeding 2 mm diam., naked below through decay of older lvs. *Lvs* 1–2.5 cm, *usually 4 in a whorl*, sometimes 3, about equalling the internodes, simply pinnatisect with 6–18 *segments* 6–20 mm. *Spike* 1–2(–3) cm, emergent, its *tip drooping in bud.* Basal whorl usually of 3 female fls, with rudimentary petals and stamens, in the axils of lf-like pinnate bracts; then other female fls, solitary or in groups of 2–4 in the axils of short pectinate bracts; next hermaphrodite fls, and in the upper half of the spike *male fls*, *usually c. 6, solitary or in opposite pairs in the axils of entire bracts shorter than the fls*; petals 2.5 mm, yellow with red streaks; stamens 8. *Fr.* 1.5–2 mm, *longer than wide*, separating into 4 nutlets. Turions 0. Fl. 5–8. $2n = 14$. Hyd.

Var. *americanum* Pugsl. has lvs only 3–5 mm, with segments 2–4 mm.

Native. Lakes, streams, ditches, etc., to 716 m in Wales and Scotland; locally common but especially in the west and north and in base-poor and peaty water.

Throughout the British Is. Europe eastwards to C. Russia and northwards to S. Scandinavia and Finland; Iceland; Greenland; Azores. Var. *americanum* in base-rich water in Lough Neagh and Lough Beg (N.E. Ireland), and in eastern North America.

*5. M. heterophyllum Michx.

Rhizomatous, with stoutish stems bearing lvs in whorls of 4 and 6. Submerged lvs 2–5 cm, with 5–12 capillary segments; *uppermost (emergent) lvs lanceolate to elliptical, simple, entire or toothed*; lvs pinnately cut in the transitional region. Spike 3–35 cm, emergent, with fls in whorls of 4 or 6, hermaphrodite or the lower female and upper male; *bracts lanceolate*, entire or toothed. *Stamens* 4. Fr. 1–1.5 mm, subglobose, each nutlet

beaked and with the outer face 2-ridged and minutely papillose. Fl. 6–9. Hyd.

Introduced. Established in a canal near Halifax (S.W. Yorks). Eastern North America.

M. verrucosum Lindley, native in Australia but seen during the 3 years 1944–6 in a disused water-filled gravel-pit at Eaton Socon, Bedford, has not been refound in the British Is. It has submerged lvs mostly in whorls of 3, simply pinnate or pinnatisect with 8–18 capillary segments. Fl.-spike in submerged plants up to 18 cm, with bracts in whorls of 3(–2), the lowest pinnatifid or coarsely serrate, uppermost broadly ovate to elliptic, subentire. Fls 1–3 per node, solitary in axils of bracts, all hermaphrodite with 4 pinkish petals, 8 stamens and 4 styles. Fr. 4-angled, separating into four 1-seeded longitudinally ridged and minutely tuberculate mericarps. Dwarfed terrestrial plants have fls almost to the base. Perhaps introduced with Australian wool, other wool-aliens being found at Eaton Socon.

2. GUNNERA L.

Perennial, often gigantic, herbs with creeping rhizomes and stalked ovate or orbicular lvs, exclusively basal, with numerous intravaginal scales. Infl. a large racemose panicle, sometimes very dense and exceeding 2×1 m. Fls unisexual, the lower female, upper male, with or without intervening hermaphrodite fls. Sepals 2–3 or 0; petals 2, free, hooded, or 0, stamens 1–2; ovary inferior, 1-celled with 1 ovule; styles and stigmas 2. Fr. a 1-seeded drupe.

About 50 spp. in South and Central America, S. and S.E. Africa, Madagascar, East Indies, New Guinea, Tasmania and New Zealand. Some of the spp. are cultivated as waterside plants of spectacular dimensions. Frequently cultivated and occasionally escaping are:

*1. G. tinctoria (Molina) Mirbel (*G. chilensis* Lam.; *G. scabra* Ruiz & Pav.) with short rhizome, round, cordate, *palmately-lobed lvs* to 2 m diam. with hispid stalks, and a compound infl. to 1 m, with broad branches up to 50 cm. Chile, Ecuador, Colombia; naturalized in W. Europe.

*2. G. manicata Linden ex André (*G. brasiliensis* Schindler), a larger plant with a long-creeping rhizome, and orbicular, *peltate*, *pedately-lobed lvs* sometimes exceeding 2 m diam., whose stalks bear *red spiny hairs*; infl.-branches up to 120 cm, narrow. S. Brazil.

75. HIPPURIDACEAE

Glabrous aquatic herbs with whorled linear exstipulate lvs and solitary axillary fls, hermaphrodite, or unisexual, epigynous. Perianth a rim round the top of the ovary. Stamen 1, anterior, median. Ovary inferior, 1-celled; ovule solitary, pendulous, anatropous; integument single; micropyle closed, the pollen-tube passing laterally through funicle and integument; style 1, long and slender with stigmatic papillae throughout its length. Anemophilous. Fr. an achene.

Perhaps 2 spp.

Sometimes included in Haloragidaceae but differing in many features of vegetative and floral morphology and unlikely to be closely related.

1. HIPPURIS L.

The only genus, with the characters of the family.

1. H. vulgaris L.   Mare's-tail.

A perennial usually aquatic herb with a stout creeping rhizome from which arise lfy shoots 25–75(–150) cm if wholly or partly submerged but only 7–20 cm in terrestrial forms. *Lvs* 1–7.5 cm × 1–3.5 mm, 6–12 *in a whorl, linear, sessile, entire, glabrous, with a hard acute tip*; submerged shoots have longer, thinner, more flaccid and more translucent lvs than those of emerged shoots, and the internodes are longer and less rigid, so that the name mare's-tail appears very appropriate, especially to the luxuriant submerged shoots of flowing water. Fls small, greenish, in the axils only of the emerged lvs. Stamens with reddish anthers, sometimes 0. Fr. ovoid, smooth, greenish. Fl. 6–7. $2n = 32$. Hyd.

Native. In lakes, ponds and slow streams, to 550 m in Scotland; especially in base-rich water, local. Throughout the British Is. Europe; W. and N. Asia; N. Africa; southernmost South America.

76. CALLITRICHACEAE

Annual or perennial herbs, usually aquatic or subaquatic, with *filiform stems* and *opposite, entire, linear to ovate, exstipulate lvs*. Monoecious. Fls axillary, either solitary or a male and female fl. in the same axil. Bracteoles 2, crescent-shaped, or 0; *perianth* 0; *stamen* 1, with a long slender filament and reniform anther; *ovary syncarpous, 4-celled by secondary septation*, 4-lobed; styles 2, long, free, papillose; ovules solitary in each cell, anatropous, pendulous; integument 1. Fr. 4-lobed, the lobes ±keeled or winged, separating at maturity *into* (2–)4 *drupelets*; seeds with a fleshy endosperm.
One genus.

1. CALLITRICHE L.

Submerged lvs usually narrowly elliptical or linear, floating and aerial lvs narrowly elliptical-spathulate to ovate or suborbicular, floating lvs commonly in a terminal rosette; lvs with or without stomata. Plants usually annual but occasionally perennial.

Almost cosmopolitan.

A taxonomically troublesome genus owing to the dependence of lf-shape upon whether the lvs are submerged, floating or aerial and upon the depth and rate of movement of the water. For this reason a key based on lf-shape is unreliable for plants from non-typical habitats. One based on fr. is more reliable but often cannot be used owing to the sterility of some species especially when growing in deep water. The following key makes use of both vegetative and fr. characters and should be used only when healthy plants with fls and fr. are available.

1 Plants growing in water with at least some lvs submerged.   2
   Plants terrestrial, growing on mud, etc., not in flowing or standing water; lvs usually elliptical.   13
2 All lvs submerged.   3
   Uppermost lvs in a floating rosette.   8
3 Lvs translucent, all linear or narrowly linear-lanceolate; plants usually with fls and fr.; fls without bracteoles; styles deciduous.   4
   Lvs not translucent, ±linear or broader; fls with 2 bracteoles; styles deciduous or persistent; fls and fr. often 0.   5
4 Most lvs exceeding 1 cm and up to 2 cm, narrowly linear-lanceolate, distinctly tapering from base to emarginate tip; fls and fr. produced freely; fr. 1–3 mm, almost circular, its lobes usually broadly winged. Mainly in lakes in N. England and Scotland.   **8. hermaphroditica**
   Lvs up to 1 cm, deep green shot with blue, linear, imperceptibly tapering to the truncate or slightly emarginate tip; fr. (rarely seen here) 1–1.2 mm, wider than long, its lobes not or very narrowly winged. Local in lowland pools and ditches mainly in England south of the Humber.   **4. truncata**
5 All or most lvs linear to linear-lanceolate.   6

Lvs all narrowly elliptical to spathulate, or only the lowest ±linear. Probably **1** *stagnalis*, **3** *obtusangula* or **4** *cophocarpa*, but these are sterile when wholly submerged and cannot be identified with certainty.
6 Lvs narrowly linear, usually expanded abruptly at the deeply emarginate tip and thus shaped like a bicycle spanner or pair of pincers; fr. 1.2–1.5 mm, about as long as wide, its lobes narrowly winged.
                                     6. hamulata
Lvs not expanded abruptly at the tip.   7
7 Some lvs with asymmetrically emarginate apex; fr. 1–1.3 mm, often somewhat longer than wide, its lobes winged.   **7. brutia**
   Lvs not asymmetrically emarginate at apex. Probably **2** *platycarpa*, **4** *cophocarpa* or **5** *palustris*, but these are sterile when wholly submerged and cannot be identified with certainty.
8 Fls in axils of both linear submerged lvs and also of the elliptical or slightly obovate rosette-lvs; fr. with persistent styles deflexed and appressed to the sides; fr.-lobes winged. Probably **6** *hamulata* and **7** *brutia*: see 6 and 7 above.
   Fls in axils only of rosette-lvs, not of submerged lvs; styles, if persistent, not appressed to sides of fr.   9
9 Rosette-lvs rhombic-spathulate, distinctly angular in outline and usually 3-ridged on upper side; submerged lvs narrowly rhombic or the lowest linear; most pollen-grains at least twice as long as broad; fr.-lobes rounded, unwinged.   **3. obtusangula**
   Rosette-lvs elliptical to almost orbicular, not or very slightly angular in outline; no pollen-grains twice as long as broad; fr.-lobes winged or not.   10
10 Rosette-lvs broadly elliptical to almost orbicular; submerged lvs elliptical, sometimes very narrowly so but never truly linear; fr.-lobes broadly winged.
                                   1. stagnalis
   Rosette-lvs elliptical (or very slightly rhombic) to almost orbicular; at least the lower submerged lvs truly linear; fr.-lobes winged or not.   11
11 Fr. c. 1 mm, its lobes unwinged. Rare. **4. cophocarpa**
   Fr.-lobes ±narrowly winged all round or only at upper end.   12
12 Fr. c. 1.5 mm, its lobes narrowly winged all round; rosette-lvs usually 3–4 times as long as wide; submerged lvs up to 3 cm. Lowland, widespread but not common.   **2. platycarpa**
   Fr. c. 1 mm, its lobes winged only near the upper end; rosette-lvs elliptical to almost orbicular. Rare and doubtfully native.   **5. palustris**

Any of our *Callitriche* spp. except **8** and **9** can occur as terrestrial plants on mud, etc., varying in habit, lf-shape and even in features of fl. and fr. from plants of the same sp. growing in water, and it may be difficult or impossible to achieve certainty in identification. The following key-entries emphasize characters most helpful in diagnosis.

13 Fr.-lobes unwinged though sometimes weakly keeled. 14

14 Fr.-lobes winged at least at the upper end.   *15*

14 Fr. 1.5–2 mm, longer than broad, its lobes with rounded margins, neither winged nor keeled; pollen-grains at least twice as long as broad; lvs elliptical to narrowly rhombic, a terminal group commonly ±in a rosette.   **3. obtusangula**

Fr. 0.8–1.2 mm, about as long as broad, its lobes unwinged but sometimes weakly keeled; pollen-grains subglobose or shortly ellipsoid.

    4. cophocarpa

15 Fr. obovate or elliptical, winged only at apical end; styles erect, deciduous.   **5. palustris**

Fr. almost circular or slightly longer than broad, its lobes winged from apex to base; styles erect or recurved, persistent.   *16*

16 Styles erect or recurved, not appressed to sides of fr.; lvs sometimes forming a rosette.   *17*

Styles deflexed and appressed to sides of fr.; lvs elliptical, dark green, not forming a rosette.   *18*

17 Styles arcuate-recurved; fr.-lobes ±broadly winged; lvs pale green, usually small, ±broadly elliptical to almost circular, the uppermost commonly in a loose rosette.   **1. stagnalis**

Styles erect or spreading; fr.-lobes narrowly winged; lvs dark green, elliptical, no distinct rosette.

    2. platycarpa

18 Fr. subsessile or rarely short-stalked.   **6. hamulata**

Fr. with distinct stalk up to 13 mm.   **7. brutia**

Section *Callitriche*: typically with narrow, often linear, submerged and broader spathulate to obovate floating lvs, all with stomata; usually not fruiting under water; subparallel lobes of the fr. joined at least in the basal half.

1. C. stagnalis Scop.

Annual to perennial. Stem to 60 cm in water or 15 cm when prostrate on mud. *Lowest lvs ±elliptical*, 3-veined, slightly emarginate, narrowest in deep or flowing water, but not truly linear, sometimes persisting to flowering but usually lost in shallow water and on mud; *uppermost 6–8 lvs forming a floating rosette*, each 1–2 cm, with *broadly elliptical or almost circular blade narrowed rather abruptly into the stalk*; blade usually 5-veined, rounded or slightly emarginate at the tip. Fls in the axils of rosette-lvs only; bracts falcate, persistent. Stamen c. 2 mm. Styles 2–3 mm, soon becoming arcuate-recurved, only occasionally persistent in fr. *Fr. 1.6–2 mm, suborbicular, conspicuously keeled, with deep lateral grooves; broadly winged, divergent.* Fl. 5–9. Male fls in axils of upper lvs and maturing before the female. Usually anemophilous but perhaps also hydrophilous. $2n = 10^*$. Th.–Hyd.

Var. *serpyllifolia* Lönnr. is the name given to very small terrestrial forms with smaller and narrower thyme-like lvs.

Native. Ponds, ditches, streams, wet mud; to c. 900 m in Wales. Common throughout the British Is. Europe; Canary Is.; N. Africa.

2. C. platycarpa Kütz.

C. polymorpha auct., *C. verna* auct.

Perennial. Stem to 100 cm in water, 1–15 cm when pro-strate on mud. *Lowest lvs linear to narrowly elliptical, emarginate but not expanded at the tip; uppermost lvs in a floating ±convex rosette*, each 10–20 mm, with a *broadly elliptical blade ±gradually narrowed into the stalk*; blade 3(–5)-veined, slightly emarginate at the tip. Fls in the axils of rosette-lvs; bracts falcate, persistent. Stamen 3–4 mm. Styles 3–4(–8) mm, erect, at least their bases persistent in fr. *Fr. 1.5–1.8 × 1.3–1.5 mm, suborbicular to shortly elliptical, distinctly keeled and with fairly deep lateral grooves; lobes* winged, *parallel*. Fl. 4–10. Anemophilous. $2n = 20^*$. Th.–Hyd.

Native. Ponds, ditches, streams, etc. Probably common throughout lowland Great Britain but distribution imperfectly known; Ireland. Europe, northwards to S. Sweden and S. Finland.

Most British material named *C. polymorpha* Lönnr. and *C. palustris* L. belongs here.

3. C. obtusangula Le Gall

Stem 10–60 cm in water, 2–12 cm on mud. *Lowest lvs 10–40 × 0.5–2 mm, linear, deeply emarginate but not widened at the apex; middle (submerged) and floating lvs 8–20 × 3–7 mm, rhombic-spathulate, 3–7-veined, blunt or ±retuse; linear lvs* commonly not persisting; mud forms may have all lvs linear or lanceolate. Fls only in the axils of floating lvs; bracts falcate, persistent. Stamen c. 5 mm. Stigmas c. 4 mm, their erect bases persisting in fr. *Fr. c. 1.5 × 1 mm, with broadly rounded lobes and barely discernible grooves*; seeds unwinged. Fl. 5–9. Anemophilous. $2n = 10^*$. Hyd.

Native. Ponds, ditches, lakes; locally frequent in S. England, rare in the north. Great Britain northwards to Cumbria and Yorks; Wigtown and W. Inverness; scattered throughout Ireland. France, Belgium, Netherlands, W. Germany, Corsica, Italy, Sardinia, Sicily, Greece; N. Africa.

Hardly distinguishable from *C. stagnalis* when sterile, but very different in its fr.

'Var. *lachii* Warren' appears to be a hybrid between *C. obtusangula* and *C. hamulata*, with rosette-lvs like those of *C. obtusangula* and submerged lvs like those of *C. hamulata*.

Section *Pseudocallitriche* Hegelm.: lvs all similar, submerged, linear to linear-lanceolate, without stomata; fr. of 4 readily separating lobes.

4. C. cophocarpa Sendtner

C. polymorpha Lönnr.

Stem up to c. 25 cm in water but only to c. 7 cm on mud. *Floating rosette-lvs usually 10 or more, elliptical-spathulate or slightly rhombic; submerged lvs narrower, the lowest linear* with emarginate tip; lvs of terrestrial plants narrowly elliptical or spathulate. Fls in axils only of rosette-lvs; bracteoles falcate, large and persistent. *Pollen-grains shortly ellipsoid to subglobose. Styles ±erect, 4–6 mm, persistent* and finally divergent though remaining erect below. *Fr. c. 1 mm, suborbicular or slightly ellipsoid, its lobes parallel, unwinged* but sometimes weakly keeled. $2n = 10$. Th.–Hyd.

Native. A plant of slow-moving water, but few records have been authenticated and British distribution still virtually unknown. Most of Europe except the Iberian Peninsula.

5. C. palustris L.

C. verna L., nom. ambig.; *C. vernalis* Koch

Stem up to c. 25 cm in water but much less on mud. *Floating rosette-lvs with elliptical to suborbicular blade narrowing into a petiole-like base*; most *submerged lvs narrowly linear with emarginate tip*; lvs of terrestrial plants small, elliptical to ±linear, the terminal group usually in a rosette. Fls in axils of rosette-lvs, each axil usually with one staminate and one ovulate fl. *Pollengrains subglobose. Styles erect, deciduous. Fr.* c. 1 mm, *usually obovate, almost black when ripe, its lobes ±parallel, winged only near apex.* $2n = 20$. Th.–Hyd.

Doubtfully native. A plant of shallow still water. Hitherto much confused with *C. stagnalis* and other spp. and, if truly native, very rare or generally overlooked. Much of Europe, mainly in mountains except in Scandinavia and N.E. Europe.

6. C. hamulata Kütz. ex Koch

C. intermedia Hoffm., *nom. illegit.*

Stem up to 50(–80) cm in water and 15 cm on mud, robust. Juvenile lvs linear, 1-veined, succeeded in shallow still water by rather broader but ±parallel-sided 3-veined lvs and then by *elliptical to ovate-spathulate 3–5-veined floating lvs forming a lax concave rosette. In deeper or flowing water the lower submerged lvs are* $10–25(–40) \times 0.5–1.3$ mm, *1-veined, usually widened and deeply emarginate at the tip* (like a bicycle-spanner or pair of pincers), with broader lvs transitional to the elliptical rosette-lvs at the water-surface; *all lvs may be submerged and narrowly linear.* In terrestrial plants all lvs may be elliptical, dark green in colour and not forming a rosette. *Fls in axils of both submerged and floating or aerial lvs*; bracteoles of staminate fl. small, falcate, soon deciduous, of ovulate fl. 0 or rarely present. Stamen up to 2 mm; anthers pale cream or whitish; *pollengrains globose.* Styles 1.5–2 mm, soon reflexing and falling or their appressed bases persisting. *Fr.* 1.2–1.5 mm, *suborbicular or slightly broader than long, subsessile* (or sometimes very shortly stalked in terrestrial plants); lateral grooves fairly deep; *fr.-lobes parallel, sharply keeled and usually narrowly winged; ripe fr. usually black.* Fl. 4–9. Hydrophilous. $2n = 38^*$. Th.–Hyd.

Native. Lakes, reservoirs, pools, ditches and slow streams throughout the British Is.; reaches c. 1000 m in Scotland. Through much of N.W. and C. Europe eastwards to Finland and Latvia and southwards to Italy.

7. C. brutia Petagna

C. pedunculata DC.; *C. intermedia* subsp. *pedunculata* (DC.) Clapham

Like 6 but usually more slender and with *linear submerged lvs not widened at the emarginate, truncate or blunt apex and, when emarginate, often asymmetrically so,* the unequal apical teeth shorter and less claw-like than in *C. hamulata*; floating rosette-lvs elliptical or slightly obovate; lvs of terrestrial plants elliptical, dark green. *Fls in axils of both submerged and floating or aerial lvs*; bracteoles falcate, deciduous. *Pollen-grains subglobose. Styles persistent, deflexed and appressed to sides of fr. Fr.* 1–1.4 mm, *suborbicular* or slightly longer than broad, ±sessile in aquatic plants but with stalk up to 13 mm *in terrestrial plants; fr.-lobes parallel,* keeled and *mostly broadly winged.* $2n = 28$. Th.–Hyd.

Native. In still shallow water, often of pools which dry up in summer, or on marginal mud; found only in a few scattered localities, but distribution imperfectly known. W. and S. Europe eastwards to Turkey; N. Africa; Near East.

C. hamulata and *C. brutia* are often not readily distinguishable in the field and have been treated as varieties or subspp. of a single species. The differences in chromosome number and in geographical distribution are in favour of their treatment as separate spp. (as by H. D. Schotsman in *Flora Europaea,* **3,** 1972), but further investigation seems necessary.

8. C. hermaphroditica L.   Autumnal Starwort.

C. autumnalis L.

Stem 15–50 cm, branched, *yellowish, entirely submerged. Lvs* $8–18 \times 1–2$ mm, those *in middle of stem* longest, linear-lanceolate, *widest at the base* and *distinctly tapering above* to an emarginate apex, 1-veined, pale green, becoming olive to blackish on drying. *Bracteoles* 0. Filaments short, little exceeding the anthers. Styles commonly longer than the ovary, spreading or deflexed, soon falling. Fr. 1.2–3 mm diam., ±orbicular, of 4 readily separating *broadly and acutely winged lobes.* Fl. 5–9. Hydrophilous. Fruits more freely than any other British sp. $2n = 6^*$. Hyd.

Native. Lakes and streams; to 380 m in Yorks. N. Britain from Anglesey, Stafford, Cheshire and Yorks northwards (i.e. north of c. 53° N) to Caithness; Inner and Outer Hebrides; Orkney; Shetland. Throughout Ireland, but common in the north. N. and E. Europe southwards to c. 48° N in Russia; Iceland, Faroes.

9. C. truncata Guss.

Stem 8–20 cm, very slender, branched, *entirely submerged, commonly reddish. Lvs* $5–11 \times 0.8–1.3$ mm, linear, almost *parallel-sided,* imperceptibly tapering to a truncate and shallowly emarginate apex, 1-veined, dark green. *Bracteoles* 0. Stamens and styles as in *C. hermaphroditica. Fr.* c. 1 mm diam., almost orbicular, of 4 readily separating *blunt unwinged lobes*; sessile or shortly stalked (1–3 mm). Fl. 5–9. Hydrophilous. Rarely fruits in the British Is. Hyd.

British material all falls in subsp. **occidentalis** (Rouy) Schotsman, which differs from the type in being more robust and more lfy, fruiting poorly, and having subsessile frs (fr.-stalks in type 2–4(–9) mm).

Native. Pools and ditches; local and only south of c. 53° N. Somerset, Sussex, Kent, Gloucester, Nottingham; Wexford; Guernsey. S. and W. Europe from Crete, Greece and Dalmatia to Spain and Portugal and northwards to N. France, England and Belgium.

Distinguished from *C. hermaphroditica*, even when sterile, by the shorter, imperceptibly tapering lvs.

77. LORANTHACEAE

Mostly shrubs partially parasitic on trees. Lvs exstipulate, usually opposite or whorled, entire, sometimes scale-like. Fls actinomorphic, unisexual. Perianth of 2 whorls, the inner often brightly coloured, the outer sometimes suppressed and the inner then sepaloid; petals free or connate. Stamens epipetalous. Rudimentary ovary often present in male fls, staminodes in female. Ovary inferior; ovules usually not differentiated from the placenta. Style simple or 0. Seed solitary, devoid of testa; embryo large, sometimes up to 3 in a seed; endosperm usually copious.

About 36 genera and 1300 spp. in tropical and temperate regions.

1. VISCUM L.

Hemiparasites. Fls unisexual. Sepals much reduced, petals sepaloid, usually 4. Stamens sessile, opening by pores. Berries viscid.

About 60 spp. in the Old World.

1. V. album L.   Mistletoe.

A somewhat woody evergreen, parasitic on the branches of trees. Stems up to c. 1 m, green, much-branched, branching apparently dichotomous. Lvs 2–8 cm, narrowly obovate and often somewhat falcate, obtuse, rather thick and leathery, yellowish-green, narrowed at base into a short petiole. Infl. a small compact cyme of 3–5 subsessile fls. Bracts united to the short pedicels. Plant usually dioecious; male fls: calyx 0; female fls: calyx small, indistinctly 4-toothed. Fr. 6–10 mm diam., white. Fl. 2–4. Fr. 11–12. $2n = 20$.

Native. On the branches of a great variety of deciduous trees, most commonly apple, rarely on evergreens and very rarely on conifers, more abundant on calcareous soils. From Cornwall and Kent to Montgomery and N.E. Yorks, common in S. England and the W. Midlands, rather infrequent elsewhere and absent from Scotland and Ireland. Most of Europe, except the extreme north and east; Caucasus.

78. SANTALACEAE

Trees, shrubs or herbs, sometimes hemiparasitic. Lvs alternate, entire, exstipulate. Fls actinomorphic, hermaphrodite or unisexual. Perianth of one whorl, sepaloid or petaloid, often fleshy; lobes 3–6, valvate. Stamens the same number as the per.-segs and opposite them; anthers 2-celled, opening lengthwise. Ovary inferior or ½-inferior, 1-celled; ovules 1–3. Fr. indehiscent, nut-like or drupaceous. Seed solitary, with abundant endosperm.

About 30 genera and 400 spp., in tropical and temperate regions.

1. THESIUM L.

Perennial hemiparasitic herbs having haustoria on their roots by means of which they attach themselves to the roots of other plants. Lvs narrow, alternate. Fls small, greenish, hermaphrodite, usually in small dichotomous cymes. Perianth funnel-shaped or campanulate; segments 5(–4), persistent. Stigma capitate. Ovules 3.

About 300 spp., widely distributed.

1. T. humifusum D.C.   Bastard Toadflax.

A slender glabrous yellowish-green perennial 5–20 cm. Stock woody. Stems spreading or prostrate, herbaceous or woody at base, angled, the angles slightly rough. Lvs 5–15(–25) mm, linear, acute or obtuse, 1-veined, lower lvs scale-like, distant. Infl. ±branched, terminal. Bract longer than the fl.; bracteoles 2, linear-lanceolate, serrulate, inserted at the base of the short stout pedicel. Fls c. 3 mm diam., yellowish. Per.-segs triangular, acute. Fr. c. 3 mm, ovoid, ribbed, green, crowned by the persistent, inrolled per.-segs. Fl. 6–8. Fr. 7–9. Ch.

Native. Parasitic on the roots of various herbs in chalk and limestone grassland, local. Dorset to Kent northwards to E. Gloucester, S. Lincoln and W. Europe.

79. CORNACEAE

Trees or shrubs, rarely herbs. Lvs simple, rarely exstipulate. Fls small, usually in panicles, sometimes umbels or heads, regular, hermaphrodite, rarely unisexual, epigynous, 4(–5)-merous. Sepals small, sometimes obsolete. Petals usually valvate, rarely 0. Stamens equalling in number and alternate with petals. Disk usually cushion-like. Ovary inferior, 1–4- (usually 2-)celled; ovules solitary in each cell, pendent from apex, anatropous; integument 1; style simple or divided, very rarely styles free. Fr. a drupe, rarely a berry; embryo straight; endosperm copious.

Twelve genera and about 100 spp., mainly in north temperate regions and S.E. Asia, a few in Africa, South America and New Zealand. **Aucuba japonica** Thunb. a dioecious shrub, with opposite, evergreen, laurel-like lvs, small dull purple fls and red berries is commonly grown, usually as a form with small yellow spots on the lvs.

1. CORNUS L.

Trees, shrubs or rhizomatous herbs. Lvs usually opposite, entire, exstipulate. Fls in corymbose cymes or umbels, hermaphrodite, 4-merous; bracts 0 or 4. Calyx-teeth small, but always present. *Petals valvate. Ovary 2-celled*; style filiform or columnar, stigma capitate or truncate. *Fr. a drupe*; *stone* 1, 2-celled. Nectar secreted by the disk.

About 45 spp., north temperate regions (to Himalaya and Guatemala); 2 in Peru and Bolivia. Some spp. are cultivated.

Shrub; fls cream-coloured, in corymbose cymes without
  involucral bracts; fr. black.   **1. sanguinea**
Herb; fls dark purple, in umbels each with 4 large, white
  involucral bracts; fr. red.   **2. suecica**

Subgenus 1. KRANIOPSIS Rafin.

Infl. a terminal, ebracteate, corymbose cyme. Fr. ±globose.

1. C. sanguinea L.   Dogwood.

Swida sanguinea (L.) Opiz; *Thelycrania sanguinea* (L.) Fourr.

Deciduous shrub 0.25–4 m. Twigs purplish-red at least on the sunny side. Lvs opposite; blade 4–8 cm, ovate or oval, cuspidate or acuminate, rounded at base, appressed pubescent on both sides, paler beneath, usually becoming purplish-red in autumn; main veins 3–4 pairs from below the middle of the lf, curving round towards the apex; petiole 8–15 mm, grooved above. Infl. many-fld, ±flat-topped, pubescent; peduncle 2.5–3.5 cm; pedicels to 6 mm. Calyx-teeth very small. Petals 4–6 mm, oblong-lanceolate, obtuse, appressed-pubescent outside, cream-coloured. Style clavate above. Fr. black, 6–8 mm diam. Fl. 6–7. Pollinated by various insects. Fr. 9. 2*n* = 22. M. or N.

Native. Woods and scrub on calcareous soils, occasionally locally dominant in chalk scrub. From Durham and Westmorland southwards, widespread and common on suitable soils; introduced farther north, extending to Perth and Aberdeen; rather local in Ireland and now native only from Galway southwards to Limerick; Jersey. Most of Europe, except the north-east and extreme north; S.W. Asia (very rare).

C. sericea L. (*Cornus stolonifera* Michx.)
Deciduous shrub to 2.5 m with numerous decumbent suckering branches. Twigs deep blood-red. Lf-blades 6–12 cm, ovate to oblong-lanceolate, acuminate, veins c.5 pairs; petiole 1–2.5 cm. Style not dilated above, crowned by the discoid stigma. Fr. white, c.5 mm diam., stone as broad or broader than high, rounded below.

Frequently planted and perhaps naturalized in a few places. Native of eastern North America.

C. alba L. (*Thelycrania alba* (L.) Pojark.)
Differs from *C. sericea* in the acute lf-blades 4–8 cm, in not or scarcely suckering, and in the stone being higher than broad, acute at both ends.
Planted, and reported as an escape. Native of N.E. Asia.

Subgenus 2. ARCTOCRANIA (Endl.) Reichenb.

Infl. a terminal umbel, with 4 large whitish bracts. Fr. globose.

2. C. suecica L.   Dwarf Cornel.

Chamaepericlymenum suecicum (L.) Ascherson & Graebner

Rhizomatous herb. Stems annual, 6–20 cm, erect, often a few together, simple or with short axillary branches from the uppermost pair of lvs, glabrous or appressed pubescent. Lvs 1–3 cm, ovate or ovate-elliptical, subsessile, acute or very shortly acuminate, 3–5-veined from the base, green and appressed pubescent above, subglaucous and glabrous beneath. Infl. terminal, fl.-like, of 8–25 dark purple fls, surrounded by 4 white ovate involucral bracts, which are 5–8 mm long; pedicels 1–2 mm, appressed-pubescent like the receptacle. Sepals small but obvious, deltate. Petals 1–2 mm, ovate-triangular, acute. Fr. red, c. 5 mm. Fl. 7–8. 2*n* = 22. Hp.

Native. Moors on mountains, usually under heather or bilberry; ascending to 990 m. Very local and rare in England; Lancs, Yorks, Northumberland; Selkirk; more frequent in the Scottish Highlands from Dumbarton to Shetland but local and mainly in the east (not recorded from the Inner or Outer Hebrides). N. Europe, extending southwards to Holland; N. Asia and North America.

Subgenus 3. CORNUS

Infl. an axillary umbel, with yellowish-green bracts. Fr. ellipsoid.

Cornus mas L.   Cornelian Cherry.

Shrub or small tree to 8 m. Lvs 4–10 cm, ovate. Fls small, yellow, produced before the lvs expand, in axillary umbels with 4 yellowish-green bracts. Fr. ellipsoid, bright red. Frequently planted, sometimes naturalized. Native of C. and S.E. Europe, W. Asia.

80. ARALIACEAE

Trees, shrubs or woody climbers, occasionally herbs. Lvs usually alternate, simple or compound, usually stipulate. Fls small, in umbels, heads, racemes or spikes often forming compound infl., regular, hermaphrodite or unisexual, epigynous, usually 5-merous. Sepals small or obsolete. Petals free or united, 3–many, valvate or imbricate, sometimes falling off together as a cap. Stamens 3–many, usually equal in number and alternate with petals, occasionally more. Disk flat or swollen. Ovary inferior; cells usually as many as petals, sometimes fewer (2–4), very rarely 1 or many; ovules solitary in each cell, pendent from apex, anatropous; styles free or united or 0. Fr. a drupe or berry; embryo small; endosperm copious.

About 55 genera and 700 spp., mainly tropical, some temperate. Resembling Cornaceae and not separable by any single character but with tendencies to have alternate and compound and lobed lvs, more compound infls and free styles. Some genera are grown in gardens, spp. of *Fatsia* Decne. & Planchon and *Aralia* L. being the most common.

1. HEDERA L.

Evergreen *woody climbers*, climbing by adventitious roots. *Lvs simple*, coriaceous, glabrous, alternate, petiolate; stipules 0. Fls in terminal, globose umbels, often arranged in a panicle; *pedicels not jointed.* Calyx with 5 small teeth. *Petals 5, valvate, with broad base,* free. *Stamens 5. Ovary 5-celled*; styles joined into a column. Fr. a berry; *endosperm ruminate.* Nectar secreted by the disk.

About 15 spp., north temperate Old World; Queensland.

1. H. helix L. Ivy.

Woody climber sometimes climbing to 30 m or creeping along the ground and forming carpets, flowering only in sun at the top of whatever it is climbing on. Stems up to 25 cm diam., densely clothed with adventitious roots. Young twigs stellate-pubescent. Lvs glabrous, dark green above, often with pale veins, sometimes tinged purple, paler beneath; blades of those of the creeping or climbing stems 4–10 cm, palmately 3–5-lobed, with ±triangular, entire lobes; those of the fl. branches entire, ovate or rhombic. Peduncle, pedicels and receptacle stellate-tomentose. Calyx-teeth small, deltate. Petals yellowish-green, 3–4 mm, triangular-ovate, somewhat hooded at apex. Fr. black, globose, 6–8 mm. Fl. 9–11. Pollinated by flies and wasps; homogamous or protandrous. $2n = 48$. MM., M., N. or Ch.

Native. Climbing in woods, hedges or on rocks and walls or creeping in woods, on all but very acid, very dry or water-logged soils, very tolerant of shade; ascending to 610 m. Common throughout the British Is. except parts of N. Scotland. Europe from Norway (60° 32′ N) and Latvia southwards (absent from N. and E. Russia); Asia Minor to Palestine and N. Iran.

81. UMBELLIFERAE (Apiaceae)

Annual or perennial herbs, rarely shrubs. Stems often furrowed, pith wide and soft or internodes hollow. Lvs alternate, usually exstipulate and much divided, petioles sheathing at base. Infl. usually a compound umbel, sometimes a simple umbel, rarely capitate or very reduced and cymose with solitary fls surrounded by a whorl of bracteoles; bracts and bracteoles usually present, whorled. Fls hermaphrodite or unisexual, usually with nectar, often strongly protandrous. Calyx-teeth usually small, 5, sometimes unequal, often 0. Petals 5, valvate or slightly imbricate, often notched with an inflexed or incurved point, sometimes very unequal, often white, sometimes pink or yellow, rarely blue. Stamens 5, alternating with the petals, inflexed in bud. Ovary inferior, 2 (rarely 1)-celled; ovules pendent, solitary in each cell; styles 2, often with an enlarged base (*stylopodium*). Fr. dry, of 2 indehiscent dorsally or laterally compressed carpels joined by a narrow or broad commissure; carpels adnate to or suspended from a slender simple or divided axis (*carpophore*), usually separating when ripe, mostly prominently 5- or 9-ribbed, and generally with four resinous canals (*vittae*) between the primary ridges (rarely in them) and 2 on the commissural face. Pollinated by various insects, particularly small beetles and flies.

About 275 genera and 2850 spp., cosmopolitan but chiefly in the north temperate region.

For illustrations and more detailed descriptions see T. G. Tutin, *Umbellifers of the British Isles* (Botanical Society of the British Isles, 1980).

1 Lvs coriaceous, with spiny margins; fls in dense heads.
4. ERYNGIUM

Lvs usually herbaceous, without spines; fls in umbels
or rarely whorls. *2*
2 All lvs entire. 26. BUPLEURUM
Lvs toothed or more deeply divided. *3*
3 Upper cauline lvs perfoliate, ovate, denticulate or crenulate. 10. SMYRNIUM
Upper cauline lvs not perfoliate. *4*
4 Lvs ±orbicular, divided for less than half-way to base.
1. HYDROCOTYLE
Lvs divided for more than half-way to base or midrib. *5*
5 Basal lvs palmately divided. *6*
Basal lvs pinnately or ternately divided. *7*
6 Fls subsessile; infl. an irregular compound false umbel; involucre inconspicuous. 2. SANICULA
Fls with long filiform pedicels; infl. a simple umbel; involucre conspicuous, pink or white.
3. ASTRANTIA
7 Plant aquatic, with finely divided, translucent submerged lvs at flowering time. *8*
Plant usually terrestrial, sometimes aquatic but with few or no finely divided submerged lvs at flowering time. *9*
8 Umbel with 1–2(–4) rays; sepals minute or absent.
28. APIUM
Umbel with (4–)5–16 rays; sepals relatively conspicuous. 19. OENANTHE
9 Lowest aerial lvs pinnately lobed to 1-pinnate. *10*
Lowest aerial lvs at least 2-pinnate or 2-ternate. *24*
10 Plant hairy, sometimes minutely so. *11*
Plant glabrous. *15*
11 Sepals about as long as petals. 41. TORDYLIUM
Sepals minute or absent. *12*
12 Fls yellow. 39. PASTINACA
Fls white or pink. *13*
13 Bracteoles absent. 13. PIMPINELLA
Bracteoles several. *14*

14 Hairs on stem not deflexed and appressed; fr. winged.
                                                    40. HERACLEUM
- Hairs on stem deflexed and closely appressed; fr. spiny or tuberculate.   42. TORILIS
15 Petiole fistular; partial umbels subglobose in fr.
                                                    19. OENANTHE
- Petiole not fistular; partial umbels not subglobose in fr.   16
16 Lower lvs with more than 20 pairs of lobes.
                                                    34. CARUM
- Lower lvs with fewer than 10 pairs of lobes.   17
17 Longest bracts at least half as long as the shortest rays.   18
- Bracts absent or all less than half as long as rays.   20
18 Bracts subulate; stem solid.   29. PETROSELINUM
- Bracts wider, often lf-like; stem hollow.   19
19 Umbel usually with 20–30 rays; petals papillose beneath; fr. 3–4 mm.   15. SIUM
- Umbel usually with 7–14 rays; petals smooth beneath; fr. 1.5–2 mm.   16. BERULA
20 Some umbels subsessile and lf-opposed.   28. APIUM
- Umbels all long-pedunculate, not lf-opposed.   21
21 Bracts 2–4.   30. SISON
- Bracts absent.   22
22 Bracteoles present.   9. CORIANDRUM
- Bracteoles absent.   23
23 Lvs ternate; plant with long rhizomes.
                                                    14. AEGOPODIUM
- Lvs pinnate; plant without long rhizomes
                                                    .13. PIMPINELLA
24 Fls yellow or greenish-yellow.   25
- Fls white, greenish-white or pinkish.   34
25 Bracts more than 3; bracteoles numerous.   26
- Bracts 0–3; bracteoles usually few or none.   29
26 Bracteoles connate below.   (p. 291) LEVISTICUM
- Bracteoles free to the base.   27
27 Lf-lobes subulate to linear-lanceolate. 17. CRITHMUM
- Lf-lobes ovate.   28
28 Lobes of lower lvs crenate or serrate, obtuse.
                                                    10. SMYRNIUM
- Lobes of lower lvs pinnatifid, acute.
                                                    29. PETROSELINUM
29 Lf-lobes filiform, entire.   30
- Lf-lobes toothed or lobed, not filiform.   32
30 Annual.   (p. 286) ANETHUM
- Perennial.   31
31 Stock without fibrous remains of petioles; bracteoles absent.   21. FOENICULUM
- Stock with abundant fibrous remains of petioles; bracteoles present.   38. PEUCEDANUM
32 Lobes of lower lvs crenate or serrate, obtuse.
                                                    10. SMYRNIUM
- Lobes of lower lvs pinnatifid, acute.   33
33 Sheathing petiole-base without or with a narrow hyaline margin; lf-margin serrulate.   22. SILAUM
- Sheathing petiole-base with a broad hyaline margin; lf-margin entire.   29. PETROSELINUM
34 Bracts all 3-fid or pinnatisect.   35
- Bracts absent or mostly undivided.   36
35 Biennial; plant ±hispid; central fl. of umbel usually purple; fr. spiny.   43. DAUCUS
- Annual; plant glabrous; central fl. of umbel white; fr. not spiny.   32. AMMI
36 Stem tapering downwards from ground level and becoming flexuous to its junction with the tuber

from which it arises.   37
- Stem arising at or near ground level, not tapering downwards.   38
37 Stem hollow after flowering; styles suberect in fr.
                                                    12. CONOPODIUM
- Stem solid after flowering; styles recurved in fr.
                                                    11. BUNIUM
38 Stems conspicuously swollen below the nodes.   39
- Stems not conspicuously swollen below the nodes.   40
39 Rays 3–5.   6. ANTHRISCUS
- Rays (4–)8–25.   5. CHAEROPHYLLUM
40 Annual, with slender roots; lower lvs usually dead at flowering-time.   41
- Biennial or perennial with stout stock, rhizome or tap-root; lower lvs often green at flowering-time.   45
41 Bracteoles 2- to 3-fid or pinnatifid; fr. with a long linear beak, developing soon after flowering.
                                                    7. SCANDIX
- Bracteoles entire; fr. without a long beak.   42
42 Sepals conspicuous, often as long as the petals.
                                                    9. CORIANDRUM
- Sepals very small or absent.   43
43 Stems with deflexed appressed hairs, at least above.
                                                    42. TORILIS
- Stems glabrous or with sparse patent or forward-pointing hairs.   44
44 Bracteoles on the outer side of the partial umbels only, deflexed.   20. AETHUSA
- Bracteoles on all sides of the partial umbels, not deflexed.   6. ANTHRISCUS
45 Base of stem surrounded by abundant fibrous remains of petioles.   46
- Base of stem without fibrous remains of petioles.   48
46 Usually dioecious; male plants with dense umbels of numerous shortly pedicellate fls; female plants with lax umbels and few long-pedicellate fls. 27. TRINIA
- Most fls hermaphrodite; umbels all similar.   47
47 Lf-lobes lanceolate to ovate in outline.   18. SESELI
- Lf-lobes filiform.   23. MEUM
48 Basal and lower cauline lvs 1- to 3-ternate.   49
- Basal and lower cauline lvs pinnately divided.   51
49 Lf-lobes linear to linear-lanceolate, the midrib (at least near the base) with a vein on either side close to and parallel with it.   33. FALCARIA
- Lf-lobes rhombic in outline, often 3-fid at apex, without accompanying parallel veins.   50
50 Lvs puberulent on margin and veins.
                                                    24. PHYSOSPERMUM
- Lvs glabrous.   36. LIGUSTICUM
51 Bracteoles on outer side of partial umbels only; ovary and fr. usually with crenulate-undulate ridges.
                                                    25. CONIUM
- Bracteoles absent or on all sides of the partial umbels; ridges on ovary and fr. not crenulate-undulate.   52
52 Rays puberulent or papillose at least on the angles (use lens).   53
- Rays glabrous and smooth.   57
53 Stems and lvs ±hairy.   54
- Stems and lvs glabrous.   55
54 Sheathing bases of upper lvs not greatly inflated; plant strongly aromatic.   8. MYRRHIS
- Sheathing bases of upper lvs greatly inflated; plant not strongly aromatic.   37. ANGELICA
55 Bracts present, persistent, deflexed.
                                                    38. PEUCEDANUM

Bracts 0 or few, caducous, not deflexed. 56
56 Lf-lobes 5–10 cm; rays usually more than 30.
    38. PEUCEDANUM
Lf-lobes 0.3–1 cm; rays usually fewer than 25.
    35. SELINUM
57 Bracteoles 0, rarely 1. 58
Bracteoles present. 61
58 Umbels with peduncle shorter than the rays.
    28. APIUM
Umbels with peduncle longer than the rays. 59
59 Umbels mostly with fewer than 10 rays; fr. smelling
    of caraway when crushed. 34. CARUM
Umbels with more than 10 rays; fr. not smelling of
    caraway when crushed. 60
60 Rhizomes far-creeping; plant glabrous.
    14. AEGOPODIUM
Rhizomes absent; plants with at least some small cris-
    pate hairs. 13. PIMPINELLA
61 Plant glabrous; sepals conspicuous. 63
Plant hairy, the hairs sometimes small and crispate;
    sepals minute. 62
62 Lobes of lower lvs less than 3 cm, frequently and
    deeply divided. 6. ANTHRISCUS
Lobes of lower lvs at least 4 cm, not or shallowly pin-
    natifid. 40. HERACLEUM
63 Bracts present. 19. OENANTHE
Bracts 0. 64
64 Pedicels at least twice as long as fr.; stock septate.
    31. CICUTA
Pedicels mostly much shorter than fr. stock not sep-
    tate. 19. OENANTHE

Subfamily 1. HYDROCOTYLOIDEAE. Lvs often simple, usually with scarious stipules. Fls usually in simple umbels or whorls. Ovary with a ±flat disk. Fr. with woody endocarp and no vittae, at least when mature. Carpophore absent. Commonest basic chromosome number 8.

1. HYDROCOTYLE L.

Perennials with stipulate lvs. Infl. a simple umbel or fls in whorls. Sepals small or 0. Fr. suborbicular, strongly compressed laterally; inner layer of fr. wall woody. Carpophore 0.

Lvs peltate, glabrous. **1. vulgaris**
Lvs with a deep basal sinus, hirsute. **2. moschata**

1. H. vulgaris L.   Marsh Pennywort, White-rot.

A slender, *creeping* or sometimes floating perennial, rooting freely at the nodes. Petioles 1–25 cm, erect, sparsely hairy. *Lvs* 8–35 mm diam., *peltate, orbicular, crenate*. Peduncles shorter than petioles. *Umbels* 2–3 mm *diam., 3–6-fld, sometimes with 1–3 whorls of fls below*; bracts triangular. Fls c. 1 mm, pinkish-green, subsessile, usually self-pollinated. Fr. 2 mm diam., carpels with 2 ridges on each face, covered with brownish resinous dots. Fl. 6–8. Fr. 7–10. $2n = 96$. H.

Native. In bogs, fens and marshes, usually on acid soils. Generally distributed in suitable localities throughout the British Is., ascending to 530 m. W., C. and S. Europe to 60° N in Scandinavia, eastwards to the Caucasus; introduced in New Zealand.

*2. H. moschata G. Forster

Like **1** but lvs hirsute on both surfaces, with a deep basal sinus and 5–7 distinctly toothed lobes; petioles up to c. 5 cm, with dense patent to deflexed hairs; umbels usually with 10–20 fls.

Introduced. A weed in lawns and well established on grassy banks on Valencia Is. in S.W. Ireland. Native in New Zealand.

Subfamily 2. SANICULOIDEAE. Lvs often simple or palmately lobed, exstipulate. Fls in simple umbels or capitula. Ovary with a ±flat disk. Fr. with membranous endocarp. Vittae usually present in the mature fr. Carpophore absent. Commonest basic chromosome number 8.

2. SANICULA L.

Erect perennials. Stock short. Lvs palmately lobed. Umbels small, arranged in an irregular cymose inflorescence and often forming a false compound umbel. Bracts few, simple, 3-fid or sometimes lf-like; bracteoles few. Calyx-teeth longer than the inflexed petals. Fr. covered with rigid forward-pointing hooked bristles; commissure broad; ridges inconspicuous; styles long, filiform.

About 40 spp. in temperate regions and on tropical mountains, absent from Australia.

1. S. europaea L.   Sanicle.

An erect glabrous perennial 20–60 cm. *Basal lvs* 2–6 cm, *3–5-lobed, lobes cuneate, coarsely and acutely serrate*, the teeth ending in a bristle; petiole 5–25 cm. False umbels usually with three rays; bracts 3–5 mm, 2–5, simple or pinnatifid; bracteoles simple. Fls pink or white, the outer male and shortly pedicellate, the inner hermaphrodite and nearly sessile. Fr. c. 3 mm, animal dispersed. Fl. 5–9. Pollinated mainly by small flies and beetles; self-pollination also possible. $2n = 16$. Hr.

Native. In woods; forming societies in chalk beechwoods and in oakwoods on loams. Throughout the British Is. except Orkney, Shetland and Channel Is. Wooded regions of Europe, Asia and N. Africa, only in mountain woods (at some altitude) in the Mediterranean region; mountains of tropical Africa (to 2500 m in Cameroun); S. Africa; S., C. and E. Asia.

3. ASTRANTIA L.

Perennial herbs. Stock short, creeping. Lvs palmately lobed or cut. Umbels simple; bracteoles many, conspicuous, often coloured. Fls male and hermaphrodite, the smaller later umbels often male only; pedicels of male fls longer than those of hermaphrodite. Calyx-teeth triangular, acuminate, longer than the notched inflexed petals. Fr. ovoid or oblong; commissure broad; carpels somewhat dorsally compressed with scaly inflated primary ridges; vittae 1 in each groove; styles filiform.

About 10 spp. in Europe and W. Asia.

*1. A. major L.   Astrantia.

An erect glabrous perennial 30–100 cm. Basal lvs

6–17 cm across, long-petiolate, with 3–7 coarsely serrate lobes. Bracteoles 1–2 cm, lanceolate, acuminate, entire or slightly toothed at apex, whitish beneath, pale greenish-purple above, equalling or exceeding the fls. Fls whitish or pinkish, in a convex umbel; pedicels filiform; male fls usually outnumbering hermaphrodite. Fr. 6–8 mm, covered with 2-dentate vesicular scales arranged in 5 vertical rows on each mericarp. Fl. 5–7. Pollinated by various insects, especially beetles. $2n = 28$. Hs.

Introduced. Naturalized in meadows and at margins of woods in several localities chiefly in N. and W. Britain; established for over 100 years near Stokesay Castle, Shropshire. C. and E. Europe, extending to Spain and White Russia.

4. ERYNGIUM L.

Rigid, often glaucous and spiny, perennial herbs. Lvs toothed, lobed or pinnatisect. Fls bracteolate, sessile in dense heads surrounded by rigid lf-like bracts. Calyx-teeth rigid, acute or pungent, longer than the petals. Petals narrow, erect, deeply notched, the tip inflexed. Fr. ovoid, subterete, scaly or papillose; commissure broad; carpels obscurely ridged; vittae slender.

About 230 spp. in temperate and subtropical regions, especially South America; absent from S. Africa. The fls have nectar and are pollinated by various insects.

Plant glaucous; basal lvs suborbicular, 3-lobed; involucral bracts oblong-cuneate, with 1–3 pairs of broad spinescent teeth.   **1. maritimum**
Plant pale green; basal lvs pinnate; involucral bracts linear-lanceolate, entire or with 1(–2) pairs of spines.
          2. campestre

1. E. maritimum L.   Sea Holly.

An *intensely glaucous* glabrous branched perennial 15–60 cm. Basal *lvs* 5–12 cm diam., stalked, *suborbicular, 3-lobed; cauline* sessile, *palmate;* both spinous-toothed with a thickened cartilaginous margin; petiole unwinged. Heads 1.5–3 cm, becoming ovoid. *Bracts oblong-cuneate,* with 1–3 pairs of broad spinescent teeth; *bracteoles linear, spinous, 3-fid,* often purplish-blue, *somewhat exceeding the fls.* Fls bluish-white; sepals 4–5 mm, longer than the petals. Fr. covered with papillae. Fl. 7–8. $2n = 16$. Hs.

Native. On sandy and shingly shores. Round the coasts of the British Is. north to Shetland but apparently extinct in N.E. England and E. Scotland. Coasts of Europe north to c. 60° N and of N. Africa and S.W. Asia.

2. E. campestre L.   Field Eryngo.

A *pale green* glabrous branched perennial 20–70 cm. *Basal lvs* stalked, *pinnate; cauline* sessile, *subcordate* at the base and ±*clasping stem,* less spiny than in *E. maritimum;* petiole winged. Heads 1–1.5 cm, ovoid. *Bracts linear-lanceolate,* spinous, *entire or with* 1(–2) *pairs of spines; bracteoles subulate,* spinous, *entire,* 2–3 *times as long as the fls.* Fls white; sepals c. 2.5 mm. Fr. densely scaly. Fl. 7–8. $2n = 14, 28$. Hs.

Native. In dry grassy places. In a few scattered localities in S. England probably introduced in some, but certainly established at Plymouth before 1670; rare. S. and C. Europe, S.W. Asia, N. Africa; introduced in North America.

Subfamily 3. APIOIDEAE. Lvs usually much divided, exstipulate. Fls usually in compound umbels. Ovary with a prominent stylopodium. Fr. with membranous endocarp. Vittae usually present in the mature fr. Carpophore present. Commonest basic chromosome number 11.

5. CHAEROPHYLLUM L.

More or less hairy herbs. Lvs usually 2–3-pinnate. Umbels compound; bracts few or 0; bracteoles several. Fls white (in our spp.). Calyx-teeth 0; petals notched, with an inflexed point. Fr. oblong or narrowly oblong-ovoid, scarcely beaked, laterally compressed, the commissure narrow; carpels subterete, the ridges broad and rounded; vittae solitary in the furrows.

About 40 spp. in north temperate regions.

Lvs dark green; fr. 4–7 mm; style as long as stylopodium
          1. temulentum
Lvs yellow-green, shortly ciliate at margins, slightly pubescent beneath; fr. 8–12 mm; style twice as long as stylopodium.
          2. aureum

1. C. temulentum L.   Rough Chervil.

A rough erect biennial 30–100 cm. *Stem* solid, somewhat grooved, *clothed with short stiff hairs,* swollen below the nodes, purple-spotted or entirely purple. *Lvs* up to 20 cm, 2–3-pinnate, *pubescent on both surfaces;* segments ovate, the lower shortly stalked, pinnately lobed, the lobes coarsely serrate. *Umbels* 3–6 cm diam., *rather irregular,* nodding in bud, the earliest terminating the main stem and subsequently overtopped by branches arising from axils of lvs immediately below it; rays (4–)8–10(–15), 1.5–5 cm, slender; bracts 0(–2); *bracteoles* 5–8, usually shorter than pedicels, lanceolate to ovate, aristate, ciliolate, spreading in fl., *deflexed in fr.* Fls white. *Fr.* 5–6 mm, oblong-ovoid, *narrowed upwards,* often purple. Fl. 6–7. $2n = 14, 22$. Hs.

Flowers just after *Anthriscus sylvestris;* readily recognized by the rough purple-spotted stems.

Native. In hedgebanks and in grassy places. Common and generally distributed in Great Britain, except W. and N. Scotland; Ireland, frequent in the east, almost absent from the west. Most of Europe, but rare in the Mediterranean region; S.W. Asia; N.W. Africa; introduced in North America and New Zealand.

*2. C. aureum L.   Golden Chervil.

An erect somewhat hairy perennial somewhat resembling *Anthriscus sylvestris* in general appearance, but stouter. *Stem* solid, slightly grooved, ±rough, swollen below the nodes, ±purple-spotted. *Lvs* 3-pinnate, *pubescent to nearly glabrous, the margins shortly ciliate;* segments lanceolate, acuminate, lobed, the lobes serrate or subentire. *Umbels* 5–8 cm diam., *nearly regular;*

rays 12–18, 1.5–3 cm; bracts 0(–3), subulate, up to 2 cm; *bracteoles 5–8, often equalling or exceeding pedicels*, lanceolate, aristate, fringed, *spreading in fl.*, ultimately deflexed. Fls white. *Fr. usually* 9 mm, *oblong, contracted near the apex*. Fl. 7. 2*n* = 22. Hs.

Introduced. Naturalized in meadows at a few places in S. England and Scotland, particularly Callander, W. Perth. C. and S. Europe; S.W. Asia.

6. ANTHRISCUS Pers.

Annual or biennial ±hairy herbs. Lvs 2–3-pinnate. Umbels compound; bracts 0 or rarely 1, bracteoles several. Fls white. Calyx-teeth minute or 0; petals notched, with an inflexed point. Fr. ovoid or oblong, beaked, commissure constricted; carpels sub- or ½-terete, ridges confined to the beak; vittae 0 or very slender and solitary in the furrows.

About 20 spp. in Europe, temperate Asia and N. Africa.

Stems glabrous or very sparsely hairy; fls 2 mm diam.; pedicels thicker than rays in fr.; fr. 3 mm, muricate.
        1. caucalis
Stems uniformly pubescent below; fls 3–4 mm diam.; pedicels not thicker than rays; fr. 7–10 mm, smooth.
        2. sylvestris

1. A. caucalis Bieb.   Bur Chervil.

A. scandicina Mansf.; *A. vulgaris* Pers., non Bernh.; *A. neglecta* Boiss. & Reuter; *Chaerophyllum anthriscus* (L.) Crantz

A sparsely hairy, branched and spreading annual 25–50(–100) cm. *Stems* hollow, striate, *glabrous* to very sparsely hairy, often curved below and purplish towards the base, somewhat thickened below the nodes. Lvs up to 10 cm, usually smaller, 2–3-pinnate, usually glabrous above, with stiff scattered hairs beneath; segments ovate, pinnatifid, lobes obtuse. Umbels 2–4 cm diam., shortly pedunculate, lf-opposed; *rays 3–6, 0.5–1.5 cm, glabrous*, slender, often divaricate or recurved in fr.; bracts 0 or rarely 1; bracteoles usually 4–5, about 2 mm, ovate, aristate, ciliate. *Fls c. 2 mm diam.*, subsessile, *pedicels elongating and becoming thicker than the rays in fr*. Fr. 3 mm, *ovoid, with hooked spines, beak short, glabrous*; *pedicels with a ring of hairs at the top; stigmas subsessile*. Fl. 5–6. 2*n* = 14. Th.

Native. In hedgebanks, waste places and on sandy ground near the sea. Scattered throughout the British Is. rather local and absent from much of the north and west. W., C. and S. Europe; temperate Asia; N. Africa; introduced in North America and New Zealand.

2. A. sylvestris (L.) Hoffm.   Cow Parsley, Keck.

Chaerophyllum sylvestre L.

A ±downy, erect biennial often perennating by means of offsets. *Stems* 60–100 cm, hollow, furrowed, *downy below, glabrous above*. Lvs up to 30 cm, 2–3-pinnate, somewhat pubescent, at least beneath; segments commonly 10–30 mm, ovate, pinnatifid and coarsely serrate. Umbels 2–6 cm diam., terminal, the earliest somewhat

overtopped by lateral branches; *rays* (3–)6–12, 1–4 cm, *glabrous*; bracts 0; *bracteoles 4–6, 2–5 mm, ovate, aristate, ciliate, spreading or deflexed*, often pink. *Fls 3–4 mm diam.*; pedicels about equalling bracteoles, elongating but scarcely thickening in fr. *Fr. 6–9 mm, oblong-ovoid, smooth, very shortly beaked; styles slender, spreading.* Fl. 4–6. Germ. spring. 2*n* = 16. Hs.

By far the commonest of the early-flowering umbellifers in the southern half of England; ±lfless in midsummer, new lvs developing in August.

Native. By hedgerows, at edges of woods and in waste places. Generally distributed and often extremely abundant throughout most of the British Is. Europe, rare in the south; temperate Asia; N. Africa; introduced in North America.

7. SCANDIX L.

Annual herbs. Lvs 2–3-pinnate, segments narrow and short. Umbels with few rays, sometimes only 1; bracts 1 or 0, bracteoles several, entire or variously lobed. Fls white. Calyx-teeth minute or 0; petals often unequal, point inflexed or 0. Fr. subcylindric with a very long beak, commissure constricted; carpels subterete, ridges prominent, slender; vittae solitary in the furrows.

About 15 spp. mainly in the Mediterranean region.

1. S. pecten-veneris L.   Shepherd's Needle.

An erect branched nearly glabrous annual 15–50 cm. Stems becoming hollow when old, striate, ±pubescent with short scattered hairs. Lvs 2–3-pinnate, oblong or narrowly deltate; segments up to 5 mm, spathulate, margins denticulate or subentire. *Umbels of 1–2 stout rays; bracteoles 5–10 mm, bifid or pinnatifid, sometimes entire, spinous-ciliate*. Fls subsessile, the pedicels elongating and becoming thickened in fr. *Fr. 15–80 mm, scabrid,* the ridges broad; styles very short, erect. Fl. 4–7. 2*n* = 26. Th.

Native. A weed of arable land. Formerly widely distributed in S.E. England, but now rather rare; very local and mainly near the coast elsewhere. C., S. and W. Europe; W. and C. Asia; N. Africa; introduced in S. Africa, North America, Chile and New Zealand.

8. MYRRHIS Miller

Finely pubescent perennial herbs. Lvs 2–3-pinnate. Umbels compound; bracts few or 0, bracteoles several, membranous. Fls white, male or hermaphrodite. Calyx-teeth minute or 0; petals notched, with a very short inflexed point. Fr. linear-oblong, beaked, commissure broad; carpels with prominent ridges; vittae 0 at maturity.

One sp. in Europe, widely naturalized elsewhere.

*1. M. odorata (L.) Scop.   Sweet Cicely.

A stout erect puberulent perennial 60–200 cm. Stems hollow, somewhat grooved. Lvs up to c. 30 cm, 2–3-pinnate, pale beneath and usually with some whitish markings; segments oblong-ovate, pinnatisect, the lobes coarsely serrate; petioles of stem-lvs sheathing. Umbels

1–5 cm diam., terminal; *rays* 5–10, *those of partial umbels bearing hermaphrodite fls* 2–3 cm, *stout*; *those of partial umbels bearing only male fls shorter, slender*; bracteoles c. 5, lanceolate, aristate. Fls white, the petals unequal. *Fr.* 15–25 mm, strongly and sharply ridged, ridges scabrid with bristly hairs; styles slender, diverging. Fl. 5–6. 2n = 22. Hs.

The whole plant has a strong aromatic smell. The commonest spring-flowering umbellifer in many northern districts, especially near dwellings. The plant is lfless in winter (cf. *Anthriscus sylvestris*).

Introduced. In grassy places, hedges and woods. Northwards from Glamorgan and Lincoln, rare and probably casual in the southern part of its range, common in N. England and S. Scotland, becoming rarer northwards; northern Ireland. Pyrenees, Alps, Apennines, W. part of the Balkan peninsula.

9. Coriandrum L.

Slender glabrous annual herbs. Lvs 1–3-pinnate, the lowest only lobed. Umbels compound; bracts 0; bracteoles few, lanceolate. Fls white or pink. Calyx-teeth acute, unequal; petals with an inflexed point. Fr. subglobose; carpels ½-terete, primary ridges low, slender; secondary broader; vittae obscure, solitary beneath each secondary ridge.

Two sp. in S.W. Asia and N. Africa.

*1. C. sativum L.   Coriander.

An erect annual 15–70 cm. Stems solid, ridged. Lower lvs 1–2-pinnate or lobed; segments ovate, cuneate at base, pinnately lobed or toothed; upper lvs 2–3-pinnate, pinnatifid. Umbels 1–3 cm diam.; rays 3–5(–10). Petals unequal, the larger 2–3 mm. *Fr.* 2–6 mm, *hard, red-brown, carpels adhering firmly to one another*; styles slender. Variants with 3 carpels and styles occur. Fl. 6. 2n = 22. Th.

The whole plant smells strongly of bed-bugs.

Introduced. A casual in waste places, sometimes escaped from cultivation. Native of N. Africa and S.W. Asia; widespread as a weed.

10. Smyrnium L.

Erect glabrous biennial or perennial herbs. Basal lvs 2–3-ternate, the segments broad. Umbels compound; bracts and bracteoles small, few or 0. Fls yellow or greenish. Calyx-teeth minute or 0; petals with a short inflexed point. Fr. ovoid, laterally compressed, didymous, commissure narrow; carpels subterete or angular with 3 prominent sharp ridges; vittae numerous; styles short, recurved.

Seven spp. in Europe and N. Africa.

*1. S. olusatrum L.   Alexanders.

A stout biennial 50–150 cm. Stem solid, becoming hollow when old, furrowed, branches in the upper part often opposite. *Lvs dark green and shiny*, the basal c. 30 cm, 3-*ternate*; *segments rhombic, obtusely serrate or lobed, stalked*; upper stem-lvs often opposite, simply

ternate; petiole base sheathing. Umbels axillary and terminal, subglobose; rays (3–)7–15(–18), 1–5 cm; bracts and bracteoles few. *Fls yellow-green*, shortly pedicellate. *Fr.* 7–8 mm, *broadly ovoid, nearly black*. Fl. 4–6. 2n = 22. Hs.

Formerly cultivated as a pot-herb; the young stems have somewhat the taste of celery.

Introduced. Extensively naturalized in hedges, waste places and on sea-cliffs. Europe northwards to N.W. France; S.W. Asia; N. Africa; Macaronesia.

*S. perfoliatum L., a biennial with simple ovate or suborbicular denticulate or crenulate amplexicaul upper lvs, is occasionally naturalized. S.Europe to Asia Minor and the Caucasus.

11. Bunium L.

Perennial herbs. Stem arising from a solitary globose tuber. Lvs 2–3-pinnate, the segments narrow. Umbels compound; bracts and bracteoles several. Fls white. Calyx-teeth minute; petals obovate, emarginate, with inflexed apex. Fr. oblong-ellipsoid, laterally compressed; carpels with 5 slender ridges. Seedlings with only one cotyledon.

About 40 spp. in Europe and temperate Asia.

1. B. bulbocastanum L.   Great Pignut.

A glabrous erect perennial 30–100 cm. *Tuber globose*, dark brown, 1–2.5 cm diam., edible. *Stems* terete, striate, *solid*. Lvs 10–15 cm, broadly triangular, 3-pinnate, mostly from the lower part of the stem, rarely a few basal, usually dead by flowering time; primary divisions stalked; segments 0.5–1 cm, spathulate to subulate with a blunt cartilaginous point. Umbels 3–8 cm diam., flat-topped; rays 10–20, 1.5–4.5 cm; bracts and bracteoles 5–10, small, lanceolate to linear-lanceolate, acute. Fls white. *Fr.* 3.5–4.5 mm, *oblong-ellipsoid*, dark brown with paler inconspicuous ridges; *styles short, recurved*. Fl. 6–7. 2n = 22. G.

Native. In chalk grassland and on banks, very local. Hertford, Bucks, Cambridge and Bedford. Europe northwards to England and eastwards to C. Germany and N.W. Yugoslavia; N.W. Africa.

12. Conopodium Koch

Glabrous or pubescent perennial herbs. Stem arising from a solitary tuber. Lvs ternately divided, segments narrow. Umbels compound; bracts and bracteoles few or 0, membranous. Fls white; stylopodium conical. Calyx-teeth 0; petals notched with an inflexed point, those of the outer fls often unequal. Fr. ovoid-oblong often shortly beaked, commissure constricted; carpels subterete, ridges slender, inconspicuous; vittae several in each furrow.

About 20 spp. in Europe, N. Africa and temperate Asia.

1. C. majus (Gouan) Loret   Pignut, Earthnut.

C. denudatum Koch; *Bunium flexuosum* Stokes

A slender erect glabrous perennial 8–50(–90) cm. Tuber 1–3.5 cm, dark brown, irregular. *Stem hollow* after fl, finely striate, flexuous, much attenuated near its junction with the tuber. Basal lvs 5–15 cm, soon withering, broadly deltate, the petioles slender, flexuous, mostly subterranean; primary divisions stalked, 2-pinnate, the segments deeply pinnatifid and the lobes linear-lanceolate; stem-lvs smaller; petioles short and sheathing and the lobes linear, the terminal much the longest. Umbels 3–7 cm diam., terminal, nodding in bud; rays 6–12, 1–6 cm, slender; bracts 0(1–2); bracteoles 2–5, membranous, linear, variable in length. *Fr.* 4 mm, *oblong-ovoid, beaked; styles* short, usually *erect.* Fl. 5–6. G.

The basal lvs wither about flowering time and the stem-lvs by the time the fr. is ripe. Tubers edible raw or cooked. Resembles *Bunium bulbocastanum* but differs in the hollow stems, less numerous bracts and bracteoles, usually erect styles and the shapes of the fr. and stylopodium.

Native. in fields and woods on acid soils. Generally distributed throughout Great Britain; common except on chalk and in fens; less frequent in Ireland. W. Europe from Norway southward and eastward to Italy.

13. PIMPINELLA L.

Perennial, rarely annual herbs. Lvs usually pinnate. Umbels compound; bracts 0; bracteoles 0 or few. Fls white or pinkish in our spp. Calyx-teeth small or 0; petals with a long inflexed point. Fr. ovoid or oblong, laterally compressed, commissure broad; carpels with 5 slender ridges; vittae numerous in each furrow; styles long. The fls are strongly protandrous. Pollinated by insects.

About 150 spp. mainly in north temperate regions.

Stem solid, subterete, tough; lower lvs with (2–)4–7 pairs
of lobes **1. saxifraga**
Stem hollow, strongly ridged or angled, brittle; lower lvs
with 3–4 pairs of lobes **2. major**

1. P. saxifraga L. Burnet Saxifrage.

An erect usually puberulent rather slender perennial 30–100 cm. Stock rather slender, usually with fibrous remains of petioles. *Stem subterete, slightly ridged, usually rough, tough, solid.* Basal lvs and upper stem-lvs usually simply pinnate, *lower stem-lvs usually 2-pinnate,* very variable; segments 1–2.5 cm, ovate to linear-lanceolate, obtusely or acutely serrate to pinnatifid, *sessile or subsessile,* those of the basal lvs broader and less divided than those of the stem-lvs; uppermost stem-lvs small or 0, petioles sheath-like, often purplish. Umbels terminal, 2–5 cm diam., flat-topped; rays 10–22, 1–4.5 cm; bracts and bracteoles 0. Fls white; *styles much shorter than petals,* elongating later. Fr. 2–2.5 mm, broadly ovoid. Fl. 7–8. 2n = 36, 40. Hs.

Native. In dry grassy places. Scattered throughout the British Is., rarer in the north and west; mainly calcicole. Most of Europe; S.W. Asia.

2. P. major (L.) Hudson Greater Burnet Saxifrage.
P. magna L.

An erect stout nearly glabrous perennial 50–120 cm. Stock rather stout, usually without fibrous remains of petioles. *Stem prominently ridged or angled, glabrous, brittle,* often magenta towards base, *hollow. Lvs all simply pinnate* (rarely bipinnate); *segments* 2–10 cm, those *of basal lvs* ovate, subcordate, *shortly stalked,* those of stem-lvs narrower, sessile, all coarsely serrate and sometimes 3-lobed; uppermost stem-lvs small, often ternate, rarely 0; petioles sheath-like, green. Umbels terminal, 3–6 cm diam., flat-topped; rays 10–20, 1–4 cm; bracts and usually bracteoles 0. Fls white or pinkish; *styles about as long as petals.* Fr. 3–4 mm, ovoid. Fl. 6–7. 2n = 20. Hs.

Native. In grassy places at margins of woods and in hedgebanks. From Northumberland southwards, but mainly in the eastern half of England and in S.W. Ireland; local and rarer in the north. Most of Europe, except the extreme north and south.

14. AEGOPODIUM L.

Glabrous rhizomatous perennial herbs. Lvs 1–2-ternate; segments broad. Umbels compound; bracts and bracteoles few or 0. Fls white. Calyx-teeth 0; petals somewhat unequal, with inflexed apex. Fr. ovoid, laterally compressed; carpels with 5 slender ridges; vittae 0.

Five to 7 spp. in Europe and Asia.

1. A. podagraria L.
Goutweed, Bishop's Weed, Ground Elder,
Herb Gerard.

A stout erect glabrous perennial 40–100 cm. *Rhizomes far-creeping,* white when young. Stem hollow, grooved. Lvs 10–20 cm, deltate; *segments* 4–8 cm, sessile or shortly stalked, lanceolate to ovate, acuminate, often oblique at base, *irregularly serrate;* petioles much longer than blade, bluntly triquetrous. Umbels 2–6 cm diam., terminal; rays 10–21, 1–4 cm; bracts and bracteoles usually 0. Fr. 3–4 mm, ovoid; styles slender, reflexed. Fl. 5–7. 2n = ?22, 44. Hs.

?Introduced. Said to have been introduced and cultivated as a pot-herb, now very well naturalized in waste places near buildings and a persistent weed in gardens; perhaps native in some woods. Generally distributed throughout the British Is. Nearly the whole of Europe except parts of the south; temperate Asia.

15. SIUM L.

Glabrous herbs. Lvs pinnate, the segments broad. Umbels compound, terminal or lateral; bracts and bracteoles 2–6. Fls white. Calyx-teeth 5, small, acute; petals with a short inflexed point. Fr. ovoid, laterally compressed, constricted at the commissure; carpels with 5 obtuse or thickened, prominent ridges; vittae 3 or more in each furrow, superficial.

10–15 spp. Widely distributed but absent from South America and Australia.

1. S. latifolium L. Water Parsnip.

A stout erect perennial up to 2 m. *Stem* hollow, *grooved.* Submerged lvs 2–3-pinnate, with linear lobes, present only in spring. Aerial lvs c. 30 cm, long-petiolate, simply pinnate; *segments* 2–12 cm, sessile, 4–6(–8) *pairs*, oblong-lanceolate to ovate, *all regularly serrate* or when submerged finely divided; petioles sheathing at base, fistular above. *Umbels* 6–10 cm diam., *terminal, flat-topped*; *rays usually* 20–30, 1.5–4 cm; bracts and bracteoles variable, often large and lf-like. *Fr.* 3–4 mm, ovoid, *longer than broad*. Fl. 7–8. $2n = 20$. Hel.

Native. In fens and other wet places. Very local and mainly south-east of a line from the Humber to the Bristol Channel; Ireland, recently recorded only from the basins of Shannon and Erne. Apparently decreasing. Most of Europe; Siberia.

16. BERULA Koch

Like *Sium* but differing in the subglobose, nearly didymous fr. with the lateral ridges slender, not marginal; vittae deeply embedded.

Three spp. in north temperate regions.

1. B. erecta (Hudson) Coville
　　　　　　Lesser or Narrow-leaved Water-parsnip.

Sium erectum Hudson; *S. angustifolium* L.

An erect or decumbent stoloniferous perennial 30–100 cm. *Stem* hollow, *striate*. Lvs up to 30 cm, long-petiolate, simply pinnate, dull, somewhat blue-green; *segments* 2–5 cm, sessile, 5–9(–14) *pairs*, oblong-lanceolate to ovate, sharply serrate or slightly lobed; *stem-lvs* small, *usually very irregularly serrate*. *Umbels* 3–6 cm diam., *lf-opposed*, *rather irregular*; *rays usually* 7–18, 0.5–2.5 cm; bracts and bracteoles many, lf-like, often trifid. *Fr.* 1.5–2 mm, subglobose, *broader than long*. Fl. 7–9. $2n = 20$. Hel.

Native. In ditches, canals, ponds, fens and marshes. Lowland districts of England, except the south-west; rare in most of Wales and Scotland; frequent in C. Ireland. Europe; W. Asia.

The lvs of this plant are often confused with those of *Apium nodiflorum* (p. 288). They differ in colour and in the usually greater number of pairs of segments.

17. CRITHMUM L.

A glabrous herb, woody at the base. Lvs 2–3-pinnate, fleshy. Umbels compound; bracts and bracteoles numerous. Fls yellowish-green. Calyx-teeth very small; petals with a long inflexed point. Fr. ovoid-oblong, terete, commissure broad; carpels with prominent ridges; vittae several in each furrow of the spongy mesocarp.

One sp. on the Atlantic coasts of Europe, the Mediterranean, the Black Sea, and Macaronesia.

1. C. maritimum L.　　　　　　Rock Samphire.

A branched perennial 15–45 cm. Stems solid, striate. Lvs deltate, *segments* 2–5 cm, terete, *subulate to linear-lanceolate*, *fleshy*, acute; petioles short, sheaths long membranous, enfolding the stem. Umbels 3–6 cm diam.; rays 8–36, rather stout; *bracts* 5–10, lanceolate, acute, ±membranous, spreading, at length reflexed. Bracteoles 6–8, lanceolate, deflexed in fr. Fr. 5–6 mm, corky, olive-green to purplish. Fl. 6–8. $2n = 20$. Hp.

Lvs make a good pickle.

Native. On sea-cliffs and rocks or more rarely shingle or sand by the sea. Common on S. and W. coasts of England and Wales and in Ireland; rare in Scotland and only in the west. Distribution of the genus.

18. SESELI L.

Biennial or perennial herbs. Lvs usually 2–3-pinnate. Umbels compound; bracts many, few or 0; bracteoles many, entire. Fls white. Calyx-teeth prominent (in our sp.); petals notched, with a long inflexed point. Fr. ovoid or oblong, subterete, commissure broad; carpels dorsally compressed, the ridges prominent; vittae solitary, rarely 2 or 3, in the furrows.

About 80 spp. mainly in Europe and temperate Asia.

1. S. libanotis (L.) Koch　　　　　Moon Carrot

An erect somewhat pubescent to nearly glabrous biennial or monocarpic perennial 30–100 cm. *Stock* stout, *with fibrous remains of petioles*. Stems solid, somewhat ridged. Lvs c. 10 cm, 2-pinnate; segments sessile, ovate, pinnatisect, the lobes 5–15 mm, oblong, mucronate, ±ciliate. Umbels 3–6 cm diam., terminal, convex in fl.; *peduncles* long, *pubescent at top*; rays 15–30, 1–3 cm, pubescent; *bracts* and bracteoles *many*, linear, pubescent, spreading or deflexed. Calyx-teeth subulate, deciduous. Fr. 2.5–3 mm, ovoid, pubescent. Fl. 7–8. $2n = 22$. Hs.

Native. In rough grassy and bushy places on chalk hills. Sussex. Hertford, Cambridge and Bedford, very local. Most of Europe; W. Asia; N. Africa.

19. OENANTHE L.

Glabrous usually marsh or aquatic herbs. Lvs 1–3-pinnate, rarely some reduced to a fistular petiole. Umbels compound; bracts several, few or 0; bracteoles (in our spp.) many. Fls white. Calyx-teeth persistent, acute; petals notched, with a long inflexed point. Fr. ovoid, cylindrical or globose, commissure broad; carpels $\frac{1}{2}$-terete, 2 lateral ridges grooved or thickened, sometimes obscure; vittae solitary in the furrows.

About 35 spp. in the north temperate regions of the Old World.

1 Some umbels lf-opposed; peduncles shorter than rays.　*2*
　Umbels terminal; peduncles longer than rays.　　　　　*3*
2 Fr. more than 5 mm.　　　　　　　**7. fluviatilis**
　Fr. less than 4.5 mm.　　　　　　　**6. aquatica**
3 Partial umbels globose in fr.; fr. sessile.　**1. fistulosa**
　Partial umbels not globose in fr.; some fr. pedicellate.　*4*
4 Rays and pedicels not thickened in fr.　　　　　　*5*
　Rays and pedicels thickened in fr.　　　　　　　　*6*
5 Lobes of basal lvs ovate or suborbicular.　**5. crocata**
　Lobes of basal lvs spathulate to linear.　**4. lachenalii**
6 Lobes of basal lvs cuneate or ovate; stem solid; partial
　　umbels flat-topped in fr.　　　**2. pimpinelloides**
　Lobes of basal lvs linear to linear-lanceolate; stem
　　hollow; partial umbels not flat in fr.　**3. silaifolia**

1. O. fistulosa L.　　　　Tubular Water-Dropwort.

An erect perennial 30–80 cm. *Roots with fusiform*

tubers. *Stems* slender, *fistular*, often constricted at nodes, rooting at lower nodes. Lvs 1-(2)-pinnate; segments of lower lvs shortly stalked, lanceolate, lobed; of upper 0.5–2 cm, linear-lanceolate or subulate, entire, distant; *petioles* long, *fistular*. Umbels terminal, pedunculate; *rays* 2–4, 1–2 cm, stout, spreading. *Partial umbels* c. 1 cm diam., dense, flat-topped in fl., globose *in fr.*; pedicels thickening after fl. *bracts* 0(–1); *bracteoles* 7–16, *shorter than pedicels*. Fls with conspicuous sepals, petals of the outer unequal. Fr. 3–4 mm, angular; *styles* 4–5 mm, *slender, diverging*. Fl. 7–9. 2n = 22. Hel. or Hs.

Native. In marshy places and shallow water. Mainly in the eastern half of England and very local in Scotland and Wales; Ireland, very rare in the west. Most of Europe; S.W. Asia; N.W. Africa.

2. O. pimpinelloides L.
Corky-fruited Water-Dropwort.

An erect branched perennial 30–100 cm. *Roots with ovoid tubers towards their ends. Stem solid*, furrowed. Lvs 2-pinnate, segments of lower lvs c. 5 mm, stalked, ovate or lanceolate, lobed of upper 1–3 cm, 1–2-pinnate linear-lanceolate to linear, usually entire. Umbels 2–5 cm diam., terminal; *rays* 6–15, 1–2 cm, *rather stout* (0.5–1 mm *diam.*) *in fr.*; partial umbels dense, flat in fl. and fr.; pedicels thickening after fl.; *bracts* 1–5 linear to linear-lanceolate, unequal; *bracteoles* 12–20, linear to linear-lanceolate, *about equalling pedicels*. Fls 3–4 mm diam, petals of outer unequal. Fr. c. 3.5 mm, cylindrical, strongly ribbed; *styles* 3 mm, *stout, erect*. Fl. 6–8. 2n = 22. Hs.

Native. In meadows and damp grassy places. From E. Devon to Hants, very local elsewhere south of a line from Worcester to Suffolk; formerly in 1 locality in Co. Cork. W. and S. Europe; S.W. Asia.

3. O. silaifolia Bieb. Narrow-leaved Water-Dropwort.

Like *O. pimpinelloides* but larger. *Roots usually thickened and fusiform from base. Stem* hollow above, rubbed and striate. Lvs 2–4-pinnate, the lower soon withering; *lobes* linear-lanceolate, *acute*. Umbels with 4–10 rays 1.5–3 cm, *becoming very stout* (1–2 mm *diam.*) *in fr.*; *bracts few or* 0; *bracteoles* 10–17, *ovate, acuminate. Fr.* 3–3.5 mm, *subcylindrical, strongly constricted at the junction with the pedicel*; *styles* 1–2 mm, *rather slender*, spreading to suberect. Fl. 6. 2n = 22*. Hs.

Native. In damp rich meadows, usually near rivers, very local. South of a line from the Severn to the Wash with outlying stations in Worcester and Nottingham. W., C. and S. Europe; S.W. Asia; N.W. Africa.

4. O. lachenalii C. C. Gmelin
Parsley Water-Dropwort.

Like *O. pimpinelloides*. *Roots cylindrical or fusiform. Stem solid or sometimes with a small cavity when old*, ribbed and striate. Lvs (1–)2(–3)-pinnate, the lower soon withering; *lobes* spathulate or linear, *obtuse or subacute*. Umbels of 5–9(–20) *rays*, slender (c. 0.25 mm

diam.) *in fr.*, bracts up to c. 5, subulate; *bracteoles* usually 5–7 *oblong-lanceolate. Fr.* c. 2.5 mm, *obovoid*; *pedicels neither thickened nor constricted at top*, very short; styles c. 1 mm, rather slender, spreading or suberect. The partial umbels have a number of long-pedicellate male fls round the outside. Fl. 6–9. 2n = 22*. Hs.

Native. In brackish and fresh water marshes and fens. Much of the British Is. but absent from N. Scotland and many inland areas. W. Europe eastwards to Poland and Yugoslavia; Algeria (very rare).

5. O. crocata L. Hemlock Water-Dropwort.

A stout erect branched perennial 50–150 cm. *Root tubers up to c.* 6 × 1 cm, sweetish-tasting but poisonous. Stems hollow, grooved. *Lvs* 30 cm *or more, deltate*, 3–4-*pinnate*; segments ovate to suborbicular, cuneate at base, 1–2-lobed, serrate, teeth obtuse or subacute with a minute apiculus, segments of stem-lvs narrower; *petioles mostly sheathing*. Umbels 5–10 cm diam., terminal; peduncles usually longer than rays; rays (7–)12–40, 3–8 cm; *bracts* c. 5, *linear to 3-fid*; *bracteoles* 6 *or more, caducous, linear-lanceolate*. Petals of outer fls unequal. *Fr.* 4–5.5 mm, *cylindrical*; pedicels neither thickened nor constricted at top; *styles c.* 2 mm, *erect*. Fl. 6–7. 2n = 22*. Grt.

Native. In wet places, usually calcifuge. Mainly in the south and west of Great Britain; absent from much of C. and W. Ireland. W. Europe; N. W. Africa.

6. O. aquatica (L.) Poiret
Fine-leaved Water-Dropwort.

O. phellandrium Lam.; *Phellandrium aquaticum* L.

An erect annual or biennial 30–150 cm. Roots of young plant tuberous, the tubers disappearing as the plant flowers. Stem hollow, striate, often very stout. Lvs 2–3-pinnate, all aerial or *lower submerged; submerged lvs with filiform segments*; aerial lvs with deeply lobed segments c. 5 mm, the *lobes* lanceolate to ovate, *acute*. Umbels 2–4 cm diam., terminal and *lf-opposed; peduncles usually shorter than rays*; rays 4–16, 1–4 cm bracts 0 or 1; bracteoles usually 4–8, subulate. *Petals of outer fls sub-equal. Fr.* 3.5–4.5 mm, *ovoid or oblong-ovoid*. Sometimes perennating by means of offsets. Fl. 6–9. 2n = 22*. Hyd.

Native. In slow-flowing or stagnant water, sometimes attaining great size in nearly dry fen ditches. England and E. Wales, S. E. Scotland; Ireland, frequent in the centre, local elsewhere. Most of Europe extending eastwards to C. Asia.

7. O. fluviatilis (Bab.) Coleman

Like *O. aquatica*, but stems submerged and ascending, then erect. Lower lvs 2-pinnate; *submerged lvs with cuneate segments cut at the ends* into longer or shorter narrow lobes; aerial lvs 1–3-pinnate, with rather shallowly lobed *segments up to* 1 cm *or more, lobes* ovate *obtuse or subacute. Fr.* 5–6.5 mm. Fl. 7–9. 2n = 22*. Hyd.

Native, In streams and ponds. S.E. England, C. and E. Ireland, 1 locality in Wales; very local. W. Europe.

20. AETHUSA L.

A glabrous annual herb. Lvs ternately 2-pinnate. Umbels compound; bracts 0 or 1, bracteoles 1–5, deflexed. Fls white. Calyx-teeth small or 0; petals notched, with an inflexed apex. Fr. broadly ovoid; carpels dorsally compressed; ridges very broad, the lateral narrowly winged; vittae solitary in the furrows.

One sp. in Europe; S.W. Asia, N. Africa; introduced in North America. Usually contains coniine and cynapine and is consequently poisonous.

1. A. cynapium L.   Fool's Parsley.

A branched lfy annual 5–120 cm. Stems hollow, finely striate, somewhat glaucous. Lvs deltate; segments ovate, pinnatifid. Umbels 2–6 cm diam., terminal and lf-opposed; rays (4–)10–20, 0.5–3 cm; bracts usually 0; *bracteoles usually 3–4 on the outer side of the partial umbels*, subulate, up to 1 cm. Petals unequal. Fr. 2.5–3.5 mm. Fl. 7–8. $2n = 20$. Th.

Native. A weed of cultivated ground. Common and generally distributed throughout much of the British Is. but becoming rarer northwards and absent from the extreme north of Scotland. Distribution of the genus.

Subsp. **cynapium**: Stems usually 30–80 cm; outer pedicels about twice as long as the fr.; bracteoles usually several times as long as the longer pedicels. Gardens, roadsides, etc.

Subsp. **agrestis** (Wallr.) Dostál: Stems usually 5–20 cm; outer pedicels usually shorter than the fr.; bracteoles usually about as long as the longer pedicels. Arable fields.

21. FOENICULUM Miller

Tall, glabrous, biennial or perennial herbs. Lvs 3–4-pinnate, the segments narrow. Umbels compound; bracts and bracteoles usually 0. Fls yellow. Calyx-teeth 0; petals with an obtuse incurved point. Fr. ovoid-oblong, subterete, commissure broad; ridges stout; vittae solitary in the furrows.

One sp. in Europe and the Mediterranean region.

1. F. vulgare Miller   Fennel.

F. officinale All.

A stout erect rather glaucous perennial 60–250 cm. Stem solid, developing a small hollow when old, striate, polished. Lvs much divided; *segments* 1–5 cm, *not all in one plane, filiform* with cartilaginous points. Umbels 4–8 cm diam., terminal and often lf-opposed; rays 10–40, 1–6 cm, glaucous; bracts and bracteoles usually 0. *Fls yellow*. Fr. 4–6 mm, ovoid. Fl. 7–10. $2n = 22$. Hp.

The whole plant has a strong and characteristic smell. Lvs used for culinary purposes.

?Native. On sea cliffs and naturalized or as a casual in waste places inland. England south of a line from the Humber to the British Channel, mainly coastal; Ireland, mainly on the south and east coasts. Native in the Mediterranean region and S. Europe; Macaronesia. Naturalized in most temperate countries.

Anethum graveolens L., Dill, is often grown for flavouring and sometimes persists in waste places for a few years. It is an annual with yellow fls, somewhat resembling *Foeniculum* but easily uprooted and usually smaller and more slender. The stem is hollow and the fruits have distinct marginal wings. It is probably native in S.W. Asia but is widely cultivated.

22. SILAUM Miller

Glabrous perennial herbs. Lvs 1–4-pinnate. Umbels compound; bracts 0–3; bracteoles usually 5–11, small. Fls yellowish. Calyx-teeth minute; petals broad and truncate at base with an incurved tip. Fr. ovoid or oblong, the commissure broad; ridges slender, the lateral narrowly winged; vittae numerous, inconspicuous.

About 10 spp. in Europe and temperate Asia.

1. S. silaus (L.) Schinz & Thell.   ·   Pepper Saxifrage.

Silaus flavescens Bernh.; *S. pratensis* Besser

An erect branched perennial 30–100 cm. Tap-root cylindrical, stout and woody with a few fibrous remains of petioles at the top. Stem solid, striate. Lower lvs deltate, 2–4-pinnate, the *segments* 1–1.5 cm, entire or pinnatisect, *finely serrulate*; upper lvs few, small, 1–2-pinnate or reduced to sheaths. Umbels 2–6 cm diam., terminal and axillary, long pedunculate; rays 4–15, 1–3 cm, rather unequal, the top of the peduncle and the rays papillose; bracteoles about equalling pedicels, linear-lanceolate with scarious margins. *Fls yellowish*. Fr. 4–5 mm, oblong-ovoid. Fl. 6–8. $2n = 22*$. Hs.

Native. In meadows and on grassy banks. Mainly in S. and E. England, a few localities in Wales and S.E. Scotland; absent from Ireland. W., C. and E. Europe northwards to the Netherlands and Sweden but absent from Portugal.

23. MEUM Miller

A glabrous very aromatic perennial herb. Lvs 3–4-pinnate, the segments filiform. Umbels compound; bracts few or 0; bracteoles 3–8. Fls white or purplish. Calyx-teeth 0; petals acute, narrowed to base, sometimes with a short inflexed point. Fr. ovoid, commissure broad; ridges slender, acute; vittae 3–5 in each furrow.

One sp. in the mountains of western Europe.

1. M. athamanticum Jacq.   Spignel, Meu, Baldmoney.

A tufted branched perennial 7–60 cm. *Stock* with abundant *coarse fibrous remains of petioles*. Stems hollow, striate. Lvs mostly basal, *divisions* filiform ±*whorled*, the lobes c. 0.5 cm. Umbels 3–6 cm diam.; rays 6–15; bracts 0, linear; bracteoles usually 3–8 linear, shorter than pedicels. Fr. 6–10 mm. Fl. 6–7. $2n = 22$. Hr.

Native. In grassy places in mountain districts. A few localities in N. England and N. Wales; Scotland, northwards to Argyll and Aberdeen, local. W. and C. Europe, extending southwards to the Sierra Nevada and C. Bulgaria.

24. PHYSOSPERMUM Cusson

Erect perennial herbs. Lvs 2-ternate with usually ovate cuneate segments. Umbels compound; bracts usually 4–7, lanceolate to linear. Fls all hermaphrodite white. Calyx-teeth shorter than the inflexed petals, sometimes 0. Fr. inflated, didymous, broader than long, the commissure narrow; carpels smooth, the ridges inconspicuous; vittae 1 in each groove.

About 4 spp. in Europe and W. Asia.

1. P. cornubiense (L.) DC.   Bladder-seed.

Danaa cornubiensis (L.) Burnat

A nearly glabrous perennial. Stem 30–120 cm, ribbed. Lvs long-petiolate, 2-ternate; *segments stalked, cuneate, laciniate, puberulent on margins and larger veins on both surfaces.* Umbels 1.5–5 cm diam., long-pedunculate; rays (6–)10–14; bracts and bracteoles lanceolate, acute, entire. Fls long-pedicellate; styles rather stout, spreading, recurved in fr. Fr. 3–4 mm, dark brown. Fl. 7–8. Hs.

Native. In bushy places. Cornwall and S. Devon; Bucks, perhaps introduced; very local. S. Europe, S.W. Asia.

25. CONIUM L.

Glabrous biennial herbs. Lvs 2–4-pinnate. Bracts and bracteoles few small; fls white, strongly protandrous. Calyx-teeth 0; petals obtuse or tip shortly inflexed. Fr. broadly ovoid to suborbicular, laterally compressed, the commissure rather narrow; carpels with five prominent ridges; vittae 0; styles short, recurved.

Two spp. in north temperate regions and S. Africa. Very poisonous.

1. C. maculatum L.   Hemlock.

An erect branched foetid biennial up to 3 m. *Stems furrowed, smooth, purple-spotted, glaucous.* Lvs up to 30 cm, 2–4-pinnate, ovate to deltate; segments 1–2 cm, oblong-lanceolate to deltate, coarsely and deeply serrate. Umbels 2–5 cm diam., terminal and axillary, shortly pedunculate; rays 10–20, 1–3.5 cm; bracts 2–5 mm, few, reflexed; bracteoles similar but smaller and only on the outside of the partial umbel. Fls white; petals with short inflexed tips. *Fr. 2.5–3.5 mm, suborbicular, the ridges usually undulate-crenate.* Fl. 6–7. $2n = 22$. Hs.

Native. In damp places, open woods and near water; also on roadsides and rubbish tips. Throughout most of the British Is., less common in the north. Europe and temperate Asia; Macaronesia. Introduced in eastern North America, California, Mexico, West Indies, temperate South America, and New Zealand.

26. BUPLEURUM L.

Annual or perennial glabrous herbs or rarely shrubs. Lvs simple, entire. Bracts and bracteoles various. Fls yellow. Calyx-teeth 0; petals with inflexed points. Fr. laterally compressed, commissure broad; carpels usually with prominent ridges; vittae 1–5 in each groove, often disappearing as fr. ripens; styles short, reflexed.

About 150 spp. in Europe, temperate Asia, Africa and North America.

1 Perennial.   2
  Annual.   3
2 Herb; lvs parallel-veined.   **5. falcatum**
  Shrub; lvs with a distinct midrib and network of lateral veins.   **6. fruticosum**
3 Upper cauline lvs perfoliate; bracts absent.   4
  Upper cauline lvs not perfoliate; bracts present.   5
4 Fr. papillose.   **2. subovatum**
  Fr. not papillose.   **1. rotundifolium**
5 Lvs with slender, but distinct cross-veins; bracteoles subulate, not concealing the fls.   **4. tenuissimum**
  Lvs without visible cross-veins; bracteoles ovate, concealing the fls.   **3. baldense**

*1. B. rotundifolium L.   Hare's-ear, Thorow-wax.

An erect glaucous annual 15–30 cm. *Stem hollow, often purple-tinged. Lvs* 2–5 cm, *elliptic-ovate to suborbicular,* apiculate, with a cartilaginous margin, the lower attenuate into petiole, *upper perfoliate.* Umbels 1–3 cm *diam.;* rays 4–8, up to 1 cm, unequal; *bracts* 0; *bracteoles ovate,* the *margins green, connate at least at base,* longer than the shortly pedicellate fls and suberect in fr. Fr. 3–3.5 mm, blackish and somewhat pruinose, not granulate, ovoid; ridges slender. Fl. 6–7. $2n = 16$. Th.

Introduced. In cornfields. Formerly fairly widespread in S.E. England, now very scarce or extinct. C. and S. Europe; temperate Asia; N. Africa; usually a weed of arable land; introduced in North America, Australia and New Zealand.

*2. B. subovatum Link ex Sprengel

False Thorow-Wax.

B. lancifolium auct., non Hornem.

Like **1** but the lvs about 3 times as long as wide (about twice as long as wide in **1**); umbels with 2–3 subequal rays; fr. 4.5–5 mm, strongly tuberculate, the tubercles developing soon after flowering and becoming more conspicuous as the fr. ripens. Fl. 6–10. $2n = 16$.

Introduced. A casual in gardens and on disturbed ground. Introduced in bird seed and chicken food. First introduced c. 1870. S. Europe; S.W. and C. Asia; N. Africa.

3. B. baldense Turra   Small Hare's-ear.

B. aristatum sensu Coste; *B. opacum* (Cesati) Lange

A slender erect annual 2–10(–25) cm. *Stem solid,* simple or divaricately branched. *Lvs* 1–1.5(–3) cm, *spathulate,* acute. *Umbels* 5–10 mm *diam.;* rays 1–4, 2–5 mm; *bracts* c. 4, 5–10 mm, lanceolate, acute, exceeding the rays; *bracteoles free, concealing fls, ovate,* pungent, rigid, *margins whitish.* Fls subsessile. Fr. c. 1.5 mm, the ridges slender. Fl. 6–7. $2n = 16$. Th.

Native. On dry banks, rocky slopes and grey dunes near the sea. S. Devon, E. Sussex and Channel Islands. W. and S. Europe east to Italy.

4. B. tenuissimum L.   Slender Hare's-ear.

A slender erect or procumbent branched annual 15–50 cm. *Stem solid, flexuous,* wiry. *Lvs* 1–5(–7) cm, *linear to narrowly oblanceolate,* acute or acuminate, rigid. *Umbels up to 5 mm diam.,* terminal or *axillary, sessile or shortly pedunculate;* rays 1–3, unequal, often very short, partial umbels few-fld; *bracts* 3–5, 2–6 mm, *subulate;* bracteoles similar, exceeding fls but not concealing them. *Fr.* 2 mm, black, *subglobose, granulate;* ridges crenulate. Fl. 7–9. $2n = 16$. Th.

Native. In salt marshes and waste places. From Dorset to the Humber; Bristol Channel; 1 locality in Durham, usually near the coast; inland near Malvern, local and rarer in the north. Europe northwards to Gotland; S.W. Asia; N.W. Africa, inland on saline steppes. Represented in Britain by subsp. **tenuissimum.**

5. B. falcatum L.   Sickle-leaved Hare's-ear.

An erect *perennial,* with a stout stock. *Stem up to* 100 cm, *hollow.* Lvs 3–8 cm, *lower petiolate, narrowly obovate,* often ±falcate, the *upper linear-lanceolate,* $\frac{1}{2}$-*amplexicaul.* Umbels 1–4 cm diam., pedunculate; rays 5–11, 5–30 mm, unequal; bracts 2–5, 2–7 mm, lanceolate-acuminate, unequal; *bracteoles* similar, 4–5, somewhat *shorter than the pedicellate fls. Fr.* 2.5–3 mm, *oblong,* the ridges prominent. Fl. 7–10. $2n = 16$. H.

?Native. Waste places and hedgebanks. Formerly in Surrey and E. Essex. First recorded in 1831 and perhaps introduced. S., C. and E. Europe and temperate Asia.

*6. B. fruticosum L.   Shrubby Hare's-ear.

A ±*evergreen* glabrous *shrub* up to 2.5 m. *Twigs slender, purplish when young.* Lvs 5–9 cm, oblong-obovate to oblanceolate, mucronate. *Umbels* 7–10 cm *diam;* peduncles slender; bracts and bracteoles reflexed. Fr. c. 7 mm. Fl. 7–8. $2n = 14$. N.

Introduced. Established in a few localities, especially in Worcester. S. Europe; N.W. Africa.

27. TRINIA Hoffm.

Glabrous usually dioecious herbs. Lvs pinnately compound. Umbels compound; bracts 0–3; bracteoles several. Fls white. Calyx-teeth minute or 0; petals acute, the point sometimes inflexed. Fr. broadly ovoid, laterally compressed or didymous, commissure narrow; carpels subterete with thick, often prominent ridges; styles short, recurved.

About 12 spp. in S. Europe and temperate Asia.

1. T. glauca (L.) Dumort.   Honewort.

T. vulgaris DC.; *Apinella glauca* (L.) Caruel

An erect *glaucous* perennial 3–20 cm. *Root* stout, *with fibrous remains of petioles at top. Stem* solid, deeply grooved, *branched from base.* Lvs 2–3-pinnate; *segments* 5–15 mm, *linear;* petioles slender. *Male umbels* c. 1 cm diam., *flat-topped;* rays 4–8, 5–10 mm, equal; *female* c. 3 mm diam.; *rays unequal,* up to 3 cm. Bracts 0 or 1, 3-fid; bracteoles 2–3, simple or 2–3-fid. Fr. c. 3 mm,

broadly ovoid; ridges rather prominent, smooth and broad. Fl. 5–6. $2n = 18^*$. Hs.

Native. In dry limestone grassland. S. Devon, N. Somerset and Bristol. C. and S. Europe; S. W. Asia. Represented in England by subsp. **glauca.**

28. APIUM L.

Annual, biennial or perennial glabrous herbs. Lvs pinnate or ternate. Umbels often lf-opposed; bracts few or 0, bracteoles several or 0. Fls white. Calyx-teeth 0 or minute; petals entire, acute, the point sometimes shortly inflexed. Fr. broadly ovoid or elliptical-oblong, laterally compressed; commissure narrow; carpels with 5 equal or rarely somewhat unequal ridges.

About 6 spp. in Europe, temperate Asia, North America and circum-Antarctic zone.

1 Lower lvs with deltate to rhombic segments, the lower stalked; bracteoles 0.   **1. graveolens**
   Lower lvs with sessile or filiform to linear segments; bracteoles 5–7.   *2*
2 Lower lvs with deeply lobed segments, the lobes usually filiform or linear; segments of upper lvs cuneate, often 3-lobed; fr. elliptic-oblong.   **4. inundatum**
   Lvs with serrate, sometimes slightly lobed segments, not linear or filiform; fr. broadly ovoid.   *3*
3 Peduncle usually shorter than rays; bracts 0–2; fr. 2–2.5 mm, ovoid, longer than wide. **2. nodiflorum**
   Peduncle longer than rays; bracts 3–7; fr. c. 1 mm, suborbicular, wider than long.   **3. repens**

1. A. graveolens L.   Wild Celery.

An erect strong-smelling biennial 30–100 cm. Stem grooved. Lower lvs simply pinnate; *segments* 1–5 cm, *deltate to rhombic,* lobed and serrate, *lower stalked,* the upper sessile; upper stem-lvs ternate, narrowly rhombic to lanceolate, subentire. Umbels terminal and axillary, shortly pedunculate or sessile in the axil of a small ternate lf; rays unequal; *bracteoles* 0. Fls greenish-white. Fr. c. 1.5 mm, broadly ovoid. Fl. 6–8. $2n = 22$. Hs.

Native. In damp places, by rivers and ditches, especially near the sea. Mainly in the maritime counties, rather local, rarer in the north but reaching S. Scotland; Ireland, around the whole coast, local. Europe to c. 56° N; S.W. Asia; Macaronesia; N. Africa; introduced in other temperate countries.

The cultivated celery (*var. **dulce** (Miller) DC.) is sometimes found by roads; celeriac or turnip-rooted celery is another variety (var. *rapaceum* (Miller) DC.) which is often cultivated.

2. A. nodiflorum (L.) Lag.   Fool's Watercress.

Helosciadium nodiflorum (L.) Koch

A procumbent or ascending perennial 30–100 cm. Stems finely furrowed, slender or stout, the flowering ones rooting at base only. Lvs simply pinnate, bright green, shiny; *segments* 1–6(–10) cm, 2–4(–6) pairs, *lanceolate to ovate,* serrate or crenate, often slightly lobed, *sessile.* Umbels lf-opposed, sessile or shortly pedunculate; rays 3–15, usually subequal, spreading or recurved; bracts usually 0; *bracteoles* 4–7, narrowly lanceolate as long as the shortly pedicellate fls. Fls white. *Fr.* 2–2.5 mm, *ovoid, longer than broad*; ridges all equal. Fl. 7–8. $2n = 22^*$. Hel.

Sometimes mistaken for watercress, with which it often grows and is eaten, apparently without ill effects.

Native. In ditches and shallow ponds. Common and generally distributed but rare in Scotland and mainly in the south. C. and S. Europe; W. and C. Asia; N. Africa; introduced in North America and Chile.

3. A. repens (Jacq.) Lag.   Creeping Marshwort.

Like **2** but stem creeping and rooting at the nodes throughout its length; lf-lobes 5–14 mm, 5–11, ovate to suborbicular; peduncle usually 2–3 times as long as the rays; rays 3–6; bracts 3–7; fr. c. 1 mm, suborbicular, wider than long. Fl. 7. $2n = 16^*$. Hel.

Native. In damp meadows with old grassland, in ditches and shallow ponds, very local. Known from 4 localities in Oxford. Often confused with small variants of **2** and distribution therefore uncertain but apparently mainly in C. Europe.

A. nodiflorum × repens sometimes occurs where the parents grow together and may persist in the absence of *A. repens*. It is sterile.

4. A. inundatum (L.) Reichenb. fil.   Lesser Marshwort.

Helosciadium inundatum (L.) Koch

A straggling often submerged or floating perennial 10–75 cm. Stem slender, nearly smooth. Lvs pinnate; *segments of lower lvs deeply pinnatifid, the lobes filiform in submerged lvs*, linear in floating or aerial lvs; *segments of upper lvs* 0.5–1 cm, *cuneate, often 3-lobed*, sessile. Umbels lf-opposed, peduncles 1–3.5 cm; rays 2–3(–4), 0.5–1 cm, somewhat unequal; bracts 0; bracteoles 3–6, lanceolate, obtuse, unequal. Fls white, subsessile. *Fr.* 2.5–3 mm, *elliptical-oblong;* ridges inflated. Fl. 6–8. $2n = 22$. Hyd.

Native. In lakes, ponds and ditches. Local but widely distributed throughout the British Is. W. Europe, extending eastwards to Sicily and S. E. Sweden.

A. × moorei (Syme) Druce resembles *A. inundatum* but is generally larger and has the segments of the lower lvs linear or strap-shaped; it is a sterile hybrid between that sp. and *A. nodiflorum*. Frequent in Ireland, rare elsewhere.

Trachyspermum ammi (L.) Sprague, an annual with finely divided lvs, white fls and densely grey-papillose ovary and fr. is found occasionally on rubbish-tips. The aromatic fr. is used for flavouring.

29. PETROSELINUM Hill

Annual or biennial herbs, root fusiform. Lvs 1–3-pinnate, the segments broad. Umbels compound; bracts and bracteoles several. Fls white or yellowish. Calyx-teeth 0; petals scarcely notched, with a small inflexed point. Fr. ovoid, laterally compressed; carpels with 5 slender ridges; vittae solitary in the furrows.

About 2 spp. in Europe, W. Asia and N. Africa.

Lvs 3-pinnate; fls yellowish.   **1. crispum**

Lvs 1-pinnate; fls white.   **2. segetum**

*1. P. crispum (Miller) A. W. Hill   Garden Parsley.

P. sativum Hoffm.; *Carum petroselinum* (L.) Bentham

A stout erect glabrous biennial 30–75 cm. Tap-root stout, fusiform. Stem terete, solid, striate, *branches ascending, strict. Lvs deltate*, 3-pinnate, shiny; *segments* 1–3 cm, 4–12 pairs *cuneate*, lobed, often much crisped in cultivated forms, upper stem-lvs often ternate. *Umbels* 2–5 cm diam., *flat-topped*; rays 8–20, 0.3–5 cm; bracts 1–3, erect, entire or 2–3-lobed, with a broad sheath-like lower part with hyaline margins; bracteoles 5–8, linear-oblong to ovate-cuspidate, often with hyaline margins. *Fls yellowish.* Fr. 2–2.5 mm, ovoid. Fl. 6–8. $2n = 22^*$. Hs.

Introduced. In grassy waste places, on walls and rocks, escaped from cultivation and naturalized. Scattered throughout the British Is. north to Fife. Perhaps native in S.E. Europe or W. Asia. Escaped from cultivation and ±naturalized in almost all temperate regions.

2. P. segetum (L.) Koch   Corn Parsley.

Carum segetum (L.) Bentham ex Hooker fil

A slender glabrous dark green ±glaucous biennial 30–100 cm. Tap-root rather slender, fusiform. *Stem* terete, striate, solid, *divaricately branched. Lvs linear-oblong, simply pinnate; segments* 0.5–3.5 cm, *ovate*, subsessile, serrate or sometimes lobed, margins thickened, teeth with forward-curving cartilaginous points. Umbels 1–5 cm diam.; rays 3–10, 0.2–4 cm very unequal; bracts and bracteoles 2–5, subulate; partial umbels of few (often 3–5) fls. *Fls white,* subsessile to long- (15 mm) pedicellate. Fr. 2.5–3 mm, ovoid. Fl. 8–9. $2n = 18$. Hs. Stem and fr. have a smell reminiscent of parsley.

Native. In hedgerows and grassy places. S. and E. England, north to the Humber; a few coastal localities in Wales. W. and S. Europe, from the Netherlands to Portugal and eastwards to C. Italy.

30. SISON L.

Biennial herbs. Lvs pinnate segments broad. Umbels compound; bracts and bracteoles few. Fls white. Calyx-teeth 0; petals suborbicular-cordate, notched, with an inflexed point. Fr. broadly ovoid or subglobose, laterally compressed; carpels with 5 slender ridges; vittae very short, in the upper half of the carpel only.

Two spp. in Europe and the Mediterranean region.

1. S. amomum L.   Stone Parsley.

An erect glabrous biennial 50–100 cm, *with a nauseous smell somewhat resembling that of nutmeg mixed with petrol.* Stems solid, finely striate. Lvs simply pinnate, lower 10–20 cm, long-petiolate, segments 2–7 cm, 2–5 pairs sessile or subsessile, oblong-ovate, serrate and often lobed, margins thickened, teeth with forward-curving cartilaginous points; upper stem-lvs usually ternate with spathulate or linear, lobed or toothed segments. Umbels 1–4 cm diam., terminal and axillary, peduncles slender; rays 3–6, 0.5–3 cm, slender; bracts and bracteoles 2–4, subulate, the bracts usually spreading or reflexed. Fls white or greenish-white. Fr. 3 mm, subglobose; vittae c. $\frac{1}{2}$ as long as fr., conspicuous. Fl. 7–9. Hs.

Resembles *Petroselinum segetum* but differs in the smell, the bright green colour, the larger lvs, the finely divided upper stem-lvs, the larger less irregular umbels with more fls in the partial umbels, and the broader fr.

Native. On hedgebanks and roadsides. S. and E. England north to the Humber; a few coastal localities in Wales. S. and S.W. Europe; N. Africa.

31. CICUTA L.

Perennial glabrous herbs. Lvs 2–3-pinnate. Umbels compound, rays many; bracts few or 0, bracteoles many, narrow. Fls white. Calyx-teeth conspicuous, acute; petals with inflexed points. Fr. subglobose or broadly ovoid; carpels with 5 equal broad flattened ridges; vittae solitary in the furrows. Pollinated by insects.

About 10 spp. in north temperate regions. Highly poisonous.

1. C. virosa L.   Cowbane.

A stout erect perennial 30–150 cm. *Stock* ovoid or shortly cylindrical, *septate*. Stems somewhat ridged, hollow. Lvs up to 30 cm, deltate, 2–3-pinnate; segments 3–9 cm, linear-lanceolate, acutely serrate, unequal at base; petiole long, stout, hollow. Umbels terminal, and lateral and lf-opposed, 7–13 cm diam., flat-topped; rays 10–30, 1–5 cm; partial umbels many-fld, dense; bracts 0; bracteoles many, strap-shaped, longer than the pedicels. Fls white; calyx-teeth ovate. Fr. c. 2 mm, broader than long. Fl. 7–8, $2n = 22$. Hel. or Hyd.

Native. In shallow water, ditches and marshes. Scattered throughout Great Britain, local and mainly in East Anglia, Shropshire, Cheshire and S.E., Scotland; northeast and north-central Ireland. N.C. Europe, temperate Asia.

32. AMMI L.

Branched annual herbs. Lvs pinnate or pinnatifid. Umbels compound; bracts several, long, pinnatifid; bracteoles numerous. Fls white. Calyx-teeth 0; petals obovate, irregularly and unequally 2-lobed with an inflexed notched point. Fr. oblong-ovoid, laterally compressed; carpels with slender ridges; vittae solitary in the furrows.

About 10 spp. from S. Europe to Macaronesia.

*1. A. majus L.   Bullwort.

A glaucous glabrous annual 15–100 cm. Lvs 1–2-pinnate or pinnatifid, the segments 1–3 cm; basal lvs 1–2-pinnate; segments oblong-ovate or spathulate, sometimes pinnatifid; stem-lvs 2-pinnate or pinnatifid, the segments linear-lanceolate; segments of all lvs serrate, the teeth ending in fine cartilaginous points. Umbels terminal, 3–6 cm diam.; rays 9–40, 1–7 cm; bracts several, $\frac{1}{3}-\frac{2}{4}$ length of rays, mostly pinnatifid, with linear segments; bracteoles subulate, about equalling the pedicels. Fr.

1.5–2 mm, oblong or ovoid; ridges prominent, pale. Fl. 6–10. $2n = 22$. Th.

Introduced. A casual in waste places, rather rare. S. Europe, N. Africa, Macaronesia, S.W. Asia; introduced in most temperate regions.

33. FALCARIA Bernh.

Biennial or perennial usually glabrous herbs. Lvs ternate. Umbels compound; bracts and bracteoles 4–15, usually large. Fls white. Calyx-teeth small but distinct; petals nearly equal, obcordate with an inflexed point. Fr. oblong; carpels with 5 low, blunt ridges; vittae slender, solitary in the furrows.

Species 2–3 in C. and S. Europe and W. Asia.

*1. F. vulgaris Bernh.   Longleaf.

A glaucous much-branched perennial 30–50 cm. *Lvs ternate or 2-ternate; segments linear-lanceolate to linear, somewhat falcate, strongly sharply and regularly serrate, up to* 30 cm *in the lower lvs. Bracts and bracteoles linear.* Fl. 7–9. $2n = 22$. H.

Introduced. Naturalized in East Anglia, S. E. England and Guernsey. Europe from N. France and C. Russia southwards; W. Asia.

34. CARUM L.

Perennial or biennial herbs. Root fibrous or fusiform. Lvs pinnate; segments usually narrow. Umbels compound; bracts and bracteoles several, few or 0. Fls white or pinkish. Calyx-teeth minute; petals deeply notched with an inflexed point. Fr. ellipsoid or oblong, laterally compressed; carpels with 5 slender ridges.

About 30 spp. in Europe, N. Africa and temperate Asia.

Lvs 1-pinnate; segments palmately multifid, the lobes filiform; bracts and bracteoles several, reflexed.
                                              1. verticillatum
Lvs 2-pinnate; segments pinnatifid, the lobes linear-lanceolate to linear; bracts and bracteoles 0 or 1, not reflexed.   **2. carvi**

1. C. verticillatum (L.) Koch   Whorled Caraway.

An erect glabrous perennial 30–60 cm. *Root of fusiform fibres thickened downwards. Stem little branched,* striate, *solid, nearly lfless,* surrounded at base by fibrous remains of petioles. *Lvs simply pinnate,* linear-oblong; *segments usually more than 20 pairs, palmately multifid; lobes* up to 5 mm, *filiform, appearing as if whorled.* Umbels 2–5 cm diam., flat-topped; rays 8–14, 1.5–4 cm; *bracts and bracteoles several, lanceolate,* acuminate or aristate, short, *reflexed.* Fls white or pinkish. *Fr. c.* 2.5 mm, *ellipsoid, the ridges prominent, acute.* Fl. 7–8. $2n = 20$. Hr.

Native. In damp grassy places, calcifuge, very local; west of England, Wales and Scotland; N. and W. Ireland; 1 locality in Surrey and 1 in Aberdeenshire. W. Europe, northwards to Scotland and the Netherlands.

2. C. carvi L.   Caraway.

An erect *much-branched* glabrous biennial 25–60 cm. *Tap-root fusiform. Stem* striate, *hollow, lfy. Lvs* all basal in the first year, 2–3-*pinnate*, narrowly triangular to linear-oblong; *segments deeply pinnatifid, the lobes linear-lanceolate to linear*. Umbels 2–4 cm diam., irregular; rays 5–16, 0.5–6 cm; *bracts and bracteoles* 0 *or few, setaceous*, or the former rarely lf-like. Fls white or pink. *Fr.* 3–4 mm, *ellipsoid*, strong smelling when crushed; *ridges low, obtuse*. Fl. 6–7. $2n = 20$. Hs.

Seeds used for flavouring.

Perhaps native in some south-eastern counties and naturalized in waste places. Scattered throughout the British Is., rather rare. Most of Europe; temperate Asia; N.W. Africa; introduced elsewhere.

35. Selinum L.

Perennial herbs. Lvs 2–3-pinnate. Umbels compound; bracts few or 0; bracteoles several. Fls white. Calyx-teeth 0; petals notched, the point inflexed. Fr. oblong to broadly ovoid, dorsally compressed, the commissure broad; carpels ½-terete, the ridges winged, the lateral broadly so; vittae solitary in each dorsal furrow.

About 4 spp. in Europe and Asia.

1. S. carvifolia (L.) L.   Cambridge Milk-parsley

An erect glabrous perennial 30–100 cm. *Stem* solid, ridged, *ridges acutely angled or almost winged*. Lvs 2–3-pinnate *segments* 3–10 mm, entire or pinnatifid, *minutely serrulate*, lanceolate to ovate, *mucronate or aristate*. Umbels 3–7 cm diam., terminal, long-pedunculate; *rays* 15–25, *papillose on the angles*; bracteoles c. 10 subulate, equalling or exceeding pedicels. Fr. 3–4 mm, ovoid-oblong. Fl. 7–10. $2n = 22$. H.

Native. In fens and damp meadows. Cambridge; formerly in N. Lincoln and Nottingham. Much of Europe eastwards to C. Asia, but absent from most of the Mediterranean region; introduced in North America.

36. Ligusticum L.

Glabrous perennial herbs. Lvs 1–2-ternate. Umbels compound; bracts 1–5; bracteoles usually c. 7. Fls mostly hermaphrodite, usually white. Calyx-teeth small or 0, shorter than the entire, inflexed petals. Fr. oblong or ovoid, subterete or dorsally compressed; commissure broad; carpels with prominent, acute or winged primary ridges; vittae many, slender or obscure.

About 30 spp. in north temperate regions.

1. L. scoticum L.   Scots Lovage.

A glabrous, shiny, bright green perennial 15–90 cm. Stock stout. Stem terete, ribbed, little-branched, often magenta towards the base. Lvs 5–10 cm, 2-ternate; *segments* 2–5 cm, ovate-cuneate, sometimes lobed, *serrate in the distal half*. Umbels 4–6 cm diam., *rays* 8–14, 1.5–4 cm; bracts 1–5, linear, entire; bracteoles usually c. 7, linear to linear-lanceolate, entire, shorter than the pedicels. Fls greenish-white, sometimes tinged with pink. Calyx-teeth small; styles very short and stout. Fr. 4–7 mm, oblong, subterete; ridges acute. Fl. 6–7. $2n = 22$. Hs. Lvs sometimes eaten as pot-herb.

Native. On rocky coasts. Scotland and the northern half of Ireland; perhaps extinct in Northumberland. Coasts of N.W. Europe from Denmark to c. 71° N; Greenland, eastern North America.

37. Angelica L.

Tall perennial herbs. Lvs 2–3-pinnate; segments broad and large. Umbels compound, many-rayed; bracts few or 0; bracteoles usually many. Fls greenish, white or pinkish. Calyx-teeth minute or 0; petals lanceolate, incurved. Fr. ovate, dorsally compressed, the commissure broad; carpels flat, with 2 broad marginal wings and 3 dorsal ridges. Fls protandrous, insect-pollinated.

About 60 spp. in north temperate regions and New Zealand.

1. A. sylvestris L.   Wild Angelica.

A stout nearly glabrous perennial (7–)30–200 cm or more. *Stem* hollow, *usually purplish and pruinose*, striate, pubescent towards base. Lvs 30–60 cm, 2–3-pinnate, deltate, the lower primary divisions of basal lvs long-stalked; segments 1.5–8 cm, obliquely ovate to lanceolate, acutely serrate, ±hispidulous on both surfaces, at least on the veins; petioles laterally compressed, deeply channelled on upper side, dilated and sheathing at base; *upper lvs reduced to inflated sheathing petioles which* ±*enclose the fl. buds*. Umbels terminal and axillary, 3–15 cm diam.; peduncles puberulent; *rays* 15–40, 2–8 cm, *puberulent*; bracts 0 or few, caducous; bracteoles usually 6–10, setaceous, as long as the pedicels, persistent. Fls white or pink, petals suberect, incurved; calyx-teeth 0. Fr. 4–5 mm, the wings scarious. Fl. 7–9. $2n = 22$. Hs.

Native. In fens, damp meadows and woods. Generally distributed and common throughout the British Is. Almost throughout Europe, rare in the south; temperate Asia; introduced in North America.

*****A. archangelica** L. (*Archangelica officinalis* Hoffm.), Garden Angelica, is somewhat similar but has the *stems usually green*, the lf-segments somewhat decurrent, the *fls greenish-white or green*, the peduncles glabrous, and the wings of the fr. corky. $2n = 22$.

Formerly cultivated and now ±naturalized on river banks and in waste places, locally abundant. N. and E. Europe, eastwards to C. Asia; Greenland; naturalized in many other parts of Europe.

*****Levisiitcum officinale** Koch, Lovage, a large herb differing from *Angelica* in its yellowish fls and in having all the ribs of the fr. narrowly winged, is cultivated for flavouring and is occasionally naturalized.

38. Peucedanum L.

Perennial, rarely biennial or annual herbs. Lvs pinnate or ternate. Umbels compound; bracts many, few or 0, bracteoles many. Fls white, yellow or pinkish. Calyx-teeth small or 0; petals with a long inflexed point. Fr. globose, ovoid or oblong, dorsally compressed, the commissure very broad; carpels with marginal ridges forming a narrow or broad wing, dorsal ones filiform, all equidistant; vittae 1–3 in each furrow, as long as the fr.

About 100 spp. in Europe, Asia, and Africa.

1 Lf-segments linear; fls yellow. **1. officinale**
   Lf-segments broader; fls white or pinkish. 2
2 Bracts 4 or more. **2. palustre**
   Bracts 0, rarely 1–2. **3. ostruthium**

1. P. officinale L.   Hog's Fennel, Sulphur-weed.

An erect glabrous perennial 60–200 cm. Root very stout and woody. *Stems solid*, striate, with persistent fibrous remains of petioles at base. *Lvs 3–6-ternate, the segments 4–15 cm, sessile, linear, attenuate at both ends*, quite entire, spreading in all directions. Umbels 5–15 cm diam.; rays usually 15–45; *bracts 0–3, linear,* usually caducous; bracteoles several, linear, shorter or longer than the *filiform pedicels. Fls yellow.* Fr. c. 7 mm, elliptical to ovoid. Fl. 7–9. $2n = 66$. H.

Native. On banks near the sea, rare and local. E. Kent and Essex. C. and S. Europe.

2. P. palustre (L.) Moench Hog's Fennel, Milk Parsley.

A glabrous biennial 50–150 cm. Stems hollow, strongly ridged. *Lvs 2–4-pinnate; segments pinnatifid, the lobes 0.5–2 cm, oblong-lanceolate, subacute or obtuse and mucronulate, exceedingly finely serrulate.* Umbels 3–8 cm diam.; *rays 15–40, minutely papillose-pubescent on the inner side; bracts 4 or more, lanceolate to linear-lanceolate, deflexed; bracteoles lanceolate-acuminate,* equalling or shorter than the *minutely papillose pedicels.* Fls greenish-white. *Fr. 4–5 mm, elliptical, with a fine white webbing over the commissural vittae.* Fl. 7–9. $2n = 22$. Hs.

All parts of the plant have a watery-milky juice when young.

Native. In fens and marshes. Mainly in East Anglia; a few scattered localities elsewhere in the southern half of England. Most of Europe extending eastwards to C. Asia.

*3. P. ostruthium (L.) Koch   Master-wort.

An erect ±downy perennial 30–100 cm. Stem hollow, ridged. *Lvs 1–2-ternate,* the lower up to 30 × 34 cm; *segments 4–10 cm, few, broad, lobed, serrate, the margin and veins often ciliolate beneath.* Umbels 5–10 cm diam.; rays 30–60; *bracts* 0(–2); bracteoles few, linear. Fls white or pinkish. *Fr.* 3.5–5 mm, *suborbicular.* Fl. 7–8. $2n = 22$.

Introduced. Formerly cultivated and now naturalized in moist meadows and on river banks mainly in N. England and Scotland; local. Mountains of C. and S. Europe; introduced in many temperate countries.

39. PASTINACA L.

Herbs with pinnate lvs. Umbels compound; bracts and bracteoles 0 or 1–2, caducous. Fls yellow. Calyx-teeth 0; petals with a truncate involute point. Fr. broadly ovoid or orbicular, strongly dorsally compressed; carpels with a broad flattened rather narrowly winged margin; dorsal ridges filiform, distant from lateral;

vittae 1 in each furrow, as long as the fr.

About 15 spp. in Europe and temperate Asia.

1. P. sativa L.   Wild Parsnip.

Peucedanum sativum (L.) Bentham ex Hooker fil.

An erect ±pubescent strong-smelling biennial 30–180 cm. Stems hollow or solid, furrowed and ±angled. *Lvs simply pinnate,* segments usually 2–5(–10) cm, (2–)5–11 pairs, ovate, lobed and serrate. Umbels 3–10 cm diam.; rays 4–11(–17). *Fls yellow.* Fr. 5–7 mm; vittae conspicuous, tapering at the ends. Fl. 7–8. $2n = 22$. Hs.

Native. Roadsides and grassy waste places. Scattered throughout England and locally abundant, particularly on chalk and limestone in the south and east; in Wales, Scotland and Ireland only as an escape from cultivation. Most of Europe; W. Asia; introduced in North and South America, Australia and New Zealand.

The following two subspp. occur in Britain:
(a) Subsp. **sativa**: Hairs on stem, petiole, upper surface of lvs and on the rays rather sparse, short and straight. Lf-lobes often narrow and acute or acuminate, often cuneate at base and ±pinnatisect in the lower part. Probably usually, if not always, an escape from cultivation.
(b) Subsp. **sylvestris** (Miller) Rouy & Camus: Hairs on stem, petiole, upper surface of lvs and on the rays usually rather dense, long and flexuous. Lf-lobes usually broad, obtuse, subcordate at base, crenate-dentate or shallowly lobed in the lower part. Commoner than subsp. *sativa* and usually with a preference for base-rich soils.

40. HERACLEUM L.

Annual or perennial herbs, sometimes of great size. Lvs 1–3-pinnate; segments broad. Umbels compound; bracts 0 or caducous; bracteoles usually present. Fls white or pinkish. Calyx-teeth small, unequal; petals often very unequal, at least the larger notched, with an inflexed point. Fr. suborbicular, obovate or oblong, strongly dorsally compressed; commissure very broad; carpels nearly flat; marginal ridges forming a broad wing, the dorsal slender; vittae solitary in the furrows, conspicuous and usually swollen at their lower ends, shorter than the fr.

About 70 spp. in north temperate regions and mountains in the tropics.

Stems seldom more than 2 m; vittae not more than 0.4 mm wide, not or only slightly expanded at the lower end.
   1. sphondylium
Stems up to 5 m; vittae up to 1 mm wide, conspicuously expanded at the lower end.   **2. mantegazzianum**

1. H. sphondylium L.   Cow Parsnip, Hogweed, Keck.

A stout erect hispid biennial 50–200 cm. Stems hollow, ridged, hairs deflexed but not closely appressed. *Lvs 15–60 cm, simply pinnate,* very rarely 2-pinnate, hispid on both surfaces; segments 4–10(–18) cm, variously lobed or pinnatifid, serrate, ovate to linear-lanceolate, the lower stalked. Umbels 5–20 cm diam., terminal and axillary; rays usually 7–10, 2–12 cm, stout; bracts few or 0, subulate; *bracteoles several, linear, reflexed. Fls white, greenish-white or pinkish,* the petals deeply notched, those of the outer fls very unequal. Fr. (6–)7–8(–10) mm, suborbicular, whitish; styles short, erect, becoming reflexed in ripe fr. Fl. 6–9. $2n = 22^*$. Hs.

Lvs edible, much esteemed by herbivorous animals.

Native. In grassy places, roadsides, by hedges and in woods. Common and generally distributed throughout the British Is., ascending to 1000 m in Argyll. Europe southwards from c. 61° N; W. and N. Asia; western N. Africa; introduced in North America.

A very variable sp. with at least 9 subspp. in Europe alone. Two of these occur in Britain:

(a) Subsp. **sphondylium**: Petals white or rarely pinkish, the outer much larger than the inner. Throughout the British Is. and N.W. Europe.

(b) Subsp. **sibiricum**: Petals greenish-white, the outer not or scarcely larger than the inner. E. Norfolk and perhaps elsewhere in East Anglia. Mainly in N.E. and E.C. Europe, but also in C. and S.W. France.

*2. H. mantegazzianum Sommier & Levier
                                    Giant Hogweed.

Biennial or monocarpic perennial, sometimes perennating by buds in the axils of the basal lvs. Stems up to 550 × 10 cm, hollow, usually purple-spotted and ±hairy. Lvs up to c. 250 cm, ternately or pinnately divided to a varying extent, usually puberulent beneath, coarsely and irregularly serrate. Umbels up to c. 50 cm diam. Rays 50–150, 15–30 cm, somewhat unequal, hairy. Petals up to 12 mm, white or rarely pinkish, the outer largest. Calyx-teeth prominent, acute. Fr. (7–)9–11 mm, elliptical, glabrous to villous; vittae up to 1 mm wide, conspicuously expanded at the lower end. Fl. 6–7. $2n = 22$.

Naturalized in waste places, especially near rivers, in many scattered localities. Native of S.W. Asia.

The plant has a strong resinous smell and can cause dermatitis when handled in bright sunlight. Hybrids between this and *H. sphondylium* sometimes occur.

41. TORDYLIUM L.

Pubescent annual herbs. Lvs simple or pinnate. Umbels compound; bracts and bracteoles 4–7, linear. Fls white or pinkish. Calyx-teeth c. ½ as long as petals; petals with an incurved point, unequal. Fr. orbicular or elliptical, strongly dorsally compressed; carpels with the lateral ridges much thickened, the dorsal ones slender; vittae 1–3 in each furrow.

About 16–20spp. in Europe, N. Africa and S.W. Asia.

1. T. maximum L.   Hartwort.

An erect branched annual or biennial 20–125 cm. Stem hollow at base, ridged, rough with stiff deflexed, closely appressed hairs. Lowest lvs simply pinnate; *segments of lower lvs broadly ovate or almost orbicular,* lobed and crenate; of upper lanceolate, obtusely serrate,

appressed-hispid on both surfaces. Umbels flat, long-pedunculate; rays (3–)5–15, hispid with short stiff forward-pointing hairs; bracts and bracteoles 4–7, setose, the latter exceeding the subsessile fls. Fls white or pinkish; *calyx-teeth conspicuous,* about ½ as long as the petals. *Fr.* 5–6mm, broadly oblong, hispid, *margins much thickened, whitish,* glabrous; vittae 1 in each furrow. Fl. 6–7. $2n = 22$. Th.

?Introduced. Naturalized or perhaps native in a few places. Recorded from hedge- and river-banks mainly along the Thames estuary. S.C. and S. Europe; N. Africa.

42. TORILIS Adanson

Annual herbs. Lvs 1–3-pinnate. Bracts several or 0; bracteoles several, subulate. Fls white or pinkish. Calyx-teeth small, triangular-lanceolate, persistent; petals with an inflexed point. Fr. ovoid, narrowed at the commissure; carpels with slender ciliate ridges, furrows thickly beset with tubercles or spines; vittae solitary in the furrows.

Eight spp. in Europe, temperate Asia and N. Africa.

1 Umbels sessile or shortly pedunculate.	**3. nodosa**	
Umbels with peduncles usually at least 5 cm.		2
2 Bracts 0–1.	**2. arvensis**	
Bracts 4–6(–12).	**1. japonica**	

1. T. japonica (Houtt.) DC.   Upright Hedge-parsley.

T. anthriscus (L.) C. C. Gmelin; *Caucalis anthriscus* (L.) Hudson

An erect shortly appressed-hispid annual 5–125 cm. Stems solid, striate, hairs closely appressed, deflexed. Lvs 1–3-pinnate; *segments* 1–3 cm, *ovate* to lanceolate, pinnatifid. serrate. Umbels 1.5–4 cm diam.; *rays* (4–)6–12, hairs forward-pointing; *bracts* 4–6(–12), unequal, up to nearly as long as rays; bracteoles about equalling pedicels. Fls pinkish- or purplish-white, outer petals larger than inner. *Fr.* 3–4 mm, *spines curved, forward-pointing; styles glabrous, recurved in fr.* Fl. 7–8. Germ. spring and autumn. $2n = 16$. Th.

In many places the commonest roadside umbellifer flowering in July, just after *Chaerophyllum temulentum*.

Native. In hedges and grassy places. Generally distributed throughout most of the British Is. Throughout Europe, except the north and a few Mediterranean islands; temperate Asia; N. Africa; Japan; introduced in North America.

2. T. arvensis (Hudson) Link
                              Spreading Hedge-parsley.

T. infesta (L.) Sprengel

An erect annual 10–40(–100) cm. Stems solid, terete below, ridged and ±angled above, sparsely deflexed-appressed-hispid or almost glabrous. Lvs 1–2-pinnate; *segments* 0.5–3 cm, *lanceolate,* pinnatifid or coarsely serrate, sparsely appressed-hispid on both surfaces. Umbels 1–2.5 cm diam.; rays (2–)3–5(–12), hairs forward-pointing; *bracts 0 or 1;* bracteoles several, densely hispid, nearly equalling the few-fld partial umbels.

Fls white or pinkish, petals unequal. *Fr.* 3–6 mm, *spines curved, thickened near tip and with a slender hooked apex; styles ±hairy, spreading in fr.* Fl. 7–9. Germ. autumn and spring. Th.

Native or introduced. In arable fields, mainly in S.E. England, probably often introduced with seed and now less frequent than formerly. W., C. and S. Europe; S.W. Asia.

3. T. nodosa (L.) Gaertner   Knotted Hedge-parsely.

An erect or more often prostrate annual. Stems up to 50 cm, solid, striate, hairs sparse, appressed, deflexed. Lvs 1–2-pinnate; *segments ovate, deeply pinnatifid, the lobes linear-lanceolate. Umbels* 0.5–1 cm, *subsessile, lf-opposed; rays* 2–3, *very short*; bracts 0; *bracteoles exceeding the subsessile fls.* Fls pinkish, petals subequal. *Fr.* 2.5–3.5 mm, *the outer carpels with straight, rarely hooked, spreading spines, the inner tubercled.* Fl. 5–7. Germ. autumn and spring. $2n = 22$. Th.

Native. On dry rather bare banks and arable fields, local; England, Wales, S. Scotland and the southern half of Ireland. S. and W. Europe; W. Asia; N. Africa.

43. DAUCUS L.

Annual or biennial hispid herbs. Lvs 2–3-pinnate. Umbels compound; bracts pinnatisect or 3-sect bracteoles numerous; entire or 3-fid. Fls white. Calyx-teeth small or 0; petals notched, often unequal. Fr. ovoid or oblong; carpels convex primary ridges 5, filiform, the secondary 4 stouter and more prominent than primary; all or the secondary only, with rows of spines; vittae solitary under each secondary ridge.

About 20 spp. in temperate regions of the northern and southern hemispheres.

1. D. carota L.   Wild Carrot.

An erect ±hispid biennial 30–100 cm. Stem solid, striate or ridged. Lvs(1–)2–3-pinnate; segments pinnatifid, the lobes usually 5–30 mm, lanceolate to ovate and acuminate to obtuse. Umbels 3–7 cm diam.; rays usually numerous; bracts 7–13, about equalling pedicels, 3-fid or pinnatifid, conspicuous, the margins broadly scarious. Fls white, the central one of the umbel usually red or purple. Fr. 2–4 mm, oblong-ovoid, the primary ridges ciliate, the secondary spiny. Fl. 6–8. $2n = 18^*$. Hs. (biennial).

Subsp. **carota**: Root tough, not fleshy. Lvs ovate in outline; umbels strongly concave in fr.

Subsp. **gummifer** Hooker fil. (*D. gingidium* auct.): Lvs narrower in outline; umbels ±flat in fr.

*Subsp. **sativus** (Hoffm.) Arcangeli, with a thick and fleshy tap-root in the first year, is the cultivated carrot.

Native. Subsp. *carota* in fields and grassy places, particularly near the sea and on chalky soils; subsp *gummifer* on cliffs and dunes, local and mainly on the south coast. Throughout most of the British Is. Most of Europe, temperate Asia and N. Africa; naturalized in most other temperate and many tropical countries.

82. CUCURBITACEAE

Herbs, rarely shrubs or small trees, and mostly annual tendril-climbers but also prostrate annuals, climbing perennials and spiny herbs and shrubs. All usually with very sappy often hispid stems. Tendrils spirally coiled, often branched, each arising at the side of a lf-axil. Lvs spirally arranged, palmately-veined and often palmately-lobed, exstipulate. Fls usually unisexual, (the plant monoecious or dioecious), actinomorphic, epigynous, and often also perigynous, solitary or in infls of various kinds. Calyx with 5 narrow sepals; corolla of 5 free or basally united petals; androecium rarely of 5 free stamens each with 2 pollen-sacs, more usually of 2 pairs of ±completely united stamens with 1 free stamen; often the anthers are borne on a common connective-column, and frequently they are variously curved and twisted; ovary usually inferior, 1- or 3-celled, the commonly 3 parietal placentae sometimes meeting in the centre; ovules numerous, rarely few, anatropous; style 1, rarely 3; stigmas 3, commissural. Entomophilous. Fr. usually a berry or pepo. Seed non-endospermic with a straight embryo.

About 700 spp. in 110 genera mainly in tropical and subtropical regions.

Although commonly gamopetalous the Cucurbitaceae appear more closely related to the Begoniaceae and other polypetalous families than to any families of the Metachlamydeae.

A family of some economic importance since it includes the melons (*Cucumis melo*) and the cucumber (*C. sativus*), the water-melon (*Citrullus lanatus*), the various marrows, pumpkins and squashes (mainly *Cucurbite pepo* but also *C. moschata*, *C. maxima*, etc.), and other edible fruits. The hard-shelled pepos of *Lagenaria* and *Cucurbita* yield calabashes and gourds, and the fibrous tissue of the fr.-wall of *Luffa aegyptiaca* is the loofah.

1. BRYONIA L.

Perennial tendril-climbing monoecious or dioecious herbs with palmately-lobed lvs and fls in axillary cymes or clusters. Corolla of 5 petals free or united below into a short tube. Androecium of 2 pairs of stamens united by their filaments and with 2 pollen-sacs, and 1 free stamen with 1 pollen-sac; female fls with 3–5 small staminodes. Stigmas 3, each bifid. Fr. a small smooth globose berry.

Eight spp. in Europe, N. Africa, W. Asia and Canary Is.

1. B. cretica L. subsp. dioica (Jacq.) Tutin
                                   White or Red Bryony.

B. dioica Jacq.

Stock erect, tuberous, massive, branched. Stem very long, branching especially from near the base, brittle, angled, hispid with swollen-based hairs, climbing by simple spirally-coiled tendrils arising from the side of the lf-stalk. Lvs palmately (3–)5-lobed with the lobes sinuate-toothed, cordate at the base, with a curved stalk shorter than the blade. Fls in axillary cymes; those of the male plant stalked, corymbose, of 3–8 pale greenish fls 12–18 mm diam., with triangular spreading sepals, oblong hairy distinctly net-veined petals 2–3 times as long as the sepals, and yellow anthers; those of the female plant ±sessile, umbellate, of 2–5 greenish fls 10–12 mm diam., with sepals and petals as in the male fls but smaller, prominent bifid stigmas and a smooth broadly ellipsoid ovary separated by a short constriction from the perianth. Berry 6–10 mm diam., red when ripe, with 3–6 large compressed seeds yellowish with black mottling or vice versa. Fl. 5–9. visited by various insects including many bees. $2n = 20$. G.

Native. Hedgerows, scrub, copses; avoided by rabbits and common in warrens; locally common, especially on well-drained soils. S., C. and E. England and Wales northwards to N.W. Yorks and Northumberland; introduced locally in N.W. England and S. Scotland. Europe, especially central and south, absent from Scandinavia; W. Asia; N. Africa. The 2 other European subspp. are of very limited distribution.

*Ecballium elaterium** (L.) A. Richard Squirting Cucumber, is naturalized in a few places on the S. coast. It is a hispid, somewhat glaucous, perennial, monoecious herb with prostrate succulent shoots, 20–60 cm, and fleshy triangular-cordate lvs, 7–10 cm, whitish-tomentose beneath, sinuate-toothed to obscurely lobed; tendrils 0. Fls c. 2.5 cm diam., yellow; female fls usually solitary, axillary; male in racemes; stamens with free anthers. Fr. oblong, 3–5 cm, hispid, greenish, detaching itself explosively at maturity and squirting the seeds to some distance from the aperture at the proximal end. A weed throughout the Mediterranean region.

83. ARISTOLOCHIACEAE

Herbs or woody climbers. Lvs alternate, simple, entire or rarely lobed, stalked, exstipulate. Fls solitary or in racemes or axillary clusters, stalked, hermaphrodite, actinomorphic to strongly zygomorphic, usually 3-merous. Perianth usually in 1 whorl, rarely in 2 whorls, petaloid, often brown or lurid purple, united below into a tube, regularly 3- or 6-lobed or with a unilateral entire or lobed limb. Stamens 6-many in 1 or 2 whorls, free or united with the stylar column; anthers extrorse. Ovary inferior, rarely ½-inferior, 4–6-celled, with numerous axile ovules; styles short, thick, 3–many, free or united into a column with a 3–many-lobed stigma. Fr. a capsule, opening variously, rarely indehiscent; embryo very small in a fleshy endosperm.

Seven genera and about 400 spp., tropical and temperate.

Perianth regular, with short tube; fls terminal. 1. ASARUM
Perianth strongly zygomorphic, with long tube; fls axillary.
2. ARISTOLOCHIA

1. ASARUM L.

Herbs with creeping rhizome. Fls solitary, terminal. Perianth regular in 1 whorl, 3-lobed, persistent in fr. Stamens 12 in 2 whorls, free or nearly so; filaments short. Ovary 6- (rarely 4-)celled with numerous (rarely 1) ovules in 2 rows in each cell; styles 6 (rarely 4) free or united into a column. Fr. a subglobose capsule opening irregularly; seeds flat.

About 70 spp., north temperate zone, mainly E. Asia.

1. A. europaeum L.   Asarabacca.

Perennial evergreen herb with thick creeping rhizome. Stems 2–5 cm, pubescent, with usually 2 lvs and 2 brown ovate scales (1–2 cm). Lvs 2.5–10 cm, reniform, broader than long, deeply cordate at base, very obtuse, entire, dark green, pubescent on the veins above; petiole much longer than blade. Fl. solitary, terminal. Perianth brownish, c. 15 mm, pubescent outside; lobes ±deltate, acuminate, about half as long as tube. Fl. 5–8. Pollinated by small flies or probably more often self-pollinated. $2n = 26, 40$. Grh.

?Native. Rare. Perhaps native in woods etc., in a few localities from Dorset to Denbigh and Leicester; S. Northumberland; apparently largely extinct in intervening areas and in Scotland where it was an escape north to Lanark and Aberdeen; formerly much grown as a medicinal plant. From S. Fennoscandia southwards to S. France, C. Italy and Macedonia; W. Siberia.

2. ARISTOLOCHIA L.

Perennial herbs or woody climbers. Fls axillary. Perianth zygomorphic with long tube, the base swollen, the upper part narrower, ±cylindrical, straight or variously curved; limb usually unilateral and entire, rarely absent or 2- or 3-lobed or 3–6-toothed, caducous. Stamens usually 6, in 1 whorl, joined to the stylar column consisting of (5–)6(–12) styles. Ovary (4–)5–6-celled. Fr. a septicidal capsule.

About 350 spp., tropical and temperate. Several are grown in greenhouses for their fantastic fls.

*1. A. clematitis L.   Birthwort.

Glabrous foetid perennial herb 20–80 cm. Rhizome long, creeping. Stems erect, simple, numerous. Lvs 6–15 cm, broadly ovate-cordate, sinus open, obtuse, finely denticulate, with a petiole about half as long. Fls in axillary clusters of 4–8; pedicels short. Perianth 2–3 cm, dull yellow, the limb brownish, glabrous; tube somewhat curved, the basal swelling globose; limb entire, oblong or ovate, about as long as tube. Capsule 2–2.5 cm, ovoid to pyriform. Fl. 6–9. Protogynous, pollinated by small flies which are trapped in the basal swelling till the stamens mature. $2n = 14$. Grh.

Introduced. Long cultivated as a medicinal plant and naturalized in a number of places mainly in E. England from Sussex to N. Lincoln, extending west to Northampton and Oxford; Glamorgan and Anglesey. Probably native in E. and S.E. Europe; formerly cultivated over most of continent northwards to *c.* 55° N; Caucasus; N. Asia Minor.

A. rotunda L.

Stock a subglobose tuber. Lf-sinus closed or nearly so; petiole short or 0. Fls solitary. Perianth yellowish with brown limb, puberulent. Rarely naturalized. Native of the Mediterranean region.

84. EUPHORBIACEAE

Trees, shrubs or, as in all the British spp., herbs. Lvs usually alternate and stipulate, simple or compound. Infl. usually compound. Fls regular or nearly so, unisexual, usually hypogynous. Perianth of 1 whorl or 0, rarely of 2 whorls. Stamens 1-many. Ovary usually 3-celled, with 1 or 2 anatropous, pendent ovules in each cell; raphe ventral. Fr. usually a capsule separating into three parts and leaving a persistent axis; rarely a drupe. Seeds with abundant endosperm and large embryo, usually carunculate.

About 300 genera and 5000 spp., mainly tropical. The British genera are not closely related and give a very inadequate idea of this large and varied family of which the ovary and fr. and unisexual fls are the most constant features. Its relationships are obscure and it is quite likely to be polyphyletic.

Perianth present, 3-merous; male fls in clusters on long
    axillary spikes, the female solitary or 0; without milky
    juice.   1. MERCURIALIS
Perianth 0; fls very small, one female and several male
    fls borne in a 4–5-lobed perianth-like involucre and
    simulating a single fl.; juice milky.   2. EUPHORBIA

1. MERCURIALIS L.

Usually dioecious herbs with watery juice. Lvs opposite; stipules small. Male fls in clusters arranged on long pedunculate axillary spikes; per.-segs 3, sepaloid; stamens 8–15. Female fls axillary, solitary or in small clusters, pedunculate or subsessile; perianth as in the male; 2 or 3 sterile filaments present. Ovary of two 1-seeded cells; styles 2, free. Fr. dehiscing by 2 valves.

Eight spp. in Mediterranean region and temperate Eurasia eastwards to N. Thailand. Wind-pollinated.

Perennial with creeping rhizome; stems simple.
                                         1. perennis
Annual; stems branched.   **2. annua**

1. M. perennis L.   Dog's Mercury.

Perennial with long creeping rhizome, ±pubescent. *Stems erect, simple,* 15–40 cm. Lvs 3–8 cm, elliptical-ovate or elliptical-lanceolate, crenate-serrate, with a short (3–10 mm) petiole. Dioecious, rarely monoecious. Fls 4–5 mm across; *female* 1(–3) *on long peduncles.* Fr. pubescent, 6–8 mm broad. Fl. 2–4. $2n = c. 64$. Hp.

Native. Woods on good soils and shady mountain rocks, ascending to 1035 m on Ben Lawers; frequently dominant in the field layer especially in beechwoods on chalk. Common over most of Great Britain but rare from Ross northwards and absent from Orkney, Shetland and the Isle of Man; rare in Ireland and usually introduced, though probably native in Clare; Jersey.

Most of Europe northwards to *c.* 66° N in Norway; Caucasus; Algeria (very rare); S.W. Asia.

2. M. annua L.   Annual Mercury.

Glabrous or sparsely hairy *annual. Stems* erect, *branched,* 10–50 cm. Lvs 1.5–5 cm, ovate to elliptical-lanceolate, crenate-serrate, with a petiole 2–15 mm. Dioecious, occasionally (var. *ambigua* (L. fil.) Duby) monoecious. *Female fls few, subsessile* in the lf axils. Fr. hispid, 3–4 mm broad. Fl. 7–10. $2n = 16$. Th.

?Native. Waste places and as a garden weed, often only a casual, widespread but local in S. England, rare in Wales and N. England, extending to Cumberland and Northumberland; Berwick, Kintyre; introduced in S.E. Ireland, north to Dublin and Westmeath; Channel Is. Most of Europe but introduced in much of the north and west, doubtfully native in the Azores.

2. EUPHORBIA L.   Spurge.

Monoecious herbs with milky juice. Lvs alternate and exstipulate. Fls very small; perianth 0, a number of male fls and a single female fl. being grouped within a cup-shaped perianth-like involucre (cyathium) with 4–5 small teeth alternating with conspicuous glands. Cyathia in compound cymes, the primary branches forming an umbel subtended by ray-lvs; subsequent branching dichasial with conspicuous opposite bracts, the ultimate pair of which (raylet-lvs) usually differ from the lvs. Male fl. consisting of a single stamen on a jointed pedicel. Female fl. consisting of the 3-celled ovary on a pedicel which elongates in fr. Styles 3, stigmas often bifid. Ovules 1 in each cell. Fr. a 3-valved capsule.

About 2000 spp. in tropical and temperate regions.

All the spp. appear to ripen fr. about a month after fl. The chromosome numbers given by different authors (and given below) differ rather oddly and cannot be considered satisfactory. All our spp. are probably pollinated by flies.

The above description applies to the British spp. The foreign spp. include trees and shrubs and many are cactus-like without or with very reduced caducous lvs. Some of these are cultivated in greenhouses, as is *E. pulcherrima* Willd. (Poinsettia) with conspicuous red bracts. A few spp. not differing greatly from ours are sometimes grown out of doors and some of these and others have been found as escapes and casuals.

1 Lvs opposite.   2
   Lvs alternate.   3
2 Erect biennial; lvs decussate, equal at base; stipules
      absent.   **2. lathyrus**
   Procumbent annual; lvs not decussate, very unequal
      at base; subulate stipules present.   **1. peplis**
3 Involucral glands rounded on outer edge.   4

Involucral glands with concave outer edge, the tips
±prolonged into horns.   10
4 Perennials with ±numerous stems; lvs oblanceolate
to oblong or lanceolate to elliptical, cuneate at
base.   5
Annuals usually with single stems; lvs either obovate
to spathulate and very obtuse or oblong- to obovate-lanceolate and deeply cordate at base.   8
5 Capsule smooth or minutely tuberculate; lvs usually
±hairy on both sides (rarely glabrous above).   6
Capsule strongly tuberculate; lvs glabrous above,
sometimes hairy beneath.   7
6 Stems stout, scaly below; ray-lvs ovate, obtuse, mucronate; raylet-lvs yellowish; capsule glabrous or
sparsely hairy.   **3. villosa**
Stems slender, not scaly below; ray-lvs oblong to
oblanceolate, obtuse to subacute, not mucronate;
raylet-lvs green or reddish; capsule densely pubescent.   **4. corallioides**
7 Stems scaly at base; raylet-lvs truncate at base, green;
involucral glands green, soon becoming purple.
                                           6. dulcis
Stems not scaly at base; raylet-lvs subcordate or
rounded at base, yellowish; involucral glands yellowish, becoming brown.   **5. hyberna**
8 Lvs and ray-lvs acute or subacute; lvs cordate at base;
capsule tuberculate.   9
Lvs and ray-lvs very obtuse; lvs tapered to base; capsule smooth.   **9. helioscopia**
9 Capsule shallowly grooved, almost globose, its tubercles ±hemispherical; lowest raylet-lvs very different from ray-lvs.   **7. platyphyllos**
Capsule deeply grooved to trigonous, its tubercles
cylindrical or conical, as long as or longer than
broad; lowest raylet-lvs grading into ray-lvs.
                                         8. serrulata
10 Raylet-lvs free; plant glabrous.   11
Raylet-lvs connate in pairs; plant hairy.
                                       16. amygdaloides
11 Annuals with single stems; lvs few.   12
Biennials or perennials with ±numerous stems, some
of them usually sterile, tufted or from a creeping
rhizome; lvs numerous.   13
12 Lvs ovate, suborbicular or obovate, stalked, green.
                                          10. peplus
Lvs linear to oblong cuneate, sessile, glaucous.
                                          11. exigua
13 Lvs ovate to oblanceolate, all less than 2 cm, ±coria-
ceous; plant without creeping rhizome; umbel with
6 rays or fewer; maritime plants.   14
Lvs linear-lanceolate or linear, mostly over 2 cm. (or
if all less than 2 cm. then narrowly linear), thin;
plant with creeping rhizome; umbel with 6 or more
rays; inland plants.   15
14 Middle lvs obovate to oblanceolate, slightly coriaceous, the midrib prominent beneath; seeds pitted.
                                       12. portlandica
Middle lvs elliptical-oblong, very coriaceous, the midrib obscure; seeds smooth.   **13. paralias**
15 Lvs of fl. stems 4 mm broad or more.   **14. esula**
Lvs of fl. stems 2 mm broad or less.   **15. cyparissias**

Subgenus 1. CHAMAESYCE Rafin. Usually procumbent
annuals. Lvs all opposite, usually unequal at base, stipulate. Cyathia axillary or clustered, not in umbels.

1. E. peplis L.   Purple Spurge.

Glaucous often purplish annual *with* usually 4 *procumbent branches* 1–6 cm long *from the base. Lvs* 3–10 mm,
±oblong in outline, obtuse or retuse, *with a large
rounded auricle on one side at the base* (except sometimes the basal lvs), entire, shortly petiolate; *stipules*
divided into subulate segments. Branches branching
dichotomously with cyathia in the forks and axils.
Cyathia 1–2 mm, stalked; glands suborbicular, entire.
Capsule 3–5 mm, trigonous, glabrous, smooth. Seeds
c. 3 mm, pale grey, smooth, not caruncled. Fl. 7–9. Th.

Native. Shingle beaches, formerly found from the
Channel Is. and Cornwall to Somerset and Isle of Wight,
and in Kent, Cardigan and Waterford; now probably
extinct. Shores of S. and W. Europe.

Subgenus 2. ESULA Pers. Usually erect herbs. Lvs
usually alternate, equal at base, exstipulate. Cyathia
almost always in umbels.

Section 1. *Lathyrus* Dumort. Biennial. Cauline lvs
decussate.

2. E. lathyrus L.   Caper Spurge.

Glabrous glaucous biennial 30–120 cm, forming a short
erect lfy stem the first year, elongating and flowering
in the second. *Lvs* 4–20 cm, *opposite* or sub-opposite
in 4 rows, narrow-oblong, the lower smaller, obtuse or
retuse, often mucronulate, rounded at base, entire, sessile. Umbel 2–6-rayed; ray-lvs triangular-lanceolate.
Raylet-lvs 1.5–7 cm, triangular-ovate, acute, cordate at
base. Glands lunate, the horns blunt. Fr. 8–20 mm, trigonous, glabrous, smooth. Seeds c. 5 mm, brown, reticulate. Fl. 6–7. $2n = 20$. Not satisfactorily referable to
any standard life-form, appearing in winter as N., but
biennial and herbaceous.

?Native. Perhaps native in woods in a few places in
England from Somerset, Gloucester and Wilts to Huntingdon and Northampton, elsewhere extinct or as a local
garden weed and escape in waste places north to Banff
and Elgin; Wexford. Formerly cultivated for its fr. S.
Europe from Spain, N. Italy and Greece to France and
Germany but probably native only in E. and C. Mediterranean region; Morocco (rare); Azores; status often
doubtful.

Section 2. *Helioscopia* Dumort. Annuals to perennials.
Stem-lvs alternate. Involucral glands transversely ovate,
emarginate or with horns.

3. E. villosa Waldst. & Kit. ex Willd.   Hairy Spurge.
E. pilosa auct., non L.

Perennial 30–100 cm, *with stout rhizome*; stems *stout,
scaly below*, ±*numerous* with axillary sterile branches
above. *Lvs* 4–10 cm, alternate, oblong or oblong-lanceolate, obtuse to subacute, broad-cuneate at base, sessile,
serrulate near apex, entire near base, *softly hairy on
both sides* or nearly glabrous above, dense and with
the uppermost less than its own length below the umbel.

Umbel 4–6-rayed often with axillary infl. branches below, its *ray-lvs* c. 2 cm, *ovate*. Raylet-lvs 8–12 mm, ovate or suborbicular, obtuse, mucronulate, *rounded at base, yellowish*, glabrous or sparsely hairy. *Glands entire. Capsule* 4–8 mm, subglobose, scarcely grooved, *glabrous or sparsely hairy, smooth or minutely tuberculate. Seeds brown,* smooth. Fl. 5–6. $2n = 18$. Hp.

?Native. Formerly in a wood near Bath and neighbouring hedgebanks, where it had been known since 1576, but now extinct. S.E., S. and E.C. Europe, extending northwards to C. Russia and N.W. France; Algeria.

*4. E. corallioides L.   Coral Spurge.

Differs from *E. villosa* as follows; Plant *not or scarcely rhizomatous*; stems *slender, not scaly below*. Uppermost lf usually more than its own length below the umbel. Ray-lvs c. 7–8 cm, *oblong-lanceolate*, similar to the lvs. Raylet-lvs *green or tinged with red*, usually hairier, the upper oblong or ovate, the lower intermediate between the upper and those of the umbel. *Capsule densely hairy*, appearing woolly, minutely tuberculate. Fl. 6–7. Hp.

Introduced. Shady places in two stations in Sussex where it has been known since 1808, and in Oxford. Native of C. and S. Italy and Sicily.

5. E. hyberna L.   Irish Spurge.

Perennial 30–60 cm, with thick rhizome; stems ±numerous erect simple *not scaly at base. Lvs* 5–10 cm, alternate, oblong, elliptical-oblong or oblanceolate-oblong, obtuse or retuse, cuneate at base, sessile or subsessile, *entire, glabrous above*, sparsely hairy beneath especially near midrib, or more rarely glabrous, eventually turning red. Umbel 4–6-rayed, often with axillary infl. branches below; ray-lvs 3–6 cm, elliptical-oblong. Raylet-lvs 8–30 mm, ovate or ovate-elliptical, obtuse to subacute, not mucronulate, *subcordate or rounded at base, yellowish. Glands* 5, *yellowish*, finally brown, *entire*, reniform. *Capsule* 5–6 mm, subglobose, grooved, glabrous, *with prominent cylindrical tubercles.* Seeds pale brown, smooth. Fl. 4–7. Hp.

Native. Woods, hedgebanks and rough pastures on lime-free soils; locally common in S.W. Ireland northwards to Waterford, Tipperary and Limerick, ascending to 550 m; also in a few isolated localities in Galway, Clare and Donegal and in Cornwall and Somerset; W. and C. Europe eastwards to N. Italy.

*6. E. dulcis L.   Sweet Spurge.

Perennial 20–50 cm, more slender than the three preceding spp., with creeping swollen and jointed rhizome; stems ±numerous erect simple, *scaly at base.* Lvs 3–5 cm, alternate, oblanceolate-oblong, obtuse, tapered to base, subsessile, entire or finely serrulate near apex, glabrous or with a few scattered hairs beneath and occasionally above. Umbel usually 5-rayed often with axillary infl. branches below; ray-lvs 2–3 cm, oblong-elliptical. *Upper ray-lvs* 7–20 mm, ovate-deltate, subacute, not mucronulate, denticulate, *truncate at base, green.*

Glands green, *soon purple*, obovate-orbicular, *entire. Capsule* 2–3 mm, subglobose, grooved, glabrous (in Britain), *with prominent cylindrical tubercles.* Seeds brown, smooth. Fl. 5–7. $2n = 12$, 28. Hp.

Introduced. Naturalized in shady places in a few localities in Somerset and Middx to Brecon, Caernarvon and Anglesey, and in Scotland from Lanark and Peebles to Inverness and Ross. W. and C. Europe extending to C. Italy and Macedonia.

7. E. platyphyllos L.   Broad-leaved Spurge.

Annual 15–80 cm, glabrous or pubescent, with single vertical root and simple stems. *Lvs* 1–4.5 cm, alternate, obovate- or oblong-lanceolate, acute, *deeply cordate or auricled at base*, sessile, serrulate except at base. Umbel usually 5-rayed, with axillary infl. branches below, the secondary forks often 3-rayed; ray-lvs 2–3 cm, elliptical-oblong. Raylet-lvs 5–15 mm, deltate, acute or obtuse, mucronate, *the lowest not differing from the upper and markedly different from the ray-lvs. Glands* suborbicular, *entire. Capsule* 2–3 mm, subglobose, *shallowly grooved, with hemispherical tubercles. Seeds* c. 2 mm, *olive-brown, smooth.* Fl. 6–10. $2n = 28$, 36. Th.

Native. In arable land and waste places, widespread but local in S. England from S. Somerset and Kent to Derby and Lincoln; Glamorgan. S., W. and C. Europe; W. Caucasus.

8. E. serrulata Thuill.   Upright Spurge.

E. stricta L.

Differs from *E. platyphyllos* as follows: Usually more slender, always glabrous. Lvs usually smaller. Umbel 2–5-rayed *Raylet-lvs usually becoming relatively narrower downwards and passing into the ray-lvs. Capsule* 2 mm *or less, deeply grooved*, with prominent *cylindrical tubercles which are longer than broad. Seeds* c. 1.5 mm, *red-brown.* Fl. 6–9. $2n = 28$. Th.

Native. Limestone woods in W. Gloucester and Monmouth. S., C. and W. Europe; Caucasus; N. Iran, Aral–Caspia.

9. E. helioscopia L.   Sun Spurge.

Glabrous *annual* 10–50 cm, with slender vertical root and single stems, simple or with a few branches below. *Lvs* 1.5–3 cm, alternate, *obovate, very obtuse*, tapered from near apex to a narrow base, serrulate above. Umbel 5-rayed without axillary infl. branches below. Ray- and raylet-lvs all similar to lvs but less tapered below and often yellowish-tinged. *Glands* transversely ovate, *entire*, green. *Capsule* 3–5 mm, subglobose, somewhat trigonous, *smooth. Seeds* c. 2 mm, brown, *reticulate.* Fl. 5–10. $2n = 42$. Th.

Native. Common in cultivated ground throughout the British Is., ascending to 450 m. Almost throughout Europe but a casual in the extreme north; C. Asia.

Section 3. *Cymatospermum* (Prokh.) Prokh. Annuals. Stem-lvs opposite or alternate.

10. E. peplus L. Petty Spurge.

Glabrous *green annual* 10–30 cm, with slender vertical root and simple or branched stems. *Lvs* 0.5–3 cm, alternate, *ovate to obovate*, *obtuse*, shortly stalked, entire. Umbels 3-rayed. Ray- and raylet-lvs like the lvs but subsessile. *Glands lunate* with long slender horns. Capsule c. 2 mm, trigonous, each valve with 2 narrow wings on the back. Seeds pale grey, pitted. Fl. 4–11. 2n = 16. Th.

Native. Very common in cultivated and waste ground throughout the British Is., ascending to 410 m. Most of Europe northwards to c. 65° N; Siberia (to Lake Baikal).

11. E. exigua L. Dwarf Spurge.

Glabrous, *glaucous annual* 5–30 cm, with slender vertical root and single simple or branched stem. *Lvs* 0.5–3 cm, alternate, *linear*, acute or more rarely obtuse, mucronate, sessile, entire. Umbel 3(–5)-rayed. Lower ray-lvs triangular-lanceolate, the upper triangular-ovate, all subcordate at base. *Glands lunate*, with long slender horns. Capsule c. 2 mm, trigonous, smooth or slightly rough on back. Seeds pale grey, tuberculate. Fl. 6–10. 2n = 24, 28. Th.

Native. Common in arable land in S. and E. England, from Somerset to Durham, less common in S.W. England, Wales and Scotland northwards to Banff, absent further north; common in C. and E. Ireland, rare elsewhere; Channel Is. Most of Europe northwards to c. 65° N, but absent from much of the east; Azores.

Section 4. *Paralias* Dumort. Annuals to perennials. Stem-lvs alternate.

12. E. portlandica L. Portland Spurge.

Glabrous *glaucous biennial or short-lived perennial* 5–40 cm, with ±vertical root and several usually simple *stems* from the base, all flowering or some sterile. *Lvs* 0.5–2 cm, alternate, dense, numerous, often caducous on fl.-stems, *somewhat coriaceous*, *obovate to oblanceolate*, acute to apiculate. *mucronate*, *tapered to base*, sessile or subsessile, entire, *the midrib prominent beneath*. Umbel 3–6-rayed; ray-lvs oblanceolate to obovate. Raylet-lvs 5–8 mm, triangular-rhombic, broader than long, mucronate. *Glands lunate*, with long horns. Capsule c. 3 mm, trigonous, granulate on the back of the valves. *Seeds grey, pitted*. Fl. 5–9. 2n = 16, 40. ?Ch.

Native. Sea sands and young dunes on the S. and W. coasts from Hants to Wigtown, very local; all round the Irish coast but rare in the south and west, Channel Is. W. coast of Europe from France to Portugal.

13. E. paralias L. Sea Spurge.

Glabrous glaucous *perennial* 20–40 cm, with short woody stock, *vertical root and several* simple fertile and sterile *stems*. *Lvs* 0.5–2 cm, alternate, numerous, dense, often imbricate, *very thick and fleshy*, entire, somewhat concave, *ovate or oblong*, obtuse to subacute, *not mucronate, with broad sessile base*; midrib obscure. Umbel 3–6(–8)-rayed. Ray-lvs ovate. *Glands lunate*, with short horns. Raylet-lvs 5–10 mm, orbicular-rhombic, thick and fleshy, mucronulate. Capsule c. 4 mm, trigonous, granulate. *Seeds* pale grey, *smooth*, with very small caruncle. Fl. 7–10. Fr. 8–11. 2n = 16. ?Ch.

Native. Sea sands and mobile dunes from Wigtown and Norfolk southwards, rather local; all round the Irish coast but rare in the north and west; Channel Is. Coasts of W. and S. Europe from the Netherlands to Romania; Morocco.

The partially sterile hybrid between *E. paralias* and *E. portlandica*, which is morphologically more or less intermediate between the parents, occurs on fixed dunes and sandy or gravelly banks near the coast in Glamorgan, Merioneth, Anglesey and Wexford.

Section 5. *Esula*. Perennials. Stem-lvs alternate.

*14. E. esula L. Leafy Spurge.

Glabrous perennial 30–80 cm, *with long creeping rhizome* and *numerous* erect fertile and sterile *stems*, often forming large patches. Fertile stems with branches 2–10 cm, axillary, sterile or occasionally flowering. *Lvs* alternate, *numerous*, oblanceolate to broadly ovate or obovate, on the main stems 2–4.5 cm × 4–7 mm, *broadest near the subacute, mucronate apex*, gradually tapered to a narrow base, sessile, entire. Umbel 6–12-rayed with axillary infl. branches below; *ray-lvs* 5–15 mm, linear-oblong. Infl. rather narrow. Raylet-lvs 5–7 mm, rhombic or deltate, apiculate, green. *Glands lunate* with rather short horns. Capsule 2–3 mm, trigonous, slightly granulate on back of valves. Seeds brown, smooth. Fl. 5–7. 2n = 64. Hp.

Introduced. Naturalized in woodlands, waste and grassy places from Dorset and Kent to Stafford and Suffolk, extending locally to Somerset, Glamorgan, S. Lancs, Cumberland and Angus. Native of much of continental Europe but an alien in the north.

British representatives of this species belong to subsp. **esula**, but it appears that most British plants are hybrids between subsp. **esula** and subsp. **tommasiniana** (Bertol.) Nyman, a native of E. and E.C. Europe. The hybrid (**E. ×pseudovirgata** (Schur) Sóo, *E. uralensis* auct., non Fischer and Link) differs from *E. esula* in having longer (7–20 cm) axillary branches on the fl. stems, narrower lvs which are broadest near the middle, with more truncate bases, a rather wide infl. and yellowish raylet-lvs. All these hybrids are at least partly fertile. Furthermore, the relationships between these hybrids and *E. esula*, and indeed with *E. cyparissias*, which is rather poorly separable from *E. esula*, need further investigation in Britain and elsewhere.

*15. E. cyparissias L. Cypress Spurge.

Glabrous perennial 10–30 cm, *with long creeping rhizome* and numerous erect fertile and sterile stems, often forming large patches. Fertile stems usually with axillary branches above, often overtopping the infl., the stem then appearing bushy. *Lvs* alternate, very numerous, *linear*, on the main stems 1.5–3 cm × 1–2 mm, obtuse to subacute, sessile, entire. Umbel 9–15-rayed, often with axillary infl. branches below; ray-lvs ±oblong. Ray-let-lvs 3–6 mm, deltate or reniform, yellowish or becoming reddish. *Glands lunate*, with very short horns. Capsule c. 3 mm, trigonous, slightly granulate. Seeds brown, smooth. Fl. 5–8. $2n = 20, 40$. Hp.

Introduced. In calcareous grassland and scrub in a few places from Kent and Somerset to Northumberland; more commonly occurring as a garden escape or casual in waste places and as such extending to Argyll and Sutherland. Most of Europe except the extreme north and the extreme south.

16. E. amygdaloides L.   Wood Spurge.

E. ×turneri Druce

Pubescent perennial 30–80 cm, with thick stock and tufted stems, sterile the first year, elongating and flowering the second, the lvs of the first year persisting at flowering in a cluster at the top of the first year stem. Lvs of the first year 3–8 cm, oblanceolate, obtuse to subacute, gradually tapered at base into a short petiole, entire, dark green; lvs of the fl.-stems oblong or obovate-oblong, not tapered at base, subsessile. Umbel with 5–10 rays with axillary infl. branches below; ray-lvs ovate. Raylet-lvs 5–10 mm, reniform, *connate* in pairs for about half the width of their bases, yellowish. *Glands lunate* with converging horns. Capsule c. 4 mm, somewhat trigonous, deeply sulcate. Seeds grey, smooth. Fl. 3–5. $2n = 18$. Not satisfactorily referable to any standard life-form, usually given as Hp., but appearing in winter as N.

Native. Damp woods, sometimes co-dominant in recent coppice; ascending to c. 430 m. S. England and Wales northwards to Cardigan, Shropshire, Leicester and Norfolk; Channel Is.; in Ireland in Cork, probably introduced in Wexford. C., S. and W. Europe, extending eastwards to E. Ukraine; Caucasus; Algeria (in the mountains).

85. POLYGONACEAE

Herbs, shrubs, climbers, or rarely trees. Lvs usually alternate and usually with sheathing stipules (ochreae). Fls hermaphrodite or unisexual. Per.-segs 3–6, sepaloid or petaloid, free or connate, persistent, imbricate in bud. Stamens usually 6–9; anthers 2-celled, opening lengthwise. Ovary superior, syncarpous, unilocular, with a solitary, basal, orthotropous ovule. Fr indehiscent, hard, trigonous or lenticular, usually enveloped in the perianth.

About 40 genera and 800 spp. distributed throughout the world, but mainly in the temperate regions.

1	Per.-segs 3 or 4.	*2*
	Per.-segs 5 or 6.	*3*
2	Small annual; lvs 3–5 mm; per.-segs 3. 1. KOENIGIA	
	Perennial; lvs 10–30 mm; per.-segs 4. 6. OXYRIA	
3	Woody climbers.	*4*
	Herbs.	*5*
4	Lvs 3–6 cm wide, cordate. 3. FALLOPIA	
	Lvs not more than 1.5 cm wide, not cordate.	
	8. MUEHLENBECKIA	
5	Per.-segs 6, the inner much enlarged in fr. 7. RUMEX	
	Per.-segs 5, the inner not enlarged in fr.	*6*
6	Outer per.-segs winged or keeled in fr.	*7*
	Outer per.-segs not winged or keeled in fr.	*8*
7	Erect rhizomatous perennials; stigmas fimbriate.	
	4. REYNOUTRIA	
	Twining annuals; stigmas not fimbriate. 3. FALLOPIA	
8	Lvs deltate-cordate, about as wide as long.	
	5. FAGOPYRUM	
	Lvs not deltate and rarely cordate, distinctly longer than wide. 2. POLYGONUM	

For illustrations and more detailed descriptions see J. E. Lousley & D. H. Kent, *Docks and Knotweeds of the British Isles* (Botanical Society of the British Isles, 1981).

1. KOENIGIA L.

Annual herb, most lvs sub-opposite. Fls clustered; bracts small. Per.-segs 3, sepaloid. *Stamens 3, alternating with 3 gland-like staminodes*. Ovary trigonous; styles 2.

One sp., Arctic and sub-Arctic, mountains of C. Asia, Rocky Mts, Tierra del Fuego.

1. K. islandica L.   Iceland Purslane.

1–6 cm. Stems reddish, flexuous; branches short. Lvs 3–5 mm; blade broadly elliptical to suborbicular, rounded at apex, rather fleshy; petiole 0.5–1 mm, thick; stipules c. 1 mm, sheathing, hyaline. Fls in terminal and axillary clusters; pedicels 0.5 mm, jointed, pale green. Per.-segs c. 1.2 mm, pale green, rounded, joined in lower $\frac{1}{3}$. Stamens alternating with per.-segs, inserted on tube; anthers reddish; staminodes yellow. Fr. 1.5 mm. $2n = 28$. Fl. 6–8. Th.

Native. On bare stony ground and among open vegetation 460–720 m in Skye and Mull. Distribution of the genus.

2. POLYGONUM L.

Annual or perennial herbs, rarely shrubby. Stipules connate into a tube (ochrea). *Per.-segs 3–6, usually 5*, all similar, *spirally arranged, the outer sometimes enlarging a little in fr. but never tuberculate*. Stamens 4–8. Styles 2–3; *fr. triquetrous or compressed and lenticular, ±enclosed in the persistent perianth*. Insect- or self-pollinated, sometimes cleistogamous.

About 250 spp., cosmopolitan, but particularly in temperate regions.

The following hybrids, with characters ±intermediate between those of the parents, are recorded, but none seems to be particularly frequent:

P. hydropiper × minus; *×mite*; *×lapathifolium*; *×persicaria*. *P. minus × mite*; *×persicaria*. *P. mite × persicaria*.

1 Stems without deflexed prickles; lvs not sagittate.
  Stems 4-angled, with weak deflexed prickles; lvs sagittate.   **16. sagittatum**

2 Infl. 1–6-fld, all axillary.   *3*
  Infl. usually many-fld, some or all terminal   *8*

3 Fr. dull, striate, enclosed by or slightly longer than the persistent perianth. (**aviculare** group)   *5*
  Fr. smooth, distinctly shining, as long as or longer than the persistent perianth.   *4*

4 Ochreae with 4–6 unbranched veins, shorter than internodes; fr. 5–6 mm, exceeding the persistent perianth.   **5. oxyspermum**
  Ochreae with 8–12 branched veins, as long as upper internodes; fr. 4–4.5 mm, enclosed by or slightly projecting from the persistent perianth (very rare).
                          6. maritimum

5 Branch-lvs much smaller than stem-lvs; persistent perianth divided almost to base; fr. trigonous with 3 concave sides.   *6*
  Branch- and stem-lvs ±equal; persistent perianth divided for half its length; fr. with 2 sides convex, 1 concave.   **4. arenastrum**

6 Stem-lvs narrow, linear-lanceolate, 1–4 mm broad; per.-segs and fr. narrow; fr. exserted. **3. rurivagum**
  Stem-lvs 5–18 mm broad; per.-segs overlapping; fr. broad.   *7*

7 Stem-lvs ovate-lanceolate, subsessile; petioles c. 2 mm, included in the ochreae; fr. 2.5–3.5 mm
                          1. aviculare
  Stem-lvs obovate-spathulate; petioles up to 4–8 mm, projecting from ochreae; fr. 3.5–4.5 mm (Shetland).   **2. boreale**

8 Plant rarely exceeding 50 cm, if more then stems soft, ±decumbent at base, and nodes often swollen.   *9*
  Stout erect herb with somewhat woody stems 60–180 cm.   **17. polystachyum**

9 Stems unbranched, never floating; rootstock stout.   *10*
  Stems usually branched, if nearly simple generally floating; root fibrous.   *12*

10 Basal lvs narrowed at the base; infl. slender, the fls mostly replaced by bulbils.   **7. viviparum**
  Basal lvs truncate or cordate at the base; infl. stout, without bulbils.   *11*

11 Blade of basal lvs truncate; petiole winged in upper part; fls pink, rarely white.   **8. bistorta**
  Blade of basal lvs cordate; petiole not winged; fls deep purplish-red.   **9. amplexicaule**

12 Lf-base cordate or rounded; stamens exserted.
                          10. amphibium
  Lvs narrowed to base; stamens included.   *13*

13 Plant with glands, sometimes sparse, on per.-segs or peduncle.   *14*
  Infl. and peduncle entirely without glands.   *15*

14 Infl. slender, nodding; per.-segs conspicuously glandular; peduncles eglandular.   **13. hydropiper**
  Infl. stout, erect; per.-segs sparsely glandular; peduncles glandular.   **12. lapathifolium**

15 Infl. dense, stout, obtuse; lvs often with a dark blotch.   **11. persicaria**
  Infl. slender, few-fld, acute; lvs never blotched.   *17*

16 Lvs usually 10–25 mm broad; infl. slightly nodding; fr. 3–4 mm.   **14. mite**
  Lvs usually 5–8 mm broad; infl. erect; fr. 2–2.5 mm
                          15. minus

Section 1. *Polygonum.* Annual or perennial. Stem branched, ±prostrate. Ochreae ±silvery or membranous. Infl. axillary, few-fld. Perianth ±petaloid. Stamens 3–8, filaments of inner dilated at base. Styles 3, rarely 2. Fr. trigonous or subtrigonous.

P. aviculare group (spp. 1–4)   Knotgrass.

A glabrous annual 3–200 cm. Lvs up to 5 cm × 18 mm, elliptical, spathulate, lanceolate or linear, acute or subobtuse; *ochreae silvery*, lacerate; veins indistinct. *Infl. 1–6-fld, axillary.* Stamens 5–8. *Fr.* 1.5–4.5 mm, trigonous, sometimes with one side much narrower than the others, brown, usually punctate or striate, *dull or slightly shining*, usually *enclosed within the persistent perianth*. Fl. 7–10. Th.

Native. In waste places, arable land and on the sea shore. Common and generally distributed. Cosmopolitan except for the Arctic and Antarctic.

The aggregate includes the following four species which are all very variable according to habitat:

1. P. aviculare L.

P. heterophyllum Lindman

Plant robust, up to 200 cm, erect or spreading, *heterophyllous when young.* Lvs 2.5–5 × 0.5–1.5 cm, *lanceolate to ovate-lanceolate*, subacute; those of main stems 2–3 times as long as those of flowering branches; *petioles up to* 2 mm, included in the ochreae. Ochreae c. 5 mm. *Per.-segs connate at the base only*, the margins pinkish or white. *Fr.* 2.5–3.5 × c. 1.5 mm, punctate, dull brown, *with 3 equal sides.* $2n = 60^*$.

Native. Roadsides and waste places and on the coast. Very common and generally distributed. Europe, temperate Asia; introduced into North America, Australasia, South America.

2. P. boreale (Lange) Small

Plant up to 100 cm, erect or suberect, *simple or sparingly branched. Lvs oblong-obovate* to *spathulate*, those on the main stems 3–5 cm × 5–18 mm, those on the branches much smaller; *petioles 4–8 mm, projecting from ochreae.* Per.-segs *broad, inclined to be open*; margins bright pinkish. Fr. 3.5–4.5 × c. 2.5 mm, included, punctate, light brown, *with 3 broad sides.* $2n = 40^*$.

Native. N. Scotland, Orkney and Shetland, where it seems largely to replace *P. aviculare.* Greenland, Iceland, Faeroes, N. Scandinavia; northern North America.

3. P. rurivagum Jordan ex Boreau

Plant seldom more than 30 cm, very slender, flexuous, suberect. *Lvs on the main stems 1.5–3.5 cm × 1–3 mm, linear-lanceolate to linear, acute*, those on the branches much smaller. Ochreae often c. 10 mm, brownish-red below. Infl. 1–2-fld, seldom more. *Per.-segs narrow*, reddish. *Fr.* 2.5–3 × 1.5–2 mm, slightly *exserted from the persistent perianth*, scarcely shining. $2n = 60^*$.

Native. Usually in arable fields on chalky soil in the south of England, local. W. and S.C. Europe extending to Sweden.

4. P. arenastrum Boreau   Small-leaved Knotgrass.

P. aequale Lindman; incl. *P. microspermum* Jordan ex Boreau and *P. calcatum* Lindman

Plant usually 5–30 cm; forming a dense prostrate mat, rarely erect. Lvs up to 20 × 5 mm, elliptical or elliptical-lanceolate, all similar. Infl. 2–3-fld. *Per.-segs connate for half their length*, greenish-white or pink. *Fr.* 1.5–2.5 mm, dull, but sometimes shining on the edges, *with 2 broad and 1 narrow sides*, or sometimes biconvex, brown to black. $2n = 40^*$.

Native. Waste places, roadsides, common throughout the British Is., though less so than *P. aviculare* and often in drier places than the latter species. Throughout Europe and temperate Asia, introduced into North America.

5. P. oxyspermum Meyer & Bunge ex Ledeb.
                                      Ray's Knotgrass.
P. raii Bab.

A straggling, prostrate, glabrous, sometimes glaucous annual or perennial 10–100 cm. Stems ±woody at base. Lvs 1–3.5 cm, elliptical-lanceolate to linear-lanceolate, usually flat. *Ochreae with 4–6 unbranched veins.* Infl. 2–6-fld. Per.-segs c. 3 mm, with broad pink or white margins. *Fr.* 2.5–5.5 × 3–3.5 mm, *dark brown*, almost flat, *shining, exceeding the persistent perianth.* Fl. 7–10. $2n = 40^*$. Th.

Represented in Britain by subsp. **raii** (Bab.) D. A. Webb & Chater, to which the above description applies.

Native. A decreasing Atlantic species on sandy shores or fine shingle above high-water spring-tides. S. and W. coasts north to the Outer Hebrides; N.E. Yorks; formerly on the east coast from Suffolk to Angus; local round the Irish coast. Coast of Europe from N.W. France to arctic Russia; eastern North America.

6. P. maritimum L.   Sea Knotgrass.

A glabrous, glaucous, procumbent perennial 10–50 cm. Stems stout and woody at base. Lvs 0.5–2.5 cm, elliptical-lanceolate; margins revolute. *Ochreae longer than most of the internodes*, *with 8–12 conspicuous branched veins.* Infl. 1–4-fld. Per.-segs c. 2–2.5 mm, with broad pink or white margins. *Fr.* 3.5–5 × 2.5 mm, *dark reddish-brown, shining, as long as, or slightly exceeding, the persistent perianth.* Fl. 7–10. $2n = 20^*$. Chh.

Native. On sand and fine shingle at, and just above high water spring-tides; extinct in England, local in the Channel Is.; formerly in Cornwall, Devon, N. Somerset and Hants. Atlantic coast of Europe from France southwards; Mediterranean region; Black Sea; Macaronesia.

*P. cognatum Meissner, a ±prostrate perennial with the fr. perianth with a very thick urceolate tube which is at least as long as the orbicular lobes and silvery ochreae usually longer than the internodes, is naturalized in a few places. S.W. Asia.

Section 2. *Bistorta* (Miller) DC. Perennial, rhizomatous. Stem unbranched, erect. Ochreae truncate. Infl. terminal, spicate. Perianth petaloid. Stamens 6–8. Styles 3. Fr. triquetrous.

7. P. viviparum L.   Alpine Bistort.

A slender erect *glabrous* perennial 6–30 cm. *Rhizome* rather stout, *not contorted.* Lvs 1.5–7 cm, *linear-lanceolate, tapering at both ends*, sometimes oval or suborbicular, the lower lvs petiolate, the upper sessile; *petioles not winged*; ochreae obliquely truncate, ±laciniate. *Infl.* terminal, *rather lax, slender* (4–8 mm diam.), *the lower part with purple bulbils.* Fls 3–4 mm, white, in upper part of infl. only, sometimes very few. Fr. rarely produced. Fl. 6–8. $2n = 132$, c. 100. ?Hp.

Native. In mountain grassland and on wet rocks. Mountain districts of N. Wales, N. England and N. Scotland; Ireland; Kerry; formerly in Sligo, Leitrim, Donegal, rare. In Scotland from sea-level in Sutherland to 1350 m. N. Europe, mountains in C. and S. Asia and North America.

8. P. bistorta L.
          Snake-root, Easter-ledges, Common Bistort.

An erect almost glabrous perennial 25–50 cm. *Rhizome* very stout and *contorted. Basal lvs* 5–15 cm, *broadly ovate*, obtuse, *truncate or subcordate at base, puberulent on the veins beneath*, folded longitudinally in bud and showing 'creases' when mature, *petioles* broadly winged in upper part; upper lvs triangular, acuminate; petioles sheathing; ochreae obliquely truncate, ±laciniate. *Infl.* terminal, *dense*, spicate, *stout* (10–15 mm *diam.*). Fls 4–5 mm, pink, rarely white, numerous. Fl. 6–8. $2n = 44$. Hs.

The young lvs are eaten as Easter-ledge pudding in the Lake District.

Native. In meadows and grassy roadsides, commoner on siliceous soils, often forming large patches. Scattered throughout the British Is. and commonest from Stafford to Cumberland, but rare and perhaps not native in the south-east or in Ireland. Much of Europe except most of Fennoscandia; only on mountains in the south; Asia; Japan.

*9. P. amplexicaule D. Don   Red Bistort.

Stems 60–100 cm, erect. Lvs ovate, acuminate, cordate, the lower petiolate, the upper sessile and amplexicaul. Fls deep red, in dense cylindrical spikes. $2n = 40$. Fl. 8–10. Hp.

Introduced. Cultivated for ornament and naturalized in Ireland and less frequently in Great Britain. Native of the Himalaya.

Section 3. *Persicaria* (Miller) DC. Annual, rarely perennial. Stem branched, erect or decumbent. Ochreae truncate, often ciliate. Infl. spicate. Perianth ±petaloid. Stamens 5–8, usually 8. Styles 2, rarely 3. Fr. compressed and lenticular or trigonous.

10. P. amphibium L.   Amphibious Bistort.

A glabrous or pubescent perennial commonly 30–75 cm. Aquatic and terrestrial forms differ considerably in vegetative features. Aquatic: glabrous; stems floating and rooting at nodes; *lvs* 5–15 cm, floating, ovate-oblong, subacute, *truncate or subcordate at base*; petioles 2–6 cm. Terrestrial: stems ascending or erect, rooting only at lower nodes, glabrous, or slightly pubescent; lvs 5–12(–15) cm, appressed-hispid or pubescent, ciliate, oblong-lanceolate, obtuse, usually ±narrowed to the *rounded base*; ochreae hispid. Infl. 2–4 cm, usually terminal, many-fld, dense, lfless and pedunculate. Fls pink or red. Per.-segs eglandular; *stamens 5, exserted*; styles 2, connate ¼ of the way up. Fr. c. 2 mm, orbicular, biconvex, brown, ±shining. Fl. 7–9. 2*n* = 66. Hyd. or Hp.

Native. In pools, canals and slow-flowing rivers; the terrestrial form on banks by water. Generally distributed throughout the British Is. Europe, Asia, North America, N. and S. Africa.

11. P. persicaria L.

Red Shank, Willow weed, Persicaria.

A branched erect or ascending nearly glabrous annual 25–75 cm. Stems reddish, swollen above the nodes. Lvs 5–10 cm, lanceolate, ciliate, sometimes woolly beneath, often black-blotched; *ochreae* truncate, *ciliate*. *Infl. stout, obtuse, continuous* or somewhat interrupted, lfless or with a single lf at base. *Fls pink. Per.-segs and peduncle eglandular*; styles 2 or 3, connate below. *Fr.* c. 3 mm, *biconvex or bluntly trigonous* with concave faces, shining. Fl. 6–10. 2*n* = 44. Th.

Native. In waste places, cultivated land and beside ponds. Generally distributed throughout the British Is., common. Temperate regions of Europe and Asia; introduced in North America.

12. P. lapathifolium L.   Pale Persicaria.
Incl. *P. nodosum* Pers.

A branched erect ±pubescent annual up to 100 cm. Stems usually greenish, swollen above nodes. Lvs 5–20 cm lanceolate, ciliate, glandular, sometimes woolly beneath, ±hispid on the midrib, often black-blotched; petioles short, ±hispid; *ochreae* truncate, or *very shortly ciliate*. *Infl. stout, obtuse, continuous, some long-pedunculate*, lfless. *Fls greenish-white*, rarely pink. *Per.-segs and peduncle rather sparsely glandular*; styles 2, separate nearly to base. Fr. 2–3 mm, *usually suborbicular, flattened, biconcave*, sometimes ±trigonous, shining. Fl. 6–10. 2*n* = 22. Th.

Native. In waste places, cultivated ground and beside ponds. Throughout most of the British Is., common except in Scotland and Ireland. Temperate regions of the northern hemisphere; S. Africa.

13. P. hydropiper L.   Water-pepper.

A glabrous *acrid* erect annual, 25–75 cm. Lvs usually 5–10 cm, lanceolate or narrowly lanceolate, subsessile, ciliate; *ochreae* truncate, *shortly* ciliate. *Infl.* slender, *nodding*, acute, *interrupted and lfy in lower part*. *Fls usually greenish. Per.-segs covered with yellow glands*; *peduncles eglandular*; styles 2, free nearly to base. *Fr.* c. 3 mm, ovoid, biconvex or subtrigonous, punctate, dark brown or black, *not shining*. Fl. 7–9. 2*n* = 20. Th.

Native. In damp places or in shallow water in ponds and ditches; calcifuge. Generally distributed throughout the British Is., except the north of Scotland, Orkney

and Shetland; common. Europe, except the north and some islands; N. Africa; temperate Asia; introduced in North America.

14. P. mite Schrank   Tasteless Water-pepper.
P. laxiflorum Weihe

Like *P. hydropiper* in general appearance but *ochreae* with *conspicuous* long *coarse* cilia and lvs rather abruptly narrowed at base. *Infl. nearly erect*, slender, interrupted but scarcely lfy, *eglandular*. *Fls pink*, rarely white. *Fr.* 3–4.5 mm, broadly ovoid, biconvex, *shining*. Fl. 6–9. 2*n* = 40. Th. *Plant not acrid and burning*.

Native. In ditches and beside ponds and rivers. Scattered throughout the southern part of Great Britain north to Durham and N.E. Yorks; rare. Most of Europe except the north; W. Asia.

15. P. minus Hudson

A spreading branched decumbent or ascending annual 10–30(–40) cm. *Lvs usually* 2–5 × 0.5–0.8 cm, narrowly lanceolate, ±obtuse, ciliate, subsessile; ochreae conspicuously and coarsely ciliate. *Infl.* slender, *erect*, ±interrupted, eglandular. Fls pink or rarely white. *Fr.* 1.5–3 mm, ovoid, biconvex, black and shining. Fl. 8–9. 2*n* = 40. Th.

Native. In wet marshy places and beside ponds and lakes. Scattered through the British Is. north to N.E. Yorks and Kintyre; local. Europe from N. Spain to 63°N in Russia; temperate Asia; introduced in North America.

*16. P. sagittatum L.

Annual, with weak 4-angled stems up to 150 cm. *Stems, petioles and midribs with short, slender deflexed prickles.* Lvs cordate or sagittate at base. Ochreae truncate. Infl. subcapitate. Perianth petaloid. Stamens 5–8. Fr. lenticular or trigonous. Fl. 8. Th.

Introduced. Naturalized in a ditch in Kerry. Native of North America.

*P. campanulatum Hooker fil., Lesser Knotweed, a perennial with pale pink fls lvs whitish to buff tomentose beneath, is naturalized in S. England, W. Scotland and W. Ireland. Himalaya.

Section 4. *Aconogonon* Meissner. Perennial. Fls in diffuse terminal panicles. Perianth petaloid. Stamens 8. Styles 3. Nut trigonous.

*17. P. polystachyum Wall. ex Meissner
Himalayan Knotweed.

Perennial. Stems 60–180 cm, erect. Lvs up to 20 cm, lanceolate, acuminate, ±cordate at base, shortly petiolate. Infl. a lax, terminal or axillary panicle. Per.-segs usually c. 3 mm, unequal, the 2 outer much narrower than the broadly obovate to suborbicular inner segs. 2*n* = 22. Fl. 8–9.

Introduced. Grown in gardens and naturalized from outcast rhizomes in many localities throughout the British Is. Native of the Himalaya.

3. FALLOPIA Adanson
Bilderdykia Dumort.

Stems twining; lvs deltate or cordate-sagittate, petiolate. Infl. terminal or axillary, lax, spike-like or branched. Per.-segs 5(–6), the 3 outer keeled or winged. Stamens 8. Stigmas capitate, subsessile. Nut trigonous.

Nine spp. in north temperate regions.

1 Woody perennial; infl. much-branched.
       3. baldschyanica
  Annuals; infl. scarcely branched.   **2**
2 Pedicels 1–3 mm in fr.; nut 4–5 mm, finely granulate.
       1. convolvulus
  Pedicels 5–8 mm in fr.; nut 2.5–3 mm, smooth and glossy.   **2. dumetorum**

1. F. convolvulus (L.) Á. Löve   Black Bindweed.
Polygonum convolvulus L.

A somewhat mealy scrambling or climbing annual 30–120 cm. Stem angular, mealy on the angles. Lvs 2–6 cm, ovate, acuminate, cordate-sagittate at base, mealy beneath, nearly smooth above, the lamina longer than the petiole; ochreae obliquely truncate, ±laciniate. Infl. pedunculate or subsessile, interrupted; *pedicels 1–3 mm, jointed above the middle. Per.-segs 5, the 3 outer obtusely keeled or* (var. *subalatum* (Lej. & Court.) Kent) *narrowly winged in fr.*, rough on back. *Fr. 4–5 mm, dull black*, minutely granulate. Fl. 7–10. $2n = 40$. Th.

Native. In waste places, arable land, and gardens. Generally distributed throughout the British Is., except Shetland; common. Europe, N. Africa, temperate Asia; introduced in North America, Japan and S. Africa.

2. F. dumetorum (L.) J. Holub
Polygonum dumetorum L.

A somewhat mealy climbing annual up to 300 cm. Like *F. convolvulus*, but fruiting *pedicels up to 8 mm, filiform, jointed at or below the middle and deflexed; outer perianth segments broadly winged in fr.*, the wings strongly decurrent on the pedicels. *Fr. 2.5–3 mm, black and shining.* Fl. 7–9. $2n = 20$. Th.

Native. In thickets and hedges. From S. Devon and E. Kent to Oxford and S. Essex; Caernarvon; very local. Europe except the extreme north; N. and W. Asia.

*3. F. baldschmanica (Regel) J. Holub   Russian-vine.
Polygonum baldschuanicum Regel

A robust woody perennial climbing to 10 m or more. Lvs ovate, acute, ±cordate at base, bright red when young, long-petiolate. Fls c. 5 mm diam., in large diffuse freely branched panicles. Per.-segs white, turning pink in fr., the outer broadly winged. Nut 4 mm, brown, finely granulate. Fl. 8–10. $2n = 20$. MM.

Introduced. Commonly cultivated and persisting as a garden outcast in many places in lowland England and Wales. Native of China. *F. aubertii* (Louis Henry) J. Holub is probably not specifically distinct.

4. REYNOUTRIA Houtt.

Stems stout, erect, arising from strong perennial rhizomes. Lvs broad, petiolate. Plants dioecious. Fls in axillary panicles, the female with staminodes, the male with abortive ovary. Per.-segs 5, enlarging in fr., the 3 outer keeled and becoming winged in fr. Stamens 8. Styles 3. Nut trigonous, not exceeding the perianth.

Lvs usually not more than 12 cm, truncate at base; fls white.   **1. japonica**
Lvs usually more than 15 cm, cordate at base; fls greenish.   **2. sachalinensis**

*1. R. japonica Houtt.   Japanese Knotweed.
Polygonum cuspidatum Siebold & Zucc.

Stems 1–2 m, numerous, glaucous, often tinged with red, branched above. Lvs 5–12 cm, broadly ovate, cuspidate, truncate at base. Infl. 8–12 cm, with slender rather lax branches. Fls white. Nut 4 mm. $2n = 44, 88$. Fl. 8–9.

Introduced. Escaped from cultivation and widely naturalized throughout the British Is. Native of Japan, now naturalized in much of C. and N. Europe.

*2. R. sachalinensis (Friedrich Schmidt Petrop.) Nakai   Giant Knotweed.

Like **1** but stems often more than 3 m; lvs 15–30 cm, ±cordate at base; infl. with shorter and stouter branches; fls greenish. $2n = 44$. Fl. 8–9.

Introduced. Scattered throughout the British Is. but much less common than *R. japonica*. Native of E. Asia.

5. FAGOPYRUM Miller

Annual herbs similar to *Polygonum* in many respects. Stems erect, hollow. *Lvs cordate at base.* Perianth petaloid. Fls heterostylous. Stamens 8; styles 3. *Fr. much exceeding the perianth.*

About 15 spp. in Asia, introduced elsewhere and frequently cultivated.

*1. F. esculentum Moench   Buckwheat.
F. sagittatum Gilib.; *Polygonum fagopyrum* L.

An erect little-branched nearly glabrous annual 15–60 cm. Lvs cordate-sagittate, acuminate, as long as or little longer than broad. Infl. a cymose panicle. Fls greenish-white, pink-tipped. Fr. 5–6 mm, 2–3 times as long as perianth, triquetrous, the angles acute, entire. Fl. 7–8. $2n = 16$. Th.

Introduced. Cultivated as a crop, particularly in the fens, and occurring as a casual in waste ground. Native in E.C. Asia. Largely cultivated in some countries as a substitute for cereals, or as green fodder.

6. OXYRIA Hill

Perennial herbs. Infl. branched, lfless. *Per.-segs 2 + 2, sepaloid, inner enlarging in fr.* but *not tuberculate;* stamens 6; anthers versatile; stigmas 2. *Fr. lenticular, broadly winged.*

One sp. in arctic regions and on mountains in the northern hemisphere.

1. O. digyna (L.) Hill   Mountain Sorrel.
Rheum digynum (L.) Wahlenb.

A glabrous often tufted perennial 5–30 cm. Stock stout. Stem erect. *Lvs* 1–3 cm, almost all basal, *reniform*, suborbicular or retuse, rarely subhastate, rather fleshy;

petioles long. Panicle lfless. Pedicels slender, jointed about the middle and thickened towards the top. Outer per.-segs spreading or reflexed, the inner pressed to fr., spathulate or ±rhombic. Fr. 3–4 mm, broadly winged. Fl. 7–8. $2n = 14$. Hr.

Native. In damp rocky places on mountains, especially beside streams, locally common. Merioneth and Caernarvon; Westmorland, Cumberland; Dumfries and Peebles, Arran and Angus northwards; Ireland, on the mountains of the west and in S. Tipperary. The Arctic and mountains of the north temperate zone.

7. RUMEX L

Annual, biennial or perennial herbs, usually with long stout roots, sometimes rhizomatous. Lvs alternate; ochreae tubular. Fls hermaphrodite or unisexual, arranged in whorls on simple or branched infls., anemophilous. *Per.-segs* 3 + 3, the outer always small and thin, *the inner* (fr. per.-segs) *enlarging and usually becoming hard in fr.* Fr. per.-segs with or without swollen corky, globose or ovoid tubercles on their midribs; tubercles developing as the fr. ripens. Stamens 3 + 3; anthers basifixed. Fr. triquetrous with a woody pericarp.

About 200 spp. in the temperate regions of the world.

Hybrids are of frequent occurrence, although some (e.g. *R. crispus × obtusifolius*) are common, they have not been included in the following key and descriptions. The majority may be easily recognized by their fairly high degree of sterility in both pollen and fr. They usually occur in close proximity to the parents and are intermediate in character between them. Hybrid swarms do not occur.

Really ripe fr. is essential for the accurate identification of many spp.

1 Foliage acid; lvs hastate; fls usually unisexual.   2
  Foliage not or scarcely acid; lvs not hastate; fls usually hermaphrodite.   4
2 Lvs about as long as broad, upper petiolate.
                 4. scutatus
  Lvs several times as long as broad, if nearly as broad as long then upper subsessile and clasping stem.   3
3 Lobes of lvs spreading or forward-pointing, upper lvs not clasping stem.   **acetosella** (spp. 1–2)
  Lobes of lvs ±downward-directed, upper lvs clasping stem.   **3. acetosa**
4 All fr. per.-segs without tubercle.   5
  At least one fr. per.-seg. with a distinct tubercle.   7
5 Pedicels of fr. with an almost imperceptible joint.
                 7. aquaticus
  Pedicels of fr. distinctly jointed.   6
6 Rhizomatous; lvs about as long as broad; fr. per.-segs ovate, truncate at base.   **6. alpinus**
  Not rhizomatous; lvs distinctly longer than broad; fr. per.-segs reniform.   **8. longifolius**
7 Fr. per.-segs distinctly toothed (teeth more than 1 mm).   8
  Fr. per.-segs entire or denticulate (teeth not exceeding 1 mm).   11
8 Branches making a wide angle with main stem, forming an entangled mass in fr.; lvs usually panduriform, lamina rarely exceeding 10 cm.   **13. pulcher**

  Branches usually making a narrow angle with main stem, not becoming entangled in fr.; lvs very rarely panduriform, lamina often exceeding 10 cm.   9
9 Fr. per.-segs 5–6 mm, reddish or brown.
                 12. obtusifolius
  Fr. per.-segs 2–3 mm, yellow or golden.   10
10 Outer per.-segs with claw-shaped forward-curved apices; fr. per.-segs with lingulate obtuse apices; teeth rigid, setaceous, not longer than the width of the seg.; fr. 2–2.5 mm   **17. palustris**
  Outer per.-segs horizontally spreading or weakly reflexed; fr. per.-segs with triangular, ±acute apices; teeth very fine and as long as the width of the seg.; fr. 1–1.5 mm.   **18. maritimus**
11 Lvs narrowly obovate, thick and coriaceous; rhizome far-creeping; infl. little-branched (dunes).
                 19. frutescens
  Not as above.   12
12 Primary infl. after flowering overtopped by secondary ones arising from lower axils; tubercles of fr. per.-segs rugose.   **20. triangulivalvis**
  Not as above.   13
13 Infl. dense, the whorls crowded, ±confluent; fr. per.-segs (3.5–)4.5(–8) mm; tubercles $\frac{1}{2}$ as long as fr. per.-segs or less.   14
  Infl. lax, the whorls distant; fr. per.-segs up to 3 mm, or if more then tubercles $\frac{2}{3}$–$\frac{3}{4}$ length of fr. per.-segs.   17
14 Fr. per.-segs triangular, truncate or subcuneate at base.   **5. hydrolapathum**
  Fr. per.-segs broadly ovate or orbicular, cordate at base.   15
15 Lvs lanceolate, usually narrow, undulate and crisped; plant up to 100 cm; fr. per.-segs 3–5 × 2–5 mm.
                 11. crispus
  Lvs ovate- to oblong-lanceolate, not crisped; plant 100–200 cm; fr. per.-segs 5–7 mm broad.   16
16 Veins in middle of lf at an angle of 45–60° with midrib.
                 10. patientia
  Veins in middle of lf at an angle of 60–90° with midrib.
                 9. cristatus
17 Lvs glaucous; fr. per.-segs c. 4 mm, all with prominent tubercles $\frac{2}{3}$–$\frac{3}{4}$ as long (S.W. England and S. Wales, by the sea).   **16. rupestris**
          (See also **13. pulcher**)
  Lvs not glaucous; fr. per.-segs up to 3 mm.   18
18 Stems usually almost straight, branches making an angle of 20°(–40°); lowest whorls on branches subtended by lvs; tubercle (usually one) globose.
                 14. sanguineus
  Stems usually flexuous, branches making an angle of 30–90°; whorls usually subtended by lvs for $\frac{2}{3}$ length of branches; tubercles (usually 3) oblong.
                 15. conglomeratus

Subgenus 1. ACETOSELLA Rafin.

Plant usually dioecious. Fr. per.-segs of female fls not enlarged in fr. or, at most, twice as large as nut, all without tubercles. Lvs often hastate or sagittate.

R. acetosella group (spp. 1–2)   Sheep's Sorrel.

Erect or decumbent dioecious perennials up to 30 cm, producing adventitious buds on horizontal roots. *Lvs* up to 4 cm, several times as long as broad, lanceolate to linear, often hastate with *spreading or forward-point-ing lobes*; *upper lvs ±distinctly petiolate, not clasping the stem*; *ochreae hyaline, lacerate.* Infl. up to c. 15 cm, lfless or nearly so. *Outer per.-segs pressed to inner. Fr.* 0.8–1.5 mm. Fl. 5–8. Hs. or Grt. Lvs slightly acid and bitter.

Native. On heaths, in grassland and cultivated land, common on acid but infrequent on calcareous soils. Generally distributed throughout the British Is. Throughout Europe; temperate Asia; N. and S. Africa; Macaronesia; temperate America; Australia; Greenland; perhaps introduced in the southern hemisphere.

1. R. acetosella L.

Incl. *R. angiocarpus* Murb.

Plant variable in size, stems erect. Lower lvs 3–4 times as long as broad (excluding lobes); margins flat. Ripe fr. 1.3–1.5 mm. $2n = 42$.

Native. Widespread and common, avoiding the poorest soils and less strongly calcifuge than *R. tenuifolius.* Throughout Europe.

2. R. tenuifolius (Wallr.) Löve

Plant small, stems ±decumbent. Lower lvs 7–10(–40) times as long as broad (excluding lobes); margins inrolled. Ripe fr. 0.9–1.3 mm. $2n = 28$.

Native. Probably widespread but rather uncommon, occurring mainly on the poorest soils. C. and N. Europe.

Subgenus 2. ACETOSA (Miller) Rech. fil.

Dioecious or polygamous. Fr. per.-segs of female fls many times longer than nut, without tubercles or with small recurved tubercles at the base. Lvs often hastate or sagittate.

3. R. acetosa L.   Common Sorrel.

An erect dioecious nearly glabrous perennial up to 100 cm, though often less. *Lvs* up to 15 cm, usually less, 2–4 times as long as broad, oblong-lanceolate, obtuse or subacute, hastate, *lobes ±downward-directed*; *upper lvs subsessile, ±clasping stem*; *ochreae ciliate.* Infl. up to 40 cm, lfless or nearly so, the branches simple. *Outer per.-segs reflexed and appressed to pedicels after flowering*; *fr. per.-segs 3–3.5 mm, orbicular-cordate*, each with a small tubercle near the base. *Fr.* 1.8–2 mm, shining. Fl. 5–6. $2n = 14$ (female), 15 (male); triploid and hexaploid variants also occur. Hs.

Our common plant is subsp. **acetosa**.

Lvs acid, sometimes used in salads and sauces.

Native. In grassland and in open places in woods. Generally distributed and common throughout the British Is. Europe, temperate Asia, Japan, North America, Greenland.

R. hibernicus Rech. fil. is a variant with stems 10–15 cm and basal lvs 10–15 × 5–10 mm with short acute slightly divergent basal lobes. The cauline lvs are few, the upper up to 8 times as long as broad and the infl. is dense, with very few short branches. It occurs on dunes in W.

Scotland and W. Ireland and is of uncertain taxonomic status.

*4. R. scutatus L.   French Sorrel.

A glabrous perennial up to 50 cm. *Lvs* up to 5 cm, *about as broad as long, broadly ovate or panduriform*, obtuse, hastate, the lobes diverging; upper lvs petiolate; *ochreae bifid, ±toothed but not ciliate.* Infl. up to c. 15 cm, lfless or nearly so. Plant polygamous. Outer per.-segs pressed to inner; fr. per.-segs c. 5 mm, orbicular, cordate. Fl. 6–7. $2n = 20$. Hp. Lvs acid, sometimes used in salads and sauces.

Introduced. Naturalized on old walls and in pastures in a few scattered localities. C. and S. Europe; W. Asia, N. Africa.

Subgenus 3. RUMEX.

At least most of the fls hermaphrodite. Fr. per.-segs many times longer than nut, with or without tubercles; tubercles not recurved. Lower lvs cuneate, rounded, or cordate at base, hastate or sagittate.

Section Rumex. Annuals, biennials, or perennials. Primary infl. not normally overtopped by secondary infl. arising from the axils of the lower lvs.

5. R. hydrolapathum Hudson   Water Dock.

A stout erect perennial up to 200 cm. Lvs up to 110 cm, lanceolate to ovate, acute or acuminate, tapering equally at both ends; margins flat or somewhat undulate. Infl. usually much-branched, somewhat lfy below, the branches strict. Whorls ±crowded. *Fr. per.-segs* 5–7 mm, *triangular, truncate or subcuneate at base, each with a prominent elongated tubercle*, usually with a few short teeth towards the base. Fr. 4.5–5 mm. Fl. 7–9. Germ. spring. $2n = 200$. Hyd. or Hel.

Native. In wet places and shallow water. Fairly generally distributed throughout the British Is., though rare in Scotland; local in Ireland. Most of Europe.

*6. R. alpinus L.   Monk's Rhubarb.

A stout erect rhizomatous perennial 30–80(–150) cm. *Rhizome extensively creeping. Lvs* 20–40 cm, *about as broad as long, rounded*, cordate, margins ±undulate. Infl. interrupted below, little-branched. Whorls confluent, lfy in the lower part. Pedicels 5–10 mm, slender, deflexed in fr. *Fr. per.-segs* 4.5–5 mm, *ovate, truncate at base*; *margins entire, tubercles 0.* Fr. 3 mm. Fl. 7. $2n = 20$. Hs.

Introduced. Near buildings and beside streams and roads in hilly districts. Devon, E. Kent and from Stafford and Derby northwards. Mountains of C. and S. Europe; Caucasus.

7. R. aquaticus L.   Scottish Dock.

A stout perennial 80–200 cm. Basal lvs triangular, acute, deeply cordate at base $1\frac{1}{2}$–$2\frac{1}{2}$ times as long as broad and broadest near the base; petiole at least as long as blade. *Pedicels of fr. very slender with an almost imperceptible joint near base. Fr. per.-segs* 5–8.5 mm, *ovate-*

triangular, truncate at base, ±acute, *entire; tubercles* 0.
Fr. c. 3.5 mm. Fl. 7–8. 2*n* = c. 200. Hyd. or Hel.

Native. In water at the margins of alder swamps. Loch
Lomond, Stirling and Dunbarton. N., C. and E. Europe; temperate Asia; Japan.

8. R. longifolius DC.   Northern Dock.

R. domesticus Hartm.; *R. aquaticus* auct., non L.

A stout erect perennial 60–120 cm. Lvs up to 80 cm,
broadly lanceolate, obtuse or subacute, 3–4 times as
long as broad, ±cordate at base. *Infl. dense, compact,
fusiform*. Whorls confluent, lfy at base. *Fr. per.-segs*
4.5–5.5 mm, *thin, reniform*, entire; *tubercles* 0. Fr.
3–4 mm. Fl. 6–7. 2*n* = 40, 60. Hs.

Native. Beside rivers, in ditches and in damp grassy
places. S. Lancs and S.W. Yorks northwards; absent
from Ireland. N. Europe; Pyrenees; USSR; Caucasus,
C. Asia; Japan; introduced in North America and
Greenland.

*9. R. cristatus DC.   Greek Dock.

R. graecus Boiss. & Heldr.

A tall perennial like *R. patientia*. Veins in middle of lf at an
angle of 60–90° with midrib. *Fr. per.-segs* roundish-cordate,
6–7 mm broad, *with frequent irregular acute teeth up to* 1 mm,
usually all tuberculate, one tubercle 2–3 mm, the others smaller
or sometimes 0. Fl. 6–7. Hs.

Introduced. In waste places, apparently well established near
Cardiff and London. Greece, Sicily, W. Anatolia.

*10. R. patientia L.

An erect perennial 100–200 cm. Lvs ovate- or oblong-
lanceolate from a truncate or subcuneate, rarely subcordate base, the apex acute. Veins in middle of lf at
an angle of 45–60° with midrib. Whorls crowded. *Fr.
per.-segs broadly ovate- or suborbicular-cordate, one
with a large tubercle* the others without or with small
tubercles; *margins entire, crenulate or minutely denticulate.* Fl. 6–7. 2*n* = 60. Hs.
Introduced. In waste places. Well established in a few
localities in S. and C. England. From S. Czechoslovakia
to S.E. Russia and S.W. Asia. Naturalized elsewhere
in Europe and in North America.

The 2 following subspp. occur in Britain:

Subsp. **patientia**. Stem usually purplish or reddish-
brown. Fr. per.-segs 5–6(–7) mm broad, only one with
a tubercle 1.5 mm.

Subsp. **orientalis** (Bernh.) Danser. Stem usually pale.
Fr. per.-segs 8–10 mm broad, one with a tubercle
2–3 mm, the others sometimes with smaller tubercles.

11. R. crispus L.   Curled Dock.

Incl. *R. elongatus* Guss.

An erect perennial 30–150 cm. *Lvs* up to 30 cm, 4–5
times as long as broad, lanceolate or oblong-lanceolate,
narrowed or ±rounded at base, tapering from about
the middle to the apex; *margins usually undulate and
strongly crisped*; petiole usually shorter than blade. Infl.
usually nearly simple or little-branched, the branches

strict, usually subtended by linear-lanceolate much
crisped lvs. Whorls close, distinct, or confluent towards
the ends of the branches. *Fr. per.-segs* 3–5.5(–6) mm,
broadly ovate-cordate, usually all three with tubercles,
one larger than the other two or (particularly in sea-
shore plants) all equal; margin entire or minutely denticulate. Fr. 2.5–3 mm. Fl. 6–10. 2*n* = 60. Hs.

Very variable. The commonest British sp. and a serious agricultural weed.

Native. In grassy places, waste ground, cultivated
land, dune-slacks, and shingle beaches. Generally distributed and common throughout the British Is. Europe,
Macaronesia and most of Africa. Naturalized in most
other parts of the world; rare in the north.

12. R. obtusifolius L.   Broad-leaved Dock.

An erect branched perennial 50–120 cm. Underside of
lvs and veins usually hairy. Lower lvs up to c. 25 cm,
ovate-oblong, cordate at base, obtuse, margins undulate; petiole slightly longer than blade; upper lvs ovate-
lanceolate to lanceolate. Infl. branched, lfy in the lower
part, branches rather spreading. *Whorls remote. Fr.
per.-segs* 5–6 mm, *triangular*, acuminate, *one* (rarely all
three) *with a prominent tubercle*; margin with 3–5 long
teeth. Fr. 3 mm. Fl. 6–10. Germ. spring. 2*n* = 40. Hs.

The above description applies to subsp. **obtusifolius,**
which appears to be the only subsp. native in this
country. Subspp. *transiens* (Simonk.) Rech. fil. and
sylvestris (Wallr.) Rech. occur as rare aliens in the
neighbourhood of London.

Native. On waste ground, in hedgerows, and at margins of fields, generally in disturbed ground. Common
and generally distributed throughout the British Is.
Almost all Europe; N. Africa, Macaronesia, temperate
Asia, Japan; introduced in North America.

13. R. pulcher L.   Fiddle Dock.

A spreading much-branched perennial 20–60 cm.
*Branches making a wide (–90°) angle with the main stem,
rather slender and drooping in fl., forming an entangled
mass in fr.* Lower lvs very rarely exceeding 10 cm, puberulent, panduriform or oblong from a cordate base, with
an obtuse apex; upper lvs lanceolate acute. Whorls distant. Pedicels jointed about the middle. *Fr. per.-segs*
4.5–5 mm, oblong- or ovate-triangular, acute or obtuse,
all tuberculate (often unequally), *margin with* c. 4 *teeth
not longer than ½ width of per.-seg.*; tubercles verrucose.
Fl. 6–7. 2*n* = 20. Hs.

The above description applies to subsp. **pulcher,**
which is the only native subsp. Subspp. *anodontus*
(Hausskn.) Rech. fil. and *divaricatus* (L.) Murb. have
occurred as rare aliens.

Native. In dry sunny habitats on sandy soils and, less
commonly, on chalk and limestone. England south of
a line from the Humber to the Severn; Wales: coastal
districts of the south and north-west (doubtfully native);
S.E. Ireland, introduced. S. and W. Europe extending
to Hungary and N. Romania; Caucasus; Macaronesia;
introduced in North America.

14. R. sanguineus L.   Red-veined Dock.

R. condylodes Bieb.; *R. nemorosus* Schrader ex Willd.

An erect perennial up to c. 100 cm. *Stem usually almost straight, the branches making an angle of about 20°(–45°) with it.* Lvs ovate-lanceolate from a rounded or subcordate base, the upper narrowly lanceolate, all acute. *Infl.* much-branched, *branches with only the lowest whorls subtended by lvs.* Whorls rather distant. *Fr. per.-segs* 2.5–3 mm, oblong, obtuse, *one with a globose tubercle* c. 1.5 mm *diam., the others devoid of tubercles or with less developed ones*; margins entire. *Fr.* 1.25–1.75 mm. The common form is var. *viridis* Sibth., with green or occasionally rusty-red veins and usually thin lvs. Var. *sanguineus* is an uncommon plant which has lvs with purple veins and stems and panicle branches often suffused with purple. It was formerly widely cultivated for medicinal purposes and is now naturalized in several places. It is suggested that var. *sanguineus* arose as a mutant which bred true and has been spread by human agency. Fl. 6–8. Germ. throughout the year. $2n = 20$. Hs.

Native. On waste ground, in grassy places, and in woods. Common in England, Wales and Ireland, less frequent in N. England and in Scotland. Most of Europe north to c. 60°N; S.W. Asia; N. Africa; introduced in North and South America.

15. R. conglomeratus Murray   Sharp Dock.

R. glomeratus Schreber; *R. acutus* sensu Sm.

An erect biennial or perennial up to c. 100 cm. *Stem usually distinctly flexuous, the branches making an angle of 30–90° with it.* Lvs oblong or frequently panduriform from a rounded or subcordate base, the upper lanceolate, all acute. Infl. much-branched. *Whorls distant, usually subtended by lvs for $\frac{2}{3}$ length of branches. Fr. per.-segs* 2–3 mm, ovate to oblong, obtuse, *all with oblong tubercles* 1.25–1.75 mm, the tubercles often nearly covering the per.-seg.; margins entire. *Fr.* 1.75–2 mm. Fl. 7–8. Germ. 5–6. $2n = 20$. Hs.

Very variable and sometimes with difficulty distinguishable from *R. sanguineus*.

Native. In damp grassy places and less frequently in woods. Common and generally distributed throughout the lowland districts of England, Wales and Ireland; Scotland, rare. Europe north to S. Sweden and C. Russia; temperate Asia; N. Africa; Macaronesia; introduced in North America.

16. R. rupestris Le Gall   Shore Dock.

An erect branched perennial up to c. 70 cm. *Branches strict.* Lvs oblong, narrowed at base, obtuse, *glaucous, blade usually much longer than the petiole.* Whorls distinct, *usually only the lowest on each branch subtended by a lf.* Fr. per.-segs 3–4 mm, oblong, obtuse, all with oblong tubercles 2.5–3 mm; margins entire. *Fr.* c. 2 mm. Fl. 6–8. Germ. 4–6. $2n = 20$. Hs.

Native. On sea-cliffs, rocky shores, and in dune-slacks. Scilly, Cornwall, S. Devon, Glamorgan, Pembroke,

Anglesey; Channel Is. Western France; N.W. Spain.

17. R. palustris Sm.   Marsh Dock.

R. limosus auct. angl., non Thuill.

An erect branched annual, biennial or short-lived perennial up to c. 100 cm, usually much less. Plant brownish in fr. Lvs oblong-lanceolate, tapering at base, the apex subacute; upper lvs linear-lanceolate to linear, acute. Whorls distant below, crowded towards ends of branches. *Pedicels of fr. rather thick and rigid*, 1–1½ *times as long as fr. peg.-segs. Outer per.-segs herbaceous, longer than ½ diam. of fr. per.-segs, the apex claw-shaped and forward-curved. Fr. per.-segs* 3–4 mm, *with* lingulate *obtuse apices, all tuberculate, tubercles obtuse in front; margins with rigid setaceous teeth shorter than the width of the segments. Fr.* 1.5–2 mm. Fl. 6–9. $2n = 40, 60$. Hs.

Native. On bare muddy ground beside lakes and at margins of reservoirs, in dried-up ponds and, more rarely, in damp grassy places. England from the Thames to the Humber; N. Somerset; Monmouth, local; absent from Ireland. S. and C. Europe northwards to the Netherlands, S. Sweden and Latvia; temperate Asia.

18. R. maritimus L.   Golden Dock

Like *R. palustris* but plant usually golden-yellow in fr.; whorls usually more crowded; *pedicels of fr. very slender, mostly longer than the fr. per.-segs; outer per.-segs thin, shorter than ½ diam. of fr. per.-segs, horizontally spreading or weakly reflexed; fr. per.-segs with triangular ±acute apices, tubercles acute in front; margins with very fine capillary teeth some of which are longer than the width of the segments; fr.* 1–1.5 mm. Fl. 6–9. $2n = 40$. Th. or Hs.

Native. In similar situations to *R. palustris.* England, local but rather commoner and extending to S. Scotland and the Isle of Man; Ireland, very local. Most of Europe; temperate Asia, Japan; North and South America (introduced).

Section *Axillares* Rech. fil. Perennials. Primary infl. overtopped after flowering by a succession of secondary infl. arising from the axils of the lvs below the primary infl.

*19. R. frutescens Thouars.

R. magellanicus auct. angl.; *R. cuneifolius* Campd.

A rhizomatous perennial c. 25 cm. *Rhizome* c. 1 m *sending up shoots at intervals. Lvs thick and coriaceous, narrowly obovate, ±obtuse, often cuneate at base, c.* 1½ *times as long as wide, undulate, margins finely crisped; petiole short. Infl.* dense, *with very few short simple branches*, secondary infl. sometimes produced from axils of lvs below primary infl. and ultimately overtopping it. *Fr. per.-segs* 4 mm, *ovate-deltate, ±acute, coriaceous, all tuberculate; margins entire.* Fr. c. 2.5 mm. Fl. 7–8. ?Hs.

Introduced. Well naturalized in dune-slacks and near ports. S.W. England, S. Wales, S. Scotland. Native of Argentina, Uruguay, Chile, Bolivia, Peru; introduced in USA and W. Europe.

*20. R. triangulivalvis (Danser) Rech. fil.
Willow-leaved Dock.

R. salicifolius group

An erect perennial 30–50(–100) cm. Stems usually several to each plant, sometimes decumbent at base. *Lvs linear-lanceolate*, gradually narrowed at both ends, *acute*, c. 5 times as long as wide, shortly petiolate or subsessile, pale green, all cauline except in seedling stage. Infl. with simple, arcuate-ascending branches. The terminal infl. when in fr. overtopped by lateral infls. which arise in the axils of lvs below the primary infl. *Fr. per.-segs* c. 3 mm, *olive*, *deltate*, *without tubercles*; *margins entire*. Fr. 2 mm. Fl. 7–8. Germ. 4–5. ?Hp.

Introduced. On waste ground and rubbish-tips, especially near ports. Scattered throughout England, Wales and S. Scotland; Ireland only near Dublin. North America; introduced in N., W. and C. Europe.

8. MUEHLENBECKIA Meissner

Usually dioecious shrubs or climbers. Perianth deeply 5-lobed, becoming succulent in fr. Stamens 8, represented by staminodes in female fls. Stigmas 3, subsessile. Nut trigonous, partly adnate to the accrescent perianth. About 15 spp., New Guinea to western South America.

*1. M. complexa (A. Cunn.) Meissner — Wire Plant.

A spreading or climbing shrub with reddish wiry stems. Lvs 4–8 mm, broadly oblong to suborbicular, entire, coriaceous, shortly petiolate; stipules deciduous. Infl. axillary or terminal, of simple or branched spikes. Per.-segs waxy and white in fr. $2n = 20$. Fl. 7–8. N.

Introduced. More or less naturalized in the Channel Is., Scilly Is. and S.W. England. Native of New Zealand.

86. URTICACEAE

Herbs, small shrubs or rarely soft-wooded trees, often with stinging hairs, usually with cystoliths in the epidermal cells. Lvs alternate or opposite, simple. Stipules usually present. Fls small, generally unisexual and cymose. Perianth 4–5-merous, often persistent and enlarged in fr. Male fls with 4–5 stamens inserted opposite the per.-segs; rudimentary ovary usually present; stamens inflexed in bud, springing open and scattering the pollen at maturity; anthers 2-celled, opening lengthwise. Female fls often with small staminodes; ovary 1-celled, superior, free or adnate to perianth; style simple; ovule solitary, erect. Fr. an achene (in our spp.) or drupe. Seeds usually with endosperm; embryo straight.

About 45 genera and 550 spp., generally distributed. Some spp. yield valuable fibres.

1 Plant normally with stinging hairs; lvs toothed; stems ridged or 4-angled. **3. URTICA**
  Plant without stinging hairs; lvs entire; stems terete. **2**
2 Stems spreading or decumbent, not rooting at nodes; most of the lvs 1 cm or more, distinctly petiolate; fls clustered. **1. PARIETARIA**
  Stems creeping, rooting at nodes; lvs not exceeding 6 mm, subsessile; fls solitary. **2. SOLEIROLIA**

1. PARIETARIA L. — Pellitory.

Annual or perennial herbs. *Lvs alternate, entire*; stipules 0. *Infl.* cymose, axillary, the cymes clustered, dichotomous, *3–several-fld*. Fls hermaphrodite or unisexual, bracteolate, green. Perianth of female fls tubular, 4-toothed, of male and hermaphrodite fls 4-partite. Fr. enclosed in per.-segs.

About 30 spp., mainly in temperate regions.

1. P. judaica L. — Pellitory-of-the-Wall.

P. officinalis auct., non L.; *P. ramiflora* auct.; *P. diffusa* Mert. & Koch

A softly hairy perennial up to 40 cm. Stems terete, much-branched, spreading or decumbent, usually reddish. Lvs up to c. 7 cm, lanceolate to ovate, obtuse to acuminate; petiole shorter than blade, slender. Fls usually unisexual. Female fls terminal. Male fls lateral, surrounded by a calyx-like involucre of 1 bract and 2 bracteoles. Fl. 6–10. $2n = 14$. Hp.

Native. In cracks in rocks and old walls, and in hedgebanks. Widely distributed but rather local in England, Wales and Ireland, rare in Scotland and absent from the north. W. and S. Europe; Macaronesia; N. Africa.

2. SOLEIROLIA Gaud.-Beaup.
(*Helxine* Req.)

A perennial herb. *Lvs alternate, entire*; stipules 0. *Fls* unisexual, green, *solitary*, axillary, surrounded by an involucre of 1 bract and 2 bracteoles. Perianth 4-lobed. Fr. enclosed in perianth and involucre.

One sp. in the Balearic Is., Corsica and Sardinia.

*1. S. soleirolii (Req.) Dandy
Mind-your-own-business, Mother of thousands.

A slender creeping puberulent herb forming dense evergreen mats. Stems 5–20 cm, very slender, rooting freely at nodes. Lvs 2–6 mm, suborbicular, subsessile. Female fls enclosed in the ±connate involucre; perianth tubular, narrowly and shortly 4-lobed. Male fls with a 4-lobed perianth. Fl. 5–10. Ch.

Introduced. Commonly planted in rock gardens and cool greenhouses and naturalized on walls and damp banks, particularly in S.W. England and S.W. Ireland.

3. URTICA L. — Nettle.

Annual or perennial *herbs usually with stinging hairs*. Stems ridged or 4-angled. *Lvs opposite, toothed*; stipules free. Infl. lateral, arising from an often suppressed lfy branch, usually spike-like with clustered cymes. Fls green, unisexual. Perianth 4-merous. Female fls with unequal per.-segs, the larger enclosing the fr.

About 50 spp. in temperate regions.

Annual; plant monoecious; petiole of lower lvs c. $\frac{2}{3}$ as long as blade. **1. urens**
Perennial; plant dioecious; petiole of lower lvs not more than $\frac{1}{2}$ as long as blade. **2. dioica**

1. U. urens L. Small Nettle.

An *annual* herb 10–60 cm. *Lvs* 1.5–4(–6) cm, ovate or elliptical, obtuse to acuminate, incise-dentate; *petiole c. ⅔ as long as blade.* Infl. c. 1 cm, borne on usually well-developed lfy lateral branches. *Achene c.* 1.6 mm. Fl. 6–9. 2n = 24, 26, 52. Th.

Native. In cultivated ground and waste places, particularly on light soils. Not uncommon but mainly in the east and rather local throughout the British Is. North temperate regions; introduced in North America.

2. U. dioica L. Common Nettle, Stinging Nettle.

A coarse hispid *perennial* 30–150(–250) cm. Roots much-branched, very tough, yellow. Stems creeping and rooting at the nodes, giving rise to erect shoots in spring. *Lvs* 4–8(–15) cm, ovate, acuminate, coarsely serrate,

usually cordate at base *petiole, of the lower not more than ½ as long as blade.* Infl. up to c. 10 cm, lateral branches usually suppressed. *Achene* c. 1.2 mm. Fl. 6–8. 2n = 48, 52. Hp.

Occasionally devoid of stinging hairs. The young shoots are eaten like spinach.

Native. In hedgebanks, woods, grassy places, fens, and near buildings, especially where the ground is covered with litter or rubble. Abundant and generally distributed throughout the British Is.; ascending to 838 m on Ben Lawers. Temperate regions of Europe and Asia; introduced elsewhere.

*****U. pilulifera** L., the Roman Nettle, is a monoecious annual or biennial with the female fls in dense globose heads up to 1 cm diam. *Achene* c. 2.4 mm. It formerly occurred as a rare alien but appears now to be extinct in the British Is.

87. CANNABACEAE

Dioecious herbs without latex. Lvs usually lobed, stipulate. Fls axillary. Male fls pedicellate; perianth 5-partite, the segments imbricate; stamens 5, erect in bud, anthers longer than their filaments; rudimentary ovary 0. Female fls sessile; perianth entire, closely enfolding ovary; ovary sessile, unilocular; ovule solitary, pendent. Fr. an achene, enveloped in the persistent perianth. Seed with fleshy endosperm; embryo curved or spiral. Two genera, widely distributed. The infructescences of *Humulus* are used in brewing; *Cannabis* yields a valuable fibre (hemp), and a narcotic resin which is largely used as a stimulant.

1. HUMULUS L.

Climbing perennial herbs rough with deflexed hairs. Lvs opposite, palmately lobed or entire, petiolate. Fls pendent. Infl. glandular, axillary or the upper forming a lfy terminal panicle. Male infl. much-branched, the fls shortly pedicellate. Female infl. a pedunculate cone-like spike; bracts broad, membranous, imbricate, persistent in fr.

Three spp., one probably native in Europe and W.

Asia but now cultivated in all temperate countries, one in North America, and the third in E. Asia.

1. H. lupulus L. Hop.

An herbaceous perennial climber 3–6 m. Stems climbing by twisting in a clockwise direction. Lvs (4–)10–15 cm, broadly ovate, ±cordate at base, usually deeply 3–5-lobed and coarsely dentate, the lobes acuminate; petiole shorter than to about as long as blade. Male fls c. 5 mm diam. Female 'cone' 15–20 mm, enlarging to c. 30 mm in fr.; bracts c. 10 mm, ovate, acute, pale yellowish-green. Fls 2–3 in the axils of the bracts, each subtended by a bracteole. Fl. 7–8. 2n = 20. Hp.

A valuable constituent of the best beers.

Native. In hedges and thickets, extensively cultivated in certain districts. Widely distributed in England and Wales, but doubtless often an escape from cultivation – always so in Scotland and Ireland. Europe; W. Asia; North America.

*****Cannabis sativa** L., Hemp, an erect annual with usually alternate lvs which are palmately 5–7-lobed almost to the base, was formerly cultivated for its fibre and occurs rarely as a casual in waste places.

88. ULMACEAE

Trees without latex. Lvs alternate, simple, often asymmetrical at base; stipules caducous. Fls clustered, arising from the 1-year-old twigs, hermaphrodite. Perianth herbaceous, shortly 4–8-lobed, lobes imbricate. Stamens the same number as the perianth lobes and opposite to them, erect in bud. Ovary of 2 connate carpels, 1–2-celled. Ovules solitary, pendent. Fr. compressed, dry, winged all round. Endosperm 0; embryo straight.

About 15 genera and 200 spp., mostly in north temperate regions.

1. ULMUS L. Elm.

Large trees, usually with suckers. Lvs ±asymmetrical at base, serrate, hairy or glabrous above, hairy beneath,

at least in the axils of the main veins ('axillary tufts'). Fls protandrous, appearing before the lvs. Perianth campanulate, persistent, usually 4–5-lobed. Anthers reddish. Ovary usually 1-locular. Fr. compressed, broadly winged, wing notched at top.

About 30 spp. in north temperate regions and on mountains of tropical Asia.

As in most trees, lvs of sucker shoots or those on rapidly growing branches and on young trees may differ greatly from those on slow-growing laterals on mature trees ('short shoots'). The latter show less variation within a sp. than the former and are therefore more useful in identification. The description of lvs in the following account refers only to the distal and subdistal lvs of short shoots of mature trees.

Lvs 7–16 cm, usually rough above; petiole usually less than
3 mm.   **1. glabra**
Lvs usually less than 7 cm, rough or smooth above; petiole
more than 3 mm.   **2. minor**

1. U. glabra Hudson   Wych Elm.

U. montana Stokes, *U. scabra* Miller

A rounded tree up to 40 m, often with 2 or 3 large
branches from near the base. Branches ascending and
spreading and forming a closed canopy; suckers 0. Twigs
stout, coarsely pubescent, becoming smooth and ashy-
grey by the third year. Winter buds with rufous hairs
on the scales. *Lvs* (Fig. 47) 7–16 cm, *on average*

2. U. minor Miller   Elm.

Incl. *U. carpinifolia* G. Suckow, *U. procera* Salisb.

Tree up to 40 m, very variable in habit and lf-shape,
suckering freely. Twigs usually slender, often long and
±pendent. *Lvs* (Figs 47, 48) *usually less than* 7 cm,

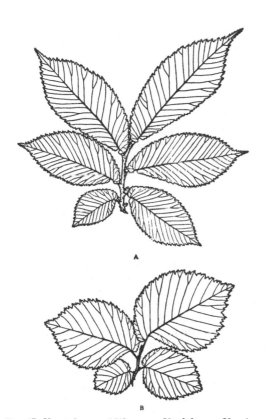

Fig. 47. Short shoots of *Ulmus*. A, *U. glabra*; B, *U. minor*.

Fig. 48. Short shoots of *Ulmus minor*.

10–11 cm, suborbicular or broadly ovate to elliptical,
long-cuspidate, usually scabrid above, coarsely to finely
pubescent beneath; base unequal, the *long side forming
a rounded auricle which overlaps and often hides the
short petiole*; total number of teeth usually more than
130. Fr. 15–20 mm, broadly obovate to elliptical; *seed
central*. Fl. 2–3. Fr. 5–6. $2n = 28$. MM.

Native. In woods, hedges and beside streams. Widely
distributed throughout the British Is., especially in hill
districts and mainly in the west and north. Most of Europe, N. and W. Asia; N. Africa (? introduced).

often smooth above; *petiole more than* 3 mm, usually
not concealed by the lf-base; total number of teeth
usually less than 110. *Seed centred $\frac{2}{3}$ of the way from
base to apex of fr.* Fl. 2–3. Fr. 5–6. $2n = 28$. MM.

Probably introduced. Formerly abundant in the
southern half of England but now largely destroyed by
a virulent strain of the fungus *Ceratocystis ulmi*, which
is spread by elm bark beetles, *Scolytus* spp.

Probably a native of S. Europe and N. Africa.

Rarely reproducing by seed but suckering freely and
so giving rise to distinctive local populations, a number
of which have been given specific status. Among the
better marked of these are the following: English Elm
(*U. procera* Salisb.), a large tree with stout, straight
trunk and suborbicular lvs; the East Anglian Elm (*U.
carpinifolia* G. Suckow), with arching branches and long
slender pendent twigs; the Lock Elm (*U. plotii* Druce)
with a scanty crown and with main stem and larger
branches curving to one side at the top – this occurs
rather locally in the Midlands; the Jersey Elm

(*U. angustifolia* (Weston) Weston) with a strong main stem and mostly rather short ascending branches giving a narrowly conical outline – this was widely planted along streets in towns. All these, plus *U. glabra*, are inter-fertile (see R. Melville in C. A. Stace (1975), (pp. 292–299). The most recent account of the British members of the genus is *Elm* by R. H. Richens (Cambridge University Press, 1983).

89. MORACEAE

Trees or shrubs, rarely herbs, with milky latex. Lvs usually alternate, simple stipulate. Fls small, unisexual, in dense infls often with an expanded axis; per.-segs usually 4; stamens usually equalling in number and opposite per.-segs; carpels 2, 1 usually aborted; ovule 1, usually pendent; styles usually 2. Fr. a drupe or achene; frequently a multiple fr. is formed from the infl. by the axis becoming fleshy. Endosperm present or 0.

About 53 genera and 1400 spp., tropical and subtropical, a few temperate.

1. Ficus L.

Usually monoecious trees or shrubs. Lvs convolute in bud; stipules intrapetiolar, at first forming a cup, soon falling and leaving a circular scar. Fls borne inside a fleshy hollow axis with a small opening. Per.-segs 2–6 in male, more in female. Stamens usually 1–2. Multiple fr. consisting of the fleshy axis enclosing the individual frs.

About 800 spp., mainly tropical.

*F. carica L.   Fig.

Deciduous shrub or small tree to 10 m. Lvs 7–20 cm, palmately lobed into 3–5 obovate lobes, scabrid above, tomentose beneath. Infl. pear-shaped, in fr. 5–8 cm, green, brownish or purple. Pollinated by a gall-wasp. Fr. 7–10. MM.

Cultivated for its fr., occasionally self-sown and ±naturalized.

90. JUGLANDACEAE

Monoecious trees. Lvs alternate, pinnate, exstipulate. Fls solitary in the axils of bracts. Male fls in lateral, usually drooping catkins; bracteoles 2, rarely 0; perianth 0 or small and 1–5-lobed; stamens 3–40, filament short; rudimentary ovary sometimes present. Female fls solitary or in spikes, terminal; bracteoles and perianth as in the male. Ovary inferior, of 2 carpels, 1-celled; ovule 1, basal, orthotropous; style short, stigmas 2. Fr. a drupe or small nut; endosperm 0.

Seven genera and over 50 spp., northern hemisphere, temperate and on mountains in the tropics and in South America. Spp. of *Caryx* Nutt. (Hickory, Pecan) and *Pterocarya* Kunth ('Wing-nut') are sometimes planted.

1. Juglans L.

Deciduous trees. Pith chambered. Buds with few scales. *Male catkins drooping*. Female fls in erect terminal few(sometimes 1)-fld spikes. Bracteoles and *perianth present* in both sexes; perianth 1–5-lobed in the male, 4-lobed in the female. Bracteoles united with the ovary, not persistent in fr. Stamens 8–40. *Stigmas simple*; car-pels median. *Fr. a large indehiscent drupe*; stone (walnut) incompletely 2–4-celled; cotyledons ruminate.

About 15 spp., north temperate regions. S.E. Asia, Jamaica, Central and South America.

*1. J. regia L.   Walnut.

Large tree to 30 m, with spreading crown; main branches tortuous. Bark grey, smooth for many years, finally fissured, not scaling. Buds c. 6 mm, broadly ovoid, blackish, glabrous or the terminal one greyish and tomentose. Twigs greenish or grey, glabrous; lf-scars Y-shaped. Lflets (2–)3–4(–6) pairs, 6–12(–15) cm, obovate or elliptical, acute or acuminate, pubescent when young becoming glabrous except in the axils of the veins beneath. Male catkins 5–15 cm. Fr. 4–5 cm, subglobose, green, gland-dotted, aromatic; stone ovoid, acute, somewhat wrinkled, 4-celled below, easily splitting into halves. Fl. 6. Wind-pollinated, some plants protandrous, others protogynous. $2n = 32$. MM.

Introduced. Planted for its fr., sometimes in wild situations and sometimes ±naturalized in S. England. Native of S.E. Europe and W. and C. Asia to China.

91. MYRICACEAE

Monoecious or dioecious trees or shrubs. Lvs alternate, simple, dotted with resinous glands, aromatic, exstipulate. Fls solitary in the axils of bracts, forming catkins. Perianth 0. Male fl. usually without bracteoles; stamens 2–16, filaments free or united below. Female fl. with 2 or more small bracteoles; ovary 1-celled; ovule 1, basal, orthotropous; style short; stigmas 2, filiform. Fr. a drupe or nut; embryo straight; endosperm 0.

Four genera and about 40 spp., cosmopolitan.

1. Myrica L.

About 35 spp., almost cosmopolitan (except C. and S. Europe, N. Africa, S.W. Asia and Australia).

1. M. gale L.   Bog Myrtle, Sweet Gale.

Usually dioecious, but plants may change sex from year to year. Deciduous shrub 60–250 cm, spreading by suckers. Twigs red-brown, with scattered yellowish glands, becoming dark, ascending at a narrow angle. Buds small, ovoid, obtuse, reddish-brown, with several scales. Lvs 2–6 cm, oblanceolate, cuneate at base, sub-sessile, obtuse or acute, serrate near the apex or sub-entire, grey-green and usually glabrous above at maturity, usually ±pubescent beneath, with conspicuous scattered shining yellowish sessile glands on both sides. Catkins lateral, on the last year's shoots, ascending, forming panicles at the ends of the shoots, appearing before the lvs; male 7–15(–30) mm, bracts glabrous, red-brown, bracteoles 0, stamens c. 4, anthers red; female 5–10 mm in fr., bracts like the male, bracteoles 2, styles red. Fr. dry, compressed, gland-dotted, 2-winged by the adnate, accrescent bracteoles. Fl. 4–5, wind-pollinated. Fr. 8–9. $2n = 48$. N.

Native. Bogs, wet heaths and fens ascending to 550 m, often abundant and sometimes locally dominant. Throughout the British Is. except Shetland and the Channel Is., but local in or absent from many counties in the Midlands, E. England, E. Wales, S.E. Scotland and E. Ireland. N.W. Europe from Scandinavia (69° N) to N.W. Spain, C. Germany and N.W. Russia (Ladoga–Ilmen region); North America from Labrador and Alaska southwards to N. Carolina and Oregon; a subsp. or allied sp. in E. Asia.

*M. caroliniensis Miller   Bayberry.
M. pennsylvanica Loisel.

Deciduous shrub to 3 m. Lvs 4–10 cm, obovate or oblong, obtuse or subacute, shallowly and obtusely dentate near the apex or subentire, pubescent on both sides, gland-dotted. Catkins on the old wood, appearing with but below the lvs. Fr. 3.5–4.5 mm, subglobose, coated with greyish-white wax. $2n = 16$.

Sometimes cultivated; naturalized in the New Forest (Hants). Native of eastern North America from Newfoundland to N. Carolina.

92. BETULACEAE

Dioecious, deciduous trees or shrubs. Buds scaly. Lvs alternate, simple; stipules caducous. Male fls in pendent catkins, three together in the axil of each bract; bracteoles 2–4 to each group of fls; perianth present; stamens 2 or 4. Female fls in erect cylindrical or ovoid catkins, 2–3 in the axil of each bract; bracteoles 2–4 to each group of fls; perianth 0; ovary 2-celled, ovules 1 in each cell, anatropous, pendent; styles 2, free. Fr. a flattened nutlet, 1–3 on the surface of a scale formed from the accrescent fused bract and bracteoles, in a dense cylindrical or cone-like fr. catkin; seed 1; endosperm 0; embryo straight; cotyledons flat. Wind-pollinated.

Two genera and 95 spp., north temperate regions and tropical mountains, C. Andes.

Fls with the lvs; stamens 2, bifid below the anthers; fr. catkin cylindrical, its scales 3-lobed, falling with the fr.
                                                        1. BETULA
Fls before the lvs; stamens 4, entire; fr. catkin cone-like, its scales 5-lobed, persistent.   2. ALNUS

1. BETULA L.

Trees or shrubs. Buds with several scales. Bracteoles 2 to each group of fls. *Stamens 2, bifid below the anthers*; perianth minute. *Female fls 3 to each bract. Fr. catkins cylindrical, the scales 3-lobed, relatively thin, falling with the fr.* Fr. winged, styles persistent.

About 60 spp., north temperate regions and Arctic.

1 Lvs ±ovate, subacute to acuminate; serrate or serrate-dentate; petiole 7 mm or more.   *2*
   Lvs ±orbicular, rounded at apex, crenate; petiole 3 mm or less.   **3. nana**
2 Lvs acuminate, sharply doubly serrate with prominent primary teeth, glabrous; trunk black and fissured into ±rectangular bosses at the base, white above.
                                                      1. pendula
   Lvs acute or subacute, irregularly or evenly serrate or serrate-dentate without markedly projecting primary teeth; trunk ±smooth throughout.
                                                    2. pubescens

1. B. pendula Roth   Silver Birch.
B. verrucosa Ehrh.; *B. alba* sensu Coste

Tree to 25 m with single stem; *bark smooth and silvery-white above, peeling, ±abruptly changing near the base of the trunk to black and fissured into rectangular bosses. Branches ±pendent. Twigs glabrous*, brown, somewhat shining, with ±conspicuous pale warts, especially well developed on vigorous shoots. Buds long, acute, not viscid. *Lf-blades* (2–)2.5–5(–7) cm ovate-deltate, *acuminate*, truncate or broadly cuneate at base, *sharply doubly serrate with the primary teeth very prominent and somewhat curved towards the apex of the lf, glabrous*; petiole 1–2 cm, glabrous. Male infl. 3–6 cm, ±pendent. Female infl. in fr. 1.5–3(–3.5) × c. 1 cm; scales with short broad cuneate base, the lateral lobes broad, spreading and curving downwards, the middle lobe deltate, obtuse; *wings of fr. 2–3 times as broad as achene, the upper edge surpassing the stigmas.* Fl. 4–5. Fr. 7–8. $2n = 28$ (42). MM.

Native. Woods especially on the lighter soils, rare on chalk, colonizing heathland and often forming woods there as a successional stage to sessile oak woodland, sometimes also forming pure woods in the Scottish Highlands, apparently more tolerant of dry conditions than *B. pubescens*. Almost throughout the British Is., but less common in Ireland and absent in Shetland; introduced in the Channel Is. Most of Europe from 69° N in Norway, but only on mountains in the south; W. Siberia, N.E. and C. Asia Minor; Morocco (Rif).

2. B. pubescens Ehrh.   Downy Birch.

Tree to 20 m with single stem, or shrub with several stems; *bark smooth, brown or grey, rarely white, not markedly different at the base of the trunk, sometimes with deep grooves but never broken up into rectangular bosses. Branches spreading or ascending* (sometimes pendent in mountain forms). *Twigs ±pubescent or glabrous, becoming dark brown or blackish, not or scarcely shining, with or without brown resinous warts (?sucker shoots always pubescent). Buds viscid or not. *Lf-blades* 1.5–5.5 cm, very variable in shape, ovate, orbicular-ovate or rhombic-ovate, *subacute or acute*, rounded or cuneate at base, *coarsely and sometimes irregularly serrate or serrate-dentate, the teeth not curved towards the apex of the lf*, usually pubescent at least on the veins beneath or in their axils, of a duller green than *B. pendula*; petiole 7–15(–20) mm. Male infl. 3–6 cm, pendent. Female infl. in fr. 1–4 × 0.5–1 cm; scales with short or long cuneate base, the lateral lobes rounded or nearly square, ±spreading or ascending, the terminal lobe long or short, narrow, oblong- or triangular-lanceolate; *wings of fr.* 1–1½ *times as broad as achene, the upper edge not surpassing the stigmas, often not projecting beyond the achene*. Fl. 4–5. Fr. 7–8. 2n = 56. MM. or M.

Native. In similar places to *B. pendula* but more tolerant of wet and cold conditions, ascending to 760 m. Throughout the British Is., but less common than *B. pendula* in S. England and there only abundant on wet soils, forming pure woods above the limit of oakwood in the north and west. N. and C. Europe northwards to Iceland and N. Russia, locally on mountains southwards from 47° N to C. Spain and Yugoslavia; Caucasus; Siberia (to Yenisei region); N.E. Asia Minor.

A very variable sp. The two following subspp. may be recognized but intermediates occur.

Subsp. **pubescens**

Tree with single stem; branches never pendent. Twigs usually conspicuously pubescent; warts few or 0. Buds not viscid. Lvs relatively large (c. 3–4 cm).

Mainly in southern and lowland areas.

Subsp. **carpatica** (Willd.) Ascherson & Graebner

B. odorata Bechst.; *B. alba* subsp. *odorata* (Bechst.) Dippel; *B. tortuosa* auct., non Ledeb.

Often shrubby or with several stems. Twigs not or sparsely pubescent, covered when young with brown resinous warts as are the young lvs. Buds viscid. Plant with a resinous smell as the lvs unfold. Lvs relatively small (usually less than 3 cm).

Northern and mountain areas; perhaps the only subsp. in the Scottish Highlands.

B. pendula × *pubescens* (*B.* × *aurata* Borkh.), comprising trees variously intermediate between the parents, is widespread in the British Is., often in the absence of the parents. Some are highly sterile, but most are fertile and their status requires further study.

3. B. nana L.   Dwarf Birch.

Shrub to 1 m, the stems mostly procumbent or spreading widely, stiff. Twigs pubescent, not warty, dull dark brown. *Lf-blades* 5–15(–20) mm, *orbicular or obovate-orbicular, rounded at both ends, deeply crenate*, glabrous at maturity; *petiole up to* 3 mm. Male infl. c. 8 mm. Female infl. in fr. 5–10 mm; *scales with* cuneate base

and 3 ±*equal narrow erect lobes at the apex*; *wings of fr. very narrow*. Fl. 5. Fr. 7. 2n = 28. N.

Native. Mountain moors from 245 to 855 m, very local; Perth and Argyll to Sutherland (on the mainland); N. Yorks, S. Northumberland; probably extinct in Lanark. N. and C. Europe southwards to the N. Alps, E. Carpathians and C. Russia, but only on mountains in the south; N. Asia; Greenland.

B. nana × *pubescens* (*B.* × *intermedia* Thomas ex Gaudin) occurs rarely within the Scottish range of *B. nana*.

2. ALNUS Miller

Trees or shrubs. Bracteoles 4 to each group of fls. *Stamens 4, not bifid. Female fls 2 to each bract. Fr. catkins ovoid or ellipsoid, cone-like, the scales 5-lobed, thick and woody, long persistent after the fall of the fr.* Fr. usually winged.

About 35 spp. north temperate regions, south to Assam, Indochina and C. Andes.

Lvs truncate or retuse at apex, bright green on both sides,
   the veins 4–7 pairs; young twigs viscid; bark dark brown;
   female catkins pedunculate.   **1. glutinosa**
Lvs subacute to acuminate, pale or glaucous beneath; the
   veins 10–15 pairs; young twigs not viscid; bark pale grey;
   female catkins ±sessile.   **2. incana**

1. A. glutinosa (L.) Gaertner   Alder.

A. rotundifolia Stokes

Tree to 20(–26) m with ±oblong crown. *Bark dark brown, fissured*. Twigs glabrous, *viscid when young*. Buds purplish, obtuse, stalked, with a large outer scale almost hiding the inner ones. *Lf-blades* 3–9 cm suborbicular or broadly obovate, *truncate or retuse*, rarely rounded, at apex, cuneate at base, irregularly and often doubly serrate-dentate, *bright green and glabrous on both sides* except for tufts in the axils of the veins beneath, glutinous when young; *veins 4–7 pairs*; petiole 1–3 cm. Fls before the lvs. Male catkins 2–6 cm, 3–6 together at the ends of the twigs. Female catkins in fl. 1 cm or less, in fr. 7–28 mm, ovoid or ellipsoid, *pedunculate*. Wing of fr. narrower than the achene. Fl. 2–3. 2n = 28*, 56. MM.

Native. Wet places in woods and by lakes and streams, ascending to 490 m, often forming pure woods in succession to fen or marsh; common almost throughout the British Is. (but planted in Shetland). Europe, except the extreme north and south; W. Asia from W. Siberia to N. and W. Asia Minor; N. Africa.

*2. A. incana (L.) Moench   Grey Alder.

Tree to 20 m, often shrubby. *Bark pale grey, smooth.* Twigs pubescent, *not viscid*. *Lf-blades* 3–10 cm, ovate to elliptical, *acute or acuminate*, usually rounded at base, regularly doubly serrate, dull green above, *grey-green or glaucous* and ±pubescent *beneath*, not glutinous when young; *veins 10–15 pairs*; petiole 1–2.5 cm. Fls before the lvs. Male catkins 2–4 together. Female catkins in fr. c. 2 cm, ±sessile. Wings of fr. about as broad as achene. Fl. 2–3. 2n = 28. MM. or M.

Introduced. Sometimes planted, especially for shelter or on poor soils, mainly in Scotland and sometimes ±naturalized. N. Europe from Scandinavia (70° 30′ N) to N. Russia south in the mountains to the French Alps, Apennines, Albania, Bulgaria and the Caucasus; W. Siberia; a subsp. in eastern North America.

93. CORYLACEAE

Monoecious, deciduous trees or shrubs. Buds scaly. Lvs alternate, simple; stipules caducous. Male fls in pendent catkins, solitary in the axil of each bract; bracteoles 2, united to the bract, or 0; perianth 0; stamens 3–14. Female fls in pairs in the axil of each bract; bracteoles present; perianth present, small, irregularly lobed; ovary inferior 2-celled; ovules 1 in each cell, anatropous, pendent; styles 2, filiform, free or joined at base. Fr. a nut, surrounded or subtended by a lf-like involucre formed from the fusion of the accrescent bract and bracteoles; seed 1; endosperm 0; cotyledons large and fleshy; embryo straight.

Four genera and about 52 spp., north temperate regions and Central America.

Female fls numerous, in pendent catkins; fr. involucre 3-lobed, unilateral; lvs with 9 or more pairs of veins.
                                                    1. CARPINUS
Female fls few, in short erect bud-like spikes; fr. involucre irregularly lobed, surrounding the nut; lvs usually with fewer than 8 pairs of veins.   2. CORYLUS

1. CARPINUS L.

Trees, rarely shrubs. *Male fls without bracteoles. Stamens c. 10, bifid below the anthers; anthers with hair-tufts at the apex. Female fls in terminal pendent catkins;* carpels median. *Fr. a small nut subtended by a large, unilateral, 3-lobed, bract-like involucre.* Cotyledons folded.

Over 25 spp., north temperate regions and Central America.

1. C. betulus L.   Hornbeam.

Tree to 30 m but usually less, with ovoid bushy crown, often coppiced as a shrub. Trunk fluted; bark smooth, grey; branches ascending at an angle of 20–30°. Buds 5–10 mm, narrowly oblong, pointed, pale brown, with numerous scales. Twigs brown, sparsely pubescent. Lvs 3–10(–12) cm, ovate, acute or acuminate, cordate or rounded at base, sharply and doubly serrate, folded along the veins in bud, glabrous except for long appressed hairs on the main *veins (9–15 pairs)* below; petiole 5–15 mm. Male catkins 2.5–5 cm; bracts ovate, greenish. Female catkins c. 2 cm in fl., 5–14 cm in fr.; involucre 2.5–4 cm, 3-lobed, the middle lobe much longer than the laterals, the lobes entire or serrate. Fr. 5–10 mm, greenish, ovoid, compressed, strongly veined, crowned by the persistent perianth. Fl. 4–5. Wind-pollinated. 2n = 64. MM. or M.

Native. Woods and hedgerows, often dominant as a coppiced shrub in oakwoods on sandy or loamy clays in S.E. England; extending as a native to Sussex, Oxford and Cambridge with isolated occurrences in Somerset and Monmouth; planted elsewhere (north to Sutherland). Europe from S. Sweden (57° 11′ N) and White Russia to W. France, S. Italy, Greece and the Caucasus; N. Asia Minor, Iran.

2. CORYLUS L.

Shrubs, rarely trees. *Male fls with 2 bracteoles.* Stamens c. 4, bifid below the anthers; anthers with hair-tufts at the apex. *Female fls few, in erect short bud-like spikes. Carpels transverse. Fr. a large nut surrounded or enclosed by a lobed involucre.* Cotyledons not folded.

About 15 spp., north temperate regions.

1. C. avellana L.   Hazel, Cob-nut.

Shrub 1–6 m with several stems, usually seen coppiced, rarely a small tree. Bark smooth, coppery-brown, peeling in thin papery strips. Buds c. 4 mm ovoid, obtuse; scales several, ciliate. Twigs thickly clothed with reddish glandular hairs. Lvs 5–12 cm, suborbicular, cuspidate, cordate at base, sharply doubly serrate or lobulate, slightly pubescent above, more so beneath, becoming nearly glabrous, *usually with fewer than 8 pairs of main veins*; petiole 8–15 mm, glandular-hispid. Fls appearing before the lvs. Male catkins 1–4 together, 2–8 cm; bracts ovate; anthers bright yellow. Female spikes 5 mm or less; styles red. Fr. in clusters of 1–4, 1.5–2 cm, globose or ovoid, brown, with hard woody shell, surrounded by an involucre about as long as or rather longer than itself; involucre deeply divided into usually toothed lobes. Fl. 1–4. Wind pollinated, protandrous, homogamous or protogynous. Fr. 9–10. 2n = 22. M.

Native. Woods, scrub and hedges, on damp or dry basic and damp neutral or moderately acid soils; the common shrub-layer dominant (as coppice) of lowland oakwoods, sometimes also dominant in the shrub-layer of ashwoods or forming scrub on exposed limestone. Ascends to over 610 m. Common almost throughout the British Is. but absent from Shetland. Europe, except the extreme north and north-east; Asia Minor.

*C. maxima Miller   Filbert.

A larger shrub (to 10 m). Involucre tubular, about twice as long as fr., lobed at the apex. Sometimes planted for its nuts.
Native of Italy and the Balkan Peninsula and W. Asia.

94. FAGACEAE

Monoecious trees or shrubs. Buds scaly. Lvs alternate, simple; stipules usually caducous. Male fls in catkins or in many-fld tassel-like heads (or in *Nothofagus* solitary or in threes); perianth 4–6-lobed; stamens usually twice as many as perianth lobes; rudimentary ovary sometimes present. Female fls in groups of 1 or 3, arranged in spikes or at the base of the male infl.; each group surrounded at the base by an involucre bearing numerous scales; perianth 4–6-lobed; ovary 3- or 6-celled; ovules 2 in each cell, pendent, anatropous; styles 3 or 6. Fr. a 1-seeded nut, in groups of 1–3, surrounded or enclosed by accrescent scaly or spiny cupule formed from the involucre; endosperm 0; cotyledons fleshy.

Eight genera and about 900 spp., absent from tropical and S. Africa, Australia (except S.E.), E. tropical South America, etc.

1 Male fls in pendent tassel-like heads; lvs nearly entire; buds fusiform; fr. completely enclosed by a woody 4-valved cupule.   1. FAGUS
  Male fls in catkins; lvs toothed or lobed, if entire then evergreen and tomentose beneath; buds ovoid.   2
2 Lvs serrate with aristate teeth, not lobed or evergreen; male catkins erect, female fls in lower part of male catkin; cupule prickly, completely enclosing the fr., splitting into 2–4 valves. 2. CASTANEA
  Lvs lobed, or if entire or serrate, evergreen and densely tomentose beneath; male catkins drooping, female fls in separate spikes; cupule cup-like enclosing only the lower half of the fr., not prickly or splitting into valves.   3. QUERCUS

1. FAGUS L.

Deciduous trees. Buds fusiform, acute. *Male fls in tassel-like heads on long peduncles*; perianth campanulate, 4–7-lobed; stamens 8–16. Female fls usually paired, surrounded below by the pedunculate cupule bearing numerous long scales; styles 3, long. *Nuts triquetrous*, 1 or 2, *enclosed in a woody regularly 4-valved cupule* covered with projecting scales or prickles. Germination epigeal.

Ten spp., north temperate regions, Mexico.

1. F. sylvatica L.   Beech.

Large tree to 30(–40) m with broad dense crown. Bark grey, smooth. Twigs brownish-grey in second year. Buds 1–2 cm, fusiform, reddish-brown. Lvs 4–9 cm, ovate-elliptical, acute, broadly cuneate or rounded at base, obscurely sinuate, glabrous except for the long-ciliate margins and silky hairs on the veins and axils beneath; veins 5–9 pairs, prominent beneath; petiole 5–15 mm, hairy. Male fls numerous, peduncles 5–6 cm. Fr. 12–18 mm, brown; cupule c. 2.5 cm, brown, with spreading subulate scales. Fl. 4–5. Fr. 9–10. Wind-pollinated, protogynous. $2n = 24$. MM.

Native. Woods, etc., native in S.E. England north and west to about Hertford, Gloucester, Brecon, Glamorgan, Somerset and Dorset; elsewhere (north to Caithness, throughout Ireland, Channel Is.) planted and often naturalized (regenerating freely in Aberdeen); the characteristic dominant of chalk and soft limestone in S.E. England, also frequently dominant on well-drained loams and sands. W. and C. Europe from S. Scandinavia (60° 30′ N in Norway) to the mountains of C. Spain, Corsica, Sicily and Greece, eastwards to W. Russia (Upper Dnieper region); Crimea.

Increasingly over the past 50 years or so spp. of **Nothofagus** Blume (Southern Beech) have been planted throughout Great Britain on a large enough scale to test their commercial value. The most successful species, *N. obliqua* (Mirbel) Oersted and *N. procera* Oersted, both deciduous trees from Chile, are distinguished from *Fagus* by their sessile or subsessile, 1–3-fld male infls and short styles. *N. procera* seeds abundantly and may become naturalized.

2. CASTANEA Miller

Deciduous trees, rarely shrubs. Buds with 3–4 scales. *Fls in erect* terminal catkins, the female fls at the base, the rest male. Male fls with 6-lobed perianth and 10–20 stamens. Female fls usually 3 in each cupule; ovary 6-celled; styles 7–9, cylindrical; stigma small. *Fr.* large, brown, 1–3, rarely more, *enclosed, except for the styles, in a symmetrical prickly cupule which dehisces rather irregularly by 2–4 valves*. Germination hypogeal.

About 12 spp., north temperate regions.

*1. C. sativa Miller Sweet Chestnut, Spanish Chestnut.
C. vesca Gaertner

Large tree to 30 m with wide crown. Bark dark brownish-grey, fissured, the longitudinal fissures often spirally curved. Twigs olive-brown, glabrous or slightly pubescent, with prominent lenticels. Buds c. 4–5 mm, yellowish-green tinged brownish, ovoid, obtuse. Lvs 10–25 cm, oblong-lanceolate, acute or acuminate, broad-cuneate to subcordate at base, coarsely and regularly serrate-dentate with aristate teeth, glabrous above, pubescent below when young, finally glabrous; petiole 0.5–3 cm. Catkins 12–20 cm, conspicuous because of the yellowish-white anthers; male fls numerous; female fls few. Fr. 2–3.5 cm, deep brown, shining; cupule green, densely covered with long branched spines. Fl. 7. Fr. 10. Pollenfls, visited by various insects. $2n = 24$. MM.

Introduced. Commonly planted (often in pure stands) throughout the British Is.; extensively naturalized in S.E. England. S. Europe from Italy eastwards, extending northwards to Hungary; planted and naturalized in many parts of W., C. and N. Europe; Algeria; Asia Minor, Caucasus, W. Iran.

3. QUERCUS L.

Deciduous or evergreen trees or sometimes shrubs. Buds with numerous scales. Male and female fls in separate infl. *Male in pendent catkins*; perianth 4–7-lobed; stamens 4–12 usually 6. Female fls in each involucre solitary or in spikes; ovary usually 3-celled; styles variously shaped, the stigma on the ±flat inner surface. *Fr.* (acorn) large, solitary, *surrounded below by a cup-like indehiscent symmetrical cupule*. Germination hypogeal. Wind-pollinated.

About 450 spp., northern hemisphere, temperate, subtropical and mountains of the tropics, extending south to the E. Indies and the Andes of Colombia.

1 Lvs evergreen, dark green above, grey-tomentose beneath, entire or remotely serrate or dentate.
                                                            2. ilex
  Lvs deciduous, green and glabrous or pubescent beneath, lobed.   2

2 Cup-scales subulate, spreading; buds surrounded by long subulate persistent stipules; lvs rough above, the lobes acute. **1. cerris**
  Cup-scales ovate, appressed; buds without subulate stipules; lvs smooth above, the lobes obtuse. 3
3 Petiole short (less than 1 cm) or 0; lvs glabrous beneath, with small reflexed auricles at the base; peduncle 2–8 cm. **3. robur**
  Petiole 1 cm or more; lvs stellate-pubescent along either side of the midrib beneath, without reflexed auricles; peduncle short (1 cm) or 0. **4. petraea**

Subgenus 1. CERRIS (Spach) Oersted.

Fr. ripening in second year. Shell of acorn usually tomentose within, with abortive ovules in the lower part; cup-scales usually long, linear, spreading or reflexed.

*1. Q. cerris L.   Turkey Oak.

Deciduous tree to 35 m; bark dark grey, fissured; branches ascending, ultimately forming a wide crown. Twigs pubescent at first, brown and subglabrous in the second year. *Buds* small, ovoid, pubescent, *with long* (to 25 mm) *subulate stipules below. Lvs* 5–10(–18) cm, very variable in shape and lobing, ±oblong in outline, cuneate to rounded at base, usually acute, pinnately lobed to lyrate-pinnatifid, greyish pubescent on both sides when young, later *dull green and somewhat rough with very small scattered hairs above*, somewhat paler beneath and finely stellate-pubescent especially on the veins; *lobes* 7–8 pairs, unequal, ovate-triangular, *acute or subobtuse*, mucronulate, some of them often lobed; petiole 1–2.5 cm; stipules long, subulate, persistent. Male catkins 5–8 cm; stamens usually 4. Female spikes 1–5-fld; styles 4. Fr. spike short (0–2 cm). Fr. 20–35(–50) mm, ±oblong, reddish-brown; *cup* covering c. ½ the fr., 10–15(–30) mm diam., *clothed with long* (to 1 cm) *spreading or somewhat reflexed subulate scales.* Fl. 5. Fr. 9. $2n = 24$. MM.

Introduced. Commonly planted and quite naturalized in many places, at least in S. England on acid soils, sporadically present northwards to W. Inverness. Native of S. and S.C. Europe and S.W. Asia (but not Iberian Peninsula); Syria and eastwards throughout Asia Minor.

Q. × hispanica Lam. (*Q. cerris × Q. suber* L.), differing in the half-evergreen, smaller, less deeply lobed lvs and somewhat corky bark, is rather frequently planted in parks, etc.

Subgenus 2. SCLEROPHYLLODRYS O. Schwarz.

Shell of acorn tomentose within, with abortive ovules in lower part; cup scales usually acuminate, flat, appressed; fr. ripening in first year after flowering.

*2. Q. ilex L.   Evergreen Oak, Holm Oak.

Evergreen tree to 30 m with broad dense crown; bark grey, scaly. Twigs densely tomentose when young, grey and subglabrous in the second year. Buds small, ovoid, tomentose, without subulate stipules below. *Lvs* 2–6(–9) cm, very variable in shape and toothing even on the same tree, ovate, oblong, elliptical or lanceolate, cuneate to subcordate at base, obtuse or acute, *entire or remotely serrate or dentate*, the teeth often mucronate or aristate, *dark green and glabrous above, densely grey-tomentose beneath*; petiole 5–10(–15) mm; stipules linear, caducous; lvs of young trees or on sucker shoots often sinuately spine-toothed, ±holly-like, green and stellate-pubescent beneath. *Male catkins* c. 3–5 cm. *Female spikes* 1–4-*fld*; styles 3–4. Fr. spike short. Fr. 2–4 cm, ovoid or ellipsoid; *cup* usually covering less than ½ the fr., grey-tomentose, *clothed with small ovate appressed imbricate scales.* Fl. 5. Fr. 9. $2n = 24$. MM.

Introduced. Commonly planted, especially in coastal areas, northwards to S. Lancs and sporadically throughout Ireland, sometimes naturalized in S. England. Native of the Mediterranean region and Portugal, extending north in W. Europe to Brittany.

Subgenus 3. QUERCUS.

Shell of acorn subglabrous within, with abortive ovules in lower part; cup scales usually short, appressed; fr. ripening the first year after flowering.

3. Q. robur L.   Common Oak, Pedunculate Oak.

Q. pedunculata Ehrh.

Large *deciduous* tree to 30(–40) m with broad crown; bark brownish-grey, fissured. Twigs glabrous, greyish-brown. Buds 2–5 mm, ovoid, obtuse, glabrous, without subulate stipules below. *Lvs* 5–12 cm, obovate-oblong, obtuse, rounded to cordate at base *with a reflexed auricle on either side* (at least on most of the lvs), pinnately lobed, *dull* green and glabrous *above*, paler beneath and sometimes with simple hairs when young, soon *glabrous*; lobes 3–5(–6) pairs, obtuse, rather unequal; stipules linear, caducous; *petiole to* 5 (rarely 10) *mm, often almost* 0. Male catkins 2–4 cm; stamens (4–)6–8(–12). Female spikes 1–5-fld, pedunculate; styles 3, obvious. *Fr. peduncles* 2–8 cm. Fr. 1.5–4 cm, ellipsoid or oblong, brown. *Cup* 1.5–2 cm diam., covering ⅓–½ of the fr., *clothed with small ovate appressed imbricate scales.* Fl. 4–5. Fr. 9–10. $2n = 24$. MM.

Native. Woods, hedgerows, etc.; reaching nearly 460 m in Derry and on Dartmoor, but rarely occurring over 300 m. From Caithness southwards and throughout Ireland; the characteristic dominant tree of heavy and especially basic soils (clays and loams) and thus of most of S., E. and C. England, sometimes also dominant or co-dominant with *Q. petraea* on the damper acid sands; in other parts of the British Is. mainly on alluvium; not thriving on acid peat or shallow limestone soils; often planted and perhaps not native in some northern and western areas. Most of Europe, except the extreme north and parts of the Mediterranean region; Caucasus, east to the Urals.

Q. petraea × robur (*Q. × rosacea* Bechst.) is rather frequent where the parents occur together and is most common in Scotland and N. England; in N. Ireland (Antrim). Variably fertile due to introgressive hybridization.

4. Q. petraea (Mattuschka) Liebl.

Durmast Oak, Sessile Oak.

Q. sessiliflora Salisb.

Differs from *Q. robur* as follows: usually branching higher; crown rather narrower. Buds often slightly larger (to 6 mm) with ciliate scales. *Lvs* cuneate to cordate at base, *without reflexed auricles, somewhat shining above*, persistently *stellate-pubescent with large hairs along either side of the base of the midrib* and main veins *beneath*, sometimes with smaller stellate hairs on the surface also; lobes more regular, 4–6 pairs; *petiole* 1–2.5 cm. Stigmas subsessile. *Fr. peduncle* 0–1 cm. Fl. 4–5. Fr. 9–10. 2n = 24. MM.

Native. Woods from Sutherland and Caithness southwards (not Channel Is.) and throughout Ireland mainly on acid soils; the characteristic dominant tree of the siliceous soils of N. and W. England, Wales, Scotland and Ireland, sometimes dominant or co-dominant with *Q. robur* on acid soils in southern England; forming woods up to 460 m in Cumberland. W., C. and S.E. Europe from nearly 62° N in Norway to the mountains of Sicily and C. Albania.

Subgenus ERYTHROBALANOS (Spach) Oersted.

Shell of acorn tomentose within; abortive ovules apical; cup-scales appressed. Fr. usually ripening in the 2nd year after flowering.

*Q. borealis Michx. fil. var. *maxima* (Marshall) Ashe (*Q. rubra* L. sec. Duroi, *Q. maxima* (Marshall) Ashe)   Red Oak.

Deciduous tree to 25(–50) m. Lvs 12–22 cm, oblong, with 3–5 pairs of triangular acute bristle-pointed lobes reaching about half-way to midrib, glabrous except for brownish tufts of hairs in the axils of the main veins beneath, often turning dark red in autumn. Fr. 2.5–3 cm; cup saucer-shaped, enclosing only base of acorn.

Often planted in parks, etc., for ornament and occasionally for forestry. Rarely fruiting. Native of North America from Nova Scotia to Florida and Texas.

95. SALICACEAE

Dioecious deciduous trees or shrubs. Lvs alternate, simple, stipulate. Fls in catkins, each fl. solitary in the axil of a bract (scale). Bracteoles 0. Perianth 0, but fls (of both sexes) usually with a cup-like disk or 1 or 2 small nectaries which possibly represent a perianth. Stamens 2–many; filaments long and slender, occasionally connate. Ovary of 2 carpels, 1-celled, with 2 or 4 parietal placentae; ovules numerous (rarely 4), anatropous; styles 1 or rarely 2; stigmas 2, often bifid. Fr. a 2-valved capsule; seeds enveloped by long silky hairs arising from the funicle; endosperm 0; embryo straight.

Three genera, the third (*Chosenia*) with 1 sp. in N.E. Asia.

Catkin scales toothed or laciniate; stamens numerous; fls with a cup-like disk; buds with several outer scales
1. POPULUS
Catkin scales entire; stamens 5 or fewer; fls with 1 or 2 nectaries, without disk; buds with 1 outer scale. 2. SALIX

1. POPULUS L.

Trees. Buds with several outer scales; terminal bud usually present. Fls appearing before the lvs, in pendent catkins, *scales toothed or laciniate*; each fl. *with a cup-like disk*; *stamens 4–many*, anthers red or purple. Wind pollinated.

About 35 spp., north temperate regions. Some spp. and hybrids, besides the following, are sometimes grown and some modern hybrids are now being grown commercially and are likely to become commoner.

1 Young lvs white- or greyish-tomentose on one or both surfaces.   2
  All lvs glabrous or subglabrous.   3
2 Mature lvs (except of sucker shoots) tomentose beneath; blade distinctly palmately lobed.   1. alba
  Mature lvs glabrous or subglabrous; blade not or usually only shallowly lobed, with coarse, obtuse sinuate teeth.   3. ×canescens

3 Lvs (except of sucker shoots) broadly ovate to suborbicular, the margin coarsely and obtusely sinuate-dentate.   2. tremula
  Lvs (except of sucker shoots) ovate to ovate-deltate or cordate, the margin shallowly serrate or serrulate.   4
4 Mature lvs whitish beneath; young lvs with strong balsam scent.   5
  Mature lvs pale green beneath; young lvs without balsam scent.   6
5 Lvs broadly ovate, abruptly narrowing to cuspidate apex, the base conspicuously cordate. 6. candicans
  Lvs ovate, gradually narrowing to acute apex, the base cuneate or truncate, rarely subcordate.
                                            7. trichocarpa
6 Branches erect or suberect to form a conspicuously columnar crown.   4. nigra var. *italica*
  Branches somewhat ascending to arching downwards, to form a wide crown.   7
7 Branches arching downwards; trunk usually with conspicuous swellings (bosses); lvs usually with cuneate to truncate base.   4. nigra
  Branches spreading or somewhat ascending; trunk without bosses; lvs usually with truncate to cordate base.   5. ×canadensis

Section 1. *Populus.* Bark smooth, grey, with conspicuous rhombic lenticels, rough only at the base of old trunks. Buds not scented when unfolding. Lvs lobed or coarsely toothed, glabrous or tomentose; petiole laterally compressed or scarcely compressed. Catkin scales ciliate with long hairs. Stamens 5–12(–20).

*1. P. alba L.   White Poplar, Abele.

Robust tree 15–20(–25) m, freely suckering, with wide, spreading crown. *Young twigs densely white-tomentose*, becoming glabrous, glossy, dark brown after a year or more; buds c. 5 mm, ovoid, subacute, the scales tomentose at base, dark brown above. Lvs of short lateral shoots and at base of long, leading shoots broadly ovate, obtuse, with irregular, obtuse, sinuate lobes, sub-glabrous or sparsely tomentose; *lvs towards apex of long shoots* palmately lobed, the blade 3–9 × 3–10 cm, *densely white-tomentose beneath*, ultimately dark green and glabrous above, the lobes acute, irregularly and indistinctly dentate; petiole 5–6 cm, subterete, tomentose. Catkin-scales narrowly ovate to obovate-cuneate, with irregular, rounded or ±deltate, long-ciliate crenations or teeth at apex. Male catkins 4–7(–10) × 0.8–1 cm; stamens 5–10; anthers purple. Female catkins 2–5(–6) × c. 0.6 cm; stigmas greenish, divided almost to base into 4 linear lobes; ovary ovoid, glabrous. Fl. 2–3, well before lvs appear. 2n = 38. MM.

Introduced. Rather frequently planted, especially in S. Britain, but apparently not truly naturalized. Native of C., E. and S.E. Europe to C. Asia.

Male trees are extremely rare in the British Is.

2. P. tremula L.   Aspen.

Tree up to 20 m, with wide, much-branched crown, but often a shrub in exposed or poor sites, freely suckering. *Twigs usually glabrous, dull greyish-brown;* buds 5–10 mm, ovoid, acute, the scales shiny, brown, sticky as buds open. Lvs of normal shoots with blade 1.5–8 cm, broadly ovate to suborbicular or oblate-orbicular, truncate or shallowly cordate at base, with irregular, coarse, obtuse, marginal serrations, dark green above, paler beneath; petiole c. 4–7 cm, strongly laterally compressed, glabrous. Lvs of sucker shoots with blade 5–12 cm, ovate, acute, cordate at base, irregularly serrate, sparsely pubescent; petiole 1.5–3 cm, subterete, pubescent. Catkins cylindrical, 5–8 × 1–1.5 cm; scales oblong, irregularly and deeply laciniate at apex, with dense, long, white hairs. Male fls with 6–12(–15) stamens; anthers reddish-purple. Female fls with pinkish or purplish stigmas divided into 2 or more irregular, short lobes; ovary flask-shaped, scabridulous. Fl. 2–3, well before lvs appear. 2n = 38. MM.

Native. Throughout the British Is., ascending to 505 m. Most of Europe, but only on the mountains in the south; temperate Asia to China and Japan.

*3. P. × canescens (Aiton) Sm.   Grey Poplar.

P. alba × tremula

Robust tree 20–30(–50) m, with wide, spreading crown. Twigs at first densely white-tomentose, later glabrous and then dark grey-brown; buds ovoid, subacute, the scales tomentose at base. Lvs of short lateral shoots and at base of long, leading shoots 3.5–6(–9) cm, suborbicular, sinuate-dentate, glabrous or subglabrous; petiole often distinctly compressed, glabrous. Lvs of upper parts of long shoots 6–8(–10) cm, broadly ovate, coarsely and irregularly erose-dentate, sometimes shallowly lobed, subglabrous and dark green above, initially sparsely greyish-tomentose beneath, later becoming ±subglabrous; petiole subterete, floccose-tomentose. Male catkins 4–9 × 1–1.2 cm, cylindrical-caudate; scales flabellate, coarsely and irregularly erose-dentate, long-ciliate at apex, brown; stamens 8–15; anthers reddish.

Female catkins 4–6 × c. 0.5 cm, narrowly cylindrical; scales narrowly oblong, irregularly laciniate, sparsely ciliate; stigmas yellowish or pinkish, deeply divided into 2 lobes or with at least 4 shorter, linear lobes. Fl. 3, well before lvs appear. 2n = 38. MM.

Introduced, probably from Holland before 1641. Common in S. England, rather infrequent elsewhere. Europe from France, Belgium and Holland eastwards to the Caucasus.

Male plants predominate. Usually propagated clonally, but some plants produce viable seed.

Section 2. Aigeiros Duby. Bark furrowed. Buds viscid but not scented when unfolding. Lvs green on both sides, glabrous or pubescent when young, toothed; teeth numerous, with a narrow translucent border; petiole laterally compressed. Catkin scales laciniate, glabrous. Stamens 8–60. Stigmas greenish, stout.

4. P. nigra L.   Black Poplar.

Robust tree up to 30(–35) m, with wide, rounded crown and stout branches *often arching downwards*, rarely suckering; trunk usually with *conspicuous swellings (bosses)*. Twigs glabrous or subglabrous, bright yellowish-brown; buds up to 10 mm, narrowly ovoid, acuminate, the scales shiny, dark brown. Lf-blade 5–10 cm (sometimes larger on strong sucker or coppice shoots), deltate-ovate, acute to acuminate, *broadly cuneate to ±truncate at base*, obtusely serrate, often indistinctly so, dark green above, paler beneath, glabrous; petiole 3–7 cm, laterally compressed, pubescent when young, later glabrous. Catkins 3–5 × 0.6–0.7 cm, the female to 10 cm in fr., caudate-cylindrical; scales oblong, deeply laciniate at apex, membranous, greenish or brownish, soon deciduous. Male fls with 12–15(–20) stamens; anthers crimson. Female fls with 2 deeply bifid, greenish stigmas; ovary subglobose, glabrous. Fl. 3–4, before lvs appear. 2n = 38. MM.

Native. Along river valleys in Britain south of the Mersey and Humber, but only introduced in Cornwall and W. Wales, planted elsewhere in the British Is. S., C. and E. Europe, eastwards to C. Asia (Kazakhstan).

Var. *italica* Moench, the Lombardy Poplar, differs principally in its erect or suberect branches, which form a conspicuously columnar crown, and in the absence of bosses on the trunk. Frequently planted it was introduced to the British Is. from N. Italy about 1758. Plants are male and propagated clonally.

*5. P. × canadensis Moench   Black Italian Poplar.

P. deltoides Marshall × nigra

Large tree up to 30(–40) m, with spreading to ascending branches forming a wide, fan-shaped crown; trunk without bosses. Twigs glabrous, shiny olive-grey; buds 10–20 mm, narrowly ovoid, acuminate, greenish-brown, viscid, the scales rather few and large, acute. Lf-blade 6–10 cm, deltate-ovate, long-acute, truncate or shallowly cordate at base, sometimes rounded, prominently and obtusely serrate, bright shiny green above, somewhat paler beneath, glabrous; petiole 4–10 cm, strongly laterally compressed, glabrous, often with 1–2 small glands at junction with blade. Catkins 3–6 × 0.8–1 cm, ovoid-cylindrical, shortly stalked; scales broadly flabellate, deeply fimbriate-laciniate at apex, membranous, purplish distally. Male fls with 20–25 or more stamens; anthers crimson. Fl. 3–4, well before lvs appear. $2n = 38$. MM.

Introduced. Widely planted throughout the British Is. northwards to Sutherland. Generally supposed to have arisen in France about the middle of the 18th century and to have been brought to Britain somewhat later.

The above description applies to var. *serotina* (Hartig) Rehder, which is always male. Var. *marilandica* (Box ex Poiret) Rehder, of which only female plants are known, was probably introduced from Holland about 1800 but is much less frequently planted in the British Is. than var. *serotina*. It differs most obviously from var. *serotina* in its broadly domed crown, spreading branches, yellowish-grey twigs, pale brown or greenish catkin-scales and female fls with a short, stout style, 2–4 yellowish-green stigmas and a subglobose, glabrous ovary.

Section 3. *Tacamahaca* Spach. Bark furrowed. Buds very viscid, strongly balsam-scented when unfolding. Lvs usually very pale beneath, glabrous or slightly pubescent, toothed, teeth numerous, without a translucent border; petiole not flattened. Catkin scales laciniate, glabrous. Stamens c. 20–60. Stigmas stout.

*6. P. candicans Aiton　　　　　　Balm of Gilead.
P. × gileadensis Rouleau

Tree up to 25 m, with spreading and suberect branches giving a rather wide crown, suckering prolifically. Twigs at first angular and sparsely pubescent, becoming subterete and shiny dark brown; buds 10–13 mm, narrowly ovoid, acuminate, shiny, viscid, the scales acute, puberulent. Lf-blade 5–15 cm (longer on sucker shoots), broadly cordate, acute, cuspidate, serrate, with obtuse, finely ciliate teeth; petiole 3–7 cm, slightly flattened on upper surface, pubescent, with 2 glands at or near junction with blade. Female catkins 4–6 cm, to 16 cm in fruit, cylindrical, very shortly stalked; scales broadly flabellate, deeply pectinate-laciniate, pale, glabrous, soon falling; style very short; stigmas 2, wide, deflexed, irregularly lobed, at first yellowish, later pink. Fl. 3, just before lvs appear. $2n = 38$. MM.

Introduced. Formerly commonly planted, and now persisting by moist roadsides and stream banks in most of the British Is.

Origin uncertain. Possibly a hybrid between the North American *P. balsamifera* L. (*P. tacamahaca* Miller) and *P. deltoides* Marshall. Only female plants known in the British Is. and in North America, whence it was apparently introduced about 1772.

*7. P. trichocarpa Torrey & A. Gray ex Hooker

Tree up to 35 m, with suberect or ascending branches and rather narrow, conical crown, infrequently suckering. Twigs at first angular and puberulent, becoming ±terete, yellowish-grey, glabrous; buds 10–15 mm, narrowly ovoid, acuminate, shiny, viscid, with few scales. Lf-blade 5–15(–23) cm, ovate, acute, cuneate to slightly cordate at base, shallowly serrate with obtuse, glandular teeth; petiole up to 4 cm, subterete or narrowly channelled above, puberulent or glabrous, sometimes with 2 inconspicuous glands at or near junction with blade. Male catkins 5–9 cm, cylindrical, very shortly stalked; scales broadly flabellate, deeply and irregularly laciniate, pale brown, sparsely hairy, soon falling; stamens 30–60; anthers crimson. Fl. 3–4, just before lvs appear. MM.

Introduced. Widely planted in the British Is. but not naturalized. Native of western North America from California to Alaska.

Only male trees are known in the British Is.

2. SALIX L.

Trees or shrubs. Buds with 1 outer scale, terminal bud 0. Fls appearing before or after the lvs, in usually erect catkins; *scales entire; each fl. with 1 or 2 small nectaries; stamens 2–5(–12)*, usually 2. Insect pollinated.

About 500 spp., mainly north temperate and arctic regions, some tropical and south temperate but absent from Australasia and the E. Indies.

In general, the British spp. of this genus are fairly clearly circumscribed. However, the identification of plants is often difficult because of various factors – the frequency of hybrids, a number of which are planted away from their parents, propagation by cuttings being easy, and the dioecism, with catkins often being produced before the lvs appear or develop fully, being the most significant. Nevertheless, over the last 100 years the extent of hybridization has been overemphasized. With relatively few exceptions there are no hybrids between spp. belonging to subgen. *Salix* and those belonging to either subgen. *Chamaetia* (dwarf alpine willows) or subgen. *Caprisalix* (sallows and osiers). Many of the hybrids are very local and most are not spontaneous, being introduced for various reasons.

The following account considers the species present in the British Is. and the most conspicuous hybrids. In addition to the usual general key, using all available characters to identify the taxa, two supplementary keys are provided to assist in the identification of male and female plants of those species in which catkins are produced before the lvs. For more detailed information see R. D. Meikle, *Willows and Poplars of the British Isles* (Bot. Soc. Brit. Is., London, 1984).

General key to species

1 Small spreading or creeping shrubs usually less than 1 m high.　　　　　2
　Trees and spreading to erect shrubs more than 1 m high.　　　　　9

2 Lvs entire or subentire.   3
Lvs serrate or crenate.   7
3 Stipules large, persistent.   **19. lanata**
Stipules small, caducous, or absent.   4
4 Lf-veins conspicuously reticulate; petioles usually at least 10 mm.   **23. reticulata**
Lf-veins not conspicuously reticulate; petioles usually less than 5(–10) mm.   5
5 Lvs pale grey or whitish beneath, usually softly tomentose or lanuginose.   **18. lapponum**
Lvs pubescent or silvery sericeous beneath, not pale grey or whitish.   6
6 Lf-veins indistinct; catkins lateral, sessile, appearing slightly before lvs.   **17. repens**
Lf-veins rather prominent; catkins terminal on short lfy shoots.   **22 × 17. ×cernua**
7 Creeping dwarf shrub, rarely more than 6 cm high; catkins terminal on short, leafy shoots.   **22. herbacea**
Spreading or decumbent shrub, usually more than 6 cm high; catkins lateral on the shoots.   8
8 Stipules conspicuous, persistent.   **21. myrsinites**
Stipules small, caducous, or absent.   **20. arbuscula**
9 Lvs ovate or obovate to oblong, elliptical or suborbicular, not more than 3 times as long as wide.   10
Lvs linear to lanceolate, oblanceolate or narrowly oblong-elliptical, more than 3 times as long as wide.   23
10 Twigs with a whitish waxy bloom   11
Twigs without bloom   12
11 Twigs reddish-brown; lvs oblong or narrowly obovate; petiole hairy above.   **7. daphnoides**
Twigs violet in winter; lvs linear-lanceolate; petiole glabrous.   **8. acutifolia**
12 Catkins developing before lvs.   13
Catkins developing with lvs.   21
13 Lvs regularly acute or shortly acuminate; catkins 3–5(–7) cm, cylindrical.   **10. ×calodendron**
Lvs not regularly acute or shortly acuminate; catkins 1–3(–5) cm, ovoid, obovoid or cylindrical.   14
14 Stipules conspicuous, persistent.   15
Stipules caducous or absent except on very robust growths.   16
15 Lvs rugose, with undulate margin; twigs slender, divaricate.   **14. aurita**
Lvs smooth or slightly rugose, with ±flat margin; twigs stout, not or little divaricate.   **13. cinerea**
16 Lvs persistently pubescent, sericeous or tomentose beneath.   17
Lvs becoming sparsely pubescent to subglabrous (rarely glabrous) beneath at maturity.   19
17 Lf-margin undulate, irregularly serrate; lvs green above, densely pubescent beneath, the veins prominent.   **12. caprea**
Lf-margin flat or recurved, entire or subentire; lvs at first silvery-pubescent on both surfaces, remaining sericeous beneath at maturity.   18
18 Lvs 3–7 × 1.5–4.5 cm; robust shrub or small tree.   **12. caprea** var. **sphacelata**
Lvs up to 5 × 0.4–2.5 cm; small slender shrub.   **17. repens**
19 Mature lvs with at least some reddish-brown hairs beneath.   **13. cinerea** subsp. **oleifolia**
Mature lvs without reddish-brown hairs.   20
20 Twigs yellowish, greyish, reddish or purplish; lvs often opposite or subopposite distally on twigs; cat-

kins sessile.   **6. purpurea**
Twigs shiny brown; lvs alternate; catkins with densely villous stalk up to 1 cm.   **13. × 16. ×laurina**
21 Mature lvs thin and papery, dull green above, pubescent, rarely subglabrous beneath; twigs dull brown, usually pubescent.   **15. myrsinifolia**
Mature lvs subcoriaceous, shiny green above, glabrous or subglabrous; twigs shiny brown, glabrous.   22
22 Lf-margins and petiole eglandular; lf-blade 2–6 × 1–5 cm; catkin-scales dark brownish; stamens 2; ovary densely hairy.   **16. phylicifolia**
Lf-margins and top of petiole glandular; lf-blade 5–12 × 2–5 cm; catkin-scales pale yellow; stamens (4–)5–8(–12); ovary glabrous.   **1. pentandra**
23 Lvs distinctly and closely serrate or serrulate.   31
Lvs entire or subentire, sometimes remotely and irregularly toothed or serrulate only towards apex.   24
24 Twigs and lvs glabrous; lvs frequently opposite or subopposite; male fls with 2 connate stamens appearing as 1.   **6. purpurea**
Twigs and lvs pubescent, at least at first; lvs usually alternate; male fls with 2 free or partly connate stamens, or male fls absent.   25
25 Lvs becoming glabrous or sparsely puberulent on both surfaces at maturity.   **6. × 9. ×rubra**
Lvs remaining pubescent, hairy or tomentose, at least beneath, at maturity.   26
26 Lvs linear or narrowly linear-lanceolate, the margins almost parallel.   27
Lvs lanceolate or narrowly oblong to oblong-elliptical, the margins distinctly convex, not parallel or subparallel.   28
27 Lvs with appressed, silvery, silky hairs beneath; ovary hairy.   **9. viminalis**
Lvs white-tomentose beneath, not silky; ovary glabrous.   **11. elaeagnos**
28 Lvs pubescent beneath, with prominent veins.   29
Lvs with sparse appressed pubescence beneath, the veins rather indistinct.   **13. × 9. ×smithiana**
29 Stipules caducous; lf-margin not recurved or very undulate.   **12. × 9. ×sericans**
Stipules prominent, persistent; lf-margin strongly undulate or narrowly recurved.   30
30 Lvs oblong-elliptical, c. 10–15 × 3–5 cm; lf-margin narrowly recurved; bracts 10–12 mm.   **10. ×calodendron**
Lvs lanceolate, 4–10 × 0.7–2 cm; lf-margin strongly undulate; bracts 5–8 mm.   **14. × 9. ×fruticosa**
31 Stipules conspicuous, persistent; bark smooth, shed in large flakes.   **5. triandra**
Stipules inconspicuous, caducous, or absent; bark fissured, not shed in flakes.   32
32 Branches pendent.   **4. babylonica**
Branches erect or spreading.   33
33 Mature lvs glossy green above.   **2. fragilis**
Mature lvs dull green or sericeous-pubescent above.   **3. alba**

Key to male plants with catkins before leaves

1 Twigs dark purple, with a whitish, waxy bloom.   **7. daphnoides**
Twigs without a waxy bloom.   2
2 Catkin-scales clothed with golden-yellow hairs.   **19. lanata**
Catkin-scales without yellowish hairs.   3

3 Anthers yellow.   6
   Anthers red or purple.   4
4 Stamens apparently 1 (really 2, completely connate).
                                               6. purpurea
   Stamens 2, free.   5
5 Low shrub up to 1(–1.5) m.   **18. lapponum**
   Shrub or small tree at least 4 m.   **13. cinerea**
6 Catkin-scales reddish or brownish at apex. **17. repens**
   Catkin-scales black at apex.   7
7 Wood of peeled twigs with longitudinal striae; twigs
      divaricate.   **14. aurita**
   Wood of peeled twigs smooth; twigs not divaricate.   8
8 Twigs pubescent.   **9. viminalis and hybrids**
   Twigs glabrous.   **12. caprea**

Key to female plants with catkins before leaves

1 Twigs dark purple, with a whitish, waxy bloom.
                                            7. daphnoides
   Twigs without waxy bloom.   2
2 Catkin-scales clothed with golden-yellow hairs.
                                              19. lanata
   Catkin-scales without yellowish hairs.   3
3 Styles long.   **9. viminalis and hybrids**
   Styles very short or absent.   4
4 Catkin-scales reddish or brownish at apex. **17. repens**
   Catkin-scales black at apex.   5
5 Low shrub up to 1(–1.5) m.   **18. lapponum**
   Shrub or small tree at least 4 m.   6
6 Catkins sessile.   **6. purpurea**
   Catkins pedunculate, sometimes shortly so.   7
7 Wood of peeled twigs with conspicuous longitudinal
      striae.   8
   Wood of peeled twigs smooth or with very few stria-
      tions.   **12. caprea**
8 Twigs usually divaricate, angular, glabrous; catkins
      1–2.5 cm at maturity.   **14. aurita**
   Twigs not divaricate, not or scarcely angular, glabrous
      or pubescent; catkins 3–5.5(–7) cm at maturity.   9
9 Bracts lf-like, 10–12 mm.   **10. ×calodendron**
   Bracts short, inconspicuous.   **13. cinerea**

Subgenus 1. SALIX. Trees or tall shrubs. Lvs ±lanceo-
late and tapered to the apex. Catkins appearing with
or after the lvs, on lfy peduncles, from lateral buds
of the preceding year. Scales yellowish, uniformly
coloured. Stamens 2 or more, free. Male fls with 2
nectaries; female fls with 1 or 2 nectaries.

1. S. pentandra L.   Bay Willow.

Shrub or small tree (2–)5–7(–17) m; bark grey or brownish, lightly fissured; branches spreading at maturity to
form a wide, rounded crown; *twigs* glabrous, green to
brown or reddish, shiny as if varnished, *not fragile*; buds
ovoid, viscid, shiny, dark brown. *Lf-blade* 5–12 ×
2–5 cm, *ovate or ovate-elliptical to elliptical- or obovate-
lanceolate*, acute to shortly acuminate, scarcely asymmetrical at apex, rounded or broadly cuneate at base,
glandular-serrulate, dark shiny green above, paler
beneath, somewhat coriaceous at maturity, ±viscid and
faintly fragrant when young; *petiole* up to c. 1 cm, stout,
with several small sessile glands near the top; *stipules
minute*, ±ovate, *caducous*. Catkins cylindrical, the rhachis and top of peduncle densely pubescent; male 2–6 ×

1–1.5 cm, dense; female 2–5 × c. 1 cm; scales c. 2 mm,
oblong, obtuse to subacute, pale yellow, pubescent towards base, glabrous above. Nectaries 2. *Stamens*
(4–)5–8(–12); anthers golden-yellow; filaments with
long hairs in basal half. Ovary 5–6 mm, narrowly ovoid,
glabrous, shortly stalked; style short, indistinct; stigmas
shortly 2-lobed to subentire. Fl. 5–6, with the lvs. Fr.
6–7. $2n = 76$. M.

Native. Streamsides, marshes, fens and wet woods,
ascending to 460 m; N. Wales, Stafford and S. Lincoln
northwards to Westerness and Moray, N. Ireland, frequent, planted elsewhere. Most of Europe except the
extreme north and south and most of the islands; Caucasus, Siberia; N.E. Asia Minor.

2. S. fragilis L.   Crack Willow.

Robust tree usually 10–15 m; *bark* greyish, *deeply fissured*; branches spreading to form a wide, rounded
crown, often pollarded; *twigs* sparsely pubescent when
young, soon glabrous, olive-brown, *fragile at junction
with branch*; buds ovoid, compressed, subglabrous,
±viscid, dull brown. *Lf-blade* 9–15 × 1.5–3 cm, lanceolate, long-acuminate, usually asymmetrical at apex,
cuneate at base, rather coarsely glandular-serrate, *dark
shiny green above*, glaucous beneath, sparsely
appressed-sericeous when young, soon glabrous; petiole
0.5–1.5 cm, rather stout, channelled above, glabrous or
sparsely hairy, usually with several distinct glands near
the top; *stipules* 3–8 mm, ±lanceolate, usually caducous,
but larger and persistent on sucker shoots. Catkins cylindrical, 4–6 × 1–1.3 cm, rather dense, the rhachis and
peduncle densely pubescent. Male catkin-scales
c. 2 mm, oblong, obtuse, pale yellow, sparsely hairy on
margin and above; *stamens* 2(–3); anthers golden
yellow; filaments hairy near base; nectaries 2. Female
catkin-scales c. 3 mm, oblong, obtuse, pale yellow,
caducous; ovary 2.5–3 mm, narrowly ovoid, glabrous,
subsessile or very shortly stalked; styles short; stigmas
2-lobed; nectaries usually 2. Fl. 4–5, with the lvs.
$2n = 76$. MM.

?Native. By streams and rivers, marshes, fens and
wet woods. Throughout the British Is., but rather local
in W. and N. Britain and in Ireland. Most of Europe
except the Arctic; local in the Mediterranean region;
W. Siberia to Iran.

A number of rather uniform variants can be recognized within this species. Most are of one sex only and
appear to be clones derived from planted material,
indeed the native status of the whole species in the British Is. has been called into question. These infraspecific
groups seem most satisfactorily treated as varieties, as
follows.

Var. **fragilis**. Description as for the species. Seeds generally infertile. Lowlands. Frequent in S.E. England and
the Midlands, extending more locally to S.W. England,
Wales and S. Scotland; perhaps native in S. England.

Var. **furcata** Ser. ex Gaudin (*S. fragilior* Host). Differs
from var. *fragilis* in having brighter brown, rather shiny

twigs; lvs 3–5 cm wide; male catkins usually more than 1 cm wide and often bifurcate, and fls commonly with 3 stamens. Only male trees known. Scattered from S. England northwards to Easter Ross and Angus.

Var. **russelliana** (Sm.) Koch, Bedford Willow. Differs from var. *fragilis* in being generally taller, up to 25 m, with less fragile, olive-brown twigs; lf-blade proportionately longer and narrow, c. 13–15 × 2–2.5(–3) cm, indistinctly glaucous beneath, the teeth markedly unequal; glands at top of petiole often somewhat lf-like; female catkins c. 4 × 0.8 cm, later up to 6 cm or more; ovary 6–7 mm, distinctly stalked. Only female trees known. Common throughout the range given for the species.

Var. **decipiens** (Hoffm.) Koch (*S. decipiens* Hoffm.), White Welsh Willow. Differs from var. *fragilis* in usually being a robust shrub 5–7 m, occasionally a tree 10–15 m; 1-yr-old twigs a distinct shiny, pale yellow-brown; lvs glabrous when young, proportionately shorter and wider, usually up to 9 × 2–3 cm; male catkins usually not more than 3 × 0.7 cm; catkins-scales c. 6 mm, rather densely hairy; stamens 2–3. Only male trees known. Rather local in Britain northwards to Perth; Derry. Sometimes grown as an osier.

*S. × **meyeriana** Rostk. ex Willd.
(*S. fragilis × pentandra*)
Shrub or small tree to 15 m; bark fissured like *S. fragilis*; twigs shiny brown like *S. pentandra*. Lvs similar to *S. pentandra*, 5–12 × 1.5–4 cm, long-acuminate, minutely and regularly glandular-serrate. Catkins 4–5 × 0.8 cm, longer and narrower than in *S. pentandra*. Male fls with (2–)3–4(–5) stamens. Female fls with ovary c. 4 mm. Fl. 5, with the lvs.

Introduced. Usually as solitary individuals or small clumps, by riversides and streamsides at lower altitudes. Rare throughout much of England.

3. S. alba L.   White Willow.

Tree 10–25(–30) m; *bark* greyish-brown, *deeply fissured*; branches ascending at 30–50° to form a narrow crown, often pollarded; *twigs* with dense, appressed silky hairs when young, becoming glabrous and shiny olive-brown, *not fragile*; buds ovoid, appressed-pubescent, dark brown. *Lf-blade* 5–10(–12) × 0.5–1.5 cm, *lanceolate*, acuminate, narrowly cuneate at base, *minutely and regularly serrulate, at first with dense, silvery-grey, appressed pubescence, usually becoming dull green and subglabrous above*; petiole up to 1 cm, pubescent, usually with a few small, dark, sessile glands at top; *stipules up to 5* mm, linear-subulate, glandular-serrate, *caducous*. Catkins spreading or erect; rhachis and peduncle densely pubescent. Male catkins 4–5 × c. 0.8 cm, dense; scales 2–3 mm, oblong, obtuse, pale yellow, pubescent towards base and on margins above; stamens 2; anthers yellow; filaments free, glabrous or sparsely pubescent towards base. Female catkins 3–4 × 0.4 cm; ovary 3–4 mm, narrowly ovoid, glabrous, subsessile or very

shortly stalked; style indistinct; stigmas 2-lobed; nectary 1. Fl. 4–5, with the lvs. Fr. 7. $2n = 76$. MM.

Native. Locally common by streams and rivers throughout the British Is. northwards to Caithness, but generally less frequent in S.W. and N.W. England, Wales and W. and N. Scotland. Most of Europe except the Arctic; local in the Mediterranean region; extending eastwards to C. Asia.

The above description refers to var. *alba*. Two further varieties are introduced and widely cultivated in Britain.

Var. **vitellina** (L.) Stokes, Golden Willow. Differs from var. *alba* in usually being less robust; 1-yr-old twigs bright yellow or orange; lf-blade soon becoming bright, rather shiny green; male catkin-scales 3–3.5 mm, sparsely pubescent; female catkin-scales as long as or longer than ovary. Formerly used in basket-work and still frequent in osier-beds, though nowadays much planted as a garden ornamental.

Var. **caerulea** (Sm.) Sm., Cricket-bat Willow. Differs from var. *alba* in the larger lf-blade, up to 10–11 × 1.5–2 cm, which is dull bluish-green above and sparsely pubescent beneath at maturity, and often more conspicuously serrate. $2n = 76$. Common and widespread in S.E. England from Hants to Norfolk, apparently infrequent elsewhere.

S. × **ehrhartiana** Sm. (*S. alba × pentandra*)
Small tree 10–15(–25) m; bark deeply fissured; twigs somewhat shiny brown or olive-brown. Lf-blade lanceolate or narrowly oblong-elliptical, intermediate between the parents. Male catkins 3–6 × up to 1 cm, the scales as in *S. alba*; fls with (2–)3–4(–5) stamens. Fl. 4–5, with the lvs. Only males known in Britain. Planted or a garden escape locally in S.E. England and Cumbria.

S. × **rubens** Schrank (*S. alba × fragilis*, *S. viridis* auct., non Fries), comprises a series of cultivated variants (nothovars) which join the two parents. It seems to be widely distributed in Britain, either planted or as an escape from cultivation.

*4. **S. babylonica** L.   Weeping Willow.

Tree up to 20 m, *with long*, glabrous, *pendent branches* almost to the ground. *Lvs* 8–16 × 0.8–1.5 cm, *linear-lanceolate*, acuminate, *serrulate*, glabrous at maturity; petiole 3–5 mm; *stipules inconspicuous*, caudate-acuminate, caducous. Female catkins short, subsessile, the fls with 1 nectary.

Introduced. Formerly planted by lakes and rivers in S. England but now very rare or perhaps extinct. Native of China.

This species is now represented in the British Is. by a great diversity of hybrids, although none is truly naturalized, these hybrids are so widely cultivated in gardens and parks that they merit brief mention here. The two most common of these 'Weeping Willow' hybrids are as follows.

S. × **sepulcralis** Simonkai (*S. alba* var. *vitellina* × *babylonica*, *S. chrysocoma* Dode) rarely exceeds 12 m, has deeply and coarsely fissured, greyish-brown bark, slender, glabrous, golden- or greenish-yellow mature twigs, finely serrulate lvs, which are pubescent or silky when young, and distinctly stalked catkins usually more than 2 cm.

S. × **pendulina** Wenderoth (*S. babylonica × fragilis*) is up to c. 12 m, has deeply and coarsely fissured bark, glabrous, olive-brown twigs, distinctly and rather irregularly serrate, glabrous or sparsely hairy lvs, and distinctly stalked catkins usually more than 2 cm.

5. S. triandra L.   Almond Willow.
S. amygdalina L.

Robust shrub or small tree 4–10 m; *bark* dark greyish, *smooth, shed in large, irregular flakes*; twigs glabrous, rather shiny olive-brown, often conspicuously angled or ridged at first, later subterete, rather fragile at junction with branch; buds ovoid, compressed dorsally, sparsely puberulent at first, later glabrous. *Lf-blade* 4–11(–15) × 1–3(–4) cm, *lanceolate to oblong-lanceolate or narrowly elliptical*, acute to long-acuminate, rounded or cuneate at base, *conspicuously* and regularly *serrate*, glabrous, rather dull dark green above, greenish or glaucous beneath; petiole usually not more than 2 cm, with often conspicuous glands at top; *stipules 5–10 mm*, ±ovate, *usually persistent*. Catkins erect, with puberulent rhachis and peduncle. Male catkins 2.5–5(–7) × 0.3–1.2 cm, cylindrical, dense or lax, fragrant; scales 1.5–2.5 mm, obovate or oblong, obtuse, pale yellow, sparsely hairy or pubescent towards base; stamens 3; anthers bright yellow; filaments pubescent towards base; nectaries 2. Female catkins usually shorter, thicker and denser than the male, with similar scales; ovary c. 2.5 mm, broadly ovoid, glabrous, distinctly stalked; styles very short, inconspicuous; stigmas short, emarginate or bifid; nectary 1. Fl. 4–5, with the lvs. Fr. 6. 2n = 38, 44, 88. M.

Native. By rivers, streams and ponds, marshes, etc., widely planted as an osier. Common in S. and E. England from Somerset and Hants northwards to Yorks, Durham and Northumberland, local in Devon and Cornwall, absent from much of Wales, N.W. England and W. Scotland, in E. Scotland northwards to Moray, but probably planted, as it is in Ireland. Most of Europe, except the Arctic, local in the Mediterranean region; N. Iran, N. and E. Asia Minor, temperate Asia to Japan; Algeria.

Var. **hoffmanniana** Bab., a much-branched shrub up to c. 4 m, with lvs 2–6(–7) × 1–1.5(–2.5) cm, lanceolate, not glaucous beneath, the petiole usually less than 1 cm, and the catkins (usually male) rarely exceeding 7 × 0.7 cm, appears to be a cultivar which is locally common in Dorset, Sussex and Surrey and occasionally found elsewhere northwards to the Outer Hebrides (S. Uist).

Subgenus 2. CAPRISALIX D..mort. From low shrubs to small trees. Lvs variously shaped. Catkins before or with the lvs on lfy peduncles or subsessile with a few

bracts at the base, from lateral buds of the preceding year. Scales blackish or dark brown at the apex, markedly 2-coloured. Stamens 2, free or united. Male and female fls with 1 nectary.

6. S. purpurea L.   Purple Willow.
Shrub 1.5–3(–5) m; bark greyish, smooth; *twigs glabrous*, yellowish or greyish, sometimes with red or purple tint, slender, flexible; buds cylindrical-ovoid, glabrous, yellowish or reddish. *Lvs often opposite or subopposite*, at least distally on twigs; blade 2–8 (–10) × 0.5–3 cm, linear-oblong to oblanceolate or narrowly obovate, acute or subacute, cuneate at base, *subentire or serrulate in apical half*, *glabrous*, or sparsely tomentose when young, dark green and dull to somewhat shiny above, paler or ±glaucous beneath, often becoming black on drying; petiole up to 1 cm; stipules small, narrowly oblong, soon caducous. Catkins 1.5–3 × 0.3–0.7 cm, cylindrical, densely fld, suberect or erect, sometimes curved, sessile, usually with 2–3 small, narrow, lf-like bracts; rhachis densely hairy; scales 1–1.5 mm, ovate or suborbicular, obtuse, blackish, reddish at base, hairy. *Male fls with stamens united* to appear 1; anthers reddish or purple; filaments glabrous. Female fls with sessile, hairy ovary c. 1–1.5 mm; style indistinct; stigmas short, ovate, often obtuse, spreading, sometimes bifid. Fl. 3–4(–5), before the lvs. Fr. 5(–6). 2n = 38. M.

Native. River margins, fens, marshes and other wet places, sometimes planted as an osier, locally common throughout the British Is. up to 440 m, northwards to Sutherland. Most of Europe except Fennoscandia and much of the USSR.

A rather variable sp., within which narrower, subentire lvd (*S. helix* L.) and wider, serrate lvd (*S. lambertiana* Sm., *S. woolgariana* Borrer) forms have been recognized at various taxonomic levels, but the numerous intermediates make any formal treatment untenable.

*7. S. daphnoides Vill.

Large shrub or small tree 6–8(–10) m; bark greyish, ±smooth; branches erect or spreading to form a rounded crown; *twigs* glabrous, dark reddish-brown, shiny, *with whitish waxy bloom*; buds compressed dorsally, glabrous or with rigid, deciduous hairs at base, dark red. *Lf-blade* (4–)7–12(–14) × (1–)2–3(–4) cm, *oblong or narrowly obovate*, acuminate, cuneate at base, regularly glandular-serrate, dark shiny green above, glaucous beneath; petiole 0.7–2 cm, channelled and hairy above; stipules up to 12 mm, narrowly ovate, often persistent. Catkins cylindrical, 2–4 × 0.8–1.8 cm, dense, erect or suberect, sessile, sometimes with a few small, densely hairy bracts; rhachis hairy; scales c. 2 mm, ovate-elliptical, acute, dark brown, hairy. Male fls with 2 free stamens; anthers yellow; filaments glabrous. Female fls with shortly stalked, narrowly ovoid ovary c. 4 mm; style indistinct; stigmas short, entire or bifid. Fl. 2–3, before the lvs. Fr. 5–6. 2n = 38. MM.

Introduced. Widely planted in wet places but probably nowhere truly naturalized in the British Is. Native of C. Europe.

*8. S. acutifolia Willd.   Violet Willow.

S. daphnoides subsp. *acutifolia* (Willd.) Dahl

Differs from *S. daphnoides* by the violet winter twigs, the linear-lanceolate lf-blade up to 15×2 cm and glabrous petiole up to 3 cm. Fl. 3–4, with or just before the lvs. Females apparently not known in the British Is. $2n = 38$.

Introduced. Sometimes planted in gardens and by rivers in S. England. Native of USSR.

9. S. viminalis L.   Osier.

Shrub or small tree 3–6(–10) m; bark greyish-brown, fissured; *branches* erect or suberect to form a narrow, truncate or rounded crown; *twigs whitish-pubescent at first*, later smooth and rather shiny yellow- or olive-brown, long, straight, very flexible; buds ovoid, shortly pubescent at first, later glabrous or subglabrous, yellowish or reddish-brown. *Lf-blade* 10–15(–25) $\times$ 0.5–1.5(–2.5) cm, *linear to linear-lanceolate*, acuminate, narrowly cuneate at base, *entire*, narrowly revolute or recurved and often distinctly undulate at margin, sparsely puberulent and dull green above, *with dense, silvery appressed hairs beneath*; petiole up to 1 cm, channelled above, pubescent; stipules up to 10 mm, linear to linear-lanceolate, often falcate, soon caducous. Catkins 1.5–3 $\times$ 0.5–1 cm, narrowly ovoid to cylindrical, densely fld, erect or somewhat recurved, sessile or subsessile, with 2–3 small, densely sericeous, oblong-lanceolate bracts; rhachis densely hairy; scales c. 2 mm, narrowly ovate-elliptical, obtuse to subacute, reddish-brown, sometimes blackish, densely hairy. Male fls with 2 free stamens; anthers yellow; filaments glabrous. Female fls with subsessile, ovoid, *densely tomentose ovary* c. 2 mm; style c. 0.8 mm; stigmas linear, undivided or 2-fid, spreading. Fl. 2–4, before the lvs. Fr. 4–5. $2n = 38$. M.

?Native. By streams and ponds, in marshes and fens, and commonly planted as an osier; ascending to 400 m. Common throughout the British Is., but less frequent in the hilly areas of W. Great Britain. Apparently always planted in Ireland and Scotland, and probably not native elsewhere except perhaps in E. and C. England. Throughout much of Europe, but perhaps planted or naturalized outside the USSR; Himalaya; Japan.

S. × mollissima Hoffm. ex Elwert
(*S. triandra* × *viminalis*)

This hybrid includes 2 rather different varieties in the British Is.

Var. **undulata** (Ehrh.) Wimmer (*S. lanceolata* Sm.). Shrub to 5 m; bark flaking; twigs olive-brown or reddish, glabrous or subglabrous; stipules often conspicuous and persistent. Lf-blade up to 10–12 $\times$ c. 1.5 cm, lanceolate, longer acuminate at apex than *S. triandra*, but with similar distinctly and often coarsely serrate margins, and a rather shiny dark green above, paler or somewhat

glaucous beneath. Female catkins 3–4 $\times$ c. 0.5 cm, narrowly cylindrical; scales with sparse, long hairs, pale yellowish; ovary very shortly stalked, glabrous or subglabrous; style short. Fl. 4–5, with the lvs. Only female plants known.

Grown as an osier and always planted or an escape. Common in S. England, extending locally northwards to Perth and found in a few places in Ireland.

Var. **hippophaifolia** (Thuill.) Wimmer. Shrub 3–5 m; twigs subglabrous; buds acuminate, puberulent. Lf-blade like that of *S. viminalis* but glabrous or soon so, dark shiny green above, duller and paler beneath. Male catkins c. 3.5 $\times$ 1 cm, subsessile or shortly stalked; scales densely hairy, yellowish; stamens 2–3, the anthers yellow; a few female fls sometimes intermixed. Female catkins rarely more than 5 mm wide, distinctly stalked, with lfy bracts; ovaries grey-pubescent when young, later sparsely so; styles rather short. Fl. 4–5, with the lvs.

Locally planted or as an escape in S. England.

S. × rubra Hudson (*S. purpurea* × *viminalis*)

Shrub or small tree (1–)3–6 m; bark greyish, fissured; branches spreading; twigs glabrous and rather shiny yellowish-brown, flexible. Lf-blade 4–12(–15) $\times$ 0.8–1(–1.5) cm, as in *S. viminalis* but margin often remotely serrulate, especially towards apex, and pale green and subglabrous to puberulent, never with appressed silvery hairs, beneath. Catkins as in *S. viminalis*, but the scales usually black and hairy as in *S. purpurea*; stamens 2, with free or partly united filaments; anthers red or yellow; ovary broadly ovoid to subglobose, densely grey-hairy. Fl. 3–4, before the lvs. $2n = 38, 57$.

Scattered throughout the British Is., often on the sites of osier-beds, but probably also occurring naturally. Widespread in Europe.

S. × forbyana Sm. (*S. purpurea* × *viminalis* × ?*cinerea*),
is similar to *S.* × *rubra* but has narrowly oblong-lanceolate to oblanceolate lvs which are sparsely whitish or reddish tomentose when very young suggesting some influence of *S. cinerea*. Formerly grown as an osier for fine basketwork, it is scattered throughout much of England, S. Scotland and N.W. Ireland.

10. S. × calodendron Wimmer

?*S. caprea* × *cinerea* × *viminalis*

Erect shrub or small tree up to c. 10 cm, of similar habit to *S. viminalis*; twigs densely greyish-white-pubescent or -tomentose, becoming purplish brown and subglabrous in their 2nd year, when peeled the wood with prominent longitudinal striae; buds ovoid, densely pubescent, dull purplish-brown. *Lf-blade* (7–)10–15 $\times$ 3–5 cm, *oblong-elliptical*, *acute or shortly acuminate*, rounded or broadly cuneate at base, the margin narrowly recurved and shallowly and distantly glandular-serrate, dull green and sparsely pubescent above, with denser, longer, whitish-grey pubescence beneath; petiole 1–1.5 cm, channelled above, densely pubescent;

stipules 8–15 mm, ±lanceolate, acute or shortly acuminate, often falcate, persistent. *Catkins* 3.5–5(–7) × 0.8–1 cm, *cylindrical*, erect or somewhat spreading, subsessile or shortly stalked; bracts 10–12 mm, lf-like; peduncle and rhachis densely hairy; scales c. 2 mm, broadly ovate, acute to obtuse, blackish-brown, densely hairy. Female fls with subsessile or shortly stalked, ovoid, densely white-tomentose ovary 3–4 mm; style c. 0.8 mm; stigmas linear-oblong, usually undivided. Fl. 3–4, before the lvs. Male plants unknown. M.

Origin uncertain. Widely but locally distributed from Surrey to N. Aberdeen, and apparently also in S. and E. Ireland. Germany, Denmark and Sweden.

*11. S. elaeagnos Scop.

S. incana Schrank

Erect shrub or slender, much-branched *tree* to 6 m; *twigs* densely *whitish- or greyish-pubescent or tomentose*, becoming glabrous and yellowish-brown or reddish after 1 yr; buds subacute or obtuse, sparsely hairy. *Lf-blade* 5–15 × c. 0.8 cm, *linear*, acuminate, narrowly cuneate at base, *entire*, subcoriaceous, dark shiny green above, *white-tomentose beneath*; petiole less than 5 mm; stipules usually absent. Catkins subsessile, crowded; bracts 2–4, linear, sparsely pubescent beneath; scales 3–4 mm, narrowly oblong, reddish or purplish at apex, subglabrous to sparsely hairy; *ovary* 4–5 mm, sessile or subsessile, narrowly ovoid, *glabrous*; style indistinct; stigmas slender, usually 4-lobed. Fl. 4–5, just before the lvs. M.

Introduced. Cultivated and an occasional garden-escape in C. and S. England. Native of C. and S. Europe, W. Asia and western N. Africa. Male plants not known in Britain in recent years.

12. S. caprea L.   Goat Willow.

Shrub or small tree 3–10 m; bark greyish-brown, irregularly fissured; branches ascending or spreading to form a rather open crown; twigs sparsely pubescent at first, soon glabrous, yellowish or greyish-brown; buds ovoid, soon glabrous and shiny, often yellow or reddish. *Lf-blade* 5–12 × 2.5–8 cm, *obovate to oblong or elliptical, sometimes suborbicular, acute to obtuse*, broadly cuneate to shallowly cordate at base, irregularly *undulate-glandular-serrate*, dull *green* and sparsely pubescent to subglabrous *above*, whitish-grey and *tomentose to pubescent beneath*; petiole 0.8–2.5 cm, channelled above, pubescent; *stipules* 8–12 mm, subcordate, *usually caducous*. Catkins 1.5–2.5 × 0.8–1.8 cm, ovoid or cylindrical, dense, erect, subsessile, with a few inconspicuous, densely villous bracts; scales c. 2 mm, broadly ovate, obtuse to acute, blackish, densely hairy. Male fls with 2 stamens; anthers golden-yellow; filaments glabrous. Female fls with distinctly stalked, narrowly ovoid, densely silver-sericeous ovary 4–5 mm; style absent or very short; stigmas oblong, shortly bifid, erect or suberect. Fl. 3–4, before the lvs. Fr. 5. $2n = 38^*$, 76^*. M.

Native. Wood-margins, hedges and rocky lake-mar-

gins, slightly calcicole, usually in the lowlands. Locally common throughout the British Is. Most of Europe, except the extreme north and south, eastwards to C. Asia.

Var. **sphacelata** (Sm.) Wahlenb. (*S. caprea* subsp. *sericea* (Andersson) Flod.), usually a small, gnarled shrub, has entire, exstipulate lvs which are at first densely appressed-sericeous. It occurs in montane areas of Scotland at altitudes of 305–825 m. It is locally common in Scandinavia, the Alps of C. Europe and in Russia.

S. × sericans Tausch ex A. Kerner
(*S. caprea* × *viminalis*)

Bush or tree 3–9 m; bark shallowly fissured; branches spreading; twigs densely greyish-white tomentose at first, soon glabrous or subglabrous and rather shiny; buds ovoid, pubescent. *Lf-blade* 5–12 × 1.3–3 cm, *lanceolate to ovate-lanceolate*, gradually acuminate, rounded or broadly cuneate at base, the margin usually somewhat undulate and *subentire or distantly crenulate*, green and subglabrous above, *densely grey-tomentose beneath*, the veins prominent; petiole 1–1.5 cm; *stipules* usually small, semicordate or ±ovate, *caducous*. Male catkins 2–3 × c. 1 cm, cylindrical, sessile or subsessile, with a few small, sericeous bracts; scales c. 2 mm, ovate-cylindrical, brownish with a darker apex, densely greyish-hairy; anthers golden yellow; filaments glabrous. Female catkins up to 5 × c. 1 cm, with shortly stalked, narrowly ovoid, densely grey-sericeous ovary c. 3 mm; style rather conspicuous, stigmas usually longer than style, narrowly oblong, bifid. Fl. 3–4, well before the lvs. Fr. 5.

Native. Scattered throughout the British Is., but a relict of osier plantings in many areas.

13. S. cinerea L.   Grey Willow, Common Sallow.

Shrub or small tree 2–10(–15) m, usually much-branched from the base; bark dark grey-brown, fissured with age; branches spreading to form a wide, rounded or flattened crown; twigs at first densely pubescent, usually becoming glabrous or subglabrous, dark reddish-brown, the wood with long, scattered striae when peeled; buds ovoid. *Lf-blade* 2–9(–16) × 1–3(–5) cm, *obovate to broadly oblanceolate or*, sometimes, *oblong to elliptical or suborbicular*, acute to obtuse, sometimes obliquely twisted at apex, cuneate at base; petiole up to 1 cm. Catkins 2–3(–5) × 0.6–1 cm, cylindrical or narrowly ovoid, erect, sessile or subsessile, becoming shortly stalked, dense, with few, small, densely hairy bracts; scales 2–2.5 mm, oblong, acute or obtuse, blackish, densely villous. Male fls with 2 free stamens; anthers yellow; filaments glabrous. Female catkins often smaller than the male; ovary 2.5–5 mm, ovoid, grey-tomentose; style usually short; stigmas narrowly oblong, erect or spreading, subentire to ±deeply 2-fid. Fl. 3–4, before the lvs. Fr. 5–6. m.

Subsp. cinerea

S. aquatica Sm.

Usually a much-branched shrub 4–6(–10) m. Twigs often

pubescent into second year; peeled wood conspicuously striate. Lf-blade distinctly undulate-serrate, dull green and puberulent above, densely pubescent, becoming sparsely pubescent and greyish-white beneath; stipules usually about as long as petiole, persistent. $2n = 76$.

Native. Usually in base-rich fens and marshes at low altitudes in S. and E. England, locally extending to the Isle of Man, Perth and N. Ireland (Down). Most of Europe from Scandinavia, Holland, E. France and Italy eastwards to W. Asia.

Subsp. oleifolia Macreight   Rusty Sallow.

S. atrocinerea Brot.; *S. cinerea* subsp. *atrocinerea* (Brot.) Silva & Sobr.; *S. oleifolia* Sm., non Vill.

Tall shrub or small tree up to 15 m. Twigs soon sub-glabrous and dark reddish-brown, the peeled wood usually rather weakly striate. Lf-blade entire or remotely and obscurely undulate-serrate, rather shiny dark green above, greyish-white and usually with sparse, stiff reddish hairs and minute blackish glands beneath; stipules usually small and caducous. $2n = 76$.

Native. On acid or base-rich soils by streams, bog- and marsh-margins, hedgerows and edges of moist woodlands, ascending to 610 m in Perth. The commonest British willow, throughout the British Is. except in parts of East Anglia. W. France, Spain and Portugal.

S. × reichardtii A. Kerner (*S. caprea × cinerea*) constitutes a series of populations which link the parents, usually with the slender, puberulent, dark reddish-brown twigs of *S. cinerea* and the wide, rather irregularly undulate-glandular-serrate lvs, softly pubescent beneath, of *S. caprea*. Locally common throughout the British Is., and perhaps more abundant than generally thought.

S. × sordida A. Kerner (*S. cinerea × purpurea*), a much-branched shrub up to 5 m, is generally intermediate between the parents and often indistinguishable from *S. cinerea* except for the male catkins with fls in which the filaments are free, shortly united and more distinctly united. Rather local in Gloucester, Northampton, Westmorland, Dumfries and Perth. Widespread in Europe.

S. × smithiana Willd.

S. cinerea × viminalis

Shrub or small tree 2–9 m; bark shallowly fissured; branches spreading; twigs densely pubescent in 1st year, dark reddish-brown, the peeled wood usually sparingly striate. Lf-blade $6-11 \times 0.8-2.5$ cm, narrowly lanceolate, acuminate, cuneate at base, usually serrulate, dull green and subglabrous above, grey-sericeous, becoming subglabrous and greenish, beneath, the veins usually prominent and often reddish; petiole 0.5–1.3 cm; stipules rather small, ovate-lanceolate, usually caducous. Male catkins $2-3 \times 1-1.2$ cm, ovoid-cylindrical, sessile or subsessile, with a few inconspicuous, sericeous bracts; scales 2–2.5 mm, ovate-elliptical, acute, brownish with a darker apex, densely greyish-hairy; anthers

golden-yellow; filaments glabrous. Female catkins usually $3-5 \times$ c. 1 cm; ovary densely grey-sericeous, shortly stalked; style usually present; stigmas ±as long as style, often with 2 lobes when mature. Fl. 3–4, before the lvs. Fr. 5.

Native. Throughout the British Isles., though clearly introduced in many areas, and generally less frequent than *S. × sericans*. Common in Europe.

14. S. aurita L.   Eared Willow.

Shrub 1–2.5(–3) m, with numerous branches; twigs puberulent but soon glabrous or subglabrous, dark reddish-brown, the peeled wood with numerous prominent striae; buds ovoid, glabrous or puberulent. *Lf-blade* $2-6 \times 1.2-2.5$ cm, *obovate to oblong-obovate*, rounded or acute, often with a twisted apiculate apex, cuneate at base, irregularly *undulate-serrate, rugose*, dull dark green above, whitish-grey and usually pubescent beneath, the veins prominent; petiole 0.3–0.8 cm; *stipules large* and conspicuous, subcordate to reniform, *persistent*. Catkins $1-2 \times 0.7-0.8$ cm, cylindrical, erect, sessile, with short, densely silvery-hairy bracts; scales 1.5–2 mm, oblong, acute to obtuse, dark purplish or reddish, sparsely to densely hairy. Male fls with 2 free stamens; anthers yellow; filaments glabrous or hairy near the base. Female fls with distinctly stalked, densely grey-tomentose, narrowly ovoid ovary 1.5–2 mm; style very short or absent; stigmas oblong, erect or somewhat spreading, entire. Fl. 4–5, before the lvs. Fr. 5–6. $2n = 38, 76$. N.

Native. Acid heaths, woods and moorlands, ascending to 730 m. Throughout the British Is., but uncommon in much of the English Midlands and East Anglia. Most of Europe, except the Arctic and the Mediterranean region, perhaps also in Turkey and N. Iran.

S. × multinervis Doell (*S. aurita × cinerea*) is a rather common hybrid which often constitutes a series of intermediates between the parents. It is normally distinguished from *S. cinerea* by its conspicuous, persistent stipules and rather wrinkled, dull dark green lvs which are sparsely pubescent beneath. It occurs wherever the parents occur together.

S. × fruticosa Doell

S. aurita × viminalis

Erect, much-branched *shrub or small tree* 1.5–3(–5) m; *twigs at first* densely or sparsely *grey-pubescent*, usually becoming glabrous and dark reddish-brown, the wood with scattered, prominent striae when peeled. *Lf-blade* $4-10 \times 0.7-2$ cm, *lanceolate*, acute or acuminate, rather narrowly cuneate at base, *subentire or* irregularly and *obtusely denticulate*, dark green and subglabrous to sparsely pubescent above, whitish-grey and softly *lanuginose or tomentellous beneath*; petiole c. 0.6 cm; *stipules conspicuous*, ±lanceolate, acuminate, *persistent*. Catkins $1-3 \times 0.5-0.7$ cm, cylindrical, erect, subsessile, with a few small, lanceolate, lf-like bracts which are densely silvery-hairy beneath; scales c. 2 mm, narrowly oblong, acute or ±acuminate, pale reddish-brown, densely hairy. Male fls with 2 free stamens; anthers yellow; filaments glabrous. Female fls with distinctly stalked, densely grey-tomentose, ovoid ovary 2–2.5 mm; style distinct; stigmas longer than style, linear or narrowly oblong, spreading, usually entire. Fl. 4, before the lvs. Fr. 5.

Native. Uncommon, but widely distributed in the British Is. Perhaps mostly the result of osier-plantings in England, where it is rather uniform, but almost certainly spontaneous in Scotland and Ireland.

15. S. myrsinifolia Salisb.   Dark-leaved Willow.
S. nigricans Sm.; *S. andersoniana* Sm.

Low and spreading to robust *shrub or small tree* 1–3(–4) m; bark dark greyish, shallowly fissured; *twigs* at first densely pubescent, later glabrous or subglabrous, *dull brown* or greenish, the wood sometimes with a few distinct striae when peeled; buds ovoid, usually pubescent. *Lf-blade* 2–6.5 × 1.5–3.5 cm, *obovate to elliptical or oblong*, usually acute, cuneate or rounded at base, the margin rather irregularly serrate or, rarely, subentire, rather *thin and papery*, at first sparsely pubescent, later somewhat shiny dark green above, glaucous beneath, turning *blackish when dried*; petiole usually less than 1 cm; stipules conspicuous, ovate, persistent, sometimes small, caducous, or absent. Catkins 1.5–4 × 1–1.5 cm, cylindrical, suberect, shortly stalked to subsessile, with 2–4 appressed-pubescent bracts c. 1 cm; scales 1–2.5 mm, oblong to suborbicular, blackish at apex, rather sparsely hairy. Male fls with 2 free stamens; anthers yellow; filaments glabrous, or hairy towards base. Female fls with shortly stalked, glabrous to pubescent, ovoid ovary 2.5–3 mm; style usually distinct; stigmas almost as long as style, suberect or spreading, entire and oblong to divided into 4 linear lobes. Fl. 4–5, with the lvs. Fr. 5–6. $2n = 114$. N.

Native. Usually on riversides, lake shores and damp rock-ledges; ascending to 915 m. Rather local, from C. Lancs and Yorks northwards to Sutherland and the Outer Hebrides, probably also in Suffolk, Cambridge and Lincoln, and in Antrim; introduced elsewhere. N. and C. Europe, Siberia.

16. S. phylicifolia L.   Tea-leaved Willow.
Incl. *S. hibernica* Rech. fil.

Similar to *S. myrsinifolia*, but differs as follows: twigs glossy brown, usually glabrous. Lvs rather rigid and coriaceous, shiny bright green above, not turning blackish when dried; stipules usually very small and caducous, or absent. Ovary usually densely pubescent.

Native. Moist rocky ground, commonly on carboniferous limestone; ascending to 700 m. Locally abundant from Lancs and Yorks northwards to Orkney, in Ireland confined to the Ben Bulben range (Leitrim, Sligo). N. Europe, extending locally to S.E. Germany, Austria and Czechoslovakia.

S. × laurina Sm.

S. cinerea × *phylicifolia*; *S.* × *wardiana* Leefe ex F. B. White

Erect, lax *shrub* 2–6 m; twigs sparsely hairy at first, becoming shiny brown. *Lf-blade* 4–10 × 1.8–3.5 cm, *elliptical to oblong or narrowly obovate*, acute, sometimes twisted at apex, cuneate at base, entire to shallowly and obtusely serrate, glabrous or subglabrous and rather shiny dark-green above, *glaucous and sparsely pubescent along the veins beneath*; petiole usually less than 1 cm; *stipules small*, ovate, *caducous*. Catkins 3–5 × 0.8–1 cm, cylindrical, with densely villous stalk up to 1 cm, with a few lanceolate bracts 6–10 mm, subglabrous above, densely sericeous beneath; scales 2–3 mm, ovate-oblong, obtuse to subacute, brownish, usually darker at apex. Male fls with 2 stamens; anthers yellow; filaments glabrous. Female fls with white-tomentose, narrowly ovoid ovary 3–4 mm; style c. 1 mm; stigmas linear-oblong, entire or divided to base. Fl. 4–5, just before the lvs.

Native. Locally abundant from Yorks northwards, but planted in Warwick and Shetland, and probably elsewhere. Male plants very uncommon. Scandinavia.

17. S. repens L.   Creeping Willow.

Prostrate to erect shrub up to 1.5(–2) m; twigs slender to rather robust, glabrous to densely appressed-sericeous, dark grey to reddish- or yellowish-brown. *Lf-blade* 1–3.5 × 0.4–2.5 cm, lanceolate to ovate-oblong or oblong, obtuse to acute or shortly mucronate, sometimes twisted at apex, cuneate to rounded at base, *entire or obscurely glandular-serrulate*; petiole usually less than 4 mm; *stipules absent* or conspicuous, lanceolate to narrowly ovate, persistent. *Catkins* 1–2.5 × 0.4–0.8 cm, obovoid or ovoid to cylindrical, erect or suberect, *sessile* or subsessile, with inconspicuous or well-developed, often pubescent or sericeous, bracts up to 5 mm; scales c. 2 mm, narrowly obovate, pale or reddish-brown towards apex, subglabrous or sparsely hairy, sometimes densely sericeous-villous. Male fls with 2 free stamens; anthers yellow; filaments glabrous or sometimes hairy at base. Female fls with subsessile or shortly stalked, glabrous to sericeous, narrowly ovoid ovary 2–2.5 mm; style conspicuous; stigmas entire or 2-fid. Fl. 4–5, with or before the lvs. $2n = 38$. N.

Native. Throughout the British Is., but rather local and absent from much of the English Midlands and C. Ireland. N., W. and C. Europe, extending to Italy and the Ukrainian Carpathians.

A very variable species in which it has proved difficult to circumscribe infraspecific taxa. However, the following 3 varieties seem worthy of recognition.

Var. **repens** (var. *ericetorum* Wimmer & Grab.). Procumbent or decumbent, often with small, subglabrous lvs, and subglabrous ovaries. Widespread throughout the range of the species on heaths and moorlands, ascending to 855 m.

Var. **fusca** Wimmer & Grab. Erect or ascending, with ovaries and underside of lvs usually densely sericeous. Locally common in the East Anglian fens.

Var. **argentea** (Sm.) Wimmer & Grab. (*S. arenaria* L.). Ascending, with pubescent twigs and large lvs with dense silky, silvery hairs on both surfaces, or only beneath. Usually on maritime dunes.

S. × ambigua Ehrh. (*S. aurita × repens*), widely distributed on acid heaths in the British Is., is generally intermediate between the parents in most characters; it has a characteristic shiny silky indumentum and the lvs, like those of *S. repens*, become blackish on drying.

S. × subsericea Doell (*S. cinerea × repens*), an uncommon hybrid known with certainty only from dunes in Merioneth and Fife, is very similar to *S. × ambigua* from which it may be distinguished by its small, caducous stipules, which are usually rather less conspicuous and persistent than in *S. × ambigua*.

S. × friesiana Andersson (*S. repens × viminalis*) is known from sand-dunes near Southport (S. Lancs).

18. S. lapponum L.   Downy Willow.
Much-branched *shrub* (0.15–)0.2–1(–1.5) m; twigs rather rigid with prominent bud-scars, at first pubescent or lanuginose, soon glabrous and rather shiny, dark reddish-brown; buds dark-brown, glabrous or shiny. *Lf-blade* 1.5–7 × 1–2.5 cm, lanceolate to narrowly obovate, sometimes ovate or oblong, acute or acuminate, sometimes twisted at apex, usually cuneate at base, *entire or subentire*, sometimes somewhat undulate, grey-green and sparsely to ±densely appressed-villous above, *pale greyish and densely tomentose or lanuginose beneath*, sometimes subglabrous above and beneath; petiole up to 5(–10) mm; *stipules absent, or small*, narrowly ovate, often falcate, soon caducous. Catkins 2–4 × 1–1.5 cm, cylindrical, erect or suberect, sessile or subsessile, with inconspicuous densely sericeous-lanuginose bracts c. 5 mm; scales 2–4 mm, elliptical or ovate, dark brown, sometimes paler towards base, sericeous-villous. Male fls with 2 free stamens; anthers yellowish or tinted purplish or reddish; filaments glabrous. Female fls with sessile or subsessile, densely whitish-tomentose, narrowly ovoid ovary c. 3–4 mm; style conspicuous; stigmas divided to base into 4 filiform lobes. Fl. 5–7. Fr. 7–8. 2*n* = 38, 76. N.

Native. Locally common on wet rocks on mountains between 220 m and 1070 m; Westmorland, Dumfries, and Perth and Angus northwards to Sutherland, not in the islands. N. Europe, extending southwards in the mountains to the Pyrenees and Bulgaria, and eastwards to the Altai and W. Siberia.

19. S. lanata L.   Woolly Willow.
Gnarled, much-branched *shrub* up to 1 m; twigs rather stout and rigid, with prominent lf-scars, at first sparsely lanuginose, soon glabrous, somewhat shiny brown; buds dark reddish-brown, at first hairy, later glabrous. *Lf-blade* 3.5–7 × 3–6.5 cm, suborbicular or broadly ovate to obovate or elliptical, obtuse to acute, sometimes shortly mucronate, broadly cuneate to rounded or cordate at base, *entire or subentire*, sometimes with a few

irregular teeth, at first silky tomentose or lanuginose, often becoming subglabrous, dark or greyish-green above, glaucous beneath, the reticulate veins usually rather conspicuous; petiole up to 1 cm, villous or subglabrous; *stipules conspicuous*, broadly ovate to obovate, often *persistent*. Catkins 2.5–3.5 × 1.3–1.5 cm (the female up to 7 cm in fr.), cylindrical, dense, erect, sessile, with inconspicuous, densely villous bracts; scales 2–3 mm, ovate or elliptical, dark purplish-brown, with dense, silky, golden yellow (rarely silver-grey) hairs. Male fls with 2 free stamens; anthers yellow; filaments glabrous. Female fls with sessile or subsessile, glabrous, narrowly ovoid ovary c. 3 mm; style distinct, often bifid at apex; stigmas linear to narrowly oblong, entire at first, becoming deeply bifid. Fl. 5–7. Fr. 7. 2*n* = 38. N.

Native. Damp mountain ledges and rocky slopes between 550 m and 915 m in Perth, Angus and Aberdeen, rare. Arctic and subarctic Europe and E. Siberia.

20. S. arbuscula L.   Mountain Willow.
Shrub up to 0.7 m, with ascending to spreading, much-divided branches; twigs sparsely puberulent at first, soon glabrous, dark, somewhat shiny reddish-brown. *Lf-blade* 1.5–3(–5) × 1–1.5(–3) cm, *ovate or oblong to elliptical*, usually acute, cuneate at base, obtusely *glundular-serrate*, glabrous and bright shiny green above, at first densely appressed-pubescent but soon glabrous and ±glaucous beneath, with rather obscure veins; petiole up to 5(–8) mm; *stipules* usually *absent*, or ovate and soon *caducous*. Catkins 1–2 × 0.4–0.5 cm, cylindrical, erect or sub-erect, the male subsessile, the female on long lfy stalks, with leaf-like, glabrous or subglabrous (at first sericeous beneath) bracts 8–20 mm; scales c. 1.3 mm, ovate to obovate, brownish or tinged with dark purple, densely hairy. Male fls with 2 free stamens; anthers dark reddish or purple; filaments glabrous. Female fls with sessile, densely grey-hairy, ovoid ovary c. 1.5 mm; style distinct; stigmas narrowly oblong, emarginate or deeply bifid. Fl. 5–6, with the lvs. Fr. 6. 2*n* = 38. N.

Native. Damp rocky mountain slopes and ledges between 460 m and 825 m. Locally abundant in Perth, Argyll and Peebles, doubtful in Dumfries, Inverness and Orkney. Fennoscandia eastwards to E. Siberia.

21. S. myrsinites L.   Whortle-leaved Willow.
Shrub up to 0.5 m, with spreading or decumbent branches, and rhizomes; twigs sparsely hairy at first, soon glabrous, dark, shiny reddish-brown. *Lf-blade* 1.5–7 × 0.5–2.5(–3) cm, *oblong or ovate* to obovate or, rarely, lanceolate, acute to obtuse, rounded or cuneate at base, *glandular-serrulate*, sparsely hairy at first, especially beneath, soon bright, shiny green on both surfaces, with prominent, reticulate veins; petiole up to 1 cm, channelled above; *stipules usually conspicuous*, ovate to oblong, ±*persistent*, sometimes caducous or absent. Catkins cylindrical, erect, usually on well-developed, sparsely hairy stalk, with leaf-like, subglabrous

or sparsely hairy bracts 10–20 mm; scales 1.5–2 mm, oblong to obovate, dark reddish-purple, villous. Male catkins 1.5–2.5 cm; fls with 2 free stamens; anthers purple or reddish; filaments glabrous or sparsely hairy near base. Female catkins 3–5 cm, to 7 cm in fr.; fls with ±hairy, ovoid ovary 4–5 mm; style distinct but rather short; stigmas oblong, subentire. Fl. 5–6, with the lvs. Fr. 6–7. 2*n* = 38, 152, 190. N.

Native. Wet ledges and mountain slopes between (90–)305 m and 1160 m, rare and local. From Argyll and Perth northwards to Orkney. Fennoscandia eastwards to the Urals.

Subgenus 3. CHAMAETIA Dumort.

Prostrate creeping shrubs. Lvs orbicular or ovate. Catkins on lfless peduncles from terminal buds of the preceding year. Scales not dark at apex, concolorous. Stamens 2, free. Fls with 1 or 2 nectaries.

22. S. herbacea L.   Dwarf Willow.

Shrub with long, creeping, branched rhizome; *branches* few up to 6 cm, sparsely hairy at first, soon glabrous, dark shiny brown or reddish; buds ovoid, sparsely hairy or subglabrous at first, soon glabrous. *Lf-blade* 0.3–2(–3) × 0.3–2(–3) *cm*, obovate to suborbicular, rounded or shallowly emarginate, rarely subacute, broadly cuneate to rounded or subcordate at base, *crenulate-serrate* or, rarely, subentire, sparsely white-hairy, soon glabrous and dark, shiny green, with prominent, reticulate veins; petiole up to 0.4 cm, channelled above, white-hairy at first, but soon subglabrous; stipules inconspicuous, membranous, caducous, or absent. Catkins 0.5–1.5 cm, 2–12-fld, subsessile or shortly stalked, without bracts; scales 1–2 mm, oblong to obovate, yellowish, often tinged reddish, subglabrous or sparsely hairy towards apex. Male fls with 2 free stamens; anthers yellow or reddish; filaments glabrous; nectaries 1 or 2, free or united in lobed cup surrounding bases of filaments. Female fls with sessile or subsessile, glabrous or sparsely hairy, narrowly ovoid ovary 2–3 mm, often reddish at maturity; style short; stigmas short, spreading, deeply bifid; nectaries 1–2, entire or irregularly lobed. Fl. 6–7(–8), with the lvs. Fr. 7–8. 2*n* = 38. Ch.

Native. Mountain summits, with grasses and sedges, and moist rocky ledges, usually from 610 m to 1310 m, but found at 150 m in Shetland. Brecon, Caernarvon, Yorkshire, Westmorland, Cumberland, Kirkcudbright, Peebles, from Kintyre and Angus northwards to Shetland. Kerry, Limerick, Mayo, Cavan, Down, Tyrone and Donegal. Arctic Eurasia and North America, extending southwards in the mountains to C. Europe, the Pyrenees, Apennines and New Hampshire.

S. × margarita F. B. White (*S. aurita × herbacea*), of which only female plants are known, occurs rarely on rock-ledges of mountains in Perth, Angus, Kintyre and Skye.

S. × cernua E. F. Linton

S. herbacea × repens

Prostrate or ascending *shrub up to* 10 cm; twigs slender, glabrous or subglabrous, reddish. *Lf-blade* 0.5–1(–2) × 0.3–0.9(–1.2) cm, ovate- to obovate-oblong, acute to obtuse or shallowly emarginate, usually rounded at base, *entire or crenulate-serrulate*, glabrous and green above, pale green to subglaucous and *appressed-hairy beneath*; *petiole up to* 0.3 cm; *stipules minute or absent.* *Catkins* 1.5–2 × 0.5–0.7 cm, *with short stalks*; scales 1–2 mm, oblong, yellowish, sometimes reddish towards apex, subglabrous and ciliate. Male fls with 2 free stamens; anthers yellow, often reddish-tinged; filaments glabrous; nectaries 2. Female fls with shortly stalked, glabrous or appressed-pubescent, narrowly ovoid ovary 2–3 mm; style distinct but short; stigmas usually divided into 4 narrow lobes. Fl. 5–6, with the lvs.

Native. Amongst *Calluna*-moorland on the mountains of Perth, Aberdeen, Rhum, Argyll, Sutherland and Caithness.

23. S. reticulata L.   Net-leaved Willow.

Shrub with much-branched rhizome or creeping, rooted stems, the *aerial branches* 5–15(–20) cm; twigs with sparse, long, silky hairs at first, soon glabrous and dark reddish-brown; buds cylindrical-ovoid, densely villous at first, becoming glabrous and reddish-brown. *Lf-blade* 1.2–4(–5) × 1–2.5(–3.5) cm, ovate to suborbicular, obtuse, rounded to broadly cuneate at base, with entire, minutely glandular and sharply recurved margins, densely villous at first, soon glabrous and dull green above, glabrous or subglabrous and whitish-grey beneath, the veins *conspicuously reticulate*; *petiole usually* 1–4 cm, channelled above, reddish, sparsely hairy but soon glabrous; stipules absent. Catkins 2–3.5 × 0.4–0.5 cm, cylindrical, erect, with long stalks; scales c. 1 mm, broadly obovate or suborbicular, brownish or purplish, with ±dense greyish hairs, especially towards the base. Male fls with 2 free stamens; anthers reddish or purplish; filaments hairy in the basal half. Female fls with densely white-lanate, broadly ovoid ovary c. 2 mm; style distinct; stigmas shortly bifid, spreading. Fl. 6–7, with the lvs. Fr. 7–8. 2*n* = 38. Ch.

Native. Wet, slightly calcareous, mountain rocks and ledges between 610 m and 1100 m. Aberdeen, Angus, Sutherland, Perth and Inverness. Arctic and subarctic Eurasia and North America, southwards to the Pyrenees, Italian Alps, Macedonia, Urals, Altai and the Rocky Mts.

METACHLAMYDEAE

96. ERICACEAE

Shrubs or rarely trees with simple leaves, stipules 0. Fls hermaphrodite, regular or rarely slightly zygomorphic, (3–)4–6(–7)-merous. Calyx often small, persistent. Corolla usually gamopetalous, rarely petals free, on the edge of a fleshy disk. Stamens free from corolla, usually twice as many as corolla-lobes and then obdiplostemonous, rarely more or fewer; anthers usually dehiscing by apical pores, often with awn-like appendages; pollen in tetrads. Carpels usually as many as the corolla-lobes, united to form a usually 4–5-celled superior or inferior ovary; placentation axile; ovules 1–many, anatropous; style simple, stigma capitate. Fr. a capsule, berry or rarely a drupe. Seeds small with copious endosperm and a usually small straight embryo.

About 50 genera and 1350 spp., arctic, temperate and mountains in the tropics (very few in Australia and only in the south-east).

A very natural family, often of characteristic heath-like aspect with corollas appearing waxy in texture. Apart from this, usually easily recognized among the Gamopetalae by the obdiplostemonous free stamens. The dehiscence of the anthers by pores is also nearly constant. All European species investigated have endotrophic mycorrhiza. All the British spp., except *Arbutus unedo*, are calcifuge.

Spp. of a number of non-British genera are sometimes cultivated as ornamental plants.

1 Corolla persistent in fruit.   2
   Corolla deciduous.   3
2 Calyx longer than and coloured like the corolla; lvs
     opposite.   13. CALLUNA
   Calyx much shorter than corolla; lvs whorled.
                                                    14. ERICA
3 Ovary inferior.   12. VACCINIUM
   Ovary superior.   4
4 Petals free; lvs rusty tomentose beneath.   1. LEDUM
   Petals ±united; lvs not rusty tomentose.   5
5 Lvs mostly opposite or whorled.   6
   Lvs alternate.   7
6 Creeping dwarf shrub; lvs not more than 7 mm.
                                                    3. LOISELEURIA
   Erect or ascending shrub; lvs at least 10 mm.
                                                    4. KALMIA
7 Corolla campanulate, slightly zygomorphic.
                                                    2. RHODODENDRON
   Corolla urceolate, regular.   8
8 Fls 4-merous in lax racemes; lvs white tomentose
     beneath. W. Ireland.   6. DABOECIA
   Fls normally 5-merous; lvs at most glaucous beneath
     (and then fls in clusters).   9
9 Lvs linear or linear-elliptical, entire or serrulate, at
     most 5 mm broad; fr. dry; fls in clusters.   10
   Lvs ovate, obovate or elliptical, if less than 5 mm
     broad, clearly though remotely serrate; fr. fleshy.   11
10 Calyx and pedicels glandular; corolla ovoid, purple
     (Perth and W. Inverness).   5. PHYLLODOCE

Calyx and pedicels glabrous; corolla subglobose,
     pink. Bogs.   7. ANDROMEDA
11 Lvs rounded to cordate at base; fr. a capsule enclosed
     in the fleshy calyx; suckering shrub. 8. GAULTHERIA
   Lvs cuneate at base; fr. a drupe or berry; not sucker-
     ing   12
12 Tree or tall shrub; lvs 4 cm or more; fr. warty (Gal-
     way, Mayo).   10. ARBUTUS
   Shrub to 2 m, but usually prostrate; lvs 2(–2.5) cm
     or less; fr. smooth.   13
13 Erect shrub; lvs ovate or lanceolate, acute, aristate;
     anthers with short erect awns.   9. PERNETTYA
   Prostrate shrub; lvs ±obovate, obtuse or subacute,
     not aristate; anthers with long reflexed appendages.
                                                    11. ARCTOSTAPHYLOS

Subfamily 1. RHODODENDROIDEAE. Ovary superior. Corolla caducous after fl. Anthers without appendages. Fr. a septicidal capsule. Seeds with or without marginal wing.

1. LEDUM L.

Evergreen shrubs. Buds scaly, conspicuous. Lvs alternate, entire, shortly petiolate. *Fls on long slender pedicels in umbel-like terminal racemes*, 5-merous. Calyx-lobes short. *Petals free.* Stamens 5–10; anthers opening by pores. Capsule opening from base. Seeds narrow, flat, with a wide wing.

About 10 spp., cold north temperate zone.

1. L. palustre L.   Labrador-tea.

Shrub to 1 m. Twigs rusty tomentose. Lvs 1–4.5 cm, linear to elliptical-oblong, dark green above, rusty-tomentose beneath; margins revolute. Pedicels glandular. Fls 1–1.5 cm across. Petals cream, obovate. Stamens (5–)10(–14). Capsule 4–5 mm, oblong. Fl. 6–7. N.

Possibly native in bogs near Bridge of Allan (Stirling and Perth), naturalized in scattered localities from Kirkcudbright to Derby and in Surrey. N. and C. Europe and N. Asia from Scandinavia to Sakhalin and N. Japan, south to C. Germany, N.E. Austria, C. Russia and the Altai; North America.

Subsp. **palustre**. Lvs linear to oblong, 4–12 times as long as broad; midrib visible from beneath. Stamens (7–)10(–11). $2n = 52$.

Perhaps only the Stirling and Perth populations belong to this subsp.

Subsp. **groenlandicum** (Oeder) Hultén. Lvs elliptical-oblong, $2\frac{1}{2}$–5 times as long as broad; midrib usually concealed by tomentum on undersurface. Stamens (5–)8(–15). $2n = 26$.

Most records seem to refer to this subsp.

2. RHODODENDRON L.

Shrubs, rarely trees. Buds scaly, conspicuous. Lvs alternate, usually entire, shortly petiolate. Fls usually in terminal racemes, usually 5-merous. Calyx usually small. *Corolla campanulate or funnel-shaped, slightly zygomorphic.* Stamens from as many to twice as many as corolla-lobes; anthers opening by pores. *Seeds* small, numerous, flat, *winged.*

About 500–600 spp., mainly E. Asia (to New Guinea and N. Australia), a few in W. Asia, Europe and North America.

Many spp. and especially hybrids are ±commonly cultivated. *Azalea*, formerly regarded as a distinct genus, is now included.

*1. R. ponticum L.   Rhododendron.

Evergreen shrub to 3 m, nearly glabrous. Lvs 6–12 cm, elliptical to oblong, dark green above, paler beneath, acute, cuneate at base. Fls numerous. Corolla widely campanulate, c. 5 cm across, dull purple spotted with brown. Stamens 10. Fl. 5–6. $2n = 26$. M.

Introduced in the 18th century. Commonly cultivated and often planted in woods, etc., and has become thoroughly naturalized in many places on sandy and peaty soils, sometimes becoming locally dominant both as the shrub layer of woods and in the open. Native of C. and S. Portugal, S. Spain, E. part of Balkan peninsula; Asia Minor and Lebanon.

Some, at least, of the naturalized populations contain plants with the larger calyx and very glandular pedicels of *R. maximum* L. and/or the hairy ovary of *R. catawbiense* Michx, reflecting the use of these eastern USA species in hybridization programmes with *R. ponticum* by early breeders in Britain.

*R. luteum Sweet (*Azalea pontica* L.).

Deciduous shrub. Twigs and infl. glandular. Lvs ±oblong, hairy. Corolla funnel-shaped, yellow. Stamens 5.

Often cultivated and naturalized in one or two places. Native of Asia Minor and the Caucasus.

3. LOISELEURIA Desv.

Creeping evergreen dwarf shrub. Buds small. *Lvs opposite*, entire, shortly petiolate. Fls 1–5 in terminal clusters, 5-merous. *Corolla widely campanulate* with spreading lobes. *Stamens 5; anthers opening by slits. Ovary 2–3-celled.* Seeds numerous small, not winged. Nectar secreted by ring at base of ovary.
One species.

1. L. procumbens (L.) Desv.   Trailing Azalea.

Intricately branched, glabrous procumbent shrub. Lvs dense, 3–8 mm, oval or oblong, obtuse, coriaceous, convex above; margins revolute. Pedicels shorter than fl. Calyx-lobes ovate-lanceolate, reddish, about half as long as corolla. Corolla 4–5 mm, pink, paler inside. Capsule 3–4 mm, ovoid. Fl. 5–7. Pollinated by various insects, but probably more often selfed; weakly protogynous. $2n = 24$. Chw.

Native. Mountain tops and moors at high altitudes from 395 m in Orkney to over 1215 m. From Dunbarton northwards, but not in Hebrides. N. Europe (to

71° 10′ N. in Norway) and Asia from Iceland and the Faeroes to N. Japan; mountains of C. Europe to Pyrenees, S. Alps and Carpathians; North America from Newfoundland and Alaska, south to the mountains of New Hampshire; Greenland.

4. KALMIA L.

Evergreen shrubs. Lvs alternate to whorled, entire, subsessile or petiolate. Fls in short, umbel-like corymbs, 5-merous. Corolla cup-shaped, with 10 pouches towards the base in which the anthers lie before dehiscence. Stamens 10; anthers opening by pores. Ovary 5-celled. Seeds numerous, small, sometimes winged.

Eight spp., North America and Cuba. Others, in addition to the following, are cultivated.

Corymbs axillary; lvs ±flat, green or brownish beneath.
                                1. angustifolia
Corymbs terminal; lvs with revolute margins, glaucous or whitish beneath.   **2. polifolia**

*1. K. angustifolia L.

Upright shrub to 150 cm; young twigs terete, glabrous. Lvs 2–6 cm, opposite or in whorls of 3, elliptical to oblong, entire, sparsely puberulent, with brownish hairs beneath when young, petiolate; margins flat or slightly revolute. Corymbs axillary; pedicels glandular-pubescent. Sepals herbaceous, acute. Corolla c. 10 mm, bright reddish-pink.

Often cultivated and apparently naturalized on Ellerside Moss, Cumbria.
Native of eastern North America.

*2. K. polifolia Wangenh.

Upright, somewhat straggling shrub to 70 cm; young twigs with 2 angular ridges, pubescent. Lvs 1–3 cm, mostly opposite, linear- to elliptical-oblong, entire, glabrous, subsessile; margins strongly revolute. Corymbs terminal; pedicels glabrous. Sepals scarious, obtuse. Corolla c. 15 cm, purplish-pink.

Often cultivated and naturalized in a bog in Surrey. Native of eastern North America.

5. PHYLLODOCE Salisb.

Low heath-like evergreen shrubs. Lvs alternate, linear, usually serrulate, shortly petiolate. *Fls on slender pedicels, in terminal clusters, 5-merous. Corolla urceolate* or campanulate. Stamens 10; anthers opening by pores. Seeds ovate, with a narrow wing. Nectar secreted by a ring at base of ovary.

About 7 spp., arctic regions and north temperate mountains.

1. P. caerulea (L.) Bab.   Blue Heath.

Menziesia caerulea (L.) Swartz; *Bryanthus caeruleus* (L.) Dippel

Low bushy shrub to 25 cm. Branches ascending. Lvs 4–10 mm, dense, linear to linear-oblong, obtuse, serrulate. Fls 2–6, nodding. Pedicels and calyx reddish, glandular-pubescent. Calyx-lobes nearly free, triangular-lanceolate. Corolla 7–8 mm, urceolate, pinkish-purple. Capsule ovoid, glandular-pubescent. Fl. 6–7. $2n = 24$. Chw.

Native. Free-draining rocky and peaty *Vaccinium* heath between 680 m and 840 m in Perth (Sow of Atholl) and Inverness (Ben Alder, Loch an Sgòir). N. Europe and Asia from Iceland to Japan; very rare in Pyrenees; Greenland to Quebec and the mountains of Vermont and New Hampshire.

6. DABOECIA D. Don

Low heath-like evergreen shrubs. Lvs alternate, entire, subsessile. *Fls in terminal racemes, 4-merous.* Calyx small. *Corolla urceolate*; lobes very short, recurved. Stamens 8; anthers opening by pores. Seeds numerous, small, tuberculate, not winged.

Two spp., the other in the Azores.

1. D. cantabrica (Hudson) C. Koch
St Dabeoc's Heath.

D. polifolia D. Don; *Menziesia polyfolia* Juss.; *Boretta cantabrica* (Hudson) O. Kuntze

Straggling shrub to 50 cm. Twigs, pedicels and calyx glandular-hairy. Lvs 5–10 mm, elliptical or elliptical-linear, acute, revolute, dark green above with scattered glandular hairs, white-tomentose beneath. Racemes lax, 3–10-fld. Pedicels c. 5 mm; fls nodding. Corolla 8–12 mm, reddish-purple. Capsule oblong, glandular-hairy. Fl. 7–9. $2n = 24$. Chw.

Native. Heaths and rocky ground in Mayo and W. Galway, ascending to 580 m, locally common. Commonly cultivated in several colour forms and occasionally found as an escape in England. W. France, N., W. and C. Spain, N.W. Portugal.

Subfamily 2. VACCINIOIDEAE. Ovary usually superior (except *Vaccinium*). Corolla caducous after fl. Anthers usually with appendages. Fr. a loculicidal capsule, drupe or berry. Seeds without marginal wing.

7. ANDROMEDA L.

Low evergreen shrubs. Lvs oblong or linear, alternate, entire, shortly petiolate. Fls in terminal clusters, 5-merous. *Calyx small, not becoming fleshy.* Corolla urceolate, lobes short. Stamens 10. Anthers opening by pores. *Fr. a capsule.* Seeds ovate, smooth. Nectar secreted by swellings at the base of the ovary.

Two spp., cold north temperate and Arctic.

1. A. polifolia L.
Bog Rosemary.

Glabrous shrub to 30 cm, with creeping rhizome and scattered erect little-branched stems. Lvs 1.5–3.5 cm, linear or elliptical-linear, acute, revolute, dark green above, glaucous beneath. Fls 2–8, nodding, on pedicels 2–4 times as long as corolla. Corolla 5–7 mm, pink fading to whitish. Capsule subglobose, glaucous. Fl. 5–9. Pollinated by humble-bees and butterflies, also selfed, homogamous. $2n = 48$. Chw.

Native. Bogs, rarely wet heaths; ascending to 535 m. From Cardigan and Stafford to Perth, local and decreasing at least in the south; apparently extinct in Huntingdon, Norfolk and Somerset; in Ireland, rather common in the Central Plain, and extending as far as Armagh, W. Galway, Kilkenny and Wicklow. N. and N.C. Europe extending locally to the S. Alps, E. Carpathians and S.C. Russia, mainly in the mountains in the southern part of its range; N. Asia to Japan; Greenland; North America south to New York and Idaho.

8. GAULTHERIA L.

Evergreen shrubs. Lvs usually alternate, shortly petiolate. Fls 5-merous. *Calyx accrescent and fleshy in fr.* Corolla urceolate or campanulate. Stamens 10; appendages short, erect, or 0. Fr. a capsule surrounded by the fleshy calyx and thus appearing berry-like. Seeds small, numerous.

About 200 spp. America, E. Asia, Australia and New Zealand. Several others are cultivated.

*1. G. shallon Pursh
Shallon.

Shrub to 1 m, with underground rhizomes and forming dense patches. Stems ascending, glandular-hairy when young. Lvs 5–12 cm, alternate, broadly ovate, acute, rounded to cordate at base, serrulate, glabrous. Fls in terminal glandular panicles. Calyx-lobes triangular. Corolla c. 1 cm, urceolate, white tinged with pink Fr. c. 1 cm, dark purple, hairy. Fl. 5–6. Fr. 9–10. $2n = 88$. N.

Introduced. Commonly planted for ornament and as game-cover, and naturalized on sand or peat in a number of places. Native of western North America from Alaska to California.

*G. procumbens L.

With underground rhizomes. Aerial stems to 15 cm. Lvs 1.5–4 cm, obovate, crenate-serrate. Fls solitary. Fr. red. One patch in a pinewood in W. Inverness. Native of eastern North America. Formerly the source of Oil of Wintergreen.

9. PERNETTYA Gaudich.

Evergreen shrubs. Lvs alternate. Fls 5-merous. Corolla urceolate. Stamens 10; appendages *short, erect,* rarely 0. *Fr. a berry* with dry, rarely fleshy, calyx at base.

About 20 spp. America from Mexico southwards, Tasmania and New Zealand.

*1. P. mucronata (L. fil.) Gaudich. ex Sprengel
Prickly Heath.

Branched suckering erect shrub to 2 m forming a dense thicket. Lvs 1–2 cm, ovate to elliptic-lanceolate, acute and aristate at apex, remotely serrate with mucronate teeth. Fls solitary, axillary, white; pedicels c. 5 mm. Fr. c. 1 cm, crimson, purple, pink or white. Fl. 5–6. N.

Introduced. Commonly planted in gardens and naturalized in a number of places, especially in Ireland. Native of S. Chile and Argentina.

10. ARBUTUS L.

Evergreen trees or shrubs. Lvs alternate, petiolate. Fls in terminal panicles, 5-merous. Calyx deeply lobed. Corolla urceolate. Stamens 10; anthers with long reflexed appendages, opening by pores. Ovules numerous in each cell. *Fr. a ±warty, globose berry.*

About 20 spp. in western North America from British Columbia to California, W. Asia and in the Mediterranean region to W. Ireland and the Canary Is. Two spp., in addition to the following, are sometimes cultivated.

1. A. unedo L. Strawberry-tree, Arbutus.

Erect shrub or tree to 12 m, glabrous or nearly so. Bark reddish-brown, thin, somewhat rough. Lvs 4–10 cm, elliptical-oblong or elliptical-obovate, acute, cuneate at base, serrate, dark green and shining above, paler beneath. Petiole c. 6 mm. Infl. c. 5 cm, many-fld. Calyx-lobes triangular. Corolla c. 7 mm, ovoid or subglobose, creamy-white sometimes tinged pink. Fr. 1.5–2 cm, red, warty. Fl. 9–12. Fr. 9–12 (the following year). $2n = 26$. M.

Native. Rock crevices and between boulders, both sandstone and limestone, persisting in scrub and developing oakwood but not in mature oakwood. Kerry, W. Cork and Sligo (Lough Gill) but formerly more widespread; locally abundant. Mediterranean region to S.W. France; Brittany.

11. ARCTOSTAPHYLOS Adanson

Deciduous or *evergreen* dwarf shrubs. Lvs alternate, usually entire. Fls in terminal racemes, panicles or clusters, normally 5-merous. Calyx deeply lobed. Corolla urceolate. Stamens 10; anthers with long reflexed appendages, opening by pores. Ovary 5–10-celled, each cell with 1 ovule. *Fr. a drupe.* Nectar secreted by a ring surrounding the ovary.

About 70 spp., confined to western North and Central America except for the following.

Lvs entire, evergreen; ripe fr. red.	**1. uva-ursi**
Lvs serrate, deciduous; ripe fr. black.	**2. alpinus**

1. A. uva-ursi (L.) Sprengel Bearberry.

Evergreen prostrate shrub with long rooting branches often forming mats. Twigs glabrous or nearly so. Lvs 1–2 cm, obovate or obovate-elliptical, usually obtuse (to subacute), cuneate at base, dark green above, paler beneath, conspicuously reticulately veined, entire. Fls 5–12 in short dense racemes; pedicels 3–4 mm. Corolla 4–6 mm, white tinged with pink or green. Fr. 6–8 mm, red, globose, glossy, with rather dry flesh. Fl. 5–7. Pollinated by humble-bees or selfed, weakly protogynous. Fr. 7–9. $2n = 52$. Chw.

Native. Moors, often covering rocks or banks; ascending to 915 m. Common in the Scottish Highlands, extending south to Northumberland, Westmorland, Derby and Cheshire; N. and W. Ireland from Donegal to Antrim and Clare, Galway and W. Mayo, local. Most of Europe except the extreme south; N. Asia to N. Japan; Caucasus, Himalaya; North America from

Labrador and Alaska to the mountains of Virginia, New Mexico and N. California.

2. A. alpinus (L.) Sprengel Alpine Bearberry.

Arctous alpinus (L.) Niedenzu

Deciduous prostrate intricately branched shrub. Twigs glabrous. Lvs 1–2(–2.5) cm, obovate, obtuse to subacute, cuneate at base, bright green, conspicuously reticulately veined, serrulate, usually ciliate. Infl. 2–4-fld. Corolla c. 4 mm, white. Fr. 6–10 mm, globose, black, juicy. Fl. 5–8. Fr. 8–10. $2n = 26$. Chw.

Native. Mountain moors, ascending to 915 m. From Inverness to Shetland (not in Hebrides), very local. N. Europe and Asia from Scandinavia to Japan; high mountains of C. and S. Europe southwards to Pyrenees, C. Apennines and N. Albania; Greenland to the mountains of New Hampshire.

12. VACCINIUM L.

Shrubs. Lvs alternate, shortly petiolate. Fls 4–5-merous, in racemes, or axillary and solitary, or clustered. Calyx-lobes short. Stamens 8 or 10. Anthers opening by pores with or without appendages. Fr. a berry. Nectar secreted by a swelling at the base of the style. *Ovary inferior.*

About 300–400 spp., arctic and north temperate regions, tropical mountains (not Africa), Andes, S. Africa, Madagascar. A number of spp. are sometimes cultivated.

1 Corolla campanulate or urceolate, with small lobes; aerial stems neither creeping nor filiform. 2
  Corolla divided nearly to base, the lobes reflexed; stems creeping and filiform. 4
2 Lvs persistent, dark green and glossy; fls in racemes; corolla campanulate. **1. vitis-idaea**
  Lvs deciduous; fls 1–4, axillary; corolla urceolate. 3
3 Lvs serrulate, acute, bright green; twigs angled. **2. myrtillus**
  Lvs entire, obtuse, blue-green; twigs terete. **3. uliginosum**
4 Lvs ±ovate, acute, to 8 mm; infl. terminal. 5
  Lvs oblong, obtuse, 6–18 mm; infl. rhachis terminated by a lfy shoot. **6. macrocarpon**
5 Pedicels puberulent; lvs ±oblong-ovate, equally wide for some distance at the base. **4. oxycoccos**
  Pedicels glabrous; lvs ±triangular-ovate, widest near base. **5. microcarpum**

Subgenus 1. VACCINIUM.

Stems ±erect. Corolla lobes short. Ovary 4–5-celled.

Section 1. *Vitidaea* Dumort. Fls in racemes. Corolla campanulate. Filaments pubescent; anthers without appendages. Lvs persistent.

1. V. vitis-idaea L. Cowberry.

Evergreen shrub to 30 cm, with creeping rhizome and numerous ±erect, often arching, much-branched stems. Twigs terete, puberulent when young. *Lvs 1–3 cm, elliptical to obovate, obtuse or emarginate, dark green and glossy above,* paler and *gland-dotted beneath,* glabrous, coriaceous, ±2-ranked; margins somewhat revolute, obscurely crenulate. Fls c. 4 in short terminal pendent racemes. Calyx-lobes ovate-orbicular, reddish. *Corolla* c. 6 mm, *campanulate*, white, tinged with pink; lobes ±revolute, c. half as long as tube. *Fr.* 8–10 mm, *red*, globose, acid, edible. Fl. 6–8. Pollinated by bees but often selfed; homogamous. Fr. 8–10. $2n = 24$. N. or Chw.

Native. Moors and woods on acid soils, occasionally locally dominant; ascending to over 1070 m. Common in Scotland, extending south to Yorks, Leicester and through Wales to Somerset; Ireland except the south-west, rare in centre. N. Europe, and Asia from Iceland and the Faeroes to Japan and in the mountains of C. Europe to the Pyrenees, N. Apennines and Macedonia; a subsp. in North America.

V. myrtillus × vitis-idaea = *V. × intermedium* Ruthe. Evergreen or nearly. Lvs elliptical, serrulate, with inconspicuous glands beneath, intermediate in colour. Fr. rare, purple.

In some quantity with the parents on Cannock Chase (Stafford) and in N. Derby, rare elsewhere.

Section 2. *Vaccinium*. Fls 1–4, axillary. Corolla urceolate. Filaments glabrous; anthers with appendages. Lvs deciduous.

2. V. myrtillus L.　　　Bilberry, Blaeberry, Whortleberry, Huckleberry.

Glabrous *deciduous* shrub to 60 cm, with creeping rhizome and numerous erect stems and branches. *Twigs angled, green.* Lvs 1–3 cm, ovate, *acute, serrulate, bright green*, conspicuously reticulately veined. Fls 1(–2) in each axil. Calyx limb scarcely more than sinuate. Corolla 4–6 mm, globose, green tinged with pink; lobes very short, reflexed. *Fr.* c. 8 mm, *black with a glaucous bloom*, globose, sweet, edible. Fl. 4–6. Pollinated mainly by bees, more rarely selfed; weakly protandrous. Fr. 7–9. $2n = 24$. N. or Chw.

Native. Heaths, moors and woods on acid soils, more tolerant of exposure and of shade than *Calluna* and becoming dominant in a zone higher on the mountains and in the field layer in more shaded woods; ascending to over 1215 m. Common throughout most of the British Is. but becoming local in England towards the south-east and absent from several counties in the east and E. Midlands. Most of Europe but only on mountains in the south; Caucasus; N. Asia.

3. V. uliginosum L.　　　Bog Bilberry.

Deciduous shrub to 50 cm, with creeping rhizome and numerous bushy stems with ±spreading branches. *Twigs terete, brownish*, glabrous or puberulent. Lvs 1–2.5 cm, obovate or ovate, *entire, obtuse, blue-green*, conspicuously reticulately veined, glabrous to somewhat pubescent. Fls 1–4 in each axil. Calyx-lobes short, broad, obtuse. Corolla c. 4 mm, subglobose or ovoid, white usually tinged with pink; lobes very short, reflexed. *Fr.* c. 6 mm, *black with a glaucous bloom*, globose, sweet, edible in small quantity. Fl. 5–6. Pollinated

by bees; weakly protandrous. Fr. 8–9. $2n = 48$. N. or Chw.

Native. Bilberry moors, ascending to 1070 m. From Shetland (absent from Hebrides), south to Angus and Kintyre, also in Selkirk, Cumberland and N. Northumberland, local but sometimes abundant. N. Europe and Asia from Iceland to Japan and in the high mountains to the Sierra Nevada, N. Apennines, Albania and Bulgaria, Caucasus and Altai; Greenland; North America, south to the mountains of New York.

Subgenus 2. OXYCOCCUS (Hill) A. Gray

Evergreen prostrate dwarf shrubs with filiform stems. Fls 4-merous, on long slender pedicels. Corolla lobed nearly to base; lobes reflexed. Ovary 4-celled.

4. V. oxycoccos L.　　　Cranberry.

Oxycoccus palustris Pers.; *O. quadripetalus* Gilib.

Stems prostrate, rooting, filiform, usually widely separated from each other. *Lvs* 4–8 mm, distant, *oblong-ovate, equally broad for some distance from the base or broader towards the middle, acute*, dark green above, glaucous beneath, *strongly revolute*. *Fls* 1–4, *in a terminal raceme*. *Pedicels* 1.5–3 cm, *puberulent*, 2-bracteolate about or below the middle; bracteoles pubescent. Calyx-lobes ciliate. Corolla pink; lobes 5–6 mm. Filaments pubescent on the edges, glabrous or slightly pubescent outside. Fr. 6–8 mm, globose to pyriform, red- or brown-spotted, edible. Fl. 6–8. Fr. 8–10. $2n = 48$. Ch.

Native. Bogs, more rarely wet heaths, local, but occurring in much of Scotland northwards to Sutherland (absent from Orkney, Shetland and Outer Hebrides) and southwards to S.W. Yorks and Gloucester, and in Norfolk, Surrey, Sussex, Hants and Devon; much of Ireland but only occasional in the south-west. N. and C. Europe extending locally to S.C. France, N. Italy and S.E. Russia; N. Asia to Japan; Greenland; North America to N. Carolina, Wisconsin and British Columbia.

5. V. microcarpum (Turcz. ex Rupr.) Hooker fil.
　　　　　　　　Small Cranberry.

Oxycoccus microcarpus Turcz. ex Rupr.

Differs from *V. oxycoccos* as follows: *Lvs* 3–5 mm, *triangular-ovate.* Fls 1–2. *Pedicels* and usually bracteoles and calyx *glabrous*. Corolla deeper pink. Filaments often pubescent outside. Fr. ellipsoid to pyriform. Fl. 7. $2n = 24$. Ch.

Native. Bogs in the Scottish Highlands from Argyll to Moray, and in Caithness; ascending to 670 m. N. and N.C. Europe, southwards to N.W. Ukraine and east to Sakhalin; Alps and Carpathians; Rocky Mountains to British Columbia and Alberta (distribution imperfectly known); absent from Greenland; extending further north in general than *V. oxycoccos*.

*6. V. macrocarpon Aiton　　American Cranberry.

Oxycoccus macrocarpos (Aiton) Pursh

Like *V. oxycoccus* but more robust. Stems prostrate,

rooting, ascending at ends, slender but scarcely filiform. *Lvs* 6–18 mm, *oblong or elliptical-oblong, obtuse,* slightly glaucous beneath, *flat or slightly revolute.* Fls 1–10 in a raceme, *the rhachis continuing as a lfy shoot.* Pedicels 2-bracteolate near apex. Corolla pink; lobes 6–10 mm. Filaments much shorter than anthers. Fr. 10–20 mm, red, edible. Fl. 6–8. Fr. 9–11. $2n = 24$. Ch.

Introduced. Sometimes grown for its fr. and naturalized in a few places. Native of North America from Newfoundland to Saskatchewan, N. Carolina and Minnesota.

Subfamily 3. ERICOIDEAE. Ovary superior. Corolla persistent in fr. Anthers with or without appendages. Fr. a capsule. Seeds without marginal wing.

13. CALLUNA Salisb.

Evergreen shrub. Lvs opposite, very small. Fls 4-merous, axillary, forming a lax terminal raceme-like infl. *Calyx large, deeply lobed, of the same colour and texture as the corolla.* Corolla smaller, campanulate. Stamens 8, with appendages; anthers opening by pores. *Capsule septicidal*, few-seeded. Nectar secreted by 8 swellings between the filament bases.

1. C. vulgaris (L.) Hull   Ling, Heather.

Diffuse evergreen shrub to 60 cm (rarely to 1 m), with numerous tortuous, decumbent or ascending branched stems, rooting at the base and bearing numerous axillary short shoots. Lvs 1–2 mm, linear, sessile with two short projections at base, the margins strongly revolute making the lf trigonous, glabrous, pubescent or (var. *hirsuta* S. F. Gray) densely grey-tomentose, those of the main stems distant, those of the short shoots densely imbricate in 4 rows. Fls solitary, axillary, on the main axis and on the short shoots, with 4 ovate bracteoles forming a calyx-like involucre under each fl.; forming a raceme- or panicle-like infl. 3–15 cm. Calyx c. 4 mm, somewhat scarious; lobes ovate-oblong, pale pinkish-purple. Capsule 2–2.5 mm, globose. Fl. 7–9. Pollinated by various insects and wind, weakly protandrous. Fr. 11. $2n = 16$. N. or Chw.

Native. Heaths, moors, bogs and open woods on acid soils; ascending to 730 m. Common throughout British Is. Dominant over large areas on well-drained acid soils (heaths and moors), though local in E. Midlands, also becoming dominant in the field layer of open woods on similar soils; var. *hirsuta* abundant in some places near the sea (as on the Culbin Sands, Moray, where it is dominant over a large area to the exclusion of the type), scarce inland. Most of Europe but rare in much of the Mediterranean region and in the south-east; common in the west, much decreasing in abundance eastwards; N.W. Morocco; Azores; eastern North America (rare, probably introduced).

14. ERICA L.

Evergreen shrubs. Lvs whorled, small, entire, shortly petiolate, revolute, dense, Infl. various. Fls 4-merous. *Calyx much shorter than corolla*, deeply lobed, not peta-

loid. Corolla urceolate, campanulate or cylindrical; lobes short. *Stamens* 8; anthers opening by pores. *Capsule loculicidal*, many-seeded. Nectar secreted by a ring round the base of the ovary.

Over 500 spp., mainly S. Africa, a few in Europe and the Mediterranean region. The British spp. and hybrids are all ±commonly cultivated, often in forms varying in fl. colour, as are two or three other European spp. Some of the S. African spp. are grown in greenhouses and often sold in pots. White-fld forms of all our native spp., except *E. mackaiana*, occur.

1 Stamens included in corolla-tube.   2
  Stamens exserted, at least partly; lvs glabrous.   7
2 Lvs and sepals ciliate with long usually glandular hairs.   3
  Lvs and sepals glabrous.   5
3 Fls in terminal umbel-like clusters; anthers with basal
    appendages   4
  Fls in racemes; anthers without appendages (Dorset
    to Cornwall).   **3. ciliaris**
4 Lvs grey-pubescent above as well as ciliate, revolute
    nearly to the midrib, shorter and more distant
    below infl. (common).   **1. tetralix**
  Lvs dark green and glabrous above, somewhat revo-
    lute but leaving much of the white under surface
    exposed, not different below infl. (Galway and
    Donegal).   **2. mackaiana**
5 Corolla urceolate, pink or purple, rarely white (and
    then white in bud).   6
  Corolla narrowly campanulate, pink in bud, white
    when open (naturalized in Dorset and Cornwall).
                                                            lusitanica
6 Low shrub with short axillary lfy shoots; infl. long,
    raceme-like (common native).   **4. cinerea**
  Erect shrub without short shoots; fls in terminal
    umbel-like clusters (naturalized in Derry).
                                                            terminalis
7 Pedicels shorter than fls; corolla ±tubular; stamens
    about half exserted (Galway and Mayo). **5. erigena**
  Pedicels longer than fls; corolla widely campanulate;
    stamens long exserted (Cornwall and Fermanagh).
                                                          6. vagans

1. E. tetralix L.   Cross-leaved Heath, Bog Heather.

Diffuse dwarf shrub to 60 cm, with numerous tortuous, ascending, branched stems, rooting at base, without short axillary shoots. Twigs pubescent, purplish, often glandular-hairy. *Lvs* 4 *in a whorl*, 2–4 mm, linear, *glandular-ciliate, grey-pubescent above, the margins revolute nearly to midrib hiding under surface, below the infl. more distant and usually shorter. Fls 4–12 in terminal, umbel-like clusters*, nodding in fl., becoming erect in fr. Pedicels c. 2 mm, pubescent, with bracteoles about the middle. *Calyx-lobes* c. 2 mm, oblong-lanceolate, *pubescent on the surface and ciliate with long wavy glandular hairs. Corolla* 6–7 mm, *urceolate, rose-pink*, sometimes somewhat paler beneath; lobes very short, patent or revolute. *Anthers included*, with appendages. Capsule pubescent. Fl. 7–9. Pollinated by small insects or selfed. Fr. 10. $2n = 24$. N. or Chw.

Native. Bogs and wet heaths and moors, rarely on

drier heaths; ascending to 730 m. Common in suitable habitats throughout the British Is., but local in the English Midlands and in parts of S.E. Ireland, sometimes locally dominant. W. and N. Europe eastwards to Latvia and C. Finland.

2. E. mackaiana Bab.   Mackay's Heath.

E. mackaii Hooker

Differs from *E. tetralix* as follows: Habit denser. Twigs almost hispid when young, soon glabrous. *Lvs oblong-lanceolate, dark green and glabrous above, the margins somewhat revolute but leaving white under-surface exposed, not different below infl.* Pedicels with scattered long hairs. *Calyx-lobes oblong-ovate, glabrous except for the shorter straight glandular cilia on the upper half. Corolla purplish-pink*, usually white beneath. Fl. 8–9. $2n = 24$. Seed not produced in Ireland. N. or Chw.

Native. In blanket bog in W. Galway and W. Donegal, very local. N.W. Spain.

3. E. ciliaris L.   Dorset Heath.

Diffuse dwarf shrub to 60 cm, with numerous ascending branched stems rooting at the base, with some axillary shoots. Twigs pubescent. *Lvs 3 in a whorl*, 1–3 mm, *ovate, glandular-ciliate, otherwise glabrous above*, the margins somewhat revolute but leaving white under-surface exposed, distant below infl. *Fls axillary, forming unilateral terminal racemes*, 5–12 cm. Pedicels 1–2 mm, bracteolate about the middle. Calyx-lobes oblong-lanceolate, ciliate, slightly pubescent. *Corolla 8–10 mm, urceolate, somewhat curved above and inflated below*, the mouth oblique, deep pink. *Anthers included, without appendages.* Ovary glabrous. Fl. 6–9. N. or Chw.

Native. Heaths in Dorset, S. Devon, W. Cornwall and Connemara, very local but sometimes abundant. W. and C. France, Spain (except east), Portugal, N.W. Morocco.

4. E. cinerea L.   Bell-heather.

Diffuse shrub to 60 cm, without numerous ascending branched stems rooting at base, *with numerous short lfy axillary shoots*, often appearing as bunches of lvs. *Lvs 3 in a whorl*, 5–7 mm, linear, *glabrous*, dark green, the margins strongly revolute. Fls in short terminal racemes and on the upper axillary shoots, the whole forming a terminal infl. 1–7 cm. Pedicels c. 3 mm, puberulent; bracteolate immediately below fl. Calyx-lobes lanceolate, glabrous, usually purple, keeled. *Corolla 5–6 mm, urceolate*, usually reddish-purple; lobes very short, reflexed. *Anthers included*, with appendages. Ovary glabrous. Fl. 7–9. Pollinated by humble-bees or selfed. Fr. 10. $2n = 24$. N. or Chw.

Native. Heaths and moors, usually dry; ascending to 670 m. Common throughout the British Is. but local in the English and Irish Midlands. W. Europe northwards to Norway (62° 20′ N) eastwards to N. Italy.

*E. terminalis Salisb.   Corsican Heath.

E. stricta Donn ex Willd.

Erect shrub 50–100 cm, with suberect branches. Lvs 4 in a whorl, 4–6 mm, linear, glabrous. Fls in terminal 3–8-fld clusters. Corolla 6–7 mm, urceolate, ovoid, bright pink. Stamens included.

Sometimes cultivated; naturalized on Magilligan dunes (Derry). Native of Corsica, Sardinia, S. Italy, S.W. Spain, N.W. Morocco.

*E. lusitanica Rudolphi

Erect shrub 1–2 m, with erect branches. Lvs 3–4 in a whorl, 5–6 mm, linear, glabrous. Fls axillary, forming long pyramidal panicles. Corolla c. 4 mm, narrowly campanulate, white tinged with pink. Stamens included.

Sometimes cultivated; naturalized in Dorset and on railway banks in Cornwall. Native of S.W. France, N.W. Spain, S. Portugal.

5. E. erigena R. Ross   Irish Heath.

E. hibernica (Hooker & Arn.) Syme, *E. mediterranea* auct., non L.

Glabrous shrub 50–200 cm. Stems several, erect, with ascending branches without short axillary shoots. Lvs 4 in a whorl, 5–8 mm, linear, dark green; margins strongly revolute. Fls axillary, pendent, forming a dense unilateral lfy raceme. *Pedicels much shorter than fls*, bracteolate about the middle. Calyx-lobes lanceolate. *Corolla 5–7 mm, ±tubular*, dull purplish-pink; lobes broad, obtuse, erect. *Anthers c. ½-exserted*, deep purple, without appendages, the cells not separated. Fl. 3–5. N.

Native. Bogs, especially when relatively well-drained as at the edges of lakes and streams. W. Galway and W. Mayo, locally plentiful. W. France (Gironde), N., W. and S. Spain, Portugal.

6. E. vagans L.   Cornish Heath.

Glabrous diffuse shrub 30–80 cm, with numerous ascending stems with erect branches, without short axillary shoots. Lvs 4–5 in a whorl, 7–10 mm, linear, bright green; margins strongly revolute. Fls axillary, forming a dense cylindrical lfy raceme 8–16 cm, often terminated by lvs. *Pedicels 3–4 times as long as fls*, bracteolate about or below the middle. Calyx-lobes triangular-ovate. *Corolla 3–4 mm, widely campanulate*, pale lilac; lobes deltate, erect. *Anthers fully exserted*, deep purple, without appendages; cells separate except at the base. Fl. 7–8. $2n = 24$. N.

Native. Heaths round the Lizard (Cornwall), where it is often dominant; probably also native in Fermanagh; commonly cultivated and an occasional escape elsewhere. W. and C. France, N. Spain.

The following hybrids occur: *E. ciliaris* × *tetralix* = *E.* × *watsonii* Bentham, common with the parents; *E. mackaiana* × *tetralix* = *E.* × *praegeri* Ostenf., common with the parents; *E. tetralix* × *vagans* = *E.* × *williamsii* Druce, near the Lizard.

97. PYROLACEAE

Evergreen perennial herbs with creeping rhizome. Fls hermaphrodite, regular, 5-merous. Petals free. Stamens obdiplostemonous, free; anthers opening by pores; pollen usually in tetrads. Ovary incompletely 5-celled, with thick fleshy axile placentae with numerous small anatropous ovules. Style simple; stigma capitate. Fr. a loculicidal capsule. Seeds very small, numerous; endosperm copious; embryo undifferentiated.

Four genera and about 35 spp., north temperate and arctic regions. A small family distinguished from Ericaceae, with which it is often united, by the herbaceous habit, incompletely septate ovary and undifferentiated embryo.

All the spp. are partial saprophytes and tend to be associated with raw humus and are especially characteristic of pine woods.

1 Fls in terminal racemes; lvs spirally arranged.   2
  Fls solitary; lvs opposite.   3. MONESES
2 Fls secund, greenish-white; lvs acute; petiole c. 1 cm.
    2. ORTHILIA
  Fls not secund, pinkish or pure white; lvs obtuse, rarely subacute; petiole 2 cm or more.   1. PYROLA.

1. PYROLA L.   Wintergreen

Herbs with a slender creeping rhizome and short, often distant, aerial stems frequently reduced to a basal rosette of lvs. *Lvs spirally arranged. Fls in racemes, not secund*, on scapes usually bearing a few scales. *Disk* 0. *Anthers with very short tubes bearing the pores. Pollen in tetrads. Valves of the capsule webbed at the edges.* Nectar secreted by base of petals.

About 20 spp., north temperate and arctic regions.

1 Style straight; fls ±globose.   2
  Style strongly curved; fls campanulate.
    3. rotundifolia
2 Style 1–2 mm, not thickened below the stigma.
    1. minor
  Style c. 5 mm, thickened into a ring below the stigma.
    2. media

1. P. minor L.   Common Wintergreen.

Stem very short or lvs all basal. Lvs 2.5–4 cm, ovate or broadly elliptical, obtuse or subacute, crenulate, light green; petioles 2.5–3 cm, shorter than blade. Scape 10–30 cm. Racemes rather dense. *Fls* c. 6 mm, ±globose. Calyx-lobes deltate, acute. *Petals white to pinkish. Style* 1–2 mm, *straight, included, shorter than stamens and ovary, without a ring below the stigma*; stigma with 5 *large spreading lobes.* Fl. 6–8. Pollinated by insects or selfed; homogamous. $2n = 46$. Hr. or Chh.

Native. Woods, moors, damp rock-ledges and dunes; ascending to 535 m in Perth. Rather local in Scotland, becoming more local and more confined to woods southwards but extending over most of England; absent from Channel Is., Devon and Cornwall, Isle of Man, Outer Hebrides, Orkney and Shetland; in Wales only in Glamorgan and Caernarvon; scattered over Ireland but very local (mainly north). N. Europe and Asia from Iceland to N. Japan, south to the mountains of C. Spain, C.

Italy and Thessaly; North America from Labrador to Alaska, New England and California.

2. P. media Sw.   Intermediate Wintergreen.

Lvs all basal, 3–5 cm, orbicular or ovate-orbicular, obtuse, obscurely crenulate, dark green; petiole 2.5–5.5 cm, about equalling blade or longer. Scape 15–30 cm. Racemes rather lax. *Fls* c. 10 mm, ±globose. Calyx-lobes triangular-ovate. *Petals white to pinkish. Style* c. 5 mm, *straight, exserted, longer than stamens and ovary, expanded into a ring below the stigma*; stigma with 5, *moderate, erect lobes.* Fl. 6–8. $2n = 92$. Hr.

Native. Woods and moors; ascending to 975 m. Sussex; S. Wilts; Worcester and Shropshire to S. Lancs; Westmorland and Yorks to Shetland (not Orkney or Outer Hebrides); N. and W. Ireland to Down and Clare; very local, commonest in the Scottish pinewoods. Europe from Scandinavia (70° 41′ N) to the mountains of E. France, N. Apennines, Serbia, Bulgaria and the Caucasus; Asia Minor.

3. P. rotundifolia L.   Round-leaved Wintergreen.

Lvs all basal, 2.5–5 cm, orbicular or ovate, obscurely crenulate, dark green and glossy; petiole 3–7 cm, longer than blade. Scape 10–40 cm. Racemes lax. Fls c. 12 mm diam., campanulate. *Petals pure white. Style decurved then curving back, longer than the petals, stamens and ovary*, expanded into a ring below the stigma; stigma with five small erect lobes. Fl. 7–9. Pollinated by various insects or selfed; homogamous. $2n = 46$. Hr.

Subsp. **rotundifolia**. Scales on scape 1–2. Pedicels 4–6 mm. Calyx-lobes usually linear-lanceolate, acute, 2–3 times as long as broad. Style 7–8 mm.

Native. Bogs, fens, damp rock-ledges and woods; ascending to 760 m. Devon and Somerset to Kent and then northwards to Orkney, very local and with a distinct eastern tendency (absent from Wales except Pembroke, N.W. England and the Hebrides and very rare in W. Scotland); Westmeath; Channel Is. Europe from Iceland to the mountains of N. Spain, N. Italy, Bulgaria and Crimea; N. Asia to the R. Lena and the Altai Mountains; Asia Minor.

Subsp. **maritima** (Kenyon) E. F. Warb. Scales on scape 2–5. Pedicels 2–5 mm. Calyx-lobes ovate, subobtuse, scarcely twice as long as broad. Style 4–7 mm, usually more thickened at apex and more curved than in subsp. *rotundifolia.*

Native. Dune-slacks in E. Norfolk, Glamorgan, Denbigh, Anglesey and Lancs. Atlantic coast from W. Germany to N.W. France.

2. ORTHILIA Rafin. (*Ramischia* Opiz ex Garcke)

Differs from *Pyrola* as follows: *Racemes secund. Disk consisting of* 10 *small glands. Anthers without tubes. Pollen grains free.*

One or two spp., north temperate and arctic regions.

1. O. secunda (L.) House   Serrated Wintergreen.
Pyrola secunda L., *Ramischia secunda* (L.) Garcke
Stems short, 2–10 cm. *Lvs 2–4 cm, ovate to ovate-elliptical, acute,* serrulate, light green; petioles c. 1 cm, shorter than blade. Scape 5–12 cm, with 1–5 scales. *Racemes secund,* dense. Fls c. 5 mm. Calyx-lobes ovate, obtuse. *Petals greenish-white,* ±erect. Style c. 5 mm, exserted, without a ring, but somewhat thickened below the stigma; stigma with five rather small spreading lobes. Fl. 7–8. Pollinated by insects or selfed. $2n = 38$. Chh?

Native. Woods and damp rock-ledges, ascending to 730 m. Orkney, Mull and W. Ross to Dunbarton and Argyll, local; Peebles, Cumberland and Westmorland; Brecon; Fermanagh and King's Co. N. Europe and Asia from Iceland to Japan and in the mountains south to the Pyrenees, Sicily and Thessaly; North America from Newfoundland to Alaska, New Jersey and California.

3. MONESES Salisb.

Differs from *Pyrola* as follows; *Lvs opposite. Fls solitary. Disk obvious,* 10-*lobed. Anthers with relatively long*

tubes. *Valves of capsule not webbed.* No nectar.
Two spp., north temperate regions.

1. M. uniflora (L.) A. Gray
                    One-flowered Wintergreen.
Pyrola uniflora L.
Stem 1–5 cm. Lvs 1–2.5 cm, orbicular, serrate, light green, decurrent down the petiole which is shorter than blade. Scape 5–15 cm, usually with 1 scale and a bract below the fl. Fls wide open, c. 15 mm diam. Petals white. Pollen in tetrads. Style without a ring below the stigma; stigma with five large spreading lobes. Fl. 6–8. $2n = 26$. Pollinated by insects and ?selfed. H. or Chh.

Native. Pinewoods, etc.; E. Scotland from Sutherland to Banff; Orkney, very local and rare. N. Europe and Asia from Iceland to N. Japan, south to the mountains of N.E. Spain, Corsica, Italy and Bulgaria; North America from Newfoundland to Alaska, Pennsylvania and New Mexico.

98. MONOTROPACEAE

Differ from Pyrolaceae in being saprophytic herbs without chlorophyll; anthers opening by longitudinal slits and the free pollen grains. They are occasionally (but not in our sp.) gamopetalous and with the ovary 1-celled with parietal placentae.

Twelve genera and 21 spp. in north temperate zone, mainly North America extending to the mountains of Colombia and Malaysia.

1. MONOTROPA L.

Fls in short racemes, 4–5-merous. Sepals free, large, oblong-spathulate. Petals free, saccate at base. Disk of 8–10 glands. Style columnar; stigma ±lobed.

About 5 spp., in north temperate zone to the mountains of Colombia.

1. M. hypopitys L.   Yellow Bird's-nest.
Whole plant uniformly yellowish or ivory-white, of waxy appearance, with simple stems 8–30 cm. Lvs 5–10 mm, scale-like, ovate-oblong, entire, ±erect, numerous, especially at the base of the stem. Infl. pendent in fl., erect in fr. Fls 10–15 mm, on short pedicels, 4–5-merous. Fl. 6–8. Pollinated by insects; homogamous. Grh.

Most of Europe but rare in the extreme south; temperate Asia to Japan, the Himalaya and N. Syria; North America (to Mexico).

The distribution and ecology of the two following are little known in this country and further observations of these and of the differential characters are needed. Both have a wide European range.

Subsp. **hypopitys** (*Hypopitys multiflora* Scop.). Infl. dense, up to 11-fld. Upper part of stem, sepals and outside of petals pubescent or glabrous. Petals 9–12(–13) mm, somewhat spreading at their apices. *Inside of petals, filaments, style* and sometimes ovary *covered with rather stiff hairs.* Ovary and capsule of lateral fls often longer than broad.

Native. Woods, especially beech and pine, and on dunes of east coast of Great Britain. Hants and Berks to Kent and northwards to Yorks and Fife. Extra-British distribution poorly known but probably most of C. and E. European populations belong here.

Subsp. **hypophegea** (Wallr.) Sóo (*M. hypopitys* var. *glabra* Roth). Infl. laxer and usually with fewer fls (up to 6) than in subsp. *hypopitys.* Upper part of stem, sepals and outside of petals glabrous. Petals 8–10 mm, straight. *Inside of petals, filaments, ovary and style glabrous.* Ovary and capsule subglobose. $2n = 16$.

Native. Woods, especially beech and pine, and on dunes among *Salix repens,* etc. Westmorland southwards to Derby, W. Norfolk and Kent in the east, and to Anglesey, Glamorgan, N. Somerset and Dorset in the west; W. Ireland from Sligo and Leitrim to N. Kerry and E. Cork. Extra-British distribution poorly known but probably only in W. Europe.

Intermediates, perhaps hybrids, occur in S.E. England westwards to N. Somerset and Glamorgan and northwards to N.E. Yorks, S. Lancs and in Moray; they are also known from Fermanagh.

99. EMPETRACEAE

Small evergreen heath-like shrubs. Lvs alternate, entire, margins strongly revolute, exstipulate. Fls axillary or in terminal heads, small, unisexual or hermaphrodite, regular, hypogynous, 2–3-merous. Per.-segs 4–6, in two ±similar whorls. Stamens half as many. Ovary 2–9-celled; ovules 1 in each cell, basal, anatropous; integument 1; raphe ventral; style short; stigmas as many as carpels. Fr. a drupe with 2–9 stones; endosperm copious, fleshy; embryo straight; cotyledons small.

Three genera and 3–9 spp., north temperate and arctic regions, southern South America and Tristan da Cunha.

1. EMPETRUM L.

Low shrubs. *Fls* 1–3, *axillary*, with scale-like bracts below. *Per.-segs* 6. *Stamens* 3; anthers introrse. *Carpels* 6–9. Stigmas toothed. Fr. juicy.

One to six closely allied spp., distribution of the family but absent from the warmer parts of the north temperate zone.

1. E. nigrum L.   Crowberry.

Low shrub 15–45 cm, with numerous procumbent and ascending stems; young twigs minutely glandular-puberulent, becoming glabrous. Lvs dense, 4–7 mm, oblong or linear-oblong, obtuse, shortly stalked, glandular on the margin when young, otherwise glabrous. Fls c. 1–2 mm diam., pinkish or purplish. Fr. c. 5 mm, black, subglobose. Fl. 5–6. Pollinated by wind or rain-splash.

Subsp. **nigrum.** Plant dioecious. Stems relatively long and slender, prostrate and rooting round the edge of the tuft. *Young twigs reddish*, becoming red-brown. *Lvs* oblong or oblong-linear, *parallel-sided, c. 3–4 times as* *long as broad.* $2n = 26^*$. Chw. or N.

Native. Moors, mountain-tops and the drier parts of blanket bogs, extending to at least 760 m (upper limit uncertain because of confusion with subsp. *hermaphroditum*), often abundant and sometimes locally subdominant. Common in Scotland, N. England and Wales, rare in S.W. England (absent from Cornwall) and absent south and east of a line from Somerset to Leicester and E. Yorks (extinct in Sussex and N. Lincoln); throughout Ireland but absent from a few counties in the centre and south. Europe from Iceland and Scandinavia (63° N) to the Pyrenees, C. Apennines, Bulgaria and S. Urals; W. Siberia (subspp. or allied spp. in E. Asia); North America.

Subsp. **hermaphroditum** (Hagerup) Böcher. Stems not prostrate nor rooting round the edge of the tuft, so that the tufts are more rounded and taller than in subsp. *nigrum*. Internodes shorter. *Young twigs green, becoming brown. Lvs* oblong or elliptical-oblong; *margins somewhat rounded, c. 2–3 times as long as broad. Fls hermaphrodite* (the stamens on some of them often persist round the fr. for some time). $2n = 52$. Chw. or N.

Native. Mountain tops and moors mostly at high altitudes, but at sea level in Shetland, dominant in the Cairngorms and elsewhere in a zone at varying heights 610–1070 m. Caernarvon; Lake District; Dumfries; Scottish Highlands from the Clyde Is. to Aberdeen northwards and westwards. N. Europe southwards to c. 60° N, Alps of France and Switzerland, Urals; arctic Siberia; W. Greenland (to 79° N); Canada; (incomplete).

100. DIAPENSIACEAE

Perennial herbs or dwarf shrubs. Fls hermaphrodite, regular. Calyx deeply 5-lobed. Disk 0. Corolla gamopetalous, deeply 5-lobed. Fertile stamens 5, alternating with the corolla-lobes, inserted on the corolla; staminodes 5 or 0. Ovary superior, 3-celled, deeply 3-lobed; ovules anatropous, on axile placentas; style simple with a 3-lobed stigma or divided above into 3. Fr. a loculicidal capsule. Seeds small; endosperm copious, fleshy; embryo small, cylindrical.

Six genera and about 20 spp., north temperate and arctic regions.

1. DIAPENSIA L.

Cushion-like evergreen *dwarf shrubs* with dense rosettes of leaves. Fls solitary ±stalked. Calyx leathery. *Anthers broadly ovate, diverging, opening lengthwise.* Style slender; stigma small, 3-lobed.

Four spp., the following and 3 in the Himalaya and W. China.

1. D. lapponica L.   Diapensia.

Cushion-like dwarf shrub up to c. 5 cm. Lvs 5–10 mm, obovate-spathulate, entire, obtuse, tapered to the base, leathery, densely crowded. Peduncles 1–3 cm, usually with 1 bract near the middle and 2 bracteoles close below the fl. Sepals c 5 mm. Petals up to 1 cm, white, obovate. Filaments broad and flat. Staminodes 0. Fl. 5–6. Protogynous. Visited by flies. $2n = 12$. Chw.

Native. In 2 localities on broken quartzite of exposed mountain ridges at about 760 m and 850 m in W. Inverness, where it was first found in 1951. Arctic regions and extending south to c. 60° N in Norway to the Urals and to New York; replaced by a subsp. in N.E. Asia (S. to Japan) and western North America.

101. PLUMBAGINACEAE

Perennial or rarely annual herbs or shrubs with simple exstipulate usually spirally arranged lvs often confined to a basal rosette. Infl. usually cymose, the cymes sometimes closely aggregated into heads. Bracts scarious. Fls hermaphrodite, actinomorphic, hypogynous, 5-merous. Calyx tubular below, scarious and often pleated above, persistent. Petals free, slightly joined at the base, or with a long basal tube. Stamens 5, free or epipetalous, opposite the petals. Ovary superior, 1-celled, with 1 basal anatropous ovule. Styles 5, or 1 with 5 stigma lobes, opposite the sepals. Fr. dry with a thin papery wall, opening with a lid or irregularly, or remaining closed. Seed with a straight embryo in mealy endosperm.

About 19 genera and 780 spp., cosmopolitan. Chiefly plants of seashores and salt-steppes, but there are some arctic and alpine species. A very natural family distinguishable from the Primulaceae by the 5 styles or stigma-lobes and the single basal ovule, and usually recognizable by the scarious and persistent brightly coloured calyx. The British members belong to the tribe Staticeae, with complex infls, a spreading coloured scarious calyx, epipetalous stamens and styles free almost to the base. The tribe Plumbagineac includes the cultivated genera *Plumbago* and *Ceratostigma* and differs in the simple infl., only slightly scarious calyx, longer corolla-tube, hypogynous stamens, and styles free only in the upper part.

Fls in terminal panicles; styles glabrous throughout.
　　　　　　　　　　　　　　　　　　　　1. LIMONIUM
Fls in dense ±globose heads; styles hairy below.
　　　　　　　　　　　　　　　　　　　　2. ARMERIA

1. LIMONIUM Miller

Perennial herbs, rarely annuals, with woody stocks and with lvs confined to a basal rosette. Fls shortly stalked in 1–5-fld cymose spikelets, each with three scale-like bracts, the *spikelets further aggregated into spikes of varying length and compactness which terminate the branches of the infl.* Calyx funnel-shaped, 5–10-ribbed at the base, expanding above into a scarious, usually coloured, 5-lobed limb, with or without smaller teeth between the lobes. Corolla with a very short tube. Stamens inserted at the base of the corolla. *Styles glabrous,* free quite or nearly to the base. Fr. opening transversely above or irregularly below. The fls secrete nectar and are visited by various insects. Some species are heterostylous and many have dimorphic pollen and stigmas. In such dimorphic species, which include *L. vulgare* and *L. bellidifolium*, plants with 'cob' stigmas have pollen of type A, while those with 'papillate' stigmas have type B pollen (Fig. 49). These species are self-incompatible, effective pollination being of cob stigmas by B pollen or of papillate stigmas by A pollen. Other species are monomorphic and self-compatible, all plants being A/papillate (e.g. *L. humile*), or, more rarely, all B/cob.

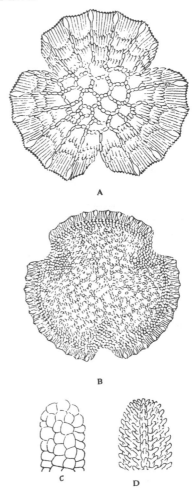

Fig. 49. Pollen and stigmas of dimorphic *Limonium*. A, B, the two pollen types; C, 'cob' stigma; D, *'papillate' stigma*. A and B × c. 800, C and D ×c. 200. (After H. G. Baker.)

There are finally some species which are monomorphic and apomictic. These have cob stigmas with either A pollen or empty anthers (*L. binervosum*), or papillate stigmas with B pollen (*L. dodartii*, a non-British species related to *L. binervosum*), or there may be some populations with the one and some with the other combination (*L. auriculae-ursifolium*). These features are observable only with the microscope, but they are often valuable in confirming identification.

About 300 maritime and salt-steppe spp. all over the world, but especially numerous in W. Asia.

Several non-British species are grown in gardens, and their infls are sold for house decoration in winter on account of the persistent coloured calyx ('Statice').

1 Lvs pinnately veined; calyx-lobes with intermediate
　　teeth.　　　　　　　　　　　　　　　　　　　　　　　*2*

Lvs not pinnately veined; calyx-lobes without intermediate teeth.   **3**

2 Stem usually not branched below the middle; infl. ±corymbose, the spikelets crowded into short spreading spikes; outer bract rounded on the back; dimorphic (A/cob and B/papillate).   **2. vulgare**

Stem usually branched below the middle; infl. not corymbose, the spikelets distant in long ±erect spikes; outer bract usually keeled; monomorphic (A/papillate).   **2. humile**

3 Infl. with very numerous slender zig-zag barren branches below; outermost bract entirely scarious; dimorphic.   **3. bellidifolium**

Infl. with few or no barren branches below; outermost bract scarious only at the margin.   **4**

4 Lvs broadly obovate-spathulate, 5- or more veined, narrowed below into a 5–9-veined stalk; monomorphic.   **4. auriculae-ursifolium**

Lvs narrowly obovate- or lanceolate-spathulate or oblong-linear, with a 1–3-veined stalk; monomorphic (**binervosum** group).   **5**

5 Spikes subcapitate, each of only 2(1–3) spikelets of which the upper has 2 or fewer bracts.   **8. paradoxum**

Spikes elongated, of numerous spikelets each with 3 bracts; spikelets arranged in 2 rows on the upperside of the spike.   **6**

6 Spikes straight, usually ±erect; spikelets distant or at least far enough apart for their outer and intermediate bracts not to overlap those of the next spike in the same row.   **5. binervosum**

Spikes straight or recurved, spreading; spikelets so closely set that their outer and intermediate bracts may overlap those of the next spike in the same row.   **7**

7 Lvs narrowly obovate-spathulate with a winged 3-veined stalk; petals contiguous or imbricating.   **6. recurvum**

Lvs oblanceolate or linear-oblong, narrowed gradually into a 1-veined stalk-like base; petals narrow, distant.   **7. transwallianum**

1. L. vulgare Miller   Sea Lavender.

Statice limonium L.

A perennial herb with deep tap-root and branched stout woody stock. *Lvs* 4–12(–25) cm *strongly pinnate-veined*, variable in shape from broadly elliptic to oblong-lanceolate, acute or obtuse, usually mucronate, narrowing gradually into a long slender stalk. *Flowering stems* 8–30(–40) cm, erect, somewhat angular, *corymbosely branched usually well above the middle*; lowest branches sometimes barren; *spikes short, dense, spreading, ±recurved*; *spikelets closely set* in 2 rows on the upper side of the spike; length from base of spikelet to apex of innermost bract 3.5–5 mm; *outer bract rounded on the back*. Calyx-teeth acute, entire or jagged, with small intermediate teeth. Corolla 8 mm diam., blue-purple, the petals broad and rounded. Anthers yellow. Fl. 7–10. Protandrous. Visited for nectar by various bees, flies and beetles. Heterostylous, *dimorphic* and self-incompatible, A/cob plants having styles exceeding the stamens, and B/papillate plants styles equalling or falling short of the stamens. $2n = 36^*$. Hr.

Native. Muddy salt-marshes; often an abundant or dominant sp. in intermediate zones. Great Britain northwards to Fife and Dumfries. S. and W. Europe northwards to S. Sweden; N. Africa; North America.

Plants with the flowering stem branching near or below the middle and with the infl. lax and less corymbose than in the type have been placed in f. *pyramidale* C. E. Salmon. There appear, however, to be all intermediates between this and the type, and the variation is probably determined environmentally, not through genetic differences; f. *pyramidale* is characteristic of the higher and drier zones of salt-marshes.

2. L. humile Miller   Lax-flowered Sea Lavender.

Statice bahusiensis Fries; *S. rariflora* Drejer

A perennial herb resembling *L. vulgare* but with oblong-lanceolate and obscurely-veined lvs. *Flowering stems branched from below the middle*; *infl. not corymbose*; *spikes long, lax, erect or somewhat incurved*; *spikelets distant*; length from base of spikelet to apex of innermost bract 5–7.5 mm; *outer bract usually keeled*. Calyx-teeth acute, denticulate, with small intermediate teeth. Anthers short, reddish. Fl. 7–8. *Monomorphic* (A/papillate, with styles shorter than stamens) and self-incompatible. $2n = 36$. Hr.

Native. Muddy salt-marshes. Great Britain northwards to Dumfries and Northumberland; all round the Irish coast, but rare in the north. W. Europe from Brittany to Norway and Sweden.

Fertile hybrids between *L. humile* and *L. vulgare* (*L × neumanii* C. E. Salmon), with intermediate characters, have been found where both parents are present.

3. L. bellidifolium (Gouan) Dumort.

Matted Sea Lavender.

Statice reticulata auct. angl., non L.; *S. caspia* Willd.

A perennial herb with deep tap-root and much-branched woody stock. Lvs 1.5–4 cm, few in each rosette, obovate- or lanceolate-spathulate, blunt, narrowed into a slender stalk, dying before flowering is over. *Flowering stems* 7–30 cm, *scabrid, decumbent*, spreading in a circle, much branched from near the base, with *numerous repeatedly forked barren branches below*; *fertile spikes* only on the uppermost branches, *dense, spreading, recurved*; spikelets closely set in 2 rows on the upper side of the spike; bracts with very broad membranous margin, the outermost almost entirely scarious. Calyx-teeth ovate, cuspidate, denticulate, with no intermediate teeth. Corolla small, 5 mm. diam., pale lilac. Fl. 7–8. Dimorphic; self-incompatible. $2n = 18$. Hr.

Native. Drier parts of sandy salt-marshes. Coasts of Norfolk and Suffolk, and formerly also of Lincoln. Shores of Mediterranean, Black Sea, Caspian Sea, and in E. Asia.

4. L. auriculae-ursifolium (Pourret) Druce

Statice lychnidifolia Girard

A *robust* plant with *large broadly obovate-spathulate* usually apiculate *lvs, glaucous and often viscid, 5–9-veined*, narrowed into a broadly winged *5–9-veined* *stalk.* Flowering stems 10–45 cm, stout, branched from near the base, rarely with barren branches below; spikes stout, dense, spreading or nearly horizontal, recurved; spikelets crowded; *inner bract* more than twice as long as the outer, with a membranous *bright red margin.* Calyx-teeth shallow, blunt. Corolla large, violet-blue. Fl. 6–9. Monomorphic (A/cob or B/papillate in different populations) and apomictic. $2n = 25^*$. Hr.

Native. Maritime rocks. Jersey and Alderney. France, Spain, Portugal, N. Africa.

5. L. binervosum (G. E. Sm.) C. E. Salmon
Rock Sea Lavender.

Statice binervosa G. E. Sm.; *S. auriculaefolia* auct.; *S. occidentalis* Lloyd?

A perennial herb with ascending branched woody stock and erect glabrous flowering stems. Lvs 2–12.5 cm, numerous, very variable in shape from obovate-spathulate to narrowly oblanceolate, 3-*veined below*, acute or blunt, apiculate or not; narrowing below into a winged *obscurely 3-veined stalk* about the same length as the blade. *Flowering stems* 5–30(–50) cm, slender, wavy, *branched from near the base*, with a *few barren branches below*, spikes usually *slender, ±straight and erect; spikelets* in 2 rows on the upperside of the spike, usually *not closely set*, so that their *outermost and intermediate bracts never overlap those of the next spikelet in the same row*; *innermost bract* about twice as long as the outermost, with a membranous *pink-tinged margin.* Calyx-teeth blunt, entire. Corolla c. 8 mm diam., violet-blue, the petals imbricating, obovate, emarginate. Fr. narrow, with a smooth reddish seed. Fl. 7–9. Monomorphic, but some plants are male-sterile; some races may be apomictic. $2n = 35^*$, 36^*. Hr.

Native. Maritime cliffs, rocks and stabilized shingle. Great Britain northwards to Wigtown and Lincoln. France, Spain and Portugal.

A very variable species with many named varieties of which the most distinct is var. *intermedium* (Syme) Druce, with obovate-spathulate lvs, *no barren branches*, *spikes short, dense, somewhat spreading*, with the *spikelets closely set* but not so closely that their bracts imbricate with those of the next in the same row.

Certain broad-leaved forms which show some failure of chromosome pairing may be segregates from a cross with *L. bellidifolium.*

The following closely related local types, all monomorphic (A/cob) or male-sterile and all probably apomictic, have been given specific rank because of striking morphological features.

6. L. recurvum C. E. Salmon
Lvs narrowly obovate-spathulate, blunt, narrowed into a broadly winged 3-*veined stalk. Flowering stem* 5–20 cm, asperous, rigid, *branched only above the middle; barren branches* 0; *spikes suberect or spreading, ±recurved*, rather *short and very dense*, terminating the short branches; *spikelets closely set* so that their *outermost and intermediate bracts overlap those of neighbour-

ing spikelets in the same row. Calyx-teeth shallow*, blunt or truncate. Corolla 6 mm diam.; petals narrow, contiguous, emarginate. $2n = 27^*$. Fl. 7–9.

Native. Apparently endemic in the British Is., occurring only at Portland, Dorset.

7. L. transwallianum (Pugsley) Pugsley
Lvs 2.5–4.5 cm, *narrowly oblanceolate or linear-oblong*, blunt, minutely mucronate, *narrowing gradually into a* 1-*veined stalk-like base. Flowering stems* 8–15(–30) cm, smooth, *branched from near the base* in large plants; *usually no barren branches; spikes ±spreading or recurved, short and very dense, usually crowded at the ends of branches; spikelets closely set* so that their *outer and intermediate bracts often overlap those of the next spikelet in the same row. Calyx-teeth deep*, triangular, obtuse or subacute. Corolla 4 mm diam., violet-blue; petals distant, narrow, oblong, emarginate. $2n = 35^*$ (Pembroke), 27^* (Clare). Fl. 7–9.

Native. Apparently endemic in the British Is., occurring only on maritime cliffs in Pembroke. A population at Poulsallagh, Co. Clare, consists of similar but more robust plants differing in chromosome number.

8. L. paradoxum Pugsley
Lvs 3–4.5 cm, oblanceolate or oblong-spathulate, blunt, mucronate or muticous, narrowed below to a winged 1-*veined stalk. Flowering stems* 5–15(–20) cm, slightly asperous, usually *branched from near the base*, with *short erect branches which do not reach the apex of the stem; barren branches* 0; fls in *subglobose heads* each consisting of 2 (rarely 1 or 3) irregular spikelets, of which the lower has 3 bracts while the upper usually has 1, but may have 2 or none. Calyx-teeth deep, triangular, obtuse. Corolla 4 mm diam., clear violet with a pink mid-vein to each of the imbricating petals. $2n = 33^*$ (St David's). Fl. 7–9. Probably male sterile and apomictic.

Native. Apparently endemic in the British Is., occurring only on basic igneous rocks at St David's Head, and doubtfully also near Malin Head, Donegal (Ireland).

In 1979 a colony of a strange *Limonium* was found growing on a coastal chalk cliff near Rottingdean in E. Sussex. It was with plants of *L. binervosum* from which it differed in its broadly obovate lvs, narrowed abruptly into a longish petiole and with rounded or retuse apex (lvs narrowly obovate-spathulate and with acute or obtuse apex in *binervosum*); in being much-branched from low down with the lower branches sterile; in its fl.-spikes up to 6 cm long with only 1–2 spikelets per cm (c. 3 cm long with 3 or more spikelets per cm in *binervosum*); and in having the calyx-teeth recurved in fr. (calyx remains tubular in *binervosum*). The fls were monomorphic and A/cob, and chromosome counts gave $2n = 35$. The best match was with *L. companyonis* (Gren. & Billot) O. Kuntze, an apomict of the *L. duriusculum* group previously known only from a short stretch of the Mediterranean coast of France near Narbonne. Further investigations are in progress (see M. J. Ingrouille, *Watsonia*, 1981, **13**, 181–4).

For a monograph covering spp. 5–8 see M. J. Ingrouille and C. A. Stace, *Bot. Jour. Linn. Soc.* **92**; 177–217 (1986).

2. ARMERIA Willd.

Perennial herbs with branched woody stocks and *basal rosettes of long very narrow entire lvs*. Fls stalked in a *solitary terminal hemispherical head* consisting of a close aggregate of bracteate cymose spikelets with a *scarious involucre*; top of flowering stem enclosed by a downward prolongation of the connate bases of the outer involucral bracts, so forming a *tubular sheath* round the growing zone *just beneath the head*. Calyx with a funnel-shaped 5–10-ribbed tube expanding upwards into a spreading scarious pleated limb; *petals free except at the extreme base, persistent*; stamens inserted at the base of the petals, their filaments broadened below; *styles* free to near the base, *hairy below*. Some spp. are dimorphic in respect of the size of the stigmatic papillae and the sculpturing of the pollen grains. Fr. enclosed in the persistent corolla, with 5 radiating ribs at the top, dehiscing transversely above or irregularly below.

About 80 spp. of maritime, arctic and alpine habitats in temperate Europe, Asia, N. Africa, North America and southern South America.

Lvs linear, 1–3-veined; calyx-teeth acute or shortly awned.
                                     1. maritima
Lvs linear-lanceolate, 3–5-veined; calyx-teeth with awns half their length.
                                     2. alliacea

1. A. maritima (Miller) Willd.   Thrift, Sea Pink.

Statice armeria L.; *S. maritima* Miller

Rootstock erect, stout, woody, branched. *Lvs* 2–15 cm, *linear*, acute or blunt, 1- (*rarely* 3-)*veined*, somewhat fleshy, punctate, glabrous, ciliate or pubescent. *Scapes* 5–30 cm, erect, lfless, usually *shortly pubescent*, rarely glabrous. Involucral sheath 8–14 mm; *outermost involucral bracts* ±green on the back, mucronate or not, *shorter than the head*; innermost wholly scarious, blunt; heads 1.5–2.5 cm diam. Fl.-stalks ±equalling the *calyx-tube* which has 5 *hairy ribs* and may or may not be hairy also between the ribs; *calyx-teeth acute or with awns less than half their overall length*. Corolla 8 mm diam., rose-pink or white. Fr. exceeding the calyx-tube. Fl. 4–10. Fls fragrant, slightly protandrous or homogamous; visited for nectar and pollen by various insects. Dimorphic (A/cob and B/papillate: see under *Limonium*, Fig. 49) and self-incompatible. $2n = 18$. Hr.–Ch.

Salt-marsh forms seem ecotypically different from those on hard substrata, whether maritime or inland. Forms with broadly linear, often 3-veined, lvs (var. *planifolia* Syme) occur in pure populations or together with narrow-leaved forms on mountains in Scotland and the Hebrides and also, less frequently,

in maritime localities.

It does not seem possible to uphold the distinction between *A. maritima* (Miller) Willd., with calyx-tube hairy only on the ribs, and *A. pubescens* Link, with hairs also between the ribs, since both types may be found on the same plant, in the same head, or even on the same calyx-tube.

The above description is of subsp. **maritima**, which is the widely distributed native subsp. in coastal habitats and inland. Subsp. **elongata** (Hoffm.) Bonnier, a more robust plant with *quite glabrous scapes* 20–55 cm and often with acuminate outer involucral bracts, occurs inland in a few localities in one small area and may perhaps be native there.

Native. Subsp. *maritima* in coastal salt-marshes and pastures, on maritime rocks and cliffs and also to 1280 m on mountains inland. In suitable localities throughout the British Is. Subsp. *elongata* only on sandy soil inland near Ancaster (Lincoln). *A. maritima* sensu lato is widely distributed in the northern hemisphere. Subsp *maritima* has a restricted W. European range from Iceland and N. Norway to N. Spain. Subsp. *elongata* is still more restricted, ranging from S. Denmark, N.W. Germany, Holland and Bavaria eastwards to S. Sweden, S. Finland, western C. Russia and Poland.

2. A. alliacea (Cav.) Hoffmans. & Link   Jersey Thrift.

A. plantaginea Willd.; *A. arenaria* (Pers.) Schultes

A more robust and more rigid plant than *A. maritima*, more densely tufted, with broader, flat, *linear-lanceolate*, acuminate, 3–5(–7)-*veined lvs*. Scapes 20–60 cm, erect, *glabrous*, rough. Involucral sheath 20–40 mm; *outermost involucral bracts acuminate*, wholly scarious or with a long herbaceous point, ±*equalling or exceeding the head*; innermost with broad scarious margins. Heads c. 2 cm diam. Calyx-tube with hairy ribs; *calyx-teeth with very long awns about half their overall length*. Fl.-stalks shorter than the calyx-tube. Corolla c. 1 cm diam., deep rose. Fl 6–9. Visited for nectar by various insects. Dimorphic. H.–Ch.

Native. On stable dunes in Jersey. C. and S. Europe.

A hybrid between *A. maritima* and *A. alliacea* has been described by Syme. It resembles *A. alliacea* but has the lvs 1-veined or indistinctly 3-veined and ciliate, and its flowering stems are densely and shortly pubescent. It flowers earlier than *A. alliacea*.

*****A. pseudarmeria** (Murray) Mansfeld (*A. latifolia* Willd.), with limp broadly oblanceolate to oblong-lanceolate 5–7-veined lvs up to 20 mm wide and stout scapes bearing heads up to 50 mm across with bright pink fls (sometimes deep rose or white), is often grown in gardens and has become established locally. Portugal.

102. PRIMULACEAE

Perennial or sometimes annual herbs, rarely dwarf shrubs. Lvs exstipulate. Fls actinomorphic, bracteate, 5(4–9)-merous, often heterostylous. Corolla present (0 in *Glaux*), rotate, campanulate, or funnel-shaped. Stamens inserted in the corolla-tube and opposite its lobes, sometimes alternating with staminodes. Ovary superior ($\frac{1}{2}$-inferior in *Samolus*), 1-celled, with a free-central placenta; style simple. Ovules numerous. Fr. a capsule dehiscing by valves or transversely. Seeds endospermic, the embryo small, straight.

About 20 genera and 1000 spp., cosmopolitan but mainly in the northern hemisphere.

1 Lvs all basal. 2
   Cauline lvs present. 3
2 Corolla-lobes incurved or spreading; stock a rhizome
   1. PRIMULA
   Corolla-lobes strongly reflexed; stock a subglobose
   corm. 3. CYCLAMEN
3 Water plant; lvs submerged, pinnatisect; fls lilac.
   2. HOTTONIA
   Land plants; lvs not pinnatisect; fls not lilac. 4
4 Fls yellow. 4. LYSIMACHIA
   Fls not yellow. 5
5 Fls white; stems erect. 6
   Stems prostrate to ascending; fls not white (except
   rarely in *Glaux*, and then apetalous). 7
6 Lvs mostly in 1 whorl; fls solitary or few; ovary superior. 5. TRIENTALIS
   Lvs not whorled; fls numerous in a lfy raceme; ovary
   ½-inferior. 8. SAMOLUS
7 Capsule dehiscing transversely; corolla present.
   6. ANAGALLIS
   Capsule dehiscing by 5 valves; corolla 0; calyx petaloid. 7. GLAUX

1. PRIMULA L.

Perennial scapigerous herbs. Fls bracteolate, in umbels or whorls, rarely apparently solitary, white, yellow, pink or purple. *Calyx 5-toothed.* Corolla funnel- or salver-shaped, 5-lobed, lobes incurved or spreading. Stamens included. Style filiform; stigma capitate. *Capsule dehiscing by valves.* Fls usually heterostylous and cross-pollinated by bees or Lepidoptera.

About 500 spp., mainly in temperate or mountainous districts of the northern hemisphere, a few in temperate South America.

1 Lvs mealy beneath; fls lilac or purple. 2
   Lvs not mealy beneath; fls yellow, rarely pink. 3
2 Lvs crenulate; fls lilac, heterostylous; corolla-lobes
   distant; capsule cylindrical, usually much exceeding
   calyx. (N. England and S. Scotland.) 1. farinosa
   Lvs not crenulate; fls purple, not heterostylous;
   corolla-lobes contiguous; capsule ovoid, slightly
   exceeding calyx. (N. Scotland and Orkney.)
   2. scotica
3 Scape distinct; pedicels finely pubescent; limb of corolla concave, rarely exceeding 20 mm diam. 4
   Scape 0 or nearly so; pedicels with shaggy hairs; limb
   of corolla flat, usually more than 30 mm diam.
   5. vulgaris
4 Calyx-teeth acute; mature capsule shorter than calyx.
   3. veris
   Calyx-teeth acuminate; mature capsule usually
   exceeding calyx. 4. elatior

Subgenus ALEURITIA (Duby) Wendelbo

Lvs usually mealy beneath, not wrinkled. Calyx-tube terete. Fls lilac or purple in our spp.

1. P. farinosa L.　　　　　Bird's-eye Primrose.

Lvs 1–5 cm, obovate-spathulate, obtuse or subacute, *crenulate*, glabrous above, usually with white or sulphur-coloured meal beneath. Scape up to 20 cm, mealy when young. *Pedicels* slender, c. 5 mm, *elongating in fr. Fls* c. 1 cm diam., *rosy lilac*, erect or spreading, *heterostylous. Calyx* 3–6 × 2 mm, mealy. *Corolla-tube* 5–6 mm, *the limb* 8–16 mm diam., the throat yellow; *lobes* flat, *distant from each other*, cuneate, bifid, *segments linear-oblong*, obtuse. *Fr.* 5–9 mm, *cylindrical*, usually *much exceeding the calyx.* Fl. 5–6. $2n = 18^*$. Hr.

Native. In damp grassy and peaty places on basic soils. Lancs and Yorks to Cumberland and Durham; formerly in S. Northumberland; Peebles, E. Lothian and Mid-lothian, locally abundant. Europe, from C. Sweden to C. Spain and Bulgaria; N. Asia to the shores of the N. Pacific and south to the Altai and Tien Shan. Other records are incorrect.

Represented in Britain by subsp. **farinosa.**

2. P. scotica Hooker

Like *P. farinosa* in general appearance. *Lvs* obovate-spathulate, usually broader than those of *P. farinosa*, *quite entire.* Scape rarely exceeding 10 cm. *Pedicels* stouter and rather shorter, *not or scarcely elongating in fr. Fls purple, homostylous. Calyx* 4–6 × 3 mm. *Cor-olla-tube* 7–10 mm, *lobes contiguous*, bifid, *segments broadly ovate or suborbicular, the limb* 5–8(–10) mm diam. *Fr.* 5–6 mm, *ovoid, equalling or slightly exceeding calyx.* Fl. 6–9. $2n = 54^+$. Hr.

Native. In damp pastures, locally abundant. W. Suth-erland, Caithness and Orkney. Endemic. The Norwe-gian plants referred to this sp. are usually *P. scandinavica* Bruun.

Subgenus PRIMULA.

Lvs not mealy beneath, pilose or pubescent, wrinkled. Calyx-tube 5-angled. Fls yellow and normally heterosty-lous in our spp.

3. P. veris L.　　　　　Cowslip, Paigle.

A ±*glandular-pubescent* perennial. Rhizome short, stout, ascending, covered with swollen persistent lf-bases and bearing many stout roots. *Lvs* 5–15(–20)) cm, ovate-oblong, obtuse, crenulate or ±serrate, *finely pubescent* on both surfaces, abruptly contracted at base and sometimes cordate; petiole about as long as blade, winged. *Scape* 10–30 cm, *finely pubescent*, 1–30-fld. Pedicels 1 cm or more. *Fls* 9–12 mm *diam., deep yellow or buff* with *distinct orange spots* at base of lobes, nod-ding but *scarcely secund. Calyx* 8–15 × 6–8 mm, *finely pubescent*, the *teeth* 2–3 mm, *ovate, obtuse and apiculate.* Corolla 9–12 mm diam., mouth with folds; lobes strongly concave, notched. *Fr.* c. 10 mm, *ovoid, enclosed by the enlarged calyx*; pedicels erect in fr. Fl. 4–5. Fr. 8–9. $2n = 22^*$. Hr.

Native. In meadows and pastures on basic and espe-cially calcareous soils, locally abundant. Generally dis-tributed north to Orkney but absent from much of Scotland; in Ireland mainly in the centre.

Subsp. **veris** in Britain and in Europe north of the Alps; other subspp. in the rest of Europe and in temper-ate Asia.

P. veris L. × **vulgaris** Hudson   Common Oxlip.

Intermediate between the parents and variable owing to back-crossing. The commonest form (presumably the F_1 cross) differs from *P. veris* in having the lvs not abruptly contracted at base; larger, paler yellow fls with a less concave limb; and longer and more shaggy pubescence. From *P. vulgaris* it differs in having a distinct scape and shorter pedicels; smaller, deeper yellow fls with a more concave limb; and shorter less shaggy pubescence. It occurs, though not usually in great quantities, where the parents grow together.

4. P. elatior (L.) Hill   Oxlip, Paigle.

Rhizome similar to that of *P. veris*. Lvs 5–20 cm, similar in shape but usually longer than those of *P. veris*, not cordate, irregularly serrate, *crisped-pubescent*, particularly beneath; petiole about as long as blade, winged. *Scape* 10–30 cm, crisped-pubescent, 1–20-fld. Pedicels c. 1 cm. *Fls 20–25 mm diam., pale yellow with diffuse orange markings* in the throat, *secund*, somewhat nodding. Calyx c. 15 × 4–5 mm, *narrower at base*, crisped-pubescent, the *teeth* c. 4 mm, lanceolate, ±acuminate. Corolla-tube c. 18 mm, the *throat without folds*; lobes concave, shallowly notched. *Fr.* up to 15 mm, *oblong-ovoid, exceeding calyx*; pedicels erect in fr. Fl. 4–5. Fr. 7. $2n = 22^*$. Hr. The total absence of folds in the throat of the corolla, the secund infl., the lvs abruptly contracted into the petiole and the smaller, paler fls distinguish this sp. from *P. veris × vulgaris* which is sometimes confused with it.

Native. In woods on chalky boulder clay. Essex, Hertford, Suffolk, Cambridge, Bedford and Huntingdon, abundant in a small area from most of which *P. vulgaris* is absent. S., W. and C. Europe eastward to C. and S.W. Asia. Subsp. **elatior** throughout most of Europe but rare in the south.

Hybrid swarms with *P. vulgaris* occur in woods where the two spp. grow together.

5. P. vulgaris Hudson   Primrose.

P. acaulis (L.) Hill

Rhizome similar to that of *P. veris*. Lvs 5–25 cm, oblanceolate to obovate, obtuse, irregularly serrate, pubescent beneath, *glabrous above* except on the veins, *narrowed gradually at base*; petiole short or 0. *Scape 0 or very short.* Pedicels 5–10 cm, with shaggy indumentum. *Fls 20–40 mm diam.*, yellow (rarely pink), ascending. Calyx 10–20 mm, *nearly cylindrical*, with shaggy hairs, teeth 4–6 mm, *narrowly triangular, acuminate*. Corolla-tube c. 15 mm, the *mouth contracted with thickened folds* and with greenish stellate markings; lobes flat, shallowly notched. *Fr.* ovoid, *shorter than calyx*; *pedicels recurved in fr.* Fl. 12–5. Seeds arillate, sticky when fresh. Fr. 3–8. $2n = 22^*$. Hr.

The roots are stated to be a strong and safe emetic. Cultivated variants with pink or white fls are sometimes found growing in hedges, and the pink in woods in Wales, where it may be native.

Native. In woods and hedgebanks and, in the west, in open grassy places, common. Throughout most of the British Is. but local in C. and S. Ireland. Now much less common than formerly in some areas (e.g. the Home Counties and the Chilterns) owing to the depredations of 'flower-lovers'. Subsp. **vulgaris**: S., W. and S.C. Europe extending to E. Denmark and N. Ukraine; other subspp. in S. Europe, N. Africa and the Caucasus.

2. Hottonia L.

Floating herbs. Lvs submerged. Infl. racemose, the fls whorled, heterostylous, white or lilac. *Calyx 5-partite.* Corolla salver-shaped, throat thickened; lobes 5. Stamens 5, included. Style filiform, stigma capitate. *Capsule dehiscing by valves which adhere at the top.* Fls contain nectar and are pollinated by various insects.

Two spp. in Europe, Siberia, and North America.

1. H. palustris L.   Water Violet.

A pale green glabrous perennial. Stems floating and rooting, nodes not inflated. Lvs up to c. 10 cm, apparently whorled, 1–2-pinnate, lobes linear. Scape up to 40 cm, or more, erect, subaerial. Fls 20–25 mm diam., lilac with a yellow throat, 3–8 in a whorl, sometimes cleistogamous. Pedicels 10–30 mm, finely glandular-pubescent, ascending in fl., deflexed in fr.; bracteoles 5–10 mm, subulate. Calyx 4–6 mm, divided almost to base; teeth linear or linear-oblong, subacute, equalling corolla-tube. Fr. c. 5 mm, globose, splitting into 5 teeth or irregularly. Fl. 5–6. $2n = 20$. Hyd.

Native. In ponds and ditches. From N. Somerset and Kent to S.E. Lancs and N.E. Yorks but local except in the east; Moray; Down, introduced elsewhere in Ireland. N. and C. Europe from C. Sweden, southwards to C. Italy and Romania; W. Asia.

3. Cyclamen L.

Herbs with large *corms* from which the lvs and fls arise. Fls nodding, solitary, on long erect lfless stems which are usually spirally coiled in fr. Corolla-tube short, the throat thickened; *lobes* large, *reflexed*. Stamens inserted at base of corolla-tube, included; anthers cuspidate, connivent. Style short; stigma simple. Capsule dehiscing by valves which become reflexed.

About 15 spp. in S. Europe, N. Africa, and W. Asia, mainly in mountains. *C. persicum* Miller is much cultivated in pots and varies greatly in colour, etc.

*1. C. hederifolium Aiton   Cyclamen, Sowbread.

C. europaeum auct.; *C. neopolitanum* Ten.

A perennial, glabrous except for the puberulent infl. Corm 3–15 cm diam., rooting mainly from the upper surface. Lvs 3–14 cm, ovate-cordate, ±strongly 5–9-angled, dark green with a whitish border above, often purplish beneath, appearing in autumn after the fls; petioles long. Fl. stalks 10–30 cm. Calyx-lobes ovate-lanceolate, equalling corolla-tube. Corolla pink or rarely white; tube 5-angled; lobes c. 2 cm. Fl. 8–9. $2n = 24$.

Introduced. Naturalized or possibly native in hedge-banks and woods in E. Sussex and E. Kent. Introduced in a number of other scattered localities. S. Europe.

4. LYSIMACHIA L.

Herbs with opposite or whorled, rarely alternate, entire, sometimes gland-dotted lvs. Fls usually 5-merous, axillary or in axillary racemes or terminal panicles, yellow (in our spp.). *Corolla rotate*, lobes spreading or connivent. Stamens included. *Staminodes usually 0. Capsule* subglobose, *5-valved, many-seeded. Seeds* rugose, ±*margined*. The fls are visited mainly by bees, for the sake of their abundant pollen.

About 200 spp. in temperate and subtropical regions of both hemispheres.

```
1 Plant prostrate or procumbent.                    2
  Plant erect.                                       3
2 Lvs not gland-dotted; calyx-teeth subulate.
                                          1. nemorum
  Lvs gland-dotted; calyx-teeth ovate.  2. nummularia
3 Fls in dense axillary racemes; corolla 7-lobed.
                                        7. thyrsiflora
  Fls not in dense axillary racemes; corolla 5-lobed. 4
4 Fls in a terminal raceme or panicle.              5
  Fls axillary.                                      6
5 Fls in a raceme, rarely produced; elongate axillary
    bulbils usually present.            6. terrestris
  Fls in a panicle; axillary bulbils absent.  3. vulgaris
6 Fls 1(–2) in a lf-axil; corolla with red basal blotches;
    lvs glabrous on the surfaces.         4. ciliata
  Fls 2 or more in a lf-axil; corolla without red blotches;
    lvs puberulent on the surfaces.      5. punctata
```

1. L. nemorum L.   Yellow Pimpernel.

A slender glabrous *procumbent* perennial up to 40 cm. *Lvs* 2–4 cm, opposite, evergreen, *ovate*, acute, rounded at base; petioles short. *Fls* c. 12 mm diam., axillary, solitary, on *filiform pedicels* which equal or exceed the lvs. *Calyx-teeth* c. 5 mm, *subulate. Corolla rotate, lobes spreading, not ciliate.* Fr. c. 3 mm diam., globose. Fl. 5–9. 2n = 16, 18, 28. Chh.

Native. In woods and shady hedgebanks. Throughout almost the whole of the British Is., though rare in the drier parts. W. and C. Europe.

2. L. nummularia L.   Creeping Jenny.

A glabrous *creeping* perennial up to 60 cm. *Lvs* 1.5–3 cm, opposite, evergreen, *broadly ovate or suborbicular*, obtuse, rounded or almost truncate at base, *gland-dotted*; petioles short. *Fls* 15–25 mm *diam.*, axillary, solitary on rather *stout pedicels*, usually shorter than lvs. *Calyx-teeth* 8–10 mm, *ovate*, acuminate. *Corolla subcampanulate*, the *lobes gland-dotted, minutely* glandular-puberulent. Fr. very rarely produced in Britain. Fl. 6–8. 2n = 32, 36, 43, 45. Chh.

Native. In usually moist hedgebanks and grassy places, though this sp. is less intolerant of drought and full sunlight than *L. nemorum*. Scattered throughout England, Wales and S. Scotland, local in Ireland; rare and usually an escape from cultivation in the north. Most of Europe eastwards to the Caucasus.

3. L. vulgaris L.   Yellow Loosestrife.

An *erect* pubescent rhizomatous perennial 60–150 cm. *Lvs* 5–12 cm, opposite or in whorls of 3–4, lanceolate to ovate-lanceolate, acute, subsessile, *dotted with orange or black glands. Fls* c. 15 mm diam. in a terminal panicle; *pedicels* 1 cm. *or less*, slender. *Calyx-teeth* triangular-lanceolate, ciliate, the *margin orange*. Corolla subcampanulate, the lobes not ciliate. Filaments connate in lower ⅓. Fr. globose. Fl. 7–8. 2n = 56, 84. Hel. or Hp.

Native. In fens and beside rivers and lakes, locally common. Scattered throughout most of the British Is., north to Argyll. Europe and Asia except the extreme north and south.

*4. L. ciliata L.

An almost glabrous *erect* rhizomatous perennial 30–100 cm. *Lvs* up to c. 10 cm, ovate, acuminate, rounded or subcordate at base, *finely ciliate* but otherwise glabrous, not gland-dotted; *petioles* c. 1 cm. Fls c. 25 mm diam., solitary (–2) in lf-axils; *pedicels* 2–4 cm, slender. Calyx-teeth linear-lanceolate, acute, glabrous, the margin green. Corolla rotate, the *lobes* suborbicular, usually shortly cuspidate, *with a red blotch towards the base*. Filaments free, alternating with small staminodes. Fl. 6–7. Hp.

Introduced. Naturalized near buildings in a few widely scattered places. North America.

*5. L. punctata L.   Dotted Loosestrife.

A somewhat pubescent perennial 30–90 cm. *Lvs* up to 10 cm, opposite or whorled, ovate, subacute, *shortly petiolate, margins puberulent. Fls* axillary, usually in pairs; pedicels 15–20 mm, shorter than the subtending lvs except at top of infl. *Calyx-teeth* 5–8 mm, oblong-lanceolate, acute or acuminate, *glandular-pubescent. Corolla* (15–)20–25(–28) mm diam., *glandular-ciliate.* Fl. 7–10. 2n = 30. Hp.

Introduced. More or less naturalized in marshy fields and beside rivers in a few scattered localities. S.E. and E.C. Europe.

*6. L. terrestris (L.) Britton, Stearns & Poggenb.
(*L. stricta* Aiton)   Lake Loosestrife.

An erect glabrous herb *usually bearing long bulbils in the axils of the narrow lanceolate lvs. Fls* (infrequently produced) *in a terminal raceme*. Calyx-teeth c. 2 mm, ovate, acute. Corolla c. 10 mm diam., the lobes lanceolate, yellow, streaked or dotted with purple. Hp.

Introduced. Well naturalized in damp places on the shores of lakes in N.W. England. North America.

7. L. thyrsiflora L.   Tufted Loosestrife.
Naumbergia thyrsiflora (L.) Reichenb.

An erect glabrous or pubescent rhizomatous perennial, 30–60 cm. Lvs 5–10 cm, opposite, rarely whorled, oblong-lanceolate, obtuse, sessile, ½-amplexicaul, densely dotted with black glands; lower lvs small and ±scale-like. *Racemes* 5–10 cm, 2–3 *in the axils of the* lvs about the middle of the stem, *many-fld, dense*, bracteate. *Fls* c. 5 mm diam., *7-merous*, on slender pedicels 2–3 mm. Calyx-teeth narrowly lanceolate, subacute. Corolla campanulate, the lobes subobtuse, erect, with black glands. Stamens somewhat exserted. Fr. ovoid, gland-dotted, shorter than calyx. Fl. 6–7. $2n = 54$. Hel.

Native. In wet marshes and shallow water by ditches and canals, rare. N.E. Yorks and S. Scotland. Europe, southwards to C. France, C. Romania and S.E. Russia; temperate Asia; North America.

5. TRIENTALIS L.

Erect, unbranched, glabrous herbs with slender rhizomes. Lvs in 1 whorl of 5–6 at the top of the stem, with a few small alternate lvs below. *Fls* white, *solitary, ebracteolate*. Calyx 5–9-partite. Corolla rotate, 5–9-partite. Stamens 5–9. *Capsule* globose, *5-valved, few-seeded*. Probably cross-pollinated by insects.

Three or 4 species in north temperate regions.

1. T. europaea L.   Chickweed Wintergreen.

A slender erect perennial 10–25 cm. Lvs 1–8 cm, obovate to obovate-lanceolate, stiff and shining, acute or obtuse, entire or finely serrulate in the upper part, with cuneate base; petiole short or almost 0. Fls erect, few, usually 1, 15–18 mm diam.; pedicel 2–7 cm, filiform. Calyx-teeth 4–7 mm, usually 7, linear-acuminate. Corolla-lobes usually 7, ovate, acute or apiculate. Fr. c. 6 mm, valves deciduous, leaving the seeds attached to the placenta. Fl. 6–7. $2n = $ c. 160. Grh.

Native. In pine woods, among moss in grassy places, usually rooting in humus, locally common; ascending to 1070 m. Very local in England; E. Suffolk and Derby and from Yorks northwards; mainly in N. and E. Scotland. N. Europe, N. Asia.

6. ANAGALLIS L.

Annual or perennial herbs. Lvs opposite, entire. Fls axillary, solitary, ebracteolate. *Corolla rotate or funnel-shaped*, 5-lobed. *Stamens inserted at base of corolla-tube*; filaments pubescent. *Capsule* globose, *dehiscing transversely*.

About 28 spp. in Europe, Asia, Africa, and America.

1 Plant 1–4(–10) cm; corolla greatly exceeded by calyx-teeth.   **3. minima**
  Plant 5–30 cm; corolla not exceeded by calyx-teeth.   *2*
2 Stems subterete, rooting at nodes; lvs obtuse; corolla 2–3 times as long as calyx.   **1. tenella**
  Stems quadrangular, not rooting at nodes; lvs acute; corolla less than twice as long as calyx. **2. arvensis**

1. A. tenella (L.) L.   Bog Pimpernel.

A slender prostrate glabrous perennial 5–15 cm. *Stems subterete, rooting at nodes. Lvs* c. 5 cm, ovate or suborbicular, *shortly petiolate*. Fls on filiform pedicels much exceeding the lvs. Calyx-teeth linear-lanceolate, acuminate. *Corolla* up to 14 mm diam., *funnel-shaped*, 2–3 *times as long as the calyx*, pink. Capsule 1.5 mm. Fl. 6–8. $2n = 22*$. Chh.

Native. In damp peaty and grassy places and in bogs. Throughout most of the British Is., though rare in S.E. England and the Midlands, and absent from much of S. and E. Scotland. Europe from the Faeroes southwards; a few isolated stations in N. Italy; Crete and Greece; N. Africa: Morocco to Algeria; Azores.

2. A. arvensis L.   Scarlet Pimpernel, Shepherd's Weather-glass.

A procumbent or ascending glabrous annual or perennial 6–30 cm. *Stems quadrangular*, gland-dotted. *Lvs* 8–28 mm, ovate to lanceolate, *sessile*, dotted with black glands beneath. Fls on slender pedicels. *Calyx-teeth* narrowly lanceolate, apiculate, *not much shorter than the corolla. Corolla rotate*, up to 14 mm diam.; lobes entire, crenulate, or denticulate. Capsule 4–6 mm diam. Fl. 6–8. Th. or Chh. Throughout the greater part of the world with the exception of the tropics.

Subsp. **arvensis**. Pedicels in fl. usually longer than the lvs. Fls usually red or pink, more rarely blue or lilac. Corolla-lobes broadly obovate, overlapping, the margins densely fringed with 3-celled glandular hairs. Calyx-teeth not concealing corolla in bud. Capsule 5-veined. $2n = 40*$.

Native. On cultivated land, by roadsides, and on dunes. Common and widely distributed throughout the British Is. but rare and mostly coastal in Scotland.

Subsp. **foemina** (Miller) Schinz & Thell., Blue Pimpernel. Pedicels in fl. usually shorter than or equalling lvs. Fls blue. Corolla-lobes narrowly obovate, not overlapping, the margins very sparingly fringed with 4-celled glandular hairs. Calyx-teeth concealing corolla in bud. Capsule more than 5-veined. $2n = 40$.

Native. In arable fields in the south and west of England, rare; elsewhere as a casual.

3. A minima (L.) E. H. L. Krause   Chaffweed.

Centunculus minimus L.

A glabrous annual 1–10 cm. Lvs 3–5 mm, alternate, ovate, subsessile, obtuse or apiculate, entire. Fls subsessile in axils of upper lvs. *Calyx* divided nearly to base, the *teeth* lanceolate, acute, *longer than corolla*. Corolla white or pink. Fr. c. 1.5 mm, exceeded by calyx-teeth. Fl. 6–7. $2n = 22$. Th.

Native. In damp sandy places in rather open communities on heaths and by the sea, often on somewhat disturbed ground, local. In suitable habitats throughout Great Britain north to Lewis; local in Ireland. Most of Europe; S.W. Asia; N. Africa; North and South America.

7. GLAUX L.

A small glabrous succulent herb. Stems creeping and rooting. Lvs entire, decussate, the lower opposite, the upper alternate. *Fls axillary, subsessile. Calyx* 5-partite, *white or pink. Corolla* 0. Stamens 5, hypogynous, alternating with calyx-lobes. *Capsule* globose, *5-valved, few-seeded*. Self-pollinated and, perhaps, sometimes cross-pollinated by Diptera.

One species, coasts of the north temperate regions and saline districts inland.

1. G. maritima L.   Sea Milkwort, Black Saltwort.

A small procumbent or suberect perennial 10–30 cm. Lvs 4–12 mm, elliptical-oblong to obovate, obtuse or acute, subsessile. Fls 3–6 mm diam. Calyx with obtuse, usually pink lobes with hyaline margins. Fr. 2–5 mm. Fl. 6–8. $2n = 30$. Hp.

Native. In grassy salt-marshes, in crevices of rocks or at the foot of cliffs by the sea or estuaries, also in saline districts inland, locally common. Coasts of the British Is. and inland in Worcester, and Stafford. North temperate regions.

8. SAMOLUS L.

Herbs. Lvs alternate or in a basal rosette. *Fls white, bracteate.* Calyx 5-toothed. Corolla 5-lobed, campanulate. *Stamens 5, alternating with staminodes; filaments very short. Ovary ½-inferior. Capsule ovoid, 5-valved.* Usually self-pollinated.

About 10–15 spp., cosmopolitan but mainly in the southern hemisphere.

1. S. valerandi L.   Brookweed.

A glabrous perennial 5–60 cm. Stem simple or little-branched, lfy. Lvs 1–8 cm, obovate to spathulate, entire, obtuse. Infl. racemose, simple or branched. Pedicels 5–15 mm, with small lanceolate bracts adnate to about the middle, straight in fl., often geniculate in fr. Calyx campanulate, the teeth triangular, obtuse. Corolla 2–3 mm diam., lobes short, obtuse. Stamens included. Fr. 2–3 mm diam. Fl 6–8. $2n = $ c. 24, 26. Hs.

Native. In wet places, especially near the sea, locally common. Scattered throughout the British Is., north to the Outer Hebrides. Cosmopolitan, though usually near the sea.

103. BUDDLEJACEAE

Shrubs or trees, rarely herbs, with simple usually opposite stipulate or exstipulate lvs; glandular hairs present. Fls hermaphrodite, actinomorphic, usually 4-merous. Disk small or 0. Stamens alternate with corolla-lobes, inserted on tube. Ovary superior (very rarely ½-inferior), of 2 united carpels with usually numerous ovules on axile placentas. Style 1. Fr. a capsule or berry. Seeds with endosperm. Vascular bundles collateral.

About 10 genera and 150 spp., mainly tropical, extending to temperate America, China, S. Japan and S. Africa.

1. BUDDLEJA L.

Trees or shrubs, rarely herbs, with interpetiolar stipules often reduced to a ridge. Indumentum of stellate and glandular hairs. Infl. cymose, often capitate or forming long terminal panicles. Calyx campanulate. Corolla with ±cylindrical usually straight tube and ±spreading limb. Stamens included or slightly exserted. Fr. a capsule with numerous very small seeds.

About 100 spp., tropical Asia to China, Japan and New Guinea, E. and S. Africa, tropical and warm temperate America.

Several other spp. are cultivated, *B. globosa* Lam. from Chile, with orange fls in globose heads, rather commonly.

*1. B. davidii Franchet

B. variabilis Hemsley

Shrub 1–5 m. Twigs somewhat angled, pubescent, pithy. Lvs 10–25 cm, ovate-lanceolate or lanceolate, acuminate, serrate, dark green and sparsely pubescent to subglabrous above, white-tomentose beneath. Fls in dense many-fld cymes forming a somewhat interrupted narrow terminal panicle, 10–30 cm. Rhachis, peduncles, pedicels and calyx ±tomentose. Bracts and bracteoles linear, inconspicuous. Corolla lilac or violet with orange ring at mouth; tube cylindrical, c. 1 cm, somewhat pubescent outside; lobes 1–2 mm. Fl 6–10. Commonly visited by butterflies. M.

Introduced. Commonly grown in gardens (introduced c. 1890), now naturalized in waste places, rarely in woods, in a number of places in S. England and sporadically northwards to W. Inverness and throughout Ireland. Native of China.

104. OLEACEAE

Trees or shrubs, usually glabrous. Lvs usually opposite, exstipulate. Infl. usually cymose. Fls hermaphrodite or rarely unisexual, actinomorphic, usually 4-merous. Calyx small. Corolla gamopetalous, rarely polypetalous or 0, lobes valvate or imbricate, not twisted in bud. Stamens 2, very rarely more, usually adnate to corolla-tube. Ovary superior, 2-celled, with usually 2 ovules in each cell. Seeds with or without endosperm; embryo straight.

About 29 genera and 600 spp., widely distributed but mainly in Asia.

1 Tree; lvs pinnate; bark grey; buds black; fr. winged.
    1. FRAXINUS
  Shrubs; lvs simple; bark and buds brown or greenish; fr. not winged.   2
2 Lvs 4–12 cm, ovate; fls usually lilac; fr. a capsule.
    2. SYRINGA
  Lvs 3–6 cm, lanceolate; fls white; fr. a berry.
    3. LIGUSTRUM

1. FRAXINUS L.

Trees with pinnate or rarely simple lvs. *Fls hermaphrodite or unisexual.* Calyx 4-lobed or 0. Petals 2–6,

±connate at base or 0. Stamens 2. *Fr. a compressed 1–2-celled samara winged at tip*; cells 1-seeded.

About 70 spp. in Europe, N. Africa, temperate Asia and North America.

1. F. excelsior L.   Ash.

A deciduous tree 15–25 m. Bark grey, smooth, becoming fissured on old trunks. Twigs flattened at the nodes. Buds large (terminal ones 5–10 mm), black. Lvs up to 30 cm, opposite, decussate, imparipinnate or rarely simple; lflets 5–11 cm, 7–13, lanceolate to ovate, apiculate to acuminate, serrate. Fls in axillary panicles, appearing before the lvs, purplish; perianth absent. Fr. c. 3 cm, pale brown. Fl. 4–5. Wind-pollinated. Fr. 10–11. Germ. spring $2n = 46$. MM.

Native. Forming woods on calcareous soils in the wetter parts of the British Is.; in oakwoods, scrub and hedges; common, but less frequent on acid soils. Generally distributed throughout the British Is. except for Orkney and Shetland, perhaps introduced in some northern localities. Most of Europe N. Africa; W. Asia.

2. SYRINGA L.

Deciduous shrubs or small trees. Lvs opposite, usually entire. Fls hermaphrodite, sweet-scented, in terminal or lateral panicles. *Fr. a coriaceous, 2-celled, loculicidal capsule with 2 winged seeds in each cell.*

About 30 spp. in Asia and Europe.

*1. S. vulgaris L.   Lilac.

A shrub or small tree up to 7 m, suckering freely. Bark fibrous. Lvs 4–12 cm, ovate. Fls lilac or rarely white, in pyramidal panicles, 10–20 cm. Corolla-tube 8–12 mm.

Fr. 8–12 mm, acute, smooth. Fl. 5. Pollinated mainly by bees. $2n = 46$. N. or M.

Introduced. Much planted in gardens and ±naturalized in hedges, thickets and shrubberies. S.E. Europe.

3. LIGUSTRUM L.

Shrubs or small trees. Lvs often evergreen, entire, opposite. Fls hermaphrodite, in terminal panicles, strong-scented. Calyx caducous. Corolla funnel-shaped. *Berry with oily flesh*, 2-celled, cells 1–2-seeded.

About 50 spp., E. Asia, Indo-Malaya to Vanuatu, one in Europe.

1. L. vulgare L.   Common Privet.

A tardily deciduous shrub up to 5 m. Bark smooth. Branches slender, the *young twigs puberulent*. Lvs 3–6 cm, *lanceolate*, obtuse to acute, shortly petiolate. Panicle 3–6 cm, *puberulent*. Fls white, shortly pedicellate, 4–6 mm diam.; *corolla-tube as long as limb*; *anthers exceeding tube, shorter than limb*. Fr. 6–8 mm, black, shining. Fl 6–7. Pollinated by various insects. Fr. 9–10. $2n = 46$. N. or M.

Native. Common in hedges and scrub, particularly on calcareous soils, throughout the British Is., but naturalized in Scotland, most of Ireland and many localities in England and Wales. S., C. and W. Europe; N. Africa.

L. vulgare is now nearly supplanted for hedging by *L. ovalifolium* Hassk., Garden Privet, a shrub with elliptical-oval to elliptical-oblong lvs which persist longer than those of *L. vulgare* except in dirty industrial districts. Young twigs and panicle branches glabrous. Corolla-tube 2–3 times as long as limb; anthers as long as limb. Fl 7. A variety with golden-yellow lvs is commonly cultivated.

Introduced. Planted, usually near houses. Native of Japan.

105. APOCYNACEAE

Woody plants, often climbing, rarely herbs, with milky latex and internal phloem. Lvs usually opposite, rarely whorled or spiral, entire, usually exstipulate. Fls hermaphrodite, actinomorphic, hypogynous, 5-merous, solitary or in cymose infls. Corolla gamopetalous, contorted in bud. Stamens as many as the corolla-lobes and alternating with them, epipetalous, with short filaments; anthers convergent on the stylar head. Ovary of 2 carpels, free below but with a common style; ovules numerous on the ventral sutures of the carpels. Fr. various; seeds often winged or plumed, usually endospermic.

About 180 genera and 1500 spp., chiefly in tropical and subtropical regions.

Funtumia elastica yields the Lagos Silk Rubber and lianas of the genus *Landolphia* the Landolphia Rubber of commerce. Seeds of *Strophanthus* spp. (also lianas) yield the cardiac stimulant strophanthine. Various spp. with poisonous latex, and especially *Acokanthera venenata*, have been used for arrow poisons. *Nerium oleander*, the Oleander, a native of the Mediterranean region, is often seen in conservatories.

1. VINCA L.

Creeping shrubs or perennial herbs with evergreen lvs in opposite pairs. *Fls solitary in the axils of the lvs.* Corolla blue or white, salver-shaped, with 5 broad asymmetric lobes and an obconic tube fluted and hairy within; stamens 5, with short sharply kneed filaments and introrse anthers terminating in *broadly triangular hairy connective flaps* which meet over the stylar head; *styles united to a column* which is slender below but with an *enlarged head* tapering upwards and *surmounted by a plume of white hairs*; stigmatic surface as a band round the broadest part of the stylar head; ovary of 2 free carpels united only by their styles. Fr. of 2 follicles each with several long narrow seeds. There are two fleshy nectaries at the base of the gynoecium, alternating with the carpels, and the fls are insect-pollinated.

Five spp. in Europe, N. Africa and W. Asia.

Fls 25–30 mm diam.; calyx-lobes glabrous.	**1. minor**
Fls 40–50 mm diam., calyx-lobes ciliate.	**2. major**

1. V. minor L.   Lesser Periwinkle.

A procumbent shrub with trailing stems 30–60 cm, *rooting at intervals*, and short erect flowing stems. *Lvs 25–40 mm, very shortly stalked, lanceolate-elliptic, quite glabrous. Flowering stems each with 1 axillary fl., rarely 2.* Fl. 25–30 mm diam. *Calyx-lobes lanceolate, glabrous.* Corolla blue-purple, mauve or white. Ripe follicles (rarely seen in Britain) 25 mm, divergent, each with 1–4 blackish seeds. Fl. 3–5. Pollinated by long-tongued bees and bee-flies. $2n = 46*$. Chw.

Doubtfully native. Found locally in woods, copses and hedgebanks throughout Great Britain northwards to Caithness. Widespread in Europe from Denmark southwards, and in W. Asia. In C. Europe occurs in ash and oak-hornbeam woods on the better soils.

*2. V. major L.   Greater Periwinkle.

A semi-procumbent shrub with trailing or somewhat ascending stems 30–100 cm, *rooting only at their tips*, and short erect flowering stems. *Lvs 20–70 mm, with stalks c. 1 cm, ovate, ±acute, somewhat cordate at the base, ciliate.* Flowering stems erect, to 25 cm, each with 1–4 axillary fls at successive nodes. Fl. 40–50 mm diam. *Calyx-lobes long, subulate, ciliate.* Corolla blue-purple, rarely white. Ripe follicles (rarely seen in Britain) 4–5 cm, each with 1–2 dark brown seeds. Fl. 4–6. Pollinated by long-tongued bees, especially *Anthophora pilipes*. $2n = 92$. Ch.

Introduced. Copses and hedgerows. S. England. Occasionally in Ireland. Native in C. and S. Europe and N. Africa.

Variants with narrowly elliptical to linear-lanceolate lvs and violet fls with narrower and more pointed corolla-lobes have been mistakenly referred to *V. herbacea* Waldst. & Kit., native in E. and S. E. Europe, or to *V. major* subsp. *hirsuta* (Boiss.) Stearn, native in W. Asia.

106. GENTIANACEAE

Herbs, usually glabrous. Lvs opposite, entire, exstipulate, usually sessile. Infl. usually a dichasial cyme. Fls hermaphrodite, regular, usually 4- or 5-merous. Calyx-lobes imbricate. Corolla persisting round the capsule; lobes contorted in bud. Stamens epipetalous, equalling in number and alternating with the corolla-lobes. Ovary superior, unilocular with 2 parietal placentas, each with numerous anatropous ovules, sometimes 2-celled by the intrusion of the placentas; style simple with a simple or bilobed stigma or two stigmas. Fr. usually a septicidal capsule; seeds small, numerous, with copious endosperm and small embryo.

About 80 genera and 900 spp., cosmopolitan, mainly temperate.

A natural family, unlikely to be confused with any other. The opposite lvs and unilocular ovary with 2 many-ovuled parietal placentae and actinomorphic fls are sufficient to distinguish the British members from the other Gamopetalae. Many species contain bitter principles. Vascular bundles bicollateral.

1 Fls 6–8-merous, yellow; cauline lvs connate in pairs.
                                      4. BLACKSTONIA
  Fls 4–5-merous; lvs not connate.   2
2 Corolla pink or yellow, rarely white; style distinct, filiform, caducous.   3
  Corolla blue or purple, rarely white; style 0 or ovary gradually tapering into style; stigmas persistent.   5
3 Calyx-lobes deltate, less than half as long as tube; corolla yellow; stigma peltate.   1. CICENDIA
  Calyx divided nearly to base, the lobes linear; corolla pink, rarely white; stigma bifid or stigmas 2.   4
4 Anthers ovate, not twisted; calyx-lobes flat; fls 4-merous. (Guernsey.)   2. EXACULUM
  Anthers linear, twisting after fl.; calyx-lobes keeled; fls usually 5-merous. (Widespread.) 3. CENTAURIUM
5 Corolla with small lobes between the large ones, not ciliate at the throat, blue.   5. GENTIANA
  Corolla, without small lobes, usually ciliate at throat, purple, rarely white.   6. GENTIANELLA

1. CICENDIA Adanson

Small annuals. Fls 4-merous. *Calyx campanulate with short deltate lobes.* Corolla with ovoid tube and short spreading lobes, yellow. *Anthers cordate, not twisted. Style filiform with peltate stigma, caducous.*

Two spp., the second in California and temperate South America.

1. C. filiformis (L.) Delarbre   Yellow Centaury.

Microcala filiformis (L.) Hoffmanns. & Link

A slender annual 3–12 cm, simple or somewhat branched. Branches strict. Lvs 2–6 mm, linear few, soon withering. Pedicels 1–5 cm. Fls 3–5 mm. Capsule ovoid, c. 5 mm. $2n = 26$. Fl. 6–10. Fr. 8–10. Th.

Native. Damp sandy and peaty places not far from the sea, very local; Cornwall to Sussex; Pembroke, Caernarvon; probably extinct in W. Norfolk and Lincoln; Kerry, W. Cork and W. Mayo; Channel Is. W. and S. Europe; Asia Minor; N. Africa; Azores.

2. EXACULUM Caruel

Small annual. Fls 4-merous. *Calyx deeply divided into 4 flat linear lobes.* Corolla with cylindrical tube and spreading lobes, pink (in Britain). *Anthers ovate, not twisted. Style filiform with bifid stigma, caducous.*

A single sp.

1. E. pusillum (Lam.) Caruel   Guernsey Centaury.

Cicendia pusilla (Lam.) Griseb.

A slender annual 3–12 cm, with divaricate branches. Lvs linear, c. 6 mm. Pedicel slender. Capsule fusiform. $2n = 20$. Fl. 7–9. Th.

Native. Sandy commons in two spots in Guernsey. S. W. Europe, from W. France and Spain to Italy; N. Africa.

3. CENTAURIUM Hill   Centaury.
(*Erythraea* Borkh.)

Annual, biennial or, rarely, perennial herbs. Fls usually 5- (rarely 4-)merous. Calyx deeply divided into keeled linear lobes. Corolla ±funnel-shaped, pink, rarely white. *Anthers linear or oblong-linear, twisting spirally after dehiscence.* Placentae intrusive, nearly meeting. *Style filiform with 2 stigmas, caducous.*

About 40 spp., cosmopolitan except tropical and S. Africa.

1 Erect annuals or biennials without decumbent sterile
    shoots; lvs sessile; corolla-lobes 7 mm. or less.   2
   Perennial with decumbent sterile shoots with orbicu-
    lar shortly petiolate lvs; corolla-lobes 8–9 mm
    (Pembroke, Cornwall and N. Devon). **5. scilloides**
2 Fls pedicellate, not clustered; corolla-lobes 3–4 mm;
    plants without basal rosette.   3
   Fls sessile or subsessile, ±clustered; corolla-lobes
    5–7 mm; plants with basal rosette.   4
3 Branches spreading at a wide angle; infl. lax; inter-
    nodes 2–4.   **1. pulchellum**
   Branches strict; infl. rather dense; internodes 5–9.
    2. tenuiflorum
4 Lvs lingulate or linear-spathulate; basal 5 mm. broad
    or less.   **4. littorale**
   Lvs ovate, obovate or oblong; basal more than 5 mm
    broad (rarely only 4 mm).   **3. erythraea**

1. C. pulchellum (Swartz) Druce   Lesser Centaury.
Erythraea pulchella (Swartz) Fries

A glabrous erect *annual; without a basal rosette*, varying in habit from very slender, 2–3 cm, unbranched and 1-fld, to 15 cm, much branched above with dichotomous *widely spreading branches. Stems with 2–4 internodes.* Lvs 2–15 × 1–10 mm, ovate or ovate-lanceolate, acute, the upper usually longer than the lower sessile. *Fls in a lax dichasial cyme on pedicels* c. 2 mm, *sometimes 4-merous in small plants.* Corolla-tube usually exceeding calyx; *lobes* 2–4 mm. Stamens inserted at top of corolla-tube. Capsule about equalling calyx. Variable. Fl 6–9. Pollinated by insects and self. Germ. spring. $2n = 36*$, 42. Th.

Native. Damp grassy places usually in rather open habitats. Common near the sea in S. and S.C. England, more local inland, extending northwards to N. Lancs and N.E. Yorks and almost confined to the coast in the northern part of its range and in Wales; apparently extinct in Scotland; S. and E. coasts of Ireland but only recorded recently from Dublin. Most of Europe northwards to c. 61°N in Finland; W. and C. Asia to the Punjab and Tien Shan; Madeira; Naturalized in North America.

2. C. tenuiflorum (Hoffmanns. & Link) Fritsch
Slender Centaury.

Erythraea tenuiflora Hoffmanns. & Link

A glabrous erect *annual* 10–35 cm, *without an obvious basal rosette, with strict branches above. Stems with 5–9 internodes.* Lvs 10–25 × 8–12 mm, ovate or elliptical, obtuse or subacute, sessile, the upper usually longer

than the lower. *Fls in a rather dense dichasial cyme on pedicels* c. 2 mm. Corolla-tube slightly exceeding calyx; *lobes* 3–4 mm. Stamens inserted at top of corolla-tube. Capsule about equalling calyx. Fl. 7–9. Germ. spring. Th.

Native. Damp grassy places near the sea in the Isle of Wight and Dorset; very rare. Coasts of S. and W. Europe from N. France southwards and eastwards.

C. latifolium (Sm.) Druce   Broad-leaved Centaury.

Fls 7–9 mm, with corolla-lobes 3–4 mm as in the preceding spp., but sessile and in a dense infl. as in the following. Lvs broadly ovate. Formerly found on sand dunes in Lancs but now extinct. Endemic.

3. C. erythraea Rafn   Common Centaury.

Erythraea centaurium auct.; *Centaurium umbellatum* auct.; *C. minus* auct.; *C. capitatum* (Willd.) Borbás

A glabrous biennial 2–50 cm, *with a basal rosette of lvs* and usually solitary but sometimes several erect stems, branched above. *Basal lvs* 1–5 cm × (4–)8–20 mm, *obovate or elliptical*, often somewhat spathulate, usually obtuse, prominently 3–7-veined; cauline shorter, sometimes narrower and acute, but *never parallel-sided. Fls sessile or subsessile*, often clustered, *forming a ±dense* corymb-like *cyme.* Corolla-tube longer than, rarely as long as, calyx; limb ±flat; *lobes* 5–6 mm. Stamens inserted at top or at the base of the corolla-tube. Capsule exceeding calyx. Fl. 6–10. Pollinated by insects and self. Germ. autumn. $2n = 40$. Hs.

Native. Dry grassland, dunes, wood margins, etc.; common in England, Wales and Ireland, less so in Scotland but quite common in the west to the Outer Hebrides, more local in the east to Moray. S. Europe from Sweden southwards; Mediterranean region; Azores; S.W. Asia to the Pamir-Alai region; naturalized in North America.

Very variable in such characters as stem-branching, lf-shape and -size, indumentum, infl.-density, relative lengths of calyx and corolla, etc. Populations showing various combinations of these characters have been treated as spp., subspp. or vars. British plants belong to subsp. *erythraea*, within which *C. capitatum* (Willd.) Borbás (Tufted Centaury), with a dense infl. and stamens at the base of the corolla-tube, at most merits varietal status (*C. erythraea* var. *capitatum* (Willd.) Melderis).

Hybrids between *C. erythraea* and *C. pulchellum* are generally intermediate between the parents and are highly fertile. They are recorded from the coasts of Somerset, Essex and W. Lancs.

Hybrids between *C. erythraea* and *C. littorale* are generally intermediate between the parents and of low fertility. They occur on the coast of Lancs and are reported from Merioneth. Hybrids which appear morphologically to be back-crosses to *C. littorale* are highly fertile and seem to be stabilized hexaploids ($2n = 60$); they occur at Ainsdale (Lancs).

4. C. littorale (D. Turner) Gilmour   Seaside Centaury.

C. turneri Druce; *C. vulgare* Rafn; *Erythraea compressa* Kunth; *E. littoralis* (D. Turner) Fries; *E. turneri* Wheldon & C. E. Salmon; *C. × intermedium* (Wheldon) Druce

An erect *biennial* 2–25 cm, usually scaberulous, *with a basal rosette of lvs* and solitary or frequently several erect stems, branched above. *Basal lvs* 1–2 cm × 3–5 mm, *linear-spathulate*, obtuse, indistinctly 3-veined; *cauline shorter, lingulate*, obtuse. *Fls sessile, clustered in a ±dense corymb-like cyme*, relatively few. Corolla-tube not longer than calyx; limb concave; *lobes* 6–7 mm. Stamens inserted at top of corolla-tube. Capsule much exceeding calyx. Variable. Fl. 7–8. Germ. autumn, $2n = 38^*$, c. 56^*. Hs.

Native. Dunes and sandy places near sea, local, almost confined to the coasts of Wales, N.W. and N. England, and Scotland, from Carmarthen to W. Ross and Northumberland to E. Sutherland, one isolated locality in Hants; Derry. Narrow-lvd forms of *C. erythraea* have frequently been erroneously recorded as this species. Coasts of W. Europe from C. Fennoscandia to N.W. France; inland in C. Europe from Austria to S.E. Russia (Upper Dnieper region).

5. C. scilloides (L.fil.) Samp.   Perennial Centaury.

C. portense (Brot.) Butcher; *Erythraea portensis* (Brot.) Hoffmanns. & Link

A glabrous perennial herb with numerous decumbent sterile stems and ascending fl.-stems. Lvs of sterile stems to 1 cm, suborbicular to rhombic, narrowed into a short petiole; of the fl. stems oblong to lanceolate, sessile. Fls 1–6, pedicellate. *Corolla-lobes* 8–9 mm. Stamens inserted at top of corolla-tube. Capsule rather longer than calyx. Fl. 7–8. Chh.

Native. Grassy cliffs near Newport (Pembroke); W. Cornwall and N. Devon. Atlantic coast from N.W. France to N.W. Portugal; Azores.

4. BLACKSTONIA Hudson (*Chlora* Adanson)

Annuals. *Fls 6–8-merous.* Calyx deeply divided into linear lobes. *Corolla rotate with short tube*, yellow. Anthers oblong or linear, sometimes slightly twisted after fl. *Style filiform with 2 deeply bilobed stigmas, caducous.*

Four spp. in Europe and Mediterranean region.

1. B. perfoliata (L.) Hudson   Yellow-wort.

Chlora perfoliata (L.) L.

An erect glaucous annual 15–45 cm, with basal rosette of lvs. Stems simple or branched above. Lvs ovate to triangular; basal 1–2 cm, obtuse, free; cauline acute, each pair connate by nearly the whole base. Fls in a lax dichasial cyme. Corolla 10–15 mm diam. Capsule ovoid Fl. 6–10. Self-pollinated. $2n = 44^*$. Th.

Native. Calcareous grassland and dunes, rather common in C. and S. England, extending north to Northumberland and Lancs but local or absent in some Midland counties, much of Wales and inland Devon and Cornwall; S. Ireland to Meath and Sligo. W., C. and S. Europe; S.W. Asia; Morocco.

5. GENTIANA L.

Perennial, rarely annual, herbs. Glabrous. Fls 5-merous. Calyx ±tubular; *teeth joined by a membrane* which forms the upper part of the tube. *Corolla-*tube of varied shape; limb spreading, not fringed, usually blue, *with small lobes between the 3-veined large ones.* Anthers not twisted, *not versatile.* Ovary gradually tapering into the style or style 0. Stigmas 2, persistent on the capsule. Nectaries at the base of the ovary.

About 400 spp., northern hemisphere, mainly in the mountains, a few in the Andes.

1 Corolla sky-blue, the tube obconical; lvs linear, 1.5–cm. **1. pneumonanthe**
  Corolla brilliant deep blue, the tube cylindrical; lvs ovate or ovate-oblong, 1.5 cm or less   2
2 Perennial with numerous rosettes of lvs; corolla c. 15 mm across. (Teesdale and W. Ireland.) **2. verna**
  Erect annual; corolla c. 8 mm across (Scotland.)
    3. nivalis

1. G. pneumonanthe L.   Marsh Gentian.

A perennial 10–40 cm, *with a few suberect simple stems. Lowest lvs scale-like; the rest* 1.5–4 cm, *linear*, obtuse. Fls 1–7, terminal and axillary, forming a rather dense infl; pedicels to 2 cm, but usually much less. Calyx-tube c. 5 mm, obconical, not angled; lobes linear, acute, about as long as tube. *Corolla sky-blue* with 5 green lines outside; *tube obconical*, 2.5–4 cm; lobes broadly ovate, ascending. Capsule ellipsoid, stipitate. Fl 8–9. Pollinated by humble-bees; protandrous. $2n = 26$. Hp.

Native. Wet heaths from Dorset to Sussex and Essex, from Norfolk to Yorks and in Shropshire, Cheshire, Lancs, Westmorland and Anglesey, very local, decreasing. Europe from 60° N in Fennoscandia to about 42° N in Spain, Portugal, Italy and the Balkans; Caucasus; Siberia.

2. G. verna L.   Spring Gentian.

A perennial 2–6 cm, with few or many underground stems from a short stock, each ending in a rosette of persistent lvs, *the rosettes forming a ±dense tuft or cushion. Rosette lvs* 5–15 mm, *ovate or ovate-oblong*, obtuse to subacute; cauline few, smaller, elliptical or oblong. Fls solitary, terminal. Calyx-tube c. 1 cm, ±cylindrical, strongly 5-angled; lobes triangular-lanceolate, much shorter than tube. *Corolla* 1.5–2.5 cm, *brilliant deep blue; tube subcylindrical*; limb 1.5–2 cm across, spreading; lobes ovate, obtuse. Stigma white. Capsule oblong, subsessile. Fl. 4–6. Pollinated by Lepidoptera. $2n = 26, 28$. Chh. or Chc.

Native. Stony grassy places on limestone; Yorks, Durham, Westmorland and Cumberland, 460–730 m; Clare, Galway, ?Mayo, 0–305 m; very local. Mountains of C.

and S. Europe, Arctic Russia; Caucasus, N. and C. Asia (subspp.); Morocco (a subsp.).

3. G. nivalis L.	Alpine Gentian.

An erect slender annual 3–15 cm, simple or branched. *Lvs* 2–5 mm, *ovate to elliptical,* lower sometimes forming a rosette. Fls terminal on stem and branches. Calyx-tube c. 8 mm, ±cylindrical, 5-angled; lobes triangular-lanceolate, shorter than tube. *Corolla* 1–1.5 cm, *brilliant deep blue; tube subcylindrical;* limb c. 8 mm across, spreading; lobes ovate. Capsule ellipsoid, subsessile. Fl. 7–9. Probably usually self-pollinated. $2n = 14$. Th.

Native. Rock ledges in the mountains of Perth and Angus, 730–1052 m. N. Europe; mountains of C. and S. Europe southwards to the Pyrenees, S. Apennines and Bulgaria; N. Asia Minor, Caucasus; arctic North America; Greenland.

6. GENTIANELLA Moench

Annual, biennial or (in foreign spp.) perennial herbs. Glabrous. Fls 4- or 5-merous. *Calyx* ±tubular; *membrane* 0. Corolla-tube cylindrical or obconical; limb spreading, fringed at the throat in the British spp., purple or whitish (blue in some foreign spp.); *lobes* 5–9-*veined, without small lobes between.* Anthers not twisted, *versatile. Ovary gradually tapering into the style or style* 0. *Stigmas* 2, *persistent on the capsule. Nectaries on the corolla.*

About 125 spp., northern hemisphere, South America, S.E. Australia, Tasmania, New Zealand.

Individual plants of *G. amarella* group may show unusual features and determinations should be based on small samples of a population.

1 Calyx-lobes 4, the two outer much larger than the inner and overlapping and enclosing them.
	1. campestris
Calyx-lobes 4 or 5 (often on the same plant), equal or somewhat unequal, not (or rarely slightly) overlapping.	*2*
2 Corolla 25–35 mm, twice as long as calyx or more; internodes 9–15 (small annual plants with smaller corollas may occur mixed with the others).
	2. germanica
Corolla (12–)13–20(–23) mm, less than twice as long as the calyx; internodes 0–9(–11). **(amarella group)**	*3*
3 Uppermost internode and terminal pedicel together forming at most ⅙ the total height of the plant; internodes usually more than 3.	**3. amarella**
Uppermost internode and terminal pedicel together forming at least ½ the total height of the plant; internodes usually 3 or fewer.	*4*
4 Upper lvs lanceolate; calyx-teeth usually subequal, up to 1.5 mm broad, ±appressed to corolla.
	4. anglica
Upper lvs ovate to ovate-lanceolate; calyx-teeth unequal, the largest up to 3 mm broad, ±spreading.
	5. uliginosa

1. G. campestris (L.) Börner	Field Gentian.
Gentiana campestris L.; incl. *G. baltica* auct.

Usually biennial, 10–30 cm, simple or branched. Basal lvs 1–2.5 cm, ovate, lanceolate or spathulate, obtuse or subacute; cauline lvs 2–3 cm, lingulate to oblong, or lanceolate, obtuse to acute. Fls 4-merous. *Calyx divided nearly to base into* 4 *lobes*; 2 *outer lobes ovate to ovate-lanceolate, acute to acuminate, overlapping and hiding most of the* 2 *lanceolate inner ones,* one or both of which may be absent. Corolla 15–25(–30) mm, bluish-lilac, rarely white; tube equalling or longer than calyx, the lobes oblong. Fl. 7–10, visited by humble-bees and Lepidoptera, sometimes selfed. Hs. or Th. $2n = 36$.

Native. Pastures and dunes, usually acid or neutral, ascending to 790 m. Common in Scotland, N. England and Wales (except the east), very local in C., S. and E. England and there absent from several counties; throughout Ireland but rare in the centre and south. N. and C. Europe eastwards to N.W. Russia and C. Austria, and southwards to E. Spain and C. Italy.

2. G. germanica (Willd.) E. F. Warb.
Gentiana germanica Willd.	Chiltern Gentian.

Biennial, (5–)7–35(–50) cm, simple or branched, producing in the first year a rosette of lvs dying in autumn. Stems usually branched above the middle; *internodes* 9–15, *all equal or the uppermost shorter.* Basal lvs spathulate, obtuse, usually dead at fl. time; *cauline lvs* 1–2.5 cm, *ovate or ovate-lanceolate,* subacute or acute, subcordate at base and *tapering from a wide base to the apex.* Fls 5-merous. Calyx-teeth ±equal, ±spreading. *Corolla* 15–35 mm, *bright bluish-purple,* at least twice as long as calyx; tube obconical. Fl. (8–)9–10. Pollinated by humble-bees, perhaps sometimes selfed. Hs.(–Th).

Native. Chalk grassland and scrub in open habitats often among tall grasses. From Hants and Surrey to Oxford and Bedford, very local. Belgium and N.E. France to the S. Alps and E. Carpathians.

3. G. amarella (L.) Börner Autumn Gentian, Felwort.
Gentiana amarella L.; *G. axillaris* (Schmidt) Reichenb.

Biennial (3–)5–30(–50) cm, simple or branched, producing in the first year a rosette of lanceolate to lingulate lvs dying in autumn. *Internodes* (2–)4–9(–12), *the terminal one and the terminal pedicel together forming not more than ⅙ the total height of the plant.* Basal lvs (of second year) obovate or spathulate; *cauline lvs* 1–2 cm, *ovate to linear-lanceolate,* ±acute. Fls 4- or 5-merous, often on same plant. *Calyx-teeth* unequal to subequal, ±*appressed to corolla. Corolla* (12–)14–22 mm, *twice as long as calyx or less,* narrower than in *G. germanica.* Fl. (late 7–)8–9(–10). Pollinated by humble-bees. Fr. 9–10. $2n = 36$. Hs.

Native. Basic pastures, usually among short grass, and dunes, rather common over most of the British Is. but absent from S.W. Scotland, C. Wales, N.E. and S.W. Ireland, Isle of Man and Channel Is. N. and C. Europe eastwards to E. Ukraine, the Caucasus and the Yenisei region.

Subsp. **amarella** (incl. subsp. *hibernica* Pritchard). Stem with 4–9(–11) internodes. Cauline lvs lanceolate or ovate-lanceolate. Corolla (14–)16–18(–22) mm, dull purple, blue, pink or whitish; lobes usually 5, ±spreading at anthesis.

Basic pastures and dunes, from Islay, Perth and Angus southwards; Ireland. Outside the British Is. as for the species, but absent from Iceland.

Subsp. **septentrionalis** (Druce) Pritchard (*G. septentrionalis* (Druce) E. F. Warb.; *Gentiana septentrionalis* (Druce) Druce, incl. subsp. *druceana* Pritchard). Stem with 2–7 internodes. Middle and upper cauline lvs ovate (or ovate-lanceolate), acute, wide at the base. Corolla (12–)14–17 mm, creamy-white within, purplish-red outside; lobes 4–5, ±erect at anthesis.

Dunes and limestone pastures. From Perth and Aberdeen to the Outer Hebrides and Shetland. Iceland.

G. amarella subsp. *amarella* × *G. germanica* (*G.* × *pamplinii* (Druce) E. F. Warb.), showing varying degrees of morphological intermediacy between the parents and reduced but significant fertility, occurs in wood margins and light scrub on the chalk of the Chilterns and Berks Downs and on the edge of Salisbury Plain. Introgression seems to be almost always towards *G. amarella*, from which hybrids can scarcely be distinguished in some of the above localities, as well as in Kent and Surrey.

G. amarella subsp. *amarella* × *G. uliginosa*, showing morphological intermediacy between the parents and also introgression towards *G. amarella*, of good fertility, occurs in dune-slacks in S. Wales.

G. amarella subsp. *amarella* × *G. anglica* subsp. *cornubiensis* forms hybrid swarms with all grades of morphological intermediacy in dune-slacks in N.W. Cornwall.

4. G. anglica (Pugsley) E. F. Warb.   Early Gentian.
Gentiana anglica Pugsley; *G. amarella* var. *praecox* Towns.; *G. lingulata* var. *praecox* (Towns.) Wettst.

Biennial 4–20 cm. *Internodes 2–3(–5), the uppermost and terminal pedicel together forming at least ½ the total height of the plant.* Basal lvs (of second year) spathulate, obtuse. Lower stem-lvs linear or linear-lanceolate; upper lanceolate. Fls 4–5-merous. *Calyx-teeth ±appressed*, not more than 1.5 mm broad. Corolla 13–20 mm, dull purple, about 1½ times as long as calyx. Fl. (3–)4–6(–early 7). Hs.

Subsp. **anglica**. Usually branched from base. Uppermost internode c. 1½ times as long as others (rarely short and then the next long). Basal lvs narrow. Terminal pedicel forming about ½ total height of plant. Calyx-teeth somewhat unequal. Corolla 13–16 mm. Fl. (4–)5–6(–7).

Native. Chalk grassland from S.E. Devon and W. Sussex to E. Gloucester and S. Lincoln, local; probably extinct on dunes in N. Devon. Endemic.

Subsp. **cornubiensis** Pritchard. Branched from middle. All internodes ±equal. Basal lvs broad. Terminal pedicel less than ⅓ total height of plant. Calyx-teeth subequal. Corolla (15–)17–20 mm. Fl (3–)4(–7).

Native. Cliffs on the N. coast of W. Cornwall. Endemic.

5. G. uliginosa (Willd.) Börner   Dune Gentian.
Gentiana uliginosa Willd.

Annual or biennial 1–15 cm, the *biennials with long branches from the base giving a pyramidal habit*, the annuals (found mixed with the biennials) small and consisting of 1 or 2 fls from the basal rosette. *Internodes 0–2(–4), the uppermost and terminal pedicel together forming at least ⅓ the total height of the plant.* Basal lvs of annuals (and first year rosettes of biennials) lanceolate, of second year rosettes obovate or spathulate. *Cauline lvs ovate or ovate-lanceolate*, acute or subacute, wide at the base. *Terminal pedicel very long, together with the terminal internode forming at least ½ total height of plant.* Fls 4–5-merous. *Calyx-teeth ±spreading, very unequal*, the largest up to 3 mm broad. Corolla 10–20(–23) mm (smaller in annuals than biennials), dull purple. Fl (7–)8–11. Hs. or Th.

Native. Dune-slacks in Glamorgan, Carmarthen and Pembroke, very local. N. and N.C. Europe.

G. ciliata (L.) Borkh. Was reported in Bucks in 1847 and has recently been found again. It has a blue corolla, not fringed at the throat and with long-fimbriate lobes. Native in much of Europe.

107. MENYANTHACEAE

Aquatic or bog plants. Lvs alternate, except sometimes on fl. stems. Fls 5-merous, hermaphrodite, regular, heterostylous. Calyx deeply divided. Corolla deeply divided, caducous; lobes valvate in bud. Stamens inserted on corolla-tube or between the lobes. Ovary superior to semi-inferior, 1-celled with 2 parietal placentae; ovules numerous, anatropous. Fr. a capsule.

About 5 genera and 33 spp., cosmopolitan.

Often united with the Gentianaceae but a small natural group of distinct habit.

Lvs ternate; fls pink or white.   1. MENYANTHES
Lvs simple, suborbicular; fls yellow.   2. NYMPHOIDES

1. MENYANTHES L.

Aquatic or bog plant. *Lvs ternate*, all alternate. *Fls in a raceme* on a lfless scape. Capsule opening by 2 valves, subglobose.

One species.

1. M. trifoliata L.   Buckbean, Bogbean.

A glabrous aquatic or bog plant with the lvs and fls raised above surface of water; rhizome creeping. Lflets obovate or elliptical, 3.5–7 cm, obtuse to subacute, entire; petioles 7–20 cm, with long sheathing base. Scape 12–30 cm, c. 10–20-fld; pedicels 5–10 mm, longer than the ovate bracts. Calyx-lobes ovate, somewhat recurved. Corolla pink outside, paler or white within,

c. 15 mm across; lobes fimbriate. Fl. 5–7. Pollinated by various insects, heterostylous. Fr. 8. $2n = 54$. Hel.

Native. Ponds, edges of lakes and in the wetter parts of bogs and fens, ascending to 915 m, sometimes locally dominant in shallow water. Rather common throughout the British Is. Most of Europe, but rare in the Mediterranean region; N. and C. Asia; N. Morocco; Greenland; North America.

2. NYMPHOIDES Séguier

Aquatic. *Lvs simple*, suborbicular, those of the fl. stems opposite. *Fls* on long pedicels *in clusters* in the lf-axils. Capsule opening irregularly, ovoid, beaked.

About 20 spp., mainly tropics and subtropics.

1. N. peltata (S. G. Gmelin) O. Kuntze

Fringed Water-lily.

Limnanthemum nymphoides (L.) Hoffmanns. & Link; *L. peltatum* S. G. Gmelin

A glabrous aquatic with floating lvs and fls; rhizome creeping, its lvs alternate. Fl. stems long, floating, their lvs opposite. Lvs suborbicular, 3–10 cm, deeply cordate at base, entire or sinuate, purplish beneath and purple spotted above, long-petiolate. Fls in 2–5-fld axillary fascicles; pedicels 3–7 cm. Calyx 5-partite; lobes oblong-lanceolate, c. 1 cm. Corolla yellow, c. 3 cm across; lobes fimbriate-ciliate. Fl. 7–8. Pollinated by various insects, heterostylous. $2n = 54$. Hyd.

Native. Ponds and slow rivers in Surrey, Berks and Oxford, and from W. Norfolk and Cambridge to S. Lincoln, local; introduced elsewhere. Most of Europe but not in Fennoscandia or northwards; N. and W. Asia to the Caucasus, Himalaya and Japan; naturalized in North America (District of Columbia).

Both fl. forms are found only in the southernmost localities. Plants in East Anglia are all of the 'pin' form suggesting an early introduction into that area and primarily vegetative reproduction since.

108. POLEMONIACEAE

Annual or perennial herbs and a few shrubs. Lvs alternate or opposite, exstipulate. Fls hermaphrodite, actinomorphic, hypogynous. Sepals 5. Corolla with basal tube and 5 free lobes above. Stamens 5, epipetalous, alternating with the corolla-lobes. Ovary superior, 3-celled with a single style and usually 3 stigmas. Fr. a capsule. Seeds endospermic.

About 15 genera with 300 spp., chiefly North American.

1. POLEMONIUM L.

Usually perennial herbs with *alternate pinnate lvs* and showy fls. Corolla rotate; *stamens inserted all at the same height* at the top of the short corolla-tube, the *broadened downwardly-curved hairy bases of the filaments almost closing the throat of the corolla.*

About 50 spp., chiefly in America but some in Europe and Asia.

1. P. caeruleum L.   Jacob's Ladder.

A perennial herb with short creeping rhizome and erect simple lfy stem 30–90 cm, hollow, angled, ±glandular-pubescent above. Lvs 10–40 cm, pinnate, with terminal and 6–12 pairs of lateral lflets; lower lvs with long slender winged stalks; upper lvs smaller, subsessile; lflets 2–4 cm, ovate-lanceolate or oblong, acuminate, entire, glabrous. Infl. corymbose, ±bractless. Fls 2–3 cm diam., many, drooping. Calyx campanulate with acute teeth. Corolla blue or white, shortly tubular (2 mm), with spreading ovate subacute lobes. Stamens exserted. Style ultimately exceeding the stamens, with 3 slender stigmas. Nectar secreted from a fleshy ring round the base of the ovary. Capsule erect, included in the calyx-tube. Seeds 4–6 in each cell, angular, rugose, shortly winged. Fl. 6–7. Fls protandrous, visited by hover-flies and humble-bees. The population at Malham Cove is gynodioecious. $2n = 18$. Hs.

Native. Locally on grassy slopes, screes and rock-ledges on limestone hills; to 580 m in N. England. England from Stafford and Derby northwards to the Cheviots. Also widely introduced as a garden-escape. N. and C. Europe. Caucasus; Siberia; North America.

Species of three mainly North American genera, differing from *Polemonium* in having stamen-filaments neither hairy nor downwardly curved, are grown in gardens and occasionally escape. *Phlox* spp., with most lvs opposite and entire and stamens inserted at different levels in the corolla-tube, are much grown for their showy fls, especially the tall perennial *P. paniculata* L. and the annual *P. drummondii* Hooker, and mat-forming perennials such as *P. subulata* L. are commonly seen on rock-gardens and walls. Hardy annual spp. of *Gilia* and *Collomia*, their opposite or alternate lvs often attractively cut or divided and their stamens inserted at the same level in *Gilia* but at different levels in *Collomia*, are also grown for their variously coloured fls. The half-hardy lf-tendril climber *Cobaea scandens* Cav. is often seen in greenhouses, less commonly on outdoor pergolas and trellises, grown for its large campanulate purple fls.

109. BORAGINACEAE

Herbs or sometimes shrubs, often hispid or scabrid. Stems usually terete. Lvs alternate, very rarely opposite, exstipulate, entire, sometimes sinuate. Fls often in scorpioid cymes, actinomorphic or sometimes zygomorphic. Calyx 5-toothed, sometimes deeply so. Corolla 5-lobed, rotate, funnel-shaped, cylindrical or campanulate, often pink in bud then bright blue in our spp.; throat often closed by scales or hairs. Stamens 5, inserted on the corolla and alternating with its lobes. Ovary superior, 2-celled or (in our spp.) 4-celled by false septa and deeply 4-lobed; style simple, terminal or from the middle of the 4 lobes (gynobasic). Fr. of 2 or 4 nutlets, rarely a drupe. Seeds usually without endosperm; embryo straight or curved.

About 100 genera with 2000 spp., cosmopolitan but specially abundant in the Mediterranean region and E. Asia.

1 Lvs all petiolate, rounded or cordate at base.
                                        2. OMPHALODES
  At least the upper lvs sessile, though sometimes narrowed towards the base.   2
2 Calyx-lobes toothed, enlarging considerably in fr. and
  forming a compressed 2-lipped covering round the
  fr.; plant procumbent   3. ASPERUGO
  Calyx not or only slightly enlarging in fr., not compressed and 2-lipped, lobes not toothed.   3
3 Nutlets covered with hooked or barbed bristles; calyx-
  teeth spreading nearly horizontally, not concealing
  fr.   1. CYNOGLOSSUM
  Nutlets without bristles; calyx ±concealing fr.   4
4 Some or all the stamens long exserted; plant hispid. 5
  Stamens all included or if slightly exserted then plant
  quite glabrous.   7
5 Stamens all exserted; corolla regular.   6
  Stamens not all exserted; corolla somewhat zygomorphic.   14. ECHIUM
6 Stamens glabrous; anthers 8–10 times as long as
  broad; connective prolonged; plant annual.
                                          5. BORAGO
  Stamens hairy; anthers 2–3 times as long as broad;
  connective not prolonged; plant perennial.
                                      6. TRACHYSTEMON
7 Plant glabrous, very glaucous; lvs punctate (sea-
  shores in the north).   13. MERTENSIA
  Plant ±pubescent or hispid, not glaucous; lvs not
  punctate.   8
8 Fls nodding.   4. SYMPHYTUM
  Fls erect.   9
9 Corolla without scales or folds in the throat, deep
  yellow.   10. AMSINCKIA
  Corolla with scales or folds in the throat, blue, rarely
  pale yellow or white.   10
10 Corolla with hairy or papillose oblong scales or folds
  in the throat.   11
  Corolla with glabrous rounded scales in the throat.
                                        11. MYOSOTIS
11 Calyx divided for ⅓–½ its length; fl. spring.
                                        9. PULMONARIA
  Calyx divided almost to base; fl. summer.   12
12 Corolla with long hairy folds in the throat.
                                      12. LITHOSPERMUM
  Corolla with conspicuous scales in the throat.   13
13 Perennial; lvs ovate; corolla-tube straight.
                                      7. PENTAGLOTTIS
  Annual; lvs linear-lanceolate to oblong; corolla-tube
  curved.   8. ANCHUSA

1. CYNOGLOSSUM L.

Biennial hispid or silky herbs. Fls in cymes, usually ebracteate. Calyx 5-partite, accrescent. Corolla with short tube and rotate limb; mouth closed with prominent scales. Stamens included. *Nutlets 4, flat or convex, covered with hooked or barbed bristles, attached to the*

conical receptacle by a narrow outgrowth of the lower surface. The fls have nectar and are pollinated mainly by bees; self-pollination also occurs.

About 60 spp. in temperate and subtropical regions, especially Asia.

Lvs grey with silky ±appressed hairs; fr. with a thickened
  border.   **1. officinale**
Lvs green, sparsely hispid, upper surface nearly glabrous;
  fr. without a thickened border.   **2. germanicum**

1. C. officinale L.   Hound's-tongue.

An erect *grey-hirsute* biennial 30–90 cm. Basal lvs up to c. 30 cm, lanceolate to ovate, usually acute, petiolate; upper stem-lvs sessile, lanceolate, usually acute; *all with rather silky ±appressed hairs on both surfaces*, upper surface of old lvs sometimes rough with papillae. Cymes usually branched, lengthening to 10–25 cm after flowering. *Pedicels* c. 1 cm, stout, recurved in fr. Calyx-lobes up to 4 mm, oblong or ovate, obtuse. Corolla 5–6 mm, dull red-purple, rarely whitish. *Nutlets* 5–6 mm, flattened, ovate, *surrounded by a thickened border* and covered with short barbed *spines, all of about the same length*. Fl. 6–8. 2n = 24, 48. Hs. (biennial). Plant smells of mice.

Native. In grassy places and borders of woods on rather dry soils on sand, gravel, chalk or limestone, particularly near the sea. Widely distributed but mainly south of a line from the Humber to the Severn, local, north to N. Aberdeen; E. Ireland from Wexford to Down. Europe, except the extreme north and south; Asia; North America.

2. C. germanicum Jacq.   Green Hound's-tongue.

C. montanum auct. non L.

Like *C. officinale*, but *green*, rough and usually more slender. *Lvs* sparsely hispid with short spreading hairs beneath, *nearly glabrous above. Pedicels* 5 mm. Calyx-lobes 4–5 mm, oblong, obtuse. Corolla 5–6 mm. *Fr. without a thickened border but with marginal spines longer than the others*. Fl. 5–7. Hs.

Native. In woods and hedgebanks, rare and apparently diminishing. Recorded recently only from Surrey, Oxford, Gloucester and W. Suffolk. W. and C. Europe.

2. OMPHALODES Miller

Annual or perennial, glabrous or nearly glabrous herbs. Calyx deeply 5-toothed, accrescent. Corolla rotate, blue or white, tube short; lobes obtuse, *the throat closed by 5 saccate invaginations.* Stamens included. *Nutlets smooth, compressed, margins membranous, inrolled leaving a groove round the edge of the nutlet; margin ciliate or dentate; nutlet attached to receptacle by the inner margin.*

About 28 spp. in Europe, Asia, Algeria and Mexico.

*1. O. verna Moench   Blue-eyed Mary.

A shortly pubescent far-creeping stoloniferous perennial 10–30 cm high. Lvs up to 20 cm, ovate, acuminate, rounded or (the lower) cordate at base; petioles

c. 10 cm. Flowering stems erect or ascending, with few lvs. Cymes short, very lax, bracteate at base. Fr. pedicels recurved. Calyx 4 mm, with appressed hairs. Corolla 10–15 mm diam., bright blue, 1–2 times as long as calyx. Nutlets with ciliate margins. Fl. 3–5. $2n = 48$. Hs.

Introduced. Frequently cultivated and sometimes naturalized in woods near dwellings. From the S.E. Alps to the N. Apennines and C. Romania; naturalized in many other countries.

3. ASPERUGO L.

A hispid procumbent annual. Fls 1–2, axillary, small, blue, on short recurved pedicels. *Calyx deeply 5-lobed, the lobes lf-like, enlarging and forming a compressed, 2-lipped covering round the fr.* Corolla funnel-shaped, throat closed by scales, the lobes rounded. Stamens included. *Nutlets 4, laterally compressed, ovate, finely warty, attached to the convex receptacle by the margin.* Fls contain nectar but are usually self-pollinated.

One sp. in Europe and Asia.

*1. A. procumbens L.   Madwort.

The only sp., found rather rarely as a casual and occasionally persisting for a few years in waste places, margins of arable fields, etc. Native of N., E. and E.C. Europe, temperate Asia and N. Africa.

4. SYMPHYTUM L.

Hispid perennial herbs. Basal lvs petiolate, the cauline usually sessile or decurrent. Fls in ebracteate terminal scorpioid cymes, nodding. Calyx campanulate or tubular, 5-toothed, accrescent. *Corolla funnel-shaped or subcylindrical, shortly and broadly 5-lobed. Scales 5, linear or subulate, ciliate, connivent,* included, rarely exserted. Stamens included. *Nutlets 4, ovoid, smooth or granulate, with a collar-like rim at the base.*

About 25 spp. in Europe, extending to the Caucasus.

1 Rhizome slender, far-creeping; fl. stems rarely more
   than 20 cm, unbranched.   **6. ibericum**
   Rhizome short, stout; fl. stems taller and nearly
   always branched.   2
2 Rhizome swollen and tuberous; roots fibrous; stems
   simple or with 1 or 2 short branches near the top;
   middle cauline lvs considerably larger than lower
   ones.   **5. tuberosum**
   Roots thick and tuberous; stems much-branched;
   lower lvs largest.   3
3 Calyx-teeth at least as long as tube.   4
   Calyx-teeth less than ½ length of tube.   **4. orientale**
4 Upper lvs shortly petiolate; calyx-teeth obtuse.
    2. asperum
   Upper lvs sessile, often decurrent; calyx-teeth acute.   5
5 Cauline lvs strongly decurrent, stem broadly winged;
   fls yellowish-white, sometimes purplish or pinkish.
    1. officinale
   Cauline lvs slightly decurrent, stem narrowly winged;
   fls blue or purplish-blue.   **3. ×uplandicum**

1. S. officinale L.   Common Comfrey.

An erect hispid perennial 30–120 cm. Root thick, fleshy, fusiform, branched. *Stem* clothed with long deflexed conical hairs, often branched, *winged with decurrent lf-bases.* Lower lvs 15–25 cm, ovate-lanceolate, petiolate; upper oblong-lanceolate, *broadly decurrent.* Calyx 7–8 mm, teeth lanceolate-subulate, twice as long as tube. Corolla 12–18 mm, whitish, yellowish-white, purplish or pink. Scales triangular-subulate, scarcely longer than the stamens. Nutlets shining, black. Fl. 5–6. $2n = 24, 48$. Hs. Still used in country districts as a poultice.

Native. In damp places, especially beside rivers and streams. Generally distributed throughout Great Britain, though less common in the north and not native there; local and nowhere native in Ireland. Most of Europe and temperate Asia. Represented in Britain by subsp. **officinale**.

*2. S. asperum Lepechin   Rough Comfrey.

A scabrid perennial up to 180 cm. Root thick, branched. *Stems* much-branched, *covered with short stout hooked bristles.* Lvs ovate or elliptical, scabrid or with tuberculate bristles; lower 15–19 cm, cordate or rounded at base, petiolate; *upper very shortly petiolate,* cuneate at base. *Calyx* 3–5 mm, accrescent, covered with short stout bristles; *teeth* linear-oblong, *obtuse,* becoming ±triangular in fr., 1–2 times as long as tube. Corolla 11–17 mm, at first pink, becoming clear blue. Scales lanceolate, about equalling stamens. Nutlets granulate. Fl. 6–7. $2n = 32$. Hs.

Introduced. Formerly cultivated for fodder, now occasionally naturalized in waste places. S.W. Asia.

3. S. ×uplandicum Nyman (*S. asperum × officinale*)
    Russian Comfrey.

Incl. *S. peregrinum* auct., non Ledeb.

Like the two preceding and showing various combinations of their characters. Stems hispid to scabrid. Upper lvs usually shortly and narrowly decurrent. Calyx sometimes accrescent, the teeth acute or subacute. Corolla blue or purplish-blue. Fl. 6–8. $2n = 36^*, 40$. Hs.

The commonest *Symphytum* of roadsides, hedgebanks, woods, etc., but usually absent from the waterside habitats occupied by *S. officinale*. Distribution imperfectly known through confusion with *S. officinale*, which appears to be much less common.

*4. S. orientale L.   White Comfrey.

A *softly pubescent* perennial up to c. 70 cm. Root fusiform, branched. Stems sparsely puberulent and pilose, much-branched. Lvs softly pubescent, ovate or oblong, subacute, base cordate, truncate or rounded; lower up to 14 cm, often less, petiolate, the petiole narrowly winged at top; upper sessile, not decurrent. *Calyx* 6–9 mm, *the teeth* c. ½ *length of tube,* ovate or oblong, obtuse. Corolla 14–18 mm, white. Scales lingulate, slightly exceeding the stamens. Nutlets tuberculate, dark brown. Fl. 4–5.

Introduced. Naturalized in hedgebanks and grassy places, not uncommon south of a line from the Humber to the Severn and near Edinburgh. N.W. Anatolia.

5. S. tuberosum L. Tuberous Comfrey.

A hispid perennial 20–50 cm. Root fibrous; *rhizome creeping, with alternate thick tuberous and thin portions. Stems covered with reflexed bristles, simple or with one or two short branches near the top.* Lvs puberulent and densely hispid; lower small, ovate or spathulate, narrowed at base, petiolate; *middle cauline 10–14 cm, considerably larger than the lowest,* ovate-lanceolate or elliptic, shortly petiolate; upper sessile. Calyx 5–8 mm, teeth lanceolate, acute, 3 times as long as tube. Corolla 13–19 mm, yellowish-white. Scales broadly triangular-subulate, acuminate, somewhat exceeding the stamens. Nutlets minutely tuberculate. Fl. 6–7. $2n = 144$.

Native. In damp woods and hedgebanks. Scattered in England and Wales; S. and E. Scotland, local; an escape from cultivation in Ireland. W., C. and S. Europe; N.W. Anatolia. Represented in Britain by subsp. **tuberosum.**

*6. S. ibericum Steven Creeping Comfrey.

S. grandiflorum auct., non DC.

Hispid perennial with *long, slender rhizomes.* Lower lvs elliptic to ovate, long-petiolate; blade 5–10 cm; cauline lvs smaller, the uppermost sessile. *Calyx 4–7 mm,* accrescent, *divided nearly to base,* the teeth linear-lanceolate, obtuse. Corolla 15–18 mm, yellowish-white. Style c. 20 mm, persistent. Fruiting calyx c. 10 mm. Hs. Fl. 4–5 and sporadically later.

Introduced. Not infrequently naturalized in hedges and woods in S. England and the Midlands. Sometimes mistaken for *S. tuberosum.* Native of the Caucasus.

5. BORAGO L.

Annual or perennial hispid herbs. Fls blue, in lax forked bracteate cymes. Calyx 5-partite. Corolla rotate, with notched scales in the throat; lobes acute. Stamens inserted in throat of corolla; *filaments broad and flattened with a narrower obtuse prolongation parallel to the anthers; anthers oblong, mucronate, exserted, connivent. Nutlets 4, rugose, base concave, with a collar-like ring.* The fls have nectar and are pollinated by bees.

Three spp. in the Mediterranean region, one extending into C. Europe.

*1. B. officinalis L. Borage.

A stout erect hispid annual 15–70 cm. Lvs 5–20 cm, ovate, obtuse or acute, lower petiolate, the upper sessile. Cymes axillary and terminal, few-fld; pedicels 0.5–3 cm; bracts linear or lanceolate, the lower lf-like. Calyx-teeth c. 1 cm, linear-lanceolate, very hispid. Corolla 2 cm diam., bright blue. Anthers purple-black. Fl. 6–8. $2n = 16$. Th.

Introduced. A garden-escape on waste ground near houses. C, S. and E. Europe; introduced in America. Widely naturalized.

6. TRACHYSTEMON D. Don

Like *Borago* but fls with a longer corolla-tube and anthers much shorter than filaments and without a pro-

longation of the connective.

Two spp. in the eastern Mediterranean region.

*1. T. orientalis (L.) G. Don Abraham-Isaac-Jacob.

A hispid perennial herb with a stout mucilaginous rhizome up to c. 5 cm diam., clothed with blackish, persistent lf-bases. Basal lvs long-stalked with an ovate, ±obtuse blade 15–50 cm. Fls similar to those of *Borago* but rather smaller. Stamens hairy. Fl. 4–5. $2n = 56$. Hs.

Introduced. Naturalized in damp woods in Devon, Kent, Yorks and probably elsewhere. Caucasus; Turkey.

7. PENTAGLOTTIS Tausch

Perennial hispid herbs. Lvs ovate. Fls actinomorphic, blue, in bracteate scorpioid cymes. Calyx 5-lobed almost to the base. *Corolla rotate; tube straight,* shorter than or as long as the obtuse lobes; *throat closed by scales.* Stamens included. *Nutlets concave at base,* with a small stalked attachment.

One sp. in Europe.

*1. P. sempervirens (L.) Tausch Alkanet.

Anchusa sempervirens L.; *Caryolopha sempervirens* (L.) Fischer & Trautv.

A somewhat hispid perennial with the habit of a *Symphytum,* 30–100 cm. Lvs ovate, acute or acuminate, entire and scarcely undulate, the lower 10–40 cm, petiolate. Cymes very hispid, in long-pedunculate, axillary, subcapitate pairs, each cyme subtended by a lf-like bract 1.5–3 cm. Fls subsessile. Calyx 2.5–3 mm; teeth linear. Corolla 8–10 mm diam., bright blue, the tube shorter than lobes; throat with white scales. Nutlets reticulate. Fl. 5–6. $2n = 22$. Hs.

Introduced. Naturalized in hedgerows and at borders of woods near buildings. Widely distributed but rather local, except in S. England. S.W. Europe from C. Portugal to S.W. France; often naturalized.

8. ANCHUSA L.

Annual, biennial or perennial, usually hispid, herbs. Fls in usually bracteate scorpioid cymes. Calyx ±deeply divided into 5 lobes. Corolla-tube straight or curved, at least as long as the regular or oblique limb; throat closed by scales or hairs. Stamens included. *Nutlets deeply concave at base, not stalked.* About 50 spp. in Europe and Asia.

1. A. arvensis (L.) Bieb. Bugloss.

Lycopsis arvensis L.

An erect very hispid annual or biennial 15–20 cm. Hairs with swollen bulbous bases. Lvs up to c. 15 cm, obovate-lanceolate to linear-oblong, obtuse or apiculate, strongly undulate, distantly and irregularly toothed; lower lvs narrowed into a petiole, the upper sessile, ½-amplexicaul. Cymes simple or forked, at first subcapitate, elongating somewhat after flowering. Fls subsessile, bracts lf-like. Calyx-teeth linear-lanceolate, enlarging in fr. Corolla 4–6 mm diam., bright blue, the tube abruptly curved about the middle; scales white. Nutlets 3–4 mm, reticulate. Fl. 6–9. $2n = 48$. Th. or Hs.

Native. On light sandy and chalky soils in arable fields, sandy heaths and near the sea; probably introduced in the north and perhaps elsewhere. Widely distributed and locally common in Great Britain; Ireland, chiefly near the sea in the north and east. Throughout the greater part of Europe; subsp. *orientalis* in Asia.

9. PULMONARIA L   Lungwort.

Perennial herbs with creeping rhizome usually ending in non-flowering shoots. Flowering stems simple; cymes terminal. Fls usually heterostylous, purple or blue, often pink in bud. Calyx 5-angled at base, cylindrical or campanulate in fl., enlarging somewhat and becoming strongly campanulate in fr.; lobes erect. *Corolla* funnel-shaped, *with 5 tufts of hair alternating with the stamens.* Stamens included. *Nutlets with a raised ring round the base; receptacle flat.* Fls have nectar and are pollinated chiefly by humble-bees.

About 20 spp. in Europe and W. Asia.

1 Basal lvs lanceolate, gradually narrowed at base; corolla 5–6 mm diam.   **1. longifolia**
   Basal lvs ovate, abruptly contracted and cordate at base; corolla c. 10 mm diam.   2
2 Summer lvs white-spotted; blade longer than petiole.   **2. officinalis**
   Summer lvs unspotted or with faint green spots; blade shorter than petiole.   **3. obscura**

1. P. longifolia (Bast.) Boreau

Narrow-leaved Lungwort.
P. angustifolia auct., non L.

A pubescent perennial 20–40 cm. Stems not scaly at base. *Basal lvs in autumn reaching* 50 cm, 6–9 *times as long as broad, lanceolate, gradually attenuate at base,* usually spotted with white; cauline lvs lanceolate or ovate-lanceolate, sessile, ½-amplexicaul. Cymes short, scarcely elongating after flowering. *Corolla* 5–6 mm *diam.*, pink then blue-violet. *Nutlets* 4×3 mm, *strongly compressed*, shining. Fl. 4–5. $2n = 14$. Hs.

Native. In woods and thickets on clay soils, very local. Dorset, S. Hants and Isle of Wight. W. Europe northwards to England.

*2. P. officinalis L.   Lungwort.

A pubescent perennial 10–30 cm. *Basal lvs up to 16 cm,* 1½ *times as long as broad, ovate, cuspidate or shortly acuminate, longer than the petiole, often cordate at base, abruptly narrowed into a winged petiole,* white-spotted; cauline lvs ovate, sessile, ½-amplexicaul Cymes short, scarcely elongating after flowering. *Corolla* c. 10 mm *diam.*, pink then blue. *Nutlets* 4×3 mm, *ovoid, acute.* Fl. 3–5. $2n = 16$ (14). Hs.

Introduced. Naturalized or possibly native in woods and on hedgebanks in a number of scattered localities in Great Britain. From the Netherlands and S. Sweden to N. Italy and Bulgaria.

*3. P. obscura Dumort.   Lungwort.

Like *P. officinalis* but summer lvs unspotted or with faint green spots, the blade shorter than the petiole. $2n = 14$.

Naturalized in scattered localities. N. and C. Europe, extending to S.E. France, Bulgaria and Crimea.

10. AMSINCKIA Lehm.

Annuals. Fls in terminal cymes. Calyx divided almost to the base. accrescent. Corolla with cylindrical tube and infundibuliform or campanulate limb, orange or yellow, without scales in the throat. Stamens included. Style included; stigma capitate. Nutlets ovoid-trigonous, rugose, attached to the receptacle for only part of their length.

About 50 spp. in temperate America.

*1. A. lycopsoides (Lehm.) Lehm.

Stem 20–50 cm, often branched, strongly hispid. Lvs 3–8 cm, linear to oblanceolate, hairy on both surfaces. Infl. ebracteate or with one bract at base. Fls subsessile; calyx 3–5 mm in fl., 6–11 mm in fr.; corolla 5–8 mm, deep yellow, hairy at the throat. Nutlets 2–3 mm.

Introduced. Naturalized on the Farne Islands and sometimes a casual elsewhere. Native of eastern North America.

11. MYOSOTIS L.

Forget-me-not, Scorpion Grass.

Annual or perennial ±hairy herbs. Cymes terminal, scorpioid, sometimes bracteate at base. Calyx 5-toothed, sometimes divided nearly to base. Corolla rotate, usually pink in bud, ultimately blue, rarely yellow; *throat closed by 5 short notched scales*; lobes emarginate or entire, flat or concave, contorted in bud. Stamens included. Style short; stigma capitate. *Nutlets small, shining, lenticular or subtrigonous*, often with a distinct rim; *attachment-area usually small.* The fls contain nectar and may be cross-pollinated by insects though self-pollination is also possible and probably more frequent; it is known to result in good seed production in several spp.

About 50 spp. in the temperate regions of both hemispheres.

1 Hairs on calyx-tube appressed, rarely almost 0.   2
   At least some hairs on calyx-tube short, stiff, hooked or crisped, not appressed, calyx always hairy.   7
2 Calyx persistent in fr; teeth at fl. broadly triangular, shorter than the tube.   3
   Calyx usually caducous; teeth at fl. narrowly triangular, longer than the tube.   4
3 Annual or biennial, without stolons; corolla-lobes concave; nutlets not more than 1 mm, brown (Jersey).   **5. sicula**
   Biennial or perennial, with stolons; corolla-lobes flat; nutlets more than 1.5 mm, black.   **1. scorpioides**
4 Stems with spreading hairs, at least at base.   **2. secunda**
   Stems with appressed hairs.   5
5 Stolons present.   **3. stolonifera**
   Stolons absent.   6

6 Pedicels of almost all fls not longer than calyx in fr.
  (Jersey).   **5. sicula**
  Pedicels of most fls longer than calyx in fr.   **4. laxa**
7 Perennial or biennial.   8
  Annual.   10
8 Calyx persistent, narrowed at the base in fr.; nutlets
  obtuse, the attachment-area large (basic rocks,
  730–1190 m).   **6. alpestris**
  Calyx deciduous, rounded at the base in fr.; nutlets
  ±acute, the attachment-area small (lowland).   9
9 Calyx closed in fr.; corolla-lobes concave. **8. arvensis**
  Calyx open in fr.; corolla-lobes flat.   **7. sylvatica**
10 Fls yellow, at least in bud; corolla-tube often lengthening with age.   **9. discolor**
  Fls blue, rarely white; corolla-tube not lengthening. 11
11 Calyx closed in fr.; lowermost pedicels longer than
  the calyx, forward-pointing; nutlets black.
                                          8. arvensis
  Calyx half open in fr.; lowermost pedicels usually
  shorter than the calyx, spreading; nutlets brown.
                                       10. ramosissima

1. M. scorpioides L.   Water Forget-me-not.

M. palustris (L.) Hill

A ±rhizomatous often stoloniferous perennial
15–45(–100) cm. Stem decumbent or erect, angular,
±hairy. Lower lvs up to 10 cm, often less, oblong to
oblong-lanceolate, usually obtuse, attenuate at base but
scarcely petiolate, subglabrous or with short appressed
hairs; upper lvs narrower and often apiculate. *Infl.
ebracteate. Fr. pedicels* up to 10 mm, spreading or reflexed. *Calyx* up to 6 mm, campanulate, hairs appressed;
teeth triangular, ¼–⅓ length of calyx. Corolla (3–)4–8 mm
diam., sky-blue, rarely white; lobes flat, emarginate.
Style equalling calyx-tube to longer than calyx. *Nutlets
up to* 1.8 × 1.2 mm, *narrowly ovoid*, obtuse, slightly bordered, not keeled, *black* and shining. Fl. 5–9. 2n = ?64,
66. Hel. or Hs.

Native. In wet places by streams and ponds. Common
and generally distributed throughout the British Is., to
500 m in Perth. C. and N. Europe; Asia, south to N.
India; N. Africa; naturalized in North America.

2. M. secunda A. Murray   Creeping Forget-me-not.

M. repens auct.

An erect ±pubescent annual to biennial 20–60 cm, perennating by means of stolons. Rhizome short, scarcely
creeping. Stem erect, with spreading or upward-pointing hairs below; *decumbent or prostrate stems arising
from its base, the non-flowering ones rooting at the nodes.*
Lower lvs c. 4 × 1.5 cm ovate-spathulate, obtuse, sparsely hairy, ciliate below; upper lvs oblong-lanceolate,
obtuse or subacute. Infl. lax, bracteate below. *Fr. pedicels* 3–5 *times as long as calyx*, reflexed. Calyx with
appressed hairs, campanulate in fr.; *teeth* lanceolate,
acute, *rather more than ½ length of calyx.* Corolla 4–8 mm
diam., blue, lobes slightly emarginate. Style equalling
or slightly exceeding calyx-tube. Nutlets 1.8 × 1.2 mm,
black, ovoid, acute, with rim, rhombic attachment-area
and spongy appendage. 2n = 24, 28. Fl. 5–8. Hel. or
Hs.

Native. In wet, often peaty, places, usually avoiding
calcareous soils. Fairly generally distributed except in
E. England, and commoner in mountain districts than
M. scorpioides. Ascends to 800 m on Carnedd Llewelyn
(Caernarvon). W. Europe; Azores; Madeira.

3. M. stolonifera (DC.) Gay ex Leresche & Levier
                              Pale Forget-me-not.

M. brevifolia C. E. Salmon

An erect ±pubescent *dark bluish-green* perennial
12–20(–30) cm. *Erect stems producing roots and stolons
from their lower nodes, the stolons bearing small lvs;
hairs appressed*, slightly spreading-ascending towards
base. *Lvs* 2 × 1 cm rounded, obtuse or emarginate,
*scarcely more than twice as long as broad, only the lowest
tapering at base.* Branches of infl. from below the middle
of the stem, usually ebracteate. Fr. pedicels spreading
or recurved, 1–2 times as long as calyx up to 3 mm in
fr., narrowly campanulate, with appressed hairs; teeth
oblong, rounded or obtuse, ½–⅔ lengths of calyx. Corolla
4–6 mm diam., pale blue; lobes entire or emarginate. Style slightly exceeding calyx-tube. Nutlets
1.2 × 0.7 mm, ovoid, obtuse. Fl. 6–8. 2n = 24. Hel. or
Hs.

Native. In wet places in mountainous districts. Lake
District, Yorks, Durham, S. Scotland. Spain and
Portugal.

4. M. laxa Lehm. subsp. **caespitosa** (C. F. Schultz)
Hyl. ex Nordh.   Tufted Forget-me-not..

M. caespitosa C. F. Schultz

An erect ±pubescent annual or biennial 20–40 cm.
Stems simple or branched from the base, *terete*, faintly
ribbed, *with straight appressed hairs*. Lvs up to 8 cm,
lanceolate, narrowed at base, obtuse. Infl. usually bracteate at base. Fr. pedicels spreading. Calyx not more
than 5 mm, campanulate in fr., hairs appressed; teeth
triangular-ovate, subacute, ½ as long as calyx. *Corolla
up to 5 mm diam.*, sky-blue, rarely white; *lobes rounded.
Nutlets* not more than 1.5 × 1 mm, broadly ovoid, *truncate at base*, obtuse, with a spongy attachment-area,
dark brown, shining. Fl. 5–8. 2n = c. 80. Hel. or Hs.

Native. In marshes and beside streams and ponds.
Common and generally distributed throughout the
British Is. Most of Europe; Asia north to Siberia and
east to the Himalaya; N. Africa; subsp. *laxa* in North
America.

5. M. sicula Guss.   Jersey Forget-me-not.

An erect or decumbent ±appressed-pubescent annual
or biennial 5–10(–30) cm. *Stems* solitary or several, simple or with *divaricate flexuous branches*; stems and lvs
with straight, appressed forward-pointing hairs. *Lower
lvs* 6 × 0.8 cm, oblong-spathulate, obtuse, glabrous or
subglabrous beneath, with scattered hairs above; upper
lvs 1–2 × 0.3–0.4 cm, linear-oblong, obtuse, hairy on
both surfaces. Infl. elongate, flexuous, rarely short,
lower branches usually divaricate; bracts sometimes present. *Lower pedicels* 1–3 times as long as calyx,

spreading or reflexed, *thickened above. Calyx up to 6 mm in fr.*, *usually with a few appressed hairs at base*, oblong-campanulate; *teeth oblong, obtuse, c. ½ length of calyx, subconnivent in fr. Corolla up to 3 mm diam.*, blue; *lobes concave*, entire. Style equalling calyx-tube. Nutlets 1 mm, narrowly ovoid, obtuse, slightly bordered, not keeled, pale or dark brown, shining. Fl. 4–6. Th.

Native. In damp places on fixed dunes. Jersey, very local. S. and W. Europe.

6. M. alpestris F. W. Schmidt   Alpine Forget-me-not.

A pubescent rhizomatous perennial 5–20 cm. Stems stiff, hairs spreading. Lvs. oblong-lanceolate, acute or subacute, the lower long-petiolate, the upper sessile, all with ±spreading hairs on both surfaces. *Infl.* ebracteate, *rather short. Fr. pedicels ascending, not longer than the calyx. Calyx up to 7 mm, campanulate, rather silvery*, hairs ±spreading, *with or without a few short stiff hooked bristles on the tube*; teeth narrow; lanceolate, ½–¾ length of tube, erect or slightly spreading in fr. Corolla 4–9 mm diam., blue; lobes flat, rounded. Style longer than calyx-tube. *Nutlets* up to 2.5 mm, roundish-ovoid, with rim only in the distal half, *black*. Fl. 7–9. 2n = 24. Hs.

Native. On basic mountain rocks, rare and local. Mickle Fell, Westmorland; Ben Lawers, Mid Perth, 730–1190 m. Mountains of Europe, from Scotland and the Carpathians to N. Spain and Bulgaria; ?Caucasus.

7. M. sylvatica Hoffm. subsp. sylvatica
Wood Forget-me-not.

An erect pubescent biennial to perennial 15–45 cm. Stems simple or branched, hairs spreading, except in the infl. Lower lvs ovate-spathulate, obtuse, forming a rosette; stem-lvs lanceolate to oblong, ±acute, sessile, all with ±spreading hairs on both surfaces. *Infl.* ebracteate, *lax, much elongated after flowering. Fr. pedicels up to 7(–15) mm, spreading, 1½–2 times length of calyx. Calyx up to 5 mm, with short crisped or hooked hairs on the tube*; teeth linear to triangular, acute, ⅔–¾ length of calyx, spreading in fr. *Corolla 6–8 mm diam.*, rarely less, bright blue rarely white; *lobes flat*, rounded; *tube equalling calyx. Style longer than calyx-tube. Nutlets 1.7 × 1.2 mm, ovoid, acute, with rim, dark brown*, shining. Fl. 5–6(–9). 2n = 18. Hs.

Native. In damp woods, locally abundant. Scattered throughout Great Britain, rare in the north and west. Most of Europe; other subspp. or closely allied spp. in Asia; introduced in North America.

8. M. arvensis (L.) Hill   Field Forget-me-not.

An erect biennial with branched stem 15–60 cm. Lfy part of stem with spreading hairs, the lfless part with straight, appressed hairs. Lower lvs up to 8 cm, oblanceolate, not distinctly stalked, the upper narrower, sessile; all with ±spreading hairs on both surfaces. Infl. ebracteate; pedicels in fr. up to 1 cm, forward-pointing. Calyx closed, with many spreading hooked hairs, deci-

duous. Corolla c. 3 mm diam.; lobes concave. Nutlets up to 2.5 × 1.2 mm, with a rim, greenish-black to black; attachment-area small. Fl. 4–9. Hs.

Native. In cultivated ground, by roads, in woods and on dunes. Generally distributed throughout the British Is., Europe, N. Africa, temperate Asia; naturalized in North America.

Two subspp. occur:
(a) Subsp. **arvensis**: Calyx not more than 5 mm in fr. Hooked hairs not more than 0.4 mm. Nutlets not more than 2 mm. 2n = 52. Common.
(b) Subsp. **umbrata** (Rouy) O. Schwarz; Calyx up to 7 mm in fr. Hooked hairs up to 0.6 mm. Nutlets up to 2.5 mm. 2n = 66. Apparently less common than (a) and confined to W. Europe.

9. M. discolor Pers.   Changing Forget-me-not.

M. versicolor Sm.

A slender erect pubescent annual up to 30 cm. Stem with spreading hairs below and appressed hairs above. Lvs oblong-lanceolate, the lower up to 4 cm, obtuse, narrowed below, the upper acute, sessile, all hairy on both surfaces. *Infl.* ebracteate, lax, *in fr. not much longer than lfy part of stem. Fr. pedicels* ascending, *shorter than calyx. Calyx up to 4.5 mm, campanulate, closed or nearly so in fr.*; tube covered with soft, deflexed hooked hairs; *teeth oblong-lanceolate, c. ½ length of calyx, ultimately nearly erect. Corolla at first yellow or white*, usually becoming pink or blue; lobes concave; *tube at length somewhat longer than calyx.* Style equalling or exceeding calyx. *Nutlets* c. 1.2 × 0.8 mm, ovoid, obtuse, with a wide rim, *dark brown or almost black*, shining. Fl. 5–9. Th.

Native. In grassy places, usually on fairly light soils and in open communities. Generally distributed and locally common throughout the British Is. Most of Europe from Iceland southward and east to Latvia and C. Yugoslavia; Azores.

Two subspp. occur:
(a) Subsp. **discolor**: Stems with at least one pair of opposite lvs; fls cream or yellow at first; corolla up to 4 mm. 2n = 72, ?64. Throughout the range of the species.
(b) Subsp. **dubia** (Arrondeau) Blaise: Stems with no opposite lvs; fls cream at first; corolla not more than 2 mm. 2n = 24. W. Europe.

10. M. ramosissima Rochel   Early Forget-me-not.

M. collina auct., non Hoffm; *M. hispida* Schlecht.

A slender erect or decumbent pubescent annual 2–40 cm. Stems with spreading hairs below and appressed hairs above. Lower lvs up to 4 cm, ovate-spathulate, obtuse, forming a rosette, the upper oblong, sessile, all hairy on both surfaces. *Infl.* ebracteate, lax, *in fr. much longer than the lfy part of stem.* Fr. pedicels ascending, shorter than to slightly longer than the calyx. *Calyx half open in fr.*, tube covered with hooked hairs. *Corolla c. 3 mm diam., blue*, rarely white; lobes concave; *tube shorter than calyx. Nutlets* 1.2 × 0.7 mm, ovoid, obtuse, without rim, *pale brown*, shining. Fl. 4–6. 2*n* = 48. Th.

Native. On dry shallow soils, locally common. Fairly generally distributed in the drier parts of the British Is., absent from much of Wales, N.W. England, Scotland and W. Ireland. Most of Europe; S.W. Asia; N. Africa.

Two subspp. occur:

(a) Subsp. **ramosissima**: Fls only in upper part of the stem. Calyx up to 4 mm in fr., the teeth narrowly triangular; corolla scarcely exceeding the calyx. Nutlets smooth. 2*n* = 48. Widely distributed.

(b) Subsp. **globularis** (Samp.) Grau: Fls almost to the base of the stem. Calyx scarcely more than 2 mm in fr., almost globose, the teeth broadly triangular; corolla distinctly exceeding the calyx. Nutlets with an indistinct rim at the apex. Sandy places near the sea. N. Portugal to S. England.

12. LITHOSPERMUM L.

Annual or perennial herbs or small shrubs, hispid or softly hairy. Fls subsessile (in our spp.) in bracteate cymes. Calyx 5-partite. Corolla rotate or funnel-shaped, *throat with hairy longitudinal folds or small scales*. Stamens included. *Nutlets very hard*, smooth or warty, *base truncate, receptacle flat*. Fls pollinated mainly by bees.

About 60 spp. in temperate regions.

1 Fls 12–15 mm diam., reddish-purple, then blue; non-flowering stems creeping, the flowering erect.
        3. purpurocaeruleum
Fls 3–4 mm diam., white, very rarely blue; all stems erect.   2
2 Lvs with prominent lateral veins; nutlets smooth, shining, white; plant perennial.   **1. officinale**
Lateral veins not apparent; nutlets warty, greyish-brown; plant annual.   **2. arvense**

1. L. officinale L.   Gromwell (Grummel).

An *erect* pubescent slightly rough perennial 30–80 cm, much branched above. Lvs up to 8 cm, lanceolate (rarely ovate), acute, sessile, *lateral veins conspicuous*. Cymes terminal and axillary, elongating after flowering. *Corolla* not much exceeding calyx, 3–4 mm diam., *yellowish or greenish-white*. Nutlets ovoid, obtuse, white, shining. Fl. 6–7. 2*n* = 28. Hp.

Native. In hedges, bushy places and borders of woods mainly on basic soils. Mainly in S. and E. England; rare in Wales and Ireland; probably introduced in Scotland. Most of Europe, eastwards to the Caucasus and Baikal.

2. L. arvense L.   Corn Gromwell, Bastard Alkanet.
Buglossoides arvensis (L.) I. M. Johnston

An erect pubescent somewhat rough *annual* 10–50(–90) cm. Stems usually simple or little branched. *Lvs* up to 3(–5) cm, the *lower obovate, obtuse, narrowed into a petiole, the upper linear-lanceolate or oblong-lanceolate*, acute or subacute, sessile, *lateral veins not apparent*. Cymes terminal, short. *Corolla* not much

exceeding calyx, 3–4 mm diam., *white, rarely bluish, tube violet or rarely blue. Nutlets trigonous-conical, greyish-brown*, warty. Fl. 5–7. 2*n* = 28. Th.

Native. Mainly in arable fields. Fairly common south of a line from the Humber to the Severn, rare and probably not native in the rest of the British Is. Most of Europe. N. Africa; temperate Asia.

Subsp. **arvense** in the British Is.

3. L. purpurocaeruleum L.   Blue Gromwell.
Buglossoides purpurocaerulea (L.) I. M. Johnston

A pubescent perennial with a *creeping woody stem from which spring long creeping non-flowering shoots and* shorter erect flowering ones up to 60 cm. Lvs up to 8 cm, narrow-lanceolate, acute, subsessile, rough and dark green above, light green beneath; lateral veins not apparent. Cymes terminal, elongating after flowering. *Corolla twice as long as calyx*, 12–15 mm *diam., at first reddish-purple, then bright blue*. Nutlets ovoid, obtuse, white, somewhat shining. Fl. 5–6. 2*n* = 16. Chh.

Native. In bushy places and at margins of woods on chalk and limestone. S. Devon, N. Somerset, Monmouth, Glamorgan, and Denbigh, very local; introduced elsewhere. S. and C. Europe; Caucasus; S.W. Asia.

13. MERTENSIA Roth

Perennial, often glabrous and glaucous herbs. Cymes terminal. Fls blue-purple or blue and pink, heterostylous. Calyx 5-lobed. Corolla-tube cylindric, the limb campanulate, the throat without scales, sometimes with 5 folds. *Stamens* inserted towards the top of the corolla-tube, *slightly exserted*. Style filiform. *Nutlets rather fleshy*, smooth or rough, *narrowed towards the flat base; receptacle small*, 2–4-*lobed, flat*.

About 50 spp. in north temperate regions.

1. M. maritima (L.) S. F. Gray   Oyster plant.

A decumbent glabrous glaucous rather fleshy perennial up to 60 cm. Stems purple-pruinose, lfy. Lvs 0.5–6 cm in 2 rows, ovate or obovate, obtuse or apiculate, lower attenuate into a petiole, upper sessile, all punctate on the upper surface. Cymes bracteate, often branched, rather short. Pedicels 2–10 mm, elongating somewhat and becoming ±recurved in fr. Calyx-lobes c. 6 mm, ovate. Corolla c. 6 mm diam., pink then blue and pink; throat with 5 folds. Nutlets c. 6 mm, flattened, fleshy, outer coat becoming inflated and papery. Fl. 6–8. Probably self-pollinated. 2*n* = 24. Hp.

Native. On sand and shingle beside the sea. Formerly in Norfolk, Caernarvon, Anglesey, N. Lancs.; W. coast of Scotland from Wigtown to Shetland; N. Aberdeen to E. Ross; Ireland, N.E. and N. coasts, rare and local; apparently decreasing everywhere. Atlantic coast of Europe from Jutland northwards, Iceland, Greenland, North America.

14. ECHIUM L.

Usually stout hispid or scabrid herbs, sometimes shrubby. Fls in paniculate unilateral cymes. Calyx

5-partite. *Corolla funnel-shaped with a straight tube and open throat, the lobes unequal. Stamens unequal, at least some exserted. Nutlets 4, attached to the flat receptacle by their flat nearly triangular bases.*

About 40 spp. in Europe, the Mediterranean region, Macaronesia.

Lvs rough and harsh, the basal with no apparent lateral veins; corolla hairy all over, blue; 4–5 stamens long-exserted. **1. vulgare**
Lvs not particularly rough, the basal with prominent lateral veins; corolla hairy on veins and margin only, pink becoming blue; 2 stamens long-exserted.
                                                    2. plantagineum

1. E. vulgare L.   Viper's Bugloss.

An erect *very rough* hispid biennial 20–90 cm. *Lvs* up to 15 cm; basal petiolate, *with a prominent midrib and no apparent lateral* veins; cauline sessile, lanceolate or oblong, rarely ovate, acute, *rounded at base. Fls* 10–19 mm, subsessile. Cymes short, dense, elongating after flowering. Calyx-teeth 5–7 mm, linear-lanceolate, shorter than corolla-tube. *Corolla* hairy all over, pinkish-purple in bud, *becoming bright blue*, rarely white. *Four to 5 stamens long-exserted*. Nutlets angular, rugose. Fl. 6–9. Fls protandrous and visited by a great variety of insects. $2n = 16, 32$. Hs.

Native. In grassy places on light dry soils, sea cliffs and dunes, locally common. Scattered throughout England and Wales, mainly in the south and east; Scotland, rare and perhaps not native, absent from the north; Ireland, mainly coastal and native only in the east. Most of Europe and temperate Asia.

2. E. plantagineum L.   Purple Viper's Bugloss.

E. lycopsis L. p.p.

An erect annual or biennial 20–60 cm. Like *E. vulgare* in general appearance but much softer. Basal lvs 5–14 cm, ovate, *lateral veins distinct*; cauline lvs oblong to lanceolate, *the upper cordate at base*; all with appressed hairs, *not very rough to the touch*. Cymes becoming very long (up to 25–30 cm) after flowering. *Corolla* 18–30 mm and nearly as broad, *pink becoming blue, hairy only on veins and margin. Two stamens long-exserted*, the others included or only slightly exserted. Fl. 6–8. $2n = 16$. Hs.

Native. On cliffs and sandy ground near the sea. Jersey, Cornwall and Scilly Is.; occasionally elsewhere as a garden-escape. S. and W. Europe; N. Africa; Caucasus; Macaronesia.

110. CONVOLVULACEAE

Herbs or shrubs; stems often twining; juice usually milky. Lvs alternate, exstipulate. Fls actinomorphic, hermaphrodite, often large and showy, in terminal or axillary racemes or heads, sometimes solitary. Sepals usually free, 5. Corolla funnel-shaped or sometimes campanulate, usually shallowly 5-lobed or -angled. Stamens 5, inserted towards the base of the corolla-tube and alternating with the lobes. Ovary 1–4-celled; ovules solitary or in pairs in each cell; style terminal. Fr. a capsule, 2–4-valved or splitting transversely near the base, sometimes fleshy and indehiscent. Seeds with endosperm which surrounds the usually folded or bent embryo.

About 55 genera and 1650 spp, mainly tropical.

1 Green plants with large lvs; fls more than 1 cm diam., solitary or few together, pedicellate.   2
  Reddish or yellowish parasites with minute scale-like lvs; fls less than 5 mm diam., sessile or subsessile in dense axillary heads.   **3. Cuscuta**
2 Bracteoles small, distant from the calyx.
                                          1. Convolvulus
  Bracteoles large, overlapping the calyx.
                                          2. Calystegia

*Dichondra micrantha Urban (*D. repens* auct., non R. & G. Forster), a small softly pubescent plant rooting at the nodes, with ±reniform entire lvs 5–30 × 5–30 mm, and small solitary greenish-white fls with silky calyces is established in Cornwall.

1. Convolvulus L.

Annual or perennial herbs. Stems often twining, sometimes woody at base. Lvs alternate, large, green. Fls large, in 1–few-fld axillary or terminal corymbs; *bracteoles small, distant from the fl.* Calyx 5-lobed, divided nearly to base. Corolla funnel-shaped or campanulate, 5-angled, entire or sinuate-lobed. Stamens 5, inserted at bottom of corolla-tube, included; anthers dilated near base. Style filiform; *stigmas 2, linear, slender.* Capsule 1–2-celled, usually with 4 seeds.

About 250 spp., chiefly in temperate regions.

1. C. arvensis L.   Field Bindweed, Cornbine.

A scrambling or climbing perennial 20–75 cm. Rhizomes stout, often ±spirally twisted, penetrating the earth to depths of 2 m or more. Stems slender, climbing by twisting in a counter-clockwise direction round the stems of other plants. Lvs 2–5(–10) cm, oblong or ovate, hastate or sagittate, obtuse, often mucronate, ±pubescent when young; petiole shorter than blade. Fls 10–25 mm, solitary or 2–3 together; peduncles exceeding lvs; pedicels with 2 subulate bracteoles c. 2 mm, not overlapping the calyx. Calyx c. 4 mm. Corolla up to c. 3 cm diam., white or pink, soon withering. Capsule c. 3 mm diam. Fl. 6–9. The fls have nectar and are scented; they are visited by many different insects which may bring about cross-pollination, though self-pollination also occurs. $2n = 50$. Hp. or Grh.

An exceedingly persistent and noxious weed.

Native. In cultivated land, waste places, beside roads and railways, and in short turf especially near the sea; commonest on the lighter basic soils. Generally distributed and common throughout England and Wales, becoming rare in Scotland and absent from Orkney and

Shetland; generally distributed in Ireland, but probably introduced in some localities. Throughout the temperate regions of both hemispheres.

2. CALYSTEGIA R.Br.

Characters of *Convolvulus* but *bracteoles large, arising immediately below the calyx and ±enclosing it, and stigmas broad.*

About 25 spp. in temperate and subtropical regions.

Stems 100–300 cm, climbing; lvs ovate-cordate or -sagittate; bracteoles longer than calyx.   **1. sepium**
Stems 10–50(–100) cm, not climbing; lvs reniform; bracteoles shorter than calyx.   **2. soldanella**

1. C. sepium (L.) R.Br.   Belbine, Hedge Bindweed.

A *climbing* rhizomatous and stoloniferous glabrous or slightly pubescent perennial 100–300 cm. Rhizomes far-creeping, seldom more than 30 cm deep. Stolons penetrating the soil and rooting at their ends. Stems climbing by twisting in a counter-clockwise direction. *Lvs up to* 15 cm, *ovate-cordate to -sagittate*, obtuse and apiculate, petiole usually shorter than blade. Fls solitary, 3–9 cm, white or pink; *bracteoles longer than calyx*, nearly flat to strongly inflated, margins overlapping or not. Calyx c. 1 cm. Corolla funnel-shaped, odourless, open by day, closed at night. Capsule 7–12 mm; seeds 4–7 mm, dark brown. Fl. 7–9. Hp. Self-sterile. Widely distributed in temperate regions.

(a) Subsp. sepium: glabrous or pubescent. Sinus of lf-base with divergent sides. Peduncle unwinged. Bracteoles usually 10–15 mm wide, not or scarcely overlapping, not closely investing the calyx, acute or rarely subobtuse, flat or keeled at the base. Corolla 30–70 mm, white or less commonly pink. $2n = 22$.

Native. In hedges, bushy places, edges of woods and in fens; introduced in gardens. Common in southern England and throughout Ireland, becoming local northwards and usually introduced in Scotland. Europe (except the north); W. Asia; N. Africa; North America.

*(b) Subsp. pulchra (Brummitt & Heywood) Tutin: Stems, petioles and peduncles pubescent, at least when young. Sinus of lf-base oblong, with ±parallel sides. Peduncle usually narrowly winged. Bracteoles usually 15–25 mm wide, overlapping, closely investing and almost concealing the calyx, ±saccate at the base, rounded or emarginate. Corolla 50–75 mm, pink. $2n = 22$.

Introduced. Rather uncommonly naturalized in the neighbourhood of gardens. E. Siberia; similar, or possibly identical, plants in North America.

*(c) Subsp. silvatica (Kit.) Griseb.: Glabrous. Sinus of lf-base rounded, with divergent or sometimes parallel sides. Peduncle unwinged. Bracteoles 14–32(–48) mm wide, overlapping, closely investing the calyx and almost concealing it, weakly to strongly saccate at the base, subacute to rounded or emarginate. Corolla usually 50–90 mm, white or occasionally with 5 pale pink stripes. $2n = 22$.

Introduced. Widely naturalized in hedges and waste places, commonest in S.E. England and the Midlands. S.E. Europe, N. Africa, Caucasus.

These 3 subspp. are sometimes given specific rank.

2. C. soldanella (L.) R.Br.   Sea Bindweed.

A procumbent glabrous perennial 10–50(–100) cm. Rhizomes slender, far-creeping. *Stems not climbing. Lf blade up to twice as wide as long, reniform; petiole usually longer than blade.* Fls solitary; *peduncles longer than lvs, sharply quadrangular; bracteoles* 10–15 mm, ovate to suborbicular, *rounded at tip, shorter than calyx.* Corolla 32–52 mm, funnel-shaped, pink or pale purple. *Capsule ovoid, acute,* incompletely 2-celled. Fl. 6–8. Pollinated mainly by humble-bees, but also sometimes self-pollinated. $2n = 22$. Hp.

Native. On sandy and shingly sea shores and dunes. Locally common in suitable habitats around the shores of the British Is., north to E. Lothian and W. Inverness; E., S. and W. Ireland from Down to Roscommon. Atlantic and Mediterranean coasts of Europe from Denmark southwards; N. Africa; Asia; North and South America; Australia and New Zealand.

3. CUSCUTA L.   Dodder.

Slender, *twining*, pink, yellow or white, annual *parasites*, sometimes perennating, attached to the host plant by suckers. *Lvs small and scale-like. Fls in lateral*, often bracteate, *heads or short spikes.* Calyx 5 (rarely 4)-lobed. Corolla urceolate or campanulate, 5- (rarely 4-) lobed; throat usually with a ring of 4–5 small dentate or laciniate petaloid scales below the insertion of the stamens. Stamens 4–5, inserted on the corolla-tube. Styles 2, free or connate; stigmas linear or capitate. Capsule 2-celled, splitting transversely near the base, cells 2-seeded; seeds angled, cotyledons rudimentary or 0.

About 100 spp. in tropical and temperate regions.

1 Stigmas capitate.   **3. campestris**
  Stigmas elongate.   2
2 Styles shorter than ovary; stamens included; scales small, applied to the corolla-tube and not closing it.   **1. europaea**
  Styles longer than ovary; stamens somewhat exserted; scales large, connivent, ±closing corolla-tube.   **2. epithymum**

1. C. europaea L.   Greater Dodder.

A reddish parasite. *Stems up to c.* 1 mm *diam., twisting in a counter-clockwise direction. Fls* 3–4(–5) mm, in dense *bracteate* heads 10–15 mm diam.; pedicels very short, fleshy. *Calyx-teeth obtuse. Corolla pinkish-white, lobes about ½ as long as tube, spreading,* rather obtuse; *scales very small, applied to the corolla-tube and not closing it. Stamens often included. Styles shorter than ovary*; stigmas linear. Fl. 8–9. $2n = 14$. Th.

Native. On *Urtica dioica, Humulus*, and rarely various other plants. England, mainly south of a line from the Wash to the Severn; Scotland and Ireland, very rare and not native. Rare and probably decreasing. Most of Europe; N. Africa; temperate Asia; introduced in North America.

*C. epilinum Weihe, Flax Dodder, which is similar to *C. europaea*, parasitizes flax (*Linum usitatissimum*) and has yellowish fls in ebracteate heads and incurved corolla-lobes c. ½ length of tube; it is sometimes introduced with flax seed. $2n = 42$.

2. C. epithymum (L.) L.   Common Dodder..

Incl. *C. trifolii* Bab.

Like *C. europaea* but much smaller. *Stems* reddish, *very slender* (c. 0.1 mm diam.). Fls 3–4 mm, pinkish, scented, sessile in dense heads 5–10 mm diam. *Calyx open-campanulate, divided for ¾ of its length, lobes acute.* Corolla-lobes acuminate, somewhat spreading; *scales connivent, closing the corolla-tube, the spaces between their bases narrow and acute. Stamens exserted* but shorter than

corolla. *Styles longer than ovary*; stigmas linear. Fl. 7–9. $2n = 14$. Th. (often perennating).

Native. On *Ulex*, *Calluna* and various other plants. Locally common in England mainly south of the Thames; probably introduced elsewhere. Most of Europe; N. Africa; temperate Asia. Introduced in North America and S. Africa.

*3. C. campestris Yuncker

Stems yellowish. Fls 2–3 mm, subsessile in dense clusters 10–12 mm diam. Calyx-teeth ovate or orbicular, obtuse. Corolla with acute, triangular lobes about as long as the shortly campanulate tube. Scales long, fimbriate, exserted. Stamens exserted. Styles about as long as the globose ovary.

Introduced. On cultivated *Trifolium* and *Medicago*, less frequently on other herbaceous plants. Native of North America, naturalized in S., C. and W. Europe.

111. SOLANACEAE

Lvs alternate, sometimes paired through adnation; stipules 0. Fls usually actinomorphic. Calyx (3–)5–(–6)-lobed, usually persistent. Corolla usually 5-lobed. Stamens inserted on the corolla-tube, alternating with corolla-lobes. Ovary superior, 2-locular, cells sometimes divided by false septa; ovules numerous; placentation axile. Fr. a capsule or berry. Seeds with endosperm; embryo straight or bent.

About 90 genera and more than 2000 spp., widely distributed but chiefly tropical.

Closely related to Scrophulariaceae which, however, usually have zygomorphic fls. The two families are best distinguished by the presence of phloem on both sides of the xylem in the vascular bundles of the Solanaceae.

1 Anthers much longer than filaments, opening by apical pores, exserted and forming a cone.
                                                   4. SOLANUM
   Anthers shorter than filaments, opening by slits, usually included, not forming a cone.   2
2 Shrubs, often spiny.   1. LYCIUM
   Herbs.   3
3 Plant viscid-pubescent; fls in a scorpioid cyme.
                                                3. HYOSCYAMUS
   Plant not viscid; fls not in a scorpioid cyme.   4
4 Fls up to 3 cm., drooping; lvs entire.   2. ATROPA
   Fls c. 7 cm., erect; lvs ±toothed.   5. DATURA

1. LYCIUM L.

Erect or pendent spinous or unarmed *shrubs*. Lvs alternate or fascicled, entire. Calyx campanulate, (3–)5-lobed. Corolla funnel-shaped, 5-lobed. *Stamens inserted at mouth of corolla-tube, long-exserted, anthers short. Fr. a berry.*

About 80–90 spp. in temperate and subtropical regions.

Lvs usually widest at the middle; corolla-tube narrowly cylindrical for 2.5–3 mm at base.   **1. barbarum**
Lvs usually widest below the middle; corolla-tube narrowly cylindrical for c. 1.5 mm at base.   **2. chinense**

*1. L. barbarum L.   Duke of Argyll's Tea-plant.

L. halimifolium Miller

Stems up to 2.5 m, arching, greyish-white, often spiny. *Lvs* up to 10 cm, very narrowly elliptical to narrowly lanceolate, *usually widest at the middle*. Fls solitary or few on short axillary shoots; *corolla* c. 9 mm, funnel-shaped, rose-purple turning pale brown, the *tube narrowly cylindrical for* 2.5–3 mm *at base*. Fr. 1–2 cm, scarlet. Fl. 6–9. Fr. 8–10. $2n = 24$. N.

Introduced. Naturalized in hedges and on walls and waste ground, mainly in England. Native of China, widely naturalized in Europe.

*2. L. chinense Miller

L. rhombifolium (Moench) Dippel

Like **1** but less spiny; *lvs* up to 14 cm, the lower much larger than the upper, lanceolate to ovate, *usually widest below the middle*; *corolla* 10–15 mm, the *tube narrowly cylindrical for* c. 1.5 mm *at base*; fr. 1.5–2.5 cm. Fl. 6–9. Fr. 8–10. $2n = 24$. N.

Introduced. Naturalized locally, mainly near the sea. Native of China, locally naturalized in W., C. and S. Europe.

2. ATROPA L.

Tall much-branched *herbs* with entire lvs. Fls solitary, axillary. Calyx 5-partite. Corolla campanulate, slightly zygomorphic. *Stamens inserted at base of corolla-tube.* Ovary 2-celled; stigma peltate. Fr. a many-seeded, 2-celled *berry* subtended by the spreading calyx.

Four spp., in Europe, W. Asia and N. Africa.

1. A. bella-donna L.   Dwale, Deadly Nightshade.

A glabrous or pubescent and glandular perennial up to 150(–200) cm. Lvs up to 20 cm, alternate or in unequal pairs, ovate-acuminate, narrowed into the petiole. Fls 25–30 mm; pedicels drooping, axillary or in the forks of the branches. Calyx somewhat accrescent. Corolla brownish-violet or greenish, lobes obtuse. Anthers pale. Berry 15–20 mm diam., black. Fl. 6–8. Fr. 8–10. 2n = 50, 72. Hp.

A powerful narcotic and very poisonous. Contains the alkaloids atropine and hyoscyamine.

Native. In woods and thickets on calcareous soils; naturalized near old buildings and in hedges, rather rare. England and Wales, from Westmorland southwards, local; introduced elsewhere. W., C. and S. Europe; W. Asia; N. Africa.

3. HYOSCYAMUS L.

Annual or biennial, often viscid, herbs. Fls axillary or in bracteate scorpioid cymes. *Calyx urceolate*, 5-toothed. *Corolla somewhat zygomorphic*, campanulate or funnel-shaped; lobes obtuse. Stamens inserted at base of corolla-tube. Ovary 2-celled; stigma capitate. *Fr. a many-seeded capsule enclosed in the calyx and constricted in the middle, circumscissile near the top.*

About 20 spp., in Europe, Asia and N. Africa.

1. H. niger L.   Henbane.

.A viscid pubescent strong-smelling annual or, more usually, biennial up to 80 cm. Hairs soft, glandular. Stem stout, rather woody at base. Lvs with a few large teeth or nearly entire; lower 15–20 cm, petiolate, oblong-ovate; cauline smaller, amplexicaul, oblong. Fls c. 2 cm, subsessile in 2 rows in a scorpioid cyme; bracts lf-like. Calyx-teeth triangular-acuminate, points becoming rigid in fr. Corolla 2–3 cm diam., pale yellow usually veined with purple. Anthers purple. Calyx-tube subglobose in fr., 15–20 mm diam., strongly veined. Fl. 6–8. 2n = 34. Hs.

Poisonous and narcotic. Contains the alkaloids hyoscyamine and scopolamine.

Native. In sandy places, specially near the sea, elsewhere usually on disturbed ground in farmyards, etc. Native in S. England and widely scattered throughout the British Is. north to Orkney as a casual, local. Almost all Europe; W. Asia; N. Africa.

*Nicandra physalodes** (L.) Gaertner, an annual with a strongly accrescent bladdery calyx which is cordate at base and with bright blue fls 2–4 cm, occurs as a casual. Native of Peru, widely naturalized in north temperate regions.

*Physalis alkekengi** L., Cape Gooseberry, a somewhat pubescent perennial with dirty white fls and a calyx similar to that of *Nicandra* enclosing an orange or red berry the size of a cherry, is occasionally naturalized. C. and S. Europe.

4. SOLANUM L.

Herbs or shrubs. Lvs alternate or in pairs. Fls in cymes, rarely reduced to 1 fl., white, purple or blue. Calyx usually 5-fid. *Corolla rotate*; lobes usually 5. *Stamens inserted on the throat of the corolla-tube; filaments short; anthers long, exserted, connivent or cohering in a cone, opening by apical pores.* Ovary 2-(3-4)-celled. Fr. a many-seeded berry.

About 1500 spp. mainly in the tropics.

S. tuberosum, the potato, is widely cultivated in temperate countries. The tomato, *Lycopersicon esculentum*, belongs to a closely related genus.

1 Scrambling perennial, woody at the base; corolla usually deep purple; cymes much-branched, usually with at least 10 fls.   **1. dulcamara**
 Annual; corolla white; cymes usually unbranched, with 2–10 fls.   2
2 Lvs deeply pinnatisect.   **4. triflorum**
 Lvs entire to sinuate-dentate.   3
3 Calyx strongly accrescent, enclosing at least the lower half of the berry.   **3. sarrachoides**
 Calyx slightly accrescent, enclosing at most the base of the berry.   **2. nigrum**

1. S. dulcamara L.   Bittersweet, Woody Nightshade.

A glabrous, pubescent or sometimes tomentose *scrambling woody perennial* 30–200(–700) cm. Lvs up to c. 8 cm, ovate, entire, or with 1–4 deep lobes or stalked pinnae at base, *apex acuminate, base rounded, cordate or sometimes hastate*; petiole shorter than blade. Cymes lf-opposed, pedunculate, branched, umbellate, with 10–25 or more fls; pedicels erect in fl., recurved in fr. *Fls* 10–15 mm diam. Calyx with broad, shallow, rounded lobes. *Corolla purple* or very rarely white; *lobes* 3–4 times as long as calyx, at first spreading then revolute. *Anthers* yellow, *cohering in a cone. Fr.* 10–15 mm, *ovoid, red.* Fl. 6–9, 2n = 24. N. or Chw.

A variant which may be genetically distinct (var. *marinum* Bab.) with fleshy lvs and prostrate stems occurs on the coast.

Native. In hedges, woods, waste ground and on shingle beaches, common. Throughout the British Is., rare in Scotland and Ireland and often introduced. Most of Europe; Asia; N. Africa.

2. S. nigrum L.   Black Nightshade.

A subglabrous or pubescent annual up to c. 60 cm. Lvs ovate or rhombic to lanceolate, entire or sinuate-dentate, acute, cuneate at base; petiole shorter than the *decurrent blade.* Cymes (3–)5–10-fld, scarcely branched, umbellate; peduncles (10–)14–30 mm; pedicels erect in fl., deflexed in fr. *Fls* 10–14 mm diam. Calyx scarcely accrescent, lobes obtuse. *Corolla white*, lobes about twice as long as calyx, at first spreading, then revolute. Anthers yellow. *Fr.* 6–10 mm, usually wider than long, *black or green.* Fl. 7–9. 2n = 72. Th.

Native. In waste places and a weed in gardens. Throughout most of England, becoming rarer northwards; very local in Wales, and introduced in Scotland and Ireland. Throughout most of Europe and widely distributed as a weed of cultivation.

Represented in Britain by subsp. **nigrum**.

*3. S. sarrachoides Sendtner

A glandular-pubescent annual up to 40 cm. Lvs ovate-lanceolate to rhombic, usually sinuate-dentate. Cymes 3–8(–10)-fld; peduncles 6–15 mm. *Calyx strongly accrescent* and enclosing at least the lower half of the berry. Corolla white. Berry green or black. Fl. 7–9, 2n = 24. Th.

Introduced. A weed of arable land in East Anglia and the Scilly Is. and a casual on rubbish dumps. Native of Brazil.

*4. S. triflorum Nutt.

Annual up to 100 cm. *Lvs* almost glabrous, ovate to elliptical, *deeply pinnatisect*, the lobes ±linear. Cymes usually 2–3-fld. *Calyx scarcely accrescent.* Corolla white. Berry marbled whitish and green. Fl. 7–9. Th.

Introduced. Naturalized in East Anglia. Native of western North America.

*S. pseudocapsicum L., Jerusalem Cherry, is often cultivated and is sometimes found as a casual. Glabrous, ±shrubby (though usually treated as an annual); lvs oblanceolate, petioles short. Fls white, solitary or in small axillary clusters. Fr. red, cherry-like, c. 1.5 cm diam. Probably South American, though known from Madeira for c. 300 years.

5. DATURA L.

Erect annuals. Lvs alternate, simple. Fls large, axillary, pendent or erect. *Calyx long, tubular, often 5-angled,* splitting transversely after flowering, the upper part deciduous, the lower persistent and accrescent. *Corolla funnel-shaped, with long cylindrical tube* and 5(–10), ±spreading, sometimes unequal lobes. Stamens included or exserted. *Fr. a capsule, dehiscing by 4 valves or irregularly.*

About 10 spp. in warm temperate and tropical regions, mostly in Central America.

*1. D. stramonium L.   Thorn-apple.

A stout erect dichotomously branched herb up to 200 cm. Lvs up to c. 20 cm, ovate to elliptical, sinuate-dentate or coarsely toothed, acute, often unequal at base; petiole up to c. 7 cm. Fls 5–10 cm, erect, solitary; pedicel short. Calyx 3–5 cm, pale green, 5-angled, splitting transversely after flowering; teeth (3–)5–10 mm, unequal, narrow-triangular. Corolla white or sometimes purple; lobes c. 1 cm, narrow, acuminate, ±erect. Fr. a many-seeded capsule (2.5–)4–7 cm, ovoid, densely clothed with long sharp spines or rarely unarmed, dehiscing by 4 valves. Fl. 7–10. $2n = 24$. Th. Narcotic and very poisonous. Contains the alkaloids hyoscyamine, hyoscine and scopolamine.

Introduced. An uncommon casual or ±naturalized in waste places and cultivated ground. England, Wales and S. Scotland; absent from Ireland. Throughout most of the temperate and subtropical parts of the northern hemisphere. Native in America.

*Salpichroa origanifolia (Lam.) Baillon is perhaps naturalized in the Channel Is. and along the south coast of England. Somewhat woody perennial. Lvs 1.5–2.5 × 1–2 cm, ovate-rhombic. Fls less than 1 cm, solitary, nodding; corolla urceolate, white, with a ring of hairs above the insertion of the stamens within. E. temperate South America.

112. SCROPHULARIACEAE

Herbs, sometimes hemiparasites or parasites, sometimes shrubs, rarely trees, with exstipulate lvs. Fls hermaphrodite, zygomorphic, hypogynous, basically 5-merous (except gynoecium). Calyx 5-lobed or sometimes 4-lobed, the top lobe being absent. Corolla gamopetalous, imbricate in bud, not plicate, very variable, from regularly 5- (rarely to 8-)lobed to strongly 2-lipped with the lobes obscure. Stamens 5, epipetalous, alternating with corolla-lobes or more frequently 4, the upper being absent or represented by a staminode, sometimes only 2; anthers introrse. Ovary 2-celled with numerous anatropous ovules on axile placentae, the septum transverse, rarely 1-celled with 2 large, 2-lobed, parietal placentae; style terminal, simple or bilobed. Fr. a capsule (rarely a berry, but not in British spp.). Seeds usually numerous, with fleshy endosperm and straight or slightly curved embryo. Vascular bundles collateral.

About 220 genera and 3000 spp., cosmopolitan.

A family of very varied aspect, distinguishable from allied families by the zygomorphic fls and structure of the ovary. Some foreign Solanaceae can only be distinguished by the bicollateral bundles and either the oblique septum of the ovary or the plicate aestivation of the corolla, but the British spp. are all nearly actinomorphic (see also Acanthaceae).

Species of *Alonsoa, Calceolaria, Nemesia, Phygelius, Paulownia, Penstemon,* etc., are ±commonly grown in gardens for their fls.

1 Parasitic herb, without chlorophyll; lvs fleshy, scale-like, whitish or brownish.   23. LATHRAEA
  Not parasitic, sometimes hemiparasitic, with chlorophyll (rarely masked by anthocyanin); lvs not markedly fleshy, not scale-like.   2
2 Stamens 5; fls many in a large erect terminal raceme or panicle, yellow or white.   1. VERBASCUM
  Stamens 4 or 2.   3
3 Stamens 2; fls blue, more rarely white or pinkish.   4
  Stamens 4; fls never a true blue.   5
4 Herb, sometimes woody at base; capsule loculicidal (sometimes also septicidal), ±compressed at right angles to septum.   14. VERONICA
  Shrub; capsule septicidal, compressed parallel to septum.   15. HEBE
5 Corolla-tube saccate or spurred at base.   6
  Corolla-tube not saccate or spurred.   11
6 Lvs lanceolate to linear or oblong, cuneate or tapered at base, not or shortly stalked   7
  Lvs ovate, orbicular, reniform or hastate, rounded to cordate at base, stalked.   10
7 Corolla saccate at base.   8
  Corolla spurred.   9
8 Annual; calyx unequally lobed, longer than corolla-tube.   3. MISOPATES
  Perennial; calyx ±equally lobed, shorter than corolla-tube.   2. ANTIRRHINUM
9 Fls in a terminal raceme, the bracts much shorter than the lvs; mouth of corolla closed; capsule opening by valves.   4. LINARIA
  Fls axillary, the bracts scarcely differing from lvs;

mouth of corolla slightly open; capsule opening by pores.   5. CHAENORHINUM

10 Lvs entire or hastate, pinnately veined; decumbent annuals; fls yellow, with purple upper lip.
    6. KICKXIA

Lvs lobed, palmately veined; creeping perennials; fls lilac, with orange spot on palate.   7. CYMBALARIA

11 Plants creeping and rooting at nodes, or lvs all basal; fls solitary; corolla rotate with very short tube, 5 mm diam. or less.   12

Plants not creeping and rooting, lfy; fls in infls.; corolla with distinct tube, strongly zygomorphic (except *Erinus*).   13

12 Lvs oblong to subulate, entire; anthers 1-celled.
    10. LIMOSELLA

Lvs reniform, dentate; anthers 2-celled.
    11. SIBTHORPIA

13 Calyx 5-lobed, not inflated.   14

Calyx 4-lobed or if 5-lobed strongly inflated after flowering and lobes usually lf-like.   17

14 Lvs opposite.   15

Lvs alternate.   16

15 Corolla with nearly globose tube and small lobes.
    18. SCROPHULARIA

Corolla with broad straight tube and large lobes.
    9. MIMULUS

16 Corolla-tube narrow, scarcely longer than calyx, the lobes 5, nearly equal; low tufted plant. 12. ERINUS

Corolla bell-shaped, the tube several times as long as calyx, the lobes small; tall erect plant.
    13. DIGITALIS

17 Calyx inflated, at least after flowering.   18

Calyx not inflated.   19

18 Lvs alternate, pinnatisect; fls pink. 16. PEDICULARIS

Lvs opposite, crenate or dentate; fls yellow or brown.
    17. RHINANTHUS

19 Upper lip of corolla laterally compressed, the mouth nearly closed; fls ±yellow.   18. MELAMPYRUM

Upper lip of corolla not compressed, the mouth open.   20

20 Upper lip of corolla with 2 recurved lobes; corolla white or lilac with violet lines and usually yellow spot on lower lip, rarely purple or yellow.
    19. EUPHRASIA

Upper lip of corolla entire or with flat lobes.   21

21 Fls small, 4–8 mm, in unilateral spikes, pink.
    20. ODONTITES

Fls larger, more than 1 cm; spikes not one-sided.   22

22 Seeds large, winged or ribbed; plant perennial; fls dull purple.   22. BARTSIA

Seeds minute; plant annual, sticky; fls yellow.
    21. PARENTUCELLIA

Subfamily 1. PSEUDOSOLANOIDEAE. Autotrophic. Lvs alternate. Fls only slightly zygomorphic. Stamens usually 5. Nectaries 0 or on the corolla.

1. VERBASCUM L.   Mullein.

Herbs (rarely shrubs), usually biennial, with basal rosette of lvs and tall erect stems with alternate lvs. *Fls in terminal racemes or panicles.* Calyx deeply 5-lobed, nearly regular. *Corolla* usually *yellow, rotate, with 5 nearly equal lobes and very short tube.* Stamens 5; filaments all, or at least the three upper, hairy, the two lower longer than upper. Capsule septicidal. Seeds small, numerous.

About 360 spp., mainly in the Mediterranean region but extending to N. Europe and N. and C. Asia.

Besides those given below, several other spp. have been recorded as casuals.

1 Hairs on the filaments white or yellow.   2
  Hairs on the filaments purple.   5
2 Lower filaments glabrous or much less hairy than the upper, with adnate or obliquely inserted anthers; fls 15–55 mm diam., in a dense spike-like raceme.   3
  Filaments all equally hairy, the anthers transversely inserted; fls less than 20 mm diam., in a laxer raceme or panicle.   7
3 Upper cauline lvs distinctly decurrent; pedicels 0–2 mm; calyx with some stellate hairs. 1. thapsus
  Upper cauline lvs not or scarcely decurrent; longer pedicels 3–15 mm; calyx without stellate hairs.
    2. phlomoides
4 Stem angled; lvs nearly glabrous above, mealy beneath.   **3. lychnitis**
  Stem terete; lvs mealy on both sides.
    4. pulverulentum
5 Anthers all transversely inserted, equal; lvs and stems hairy.   **5. nigrum**
  Lower anthers obliquely inserted, larger than the upper; lvs and stems glabrous or nearly so.   6
6 Pedicels longer than the calyx; fls always solitary.
    6. blattaria
  Pedicels shorter than the calyx; fls 1–5 together.
    7. virgatum

1. V. thapsus L.   Great Mullein, Aaron's Rod.

An erect biennial or rarely annual 30–200 cm, ± *densely greyish- or whitish-tomentose.* Lvs obovate-lanceolate to oblong, acute or subacute, crenate; basal 15–45 cm, attenuate into a narrowly winged petiole; cauline smaller, the *base decurrent* nearly to the next lf below. *Fls in a dense terminal spike-like raceme,* rarely with axillary racemes from the upper lvs. Bracts 12–18 mm, triangular-lanceolate, the apex with a long acuminate point, usually rather longer than the fls. Pedicels 0–2 mm. Sepals triangular-ovate, acuminate, with stellate hairs. Corolla yellow, 1.5–3 cm diam., concave. *Three upper filaments clothed with whitish or yellowish hairs, the two lower glabrous or sparingly hairy, their anthers obliquely inserted and partially adnate to the filaments.* Stigma capitate. Capsule 7–10 mm, elliptical-ovoid, rather longer than calyx. Fl. 6–8. Pollinated by various insects or selfed, self-fertile. $2n = 36$. Hs.

Native. Rather common on sunny banks and waste places usually on a dry soil. Throughout much of England, Wales and S. Scotland, more locally northwards to Argyll, Ross and Caithness, and perhaps not native. Most of Europe except the extreme north and much of the Balkan peninsula; Asia, south to Caucasus and Himalaya and east to W. China; naturalized in North America.

*2. V. phlomoides L.   Orange Mullein.

Similar to *V. thapsus* but distinguished as follows: plant

greyish- or whitish- to yellowish-tomentose; upper cauline lvs not or only shortly decurrent; longer pedicels 3–15 mm; bracts 9–15 mm, sepals without stellate hairs; corolla flat or almost so, brighter yellow or rarely white; stigma spathulate, decurrent on style. $2n = 34$.

Introduced. Roadsides, waste ground and rubbish tips. Locally naturalized, though usually a casual, in S.E. England, extending to Lincoln, Nottingham, Derby and Cardigan; Ayr, Channel Is. Native of most of Europe except the north.

*V. densiflorum Bertol. (*V. thapsiforme* Schrader), which has long-decurrent upper cauline lvs, long-acuminate bracts 15–40 mm, yellowish hairs on the inflorescence and a spathulate stigma, is sometimes found as a casual. It occurs through much of Europe northwards to the Netherlands, S.C. Sweden and C. Russia.

V. × kerneri Fritsch (*V. phlomoides × thapsus*), which is morphologically intermediate between the parents except for the absence of stellate calyx-hairs, and virtually sterile, has been reported from Middlesex.

3. V. lychnitis L.　　　　　　　White Mullein.

An erect biennial 50–150 cm. *Stem angled, with a short powdery stellate pubescence. Lvs dark green above* and subglabrous, *with a dense white powdery stellate-pubescence beneath*, crenate; the basal 10–30 cm, oblong or lanceolate, narrowed into a short petiole; cauline smaller, ovate, acuminate; the upper sessile, often narrower. Fls in a narrow panicle, 2–7 in the axil of each bract. Bracts and bracteoles linear-lanceolate or linear, shorter than the fls. Pedicels varying from as long as to twice as long as calyx. Sepals linear-lanceolate, acute, very woolly. Corolla white or yellow, 15–20 mm diam. *Filaments all clothed with yellowish or whitish hairs. Anthers equal, transversely inserted.* Capsule ovoid or pyramidal, longer than calyx. Fl. 7–8. Pollinated by various insects or selfed, homogamous. $2n = 32$. Hs.

Native. Waste places and calcareous banks. Very local from Kent and Essex to Sussex and Berks, Devon, Somerset and Denbigh; extinct or casual further north to Yorks and Edinburgh to Perth. Europe southwards from C. Russia, but rare in the Mediterranean region; Caucasus; W. Siberia (Upper Tobol region); Morocco; naturalized in North America.

The common British form is white-fld (var. *album* (Miller) Druce), the yellow-fld form being known only from Somerset.

V. × thapsi L. (*V. lychnitis × thapsus*), which is morphologically intermediate to varying degrees between the parents, and of very low fertility, is recorded from Sussex, Kent, Surrey, Hertford, Suffolk, Denbigh and Anglesey.

4. V. pulverulentum Vill.　　　　Hoary Mullein.

An erect biennial 50–120 cm, *thickly clothed with a mealy white wool which is easily rubbed off. Stem terete*, weakly striate. *Lvs mealy on both sides*, crenate; basal 20–50 cm, broadly oblong, narrowed to a very short petiole; cauline smaller, sessile, the upper cordate. Fls in a pyramidal panicle, 4–10 in the axil of each bract. Pedicels at first all short, some becoming longer later.

Sepals linear-lanceolate, acute. Corolla yellow, less than 2 cm diam. *Filaments all clothed with white hairs. Anthers equal, transversely inserted.* Capsule ovoid, rather longer than calyx. Fl. 7–8. $2n = 32$. Hs.

Native. Roadsides in Norfolk, Suffolk and Essex, very local; a rare casual elsewhere from Cornwall to Nottingham. W., S. and S.C. Europe.

*V. speciosum Schrader
Whole plant thickly clothed with persistent white wool. Stem angled. Lvs entire; basal oblanceolate, obtuse, gradually tapered to base; cauline ovate, acute, cordate at base. Infl. a panicle. Corolla 2–2.5 cm diam., yellow. Stamens as in *V. lychnitis* and *V. pulverulentum*.

Occasionally ±naturalized. Native of S.E. and E.C. Europe, Asia Minor, Caucasus.

V. × godronii Boreau (*V. pulverulentum × thapsus*), which is generally intermediate between the parents in most morphological characters, and usually completely sterile, is rarely recorded where the parents occur together.

V. × regelianum Wirtgen (*V. lychnitis × pulverulentum*), similarly sterile and morphologically intermediate between the parents, is recorded from Devon.

5. V. nigrum L.　　　　　　　　Dark Mullein.

An erect biennial 50–120 cm. *Stem angled, stellate-pubescent. Lvs dark green above, sparsely pubescent, pale beneath* and more *conspicuously stellate-pubescent*, crenate; the basal 10–30 cm, long-petiolate, ovate to lanceolate, cordate at base; cauline smaller, broadly cuneate at base, the upper nearly sessile. *Fls in a terminal raceme*, sometimes with other racemes from the axils of the upper lvs, 5–10 *in the axil of each bract*. Upper bracts linear, acute, the lower somewhat lf-like. Pedicels 5–12(–15) mm. Sepals linear, acute. Corolla yellow, sometimes cream, with small purple spots at the base of each lobe, 12–22 mm diam. *Filaments all clothed with purple hairs. Anthers equal, transversely inserted.* Capsule ovoid, truncate at the apex, rather longer than calyx. Fl. 6–10. Pollinated by various insects, self-sterile, homogamous. $2n = 30$. Hs.

Native. Waysides and open habitats on banks, etc., usually on calcareous soil. Rather common in S.C. and S.E. England, extending locally to Cornwall, Pembroke, Flint and N. Lincoln; naturalized northwards to Angus and Easterness, as well as in Waterford and Wicklow. Europe from Scandinavia (c. 65° N) and N. Russia to N. Spain, N. Italy and Macedonia and the Caucasus; Siberia (east to the Yenisei region).

V. × semialbum (*V. nigrum × thapsus*), which is variably intermediate between the parents and virtually sterile, is rather common in S. and E. England northwards to Derby, and in Kirkcudbright.

V. × schiedeanum Koch (*V. lychnitis × nigrum*) is sterile and intermediate between its parents in leaf characters. The petiole length and stem are variable, while the filament-hairs are usually purple. It is recorded from Cornwall, Sussex, Surrey and Kent.

V. × wirtgenii Franchet (*V. nigrum × pulverulentum*), which is also sterile and generally intermediate between its parents, is recorded from Cornwall, Suffolk and Norfolk.

*V. sinuatum L.

Basal lvs sinuate-pinnatifid, tapered to base. Infl. a panicle. Corolla c. 2 cm diam., yellow. Stamens as in *V. nigrum*. A rather frequent casual. Native of the Mediterranean region.

*6. V. blattaria L.   Moth Mullein.

An erect biennial to 1 m, *glabrous below, the infl. glandular*. Stem angled. *Lvs glabrous*; basal 10–25 cm, oblong, gradually narrowed to the base, crenately or sinuately lobed, the lobes often toothed; cauline smaller, the upper triangular, cordate at base, sessile, with acute triangular teeth. *Fls in a terminal raceme occasionally with axillary racemes in addition, solitary in the axil of each bract*. Bracts lanceolate, acuminate, the lower somewhat lf-like. *Pedicels much longer than calyx*. Sepals linear-elliptical, subacute. Corolla yellow, rarely whitish, 2–3 cm diam. *Filaments all clothed with purple hairs. Anthers of the three upper filaments small, of the lower very obliquely inserted and adnate to the filament*. Capsule subglobose, as long as or rather longer than calyx. Fl. 6–10. Pollinated by various insects or selfed, homogamous. 2n = 30, 32. Hs.

Introduced. Rather rare, in waste places. England northwards to Yorks, in Glamorgan, Pembroke, Flint and Berwick, apparently extinct in intervening areas and in Ireland, rarely persisting in the same spot for more than a few years. Europe northwards to the Netherlands and C. Russia; W. and C. Asia, from Upper Tobol region south to Palestine and Afghanistan; N. Africa; naturalized in North America.

7. V. virgatum Stokes   Twiggy Mullein.

Differs from *V. blattaria* as follows: More glandular. *Fls 1–5 in the axil of each bract. Pedicels much shorter than calyx*. Fl. 6–8. 2n = 32. Hs.

Native. Native in Cornwall, Devon, Scilly and Channel Is. Elsewhere naturalized or casual northwards to Fife and Moray; casual in Kerry, Cork, Mayo and Antrim. W. Europe; widely naturalized elsewhere.

*V. phoeniceum L., similar in general aspect to the last two species but with purple solitary fls, on long pedicels and anthers all equal, is a rare escape from cultivation. Native in E. Europe and W. Asia.

*V. pyramidatum Bieb., a native of the Caucasus and the Crimea which is a perennial up to 1 m, with obovate-oblong, crenately toothed and weakly lobed basal lvs 12–40 cm, glandular hairy above and with simple eglandular and stellate hairs beneath, with fls solitary in axils of bracts, and with purple hairs on filaments, is casual or locally naturalized in Wilts, Surrey and Cambridge.

Subfamily 2. ANTIRRHINOIDEAE.. Autotrophic. Stamens 4 or 2. Nectaries at the base of the ovary, usually ring-shaped.

Tribe 1. ANTIRRHINEAE. At least the lower lvs opposite. Fls axillary or in terminal spikes or racemes. Corolla spurred or saccate at base, the two upper lobes outside the lateral ones in bud. Stamens 4.

2. ANTIRRHINUM L.

Perennial herbs with lower lvs opposite, the upper alternate. Fls in terminal racemes. *Calyx equally 5-lobed. Corolla strongly 2-lipped, the lower lip 3-lobed with a projecting 'palate' closing the mouth of the corolla; the upper 2-lobed; tube broad, gibbous or saccate at base. Capsule with 2 unequal cells, opening by 3 pores*.

About 42 spp., mainly in the Mediterranean region, a few extending farther north, and in western North America.

*1. A. majus L.   Snapdragon.

A *perennial* 30–80 cm, often becoming woody at base but usually not of long duration, branched at the base and of rather bushy habit, glabrous below and glandular-pubescent above. Lvs 3–5 cm, lanceolate or linear-lanceolate, entire, attenuate at base. *Fls in a terminal raceme, in the axils of short, sessile, ovate bracts. Calyx-lobes ovate. Corolla* 3–4 cm, *several times as long as calyx*, reddish-purple, more rarely yellowish-white. Capsule longer than calyx. Fl. 7–9. Pollinated by humble-bees, self-fertile or self-sterile, homogamous. 2n = 16. Hp.

Introduced. Naturalized, usually on old walls, in many parts of Great Britain northwards to Perth and Angus, and in Galway, Wexford, Mayo, Dublin and Louth in Ireland. S.W. Europe, extending to Sicily, naturalized elsewhere. Commonly cultivated in gardens in many colour and habit varieties.

3. MISOPATES Rafin.

Differs from *Antirrhinum* as follows: Annual. Calyx-lobes distinctly unequal, all longer than corolla-tube. Seeds somewhat flattened, with one face smooth, keeled and produced into a narrow wing, the other finely tuberculate and with a wide, raised, sinuate border.

1. M. orontium (L.) Rafin.

Lesser Snapdragon, Weasel's Snout, Calf's Snout.

Antirrhinum orontium L.

An *erect annual* 20–50 cm, simple or branched, usually glandular-pubescent above. Lvs 3–5 cm, linear or narrowly elliptical, entire, attenuate at base. *Fls subsessile in axils of upper lvs forming a lfy terminal raceme. Calyx-lobes linear, unequal, as long as or longer than corolla. Corolla* 10–15 mm, pinkish, rarely white. Capsule shorter than calyx. Fl. 7–10. Pollinated by bees or selfed, homogamous. 2n = 16. Th.

?Native. In cultivated ground. S. England and Wales northwards to Radnor and Norfolk, then very infrequently to Cumberland, local; Cork. S., W. and C. Europe, extending east to the Himalaya and south to the Canaries and Abyssinia.

ASARINA Miller

Differs from *Antirrhinum* as follows: Lvs opposite, palmately veined. Capsule with two equal cells, opening by 2 pores. One sp. in Europe, 15 in North America.

*A. procumbens Miller. Viscid perennial with long procumbent stems. Lvs 2–7 cm, orbicular, crenately lobed. Fls c. 3 cm, pale yellow, solitary, axillary. Rarely naturalized on walls, etc. S. France, N. Spain.

4. LINARIA Miller

Herbs. *Lvs* all opposite or whorled or, more commonly, the upper alternate, *pinnately veined. Fls in terminal racemes*; bracts small. Fls as in *Antirrhinum* but *corolla spurred. Capsule opening by* 4–10 *apical valves* of varying length.

About 150 spp., mainly Mediterranean, a few in C. Europe, temperate Asia, temperate America.

In addition to the following, a number of spp. are cultivated and several are from time to time found as garden-escapes or among ballast.

1 Plant glabrous or only infl. glandular; corolla (excluding spur) 8 mm or more.   2
  Whole plant viscid-pubescent; corolla 4–7 mm.
                                                6. arenaria
2 Corolla either violet or whitish striped with purple.   3
  Corolla yellow.   5
3 Annual; fls 10 or fewer; spur nearly as long as rest of corolla, straight (Jersey).   **1. pelisseriana**
  Perennial; fls 15–40; corolla-spur short or strongly curved.   4
4 Infl. dense; fls violet; spur c. ½ as long as rest of corolla, curved.   **2. purpurea**
  Infl. lax; fls whitish striped with purple; spur c. ¼ as long as rest of corolla, straight.   **3. repens**
5 Erect perennial; sepals ovate or lanceolate, acute.
                                                4. vulgaris
  Decumbent annual; sepals linear, obtuse.   **5. supina**

1. L. pelisseriana (L.) Miller   Jersey Toadflax.

A glabrous, glaucous *annual* 15–30 cm; stems erect, usually simple, sometimes several, often with short sterile branches at base. Lower lvs and those of sterile stems in whorls of 3, elliptical; upper lvs 1–3 cm, alternate, linear. *Fls* 10 or *fewer* in a raceme which is at first short and dense, later elongating. Bracts linear. Pedicels longer than or equalling bracts and calyx. Sepals linear-lanceolate, acute. *Corolla* (10–)15–20 mm, *violet*, with a whitish palate; *spur slender, straight, nearly as long as corolla. Capsule* wider than long, 2-lobed, *shorter than calyx. Seeds flat, tuberculate on one face only, winged.* Fl. 5–7. Th.

Native. Heathy places in Jersey only. Rarely occurring as a casual elsewhere. S. and W. Europe from W. and C. France (not Portugal or W. Spain), eastwards to Transcaucasia and Palestine; Algeria.

*2. L. purpurea (L.) Miller   Purple Toadflax.

A glabrous, glaucous *perennial* 30–90 cm; stems often branched above. Lvs linear or linear-lanceolate, 2–5 cm. *Fls* 15–40 in dense racemes terminal on the stem and branches. Pedicels shorter than the linear bracts. Sepals linear, acute. *Corolla* 9–12 mm, *purplish-violet*, rarely bright pink; *spur* c. 5 mm, *curved, more than half as long as corolla. Capsule* longer than calyx. *Seeds angled, wingless, reticulate.* Fl. 6–8. Pollinated by bees, self-sterile. $2n = 12$. Hp.

Introduced. Much cultivated in gardens and sometimes naturalized on old walls and in waste places in Great Britain northwards to Lanark and Sutherland, though more common in S. England and in parts of Ireland (Kerry, Carlow, Wicklow, Roscommon). C. and S. Italy, Sicily.

*3. L. repens (L.) Miller   Pale Toadflax.

A glabrous glaucous *perennial* 30–80 cm, with a creeping underground rhizome and numerous *erect stems*, usually branched above. Lvs 1–4 cm, linear, whorled below, alternate above. Fls 10–30 in rather long terminal racemes, which are dense in fl. and lax in fr. Fr. pedicels usually rather longer than calyx and linear bracts. Sepals linear-lanceolate, acute. *Corolla* 8–15 mm, *white to pale lilac, with violet veins*, the palate with orange spot; *spur* short, *straight, about a quarter as long as corolla. Capsule* subglobose, *longer than calyx. Seeds angled, wrinkled, wingless.* Fl. 6–9. Pollinated by bees, self-incompatible. $2n = 12$. Hp.

Probably introduced. Local in dry stony fields and waste places, usually calcareous. Great Britain northwards to Roxburgh and Easterness, though very sporadic north of the English Midlands, and Carlow, Wicklow and Louth in Ireland; casual elsewhere in the British Is. Native from N. Spain and N.W. Italy to N.W. Germany; naturalized elsewhere in N.W. and C. Europe.

L. × dominii Druce (*L. purpurea* × *repens*), intermediate between the parents in most characters and apparently sometimes producing fertile seed, has been recorded on waste ground and along railway lines in Berks, Bedford, Derby and Lancs.

4. L. vulgaris Miller   Common Toadflax.

A glaucous *perennial* 30–80 cm, glabrous except for the sometimes glandular infl.; rhizome creeping; *stems* numerous, *erect*, often branched above. Lvs 3–8 cm, linear to narrowly elliptical or linear-oblanceolate. *Fls* 5–30 *in a long dense raceme.* Pedicels usually rather longer than calyx and linear bracts. *Sepals ovate or lanceolate, acute. Corolla* 15–33 mm, *yellow* with orange palate; spur ±straight, about half as long as corolla. *Capsule* ovoid, *more than twice as long as calyx. Seeds tuberculate, winged.* Fl. 7–10. Pollinated by large bees, self-incompatible. $2n = 12$. Hp.

A 'peloric' form of this species, with a regular 5-spurred corolla, sometimes occurs.

Native. Common in grassy and cultivated fields, hedgebanks and waste places in England, Wales and S. and C. Scotland, then less frequent northwards to the Hebrides and Sutherland; widespread in Ireland but less common. Most of Europe except the extreme north and much of the Mediterranean region; W. Asia, east to the Altai; naturalized in North America.

*L. dalmatica (L.) Miller, similar to *L. vulgaris* but with broadly lanceolate, subcordate lvs and wingless seeds, is rarely naturalized. S.W. Europe, etc.

L. × sepium Allman (*L. repens* × *vulgaris*) is very variable, having a usually yellowish corolla striped with violet

12–21 mm, and lvs mostly wider than in *L. repens*, and is highly fertile so that backcrossing produces a more or less complete range of variation between the parents. It is scattered throughout England and Wales northwards to Merioneth and Lincoln, with isolated localities in Cheshire, Lancs and Durham, and is known in Ireland from Waterford.

5. L. supina (L.) Chaz.   Prostrate Toadflax.

A glaucous *annual* 5–20 cm, glabrous except for the glandular infl., sometimes branching at the base, *the branches decumbent*, the ends ascending. Lvs 1–3 cm, linear to linear-oblanceolate. *Fls usually 2–10 in a short dense raceme.* Pedicels shorter than calyx and linear bracts. *Sepals linear, obtuse. Corolla 10–15 mm, yellow*, sometimes purple-tinged, with orange palate; spur almost straight, nearly as long as corolla. *Capsule sub-globose, not much longer than calyx. Seeds smooth, winged.* Fl. 6–9. Self-incompatible. $2n = 12$. Th.

?Native. In sandy places near Par, Cornwall; naturalized in waste places round Plymouth (Devon), elsewhere in Cornwall, and in Carmarthen. France, Spain, Portugal, N.W. Italy; Morocco.

*6. L. arenaria DC.   Sandy Toadflax.

A *viscid-pubescent annual* 5–15 cm with a bushy habit. *Lvs* 3–10 mm, *oblanceolate-elliptical*, attenuate at base into a short petiole. Fls few in a short, ultimately lax, raceme. Pedicels shorter than calyx. Sepals linear-oblanceolate. *Corolla* 4–6 mm, *yellow*; spur rather shorter than corolla, slender, often violet. *Capsule obovoid, about as long as calyx.* Seeds black, smooth, narrowly winged. Fl. 5–9. Th.

Introduced. Naturalized on dunes at Braunton Burrows (N. Devon). Native of W. and N.W. France from Dunkirk to the Gironde.

5 CHAENORHINUM (DC.) Reichenb.

Differs from *Linaria* in having *axillary fls, the mouth of the corolla somewhat open* and the unequal cells of the *capsule, which open by pores.*

About 20 spp., all except the following confined to the Mediterranean region.

1. C. minus (L.) Lange   Small Toadflax.

Linaria minor (L.) Desf.

An erect annual 8–25 cm, usually glandular-pubescent, rarely glabrous; branches ascending. Lvs 1–2.5 cm, alternate, linear to oblong-lanceolate, entire, obtuse, narrowed at base into a short petiole. Fls solitary in the axils of the rather smaller upper lvs. Pedicels 3–20 mm in fr. Sepals ±linear, obtuse. Corolla 6–9 mm, purple outside, paler within; spur short, obtuse. Capsule ovoid, shorter than calyx. Seeds ovoid, ridged longitudinally. Fl. 5–10. Probably self-pollinated. $2n = 14$. Th.

Native. In arable land and waste places, particularly common along railways. Common in S. England becoming rarer northwards to Kincardine and Argyll, and in Banff to Easterness; widespread, but local, in Ireland. Europe (except Mediterranean islands, Russia, Iceland and Faeroes); W. Asia to the Punjab.

*C. origanifolium (L.) Fourr.

Biennial or perennial with ±numerous ascending stems. Lvs thick; lower ovate, 1 cm or less, close together; upper narrower. Fls bright violet, 8–15 mm. Usually casual, but established on walls in W. Kent for over 100 yrs. Native of S. W. Europe.

6. KICKXIA Dumort.   Fluellen.

Annual herbs with mostly alternate *pinnately veined petiolate lvs. Fls axillary*, the bracts like the lvs but rather smaller, otherwise as in *Linaria. Capsule opening by pores with deciduous lids.*

About 25 spp. in Europe, Africa and W. Asia to India.

Lvs ovate or orbicular; pedicels villous.	**1. spuria**
Lvs hastate; pedicels glabrous.	**2. elatine**

1. K. spuria (L.) Dumort.   Round-leaved Fluellen.

Linaria spuria (L.) Miller

A hairy and glandular annual, decumbent and branched from base; stems 20–50 cm. *Lvs* shortly petiolate, *ovate to suborbicular*, entire or the lower slightly toothed, *rounded or subcordate at base*, the lowest stem-lvs to 6 cm, the upper less than 1 cm. *Pedicels* 1–2(–2.5) cm, *villous.* Sepals ovate, acute. Corolla 8–11 mm, yellow with deep purple upper lip; spur curved, about as long as the corolla. Capsule globose, glabrous, shorter than calyx. Seeds pitted. Fl. 7–10. Self-compatible, cleistogamous fls occur. Th.

Native. Rather local in arable land, usually cornfields, on light soils. S. England and Wales northwards to Glamorgan, Hereford and Lincoln, casual northwards to Durham. S., W. and C. Europe; N. Africa; Macaronesia.

2. K. elatine (L.) Dumort.   Sharp-leaved Fluellen.

Linaria elatine (L.) Miller

Differs from *K. spuria* as follows: Usually more slender, less hairy and scarcely glandular. *Upper and middle lvs hastate. Pedicels glabrous.* Sepals narrower. Corolla with upper lip paler purple; spur straight. Fl. 7–10. Probably self-incompatible. $2n = 18, 36$. Th.

Native. Rather local in the same habitat as *K. spuria* but extending northwards to Denbigh, Cheshire and Lincoln, casual northwards to Durham and Cumberland; S. and W. Ireland, from Galway to Wexford, probably now present and native only in Cork. S., W. and C. Europe; Macaronesia; naturalized in North America.

7. CYMBALARIA Hill

Creeping perennial herbs with *palmately veined* stalked lvs which are alternate above, opposite below. *Fls axillary* (bracts not differentiated), otherwise as in *Linaria. Capsule opening by 2 lateral pores, each pore with 3 valves.*

About 15 spp. in the Mediterranean region and W. Europe.

*1. C. muralis Gaertner, Meyer & Scherb.

        Ivy-leaved Toadflax..

Linaria cymbalaria (L.) Miller

A glabrous perennial with trailing or drooping, rooting,

often purplish stems 10–80 cm. *Lvs* 2.5 cm, nearly all alternate, (3–)5(–7)-*lobed*, thick, sometimes purplish beneath; petiole longer than blade. Pedicels recurved, rather long (c. 2 cm). Sepals linear-lanceolate, about half the length of corolla-tube. Corolla 8–10 mm, lilac to violet (rarely white) with a yellowish or white palate with yellow spot at the mouth and darker lines on the upper lip; spur curved, about ⅓ as long as corolla. Capsule globose. Seeds ovoid, with thick flexuous ridges. Fl. 5–9. Pollinated by bees, self-compatible. $2n = 14$. Chh.

Introduced. First recorded in 1640, now common on old walls, rarely on rocks, almost throughout the British Is. but very local between Argyll, Kincardine and Caithness. Native in S. Alps, W. Yugoslavia, C. and S. Italy and Sicily; naturalized elsewhere.

*C. pallida (Ten.) Wettst.

Linaria pallida (Ten.) Guss.

Differs from *C. muralis* in being pubescent. Lvs mostly opposite; lobes more obtuse. Corolla larger, 10–15 mm; spur relatively longer (6–9 mm). $2n = 14$. Naturalized near Loweswater (Cumbria), apparently extinct at Bardsea. Native of C. Italy.

Tribe 2. CHELONEAE. Lvs usually opposite. Fls in cymes, often forming a terminal panicle. Corolla with well-developed tube, not saccate or spurred at base, the two upper lobes outside the lateral ones in bud. Stamens usually 4.

8. SCROPHULARIA L.

Herbs with square stems and opposite lvs. *Fls in cymes* in the axils of the upper lvs or of bracts and then forming a terminal panicle. Calyx 5-lobed. *Corolla* usually dingy in colour, *with a nearly globose tube and 5 small lobes*, the two upper united at the base. *Fertile stamens* 4, bent downwards, the fifth usually represented by a scale-like staminode inserted at the base of the upper lip of the corolla, sometimes 0. Stigma capitate. *Capsule septicidal*. Seeds small, ovoid, rugose.

About 300 spp. in temperate Eurasia; 12 in North and tropical America. All spp. protogynous, mostly pollinated by wasps.

1 Sepals obtuse, with a scarious border; fls brownish
   or purplish; staminode present.   2
   Sepals acute, without a scarious border; fls yellowish;
   staminode 0.   **5. vernalis**
2 Lvs and stems glabrous or nearly so.   3
   Lvs (on both surfaces) and stems downy.
                         4. scorodonia
3 Stems 4-angled, not winged; sepals with narrow scarious border.   **1. nodosa**
   Stems 4-winged; sepals with broad (0.5–1 mm) scarious border.   4
4 Lvs crenate, often cordate at base; staminode orbicular or reniform.   **2. auriculata**
   Lvs serrate, not cordate at base; staminode ±2-lobed with diverging lobes.   **3. umbrosa**

Section 1. *Scrophularia*. Staminode present, scale-like.

1. S. nodosa L.   Common Figwort.

A perennial 40–80 cm, *glabrous* except for the glandular infl. (rhachis and pedicels); rhizome short, swollen and nodular. *Stem sharply quadrangular but not winged. Lvs* 6–13 cm, ±ovate, *acute*, coarsely and unequally *serrate*, ±truncate at base but usually slightly and often unequally decurrent on the *petiole, which is not winged*. Infl. a panicle made up of cymes borne in the axils of the bracts. Lowest pair of bracts like the lvs; 1 or 2 more pairs often slightly lf-like; upper small, linear, alternate (rarely all lf-like). Pedicels 2 or 3 times as long as the fl. *Calyx-lobes* ovate, obtuse, *with very narrow*, often scarcely visible, *scarious border*. Corolla 7–10 mm, with greenish tube and purplish-brown upper lip (rarely wholly green). *Staminode* obovate, broader than long, *retuse*. Capsule 6–10 mm, ovoid, acuminate. Fl. 6–9. Pollinated by wasps, less frequently by bees. $2n = 36$. Hp.

Native. Common in damp and wet woods and hedgebanks throughout the British Is., except Shetland; ascending to 460 m in Yorks. Most of Europe southwards from 69° 48′ N in Norway; temperate Asia, eastwards to the Yenisei region.

2. S. auriculata L.   Water Figwort.

A perennial 50–100 cm, *glabrous* except for the somewhat glandular infl. and the sometimes puberulent lvs. Rhizome not nodular. *Stem 4-winged. Lvs* 6–12 cm, ovate to elliptical, *obtuse*, subcordate to broadly cuneate at base, *crenate*, often with 1 or 2 small basal lobes; *petiole winged*. Infl. a panicle, made up of cymes in the axils of bracts. Upper bracts oblong, the lower somewhat lf-like but all differing markedly from the lvs. Pedicels about as long as the fls. *Calyx-lobes* rounded *with broad* (0.5–1 mm) *scarious border*. Corolla 5–9 mm, greenish, with brownish-purple upper lip. *Staminode suborbicular* or rather broader than long, *entire*. Capsule 4–6 mm, subglobose, apiculate. Fl. 6–9. Pollinated by wasps, less frequently by other insects. $2n = 80^*$. Hp. or Hs.

Native. Common on the edges of ponds, streams, and in wet woods and meadows in the south of England extending north to Ayr and Fife, and in Ross, but rare in Scotland and W. Wales; over the whole of Ireland, but rather local in the centre and north. W. Europe, northwards to the Netherlands; Italy, Sicily and Crete; Morocco and Tunisia; Azores.

3. S. umbrosa Dumort.   Green Figwort.

S. ehrhartii Stevens; *S. alata* Gilib.

A *glabrous* perennial 40–100 cm. Differs from *S. aquatica* as follows: Stem more broadly winged. *Lvs serrate, acute or subobtuse*, never cordate at base, without basal lobes. Cymes laxer. Bracts larger, lf-like. *Staminode with two diverging lobes*, so that it is much broader than long. Fl. 7–9. $2n = 52$. Hp.

Native. Rare, in damp shady places in England and S. Scotland, absent west of Wilts, Gloucester and Montgomery and north of Renfrew and Moray; Limerick, Dublin, Kildare, Fermanagh, Derry. Europe, from Denmark and Latvia southwards, but absent from much of the west; Asia, east to Tibet and south to Palestine.

4. S. scorodonia L. — Balm-leaved Figwort.

A perennial 60–100 cm, *whole plant greyish pubescent. Stem quadrangular, not winged.* Lvs 4–10 cm, ovate, petiolate, cordate at base, obtuse to acute, *doubly dentate, with mucronate teeth, rugose; petiole not winged.* Infl. a panicle of lax few-fld cymes in the axils of ±lf-like bracts. Pedicels 2 or 3 times as long as the fls, arcuate, divaricate. *Sepals* rounded, *with a broad scarious border.* Corolla 8–11 mm, dull purple. *Staminode suborbicular, entire.* Capsule 6–8 mm, subglobose or ovoid, apiculate. Fl. 6–8. Hp.

Native. Hedgebanks, etc., in Cornwall, Devon, Guernsey, Jersey and the Scilly Is., locally frequent; naturalized in Glamorgan. W. France, Spain, Portugal, Madeira, Azores; N.W. Morocco.

S. canina L. Differs from all the above in its pinnatifid lvs. An infrequent casual. $2n = 26$. Native of S. and S.C. Europe.

Section 2. *Venilia* G. Don. Staminode 0.

*5. S. vernalis L. — Yellow Figwort.

A biennial or perennial 30–80 cm, *softly glandular-hairy.* Stem obscurely quadrangular. Lvs 4–15 cm, thin, broadly ovate, cordate, petiolate, acute, deeply toothed. *Fls in cymes in the axils of the upper lvs.* Cymes long-pedunculate, many-fld and compact, with lf-like bracts. Pedicels shorter than calyx. Sepals oblong-lanceolate, *subacute, without a scarious border.* Corolla 6–8 mm, *greenish-yellow, not 2-lipped, contracted at the mouth,* the lobes very small, nearly equal. *Staminode* 0. Stamens finally protruding. Capsule 8–10 mm, ovoid-conical. Fl. 4–6. Visited by bees. $2n = 40$. Hp.

Introduced. In plantations and waste places, usually in shade. Very local but found in a number of places in Great Britain, mainly in the south-east but extending north to Aberdeen and west to Cornwall and Cardigan. Native of C. and S. Europe in the mountains from S.C. Germany and C. Russia southwards to the Pyrenees, Sicily and C. Yugoslavia; naturalized in the plains.

Tribe 3. GRATIOLEAE. Lvs opposite, at least below. Fls solitary or in spikes or racemes. Corolla not saccate or spurred at base, the two upper lobes outside the lateral in bud. Stamens 4 or 2.

9. MIMULUS L.

Herbs with opposite lvs. Fls on peduncles from the axils of the lvs or bracts. Calyx tubular, 5-angled and 5-toothed. *Corolla with a long tube, hairy in the throat, 2-lipped;* upper lip 2-lobed; lower longer, 3-lobed, the lobes all flat. Stamens 4; anthers 2-celled. Stigma with 2 flat lobes. *Capsule included in the calyx, loculicidal.* Seeds small, numerous.

About 100 spp., mostly in temperate America (mainly California), a few in the Old World in E. Asia, Australia, New Zealand and S. Africa.

1 Plant glabrous or pubescent only above; calyx-teeth conspicuously unequal; corolla c. 4×3 cm. — 2
  Plant viscid-hairy all over; calyx-teeth ±equal; corolla c. 2×1 cm. — **3. moschatus**
2 Calyx and pedicels pubescent; pedicels mostly 1.5–3 cm; corolla markedly 2-lipped, the throat nearly closed. — **1. guttatus**
  Calyx and pedicels glabrous; pedicels mostly 3.5 cm or more; corolla only slightly 2-lipped, the throat wide open. — **2. luteus**

*1. M. guttatus DC. — Monkeyflower.

M. luteus auct. angl, pro parte; *M. langsdorffii* Donn ex Greene

A perennial (5–)20–50 cm, *glabrous below, ±pubescent* (often glandular-pubescent) *above* at least on the pedicels or calyx (in America sometimes entirely glabrous). Fl. stems ascending or decumbent. Lvs 1–7 cm, irregularly dentate; lower petiolate, ovate or oblong; upper sessile, ovate or orbicular-ovate. Infl. on well-developed plants many-fld, the upper bracts much smaller than the lvs but the lower passing into the lvs, but sometimes few fld and with the bracts more like the lvs. *Pedicels* (1–)1.5–3 (the lower sometimes up to 5.5) cm. Calyx becoming inflated in fr.; *teeth ±deltate, the upper much longer than the others. Corolla* 2.5–4.5 × c. 3 cm, *yellow usually with small red spots in the throat, markedly 2-lipped with the lower lip much longer than the upper and with a prominent palate nearly closing the throat.* Capsule oblong, obtuse; seeds finely striate. Fl. 7–9. Pollinated by bees, homogamous. $2n = 28, 48*$. Hp., Hs. or Hel.

Introduced. First recorded in 1830, now rather common on the banks of streams, etc., through nearly the whole of the British Is., local or absent in parts of N. Scotland, East Anglia and the E. Midlands, and much of C. Ireland. Native of western North America (Alaska and Montana to N.W. Mexico). Cultivated for ornament and naturalized in much of Europe and in the eastern USA.

*2. M. luteus L. — Blood-drop-emlets.

Differs from *M. guttatus* as follows: *Glabrous all over* (except inside the calyx and corolla). Fl. stems usually decumbent. Lvs narrower, more acuminate, usually with fewer and more regular teeth. Fls always few (up to c. 10). *Pedicels* (3–)3.5–6(–10) cm. *Corolla yellow with small red spots in the throat and usually with large red spots on the lobes or with the lobes ±variegated with pinkish-purple* (sometimes, but apparently not in Britain, coloured as *M. guttatus); lower lip little longer than the upper; throat open.* Fl. 6–9. Hp. or Hel.

In similar places to *M. guttatus* but much less common; mainly in C. and S. Scotland but extending locally to Shetland, and southwards to Cornwall, in the west except for Norfolk, Hertford and Kent. Native of Chile;

perhaps naturalized also in C. Europe.

M. guttatus × luteus, intermediate between the parents in lf- and corolla-shapes and with very variable colour patterns on the corolla, is a highly sterile hybrid which is widespread and locally abundant by rivers and streams in N. and W. Britain. It is much the commonest taxon at higher elevations and is often mistaken for one or other of the parents.

M. guttatus × M. variegatus Lodd., which is difficult to separate from the above, has the lvs as long as wide, and a very large corolla with lobes heavily blotched with orange, red or purple. It is cultivated as a garden-ornamental in Scotland and is locally naturalized from Dumbarton northwards. *M. variegatus* is a native of Chile.

M. × burnetii S. Arnott (*M. cupreus* Dombrain × *guttatus*), a usually sterile hybrid distinguished by its predominantly copper-coloured corolla, which is sometimes double, is a garden-escape which is locally established in N. England and Scotland. *M. cupreus* is a native of Chile.

M. cupreus × luteus, with a usually yellow corolla having confluent small blotches of coppery orange, is a garden hybrid very locally established from Peebles northwards.

*3. M. moschatus Douglas ex Lindley   Musk.

A viscid-hairy perennial with decumbent stems 10–40 cm. Lvs 1–4 cm, all alike, shortly petiolate, ovate or elliptical with small distant teeth. Pedicels equalling or shorter than lvs. *Calyx-teeth triangular-lanceolate, subequal. Corolla* 1–2 cm, yellow, not blotched. Capsule acute. Fl. 7–8. Hp.

Introduced. Commonly cultivated in gardens and casual or occasionally naturalized in wet places in Great Britain and in Ireland (Wicklow, Down, Tyrone). Native of North America (British Columbia to Montana and California). Naturalized also in C. Europe and eastern USA. Formerly much cultivated for the musky scent of all parts of the plant. All the plants in this country today, however, appear to be scentless.

10. LIMOSELLA L.

Annual herbs, creeping by runners, *the lvs all basal.* Fls small, axillary, solitary. Calyx 5-toothed. *Corolla rotate, nearly regular, 4–5-lobed, with short tube.* Stamens 4, with 1-*celled anthers. Capsule septicidal*, the septum incomplete.

Fifteen spp., almost cosmopolitan.

Lvs of mature plants elliptical; calyx longer than corolla-
tube.   **1. aquatica**
Lvs all linear-subulate; calyx shorter than corolla-tube.
                2. australis

1. L. aquatica L.   Mudwort.

A glabrous annual, creeping by runners which are at first upright then become horizontal and produce fresh rosettes at the nodes. *Upper lvs* 5–15 mm, *elliptical, with petiole several times as long as blade*, dark green, the

lower lanceolate-spathulate, the lowest subulate. *Calyx longer than corolla-tube.* Corolla 2–5 mm diam., white or lavender sometimes splashed purple on the back; *tube* 1.5 mm, *campanulate*; lobes triangular, acute, with few long hairs. Style short; stigma medium-sized. Capsule globose-ovoid to ellipsoid. Fl 6–10. $2n = 40^*$. Th.

Native. In wet mud at the edges of pools or where water has stood. Local in England and Wales, from Glamorgan, Hants and Surrey to Yorks and Cumberland, ascending to 460 m in Yorks; apparently now extinct in S. Scotland; very local in Ireland, only known from the limestone in Clare and S. Galway, and in Fermanagh and Cork. Most of Europe except the Mediterranean region, but often rare and local; Egypt; arctic and temperate Asia to Japan and India; Greenland; North America from Labrador to North-West Territory and south in the mountains to Colorado and California.

2. L. australis R. Br.   Welsh Mudwort.

L. subulata Ives, *L. aquatica* var. *tenuifolia* auct.

Runners arching, often below the soil surface, occasionally perennating. *Lvs* 1–2.5 cm, *all linear-subulate*, light green. *Calyx* c. $\frac{2}{3}$ *as long as corolla-tube.* Corolla up to 4 mm diam., white with orange tube; *tube* c. 3 mm, *contracted below the insertion of the stamens*; lobes ovate-ligulate with numerous short hairs. Style long; stigma small. Fr. nearly globose. Fl. 6–10. $2n = 20^*$. Th.

Native. In similar habitats to *L. aquatica*, very rare. Glamorgan (3 stations), near the River Glaslyn in Caernarvon and Merioneth. Eastern North America from Labrador to Maryland.

L. aquatica × australis, hybrids vegetatively more vigorous and producing more fls than either parent, are intermediate in lf-shape and in many flower characters, apparently sterile and regularly surviving the winter better than the parents. $2n = 30^*$. Only known from Morfa Pools, Glamorgan, where it is often abundant.

Tribe 4. DIGITALEAE. Corolla not saccate or spurred, its lobes ±spreading, the two upper inside the lateral ones in bud. Stamens 2–4(–8).

11. SIBTHORPIA L.

Creeping herbs with reniform basal and alternate lvs. Fls small, axillary, solitary. Calyx 4–8-lobed. *Corolla* 5–8-fid, nearly regular, *rotate; tube very short. Stamens as many as or one fewer than corolla lobes*, with 2-celled anthers. Capsule loculicidal.

About 5 spp. in S. and W. Europe, Azores, Madeira, mountains of tropical Africa, Central America and the Andes of South America.

1. S. europaea L.   Cornish Moneywort.

A creeping hairy perennial, rooting at the nodes. Lvs 0.5–2 cm diam., in clusters at the nodes, alternate from the upper part of the filiform stem, reniform, long-stalked, crenately 5–7-lobed. Calyx 4- or 5-lobed. Corolla 1–2.5 mm diam., 5-lobed, the two upper lobes whitish or cream, the three lower broader, pinkish. Stamens 4. Fl. 7–10. Chh.

Native. Very local in moist shady places. Sussex, Somerset to Cornwall, S. Wales (Glamorgan, Carmarthen and Cardigan); Lewis, probably introduced; Dingle Peninsula (Kerry) ascending to 520 m; Channel Islands. W. France, W. Spain, Portugal and the mountains of Greece and Crete.

12. ERINUS L.

Low tufted perennial with alternate lvs. Fls in terminal bracteate racemes. Calyx deeply 5-lobed. *Corolla with a slender narrow cylindrical tube* about as long as the calyx; lobes 5, spreading, nearly equal, deeply emarginate. *Stamens* 4. Capsule loculicidal.

One sp.

*1. E. alpinus L.   Fairy Foxglove.

A tufted perennial 5–15 cm; stems numerous, simple, hairy, ascending. Lvs c. 1.5 cm, oblanceolate to spathulate, attenuate into a petiole, crenate or dentate, green, glabrous or hairy, the lower forming a basal rosette. Raceme corymbiform. Bracts entire. Calyx-lobes linearoblong. Corolla purple (rarely white). Capsule ovoid, shorter than calyx. Fl. 5–10. Pollinated probably mainly by Lepidoptera, self-pollination possible. Chh.

Introduced. Naturalized on walls and in rocky woods in a few places scattered from Devon and Kent to Aberdeen and Sutherland, and from Clare and Wexford to Galway and Derry. Mountains of S. and S.C. Europe, from N. and E. Spain to C. Italy and W. Austria. Algeria and Morocco.

13. DIGITALIS L.

Tall biennial or perennial herbs with alternate lvs, the lowest forming a rosette. Fls nodding, in terminal, unilateral, bracteate racemes. Calyx deeply 5-lobed. *Corolla with a long campanulate tube,* constricted near the base, shortly 5-lobed, the two upper lobes forming an emarginate upper lip, the lowest usually longer and more prominent. *Stamens* 4. Capsule septicidal, with many small seeds.

About 20 spp. in Europe and the Mediterranean region to C. Asia.

Several spp. besides our own are cultivated and *D. lutea* L., with a yellow corolla to 2 cm., and *D. grandiflora* Miller, with yellow corolla 3–4 cm, have been recorded as escapes.

*D. lanata Ehrh.

Biennial or perennial. Lvs glabrous or ciliate, ±lanceolate. Infl. woolly. Corolla 20–25 mm, yellowish-white with purplish network; lowest lobe very long, whitish. Perhaps sometimes established outside gardens. Native of S.E. Europe.

1. D. purpurea L.   Foxglove.

An erect biennial, rarely perennial, 50–150 cm. Stem usually simple, greyish tomentose or the lower part glabrous. Lvs 15–30 cm, ovate to lanceolate, crenate, green and softly pubescent above, grey-tomentose beneath attenuate at the base into a winged petiole. Raceme

20–80-fld. Bracts lanceolate, sessile, entire, decreasing in size upwards. Pedicels tomentose, longer than calyx. Lower sepals ovate, the upper lanceolate, all acute. Corolla 4–5 cm, 3 or 4 times as long as calyx, pinkish-purple with deeper purple spots on a white ground inside the lower part of the tube, rarely white and spotted or unspotted, shortly ciliate and with a few long hairs within. Stamens and filiform style included. Capsule ovoid, rather longer than calyx. Fl. 6–9. Pollinated by humble-bees, protandrous. $2n = 56^*$. Hs.

Native. Common in open places in woods, etc., and on heaths and mountain rocks, ascending to 885 m. On acid soils throughout the British Is. (not Shetland), often becoming dominant in clearings and burnt areas in woods on light dry soils. W., S.W. and W.C. Europe, extensively cultivated for ornament and as a medicinal plant elsewhere; Morocco. The drug digitalin, obtained from this plant, is still extensively used for heart complaints.

14. VERONICA L.

Annual or perennial herbs or low shrubs with opposite lvs. Fls blue, rarely white or pinkish, in axillary or terminal racemes or solitary in the axils of lvs similar to the stem-lvs but alternate. Calyx with 4 lobes, the upper being absent (or rarely 5-lobed, the upper much smaller). *Corolla with a very short tube and rotate 4-cleft limb,* the upper lobe the largest (representing 2 lobes of the 5-lobed genera), the lower the smallest. *Stamens* 2. Capsule ±laterally compressed. Seeds few.

About 300 spp. in temperate regions of both hemispheres, but most in the north. A number of spp. are cultivated in gardens.

1	Fls in axillary racemes.	2
	Fls in terminal racemes or solitary.	8
2	Plant glabrous or, if hairy, lvs linear-lanceolate; growing in wet places.	3
	Plant ±hairy, lvs ovate or oblong; growing in drier places.	6
3	Racemes alternate (i.e. from one only of a pair of lvs).	**4. scutellata**
	Racemes opposite.	4
4	Stems decumbent; lvs all petiolate.	**1. beccabunga**
	Stems ±erect, at least upper lvs sessile.	5
5	Corolla pale blue; pedicels ascending after flowering; capsule not wider than long.	**2. anagallis-aquatica**
	Corolla pinkish; pedicels spreading after flowering; capsule usually much wider than long.	**3. catenata**
6	Pedicels 2 mm or less, shorter than bract and calyx; racemes rather dense, pyramidal.	**5. officinalis**
	Pedicels 4 mm or more, as long as or longer than bract and calyx; raceme lax.	7
7	Stem hairy all round; capsule c. 8 mm wide, longer than calyx; lvs with petiole 5–15 mm.	**6. montana**
	Stem hairy on two opposite sides only; capsule not more than 6 mm wide, shorter than calyx; lvs sessile or with petiole less than 5 mm.	**7. chamaedrys**
8	Fls in dense many-fld terminal racemes; corolla-tube longer than wide.	**8. spicata**
	Fls in lax racemes, or few-fld head-like racemes, or solitary; corolla-tube very short, much wider than	

long.

9 Fls in racemes, the bracts often passing gradually into the lvs but the upper, at least, very different from them.   *10*

Fls solitary in the axils of lvs resembling the cauline lvs, though the upper sometimes rather smaller.   *19*

10 Lvs glabrous or finely puberulent, entire or obscurely crenulate; perennial (except *peregrina*).   *11*

Lvs conspicuously pubescent, often glandular, dentate or lobed (but only obscurely crenulate in *acinifolia*); annual.   *15*

11 Perennial; bracts (at least the upper) shorter or scarcely longer than fls.   *12*

Annual; bracts all much longer than fls. **13. peregrina**

12 Shrubby at base; corolla bright blue with reddish centre, c. 1 cm diam.   **9. fruticans**

Herbs; corolla white, pale or dull blue or pink, smaller.   *13*

13 Pedicels much longer than bracts; fls pink. **10. repens**

Pedicels shorter than bracts; fls not pink.   *14*

14 Pedicels longer than calyx; fr. wider than long, scarcely exceeding calyx; fls white or pale blue.   **12. serpyllifolia**

Pedicels shorter than calyx; fr. longer than wide; much exceeding calyx; fls dull blue.   **11. alpina**

15 Pedicels much shorter than calyx.   *16*

Pedicels longer than calyx.   *17*

16 Upper cauline lvs toothed.   **14. arvensis**

Upper cauline lvs pinnatifid, the lobes longer than the entire portion.   **15. verna**

17 Lower bracts and upper cauline lvs ±toothed or nearly entire; fr. longer than calyx.   *18*

Bracts and lvs deeply digitately 3–7-lobed; fr. shorter than calyx.   **18. triphyllos**

18 Lvs obscurely crenate; pedicels 2–3 times as long as calyx; fr. wider than long.   **16. acinifolia**

Lvs conspicuously dentate; pedicels not twice as long as calyx; fr. longer than wide.   **17. praecox**

19 Lvs with 5–7 large teeth or small lobes near base; sepals cordate at base.   **19. hederifolia**

Lvs regularly crenate-serrate; sepals narrowed at base.   *20*

20 Decumbent annuals; lvs not reniform; pedicels not twice as long as lvs.   *21*

Creeping perennial often forming mats; lvs reniform; pedicels several times as long as lvs.   **23. filiformis**

21 Lobes of fr. divergent; fls 8–12 mm diam. **20. persica**

Lobes of fr. not divergent; fls 4–8 mm diam.   *22*

22 Sepals ovate, acute or subacute; fr. with short crisped eglandular hairs and a few longer glandular hairs; corolla usually uniformly blue.   **21. polita**

Sepals oblong, obtuse or subobtuse; fr. with sparse, long glandular hairs only; corolla usually whitish.   **22. agrestis**

Section 1. *Beccabunga* (Hill) Dumort. Fls in opposite axillary racemes. Capsule loculicidal. Normally perennial.

1. V. beccabunga L.   Brooklime.

A glabrous perennial 20–60 cm; stems creeping and rooting at base, then ascending, fleshy. *Lvs 3–6 cm*, rather thick and fleshy, *ovate or oblong*, *obtuse*, the base rounded, shallowly crenate-serrate, *shortly petiolate*. *Racemes opposite*, rather lax, 10–30-fld. Bracts linear-lanceolate, ±equalling the slender pedicels.

Calyx-lobes narrowly ovate, acute. Corolla 5–7 mm diam., blue. Capsule subglobose, retuse, shorter than calyx. Fl. 5–9. Pollinated by various Diptera and Hymenoptera, protogynous, often selfed. $2n = 18$. Hel. or Hp.

Native. *In streams, ponds, marshes and wet places in meadows*. Common throughout the British Is. but local in N.W. Scotland and not in Shetland. Most of Europe southwards from 65° N.; N. Africa; temperate Asia extending to Japan and the Himalaya.

2. V. anagallis-aquatica L.   Blue Water-speedwell.

A perennial or sometimes annual herb, *glabrous* except for the sometimes glandular infl. Stems 20–30 cm high, shortly creeping and rooting at base, then ascending, fleshy, usually branched, green. *Lvs 5–12 cm*, the lower ovate, subentire, often petiolate, the upper *ovate-lanceolate or lanceolate*, *acute*, *±amplexicaul*, remotely serrulate. *Racemes opposite*, rather lax, 10–50-fld, ascending. *Bracts linear*, *shorter than or equalling pedicels at flowering*. Calyx-lobes ovate-lanceolate, acute. *Corolla 5–10 mm diam.*, *pale blue with violet veins. Pedicels ascending after flowering*. Capsule ±orbicular, slightly emarginate, usually slightly longer than broad. Fl. 6–8. Visited by Diptera, easily self-pollinated. $2n = 36$. Hp. or Hel.

Native. In ponds, streams, wet meadows and wet mud, rather common. Throughout the British Is. but absent from much of S.W. England, Wales and N. Scotland. All Europe except the extreme north; Asia to Japan and the Himalaya; N.E. and S. Africa; North and South America; New Zealand.

3. V. catenata Pennell   Pink Water-speedwell.

V. aquatica Bernh., non S. F. Gray; *V. comosa* auct.

Differs from *V. anagallis-aquatica* as follows: Stem usually purplish-tinged. Lvs linear to linear-lanceolate. Racemes laxer, more spreading. *Bracts wider*, c. 1.5 mm, usually lanceolate, less acute, *longer than pedicels at flowering*. Calyx-lobes elliptical or oblong, widely spreading after flowering. *Corolla usually pink with darker lines. Pedicels spreading at right angles after flowering*. Capsule more definitely emarginate, usually wider than long. Fl. 6–8. $2n = 36$. Hp. or Hel.

Native. In similar places to the last, with which it often grows. E., C. and S. England, uncommon in the southwest, the north and in Wales, rare in Scotland (Berwick, Inner Hebrides, Aberdeen); scattered throughout C. and C.S. Ireland, isolated localities near the north coast. Most of Europe from c. 58° N southwards.

V. × lackschewitzii Keller (*V. anagallis-aquatica* × *catenata*) is most readily recognized by its usual sterility and its robust and vegetatively vigorous habit. It is common in England and Wales from Wilts and Surrey northwards to Northumberland and the Isle of Man, but very rare in Scotland; in Ireland recorded from Kerry and Waterford.

Section 2. *Veronica*. Fls usually in alternate axillary

racemes. Capsule septicidal. Perennial.

4. V. scutellata L.   Marsh Speedwell.

A *glabrous perennial*, sometimes glandular or (var. *villosa* Schumach.) densely pubescent. Stems creeping below and then ascending, simple or slightly branched, 10–15 cm high. *Lvs 2–4 cm, linear-lanceolate or -oblong*, acute, semi-amplexicaul, entire or remotely denticulate, yellowish-green often tinged with purple; midrib impressed and conspicuous above. Fls in very lax, slender, few (up to 10)-fld *alternate racemes*. Bracts linear. Pedicels 7–10 mm, slender, more than twice as long as the bracts, spreading at right angles in fr. Sepals ovate. Corolla 5–6 mm diam., white or pale blue with purple lines, sometimes pinkish. *Capsule flat*, wider than long, deeply emarginate, much longer than calyx. Fl. 6–8. $2n = 18$. Hp. or Hel.

Native. In ponds, bogs, wet meadows, etc., often on acid soils. Rather common throughout the British Is.; var. *villosa* less common, often in rather drier habitats. Most of Europe but rare in the Mediterranean region; N. Asia, east to Kamchatka.

5. V. officinalis L.   Heath Speedwell.

A perennial herb with *stems* 10–40 cm, creeping and rooting and often forming large mats, ascending above, *hairy all round*. Lvs 2–3 cm, ovate to elliptical, crenate, subacute, cuneate at base, subsessile, hairy on both sides. *Racemes* from the axils of one or, more rarely, both of a pair of lvs, long-stalked, *rather dense and pyramidal*, 15–25-fld. *Bracts* linear, *about twice as long as pedicels* which are 1–2 mm. Calyx-lobes lanceolate. *Corolla* c. 8 mm diam., *lilac*. Filaments, anthers and style ±lilac. Capsule obovate or obcordate, longer than calyx. Fl. 5–8. Pollinated by various Diptera and Hymenoptera, homogamous, protandrous or protogynous. $2n = 18$ (rare), 36. Chh.

Native. In grassland, heaths and open woods, often on dry soils. Common throughout the British Is. Almost throughout Europe; Asia Minor and Caucasus; Azores; eastern North America (probably introduced).

6. V. montana L.   Wood Speedwell.

A perennial herb; *stems* 20–40 cm, *hairy all round*, creeping and rooting, the ends of the branches ascending. Lvs 2–3 cm, ovate or orbicular-ovate, subobtuse, truncate or broadly cuneate at base, coarsely crenate-serrate, light green, hairy on both sides; *petiole* 5–15 mm. Racemes from the axil of one or both of a pair of lvs, long-stalked, *lax*, 2–5-fld. *Bracts* small, linear, *much shorter than pedicels* which are 4–8 mm. Calyx-lobes obovate. *Corolla* 8–10 mm diam., *lilac-blue*. Filaments and anthers paler. *Capsule* nearly flat, ±orbicular, retuse or emarginate, *longer than calyx*, glandular-ciliate. Fl. 4–7. Pollinated by various Diptera and Hymenoptera. $2n = 18*$. Chh.

Native. In damp woods. Throughout England and Wales, but rather local, less common in Scotland and absent from much of the north; scattered throughout

Ireland. W., C. and S. Europe northwards to N. Denmark and eastwards to Latvia and W. Ukraine; Caucasus; mountains of Algeria and Tunisia.

7. V. chamaedrys L.   Germander Speedwell.

A perennial herb 20–40 cm; *stems* prostrate and rooting at the nodes, ascending above, *with* long white *hairs in 2 lines on opposite sides*, glabrous between. *Lvs* 1–2.5 cm, triangular-ovate, *sessile or shortly stalked* (petiole to 5 mm), subobtuse, subcordate to cordate at base, coarsely crenate-serrate, dull green, hairy, especially on margins and veins beneath. *Racemes* from the axils of one or, more rarely, both of a pair of lvs, long-stalked, lax, 10–20-*fld. Bracts* lanceolate, *about as long as or shorter than pedicel*. Pedicels 4–8 mm. Calyx-lobes ±lanceolate, hairy. *Corolla* c. 10 mm diam., deep *bright blue* with white eye. Filaments and style blue; anthers pale. *Capsule obcordate, shorter than calyx*, wider than long, ciliate and pubescent. Fl. 3–7. Pollinated by various Diptera and Hymenoptera, homogamous. $2n = 32$. Chh.

Native. In grassland, woods, hedges, etc. Very common throughout the British Is. Throughout Europe except for some islands and much of the Arctic; N. and W. Asia; naturalized in North America.

Section 3. *Pseudolysimachium* Koch. Fls in terminal racemes. Corolla-tube longer than wide. Capsule septicidal. Perennial.

8. V. spicata L.   Spiked Speedwell.

A pubescent perennial 8–60 cm, with shortly creeping, somewhat woody rhizome and erect flowering stems. Lowest lvs ovate, petiolate, passing gradually into the lanceolate or linear, sessile upper ones, all ±crenate or crenate-serrate or the upper entire. *Fls in a many-fld terminal spike-like raceme*. Pedicels usually less than 1 mm. Bracts and calyx-teeth ±lanceolate. Corolla 4–8 mm diam., violet-blue, with a rather long tube. Capsule ±orbicular, about as long as calyx-lobes, retuse or emarginate. Fl. 7–9. Pollinated by various insects. $2n = 34, 68*$. Ch.

Europe from S. Scandinavia to C. Spain, Italy and Greece; W. and C. Asia to China and Japan.

Subsp. **spicata**. Plant 8–30 cm. Lowest lvs 1.5–3 cm × 8–12 mm, sparingly crenate mostly near the middle, usually widest near the middle, gradually narrowed into a narrow petiole.

Native. In dry grassland on basic soils in the breckland of East Anglia and even there rare.

Subsp. **hybrida** (L.) E. F. Warb. Plant 15–60 cm, more robust. Lvs 2–4 × 1–2 cm, more deeply crenate or crenate-serrate nearly all round, usually widest below the middle, abruptly narrowed into a broad petiole.

Native. On limestone rocks; Avon Gorge (Bristol), Wales, W. Yorks, W. Lancs and Westmorland; very local.

The two following rare garden-escapes are sometimes confused with *V. spicata* from which they can be distinguished as follows:

*V. paniculata L. (*V. spuria* auct., non L.) Taller (40–80 cm). Lvs sharply serrate. Pedicels about as long as bracts and calyx. Fr. glabrous. C.E. and S.E. Europe.

*V. longifolia L. Lvs sharply serrate. Racemes dense, usually arranged in terminal panicles. Pedicels much shorter than bracts and calyx. Rarely naturalized. N., E. and E.C. Europe.

The above perhaps correctly *V. longifolia* × *spicata*, the parentage attributed to cultivated spp. which are sometimes ±naturalized as garden-escapes.

Section 4. *Veronicastrum* Koch. Fls in terminal racemes, the lower bracts often lf-like. Capsule septicidal. Seeds flat, convex or cup-shaped. Annual or perennial.

9. V. fruticans Jacquin   Rock Speedwell.

V. saxatilis Scop.

A perennial 5–20 cm, *woody at base*, glabrous below, puberulent above; branches ascending, numerous. Lvs c. 1–2 cm, obovate or oblong, entire or slightly crenulate, cuneate at subsessile base, coriaceous. Raceme lax, up to 10-fld. Bracts narrow, not passing into lvs, shorter than pedicels. Sepals narrow, oblong, obtuse. *Corolla* 11–15 mm *diam.*, *deep bright blue*, reddish in the middle. Capsule ovate or elliptical, longer than sepals, about as long as style; apex entire. Fl. 7–8. Visited by various insects but apparently often selfed, homogamous. $2n = 16$. Chw.

Native. On alpine rocks from 490 to 915 m. Perth, Angus, S. Aberdeen, Argyll, Inverness; very local. Arctic Europe from Iceland to Russia; alpine Europe south to the Pyrenees, Corsica, the Apennines and Bosnia; Greenland.

*10. V. repens Clarion ex DC.

A *perennial herb* 4–10 cm, glabrous below, glandular-puberulent above; stems slender, creeping and rooting, with short lateral flowering branches. Lvs ovate-orbicular to elliptical, shortly petiolate, entire or obscurely crenulate. *Racemes* terminal, *lax*, *3–6-fld*. *Pedicels longer than bracts and calyx*. Calyx-lobes elliptical. *Corolla* 10 mm diam., *pink*. Capsule obovate, emarginate; style nearly 3 times as long. Fl. 4–5. $2n = 14$. Chh?

Introduced. Sometimes found as an escape and ±naturalized in a few places in N. England and S. Scotland. Native of mountains of Corsica and S. Spain.

11. V. alpina L.   Alpine Speedwell.

A *perennial herb* 5–15 cm, glabrous below, glandular-pubescent above; stems shortly creeping at base, with very few branches, the flowering ascending. Lvs ovate, entire or serrulate, cuneate at base, subsessile. *Racemes* 4–12-fld, *dense*, head-like. *Bracts* not passing into the lvs, *longer than the pedicels*. Calyx-lobes elliptical, subacute. *Corolla* c. 7 mm diam., *dull blue*, about twice as long as calyx. *Capsule* obovate, glabrous, slightly emarginate, *longer than calyx and much longer than the very short* (1 mm) *style*. Fl. 7–8. Probably usually self-pollinated; insect visits few (flies). $2n = 18$. Chh?

Native. Damp alpine rocks in Scotland from Perth and Argyll to Aberdeen and Inverness, 490–1130 m,

local. Arctic Europe and North America; high mountains of Europe southwards to S. Spain, S. Italy and S. Bulgaria; Siberia, Manchuria, Korea; mountains of North America southwards to New England and Colorado.

12. V. serpyllifolia L.   Thyme-leaved Speedwell.

A *perennial herb* 10–30 cm; stems creeping and rooting at nodes, the flowering usually ascending, puberulent. Lvs 1–2 cm, ovate or oblong, entire or weakly crenulate, rounded at both ends, subsessile or shortly petiolate, glabrous, light green. *Racemes* terminal, up to 30-fld, often long (10 cm or more), *lax*. Upper *bracts* narrowly oblong, the lower larger and broader, passing into the lvs, *longer than the pedicels*. Calyx-lobes oblong. *Corolla* 6–10 mm diam., *white or pale blue* with darker lines. Filaments and style white; anthers slaty-violet. *Capsule* obcordate, ciliate, wider than long, *about as long as calyx and style*. Infl. sometimes glandular. Fl. 3–10. Visited by flies.

Almost throughout Europe; mountains of N. Africa; Madeira and Azores; temperate Asia; North and South America.

Subsp. **serpyllifolia.** Shortly creeping; flowering branches ascending. Racemes elongate, with 20–40 fls, subglabrous to eglandular-pubescent; pedicels about as long as calyx. Corolla 6–8 mm diam., white or pale blue with slaty-violet lines. $2n = 14$.

Native. In grassland, heaths, waste places, etc., sometimes as a garden weed, often on rather moist ground. Common throughout the British Is.

Subsp. **humifusa** (Dickson) Syme. *V. tenella* All.; *V. humifusa* Dickson; *V. borealis* (Laest.) Hooker fil. Decumbent; stems rooting for most of their length. Racemes usually short, with 8–15 fls, glandular-pubescent; pedicels much longer than calyx. Corolla 7–10 mm diam., blue.

Native. Damp places in mountainous districts, rather local. Monmouth, Brecon, Caernarvon, W. Yorks, N. Northumberland, Cumbria, Selkirk, Stirling, and Kincardine to Sutherland. On the mountains throughout much of Europe.

Section 5. *Pocilla* Dumort. Fls solitary in lf-axils, or in terminal racemes. Corolla-tube wider than long. Capsule loculicidal and also to some extent septicidal. Seeds ±compressed. Annual (except *V. filiformis*).

*13. V. peregrina L.   American Speedwell.

A *glabrous annual* 5–25 cm; stems erect, simple or with spreading branches. Lvs ovate or oblong, entire or weakly toothed, tapering at base to a short petiole. Racemes terminal, long and lax. Upper *bracts* lanceolate, entire, *much longer than the fls*. Pedicels very short. Calyx-lobes lanceolate, about 6 times as long as the pedicel. Corolla 2–3 mm diam., whitish, shorter than calyx. *Style almost* 0. Capsule glabrous, scarcely emarginate, about as wide as long. Fl. 4–7. Self-pollinated. $2n = 52$. Th.

Introduced. Locally naturalized on cultivated ground from Cornwall and Sussex to Cambridge and Worcester, and from Lancs to Perth; in Ireland from Mayo, Cavan and Down northwards; Jersey. Native of North and South America; widely naturalized in W. and C. Europe.

14. V. arvensis L.   Wall Speedwell.

An *erect annual* 5–25 cm, but very variable in size, simple or branched at base with ascending branches, pubescent and sometimes glandular. *Lvs* to 1.5 cm but usually less, triangular-ovate, *coarsely crenate-serrate*, the lowest petiolate, the upper sessile. Racemes long, lax, occupying the greater part of the length of the stem. *Upper bracts* 4–7 mm, lanceolate, entire, ciliate, *longer than the fls*, lower gradually passing into the lvs. *Pedicels very short* (less than 1 mm). Calyx-lobes like the upper bracts but smaller. Corolla 2–3 mm diam., blue, shorter than calyx. Style c. 1 mm or less. Capsule about as long as wide, obcordate, ciliate, shorter than calyx. *Seeds flat.* Fl. 3–10. Visited by small bees, probably often selfed. $2n = 16$. Th.

Native. In cultivated ground and grassland and on heaths in ±open habitats, usually on dry soils. Common throughout the British Is. Almost throughout C. and W. Asia; N. Africa; Macaronesia; naturalized in North America.

15. V. verna L.   Spring Speedwell.

An *erect annual* 3–15 cm. Differs from *V. arvensis* as follows: *Lvs pinnatifid* with 3–7 lobes. Raceme denser, always glandular. Capsule wider than long. Fl. 5–6. $2n = 16$. Th.

Native. In open habitats in dry grassland in the breckland of Norfolk and Suffolk and local there. Much of Europe but absent from much of the north and parts of the west and of the Mediterranean region; W. Asia, east to the Altai; Morocco.

*16. V. acinifolia L.

An *annual* 5–15 cm, usually with erect or ascending branches from the base, ±glandular-pubescent. *Lvs* 1 cm or less, ovate, *obscurely crenate* or almost entire. Infl. lax. Upper *bracts* elliptic-oblong, entire, the lower gradually passing into lvs, *as long as or shorter than pedicels. Calyx-lobes* ovate-oblong, ⅓ or ½ *as long as pedicel.* Corolla blue, rather longer than calyx. *Capsule wider than long*, 2-lobed with a deep sinus, slightly longer than calyx. *Seeds flat.* Fl. 4–6. $2n = 14$. Th.

Introduced. Sometimes occurring as a casual in cultivated ground in England and perhaps naturalized in one or two places in Dorset and adjacent counties. Native of S., S.C. and W. Europe northwards to N.C. France; Asia Minor.

17. V. praecox All.   Breckland Speedwell.

An *erect annual* 5–20 cm, simple or with erect or ascending branches, glandular-pubescent. *Lvs* 1.2 cm or less, ovate, *deeply toothed*, shortly petiolate. Racemes lax.

Upper bracts elliptic-oblong, entire, the lower passing into the lvs, *slightly shorter than the pedicels. Calyx-lobes* oblong, hairy, *slightly shorter than the pedicel.* Corolla c. 3 mm diam., blue, rather longer than calyx. Style c. 2 mm. Capsule obovate, longer than wide, emarginate, rather longer than calyx. *Seeds cup-shaped.* Fl. 3–6. $2n = 18$. Th.

?Introduced. Known only from a few cultivated fields in W. Norfolk and Suffolk, where it was first found in 1933. S., C. and W. Europe, extending northwards locally to S.E. Sweden; Asia Minor, Caucasus; N. Africa.

18. V. triphyllos L.   Fingered Speedwell.

An *annual* 5–20 cm, suberect with spreading or decumbent branches, glandular-pubescent. *Lvs digitately* 3–7-*lobed*, the lobes spathulate or oblong, to c. 1 cm, the lower petiolate, the upper sessile. *Uppermost bracts* usually entire, *shorter than the slender pedicels*, passing gradually but rather soon into the lower lf-like ones. Calyx-lobes spathulate, obtuse, shorter than pedicel. Corolla 3–4 mm diam., deep blue, shorter than calyx. Style 1 mm. Capsule about as long as wide, deeply lobed, shorter than calyx. *Seeds cup-shaped.* Fl. 4–6. Visited by small bees, often selfed. $2n = 14$. Th.

Native. In sandy arable fields in Norfolk and Suffolk, and rare even there; Yorks (extinct); doubtfully naturalized in Surrey. Europe from S. Sweden and Latvia southwards, but rather rare in the Mediterranean region; Asia Minor, Caucasus; N. Africa.

19. V. hederifolia L.   Ivy-leaved Speedwell.

An annual, branched at the base, with decumbent branches 10–60 cm; stem hairy. *Lvs* to c. 1.5 cm, ±reniform, *with 2 or 3 large teeth or small lobes on each side near the base*, rather thick, light green, petiolate, obtuse at apex, ±truncate at base, 3-veined, ciliate and with a few scattered short hairs beneath. Upper lvs somewhat smaller than lower. Pedicels usually rather shorter than lvs. *Sepals* ovate, strongly *cordate at base. Corolla* 4–9 mm diam., shorter than calyx. Capsule glabrous, scarcely compressed, scarcely emarginate, wider than long. Fl. (3–)4–5(–8). Visited by various insects, often selfed. Th.

Native. Common in cultivated ground throughout the British Is., though local in Ireland and restricted or absent in much of W. and N. Scotland. Europe; temperate Asia to Japan; N. Africa; Madeira; naturalized in North America.

Subsp. **lucorum** (Klett & Richter) Hartl (*V. sublobata* Fischer). Lvs thin, shallowly 5–7-lobed; middle lobe as long as or longer than wide. Pedicels 3½–7 times as long as calyx. *Calyx* glabrous or sparsely pubescent, *shortly ciliate.* Corolla 4–6 mm diam., pale lilac. *Anthers* 0.6 mm, *whitish.* Style 0.3–0.5 mm. $2n = 36$.

Woodland, shady places, gardens. Widespread and common in the British Is. N., N.W. and C. Europe, extending to northern part of Balkan peninsula.

Subsp. **hederifolia.** Lvs rather thick, 3–5-lobed; middle lobe wider than long. Pedicels (2)3–4 times as long as calyx. *Calyx long-ciliate*, otherwise glabrous. Corolla 6–9 mm diam., pale blue with white centre. *Anthers* 1 mm, *blue*. Style 0.7–1 mm. $2n = 54$.

Cultivated ground, roadsides. Widespread, but more local than subsp. *lucorum* in the British Is. Throughout the European range of the species.

*20. V. persica Poiret   Common Speedwell.

V. buxbaumii Ten., non Schmidt; *V. tournefortii* auct.

An annual, branched at the base, with decumbent branches 10–40 cm; stem hairy. Lvs 1–3 cm, triangular-ovate, shortly petiolate, coarsely crenate-serrate, light green, hairy on the veins beneath and minutely ciliate. *Pedicels up to* 40 mm, *longer than lvs* but not twice as long, decurved in fr. Calyx-lobes 5–6 mm at flowering time, ovate, ciliate, accrescent and strongly divaricate in fr. *Corolla 8–12 mm diam.*, *bright blue*, the lower lobe often paler or white. *Capsule* 2-lobed, *lobes* sharply keeled and *divergent*, so that the capsule is nearly twice as wide as long, ciliate. Fl. 1–12. Visited by various insects, often selfed. $2n = 28$. Th.

Introduced. First recorded 1825. Now common in cultivated land throughout the British Is. and the commonest species of the genus in this habitat. Naturalized almost throughout Europe. Native of S.W. Asia.

21. V. polita Fr.   Grey Speedwell.

V. didyma auct.

A *pubescent annual*, branched at base, with decumbent branches. *Lvs 5–15 mm*, *ovate*, shortly petiolate, the lower wider than long, upper longer than wide, dull green, obtuse, ±truncate at base, coarsely and regularly crenate-serrate. *Pedicels 5–15 mm*, *as long as or shorter than lvs*, decurved in fr. *Sepals ovate*, *acute or subacute*, accrescent, conspicuously veined. *Corolla 4–8 mm diam.*, *usually uniformly bright blue*, rarely the lower lobe paler. *Capsule* rather wider than long; *lobes erect*, not keeled, *clothed with short, crispate, eglandular hairs* and some longer glandular ones. Fl. 1–12. Usually self-pollinated. $2n = 14^*$. Th.

Native. In cultivated ground throughout the British Is., common S.E. of a line from the Humber to the Severn, more local northwards to S. Scotland and in Ireland, very local or absent in N. Scotland. Most of Europe, except the Arctic; temperate Asia; N. Africa.

22. V. agrestis L.   Green Field-speedwell.

Differs from *V. polita* as follows: Lvs all longer than wide, rather lighter green, irregularly crenate-serrate. *Sepals oblong or ovate-oblong*, *obtuse or subobtuse*, faintly veined. *Corolla* usually *pale blue with the lower lobe or 3 lobes white or very pale*, less frequently all white, or pink above. *Capsule* lobes obscurely keeled, with long glandular hairs, often with rather shorter glandless ones but *without short crispate hairs*. Fl. 1–12. Visited by Diptera and Hymenoptera, homogamous, often selfed. $2n = 28$. Th.

Native. In cultivated ground throughout much of the British Is. Most of Europe except the extreme north and parts of the south-east; mountains of N. Africa; Asia Minor.

*23. V. filiformis Sm.   Slender Speedwell.

A pubescent *perennial with numerous creeping stems*, often forming large patches. *Lvs 5–10 mm*, *reniform*, shortly petiolate, crenate. *Pedicels* up to 40 mm, *filiform, several times as long as lvs*. Calyx-lobes oblong, obtuse. Corolla 10–15 mm diam., pale lilac blue. Fr. 2-lobed, with subparallel lobes and rather wide sinus, very rare in Britain. Fl. 4–6. $2n = 14$. Chh.

Introduced. Grown in gardens and now extensively naturalized in grassy ground by streams, roads, on lawns, etc., throughout the British Is., especially in S. England; first recorded as an escape in 1927. Native of N. Anatolia and the Caucasus.

*V. crista-galli Steven

Annual resembling *V. persica* in habit. Lvs 1–2 cm, ovate, the lower petiolate. Sepals connate in pairs, each pair wider than long, divided near apex into two acuminate lobes. Corolla small, pale blue. Naturalized in Sussex for about 80 years, probably now extinct. Native of Caucasus, etc.

15. HEBE Comm.

Like *Veronica* but erect, evergreen shrubs; lvs opposite, entire or subentire; fls in opposite axillary racemes; capsule septicidal, compressed parallel to septum.

Lvs narrowly lanceolate, 5–8 times as long as wide; corolla
          white, sometimes pinkish.   **1. salicifolia**
Lvs obovate-oblong, 2–4 times as long as wide; corolla
          violet to reddish-purple.   **2. speciosa**

*1. H. salicifolia (Forster fil.) Pennell

Rather straggling shrub up to 2.5 m. Twigs glabrous. Lvs 4–15 cm, narrowly lanceolate, 5–8 times as long as wide, puberulent on midrib and petiole. Racemes 10–20 cm, slender, rather dense. Corolla 3–5 mm diam., white sometimes tinged with pink. Capsule c. 3.5 × 2.5 mm.

Introduced. Locally naturalized on walls and sea-cliffs in S.W. England and in Ireland, from Connemara and Dublin and probably elsewhere. Native of New Zealand and S. Chile.

*2. H. speciosa (R. Cunn. ex A. Cunn.) Andersen

Compact shrub up to 1.5 m. Twigs glabrous. Lvs 5–10 cm, obovate-oblong, 2–4 times as long as wide, glabrous. Racemes 7–10 cm, stout, very dense. Corolla 8–10 mm diam., violet to reddish-purple. Capsule 5–6 × 3–4 mm.

Introduced. Locally naturalized on sea-cliffs in W. Ireland, and perhaps also in S.W. England. Native of New Zealand.

H. elliptica (Forster fil.) Pennell, which differs from *H. speciosa* in having pubescent young twigs, ciliate mature lvs and lax racemes with only c. 10 fls, is reported as naturalized in Clare (nr. Hag's Head). It is a native of New Zealand, the Falkland Islands and S. Chile.

H. × franciscana (Eastw.) Souster (*H. elliptica × speciosa*) (*H. lewisii* auct.) is a shrub up to c. 1 m, with narrowly obovate-elliptical, obtuse lvs 4–5 cm, and terminal, axillary racemes of fls having a bluish or purplish corolla 10–12 mm diam. It is fertile.

This widely cultivated hybrid, synthesized in an Edinburgh garden about 100 years ago, is naturalized in coastal areas of Cornwall, S. Devon and the Scilly and Channel Is., where it is a garden-escape or, sometimes, self-sown.

Subfamily 3. RHINANTHOIDEAE. Hemiparasitic. Nectaries at base of ovary, unilateral. Stamens (2–)4. Fls in spikes or racemes. Corolla 2-lipped, the upper lobes inside the lateral in bud. Capsule loculicidal.

16. PEDICULARIS L.

Herbs with alternate pinnatisect lvs. Fls in terminal lfy-bracted spikes or racemes. *Calyx tubular* in fl., soon *becoming inflated*, with 2–5 lf-like lobes. *Upper lip of corolla laterally compressed*, entire or with 2 or 4 small teeth near the end; lower lip 3-lobed. Stamens 4, included in the upper lip. Capsule compressed, with a few large seeds in the lower part. Nectar secreted by a swelling at the base of the ovary; pollinated by humble-bees.

About 500 spp. in north temperate regions, especially the mountains of C. and E. Asia, and the Andes.

Calyx clearly 2-lipped; upper lip of corolla with a small
   tooth on each side some distance below apex.
                                1. palustris
Calyx entire or indistinctly 2-lipped; upper lip of corolla
   with entire lateral margins.   **2. sylvatica**

1. P. palustris L.   Marsh Lousewort, Red-rattle.

Biennial, sometimes annual 8–60 cm, glabrous or sparsely hairy. Stem branching from near the base to the middle. Lvs 2–4 cm, triangular-lanceolate to oblong in outline, pinnatisect, the lobes dentate to pinnatifid. Bracts similar but smaller. Racemes lax. Pedicels short. *Calyx pubescent*, at least near the upper edge, often reddish, strongly 2-lipped, without distinct teeth. Corolla 2–2.5 cm, reddish-pink; *upper lip with a tooth on each side some distance below the apex.* Capsule ovoid, curved, longer than the calyx. Fl. 5–9. $2n = 16^*$. Th.

Native. Wet heaths and meadows, rather common throughout the British Is., but less so in C. and S.E. England; ascending to 855 m. Europe southward to the Pyrenees, N. Italy, S. Bulgaria and S. Ural; Caucasus.

2. P. sylvatica L.   Lousewort.

Nearly glabrous *perennial or biennial* 8–25 cm, *with many decumbent branches from the base*, and with thick tap-root. Lvs to 2 cm, pinnatisect, oblong or linear in outline, the divisions deeply dentate. Racemes terminal, 3–10-fld, rather lax. Bracts trisect, lf-like, passing into the lvs below. Pedicels short, stout. Calyx ±cylindrical in fl., 5-angled with 4 small lf-like 2–3-lobed teeth, the 5th (upper) tooth small and linear, arising at a lower level, pubescent on the lobes within. Corolla 2–2.5 cm,

pink or red (rarely white); *upper lip with entire lateral margins*; lower lip with 3 ±orbicular lobes. Capsule ovoid, obliquely truncate, about as long as the calyx. Fl. 4–7. $2n = 16$. Hp.

Native. Damp heaths, bogs and marshes, rather common throughout the British Is., ascending to 915 m. W. and C. Europe extending northwards to C. Sweden and eastwards to Lithuania and W. Russia.

Subsp. **sylvatica**. Stem glabrous or with 2 lines of hairs; central stem up to 15 cm. Calyx and pedicels glabrous. Throughout most of the British Is., and the range of the sp. elsewhere, except Portugal.

Subsp. **hibernica** D. A. Webb. Stems pubescent; central stem not more than 10 cm. Calyx and pedicels villous. Moors and bogs. W. Ireland from Cork to W. Donegal, also in Waterford and Wicklow, and in W. Scotland from Kintyre and Lewis. N.W. Europe northwards to Norway.

17. RHINANTHUS L.

Annual, hemiparasitic herbs with opposite toothed lvs. Fls in terminal lfy-bracted spikes. *Calyx flattened and inflated* in fl., accrescent, with 4 entire teeth. *Upper lip of corolla laterally compressed* with 2 teeth at the end; lower lip 3 lobed. Stamens 4, included in the upper lip. Capsule compressed, with a few large, usually winged seeds. Nectar secreted at the base of the ovary.

About 50 spp. in north temperate regions.

The spp. of this genus exhibit a reticulate pattern of variation, with which is associated a form of seasonally correlated ecotypic differentiation, so that the delimitation of taxa is still the subject of much uncertainty.

'Intercalary lvs' are the lvs on the main stem between the topmost branches and the lowest fls of the terminal spike. They are often transitional between the lower lvs and bracts in shape and toothing. 'Bracts' in the following descriptions exclude the two lowest pairs which are often transitional to the lvs.

1 Teeth of upper lip of corolla long (c. 2 mm), twice
   as long as wide; corolla-tube curved upwards.
                     1. angustifolius
Teeth of upper lip of corolla less than 1 mm, not
   longer than wide; corolla-tube straight.   **2. minor**

1. R. angustifolius C. C. Gmelin
                  Greater Yellow Rattle.
R. major auct., non L.; *R. serotinus* (Schönheit) Oborny

Plant ±robust 20–60 cm; stem black-spotted, simple or more frequently with short or long branches from about the middle. Stem-lvs 2.5–7 cm × 4–15 mm, lanceolate or linear-lanceolate, often strongly crenate-serrate with prominent teeth but sometimes the teeth less prominent, scabrid; branch lvs usually smaller. Intercalary lvs 0–1(–2) pairs. Main inflorescence 8–18-fld. Bracts yellowish-green, ovate or rhombic-ovate, usually acuminate, at least the basal teeth usually deep and sharp. Calyx glabrous except for the pubescent margin. *Corolla*

15–20 mm, yellow with violet teeth; *tube 2–3 mm wide, curved upwards, mouth closed*; *teeth* c. 2 mm, *conical, twice as long as wide*. Seeds varying from broadly winged to wingless (*R. apterus* (Fries) Ostenf.). Fl. 6–9. Pollinated by humble-bees, self-pollination not possible. $2n = 22^*$. Th.

Native. In cornfields, less commonly in meadows and on sandhills. Scattered localities in England (Surrey, Gloucester, Cheshire, S. Yorks) and Scotland (Angus, Aberdeen, Barra, Coll, Tiree). Most of Europe, but absent from the Mediterranean region, the south-west and most islands; Siberia.

2. R. minor L.   Yellow Rattle.

Stem up to 50 cm, simple or branched, usually black-spotted. Lvs oblong to linear, crenate-dentate, scabrid. Bracts triangular-ovate to lanceolate, with triangular acute or shortly aristate teeth. *Calyx* shortly hairy on margin, otherwise *glabrous*. *Corolla* c. 12–15 mm; yellow or brown with violet, rarely white teeth; *tube straight*, mouth somewhat open; *teeth rounded, short, not longer than wide*. Seeds always winged. Fl. 5–8, pollinated by humble-bees, selfed if not visited. $2n = 22^*$. Th.

Grassland, etc., common throughout the British Is. Most of Europe, but rare in the Mediterranean region; Caucasus; W. Siberia; S. Greenland, Newfoundland.

This species exhibits a great deal of only partially correlated variation in several characters as well as that resulting from the seasonal ecotypic differentiation prevalent in the genus. None of the formal taxonomic treatments available seems to reflect this variation adequately. However, since most recent British studies have attempted to express the variation at the subspecific level, this is adopted in the following account, though varietal status may seem more suitable, and in each case the appropriate seasonal ecotype is indicated.

Subsp. **minor**. Stem 12–40 cm, usually with only *very short non-flowering branches but sometimes with longer suberect flowering branches from the middle and upper part of stem*. Internodes (except the lowest) ±equal. *Lvs* of main stem (10–)20–40(–50) x (3–)5–7 mm, *usually oblong* (less often oblong-linear) parallel-sided for the greater part of their length. *Intercalary lvs* 0(–1) pairs (rarely more). *Lowest fls usually from 6th–9th node*. Calyx hairy only on the margin. Corolla yellow; teeth violet (rarely white). Fl. 5–7. Vernal ecotype.

Pastures, especially on dry basic soils; throughout the British Is. but less common in Scotland and there often tending towards subsp. *stenophyllus*. Almost throughout the range of the sp.

Subsp. **stenophyllus** (Schur) O. Schwarz (*R. stenophyllus* (Schur) Druce). *Stem* (15–)25–50 cm, usually with *long arcuate-ascending flowering branches* and shorter non-flowering branches from the lower and middle part of the stem. Lower internodes usually short, as long as or shorter than lvs, the upper much longer. *Lvs of*

main stems 15–45 × 2–5(–7) mm, *narrowly lanceolate or linear-lanceolate*, ±tapering from near the base. *Intercalary lvs* (0–)1–2(–4) pairs. *Lowest fls usually from* (8th–)10th–13th(–15th) node. Calyx hairy only on the margin. Corolla yellow, sometimes becoming brown; teeth violet. Fl. 7–8. Autumnal ecotype.

Damp grassland, fens, etc., common from the Isle of Man and Northumberland northwards, less so elsewhere and absent from much of S.E. England and S. Ireland. Much of Europe, especially in the uplands.

Subsp. **monticola** (Sterneck) O. Schwarz (*R. monticola* (Sterneck) Druce; *R. spadiceus* Wilmott). *Plant often tinged purple. Stem* (5–)10–20(–25) cm, *usually with short or medium length non-flowering branches from near the base, sometimes with 1–3 longer flowering branches in addition*. Lower internodes usually very short, the upper much longer. *Lvs* of main stems 10–20(–25) × 2–4 mm, *linear-lanceolate*, ±tapering from near base, tending to be more erect than in subsp. *stenophyllus*. *Intercalary lvs* usually 1–2(–3) pairs. *Lowest fls usually from* (7th–)8th–11th(–12th) node. Calyx hairy only on the margins. Corolla dull yellow *becoming light brown, or constantly light brown*; teeth violet. Fl. 7–8. Montane ecotype.

Grassy places in mountain districts in Yorkshire and from Selkirk to Shetland; Kerry, Derry. Alps.

Subsp. **calcareus** (Wilmott) E. F. Warb. (*R. calcareus* Wilmott). Stem 25–50 cm, usually with *long arcuate-ascending flowering branches from about the middle*. Lower internodes short; upper very long. *Lvs of main stems* 10–25 × 1.5–3 mm, *linear*, ±spreading. *Intercalary lvs usually* (2–)3–6 pairs. *Lowest fls usually from* 14th–19th node. Calyx hairy only on margin. Corolla yellow; tooth violet. Autumnal ecotype.

Chalk and limestone downs from Dorset and Kent to Gloucester and Bucks, Northampton, local.

Subsp. **borealis** (Sterneck) Sell (*R. borealis* (Sterneck) Druce)

Stem (5–)9–20(–28) cm, *unbranched* or occasionally with short axillary non-flowering (very rarely flowering) branches and then intercalary lvs 0. Stem-lvs 10–30 × 3–7 mm, oblong or oblong-linear, ±parallel-sided. Lowest fls from 5th–7th(–8th) node. *Calyx hairy all over*. Corolla bright yellow; teeth violet or white. Fl. 7–8. Th. Montane ecotype.

Native. Grassy places on mountains, local, ascending to 975 m. Dumfries and Angus to Shetland and the Outer Hebrides; Kerry. Alaska, Greenland, Iceland, ?Scandinavia.

Scottish populations combining features of two pairs of subspecies have been considered to be fertile hybrids, *R. minor* subsp. *borealis* x subsp. *monticola* (**R. x gardineri** Druce) and *R. minor* subsp. *borealis* x subsp. *stenophyllus*. It has been suggested that together these constitute a distinct taxon, **R. minor** subsp. **lintonii** (Wilmott) Sell.

18. MELAMPYRUM L. Cow-wheat.

Annual, hemiparasitic herbs with opposite mostly entire lvs. Fls in terminal bracteate racemes or spikes, the bracts not inflated, often toothed. *Calyx tubular*, not inflated, 4-toothed. Corolla 2-lipped; *upper lip laterally compressed*; lower lip 3-lobed, shorter, with prominent palate nearly closing the mouth. Stamens 4, included under the upper lip of corolla. *Capsule compressed, with 1–4 ovoid seeds.* Nectary at the base of the ovary. All our spp. are pollinated by humble-bees but can be selfed if not visited.

About 35 spp., in Europe, temperate Asia and eastern North America.

The spp. of this genus are very variable though *M. cristatum* and *M. arvense* show little variation in Britain.

1 Fls in ±dense spikes; bracts densely imbricate, concealing spike-axis. ... 2
   Fls in pairs in the axils of lf-like bracts which do not conceal infl. axis, forming a very lax, secund, interrupted raceme. ... 3
2 Spikes 4-sided, very dense; bracts cordate, folded along midrib and with apex recurved. **1. cristatum**
   Spikes conical or cylindrical, laxer; bracts not cordate, ±flat, with apex rarely recurved. **2. arvense**
3 Corolla 10–18 mm, the tube much longer than the calyx, the lower lip straight. **3. pratense**
   Corolla 6–10(–11) mm, the tube as long as or shorter than the calyx, the lower lip deflexed. **4. sylvaticum**

Section 1. *Spicata* Wettst. Fls in dense spikes, which are not secund; bracts densely imbricate, concealing the spike-axis. Lvs sessile.

1. M. cristatum L. Crested Cow-wheat.

Puberulent annual 20–50 cm, simple or with a few spreading branches. Lvs 5–10 cm, sessile, linear-lanceolate to lanceolate. *Fls in a dense 4-sided spike. Bracts cordate, folded along midrib, recurved at apex, the base bright rosy-purple and finely pectinate* (teeth less than 2 mm); *lower with long lf-like green entire points*, the upper with short points. Calyx-tube with two lines of hairs; *teeth shorter than tube, unequal.* Corolla 12–16 mm, pale yellow, tinged at least on lower lip with purple; palate deeper yellow; tube longer than the calyx. Capsule normally 4-seeded, dehiscing along one margin only. Fl. 6–9. $2n = 18^*$. Th.

Native. Very local, at the edges of woods. Essex, Bedford, Huntingdon, Norfolk, Northampton and Leicester, apparently extinct northwards to Lincoln and southwards to Hants. Much of Europe except parts of the north and south; W. Asia, east to the Yenisei region.

2. M. arvense L. Field Cow-wheat.

Pubescent annual 20–60 cm, with spreading branches. Lvs 3–8 cm, sessile, lanceolate, the upper usually with a few, long teeth at the base. *Fls in a rather lax cylindrical or conical, flat spike. Bracts lanceolate, ±flat, erect, pink at first, pinnatifid, with long* (to 8 mm) *slender teeth, without lf-like points.* Calyx-tube pubescent; *teeth longer than tube, nearly equal.* Corolla 20–25 mm, purplish-

pink, yellow at the throat and sometimes on lower lip; tube about as long as the calyx. Capsule normally 2-seeded. Fl. 6–9. $2n = 18$. Th.

Native. Cornfields. Isle of Wight, Bedford, Wilts, Essex, Hertford, and S. Yorks, very local and rare. Europe from S.W. Finland southwards to N. Spain, Italy and Turkey; Caucasus; W. Siberia (upper Tobol region).

Section 2. *Melampyrum.* Fls in lax, ±secund spikes or racemes; bracts not concealing infl.-axis. Lvs often shortly petiolate.

3. M. pratense L. Common Cow-wheat.

Very variable. Annual 8–60 cm, glabrous or somewhat hispid, with suberect to spreading branches. Lvs 1.5–10 cm, sessile or very shortly petiolate, linear to ovate, entire. *Fls in the axils of each of distant opposite green lf-like bracts, both fls turned to the same side of the stem* (secund), ±horizontal. *Upper bracts pectinate or dentate*, except in small plants with only 2–3 pairs of bracts; lower often lf-like. *Calyx with linear-setaceous appressed lobes*, usually rather longer than the tube. Corolla 11–17 mm to tip of upper lip, deep yellow to whitish, sometimes tube and upper lip tinged with red or purple; *tube about twice as long as calyx*, the mouth ±closed; *lower lip straight.* Capsule normally 4-seeded, dehiscing along one margin. Fl. 5–10. $2n = 18^*$. Th.

Native. In woods, heaths, etc.; ascending to over 915 m, common in many districts, local in others, especially E.C. England and S.C. Ireland. Most of Europe, W. Asia, east to the Yenisei region.

A variable sp. The characters which have been relied upon to distinguish infraspecific taxa are the habit, whether glabrous or hispid, shape of lvs, toothing of the bracts, colour of the fls both when fresh and when fading and the relative length of the anther-hairs and anther-appendages. The variation results from a combination of climatic and ecological selection, together with random isolation of particular genotypes in individual populations. Populations confined to calcareous habitats in S.E. England have been distinguished as subsp. **commutatum** (Tausch ex Kerner) C.E. Britton, while those with golden yellow fls, apparently predominating on the palaeogenic areas of W. Britain and W. Ireland, have been called subsp. **pratense** var. **hians** Druce. On present evidence there seems little justification for maintaining these.

4. M. sylvaticum L. Small Cow-wheat.

Very similar to some variants of *M. pratense* but may be distinguished as follows: Lvs linear-lanceolate. Upper bracts entire except in some large plants which may have a few teeth on the uppermost bracts. Calyx-lobes spreading, not appressed to corolla, longer than the tube. Corolla 6–10(–11) mm; tube as long as or shorter than the calyx; lower lip deflexed. Capsule 2-seeded, dehiscing along 2 margins. Fl. 6–8. $2n = 18$. Th.

Native. Mountain woods, ascending to 395 m. N. Yorks and Durham (Teesdale); Mull, W. Perth and Angus northwards to W. Sutherland; Derry (? extinct in Antrim). N. Europe, extending southwards in the mountains to the Pyrenees, C. Italy and S. Bulgaria.

19. EUPHRASIA L.

Annual herbs. Lvs opposite or the upper alternate, rather small. Fls sessile in the axils of the bracts (floral lvs) which usually differ somewhat from the cauline lvs, thus forming a terminal spike. Calyx campanulate, 4-toothed. *Upper lip of corolla slightly concave with* 2 *porrect or reflexed lobes*; lower lip flat, 3-lobed, the lobes emarginate. Stamens 4; anther cells pointed at the base. Capsule loculicidal. Seeds small, numerous, oblong or fusiform, furrowed.

About 200 spp., temperate northern hemisphere, temperate South America, Australia and New Zealand. All the north temperate spp. belong to the section *Euphrasia*.

The determination of the spp. is not easy. They have, however, definite habitats and geographical distributions. The best time for determinations is when the fls are well out and some capsules have formed. Samples of at least 6 plants should be used and depauperate or damaged plants neglected. Very dwarf, compact, unbranched forms of many spp. occur, especially in N. Scotland. These are very difficult to determine and no allowance has been made for them in the key.

There is still debate as to the most useful, and useable, taxonomic treatment of *Euphrasia*. It has been found, for example, that certain characters used here for separating taxa, e.g. length of internodes, node at which the lowest fl. is situated, number of branches, shape of the lvs, size and shape of capsule, are often habit- or habitat-correlated, and are thus unsatisfactory characters on which to base a classification. However, the tradition of British students of the genus has been to recognize the major variants at the species level, and this approach is followed here.

Certain characters, not included in the generic diagnosis, are common to all the British spp. These have not been repeated in the detailed descriptions below. They are as follows:

Stem clothed with crispate whitish hairs. Lower cauline lvs with 1–2 teeth on each side; bracts usually with 5–6 teeth. Indumentum of the calyx like that of the lvs. Corolla (which may be white, of various shades of blue or purple, or rarely yellow) with a yellow blotch and purple lines on the lower lip. In all species, especially the less well differentiated where individual characters may vary widely, it is essential that *all* characters be evaluated. Correct determination lies in taking the characters as a whole.

Nodes are counted from the base, excluding the cotyledonary node. Relative lengths of nodes and their subtending lvs refer to the main stem. Corolla length is measured from the base of the tube to the apex of the upper lip in its normal position (drying and pressing adds 0.5–1 mm to the length).

Hybrids and hybrid swarms seem to occur commonly, sterile hybrids being comparatively rare. The swarms sometimes occur where one of the parents is not present, though it can usually be found in the district. Hybridization is particularly frequent between spp. with the same chromosome number (either $2n = 22$ or $2n = 44$) within the subsection *Ciliatae*, which includes all but one of the British spp. However, several hybrids involving subsections *Ciliatae* and *Angustifoliae* are known. Only the more frequent hybrids are mentioned in the text and none is included in the key; for more information reference should be made to Yeo, in C. A. Stace (1975) (p. 373).

For a full account of the British spp. see Pugsley, *J. Linn. Soc.*, **48** 467 (1930), from which grew the modern monograph of the genus in Europe by Yeo, *Bot. J. Linn. Soc.*, **77**, 223 (1978), on which this account is largely based. The larger-fld spp. appear to be pollinated by bees or other insects, the small-fld ones are usually selfed.

1 Middle and upper lvs bearing glandular hairs with a stalk 10–12 times as long as the gland. 2
  Middle and upper lvs without glandular hairs, or glandular hairs with a stalk not more than 6 times as long as the gland. 9
2 Capsule more than twice as long as wide. **5. arctica**
  Capsule not more than twice as long as wide. 3
3 Corolla not more than 7 mm. 4
  Corolla more than 7 mm. 5
4 Lowest fl. at node 5–8; branches usually 1–5 pairs. **3. anglica**
  Lowest fl. at node 2–6(–7); branches usually 0–2 pairs. **2. rivularis**
5 Lowest fl. at node 2–5(–6). 6
  Lowest fl. at node 5 or above. 7
6 Corolla 9–12.5 mm; lower floral lvs 5–12(–20) mm. **1. rostkoviana**
  Corolla not more than 9 mm; lower floral lvs not more than 6(–7) mm. **2. rivularis**
7 Lvs dull greyish-green, ±strongly tinted with dull violet or black; corolla usually lilac to deep reddish-purple. **4. vigursii**
  Lvs light or dark green, usually with much purplish tinting; corolla usually with at least the lower lip white. 8
8 Stem usually erect, with erect or divergent branches; lower floral internodes mostly 1.5–3 times as long as lvs; corolla 8–12 mm. **1. rostkoviana**
  Stem usually flexuous, with flexuous or arcuate branches; lower floral internodes mostly less than 1.5 times as long as lvs; corolla usually 6.5–8 mm. **3. anglica**
9 Capsule glabrous, or with a few small cilia; at least some lf-teeth distant. **20. salisburgensis**
  Capsule ciliate with long, fairly numerous hairs; all lf-teeth usually contiguous. 10
10 Corolla more than 7.5 mm. 11
  Corolla not more than 7.5 mm. 15
11 Basal pairs of teeth of lower floral lvs directed towards apex. **9. confusa**
  Basal pairs of teeth of lower floral lvs patent. 12
12 Lowest fl. at node 8 or lower; capsule usually elliptical to obovate. **5. arctica**
  Lowest fl. at node 9 or higher; capsule oblong to elliptical-oblong. 13
13 Stems and branches usually flexuous; lvs near base of branches usually very small. **9. confusa**
  Stems and branches usually not flexuous; lvs near base

of branches not much smaller than the others. 14

14 Teeth of floral lvs acute to acuminate; capsule usually
7. **nemorosa**
Teeth of floral lvs mostly aristate; capsule much
shorter than calyx; stem and branches relatively
slender. 8. **pseudokerneri**

15 Calyx-tube whitish and membranous, with prominent
green veins. 16. **campbelliae**
Calyx-tube green and herbaceous. 16

16 Lowest fl. at node 6 or higher. 17
Lowest fl. at node 5(–6) or lower. 41

17 Cauline internodes mostly 2–6 times as long as lvs. 18
Cauline internodes mostly not more than twice as long
as lvs. 29

18 Basal pairs of teeth of lower floral lvs directed towards
apex. 19
Basal pairs of teeth of lower floral lvs patent. 20

19 Teeth of lower floral lvs obtuse to acute, fairly short;
corolla not more than 6.5 mm. 17. **micrantha**
Teeth of lower floral lvs acute to aristate, long; corolla usually at least 7 mm. 5. **arctica**

20 Corolla at least 6.5 mm. 21
Corolla not more than 6.5 mm. 22

21 Lowest fl. at node 9 or higher; lvs usually without
glandular hairs; lower floral lvs smaller than upper
cauline. 7. **nemorosa**
Lowest fl. at node 8 or lower; lvs usually with glandular hairs; lower floral lvs larger than upper cauline.
5. **arctica**

22 Lvs sparsely hairy to subglabrous. 23
Lvs densely hairy. 25

23 Stem and branches very slender, blackish; lvs strongly
tinged with purple, not darker beneath than above;
corolla usually lilac to purple. 17. **micrantha**
Stem and branches either stout or lightly pigmented;
lvs weakly or moderately tinged with purple; corolla usually white. 24

24 Lowest fl. at node 8 or higher; stem stout; lvs not
darker beneath than above; capsule usually shorter
than calyx. 7. **nemorosa**
Lowest fl. at node 7 or lower; lvs usually light green
above and purplish beneath; capsule usually longer
than calyx. 18. **scottica**

25 Lowest fl. at node 9 or higher; stem up to 40 cm;
lower floral lvs often longer than wide. 7. **nemorosa**
Lowest fl. at node 8 or lower; stem not more than
15 cm; lower floral lvs about as wide as long. 26

26 Lvs hairy mainly towards the apex, the cauline obovate to narrowly ovate or elliptical. 16. **campbelliae**
Lvs ±uniformly hairy, usually suborbicular, ovate or
ovate-oblong. 27

27 Teeth of lower floral lvs mostly wider than long;
branches not more than 3 pairs. 15. **rotundifolia**
Teeth of lower floral lvs mostly as long as wide;
branches up to 5 pairs. 28

28 Corolla 5.5–7 mm; capsule usually more than twice
as long as wide. 14. **marshallii**
Corolla 4.5–6 mm; capsule not more than twice as
long as wide. 13. **ostenfeldii**

29 Basal pairs of teeth of lower floral lvs directed towards
apex. 30
Basal pairs of teeth of lower floral lvs patent. 34

30 Teeth of lower floral lvs much longer than wide.
9. **confusa**

Teeth of lower floral lvs not much longer than wide. 31

31 Lvs hairy mainly towards the apex, the cauline obovate to narrowly ovate or elliptical. 16. **campbelliae**
Lvs ±uniformly hairy, usually suborbicular, ovate or
ovate-oblong. 32

32 Teeth of lower floral lvs mostly wider than long;
branches not more than 3 pairs. 15. **rotundifolia**
Teeth of lower floral lvs mostly as long as wide;
branches up to 5 pairs. 33

33 Corolla 5.5–7 mm; capsule usually more than twice
as long as wide. 14. **marshallii**
Corolla 4.5–6 mm; capsule not more than twice as
long as wide. 13. **ostenfeldii**

34 Lowest fl. at node 10 or higher. 35
Lowest fl. at node 9 or lower. 36

35 Stem erect, stout, with stout ascending branches;
lower floral lvs mostly opposite. 7. **nemorosa**
Stem and branches slender, flexuous; lower floral lvs
mostly alternate. 9. **confusa**

36 Lvs with few eglandular hairs. 37
Lvs with numerous eglandular hairs. 38

37 Capsule 5.5–7 mm, often slightly curved.
19. **heslop-harrisonii**
Capsule usually not more than 5.5 mm, straight.
6. **tetraquetra**

38 Lvs hairy mainly towards the apex, the cauline obovate to narrowly ovate or elliptical. 16. **campbelliae**
Lvs ±uniformly hairy, usually suborbicular, ovate or
ovate-oblong. 39

39 Teeth of lower floral lvs mostly wider than long;
branches not more than 3 pairs. 15. **rotundifolia**
Teeth of lower floral lvs mostly as long as wide;
branches up to 5 pairs. 40

40 Corolla 5.5–7 mm; capsule usually more than twice
as long as wide. 14. **marshallii**
Corolla 4.5–6 mm; capsule not more than twice as
long as wide. 13. **ostenfeldii**

41 Cauline internodes mostly at least 2½ times as long
as lvs. 42
Cauline internodes mostly less than 2½ times as long
as lvs. 49

42 Capsule broadly elliptical to obovate-elliptical. 43
Capsule oblong to narrowly elliptical. 44

43 Teeth of lower floral lvs mostly subacute and not
longer than wide; corolla 4.5–7 lowest fl. at
node 2–4(–5). 10. **frigida**
Teeth of lower floral lvs usually acute or acuminate
and longer than wide; corolla at least 6.5 mm; lowest fl. usually at node 4 or higher. 5. **arctica**

44 Upper cauline lvs elliptic-ovate to narrowly obovate.
18. **scottica**
Upper cauline lvs suborbicular to broadly ovate or
broadly obovate. 45

45 Lowest fl. at node 4 or lower; lower floral lvs often
much larger than the upper cauline. 10. **frigida**
**Lowest fl. at node 4 or higher; lower floral lvs scarcely
larger than the upper cauline.** 46

46 Lvs hairy mainly towards the apex, the cauline obovate to narrowly ovate or elliptical. 16. **campbelliae**
Lvs ±uniformly hairy, usually suborbicular, ovate or
ovate-oblong. 47

47 Teeth of lower floral lvs mostly wider than long;
branches not more than 3 pairs. 15. **rotundifolia**
Teeth of lower floral lvs mostly as long as wide;
branches up to 5 pairs. 48

48 Corolla 5.5–7 mm; capsule usually more than twice
as long as wide.   **14. marshallii**
Corolla 4.5–6 mm; capsule not more than twice as
long as wide.   **13. ostenfeldii**
49 Corolla at least 6 mm.   50
Corolla not more than 6 mm.   52
50 Teeth of lower floral lvs usually very acute, all directed towards apex.   **9. confusa**
Teeth of lower floral lvs subacute to acute, the basal
pairs patent.   51
51 Capsule at least as long as calyx, usually emarginate.
                                       10. frigida
Capsule shorter than calyx, truncate or slightly emarginate.   **6. tetraquetra**
52 Lower floral lvs ovate to rhombic, with acute to aristate teeth, the basal pairs directed towards apex.
                                       9. confusa
Lower floral lvs broadly ovate or deltate to suborbicular, with obtuse to subacute teeth, the basal pairs
patent.   53
53 Lvs with numerous hairs, all eglandular.   54
Lvs with few eglandular hairs, short glandular hairs
sometimes present.   58
54 Lower floral lvs much larger than the upper cauline.
                                       10. frigida
Lower floral lvs scarcely larger than the upper cauline. 55
55 Lvs hairy mainly towards the apex, the cauline obovate to narrowly ovate or elliptical. **16. campbelliae**
Lvs ±uniformly hairy, usually suborbicular, ovate or
ovate-oblong.   56
56 Teeth of lower floral lvs mostly wider than long;
branches not more than 3 pairs. **15. rotundifolia**
Teeth of lower floral lvs mostly as long as wide;
branches up to 5 pairs.   57
57 Corolla 5.5–7 mm; capsule usually more than twice
as long as wide.   **14. marshallii**
Corolla 4.5–6 mm; capsule not more than twice as
long as wide.   **13. ostenfeldii**
58 Capsule elliptical to obovate, emarginate.
                                       12. cambrica
Capsule oblong to elliptic-oblong, usually truncate.   59
59 Capsule usually shorter than calyx; distal teeth of
lower floral lvs not incurved.   **6. tetraquetra**
Capsule as long as or longer than calyx; distal teeth
of lower floral lvs ±incurved.   60
60 Capsule 4.5–5.5(–7) mm, about twice as long as wide,
straight; upper cauline lvs only obscurely petiolate,
the teeth without sinuate margins.   **11. foulaensis**
Capsule (4.5–)5.5–7 mm, 2–3 times as long as wide,
often slightly curved; upper cauline lvs ±distinctly
petiolate, the teeth with sinuate margins.
                                  19. heslop-harrisonii

All British spp. belong to Sect. *Euphrasia*.

Subsection *Ciliatae* Joerg. (*E. officinalis* sens. lat.)
                                       Eyebright.
Floral lvs generally more than half as long as wide. Capsule ciliate, with long, fine hairs.

1. E. rostkoviana Hayne

E. officinalis L., nom. ambig.
Stem erect, up to 35(–50) cm, light green or purplish,
usually branched; branches usually again branched.

Cauline and lower floral internodes shorter than or
1.5–3(–4) times as long as lvs. Lowest fl. at node
(3–)6–10(–14). Lvs usually light green, or the lower
sometimes purplish above, setose, all but the lowest
also more or less densely clothed with glandular hairs
having the stalk c. 10 times as long as gland. Cauline
lvs 3–17 mm, orbicular-ovate or narrowly ovate,
rounded or truncate at base, rarely abruptly contracted
into a very short petiole, crenate to crenate-serrate with
1–7 pairs of obtuse to subacute teeth and an obtuse
terminal lobe, teeth of middle lvs usually with both margins convex. Lower floral lvs 6–15 mm, deltate or
broadly ovate to oblong-ovate, base rounded, truncate
or subcordate, margin serrate with 4–9 pairs of subacute
to acute or acuminate teeth; teeth usually not very much
longer than broad, the basal pairs patent or retrorse;
terminal lobe obtuse to acute. Calyx with the teeth
narrowly deltate and acuminate, or linear and aristate.
Corolla 6.5–12.5 mm; lower lip usually porrect,
white. Capsule usually less than twice as long as wide.
Fl. 6–9. Th.

Subsp. rostkoviana

E. hirtella auct. angl., non Jordan ex Reuter, pro parte
Stems with 1–5(–12) pairs of ascending to erect
branches. Cauline internodes mostly not more than 3
times as long as lvs. Lowest fl. at node 6–10. Upper
corolla lip usually lilac. Capsule 4–5 mm, shorter than
to about as long as calyx, oblong, truncate or retuse.
$2n = 22$.
Native. Moist meadows. Wales and W. England and
Scotland from Gloucester to Argyll, locally eastwards
to Northumberland and Perth; Ireland, but local in the
north-east. N. and C. Europe, except N. Fennoscandia
and the northern islands, extending southwards to the
Pyrenees, and the N. Balkan peninsula; Urals; Caucasus, N.E. Turkey.

E. anglica introgressed with *E. nemorosa*, which occurs
in southern England, and forms of *E. arctica* subsp.
borealis with long, glandular hairs, which occur in Scotland, are liable to be confused with *E. rostkoviana*
subsp. *rostkoviana*.

Subsp. montana (Jordan) Wettst.

Stems unbranched, or with 3(4) pairs of erect branches.
Cauline internodes 2–6(–10) times as long as lvs.
Lowest fl. at node 2–6. Upper corolla-lip often lilac,
rarely whole corolla lilac or purple. Capsule
(4–)4.5–5.5(–6.5) mm, often longer than calyx, oblong
or elliptical, truncate to retuse or emarginate and with
rounded lobes. $2n = 22$.
Native. Montane grasslands. Brecon, Caernarvon,
W. Yorks and Westmorland to Dumfries and Selkirk.
From the Baltic states and Belgium southwards to the
mountains of the Balkan peninsula, the Carpathians and
the Urals.

2. E. rivularis Pugsley

Stem slender, flexuous, up to c. 10(–15) cm, light green
or purplish, with 0–2 pairs of short branches. Cauline

internodes shorter than or up to 4 times as long as lvs; lower floral internodes c. 1–2.5 times as long as lvs. Lowest fl. at node (2–)3–5(–6). Lvs usually light green, sometimes purplish above, ±setose; upper with sparse multicellular glandular hairs having stalk 6–10 times as long as gland. Cauline lvs 2–5.5 mm, orbicular to elliptical, rounded at base, contracted into a very short petiole, crenate, with 1–4 pairs of very obtuse teeth and an obtuse terminal lobe. Lower floral lvs 3–6(–7) mm, orbicular, broadly ovate or elliptical, truncate, rounded or broadly cuneate at base, crenate or serrate, with 3–5(–6) pairs of obtuse to acute and slightly curved teeth on each side, the basal pairs patent, or directed towards apex and then sometimes distant from lf-base; terminal lobe obtuse to acute, sometimes nearly equalled by the distal pair of teeth. Calyx sometimes blotched with black; teeth deltate, acute or acuminate. Corolla 6.5–9 mm; lower lip relatively large, porrect, white or lilac; upper lip lilac. Capsule 3.5–5 mm, longer or shorter than the calyx, usually much less than twice as long as wide, broadly obovate-elliptical, rounded to truncate or emarginate. Fl. 5–7. 2n = 22. Th.

Damp mountain pastures. N. Wales (Merioneth, Caernarvon), Cumberland. Endemic.

3. E. anglica Pugsley

E. hirtella auct. angl., non Jordan ex Reuter, p.p.

Stem usually flexuous, 5–16(–30) cm, green or purplish; branches (0–)1–4(–6) pairs, flexuous or arcuate, usually again branched. Cauline internodes shorter than or up to 2.5(–3) times as long as lvs (the lower often very short); lower floral internodes shorter than or up to 1.5(–2.5) times as long as lvs. Lowest fl. at node 5–8. Lvs green, sometimes brownish or purplish above, setose; upper also with usually rather dense glandular hairs, having a stalk c. 10 times as long as gland. Cauline lvs 2–9 mm, oblong, ovate or deltate, rounded to cuneate or subcordate at base, usually contracted into a very short petiole, crenate or crenate-serrate, with 1–6 pairs of obtuse to subacute teeth and an obtuse terminal lobe. Lower floral lvs 5–12 mm, suborbicular, deltate, ovate or oblong-ovate, cordate to rounded at base, crenate or serrate, with 4–7 pairs of obtuse to acute teeth, the teeth usually not much longer than wide, the basal pairs patent to retrorse; terminal lobe obtuse to acute. Calyx-teeth deltate to acuminate or aristate. Corolla (5–)6.5–8(–10) mm; lower lip usually porrect, white or lilac; upper usually lilac. Capsule 4–5.5 mm, usually slightly shorter than the calyx and not more than twice as long as wide, oblong, truncate or retuse. Fl. 5–9. 2n = 22. Th.

Pastures, heathland. England and Wales, northwards to Lincoln, Westmorland and Isle of Man; Kirkcudbright; local in Ireland from Waterford and Clare to Cavan and Antrim. Endemic.

Distinguished from *E. rostkoviana* chiefly by its shorter lower internodes and usually by its flexuous lower stem and branches, and smaller corolla.

4. E. vigursii Davey

Stem erect, 6–18(–25) cm, purplish; branches 0–5(–7) pairs, erect, often branched. Cauline internodes 1–3 times as long as lvs; lower floral internodes 1–2(–3) times as long as lvs. Lowest fls at node 7–10(–12). Lvs greyish-green, ±dull violet at margins and on veins beneath, or dull violet to blackish purple all over upper surface, ±black-spotted, slightly scabrid to setose, the upper also sparsely to densely glandular-hairy, the longer hairs with stalks 6–10 times as long as gland, some plants with all lvs scabrid. Cauline lvs 3–11 mm, the lower broadly elliptical or oblong-elliptical, the upper ovate, oblong-ovate or orbicular, rounded or truncate at base, ±contracted into a very short petiole, crenate or serrate, with 1–6 pairs of very obtuse to acute teeth and an obtuse terminal lobe which is broadly rounded in the lower leaves. Lower floral lvs 5–10 mm, ovate, rounded to subcordate at base, crenate-serrate, with 4–6 pairs of subacute to acute teeth; teeth usually not much longer than wide, the basal pairs patent; terminal lobe subacute to acute. Calyx sometimes spotted or blotched with black; teeth ±narrowly deltate, acute or acuminate. Corolla (6–)7–8.5 mm; lower lip relatively large, occasionally white but usually lilac to deep reddish purple; upper lip lilac to reddish purple. Capsule (3–)4–5.5 mm, shorter than to about as long as calyx, not more than twice as long as wide, oblong or obovate-oblong, truncate to emarginate. Fl. 6–9. 2n = 22. Th.

Native. *Agrostis curtisii–Ulex gallii* heaths. Cornwall and Devon. Perhaps originating from hybridization between *E. anglica* and *E. micrantha*.

5. E. arctica Lange ex Rostrup

Stem erect from a flexuous or decumbent base, stout, up to 30(–35) cm, green or purplish; branches ascending or erect, sometimes again branched. Cauline internodes up to 4(–5) times as long as lvs; lower floral internodes usually 1–4 times as long as lvs. Lvs bright or dark green, the lower often tinged with brown or purple above. Upper cauline lvs 12(–16) mm, usually broadly ovate. Lower floral lvs 4–10(–14) mm, rounded, truncate or subcordate at base, usually contracted into a very short petiole, crenate to serrate. Calyx-tube occasionally blotched with black; veins and margins frequently blackish. Corolla-tube relatively short; lower lip relatively large, white to purple, but usually with lower lip white and the upper lilac. Fl. 6–8. Th.

Native. N. Europe extending southwards to the E. Carpathians.

Subsp. arctica

Stem with 0–2(–5) pairs of long, often flexuous, branches. Lowest fl. at node (3)4–6(7). Lvs glabrous or scabridulous. Cauline lvs 4–11(–16) mm, orbicular to broadly ovate or oblong, rounded or broadly cuneate at base, sessile or shortly petiolate, crenate or crenate-serrate, with 1–4 pairs of obtuse to subacute teeth, the proximal margins of distal teeth usually with a distinct obtuse angle; terminal lobe obtuse. Lower floral lvs suborbicular to broadly ovate, with 3–5(–6) pairs of obtuse to acuminate teeth; teeth usually not much longer than wide, the basal pairs patent. Calyx-teeth elliptical and obtuse to deltate, acute or acuminate. Corolla 7–11(–13) mm. Capsule (5.5–)6–7.5(–8) mm, about as long as calyx, elliptical to obovate, emarginate or retuse. Early flowering.

Native. Rough pastures in Orkney and Shetland; Faeroes.

Subsp. **borealis** (Towns.) P. F. Yeo
E. brevipila sensu Wettst., Pugsley, non Burnat & Gremli.
Stem with 0–5(–6) pairs of straight, often short branches. Lowest fl. at node 4–8(–10). Lvs scabridulous or shortly setose, the upper usually also with short glandular hairs having stalk 2–6(–10) times as long as gland. Cauline lvs 3–14 mm, ovate or oblong, cuneate to truncate at base, crenate to serrate, with 1–5(–6) pairs of obtuse to acute teeth; terminal lobe ±acute. Lower floral lvs ovate to deltate, with (3–)4–6(–7) pairs of subacute to aristate teeth; teeth usually much longer than wide, the basal pairs usually patent but sometimes incurved or directed to apex. Calyx-teeth deltate, acute to aristate. Corolla 6–9(–10) mm. Capsule (4–)4.5–6.5(–7) mm, shorter than to about as long as calyx, elliptical to oblong, truncate to emarginate. $2n = 44$.

Native. Meadows, pastures, roadsides and disturbed ground. Rather local from Cornwall to Isle of Wight and Bucks northwards to Yorks, more widespread and rather common further north (except Shetland) and in N. Wales; throughout Ireland but more frequent in the west. N. France and Norway.

*E. stricta J. P. Wolff ex J. F. Lehm, native over much of Europe, which differs most obviously from *E. arctica* in its proportionately longer capsule and with the lower floral lvs having the basal teeth markedly directed to the apex, is probably locally introduced in some meadows in Scotland.

6. E. tetraquetra (Breb.) Arrond.

E. occidentalis Wettst.

Stem erect, stout, up to 15(–20) cm, usually purplish; branches 0–5(–8) pairs, frequently rather short, erect or ascending, sometimes branched. Cauline internodes up to 1.5(–3) times as long as lvs; lower floral internodes up to 1.5(–2) times as long as lvs, frequently all the internodes shorter than the lvs. Lowest fl. at node (3–)5–7(–9). Lvs light to dark green, glossy, rugose, sometimes tinged with dull violet or purple, the lower setose, setulose or scabridulous, the upper also often with glandular hairs having stalks up to c. 4 times as long as gland, or glabrous. Cauline lvs 2–9(–11) mm, ovate to obovate or suborbicular, cuneate or rounded at base, with a short petiole, crenate or crenate-serrate, with 1–3(–4) pairs of obtuse to subacute teeth, the basal pairs usually distant from the leaf-base; terminal lobe broadly obtuse. Lower floral lvs 5–10(–14) mm, ovate,

deltate to occasionally oblong-ovate or trullate, truncate or occasionally broadly cuneate, rounded or subcordate at base, crenate-serrate to serrate, with 3–5(–6) pairs of subacute to acute or sometimes narrowly acuminate teeth; teeth as long as or longer than wide, the basal pairs patent or slightly retrorse; terminal lobe obtuse to acute. Floral lvs usually forming a dense, sometimes 4-angled, spike. Calyx-teeth deltate, acute or aristate. Corolla (4–)5–7(–8) mm; lower lip deflexed, white or sometimes lilac; upper lip white or frequently lilac. Capsule 4.5–5.5(–6) mm, usually shorter than calyx, at least twice as long as wide, oblong or elliptical-oblong, retuse or truncate. Fl. 5–8. $2n = 44$. Th.

Native. Short turf on sea-cliffs, maritime dunes and inland limestone pastures. Coasts of the British Is. except E. England from Essex to Yorks, and Ross, Sutherland and the Outer Hebrides; inland in Devon, Somerset, Dorset, Wilts, Gloucester and Oxford, and in King's Co., Sligo and Antrim in Ireland. Coasts of N.W. France, perhaps Germany.

An early-flowering species.

Hybrids between *E. tetraquetra* and *E. confusa* are frequent in Britain.

7. E. nemorosa (Pers.) Wallr.

E. curta (Fries) Wettst.

Stem erect, up to 35(–40) cm, usually purplish, usually with 1–9 pairs of ascending branches; branches often again branched. Cauline internodes shorter than to up to 4(–7) times as long as lvs; lower floral internodes c. 1–2(–3) times as long as lvs, occasionally all internodes shorter than lvs. Lowest fls at node (5–)10–14. Lvs ±dark green, glossy, rugose, the veins being rather deeply impressed; lower lvs often dark violet above, the upper sometimes tinged blackish violet towards the margins; lower lvs shortly setose to hirsute, the upper glabrous or scabridulous to densely hirsute, rarely with some glandular hairs having stalk up to c. 4 times as long as gland. Cauline lvs 2–12 mm, elliptical or oblong to ovate or deltate, the lower cuneate at base and with short petiole, the upper with rounded or truncate base, sometimes shortly petiolate, crenate or crenate-serrate, with 1–5(–7) pairs of obtuse to acute teeth; terminal lobe obtuse to acute. Lower floral lvs 4–9(–11) mm, ovate or deltate, rounded, truncate or occasionally subcordate at base, crenate-serrate to serrate, with (3–)4–6(–9) pairs of subacute to acuminate or aristate teeth; teeth usually slightly longer than wide, the basal pairs patent or slightly retrorse; terminal lobe subacute to acute. Calyx with veins sometimes blackish; teeth narrowly deltate, acuminate to aristate. Corolla 5–7.5(–8.5) mm; lower lip deflexed, white or sometimes lilac; upper lip white or lilac. Capsule 4–5.5(–6) mm, usually slightly shorter than calyx, more than twice as long as wide, oblong or elliptical-oblong, truncate or retuse. Fl. 7–9. $2n = 44$. Th.

Native. Pastures, heaths, wood margins and scrub, dunes. Common in England and Wales but more local in Scotland and absent from much of the area from Fife and Inverness to Aberdeen; widespread but very local in Ireland. N. and C. Europe, extending southwards to N.E. Spain. North America.

A late-flowering species. On dunes in N. Scotland and W. Ireland *E. nemorosa* tends to have larger and wider lvs, and depauperate plants may resemble *E. tetraquetra*, from which they differ in the more profuse branching, long fine basal teeth of the lower floral lvs and larger corollas.

The hybrid *E. nemorosa × confusa* is widespread from Cornwall and East Anglia to N. Scotland.

8. E. pseudokerneri Pugsley

Stem erect or flexuous, up to 20(–30) cm, usually purplish; branches (0–)3–8(–10) pairs, spreading or ascending, often again branched. Cauline internodes shorter than to 2.5(–3) times as long as lvs; lower floral internodes shorter than to 1.5(–2) times as long as lvs. Lowest fl. at node (5–)10–16(–18). Lvs green, usually tinged brown, deep reddish or blackish purple above, especially in the cauline lvs; lower lvs usually scabridulous or setulose, the upper glabrous. Cauline lvs 2–10(–15) mm; lower oblong-ovate, rounded or cuneate at base, with obtuse to subacute teeth and obtuse terminal lobe; upper ovate, rounded or truncate at base, serrate, with up to 5(–7) pairs of acute to acuminate teeth; terminal lobe subacute to acute. Lower floral lvs 3–9(–11) mm, deltate, ovate or oblong-ovate, truncate, rounded or broadly cuneate at base, serrate, with 3–6(–7) pairs of acute to aristate teeth; teeth about as long as or much longer than wide, the basal pairs patent and usually aristate; terminal lobe acute or acuminate. Calyx with the veins and margins usually blackish; teeth narrowly deltate to sublinear, acuminate or aristate. Corolla (6–)7–9(–11) mm, white or lilac; lower lip deflexed. Capsule 3.5–5(–6) mm, much shorter than calyx, c. 2.5 times as long as wide, oblong or elliptical-oblong, truncate or retuse. Fl. 7–9. $2n = 44$. Th.

Native. Dry calcareous grassland, especially on chalk, and calcareous fens. S. England from Somerset, Gloucester and Lincoln eastwards; on coastal limestone in W. Ireland from Limerick to Donegal. Endemic.

9. E. confusa Pugsley

E. atroviolacea Druce & Lumb

Stem flexuous, sometimes decumbent at the base, up to 20(–45) cm; branches (0–)2–8(–10) pairs, usually long, slender, flexuous, spreading or ascending, or sometimes short and erect, usually branched. Cauline internodes shorter than or up to 2.5 times as long as lvs; lower floral internodes shorter than or up to 2(–3) times as long as lvs, decreasing upwards gradually; sometimes all the internodes shorter than the lvs. Lowest fl. at node (2–)5–12(–14). Lvs deep or greyish-green, often flushed brown or dark violet, especially towards the margins and on the upper surface, glabrous, or the lower scabridulous, or all scabridulous or finely and sparsely setulose, occasionally with a few long bristles, sometimes also with more or less sparse short glandular

hairs having the stalk about 5 times as long as gland. Cauline lvs 2–10 mm, ovate, oblong- to lanceolate-ovate, cuneate or rounded at base, usually shortly petiolate, crenate to serrate, with 1–5 pairs of obtuse to acute teeth; terminal lobe broadly obtuse to subacute. Lower floral lvs 3.5–9(–10) mm, often alternate and with a flower only in every other axil, ovate to oblong-ovate, rounded or cuneate at base, occasionally truncate, usually very shortly petiolate, crenate-serrate to serrate, with 2–6 pairs of subacute to acute teeth; teeth about as long as or much longer than wide, often incurved, the basal pairs patent or directed towards apex and distant from the leaf-base; terminal lobe obtuse to acute. Calyx-tube sometimes blotched with black, the veins and margins sometimes blackish; teeth narrowly deltate, acuminate. Corolla 5–9 mm; upper lip white or more usually lilac, the lower white or lilac, or both more or less deep reddish-purple or, very rarely, yellowish. Capsule 3.5–5.5(–6.5) mm, usually about as long as calyx, c. 2–3 times as long as wide, oblong or elliptical-oblong, truncate, retuse or emarginate. Fl. 7–9. $2n = 44$. Th.

Native. Rough grassland, moorland, heaths and dunes. Throughout the British Is. but absent from much of C. and S.E. England; mostly coastal in Ireland from Wexford and Clare northwards, Kerry and Cork. Faeroes.

Irish plants are always unbranched.

10. E. frigida Pugsley

Stem flexuous or erect, up to 20(–30) cm, green or purplish; branches 0–2 pairs, erect, occasionally branched. Cauline internodes c. 1–5(–10) times as long as lvs; lower floral internodes 1–4(–5) times as long as lvs, usually decreasing upwards gradually. Lowest fl. at node 2–4(–5). Lvs light green, the lower sometimes tinged with brown or purple above, occasionally green above and purple beneath, glabrous or usually pilose or setose with sparse to dense, straight, slender bristles, some hairs occasionally glandular and having stalk up to c. 5 times as long as gland. Cauline lvs 3–11(–13) mm, suborbicular to elliptical, ovate, oblong or obovate, ±broadly cuneate at base, shortly petiolate, shallowly to deeply crenate or crenate-serrate, with 1–4(–5) pairs of obtuse to subacute or, rarely, acute teeth; terminal lobe truncate, rounded or subacute. Lower floral lvs (3–)5–12(–18) mm, frequently alternate, elliptical to ovate or deltate to suborbicular, cuneate to truncate or subcordate at base, crenate or crenate-serrate, with 2–5(–8) pairs of obtuse to acute teeth; teeth usually not longer than wide, having the proximal margin at least strongly curved, the basal pairs ±patent, or sometimes directed to apex; terminal lobe truncate to subacute. Calyx-tube occasionally blotched with black; veins sometimes black; teeth deltate, acute or acuminate. Corolla 4.5–7(–8) mm; upper lip white or lilac, rarely purple; lower lip white, rarely lilac. Capsule (4–)5–7 mm, about as long as or longer than calyx, less than to slightly more than twice as long as wide, obovate, elliptical to broadly oblong, emarginate to retuse or truncate. Fl. 7–8. $2n = 44$. Th.

Native. Damp alpine rock-ledges, grassland and mountain tops, usually above 610 m and ascending to 1070 m. Cumberland, Dumfries, Peebles, Roxburgh and from Islay and Angus northwards to Sutherland and Outer Hebrides; very local in Kerry, Galway, Sligo and S. Donegal. N. Europe; North America.

11. E. foulaensis Towns. ex Wettst.

Stem erect, rather stout, up to 6(–9) cm, purplish; branches up to 3(–4) pairs, short, ascending, occasionally branched. Cauline internodes up to 1–2(–4) times as long as lvs; lower floral internodes usually slightly shorter than lvs, the upper much shorter, sometimes all the internodes shorter than the lvs. Lowest fl. at node (2–)4–6. Lvs with thickened margins, deep green, often tinted dull violet, easily blackening on drying, glabrous or sparsely hairy. Cauline lvs 2–10 mm, suborbicular, broadly obovate or oblong, broadly cuneate or rounded at base, contracted into a very short petiole, crenate with 1–4 pairs of shallow, obtuse teeth; terminal lobe obtuse. Lower floral lvs (3–)5–9 mm, suborbicular to broadly ovate, broadly cuneate or rounded at base, contracted into a very short petiole, crenate to serrate, with 2–5(–6) pairs of obtuse to acute or sometimes acuminate teeth; teeth usually not much longer than wide, with both margins usually convex, the basal pairs patent, the distal usually strongly incurved; terminal lobe obtuse to subacute or cuspidate. Calyx-tube sometimes blotched or striated with black; veins and margins sometimes blackish; teeth deltate, acute. Corolla 4–6 mm, white to purple. Capsule 4.5–5.5(–7) mm, usually as long as or longer than the calyx, about twice as long as wide, oblong or elliptic-oblong, sometimes blotched with black, retuse or emarginate. Fl. 7–8. $2n = 44$. Th.

Native. Sea-cliffs, coastal pastures and edges of salt-marshes from Moray and Elgin to Shetland and Outer Hebrides. Faeroes.

12. E. cambrica Pugsley

Stem flexuous, up to c. 8cm, with 0–2 pairs of flexuous branches. Cauline internodes much shorter than or up to 1.5(–2) times as long as lvs; lower floral internodes shorter than or up to 1(–1.5) times as long as lvs. Lowest fl. at node 2–4. Lvs light green, sometimes purplish, with rather sparse long eglandular hairs. Cauline lvs 2–7 mm, broadly ovate or oblong to obovate, rounded at base, shortly petiolate, crenate, with 1–3 pairs of obtuse teeth and an obtuse terminal lobe. Lower floral lvs 3.5–9 mm, broadly ovate or suborbicular to deltate, truncate or rounded at base, crenate or crenate-serrate, with 2–4 pairs of obtuse to subacute teeth; teeth about as long as wide, the basal pairs patent; terminal lobe obtuse. Calyx teeth deltate, acute or acuminate. Corolla 4–5.5 mm; lower lip white or yellowish white, the lobes usually very small and sometimes incurved; upper lip white or lilac. Capsule (4–)5–7 mm, longer than the calyx, up to twice as long as wide, elliptical or obovate, emarginate. Fl. 7. $2n = 44$. Th.

Mountain grassland and rocky ledges in Caernarvon and Merioneth. Endemic.

13. E. ostenfeldii (Pugsley) P. F. Yeo

E. curta var. *ostenfeldii* Pugsley, *E. curta* auct., non (Fries) Wettst., pro parte

Stem erect or flexuous at base, up to 12(–15) cm; branches 0–4(–6) pairs, erect or ascending, sometimes branched. Cauline internodes shorter than or up to 3(–5) times as long as lvs; lower floral internodes shorter than or up to 1.5(–2.5) times as long as lvs, the upper often very short. Lowest fl. at node (3–)4–7(–9). Lvs with recurved margins, ±strongly purplish on upper or both surfaces, ±densely clothed with long eglandular hairs. Cauline lvs 2–10(–14) mm, suborbicular or ovate to oblong-ovate or obovate, rounded or cuneate at base, crenate to crenate-serrate, with 1–4(–5) pairs of obtuse to subacute teeth; terminal lobe obtuse. Lower floral lvs 4–8(–10) mm, suborbicular, broadly ovate or oblong-ovate, rounded, truncate or broadly cuneate at base, crenate-serrate to serrate, with 3–5 pairs of subacute to acute teeth; teeth usually about as long as wide, the basal pairs more or less patent; terminal lobe obtuse or subacute. Calyx-tube often tinged with purple, sometimes with black spots, sometimes blackish on veins; teeth deltate, acute or acuminate. Corolla (3.5–)4.5–6 mm; lower lip ±hairy beneath, white; upper white or lilac. Capsule 4–5.5(–6) mm, usually about as long as or longer than calyx, up to twice as long as wide, elliptical-oblong or oblong, retuse or emarginate. Fl. 7–9. $2n = 44$. Th.

Native. Grassland, stony and sandy places, rock-ledges, often near the sea. N.W. of Great Britain, from N. Wales to Shetland. Faeroes; Iceland.

E. eurycarpa Pugsley, described from montane slopes at 610 m on the Isle of Rhum (Inner Hebrides), is a dwarf hairy plant with rounded lvs and very wide, conspicuously emarginate capsules. It is probably a local hybrid derivative of *E. ostenfeldii*.

14. E. marshallii Pugsley

Stem erect, up to c. 12 cm, with (0–)1–5 pairs of rather long erect branches; branches sometimes again branched. Cauline internodes shorter than or up to 2.5(–3) times as long as lvs; lower floral internodes shorter than or up to 1.5 times as long as lvs, the upper very short. Lowest fl. at node (5–)7–9. Lvs grey-green, densely hairy. Cauline lvs 2–11 mm, ovate or ovate-oblong to elliptical, rounded or cuneate at base, crenate to serrate, with 1–5 pairs of obtuse to acute teeth; terminal lobe obtuse to subacute. Lower floral lvs 5–10 mm, broadly ovate or rhombic truncate to cuneate at base, serrate, with 4–5 pairs of acute teeth; teeth mostly about as wide as long, the basal pairs patent or directed towards apex; terminal lobe obtuse to acute. Calyx-teeth deltate, acute to aristate. Corolla 5.5–7 mm, white or lilac; lower lip hairy beneath, often large. Capsule 4.5–5.5(–6.5) mm, usually slightly shorter than to about as long as calyx, rarely longer, about twice or more as long as wide, oblong or elliptic-oblong, retuse or truncate. Fl. 7–8. $2n = 44$. Th.

Native. Very local in grassland on sea-cliffs or stabilized dunes. West Ross, Sutherland, Caithness, Outer Hebrides, Orkney and Shetland. Endemic.

Apparently hybridizes with *E. nemorosa* and *E. arctica* subsp. *borealis*.

15. E. rotundifolia Pugsley

Stem erect, up to c. 10 cm, with 0–3 pairs of short erect branches. Cauline internodes 1–2(–2.5) times as long as lvs; lower floral internodes shorter than to up to 1.5 times as long as lvs, the upper very short. Lowest fl. at node 6–8(–9). Lvs grey-green, densely hairy. Cauline lvs 2–12 mm, orbicular or broadly to oblong-ovate, rounded or truncate to broadly cuneate at base, crenate, with 1–4(–5) pairs of obtuse teeth and an obtuse terminal lobe. Lower floral lvs 5–8 mm, suborbicular to broadly ovate, rounded at base, crenate or crenate-serrate, with 3–5 pairs of obtuse to subacute teeth; teeth mostly wider than long, the basal pairs patent; terminal lobe obtuse. Calyx-teeth rather broadly deltate, subacute to acute. Corolla 5–6 mm; lower lip hairy beneath, white or lilac; upper lip white or purplish. Capsule 5.5–6 mm, slightly longer than calyx, elliptic-oblong, about twice as long as wide, retuse. Fl. 7. Th.

Native. Sea-cliffs or stabilized dunes. Sutherland, Caithness, Outer Hebrides, Orkney and Shetland. Endemic.

16. E. campbelliae Pugsley

Stem erect, up to c. 10 cm, green or purplish, with 0–2 pairs of short erect branches. Cauline internodes 1–2.5(–4) times as long as lvs; lower floral internodes up to about as long as lvs. Lowest fl. at node 5–7. Lvs usually ±tinged with purple or brown towards the edges, sometimes purple beneath and green above; veins beneath sometimes blackish, with sparse to moderately dense long bristles, especially distally. Cauline lvs 2–7 mm, obovate to rather narrowly ovate or elliptical, cuneate at base, crenate or crenate-serrate, with 1–3 pairs of obtuse to subacute teeth and an obtuse terminal lobe. Lower floral lvs 4–8 mm, ovate, rounded or cuneate at base, tapered into a short petiole, crenate to serrate with 2–5(–6) pairs of obtuse to acute teeth; teeth about as long as or longer than wide, the basal pairs directed to apex or patent; terminal lobe obtuse to subacute. Calyx-tube often papery and whitish; veins and margins usually blackish; teeth deltate, acute or acuminate. Corolla 5.5–7 mm; lower lip white or tinged with lilac; upper lip lilac or sometimes white. Capsule 4.5–7(–7.5) mm, shorter than the calyx, more than twice as long as wide, oblong or elliptic-oblong, retuse or emarginate. Fl. 7. Th.

Native. Grassy moors near the sea. Isle of Lewis (Outer Hebrides), perhaps also in Shetland. Endemic.

Perhaps originating from hybridization between *E. marshallii* and *E. micrantha* or *E. scottica*.

17. E. micrantha Reichenb.

Stem erect, slender, up to c. 25 cm, usually dark purple; branches (0–)2–7(–10) pairs, slender, erect, usually branched. Cauline internodes shorter than to 2–3(–4) times as long as lvs; lower floral internodes 1.5–2.5(–3) times as long as lvs, decreasing upwards gradually. Lowest fl. at node (4–)6–14(–16). Lvs usually glossy, green and tinged with purple or blackish purple, glabrous or scabridulous. Cauline lvs 2–8(–11) mm, rather narrowly ovate or ovate-oblong to obovate, cuneate at base, often tapering into a short petiole, crenate to serrate, with 1–6 pairs of obtuse to acute teeth and a large obtuse to acuminate terminal lobe. Lower floral lvs 3.5–7(–8) mm, ovate, deltate or rhombic, rounded, truncate or cuneate at base, serrate, with 3–6 pairs of acute or acuminate teeth; teeth about as long as or longer than wide, the basal pairs patent or directed to apex; terminal lobe subacute to acute or acuminate. Calyx-teeth narrowly deltate, acuminate or aristate. Corolla 4.5–6.5 mm, lilac to purple or white with upper lip lilac, or occasionally entirely white; lower lip usually small. Capsule 3–5(–6) mm, shorter than or sometimes longer than the calyx, more than twice as long as wide, oblong or occasionally elliptic-oblong, truncate, rounded or retuse. Fl. 7–9. $2n = 44$. Th.

Native. Heaths, normally associated with *Calluna*, sometimes in small bogs. S.W. England, Wales and from Lancs and Yorks northwards, also in scattered localities in Warwick, Wilts and Dorset to Sussex but perhaps extinct in the south-east; throughout Ireland but not reported from much of the centre and south. N. and C. Europe, extending to N. Spain and N. Italy; North America.

E. rhumica Pugsley, described from the Isle of Rhum (Inner Hebrides), which has very small lvs and corolla, and a retuse or emarginate capsule, appears to be a local hybrid segregate of *E. micrantha*.

E. micrantha hybridizes especially with *E. arctica* subsp. *borealis* and *E. confusa*.

18. E. scottica Wettst.

E. paludosa Towns., non R. Br.

Stem erect, up to c. 25 cm, green or purplish, with 0–4 pairs of long arcuate-erect branches. Cauline internodes (1–)2–5 times as long as lvs; lower floral internodes c. 1.5–3(–5) times as long as lvs, decreasing upwards gradually. Lowest fl. at node (2–)3–6(–8). Lvs light green, ±tinged with purple or brown above, frequently purple beneath and green above, glabrous to setulose or sometimes moderately hairy. Cauline lvs 2–13(–17) mm, elliptical to ovate-elliptical or obovate, cuneate or rounded at base, sometimes contracted into a distinct short petiole, crenate or crenate-serrate, with 1–5(–6) pairs of shallow, obtuse to subacute teeth and an obtuse terminal lobe. Lower floral lvs 4–10(–17) mm, some often alternate, ovate to oblong or rhombic, rounded, truncate or cuneate at base, sometimes contracted into a distinct short petiole, crenate-serrate or serrate, with 3–5(–6) pairs of subacute to acute teeth; teeth usually not longer than wide, the basal pairs directed to apex or sometimes patent; terminal lobe obtuse to acute. Calyx with the veins and margins often blackish; teeth deltate, usually narrowly so, acute to aristate. Corolla (3.5–)4.5–6.5 mm; lower lip usually small, narrowly lobed and white, rarely tinged with lilac; upper lip white or lilac. Capsule 4.5–5.5(–9) mm, as long as or longer than calyx, usually more than twice as long as wide, oblong or elliptical-oblong, retuse or emarginate. Fl. 7–8. $2n = 44$. Th.

Native. Wet moorland and flushes, ascending to 915 m. N. Devon, Brecon to Anglesey, Lancs and Yorks northwards to Shetland; in Ireland usually in *Schoenus* fens from Kerry, Cork and Carlow to Fermanagh and W. Donegal. Fennoscandia, Faeroes, ? C. Russia.

The hybrid *E. confusa* × *scottica*, which differs from *E. scottica* in its more flexuous stems and branches, its longer, finer lf-teeth and sometimes its shorter, wider lvs and larger corolla, occurs throughout much of the range of *E. scottica* except the C. and N. Highlands. It is usually found in flushes or damp stream-margins in hill country and may be locally more common than *E. scottica*.

19. E. heslop-harrisonii Pugsley

Stem usually flexuous at base, up to 15 cm; branches 0–4(–5) pairs, erect or spreading, sometimes branched. Cauline internodes shorter than or 1–2(–3) times as long as lvs; lower floral internodes shorter than or up to 2(–2.5) times as long as lvs, sometimes all internodes shorter than lvs. Lowest fl. at node 4–7(–8). Lvs light green, occasionally tinged with brown or purple, scabridulous or sparsely setose, or the upper often glabrous. Cauline lvs 2–11.5 mm, ovate or elliptical to elliptical-oblong or narrowly obovate, rounded or cuneate at base, usually contracted into a distinct short or rather long petiole, crenate or crenate-serrate, with 1–4 pairs of obtuse to subacute teeth; teeth with the proximal margin usually sinuous, the basal pair usually distant from lf-base; terminal lobe obtuse. Lower floral lvs 4–10.5 mm, broadly ovate to oblong-ovate, rounded, truncate or broadly cuneate at base, sometimes contracted into a short or very short petiole, crenate to serrate, with 2–5 pairs of obtuse to acute or acuminate teeth; teeth about as long as wide, the basal pairs patent or, if the lf-base is rounded, directed to apex, but then frequently with patent tips. Upper floral lvs with teeth not aristate. Calyx with veins occasionally purplish; teeth deltate, acute. Corolla 4.5–6(–6.5) mm, white or occasionally lilac; lower lip relatively small. Capsule (4.5–)5.5–7 mm, as long as or longer than calyx, 2–3 times as long as wide, oblong or elliptical, often slightly curved, truncate, retuse or occasionally emarginate. Fl. 8. Th.

Native. Coastal grassland and marshy areas. Isle of Rhum (Inner Hebrides), W. Ross, Sutherland, Orkney and Shetland. Endemic.

Perhaps a hybrid derivative of *E. scottica*.

Subsection *Angustifoliae* (Wettst.) Joerg. Floral lvs generally less than half as wide as long. Capsule glabrous, or ciliate with weak marginal hairs.

20. E. salisburgensis Funck   Irish Eyebright.

Stem erect or flexuous, up to c. 12 cm, usually purplish; branches (0–)1–7 pairs, slender, erect or spreading, often branched. Cauline and lower floral internodes shorter than or up to twice as long as lvs, sometimes all internodes shorter than lvs. Lowest fl. at node 5–13. Lvs light green, glossy, often strongly tinged brown or purple, glabrous or scabridulous. Cauline lvs 2–8(–10) mm, up to 5 mm wide overall, mostly oblanceolate to oblong or ovate-lanceolate, with 1–3 distant pairs of obtuse, deltate to narrowly falcate, acuminate teeth, the basal distant from the leaf-base; terminal lobe obtuse to acuminate. Lower floral lvs 3.5–8(–9) mm, up to 5 mm wide overall, lanceolate to ovate-lanceolate or oblong-ovate, with 2–4 pairs of deltate, subacute to sublinear, aristate teeth, the apical pair distant from the others, the basal pair distant from the leaf-base; terminal lobe obtuse to acuminate. Calyx with veins and margins sometimes blackish; teeth deltate, acute to acuminate. Corolla 4.5–6.5 mm, white, or the upper lip sometimes lilac. Capsule 3.5–6 mm, much shorter than to much longer than calyx, 2–4 times as long as wide, oblong or elliptical-oblong to elliptical-obovate or cuneate-obovate, truncate, retuse or emarginate. Fl. 7–8. $2n = 44$. Th.

Native. Among limestone pavement and on dunes, usually associated with *Thymus arcticus*. Widespread in W. Ireland from Limerick to Donegal. Mountains of C. and S. Europe, Norway, Sweden (Gotland); N.W. Turkey.

The above description refers to var. **hibernica** Pugsley, which is endemic. Hybridizes with *E. micrantha*, *E. nemorosa* and *E. arctica* subsp. *borealis*.

20. ODONTITES Ludwig

Annual herbs with opposite lvs. Fls in a secund, terminal, spike-like raceme, in the axils of bracts resembling the lvs but smaller. Calyx 4-toothed, tubular-campanulate. Corolla 2-lipped; *upper lip concave, entire or emarginate*, the lower 3-lobed with entire lobes. Stamens 4; anther cells mucronate. *Seeds* small, rather few, oblong or fusiform, *furrowed, hilum basal*.

About 30 spp., in W. and S. Europe, W. Asia and N. Africa.

1. O. verna (Bell.) Dumort.   Red Bartsia.

Bartsia odontites (L.) Hudson; *O. rubra* Gilib.

Erect branching pubescent annual up to 50 cm, often purple-tinted. Lvs sessile, 12–40 mm, lanceolate to linear-lanceolate, remotely dentate. Racemes terminal on the stem and upper branches. Bracts like the lvs but smaller. Calyx campanulate; teeth triangular, obtuse. Corolla 8–10 mm, purplish-pink; tube 5–6 mm. Anthers coherent, slightly exserted. Style long, filiform.

Capsule 6–8 mm, oblong, pubescent. Pollinated by bees. Th.

Common in cultivated and uncultivated fields and waste places throughout the British Is., ascending to over 395 m. Most of Europe except some of the islands; N. and W. Asia.

A very variable species difficult to separate adequately into infraspecific categories. There are local populations with distinct facies, and some fairly well-marked ecotypes, and the species exhibits seasonal dimorphism. The three reasonably most distinct variants are indicated here, but intermediates are fairly common, and neither their circumscriptions nor their distributions are given with much confidence.

Subsp. **verna**. Stem 10–30 cm. *Branches fairly short, coming off at an angle of less than 45°, ±straight. Lvs lanceolate*, rounded at base, distinctly crenate to serrate. *Bracts longer than fls*. Fl. 6–7. $2n = 40$.

Apparently less common in S. England than subsp. *serotina*. N. and C. Europe, eastwards to N.W. Russia, and southwards in the mountains to N. Portugal, C. Italy and N. Greece; N. Asia.

Subsp. **pumila** (Nordstedt) A. Pedersen, on the coasts of the Outer Hebrides, Sutherland and Caithness, similar to the above but with very short internodes and later flowering, is apparently transitional to subsp. *serotina*.

Subsp. **serotina** (Dumort.) Corb. Stem 20–25 cm. *Branches long, spreading at a wide angle*, sometimes nearly at right angles, their tips often upcurved. *Lvs linear-lanceolate*, somewhat narrowed at the base, obscurely crenate to subentire. *Bracts shorter than to as long as the fls*. Fl. 7–8. $2n = 20$.

Common in S. England, rare or absent in N. Scotland. Almost throughout the range of the species, but rare in parts of the north; W. Asia.

21. PARENTUCELLIA Viv.

Annual herbs with opposite lvs. Fls in the axils of bracts similar to the lvs, forming terminal spike-like racemes. Calyx tubular, 4-dentate. Corolla 2-lipped; *upper lip entire or emarginate, forming a hood*; lower lip with 3 entire lobes. Stamens 4; anthers aristate at the base. Capsule lanceolate. *Seeds many, minute, smooth; hilum basal.*

Four spp. in W. Europe, the Mediterranean region, C. and W. Asia, Iran.

1. P. viscosa (L.) Caruel   Yellow Bartsia.

Bartsia viscosa L.; *Eufragia viscosa* (L.) Bentham

Erect viscid-hairy annual 10–50 cm, usually unbranched. Lvs 1.5–4 cm, oblong to lanceolate, coarsely serrate, acute or subacute, sessile. Calyx-teeth linear-lanceolate, subacute, about as long as the tube. Corolla 16–24 mm, yellow (rarely white), the lower lip much longer than the upper. Stamens included in the upper lip, anthers hairy. Capsule 7–9 mm, slightly longer than the calyx-tube. Fl. 6–10. $2n = 48^*$. Th.

Native. Damp grassy places usually near the south and west coasts. From Cornwall to Pembroke, Glamorgan, Wilts, Sussex and Kent, Channel and Scilly Is., scattered localities from S. Lancs to Dunbarton, casual further east and north in England; Cork, Kerry and N.W. Ireland from Sligo to Derry. S. and W. Europe, northwards to N. France; N. Africa; Azores, Canary Is.

22. BARTSIA L.

Perennial herbs with opposite lvs. Infls and fls as in *Parentucellia*. Capsule wide. *Seeds few, large, with strong longitudinal ribs or wings and a lateral hilum.*

About 30 spp., in Europe, N. Africa and tropical mountains; only the following more widespread.

1. B. alpina L.   Alpine Bartsia.

Glandular-hairy; stems 10–20 cm, erect, simple; rhizome short. Lvs 10–25 mm, sessile, ovate, crenate-serrate, obtuse to subacute. Infl. short, few-fld; bracts purplish. Calyx-teeth ovate-lanceolate, obtuse, about as long as the tube. Corolla 15–20 mm, dull purple; upper lip much longer than the lower. Anthers hairy. Capsule about twice as long as the calyx. Fl. 6–8, pollinated by humble-bees. $2n = 24$. Hp.

Native. Mountain meadows and rock-ledges on basic soils; from 295 to 915 m. Yorks, Durham, Westmorland, Perth and Argyll. N. Europe, southwards in the mountains to the Pyrenees, S. Alps and S.W. Bulgaria; Greenland; eastern North America southwards to Labrador.

23. LATHRAEA L.

Herbaceous *root-parasites with* branched creeping *rhizomes* covered *with* wide whitish fleshy *imbricate scales* and bearing rootlets which are swollen where they are attached to roots of the host plant. Fls borne singly in the axils of scales. *Calyx campanulate, equally 4-lobed* above; corolla 2-lipped, the upper lip strongly concave, entire, the lower smaller and 3-lobed; stamens subexserted; style exserted, curved downwards at its tip. Nectary at base of ovary, crescent-shaped. Capsule opening elastically from the apex into 2 valves. Plants blackening on drying.

Seven spp., in temperate Europe and Asia.

Aerial shoot 8–30 cm; fls white or dull purple, short-stalked.   **1. squamaria**

Aerial shoot 0; fls bright purple, long-stalked.   **2. clandestina**

1. L. squamaria L.   Toothwort.

A perennial herb with a *stout simple erect flowering shoot* 8–30 cm, white or pale pink, slightly pubescent above and with a few whitish ovate scales below. Infl. a *one-sided raceme* with scaly bracts, at first drooping, but straightening later. Fls shortly stalked, each in the axil of a broadly ovate bract. *Calyx glandular-hairy*, tubular, with 4 broadly triangular teeth. *Corolla white ±tinged with dull purple*, slightly longer than the calyx. *Capsule ovoid*, acuminate, with *numerous seeds*. Fl. 4–5. Visited by humble-bees. $2n = 42^*$; 36. Grh.

Native. On roots of various woody plants, especially of *Corylus* and *Ulmus*, in moist woods and hedgerows on good soils, and locally common in some limestone areas; to 305 m in N. England. Great Britain northwards to Perth and Inverness; formerly scattered throughout much of Ireland, but perhaps now only in the east. Throughout much of Europe except parts of the north and south; W. Asia to the Himalaya.

*2. L. clandestina L.

A perennial herb with *no aerial shoot. Fls in corymbose clusters, long-stalked, arising* singly in the axils of fleshy rhizome-scales *at or just beneath the soil surface. Calyx glabrous*, campanulate, with 4 short triangular lobes. *Corolla bright purple, twice or more as long as the calyx. Capsule globose*, with 4–5 *seeds*. Fl. 4–5. Visited by humble-bees. $2n = 42$. Grh.

Introduced. Naturalized in a few localities in England and S. Scotland on roots of *Populus* and *Salix* in damp shady places. Native in N. Spain, Italy, W. and C. France, and Belgium.

113. OROBANCHACEAE

Perennial, rarely annual, *herbs devoid of chlorophyll and completely parasitic on the roots of other phanerogams. Aerial stems* erect, usually simple and often brownish, *bearing small alternate scale-like and ±succulent lvs* and *fls in a terminal spike or raceme*, rarely in a panicle or solitary. Calyx tubular, campanulate or 2-lipped. *Corolla* with 5 free lobes, *±distinctly 2-lipped*. Stamens 4, 2 long and 2 short. *Ovary superior, unilocular, with numerous ovules on 4 (rarely 2) often deeply lobed parietal placentae*; style single with a 2-lobed stigma. *Fr. a loculicidal capsule* with numerous small seeds.

The genus *Lathraea* (p. 395) has commonly been included in the family as a root-parasite devoid of chlorophyll and with a unilocular ovary. There are, however, reasons for doubting any very close relationship between *Lathraea* and *Orobanche* (see J. Kuijt, *Beitr. Biol. Pflanzen*, 1973, **49**, 137–46), and the former was transferred to Scrophulariaceae in *Flora Europaea*, **3** (1972), as earlier in the 3rd Edn of Hooker's *The Student's Flora of the British Islands* (London: Macmillan, 1884). This course has been adopted in the present volume. Further investigation may well lead to a merging of the whole family in the Scrophulariaceae.

The family, excluding *Lathraea*, comprises about 10 genera mainly in warm temperate regions of the Old World with a few spp. in North America. There is a single British genus.

1. OROBANCHE L.

Herbaceous root-parasites ('broomrapes'), usually short-lived perennials, *which develop underground tubers* where the tips of seedlings become attached to roots of a host plant. From these tubers arise erect scaly *fl.-stems, slightly to strongly swollen at base* and with terminal spikes or racemes of sessile or short-stalked fls; bracteoles, adnate to the calyx, present or 0. Calyx with 4 subequal teeth (rarely with a smaller posterior 5th tooth) or divided ±to the base into 2 lateral entire or bifid segments. Corolla with cylindrical to campanulate tube and 2 distinct lips above, the upper erect and 2-lobed to entire, lower spreading, 3-lobed. *Stamens included* in corolla-tube. Ovary with 4 variably lobed placentae; style down-curved near tip; stigma-lobes

fleshy, variously coloured. *Capsule with valves* free or more *commonly remaining attached at their tips below the persistent style.*

About 100 spp., chiefly north temperate and subtropical.

1 Fls each with 1 bract on the anterior side and 2 lateral bracteoles ±adnate to the calyx; valves of ripe fr. free at their tips below the persistent style. 2

  Fls each with 1 bract on the anterior side but lacking bracteoles; valves of ripe fr. united at their tips below the persistent style. 3

2 Stem usually unbranched; corolla 18–30 mm, whitish below, bluish-violet above with dark violet veins; parasitic chiefly on *Achillea millefolium*. **1. purpurea**

  Stem usually branched; corolla 10–16 mm, whitish below (rarely throughout), creamy above and usually tinged with blue or violet; parasitic chiefly on hemp, tobacco or *Solanum* spp. and now extinct or nearly so. **O. ramosa** (p. 397)

3 Lobes of lower lip of corolla with distinctly glandular-ciliate margins. 4

  Lobes of lower lip of corolla not or very slightly glandular-ciliate. 6

4 Robust plant 20–80 cm, parasitic on gorse, broom and some other Leguminous shrubs; stigma-lobes pale yellow at fl.-opening. **2. rapum-genistae**

  Not parasitic on Leguminous shrubs; stigma-lobes red or purple at fl.-opening. 5

5 Stem and corolla usually purplish-red, rarely paler; corolla usually 15–25 mm; stigma-lobes contiguous; parasitic on *Thymus* and other Labiatae. **3. alba**

  Stem and corolla usually yellowish, sometimes tinged with purple; corolla 20–35 mm; stigma-lobes distant; parasitic chiefly on *Galium* spp. and other Rubiaceae. **4. caryophyllacea**

6 Stigma-lobes yellow at fl.-opening. 7

  Stigma-lobes purple, dark red or orange at fl.-opening. 9

7 Lateral lips of calyx connate at base on anterior side, each ±unequally bifid; corolla usually 18–25 mm; stigma-lobes contiguous but free; parasitic on *Centaurea scabiosa*. **5. elatior**

  Lateral lips of calyx free to base, each entire or ±bifid; stigma-lobes free or partly united. 8

8 Corolla cream-coloured, tinged distally with reddish-purple, its tube inflated below then gradually narrowing to the mouth; stigma-lobes contiguous and partly united; parasitic on ivy.   **9. hederae**
Corolla yellow throughout, not narrowing upwards to the mouth; stigma-lobes free; parasitic on cultivated clovers and a wide variety of other plants.
7. minor
See also *O. crenata* (p. 399)

9 Corolla broadly cylindrical-campanulate; stamen-filaments ±glabrous below, usually glandular-puberulent above; parasitic on spp. of *Cirsium* and *Carduus* in Yorks, rare.   **6. reticulata**
Corolla tubular or narrowly campanulate; stamen-filaments ±hairy below, subglabrous above.   10

10 Bracts 10–20 mm; corolla 14–22 mm, the stamen-filaments inserted 3–5 mm above its base; parasitic on *Picris*, *Crepis* and *Artemisia* spp. and some other Compositae; rare and local.   **8. loricata**
Bracts 7–15 mm; corolla 10–18 mm, the stamen-filaments inserted 2–3 mm above its base; parasitic on cultivated clovers and a wide variety of other plants.
7. minor
See also *O. maritima* (p. 399)

Section 1. *Trionychon* Wallr. Stem simple or branched; fls each with 1 bract and 2 bracteoles; capsule with the valves free above.

1. O. purpurea Jacq.
Purple Broomrape, Yarrow Broomrape
O. caerulea Vill.; *O. arenaria* auct.; *Phelypaea caerulea* (Vill.) C. A. Meyer

Flowering *stem* 15–45 cm, *simple*, rarely branched, *bluish*, fairly stout, with distant narrow scales towards the base, glandular-pubescent especially above. Fls in a lax spike; *bracts* somewhat shorter than calyx, *lanceolate*. Calyx hairy, tubular, with 4 lanceolate-acute teeth usually shorter than the tube. *Corolla* 18–30 mm, twice as long as the calyx, *dull bluish-purple* suffused with yellow at the base; corolla-tube constricted just below the middle, slightly curved in the upper half; *upper lip of corolla with 2 acute lobes*, lower with 3 subacute lobes. Stamens inserted just below the constriction in the corolla; anthers ±glabrous. Stigma-lobes white. Fl. 6–7. Apparently self-pollinated. $2n = 24$. G.

Native. Parasitic on *Achillea millefolium* and a few other Compositae. Locally frequent in the Channel Is., especially on walls; formerly in a few scattered localities in S. and E. England and S. Wales but seen recently only in N. Hants, E. Kent and N. Norfolk. C. and S. Europe; N. Africa; W. Asia to the Himalaya.

O. arenaria Borkh. with fls 25–35 mm long and hairy anthers, parasitic on *Artemisia campestris*, has been reported in error from Alderney. Its fls are pale purple or lavender with no yellow at the base. The Alderney plant is a form of *O. purpurea*.

O. ramosa L. (*Phelypaea ramosa* (L.) C. A. Meyer) differs from **1.** in the *usually branched* fl.-stem and the *smaller corolla*, only 10–17 mm. The slender stem, 5–30(–40) cm, is yellowish-white or tinged with blue, and the fls have a yellowish-white corolla usually pale blue or lilac at least round the mouth, the upper lip of 2 rounded lobes and lower of 3 blunt ciliate

lobes. Fl. 7–9. $2n = 24$. Usually parasitic on cultivated hemp, tobacco and *Solanum* spp. but also on various other hosts. Perhaps native, at least until recently, in the Channel Is. and possibly also in S. England from Devon to Kent where it once grew in a few scattered localities, but almost certainly introduced with its host, *Cannabis*, in the former hemp-growing areas of Norfolk and Suffolk. It now seems extinct in all its English localities. S. and S.C. Europe; naturalized further north.

Section 2. *Orobanche*. Stem simple; fls with 1 bract and no bracteoles; capsule with the valves coherent above.

2. O. rapum-genistae Thuill.   Greater Broomrape.
O. major L., p.p.; *O. rapum* auct.

Flowering *stem* 20–80 cm, simple, *yellowish*, stout, swollen near the base, glandular-hairy throughout. Fls in a long compact spike; bracts linear-lanceolate, shorter to slightly longer than the fls. Lateral calyx-lips usually equally bifid and nearly equalling the corolla-tube. Corolla 20–25 mm, yellowish tinged with purple, glandular-pubescent; corolla-tube campanulate, inflated at the base in front, its back curved throughout; *upper lip of corolla almost entire* with spreading margins, lower 3-lobed, the middle lobe much larger than the lateral, all ciliate; both lips waved and indistinctly denticulate. *Stamens inserted near the base of the corolla-tube*; filaments glabrous at base, glandular above. *Stigma-lobes distant, pale yellow*. Fl. 5–7. Visited by bees. Corolla often perforated at the base by *Bombus terrestris*. $2n = 38$. G.

Native. Parasitic on the roots of shrubby Leguminoae, chiefly on *Ulex* and *Cytisus*, occasionally on *Genista tinctoria*, etc. Formerly through much of England and Wales but now restricted to scattered localities mainly in S. England and Wales but with a few outliers northwards to Northumberland; also in single localities in Kirkcudbright and Sutherland; now very rare in Ireland. W. Europe eastwards to C. Germany, Italy and Sicily; N. Africa.

3. O. alba Stephan ex Willd.   Red Broomrape.
Inc. *O. rubra* Sm., *O. epithymum* DC.

Flowering *stem* 8–25(–35) cm, simple, *purplish-red*, rather stout, *with numerous reddish scales* near the base, glandular-hairy throughout. Fls rather few in a lax spike; *bracts* ovate-lanceolate, acuminate, *shorter than the fls*. Calyx 2-lipped, the *lips entire*, subulate, about equalling corolla-tube. *Corolla* 15–20 mm, *dull purplish-red*, sparingly glandular-pubescent; corolla-tube campanulate, back strongly curved only at the basal and distal ends; *upper lip of corolla notched* with somewhat spreading margins, lower 3-lobed with the lobes almost equal; both lips crisped and distinctly denticulate. *Stamens inserted 1–2 mm above base of corolla-tube*; filaments slightly hairy below, glandular above. *Stigma-lobes contiguous, reddish*. Fl. 6–8. Fls fragrant, visited by humble-bees. G.

Native. Parasitic on *Thymus* (rarely also other genera of Labiatae) on rocky slopes, cliff-ledges, screes, etc., especially near the sea; usually on calcareous substrata. Very local in W. Cornwall, W. Yorks and W. Scotland from Wigtown to Sutherland; Inner and Outer Hebrides; very local near N. and W. coasts of Ireland. Most of Europe northwards to Scotland, Belgium and C. Russia, with outliers in the Baltic islands of Öland and Gotland; W. Asia to the Himalaya.

Most British specimens have deep reddish fls (*O. rubra* Sm.), but the form common on the Continent, with much paler fls, is sometimes found.

4. O. caryophyllacea Sm.   Clove-scented Broomrape.

O. vulgaris Poiret

Flowering *stem* 15–40 cm, simple, yellowish usually tinged with purple, *with numerous* yellow or purple-brown *scales* below, glandular-hairy. Fls rather few in a lax spike; *bracts* ovate-lanceolate, acuminate, *shorter than the fls*. Lateral lips of calyx free, entire to bifid, *shorter than the corolla-tube*. Corolla 20–30 mm, yellowish or tinged with reddish-brown or purplish, densely glandular-pubescent; corolla-tube campanulate, back curved throughout; upper lip notched with the lobes at first erect; lower 3-lobed with the lobes nearly equal; all lobes crisped and denticulate, those of the lower lip almost fimbriate. *Stamens inserted* 1–3 mm *above base of corolla-tube*; *filaments hairy below*, glandular above. *Stigma-lobes distant, purple*. Fl. 6–7. Fls said to smell of cloves; visited for nectar by humble-bees. G.

Native. Parasitic on *Galium mollugo*, but also on other *Galium spp.* and other genera of *Rubiaceae* in continental Europe. Extremely rare; only near the sea-coast of S.E. Kent. Much of Europe northwards to Norway, Poland and C. Russia; Caucasus, Siberia; N. Africa.

5. O. elatior Sutton   Tall Broomrape.

O. major auct.; incl. *O. ritro* Gren. & Godron

Flowering stem 15–70 cm, simple, yellowish or reddish, stout, with numerous acuminate scales below, glandular-hairy throughout. Fls numerous, in a rather dense spike. Bracts lanceolate-acuminate, as long as the fls. *Lateral lips of calyx* c. 10 mm ±unequally bifid, *connate at base* on anterior side. Corolla 18–25 mm, pale yellow usually tinged with purple, glandular-pubescent, rather widely tubular, back curved throughout; upper lip ±2-lobed usually spreading; lower lip 3-lobed, the lobes nearly equal; all *lobes crisped, denticulate, not ciliate*. Stamens *inserted* 4–6 mm *above base of corolla-tube*; *filaments hairy below*, glandular throughout. *Stigma-lobes yellow*. Fl. 6–7. Nectar-secreting but scentless. $2n = 38$. G.

Native. Parasitic, perhaps exclusively, on *Centaurea scabiosa* in rough pastures over chalk or limestone, in and round chalk-pits and limestone quarries, on way-sides, etc., mainly on calcareous substrata. Rare but widely distributed south and east of a line from near Scarborough (N. Yorks) to the Severn estuary and C. Somerset, also in Glamorgan; not in Ireland. Much of Europe from England, Denmark, S. Sweden, Estonia and C. Russia southwards to E. Spain, N. Italy, Greece

and Bulgaria; Asia Minor and Caucasus eastwards to India.

6. O. reticulata Wallr.   Thistle Broomrape.

Flowering stem 15–50 cm, simple, yellowish to purplish, glandular-hairy, with a few lanceolate scales mostly below. Fls numerous, in a spike compact above but lax below; bracts narrowly triangular, about equalling the fls. Lateral lips of calyx entire or unequally bifid, somewhat shorter than the corolla-tube. *Corolla* 15–22 mm, yellowish, ±purple-tinged marginally, *with sparse dark glands*; tube somewhat widened above the insertion of the stamens, its back strongly curved near the base and into the upper lip; upper lip notched, with spreading lobes; lower with 3 ±equal lobes, the central truncate or squarish; all lobes denticulate. Stamens inserted 2–4 mm above the base of the corolla, their filaments glabrous or sparsely hairy below and glabrous or sparsely glandular above. *Stigma-lobes dark purple*. Fl. 6–8. Fls unscented. $2n = 38$. G.

The description is of subsp. **pallidiflora** (Wimmer & Grab.) Hegi, which in its var. *procera* (Koch) Hegi includes all British specimens. Subsp. *reticulata* has the corolla yellowish only at the base (elsewhere bright reddish-purple with darker veins) and densely dark-glandular, and the stamen-filaments densely glandular above. It is found on a wider range of hosts than subsp. *pallidiflora*.

Native. On *Cirsium arvense*, *C. eriophorum* and other spp. of *Cirsium* and *Carduus*; very local. Reported only from a few stations in Yorks. C. and S. Europe northwards to England, Denmark, S. Sweden and Estonia.

7. O. minor Sm.   Lesser Broomrape.

O. apiculata Wallr.; *O. trifolii pratensis* Schultz

Fl.-stem 10–50 cm; simple, yellowish but usually tinged reddish-purple, ±glandular-pubescent; scales 10–25 mm, ovate to linear-lanceolate, brownish, crowded at base of stem but sparse above. Fl.-spike usually lax below; bracts 7–15 mm, ovate-lanceolate, acuminate, about equalling fls or longer. *Lateral lips of calyx* 7–12 mm, *free, ±unequally bifid to entire*. Corolla 10–18 mm, *pale yellow usually tinged and veined above with violet*, glandular-pubescent to subglabrous, *tubular or narrowly campanulate* (c. 5 mm diam.), with *tube curving regularly throughout its length*; upper lip notched to 2-lobed, lower ±equally 3-lobed or the reni-form middle lobe largest; *lower lip* denticulate but *not ciliate*. Stamen-filaments inserted 2–3 mm above base of corolla, somewhat hairy below but subglabrous above. *Stigma-lobes not contiguous, purple*, rarely yellow, at fl.-opening. Fl. 6–9. $2n = 38$. Th.-G.

Variable. Plants having both corolla and stigma-lobes yellow with no tinge of purple or red have been named var. *flava* Regel. Also paler-fld than typical *O. minor* is var. *composit-arum* Pugsley, with rather slender ±reddish stem and a many- and usually dense-fld spike with fls less curved than in the type and with suberect corolla 12–18 mm, yellowish-white tinted with pale or dull violet, its tube only 3–4 mm diam. and often only slightly glandular-pubescent; stigma-lobes dull or pale purple.

In the mid-19th century plants growing on *Daucus carota* on the coast of S.E. Cornwall were mistakenly identified as *O. amethystea*, a continental species related to *O. minor* but with longer scales, bracts and calyx-lips and a larger corolla, 15–25 mm, straight-backed and white or cream tinged distally with violet. Plants from other coastal localities were similarly named, but doubts as to their specific distinctness from *O. minor* were raised by some botanists. Pugsley realized that our plants were certainly not *O. amethystea* but thought they merited treatment as a separate species which he named *O. maritima*. They are often larger than typical *O. minor*, with laxer spikes and deep purple fls having ±straight-backed corollas. More detailed recent studies of one of these coastal populations, in which several different host-species are parasitized, shows a full range of intermediates between typical *minor* and Pugsley's *maritima*, so that the latter can be regarded at most as a variety of *minor*. (F. J. Rumscy, unpublished).

Plants parasitic on *Eryngium maritimum* in coastal sites of S. England and the Channel Is., also referred earlier to *O. amethystea* (which is commonly on *E. campestre* in mainland Europe), seem to belong to other infraspecific variants of *O. minor*.

Probably native. Parasitic mainly on *Trifolium* spp. and other herbaceous Leguminosae in fields, on waysides and railway-banks, in and round pits and quarries, on fixed sand dunes, etc.; often to be seen earlier in fields of rotational red clover; occurs also on a wide range of other hosts including many garden plants, both herbaceous and woody; var. *compositarum* on *Crepis capillaris*, *Hypochaeris radicata* and other Compositae, and 'var. *maritima*' on *Daucus carota* and less commonly on *Ononis repens*, *Plantago coronopus*, *Calystegia soldanella*, etc. Largely to S. and E. of a line from N. Yorks to St David's Head, with a few outliers northwards to N. Wales, Lancs, Isle of Man and Cumbria, but no longer in Scotland; frequent in S. and E. Ireland but rather rare elsewhere. W. and S. Europe eastwards to S.W. Russia and Turkey and northwards to Belgium and Czechoslovakia; probably introduced in S. Scandinavia and Poland; Asia Minor; N. Africa.

8. O. loricata Reichenb.

Ox-tongue Broomrape, Picris Broomrape.

Inc. *O. picridis* F. W. Schultz ex Koch

Flowering stem 10–60 cm, simple, pale yellowish often tinged with purple, glandular-pubescent or nearly glabrous, with a few brownish acuminate scales below. Fls numerous, in a spike lax below; bracts narrowly ovate-acuminate, equalling or exceeding the fls. Lateral lips of calyx entire or unequally bifid, equalling or somewhat exceeding the corolla-tube. *Corolla* 15–20 mm, yellowish-white tinged and veined with pale purple, ±glandular-pubescent; *tube* narrowly campanulate, its back *curved at the base then nearly straight*; upper lip retuse to notched, folded in the middle, erect; lower lip ±equally 3-lobed, the middle lobe squarish; all lobes crisped and denticulate but not ciliate. *Stamens inserted 3–5 mm above the base of the corolla-tube; filaments densely hairy below*, ±glabrous above. *Stigma-lobes just touching, purple*. Fl. 6–7. G.

Native. Rare and decreasing; now only in S.E. England in a very few localities in E. Kent, Bucks, Hertford and Cambridge. S., W. and C. Europe extending northwards to Denmark, Poland and S.W. Russia; Palestine; N. Africa.

9. O. hederae Duby   Ivy Broomrape.

Flowering stem 10–60 cm, simple, purplish, glandular-pubescent, with a few brownish acuminate scales below. *Fls* rather few, *in a long lax spike*; bracts ovate-lanceolate acuminate ±serrate near the tip, equalling or exceeding the fls. Lateral lips of calyx entire or unequally bifid, 1-veined, about equalling the corolla-tube. Corolla 12–20 mm, cream strongly veined with purple, sparsely glandular; tube inflated near the base then gradually narrowed upwards, its back straight except at the base; upper lip entire to notched, the lobes forwardly directed or spreading, lower 3-lobed with the middle lobe truncate; all lobes crisped and denticulate, not ciliate. *Stamens inserted 3–4 mm above the base of the corolla-tube; filaments almost glabrous below*, glabrous or sparsely glandular above. *Stigma-lobes partly united*, yellow. Fl. 6–7. $2n = 38$. G.

Native. On *Hedera helix*. A local plant of S. England and Wales to the S. and W. of a line from E. Kent to N. Wales and commonest in coastal districts; introduced in a few midland sites northwards to Leicester; occasional in S. Ireland, much rarer in N. W. and S. Europe northwards to N. Ireland, Belgium, Hungary and the Crimea; Asia Minor; N. Africa.

*O. crenata Forskål, native in S. Europe and usually parasitic on leguminous crops, has recently been reported from S. Essex. It has yellowish fl.-stems up to 50 cm or more, and a spike, dense above, of large fls with corolla 20–30 mm, white with lilac veins; stigma-lobes white, yellow or pinkish.

114. LENTIBULARIACEAE

Carnivorous aquatic, bog or terrestrial herbs (or epiphytic), scapigerous. Lvs alternate or all basal. Fls solitary or in a raceme. Calyx 5-lobed or 2-lipped with the lobes obscure. Corolla gamopetalous, 2-lipped, spurred, imbricate; upper lip 2-lobed, the lower 3-lobed or the lobes obscure. Stamens 2, inserted on base of corolla, 1- or 2-celled, introrse. Carpels 2, forming a unilocular ovary with numerous ovules on a free-central placenta. Stigma often sessile, 2-lobed with 1 lobe much reduced. Fr. a capsule opening irregularly or by 2 or 4 valves. Seeds small, numerous, without endosperm.

Four genera and about 170 spp., cosmopolitan.

A small family, clearly distinguished from the Scrophulariaceae by being carnivorous and by the placentation

Lvs entire, in a basal rosette; carnivorous by sticky glands covering the whole plant. 　　1. Pinguicula

Lvs divided into filiform segments, alternate; carnivorous by special bladders borne on the lvs. 　2. Utricularia

1. Pinguicula L.

Perennial scapigerous herbs with lvs all forming a basal rosette, clothed all over (except corolla) with sticky glands which catch insects. Lvs entire, sessile; margins involute. Fls solitary on naked scapes. Calyx 5-lobed, the lobes unequal. Corolla 2-lipped; upper lip 2-, the lower 3-lobed, spurred, open at mouth. Capsule opening by 2 valves.

About 46 spp., northern hemisphere and South America.

1 Fls white, with 1 or more yellow spots; roots thick.
　　　　　　　　　　　　　　　2. alpina
　Fls pinkish to violet; roots thin. 　　　　　2
2 Corolla not more than 10 mm, the lobes of the lower lip emarginate. 　　　　**1. lusitanica**
　Corolla at least 15 mm, the lobes of the lower lip entire. 　　　　　　　　　3
3 Corolla 15–22 mm, the lobes of the lower lip widely separated; spur 3–6 mm. 　　**3. vulgaris**
　Corolla 25–35 mm, the lobes of the lower lip contiguous or overlapping; spur 10–12 mm.
　　　　　　　　　　　　　4. grandiflora

1. P. lusitanica L. 　　　　Pale Butterwort.

Overwintering as a rosette. Lvs 1–2 cm, oblong, yellowish, tinged purple. Scapes 3–15 cm, very slender. Calyx-lobes suborbicular or ovate, obtuse. *Corolla* 7–9 mm, *pinkish to pale lilac*, yellow in the throat; lobes of upper lip suborbicular, of the lower emarginate; *spur* 2–4 mm, bent downwards, ±*cylindrical*, obtuse. Capsule globose. Fl. 6–10. Self-pollinated. Hr.

Native. Bogs and wet heaths; ascending to 490 m; local. Cornwall to Somerset and Hants; Pembroke; Isle of Man; W. Scotland from Kirkcudbright to Orkney and Outer Hebrides; throughout Ireland but rare in the centre. W. France, W. Spain, Portugal; N.W. Morocco.

2. P. alpina L.

Roots long, thick. Lvs 1–2 cm, elliptical. Scapes 5–10 cm. *Corolla* 8–10 mm, *white* with 1 or more yellow spots at mouth; *spur* 2–3 mm, bent downwards, *conical*. Capsule ovoid. Fl. 5–8. $2n = 32$. Hr.

Formerly occurred in E. Ross, now believed extinct. N. Europe from Iceland to Finland and Baltic States; Pyrenees, mountains and uplands of C. Europe; Himalaya.

3. P. vulgaris L. 　　　Common Butterwort.

Overwintering as a rootless bud. Lvs 2–8 cm, ovate-oblong, bright yellow-green. Scapes 5–15 cm. Calyx-lobes ovate or oblong, broad-based, obtuse or subacute, the upper lip divided to middle or less. *Corolla violet*, 15–22 × c. 12 mm; usually white in the throat; *lobes of lower lip* deep, *much longer than broad*, flat, entire,

divergent; spur 3–6 mm, directed backwards or somewhat downwards, straight or somewhat curved, slender, *acute*. Capsule ovoid. Fl. 5–7. Pollinated by small bees. $2n = 64$. Hr.

Native. Bogs, wet heaths and among wet rocks; ascending to 1040 m. Common throughout the British Is. except C. and S. England and S. Ireland where it is rare and absent from several counties. N., W. and C. Europe, extending eastwards to W. Ukraine; N. Asia; North America south to New York and British Columbia; N. Morocco (Rif).

4. P. grandiflora Lam. 　Large-flowered Butterwort.

Overwintering as a rootless bud. Lvs 2–8 cm, ovate-oblong, bright yellow-green. Scapes 8–20 cm. Calyx-lobes ovate or oblong, upper lip divided nearly to base. *Corolla* 25–35 × 25–30 mm, *violet to pinkish or pale lilac, with a long cuneate white patch in the throat, the lobes of lower lip* shallow, *broader than long*, somewhat undulate, *contiguous or overlapping*; spur 10–12 mm, directed backwards, straight, rather stout, sometimes slightly bifid. Capsule subglobose. Fl. 5–6. $2n = 32$. Hr.

Native. Bogs and wet rocks, locally common in Kerry and W. Cork, extending to Clare, ascending to 855 m; naturalized in Cornwall and S. Somerset. Jura, French Alps, Pyrenees and mountains of N. Spain.

P. grandiflora × vulgaris = P. × scullyi Druce is intermediate between the parents, especially in floral characters; it occurs in S. Kerry, W. Cork and Clare, perhaps elsewhere in S. and W. Ireland.

2. Utricularia L.

Perennial rootless herbs overwintering by turions. Stems long, lfy. Lvs divided into filiform segments, bearing small bladders which trap animals. Fls in short racemes on lfless scapes. Calyx 2-lipped, divided nearly to base; lips entire or obscurely toothed. Corolla 2-lipped, spurred; lips entire, lower with a projecting palate, larger than upper. Capsule globose, opening irregularly.

About 120 spp., cosmopolitan, mainly tropical.

1 Stems of one kind, all bearing green lvs furnished with numerous bladders. 　(**vulgaris** group) 　2
　Stems of two kinds (*a*) bearing green lvs with few or no bladders, and (*b*) colourless, bearing bladders on much reduced lvs, often beneath surface of substratum. 　　　　　　　　3
2 Lower lip of corolla with deflexed margins; pedicels 6–15 mm; fr. freely produced. 　　**1. vulgaris**
　Lower lip of corolla ±flat; pedicels 10–25 mm; fr. very rare. 　　　　　　　**2. australis**
3 Lf-segments denticulate, with bristles on the teeth; bladders confined or nearly confined to colourless stems. 　　　　　　　**3. intermedia**
　Lf-segments entire, without bristles; bladders present on green lvs. 　　　　　**4. minor**

(1–2). U. vulgaris group 　　Greater Bladderwort.

Free floating. *Stems* 15–45 cm, *all alike, bearing green lvs with numerous bladders*. Lvs 2–2.5 cm, broadly ovate in outline, pinnately divided; segments denticulate, with small bristles on the teeth; bladders c. 3 mm. Scape 10–20 cm, 2–8-fld. *Corolla* 12–18 mm, *bright yellow*; spur conical; upper lip ±ovate.

Native. Lakes, ponds and ditches usually in relatively deep water; ascending to 610 m. Throughout the British Is. (except Channel Is.), local; flowering sporadically and fls unknown from the north part of its range and, as the two following are not certainly determinable without fls, their complete distribution is unknown. Europe; temperate Asia; N. Africa; North America.

1. U. vulgaris L.

Lf-segments with groups of bristles. *Pedicels* 6–15 mm, rather stout. *Upper lip of corolla about as long as palate; lower lip with margins deflexed ±at right angles. Fr. freely produced* (if the plants fl.). Fl. 7–8. Pollinated by bees. $2n = 36$–40. Hyd.

Native. Base-rich waters. North to S.E. and E. England extending to S. Lancs, Durham and Edinburgh, rare in the west; Ireland. Most of Europe but rare in the south; temperate Asia.

2. U. australis R.Br.

U. major auct., *U. neglecta* Lehm.

Lf-segments with solitary, rarely grouped bristles. *Pedicels* 10–25 mm, rather slender. *Upper lip of corolla about twice as long as the less projecting palate; lower lip ±flat, somewhat undulate. Fr. very rare.* Fl. 7–8. Pollinated by bees. $2n = 36$–40. Hyd.

Native. Acid waters. North to Perth, rare in the east; Ireland. Most of Europe except the extreme north and most of the USSR.

3. U. intermedia Hayne   Intermediate Bladderwort.

Incl. *U. ochroleuca* auct. angl., non R. W. Hartman

Stems slender, 10–25 cm, *of two kinds* (*a*) bearing green lvs without or occasionally with very few bladders, and (*b*) colourless, often buried in substratum, bearing bladders on very much reduced lvs. Lvs distichous, 4–12 mm, orbicular in outline, palmately divided; *segments denticulate, with 1–2 small bristles on the teeth.* Bladders c. 3 mm. Fls very rarely produced, 2–4, on a scape 9–16 cm. *Corolla* c. 8–12 mm, *bright yellow*, marked with reddish-brown lines; spur 5–6.5 mm, conical. Fl. 7–9. Hyd.

Native. Lakes, pools and ditches, usually in shallow peaty water, very local; ascending to 990 m (but rare at high altitudes); commonest in N. and W. Scotland and Ireland. Dorset, Hants, Norfolk; Westmorland to Wigtown; Fife and Kintyre to Shetland; widespread in Ireland (rare in the east). Europe from Scandinavia to W. and C. France, N. Italy and S.C. Russia; N. Asia; North America from Newfoundland and British Columbia to New Jersey and California.

4. U. minor L.   Lesser Bladderwort.

Stems 7–25 cm, slender, *of two kinds* (*a*) bearing green lvs with a few bladders, and (*b*) colourless, often buried in substratum, bearing bladders on very much reduced lvs. Lvs 3–6 mm, orbicular in outline, palmately divided; *segments entire, without bristles.* Bladders c. 2 mm. Scapes 4–15 cm, 2–6-fld. *Corolla* 6–8 mm, *pale yellow*; spur very short, obtuse. Fl. 6–9. Hyd.

Native. Ponds and ditches, bog- and fen-pools; ascending to 685 m. Throughout the British Is. but local and extinct in or absent from many counties in the English Midlands and E. Scotland. Most of Europe, but rare in the Mediterranean region; Morocco (Atlas Mountains); N. Asia; North America south to Pennsylvania and California.

115. ACANTHACEAE

Differs from the Scrophulariaceae chiefly in the indurated funicle, which often develops into a 'jaculator' for ejecting the seeds, and almost always in the absence of endosperm. In addition the lvs are always opposite, coloured bracts are often present, the corolla is strongly zygomorphic, the stamens often unequally 2-celled or only 1-celled and the style unequally bilobed.

About 250 genera and 2500 spp., tropical and warm temperate regions.

1. ACANTHUS L.

Herbs (or shrubs) usually with a large basal rosette of lvs. Fls in large terminal bracteate spikes. Calyx 4-partite into large upper and lower lips and small lateral lobes. Corolla 1-lipped (upper lip 0), 3-lobed; tube very short. Stamens 4, shorter than the corolla; filaments thick, bent near the top; anthers 1-celled, connivent in pairs. Style filiform; stigma bifid. Jaculators well developed.

About 50 spp. in S. Europe, tropical and subtropical Asia and Africa. Several, all similar in general appearance, are grown in gardens.

*1. A. mollis L.   Bear's-breech.

Perennial herb 30–100 cm. Stems several, stout, simple. Basal lvs 25–60 cm, oblong-ovate, pinnatifid into oblong, lobed divisions, glabrous to puberulent, petiolate. Bracts ovate, spinose-dentate, tinged purple. Calyx glabrous. Corolla 3–5 cm, whitish with purple veins. Capsule glabrous, ovoid, with 2–4 seeds. Fl. 6–8. Pollinated by humble-bees; protandrous. $2n = 56$. Hs.

Introduced. Naturalized in waste places from W. Cornwall (including Scilly Is.) to Wilts, and in Huntingdon, Pembroke and Nottingham, first recorded 1820. Native of W. and C. Mediterranean region, and Portugal.

116. VERBENACEAE

Herbs, shrubs, trees and woody climbers usually with opposite or whorled exstipulate lvs. Fls solitary, axillary or more usually in many-fld infl. of various kinds; hermaphrodite, ±zygomorphic, hypogynous. Calyx of 5 or rarely 4 sepals, sometimes 2-lipped, often enlarging in fr.; corolla gamopetalous, commonly 2-lipped; stamens antesepalous, epipetalous, rarely 5, usually 4, sometimes 2, the missing stamens sometimes represented by staminodes; *ovary* syncarpous, initially 1-celled but becoming 2-, 4- or 5-celled by ingrowth of the placentae, and *most commonly 4- (8- or 10-)celled through formation of false septa between the placentae* (as in Labiatae); *style terminal*, 1 or dividing above into 2, 4 or 5 stigmatic lobes; ovules anatropous with the micropyle directed downwards, usually 1 per cell. *Fr. usually a drupe* with 1, 2 or 4 stones each 4–1-celled with 1 seed per cell; sometimes a capsule (*Avicennia*) and rarely of four 1-seeded nutlets (*Verbena*). Seeds usually non-endospermic, with a straight embryo.

About 3000 spp. in 75 genera, chiefly tropical and subtropical.

Teak (*Tectona grandis* L.f.), a large timber tree of Burma, Malaya, the East Indies, etc., is by far the most important member of the family commercially. *Clerodendrum* includes several small shrubs grown for their attractive fls whose corolla and persistent calyx have contrasting colours. *Lantana* spp., especially *L. camara* L., are pestilential weed-shrubs in many tropical and subtropical countries. *Avicennia* is a genus of mangroves with pneumatophores and viviparous fr., and *Verbena* includes many herbs grown for their brightly coloured fls.

1. VERBENA L.

Annual to perennial herbs or dwarf shrubs usually with opposite or whorled lvs and spikes, corymbs or panicles of smallish fls. Calyx tubular below, unequally 5-toothed above; corolla-tube straight or curved, downy within, the limb ±2-lipped with 5 spreading lobes; stamens 4, in pairs, rarely 2 or 5, not exserted, ovary 4-celled with 1 ovule in each cell; style slender, somewhat 2-lobed above. Fr. of 4 nutlets separating at maturity.

About 250 spp. chiefly in America, many of them weeds of arable land and waste places in North America.

The commonly cultivated *Verbena* is *V.* × *hybrida* Voss, probably of mixed hybrid origin with *V. peruviana* (L.) Druce as one of the parents.

1. V. officinalis L.　　　　　　　　　　　　Vervain.

A perennial herb with a woody stock and several stiffly erect tough stems 30–60 cm, ±hispid, paniculately branched above. Lvs 2–7.5 cm, oblanceolate or rhombic in outline, pinnatifid, with acute or blunt, ovate to oblong lobes, the lowest sometimes much the largest; upper lvs narrower, less divided, sometimes ±entire; all dull green, hispid. Fls in slender terminal spikes, at first dense, then elongating to become lax in fr. Bracts c. 2 mm, ovate-acuminate, ciliate, reaching half-way up the calyx. Fls 4 mm diam., subsessile. Calyx 2–3 mm, ribbed, shortly hairy. Corolla-tube almost twice as long as the calyx; limb pale lilac. Some fls have only 2 fertile anthers. Nutlets 4, reddish-brown, truncate, granulate on the inner face. Fl. 7–9. Homogamous; visited by small bees, hoverflies and butterflies, and automatically self-pollinated. $2n = 14$. H.

Native. Waysides and waste places; local. England and Wales, northwards to Cumbria and N. Yorks. Europe northwards to Denmark; N. Africa; W. Asia to Himalaya. Introduced in North America.

117. LABIATAE (Lamiaceae)

Herbs, less often shrubs, with ±quadrangular stems and opposite simple lvs; stipules 0. Infl. basically of cymose type, borne in the axils of opposite bracts which vary from being like the cauline lvs to very different from them, usually much contracted so that the 2 opposite cymes form a whorl-like infl. (referred to in the following account as a whorl), the whorls themselves often close together at the apex of the stem and forming a spike-like or head-like infl. (referred to as spikes or heads in the following account). Fls sometimes solitary in the axils of each bract. Bracteoles usually small, sometimes 0. Fls hypogynous. Calyx usually 5-toothed, often 2-lipped with the upper lip 3-, the lower 2-toothed, rarely with the lips entire. Corolla with a ±well-developed tube, basically 5-lobed, but with the 2 upper lobes nearly always closely united to form a single lip, its double origin sometimes apparent only from the venation; lower 3 lobes often forming a ±distinct lower lip but never so closely united as the upper, the middle lobe usually larger than the 2 lateral; rarely (*Teucrium*) the 5 corolla lobes form a single lower lip; aestivation imbricate. Stamens 4 (the upper absent), didynamous, more rarely reduced to 2, epipetalous, alternating with the corolla-lobes; anthers introrse. Carpels 2, each with 2 ovules but ovary apparently equally 4-lobed because of a secondary division developing later. Ovules anatropous, usually erect and basal. Style simple below, branched into 2 above, usually gynobasic. Fr. of four 1-seeded nutlets; endosperm 0 or scanty. Nectar secreted at the base of the ovary.

About 180 genera and 3500 spp. in tropical and temperate regions, rare in the Arctic.

A very natural family, usually recognizable at sight. The 4 nutlets are characteristic, but are found also in the Boraginaceae; these usually have alternate lvs and actinomorphic fls and the infl. and appearance are very different. The quadrangular stems and opposite lvs are found also in certain Scrophulariaceae, e.g. *Scrophularia*, *Rhinanthus*, the latter especially somewhat resembling the Labiatae in appearance but with a very different ovary. The Labiatae are frequently aromatic and have a characteristic smell (which may be either pleasant or unpleasant); many common pot-herbs belong to this family.

The fls are normally hermaphrodite but in several genera female fls, which normally have a smaller corolla, occur either on the same or more commonly on different plants.

Commonly cultivated non-British genera are: *Lavandula* (Lavender), *Rosmarinus* (Rosemary), *Physostegia*, *Coleus* and *Monarda*.

1 Corolla with 4 nearly equal lobes.   2
   Corolla 2-lipped or ±1-lipped.   3
2 Stamens 4; lvs entire or toothed.   1. MENTHA
   Stamens 2; lvs pinnately lobed.   2. LYCOPUS
3 Corolla ±1-lipped, the upper lip absent or represented by 2 short teeth.   4
   Corolla 2-lipped.   5
4 Corolla of a single 5-lobed lip; corolla-tube glabrous inside.   24. TEUCRIUM
   Corolla with conspicuous, 3-lobed lower lip and upper lip of 2 short teeth; corolla-tube with a ring of hairs inside.   25. AJUGA
5 Calyx 2-lipped; lips entire, the upper with an erect dorsal projection.   23. SCUTELLARIA
   Calyx ±equally 5-(rarely 10-)toothed, or 2-lipped with toothed lips and no dorsal projection.   6
6 Fertile stamens 2, each with a unilocular anther.   11. SALVIA
   Fertile stamens 4, each with a bilocular anther.   7
7 Stamens, at least the longer pair, longer than the upper lip of the corolla, diverging.   8
   Stamens included in the corolla-tube, or the anthers placed under the upper lip of the corolla, parallel or connivent.   10
8 Fls in a lax terminal panicle; bracts ovate, conspicuous.   3. ORIGANUM
   Fls in a dense head or spike, which may be interrupted below; bracts linear.   9
9 Calyx 2-lipped; fls purple or pink-purple.   4. THYMUS
   Calyx ±equally 5-toothed; fls blue or violet.   5. HYSSOPUS
10 Calyx 2-lipped.   11
   Calyx ±equally 5- or 10-toothed.   13
11 Corolla-tube curved upwards.   10. MELISSA
   Corolla-tube straight.   12
12 Lower lip of calyx with acute lobes; corolla 1.5 cm or less.   13. PRUNELLA
   Lower lip of calyx with obtuse lobes; corolla 2.5–4 cm.   12. MELITTIS
13 Upper lip of corolla distinctly hooded (concave), ±as long as to much longer than lower lip.   14
   Upper lip of corolla ±flat or convex, usually shorter than lower lip.   20
14 Lower lvs deeply lobed.   18. LEONURUS
   Lvs not lobed.   15
15 Calyx funnel-shaped (i.e. with a spreading rim below the base of the teeth); calyx-teeth 1–2 mm, broadly ovate.   15. BALLOTA
   Calyx tubular or campanulate; calyx-teeth usually at least 2.5 mm, narrow and often mucronate.   16
16 Lateral lobes of lower lip of corolla short, obscure, with one or more small teeth.   17. LAMIUM
   Lateral lobes of lower lip of corolla well developed.   17
17 Corolla deep yellow (see also *Phlomis*); stoloniferous perennial. In woods.   16. LAMIASTRUM
   Perennial with purple fls or annual of cultivated ground with purple or pale yellow fls.   18

18 Corolla with 2 bosses at base of lower lip; upper lip of corolla laterally compressed, helmet-shaped; annual.   19. GALEOPSIS
   Corolla without bosses; upper lip of corolla concave but not laterally compressed; annual or perennial.   19
19 Outer pair of stamens longer than the inner (sometimes curving downwards after dehiscence and then appearing shorter).   14. STACHYS
   Outer pair of stamens shorter than the inner.   20. NEPETA
20 Stamens and style included in corolla-tube.   22. MARRUBIUM
   Anthers placed under the upper lip of the corolla.   21
21 Outer pair of stamens shorter than the inner; corolla-tube hairy within at base of lower lip.   21. GLECHOMA
   Outer pair of stamens longer than the inner; corolla-tube glabrous or with sparse scattered hairs within.   22
22 Stamens straight.   14. STACHYS
   Stamens curved, connivent.   23
23 Dwarf shrub; lvs linear-lanceolate.   6. SATUREJA
   Herb; lvs elliptical to ovate or ovate-orbicular.   24
24 Calyx-tube straight; fls in opposite, axillary pedunculate cymes.   7. CALAMINTHA
   Calyx-tube curved; fls in opposite, axillary whorls without a common peduncle.   25
25 Whorls many-fld, dense; calyx-tube straight; fls pinkish-purple.   9. CLINOPODIUM
   Whorls 3–8-fld; calyx-tube gibbous at base; fls violet.   8. ACINOS

Subfamily 1. STACHYOIDEAE. Style gynobasic. Nutlets with small basal attachment. Seeds erect. Radicle short, straight. Calyx 5-(rarely 10-)toothed.

Tribe 1. SATUREJEAE. Calyx 10–15-veined. Corolla of 4 nearly equal lobes or 2-lipped, the upper lip flat or nearly so. Stamens 4, the outer pair longer than the inner, diverging or curved and connivent under the upper lip, rarely 2. (Genera 1–10).

1. MENTHA L.   Mint.

Perennial (rarely annual) herbs with a characteristic pleasing smell and creeping rhizome. Fls small, purple, pink or white, in axillary whorls, often forming a terminal spike or head. Bracteoles small or 0. Calyx tubular or campanulate, 10–13-veined, with 5 nearly equal teeth. Corolla-tube shorter than the calyx; lobes 4, nearly equal but the upper usually broader and often emarginate. Stamens 4, diverging, ±equal in length, usually exserted, but sometimes, especially in hybrids, included in the corolla; anther cells parallel. Nutlets ovoid, rounded at apex, smooth.

About 25 spp. in north and south temperate regions of the Old World. The species are very variable and hybridize freely. Several are commonly cultivated for flavouring and often escape. Propagation is largely vegetative and the hybrids are thus often found away from their parents and the parentage of many of them is in doubt. Small-fld female plants of most spp. occur; hermaphrodite fls protandrous; pollinated by various insects.

The following account is confined to the spp. and hybrids, the numerous varieties being mostly omitted. Hybrids are included in the key.

1 Lvs less than 1.5 cm; calyx hairy in throat; calyx-teeth unequal, the two lower narrower and longer than the three upper.
  Lvs more than 1.5 cm; calyx glabrous in throat; calyx-teeth ±equal.   3
2 Lvs ovate-orbicular, 2–7 mm; stems filiform, creeping; whorls 2–6-fld; corolla-tube straight.
                                                            1. requienii
  Lvs narrowly elliptical or suborbicular 8–20(–30) mm; stem not filiform; whorls many-fld; corolla-tube somewhat gibbous below the mouth.   2. pulegium
3 Whorls distant, in the axils of bracts like the lvs but smaller, with no or very few fls in the axils of the uppermost pair.   4
  Whorls in the axils of narrow, inconspicuous bracts, forming terminal spikes or heads.   7
4 Calyx broadly campanulate, with teeth about as long as wide; plant fertile.   3. arvensis
  Calyx narrowly campanulate or tubular, with teeth about twice as long as wide; plant usually sterile.   5
5 Calyx 2–3.5 mm, campanulate, the teeth rarely more than 1 mm; plant usually glabrous.   5. ×gentilis
  Calyx 3.5–4 mm, tubular, or, if shorter, the teeth usually 1–1.5 mm and plant distinctly hairy.   6
6 Plant distinctly hairy; upper bracts ovate to ovate-lanceolate, not cuspidate.   4. ×verticillata
  Plant subglabrous; upper bracts usually suborbicular, cuspidate.   6. ×smithiana
7 Fls in a head or oblong spike 12–20 mm diam.; lvs distinctly petiolate; calyx tubular.   8
  Fls in a ±slender spike 5–10(–15) mm diam.; lvs sessile or with very short petiole; calyx campanulate.   10
8 Fls in a head, sometimes with 1–3 whorls below; lvs usually ovate.   9
  Fls in an oblong spike; lvs usually lanceolate; plant sterile.   8. ×piperita
9 Lvs and calyx-tube hairy; plant fertile.   7. aquatica
  Lvs and calyx-tube glabrous or subglabrous; plant sterile.   8. ×piperita
10 Stamens not longer than corolla. Plant sterile.
                                                           10. ×villosa
  Stamens longer than corolla. Plant fertile.   11
11 Lvs not more than 45 mm, with numerous branched hairs beneath; spike 4–9 cm.   11. suaveolens
  Lvs (30–)50–90 mm, glabrous to hairy with some branched hairs beneath; spike 3–6 cm.   9. spicata

Section 1. *Audibertia* (Bentham) Briq. Bracts like lvs. Whorls 2–6-fld. Calyx turbinate-campanulate, weakly 2-lipped, the 2 lower teeth narrower than the 3 upper; throat hairy within. Corolla-tube straight.

*1. M. requienii Bentham   Corsican Mint.

Low perennial 3–12 cm, very strongly scented, glabrous or slightly hairy; *stems* creeping and rooting at the nodes, *filiform*, forming mats, the fl. stems ascending. *Lvs* 2–7 mm, *ovate-orbicular*, entire, petiolate. Corolla pale lilac, little exserted. Fl. 6–8. $2n = 18$. Chh.

Introduced. Grown in gardens for its scent and sometimes naturalized as in E. Sussex and by numerous rills

at c. 305 m on Slieve Gullion (Armagh). Native of Corsica and Sardinia.

Section 2. *Pulegium* (Miller) DC. Bracts like lvs. Whorls many-fld. Calyx tubular, weakly 2-lipped, the 2 lower teeth narrower than the 3 upper; throat hairy within. Corolla-tube gibbous on the lower side.

2. M. pulegium L.   Pennyroyal.

Perennial 10–30 cm. *Stems* prostrate, less often erect (var. *erecta* Martyn), pubescent, often red, *relatively stout*. *Lvs* 8–20(–30) × 6–10 mm, *narrowly elliptical*, rarely suborbicular, obtuse, ±cuneate at the base, minutely puberulent, gland-dotted, obscurely crenate-serrate with 1–6 teeth on each side, shortly petiolate. Whorls distant. Bracts like the lvs, but becoming smaller upwards, all longer than the fls. Pedicels and calyx shortly pubescent. Corolla hairy outside, glabrous within, lilac. Fl. 8–10. $2n = 20^*$. Hp.

Native. Wet places on sandy soil. Very local but widespread in S. England, from Cornwall to E. Kent northwards to Hereford and W. Norfolk, S. Wales, Derby, S. Lancs and Yorks, extinct elsewhere; introduced in S. Scotland; Ireland, Kerry, Wexford and Down, extinct or introduced elsewhere; Channel Is. W., C. and S. Europe; N. Africa; Macaronesia.

Section 3. *Mentha*. Calyx tubular or campanulate, equally or nearly equally 5-toothed, not hairy in the throat. Corolla-tube straight.

3. M. arvensis L.   Corn Mint.

Very variable perennial or rarely annual 10–60 cm, with erect or ascending, simple or branched, ±hairy stems, with sickly scent. Lvs (13–)20–65 × 10–20(–32) mm, often elliptical but varying from lanceolate to suborbicular, usually obtuse, cuneate or rounded at the base, shallowly crenate or serrate, petiolate, ±hairy on both sides. *Fls in distant axillary compact whorls; the bracts like the lvs*, gradually decreasing in size upwards but *always much longer than the fls*. Pedicels glabrous or hairy. *Calyx* broadly *campanulate, hairy all over; teeth about as long as wide, broadly triangular*, often deltate, obtuse to acuminate. Corolla lilac, white or pinkish, hairy outside. *Stamens normally exserted*. Fl. 5–10. $2n = 72^*$. Usually sterile.

Native. Common in arable fields, etc., in rides in woods and in damp places, ascending to 365 m; throughout the British Is. but rarer in N. Scotland, and absent from Orkney and Shetland. Most of Europe; N. Asia to the Himalaya.

4. M. × verticillata L. (*M. aquatica × arvensis*)

M. sativa L.

Very variable. Perennial 30–90 cm, often more robust than *M. arvensis*, and often hairier, with sickly scent. Lvs 20–65(–80) × (10–)15–40 mm, usually ovate to elliptical, oblong or lanceolate, acute or obtuse, cuneate to subcordate at base, serrate, petiolate, ±hairy on both sides. *Fls in whorls*, all distant or the upper crowded;

whorls less compact than in *M. arvensis* and often stalked; bracts like the lvs but often decreasing in size more than in *M. arvensis*, the upper sometimes shorter than the fls. *Pedicels and calyx hairy. Calyx* 3.5–4 mm *tubular*; teeth usually about twice as long as wide, ±triangular, the apex acuminate sometimes with a subulate point. Corolla lilac, hairy. *Stamens included*, very rarely exserted. $2n = 42*, 84*, 120*, 132*$. Rather common throughout the British Is., but local in N. Scotland and absent from the Outer Hebrides, Orkney and Shetland, often occurring where only one or neither parent is present.

5. M. × gentilis L. (*M. arvensis × spicata*)

Incl. *M. gracilis* Sole and *M. cardiaca* (S. F. Gray) Baker
Variable. Perennial 30–90 cm, *with pungent smell*. Stem erect, glabrous or hairy, sometimes red. *Lvs* (20–)30–60 (–75) × 10–25 mm, *ovate-lanceolate, lanceolate, elliptical or oblong*, usually thinly hairy on both faces, sometimes subglabrous above and with the hairs restricted to the veins beneath, obtuse or shortly acuminate, cuneate or rarely rounded at the base, shortly petiolate. *Whorls* usually *all separate*; the bracts ±like the lvs, gradually becoming smaller upwards in a variable degree. *Pedicels and calyx-tube* usually *glabrous*, but glandular. *Calyx* 2–3.5 mm, *campanulate*; teeth about twice as long as broad, subulate, acuminate or with a setaceous point. Corolla lilac, c. 3 mm. *Stamens included*. $2n = 54*, 60*, 84*, 96*, 108*, 120*$. Usually sterile.

A very variable plant apparently including amphidiploids as well as primary hybrids and back-crosses. There is insufficient evidence at present to separate the various types morphologically.

Sides of ditches, waste ground, etc.; throughout Great Britain but rather local; absent from much of N. Scotland, Outer Hebrides, Orkney and Shetland. Scattered from N. to S.E. Ireland.

6. M. × smithiana R. A. Graham (*M. aquatica × arvensis × spicata*)

M. rubra Sm., non Miller

Stem 30–150 cm, erect, simple or branched above, usually red or purple, glabrous or more rarely sparsely hairy; *smell sweetly pungent. Lvs* (20–)30–60 × 15–35 mm, *ovate*, usually glabrous or nearly so, sometimes thinly hairy on both faces, obtuse or acute, rounded or often ±cuneate at base, serrate, shortly petiolate, often purple-veined. *Whorls separate; bracts like the lvs, but gradually smaller, usually suborbicular, cuspidate, the uppermost usually longer* but occasionally shorter *than the fls. Pedicels and calyx glabrous*, except for the shortly ciliate teeth, but glandular. *Calyx* 3.5–4 mm *tubular*; teeth about twice as long as broad. Corolla lilac, c. 5 mm. Stamens usually ±exserted. Fl. 8–10. $2n = 120*$. Hp. Usually sterile, but occasionally producing a few viable seeds.

Probably originating as a hybrid between *M. × verticillata* and *M. spicata*, but with a double chromosome number.

Native? Sides of ditches, damp hedgebanks, waste ground, etc., often in the absence of *M. spicata*; in scattered localities throughout the British Is., local; absent from much of W. and N. Scotland and from C. and W. Ireland.

M. × muellerana F. W. Schultz (*M. arvensis × rotundifolia*)
Like *M. arvensis* but much more robust, lvs broader, bracts suborbicular or broadly ovate, markedly decreasing in size upwards. Calyx-teeth lanceolate. Stamens included. $2n = 60*$. Sterile.
In 1 locality in S. Devon.

7. M. aquatica L. Water Mint.

Variable. Perennial 15–90 cm, strongly but not pungently scented. Stems ±erect, simple or branched, often purplish in exposure. Lvs 20–60(–90) × (12–)15–40 mm, usually ovate, ±hairy on both faces, or sometimes subglabrous, obtuse or acute, cuneate to subcordate at base, serrate or crenate-serrate, petiolate. *Infl. consisting of a terminal head* c. 2 cm across, composed of 2–3 congested whorls, and *usually 1–3 distant axillary whorls below*. Lower bracts lf-like, the upper lanceolate or linear-lanceolate, hidden by the fls. *Pedicels and calyx hairy*. Corolla lilac. *Stamens* normally *exserted*. Fl. 7–10. $2n = 96*$. Hp. or Hel.

Native. In swamps, marshes, fens and wet woods and by rivers and ponds, ascending to 460 m. Common throughout the British Is. Europe except the extreme north; S.W. Asia; N. and S. Africa; Madeira.

8. M. × piperita L. (*M. aquatica × spicata*)

Peppermint.

Subglabrous, or sometimes hairy to grey-tomentose perennial 30–80 cm, with a pungent smell and taste (of peppermint). Stem erect, usually branched, reddish or purple. *Lvs* (20–)35–85 × 15–38 mm, usually lanceolate, less often ovate, acute, usually cuneate to (rarely) subcordate at base, sharply serrate, *petiolate. Infl. a terminal oblong spike* (2–)3.5–6 cm, usually interrupted at the base, or a head; bracts lanceolate, about as long as the fls, the lowest 1–2 pairs somewhat lf-like. *Pedicels and calyx-tube usually glabrous*, ±glandular. Calyx tubular, often strongly tinged with red; teeth subulate, long-ciliate. *Corolla lilac-pink. Stamens included.* $2n = 66*, 72*$. Sterile.

Sides of ditches, etc., and damp roadsides; throughout the British Is. but rather local. Cultivated as the source of peppermint. A naturalized alien throughout much of Europe but absent from Fennoscandia and N. Russia.

nm. **citrata** (Ehrh.) Boivin. A sterile cultivar resembling *M. aquatica* in infl. but with glabrous pedicels and calyx, glabrous or very sparsely hairy, ovate, subcordate lvs and included stamens and a characteristic lemony scent. $2n = 84*, 120*$.

Widely cultivated and occasionally naturalized in scattered localities from Cornwall to Surrey, Hertford, and

Worcester, Pembroke, Montgomery, Stafford, Yorks and Northumberland.

M. × *dumetorum* Schultes (*M. palustris* Sole, non Miller; *M. pubescens* auct.) (*M. aquatica* × *longifolia*) has been reported as naturalized in S. and W. England, but these records almost certainly refer to hairy forms of *M.* × *piperita*.

M. × *maximilianea* F. W. Schultz (*M. aquatica* × *suaveolens*), with $2n = 60, 72–78, 120$, occurs occasionally in the vicinity of its parents, between which it is intermediate in morphology and usually sterile, though fertile backcrosses are reported. It is recorded from W. Cornwall, N. Devon, S. Somerset and Jersey.

*9. M. spicata L.   Spear Mint.

M. viridis (L.) L.; *M. longifolia* auct., non (L.) Hudson; *M. cordifolia* auct., non Opiz; *M. niliaca* auct., non Juss. ex Jacq.; *M. sylvestris* auct., non L.

Perennial 30–90 cm, with a pungent smell. *Stem* erect, usually branched, *glabrous. Lvs* 40–90 × 13–30 mm, *lanceolate* or ovate-lanceolate, acute or acuminate, sharply serrate, green and *glabrous* to densely hairy, *sessile or* the lowest *subsessile* (petiole not more than 3 mm), smooth or rugose. *Infl. a terminal cylindrical spike* 3–6 cm, 5–10(–15) mm diam.; whorls often becoming ±separate. Spikes often clustered at the top of the main axis. Bracts linear-setaceous, longer than fls, the lowest pair sometimes lf-like. *Pedicels and calyx glabrous or hairy.* Corolla lilac, pink or white, glabrous. Stamens exserted. Fl. 8–9. $2n = 36^*, 48^*, 84$. Hp.

Origin unknown, but probably arose in cultivation. Widely cultivated as a pot-herb and naturalized in usually damp areas by roads and in watery places throughout the British Is., though only occasionally so in Ireland. Naturalized throughout much of Europe.

10. M. × villosa Hudson (*M. spicata* × *suaveolens*)

M. × *niliaca* auct., non Juss. ex Jacq.; *M. cordifolia* auct.; *M. scotica* R. A. Graham.

Very variable. May be intermediate between parents in morphological characters or resemble one or the other (usually *M. spicata*) very closely, though it usually has wider, more rugose lvs and the stamens are shorter than the corolla. $2n = 36$. Sterile.

Widely naturalized throughout Great Britain and probably often arising spontaneously. Many variants have spread vegetatively to form locally common clones which have been given taxonomic recognition. Such an example is *M. scotica* R. A. Graham ($2n = 44$), found by streams in E. Scotland from Berwick to Moray, which has oblong-ovate lvs which are grey above and densely white-tomentose beneath. The most widespread form is nm. *alopecuroides* (Hull) Briq., which is more robust than *M. spicata*, with broadly ovate to suborbicular lvs with patent teeth. It has $2n = 36$.

Hairy variants of *M. spicata* have, in the past, been erroneously attributed to *M. longifolia* (L.) Hudson, which does not occur in Britain. However, the hybrid *M. longifolia* × *spicata* (*M.* × *villosonervata* auct.) is widely cultivated and has become locally naturalized.

It is difficult to separate from *M. spicata* and *M.* × *villosa*, though tending to have narrower, patently toothed lvs with few or no branched hairs, than the latter.

11. M. suaveolens Ehrh.   Round-leaved Mint.

M. rotundifolia auct., non (L.) Hudson

Perennial 60–90 cm, strongly fragrant, stoloniferous. *Stem* erect, usually branched above the middle, sparsely to *densely clothed with white hairs. Lvs* 20–40 × 15–30 mm, *oblong, ovate or suborbicular* on the same stem, *rounded or minutely cuspidate at apex, crenate-serrate*, the teeth sometimes cuspidate, pubescent above, *grey or white tomentose beneath*, subcordate at base, *sessile* or very shortly petiolate, *rugose. Fls in dense spikes*, 3–5 cm, falcate when young, ±interrupted below, often forming a panicle. Bracts lanceolate-subulate, longer than the fls. *Pedicels and calyx hairy.* Corolla pink or whitish, hairy outside. Stamens usually exserted. Fl. 8–9. $2n = 24^*$. Hp.

Native from Cornwall to Hants and Anglesey; naturalized elsewhere in S. England, Warwick, Yorks and Ross, and locally so throughout much of Ireland except the centre. S. and W. Europe; N. Africa, Azores.

2. LYCOPUS L.

Perennial odourless herbs with creeping rhizome. Fls in many-fld distant axillary whorls; bracts not differentiated from the lvs. Bracteoles small. Fls small. Calyx campanulate, 13-veined, with 5 equal teeth. *Corolla-tube shorter than calyx, with 4 subequal lobes, the uppermost usually widest. Stamens* 2, diverging, longer than the corolla. Anther cells parallel. *Nutlets tetrahedral; apex truncate*, bordered.

About 14 spp. in temperate regions of the northern hemisphere.

1. L. europaeus L.   Gipsywort.

Stems 30–100 cm, erect, somewhat hairy, usually with ascending branches. Lvs to 10 cm, ovate-lanceolate or elliptical, acute, shortly petiolate, pinnately lobed, with numerous triangular, acute lobes, the lower often reaching to the midrib, but sometimes all shallow. Calyx-teeth lanceolate, hairy, spiny-pointed. Corolla white with a few small purple dots on the lower lip, c. 3 mm long and across. Fl. 6–9. Pollinated by various insects; protandrous; small female fls occur. $2n = 22$. Hp. or Hel.

Native. On the banks of rivers and ditches and in marshes and fens. Common throughout much of England and Wales, in Scotland mainly in the west and extending to Sutherland and the Outer Hebrides, in the east on the coast northwards to Ross; scattered throughout Ireland; Channel Is. Europe northwards to 64° 30′ N in Fennoscandia; N. Africa; N. and C. Asia; naturalized in North America.

3. ORIGANUM L.

Aromatic perennial. Fls in whorls grouped into short spicules; spicules arranged in a corymb or panicle. Bracts ovate, exceeding the calyx, imbricate, distinct from the lvs. *Calyx* campanulate, 13-veined, *with 5 nearly equal teeth. Corolla 2-lipped. Stamens 4, straight, diverging, longer than the corolla* (at least the longer pair). Anther cells divergent. Nutlets ovoid, smooth.

About 15 spp., north temperate Old World, mainly Mediterranean.

1. O. vulgare L.   Marjoram.

Woody rhizomatous perennial. Stems 30–80 cm, erect, branched above, and often with very short axillary sterile branches below, somewhat hairy. Lvs 1.5–4.5 cm, ovate, petiolate, entire or obscurely crenate-serrate, glabrous or with scattered appressed hairs on both sides. Infl. lax or dense. Bracts conspicuous, ovate, purple, longer than the calyx. Calyx yellow-glandular-punctate, hairy within; teeth short. Corolla 6–8 mm, violet-purple; tube longer than the calyx. Fl. 7–9. Pollinated by various insects; protandrous; small female fls occur (gynodioecious). $2n = 30^*$, 32. Chh.

Native. Dry pastures, hedge-banks and scrub, usually on a calcareous soil. Common in much of England and Wales; local in Scotland, extending to Argyll and Moray; throughout Ireland, but less common in the south-west and north-east. Most of Europe; N. and W. Asia.

4. THYMUS L.   Thyme.

Small aromatic shrubs. Fls in few-fld whorls, forming a terminal spike or head. Bracts ±differentiated. Bracteoles minute. *Calyx 2-lipped*, the upper lip with 3 ±equal teeth, the lower 2-lobed. *Corolla, stamens* and fr. *as in Origanum*.

About 300 spp. in temperate Eurasia.

T. vulgaris L., 'Garden Thyme', an erect greyish shrublet, is the sp. usually cultivated as a pot-herb. Some other spp. are also cultivated for their aromatic lvs.

Gynodioecism is common in the British spp.; the female fls are normally smaller than the hermaphrodite fls. Pollinated by various insects, protandrous.

1 Fl.-stem sharply angled, with long hairs on the angles only (2 faces shortly pubescent); plant tufted with the branches ascending; lateral veins of lvs not prominent beneath when dry; infl. elongated on normally developed plants.   **1. pulegioides**
   Fl.-stem obscurely angled, hairy on at least two opposite sides; plant with long creeping branches, forming a mat; lateral veins of leaves prominent beneath when dry; infl. usually capitate.   2
2 Fl.-stem hairy on two opposite sides only or more hairy on two opposite sides than on the other two; lateral veins of lvs curved along margin and anastomosing at apex.   **2. praecox**
   Fl.-stem equally hairy all round; lateral veins of lvs disappearing towards margin.   **3. serpyllum**

1. T. pulegioides L.   Large Thyme.

T. chamaedrys Fr.; *T. ovatus* Miller; *T. glaber* Miller

Tufted, *strongly aromatic* dwarf shrub to 25 cm, *with ascending fl.-branches* and short creeping branches (more prostrate if heavily grazed). *Fl.-stems*, below the

infl., *sharply 4-sided with long hairs only on the angles*, two opposite faces narrow and shortly pubescent, the other two wider and glabrous. *Lvs* 6–10(–17) mm × 3–6 mm, ovate to elliptical, obtuse, shortly stalked, ciliate at base, otherwise glabrous, *slightly folded upwards about the midrib; lateral veins slender, not prominent beneath when dry; stomata* 21–23 μm, very few on upper surface. *Infl.* elongated and interrupted below when normally developed, but capitate on grazed plants. Calyx 3–4 mm, ±hairy, the teeth longciliate. Corolla pink-purple. Fl. 7–8. Fr. 9. $2n = 28^*$, 30. Chw.

Native. Dry grassland, usually calcareous, common in S.E. England, extending to Devon, Monmouth, Shropshire and Yorks; Wigtown, Angus; Cork, Cavan, but doubtfully native. Europe, except parts of the north and east and many of the islands.

2. T. praecox Opiz   Wild Thyme.

T. drucei Ronn., *T. arcticus* (E. Durand) Ronn., *T. britannicus* Ronn., *T. neglectus* Ronn., *T. zetlandicus* Ronn., *T. pseudolanuginosus* Ronn., *T. carniolicus* auct. brit., *T. pycnotrichus* auct., *T. serpyllum* auct. brit., pro parte.

Mat-like, faintly (to moderately) *aromatic dwarf shrub*, rarely more than 7 cm, *with long creeping branches*, the fl.-stems in rows on the branches. *Fl.-stems*, below infl. 4-*angled with 2 opposite sides densely hairy with hairs of varying length, the other 2 sides less hairy or glabrous* (very rarely all equally hairy). *Lvs* 4–8(–11.5) mm × 1.5–4 mm, usually borne horizontally (rarely upright), suborbicular to elliptical or oblanceolate, obtuse, shortly petiolate, glabrous or hairy as well as ciliate, *flat, lateral veins prominent beneath when dry, curved along lf-margin and anastomosing at apex; stomata* 24.5–28 μm, numerous on both surfaces (though more so beneath). *Infl.* capitate, rarely somewhat elongated. Calyx 3.5–4 mm, ±hairy, the upper teeth long-ciliate. Corolla rose-purple. Very variable, several ecotypes occurring. Fl. 5–8, earlier than the other spp. when growing together. $2n = c. 50$, 51, 54. Chw.

Native. Dry grassland, heaths, dunes, screes and among rocks. Common throughout the British Is., but less frequent in East Anglia, parts of the English Midlands and C. Ireland, ascending to 1130 m in Perth. Greenland, S., W. and C. Europe.

British plants belong to subsp. **arcticus** (E. Durand) Jalas; which is restricted to W. Europe.

3. T. serpyllum L.   Breckland Thyme.

Similar to *T. praecox* but much less variable in Britain. Small. Faintly aromatic. *Fl.-stems* below infl. scarcely angled, *with short white hairs evenly distributed all round*. Lvs 4–5(–7) mm × 1.5 mm, borne upright, oblanceolate-spathulate, ciliate, usually otherwise glabrous; *lateral veins disappearing towards lf-margins; stomata* 22–24 μm. Infl. capitate. Fl. 7–8. $2n = 24^*$. Chw.

Native. Open sandy heath and grassland in the Breckland of East Anglia. Europe northwards from N.E. France, N. Austria and N. Ukraine.

T. × *oblongifolius* Opiz (*T. pulegioides* × *serpyllum*) has been erroneously reported from Norfolk (Swaffham), the material being *T. praecox*. The parental species occur together in East Anglia but no hybrids have yet been found.

5. HYSSOPUS L.

Aromatic perennials, woody at the base. Lvs lanceolate or linear, entire. Fls in 4–16-fld whorls, forming a terminal unilateral spike-like infl. Bracteoles, small, linear. *Calyx* tubular, 15-*veined*, with 5 nearly equal teeth. *Corolla 2-lipped. Stamens 4, ascending and curved at the base, then suddenly diverging,* longer than the upper lip of the corolla; anther cells divergent. Nutlets ovoid-trigonous, smooth.

About 15 spp. in S. Europe and N. Africa to C. Asia.

*1. H. officinalis L.　　　　　　　　　　　Hyssop.

Stems 20–60 cm high, nearly glabrous, green. Branches erect, glabrous or puberulent. Lvs 1.5–2.5 cm, oblong-lanceolate to linear, suboblique. Infl. rather dense and long. Bracts about as long as the fls, linear, acuminate. Calyx 4–5 mm, with ovate, shortly aristate teeth 2–3 mm long. Corolla 10–12 mm, violet or blue, rarely white. Fl. 7–9. Pollinated by bees; protandrous. $2n = 12$. N.

Introduced. Formerly much cultivated as a herb and still sometimes grown for ornament. Has become naturalized on old walls in a few places as at Beaulieu Abbey. S., S.C. and E. Europe; W. Asia; Morocco.

6. SATUREJA L.

Aromatic herbs or dwarf shrubs. *Fls in few-fld axillary whorls,* forming a crowded terminal unilateral infl. Bracteoles small. *Calyx* 10-veined, ±equally 5-toothed. *Corolla 2-lipped; tube straight,* naked within. *Stamens 4, shorter than the corolla, converging.* Style-branches nearly equal, subulate. Nutlets ovoid, smooth.

About 30 spp. in temperate and warm regions.

*1. S. montana L.　　　　　　　　　Winter Savory.

Dwarf shrub 15–40 cm high, with erect or ascending stems having stiff erect branches. Lvs 1–2 cm, lanceolate-linear, coriaceous, mucronate, ciliate, acute, longer than the internodes, entire. Fls in small shortly pedunculate 2–4-fld cymes, forming a long terminal unilateral infl. Bracts like the lvs. Bracteoles shorter than the fls, mucronate. Calyx-teeth triangular-lanceolate, acuminate. Corolla white or pink; tube 6–7 mm, much longer than the calyx. Fl. 7–10. $2n = 30$. N.

Introduced. Formerly much cultivated as a pot-herb, rarely naturalized on old walls, as at Beaulieu Abbey. S. Europe; Algeria.

7. CALAMINTHA Miller

Perennial herbs. *Fls in opposite axillary pedunculate cymes*; bracts similar to cauline lvs but becoming much smaller above. *Calyx* tubular, 13-*veined*, 5-toothed; tube straight, *hairy within. Corolla 2-lipped; tube straight,* naked within. *Stamens 4, shorter than corolla, curved and converging.* Style-branches unequal, the upper subulate, the lower longer and wider. Nutlets ovoid, smooth.

About 70 spp. in north temperate regions from W. Europe to C. Asia.

Forms with small female fls only occur. The measurements given apply to the hermaphrodite fls. Measurements of the lvs apply to the cauline lvs of the main stem.

Lower calyx-teeth 2–4 mm, usually densely long-ciliate; hairs in throat of calyx ±included.　　**1. sylvatica**
Lower calyx-teeth 1–2 mm, with no or very few long cilia; hairs in throat of calyx somewhat exserted.　　**2. nepeta**

*C. grandiflora (L.) Moench with larger (3–4 cm) fls and more deeply toothed lvs than our native spp. has occurred as a garden-escape. Native of W., S. and S.C. Europe, S.W. Asia and Algeria.

1. C. sylvatica Bromf.

C. intermedia auct.; *Satureja sylvatica* (Bromf.) Maly

Perennial with creeping rhizome. *Stems* erect, 30–60 cm, *little branched,* with long spreading hairs. Lvs ovate, obtuse or subobtuse, broad-cuneate or rounded at base and often somewhat decurrent down the petiole (5–15 mm long), *serrate with 5–10 teeth on each side,* green, with long ±appressed hairs on both surfaces. Peduncles up to 9(–12)-fld, upper spreading after fl.; partial peduncles short or 0. Calyx 7–10 mm, with long hairs on the nerves and sessile shining glands between; *upper teeth spreading; lower teeth much longer, long-ciliate* (cilia longer than width of tooth), curved upwards; *throat with ring of hairs included. Corolla* 17–22 mm, pink or lilac, much variegated with purple on lower lip. Fl. 7–9. Hp.

Subsp. **sylvatica**　　　　　　　　Wood Calamint.

C. intermedia auct.; *C. baetica* Boiss. & Heldr.; *Satureja sylvatica* (Bromf.) Maly

Lvs 25–70 × 10–45 mm, the middle and upper ovate, coarsely dentate or crenate-serrate, with 6–10 teeth on each side, the lower suborbicular. Peduncles up to 10 mm. Lower calyx-teeth 3–4 mm. Corolla 17–22 mm. $2n = 24$.

Native. Chalky banks in Isle of Wight. From France eastwards to S.W. Ukraine.

Subsp. **ascendens** (Jordan) P. W. Ball
　　　　　　　　　　　　　　Common Calamint.

C. ascendens Jordan; *C. officinalis* auct. angl., non Moench; *Satureja ascendens* (Jordan) Maly

Lvs (15–)20–40(–50) × 15–35 mm, ovate or orbicular-ovate, subentire or shallowly crenate-serrate, with 5–8 teeth on each side. Peduncles 0.5(–10) mm. Corolla 10–16 mm. $2n = 48$.

Native. Dry banks, usually calcareous; from Durham, Yorks and Isle of Man southwards and from Galway, Roscommon and Dublin southwards. W., S. and S.C. Europe; Caucasus; N. Africa.

2. C. nepeta (L.) Savi　　　　　　Lesser Calamint.

Satureja nepeta (L.) Scheele

Perennial, appearing more greyish than the preceding

sp., with long creeping rhizome. *Stems* erect, 30–60 cm, *much-branched*, of rather bushy appearance, grey with long soft spreading hairs. *Lvs* 10–20 mm, ovate, obtuse, usually broad-cuneate at base, *shallowly crenate or crenate-serrate with 5 or fewer teeth on each* side, greyish, above with short scurfy and few long hairs, beneath with more numerous long hairs; petiole usually 5 mm or less. Peduncles short, up to 15-fld, ascending after fl.; *partial peduncles present. Calyx* 4–6 mm, with short hairs often obscuring the shining sessile glands; *upper teeth nearly straight* or somewhat spreading; *lower teeth* somewhat *longer*, puberulent, rarely with a few cilia less than width of tooth at its base, straight or nearly so; *throat with rings of hairs protruding. Corolla* 10–15 mm, lilac, scarcely spotted. Fl. 7–9. Visited by various insects, mainly bees. Hp.

Native. Dry banks, usually calcareous; from Kent to Sussex, Gloucester, Monmouth, Northampton and Suffolk; local. S., W. and S.C. Europe from W. and C. France to S. Russia; N. Africa, N. Syria and N. Iran.

British material belongs to subsp. **glandulosa** (Req.) P. W. Ball, to which the above description refers. Subsp. *nepeta*, with longer peduncles and lvs with more teeth, is found in the mountains of S. and S.C. Europe.

8. ACINOS Miller

Herbs. *Fls in c. 6-fld axillary whorls.* Bracts not differentiated. Bracteoles minute. *Calyx* tubular, 13-veined; *tube curved, gibbous at base, hairy within.* Corolla, *stamens and style as in Calamintha.*

About 10 spp., in Europe and Mediterranean region to C. Asia and Iran.

1. A. arvensis (Lam.) Dandy   Basil Thyme.
Calamintha acinos (L.) Clairv.; *Acinos thymoides* Moench; *Satureja acinos* (L.) Scheele

Usually annual, occasionally lasting for more than one year, branched from the base. Stems 10–20 mm, ascending, pilose. Lf-blades 5–15 mm, ovate to elliptical, petiolate, subacute, cuneate at base, obscurely crenate, glabrescent. Fls in 3–8-fld axillary whorls, forming a lax terminal infl. Calyx pilose; tube contracted in the middle; teeth subulate, the upper one broad-based. Corolla 7–10 mm, violet with white markings on the lower lip. Fl. 5–9, pollinated by bees. $2n = 18$. Th.

Native. Arable fields, open habitats in grassland or rocks on dry usually calcareous soils, rather local; Devon and Somerset to Kent and northwards to Cumberland and Northumberland, Moray, very local in N. England and Wales, decreasing; Wexford, Carlow, Kildare, Tipperary and Leix, apparently extinct elsewhere in S.E. Ireland. Most of Europe except the extreme north and parts of the south; Asia Minor, Caucasus.

9. CLINOPODIUM L.

Herbs. *Fls in remote many-fld terminal and axillary whorls.* Bracts not differentiated. Bracteoles subulate, conspicuous. *Calyx* cylindrical, 13-veined; *tube curved,*

not gibbous, glabrous or weakly hairy within. Corolla, stamens and style as in Calamintha.

About 10 spp., in north temperate zone.

1. C. vulgare L.   Wild Basil.
Calamintha vulgaris (L.) Druce; *Calamintha clinopodium* Bentham

Pubescent to densely villous perennial; scent weak. Rhizome shortly creeping. Stems 30–80 cm, erect, simple or sparingly branched. Lf-blades 20–50 mm, ovate, petiolate, subobtuse, rounded or broad-cuneate at base, shallowly and remotely crenate-serrate. Bracteoles numerous, long-ciliate. Calyx somewhat 2-lipped, the 3 upper teeth broad-based, up to 2.5 mm, the 2 lower 2.5–4 mm. Corolla 15–20 mm, pinkish-purple. Fl. 7–9. Pollinated by bees and Lepidoptera; protandrous; small female fls occur. $2n = 20$. Hp.

Native. Hedges, wood borders and scrub; less often in grassland on dry, usually calcareous soils; ascending to 395 m. Common in England, becoming more local northwards, extending to Perth and Ross; only as a rare casual in Ireland, mainly in the south and east. Most of Europe; C. and W. Asia; Siberia; N. Africa, Azores; Madeira; North America.

10. MELISSA L.

Perennial herbs. Fls in axillary whorls. Bracteoles ovate or obovate. Calyx campanulate, 2-lipped, 13-veined. *Corolla 2-lipped*; upper lip sometimes slightly concave; *tube curved upwards and dilated above the middle. Stamens 4, shorter than the corolla, curved, converging.* Style-branches equal. Nutlets obovoid, smooth.

Three spp. in Europe to C. Asia, N. Africa, Iran.

*1. M. officinalis L.   Balm.

Sweet-scented perennial herb with short rhizome and erect, branched, somewhat hairy stems 30–60 cm high. Lf-blades 3–7 cm, ovate, with a petiole more than ½ as long, deeply crenate or serrate, glabrescent, passing into the smaller, more sharply toothed bracts. Calyx with long spreading hairs; upper teeth broadly triangular, the lower triangular-lanceolate, all with subulate points. Corolla c. 12 mm, pale yellow, becoming white or pinkish. Fl. 8–9, pollinated mainly by bees; protandrous to protogynous; small female fls occur. $2n = 32$. Hp.

Introduced. Cultivated for its sweet scent; a not uncommon garden-escape in S. England and naturalized in some places northwards to Norfolk and Flint, in Westmorland and Isle of Man; scattered localities in S. and E. Ireland. Native of S. Europe, W. Asia and N. Africa.

Tribe 2. SALVIEAE. Corolla 2-lipped; lobes usually very unequal. Upper lip often concave. Stamens 2; connective much elongated, with a single linear anther-cell.

11. SALVIA L.

Annual or perennial herbs or undershrubs. Fls in axillary whorls forming a ±interrupted terminal spike, the bracts differentiated. Bracteoles usually small. Calyx tubular or campanulate, 2-lipped, the upper with 3 teeth, lower with 2. Stamens 2, the filaments short, the connective much elongated, one anther cell aborted so that the stamen appears branched with an anther cell on the end of the longer branch. Nutlets ovoid-trigonous, smooth.

About 700 spp., in tropics and temperate zones.

A number of spp. are cultivated, including *S. officinalis* L. (Sage) from S. Europe, a greyish-leaved undershrub much cultivated as a pot-herb, and *S. splendens* Sellow ex Nees from Brazil with brilliant scarlet fls much used as a bedding plant. About 15 other spp. have been recorded as garden-escapes or casuals.

> *1* Upper lip of calyx with conspicuous, ±equal teeth; corolla-tube with a ring of hairs within; fls usually in whorls of 15–30.   **1. verticillata**
> Upper lip of calyx with very short teeth, the two lateral connivent over the middle one; corolla-tube without a ring of hairs; fls usually in whorls of 1–10.   *2*
> *2* Calyx pubescent and glandular but without long white hairs; corolla 10–25 mm, the smaller fls female only; upper lip glandular outside.   **2. pratensis**
> Calyx glandular and pilose with long white hairs; corolla 6–15 mm; fls all hermaphrodite, the smaller cleistogamous; upper lip not glandular outside.
> **3. verbenaca**

Section 1. *Hemisphace* Bentham. Calyx tubular; upper lip with 3 nearly equal teeth. Corolla-tube with a ring of hairs within; upper lip weakly convex. Whorls many-fld.

*1. S. verticillata L.   Whorled Clary.

Perennial 30–80 cm, hairy, foetid. Lf-blades 5–15 cm, broadly ovate, obtuse, cordate at base, irregularly crenate-dentate, sometimes pinnatifid, petiolate. Bracts small, brown, reflexed. Pedicels 3–5 mm. Calyx c. 5.5 mm, tubular-campanulate, 12-veined, *the upper lip with conspicuous teeth, the middle tooth slightly broader and shorter than the lateral.* Corolla 10–15 mm, violet; tube exserted. Fl. 6–8. Pollinated by bees; protandrous; small female fls occur. $2n = 16$. Hs.

Introduced. In waste places, recorded from many localities, from Kent to Somerset, and northwards to Norfolk, Cheshire and Denbigh; in Ireland naturalized in N. Tipperary. Mountains of S., E. and E.C. Europe and of S.W. Asia.

Section 2. *Plethiosphace* Bentham. Calyx campanulate; upper lip with 3 short teeth, the middle one very small (less than 0.3 mm), the two lateral connivent over it. Corolla-tube without a ring of hairs within; upper lip concave. Whorls c. 6-fld.

2. S. pratensis L.   Meadow Clary.

Perennial 30–100 cm, pubescent, glandular above, aromatic. Basal lf-blades 7–15 cm, ovate or oblong, obtuse, cordate at base, irregularly doubly crenate or occasionally lobed, rugose, with long petioles; cauline 2–3 pairs, smaller, the upper sessile. Bracts green, ovate, acuminate, entire, shorter than the calyces. Fls hermaphrodite or female, usually on separate plants, never cleistogamous. *Calyx* pubescent and glandular but *without long*

white hairs, c. 6.5 mm in the hermaphrodite fls. *Corolla of the hermaphrodite fls* 15–25 mm, violet-blue; the tube exserted; *upper lip laterally compressed, forming a hood, falcate, glandular outside; style long-exserted.* Female fls much smaller, sometimes only 10 mm. Fl. 6–7. Pollinated by long-tongued bees; protandrous or homogamous. $2n = 18$. Hs.

Native. On calcareous grassland from Kent and Surrey to Monmouth and Worcester, naturalized or casual elsewhere in S. England and northwards to S. Lincoln and locally to Cumberland; rare. Most of Europe northwards to N. Germany and N.C. Russia; Morocco.

*S. nemorosa L. (*S. sylvestris* auct., non L.) differing in its more numerous, oblong-lanceolate stem-lvs, purple bracts and numerous whorls of smaller (10–14 mm) fls is naturalized in one place at Barry Docks (Glamorgan), and occurs as a casual elsewhere. Native of C., S.E. and E. Europe and W. Asia.

3. S. verbenaca L.   Wild Clary, Guernsey Clary.

S. clandestina L.; *S. horminioides* Pourret; *S. marquandii* Druce

Perennial 30–80 cm, little-branched, pubescent, glandular above, slightly aromatic. Basal *lf-blades* 4–12 cm, oblong or ovate, obtuse, crenate-serrate or sinuate to incised or pinnately lobed with the lobes crenate-serrate, rugose, petiolate. Stem-lvs usually 2–3 pairs, smaller, the upper sessile. Upper part of stem and calyces often strongly tinted with dull blue-purple. Bracts as in *S. pratensis.* Fls 12–15 mm, often cleistogamous and much smaller. *Calyx* c. 7 mm, pubescent, glandular and *pilose with long white hairs* which occur especially near the base and round the sinuses, with more prominent veins and longer points than in the preceding sp. *Corolla up to* 15 mm in the largest open-fld forms, blue, lilac or *with 2 white spots at the base of the lower lip,* and then with the upper lip somewhat compressed and falcate and the style slightly exserted; *in the cleistogamous fls as little as* 6 mm, the lips nearly equal and connivent, the upper nearly straight and the style included. Fl. 5–8. $2n = 64$. Hs.

Native. Local in dry pastures, roadsides and occasionally coastal dunes. Widespread in S. England to Worcester, Leicester and Norfolk; then northwards in the east to Yorks and in N. Northumberland and Fife; Monmouth to Pembroke, Caernarvon and Flint; Ireland, locally on the south coast and in Dublin. S. and W. Europe; Algeria.

S. reflexa Hornem., a native of S.C. USA and Mexico, is apparently brought into Britain in bird-seed and is a frequent casual, though nowhere naturalized, from Kent and Somerset to Northampton and Lancs, and in S. Scotland. It is an annual 30–40 cm high, with a glabrous or puberulent stem bearing linear- to oblong-lanceolate lvs 30–80 × 4–12 mm. Fls 2–4 in each whorl, forming a slender interrupted spike; calyx deeply 2-lipped, 6–8 mm, the upper lip entire, the lower 2-lobed; corolla 7–11 mm, pale blue, the tube not exceeding the calyx, the lower lip twice as long as the upper. Fl. 8–10.

Tribe 3. STACHYEAE. Calyx 5–10-veined. Corolla strongly 2-lipped; upper lip concave, often falcate or forming a hood. Stamens 4, the outer pair longer than the inner; filaments parallel; anthers ovate, connivent in pairs under the upper lip of the corolla.

12. MELITTIS L.

Perennial herb. Fls in 2–6-fld axillary whorls; bracts not differentiated. *Calyx open in fr., 2-lipped; upper lip with 2–3 small irregular teeth, the lower with 2 rounded lobes. Upper lip of corolla only slightly concave;* tube broad, naked within. Anther cells diverging. Nutlets ovoid, smooth.

One sp.

1. M. melissophyllum L.   Bastard Balm.

Strong smelling. Stems erect 20–50 cm, hairy. Lf-blades 5–8 cm, ovate, petiolate, crenate. Corolla 2.5–4 cm; tube much exceeding the calyx, pink, or white spotted with pink. Fl. 5–7. Pollinated by humble-bees and hawkmoths; protandrous. Hp.

Native. In woods and hedgebanks, very local. Cornwall, Somerset, Devon, Wilts, Hants, Sussex, Pembroke and Cardigan. W., C. and S. Europe.

13. PRUNELLA L.

Perennial herbs. Fls in few- (c. 6-)fld whorls forming a dense terminal cylindrical spike. Bracts orbicular, sessile, differing markedly from lvs. *Calyx strongly 2-lipped, closed in fr.; upper lip truncate with 3 very short aristate teeth, the lower with 2 long teeth. Upper lip of corolla very concave;* tube straight, obconical. Anther cells diverging; filaments with a subulate appendage below the apex. Nutlets oblong, smooth.

Seven spp., *P. vulgaris* almost cosmopolitan, the remainder in Europe and the Mediterranean region.

Upper lvs entire or shallowly toothed; fls normally violet; sinus between upper calyx-teeth gradually rounded.
                                                    1. vulgaris
Upper lvs lyrate or pinnatifid; fls normally cream; sinus between upper calyx-teeth parallel-sided. **2. laciniata**

1. P. vulgaris L.   Selfheal.

Sparingly pubescent perennial with short rhizome and ascending or erect stems, 5–30 cm high. *Lf-blades* 2–5 cm, ovate, *entire or shallowly dentate*, petiolate, cuneate or rounded at base. Bracts and calyx with long white hairs, usually purplish-tinged. Upper lip of calyx with 2 lateral teeth ill-developed, their points diverging, the *sinus between them and the middle tooth rounded*, sometimes obsolete; teeth of lower lip shortly ciliate. *Corolla* 10–14 mm, *violet*, rarely pink or white. Appendages of longer stamens nearly straight. Fl. 6–9. Pollinated mainly by bees; protandrous or homogamous; small female fls occur. $2n = 32$. Hs.

Native. Very common in grassland, clearings in woods and waste places, mainly on basic and neutral soils; ascending to 760 m. Throughout the British Is. Almost throughout Europe; temperate Asia; N. Africa; North America; Australia.

*2. P. laciniata (L.) L.   Cut-leaved Selfheal.

Differs from *P. vulgaris* as follows: *Lvs with more abun-dant and softer hairs*, somewhat whitened beneath, narrower, the lower oblong-lanceolate, often entire, *the upper very variable, from deeply pinnatifid to toothed with a single lobe on each side near the base.* Lateral teeth of upper lip of calyx better developed, their points nearly erect, the *sinus between them and the middle tooth narrow, parallel-sided and well marked;* teeth of lower lip longer and long-ciliate. *Corolla larger*, c. 15 mm, *cream-white*, rarely pink or pale blue-violet. Appendages of longer stamens curved. Fl. 6–8. Pollinated by bees; protandrous; small female fls occur. $2n = 32$. Hs.

Probably introduced. First recorded 1887. In dry calcareous grassland from Dorset, Somerset, Gloucester and Hereford to Kent and Cambridge, and in Leicester and Lincoln, very local; appearing quite native; Alderney. S., W. and C. Europe; W. Asia; N. Africa.

P. laciniata × *vulgaris* (*P.* × *hybrida* Knaf) is often found where the parents grow together. It appears to be sterile and is morphologically intermediate between the parents, usually being noticeable because of varied colours of the corolla, as by a combination of large, creamy-white fls with entire lvs, or of purple fls with pinnatifid or deeply toothed lvs.

14. STACHYS L.

Annual or perennial herbs with or without well-marked basal rosette of lvs. *Stem-lvs ±numerous. Calyx tubular or campanulate, with 5 narrow equal teeth. Upper lip of corolla concave, but not helmet-shaped,* or nearly flat. Corolla-tube usually with a ring of hairs within. *Outer stamens often diverging laterally from corolla after fl. Anther-cells diverging or parallel. Nutlets obovoid, rounded at apex.*

About 300 spp., cosmopolitan, except Australia and New Zealand.

1 Fls white or pale yellow; annual   **1. annua**
  Fls purple.   2
2 Perennial or biennial; corolla 12 mm or more.   3
  Annual; corolla 7 mm or less.   **2. arvensis**
3 Stem and lvs densely clothed with long white silky hairs.   **3. germanica**
  Lvs green.   4
4 Most lvs in a large basal rosette; outer stamens not diverging laterally from the corolla after fl.; anther-lobes almost parallel.   **7. officinalis**
  No well-marked basal lf-rosette; outer stamens diverging laterally from the corolla after fl.; anther-lobes widely diverging.   5
5 Bracteoles nearly as long as the calyx-tube; calyx-teeth less than half as long as tube (very rare).
                                                    4. alpina
  Bracteoles very short, scarcely longer than the pedicel; calyx-teeth more than half as long as tube (common).   6
6 Lvs lanceolate, with short petioles or subsessile; fls dull purple.   **5. palustris**
  Lvs ovate, with long petioles; fls dull reddish-purple, white or pale pink.   **6. sylvatica**

*1. S. annua (L.) L.

Annual 10–30 cm, much-branched, pubescent and sometimes glandular. *Lvs* 2–6 cm, *oblong*, obtuse, *cuneate at base*, shallowly crenate, the lower shortly petiolate, the upper subsessile. Whorls 3–6-fld, in the axils of lanceolate acute bracts, passing into the lvs below. Bracteoles linear, very small. Calyx c. 8 mm, tubular-campanulate, hirsute; teeth narrowly triangular-lanceolate, mucronate. *Corolla* 11–13 mm, *white or pale yellow*. Fl. 6–10. Visited by humble-bees. $2n = 34$. Th.

Introduced. A casual recorded from many localities in waste places and formerly an abundant weed of cornfields, now much reduced; rarely naturalized. C. and S. Europe; W. Siberia; Orient.

*S. recta L., with white or pale yellow fls like the preceding but differing in the perennial habit and broader calyx-teeth, is naturalized at Barry Docks (Glamorgan). Native of C. and S. Europe, Asia Minor and the Caucasus.

2. S. arvensis (L.) L.   Field Woundwort.

Annual 10–25 cm with slender ascending stems usually branched at the base, hirsute. *Lf-blades* 1.5–3 cm, *ovate*, *obtuse*, *truncate* or cordate at base, crenate-serrate, petiolate. Whorls 2–6-fld, in the axils of bracts resembling the lvs but becoming much smaller and subsessile above, forming a very lax spike, much interrupted below. Bracteoles linear, very small. Calyx 4–6 mm, tubular-campanulate, hirsute; teeth triangular-lanceolate or triangular-ovate, mucronulate. *Corolla* 6–7 mm, *pale purple*. Fl. 4–11. Self-pollinated, insect visits few. $2n = 10$. Th.

Native. Arable fields on non-calcareous soils; ascending to 380 m. Throughout Great Britain but local or absent in the Midlands, parts of N. England and much of Scotland; S.E. Ireland, local or absent elsewhere. S., W. and C. Europe, northwards to S. Sweden, and eastwards to W. Poland, W. Yugoslavia and Crete; N. Africa; Palestine. Naturalized in America.

3. S. germanica L.   Downy Woundwort.

Perennial or biennial 30–80 cm. *Whole plant densely covered with long white silky hairs*, giving it a whitish appearance. Lf-blades 5–12 cm, ovate-oblong to lanceolate, cuneate to cordate at base, crenate, the lower with long petioles, the upper sessile or nearly so; venation reticulate, conspicuous. Whorls many-fld, forming a dense terminal spike interrupted below; bracts lanceolate, passing into the lvs. *Bracteoles linear, nearly as long as the calyx.* Calyx 9–11 mm, tubular, very silky, with triangular mucronate, somewhat unequal *teeth less than half as long as the tube*. Corolla c. *twice as long as the calyx*, pale pinkish-purple, hairy outside. Fl. 7–8. Pollinated by bees; protandrous; small female fls occur. $2n = 30$. Hs.

Native. Pastures and hedgebanks, very rare; now believed extinct except in Oxford but formerly also in Hants, Leicester and Lincoln; introduced elsewhere. W., C. and S. Europe, extending eastwards to S.C. Russia; N. Africa; Orient.

4. S. alpina L.   Limestone Woundwort.

Perennial 40–100 cm, *green*, hairy, glandular above.

Lf-blades 4–16 cm, ovate, cordate at base, crenate-serrate; petioles 3–10 cm. Whorls many-fld, distant in the axils of subsessile ovate or lanceolate bracts, the lower crenate-serrate, lf-like, the uppermost smaller, entire. *Bracteoles entire, nearly as long as the calyx.* Calyx c. 8 mm, tubular, glandular-hairy; *teeth less than half as long as the tube*, triangular-ovate, mucronate, somewhat unequal. *Corolla* 15–20 mm, *dull purple*, rarely tinged with yellow, hairy. Fl. 6–8. $2n = 30$. Hp.

Native. Very rare in open woods in Gloucester and Denbigh. W., C. and S. Europe.

5. S. palustris L.   Marsh Woundwort.

Perennial herb with lông creeping rhizome producing small tubers at the apex in autumn, *green*, hairy, almost hispid, *odourless*. Stems 40–100 cm, simple or slightly branched, hollow. *Lvs* 5–12 cm, *oblong-lanceolate* or linear-lanceolate, acute, rounded or subcordate at base, crenate-serrate, *the lower very shortly petiolate* (petiole 5 mm or less), *the upper sessile*. Whorls c. 6-fld, forming a terminal spike, dense above, interrupted below. Lower bracts resembling the lvs but smaller, the upper small, shorter than the fls, entire. *Bracteoles* linear, *scarcely reaching the base of the calyx*. Calyx c. 8 mm, tubular-campanulate, pilose, *eglandular* or sparingly glandular, *the teeth triangular-subulate, more than half as long as the tube*. Corolla 12–15 mm, *dull purple*, pubescent outside. Fl. 7–9. Pollinated mainly by bees; protandrous. $2n = $ c. 64, 102. Gt.

Native. Common by streams and ditches and in swamps and fens, sometimes also in arable land; ascending to 460 m. Throughout the British Is., but absent from parts of N.E. England and W. Scotland. Most of Europe, but rare in the Mediterranean region; temperate Asia (to Japan); North America.

6. S. sylvatica L.   Hedge Woundwort.

Perennial herb with long creeping rhizome not producing tubers, *green*, almost hispid, *foetid* when bruised. Stems 30–100 cm, often branched, solid. *Lf-blades* 4–9 cm, *ovate*, acuminate, cordate at base, coarsely crenate-serrate, *all petiolate* (petioles 1.5–7 cm). Whorls c. 6-fld, forming an interrupted terminal spike. Bracts shortly petiolate, the lower ovate-lanceolate, toothed, the upper lanceolate, entire. *Bracteoles* linear, *scarcely reaching the base of the calyx*. Calyx c. 7 mm, campanulate, hairy and *glandular, the teeth triangular-lanceolate, more than half as long as the tube*. Corolla 13–15 mm, dull reddish-purple with white markings, pubescent outside. Fl. 7–8, pollinated by bees; protandrous. $2n = 66$. Hp.

Native. Common in woods, hedgebanks and shady waste places on the richer soils; ascending to 460 m. Throughout the British Is. Most of Europe but rare in the Mediterranean region; Caucasus, Kashmir and the Altai.

S. palustris × *sylvatica* = *S.* × *ambigua* Sm. has oblong shortly-petiolate lvs, bright red fls, and a foetid smell and is normally sterile. It is widespread and not uncommon with the parents, or in the absence of one or both of them.

7. S. officinalis (L.) Trev.   Betony.

Stachys betonica Bentham, *Betonica officinalis* L.

Sparingly hairy perennial 15–60 cm, with a short woody rhizome, a well-marked basal rosette of lvs, and erect, sparsely lfy stems, simple or somewhat branched below. Lf-blades 3–7 cm, oblong or ovate-oblong, cordate at base, obtuse, coarsely crenate; the basal numerous on very long (to 7 cm) petioles; cauline 2–4 pairs, distant, several times their own length apart; petioles becoming shorter upwards, the uppermost subsessile. Infl. often interrupted below. Bracts ovate or lanceolate, entire; the lowest pair crenate-serrate, rather lf-like. Bracteoles lanceolate, aristate, about equalling the calyx. Calyx 7–9 mm, the teeth triangular-lanceolate, aristate. Corolla c. 15 mm, bright reddish-purple, the tube without or with scattered hairs within, longer than the calyx; upper lip nearly flat. Fl. 6–9. Pollinated mainly by bees; protandrous or homogamous. $2n = 16$. Hs.

Native. Open woods, hedgebanks, grassland and heaths, usually on the lighter soils; ascending to 460 m. Common in England and Wales, local in Scotland in Dumfries, Kirkcudbright, Roxburgh, Haddington, Perth and the Inner Hebrides, and in Ireland (Kerry, Cork, Wexford, Galway and Meath); Jersey. Most of Europe northwards to S. Sweden and N.W. Russia; Caucasus; Algeria.

15. BALLOTA L.

Perennial herbs. Fls in many-fld axillary whorls; bracts not differentiated. *Calyx funnel-shaped, 10-veined, with 5 broadly ovate acuminate, mucronate ±equal teeth. Upper lip of corolla somewhat concave*; tube shorter than the calyx, with a ring of hairs within. Stamens 4, parallel, the outer pair the longer; anther cells diverging. *Nutlets oblong, rounded at apex.*

About 35 spp., mainly Mediterranean, the following more widespread, and 1 in S. Africa.

1. B. nigra L. subsp. **foetida** Hayek Black Horehound.

Hairy perennial with an unpleasant smell. Rhizome short, stout. Stems 40–100 cm, branched. Lf-blades 2–5 cm, ovate or orbicular, broad-cuneate to cordate at base, petiolate, coarsely crenate. Whorls numerous, many-fld, in the axils of bracts resembling the cauline lvs but smaller. Bracteoles subulate. Fls subsessile. Calyx c. 1 cm; teeth broadly ovate, suddenly acuminate, 1–2 mm. Corolla 12–18 mm, purple, hairy. Fl. 6–10. Pollinated mainly by bees; protandrous. $2n = 22*$. Hp.

Native. Common on roadsides and hedgebanks in much of England and Wales, but local in E.C. Wales, N.W. England and Scotland from Roxburgh and Berwick to Forfar and in Wigtown, Ayr, Lanark and Moray (introduced); local in E. Ireland and not native. The subsp. from Germany, Italy and Albania westwards,

but naturalized in S. Scandinavia; Palestine, N. Iran; Azores.

*Subsp. **nigra** (*B. nigra* var. *ruderalis* (Swartz) Koch) differs in the shorter calyx-tube and longer (2–4 mm) lanceolate, gradually tapered calyx-teeth. It is rare and not native.

16. LAMIASTRUM Heister ex Fabr.

Stoloniferous perennial herbs. Fls in dense axillary whorls, the bracts like the lvs. *Calyx tubular-campanulate with 5 nearly equal mucronate teeth*. Corolla 2-lipped; *upper lip laterally compressed, helmet-shaped*; lower lip 3-lobed, *the middle lobe only slight larger than the lateral*; *tube* longer than the calyx, straight, *dilated above*, with a ring of hairs within. *Anther cells divaricate, glabrous. Nutlets trigonous, truncate at apex.*

One sp.

1. L. galeobdolon (L.) Ehrend. & Polatschek
    Yellow Archangel.

Lamium galeobdolon (L.) L., *Galeobdolon luteum* Hudson

Perennial herb with long lfy stolons sometimes not produced till after fl., sparingly hairy. Stems 20–60 cm. Lf-blades 4–7 cm, ovate, acute or acuminate, truncate or rounded at base, petiolate, irregularly crenate-serrate or serrate, those of the stolons shorter and broader than those of the fl.-stems. Calyx c. 10 mm. Corolla c. 2 cm, yellow with brownish markings. Fl. 5–6. Pollinated by bees; homogamous. $2n = 18$. Chh.

Native. In woods, usually on the heavier soils, sometimes becoming locally dominant, especially after coppicing. Common in England and Wales; rare in the north (Lancs, Cumbria, Yorks and Durham) and in Scotland, only in Ayr; in Ireland only in the southeast in Wexford, Wicklow and Dublin. Most of Europe but rare in the Mediterranean region and the north; Iran.

17. LAMIUM L.

Annual or perennial herbs without aerial stolons. Fls in dense axillary whorls, the bracts ±like the lvs. Calyx tubular or tubular-campanulate with 5 nearly equal mucronate teeth. Corolla 2-lipped, the upper lip laterally compressed forming a hood; lower lip 3-lobed, *the lateral lobes very small, each with a small tooth* (in all the other genera the lateral lobes are well developed); *tube dilated above. Anther cells divaricate, hairy. Nutlets trigonous, truncate at apex.*

About 40 spp., in Europe, temperate Asia, N. Africa.

1 Annuals; corolla not more than 15 mm; tube not suddenly contracted near the base. 2
  Perennials; corolla 2 cm or more; tube suddenly contracted near the base. 5
2 Calyx 8–12 mm, the teeth usually longer than the tube; lower lip of corolla c. 4 mm.
    2. moluccellifolium
  Calyx 5–7 mm, the teeth not longer than the tube; lower lip of corolla 1.5–2.5 mm. 3

3 Bracts ±amplexicaul, sessile, usually wider than long; corolla usually much longer than calyx.
1. amplexicaule

Bracts not amplexicaul, most at least shortly petiolate, longer than wide; corolla not or only slightly longer than calyx. 4

4 Lvs and bracts ±regularly crenate or crenate-serrate, not decurrent along petiole. **4. purpureum**

Lvs and bracts ±irregularly incised dentate, the upper ±decurrent along petiole. **3. hybridum**

5 Corolla white; tube with an oblique ring of hairs towards the base. **5. album**

Corolla purple; tube with a transverse ring of hairs towards the base. **6. maculatum**

Section 1. *Lamiopsis* Dumort. Annuals. Corolla-tube regularly cylindrical below, enlarged above into a wide throat, with or without a ring of hairs.

1. L. amplexicaule L. Henbit Dead-nettle.

Finely pubescent annual 5–25 cm, with ascending branches from the base. *Lf-blades* 1–2.5 cm, *orbicular or ovate-orbicular*, obtuse, truncate, rounded or subcordate at base, *crenate-lobulate*, with a long petiole (3–5 cm in the lower lvs). *Bracts* similar to lvs, often larger and lobed, but *sessile* (rarely lowest stalked), ±amplexicaul, usually broader than long. Whorls few, rather distant. *Calyx* 5–7 mm, tubular, densely *clothed with white, ±spreading hairs*; the *teeth rather shorter than to as long as the tube, connivent in fr. Corolla* when well developed c. 15 mm, pinkish-purple, long-exserted, or small, included and cleistogamous; lower lip 1.5–2.5 mm; *tube glabrous within*. Fl. 4–8. Insect visitors rare (bees), mainly self-pollinated. $2n = 18$. Th.

Native. Cultivated ground, usually on light dry soils through nearly the whole of the British Is., rather common in Great Britain, though local in the west, local in Ireland (Kilkenny and Wexford to Dublin, Cork, Offaly and Donegal). All Europe, except the extreme north, extending to the Azores, Madeira, Canaries, N. Africa, Palestine, Iran and the Usuri region. Naturalized in North America.

2. L. moluccellifolium Fries Northern Dead-nettle.

L. intermedium Fries

Differs from *L. amplexicaule* as follows: usually more robust. Lvs somewhat deltate. Bracts not amplexicaul, the lowest shortly stalked. Infl. denser. *Calyx* 8–12 mm in fl., *with appressed hairs; the teeth longer than the tube, spreading in fr.* Lower lip of corolla c. 4 mm. Corolla-tube scarcely longer than calyx, with a faint ring of hairs within. Fl. 5–9. $2n = 36$. Th.

Native. In cultivated ground. Widespread in lowland areas of Scotland, rare in England, native only in N. Northumberland, Isle of Man, occasional casual southwards to Derby; scattered localities in Ireland southwards to Tipperary and Wexford. N. Europe to c. 67° N in Fennoscandia and southwards to 52° 30′ in Germany.

3. L. hybridum Vill. Cut-leaved Dead-nettle.

L. incisum Willd.; *L. dissectum* With.

Differs from *L. purpureum* as follows: more slender, less pubescent. Lvs often smaller, *the upper truncate at the base and ±decurrent down the petiole*, all ±*irregularly incised-dentate. Corolla-tube less exserted without or with a faint ring of hairs* towards the base. Fl. 3–10. Visited by bees. $2n = 36$. Th.

Native. In cultivated ground, scattered over the whole of the British Is. but local, commonest in parts of East Anglia and Nottingham. Most of Europe except the southeast; Morocco, Algeria (rare).

4. L. purpureum L. Red Dead-nettle.

Pubescent annual 10–45 cm, branched from the base, often somewhat purple-tinted. *Lf-blades* 1–5 cm, *ovate, obtuse, cordate at base, ±regularly crenate or crenate-serrate*, petiolate. *Bracts* similar, *rounded or truncate at base, petiolate*, or the upper subsessile. Infl. rather dense. Calyx 5–7 mm, tubular-campanulate, pubescent, the teeth about as long as the tube, spreading in fr. *Corolla* 10–15 mm, pinkish-purple, lower lip of corolla 1.5–2.5 mm, *tube* longer than the calyx, *with a ring of hairs near the base.* Fl. 3–10. Pollinated by bees (rarely other insects) or selfed; homogamous. $2n = 18$. Th.

Native. Very common in cultivated ground and waste places throughout the British Is.; ascending to 610 m. Most of Europe, in the south only in the mountains and absent from several of the Mediterranean islands, to Palestine and the Yenisei region.

Section 2. *Lamium.* Perennials. Corolla-tube cylindrical at the base for a very short distance, then suddenly enlarged and gibbous, the upper part curved, with a ring of hairs within at the enlargement.

5. L. album L. White Dead-nettle.

Hairy perennial 20–60 cm, with a creeping rhizome and erect stems. Lf-blades 3–7 cm, ovate, acuminate, cordate at base, coarsely and often doubly serrate or crenate-serrate, petiolate. Bracts similar. Whorls mostly distant. Calyx 9–13 mm, tubular-campanulate, the teeth about as long as or slightly longer than the tube. *Corolla* 20–25 mm, *white*, the *tube with an oblique ring of hairs near base*, lateral lobes of lower lip with 2–3 small teeth; upper lip long-ciliate. Fl. 5–12. Pollinated by long-tongued Hymenoptera, mainly humble-bees; homogamous. $2n = 18$. Hp.

Native. Hedgebanks, roadsides and waste places. Common in England, absent from the Scottish islands and very rare in Scotland north of the Caledonian Canal; in Ireland mainly in the east, also in Kerry and Clare; local and not native. Most of Europe but rare in the south and absent from many islands; Himalaya and Japan; only in the mountains in the southern part of its range.

*6. L. maculatum L. Spotted Dead-nettle.

Differs from *L. album* as follows: lf-blades 2–5 cm, usually acute, often with a large whitish blotch. Calyx-teeth relatively rather shorter. *Corolla pinkish-purple*, the *tube with a transverse ring of hairs within*; lateral lobes of lower lip with a single tooth; upper lip shortly ciliate. Fl. 5–10. Pollinated by humble-bees; homogamous. $2n = 18$. Hp.

Introduced. Commonly cultivated in gardens in the form with a whitish blotch on the lf and sometimes found as an escape. Europe northwards to c. 54° N in Germany and c. 59° N in N.C. Russia, but absent from most of the islands; Iran.

18. LEONURUS L.

Perennial herbs. Fls in dense-fld, distant, axillary whorls. *Calyx campanulate, 5-veined with 5 nearly equal spiny-pointed teeth.* Corolla 2-lipped; *tube shorter than the calyx, not widened at the throat.* Anther cells opposite, opening by a common slit. *Nutlets trigonous, truncate at apex.*

About 14 spp., in Europe and temperate Asia, 1 widespread in the tropics.

*1. L. cardiaca L.   Motherwort.

Pubescent perennial 60–120 cm with stout rhizome and branched stems. Lf-blades 6–12 cm, cordate at base; the lower ovate-orbicular, palmately 5–7-lobed, the lobes irregularly dentate; the upper trifid, lanceolate, cuneate at base, passing into the lf-like bracts; all petiolate. Whorls numerous, distant, many-fld. Calyx 5-angled; teeth ovate-lanceolate, the 2 lower deflexed. Corolla 8–12 mm, white or pale pink, sometimes with small purple spots. Fl. 7–9. Pollinated by bees. $2n = 18*$. Hp.

Introduced. Waste places and hedgebanks, rare, scattered over England and Wales north to Yorks and Isle of Man; Offaly and Armagh in Ireland. Most of Europe, but absent from the extreme north, the islands and much of the Mediterranean region.

PHLOMIS L.

Small shrubs or large herbs with large fls, many in a whorl. Differs from all the British members of the tribe in the unequal style branches.

*P. fruticosa L.   Jerusalem Sage.

Whitish-tomentose shrub to 1 m. Lvs 3–6 cm, ovate or oblong, entire, shortly stalked. Bracts ovate or lanceolate. Corolla 2–3 cm, bright yellow. Often cultivated, and naturalized in Somerset and Devon. Mediterranean region westwards to Sardinia.

*P. samia L.

Herb c. 1 m. Lvs 8–15 cm, ovate, crenate, green above, grey-tomentose beneath; lower long-stalked. Bracts subulate. Corolla c. 3 cm, purple. Probably only casual. Greece, S. Yugoslavia; Asia Minor.

19. GALEOPSIS L.   Hemp-nettle.

Annual herbs. Fls in dense whorls, terminal and axillary, the bracts like the cauline lvs, but often smaller. *Calyx tubular or campanulate, with 5 somewhat unequal spiny-pointed teeth.* Corolla 2-lipped; *upper lip laterally compressed, helmet-shaped; lower lip 3-lobed, with two conical projections at the base* (not found in the other genera of the tribe); tube longer than the calyx, straight, dilated above, with a ring of hairs within. *Anther cells parallel, superposed, opening separately,* ciliate. Nutlets trigonous, rounded at the apex.

Ten spp., in Europe and temperate Asia.

1 Stem softly pubescent or subglabrous, not swollen at
     the nodes.   2
   Stem hispid, swollen at the nodes.   4
2 Lvs and calyx hairy but not silky; fls pinkish-purple.   3
   Lvs beneath and calyx velvet-silky; fls pale yellow.
                                                               3. segetum
3 Lvs linear-lanceolate to oblong-lanceolate with 1–4
     teeth on each side.   **1. angustifolia**
   Lvs ovate or ovate-lanceolate with 3–8 teeth on each
     side.   **2. ladanum**
4 Corolla 13–20 mm, pink, purple or white, rarely pale
     yellow with violet spot, the tube nearly always scarcely exceeding the calyx.   5
   Corolla 27–34 mm, pale yellow usually with a violet
     spot on the lower lip, the tube much exceeding the
     calyx.   **6. speciosa**
5 Middle lobe of lower lip of the corolla broad and
     flat, entire.   **4. tetrahit**
   Middle lobe of lower lip of corolla narrower, convex,
     with the margins ±deflexed, emarginate.   **5. bifida**

Subgenus 1. LADANUM Reichenb.

Stem softly pubescent or subglabrous, not swollen at the nodes.

1. G. angustifolia Ehrh. ex Hoffman.

Red Hemp-nettle.

G. ladanum auct., pro parte

Stem 10–80 cm, pubescent or subglabrous, not swollen at the nodes. Lvs 1.5–8 cm, linear-lanceolate to oblonglanceolate, less than 1 cm broad, acute, attenuate at base into a short petiole, often abundantly appressed-hairy, *with 1–4 small serrations on each side.* Bracteoles often as long as the calyx. Calyx tubular, ±hairy, often with appressed whitish hairs, usually eglandular or with few glands. Corolla 1.5–2.5 cm, pinkish-purple, the tube usually much longer than the calyx. Fl. 7–10. Insect or self-pollinated. $2n = 16$. Th.

Native. Rather common in arable land in S. and E. England extending to Yorks, Derby and Cheshire, and in Lancs, Cumbria and Berwick, S. Wales, Montgomery, Carmarthen and Denbigh; in Ireland only in Offaly, Kildare and Westmeath. W., C. and S. Europe eastwards to Poland and Bulgaria.

*2. G. ladanum L.

Differs from *G. angustifolia* as follows: *Lvs* 1–3 cm broad, not silky, *ovate-oblong or ovate-lanceolate,* cuneate at base, serrate, with 3–7 *prominent teeth on each side.* Bracteoles shorter. Calyx green, hirsute and glandular, usually longer. Corolla-tube often scarcely exceeding calyx. Fl. 7–10. Insect or self-pollinated. $2n = 16$. Th.

Introduced. A rare casual of waste or cultivated ground. Most of Europe except the islands and the extreme south; Soviet Asia.

3. G. segetum Necker   Downy Hemp-nettle.

G. ochroleuca Lam.; *G. dubia* Leers

Differs from *G. ladanum* as follows: *lvs velvety silky, especially beneath*, with more prominent veins. Bracteoles much smaller. Whorls often fewer-fld. Calyx velvety silky and glandular. *Corolla* 2–3 cm, *pale yellow*, c. 4 *times as long as the calyx*. Fl. 7–10. Insect pollinated (but not self-sterile). $2n = 16$. Th.

Native. Formerly in arable land in Caernarvon and a casual elsewhere, principally from Lincoln to W. Yorks; extinct. Denmark to France and N. and E. Spain, eastwards to N.E. Italy.

Subgenus 2. GALEOPSIS

Stem hispid, swollen at the nodes.

4. G. tetrahit L.   Common Hemp-nettle.

Stems 10–100 cm, *with ascending branches, hispid, especially below the nodes, and with red-tipped glandular hairs in a group below the nodes*. Lvs 2.5–100 cm, ovate to ovate-lanceolate, acuminate, cuneate at base, crenate-serrate, ±hairy. Calyx ±hispid, with rather prominent veins. Corolla 15–20 mm, purple, pink, or white with darker markings; *middle lobe of lower lip of corolla flat, entire*, often nearly as broad as long, *network of dark markings restricted to base, never reaching margin*; tube usually scarcely longer than calyx. Fl. 7–9. Usually self-pollinated. $2n = 32$. Th.

Native. Arable land, less often in woods, fens and wet heaths; ascending to 450 m. Common throughout the British Is. Most of Europe but rare in the southeast. Naturalized in North America.

A plant indistinguishable from this sp. has been produced artificially by crossing *G. speciosa* and the continental *G. pubescens* Besser and back-crossing a triploid raised from this hybrid with *G. pubescens* (see Müntzing, *Hereditas*, 1938).

5. G. bifida Boenn.

G. tetrahit var. *bifida* (Boenn.) Lej. & Court.

Differs from *G. tetrahit* in the narrower lower lip of the corolla, which is convex with the margins ±deflexed, and which is distinctly emarginate. Fl. 7–9. Usually self-pollinated. $2n = 32$. Th.

Scattered throughout the range of *G. tetrahit*, but distribution not accurately known. Most of Europe, but absent from many of the islands and the south-west. N. Asia. Probably originated in a similar way to *G. tetrahit*.

G. × *ludwigii* Hausskn. (*G. tetrahit* × *G. bifida*) can be distinguished by its intermediate corolla, high pollen sterility and poor nut-production. It has been recorded as single plants in Bedford and Merioneth.

6. G. speciosa Miller   Large-flowered Hemp-nettle.

Differs from *G. tetrahit* as follows: usually more robust. Stem uniformly hispid, glandular hairs yellow-tipped, mainly on the upper half of the internodes. Veins of calyx less prominent. *Corolla* 22–34 mm, *pale yellow*

with lower lip mostly violet, occasionally all yellow; *tube about twice as long as the calyx*; lower lip entire or slightly emarginate. Fl. 7–9. Insect pollinated (but not self-sterile). $2n = 16$. Th.

Native. In arable land, often on black peaty soil; ascending to 450 m. Scattered over Great Britain; but rather local from Kent and Surrey to Glamorgan and Pembroke northwards to Sutherland; absent from Outer Hebrides, Orkney and Shetland; N. Ireland extending to Clare and Wicklow. Europe, except many of the islands and much of the south; Siberia (Yenisei region).

Tribe 4. NEPETEAE Calyx 15-veined. Corolla as Stachydeae, but upper lip sometimes flat. Stamens 4, the outer pair shorter than inner, otherwise as in Stachydeae.

20. NEPETA L.

Perennial herbs. *Fls in many-fld axillary whorls forming a terminal spike-like infl. Calyx tubular, 5-toothed. Upper lip of corolla flat; tube rather suddenly curved and dilated at the middle, glabrous within. Anther cells diverging, opening by a common slit.* Nutlets obovoid.

About 250 spp., in Europe, Asia, N. Africa and mountains of tropical Africa.

N. × *faassenii* Bergmans ex Stearn (*N. mussinii* hort. pro parte) and *N. mussinii* Sprengel are used extensively as edging-plants.

1. N. cataria L.   Cat-mint.

Strongly scented. Stems 40–100 cm, erect, branched, hoary pubescent. Lf-blades 3–7 cm, ovate, petiolate, cordate at base, coarsely serrate, white-tomentose beneath; bracts similar but much smaller, mostly shorter than fls. Upper whorls crowded, the lower more widely spaced. Calyx 5–6.5 mm, pubescent; teeth lanceolate-subulate, the upper the longest, the two lowest the shortest; tube ovoid. Corolla 7–12 mm, white with small purple spots. Nutlets smooth. Fl. 7–9. Pollinated by bees; protandrous. $2n = 32, 36$. Hp.

?Native. Hedgebanks and roadsides usually on calcareous soil from Yorks and Anglesey southwards, rather local; introduced elsewhere in N. England, S. Scotland and Ireland; Jersey. S., E. and perhaps W. Europe; W. and C. Asia to Kashmir. Naturalized in North America and S. Africa.

21. GLECHOMA L.

Perennial herbs. *Fls in few (2–4)-fld secund axillary whorls, the bracts not differing from the foliage lvs.* Calyx tubular, somewhat 2-lipped. Upper lip of *corolla* flat; *tube narrowly obconical, straight*, hairy within at base of lower lip. *Anther cells at right angles, each opening by a separate slit.* Nutlets obovoid, smooth.

About 10 spp., in Europe and temperate Asia.

1. G. hederacea L.   Ground-ivy.

Nepeta hederacea (L.) Trev.; *N. glechoma* Bentham

Softly hairy or nearly glabrous perennial with stems creeping and rooting. Flowering branches 10–30 cm, ascending. Lf-blades 1–3 cm diam., reniform to ovate-cordate, obtuse, crenate; petiole long. Corolla 15–20 mm, violet with purple spots on lower lip. Nutlets smooth. Fl. 3–5. Pollinated mainly by bees; protandrous; small female fls occur commonly. $2n = 18, 24, 36^*$. Hp.

Native. Woods, grassland and waste places, usually on the damper and heavier soils, sometimes becoming locally dominant in damp oakwoods, especially after coppicing; ascending to 395 m. Common throughout the British Is. except in N. Scotland where it is rare (absent from Orkney, Shetland and the Outer Hebrides). Almost throughout Europe; W. and N. Asia to Japan.

Tribe 5. MARRUBIEAE. Corolla 2-lipped. Stamens 4, like the style included in the corolla-tube.

22. MARRUBIUM L.

Perennial herbs. Fls in many-fld axillary whorls; bracts not differentiated. Bracteoles linear. Calyx tubular, 10-veined, with 5 or 10 teeth. Upper lip of corolla nearly flat; tube shorter than the calyx, naked or with a poorly developed ring of hairs within. Anther cells diverging. Nutlets ovoid, smooth.

About 40 spp., in Europe, temperate Asia and North Africa.

1. M. vulgare L.   White Horehound.

White-tomentose perennial 30–60 cm, with short stout rhizome and branched stems. Lf-blades 1.5–4 cm, orbicular or ovate-orbicular, crenate, cordate or cuneate at base, the lower long-, the upper short-petiolate, obtuse, rugose, green above. Whorls many-fld, broader than high. Calyx with 10 small hooked teeth. Corolla c. 1.5 cm, whitish. Fl. 6–11. Pollinated mainly by bees or selfed; homogamous or weakly protandrous; small female fls occur. $2n = 34^*$, 36. Hp.

Native. Local on downs, in waste places and by roadsides in Somerset, Isle of Wight, E. Sussex, Glamorgan, Carmarthen and Caernarvon; naturalized northwards to Haddington, and in Argyll; in Ireland naturalized in Leix, apparently extinct elsewhere. Europe, from S. Sweden and C. Russia southwards; C. and W. Asia; N. Africa; Macaronesia.

Subfamily 2. SCUTELLARIOIDEAE. Style gynobasic. Nutlets with small basal attachment. Seeds ±transverse. Radicle bent, lying along one of the cotyledons. Calyx 2-lipped with entire lips. Corolla 2-lipped, the upper helmet-shaped. Stamens 4, the outer pair longer than the inner.

23. SCUTELLARIA L.

Perennial herbs. Fls in axillary pairs, sometimes forming a terminal raceme. Calyx campanulate, 2-lipped, closed after flowering; the lips entire, the upper bearing a small scale on the back. Corolla 2-lipped, the lateral lobes small, ±free from the lips; tube naked within, dilated above, straight or curved below. Anthers of outer pair of stamens with 2 nearly parallel cells, of the inner with 1 cell. Nutlets subglobose, smooth or tuberculate.

About 300 spp., cosmopolitan except S. Africa.

1 Fls blue-violet, 10–20 mm.   2
  Fls pale pinkish-purple, 6–10 mm.   **2. minor**
2 Lvs cordate at base, toothed; fls not forming a well-marked raceme; corolla-tube slightly curved; common.   **1. galericulata**
  Lvs hastate at base, otherwise entire; fls forming a short terminal raceme; corolla-tube strongly curved; Norfolk.   **3. hastifolia**

1. S. galericulata L.   Skullcap.

Pubescent or subglabrous perennial 15–50 cm high, with simple or branched stems from a slender creeping rhizome. *Lvs 2–5 cm*, ovate-lanceolate or oblong-lanceolate, obtuse or subacute, cordate at base, *remotely and shallowly crenate*; petiole short. Bracts like the lvs but gradually decreasing in size upwards; *fls not in a well-marked raceme*. Pedicels very short. Calyx glabrous or pubescent. *Corolla 10–20 mm, blue-violet*, several times as long as the calyx, the tube somewhat curved below. Fl. 6–9. Visited by various insects; homogamous; small female fls rare. $2n = $ c. 32. Hp.

Native. On the edges of streams and in fens and water meadows; ascending to 365 m. Throughout the British Is., common, local in N.E. England and E. Scotland (not in Shetland), and scattered throughout Ireland. Almost throughout Europe, but absent from some islands; N. and W. Asia; Algeria; North America.

2. S. minor Hudson   Lesser Skullcap.

Smaller in all its parts than *S. galericulata*. Stems 10–15(–30) cm, usually more glabrous. Lvs 1–3 cm, entire except near the base, the lower sometimes ovate. *Bracts* often truncate at the base, *quite entire. Corolla 6–10 mm, pale pinkish-purple with darker spots*, 2–4 times as long as the calyx; tube nearly straight. Fl. 7–10. Hp.

Native. On wet heaths, etc.; ascending to 460 m. Wales and S. England northwards to Essex, Hertford and Hereford, scattered localities from Shropshire and Norfolk northwards to Ayr and Durham, W. Scotland from Kintyre to the Outer Hebrides; S.E., S.W. and C.W. Ireland northwards to Sligo. W. Europe, eastwards to E. Germany and W. Italy; S.W. Sweden; Azores.

S. galericulata × minor = S. × hybrida Strail (*S. × nicholsonii* Taub.), which is intermediate between the parents, occurs from Cornwall to Kent and in Kerry and Waterford.

*3. S. hastifolia L.

A subglabrous ±erect perennial herb 20–40 cm; rhizome slender, creeping. *Lvs 1–2.5 cm*, ovate-lanceolate, truncate and *hastate* at the base, *otherwise entire*, acute to suboctuse, shortly petiolate. Bracts smaller than the lvs, the uppermost c. 5 mm; *fls forming a well-marked* short and rather dense *terminal raceme*. Pedicels very short. Calyx glandular-hairy. *Corolla blue-violet, 15–25 mm; tube strongly curved* near the base. Fl. 6–9. $2n = $ c. 32. Hp.

Introduced (first found 1948). In semi-natural oak-wood near Brandon, Norfolk. Much of Europe but. absent from the south-west, the extreme north and the islands; Asia Minor.

*S. altissima L. (*S. columnae* auct.). Lvs large, ovate, long-petiolate, strongly crenate, hairy only on the veins beneath. Bracts small, lanceolate. Corolla 15–18 mm; upper lip bluish-purple, the lower whitish. Naturalized in N. Somerset and Surrey. Native of S.E. Europe and S.W. Asia.

Subfamily 3. AJUGOIDEAE. Style not gynobasic. Nutlets with large lateral-ventral attachment. Calyx 10-veined, 5-toothed. Corolla 1-lipped or nearly so. Stamens 4, exserted, the outer pair longer than the inner.

24. TEUCRIUM L.

Herbs or dwarf shrubs. Infl. various. Calyx tubular or campanulate, equally 5-toothed or the upper tooth larger. *Corolla of one 5-lobed lip* (the lower), the 4 upper lobes short, the middle lobe large; tube usually included, without a ring of hairs within. Nutlets obovoid, smooth or reticulate.

About 300 spp., cosmopolitan, mainly Mediterranean.

  1 Fls in terminal racemes, the bracts very different from the lvs; upper tooth of calyx much larger than the others; corolla pale yellowish-green. **4. scorodonia**
  Fls in axillary whorls, sometimes forming a terminal spike, the bracts similar to the lvs though sometimes smaller; calyx-teeth equal or nearly so; corolla purple. 　　　　　　　　　　　　　　　　2
  2 Lvs pinnatifid. 　　　　　　　　**3. botrys**
  Lvs toothed. 　　　　　　　　　　　　　3
  3 Whorls forming a terminal spike; upper bracts shorter than the fls though somewhat gibbous (on walls).
  　　　　　　　　　　　　　1. chamaedrys
  Whorls distant; all the bracts longer than the fls (in wet places). 　　　　　　　**2. scordium**

Section 1. *Chamaedrys* (Miller) Schreber. Whorls 2–6-fld, forming terminal spikes. Calyx tubular-campanulate, teeth subequal, tube not saccate.

*1. T. chamaedrys L. 　　　　Wall Germander.

Perennial, almost woody at base, without a creeping rhizome, forming low tufts. Stems 10–30 cm, many, ascending, hairy, rooting at the base. Lvs 1–3 cm, ovate, obtuse, attenuate at base into a short petiole, *deeply crenate* or lobulate, dark green and shining above, ±hairy. *Bracts similar but smaller, the upper shorter than the fls*, subsessile. Infl. somewhat secund, short. Pedicels c. 2 mm. Calyx 5–8 mm. *Corolla* 9–16 mm, *pinkish-purple*; the tube c. 5 mm; lip c. 8 mm, the upper 4 lobes acute, ciliate, the middle broad, obovate-cuneate, crenulate. Fl. 7–9. Pollinated by bees, self-pollination possible; protandrous. $2n = 64$. Chw.

Introduced. Grown in gardens, and sometimes naturalized on old walls in Cornwall, E. Sussex, Kent, Oxford, Glamorgan, Lincoln, Durham and Edinburgh; apparently no longer established in Ireland. W., C. and S. Europe; S.W. Asia; N. Africa.

Section 2. *Scordium* (Miller) Bentham. Whorls 2–6-fld, axillary. Calyx tubular; teeth subequal; tube saccate at the base on the lower side.

2. T. scordium L. 　　　　Water Germander.

Softly hairy *perennial* with creeping rhizome or lfy stolons. Stems 10–60 cm, rooting at the base, ascending. Lvs 1–5 cm, oblong, sessile or subsessile, usually rounded at base, *coarsely serrate*, not shining above. *Bracts scarcely different*. Whorls distant, secund. Pedicels short. *Corolla* 7–12 mm, *purple*. Fl. 7–10. Pollinated by bees, self-pollination possible; protandrous. Hp.

Native. Banks of rivers and ditches on calcareous soil, and dune-slacks. Rare in England and only recorded from Devon, Cambridge and Huntingdon since 1930; in Ireland only in Clare, Galway and Roscommon since 1930. Europe from Sweden and Estonia southwards; W. Siberia and the Aral-Caspian region.

3. T. botrys L. 　　　　Cut-leaved Germander.

Annual 10–30 cm, softly hairy. *Lf-blades* 1–2.5 cm, ±ovate in outline, petiolate, *pinnatifid*; the segments oblong, often lobed, obtuse, 1–2 mm broad. *Bracts smaller but longer than the fls*, once-pinnatifid. Whorls all along the branches, secund. Pedicels 3–4 mm. Calyx c. 7 mm, reticulately veined. *Corolla* 15–20 mm, *pinkish-purple*; the tube c. 6 mm; lip deflexed, c. 8 mm. Fl. 7–9. Pollinated by bees, self-pollination possible. $2n = 10$. Th.

Native. Chalky fallow fields and open habitats in chalk grassland. Wilts, Hants, Kent, Surrey, Gloucester; very rare. W., C. and S. Europe, northwards to S. Poland and eastwards to Romania; Algeria.

Section 3. *Scorodonia* (Hill) Bentham. Fls in pairs in terminal racemes, the bracts very different from the lvs. Calyx campanulate, the upper tooth orbicular-ovate, much broader than the others; tube gibbous.

4. T. scorodonia L. 　　　　Wood Sage.

Pubescent perennial with creeping rhizome. Stems 15–30 cm, erect, branched. Lf-blades 3–7 cm, ovate, cordate at base, petiolate, subacute, crenate, rugose. *Bracts ovate-lanceolate, entire, shorter than the fls*. Pedicels short. Calyx c. 5 mm. *Corolla pale yellowish-green*; tube c. 8 mm; lip 5–6 mm, deflexed. Fl. 7–9. Pollinated by bees; protandrous. $2n = 32, 34*$. Hp.

Native. Woods, grassland, heaths and dunes usually on dry not strongly calcareous soils; ascending to 550 m. Common in Great Britain, but scattered in parts of the E. Midlands; in Ireland, local or absent from much of the centre. Channel Is. S., W. and C. Europe, northwards to S. Norway and eastwards to W. Poland and N.W. Yugoslavia.

25. AJUGA L.

Herbs with 2–many-fld whorls, sometimes forming a terminal infl. Bracts differentiated or not. Bracteoles small or 0. Calyx tubular-campanulate, nearly equally

5-toothed. *Corolla with a very short upper lip and conspicuous 3-lobed lower lip*; tube ±exserted, with a ring of hairs within. Nutlets obovoid, reticulate.

About 40 spp., in temperate regions of the Old World.

1 Lvs divided into 3 linear lobes; corolla yellow.
1. chamaepitys
　　Lvs not divided; entire or toothed; corolla blue or violet. 2
2 With aerial stolons; stem usually hairy on 2 opposite sides. **2. reptans**
　　Without aerial stolons; stem usually hairy all round. **3. pyramidalis**

Section 1. *Chamaepitys* (Hill) Bentham. Whorls 2-fld, axillary, very rarely 4-fld. Corolla often yellow.

1. A. chamaepitys (L.) Schreber　　Ground-pine.

Hairy annual 5–20 cm, smelling of pine when crushed, branched below, the branches ascending. *Lf-blades 2–4 cm, divided into 3 linear, obtuse lobes*, attenuate into a petiole; basal lvs withering early, entire or toothed. Bracts not differentiated. Whorls 2-fld, axillary, many, much shorter than the lvs. Calyx c. 8 mm, campanulate, hairy. *Corolla yellow*; lower lip red-spotted; tube included. Fl. 5–9. Visited by bees. $2n = 28^*$. Hs.

Native. Very local in chalky arable fields and open habitats in chalk grassland in Hants, Sussex, Surrey, Kent, Bedford, Hertford and Suffolk. Most of Europe, except the north; Orient; N. Africa.

Section 2. *Ajuga*. Whorls 6–many-fld, forming a terminal spike. Fls never yellow.

2. A. reptans L.　　Bugle.

Perennial 10–30 cm, with short rhizome and *long lfy and rooting stolons*. Stems simple, ±hairy, often on two opposite sides only. Basal *lvs* forming a rosette; blades 4–7 cm, *subglabrous*, obovate or oblong, *entire to obscurely crenate*, obtuse, attenuate at base into a long petiole; cauline shorter, subsessile, few. Upper bracts shorter than their fls, ±blue-tinted, ovate, entire. Calyx 4–6 mm, campanulate; teeth rather shorter than the tube. *Corolla* 10–18 mm, *blue*, rarely pink or white; tube exserted. Fl. 5–7. Pollinated by bees, self-pollination possible; usually homogamous. $2n = 32^*$. Hs.

Native. Common in woods, usually damp, and in damp meadows and pastures; ascending to 670 m. Throughout the British Is. Most of Europe northwards to c. 61°N; Caucasus; S.W. Asia; E. Algeria, Tunisia.

3. A. pyramidalis L.　　Pyramidal Bugle.

Perennial 10–30 cm, with *short rhizome, without stolons*. Stems simple, hairy all round. Basal *lvs* hairy or subglabrous, obovate, obtuse, obscurely crenate, attenuate at base into a *short petiole, persistent at fl. time*; cauline obovate-oblong, very shortly petiolate or sessile, much shorter than the basal. *Bracts all much longer than fls*, sometimes lobed, entire, tinged violet or blue. Calyx c. 8 mm; teeth longer than the tube. *Corolla pale violet-blue*. Fl. 5–7. Pollinated by bees, self-pollination possible; protandrous or homogamous. $2n = 32$. Hs.

Native. Crevices of usually basic rocks; ascending to 535 m. Westmorland, Dumfries to the Outer Hebrides and Caithness, apparently extinct in Orkney; Clare, Galway, Armagh. Europe, southwards to N. Portugal, N. Italy and Bulgaria.

A. pyramidalis × *reptans* (*A.* × *hampeana* Braun & Vatke) is found in Clare and Sutherland, and there are older records from Ross and Orkney. The fls, lvs and inflorescence are ±intermediate between the parents.

A. genevensis L., a native of much of Europe except the south-west, the islands and most of the north, and S.W. Asia, was formerly naturalized in chalk grassland in Berks and on dunes near Hayle (Cornwall). It differs from *A. pyramidalis* in having the upper bracts shorter than the fls and the basal lvs withering before fl. time. It now seems to be extinct in Britain.

118. PLANTAGINACEAE

Annual or perennial, usually scapigerous herbs. Lvs usually all basal and spirally arranged, rarely cauline and spiral or opposite. Scapes axillary. Fls bracteate, small, usually in racemose heads or spikes, actinomorphic, hermaphrodite or rarely unisexual, usually 4-merous. Perianth green or ±scarious, lobes imbricate. Sepals persistent, connate at base. Corolla gamopetalous, scarious. Stamens inserted on the corolla-tube, rarely hypogynous, filaments usually long, anthers large. Ovary superior, 1–4-celled; ovules 1–many in each cell, axile or basal. Fr. a circumscissile capsule or 1-seeded, hard and indehiscent. Seeds often mucilaginous when wet.

Three genera and about 270 spp., cosmopolitan, but mainly in temperate regions.

Terrestrial; stolons 0; fls hermaphrodite, many together; fr. dehiscent. 1. PLANTAGO
Aquatic; stolons present; fls unisexual, male solitary, female few; fr. indehiscent. 2. LITTORELLA

1. PLANTAGO L.　　Plantain.

Terrestrial herbs. Fls 4-merous, in heads or cylindrical spikes, mostly hermaphrodite. Stamens inserted on the corolla-tube. Ovary 2–4-celled; ovules 2–many. *Capsule circumscissile*. Fls mostly protogynous and wind pollinated.

About 260 spp., in the temperate regions of both hemispheres, a few in the tropics.

1 Stems long; lvs opposite; lower bracts with lf-like tips.
6. arenaria
　　Stems short; lvs spirally arranged, forming a basal rosette; bracts not normally lf-like. 2
2 Corolla-tube glabrous; lvs not linear or pinnatifid. 3
　　Corolla-tube pubescent; lvs linear or pinnatifid. 5
3 Scape deeply furrowed; corolla-lobes with prominent brown midrib. **3. lanceolata**
　　Scape not furrowed; corolla-lobes without a midrib. 4
4 Lvs abruptly contracted at base; petiole usually as long as blade; scape scarcely exceeding lvs; capsule 6–34-seeded **1. major**

Lvs gradually narrowed at base; petiole much shorter than blade; scape much exceeding lvs; capsule c. 4-seeded. **2. media**

5 Lvs 3–5(–7)-veined, never pinnatifid; bracts obtuse or subacute; corolla-lobes with a brown midrib; capsule 2-celled, 2–4-seeded. **4. maritima**

Lvs 1-veined, often pinnatifid; bracts often acuminate; corolla-lobes without a midrib; capsule 3-celled, 3–6-seeded. **5. coronopus**

Subgenus 1. PLANTAGO.

Lvs spirally arranged, usually in a basal rosette.

1. P. major L.

A glabrous or pubescent perennial. Lvs (1.5–)5–15(–40) cm, ovate or elliptical, entire or irregularly toothed, 3–9-veined; *petiole usually about as long as the blade.* Infl. (1–)10–15(–50) cm; scape equalling or exceeding the lvs, not furrowed. Bracts 1–2 mm, ovate, glabrous, brownish with a green keel. Sepals 1.5–2.5 mm, green. Corolla-tube c. 2 mm, glabrous, the lobes c. 1 mm, yellowish-white, subobtuse, glabrous. Stamens exserted 2–3 mm; anthers at first lilac, later dirty yellow. Fr. 2–4 mm, 2-celled; *seeds* (4–)6–34. Fl. 5–9. Wind-pollinated. Hr (lfless in winter).

Native. In ±disturbed habitats, usually with rather open vegetation, such as farmyards, roadsides and cultivated ground. Generally distributed throughout the British Is. Europe; N. Africa; N. and C. Asia; naturalized throughout most of the world in temperate climates. Two subspp. occur in Britain.

Subsp. **major**: Lvs 5–9-veined, abruptly contracted into the petiole and ±cordate at base, usually subglabrous. Spikes tapering upwards. Seeds (4–)6–10(–13). 2n = 12. Usually in dry non-saline habitats, widespread.

Subsp. **intermedia** (DC.) Arcangeli. Lvs 3–5-veined, gradually narrowed into the petiole, usually puberulent. Spikes cylindrical. Seeds 14–34. 2n = 12. In damp, usually somewhat saline habitats, rather local.

2. P. media L.   Hoary Plantain.

A finely pubescent perennial. Lvs (2–)4–6(–30) cm, elliptical to ovate, entire or remotely dentate or crenate, 5–9-veined, gradually narrowed into a *petiole usually less than half as long as the blade.* Infl. (1–)2–6(–10) cm; scapes much exceeding the lvs, often 30 cm, not furrowed. Bracts 2–3 mm, ovate, acute, with membranous margins. Fls scented; sepals c. 2 mm, almost free, glabrous. Corolla-tube c. 2 mm, glabrous, the lobes 1.5–2 mm, ovate-lanceolate, subacute, glabrous. Stamens exserted 8–13 mm; filaments purple, the anthers lilac or white. Fr. 2–4 mm; *seeds* 2–4(–6). Fl. 5–8. Insect-pollinated. 2n = 12, 24. Hr.

Native. In grassy places on neutral and basic soils. Fairly generally distributed in S. England and the Midlands, becoming rarer northwards and westwards; often introduced in Scotland; introduced in Ireland. Most of Europe; temperate Asia.

3. P. lanceolata L.   Ribwort Plantain.

Perennial with several rosettes. Lvs 2–30 × 0.5–3.5 cm, linear-lanceolate to lanceolate, entire or with few shallow teeth, 3–5(–7)-veined, sessile or with a petiole up to as long as the blade. Infl. 0.5–5(–8) cm; *scapes about twice as long as lvs, with 5 deep furrows.* Bracts 2.5–3.5 mm, the anterior connate for most of their length but their midribs separate, often shortly hairy. Corolla-tube 2–3 mm, glabrous, the lobes 1.5–2.5 mm, lanceolate to ovate, acute or acuminate, glabrous. Stamens exserted 3–5 mm, the anthers yellowish. Fr. 3–4 mm; seeds 2. Fl. 4–8. Wind-pollinated. 2n = 12 + 0–1B. Hr.

Native. In grassy places on neutral or basic soils. Generally distributed throughout the British Is. Europe, except the extreme north; N. Africa; N. and C. Asia; introduced in most other temperate countries.

4. P. maritima L.   Sea Plantain.

A nearly or quite glabrous perennial. Stock stout, woody, sometimes with long silky hairs. *Lvs* 3–15(–25) cm, *narrowly linear*, somewhat fleshy, entire or slightly toothed, faintly 3–5-veined, rarely up to 15 mm broad and 7-veined. Infl. 2–6(–12) cm; scape equalling or exceeding lvs, not furrowed. Bracts ovate, obtuse or subacute, strongly keeled; *sepals free, the posterior keeled, not winged; corolla brownish, the lobes* ovate, acute, *with a rather broad indistinct brown midrib* reaching the apex; tube pubescent. Stamens pale yellow. Fr. 2-celled; seeds 2–4. Fl. 6–8. Wind-pollinated. 2n = 12*, 18, 24. Hr.

Native. In salt marshes, in short turf near the sea and beside streams on mountains. Round most of the coasts of the British Is. and on the Snowdon range, the Pennines and the higher mountains of Scotland; inland in N.W. Scotland and on limestone in W. Ireland. Most of Europe but rare in the south; N. Africa; Greenland; North and southern South America.

5. P. coronopus L.   Buck's-horn Plantain.

A pubescent or occasionally glabrous biennial or perhaps sometimes annual or perennial. *Lvs* usually 2–6 cm, very variable, linear, nearly entire, toothed or (*most often*) 1–2-*pinnatifid*, 1(–3)-*veined*. Infl. 0.5–4 cm; scape shorter than to somewhat longer than lvs, often curved below, not furrowed. *Bracts ovate, often long-acuminate with spreading points*, sometimes obtuse and appressed. Sepals free, the posterior *winged; corolla* brownish, *lobes* ovate, acute or acuminate, *the tube pubescent.* Stamens pale yellow. *Fr. 3-celled*; seeds 3–6. Fl. 5–7. Wind pollinated. 2n = 10 + 0–1B. Th. or Hr.

Native. In dry and ±open habitats on sandy and gravelly soils and in cracks in rocks, most common near the sea. Through most of England and Wales in suitable habitats but absent from some inland counties; strictly maritime in Scotland and Ireland. Coasts of C. and S. Europe from S. Sweden southwards; N. Africa; W. Asia; Macaronesia; introduced in North America, Australia, New Zealand.

Subgenus 2. PSYLLIUM (Miller) Harms

Lvs opposite; stem long, lfy, often branched.

*6. P. arenaria Waldst. & Kit.

P. psyllium L., nom. ambig.; *P. ramosa* Ascherson; *P. indica* L., nom. illegit.

An erect or spreading ±pubescent *usually much-branched* annual 15–50 cm. Lvs 3–8 cm, narrow-linear, entire or obscurely toothed, *the lower with short very lfy shoots in their axils.* Infl. 0.5–1.5 cm; scape slender, exceeding the subtending lf. Lowest 2 bracts 6–10 mm, with herbaceous midrib and wide scarious margin, ovate-orbicular with linear-subulate apex, straight and suberect with divergent lateral veins at the base; upper bracts 3.5–4.5 mm. Sepals unequal. Corolla-tube 3.5–4 mm, the lobes c. 2 mm. Fl. 7–8. $2n = 12$. Th.

Introduced. In disturbed ground and sometimes on dunes, widely distributed but infrequently as a casual, ±naturalized in a few localities. E., S. and C. Europe; S.W. Asia; N. Africa.

2. LITTORELLA Bergius

Perennial scapigerous *aquatic herbs. Fls unisexual. Male fls* 4-merous, *solitary on short scapes; stamens hypo-gynous,* rudimentary ovary small. Female fls 2–4-merous, 2–8 at base of male scape; *ovary 1-celled;* style long, rigid; ovule 1, rarely 2, erect, campylotropous. *Fr. indehiscent, hard.*

Three spp., one in Europe (excluding the Mediterranean region), and the Azores, one in North America and one in temperate South America.

1. L. uniflora (L.) Ascherson   Shore-weed.

L. lacustris L.

An abundantly stoloniferous perennial often forming an extensive turf in shallow water. Stolons slender, far-creeping, rooting and producing rosettes of lvs at the nodes. Lvs 2–10(–25) cm, ½-cylindrical and linear-subulate or sometimes flattened and broader, sheathing at base, not septate. Scape shorter than lvs, rarely equalling them, slender, bracteate below the middle. Male fl. 5–6 mm; stamens 1–2 cm, Female fls 4–5 mm, subsessile; style c. 1 cm. Fl. 6–8. Wind-pollinated. $2n = 24^*$. Hyd. Fls are produced only when the plant is exposed.

Native. In shallow water down to about 4 m or just exposed on sandy and gravelly shores of non-calcareous lakes and ponds. In suitable habitats throughout the British Is., but commoner in the north. W., C. and N. Europe to c. 68° N in Norway; Azores.

119. CAMPANULACEAE
(incl. Lobeliaceae)

Mostly herbs, nearly always with latex. Lvs usually alternate, simple, exstipulate. Fls usually hermaphrodite, actinomorphic or zygomorphic, often showy. Calyx-tube adnate to ovary. Corolla gamopetalous, campanulate or sometimes 1–2-lipped; lobes valvate. Stamens as many as corolla-lobes and alternate with them, free, inserted towards base of corolla or on the disk. Ovary inferior or rarely superior, 2–10-celled; placentation usually axile. Ovules numerous, rarely few. Fr. a capsule or fleshy.

About 60 genera and 2000 spp., throughout most of the world.

Certain genera (e.g. *Phyteuma* and *Jasione*) approach the Compositae in having numerous small fls arranged in compact bracteate heads, and in their pollination mechanism.

1 Fls solitary or in racemes or panicles; corolla-lobes
    usually shorter than tube, ovate or broadly triangular; style not or only shortly exserted.   2
  Fls capitate or spicate, very numerous; corolla-lobes
    much longer than tube, linear; style long-exserted.   5
2 Stem creeping, slender; lvs all similar, long-petiolate.
                                               1. WAHLENBERGIA
  Stem erect or ascending; upper lvs much smaller or
    narrower than basal, sessile or nearly so.   3
3 Perennials or biennials; ovary and capsule ovoid or
    subglobose.   4
  Annual; ovary and capsule subcylindrical.
                                               3. LEGOUSIA

4 Fls actinomorphic; corolla-tube not split dorsally.
                                               2. CAMPANULA
  Fls zygomorphic; corolla-tube deeply split dorsally
    and corolla 2-lipped.   6. LOBELIA
5 Plant glabrous or very nearly so; fl. buds curved; stig-
    mas linear (on chalk).   4. PHYTEUMA
  Plant pubescent or hispid; fl. buds straight; stigmas
    short, stout (on lime-free soils).   5. JASIONE

1. WAHLENBERGIA Schrader ex Roth

Annual or perennial herbs of varied habit. Fls 3–5-merous, usually blue and nodding, solitary or in lax panicles. *Calyx-tube hemispherical or oblong-obconic.* Corolla campanulate or subrotate. Anthers free. Ovary 2–5-celled; stigmas 2–5, filiform. *Capsule dehiscing by 2–5 apical loculicidal valves,* alternating with the persistent calyx-teeth.

More than 150 spp., mostly south temperate particularly in South Africa, a few in tropical America and the temperate regions of the Old World.

1. W. hederacea (L.) Reichenb. Ivy-leaved Bellflower.

Campanula hederacea L.

A slender glabrous creeping perennial. Stems up to c. 30 cm, little-branched, weak. Lvs all petiolate, the upper often subopposite; blade 5–15 mm, suborbicular, angled or obscurely lobed, ±cordate. Peduncles up to c. 4 cm, much longer than petioles, filiform, 1-fld. Fls ±nodding. Calyx 2–3 mm, teeth subulate, erect, much longer than tube. Corolla 6–10 mm, campanulate, pale blue; lobes equalling or shorter than tube, ovate, ±acute. Capsule c. 3 mm, turbinate, erect. Fl. 7–8. H. or perhaps Chh.

Native. In damp acid peaty places on moors, heaths and by streams. S. and W. England and Wales; Kintyre and Argyll; S. and E. Ireland, north to Dublin; Mayo; local everywhere. W. Europe from Scotland and Belgium to Spain and Portugal.

2. CAMPANULA L.   Bellflower.

Herbs of varied habit, usually perennial, rarely annual or biennial. Fls 5-merous, blue or purplish, rarely white, usually in racemes or panicles, sometimes cymose. *Calyx-tube ovoid or subglobose*; lobes flat or folded at the sinus. Corolla rotate or campanulate. Anthers free. Ovary 3 or 5-celled; *style clavate, hairy opposite the anther cells*, stigmas 3 or 5, filiform. *Capsule* ovoid or turbinate, 3- or 5-celled, *dehiscing by lateral or basal pores or valves*.

About 300 spp., in north temperate regions and on mountains in the tropics. Several are cultivated.

The fls are strongly protandrous, the pollen being shed in bud and deposited on the hairs of the style. As the fl. opens the stamens wither and the style presents the pollen to insects which come for the nectar, which is protected by the persistent triangular bases of the stamens. The stigmas eventually separate and finally curl right back so that, if cross-pollination fails, self-pollination may occur (cf. Compositae).

1 Hispid biennial; calyx with broad-cordate, reflexed appendages between the teeth; stigmas 5.
        9. medium
  Calyx without appendages; stigmas 3.   2
2 Fls sessile.   **5. glomerata**
  Fls distinctly pedicellate   3
3 Middle stem-lvs ovate (2–4 times as long as broad).   4
  Middle stem-lvs linear, linear-lanceolate or oblong.   6
4 Calyx-teeth spreading in fl.; fls nodding; plant with spreading roots producing numerous adventitious buds.   **3. rapunculoides**
  Calyx-teeth ±erect in fl.; fls erect or inclined; no root buds.   5
5 Stem obscurely and bluntly angled; basal lvs decurrent on petiole, rarely somewhat cordate, blade 10 cm or more; stem-lvs sessile; corolla 40–55 mm; plant softly hairy.   **1. latifolia**
  Stem sharply angled; basal lvs broadly ovate, deeply cordate, blade 10 cm or more; stem-lvs petiolate; corolla 25–35 mm; plant ±hispid.   **2. trachelium**
6 Fls nodding; basal lvs orbicular, cordate; lower stem-lvs stalked.   **6. rotundifolia**
  Fls erect or suberect; basal lvs not orbicular or cordate; lower stem-lvs sessile.   7
7 Plant glabrous; infl. 1–8-fld; corolla 25–35 mm.   **4. persicifolia**
  Plant pubescent or scabrid; infl. typically many-fld; corolla not more than 25 mm.   8
8 Root slender, fibrous; basal lvs narrowed at base and decurrent on petiole; infl. with many long branches.   **7. patula**
  Root swollen, fleshy; basal lvs abruptly contracted at base; infl. simple or nearly so.   **8. rapunculus**

1. C. latifolia L.   Giant Bellflower.

A stout erect perennial 50–120 cm. *Stem simple, obscurely and bluntly angled.* Lvs (Fig. 50A) ovate to ovate-

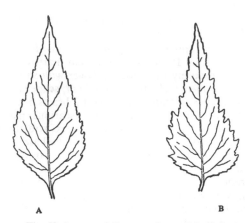

Fig. 50. Leaves of *Campanula*. A, *C. latifolia*; B, *C. trachelium*. ×⅓.

oblong, acuminate, rounded or narrowed at base, margins irregularly 1–2-serrate; *basal lvs* petiolate, the blade 10–20 cm, *usually decurrent on the petiole*, rarely cordate at base; *upper lvs sessile*. Infl. a raceme, sometimes with short branches below, lower bracts lf-like. Fls suberect or inclined; pedicels c. 2 cm. Calyx-tube 5–7 mm, 5-ribbed, glabrous or nearly so; *teeth* 15–25 mm, narrowly triangular, often serrate, acuminate, *erect. Corolla* 40–55 mm, blue-purple, rarely white; lobes slightly shorter than tube, suberect, acute or acuminate, ciliate. Stigmas 3. Capsule 12–15 mm, ovoid, nodding, opening by basal pores. Fl. 7–8. $2n = 34 + 0$–5B. Hs.

Native. In woods and hedgebanks. Widely distributed but local in Great Britain and much commoner in N. England and S. Scotland than elsewhere. Most of Europe except the Mediterranean region and much of the S.W.; W. Asia; Siberia.

2. C. trachelium L.   Bats-in-the-Belfry.

An erect ±hispid perennial 50–100 cm. *Stem simple, sharply angled. Basal lvs* long-petiolate, the blade c. 10 cm, *broadly ovate* or almost an equilateral triangle in outline, *deeply cordate* at base, tapering to a rounded or subacute apex, coarsely dentate or irregularly crenate-serrate; *stem-lvs* (Fig. 50B) *shortly petiolate*, smaller, ovate to ovate-oblong, ±truncate at base, ±acuminate, irregularly and coarsely 1–2-serrate. Infl. a lfy panicle with short branches bearing 1–4 fls. Fls suberect or inclined; pedicels up to c. 1 cm. Calyx-tube 3–5 mm, usually hispid; *teeth* c. 10 mm, triangular, acute, *erect. Corolla* 25–35 mm, blue-purple; lobes shorter than tube, suberect, acute, ciliate or sparsely hispid. Stigmas 3. Capsule c. 7 mm, hemispherical, nodding, opening by basal pores. Fl. 7–9. $2n = 34$. Hs.

Native. In woods and hedgebanks, usually on clayey soils. England, north to the Humber and Dee, but absent from the south-west; E. Wales; S.E. Ireland. Most of Europe to c. 62° 30' N in Sweden; Siberia; N. Africa.

Represented in Britain by subsp. **trachelium**.

*3. C. rapunculoides L.   Creeping Bellflower.

A puberulent or subglabrous perennial 30–100 cm, producing numerous adventitious buds from slender, far-spreading roots. *Stems subterete. Basal lvs* petiolate, the blade 5–8 cm, *ovate, cordate or rounded at base,* subacute, serrate; *stem-lvs sessile,* narrower and ±acuminate. Infl. a secund raceme or panicle. Fls nodding; pedicels c. 5 mm. *Calyx-tube* 3–4 mm, *covered with short, stiff, appressed, downward-directed hairs; teeth* c. 8 mm, linear, acute, *spreading or reflexed at flowering time. Corolla* 20–30 mm, funnel-shaped, blue-purple; lobes about as long as tube, spreading, acute, ciliate. Stigmas 3. Capsule c. 7 mm, hemispherical, nodding, opening by basal pores. Fl. 7–9. $2n = 102$. Hs.

Introduced. In fields and ±disturbed grassy places (e.g. railway banks), and occasionally in open woods. Widely distributed throughout the British Is., but rather local and chiefly near houses. Sometimes grown in gardens where it speedily becomes a weed. Most of Europe except the islands; S.W. Asia; Caucasus.

*C. lactiflora Bieb., a tall (up to 1.5 m) perennial with large lax panicles and generally 3-fld peduncles, fls 25–35 mm diam., and ovate hispid calyx-lobes, is established in S. Aberdeen. Native of S.W. Asia.

*4. C. persicifolia L.   Peach-leaved Bellflower.

A glabrous perennial up to 70 cm, with linear-lanceolate to subulate, finely and remotely crenulate stem-lvs, is often cultivated and is naturalized in a few places. *Raceme few-*(1–8-)*fld.* Fls suberect. *Corolla* 25–35 mm and about as broad, blue, rarely white; lobes broadly ovate, acute, about ½ as long as tube. Fl. 6–8. $2n = 16$. Hs.

Introduced. In open woods and on commons. Well established or possibly native in a few localities in S. England; introduced elsewhere in Great Britain. Most of Europe; western and northern Asia.

5. C. glomerata L.   Clustered Bellflower.

An erect downy perennial 3–20(–75) cm. Basal lvs long-petiolate, the blade 2–4(–8) cm, ovate, obtuse, cordate or rounded at base, serrulate, teeth obtuse; stem-lvs sessile, ½-amplexicaul, or the lower petiolate, ovate to lanceolate, obtuse or subacute, rounded or narrowed at base, entire or serrulate. Infl. subcapitate, often with several ±distant fls or short few-fld branches below the terminal head. *Fls erect, sessile.* Calyx-tube c. 3 mm, obconical, 5-ribbed; teeth rather longer than tube, triangular. Corolla 15–25 mm, bright blue-purple, rarely white; lobes nearly as long as tube, suberect, eventually spreading, acute. Stigmas 3. Capsule c. 3 mm, erect, opening by apical pores. Fl. 5–9. $2n = 30$. Hs.

Native. In grassy places usually on calcareous soils, particularly in chalk grassland, less commonly on sea-cliffs or in woods. From Dorset and Kent to Cumberland and Kincardine, locally common, but mainly in the south and east. Most of Europe; temperate Asia.

Represented in Britain by subsp. **glomerata**.

6. C. rotundifolia L.   Harebell; Bluebell (in Scotland).

A slender nearly or quite glabrous perennial 15–40(–60) cm, producing slender underground stolons. Stems decumbent at base. *Basal lvs* long-petiolate, the blade 5–15 mm, ovate or *suborbicular,* crenate, *±cordate* at base; *lvs on decumbent part of stem petiolate,* lanceolate, entire or serrate, acute; lvs on erect part of stem sessile, narrowly linear, entire, acute. Infl. a ±branched panicle, or reduced to a solitary terminal fl.; bracts small, linear. Buds erect, the *fls nodding;* pedicels very slender. Calyx-tube c. 2 mm; teeth c. 5 mm, setaceous, spreading. Corolla c. 15 mm, broadly campanulate, narrowed at base, blue, rarely white; lobes about ½ as long as tube, broadly ovate, subacute. Stigmas 3. Capsule 3–5 mm, subglobose, nodding, opening by basal pores. Fl. 7–9. $2n = 34, 68$. Hs.

Native. In dry grassy places and on fixed dunes, often in poor shallow soils, locally common. Throughout the British Is., except for the Channel Is., Orkney and much of C. and S. Ireland. North temperate regions to over 70° N. in Norway.

7. C. patula L.   Spreading Bellflower.

A *scabrid* biennial 20–60 cm. *Root slender, fibrous.* Stems slender, angled. *Lower lvs* c. 4 cm, obovate-oblong, narrowed below into a short petiole and *decurrent* on it, crenate; stem-lvs smaller and narrower, sessile, acute, serrate or nearly entire. *Infl.* cymose, *much-branched,* branches up to c. 20 cm, spreading. Fls erect; stalks 2–5 cm, slender. Calyx-tube c. 4 mm, funnel-shaped; teeth c. 1 cm, triangular to setaceous, usually with 1–2 teeth at base. Corolla 20–25 mm, broadly campanulate, narrowed at base, purple; lobes about as long as tube, ovate, acute, spreading. Capsule c. 1 cm, obconical, erect, opening by pores at top; calyx-teeth erect in fr. Fl. 7–9. $2n = 20*$. Hs.

Native. In shady woods and hedgebanks. From Hants and Surrey to Shropshire, local, sometimes naturalized elsewhere. Most of Europe.

*8. C. rapunculus L.   Rampion.

A biennial somewhat similar to *C. patula,* is naturalized in a number of places. Root swollen, fleshy. Basal lvs abruptly contracted at base. Infl. simple or with a few short branches in the lower part. Corolla 10–20 mm, the lobes about ½ as long as tube. Fl. 7–8. $2n = 20$. Hs.

Introduced. In fields and hedgebanks, usually on gravelly soils. Hants and E. Gloucester to Hertford and Surrey; Fife, rare and local. Europe from the Netherlands southwards; N. Africa.

*9. C. medium L.   Canterbury Bell.

A stout erect very hispid biennial up to c. 60 cm. Lvs

crenate-dentate or serrate, the basal ovate-oblong, petiolate, the cauline lanceolate, sessile. Fls terminal and axillary. Calyx with ovate-lanceolate, acuminate teeth shorter than the corolla, and broadly ovate, cordate, reflexed appendages between the teeth. Corolla 40–50 mm, campanulate, inflated in the middle. *Stigmas* 5. Fl. 5–6. $2n = 34$. Hs. Many colour and other garden forms exist.

Introduced. Naturalized, particularly on railway banks in S.E. England and the E. Midlands, and an impermanent casual elsewhere. N. and C. Italy, S.E. France, naturalized elsewhere in Europe.

*C. alliariifolia Willd., a perennial with nodding creamy fls, appendages between the calyx-teeth, and 3 stigmas, is naturalized along railway lines in Cornwall. Native of Caucasus and Anatolia.

3. LEGOUSIA Durande

Characters of *Campanula* but ovary and capsule much elongated and the corolla ±rotate.

About 15 spp., in north temperate regions and South America.

1. L. hybrida (L.) Delarbre   Venus's Looking-glass.
Specularia hybrida (L.) A.DC.

An erect hispid annual 5–30 cm. Stems simple or ±branched. Lvs 1–3 cm, sessile, oblong or oblong-obovate, strongly undulate. Fls erect, mostly in terminal few-fld cymes. Calyx-teeth elliptic-lanceolate to linear-lanceolate, acute or obtuse, c. ½ as long as ovary. Corolla 8–15 mm diam., scarcely ½ as long as calyx, reddish-purple or lilac. Capsule 15–30 mm, subcylindrical, opening by valves just below the calyx-teeth. Fl. 5–8. $2n = 20^*$. Th.

Native. In arable fields, locally common, becoming rarer in the north. Mainly south and east of a line from S.E. Yorks to Dorset; absent from Wales, Scotland and Ireland. W. and S. Europe; W. Asia; N. Africa; Macaronesia.

4. PHYTEUMA L.

Perennial herbs with a fleshy stock. *Fls 5-merous, small, numerous, usually sessile in terminal heads or dense spikes. Corolla divided nearly to base, lobes linear*, at first cohering near the top to form a tube, later spreading. *Anthers free.* Ovary 2–3-celled. *Stigmas 2–3, linear. Capsule opening by lateral pores or small valves between the ribs.*

About 40 spp., in Europe (especially the Mediterranean region), and temperate Asia. Pollination mechanism similar to *Campanula* except that the pollen is held in the tube formed by the corolla-lobes and pushed out by the elongating style.

Infl. capitate; fls violet.   **1. orbiculare**
Infl. spicate; fls yellowish-white.   **2. spicatum**

1. P. orbiculare L.   Round-headed Rampion.
Incl. *P. tenerum* R. Schulz

A glabrous or slightly hairy erect perennial 5–50 cm.

Root with a deeply-buried fusiform enlargement. *Basal lvs* long-petiolate, the blade 2–4 cm, lanceolate to ovate-oblong, acute or obtuse, crenate, *narrowed, rounded or subcordate* at base; stem-lvs much smaller, sessile, linear-lanceolate, few and scattered. *Infl.* 1–2.5 cm diam., *depressed-globose*. Bracts narrowly triangular, shorter to longer than infl. *Corolla* c. 8 mm, *deep violet*, curved and cylindrical in bud; lobes narrow, eventually free nearly to base and spreading or reflexed. Style c. 1 cm; stigmas 2–3. Fruiting head c. 2 cm, ovoid or cylindrical. Capsule c. 5 mm, ovoid, crowned by the short, stiff, triangular, erect calyx-teeth. Fl. 7–8. $2n = 22 + 0$–2B. Hs.

Native. In chalk grassland, locally abundant. Wilts to E. Sussex. From S. England and Latvia to S. Spain and Albania.

2. P. spicatum L.   Spiked Rampion.

A robust glabrous perennial 30–80 cm. Root swollen, fleshy, fusiform. *Basal lvs* long-petiolate, blade 3–5 cm, ovate, obtuse, crenate or serrate, *deeply cordate at base*; lower stem-lvs petiolate, as large as or even larger than the basal, the upper smaller, sessile, lanceolate to linear. *Infl.* 3–8 cm, *cylindrical*. Bracts subulate, about as long as diam. of infl. *Corolla* c. 1 cm, *yellowish-white*, somewhat curved and cylindrical in bud; lobes eventually separating nearly to base. Style c. 1 cm; stigmas 2. Fruiting head 5–12 cm, cylindrical. Capsule similar to that of *P. tenerum*, but calyx-teeth subulate, spreading. Fl. 7–8. $2n = 22 + 0$–4B. Hs.

Native. In woods and thickets, very local. E. Sussex. Europe, from S. Norway and Estonia to N. Spain and Crna Gora.

In Britain the sp. is represented by subsp. **spicatum**.

5. JASIONE L.

Perennial, biennial, or annual herbs. *Fls small, numerous, sessile or subsessile in a terminal head* surrounded by 1 or more rows of bracts. Calyx-tube ovoid or turbinate; teeth 5. *Corolla blue, rarely white, 5-partite nearly to base, lobes narrow*, spreading. *Anthers* usually *shortly connate at base*. Ovary 2-celled. *Stigmas 2, short, stout. Capsule dehiscing loculicidally by 2 short valves within the persistent calyx-teeth.*

About 10 spp., in Europe and the Mediterranean region. Pollination mechanism similar to *Phyteuma* except that the anthers cohere and form a tube while the corolla-lobes spread out soon after the fl. opens. (cf. Compositae.)

1. J. montana L.   Sheep's-bit.

A pubescent spreading or ascending herb, usually biennial but sometimes annual. Stems 5–50 cm, usually decumbent at base, stout or slender, simple or branched, lfless in the upper half. Lvs up to c. 5 cm, linear-oblong to linear-lanceolate, obtuse or subacute, undulate or crenate, ciliate, the basal narrowed into a short petiole, the cauline sessile. Infl. 5–35 mm diam., depressed-globose. Bracts numerous, shorter than infl., triangular-acute to ovate-cuspidate. Corolla c. 5 mm, persistent, blue, rarely white. Calyx-teeth subulate, about as long as the unopened corolla. Fl. 5–8. $2n = 12$. Hs.

Native. In grassy places on light sandy or stony lime-free soils, in rough pastures, on heaths, cliffs and banks. Great Britain north to Moray and W. Inverness, mainly in the west. Orkney, Shetland; Ireland, but absent from most of the Centre; generally local, but abundant in some areas (e.g. Cornwall and Shetland). Distribution of the genus.

6. LOBELIA L.

Herbs or rarely shrubs. Fls solitary in the axils of lvs or arranged in terminal bracteate racemes. *Corolla*-tube oblique and curved, split to base along the back, 2-*lipped*, rarely with subequal connivent lobes. Stamens not epipetalous; 2 or all of the anthers bearded at the tip. Stigma shortly 2-lobed. Capsule dehiscing loculicidally by 2 valves within the calyx-teeth.

About 250 spp., in all the warm and temperate parts of the world except C. and E. Europe and W. Asia.

A number of spp. are cultivated for ornament, *L. erinus* L., the annual blue-fld sp. used for bedding-out, being the commonest.

Terrestrial; stems lfy; lvs toothed; fls erect or spreading, blue.   **1. urens**
Aquatic; stems lfless except for a few small scales; lvs entire; fls nodding.   **2. dortmanna**

1. L. urens L.   Heath Lobelia.

A nearly or quite glabrous erect perennial 20–60 cm. Juice very acrid. *Stems* slender, solid, angular, lfy. *Lvs* up to c. 7 cm, obovate or obovate-oblong, upper linear-oblong, all acute or obtuse, irregularly *serrate*. Raceme up to c. 20 cm, lax. *Fls erect or spreading*, shortly pedicellate, bracts linear, ±equalling calyx. Calyx c. 8 mm, the teeth rather longer than or about equalling the narrowly obconic tube, subulate, spreading. *Corolla* 10–15 mm, *blue* or purplish. Anthers shortly exserted. Capsule erect. Fl. 8–9. $2n = 14$. Hs.

Native. In rough pasture, on grassy heaths and at margins of woods, on damp acid soils, very local but apparently increasing. E. Cornwall to S. Hants; E. Sussex; Waterford. W. Europe from Belgium southwards; Madeira, Azores; N. Africa.

2. L. dortmanna L.   Water Lobelia.

A glabrous erect stoloniferous perennial 20–60 cm. *Stems* slender, terete, fistular, *lfless* except for a few small scales. *Lvs* 2–4(–8) cm, in a basal rosette, linear, obtuse, ±recurved, *entire*, submerged. Racemes up to c. 10 cm, very lax and few-fld, emersed. *Fls nodding*, pedicels up to c. 1 cm, bracts oblong, obtuse, much shorter than pedicels. Calyx c. 5 mm, teeth shorter than tube, ovate, obtuse, erect. *Corolla* 15–20 mm, *pale lilac*. Anthers included. Capsule nodding. Fl. 7–8. $2n = 16$. Hyd.

Native. In stony lakes and tarns with acid water. Locally common in Wales, the Lake District and most of Scotland and Ireland, except the centre and south. N. and N.C. Europe extending locally to S.W. France and White Russia.

120. RUBIACEAE

Woody plants or herbs with opposite and decussate or whorled simple entire lvs whose stipules may stand between the lvs (interpetiolar) or between lf and stem (intrapetiolar). Fls usually small in terminal and axillary cymes, sometimes in heads; usually hermaphrodite, actinomorphic, epigynous, 4–5-merous. Sepals usually free, often very small or represented only by an annular ridge; corolla funnel-shaped or rotate with the petals joined below into a longer or shorter tube; stamens isomerous, epipetalous, alternating with the petals; style simple or bifid; ovary inferior, usually 2-celled, with 1–many anatropous ovules on an axile placenta in each cell. Fr. a capsule, berry or drupe, or dry and schizocarpic; seeds endospermic with a usually straight embryo.

About 6000 spp. and 500 genera, cosmopolitan but chiefly tropical.

The British representatives all belong to the tribe Galieae with 2 or more lf-like interpetiolar stipules so that there appear to be 4 or more lvs in a whorl, of which usually only 2 have axillary buds. The fls are very small and the fr. is usually of 2 separating indehiscent 1-seeded parts.

1 Fls in small heads terminating main stem and branches, usually with a basal involucre of lf-like bracts.   2
  Fls in panicles or cymes, not in terminal heads.   4
2 Calyx of (4–)6 distinct teeth, enlarging and persistent in fr.; corolla pink, with tube at least twice as long as free lobes.   **1. SHERARDIA**
  Calyx an inconspicuous rim (sometimes with 4–5 very short teeth) or 0.   3
  Stem erect or ascending; fls 4-merous, whitish or blue.
                                                         2. ASPERULA
  (Stem procumbent; fl.-heads lacking basal involucre; fls 5-merous, deep pink. Garden escape
                                          PHUOPSIS, (p. 427))
4 At least the lower lvs in whorls of 4.   5
  Most lvs in whorls of more than 4.   8
5 Lvs usually in whorls of 4 throughout.   6
  Lvs in whorls of 4–6 or more.   8
6 Corolla pink, with tube at least as long as free lobes.
                                                         2. ASPERULA
  Corolla white, yellowish-green or yellow.   7
7 Lvs conspicuously 3-veined; fls yellow, in axillary cymes shorter than axillant lvs, with terminal fl. hermaphrodite, laterals mostly male; fr. smooth, glabrous.   **4. CRUCIATA**
  Lvs indistinctly 3-veined; fls white in a terminal lfy panicle with branches exceeding their axillant bracts; all fls hermaphrodite; fr. rough with hooked hairs.   *Galium boreale*

8 Corolla funnel-shaped, white, with tube slightly longer to c. twice as long as free lobes; fr. rough with hooked bristles.   *Galium odoratum*
    Corolla rotate, yellow or white, with lobes spreading horizontally from top of very short tube.   9
9 Corolla with 4(–3) free lobes; fr. dry.   3. GALIUM
    Corolla with 5 free lobes; fr. a berry.   5. RUBIA

1. SHERARDIA L.

Lvs (and lf-like stipules) 4–6 in a whorl. *Fls few in small terminal head-like clusters with a stellate involucre of lf-like basally concrescent bracts. Calyx with 4–6 distinct sepals; corolla funnel-shaped* with 4 lobes; stamens 4; style bifid with capitate stigmas.
    One sp.

1. S. arvensis L.   Field Madder.

An annual herb with slender reddish roots and numerous prostrate or decumbent spreading stems 5–40 cm, simple or branched, ±glabrous, with 4 rough angles. Lower lvs 4 in a whorl, obovate-cuspidate, soon withering, upper 5–18 mm, 5–6 in a whorl, elliptical-acute, ±glabrous but the margins and underside of the midrib scabrid with forwardly directed prickles. Fls 3 mm diam., subsessile, 4–8 in terminal heads. Involucre of 8–10 lanceolate lf-like bracts longer than the fls. *Sepals* 4–6, green, triangular-lanceolate, at first small but persistent and usually *enlarging in fr.* Corolla 4–5 mm, pale lilac, funnel-shaped, with a long slender tube about twice as long as the 4 lobes. Fr. 4 mm, crowned by the sepals, of 2 obovoid mericarps rough with short appressed bristles. Fl. 5–10. Said to be gynomonoecious or -dioecious in Continental Europe. Visited chiefly by flies. $2n = 22$. Th.
    Native (but see below). Mainly in arable fields and open waste ground including road-verges, on hedge-banks and occasionally in rough dry grassland (especially on sandy soil and near the sea-coast); common in lowlands throughout British Is. and reaching 370 m in Scotland; on all types of fairly dry soil but commonest over chalk or limestone. Through most of Europe to c. 67° N in Norway; W. Asia; N. Africa. Introduced in North America, Australia and other temperate countries, and probably also, in prehistoric times, in C. and N. Europe.

2. ASPERULA L.

Usually perennial herbs with lvs and lf-like stipules in whorls. Fls in heads or panicles, 4-merous. *Calyx an indistinct annular ridge* or of 4–5 small teeth, not persistent in fr.; *corolla funnel-shaped,* its tube longer than the 4(–3) lobes; styles ±connate; stigmas capitate. Fr. of two 1-seeded dry mericarps, with the testa of the seed adherent to the fr. wall.
    About 200 spp. in Europe, Asia and Australia.

1 Plant annual; lvs linear, mostly 6–8 in a whorl; fls bright blue in terminal involucrate heads. A casual.
                                        arvensis (p. 427)

Plant perennial; lvs mostly 4 in a whorl; fls pink-lilac or white tinged with pink.   2
2 Upper lvs ovate-lanceolate; fls in involucrate heads.
                                        3. taurina
    Upper lvs linear to linear-lanceolate; fls not in involucrate heads, pale pink-lilac outside, white within.   3
3 Middle stem-lvs ±linear, commonly 10–12(–25) mm long and less than 1 mm wide; fls usually short-stalked; corolla-tube usually somewhat longer than the free lobes.   **1. cynanchica**
    Middle stem-lvs narrowly oblanceolate, commonly 3–10(–15) mm long and 1–2 mm wide; fls sessile; corolla-tube about as long as the free lobes.
                                        2. occidentalis

1. A. cynanchica L.   Squinancy Wort.

A *perennial,* loosely to densely caespitose herb with branching but *non-stoloniferous* stock producing ±numerous slender prostrate to ascending sterile shoots and ascending to erect flowering shoots, 10–50 cm; all shoots branched, 4-angled, usually rough with short hairs below but ±glabrous above. *Lvs* glabrous, in whorls of 4 (rarely 6), the basal broadest, soon withering, the middle longest, (6–)10–12(–25) mm, *usually less than 1 mm wide,* linear to linear-lanceolate; *true lvs often distinctly longer than stipular members,* especially in upper whorls; all lvs acute to mucronate, strongly 1-veined with flat or slightly revolute margins, firm but *not fleshy,* stiffly recurved. Fls 3–4 mm diam., with pedicels 0–1 mm, in lax, long-stalked, terminal and axillary few-fld corymbs; bracts small, subulate. Calyx of 4 small teeth. *Corolla* funnel-shaped, white within, pale pink-lilac and roughly papillose outside, its *tube usually somewhat longer than the acute ovate-lanceolate free lobes.* Fr. 2–3 mm, densely papillose to rugose, usually glabrous. Fl. 6–7. Homogamous; vanilla-scented and visited by various small insects. $2n = 40^*$. Hp.
    Native. Dry calcareous grassland and calcareous sand dunes, to 305 m in Ireland. Locally abundant in S. England and S. Wales and extending northwards to N. Yorks and Cumbria; S.W. Ireland from Kerry to Galway; Channel Is. Most of Europe to c. 54° 30′ N. in England and to C. Russia; Caucasus.

2. A. occidentalis Rouy   Western Squinancy Wort.

A perennial, loosely caespitose herb closely resembling **1** but with the stock *spreading by slender orange-coloured hypogeal stolons.* The procumbent to ascending lfy stems are only 3–16(–35) cm and the *middle stem-lvs only* 3–10(15) mm *long but* 1–2 mm *wide, narrowly oblanceolate and rather fleshy.* The infl. is more compact and the *fls are sessile,* otherwise much as in **1** but with the *corolla-tube about equalling the free lobes.*
    Native. Recognized in Britain as distinct from **1** in 1973 and so far known only from sand dunes in S. Wales (Glamorgan and Pembroke) and in S.W. Ireland (N. Kerry and W. Galway). Elsewhere known only from coastal sand dunes between Santander in N. Spain and Biarritz in S.W. France.
    Further study of the very variable *A. cynanchica* com-

plex is needed and may lead to treatment of *A. occiden-talis* as a subsp. of *cynanchica*.

*3. A. taurina L.   Pink Woodruff; Turin Woodruff.

A *perennial* herb *spreading by slender orange-red hypo-geal stolons* and with stout, erect, simple or sparingly branched fl.-stems (10–)20–50 cm, 4-angled, ±sparsely pubescent with patent hairs. *Lvs* 4 *in a whorl*, 3–6 × 1–2.5 cm, *elliptical-lanceolate to -ovate*, very shortly petiolate, somewhat acuminate, *with* 3(–5) *main veins* and a distinct reticulum of subsidiary veins, and with patent hairs especially on veins and margins; *pale beneath*. *Fls* 5 mm diam. *in dense terminal heads sur-rounded by an involucre of lf-like ciliate bracts*, *the outer exceeding the fls*, the inner shorter and narrower. *Cor-olla* 10–14 mm, *white* tinged with yellowish-pink, *the slender tube up to* 3 *times as long as the free lobes*. Sta-mens exserted, with violet anthers. Style included. Fr. 1–3 mm, glabrous, ±smooth. Fl. 5–6. Protandrous and visited by flies and bees. 2*n* = 22. Grh.–Hp.

Introduced. Established very locally in woods and grasslands and on roadsides where escaped from culti-vation or perhaps planted; covers a wide area of stream-bank sand and shingle under trees in Mid-Perth. Subsp. *taurina*, described above, is native in S. and S.C. Europe; a different subsp. in the N. Balkan Peninsula, Hungary and Romania.

*A. arvensis L., an annual herb with an erect 4-angled glabrous branching stem 8–30(–50) cm, bearing persistent cotyledons and distant whorls of linear blunt ciliate lvs, 12–25 cm, 6–8 in a whorl; bright blue fls, 4 mm diam., in involucrate sessile terminal heads, the ciliate bracts exceeding the fls, and brown glabrous ±smooth fr. 2 mm, has occurred as a casual in arable land and waste places in various localities in England. S. and C. Europe; N. Africa; Asia Minor.

*Phuopsis stylosa (Trin.) B. D. Jackson (*Crucianella stylosa* Trin.) is much grown in gardens and frequently escapes. It is a procumbent annual (rarely longer-lived) 15–25 cm, with lanceolate hispid lvs, 12–18 mm, 8–9 in a whorl, and small 5-merous deep pink fls in rounded terminal heads, 10–15 mm diam.; style long-exserted, 2-cleft; ovules 1 per cell. Iran.

3. GALIUM L.

Annual to perennial herbs with whorls of 4–10 lvs and lf-like stipules. Fls in terminal and axillary cymes, rarely solitary; small, hermaphrodite or polygamous. Calyx a minute annular ridge; corolla usually rotate with a very short tube and 4 (3–5) lobes, sometimes funnel-shaped with the tube almost equalling the lobes; stamens 4, exserted; styles 2, short, connate below; stigmas capi-tate. Nectar-secreting and insect pollinated. Fr. di-dymous, of two 1-seeded mericarps, glabrous, hairy or with hooked bristles, with the testa of the seed adherent to the pericarp.

About 400 spp., almost cosmopolitan.

1 Lvs in whorls of 4 throughout, each with 3 ±parallel longitudinal veins from the base; fls white.
　　　　　　2. boreale

Lvs neither in whorls of 4 throughout nor distinctly 3-veined, though sometimes with strong midrib and weaker diverging laterals.　　2

2 Corolla funnel-shaped, white and fragrant, with tube at most about equalling the free lobes, usually shorter.　**1. odoratum**

Corolla rotate, yellow or white, with lobes spreading horizontally from top of very short tube.　　3

3 Lvs blunt to acute but never cuspidate or mucronate.　4
Lvs cuspidate or mucronate.　　6

4 Lvs narrowly linear; submerged lvs very narrow, flac-cid, up to 2 cm; infl. obconical, its branches erect-ascending throughout. Very rare in ponds in S. England.　**11. debile**

Lvs linear-lanceolate to broadly oblanceolate; sub-merged autumn lvs like those on aerial shoots; infl. pyramidal or oblong, its branches soon spreading.
　　　　　(**palustre** sens. lat.)　5

5 Stems to 120 cm, stout but weak; lvs usually 1.5–2 cm, oblanceolate, blunt; fls c. 4 mm diam. in large pyra-midal panicles with branches spreading in fr.; meri-carps 1.6 mm diam., coarsely rugose. Usually in standing water.　**10. elongatum**

Stems 15–30(–50) cm, slender but ±erect; lvs usually 0.5–1 cm, linear-oblanceolate, blunt or rarely sub-acute; fls c. 3 mm diam. in oblong panicles with branches ±reflexed in fr.; mericarps 1.2 mm diam., finely rugose. Commonly where water stands only in winter.　**9. palustre**

6 Stems markedly rough with recurved prickles on angles.　　7
Stems smooth or only slightly rough on angles.　12

7 Perennial plant of wet peaty places; fls white.
　　　　　　12. uliginosum

Annual; fls greenish, cream or reddish, not pure white.　　8

8 Fls commonly 1-3-fld axillary cymes shorter or barely longer than axillant lvs; lateral fls often male; fr.-pedicels down-curved; fr. verrucose; stem rough with downwardly-directed prickles.　9

Fls in cymes longer than axillant lvs, usually all herm-aphrodite; fr.-pedicels straight; stem rough or not.　10

9 Lvs in whorls of 6–8, their margins with backwardly-directed hooked prickles; fls less than 2 mm diam., white.　**13. tricornutum**

Lvs in whorls of 5–6(–7), their margins with for-wardly-directed prickles; fls usually 2–2.5 mm diam., often greenish. Casual. **verrucosum** (p. 431)

10 Lvs 3–6 cm; fls at least 1.5 mm diam.; mericarps 3–6 mm diam., with dense hooked swollen-based bristles.　**14. aparine**

Lvs up to 3.5 cm, often much shorter; fls up to 1.3 mm diam.; mericarps c. 1–3 mm diam., smooth or with hooked hairs lacking swollen bases.　11

11 Lvs in whorls of 6–10, with backwardly-directed mar-ginal prickles and long-awned apex; fls c. 1 mm diam., greenish; mericarps 2–3 mm　**15. spurium**

Lvs in whorls of 5–7, with forwardly-directed marginal prickles and short-awned apex; fls 0.5–1 mm diam., greenish inside, reddish outside; mericarp c. 1 mm.　　**16. parisiense**

12 Fls yellow.　　13
Fls white.　　14

13 Stem ±terete with 4 raised lines; lvs linear, usually less than 2 mm wide; infl. usually densely hairy;

fls golden yellow, corolla-lobes acute but hardly apiculate.   **4. verum**

Stem 4-angled below; lvs narrowly lanceolate, usually more than 1.5 mm wide; infl. hairy but not densely so; fls variable in colour but usually paler than in *verum*.   ×**pomeranicum** (p. 429)

14 Stem robust, 4-angled, decumbent to erect, with fls in large panicles; corolla-lobes apiculate to long-cuspidate; fr. rugose.   **3. mollugo** group

Stem slender, ±decumbent, with fls in small axillary and terminal corymbs; corolla-lobes usually acute; fr. papillose or tuberculate.   15

15 Lvs on flowering shoots obovate to oblanceolate, their margins with forwardly-directed prickles; axillary cymes shorter than stem-internodes.   **5. saxatile**

Lvs on flowering shoots oblanceolate to linear, with marginal prickles backwardly-directed or 0; axillary cymes longer than stem-internodes.   16

16 Fr. covered with acute tubercles. Only north of line from Severn estuary to Humber.   **8. sterneri**

Fr. smooth or with low rounded papillae. Mostly south of line from Severn estuary to Humber.   17

17 Loosely caespitose; lvs commonly 1–2 cm, ±falcate; pedicels 1–1.5 mm.   **6. pumilum**

Densely caespitose; lvs commonly 0.6–1 cm, straight; pedicels usually not exceeding 1 mm; Cheddar cliffs only.   **7. fleurotii**

1. G. odoratum (L.) Scop.   Sweet Woodruff.

Asperula odorata L.

A perennial herb, *hay-scented* when dried, with slender branched far-creeping rhizomes and *erect simple* 4-angled *stems* 15–45 cm, hairy beneath the nodes, otherwise glabrous. Lvs firm, in distant whorls, lanceolate or elliptical, ±cuspidate, glabrous but with forwardly-directed marginal prickles; lowest lvs small, 6 in a whorl; middle lvs 2.5–4 × 0.6–1.5 cm, 6–8(–9) in a whorl. Fls c. 6 mm diam., short-stalked, in long-stalked terminal and 1–2(–4) lateral cymes, together forming a ±umbellate infl. Bracts small, linear-lanceolate. *Corolla* 4–6 mm, *4-lobed to* c. *halfway*, pure white, the lobes blunt, slightly recurved, downy within. Fr. 2–3 mm, rough with hooked black-tipped bristles. Fl. 5–6. Homogamous. Fragrant and visited chiefly by flies and bees. $2n = 44$. Grh.–Hp.

Native. Locally abundant in woods on damp calcareous or base-rich soils, up to 640 m in Scotland. Throughout the British Is., except the Outer Hebrides and Orkney. N. and C. Europe and montane woods in Italy and the Balkans; N. Africa; Siberia.

There is difficulty in drawing a line between *Galium* and *Asperula*. *G. odoratum* has funnel-shaped fls but with a corolla-tube hardly longer than the lobes. The closely related *G. glaucum* L., commonly placed in *Asperula*, has an even shorter corolla-tube and hybridizes with *G. verum*. It seems desirable to transfer both spp. to *Galium*, although they lack the rotate corolla typical of the genus.

2. G. boreale L.   Northern Bedstraw.

A perennial herb with creeping stock and erect, rigid, 4-angled, glabrous or pubescent stems 20–45 cm, with erect-ascending branches. *Lvs* 1–4 cm, lanceolate or

elliptical, 3-*veined*, *rough on the margins and the underside of the midrib*, 4 *in a whorl*, *bright green*, turning black when dried. Fls 3 mm diam., in a ±pyramidal terminal lfy panicle whose ascending branches exceed their bracts. *Corolla white* with 4 apiculate lobes. *Fr.* 1.5–2 mm, olive-brown, *densely hispid with hooked bristles*. Fl. 7–8. Hermaphrodite, slightly protandrous. Visited by various small insects. $2n = 44^*$, 66. Hp.

Native. Rocky slopes, moraine, scree and shingle, sides of streams and lakes, stable sand dunes, etc., to 1060 m in Scotland. Rare in Wales, locally common in N. Britain northwards from Lancs and Yorks; Inner Hebrides; not in Outer Hebrides or Orkney but ?introduced in Shetland; locally frequent in W. Ireland but no longer in the north. N. and C. Europe southwards to N. Italy, Yugoslavia and Bulgaria; Caucasus; N. and C. Asia eastwards to Japan; North America; Greenland.

3. G. mollugo L.   Hedge Bedstraw.

Incl. *G. album* Miller (=*G. erectum* Hudson 1778, non 1762)

Perennial herb with ±woody stock from which arise hypogeal stolons and aerial *stems* 30–150 cm, *procumbent or weakly ascending to ±erect*, often much branched below, *4-angled*, glabrous or pubescent, ±*swollen at the nodes, not blackening when dried*. Lvs c. 8–30 × 2–7 mm, usually 6 or 8 in a whorl, *elliptical-obovate or oblong-oblanceolate to linear-oblanceolate*, cuspidate to acuminate, often mucronate; *margins usually rough* with short forwardly-directed prickles, faces wholly glabrous or finely pubescent beneath. Infl. a broadly pyramidal to oblong panicle of spreading or ascending cymose branches. *Fls* 2–5 mm diam., ±white, usually *with* 4 (3–5) *cuspidate corolla-lobes*. Ripe fr. 1–2 mm, glabrous, rugulose, brown or blackish; fr.-pedicels ±divaricate. Fl. 6–9; fls hermaphrodite, protandrous, visited chiefly by small flies. $2n = 44^*$. Hp.

Native. Hedgebanks, waysides, pastures, grassy slopes, waste land, also woodland and scrub, on nutrient-rich and calcareous soils, to 305 m in Scotland. Common in S. England; diminishing northwards but reaching northernmost Scotland and Orkney, though perhaps introduced in some areas; infrequent in Wales and in Ireland.

Very variable. The *G. mollugo* group, with representatives throughout Europe and in N. Africa and W. Asia, comprises a range of variants, the taxonomic treatment of which has proved difficult and controversial. Problems raised by British members led Miller in 1768 to separate his *G. album*, later usually named *G. erectum* Hudson. It was soon realized that the two are connected by many intermediates and they have therefore often been treated as subsp. or lesser variants of a single sp. Over its whole distribution-range the group shows cytological as well as morphological variation, and the account by F. Krendl (*Österr. Bot. Zeitschr.*, 1967, **114**, 508–49) of C. European representatives distinguishes *G. mollugo* L. and *G. album* Miller as exclusively diploid ($2n = 22$) and tetraploid ($2n = 44$) respectively. He states that the former appears to occur in S. England, but K. M. Goodway (personal communication) reports only tetraploid counts for all British material so far investigated, whether identified as *G. mollugo* or as *G. album*. Plants collected as *G. mollugo* L. in Switzerland, France, Germany, the Low Countries and Denmark have also yielded exclusively tetraploid counts (Kliphuis, 1962). If Krendl's conclusions are accepted, all these tetraploids should have been referred to *G. album*, a course already taken by some British botanists. It may, alternatively, be judged that *G. mollugo* L., like other *Galium* spp., reasonably comprises both diploid and tetraploid plants. It is clear, however, that much more morphological and cytological information, and a deeper study of the applicability of the two names, is needed before British members of the group can be dealt with satisfactorily.

The account below is based on published descriptions of the two taxa into which British plants have commonly been divided, although the distinction between them is very far from clear cut.

3a. G. mollugo L. (or subsp. *mollugo*)

Fl.-stems weak, decumbent to ascending, diffusely branched, markedly swollen at the nodes. *Lvs* up to c. 25 mm, *narrowly obovate to oblong-oblanceolate*, rarely narrower, those *on main stem commonly ±horizontal or deflexed. Infl. with spreading branches, ±broadly ovoid or pyramidal*, the lower branches divaricate in fr.

3b. G. album Miller (or *G. mollugo* L. subsp. *album* (Miller))

Fl.-stems decumbent to ±erect, with usually erect branches, less swollen at the nodes than in *mollugo*. *Lvs* up to 30 mm, *oblong to linear-oblanceolate*, mucronate, those *on main stem commonly obliquely erect. Infl. with ascending branches, oblong and usually narrower than in mollugo*, the lower branches less divaricate in fr.

The fls of *mollugo* are commonly stated to be only 2–3 mm diam., those of *album* 3–5 mm, but this seems an unreliable character, as does the greater pedicel-length in *mollugo* (2–4 mm) than in *album* (1.5–3 mm). The distributions of the two are inadequately known, but *album* seems to become relatively more frequent northwards. Some British botanists, however, now regard all our members of the *mollugo* group as variants of *G. album* subsp. *album*, widespread in Continental Europe.

G. × pomeranicum Retz. (*G. ochroleucum* Wolf ex Schweiger), the hybrid between *G. mollugo* sens. lat. and *G. verum*, is intermediate between the parents in fl.-shape, form of infl. and the pale yellow colour of its fls but resembles *G. mollugo* in not blackening when dried. It is moderately fertile and can also backcross with the parent spp. giving rise to a series of connecting variants. Native. Most frequent along the coast of S. England but also occurs inland on chalk and limestone through much of S. and E. England and to a smaller extent in N. England from the Humber estuary to Solway Firth, but is virtually absent from Scotland, including the Hebrides, Orkney and Shetland, and from Ireland. It is found in most of Europe except the Mediterranean islands.

4. G. verum L.   Lady's Bedstraw.

A perennial stoloniferous herb with a slender creeping stock and erect to decumbent, glabrous or sparsely pubescent, bluntly 4-angled stems 15–100 cm, with numerous ascending branches; *blackening when dried*. *Lvs* 6–25 × 0.5–2 mm, *linear*, mucronate, 1-veined, dark green and rough above, pale and pubescent beneath, *with revolute margins*, 8–12 in a whorl. Fls 2–4 mm diam., in a terminal lfy compound panicle. *Corolla bright yellow*, its 4 lobes apiculate. Fr. 1.5 mm smooth, glabrous, ultimately black. Fl. 7–8. Hermaphrodite, protandrous. Coumarin-scented and visited by various small insects, especially flies. $2n = 22, 44$. Hp.

Native. In grassland on all but the most acid soils, hedge-banks, stable dunes, etc, to 655 m in Scotland. Abundant throughout the British Is. All Europe except Russia, north to Iceland; W. Asia. The fls are used for coagulating milk and the stolons yield a red dye.

5. G. saxatile L.   Heath Bedstraw.

G. harcynicum Weigel; *G. hercynicum* auct.; *G. montanum* Hudson

A perennial mat-forming herb with numerous prostrate non-flowering branches and decumbent or ascending flowering shoots (5–)10–20(–30) cm, 4-angled, glabrous, smooth, much branched. *Plant turning black when dried*. Lvs (5–)7–10(–11) mm, 6–8 in a whorl, obovate on non-flowering and obovate to oblanceolate on flowering shoots, mucronate, with small straight *marginal prickles pointing obliquely forwards*. Fls 3 mm diam., in *few-fld ascending cymes* longer than their subtending lvs but *shorter than the stem internodes* and forming a cylindrical panicle. Corolla pure white, its 4 lobes acute, not cuspidate. Fr. 1.75–2 mm, glabrous, *covered with high acute tubercles*. Fl. 6–8. Hermaphrodite, protandrous, self-incompatible. Visited by flies and other small insects. $2n = 22, 44^*$. Hp.

Native. Heaths, moors, grasslands and woods on acid soils; to 1310 m in Scotland. Common throughout the British Is.; strictly calcifuge. W. Europe from c. 63° N. in Norway to N. Spain and Portugal and eastwards to the Carpathians and N. W. Russia; Newfoundland. An Atlantic species.

Plants intermediate between this species and *G. sterneri* occur on some Scottish mountains and seem to be hybrids, they are highly sterile but may occasionally backcross with the parent spp.

6. G. pumilum Murray   Slender Bedstraw.

G. sylvestre Pollich, non Scop.

A perennial *loosely caespitose herb with ascending to erect flowering shoots* 10–30(–60) cm and a few ±prostrate non-flowering shoots, all little branched, glabrous or hairy, 4-angled, *remaining greenish when dried*. Lvs 10–16(—30) × c. 1.5 mm, usually *narrowly oblanceolate*, uppermost narrower, all mucronate, ±*falcate*, in whorls of 5–9; *lf-margins with rather few backwardly-directed prickles*; basal stem-internodes shorter than the early-falling lowest lvs, but *middle internodes 2–4 times as long as lvs*. Infl. a lax panicle of open flat-topped axillary cymes exceeding the stem-internodes. Fls 2–3 mm diam., creamy white. *Ripe fr.* to 1.5 mm, glabrous and almost smooth but *covered with tiny dome-shaped papillae.* Fl. 6–8. Hermaphrodite, protandrous, self-incompatible; visited by flies and other small insects. $2n = 88^*$. Hp.

Native, and introduced in a few localities: mainly in chalk and limestone grassland. Occurs in a number of small isolated populations mostly in S.E. England but extending to Cornwall and Gloucester and northwards to Lincoln. Widespread and variable in W. and C. Europe from France northwards to Denmark and eastwards to the Baltic States, Poland and Romania.

7. G. fleurotii Jordan   Cheddar Bedstraw.

A perennial herb resembling *G. pumilum* but more densely caespitose, with *stems usually reddish at base* and with linear-lanceolate or narrowly oblanceolate *straight lvs*, commonly $6–10 \times 0.5–1$ mm, their margins with spreading hairs or backwardly-directed prickles. *Infl. rather dense*, with *pedicels usually not exceeding* 1 mm (in *pumilum* 1–1.5 mm). Fls c. 2.5 mm diam. Ripe fr. c. 1.4 mm, ±papillose. Fl. 6–7. $2n = 44, 88$. Hp.

Native. Known only on limestone cliffs at Cheddar, Somerset. Elsewhere only on calcareous screes and cliffs in France.

8. G. sterneri Ehrend.   Limestone Bedstraw.

G. pumilum subsp. *septentrionale* Sterner

A perennial herb overwintering as a mat (more prostrate and compact than *G. pumilum*) and later with many prostrate non-flowering and erect or ascending flowering shoots (10–)15–25(–30) cm, glabrous or hairy, much branched at the base, giving a tangled new growth. Lower stem-internodes short, so that lf-whorls are crowded. *Plant turning dark green when dried.* Lvs on flowering shoots (8–)10–14(–16) mm, 6–8 in a whorl, oblanceolate to linear, mucronate, their *margins with many curved backwardly-directed prickles.* Fls 2.5–3.5 mm diam., in compact ascending *cymes longer than the stem-internodes* and forming a compact pyramidal panicle. Corolla creamy white. *Fr.* 1.25 mm, glabrous, *covered with acute tubercles.* Fl. 6–7. Hermaphrodite, protandrous, self-incompatible. Visited by flies and other small insects. $2n = 44^*$ (and 22, see below). Hp.

Native. On grassy calcareous slopes and on calcareous or basic igneous rocks; to 914 m in Scotland. Abundant on Carboniferous Limestone in Brecon, Derby, Yorks, Durham and W. Ireland; occasional on other basic outcrops from Caernarvon and Northumberland to Sutherland and Orkney, and in N.E. Ireland. The north British calcicole counterpart of the calcifuge *G. saxatile.*

The plants from Caernarvon, from the Durness Limestone in Sutherland and from W. Ireland are diploid ($2n = 22^*$). They are more slender and smaller in all parts (flowering shoots 10–15(–20) cm; lvs 7–9 mm; fls 3.25 mm. diam.; fr. 1–1.15 mm),

the infl. is more open and the cymes smaller and more compact; but there is overlap with the tetraploid in all these features.

G. sterneri is restricted to N.W. Europe, like its close allies *G. normanii* O. C. Dahl (Iceland and W. Norway), *G. suecicum* (Sterner) Ehrend. (Sweden and N.E. Germany) and *G. oelandicum* (Sterner & Hyl.) Ehrend. (Öland only).

9–11 G. palustre group   Marsh Bedstraws.

Perennial herbs with slender creeping hypogeal or epigeal stocks and decumbent or ascending to ±erect *lfy shoots* to 120 cm or more, 4-*angled*, glabrous, smooth or more usually *±scabrid with downwardly directed prickles on the angles*, blackening when dried. Lvs 4–6 in a whorl, 1-*veined*, *broadly oblanceolate to linear, blunt or subacute*, rarely acute, never mucronate, margins commonly rough. *Fls* usually 2.5–5 mm diam., *±white*, in lax cymose panicles. *Ripe fr.* glabrous, *granulate to coarsely rugose.* Fl. 5–7; fls hermaphrodite, protandrous; visited chiefly by bees. $2n = 24^*, 48^*, 96^*$. H.–Hel.

A morphologically and cytologically variable group presenting taxonomic problems. The present treatment follows that in *Flora Europaea* 4 (1976).

9. G. palustre L.   Lesser Marsh Bedstraw.

Incl. *G. witheringii* Sm.

Stems slender, 15–30(–50) cm, ±erect, somewhat rough on the angles, rarely smooth. *Lvs* usually 0.5–1 cm, *linear-oblanceolate*, blunt or rarely subacute, often rough on the margins. *Fls* usually 2.5–4 mm diam., averaging c. 3 mm, *in lax oblong panicles with branches erect-ascending in fl. but ±reflexed in fr.* Mericarps averaging 1.2 mm diam., finely rugose; *fr.-pedicels markedly divaricate.* Fl. 6–7, 2–3 weeks earlier than *G. elongatum.* $2n = 24^*$, very rarely 48^*.

Native. The diploid ($2n = 24$) is locally common throughout the British Is. in mowing meadows and fens where water stands only intermittently. The tetraploid ($2n = 48$), so far found only in single localities in Devon and Cheshire, seems a somewhat larger plant but is not certainly separable unless perhaps by stomatal dimensions: modal length of guard-cells of fully-grown lvs 29–33 μm; in diploid 25–30 μm, in octoploid *G. elongatum* 35–39 μm. The tetraploid is therefore treated as conspecific with *G. palustre*, following *Flora Europaea*, 4. The diploid is native through most of Europe but the tetraploid has been reported only from Portugal, Spain, Yugoslavia and Romania.

10. G. elongatum C. Presl   Great Marsh Bedstraw.

Stems 50–120 cm or more, *stout but weak* and commonly supported by surrounding vegetation, diffusely branched, usually *±rough on the angles.* Lvs in midstem c. $20–40 \times 2.5–5$ mm, varying from lanceolate to elliptical or near-ovate but most often *broadly oblanceolate*, usually *blunt* and usually rough on margins and midrib. *Fls* usually 3–5 mm diam. and averaging c. 4 mm, *in a large lax panicle* with *branches erect-ascending in* *fl. and spreading*, but *not strongly reflexed, in fr.* Mericarps averaging 1.6 mm diam., *coarsely rugose*; fr.-pedicels divaricate. Fl. 6–7, somewhat later than *G. palustre.* $2n = 96^*$, 144.

Native. Common throughout the British Is. in permanently wet habitats and usually in standing water, often at the landward edge of the reed-swamp zone of streams and ponds. It is typically a plant much larger than the diploid in all its parts and the two are readily distinguishable. The very rare tetraploid is intermediate but closer to the diploid (see under **9**, above). Native in most of Europe eastwards to Finland, W. Ukraine and Turkey but not in Faeroes, Iceland or Spitsbergen; also in N. Africa and W. Asia.

11. G. debile Desv.   Slender Marsh Bedstraw.

G. constrictum Chaub.

A perennial herb with slender 4-angled glabrous stems 15–40 cm, prostrate below then ascending, smooth or slightly rough on the angles, not blackening when dried. *Lvs* 0.5–1 cm, 4–6 in a whorl, *linear*, broadest beyond the middle, subacute, sometimes minutely apiculate but never mucronate, glabrous but the margins rough with forwardly-directed prickles; *autumn lvs of submerged shoots* 1–2(–3) cm, *very narrow, flaccid.* Fls 2.5 mm diam., in long-stalked erect-ascending axillary corymbs forming an *obconical panicle* whose branches do not spread in fr. Corolla pinkish-white, its 4 lobes acute. Fr. 1.2 mm the mericarps 1 mm diam., glabrous, granulate; fr.-pedicels not divaricate. Fl. 5–7. $2n = 22$. Hel.–Hyd.

Native. Pond margins and pools drying out in summer. Devon, the New Forest (Hants) and Humberside; Channel Is. S. W. and S. Europe from France, Spain and Portugal to S. Balkans and Crete.

12. G. uliginosum L.   Fen Bedstraw.

A perennial herb with slender creeping stock and weak decumbent or ascending glabrous 4-angled stems 10–60 cm, very rough on the angles with downwardly directed prickles, *not blackening when dried.* Lvs 0.5–1(–1.5) cm, (4–)6–8 *in a whorl, linear-oblanceolate, mucronate*, 1-veined, glabrous, the margins rough with backwardly-directed prickles. Fls 2.5–3 mm diam., in small axillary corymbs forming a *narrow panicle.* Corolla white, the 4 lobes acute. Anthers yellow. Fr. 1 mm, glabrous, rugulose, ultimately dark brown; stalks deflexed in fr. Fl. 7–8. Hermaphrodite, protandrous; coumarin-scented and visited by small insects. $2n = 22^*$, 44. Hel.

Resembles small types of *G. palustre* but is readily distinguishable by the 6–8 mucronate lvs in each whorl, and by remaining green when dried.

Native. Fens; to 500 m in N. England. Locally frequent throughout Great Britain northwards to Ross and Banff; Inner Hebrides; Ireland northwards to Mayo and Meath, mostly near the coast. Europe from Portugal and N. Spain, N. Italy and Greece to c. 70° N. in Scandinavia.

13. G. tricornutum Dandy   Rough Corn Bedstraw.

G. tricorne Stokes, p.p.

An annual herb with decumbent or ascending-scrambling stems 10–40–(60) cm, glabrous, sharply 4-angled, the *angles very rough* with downwardly directed prickles. Lvs 2–3 cm, 6–8 in a whorl, linear-lanceolate, mucronate, 1-veined, glabrous, the margins with backwardly directed strong hooked prickles. Fls 1–1.5 mm diam., in 1–3-fld stalked axillary cymes. Corolla cream-coloured, the 4 lobes long, acute. Fr. 3–4 mm, wider than the corolla, pale-coloured, often of a single 1-seeded ±spherical mericarp by abortion of the second, *granulate* with large papillae, not bristly; their *stalks strongly recurved* (Fig. 51A). Fl. 6–9. Hermaphrodite,

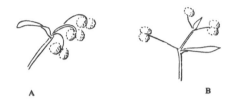

Fig. 51. Partial infructescences of *Galium tricornutum* (A) and *G. aparine* (B). ×1.

or lateral fls often male, ±homogamous. Little visited by insects. $2n = 44$. Th.

Very doubtfully native. In cultivated fields, especially with cereal crops, and on waste land, chiefly on calcareous soil. Formerly widespread in S. and E. England but now very much decreased and recently (1977) in only a very few localities from Somerset to Kent and northwards to W. Suffolk and Cambridge with an outlier east of Leeds, Yorks. Through much of Europe but casual in the north and east.

*G. verrucosum Hudson (*G. saccharatum* All., *G. valantia* Weber), an annual herb resembling *G. tricornutum*, is a rare casual. It is distinguishable by its smaller *lvs only* 0.5–1.5 cm and borne in whorls of 5–6(–7), *their margins rough with forwardly-directed prickles*; the *3-fld cymes with central fl. hermaphrodite and laterals male*, and the *large fr.*, 4–6 mm, *covered with long conical papillae.* A weed of arable fields and other open habitats in S. Europe, naturalized in C. Europe but casual further north.

14. G. aparine L.
          Goosegrass, Cleavers, Hairif, Sticky Willie.

An annual herb with prostrate or, more usually, scrambling-ascending diffusely branched stems 15–120 cm, glabrous or hairy above the nodes, the 4 *angles very rough with downwardly-directed prickles.* Lvs in whorls of 6–9, 12–60 × 3–8 mm, *narrowly to broadly oblanceolate or elliptical, mucronate*, 1-veined, glabrous or with hooked bristles above, *margins rough with prickles backwardly directed* except near lf-tip. Fls 1.5 mm diam. in 2–5-fld axillary cymes, their *peduncles topped by a whorl of 4–8 bracts.* Corolla whitish, with 4 acute lobes.

Fr. 4–6 mm, wider than the corolla, olive or purplish, covered *with white hooked bristles with tuberculate bases*, their stalks divaricate (Fig. 51B). Fl. 6–8. Hermaphrodite, protandrous. Sparingly visited by small insects. $2n = 42, 44, c. 66, c. 88$. Th.

Native. Hedges, waste places, drained fen peat, limestone scree, maritime shingle, etc. to 366 m in Yorks. Abundant throughout the British Is. Europe to 69° N in Scandinavia; N. and W. Asia. Widely introduced.

15. G. spurium L.   False Cleavers.

Incl. *G. vaillantii* DC.

An annual herb resembling *G. aparine* but *with somewhat narrower* linear-oblanceolate acuminate and long-awned *lvs* (their margins rough with backwardly directed prickles as in *aparine*) and *greenish fls* c. 1 mm diam. *in axillary 3–9-fld cymes with peduncles topped by only 2(–3) lf-like bracts*. Fr. 2–3 mm, blackish, rugulose, glabrous or (*G. vaillantii* DC.) *thickly covered with white hooked hairs having non-tuberculate bases*. Fl. 7–9. $2n = 20$. Th.

Doubtfully native, but recorded from a Late Bronze Age site in Sussex. Formerly established or casual as a weed of arable land in scattered localities mainly in the southern half of England, but recently much decreased and believed extinct until refound in Cambridge in 1974. In much of Europe northwards to arctic Scandinavia but naturalized or merely casual in parts of the north.

16. G. parisiense L.   Wall Bedstraw.

G. anglicum Hudson

An annual herb with decumbent or ascending, usually much-branched, weak, slender, glabrous stems 10–30 cm, 4-angled, the angles rough with small downwardly directed prickles, blackening when dried. *Lvs* 3–12 mm, 5–7 in a whorl, linear-oblong, shortly mucronate, 1-veined, at first spreading then *reflexed*, glabrous, the margin with *forwardly-directed prickles*. Fls c. 0.5 mm diam., in few-fld axillary stalked corymbs forming a long narrow panicle. Corolla greenish inside, reddish outside, its 4 lobes acute. Fr. 1 mm, wider than the corolla, glabrous, granulate, blackish (Fig. 52B). Fl. 6–7. $2n = 44, 66$. Th.

Native. On walls and in open vegetation on dry calcareous sand, the sides of chalk- and sand-pits and other disturbed areas on calcareous or circumneutral substrata. Rare and local in S.E. England from Kent and Sussex northwards to Norfolk and Cambridge; also in N. Hants and N. Devon. S., W. and C. Europe northwards to E. England and eastwards to Czechoslovakia and Bulgaria; N. Africa; W. Asia. Variants with fr. covered with hooked bristles occur in Continental Europe.

4. CRUCIATA Miller

Annual to perennial herbs, sometimes suffruticose, with *lvs in whorls of 4*, 1- or 3-veined. *Fls in opposite pairs of short axillary cymes shorter than or at most equalling the axillant lvs* and with strong basal branches so that there may appear to be whorls rather than pairs of cymes at each node, the whole forming a narrow panicle; *central fls hermaphrodite, laterals male or 0; peduncles and pedicels ±deflexed under the lvs*, but not coalescing or encircling the fr. Calyx 0; *corolla rotate, 4-lobed, yellow*; stigmas capitate. *Fr. dry*, with 1–2 glabrous or hairy mericarps.

A small genus in Europe and W. Asia, very close to *Galium*, and not separated by some authors, but differing in the combination of *lvs in whorls of 4*, a *yellow corolla*, and *peduncles and pedicels ±deflexed under the lvs* (in the related *Valantia* L. similarly deflexed peduncles and pedicels enlarge and coalesce during development and encircle the ripe fr.). There is a single British sp.

1. C. laevipes Opiz   Crosswort, Mugwort.

Valantia cruciata L.; *Galium cruciata* (L.) Scop.; *Cruciata chersonensis* auct.

A perennial herb with slender rhizome from which arise hypogeal stolons and slender, 4-angled, softly hairy aerial stems 15–60 cm, *decumbent but usually with several ascending branches* from near the base. *Lvs* 12–25 × 4–10 mm, largest in mid-stem but *mostly shorter than adjacent internodes*, *elliptical to ovate-elliptical* and ±blunt, *with 3 longitudinal veins* of which the midrib is more distinct than the laterals; all *softly hairy on both sides* but more so beneath, *yellowish-green*, especially in upper part of stem. Fls 2–3 mm diam., usually in 5–9-fld partial cymes; peduncles and pedicels usually hairy, short but elongating in fr., the pedicels becoming recurved; peduncles with 2 bracteoles. Corolla pale yellow with acuminate lobes; styles free to base. Fr. with 1–2 mericarps 2–2.5 mm diam., ±globose, glabrous and smooth, ultimately blackish. Fl. 4–7. Andromonoecious, the central hermaphrodite fls protandrous, laterals male; honey-scented and visited by bees and flies. $2n = 22$. Hp.

Native. Open woodlands and scrub, hedges, waysides and pastures especially on calcareous soils; to 470 m in Wales. Through most of Great Britain but decreasing northwards to Moray and Inverness; very local in Outer and Inner Hebrides; naturalized in N. Ireland (Co. Down). In most of Europe northwards to the Netherlands, N. Germany, N. Poland and C. Russia; Caucasus; W. Asia.

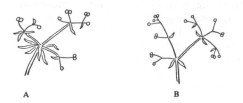

A   B

Fig. 52. Partial infructescences of *Galium spurium* (A) and *G. parisiense* (B). ×1.

5. RUBIA L.

Perennials, often woody below, with whorls of lvs and lf-like stipules. Fls yellowish, in terminal and axillary dichasia. *Calyx represented by an annular ridge; corolla rotate with a very short tube and usually 5 lobes*; stamens 5; styles 2, connate below; stigmas capitate. *Fr. succulent*, usually globose and 1-seeded and derived from only 1 cell of the ovary, rarely didymous with the parts not separating.

About 38 spp., chiefly in temperate regions of the Old and New Worlds.

1. R. peregrina L.   Wild Madder.

A perennial evergreen plant with a long creeping stock and trailing or scrambling-ascending glabrous stems 30–120 cm, woody and terete below, sharply 4-angled above, the angles rough with downwardly directed prickles. Lvs 1.5–6 cm, ovate- to elliptical-lanceolate, rigid, leathery, shining above, the cartilaginous margins and the underside of the midrib rough with curved prickles, 1-veined, 4–6 in a whorl. Fls 5 mm diam. in terminal and axillary cymes forming a lfy panicle. Corolla pale yellowish-green, its 5 lobes long-cuspidate. Anthers not more than twice as long as wide. Fr. 4–6 mm diam., subglobose, black. Fl. 6–8. $2n = 44$, ?66. H.–Ch.

Native. In hedgerows, scrub and copses and on cliffs, rocks, cuttings and walls near the sea-coast and very locally further inland, usually on chalk or limestone. Restricted to coastal counties from Kent to Gloucester and round Wales and in Scilly and Channel Is.; local near coasts of the southern half of Ireland. S. and W. Europe northwards to 53° 30′ N. in W. Ireland and eastwards to Greece and Turkey; N. Africa.

*****R. tinctorum** L., Madder, was formerly grown in England and still occurs as a casual. It has longer and narrower lanceolate lvs with a conspicuous network of lateral veins on the underside, brighter yellow fls with the corolla-lobes acute but not cuspidate, anthers linear-oblong and 5–6 times as long as wide, and a reddish-brown berry.

Native in the Mediterranean region and Near East. Madder is obtained from the roots.

121. CAPRIFOLIACEAE

Shrubs, rarely herbs. Lvs opposite. Stipules 0 or small and adnate to petiole, very rarely conspicuous. Infl. of a cymose type. Fls hermaphrodite, usually 5-merous (except ovary). Calyx often small. Corolla gamopetalous; lobes imbricate, sometimes 2-lipped. Stamens inserted on corolla-tube, equalling in number and alternate with corolla-lobes, rarely 1 stamen suppressed; anthers introrse (except *Sambucus*). Ovary inferior. Carpels 2–5 (sometimes two of them sterile), with 1–many axile ovules. Fr. a berry, drupe or achene, rarely a capsule. Seeds with fleshy endosperm and usually small straight embryo.

Thirteen genera and about 490 spp., cosmopolitan, but mostly north temperate regions and tropical mountains.

Very near Rubiaceae and chiefly distinguished by the stipules. There is, however, no close resemblance between the British members of the two families. The woody habit, inferior ovary and stamens equalling in number the corolla-lobes are not found in combination in any other British member of the Metachlamydeae.

Besides members of the British genera, species and hybrids of *Weigela* are commonly cultivated. They have conspicuous large usually pink or white fls and differ from all the other genera except *Diervilla* in the capsular fr. *Abelia* and *Dipelta* are less frequently cultivated.

1 Lvs pinnate.	1. SAMBUCUS	
Lvs simple (sometimes lobed).		2
2 Dwarf shrub with slender procumbent stems.	4. LINNAEA	
Erect shrub or woody climber.		3
3 Fls in compound cymes.	2. VIBURNUM	
Fls in racemes, heads, whorls or in pairs.		4
4 Fls in spike-like racemes or clusters; fr. a white drupe with 2 pyrenes.	3. SYMPHORICARPOS	

Fls in pairs, heads or whorls; fr. a red, black or blue, few-seeded berry.   4. LONICERA

1. SAMBUCUS L.

Deciduous shrubs or small trees, rarely herbs, with large pith. Buds with several pairs of scales. *Lvs pinnate*. Stipules present or not. Fls regular, in compound umbel-like or panicle-like cymes, usually 5-merous. Calyx-limb very small. Corolla rotate with short tube and flat spreading limb. Anthers extrorse. *Ovary 3–5-celled*, with 1 pendent ovule in each cell. Style short, with 3–5 stigmas or branches. Fr. a drupe. Seeds compressed.

About 40 spp., temperate and subtropical regions (not C. and S. Africa).

1 Herb; stipules conspicuous.	**1. ebulus**	
Shrub; stipules 0 or very small.		2
2 Cymes flat-topped; fr. black.	**2. nigra**	
Cymes panicle-like; fr. red.	**3. racemosa**	

1. S. ebulus L.   Danewort, Dwarf Elder.

Perennial, foetid, glabrous *herb* 60–120 cm, with creeping rhizome. Stems numerous, stout, erect, grooved, simple or little–branched. Lflets 5–15 cm, 7–13, oblong or oblong-lanceolate, acuminate, sharply serrate. *Stipules conspicuous*, ±ovate. *Infl. flat-topped*, 7–10 cm diam., *with 3 primary rays*. Corolla white, sometimes pink-tinged outside. *Anthers purple. Fr. globose, black*. Fl. 7–8. Pollinated by insects. $2n = 36$. Hp.

?Native. Local by roadsides and in waste places, scattered over the greater part of the British Is., but very sporadically north of Kirkcudbright and Renfrew. Most of Europe from the Netherlands and N. Ukraine southwards; W. Asia to Himalaya; Madeira (probably introduced).

2. S. nigra L.   Elder.

Shrub or more rarely a small tree to 10 m often with straight vigorous erect shoots from the base; branches often arching. Bark brownish-grey, deeply furrowed, corky. Twigs stout, greyish, with prominent lenticels. Lflets 3–9 cm, (3–)5–7(–9), ovate-lanceolate or ovate-elliptical, acuminate, rarely (var. *rotundifolia* Endl.) orbicular or (var. *laciniata* L.) deeply dissected, sparingly hairy on veins beneath, serrate. *Stipules* 0 *or very small and subulate. Infl. flat-topped,* 10–20 cm diam., *with 5 primary rays.* Corolla c. 5 mm diam., cream-white. *Anthers cream. Fr. 6–8 mm, globose, black,* rarely greenish. Fl. 6–7. Pollinated by small flies, etc. Fr. 8–9. $2n = 36$. M.

Native. Woods, scrub, roadsides and waste places, especially characteristic of disturbed, base-rich and nitrogen-rich soils; very resistant to rabbits, comparatively infrequent in closed communities; ascending to over 460 m. Throughout the British Is., common except in N. Scotland; introduced in Orkney and Shetland. Most of Europe except the extreme north; W. Asia; N. Africa; Azores.

S. canadensis L., a native of eastern North America from Nova Scotia to Florida and Texas, has been widely planted on railway property and may be confused with *S. nigra*. It may be distinguished by its smaller, sharper lf-teeth, with attenuate lf-apices; its fl-period, 7–9, is generally later than *S. nigra*.

*3. S. racemosa L.   Red-berried Elder.

Shrub to 4 m, glabrous. Lflets 4–8 cm, 5–7, ovate, ovate-lanceolate or elliptical, acuminate, serrate. *Stipules represented by large glands. Infl. a dense ovoid panicle,* 3–6 cm. Corolla cream-white. Anthers cream. *Fr.* c. 5 mm, globose, *scarlet.* Fl. 4–5. Pollinated by insects. Fr. 6–7. $2n = 36$. M.

Introduced. Commonly planted and sometimes naturalized, especially in Scotland. Native of W., C. and S. Europe and W. Asia.

2. VIBURNUM L.

Shrubs or small trees. *Lvs simple.* Stipules 0 or small. *Fls regular, in compound umbel-like, rarely panicle-like cymes,* 5-merous (except ovary). Calyx-teeth very small. Corolla rotate, funnel-shaped or campanulate. *Ovary* 1-*celled, with 1 pendent ovule. Stigmas 3, sessile. Fr. a drupe* the stone usually compressed.

About 200 spp., north temperate zone extending to Central America and Java. A number of spp. are cultivated.

Fls all alike; lvs serrulate; buds naked.   **1. lantana**
Outer fls sterile and much larger than inner; lvs lobed; buds scaly.   **2. opulus**

1. V. lantana L.   Wayfaring-tree.

Deciduous shrub 2–6 m. *Twigs and naked buds* greyish *scurfily stellate-pubescent. Lf-blades* 5–10 cm, oval, ovate or obovate, usually acute, cordate at base, *serrulate,* rugose above and sparingly stellate-pubescent, densely stellate-tomentose beneath; petiole 1–3 cm, exstipulate. Infl. 6–10 cm diam., umbel-like, dense, on a short peduncle. *Fls all alike and fertile.* Corolla c. 6 mm, cream-white. Fr. c. 8 mm, ovoid, compressed, at first red, finally changing quickly to black. Fl. 5–6. Pollinated by insects or self-pollinated. Fr. 7–9. $2n = 18$. M.

Native. Scrub, woods and hedges on calcareous soils; common in S. England, becoming rarer northwards to Nottingham and Derby, and in S. Wales and Denbigh; introduced in places northwards to Moray and occasionally in Ireland. C. and S. Europe extending to Belgium and N. Spain and eastwards to C. Ukraine; Caucasus; N. Asia Minor; Morocco and Algeria (very rare).

2. V. opulus L.   Guelder-rose.

Deciduous shrub 2–4 m. *Twigs* greyish, *glabrous,* slightly angled. Buds scaly. *Lf-blades* 5–8 cm, *with* 3(–5) *acuminate, irregularly dentate lobes,* glabrous above, sparingly pubescent or glabrescent beneath, usually reddening in autumn; petiole 1–2.5 cm, with subulate stipules and discoid glands. Infl. 5–10 cm diam., rather lax; peduncle 1–4 cm. *Inner fls fertile,* c. 6 mm diam.; *outer* 15–20 mm, *sterile,* white. Fr. c. 8 mm, subglobose, red. Fl. 6–7. Pollinated by insects or selfed. Fr. 9–10. $2n = 18$. M.

Native. Woods, scrub and hedges especially on damp soils; rather common in England, Wales and Ireland, less common in Scotland but extending to Caithness. Europe, except for parts of the north and most of the Mediterranean region; N. and W. Asia; Algeria (very rare).

A form with all fls sterile (var. *roseum* L., Snowball Tree) is commonly grown in gardens.

*V. tinus L. (Laurustinus), with evergreen entire lvs, fls pinkish outside, white within, and metallic blue fr., is frequently planted and occasionally self-sown. Native of S. Europe.

3. SYMPHORICARPOS Duh.

Deciduous shrubs. Buds scaly. Lvs entire or lobed, exstipulate. *Fls regular, in terminal spike-like racemes or clusters,* 4–5-merous. Calyx-teeth small. Corolla tubular or campanulate. *Ovary with 2 fertile cells,* each with 1 ovule *and 2 sterile cells* with numerous ovules. Style slender, with capitate stigma. *Fr. a drupe* with 2 pyrenes.

About 17 spp., in North America and 1 in China.

*1. S. albus (L.) S. F. Blake   Snowberry.

S. racemosus Michx; *S. rivularis* Suksdorf

Shrub 1–3 m, spreading underground, with numerous rather slender erect stems. Twigs slender, yellowish-brown, ascending, glabrous. Lvs of the twigs 2–4 cm, oval or ovate, obtuse, cuneate or rounded at base, entire or a few sinuately lobed, dull green and glabrous or sparsely pilose beneath; petiole c. 5 mm; lvs of the sucker shoots often conspicuously lobed. Fls 3–7, in terminal spike-like racemes, 2 cm or less. Bracts and bracteoles ovate, small. Corolla 5–6 mm, campanulate, pink, hairy at the throat within. Fr. 1–1.5 cm, globose, white. Fl. 6–9. Pollinated by bees, wasps and syrphids. Fr. 9–11. $2n = $ c. 54. N.

Introduced. Commonly planted and ±naturalized in many places but absent from N. Scotland; often spreading by suckers and forming large thickets, apparently rarely spreading by seed. Native of western North America from Alaska and Alberta to California and Colorado.

British and other European material belongs to var. *laevigatus* (Fernald) S. F. Blake, which differs from the typical variety in its glabrous twigs and leaves, and its larger fruit.

4. LINNAEA L.

Evergreen creeping dwarf shrub. Lvs small, exstipulate. *Fls in pairs on long peduncles, terminal on short lateral branches.* Calyx lobes 5, narrow-lanceolate, caducous. Corolla campanulate, 5-lobed. Stamens 4. *Ovary with 1 fertile cell* with 1 ovule *and 2 sterile cells.* Style filiform, with capitate stigma. *Fr. an achene.*

One sp.

1. L. borealis L.   Twinflower.

Prostrate plant with slender pubescent stems, often forming large mats. Lf-blades 5–15 mm, broadly ovate to orbicular, subobtuse, crenate-dentate in the upper half, sparingly hairy, tapered at the base into a petiole 2–3 mm. Peduncles 3–7 cm, pubescent and glandular, with 2 small lanceolate membranous bracts at apex. Pedicels 1–2 cm, pubescent and glandular, with 2 small lanceolate bracteoles near the fl. Corolla pink, often beautifully marked, c. 8 mm, hairy within. Fr. c. 3 mm, densely glandular-hairy, but now very rare. Fl. 6–8. Pollinated by insects. $2n = 32$. Chw.

Native. Woods, especially pine, and in the shade of rocks; ascending to 730 m. From Cumberland to Sutherland, almost confined to the east, very local and rare, especially in the southern part of its range. N. Europe and Asia (from 71° 10′ N. in Norway) extending locally southwards to the Alps, and E. Carpathians, Caucasus; subspp. in North America.

5. LONICERA L.

Deciduous, rarely evergreen, shrubs or woody climbers. Buds scaly. Lvs usually entire, exstipulate. Fls sessile, either in axillary pairs on long peduncles or in terminal heads or whorls. Calyx with 5 small teeth. *Corolla* either (as in all the British spp.) *strongly 2-lipped* with 4-lobed upper and entire lower lip or with a nearly regular 5-lobed limb; tube long or short. Stamens 5. *Ovary 2-celled* with numerous axile ovules. Style slender; stigma capitate. *Fr. a few-seeded berry.*

About 200 spp., north temperate zone, extending to Mexico and Java. Many spp. are grown in gardens, including *L. nitida* Wilson from W. China, with evergreen lvs, c. 1 cm, now much used for hedges.

1 Upright shrub; fls in pairs.   **1. xylosteum**
　Woody climbers; fls in heads or whorls.   *2*
2 Lvs all free; bracts small.   **2. periclymenum**
　Upper lvs and bracts connate in pairs; bracts large.
          3. caprifolium

Section 1. *Lonicera.* Upright shrubs. Fls in axillary pairs. Corolla-tube short.

1. L. xylosteum L.   Fly Honeysuckle.

Deciduous *bushy shrub* 1–2 m. Twigs grey, somewhat pubescent. Lvs 3–6 cm, ovate to obovate, acute, broad-cuneate or rounded at base, greyish-green, ±pubescent, especially beneath; petiole 3–8 mm. *Fls in pairs*, sessile in the axils of subulate bracts about as long as ovaries, on pubescent peduncles 1–2 cm. Bracteoles ovate, about $\frac{1}{2}$ as long as ovaries. Ovaries glandular. *Corolla* c. 1 cm, yellowish, often tinged reddish, pubescent outside. Fr. red, globose. Fl. 5–6. Pollinated by humble-bees. Fr. 8–9. $2n = 18$. N.

Native. Woods and hedges in one or two places in W. Sussex; elsewhere probably introduced, occurring in a number of places in England and Wales and a very few in E. Scotland (to Ross) and Ireland. Europe from Scandinavia (c. 64° N.) to Spain, Sicily, Macedonia and the Caucasus (rare in W. France); N. and W. Asia.

Section 2. *Nintooa* (Spach) Maxim. Woody climbers. Fls in pairs. Corolla-tube long.

*L. japonica Thunb.

A half-evergreen twining shrub with white fls 3–4 cm, in pedunculate axillary pairs and black fr. Perhaps naturalized in Devon. Native of E. Asia.

Section 3. *Caprifolium* (Miller) DC. Fls in heads or whorls. Corolla-tube long.

2. L. periclymenum L.   Honeysuckle.

Twining shrub reaching 6 m but often low and trailing or scrambling, glabrous or somewhat pubescent. *Lvs* 3–7 cm, ovate, elliptical or oblong, dark green above, glaucous beneath, usually acute, lower shortly petiolate, upper subsessile, smaller, *all free. Fls in terminal heads. Bracts small, not exceeding ovary. Corolla* 4–5 cm, with long tube and spreading limb, cream-white within, turning darker after pollination, purplish or yellowish and glandular outside. Fr. red, globose. Fl. 6–9. Pollinated by hawk-moths, also humble-bees, etc. Fr. 8-9. $2n = 18*, 36*$. M. or N.

Native. Woods, hedges, scrub and shady rocks; ascending to 610 m, common throughout the British Is. W., C. and S. Europe extending northwards to S. Sweden; N. and C. Morocco.

*3. L. caprifolium L.   Perfoliate Honeysuckle.

Twining glaucous shrub. Lvs 4–10 cm, ovate or oblong, dark green above, glaucous beneath, obtuse, lower shortly petiolate, *2 or 3 pairs below infl. connate by the greater part of their bases. Fls in terminal heads*, often *with axillary whorls in addition. Bracts like the upper lvs but smaller.* Corolla like the last sp., but not glandular. Fr. red, globose. Fl. 5–6. Pollinated by hawk-moths. Fr. 8–9. $2n = 18$. M.

Introduced. Hedges, etc., in a number of places from Dorset and Sussex to Hereford and Suffolk then scattered from Yorks to Inverness, apparently decreasing. E.C. and S. Europe westwards to Italy; Caucasus, Asia Minor.

LEYCESTERIA Wall.

Deciduous shrubs. Fls in whorls in the axils of lf-like bracts forming drooping terminal spikes. Fls regular, 5-merous. Corolla funnel-shaped. Ovary 5-celled. Six spp. in Himalaya and China.

*L. formosa Wall. (Himalayan Honeysuckle). Lvs ±ovate, acuminate, 5–18 cm. Bracts purplish. Corolla 1.5–2 cm, purplish. Fr. deep brownish-purple. Commonly cultivated and sometimes escaping. Native of the Himalaya and S.W. China.

122. ADOXACEAE

Perennial rhizomatous herb. Lvs ternate, exstipulate. Fls c. 5 in terminal heads. Terminal fl. with 2-lobed calyx (or bracts) and 4-lobed corolla (or perianth); stamens 4, epipetalous, alternating with the corolla-lobes but divided to the base and appearing as 8 but each bearing $\frac{1}{2}$-anther only; lateral fls with 3-lobed calyx (or bracts), 5-lobed corolla (or perianth) and 5 (apparently 10) stamens. Ovary 3–5-celled, $\frac{1}{2}$-inferior, tapered above into the styles; ovule 1 in each cell, anatropous, pendent from the inner angle. Fr. a drupe; endosperm copious; embryo small.

3 genera of obscure relationships.

1. ADOXA L.

One sp. Nectar secreted by a ring round the base of the stamens.

1. A. moschatellina L.   Moschatel, Townhall Clock.

Glabrous perennial herb 5–10 cm. Rhizome far-creeping with fleshy whitish scales at the apex. Basal lvs ternate,

on long petioles, light green, dull above, rather glossy beneath; lflets 1–3 cm, long-stalked, ternate or trisect, the divisions often 2–3-lobed, the lobes oval or oblong, obtuse, mucronate, Fl.-stems erect, unbranched; cauline lvs 2, opposite, 8–15 mm, shortly stalked. ternate or trisect, the terminal lflet often 3-lobed. Infl. c. 6 mm diam.; calyx and corolla light green, like the lvs. Anthers yellow. Fr. green, rarely produced. Fl. 4–5. Visited by various small insects; homogamous or slightly protogynous, self-pollination possible. $2n = 36$, 54, 72. Grh.

Native. Woods, hedge-banks and mountain rocks; ascending to 1100 m. From Sutherland southwards, widespread but rather local; absent from the Isle of Man and Channel Is.; Antrim only in Ireland. Europe from Scandinavia to the mountains of C. Spain, Italy, Montenegro and Bulgaria; Morocco (mountains of N.W.); N. and C. Asia to the Caucasus, Himalaya and Kamchatka; North America.

123. VALERIANACEAE

Herbs, sometimes woody at the base, often with strong-smelling rhizomes. Lvs opposite or basal, entire or pinnatifid, exstipulate. Infl. cymose, often capitate. Fls generally small, hermaphrodite or unisexual, often somewhat zygomorphic. Calyx annular or variously toothed, often inrolled in fl. and forming a feathery pappus in fr. Corolla funnel-shaped, the base equal, saccate, or spurred on one side; lobes 5 (3–4), imbricate, obtuse. Stamens 1–4 (1–3 in our spp.), inserted towards the base of the corolla-tube. Ovary 3-celled, one cell fertile, the two sterile cells usually small or almost 0; ovule solitary, pendent. Fr. dry, indehiscent.

About 13 genera and 400 spp., generally distributed except for Australia.

1. VALERIANELLA Miller
Lamb's Lettuce, Corn Salad.

Small *annual* herbs with apparently dichotomous branching. Fls solitary in the forks of the branches and in terminal bracteate cymose heads. *Calyx neither inrolled in fl. nor forming a pappus in fr., but a toothed or funnel-shaped rim, or sometimes almost 0. Corolla funnel-shaped, regular, not spurred or saccate at base; lobes 5. Stamens 3. Fr. of one 1-seeded cell and 2 distant but sometimes small sterile empty cells.* Self-pollination is probably the rule.

About 80 spp., in the temperate parts of the northern hemisphere.

Ripe fr. is essential for the determination of the spp.

1 Fr. compressed or nearly quadrangular in section; calyx in fr. very small and inconspicuous.   *2*
  Fr. ovoid or oblong, flat on one face, convex on the other; calyx in fr. distinct.   *3*
2 Fr compressed, suborbicular in side view, c. 2.5 × 2 mm; fertile cell corky on back.   **1. locusta**
  Fr. oblong, nearly quadrangular in section, c. 2 × 0.75 mm; fertile cell not corky on back.   **2. carinata**
3 Calyx in fr. minutely toothed; sterile cells of fr. together larger than the fertile.   **3. rimosa**

1 Annual; corolla not spurred or saccate at base; calyx not forming a feathery pappus in fr.
    1. VALERIANELLA
  Perennial; corolla spurred or saccate at base; calyx forming a feathery pappus in fr.   *2*
2 Corolla saccate at base; stamens 3.   2. VALERIANA
  Corolla spurred; stamen 1.   3. CENTRANTHUS

Calyx in fr. strongly toothed, sterile cells of fr. together much smaller than the fertile.

4 Calyx in fr. as broad as the oblong fr., strongly veined, with 5–6 subequal deep teeth. **4. eriocarpa**

Calyx in fr. ½ as broad as the ovoid fr., weakly veined, with one tooth much larger than the others. **5. dentata**

1. V. locusta (L.) Laterrade

Lamb's Lettuce, Cornsalad.

V. olitoria (L.) Pollich

A slender erect nearly glabrous annual 7–40 cm. Stems rather brittle, much-branched, weakly angled, slightly pubescent below. Lvs 2–7 cm, entire or sometimes dentate, lower spathulate, obtuse, upper oblong, obtuse or subacute. Cymes capitate; bracts shortly ciliate. Fls small, pale lilac. Calyx indistinct, 1-toothed. *Fr.* (Fig. 53A) c. 2.5 × 2 mm, *compressed, suborbicular in side view; fertile cell corky on the back*; the sterile cells confluent. Fl. 4–6. $2n = 14$. Th.

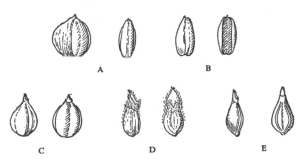

Fig. 53. Fruits of *Valerianella*. A, *V. locusta*; B, *V. carinata*; C, *V. rimosa*; D, *V. eriocarpa*; E, *V. dentata*. ×5.

Native. On arable land, hedgebanks, rocky outcrops and dunes, usually on dry soils; much the commonest British sp. Throughout nearly the whole of the British Is., though rather local; rare in most of Scotland. Most of Europe, but rare in the north; Madeira; N. Africa; W. Asia; introduced in North America.

2. V. carinata Loisel.   Keel-fruited Cornsalad.

An annual very similar in general appearance to *V. locusta*. Calyx indistinctly 1-toothed. *Fr.* (Fig. 53B) 2 × 0.75 mm, *oblong, nearly quadrangular in section, fertile cell not corky on the back*, sterile cells nearly confluent. Fl. 4–6. $2n = 18$. Th.

Native. On arable land, banks, old walls and rocky outcrops. Local in England and mainly in the south; Brecon, Pembroke and Caernarvon; local in E. Ireland. C. and S. Europe, extending north-eastwards to N. Ukraine; N. Africa.

3. V. rimosa Bast.   Broad-fruited Cornsalad.

V. auricula DC.

An annual similar in general appearance to *V. locusta*. Stems rather rough on the angles. Upper lvs and bracts

usually linear. *Cymes rather lax. Calyx in fr. distinct, minutely toothed,* c. ⅓ *as broad as the fr. Fr.* (Fig. 53C) c. 2 mm, *ovoid, the sterile cells together larger than the fertile one,* confluent. Fl. 7–8. $2n = 14$. Th.

Native. In cornfields, local C. and S. England, S. Wales and C. and S. Ireland, mainly in the south and east; not recorded recently from Scotland. C. and S. Europe from Denmark southward; N. Africa.

*4. V. eriocarpa Desv.   Hairy-fruited Cornsalad.

An annual similar in general appearance to *V. locusta*. Stems rather rough. *Cymes very dense. Calyx in fr. distinct, as broad as the fr., oblique, strongly net-veined, deeply 5–6-toothed, teeth subequal. Fr.* (Fig. 53D) c. 1 mm, *oblong, ±hispid, fertile cell many times larger than the distant sterile ones.* Fl. 6–7. $2n = 14$. Th.

Introduced. Naturalized on banks and old walls in a few places, mainly in southern England; Scotland: Moray. S. Europe; N. Africa.

5. V. dentata (L.) Pollich   Narrow-fruited Cornsalad.

Differs from *V. eriocarpa* in its laxer cymes, the calyx ½ as broad as the ovoid glabrous fr., scarcely veined, with one tooth much longer than the others (Fig. 53E). Fl. 6–7. $2n = 14$. Th.

Native. Locally common in cornfields. Fairly generally distributed throughout the British Is., but infrequent in the west and north. Europe, northwards to Scotland and S.E. Sweden; Macaronesia; N. Africa; W. Asia.

2. VALERIANA L.

Perennial herbs, mostly with a bitter taste and peculiar smell, particularly evident when dry and very attractive to cats. Lvs entire, pinnate or pinnatifid. Infl. cymose, terminal and usually subcapitate in fl. Fls hermaphrodite or unisexual, bracteolate. Calyx inrolled in fl., enlarging and forming a pappus in fr. *Corolla funnel-shaped; tube slightly saccate at base;* lobes 5, rarely 3 or 4, unequal. *Stamens* 3. *Ovary apparently of 1 carpel.* Fr. a unilocular, 1-seeded nut.

More than 200 spp., in Europe, Asia, Africa and America.

1 Fls unisexual; stolons long.   **3. dioica**
  Fls hermaphrodite; stolons 0 or very short.   *2*
2 All lvs pinnate.   **1. officinalis**
  Lower lvs simple, the upper simple or with 1–2 pairs of small lflets.   **2. pyrenaica**

1. V. officinalis L.   Common Valerian.

An erect nearly glabrous perennial 20–150 cm, *rarely producing short stolons.* Stems hairy below. *Lvs* up to c. 20 cm, *all imparipinnate,* the lower long-petiolate, the upper nearly sessile; *lflets lanceolate, entire or distantly and irregularly toothed.* Fls 4–5 mm, hermaphrodite. Corolla c. 5 mm diam., pale pink. Fr. 4 mm, ovate-oblong. Fl. 6–8. $2n = 28^*, 56^*$. Hs.

A variable sp. the extreme forms of which have been named *V. mikanii* Syme and *V. sambucifolia* Mikan fil. The tetraploid and octoploid forms differ somewhat in, but appear not to be certainly distinguishable by, morphological features.

Native. In rough grassy and bushy places, usually on damp soils but also in dry habitats. The tetraploid apparently confined to the south and parts of the Midlands, on chalk, oolite and limestone; the octoploid throughout the British Is., in damp valleys in the south and both damp and dry habitats in the north. Most of Europe; temperate Asia to Japan.

*2. V. pyrenaica L.   Pyrenean Valerian.

An erect dark green ±pubescent perennial c. 100 cm. Stolons 0. Stem pubescent at nodes. *Lvs 8–20 cm broad, simple or the upper with 1–2 pairs of small lateral lflets and a large terminal one, ovate or suborbicular, cordate, deeply and irregularly dentate*, the lower obtuse, the upper acuminate. Fls 5–6 mm, hermaphrodite. Corolla c. 5 mm diam., pale pink. Fr. 5–6 mm, linear-oblong. Fl. 6–7. Hs.

Introduced. Naturalized in woods in a number of localities in the western parts of the British Is., mainly in Scotland. Cordillera Cantábrica, Pyrenees, one station in S.E. Spain.

3. V. dioica L.   Marsh Valerian.

An erect nearly glabrous bright green dioecious perennial 15–30(–50) cm. *Stolons present, long.* Stem slightly pubescent at nodes. Basal lvs long-petiolate, *blade 2–3 cm, elliptical to ovate, obtuse and quite entire*; stem-lvs sessile or subsessile, *pinnatifid*. Male fls c. 5 mm diam., the female c. 2 mm diam., pinkish. Fr. c. 3 mm, elliptical. Fl. 5–6. $2n = 16$. Hs.

Native. In marshy meadows, fens and bogs. Scattered throughout most of England, Wales and S. Scotland, local. From S. Norway (1 locality) and S. Sweden to N. Spain, eastwards to Macedonia and the western borders of the USSR.

3. Centranthus DC.

Perennial or annual glabrous herbs. Fls in terminal panicled cymes, red, pink or white, bracteolate. Calyx inrolled in fl., enlarging and forming a pappus in fr. *Corolla-tube with a spur or small conical projection at base*, and with a longitudinal septum, one compartment containing the style, the other leading to the spur; lobes 5. *Stamen 1. Ovary apparently of 1 carpel.* Fr. a 1-seeded nut.

About 12 spp., in Europe and the Mediterranean region.

*1. C. ruber (L.) DC.   Red Valerian.

An erect somewhat glaucous perennial 30–80 cm. Lvs c. 10 cm, ovate or ovate-lanceolate, entire or the upper sinuate-dentate, the lower narrowed at base into a petiole, the upper sessile. Fls c. 5 mm diam., red or less commonly white, scented; corolla-tube 7–10 mm, slender; spur 4–7 mm. Stamen exserted. Fl. 6–8. The fls are protandrous and are cross-pollinated by long-tongued insects, mainly Lepidoptera. $2n = 32$. Hp.

Introduced. Frequently cultivated and well naturalized on old walls, cliffs, dry banks and in waste places, locally abundant, particularly in S. and W. England, Wales and S. and E. Ireland. Mediterranean region; W. Asia; introduced in many places.

124. DIPSACACEAE

Herbs, rarely shrubs, with opposite or whorled exstipulate lvs. Infl. usually of heads of fls (*capitula*) with a calyx-like involucre of bracts, but in *Morina* an interrupted spike of fls in axillary 'false whorls' (*verticillasters*) as in many Labiatae. Fls hermaphrodite, zygomorphic, epigynous, each surrounded at its base by an epicalyx or 'involucel' of united bracteoles, and often subtended by a receptacular bract. Calyx small, cup-shaped or ±deeply cut into 4–5 segments or into numerous teeth or hairs; corolla sympetalous with the tube often curved and the 4–5 lobes ±equal or forming 2 lips; stamens 4 or 2, epipetalous, alternating with the corolla-lobes, exserted, anthers free; ovary inferior, 1-celled; ovule solitary, anatropous, pendulous; style slender with a simple or 2-lobed stigma. Insect-pollinated. Fr. dry, indehiscent, 1-seeded, enclosed in the involucel and often surmounted by the persistent calyx; endosperm present; embryo straight.

About 155 spp. in 9 genera, chiefly in the Mediterranean region and Near East.

Readily distinguished from Compositae by the exserted stamens with free anthers, and by the fr. enclosed in an involucel.

1 Stem spiny or prickly; involucral and receptacular bracts spine-tipped.   1. **Dipsacus**
  Stem not spiny or prickly; involucral and receptacular bracts not spine-tipped.   2
2 Receptacle hemispherical, hairy; receptacular bracts 0; calyx of 8(–16) teeth or bristles.   2. **Knautia**
  Receptacle elongated, not hairy; receptacular bracts present; calyx of 4–5 teeth or bristles.   3
3 Stem-lvs pinnatifid or pinnate; marginal fls larger than the central; corolla 5-lobed; involucel ending above in a pleated scarious funnel-shaped cup.   3. **Scabiosa**
  Stem-lvs entire or faintly toothed; fls equal; corolla 4-lobed; involucel ending above in 4 herbaceous teeth.   4. **Succisa**

1. Dipsacus L.

Usually *large biennial herbs with prickly stems*, lvs in opposite pairs and fls in subglobose, ovoid or cylindrical heads. Basal lvs often petiolate, though sometimes very shortly and indistinctly so; *stem-lvs shortly petiolate or sessile*, members of a pair *often connate into a basal cup*. Fl.-heads with 1–2 rows of linear to lanceolate, erect or ±spreading, spine-tipped involucral bracts often considerably longer than the also spine-tipped receptacular bracts. *Involucel 4-angled in section, 8-ribbed,* with strong ribs at the corners and a weaker rib up the centre of each face united with the ovary below and *ending above in a very short 4- or 8-lobed or many-toothed cup,* initially pubescent but sometimes becoming glabrous. *Calyx* with short and slender basal tube surmounted by *a wide 4-angled pubescent cup* with ciliate and ±shallowly lobed margin, commonly falling before the fr. matures. Corolla with long tube and 4 unequal lobes. Stamens 4. Stigma oblique, entire. *Fr. 4-angled,* appressed-hairy, *crowned by the calyx at least in early stages of maturation.*

About 15 spp. in Europe, W. Asia and N. Africa.

1 Stem-lvs sessile; fl.-heads 3–9 cm, ovoid to ovoid-cylindrical; involucral bracts much longer than receptacular bracts.   2

  Stem-lvs shortly petiolate; fl.-heads 1.5–4.5 cm, sub-globose; involucral and receptacular bracts about equally long.   3

2 Involucral bracts curving upwards, the longest equalling or overtopping the head; receptacular bracts exceeding fls and ending in a long straight but flexible spine.   **1. fullonum**

  Involucral bracts spreading ±horizontally with none equalling the head; receptacular bracts almost equalling the fls and ending in a stiff recurved spine.
    2. sativus

3 Fl.-heads 1.5–2.5 cm diam.; involucral bracts 10–12 mm, abruptly narrowed into a spiny tip; corolla whitish.   **3. pilosus**

  Fl.-heads 3–4.5 cm diam.; involucral bracts 15–25 mm, gradually tapered into a spiny tip; corolla pale yellow.   **4. strigosus**

1. D. fullonum L.   Wild Teasel.

D. sylvestris Hudson; *D. fullonum* subsp. *fullonum*

A robust biennial herb with stout yellowish tap-root and erect angled and furrowed stem 50–150(–200) cm, glabrous, prickly on the angles, branching above. *Basal lvs* in a rosette *dying early in the 2nd season,* oblong to elliptical-oblanceolate, obtuse or acute, *entire to somewhat crenate,* glabrous but with scattered swollen-based prickles, *very shortly and indistinctly petiolate* and lying flat on the ground; *stem-lvs* narrowly lanceolate, *sessile and with each pair connate at the base into a water-collecting cup,* prickly only on the midrib beneath, *distinctly crenate-serrate* or (especially the uppermost) entire, often ciliate. *Fl.-heads 5–8 cm, erect throughout,* ovoid-cylindrical. *Involucral bracts* linear, prickly, *strongly ascending,* of unequal length with *the longest equalling or exceeding the head. Receptacular bracts exceeding fls,* oblong-obovate, ciliate, *ending in a long straight but flexible spine. Corolla purplish rose,* its *tube* 9–11 mm. Fr. 3–5 mm. Fl. 7–8. Visited by bees and long-tongued flies. $2n = 18$. Hs (biennial).

Native. Damp wood-margins, copses, thickets and hedgerows, stream-banks, field-borders, roadsides, banks by the sea, etc., especially where heavy soil has been disturbed. Locally frequent, south of a line from the Humber to the Severn Estuary and in more scattered localities westwards to Cornwall and Wales and northwards to Cumbria, the Forth Estuary and Fife; local in Ireland; Channel Is. S., W. and C. Europe eastwards to C. Russia and Turkey; W. Asia; N. Africa.

This is the wild plant whose receptacular bracts are too flexible for use in combing woollen cloth.

*2. D. sativus (L.) Honckeny   Fuller's Teasel.

D. fullonum subsp. *sativus* (L.) Thell.

A biennial herb closely resembling **1.** but with *stem-lvs usually distinctly and irregularly crenate-toothed* though sometimes ±entire and with fewer prickles on the mid-rib beneath. Fl.-heads ovoid. *Involucral bracts spreading ±horizontally so that none equals or overtops the head. Receptacular bracts about equalling the fls,* spinose-ciliate, *ending in a stiff recurved spine. Corolla-tube* 13 mm. Fl. 7–8. $2n = 18$.

Introduced. Formerly cultivated extensively, the heads, with receptacular bracts tipped by a rigidly recurved spine, being used for raising the nap of woollen cloth after fulling. Still seen occasionally as a casual on rubbish tips, railway sidings and other waste places. Of uncertain origin but naturalized in parts of S., W. and C. Europe.

3. D. pilosus L.   Small Teasel.

Cephalaria pilosa (L.) Gren.

A biennial herb with an erect angled and furrowed stem 30–120 cm, with sparse weak prickles on the angles. *Basal lvs* in a rosette, *ovate,* acute or shortly acuminate, narrowed into a *long stalk,* crenate-toothed, hairy, sometimes prickly on the midrib beneath; *stem-lvs short-stalked,* ovate to narrowly elliptical, simple or the upper lvs usually with a basal pair of small, often unequal, ±free, elliptical lflets, entire to crenate-toothed, sparsely hairy. *Heads* 1.5–2.5 cm diam., *spherical, at first drooping,* later erect, on long weakly prickly stalks. Involucral bracts c. 1 cm, narrowly triangular, spine-tipped, sparsely covered with long silky hairs, spreading or slightly reflexed, falling short of the fls. Receptacular bracts equalling the fls, obovate, abruptly contracted into a straight spiny point, ciliate with long silky hairs. Corolla 6–9 mm, whitish. Anthers dark violet. Fr. 4–5 mm, brown. Fl. 8. Sparingly visited by small bees and flies. $2n = 18$. Hs. (biennial).

Native. Damp woods, hedgebanks, ditch-sides, etc., especially on chalk or limestone, to 260 m in Derby; local and infrequent. England and E. Wales northwards to Lancs and Durham. C. Europe from Spain, N. Italy, N. Balkans and S. Russia northwards to the Netherlands, Denmark, Sweden and Hungary, W. Asia; Japan.

*4. D. strigosus Willd.

Closely resembling **3** but usually larger in all its parts, the stem 50–200 cm, angled and deeply furrowed, the angles with distinctly upward-pointing prickles; *heads* 2.5–4.5 cm *diam.* with *involucral bracts* 15–25 mm, *tapering gradually into a spiny point* (abruptly contracted in **3**); *receptacular bracts* 15–20 mm, broadly obovate, *tapering ±gradually into a long spine-like* *appendage*; *corolla pale yellow*; fr. 4–4.5 mm, grey-brown with black streaks.

Introduced. Collected and recorded as *D. pilosus* at intervals over the past 150 years and almost certainly an escape from the Cambridge Botanic Garden and later from the Botanic Gardens at Oxford and Kew, but also found near Bristol; still near Cambridge in 1975. Native in S. Russia and the Ukraine but introduced and naturalized elsewhere in Europe.

The genus *Cephalaria* Schrader, to which *D. pilosus* was often formerly assigned, is now defined as differing from *Dipsacus* in having non-prickly stems and involucral bracts in more than 3 rows. Some spp. are grown in gardens and occasionally escape, in particular *C. gigantea* (Ledeb.) Bobrov (*C. tatarica* auct., non Roemer & Schultes), native in the Caucasus and up to 2 m high with scabious-like yellow-fld heads.

2. KNAUTIA L.

Annual to perennial herbs with simple to pinnate lvs in opposite pairs. *Fl.-heads long-stalked*, hemispherical to almost flat; *involucral bracts numerous, non-spinous*; receptacle hairy; *receptacular bracts* 0. Involucel somewhat compressed, bluntly 4-angled, not furrowed, surmounted by a very short entire or obscurely toothed cup. *Calyx* a short stipitate cup *prolonged upwards as* (6–)8–16(–24) *suberect teeth or bristles*, deciduous in fr. *Corolla ±unequally 4-lobed*, often larger in the marginal than the central fls. Stamens 4, free. Stigma emarginate or shortly lobed. Fr. ovoid to cylindrical, ±hairy, with basal elaiosome favouring dispersal by ants.

Fifty or more spp. in Europe, W. Asia and N. Africa.

1. K. arvensis (L.) Coulter   Field Scabious.

Scabiosa arvensis L.

A perennial herb with loosely caespitose and usually stoloniferous stock giving rise to lf-rosettes and erect fl.-stems 25–100 cm, ±terete, simple or branched, rough at least below with stiff downwardly directed hairs. *Basal lvs in an overwintering rosette* from the stock or the base of an old fl.-stem, oblanceolate, shortly petiolate, *commonly simple* and entire or crenate-toothed but *sometimes lyrate-pinnatifid*; *stem-lvs usually deeply pinnatifid* with elliptical terminal segment and linear-oblong lateral segments; some of the uppermost usually ±entire; all dull green, ±hairy. *Gynodioecious*. Heads of hermaphrodite plants 3–4 cm diam., of female 2–3 cm diam.; peduncles hairy. *Involucral bracts* in 2 rows, *shorter than the fls*, ovate-lanceolate, hairy. *Corolla of marginal fls larger and more unequally 4-lobed than of the central fls*; all pale bluish-lilac, rarely purple or pink. *Calyx with c. 8 suberect setaceous teeth*, 2–3 mm, ciliate below. Fr. 5–6 mm (without calyx), densely hairy. Fl. 7–9. Protandrous and visited by bees and butterflies. $2n = 40, 20$. Hs.

Very variable, especially in the form of the lvs.

Native. Dryish rough pasture, especially over limestone or chalk but also on sand; dry hedgebanks, field-borders, roadsides and waste places, and locally on limestone rock-ledges; to 366 m in Derby. Throughout the British Is. except the Hebrides and Shetland, but infrequent in W. and N. Scotland. Europe northwards to 68° 50′ N. in Norway; W. Siberia.

3. SCABIOSA L.

Annual to perennial herbs, rarely suffruticose, with simple or pinnate lvs in opposite pairs. Heads long-stalked, convex, with an involucre of numerous herbaceous bracts in 1–3 rows. *Receptacle elongated*, with pale linear-lanceolate *non-spinous receptacular bracts*. *Involucel* with 8 vertical furrows or 8 pits above, *ending upwards in a ±scarious pleated funnel-shaped cup*. *Calyx* cup-shaped below, usually *prolonged upwards as 5 ±spreading setaceous teeth*. Corolla with a short tube and (4–)5 *unequal lobes*, the marginal fls usually larger than the central. Stigma ±2-lobed. Fr. crowned by the persistent calyx, wind-dispersed.

About 100 spp. in Europe, N. Africa and W. Asia, chiefly in the Mediterranean region.

1. S. columbaria L.   Small Scabious.

A perennial herb with a long tap-root prolonged upwards into an erect branching stock. Flowering stems 15–70 cm, lfy, slender, terete, sparsely hairy, branching at or below the middle. Basal lvs ±long-stalked, obovate to oblanceolate, simple and crenate or ±lyrate-pinnatifid; stem-lvs successively shorter-stalked and more deeply pinnatifid with narrower segments, the uppermost with linear segments; all subglabrous to pubescent. Heads 1.5–3.5 cm diam., their long slender stalks usually pubescent. Involucral bracts c. 10, in 1 row, linear-lanceolate, shorter than the fls. *Corolla* bluish-lilac, rarely pinkish or white; those *of outer fls much larger than of central fls*. Fruiting head ovoid. Fr. 3 mm, its involucel deeply 8-furrowed, hairy, with the scarious cup 1.5 mm; calyx-teeth 3–6 mm, setaceous, blackish. Fl. 7–8. Homogamous to protandrous and visited by bees and Lepidoptera. $2n = 16$. Hs.

Native. Dry calcareous pastures, banks, etc., to 610 m in N. England; locally common. Great Britain northwards to Berwick and Angus. Europe to the Arctic Circle; W. Asia and Siberia; N. Africa.

*S. atropurpurea L. (S. Europe), an annual herb with branching stems 30–60 cm, lyrate-pinnatifid coarsely toothed basal lvs and pinnatifid stem-lvs, and long-stalked heads of dark purple, rose or white fls, is much grown in gardens and is naturalized near Folkestone and Newquay (Cornwall).

S. caucasica Bieb. (Caucasus) is the large blue scabious much grown in gardens and for cutting.

4. SUCCISA Haller

A perennial herb with lvs in opposite pairs. Heads long-stalked, hemispherical, with an involucre of numerous herbaceous bracts in 2–3 rows. *Receptacle elongated, not hairy*, with *non-spinous receptacular bracts*. *Involucel* 4-angled with 2 furrows in each face, *ending above in 4 erect triangular herbaceous lobes*. *Calyx* a short cup *prolonged upwards as 5(–4) setaceous teeth*. *Corolla almost equally 4-lobed*, those of the *marginal fls not appreciably larger than* those of the *central*. Calyx persistent in fr.

One sp.

1. S. pratensis Moench   Devil's-bit Scabious.
Scabiosa succisa L.

Rootstock short, erect, *premorse*, with long stout roots. Stem 15–100 cm, erect or ascending, ±terete, subglabrous or appressed-hairy above. Basal lvs 5–30 cm, in a rosette, obovate-lanceolate to narrowly elliptical, blunt or acute, narrowed into a short stalk, firm, reticulate beneath, usually entire, sparsely hairy, rarely glabrous; stem-lvs few, resembling the basal but narrower, entire or distantly toothed, sometimes all ±bract-like. Gynodioecious. Heads 1.5–2.5 cm diam., the female usually smaller than the hermaphrodite. Involucral bracts broadly lanceolate, pubescent, ciliate. Receptacular bracts elliptical, herbaceous, purple-tipped, much

exceeding the calyx-teeth. Corolla 4–7 mm, usually smaller in female heads, mauve to dark blue-purple, rarely white or pink. Anthers 2–2.5 mm, red-purple and much exserted in hermaphrodite heads; smaller, whitish and not or hardly exserted in female heads. Fr. 5 mm, downy, surmounted by the involucel and the 5(–4) persistent reddish-black calyx-teeth which fall short of the lf-like bracts. Fl. 6–10. Protandrous; visited by butterflies and various bees. There is a variable degree of abortion of the anthers in 'female' heads. $2n = 20$. H.

Native. Marshes, fens, meadows and pastures, damp woods; common. Throughout the British Is. Europe northwards to 68°21″N.; Caucasus; W. Siberia; N. Africa.

Morina longifolia Wall., Whorl-Flower, a thistle-like perennial with pinkish fls in a spike of axillary verticillasters, is sometimes grown in gardens as a border plant.

125. COMPOSITAE (Asteraceae)

Herbaceous or sometimes woody plants of very diverse habit; often with latex or oil-canals, rarely with both. Lvs exstipulate. Fls small (*florets*), aggregated into heads (*capitula*) simulating a single larger fl. and surrounded by a calyx-like *involucre* of one or more rows of bracts which vary greatly in number, arrangement, form and consistency. Receptacle of the head expanded and concave, flat, convex or elongated-conical, with or without *receptacular scales* each subtending a floret. Florets all similar (head *homogamous*) or central and marginal florets differing (head *heterogamous*), and then the central florets usually hermaphrodite or rarely male, the outer female or rarely neuter, but there are many variations. Calyx never typically herbaceous but represented by a *pappus* either of numerous simple or feathery hairs in one or more rows, or of a small number of membranous scales, or of teeth or bristles, or of a continuous membranous ring; sometimes 0. Corolla gamopetalous, variable in form but of three main types: (*a*) *tubular*, actinomorphic, the corolla-tube being surmounted by 5 ±short equal teeth; (*b*) tubular, 2-lipped; (*c*) *ligulate*, the corolla-tube being prolonged only along one side as a strap-shaped '*ligule*', usually 3- or 5-toothed at its tip. In homogamous heads the florets may all be of any of these types, but in heterogamous heads the central or disk-florets are usually tubular and actinomorphic and the marginal or ray-florets usually ligulate or, if tubular, distinctly larger and more conspicuous than the disk-florets. Stamens 5, epipetalous; their anthers, often sagittate or tailed below and often with terminal appendages, are usually united laterally so that they form a closed cylinder round the style; dehiscence introrse, into the interior of the cylinder. Ovary inferior, 1-celled, with 1 basal anatropous ovule having a single integument; style single below but branching above into 2 stigmatic arms of variable length and shape. Pollination entomophilous, rarely anemophilous (*Artemisia*, *Xanthium*, etc.); some genera have apomictic spp. (*Taraxacum*, *Hieracium*). Typically the hermaphrodite floret is protandrous, pollen being shed into the anther-tube either above the closed style-arms or upon 'collecting hairs' on their outer side or on the stylar shaft below them. Pollen is thus presented to a visiting insect during or after emergence of the style and before the style-arms diverge, so that cross-pollination is favoured. Self-pollination is often also possible when, at a later stage, the style-arms curl downwards and bring the stigmatic surfaces into contact with residual pollen on the collecting hairs, but self-incompatibility is common. Fr. an inferior achene ('cypsela') crowned by the pappus, sometimes with a slender 'beak' between them. Dispersal usually by wind, but achenes with spiny or barbed bristles (*Bidens*, etc.) or whole heads with hooked bristles or spines on the involucre (*Arctium*, *Xanthium*, etc.) may be carried by animals. Seeds non-endospermic, usually oily.

Over 900 genera and 14 000 spp., throughout the world.

The largest family of flowering plants, comprising examples of almost every ecological type and life-form; trees, shrubs, perennial and annual herbs; aquatic, alpine and desert plants; climbers, succulents and spiny shrubs. There is nevertheless a striking constancy in characters of the fl. and infl., though dioecism, anemophily or reduction of the head to a single fl. are encountered in a small minority of genera.

Many spp. are cultivated as vegetables, the most familiar being: *Lactuca sativa* L. (lettuce), *Cichorium endivia* L. (endive), *Helianthus tuberosus* L. (Jerusalem artichoke), *Cynara cardunculus* L. (cardoon), *C. scolymus* L. (globe artichoke), *Tragopogon porrifolius* L. (salsify), *Scorzonera hispanica* L. (scorzonera). *Helianthus annuus* L. (annual sunflower), *Madia sativa* Mol. and *Sigesbeckia orientalis* L. are grown for their oily seeds, *Cichorium intybus* L. yields chicory, *Carthamus tinctorius* L. (safflower) an orange or red dye, and *Taraxacum bicorne* Dahlst. (*T. kok-saghyz* Rodin) is being grown extensively for its rubber-yielding latex. Many genera, including *Achillea*, *Aster*, *Calendula*, *Callistephus*, *Centaurea*, *Cosmos*, *Dahlia*, *Echinops*, *Erigeron*, *Gaillardia*, *Helenium*, *Helianthus*, *Rudbeckia*, *Tagetes*, etc., include valuable garden plants.

A. *Subfamily* ASTEROIDEAE: plants usually without latex;

at least some disk-fls not ligulate; pollen-grains usually with uniformly distributed spines.

Tribe 1. HELIANTHEAE: lvs usually opposite; involucral bracts not broadly scarious-margined; receptacular scales present; heads usually with tubular hermaphrodite or male disk-florets and ligulate female or neuter ray-florets, sometimes of tubular florets only and then commonly dioecious; anthers rounded to acute at the base, not tailed; style-arms flat, with hairs on the outside which rarely extend below the bifurcation, each arm with 2 marginal stigmatic strips; pappus various, never of hairs, commonly of 2–3 scales or awns.

All disk-florets subtended by receptacular scales.
  Ray-florets usually present, their corollas ligulate, rarely tubular; entomophilous.
    Both disk- and ray-florets setting fr.
      Corolla of ray-florets deciduous.
        Achenes of disk-florets 4–5-angled or laterally compressed.
          Ray-florets ligulate; receptacle flat or somewhat convex.   1. HELIANTHUS
          Ray-florets ligulate; receptacle markedly conical.   2. RUDBECKIA
          Ray-florets tubular; outer involucral bracts linear-spathulate, glandular.   (SIGESBECKIA)
          Ray-florets 0; outer involucral bracts ovate-lanceolate, not glandular.   (SPILANTHES)
        Achenes of disk-florets dorsiventrally compressed.
          Corolla hairy outside at the base.   (GUIZOTIA)
          Corolla glabrous outside.
            Disk-florets with pappus 0 or of 2–3 bristles with no downwardly directed barbs.
              Perennial herbs with root-tubers; outer involucral bracts ±lf-like; heads inclined or drooping.   (DAHLIA)
               Roots not tuberous; outer involucral bracts not lf-like; heads erect.   (COREOPSIS)
             Disk-florets with a pappus of 2–4 bristles with downwardly directed barbs.
              Ray-florets yellow, white or 0.   3. BIDENS
              Ray-florets red or purple.   (COSMOS)
            Disk-florets with a pappus of several fimbriate scarious scales.   4. GALINSOGA
      Corolla of ray-florets persistent in fr.
        Lvs entire, sessile.   (ZINNIA)
        Lvs entire, stalked.   (SANVITALIA)
        Lvs toothed.   (HELIOPSIS)
    Disk-florets not setting fr.   (SILPHIUM)
  Corolla of ray-florets rudimentary or 0; anemophilous.
    Male and female florets in the same head.   (IVA)
    Male and female florets in different heads.
      Involucral bracts of male head joined below; fruiting head 1-seeded with a ring of 4–8 spines. 5. AMBROSIA
      Involucral bracts of male head free; fruiting head 2-seeded with 2 distal beak-like appendages.
                               6. XANTHIUM
Only the outermost row of disk-florets subtended by receptacular scales.   (MADIA)

Tribe 2. HELENIEAE: like Heliantheae but lvs often alternate and receptacular scales 0; conspicuous oil-glands commonly present.

Genera with names in brackets include species found as infrequent casuals or garden-escapes. Brief descriptions of many of these spp., but usually not of the genera, will be found after the accounts of native members of their tribes.

Conspicuous ligulate ray-florets present.
  Lvs opposite; plants strongly aromatic.   (TAGETES)
  Lvs alternate.
    Receptacle naked; involucral bracts herbaceous, not coriaceous at the base.   (HELENIUM)
    Receptacle bristly; involucral bracts herbaceous or lf-like, coriaceous at the base.   (GAILLARDIA)
Ray-florets 0 or few and very shortly ligulate. (SCHKUHRIA)

Tribe 3. SENECIONEAE: lvs alternate, rarely opposite; involucre usually of a single row of equal bracts with or without a few shorter outer bracts; margins of bracts not scarious; receptacular scales usually 0; heads with tubular hermaphrodite disk-florets, sometimes not setting fr.; marginal female florets ligulate, tubular or 0; anthers and style-arms as in Heliantheae; pappus of many simple hairs.
Disk-florets not setting fr.
  Heads all similar.   9. TUSSILAGO
  Heads subdioecious.   10. PETASITES
Disk-florets fertile.
  Marginal florets with obliquely truncate filiform tube.
                          11. HOMOGYNE
  Marginal florets ligulate or heads homogamous.
    Receptacle convex; involucral bracts in several equal rows.   8. DORONICUM
    Receptacle flat; involucral bracts in 1 row with or without smaller ones at the base.   7. SENECIO

Tribe 4. CALENDULEAE: lvs usually alternate, simple; receptacular scales 0; heads with tubular hermaphrodite disk-florets which commonly do not set fr. and usually with ligulate female ray-florets; anthers acute or shortly tailed at the base; style-arms usually truncate at the ends, otherwise as in Heliantheae; achenes large, often spiny and irregularly shaped; pappus 0.
Disk-florets not setting fr.   (CALENDULA)
Disk-florets setting fr.   (DIMORPHOTHECA)

Tribe 5. INULEAE: lvs usually alternate, simple, often with white woolly hairs; fl.-heads usually with inner fls hermaphrodite and tubular, marginal fls female, ligulate or tubular-filiform; sometimes ±dioecious; anthers with basal tail-like appendages; style-arms various; pappus usually present and of simple or feathery hairs, sometimes at least partly of scales or represented by a scarious corona or rim, sometimes 0 in some or all fls.
Marginal fls of fl.-heads with ligule, sometimes very short.
  Receptacular scales 0.
    Achenes ribbed or 4-sided, not abruptly narrowed just below pappus.
      Pappus entirely of simple hairs free to their base.
                            12. INULA
      Pappus with inner row of hairs and outer row of small ±connate scales.   13. PULICARIA
    Achenes cylindrical, not ribbed, abruptly narrowed into a neck just below the pappus.   (DITTRICHIA)
  Receptacle with numerous scales folded round the achenes; pappus a scarious denticulate rim. (TELEKIA)
Marginal fls of fl.-heads tubular-filiform, eligulate.
  Plants ±dioecious; all involucral bracts scarious and usually shining white or pink; pappus of functionally male fls clavate like the antennae of a butterfly.
    Fl.-heads few in a close terminal cluster; plant 5–20 cm.
                      17. ANTENNARIA
    Fl.-heads numerous in dense terminal corymbs; plants 30–100 cm.   16. ANAPHALIS
  Plants not ±dioccious; fl.-heads small; involucral bracts

all scarious or the outer herbaceous, variously coloured and often with darker central stripe, the outer often woolly at least below.

Marginal (female) and often some inner fls subtended by involucral bracts or receptacular scales resembling the innermost bracts; pappus of marginal fls a single row of simple hairs or 0, of central fls several rows of simple hairs.   14. FILAGO

Marginal fls not subtended by bracts or scales; all fls with pappus of a single row of simple hairs.
  15. GNAPHALIUM

Tribe 6. ASTEREAE: lvs alternate; involucre usually of numerous herbaceous bracts in several imbricating rows, progressively shorter towards the outside; receptacular scales 0; heads heterogamous with the marginal florets ligulate or filiform with short or no ligule, or homogamous with all florets tubular; anthers usually blunt at the base with basally inserted filaments; style-arms broad and flat, densely hairy outside towards their tips, each with 2 well-marked marginal stigmatic strips; pappus various or 0.

Pappus present.

Pappus of 2–8 deciduous awn-like bristles; ray-florets yellow.   (GRINDELIA)

Pappus of persistent hairs or bristles, sometimes with an outer row of scales.

Shrubs; pappus of many unequal bristles; ray-florets whitish.   (OLEARIA)

Herbs.

Ray-florets yellow, ligulate; pappus-hairs equal.
  18. SOLIDAGO

Ray-florets usually blue, red or white, ligulate, or if filiform then with definite ligules at least equalling the pappus.

Ray-florets in 1 row, not filiform; pappus of hairs only.
  19. ASTER

Ray-florets in 1 row, not filiform; like *Aster* but with lf-like involucral bracts and a row of concrescent scales outside the pappus-hairs.   (CALLISTEPHUS)

Ray-florets in 2 or more rows, ligulate, narrow, or if filiform at least the outermost with long ligules; pappus of hairs.   20. ERIGERON

Ray-florets filiform with or without tiny ligules (whitish in *C. canadensis*).   21. CONYZA

Ray-florets 0; all florets tubular, yellow.   19. ASTER

Pappus 0.   22. BELLIS

Tribe 7. EUPATORIEAE: lvs often opposite; receptacular scales 0; florets all tubular and hermaphrodite, never pure yellow; anthers blunt at the base with basally inserted filaments; style-arms hairy outside, each with 2 short marginal stigmatic strips.

Pappus of many simple hairs.   23. EUPATORIUM

Pappus of 5–20 scales.   (AGERATUM)

Tribe 8. ANTHEMIDEAE: lvs usually alternate and commonly pinnatifid; involucral bracts usually with a broadly scarious margin; receptacular scales present or 0; heads with tubular, hermaphrodite, usually fertile, disk-florets; ray-florets ligulate, and female or neuter, sometimes 0; anthers blunt or rounded at the base, not tailed; style-arms of the disk-florets usually truncate at the end, otherwise as in Heliantheae; pappus rudimentary or 0.

Receptacular scales present.

Ray-florets tubular or 0; corolla-tube with a basal appendage covering the top of the achene.

Woolly maritime herb with simple lvs.   27. OTANTHUS

Strongly aromatic dwarf shrubs.   28. SANTOLINA

Ray-florets ligulate, rarely 0 (and then corolla-tube lacking basal appendage).

Achenes not strongly compressed.

Achenes slightly compressed, weakly 3-veined on the posterior face; corolla-tube of disk-florets saccate at base (or spurred), so that it covers the top of the achene.   25. CHAMAEMELUM

Achenes not or scarcely compressed, distinctly ribbed all round; corolla-tube not saccate at base.
  24. ANTHEMIS

Achenes strongly compressed.

Achenes unwinged; ligules short and broad.
  26. ACHILLEA

At least the marginal achenes strongly winged; ligules longish.   (ANACYCLUS)

Receptacular scales 0.

Heads small, drooping, inconspicuous, anemophilous, usually in elongated racemose panicles (solitary or 2 in a very rare Scottish alpine); florets all tubular.
  35. ARTEMISIA

Heads ±conspicuous, not drooping, entomophilous, not in elongated racemose panicles.

Ray-fls with conspicuous ligules.

Receptacle hemispherical to conical; lvs repeatedly pinnatisect with ultimate segments linear.

Achenes horizontally truncate above and with truly basal and horizontal insertion-scar, 3-ribbed on adaxial (ventral) face and with 2 conspicuous oil-glands on abaxial (dorsal) face.
  29. TRIPLEUROSPERMUM

Achenes obliquely truncate or rounded above and with oblique (basal-lateral) insertion-scar, 3–5-ribbed on adaxial (ventral) face; oil-glands 0.   30. MATRICARIA

Receptacle flat to slightly convex; lvs simple to pinnatifid or pinnate but ultimate segments not linear.

Strongly aromatic herbs with usually many fl.-heads in dense corymbs; disk-fls with unwinged corolla-tube; achenes all similar; ligule of ray-fls white in native species.   33. TANACETUM

Not strongly aromatic; fl.-heads solitary at end of primary stem and main branches.

Annual; corolla-tube of disk-fls 2-winged; achenes dimorphic, of ray-fls 3-angled with angles often winged, of disk-fls 10-ribbed, usually unwinged; ligule of ray-fls yellow in the single native species.
  31. CHRYSANTHEMUM

Perennial, rarely annual; corolla-tube of disk-fls unwinged; achenes all similar, 10-ribbed; ligule of ray-fls white in the single native species.
  32. LEUCANTHEMUM

Ray-fls very inconspicuously ligulate or ligule 0.

Fls 4-merous; heads solitary, hemispherical.
  34. COTULA

Fls 5-merous.

Receptacle flat or slightly convex; fl.-heads in a dense almost flat corymb.   33. TANACETUM

Receptacle conical; fl.-heads not in a dense corymb.
  30. MATRICARIA

Tribe 9. ARCTOTIDEAE: lvs usually alternate; involucral bracts in many rows, the outer usually shorter, commonly spiny and often scarious-margined; receptacle often bristly, but scales 0; heads with hermaphrodite tubular disk-florets and usually neuter ligulate ray-florets; anthers blunt to

acuminate at the base, but not tailed; style commonly with very short arms and with a ring of hairs well below the bifurcation; pappus of scales, a membranous ring or 0.
Involucral bracts free; rays usually white or pink; lvs cottony. (ARCTOTIS)
Involucral bracts united below; rays yellow or orange.
(GAZANIA)

Tribe 10. CYNAREAE: lvs alternate, often spinous; involucral bracts in many rows, often spinous; receptacle often bristly but subtending scales 0; florets usually all tubular and all hermaphrodite or the outermost female or neuter; anthers usually tailed at the base; style with a ring of hairs or swelling below the bifurcation; pappus usually of many hairs or scales.
Heads 1-fld, aggregated into globose secondary heads.
(ECHINOPS)
Heads with several florets, not aggregated into secondary heads.
 Achene with a straight basal insertion.
  Achene covered with silky hairs, not bordered round its upper end. 36. CARLINA
  Achene usually glabrous, bordered round its upper end.
   Receptacle bristly.
    Outer involucral bracts hooked at the tip.
37. ARCTIUM
    Outer involucral bracts not hooked.
     Stamen-filaments free.
      Receptacle not fleshy.
       Pappus of simple hairs. 38. CARDUUS
       Pappus of feathery hairs. 39. CIRSIUM
      Receptacle fleshy. (CYNARA)
     Stamen-filaments united. 40. SILYBUM
   Receptacle not bristly.
    Lvs spiny. 41. ONOPORDUM
    Lvs not spiny. 42. SAUSSUREA
 Achene with an oblique lateral insertion.
  Heads surrounded at the base by spinous lf-like bracts.
   Pappus of 1 row of scales or bristles. (CARTHAMUS)
   Pappus of 2 unequal rows of bristles. (CNICUS)
  Heads not surrounded by lf-like bracts.
   Involucral bracts with a terminal scarious or spinous appendage. 42. CENTAUREA
   Involucral bracts without terminal appendages.
44. SERRATULA

B. *Subfamily* CICHORIOIDEAE: latex present; all fls ligulate, the ligules 5-toothed at their tips; pollen-grains usually with spines in rows in a polygonal pattern.

Tribe 11. CICHORIEAE: lvs basal or alternate; receptacular scales present or 0; all florets hermaphrodite; anthers acute at the base, sometimes tailed; style-arms long, flattened and stigmatic above, rounded and hairy beneath; pappus various but commonly of many hairs.
Thistle-like plants. (SCOLYMUS)
Not thistle-like.
 Long filiform bracts borne beneath the involucre.
(TOLPIS)
 Filiform extra-involucral bracts 0.
  Pappus 0.
   Florets blue, rarely pink or white. 45. CICHORIUM
   Florets yellow, orange or reddish.
    Stems lfy. 41. LAPSANA
    Lvs all basal; stems markedly swollen beneath the heads. 47. ARNOSERIS
  Pappus of hairs or scales present.
   Pappus-hairs feathery.

Receptacular scales present. 48. HYPOCHOERIS
Receptacular scales 0.
 Involucral bracts in 1 row. 51. TRAGOPOGON
 Involucral bracts in many rows.
  Lvs all basal or stems only with a few small bracts.
49. LEONTODON
  Stems lfy.
   Stems rough with hook-tipped bristles; pappus-hairs not interlocking, readily deciduous.
50. PICRIS
   Stems not rough; pappus-hairs interlocking, not deciduous. 52. SCORZONERA
Pappus-hairs simple or rough, not feathery.
 Achenes ±strongly compressed.
  Achenes beaked or at least strongly narrowed upwards.
   Involucral bracts in many unequal rows; pappus-hairs in 2 equal rows. 53. LACTUCA
   Involucral bracts and pappus-hairs both in 2 unequal rows, inner long, outer short. 54. MYCELIS
  Achenes neither beaked nor much narrowed upwards.
   Florets yellow; pappus hairs in 2 equal rows.
55. SONCHUS
   Florets blue; pappus-hairs in 2 unequal rows, inner long, outer short. 56. CICERBITA
 Achenes not strongly compressed.
  Achenes truncate above, neither narrowed nor beaked; pappus brown, brittle. 57. HIERACIUM
  Achenes narrowed above or beaked.
   Achenes beaked and conspicuously muricate below the beak. 59. TARAXACUM
   Achenes beaked or not, not conspicuously muricate. 58. CREPIS

Key to Native and Naturalized Genera

1 Plant with milky latex; ; florets all ligulate (subfamily Cichorioideae). 59
  Plant without milky latex; at least the central florets tubular (subfamily Asteroideae). 2
2 Heads with distinct ligulate ray-florets. 3
  Florets all tubular (or the outermost very slender and minutely ligulate). 29
3 Ray-florets yellow. 4
  Ray-florets not yellow. 16
4 Stem-lvs all or mostly in opposite pairs. 5
  Stem-lvs all or mostly alternate or lvs all basal. 6
5 Heads not exceeding 2 cm diam.; pappus of 2–4 persistent rigid barbed bristles. 3. BIDENS
  Heads at least 5 cm diam., pappus of 2 deciduous scale-like awns. 1. HELIANTHUS
6 Pappus of hairs, with or without an outer row of scales. 7
  Pappus not of hairs, or 0. 12
7 Head solitary, terminal on a scaly scape and appearing before the exclusively basal true lvs., 9. TUSSILAGO
  Head not terminal on a scaly scape. 8
8 Involucral bracts in 2 or more ±imbricating rows, progressively shorter towards the outside. 9
  Involucral bracts in 1 or 2 rows, all of equal length or with a few much shorter bracts at the base. 11
9 Heads 5–10(–20) mm diam., usually in elongated racemes or racemose panicles. 18. SOLIDAGO
  Heads 10 mm diam. or more, usually solitary or in cymose infls. 10
10 Pappus of an inner row of hairs and an outer row of scales. 13. PULICARIA

Pappus of 1 row of hairs.   12. INULA
11 Heads 4 cm or more in diam.; involucral bracts in 2 equal rows.   8. DORONICUM
Heads not exceeding 3.5 cm diam.; involucral bracts equal in 1 row or with a few much shorter basal bracts.   7. SENECIO
12 Heads tiny (c. 4 mm diam.) in dense compound corymbs; ray-florets 4–6.   26. ACHILLEA
Heads at least 2.5 cm diam.   13
13 Heads at least 6 cm diam.; receptacular scales present.   14
Heads less than 6 cm diam. (or, if as large, with no receptacular scales).   15
14 Lvs simple; receptacle not strongly conical.   1. HELIANTHUS
Lower lvs commonly pinnatifid; receptacle strongly conical.   2. RUDBECKIA
15 Lvs green, deeply pinnatifid with toothed lobes, ±woolly beneath; receptacular scales present; achenes all similar.   24. ANTHEMIS
Lvs glaucous, toothed or pinnately lobed, glabrous; receptacular scales 0; achenes dimorphic.
  31. CHRYSANTHEMUM
16 Lvs opposite; heads less than 1 cm diam.; ray-florets usually only 5, very small, white.   4. GALINSOGA
Lvs alternate or all basal.   17
17 Pappus of hairs.   18
Pappus not of hairs, or 0.   22
18 Shrub with coarsely toothed lvs and small heads with whitish ray-florets.   *Olearia macrodonta* (p. 470)
Herbs.   19
19 Ray-florets not extremely narrow, in 1 row.   20
Ray-florets very narrow, usually in 2 or more rows.   21
20 Ray-florets blue, reddish or white; stem lfy.
  19. ASTER
Ray-florets yellowish-white; lvs mostly basal.
  7. SENECIO
21 Heads solitary (–3) or in ±lax corymbs; ray-florets with evident and often spreading ligules.
  20. ERIGERON
Heads small, numerous, in elongated panicles; ray-florets with tiny erect whitish ligules.   21. CONYZA
22 Lvs all basal; heads solitary.   22. BELLIS
Lvs not all basal.   23
23 Receptacular scales present.   24
Receptacular scales 0.   26
24 Heads 4–6 mm diam. or, if larger, then lvs simple, narrow, finely toothed; ray-florets with short broad ligules; achenes strongly compressed.
  26. ACHILLEA
Heads exceeding 1 cm diam.; lvs finely divided into narrow segments; ray-florets with long narrow ligules; achenes not or little compressed.   25
25 Annual; tube of disk-florets flattened and somewhat winged below; achenes ribbed on both faces; truncate above.   24. ANTHEMIS
Perennial; tube of disk-florets not flattened or winged; achenes ribbed only on 1 face, rounded above.
  25. CHAMAEMELUM
26 Lvs finely divided into ±narrowly linear segments.   28
Lvs not divided into narrowly linear segments.   27
27 Perennial strongly aromatic herb; lvs yellowish-green, mostly pinnatisect to pinnate with deeply lobed segments or lflets; fl.-heads 1–2 cm diam., in corymbs.
  33. TANACETUM
Perennial herbs with simple toothed or lobed lvs;

fl.-heads usually at least 2.5 cm diam., solitary.
  32. LEUCANTHEMUM
28 Heads with ray-florets soon deflexed; receptacle conical from the first, hollow; achenes with 3–5 slender ribs on one face; oil-glands 0.   30. MATRICARIA
Heads with ray-florets spreading until near end of flowering; receptacle becoming conical in fr., solid; achenes with 3 broad corky ribs on one face and 2 conspicuous dark oil-glands on the other.
  29. TRIPLEUROSPERMUM
29 Heads unisexual, greenish, made conspicuous only by the yellow anthers; female heads axillary, solitary or clustered, their involucres either densely covered with hooked spines or with a single ring of straight spines.   30
Heads not both unisexual and spiny.   31
30 Fruiting heads covered densely with hooked spines and with 2 prominent terminal processes.
  6. XANTHIUM
Fruiting heads with a single ring of straight spines.
  5. AMBROSIA
31 Heads small, inconspicuous, campanulate to ellipsoid, usually numerous in long racemose panicles; lvs in native and naturalized spp. pinnatifid or pinnate (rarely heads 1(–3), and then a very dwarf Scottish alpine with silky white subpalmate lvs).
  35. ARTEMISIA
Heads not as above, or, if so, then lvs not pinnatifid or pinnate.   32
32 Stem-lvs all or mostly in opposite pairs.   33
Stem-lvs all or mostly alternate, or all lvs basal.   34
33 Florets pale pink; pappus of hairs.   23. EUPATORIUM
Florets yellow; pappus of 2–4 stiff barbed awns.
  3. BIDENS
34 Heads purplish or white in stout racemes appearing early in the year, before or with the large cordate basal lvs.   10. PETASITES
Not as above.   35
35 Florets red, purplish or blue, rarely white.   36
Florets not red, purplish or blue; commonly yellow.   45
36 Basal lvs long-stalked, reniform, to 4 cm across; stem with c. 2 distant ±scale-like lvs and a solitary pale violet head; a rare Scottish alpine.   11. HOMOGYNE
Not as above.   37
37 Involucral bracts slender, spreading, hook-tipped, the fruiting head forming a ±globose adhesive bur containing several achenes.   37. ARCTIUM
Involucral bracts not slender, spreading and hook-tipped.   38
38 Involucral bracts with a scarious or spinous terminal appendage, differing in colour and texture from the basal part; pappus 0, or of scales, or of hairs not exceeding the length of the achenes.
  43. CENTAUREA
Involucral bracts spinous or not, but not appendaged; pappus of hairs exceeding the length of the achenes.   39
39 Pappus-hairs feathery, i.e. with slender lateral hairs readily visible (when dry) to the naked eye.   40
Pappus-hairs simple or toothed but not feathery to the naked eye.   42
40 Lvs spinous at least at the margins; involucral bracts usually distinctly spine-tipped (or sometimes appressed and mucronate but not evidently spine-tipped).   39. CIRSIUM
Neither lvs nor involucral bracts spinous.   41

41 Heads usually 4–10 in a close corymb; lvs entire or distantly toothed; a rare alpine.   42. SAUSSUREA
    Heads usually solitary, rarely 2–3; lvs simple (rarely pinnate-lobed), fringed with fine prickles.
                                        39. CIRSIUM
42 Neither lvs nor involucral bracts spinous.
                                        44. SERRATULA
    Lvs and involucral bracts spinous; thistles.   *43*
43 Lvs glabrous, conspicuously white-veined above; stem not spinous- winged.   40. SILYBUM
    Lvs not conspicuously white-veined above, decurrent down the stem in spinous wings.   *44*
44 Lvs white with dense cottony hairs above and beneath; stem cottony-white; achenes 4-angled.
                                        41. ONOPORDUM
    Lvs not white on both sides; achenes not 4-angled.
                                        38. CARDUUS
45 Involucre spinous.   *46*
    Involucre not spinous   *47*
46 Lvs spinous; inner involucral bracts yellow, scarious, spreading in fr.   36. CARLINA
    Lvs not spinous; stem with wavy non-spinous wings; involucral bracts with a terminal spinous appendage.   43. CENTAUREA
47 Lvs simple, entire or toothed.   *48*
    Lvs pinnately lobed or divided.   *55*
48 Lvs glabrous.   19. ASTER
    Lvs not glabrous.   *49*
49 Pappus 0; a very rare white-woolly maritime plant.
                                        27. OTANTHUS
    Pappus of hairs.   *50*
50 Lvs hairy, at least beneath, but not woolly or cottony beneath; involucral bracts neither woolly nor wholly scarious (though they may have scarious margins).   *51*
    Stem-lvs linear to broadly lanceolate or elliptical, cottony or woolly at least beneath; heads usually small and inconspicuous, but if as much as 1 cm diam., then with scarious or woolly bracts.   *52*
51 Heads c. 1 cm diam., in a corymb; lvs ovate to ovate-oblong, pubescent especially beneath.
                                        12. INULA
    Heads 3–5 mm diam., in a long panicle; minute whitish ray-florets present; lvs narrowly lanceolate to linear; hairy and ciliate.   21. CONYZA
52 Annual woolly herbs with usually erect stem-lvs and cymosely branching stems, the main stem and the successively overtopping branches each ending in a small roundish cluster of heads; outer involucral bracts herbaceous, ±woolly; receptacle with marginal scales.   14. FILAGO
    Perennial or rarely annual herbs; heads variously arranged, but if as in *Filago* then with spreading stem-lvs; all involucral bracts scarious; receptacular scales 0.   *53*
53 Perennial herbs, 5–25 cm, with basal rosettes of obovate-spathulate lvs and long lfy stolons; stem-lvs erect, 1–2 cm; heads in a close terminal cluster; involucral bracts white or pink; dioecious.   17. ANTENNARIA
    Not as above.   *54*
54 Perennial herb, 25–100 cm, with short stolons; stem-lvs spreading, 5–10 cm; heads in a cymose panicle; involucral bracts shining white; subdioecious.
                                        16. ANAPHALIS
    Perennial herbs with heads in spike-like racemes, or

annuals with heads in roundish clusters; involucral bracts straw-coloured or brownish; not dioecious.
                                        15. GNAPHALIUM
55 Aromatic shrub with whitish-tomentose pinnatifid lvs.
                                        28. SANTOLINA
    Herbs.   *56*
56 Pappus of hairs.   7. SENECIO
    Pappus not of hairs.   *57*
57 Florets 4-merous; heads solitary, hemispherical.
                                        34. COTULA
    Florets 5-merous.   *58*
58 Lvs repeatedly divided into linear segments; receptacle conical.   30. MATRICARIA
    Lvs not divided into linear segments; receptacle flat; heads numerous, in a flat-topped corymb.
                                        33. TANACETUM
59 Pappus of scales or 0.   *60*
    Pappus, at least of the central achenes, of hairs.   *62*
60 Heads blue; pappus of scales.   45. CICHORIUM
    Heads yellow; pappus not of scales.   *61*
61 Stem lfy.   46. LAPSANA
    Lvs all basal.   47. ARNOSERIS
62 All pappus-hairs, or the inner of 2 rows, feathery, i.e. with slender lateral hairs visible (when dry) to the naked eye.   *63*
    Pappus-hairs simple or shortly toothed but not feathery to the naked eye.   *67*
63 Involucral bracts in 1 row; stem-lvs markedly sheathing; heads yellow or purple.   51. TRAGOPOGON
    Involucral bracts in more than 1 row; heads yellow.   *64*
64 Outer involucral bracts either narrow and lax or broadly cordate; stems rough with hooked bristles.
                                        50. PICRIS
    Outer involucral bracts appressed, not broadly cordate.   *65*
65 Stem-lvs linear-lanceolate to linear, entire; lateral hairs of pappus interlocking. Very rare marsh plant.
                                        52. SCORZONERA
    Stem-lvs minute or 0.   *66*
66 Receptacular scales present.   48. HYPOCHOERIS
    Receptacular scales 0.   49. LEONTODON
67 Achenes strongly compressed.   *68*
    Achenes not strongly compressed.   *71*
68 Achenes beaked or at least markedly narrowed upwards.   *69*
    Achenes neither beaked nor markedly narrowed upwards.   *70*
69 Involucre of many unequal rows of bracts; pappus-hairs in 2 unequal rows.   53. LACTUCA
    Involucre of an inner row of equal long bracts and an outer row of much shorter bracts; pappus-hairs in 2 rows, the inner long, the outer short.
                                        54. MYCELIS
70 Heads yellow.   55. SONCHUS
    Heads blue or lilac.   56. CICERBITA
71 Lvs all basal; achenes strongly muricate at the base of the beak.   59. TARAXACUM
    Stem with at least 1–few small lvs; achenes not muricate.   *72*
72 Involucre commonly of many unequal rows of bracts; achene not or scarcely narrowed upwards; pappus-hairs brownish, brittle (but see also *Crepis paludosa*).   57. HIERACIUM
    Involucre usually of bracts of equal length except for some shorter ones near the base; achene usually distinctly narrowed upwards; pappus usually white

and soft (but *C. paludosa* has the achenes not or scarcely narrowed above and a yellowish-white pappus: it differs from most *Hieracium* spp. in being quite glabrous). 58. CREPIS

Subfamily ASTEROIDEAE

Tribe 1. HELIANTHEAE

1. HELIANTHUS L.

Tall annual to perennial herbs with simple opposite or alternate lvs and large heads. Involucral bracts herbaceous, often with lf-like tips, in 2 to several imbricating rows. *Receptacle* flat to conical, *with scales which partly enclose the achenes at maturity*. Disk-florets hermaphrodite, tubular, yellow to brown; ray-florets neuter, ligulate, yellow, deciduous. Achenes somewhat flattened, slightly angled; pappus of 2(–4) deciduous bristles or setae, sometimes with or replaced by small scales.

About 110 spp. in North America.

The following are found as escapes from cultivation:

1 Annual; lvs alternate, broadly ovate, the lower cordate; heads 10–30 cm diam. **1. annuus**
 Perennial; lvs mostly opposite, not cordate; heads not exceeding 10 cm. 2
2 Stolons tuberous; lvs ovate-acuminate with winged stalks; bracts lanceolate-acuminate; rarely flowering. **2. tuberosus**
 Stolons not tuberous; free-flowering. **Perennial Sunflowers** (see below)

*1. H. annuus L. Common Sunflower.

A large *annual* herb with a stout erect stem 75–300 cm, often unbranched. *Lvs* 10–40 cm, alternate, petiolate, broadly ovate, the lower *cordate* at the base, 3-veined, sinuate-toothed, hispid with stiff appressed hairs above and beneath. Heads 10–30 cm diam., ±drooping. Involucral scales ovate-acute, ciliate. Disk-florets brownish; ray-florets golden-yellow. Achenes 7–17 mm, obovoid-compressed, pubescent, variable in colour but often white with black streaks; pappus falling early. Fl. 8–10. Visited for nectar and pollen by many bees and long-tongued flies. $2n = 34$. Th.

Introduced. Commonly cultivated for its oil-yielding achenes or as a garden plant, sometimes escaping; also a bird-seed adventive. The many cultivated races differ in stature, diam. and colour of head, etc. Probably native of Mexico.

*2. H. tuberosus L. Jerusalem Artichoke.

A large perennial herb overwintering by stolons irregularly tuberized at their tips. Stems 100–250 cm, usually branched above, scabrid below, ±pubescent above. *Lvs* mostly opposite, ovate-acuminate, narrowed below into the *winged stalk*, *coarsely toothed*, scabrid above and scabrid-pubescent beneath. Heads 4–8 cm diam., erect. Bracts lanceolate-acuminate, ciliate. All florets yellow. Achenes with 1–4 ciliate awns. Fl. 9–11. Visited by many bees and Diptera. $2n = 102$. Gr.

Introduced. Much cultivated for its inulin-containing tubers, called 'Jerusalem artichokes'. Frequently escaping from cultivation. Native throughout much of Canada and USA, and introduced into England in 1616. Flowers only after a long hot summer.

Some of the North American spp. and their hybrids have given rise to the numerous horticultural races of 'perennial sunflowers' which are much grown in gardens in this country and sometimes escape. The best known of these are *H. rigidus* (Cass.) Desf. and its hybrid with *H. tuberosus* (*H.* × *laetiflorus* Pers.), *H. decapetalus* L. and *H. strumosus* L. They all differ from *H. tuberosus* in having non-tuberous stolons, but hybridization and horticultural selection have led to taxonomic difficulties.

2. RUDBECKIA L.

Annual or perennial herbs with 1 or few large long-stalked heads. Involucre of many imbricating herbaceous bracts. *Receptacle strongly convex to elongated-conical, covered with rigid acute scales*. Disk-florets hermaphrodite, tubular, brown or purple; ray-florets neuter, ligulate, yellow, orange or red. Achenes prismatic, ±angled; pappus a short cup or 0.

About 25 spp. in North America.

Lvs variable, some pinnatifid or pinnate; pappus a short cup. **1. laciniata**
Lvs all simple, ±entire; pappus 0. **2. hirta**

*1. R. laciniata L.

A tall perennial herb with a branched creeping rhizome and erect glabrous stems 50–250 cm. *Lvs* alternate, the lower petiolate, simple or 1–2-pinnatifid with acute lobes, the middle *divided into* 2–3 ±*pinnatifid segments*, the uppermost ovate, simple, sessile; all entire or coarsely toothed, glabrous or sparsely hairy. Heads 7–12 cm diam., long-stalked. Bracts ovate-oblong, with reflexed acute tips. Receptacle conical. Disk-florets brownish-black. Ray-florets golden-yellow, twice as long as the involucre. Receptacular scales equalling the achenes. Achenes 5 mm, ±4-angled, glabrous; pappus a small 4-toothed cup. Fl. 7–10. Visited by various bees and hoverflies. $2n = 38, 76$. G.

Introduced. Grown in gardens, sometimes escaping and locally naturalized. Introduced into Europe early in the seventeenth century and now established in many parts of C. Europe.

*2. R. hirta L. Black-eyed Susan.

An annual to perennial herb with an erect hispid stem 5–60 cm. *Lvs* alternate, simple, oblong-lanceolate to lanceolate, the lowest narrowed into a stalk-like base, the middle and upper sessile; all hispid, *entire to remotely toothed*, distinctly 3-veined. Heads 6–8 cm, long-stalked. Disk-florets brownish-black; ray-florets bright yellow, much exceeding the involucre. Achenes 4-angled; pappus 0. Fl. 7–9. $2n = 38$. Th.–H.

Introduced. Grown in gardens and occasionally escaping but not naturalized. Established in Germany and elsewhere in C. Europe.

Cultivated races derived from other spp. are also grown in gardens, as is the related *Echinacea purpurea* (L.) Moench, with rose-purple ray-florets.

3. Bidens L.

Annual to perennial herbs with opposite simple to bipinnate lvs. *Fl.-heads usually solitary* at the ends of stem and main branches. Involucral bracts in 2 rows, outer usually large and lf-like, inner shorter, ±scarious. *Receptacle* flat or slightly convex, *with scales. Ligulate ray-florets usually 0,* neuter if present; disk-florets tubular, hermaphrodite. *Achenes* unbeaked, *compressed tangentially and broadened upwards,* the broad faces with or without prominent midribs and the resulting 2–4 *angles usually prolonged upwards as stiffly barbed awns;* more rarely not compressed but narrowly tetragonal and tapering to both ends, with 2–4 barbed awns above.

About 120 spp., cosmopolitan but mostly native in the New World though widely introduced as weeds. The shape, size and surface features of ripe achenes are of great importance for identification and should be examined in achenes of central disk-florets.

1 Most lvs simple, ±lanceolate, coarsely serrate (lowest sometimes with 1–2 pairs of basal lateral lobes); central achenes usually with 4 strong awns prolonging the lateral angles and the stout midribs of the faces. 2

Most lvs deeply lobed or compound; central achenes usually with 2 strong awns above the lateral angles and often with 1–2 shorter awns above the weak midribs of the faces. 3

2 Fl.-heads nodding; central achenes convex above, lateral angles and awns with downwardly directed bristles. **1. cernua**

Fl.-heads ±erect; central achenes truncate above, lateral angles non-bristly or with ascending cilia above, and awns in British material with upwardly directed bristles (downwardly in some North American and European races). **2. connata**

3 Most lvs 3(–5)-partite; ripe achenes 5–6 mm, convex and non-ciliate above. **3. tripartita**

Most lvs imparipinnate with 1(–2) pairs of stalked lateral lflets and larger terminal lflet; ripe achenes truncate and ciliate above. **4. frondosa**

1. B. cernua L.   Nodding Bur-Marigold.

An annual herb with erect glabrous or sparsely hairy stems 8–60 cm, simple or branched above. *Lvs* 4–15 cm, *simple, lanceolate-acuminate, sessile,* the members of an opposite pair slightly connate, coarsely serrate, glabrous or slightly hairy, pale green. *Heads* 15–25 mm diam. (without ray-florets), long-stalked, *drooping,* usually solitary at the ends of the stem and main branches. Outer bracts 5–8, lanceolate, lf-like, much longer than the inner, spreading; inner broadly ovate, dark-streaked. Ray-florets usually 0; when present (var. *radiata* DC.) c. 12 mm, spreading, yellow; disk-florets numerous, yellow; receptacular scales c. 8 mm, oblanceolate, scarious. *Achenes* 5–6 mm, straight-sided, broadening upwards, compressed, the outer with 3 and the inner with 4 barbed angles which *continue upwards as* 4 (rarely 3) *downwardly barbed awns.* Fl. 7–9. Visited sparingly by hive-bees and flies. $2n = 24*$. Th.

Native. A locally common plant of ponds and streamsides, and especially of places with standing water in winter but not during the growing season. Throughout the British Is. northwards to Angus and Argyll. Europe northwards to 60° 40′ N; Caucasus; N. Asia. Introduced in North America.

*2. B. connata Muhl. ex Willd.

An annual herb with erect usually glabrous stem 15–100 cm or more. Lvs lanceolate-elliptical, *unlobed,* long-acuminate, coarsely serrate, *narrowed below into a winged petiole. Fl. heads* c. 15 mm diam., ±erect. Outer involucral bracts 4–6, much longer than inner, linear to spathulate, lf-like and spreading or somewhat ascending; c. 8 inner bracts 6–10 mm, broadly elliptical, blunt, brown. Ligulate ray-florets 0; disk-florets orange- or reddish-yellow. Ripe central *achenes* 5–8 mm, compressed but made *strongly 4-angled* by the stout midribs of the faces, broadened evenly to near the *truncate top,* the (3–)4 *awns upwardly bristly* as are the lateral margins towards the top; *faces* verrucose and *with large roundish raised areas.* Fl. 7–9. Th.

Introduced. First recognized in the British Is. in 1977 but established in many areas by the Grand Union Canal in Middx, Bucks and Hertford. Native in North America and naturalized in W. and C. Europe.

Very variable. The above description is based on British material. Some North American and Continental European material differs in having lvs with 1–2(–3) pairs of basal lateral lobes, heads with ligulate ray-florets or achenes with downwardly bristly awns or lateral margins, and plants with these features may turn up here at any time.

3. B. tripartita L.   Tripartite Bur-Marigold.

An annual herb with an erect usually much-branched glabrous or somewhat downy stem 15–60 cm. *Lvs* 5–15 cm, usually 3(–5)-*partite,* sometimes simple, narrowed below into a *short-winged stalk;* lflets lanceolate, acute, coarsely serrate, the terminal lflet sometimes broader and 3-lobed. *Heads* 15–25 mm diam., solitary, *suberect.* Outer bracts 5–8, oblong, lf-like, spreading, inner broadly ovate, brownish. Receptacular scales broadly linear, equalling the achenes. Ray-florets usually 0; disk-florets yellow. *Achenes* 7–8 mm, obovoid, oblong, much compressed, glabrous, with barbed angles, 2 *of which continue upwards as downwardly barbed bristles* with or without 1–2 shorter ones. Fl. 7–9. Visited by bees and hover-flies. $2n = 48*$. Th.

Native. A locally common plant of ditches, pond and lake margins, stream-sides, etc. Throughout the British Is. northwards to Perth and Colonsy. Throughout Europe to 63° 45′ N. and in W. Asia. Introduced in Australia and North America.

*4. B. frondosa L.   Beggar-Ticks, Stick-tight.

An annual herb with erect glabrous stems 10–80 cm or more. *Lvs* 5–15 cm, *pinnate,* with terminal lflet and 1–2 pairs of somewhat smaller lateral lflets, all lanceolate to ovate-lanceolate, sharply and irregularly toothed, stalked. Heads 15–20 mm diam., solitary, erect, long-stalked. Outer involucral bracts 5–8, linear-lanceolate, green, lf-like; inner 6–12, oblong, blunt, brown or livid. Ray-florets usually 0; when present yellow; disk-florets orange-yellow; receptacular scales equalling the florets. *Achenes* (5–)7–10 mm, obovoid-compressed, *blackish*, their *faces papillose* and their *angles upwardly ciliate* (but *not downwardly barbed*), 2 of them prolonged into downwardly barbed bristles. Fl. 7–10. Th.

Introduced. Established locally in wet places and waste ground. North, Central and South America. Widely introduced.

Several other *Bidens* spp. have been recorded as casuals in the British Is. but have not so far established themselves. Two of these are wool-aliens which have also been reported from continental Europe. **B. vulgata** E. L. Greene (North America) resembles **4** but differs in having fl.-heads with usually 10–16 outer involucral bracts, densely hispid-ciliate instead of sparingly ciliate, and well over 60 yellowish achenes (c. 50 and blackish in **4**), their faces almost glabrous with few or no papillae and with stiff upwardly directed cilia on the lower $\frac{3}{4}$ of the margins but downwardly directed near the top. It has been found in a few widely separated areas in S. and C. England. More widespread is **B. pilosa** L. (South America), belonging to a section of the genus characterized by slender fusiform achenes, tetragonal and not or little compressed. Its white-rayed fl.-heads are only 5–15 mm diam. and its achenes are glabrous except for a few upwardly directed setae on the angles; the 2–3 awns are downwardly barbed. It has been found in several scattered localities in S. and C. England and northwards to S. Lancs and N. Yorks.

4. GALINSOGA Ruiz & Pavón

Annual herbs with simple 3-veined lvs in opposite pairs and small fl.-heads in dichasia. *Involucral bracts ovate, in 1–2 rows with 1–4 shorter outer bracts. Receptacular scales present.* Disk-florets yellow, tubular, hermaphrodite; *ray-florets 4–8, commonly 5, white* (or purplish), *shortly ligulate*, female. Achenes obovoid-prismatic; pappus of fimbriate to laciniate scales or 0.

About 4 spp. in tropical and North America.

Stem ±glabrous to moderately pubescent with very short hairs; peduncles with a few glandular hairs; outer involucral bracts 2–4, persistent; receptacular scales trifid; pappus-scales of disk-florets not awned.　**1. parviflora**
Stem and peduncles moderately to densely pilose; peduncles with many long glandular hairs; outer involucral bracts 1–2, deciduous; receptacular scales not trifid; pappus-scales of disk-florets distinctly awned.　**2. ciliata**

*1. G. parviflora Cav.　Gallant Soldier, Joey Hooker.

Stem 10–80 cm, erect, much branched, *almost glabrous to moderately pubescent* (especially above) with ±appressed hairs only about 0.5 mm. Lvs usually 2–5 cm, ovate, acute to acuminate, shortly petiolate, their margins undulate, ±ciliate and with blunt shallow teeth. *Fl.-heads with 2–4, commonly 3, short scarious-margined outer involucral bracts* which, like the inner bracts, are *persistent. Receptacle shortly conical with trifid scales* finely toothed above, those subtending ray-florets only c. 0.4 mm long, those of the disk at least

1 mm; scales sometimes 0. Ray-florets usually 5, with inconspicuous dirty-white ligule up to c. 1.5(–2) mm, sometimes 0. Disk-florets c. 50. *Ripe achenes of disk-florets* 1.5–2 mm, narrowly obovoid-prismatic, their *faces with ±sparse short ascending setae; pappus-scales* 15–20, c. 1.5 mm, silvery, lanceolate to spathulate, *deeply fimbriate*, blunt to acute but *not awned*; achenes of ray-florets flattened, 3-angled, ±curved, setose only above, their pappus-scales only c. 0.4 mm, often 0. Fl. 5–10. $2n = 16^*$. Th.

Introduced. Well established as a weed of arable land and waste places especially in S.E. England, also in scattered localities westwards to Dorset and S. Wales and northwards to Northumberland, Edinburgh and E. Perth. Native in tropical South America but now a cosmopolitan weed.

*2. G. ciliata (Rafin.) S. F. Blake　Shaggy Soldier.

G. quadriradiata auct, ?Ruiz & Pavón

Like **1** but rather more robust and with *stem moderately to densely whitish-pilose with spreading flexuous hairs* c. 1.5 mm long *and with numerous spreading glandular hairs above. Larger stem-lvs* usually 6–7 cm, *cuneate* rather than rounded at base and with longer and more acute marginal teeth. *Outer involucral bracts* 1(–2), *not scarious-margined, deciduous* like the inner bracts. *Receptacular scales usually entire*, rarely weakly 2–3-fid. *Ligule of ray-florets* up to 2.5 mm *long, pure white* (or purplish-red). *Disk-florets* 15–30, *the faces of their ripe achenes hispid with abundant long spreading hairs* and the 8–20 *pappus-scales* white, laciniate, acute and usually *distinctly awned*; ripe achenes of ray-florets similar. Fl. 5–10. $2n = 32^*$. Th.

Introduced. First recorded for British Is. in 1919, now spreading fast as a weed of cultivated ground and waste places in scattered localities from S.E. England westwards to Dorset and S. Wales and northwards to Cumbria and Northumberland and, more thinly, to Easterness and Moray. Native in America from Mexico to Chile, but widely established as a weed.

5. AMBROSIA L.

Annual to perennial herbs, some aromatic, with opposite or alternate lvs which are often pinnately or palmately lobed or dissected. *Fls greenish, inconspicuous and anemophilous, in unisexual heads*; plants *usually monoecious*, sometimes dioecious. *Male heads in terminal ebracteate racemes or spikes*, hemispherical, drooping, each with 7–12 connate involucral bracts enclosing 5–20 fls with or without receptacular scales, stamens with anthers separate or nearly so being borne on the tubular corolla of each fl. *Female heads usually in the axils of lvs or bracts at the base of the male racemes, each with a single female fl. lacking pappus, corolla and stamens but enclosed in an involucre which is united with the ovary below, bears a ring of 4–8 spines, setae or tubercles at or above midway and then tapers into a weakly bifid beak* through the top of which the stigmatic arms protrude; the ripe fr. is nut-like but with the involucre assuming the role of sclerocarp.

About 40 spp., chiefly in the New World but including several cosmopolitan weeds.

1 Lvs usually 1–3 times pinnatifid or pinnatisect, some opposite and some alternate.   2
  Lvs some or all palmately 3(–5)-cleft, all in opposite pairs.   **3. trifida**
2 Annual; lvs ±glabrous and smoothish above; fr.-involucre with 4–7 short spinose teeth or sharp tubercles well below the top of the slender beak.
                                                    1. artemisiifolia
  Usually perennial with creeping rootstock and roots; lvs scabrid above; fr.-involucre unarmed or with 4(–6) short blunt tubercles.   **2. coronopifolia**

*1. A. artemisiifolia L.

Common Ragweed, Roman Wormwood

Incl. *A. elatior* L.

An annual usually monoecious non-aromatic herb with erect furrowed and bluntly ridged stem 25–125 cm, branching above, ±densely covered with short appressed hairs. *Lvs* mostly in opposite pairs but uppermost often alternate, *ovate-lanceolate in outline but deeply pinnatisect into narrow acuminate segments which are mostly pinnatifid to coarsely and sharply toothed*, the uppermost sometimes entire; all ±glabrous and dark green above, shortly hairy and somewhat grey-green beneath. Male heads short-stalked, 4–5 mm diam., in terminal racemes, each with 10–15 fls. Female heads borne singly or in clusters of 2–4 in the axils of lvs immediately below the male racemes. *Fr.* 4–5 mm *with a ring of 4–7 spines or sharp tubercles* overtopped by the slender apical beak. Fl. 8–10. Anemophilous. $2n = 36^*$. Th.

Introduced. A not infrequent casual on rubbish-tips and waste places in scattered localities northwards to Cumbria and sometimes persisting; perhaps entering commonly with imported animal food or bird-seed. Native in North America, where it is an important cause of hay-fever, but now a cosmopolitan weed.

*2. A. coronopifolia Torrey and A. Gray

A. psilostachya auct., non DC.

A herb resembling **1** but *usually perennial* with creeping stock and roots, sometimes annual. Stem erect, branching above, scabrid and with ±dense short stiff hairs. *Lvs once pinnatisect, sessile or nearly so, the oblong acute segments ±entire to incise-serrate,* scabrid above. *Male heads with involucres hispid to long-pilose. Fr. unarmed or with* c. 4 *short blunt tubercles.* Fl. 8–10. $2n = 36, 72, 108$. Hp.–Th.

Introduced. A casual until recently often mistaken for **1** but now recognized as well established in N. Wilts, S. Lancs and Ayr. Native in North America but widely naturalized in Europe and elsewhere.

*3. A. trifida L.   Great Ragweed, Buffalo-Weed.

A *robust annual herb* with stems up to 200 cm or more, glabrous below, ±roughly hairy above. *Lvs* all in opposite pairs, *some or all deeply and palmately 3(–5)-cleft*

into ovate-lanceolate serrate lobes; all petiolate, scabrid. Male heads in one or more racemes, each with cup-shaped 3–5-ribbed involucre c. 1 mm enclosing 10–15 male fls with tubular corolla. Female involucre 6–10 mm in fr., each of its 4–10 ribs ending above in a short spiny tubercle, the ring of spines overtopped by the conical beak 2–4 mm. Fl. 7–9. $2n = 24$. Th.

Introduced. An infrequent casual on rubbish-tips, etc. Native in North America and widely naturalized in Europe.

*Iva xanthifolia Nutt., an annual anemophilous herb, 1–2 m high, native in North America (Prairie Ragweed) and a serious cause of hay-fever, differs from spp. of *Ambrosia* in that the small greenish fl.-heads are all similar, with 8–20 male florets surrounded by 1–5 female florets, the latter with abortive corolla and no pappus. Lvs ovate-deltate, petiolate, coarsely and irregularly dentate, mostly opposite. An infrequent casual.

6. Xanthium L.

Annual monoecious herbs with alternate petiolate lvs, often with trifid spines at base of petiole. *Male heads subglobose in short terminal or axillary ebracteate racemes or panicles, each of many male fls and their axillant receptacular scales with an involucre of free bracts; male fl. with tubular 5-toothed corolla bearing 5 stamens with free anthers;* ovary abortive but the undivided style extending upwards to mid-anther level. *Female heads in axillary subsessile clusters beneath the racemes of male heads, each with (1–)2 fls in a leathery ovoid to ellipsoid 1(–)2-locular involucre of connate bracts bearing hooked spines* and ending above in (1–)2 beak-like processes; *female fl. consisting only of ovary with style and 2 stigmatic arms* which emerge through a hole on the inner side of the beak; pappus, corolla and stamens all 0; anemophilous. *Fr. a hard-walled animal-dispersed beaked bur* with (1–)2 flattened achenes in separate cells.

About 30 spp. in temperate and warmer parts of the world, several of the cosmopolitan weeds carried in wool. The following regularly occur in the British Is. as casuals and may persist for some time.

1 Plant with 1–2 large yellow trifid spines at the base of most petioles; lvs white-tomentose beneath.   2
  Plant lacking trifid spines; lvs green beneath.
                                                    3. strumarium
2 Plant ±erect; lvs short-stalked, narrowly rhombic, usually with 3–5 narrow triangular lobes; fr. 10–15 mm.   **1. spinosum**
  Plant decumbent; lvs long-stalked, usually twice pinnatifid; fr. 7–8 mm.   **2. ambrosioides**

Section 1. *Acanthoxanthium* DC.: lvs narrowed at both ends, white-tomentose beneath, usually with trifid spines at base of petiole; fr.-bur with 1 or 2 short straight beaks or beakless.

*1. X. spinosum L.

Spiny Cocklebur, Barbed-wire Weed.

Stem 15–100 cm, slender, ±*erect*, much branched, shortly pubescent to glabrous, *with strong yellow trifid spines,* 1–3 cm, *at base of most petioles. Lvs narrowly* rhombic, cuneate-based, shortly petiolate, usually with 3–5 narrow triangular lobes, dark green above, whitish-tomentose beneath. Male heads 4–5 mm diam., sub-globose. Fr.-burs 8–12 × 5–8 mm, ellipsoid, with single short straight beak or beakless (rarely with 2 beaks), somewhat pubescent and densely clothed throughout with hooked spines. Fl. 7–10. 2n = 36. Th.

Introduced. A wool-adventive on railway-sidings, rubbish-tips, sewage-works and waste ground near woollen mills and factories using wool, also on farmland manured with shoddy; frequent in suitable areas in England and S. Scotland and perhaps able to persist for a few years. Thought to be native in South America but naturalized in C. and S. Europe and widely elsewhere.

*2. X. ambrosioides Hooker & Arnott
Ragweed Cocklebur.

Like **1** in bearing trifid spines at base of most petioles but decumbent with the weak main stem and branches ascending only to 20–30 cm and with bipinnatifid blunt-lobed lvs narrowly elliptical to triangular-rhombic in outline and about twice as long as broad. Fr.-bur small, 7–8 × 3–5 mm, ±sparsely covered with hooked spines and usually with a single straight beak. Fl. 7–10. Th.

Introduced. A wool-adventive so far known only from a few scattered localities in England northwards to S.W. Yorks, always on wool-waste; has shown signs of becoming established in Bedford. Endemic to S. Argentina but introduced in North America and Europe.

Section 2. Xanthium: lvs ±deltate, ovate or sub-orbicular, often cordate-based, green beneath, lacking trifid spines at base of petiole; fr.-bur with usually 2 strong beaks which are often curved or hooked but sometimes straight.

*3. X. strumarium L., sensu lato Common Cocklebur,
Common Bur-Marigold.

Incl. X. echinatum Murray, X. italicum Moretti, X. orientale L., X. macrocarpum DC., X. riparium Itz. & Hertsch, X. saccharatum Wallr., etc.

Stem 20–120 cm, erect, much branched, variously hairy. Lvs broadly ovate to deltate, coarsely and irregularly toothed and usually palmately 3–5-lobed, ±cordate-based or rarely cuneate, long-petiolate, variously hairy and scabrid, sometimes glandular. Male heads 5–8 mm diam., subglobose to conical. Fr.-bur with straight or hooked spines and straight or curved beaks. Fl. 7–10. Predominantly inbreeding but outcrossing has been shown to be readily possible. 2n = 36. Th.

A polymorphic complex within which several spp. or subspp. have been claimed distinguishable. A rich gene-pool and predominant inbreeding have undoubtedly favoured the formation of local races, but many of the previously recognized taxa are not only very variable and connected by intermediates but have also been observed to undergo considerable changes, presumably consequences of outcrossing with individuals from

fr.-burs originating elsewhere (see D. Löve, 1974). The following is based on the treatment by D. Löve in Flora Europaea 4 (1976).

Subsp. **strumarium**. Not aromatic. Stems and branches green. Fr.-bur 12–15 × 6–10 mm, green to greyish-green when ripe, covered with dense but slender spines. Regarded as native in E., C. and S. Europe; casual in North America and elsewhere.

Subsp. **italicum** (Moretti) D. Löve (X. echinatum Murray, X. italicum Moretti, X. californicum E. L. Greene, X. strumarium subsp. cavanillesii (Schouw ex Didr.) Löve & Dansereau). Aromatic. Stem and branches often with violet or brownish lines or dots. Fr.-bur 15–35 × 6–25 mm, yellow or brown when ripe, covered with stout spines. Native in North and South America; naturalized in S. Europe, probably the outcome of early introduction from the New World; casual in other parts of Europe.

Very variable populations in Eurasia and in North America, believed derivative from hybrids between these 2 subsp., have been named 'X. albinum (Widder) H. Scholz' (incl. X. orientale L. and X. macrocarpum DC., X. brasilicum Velloso, X. saccharatum Wallr., X. riparium Itz. & Hertsch, etc.).

Introduced, mainly as wool-adventives. Subsp. strumarium is an occasional casual in suitable localities and especially on rubbish-tips in S. England. Subsp. italicum is more infrequent but has been recorded (e.g. as X. echinatum Murray) from time to time, as also have putative hybrids between the 2 subspp.

SIGESBECKIA L.

Annual herbs with erect usually much branched and sometimes glandular stems; lvs opposite. Fl.-heads small, heterogamous, in loose panicles or rarely solitary; involucral bracts, often glandular, in 2 rows, outer linear to linear-spathulate, spreading, usually much longer than inner. Ray-florets yellow, shortly ligulate or irregularly 2–3-lobed, female; disk-florets with axillant receptacular scale, yellow, tubular, hermaphrodite. Fr. of ray-florets clasped by involucral bract which may remain adherent after shedding, of disk-florets ±enclosed by their receptacular scale; pappus 0.

Six spp. native in warmer parts of the world, some widespread as introductions, often as wool-adventives. Three spp. have been recorded from the British Is.:

1 Lvs narrowly elliptical or lanceolate, up to 2 mm wide, ±sessile; involucral bracts not exceeding fls and frs. A very rare casual, native in Australia.
S. microcephala DC.

Lvs triangular-hastate to ovate or cordate, the larger 2–14 cm wide, narrowed below into a winged petiole; outer involucral bracts conspicuous and up to 5 times as long as fls and frs.   2

2 Lvs broadly ovate to cordate, shallowly and regularly crenate to serrate, with petiole winged to base and ±amplexicaul.   **1. jorullensis**

Lvs ±triangular-hastate, irregularly dentate or lobed, with petiole tapering downwards and unwinged in lower part, not amplexicaul. An infrequent casual, native in warm temperate and tropical regions of the Old World.   *S. orientalis* L.

*1. S. jorullensis Kunth   Indian-weed.
S. cordifolia Kunth

Stem up to 120 cm, pubescent, often glandular above. *Lvs broadly ovate to cordate, shallowly and regularly crenate to serrate*, acute or subobtuse at apex, *cuneate at base and narrowing into a petiole broadly winged throughout, ±amplexicaul.* Fl.-heads 5–8 mm diam. (excl. the linear-spathulate outer involucral bracts which are usually 10–20 mm long); all involucral bracts with stalked glands.
Introduced. An occasional casual of rubbish-tips, roadsides, waste places and gardens; locally established in S. Lancs. Formerly confused with *S. orientalis* L., a less frequent casual. Native in Central and South America but widespread as a casual and established also in Germany.

Several other members of the Heliantheae occur as casuals, the most frequently encountered being:

*Guizotia abyssinica (L. fil.) Cass. An annual herb with erect divaricately branched usually purple-streaked stems to 1–2 m, glabrous below, glandular-hairy above. *Lvs opposite* or the uppermost alternate, *ovate-lanceolate*, distantly toothed. *sessile and amplexicaul.* Fl.-heads numerous, 2–3 cm diam.; outer involucral bracts 5, ovate-lanceolate, herbaceous, the inner scarious. *Ray-florets c. 8, shortly and broadly ligulate,* female; disk-florets tubular, hermaphrodite; *all with yellow corollas hairy outside at the base.* Achenes 3.5–5 mm, *widening upwards, 3–4-angled; pappus 0.* A bird-seed alien frequent on rubbish-tips, in gardens, etc. Native in E. Africa and cultivated there and in India, the West Indies, etc. for the 'Niger-seed oil' from the achenes, used as a food, as burning oil and in soap manufacture; the achenes themselves are sold as bird-seed.

Several garden plants which occasionally escape are also members of the Heliantheae: *Coreopsis tinctoria* Nutt. and other *Coreopsis* spp., *Heliopsis scabra* Dunal, *Cosmos bipinnatus* Cav., *Dahlia pinnata* Cav. and *D. merckii* Lehm., *Zinnia elegans* Jacq., *Sanvitalia procumbens* Lam., etc.

*Schkuhria pinnata (Lam.) Thell., *sens. lat.,* also native in South America, is an annual herb with erect freely branched stem 25–75 cm, appressed-pubescent at least above, alternate 1–2-pinnatisect lvs with linear segments, and numerous small long-stalked fl.-heads in a loose ±corymbose panicle. Involucral bracts few, gland-dotted, purplish, with scarious yellow margins in fr. Ray-florets 0–2, very shortly ligulate, yellow; disk-florets commonly 4. Achenes obovoid, appressed-pubescent and often with a basal tuft of hairs; pappus of 8 scales usually either all short and awnless or 4 short and awnless alternating with 4 longer and awned, rarely all 8 awned; intermediates also occur. A not uncommon wool-alien recorded as a casual in a number of localities northwards to S. Scotland.

The related tribe HELENIEAE has no native British members but includes familiar garden plants and some frequently encountered casuals. Amongst the former are spp. of 2 American genera: *Helenium* L., in particular the much-grown *H. autumnale* L., a tall late-flowering perennial with rays ranging in colour from plain yellow to mahogany-red; and *Gaillardia* Foug., the annual *G. amblyodon* Gay and *G. pulchella* Foug. being commonly seen as well as the perennial *G. aristata* Pursh.

Tagetes erecta L., *T. patula* L. and *T. lucida* Cav. are the African, French and Mexican marigolds, respectively, while the Dwarf Marigold is a horticultural race of *T. tenuifolia* Cav. (*T. signata* Bartl.) commonly known as *T. signata* 'pumila'.

*Tagetes minuta L., a wool-adventive native in South America, is the most commonly occurring casual of the genus *Tagetes*. It is an annual herb with erect somewhat flexuous stem to 1 m or more, mostly opposite pinnatisect lvs with sharply serrate linear-lanceolate segments, and numerous small fl.-heads in dense corymbs c. 1 cm high. The minute lemon-yellow ligules of the ray-florets are only 1–2 mm long, and the black fr. has a pappus of 5–6 setae with one stiffer and much longer than the rest. A common weed in S. Africa, Australia, etc. and one of our most widespread wool-adventives northwards to Lancs and N.E. Yorks.

Tribe 3. SENECIONEAE

7. SENECIO L.

Annual or perennial herbs, shrubs or trees, some climbers and many succulents; lvs alternate. Heads solitary or in corymbs, usually heterogamous; *bracts mostly in 1 row* with a few outer short ones; *receptacle naked.* Ray-florets ligulate, female, or 0; disk-florets tubular, hermaphrodite, yellow. Achenes cylindrical, ribbed; *pappus of simple hairs,* rarely 0.

About 2000 spp., cosmopolitan: many in S. Africa, the Mediterranean region, temperate Asia and America.

1 Plant with elongated twining stem; lvs ivy-shaped with 5–7 angles or deltate lobes and cordate or hastate base; ray-florets 0.   **16. mikanioides**
  Plant not twining; lvs not angled or deltate-lobed.   **2**
2 Lvs ±deeply pinnatifid or pinnatisect, or some linear and some pinnatifid.   **3**
  Lvs simple, entire or toothed, not linear.   **13**
3 Lvs 3–4 cm, linear, entire, toothed or with short spreading lobes, their margins revolute; plant ±glabrous.   **17. inaequidens**
  Lvs not linear with revolute margins.   **4**
4 Robust glabrous perennial, to 2 m; lvs broad in outline, pinnatifid, pale beneath, the stalks of the lower with broadened clasping base; heads numerous, small, with only 3–4 ray-florets. **tanguticus** (*see* **17**)
  Not as above.   **5**
5 Lvs densely white-felted beneath.   **14. cineraria**
  Lvs glabrous, pubescent or cottony beneath.   **6**
6 Heads with ligule of ray-florets short (5 mm or less) and soon becoming revolute, or 0.   **7**
  Heads with ray-florets conspicuous and spreading.   **10**
7 Heads at first subsessile in dense clusters, later stalked; ray-florets usually 0 (but short in var. *radiatus*).   **8. vulgaris**
  Heads stalked from the first, in loose corymbs; ray-florets present.   **8**
8 Involucre conical; ligule of ray-florets very short and strongly revolute from the first.   **9**
  Involucre broadly cylindrical; ligule of ray-florets 4–5 mm, at first erect or spreading, then revolute; achene 3–3.5 mm.   **5. cambrensis**
9 Stem and lvs very viscid with glandular hairs; involucral bracts c. 20, the outermost 2–3 more than one-third as long as the rest; achenes glabrous.
       7. viscosus

Stem and lvs not or slightly glandular; involucral bracts c. 13, the outermost very short; achenes pubescent. **6. sylvaticus**

10 Heads in dense flat-topped corymbs. *11*
  Heads in irregular loose corymbs or cymes. *12*

11 Lower stem-lvs with small narrow acute terminal lobe; small outer involucral bracts about half as long as the rest; all achenes hairy. **3. erucifolius**
  Lower stem-lvs with blunt terminal lobe; small outer involucral bracts about ¼ as long as as the rest; achenes of ray glabrous, of disk hairy. **1. jacobaea**

12 Weed of waste places, banks, walls, etc.; 5–13 small outer involucral bracts; all involucral bracts conspicuously black-tipped. **4. squalidus**
  Biennial marsh plant; 2–5 small outer involucral bracts; bracts not black-tipped. **2. aquaticus**

13 Ray-florets whitish. **15. smithii**
  Ray-florets yellow. *14*

14 Involucre with a few short outer bracts forming a kind of epicalyx at its base. *15*
  Involucre with no short outer bracts. *17*

15 Lvs cottony beneath; ray-florets 10–20. **9. paludosus**
  Lvs glabrous beneath; ray-florets fewer than 10. *16*

16 Stem-lvs thick, fleshy, decurrent; ray-florets 4–6. **10. doria**
  Stem-lvs thin, not decurrent; ray-florets 6–8. **11. fluviatilis**

17 Infl. a simple terminal corymb; ray-florets c. 13; achenes hairy. **13. integrifolius**
  Infl. compound, of a terminal and several lateral corymbs; ray-florets c. 21; achenes glabrous. **12. congestus**

1. S. jacobaea L.   Ragwort.

A biennial to perennial *non-stoloniferous* herb with a short erect stock and erect flowering stems 30–150 cm, furrowed, glabrous or cottony, branched above the middle. Basal lvs in a rosette 7–15 cm diam., usually dying before flowering, petiolate, lyrate-pinnatifid with a large ovate blunt terminal lobe and 0–6 pairs of much smaller oblong lateral lobes, all sinuate-toothed or pinnatifid; lower *stem-lvs* stalked, upper semi-amplexicaul, *pinnatifid to bipinnatifid*, the *terminal lobe* not much longer than the laterals, *blunt*; all lvs ±glabrous or sparsely cottony beneath, crisped, firm, dark green. *Heads* 15–25 mm diam. *in a large flat-topped dense compound corymb*. Bracts oblong-lanceolate acute ±glabrous with 2–5 subulate outer bracts less than one-quarter as long. Ray-florets 12–15, rarely 0 (var. *flosculosus* DC.) bright golden yellow like the disk-florets. *Achenes* 2 mm, c. 8-ribbed; those *of the ray glabrous, of the disk hairy*; pappus twice as long as the achenes, readily falling. Fl. 6–10. Visited by various bees and flies. $2n = 40^*, 32^*, 80$.

Native. Sand-dunes and some open woodland, also a weed of waste land, waysides and neglected or overgrazed pastures on all but the poorest soils. Abundant throughout the British Is., reaching 670 m in Scotland. Europe to c. 66° N. in Sweden and Finland; Caucasus; W. Asia. Introduced in New Zealand (1874), Australia, S. Africa, North and South America.

Poisonous to cattle when eaten in quantity either fresh or dried; not eaten by rabbits.

2. S. aquaticus Hill   Marsh Ragwort.

A usually biennial *non-stoloniferous* herb with a short ±erect premorse stock and erect flowering stems 25–80 cm, glabrous or cottony above, often reddish, with *ascending branches* above. Basal lvs long-stalked, elliptical to ovate, undivided, or lyrate-pinnatifid with a large ovate to ovate-oblong terminal lobe and 1–several pairs of much smaller oblong lateral lobes; lower *stem-lvs* stalked, ±lyrate-pinnatifid, middle and upper semi-amplexicaul, pinnatifid, *with the lateral lobes directed forwards*; all crenate to coarsely serrate, ±glabrous, firm, slightly waved, often purplish below. *Heads* 2.5–4 cm diam. *in irregular lax corymbs*. Involucral bracts narrowly acuminate, green with white margins, with a few narrower and much shorter bracts at the base. Ray-florets 12–15, golden yellow. *Achenes* 2.5–3 mm, *all ±glabrous*; pappus about twice as long as the achene, readily falling. Fl. 7–8. Visited by flies. $2n = $ c. 40^*. Hs. Very variable in the shape of the lvs and especially of the basal lvs.

Native. Marshes, wet meadow and ditches. Common throughout the British Is. Reaches 460 m in England and Ireland. W. and C. Europe from N. Italy northwards to 62°47′ N. in Scandinavia and eastwards to Posen and lower Silesia.

S. erraticus Bertol., close to *S. aquaticus* but with divaricate branching, basal lvs always lyrate-pinnatifid and lateral lobes of stem-lvs spreading at right-angles to the lf-axis, is very doubtfully claimed for S. England and the Channel Is.

3. S. erucifolius L.   Hoary Ragwort.

A perennial *stoloniferous* herb with short creeping stock and erect sparsely cottony furrowed stems 30–120 cm, with ascending branches above. Basal lvs stalked, obovate-lanceolate, ±pinnately lobed; *stem-lvs* ovate-oblong in outline, deeply pinnatifid with a *small narrow acute terminal lobe* and long parallel *linear-oblong lateral lobes*, entire or with a few teeth; lower stem-lvs stalked, upper sessile, clasping the stem with their small entire basal lobes; all firm, somewhat revolute at the margins, *cottony, especially beneath*. Heads 12–20 mm diam., in terminal and axillary corymbs. Bracts lanceolate-acute, cottony, the *4–6 outer about half as long as the rest*. Ray-florets 12–14, bright yellow. *Achenes* 2 mm, *all with hairy ribs*; *pappus* about three times as long as the achenes, persistent. Fl. 7–8. Visited by flies and bees. $2n = 40^*$. Hs.

Native. Roadsides, field-borders, shingle-banks, grassy slopes, etc., chiefly on lowland calcareous and heavy soils. Locally common through most of England and Wales but rare in S. Scotland northwards to Fife; Ireland, only in the east. C. and S. Europe, northwards to Denmark, S. Sweden and Lithuania; W. Asia.

*4. S. squalidus L.   Oxford Ragwort.

An annual herb, rarely biennial or perennial, with ±glabrous stems 20–30 cm, decumbent at the base then erect, tough, almost woody below, flexuous, branched. Lower lvs narrowed into a winged stalk, upper semi-amplexicaul; all ±glabrous, usually deeply pinnatifid, the lobes oblong, entire or toothed, rather distant, the basal lobes of the upper lvs clasping the stem. Heads 16–20 mm diam. in an irregular lax simple or compound corymb. *Involucre campanulate*, broadening below but remaining constricted above in older heads; inner *bracts* usually 21; outer 5–13, very short; *all conspicuously black-tipped*. Ray-florets usually 13, bright yellow, broad. Achenes 1.5–2 mm, brownish, usually hairy in the grooves; pappus long. Fl. 5–12. $2n = 20^*$. Th.–H. Very variable in degree of dissection of the lvs, which may be merely toothed.

Introduced. On old walls, waste ground, railway embankments, waysides, bombed sites, etc., throughout lowland England, very locally in mainland Scotland north to Moray and in S. Ireland. Probably introduced from the Oxford Botanic Garden. First recorded on walls in Oxford in 1794 and then spreading very rapidly. Native in C. and S. Europe, mainly in the mountains.

For hybrids with *S. vulgaris* see **5** and **8** below.

5. S. cambrensis Rosser

Intermediate between *S. squalidus* and *S. vulgaris* but more robust than either (to 50 cm high). *Heads broadly cylindrical*, 6 mm diam. (10 mm with rays expanded). *Ray-florets* 8–15, bright yellow, short (c. 5 mm) and broad, *soon becoming revolute. Achenes* 3–3.5 mm. Fl. 5–10. $2n = 60^*$. Hs.

Native. Local on roadsides in N. Wales and at Leith, S. Scotland. A naturally arisen alloploid from *S. squalidus × vulgaris*; the F_1 hybrid has low fertility and achenes less than 2.5 mm.

6. S. sylvaticus L.   Wood Groundsel.

An annual herb with an erect slender furrowed stem 30–70 cm, ±cottony or pubescent, ±*glandular but not viscid*; branches ascending. *Lvs yellow-green*, deeply and irregularly pinnatifid, the lobes unequal, ±cut, or toothed; lower lvs oblanceolate or obovate, narrowed into a short stalk, upper oblong, sessile or clasping the stem with enlarged auricle-like basal lobes; all cottony at first but *becoming ±glabrous*. Heads 7–9 × c. 5 mm in a large flat-topped terminal corymb. Involucre conical, the inner bracts narrow, green, the *outer less than ¼ their length*; all glandular-hairy. Ray-florets 8–14, bright yellow, very short and revolute. Achenes 2.5 mm, dark green, *stiffly hairy* on the ribs; pappus whitish, nearly twice as long as the achenes. Fl. 7–9. Visited by flies, etc. $2n = 40^*$. Th.

Native. In open vegetation on sandy non-calcareous substrata. Locally common throughout the British Is. except the Outer Hebrides and Shetland. C. Europe from Portugal, C. Italy and N. Balkans to 66° N. in Scandinavia; W. Asia.

7. S. viscosus L.   Stinking Groundsel.

An annual *foetid* herb with 1 or more erect *very viscid* glandular-hairy stems 10–60 cm, somewhat flexuous, usually with many spreading branches. *Lvs dark green*, glandular-pubescent and *very viscid*, deeply pinnatifid with nearly equal toothed or pinnatifid lobes; lower lvs obovate in outline, narrowed into a short stalk; upper oblong, sessile, not or slightly amplexicaul. Heads 8–11 × 6–10 mm, long-stalked, in a large irregular compound corymb. Involucre, ovoid-conical, densely glandular, the inner bracts green, linear, acute, *the 3 or 4 outer only* 2–4 mm; all concolorous. Ray-florets about 13, yellow, short and revolute. *Achenes* 3–4 mm, yellowish, strongly ribbed, *glabrous*; pappus white, very long. Fl. 7–9. Visited by flies and bees. $2n = 40$. Th.

Probably native. Waste ground, railway banks and tracks, sea-shores, etc. Locally common and scattered throughout lowland Great Britain; increasing. Very rare in Ireland. Europe to 66° N. in Scandinavia and eastwards to Karelia; Asia Minor. Introduced in North America.

8. S. vulgaris L.   Groundsel.

An annual or overwintering herb with weak, irregularly branched, erect or ascending stems 8–45 cm, rather succulent, subglabrous or somewhat floccose, *non-glandular. Lvs coarsely pinnatifid with distant obtuse and irregularly toothed lobes*; basal and lower stem-lvs ±oblanceolate in outline, narrowed into a short petiole; middle and upper stem-lvs oblong in outline, auricled and semi-amplexicaul; all subglabrous or somewhat floccose. Fl.-heads 8–10 × 4–5 mm, subsessile but peduncles elongating in fr., in dense subcorymbose clusters which later loosen. Involucre ±cylindrical; inner bracts 5–8 mm, usually glabrous and often black-tipped; *outer bracts* 8–10, 1–2 mm, *usually black-tipped* and sometimes black throughout. *Ligulate ray-florets usually* 0 (but see below). Achenes 1.5–2(–2.5) mm, appressed-hairy between ribs; pappus white, long. Fl. 1–12. Little visited by insects and normally self-pollinated. $2n = 40^*$. Th.

Very variable in habit, lf-shape and hairiness. Plants with ligulate ray-florets (f. *radiatus* Hegi) fall into two main categories. Those in natural maritime habitats such as sea-cliffs and sand-dunes are typically not or weakly branched, ±strongly floccose and with only 1–8 fl.-heads having 6–12 short yellow ligules which become revolute after flowering. They have been named var. *denticulatus* (O. F. Mueller) Hyl. or, in view of their ecological as well as morphological distinctness, subsp. *denticulatus* (O. F. Mueller) P. D. Sell. Rayed fl.-heads are also found in scattered inland areas where populations commonly include some plants with long and some with short ligules, usually with a majority lacking ligules. Such populations, also found near some sea-ports, are mostly in areas where *S. squalidus* has been recorded, and where it is frequent the proportion of radiate plants has recently been increasing. The triploid hybrid between *S. squalidus* and *S. vulgaris* is highly sterile, but these populations yield seeds giving rise to fertile tetraploids which, selfed or backcrossed with *S. vulgaris*, show some segregation of *squalidus* characters in the otherwise *vulgaris*-like offspring. The mode of origin of these tetraploids is not yet known, but it seems that introgression of genes from *squalidus* into *vulgaris*, for which there is also biochemical evidence, is progressing quite rapidly in some places (see D. H. Kent (*Glasgow Nat.*, 1966, **18**, 407–8), D. E. Allen (*Proc. Bot. Soc. Br. Is.*, 1967, **6**, 362–3), P. Crisp & B. M. G. Jones (*Watsonia*, 1970, **8**, 47–8), P. Hull (*Ann. Bot.*, *London*, N.S., 1974, **38**, 697–700; *Watsonia*, 1974, **10**, 69–75).

Native. Cultivated ground and waste places; abundant throughout the British Is., reaching 530 m in Scotland. Subsp. *denticulatus* very locally on south-western and western coast of Great Britain (Devon, Cornwall, Cheshire, S. Lancs) and also in Channel Is., Anglesey and Isle of Man. Other ligulate plants scattered through S. England and Wales northwards to Durham and in a few places in S. Scotland, especially near Edinburgh. Europe to 71°N; Asia; N. Africa. Widely introduced. Often a troublesome weed of arable land and gardens.

*S. vernalis Waldst. & Kit., until recently a rare casual of cultivated and waste ground, now shows signs of becoming established locally. An annual herb with erect stem up to 50 cm, unbranched or branching only above; lvs narrowly oblong to oblong-spathulate in outline, pinnatifid with shortly deltate broad and ±rounded lobes and with the whole lf-margin sharply dentate, lowest lvs petiolate, middle and upper sessile and amplexicaul; stem and both sides of lvs initially rather densely arachnoid-woolly but often glabrescent; non-glandular. Fl.-heads 2–2.5 cm diam. in a lax terminal subcorymbose infl., their involucral bracts 8–10 mm with 5–13 outer bracts c. 3 mm and usually black-tipped. Ray-florets c. 13 with yellow ligules c. 10 mm. Achenes 2–3 mm, appressed-hairy. Fl. 5–11. $2n = 20, 40$. Th. Native in W. Asia, E. Europe and N. Africa, but has been spreading westwards during the past century and is now a troublesome weed in parts of C. and W. Europe. *S. vernalis* is distinguishable from typical members of the three preceding spp. by the conspicuous non-revolute ligules of the ray-florets, by the short broad lf-lobes and by the usually arachnoid-woolly and non-glandular young shoots. The fl.-heads are larger than in *S. squalidus* or in radiate plants otherwise resembling *S. vulgaris* which are commonly but not invariably subglabrous. Robust plants seen near Market Harborough in 1968–9 with the presumed parents were intermediate in lf-shape and had deformed pollen and seem likely to have been hybrids between *S. vernalis* and *S. vulgaris* (see C. A. Stace, 1975).

9. S. paludosus L.   Great Fen Ragwort.

A perennial herb with shortly creeping stock and erect cottony stems 80–200 cm. *Lvs* 7–15 × 1.3–2 cm, narrowly elliptical, acute, sharply serrate; the basal lvs narrowed into a short broad stalk-like base, upper sessile, very slightly amplexicaul; all ±*cottony beneath.*, Heads 3–4 cm diam., in a usually simple terminal corymb. Involucre campanulate, the inner bracts lanceolate, the outer numerous, almost half as long as the inner, all glabrous or cottony. Ray-florets 10–16, spreading bright yellow. Achenes 3 mm, all glabrous; pappus three times as long as the achene. Fl. 5–7. Visited by various insects. $2n = 40$. Hel.

Native. Formerly in fen-ditches in Lincoln, Cambridge, Norfolk and Suffolk, but thought extinct until rediscovered in Cambridge in 1972. Much of C. Europe from Spain, N. Italy and the Balkan peninsula northwards to the Netherlands, southernmost Sweden and N. Russia; N. Asia.

*10. S. doria L.

A perennial herb with an oblique premorse stock and erect glabrous stems 40–150 cm, branched above. Lvs narrowly elliptical, blunt or acutish, *entire or denticulate*, *glabrous*, *glaucous*, *coriaceous*; basal lvs narrowed into a short stalk; *stem-lvs* sessile, ±*decurrent*. Heads 1.5–2 cm diam., numerous, in a large compound corymb. *Involucre* narrowly campanulate, ±*glabrous*, the bracts 6–8 mm, lanceolate, the 4 outer bracts about half the length of the inner. *Ray-florets* 4–6, bright golden-yellow, short, broad, spreading. Achenes 3 mm, ribbed, all glabrous; pappus at least three times as long as the achenes, whitish. Fl. 7–9. $2n = 40$. Hel.

Introduced. Wet meadows and stream-sides. Established in a few scattered localities. S. and S.E. Europe; N. Africa.

*11. S. fluviatilis Wallr.   Broad-leaved Ragwort.

S. sarracenicus L., p.p.; *S. salicetorum* Godr.

A perennial herb with creeping stock and *long stolons*. Flowering stems 80–150 cm, erect, very lfy, glabrous below, slightly downy and sometimes glandular above, corymbosely branched above. *Lvs* 10–20 × 2–6 cm, elliptical, acute, sessile, hardly amplexicaul, ±*glabrous*, the margins with small rather irregular forwardly directed and somewhat incurved cartilaginous teeth. Heads 3 cm diam., numerous, in a large compound corymb. Involucre ovoid-campanulate, about as long as broad, the 12–15 inner bracts 6–8 mm, with about 5 outer *bracts* half as long, all *pubescent*. *Ray-florets* 6–8, bright yellow, spreading. Achenes 3–4 mm, ribbed, glabrous; pappus three times as long as the achenes. Fl. 7–9. Visited by bees and flies. Hel.

Introduced. Stream-sides, fens and fen-woods. Naturalized in several localities scattered throughout Great Britain northwards to Angus, and in Ireland; decreasing. C. and S. Europe from Spain, N. Italy, N. Balkan peninsula and S. Russia northwards to the Netherlands and Estonia; Siberia.

12. S. congestus (R.Br.) DC.   Marsh Fleawort.

Cineraria palustris (L.) L.; *S. palustris* (L.) Hooker, non Velloso

A biennial or sometimes perennial herb with a short, stout, erect stock and stout erect *woolly* flowering stems 30–100 cm, furrowed, hollow, very lfy, branched only above. Lvs 7–12 cm, broadly lanceolate-acute, entire or sinuate-toothed, pale yellowish-green, glandular-woolly; rosette lvs of the first season narrowed into a short stalk-like base, dying before flowering; stem-lvs sessile with a broad semi-amplexicaul base. Heads 2–3 cm diam., short-stalked in small dense terminal and axillary corymbs. Involucral bracts all equal, about 10 mm, oblong-lanceolate acuminate, ±glandular-woolly. Ray-florets about 20, sulphur-yellow, spreading. Achenes glabrous, 10-ribbed. Fl. 6–7. $2n = 48$. Hel.

Formerly native. Fen ditches. East Anglia, Lincoln, Sussex; but now apparently extinct. C. Europe from

N. France, Denmark, S. Sweden and N. Russia to Alsace, Bohemia, Hungary and Ukraine; Siberia.

13. S. integrifolius (L.) Clairv.   Field Fleawort.

S. campestris (Retz.) DC., pro parte; incl. *S. spathulifolius* auct. angl., non Griesselich

Perennial herb with short ±erect stock and erect stem usually 5–60 cm, arachnoid-lanate but becoming ±glabrous, branching only in the infl. *Basal lvs in a rosette ±appressed to the ground*, usually 2–10 × 2–5 cm, *broadly lanceolate or ovate-spathulate to suborbicular, entire or toothed*, abruptly narrowed into a winged petiole shorter than the blade, persistent at flowering; *stem-lvs ±lanceolate*, narrowing upwards, *sessile and semi-amplexicaul or the lowest with short winged petiole*; all lvs grey or whitish on both surfaces with arachnoid woolly hairs. Fl.-heads 15–25 mm diam., (1–)3–15 in a simple corymb. Involucral bracts 6–12 mm, all equal, arachnoid-woolly at base. Ray-florets 12–15, bright yellow. Achenes 2.5–4 mm, shortly and usually densely hairy. Fl. 6–7. $2n = 48^*$, 64, c. 96. Hs.

Two subspp. are recognized as native in the British Is.:

Subsp. **integrifolius** has stems commonly 8–30 cm; basal lvs oblong to ovate-lanceolate with ±rounded base, entire or remotely denticulate; stem-lvs few, oblong-lanceolate to narrowly lanceolate, and involucre 6–8 mm, often glabrescent. The widespread variant in the British Is.

Subsp. **maritimus** (Syme) Chater (*S. spathulifolius* auct. angl., non Griesselich) is more robust with stems commonly 25–60(–90) cm; basal lvs broadly oblong or ovate-spathulate, rather coarsely crenate-dentate, with truncate or subcordate base and broadly winged petiole; stem-lvs several, broadly lanceolate, at least the lowest distinctly amplexicaul; fl.-heads 3–12, with involucre 8–12 mm; fls larger and opening somewhat earlier than in subsp. *integrifolius*. Known in only one locality in Wales.

Plants on Carboniferous Limestone above Appleby, Cumbria, now very rare, are in some respects intermediate between these 2 subspp. and have been referred to both.

Native. Subsp. *integrifolius* is a local plant of chalk downs in S. England from Cambridge and W. Kent westwards to Wilts and Dorset and is on limestone in Gloucester. Subsp. *maritimus* is known only on submaritime cliffs near South Stack, Holyhead Island. The sp. as a whole is native in C. and E. Europe from the British Is. and France eastwards to the Ural Mts, Romania and Greece, northwards to c. 70° N. in Norway and to Arctic Russia; temperate Asia eastwards to Japan. Subsp. *integrifolius* occurs in most of the range of the species, but subsp. *maritimus* seems endemic to N. Wales, though variants of *S. helenitis* (L.) Schinz & Thell. resemble it.

*14. S. cineraria DC., *sens. lat.*   Silver Ragwort, Dusty Miller.

S. bicolor (Willd.) Tod., non Viv.; incl. *Cineraria maritima* L.

The Silver Ragwort of gardens is a ±hardy suffruticose perennial with stout white-tomentose stem 25–60 cm or more, branched especially below. Lvs 4–15 cm, ovate-oblong in outline and white-tomentose beneath, may range, from below upwards, from crenate-dentate with petiole about equalling the blade to deeply pinnatisect and almost sessile, and between different races from silvery-hairy above to somewhat cottony but green and glabrescent. Fl.-heads 8–15(–25) mm diam., in dense compound corymbs; involucre white-tomentose, 5–8 mm, with up to 5 short outer bracts; ray-florets 10–13, 3–6 mm, yellow. Fl. 6–8. $2n = 40$. Ch.–N.

Introduced. Frequently grown in gardens and sometimes escaping: plants established in several localities, mainly on sea-cliffs in S. England and formerly also in S. Wales, and in one locality near Dublin in E. Ireland, are almost certainly garden-escapes. The wild *S. cineraria* is native in rocky and sandy places in the Mediterranean region, usually near the coast. It varies considerably, especially in the degree of dissection of the lvs and the shape of their ultimate lobes, the density and persistence of white tomentum on the involucre, and peduncle-length. The plants cultivated in this country and locally naturalized most closely resemble subsp. *cineraria* of the W. and C. Mediterranean region but with certain differences, and further taxonomic research is needed to clarify the position. Meanwhile the British coastal plant is provisionally referred to *S. cineraria* DC. subsp. *cineraria*.

*15. S. smithii DC.

A robust perennial 60–120 cm, with erect ±woolly stem branched only in the infl. Basal lvs with blade 15–35 cm, oblong, acute, cordate at base, long-stalked, irregularly toothed; stem-lvs acuminate, sessile, crenate-toothed; all loosely woolly at least beneath. Heads 6–12 or more in an irregular corymb. Involucre 10–12 mm, broadly campanulate. Ray-florets c. 15, yellowish-white. Achenes hairy, ribbed.

Introduced. Established in a few places on the coast in Scotland. Chile and Patagonia.

*16. S. mikanioides Otto ex Walpers   German Ivy.

A slender much-branched woody twiner reaching 3(–6) m; glabrous. The somewhat fleshy lvs, 3–10 cm, suborbicular to deltate with (3–)5–7(–11) acute angles or triangular lobes and hastate or cordate base, resembling lvs of ivy; petiole usually longer than blade, often with small basal auricles. Fl.-heads 5–7 mm diam., many, in terminal and axillary corymbs. Involucre 3–4 mm, cylindrical, of c. 8 narrow bracts with 2–4 shorter outer bracts. Ray-florets 0; disk-florets 8–12, tubular, yellow. Achenes 1–2 mm, ±cylindrical, glabrous; pappus 3–4 mm. Winter-flowering. $2n = 20$. N.–M.

Introduced. Grown in gardens; sometimes escaping and occasionally becoming established. Native in S. Africa.

*17. S. inaequidens DC.

A suffruticose sparsely hairy or glabrescent perennial with erect or ±widely spreading much-branched stems to 60 cm or more. *Lvs usually* 2.5 cm × c. 1 mm, *linear, acute, mostly entire* but sometimes irregularly denticulate to shortly pinnatifid, often broadened below to an amplexicaul base. *Fl.-heads 2–2.5 cm diam.* in a lax corymb. Involucre with 10–20 short outer bracts with white scarious fimbriate margins. Ray-florets 10–15, yellow; disk-florets c. 60, yellow. Achenes shortly pubescent. Fl. 6–10. Ch.

Introduced. Casuals, mainly wool-aliens, conforming ±closely to the above description and clearly belonging to Section Fruticulosi DC., have been recorded from several localities in England and S. Scotland and seem able to persist in some places, as near Dover and by the R. Tweed. They were formerly misidentified as the Australian *S. lautus* Solander ex Willd. but seem for the most part to be *S. inaequidens*, native in S. Africa.

Several other *Senecio* spp. occur as rare casuals and some grown in gardens escape and may persist for a time, amongst them **S. tanguticus** Maxim., of problematic taxonomy and sometimes placed in *Ligularia* (see below) or in any of three other genera. It is a robust perennial herb with branching stems up to 2 m or more. *Lvs 12–18 cm, broadly ovate or deltate in outline, pinnately divided* into narrow coarsely toothed segments, *dark green above, pale beneath*, lower lvs with petioles broadened into a ±amplexicaul base. *Fl.-heads small, very numerous, in terminal pyramidal panicles.* Ray-florets 3–4, yellow; disk-florets 3–4, with revolute corolla-lobes. Fl. 7–9. Native in W. China. Widely and thinly scattered through Great Britain as a garden-escape and persisting for some time but doubtfully naturalized.

S. cruentus DC., 'Cineraria', is much grown in greenhouses for its red, violet, purple or blue fl.-heads.

LIGULARIA Cass.

Perennial and often very robust herbs not readily distinguishable from *Senecio* but usually with alternate stem-lvs of which at least the lower have broadly sheathing petioles, rarely with all lvs basal. The yellow fl.-heads have involucral bracts originating in a single row but appearing 2-rowed because of marginal overlapping. The style-arms are rounded at the ends (truncate in *Senecio*) and are stigmatic above and hairy beneath over their whole length. Several spp. are grown in gardens and *L. dentata* (A. Gray) Hara (*S. clivorum* Maxim.), reaching 1.2 m or more, with large long-stalked reniform to suborbicular-cordate sharply dentate basal lvs which may be 50 cm across and fl.-heads up to 10 cm diam. with orange-yellow rays and dark brown disk, has escaped or been planted in a few scattered localities where it seems ±naturalized. Native in China and Japan.

For *L. tangutica* (*Senecio tanguticus*) see above.

8. DORONICUM L.

Perennial herbs usually with *locally tuberized stolons*; lvs alternate, simple, and fl.-heads large, long-stalked, yellow. Bracts herbaceous in 2–3 almost equal rows. Receptacle convex, often hairy. Heads heterogamous; disk-florets hermaphrodite, tubular, yellow; ray-florets in 1 row, female, ligulate, yellow. Achenes ±cylindrical, ribbed; pappus of 1–2 rows of simple hairs, or 0 in marginal florets.

About 34 spp. in Europe, Asia and Africa and especially in the mountains of temperate Asia.

The following key is to plants naturalized in the British Is.:

1 Basal lvs all gradually narrowed into the petiole, sparsely hairy to glabrescent; peduncles with short glandular hairs only; fl.-heads commonly solitary; receptacle glabrous. **2. plantagineum**

Basal lvs mostly truncate to deeply cordate at base, sparsely to densely hairy; peduncles usually with some long flexuous eglandular as well as shorter glandular hairs; fl.-heads 1-several; receptacle hairy. 2

2 Basal lvs deeply cordate; fl.-heads usually 3–8, mostly less than 4.5 cm diam. **1. pardalianches**

Basal lvs mostly truncate to shallowly cordate at base; fl.-heads usually 1–3 and mostly more than 4.5 cm diam. Probable hybrids (see below)

*1. D. pardalianches L.   Great Leopard's-bane.

Herb perennating by numerous stout hypogeal stolons which are tuberized at their tips and give rise in spring to rosettes of long-stalked *broadly ovate-cordate* ±entire *lvs*, ciliate and hairy on both sides. Flowering rosettes produce erect lfy stems 30–90 cm, ±woolly with spreading hairs. Lower stem-lvs with a long stalk winged above and amplexicaul below; middle stem-lvs panduriform, amplexicaul; uppermost *ovate-amplexicaul*; all pale green, thin, hairy, with entire, crenate or sinuate-toothed margins. *Heads 4–6 cm diam., usually several.* Involucre saucer-shaped, its bracts triangular-subulate, glandular-ciliate. Receptacle pubescent. Florets bright yellow. Achenes black, 10-ribbed; those of the disk pubescent and with a pappus; those of the ray glabrous, with no pappus. Fl. 5–7., Visited by flies, beetles and Lepidoptera. $2n = 60$. Hs.

Introduced. Woods, plantations; local. Scattered throughout Great Britain northwards to W. Ross and Moray. W. Europe, eastwards to Italy, W. Switzerland and W. Germany.

Formerly much cultivated as a medicinal drug.

*2. D. plantagineum L.   Leopard's-bane.

Like *D. pardalianches* but stems more slender and less woolly, glandular-pubescent above. *Basal lvs ovate-elliptical narrowed gradually* into the long stalk; *uppermost lvs elliptical to lanceolate, sessile decurrent*; all with prominent curving lateral veins rather as in *Plantago major*. *Heads 5–8 cm diam.*, commonly *solitary*; receptacle glabrous. Fl. 6–7. $2n = c. 120$. Hs.

Introduced. Grown in gardens and rarely naturalized. Scattered throughout Great Britain northwards and rarely naturalized to Banff. S.W. Europe (Portugal, Spain, S. France, Italy).

D. plantagineum var. *willdenowii* (Rouy) A. B. Jackson, with broadly ovate basal lvs truncate or somewhat rounded below, often with blunt apex and with margins inconspicuously dentate or crenate-dentate, young lvs often densely hairy, probably originated from a cross between *D. plantagineum* and *D. pardalianches*. *D. plantagineum* var. *excelsum* N. E. Brown, often sold as 'Harpur Crewe', having basal lvs usually shallowly cordate (but late-season lvs may be rounded below), with acute apex and prominently dentate margins, young lvs less hairy than in var. *willdenowii*, has been regarded as derived from the same interspecific cross but the strongly toothed lf-margin suggests that a third sp., perhaps the S.E. European *D. columnae* Ten., may also have been implicated. Both 'varieties' are grown in gardens and both have been recorded as ±persistent escapes (see A. C. Leslie, *B.S.B.I. News*, 1981, **27**, 22–3 for descriptions and outlines of basal lvs).

9. TUSSILAGO L.

Heads solitary, heterogamous; involucral bracts numerous, mostly in 1 row; receptacle slightly concave, naked. Ray-florets ligulate, numerous (up to 300), in many rows, female; disk-florets few, male. Pappus of many rows of long simple hairs. One sp.

1. T. farfara L   Coltsfoot.

A perennial with long stoutish white scaly stolons, these and their short branches terminating in rosettes of lvs in whose axils arise the flowering shoots of the following season. Lvs 10–20(–30) cm across, all basal, roundish-polygonal, very shallowly 5–12-lobed, the lobes acute and with small distant blackish teeth; ±deeply cordate at the base, with a stalk broadly furrowed above; at first white-felted above and beneath, later only beneath. *Heads* 15–35 mm diam., *solitary, terminal on purplish scaly and woolly flowering shoots* 5–15 cm, *opening long before lvs appear*, erect in bud. Bracts numerous linear, blunt, green or purplish, somewhat hairy, mostly in 1 row but with a few broader basal scales transitional to the scale lvs of the flowering stem. Florets bright pale yellow. After flowering the head first droops and then re-erects when the fr. is ripe, the stem meanwhile lengthening to c. 30 cm. Achenes 5–10 mm, glabrous, pale; pappus white, much longer than the achenes. Fl. 3–4. Visited chiefly by flies and bees. The fls close at night, $2n = 60$. Grh.–Hr.

Native. Abundant, especially on stiff soils, in arable fields (often a troublesome weed), waste places, banks, landslides, boulder-clay cliffs, etc.; also on dunes, screes and stream-side shingle and in seepage-fens on hillsides. Throughout the British Is. Reaches 1070 m on Ben Lawers. Europe northwards to 71° N. in Norway; W. and N. Asia; N. Africa. Introduced in North America.

The dried lvs were formerly smoked as a remedy for asthma and coughs.

10. PETASITES Miller

Perennial rhizomatous polygamodioecious herbs with large basal lvs produced with or after the fls. Fl.-stems scapose with few to many bract-like scale-lvs with or without rudimentary lamina and 1-many heterogamous fl.-heads in spike-like racemes or panicles which in 'female' plants greatly elongate after flowering. Involucral bracts in 1 row but commonly with additional outer bracts and then irregularly 2–3-rowed; receptacle flat, lacking scales. Fl.-heads of 'male' plants with numerous hermaphrodite but functionally male tubular fls, usually with 1–5 peripheral female but sterile fls, tubular or shortly ligulate; of 'female' plants with numerous fertile female fls. tubular or ligulate, surrounding 0–5(–8) hermaphrodite but sterile central tubular fls. Achenes cylindrical, glabrous; pappus of long slender simple hairs, numerous in female, few in male or sterile fls.

About 15 spp. in the north temperate zone of the Old World including a few in North America.

1 Lvs orbicular-cordate, dentate, with shallow lobes each centred on a large tooth marking the end of a major vein; bract subtending lowest infl.-branch usually much shorter than scale-lvs near base of stem and at most 1.6 cm wide.   2

Lvs reniform-cordate, dentate, not shallowly lobed; bract subtending lowest infl.-branch almost as long as scale-lvs near base of stem and 1.7–4 cm wide.   3

2 Lf-blade with 2–5 lateral veins bordering the basal sinus and with basal lobes convergent; fls usually pale lilac-pink, sometimes yellowish.   **1. hybridus**

Lf-blade with no vein, or only 1, bordering the basal sinus and with basal lobes usually divergent; fls yellowish-white.   **2. albus**

3 Lvs irregularly dentate, their basal lobes convergent; stem with 15–25 broad and crowded scale-lvs; fl.-heads more than 20; fls creamy-white, all tubular.   **3. japonicus**

Lvs regularly dentate, their basal lobes slightly convergent to divergent; stem with 2–7 scale-lvs; fl.-heads 6–20; fls pale lilac, vanilla-scented; marginal fls ligulate.   **4. fragrans**

Section 1. *Petasites*. Corolla of the marginal (female) fls filiform, obliquely truncate, not ligulate.

1. P. hybridus (L.) P. Gaertner, B. Meyer & Scherb.
Butterbur.

Tussilago petasites L. (male) and *T. hybrida* L. (female); *P. ovatus* Hill; *P. officinalis* Moench; *P. vulgaris* Desf.

Rhizome stout, ±horizontal, with branches up to 150 cm. *Lvs* 10–90 cm across, mostly basal, long-petiolate, roundish, deeply cordate, at first downy on both sides but later green above and *greyish beneath*; petioles stout, hollow, channelled above; blade with larger distant teeth and smaller intervening teeth; *lower part of each basal lobe bordered by 2–5 lateral veins*. Flowering stems 10–40 cm (–80 cm in fr.), appearing before the lvs, stout, purplish below, covered with greenish lanceolate scales, some with a rudimentary blade. Heads 1–3 in the axils of linear acute bracts, pale reddish-violet; 'male' heads 7–12 mm, very short-stalked, with 0–3 female and 20–40 sterile 'hermaphrodite' fls; 'female' heads 3–6 mm, lengthening in fr., longer-stalked, with c. 100 female and 1–3 sterile fls. Involucral bracts narrow, blunt, glabrous, purplish. Achenes 2–3 mm, yellowish-brown, cylindrical; pappus whitish. Fl. 3–5. Visited chiefly by bees; only the sterile fls secrete nectar. $2n = $ c. 60*. G.

Native. In wet meadows and copses and by streams to 460 m in Scotland. The male plant is locally common throughout the British Is. The female plant is not uncommon in Lancs, Yorks, Cheshire, Derby and Lincoln, but rare or absent elsewhere, though extending north to Perth. Europe to 63° 26′ N. in Scandinavia; N. and W. Asia. Introduced in North America.

*2. P. albus (L.) Gaertner   White Butterbur.

Tussilago alba L.

Like *P. hybridus* but smaller. *Lvs* 15–30 cm across, roundish, deeply cordate, long-petiolate, glabrous above and *white-woolly* *beneath* when mature; margin with conspicuous large teeth (more prominent than in *P. hybridus*), the intervening spaces sharply denticulate; *basal lobes not bordered by lateral veins.* Flowering stems 10–30 cm (–70 cm in fr.), appearing before the lvs, becoming ±glabrous, covered with pale green lanceolate scales. Infl. not much longer than broad. Heads whitish, the male larger and more crowded. Bracts linear, ±acute, pale green, glandular. Achenes 2–3 mm; pappus white. Fl. 3–5. Visited by various insects. $2n = 60$. G.

Introduced. Locally in waste places, roadsides, plantations and woods from Leicester, Warwick and Merioneth northwards to Aberdeen; Ireland (Down); C. and N. Europe from Spain, Italy and the N. Balkan peninsula to S. Scandinavia and C. Russia; Caucasus; W. Asia.

*3. P. japonicus (Siebold & Zucc.) Maxim.

Creamy Butterbur.

Nardosmia japonica Siebold & Zucc.

A robust perennial herb with long-petiolate reniform-cordate basal lvs up to 1 m across, sharply but irregularly dentate, glabrous above, hairy beneath; basal sinus of lf bordered by lateral veins, basal lobes convergent. Fl.-stem, appearing before basal lvs, up to 1 m or more bearing 15–25 broadly oblong-lanceolate pale green and ±lf-like scale-lvs, those in mid-stem 5–7 cm, crowded and overlapping at least above. Fl.-heads short-stalked in dense creamy-white corymbose clusters so that top of fl.-stem resembles a small cauliflower; involucral bracts 8–10 mm, pale green, broadly lanceolate, united below; fls all tubular, fragrant. Fl. 3–4. $2n = 84$–87. G.

Introduced. The male plant has escaped from gardens and is established in a few places northwards to Fife, chiefly on stream-banks. Native in Japan and Sakhalin but widely grown in gardens and locally naturalized in N.W. and C. Europe.

Section 2. *Nardosmia* (Cass.) Fiori. Corolla of the marginal female fls with a distinct but short ligule.

*4. P. fragrans (Vill.) C. Presl   Winter Heliotrope.

A perennial herb with far-creeping rhizome and long-stalked reniform-cordate basal lvs 10–20 cm across, equally denticulate, glabrescent above, somewhat hairy beneath; basal lf-lobes usually sub-parallel near base then divergent, with 2–5 lateral veins bordering the deep sinus; lvs appearing in spring and often remaining green through the following winter until the new lvs appear. Fl-stem 10–25 cm, bearing 2–7 scale-lvs, commonly with rudimentary lamina, those in mid-stem 3–7 cm. Fl.-heads 6–20 in a short lax cluster, pale lilac, vanilla-scented; involucral bracts 7–10.5 mm, narrow, acute, pale green or purplish, ±glabrous; 'male' heads chiefly of tubular fls, 'female' of slender fls with broad ligule c. 5 mm long. Fl. 1–3. Visited by flies and hive-bees. $2n = $ c. 60. G.

Introduced. An escape from gardens on streamsides

and banks and in waste places and naturalized in scattered localities throughout Great Britain, Isle of Man and Ireland; Scilly, Channel Is.; Mull, Orkney and Shetland. Native in Italy, Sardinia and Sicily but much grown in gardens and widely naturalized.

11. HOMOGYNE Cass.

Perennial rhizomatous herbs with *reniform basal lvs* and scapes bearing a few small stem-lvs and terminating in a *solitary head* (rarely 2 or more). Involucre of a single row of bracts. *Heads heterogamous: central fls tubular,* hermaphrodite, fertile; *marginal with short slender obliquely truncate tube,* female, fertile. Receptacle naked. Achenes cylindrical, slightly ribbed; pappus of several rows of simple hairs.

Three spp. in the mountains of C. Europe and the Balkan peninsula.

*1. H. alpina (L.) Cass.

Tussilago alpina L.

Rhizome slender, covered with woolly scales, emitting lfy stolons. Basal lvs to 4 cm across, long-stalked, orbicular to reniform, deeply cordate at base, shallowly sinuate-toothed, dark green and glabrous above, paler and often purplish beneath, hairy on the petiole and on the veins beneath. Scape 10–30 cm, cottony below and with crisped and glandular hairs above, bearing a few ovate-lanceolate scales of which the lowest may have a small blade and a sheathing base. Head 10–15 mm, usually solitary. Involucral bracts linear-lanceolate, cottony below, purplish at margins and tip. Fls somewhat exceeding the involucre, pale violet. Achenes 4–5 mm, with pure white pappus. Fl. 5–8.

Almost certainly introduced. Clova (where it was recently rediscovered); 1 locality in Outer Hebrides. Mountains of C. Europe from Pyrenees to N. Balkan peninsula.

Tribe. 4. CALENDULEAE

There are no native members of the tribe Calenduleae, but several *Calendula* spp. occur as garden-escapes or casuals.

*C. officinalis L. (Pot Marigold) has the lower lvs oblong-ovate, the fruiting heads erect, and all or most of the achenes boat-shaped. This is the familiar garden plant, often escaping and well-naturalized in the Scilly Is. ***C. arvensis** L. has all the lvs oblong-lanceolate and ±toothed, the fruiting heads drooping, the outer achenes narrow and not greatly curved, the middle boat-shaped and the innermost curved into a complete ring. A frequent casual. Both spp. are apparently native in C. Europe and the Mediterranean region.

Dimorphotheca sinuata DC. (*D. aurantiaca* hort.) is a related sp., often grown in gardens and differing from *Calendula* in that the disk-florets set fr.

Tribe 5. INULEAE

12. INULA L.

Perennial, rarely biennial, herbs or small shrubs with *alternate usually simple lvs.* Fl.-heads rather large, solitary or in corymbs or panicles, heterogamous; involucral bracts herbaceous, imbricate, many-rowed; *receptacle* flat or slightly convex, *without scales*. Peripheral fls female, typically ligulate but sometimes with ligule very short or almost 0 and therefore ±tubular; disk-fls hermaphrodite, tubular; all fls yellow, rarely reddish. *Achenes distinctly ribbed or angled, not abruptly contracted into a neck just below the apex* (cf. PULICARIA below); *pappus entirely of shortly ciliate and therefore rough hairs which are free to their base or nearly so.*

About 200 spp. in temperate and subtropical regions of the Old World.

1 Lower stem-lvs 25–70 cm, ovate-elliptical; fl.-heads 6–8 cm diam.; outer involucral bracts ovate, 4–5 mm wide; achenes 3–5 mm.   **1. helenium**
  Lower stem-lvs rarely exceeding 15 cm; fl.-heads less than 6 cm diam.; outer involucral bracts less than 4 mm wide; achenes 1–3 mm.   2
2 Ligules of peripheral fls at most equalling the inner involucral bracts and commonly shorter or almost 0; lvs resembling those of *Digitalis*.   **3. conyza**
  Ligules of peripheral fls exceeding inner involucral bracts.   3
3 Maritime plant with ±linear succulent lvs, often 3-toothed at apex; fl.-heads up to 2.5 cm diam.
                                                          4. crithmoides
  Not maritime; lvs lanceolate or broader, not succulent; fl.-heads usually exceeding 2.5 cm diam.   4
4 Lvs glabrous above, sparingly hairy on veins beneath; involucral bracts glabrous with ciliate margins, outer lanceolate, inner linear. W. Ireland only.
                                                               2. salicina
  Lvs sparsely hairy above, very sparsely to densely beneath; involucral bracts all linear, softly hairy. Very rare casual, formerly naturalized in a single locality near Leicester.   *I. britannica* (p. 460)

Section 1. *Inula*. Outer involucral bracts broad, herbaceous. Achenes 4-sided.

*1. I. helenium L.   Elecampane.

A *robust perennial herb* with branching tuberous stock and stout erect furrowed stems 60–150(–250) cm, simple or corymbosely branched above, tomentose. Basal lvs with ovate-elliptical lamina 25–60 cm, narrowed below into a long petiole, the lowest stem-lvs similar but short-stalked or subsessile; *middle and upper stem-lvs smaller and diminishing upwards*, ±narrowly ovate-cordate, acute, sessile and amplexicaul; all with margins finely and irregularly toothed and somewhat undulate and all ±glabrous above, softly tomentose beneath. *Fl.-heads 6–8 cm diam.*, rather shortly pedunculate, solitary or in few-headed corymbs at ends of main stem and branches. *Involucre 1.5–2 cm, hemispherical*, its *outer bracts herbaceous* and softly hairy, *ovate with spreading or recurved tips, exceeded by the scarious inner bracts which are lanceolate-acuminate and erect*. Peripheral fls *with narrow spreading ligules 3–4 cm*, bright yellow like the disk-fls. *Achenes 3–5 mm, 4-sided with strong ribs at the angles*, glabrous; pappus longer than achene, often reddish. Fl. 7–8. Visited by many bees and hoverflies. $2n = 20^*$. H.

Introduced. Uncommon but widely scattered on waysides and waste places and in hedgerows, rough pastures, copses, etc., as an established escape from cultivation, throughout Great Britain and Ireland and in the Inner and Outer Hebrides and Orkney but no longer in Scilly or the Channel Is. Thought to be native in S.E. Europe and W. Asia but widely naturalized elsewhere in Europe and in North America and Japan.

Formerly much grown, especially for medicinal purposes, the root-stock being used as a remedy for coughs and colds and as a tonic, and candied with sugar as a popular sweetmeat. The Romans are said to have eaten the lvs as a pot-herb.

Section 2. *Enula* Duby. Outer involucral bracts linear to lanceolate, herbaceous. Achenes with 5–12 ±distinct ribs; pappus of 1 row of shortly ciliate hairs.

2. I. salicina L.   Irish Fleabane.

A perennial herb with *slender white hypogeal stolons* and stiffly *erect very lfy stems* 25–50(–75) cm, simple or corymbosely branched above, ±glabrous. *Lvs 2–6(–10) cm*, lowest oblanceolate, *middle and upper ±broadly lanceolate-elliptical, cordate and semi-amplexicaul at the base*, all firm, entire or remotely denticulate, stiffly ciliate, glabrous or somewhat hairy on the veins beneath, prominently reticulate-veined above. *Fl.-heads 2–3 cm diam., solitary or 2–5 in a corymb*. Involucre 8–12 mm, hemispherical; outer bracts 5–7 mm, lanceolate with ±spreading tips, inner longer and narrower. *Peripheral fls with ligule 1.5–2.5 cm, golden-yellow* like the disk-fls. Achenes 1.5–2 mm, cylindrical, rather weakly ribbed, glabrous; pappus whitish. $2n = 16^*$. Fl. 7–8. H.

Native. Confined to the limestone shores of Lough Derg in N. Tipperary and S.E. Galway, close to winter flood-level on rough grassy stony ground with *Schoenus nigricans*, *Molinia caerulea*, *Sesleria albicans*, etc. Native in most of Europe and in W. Asia, especially in *Molinia*-rich vegetation or moist calcareous grassland, and reaching the Arctic Circle in Finland.

I. britannica L. was recorded in 1894 as established by the Cropstone Reservoir near Leicester, presumably introduced by waterfowl. It persisted there for perhaps half a century but could not be found in 1947. It is a perennial herb with erect usually simple appressed-hairy stem 15–75 cm and *narrowly lanceolate to oblong entire or denticulate lvs*, the lower 4–15 cm, narrowed into a short petiole, the upper rounded to subcordate at the sessile and somewhat amplexicaul base; all sparsely hairy above, ±densely pubescent beneath. *Fl.-heads 2–4(–5.5) cm diam.*, solitary or in few-headed corymbs. Involucre 7–9 mm, hemispherical, with 2 rows of *herbaceous and softly hairy linear bracts*. Peripheral fls with yellow ligules 15–25 mm. Achenes 1–1.5 mm, hairy. Fl. 7–8. $2n = 32$. H. Native in damp pastures and by streams and ditches in much of Europe, but now at most a very rare casual in the British Is.

3. I. conyza DC.   Ploughman's Spikenard.

Conyza squarrosa L.; *I. squarrosa* (L.) Bernh., non L.

A biennial to perennial herb with oblique irregularly thickened stock and erect or ascending stems 30–120 cm, branched above, often reddish, softly pubescent to tomentose. *Lower lvs ovate-oblong*, narrowed into a flattened petiole; *upper elliptical to lanceolate, sub-sessile*, acute; all 3–5 cm, irregularly denticulate, downy, especially beneath. *Fl.-heads c.* 1 *cm diam., numerous in corymbs. Involucre* 9–15 mm, *broadly cylindrical; outer bracts* lanceolate, green, pubescent, with spreading or recurved tips; *inner longer, narrower, ±scarious*, ciliate, often purple. *Peripheral fls tubular or with ligule shorter than inner bracts;* all fls yellowish. *Achenes c.* 2 mm, *dark brown, strongly ribbed*, sparsely hairy above; pappus reddish-white. Fl. 7–9. 2n = 32*. H.

Native. A calcicolous plant of dry or rocky slopes and cliffs and of open scrub-woodland, locally common in England and Wales northwards to Westmorland and Durham; Channel Is. C. and S.E. Europe northwards to Denmark; Near East; Algeria. The basal lvs are sometimes mistaken for those of the Foxglove.

4. I. crithmoides L.   Golden Samphire.

A *perennial suffruticose maritime* plant with branched woody stock and ascending *very fleshy glabrous stems* 15–60(–90) cm, branched above, rarely somewhat glandular. *Lvs* 2–4.5(–6) cm, glabrous, *linear to oblanceolate, narrowed below to the sessile base*, those on main stem *often 3-toothed at the apex*, otherwise entire; *upper lvs with axillary lf-clusters. Fl.*-heads c. 2.5 cm diam. in a few-headed corymb, the lvs beneath them small and bract-like. Involucre 5–10 mm, hemispherical; outer bracts linear-lanceolate, inner longer, narrower and with scarious margins; all acuminate. *Golden-yellow ligules* of the numerous marginal fls 14–25 mm, *up to twice as long as inner involucral bracts*; disk-fls orange-yellow. Achenes 2–3 mm, grey-brown, cylindrical, faintly ribbed, pubescent; pappus somewhat reddish. Fl. 7–8. 2n = 18*. H–Ch.

Native. Salt-marshes, shingle-banks and maritime cliffs and rocks on the coasts of southern Great Britain from N. Essex to Anglesey and near the Mull of Galloway; E. and S. Ireland from N. Dublin to W. Cork; Channel Is. Atlantic and Mediterranean coasts of W. and S. Europe northwards to c. 55° 30′ N. in Scotland and eastwards to Yugoslavia and Greece; inland in E. Spain.

The closely related genus *Dittrichia* W. Greuter differs from *Inula* in having medium to small fl.-heads, *cylindrical unribbed achenes narrowed above into a neck and then expanded into a rim or cupule upon which the pappus-hairs are inserted*, and *pappus-hairs connate near their base.* There are no native spp., but 2, both densely glandular and viscid, occur as infrequent casuals. **D. viscosa** (L.) W. Greuter (*Inula viscosa* (L.) Aiton), native in S. Europe, N. Africa and the Near East, is a resin-scented perennial, 40–130 cm, woody at base, with linear to oblong-lanceolate distantly denticulate lvs, the upper sessile and semi-amplexicaul, and medium-sized fl.-heads in long racemes, their marginal fls with ligules 10–12 mm, distinctly exceeding the involucre (6–8 mm). It cannot be regarded as truly naturalized but individual plants seem able to persist for several years. **D. graveolens** (L.) W. Greuter (*Inula graveolens* (L.) Desf.), native in the Mediterranean region and in W. Europe northwards to near Paris, is a narrow-leaved annual, 20–50 cm, smelling of camphor. Its fl.-heads are small with involucre only 4–7 mm, not or barely exceeded by ligules of

the marginal fls. An occasional wool-alien, said to be incapable of setting seed in this country.

13. PULICARIA Gaertner

Annual to perennial herbs resembling *Inula* spp. but *achenes with a 2-rowed pappus*, the *inner row of very shortly ciliate hairs* or narrow scales, the *outer a denticulate to ±deeply laciniate cup.* Involucral bracts numerous, narrow. Receptacle naked. Ligules of marginal fls equalling or exceeding the involucre.

Upwards of 50 spp. in Europe, Africa and W. Asia eastwards to the Altai Mountains and India.

Stoloniferous perennial; middle and upper stem-lvs sessile, cordate, clasping the stem with large basal auricles; fl.-heads usually 1.5–2.5 cm diam. with ligules of marginal fls spreading and much exceeding the involucre; cup of outer pappus with denticulate margin. **1. dysenterica**
Annual; middle and upper stem-lvs narrowed into a sessile base, neither cordate nor auricled but somewhat amplexicaul; fl.-heads usually c. 1 cm diam. with ligules of marginal fls erect and about equalling the involucre; cup of outer pappus laciniate to half-way or more.
                                                2. vulgaris

1. P. dysenterica (L.) Bernh.   Fleabane.
Inula dysenterica L.

A perennial stoloniferous herb with erect ±tomentose stems 20–60 cm, corymbosely branched above. Lowest lvs oblong, narrowed into the petiole; middle and upper *stem-lvs oblong-lanceolate to lanceolate, cordate at base, amplexicaul*; all 3–8 cm, undulate, entire or distantly sinuate-toothed, green above, greyish-tomentose beneath. Heads 1.5–3 cm diam., c. 2–12 in a loose corymb; bracts linear, herbaceous with long fine scarious tips, covered with long hairs, glandular. *Ray-florets* numerous, ligulate, linear, *almost twice as long as the involucre and disk-florets*, golden yellow like the disk. Achenes 1.5 mm, hairy; *outer pappus a small denticulate or crenate cup*, inner of long hairs. Fl. 8–9. Visited by many insects, chiefly flies. 2n = 18*. H.

Native. A common plant of marshes, wet meadows, ditches, etc., throughout the British Is. northwards to Islay and Fife; Channel Is. Europe northwards to Denmark and C. Russia; N. Africa; Caucasus; Asia Minor.

2. P. vulgaris Gaertner   Small Fleabane.
Inula pulicaria L.

An annual herb with much-branched slightly glandular-pubescent stems 8–45 cm, the branches overtopping the main stem. Basal lvs oblanceolate narrowed into a stalk-like base; middle and upper *lvs* elliptical or lanceolate, rounded at the base, half-amplexicaul but *not or hardly cordate*; all 2.5–4 cm, undulate at the margin, entire or distantly sinuate-toothed, glandular-pubescent or ±glabrous. Heads c. 1 cm diam., numerous, in a lax subcorymbose panicle. Bracts linear, herbaceous with long fine scarious points, glandular-pubescent. *Ray-florets* in 1 row, shortly ligulate, ±erect, *hardly exceeding the involucre* and pale yellow disk-florets. Achenes 1.5 mm, hairy; outer pappus a deeply laciniate cup; inner of hairs little longer than the achenes. Fl. 8–9. $2n = 18^*$. Th.

Native. A very rare plant of moist sandy places, pond-margins, etc., where water stands in winter but not during the growing season. Formerly widespread in C. and S.E. England but now restricted to a very few localities from S. Wilts to Surrey and near the south coast from Hants to W. Sussex. Europe north to Denmark and S. Sweden; Caucasus; W. Asia; N. and E. Africa.

*__Telekia speciosa__ (Schreber) Baumg. (_Buphthalmum speciosum_ Schreber), Large Yellow Ox-eye, is a robust aromatic perennial herb with erect fl.-stems 60–200 cm, ±pubescent and with small sessile glands. _Basal lvs_ c. 30 cm, _broadly ovate- to triangular-cordate_, petiolate; stem-lvs smaller upwards, becoming rounded to cuneate at base and more shortly petiolate to sessile; all ±coarsely dentate, ±glabrous above and pubescent beneath. Fl.-heads 5–8 cm diam., long-stalked in a terminal corymb of 2–8 heads; _involucre hemispherical_, c. 1.5 cm diam., _its bracts_ coriaceous below, _ovate to ovate-lanceolate_, blunt, the _outer with deflexed herbaceous tips_; _receptacle_ somewhat convex _with long filiform scales equalling the subtended fls. Ray-fls with deep yellow ligules only_ 1 mm _wide_ but c. twice as long as involucre; disk-fls brownish-yellow. Achenes of ray- and disk-fls similar, 6 mm, ±cylindrical, many-ribbed, glabrous; _pappus a short membranous cup with irregularly toothed margin_. Resembles _Inula helenium_ but has somewhat smaller fl.-heads with ligules only about half as long and differs also in the scaly receptacle and lack of pappus-hairs.

A native of S.E. Europe and the Near East, first introduced in 1739 and since then much grown in gardens. Following escape from gardens or spread from plantings on large estates it has become established in a few scattered localities in Great Britain from the south Midlands of England and mid-Wales to Argyll and Aberdeen, mostly by streams and lakes and on waysides and waste ground; occasionally a grain-adventive.

14. FILAGO L.

Annual herbs with stems and alternate lvs ±densely hairy, _often tomentose and greyish-white_, less commonly yellowish or greenish. _Fl.-heads_ heterogamous, _small, mostly in roundish terminal and axillary clusters_, rarely solitary; involucre ±5-angled, of numerous imbricating bracts, the outer commonly hairy on the back; _receptacle_ ranging from ±flat to conical or obconical and to cylindrical, _naked except for scales subtending outer fls, often doubtfully distinguishable from inner involucral bracts. Fls all tubular, outer female with filiform corolla, inner hermaphrodite_ (or functionally male) _with broader corolla_, often with female fls intermixed. Achenes slightly compressed laterally, the outer usually larger and sometimes enclosed by involucral bracts and shed with them; pappus of central fls of many simple hairs, of outer fls of few hairs or 0, sometimes 0 on all achenes.

About 25 spp. mainly in Europe, Asia and N. Africa

with a few in North and South America.

1　Fl.-heads usually 8–40 in each cluster, rarely fewer; involucral bracts cuspidate or awned, erect to somewhat divergent in fr.; marginal fls subtended but not enclosed by involucral bracts or receptacular scales.　(_Filago_ sens. str.)　2

　Fl.-heads usually fewer than 8 in each cluster, sometimes solitary, rarely up to 14; involucral bracts not cuspidate or awned, spreading star-wise in fr.; marginal fls ±enclosed by involucral bracts or by receptacular scales resembling them.

　　　　　　　(_Logfia_ Cass., emend.)　4

2　Lvs linear to oblong-lanceolate, widest below the middle, usually distinctly undulate; fl.-heads in dense clusters usually of 20–35, not overtopped by subtending lvs; involucre not or obscurely 5-angled.

　　　　　　　　　　　　　　__1. vulgaris__

　Lvs linear-oblong to spathulate, widest above the middle, not or hardly undulate; fl.-heads in loose clusters usually of 8–25, often overtopped by 1 or more subtending lvs; involucre ±clearly 5-angled with bracts in 5 distinct vertical rows.　3

3　Plant greyish-white; involucre sharply 5-angled with 4–6 bracts in each vertical row; bracts divergent in fr. with apical awn yellowish and recurved.

　　　　　　　　　　　　　__3. pyramidata__

　Plant yellowish-green; involucre weakly 5-angled with 3(–4) bracts in each vertical row; bracts with apical awn reddish, straight.　__2. lutescens__

4　Clusters of fl.-heads usually much overtopped by lvs at their base; lvs 8–25 mm, mostly linear-subulate; marginal achenes completely enclosed by saccate-based involucral bracts.　__4. gallica__

　Clusters of fl.-heads usually not overtopped by subtending lvs; lvs up to 20 mm but often much shorter; marginal achenes incompletely enclosed by scales or bracts not or only slightly saccate-based.　5

5　Plant greyish with short silky hairs; lvs narrowly oblong to linear-lanceolate, erect and closely stem-appressed; involucre distinctly 5-angled, its blunt bracts woolly below but glabrous above, the outer keeled.　__5. minima__

　Plant densely white-woolly; lvs linear-lanceolate, ±erect but not closely stem-appressed; involucre not or obscurely angled, its acute bracts woolly to the tip.　_F. arvensis_ (p. 464)

1. F. vulgaris Lam.　　　　　Cudweed.

F. canescens Jordan, _F. germanica_ L., non Hudson

An annual herb with erect or ascending densely woolly stems 5–30(–45) cm, simple or branched at the base, main stem and branches with further branches immediately beneath the terminal clusters of heads. _Lvs_ 1–2(–3) cm, erect, lanceolate, blunt or tapering to an acute apex, entire and usually _undulate_, covered with _white_ woolly hairs. Heads 20–40 in ±sessile clusters c. 12 mm diam. terminating main stem and branches and half sunk in white woolly hairs, each head c. 5 mm. _Bracts_ linear, longitudinally folded, cuspidate, erect, in 5 rows; the outer bracts short, densely woolly, straight-pointed, the inner longer, scarious, yellowish, _with a yellow awn-like point_. Florets small, yellow, several rows of female surrounding the central hermaphrodite florets. Achenes 0.6 mm, somewhat compressed, papillose; pappus of inner achenes scabrid, longer than the achenes; of outer 0. Fl. 7–8. $2n = 28*$. Th.

Native. A fairly common plant of heaths, dry pastures, fields and waysides, usually on acid sandy soils, throughout England, Wales, Ireland and S. Scotland and reaching Sutherland; rare in N. Scotland and not in the Hebrides, Orkney or Shetland; occasional in Ireland; Channel Is. C. and S. Europe northwards to Denmark and Poland; Siberia and W. Asia; N. Africa; Canary Is. Introduced in North America.

2. F. lutescens Jordan   Red-tipped Cudweed.

F. apiculata G.E. Sm. ex Bab.; *F. germanica* auct., non L. nec Hudson

An annual herb 10–25 cm, resembling **1** but *with broader* (3–6 mm) *oblong-lanceolate to spathulate, apiculate, hardly undulate lvs*; stem and lvs *covered with yellowish woolly hairs.* Fl.-heads c. 5 mm, 10–25 in each cluster, half-sunk in woolly hairs and *overtopped by 1–2 lvs at their base. Involucre weakly 5-angled,* its *bracts* 3(–4) in each vertical row, oblong-ovate to lanceolate, *erect,* the outer yellowish-woolly on the back, all *with straight usually reddish awns.* Inner fls mainly female. Achenes oblong-cylindrical; pappus of outer female fls 0. Fl. 7–8. $2n = 28$. Th.

Native. A rare and decreasing plant of sandy fields and waysides in a few localities in S. E. England from N.W. Norfolk to Hants and W. Sussex. Much of Europe from Denmark and S. Sweden southwards to Portugal, C. Spain and Sicily and eastwards to Poland and Bulgaria; W. Asia; N. Africa.

3. F. pyramidata L.   Broad-leaved Cudweed.

F. spathulata C. Presl; *F. germanica* Hudson

An annual herb 5–30(–40) cm, usually with *several decumbent or ascending branches from near the base* of the rather short main stem, *all stems with divaricate branches above.* Lvs (5–)10–15 mm, *narrowly oblong to ±broadly spathulate, widest above the middle, entire, usually apiculate, hardly undulate.* Stems and lvs greyish-white with woolly hairs. Fl.-heads c. 5 mm *in sessile clusters* 5–12 mm diam., each with (5–)10–20 heads and *commonly overtopped by 3–5 lvs at their base. Involucre sharply 5-angled,* its scarious cuspidate *bracts* 4–6 *in each vertical row, the outer* softly woolly on the back and *distinctly divergent in fr. with recurved yellowish awn,* the inner longer, awnless, whitish and shining. Achenes c. 0.6 mm; pappus of inner fls longer than achenes, of outer 0. Fl. 7–8. $2n = 28$. Th.

Native. A rare and decreasing plant of sandy and chalky fields and waysides in S. England from E. Kent to Devon and Cornwall and northwards to Cambridge, Bedford and Berks; also S.E. Yorks (Spurn Pt). S. and W. Europe northwards to E. England and Netherlands and eastwards to Bulgaria and the Crimea; W. Asia; N. Africa; Canary Is.

4. F. gallica L.   Narrow-leaved Cudweed.

Logfia gallica (L.) Cosson & Germ.

An annual herb, 5–20(–25) cm, with erect or ascending slender stems commonly much *branched below and also, dichasially, above.* Lvs 8–25 mm, mostly *linear-subulate,* held *loosely erect* but with the long acute *tip somewhat spreading. Stem and lvs ±grey-green* with short silky hairs. Fl.-heads 4 mm *in clusters of 2–7(–14), usually much overtopped by linear-lanceolate lvs at their base. Involucre 3–4 mm, distinctly but bluntly 5-angled,* with a few small outermost ovate bracts much shorter than the *main outer set of broadly lanceolate-triangular bracts with strongly saccate base completely enclosing achenes of marginal female fls, woolly on the back* except for the infolded margins and long subacute tip; *inner bracts and receptacular scales* becoming more narrowly lanceolate, less hairy on the back and less concave-based from outside inwards, *subtending but not enclosing some achenes of central* female and hermaphrodite *fls;* all spreading star-wise in fr. Achenes of outer fls 0.9 mm, falling with the enclosing bracts and with pappus 0; of central fls c. 0.6 mm, covered with transparent papillae and with pappus 2–2.7 mm. Fl. 7–9. $2n = 28$. Th.

?Native. Channel Is. Formerly in a very few dry sandy and gravelly places in S.E. England but perhaps now extinct. W. and S. Europe northwards to Belgium and eastwards to Bulgaria and Turkey; W. Asia; N. Africa; Azores.

5. F. minima (Sm.) Pers.   Small Cudweed.

Logfia minima (Sm.) Dumort.

An annual herb, 2–15(–30) cm, with slender erect or ascending stems *branched* below and also ±*dichasially above.* Lvs 4–10 mm, *narrowly oblong to linear-lanceolate, acute, erect and closely stem-appressed. Stem and lvs greyish* with silky hairs. *Fl.-heads* c. 3 mm *in small clusters of 3–7* at ends and in forks of main stem and branches; *clusters not overtopped by lvs* at their base. Involucre distinctly 5-angled with a few narrowly lanceolate outermost bracts less than half as long as the main *outer* set of *broadly lanceolate bracts* 2.5–3.5 mm, *keeled and with slightly saccate base and long-tapering but blunt tip, woolly on the back but the upper end glabrous, scarious and shining;* inner bracts similar but somewhat shorter and narrower with non-saccate base and rounded tip; innermost bracts and receptacular scales narrowly oblong, scarious and quite glabrous; all spreading star-wise in fr. Central fls all female apart from 3–5 hermaphrodite. Achenes of outer fls c. 0.9 mm, smooth and lacking pappus, of central fls c. 0.5 mm, papillose and with many-rowed deciduous pappus. Fl. 6–9. $2n = 28$. Th.

Native. Widespread and locally frequent on circum-neutral to moderately acid sands and gravels in open vegetation of heaths, grassland and sand dunes, dry fields, waysides and waste ground, also in sand-pits and stone-quarries and on walls, mainly in eastern Great Britain northwards to Caithness but extending more sparsely westwards to Cornwall, W. Wales, Isle of Man and S.W. Scotland with Arran and Islay; rare in Ireland, chiefly in the east and north; Channel Is. Through much of Europe northwards to c. 60° N. in Scandinavia with a few scattered localities further north; Siberia.

F. arvensis L. (*Logfia arvensis* (L.) J. Holub), now an infrequent casual, resembles *F. minima* but has longer (10–20 mm) and less stem-appressed lvs which, like the stem, are densely white-tomentose. The short ±simple upper branches arise singly (seldom dichasially) and bear axillary as well as terminal clusters of fl.-heads which are sometimes overtopped by lvs at their base. The involucre is obscurely angled, its outer bracts unkeeled and woolly to the short glabrous tip. Fl. 7–9. $2n = 28$. Native in most of Europe to c. 65° N. in Scandinavia; W. Asia; Canary Is.

15. GNAPHALIUM L.

Annual to perennial herbs with *stems and alternate simple lvs usually covered with whitish woolly hairs. Flheads* heterogamous, small, *borne without subtending lvs in terminal clusters or corymbose groups of clusters or in lfy spikes or racemes* of either single or clustered heads, rarely solitary; *involucral bracts* imbricate in several rows, *scarious, spreading in fr.; receptacle* flat, *naked. Fls all tubular*, the *outer female with filiform corolla* and lacking anthers, the *inner hermaphrodite with broader corollas*, often with female intermixed. Achenes not or little compressed, all with pappus of 1 row of simple hairs, sometimes ciliate below, free to the base or connate in a basal ring.

About 200 spp. mainly in temperate regions of both Old and New Worlds and on tropical mountains, a few in the Arctic.

Gnaphalium as defined above is a large genus which taxonomists have divided in various ways, but it has seemed best to retain it intact until a generally agreed scheme becomes available.

1 Perennials with single or clustered fl.-heads in a terminal spike or raceme; non-flowering shoots commonly present.   (*Omalotheca* Cass.) **2**
   Annuals with fl.-heads in dense terminal clusters or in corymbose groups of clusters; non-flowering shoots 0.   **4**
2 Densely tufted Scottish mountain plant usually only 2–12 cm; fl.-heads 2–10 in a short compact spike, rarely solitary; pappus-hairs free, falling separately.   **3. supinum**
   Plant 8–60 cm, not densely tufted, lowland or alpine; terminal spike ±elongated with at least 10 and often very many fl.-heads; pappus-hairs united in a basal ring and falling together.   **3**
3 Lvs narrowly lanceolate to linear-oblong, mostly 1-veined, diminishing upwards with no abrupt change in size near base of spike occupying at least upper $\frac{1}{3}$ of stem. Widespread lowland plant. **1. sylvaticum**
   Lvs elliptic- to oblong-lanceolate, mostly 3-veined, diminishing abruptly in size near base of spike occupying no more than $\frac{1}{4}$ of stem. Rare plant of Scottish mountains.   **2. norvegicum**
4 Clusters of fl.-heads with conspicuous lvs at their base, some erect and overtopping the clusters; outer

involucral bracts mottled brown and paler; pappushairs free, falling separately. (*Filaginella* Opiz). Stem much branched below.   **4. uliginosum**
   Clusters of fl.-heads lacking lvs at their base; involucral bracts scarious and shining, entirely white or yellowish; pappus-hairs ciliate below, free, falling ±separately.   (*Gnaphalium* L., sens. str.) **5**
5 Stem 8–40 cm; stem-lvs not decurrent, white-tomentose above and beneath; involucral bracts strawcoloured.   **5. luteo-album**
   Stem 20–80 cm; stem-lvs decurrent, green and asperous above, white-tomentose beneath; involucral bracts a slightly yellowish white.   ***G. undulatum** (p. 465)

Section 1. Tomentose perennials with non-flowering shoots. Achenes with pappus-hairs united in a basal ring and falling together (*Omalotheca* Cass.)

1. G. sylvaticum L.   Heath Cudweed.
Omalotheca sylvatica (L.) Schultz Bip. & F. W. Schultz
A perennial herb with woody stock bearing short ascending non-flowering shoots and erect usually simple flowering shoots (8–)20–60 cm, all *whitish-tomentose*. Rosette-lvs and lower stem-lvs 2–8 cm, lanceolate, narrowed into a long petiole-like base; *middle and upper stem-lvs* linear-lanceolate to linear, sessile, *diminishing in size ±steadily upwards; all acute, 1-veined* (rarely indistinctly 3-veined), *whitish-tomentose beneath, glabrescent above*. Fl.-heads c. 6 mm, *in clusters of 2–8, or sometimes solitary, forming a lfy spike*, often interrupted below, *occupying at least the upper third and commonly more than half of the stem*. Outer involucral bracts broadly oblong with blunt ±emarginate tip, brown but with hyaline upper end and margins, woolly on the back below; inner longer, the innermost about equalling the fls, usually with pale green central stripe and pink or silvery upper end and margins. Fls yellowish, female except for only 3–4 hermaphrodite in the centre. Achenes 1.5 mm, cylindrical, hispid; pappus reddish. Fl. 7–9. Little visited by insects. $2n = 56*$. H.

Native. Locally common in dry open heathy woods and on heaths, dry pastures and waysides on acid soils throughout the British Is. but less common in the west. Europe to 70° N. in Scandinavia; Caucasus; North America.

2. G. norvegicum Gunn.   Highland Cudweed.

A perennial herb resembling *G. sylvaticum* in habit but only 8–30 cm and with broader *lanceolate-acuminate 3-veined stem-lvs* which diminish abruptly in length only in the shorter more *compact spike*, continuous or interrupted below, occupying no more than the upper $\frac{1}{4}$ *of the length of the stem*. Heads 6–7 mm, solitary or 2–3 in the axils of the upper lvs. Bracts with an olive central stripe and *dark brown* scarious margins, the *inner shorter than the fls*. Achenes 1.5 mm, cylindrical, hispid; pappus white. Fl. 8. $2n = 56$. H.

Native. A rare plant of alpine rocks to 1100 m on mountains in Perth, Angus, Aberdeen, Inverness and Ross; also reported from Caithness. Mountains of

C. Europe, Balkan peninsula and Caucasus; arctic and subarctic Europe; W. Asia; eastern North America; Greenland.

3. G. supinum L. Dwarf Cudweed.

A *dwarf perennial tufted herb* with a slender creeping branched stock producing numerous short lfy non-flowering and simple erect or ascending flowering stems 2–12(–20) cm. Basal lvs linear-oblanceolate, stem-lvs linear acute; all 0.5–1.5(–2) cm, entire, woolly above and below. *Heads 1–7 in a short ±compact terminal spike* somewhat lengthening in fr., each campanulate, c. 6 mm long and 8 mm diam. Bracts in 3–4 rows, broadly ellipti-cal with a woolly olive-coloured central stripe and broad brown scarious margins, the inner bracts almost equal-ling the fls. Female fls in 1 marginal row. Achenes 1.5 mm, spindle-shaped, compressed, shortly hairy. Fl. 7. 2*n* = 28*. Chh.

Native. An alpine plant of cliffs and moraines of Scot-tish mountains from Stirling to Sutherland, and in Skye. Mountains of C. Europe and W. Asia; arctic Europe; Greenland; North America.

Section 2. Tomentose annuals without non-flowering shoots. Clusters of fl.-heads subtended by lvs. Achenes with free filiform pappus-hairs not ciliate at base, falling separately (*Filaginella* Opiz).

4. G. uliginosum L.
Marsh Cudweed, Wayside Cudweed.
Filaginella uliginosa (L.) Opiz

An annual grey to whitish tomentose herb with erect or ascending main stem 4–20 cm, usually with many decumbent to ascending branches from near the base. Lvs 10–50 × 2–5 mm, *linear-oblong to oblanceolate*, nar-rowed below to a sessile base, ±acute, with entire but often undulate margins, ±densely *cottony above and beneath. Fl.-heads 3–4 mm, in dense terminal clusters* of 3–10 on main stems and short lateral branches, *each cluster with conspicuous subtending lvs some of which overtop it. Involucral bracts longer inwards*, the inner-most at least equalling fls, outermost less than half as long, *all brown and glabrous above, paler and ±hairy below*, spreading in fr. *Fls yellowish*, all female apart from 5–8 hermaphrodite in centre. Achenes olive-brown, *tiny* (0.5 mm), ±glabrous; *pappus* 1.5 mm, *deci-duous.* Fl. 7–9. Little visited by insects. 2*n* = 14. Th.

Very variable. Our native plant is placed in the wide-spread subsp. **uliginosum.**

Native. Common throughout the British Is. in open vegetation of periodically damp places on a wide range of moderately acid soils including banks of streams, ditches and ponds, woodland rides, waysides, arable fields and gardens, etc. Most of Europe to c. 68.5° N. in Scandinavia; W. Asia; North America.

Section 3. Tomentose annuals without non-flowering shoots. Clusters of fl.-heads without subtending lvs. Achenes with free filiform pappus-hairs ciliate at their base and falling in small groups. (*Gnaphalium* L.)

5. G. luteo-album L. Jersey Cudweed.

An annual herb with erect stem 8–45 cm, usually with decumbent then *erect branches* from near the base, the main stem and branches simple below but *corymbosely branched above*, all very densely covered with white woolly hairs. Basal lvs broadly oblanceolate, usually blunt; stem-lvs oblong-amplexicaul, ±undulate, acute; all 1.5–3(–7) cm, woolly on both sides. *Heads* in dense terminal lfless clusters of 4–12, *not overtopped by basal lvs*; each head 4–5 mm, ovoid. *Bracts* elliptical, largely scarious, *shining, straw-coloured*, only the outermost woolly below. Fls yellowish with red stigmas. Achenes 0.5 mm, brown, tubercled. Fl. 6–8. Visited by various flies and bees. 2*n* = 14.

Native in the Channel Is.; probably introduced in Great Britain. A rare plant of sandy fields and waste places in Jersey and Guernsey and formerly established in a few localities in S. and S.E. England where it may have been native; reappeared on Norfolk coast in 1977; casual elsewhere. Europe northwards to Holstein, S. Sweden and Latvia; temperate W. and C. Asia.

***G. undulatum** L. (S. Africa), 20–80 cm, more robust than *G. luteo-album*, and with the *lvs decurrent, green and asperous above*, is established in the Channel Is and is casual in E. Cornwall.

16. ANAPHALIS DC.

Perennial white-woolly herbs, some suffruticose, differ-ing from *Gnaphalium* in having *individual fl.-heads in terminal corymbs*, in being ±*dioecious*, and in the some-what *clavate pappus-hairs* of the abortive ovaries in fls of male plants. From the closely related *Antennaria* they differ in being *less completely dioecious*, in the *less dis-tinctly clavate pappus-hairs of male plants* and in that the *pappus-hairs of fertile fls are free to the base*, falling separately.

About 30 spp., chiefly in temperate and tropical Asia but also in North America.

*1. A. margaritacea (L.) Bentham Pearly Everlasting.

Gnaphalium margaritaceum L.; *Antennaria margarita-cea* (L.) S. F. Gray

A perennial subdioecious herb with erect robust lfy flowering stems 30–100 cm, woolly with white hairs. Lvs 6–10 × 1–1.5 cm, elliptical, acute, ±entire, woolly beneath, becoming ±glabrous above. Heads 9–12 mm diam., numerous in terminal corymbs. *Bracts* oblong, brown below, the outer woolly, the inner glabrous, all *with a shining white scarious rounded apex*. All fls of male plants with abortive ovaries; those of the female without anthers except for a few central herm-aphrodite fls; corolla yellowish. Achenes spindle-shaped, papillose; pappus-hairs slender in fertile fls but thickened above (as in *Antennaria*) in male fls. Fl. 8. 2*n* = 28*. H.

Introduced. Long cultivated in gardens for its 'ever-lasting' fls and naturalized in moist meadows, by rivers, on wall-tops and in sandy and waste ground in several scattered localities northwards to Skye and Caithness but especially in S. Wales. Native in E. Asia and in North America; introduced early into Europe.

17. ANTENNARIA Gaertner

Perennial *dioecious or apomictic herbs*, sometimes suffruticose, *commonly stoloniferous*, with alternate simple entire lvs; *basal lvs usually in well-marked rosettes*, broader than the ±*erect stem-lvs* of the simple erect or ascending flowering shoots; *stem and lvs usually ±densely covered with whitish woolly or silky hairs. Fl.-heads few (up to 8(–12)) in terminal usually compact corymbs or racemes*, rarely solitary; *involucral bracts closely imbricate, scarious* white or coloured, often perianth-like, *not spreading in fr.*; *receptacle* flat or convex, *naked. Fls* all tubular, those *of female plants filiform, lacking anthers and with pappus a single row of slender hairs united below* and falling together; *fls of* the usually smaller *male plants with tailed anthers and abortive ovary bearing clavate pappus-hairs free to the base* and falling separately (the name *Antennaria* derives from their resemblance to the *antennae* of butterflies). Achenes cylindrical or somewhat compressed.

Many spp. in alpine and arctic areas especially of North America but also of Europe, Asia, Australia and extratropical South America.

Close to *Gnaphalium* but dioecious or nearly so in sexually reproducing spp. Male plants unknown in the many spp. shown or presumed to be apomictic.

1. A. dioica (L.) Gaertner
Mountain Everlasting, Cat's-foot.

Gnaphalium dioicum L.

A perennial with above-ground creeping woody stock producing lfy *stolons rooting at the nodes* and erect simple woolly flowering shoots 5–20 cm. *Lvs* 1–4 cm, mostly in rosettes at the ends of the stock and stolons, obovate-spathulate, blunt or apiculate; upper stem-lvs erect and appressed, lanceolate to linear, acute; all green and glabrous or sparsely hairy above, *white-woolly beneath.* Fl.-heads short-stalked, 2–8 in a close terminal corymb; those on female plants c. 12 mm diam., on functionally male plants c. 6 mm diam. Outer involucral bracts woolly below, scarious and glabrous above; of the male heads obovate-spathulate blunt, usually white, sometimes pink, spreading above like ray-florets; of the female heads linear-lanceolate ±acute, usually rose-pink, erect. Achenes 1 mm.; pappus white. Fl. 6–7. Visited by various insects and reproducing sexually. $2n = 28*$. Chh.

In var. *hyperborea* (D. Don) DC. the lvs are broader and white with hairs on both sides. The status of this variety is not yet clear, but it may merit subspecific rank.

A. hibernica Br.-Bl., described as having long-stalked heads (stalks 0.5–2 cm) with white, not pink, involucral bracts which are rounded or emarginate at the apex, is reported from limestone areas in Ireland. It seems insufficiently distinct to warrant specific rank.

Native. On heaths, dry pastures and dry mountain slopes throughout the British Is., but rare in the south; usually over limestone, basic igneous rock or base-rich boulder-clay, perhaps most commonly where the surface-soil is somewhat leached. Reaches 914 m in Scotland. Var. *hyperborea* in the Inner and Outer Hebrides. N. and C. Europe (not the Mediterranean region), Siberia, W. Asia, North America.

Tribe 6. ASTEREAE.

18. SOLIDAGO L.

Perennial herbs with ±sessile simple lvs. Fl.-heads small, yellow, usually in racemose panicles of scorpioid cymes; involucre ±cylindrical with many rows of imbricating ±lf-like bracts; receptacle flat, naked, often pitted. Ray-florets in 1 row, female or neuter; disk-florets hermaphrodite. Style-arms with terminal papillose cones. *Achenes many-ribbed, not compressed; pappus-hairs* shortly ciliate, *in* 1(–2) *row*.

About 120 spp., chiefly American.

1. S. virgaurea L.   Golden-rod.

A perennial herb with a stout obliquely ascending stock and an erect simple or somewhat branched lfy stem 5–75 cm, terete, glabrous or pubescent. Basal lvs 2–10 cm, obovate or oblanceolate, narrowing to a short stalk-like base, usually toothed; stem-lvs elliptical or oblong-lanceolate, ±acute, entire or obscurely toothed. Heads 6–10 mm in fl., short-stalked, in a panicle with straight erect branches or in a raceme. Involucral bracts greenish-yellow, linear, acute, glabrous or slightly downy, with scarious margins. Florets all yellow; ray-florets 6–12, spreading. Achenes c. 3 mm, brown, pubescent; pappus whitish. Fl. 7–9. Much visited by various bees and flies, and automatically self-pollinated. $2n = 18*$. H.

Very polymorphic, with many named varieties differing in stature, pubescence, size, shape and serration of lvs, branching of infl. and size of individual heads. Mountain forms only 5–20 cm with simple racemes of large heads have been named var. *cambrica* (Hudson) Sm. They seem to grade into the type but closer investigation may lead to the recognition of 2 or more subspp.

Native. A common plant of dry woods and grassland, rocks, cliffs and hedge-banks, dunes, etc., on acid and some calcareous substrata, throughout the British Is., but infrequent in midland England. Reaches 1080 m in Scotland. Europe; Asia; North America.

The common Golden-rod of gardens is *S. canadensis* L., a tall rhizomatous herb with stems 30–250 cm, pubescent throughout, and lanceolate 3-veined lvs roughly hairy on both sides or only beneath, toothed except sometimes at the base. Heads 5 mm. diam., golden-yellow, in dense one-sided recurved axillary partial infls to form a pyramidal panicle. Ray-florets barely exceeding the involucre and about equalling the disk-florets. Fl. 8–10. $2n = 18$. Much cultivated and often escaping. Native of North America.

S. gigantea Aiton (*S. serotina* Aiton), like *S. canadensis* but more robust, with stems glabrous below and ray-florets distinctly exceeding the involucre and disk-florets, and *S. graminifolia (L.) Salisb. (*S. lanceolata* L.) with linear-lanceolate 3–5-veined entire lvs, glabrous, scabrid on the veins beneath, and erect corymbose panicles of golden yellow heads whose ray-florets do not exceed the disk-florets, also occur as garden-escapes or casuals, the former naturalized locally in mid-Wales. North America.

19. ASTER L.

Perennial, rarely annual herbs with alternate *simple lvs.* Fl.-heads heterogamous or homogamous; *involucre of many rows of imbricating herbaceous* or scarious *bracts; receptacle flat, naked,* pitted, the pits with toothed membranous borders. *Ray-florets ligulate, blue, red or white,* 1-rowed, female or neuter, sometimes 0; disk-florets tubular, yellow, hermaphrodite. Style-arms of ray-florets linear, of disk-florets short with terminal papillose cones. *Achenes compressed, not ribbed; pappus of 1, 2 or several rows of shortly ciliate hairs.*

About 500 spp. in America, Asia, Africa and Europe.

Crinitaria Cass. (*Linosyris* Cass., non Ludwig) with ray-florets 0 and 2-rowed pappus, and *Galatella* Cass., with neuter ray-florets and many-rowed pappus, seem insufficiently distinct to warrant generic separation.

1 Maritime plant, glabrous or nearly so with markedly succulent faintly 3-veined lvs; ray-fls blue or whitish, sometimes 0. **1. tripolium**
  Lvs not markedly succulent. 2
2 Ray-fls 0; all lvs linear to linear-lanceolate, gland-dotted, sessile. Limestone rocks near the sea, rare. **2. linosyris**
  Ray-fls present; lvs broadly cordate to linear-lanceolate, all sessile or at least the lowest petiolate. Michaelmas Daisies, garden plants locally established on banks of streams, railways, etc., in fens and in waste places. 3
3 Lvs of basal tufts and basal and lower stem-lvs petiolate, the broadly cordate lamina 6–15 cm across with ±rectangular basal sinus; ray-fls white. **10. schreberi**
  Basal lvs, if any, and lower stem-lvs sessile or petiolate, often auricled at base but never both petiolate and broadly cordate; ray-fls usually bluish, sometimes reddish or white. 4
4 Infl.-branches and involucre of fl.-heads densely glandular-hairy; ray-fls reddish or pink. **3. novae-angliae**
  Infl.-branches and involucre of fl-heads not glandular-hairy; ray-fls usually bluish or white. 5
5 Fl.-stem purplish-red, usually rough all round with stiff swollen-based hairs, not glandular. **4. puniceus**
  Fl.-stem green or purplish, ±glabrous or with hairs restricted to narrow vertical bands. 6
6 Lower lvs tapering into a long distinctly winged petiole, middle and upper sessile, ±amplexicaul, all glaucous (pruinose) on both faces; involucral bracts whitish, closely appressed; ray-fls violet-blue. **5. laevis**
  Lower lvs sessile or with unwinged petiole; lvs at most somewhat pruinose beneath. 7
7 Upper lvs ±distinctly auricled and amplexicaul. 8
  Upper lvs not or very slightly auricled and amplexicaul. 9

8 Middle stem-lvs mostly 2½–5 times as long as wide; all involucral bracts erect and ±appressed; ray-fls finally bluish, sometimes white initially. **6. ×versicolor**
  Middle stem-lvs mostly 4–10 times as long as wide; outer involucral bracts loosely spreading or recurved; ray-fls usually violet-blue. **7. novi-belgii**
9 Lvs usually indistinctly auricled; involucral bracts up to 7 mm, somewhat loosely appressed; ray-fls initially white then violet-blue. **8. ×salignus**
  Lvs not at all auricled; involucral bracts not exceeding 5.5 mm, all appressed; ray-fls white. **9. lanceolatus**

1. A. tripolium L. Sea Aster.

A short-lived perennial *maritime* herb with a short swollen suberect rhizome and stout erect stems 15–100 cm, glabrous. *Lvs* 7–12 cm, *fleshy, glabrous,* faintly 3-veined, entire or obscurely toothed; basal lvs oblanceolate to obovate, narrowing into a long stalk; stem-lvs narrowly oblong to linear. Heads 8–20 mm diam., in corymbs, their stalks with 1–2 small bracts. Involucre of not many appressed narrow blunt bracts, the outer scarious-tipped, the inner longer, largely scarious. Ray-florets spreading, blue-purple or whitish, or 0 (var. *discoideus* Reichenb. fil); disk-florets yellow. Achene 5–6 mm, brownish, hairy; pappus 10 mm, brownish. Fl. 7–10. Visited by many flies and bees, etc.; self-pollination possible. $2n = 18^*$. Hel.

Native. A common salt-marsh plant occurring also on maritime cliffs and rocks all round the coasts of the British Is., but almost confined to estuaries in N. England and Scotland. Rarely found inland at saltworks. Var. *discoideus* is frequent southwards from N. Lancs and N. Yorks and occurs in two isolated localities in Scotland. Most European coasts but not in Iceland and the Faeroes; N. Africa; Caspian Sea; Lake Baikal; inland saline areas of Europe and C. Asia.

2. A. linosyris (L.) Bernh. Goldilocks.

Chrysocoma linosyris L.; *Linosyris vulgaris* DC.; *Crinitaria linosyris* (L.) Less.

A perennial herb with a woody stock and erect slender glabrous stems 10–50 cm, wiry, very lfy. *Lvs* (1–)2–5 cm, *very numerous, linear,* acute, narrowing to the base, 1-veined, entire and glabrous but rough at the margins and punctate above. Heads 12–18 mm diam., in dense corymbs. Involucre lax, of many acute bracts; the outer linear, spreading at the tip, lf-like; the inner oblong, yellowish, scarious-margined. Florets exceeding the bracts, bright yellow. Achenes 5 mm, brown, pubescent; pappus about equalling the achene, reddish. Fl. 8–9. $2n = 18^*, 36$. Hp.

Native. A rare plant of limestone cliffs in S. Devon, N. Somerset, Glamorgan, Caernarvon, Pembroke and on Humphrey Head, Cumbria. Most of Europe northwards to England and S.E. Sweden.

The following are Michaelmas Daisies, mostly native in North America, which are grown in gardens and have escaped and established themselves in at least one locality. Establishment is most commonly on banks of streams and ditches and in fens or marshes but often on railway banks or waste ground.

*3. A. novae-angliae L. *Fl.-stems* robust, to 2 m or more, ±rough with spreading hairs, *glandular in infl.* Stem-lvs 5–10 cm, lanceolate-acuminate, *entire*, the *sessile* base *auricled and semi-amplexicaul*; hairy or shortly hispid. *Fl.-heads* 2–5 cm diam. in a corymbose panicle with *short densely glandular peduncles*; *involucre* hemispherical, *glandular*, its linear, often purplish, *viscid bracts* held *loosely erect* but *with acute to awned tips spreading or recurved*. *Ray-fls purplish-red*, more rarely pink or white. Much grown in gardens and widely established northwards to Argyll. Distinctive in the glandular infl.-branches and involucres and the purplish-red ray-fls.

*4. A. puniceus L. *Fl.-stems* to c. 180 cm, commonly *purplish-red* and *hispid all round with stiff spreading swollen-based hairs*, but sometimes more softly hairy to glabrous, *eglandular*. Stem-lvs up to 20 cm, broadly to narrowly lanceolate-acuminate, entire to coarsely serrate, the *sessile base strongly auricled* and ±amplexicaul; *±scabrid*. Fl.-heads in a corymbose panicle; *involucre* 7–12 mm, hemispherical, its subequal herbaceous but *scarious-margined linear bracts* held *loosely erect with ±recurved attenuate* tips. Ray-fls violet-blue, more rarely lilac, pink or white. Established in a few places in E. Scotland northwards to Perth.

*5. A. laevis L. *Fl.-stems* to c. 120 cm, glabrous, *±glaucous* (pruinose). Stem-lvs ovate to narrowly lanceolate, the *broad basal lvs tapering into winged petioles*, the narrower *middle and upper sessile, auricled and ±amplexicaul*; all *thick*, entire or obscurely dentate, somewhat *pruinose*. Fl-heads in a long panicle; *involucre* 5–8 mm, campanulate or turbinate, the *very unequal bracts* linear-oblong to linear, coriaceous and *appressed*, *whitish* but with green deltate or rhombic patches above. Ray-fls violet-blue, more rarely white. Established in a very few places in Great Britain northwards to Nairn.

*6. A. ×versicolor Willd. (probably *A. laevis × A. novi-belgii*). *Fl. stems* 1–2 m, *glabrous or with hairs in narrow vertical bands*. Resembles *A. laevis* in its *broadish lvs* 2½–5 *times as along as wide*, the middle and upper ±clearly auricled and somewhat amplexicaul, but differs in that the *lowest lvs have short and unwinged petioles* and *no lvs are more than slightly pruinose beneath*. Fl.-heads in a corymbose panicle; *involucre* c. 6 mm, campanulate, its *bracts* ±appressed but less regularly imbricate and with longer green tips than in *A. laevis* and not as spreading or recurved as in *A. novi-belgii*. *Ray-fls* commonly *white initially but becoming violet-blue* or pink. Established in a very few places in S. and C. England, Isle of Man and S. Scotland (Midlothian). Very rare as a wild plant in North America but grown in gardens in many European countries and locally naturalized.

*7. A. novi-belgii L. Fl.-stems up to 120 cm, ±erect, *glabrous or with hairs in narrow vertical bands*. Stem-lvs oblong- to linear-lanceolate, acuminate, 4–10 *times as*

long as wide, firm, *auricled and somewhat amplexicaul* at the *sessile* base, entire or distantly toothed. Fl.-heads 2.5–4 cm diam. in a usually corymbose panicle, their *peduncles with small recurved bract-like lvs*; involucre 6 mm or more, campanulate, its herbaceous *bracts* linear-oblong c. 1 mm wide in several *±equal* rows, at least *the outer loosely spreading or recurved*. Ray-fls violet-blue, white or purplish in some cultivars. Much grown in gardens and the most widely and abundantly established Michaelmas Daisy in many localities through much of Great Britain; recorded also from Ireland and from Islay and Orkney. Very variable as a wild species in North America and now a complex of cultivars not always clearly distinguishable from **6** and **8**. Plants with a close corymbose panicle of numerous small heads have been named *A. floribundus* Willd., those with a racemose panicle having short branches with only one or a few fl.-heads *A. laevigatus* Lam., and those with narrowly lanceolate to linear-lanceolate lvs up to 18 cm, attenuate at each end, *A. longifolius* Lam. (a component of the shingle-flora of the R. Tay in Mid-Perth, Scotland). These variants are doubtfully worthy even of subspecific rank within *A. novi-belgii*.

*8. A. ×salignus Willd. (probably *A. novi-belgii × A. lanceolatus*) is intermediate between its putative parents, with *stem-lvs commonly auricled* but less distinctly than in *A. novi-belgii* and sometimes with auricles 0 as in *A. lanceolatus*; lvs are usually broader than those of *A. lanceolatus*. The *involucral bracts* are *herbaceous* and typically *appressed* but more loosely than in *A. lanceolatus* though much less spreading or recurved than is usual in *A. novi-belgii*. The ray-fls are pale bluish but may be initially white. Grown in gardens and now established in scattered localities in Great Britain northwards to E. Ross and Caithness. The distinction from *A. novi-belgii* may be difficult and some earlier records, like that for Wicken Fen, are now thought erroneous.

*9. A. lanceolatus Willd. (incl. *A. simplex* Willd.; *A. paniculatus* Lam., non Miller, and *A. tradescanti* auct. eur., non L.). Fl.-stems 50–130 cm, glabrous throughout or pubescent above and then often with vertical bands of hairs; sometimes purple-tinged. *Stem-lvs* linear- to oblong-lanceolate, acute to acuminate, *narrowed below to a sessile but not or barely auricled base*, ±glabrous, entire or obscurely toothed. *Fl.-heads small*, c. 1.5 cm diam., in a rather narrow panicle; *involucre* to 5 mm, its *bracts narrow* (to 0.5 mm wide) and *very unequal*, innermost much the longest, *all appressed*. Ray-fls with narrow white or violet-blue ligules. Variable and taxonomically problematical. Established, and recorded under various names, in scattered localities in England and S. Scotland. Said to have been grown in gardens from the early 17th century and to be the original 'Michaelmas Daisy'.

*10. A. schreberi Nees. *Fl.-stems* 30–80 cm, glabrous or ±hairy above, *eglandular*. *Basal lvs of non-flowering tufts and lowest lvs of fl.-stems long-petiolate and with* *large cordate lamina* 6–15 cm wide, the basal sinus ±rectangular. Fl.-heads in a roundish-topped corymbose panicle; *involucre* 8–9 mm, narrowly cylindrical, its *loosely erect scarious bracts* c. 1 mm wide, *very unequal*, the *innermost much the longest. Ray-fls white.* Established at Lochside railway station, Renfrew. Formerly recorded as *A. macrophyllus* L., but this has glandular-viscid infl.-axes and usually violet or pale blue ray-fls.

A few other North American spp. are grown in gardens and occasionally escape but so far without becoming established. Also grown are the S. and E. European *A. amellus* L., 'Italian Starwort', and *A. sedifolius* L. (*A. acris* L.), both earlier-flowering than the North American Michaelmas Daisies. *A. amellus* is up to 50 cm high with erect shortly hairy fl.-stem, lower lvs ovate-oblong to elliptical, petiolate, middle and upper narrower, sessile; all roughly hairy above and beneath, ±3-veined. *A. sedifolius* is a plant of much the same size but is ±glabrous and has gland-dotted mostly ±linear lvs. Both have blue ray-fls. The China Aster of gardens is *Callistephus chinensis* (L.) Nees.

20. ERIGERON L.

Perennial, rarely annual, herbs with spirally arranged simple lvs. Heads heterogamous; involucre of narrow, equal, scarcely imbricating herbaceous bracts; *ray-florets* female, usually *numerous and in 2 or more rows, with narrow ligules*; disk-florets hermaphrodite. Achenes compressed, usually pubescent and 2-veined; pappus usually of 1 row of simple hairs with much shorter ones intermixed, or with an outer row of small bristles or chaffy scales. In section *Trimorpha* (Cass.) Reichenb. The female florets have filiform tubes but at least the outermost have definite ligules equalling or exceeding the pappus. (In *Conyza* ligules, if present, are inconspicuous, shorter than the filiform tubes and scarcely exceeding the pappus.)

Close to *Aster* but differing in the more numerous and narrower ligulate florets.

About 200 spp., chiefly American but with some in the Old World, including several arctic–alpines.

1 Heads usually several in a panicle or corymb. 2
  Heads solitary, terminal, rarely 2–3. 3
2 Annual or biennial; lvs all entire, softly hairy; ray-florets pale purple-blue. **1. acer**
  Slender branching perennial; some of the lower lvs usually 3-lobed or 3(–5)-toothed at the apex; lvs sparsely hairy or ±glabrous; ray-florets white above, purplish beneath. **3. karvinskianus**
3 Heads 1.5–2 cm diam.; rare alpine plant. **2. borealis**
  Heads 3–4 cm diam.; lowland alien. **4. glaucus**

1. E. acer L. Blue Fleabane.

An annual or biennial herb with an erect slender stem 8–40 cm, usually branched above, rough with long hairs, reddish. Basal lvs 3–7.5 cm, obovate-lanceolate, stalked; stem-*lvs* numerous, linear-lanceolate, semi-amplexicaul; all *entire*, hairy. Heads 12–18 mm diam., 1–several in a corymbose panicle; bracts linear, glabrous or hairy, not glandular, red-tipped. *Ray-florets* pale purple, very slender, *erect, not much exceeding the yellow disk-florets*; outer disk-florets filiform, female; inner

broader, hermaphrodite. Achenes 2–3 mm, yellowish, hairy; pappus much longer than the achenes, reddish-white. Fl. 7–8. 2n = 18*. Th.

Native. A locally common plant of dry grassland, dunes, banks and walls, especially on calcareous substrata, throughout England and Wales but in Scotland only in Angus and Banff; local in Ireland. Reaches 430 m in Banff. Temperate regions of the northern hemisphere.

2. E. borealis (Vierh.) Simmons Boreal Fleabane.

E. alpinus auct. angl., non L.

A perennial alpine herb with a short creeping woody stock and erect flowering stems 7–20 cm, usually unbranched, hairy. Lvs mostly in a basal rosette, 1.5–3 cm, narrowly oblanceolate, narrowing to a long winged stalk; stem-*lvs* few, linear-oblong, sessile, somewhat amplexicaul; all *very hairy, ciliate towards the base*, entire. Heads c. 18 mm diam., usually solitary, rarely 2–3; *bracts* linear-acute, *hairy, not glandular. Ray-florets* numerous, *purple*, slender, *spreading*, much exceeding the yellow disk-florets; *outer disk-florets very slender, female*, inner broader, hermaphrodite. Achenes yellowish, downy; pappus about equalling the achene, reddish. Fl. 7–8. 2n = 18. Ch.

Native. A rare and local plant of alpine rock-ledges between 730 m and 1070 m on Ben Lawers and other mountains in Perth, Angus and S. Aberdeen. Scandinavia, Iceland, Greenland.

E. uniflorus L., a perennial alpine herb closely resembling *E. borealis* but rather less tall (5–15 cm), with glabrescent basal lvs, fewer (2–5) and smaller ±ciliate stem-lvs, the heads always solitary with *whitish ray-florets which later turn pale lilac*, and *disk-florets all similar*, yellow, has been reported from rock-ledges on Rhum (Inner Hebrides) but needs confirmation. 2n = 18. Mountains of C. Europe; Scandinavia, Iceland; Greenland; arctic America; arctic W. Asia.

*3. E. karvinskianus DC.

E. mucronatus DC

A perennial with branched slender lfy stems 10–25 cm, somewhat woody below, sparsely hairy. Lower lvs obovate-cuneate, often 3-lobed or coarsely 3(–5)-toothed at the apex, the teeth mucronate; upper lvs linear-lanceolate, ±entire; all sparsely hairy to nearly glabrous, often ciliate towards the base. Heads c. 1.5 cm diam. in a lax corymb, bracts linear-acute, hairy. Ray-florets in 2 rows, white above, purple beneath. Achenes 1.5 mm, reddish-brown, shining, somewhat hairy; pappus whitish, longer than the achene. Fl. 7–8. Ch.

Introduced. Naturalized for over 80 years on old walls at St Peter Port, Guernsey and now well established in the Channel Is., Scilly Is. and S. England and in a few scattered localities northwards to Warwick and N. Norfolk. Native of Mexico.

*4. E. glaucus Ker-Gawler Beach Aster.

A perennial herb with a stout woody stock bearing a rosette of somewhat fleshy finely puberulent or glabrescent lvs, 2.5–10 cm, obovate-spathulate, blunt, entire or with a tooth on each side of the apex; other rosettes terminating prostrate woody offsets. Stems erect, 10–20(–30) cm, pubescent to villous, bearing a few small stem-lvs and terminating in a solitary head, 3–4 cm diam., or less commonly branching and bearing 2 or more heads. Involucre shaggy and somewhat glandular; ray-florets broadish, lilac or violet; disk-florets yellow. Fl. 5–8.

Introduced. Established in a few places. Coast of Pacific North America.

The lf-rosettes resemble those of *Limonium vulgare*

*E. annuus (L.) Pers. (*Stenactis annuus* (L.) Nees), sens lat., native in North America and formerly grown here as a garden plant, is an annual to perennial herb, 20–100 cm or more, somewhat resembling a Michaelmas Daisy. *Ovate to obovate rosette-lvs, long-petiolate*, wither early; lower stem-lvs ovate-lanceolate, shortly petiolate to ±sessile, distantly toothed; upper lanceolate, sessile, usually entire. Fl.-heads 1–2 cm diam. in a corymbose panicle; ray-fls white or pale blue. Achenes 1.5 mm, hairy, *marginal with short 1-rowed pappus-hairs, central with* 2-rowed pappus, the *inner of long silky hairs, outer of quite short bristles*. Apomictic and very variable. Now a common weed over much of Europe. An infrequent casual here but recently 'subsp. *strigosus* (Muhl. ex Willd.) Wagenitz' has become well established on an air-strip in N. Hants.

21. CONYZA Less.

Perennial or annual herbs with spirally arranged simple lvs. *Heads with a few central hermaphrodite florets surrounded by numerous female florets with filiform tubular corollas; ligulate florets 0 or with inconspicuous ligules* shorter than the filiform corolla-tube and scarcely if at all exceeding the pappus. Otherwise like *Erigeron*.

It is difficult to draw a distinct line between *Erigeron* and *Conyza*: restriction of the latter to spp. with no ligulate florets separates obviously close relatives. *C. canadensis* does not seem closely related to any species of undoubted *Erigeron* and has the habit of many other North American weeds falling clearly into *Conyza*. It can, however, produce sterile hybrids with *E. acer*.

*1. C. canadensis (L.) Cronq.   Canadian Fleabane.
Erigeron canadensis L.

An annual herb with stiffly erect very lfy stems 8–100 cm, much branched, sparsely hairy or ±glabrous. Basal lvs obovate-lanceolate, stalked, entire or ±toothed, soon dying; *stem-lvs* 1–4 cm numerous, *narrowly lanceolate or linear*, acute, entire or obscurely toothed, hairy and bristly-ciliate. Fl.-heads (3–5 mm diam.), numerous, in long panicles; involucral bracts linear-attenuate, glabrous, with broad scarious margins. *Ray-florets whitish* to pale lavender, in several rows, narrow, erect, female; disk-florets pale yellow, hermaphrodite. Achenes 1.5 mm, pale yellow, downy; pappus yellowish, longer than the achene. Fl. 8–9. Visited by small insects and said not to be self-pollinated. $2n = 18$. Th.

Introduced. A local weed of waste ground, waysides, cultivated land on light soils, dunes, walls, etc., throughout England and Wales but rare in the north and in only a few localities in S. Scotland. North America, but widely naturalized.

*C. bonariensis (L.) Cronq. is an occasional casual resembling *C. canadensis* but more robust and more densely hairy, differing also in the larger fl.-heads, over 1 cm diam. and usually with 50–100 female fls (fewer than 50 in *C. canadensis*), and in the non-ciliate though sometimes minutely hooked lf-margins. Native in tropical America and well established as a weed in S. and S.W. Europe.

*Olearia macrodonta Baker, a shrub or small tree with a strong musky scent, native in mountainous areas of New Zealand, is much grown in gardens and has established itself locally. Its branchlets are tomentose and the spirally arranged, ovate to narrowly oblong, acute, rounded-based, rigid lvs 5–10 × 2.5–4 cm, coarsely and sharply toothed on the waved margins, are glabrescent above and white-tomentose beneath. Heads 6–8 mm, numerous, in large compound corymbs; involucral bracts pubescent; ray-florets 3–5, whitish, their ligules short and narrow; disk-florets 4–7. Achenes pubescent, grooved; pappus dirty-white or reddish.

22. BELLIS L.

Annual to perennial herbs with spirally arranged lvs often confined to a basal rosette. Heads solitary; involucre of many lf-like bracts in (1–)2 rows; receptacle conical, pitted. Ray-florets in 1 row, white or pink, female; disk-florets hermaphrodite. Style arms short, thick, with terminal papillose cones. *Achenes obovate, compressed, bordered, not ribbed*; pappus 0.

Fifteen spp., in Europe and the Mediterranean region.

1. B. perennis L.   Daisy.

A perennial herb with a short erect stock and stout fibrous roots. Lvs 2–4(–8) cm, confined to a basal rosette, obovate-spathulate, broad and rounded at the end, crenate-toothed, narrowed abruptly into a short broad stalk, sparsely hairy. Scapes 3–12(–20) cm, naked, hairy. Head 16–25 mm diam.; bracts oblong, blunt, green or black-tipped, hairy. Ray-florets numerous, narrow, spreading, white or pink; disk bright yellow. Achenes 1.5–2 mm, pale, strongly compressed, distinctly bordered, ±downy. Fl. 3–10. Visited by many small insects. $2n = 18^*$. Hr.

Native. An abundant plant of short grassland throughout the British Is. Reaches 915 m in Scotland. Europe, W. Asia.

The following introduced species also belong to the Astereae:

*Grindelia squarrosa (Pursh) Dunal, the Broadleaf Gumplant, native of North America, a perennial viscid herb to 100 cm, corymbosely branched, with lvs 3–7 cm, ovate-oblong, ±amplexicaul, serrate-crenate or entire, strongly gland-dotted. Fl.-heads 2–3 cm diam. terminating main stem and branches; involucre very viscid, its narrow bracts with their distal halves squarrose to recurved; ray-fls numerous, yellow. Achenes 2–3 mm, striate, with pappus of 2–8 caducous awns 3–5 mm long. A weed of open grassland and waste ground and a frequent casual. Widely naturalized in Europe, Australia and other temperate countries

*G. rubricaulis DC. var. *robusta* (Nutt.) Steyermark (*G. robusta* Nutt.), Californian Gumplant, sometimes grown in gardens, has become established on sandstone cliffs near Whitby harbour, N. Yorks. It has yellow ray-fls, oblanceolate stem-lvs and its young parts are highly viscid.

*Calotis cuneifolia R.Br., native in Australia, is a frequent wool-alien. It is a perennial herb with ±ascending stiffly hairy stems up to 60 cm and simple broadly spathulate stem-lvs, commonly 2–3 cm including the petiole-like base, rounded and toothed above, cuneate and entire below. Fl.-heads 1–1.5 cm diam., terminating main stem and branches; involucral bracts up to 3.5 mm, ±narrow. Ray-fls numerous with pale lilac or whitish ligules up to 1 cm long. Achenes to 1.5 mm, cuneate, flattened; pappus of 2(–4) awns c. 3 mm with distal barbs alternating with as many broad scales. Other spp. of *Calotis* are less frequent wool-aliens.

Tribe 7. EUPATORIEAE.

23. EUPATORIUM L.

Perennial herbs or shrubs with usually opposite lvs. *Heads few-fld white, pink* or *purplish*, in terminal corymbs or panicles. Bracts few, loosely imbricate in 2–3 rows; *receptacle naked, flat. Fls all tubular*, hermaphrodite, actinomorphic, 5-merous. Achenes 5-angled; *pappus-hairs in* 1 *row*, denticulate.

About 1200 spp., chiefly American.

1. E. cannabinum L.   Hemp Agrimony.

A large perennial herb with a woody rootstock and erect downy striate shoots 30–120 cm, simple or with short branches. Basal lvs oblanceolate, stalked; stem-lvs subsessile, 3(–5)-partite with elliptical-acuminate toothed segments 5–10 cm; branch lvs simple, ovate or lanceolate; all lvs opposite, shortly hairy and gland-dotted. Heads in dense terminal corymbs; each head with 5–6 reddish-mauve or whitish florets and c. 10 oblong purple-tipped involucral bracts, the inner c. 6 mm, narrow, ±scarious, the outer much shorter. Styles white, long. Achenes blackish, 5-angled, gland-dotted; pappus whitish. Fl. 7–9. Protandrous. Visited chiefly by Lepidoptera and by some flies and bees; automatic cross-pollination between different fls in the same head occurs. 2n = 20. H.–Hel.

Native. A common gregarious plant of marshes and fens, stream banks and moist woods throughout most of the British Is. but less common in Scotland and not in the Outer Hebrides, Orkney or Shetland; Channel Is. Europe; W. and C. Asia; N. Africa.

Some species of the related genus *Ageratum*, differing from *Eupatorium* in having the pappus of free or basally united scarious scales, are often cultivated. Best known is **A. houstonianum** Miller (*A. mexicanum* Sims), with broadly ovate ±cordate lvs and blue (sometimes pink or white) fls, much grown as an edging-plant and in window-boxes and occasionally escaping.

Tribe 8. ANTHEMIDEAE.

24. ANTHEMIS L.

Annual to perennial usually strongly scented herbs with alternate 1–3 *times pinnately divided lvs* whose *ultimate segments* are *linear*. Fl.-heads solitary, usually heterogamous; bracts imbricate, blunt, usually scarious and often dark-coloured at the margins; *receptacle* flat to conical, *with narrow flat scarious scales* subtending some or all of the florets. Ray-florets ligulate, female or

neuter, yellow or white, sometimes 0; disk-florets tubular, hermaphrodite, yellow, their corolla-tubes compressed and winged. *Achenes with at least* 10 *ribs*, not or somewhat compressed in the anterior-posterior plane; *pappus* represented by a *small often oblique membranous rim, or* 0.

About 100 spp., chiefly in the Mediterranean region and Near East.

1 Ray-florets yellow.   **1. tinctoria**
  Ray-florets white.   2
2 Plant glabrous or slightly hairy, foetid; receptacle-scales linear-acute; ray-florets usually without styles; achenes tubercled.   **3. cotula**
  Plant usually pubescent or woolly, aromatic; receptacle-scales lanceolate-cuspidate; ray-florets with styles; achenes strongly ribbed, not tubercled.   **2. arvensis**

Subgenus COTA (Gay ex Guss.) Rouy.

Receptacle with persistent rigid scales over its whole surface, ±hemispherical in fr. Achenes narrow below, truncate above, ±compressed, rhombic in cross-section, slightly striate on the faces, smooth.

*1. A. tinctoria L.   Yellow Chamomile.

A biennial to perennial herb with erect or ascending woolly stems 20–60 cm, usually branched. Lvs 4–7 cm, deeply pinnatisect with pinnately lobed or toothed segments united only by a narrow wing, ±glabrous above, sparsely woolly beneath. Heads 2.5–4 cm diam., solitary, long-stalked. Bracts lanceolate, acute, scarioustipped, ±woolly at the back, with brown ciliate margins. *Receptacle hemispherical*, its scales lanceolate, cuspidate, barely exceeding the disk-florets. *Ray-florets* female, *golden-yellow* like the disk-florets, rarely 0 (var. *discoidea* Willd.). Achenes c. 2 mm, 4-angled, faintly ribbed on each face, glabrous; pappus represented by a very short border. Fl. 7–8. Much visited by a great variety of insects. 2n = 18. H.

Introduced. A garden plant naturalized in Bucks and found as a casual in waste places, waysides, banks, etc., in many English and a few Scottish counties. S. and C. Europe northwards to Scandinavia and Finland; W. Asia. Introduced in North America. The fls yield a yellow dye.

Subgenus ANTHEMIS.

Receptacle with scales over its whole surface or only its upper part, the scales persistent or not, membranous or slightly stiff; receptacle often conical in fr. Achenes cylindrical to obconical, truncate above, not compressed, rhombic in cross-section, usually with distinct ribs which may be smooth, granulate or tuberculate.

2. A. arvensis L.   Corn Chamomile.

An annual aromatic herb with decumbent or ascending pubescent stems 10–50 cm, much-branched below, the branches simple or irregularly branched above. Lvs 1.5–5 cm, 1–3 times pinnate, the *ultimate segments* short, *oblong*, acute, *hairy or even ±woolly beneath*, especially when young. Heads 20–40 mm diam., solitary, long-stalked. *Bracts* oblong, scarious-tipped, pale, *downy*. *Receptacle conical, with lanceolate-cuspidate scales* just exceeding the disk-florets. *Ray-florets* female (*with styles*), white, spreading; disk-florets yellow. *Achenes* 1.5–3 mm, whitish, c. 10–ribbed, glabrous, *ribs not tubercled*, rugose on top; border crenate. Fl. 6–7. Fragrant and much visited by bees and flies. $2n = 18^*$. Th.

Var. *anglica* (Sprengel) Syme, with fleshy bristle-pointed lf-segments and a 'flat receptacle' was probably a maritime form or ecotype of the Durham coast, now apparently extinct.

Native. A locally common calcicolous plant of arable land and waste places throughout the British Is. to E. Ross; Orkney; introduced in E. Ireland. Europe northwards to S. Norway, C. Sweden and Lake Onega; Asia Minor; N. Africa. Introduced elsewhere.

3. A. cotula L.   Stinking Mayweed.

An annual *foetid* herb with erect *sparsely hairy* stems 20–60 cm, usually branched below and corymbosely branched above. *Lvs* 1.5–5 cm, 1–3 times pinnate, the ultimate segments narrowly linear, acute, ±*glabrous*. Heads 12–25 mm diam., solitary, rather short-stalked. Bracts oblong, blunt, ±glabrous, with a green central stripe and broad scarious margins. *Receptacle long-conical* with *linear-subulate scales* only near the apex. *Ray-florets* usually neuter (*without styles*), white, at first spreading, later *reflexed*; disk-florets yellow. *Achenes* c. 1.5 mm, yellowish-white, c. 10-ribbed, *the ribs tubercled*; membranous border crenate, inconspicuous. Fl. 7–9. $2n = 18^*$. Th.

In var. *maritima* Bromf, the stems are prostrate and the lvs fleshy.

Native. A locally common weed of arable land and waste places especially in S. and C. England and on heavy soils, rarer in the north but reaching S.W. Inverness and Moray and introduced in the Hebrides, Orkney and Shetland. S. Europe northwards to S. Norway; N. and W. Asia. Introduced elsewhere.

Several spp. of *Anthemis* native in S. and E. Europe occur as casuals.

25. CHAMAEMELUM Miller

Like *Anthemis* but *disk-florets with corolla-tube saccate at base*, especially in the anterior-posterior plane, *so that it covers the top of the achene* (in *C. mixtum* prolonged into a posterior spur); and *achenes laterally compressed. unribbed* (but with 3 slender vascular bundles making the inner face weakly striate).

Three spp. in W. Europe and Mediterranean region.

1. C. nobile (L.) All.   Chamomile.

Anthemis nobilis L.

A perennial *pleasantly scented* herb with a short much-branched creeping stock and decumbent or ascending branched hairy stems 10–30 cm. Lvs 1.5–5 cm, 2–3 times pinnate, the ultimate segments short, linear-subulate, sparsely hairy. Heads 18–25 mm diam., solitary, long-

stalked. Bracts oblong, downy, with broadly scarious and laciniate white margins. *Receptacle* conical, *scales oblong, concave, blunt*; often laciniate at the apex. *Ray-florets* female (*with styles*), broad, white, spreading, rarely 0; disk-florets yellow, with the *enlarged and persistent base of the corolla-tube enveloping the apex of the achene*. *Achene* 1–1.5 mm, obovoid, *narrowed and rounded above*, smooth except for 3 faint striae on the inner face; border inconspicuous. Fl. 6–7. $2n = 18^*$. H.

Native. A local plant of sandy commons and pastures and grassy roadsides in S. England and in Wales and Ireland; introduced in N. England and in Scotland but casual there and in the Hebrides, Orkney and Shetland; decreasing. W. Europe from Belgium southwards; N. Africa; Azores.

26. ACHILLEA L.

Usually perennial herbs with spirally arranged usually pinnatisect, rarely simple, lvs. *Heads in corymbs*, rarely solitary, heterogamous; bracts imbricate, many-rowed, scarious-margined; *receptacle* small, flat or slightly convex, *with narrow scarious scales*. Ray-florets female, ligulate, usually short and broad, white, yellow or reddish; disk-florets hermaphrodite, tubular, white or yellow. *Achenes strongly compressed, truncate above, not ribbed; pappus* 0.

About 200 spp., chiefly in temperate regions of the Old World; 1 in North America.

Close to *Anthemis* but with the achenes strongly compressed and the smaller heads usually in corymbs.

> *1* Ray florets yellow   **3. tomentosa**
>   Ray florets white.   2
> *2* Lvs finely pinnatisect; heads 4–6 mm diam., numerous, in a dense corymb.   **2. millefolium**
>   Lvs simple, broadly linear, serrate; heads 12–18 mm diam., few, in a lax corymb.   **1. ptarmica**

Section 1. *Ptarmica* (Miller) Dumort. Ray-florets 6–25, female, white, almost as long as the involucre; receptacle convex or hemispherical, not much elongated in fr.

1. A. ptarmica L.   Sneezewort.

A perennial herb with a creeping woody stock and erect angular stems 20–60 cm, simple or branched above, glabrous below but hairy above. *Lvs* 1.5–8 cm, *linear-lanceolate*, sessile, acute, ±glabrous, sharply *serrulate*, the serrations with a cartilaginous and denticulate margin. *Heads* 12–18 mm *diam., not numerous, in a rather lax corymb*. Involucre hemispherical, its bracts lanceolate to oblong, blunt, green-centred, ±woolly, with reddish-brown scarious margins. Ray-florets 8–13, ovate, white, as long as the involucre; disk-florets greenish-white. Achenes 1.5 mm, pale grey. Fl. 7–8. Freely visited by bees and flies. $2n = 18^*$. H.

Native. Common throughout the British Is. in damp meadows and marshes, and by streams, reaching 730 m in the Lake District. Europe except the southern Mediterranean region; Asia Minor; Caucasus; Siberia. Introduced in North America.

Double forms are much grown in gardens as Bachelors' Buttons. Formerly used as a salad plant. Infusions of the lvs and fls were also used medicinally.

Section 2. *Achillea*. Involucre ovoid; ray-florets 4–6, white, shorter than the involucre; receptacle long-conical in fr.

2. A. millefolium L.   Yarrow, Milfoil.

A perennial strongly scented far-creeping stoloniferous herb with erect, furrowed, usually simple, ±woolly stems 8–45(–60) cm. *Lvs* 5–15 cm, lanceolate in outline, *2–3 times pinnate*, the ultimate segments linear-subulate; basal lvs long, stalked; upper shorter, sessile, often with 2–3 small axillary lvs. *Heads 4–6 mm diam., numerous, in dense terminal corymbs*. Involucre ovoid, bracts rigid, oblong, blunt, keeled, ±glabrous, with a broad brown or blackish scarious margin. Ray-florets usually 5, about half as long as the involucre and as broad as long, 3-toothed at the apex, white, rarely pink or reddish; disk-florets white or cream-coloured. Achenes c. 2 mm, shining greyish, somewhat winged. Fl. 6–8. Much visited by a great variety of insects. $2n = 54^*$. Chh. Very variable in hairiness and in the colour of the bracts.

Native. A common plant of meadows and pastures, grassy banks, hedgerows and waysides on all but the poorest soils throughout the British Is. Europe to 71° 10′ N.; W. Asia. Introduced in North America, Australia and New Zealand.

Used medicinally for a great variety of purposes from early times, and still officinal in Austria and Switzerland as *Herba Millefolii* or *Flores Millefolii*. It contains the two alkaloids achillein and moschatin and an ethereal oil which is responsible for its aromatic scent.

A. ligustica All., native in dry grassland and scrub of the Mediterranean region, has persisted for c. 40 years on railway ballast near the entrance to the Newport docks, Gwent. Its erect stems commonly reach 50–100 cm, the twice pinnatifid to pinnatisect lvs differ from those of *A. millefolium* in being more finely divided and having a ±toothed rhachis, and the greenish fls open earlier. Some other *Achillea* spp. have been reported as infrequent casuals.

*3. A. tomentosa L.

A perennial herb with very short stolons and erect or ascending *densely woolly stems* 8–30 cm. *Lvs linear-lanceolate* in outline, *twice-pinnatisect*, *with* crowded pinnatifid or entire *mucronate ±woolly segments*. Heads c. 3 mm diam., numerous, in dense compound corymbs, with 4–6 *bright yellow ray-florets* half as long as the involucre, and yellow disk-florets.

Introduced. Grown in gardens and established in a few localities in Scotland and N.E. Ireland. S. Europe northwards to C. France and Tirol; W. Asia.

27. Otanthus Hoffmanns. & Link

A maritime herb with corymbs of yellow heads. Involucre campanulate-hemispherical, of *numerous imbricating woolly bracts*. Receptacle shortly conical, with ovate-acuminate scales. *Florets all hermaphrodite and tubular*, the *tube prolonged downwards into 2 auricle-like spurs which almost enclose the ovary*. Achenes compressed, longitudinally ribbed; pappus 0.

One sp.

1. O. maritimus (L.) Hoffmanns. & Link Cotton-weed.

Athanasia maritima (L.) L.; *Diotis maritima* (L.) Desf. ex Cass.

A perennial maritime herb with long branching woody stock and *white-woolly* decumbent to ascending *fl.-stems* 15–40 cm, stout, branched only above. *Lvs* alternate, oblong to oblanceolate or spathulate, *sessile*, with a rounded tip and ±upturned *entire or crenate-dentate margins, densely white-felted on both faces*. Fl.-heads almost globose, 6–9 mm diam., short-stalked, several in a dense corymb. *Involucre* of rather few *ovate bracts*, the *outer densely white-felted over the whole outer face*, the inner only near the tip. Fls yellow, little exceeding the involucre. *Achenes* curved, longitudinally 4–5-ribbed but smooth and glabrous, *partly enclosed by the persistent enlarged and spongy base of the corolla-tube* and by its spurs.

Native. Sandy sea-shores and stable shingle. Formerly in a few localities in S. England and the Channel Is. but now only in the south of Co. Wexford, Ireland. S. and W. Europe, including the Mediterranean islands, and northwards to S. Ireland, eastwards to Greece and Turkey; N. Africa; Near East.

28. Santolina L.

Suffruticose aromatic plants with pinnatifid lvs. Heads long-stalked, ±homogamous, the marginal florets tubular or very shortly ligulate; receptacular scales present. *Corolla-tube* compressed, winged, and *with a basal one-sided appendage enclosing the apex of the achene*. Achene ±3–5-angled, glabrous; pappus 0.

About 8 spp., in S.W. Europe, eastwards to Dalmatia, and N. Africa.

*1. S. chamaecyparissus L.   Lavender Cotton.

A small evergreen strongly aromatic shrub with pubescent branching woody stems to 50 cm. Lvs only 2–3 mm wide, linear in outline, whitish-tomentose, pinnatifid with 4 rows of ±crowded fleshy blunt lobes, 1–2 mm. Heads 10–15 mm diam., solitary, terminal, globular; involucral bracts ±glabrous, somewhat scarious-tipped. Florets all tubular, yellow. Fl. 7–8. $2n = 18$. Ch.–N.

Introduced. A garden-escape, sometimes establishing itself. Mediterranean region from Spain to Dalmatia; N. Africa.

The taxonomy and nomenclature of the five following genera are still under discussion and the treatment here must be regarded as provisional.

29. Tripleurospermum Schultz Bip.

Annual to perennial *herbs with alternate 2–3 times pinnatisect lvs*, the *ultimate segments narrowly linear*. Fl-heads solitary, usually 2–5 cm diam., pedunculate, usually heterogamous; involucral bracts in 2 or more rows, scarious-margined; *receptacle* hemispherical to conical, *solid; receptacular scales 0. Ray-fls* female, *with ligule white*, sometimes 0; disk-fls tubular, yellow, with corolla-tube laterally compressed. *Achene with broad truncate apex and narrower base, both at right angles to the long axis*, broader tangentially than dorsiventrally and ±trigonous in cross-section because of the 3 *strong ribs on the adaxial (ventral) face*; with usually 2 *conspicuous oil-glands high up on the abaxial (dorsal) face*; pappus represented by a small membranous border or corona, sometimes 0.

Perhaps c. 30 spp., chiefly in Europe and W. Asia.

Perennial or biennial herb of open coastal habitats; stem usually prostrate or decumbent; lf-segments short and fleshy; ribs of adaxial face of mature achene contiguous or separated by less than ¼ of their width; oil-glands on abaxial face usually at least twice as long as wide.
1. maritimum

Annual herb of cultivated and waste ground; stem erect or ascending; ultimate lf-segments usually fine and non-fleshy; ribs on adaxial face of mature achene clearly separated by at least ⅓ of their width; oil-glands on abaxial face ±circular, at most 1.5 times as long as wide.
2. inodorum

1. T. maritimum (L.) Koch   Sea Mayweed.
Matricaria maritima L.

A *perennial* maritime herb with branching woody stock and usually *prostrate or decumbent* branching stem 10–50(–80) cm, rarely ascending to erect. *Lvs* oblong in outline with *ultimate segments short, blunt, ±cylindrical* and *fleshy*; stems and lvs glabrous or with a few scattered hairs. Fl-heads 3–5 cm diam., long-stalked; *involucre* hemispherical, its *bracts oblong to broadly triangular* with pale to blackish-brown scarious margins, *outer bracts much shorter than inner*. Ray-fls with ligule 1–1.6 cm. Achenes 1.8–3.5 mm; *ribs on adaxial face contiguous or separated by less than ¼ of their width*; abaxial face blackish-brown and transversely rugose, the *oil-glands* near its upper end *vertically elongated* and *usually at least twice as long as wide*; pappus a short membranous border. Fl. 7–9. Visited and cross-pollinated mainly by flies; normally self-incompatible. $2n = 18^*$, 36. H.–Ch.

British representatives have been allocated to the following subspp.:

Subsp. **maritimum**: involucral bracts oblong or narrowly triangular with pale to dark brown scarious margin up to 0.3 mm wide; oil-glands usually much more than twice as long as wide.

Subsp. **phaeocephalum** (Rupr.) Hämet-Ahti: involucral bracts broadly triangular with blackish-brown scarious margin 0.4–1 mm wide; oil-glands usually more than twice as long as wide.

Native. Subsp. *maritimum* is a locally common maritime plant of the drift-line at the foot of sand dunes, of shingle-beaches, coastal rocks and cliffs, walls, etc., all round the British Is. but less frequent on the east coast than the west. Subsp. *phaeocephalum* is restricted to coasts of northern Scotland and Shetland. Coasts of W. and N. Europe; subsp. *maritimum* in W. and N.W.

Europe, subsp. *phaeocephalum* in arctic Europe, incl. Iceland and Faeroes, and in arctic North America, incl. Greenland.

2. T. inodorum (L.) Schultz Bip.   Scentless Mayweed.
Matricaria inodora L.; *M. perforata* Mérat; *T. maritimum* subsp. *inodorum* (L.) Hyl. ex Vaarama

A usually *annual* almost scentless herb with *erect or ascending stem* 15–60 cm or more, usually branched above, sometimes also near base. *Lf-segments long, slender, acute or bristle-pointed, not fleshy*. Stem and lvs glabrous at maturity, sometimes sparsely pubescent when young. Fl.-heads few to many, 3–4.5 cm diam.; involucre hemispherical, its *subequal bracts* oblong or narrowly triangular with narrow usually brownish scarious margins. Ray-fls with ligule 1–1.8 cm. Achenes 1.5–2 mm; *ribs on adaxial face clearly separated* by at least ⅓ of their width; abaxial face blackish-brown, transversely rugose, the *oil-glands* near its upper end *±circular*, at most 1.5 times as long as wide; pappus a short membranous border. Fl. 7–9. Visited mainly by flies; normally self-incompatible. $2n = 18^*$ (a single plant with $2n = 36^*$ may have been introduced from continental Europe). Th.

Native. Abundant throughout the British Is. as a weed of cultivated and waste land on all kinds of soil. Native in most of continental Europe; W. Asia. Widely introduced as a weed or ruderal.

Hybrids between **1** and **2**, vigorous and usually at least moderately fertile, occur not infrequently in places where both parents are nearby and where there are suitably intermediate habitats. Populations showing presumed introgression of *inodorum* into *maritimum* have also been reported. For such reasons the two are often treated as subspp. of a single species, but the marked morphological and ecological differences between them, the relative infrequency of hybrid and intermediate populations and the near or complete sterility of some hybrid individuals are held to justify their maintenance as separate spp.

30. MATRICARIA L.

Usually annual herbs with finely pinnatisect lvs, the ultimate segments narrowly linear. Closely resembling *Tripleurospermum* but heads sometimes homogamous (lacking ray-florets), *receptacle* often *hollow* and *achenes* strikingly different, these being *obovoid*, somewhat compressed laterally, *obliquely truncate* above and with the *basal insertion-scar oblique and distinctly lateral* (posterior), *rather inconspicuously 3–5-ribbed* on the posterior face and *lacking oil-glands* on the anterior face; pappus represented by a membranous rim or 0.

About 40 spp. in the Mediterranean region and W. Asia, some in S. Africa and Pacific North America.

Matricaria and *Tripleurospermum* differ in embryo-sac development, achene anatomy and orientation of the cotyledons within the seed, in addition to the differences noted above.

Ray-florets 0; plant strongly aromatic.   **2. matricarioides**
Ray-florets white, soon deflexed; plant pleasantly aromatic.   **1. recutita**

1. M. recutita L.   Wild Chamomile.

M. chamomilla L. 1755, non L. 1753; *Chamomilla recutita* (L.) Rauschert

An annual *pleasantly aromatic* herb resembling *Tripleurospermum inodorum* with erect glabrous stems 15–60 cm, usually much branched, and 2–3 times pinnatisect lvs with ultimate segments narrowly linear and bristle-pointed. Heads 1–2.5 cm diam. Involucral bracts linear, blunt, yellowish-green with the *narrow scarious margins much the same colour* (not brown as in *T. inodorum*). *Receptacle* markedly conical from the first, hollow. Ray-florets c. 15, white, 6–9 × 2–3 mm, *deflexed soon after flowering begins*, sometimes 0; disk-florets yellow. *Achenes* c. 1 mm, pale grey, slender, obliquely truncate and *unbordered* above, with 4–5 ribs on the inner face, and *lacking oil-glands* on the outer face. Fl. 6–7. Freely visited by flies and some small bees. $2n = 18^*$. Th.

Native. Locally abundant as a weed of sandy or loamy arable soil or waste places throughout England and Wales; only casual in Scotland northwards to Moray and in Ireland. Europe (probably introduced in the north); W. Asia to India. Introduced in North America and Australia. Used as substitute for true Chamomile, *Chamaemelum nobile*, Oil of Chamomile being distilled from the flower heads.

*2. M. matricarioides (Less.) Porter
                    Pineapple Weed, Rayless Mayweed.

M. discoidea DC.; *M. suaveolens* (Pursh) Buchenau, non L.; *Chamomilla suaveolens* (L.) Rauschert

An annual *strongly aromatic* herb with erect glabrous stems 5–30(–40) cm, much branched above, the branches rigid. Lvs 2–3 times pinnatisect with the ultimate segments linear, bristle-pointed. *Heads* 5–10 mm diam., solitary, *short-stalked*. Involucral *bracts* oblong, blunt, *with broad scarious margins. Receptacle conical*, hollow. Ray-florets 0; disk-florets dull greenish-yellow. Achenes 1.5 mm, with 3–4 inconspicuous ribs on the inner face and an obscure rim at the apex. Fl. 6–7. Little visited by insects. $2n = 18^*$. Th.

Robust plants from W. Ireland with larger heads (to 12 mm diam.) and achenes having a large, often unilateral, toothed or lobed crown at the apex, have been named *M. occidentalis* Greene (California), but they seem to grade into typical *M. matricarioides*.

Introduced. An abundant and increasing weed of waysides and waste places, and especially of tracks, paths and trampled gateways, throughout the British Is. Probably native in N.E. Asia but established in North America, throughout Europe, in Chile and in New Zealand.

*M. disciformis DC. (Asia Minor), like *M. matricarioides* but with *long-stalked* rayless heads, is sometimes found as a casual.

31. Chrysanthemum L.

Annual usually glabrous herbs with alternate lvs ranging from subentire to twice pinnatisect. Fl.-heads solitary at ends of main stem and branches, pedunculate, heterogamous; involucral bracts in 2–3 rows, scarious-margined; receptacle convex; *receptacular scales* 0. *Outer fls female*, usually ligulate, the *ligules yellow*, sometimes differently coloured near base; *inner fls hermaphrodite*, tubular, the tube 2-winged. *Achenes* usually ±*distinctly ribbed*, dimorphic, those of ray-fls ±3-*angled with at least the 2 lateral angles winged*, of disk-fls not or hardly angled and unwinged or with a single adaxial wing; conspicuous oil-glands 0; *pappus* 0.

A few spp. in S. and S.W. Europe, N. Africa and W. Asia; some widely naturalized in W., C. and N. Europe and in North and South America.

1. C. segetum L.   Corn Marigold.

An *annual* glabrous and *glaucous* herb with erect usually branching stems 20–60 cm or more. *Lvs* 2–10 cm, simple, ±oblong in outline, the *lowest narrowed into a winged petiole-like base*, all *others sessile and semiamplexicaul*; lower and middle lvs *sharply toothed below but broadening* and becoming more *coarsely toothed to ±deeply pinnatifid distally*, the *uppermost* shorter and *narrowly* lanceolate, toothed below or subentire. *Fl.-heads* 3–6 cm diam., *solitary, with long peduncle thickened upwards*; *involucre* hemispherical, its *bracts broadly ovate*, *glaucous, with hyaline scarious margin* much broadened at the blunt apex. *Ligule of ray-fls* 1–1.5 cm, *golden-yellow*, c. ½ as wide as long, *with 2 lateral wings and 3 prominent whitish ribs* on adaxial face; *inner achenes* ±*cylindrical*, *10-ribbed, unwinged*. Fl. 6–8: freely visited by various insects, especially flies. $2n = 18^*$. Th.

Probably introduced. A locally common but decreasing weed of acid arable soils and waysides throughout the British Is. Often a troublesome weed on loamy and sandy soils, but much reduced in abundance by liming. Probably a native of the E. Mediterranean region and W. Asia, but well established throughout Europe to 70° N. in Norway. Introduced in North and South America and N. Africa.

The most commonly grown of the annual Summer Chrysanthemums of gardens are **C. coronarium** L. (Mediterranean region and S.W. Europe), with uniformly golden-yellow to white ray-fls and yellow disk-fls, and **C. carinatum** Schousboe (Morocco), with ligules of ray-fls having bands near their base coloured differently from the main distal part and with purplish disk-fls; both have twice-pinnatisect lvs. *C. coronarium* has been reported as a garden-escape.

The perennial Florist's Chrysanthemums, autumn-flowering and half-hardy, are plants of uncertain and probably E. Asian origin. Most were formerly referred to *C. morifolium* Ramat., now *Dendranthema morifolium* (Ramat.) Tzvelev. Costmary, formerly much grown for the mint-like smell of its lvs, is *Balsamita major* Desf. (*Chrysanthemum balsamita* (L.) Baillon), with simple glandular-punctate lvs and white ray-fls.

32. Leucanthemum Miller

Perennial herbs, rarely annual. Lvs alternate, entire to twice pinnatisect. *Fl.-heads* usually *solitary*, terminal; involucral bracts in 2–3 rows; receptacle usually convex; *receptacular scales* 0. *Outer fls* female, *ligulate*, *white or*

pinkish, rarely yellow and then either hermaphrodite or female; *inner fls* hermaphrodite, *tubular, yellow. Achenes all similar*, obconical-cylindrical, *usually with* 10 *prominent ribs, unwinged; pappus a membranous border* or corona, sometimes incomplete or 0.

Several spp. in Europe and Asia, some widely introduced elsewhere.

1. L. vulgare Lam.

Ox-eye Daisy, Moon Daisy, Marguerite.

Chrysanthemum leucanthemum L.

A perennial herb with slender branched oblique woody stock producing non-flowering lf-rosettes and erect simple or branched flowering stems 10–100 cm, sparsely hairy or almost glabrous. Basal and lower stem-lvs roundish to obovate-spathulate, crenate to dentate, long-stalked; upper stem-lvs oblong, blunt, toothed to pinnatifid, sessile, half-amplexicaul; all dark green, very sparsely hairy. *Heads* 2.5–5 cm diam., *solitary*, long-stalked. Bracts lanceolate to oblong, green, with narrow dark purplish scarious margins and tips. Ligules of ray-fls long, *white*, rarely very short or 0; disk-fls yellow. Achenes 2–3 mm, pale grey, with 5–10 strong ribs; ray-achenes with a border, or, like the disk-achenes, not bordered. Fl. 6–8. Freely visited by a great variety of bees, flies, beetles, butterflies and moths. $2n = 18$*. H.

Native. A common plant of grassland on all the better types of soil. Throughout the British Is., though less common in Scotland. Throughout Europe to Lapland; Siberia. Introduced in North America and New Zealand.

Over its whole range *L. vulgare* is a morphologically and cytologically very variable species or species-complex posing serious problems for the taxonomist. One member of the complex, native in the Pyrenees and with $2n = 90$ or 108, is much grown in gardens as 'Shasta Daisy' and often escapes. Treated as a distinct species it is **L. maximum** (Ramond) DC., distinguished from our native *L. vulgare* by its greater robustness, by the elliptical-lanceolate blades of its lower lvs (obovate-spathulate in *vulgare*) and by its larger fl.-heads, 6–9 cm diam.

33. TANACETUM L.

Annual to perennial herbs, *often aromatic. Lvs* alternate, *pinnately lobed or cut. Fl.-heads in terminal corymbs*, rarely solitary; involucral bracts in 3 rows; receptacle convex to subglobose, punctate-tuberculate; *receptacular scales* 0. *Outer fls usually* female *with white or yellow ligule, sometimes* hermaphrodite or female and tubular *with ligule* 0; *inner fls* hermaphrodite, *tubular and yellow. Achenes* usually 3–10-ribbed, *all similar*; pappus usually a membranous border.

Several species in Europe and Asia.

Section 1. *Pyrethrum* (Zinn) Reichenb. fil. Ligules white, always present. Achenes lacking secretory lacunae.

*1. T. parthenium (L.) Schultz Bip.   Feverfew.

Matricaria parthenium L.; *Chrysanthemum parthenium* (L.) Bernh.; *Pyrethrum parthenium* (L.) Sm.

A perennial *strongly aromatic* herb with ±vertical rootstock and erect stem 25–60 cm, ridged, ± downy, corymbosely branched above. Lvs 2.5–8 cm, yellowish-green, downy to subglabrous; lower lvs long-stalked, ovate in outline, pinnatisect to pinnate, the narrowly ovate segments toothed or lobed, upper lvs shorter-stalked and less divided. *Heads* 1.5–2 cm diam., long-stalked, *in ±lax corymbs*. Involucre hemispherical; bracts lanceolate to oblong, bluntly keeled, downy, with narrow pale scarious margins and laciniate tips. *Receptacle hemispherical*. Ligule of ray-fls short and broad; disk-fls yellow. Achenes 1.5 mm, 8–10-ribbed, all with a short membranous border. Fl. 7–8. Visited by bees and flies. $2n = 18$*. H.

Probably introduced. A frequent plant of walls, waste places, waysides etc., throughout Great Britain, and in several localities in Ireland but not in the Outer Hebrides, Orkney and Shetland. Formerly cultivated as a medicinal herb and used as a febrifuge, whence the common name Feverfew. Probably native in S.E. Europe, Asia Minor and the Caucasus, but now established throughout Europe and in North and South America.

Section 2. *Tanacetum.* Ligules of ray-fls yellow or 0.

2. T. vulgare L.   Tansy.

Chrysanthemum vulgare (L.) Bernh.; *C. tanacetum* Karsch, non Vis.

A *perennial aromatic herb* with creeping stoloniferous stock and stiffly erect lfy *stem* 30–120 cm, *angled, usually purplish*, ±glabrous, corymbosely branched above. *Lvs* up to 25 cm, broadly elliptical to ovate-oblong in outline, *pinnatisect* with up to 12 pairs of *oblong-lanceolate segments* which are ±*deeply pinnatifid* or at least *coarsely and sharply toothed*, the rhachis between them sharply toothed distally or throughout; uppermost lvs smaller and narrower with fewer segments or merely toothed; *lower lvs shortly petiolate, upper sessile and semi-amplexicaul*; all lvs gland-dotted and strongly aromatic, glabrous or sparsely pubescent. *Fl-heads* 7–12 mm diam., vertically flattened and ±button-shaped, 10–70 or more *in a dense compound corymb; involucre* hemispherical, its *outer bracts ovate-lanceolate, inner longer, broader and more rounded distally, all with scarious margins* broadest near the apex. *Fls golden-yellow*, the marginal female fls usually tubular, 3-toothed and obliquely truncate, rarely shortly ligulate, sometimes 0; inner fls hermaphrodite, tubular, 5-toothed. *Achenes* c. 1.5 mm, *cylindrical-obovoid*, greenish-white, 5-ribbed, with scattered sessile glands; pappus a short unevenly lobed and toothed membranous border. Fl. 7–9; freely visited by a great variety of small insects. $2n = 18$*. H.

Native. Roadsides, hedgerows, waste places, etc., to 365 m in Scotland; common. Throughout the British Is. Europe. Caucasus; Armenia; Siberia. Formerly much cultivated as a medicinal and pot-herb.

The Pyrethrums of gardens are hardy perennials now placed in the genus *Tanacetum*, most of them derived from *T. cocci-*

neum (Willd.) Schultz Bip. (*Pyrethrum roseum* Bieb.), native in the Caucasus and S.W. Asia. Its dried fl.-heads are used in the preparation of 'pyrethrum powder', a valuable contact-insecticide the main source of which is *T. cinerariifolium* (Trev.) Schultz Bip., native in the N. Balkan peninsula but widely cultivated on a commercial scale.

34. COTULA L.

Annual to perennial *marsh and water plants usually with pinnatifid lvs* and small long-stalked *rayless heads of yellow fls*. Receptacular scales 0. Central fls tubular, hermaphrodite; *marginal fls* female, usually *with so short a ligule as to appear tubular, or with corolla ±abortive*. Achenes of outer fls compressed; pappus 0.

About 75 spp. chiefly in Africa, Australia and South America.

*1. C. coronopifolia L.

An annual to perennial strongly aromatic glabrous herb with a procumbent or ascending branched stem 8–30 cm. *Lvs* 2–5 cm, alternate, sessile, oblong-lanceolate in outline, deeply toothed to irregularly pinnatifid with long narrow lobes, rarely entire, *broadening below into a whitish sheathing base*. *Fl.-heads* 6–10 mm diam., *hemispherical* on stalks exceeding the lvs; involucre of 2 rows of bracts with scarious margins and rounded ends. Disk-florets with a white corolla-tube and 4 short yellow lobes; *marginal fls* in 1 row *with* a short style and *±abortive corolla*. Achenes stalked, papillose on the inner face; those of the disk narrowly, of the marginal row broadly winged. Fl. 7–8. $2n = 20$. Th.–H.

Introduced. A casual, locally established in damp saline habitats, as in coastal areas of N.W. England. Probably native in S. Africa but now widely distributed in both hemispheres.

35. ARTEMISIA L.

Suffruticose or herbaceous perennials, rarely shrubs or annual herbs, with alternate pinnatisect or palmatisect lvs. *Fl.-heads small, numerous, in racemes or racemose panicles*, rarely few or only 1, homogamous or heterogamous; involucre cylindrical or globular, of many imbricating bracts with narrow scarious margins; *receptacle ±flat, naked*, glabrous or hairy. *Marginal fls tubular, female or* 0; disk-fls tubular, hermaphrodite. Achenes cylindrical or somewhat compressed, not strongly ribbed; pappus usually 0.

About 400 spp., chiefly in the steppes and prairies of E. Europe, Asia and North America.

1 Plant only a few cm. high with a single head (rarely 2) c. 12 mm diam.; lvs ±palmately cut; a rare Scottish alpine. **4. norvegica**
  Plant usually more than 20 cm high with numerous heads in a racemose panicle. 2
2 Ultimate segments of lvs broadly linear or lanceolate, at least 2 mm wide. 3
  Ultimate segments of lvs narrowly linear or subulate, c. 1 mm wide or less. 6
3 Lf-segments glabrous or nearly so above, whitish-tomentose beneath; heads ovoid to ellipsoid. 4

Lf-segments whitish, with hairs on both sides; heads broadly campanulate. 5
4 Plant caespitose or with short underground shoots; stems usually glabrescent, with large central pith; lvs with only the larger veins translucent; infl.-branches strict, straightish; fl. 7–9. **1. vulgaris**
  Plant with long rhizomes; stems persistently pubescent, with narrow central pith; lvs with even the smaller veins translucent; infl.-branches arcuate-divaricate; fl. 10–11. **2. verlotiorum**
5 Stems, lvs and heads densely white-felted; heads 5–7 mm diam. longer than wide. **3. stellerana**
  Stems, lvs and heads covered with silky white hairs but not densely felted; heads 3–5 mm diam., wider than long. **5. absinthium**
6 Plant hardly scented, ±glabrous or sparsely silky. **7. campestris**
  Plant strongly aromatic; lvs whitish-woolly beneath. **6. maritima**

Section 1. *Artemisia*. Marginal fls female, rarely 0, central hermaphrodite, all fertile; receptacle glabrous or hairy.

1. A. vulgaris L. Mugwort.

A perennial ±caespitose aromatic herb with branching root-stock and erect, *glabrescent or sparsely pubescent*, reddish, grooved and angled *stems* 60–120 cm, with large *white central pith and narrow green peripheral tissues*. *Lvs* 5–8 × 2.5–5 cm, *dark green and ±glabrous above*, *whitish-tomentose beneath*; basal lvs short-stalked, ±lyrate-pinnatifid, auricled; stem-lvs ±sessile, amplexicaul, bipinnate, the uppermost simply pinnate; *ultimate segments shortish, lanceolate* to oblong, 3–6 mm wide, toothed or entire; *only the main veins translucent*. Heads 3–4 mm, 2–2.5(–3.5) mm diam., narrowly campanulate to ovoid, ±erect, numerous, in dense sparsely lfy racemose panicles, the *infl.-branches strict and nearly straight*. Involucral *bracts* lanceolate to oblong, ±*densely arachnoid-pubescent* with broad scarious margins. Florets usually reddish-brown, all fertile. Achenes c. 1 mm, glabrous. Fl. 7–9. Wind-pollinated. $2n = 16*, 18$. Hp.

Variable in the degree of dissection of the lvs and branching of the infl.

Native. Common in waste places, waysides, hedgerows, etc., throughout the British Is. Most of the temperate regions of the northern hemisphere to 70° N. in Norway and 74° N. in Siberia.

*2. A. verlotiorum Lamotte Verlot's Mugwort.

A perennial herb resembling *A. vulgaris* but plant more strongly and pleasantly aromatic, not caespitose but with *long rhizomes and overwintering lf-rosettes*; *stem* usually *more densely and persistently pubescent* and with a *small central pith* and relatively broader green peripheral tissues; *ultimate lf-segments* conspicuously *elongated, linear-lanceolate to linear*, even the *smaller veins clearly translucent*; heads 3.5–5 mm, 2.5–3(–4) mm diam., ellipsoid, the *infl.-branches arcuate-divaricate* and the whole infl. very lfy; involucral *bracts* only *thinly arachnoid-pubescent*; achenes not as yet seen in Britain. Fl. 10–11. $2n = 54$. Hp.

Introduced. Established on waysides and waste places in the London area and a few scattered places northwards to Lancs and Northumberland. S.W. China.

*3. A. stellerana Besser
Old Woman, Dusty Miller, Beach Sagewort.

A. ludoviciana Hort.

A perennial *non-aromatic* herb with a creeping woody stock and *densely white-felted stems* 30–60 cm. *Lvs* pinnate or pinnatifid, the uppermost ±entire, the lobes broad, blunt, *densely white-felted above and beneath.* Heads 5–9 mm diam., broadly campanulate, somewhat longer than broad, numerous, in a racemose panicle. Bracts oblong to ovate, densely felted. Receptacle glabrous. Marginal fls female, central hermaphrodite, all yellow, fertile. Fl. 7–9. Wind-pollinated. H.

Introduced. A common garden plant grown for its white foliage, occasionally escaping and established in a few localities in Cornwall, Devon, Hants, S.W. Scotland etc., and near Dublin. N.E. Asia and the Atlantic coast of North America from Massachusetts to Delaware; Kamchatka.

4. A. norvegica Fr. var. *scotica* Hultén
Norwegian Mugwort.

A dwarf tufted aromatic perennial herb with flowering stems 3–10 cm, hairy. *Basal lvs* c. 2 cm, stalked, *subpalmate*, with deeply 3–5-toothed *cuneate lobes*; stem-lvs ±sessile; all *silky-white. Heads usually solitary* (–3), c. 10 mm diam., nodding. Involucral bracts with green midrib and broad dark brown scarious margins. Fl. 7–9. $2n = 18$. Hs.

Native. At 550–760 m in *Rhacomitrium* heath on two mountains in W. Ross. Norway and the Ural Mts.

5. A. absinthium L.
Wormwood.

A perennial aromatic plant with non-flowering rosettes and erect silky-hairy grooved and angled stems 30–90 cm, ±woody below. Lvs 2.5–5(–10) cm, those of the barren rosettes and the lower stem-lvs tripinnate, middle stem *lvs* bipinnate, and the uppermost simply pinnate or undivided; ultimate segments lanceolate or linear-oblong, c. 2–3 mm wide, usually blunt, *punctate, silky-hairy on both sides. Heads* 3–4 mm diam., *broadly campanulate to globose*, rather broader than long, *drooping, numerous*, in a much-branched racemose panicle. Involucre silky-hairy, the outer bracts linear, inner ovate, all blunt and broadly scarious-margined. *Receptacle with long hairs.* Marginal fls female, central hermaphrodite, all fertile, yellow. Achenes 1.5 mm, glabrous. Fl. 7–8. Wind-pollinated. $2n = 18^*$. H.–Ch.

Native. A not infrequent plant of waste places throughout the British Is. to Aberdeen and Ross; Orkney; introduced in Ireland. Temperate Europe and Asia northwards to Lapland, Karelia and S. Siberia. Introduced in North and South America and New Zealand.

6. A. maritima L.
Sea Wormwood.

A perennial strongly aromatic herb with a short usually branching vertical woody stock producing non-flowering rosettes and decumbent then erect flowering shoots 20–50 cm, usually downy, branched above. Lvs 2–5 cm, mostly bipinnate, the ultimate segments linear, c. 1 mm wide, blunt; lower lvs stalked, auricled; upper *lvs* simply pinnate, sessile; uppermost pinnatifid or entire; all ±*woolly on both sides*, not punctate. Heads 1–2 mm diam., ovoid, longer than wide, erect or drooping, numerous, in lfy racemose panicles with short branches. Bracts oblong, outer herbaceous, downy, inner with broad scarious margins. Receptacle glabrous. Fls yellowish or reddish, all hermaphrodite, but the central ones sometimes sterile. Achenes apparently very rarely produced. Fl. 8–9. Wind-pollinated. $2n = 36, 54^*$. H.–Ch.

All British plants belong to subsp. **maritima**, with rather few long non-flowering rosettes, ascending stems scarcely woody at base, and usually a broad panicle. Forms with the panicle-branches short, erect, crowded and the individual heads erect and crowded are placed in var. subsp. **subgallica** Rouy (*A. gallica* auct., non Willd.), but no clear distinction can be made.

Native. Locally common on the drier parts of salt-marshes and sea-walls. All round the coast of Great Britain northwards to N. Aberdeen and Cumbria, but absent in the extreme west and north; very local on the east and west coasts of Ireland. Subsp. *maritima* is confined to the coasts of N.W. Europe from W. France to Denmark and S. Sweden, but other subspp. occur in the W. Mediterranean littoral from Spain to Dalmatia, in W. Switzerland (on dry calcareous substrata), in the inland saline areas of Germany, Hungary, etc., and from the Black Sea coasts of S. Russia, and the Caucasus, Georgia and Armenia across C. Asia to Lake Baikal.

Section 2. *Dracunculus* Besser. Marginal florets female, fertile; central hermaphrodite but mostly sterile: receptacle glabrous.

7. A. campestris L.
Field Southernwood.

A perennial *scentless* plant with a branched creeping woody stock producing tufts of short non-flowering shoots and decumbent then ascending ±glabrous flowering shoots 20–60 cm, ±woody below, paniculately branched above. Basal and lower stem-lvs 2–3 times pinnate, stalked, auricled; the upper stem-lvs less divided and shorter stalked upwards; the uppermost linear, entire, sessile; ultimate segments long, linear, c. 1 mm wide, mucronulate, at first silky-hairy on both sides, later glabrous. Heads 3–4 mm diam., broadly ovoid, ±erect, numerous in a narrow elongated racemose panicle. Bracts ovate to oblong, green or reddish, glabrous, with scarious margins. Receptacle glabrous. Fls yellow or reddish; the marginal female, fertile; the central hermaphrodite but mostly sterile. Achenes glabrous. Fl. 8–9. Wind-pollinated, $2n = 18, 36$. Ch.

The British plant belongs to subsp. **campestris**.

Native. A very local plant confined to the 'breckland'

heaths of S.W. Norfolk, N.W. Suffolk and E. Cambridge; naturalized on sand dunes at one locality in S. Wales, casual elsewhere. Casual near Belfast. The various subspp. are distributed throughout the northern hemisphere to 75° N. in Novaya Zemlya and to 2700 m in the Alps.

Several spp. of *Artemisia* occur as infrequent casuals or garden-escapes. Amongst them are: **A. abrotanum** L. (section *Artemisia*), Southernwood or Lad's Love, a strongly aromatic shrubby plant, up to 1 m high, with 1–3 times pinnatifid lvs (uppermost simple) whose filiform gland-dotted segments are glabrous above and ±hairy beneath, and very small drooping heads, much grown in gardens; **A. biennis** Willd. (section *Artemisia*), with 2–3-pinnatisect lvs, ultimate segments linear-acute, entire or sharply toothed, ±glabrous, and erect fl.-heads on short branchlets, the whole infl. very narrow, a rare casual reported established in one locality in Somerset; **A. pontica** L. (section *Artemisia*), Roman Wormwood, shrubby, with finely dissected lvs, white-woolly beneath, and numerous hemispherical short-stalked drooping fl.-heads in a narrow racemose panicle; and **A. scoparia** Waldst. & Kit. (section *Dracunculus*), close to *A. campestris* but biennial with a slender stock, a single fl.-stem 30–60 cm and drooping subglobose fl.-heads, a rare casual.

Also belonging to the Anthemidae is the genus **Anacyclus** which closely resembles *Anthemis* in habit and in having receptacular scales but differs in the broadly membranous-winged achenes. Occasionally found as casuals are *A. radiatus* Loisel., with yellow ray-florets, *A. clavatus* (Dcsf.) Pers., with white ray-florets, and *A. valentinus* L. with ray-florets 0 or so shortly ligulate as to be inconspicuous.

Tribe 9. ARCTOTIDEAE.

The tribe Arctotideae resembles Anthemidae in that the heads have ligulate ray-florets, but has the style of Cynareae, with a ring of hairs below the bifurcation. There are no native members, but *Gazania* spp. and *Arctotis stoechadifolium* are commonly grown in gardens and sometimes escape and become established (as has *G.* × *splendens* Lemaire in Scilly).

Tribe 10. CYNAREAE.

*Echinops sphaerocephalus L. (Globe Thistle), a native of C. and S. Europe and W. Asia, is much grown in gardens and occasionally establishes itself. It is a rigidly erect thistle-like perennial herb 50–200 cm, with ovate-oblong pinnatifid lvs, green, densely glandular and sparsely bristly above, whitish-tomentose beneath, the ±triangular lobes ending in a strong spine and having spinous-toothed margins. Heads 1-flowered, aggregated into globose secondary heads 4–6 cm diam. General involucre of laciniate straw-coloured scales hidden by the downwardly-directed heads; individual involucre of outer bristle-like bracts only half as long as the broader fimbriate inner ones which are glandular on the back. Florets tubular, pale blue or whitish; anthers blue-grey. Achenes 7–8 mm, silky; pappus represented by a cupule of partially joined hairs. Also sometimes established are the eglandular *E. banaticus* Rochel ex Schrader (with blue fls) and *E. exaltatus* Schrader (with whitish fls).

36. CARLINA L.

Thistle-like herbaceous or suffruticose plants with ±pinnatifid *spinous lvs*. Heads homogamous, solitary or in corymbs; outer bracts lf-like, *inner longer*, *scarious*, *coloured*, *shining*, *spreading* in dry weather; receptacle flat,

pitted, with ±divided scales and often bristles. Fls all tubular, hermaphrodite; style-arms short and almost closed, forming a pubescent cone. Achenes cylindrical, covered with appressed forked hairs; *pappus of feathery hairs in 1 row*, *united at the base in groups of 2 to 4*, deciduous.

About 20 spp., in Europe and W. Asia; Canary Is.

1. C. vulgaris L.   Carline Thistle.

A biennial herb with a tap-root and stiffly erect flowering stems 10–60 cm, usually purplish, somewhat cottony, usually branched above. Rosette lvs of the first season dying before flowering, 7–13 cm, oblong-oblanceolate, acute, narrowing to a short petiole, cottony especially beneath; stem-lvs shorter and broadly sessile, semi-amplexicaul, ±glabrous; all with undulate and somewhat lobed margins carrying numerous short weak spines. Heads 1.5–3 cm diam., 2–5 or more in a corymb, rarely solitary. Outer bracts broadly lanceolate, cottony, green or purple-tinged, spiny at the tips and margins, shorter than the numerous long linear *straw-yellow inner bracts which spread when dry and simulate ray-florets*. Achenes 2–4 mm, covered with rusty hairs; pappus two or three times as long as the achenes. Fl. 7–10. Visited chiefly by bees and hoverflies. $2n = 20^*$. H. (biennial).

Native. Almost always in calcareous grassland. Locally common throughout lowland Great Britain northwards to Ross; Inner Hebrides; frequent in C. Ireland; reaches 460 m in Westmorland. Europe northwards to c. 62° N. in Scandinavia; Siberia; Caucasus; Asia Minor.

37. ARCTIUM L.

Tall *biennial* herbs with long stout tap-root and *alternate non-spiny lvs, basal large, ovate-cordate*. Fl.-heads solitary or in racemes or corymbs; involucre ovoid-conical to hemispherical or ±globose, its *bracts* numerous, flat and imbricate below, *narrowing above into a long rigid subulate ±spreading point*, all, or only the outer, *hooked at apex; receptacle* flat, *with rigid subulate scales. Fls all tubular, hermaphrodite*, reddish-purple or white; anthers acuminate above, prolonged into filiform tails below; style swollen below, its branches cuneate. Achenes obovoid-oblong, compressed, rugose; *pappus of scabrid golden-yellow hairs* in several rows, free to the base. Fls normally self-pollinated but with occasional outbreeding. *Fr.-heads animal-dispersed.*

A few closely related spp. in Europe and Asia, some of them naturalized in North America.

Petioles solid; main infl.-branches with few fl.-heads in a ±corymbose cluster; peduncles 3–10 cm; fr.-involucre usually at least 3.5 cm diam., yellow-green and shining.
                                                        1. lappa
Petioles with a central hollow; main infl.-branches with sessile or pedunculate fl.-heads in a raceme ending in a solitary head or a cluster of 2–4; fr.-involucre up to 3.5 cm diam., often much smaller. **(2.–4.) minus** group

1. A. lappa L. Great Burdock.

A. majus Bernh.; *Lappa officinalis* All.; *L. major* Gaertner

Fl.-stems 90–150 cm or more, stout, furrowed, often reddish, ±cottony, with many spreading-ascending branches. *Basal and lower stem-lvs* up to 50 cm, *broadly ovate-cordate* with ±rounded blunt or apiculate apex, *entire or distantly denticulate* and often undulate margin, and *petiole* up to 30 cm, grooved above, *solid*; middle and upper lvs smaller, narrower, less cordate and more shortly petiolate upwards, uppermost often ovate-lanceolate and not at all cordate; all ±sparsely cottony and green above, grey-cottony beneath. Fl.-heads 3–4 cm diam. overall, on peduncles 3–10 cm, borne in axillary clusters of 1–4, together forming *loosely corymbose groups terminating the main infl.-branches*; heads globular in bud, hemispherical and *widely open above in fr*. Involucral bracts glabrous or slightly ciliate at base (cottony in var. *subtomentosum* Lange), *tips of the inner bracts about equalling the open fls*. Fls reddish-purple, the *wider upper part of the corolla distinctly shorter than the filiform lower part*. *Fr.-involucre* 3.5–4.2 cm diam., *yellow-green to straw-coloured*. Achenes 6–7 mm, ±pale fawn, often with dark blotches, somewhat rugose above. Fl. 7–9. Visited by some bees and Lepidoptera but normally self-pollinated before fls open. $2n = 36^*$. H.

Native. Waysides, field borders and waste places, rarely in woods. Scattered through lowland Great Britain northwards to N. Wales and Humberside, rare in Ireland. Most of Europe to c. 63° N. in Scandinavia; Asia Minor.

(2.–4.) A. minus group

Fl.-stems 50–150 cm, furrowed, often reddish, somewhat cottony, with many spreading or down-curving branches. Basal lvs up to 40 cm or more, ±broadly ovate-cordate, much as in **1**, but somewhat narrower and more acute; stem-lvs becoming smaller upwards, more shortly petiolate and less or not at all cordate, uppermost often ±broadly lanceolate; all nearly glabrous above, *paler* and somewhat cottony *beneath*; *petiole with a central hollow*. *Fl.-heads in racemes* or the top 2–4 subcorymbose; fls red-purple, the *wide upper part of the corolla about equalling the filiform lower part*. *Achenes* 5–9 mm, pale to dark *brown*, commonly with *blackish blotches*. Fl. 7–9. H.

The following 3 spp. are variable and their delimitation presents some difficulties. All seem normally to be self-pollinated and, as would then be expected, isolated wild populations tend to be morphologically uniform and distinguishable from other populations. Fertile intermediates nevertheless occur, both within the group and between each member and *A. lappa*, all four appearing to constitute a potentially interbreeding complex. Complete merging has presumably been prevented by some geographical and ecological separation of the recognized taxa and by the fact that selfing usually occurs before the fls open. The not infrequent occurrence of intermediates and the approach, at least locally, to a continuum of variants between any two, has led some to treat all three as subspp. of *A. minus*.

Others claim that most plants are assignable without serious difficulty to one or other of the three and, like *Flora Europaea*, 4 (1976), treat them as separate spp., the course provisionally adopted here.

> 1 Fl.-heads small, 1.5–2.5 cm diam. in fr. (incl. involucral bracts), shortly pedunculate or subsessile; corollas of open fls exceeding surrounding involucral bracts; fr.-involucre usually closed above. **2. minus**
>
> Fl.-heads 3–4 cm diam. in fr. (incl. involucral bracts); corollas of open fls at most equalling surrounding involucral bracts; fr.-involucre open or closed above. 2
>
> 2 Peduncles usually 1–4 cm but sometimes much longer; fr.-involucre ±straw-coloured, usually open above. **4. pubens**
>
> Peduncles usually less than 1 cm; fr.-involucre green or tinged with dark purple. **3. nemorosum**

2. A. minus Bernh. Lesser Burdock.

Lappa minor Hill

Fl.-stems 50–120 cm. Most lvs rather sharply dentate or crenate-denticulate, uppermost often ±entire. *Fl.-heads small*, 1.5–2.5 cm overall diam. in fr., *subsessile or with short peduncles* to 1 cm, *terminal head usually solitary*; *corollas of open fls overtopping surrounding involucral bracts*; *involucre green or purple-tinged*, often cottony when young but becoming ±glabrous, *usually closed above in fr*.; *all bracts with hooked tips*. Achenes 5–7 mm, brownish, often with darker blotches. $2n = 36^*$.

Native. Open woods and wood-margins, waysides, waste ground, etc., on a wide range of soils. Local in England and Wales northwards to N. Lancs and N. Yorks with outliers in Northumberland and E. Lothian; rare in Ireland. Most of Europe to c. 66° N. in Scandinavia and Finland; Caucasus; N. Africa.

3. A. nemorosum Lej. Wood Burdock.

Lappa vulgaris Hill

Fl.-stems robust, commonly 90–200 cm. Most lvs rather distantly and not very sharply crenate-dentate, often crenate-denticulate towards apex, uppermost ±entire. *Fl.-heads* 3–4 cm overall diam. in fr., *subsessile or very shortly pedunculate, in racemes* but with *the top 2–4 usually in a subcorymbose cluster; corollas of open fls at most equalling the surrounding involucral bracts; involucre green or purple-tinged*, sparsely cottony, usually closed in fr., sometimes ±open; *all bracts with hooked tips*. Achenes 6–9 mm, brown, usually with blackish blotches. $2n = 36^*$.

Native. Woodland margins and clearings, scrub, hedgebanks and waysides, waste places, etc. Locally frequent in S.E. England and north of a line from S.W. Wales to the Wash and including S.E. Scotland, but rare elsewhere in Scotland though reaching Caithness; Inner and Outer Hebrides, Orkney; locally frequent in much of Ireland except the south. Europe to 64° N. in Scandinavia; Caucasus; N. Africa.

It will be noted that the distributions of **2.** and **3.** appear to be largely complementary but overlap in S.E. England and in a broadish belt including much of W. and N. Wales and across to Lincoln and Humberside. There is, however, need for more precise information.

4. A. pubens Bab.

Much like **2.** but with *larger fl.-heads*, 2–3.5 cm overall diam. in fr., *longer peduncles*, 1–4(–12 or more) cm, *corollas of open fls about equalling surrounding involucral bracts*, and *involucre densely cottony at first* but *in fr. subglabrous, straw-coloured and open above.* Achenes much like those of **2.**

Native. Probably widely but thinly scattered in habitats similar to those of **1.–3.**, but many records are doubtful and the detailed distribution is still very uncertain. Recorded from much of W., C. and S. Europe.

It seems likely that *A. pubens* originated from crosses between *A. lappa* and *A. minus* and may have been made more variable through backcrossing with the parent spp.

*****A. tomentosum** Miller, with solid petioles and long-stalked fl.-heads in corymbs, their involucres densely and persistently cottony and with the reddish tips of the broad inner involucral bracts not hooked and falling short of the corollas of open fls, is native in most of Europe but not in the British Is., where it is an occasional casual which may persist for a time.

38. Carduus L.

Annual to perennial herbs (thistles) whose alternate, ±entire to pinnatisect lvs have ±spiny margins. Heads solitary or in clusters at the ends of the main stem and branches, homogamous; bracts imbricating, many-rowed, narrow, usually spine-tipped; *receptacle* deeply pitted, *densely bristly.* Fls all tubular, hermaphrodite; anthers tailed below; style-arms united below, forming a shortly 2-lobed column with a ring of hairs at the base. Achenes obovate-oblong, grooved, glabrous; *pappus of many rows of simple* but scabrid (not plumose) *hairs*, united below, and deciduous.

About 120 spp., in Europe, N. Africa, and Asia, with a few in the Canary Is. and tropical Africa.

1 Heads oblong-cylindrical, falling when the achenes are ripe; corolla equally 5-lobed.   2

Heads ovoid or hemispherical, not falling when the achenes are ripe; corolla distinctly 2-lipped, with one entire and one 4-lobed lip.   3

2 Stem winged to close beneath the fl.-heads, with continuous broad spinous wings; lvs somewhat cottony beneath; heads 3–10 or more in a cluster; inner bracts equalling or exceeding the open fls.

     1. tenuiflorus

Stem naked close beneath some fl.-heads, below with interrupted narrow spinous wings; lvs densely cottony beneath; heads solitary or 2–3 in a cluster; inner bracts falling short of the open fls.

     2. pycnocephalus

3 Heads 3–5 cm diam., usually solitary, drooping; bracts lanceolate above then contracted abruptly into an oblong base; middle and outer strongly reflexed, inner erect.   **3. nutans**

Heads 1–3 cm diam., solitary or clustered, erect; bracts linear-subulate, straight or the outermost recurved at the tip, not contracted above the base.

     4. acanthoides

1. C. tenuiflorus Curtis   Slender Thistle.

Annual or biennial herb with stout tap-root and erect broadly (to 10 mm) and *continuously spinous-winged* stems 15–120 cm, ±cottony, branched above, the branches erect-ascending. Basal lvs oblanceolate in outline, blunt, narrowed to the base; stem-lvs decurrent, acute; all sinuate-pinnatifid, spinous-margined, ±cottony beneath. *Heads* about 15 × 8 mm, *cylindrical*, sessile, in dense terminal clusters of 3–10(–20), with the stem winged close beneath them. Bracts ovate-lanceolate, acuminate, glabrous, broadly scarious-margined, 1-veined, terminating in a ±outwardly curved spine; inner bracts scarious, equalling or exceeding the florets. Florets pale purple-red, rarely white. Achenes 3.5 mm, fawn, shining, with fine transverse wrinkles, prominently tubercled above; pappus 1–1.5 cm, white. Fl. 6–8. Visited by bees. $2n = 54^*$. Th.–H.

Native. Waysides and waste places, especially near the sea. Locally common throughout lowland Great Britain, except in parts of the W. Midlands, northwards to the Clyde Is. and Moray; throughout Ireland. W. Europe eastwards to the Netherlands and Italy, incl. islands of W. Mediterranean; naturalized in Scandinavia.

*2. C. pycnocephalus L.

Annual or biennial herb closely resembling *C. tenuiflorus* but with *stem narrowly* (to 5 mm) *and discontinuously spinous-winged, naked beneath some fl.-heads; lvs densely cottony beneath;* heads 2 × 1 cm, ±distinctly stalked, *solitary* or 2–3 in a cluster; bracts cottony, tips spiny, straight and erect, *inner bracts falling short of the open fls;* pappus 1.5–2 cm. Fl. 6–8. Th.–H. (biennial).

Introduced. Waste places. Established at Plymouth and a rare casual elsewhere. S. Europe from Spain to the Balkans; Canary Is.; N. Africa; W. Asia.

3. C. nutans L.   Musk Thistle.

A usually biennial herb with an erect interruptedly spinous-winged *stem* 20–100 cm, *naked for some distance beneath the heads*, ±cottony throughout, simple or with spreading ascending branches above. Basal lvs elliptical, narrowing into a stalk-like base, sinuate; stem-lvs oblong-lanceolate, decurrent, deeply pinnatifid with triangular 2–5-lobed spine-tipped segments; all with undulate and spinous margins, sparsely hairy on both sides, woolly on the veins beneath. *Heads* 3–5 cm *diam.*, *solitary*, or 2–4 in a loose corymb, hemispherical, ±*drooping.* Involucre cottony, often purplish, the outer and middle *bracts* ±strongly reflexed, the inner erect; all spine-tipped, *lanceolate-acuminate, contracted abruptly into an oblong base.* Florets red-purple, distinctly 2-lipped. Achene 3–4 mm, fawn, with fine transverse wrinkles; pappus long, whitish. Fl. 5–8. Slightly musky and visited by many bees, hover-flies and Lepidoptera. $2n = 16^*$. H.

Native. Pastures, waysides, arable fields and waste places on calcareous soils, up to 500 m in Yorks. Locally common through much of England, Wales and S.E. Scotland with a few outliers northwards to Moray and E. Ross; seems extinct in Ireland. Europe northwards to c. 67° N. in Fennoscandia; Siberia; Caucasus; Asia Minor; N. Africa.

4. C. acanthoides L.   Welted Thistle.

A biennial herb with a slender tap-root and an erect cottony *stem* 30–120 cm, usually branched above, *usually naked just beneath the fl.-heads* but otherwise with a continuous narrow undulate spinous-margined wing. Basal lvs elliptical in outline, sinuate-pinnatifid, narrowed into a stalk-like base; stem-lvs lanceolate in outline, decurrent, deeply pinnatifid with 3-lobed ovate segments whose terminal lobe is longest; all dull green, cottony beneath, with weakly spinous margins. *Heads* c. 2.5–3 cm *diam.*, erect, ±globose, usually in dense *clusters of* 3–5, rarely solitary. Involucre roundish-ovoid, slightly cottony, the *bracts linear-subulate, not contracted above the base*, ending in a weak slender spine, the outermost somewhat spreading, green, the inner erect, purplish, shorter than the florets. Florets red-purple or white, 2-lipped. *Achenes* 3–4 mm, fawn, *with* fine transverse wrinkles and *a prominent terminal tubercle which is not 5-angled*; pappus 11–13 mm, whitish. Fl. 6–8. Visited by many bees, hover-flies and Lepidoptera. $2n = 16$. H.

Native. Damp grass verges and stream-sides, hedgerows, waste places. A lowland plant common in the south but becoming rarer in the north though reaching E. Ross; Inner Hebrides; Ireland. Guernsey. Often with *Urtica dioica*, *Galium aparine* and *Poa trivialis*. Europe northwards to c. 71° N. in Scandinavia; Siberia; Caucasus. Introduced elsewhere.

The hybrid of *C. acanthoides* and *C. nutans*, intermediate between the parents, is frequently found.

*C. crispus L. closely resembles *C. acanthoides* L. but may be distinguished by the stronger spines on the margins of the stem-wings, and by the wings usually extending up to the heads, not falling short of them as is more usual in *C. acanthoides*. The lvs are green and ±glabrous beneath; when pinnatifid the terminal segment is smaller than the laterals, and the marginal spines are stiffer than in *C. acanthoides*. The heads are 1.5–2.5 cm diam. and usually in clusters; the involucre is subglobose with linear-lanceolate bracts ending in a stout spine the outer often with ±recurved tip. The achenes are olive-green, with the terminal tubercle 5-angled. Fl. 6–10. $2n = 16$. H.

Introduced and at most a rare casual. Further investigation is needed; many of the records are of forms of *C. acanthoides* with large solitary heads. Europe, especially the south and south-east; Caucasus.

There has been doubt about the allocation of the two names *C. crispus* L. and *C. acanthoides* L.

39. CIRSIUM Miller

Annual to perennial herbs (thistles) with alternate simple or pinnately lobed or divided lvs, usually with prickly margins. Heads solitary or in corymbs or dense clusters, homogamous; the florets all tubular, hermaphrodite or female. Involucre, receptacle and florets as in *Carduus* but achenes with a *pappus of* many rows of *feathery hairs* united at the base.

About 150 spp. throughout the northern hemisphere.

1 Plant stemless, with 1–3 usually sessile heads at the centre of the lf-rosette.   **6. acaule**
   Plant with elongated stem.   2
2 Lvs roughly hairy and prickly on the upper surface, and therefore dull, not shining.   3
   Lvs not prickly (though sometimes hairy) on the upper surface and often ±glossy.   4
3 Stem not winged; involucre very cottony.   **1. eriophorum**
   Stem interruptedly spinous-winged; involucre not or scarcely cottony.   **2. vulgare**
4 Stem continuously spinous-winged.   **3. palustre**
   Stem not winged, or, if so, then only for short distances.   5
5 Heads yellow, overtopped by large pale ovate bract-like uppermost lvs; most stem-lvs ovate-acuminate, sessile and clasping, hardly spinous on the margins.   **5. oleraceum**
   Heads red-purple or pale purple, rarely white.   6
6 Heads 1.5–2.5 cm; florets with the slender basal tube of the corolla longer than the broad upper part which is 5-cleft almost to its base.   **4. arvense**
   Heads 2.5–5 cm; florets with the slender basal tube of the corolla about equalling the broad upper part which is 5-cleft to about the middle.   7
7 Lvs densely white-felted beneath with softly prickly-ciliate margins; head 3.5–5 cm.   **7. helenioides**
   Lvs at most whitish-cottony beneath, not densely felted; head 2.5–3 cm.   8
8 Lvs deeply pinnatifid, green beneath; heads solitary or 2–4 in a cluster; roots swollen, spindle-shaped; rare plant of calcareous pastures.   **9. tuberosum**
   Lvs sinuate-toothed, sometimes ±lobed, whitish-cottony beneath; head usually solitary; roots not swollen; fens and bogs.   **8. dissectum**

1. C. eriophorum (L.) Scop.   Woolly Thistle.

Carduus eriophorus L.; *Cnicus eriophorus* (L.) Roth

A biennial herb with a thick tap-root and a stout erect *unwinged* furrowed *stem* 60–150 cm, corymbosely branched above, cottony, not prickly. Basal lvs up to 60 cm, ovate-oblong in outline, narrowing into a short stalk, deeply pinnatifid and strongly undulate, the narrowly lanceolate distant spine-tipped *segments* usually *2-lobed with one lobe directed upwards and one downwards*; stem-lvs similar but sessile, auricled, *semi-amplexicaul, not decurrent*; all prickly-hairy but green above, white-cottony beneath, spiny and ciliate on the margins. *Heads* 3–3.5 × 4–7 cm, usually *solitary* ±erect, usually with a few small lvs close below. *Involucre very cottony*, its bracts lanceolate-acuminate ending in a long narrow ±spreading reddish-ciliate point usually with a slight dilation just below its apex; the outermost spine-tipped. Florets pale red-purple. Anthers blue-purple. Achenes 6 mm, buff mottled with black, smooth, shining; pappus very long, shining white. Fl. 7–9. Visited by long-tongued bees and Lepidoptera to whom alone the nectar is accessible. $2n = 34^*$. Hs. (biennial).

Native. Grassland, open scrub and roadsides on calcareous soil, to 260 m in Yorks. Local in England and S.E. Wales northwards to Durham, but not in W. and N. Wales or in S.W. and N.W. England; not in Ireland. C. Europe from France, Belgium and the Netherlands to N. Balkan peninsula and Upper Volga.

2. C. vulgare (Savi) Ten.   Spear Thistle.

Carduus lanceolatus L.; *Cirsium lanceolatum* (L.) Scop., non Hill

A biennial herb with a long tap-root and erect *interruptedly spiny-winged* cottony furrowed *stems* 30–150 cm, branched above. *Basal lvs* 15–30 cm, obovate-lanceolate in outline narrowed into a short stalk-like base, ±*deeply pinnatifid and undulate* with the segments usually 2-lobed, the upper lobe toothed near the base, the lower entire, lobes and teeth tipped with long stout spines; *stem-lvs* similar but sessile and *decurrent, with a long narrow terminal segment; all prickly-hairy above*, rough or cottony beneath; earliest lvs not undulate and only slightly pinnatifid. Heads 3–5 × 2–4 cm, ovoid-oblong, solitary or 2–3 in a pedunculate cluster. Involucre slightly cottony, its bracts green, lanceolate-acuminate with the long neither ciliate nor dilated point recurved and spine-tipped in the outer, erect and scarious in the inner bracts. Fls pale red-purple. Achenes 3.5–5 mm, yellow streaked with black; pappus 2–3 cm, white. Fl. 7–10. Visited chiefly by long-tongued bees, hover-flies and butterflies. $2n = 68^*$, 102. Hs. (biennial).

Plants with lvs hardly undulate, densely cottony beneath; branches few, strict, or 0; heads few, ovoid-globose, are of uncertain taxonomic status.

Native. Fields, waysides, gardens, waste places, to 625 m in N. England. Common throughout the British Is. Europe to 67° 50′ N. in Scandinavia; W. Asia; N. Africa. Introduced in North America and Chile.

3. C. palustre (L.) Scop.   Marsh Thistle.

Carduus palustris L.

A biennial herb with a short erect premorse stock and an erect, furrowed, narrowly but continuously spiny-winged, hairy and cottony stem 30–150 cm, usually with short ascending branches. Basal lvs narrowly oblanceolate in outline, stalked, pinnatifid, the lobes shallow with spinous margins; *stem-lvs* sessile, *long-decurrent*, deeply pinnatifid and undulate, the segments each with 2–3 spine-tipped and spiny-ciliate lobes; all *hairy above*, slightly cottony beneath; earliest rosette lvs flat, hardly pinnatifid, cottony above and beneath. Heads 1.5–2 × 1–1.5 cm, subsessile in crowded lfy clusters of 2–8 at ends of main stem and branches. Involucre ovoid, slightly cottony, its bracts purplish, lanceolate, appressed, the outer mucronate, inner acuminate. Florets dark red-purple, rarely white. Achenes 3 mm, pale

fawn, smooth; pappus 2–3 times as long as the achenes, dirty white. Fl. 7–9. Visited by many bees, flies and Lepidoptera, the nectar being more readily accessible than in *C. vulgare* and *C. eriophorum*; said to be gynodioecious in continental Europe. $2n = 34^*$. Hs. (biennial).

Native. Marshes, moist grassland, hedgerows, woods, to 760 m in Scotland. Abundant throughout the British Is. Europe to 67° 50′ N. in Scandinavia; W. Asia; N. Africa.

4. C. arvense (L.) Scop.   Creeping Thistle.

Serratula arvense L.; *Carduus arvensis* (L.) Hill

A *perennial ±dioecious* herb initially with a slender tap-root producing *far-creeping whitish lateral roots which bear numerous adventitious* non-flowering and flowering shoots, the latter 30–90(–150) cm, erect, furrowed, *unwinged*, glabrous or cottony above, usually branched. Basal lvs not in a compact rosette, oblong-lanceolate in outline, narrowed to a short stalk-like base, usually ±pinnatifid and undulate with triangular toothed and spiny-ciliate lobes ending in strong spines; middle and upper lvs similar but sessile, semi-amplexicaul, more deeply pinnatifid, not or very slightly decurrent; all ±glabrous on both sides or cottony beneath. Heads 1.5–2.5 cm, short-stalked, solitary or in terminal clusters of 2–4, together forming an irregular corymb. Involucre purplish, glabrous or somewhat cottony, of male heads ±globose, of female ovoid; *bracts* numerous, appressed, the outer short, ovate-mucronate with ±spreading spiny points, the inner longer lanceolate-acuminate with erect scarious tips. Fls dull pale purple or whitish, *the broad upper part of the corolla shorter than the slender basal tube and 5-cleft to its base.* Achenes 4 mm, dark brown, smooth; pappus very long, brownish. Fl. 7–9. The male heads have abortive ovaries but sometimes ripen a few fr.; in the female heads the anthers are abortive. Strongly honey-scented and visited freely by a great variety of insects. $2n = 34^*$. Gr.

Very variable, especially in the lvs. Plants with broad, flat, hardly lobed and weakly spiny lvs, green and ±glabrous beneath, have been placed in var. *setosum* C. A. Meyer or var. *mite* Winmer & Grab., but intermediates are found. Var. *incanum* (Fischer) Ledeb. with similar but narrower lvs, densely white-cottony beneath, is a S. European form which may merit subspecific rank. It occurs in Britain as a casual.

Native. Fields, waysides and waste places to 640 m in N. England. Abundant throughout the British Is. and a very troublesome weed of cultivated land because of its capacity for regenerating from fragments of root. Europe to 68° 50′ N. in Scandinavia; Asia; N. Africa. Introduced in North America.

*5. C. oleraceum (L.) Scop.

Cnicus oleraceus L.

A perennial herb with an obliquely ascending stock and erect furrowed unwinged stems 50–120 cm, simple or rarely branched above, ±glabrous. Basal lvs ovate to broadly elliptical in outline, narrowed to a stalk-like base, simple or ±deeply pinnatifid with triangular-acuminate lobes; *middle and upper lvs* usually *unlobed, ovate-acuminate, sessile and amplexicaul* with large rounded basal auricles; all green and flaccid, ±glabrous, sharply toothed, with ciliate but hardly spinous margins, not decurrent. *Heads* 2.5–4 cm, ovoid, erect, clustered on short cottony stalks, *exceeded by the yellowish ovate-acuminate ciliate bract-like uppermost lvs.* Involucre slightly cottony, its bracts linear-lanceolate, erect, but with ±spreading tips, the outer ending in a short spine. *Fls yellowish-white*, rarely reddish. Achenes 4 mm, pale grey, angled. Fl. 7–9. The usually hermaphrodite heads are freely visited by bees and butterflies. $2n = 34$. H.

Introduced. Established in a few localities in Great Britain and Ireland. In marshes, fens, flushes, stream-sides and wet woods in C. Europe from C. France, N. Italy and N. Balkan peninsula northwards to 61° 15′ N. in Norway and to C. Russia; Siberia.

6. C. acaule Scop.   Stemless Thistle.

Carduus acaulos L.

Perennial with short rhizome, long stout roots, and a rosette of spiny lvs usually with a *few sessile heads* at its centre, there being only rarely (f. *caulescens* Rchb.) an elongated unwinged aerial stem to 30 cm, simple or branched. Lvs 10–15 × 2–3 cm, oblong-lanceolate in outline, stalked, ±deeply pinnatifid, strongly and stiffly undulate, the segments with 3–4 triangular teeth or lobes, stoutly spine-tipped and spiny-ciliate, ±glabrous above, hairy on the veins beneath. Heads 3–4 × 2.5–5 cm, ovoid, 1–3, usually sessile on the rosette. Involucre glabrous, purplish, the outer bracts ovate with a short spiny mucro, the inner oblong-lanceolate blunt; all appressed. Fls bright red-purple with the broad upper part of the corolla somewhat shorter than the slender basal tube, 5-cleft to about the middle. Achenes 3–4 mm, smooth; pappus very long, whitish. Fl. 7–9. Gynodioecious, the hermaphrodite heads somewhat larger. Visited by hover-flies, bees and Lepidoptera. $2n = 34$. Hr.

Native. In closely grazed pastures, especially on chalk or limestone, to 390 m in Derby. Locally common in Great Britain northwards to N. Yorks, Derby and Clwyd. N. Spain, N. Italy and N.W. Balkan peninsula to S. Scandinavia and Estonia; W. Asia.

7. C. helenioides (L.) Hill   Melancholy Thistle.

C. heterophyllum (L.) Hill

A perennial *stoloniferous* herb with obliquely ascending stock and erect usually simple grooved *cottony unwinged stems* 45–120 cm. Basal lvs 20–40 × 4–8 cm, elliptical-lanceolate, long-stalked, finely toothed; lowest *stem-lvs* narrowed to the base, sometimes ±pinnatifid with forwardly directed lobes, the remainder usually unlobed, lanceolate-acuminate, broad-based, *amplexicaul, with rounded auricles*, entire or ±toothed; all *flat and flaccid, green and glabrous above, white-felted beneath with softly prickly-ciliate margins*, not decurrent. Heads

3.5–5 × 3–5 cm, solitary or rarely 2–3 in a terminal cluster. Involucre broadly ovoid, its bracts ovate-lanceolate, the outer mucronate, innermost blunt, all appressed, glabrous or finely pubescent, purplish-tipped. Fls red-purple, rarely white, the broad upper part of the corolla longer than the slender basal tube, 5-cleft to about the middle. Achenes 4–5 mm, fawn, smooth; pappus very long, whitish. Fl. 7–8. Visited chiefly by bees. $2n = 34$*. H.

Native. Hilly pastures and stream-sides, upland scrub and open woodland, from 90 m in Yorks to 975 m in Scotland. From Sutherland and Caithness southwards to Powys, Stafford, Derby and N. Yorks; Inner and Outer Hebrides; very rare in N.W. Ireland. N. Europe from 71° 10′ N. in Scandinavia southwards to Schleswig and Pomerania, and in the mountains of C. Europe from the Pyrenees to Romania; C. Russia; Siberia. Introduced in North America.

8. C. dissectum (L.) Hill
                    Meadow Thistle, Marsh Plume Thistle.

Carduus dissectus L.; *C. pratensis* Hudson; *C. anglicus* Lam.

A perennial shortly *stoloniferous* herb with short obliquely ascending stock, *cylindrical roots* and an erect, usually simple, terete cottony unwinged *stem* 15–18 cm, usually *with few small bract-like lvs above the middle*. Basal lvs 12–25 × 1.5–3 cm, elliptical-lanceolate, long-stalked, sinuate-toothed or slightly pinnatifid; stem-*lvs* usually only 3–5, like the basal but oblong-lanceolate and semi-amplexicaul with basal auricles; all green and hairy above, *whitish cottony beneath*, the *margins with soft prickles*, longest on the teeth or lobes; not decurrent. Heads 2.5–3 × 2–2.5 cm, usually solitary. Involucre ovoid, purplish, cottony, bracts lanceolate, appressed, the outer spine-tipped, the inner acuminate. Fls dark red-purple. Achenes 3–4 mm, pale fawn, smooth; pappus long, pure white. Fl. 6–8. Fls hermaphrodite. $2n = 34$*. Hel.

Native. Fens and bog-margins, always on wet peat, to 500 m in W. Ireland. Local in England and Wales northwards to N. Yorks, Derby, Shropshire and Dyfed; throughout Ireland, common in the west. W. Europe from Spain to the Netherlands and N.W. Germany.

9. C. tuberosum (L.) All.   Tuberous Thistle.

Carduus tuberosus L.

A perennial *non-stoloniferous* herb with an obliquely ascending stock, *swollen fusiform roots* and an erect grooved cottony unwinged stem 20–60 cm, lfy chiefly below the middle, simple or with very long erect branches. Basal *lvs* long-stalked broadly elliptical in outline, usually *deeply pinnatifid*, the segments distant, each with 2–5 spreading oblong lobes; stem-lvs few, less deeply divided, ±sessile, semi-amplexicaul but with auricles small or 0; all *green on both sides*, slightly cottony beneath, undulate and spinous-ciliate, not decurrent. Heads 2.5–3 cm, usually solitary. Involucre subglobose, cottony below, its bracts appressed, oblong-lanceolate, the outer shortly mucronate. Fls dark red-purple. Achenes 3–4 mm, pale fawn, smooth; pappus long, white. $2n = 34^*$. G.

Native. On chalk downs and other calcareous pastures. Very rare and local in England and S. Wales (Cambridge, Wilts, E. Gloucester and Glamorgan). W. and C. Europe eastwards to S. Sweden, Saxony, Bohemia, Tirol and N. Italy.

Closely related to *C. dissectum* but readily distinguishable by the tuberous roots and absence of stolons and by the deeply pinnatifid lvs.

The following hybrids have been reported from Britain: *C. acaule × arvense, C. acaule × dissectum, C. acaule × tuberosum, C. acaule × vulgare, C. arvense × palustre, C. arvense × vulgare, C. dissectum × palustre, C. helenioides × palustre, C. palustre × vulgare.*

40. SILYBUM Adanson

Annual or biennial thistles with *white-veined or otherwise variegated lvs*. Heads homogamous; involucre of many rows of spiny bracts; *receptacle hairy*, not pitted. Fls all tubular and hermaphrodite, red-purple; *stamen-filaments united at their base into a tube*; anthers with short terminal points; style-arms connate. Achenes obovoid compressed, crowned with a membranous border; pappus of many rows of rough hairs united below into a basal ring.

Two spp. in the Mediterranean region, C. Europe, and the Near East. Differs from *Carduus* in the connate stamen-filaments.

*1. S. marianum (L.) Gaertner   Milk-Thistle.
Carduus marianus L.

An annual to biennial herb with an erect grooved slightly cottony unwinged stem 40–120 cm, simple or branched above. *Lvs* oblong, sinuate-lobed or pinnatifid with strongly spinous margins, glabrous, *pale shining green variegated with white along the veins above*; basal lvs narrowed into a sessile base, stem-lvs amplexicaul with rounded spinous-ciliate auricles, hardly decurrent. Heads 4–5 × 1–2 cm, solitary, erect or ±drooping. Involucre ovoid, glabrous, the outer bracts with an ovate-oblong base surmounted by a triangular spinous ciliate lf-like appendage which, in all but the basal bracts, ends in a stout *yellowish spine, long and spreading or recurved in the middle bracts*, shorter and erect in the innermost. Fls red-purple. Achenes 6–7 mm, blackish, grey-flecked and with a yellow ring near the apex, transversely wrinkled; pappus long, pure white. Fl. 6–8. Visited by various bees. $2n = 34$. H. (biennial).

Introduced. Naturalized in waste places locally and a frequent casual throughout lowland Great Britain, northwards to Nairn; decreasing; rare in Ireland. S. Europe from Spain to S. Russia; N. Africa; Caucasus; Near East. Introduced in C. Europe to Denmark, and in North and South America and S. Australia.

41. ONOPORDUM L

Biennials with sinuate to pinnatisect spinose lvs. Heads homogamous; involucre of many rows of coriaceous spiny bracts; *receptacle naked, deeply pitted, the pits with toothed membranous borders*. Fls all tubular and hermaphrodite, usually red-purple: anthers with terminal subulate appendages and short basal tails; style-arms connate. Achenes obovoid, compressed or 4-angled; pappus of many rows of rough hairs united into a basal ring.

About 40 spp. in the Mediterranean region and Near East. Close to *Carduus*, but lacking receptacular scales.

1. O. acanthium L.   Scotch Thistle, Cotton Thistle.

A large biennial thistle with a stout tap-root and a stiffly erect *continuously and broadly spinous-winged white-woolly stem* 45–250 cm, shortly branched above. Lvs sessile, elliptic-oblong, with sinuate teeth or triangular lobes ending in strong spines, cottony above and beneath; stem-lvs decurrent. Heads 3–5 × 3–5 cm, usually solitary. Involucre subglobose, cottony, bracts lanceolate-subulate, green, tipped with yellowish spines, spreading or reflexed. Fls pale purple, rarely white. Achenes 4–5 mm, grey-brown with darker mottling, transversely wrinkled; pappus pale reddish, up to twice as long as the achene, its hairs strongly toothed. Fl. 7–9. Visited chiefly by bees. $2n = 34^*$. H. (biennial).

Doubtfully native. Fields, roadsides and waste places. Scattered throughout Great Britain northwards to Ross, but rare in Scotland. Europe northwards to S. Scandinavia and C. Russia; W. Asia. Introduced in North America.

42. SAUSSUREA DC.

Perennial herbs with alternate *non-spinous lvs*. Heads solitary or corymbose, homogamous; involucre with imbricating non-spinous bracts in many rows; *receptacle flat, with dense chaffy bristles*. Fls all tubular, hermaphrodite; *anthers with* long acute terminal appendages and *basal feathery tails*; style-arms connate. Achenes cylindrical, 4-ribbed, glabrous, smooth or wrinkled; *pappus of an outer row of persistent rough hairs and an inner row of deciduous feathery hairs united below into a basal ring*.

About 130 spp., chiefly alpine herbs of C. and E. Asia, others in Europe and North America, 1 in Australia.

Differs from *Serratula* in the long anther-tails and the inner row of feathery hairs of the pappus.

1. S. alpina (L.) DC.   Alpine Saussurea.
Serratula alpina L.

A perennial herb with a ±horizontal scaly stock producing *short stolons* ending in lf-rosettes and an erect lfy grooved somewhat cottony simple flowering stem, 7–45 cm. Basal lvs 10–18 cm, ovate to lanceolate, stalked; stem-lvs diminishing upwards, the lower short-stalked, the uppermost narrow, sessile; all sharply toothed or ±entire, becoming glabrous above, white-cottony beneath. Heads 1.5–2 cm, ±sessile in a small dense terminal corymb; involucre ovoid-cylindrical, the outer bracts ovate, concave, sparsely hairy, the inner ovate-lanceolate covered with long dense grey hairs, all blunt, purplish. Florets exceeding the bracts, white below purple above. Anthers dark purple. Achenes 4 mm, brown, with pale ribs; pappus very long, whitish. Fl. 8–9. Protandrous. Fragrant and visited by flies and bees. $2n = 54^*$. Chh.

Native. Alpine and maritime cliffs, from 46 to 1190 m in Scotland. N. Wales, Craven Pennines, Lake District, W. and N. Scotland, Inner and Outer Hebrides, Orkney, Shetland; higher mountains of Ireland. N. Europe and the mountains of C. Europe; reaching 75° N. in Siberia; North America.

CNICUS L.

One sp. in the Mediterranean region.

*C. benedictus L. (Blessed Thistle.) An annual thistle-like herb with an erect *pubescent* stem 10–50 cm, branched above. Basal lvs stalked; stem-lvs sessile, ±amplexicaul, shortly decurrent; all oblong, toothed or pinnatifid with spreading or backwardly directed lobes, pubescent, spinous-ciliate, with the *veins white and prominent on the underside.* Heads up to 4 × 2 cm, solitary at the end of the main stem with or without others on branches overtopping the primary head. *Involucre ovoid, woolly, equalled or exceeded by a rosette of lvs just beneath the head;* bracts ovate-acuminate, the point spiny and with spreading lateral spines. Receptacle with shining filiform scales. *Fls yellow,* the marginal very small, sterile. Achenes yellow-brown, shining, ribbed; pappus longer than the achenes. Fl. 5–9. Chh.

Introduced. A casual. Native of the Mediterranean region and Near East. Contains a bitter glucoside, cnicin, and was formerly much used as a tonic and as a cure for gout.

43. CENTAUREA L.

Annual to perennial herbs, rarely suffruticose, with alternate usually non-spiny lvs. Heads homogamous or heterogamous; *involucral bracts* imbricate, *usually with membranous or scarious terminal appendages* which are laciniate, fimbriate, ciliate, dentate, pectinate or spiny, rarely entire; receptacle ±flat, bristly. *Fls tubular,* all similar and hermaphrodite or the marginal fls larger and spreading, neuter; anthers with a long terminal appendage and with or without basal tails; style-arms connate below. *Achenes* ±compressed, smooth, *with oblique attachment-scar;* pappus of 2-many rows of rough to feathery hairs or of scales, the innermost shortest and sometimes connate at base, or pappus 0.

A very large genus (c. 600 spp.) with representatives chiefly in the Mediterranean region and Near East but extending into C. and N. Eurasia and with a few species in the New World. Widely introduced and naturalized.

1 Involucral bracts with a terminal appendage which is non-spiny or at most minutely prickly.   2
　Involucral bracts with a terminal appendage bearing 1 or more distinct spines.   7
2 Appendages of bracts clearly decurrent, i.e. extending some way down the sides of the basal part of the bract.   3
　Appendages of bracts not or very slightly decurrent.   5

3 Most lvs 1–2-pinnatisect with much variation in degree of division, uppermost pinnatifid at least near base.   **1. scabiosa**
　All lvs, or all upper lvs, simple and ±entire.   4
4 Annual weed usually of corn or flax (now rare) with upper lvs simple, linear-lanceolate, lower usually pinnatifid with distant narrow lateral segments and a much larger terminal segment; marginal fls bright blue.   **2. cyanus**
　Perennial garden-escape with all lvs simple, ovate to broadly lanceolate, usually entire; marginal fls commonly blue but sometimes white, pink or dark purple.   **3. montana**
5 Involucre 3–6 mm diam.; involucral bract-appendages small, slightly decurrent; middle stem-lvs deeply pinnatifid with ±linear segments.   **4. paniculata**
　Involucre at least 10 mm diam.; involucral bract-appendages not decurrent; middle stem-lvs entire, sinuate-toothed or shallowly lobed, not deeply pinnatifid.   6
6 Involucral bract-appendages pale brown, scarious, the broad pale margins entire to laciniate but not pectinate; pappus 0.   **5. jacea**
　Involucral bract-appendages dark brown or blackish, those of outer bracts deeply pectinate; pappus present.   **6. nigra**
7 Involucral bract-appendages spreading or reflexed with palmately-arranged short and ±equal spines.   **7. aspera**
　Involucral bract-appendages with a longer terminal spine, lateral spines much shorter or 0.   8
8 Fls reddish-purple to pink; middle stem-lvs not decurrent.   9
　Fls yellow; middle stem-lvs strongly decurrent.   10
9 Terminal spine of bract-appendage 1–2 cm or longer, stout and spreading, with 2–6 shorter lateral spines mostly near its base; pappus 0.   **8. calcitrapa**
　Terminal spines of bract-appendage slender but rigid; shorter lateral spines 0; pappus present, of inner fls as long as achene, of outer very short. **11. diluta**
10 Lvs white-tomentose; spines of bract-appendage palmately-arranged, the short lateral spines confined to near base of long terminal spine; corolla not glandular.   **9. solstitialis**
　Lvs green but somewhat cottony, gland-dotted; spines of bract-appendage pinnately-arranged, the short lateral spines extending c. ½-way up the long terminal; corolla glandular.   **10. melitensis**

Subgenus LOPHOLOMA (Cass.) Dobrocz.

Lower lvs usually pinnatisect, the uppermost ±pinnatifid or almost entire. Fl.-heads rather large, the involucral bract-appendages horseshoe-shaped and decurrent along the sides but not reaching the base of the bract. Pappus present.

1. C. scabiosa L.   Greater Knapweed.

A perennial herb with a stout woody oblique branching stock, enclosed above in fibrous scales, and erect grooved ±pubescent stems 30–90 cm, usually branched above the middle. *Lower lvs* 10–25 cm, oblanceolate in outline, petiolate, 1–2-pinnatisect, sometimes merely dentate or ±entire; uppermost smaller, sessile, ±pinnatifid to entire; all firm, usually sparsely hispid above and beneath, often shining above. Heads 3–5 cm diam., solitary on long ±glabrous stalks. Involucre ovoid-globose, bracts with *blackish-brown horseshoe-shaped pectinate decurrent appendages*, which do not completely conceal the green basal parts. Fls red-purple, with or rarely without a large neuter marginal row. Achenes 4–5 mm, greyish, pubescent; pappus stiff, whitish, about equalling the achene. Fl. 7–9. Freely visited by various bees and flies. $2n = 14^*, 20^*, 24$. Hs.

A very variable plant split by Continental taxonomists into several subspp.

Native. Dry grassland, hedgebanks, roadsides, cliffs, etc., especially on calcareous substrata, to 320 m in Derby. Throughout lowland Great Britain to Sutherland, common in the south but rare in Scotland. Europe northwards to 67° 56′ in Scandinavia, Finland and Karelia; Caucasus.

Subgenus CYANUS (Miller) Hayek

Lvs undivided to pinnatisect. Fl-heads rather small, the involucral bracts with horseshoe-shaped appendages narrowly decurrent to base of bract, denticulate to fimbriate, muticous. Pappus usually present.

2. C. cyanus L.   Cornflower, Bluebottle.

An annual or overwintering herb with an erect wiry grooved cottony stem 20–90 cm, usually with many slender ascending branches. Lower lvs 10–20 cm, stalked, usually lyrate-pinnatifid with narrow distant lobes, rarely oblanceolate, distantly toothed or entire; *upper lvs smaller*; *sessile, linear-lanceolate*; all *greyish with cottony hairs*. Heads 1.5–3 cm diam., solitary on main stem and branches, on long cottony stalks. Involucre ovoid, cottony, its *bracts* green below *with narrow decurrent appendages* cut half-way into long spreading narrowly triangular teeth, those of the outer bracts usually *silvery white*, of the middle bracts brown with white-edged teeth. *Fls of the marginal row large, bright blue*, of the centre red-purple. Anthers purple. Achenes 3 mm, silvery-grey, finely pubescent; pappus reddish, shorter than the achene. Fl. 6–8. Freely visited, especially by flies and bees. $2n = 24$. Th.

Native, *fide* Godwin. Cornfields and waste places to 380 m in Scotland. Formerly common throughout Great Britain but now rare owing to greater care in cleaning seed-grain. Probably native in most of Europe and the Near East, but widely introduced as a cornfield weed.

*3. C. montana L.   Perennial Cornflower.

A perennial often stoloniferous herb with stout creeping rhizome and erect or ascending, broadly winged, usually unbranched fl.-stems to 70 cm. Lvs ovate to oblong or broadly lanceolate, ±entire, the middle and upper narrowed into the sessile base and long-decurrent as a wing down the stem, the lowest shortly petiolate; all initially floccose-tomentose at least beneath but glabrescent later. Fl.-heads usually solitary; involucre 10–15 mm diam., ovoid-cylindrical, its bracts with decurrent appendages which are blackish below and with deeply

fimbriate dark brown margin. Inner fls blue-violet, marginal commonly blue but sometimes white, pink or dark purple. Achenes 5–6 mm; pappus c. 1.5 mm. $2n = 44$. Hs.

Introduced. Much grown in gardens, frequently escaping, and established in several places, especially in Scotland. Native in mountains of C. and S. Europe from the Pyrenees and Ardennes eastwards to the Carpathians and the Caucasus.

Subgenus ACROLOPHUS (Cass.) Dobrocz.

Lvs usually pinnatisect with narrow segments. Fl-heads rather small, the involucral bracts prominently veined, their appendages (not always present) small, shortly decurrent, usually fimbriate and often, but not invariably, spiny at apex. Pappus usually present.

*4. C. paniculata L. subsp. **paniculata**
                            Jersey Knapweed.

A biennial herb with erect, slender, sharply angled, cottony fl.-stems 20–70 cm, much branched below. Lvs up to 8 cm, the lower 1–2-pinnatisect with ±linear acute lobes, cottony above and beneath. Fl.-heads c. 2 × 1.5 cm, in a long panicle. Involucre ovoid-oblong, 3–6 mm diam., its bracts striate with small pale brown pectinate appendages, acuminate but not spiny. Fls purple, the marginal row larger and neuter. Achenes 2 mm, silvery-white, glabrous; pappus of very short scale-like bristles. Fl. 7–8. H. (biennial).

Introduced. Established in Jersey and a rare casual elsewhere in the British Is. Native in the W. Mediterranean region from Spain to Italy.

Subgenus JACEA (Miller) Hayek

Lower lvs entire to pinnatifid, upper entire or dentate, sessile. Fl-heads varying widely in size; involucral bracts often ±completely concealed by the large imbricating terminal appendages which are not or very slightly decurrent and entire to fimbriate, blunt at apex or with a small mucro but never spiny. Pappus present, often very short, or 0.

*5. C. jacea L.   Brown Knapweed.

A perennial herb with an oblique branching stock and erect or ascending grooved glabrous or cottony stems 15–60 cm, usually with a few long slender subcorymbose branches above. Basal and lower stem-lvs oblanceolate narrowing into a stalk-like base, entire, coarsely toothed or somewhat pinnatifid; upper stem-lvs lanceolate, sessile, entire or with 1–2 teeth towards the base; all roughly hairy. Heads 10–20 cm diam., solitary and subsessile; peduncles not or only slightly thickened immediately beneath heads. Involucre ovoid-globose, bracts ovate to oblong, bases quite concealed by the border, with *rounded pale brown scarious non-decurrent appendages* whose margins are *irregularly but usually not deeply laciniate*, those of the innermost entire. Fls red-purple, the marginal row usually larger. Achenes 3 mm, greyish, shining; *pappus* 0. Fl. 8–9. Male and female as well as hermaphrodite heads have been reported from Continental Europe. Freely visited by bees, flies and Lepidoptera. $2n = 44^*$. H.

Very variable. The usual British form is var. *angustifolia* Reichenb., described above, but forms with broader basal lvs occur as casuals.

Introduced. Grassland and waste places. Established but rare in S. England and casual elsewhere. Formerly near Belfast. Europe northwards to 63° 41′ N. in Scandinavia and Karelia; N. Africa; N. and W. Asia.

C. jungens (Gugler) C. E. Britton is the 1st generation hybrid of *C. jacea* and *C. nigra*; and *C. pratensis* Thuill., *C. drucei* C. E. Britton and probably *C. surrejana* C. E. Britton are segregants from this cross. All have characters intermediate between those of the parent spp., and in particular have pale but deeply laciniate or pectinate scarious bract-appendages.

6. C. nigra L.   Lesser Knapweed, Hardheads.

A perennial herb with a stout branching oblique stock and erect tough rigid stems 15–60(–90) cm, grooved, usually ±roughly hairy, branched above. Basal lvs stalked, entire, sinuate-toothed or somewhat pinnatifid; stem-lvs sessile, entire or with a few teeth towards the base; all softly or roughly hairy, often cottony beneath when young. Heads 2–4 cm diam., subsessile, solitary. Involucre ovoid-globose, bracts with *brown or blackish triangular non-decurrent ±deeply pectinate appendages*. Fls red-purple, those of the marginal row sometimes larger and neuter. Achenes 3 mm, pale brown, ±pubescent; *pappus of short bristly hairs*. Fl. 6–9. Freely visited by a great variety of insects. $2n = 44^*, 22^*$. H.

Native. Grassland, waysides, cliffs, etc., to 580 m in Wales. Throughout the British Is. but introduced in Shetland.

Very variable. Commonly treated as comprising the following subspp. but these are not clearly distinct:

Subsp. **nigra** (*C. obscura* Jordan)
Plant hispid, with broadly lanceolate usually toothed or shallowly pinnatifid lvs. Stem stout, *conspicuously swollen beneath the heads*. Appendages of bracts blackish-brown, ±*completely concealing the pale basal parts of the bracts*; those of the outer bracts very broadly triangular, *teeth about equalling the undivided central portion*. Heads dark purple, rather rarely with enlarged marginal florets.

Common in England, Wales and S. Scotland and extending more locally to northernmost Scotland, Inner Hebrides and Orkney; scattered through much of Ireland; Channel Is. Tolerant of a wider range of soil-type than subsp. *nemoralis*.

Subsp. **nemoralis** (Jordan) Gugler (*C. nemoralis* Jordan)
Plant pubescent, more branching than subsp. *nigra*, with narrowly lanceolate, entire or sinuate-toothed, rather softly hairy lvs, diminishing rapidly up the stem. Stem slender, *not much swollen beneath the heads*. Appendages of bracts brown or brownish-black, *not completely concealing the pale basal parts of the bracts*; those of

the outer bracts about equilaterally triangular; *teeth longer than the undivided portion*. Heads usually pale purplish-red, not uncommonly with enlarged marginal fls.

Restricted to England and Wales and especially common on light calcareous soils. Mostly south of a line from the Humber to N. Wales, only in a few scattered localities northwards to N. Yorks and Cumbria and a rare casual elsewhere; not in Ireland or the Channel Is.; largely coastal in S.W. England and Wales.

Intermediates between the 2 subspp. have been reported from a number of localities within the area where both occur but also in Ireland (especially the south-west) and the Channel Is., where subsp. *nemoralis* does not now occur.

The aggregate species is native in W. Europe from Portugal, Spain and Italy northwards to c. 64° N. in Scandinavia, and eastwards to the Netherlands, W. Germany and Switzerland. Introduced in North America and New Zealand.

C. microptilon Gren. appears to be a segregant from *C. jacea* × *nigra* and shows intermediate characters, with narrow acute dark-coloured bract-appendages.

Subgenus SERIDIA (Juss.) Czerep.

Lower lvs entire to pinnatisect, upper entire to pinnatifid and usually decurrent. Fl.-heads often large and solitary; involucral bract-appendages non-decurrent, palmately spiny above, the apical spine not or slightly longer than the others. Pappus usually present.

7. C. aspera L.   Rough Star-Thistle.

A perennial herb with a slender stock and ascending stems 20–60 cm, sparsely hairy below, cottony above, with many slender spreading branches. Basal lvs usually lyrate, narrowed to a petiole-like base; stem-lvs 2–4 cm, narrowly oblong, pinnatifid, sinuate-lobed, or more usually toothed or entire, sessile or ±amplexicaul; all sparsely hairy. Heads 2.5 cm diam., solitary, subsessile. Involucre ovoid-globose, glabrous or slightly cottony, its *bracts yellowish, leathery, with spreading or reflexed reddish-brown appendages* having 3–5 subequal palmately-arranged short spines about 3 mm. Receptacular scales white. Fls pale red-purple, the marginal neuter but little larger than the central. Achenes 3–5 mm, greyish-white, pubescent; pappus reddish, shorter than achene. Fl. 7–9. $2n = 20, 22^*$. H.

Doubtfully native in Channel Is. Dunes and waste places. A rare plant of Jersey and Guernsey, naturalized in S. Wales (Glamorgan) and casual elsewhere. S. and W. France, Spain, Portugal, Corsica and Sardinia, Italy.

Subgenus CALCITRAPA (Heister ex Fabr.) Hayek

Biennials with non-decurrent pinnatisect lvs. Fl.-heads usually ±sessile; involucral bract-appendages non-decurrent, palmate- or pinnate-spiny at apex, the apical spine much the longest. Pappus usually present.

*8. C. calcitrapa L.   Star-Thistle.

A biennial herb with a stout tap-root and branched erect

stock producing erect or ascending grooved ±glabrous flowering stems 15–60 cm, with stiffly flexuous divaricate-ascending branches arising from just beneath the heads. Basal and lower stem-lvs to 8 cm, deeply pinnatifid with narrow distant entire or toothed lobes; upper stem-lvs sessile, irregularly toothed or entire; all sparsely hairy above and beneath, the lobes and teeth bristle-pointed. Heads 8–10 mm diam., subsessile; branches successively overtopping the older heads so that these appear lateral. Involucre ovoid, glabrous, its bracts with the *appendage ending in a stout spreading spine* 2–2.5 cm, yellow and channelled above, *with shorter spines at its base. Fls pale red-purple, glandular*, the marginal no larger than the central. Achene 3–7 mm, ovoid, whitish with or without brown mottling, glabrous; pappus 0. Fl. 7–9. Visited by bees and flies. $2n = 20$. H. (biennial).

Probably introduced. Waysides and waste places on sandy and gravelly soils and on chalk (especially near the Sussex coast). A rare plant of S. England from Cornwall to Northampton and Kent, and in S. Wales (Glamorgan); East Anglia; a casual near ports and elsewhere over a wider area. S. and S.C. Europe north to Switzerland and Czechoslovakia, but more widely naturalized; N. Africa and Canary Is.; W. Asia.

Subgenus SOLSTITIARIA (Hill) Dobrocz.

Annual to perennial herbs with lvs usually decurrent, lower sinuate-dentate to pinnatifid, upper entire to dentate. Fl.-heads usually solitary, sometimes in groups of 2–3; involucral bract-appendages not or shortly decurrent, palmate- or pinnate-spiny above, the apical spine usually much the longest; rarely with a single apical spine. Pappus usually present.

*9. C. solstitialis L.
                    Yellow Star-Thistle, St Barnaby's Thistle.

An annual or rarely biennial herb whose erect or ascending stiff cottony stems 20–60 cm, have *broad and continuous wavy wings* as do the many slender ascending branches. Basal lvs deeply lyrate-pinnatifid with distant, narrow, toothed or entire lobes; middle and upper lvs lanceolate, ±entire, sessile and decurrent into the wings of the stem; all cottony above and beneath. Heads 12 mm diam., solitary, stalked. Involucre ovoid-globose, cottony, rarely ±glabrous, all but the innermost bracts with a *palmately spinous appendage*, the *terminal spine spreading, yellow*, channelled above, short in the lowest bracts but 10–20 mm in the middle ones. *Fls pale yellow, eglandular*, the marginal no larger than the central. Achenes 2.5–3.5 mm, obovoid, the central yellowish mottled with brown, and with a white pappus equalling the achene; the marginal dark brown with no pappus. Fl. 7–9. Visited chiefly by bees. $2n = 16$. Th.

Introduced. Cultivated land, especially in lucerne and sainfoin fields. Rare in S. and E. England where it may persist for many years; casual elsewhere. S. and S.E. Europe; W. Asia. Introduced in C. and N. Europe.

*10. C. melitensis L.   Maltese Star-Thistle.

An annual herb resembling *C. solstitialis* but with involucral bract-appendages whose slender brownish 5–8 mm terminal spine bears *pinnately arranged short lateral spines in its lower half*. Heads subsessile. Fls with yellow *glandular corolla*, the marginal no larger than the central. Achenes 2–2.3 mm, pale greenish-grey with whitish stripes; pappus shorter than the achene, whitish. Fl. 7–9. $2n = 22$. Th.

Introduced. Waste places and roadsides. A not infrequent casual. Mediterranean region eastwards to Greece and Tunis; Madeira; Canary Is. Widely introduced in C. and N. Europe, North and South America, S. Africa and Australia.

*11. C. diluta Aiton

A *perennial* herb with erect branching fl.-stems to 50 cm. Lower lvs dentate, the lowest lyrate, upper entire, semi-amplexicaul. Fl.-heads solitary, stalked; *involucre* 8–12 mm diam., ovoid, its appressed *bracts* having shortly decurrent orbicular-ovate *terminal appendages with membranous and irregularly lacerate margin* and with *a single very slender but rigid spine* from the notch of the emarginate apex. *Fls reddish-purple, the marginal paler, long and spreading*, so that the head resembles that of *C. scabiosa* but is smaller. Pappus of inner achenes about equally long, of outer very short.

Introduced. A bird-seed alien and a frequent casual in gardens and waste ground near houses, also on rubbish-tips, etc. Native in Algeria, Morocco and S.W. Spain, casual elsewhere in Europe.

Several other *Centaurea* spp., especially natives of the Mediterranean region and Near East, occur as casuals.

44. SERRATULA L.

Perennial herbs with alternate *non-spinous lvs*. Heads solitary or in corymbs, homogamous, often gynodioecious or ±dioecious; involucre with imbricating acute but not spinous bracts in many rows; receptacle flat, with *dense chaffy scales*. Fls all tubular, similar or sometimes the central hermaphrodite and the marginal female; *anther-tails short or* 0; style-arms connate or free. Achenes oblong, slightly compressed, glabrous; *pappus of many rows of stiff, rough, deciduous, simple hairs, all free to the base*, the outermost shortest.

About 70 spp. in Europe, N. Africa and temperate Asia. Like *Saussurea* but with the anthers hardly tailed and the pappus-hairs non-feathery.

1. S. tinctoria L.   Saw-wort.

A perennial glabrous subdioecious herb with a short stout ±erect stock and an erect, slender, wiry, grooved stem 10–90 cm, with a few ascending branches above. Lvs 12–25 cm, ovate-lanceolate to lanceolate in outline, glabrous or with a few scattered hairs, their *margins with fine bristle-tipped teeth*, very variable in degree of dissection from undivided to lyrate-pinnatifid or almost pinnate with narrow lateral lobes and a larger ±narrowly elliptical terminal lobe; basal and lower stem-lvs petiolate, upper sessile. Heads 1.5–2 cm, pedunculate and loosely corymbose, or subsessile and crowded ('var. *monticola* (Boreau) Syme'), the female heads larger than the male; involucre of female heads ovoid-cylindrical, the florets exceeding the bracts; of male heads oblong-cylindrical, the florets about equalling the bracts; outer bracts ovate-acute, downy on the margins, inner narrower, rough on the margins; all appressed, purplish where exposed. Fls reddish-purple, rarely white; the female with corolla swollen in the middle, white abortive anthers and spreading style-arms; the male with corolla not swollen, dark-blue anthers and appressed style-arms. Achenes 5 mm, fawn, slightly rough; pappus yellowish. Fl. 7–9. Visited by flies and bees. $2n = 22*$. H.

Native. Wood margins, clearings and rides and open grassland on moist basic soils over limestone or chalk; to 380 m in Wales. Local throughout England and Wales, and in Dumfries and Kirkcudbright; in Ireland in a single locality in Co. Wexford. C. Europe from N. Spain, C. Italy and N. Balkans northwards to C. Scandinavia, Estonia and C. Russia; Siberia; Algeria.

Amongst other members of the Cynareae which occur as casuals the most frequently encountered are:

*Carthamus tinctorius L. An annual to biennial herb with an erect simple or little branched stem 10–60(–100) cm, furrowed, pale yellow, glabrous. Lower lvs ovate-oblong narrowed into a short stalk; upper lvs ovate-lanceolate, sessile, with a ±cordate amplexicaul base; all glabrous, finely and softly spinous-toothed. Heads 2–3 cm diam., surrounded by a cluster of spreading lf-like bracts passing over into ±appressed involucral bracts with green spinous-toothed terminal appendages. Fls bright reddish-orange. Achenes obovoid, shining white; pappus of numerous narrow scales. Not known as a wild plant, but still cultivated for the red and saffron dyes from its flowers (Safflower, False Saffron) and for the oil from its achenes.

*C. lanatus L. An annual herb 20–60 cm, with a terete, pale yellow, ±woolly stem and sessile to amplexicaul lanceolate coarsely sinuate-toothed or pinnatifid lvs with long stout marginal spines. Heads surrounded by narrow spreading spinous bracts; involucre of narrow appressed bracts with spinous terminal appendage. Fls yellow. Achenes 4-angled, rugose, dark brown; pappus of linear scales.

C., S. and S.E. Europe, Mediterranean region and Canary Is.

Subfamily CICHORIOIDEAE

Tribe 11. CICHORIEAE.

45. CICHORIUM L.

Usually perennial herbs with alternate runcinate or dentate basal lvs. Heads terminal and axillary; involucre cylindrical, its *bracts in 2 rows*, the inner longer; receptacle ±flat, usually naked. Ligules usually *blue*; anthers without basal tails; style-arms hairy. *Achenes* obovoid, ±angled, flat-topped; *pappus of 1–2 rows of short blunt scales*. Nine spp. in the Mediterranean region with 1 reaching N. Europe and another in Ethiopia.

1. C. intybus L.   Chicory, Wild Succory.

A perennial herb with a long stout tap-root, a short vertical stock and stiffly erect, tough, grooved stems 30–120 cm, with stiff spreading-ascending branches, all roughly hairy or ±glabrous. Basal lvs short-stalked, oblanceolate in outline, runcinate-pinnatifid or toothed;

lower stem-lvs similar but sessile and ±amplexicaul; upper stem-lvs lanceolate, entire or distantly toothed, sessile, clasping the stem with the pointed auricles of their broadened base; all ±glabrous or roughly hairy beneath, glandular-ciliate. *Heads* 2.5–4 cm diam., solitary, terminal on a somewhat thickened stalk and in *subsessile clusters of 2–3 in the axils of upper lvs*. Outer bracts about 8, broadly lanceolate, spreading above; inner about 5, twice as long, narrower, erect; all green, herbaceous. *Fls large, bright blue*, rarely pink or white. Achenes 2–3 mm, irregularly angular, pale brown often with darker mottling; pappus of fimbriate scales $\frac{1}{10}-\frac{1}{8}$ as long as the achene. Fl. 7–10. The heads open in early morning and close soon after midday. Visited chiefly by bees and hover-flies. $2n = 18$. H.

Probably native. Roadsides and pastures to 275 m in Scotland. Locally common especially on calcareous soils in England and Wales, more local and probably introduced in Scotland and Ireland. Europe northwards to C. Scandinavia, Finland and C. Russia; W. Asia; N. Africa. Introduced in E. Asia, North and South America, S. Africa, Australia and New Zealand.

The dried and ground roots yield the chicory of commerce. In some countries the roots are boiled and eaten as a vegetable. The plant is sometimes included in seed mixtures on shallow chalky soils in England, both because cattle readily eat its lvs and for the effect of its deep tap-root in breaking up the subsoil.

The cultivated endive, **C. endivia** L. subsp. *endivia*, closely resembles C. intybus but its lvs are glabrous, the basal lvs are less deeply lobed and often merely sinuate-toothed, the stalk of the terminal head is conspicuously thickened, the bracts are eglandular, and the larger achenes (2.5–3.5 mm) have a pappus of scales $\frac{1}{6}-\frac{1}{2}$ as long as the achene. It is widely grown as a salad plant, often in varieties with strongly crisped lvs.

The wild subsp. *divaricatum* (Schousboe) P. D. Sell (Mediterranean region) is a low-growing herb with runcinate and hairy basal lvs but with the long pappus-scales of the cultivated form.

46. LAPSANA L.

Annual to perennial herbs with *lfy flowering stems* and *small yellow fl.-heads* in loose corymbose panicles. Involucral bracts erect, in 1 row with a few very small basal scales; *receptacle flat, naked*. Anther-lobes without basal tails; style-arms slender, hairy. *Achenes somewhat compressed, about 20-ribbed, rounded above, glabrous*, the outermost longer than the central and distinctly curved; *pappus 0*.

About 9 spp. in temperate Europe and Asia.

1. L. communis L.   Nipplewort.

Lampsana communis auct.

Annual to perennial herbs with stem 15–125 cm, erect, usually hairy below but ±glabrous above, paniculately branched in upper half, the branches ascending. *Lvs* to 15 cm, *lower long-petiolate, usually lyrate-pinnatifid with large* often ±cordate *terminal lobe*, upper shortly petiolate or sessile, ovate to linear-lanceolate; all ±dentate or the upper entire. *Fl.-heads in a ±corymbose pan-*

icle; involucre 5–10 × 2–5 mm, the inner bracts linear-oblong, the few outer only 0.5–1 mm, ovate-lanceolate; *peduncles usually more than twice as long as involucre.* Fls with ligules exceeding involucre. Achenes 2.5–9 mm, outer much longer than inner. $2n = 14$, rarely 12 or 16.

Very variable. Of the several not very distinct subspp. that have been recognized in Europe the following are recorded for the British Is.:

Subsp. **communis.** *Annual.* Stem with eglandular hairs below. *Lvs,* if pinnatifid, *with ovate terminal segment wider than laterals; upper lvs ovate-rhombic, broadly elliptical or lanceolate. Fl.-heads* 1.5–2 cm *diam.* with *involucre* 5–7(–8) mm and slender peduncles more than twice as long. *Fls with ligules up to 1½ times as long as involucre.*

Native. Waysides, hedgerows, wood-margins, cultivated ground, walls, waste places, etc. Common throughout the British Is. Europe to 70° N. in Scandinavia; N. Africa; W. and C. Asia; introduced in North America. Formerly used as a salad plant.

*Subsp. **intermedia** (Bieb.) Hayek (*L. intermedia* Bieb.). *Annual to perennial.* Stem with eglandular hairs below. *Lvs,* if pinnatifid, *with lateral segments often as wide as the terminal;* upper lvs lanceolate to linear-lanceolate. *Fl.-heads* 2.5–3 cm *diam.* with *involucre* 7–10 mm and peduncles mostly more than twice as long. *Fls with ligules* 2–2½ *times as long as involucre,* so that the plant might be mistaken for a *Crepis.*

Almost certainly introduced. First recorded in 1945 on a steep railway embankment on the Bedford chalk and more recently in semi-natural limestone grassland on the Great Orme, Caernarvon, and seems established in both sites; also seen on a roadside in Flint, Wales. Native in S.E. Europe.

47. Arnoseris Gaertner

A small annual scapigerous herb with yellow fl.-heads whose *involucral bracts* are *in 1 row with a few tiny basal scales;* receptacle flat, naked, pitted marginally. Anther-lobes without basal tails; style-arms short, blunt, hairy. *Achenes obovoid,* strongly ribbed; *pappus a very short membranous border.* One sp.

Differs from *Lapsana* in habit and in the membranous border of the achenes.

1. A. minima (L.) Schweigg. & Koerte
Lamb's or Swine's Succory.

Hyoseris minima L.; *Arnoseris pusilla* Gaertner

Lvs 5–10 cm, *all* basal, spathulate or oblanceolate, narrowed into a short stalk, distantly and rather coarsely toothed, ciliate, glabrous or ±hairy on both sides. *Scapes* 7–30 cm, many, ±glabrous, simple or sparingly branched above, the *branches,* in the axils of minute bracts, *curving upwards and finally overtopping the main stem;* all fistular, enlarging upwards and so *markedly clavate* beneath the heads. Heads 7–11 mm diam., solitary, terminal. Involucre campanulate, ±glabrous, its

15–20 *bracts* narrowly triangular-acuminate with a *prominent pale keel, connivent in fr.* with the extreme tips slightly spreading; basal scales tiny, subulate. Fls yellow, half as long again as the bracts. Achenes 1.5–2 mm, with 5 strong and 5 weaker intermediate ribs, transversely wrinkled between the ribs; border very short, sometimes obscurely toothed above the ribs. Fl. 6–8. Sparingly visited by flies. $2n = 18^*$. Th.

Probably native. Arable fields, mainly on sandy soils. Formerly in eastern England from Dorset and Sussex northwards to Humberside; now apparently extinct. W. and C. Europe from Spain and Portugal, Corsica, N. Italy, Hungary, Romania and S. Russia northwards to England, S. Sweden, N. Poland and C. Russia. Introduced in North America, Australia and New Zealand.

48. Hypochoeris L.

Annual to perennial usually scapigerous herbs, commonly with the branches of the scape thickened distally beneath the heads. Involucre of many imbricating rows of lanceolate bracts. *Receptacle flat, with numerous lanceolate scales.* Fls yellow. Anthers shortly tailed below and with rounded terminal appendages. Achenes cylindrical, ribbed, at least the *inner* usually *beaked;* pappus commonly of 1 row of feathery hairs, sometimes of 2 rows of hairs, the inner long, feathery, outer shorter and sparsely feathery or merely scabrid, rarely the marginal achenes with pappus of fimbriate scales.

About 100 spp., 12 in Europe and the Mediterranean region, the remainder in South America.

1 Annual; lvs ±glabrous; fls about equalling the involucre, their ligules only about twice as long as broad; heads opening only in full sun.   **2. glabra**
  Perennial; lvs hispid; fls exceeding the involucre, their ligules four times as long as broad; heads opening in dull weather.   2
2 Heads commonly solitary, sometimes 2–4; scape unthickened above or thickened only immediately beneath the heads; lvs obovate-oblong, often purple-spotted; pappus of 1 row of feathery hairs.
     3. maculata
  Heads usually several; scape thickened for some distance beneath the heads; lvs ±broadly oblanceolate-oblong, not spotted; pappus hairs in 2 rows, the inner feathery, the outer shorter, ±simple.
     1. radicata

Section 1. *Hypochoeris*
Pappus hairs in 2 rows, the inner feathery, the outer shorter, ±simple.

1. H. radicata L.   Cat's Ear.

A perennial scapigerous herb with a short ±erect branching premorse stock and fleshy roots. *Lvs* 7–25 cm, in a basal rosette, ±broadly oblong-lanceolate, narrowed gradually into the broad stalk-like base, sinuate-toothed to sinuate-pinnatifid, usually *hispid with simple hairs,* dull green above and somewhat glaucous beneath. *Scapes* 20–60 cm, usually several from each rosette, erect or ascending, lfless or with 1–2 small lvs below, usually *forking, enlarged below the heads and bearing* *numerous small scale-like bracts*. Heads 2–4 cm diam. Involucre 18–25 mm, cylindrical-campanulate; bracts lanceolate-acuminate, dull green, bristly on the midrib, the outermost somewhat lax. Fls bright yellow, exceeding the involucre, the *ligules about 4 times as long as broad*, the *outer* ones *greenish or grey-violet beneath*. Achenes 4–7 mm (excl. beak), orange, strongly muricate, narrowed above into a beak, that of the central achenes exceeding, of the outer equalling or falling short of, the achenes or sometimes 0. Fl. 6–9. Visited freely by many kinds of insects, especially bees, and not automatically self-pollinated. $2n = 8^*$. Hr.

Native. Meadows and pastures, grassy dunes, waysides, etc., to 600 m in Ireland; common throughout the British Is. Europe northwards c. 63° N. in Scandinavia and to N. Russia; Asia Minor; N. Africa.

2. H. glabra L.   Smooth Cat's Ear.

An *annual* herb with a basal rosette of oblanceolate *lvs* 1–20 cm, narrowed gradually to the stalk-like base, sinuate-toothed to sinuate-pinnatifid, ±*glabrous*, pale green and sometimes reddish near the margin. Scapes 10–40 cm, usually several from each rosette, erect, ascending or decumbent, lfless or with 1–2 small lvs, ±branched, the branches slightly enlarged beneath the heads and with a few scale-like bracts. *Heads* 0.5–1.5 cm diam., *opening widely only in full sunlight*. Involucre 12–15 mm, cylindrical; its bracts very unequal, lanceolate-acuminate, with whitish margins and dark points. *Fls* bright yellow, *about equalling the involucre*, their *ligules only about twice as long as broad*. Achenes 4–5 mm (excl. beak), reddish-brown, muricate; those of central fls with long slender beak, of outer usually unbeaked (all unbeaked in var. *erostris* Cosson & Germ.). Fl. 6–10. Visited by bees, etc. $2n = 10^*$. Th.

Native. Grassy fields, derelict arable land, heaths, fixed dunes, etc., on sandy soils; locally frequent. Great Britain north to Cumbria and N. Yorks, also in Fife, Moray and Inverness; Isle of Man, Channel Is. In Ireland confined to Derry. Europe northwards to S. Scandinavia and N. Poland; Asia Minor and Syria; N. Africa. A characteristic plant of open communities on non-calcareous sand, on dunes with *Corynephorus canescens* and *Jasione montana*, and on arable land with *Scleranthus* spp., *Teesdalia nudicaulis*, etc.

Section 2. *Achyrophorus* Duby
Hairs of pappus in 1 row, all feathery.

3. H. maculata L.   Spotted Cat's Ear.

A perennial herb with a stout cylindrical blackish stock and a basal rosette of obovate-oblong *lvs* 4–15(–30) cm, *usually spotted with dark purple* and with reddish midribs, narrowed to the stalk-like base, ±sinuate-toothed, hispid. *Scapes* 20–60 cm, 1 or a few from each rosette, erect, *simple* or less commonly with 1–3 branches, lfless or with 1–2 small lvs, not enlarged above or enlarged only immediately below the heads and with *scale-like bracts few or 0*. Heads 3–4.5 cm diam., solitary or 2–4.

Involucre 18–23 mm, campanulate, blackish-green; its outer bracts lanceolate, hispid, the middle and inner linear-lanceolate with woolly margins. Fls lemon-yellow, twice as long as the involucre. Achenes 5–7 mm (excl. beak), slightly muricate, transversely ridged, all rather shortly beaked. Fl. 6–8. $2n = 10^*$. Hr.

Native. Calcareous pastures and grassy cliffs; rare and decreasing. E. England from Bedford and Hertford to Northampton and Lincoln; Cornwall, Caernarvon, N. Lancs, Westmorland. Europe from the Pyrenees, S. France, N. Italy, N. Balkans and S. Russia northwards to c. 66° N. in Scandinavia and to Karelia and C. Russia. A 'pontic' plant in C. Europe, growing commonly with *Filipendula vulgaris*, *Peucedanum oreoselinum*, *Asperula cynanchica*, *Orchis ustulata*, etc.; usually calcicolous but sometimes found in *Calluna* heath.

49. LEONTODON L.

Usually perennial *scapigerous* herbs with rosettes of entire or pinnatifid lvs, the lvs and scapes commonly with forked hairs. Involucral bracts in several imbricating rows. *Receptacle naked*, pitted, the pits often with toothed or ciliate margins. Anthers not tailed, with blunt terminal appendages. Achenes little compressed, narrowed above and sometimes beaked, strongly ribbed. *Pappus usually of 2 rows of hairs, the inner feathery, the outer simple* or sometimes 0; that of the outermost row of achenes sometimes represented by a small cup of scarious scales.

About 50 spp. in Europe, C. Asia, N. Africa and the Azores.

1 Lvs glabrous or with simple hairs; scape usually branched and bearing 2 or more heads; pappus of a single row of feathery hairs.   **1. autumnalis**
  Lvs usually with forked hairs; scape simple with a single terminal head; pappus of central achenes consisting of an inner row of feathery and an outer row of shorter scabrid simple hairs.   2
2 Scape usually densely hairy above; involucre exceeding 10 mm; outer fls orange or reddish beneath, rarely pale grey-violet; achenes all with feathery hairs.   **2. hispidus**
  Scape sparsely hairy, especially below; involucre not exceeding 10 mm; outer fls grey-violet beneath; outermost achenes surmounted by a cup of scarious scales.   **3. taraxacoides**

Section 1. *Oporinia* (D. Don) Koch
All achenes with a single row of feathery hairs, dilated at the base; lvs glabrous or with simple hairs.

1. L. autumnalis L.   Autumnal Hawkbit.

A perennial herb with an oblique usually branched premorse stock, each branch terminating in a rosette of ±oblanceolate *lvs* varying from distantly sinuate-toothed to deeply pinnatifid, and *glabrous or with simple hairs*. Scapes 5–60 cm, decumbent below then erect or ascending, usually *branched*, rarely simple, somewhat enlarged and hollow above and bearing *numerous scale-like bracts just beneath the heads*, glabrous or sparsely hairy, especially below. Head 12–35 mm diam., erect in bud. Involucre c. 8 mm, ovoid-cylindrical, dark green, glabrous to woolly; its bracts linear-lanceolate, acute. *Fls* golden yellow, the outer *with reddish streaks beneath.* Achenes 3.5–5 mm, reddish-brown, slightly narrowed above, longitudinally ribbed and with numerous small transverse ridges; beak 0; pappus of a single row of feathery hairs. Fl. 6–10. Freely visited by a great variety of insects, and automatically self-pollinated. $2n = 12^*$, 24. Hr.

Very variable. Var. *pratensis* (Link) Koch, with the involucre thickly covered with usually blackish woolly hairs, is specially characteristic of mountainous districts, though not confined to them; but it is not clearly separable from the type. Its alpine forms have commonly a single head but in cultivation the scapes branch.

Native. Meadows, pastures, waysides, screes; to 975 m in Scotland and Ireland. Abundant throughout the British Is. Throughout Europe except Greece; N. and W. Asia; N.W. Africa; Greenland. Introduced in North America.

Section 2. *Leontodon*

All achenes with 2 rows of hairs, the inner feathery, the outer shorter, simple, scabrid; lvs usually with forked hairs.

2. L. hispidus L.   Rough Hawkbit.

A perennial herb with an erect or oblique usually branched premorse stock, each branch terminating in a rosette of ±oblanceolate *lvs* varying from distantly sinuate-toothed to runcinate-pinnatifid, narrowed into a long stalk-like base, *usually hispid with forked hairs,* rarely subglabrous. *Scapes* 10–60 cm, 1 or a few from each rosette, erect or ascending, *simple,* slightly enlarged above and with 0–2 small bracts beneath the *solitary head,* usually densely hairy at least above. Head 25–40 mm diam., drooping in bud. Involucre 10–17 mm, ovoid, dark to blackish green, hispid to nearly glabrous; its bracts linear-lanceolate with the outermost lanceolate and somewhat spreading. Fls golden yellow, the outermost orange or reddish, rarely grey-violet, beneath. Achenes 5–8 mm, pale brown, fusiform, narrowing for the upper ⅔ of their length but not beaked, with muricate longitudinal ribs and numerous distinct transverse ridges; pappus dirty white, of 2 rows of hairs, the inner feathery, the outer shorter, simple, scabrid. Fl. 6–9. Freely visited by a variety of insects, especially bees and flies. $2n = 14^*$. Hr. Very variable, especially in hairiness. Almost or wholly glabrous plants corresponding with var. *glabratus* (Koch) Bischoff (*L. hastilis* L.) are occasionally found in this country.

Native. Meadows, pastures, grassy slopes, etc., especially on calcareous soils, to 600 m in N. England. Common in England, Wales and much of S. Scotland, but very local further north to Caithness; Ireland northwards to Meath and Mayo; Inner Hebrides, not in Orkney and Shetland. Europe from N.E. Spain, S. Italy and Greece northwards to Norway, Sweden and Karelia; Asia Minor; Caucasus; N. Iran.

Section 3. *Thrincia* (Roth) Dumort.

Pappus of marginal achenes represented by a cup of small scarious scales; that of the central achenes usually of 2 rows of hairs, the inner feathery with dilated bases, the outer shorter, simple, scabrid; lvs with simple or forked hairs.

3. L. taraxacoides (Vill.) Mérat   Hairy Hawkbit.
L. leysseri Beck; *Crepis nudicaulis* auct.; *Thrincia hirta* Roth; *Leontodon hirtus* auct., non L.

A perennial, rarely biennial, herb with a short erect premorse stock and a basal rosette of narrowly oblanceolate *lvs* gradually narrowed into the long stalk-like base, remotely sinuate-toothed to runcinate-pinnatifid ±glabrous or *with ciliate margins and rather dense forked hairs.* Scapes 2.5–30 cm, 1 to several from each rosette, ascending from a decumbent base, slender, bractless, hardly thickened beneath the *solitary head,* ±glabrous or sparsely hairy below with forked hairs. Head 12–20 mm diam., drooping in bud. Involucre 7–9 mm, its bracts narrowly lanceolate, the inner equal, the outer shorter and imbricating; all glabrous or with bristly midribs and ciliate margins. *Fls* golden yellow, the *outermost grey-violet beneath.* Achenes c. 5 mm, those of the central florets chestnut, attenuate above and ±short-beaked, straight, with strongly muricate longitudinal ribs and a brownish-white pappus of feathery hairs; those of the outer row pale brown, ±cylindrical, curved, with fainter transversely wrinkled longitudinal ribs, and surmounted by a cup of small scarious scales. Fl. 6–9. Visited by many bees and syrphids, etc. $2n = 8^*$, 10. Hr.

The above description is of subsp. **taraxacoides** (*Thrincia hirta* Roth, sens. str.), widespread in Europe. Subsp. **longirostris** Finch & P. D. Sell, usually annual and with long-beaked achenes (beak 2–3 mm; in subsp. *taraxacoides* c. 1 mm), is restricted to S. Europe.

Native. Dry grassland, especially on base-rich soils; fixed dunes; reaching 400 m in Ireland. Throughout the British Is. northwards to Inverness and W. Ross, Inner Hebrides. Europe northwards to Scotland, Denmark, Gotland and C. Russia.

50. PICRIS L.

Annual to perennial *stiffly hairy* herbs with flowering *stems bearing* alternate entire or sinuate-toothed to pinnatisect *lvs.* Heads solitary or in corymbs; bracts imbricate in many rows, those of the innermost row longest, erect and equal, the outermost sometimes very broad and resembling an epicalyx; receptacle flat, naked, pitted. Fls yellow; anther-lobes with short basal tails; style-arms hairy. Achenes curved, ribbed with transverse wrinkles between the ribs, beaked or not; *pappus* of 2 rows, the inner always of *feathery deciduous* hairs, the outer similar or of rough simple hairs.

About 50 spp., in the Mediterranean region and in temperate Europe and Asia, with 4 spp. in Abyssinia.

1 Outer bracts 3–5, large, ovate-cordate, resembling an
    epicalyx; achenes long-beaked.   **1. echioides**
  Outer bracts small, narrow, ±spreading; achenes not
    or very shortly beaked.   2
2 Heads distinctly stalked; bracts obscurely keeled,
    densely covered with mostly simple bristles, with
    a few forked and hooked bristles.   **2. hieracioides**
  Heads ±sessile, in dense terminal and axillary clus-

ters; bracts distinctly keeled in fr., the keel almost spinous with whitish forked and hooked bristles.

3. spinulosa

Section 1. *Helmintia* (Juss.) O. Hoffm.

Outer bracts large, ovate-cordate, resembling an epicalyx; achenes beaked, the upper part of the beak falling with the pappus; central achenes straight, marginal curved.

1. P. echioides L.   Bristly Ox-Tongue.

Helmintia echioides (L.) Gaertner

An annual or biennial herb with a stout furrowed erect irregularly forked stem 30–90 cm, covered with short rigid hairs which are tuberculate at the base, trifid and minutely hooked at the apex. Basal and lower stem-lvs oblanceolate, narrowed into a stalk-like base; *middle and upper lvs* lanceolate to oblong, *sessile, ±cordate, amplexicaul* or shortly decurrent; all coarsely toothed or sinuate, bristly-ciliate, and *very rough with scattered bristles on white tuberculate bases*. Heads 2–2.5 cm diam., somewhat crowded on short lateral stalks in an irregular corymb. *Outer bracts 3–5, lf-like, broadly cordate-acuminate*, bristly-ciliate and rough like the lvs, not quite equalling the lanceolate, awned, bristly inner bracts. Fls almost twice as long as the involucre, yellow, the outermost purplish beneath. Achenes 2.5–3.5 mm (excluding the beak), the central red-brown, ±straight, glabrous, the marginal whitish, curved, downy on the ventral side; all transversely wrinkled, with a *slender beak about as long as the achene; pappus of pure white feathery hairs, falling with the end of the beak*. Fl. 6–10. Visited by hive-bees, but said to be apomictic. $2n = 10^*$. Th.–H. (biennial).

Doubtfully native. Roadsides, hedge-banks, field margins and waste places, especially on stiff and calcareous soils. Locally common in lowland England and Wales but now only casual in S.E. Scotland and perhaps recently introduced further north to Aberdeen. S. and E. Ireland. Native in the Mediterranean region, Canary Is. and S.W. Asia; perhaps introduced in C. Europe northwards to the Netherlands and Denmark. Introduced in North America.

Section 2. *Picris*

Outer bracts short, narrow; achenes all similar, not or very shortly beaked.

2. P. hieracioides L.   Hawkweed Ox-Tongue.

A biennial to perennial herb with a stout furrowed erect stem 15–90 cm, rough with short forked and hooked bristles especially below, usually branched above, the branches spreading, corymbose. Basal and lower stem-lvs 10–20 cm, oblanceolate, narrowed into a stalk-like base; middle and upper *lvs* lanceolate, usually *broadened at the base and ±amplexicaul*, sometimes narrowed; all *±sinuate-toothed and undulate*, bristly-ciliate and bristly at least on the veins beneath. Heads 2–3.5 cm diam., solitary, terminal on the main stem and branches, which are bracteate, roughly hairy, and somewhat thickened distally. Involucre ovoid, its inner bracts 8–15 mm, lanceolate, obscurely keeled, with bristles and short

white hairs down a central strip, the margins ±glabrous; *outer bracts short, narrow*, usually spreading or recurved, with blackish mostly simple hairs. Fls bright yellow. Achenes 3–5 mm, fusiform, slightly curved, reddish-brown, with fine interrupted transverse wrinkles; *beak very short*; pappus of cream-coloured, deciduous, feathery hairs. Fl. 7–9. Freely visited by flies and bees, but said to be apomictic. $2n = 10^*$. H. Very variable, especially in hairiness and mode of branching.

Native. Grassland, especially on calcareous slopes, waysides, etc. Locally common in lowland England and Wales and reaching Cumbria and Northumberland. Introduced on railway banks in Ireland. Europe, from N. Spain, Italy, N. Balkan peninsula and S. Russia northwards to Denmark, S. Sweden and Karelia; W. and C. Asia. Introduced in North America, Australia and New Zealand.

*3. P. spinulosa Bertol. ex Guss.

Very closely related to *P. hieracioides* and often treated as a subsp., may be distinguished by its subsessile heads crowded in terminal and axillary clusters, and by the distinct and almost spinous keel of its bracts whose bristles are very stiff and mostly forked and hooked. Fl. 7–9. H. (biennial).

Introduced. A Mediterranean species formerly? established in a few villages in W. Kent, now a casual.

51. TRAGOPOGON L.

Herbs with copious latex and *linear or linear-lanceolate entire long-pointed sheathing lvs* resembling those of leeks. Heads large, yellow or purple. *Involucre conical in bud, of 1 row of lanceolate-acuminate bracts* united at their base. Receptacle naked. Anthers shortly tailed below and with short terminal appendages. Achene fusiform, 5–10-ribbed; *beak long*, ending upwards in a hairy ring; *pappus*, at least of the central achenes, of 1 *row of hairs simple below and densely feathery at the tips* except for 5 which exceed the remainder and are simple throughout; marginal achenes sometimes with a pappus of stiff bristles.

About 50 spp., in Europe and W. Asia.

Fls yellow   **1. pratensis**
Fls lilac to deep reddish-purple (see also *T. hybridus*, p. 495).   **2. porrifolius**

1. T. pratensis L.

Goat's-Beard, Jack-go-to-bed-at-noon.

An annual to perennial herb with a long brownish cylindrical tap-root surmounted by the remains of old lvs. Stem 30–70 cm, erect, simple or little branched above, glabrous or slightly woolly when young, somewhat glaucous. Basal lvs linear-lanceolate, long-pointed, entire, glabrous, broadened and somewhat sheathing at the base, with conspicuous white veins; stem-lvs similar but more abruptly narrowed into the long acumen from a broad semi-amplexicaul base. *Heads large, yellow*, terminal on the main stem and its few branches which are slightly enlarged just beneath the heads. Involucre 2.5–3 cm, of 8 or more equal lanceolate-acuminate bracts, glabrous or with some woolly hairs at their base. *Fls falling short of, rarely equalling, the spreading involucre*. Achenes 10–22 mm, yellowish, those in the centre commonly smooth and the outer ones scaly-muricate on the ribs and sometimes also tuberculate between the ribs, but sometimes all ±smooth or all ±muricate; *beak usually about equalling achene*; pappus *very large*, the feathery hairs interwoven. Fl. 6–7. Visited by various insects and ultimately self-pollinated; the heads close round noon ('Jack-go-to-bed-at-noon'). $2n = 12*$. H.–G.

The species has been divided into 3 subspp. of which subsp. *minor* is the commonest in this country:

Subsp. **minor** (Miller) Wahlenb., with *bright yellow fls only about half as long as the red-bordered involucral bracts*, anthers brownish above, achenes 10–12 mm, the outer scaly-muricate on the ridges, tuberculate between. Fls closing in dull weather. This has the most westerly continental distribution and may be the only subsp. native in the British Is.

Subsp. **pratensis** has the *pale yellow fls almost or quite equalling the pale-bordered involucral bracts*, the anthers are yellow below and dark violet at their tips, the achenes are 15–20 mm, the outer smooth or slightly scalymuricate. Fls usually remaining open in dull weather. This is the main subsp. in C. Europe north of the Alps and the Danube. In Britain it is much less frequent than subsp. *minor*, if indeed it occurs at all as a native.

*Subsp. **orientalis** (L.) Čelak has *golden yellow fls* which *equal or exceed the whitish-bordered involucral bracts*, the anthers are yellow with dark brown lines, and the *achenes are large, to *twice as long as beak*, the marginal ones muricate with cartilaginous scales. It is more easterly in distribution and is known only as a casual in this country.

These three subspp. differ in several morphological features and in geographical range, but their ranges appear to overlap and individuals are not infrequently found which combine features of different subspp. It seems inadvisable, therefore, to treat them as distinct species as has been done in the past.

Probably native. Meadows, pastures, dunes, roadsides, waste places, etc., to 370 m in Derby. Locally common in lowland Great Britain northwards to northernmost Scotland, but not in Hebrides, Orkney or Shetland; mainly in C. Ireland; Channel Is. Most of Europe northwards to c. 64° N.; Caucasus, Armenia, Iran, C. Asia.

*2. T. porrifolius L.   Salsify.
An annual or biennial glabrous and ±glaucous plant, with a branched irregularly cylindrical tap-root and an erect branching stem 40–120 cm. Resembles *T. pratensis* but the lvs taper more gradually and the stem and *branches are conspicuously enlarged just beneath the heads*. Involucre 3–5 cm, usually of 8 bracts. *Fls purple*, varying from half as long to as long as the involucre. Achene c. 4 mm, faintly 10-ribbed, scaly, gradually narrowed upwards into a beak somewhat exceeding the achene. Fl. 6–8. Visited by various insects. $2n = 12*$. Th.–H. (biennial).

Introduced. Cultivated for its tap-roots and occasionally escaping. Native in the Mediterranean region but widely cultivated. Hybrids between *T. porrifolius* and *T. pratensis* are found in the British Is.

*T. hybridus L. (*Geropogon glaber* L.), native of the Mediterranean region introduced here in seed for wild birds, is an occasional casual in and near gardens. It has pinkish-lilac fls no more than half as long as the involucral bracts, and its achenes differ from those of other *Tragopogon* spp. in lacking the usual thickened ring (annulus) between beak and pappus, and its marginal achenes in having a pappus of c. 5 rigid and scabrid bristles.

52. SCORZONERA L.

Usually perennial herbs with copious latex and mostly with simple entire or dentate linear to ovate-lanceolate lvs (but 1–2-pinnatisect in one section). Fl.-heads, 1–many, with involucral bracts imbricate and manyrowed; receptacle without scales. Fls yellow, whitish or purple. Achenes not or obscurely beaked; pappus of several rows of usually feathery hairs or the outermost simple and scabrid. Differs from *Tragopogon* in the many rows both of involucral bracts and pappus-hairs.

About 150 spp. in Europe and Asia, chiefly in the Mediterranean region and the Near East. A single sp. in the British Is.

1. S. humilis L.   Viper's Grass.
A perennial herb with a black often branched cylindrical stock, scaly beneath the basal lvs of the current season. Stem 7–50 cm, erect or ascending, usually simple, woolly when young but becoming ±glabrous. Basal lvs 10–20 cm, narrowly lanceolate to elliptical, long-acuminate, narrowed into a long stalk-like half-sheathing base, at first woolly then becoming green and glabrous; stem-lvs narrower but more abruptly broadened below into the semi-amplexicaul base. Head 2.5–3 cm diam., usually solitary. Involucre 2–2.5 cm, woolly below; its outer bracts ovate, inner oblong-lanceolate, all blunt. Fls pale yellow, twice as long as the involucre. Achenes 7–9 mm, with smooth longitudinal ribs; pappus dirty white. Fl. 5–7. Visited by various bees and other insects, and automatically self-pollinated. $2n = 14*$. H.

?Native. Marshy fields in Dorset and Warwick, very local. Europe from Portugal, C. Spain, S. France, N. Italy, N. Balkan peninsula and S. Russia northwards to Denmark, S. Sweden, Karelia and C. Russia; Caucasus.

53. LACTUCA L.

Annual to perennial herbs usually with overwintering lf-rosettes and lfy fl-stems; lvs unlobed to deeply pinnatifid, with margins and underside of midrib often prickly. Fl-heads rather small, solitary or in corymbose to spike-like panicles; *involucre cylindrical, with many rows of imbricating bracts*, the outermost sometimes resembling an epicalyx. *Receptacle without scales*, pitted. *Fls yellow or bluish*; anthers short-tailed below and with short terminal appendages. *Achenes strongly compressed*, usually ribbed, *abruptly contracted into a beak*; *pappus of 2 equal rows of soft white simple hairs*.

About 100 spp., ±cosmopolitan but chiefly in drier temperate and subtropical regions.

1 Fls blue; achenes with body and beak much the same colour and with beak at most half as long as body.
 5. tatarica
 Fls yellow; achenes with beak paler in colour and from somewhat shorter to much longer than body. 2
2 At least the upper stem-lvs very narrowly oblong to linear, long-tapering to tip and sagittate at base; underside of midrib at most sparsely hispid or muricate, not prickly; fl.-heads sessile or on very short branches in a long spike-like panicle. **4. saligna**
 Upper stem-lvs variously shaped but not approaching linear; underside of midrib prickly or not; fl.-heads in a ±spreading, not spike-like, panicle. 3
3 Middle and upper stem-lvs ovate-oblong to orbicular, not lobed, sessile and cordate-amplexicaul, margins not spinose, midrib smooth beneath; stem-lvs not held vertically. **1. sativa**
 Middle and upper stem-lvs ±oblong in outline, unlobed to pinnatifid, usually with spinose-ciliate margins and midribs prickly beneath; stem-lvs often held vertically. 4
4 Stem whitish; stem-lvs with white midrib and flat margins; ripe achenes 3–4 × 0.8–1.3 mm, olive-grey, mottled, with many simple bristles near apex.
 2. serriola
 Stem maroon-coloured; stem-lvs with undulate margins; ripe achenes 4–5 × 1.3–1.6 mm, blackish-red, with a few palmate bristles near apex. **3. virosa**

*1. L. sativa L.   Garden Lettuce.

An annual or biennial herb with a slender tap-root, a dense basal rosette and a tall erect flowering stem 30–100 cm, whitish, glabrous. Rosette lvs entire or runcinate-pinnatifid, very short-stalked; *stem-lvs ovate to orbicular*, cordate-amplexicaul, sessile; all glabrous *with entire margins, smooth on the main veins beneath*. Heads numerous, in a *dense corymbose panicle* with small sagittate scale-like bracts. Involucre 10–15 mm, bracts ovate-lanceolate, blunt, brownish-green with pale margins. Fls few, exceeding the involucre, pale yellow, often violet-streaked. Achene 3.2–4(–5) mm, narrowly obovate, 5–7-ribbed on each face, with 0–few simple apical bristles; beak white, equalling the achene. Fl. 7–8. Visited by flies and automatically selfed. $2n = 18$, $36 + 1$. Th.–H. (biennial).

There are many cultivated races of which most are included in var. *capitata* L., the cabbage lettuce. Var. *crispa* L., the cos lettuce, has long, erect, crisped and ±lobed lvs.

Introduced. Long cultivated as a salad plant and frequently escaping on waste ground. Origin unknown but probably from S.W. Asia or Siberia.

2. L. serriola L.   Prickly Lettuce.

L. scariola L.

An overwintering or spring-germinating annual, rarely biennial, with lfy fl-stems (15–)30–200 cm, stiffly erect, glabrous or somewhat prickly below, whitish or red-tinged. Lvs of overwintering rosette obovate-oblong,

unlobed to sinuate; stem-lvs obovate-oblong in outline, unlobed or sinuate-lobed, less commonly runcinate-pinnatifid with acuminate terminal lobe and a few distant pairs of narrowish acute lateral lobes which are back-curved distally and shaped like a curved bill-hook; upper lvs less lobed and the uppermost ±simply hastate or sagittate; all rigid, ±glaucous, glabrous but *spinous-ciliate on the margins and prickly on the underside of the white main veins*. The *stem-lvs* of plants fully exposed to the sun are all *held vertically* in the north–south plane ('compass plant'). Infl. an elongated *pyramidal panicle*; *bracts* sagittate *with spreading auricles*. Heads 11–13 mm diam., closely spaced along the distal halves of the panicle branches. Involucre 8–12 mm, narrowly cylindrical; its bracts lanceolate, glabrous, glaucous and often violet-tipped. Fls few, pale yellow, often mauve-tinged, exceeding involucre. Ripe achenes 3–4 mm, elliptical, olive-grey and mottled, 5–7-ribbed on each face, narrowly bordered with many simple bristles near the apex; beak white, equalling the achene. Fl. 7–9. Little visited by insects and automatically selfed. $2n = 18^*$. Th.–H. (biennial).

Probably native. In waste places and on walls, sometimes on ±stable dunes. S. England and Wales. S. and C. Europe, but established northwards to C. England, the Netherlands, N.W. Germany, Denmark, Gotland and C. Russia; W. Asia to the Altai and Himalaya; N. Africa from the Canary Is. to Ethiopia. Introduced in North America. In C. Europe associated with steppe species such as *Stipa pennata* and *Artemisia campestris*.

3. L. virosa L.

An overwintering annual or biennial herb with a branched tap-root and an erect white or reddish lfy stem 60–250 cm, glabrous or prickly below. Lvs obovate-oblong in outline (broader than in *L. serriola*), undivided or ±deeply pinnatifid, the basal lvs narrowed into a stalk-like base, the *stem-lvs* sessile and *cordate-amplexicaul with appressed auricles*; all rigid, ±glaucous, glabrous but spinous-ciliate and prickly on the underside of the main veins. Infl. an elongated *pyramidal panicle*; *bracts* amplexicaul with *appressed ±rounded auricles*. Heads 10 mm diam. Involucre 8–12 mm, cylindrical-ovoid, glabrous; its numerous imbricating bracts glaucous with a white margin and crimson tip. Fls pale greenish-yellow, exceeding the involucre. *Achenes* 3 mm, narrowly elliptical, with a narrow wing-like border (rather broader than in *L. serriola*), *blackish* when ripe, 5-ribbed on each face, ±*glabrous at the apex*; beak white, equalling the achene. Fl. 7–9. $2n = 18^*$. Th.–H. (biennial).

Probably native. Local in naturally unstable habitats such as wood-margins, sand dunes and cliff-ledges especially on calcareous substrata, but also on grassy wayside and streamside banks and verges, on walls, in gravel-pits and quarries, on industrial estates, etc.; mainly south and east of a line from Durham to S. Devon with a few outliers in Wales, W. Midlands, Cumberland and

S.E. Scotland. S. and C. Europe from Portugal to Greece and Turkey and northwards to S. Scotland, Belgium, Austria and Romania. A sub-Mediterranean species associated in C. Europe with such thermophilous plants as *Acer monspessulanum*, *Seseli libanotis* and *Aster linosyris*.

4. L. saligna L.   Least Lettuce.

An annual rarely biennial herb with an erect stem 30–100 cm, whitish, glabrous or bristly below, with long slender steeply ascending branches, the lowest often arising from near the base of the stem. Basal lvs oblong, entire, sinuate-pinnatifid, or sometimes runcinate-pinnatifid with narrow distant acute lateral lobes and a long slender terminal lobe, narrowed below into a petiole-like base, withered at flowering; *stem-lvs linear-lanceolate, entire*, with a sagittate amplexicaul base, or pinnatifid with a few distant narrow lobes, commonly *held vertically* and all ±in one plane; all lvs glabrous and ±glaucous with a conspicuous broad white midrib and *±entire margins*. Heads borne singly or in small clusters in the axils of sagittate bracts on the long branches of the *narrow strict panicle*. Involucre 15 mm, narrowly cylindrical; its bracts linear-lanceolate, blunt, greenish with a narrow white margin. Fls few, pale yellow, often reddish beneath, exceeding the involucre, becoming deep blue when dry. Achenes 3–4 mm, ±ribbed on each face, very narrowly bordered, finely muricate above; beak white, twice as long as the achene. Fl. 7–8. $2n = 18$. Th. (H. biennial).

Probably native, but decreasing. Formerly local in disturbed habitats on calcareous substrata, especially near the sea, in E. England from E. Sussex and Kent northwards to Bedford, Huntingdon and N. Norfolk, but now seems extinct inland and almost restricted to sea-walls of the Thames estuary. Much of Europe northwards to S.E. England, C. Germany and S.C. Russia; introduced in Australia.

*5. L. tatarica (L.) C. A. Meyer   Russian Blue Sowthistle..
Sonchus tataricus L.

A perennial herb with vertical stock giving rise to hypogeal stolons and erect fl.-stems 30–90 cm, branching above. In British plants all or most *lvs elliptic-lanceolate, unlobed* (elsewhere at least some lvs may be runcinate-pinnatifid with long terminal lobe), attenuate below into a petiole-like base, *margins ±sinuate-dentate but not spinose-ciliate*; bracts smaller, widened below into a sessile semi-amplexicaul base and with ±entire margins; stem and lvs subglabrous. Fl-heads in a ±corymbose panicle; involucre c. 1.5 cm, cylindrical. *Fls* c. 20 in each head, *lilac-blue*. *Ripe achenes* 4.5–6.5 mm, *almost black*, with *beak varying in length from very short to half as long as body*; pappus white. $2n = 18$. Hr.

Introduced. An infrequent casual in England, Wales, Isle of Man and Channel Is. which has persisted in a few spots for some years, especially on sand near the sea (as at Llandudno, Caernarvon), and also on stony sea-shores in W. Ireland. Native in steppe and coastal vegetation in E. Europe and Asia; now widely naturalized in N. and C. Europe.

54. MYCELIS Cass.

Annual to perennial herbs with small heads in ±spreading panicles. Involucre cylindrical, of 2 rows of bracts, the outer row short and resembling an epicalyx. Receptacle naked. Fls usually 3–5, yellow. *Achenes somewhat flattened, abruptly beaked. Pappus of an inner row of long simple hairs and an outer row of shorter hairs*. The genus therefore differs from *Lactuca* in pappus structure in the same way that *Cicerbita* differs from *Sonchus*.

About 30 spp. in Europe, Asia and Africa.

1. M. muralis (L.) Dumort.   Wall Lettuce.

Prenanthes muralis L.; *Lactuca muralis* (L.) Gaertner

A perennial herb with a short premorse stock and an erect stem 25–100 cm, glabrous, paniculately branched above. Lower lvs lyrate-pinnatifid with long winged stalks, the lobes rhombic or hastate, the terminal lobe much larger than the laterals and itself often hastately 3-lobed; the middle and upper lvs sessile and ±amplexicaul, becoming succesively smaller and less divided; all lvs thin, glabrous, often reddish, their lobes triangular-toothed. Heads in a large open panicle. Involucre 7–10 mm, narrowly cylindrical, the inner bracts linear, the outer very small and spreading, all blunt and often reddish. Fls usually 5, yellow, slightly exceeding the involucre. Achene 3–4 mm (including the short pale beak), fusiform, blackish. Fl. 7–9. $2n = 18^*$. Hp.

Native. On walls and rocks, in woods particularly on beech on chalk, and in cultivated ground and waste places, usually on nutrient-rich soils. Throughout Great Britain northwards to E. Ross but local and infrequent in upland Scotland; in Skye but not in Outer Hebrides, Orkney or Shetland; local in Ireland; not in Channel Is. Europe northwards to c. 68.5° N. in Norway; N.W. Africa; Asia Minor; Caucasus.

55. SONCHUS L

Annual to perennial herbs with copious latex, sometimes suffruticose, with amplexicaul often spinous-ciliate stem-lvs and large fl.-heads. Involucre of several rows of imbricating bracts. *Receptacle naked*, pitted. Anthers not tailed at the base but with a short blunt distal appendage. Achene flattened, somewhat narrowed above and below, truncate above, ±strongly ribbed; *beak* 0; *pappus white, of two equal rows of simple hairs*, the outermost thickened near the base.

About 70 spp. throughout the Old World but especially in the Mediterranean region and Africa.

1 Annual to biennial; achenes with 3 longitudinal ribs
  on each face.   2
  Perennial; achenes with 5–8 often unequally strong
  ribs on each face.   3
2 Stem-lvs with rounded auricles; achene smooth.
                                                          4. asper
  Stem-lvs with pointed auricles; achene rugose.
                                                      3. oleraceus
3 Plant with creeping rhizome; stem-lvs with rounded
  auricles; glands on involucre usually yellow;
  achenes brown, compressed.   **2. arvensis**

Plant with ±erect stock; stem-lvs with pointed auricles; glands on involucre usually blackish-green;
achenes yellow, ± tetragonal.   **1. palustris**

1. S. palustris L.   Marsh Sow-Thistle.

A tall perennial herb with a *short ±erect tuberous stock*
and a stout erect stem 90–300 cm, 4-angled, hollow, with
the large central cavity square in cross-section, glabrous
below but glandular-hairy above. Basal *lvs* lanceolate-
oblong in outline with a *deeply and acutely sagittate* sessile *base*, pinnatifid with a few distant lanceolate lateral
lobes and a larger lanceolate-acute terminal lobe, all
spinous-ciliate and with spine-tipped teeth; stem-lvs
becoming less pinnatifid, the uppermost simply linear-
lanceolate with a deeply sagittate amplexicaul base, the
auricles long, narrow, acute. Heads to 4 cm diam., in
a dense corymbose panicle whose branches, like the
stalks and *involucres* of the heads, are densely *covered
with blackish-green* (rarely yellow) *glandular hairs.*
Involucre 12–15 mm, ovoid-cylindrical; its outer bracts
ovate-acuminate, blunt. Fls pale yellow, exceeding the
involucre. *Achene* 4 mm, yellowish, slightly flattened
but made ±tetragonal by the 4 strongest of the longitudinal ribs; pappus c. twice as long as achene. Fl. 7–9.
$2n = 18^*$. Hp.

Native. Marshes, fens, stream-sides; rare and decreasing. S.E. England from Kent to Norfolk and Huntingdon; formerly in Oxford and Bucks. C. Europe from
Spain, Corsica, N. Italy, Serbia and S. Russia northwards to S. England, the Netherlands, Denmark, S.
Sweden and C. Russia; Caucasus; Armenia.

Differs from *S. arvensis* in being non-stoloniferous and in the
pointed (not rounded) auricles of the amplexicaul lvs.

2. S. arvensis L.   Field Milk-Thistle.

A perennial herb with *creeping underground stolons* and
erect or ascending stems 60–150 cm, furrowed, hollow
with the small central cavity elliptical in cross-section,
glabrous or at first cottony below, glandular-hairy
above. Basal lvs oblong or lanceolate in outline, narrowed into a winged stalk, runcinate-pinnatifid with
short triangular-oblong spinous-ciliate and spine-
toothed lobes; *stem-lvs* similar but less divided and sessile, the *cordate-amplexicaul base* having *rounded
appressed auricles*. Heads 4–5 cm diam. in a loose
corymb whose branches like the *involucres* are usually
densely covered with yellowish glandular hairs. Involucre 13–20 mm, campanulate; bracts oblong-lanceolate, blunt. Fls golden yellow. *Achene* 3–3.5 mm, *dark
brown*, narrowly ellipsoid with 5 strong ribs on each
face. Fl. 7–10. Visited freely by many kinds of insects,
especially bees. $2n = 36, 54^*$. H.

Native. Stream-sides, drift-lines on salt- and brackish-
marshes, banks, arable land, etc., to 380 m in Wales;
common. Throughout the British Is. Throughout.
Europe to 70° 33′ N. in Scandinavia; W. Asia. Widely
introduced in Asia, America, Africa and Australia.

3. S. oleraceus L.   Milk- or Sow-Thistle.

Annual or overwintering herb with a long slender pale
tap-root and stout erect glabrous stems 20–150 cm, ±5-
angled, hollow except at the nodes, branched above.
Lvs very variable; basal lvs usually ovate, stalked; lower
stem *lvs runcinate-pinnatifid* with the terminal lobe
usually wider than the uppermost pair of laterals and
with a short winged stalk and enlarged *acute spreading
auricles*; uppermost with a reduced blade and more
broadly winged stalk; all glabrous (or cottony only when
young), ±glaucous, *dull, never spinous*. Infl. an irregular cymose umbel, its branches sometimes glandular-
hairy. Heads 2–2.5 cm diam. Involucre 1–1.5 cm, glabrous (or cottony in young bud), rarely glandular-hairy,
its outer bracts broadly lanceolate, shorter and more
acute than the inner. Florets yellow, the outer purple-
tinged below. *Achenes* 3 mm, *oblanceolate*, compressed,
never winged, first yellow then brown, longitudinally
3-ribbed on each face, *transversely rugose*. Fl. 6–8.
Visited by various insects, especially bees and hover-
flies. $2n = 32^*$. H.

Native. Cultivated soil, waysides, waste places, etc.,
throughout the British Is. Europe to 66° 13′ N. in Scandinavia; N. and W. Asia; N. Africa and the Canary Is.
Widely introduced as a weed of cultivation.

4. S. asper (L.) Hill   Spiny Milk- or Sow-Thistle.

An annual or overwintering herb closely resembling *S.
oleraceus* but differing in the form and appearance of
the lvs, which are less often pinnatifid and then have
the terminal lobe narrower than the uppermost pair of
laterals, show a less clear distinction between blade and
stalk, have *rounded appressed auricles*, are usually *dark
glossy green* above, and are commonly *crisped and spinous-ciliate* at the margin. In var. *inermis* Bisch. the lvs
are all simple, obovate or ovate-lanceolate, with flat
and softly spinous-ciliate margins. Florets usually golden-yellow, and so a deeper colour than is usual in *S.
oleraceus*. *Achenes* 2.5 mm, obovate, compressed,
sometimes winged (Scotland), usually brown but variable in colour, longitudinally 3-ribbed on each face,
otherwise *smooth*. Fl. 6–8. Visited by bees and hover-
flies, etc. $2n = 18^*$. Th.

Native. Cultivated soil, waste places, etc., to 395 m
in England; common. Throughout the British Is.
Europe northwards to 64° 5′ N. in Scandinavia; N. and
W. Asia; N. Africa. Widely introduced as a weed of
cultivated land.

A sterile hybrid between *S. oleraceus* and *S. asper* has been
reported from Great Britain, but appears to be rare.

56. CICERBITA Wallr.

Perennial herbs usually with panicles of large heads
whose involucres consist of an inner row of larger and
an outer row of shorter bracts. Receptacle naked. Fls
all ligulate, blue or yellow. Anthers tailed below and
with short terminal appendages. Achenes flattened, narrowed above and below; *beak 0; pappus of 2 rows of
simple hairs, the outer shorter.*

Eighteen spp., of which 14 are native in the mountains of Europe, Asia and N. Africa, and 4 in America. Differs from *Sonchus* in the outer row of shorter pappus hairs.

Plant with erect branched stock; lower lvs with triangular terminal lobe and a few pairs of small lateral lobes; infl. an elongated simple or compound raceme; heads blue.　　　　　　　　　　　　　　　**1. alpina**
Plant with far-creeping rhizome; lower lvs with cordate terminal lobe and usually a single pair of lateral lobes; infl. corymbose; heads lilac.　　　　　**2. macrophylla**

1. C. alpina (L.) Wallr.　　　　Blue Sow-Thistle.

Sonchus alpinus L.; *Mulgedium alpinum* (L.) Less.; *Lactuca alpina* (L.) A. Gray

A tall perennial herb with a ±erect cylindrical rootstock and a stout erect furrowed stem 50–200 cm, simple or branched, bristly below and with dense reddish glandular hairs above. Lowest lvs stalked, lyrate- or runcinate-pinnatifid with a large broadly triangular-acuminate terminal lobe and a few pairs of much smaller ±triangular denticulate lateral lobes, narrowed into a winged stalk; the succeeding lvs becoming smaller and less divided, the *winged stalk* broadened into a *cordate-amplexicaul* base, the uppermost ±lanceolate; all lvs glabrous, somewhat glaucous beneath. *Heads* c. 2 cm diam., *pale blue*, in a simple or compound raceme, their stalks, like the infl. axis, densely glandular-hairy. Involucre 1–1.5 cm, purplish-green, with long glandular hairs. Achenes 4.5–5 mm, linear-oblong, with 5 strong and several weaker ribs; pappus 7 mm. Fl. 7–9. $2n = 18$. H.

Native. Moist places on alpine rocks; very rare. Angus and S. Aberdeen. Mountains of C. Europe from the Pyrenees to the Carpathians; and in Scandinavia and Karelia, to 71° 7′ N. in Norway. Reaches over 2100 m in Switzerland, where it is associated with *Peucedanum ostruthium, Senecio nemorensis, Adenostyles alliariae, Achillea macrophylla*, etc.

*2. C. macrophylla (Willd.) Wallr.

A tall perennial herb with *far-creeping rhizome* and erect stems 60–150 cm, *glabrous below*, glandular-hispid above. Lower lvs large, ±lyrate, with large *cordate terminal lobe* and usually a *single pair* of small lateral segments decurrent into a winged stalk with broadened cordate-amplexicaul base; upper lvs smaller and ±sessile but with cordate-amplexicaul base, the uppermost lanceolate to linear; all glandular-bristly, somewhat glaucous beneath. Infl. a loose ±*corymbose panicle* with glandular-bristly branches. Heads c. 30 mm diam.; involucral bracts glandular-hairy; *florets lilac-coloured*. Achenes 5 mm, glabrous, narrowly winged, 3-ribbed on each face. Fl. 7–9.

Introduced. A garden escape established in hedge-banks, waysides, etc., in a few places in Great Britain and W. Ireland. Caucasus.

C. plumieri (L.) Kirschl., a smaller plant, 60–130 cm, *wholly glabrous*, non-creeping, with lyrate-pinnatifid amplexicaul lvs,

blue heads in a loose corymb and achenes c. 6 mm, is often cultivated and sometimes escapes. Mountains of C. Europe.

57. HIERACIUM L.　　　　Hawkweed.

Perennial herbs, sometimes stoloniferous (subgenus *Pilosella*), with vertical to horizontal stocks and stout fibrous roots. Lvs spirally arranged on the flowering stems or some or all in a basal rosette. Infl. a cymose and often corymbose panicle or a few-fld forking cyme or the heads solitary, terminal. Involucral bracts erect or incurved in bud, imbricate in few to several irregular rows, the outermost shortest. Stem, lvs, infl. and involucre glabrous or clothed with simple hairs, toothed hairs, glandular hairs, sessile micro-glands and soft white stellate hairs ('floccose') in varying proportions. Receptacle pitted, the scarious margins of the pits variously toothed or fimbriate; *receptacular scales* 0. Florets usually yellow, glabrous or sometimes hairy at the tips ('ciliate-tipped') and less commonly also on the backs of the ligules. *Achenes* 1.5–5 mm, *cylindrical*, 10-ribbed, *truncate above*, neither appreciably narrowed upwards nor beaked; *pappus-hairs* in 1 or 2 rows, *simple, rigid*, brittle, whitish to tawny, usually *pale brownish*.

Perhaps 10 000–20 000 'species', chiefly in temperate, alpine and arctic regions of the northern hemisphere but some in South America (subgenera *Stenotheca* and *Mandonia*) and some in S. Africa, Madagascar, S. India and Ceylon (subgenus *Stenotheca*).

Hieracium is taxonomically one of the most difficult genera of the British flora. This is because seeds are usually produced apomictically, without fertilization of the ovum by a male nucleus from the pollen. Plants of subgenus *Hieracium* produce little or no viable pollen and are almost invariably apomictic, so that each individual and its descendants become more or less completely closed units. Genetic variation must depend on mutation or on the occasional crossing that appears to take place, and the resultant innumerable true-breeding lines are not at all closely comparable with normally outbreeding species. In subgenus *Pilosella* sexual reproduction is much more frequent, but plants are commonly capable also of apomictic seed-production and many can multiply vegetatively, so that taxonomic problems are still formidable.

The following is a key to the two subgenera and the several sections into which the British Hieracia have customarily been divided, though the sections are by no means clearly distinct units. There follow brief accounts of the sections with notes on the more widespread and abundant of the component 'micro-species' or groups of microspecies. It is hoped that this will prove of some use to those who wish to make a start on identifying hawk-weeds, but more serious students are recommended to consult the monograph by H. W. Pugsley (*A Prodromus of the British Hieracia, J. Linn. Soc., Botany*, **54**, 1948), the account by P. D. Sell & C. West in *Flora Europaea*, and the maps and explanatory text, also by Sell & West, in the *Critical Supplement to the Atlas of the British Flora* (Ed. F. H. Perring; Bot. Soc. Br. Is., 1968).

In the key the term 'inner rosette-lvs' refers to those that succeed the earliest-formed and outermost, which are usually smaller and broader than the later lvs.

Key to Subgenera and Sections

1 Pappus hairs ±equal, in 1 row; achenes 1.5–2 mm, their 10 longitudinal ridges each ending above in a projecting tooth; plants overwintering as rosettes

and often stoloniferous (as in the commonest British species). Subgenus PILOSELLA   13
Pappus hairs of various lengths, in 2 rows; achenes 2.5–4.5 mm, their 10 longitudinal ridges merging above into a swollen ring round the top; plants overwintering as rosettes or by lateral buds, never stoloniferous. Subgenus HIERACIUM   2

Subgenus HIERACIUM

2 No rosette formed, or rosette withered at time of flowering, rarely a few (often moribund) rosette-lvs persisting; stem-lvs 8–many, rarely fewer in small plants.   3
Rosette-lvs present at flowering; stem-lvs 0–8(–10).   8

3 Middle stem-lvs somewhat constricted just above the broad amplexicaul base; all lvs strongly reticulate and somewhat glaucous beneath; involucre and peduncles floccose and densely glandular; achenes tawny or pale yellowish-brown.   5. *Prenanthoidea*
Lvs not as above; involucre not or slightly floccose; achenes purplish- or blackish-brown.   4

4 Middle stem-lvs not or hardly amplexicaul.   5
Middle stem-lvs distinctly though not broadly amplexicaul; involucre not or slightly glandular.   7

5 Lvs numerous, rather crowded at least below, all of similar shape, commonly linear to narrowly lanceolate, rarely broader, narrowed into a sessile base, their margins revolute and scabrid; heads in a ±umbellate panicle; involucral bracts ±glabrous, all except the inner with recurved tips; styles yellow.   1. *Umbellata*
Lower lvs petiolate; middle and upper lvs shortly petiolate or sessile, lanceolate to broadly ovate, their margins not revolute; involucral bracts glabrous or somewhat but not densely hairy or glandular, their tips very rarely recurved; styles usually dark or with short dark hairs, sometimes yellow.   6

6 No rosette-lvs; stem-lvs numerous, crowded at least below; upper stem-lvs short with broad rounded bases; involucral bracts olive- or blackish-green.   2. *Sabauda*
Rosette-lvs withering early or a few persisting until flowering; stem-lvs usually not very numerous, rather widely spaced in the upper half or sometimes confined to the lower half of the stem; all lvs narrowing to the base, usually toothed in the middle.   4. *Tridentata*

7 No rosette-lvs; stem-lvs fairly numerous, all ±amplexicaul or the lowest merely sessile, paler and ±glaucous and reticulate beneath; involucre sparsely hairy and glandular; ligules glabrous-tipped.   3. *Foliosa*
Rosette-lvs petiolate, withering early or a few persisting; stem-lvs not numerous, rather widely spaced, often confined to lower half of stem, the lowest often petiolate; panicle lax, somewhat lfy; involucral bracts moderately glandular and hairy; ligules glabrous- or ciliate-tipped.   6. *Alpestria*

8 Stem-lvs 1–6, large, usually yellowish-green, the upper amplexicaul with large rounded auricles; whole plant glandular-viscid.   9. *Amplexicaulia*
Plant not viscid; stem-lvs if amplexicaul not yellowish-green.   9

9 Stem-lvs 1–7, ±amplexicaul; all lvs glaucous with long stout hairs on both sides; heads c. 5 cm diam.; invo-

lucral bracts with tips incurved in bud; florets often pale yellow; styles usually dark; robust plants of N. Britain, Hebrides and Ireland.   10. *Cerinthoidea*
Stem-lvs 0, or if present not amplexicaul.   10

10 Lvs with at least some glandular hairs, sometimes few and minute, especially along their margins; stem-lvs 0 or few, bract-like or like the inner rosette-lvs; heads solitary or few; involucre often shaggy with long hairs or densely floccose; ligules usually ciliate-tipped; mountain plants.   11
Lvs not glandular nor heads shaggy with long hairs; ligules usually glabrous-tipped.   12

11 Rosette-lvs numerous, small; stem-lvs 0–3; heads usually solitary and rather large, sometimes 2–5; plants commonly 5–30 cm; at altitudes above 600 m.   11. *Alpina*
Rosette-lvs not numerous; stem-lvs (0–)1–4; heads normally 2–6, sometimes more, rarely solitary; plants comonly 20–45 cm; at altitudes above 450 m.   12. *Subalpina*

12 Lvs glaucous, firm, often purple-spotted or purplish beneath, usually bristly with stiff stout-based hairs, especially on and near margins; stem forking above; heads usually few, large; involucral bracts erect in bud, with glandular and stout black-based hairs, not densely floccose; styles usually yellow.   8. *Oreadea*
Lvs green or glaucous, hairy but not truly bristly on margins and underside; heads various in size and number; involucral bracts incurved in bud, usually densely floccose at least on margins, usually ±glandular, often with slender and sometimes with stout black-based hairs, but never shaggy as in *Alpina*.   7. *Vulgata*

Subgenus PILOSELLA

13 Stem-lvs 0; head 1; rosette-lvs white- or grey-felted beneath; ligules reddish beneath.   13. *Pilosella*
Stem-lvs 1–several; heads 4–many.   14

14 Rosette-lvs yellow- or grass-green, elliptic to lanceolate, flaccid; stem-lvs usually 1–2; heads often in a compact, sometimes subumbellate corymb; ligules yellow, brownish or brick-red.   14. *Pratensina*
Rosette-lvs glaucous, lanceolate to linear, rigid; stem-lvs 1–3(–7); heads panicled; stolons sometimes 0; ligules yellow.   15. *Praealtina*

Subgenus HIERACIUM

Section 1. *Umbellata* F. N. Williams (Fig. 54A)
British representatives are now placed in a single variable but not readily divisible species which includes both sexual and apomictic lines.

H. umbellatum L.

Stems commonly 30–80 cm, slender, usually hairy below, eglandular. Lvs numerous (up to 50), dark green, typically linear to linear-lanceolate with revolute scabrid margins, subentire or with a few distant teeth, narrowed into a sessile base, usually subglabrous above, hairy and floccose beneath. Heads few to many in an umbellate panicle, their peduncles floccose but neither hairy nor glandular. Involucral bracts glabrous or nearly so, uniformly green or blackish, blunt, all but the inner

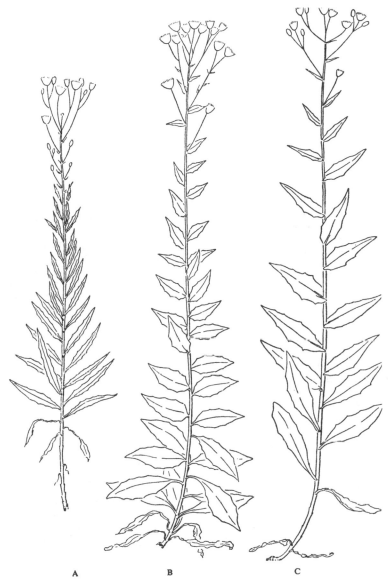

Fig. 54. *Hieracium.* A, *Umbellata* (*H. umbellatum*); B, *Sabauda* (*H. sabaudum*); C, *Foliosa* (*H. subcrocatum*). ×⅕.

with spreading or recurved tips. Ligules glabrous-tipped. Styles usually yellow. Achenes brownish-black. Locally common in open woods and copses, roadsides, heaths, etc., through much of lowland England and Wales, more scattered in Scotland and Ireland. A variant with ovate-lanceolate to oblong lower lvs, all lvs clear green and involucre usually olive-green, restricted to Wales, S.W. England, W. Ireland and Channel Is., has been named ssp. **bichlorophyllum** (Druce & Zahn) P. D. Sell & C. West, but it grades into the more wide-spread type.

Section 2. *Sabauda* F. N. Williams (Fig. 54B)

Stem 30–120 cm, robust, usually hairy at least below, eglandular. Rosette-lvs 0; stem-lvs up to 50 or more, crowded below, rapidly smaller upwards; lowest ovate-lanceolate to lanceolate, acute, narrowing gradually into the long petiole; middle lanceolate to oval-elliptical or broadly elliptical; middle and upper rounded or cuneate at the ±sessile base but never truly amplexicaul; all ±toothed, with stout-based hairs on both surfaces and on margins. Heads few to numerous in a subcorymbose panicle. Involucre olive- or blackish-green, often hairy and glandular; bracts linear-lanceolate, usually with appressed tips. Ligules glabrous-tipped. Styles usually dark. Achenes purplish- or blackish-brown. Chiefly low-land, especially on woodland margins and in clearings, hedge-banks, etc., in England and Wales, more local in Scotland and Ireland. Of the 5 native spp., 2 are widespread:

Stem densely covered below with long simple hairs, fewer above; peduncles floccose and with spreading simple hairs; involucre with many fine glandular and numerous long simple hairs: **H. sabaudum** L. (incl. *H. perpropin-quum* (Zahn) Druce & *H. bladonii* Pugsley). Common in England and Wales and extending northwards to Easter Ross; E. Ireland.

Stem ±glabrous above; peduncles usually floccose only but sometimes with scattered simple hairs; involucre gla-brous or nearly so, eglandular: **H. vagum** Jordan. Chiefly in C. and N. England and Wales but extending from Somerset and Kent to Angus; not in Ireland. **H. rigens** Jordan differs in the long slender infl.-branches (short in *H. vagum*) and the densely glandular involucre. It is locally common in open woodland and on heaths, roadside- and railway-banks, etc., in S.E. England.

Section 3. *Foliosa* Fries (Fig. 54C)

Stems up to 120 cm, often reddish. Rosette-lvs 0; stem-lvs 10–30, lower ±oblanceolate, long-attenuate below but only the lowest sometimes shortly petiolate; middle and upper narrowly to broadly lanceolate or elliptic, sessile and at least the upper semi-amplexicaul, all paler and reticulate beneath. Heads in corymbs. Involucral bracts with tips incurved in bud and with numerous glan-dular but few simple or stellate hairs. Ligules glabrous-tipped. Achenes purplish- or blackish-brown. Hills and coastal cliffs, at no great altitude, from Wales, Stafford and Yorks northwards to Sutherland, Outer Hebrides and Shetland.

Distinguishable from *Umbellata* and *Sabauda* by the semi-amplexicaul lvs, reticulate beneath, and from *Pre-nanthoidea* by the lvs not being fully amplexicaul, the much less glandular infl. and involucre, the glabrous-tipped ligules and the dark achenes.

Of the 10 native spp., the most widespread may be distinguished thus:

1 Lvs in mid-stem elliptic- to rhombic-lanceolate,
  cuneate at base.   2
  Lvs in mid-stem oblong-lanceolate or lingulate,
  rounded at base.   3
2 Heads truncate below; style blackish.
              H. subcrocatum (E. F. Linton) Roffey
  The most widespread species of the Section. On riverside rocks and sea-cliffs from Wales and Staf-ford northwards to Sutherland; Inner and Outer Hebrides; E. Ireland.
  Heads rounded below; styles pure yellow.
              H. latobrigorum (Zahn) Roffey
  N. Wales and N. England and northwards to Ork-ney and Inner Hebrides; E. Ireland.
3 Heads truncate below; styles dark.
              H. strictiforme (Zahn) Roffey
  Very common in C. and N. Scotland and N.E. Ire-land but with a few scattered localities in Wales, N. England and S. Scotland.

Heads narrowed below; styles yellow.
              H. reticulatum Lindeb.
  Widespread in C. and N. Scotland and a very few localities further south.

Section 4. *Tridentata* F. N. Williams (Fig. 55A)

Stems up to 100 cm, rigid, often reddish below, subgla-brous to densely hairy. True rosette-lvs 0 or a few wither-ing early and usually dead before flowering; stem-lvs 3–15 or more, often confined to lower half of stem and the lowest sometimes simulating a rosette, linear- or ovate-lanceolate, narrowed to the base and at least the lowest petiolate, all acuminate and usually with 3–5 teeth in the basal two-thirds of each side; often ±glabrous above, paler and commonly somewhat floccose and hairy beneath, the margins usually thickened and ciliate or scabrid. Heads few to many in a corymbose panicle. Involucral bracts incurved in bud but later appressed, olive- to blackish-green, variously clothed with stellate, simple and glandular hairs but not often densely glandu-lar. Ligules glabrous-tipped. Achenes blackish-brown.

There are 21 British spp. in woods and copses and on rocky slopes, walls, waste places, etc., from sea-level to subalpine altitudes. The following key will assist in distin-guishing the 4 most widespread or locally most abundant spp.

1 Lvs linear-oblong to linear-lanceolate or lanceolate, often blotched with purple-brown, usually suben-tire.   **H. sparsifolium** Lindeb.
  Stem 20–60 cm, usually purplish below. Lvs 10–16, the lowest often simulating a rosette. Infl. with 2–10 or more heads on long slender branches. Involucre with many dark-based simple hairs and some short glandular hairs. Styles yellow or darkish. Wales, N. England and Scotland, with a few scattered localities in W. Ireland.
  Lvs in mid-stem elliptic- or rhombic-lanceolate,
  ±toothed.   2
2 Slender plant 25–35 cm, with lowest lvs ±simulating a rosette; all lvs small, suberect, remotely serrate; involucre blackish-green; styles dark.
              H. calcaricola (F. J. Hanb.) Roffey
  Variable, but typically slender and small-leaved and with few large dark heads. In grassy places and heath-land over much of lowland England and Wales.
  Robust plants, commonly 50–100 cm; lowest lvs some-times simulating a rosette; lvs not markedly small, distinctly toothed; involucral bracts olive-green; styles yellow or darkish.   3
3 Lvs 8–15, ±coarsely toothed, those in mid-stem lan-ceolate to elliptical-lanceolate, all ±glabrous above, hairy beneath; heads usually 6–15; styles yellowish to discoloured.
              H. trichocaulon (Dahlst.) K. Joh.
  Common in open woodland and on sandy heaths and roadsides in S.E. England and S. England with a few outlying localities.
  Lvs 10–25(–30), sharply and irregularly toothed, those in mid-stem elliptical- to rhombic-lanceolate, all hairy on both sides; heads usually 10–30 or more; styles discoloured to dark.   **H. eboracense** Pugsley
  Frequent in open woodland, grassy places and heaths over much of lowland England and Wales.

Fig. 55. *Hieracium*. A, *Tridentata* (*H. trichocaulon*); B, *Prenanthoidea* (*H. prenanthoides*); C, *Alpestria* (*H. dewarii*). ×⅕.

Section 5. *Prenanthoidea* Koch (Fig. 55B)

Stems up to 120 cm. Rosette-lvs 0; stem-lvs 6–30(–50), diminishing gradually upwards; lowest oblanceolate, narrowed into a short broadly winged petiole, the middle elliptical-oblong, usually constricted just above the broad base, the upper ovate-lanceolate; all amplexicaul, subentire or with small distant teeth, green and ±glabrous above, glaucous, distinctly net-veined and somewhat hairy beneath. Heads small, numerous, in a subcorymbose panicle, the peduncles densely floccose and glandular. Involucre dark brownish-green, densely glandular; bracts few, incurved in bud, the outer short and acute, the inner long blunt. Ligules ciliate-tipped. Styles dark. Achenes tawny to yellow-brown. Only one sp. now native in the British Is.

H. prenanthoides Vill., with stem commonly 40–100 cm, reddish below; lowest lvs often withered at flowering; inner involucral bracts linear-oblong, sparsely floccose. Streamsides and steep-sided valleys in mountain dis-

tricts. Locally frequent in S. Wales and N. Britain from Derby to Aberdeen and Inverness.

Section 6. *Alpestria* (Fries) F. N. Williams (Fig. 55C)

Stem 20–100 cm, rather slender, usually with white stout-based hairs below, floccose and often also with black-based hairs above. Rosette-lvs few, usually withered by flowering, sometimes persistent; stem-lvs 2–10(–15), distant, the lowest narrowed into a petiole-like base, middle and upper semi-amplexicaul; all dark- or grass-green, paler but not clearly reticulate beneath, subentire or distantly toothed, ciliate and often with stout-based hairs at least beneath. Heads 2–6(–17) in a corymbose panicle, their peduncles floccose. Involucral bracts blackish-green, broad and blunt, sparsely or not at all floccose, with or without glandular and simple hairs. Ligules slightly hairy-tipped. Styles yellow or dark. Achenes dark brown. Rocky cliffs, streamsides and grassy slopes. Of the 18 native spp., 13 are confined to Shetland, 3 have very restricted distributions in Scotland and 1 in N. England. The only sp. at all widespread is **H. dewarii** Syme, of stream-banks and rocky ledges in W. Central Scotland. Its rosette-lvs wither at or before flowering, and the lower and middle stem-lvs are broadly elliptic, the upper being ovate to ovate-lanceolate, all entire or distantly denticulate, very hairy above and beneath. The inner involucral bracts are very slightly floccose below with a few glandular and white black-based simple hairs. The ligules are shortly ciliate-tipped and the styles dark. **H. mirandum** P. D. Sell & C. West, very rare in N. England, differs in its ovate middle stem-lvs, eglandular involucre and glabrous ligules.

Section 7. *Vulgata* F. N. Williams

Very varied in robustness, in number, shape and colour of rosette- and stem-lvs and in number and size of heads. Lvs usually softly hairy and sometimes floccose but lacking glandular hairs and marginal bristles. Involucral bracts variously floccose, glandular and simply hairy, their tips usually incurved in bud. Ligules usually glabrous-tipped. Styles usually dark, rarely yellow. Achenes dark red or blackish.

A large and complex section often split into the more or less distinct subsections included in the following key:

1 Stem-lvs 0–1(–3); rosette-lvs several to many, ±broad, often with concave-sided spreading or reflexed basal teeth, the base of the blade usually cordate, truncate or rounded and so clearly demarcated from the often shaggy petiole; styles commonly yellow or discoloured.   2

  Stem-lvs 2–5(–12); rosette-lvs not numerous, often very few, the blade broadly to narrowly lanceolate, narrowed gradually into the petiole and with outwardly or forwardly directed (not reflexed) teeth; styles commonly discoloured or dark.   4

2 Lvs dull green or glaucous, usually ±glabrous (rarely floccose) above; stem-lvs 0–1, bract-like or narrowly lanceolate; stem glabrous or slightly hairy below, ±forked above, commonly with 2–8 medium-sized or large heads.   Subsection *Bifida*

  Lvs grass- or yellow-green, ±hairy above, more so beneath; stem-lvs (0–)1–2(–3), the one in mid-stem usually fairly large and petiolate; stem usually with many long hairs, glandular above; heads 2–10 in a corymbose panicle.   3

3 Clothing of involucre predominantly glandular, with or without hairs.   Subsection *Glandulosa*

  Involucre usually densely clothed with simple hairs, glandular hairs sparse or 0.   Subsection *Sagittata*

4 Lvs glaucous or pale green; stem-lvs 2–5(–8); stem subglabrous or hairy below, ±floccose above, not or slightly glandular; heads few, loosely panicled or stem forked above, with floccose peduncles but sparse hairs and glands; involucre hairy and floccose but not or sparsely glandular.

                Subsection *Caesia*

  Lvs grass- or yellow-green; stem densely hairy below, floccose and glandular above; heads small to medium, panicled, few to many, peduncles and involucres usually densely glandular, hairy and floccose.   Subsection *Vulgata*

Subsection *Bifida* Pugsley (Fig. 57B)

Plants usually with oval to oblong rosette-lvs, often toothed near the cordate to cuneate base; stem-lvs usually 0–1; all nearly glabrous above (but floccose on both surfaces in *H. cymbifolium*). Heads commonly 2–8 in a corymbose, often forking, infl., medium-sized to large; involucre floccose and with glandular and simple hairs in varying proportions but not as glandular as in subsection *Glandulosa*. Most species are of very restricted distribution.

Lvs floccose on both surfaces, their margins often incurved making the lf boat-shaped; involucre with many dark glandular and longer simple hairs. A robust plant, 25–50 cm, with thick rosette-lvs, cordate or subtruncate at base, and 3–12 heads, their peduncles and involucres densely floccose and with many glandular and simple hairs. Common on limestone cliffs and rocks, on grassy slopes and on walls and railway banks in Stafford, Derby and N.W. Yorks and in a few other scattered localities.

           H. cymbifolium Purchas

Lvs not appreciably floccose above and not laterally incurved; involucre with a few fine glandular hairs and some dark-based simple hairs. Stem 30–40 cm, with thin deep green oval to elliptic-ovate inner rosette-lvs, subentire or minutely toothed, with rounded or subcordate base, ±glabrous above. Heads 5–8; styles dark. On limestone in N.W. Yorks and Cumberland, with a few scattered localities in Scotland and N. Ireland.

           H. silvaticoides Pugsley

Subsection *Glandulosa* Pugsley (Fig. 57A)

Plants variable but typically with hairy stems and lvs, rosette-lvs green and truncate-based, stem-lf 0 or 1, often well developed and petiolate. Heads small- to medium-sized. Involucral bracts blackish-green, incurved in bud, always with numerous glandular hairs, with or without stellate and simple hairs. Ligules usually glabrous-tipped and styles dark. Widespread in lowland as well as upland areas throughout Great Britain and more locally in Ireland.

The following are more or less widespread.

1 Heads small or very small, narrow; involucre with many dark glandular hairs and with or without simple hairs as well.   2

  Heads medium-sized, rounded below; involucre with dark glandular and some simple hairs. Plants of cliff-ledges and rocky streamsides in C. and N. Scotland, with bright green rosette-lvs, often tinged or spotted with purple, and 0–1 ovate-lanceolate stem-lf.   5

2 Involucre with many long dark simple hairs as well as black glandular hairs; rosette-lvs cuneate-based. Stem 30–50 cm. Inner rosette-lvs dull green, sometimes purple-spotted, oblong to lanceolate; stem-lvs 0–2; all stiffly hairy on both sides. Heads 2–6, very small; ligules hairy-tipped; styles dark. Cliffs and rocky streamsides from N. England to Sutherland; Hebrides; E. Ireland.   **H. duriceps** F. J. Hanb.

  Involucre with dense dark glandular but no simple hairs; rosette-lvs rounded, truncate or cordate at base.   3

3 Lvs dark green, rarely purple-spotted, broadly oval, nearly glabrous above. Stem 30–55 cm. Stem-lf 0–1,

Fig. 56. *Hieracium. Vulgata*: A, B, Subsection *Vulgata* (A, *H. acuminatum*; B, *H. vulgatum*); C, Subsection *Caesia* (*H. cravoniense*); D, Subsection *Sagittata* (*H. oistophyllum*). ×⅕.

ovate-cordate or bract-like. Heads 3–20; involucre blackish-green; styles dark. Common on limestone in N. England, Peak District, S. Wales, Gloucester.
H. pellucidum Laest.

Lvs pale or yellowish-green, oval to oblong-lanceolate, varying in hairiness.   4

4 Lvs with broad shallow teeth, almost glabrous above; styles yellowish. Frequent on roadsides and railway banks and occasional in open woodland south and west of a line from S. Lancs to Essex, with a few localities in W. Scotland and S.E. Ireland. Doubtfully native.   **H. sublepistoides** (Zahn) Druce
(**H. cinderella** (A. Ley) A. Ley, locally frequent in similar places in C. and S. Wales and the English border area, differs in its pale green heads and more sharply toothed lvs. It is also doubtfully native.)

Lvs usually deeply toothed, distinctly hairy above; styles dark. Naturalized in scattered localities, by roads and railways, etc., all over Great Britain; N.E. Ireland.   ***H. grandidens** Dahlst.
(The above 3 spp. were formerly included in *H. exotericum* auct.)

5 Inner rosette-lvs often purplish, rarely purple-spotted, oblong to ovate-lanceolate, denticulate to serrate, ±abruptly narrowed at base, nearly glabrous above, hairy beneath; infl. compact, usually with 2–6 heads; styles yellow or discoloured. Common in the Scottish Highlands. **H. pictorum** E. F. Linton

Inner rosette-lvs spotted with purplish-brown, oval to oblong, denticulate, hairy above and beneath; infl. with 2–6 heads with arching peduncles; ligules hairy-tipped; styles dark. Locally frequent in the Scottish Highlands.   4

H. piligerum (Pugsley) P. D. Sell & C. West

Subsection *Sagittata* Pugsley, emend. (Fig. 56D)

Plants typically 20–50 cm, with hairy stems and lvs; rosette-lvs green, oval to oblong, often with truncate or sagittate base but sometimes abruptly narrowed or rounded; stem-lvs 0–2, the lower, if 2, well-developed and petiolate. Heads medium-sized to rather large. Involucral bracts usually incurved in bud, with many simple and fewer glandular hairs. Mostly upland plants of cliffs and rocky streamsides in N. England and Scotland and often local and uncommon, but the following are more widespread.

1 Heads medium-sized; inner rosette-lvs sagittate or truncate at base.   2
  Heads rather large; inner rosette-lvs abruptly narrowed or rounded at base. N.W. England only.   3
2 Lvs sagittate at base; heads 3–12(–20); involucral bracts olive-green, broad and blunt, with numerous dark and simple and fewer fine short glandular hairs; ligules shortly hairy-tipped; styles darkish.

Fig. 57. *Hieracium.* A, B, *Vulgata*: A, Subsection *Glandulosa* (*H. exotericum sens. lat.*); B, Subsection *Bifida* (*H. sanguineum*); C, *Oreadea* (*H. hypochaeroides*); D, *Amplexicaulia* (*H. amplexicaule*). ×⅕.

Cliffs and grassy slopes in N. England and S. and C. Scotland.   **H. oistophyllum** Pugsley

Lvs shortly cuneate to truncate at base, often with large horizontal teeth; heads 2–6; involucral bracts grey-green, narrow, usually acute, with dense simple and sometimes a few glandular hairs; ligules sparingly hair-tipped; styles darkish. Cliffs and rocky streamsides in C. and N. Scotland.

H. subhirtum (F. J. Hanb.) Pugsley

3 Lvs bright green, often purple beneath, oval to oblong-lanceolate, with sharp spreading teeth near the abruptly contracted base, hairy above and beneath; heads 2–10 in a forking corymb; involucral

bracts erect in bud, dark green, with many dark simple and fewer glandular hairs; ligules sparsely hairy-tipped; styles darkish.

H. crebridentiforme Pugsley

Lvs dark green, oval to elliptical with the lowest teeth often deflexed about the rounded or subcordate base, with short stiff hairs above, longer beneath; heads 2–12 in a lax corymbose panicle; involucral bracts grey-green with many long simple and shorter glandular hairs; ligules golden-yellow, glabrous-tipped; styles dull yellow.

H. auratiflorum Pugsley

Subsection *Caesia* W. R. Linton, emend. (Fig. 56C)

Plants typically with pale green or glaucous basal lvs which are ±cuneate-based, never truly truncate and never with the lowest marginal teeth parallel to the petiole or curved towards the lf-base; stem-lvs 1–3(–5), the lowest usually large and petiolate. Heads 1–12 or more, of varying size; involucral bracts usually incurved in bud, somewhat floccose and with many simple hairs but with glandular hairs fewer and shorter, sometimes 0. Achenes blackish.

Most of the 20 or more British spp. are rare and local, chiefly in Scotland.

1 Heads rather small, the terminal pair very short-stalked and therefore appearing geminate; glandular hairs on involucre very few or 0; lvs denticulate or shallowly sinuate-dentate.   2

Heads rather large, the terminal pair not appearing geminate; glandular hairs present on involucre, but not abundant; lvs with long spreading or ascending teeth, at least near base of lf.   3

2 Inner rosette-lvs pale green to glaucous, narrowly oblong to lanceolate, narrowed below into the short winged petiole; heads commonly stylose, i.e. with undeveloped ligules; involucral bracts rather broad, blunt, floccose and shaggy with long pale simple hairs, almost eglandular; styles dark. Slender plants, 25–60 cm, with narrow rosette-lvs and 2–4 stem-lvs; lvs rather stiffly hairy on both sides, ciliate and with shaggy petioles. The narrow lvs and the lax panicle of small geminate and frequently stylose heads are distinctive. Common in rocky and grassy places in N. England and Scotland.

H. cravoniense (F. J. Hanb.) Roffey

Inner rosette-lvs dull to deep green, elliptic to oblong or ovate-lanceolate, cuneate or abruptly narrowed below; heads rarely stylose; involucral bracts few, broad, blunt, with numerous dark-based simple and few or no glandular hairs; styles yellowish to dark. Variable. Usually robust plants, 20–50 cm, with 1–2 stem-lvs, the upper bract-like; lvs shortly and stiffly hairy on both sides, ciliate and with shaggy petioles. Rocky and grassy places in Wales and from S. Scotland to Orkney; Ireland.

H. orcadense W. R. Linton (incl. *H. euprepes* F. J. Hanb.)

3 Inner rosette-lvs dull green, ovate-lanceolate to oblong, denticulate or with sharp ascending teeth near the cuneate or rounded base; stem-lvs 1–3(–5), the lowest large, ovate-lanceolate, irregularly toothed below; all lvs usually roughly hairy on both sides, ciliate and with shaggy petioles; involucre dark olive-green, sparingly floccose and with dense long dark-based simple hairs and a few shorter glandular hairs; styles usually yellow. Abundant on limestone in the Peak District and N. England, and in scattered localities in Scotland, Hebrides and N. Ireland.   **H. rubiginosum** F. J. Hanb.

Inner rosette-lvs bright green or glaucous, oblong-lanceolate to lanceolate with long spreading teeth near the abruptly narrowed or sagittate base; stem-lvs 1–2; all ±hairy above and beneath, floccose beneath and the long petioles shaggy; involucre greyish- or olive-green, floccose, with dense simple and some shorter glandular hairs; styles yellowish. Common in rocky and grassy places in C. and N. Scotland.   **H. caesiomurorúm** Lindeb.

Subsection *Vulgata* (Fig. 56A, B)

Stems 15–100 cm, usually slender and wiry, hairy below, ±glandular and floccose above. Rosette-lvs 2 to several, petiolate, ±persistent until flowering; stem-lvs 2–8 (–12), diminishing upwards, the lowest petiolate, upper ±sessile; all grass- or yellow-green, lanceolate to elliptic, narrowed at both ends and passing gradually into the petiole, usually with spreading or forwardly directed (not reflexed) teeth, hairy or almost glabrous above, densely hairy on margins and petioles and at least on the veins beneath. Heads small to medium-sized, panicled, the peduncles and involucre usually floccose and ±densely glandular-hairy but usually with few or no simple hairs (though with many in *H. vulgatum*). Ligules glabrous-tipped. Styles usually dark. Achenes dark brown or black. Distinguishable from *Caesia* by the non-glaucous lvs and the densely glandular involucres with few or no simple hairs; and from the remaining subsections by the gradual narrowing of lf-blade into petiole and the non-reflexed basal teeth.

Some of the commonest and most widespread of our lowland hawkweeds are placed in this subsection.

1 Rosette-lvs dark green spotted with purple-brown. Stem 30–50 cm, densely hairy below; rosette-lvs few, oblong to elliptic-lanceolate, acute, shortly narrowed into long petioles; stem-lvs 3–5; all rather stiffly hairy on both sides. Infl. with 6–20 medium-sized heads; involucre grey-green, floccose, with numerous short dark glandular hairs and a few longer simple hairs. Styles dark. Scattered throughout England and in a few places in Wales, Scotland and Ireland. Introduced and mainly on waste ground, slag-heaps, walls, quarry slopes, etc.   ***H. maculatum** Sm.

Rosette-lvs not spotted.   2

2 Involucral bracts with both glandular and simple hairs ±numerous. Stem 20–80 cm, hairy, especially below; rosette-lvs dull or pale green, often purplish beneath, narrow-oblong to lanceolate, commonly with sharp ascending teeth below, cuneate at base; stem-lvs 2–4(–5), the lowest petiolate; all sparsely hairy above and beneath. Heads 1–20, smallish; involucre grey-green with numerous whitish dark-based simple hairs and some fine short glandular hairs. Styles darkish. The commonest hawkweed throughout N. England and Scotland, with scattered localities southwards to a line joining the Severn and Thames estuaries and in N. Ireland. In rocky and grassy places at a wide range of altitudes; sometimes on roadsides and railway banks in S. England.   **H. vulgatum** Fries

Involucral bracts ±densely glandular but with simple hairs 0 or very few.   3

3 Rosette-lvs ovate- or elliptical-lanceolate; stem-lvs 3–10(–20); heads 2–50 or more, medium-sized with rounded base; involucre distinctly floccose. Stem 30–100 cm, hairy especially below, floccose and glandular above. Rosette-lvs few, deep or light

green, sharply dentate towards the cuneate base; lowest stem-lf petiolate and with sharp spreading teeth in the lower half; all lvs usually hairy on both sides. Involucre with many usually dark glandular hairs, rarely with a few simple hairs. Styles yellow or darkish. The commonest hawkweed of lowland England and Wales, extending northwards to the Lake District, Craven Pennines and N.E. Yorks, where it is in upland rocky places. Elsewhere it grows on roadsides, railway banks, walls and grassy places.   **H. acuminatum** Jordan (incl. *H. lachenalii* auct. & *H. strumosum* auct.)

Rosette-lvs oblong to lanceolate; stem-lvs 2–5(–9); heads 3–20, rather small, sometimes medium-sized; involucre not or sparingly floccose. Very variable. Stem 30–80 cm, hairy at least below. Rosette-lvs green, sometimes bluish- or greyish-green, denticulate or more deeply and sharply toothed especially towards the base; all lvs hairy on both sides or almost glabrous above. Involucre with numerous unequal dark glandular hairs but simple hairs 0 or very few; styles yellow to darkish. A common lowland hawkweed of C. England and Wales, extending to N. England and S.W. Scotland and southwards to Kent and Cornwall, and usually on roadsides, banks of railways and canals, waste ground, etc., where it usually has thin green lvs and many heads and is likely to be introduced. In Wales the bluish-green rigid-leaved variant grows on cliff-ledges and appears native. **H. diaphanum** Fries (including *H. anglorum* (A. Ley) Pugsley and *H. daedalolepioides* (Zahn) Roffey).

H. diaphanoides Lindeb., which differs in its bright green lvs and narrow velvety-black involucre, grows on cliff-ledges and by rocky streams chiefly in N. England and C. and S.W. Scotland.

Section 8. *Oreadea* Fries, emend. (Fig. 57C)

Plants with firm, commonly glaucous and sometimes purple-spotted lvs; rosette-lvs varying from cuneate to truncate at base, and usually 0–1 stem-lf but occasionally 2 or more; all usually ±hairy, the hairs often, though not invariably, long and bristly, especially along the margins, in species with cuneate-based lvs but softer, shorter and often fewer in those with rounded- to truncate-based lvs (which probably arose through hybridization with members of section *Vulgata*). Stem forking above into a corymb of rather large heads. Involucral bracts with stout black-based simple and some glandular hairs, not densely floccose. Ligules usually glabrous-tipped. Styles usually yellow. Achenes blackish. Mainly on limestone and other basic rocks.

There are more than 40 British spp., many of very restricted distribution. The following key includes only a few of the more widespread or abundant spp.

1 Rosette-lvs linear-lanceolate to oblong or oval, glaucous, not spotted, cuneate or less commonly rounded at base; styles yellow.   2
  Rosette-lvs oval or elliptic to ovate-lanceolate, green or glaucous, often spotted, with base rounded, subcordate or truncate; styles yellow or darkish.   4
2 Inner rosette-lvs narrow, lanceolate or linear-lanceolate, subentire or sinuate-toothed, narrowed into a short winged petiole; stem-lvs usually 2, spreading or bract-like; all ±glabrous or with long and scarcely bristly hairs; heads usually 2–4. Stems 15–40 cm, slender. Common on rocky ledges, streamsides and grassy banks, from sea-level to 650 m, in Wales, N. England and much of Scotland and the Scottish islands.   **H. argenteum** Fries
  Inner rosette-lvs broader, oblong or oval to lanceolate, subentire or toothed below the middle; stem-lf 0–1; all with some bristly hairs; heads 1–6.   3
3 Inner rosette-lvs oblong, blunt and rounded at apex and usually rounded at base; stem-lf 0(–1); all with long bristly hairs above, more softly hairy beneath and densely fringed with long hairs. Small slender plants, usually 10–20 cm, with very glaucous lvs and medium-sized heads, involucral bracts erect in bud, grey-green, floccose and with many dark glandular and dark-based simple hairs. Scattered through Wales, Lake District and Scotland; N. Ireland.   **H. lasiophyllum** Koch
  Inner rosette-lvs oval, elliptic or lanceolate, blunt to acute at apex, gradually or more rapidly narrowed into long petioles; stem-lf usually 1, narrow and sessile or bract-like; all bristly on margins, less so on surfaces. Stem 15–40 cm, slender, with glaucous lvs and rather large heads, the involucral bracts erect in bud, dark green, sparingly floccose with many black-based simple and fine glandular hairs. Scattered through western Britain from Devon to Sutherland and in E. Highlands.   **H. schmidtii** Tausch
4 Stem-lvs 1–3, usually 2; involucral bracts rather broad, blunt or subacute, their tips incurved in bud. Lvs dull or yellowish-green, often purplish beneath; inner rosette-lvs ovate-lanceolate to oblong-elliptic, distantly toothed, ±abruptly narrowed into the petiole; lowest stem-lf often large; all with soft white hairs on both sides and on margins and petioles. Heads 1–4(–6); styles brownish-yellow. From Wales and the Peak District to Outer Hebrides and Shetland; N. and E. Ireland.   **H. caledonicum** F. J. Hanb.
  Stem-lvs 0–1; involucral bracts with tips erect in bud.   5
5 Inner rosette-lvs broadly oval to elliptic, pale green, spotted or marbled with purple-brown, subcordate at base. Stem 20–40 cm, ±hairy. Lvs subentire or with a few teeth near the base, almost glabrous above, sparsely hairy beneath and with bristly margins and shaggy petioles. Heads 2–4 or more, with long glandular-hairy peduncles; involucral bracts dark green, broad and blunt, floccose and with many short dark glandular and longer simple hairs; styles yellowish. On limestone in S. and N. Wales, N.W. England, Antrim and Co. Clare. A very beautiful hawkweed.   **H. hypochaeroides** Gibson
  Inner rosette-lvs often glaucous, sometimes spotted, oval to ovate-lanceolate, ±truncate at base and usually with some sharp and often coarse and spreading teeth near the base which may thus become somewhat sagittate. Heads usually 2–7.   **H. britannicum** group

This group comprises several species, mostly very local, which are not easily distinguishable. The most widespread is **H. dicella** P. D. Sell & C. West,

Fig. 58. *Hieracium*. A, *Cerinthoidea* (*H. iricum*); B, *Subalpina* (*H. gracilifolium*); C, D, *Alpina* (C, *H. hanburyi*; D, *H. holosericeum*). ×⅕.

with dark green red-tinged lvs, the inner rosette-lvs ovate to ovate-cordate, their petioles shaggy with long and dense brownish-red hairs; lvs shortly toothed below, almost glabrous above but densely hairy on the midrib beneath. Involucral bracts narrow, the inner with a long acute tip, floccose especially on the margins and with dark simple and a few glandular hairs. Chiefly on limestone in Wales, Peak District, N. England and scattered through Scotland. **H. britannicum** F. J. Hanb. sens. str., confined to the Peak District and more abundant there than *H. dicella*, has slightly glaucous and rarely spotted lvs, mostly broadly oval to elliptic-lanceolate, with sharp teeth increasing in size and depth towards the base. Other local species have more definitely glaucous lvs, and 2 of them, **H. stenolepiforme** (Pugsley) P. D. Sell & C. West (Cheddar Gorge only) and **H. britanniciforme** Pugsley (on and near Great Orme's Head and in Kintyre) have numerous glandular hairs on the involucre, the latter also having bristly-hairy and usually purple-spotted lvs.

Section 9. *Amplexicaulia* Fries (Fig. 57D)

Whole plant covered with viscid glandular hairs. Rosette-lvs numerous, narrowed at base; stem-lvs 3–6, often semi-amplexicaul; all pale bluish- or yellowish-green. Heads large; involucral bracts with tips incurved in bud, densely yellow-glandular but with no simple hairs. Ligules densely hairy- or ciliate-tipped. Achenes blackish.

No native species but 3 are naturalized, mainly on old walls, in a few scattered localities. ***H. amplexicaule** L., still on old walls in Oxford, has dense glandular but no simple hairs all over stems and leaves, and its styles are yellow. The other 2 have simple as well as glandular hairs and dark styles. ***H. speluncarum** Arvet-Touvet has pale green, often somewhat glaucous, lvs and involucre with blackish glandular hairs, and its localities range from S. England to C. Scotland.

Section 10. *Cerinthoidea* Fries (Fig. 58A)

Usually robust non-glandular plants, commonly 15–40 cm, with ±glaucous lvs, the stem-lvs usually semi-amplexicaul. Heads large, usually with pale yellow florets. Involucral bracts with tips incurved in bud, densely clothed with simple hairs and sometimes also with glandular hairs. Styles usually dark. Achenes reddish-black to black.

A group of 10 distinctive and closely allied species of cliff-ledges and rocky streamsides in coastal and upland areas of N. England, Scotland, the Scottish islands, and Ireland. The 3 included in the following key are the commonest and most widespread, all being spp. found in N. England, C. and N. Scotland, Inner Hebrides and Ireland, with *H. iricum* and *H. anglicum* also in the Outer Hebrides and Orkney.

1 Stem-lvs 0–2; ligules pale yellow.   2
  Stem-lvs 3–6(–9); ligules bright yellow. Stem usually 30–40 cm, hairy throughout with long waved hairs and ±floccose above. Rosette-lvs firm, elliptic to oblong-lanceolate, acute, narrowed into winged shaggy petioles; stem-lvs smaller upwards, lowest shortly petiolate. others semi-amplexicaul; all long-hairy, especially beneath. Styles dark.

                            H. iricum Fries

2 Lvs ±cuneate-based; peduncles long, arching; involucre with dense long simple hairs. Stem usually 20–30 cm; rosette-lvs thinnish, the inner large, ovate to ovate-lanceolate, toothed towards base or subentire, narrowed below into rather long petioles; stem-lvs usually 2; all with long waved simple hairs on both sides and on margins and petioles. Heads 1–4, large and handsome; involucre densely clothed with long black-based simple hairs with or without a few glandular hairs. Ligules lemon-yellow. Styles dark.   **H. anglicum** Fries
  Lvs abruptly narrowed below; peduncles shorter than in *H. anglicum*; involucre with shorter simple hairs. Stem 20–50 cm; inner rosette-lvs oval or oblong, often with a few coarse teeth near the subtruncate base; stem-lf usually 0–1, petiolate, sometimes bract-like; all hairy above and beneath, on margins and on petioles. Heads 2–10, large. Styles dark. Abundant on basic rocks in N.W. England and scattered throughout W. Scotland and N.E. Ireland.

                 H. ampliatum (W. R. Linton) A. Ley

Section 11. *Alpina* F. N. Williams (Fig. 58C, D) and **Section 12.** *Subalpina* Pugsley (Fig. 58B)

These two sections have it in common that they are plants of upland areas, usually at more than 450 m above sea-level, that all or most of their lvs are in a basal rosette, and that the lvs have at least a few stalked glands on their margins. The *Subalpina* probably arose from crosses between *Alpina* and *Vulgata* or *Cerinthoidea*.

The *Alpina* are small plants, usually 5–30 cm, of altitudes above 600 m. Their rosette-lvs are typically numerous, the inner obovate-lanceolate to linear-oblong, cuneate to subtruncate at base; stem-lvs 0–3, small, non-amplexicaul; all with simple and often stiff hairs and fewer very fine yellow glandular hairs. Heads fairly large, usually solitary, sometimes 2–5. Involucral bracts with tips incurved in bud, densely covered with long whitish simple hairs and usually a few glandular hairs, not floccose. Ligules hairy-tipped. Styles yellow or dark. Achenes blackish.

Most species are restricted to Scotland and local there, but one, **H. holosericeum** Backhouse, is found in the mountains of N. Wales, the Lake District, S. Scotland and in many localities in C. and N. Scotland. It is a very distinctive plant, usually 5–15 cm, covered throughout with long spreading whitish dark-based hairs with some fine glandular hairs above. The rosette-lvs

are light green, narrowly linear-oblong, subentire, narrowed into a long petiole; stem-lvs usually 0–2, narrow, sessile; all covered with long stiff silky simple hairs. Heads solitary, smallish; involucre very densely covered with very long silky, whitish, dark-based simple hairs. Styles yellow.

Also in the Lake District, but nowhere else, is **H. subgracilentipes** (Zahn) Roffey, 12–35 cm, differing in its broader elliptic-lanceolate lvs, dusky-hairy involucre and dark styles. Of the several species confined to Scotland **H. eximium** Backhouse is locally abundant in the mountains of C. Scotland, differing from *H. holosericeum* in its longer, broader and more toothed oblong to lanceolate lvs and long winged petioles, its 1–3 larger heads and the usually dark styles, and in general in being less densely silky-haired. **H. calenduliflorum** Backhouse, also restricted to the C. Highlands, has obovate to lanceolate distinctly toothed lvs and strikingly large heads, often 5–7 cm across, with very dark styles.

H. hanburyi Pugsley has irregularly and deeply toothed lvs and large usually solitary heads with velvety black glandular involucres lacking the long silky or shaggy hairs of most of the *Alpina*. It is frequent in high glens of C. and N. Scotland.

The *Subalpina* comprise taller plants with lvs having fewer glandular hairs than in *Alpina* and normally with 2–5 medium-sized to large heads. Involucral bracts usually incurved in bud. Ligules usually hairy-tipped. Styles yellow to dark.

Of the 30 native spp., 25 are restricted to the mountains of Scotland. The most widely distributed of these is **H. lingulatum** Backhouse, commonly 25–40 cm, with inner rosette-lvs to 20 cm long, oblong to lingulate, narrowed into short petioles; stem-lvs 2–3; all stiffly hairy on both sides. Heads 2–4, with numerous dusky simple hairs and a few glandular ones. Not uncommon on rock-ledges and rocky streamsides in the higher glens of C. and N.W. Scotland. **H. senescens** Backhouse has toothed lvs and strongly glandular involucres; also widespread in Scotland. **H. gracilifolium** (F. J. Hanb.) Pugsley, common in C. Scotland, has the involucral bracts ±densely glandular but lacking simple hairs and very sparsely floccose.

Subgenus PILOSELLA (Hill) S. F. Gray

Section 13. *Pilosella* Fries

Stoloniferous herbs with rosette-lvs linear-oblong to elliptic, entire or nearly so, white or grey with dense stellate hairs beneath, sometimes also above but usually with scattered simple hairs above; stem-lvs 0 or 1–2, very small and bract-like. Head solitary. Ligules yellow, the outer red-striped beneath. Styles yellow.

H. pilosella L., Mouse-ear Hawkweed, has a long slender rhizome and numerous long epigeal stolons bearing small distant lvs and ending in overwintering rosettes. Flowering stem 5–30 cm, floccose especially above, with some black-based simple and glandular hairs. Rosette-lvs linear- to obovate-oblong, blunt, narrowed into a long petiole-like base, white-felted beneath and with sparse long stiff hairs on both sides; stem-lvs usually 0. Involucre of the solitary head floccose and with dense glandular black-based and simple hairs. Ligules pale yellow. Variable. Grassy pastures and heaths, banks, rocks, walls, etc. Locally common throughout the British Is. except Shetland and parts of the Fenland.

H. peleteranum Mérat, in a few scattered localities in S. and C. England, Wales and the Channel Is., differs in its short stolons with large crowded lvs, and its ±acute lanceolate to elliptical rosette-lvs.

Section 14. *Pratensina* Zahn

Stoloniferous herbs with densely long-hairy stems and green lanceolate to elliptical rosette-lvs, not rigid, stiffly hairy on both sides; stem-lvs (0–)1–3, like the rosette-lvs or bract-like; lvs with or without stellate hairs beneath. Heads (1–)2–30 or more, smallish in a compact and sometimes subumbellate corymb; involucre ±floccose and with often long simple and numerous shorter glandular hairs. Ligules yellow to brownish or deep red. Styles yellow or dark. Of the 4 spp. occurring in the British Is., only one is rather doubtfully native, the others certainly introduced.

1 Ligules orange-brown to brick-red, turning purplish when dry.   2
  Ligules yellow, the outer often red-streaked beneath.   3
2 Ligules orange-brown; stolons long and lfy, mostly epigeal; lvs oblong to oblanceolate. Widely naturalized on roadsides, railway banks, walls, grassy and waste places and an aggressive garden weed throughout England, Wales and much of Scotland.
                      *H. brunneocroceum Pugsley
  Ligules brick-red; stolons rather short and mostly hypogeal; lvs obovate to elliptical. In a few scattered localities throughout Great Britain, but much less common than *H. brunneocroceum*.
                          *H. aurantiacum L.
  (In *Flora Europaea* both are treated as subspp. of *H. aurantiacum*, brunneocroceum as subsp. *carpathicola* Naegeli & Peter. Intermediates occur not infrequently.)
3 Heads 2–6 in a ±racemose panicle with long peduncles; styles yellow. Plants with green lvs and epigeal large-lvd stolons. On roadsides and railway banks, etc., in a few places in S. and C. England and near the Forth Estuary, and almost certainly introduced, but a slight variant in one locality in Shetland may be native.   **H. flagellare** Willd.
  Heads usually 15–30 in a compact umbel-like corymb; styles dark. Plants with slightly glaucous lvs and epigeal stolons. In a few scattered localities, mainly on railway banks and old walls, in England and Scotland with one in Ireland.
    *H. caespitosum Dumort. subsp. **colliniforme**
(Peter) P. D. Sell (*H. pratense* Tausch; *H. colliniforme*   (Naegeli & Peter) Roffey)

Section 15. *Praealtina* Zahn

Tall herbs, stoloniferous or not. Rosette-lvs linear to lanceolate, glaucous, rigid; stem-lvs usually 2–3. Infl.

corymbose, compact, of 10–30 or more small heads with involucre 6–8 mm, dark green, floccose and ±densely hairy and glandular; simple hairs sometimes 0. Ligules yellow. Styles yellow.

The 3 spp., 2 with and 1 without stolons, that have been recorded for the British Is., mainly on railway-banks in S. England where they are certainly introduced, are now treated as subspp. of ***H. praealtum** Vill. ex Gochnat. All are very rare.

58. Crepis L.

Usually herbaceous plants with alternate, entire to pinnatisect often runcinate lvs and branching lfy flowering stems. Heads in panicles or corymbs, rarely solitary; involucral bracts many, in 2 rows, the outer shorter than the inner; receptacle flat, naked, pitted, the pits with toothed or hairy margins. Florets usually yellow, sometimes pink or white; anther-lobes without basal tails; style-arms slender, hairy. *Achenes* ±cylindrical, ribbed, *narrowed* above, with or without a beak; *pappus of simple hairs in 1–many rows, usually white*, sometimes brownish; marginal achenes sometimes without pappus.

About 200 spp. chiefly in temperate and subtropical Europe and Asia, some in N. and C. Africa and North America, and 1 each in S. Africa and South America.

Close to *Hieracium*, but in that genus the achenes are not or scarcely narrowed to the truncate apex and the pappus is brownish and brittle. *C. paludosa*, however, combines features of both genera, resembling *Crepis* in habit and in the distinct though slight narrowing of the achene towards its apex, but having the discoloured brittle pappus of *Hieracium*.

1 Perennial herbs with short oblique to erect fibrous-rooted stock, entire to sinuate- or coarsely runcinate-dentate (but not pinnatifid) lvs and unbeaked achenes; plants of moist upland habitats.   2
  Annual to biennial (rarely perennial) herbs with taproots, usually with some pinnatifid lvs and with mature achenes beaked or not; commonly plants of lowland pastures, cultivated ground, walls, waste places, etc.   3
2 Stem-lvs oblong-spathulate with entire or minutely sinuate-toothed margins; involucral bracts sparsely pubescent and glandular-hairy; mature achenes reddish-brown, distinctly narrowed upwards; pappus soft, white.   **8. mollis**
  Stem-lvs ovate to lanceolate, their margins with well-developed often runcinate teeth; involucral bracts with long black bristles and shorter glandular hairs; mature achenes straw-coloured, very slightly narrowed to the truncate top; pappus brittle, brownish.   **9. paludosa**
3 Involucral bracts and upper infl.-branches ±prickly with yellow non-glandular bristles; mature achenes beaked.   **3. setosa**
  Involucre not prickly with yellow bristles; mature achenes prickly or not.   4
4 Involucral bracts with their inner faces hairy, their scarious margins distinctly pale-coloured; outermost bracts short, narrow, ±spreading.   5
  Involucral bracts with their inner faces glabrous, their scarious margins not conspicuously pale-coloured;

outermost bracts remaining appressed or slightly
  spreading.   7
5 Mature achenes beaked.
                                2. vesicaria subsp. **haenseleri**
  Mature achenes somewhat narrowed upwards but not
  beaked.   6
6 Stem-lvs with down-rolled margins; mature achenes
  dark purplish-brown, 10-ribbed.   **6. tectorum**
  Stem-lvs with flat margins; mature achenes yellowish-
  to cinnamon-brown, usually with 13 or more ribs.
                                             4. biennis
7 Fl.-heads commonly drooping in bud; involucre
  pubescent, sometimes also with short glandular
  hairs; mature central achenes long-beaked, marginal very short-beaked and closely clasped by involucral bracts; pappus persistent.   **1. foetida**
  Fl.-heads erect in bud; involucre not pubescent but
  sometimes with blackish glandular hairs or bristles;
  mature achenes all without beak and the marginal
  not clasped by involucral bracts; pappus deciduous.   8
8 Stem ±glabrous below; lvs smooth, ±glabrous; outer-
  most involucral bracts mostly remaining appressed;
  receptacle usually with few short cilia round pits;
  achenes 1.5–2.5 mm, palish brown.   **5. capillaris**
  Stem shortly hispid-pubescent below; lvs scabrid and
  ±hispid; outermost involucral bracts lax and
  slightly spreading; receptacle with membranous
  fringes round pits; achenes 2.5–4 mm, golden-
  brown.   **7. nicaeensis**

Section 1. *Barkhausia* (Moench) Gaudin
Annual to biennial herbs usually with some pinnatifid
lvs; achenes at least of central fls prolonged upwards
into a distinct beak.

1. C. foetida L.   Stinking Hawk's-beard.

An annual or biennial *foetid* herb with erect main stem
20–60 cm, and usually several spreading-ascending
corymbose branches from its base and lower half, all
slightly furrowed, hairy. Lvs chiefly in a basal rosette,
runcinate-pinnatifid, petiolate; middle and upper stem-
lvs few, small, lanceolate, cut or toothed below, sessile,
semi-amplexicaul; all densely hairy on both sides. *Heads*
1.5–2 cm diam., *drooping in bud*, solitary, terminal on
long incurved bracteolate stalks somewhat thickened
upwards. Involucre 10–13 mm, grey-downy and usually
with some longer glandular hairs; *inner bracts* linear-
lanceolate, *later hardening and enclosing the outer
achenes*, finally spreading starwise; outer bracts short,
narrow, ±spreading. Florets pale yellow, exceeding the
bracts, the outermost purplish beneath. Styles yellow.
Achenes 3–4 mm (6–15 mm including the beak); *marginal short-beaked, central long-beaked*; pappus
6–8 mm, white. Fl. 6–8. 2n = 10. Th.–H. (biennial).
  Native. Waysides and waste places, especially on
chalk and shingle. A rare plant of S.E. England from
Sussex and Kent to Cambridge and Norfolk; probably
introduced in Worcester, Hereford, etc. W. and S.
Europe from Spain and Portugal to Belgium, Germany,
and Yugoslavia. The closely related *C. rhoeadifolia*
Bieb. occurs in E. Europe and the Near East.

*2. C. vesicaria L. subsp. **haenseleri** (Boiss. ex DC.)
P.D. Sell   Beaked Hawk's-beard.

C. taraxacifolia Thuill.

A biennial herb, sometimes annual or perennial, with
1 or more erect ±downy stems 15–80 cm, hispid and
purplish below, branched above or from near the base.
Basal lvs petiolate, usually oblanceolate, blunt or acute,
lyrate- or runcinate-pinnatifid with the lobes very variable in length and width, sometimes merely toothed;
stem-lvs shortly petiolate or sessile, pinnatifid to
±entire, the middle ones amplexicaul; all finely pubescent on both sides. *Heads* 1.5–2.5 cm diam., *erect in
bud*, on slender, ±straight, non-bracteolate stalks in
corymbs terminating the main stems and branches.
Involucre 8–12 mm, cylindric-campanulate, tomentose
and often glandular; *inner bracts* lanceolate blunt, hardening in fr. but *not enclosing the marginal achenes*;
outer bracts narrower, *spreading*, scarious-margined.
Florets yellow, the outermost brown-striped beneath,
exceeding the involucre. Styles greenish-brown.
Achenes 4–5 mm, pale brown, with 10 narrow ±rough
ribs, *all narrowed when mature into a beak about as
long as the achene*; pappus 4–6 mm, white, soft, exceeding the involucre. Fl. 5–7. 2n = 8*, 16. Th.–H.
(biennial).
  Introduced. Waysides, walls, railway banks, waste
places, especially on calcareous soils, to 150 m in England. First recorded in 1713 and now locally common
in England and Wales northwards to Durham and Lancs
and in C. and S. Ireland; rapidly spreading. W. and
S. Europe from Spain and Portugal to N.W. Germany
and the Balkan peninsula; N.W. Africa. Part of a large
complex of forms placed under *C. vesicaria* L.

*3. C. setosa Haller fil.   Bristly Hawk's-beard.

Annual or biennial herb with erect usually hispid stem
20–70 cm, branched from near the base or throughout,
the branches long, spreading-ascending. Basal lvs petiolate, oblanceolate, blunt or acute, runcinate-pinnatifid
with lobes very variable in size, or merely toothed; stem-
lvs lanceolate-acuminate, entire or toothed or cut near
the base, amplexicaul with pointed auricles; all hispid.
Heads 1–1.4 cm diam., *erect in bud*, on slender deeply
grooved, usually hispid or prickly stalks in few-headed
corymbs terminating the main stem and branches. *Involucre* 8–10 mm, cylindrical-campanulate, *contracted
above in fr.*, ±*prickly with yellow non-glandular bristles*;
inner bracts lanceolate-acuminate, later hardening but
not enclosing the outer achenes; outer bracts linear,
keeled, up to half as long as the inner, spreading. Florets
pale yellow, reddish beneath, exceeding the involucre.
Styles blackish-green. *Achenes* 3.5–5 mm, pale brown,
with 10 prominent ribs rough above, *all narrowed into
a slender beak* up to half as long as the achene; *pappus*
2.5–5 mm, white, soft, *slightly exceeding the involucre*.
Fl. 7–9. 2n = 8. Th.–H. (biennial).
  Introduced. A casual in arable fields, especially of
clover and newly sown grass; local northwards to Cum-

berland. S. and S.E. Europe from S. France to the Crimea; W. Asia. Introduced northwards to Denmark and C. Russia.

Section 2. *Crepis*

Annual to biennial herbs usually with some pinnatifid lvs; achenes not or very shortly beaked; pappus white, soft.

4. C. biennis L.   Rough Hawk's-beard.

A biennial herb with an erect stout grooved ±hispid stem 30–120 cm, often purplish below, corymbosely branched above. Basal lvs 15–30 cm, irregularly lyrate-pinnatifid, runcinate, petiolate; *stem-lvs* mostly similar but smaller and sessile, *semi-amplexicaul but not or scarcely sagittate* at the base; all ±rough with scattered hairs. Heads 2–3.5 cm diam., at first crowded, then with elongation of the stout hairy stalks forming corymbs terminating the main stem and branches. Involucre 10–13 mm, cylindrical-campanulate, pubescent and often glandular; *inner bracts* linear-lanceolate, blunt, *downy within*; *outer* narrow, *spreading*, without scarious margins. Florets golden yellow, not reddish beneath, exceeding the involucre. Styles yellow. *Achenes* 7–12 mm, narrowed above but not beaked, reddish-brown, usually with 13 ±smooth ribs and numerous fine transverse wrinkles; *pappus equalling or slightly exceeding the involucre*, white, soft. Fl. 6–7. Visited freely, chiefly by bees and hover-flies, but probably often apomictic. 2n = 40. H. (biennial).

Probably native. Pastures, waysides, clover and lucerne fields, waste places. Locally frequent in lowland Great Britain northwards to E. Lothian, often on calcareous soils; decreasing. Introduced in Ireland. C. Europe from C. Spain, N. Italy, C. Balkan peninsula and C. Russia northwards to Denmark, S. Scandinavia and Estonia.

5. C. capillaris (L.) Wallr.   Smooth Hawks-beard.
C. virens L.

A usually annual herb with one or more erect or ascending stems 20–90 cm, glabrous or ±hairy below, branched above, or with several spreading-ascending branches from the base upwards. Basal and lower stem-lvs 5–25 cm; very variable, usually oblanceolate or lanceolate, narrowed into a stalk-like base, lyrate- or runcinate-pinnatifid with distant often toothed lobes, or merely toothed; middle and upper *stem-lvs* lanceolate, acute, sessile, amplexicaul with a *sagittate base*; all glabrous or somewhat hairy on one or both sides. Fl. heads 1–1.3(–2.5) cm diam., erect in bud, on slender glabrous or hairy stalks in lax terminal corymbs. Involucre 5–8 mm, cylindrical, contracted above in fr., usually shortly downy and often with black glandular bristles; *inner bracts* linear-lanceolate, acute, *glabrous within*; *outer bracts* ⅓ as long, *appressed*. Receptacle with shortly and sparsely ciliate pit-borders. *Florets* exceeding the involucre, bright yellow, the outermost often *reddish beneath*. Styles yellow. *Achenes* 1.5–2.5 mm, pale

brown, with 10 ±smooth ribs; pappus equalling or slightly exceeding the involucre, snow-white, soft. Fl. 6–9. Freely visited by flies and bees. 2n = 6. Th.–H. (biennial).

Very variable in stature and robustness, degree of glandulosity of peduncles and involucral bracts, etc., but the variation seems ±continuous and no satisfactory infra- taxonomic taxa have been recognized here.

Native. Grassland, heaths, walls, waste places, etc., to 445 m. Common throughout the British Is. Most of Europe northwards to Denmark and S. Sweden; Canary Is. Introduced in North America.

*6. C. tectorum L.

An annual or overwintering herb with slender tap-root and erect fl.-stems to 65 cm or more, ±pubescent above. *Lvs of basal rosette* to 15 × 4 cm, lanceolate to oblanceolate in outline, acute, narrowed below into a petiole-like base, *varying from sinuate-denticulate to runcinate-pinnatifid* or 1–2-pinnatisect with distant narrow acute lobes; stem-lvs smaller, narrower and less divided upwards, the *upper ±linear and entire with sessile sagittate base and down-rolled margins*; all lvs glabrous to sparsely tomentose, sometimes glandular. Fl.-heads usually numerous; *involucral bracts* linear-lanceolate, 7–10 mm, ±tomentose on outer and *appressed-hairy on inner face*, the *outermost* short, lax, *spreading*. Ligules 12–15 mm, yellow. Mature achenes rugulose and somewhat spinulose near apex, 10-ribbed; pappus slightly longer than achene. 2n = 8. Th.

Introduced. First recorded in S. Scotland in 1872 and now a not infrequent casual, perhaps introduced in grass-seed, which has been reported recently from a number of localities in England. Native in much of Europe except the Mediterranean region.

Section 3. *Catonia* (Moench) Bentham & Hooker fil.

Perennial herbs with lvs entire to coarsely dentate but not pinnatifid; achenes not beaked; pappus of soft white or sometimes brittle brownish hairs.

*7. C. nicaeensis Balbis   French Hawk's-beard.

A usually biennial herb with an erect strongly ribbed stem, hispid-pubescent at least below and often reddish below, branched above. Basal and lower stem-lvs oblanceolate, blunt, shortly petiolate, runcinate-pinnatifid; middle and upper *lvs* lanceolate, ±deeply toothed below, sessile, amplexicaul with a *sagittate base*; uppermost narrow, entire; all roughly hairy on both sides. Heads 2.5 cm diam., erect in bud, on slender ±glandular-pubescent stalks in a terminal corymb. Involucre 8–10 mm, campanulate, ±glandular-pubescent and grey-tomentose; *inner bracts* lanceolate, blunt at the ciliate tip, *glabrous within*, strongly keeled in fr.; *outer bracts* about half as long, ±spreading. *Receptacle with elevated membranous fimbriate-ciliate pit-borders*. Florets yellow often red-tipped, exceeding the involucre. Styles greenish-brown. Achenes 3.5–4.5 mm, yellowish-brown, much narrowed above, with 10 smooth or ±rough ribs; pappus 4–5 mm, white, hardly exceeding the involucre. Fl. 6–7. 2n = 8. H. (biennial).

Introduced. Arable fields and leys. A casual introduced as a seed-impurity. S. Europe from S. France to N. Balkan peninsula; Caucasus. Introduced in C. Europe and North America.

8. C. mollis (Jacq.) Ascherson   Soft Hawk's-beard.

C. succisifolia (All.) Tausch; *C. hieracioides* Waldst. & Kit., non Lam.

A perennial herb with a short erect blackish premorse stock and an erect glabrous or hairy stem 30–60 cm, branched above. Basal lvs 5–10 cm, oblanceolate narrowed into a long winged petiole, blunt; middle and upper lvs oblong, sessile, *semi-amplexicaul with rounded auricles*; all sinuate-toothed or entire, glabrous or hairy. Heads 2–3 cm diam., in a terminal corymb, erect in bud, on rather long, slightly incurved, slender stalks. *Involucre* 8–13 mm, subcylindric, ±sparsely *glandular-hairy*; inner bracts linear-lanceolate, acute, outer very short, appressed. Florets yellow, almost twice as long as the involucre. Achenes 4.5 mm, yellow, with 20 smooth ribs; pappus 6 mm, pure white, soft. Fl. 7–8. Visited by bees. 2*n* = 12. H.

Native. Streamsides and woodlands in mountainous districts, very local; reaches 370 m in N. England. N. Wales and N. Britain from Yorks and N. Lancs to Aberdeen. C. Europe from the Pyrenees, N. Italy and S. Russia northwards to Germany, Poland and C. Russia.

9. C. paludosa (L.) Moench   Marsh Hawk's-beard.

Hieracium paludosum L.; *Aracium paludosum* (L.) Monnier.

A perennial herb with an oblique blackish premorse stock and an erect striate glabrous stem 30–90 cm, often reddish below, branched only above. Basal and lower stem-lvs obovate-lanceolate narrowed into a short winged petiole; *middle lvs panduriform-acuminate, sessile, amplexicaul*; upper lvs ovate- to lanceolate-acuminate, amplexicaul with long-pointed auricles; all thin, sinuate- or runcinate-toothed, glabrous. Heads 15–25 mm diam., few, on ±straight slender stalks in a terminal corymb. *Involucre* 8–12 mm, campanulate, *woolly and with many black glandular hairs*; inner bracts linear-lanceolate, acute, blackish-green; outer bracts hardly half as long, ±appressed. Florets yellow. Styles greenish-black. *Achenes* 4–5 mm, brownish, cylindrical, *hardly narrowed at the ends*, 10-ribbed, smooth; pappus about as long as the achene, of *stiff brittle brownish hairs*. Fl. 7–9. Visited by bees and flies. 2*n* = 12*. Hel.

Native. Streamsides, wet copses, wet meadows and fens, to 915 m in Scotland. Locally common in N. Britain to Caithness and extending southwards to Glamorgan and Worcester; throughout Ireland. Europe from Spain, C. France, N. Italy and N. Balkan peninsula northwards to 70° 30′ N. in Norway; Russia; W. Siberia.

C. rubra L. (Greece), a branching annual 15–45 cm, with glabrous runcinate-toothed lvs and red or whitish heads with a hispid involucre, is grown in gardens.

59. TARAXACUM Weber   Dandelion.

Perennial herbs with simple or branched tap-root and alternate entire to pinnate-lobed, commonly runcinate lvs, confined to a basal rosette. Scape unbranched with a solitary terminal head. Inner involucral bracts erect, equal; outer shorter, often spreading or reflexed. Receptacular scales 0. Fls yellow, the ligules often dark-striped on the outer side. Achenes narrowly obovoid, less commonly cylindrical, ribbed, the ribs usually muricate above; prolonged above first into a conical cusp or short narrow cylinder ('cone') and then into a long beak surmounted by the pappus of many rows of simple rough white hairs.

Most British dandelions are apomictic (fruit being formed independently of fertilization) and, as in such other largely apomictic genera as *Hieracium* and *Rubus*, considerable problems arise for the taxonomist. The following key and notes deal only with the *Sections* represented in the British Is. and with the most frequently encountered *groups* of microspp. Those wishing to carry identification to the level of individual microspp. (of which 132 had been recorded by 1972) should consult *The Taraxacum Flora of the British Isles* by A. J. Richards (*Watsonia*, **9**, Suppl., 1972), on which this account is largely based.

1 Outer involucral bracts up to 7 mm, appressed (rarely erect), with a wide pale or scarious margin amounting to at least ¼ of the total area of the bract; lvs often linear, ±unlobed. Rare plants of wet places.
    Section B. *Palustria*
  Outer bracts with at most a narrow pale or scarious margin, its area less than ¼ of the total area of the bract; lvs never linear.   2
2 Outer involucral bracts with a small appendage on the outer side near the tip, which thus appears double. Small delicate plants of dry places, with strongly dissected lvs.   3
  Outer involucral bracts not appendaged. Robust plants.   4
3 Achenes grey-brown with 'cone' shortly conical. Plants of sand-dunes in Scotland, very rare elsewhere.   Section C. *Obliqua*
  Achenes usually with a reddish or purplish tint and with the 'cone' cylindrical.
    Section D. *Erythrosperma*
4 Lvs often dark-spotted all over and often with red petiole and midrib; outer involucral bracts (5–)7–9 (–12) mm, ovate-lanceolate to lanceolate, appressed to spreading; ligules often striped carmine or purple. Plants of wet places, especially in upland Britain.   Section A. *Spectabilia*
  Lvs sometimes with localized heavy blotching, especially between lobes, but in only one microspecies of doubtful status are there spots all over and petioles and midribs are very rarely red; outer bracts (7–)9–15 mm lanceolate to linear, erect to recurved; ligules usually striped grey-violet, sometimes purple, never carmine.
    Section E. *Taraxacum* (*Vulgaria*)

Section A. *Spectabilia* Dahlst.

Moderately robust plants with usually shallow-lobed, dull-green and hairy lvs, often dark-spotted all over; petioles and midribs often red or purple. Heads 3.5–5.5 cm diam.; outer involucral bracts not double-tipped, with a narrow pale margin. British representatives have been placed in the following 5 groups of microspp.:

1 Achenes reddish; lvs unspotted; outer involucral bracts spreading. Two microspecies, chiefly in N. and W. Britain, the more widespread being very distinctive in the curving of the lf-lobes towards the lf-base and the pink-tipped and white-bordered outer bracts. *T. unguilobum* group

Achenes very rarely reddish. 2

2 Outer involucral bracts erect, appressed or not. 3

Outer bracts spreading. 4

3 Body of achene 4 mm or longer, straw-coloured or brown; lvs often spotted; ligules striped carmine. Common plants especially of hilly districts but some also in the south. *T. spectabile* group

Body of achene not more than 4 mm, brown; lvs rarely spotted; ligules striped grey, red or purple. Includes the only members of the Section common in the southern half of Britian. *T. adamii* group

4 Lvs commonly spotted; outer involucral bracts ovate-lanceolate; 'cone' of achene conical; ligules rather pale yellow with grey or brown striping. Wet grassland, especially calcareous, stream banks, wet cliffs, waysides, etc., especially in the north. *T. praestans* group

Lvs usually unspotted; outer bracts lanceolate; 'cone' of achene cylindrical; ligules deep or orange yellow with grey or brown striping. Wet ledges of calcareous or base-rich but acidic rocks, chiefly in Scotland. *T. croceum* group

Section B. *Palustria* Dahlst.

Distinctive because of the very broad pale or scarious border to the usually appressed ovate outer bracts and the linear to linear-lanceolate often unlobed lvs. Represented in the British Is. only by 3 microspp. of the *T. palustre* group, all now rare and local through drainage of the wet fens and seasonally flooded hay-meadows in which they formerly grew, and therefore in serious danger of extinction.

Section C. *Obliqua* Dahlst.

Comprises only the 2 microspp. of the *T. obliquum* group, both native in the British Is., chiefly on grey dunes and in dune-slacks and machair in Scotland. Their lvs are highly dissected, never spotted; the outer involucral bracts are ovate-lanceolate, bordered, double-tipped, erect or appressed; the ligules are striped red, and the achenes are c. 3 mm, grey-brown, with the 'cone' shortly conical.

Section D. *Erythrosperma* Dahlst.

These are our most frequently encountered dandelions apart from members of the next Section. They are typically small plants with strongly dissected narrow-lobed unspotted lvs, have small heads, 1.5–3.5 cm diam., with small double-tipped outer involucral bracts, and commonly reddish or purplish achenes with the 'cone' c. 1 mm, cylindrical. The 27 or so British representatives fall into 3 groups of microspp.:

1 Achenes distinctly reddish or purplish. 2

Achenes brown, yellow, cream or grey without any red or pink tinge. Outer involucral bracts 6–9 mm, erect, spreading or recurved; ligules striped violet to grey. Local plants of dry grassland and rocky places. *T. simile* group

2 Achenes red, purple or violet. Outer involucral bracts up to 6 mm, bordered, appressed to spreading; ligules pale yellow striped grey, brown or purple. Includes several common microspecies of calcareous downs, sand dunes, sandy heaths, walls, paths, etc. *T. erythrospermum* group

Achenes reddish- or pinkish-brown (cinnamon-coloured), otherwise like *T. erythrospermum* group. Widespread and locally common plants of dry grassland of downs, dunes and cliff tops, less commonly on sandy heaths, paths and walls. *T. fulvum* group

Section E. *Taraxacum* (*Vulgaria* Dahlst.)

The numerous microspp. of this Section are all placed in the single group *T. officinale*. They are robust plants, chiefly of waysides, paths, gardens, etc., and of managed grasslands. Their lvs are usually triangular-lobed and may be dark-blotched, especially between the lobes. The heads are large, 2.5–7.5 cm diam., and the outer involucral bracts are up to 17 mm long, linear-lanceolate, indistinctly bordered and erect to recurved. The achenes are brownish, with the 'cone' conical.

Of the several alien members of the Cichorieae occasionally found as casuals, the most frequent are:

*Scolymus hispanicus L. A biennial to perennial thistle-like herb with an erect branching stem 20–80 cm, pubescent, interruptedly spinous-winged. Lvs oblong-lanceolate in outline, white-veined, sinuate-pinnatifid with spinous lobes and teeth; stem-lvs decurrent. Heads axillary, subsessile, each surrounded by 3 long ascending lf-like spiny bracts; involucral bracts linear-lanceolate, acuminate. Florets yellow. Achene enclosed by the laterally winged receptacular scale; pappus of 2–3 deciduous bristles. Mediterranean region, Madeira, Canary Is. Cultivated for its salsify-like tap-root.

*S. maculatus L. An annual herb with strongly spiny white-bordered lvs, yellow florets and no pappus. Mediterranean region.

*Tolpis barbata (L.) Gaertner. An annual herb with an erect ±pubescent branching stem 20–50 cm. Lower lvs oblong-lanceolate shortly petiolate; upper lvs narrower, sessile; all toothed, ±pubescent. Heads small, in a cyme-like panicle; outer involucral bracts setaceous, spreading, exceeding the inner bracts and the florets. Florets all yellow or those in the centre reddish-brown. Marginal achenes with numerous pappus-bristles, those of the centre with 4–5 long hairs. Mediterranean region.

MONOCOTYLEDONES

126. ALISMATACEAE

Annual or perennial herbs, erect or sometimes with floating lvs, living in water or wet places. Lvs petiolate, the blades linear-lanceolate to ±orbicular. Fls actinomorphic, usually hermaphrodite, often whorled, arranged in panicles or racemes. Perianth in 2 whorls, heterochlamydeous, the outer whorl persistent. Stamens (3–)6 or more, free; anthers 2-celled. Ovary superior. Carpels free or rarely connate at base. Ovules 1 or several, basal or in the inner angle of the carpel. Fr. usually a head or whorl of achenes, rarely dehiscing at base. Seeds small, endosperm 0; embryo horseshoe-shaped.

Thirteen genera and about 90 spp. in temperate and tropical regions, mainly in the northern hemisphere.

An interesting family, showing some striking similarities with the Ranunculaceae and perhaps derived from them.

1 Fls unisexual; stamens numerous; lvs often sagittate.
　　　　　　　　　　　　　　　　　5. SAGITTARIA
　Fls hermaphrodite; stamens 6; lvs not sagittate. 　2
2 Fls in more than 2 whorls. 　　　　　　　　　　3
　Fls in not more than 2 whorls. 　　　　　　　　4
3 Lvs not cordate at base; ripe carpels 2–3 mm, numerous. 　　　　　　　　　　　　　　3. ALISMA
　Lvs cordate at base; ripe carpels 5–12 mm, 6(–9).
　　　　　　　　　　　　　　　4. DAMASONIUM
4 Lvs lanceolate to linear-lanceolate, acute.
　　　　　　　　　　　　　　　　　1. BALDELLIA
　Lvs elliptical to ovate, rounded at apex. 2. LURONIUM

1. BALDELLIA Parl.

Glabrous perennial scapigerous herbs, sometimes stoloniferous. Lvs in a basal rosette. *Fls long-pedicellate, umbellate* or (rarely) in 2 simple whorls, sometimes solitary. Stamens 6. *Carpels numerous*, free, *forming a crowded head. Ripe carpels ovoid.*

Two spp. in Europe and N. Africa.

1. B. ranunculoides (L.) Parl. 　Lesser Water-Plantain.

Alisma ranunculoides L.; *Echinodorus ranunculoides* (L.) Engelm.

A variable plant, usually between 5 and 20 cm. Lvs long-petiolate, the blade up to 10 cm, lanceolate to linear-lanceolate, acute, narrowed gradually into petiole. Stems usually erect or spreading, sometimes decumbent and then rooting and producing tufts of lvs and (often solitary) fls at nodes. Petals 7–10 mm, white or pale purplish; fls opening in succession. Pedicels unequal, up to c. 10 cm. Bracts small, scarious. Heads of achenes ±globose. Achenes 2–3.5 mm, ovoid, curved, strongly 3-ribbed on the back with 2 ventral ribs. Fl. 5–8. $2n = 16^*, 30^*$. Hel.

Native. In damp places beside streams, ponds and lakes and in fen ditches, locally common. Scattered throughout the British Is., north to E. Ross and the Outer Hebrides. Europe north to S. Norway and east to Lithuania and W. Greece; N. Africa.

2. LURONIUM Rafin.

A slender herb. *Stems floating* and rooting at nodes. *Fls* 1(–5) *in the axils of the lvs*, long-pedicellate. Stamens 6. *Carpels* 6–15 in an irregular whorl, free. *Ripe carpels oblong-ovoid.*

One sp. in W. and C. Europe.

1. L. natans (L.) Rafin. 　　Floating Water-Plantain.

Alisma natans L.; *Elisma natans* (L.) Buchenau

Stems 50 cm or more, slender, floating but rooted at base. Lower lvs submerged, reduced to linear, flattened, translucent petioles up to c. 10 cm × 2 mm; other lvs floating, long-petiolate, the blade up to 4 cm, ovate or elliptical, rounded and obtuse at apex. Petals 7–10 mm, white with a yellow spot at base. Head of achenes hemispherical. Fl. 7–8. $2n = 42$. Hyd.

Native. In lakes, tarns and canals with acid water, very local but apparently increasing. Pembroke to Caernarvon, Stafford and N.W. Yorks; also here and there in canals in several other English counties, where it is of recent introduction. W. and C. Europe to S. Norway and Bulgaria.

3. ALISMA L.

Glabrous perennial scapigerous herbs with acrid juice. *Infl. usually much-branched. Carpels numerous, in one whorl;* style lateral. *Ripe carpels strongly compressed.*

Ten spp. in north temperate regions and Australia.

1 Style shorter than ovary, recurved; anthers suborbicular; lvs usually linear. 　　　**3. gramineum**
　Style as long as or longer than ovary, ±erect; anthers elliptical; lvs lanceolate to ovate. 　　2
2 Lvs ovate, rounded to subcordate at base; styles stigmatose in upper $\frac{1}{8}$–$\frac{1}{5}$ of their length.
　　　　　　　　　　　　1. plantago-aquatica
　Lvs lanceolate, gradually narrowed at base; styles stigmatose in upper $\frac{1}{2}$–$\frac{2}{3}$ of their length.
　　　　　　　　　　　　　　2. lanceolatum

1. A. plantago-aquatica L. 　　　Water-Plantain.

An erect glabrous perennial 20–100 cm. Stem stout, usually unbranched in lower half. *Lvs* long-petiolate, the blade 8–20 cm, *ovate*, subacute to acuminate, *usually rounded or subcordate at base*; the first lvs of land and water forms reduced to a $\frac{1}{2}$-cylindrical petiole with small narrow blade; floating lvs sometimes occur in the water form. Infl. branches ±straight, usually ascending. Fls usually pale lilac, open from 1 to 7 p.m. Petals 3.5–6.5 mm. Stamens (in fl.) longer than carpels (excluding style) *styles ±straight, long, stigmatose in the upper* $\frac{1}{8}$–$\frac{1}{5}$ *of their length.* Carpels c. 20, in a ±flat head. Fl. 6–8. $2n = 14^*$. Hyd. or Hel.

Native. On muddy substrata beside slow-flowing rivers, ponds, ditches and canals, in damp ground or shallow water. Throughout much of the British Is., rarer in the north and absent from most of the north of Scotland, the Outer Hebrides, Orkney and Shetland. Temperate Eurasia.

2. A. lanceolatum With.

Like *A. plantago-aquatica* in general appearance. *Lvs lanceolate, narrowed gradually into petiole.* Fls usually pink, open from 9 a.m. to 2 p.m. Petals 4.5–6.5 mm. Stamens (in fl.) somewhat longer than carpels (excluding style). *Styles stigmatose in the upper $\frac{1}{2}$–$\frac{2}{3}$ of their length.* $2n = 26^{*}$, 28. Hyd. or Hel.

Native. In similar situations to the foregoing. Distribution imperfectly known, but apparently a much less frequent plant than *A. plantago-aquatica.* Scattered in England, Wales and Ireland; Kincardine. Europe; N. Africa; W. Asia; Macaronesia.

3. A. gramineum Lejeune
Ribbon-leaved Water plantain.

A. plantago-aquatica L. var. *graminifolium* Wahl.

A perennial herb (5–)15–30(–60) cm. Stem stout usually ±curved, branched below the middle. *Lvs linear, ribbon-like,* the blade not or scarcely distinguishable from the petiole; later lvs sometimes with a short (15–40 mm) linear-lanceolate or narrowly oblong blade. Floating lvs 0. Infl. branches ±recurved particularly in fr. Fls open between 6 and 7.15 a.m.; petals 2.5–3.5 mm, caducous. Stamens as long as carpels (excluding style); anthers suborbicular. *Style shorter than ovary, recurved.* Fl. 6–8. $2n = 14$. Hyd. or Hel.

?Native. In a shallow artificial pond near Droitwich, Worcs., where it was first recorded in 1920. C. and E. Europe, northwards to the Leningrad region; North America.

4. DAMASONIUM Miller

Glabrous scapigerous perennial herbs. Fls hermaphrodite, in 1–several, usually simple whorls. Inner per.-segs entire. Stamens 6. *Carpels 6–9, in one whorl, connate at base;* style apical. Ovules 2–several in each carpel. *Ripe carpels (1–)2–several-seeded, indehiscent or tardily dehiscent at base, spreading stellately.*

Three spp., in Europe, N. Africa, W. Asia and Australia.

1. D. alisma Miller
Starfruit, Thrumwort.

D. stellatum Thuill.; *Actinocarpus damasonium* R.Br.

An erect herb 5–30(–60) cm. *Lvs* long-petiolate, floating or sometimes submerged, the blade up to 8 cm, ovate to oblong, obtuse, *cordate at base.* Fls c. 6 mm diam.; petals white with a yellow spot at the base. *Ripe carpels 5–12 mm, tapering into a long beak,* usually 2-seeded. $2n = 28$. Fl. 6–8. Hyd.

Native. In gravelly ditches and ponds, very local and

decreasing. Surrey, Bucks. W., S. and S.E. Europe; N. Africa; Asia.

5. SAGITTARIA L.

Scapigerous herbs, usually perennial. Infl. of several simple or rarely slightly branched whorls. *Fls unisexual. Stamens numerous. Carpels numerous, spirally arranged* on a large receptacle, free, strongly *compressed.*

About 20 spp. throughout the temperate and tropical regions.

Lvs typically sagittate; petals white, with a dark violet patch at base.   **1. sagittifolia**
Lvs sagittate; petals entirely white, or slightly yellowish at base.   **2. rigida**

1. S. sagittifolia L.
Arrow-head.

A monoecious herb 30–90 cm, perennating by means of turions borne at the ends of slender runners. Turions c. 3 cm, ovoid or subcylindrical, bright blue with yellow spots. Submerged lvs linear, translucent; floating lvs lanceolate to ovate; *aerial lvs long-petiolate, the blade* 5–20 cm, *sagittate,* acute or obtuse, lateral lobes about as long as main portion of blade. Fls 3–5 in a whorl, female in lower part of infl., rather smaller than male. *Scape longer than lvs.* Pedicels of male fls c. 20 mm, of female fls c. 5 mm. Petals 10–15 mm, *white with a dark violet patch at base.* Filaments linear, glabrous. Mature head of carpels c. 15 mm diam., hemispherical. Fl. 7–8. $2n = 22^{*}$. Hyd.

Native. In shallow water in ponds, canals and slow-flowing rivers on muddy substrata. Scattered throughout England, rather local and rarer in the north; Glamorgan, Pembroke and Anglesey; scattered throughout C. and N.E. Ireland, local; not native in Scotland. Europe and Asia.

*2. S. rigida Pursh
Canadian Arrowhead.

S. heterophylla Pursh, non Schreber

Like *S. sagittifolia* but *aerial lvs ovate or elliptical* (rarely sagittate in America); *scape shorter than lvs;* pedicels of male fls c. 12 mm, female fls subsessile; *petals entirely white or pale yellowish at base;* filaments dilated, hairy.

Introduced. Naturalized in the river Exe, in and near Exeter; first observed in 1898. Native of North America from Quebec to Tennessee and west to Minnesota.

S. subulata (L.) Buchenau, native in North America, is naturalized in one locality in S. England. It has lvs which are either linear and submerged or floating and elliptical to ovate-oblong, and the filaments of the stamens are dilated and glabrous.

127. BUTOMACEAE

Rhizomatous herbs, living in water or wet places. Lvs linear. Fls in long-pedunculate umbels. Perianth in 2 petaloid whorls. Stamens 9; filaments flattened; anthers basifixed, 2-celled, opening by lateral slits. Ovary superior. Carpels 6, connate at the base. Ovules numerous, scattered over the inner surface of the carpel wall. Seeds

small, endosperm 0; embryo straight.

One genus in temperate Eurasia.

1. BUTOMUS L.

A scapigerous herb devoid of latex. Lvs linear, erect. Fls hermaphrodite, umbellate. Per.-segs persistent.

Stamens 9. Carpels 6, connate at base.
One sp. in Europe and temperate Asia.

1. B. umbellatus L. Flowering Rush.

A glabrous rhizomatous perennial up to 150 cm. Lvs about as long as stems, in a basal rosette, triquetrous, twisted, acuminate, sheathing at base. Stems terete. Umbel with an involucre of acuminate bracts. Pedicels up to c. 10 cm, unequal. Fls opening in succession. Per.

-segs 10–15 mm, pink with darker veins, the outer somewhat smaller and narrower than the inner. Ripe carpels obovoid, the style persistent. Fl. 7–9. $2n = 26, 39$. Hel. or Hyd.

Native. In ditches, ponds and canals, and at margins of rivers. England and Ireland, rather local; rare in Wales and not native in Scotland. Europe, temperate Asia; naturalized in North America.

128. HYDROCHARITACEAE

Aquatic herbs, wholly or partially submerged, inhabiting fresh waters or the sea. Fls actinomorphic, hermaphrodite or unisexual, arranged in a spathe composed of 1 bract or 2 opposite free or connate bracts. Spathes sessile or pedunculate; male fls usually numerous, the female solitary. Sepals 3; petals 3 or 0. Stamens 2–15. Rudimentary ovary present in male fls. Staminodes sometimes present in female fls. Ovary inferior, unilocular, with 3–6 parietal placentae; ovules numerous.

About 16 genera and 80 spp., in the warmer parts of the world, a few in temperate regions.

1 Lvs petiolate, orbicular-reniform, floating.
      1. HYDROCHARIS
Lvs sessile, submerged or partly so.   2
2 Lvs basal, more than 10 cm.   3
Lvs cauline, not more than 4 cm.   4
3 Lvs spinous-serrate, rigid, tapering from base.
      2. STRATIOTES
Lvs denticulate at top, flaccid, ribbon-shaped.
      3. VALLISNERIA
4 Lvs spirally arranged, usually recurved; fls minute.
      7. LAGAROSIPHON
Lvs opposite or in whorls, not usually recurved.   5
5 Petals 8–11 mm; lvs up to 4 cm, linear, acute.
      4. EGERIA
Petals not more than 3.5 mm; lvs less than 2 cm.   6
6 Lvs in whorls of 3–8; nodal scales fringed with long orange-brown hairs; plant pale green. 6. HYDRILLA
Lvs opposite or in whorls of 3(–4); nodal scales entire; plant dark green.   5. ELODEA

1. HYDROCHARIS L.

A floating herb. Lvs orbicular-reniform, petiolate. Fls white, unisexual, male 1–4 in a pedunculate spathe, the female solitary, in a sessile spathe. *Sepals herbaceous, narrower and smaller than the petals.* Stamens 9–12, 3–6 outer usually sterile. Female fls with 6 staminodes; ovary 6-celled with 6 bifid styles. Fr. fleshy, indehiscent.
Six spp. in Europe, Asia, Africa and Australia.

1. H. morsus-ranae L. Frogbit.

A floating stoloniferous herb with lvs in groups at the nodes. Roots in bunches at the nodes. Turions enclosed by 2 scale-lvs, produced at ends of stolons in autumn. Lvs floating on the surface, blade c. 3 cm diam.; stipules large, scarious. Fls erect, aerial; petals c. 10 mm, broadly obovate, crumpled, white, with a yellow spot

near base. Fr. rarely if ever produced in this country. Fl. 7–8. $2n = 28*$. Hyd.

Native. In ponds and ditches, usually in calcareous districts, locally common. Scattered throughout England from Devon and Kent to S. Lancs and N.E. Yorks; Wales: Carmarthen, Glamorgan and Flint; absent from Scotland; Ireland, mainly in the N. Centre extending to Clare, Dublin and Down. Temperate Eurasia; generally local and probably diminishing.

A land-form occurs occasionally in dry seasons. The plant is probably monoecious but with male and female fls at different nodes and often separated by a long slender stem which usually breaks when removed from the water.

2. STRATIOTES L.

A submerged dioecious herb rising to the surface at flowering time. *Lvs basal, sessile, tapering from the base.* Fls white; *male several in a spathe, bracteolate, pedicellate; female solitary, sessile.* Stamens 12 fertile surrounded by numerous sterile. Fls otherwise much as in *Hydrocharis.*
One sp. in Europe and N.W. Asia.

1. S. aloides L. Water Soldier.

A stoloniferous herb with lvs in large rosettes, reproducing mainly by means of offsets. Lvs 15–20 cm, spinous-serrate, rigid, brittle, many-veined, resembling those of an aloe. Fls erect, aerial; petals 15–25 mm, suborbicular, thick. Ovary becoming deflexed after fl.; fr. never produced in this country. Fl. 6–8. $2n = 24$. Hyd.

Native. In Broads, ponds and ditches in calcareous districts, very local and probably diminishing. E. England from Cambridge to S.W. Yorks; Wales: Flint; Scotland and Ireland: introduced in a few localities. Distribution of the genus. In the north the plants are entirely female, in the south predominantly or entirely male; both sexes occur in an intermediate area.

Only the female plant occurs normally in Britain, though plants with hermaphrodite fls have been recorded. The floating and submerging of the plant is said to be due to changes in the amount of calcium carbonate on the lvs.

3. VALLISNERIA L.

Submerged perennial stoloniferous herbs. Fls unisexual, male many together in a tubular, 2-toothed, shortly pedunculate spathe, female solitary in a *spathe borne on a long filiform peduncle. Per. of male fls single, of female double but petals rudimentary.* Stamens 1–3, usually 2. Stigmas 3, bifid. Capsule cylindrical. Water-pollinated, the male fls breaking off and floating to the surface when the spathe opens.

About 8 spp. in the warmer parts of the world.

*1. V. spiralis L.

Lvs basal, ribbon-shaped, obtuse and denticulate at top. Fls pinkish-white. Peduncle of female fls spirally twisted after fl. Fl. 6–10. $2n = 20$. Hyd.

Introduced. Naturalized in W. Gloucester, Worcester, S.W. Yorks, and S. Lancs, often where water is heated by effluents from mills. Widely distributed in the warmer parts of the world, in Europe reaching as far north as N.C. France and C. Ukraine.

4. EGERIA Planchon

Dioecious. Lvs submerged, whorled, sessile, linear. Fls pollinated above the water-surface. Spathes tubular, axillary, sessile, the male 2–4-fld, the female 1-fld. Petals longer and wider than the sepals; stamens 9(–10); ovary with a long slender beak; styles 3, 2(–3)-fid. Insect-pollinated.

*1. E. densa Planchon

Lvs up to 40×5 mm, in ±densely crowded whorls of 3–5, minutely denticulate. Sepals 3–4 mm; petals 8–11 mm, obovate to suborbicular, white. Hyd.

Introduced. Grown in aquaria and locally naturalized in still or slow-flowing water. Native in South America.

5. ELODEA Michx

Dioecious submerged herbs with sessile whorled lvs. Fls solitary in tubular spathes which are sessile in the axils of the lvs. Petals narrower than sepals. *Stamens usually 9, filaments short or 0.* Ovary with a long slender beak. *Styles 3, usually lobed or notched,* free to base. Water-pollinated, the female fls reaching the surface by the elongation of the axis, the male usually breaking off and floating.

About 10 spp. in North and South America.

Lvs usually c. 1 cm, oblong, rounded at the apex; male fls not breaking free.	**1. canadensis**
Lvs usually c. 1.5 cm, linear, acute; male fls breaking free and floating.	**2. nuttallii**

*1. E. canadensis Michx. Canadian Pondweed.

Anacharis canadensis Planchon

A dark green translucent submerged herb perennating by means of winter buds. Stems up to 3 m, usually much less, very brittle. *Lvs* 5–12(–15) × 1.5–2.5(–3.2) mm, usually 3 in a whorl, ±rigid, *usually obtuse*, serrulate for about the distal ⅔; scales on upper surface of lf towards the base minute, entire, green. Fls floating, greenish-purple. *Sepals of female fls* 2–2.7 mm; petals white or pale purple. Male fls floating and remaining attached to the pedicel, very rare in Britain. Fl. 5–10. $2n = 24$; 48. Hyd.

Introduced. Naturalized in slow-flowing fresh waters throughout most of the British Is. First introduced in 1834 in Ireland and 1842 in Britain, spread rapidly and attained great abundance so as to block many waterways, then diminished; now widespread but seldom abundant. Native in North America; introduced in most of Europe.

*2. E. nuttallii (Planchon) St John

Like **1** but light green; lvs 8–17 × 1–2.5 mm, 3–4 in a whorl, flaccid, acute; sepals of female fls 1–1.8 mm; male fls becoming detached from the pedicel, not known in Britain. $2n = 48$. Hyd.

Introduced. Spreading rapidly in similar habitats to and tending to displace *E. canadensis*. Native of North America.

*E. ernstiae St John (*E. callitrichoides* auct., non (L. C. M. Richard) Caspary) is also recorded as naturalized. It is like **2** but has rather larger lvs and the male fls (which have never been seen in Britain) do not become detached from the pedicel. The sepals of the female fls are 3–3.5 mm. Native of South America.

Elodea and the other morphologically similar naturalized genera require further investigation. The variability of vegetative characters and the usual absence of flowers often makes certain identification impossible.

6. HYDRILLA L. C. M. Richard

Dioecious. Lvs submerged, whorled or the lower opposite, sessile, linear to rarely ovate. Fls pollinated at the water-surface. Spathes tubular, axillary, sessile, 1-fld. Petals and sepals subequal. Male fls becoming detached and rising to the surface; stamens 3; ovary with a long slender beak; styles 3(–5), simple.

One sp., Eurasia and Africa to Australia.

1. H. verticillata (L.fil.) Royle

Lvs usually in whorls of 3–8, up to 20(–40) × 2(–5) mm, denticulate; nodal scales fringed with orange-brown hairs. Sepals of female fls 1.5–3 mm; petals narrower than the sepals. $2n = 16$. Hyd.

Native. In 1.5–2.6 m of water on blue-green clayey mud, Esthwaite Water, Lake District (?extinct); W. Galway. Very local in Europe and often impermanent; S. and E. Asia; E. Africa; Australia.

7. LAGAROSIPHON Harvey

Dioecious. Lvs submerged, alternate, sessile, linear. Fls pollinated at the water-surface. Spathes tubular, axillary, sessile, the male with many fls, the female with 1(–3) fls. Sepals and petals subequal. Male fls becoming detached and rising to the surface; stamens 3; staminodes 3. Ovary with a long slender beak; styles 3, 2-fid.

Fifteen spp. in Africa and India.

*1. L. major (Ridley) Moss

Lvs up to 30 × 3 mm, densely crowded towards the ends of the branches, recurved, denticulate. Fls pinkish. Hyd.

Introduced. Grown in aquaria and naturalized here and there in ponds and lakes. Native of South Africa.

129. SCHEUCHZERIACEAE

Perennial herb. Lvs linear, sheathing at base; ligule present. Infl. a few-fld, terminal raceme. Bracts present. Fls hermaphrodite. Per.-segs 6, all sepaloid, persistent. Stamens 6, free. Ovary superior; carpels 3–6, shortly connate towards the base, divaricate and free or nearly so in fr.; stigmas sessile; ovules 2 or few, basal, erect, anatropous. Fr. dehiscing on the curved, adaxial side.

One genus and 1 sp. in the colder parts of the northern hemisphere.

1. SCHEUCHZERIA L.

The only genus.

1. S. palustris L.   Rannoch Rush.

An erect perennial 10–20(–40) cm. Rhizome creeping, clothed with persistent lf-bases. Stems lfy. Lvs alternate, linear, slightly grooved, obtuse, with a conspicuous pore at the tip. Infl. 3–10-fld, very lax, overtopped by the lvs. Fls yellowish-green; per.-segs 2–3 mm, lanceolate, acute. Carpels usually 3. Fl. 6–8. $2n = 22^*$. Hel.

Native. In very wet *Sphagnum* bogs, usually in pools. Perth and Argyll very rare; formerly in Shropshire, some north English counties and Offaly. Colder parts of the northern hemisphere.

130. JUNCAGINACEAE

Annual or perennial scapigerous marsh or aquatic herbs. Lvs mostly basal, linear, sheathing at the base. Fls small, in spikes or racemes, hermaphrodite or unisexual, actinomorphic or slightly oblique, 2–3-merous. Bracts 0. Per.-segs 2–6, sepaloid. Stamens 3 or 6; filaments very short. Ovary superior; carpels 6 (sometimes 3 sterile) or 3, free or ±connate; style short or 0; ovules solitary, basal, anatropous. Fr. dehiscent or not. Fls protogynous, wind pollinated.

Three genera with about 25 spp., widely distributed, particularly in the temperate and cold regions of both hemispheres.

1. TRIGLOCHIN L.

Rhizomatous herbs with fibrous roots and ±tuberous stems. Lvs erect, linear, ½-cylindrical. Infl. a raceme. Fls 3-merous. Per.-segs deciduous. Carpels all fertile or alternate ones sterile. Fr. dehiscing by the carpels separating from the central axis.

About 15 spp. in temperate regions.

Lvs deeply furrowed on upper surface towards base; fr. 7–10 × 1 mm, clavate, appressed to scape.   **1. palustris**
Lvs not furrowed; fr. 3–4 × 2 mm, oblong-ovoid, not appressed to scape.   **2. maritima**

1. T. palustris L.   Marsh Arrowgrass.

A slender perennial 15–70 cm. Rhizomes long, slender. Lvs ½-cylindrical, deeply furrowed on upper surface towards the base. Raceme elongating after fl. Fls 2–3 mm; per.-segs purple-edged. *Fr.* 7–10 × 1 mm, *clavate, appressed to scape; carpels 3 sterile and 3 fertile, remaining attached to the triquetrous axis at the top* after dehiscence. Fl. 6–8. $2n = 24$. Hel. or Hr.

Native. In marshes, usually among tall grass. Throughout the British Is. but very local in most southern and midland counties. Europe except most of the Mediterranean region; N. Africa; N. Asia; North and South America; Greenland.

2. T. maritima L.   Sea Arrowgrass.

A rather stout perennial 15–60 cm. Rhizomes short, stout. Lvs subulate or linear, up to 4 mm wide, ½-cylindrical, not furrowed. Raceme scarcely elongating after fl.; fls 3–4 mm; pedicels 1–2 mm, elongating after fl. *Fr.* 3–4 × 2 mm, *oblong-ovoid, not appressed to scape; carpels 6, all fertile, separating completely at dehiscence.* Fl. 7–9. $2n = 24, 30, 36, 48$. Hel. or Hr.

Native. In salt-marsh turf and grassy places on rocky shores. In suitable habitats around the entire coast of the British Is. Europe southwards to C. Portugal and Bulgaria; N. Africa; W. and N. Asia; North America.

131. APONOGETONACEAE

Aquatic herbs. Fls hermaphrodite or rarely unisexual, ebracteate, in pedunculate spikes. Per.-segs 1–3, sometimes petaloid, usually persistent. Stamens 6–18 free, persistent; pollen subglobose or ellipsoid. Ovary of 3–8, free, sessile carpels. Ovules 2–8 in each carpel, basal, anatropous. Seeds without endosperm and with a straight embryo.

One genus and c. 30 spp. in the tropics of the Old World and S. Africa.

1. APONOGETON L.

The only genus.

***1. A. distachyos** L.   Cape Pondweed.

A perennial herb with floating lvs and edible tuberous stock. Stems long, green, spongy. Lvs up to 25 × 7 cm, oblong-elliptical, rounded at base and apex. Infl. of two terminal, rigid, whitish spikes. Fls c. 10 on each spike, distichous, white, fragrant; per.-segs 1–2, 10–20 mm, white, ovate. Stamens 6–18. Hyd.

Introduced. Frequently planted in ponds and sometimes ±naturalized. S. Africa.

132. ZOSTERACEAE

Perennial submerged marine herbs with a creeping rhizome bearing at each node 2 or more unbranched roots and a lf with a short shoot in its axil. Short shoots with several distichous linear lvs. Flowering stems lateral or terminal; infl. enclosed in the sheathing base of a spathe. Fls with 1 dorsifixed sessile stamen and one unilocular ovary with a style and 2 filiform stigmas arranged alternately on the margin of the flattened axis; perianth absent. Pollen-grains filiform. Ovule 1.

1. ZOSTERA L.

Rhizome monopodial, bearing alternate distichous lvs. Flowering shoots annual, simple or branched. Pollination by water; reproduction mostly vegetative by the breaking up of the rhizome.

Twelve spp. in temperate seas of the world.

1   Flowering stems lateral, simple or sparingly branched; lf-sheaths open.   **3. noltii**
     Flowering stems terminal, freely branched; lf-sheaths closed, splitting when old.   **2**
2   Lvs on non-flowering shoots usually 5–10 mm wide; stigma twice as long as style; fr. 3–3.5 mm.
                                                                1. marina
     Lvs on non-flowering shoots usually 1–2 mm wide; stigma about as long as style; fr. 2.5–3 mm.
                                                                2. angustifolia

1. Z. marina L.   Eelgrass, Grass-wrack.

A rhizomatous perennial with dark green grass-like lvs. Rhizome 2–5 mm thick; internodes short; cortex with 2 bundles of fibres in its outer layers. *Lvs* of non-flowering shoots 20–50(–200) cm, (2–)5–10 *mm broad, rounded and mucronate at apex;* sheaths closed. Lvs of flowering shoots shorter and narrower, sometimes emarginate. Flowering stems up to 60 cm, much-branched. Infl. (4–)9–12(–14) cm; *stigma twice as long as style. Seed* 3–3.5 mm, ellipsoid, pale brown or bluish-grey, longitudinally ribbed. Fl. 6–9. Fr. 8–10. Germ. autumn. $2n = 12^*$. Hyd.

Native. On fine gravel, sand or mud in the sea, from low-water spring tides down to 4 m, rarely in estuaries. Local but formerly covering large areas in suitable localities; decreased markedly in abundance about 1933. Coasts of the British Is., becoming rarer northwards. Europe from Norwegian Lapland (71°N.) to the Mediterranean; W. Greenland (64°N.); Atlantic and Pacific coasts of North America from N. Carolina to Hudson Bay and California to Unalaska Bay.

2. Z. angustifolia (Hornem.) Reichenb.
                                        Narrow-leaved Eelgrass.
Z. hornemanniana Tutin; *Z. marina* L. var. *angustifolia* Hornem.

A slender rhizomatous perennial with narrow dark green lvs. Rhizome 1–2 mm thick, rooting at the black slightly inflated nodes; cortex with 2 bundles of fibres in its outer layers. *Lvs* of non-flowering shoots 15–30 cm, 2 mm wide in summer, in winter 5–12 cm, c. 1 mm wide, obtuse and rounded at apex when young, *later emarginate;* sheaths closed. Lvs of flowering shoots 4–15 cm, 2–3 mm wide. Flowering shoots 10–30 cm, compressed, c. 1 mm wide, pale green or white, branched. Infl. 8–11 cm; *style and stigmas approximately equal in length. Seed* 2.5–3 mm, ellipsoid, pale brown, longitudinally ribbed. Fl. 6–11. Fr. 7–12. Germ. autumn. $2n = 12^*$. Hyd.

Native. On mud flats in estuaries and in shallow water, from half-tide mark to low-tide mark or rarely down to 4 m, in salinities between 25 and 42 g/l. Scattered round the coasts of the British Is. north to Orkney and not uncommon in suitable habitats. Europe, so far recorded only from Denmark and Sweden.

3. Z. noltii Hornem.   Dwarf Eelgrass.
Z. nana Roth, pro parte

A slender shortly creeping rhizomatous perennial with narrow lvs. Rhizome 0.5–1 mm thick; *cortex with bundles of fibres in its innermost layers. Lvs* of non-flowering shoots (4–)6–12(–20) cm, *up to* c. 1 mm wide, emarginate; *sheaths open. Flowering stems unbranched* or occasionally with 1 or 2 branches from very near the base; peduncles 0.5–2 cm. Infl. 3–6 cm; *sheath inflated. Seed* 2 mm, *smooth,* dark brown when ripe, subcylindrical. Fl. 6–10. Fr. 7–11. Germ. autumn. $2n = 12^*$. Hyd.

Native. On mud-banks in creeks and estuaries from half-tide mark to low-tide mark. Locally common in suitable habitats around the coasts of the British Is. north to E. Sutherland; east coast of Ireland. Europe from the Mediterranean to S.W. Norway and Sweden.

133. POTAMOGETONACEAE

Aquatic herbs, chiefly of fresh water, with alternate or opposite distichous usually 'stipulate' lvs. Fls in axillary or terminal bractless spikes, inconspicuous, hermaphrodite, actinomorphic, hypogynous. Per.-segs 4; stamens 4, sessile on the claws of the per.-segs; carpels (1–3) 4, free, each with 1 campylotropous ovule near the base of the ventral margin; stigma ±sessile. Fr. a small green or brownish drupe or achene, 4 or fewer from each fl.; seed non-endospermic; embryo with a massive hypocotylar 'foot'.

Two genera.

All lvs in opposite pairs or in whorls of 3; stipules 0, except

for the involucral lvs subtending fl.-spikes.
2. GROENLANDIA
All or most lvs alternate and all stipulate. 1. POTAMOGETON

1. POTAMOGETON L.

Chiefly perennial and rhizomatous, and overwintering both by the rhizome and by specialized winter buds ('turions') which may be borne directly on the rhizome, on rhizome-stolons or on the lfy stems. In some spp. the creeping stem rarely overwinters, and in others none is formed, perennation being only by the turions. Lvs all submerged, thin and translucent, or some floating lvs which are usually ±coriaceous and opaque; submerged lvs linear or with a ±broadened sessile or stalked blade; floating lvs usually narrowly to broadly elliptical-oblong. In most spp. the lf has in its axil a ±delicate membranous sheathing scale which may be free throughout ('stipule') or may be adnate to the lf-base in its lower part ('stipular sheath') and free above ('ligule'); in either case the basal part may be open and with overlapping (convolute) margins, or tubular. Spikes ovoid to cylindrical, dense, lax or interrupted; either submerged, with pollination by water, or emergent and wind-pollinated. Perianth of 4, free, rounded, shortly clawed, valvate segments, sometimes regarded as appendages of the connectives of the anthers.

About 100 spp. Cosmopolitan.

The genus is reputed to be taxonomically difficult because most species are very plastic in their vegetative morphology, varying greatly in the size and shape of their lvs at different stages of development and in different conditions of light intensity, mineral nutrient supply, speed of water-movement, etc. Spp. which can form floating lvs in sufficiently shallow water may fail to do so in deep water, and some form lvs only of the floating type when growing subterrestrially. Moreover many hybrids are known, some of them being fairly common plants. They are usually sterile but vegetative material must often be examined very closely before a confident judgment can be made. Descriptions are given only of the three most frequently encountered hybrids; others reported for Britain are listed, and it must be understood that they are intermediate in certain features between their putative parents.

Amongst the most important diagnostic features are the venation of the lvs, including the point at which lateral veins join the midrib and the angle of the join, the shape of the lf-apex, the denticulation or otherwise of the lf-margin, and whether the stipules of the young lvs are open throughout or tubular in their lower part. These features are best examined in fresh or soaked-out material by means of a strong hand lens or, better, a binocular dissecting-microscope.

1 Floating lvs present.   2
  Floating lvs 0.   10
2 Submerged lvs (phyllodes) all linear with no expanded blade; floating lvs with the margins decurrent for a short distance down the stalk, making a flexible joint.   **1. natans**
  Submerged lvs with a ±expanded translucent blade; floating lvs not jointed.   3
3 Stems compressed; submerged lvs all linear, sessile, with a well-marked band of air tissue bordering the midrib.   **13. epihydrus**

Stems ±terete; submerged lvs not all linear, often more than 8 mm wide.   4
4 All lvs thin, translucent, distinctly and finely net-veined; floating and most of the submerged lvs ±broadly elliptical, all short-stalked. **3. coloratus**
  Floating lvs ±coriaceous, opaque, not translucent.   5
5 Submerged lvs mostly 10–16 × 3.5–5 cm, elliptical-lanceolate, transparent, very conspicuously but delicately net-veined; floating lvs of similar shape; all lvs long-stalked.   **4. nodosus**
  Submerged lvs translucent but not conspicuously net-veined unless held against the light; floating lvs usually broader than submerged lvs.   6
6 All lvs distinctly stalked; usually in very shallow water and then most or all lvs commonly of the floating type, coriaceous and opaque, broadly elliptical to elliptical-ovate.   **2. polygonifolius**
  Some or all submerged lvs sessile.   7
7 At least the lower submerged lvs rounded and semi-amplexicaul at the base; a sterile hybrid. **9. ×nitens**
  Submerged lvs narrowed to the base, sessile or short-stalked.   8
8 Submerged lvs blunt, with quite entire margins, commonly 6–12 × 1–2 cm, often reddish; stipules 2–5 cm, broad, blunt; peduncles not thickened upwards.   **10. alpinus**
  Submerged lvs acute, cuspidate or mucronate, with microscopically denticulate margins; peduncles thickened upwards.   9
9 Submerged lvs commonly 2.5–10 × 0.5–1 cm, acute or acuminate, sessile; stipules 1–2 cm; plant much branched at the base.   **8. gramineus**
  Submerged lvs commonly 8–12 × 2–3 cm, acuminate or cuspidate; stipules 2–5 cm; plant not much branched at the base.   **6. ×zizii**
10 Grass-leaved; lvs narrowly linear or filiform, not exceeding 6 mm in width, parallel-sided.   19
   Lvs usually exceeding 6 mm wide, or if narrower linear-lanceolate, not parallel-sided.   11
11 At least the lower lvs ±amplexicaul.   12
   Lvs not amplexicaul.   13
12 Lvs usually narrowly to broadly ovate, all cordate and amplexicaul; stipules small, fugacious; fertile.
                                              12. perfoliatus
   Lvs lanceolate, narrowed to a cuspidate tip; at least the lower lvs rounded and semi-amplexicaul at the base; stipules usually ±persistent; a sterile hybrid.
                                                  9. ×nitens
13 Stem compressed; lvs commonly 4–9 × 1–1.5 cm, linear-oblong, margins distinctly serrate and often strongly undulate; beak equalling the rest of the fr.   **22. crispus**
   Stem ±terete; lf-margin not distinctly serrate to the naked eye; beak shorter than the rest of the fr.   14
14 Lvs blunt; peduncles not or hardly thickened upwards.   15
   Lvs not blunt; peduncles distinctly thickened upwards.   17
15 Lvs distinctly hooded at the tip and rounded at the base, commonly 10–18 × 2–4.5 cm, with microscopically entire margins.   **11. praelongus**
   Lvs not hooded.   16
16 Lvs rounded at base and apex, sessile, commonly 6–10 × 1.5–2 cm, margins microscopically denticulate at least when young; fr. abortive.   **7. ×salicifolius**

Lvs narrowed to the sessile base, commonly 6–12 × 1–2 cm; margins microscopically entire; fertile.

        10. alpinus

17 Plant much branched at the base; lvs commonly 2.5–10 × 0.3–1 cm, acute or acuminate, sessile;stipules 1–2 cm.   **8. gramineus**

   Plant not much branched at the base; lvs 8–20 × 2–5 cm; stipules 2–6 cm.   18

18 Lvs commonly 8–12 × 2–3 cm, lanceolate to oblong, cuspidate or rounded and mucronate.   **6. ×zizii**

   Lvs commonly 12–20 × 3.5–5 cm, oblong-lanceolate, cuspidate or acuminate.   **5. lucens**

19 Lvs with 2 large air-filled longitudinal canals, one on each side of the midrib, occupying the greater part of their volume; stipules adnate below to the lf-base, forming a stipular sheath with a free ligule.   20

   Lvs not as above; stipules free from the lf-base throughout their length.   21

20 Stipular sheath open and convolute with a whitish margin; fr. 3–5 × 2–4 mm, with a short beak terminating the ventral margin.   **24. pectinatus**

   Stipular sheath tubular below when young; fr. 2–3 × 2 mm with an extremely short almost central beak.   **23. filiformis**

21 Lvs with 3(–5) principal and many faint intermediate longitudinal strands; stems strongly compressed or even winged.   22

   Lvs usually with only 3–5 longitudinal veins or apparently 1-veined especially when very narrow, but never with many faint intermediate longitudinal strands; stems not or slightly compressed.   23

22 Lvs 10–20 cm × 2–4 mm, usually rounded and cuspidate at the tip, sometimes acuminate; principal lateral veins usually distinctly joining the midrib below the tip of the lf; stipules blunt; fr. smooth, with a straight beak.   **20. compressus**

   Lvs 5–13 cm × 2–3(–4) mm, finely acuminate; principal lateral veins usually not distinctly joining the midrib; stipules acute; fr. toothed near the base of the ventral margin and tubercled along the dorsal margin; beak recurved.   **21. acutifolius**

23 Lvs mostly 5-veined, the laterals closer to the margin and to each other than to the midrib; stipules tubular below.   **14. friesii**

   Lvs mostly 3-veined or apparently 1-veined, rarely 5-veined and then with equal spacing.   24

24 Lvs 2–3 mm wide, 3(–5)-veined, the laterals joining the midrib at a wide angle close below the blunt mucronate tip; stipules open, convolute.

        17. obtusifolius

   Lvs mostly less than 2 mm wide.   25

25 Stipules tubular below.   26

   Stipules open, convolute.   27

26 Stipules tubular for at least ⅔ of their length; lf-tip gradually acute with a narrow blunt terminal cusp.

        16. pusillus

   Stipules tubular only towards the base; lf narrowed to a sharply acuminate tip (Shetland and Outer Hebrides).   **15. rutilus**

27 Lvs rarely exceeding 1 mm wide, subsetaceous, narrowing to the base and tapering to a long fine point; usually no air-filled lacunae bordering the midrib; fr. often toothed below and tubercled on the dorsal margin, usually 1 per fl.   **19. trichoides**

   Lvs usually exceeding 1 mm wide, subacute to rounded and mucronate, with air-filled lacunae bordering the midrib; rarely narrower and acute; fr. smooth, usually 4 per fl.   **18. berchtoldii**

Subgenus 1. **POTAMOGETON**

Lvs all submerged or some floating, alternate (only the involucral ones opposite), variously shaped; stipules free from the lf throughout, or adnate only at the very base; stigma with small papillae; fr. spike not or hardly interrupted, its stalk and rhachis rigid; fr. drupaceous, with fleshy exocarp and bony endocarp; wind-pollinated.

1. P. natans L.   Broad-leaved Pondweed.

Incl. *P. hibernicus* (Hagström) Druce

Rhizome extensively creeping. Lfy stems commonly to 100 cm but reaching 500 cm in deep water, ±terete, not or little branched. Submerged lvs (phyllodes) 15–30 (–80) cm × 1–3 mm, linear, channelled, with several longitudinal veins, rarely with a small blade. *Floating lvs* stalked, 15–60 cm overall, with the blade 2.5–12.5 × 0.8–7 cm, elliptical to ovate-lanceolate, ±acute, rounded to subcordate at the base, *inrolled at the base* for a time after emergence, *coriaceous*, and with 2 coriaceous wings decurrent for a short distance down the stalk which therefore appears *jointed just below the blade*; longitudinal veins c. 20–25; transverse veins indistinctly visible against the light. Stipules 5–12(–18) cm, persistent, at length fibrous. Fr. spike 3–8 cm, cylindrical, dense; stalk 5–12 cm, axillary, stout, not enlarging upwards. Fr. (Fig. 59A) 4–5 × 3 mm, olive green, obovoid, somewhat compressed; ventral margin convex, dorsal ±semi-circular and keeled when dry; beak short, straight. Fl. 5–9. Turions usually 0. $2n = 52$. Hyd.

Fig. 59. Fruits of *Potamogeton*. A, *P. natans*; B, *P. polygonifolius*; C, *P. coloratus*; D, *P. lucens*; E, *P. gramineus*; F, *P. alpinus*; G, *P. praelongus*. ×3.

Native. Lakes, ponds, rivers, ditches, especially on a highly organic substratum, and usually in water less than 1 m deep. Common throughout the British Is. Northern hemisphere.

Readily recognized by the jointed blade of the floating lf and the very long stipules.

   P. × fluitans Roth (*P. lucens × natans*) and *P. × sparganifolius* Laest. ex Fries (*P. gramineus × natans*), both with blade-bearing submerged lvs, have been reported.

2. P. polygonifolius Pourret   Bog Pondweed.

P. oblongus Viv.; *P. anglicus* Hagström

Rhizome extensively creeping. Lfy stems commonly to 20 cm, but to 50 cm in deeper water, slender, terete, unbranched. *Lvs all with blade and stalk.* Submerged lvs with blade commonly 8–20 × 1–3 cm, ±narrowly lanceolate-elliptical but very variable in size and shape, membranous, translucent. *Floating lvs* with blade commonly 2–6 × 1–4 cm and stalk one-half to twice as long, broadly elliptical to lanceolate, cuneate to cordate at the base, *not jointed below the blade,* subcoriaceous, not translucent; longitudinal veins c. 20; transverse veins plainly visible against the light. Stipules 2–4 cm, blunt. Fr. spike 1–4 cm, cylindrical, dense; stalk much exceeding the spike, slender, not widening upwards. Fr. (Fig. 59B) 2 × 1.5 mm, reddish, obovoid, slightly compressed; ventral margin convex, dorsal semicircular and slightly keeled when dry; hardly beaked. Fl. 5–10. Turions little differentiated from lfy shoots. $2n = 26$. Hyd.

Native. Ponds, bog-pools, ditches and small streams with acid and usually shallow water; to 716 m in Wales. Common throughout the British Is. Europe; N.W. Africa; eastern North America.

Very variable in size and shape of lvs with varying water depth. A state which has been called f. *cancellatus* Fryer has thin, translucent, finely reticulate-veined submerged lvs somewhat resembling those of *P. coloratus.*

3. P. coloratus Hornem.   Fen Pondweed.

P. plantagineus Du Croz ex Roemer & Schultes

Resembling *P. polygonifolius* in habit but *all lvs with stalk usually shorter than blade,* ±blunt, *thin, translucent, finely and distinctly reticulate-veined.* Submerged lvs 6–10(–18) cm, linear-lanceolate to narrowly elliptical, often subsessile. Floating lvs with blade 2–7(–10) × 1.5–5 cm, ovate-elliptical. Stipules 2–4 cm, blunt, ±persistent. Fr. spike 2.5–4 cm, cylindrical, dense; stalk 5–20 cm, slender, not widening upwards. Fr. (Fig. 59C) 1.7–2 × c. 1 mm, green, ovoid, compressed; ventral margin slightly convex, dorsal semicircular and slightly keeled when dry; beak very short, curved. Fl. 6–7. Turions little differentiated from lfy shoots. $2n = 26$. Hyd.

Native. Shallow ponds and pools in fen peat, especially in calcareous water. Local throughout the British Is. northwards to Argyll; Hebrides. Ireland. Europe.

P. × billupsii Fryer (*P. coloratus × gramineus*) and *P. × lanceolatus* Sm. (*P. berchtoldii × coloratus*) have both been recorded from a few localities in the British Is.

4. P. nodosus Poiret   Loddon Pondweed.

P. petiolatus Wolfg.; *P. drucei* Fryer; *P. × fluitans* Roth, pro parte

Rhizome creeping. Lfy stems commonly 20–30 cm but reaching 2 m, terete, stout, simple. *Lvs all with a ±long-stalked blade. Submerged lvs* with blade 10–20 × 1.5–4 cm, broadly lanceolate, or elliptical-lanceolate, ±*equally narrowed* at each end, ±*cuspidate,* thin, translucent, *very finely and beautifully reticulate-veined.* Floating lvs with blade 6–15 × 2.5–6 cm, oblong-elliptical or ovate, shortly mucronate, coriaceous; transverse veins visible against the light. Stipules 7–10 cm, lanceolate. Fr. spike 2–6 cm, cylindrical, fairly dense but often with many abortive fr.; stalk long, stout, not thickening

upwards. *Fr.* c. 3.5 × 2.5 mm, obovoid; ventral margin convex, dorsal almost semicircular and *acutely keeled;* beak short but stout (c. 2 × 1.5 mm). Fl. 8–9. Turions borne on slender stolons. Hyd.

Native. Gravelly shallows and deeper water or slow-flowing base-rich rivers. Known only in the Avon (Somerset, Gloucester and Wilts), Stour (Dorset), Thames (Oxford, Bucks and Berks) and Loddon (Berks). Mediterranean region, Canary Is., Madeira, Azores; W. and C. Europe northwards to Poland and Germany.

Easily recognizable by the ±lanceolate beautifully net-veined submerged lvs, which are quite different from those of any other British sp.

5. P. lucens L.   Shining Pondweed.

Incl. *P. acuminatus* Schumacher and *P. longifolius* Gay ex Poiret

Rhizome creeping extensively. Lfy stems 0.5–2(–6) m, stout and tough. *Submerged lvs* commonly 10–20 × 2.5–6 cm, *oblong-lanceolate,* rarely ovate, *shortly stalked* with the blade decurrent on the stalk and so appearing subsessile, rounded and cuspidate or apiculate at the tip, or acuminate, margins minutely denticulate, thin, translucent, shining; longitudinal veins c. 11–13. *Floating lvs 0.* Stipules 3–8 cm, blunt, *prominently 2-keeled.* Fr. spike 5–6 cm, cylindrical, stout, dense; *stalk* 7–25 cm, or more, stout, *thickening upwards.* Fr. (Fig. 59D) c. 3.5 × 2.2 mm, olive-green, ovoid, swollen, hardly compressed; ventral margin almost straight, dorsal ±semicircular, hardly keeled; beak short. Fl. 6–9. Turions produced on the rhizome-system. $2n = 52$. Hyd.

Native. Lakes, ponds, canals and slow streams on nutrient-rich inorganic substrata; commonly in calcareous water, where the lvs become chalk-encrusted; to 380 m in Malham Tarn (Yorks). Locally common in S. and E. England but not in Devon or Cornwall, very local in Wales and N.W. England and thinly scattered northwards to Nairn and Skye. Europe; W. Asia.

One of the largest-leaved British spp., known by the large oblong-lanceolate short-stalked submerged lvs with denticulate margins and the upwardly thickened stalk of the fr. spike. *P. × cadburyi* Dandy & Taylor (*P. crispus × lucens*) and *P. × nerviger* Wolfg. (*P. alpinus × lucens*) have been reported.

6. P. × zizii Koch ex Roth

P. angustifolius auct. mult.; *P. gramineus × lucens*

Rhizome creeping. Lfy stems 20–50(–150) cm, terete, stout, branching below. *Submerged lvs* ±sessile or very shortly stalked, their blades c. 3.5–15 × 0.7–2.5 cm *oblanceolate to oblong-lanceolate, narrowed to the base* and either cuspidate or abruptly rounded and mucronate at the apex, margins *undulate* and irregularly denticulate throughout, thin, translucent. *Floating lvs* often produced, *oblong-lanceolate,* usually *cuspidate, distinctly stalked* but the stalk never exceeding the blade, coriaceous, with denticulate margins. Stipules 1.5–4 cm, broad, blunt, 2-keeled. Fr. spike 3–6 cm, cylindrical, with *mature achenes very variable in number, sometimes*

0; *stalk* commonly 7–12 cm, but very variable, stout, *thickened upwards*. Fr. c. 3 × 2 mm, ovoid, compressed; ventral margin almost straight, dorsal ±semicircular, strongly keeled and with lateral ridges; beak very short. Fl. 6–9. Turions at the ends of stolons borne on the rhizome. Hyd.

Native. Ponds, lakes, streams, etc.; local. Widely distributed throughout Great Britain from Surrey northwards to Ross; Ireland; Orkney. Europe.

Resembles *P. lucens* but distinguished by the presence (commonly) of floating lvs, the absence of a stalk in at least some of the submerged lvs, and by the smaller fr.

7. P. × salicifolius Wolfg.

P. lucens × perfoliatus; *P. × decipens* Nolte ex Koch

Rhizome creeping. Lfy stems long, commonly to 1 m, rarely to 3 m, terete, stout, slightly branched above. *Submerged lvs* very variable; blades 3–20 × 1.5–4 cm, *oblong, ±sessile and clasping at the base and ±rounded at the non-hooded apex*, margins *at first minutely denticulate* but sometimes becoming smooth later, thin, translucent. *Floating lvs* 0. Stipules c. 2.5–3 cm, blunt, slightly 2-keeled, persistent. *Spike* 2.5 cm, dense, *always sterile*; stalk to 8 cm, not or slightly thickened upwards. Fl. 6–9. Hyd.

Native. Ponds, canals, rivers, etc., with the parents. Local, but scattered throughout Great Britain from Dorset, Surrey and Kent northwards to Perth; Ireland from Clare to Dublin and Fermanagh; Inner Hebrides. Europe.

Resembles *P. lucens* but distinguished by its sessile and smaller lvs (commonly 6–10 × 1.5–2 cm) with ±clasping base, its scarcely keeled stipules and its sterility.

8. P. gramineus L.   Various-leaved Pondweed.

P. heterophyllus Schreber.; incl. *P. lonchites* Tuckerman and *P. graminifolius* H. & J. Groves

Rhizome creeping. Lfy stems to 1 m or more, terete, slender, flexuous, with *very numerous short non-flowering branches near the base*. *Submerged lvs* 2.5–8 (–18) × 0.3–1.2(–3) cm, usually *linear-lanceolate* or elliptical-oblong, *narrowed into a sessile base*, acuminate or cuspidate, with *minutely serrate* margins, thin, translucent; longitudinal veins 7–11. *Floating lvs few or 0, long-stalked*; blade 2.5–7 × 1–2.5 cm, usually broadly *elliptical-oblong*, ±rounded at the base, coriaceous, with transverse veins visible against the light. Stipules 2–5 cm, c. half as long as the lower internodes, broadly lanceolate, acute, not keeled. Fr. spike c. 2.5–5 cm, cylindrical, dense; *stalk* 5–8(–25) cm, stout, *thickened upwards*. Fr. (Fig. 59E) 2.5–3 × c. 2 mm, green, ovoid, ±compressed; ventral margin nearly straight, dorsal semicircular and slightly 3-keeled; beak short, ±straight. Fl. 6–9. Turions produced at the ends of stolons. $2n = 52$. Hyd.

Native. Lakes, meres and ponds and occasionally in canals and slow streams; on bottom muds in water with a wide range of acidity but commonly circumneutral or somewhat acid though sometimes calcareous. Locally frequent through much of Great Britain and Ireland but not in S.W. England or Channel Is. and in only a few scattered localities in Wales; Inner and Outer Hebrides, Orkney and Shetland. Much of Europe but rare in the south; North America; Greenland, boreal Asia.

Very variable, with numerous habitat-forms to which many names have been given. Plants with small narrow submerged lvs not exceeding 5 cm have been named var. *lacustris* (Fries) Ascherson & Graebner, and others, usually without floating lvs, with submerged lvs 5–15 cm, var. *fluvialis* (Fries) Blytt.

P. × nericius Hagström (*P. alpinus × gramineus*) has been reported from Scotland.

9. P. × nitens Weber

P. gramineus × perfoliatus

Rhizome creeping. Lfy stems to 120 cm, terete, slender, *sparingly branched*. *Submerged lvs* 2–8(–15) × 0.5–1.5 (–2) cm, *oblong-lanceolate*, *rounded*, ±*cordate* and usually *semi-amplexicaul* at the sessile base, ±acute or cuspidate, undulate and at first minutely denticulate at the margins (which may later become smooth), thin, translucent; longitudinal veins 7–15. *Floating lvs few or 0*, elliptical-oblong, long-stalked to sessile or semi-amplexicaul, coriaceous. Stipules 1–2 cm, lanceolate, acute, not keeled, persistent. *Spike* 1–2 cm, cylindrical, dense, *sterile*; stalk 2.5–10(–15) cm, variable in thickness, usually widest at or above the middle. Fl. 6–8. Turions produced at the ends of stolons from the rhizome. Hyd.

Native. Lakes, ponds, streams, etc., to 610 m; rather local throughout the British Is. In Great Britain recorded from Devon and Surrey and from Suffolk, Cambridge and Anglesey northwards; Inner and Outer Hebrides, Orkney and Shetland; scattered in Ireland. Northern hemisphere.

A plant having much the habit of *P. gramineus* but lacking the dense basal branching and having the lvs rounded to semi-amplexicaul at the base; always sterile.

Very variable, and has been divided by Hagström into varieties differing in the form of the involucral lvs which subtend the spikes; in var. *subgramineus* (Raunk.) Hagström they are distinctly stalked; in var. *subperfoliatus* (Raunk.) Hagström sessile and rounded to semi-amplexicaul at the base, and in var. *subintermedius* Hagström short-stalked or sessile, sometimes with one rounded at the base.

10. P. alpinus Balbis   Reddish Pondweed.

P. rufescens Schrader

Rhizome creeping. Lfy stems 15–20 cm, terete, slender, ±simple. *Submerged lvs* usually 6–15 × 1–2 cm, often reddish, *narrowly oblong-elliptical*, *narrowed to each end*, subsessile or shortly stalked, always *blunt, entire*, thin, translucent, with a conspicuously reticulate midrib; longitudinal veins 7–11. Floating lvs 3–8 × 0.8–2 cm, oblanceolate- to obovate-elliptical, narrowing to a short stalk, blunt, entire, subcoriaceous, with the transverse veins easily visible against the light; sometimes 0. *Stipules* 2–6 cm, shorter than the internodes, *ovate, blunt,* not keeled, robust. Fr. spike 2–4 cm, ±cylindrical, dense; stalk 5–18 cm, slender, not thickened upwards. Fr. (Fig. 59F) 3 × 2 mm, becoming pale reddish, ovoid-acuminate, somewhat compressed; ventral margin very convex, dorsal semicircular and sharply keeled, narrowed above subequally into the fairly long somewhat curved beak. Fl. 6–9. $2n = 52$. Hyd.

Native. Lakes, ditches, streams, etc., especially in non-calcareous water and on substrata rich in organic matter; to 1020 m in Scotland. Throughout the British Is. except Cornwall, Orkney and Shetland, but local in S.W. England and S. Wales. Northern hemisphere.

P. × griffithii Ar. Benn. (*P. alpinus × praelongus*), *P. × prussicus* Hagström (*P. alpinus × perfoliatus*) and *P. × olivaceus* Baagöe ex G. Fischer (*P. venustus* Ar. Benn.; *P. alpinus × crispus*) have all been recorded from two or more localities.

11. P. praelongus Wulfen   Long-stalked Pondweed.

Rhizome creeping. Lfy stems 0.5–2(–6) m, terete, stout, somewhat branched above. *Submerged lvs* 6–18 × 2–4.5 cm, green, strap-shaped to oblong, usually 4–6 times as long as wide, *rounded at the sessile ±semi-amplexicaul base*, narrowing gradually to the *blunt and hooded apex*, slightly undulate, *entire*, thin, translucent, not shining; longitudinal veins c. 13–17, with 1 strong and 5–7 weaker veins on each side of the midrib. *Floating lvs 0*. Stipules 0.5–6 cm, often exceeding the internodes, blunt, thin, not ridged or keeled. Fr. spike 3–7 cm, cylindrical, dense; stalk 15–40 cm, fairly stout, not widening upwards. Fr. (Fig. 59G) 4–6 × 3–4 mm, green, asymmetrically obovoid, hardly compressed; ventral margin very slightly convex, dorsal ±semicircular and sharply keeled or winged; beak short, straight. Fl. 5–8. No specialized turions. $2n = 52$. Hyd.

Native. Lakes, East Anglian Broads, ditches, canals and slow streams; typically in deep clear water on substrata not very rich in organic matter; to 915 m in Scotland. Rather local in Great Britain northwards from Essex, Surrey, Berks and Radnor; frequent in some of the deeper English Lakes and reaching northernmost Scotland; Inner and Outer Hebrides, Orkney and Shetland; in Ireland mainly near the north and west coasts. N. and C. Europe; North America; N. Asia.

A very distinct species recognizable by its sessile, tapering, hooded lvs and its very long peduncles.

P. × cognatus Ascherson & Graebner (*P. perfoliatus × praelongus*) and *P. × undulatus* Wolfg. (*P. crispus × praelongus*) have been reported from Britain, the latter also from Ireland.

12. P. perfoliatus L.   Perfoliate Pondweed.

Rhizome extensively creeping. Lfy stems 0.5–2(–3) m, terete, usually stout, branching above. *Submerged lvs* 2–6(–10) × 1.3–4(–6) cm, all *sessile and ±completely amplexicaul at the wide cordate base*, commonly ovate, but very variable in shape from lanceolate to orbicular, blunt or rarely mucronate, irregularly and microscopically denticulate, very thin, translucent; 5–7 strong longitudinal veins with fainter intermediate veins. *Floating lvs 0*. Stipules up to 1 cm, blunt, not keeled, very delicate

and *soon disappearing* or persisting only where the lvs are opposite. Fr. spike 1–3 cm, cylindrical, stout, dense; stalk 3.5–10(–13) cm, stout, not thickened upwards. Fr. (Fig. 60A) 3.5–4 × 2.5–3 mm, olive-green, hardly compressed; ventral side concave below but convex above, dorsal semicircular, obscurely keeled with faint lateral ridges. Fl. 6–9. Turions formed on stolons from the rhizomes. $2n = 52$. Hyd.

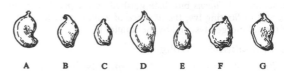

Fig. 60. Fruits of *Potamogeton*. A, *P. perfoliatus*; B, *P. friesii*; C, *P. pusillus*; D, *P. obtusifolius*; E, *P. berchtoldii*; F, *P. trichoides*; G, *P. compressus*. ×3.

Native. Lakes, ponds, streams, canals, etc., especially on substrata of moderate but not very high organic content; to 700 m in Scotland. Common throughout the British Is. Northern hemisphere; Australia.

Recognizable by the wide-based cordate and amplexicaul lvs, which are all submerged, and the fugacious stipules. The great variability in lf size and shape is largely if not wholly dependent on factors of the external environment. Narrow-leaved forms occur in low light intensity and on substrata of low calcium content, while very broad-leaved forms are characteristic of highly calcareous substrata. *P. × nitens* frequently has floating lvs and its submerged lvs are typically narrower than those of *P. perfoliatus*, and only semi-amplexicaul, with persistent stipules: the plant is, moreover, always sterile.

P. × cooperi (Fryer) Fryer (*P. crispus × perfoliatus*) has been reported from numerous scattered localities.

13. P. epihydrus Rafin.   Leafy Pondweed.

P. nuttallii Cham. & Schlecht.; *P. pennsylvanicus* Willd. ex Cham. & Schlecht.

Rhizome creeping. *Stems* slender, *compressed*, mostly simple. *Submerged lvs* 8–20 cm × 3–8 mm, *linear, sessile*, tapered to a blunt or subacute apex, thin, translucent; longitudinal veins (3–)5(–7); there are transverse veins but no fainter intermediate longitudinal veins; *midrib bordered by a well-marked band of lacunar tissue* which extends sideways towards or beyond the inner lateral veins. *Floating lvs oblong to elliptical*, blunt, tapered into the stalk, coriaceous, the blade 3.5–7 × 1–2.5 cm. *Stipules* up to 3.5 cm, *open*, broad, subtruncate or rounded. Spike 1–2.5 cm, cylindrical, dense and continuous; stalk 3–4 cm, slender, not thickened upwards. Fr. 2.5–3.5 mm, round-obovoid, 3-keeled, sides flat; beak short; *embryo subspiral (coiled more than 1 complete turn)*. Fl. 6–8. Hyd.

Native. Lakes. Known as a native plant only from S. Uist, Outer Hebrides. Also introduced in S.W. Yorks and S.E. Lancs, in the R. Calder and canals in the neighbourhood of Halifax. Widely distributed in North America.

A very distinct species, known as an alien in S.W. Yorks for more than 40 years but only recently discovered as a native in the Outer Hebrides. The distribution of the species recalls that of *Eriocaulon aquaticum*. The British plants are referable to var. *ramosus* (Peck) House.

14. P. friesii Rupr.   Flat-stalked Pondweed.

P. compressus auct. mult.; *P. mucronatus* Schrader ex Sonder

Rhizome 0. *Stem* 20–100 cm, arising from a turion, *strongly compressed*, slender, ±simple below, branching above, and producing *numerous very short lfy branches or fascicles of lvs* which later develop turions. Submerged lvs commonly 4–6.5 cm × 2–3 mm, linear, sessile, subacute or blunt and abruptly mucronate, thin, pale green, translucent; *longitudinal veins usually* 5 (occasionally 3 or 7), the distance between the midrib and the nearest lateral veins being almost twice that between the laterals and between the farther lateral and the margin; there are transverse veins but no fainter intermediate longitudinal veins. Floating lvs 0. *Stipules* 0.7–1.5 cm, *tubular below* at first but soon splitting, fibrous-persistent, whitish. Fr. spike 0.7–1.5 cm, with 3–4 remote whorls of fr.; stalk 1.5–5 cm, flattened, somewhat thickened upwards. Fr. (Fig. 60B) 2–3 × 1.5–2 mm, olive, asymmetrically obovoid, slightly compressed; ventral face strongly convex above, dorsal ±semicircular, bluntly 3-keeled; beak c. 5 mm, erect or recurved. Fl. 6–8. *Turions* terminal on short lateral branches, *prominently ribbed towards the base.* $2n = 26$. Hyd.

Native. Lakes, ponds, canals, etc., especially on a muddy substratum; lowland. Frequent in E. and C. England northwards to N. Yorks and westward to Avon and Shropshire; very rare in E. Wales and in only a few widely scattered localities in Scotland; Outer Hebrides, Orkney; rare in C. and W. Ireland. Northern hemisphere.

Distinguishable from *P. obtusifolius*, which sometimes has 5 veins with no faint intermediates, by the unequal spacing of the veins, the tubular stipules, the longer peduncles and the smaller fr.

P. × *lintonii* Fryer (*P. crispus* × *friesii*) is known from several localities. It resembles *P. crispus* in habit, and has irregularly and microscopically denticulate lvs whose stipules are tubular for a short distance. *P.* × *pseudofriesii* Dandy & Taylor (*P. acutifolius* × *friesii*) has also been reported.

15. P. rutilus Wolfg.   Shetland Pondweed.

Rhizome 0. Stem 3–60(–100) cm, arising from a turion, compressed, very slender, much branched. *Submerged lvs* 3–6 cm × 0.5–1 mm, very narrowly linear, sessile, *gradually tapering to a finely pointed apex*, bright green or reddish below, thin, translucent; longitudinal veins usually 3, sometimes 5 below, the laterals joining the midrib at some distance below the apex or vanishing without doing so; there are no faint intermediate longitudinal veins. Floating lvs 0. *Stipules* 1–2 cm, *tubular below*, acuminate, *strongly veined*, ±*fibrous-persistent*. Fr. spike 5–10 mm, shortly cylindrical, 6–8-fld; stalk

c. 2–4 cm, slightly thickened upwards. Fr. 1–2 mm, brownish-red, semi-ovoid; ventral margin slightly convex, dorsal semicircular, not keeled; beak short, stout. Fl. 8. Turions formed at the tips of lateral branches. Hyd.

Native. Lakes. Known from Easterness, the Outer Hebrides, Tiree and Shetland. Europe; W. Asia.

Distinguishable from *P. pusillus* by the stipules, which in *P. rutilus* are tubular for a shorter distance, firmer, more strongly veined and fibrous-persistent; and by the gradually tapered finely pointed lf-tip.

16. P. pusillus L.

P. panormitanus Biv.

Rhizome usually 0. Stem 20–100 cm, generally arising from a turion, slightly compressed, very slender, usually with many long branches from near the base. Submerged lvs 1–4(–7) cm × 0.3–(–3) mm, narrowly linear, sessile, narrowing gradually to a long but blunt tip, firm, translucent; longitudinal *veins* 3 (*rarely* 5), *the laterals joining the midrib, usually at a narrow angle and at a distance of* 2–3 *lf-widths below the tip*; *midrib* except of the uppermost lvs, usually *not bordered by pale bands of large-celled lacunae*, or sometimes by a single row on each side. Floating lvs 0. *Stipules* 0.6–1.7 cm, *tubular to above half-way*, splitting later, ±persistent, pale brown. Spikes 6–12 mm, terminating very short branches, cylindrical, interrupted, of c. 2–8 fls in 2–4 whorls; stalk 1.5–3 cm or more, filiform, not thickened upwards. *Fr* (Fig. 60C) 2–2.5 × 1–1.5 mm, pale olive, obovoid, *smooth*; ventral face convex, dorsal more strongly convex, broadly and obscurely keeled when dry; beak c. 0.4 mm, almost centrally placed, ±straight. Fl. 6–9. Turions 10–15 × 0.5 mm, *narrowly fusiform*, *chiefly axillary* and produced first towards the base of the branches, later upwards. $2n = 26$. Hyd.

Native. Lakes, ponds, canals, streams; especially in highly calcareous or even brackish waters; not uncommon. Throughout the British Is. but very local in N. Scotland and Ireland; Hebrides, Orkney and Shetland. Northern hemisphere; Africa.

Often confused with *P. berchtoldii* (*P. pusillus* auct., non L.) but distinguishable by the usual absence of lacunae along the midrib, the tubular stipules and the slender axillary turions. *P.* × *trinervius* G. Fischer (*P. pusillus* × *trichoides*) has been reported.

17. P. obtusifolius Mert. & Koch   Grassy Pondweed.

P. gramineus auct. mult., non L.

Rhizome 0. Stem 20–100 cm, slender, compressed (c. 2:1), frequently forked and with many short lateral lfy branches. *Submerged lvs* 3–9 cm × 2–4 mm, linear, *narrowed to the sessile base*, rounded and shortly apiculate at the tip, dark green, thin, very translucent; longitudinal veins 3, sometimes 5, the laterals joining the midrib at a wide angle, sometimes greater than a right-angle, and usually close below the tip; faint intermediate longitudinal veins 0; midrib bordered by pale bands of elongated lacunae, especially towards the base. Floating lvs 0. *Stipules* 1.3–2 cm, *open*, *broad*, *blunt*, with many faint veins. Spike 0.6–1.3 cm, ovoid to stoutly cylindrical, dense and continuous, fruiting freely; stalk 0.8–2 (–3.5) cm, slender, straight, not thickened upwards. Fr. (Fig. 60D) 3–4 × c. 2 mm, brownish-olive, oblong-ovoid, slightly compressed; ventral face convex, dorsal semicircular, 3-keeled when dry; beak c. 0.6 mm, straight. Fl. 6–9. Turions 2–4 cm × 3.5–7 mm, narrowly fan-shaped, terminal. $2n = 26$. Hyd.

Native. Lakes, ponds, streams, canals, ditches, etc.; local. Throughout Gt. Britain from S. Devon and Kent northwards to Easterness and Argyll, but very local in S.W. England and Wales; Colonsay. Orkney. Northern hemisphere.

Distinguishable from *P. compressus* and *P. acutifolius* by the absence of faint intermediate longitudinal veins, and from *P. friesii* by having usually only 3 veins (or 5 evenly spaced), by the open stipules and the shorter peduncles.

P. sturrockii (Ar. Benn.) Ar. Benn., from Marlee Loch in E. Perth, has been regarded as a distinct sp., as a subsp. of *P. berchtoldii*, and as a hybrid of *P. obtusifolius* and *P. pusillus*. It more probably represents a slender state of *P. obtusifolius*, narrow-leaved forms of which may easily be confused with *P. berchtoldii*.

18. P. berchtoldii Fieber   Small Pondweed.

P. pusillus auct. mult, non L.; incl. *P. tenuissimus* (Mert. & Koch) Reichenb., *P. lacustris* (Pearsall & Pearsall fil.) Druce and *P. millardii* Heslop-Harrison

Rhizome usually 0. Stem 10–100 cm, usually arising from a turion, very slender, very little compressed (less than 2:1), almost simple to freely branched; internodes commonly 1–5 cm. Submerged lvs 2–5.5 cm × 0.5–2 mm, linear, sessile, rounded to acute at the tip usually shortly mucronate, dark green, thin, translucent; *longitudinal veins always* 3, the laterals close to the margin, meeting the midrib almost at right angles and at ½–1 *lf-width below the tip*; *no faint intermediate longitudinal veins*, but transverse veins often present; midrib bordered at least towards the base by pale bands of lacunae. Floating lvs 0. *Stipules* 3–10 mm or more, *open*, convolute, *blunt*, faintly 6–8-veined, ±deciduous but persisting in the axils of the uppermost lvs. Fr. spike 2–8 mm, subglobose, continuous or slightly interrupted; stalk 0.5–3(–10) cm, filiform, hardly thickened upwards. Fr. (Fig. 60E) 2–2.5 × c. 1.5 mm, dark olive, ±obovoid-acuminate, hardly compressed, ±tubercled below when dry; ventral margin convex, dorsal rounded, broadly and bluntly keeled when dry; beak short. Fl. 6–9. Turions 7–15 × 0.5–2.5 mm, *terminal*, fusiform, olive. $2n = 26$. Hyd.

Native. Lakes, ponds, canals, streams, ditches, etc., in very calcareous to very acid waters. Common throughout the British Is. Northern hemisphere.

Names have been given to some of the numerous states of this sp. growing in different habitat conditions. Thus the name *P. tenuissimus* has been used for plants, usually of shallow ditches and pools, in which the lvs are very narrow, dark-coloured and with the midrib bordered by lacunae only towards the base. At the other extreme are plants found in deep, clear, nutrient-poor waters, as in the English Lakes, in which the lvs tend to be broad, very delicate and light-coloured, the apex rounded and minutely mucronate. Plants of this form were formerly confused with *P. sturrockii* and have also been treated as a distinct subspecies or species (*P. lacustris*). For differences from *P. pusillus* see under that species.

P. × *sudermanicus* Hagström (*P. acutifolius* × *berchtoldii*) is known from Dorset.

19. P. trichoides Cham. & Schlecht.

Hair-like Pondweed.

Rhizome filiform or 0. Stems 20–75 cm, filiform, terete or very slightly compressed, repeatedly and divaricately branched, most of the branches ultimately bearing spikes. *Submerged lvs* 2–4(–6.5) cm × c. 0.5–1 mm, deep dull green, *narrowly linear to subsetaceous*, narrowed to the sessile or semi-amplexicaul base and tapering to a *long fine point*, spreading, ±rigid, translucent; *longitudinal veins* 3, the midrib thick and prominent but the *laterals faint* and often indistinct even under the microscope, *midrib usually not bordered by lacunae*. Floating lvs 0. *Stipules* 7–11(–20) mm, *open* and convolute, narrow, ±acute, somewhat rigid. Spikes 1–1.5 cm, 3–6 fld, ±ovoid, interrupted, usually with only 1–3 carpels per fl. and ripening only 1 *fr. per fl.*; stalk 5–10 cm, filiform, not thickened upwards, often curved above. Fr. (Fig. 60F) c. 2.5 × 2 mm, ovoid, somewhat compressed, often with a tooth near the base of each side; ventral margin almost straight, often toothed near its base, dorsal strongly rounded, obscurely keeled and often ±tuberculate; beak short, straight. Fl. 6–9. $2n = 26$. Hyd.

Native. Ponds, canals, ditches, etc. Chiefly in S. and E. England from E. Cornwall, Somerset and Gloucester to Sussex and Kent and then northwards to Lancs and Yorks; also in Glamorgan and Anglesey, Dunbarton and Stirling. Europe; W. Asia; Africa.

Distinguished from *P. pusillus* and *P. rutilus* by the open stipules and from *P. berchtoldii* by the very narrow obscurely 3-veined finely pointed lvs with no lacunae bordering the thick midrib; and from all three by the single often tuberculate fr. from each fl.

P. × *bennettii* Fryer (*P. crispus* × *trichoides*) is known only from Stirling.

20. P. compressus L.   Grass-wrack Pondweed.

P. zosterifolius Schumacher

Rhizome terete or 0. *Stem* 0.5–2 m × 3–6 mm, *strongly flattened and ±winged*, branched. *Submerged lvs* 10–20 cm × 2–4 mm, linear, sessile, rounded and cuspidate or sometimes acuminate, thin, translucent; main longitudinal veins usually 5, *with many faint intermediate longitudinal strands* most of which approach the main laterals below the tip. Floating lvs 0. Stipules 2.5–3.5 cm, open, convolute, very blunt, 2-keeled, persistent. Spike 1–3 cm, cylindrical, rather dense, continuous; stalk 3–6 cm or more, stout, compressed, not thickened upwards. *Fr.* (Fig. 60G) 3–4.5 × 2–3 mm, ±obovoid, somewhat compressed; ventral margin convex, dorsal semicircular, smooth, bluntly 2-keeled, usually *not toothed*; beak very short, stout, almost centrally placed. Fl. 6–9. Turions with the inner lvs exceeding the outer by up to 10 mm. 2*n* = 26. Hyd.

Native. Lakes, slow streams, canals, ponds, ditches, etc., local. C. and E. England from Gloucester, Surrey and Essex northwards to Lancs and N. Yorks; Montgomery, Flint; Angus. Europe, but rare in the Mediterranean region.

Readily distinguished from all other linear-leaved spp. except *P. acutifolius* by the numerous longitudinal sclerenchyma strands. For differences from *P. acutifolius* see under that species.

21. P. acutifolius Link   Sharp-leaved Pondweed.
P. cuspidatus Schrader

Rhizome terete or 0. Stem 0.5–1 m × 2–4 mm, strongly flattened, much branched. Submerged lvs 5–13 × 2–3 (–4) mm, linear, sessile, gradually acuminate or long-cuspidate, thin, translucent; main longitudinal veins 3, the laterals sometimes hardly stronger than the main faint intermediate longitudinal strands which usually have free ends just below the lf-tip. Floating lvs 0. Stipules 1.5–2.5 cm, open, acute, many-veined, becoming fibrous and ±persistent. Spike 4–10 mm, ovoid to globose, 4–8-fld, dense; *stalk* 5–15(–35) mm, commonly *about equalling the spike*, slender, not thickened upwards. Fr. (Fig. 61A) 3–4 × 2 mm, greenish-brown, half-obovoid, compressed; *ventral margin* nearly straight, *toothed near the base and continuing upwards into the beak*, *dorsal margin* curved into more than a semicircle, somewhat keeled and ±crenulate; *beak* fairly long and *recurved*. Fl. 6–7. Turions with the inner lvs usually not protruding beyond the outer. 2*n* = 26. Hyd.

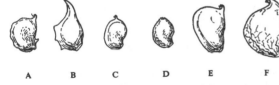

Fig. 61. Fruits of *Potamogeton*. A, *P. acutifolius*; B, *P. crispus*; C, *Groenlandia densa*; D, *P. filiformis*; E, F, *P. pectinatus*. ×3.

Native. Ponds, streams, ditches, chiefly in calcareous water; rare. S. and E. England from Dorset to Kent and northwards to Gloucester, Warwick, Northampton, Lincoln and E. Yorks. Europe.

Closely resembles *P. compressus* in vegetative structure, but distinguished by the usually narrower stem and lvs, the latter short and more gradually acuminate, with many free-ending longitudinal strands and a better-marked band of air tissue along the midrib, by the shorter and more acute stipules, and the more slender turions with the inner lvs not protruding. The spike is shorter with a stalk about equalling it, instead of about twice as long as in *P. compressus*, and the fr. differs considerably.

22. P. crispus L.   Curled Pondweed.
Incl. *P. serratus* auct. mult.

Rhizome creeping, perennating or arising from a turion. Stem 30–120 cm, slender, compressed, ±4-angled with the broader sides furrowed when mature, simple below,

repeatedly forked above. *Submerged lvs* 3–9(–10.5) cm × (5–)8–15 mm, *lanceolate to linear-lanceolate*, sessile, *rounded and blunt* or rarely acute and slightly mucronate, *serrate*, and often *strongly undulate* at the margin when mature, though commonly not so when young, often reddish, shining, translucent; longitudinal veins 3–5, the laterals close to the margin, with no faint intermediate longitudinal veins. Floating lvs 0. Stipules 10–20 mm, ±triangular, narrowed below, blunt, all but the uppermost evanescent, soon becoming torn and decayed. Fr. spike c. 1–2 cm, oblong-ovoid, rather lax; stalk 2.5–7(–10) cm, fairly slender, somewhat compressed, narrowed upwards and commonly curved. Fr. (Fig. 61B) 2–4 mm (excluding beak), dark olive, ovoid-acuminate, compressed; ventral margin convex, dorsal strongly rounded and ±keeled; *beak about equalling rest of fr.*, tapering, ±*falcate*. Fl. 5–10. Turions 1–5 cm, with thick, horny, spiny-bordered, ±squarrose lvs. 2*n* = 52. Hyd.

Native. Common but decreasing in lowland lakes, ponds, slow streams, canals, etc., throughout Great Britain and reaching 400 m in Scotland but very local in Wales and absent from much of N.W. Scotland; Outer Hebrides; Orkney; locally frequent in Ireland. Europe northwards to 61°30′ N. in Finland; N. and C. Asia eastwards to Japan; introduced in North America and Australia.

Readily recognized by its serrate and usually undulate lvs and by the highly characteristic fr. with its long curved beak. Young growths and other phases in which the lvs are flat and not crisped at the margins have often been named '*P. serratus*'.

Subgenus 2. COLEOGETON (Reichenb.) Raunk.

Lvs all submerged, alternate (only the involucral ones opposite), narrowly linear, sheathing at the base, entire, with 2 wide longitudinal air-filled canals, one on each side of the midrib, occupying the greater part of their interior; stipules adnate below to the lf-base (forming the basal sheath), but free above as a ligule; spike few-fld, interrupted; stigma with large papillae; fr. drupaceous with fleshy exocarp and bony endocarp; water-pollinated.

23. P. filiformis Pers.   Slender-leaved Pondweed.
P. marinus auct.

Rhizome extensively creeping. Lfy stem 15–30(–45) cm, filiform, ±cylindrical, sparingly forked or branched at the base. *Submerged lvs* 5–20 cm × 0.25–1 mm, *linear-setaceous*, *tapering*, *blunt* (*not mucronate*), yellowish-green, ±translucent; longitudinal veins 3, but *only midrib readily visible*, the laterals faint and actually in the margin. Floating lvs 0. *Stipular sheath* 0.5–3 cm, *tubular* below when young, its ligule 0.5–1.5 cm, acute, deciduous. Spike 4–12 cm, slender, interrupted, of 2–5 usually 2-fld whorls; stalk 5–25 cm, filiform not thickened upwards. Fr. (Fig. 61D) 2–3 × 2 mm, pale olive, ±obovoid, slightly compressed; *ventral margin convex*, dorsal semicircular, bluntly 3-keeled when dry; beak very short, almost central. Fl. 5–8. Perennation by tuberized winter buds borne on the rhizome. $2n = $ c. 78. Hyd.

Native. Lakes and streams, chiefly near the coast and sometimes in brackish water, very local. Anglesey, and Scotland from Berwick and Ayr northwards; Inner and Outer Hebrides, Orkney, Shetland; N. and W. Ireland; not in England. Northern hemisphere.

Distinguished from *P. pectinatus* by the tubular stipular sheath and the extremely short and ±central beak of the fr. and by the blunt (not mucronate) lvs.

P. × suecicus K. Richter (*P. filiformis × pectinatus*) occurs in a number of localities in Scotland, and W. Yorks. It is intermediate in many features between the parent spp., and sterile.

24. P. pectinatus L.   Fennel-leaved Pondweed.
Incl. *P. interruptus* Kit. (*P. flabellatus* Bab.)

Rhizome extensively creeping, usually arising from a tuberous winter bud. Lfy stem 0.4–2 m, very slender, ±cylindrical, usually much-branched. Submerged lvs 5–20 cm × 0.25–2(–5) mm, setaceous to linear (stem lvs always broader than branch lvs, and lower lvs than upper), blunt and mucronate or tapering and acute, dark green, ±translucent; longitudinal veins 3–5, the laterals often indistinct in narrow lvs. Floating lvs 0. *Stipular sheath* 2–5 cm, *open and convolute with a whitish margin*; ligule long, ±blunt, deciduous. Spike 2–5 cm, of 4–8 usually 2-fld whorls, ±interrupted, especially below; stalk 3–10(–25) cm, filiform, not thickened upwards. (Fig. 61E, F) 3–5 × 2–4 mm, olive tinged with orange, obovoid to semi-obovoid, ±compressed; ventral margin straight or somewhat convex, dorsal semicircular, 1–3-keeled; beak continuing the ventral margin, short. Fl. 5–9. Usually perennates as tuberized lateral buds of the rhizome. $2n = 78$. Hyd.

Native. Ponds, rivers, canals, ditches, etc. Abundant in base-rich waters of the lowland zone throughout the British Is. but absent from mountainous districts of Wales, N. England and Scotland and from much of C. Ireland. Almost cosmopolitan.

Very variable in mode of branching, size and acuteness of the lvs, and size and shape of the fr. Robust plants much branched above so as to spread in a fan-like manner, with lower stem-lvs abruptly rounded to the subacute tip and reaching 3–5 mm in width, upper stem-lvs much narrower, branch lvs filiform, acu-

minate, have been placed in var. *interruptus* (Kit.) Ascherson (*P. interruptus* Kit.; *P. flabellatus* Bab.). They have been described as having a much interrupted spike and large fr. differing from the type in having the ventral margin straight, not convex, and the dorsal gibbous above, with a strong central keel instead of the two distinct lateral keels of the type. There is, however, no close correlation between vegetative and fr. characters, nor between the various fr. characters, and the var. *interruptus* can be regarded as no more than a growth form.

2. GROENLANDIA Gay
Aquatic herbs with all lvs submerged and flowering-spike emergent. Like *Potamogeton* but *all lvs in opposite pairs* or rarely in whorls of 3, sessile and amplexicaul; *stipules* 0, except for the involucral lvs subtending spikes, where they are adnate to the lf-base and form 2 lateral auricles; spike few-fld; fr. an achene with thin (not bony or fleshy) pericarp. Wind-pollinated.

One sp. in Europe, W. Asia and N. Africa.

1. G. densa (L.) Fourr.   Opposite-leaved Pondweed.
Potamogeton densus L.; *P. serratus* L.

Rhizome creeping, much branched. Lfy stem 10–30 cm, ±cylindrical, forking above. *Submerged lvs* (0.5–)1.5–2.5(–4) × 0.5–1.5 cm, ovate-triangular to lanceolate, blunt or acute, with minutely serrate margins especially near the apex, commonly folded longitudinally and recurved, translucent; longitudinal veins 3–5, with sparse transverse connections. Floating lvs 0. *Spike usually of only 4 fls, ovoid in fr.*; stalk 5–15 mm, slender, shorter than the lvs, erect at first, then strongly recurved. Fr. (Fig. 61C) c. 3 × 2 mm, olive, orbicular-obovate, strongly compressed; ventral margin convex, dorsal semicircular, sharply keeled; beak short, ±central, recurved. Fl. 5–9. $2n = 30$. Hyd.

Native. In streams, canals, ditches and ponds with clear nutrient-rich water, often ±markedly calcareous and over a stony bottom. Locally abundant in lowland England and in a very few widely scattered localities in Wales and Scotland northwards to Easter Ross; not in Hebrides, Orkney or Shetland; local in Ireland, chiefly in the south. In much of Europe northwards to Scotland and Denmark; N.W. Africa; W. Asia.

Variable in size and shape of the lvs. In some plants they are only 5–10 × c. 3 mm.

134. RUPPIACEAE

Submerged aquatic herbs of salt or brackish pools, rarely in fresh waters. Lvs linear or setaceous, sheathing at the base. Infl. a short terminal raceme appearing sub-umbellate. Bracts 0. Fls hermaphrodite, small. Perianth 0. Stamens 2, opposite each other; filaments very short and broad. Ovary superior; carpels 4 or more, becoming long-stipitate in fr.; ovule solitary, pendent, campylotropous. Fr. indehiscent.

One genus and about 3 spp. throughout temperate and subtropical regions and rarely in mountain lakes in the tropics.

1. RUPPIA L.
The only genus.

Peduncle in fr. many times longer than pedicels, spirally coiled in fr.   **1. cirrhosa**
Peduncle in fr. about equalling pedicels, not spirally coiled in fr.   **2. maritima**

1. R. cirrhosa (Petagna) Grande   Spiral Tasselweed.
R. maritima auct. mult., non L., *R. spiralis* L. ex Dumort.

A slender perennial 30 cm or more. Stems filiform, much-branched. *Lvs* alternate or opposite, filiform (c. 1 mm wide), *rounded or obtuse at apex*, dark green; sheaths dilated, brownish. *Peduncle* 10 cm *or more, much elongated after flowering, several times as long as the pedicels of the carpels*, spirally coiled. *Fr. ovoid, nearly symmetrical, beak slightly oblique.* Fl. 7–9. $2n = 40$. Hyd.

Native. In brackish ditches near the sea, rather rare. Scattered round the coasts of England and Ireland; Scotland: Wigtown, Orkney and Shetland. Distribution of the genus.

2. R. maritima L.   Beaked Tasselweed.

R. rostellata Koch

Like *R. cirrhosa* in general appearance. *Lvs* c. 0.5 mm wide, *acute*, light green; sheaths narrow. *Peduncle 0.5 cm, shorter than to twice as long as the pedicels of the carpels*, flexuous but not spirally twisted. *Fr. strongly asymmetrical, ventral side convex ending in a long beak, dorsal side strongly gibbous at base.* Fl. 7–9. $2n = 20$. Hyd.

Native. In brackish ditches and salt-marsh pools, local. Round most of the coasts of the British Is. north to Shetland, less frequent in the north. Most of Europe; N. Africa; W. Asia; North America.

135. ZANNICHELLIACEAE

Submerged perennial aquatic herbs of fresh, brackish or sea-water, with slender creeping rhizomes. Lvs alternate, opposite or ±whorled, linear, with sheathing bases and usually with a ligule at the junction of blade and sheath. Fls very small, solitary axillary or in cymes; unisexual (monoecious or ?dioecious), hypogynous. Perianth of 3 small scales or 0; stamens 3, 2 or 1, with 1–2-celled anthers and spherical or filiform pollen grains; carpels 1–9, free, each with 1 pendent orthotropous ovule; style terminating in a capitate, peltate or spathulate stigma. Pollination hydrophilous. Fr. of 1–9 sessile or stalked achenes, each with 1 non-endospermic seed.

About 20 spp. in 6 genera, widely distributed, mainly in salt water.

1. ZANNICELLIA L.

Perennial monoecious herbs of fresh or brackish water with slender simple or branched lfy stems. Lvs mostly opposite, linear, entire, with sheathing or free axillary stipules. Fls axillary in a hyaline deciduous cup-shaped spathe, 1 male and 2–5 female fls often in the same spathe. Male fl. of 1 stamen; pollen grains spherical. Female fl. of 1 carpel with a short or long style and a flattened peltate or lingulate stigma. Perianth 0. Fr. a stalked ±curved achene, entire or toothed along one or both margins.

The only sp.

1. Z. palustris L.   Horned Pondeweed.

A submerged herb with a filiform rhizome and filiform much-branched lfy shoots commonly to 50 cm, rooting only near the base or creeping and rooting over much of their length. Lvs 1.5–5(−10) cm × 0.4–2 mm, linear to capillary, tapering to a fine point, parallel-veined, translucent; stipule amplexicaul, tubular below, scarious, soon falling. Style stout below, tapering upwards; stigma peltate with a ±waved and crenate margin. Achenes 2–6, each 2–3 mm excluding the persistent style (0.5–1.5 mm), subsessile or with a short stalk up to 1 mm; dorsal margin ±toothed. Fl. 5–8. $2n = 28$. Hyd.

Very variable. Var. *pedicellata* (Fr.) Wahlenb. & Rosen. (*Z. pedicellata* Fr., *Z. maritima* Nolte) has the 2–5 achenes distinctly stalked (1.5–2.5 mm) and often on a common peduncle, the persistent style longer (2–2.5 mm) and more slender, and the stigma lingulate instead of peltate. It is said to be a more slender plant with a preference for brackish water, and may need to be treated as a separate ssp. *Z. pedunculata* Rchb. and *Z. gibberosa* Rchb., the latter with both dorsal and ventral margins of the achenes muricate, appear to belong here.

Native. Rivers, streams, ditches and pools of fresh or brackish water; to 213 m in Ireland; locally common. Throughout the British Is though most frequent in England and E. Ireland. Cosmopolitan.

136. NAJADACEAE

Slender submerged fresh- or brackish-water herbs. Lvs opposite or apparently in whorls of 3, linear; base sheathing; two small scales (*squamulae intravaginales*) within each sheath. Fls unisexual, small; plants monoecious or rarely dioecious. Male fl. enclosed in a spathe; perianth 2-lipped; stamen 1, anther subsessile. Female fl. without perianth; carpel one, 1-celled, stigmas 2–4; ovule solitary, basal, erect, anatropous. Fr. indehiscent.

One genus and about 50 spp. in temperate and tropical regions.

1. NAJAS L.

The only genus.

Lvs with many large spinous teeth (Norfolk Broads).
                 2. marina
Lvs minutely denticulate or nearly entire (Lakes). **1. flexilis**

1. N. flexilis (Willd.) Rostk. & W. L. E. Schmidt
                 Slender Naiad.

A slender brittle annual. Stems c. 30 cm, smooth. Lvs c. 20 × 0.75 mm, 2–3 together, translucent, apiculate, very minutely and remotely denticulate. Fls 1–3 in the axils of the lvs. Fr. 2–3 mm, narrowly ovoid. Fl. 8–9. $2n = 24$. Hyd.

Native. In lakes. N. Lancs, Islay, Outer Hebrides, Kerry, Galway, Mayo and Donegal. N. and C. Europe, south to Switzerland; North America.

2. N. marina L.   Holly-leaved Naiad.

Stems with occasional teeth near the top, forked, rather stiff and brittle. Lvs strong spinous-dentate and ±toothed on the back. Fr. 3–8 mm, ellipsoid or ovoid. Fl. 7–8. $2n = 12*$. Hyd.

Native. In slightly brackish water in a few of the Norfolk Broads. Cosmopolitan, except for the colder parts of the temperate regions; in Europe extending northwards to c. 63° N. in Finland.

137. ERIOCAULACEAE

Perennial, monoecious, usually scapose herbs, sometimes woody at base. Lvs usually linear, crowded, mostly basal. Fls small, actinomorphic or weakly zygomorphic crowded in bracteate heads, male and female often mixed, or male in the middle, female surrounding them, very rarely male and female in separate heads. Perianth scarious or membranous; segments in 2 series, each 2–3-merous, the outer free or ±connate, the inner often tubular, rarely 0. Stamens as many as or double (rarely fewer than) the number of per.-segs and opposite them; staminodes usually 0 in female fls. Ovary superior, 2–3-celled, style short, terminal; stigmas 2–3; ovules solitary and pendent in each cell. Fr. membranous, loculicidal.

About 13 genera and 1150 spp., distributed throughout most of the world except the continent of Europe, few in temperate regions, particularly abundant in South America; frequently in swampy habitats.

1. ERIOCAULON L.

Roots with conspicuous transverse striations. Lvs in a basal rosette. Outer per.-segs free or connate; inner segs connate in male fls, free or absent in female fls. Stamens twice as many as inner per.-segs. About 400 spp. with the distribution of the family. The only British genus.

1. E. aquaticum (Hill) Druce   Pipe-wort.
E. septangulare With.

A slender *scapose perennial herb with a creeping stock. Roots white, soft and worm-like. Lvs 5–10 cm, subulate, laterally compressed, translucent, septate.* Scape 7–20(–150) cm, *6–8-furrowed, twisted.* Heads 6–11 mm diam.; *bracts obovate, obtuse, lead-coloured;* bracteoles black, obtuse. Fls 2-merous, the outer per.-segs lead-coloured, with a tuft of hairs at tip; inner pale with a black spot near top, ciliate. Fl. 7–9. $2n = 64^*$. Hyd. or Hel.

Native. In shallow water or bare wet ground, forming dense mats on peaty soil. Scotland: Skye and Coll; Ireland: from W. Cork to W. Donegal, locally abundant near the coast and ascending to 300 m. Widely distributed in eastern North America.

138. LILIACEAE

Herbs, rarely shrubby, sometimes climbing. Lvs alternate or in several whorls or all basal, rarely in a single-whorl. Fls usually in a raceme, usually hermaphrodite and actinomorphic, usually 3-merous (rarely 2-, 4- or 5-merous). Perianth usually petaloid, usually in two similar whorls. Stamens in two whorls, opposite per.-segs, inserted on them or free. Ovary superior, usually 3-celled with axile placentation, rarely 1-celled with parietal placentation. Ovules usually numerous, in 2 rows on each placenta, anatropous. Fr. a capsule or berry. Seeds with copious fleshy or cartilaginous endosperm.

About 240 genera and 3000 spp., cosmopolitan.

A family of very diverse habit and appearance but of uniform floral structure. It might well be divided into several families but no satisfactory way of doing this has yet been devised.

1 Lvs normally developed; herbs with the aerial stems
    unbranched or branched only in infl.   2
  Lvs scale-like, with lf-like or needle-like assimilating
    organs (cladodes) borne in their axils.   19
2 Fls axillary.   5. POLYGONATUM
  Fls or infl. terminal.   3
3 Fls arising from corm; plant lfless at fl.
                           19. COLCHICUM
  Fls with evident scape, or arising from lfy aerial stems;
    plant lfy at fl.   4
4 Fls in umbels, enclosed in bud by one or more spathe-
    like bracts; plant with onion-like smell when
    bruised.   15. ALLIUM
  Fls solitary or in panicles or racemes, rarely sub-
    umbellate and then bracts lf-like.   5
5 Lvs alike on both surfaces, vertical, iris-like.   6
  Lvs ±horizontal, the surfaces differing.   7
6 Styles 3, free throughout their length; fls whitish.
                            1. TOFIELDIA

Style single, with 3-lobed stigma; fls yellow.
                     2. NARTHECIUM
7 Per.-segs united at base for at least $\frac{1}{5}$ of their length.   8
  Per.-segs free, or united at base for less than $\frac{1}{5}$ of
    their length.   9
8 Lvs ovate-lanceolate, petiolate; stock a rhizome.
                 4. CONVALLARIA
  Lvs linear, sessile; stock a bulb.   17. MUSCARI
9 Lvs in a single whorl near top of stem   20. PARIS
  Lvs alternate or basal or in several whorls.   10
10 Fls in a panicle; filaments densely hairy. 3. SIMETHIS
  Fls solitary, in a raceme, or subumbellate; filaments
    glabrous.   11
11 Lvs all basal (sometimes 1 or 2 sheathing the stem
    and appearing cauline).   12
  Some cauline lvs present.   14
12 Fls white, marked with green outside.
              14. ORNITHOGALUM
  Fls blue, lilac or pink, rarely pure white.   13
13 Bracts 0 or 1 to each fl.; per.-segs free.   15. SCILLA
  Bracts 2 to each fl.; per.-segs united at base.
              16. HYACINTHOIDES
14 Lvs cordate; per.-segs and stamens 4; fr. a berry.
             6. MAIANTHEMUM
  Lvs not cordate; per.-segs and stamens 6; fr. a cap-
    sule.   15
15 Per.-segs recurved; fls in a raceme; anthers dorsifixed,
    ±versatile.   9. LILIUM
  Per.-segs not recurved; fls solitary or in a raceme;
    anthers basifixed, introrse.   16
16 Fls subumbellate.   13. GAGEA
  Fls solitary, very rarely in pairs.   17
17 Fls nodding, purplish, rarely whitish. 10. FRITILLARIA
  Fls erect, yellow or white.   18
18 Fls yellow; stigma sessile; lower lvs at least 6 mm wide.
             11. TULIPA
  Fls white; style well-developed; lower lvs up to 2 mm
    wide.   12. LLOYDIA

19 Cladodes linear; fls axillary; filaments free.
7. ASPARAGUS
Cladodes broadly ovate to lanceolate, bearing fls in middle of one surface; filaments united in a tube.
8. RUSCUS

1. TOFIELDIA Hudson

Stock a rhizome. Basal lvs numerous, vertical, distichous. Fls in racemes. Per.-segs free or almost so. *Anthers* ovate, *introrse*; filaments glabrous. Ovary trigonous, carpels free above. *Styles 3, free. Capsule septicidal.* Seeds oblong, numerous, very small.

About 20 spp., north temperate and Arctic regions and Andes.

1. T. pusilla (Michx) Pers.   Scottish Asphodel.

T. palustris Hudson pro parte; *T. borealis* (Wahlenb.) Wahlenb.

Glabrous. Rhizome short. Basal lvs 1.5–4 cm × 1–2 mm, rigid, 3–4-veined; stem-lvs 0 or 1–2, much smaller. Fl.-stem 5–20 cm. Raceme 5–15 mm, dense, with 5–10 fls. Bracts 3-lobed, scarious, shorter than or equalling pedicel. Pedicel 1 mm or less. Per.-segs greenish-white, obovate-oblong, c. 2 mm, obtuse. Fl. 6–8. Pollinated by insects and selfed. $2n = 30$. Hs.–Hel.

Native. By springs and streams on mountains from 215 m to 915 m, local; N.W. Yorks and Durham; Argyll and Angus to Caithness. N. Europe eastwards to the C. Urals; Alps; W. Carpathians; E. Siberia; Arctic North America.

2. NARTHECIUM Hudson

Stock a rhizome. Basal lvs numerous, vertical, distichous. Fls in racemes. Per.-segs free or almost so. *Anthers* linear, *extrorse*, versatile. Filaments woolly. Ovary trigonous, carpels completely united. *Style 1, simple*; stigma slightly 3-lobed. *Capsule loculicidal.* Seeds numerous, elongated into a tail at each end.

About 6 spp., north temperate regions.

1. N. ossifragum (L.) Hudson   Bog Asphodel.

Glabrous. Rhizome creeping; roots many, fibrous. Basal lvs 5–30 cm × 2–5 mm, rigid, often curved, usually 5-veined; stem-lvs few, 4 cm or less, sheathing. Fl.-stem 5–45 cm. Raceme 2–10 cm, with 6–20 fls, rather dense, or lax below. Bracts lanceolate, entire, about equalling pedicel. Pedicel 5–10 mm, with a subulate bracteole above the middle. Per.-segs 6–9 mm, yellow, linear-lanceolate, spreading in fl., erect in fr. Anthers orange. Stem, per.-segs and ovary becoming uniformly deep orange after fl. Capsule narrow-ovoid, 6-grooved, mucronate, c. 12 mm. Fl. 7–9. Pollinated by insects. Fr. 9. $2n = 26$. Hs.–Hel.

Native. Bogs, wet heaths and moors and wet acid places on mountains, ascending to 1000 m. Throughout most of the British Is., becoming less common towards the southeast and absent from many eastern and midland counties; often abundant, and dominant in the hollows of the raised bog cycle, sometimes also locally dominant in blanket bogs. N. and W. Europe, eastwards to S.E. Sweden and southwards to N. Portugal (in the mountains).

3. SIMETHIS Kunth

Stock a short rhizome. Lvs all basal. Infl. a panicle. Per.-segs free, spreading. *Filaments woolly*, somewhat thickened, *inserted into a pit* on the anther; *anthers* introrse, *versatile, dorsifixed*. Ovules 2 in each cell. Style simple, filiform. Fr. a loculicidal capsule.

One sp.

1. S. planifolia (L.) Gren.   Kerry Lily.

S. bicolor (Desf.) Kunth; *Pubilaria planifolia* (L.) Druce

Glabrous. Rhizomes very short, erect, clothed with the brown fibrous remains of the lf-bases; roots tufted, fusiform, fleshy. Lvs linear, 15–45 cm × c. 6 mm, glaucous. Scape about as long. Panicle lax. Bracts shorter than pedicels. Per.-segs elliptical 9–11 mm, white within, purplish outside. Capsule c. 6 mm, subglobose, obscurely angled. Seeds black, shining. Fl. 5–7. Gr.

Native. Rough rocky furzy heath near the coast, over an area of 12 or 14 × 1–3 km near Derrynane (Kerry); formerly naturalized in S. England, now extinct. S.W. Europe; Morocco, Algeria, Tunisia.

The commonly cultivated genus *Kniphofia* (Red-Hot Poker) with numerous scarlet cylindrical fls is chiefly distinguished from the preceding by the cylindrical or campanulate perianth with united segments.

PHORMIUM J. R. & G. Forster

Stock a short rhizome. Lvs mostly basal, very fibrous, distichous. Fls in terminal panicle. Perianth with short tube and erect lobes. Anthers introrse, dorsifixed. Fr. a loculicidal capsule. Seeds numerous.

Outer per.-segs brownish-red; capsule erect, not twisted.
tenax
Outer per.-segs yellow; capsule pendent, twisted. **colensoi**

*P. tenax J. R. & G. Forster   New Zealand Flax.

Lvs up to 250 × 12 cm, keeled on outer surface, margins and base often reddish. Stem, with panicle, up to 400 cm. Perianth 3–5 cm, outer lobes dark brownish-red, the inner yellow. Capsule 7–9 cm, erect, trigonous, dark brown.

Introduced. Naturalized in W. England and W. Ireland. Native of New Zealand and Norfolk Island.

*P. colensoi Hooker fil.

Lvs up to 150 × 7 cm. Stem up to 225 cm. Outer per.-lobes yellow. Capsule 10–17 cm, pendent, twisted, becoming pale with age.

Introduced. Locally naturalized on coastal rocks in Scilly Is. Native of New Zealand.

HEMEROCALLIS L.   Day Lily.

Stock a very short rhizome, with ±fleshy roots. *Lvs* long, *linear. Infl. a panicle.* Stamens inserted on the perianth-tube, curved upwards; anthers oblong-linear, introrse. Ovary trigonous; ovules many in each cell; style long, simple, curved upwards. Fr. a loculicidal capsule.

About 20 spp., temperate Asia.

Perianth dull orange, 8–10 cm, segments undulate.   **fulva**
Perianth lemon-yellow, 6–8 cm, segments flat.
liloasphodelus

*H. fulva (L.) L.

Roots thick, not swollen. Lvs 1–2 cm broad. Stems 50–100 cm, with a few scale-like lvs. Fls 8–10 cm, dull orange, scentless, per.-segs elliptic-oblong, reticulately veined, margins undulate. Sterile. Fl. 6–8. $2n = 33$.

Commonly grown in gardens and sometimes naturalized. A sterile triploid, not known wild. Allied diploids in China.

*H. liloasphodelus L. (*H. flava* L.)

Roots with swollen tubers. Lvs 5–10 mm broad. Stem 40–80 cm, naked or nearly so. Fls 6–8 cm, lemon-yellow, scented; per.-segs oblong-lanceolate, without transverse veins, margins flat. Fl. 5–6. $2n = 22$.

Commonly grown in gardens and sometimes escaping. Native probably of E. Asia.

4. CONVALLARIA L.

Rhizome long, creeping, much-branched, with clusters of roots from the nodes. Lvs arising from rhizome, with convolute, sheathing bases which simulate a lfy stem; each such 'stem' comprises 1–4 foliage lvs, with scale-lvs at the base c. 5. Scapes from the axil of one of the scale-lvs. Fls nodding *in unilateral racemes. Perianth* globose-campanulate; *segments united* for about ⅔ of their length. Stamens inserted on base of perianth, included. Anthers introrse. Ovary with 4–8 ovules in each cell. Fr. a berry.

Three spp., north temperate regions.

1. C. majalis L.   Lily-of-the-valley.

Glabrous. Lvs 8–20 × 3–5 cm, ovate-lanceolate or elliptical, acute; sheath 5–12 cm. Infl. 6–12-fld. Bracts scarious, ovate-lanceolate, shorter than pedicels. Fls sweet-scented. Perianth white, 5–9 mm. Berry globose, red. Fl. 5–6. Pollinated by insects and selfed; weakly protandrous. $2n = 38$. Grh.

Native. Dry woods, mainly calcareous; local and with a distinct eastern tendency; widespread in England; in Wales only in the south-east and north; in Scotland more local, native as far north as E. Inverness; not native in Ireland. Commonly cultivated and sometimes escaping. Most of Europe except the extreme north and extreme south; N.E. Asia.

Two spp. of *Aspidistra* Ker-Gawler, chiefly differing from the preceding in the solitary or densely spicate fls and lobed stigmas, are commonly grown in pots in houses. They have solitary brownish fls at ground level and are said to be pollinated by snails.

5. POLYGONATUM Miller

Rhizome thick, long, creeping. Lvs all cauline, numerous. *Fls* 3-merous, nodding, *axillary, solitary, or in few-fld axillary racemes. Perianth* tubular-campanulate, *united into a tube* for the greater part of its length. Stamens included, inserted on perianth-tube. Anthers oblong, 2-lobed at base, introrse. Ovules 4–6 in each cell; style filiform, with a small, 3-lobed stigma. Fr. a berry. Plants glabrous or nearly so.

About 50 spp., north temperate regions.

1 Middle and upper lvs in whorls.   **1. verticillatum**
  Lvs alternate.   2
2 Fls 1–2; perianth cylindrical, not contracted in the middle; stem angled.   **2. odoratum**
  Fls 2–6; perianth contracted in the middle; stem terete.   **3. multiflorum**

1. P. verticillatum (L.) All.   Whorled Solomon's Seal.

Stem 30–80 cm, angled. Middle and upper *lvs in whorls* of 3–6, 5–12 cm, linear-lanceolate, sessile. Fls 1–2 on a common peduncle. Perianth 6–8 mm, greenish-white, somewhat contracted in the middle. Filaments papillose. Fr. c. 6 mm, globose, red, becoming dark purple. Fl. 6–7. Pollinated by bees and selfed; homogamous. $2n = 28$ (30, 60, 86–91). Grh.

Native. Mountain woods, very rare; Northumberland, Dumfries, Perth. Arctic Norway (69° 35′ N.) southwards to the mountains of N. Spain, C. Italy and Bulgaria and eastwards to Latvia and Romania; Caucasus (not elsewhere in Russia); Asia Minor, Himalaya.

2. P. odoratum (Miller) Druce
    Angular Solomon's Seal.

P. anceps Moench; *P. officinale* All.

Stem 15–20 cm, *angled,* arching. *Lvs* 5–10 cm, *alternate,* subdistichous, ovate to elliptical-oblong, acute, sessile. *Fls scented,* 1–2 on a common peduncle. *Perianth* 18–22 mm, *cylindrical, not contracted in the middle,* greenish-white. *Filaments glabrous.* Fr. c. 6 mm, globose, blue-black. Fl. 6–7. Pollinated by humble-bees and selfed; homogamous. $2n = 20$ (26, 28, 29, 30). Grh.

Native. Limestone woods, very local N. Somerset, Gloucester, Stafford, Derby, mid-W. Yorks, Westmorland; Brecon; occasionally introduced elsewhere. Throughout most of Europe southwards from 66° N. in Finland to the mountains of Spain and Portugal, C. Italy, Greece and the Caucasus; Morocco; Siberia; W. Himalaya; China.

3. P. multiflorum (L.) All.   Solomon's Seal.

Stem 30–80 cm, *terete,* arching. *Lvs* 5–12 cm, *alternate,* subdistichous, ovate to elliptical-oblong, acute, sessile. *Fls* 2–5 in axillary racemes. *Perianth* 9–15 mm, *somewhat contracted in the middle,* greenish-white. *Filaments pubescent.* Fr. c. 8 mm, globose, blue-black. Fl. 5–6. Pollinated by humble-bees and selfed; homogamous. $2n = 18$ (20, 24, 28, 29, 30). Grh.

Native. In woods, particularly on limestone in S. England, S. Wales and Cumberland and in a few isolated localities in between, local; naturalized in many localities north to Moray. Much of Europe but absent from parts of the south-west and many islands; temperate Asia to Japan (but not in Asiatic Russia).

The hybrid *P. multiflorum × odoratum* = **P. × hybridum** Brügger is probably the commonest member of the genus in gardens. Fls 2–4 together, the size of those of *P. odoratum*, somewhat contracted in the middle. It is sometimes found ±naturalized and may occur wild.

6. MAIANTHEMUM Weber

Rhizome slender, creeping. Stems erect, with 2 scale lvs at base and 2 foliage lvs near the apex. *Fls 2-merous,* in an erect *terminal raceme.* Per.-segs free, ±spreading. Stamens inserted on the base of the per.-segs, anthers ovoid, introrse. Ovules 2 in each cell. Style short, stigma obscurely lobed. Fr. a berry.

Three spp., north temperate regions.

1. M. bifolium (L.) Schmidt   May Lily.
M. convallaria Weber; *Unifolium bifolium* (L.) Greene; *Smilacina bifolia* (L.) J. A. & J. H. Schultes
Glabrous except for the upper part of the stem which has stiff white hairs. Stem 8–20 cm, flexuous. Lvs 4–6 cm, ovate, acute or shortly acuminate, deeply cordate at base, the solitary basal lvs long-petiolate, upper cauline often sessile. Infl. 2–5 cm, (8–)15–20-fld, rather dense. Perianth white, segments 1–2 mm. Fr. c. 6 mm, red. Fl. 5–6. Pollinated by insects and selfed; protogynous. $2n = (30, 32), 36, (38, 42)$.

Native. Woods, very rare; N. Lincoln, N.E. Yorks, Durham and S. Northumberland; occasionally introduced elsewhere in England. Europe southwards from 69° 50′ N. to N. Spain, N. Apennines, Bosnia and S. Russia (Lower Don region); N. Asia to Kamchatka and Korea.

7. ASPARAGUS L.

Stock a rhizome. Stems erect or climbing, sometimes woody, much-branched. Lvs reduced to small scarious scales, bearing in their axils 1 or more green assimilating cladodes. Fls solitary or in racemes, axillary, not borne on the cladodes. Pedicels articulated. Per.-segs free or nearly so. Stamens free. Anthers introrse, dorsifixed. Ovules few. Fr. a berry.

About 300 spp., Old World.

1. A. officinalis L.   Asparagus.
Glabrous herb with short rhizome. Stems annual, erect or procumbent, much-branched. Lvs scarious, whitish, thin; those of the main stem triangular-lanceolate, to 5 mm; those of the branches sagittate, ovate, less than 1 mm. Cladodes in clusters (up to 10 in each), needle-like. Fls 1 or 2, axillary on separate pedicels, dioecious by abortion, rarely hermaphrodite. Perianth campanulate, 3–6 mm, male larger than female. Fr. red, globose. Fl. 6–9. Pollinated by insects. Grh.

Subsp. prostratus (Dumort.) Corb.
A. officinalis var. *maritimus* auct., non L.; *A. prostratus* Dumort.
Stems ±procumbent, 10–30(–40) cm. Cladodes thick, rigid, glaucous, (4–)5–10(–15) mm. Pedicels 2–6 mm. Per.-segs of male fl. 7–8.5 mm, yellow, tinged red at base, recurved at tip; of female fl. 4–4.5 mm, yellow to whitish-green, slightly recurved. Fr. 5–7 mm, seeds 1–6, most frequently 2. Fl. 6–7. $2n = 40*$.

Native. Grassy sea-cliffs, very local and rare. Dorset, Cornwall, W. Gloucester, Wales; maritime sands, Waterford, Wexford, Wicklow; Channel Is.; decreasing. Coast of Europe from N. Germany to N. Spain.

*Subsp. officinalis
Incl. *A. officinalis* var. *altilis* L.
Stems erect, 30–150 cm. Cladodes slender, often flexuous, green, 5–15(–20) mm. Pedicels 6–10 mm. Per.-segs of male fl. 5–6 mm, yellow, straight; of female fl. c. 4 mm, yellow-green, straight. Fr. 7.5–8.5 mm, seeds 5–6. Fl. 7–9. $2n = 20*$.

Introduced. Commonly cultivated as a vegetable and naturalized in many localities in waste places and dunes, etc. Most of Europe southwards from S. Denmark.

8. RUSCUS L.

Dioecious shrub. Stock a rhizome. Stems woody, erect. Lvs reduced to small scarious scales, bearing in their axils a broad green assimilating flattened lf-like cladode. Fls solitary or in clusters on the surface of the cladode. Inner per.-segs free, smaller than outer. Stamens 3, filaments connate into a column; anthers extrorse, sessile. Ovules 2 in each cell. Style short. Perianth persistent. Fr. a berry.

Seven spp. in Mediterranean region, W. Europe, Madeira, Azores.

1. R. aculeatus L.   Butcher's broom.
Glabrous. Rhizome creeping, thick, fibrous. Stems erect, 25–80 cm, stiff, much-branched, green, striate. Lvs ±triangular, less than 5 mm, scarious, thin, brownish. Cladodes 1–4 cm, ovate, entire, spine-pointed, thick, rigid, dark green, twisted at base. Fls 1–2 on the upper surface of the cladode in the axil of a small scarious bract. Perianth greenish, c. 3 mm. Female fl. with a cup representing the stamens. Fr. c. 1 cm, red, globose. Fl. 1–4. Fr. 10–5. Germ. summer. $2n = 40$. N.

Native. Dry woods and among rocks; widespread in S. England but rather local; extending north to Caernarvon and Norfolk; commonly cultivated and sometimes escaping in N. England and Scotland. Mediterranean region north to Transylvania, S. and W. Switzerland and N. France; Azores.

9. LILIUM L.   Lily.

Stock a bulb of numerous imbricate fleshy scales *without a tunic.* Lvs. all cauline, numerous, alternate or whorled. Fls large, solitary or in a terminal raceme, erect to nodding. Perianth of various shapes; segments free, spreading or revolute, caducous, with a longitudinal groove-like nectary at the base. *Anthers dorsifixed,* introrse, *versatile.* Style long; stigma ±3-lobed. Fr. a loculicidal capsule. Seeds numerous, flat.

About 90 spp., temperate regions. Many are grown in gardens.

Lvs whorled; fls purple.	**1. martagon**
Lvs alternate; fls yellow.	**2. pyrenaicum**

1. L. martagon L.   Martagon Lily.
Stem 50–100 cm, erect, scaberulous. *Lvs 7–20 cm, mostly in distant whorls* of 5–10, obovate-lanceolate, scaberulous on margins; upper alternate, smaller. Fls c. 4 cm across, alternate, nodding, 3–10 in a terminal raceme. *Perianth dull purple* with darker raised projections; segments oblong, clawed, strongly recurved. Anthers reddish-brown. Capsule obtusely 6-angled, obovoid. Fl. 8–9. Pollinated by Lepidoptera and selfed; homogamous. $2n = 24$. Gb.

?Introduced. Commonly grown in gardens and naturalized in woods in a number of places from Somerset and Kent to Monmouth, Cumberland and Yorks, mainly in the south and possibly native in Surrey and Gloucester; Fife. Mountains of Europe from N.E. France, Estonia and C. Urals southwards to C. Spain, C. Greece and the Caucasus; naturalized in Scandinavia; Siberia, N. Mongolia.

*2. L. pyrenaicum Gouan   Pyrenean Lily.
Stem 40–80 cm, erect, glabrous. *Lvs alternate*, very
dense, linear-lanceolate. Fls nodding, solitary or up to
8 in a terminal raceme. *Perianth yellow* with small black
dots; segments oblong, strongly recurved. Anthers
reddish. Capsule obtusely 6-angled, obovoid. Fl. 5–7.
$2n = 24$. Gb.

Introduced. Grown in gardens, sometimes escaping
and quite naturalized on hedgebanks in N. Devon,
Cumberland and elsewhere. Native of Pyrenees.

10. FRITILLARIA L.
Stock a small bulb with few scales and thin white tunic.
Fls usually solitary, terminal, nodding. *Perianth* campanulate, caducous, *with a large glistening nectary near
the base of each segment*, segs free. *Anthers basifixed*,
introrse. Style long, entire or with 3 stigmas. Fr. a loculicidal capsule. Seeds many, flat, often winged.

About 85 spp., north temperate regions. A number
are cultivated, including *F. imperialis* L. (Crown Imper-
ial) with numerous lvs partly in whorls and a single whorl
of yellow or orange fls.

1. F. meleagris L.   Snake's Head, Fritillary.
Glabrous, somewhat glaucous. Stem erect, 20–50 cm.
Lvs few (3–6), alternate, linear, 8–20 cm (upper
shorter) × 4–9 mm. Fls solitary, rarely paired. Perianth
3–5 cm, dull purple, chequered dark and pale, rarely
cream-white, segments oblong. Stigmas linear. Capsule
subglobose, trigonous. Fl. 4–5. Pollinated by humble-
bees and selfed. $2n = 24$. Gb.

Native. Damp meadows and pastures in W. Sussex
and from Wilts and Kent to Stafford and Suffolk; S.W.
Yorks; Radnorshire; very local; introduced elsewhere.
From C. Russia southwards to the S. Alps and C. Yugoslavia; naturalized in Fennoscandia and elsewhere.

11. TULIPA L.
Stock a bulb with numerous scales and brown tunic.
Fls usually solitary, terminal. *Perianth* ±campanulate
or rotate, erect, *segments without nectaries, caducous*,
free. *Anthers basifixed*, introrse. Style usually 0, stigma
3-lobed. Fr. a loculicidal capsule. Seeds many, flat.

About 100 spp., C. and S. Europe, temperate Asia,
N. Africa. The garden tulip (*T. gesnerana* L.) was introduced to Europe from Turkey where it was cultivated,
in the sixteenth century. Its origin is unknown, no
closely related certainly wild species having been found.

*1. T. sylvestris L.   Wild Tulip.
Glabrous, slightly glaucous. Stem 30–60 cm, flexuous,
with 2–3 lvs near the base. Lvs linear, 15–30 cm ×
6–12 mm. Fls drooping in bud, solitary or 2. Perianth
3–5 cm, yellow, greenish outside; segments elliptical,
acute, opening widely in the sun. Filaments villous at
base. Fr. oblong-trigonous c. 3 cm; rarely produced. Fl.
4–5. Pollinated by small insects and selfed; homogamous. $2n = 48$. Gb.

Introduced. Naturalized in meadows and orchards
from N. Devon and Hants to Mid-west Yorks; Berwick;
local. S. and S.E. Europe north to N.W. France and
E.C. Russia; naturalized in C. and N. Europe.

12. LLOYDIA Salisb. ex Reichenb.
Stock a small bulb, with a membranous tunic. Fls 1–2,
terminal. *Perianth* erect, *segments persistent*, free,
spreading, *with a small transverse nectary* at the base.
Anthers basifixed, introrse. Style filiform, stigma
obtuse. Fr. a loculicidal capsule. Seeds many, 3-angled.

About 20 spp. in north temperate and arctic regions.

1. L. serotina (L.) Reichenb.   Snowdon Lily.
Glabrous. Stem 5–15 cm, with 2(–4) lvs. Lvs filiform,
basal 15–25 cm, cauline much shorter, relatively
broader. Perianth white with purplish veins, segments
oblong, c. 1 cm. Fr. trigonous. Fl. 6. Pollinated by flies,
etc.; protandrous. $2n = 24$. Gb.

Native. Basic rock-ledges from 460 to 760 m in the
Snowdon range (Caernarvon). Arctic Russia, Alps,
Carpathians, Bulgaria, Urals, Caucasus; Soviet Asia
(widespread); N. Japan, Himalaya, W. China; western
North America from Alaska to Oregon and New
Mexico.

13. GAGEA Salisb.
Stock a tunicate bulb. *Infl. a raceme or subumbellate,
often subtended by lf-like bracts*. Stem with only the
bracts or with 1–4 lvs in addition. Fls erect. Per.-segs
spreading, persistent, free without nectaries. *Anthers
basifixed*, introrse. Style trigonous, stigma obscurely
lobed. Fr. a loculicidal capsule.

About 100 spp., north temperate Old World, mainly
Europe.

Bulb 1; basal lf 1, linear-lanceolate.   **1. lutea**
Bulbs 2, in common tunic; basal lvs 2–4; filiform.
                                                  2. bohemica

1. G. lutea (L.) Ker-Gawler
                                 Yellow Star-of-Bethlehem.
Bulb 1. Stem 8–25 cm, with a small lf-like bract 2–3 cm,
subtending the infl. and another much larger lf or bract
a short distance below, occasionally a third bract and
2 infl. are present. Basal lf single, (7–)15–45 cm ×
7–12 mm, linear-lanceolate, suddenly contracted to a
hooded acuminate apex, often curled. Fls 1–5, subumbellate. Pedicels somewhat unequal, 1.5–5 cm, glabrous or pubescent. Perianth yellow with a green band
outside; segments 10–15 mm, narrow-oblong. Fl. 3–5.
Pollinated by insects, weakly protogynous. $2n = 72$. Gb.

Native. Damp woods and pastures, especially on basic
soils. England and Scotland northwards to mid-Perth.
Very local, commonest in N. and C. England. Most
of Europe; temperate Asia to Kamchatka and Japan.

It may persist for years without flowering, but can
be distinguished from *Hyacinthoides* by the 2 lateral
ribs on the under surface of the lf.

2. G. bohemica (Zauschner) Schultes & Schultes fil.
Bulbs 2, in common tunic. Stem 1.5–3 cm, almost glabrous, with 1–2 bracts per flower; cauline lvs 4, alternate, long ciliate. Basal lvs 2–4, 4–9 cm × c. 1 mm,
filiform, acuminate, glabrous or shortly hairy. Fls 1(–4).
Pedicel c. 1 cm, pubescent. Perianth yellow, greenish
outside; segments 12–18 mm, narrowly oblong-lanceolate. Fl. 1–4. Gb.

Native. Pockets of shallow soil on south and east facing limestone cliffs. C.E. Wales (Radnorshire). W. France, C. and S. Europe eastwards to W. Turkey; Syria, Israel.

Species of *Camassia*, *Chionodoxa*, *Puschkinia*, *Hyacinthus* and *Galtonia* are ±commonly grown in gardens.

14. ORNITHOGALUM L.

Stock a bulb with tunic. Lvs all basal. Fls in a raceme. *Per.-segs* all alike, *free*, *obscurely veined*, persistent, usually white marked with green. Filaments flattened, hypogynous. Anthers introrse, dorsifixed. Fr. a loculicidal capsule. Seeds ovoid or globose, numerous.

About 150 spp. in Europe, Africa and W. Asia.

1 Infl. corymbiform, lower pedicels much longer than
   upper.   **1. umbellatum**
   Pedicels all ±equal.   2
2 Fls 2–3 cm, few (2–12); filaments with 2 teeth at
   apex.   **2. nutans**
   Fls 6–10 mm, many (more than 20); filaments without
   teeth.   **3. pyrenaicum**

1. O. umbellatum L.   Star-of-Bethlehem.

Glabrous. Bulb c. 2.5 cm, with numerous bulbils. Lvs 15–30 cm × c. 6 mm, linear, green, grooved, with a white stripe down the midrib. Scape 10–30 cm. *Infl. a corymbiform raceme*, 5–15-fld; pedicels ascending, lower to 10 cm, upper shorter; fls erect. Bracts linear-lanceolate, acuminate, lower 2–3 cm, thin, whitish. Per.-segs 1.5–2 cm, oblanceolate-oblong or linear-oblong, white with green band on the back. Stamens c. ½ as long, filaments lanceolate, acuminate. Capsule 1–1.5 cm, ob-ovoid, 6-angled. Fl. 4–6. Pollinated by insects and selfed. 2n = 18, 27, 36, 43, 45, 52, 54, 72. Gb.

?Native. Probably native in E. England, elsewhere naturalized in grassy places, widespread but local in England, more local in Wales and Scotland extending north to Sutherland and S. Inner Hebrides. S. and S.C. Europe, extending north to Sweden, Denmark and S.W. Russia (Middle Dnieper region) but very doubtfully native in the northern part of its range.

*2. O. nutans L.   Drooping Star-of-Bethlehem.

Glabrous. Bulb ovoid, c. 5 cm. Lvs 25–60 cm × 6–10 mm, linear, with a broad whitish stripe, channelled. Scape 25–60 cm. *Infl. a 2–12-fld unilateral raceme*; pedicels 5–12 mm, curved, all ±equal; fls drooping. Bracts ovate-lanceolate, acuminate, thin, whitish, longer than pedicels. *Per.-segs* 2–3 cm, oblong-lanceolate, white with a green band covering most of the back. Stamens shorter than perianth; *filaments* broad, *deeply bidentate at the apex* with the anther in the sinus. Capsule ovoid, pendent, 6-grooved. Fl. 4–5. Pollinated by insects and selfed; protandrous. 2n = 16, 30, 42. Gb.

Introduced. ±Naturalized in grassy places in E. and C. England, usually in small quantity. S.E. part of Balkan peninsula. Asia Minor; naturalized elsewhere in Europe.

3. O. pyrenaicum L.   Spiked Star-of-Bethlehem.

Glabrous. Bulb c. 5 cm, ovoid. Lvs 30–60 cm × 3–12 mm, linear, glaucous, withering early. Scape 50–100 cm. *Infl. a many- (more than 20-)fld* raceme; pedicels slender, spreading or ascending, all ±equal, 1–2 cm; fls ±erect. Bracts lanceolate-acuminate, thin, whitish, shorter than pedicels. *Per.-segs* 6–10 mm, oblong-linear, greenish-white with deeper band. Stamens about ¾ as long as perianth; *filaments* lanceolate, *acuminate*. Capsule c. 8 mm, ovoid, 3-grooved. Fl. 6–7. 2n = 16, 32. Gb.

Native. Woods and scrub, very local but often abundant, the young infls being sold in Bath market as 'Bath Asparagus'; N.-Somerset, Wilts, Berks and Bedford, a rare casual elsewhere. S., W. and S.C. Europe, northwards to Belgium; Asia Minor; Morocco (mountains).

15. SCILLA L.

Stock a bulb persisting several years, scales not tubular. Lvs all basal. Bracts 0 or 1 to each fl. *Per.-segs* all alike, *free*, spreading, *with a prominent midrib*, usually blue or purple. Filaments filiform or dilated at base, inserted on base of perianth. Anthers introrse, dorsifixed. Fr. a loculicidal capsule. Seeds ovoid or subglobose, sometimes angled, black, few or many.

About 80 spp., Europe, Africa and temperate Asia. Several spp. are grown in gardens, notably *S. siberica* Andr. with about 3 nodding bright blue fls.

Bracts conspicuous; fl. Apr.–May.   **1. verna**
Bracts absent; fl. July–Sept.   **2. autumnalis**

1. S. verna Hudson   Spring Squill.

Glabrous. Bulb 1.5–3 cm. Lvs 3–6, 3–20 cm × 2–4 mm, linear, produced before the fls. Scape 5–15 cm. Infl. 2–12-fld, dense, corymbose. Pedicels 3–12 mm, ascending. *Bracts solitary*, bluish, lanceolate, *usually longer than pedicels*. Per.-segs 5–8 mm, violet-blue, ovate-lanceolate, acute, ±ascending. Filaments lanceolate, anthers violet-blue. Capsule c. 4 mm, subglobose, trigonous, the cells c. 4-seeded. Fl. 4–5. Pollinated by insects. Fr. 7. 2n = 22*. Gb.

Native. Dry grassy places near the coast, very local; Cornwall and Devon, Wales, Isle of Man, Kirkcudbright to Ayr, Kintyre, Hebrides, Sutherland to Shetland, Moray to Aberdeen, Berwick and N. Northumberland; E. Ireland from Wexford to Derry. W. Europe from N. Portugal to the Faeroes.

2. S. autumnalis L.   Autumn Squill.

Almost glabrous. Bulb 1.5–3 cm. Lvs 4–15 cm × 1–2 mm, linear, produced after fl. Scape 4–25 cm. Infl. 4–20-fld, dense becoming lax, not or slightly corymbose. Pedicels 2–20 mm, ±ascending. *Bracts* 0. Per.-segs 4–6 mm, purple, oblong or oblong-lanceolate, obtuse or subacute, ±ascending. Filaments narrow, somewhat dilated below; anthers purple. Capsule c. 4 mm, ovoid-globose, obscurely trigonous; cells 2-seeded. Fl. 7–9. 2n = 14, 28, ?44. Gb.

Native. Dry short grassland usually near the sea, very local. Devon and Cornwall, Isle of Wight, Surrey, Essex, W. Gloucester, Channel Is. S. and W. Europe to N. France and Hungary.

16. Hyacinthoides Medicus
(*Endymion* Dumort.)

Differs from *Scilla* as follows: Bulb renewed annually; scales tubular. Bracts 2 to each fl. Perianth ±campanulate, the segments united at base. Stamens, at least the outer, inserted about the middle of the perianth.

Three or 4 spp. in W. Europe and N.W. Africa.

Raceme drooping at tip; per.-segs ±parallel below; anthers cream. Common.   **1. non-scripta**
Raceme erect; per.-segs not parallel below; anthers blue. Rare introduction.   **2. hispanica**

1. H. non-scripta (L.) Chouard ex Rothm.
Bluebell, Wild Hyacinth.

Endymion non-scriptus (L.) Garcke; *Scilla non-scripta* (L.) Hoffmanns. & Link; *S. nutans* Sm.

Glabrous. Bulb 2–3 cm, ovoid. Lvs linear, 20–45 cm × 7–15 mm. Scape 20–50 cm. *Raceme drooping at tip*, 4–16-fld, *unilateral*, the fls erect in bud, nodding when fully open. Pedicels c. 5 mm, afterwards elongating to c. 3 cm, and becoming erect. Bracts paired, bluish, the lower linear-lanceolate, longer than the pedicel, the upper smaller. *Per.-segs* 1.5–2 cm, violet-blue, rarely pink or white, ±*parallel* and erect, so that the lower part of the fl. appears cylindrical, the tips somewhat recurved. Filaments narrow, outer inserted about middle of perianth, inner lower. *Anthers cream.* Fr. c. 15 mm, ovoid; seeds several in each cell. Fl. 4–6. Pollinated by insects. Fr. 7. $2n = 16$. Gb.

Native. Common in woods, hedgebanks, etc., rarely in pastures, throughout the British Is., except Orkney and Shetland; ascending to 610 m; often dominant in woods on light acid soils. W. Europe from C. Spain to the Netherlands; locally naturalized in C. Europe.

*2. H. hispanica (Miller) Rothm.   Spanish Bluebell.

Scilla hispanica Miller; *S. campanulata* Aiton, *Endymion hispanicus* (Miller) Chouard

Differs from *H. non-scripta* as follows: lvs broader, 10–35 mm. *Racemes erect, not unilateral.* Pedicels ascending, not curving at fl. so that the fls are ±erect. Per.-segs paler, ±spreading, so that the *perianth is campanulate*, tips not recurved, broader and somewhat shorter. Filaments all inserted about middle of perianth. *Anthers blue.* Fl. 5. Pollinated by insects. $2n = 16$. Gb.

Introduced. Commonly grown in gardens, being usually supplied for the preceding, and naturalized in a few places. Native of Spain and Portugal; N. Africa.

Hybrids occur in gardens and may be expected as escapes.

17. Muscari Miller

Stock a bulb with tunic. Lvs all basal. Fls in a raceme. *Upper fls sterile. Perianth urceolate*, the lobes all alike, connate, very small. Stamens inserted on tube, included; filaments short, anthers introrse, dorsifixed. Ovary with 2 seeds in each cell. Style short. Fr. a triquetrous, loculicidal capsule. Seeds obovoid or globose, black.

About 60 spp., Europe, W. Asia, N. Africa. Several are commonly grown in gardens.

*1. M. neglectum Guss. ex Ten.   Grape Hyacinth.

M. racemosum auct.; *M. atlanticum* Boiss. & Reuter

Glabrous. Lvs 3–5, 15–30 cm × 1–3 mm, linear, semicylindrical narrowly grooved above. Scape 10–25 cm. Fls many, in a dense terminal raceme, drooping at fl. Pedicels shorter than fl. Bracts very small. Perianth 3–5 mm, ovoid, mouth small, dark blue, the lobes white or pale, sterile fls smaller, bright blue, never opening. Capsule c. 4 mm, broader than long, emarginate. Fl. 4–5. Pollinated by insects and selfed; protogynous. $2n = 36, 45, 54$. Gb.

Introduced. Dry grassland in Norfolk, Suffolk and Cambridge, mainly in Breckland, and in Oxford; often recorded as an escape elsewhere, but sometimes in error for other species, as it is not now commonly grown in gardens. Much of Europe, north to N. France and S.C. Russia; N. Africa.

*M. comosum (L.) Miller   Tassel Hyacinth.

Hyacinthus comosus L.

Lvs 6–15 mm broad. Scape 20–50 cm. Infl. lax; fertile fls 7–8 mm, brown, obovoid-cylindric, spreading; sterile fls purple, on long ascending pedicels. Fl. 4–7. $2n = 18$. A rather frequent casual, naturalized in Glamorgan. Much of Europe north to N. France, but doubtfully native in the northern parts of its range.

Species and hybrids of **Agapanthus**, robust plants with a rhizomatous stock, usually blue fls with per.-segs united below, are frequently cultivated and **A. praecox** Willd. subsp. **orientalis** (F. M. Leighton) F. M. Leighton is naturalized on rocky ground in the Scilly Is. S. Africa.

Species of *Brodiaea* and allied genera are sometimes cultivated, and *Ipheion uniflorum* (R. C. Graham) Rafin., with solitary bluish-white fls and with an onion-like smell, is naturalized in Jersey and elsewhere.

18. Allium L.

Stock usually a bulb with tunic. *Plant smelling of onion or garlic*. Spathe with 1 to several valves. *Perianth* campanulate or rotate; *segments free or nearly so.* Style gynobasic. Ovules usually 2, rarely more, in each cell. Seeds black. Bulbils often present between fls, sometimes completely replacing them.

About 700 spp., north temperate regions to Ethiopia, southern Africa and Mexico. Several spp. are grown as vegetables and others for their fls. With the exception of *A. schoenoprasum* and *A. ursinum*, all the species in the British Is. appear to have been introduced, inadvertently or deliberately by Man.

1 Lvs linear or cylindrical, sessile.	*2*
Lvs ±elliptical, stalked.	**3. ursinum**
2 Scape terete; perianth 3–12 mm.	*3*
Scape triquetrous; perianth 10–18 mm, white.	*12*
3 Infl. of bulbils only, without fls.	*4*
Infl. with fls.	*5*
4 Spathe 1-valved, quickly caducous, shorter than infl.	
(Common.)	**11. vineale** var. **compactum**

Spathe 2-valved, persistent, longer than infl. with long lf-like points. (Rare.) **6. oleraceum**

5 Inner filaments divided at apex into 3 long points, the middle one bearing the anther, outer entire. 6

Filaments all entire or the inner with 2 small teeth at base. 9

6 Lvs flat, solid. 7

Lvs cylindrical or semicylindrical, hollow. 8

7 Spathe 1-valved, long-beaked; bulblets yellowish; plant robust, with lvs 12–40 mm broad. (Western.) **8. ampeloprasum**

Spathe 2-valved, short-beaked; bulblets reddish-black; plant rather slender, with lvs 20 mm broad. (Northern.) **9. scorodoprasum**

8 Spathe 2–3-valved, persistent; inner stamens with lateral points shorter than anther; bulbils absent; perianth reddish-purple. (Bristol and Jersey.) **10. sphaerocephalon**

Spathe 1-valved, generally caducous; inner stamens with lateral points longer than anther; bulbils nearly always present; perianth pink or greenish-white. (Widespread.) **11. vineale**

9 Spathe shorter than fls; per.-segs 7–12 mm. 10

Spathe with long lf-like points, much longer than fls; per.-segs 5–7 mm. 11

10 Lvs cylindrical, hollow. **1. schoenoprasum**

Lvs flat. **2. roseum**

11 Anthers included. **6. oleraceum**

Anthers exserted. **7. carinatum**

12 Lvs 2–5, infl. without bulbils. **4. triquetrum**

Lf 1; infl. with bulbils. **5. paradoxum**

Section 1. *Schoenoprasum* Dumort. Bulbs narrow, elongate, clustered on a short rhizome in perennial spp. Lvs distichous, linear, hollow, their bases sheathing the lower third or so of the stem. Stem hollow. Spathe 2–3-valved, persistent, shorter than the pedicels. Per.-segs somewhat connivent. Filaments all slender, entire. Ovary with 3 distinct nectary pits but without projections. Stigma entire. Seeds angular, without aril.

1. A. schoenoprasum L.   Chives.

A. sibiricum L.

Growing in tufts. Lvs cylindrical, subglaucous, 10–25 cm × 1–3 mm. Scape 15–40 cm, cylindrical. Spathe usually 2-valved; valves ovate, shortly acuminate, scarious. Umbel subglobose, dense-fld, without bulbils; pedicels shorter than fl. *Per.-segs* 7–12 mm, *spreading, pale purple or pink.* Stamens about half as long as perianth. Fl. 6–7. $2n = 16^*$ (32 in a Siberian form). Gb.

Plants from Pembroke and Cornwall have been separated as a sp. under the name of *A. sibiricum* L. They appear to differ in their flexuous lvs but are not *A. sibiricum*, which is not itself specifically separable.

Native. Rocky pastures, usually on limestone, sometimes beside seasonal streams, very local; Cornwall, Hereford, Brecon, Carmarthen, Radnor, Pembroke, Northumberland, Berwick, Westmorland; Lough Mask (Mayo). Sometimes cultivated for flavouring and occasionally escaping elsewhere. N. Europe and Asia from Scandinavia and arctic Russia to Japan, south in the mountains to N.W. Portugal, Corsica, C. Apennines,

Greece, Asia Minor; North America from Newfoundland and Alaska to New York and Washington.

The following two species are cultivated and have numerous cultivars. Their stems are strongly inflated.

***A. cepa** L.   Onion.

Whole plant robust. Bulb usually solitary and biennial. Lvs ±semicircular in section, slightly channelled. Scape inflated and fusiform below the middle. Infl. very many-fld. Fls greenish-white. Per.-segs 4–5 mm, spreading. Inner filaments with a small tooth on each side. $2n = 16, 32$. Commonly grown as a vegetable. The Shallot (*A. ascalonicum* auct. non L.) is a perennial form, not known wild but closely allied to *A. oschaninii* O. Fedsch. from C. Asia.

***A. fistulosum** L.   Welsh Onion.

Whole plant robust. Lvs circular in section. Scape inflated near the middle. Infl. many-fld. Fls yellowish-white. Per.-segs 6–8 mm, connivent. Sometimes grown as a vegetable. Not known wild but probably originating in E. Asia, where a closely related sp., *A. altaicum* Pallas, occurs.

Section 2. *Molium* G. Don ex Koch. Bulb subglobose; rhizome 0. *Lvs spiral,* linear, flat, almost basal. Stem terete, not inflated. Spathe shorter than the pedicel. Per.-segs spreading. Filaments all slender, entire, free. Ovary with nectary pits. Stigma entire. Seeds angular, without aril.

†2. A. roseum L.

Bulb with numerous bulblets; outer tunic perforate. *Lvs 2–4, linear,* 4–12 mm broad, not keeled beneath. Scape 30–80 cm, cylindrical, sheathed by lvs at base. Spathe 2–4-valved, scarious. Pedicels 2–3 times as long as fl. Perianth broadly campanulate; *segments* 10–20 mm, *pink,* becoming scarious. Stamens included. Fl. 6. Gb.

Introduced. Native of Mediterranean region; Azores.

Var. *roseum.* Umbel without bulbils, with numerous fls. $2n = 32$. Rarely naturalized.

Var. *bulbiferum* DC. Umbel with fls and bulbils. $2n = 48$. More commonly naturalized.

Section 3. *Arctoprasum* Kirschl (*Ophioscorodon* (Wallr.) Bubani). Bulb long, narrow; rhizome short. *Lvs basal, stalked; blade* lanceolate, narrowly elliptical to narrowly ovate. Stem triquetrous, not inflated. Spathe 2-valved, shorter than the pedicels. Per.-segs spreading. Filaments slender, entire, equal, free. Ovary deeply 3-lobed; nectary pits 0. Stigma entire. Seeds globose; aril 0.

3. A. ursinum L.   Ramsons.

Bulb narrow, solitary, consisting of a single petiole-base. *Lvs 2(–3), narrowly elliptical to narrowly ovate,* 10–25 × 4–8 cm, acute, bright green; petiole 5–20 cm, strongly curved. Scape trigonous or semicylindrical and 2-angled, 10–45 cm, sheathed by petioles at base. Spathe scarious; valves ovate, acuminate, shorter than fls. Infl. 6–20-fld; flat-topped, without bulbils; pedicels longer than fls. Per.-segs 8–10 mm, white, lanceolate, acute. Stamens shorter than perianth. Stigmas obtuse. Fl. 4–6. Pollinated by insects and selfed; protandrous. $2n = 14^*$. G.

Native. Damp woods and shady places, sometimes forming local societies; ascending to 425 m. Rather common throughout British Is., but absent from Orkney, Shetland and Channel Is. Europe from Scandinavia (c. 64°N. in Norway) and C. Russia (Upper Dnieper and Volga-Don areas) to C. Spain, Corsica, Sicily, the Balkan peninsula and the Caucasus; Asia Minor.

In W. and C. Europe, and N. Italy, subsp. *ursinum*, with papillose pedicels; subsp. *ucrainicum* Kleopow & Oxner, with smooth pedicels, occurs in E. Europe, the Balkan peninsula and S. Italy.

Section 4. *Briseis* (Salisb.) Stearn. Bulb subglobose; rhizome 0. Lvs distichous or solitary, linear, so strongly keeled as to be almost 3-sided, almost basal. *Stem triquetrous*, not inflated. Spathe shorter than the pedicels. Per.-segs connivent after fl. Filaments slender, entire, unequal, connate at the base with the per.-segs. Ovary without deep nectary pits. *Stigma trifid.* Seeds angular, with aril.

*4. A. triquetrum L.   Triquetrous Garlic.

Bulb small, whitish. *Lvs* 2–5, linear, 12–20 cm × 5–10 mm. Scape 20–50 cm. Spathe 2-valved, scarious; valves lanceolate. *Infl.* 3–15-fld, lax, *without bulbils*, somewhat unilateral; fls drooping; pedicels longer than fls. Perianth ±campanulate; *segments* 12–18 mm, oblong, acute, *white with green line.* Stamens shorter than perianth. Fl. 4–6. Seeds dispersed by ants. $2n = 18^*$. Gb.

Introduced. Thoroughly naturalized in hedgebanks and waste places in Channel Is., S.W. England, S. Wales, S. Ireland and perhaps elsewhere, increasing. Native of W. Mediterranean region from S. Spain and Portugal and Morocco to W. Italy, Sicily and Tunisia.

*5. A. paradoxum (Bieb.) G. Don

Lf 1, brighter green than in *A. triquetrum. Infl.* 1–4-fld with *numerous bulbils.* Per.-segs c. 10 mm, white. Fl. 4–5. Gb.

Introduced. Naturalized in a number of places. Native of the Caucasus, N. Iran and mountains of Turkmenia.

Section 5. *Codonoprasum* Rchb. Bulb subglobose; rhizome 0. Lvs linear, their bases sheathing at least the lower ⅓ of the stem. Spathe 2-valved, persistent, *the valves unequal*, each from an ovate base drawn out into a long slender appendage *much longer than the pedicels.* Filaments all slender, entire. Ovary without conspicuous nectary pits or projections. Stigma entire. Seeds angular; aril 0.

*6. A. oleraceum L.   Field Garlic.

Bulb ovoid, usually with bulblets. Lvs semi-terete, grooved, usually hollow at least below, 15–30 cm × 2–3 mm or (var. *complanatum* Fr.) flat and to 4 mm broad. Scape 25–80 cm, cylindrical, sheathed by lf-bases in the lower half. Spathe 2-valved, with long lf-like points much longer than fls (often several times as long). Umbel lax, few-fld, or (var. *complanatum*) many-fld, with bulbils, rarely fls 0; pedicels unequal, much longer than fls. Perianth campanulate; *segments* 5–7 mm,

oblong, obtuse, pinkish, greenish or brownish. *Stamens included* in perianth. Fr. rarely or never produced. Fl. 7–8. $2n = 32$. Gb.

Probably introduced. Dry grassy places, local from Devon and Kent to Wigtown and Moray, with a distinct eastern tendency; var. *complanatum* confined to the north; in Ireland very local and only in Wexford, Dublin and Antrim. Europe from Scandinavia and N. Russia (Ladoga-Ilmen region) to N. Spain, Corsica, C. Italy, Yugoslavia, Bulgaria and the Caucasus.

*7. A. carinatum L.

Differs from *A. oleraceum* as follows: lvs always flat, somewhat grooved. Per.-segs bright pink. *Stamens conspicuously exserted.* Fr. never produced. Fl. 8. $2n = 24$ (triploid). Gb.

Introduced. Thoroughly naturalized in a number of places from Gloucester and Lincoln to Kirkcudbright and Angus and in N.E. Ireland. Scandinavia and Bornholm to E. France, Switzerland and the Balkan peninsula.

Section 6. *Allium.* Bulb subglobose or ovoid; rhizome 0. Lvs distichous, linear, their bases sheathing the lower ⅓ of the stem or more. Stem not inflated, terete. Spathe 1–2-valved, usually cast off early if longer than pedicels. Per.-segs connivent. *Filaments dimorphic*; the outer slender, entire; the inner broad, divided above into 3 points, the middle one bearing the anther. Ovary with a small shelf-like projection above each nectary pit. Stigma entire. Seeds angular, without aril.

*8. A. ampeloprasum L.   Wild Leek.

Bulb with bulblets inside the tunic. *Lvs linear*, keeled, 15–60 cm × 12–35 mm, scabrid on margins and keel, glaucous. *Scape* cylindrical, *stout*, 60–200 cm. *Spathe scarious*, 1-*valved*, with a compressed beak, *caducous before fl.* Umbels many-fld, 7–10 cm across, *globose, without or with a few small bulbils.* Perianth campanulate, pale purple or whitish, c. 8 mm. *Stamens slightly exserted*; lateral points of inner filaments much longer than the antheriferous one which about equals or is shorter than undivided lateral part of filament. Style exserted. Fl. 7–8. $2n = 32^*$. Gb.

Introduced. Rocky and waste places near the coast, very rare. S. and W. Europe; Macaronesia.

Var. *ampeloprasum.* Umbel dense, without bulbils. Cornwall, Steep Holm (Somerset), Flat Holm (Glamorgan) Pembroke; found as a casual elsewhere.

Var. *babingtonii* (Borrer) Syme (*A. babingtonii* Borrer). Differs from var. *ampeloprasum* as follows: umbel lax, irregular, few-fld, with numerous large bulbils, usually with some of the pedicels longer (5–10 cm) and bearing secondary heads. Fl. 8. $2n = 48^*$. Gb. Clefts of rocks and sandy places near the coast, Cornwall, Dorset; W. Ireland from Clare to Donegal, especially on the islands off the coast. Probably a relic of ancient cultivation.

*A. porrum L.   Leek.

Differs from *A. ampeloprasum* principally in the thick, cylindrical bulb which lacks bulblets. Commonly cultivated as a vegetable. Origin not definitely known, but evidently derived from *A. ampeloprasum* in cultivation.

*9. A. scorodoprasum L.   Sand Leek.

Bulb with reddish-black bulblets. *Lvs linear*, flat, 15–20 cm × 7–15 mm, scabrid on the margins and keel. *Scape* 30–80 cm, cylindrical, *rather slender*, sheathed by the lf bases in the lower half. *Spathe 2-valved*, shorter than umbel, with a very short beak. *Umbels with few fls and purple bulbils*. Perianth campanulate, reddish-purple; segments 5–8 mm, oblong-lanceolate, scabrid on keel. *Stamens included*; antheriferous point of inner filament much shorter than the lateral points and about half as long as undivided part. Fl. 5–8. $2n = 16, 24$. Gb.

Probably introduced. Grassland and scrub on dry soils from N. Lincoln and Cheshire to Perth and Angus, very local; Kerry and Cork. Wicklow; casual elsewhere. Europe from Scandinavia, Finland and C. Russia (Upper Dnieper region) to S.E. France, C. Italy and the Balkan peninsula.

*A. sativum L.   Garlic.

Lvs flat, not scabrid. Spathe 1-valved with long point, caducous. Fls whitish, mixed with bulbils. Stamens included; points of the inner filaments nearly equal. $2n = 16$.

Native of C. Asia, sometimes cultivated for flavouring and occasionally escaping.

*10. A. sphaerocephalon L.   Round headed Leek.

Bulb with stalked bulblets within the tunic. *Lvs subcylindrical*, grooved, *hollow*, 20–60 cm × 1–4 mm. Scape 30–80 cm, cylindrical, sheathed by the lf bases in the lower half. Spathe usually 2-valved, shorter than fls, persistent. *Umbel* dense, subglobose, 2–3 cm across, many-fld, *without bulbils*; outer pedicels about as long as fls, inner long. Perianth campanulate, reddish-purple; segments c. 5 mm, ovate, oblong, scabrid on back. Stamens exserted; *antheriferous point of inner filaments often rather longer than lateral points*. Fl. 6–8. Pollinated by insects and selfed; protandrous. $2n = 16$. Gb.

Probably introduced. Limestone rocks, St Vincent's Rocks, Bristol and at St Aubin's Bay, Jersey. Mediterranean region north to Belgium, central W. Germany and S.C. Russia (Upper Dnieper region).

*11. A. vineale L.   Crow Garlic.

Bulb with numerous bulblets. *Lvs subcylindrical*, somewhat grooved, *hollow*, 20–60 cm × c. 2 mm. Scape 30–80 cm. Spathe usually 1-valved and caducous, scarious, with a beak about as long as itself, not or scarcely exceeding fls. Umbels rather lax with fls and bulbils (var. *vineale*) or more commonly with bulbils and without fls (var. *compactum* (Thuill.) Boreau), rarely with fls only (var. *capsuliferum* Koch); pedicels several times as long as fls. Perianth campanulate, pink or greenish-white; segments ±oblong, c. 5 mm, not scabrid on back. Stamens exserted; *antheriferous point of inner filaments about half as long as lateral points*. Fl. 6–7. Pollinated by insects; strongly protandrous. $2n = 32, 40$. Gb.

Introduced. Fields and roadsides; ascending to 460 m in Yorks. Rather common in England and Wales and a serious weed in parts of E. England and the S. Midlands; local in Scotland, extending north to Aberdeen and the S. Inner Hebrides; local in Ireland and mainly in the south and centre; Channel Is. Most of Europe except the extreme north and C. and E. Russia; N. Africa (rare); Caucasus, Lebanon.

Probably originating in the Balkan peninsula or W. Asia but spread widely as a weed by human activity.

NOTHOSCORDUM Kunth

Stock a bulb with tunic. Plant without onion smell. Spathe always 2-valved. Perianth usually campanulate; segments joined at base into a short tube. Style terminal. Ovules 4–12 in each cell. Seeds black.

About 15 spp., America.

N. gracile (Aiton) Stearn

N. inodorum auct., non (Aiton) Nicholson, *N. fragrans* (Ventenat) Kunth, *Allium gracile* Aiton, *A. fragrans* Ventenat

Bulb subglobose. Lvs basal, 25–30 cm × 5–15 mm, linear. Scape 20–40 cm. Infl. many-fld, loose, fastigiate; fls scented. Per.-segs 8–14 mm, dull white with greenish base and reddish midrib outside. Stamens included.

Occurs on waste ground and as a garden weed in a number of places. Native of North America; naturalized in many parts of the Old World.

19. COLCHICUM L.

Stock a corm. Lvs all basal. Fls 1–3, from the ground. Per.-segs all alike, united below into a long tube. Anthers introrse, dorsifixed. Styles 3, filiform, free from the base. Fr. a septicidal capsule.

About 65 spp., Europe, W. and C. Asia, N. Africa. Some besides the following are sometimes grown in gardens.

1. C. autumnale L.
          Meadow Saffron, Naked Ladies, Autumn Crocus.

Glabrous. Corm 3–5 cm, large, with brown outer scales. Plant lfless at fl., lvs produced in spring as the fr. ripens. Perianth pale purple, lobes oblong, 3–4.5 cm; tube 5–20 cm. Scape elongating in fr. and appearing above the ground, sheathed by the lf-bases. Lvs 12–30 × 1.5–4 cm, oblong-lanceolate, bright glossy green. Fr. 3–5 cm, obovoid; seeds numerous. Fl. 5–10. Pollinated by insects and selfed, protogynous. Fr. 4–6. $2n = 38, 42$. Gt.

Native. Damp meadows and woods on basic and neutral soils, local but sometimes in quantity, from Devon and Sussex to Westmorland and N.E. Yorks; S.E. Ireland in Kilkenny; occasionally naturalized in Scotland. S., W. and C. Europe eastwards to White Russia and N.W. Ukraine.

20. PARIS L.

Stock a creeping rhizome. Lvs 4 or more in a whorl. Fls solitary, 4–6-merous, outer per.-segs sepaloid, inner petaloid, narrow. Filaments short, flat; anthers linear, basifixed, connective elongated. Styles free. Fr. a fleshy loculicidal capsule.

About 20 spp., Europe, temperate Asia and North America.

1. P. quadrifolia L.   Herb-Paris.

Glabrous. Stems 15–40 cm. Lvs 6–12 cm, (3–)4(–8), obovate, shortly acuminate, cuneate at base, subsessile, 3–5-veined with reticulate veins between. Pedicels 2–8 cm. Fls 4(–6)-merous. Sepals 2.5–3.5 cm, green, lanceolate, acuminate. Petals subulate, nearly as long as sepals. Ovary 4–5-celled. Fr. berry-like, black, globose, finally dehiscent. Fl. 5–8. Pollinated by flies and selfed; strongly protogynous. $2n = 20^*$. Grh.

Native. Damp woods on calcareous soils; ascending to 360 m in Westmorland. From Somerset and Kent to Caithness, local and with an eastern tendency, absent from W. Scotland north of Argyll and from Ireland and the Isle of Man. Most of Europe but rare in the Mediterranean region; Caucasus; Siberia.

Species of *Trillium*, similar to *Paris* but with 3 lvs and 3-merous fls, are frequently cultivated.

139. PONTEDERIACEAE

Aquatic herbs. Lvs sheathing. Infl. a raceme or spike, subtended by a spathe-like bract. Fls hermaphrodite, somewhat zygomorphic, hypogynous, 3-merous. Perianth corolla-like, segs usually united into a tube. Stamens 6, 3 or 1, inserted on the perianth. Ovary 3-celled with axile placentas or with only one cell fertile with a single pendent ovule; style long; stigma entire or shortly lobed. Fr. a capsule or nutlet; endosperm copious, mealy.

Seven genera and about 30 spp., mainly tropical.

1. PONTEDERIA L.

Rhizome creeping. Fl.-stems erect with 1 lf (besides the bract).

Infl. a spike. *Perianth funnel-shaped*, *2-lipped*, *with 3 lobes to each lip*, upper united c. $\frac{1}{2}$-way, lower free from each other. Stamens 6. *Ovary with 1 ovule*. Fr. a nutlet enclosed in the accrescent perianth.

About 4 spp., America.

*1. P. cordata L.

Stem erect, to 10 cm. Lvs 9–18 cm, ovate to lanceolate, cordate to broadly cuneate at base, entire. Infl. 5–15 cm. Perianth 12–17 mm, violet-blue, upper seg. with yellow markings.

Grown in gardens, rarely naturalized. Native of eastern North America from Nova Scotia to Texas.

140. JUNCACEAE

Herbs, usually perennial, frequently tufted or with creeping sympodial rhizomes. Lvs long and narrow, terete, channelled or grass-like, with sheathing base, sometimes reduced to scales, glabrous or sparsely hairy. Fls few to many, usually in numerous crowded monochasial cymes, sometimes condensed into heads (rarely solitary), regular, hermaphrodite, protogynous, wind-pollinated. Per.-segs 6, in 2 whorls, equal or subequal, usually greenish or brownish. Stamens free, in 2 whorls of 3 or with the inner whorl missing. Pollen remaining in tetrads. Ovary syncarpous, 1- or ±completely 3-locular, forming a loculicidal capsule; stigmas 3, brush-like. Seeds 3 or many, often with appendages; dispersed by various agencies. Endosperm starchy.

Nine genera and about 400 spp., cosmopolitan, but chiefly in temperate or cold climates or at high altitudes in the tropics.

A natural family, resembling the Liliaceae in fl.-structure, but wind-pollinated and with a characteristic vegetative habit.

Lvs glabrous, various, seldom flat and grass-like; capsule
  many-seeded.   1. JUNCUS
Lvs sparsely hairy, at least when young, flat or channelled
  and grass-like; capsule 3-seeded.   2. LUZULA

1. JUNCUS L.

Glabrous perennial herbs to 1 m or more, or dwarf annuals; erect, ±tufted, *often rhizomatous*. Stem-*lvs with* usually *split sheathing base which is often produced above into auricles*; lamina channelled, compressed or ±terete, sometimes wanting. Infl. a cluster of cymes, sometimes condensed into a head, terminal but in some species seeming lateral because the lowest bract appears to be a continuation of the stem; fls rarely few or solitary. *Capsule many-seeded.* Seeds sometimes with appendages; testa usually finely sculptured, often becoming mucilaginous.

About 300 spp., cosmopolitan.

In addition to *J. pallidus* (p. 545), other *Juncus* spp. of Australasian origin are sometimes found as casuals in places where wool waste has been dumped.

1 Flowering stems bearing only brown sheaths below
    the infl., which is apparently lateral and exceeded
    by a bract continuing the stem.   2
  Lvs on flowering stems (when present) with green
    lamina; infl. obviously terminal, or if apparently
    lateral, then lvs with prickly points.   7
2 Stem very slender, 1 mm or less in diam.; infl. at the
    middle of the apparent stem or lower. **3. filiformis**
  Stem stouter; infl. above the middle of the apparent
    stem.   3
3 Far-creeping, growing in straight lines; infl. usually
    5–20-fld.   **4. balticus**
  Densely tufted; infl. usually many-fld.   4
4 Pith interrupted; stem glaucous, with 12–16 promi-
    nent ridges.   **5. inflexus**
  Pith continuous, at least in the lower part of the stem;
    stem not glaucous, usually with more than 18 ridges
    or striae.   5
5 Stem with 18–45 ridges; capsule with no fertile seeds.
                                    effusus × inflexus
  Stem usually with more than 40 ridges or striae; cap-
    sule fertile.   6
6 Stem when fresh strongly ridged especially below the
    infl.; capsule shortly mucronate. **7. conglomeratus**
  Stem when fresh striate, scarcely ridged; capsule not
    mucronate.   **6. effusus**

7 Annuals (seldom perennial), rarely over 30 cm.   8
    Perennials, usually over 30 cm (except some small alpine plants).   12

8 Infl. usually a much-branched lfy panicle occupying the greater part of the plant; fls rarely in terminal heads; seeds c. 1.5 times as long as broad. **(bufonius group)**   9
    Fls in 1 or a few terminal heads; seeds twice as long as broad.   11

9 Lvs usually more than 1.5 mm wide; per.-segs usually with dark lines on either side of the midrib. **16. foliosus**
    Lvs rarely more than 1.5 mm wide; per.-segs without dark lines.   10

10 Inner per.-segs obtuse to subacute; capsule truncate, as long or longer than the inner per.-segs. **15. ambiguus**
    Inner per.-segs acute to acuminate; capsule acute, shorter than the inner per.-segs. **14. bufonius**

11 Per.-segs less than 4 mm, with fine, often recurved points; lf-sheath without auricles. **18. capitatus**
    Per.-segs 5 mm or more, gradually tapering to an acute point; lf-sheath with pointed auricles. **20. pygmaeus**

12 Lvs and lowest bracts ending in stiff prickly points; infl. many-fld, apparently lateral.   13
    Lvs and lowest bract not prickly-pointed; infl. terminal.   14

13 Fls reddish-brown; capsule much exceeding the perianth. **2. acutus**
    Fls straw-coloured; capsule not or barely exceeding the perianth. **1. maritimus**

14 Fls solitary and terminal or in capitate clusters of 6 or fewer; seeds with appendages. (Alpine plants.)   15
    Fls neither solitary nor capitate, usually in panicles; seeds without appendages. (Lowland or alpine.)   18

15 Fl.-stems densely tufted; fls 1–3 together between axils of 2–3 long filiform bracts. **9. trifidus**
    Fl.-stems solitary or in small not dense tufts; the longest bract not or only shortly exceeding the fls.   16

16 Stoloniferous; outer per.-segs acute. **27. castaneus**
    Not stoloniferous; outer per.-segs obtuse or bluntly pointed.   17

17 Fls 2–3 in a head; lvs in section of 2 tubes. **26. triglumis**
    Fls 1–2 in a head; lvs in section of 1 tube. **25. biglumis**

18 Lvs all basal.   19
    Cauline lvs 1–4.   20

19 Lvs tough and solid, not thin and grass-like; stamens 6. **10. squarrosus**
    Lvs thin, like those of a *Luzula*; stamens 3. **17. planifolius**

20 Lvs without transverse septa, usually solid, never setaceous.   21
    Lvs hollow, with more or less obvious transverse septa, sometimes setaceous.   24

21 Stem 50–100 cm, with 1–4 cauline lvs; lowest bract much less than $\frac{1}{2}$ as long as infl. **8. subulatus**
    Stem seldom exceeding 50 cm, with 1–2 cauline lvs; lowest bract at least $\frac{1}{2}$ as long as infl.   22

22 Fls greenish or straw-coloured; per.-segs very acute. **13. tenuis**
    Fls dark brown; per.-segs blunt.   23

23 Usually in salt-marshes; style as long as capsule; perianth nearly equalling fr. **12. gerardi**

Usually not on saline soils; style shorter than capsule; perianth $\frac{1}{2}-\frac{2}{3}$ as long as fr. **11. compressus**

24 Lvs setaceous, in section of 2 tubes (use lens); short shoots often proliferating from the infl. **21. bulbosus**
    Lvs not setaceous, in section of 1 or more than 2 tubes; short shoots not normally proliferating from the infl.   25

25 Lvs with longitudinal as well as transverse septa; perianth pale straw-coloured to light reddish-brown. **19. subnodulosus**
    Lvs without longitudinal septa; perianth brown to blackish, never straw-coloured.   26

26 Per.-segs obtuse, the outer often mucronate; capsule obtuse. **24. alpinoarticulatus**
    Per.-segs acute; capsule acute or acuminate.   27

27 Lvs strongly laterally compressed; capsule abruptly acuminate, fertile. **23. articulatus**
    Lvs subterete, slightly compressed; capsule acute, or if acuminate, without fertile seeds.   28

28 Fls usually less than 7 in head; capsule sterile, either much shorter than perianth or longer and abruptly acuminate. **acutiflorus × articulatus**
    Fls 6–12 in head; capsule many-seeded, gradually tapered to a very acute point. **22. acutiflorus**

Subgenus JUNCUS.

Rhizomatous perennials, halophytic. Lvs and bract very sharply pointed, terete, not septate, their pith of rounded cells with scattered vascular bundles. Infl. apparently lateral. Fls without involucral bracteoles.

1. J. maritimus Lam.   Sea Rush.

An erect, densely tufted, *very tough* perennial. Stem 30–100 cm, light green, smooth when fresh, lfy only near the base; pith continuous. Lowest stem-lvs brown glazed sheaths, upper with terete green lamina, sharply pointed. *Infl.* appearing lateral, many-fld, *forming an interrupted irregularly compound panicle*, with ascending branches, *shorter than the sharply pointed bract*, rarely a dense head or diffuse and exceeding the bracts. *Per.-segs* 3–4.5 mm, *straw-coloured, lanceolate*, outer acute, inner blunt. *Capsule ovoid-trigonous, mucronate, about equalling the perianth. Seeds* obliquely ovoid, *with a large appendage.* Fl. 7–8. $2n = 48$. Grh.

Native. On salt-marshes above high-water mark of spring tides, often dominant over large areas. Abundant on the coasts of the British Is., north to Lewis and Moray. Atlantic and Mediterranean coasts of Europe, extending into the Baltic and Black Sea and east to Sind; N. and S. Africa; North and South America; Australasia; inland in salt areas of S. Europe and C. Asia.

2. J. acutus L.   Sharp Rush.

A tall and *very robust* perennial, forming *dense prickly tussocks*, somewhat like *J. maritimus*, but with stouter and taller stems, 25–150 cm. bearing a few very sharply pointed foliage lvs, the *infl. compact, ±rounded, with more spreading branches*, much shorter than the bract. *Per.-segs* 2.5–4 mm, *reddish-brown, becoming almost woody*, ovate-lanceolate, with broad scarious margin.

Capsule turgid, *broadly ovoid, at least twice as long as the perianth.* Seeds as in *J. maritimus* but broader. Fl. 6. $2n = 48$. Hs.

Native. Local on sandy seashores and dune-slacks, less frequently on salt-marshes on the west, south and east coasts of England and Wales from Caernarvon to Norfolk; extinct in N.E. Yorks; southeast coast of Ireland from Cork to Dublin. Mediterranean region of Europe and N. Africa, extending up the Atlantic coast to N. France, east to Transcaucasia; Macaronesia; California; South America; S. Africa.

The British population belongs to subsp. **acutus**.

Subgenus GENUINI Buchenau

Perennials bearing flowering stems and short shoots with a single leaf. Lvs not sharply pointed, terete, not septate, their pith of stellate or angular cells with subepidermal vascular bundles. Infl. apparently lateral. Fls with involucral bracteoles.

3. J. filiformis L.　　　Thread Rush.

A slender wiry perennial *in not very extensive tufts. Stems* 15–30 cm × 0.75–1 mm, stiffly erect, very faintly ridged when fresh, *filiform,* bearing several *brownish lf-sheaths, of which the uppermost often has a short green lamina. Infl.* of *7 fls or fewer,* forming a ±compact head, usually *placed half-way or lower down the apparent stem.* Per.-segs 2.5–3 mm, lanceolate, becoming straw-coloured. *Capsule almost globose, very shortly mucronate,* not exceeding the perianth. Seeds oblong with an inconspicuous appendage. Fl. 6–9. $2n = 84$. Hs.

Native. On stony lake shores among *Molinia,* other *Junci,* etc. Very local and easily overlooked. Lake District; in Scotland scattered north to Inverness; also as a recent colonist by reservoirs in Caernarvon, Leicester and W. Yorks. Northern and arctic Eurasia, south to mountains of Portugal, northern Apennines and Caucasus; Iceland; North America; Patagonia.

4. J. balticus Willd.　　　Baltic Rush.

A far-creeping rhizomatous perennial, not forming large tufts. *Stem* 15–45 cm × 1–2 mm, stiffly erect, dull green, *smooth when fresh; pith continuous. Infl.* $\frac{1}{2}$ or less from apex of apparent stem, appearing lateral, *about 7–20-fld, lax, with ascending branches.* Per.-segs to 5 mm, brown, ovate-lanceolate, acute. *Capsule* dark chestnut-brown, glossy, *ovoid, abruptly mucronate.* Seeds oblong, without appendages. Fl. 6–8. $2n = ?80, 84$. Said rarely to reproduce by seed. Grh.

Native. In dune-slacks, rarely in other damp sandy places. East and north coast of Scotland from Fife to Sutherland and the Hebrides; Lancs; rare inland (to 278 m in E. Inverness). N. Europe, south to the Netherlands; Faeroes; Iceland; North America; Patagonia; New Caledonia.

British populations belong to var. *balticus.* Hybrids with *J. inflexus* and *J. effusus* have been found in Lancs.

5. J. inflexus L.　　　Hard Rush.

J. glaucus Sibth.

Perennial in *large dense grey-green tufts.* Rhizomes horizontal, matted, with very short internodes. *Stem slender,* 25–60 cm × 1–1.5 mm, *stiffly erect, with* 12–18 *prominent ridges,* dull, glaucous, pith interrupted. *Lf-sheaths glossy,* dark brown or blackish. *Infl. apparently lateral,* of many ascending branches, lax. *Per.-segs* 2.5–4 mm, *lanceolate* with subulate points, unequal. *Capsule dark chestnut-brown, glossy, ovoid-acuminate, mucronate, about equalling the perianth.* Fl. 6–8. Germ. spring. $2n = 40$. Hs.

Native. Chiefly in damp pastures, preferring heavy basic or neutral soils. North to Banff, abundant in most of England, more local in Wales and Ireland; in Scotland mainly in the southeast, rare and doubtfully native further north. Europe north to S. Sweden and C. Russia, east to India and Mongolia; N. Africa; Macaronesia. Introduced in North America and New Zealand. A subsp. in S. Africa and E. Java.

J. effusus × inflexus (*J. × diffusus* Hoppe) differs from *J. inflexus* in the green, not glaucous, scarcely grooved stems, with c. 18–45 coarse striae (cf. *J. effusus*); pith continuous or ±interrupted above. Infl. and perianth as in *J. inflexus.* Capsule much shorter than the perianth, usually with few seeds. With the parents, but not common, spreading vegetatively.

6. J. effusus L.　　　Soft Rush.

J. communis var. *effusus* (L.) E. Meyer

A *densely tufted,* stiffly erect perennial. *Stems* 30–150 cm × 1.5–3 mm (just below infl.), rather soft, *bright to yellowish green,* glossy and quite *smooth throughout their length when fresh, with* 40–90 *striae;* pith *continuous.* Lf-sheaths reddish to dark brown, not glossy. *Infl.* apparently lateral, *placed about* $\frac{1}{5}$ *the distance from the top of the apparent stem,* many-fld, lax or *condensed into a single rounded head* (var. *compactus* Hoppe), with ascending spreading and deflexed branches. Per.-segs 2–2.5 mm, lanceolate, finely pointed. Stamens 3(–6). *Capsule* yellowish to chestnut-brown, *broadly ovoid, retuse, not mucronate.* Fl. 6–8. Occasionally cleistogamous. Germ. spring. $2n = 42$. Hs.

Native. Very abundant and locally dominant in wet pastures, bogs, damp woods, etc., especially on acid soils. Throughout the British Is. Europe, to 65° 25′ N. in Norway. North and south temperate zones and on mountains in the tropics.

7. J. conglomeratus L.　　　Compact Rush.

J. communis var. *conglomeratus* (L.) E. Meyer; *J. leersii* Marsson

Closely resembling J. effusus and often confused with its var. *compactus,* but usually less robust and forming smaller tufts. *Stems* bright to greyish-green, *not glossy, with numerous ridges which are especially prominent just below the infl. Bract* with *widely expanded sheathing base* which allows it to hinge backwards when withered. *Infl.*

usually *condensed into a rounded head, more rarely of several stalked heads.* Capsule as in *J. effusus*, but with the *remains of the style on a small elevation in the hollowed top.* Fl. 5–7 (usually about a month earlier than *J. effusus*). Germ. spring. $2n = 40$. Hs.

Native. In similar habitats to *J. effusus*, but with a narrower ecological range and a more marked preference for acid soils. Throughout the British Is., but uncommon in many districts and often confused with *J. effusus* var. *compactus.* Europe, north to Faeroes and to 68° 55′ N. in Norway, extending east to Syria, Kurdistan and Dzungaria (W. China); N. Africa, Macaronesia; eastern North America.

J. conglomeratus × J. effusus. Presumed hybrids are sometimes found with the parents, but they appear to be infrequent and variants of *J. conglomeratus* are probably sometimes mistaken for them.

***J. pallidus** R. Br., an Australasian species of this subgenus, occurs in gravel pits, etc., where wool shoddy has been dumped. It is very robust (stems 120–200 cm × 3–4 mm). Infl. rather lax, straw-coloured. Per.-segs ovate-lanceolate, coriaceous. Stamens 6. Capsule much exceeding the perianth. Putative hybrids with *J. effusus* and *J. inflexus* have been found with the parents.

Subgenus SUBULATI Buchenau

Like *Genuini*, but lvs all cauline. Infl. longer than the lowest bract, not appearing lateral.

***8. J. subulatus** Forskål

A tall perennial with horizontally creeping rhizome. *Stems* to 1 m, *bearing 1–4 lvs with subterete hollow lamina.* Panicle terminal, many-fld, *narrow, interrupted, the bracts inconspicuous, much less than ½ as long as the infl.* Per.-segs finely acuminate, straw-coloured, *slightly larger than the ovoid-trigonous, mucronate capsule.* Seeds without appendage. $2n = 42$. Grh.

Probably introduced. Established for over 30 years in a salt-marsh at Berrow, N. Somerset. Mediterranean region east to Egypt and Syria. Portugal.

Subgenus PSEUDOTENAGEIA V. Krecz. & Gontsch.

Perennials with slender rhizomes, 1 sp. halophytic. Lvs all cauline, channelled. Infl. terminal. Fls with involucral bracteoles.

9. J. trifidus L.   Three-leaved Rush.

A slender, densely tufted, grass-like perennial, often forming circular patches. Rhizome horizontal, usually less than 15 cm, thickly covered with dead branches and lf-sheaths. Stems rigid, erect, terete, finely striate. *Basal lvs mostly sheaths only,* prolonged above into *laciniate auricles,* some with filiform lamina to 8 cm. *Infl. of 1 terminal and 0–3 lateral, sessile or shortly stalked fls. Stem-lvs* (bracts) 2–3, *filiform,* 2–8 cm. Per.-segs 2.5–3 mm, ovate-lanceolate, dark chestnut brown. Capsule ovoid, long beaked, longer than the perianth. Fl. 6–8. $2n = 30$. Hs.

Native. Detritus and rock-ledges on high mountain tops, where it is sometimes the most abundant plant over large areas. Scotland from Arran and Dumfries to the Outer Hebrides and Shetland. High mountains of Europe from C. Spain to Caucasus and E. Siberia. Subarctic and Arctic to 71° 30′ N.; North America. Two subsp. occur on the Continent of which only one (subsp. *trifidus*) is British.

10. J. squarrosus L.   Heath Rush.

A tough, wiry perennial, forming *dense low tufts.* Lvs 8–15 cm, *usually all basal,* from a very short upright stock, subulate, deeply channelled, rigid, *sharply deflexed above the sheathing base.* Fl. stems 15–50 cm, erect, very stiff. Lowest bract less than ½ the length of the infl., usually lf-like. Infl. lax, the branches ending in clusters of 2–3 fls. *Per.-segs* 4–7 mm, *dark chestnut brown, lanceolate, bluntly pointed* or acute, slightly exceeding the obovoid mucronate capsule. Fl. 6–7. $2n = 40$. Hr.

Native. On moors, bogs and moist heaths, confined to acid soils. Abundant throughout the British Is., except where there is a scarcity of suitable habitats. W., C. and N. Europe, from the mountains of S. Spain and N. Italy to 68° 15′ N. in Norway, east to Ukraine and 41° E. in N.W. Russia; Morocco; Iceland; S. Greenland.

11. J. compressus Jacq.   Round-fruited Rush.

A small- or medium-sized tufted perennial, *seldom forming extensive patches.* Rhizome horizontal, usually less than 5 cm, rarely far-creeping. *Lvs narrowly linear, dorsiventrally flattened.* Fl.-stems 10–30 cm, curved, not stiffly erect, compressed throughout their length, *bearing 1–2 lvs.* Panicle compound, terminal, subcymose, lax to compact, usually shorter than the lowest bract. Per.-segs 1.5–2 mm, ovate, very obtuse, light brown. *Anthers slightly shorter than the filaments. Style shorter than the ovary. Capsule subglobose, obtuse,* c. 1½ times *as long as the perianth,* very shortly mucronate, very glossy. Fl. 6–7. $2n = 44$. Grh.

Native. In marshes, alluvial meadows and grassy places where the vegetation is kept low by mowing or grazing, chiefly on non-acid soils, England and Wales, north to Westmorland, rather rare and mainly in S.E. England; in Scotland north to Argyll and Forth. Eurasia (except the Arctic); eastern North America.

Hybrids between *J. compressus* and *J. gerardi* (*J.* × *royeri* P. Fourn.) have been doubtfully reported from several localities.

12. J. gerardi Loisel.   Saltmarsh Rush.

Resembling *J. compressus* and differing in a combination of variable characters. Usually *taller, forming more extensive tufts or patches. Rhizome far-creeping.* Lvs 10–20 cm, dark green. Fl.-stems straight, stiffly erect, compressed below, triquetrous above. *Infl.* usually laxer and *with straighter, less spreading branches,* usually considerably exceeding the lowest bract. *Perianth dark brown to blackish. Anthers 3 times as long as the filaments. Style equalling or slightly longer than the ovary. Capsule acuminate, not or slightly exceeding the perianth.* Fl. 6–7. 2n = 80. Grh.

Native. In salt-marshes, from just below high-water mark of spring tides upwards, abundant and locally dominant; occasionally inland in saline sites. Great Britain north to Shetland; Ireland. Coasts and inland salt areas of Eurasia; N. Africa; North America.

*13. J. tenuis Willd.   Slender Rush.

J. macer S. F. Gray

A rather weak perennial, very variable in stature. *Lvs* 10–25 cm, usually *all basal*, on a very short, usually upright stock, curved, flexible, narrowly linear, channelled, with a broad sheathing *base* which is *produced above into obtuse scarious auricles several times as long as wide*, rarely (var. *dudleyi* (Wiegand) F. J. Hermann) brown, not scarious, broader than long. Fl.-stems 15–35 cm, erect, slightly compressed. Fls in a terminal, usually lax *panicle, much exceeded by at least one of the very narrow lf-like bracts. Per.-segs* 3–4 mm, greenish, becoming *straw-coloured, narrowly lanceolate, very acute.* Capsule ovoid, obtuse, mucronate, shorter than the perianth. *Seeds* becoming *very mucilaginous.* Fl. 6–9. Seeds dispersed on cart-wheels, etc. Germ. spring. 2n = 32. Hs.

Naturalized (according to some, native in S.W. Ireland) and still extending its range in Britain (first recorded 1883), on roadsides, waste ground and by field paths. North to Ross and the Outer Hebrides, locally abundant in parts of Wales, Scotland and Ireland uncommon in most of the Midlands and the east coast counties from the Thames to the Tees, . North and South America. Naturalized in many temperate countries.

Subgenus POIOPHYLLI Buchenau

Tufted annuals. Lvs all cauline, flat or channelled. Infl. (in British spp.) usually occupying at least the upper ½ of the plant. Fls with involucral bracteoles.

(14–16) J. bufonius group   Toad Rush.

Slender *annuals*, more rarely short-lived perennials, of very variable dimensions. *Stems* 3–25(–40) cm, simple or *much branched from the base*, usually erect, generally with 1 cauline and several basal lvs. Lvs 1–5 cm, linear from a sheathing base, channelled. *Infl.* much branched, usually *occupying more than half the plant.* Fls sessile, solitary or in groups of 2–3. *Per.-segs* 2.5–7.3 mm, outer lanceolate-acuminate, finely pointed, *green with hyaline borders*, the inner shorter, acute to obtuse. Stamens 6. Capsule equalling or shorter than the perianth, oblong, acute to truncate above. *Seeds* broadly obovoid to barrel-shaped, c. 1½ *times as long as broad, without appendages.* Fl. 5–9. Cleistogamous or wind-pollinated. Seeds mucilaginous, easily carried on wheels, feet of animals, etc.

Native. Paths, road edges, cultivated land and mud by ponds and streams. Abundant throughout the British

Is. Cosmopolitan, but chiefly in the temperate zones.

The three following taxa, differing in morphology, ecology and chromosome numbers, maintain their characters in cultivation and can be regarded as distinct spp. See T. A. Cope & C. A. Stace, *Watsonia*, **12**, 113–128 (1978) and **14**, 263–272 (1983).

14. J. bufonius L. (sens. str.).

Annual, very variable, 2–25(–35) cm. *Lvs seldom more than* 1.5 mm *wide*, dull green. *Infl. open*, seldom contracted, *the branches diverging at less than* 90°. Fls solitary (2–5 together in var. *fasciculatus* Koch). *Per.-segs* pale green, usually *without darker lines*, the outer acute, or shortly acuminate, 4.1–7.3 mm, *inner usually acute*, 3.5–5.8 mm. Anthers generally *shorter than the filaments. Capsule acute* at apex, shorter than inner per.-segs. Seeds usually *obliquely ovoid*, 340–520 μm; when ripe the cross-bars between the longitudinal ridges on the testa close together. 2n = 108 (other numbers also reported). Th.

Cosmopolitan weed. Abundant in British Is.

15. J. ambiguus Guss.

J. ranarius Song. & Perr.

Annual. Stems to 17 cm. Lvs as in *J. bufonius* sens. str. Infl. open, branches often curved. *Fls often 2–5 together at ends of branches.* Inner per.-segs ±obtuse at apex or rounded and mucronate. Anthers usually *shorter than the filaments. Capsule truncate*, equalling or longer than the inner perianth, truncate. Seeds ovoid or barrel-shaped, the testa when ripe *with inconspicuous longitudinal ridges* and cross-bars. 2n = 34 (30, 32 and 60 also reported). Th.

On damp, usually saline sand or mud near coast, occasionally inland in salt-flashes and on highly basic waste dumps. Distribution not fully known, but appears to be frequent round the coast of the British Is. Inland in Cheshire, Stafford, Worcester, and perhaps elsewhere. Europe; N. Africa; Asia; North America.

16. J. foliosus Desv

Annual or short-lived perennial. Often *densely tufted* with erect or ascending branches from a slightly procumbent base, *to* 35 cm, more robust than **14** and **15**. *Lvs bright green*, 2–5 mm *wide. Infl. open, branches* straight, often *diverging at angles of more than* 90°. Fls 1–3(–5) at ends of branches. *Per.-segs with brown or black line on either side of midrib*, the inner *subacute or acute*, 3.6–5.4 mm. *Anthers* 1.2–5 *times as long as the filaments.* Capsule usually subacute, about equalling the inner per.-segs. *Seeds large*, 430–600 × 270–400 μm, obovoid, often truncate, the testa when ripe *with conspicuous longitudinal ribs, the cross-bars closer than in J. bufonius* sens. str. 2n = 26. Th. or Chh.

Wet muddy places by streams, lakes and in damp fields and marshes. Locally abundant, especially in S. England, N. and W. Wales and W. Scotland; widespread in Ireland. Oceanic. S. and W. Europe; western N. Africa; Madeira.

Subgenus GRAMINIFOLII Buchenau

Rhizomatous or tufted perennials. Lvs basal and cauline, usually flat or involute. Infl. terminal. Fls without involucral bracteoles.

*17. J. planifolius R. Br.

Tufted perennial. *Stems* 10–30(–60) cm, *with numerous basal lvs much shorter than the flowering stems. Lvs 2–8 mm wide*, resembling those of a *Luzula*, *thin, translucent, flat*, becoming slightly involute above and gradually narrowing to the minutely mucronate apex, the sheathing base without auricles, suffused with pink or red. *Infl. with several unequal main branches and* (1–)4–12(–50) *dense rounded heads of* c. 8–10 *fls, much longer than the narrow lf-like lowest bract. Per.-segs* 1.5–3 mm, subequal, *ovate-lanceolate*, the outer acuminate, the inner obtuse, dark brown with hyaline margins. Stamens 3, almost equalling the perianth; anthers much shorter than the filaments. *Capsule broadly obovoid, trigonous, with prominent mucro*, dark brown, glossy. Seeds broadly ellipsoid, without appendages, 0.35–0.4 mm. Fl. 7–8. Hs.

Introduced. By lakes and streams and in bogs. Locally abundant over an area of c. 25 km² near Carna and Cashel, Galway; first observed 1971 (M. J. P. Scannell). Australia; New Zealand; southern South America. Naturalized in Hawaii and Oregon.

Subgenus JUNCINELLA V. Krecz. & Gontsch.

Small tufted annuals. Lvs on fl.-stems all basal, with setaceous lamina and short sheath without auricles. Fls terminal, usually clustered, without involucral bracteoles.

18. J. capitatus Weigel   Dwarf Rush.

A dwarf tufted annual 1–5 cm, becoming reddish in fr., with *setaceous, stiffly erect* stems branched only from the base. *Lvs* 0.5–4 cm all basal, *setaceous from a short sheathing base without auricles*, ±channelled. Infl. of 3–8 sessile fls in a *single terminal head*, rarely with 1–2 lateral heads. Bracts 1–2, the longer much exceeding the infl. *Outer per.-segs* 3–5 mm, greenish, becoming reddish-brown, *curved, ovate-lanceolate, with fine, ±recurved points*; inner shorter, mostly membranous. *Capsule broadly ovoid, obtuse*, mucronate, *shorter than the perianth. Seeds narrowly obovoid*, about twice as long as wide. Fl. 3–6. Fls of two types, short-styled cleistogamic and long-styled wind pollinated. 2n = 18. Th.

Native. On damp heaths, especially where water has stood during the winter and characteristically associated with *Radiola*, *Isoetes histrix*, etc. Rare; Cornwall, Anglesey; Channel Is. S. and W. Europe and sparingly through C. Europe to S. Sweden, Finland and N.W. Russia, east to the lower Don; Africa; Newfoundland (introduced?); South America; Australia.

Subgenus SEPTATI Buchenau

Perennials with far-creeping rhizomes (except *J. pygmaeus*). Lvs all on flowering stems, terete or laterally compressed, usually obviously septate. Infl. generally open, of many heads, each with many flowers, without involucral bracteoles.

19. J. subnodulosus Schrank   Blunt-flowered Rush.

J. obtusiflorus Ehrh. ex Hoffm.

A tall, erect, *rather soft* perennial, growing in extensive patches, *not densely tufted. Stems and lvs* similar, 50–120 cm, bright green, terete, *smooth*, hollow, *with longitudinal and transverse septa*. Sterile stems with 1 lf; fl.-stems with brown scale-lvs and 1–2 *foliage lvs*, each with 35–60 *transverse septa*. Infl. repeatedly compound, of many heads each with 3–12 fls, the *branches* of the second order *diverging at an angle of more than 90°*, rarely congested. *Fls pale* straw-coloured, darkening to light reddish-brown. *Per.-segs* 2–2.25 mm, *incurved, obtuse*. Stamens 6. *Capsule* light brown, *broadly ovoid*, trigonous, *acuminate, shortly beaked, slightly longer than the perianth* Fl. 7–9. Germ. spring. 2n = 40*. Hs. or Hel.

Native. Fens, marshes and dune-slacks with basic ground water, often on calcareous peat. Locally abundant and, in East Anglia, dominant over large areas. North to Angus and S.W. Scotland; in Ireland frequent in C. and W., local elsewhere. W., C. and S. Europe, northwards to S. Sweden and Estonia, eastwards to the Black Sea; Kurdistan; N. Africa.

20. J. pygmaeus L. C. M. Rich.   Pygmy Rush.

J. mutabilis Lam.

A *dwarf annual*, forming small spreading tufts, *often becoming suffused with purple*. Stems 2–8 cm, ascending or upright. Lvs mostly basal, narrowly subulate from *a sheathing base with 2 pointed auricles*, indistinctly septate. *Fls cylindrical or narrowly conical*, subsessile, in 1 or a few heads each of 1–5 fls; lowest bract much exceeding the infl. *Per.-segs* 4.5–6 mm, *subequal greenish or purplish, linear-lanceolate*, blunt, *with straight or incurved points*. Capsule 3–3.5 mm, shorter than the perianth, pale, *obclavate. Seeds ovoid-pyriform*, about twice as long as broad. Fl. 5–6. 2n = 40. Th.

Native. Damp hollows and cart-ruts on heaths. W. Cornwall (Lizard). W. and S. Europe, northwards to Denmark, eastwards to S.E. Italy and Yugoslavia; N. Africa; Asia Minor.

21. J. bulbosus L.   Bulbous Rush

J. supinus Moench

An amphibious perennial, *very variable especially in habit and stature*. Land forms erect, grass-like, densely tufted, 1–20 cm, or procumbent and rooting at the nodes; aquatic forms floating and much branched, to 30 cm or submerged with flexible stems to 1(–4) m. *Stem* slender, *often slightly swollen at the base*, not forming a creeping rhizome. *Lvs* mostly basal (in land forms), *setaceous, with* sheathing base and *numerous very indistinct septa.* Infl. simple or with few to many unequal branches and 3–20 *heads of 2–6 fls, often with small adventitious shoots in the flower heads or proliferating without flowers.* Per.-segs c. 2 mm, green or brown with hyaline borders, the inner obtuse, the outer acute, mucronate. Stamens 3–6; anthers equalling or shorter than the filaments. *Capsule* 2.5–3 mm, *oblong to obovoid, subacute to retuse, ±trigonous above.* Fl. 6–9. $2n = 40$. Hs, Hel. or Hyd.

Native. Moist heaths, bogs, cart-ruts and rides in woods, chiefly on acid soils. Abundant throughout the British Is. Europe, except the southeast, northwards to 69° 15′ N. in Norway, eastwards to W. Russia; N. Africa; Macaronesia; eastern North America.

J. kochii F. W. Schultz is often regarded as a taxon separate from *J. bulbosus.* It differs in being rather more robust, stamens 6, anthers oblong, distinctly shorter than the filaments; perianth and capsule dark brown, the latter obovoid, retuse, sharply trigonous above. Plants with these characters are frequent especially in W. and N. Britain and in W. Europe. The consensus is that *J. kochii* cannot be considered as a distinct sp. or subsp.

22. J. acutiflorus Ehrh. ex Hoffm.

Sharp-flowered Rush.

J. sylvaticus auct.

A tall, *stiffly erect* perennial with stout, far-creeping rhizome, not densely tufted. Stem 30–100 cm with 2–4 deep green, *straight, subterete lvs,* each *with 18–25 conspicuous transverse septa.* Infl. richly branched, repeatedly compound, *of many small, shortly-stalked heads,* each of 6–12 fls; *branches* of second order *diverging at an acute angle.* Fls chestnut-brown. *Per.-segs* 3–3.5 mm, lanceolate, acute, *tapering to awn-like points, the outer recurved at the tips,* the inner with narrow brownish margins. Stamens 6. *Capsule* chestnut-brown, *evenly tapered to an acute point,* longer than the perianth. Seeds about 12 per capsule. Fl. 7–9 (the latest of the common British *Junci*). Germ. spring. $2n = 40^*$. Hs. or Hel.

Native. Wet meadows, moorlands and swampy woodlands, abundant throughout the British Is., especially on acid soils. W., C. and S. Europe, north to Denmark and east to Moscow, but rare north of S. Schleswig and from W. Prussia eastwards; Newfoundland.

J. acutiflorus × articulatus (*J.* × *surreyanus* Druce ex Stace & Lambinon) is common in some districts. Intermediate between parents in most characters. Far-creeping; fl stem to 100 cm or more. Lvs subterete, often curved. Fls large, often more than 12 in head. Capsule usually with few or no seeds. $2n$ 60.

A large form of *J. articulatus* with 80 chromosomes from near Oxford (Timm & Clapham, *New Phytol.,* **39,** 1–16, 1940) and elsewhere may possibly be a hybrid of *J. acutiflorus* and *J. articulatus* derived from the fusion of gametes with 40 chromosomes from each sp. It differs from *J. articulatus* chiefly in

its size, stiffly erect habit, large heads with more numerous (14–18) fls and more numerous seeds per capsule.

23. J. articulatus L.　　Jointed Rush.

J. lampocarpus Ehrh. ex Hoffm.

Very variable, especially in size and habit. Perennial, ascending, decumbent or prostrate, seldom stiffly erect, with slender rhizome, often subcaespitose. Stem to 80 cm, terete, bearing 2–7 *laterally compressed, usually curved deep green lvs,* with 18–25 *inconspicuous transverse septa.* Infl. repeatedly compound, often sparingly branched, with few to many stalked heads, each of 4–8 fls; *branches diverging at an acute angle. Fls* c. 3 mm, *dark chestnut-brown to almost black. Per.-segs* lanceolate, *acute,* the inner with broad colourless margins. Stamens 6. *Capsule long-ovoid,* contracted above to an acumen. Seeds about 40 per capsule. Fl. 6–9. Germ. spring. $2n = 80^*$. Hs. or Hel. Exists in several ecotypes.

Native. Wet ground, especially on acid soils, preferring meadows or moors which are grazed or mown. Abundant throughout the British Is. Europe and Asia (except the Arctic); Iceland, North America; N. Africa. Introduced in S. Africa, Australia and New Zealand.

24. J. alpinoarticulatus Chaix　　Alpine Rush.

J. alpinus Vill.

An erect perennial resembling slender variants of *J. articulatus.* Rhizome shortly creeping. Stem smooth, usually terete. Lvs 2–3, subterete, distinctly septate. Infl. variable in form and size. *Per.-segs* dark brown or blackish, *incurved, the outer ovate, obtuse, often mucronate, the inner broadly obtuse. Capsule ovoid, obtuse,* shortly mucronate, equalling or slightly exceeding the perianth. Fl. 7–9. $2n = 40$. Hs.

Native. Gravelly stream beds and marshy places in mountains. Rare. Northern Pennines and Scotland (Southern uplands and Caithness). Northern Eurasia, extending southwards to the mountains of S. Europe and to the Caucasus; Greenland; North America.

Represented in Britain largely or wholly by subsp. *alpinoarticulatus.* Plants from E. Ross and Aberdeenshire described as *J. marshallii* Pugsley have been referred doubtfully to subsp. *nodulosus* (Wahlenb.) Lindmann, but further work is required.

Subgenus ALPINI Buchenau

Small to medium-sized mountain perennials. Lvs mostly basal, terete or channelled. Infl. of 1–3 heads, each of few, relatively large, flowers without involucral bracteoles.

25. J. biglumis L.　　Two-flowered rush.

A small, neatly tufted perennial, with shortly creeping rhizome. *Stem* 5–12 cm, erect, *channelled along one side. Lvs* 3–6 cm, all basal, curved, subulate from a sheathing base *with very small auricles, in section of one tube. Infl.* usually *of 2 fls, one just below the other, usually exceeded by the lowest bract. Per.-segs* oblong, obtuse, purplish-brown. *Capsule 3–4 mm, turbinate, retuse,* much longer than the perianth. *Seeds 1–1.25 mm, with a short white appendage at each end.* Fl. 6–7. $2n = 60$. Hr.

Native. Wet stony places and rock-ledges on high mountains, on base-rich soil, 600–1000 m. Scottish Highlands from Breadalbane to Argyll, Inner Hebrides and Sutherland. Throughout the Arctic (to 83° 6′ N. in Greenland), extending southwards to the Alps (one locality), S. Norway, Altai, Kamchatka and to 40° N. in the Rocky Mts.

26. J. triglumis L.   Three-flowered Rush.

A small tufted perennial, forming *larger and taller tufts than J. biglumis. Stems 5–10 cm,* slender, stiffly erect, *terete. Lvs 3–10 cm,* all basal, curved, subulate from a *sheathing base with large auricles, in section of 2 tubes. Infl. a terminal cluster of 2–3 fls, almost at one level, usually exceeding the lowest bract. Per.-segs* ovate-lanceolate, blunt, *light reddish-brown. Capsule 5 mm, ovate-oblong, obtuse, mucronate, shortly exceeding the* perianth. *Seeds 1.75–2 mm, with a long white appendage* at each end. Fl. 6–7. $2n = 50, 134$. Hr.

Native. Bogs and wet rock-ledges on high mountains, preferring non-calcareous rocks. Less alpine than *J. biglumis*, descending to 310 m in Perthshire and 60 m in Shetland; N. Wales, Lake District and northern Pennines, Scottish Highlands, Inner Hebrides, Shetland. Arctic and sub-Arctic to 80° N., extending southwards to the Pyrenees, Alps, Apennines, Balkan peninsula, and Caucasus; Himalaya; Colorado, Labrador.

27. J. castaneus Sm.   Chestnut Rush.

A medium-sized *stoloniferous* perennial, *not tufted.* Stem 8–30 cm, erect, smooth, terete, with 2–3 cauline lvs. *Lvs 5–20 cm, soft, subulate, ±channelled above,* from a *long sheathing base without auricles,* bluntly pointed. *Fls very large, dark chestnut-brown, in 2–3 ± approximated terminal capitate clusters.* Per.-segs linear-lanceolate, outer acute, inner blunter. *Capsule 6–8 mm, elliptical-oblong, obtuse, mucronate, much longer than the perianth.* Seeds 2–3 mm, with long white appendages. Fl. 6–7. $2n = 40$. Hs. or Hel.

Native. Marshes by springs and streams in high mountains. Highlands of Scotland (chiefly Clova and Breadalbane). Circumpolar, reaching 78° 30′ N. in Spitsbergen, extending south along the mountain ranges to the Alps, Urals, Altai and Shensi; in North America, southwards to Quebec and New Mexico.

2. LUZULA DC.   Woodrush

Tufted grass-like perennials, rarely more than 1 m high, sometimes stoloniferous. *Lvs mostly basal, with closed sheathing base, without auricles, lamina flat or channelled, fringed with long colourless hairs.* Cymes few- or many-fld, sometimes condensed into a head. *Capsule 1-celled;* seeds 3, smooth and shiny, usually with a conspicuous appendage (aril).

About 80 spp., cosmopolitan, but chiefly in the cold and temperate regions of the northern hemisphere.

1	Fls borne singly in the infl., rarely in pairs	2
	Fls in heads or clusters of 3 or more.	3
2	Infl.-branches drooping to one side; capsule acuminate, not suddenly contracted above the middle. **9. forsteri**	
	Infl.-branches spreading; capsule truncate, suddenly contracted above the middle. **8. pilosa**	
3	Perianth white or whitish, sometimes suffused with red. **7. luzuloides**	
	Perianth chestnut or yellowish-brown.	4
4	Robust; fl. stem often over 40 cm; lvs 6–12 mm broad. **6. sylvatica**	
	Fl. stem rarely to 40 cm; lvs less than 6 mm broad.	5
5	Infl. drooping, spike-like. **5. spicata**	
	Infl. of stalked clusters or if compact not drooping.	6
6	Dwarf alpine with channelled lvs; fls drooping in a subumbellate panicle of many clusters of 2–5 fls. **4. arcuata**	
	Lvs not channelled; fls in each cluster more numerous.	7
7	Stoloniferous; anthers 2–6 times as long as filaments. **1. campestris**	
	Not stoloniferous; anthers as long as or shorter than the filaments.	8
8	Fls 2.5–3 mm, chestnut-brown. **2. multiflora**	
	Fls 2 mm, pale yellowish-brown. **3. pallescens**	

1. L. campestris (L.) DC.

Sweep's Brush, Field Woodrush.

L. campestris (L.) DC. subsp. *vulgaris* (Gaudin) Buchenau

A compact, usually *loosely tufted* perennial, with short stock and *shortly creeping stolons,* variable in its characters. Lvs usually 2–4 mm broad, linear, grass-like, with a small truncate swelling at the apex, bright green, thinly clothed with long colourless hairs. *Fl. stem seldom more than* 15 cm, bearing a lax sometimes drooping *panicle of 1 sessile and 3–6 stalked globose obovate clusters of 3–12 fls.* Branches of infl. ±curved, reflexed in fr. Fls 3–4 mm, chestnut-brown. Per.-segs lanceolate, the outer finely pointed, subequal, longer than capsule. *Anthers 2–6 times as long as the filaments.* Capsule 2.5–3 mm, obovoid, obtuse, apiculate. *Seeds nearly globose, with a white basal appendage up to ½ their length.* Fl. 3–6. $2n = 12$. Hs.

Native. Very common in grassy places throughout the British Is. Europe to 63° 55′ N. in Norway; north and mountains of tropical Africa; North America (perhaps native in Newfoundland, naturalized elsewhere); native or introduced in various other parts of the world.

2. L. multiflora (Retz.) Lej.   Heath Woodrush.

L. campestris (L.) DC. subsp. *multiflora* (Retz.) Buchenau

A *densely tufted perennial with few or no stolons, taller* than *L. campestris.* Lvs to 6 mm broad, bright green, sparsely hairy. Fl. stem 20–40 cm, erect, wiry. *Infl. somewhat umbellate of up to 10 ovate or elongate*

8–16-*fld clusters on straight, slender erect branches* or subsessile *in a rounded or lobed head*. Fl. 2.5–3 mm, chestnut-brown. Per.-segs broadly lanceolate, equalling or longer than the capsule. *Anthers about as long as the filaments*. Capsule almost globose, apiculate. *Seeds oblong, nearly twice as long as broad*, with a white basal appendage up to $\frac{1}{2}$ their length. Fl. 4–6 (later than *L. campestris*). Hs.

The two following taxa are variously regarded as subsp. or spp.

Subsp. **multiflora**. Infl. subumbellate. Fl.-clusters pedunculate. Per.-segs about as long as capsule. $2n = 24, 36$.

Subsp. **congesta** (Thuill.) Hyl. Fl.-clusters subsessile, in a compact rounded or lobed head. Per.-segs longer than the capsule. $2n = 48$.

Intermediates between the subspp. with $2n = 42$ are sometimes found. These are usually ±sterile, but some backcrossing seems to occur (C. A. Stace, 1975).

Native. Heaths, moorland and woods, chiefly on acid and peaty soils. Throughout the British Is. Europe, becoming restricted to mountains in the south; N. Africa; Asia; North America; introduced in New Zealand. Subsp. *congesta* is more strongly calcifuge and more oceanic in its European distribution than subsp. *multiflora*.

3. L. pallescens Swartz   Fen Woodrush.

Tufted, like *L. multiflora*, but *paler in colour*. Lvs light green, almost glabrous, to 4 mm broad. Fl. stem 10–30 cm, slender. *Infl. subumbellate, of 5–10 roundish-oblong clusters* of 8–15 fls, the central larger cluster subsessile, the rest on straight erect branches. *Fls small (to 2 mm), pale yellowish-brown*. Per.-segs broadly lanceolate, the outer longer than the capsule, the inner shorter than the outer and not so long-acuminate. *Anthers slightly shorter than the filaments*. Capsule 1.5 mm, *obovoid, with a blunt apiculus*, contracted towards the base. *Seeds* about *twice as long as broad* with a broad appendage less than $\frac{1}{2}$ the length of the seed. Fl. 5–6. $2n = 12$. Hs.

Native. In open grassy places in fens, very local. Hunts. Introduced in Surrey. C. and N. Europe, east to Japan; eastern North America.

4. L. arcuata Swartz   Curved Woodrush.

A *dwarf perennial in small neat tufts* with short stolons. Rhizome creeping, clothed with the persistent lf-sheaths. *Basal lvs* 2–5 cm × less than 2 mm, narrowly linear, recurved, stiff, *deeply channelled, almost glabrous*. Fl.-stems 3–8 cm. Infl. bent to one side when young, a *subumbellate panicle of 2–5-fld clusters*, the outer on recurved stalks. Per.-segs equal, broadly lanceolate, acuminate, much longer than the broadly ovoid capsule. Seeds to 1.2 mm, oblong, with a very small basal appendage. Fl. 6–7. $2n = 36, 42$. Hs.

Native. Open stony ground (*Juncetum trifidi*, etc.) on high mountains, chiefly over 1000 m. Scottish Highlands from the Cairngorms to Sutherland. Arctic and sub-Arctic Europe, rarer in Asia and North America; Greenland.

5. L. spicata (L.) DC.   Spiked Woodrush.

A small *tufted alpine* perennial with short stolons. Stock clothed with the persistent lf-sheaths. *Basal lvs* 2–8 cm × up to 2 mm, recurved, *somewhat channelled*, sparsely hairy. Fl. stems 2–20 cm, erect, very slender, ending in the *dense drooping spike-like infl.* of chestnut-brown fls. Per.-segs equal, lanceolate, acute or sub-obtuse, with fine awn-like points, equalling or slightly longer than the broadly ellipsoid capsule. Seeds with a short appendage at the base. Fl. 6–7. $2n = 24$. Hs.

Native. Rocks, screes, alpine heaths, etc., on non-calcareous substrata, local. Lake District, Scottish Highlands, Hebrides, Shetland. Arctic and sub-Arctic, extending south into the mountains of S. Europe, Corsica, Atlas Mts, Himalaya; North America south to Arizona and New England.

British populations belong to subsp. **spicata**.

6. L. sylvatica (Hudson) Gaudin   Great Woodrush.

L. maxima (Reichard) DC.

A tall robust perennial, forming bright green mats or tussocks, with short ascending stock and numerous stolons. *Basal lvs* 10–30 cm × 6–12 mm or more, spreading, glossy, broadly linear, gradually tapering to a very acute point, sparsely hairy. Fl. stems 30–80 cm, erect, with about 4 stem-lvs, the longest about 5 cm. *Fls chestnut-brown*, 3–4 together, *in a lax terminal cyme*, the branches spreading in fr. *Per.-segs* 3–3.5 mm, lanceolate, acuminate, *about equalling the finely beaked, ovoid fr*. Seeds 1–2 mm, slightly shiny, tubercled at the tip. Var. *gracilis* Rostrup differs markedly in its much smaller dimensions. Lvs 3–7 cm × 3–4 mm. Cyme little-branched, few fld. Fl. 5–6. $2n = 12$. Hs.

Native. Woods (especially oak) on acid soil and peat and open moorlands, especially on rocky ground near streams. Throughout the British Is., but rare in East Anglia and most abundant in the west and north. Ascends to 1000 m or more. Var. *gracilis* in Shetland only. W., C. and S. Europe, to 68° 20′ N. in Norway; Caucasus; Asia Minor. Doubtfully native in South America.

*7. L. luzuloides (Lam.) Dandy & Wilmott
     White Woodrush.

L. albida (Hoffm.) DC.; *L. nemorosa* (Poll.) E. Meyer, non Hornem.

A loosely tufted *medium-sized* perennial. Basal lvs grass-like, hairy, 3–6 mm broad; *stem-lvs* 10–20 cm, *long and grass-like*. Fl. stem 30–60 cm, slender. *Infl. corymbose, lax*, shorter than the uppermost lf, with divaricate branches ending in *clusters of 2–8 fls. Fls dirty white*, sometimes tinged with pink or red. *Per.-segs* 2.5–3 mm, lanceolate, acute, very unequal, *about as long as the ovoid beaked capsule*. Fl. 6–7. $2n = 12$. Hs.

Introduced. Naturalized in various parts of England, Wales and Scotland in woods and damp places by streams, chiefly on acid soils. Established in some localities for over 100 years. C. and W. Europe to N. Germany, S. to Pyrenees, east to Moscow. Introduced with grass and other seed into Scandinavia, North America, and other temperate countries. British plants belong to subsp. **luzuloides**.

*L. nivea (L.) DC., a native of the mountains of W. and C. Europe, has been reported as planted or naturalized in some places. Differs from *L. luzuloides* in the snow-white fls, perianth twice as long as capsule (4.5–5.5 mm), and the more compact infl. with fls in more numerous clusters.

8. L. pilosa (L.) Willd.   Hairy Woodrush.

A tufted perennial with short upright stock and slender stolons. *Basal lvs 3–5 mm broad*, about ½ the length of the stem or longer, grass-like, sparsely hairy, with a small truncate swelling at the apex. Fl. stem 15–30 cm. Infl. a lax cyme, with unequal *spreading* capillary *branches, deflexed in fr*. Fls single, rarely in pairs, *dark chestnut-brown*. Per.-segs 3–4 mm, *ovate-lanceolate*, acute, with broad hyaline border, *shorter than* or almost equalling the fr. *Capsule ovoid, very broad below*, sud-denly contracted above the middle to a truncate conical top. Seeds with a long hooked appendage. Fl. 4–6 (sporadically later). 2n = 66, 72. Hs.

Native. Woods, hedgebanks, etc., throughout the British Is. Europe, excepting the sclerophyll region of the Mediterranean; east to Caucasus and Siberia.

L. forsteri × pilosa (*L. × borreri* Bromf. ex Bab.) is not uncommon with the parents. Resembles *L. pilosa* but often taller. Lvs 2.5–5 mm broad. Infl. branches fewer than in *L. pilosa*, spreading. Capsule much shorter than the perianth, generally sterile.

9. L. forsteri (Sm.) DC.   Southern Woodrush.

Tufted, like *L. pilosa*, but usually smaller and more slender. *Basal lvs 1.5–3 mm broad. Infl. branches drooping to one side, remaining erect in fr. Fls reddish chestnut-brown*. Per.-segs 3–4 mm, lanceolate, acute, narrower and more finely pointed than in *L. pilosa*, *usually longer than the ovoid acuminate mucronate capsule*, which is not suddenly contracted nor truncate. *Seeds with short straight appendage*. Fl. 4–6. 2n = 24. Hs.

Native. Woods and hedgebanks in S. England and S. Wales, local, but in some districts more abundant than *L. pilosa*; north to Radnor and Leicester. S. and W. Europe, north to Belgium, east to Crimea, Syria and Iran; N. Africa; Macaronesia.

141. AMARYLLIDACEAE

Bulbous herbs. Lvs all basal. Fls solitary or umbellate, enclosed before flowering in a usually scarious spathe, hermaphrodite, usually actinomorphic, 3-merous. Perianth petaloid, in two whorls, sometimes with a corona. Stamens in 2 whorls, opposite per.-segs, inserted on them or free; anthers introrse. Ovary inferior, 3-celled, with axile placentation. Ovules usually numerous, in 2 rows on each placenta, anatropous. Style simple; stigma capitate or 3-lobed. Fr. a capsule, rarely a berry. Seeds with copious fleshy endosperm.

About 85 genera and 1100 spp., temperate and tropical, mainly warm temperate.

A family conventionally distinguished from the Liliaceae only by the inferior ovary. It has been divided by Hutchinson into several families based mainly on habit and infl. These are probably derived from the Liliaceae along independent lines. Hutchinson includes certain Liliaceae (i.e. Allieae and 2 allied tribes) in the Amaryllidaceae on account of their essentially similar infl. They appear to link the two families and are here retained in the Liliaceae. Apart from these, the above diagnosis applies to the Amaryllidaceae in Hutchinson's sense.

Numerous genera and spp. are cultivated, among which *Amaryllis*, *Nerine*, *Crinum* and *Hippeastrum* may be mentioned.

1 Perianth without a corona.   2
   Perianth with a trumpet-like or ring-like corona inside
   the 6 segments.   3. NARCISSUS
2 Per.-segs all alike and equal.   1. LEUCOJUM
   Inner per.-segs much smaller than outer.
                                             2. GALANTHUS

1. LEUCOJUM L.

Fls solitary to several, nodding. *Perianth campanulate, segments all alike; no corona*; tube 0 or very short.

Eight spp., C. Europe and Mediterranean region.

Fls 1(–2); per.-segs 20–25 mm.	**1. vernum**
Fls (2–)3–7; per.-segs 14–18 mm.	**2. aestivum**

*1. L. vernum L.   Spring Snowflake.

Bulb rather large (1.5–3 cm), subglobose. Lvs 20–30 cm × c. 1 cm, linear, bright green. Scape 15–20 cm. Spathe 3–4 cm usually bilobed at apex, green in the middle with scarious borders. *Fls* 1(–2). *Per.-segs* 20–25 mm, white, tipped with green, obovate, bluntly acuminate. Capsule pyriform; seeds pale, appendaged. Fl. 2–4. Pollinated by bees and Lepidoptera; homogamous. Seeds distributed by ants. 2n = 22. Gb.

Introduced. Very rare. Damp scrub and hedgebanks in 2 localities in Dorset and S. Somerset; quite often cultivated but very rarely recorded as an escape. Hills of Europe from Belgium, N. and E. France and C. Germany to N. Spain, N. Italy and C. Yugoslavia.

2. L. aestivum L.   Loddon Lily, Summer Snowflake.

Bulb often larger than in **1** (2.5–4 cm), ovoid. Lvs 30–50 cm × 10–15 mm, linear, bright green. Scape 30–60 cm. Spathe 4–5 cm, green at apex. *Fls* (2–)3–7. *Per.-segs* 14–18 mm, white, tipped with green, obovate, obtuse or very bluntly acuminate. Capsule pyriform;

seeds black, not appendaged. Fl. 4–5. Pollinated by bees; homogamous. Seeds distributed by water. $2n = 22$. Gb.

Native. Wet meadows and willow thickets, very local; from Devon and Hants to Oxford and Suffolk and from Wexford and Cork to Antrim and Fermanagh; certainly native along the Thames and Shannon and probably elsewhere within the range given above. Commonly cultivated and sometimes found as an escape in other places. Europe from the Netherlands and Czechoslovakia southwards to Sardinia and Greece; Crimea, Caucasus; naturalized further north; N. Asia Minor, N. Iran.

2. GALANTHUS L.

Fls solitary, nodding. Outer per.-segs somewhat spreading, separated; *inner whorl* campanulate, *much smaller*; *no corona*; tube 0 or very short.

About 4 spp., all except the following, confined to E. Mediterranean region.

*1. G. nivalis L.   Snowdrop.

Bulb ovoid, c. 1 cm across. Lvs 10–25 cm × c. 4 mm, linear glaucous. Scape 15–25 cm. Spathe 2-fid, green in middle with broad scarious margins. Outer per.-segs 14–17 mm, pure white, obovate-oblong, obtuse; inner about half as long, obovate, deeply emarginate, white with a green spot at the incision. Capsule ovoid. Fl. 1–3. Pollinated by bees; homogamous. Fr. 6. $2n = 24$. Gb.

Probably introduced. Local in damp woods and by streams from Cornwall and Kent to Dunbarton and Moray, possibly native in some places in Wales and W. England but very commonly planted and usually only naturalized. From N.C. France and White Russia southwards to the Pyrenees, Sicily and S. Greece; N. Syria, Asia Minor, Caucasus and S.E. Russia (Lower Don region); naturalized further north.

STERNBERGIA Waldst. & Kit.

Fls solitary, erect, yellow, with distinct tube and no corona. ***S. lutea** (L.) Ker-Gawler ex Sprengel is often grown in gardens and is naturalized in Jersey.

Native of Mediterranean region.

3. NARCISSUS L.

Fls solitary or several, usually horizontal. Per.-segs all alike, usually spreading; *corona trumpet-like or ring-like*, inserted between the per.-segs and stamens; tube distinct.

About 60 spp. in Europe, W. Asia and N. Africa.

The Daffodils and Narcissi of our gardens are now mainly hybrids derived from the following and numerous other spp. A number of spp. are also cultivated. Several have been found as escapes and others, as well as garden hybrids, are likely to occur.

1 Corona about as long as perianth. **1. pseudonarcissus**
  Corona less than ½ as long as perianth.   2
2 Fls solitary; corona with a red rim.   **2. poeticus**
  Fls usually 2; corona concolorous or with pale rim.
                             3. ×medioluteus

1. N. pseudonarcissus L.   Wild Daffodil.

Bulb 2–3 cm, ovoid. Lvs 12–35 cm × 5–12 mm, linear, erect, glaucous, usually somewhat channelled. Scape 20–35 cm, ±compressed, 2-edged. Fls solitary, drooping to nearly horizontal. *Pedicels* 3–10 mm, *strongly deflexed. Perianth* 35–60 mm, *pale yellow*; tube 15–22 mm; segments oblong-lanceolate to ovate-lanceolate or elliptic, obtuse and mucronate to acuminate, ascending, wavy or spirally twisted. *Corona as long as* or slightly shorter than *per.-segs*, *deep yellow*, *scarcely expanded or spreading at mouth*, irregularly lobulate, the lobules numerous and toothed. Anthers without dark apical spot. Fr. 12–25 mm, obovoid or subglobose, very obtuse, roundly trigonous to nearly terete. Fl. 2–4. Pollinated by humble-bees and other insects; homogamous. Fr. 6. Germ. winter. $2n = 14$. Gb.

Native. Damp woods and grassland, throughout England and Wales and in Jersey but rather local in most districts; naturalized in Scotland and Ireland. Belgium, France, N. and C. Spain and Portugal, N. Italy, W. Germany (to the Rhine), Switzerland (?native); naturalized in Scandinavia and further east.

N. obvallaris Salisb. (*N. lobularis* (Haw.) A. & J. H. Schultes), the Tenby Daffodil, has both the corona and perianth deep yellow, the corona spreading at the mouth, the per.-segs not twisted and somewhat curved pedicels 10–15 mm. It seems to be a variant within *N. pseudonarcissus* close to subsp. *major*, which probably arose in cultivation. It is naturalized in pastures near Tenby, Pembroke, has been reported from Carmarthen and formerly occurred in Shropshire.

N. pseudonarcissus subsp. **major** (Curtis) Baker (*N. hispanicus* Gouan), the Spanish Daffodil, has both the corona and perianth deep yellow, the corona widely spreading at the mouth, the per.-segs spirally twisted and erect pedicels 25–35 mm. A native of S. France, N. Spain and N. Portugal, it was formerly much grown in gardens and found ±naturalized in grasslands, but is now rarely cultivated.

*N. × incomparabilis Miller

Lvs 8–15 mm broad, glaucous. Fls solitary. *Perianth* 4–6 cm across, *pale* yellow, *corona* 8–12 mm deep yellow, cup-shaped, *about half as long as perianth*. Fl. 4. $2n = 14, 21$. Gb.

A hybrid or series of hybrids between *N. pseudonarcissus* and *N. poeticus*. Grown in gardens and sometimes found as an escape. Garden hybrids of similar origin are very numerous, the perianth varying in colour from white to deep yellow and the corona from very pale yellow to reddish orange.

N. × odorus L.

*N. × infundibulum Poiret

Lvs 4–5 mm broad, linear, semi-terete, green. Scape 30–40 cm, subcylindric. Fls 1–2(–3). Perianth c. 4–5 cm across, bright yellow, tube longer than segments; segments separated, undulate, broadly elliptic. Corona c. ⅔ as long as per.-segs, funnel-shaped, deeper yellow with erect nearly truncate margin.

A hybrid between *N. jonquilla* L. and *N. pseudonarcissus*. Probably of garden origin. Naturalized in parts of S. Europe. Formerly naturalized in quantity near St. Austell (Cornwall) but now much reduced.

*2. N. poeticus L.   Pheasant's Eye.

N. majalis Curtis

Bulb 2.5–3 cm, ovoid. Lvs 9–13 mm broad, linear, erect, glaucous, grooved. Scape 40–50 cm, compressed, 2-edged. Fl. solitary. *Perianth* 5.5–7 cm across, *white*, tube about equalling segments, green; segments imbricate below, margins reflexed, oblong or obovate-oblong, obtuse or subacute, ±spreading. *Corona* c. 3 mm, *cup-shaped*, yellow with a green base and a broad white zone below the *light red margin* which is denticulate-fimbriate. Stamens unequal, three outer anthers slightly exserted, three inner included in the per.-tube and only just exceeding base of outer. Fr. c. 15 mm, triangular-obovoid, scarcely furrowed. Fl. 5. Pollinated by Lepidoptera and selfed. Gb.

Introduced. ±Naturalized in a number of places. For-merly much cultivated, now little grown. Native from E.C. France southwards to C. Spain, S. Italy and N.W. Greece.

*3. N. × medioluteus Miller   Primrose-peerless.

N. biflorus Curtis

Bulb 2.5–3.5 cm, subglobose. Lvs 25–45 cm × 7–14 mm, linear, scarcely glaucous. Scape 30–60 cm, compressed, 2-edged. *Fls* (1–)2(–3). *Perianth cream or yellowish-white*, 3–5 cm across; tube about as long as segments; segments broadly obovate, obtuse, mucronulate, imbricate. *Corona* 3–5 mm, *yellow, usually with crenate whitish margins*. Anthers aborted. Fl. 4–5. Fr. not produced. $2n = 17, 24$. Gb. A hybrid between *N. poeticus* and *N. tazetta* L.

Naturalized in grassy places in a number of localities scattered over the whole of England and Wales and in S. Ireland to Dublin and Clare; Midlothian; Jersey; after *N. pseudonarcissus* the most frequent member of the genus. Native in S. France; cultivated elsewhere for ornament and naturalized in parts of S. Europe.

142. IRIDACEAE

Perennial herbs with rhizomes, corms or bulbs. Lvs not differentiated into blade and petiole, often ensiform, sheathing at base and equitant. Fls 3-merous, hermaphrodite, usually actinomorphic, usually with 1 or 2 bracts forming a spathe at their base. Perianth in 2 series, petaloid, withering and persistent after flowering; segments usually connate at base into a longer or shorter tube; tube straight, or curved in zygomorphic fls. Stamens 3, free or partially connate. Ovary inferior, 3-celled with axile placentas or 1-celled with 3 parietal placentas; style 3-lobed, the branches entire or divided, sometimes petaloid; stigmas terminal or on the underside of the branches; ovules numerous, rarely few or 1, anatropous. Capsule loculicidal, opening by valves, usually with a conspicuous circular scar at top marking the point of attachment of the perianth.

About 60 genera and 800 spp. generally distributed throughout the world. Many cultivated as ornamental plants.

1 Outer per.-segs considerably longer than inner.   2
  Outer per.-segs not noticeably longer than inner.   3
2 Rhizomatous plants; lvs flat; ovary 3-celled.   2. IRIS
  Plant with tuberous roots; lvs quadrangular in section; ovary 1-celled.   3. HERMODACTYLUS
3 Outer per.-segs bearded.   2. IRIS
  Outer per.-segs not bearded.   4
4 Infl. spicate.   5
  Infl. not spicate.   6
5 Fls reddish-purple; stigmas widened at apex.
      7. GLADIOLUS
  Fls orange-yellow; stigmas filiform.   6. TRITONIA
6 Stock a short, slender rhizome; stems strongly compressed.   1. SISYRINCHIUM
  Stock a corm; stems not compressed.   7
7 Per.-tube more than 1.5 cm; lvs with a white stripe.
      4. CROCUS

Per.-tube not more than 1 cm; lvs entirely green.
      5. ROMULEA

Tribe 1. SISYRINCHIEAE. Rhizomatous. Perianth-tube 0 or very short. Style-branches undivided.

1. SISYRINCHIUM L.

Roots fibrous from a short rhizome. Lvs linear-ensiform. Scapes usually flattened. Spathes 2–several-fld. Fls actinomorphic, blue or yellow, very fugitive. Per.-segs in 2 series, all similar; tube very short. Stamens inserted in the throat of the perianth-tube. *Style-branches undivided*. Capsule exserted from the spathe.

About 100 spp. in North and South America, one in W. Ireland.

1 Fls yellow; capsule 9–12 mm.   **3. californicum**
  Fls blue; capsule c. 5 mm.   *2*
2 Fls 15–20 mm diam., pale blue; fruiting pedicels nodding, distinctly longer than the inner spathe-valve.
      1. bermudiana
  Fls 25–35 mm diam., deep blue; fruiting pedicels erect, scarcely longer than the inner spathe-valve.
      2. montanum

1. S. bermudiana L.   Blue-eyed Grass.

S. angustifolium Miller, *S. graminoides* Bicknell

A glabrous erect perennial 15–45 cm. Lvs 7–15 cm, 1–3 mm wide, ensiform. Scape flattened and narrowly winged. Spathe (1–)2–4(–6)-fld. *Fls* 15–20 mm diam., *pale blue*; per.-segs connate, obovate, retuse, long-mucronate; filaments connate almost to top. Fruiting pedicels clearly exceeding the inner spathe-valve. *Capsule* c. 5 mm, globular-trigonous. Fl. 6–7. $2n = 64^*, 88$. Hr.

Native. In marshy meadows and lake shores, locally abundant. W. Ireland from Cork to Donegal, but apparently absent from Mayo. Apparently endemic. A closely related sp. occurs in eastern North America.

*2. S. montanum E. L. Greene   Blue-eyed Grass.
Like **1** but fls 25–35 mm diam., deep blue; fruiting pedicels erect, scarcely exceeding the inner spathe-valve. Fl. 6–7. 2n = 96.

Introduced. Dry grassland and disturbed ground. Naturalized in scattered localities; native in eastern North America.

*3. S. californicum (Ker-Gawler) Aiton fil.
Rather similar in general appearance to *S. bermudiana* but stouter. Lvs 3–6 mm wide. Scape broadly winged. Spathe 3–5-fld. *Fls yellow becoming orange*; per.-segs oval; filaments free almost to base. *Capsule* c. 12 mm, ellipsoid-trigonous. Fl. 6–7. 2n = 34. Hr.

Introduced. Naturalized in marshy meadows, Wexford; reputed to have been introduced by wreckage. Native of California and Oregon.

Tribe 2. IRIDEAE. Rhizomatous or with corms. Per.-tube usually very short, rarely longer than ovary. Style-branches divided and ±petaloid.

2. IRIS L.

Perennial herbs with sympodial rhizomes, fleshy roots or bulbs. Lvs ensiform, sometimes very narrow, often distichous. Spathe usually with scarious margins. Fls actinomorphic, large and showy. Per.-segs in 2 series, outer ('falls') usually deflexed and larger than inner ('standards') which are often erect, usually with a well-marked limb and claw ('haft'); tube usually short. Stamens inserted at base of outer per.-segs. *Style-branches* ('crest') *broad, petaloid, bifid at tip. Ovary 3-locular, placentation axile.* Capsule trigonous.

About 300 spp. throughout the temperate regions of the northern hemisphere. Numerous spp. are cultivated; the common 'bearded' garden irises are mostly hybrids of ±complex parentage.

1   Falls bearded, nearly equalling standards; per.-tube
    longer than ovary.   **5. germanica**
    Falls not bearded, larger than standards; per.-tube
    shorter than ovary.   2
2   Growing in dry places; lvs dark green, polished, ever-
    green; seeds orange-red.   **3. foetidissima**
    Growing in wet places; lvs usually glaucous, mostly
    dying in autumn; seeds brown.   3
3   Crests yellow.   **4. pseudacorus**
    Crests violet or pale pinkish-purple.   4
4   Crests violet; claw of falls twice as long as limb.
                                              1. spuria
    Crests pale pinkish-purple; claw of falls about as long
    as limb.   **2. versicolor**

1. I. spuria L.   Blue Iris.
An erect glabrous perennial 30–90 cm. Rhizome c. 2 cm diam. Lvs linear, shorter than the unbranched cylindrical scape. Spathes green with a narrow scarious margin, 2–4-fld. Pedicels c. 2 cm. *Fls blue-violet and whitish*, not scented. *Outer per.-segs* with limb almost orbicular, not bearded, abruptly contracted below, *claw twice as long*

as limb; *inner shorter than outer, narrowly obovate, violet*; tube short. *Style-branches violet. Capsule narrowed into a long point at top with 2 ridges at each angle.* Fl. 6. 2n = 22, 44. Hel.

Native or introduced. Known for more than 100 years beside fen ditches in Lincoln and perhaps native there; Dorset; rare. S. Sweden to C. Spain, N.E. Greece, and S. Ukraine; Caucasus; N. Africa.

*2. I. versicolor L.   Purple Iris.
An erect glabrous perennial 50–100 cm. Rhizome c. 1 cm diam. Lvs ensiform shorter than the branched scape. Spathes herbaceous. 2–9-fld. Pedicels 2.5–5.5 cm. *Fls rather pale pinkish-purple.* Outer per.-segs broadly ovate, gradually contracted below; *claw about as long as limb*; inner shorter than outer, oblanceolate; tube short. *Style-branches* nearly white. Fl. 6–7 Hel.

Introduced. Naturalized in reed-swamp in Ullswater (for more than 60 years), Windermere, R. Calder (Yorks), Loch Tay (mid-Perth), and probably elsewhere. North-eastern America.

3. I. foetidissima L.   Gladdon, Stinking Iris.
A dark green glabrous perennial 30–90 cm, with a strong unpleasant smell when bruised. Lvs evergreen, ensiform, equalling or exceeding the unbranched *scape* which is *angled on one side.* Spathes with a narrow scarious margin, 1–3-fld. Pedicels 2–10 cm. *Fls* c. 8 mm diam., *purplish*, tinged with dull yellow, rarely yellow. Outer per.-segs obovate-lanceolate, not bearded; inner oblanceolate, yellowish, shorter than outer; tube very short. Style-branches spathulate, pale yellowish. *Capsule clavate; seeds orange-red.* Fl. 5–7. 2n = 40. Ch.

Native. In hedgebanks, open woods and on sea-cliffs, usually on calcareous soils. S. England and Wales, fairly widely distributed in suitable habitats; naturalized in Scotland and Ireland. S. and W. Europe from France southwards, east to N.E. Italy; N. Africa; Macaronesia.

4. I. pseudacorus L.   Yellow Iris, Yellow Flag.
An erect glabrous rather glaucous perennial 40–150 cm. Rhizome often 3–4 cm diam. Lvs 10–30 mm broad, ensiform, about equalling the *compressed terete scape.* Spathes with broadly scarious margins towards the top, 4–12-fld. Pedicels 2–5 cm. *Fls* 8–10 cm diam., *yellow*, varying from pale to almost orange. Outer per.-segs variable in form, shortly clawed, often purple-veined with an orange spot near the base, not bearded; inner spathulate, smaller than outer; tube short. Style-branches yellow. *Capsule cylindrical, apiculate; seeds brown.* Fl. 5–7. 2n = 24. 32–34. 40.

Native. In marshes, swampy woods, and in shallow water or wet ground at edges of rivers and ditches. In suitable habitats throughout the British Is. Most of Europe; N. Africa; Caucasus and W. Asia.

*5. I. germanica L.
A stout glabrous perennial 30–100 cm. Rhizome often up to 5 cm diam. Lvs up to 3.5 cm broad, ensiform,

shorter than the branched cylindrical scape. *Spathes scarious in the upper half*, usually 4-fld. Pedicels short. Fls c. 10 cm diam., bluish-violet or white tinged with blue, scented. Outer per.-segs ovate-oblong, *beard yellow*; *inner about equalling outer*, incurved, light purple; *tube longer than ovary*. Style-branches spathulate. Capsule ovoid. Fl. 5–6. H.

Introduced. Frequently cultivated and sometimes naturalized in waste places. Perhaps native in the E. Mediterranean region, but widely cultivated and naturalized in many places. Probably of hybrid origin.

3. HERMODACTYLUS Miller

Like *Iris* but roots tuberous; *lvs quadrangular in section*; *fls solitary*; outer per.-segs much larger than inner; *ovary 1-celled*; *placentation parietal*.

One sp. in the Mediterranean region.

*1. H. tuberosus (L.) Miller   Snake's-head Iris.
Iris tuberosa L.

An erect glabrous perennial 20–40 cm. Stock short, producing several fibrous and 2–4 oblong tuberous roots. Lvs linear, quadrangular in section, exceeding the slender scape. Spathes herbaceous, 1-fld. Outer per.-segs not bearded, limb smoky purple; claw green, twice as long as limb. Inner per.-segs greenish-yellow, thread-like, inconspicuous, incurved and shorter than style-branches. Anthers yellow. Capsule obovoid. Fl. 4. Grt. $2n = 20$.

Introduced. Not infrequently cultivated. Naturalized in a few localities in Cornwall and Devon. Mediterranean region eastwards from S.E. France.

Tribe 3. CROCEAE. Corm. Lvs tufted. Fls solitary and subsessile or stalked on the corm. Per.-tube distinct, often long.

4. CROCUS L.

Corm symmetrical, covered by the lf-bases of the previous year (*tunic*), which are often fibrous. *Foliage lvs tufted*, linear, channelled, surrounded by scarious sheaths, midrib white beneath. Fls actinomorphic, shortly pedicellate, large and showy, enclosed in bud by 2 spathes, one at the base, the other at the top of the pedicel, or the basal spathe 0; pedicel elongating in fr. Per.-segs in 2 series; *tube long and slender*. Ovary subterranean; style-branches cuneate or variously divided; stigmas terminal. Capsule on a longer slender pedicel.

About 75 spp., in temperate regions of the Old World, particularly the Mediterranean.

Fls appearing in autumn, long before the lvs. **1. nudiflorus**
Fls appearing in spring, with or shortly after the lvs.
                                        2. vernus

*1. C. nudiflorus Sm.   Autumnal Crocus.

Corm-tunic with parallel fibres, the corm producing stolons in spring. Lvs 2–4 mm wide, appearing in spring. Per.-tube usually 10–20 cm, white tinged with lilac or

purple at the top; per.-segs deep lilac-purple; anthers yellow. Fl. 9–10. Gt. $2n = 48$.

Introduced. Naturalized in meadows and pastures, rare. Scattered localities in England northwards to Cumberland. Native of S.W. France, N., C. and E. Spain.

*2. C. vernus (L.) Hill   Purple Crocus.
C. purpureus Weston

Corm-tunic with fine parallel or slightly reticulate fibres; corm not stoloniferous. Lvs (2–)4–8 mm wide, appearing in spring with the fls. Per.-tube 2.5–15 cm, the same colour as the rest of the fl.; per.-segs usually purple or purple and white striped; anthers yellow. Fl. 3–5. Gt. $2n = 8, 10, 12, 16, 18, 19, 20, 22, 23$.

Introduced. Subsp. *vernus* is more or less naturalized in short turf in a few localities. Native in C. and S. Europe, particularly in mountain regions.

C. sativus L., Saffron Crocus, was formerly cultivated for its styles, used as a spice. The corm-tunic has reticulate fibres, the lvs are 1–2 mm wide and appear in autumn with the fls. It is a sterile triploid, not known in the wild.

5. ROMULEA Maratti

Corm usually asymmetrical; tunic hard, smooth, brown. Foliage lvs usually 2, linear. Fls actinomorphic, solitary and pedicellate in the axils of the lf-like bracts. Per.-segs in two similar series; *tube short*. Style-branches 3, linear, bifid. Capsule ovoid, 3-lobed.

About 90 spp., mainly in the Mediterranean region and S. Africa.

1. R. columnae Sebastiani & Mauri
R. parviflora Bubani; *Trichonema columnae* (Sebastiani & Mauri) Reichenb.

Corm small, ovoid, producing offsets freely. *Lvs* 2, 5–10 cm, *almost setaceous, wiry*. *Stem shorter than the lvs, recurved after fl.* Spathe of 2 subequal blunt bracts. Fls 1–3, 9–19 mm. *Per.-segs* subacute, *greenish outside, purplish-white inside, yellow towards the base, with purple veins*. Fl. 3–5. Gt. $2n = $ c. 60.

The British plant is var. *occidentalis* Béguinot, which is the variant also found in western France.

Native. In short turf on sandy ground near the sea, very local. S. Devon and Channel Is.; apparently extinct in Cornwall. W. Europe from northern France southwards; Mediterranean region; Macaronesia.

*R. rosea (L.) Ecklon, from S. Africa, is naturalized in Guernsey. It has symmetrical subglobose corms, several basal lvs and is larger in all its parts than *R. columnae*.

Tribe 4. GLADIOLEAE. Corm. Stems lfy. Fls distinctly zygomorphic or limb oblique and tube ±curved. Tube widening gradually from base to mouth. Stamens usually bent over towards one side of the fl.

6. TRITONIA Ker-Gawler

Corm covered by a reticulate fibrous tunic. Foliage lvs ensiform, distichous, basal and cauline. Infl. long, spike-like, secund. *Spathes membranous*, the outer 3-pointed, the inner entire. Fls weakly zygomorphic, inclined or horizontal, showy. *Per.-tube straight or nearly so*, usually shorter than the per.-segs, widening gradually upwards. Capsule small, ovoid or oblong.

About 55 spp. in S. Africa.

*1. T. crocosmiflora (Lemoine) Nicholson Montbretia.

T. aurea Pappe ex Hooker × *pottsii* (Baker) Baker; *Crocosmia crocosmiflora* (Lemoine) N.E.Br.

An erect glabrous perennial 30–90 cm. Corms nearly 2 cm diam., often several together in a row, freely stoloniferous. Stems slender, simple or branched, equalling or exceeding the lvs. Lvs 5–20 mm wide. Bracts small, oblong reddish. Fls 2.5–5 cm diam., deep orange suffused with red, funnel-shaped with rather spreading per.-segs, longer or sometimes shorter than the tube. Style-branches short, simple. Fl. 7–8. Gt. 2*n* = 22, 33, 44.

Raised at Nancy, France, by Victor Lemoine by crossing *T. pottsii* (female) with *T. aurea* (male); flowered for the first time in 1880.

Introduced. Commonly cultivated and very tolerant of shade; naturalized by sides of lakes, rivers and ditches, in hedgebanks, on waste ground and in woods, particularly in the west; spreading by vegetative means and by seed. Garden origin; parents from S. Africa.

7. GLADIOLUS L.

Corm covered by fibrous tunic. Lvs ensiform, distichous, basal and cauline. Infl. long, spike-like, secund. *Spathes* usually *herbaceous*. Fls weakly zygomorphic, inclined or horizontal, showy, solitary in spathes. Per.-segs 6, in 2 similar series; *tube short, curved*. Stamens curved towards the upper side of the perianth. Style-branches undivided, obovate to obcordate. Capsule subglobose or oblong-trigonous.

About 300 spp. in Europe, W. Asia, Africa and the Mascarene Is.

1 Anthers longers than filaments.   **3. italicus**
    Anthers equalling or shorter than filaments.   2
2 Plant 25–50 cm; lvs 10–40 × 0.4–1 cm; infl. with 3–10 fls.   **1. illyricus**
    Plant 50–100 cm; lvs 30–70 × 0.5–2.2 cm; infl. with 10–20 fls.   **2. communis**

1. G. illyricus Koch   Wild Gladiolus.

An erect glabrous perennial 25–50 cm. Corm c. 1 cm diam., producing numerous offsets. Lvs 10–40 cm × 4–10 mm, acuminate. Scape usually simple, exceeding the lvs. Infl. with 3–10 fls. Spathe of 2 subequal ±herbaceous often purple-tipped bracts. Per.-segs 2.5–4 cm, obovate, long-clawed, crimson-purple; tube short, curved. Anthers shorter than their filaments. Style-branches linear at base, then abruptly dilated and ovate. Capsule obovoid, with 3 acute angles. Seeds narrowly winged. Fl. 6–8. 2*n* = 90. Gt.

Native. Among bracken on bushy heaths, very local. Dorset, S. Hants and Isle of Wight. S. and W. Europe.

*2. G. communis L.

G. byzantinus Miller

Like **1** but plant 50–100 cm; infl. with 10–20 fls; per.-segs 3–4.5 cm, pink, red or purplish-red. 2*n* = 90, 120. Gt.

Introduced. Sometimes cultivated; a garden weed or ±naturalized in waste ground near the sea, chiefly along the south coast. S. Europe; N. Africa.

*3. G. italicus Miller

G. segetum Ker-Gawler

Like **1** but plant 50–100 cm; infl. with 6–16 fls; per.-segs bright purplish-red to bright pink; anthers longer than their filaments or aborted; seeds not winged. 2*n* = 120, 171 ± 2. Gt.

Introduced. Sometimes cultivated; a weed of disturbed ground, chiefly along the south coast. S. Europe; N. Africa; S.W. Africa.

143. DIOSCOREACEAE

Usually dioecious slender herbaceous or woody twiners with tuberous rhizomes or stocks or thick woody stem-tubers (above ground in *Testudinaria*). Lvs usually spirally arranged, often cordate, entire, lobed or palmately divided, with palmate main veins and a network of smaller veins. Fls small, in spikes, racemes or panicles; unisexual, actinomorphic, epigynous. Perianth campanulate, of 3 + 3 ±equal segments united below into a short tube; stamens in male fl. borne on the base of the perianth-tube, 3 + 3 or 3 with or without staminodes replacing the missing set; in the female fl. rudimentary or 0; ovary in the female fl. inferior, 3-celled, with 2 superposed anatropous ovules on axile placentae in each cell; styles 3, or 1 divided above into 3 stigmatic lobes; ovary in the male fl. rudimentary or 0. Fr. a 3-valved capsule or berry; seeds often flattened or winged, with horny endosperm and a small embryo.

Seven hundred and fifty spp. in 8–9 genera, widespread in tropical and warm temperate regions.

The large genus *Dioscorea*, with 1 sp. in the Balkan peninsula, includes several spp. cultivated throughout the tropics for their starchy root-tubers or tuberous rhizomes (yams). The closely related *Borderea* Miégeville is endemic in the Pyrenees and is probably a Tertiary relict.

1. TAMUS L.

Perennial dioecious herbs with large hypogeal stem-tubers, slender annual twining stems, and entire cordate lvs. Fls small, in axillary racemes or spikes. Perianth campanulate; stamens 6, rudimentary in the female fl.; style 1 with three 2-lobed stigmas. Fr. a berry, incompletely 3-celled, with few globose seeds.

One sp. in the Mediterranean region, N. Africa, Caucasus and W. Asia.

1. T. communis L.   Black Bryony.

A tall herb with a large irregularly ovoid blackish tuber up to 20 (rarely to 60) cm and 10–20 cm below the soil surface. Stems 2–4 m, slender, angled, unbranched, glabrous, twining to the left. Lvs 3–15 × 2.5–10 cm, broadly ovate, deeply cordate at the base, finely acuminate, entire, dark shining green, with 3–9 curving main veins; stalk long, with 2 stipule-like emergences at its base. Fls yellowish-green in axillary racemes, the male stalked, long, erect or spreading, the female subsessile, recurved, shorter and fewer-fld than the male; bracts minute, subulate, scarious; male fls 5 mm, female fls 4 mm diam. Perianth with 6 narrow, somewhat recurved, lobes. Stigmas recurved. Berry c. 12 mm ovoid-ellipsoid, apiculate, pale red, glabrous, with 1–6 pale yellow rugose seeds. Fl. 5–7. Visited for nectar by many insects, including small bees. $2n = 48$. G.

Native. Wood-margins, scrub, hedgerows, etc., on moist well-drained fertile soils; to 244 m in Derby; common in the south. Throughout England and Wales northwards to S. Cumberland and N. Northumberland; in Ireland only near Lough Gill (Sligo and Leitrim), and probably introduced; Channel Is. S. and W. Europe northwards to England, Belgium, W. Germany, Austria, Hungary, Transsylvania and the northern shore of the Black Sea. N. Africa; Palestine and Syria; coastal regions of Asia Minor; Caucasus.

144. ORCHIDACEAE

Perennial herbs with rhizomes, vertical stocks or tuberous roots; terrestrial, epiphytic or sometimes saprophytic, and almost invariably mycorrhizic. Stems often swollen at the base ('pseudo-bulbs'), and, in epiphytic types, often with aerial roots. Lvs entire, spirally arranged or distichous, rarely opposite, often with a sheathing base and sometimes with dark spots or blotches; in saprophytic spp. reduced to scales or membranous sheaths lacking chlorophyll. Infl. a spike, raceme or racemose panicle, commonly pendent in epiphytic types. Fls zygomorphic, epigynous and usually hermaphrodite, often large and very striking in form and colour. Perianth of 6 segments in 2 whorls, usually all petaloid (though often green) or with the outer whorl sepaloid and the inner petaloid; the median (posterior) segment of the inner whorl (*labellum*) is commonly larger and different in shape from the remainder, and is usually on the lower side of the fl. and directed ±downwards owing to the inversion of the fl. in a pendent infl. or to the twisting of the ovary or its stalk through 180° in types with an erect infl.; the labellum is often spurred at its base. The anthers and stigmas are borne on a special structure, the *column*; stamens 2 (Cypripedioideae) or 1 (Orchidoideae), with ±sessile 2-celled anthers seated behind or upon the summit of the column, their pollen of single grains or more commonly of tetrads cohering in packets which are bound together with elastic threads, the contents of each cell forming 1–4 granular or waxy masses (*pollinia*) often narrowed at their apical or basal end into a sterile stalk-like *caudicle*; ovary inferior, 1-celled, with numerous minute ovules on 3 parietal placentae, or rarely 3-celled; style 0; stigmas 3 fertile (Cypripedioideae) or 2 fertile and the other sterile and forming a ±beak-like process, the *rostellum*, between the anther and fertile stigmas (Orchidoideae); in some genera the rostellum forms 1 or 2 viscid bodies (*viscidia*) to which the pollinia are attached, and it may be represented only by viscidia or may be quite lacking. Fr. a capsule opening by 3 or 6 longitudinal slits; seeds very numerous and very minute, with an undifferentiated embryo and no endosperm. Pollination usually by insects to whose bodies the pollinia become attached in such a way that they are placed on the stigmas of other fls; some spp. are automatically self-pollinated and others apomictic, although retaining the structural features of cross-pollinated types.

735 genera and 17 000 spp., and thus one of the largest families of Angiosperms. Largely tropical but spread over the whole world.

Synopsis of Classification

Subfamily CYPRIPEDIOIDEAE: fertile stamens 2; pollen granular, not united in pollinia; fertile stigmas 3.

Tribe CYPRIPEDIEAE: median stamen forming a large staminode; labellum large, deeply concave, with inrolled margins.   1. CYPRIPEDIUM

Subfamily ORCHIDOIDEAE: fertile stamen 1; pollen granular or waxy, united in pollinia; fertile stigmas 2, often confluent; median (sterile) stigma commonly extended as a beak-like rostellum.

A.   Caudicle 0 or at the apex of the pollinium, i.e. the end which is in contact with the rostellum; anther hinged to the summit or back of the column and often readily detachable.

Tribe NEOTTIEAE: pollen granular, soft; anther usually persistent; infl. always terminal.

Subtribe NEOTTIINAE: anther usually withering in situ; pollinia 2–4.

(*a*) Anther ±erect, exceeding the rostellum when present; labellum with a concave basal hypochile separated by a constriction or fold from the distal tongue-like epichile (Fig. 62).

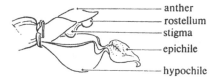

Fig. 62. Labellum and column of *Epipactis helleborine*.

Fl. erect; rostellum 0; labellum with an erect hypochile which clasps the base of the column and a forwardly directed epichile; spur 0; pollen not in tetrads.   2. CEPHALANTHERA

Fl. horizontal or pendent; rostellum ±globose or 0; labellum with a horizontal hypochile and downwardly directed epichile; spur 0; pollen in tetrads.

3. EPIPACTIS

Saprophytic; fl. not inverted, with upwardly directed labellum and spur; pollinia with caudicles attached to the rostellar viscidium.   4. EPIPOGIUM

(b) Anther erect, parallel to and equalling the conspicuous flattened rostellum; pollinia granular, not breaking into large angular masses; labellum not differentiated into hypochile and epichile.

Per.-segs all directed forwards; labellum embracing the column at its base; fls in spirally twisted rows.

5. SPIRANTHES

Per.-segs ±spreading; labellum not embracing the column; lvs 2, opposite, borne some way up the stem.   6. LISTERA

Saprophytic; stem covered with numerous brown scales; per.-segs connivent to form a helmet.

7. NEOTTIA

(c) Anthers erect or forwardly inclined, about equalling the rostellum; pollinia breaking into several large angular masses; labellum various; lvs often net-veined.

Spur 0; labellum with a saccate base and spout-like distal part which are not clearly separated.

8. GOODYERA

Tribe EPIDENDREAE pollen waxy or bony; anther usually soon deciduous; infl. terminal or lateral.

Subtribe LIPARIDINAE: infl. terminal, or axillary to the uppermost lvs; pollinia 4; caudicles 0.

Anther erect, persistent; column very short, straight; labellum directed upwards.   9. HAMMARBYA

Anther inclined, deciduous; column curved, winged at the base; labellum variously orientated. 10. LIPARIS

Saprophytic; anther inclined; column long, curved.

11. CORALLORHIZA

B. Caudicle at the base of the pollinium, attached to a viscidium; anther erect, closely adnate to the summit of the column and never deciduous; pollinia always granular.

Tribe ORCHIDEAE

(d) Viscidia 2, naked, or with a delicate skin which is removed with them; never enclosed in a pouch which remains behind after their removal.

Viscidia covered by a delicate skin; caudicles very short; viscidia large.   12. HERMINIUM

Viscidia covered by a delicate skin; caudicles fairly long; viscidium hardly wider than the caudicle.

13. COELOGLOSSUM

Viscidia naked, long and narrow, close together; caudicle attached at right angles to the viscid surface; per.-segs spreading; spur long, filiform.

14. GYMNADENIA

Like *Gymnadenia* but viscidia less elongated; per.-segs connivent; spur short, conical.

15. PSEUDORCHIS

Viscidia naked, oval or circular, distant; caudicle attached laterally so as to be parallel to the viscid surface.   16. PLATANTHERA

(e) Viscidia 2, naked; stigmas free and often long.

Stigmas crescent-shaped, diverging upwards and leaving a broad flat plate between them; per.-segs connivent into a helmet; spur short.   17. NEOTINEA

(f) Viscidia 2, enclosed in 1 or 2 pouches which are left behind when the viscidia are removed.

Viscidia in 2 separate pouches; labellum insect-like, velvety; spur 0.   18. OPHRYS

Viscidia in a single pouch.

Viscidia 2 in a common pouch.

Spur 0.   21. ACERAS

Spur present.

Labellum with a very long ribbon-shaped mid-lobe which is spirally curled in bud.

19. HIMANTOGLOSSUM

Labellum not very long and ribbon-shaped.

Root-tubers entire.   20. ORCHIS

Root-tubers palmately lobed or forked.

22. DACTYLORHIZA

Viscidium 1, strap-shaped; labellum with vertical 'guide-plates'.   23. ANACAMPTIS

Key to Genera

1 Plant lacking green lvs (saprophytic).   2
  Plant with green lvs (or with green bract-like scales on a green stem).   4
2 Labellum directed upwards, pink; spur fairly long, directed upwards; stem swollen above the base.
  4. EPIPOGIUM
  Labellum directed downwards; spur 0 or very short and adnate to the ovary; stem not swollen.   3
3 Stem covered with numerous brownish scales; fls brown; per.-segs connivent into an open hood.
  7. NEOTTIA
  Stem with 2–4 long sheathing scales; fls yellowish-green with a whitish labellum; outer lateral per.-segs curved downwards close to the labellum.
  11. CORALLORHIZA
4 Fl. spurred.   5
  Fl. not spurred.   15
5 Labellum with a long ribbon-like mid-lobe 3–5 cm × 2 mm, coiled in bud like a watch-spring; a tall plant with long spikes of unpleasantly smelling fls.   19. HIMANTOGLOSSUM
  Labellum not as above.   6
6 Fls greenish-white.   7
  Fls green, white, pink, red, etc., but not greenish-white.   8
7 Fls large, strongly fragrant, with spreading per.-segs, entire strap-shaped labellum and long slender spur 15–30 mm.   16. PLATANTHERA
  Fls in a narrow cylindrical spike, very small, ±campanulate, faintly vanilla-scented, with connivent per.-segs, a 3-lobed labellum and a short blunt conical spur.   15. PSEUDORCHIS
8 Spur very short (c. 2 mm); per.-segs connivent into a hood.   9
  Spur exceeding 5 mm; per.-segs connivent or spreading.   11
9 Fls green, often red-tinged; labellum narrowly oblong with parallel distal lobes.   13. COELOGLOSSUM
  Fls not green; labellum with 3 or 5 ±spreading lobes.   10
10 Lvs unspotted; labellum 5-lobed; per.-segs dark purplish-brown at first but becoming paler, so that the tip of the spike appears burnt or scorched; labellum white with dark spots.   20. *Orchis ustulata*
  Lvs sometimes with rows of small dots; labellum 3-lobed, the central lobe longer and forked; fls small, whitish or pink, in a dense spike 2.5–8 cm; W. Ire-

land and Isle of Man. 17. NEOTINEA

11 Spur c. 12 mm, filiform; labellum with 3 subequal lobes; 2 fertile stigmas borne on lateral lobes of the column; viscidia 1 or 2, elongated. *12*

Spur less than 12 mm, or, if as long, then not filiform; 2 fertile stigmas ±confluent on the front of the column; viscidia 2, globose. *13*

12 Spike markedly conical; fls not sweet-scented; labellum with vertical plates decurrent on its base from the lateral lobes of the column; viscidium 1, strap-shaped. 23. ANACAMPTIS

Spike ±cylindrical; fls strongly and sweetly scented; no vertical plates on labellum; viscidia 2, naked. 14. GYMNADENIA

13 All per.-segs except the labellum connivent into a helmet over the column. 20. ORCHIS

At least the outer lateral per.-segs spreading or upturned. *14*

14 Emerging fl.-spike enclosed by thin spathe-like lvs; bracts membranous; spur ±ascending. 20. ORCHIS

Emerging fl.-spike not enclosed in spathe-like lvs; bracts lf-like; spur descending. 22. DACTYLORHIZA

15 Labellum concave, slipper-shaped; fls very large, solitary (–2); per.-segs 6–9 cm. 1. CYPRIPEDIUM

Labellum not strongly concave; fls normally more than 2; per.-segs less than 6 cm. *16*

16 Labellum with an insect-like ±velvety mid-lobe. 18. OPHRYS

Labellum not insect-like and velvety. *17*

17 Labellum on the upper side of the fl. and directed upwards in some or all of the small yellowish-green fls. *18*

Labellum on the lower side of the fl. *19*

18 Lvs 0.5–1 cm, obovate, rounded and concave, the margin fringed with tiny green bulbils; labellum always on the upperside of the fl.; a small plant usually growing in *Sphagnum*. 9. HAMMARBYA

Lvs 2.5–8 cm; oblong-elliptical; labellum variously orientated; usually in fen peat. 10. LIPARIS

19 Fls small, whitish, in 1 or more spirally twisted rows; infl. narrowly cylindrical. *20*

Fls not whitish and in spirally twisted rows. *21*

20 Stoloniferous; lvs ovate, stalked, conspicuously net-veined; labellum free from the column, with a saccate base and an entire narrow spout-like distal part. 8. GOODYERA

Not stoloniferous; lvs not conspicuously net-veined; labellum ±adnate to and embracing the column at its base, its distal part entire, frilled, recurved. 5. SPIRANTHES

21 Labellum with a concave basal part (hypochile) separated by a constriction or fold from the tongue-like distal part (epichile) (Fig. 62). *22*

Labellum not as above; fls greenish. *23*

22 Fls erect, ±sessile; hypochile embracing the column; ovary straight or twisted. 2. CEPHALANTHERA

Fl. horizontal or pendent, stalked; hypochile not embracing the column; ovary straight. 3. EPIPACTIS

23 Labellum shaped like a man, with slender lobes resembling arms and legs. 21. ACERAS

Labellum not shaped like a man. *24*

24 Lvs 3–6; labellum ovate, or narrower and then much like the other per.-segs.
3. *Epipactis phyllanthes* var. *phyllanthes*

Lvs 2, with or without 1 or 2 much smaller lvs above. *25*

25 Lvs ±opposite, borne some way up the stem; labellum forked distally, sometimes with a small tooth in the sinus. 6. LISTERA

Lvs basal, with 0–2 smaller stem-lvs; labellum 3-lobed, the central lobe longer than the laterals. 12. HERMINIUM

1. CYPRIPEDIUM L.

Herbs with creeping rhizomes, green lfy stems and 1–3 large fls. Per.-segs spreading. *Labellum large*, *concave*, *inflated*, *shoe- or slipper-shaped*; spur 0. Column curving forwards so as partly to close the aperture of the labellum, and terminating in a large shield-shaped staminode with 2 *lateral fertile anthers* below and behind it; stigmas confluent, *discoid*, *stalked*, projecting downwards below the fertile anthers and just behind the staminode; rostellum 0. *Pollen not in tetrads*, *granular*. Ovary straight.

About 50 spp. in tropical and temperate regions of both Old and New Worlds.

C. calceolus L. Lady's Slipper.

Stem 15–45 cm, erect, somewhat pubescent, with basal sheathing scales. Lvs 3–4, ovate-oblong, sheathing, acute or acuminate, slightly pubescent, strongly furrowed above and ribbed beneath along the several parallel veins. Fls 1(–2); bracts large, lf-like. Per. segs 6–9 cm, maroon; upper outer segment ovate-lanccolate; ±erect, outer lateral segments narrower and usually ±connate to form a single bifid downwardly directed segment below the labellum; inner lateral segments narrowly lanceolate-acuminate, often somewhat twisted. Labellum rather shorter than the per.-segs, obovoid, rounded at the tip, pale yellow with faint darker veins, its interior and the column with reddish spots. Ovary ±pubescent. Fl. 5–6. Cross-pollinated by insects which crawl from the cavity of the labellum past the stigmatic disk and the fertile anthers. $2n = 22$.

Native. In open woods of oak, ash and hazel on steep rocky limestone slopes. Formerly in several such woods in N. England from Cumberland to Durham with one or more outliers in Derby, but now extinct or nearly so. C. and N. Europe northwards to c. 70° N. in Scandinavia and southwards to N. Greece, but not in the Mediterranean lowlands; N. Asia.

2. CEPHALANTHERA L. C. M. Richard

Herbs with shortly creeping rhizomes, lfy stems and a few large suberect white or pink fls in lax spikes. Per.-segs commonly connivent so that the *fl. never opens widely*. *Labellum with a constriction* between the suberect concave hypochile, which clasps the base of the column, and the forwardly directed epichile with several interrupted longitudinal crests along its upperside and a recurved tip; basal bosses 0; spur 0 or very short. Column long, erect; stigma large oblong; *rostellum* 0. Anther hinged to the summit of the column; pollinia 2, clavate, each ±completely divided into longitudinal halves; caudicles 0; pollen grains single, in powdery masses. Capsule erect.

Fourteen spp. in the north temperate zone.

1 Fls bright rose-pink; upper part of stem and ovaries
   pubescent.   **3. rubra**
   Fls white; upper part of stem and ovaries ±glabrous.   2
2 Lvs ovate or ovate-lanceolate; bracts exceeding the
   ovary; outer per.-segs blunt.   **1. damasonium**
   Lvs lanceolate; bracts shorter than the ovary; outer
   per.-segs acute.   **2. longifolia**

1. C. damasonium (Miller) Druce   White Helleborine.
C. latifolia Janchen; *C. grandiflora* S. F. Gray; *C. pallens*
L. C. M. Richard

Stem 15–50 cm, erect, rigid, angled, glabrous, with 2–3
brown membranous sheathing basal scales, the uppermost often green-tipped. Lvs 5–10 cm, the lowest short,
ovate-lanceolate, the middle ovate-oblong, the upper
lanceolate, passing into the bracts. *Fls* 3–12, the lowest
distant and all but the uppermost *much exceeded by
the lf-like lanceolate bracts*, *creamy-white*, scentless,
except for a brief period closed and ±tubular. *Outer
per.-segs* oblong, *blunt*; inner lateral per.-segs oblong-
lanceolate, blunt, shorter than the outer; labellum
shorter than the per.-segs, with an orange-yellow blotch
in the concavity of the hypochile and 3–5 interrupted
orange-yellow crested ridges running along the cordate,
crenate epichile. *Ovary glabrous*, not twisted. Fl. 5–6.
Usually self-pollinated, the pollen falling on the stigma.
$2n = 32; 36*$. Grh.

Native. Locally frequent in woods and shady places
on calcareous soils, especially under beech on chalk in
S. England from Kent to Dorset and northwards to
Cambridge, Northampton and Hereford & Worcester,
with a single outlier at the southern end of the Yorks
Wolds. Much of Europe northwards to Denmark and
Gotland; N. Africa; Caucasus; Asia Minor.

2. C. longifolia (L.) Fritsch   Long-leaved Helleborine.
C. ensifolia (Schmidt) L. C. M. Richard

Stem 20–60 cm, ±glabrous, slightly ridged above and
with 2–4 whitish, sometimes green-tipped, loose-topped
basal sheaths. Lvs c. 7–20 × 1.5 cm, lanceolate, longer
and narrower than in *C. damasonium*, often folded, the
uppermost linear and sometimes exceeding the spike.
Fls 3–15, *exceeding all or all but the lowest bracts*; *pure
white*, resembling those of *C. damasonium* but smaller.
Outer per.-segs lanceolate, *acute*. Ovary *glabrous*,
twisted through c. 180°. Fl. 5–7. Probably cross-pollinated by small bees. $2n = 32$. Grh.

Native. A rare and decreasing plant of woods and
shady places, especially on calcareous soils. Very local
from Essex and Kent to Hants, north-westwards to
Oxford, Powys and Anglesey, then north to Durham,
Cumberland and W. Scotland to Sutherland; Mull,
Skye; very rare in Ireland. Europe northwards to
c. 63° N. in Norway; N. Africa; W. Asia to Kashmir,
Siberia, Japan.

Hybrids between **1** and **2** have been recorded with the parents
from a wood in E. Hants. The lower lvs were ovate as in **1**,

the fls pure white as in **2**, and the ovary as long (2 cm) as
in **1** but was twisted through 180°.

3. C. rubra (L.) L. C. M. Richard   Red Helleborine.
Serapias rubra L.

Stem 20–50 cm, erect, striate and *glandular-hairy* above,
with brownish sometimes green-tipped basal sheaths.
Lvs few, the lowest oblong-lanceolate, the upper linear-
lanceolate, all very acute. *Fls* 3–10 exceeding all or all
but the lowest of the linear-lanceolate acute bracts,
bright rose-red, scentless, opening fairly wide. Outer
per.-segs lanceolate acute, spreading, glandular-hairy
outside; inner per.-segs shorter and broader, connivent;
labellum erect, white, its epichile lanceolate with red-
violet margins and tip and 7–9 narrow crested ridges.
Ovary glandular-hairy with violet ribs. Fl. 6–7. Cross-
pollinated by small bees. Grh.

Native. Confined to beech-wood and scrub on oolite
in Gloucester and on chalk in Bucks; very rare. Europe
northwards to c. 60° N. in Scandinavia; Near East.

3. Epipactis Zinn

Herbs with horizontal or vertical, often very short, rhizomes, numerous fleshy roots, and rather inconspicuous
spreading or pendent fls in ±1-sided racemes. Per.-segs
spreading or connivent, dull reddish-brown or greenish.
*Labellum usually in 2 parts separated by a narrow joint
or a fold*, the basal part (*hypochile*) forming a *nectar-
containing cup*, the distal part (*epichile*) a ±cordate or
triangular *forwardly-directed terminal lobe*; *spur* 0. Column short, with a shallow cup (*clinandrium*) at its apex;
stigma prominent, ±transversely oblong. *Rostellum*
placed centrally above the stigma, *large and globose,
persistent, evanescent and absent at fl.-opening, or* 0.
Anther free, hinged at the back of the summit of the
column, behind the stigma and rostellum; pollinia 2,
tapering towards their apices near which they are
attached to the rostellum, each ±divided longitudinally
into halves; caudicles 0; pollen grains globose, cohering
in tetrads bound loosely together by weak elastic
threads, so that the masses are very friable. Cross-pollinated by insects, or self-pollinated; perhaps sometimes
apomictic. Ovary straight, often resembling a twisted
stalk.

Several spp. in the north temperate zone, but very
imperfectly understood.

1 Fen plant with long creeping rhizome; fls brownish
   (rarely yellowish) outside; hypochile with 2 erect
   lateral lobes; epichile connected by a narrow hinge.
                                  1. palustris
   Not fen plants; rhizome horizontal or vertical, short;
   fls greenish or purple outside; hypochile without
   lateral lobes; epichile connected by 1 or more folds;
   (rarely no distinct hypochile and epichile).   2
2 Fls entirely dull reddish-purple; infl.-axis with dense
   whitish pubescence; epichile broader than long,
   with 3 large strongly rugose basal bosses; ovary
   pubescent.   **7. atrorubens**
   Fls not entirely reddish; infl.-axis glabrous to pubescent but not densely whitish-pubescent; basal

bosses of labellum smooth or nearly so.   3

3 Lvs spirally arranged; rostellum prominent and persistent unless removed by insects with the pollinia.   4
  Lvs in 2 opposite ranks (distichous); rostellum 0 in mature fls; fls sometimes opening only partly or not at all.   5

4 Stems 1–3, rarely more; lvs broadly ovate to ovate-lanceolate, largest 6–15 cm, usually dark green; per.-segs green to deep purplish; hypochile dark maroon or dark green inside.   **2. helleborine**
  Stems commonly 6–10 or more; lvs lanceolate, largest 3–8 cm, grey-green, often violet-tinged; per.-segs pale whitish-green; hypochile ±mottled with violet inside.   **3. purpurata**

5 Infl.-axis glabrous or sparsely pubescent; fls hanging almost vertically downwards; hypochile greenish-white inside, sometimes 0.   **6. phyllanthes**
  Infl.-axis markedly pubescent; fls spreading horizontally or hanging obliquely downwards; hypochile purplish inside.   6

6 Fls opening widely.   **4. leptochila**
  Fls rarely opening widely.   7

7 Fls remaining partly closed and therefore campanulate; epichile with a recurved tip; a rare plant of dunes in Anglesey and N. England.   **5. dunensis**
  Fls usually remaining closed; epichile flat; formerly in beechwoods in Gloucester, perhaps now extinct.
    4. leptochila (*E. cleistogama*)

1. E. palustris (L.) Crantz   Marsh Helleborine

Serapias palustris (L.) Miller; *Helleborine palustris* (L.) Schrank

Rhizome long-creeping with stolon-like branches. Stem 15–45(–60) cm, erect, slender, wiry, pubescent above; often purplish below and with 1 or more sheathing basal scales. Lvs 4–8, the lower 5–15 cm, oblong-ovate to oblong-lanceolate, acute, concave above, often purple beneath; uppermost narrow, acuminate, bract-like; all ±erect and folded, with 3–5 prominent veins beneath. Raceme 7–15 cm, with 7–14 fls ±turned to one side; bracts lanceolate-acuminate, lowest about equalling the fls, the rest falling short. Fls drooping in bud and in fr., almost horizontal when open, scentless. *Outer per.-segs* ovate-lanceolate, *brownish or purplish-green and hairy outside*, paler inside; inner per.-segs somewhat shorter and narrower, whitish with purple veins at the base, glabrous. Labellum c. 12 mm; *hypochile shallow with an erect triangular lobe at each side*, white with rose veins and many bright yellow raised spots down the middle; *epichile joined by a narrow hinge, white with red veins*, broadly ovate with *waved and frilled upturned sides*, and with a basal 4-toothed furrowed boss, 3–4-toothed on its forward edge. Rostellum persistent. Ovary narrowly pear-shaped, puberulent. Fl. 6–8. Cross-pollinated by hive-bees and other insects. $2n = 40, 44, 46, 48$. G.

Var. *ochroleuca* Barla has yellowish fls with a white lip.

Native. A decreasing but still locally frequent plant of fens and dune-slacks through much of England and Wales with a few localities in S.E. Scotland northwards to Fife; Ireland except the north-east; Channel Is. Europe to c. 63° N. in Scandinavia; temperate Asia; N. Africa.

2. E. helleborine (L.) Crantz   Broad Helleborine

Serapias helleborine var. *latifolia* L.; *E. latifolia* (L.) All.

Rhizome usually very short. Roots numerous, arising in a cluster from the base of the stem. Stems 1–3, rarely more, 25–80 cm, erect, solid, whitish with short hairs above, often violet-tinged below and with 2 or more basal sheathing scales. Largest *lvs* 5–17 × 2.5–10 cm, *spirally arranged*, somewhat crowded near the middle of the stem, usually *broadly ovate-elliptical or almost orbicular*, acute or shortly acuminate, with a few smaller but relatively broader lvs below, and with smaller and narrower lanceolate-acuminate lvs above, grading into the bracts; all dull green with c. 5 prominent veins beneath which are rough like the margins. Raceme 7–40 cm, of 15–50 greenish to dull purple fls, ±turned to one side, drooping, opening ±widely, scentless; bracts lanceolate-acuminate, the lowest equalling the fls, the upper the ovary. Outer lateral per.-segs c. 1 cm, ovate to ovate-lanceolate, green to dull purple. Labellum shorter than outer per.-segs; hypochile green outside and dark brown or green inside; *epichile* cordate to triangular, *broader than long*, purplish, rose or greenish-white, with a reflexed acute tip and 2(–3) *smooth or slightly rugose basal bosses. Rostellum large*, whitish, *persistent*. Ovary glabrous or with a few scattered hairs, pendent in fr. Fl. 7–10. Said to be cross-pollinated by wasps or self-pollinated. $2n = 40$, with aneuploid variants. G.

Very variable in breadth of lvs, colour of fls, shape of labellum, rugosity of basal bosses, and presence or absence of a third (central) boss.

Native. Woods, wood margins and clearings, shaded banks, scree-slopes, etc., on both calcareous and acid substrata. Frequent through much of Great Britain northwards to S. Perth and Stirling, also in a very few mainly western localities further north to Sutherland; Skye, Mull; scattered throughout Ireland. Most of Europe to c. 68° N. in Norway; N. Africa; temperate Asia eastwards to Japan. Introduced in North America.

3. E. purpurata Sm.   Violet Helleborine

E. sessilifolia Peterm.; *E. violacea* (Dur. Duq.) Boreau

Rhizome vertical, often branched above, bearing fleshy roots at successive thickened nodes. *Stems 1–many*, 20–70 cm, erect, solid, shortly hairy above, violet-tinged below, with 2–3 brown sheathing basal scales. Largest *lvs* 6–10 cm, near the middle of the stem, *spirally arranged, ovate-lanceolate to lanceolate*, acute to shortly acuminate; lower lvs smaller but relatively broader; uppermost narrow and grading into the bracts; all *grey-green, often violet-tinged*, especially beneath; main veins 3–7; margins rough. Raceme 15–25 cm, dense, 1-sided; fls numerous, pale greenish-white, slightly fragrant;

bracts narrow, acuminate, often violet-tinged, *equalling or exceeding the fls.* Outer per.-segs 10–12 mm, lanceolate, bluntish, green outside, whitish inside; inner shorter, ovate-lanceolate, whitish and sometimes rose-tinged. *Labellum* slightly *shorter than the outer per.-segs*; *hypochile* greenish outside, usually *mottled with violet within*; *epichile at least as long as broad*, triangular-cordate, dull white, with an acute reflexed tip and 2–3 ±confluent *smoothly plicate basal bosses*, the central one elongated and faintly violet-tinged. Rostellum whitish, persistent. Ovary rough with short hairs, held ±horizontally in fr. Fl. 8–9. Cross-pollinated by wasps. $2n = 40^*$. G.

Distinguishable from *E. helleborine* by the usually clustered stems, the narrower often violet-flushed lvs, longer bracts, paler fls, and longer and narrower epichile.

Native. Woods, especially of beech, usually on calcareous soil or clay-with-flints. Locally frequent in S.E. and C. England northwards to Shropshire, Leicester and S. Lincoln. N.W. and C. Europe northwards to Denmark and south-eastwards to Bulgaria; W. Siberia.

The 3 following species are characterized by having the rostellum rudimentary or lacking. The British forms whose fls are reduced in this way are taxonomically very difficult. There appear to be many local races differing in details of floral structure from their nearest relatives, a situation which may be presumed to arise from their being predominantly or exclusively self-pollinated. Some of the local forms have fls which are partly or completely closed throughout the flowering period.

4. E. leptochila (Godfery) Godfery
Green-leaved or Narrow-lipped Helleborine.

Rhizome ±vertical, branched above, bearing numerous fleshy roots at successive nodes. Stems 1–several, 20–70 cm, ±pubescent, especially above, usually green throughout; basal scales 2–3, violet, later becoming brown. Largest *lvs* 5–10 × 2–5 cm, broadly ovate to broadly lanceolate, acute, the lowest smaller, the uppermost smaller and narrower, grading into the bracts; all *arranged in 2 opposite ranks* (distichously), dull yellow-green or dark green, ±undulate at the margin, with c. 7 prominent veins beneath. Raceme long; fls many, large, yellow-green, scentless, all ±turned to one side; *lowest bracts much exceeding and the others about equalling the spreading widely open fls.* Outer per.-segs 12–15 mm, lanceolate-acuminate, pale green; inner c. 10 mm, ovate-lanceolate, acute, whitish-green. Labellum with circular *hypochile* pale green outside and *mottled with red within*; *epichile longer than broad*, *somewhat concave*, yellow-green with a white border, narrowly cordate-acuminate with the long acute tip usually spreading, not reflexed, and with 2 smooth or somewhat rugose white or rose-tinged basal bosses separated by a deep channel. Anther prominently stalked. *Rostellum evanescent.* Ovary covered with rough tubercles and sparse often blackish hairs. Fl. 6–8. Automatically self-pollinated by loosened pollen-masses falling from the pollinia on to the stigma; perhaps rarely cross-pollinated by wasps in the earliest stages

of fl.-opening, when the hypochile may hold a little nectar. $2n = 36$. G.

Very variable in size and number of lvs and fls, in shape of epichile and in length of anther-stalk, etc. *E. cleistogama* C. Thomas, a robust plant with autogamous green fls which remain closed and have a flat green epichile, seems to belong here.

Native. In deep shade of beech, in ash–hazel coppice or in birch scrub, always on calcareous or circumneutral soil; tolerant of high levels of lead, zinc and other heavy metals. Mainly in S. England from Kent to S. Devon northwards to the Chiltern and Cotswold Hills with outliers in the Wye Valley, Hereford and Worcester and Shropshire, and also in N.W. England from Southport to S. Cumberland, in N. Lincoln, S. Northumberland on substrata rich in heavy metals, and on Holy Island. *E. cleistogama* only under beech on the steep western slope of the Cotswolds near Wotton-under-Edge (Gloucester) but not seen recently. N.W. and C. Europe south-eastwards to Yugoslavia and N. Greece.

E. youngiana A. J. Richards & A. F. Porter (1982), reported recently from two sites in S. Northumberland, shares with *E. leptochila* disintegrating pollinia and automatic self-pollination and also many vegetative features including rather narrow ±distichous yellowish lvs and a pubescent rhachis, but the large open and pink-tinged fls, held ±horizontally, have a usually sessile anther about equalled by the long acute rostellum, a cordate epichile with reflexed tip, and a subglabrous ovary. *E. helleborine* grows in both sites and it is surmised that *E. youngiana* may have arisen from a cross between that species and *E. leptochila* or perhaps *E. dunensis*.

5. E. dunensis (T. & T. A. Stephenson) Godfery
Dune Helleborine.

E. leptochila var. *dunensis* (T. & T. A. Stephenson) T. & T. A. Stephenson

Rhizome short, slender, ascending, or more commonly reduced to a small irregular deeply buried mass; roots few, slender, wiry. Stems 1(–3), 20–40(–60) cm, slender green and shortly pubescent above, violet-tinged and glabrous below, with 2–3 loose sheathing basal scales. *Lvs in 2 opposite ranks*, the largest near the middle of the stem, oblong-lanceolate acute; lowest broadly ovate, blunt; uppermost bract-like; all commonly yellowish-green, rigid and plicate, with undulate margins and c. 9 main veins. Raceme 5–12(–17) cm; fls 7–20 (–25), small, pale yellowish-green; *bracts* linear-lanceolate, *lowest longer than the spreading campanulate, incompletely opening fls.* Outer per.-segs c. 7 mm, ovate, blunt, yellowish-green; inner smaller. Labellum with a ±circular *hypochile, mottled with red inside*; *epichile as broad as long*, *whitish* with a green or rosy tinge, *broadly triangular with a recurved tip*, and 2(–3) almost smooth basal bosses. *Rostellum* usually disappearing in early bud. Ovary subglabrous to glabrous. Fl. 6–7. Usually self-pollinated by pollen falling over the edge of the stigma.

Native. In somewhat peaty but not very moist hollows in coastal dunes, chiefly amongst *Salix repens* but sometimes beneath planted pines, and then taller and darker green. Known only from Anglesey and N.W. England from Southport to S. Cumberland, N. Lincoln and Northumberland (Holy Is.). Probably endemic.

Distinguishable from *E. leptochila* by the incompletely open fls and the broadly triangular epichile; from *E. phyllanthes* by the spreading, not drooping, fls.

6. E. phyllanthes G. E. Sm.
Pendulous-flowered Helleborine.

Incl. *E. vectensis* (T. & T. A. Stephenson) Brooke & Rose, and *E. pendula* C. Thomas, *nom. illegit.*

Rhizome short, horizontal or oblique, with numerous short fleshy roots. Stem (8–)20–45(–65) cm, solitary or rarely 2–3, glabrous or sparsely pubescent, green. *Lvs* 3.5–7 × 3–6 cm, few (3–6), *distichous*, orbicular to lanceolate, acuminate, the uppermost narrow and grading into bracts; all *glabrous* or slightly pubescent above, often with undulate margins. Infl. up to 15 cm with 15–35 fls, the lower bracts usually distinctly exceeding the fls. *Fls hanging ±vertically downwards, sometimes opening only slightly or not at all*. Per.-segs 8–10 mm, lanceolate to ovate-lanceolate, thick and firm, *pale yellowish-green*, the inner sometimes violet-tinged. *Labellum varying ±continuously from having hypochile and epichile clearly differentiated*, about equally long and connected by a channelled constriction, *to showing no such differentiation and closely resembling the other per.-segs*. Anther sessile or stipitate; *pollinia disintegrating in bud*. *Rostellum* not secreting a sticky fluid as in allogamous spp. and *commonly shrivelling at an early stage* of flowering. Ovary 0.9–1.3 cm, pyriform, 6-ribbed, ±glabrous. Seeds 1–1.5 mm, tapered at each end. Fl. 7–8. Self-pollinated by pollen falling on the stigma from the disintegrating pollinia. $2n = 36$. G.

A taxonomically difficult aggregate of local forms differing mainly in details of floral structure. D. P. Young (*Watsonia*, 1952, **2**, 253–76) draws attention to the ±continuous range from open fls with labellum showing full differentiation of hypochile and epichile to cleistogamous fls with labellum little different from other perianth segments, the former mainly in N. England. He nevertheless recognizes four varieties, though intermediates are not infrequent. The following key is based on one in his paper:

1 Labellum with well-formed hypochile and epichile.   2
    Hypochile rudimentary or 0.   3
2 Hypochile c. 4 mm, about equalling the strongly re-
     flexed epichile; anther sessile.   var. *pendula*
    Hypochile small, 2.5–3(–3.5) mm, shorter than the
     patent epichile; anther sessile or very shortly stipi-
     tate.   var. *vectensis*
3 Labellum with rudimentary hypochile; anther often
     distinctly stipitate.   var. *degenera*
    Labellum completely undifferentiated, sepal-like;
     anther sessile or stipitate.   var. *phyllanthes*

Var. *pendula* D. P. Young: labellum perfectly differentiated, large, spreading (i.e. not embracing the column as in var. *vectensis*), hypochile and epichile about equally long and connected by a channelled constriction, the epichile cordate, acuminate, rugose at base or with two bosses, usually strongly reflexed; anther cuneiform, sessile; fl. rarely cleistogamous.

Var. *vectensis* (T. & T. A. Stephenson) D. P. Young: labellum ±perfectly differentiated but with small hypochile, shorter than the epichile, and embracing the column ±closely, hypochile and epichile connected by a channelled constriction, the often white or pinkish epichile cordate-deltate, usually acuminate and with two basal lateral bosses, not reflexed; anther variable in shape, sessile or subsessile; fl. often cleistogamous.

Var. *degenera* D. P. Young: labellum imperfectly differentiated, broadly to narrowly ovate, with the hypochile represented by a shallow or ventricose depression and with no distinct constriction separating hypochile from epichile, the latter variable in width and often with basal lateral bosses; anther cuneiform or ovate-cylindrical, sessile to long-stipitate; fl. rarely opening widely and often cleistogamous.

Var. *phyllanthes*: labellum completely undifferentiated, ovate to lanceolate, resembling the lateral perianth segments; anther variable in shape, sessile or stipitate; fl. rarely opening widely and often cleistogamous.

E. cambrensis C. Thomas seems close to *E. phyllanthes* and especially to var. *vectensis*, but further investigation is desirable.

Native. Under trees, especially beech, usually in small woods, plantations and shelter-belts. and restricted to margins of more extensive woods; often on stream banks under willows or other trees, and on sand dunes, commonly with *Salix repens*; always on calcareous substrata. Mostly in S. England and the S. Midlands but extending to N. Wales, Lancs and Cumbria and to S. Yorks and S. Northumberland. Var. *pendula* often on sand dunes and mainly in N. England and N. Wales but southwards to Hereford, Oxford and Hertford where intermediates with var. *vectensis* are not infrequent; var. *vectensis* mainly in S. England and the S. Midlands but reaching S.E. Yorks; vars *degenera* and *phyllanthes* more exclusively in S. England. Var. *pendula*, and perhaps also var. *vectensis*, has been recorded from a few localities in Ireland. *E. cambrensis* is known only from sand dunes in S. Wales. *E. phyllanthes* has been reported from France (Atlantic coast and Pyrenees) and from Denmark.

7. E. atrorubens (Hoffm.) Besser
Dark-red Helleborine.

E. atropurpurea Rafin.; *E. rubiginosa* (Crantz) Gaudin

Rhizome ±horizontal, short, with many long slender roots. Stem 15–30(–60) cm, usually solitary, erect, ±pubescent, densely so above, *violet below*, with 1–3 loosely sheathing scales. *Lvs* c. 5–10 *in 2 opposite ranks*, the largest 5–7.5(–10) × 2.5–4.5 cm, elliptical, acute, not or hardly sheathing, with a few smaller but broader sheathing lvs below, and others above, narrower, not sheathing, grading into bracts; all keeled, folded, many-veined, rough above and beneath. Raceme spike-like; *fls 8–18, red-purple*, faintly fragrant, short-stalked (3 mm); bracts lanceolate, acute, only the lowest equalling or sometimes exceeding the fls. Outer *per.-segs* c. 8 × 4 mm, ovate-acute, the laterals asymmetrical; inner c. 7 × 5 mm; blunt; *all* spreading-incurved, *reddish-violet outside*, the outer somewhat greenish especially within. Labellum shorter than the outer per.-segs; hypochile green with a red margin and red-spotted inside; *epichile* 3 × 6 mm, deep reddish-violet with a small, acute, reflexed tip and 3 *brighter red, strongly rugose, confluent basal bosses. Rostellum an obvious whitish oval ledge. Ovary* 6–7 mm, pyriform, 6-ribbed, reddish-violet, *pubescent*; ripe capsule drooping. Fl. 6–7. Cross-pollinated by bees and wasps. $2n = 40$. G.

Native. Limestone rocks and screes, in woods or in the open; to 305 m in W. Ireland. Local and rather rare throughout Great Britain from N. Wales and Derby northwards to S. Cumberland and Durham, with a few scattered localities in N.W. Scotland to Sutherland; Skye and Raasay; rare in W. Ireland: Clare and Galway only. Europe northwards to arctic Scandinavia; Caucasus; N. Iran.

4. Epipogium R.Br.

Lfless and rootless saprophytic herbs with coralloid rhizomes. Fls in racemes; drooping, spurred, with the labellum uppermost and the *spur directed vertically upwards*. Column pointing downwards with the horseshoe-shaped stigma on the overhanging base of its upper side. Rostellum large, cordate, at the apex of the column. Anther helmet-like, sessile in the concave summit of the column; pollinia 2, pear-shaped, with caudicles attached to their base; pollen in packets of tetrads. Ovary not twisted.

Two spp. in N. and C. Europe, N. Asia and Himalaya.

1. E. aphyllum Swartz

Spurred Coral-Root, Ghost Orchid.

E. gmelinii L. C. M. Richard

Rhizome whitish with many very short bluntly 2–3-lobed fleshy branches and 1–2 long (5–7 cm) filiform whitish stolons bearing at intervals buds which give rise to new rhizomes. Stem 5–20 cm, erect, ±*translucent, white tinged with pink* and with numerous short reddish streaks, weak, turgid, much swollen below then tapering suddenly to its very weak attachment to the rhizome. *Lvs represented only by 2–3 brownish basal sheathing scales* and 1–2 long close-fitting usually dark-edged sheaths higher up the stem. Fls 1–4(–7) distant, pendent on slender stalks, the per.-segs directed downwards, the labellum and spur upwards. Bracts oblong, blunt, membranous, translucent, equalling the fl. stalk. Outer per.-segs linear, yellowish or reddish; inner per.-segs lanceolate, blunt, yellowish with a few short violet lines; all ±equally long, curving downwards and with upturned edges. Labellum with 2 short rounded basal lateral lobes and a large concave cordate terminal lobe, white with violet spots and irregularly tubercled crests which leave a deep channel down the middle. Spur about 8 × 4 mm, rounded at the tip, white tinged with yellow or reddish outside and with lines of violet spots within. The labellum is bent backwards near its middle and almost touches the spur. Capsule ±globose, drooping, opening by short slits which reach neither base nor apex. Fl.

6–8. Cross-pollinated by humble-bees and other insects. G. Saprophyte.

Native. In deep shade of oak or beech. Found, usually singly on a very few occasions, in Hereford, Shropshire, Oxford and Bucks. N. and C. Europe southwards to Pyrenees, N. Greece and Crimea; Caucasus; Siberia; Himalaya.

5. Spiranthes L. C. M. Richard

Small herbs with erect stock, 2–6 ±tuberous roots, lfy stems and *small fls in spirally twisted spike-like racemes*. Outer and inner per.-segs similar, the back outer segment cohering with the 2 inner to form the upper half of a 2-lipped trumpet-like tube round the column. *Labellum frilled and ±recurved distally, furrowed below and embracing the base of the column* to form the lower part of the perianth-tube, its edges being overlapped by the inner per.-segs; spur 0; nectar secreted by 2 bosses at the base of the labellum. Column horizontal, with the circular stigma on its underside, facing the labellum. Rostellum narrow, projecting beyond the stigma and consisting of a narrowly elliptical viscidium supported between two long narrow teeth which are left behind like the prongs of a fork when the viscidium is removed. Anther hinged to the back of the column, resting on the rostellum: pollinia 2, each of 2 plates of coherent pollen tetrads, attached to the viscidium near their summits; caudicles 0.

Twenty-five spp. in the north temperate zone and South America.

1 Flowering stem bearing only bract-like sheathing
　　appressed scales, the true lvs of the current season
　　being in a lateral basal rosette.　　　**1. spiralis**
　　Flowering stems bearing true lvs.　　　　　　2
2 Fls in a single spirally twisted row.　　**2. aestivalis**
　　Fls in 3 spirally twisted rows.　　**3. romanzoffiana**

1. S. spiralis (L.) Chevall.　　Autumn Lady's Tresses.

Ophrys spiralis L.; *S. autumnalis* L. C. M. Richard

Roots 2–3(–5), 1.5–2(–4) cm, ovoid or ovoid-oblong, radish-shaped, pale brown, hairy. Stem 7–20 cm, erect, terete, glandular-hairy especially above, with several pale green, lanceolate-acuminate, *appressed bract-like scales*, and sometimes with the withered remains of last season's rosette lvs still visible at its base. *Lvs* (4–5) of current season appearing with or after the fls *in a lateral basal rosette* which will flower next season; each c. 2.5 cm, ovate, acute, stiff, bluish-green, glabrous. Spike 3–12 cm, of 7–20 small (4–5 mm), white, day-scented *fls in a single ±spirally twisted row.* Bracts lanceolate-cuspidate, concave and hooded, sheathing the ovary and about equalling it. Outer per.-segs oblong-lanceolate, blunt, translucent, white, ±ciliate, slightly glandular outside. Inner per.-segs narrowly oblong, blunt. Labellum broadened, rounded and recurved distally, pale green with a broad white irregularly crenate or fringed margin. *Stigma ciliate below.* Ovary short, bent outwards at its apex, green, usually not twisted. Capsule 6 mm, obovoid. Fl. 8–9. Pollinated by humble-bees. $2n = 30$. Grt.

Native. Hilly pastures, downs, moist meadows and grassy coastal dunes, usually on a calcareous substratum. Throughout England and Wales northwards to Westmorland and N.E. Yorks, but common only in the west; S. Ireland, to Dublin and Sligo; Scilly and Channel Is. Europe northwards to Denmark and C. Russia; N. Africa; Asia Minor.

2. S. aestivalis (Poiret) L. C. M. Richard

Summer Lady's Tresses.

Roots 2–6, 5–8 cm, fleshy, fusiform-cylindrical. Stem 10–40 cm, erect, somewhat glandular-hairy above. *Lvs* 5–12 cm × 5–9 mm, borne ±erect *on the flowering shoot of the current season*, linear-lanceolate, blunt, the lower narrowing downwards into a sheathing base, the uppermost very short, bractlike, appressed; all bright green, glossy on both sides, glabrous. Spike 5–8 cm, of 6–18 small, night-scented *fls in a single twisted row. Bracts* as in *S. spiralis*, but *exceeding the ovary. Fls* as in *S. spiralis* but somewhat larger, *pure white* and with the *stigma not ciliate below*. Ovary 9 mm, ±glandular-hairy. Capsule oblong. Fl. 7–8. Probably pollinated by moths. Grt.

Native. Marshy ground with sedges and rushes. Known in Great Britain only in the New Forest (Hants), but perhaps now extinct. C. and S. Europe northwards to Belgium and Germany. N. Africa; Asia Minor.

3. S. romanzoffiana Cham.

Irish Lady's Tresses, Drooping Lady's Tresses.

Roots 2–6, long, fleshy, fusiform-cylindrical. Stem 12–25 cm, erect, bluntly 3-angled, sparsely pubescent above. Lower *lvs* 5–10(–15) cm × 5–10 mm, *borne erect on the flowering shoot of the current season*, linear-oblanceolate, acute, glabrous; uppermost short, bract-like, acuminate, with a loosely sheathing base. Spike 2.5–5(–8) cm, slightly twisted, stout, dense, glandular-pubescent, with many large (8–11 mm), white, hawthorn-scented *fls in 3 spirally curved rows*. Bracts lanceolate, acute or acuminate, concave, sheathing the ovary, the lowest about 3 cm long, exceeding the fls, the upper exceeding the ovary. Ovary 10 mm, cylindrical, subsessile, glandular-hairy, not twisted, turned to one side and bent at the tip so that the fl. is held horizontally. Outer lateral per.-segs 12 mm, lanceolate-acuminate, white, greenish below, glandular-hairy outside, united below and sheathing the base of the labellum. Inner per.-segs linear, blunt. Labellum white with green veins, tongue-shaped, enlarging distally to the rounded frilled denticulate apex. Stigma crescent-shaped. Fl. 7–8. Cross-pollinated by insects. Grt.

Native. Wet peaty pastures and meadows, especially where flooded in winter, species-rich upland bogs, wet stony shores of rivers and lakes, etc., occasionally on rather drier grassy slopes. Restricted to a very few areas in W. Scotland and in Ireland and one locality on Dartmoor in S. Devon: in Scotland in N.W. Argyll and W.

Inverness, Inner Hebrides (Colonsay and Coll) and Outer Hebrides (Barra and Benbecula); in Ireland in W. Cork and S. Kerry, W. Galway and around and downstream from Lough Neagh. Not elsewhere in Europe but in North America.

The S.W. Irish plants have broader lvs and a denser fl.-spike than those near Lough Neagh or in Scotland and the Hebrides, and have been treated by some as a different sp. The Devon plant, however, is intermediate in certain respects and it seems best to regard all British populations as belonging to a single variable sp.

6. LISTERA R.Br.

Herbs with short rhizomes bearing numerous slender roots, erect stems usually with 2 *sessile ±opposite green lvs*, and spike-like racemes of inconspicuous greenish or reddish fls. Per.-segs spreading or loosely connivent. Labellum long and narrow, forking distally into 2 narrow segments and sometimes with 2 lateral lobes near the base; spur 0; nectar secreted by the central furrow of the labellum. Column short, erect; stigma transversely elongated, prominent; rostellum broad, flat, blade-like, arching over the stigma and expelling its viscid contents explosively in a terminal drop when touched. Anther hinged to the back of the column; pollinia 2, club-shaped, each +divided longitudinally into halves; caudicles 0; pollen friable in loosely bound tetrads.

About 30 spp. in north temperate and sub-arctic zones.

Plant 20–60 cm; lvs large (5–20 cm), broadly ovate-elliptical; spike 7–25 cm; fls many, green. **1. ovata**

Plant 6–20 cm; lvs small (1–2.5 cm), ovate-deltate; spike 1.5–6 cm; fls reddish-green. **2. cordata**

1. L. ovata (L.) R.Br. Twayblade.

Ophrys ovata L.

Rhizome horizontal, deep, with very numerous rather fleshy but slender roots. Stem 20–60 cm, rather stout, glabrous below, pubescent above; basal sheaths 2(–3), membranous. *Lvs* 5–20 cm, in a subopposite pair rather below the middle of the stem, *broadly ovate-elliptical*, sessile, with 3–5 prominent ribs; upper part of stem with 1–2 tiny triangular bract-like lvs. Raceme 7–25 cm, rather lax; fls numerous, short-stalked, yellowish-green, bracts minute, shorter than the fl.-stalks. Per.-segs subconnivent, the outer laterals not quite contiguous with the others; outer segments ovate, bluntish, green, inner narrower, yellow-green. *Labellum* 7–15 mm, *yellow-green*, directed forwards at the cuneate furrowed base, then turning abruptly and almost vertically downwards; lateral lobes 0 or rarely represented by 2 small erect teeth near the base, *central lobe oblong*, broadening slightly downwards and *deeply forked almost to half-way into* 2 *narrowly strap-shaped parallel segments* sometimes with an intervening tooth. Capsule ±globose on an ascending stalk. Fl. 6–7. Cross-pollinated by small insects which crawl up the nectar-secreting centre furrow of the labellum. $2n = 42^*$; 32, 34*, 36, 38. G.

Var. *platyglossa* Peterm. has the labellum 6–8 mm wide, with its distal segments slightly divergent.

Native. Moist woods and pastures on base-rich soils, and on dunes; common and locally abundant. Throughout the British Is., except Shetland. Var. *platyglossa* on dunes in S. Wales. Most of Europe northwards to c. 70° N. in Norway; Caucasus; Siberia.

2. L. cordata (L.) R.Br. Lesser Twayblade.
Ophrys cordata L.

Rhizome creeping, very slender, with a few whitish filiform roots. Stem 6–20 cm, erect, slender, glabrous below, angled and slightly pubescent above; basal scales 1–2, closely sheathing. Lvs 1–2(–2.5) cm, in a subopposite pair (rarely 3–4) rather below the middle of the stem, *ovate-deltate* with rounded horny tip and basal corners and a *broadly cuneate* to subcordate sessile *base*; shining above, paler beneath, somewhat translucent. Raceme 1.5–6 cm, lax, of 4–12 fls, the 3 lowest sometimes in a whorl; bracts minute (1 mm) triangular, greenish; fl.-stalks 1–2 mm. Outer per.-segs green, the upper one broad, like a little hood, the 2 lateral narrower and forward-spreading like the inner segments, which are green outside but reddish inside. *Labellum* 3.5–4 mm, *reddish*, twice as long as the per.-segs, very narrow, with 2 linear-oblong lateral lobes close to the base, the *central lobe forking about half-way into 2 widely diverging linear tapering segments*. Capsule ±globose, strongly 6-ribbed, the ribs often reddish. Fl. 7–9. Cross-pollinated by minute flies and Hymenoptera. $2n = 42^*$; 38, 40. G.

Native. Mountain woods, especially of Scots pine, and peaty moors, especially under heather; often amongst sphagnum; to 825 m in Scotland. Very rare in Devon, Somerset, Dorset and Hants; more frequent northwards from N. Wales, Shropshire and Derby, and locally common in Scotland to Sutherland and Caithness; Hebrides; Orkney; Shetland. Ireland, but rare in the south. Europe from the Pyrenees and Apennines to Iceland, northernmost Scandinavia and the Kola Peninsula; Transcaucasus; N. Asia; North America.

7. NEOTTIA Ludwig

Perennial *saprophytic herbs* with short creeping rhizomes concealed in *a mass of short thick fleshy blunt roots* ('bird's nest'), *stems* densely *covered with brownish scales*, and spike-like racemes of pale brownish fls. Per.-segs connivent into an open hood. *Labellum saccate at the base, with 2 distal lobes; spur* 0. Column long, slender, erect; stigma large, prominent; rostellum broad, flat, blade-like, arching over the stigma, expelling its viscid contents explosively when touched. Anther hinged to the back of the column, directed forwards; pollinia 2, ±cylindrical, slender, each divided longitudinally into halves; pollen tetrads loosely united by a few weak threads, so that the pollen masses are very friable.

Nine spp. in temperate Europe and Asia.

1. N. nidus-avis (L.) L. C. M. Richard
Bird's-nest Orchid.
Ophrys nidus-avis L.

Roots cylindrical, pale fawn. Stem 20–45 cm, erect, robust, brownish, somewhat glandular above, clothed below with numerous brownish scarious sheathing scales. Green lvs 0. Raceme 5–20 cm, rather lax below; bracts falling short of the ovary, lanceolate-acuminate, scarious. Per.-segs ovate-oblong, the inner slightly shorter than the outer. Labellum c. 12 mm, twice as long as the per.-segs, darker brown, obliquely hanging, with 2 small lateral teeth near the base, and 2 ±oblong blunt distal lobes extending almost halfway and diverging, straight or outwardly curved. Capsule c. 12 mm, erect. Fl. 6–7. Cross-pollinated by small crawling insects which touch the sensitive rostellum, or self-pollinated. $2n = 36^*$.

Native. Shady woods, especially of beech and especially on humus-rich calcareous soils. Throughout Great Britain northwards to Banff and Inverness; Ireland. Europe, northwards to c. 64° N. in Norway and to S. Sweden and S. Finland; Caucasus; Siberia.

8. GOODYERA R.Br.

Small herbs with creeping rhizomes, *ovate stalked lvs* and small fls in *twisted one-sided spike-like racemes*. *Labellum in 2 parts*, a basal pocket-like hypochile and a distal narrow spout-like epichile; spur 0. Column horizontal, short, with the roundish stigma on its lower side, facing the labellum. Rostellum projecting beyond the stigma and consisting of two short curved horns enclosing the ±circular viscidium. Anther hinged to the back of the column, resting on the rostellum; pollinia 2, partially divided lengthways, attached to the rostellum just beneath their summits, usually without caudicles: pollen in tetrads cohering in packets.

Forty spp. in the north temperate zone, tropical Asia, Madagascar and New Caledonia.

1. G. repens (L.) R.Br. Creeping Lady's Tresses.
Satyrium repens L.

Rhizome creeping and giving rise to a few short roots and pale slender stolons which end in lf-rosettes. Stem 10–25 cm, stiffly erect, terete below but angled above, glandular-hairy especially above, with 1 whitish sometimes green-tipped, sheathing, basal scale. Basal lvs 1.5–2.5 cm, in a lax rosette persisting through the winter, ovate to ovate-lanceolate, firm, narrowed into a winged *stalk-like base*, dark green often mottled with lighter green, with 5 main veins and a *conspicuous network of secondary veins*; uppermost ±bract-like, appressed to the stem. Fls cream-white, fragrant. Bracts 10–15 mm, exceeding the ovary, linear-lanceolate, acute, ciliate, green outside, glossy white within. Outer per.-segs ovate, blunt, glandular-hairy, white or greenish, the lateral pair slightly spreading. Inner per.-segs narrowly lanceolate, blunt, glabrous, whitish, connivent with the back outer per.-segs. Labellum shorter than the outer per.-segs, the hypochile deeply pouched, epichile narrow furrowed, recurved. Ovary 9 mm, glandular-hairy, subsessile. Fl. 7–8. Cross-pollinated by humble-bees. $2n = 30$. Grh.

Native. Locally in pine-woods rarely under birch or on moist fixed dunes. From Cumberland, E. Yorks (?extinct) and Northumberland to Sutherland and Orkney; Norfolk and Suffolk (probably introduced with planted pines). C. and N. Europe from Pyrenees and Balkans to N. Scandinavia; Russia. Asia Minor; Afghanistan; Himalaya; Siberia; Japan; North America.

9. HAMMARBYA O. Kuntze

A small green herb with a *pseudo-bulb* covered by pale sheathing scales and connected with the daughter pseudo-bulb above it by a short vertical stolon. Lvs few, short, broad. Raceme short, of several minute yellowish-green fls which are turned through 360° so that the *labellum* is *directed upwards*. Per.-segs spreading. Labellum short, entire; *spur 0*. Column very short, stigma in a deep fold on its front; rostellum a membrane above the stigma and in front of the anther, surmounted by a small viscid mass. Anther hinged to the top of the column behind the rostellum; pollinia 2, each of 2 thin flat plates of waxy pollen, broad below and tapering upwards, standing in a clinandrium formed by 2 lateral membranous lobes of the column; pollen in tetrads which cohere firmly in the 4 plates; caudicles 0.

One sp. in central and north temperate Europe, N. Asia, and North America.

1. H. paludosa (L.) O. Kuntze   Bog Orchid.

Ophrys paludosa L.; *Malaxis paludosa* (L.) Swartz

Old pseudo-bulb buried in moss or peat, ovoid, ±angled above, tapering below into a root; daughter bulb 1–2 cm higher, enclosed by lvs. Stem 3–12 cm, slender, 3–5-angled above, glabrous. *Lvs* 0.5–1 cm, 3–5, of which the lowest 1–2 may have no blades, small, short, concave, *broadly rounded at the apex* and broadly sheathing at the base, 3–7-veined, pale green, usually *bearing a marginal fringe of tiny bulbils*. Raceme 1.5–5 cm, spike-like, becoming lax; its bracts lanceolate-acute, just exceeding the fl.-stalk. Outer lateral per.-segs ovate-lanceolate, erect; outer median segment pointing downwards, rather longer and broader. Inner lateral segments linear-lanceolate, spreading with down-turned tips. Labellum lanceolate, acute, shorter than the outer per.-segs, erect, with its base clasping the column. All floral parts greenish-yellow. Ovary turbinate, straight, its stalk twisted. Fl. 7–9. Cross-pollinated by small insects. $2n = 28$.

Native. Usually in wet sphagnum, to 490 m in Scotland. In England and Wales very rare and decreasing and now only in Hants, Dorset and S. Devon, Norfolk, W. Wales and S. Cumberland; in Scotland in a number of localities north of a line from N. Kintyre to S. Aberdeen and reaching Sutherland, Orkney and both Inner and Outer Hebrides; now only in a few places in N.E. and S.E. Ireland. C. and N. Europe from France, Switzerland and Yugoslavia to just north of the Arctic Circle

in Sweden, Finland and N.W. Russia; Siberia; North America.

10. LIPARIS L. C. M. Richard

A large genus whose European representatives are small herbs with 2 *ellipsoid pseudo-bulbs* (parent and daughter) side by side, an angled stem usually with 2 green lvs, and a raceme of small yellowish-green fls. Per.-segs narrow, spreading. *Labellum pointing* in various directions, *most commonly upwards*, broad, entire; *spur 0*. Column long, slender; stigma transversely oblong, depressed, flanked by lateral wings of the column; rostellum minute with 2 evanescent viscidia. Anther on the top of the column, deciduous; pollinia 2, each of 2 flat plates of waxy pollen and each attached to one of the viscidia; caudicles 0.

About 250 spp., some epiphytic, widely distributed in temperate and tropical regions.

1. L. loeselii (L.) L. C. M. Richard   Fen Orchid.

Ophrys loeselii L.

Parent pseudo-bulb clothed in a pale network of old lf-bases, and giving rise to a short horizontal stolon on which the daughter stock is borne. Stem 6–20 cm, erect, glabrous, strongly 3(–5)-angled above and with 2–3 sheathing basal scales enclosing the developing pseudo-bulb. *Lvs* 2.5–8 cm, *in a subopposite pair*, oblong-elliptical, acute, keeled, many-veined, shining and greasy-looking, with long sheaths. Raceme 2–10 cm, rather lax, of 1–10(–15) fls; lower bracts lanceolate, often equalling or exceeding the fls, upper ones (or all) minute, falling short of the ovary. Outer *per.-segs* linear-lanceolate, spreading; inner shorter and narrower; all *yellow-green*. *Labellum commonly directed ±upwards*, almost equalling the outer per.-segs, oblong-obovate, furrowed, undulate or ±crenate, darker green. Column slightly inflexed. Capsule fusiform, almost straight and ±erect on an ascending twisted stalk. Fl. 7. Probably cross-pollinated by insects. $2n = 32$.

Var. *ovata* Riddelsd. ex Godfery is the dune form, with broader ovate-elliptical blunt lvs.

Native. Rare and decreasing. In wet fen peat and the edges of lakes and pools in Norfolk, Suffolk and Cambridge; var. *ovata* in dune-slacks on coasts of N. Devon, Glamorgan and Carmarthen. Europe from Norfolk and c. 61° N. in Scandinavia southwards to S.W. France, S. Romania and S. Russia; North America.

11. CORALLORHIZA Chatel.

Brown saprophytic rootless herbs with coral-like much-branched fleshy rhizomes, stems with sheathing scales but *no green lvs*; fls small in spike-like racemes. Per.-segs ±spreading or the outer median and inner lateral sub-connivent. *Labellum* short, *directed downwards*, ±3-lobed, with the lateral lobes very small or 0; spur short, ±adnate to the ovary, or 0. Column long, erect; stigma discoid or triangular; rostellum small, globose. Anther terminal on the column, lid-like, deciduous; pollinia 4, subglobose; pollen waxy or powdery.

About 15 spp., in Europe (1 sp.), temperate Asia, North America and Mexico.

1. C. trifida Chatel.   Coral-root.

Ophrys corallorhiza L.; *C. innata* R.Br.

Rhizome cream-coloured or pale yellowish, with short rounded branches. Stem 7–25 cm, erect, slender, yellowish-green, glabrous, with 2–4 long, brown-veined, sheathing scales, often reaching half-way up the stem. Raceme lax, of 2–12 rather inconspicuous fls; its bracts very small, much shorter than the ovary, triangular, membranous. Outer lateral per.-segs curved downwards close to the labellum, strap-shaped, with incurved margins; outer median per.-segs ovate-lanceolate, ±connivent with the narrow inner segments; all yellowish- or olive-green, often with reddish borders or spots. Labellum c. 5 mm, about equalling the per.-segs, oblong, 3-lobed with 2 very small rounded or tooth-like lateral lobes near its base; lateral lobes sometimes almost equalling the central, or 0; the whole whitish with crimson lines or spots, and with 2 broad longitudinal ridges near the base. Ovary c. 7 mm, straight, with a short twisted stalk. Fl. 5–8. Probably cross-pollinated by insects. $2n = 42$.

Native. Damp peaty or mossy woods, especially of birch, pine or alder, and moist dune-slacks; rare. From S. Cumberland, W. Yorks and Northumberland northwards to E. Sutherland; Mull; not in Ireland; especially in upland woods and on E. Scottish dunes. Europe from the Pyrenees, S. Apennines, N. Greece and Crimea to northernmost Scandinavia and the Kola Peninsula; Siberia; North America.

12. HERMINIUM Guett.

Small herbs with 1 fully developed entire root-tuber at flowering; *daughter-tubers at the tips of slender stolons*. Lvs commonly 2. Spike slender, dense; fls small, green, ±campanulate. Per.-segs incurved, connivent. Labellum 3-lobed; spur 0. Column very short: stigma 2-lobed; rostellum represented by 2 *large viscidia* each covered by a delicate skin but not enclosed in a pouch. Anther adnate to the top of the column; pollinia 2, ovoid, diverging downwards and each attached basally by a *very short caudicle* to a viscidium which almost equals it in size; pollen in tetrads bound together by elastic threads.

About 30 spp. in Europe and Asia.

Distinguishable by the spurless fls with 2 ±exposed viscidia but no rostellar projection.

1. H. monorchis (L.) R.Br.   Musk Orchid.

Ophrys monorchis L.

Root-tubers small, globose; daughter-tubers at the tips of stolons up to 10 cm; roots slender, few. Stem 7–15 (–30) cm, erect, slender, glabrous, angled above, with 1–2 closely sheathing scales below. Lvs 2(–4), 2–7 cm, oblong or elliptical-oblong, ±acute, keeled; often with 1–2(–3) much smaller bract-like lvs higher up the stem.

Spike 1–5(–10) cm, slender, cylindrical, often 1-sided, ±dense, with small drooping ±campanulate greenish fragrant fls; bracts about equalling the ovary or the lowest equalling the fls, lanceolate, green. Outer per.-segs connivent, the upper ovate-oblong, the laterals lanceolate; inner per.-segs longer but narrower, ±laterally lobed; all pale greenish-yellow. Labellum c. 4 mm, greenish, its lateral lobes short and widely divergent, the central lobe oblong, blunt; base of labellum saccate. Pollinia white. Capsule oblong, twisted, ±erect with the apex curved outwards. Fl. 6–7. Pollinated by minute flies, beetles and Hymenoptera, the viscidia becoming attached to their legs. $2n = 40$.

Native. In short grassland over chalk or limestone; rare and local and recently much decreased in range. Now only in S. and S.E. England from Hants to Kent and northwards to Gloucester, Bucks and Beds. Much of Europe from the Pyrenees, S. Apennines, N. Greece, Crimea and Caucasus to northernmost Scandinavia and the Kola Peninsula; N. and C. Asia eastwards to N. China, Korea and Japan.

13. COELOGLOSSUM Hartman

Small herbs with palmate root tubers, ovate or oblong lvs, and small green and brown fls whose outer and inner lateral per.-segs are connivent into a hood. Labellum narrowly oblong; spur short, saccate, with free nectar. Column short, erect; stigma central, reniform, depressed; rostellum of 2 widely separated protuberances, one on each side of the upper edge of the stigma. Anthers adnate to the top of the column; pollinia club-shaped, converging above, but narrowing downwards into widely separated caudicles each attached basally to one of the 2 *oblong viscidia* which are covered by a delicate skin but only *partially enclosed in small pouches*; pollen in packets of tetrads bound by elastic threads.

Two spp. in the north temperate and arctic zones.

1. C. viride (L.) Hartman   Frog Orchid.

Satyrium viride L.; *Habenaria viridis* (L.) R.Br.

Tubers 2, ovoid, usually palmately lobed with 2–4 tapering segments. Stem 6–25(–40) cm, angled and often reddish above, with 1–2 brown sheathing basal scales. Lvs 2–5, the lower 3–8(–11) cm, broadly oblong, blunt, the lowest sometimes almost orbicular; upper lvs smaller, lanceolate, acute; all unspotted. Spike 2–10 (–15) cm, cylindrical, rather lax-fld; bracts green, the lowest about equalling the fls. Fls greenish, inconspicuous, slightly scented. Outer per.-segs c. 5 × 3.5 mm, brownish or greenish-purple, ±connivent into a hood; inner per.-segs green, linear, ±hidden by the outer. Labellum 3.5–6 × 1.5–3 mm, oblong, straight and ±parallel-sided, hanging almost vertically, 3-lobed near its tip, the outer lobes narrowly oblong and parallel, the central usually much shorter, rounded or tooth-like; colour of labellum variable, uniformly green or edged with chocolate-brown or brownish in the distal half.

Spur c. 2 mm, ovoid-conical, translucent greenish-white, containing free nectar. Fl. 6–8. Cross-pollinated by insects. $2n = 40$.

Native. Pastures and grassy hillsides, especially on calcareous soils; occasionally on dunes and rock-ledges; to 1000 m in Scotland. Throughout the British Is., though more frequent in the north. Europe southwards to C. Spain, S. Apennines, Bulgaria and Crimea, but only on mountains in the south; W. Asia; North America.

14. GYMNADENIA R.Br.

Herbs with usually palmately lobed root-tubers, lfy stems and dense spikes of small fragrant fls. Outer lateral per.-segs spreading; outer median and inner lateral per.-segs connivent into a hood. Labellum shortly 3-lobed, directed downwards; spur long and slender, with free nectar. Column short, erect, with the *2 stigmas on oval lateral lobes*; rostellum elongated, projecting between the viscidia. Anther wholly adnate to the column; pollinia 2, convergent and narrowed below into caudicles each of which is attached basally to a *long linear viscidium* about equalling the caudicle; the viscidia lie close together and are naked, i.e. *not enclosed in a pouch*; pollen in packets of tetrads bound together by elastic threads.

Ten spp. in Europe, North America and N. Asia.

1. G. conopsea (L.) R.Br. Fragrant Orchid.

Orchis conopsea L.; *Habenaria conopsea* (L.) Bentham, non Reichenb. fil.

Tubers compressed, with 3–6 tapering blunt segments. Stem 15–40(–60) cm, erect, glabrous, sometimes purplish above, with 2–3 close brown basal sheaths. Lower lvs 3–5, c. 6–15 × 0.5–2 cm, ±narrowly oblong-lanceolate, keeled and folded, slightly hooded, blunt or subacute; upper lvs c. 2–3, smaller, lanceolate-acuminate, bract-like, appressed to the stem; all unspotted, with minutely toothed margins. Spike (3–)6–10(–12) cm, ±cylindrical, rather dense-fld; bracts green or violet-flushed, about equalling the fls. *Fls* small, reddish-lilac, rarely white or magenta, *very fragrant*. Outer lateral per.-segs spreading horizontally or downwardly curved, their margins revolute. Labellum c. 3.5 × 4 mm, its 3 lobes subequal and rounded. *Spur* 11–13 mm, *very slender*, somewhat curved below, acute, *almost twice as long as the ovary*. Fl. 6–8. Pollinated by moths to whose probosces the viscidia become attached. $2n = 20^*; 40, 80$.

Native. Commonly on calcareous or circumneutral soil in ±species-rich grassland over chalk or limestone but sometimes on more acid soil with ericaceous spp., dwarf gorse, etc.; reaches 640 m in Scotland; subsp. *densiflora* characteristically and subsp. *conopsea* not infrequently in fens or on marshy or boggy ground. Subsp. *conopsea* locally frequent throughout the British Is. but not in the Channel Is.; subsp. *densiflora* in scattered localities from Devon to Kent and northwards to S. Cumbria and Northumberland with outliers in Perth,

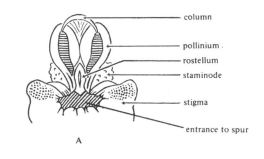

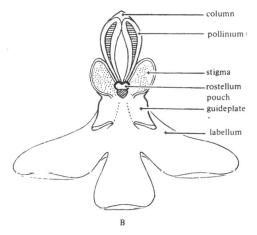

Fig. 63. Columns of A, *Gymnadenia conopsea*; B, *Anacamptis pyramidalis*.

Sutherland, Caithness and Orkney; similarly scattered through Ireland but not in the south. Most of Europe to northernmost Scandinavia; N. and W. Asia.

The above description is of the common and widespread subsp. **conopsea**. The much more local subsp. **densiflora** (Wahlenb.) G. Camus, Bergon & A. Camus is typically larger and with the lower lvs usually not less than 1.5 cm across and up to 3 cm. Its fls, more darkly coloured (deep pink to magenta) and in a longer and denser spike, are clove-scented and have the labellum much broader than long with its lateral lobes 'shouldered', the inner lateral per.-segs truncate-ended and the spur only a little longer than the ovary. It tends also to grow in marshes and fens, habitats wetter than those usual for subsp. *conopsea*. It has been treated by some authors as a distinct sp. but by others as a mere variety of *conopsea*, and its taxonomic status needs closer study.

G. odoratissima (L.) L. C. M. Richard, differing from *G. conopsea* in its narrower lvs, smaller fls, horizontally spreading (not downwardly curved) outer lateral per.-segs, narrower labellum somewhat longer than broad, and shorter spur about equalling the ovary, has been reported more than once but seems extinct, if it ever occurred. C. Europe.

15. PSEUDORCHIS Séguier

Herbs with root-tubers deeply palmate so that in our sp. the several unequal ±cylindrical segments may appear separate. Close to *Gymnadenia* but with the outer lateral *per.-segs connivent* with the outer median and the inner per.-segs *into a hood*, and the labellum ±connivent with the hood so that the *fl.* becomes almost *campanulate*. Spur short. Viscidia elliptical, less elongated and further apart than in *Gymnadenia*, but naked as in that genus.

Three spp.

1. P. albida (L.) Á. & D. Löve   Small White Orchid.

Satyrium albidum L.; *Habenaria albida* (L.) R.Br.; *Gymnadenia albida* (L.) L. C. M. Richard; *Leucorchis albida* (L.) E. H. F. Meyer.

Stem 10–30 cm, solid, stiffly erect, glabrous, with 2–3 brownish or whitish sheathing basal scales. Lower lvs c. 4, 2.5–8 × 1–1.7 cm, ±oblong-oblanceolate, the lowest blunt, all keeled, firm, glossy above, unspotted but with rows of translucent dashes between the veins; upper lvs 1–2, narrow, acute, bract-like. Spike 3–6 cm, narrowly cylindrical, dense, often slightly curved; bracts lanceolate, 3-veined, about equalling the ovary. Fls 2–2.5 mm, *half-drooping*, turned to one side, *greenish-white, faintly vanilla-scented*. Outer and inner lateral per.-segs connivent into a flattish short broad hood over the slightly longer 3-lobed labellum whose downwardly curved triangular central lobe exceeds the smaller tooth-like lateral lobes. From the front the fl. appears compressed-campanulate, its aperture being a horizontally elongated opening c. 3 × 1 mm. Spur not ½ as long as the ovary, thick, conical, blunt, downwardly curved, with free nectar. Fl. 6–7. Probably cross-pollinated by tiny insects. $2n = 42$.

The above description is of the native subsp. **albida**, widespread in continental Europe.

Native. Usually in rough but well-drained grassy meadows and pastures, especially in hilly districts; tolerant of both calcareous and non-calcareous soils and sometimes in vegetation with *Calluna* and other ericaceous spp. present. Rare and decreasing. In S. England now known only in Sussex but still in a few places in W. Wales and one in Derby, also in Cumbria and N. Yorks and a few scattered localities in S. Scotland but becoming more frequent further north, especially in N.W. Scotland and the Inner Hebrides, and reaching Caithness and Orkney. Much of W. and C. Europe from northernmost Norway southwards to the Pyrenees, S. Apennines and S. Bulgaria.

The S.E. European *P. frivaldii* (Hampe ex Griseb.) P. F. Hunt has been recorded from this country on a single occasion.

16. PLATANTHERA L. C. M. Richard

Herbs with entire tapering root-tubers, stems with 2(–3) broad unspotted lower lvs, and lax spikes of strongly scented whitish fls. Outer lateral per.-segs spreading, outer median and inner lateral pair connivent in an ovate ±erect hood. *Labellum narrowly strap-shaped, entire; spur* usually *long and slender*, with free nectar. Column

rather short; stigma transversely elongated, ±oblong, depressed; rostellum represented only by the laterally placed viscidia. Anthers adnate to the top of the column; pollinia 2, narrowed downwards into slender caudicles, each attached laterally, close to its base, either directly to one of the 2 *naked viscidia*, or indirectly through a short connecting stalk; pollen in packets of tetrads ±firmly tied by elastic threads.

Two hundred spp. in the north temperate and tropical zones.

Spike ±pyramidal, greenish; fl. 18–23 mm across; labellum 10–16 mm; pollinia 3–4 mm, divergent downwards; viscidia c. 4 mm apart, circular.   **1. chlorantha**
Spike ±cylindrical, whitish; fl. 11–18 mm across; labellum 6–10 mm; pollinia c. 2 mm, parallel; viscidia c. 1 mm apart, oval.   **2. bifolia**

1. P. chlorantha (Custer) Reichenb.

Greater Butterfly Orchid.

Orchis chlorantha Custer; *Habenaria chlorantha* (Custer) Bab.; non Sprengel.

Root tubers 2, ovoid, attenuate below. Stem 20–40 (–60) cm, glabrous, ±angled above, with 1–3 brown basal sheathing scales. Lower lvs usually 2, 5–15(–20) × 1–5(–7) cm, elliptical or elliptical-oblanceolate, blunt; upper lvs 1–5, grading into bracts; all unspotted. *Spike* 5–20 cm, ±pyramidal, lax, *greenish*; bracts ovate-lanceolate, acuminate, blunt, green, variable in length but usually about equalling the ovary. Fls 18–23 mm across, heavily fragrant, especially at night. Outer lateral per.-segs c. 10–11 × 5 mm, ovate, spreading; outer median per.-seg. broadly triangular-cordate, blunt; inner lateral per.-segs narrowly lanceolate, erect; all greenish-white. Labellum 10–16 × 2–2.5 mm, slightly tapering downwards, rounded at the tip, green below, greenish-white distally. Spur 19–28 × c. 1 mm, curved downwards and forwards, sometimes almost into a semicircle. *Pollinia* 3–4 mm in overall length, sloping forwards and outwards and so *diverging* from c. 2 mm apart above to c. 4 mm between the large *circular viscidia*. Fl. 5–7. Cross-pollinated mainly by noctuid moths to whose eyes the pollinia become attached. $2n = 42$.

Native. Woods and grassy slopes, especially but not exclusively on base-rich and calcareous soils, to 460 m in N. England; local. Throughout the British Is. except Orkney and Shetland. Europe; Caucasus; Siberia.

2. P. bifolia (L.) L. C. M. Richard

Lesser Butterfly Orchid.

Orchis bifolia L.; *O. montana* Schmidt; *Habenaria bifolia* (L.) R.Br.

Closely resembling *P. chlorantha* but somewhat smaller in all its parts. Stem 15–30(–45) cm, with 2–3 brown or whitish sheathing basal scales. Lower lvs usually 2, 3–9(–15) × 0.7–3(–4) cm, elliptical, short; upper lvs 1–5, smaller, grading into the bracts; all unspotted. *Spike* 2.5–20 cm, usually lax, ±cylindrical, *whitish*. Fls 11–18 mm across, rather less strongly night-scented than in *P. chlorantha*. Outer lateral per.-segs lanceolate.

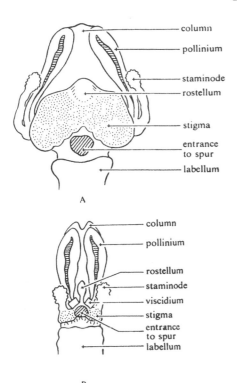

Fig. 64. Columns of A, *Platanthera chlorantha*; B, *P. bifolia*.

Labellum 6–10 mm. Spur 15–20 mm, slender, almost horizontal. *Pollinia* c. 2 mm in overall length, *vertical, parallel, their small oval viscidia* c. 1 mm apart. Fl. 5–7. Pollinated mainly by sphingid moths, to whose probosces the pollinia become attached. 2n = 42.

Distinguishable from *P. chlorantha* at a distance by the narrower, almost cylindrical spike of smaller, less distant and whiter fls. The fls are somewhat more sweetly scented, lacking the slight pungency of those of *P. chlorantha*.

Native. Heaths, moorlands and boggy or calcareous flushes, less commonly on grassy slopes or in open woodland on calcareous or circumneutral soil; widespread in S. England though nowhere abundant, still sparser in C. and E. England but becoming more frequent further west and north, especially in Cumbria and W. Scotland; Inner and Outer Hebrides, but not in Orkney or Shetland; frequent in much of Ireland but rarer in the east and north. Most of Europe to 70° N. in Norway; N. Asia; N. Africa.

P. chlorantha and *P. bifolia* are freely interfertile when artificially crossed but intermediates are quite infrequent in the wild even though the two may grow in the same area, overlap in flowering time and be visited by the same species of nocturnal moths. This arises in part from differences in peak flowering-times and because individuals of the pollinating moths tend to restrict visits to fls of similar fragrance ('odour-based constancy'), but especially from the different spatial relations between spur-entrance, viscidia and stigma-lobes in the two spp. and their F₁ hybrid, which makes out-crossing improbable.

17. Neotinea Reichenb. fil.

A small herb with entire root-tubers, usually spotted lvs and a short dense 1-sided spike of pale fls. Per.-segs all connivent. Labellum directed forwards, 3-lobed; spur very short. Column very short and small; stigmas 2, large, ±crescentic, joined below, borne on lateral wings of the column; *rostellum a broad flat plate between the stigma-lobes, its apex curving over the* 2 *otherwise naked viscidia*. Anther adnate to the top of the column; pollinia 2, club-shaped, each narrowed downwards into a short caudicle attached basally to 1 of the naked viscidia; pollen in tetrads bound by elastic threads.

One sp.

1. N. maculata (Desf.) Stearn   Dense-flowered Orchid

N. intacta (Link) Reichenb. fil.

Root-tubers 2, ovoid. Stem 10–30 cm, erect, glabrous, with brownish sheathing scales below. *Basal lvs* 2–3(–4), elliptical-oblong, blunt, mucronate, unspotted or more often *with small brownish spots in interrupted parallel lines*; upper lvs smaller, narrower, acute, grading into the bracts. Spike 2.5–8 cm, narrowly cylindrical, of numerous whitish or pink fls which do not open widely; bracts ovate-lanceolate, membranous, not exceeding the ovary. Outer per.-segs lanceolate-acute, coherent below, forming the much narrower inner segments a long almost closed helmet over the column; outer segments sometimes purple-spotted. Labellum small, equalling or somewhat exceeding the outer per.-segs, directed forwards or obliquely downwards, often with 2–3 purplish blotches at the furrowed base, 3-lobed with linear acute lateral lobes and a longer and broader central lobe, truncate or notched at the apex, sometimes with a tooth in the notch. Spur c. 2 mm, conical, blunt. Pollinia pale green. Capsule spindle-shaped, glabrous, twisted, ±erect. Fl. 4–6. Cross- or self-pollinated.

Native. Confined to an area in W. Ireland extending from the coast inland to Co. Offaly, and a single locality on the north coast of the Isle of Man. In Ireland mainly on rocky or grassy hillsides, in pastures or on grassy banks and roadsides in the limestone country near the coast from Co. Clare to Co. Mayo, including the Burren Hills, and reaching 300 m above sea-level; also on ±calcareous sand dunes in W. Ireland and in the Isle of Man, and rarely on somewhat peaty soils over acid rocks. Mediterranean region and Portugal eastwards to Greece; Madeira; Canary Is.; N. Africa; Cyprus; Asia Minor.

18. Ophrys L.

Herbs with entire ovoid to subglobose root-tubers, lfy stems and lax-fld spikes. *Per.-segs spreading*, the inner lateral segments usually smaller than the outer. *Labellum large, entire or* 3-*lobed, often convex, velvety, usually dark-coloured and conspicuously marked*; spur 0; nectaries 0, but 2 shining eye-like projections commonly stand one on each side of the base of the labellum. Column long, erect; stigma single, large, central, depressed; rostellum represented by 2 *separated pouches* above the stigma. Anther adnate to the top of the column; pollinia 2, narrowed downwards into long caudicles which are attached basally to *separate ±globose viscidia* enclosed in the distinct rostellar pouches; pollen in packets of tetrads ±firmly united by elastic threads. Ovary not twisted.

About 30 spp. in Europe, W. Asia and N. Africa.

A very distinct genus easily recognizable by the large spurless labellum, often resembling an insect, and the 2 separate rostellar pouches (Fig. 65B)

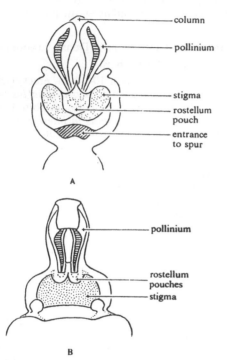

Fig. 65. Columns of *Orchia mascula* (A); *Ophrys fuciflora* (B).

1 Labellum distinctly 3-lobed, lateral lobes spreading, the long central lobe deeply emarginate or 2-lobed; forwardly-directed top of column very short and obtuse; outer per.-segs green, inner lateral dark violet, filiform.   **4. insectifera**
  Labellum entire or with 2 often indistinct basal lateral lobes; forwardly-directed top of column prolonged and narrowed into the acute or acuminate tip (from the side ±like a bird's head); inner lateral per.-segs variable in width but never filiform.   2
2 Labellum lacking terminal glabrous appendage, usually entire, with or without large basal bosses and with shining bluish H- or π-shaped marking; outer per.-segs green, inner lateral per.-segs at least ½ as long as outer, ±oblong, undulate. **3. sphegodes**
  Labellum with glabrous terminal appendage but with or without basal lobes and bosses; outer per.-segs pink, rarely white, inner lateral per.-segs less than ½ as long as outer.   3
3 Labellum strongly convex with 2 small hairy basal lobes and with the large central lobe ending in a

long narrow tooth-like appendage curved downwards and backwards so as to be ±invisible from above; labellum markings U- or W-shaped.
                                 1. apifera
  Labellum only slightly convex with basal lobes usually represented only by bosses, ending in a wide and often 3-toothed upcurving yellowish appendage; labellum markings variable.   **2. fuciflora**

1. O. apifera Hudson   Bee Orchid.

Root-tubers subglobose. Stem 15–45(–60) cm, glabrous. Lvs 3–8 × 0.5–2 cm, elliptical-oblong, subacute, diminishing rapidly up the stem and grading into the bracts; all unspotted. Spike 3–12 cm, of 2–5(–10) large, rather distant fls; bracts large and lf-like, commonly exceeding the fls. Outer per.-segs 10–15 mm, elliptical-oblong, rose-pink or whitish, greenish on the back; inner lateral per.-segs lanceolate, shorter than the outer, pinkish-green, ±downy on the inner side. *Labellum* 10–15 mm, *resembling a humble-bee*, strongly convex, semi-globose, 3-lobed, the lateral lobes small, hairy, ±rounded, with basal protuberances and a blunt forwardly and upwardly directed apex; central lobe much larger, curved downwards distally and terminating in 2 truncate lateral segments and a *long narrowly triangular central appendage curled up behind* so as to be invisible from above; the central lobe velvety, purplish- to dark brown, marked with glabrous yellow spots distally and a greenish-yellow horseshoe or H basally; lobes ±bordered with yellow. Fls 6–7. The pollinia fall forwards and downwards on to the stigma while remaining attached by their long slender caudicles, so that the fls are habitually, though not invariably, self-pollinated in this country. 2*n* = 36. Grt.

Var. *chlorantha* Godfery has the outer per.-segs white and the labellum greenish-yellow and very narrow. Var. *trollii* (Hegetschw.) Druce (*O. trollii* Hegetschw.) is a curious form in which the labellum is pale yellow with irregular brown markings, narrower than in the type, with the lateral lobes acute and ±reflexed and the terminal tooth long, acute and ±spreading.

Native. In short often open grassy turf of pastures, field-borders, banks and copses, etc., over chalk and limestone, especially in recently disturbed sites; also on nutrient-rich clays and ±stabilized sand dunes, and rarely on non-calcareous substrata as of débris from gravel-pits, etc. Local throughout England and Wales northwards to Cumberland and Durham; recorded from S. Scotland but probably now extinct; scattered through much of Ireland but decreasing. S., W. and C. Europe northwards to N. Ireland; N. Africa.

2. O. fuciflora (F. W. Schmidt) Moench
                        Late Spider Orchid.

O. arachnites (L.) Reichard

Root-tubers subglobose. Stem 10–35 cm, glabrous. Lvs 4–10 × 0.5–2.5 cm, elliptical-oblong, subacute. Spike lax, of 2–7 fls whose bracts exceed the ovaries. *Outer per.-segs* 10–12 mm, ovate-oblong, spreading, *rose-pink*; inner lateral per.-segs triangular, blunt, downy above, pinkish or with a green central stripe. *Labellum* 10–12 × 11–15 mm, *broader than long*, obovate-suborbicular or obscurely angled, *slightly convex*, 3-lobed; basal lobes broad but very short, usually reduced to bosses, sometimes 0; central lobe large, with a pair of small spreading or reflexed triangular teeth at its distal corners and terminating in a thick, wide and often 3-toothed, flat or upcurved *appendage*; the central lobe velvety, maroon to dark brown, marked with a bold symmetrical pattern in greenish-yellow lines. Fl. 6–7. Cross-pollinated by bees and other insects. 2*n* = 36. Grt.

Native. Only over chalk, mostly in short turf; now very rare and local. Kent. C. and S. Europe; Near East.

Very like *O. apifera* but distinguishable by the more obovate and less convex labellum whose terminal appendage is never reflexed as in *O. apifera*.

3. O. sphegodes Miller   Early Spider Orchid.

O. aranifera Hudson

Root-tubers ovoid to subglobose. Stem 10–35 cm, glabrous. Lvs 4–10 × 0.5–1.5 cm, elliptical-oblong, the lower spreading. Spike lax, of 2–8 fls; lower bracts exceeding their fls. *Outer per.-segs* 10–12 mm, ovate-oblong, *pale yellowish-green*; *inner per.-segs* somewhat shorter, ±narrowly oblong, blunt, greenish-yellow, usually *undulate*, glabrous, rough or downy. *Labellum* suborbicular to ovate-oblong, strongly convex, subentire or 3-lobed at the base, the lateral lobes small, with basal protuberances, the central lobe entire or emarginate, usually *with no terminal appendage*; central lobe *velvety*, *dull purplish-brown*, later turning yellowish, with various *glabrous bluish markings*, often H- or horseshoe-shaped. Fl. 4–6. Visited sparingly by bees and setting seed only if visited. 2*n* = 36. Grt.

The above description is of the only native subsp., subsp. **sphegodes**, widespread in continental Europe.

Native. In short open grassland over chalk or limestone, often on disturbed soil or quarry débris; local. Recently much decreased and now only in S. England from Cornwall and Dorset to Kent and northwards to Gloucester and Wilts. W., C. and S. Europe northwards to S. England and C. Germany and eastwards to Crimea; N. Africa.

4. O. insectifera L.   Fly Orchid.

O. muscifera Hudson

Root-tubers ovoid to subglobose. Stem 15–60 cm, ±glabrous. Lvs 4–12.5 cm, few, oblong to elliptical, subacute. Spike long, of 4–12 distant fls; the lower bracts much exceeding their fls. Outer per.-segs 6–9 mm, oblong yellowish-green; *inner lateral per.-segs filiform*, *purplish-brown*, velvety. *Labellum* c. 12 mm, obovate-oblong, not convex longitudinally, 3-*lobed near the middle*; lateral lobes without basal protuberances, narrowly oblong, spreading; middle lobe longest, obovate, ±*deeply bifid*; the whole purplish-brown, downy, with a *broad glabrous shining bluish transverse patch* or blotch just beyond the insertion of the lateral lobes.

Fl. 5–7. Rather sparingly visited by a small wasp and setting seed only if visited. 2*n* = 36. Grt.

Native. Woods, copses, field-borders, spoil-slopes, banks and grassy hillsides on chalk or limestone and in fens; not infrequent. Locally throughout lowland England from S. Devon, Somerset and Dorset to Kent and northwards to Westmorland and Durham; Glamorgan, Denbigh, Anglesey; C. Ireland. Europe northwards to c. 67° N. in Norway.

Hybrids between *O. apifera* and each of the other 3 spp. have been reported from this country, as also between *O. insectifera* and *O. sphegodes*. The record for *O. fuciflora* × *O. sphegodes* seems doubtful.

19. HIMANTOGLOSSUM Koch

Herbs with entire tubers, tall stout lfy stems, and long cylindrical spikes of large fls. All per.-segs except the labellum connivent into a hood. Labellum 3-lobed, the lateral lobes short, the *central lobe very long and narrow, spirally coiled in bud like a watch-spring*; spur very short. Column rather short, erect; stigma large, single; rostellum beak-like, projecting above the stigma. Anther adnate to the top of the column; pollinia 2, narrowed below into caudicles both of which are attached to *a single ±4-angled viscidium enclosed in a pouch* in the rostellum; pollen in packets of tetrads united by elastic threads.

Four spp. in C. Europe and the Mediterranean region.

1. H. hircinum (L.) Sprengel   Lizard Orchid.

Satyrium hircinum L.; *Orchis hircina* (L.) Crantz; *Loroglossum hircinum* (L.) L. C. M. Richard

Root-tubers ovoid or ±globose. Stem 20–40(–90) cm, stout, glabrous, faintly mottled with purple. Lower lvs 4–6, 6–15 × 3–5 cm, elliptical-oblong, ±blunt; upper narrower, acute, clasping the stem; all unspotted. Spike 10–25(–50) cm, rather lax; bracts linear-lanceolate, acute, 3–5-veined, membranous, the lowest equalling the fls, the remainder exceeding the ovary. Fls numerous, large, greenish, untidy-looking, smelling strongly of goats. Outer per.-segs cohering at the base, ovate, blunt, pale greenish often streaked and spotted with purple; inner per.-segs linear, adhering laterally to the outer. Labellum 3–5 cm × 2 mm whitish with purple spots and furry with papillae near the cuneate base, pale brownish-green and glabrous distally; lateral lobes 5–10 mm, narrow, acute, curly; central lobe narrowly strap-shaped, c. 2 mm wide, ±undulate and somewhat spirally coiled, truncate and notched or 2–4 toothed at the tip. Spur c. 4 mm, conical, blunt, slightly curved downwards. Fl. 5–7. Cross-pollinated by Hymenoptera. 2*n* = 24.

Native. Wood margins, by woodland paths, amongst bushes, on field borders and in grassland, chiefly over chalk or limestone; rare and local but now much more widespread than in the early years of the century, perhaps through climatic change. Chiefly in S. and E.

England from N. Devon, Somerset and Dorset to Kent and northwards to a line from Gloucester to E. Suffolk and in scattered localities further north to Cambridge, Norfolk, Lincoln and N.E. Yorks. S., S.C. and W. Europe northwards to C. England and the Netherlands and eastwards to the Balkan peninsula and the Crimea; N. Africa.

The above description is of the native subsp. **hircinum**, widespread in continental Europe except for the southeast.

20. ORCHIS L.

Herbs with *rounded unforked root-tubers* and sheathing lvs, the basal *lvs in a rosette*. Fls in a spike which is usually enclosed by thin spathe-like lvs during emergence. *Bracts membranous. All per.-segs except the labellum connivent into a helmet over the column, or only the outer lateral segs spreading or turned upwards.* Labellum usually 3-lobed, directed downwards, with descending or ascending spur. Column erect; stigma ±2-lobed, roofing the entrance to the spur; *rostellum overhanging the stigma as a single pouch enclosing 2 separate ±globose viscidia.* Anther adnate to the top of the column; *pollinia 2, each narrowed below to a caudicle and attached by a basal disk to one of the viscidia*; pollen in packets of tetrads united by elastic threads. Capsule erect.

About 35 spp. in Europe, temperate Asia, N. Africa, Canary Is.

1 All per.-segs except the labellum connivent to form a 'helmet' over the column; spur usually descending (horizontal or ascending in **5**).　　2
　Outer lateral per.-segs erect or spreading; spur ±ascending.　　6
2 Labellum shaped somewhat like a man, the central lobe much longer than the ±slender lateral lobes (arms) and forking distally into 2 branches (legs) often with a short tooth in the sinus between them; outer per.-segs usually ±coherent.　　3
　Labellum shallowly 3-lobed, not shaped like a man; outer per.-segs green-veined, not coherent.　　**5. morio**
3 Helmet of young fls dark reddish-brown (maroon); outer per.-segs free to the base; labellum white, c. 6 mm; spur about ¼ the length of the ovary; bracts long.　　**4. ustulata**
　Helmet not dark reddish-brown; outer per.-segs coherent below; labellum at least 10 mm; spur about ½ the length of the ovary; bracts minute.　　4
4 Helmet heavily blotched with dark reddish-purple, almost black in unopened fls; labellum usually tinged with pink or pale purple and with darker spots, its distal branches broadly rhombic.　　**1. purpurea**
　Helmet greyish, flushed, veined and spotted with pale purple or rose; distal branches of labellum linear or oblong, not broadly rhombic.　　5
5 Lobes of labellum all very narrow; helmet whitish, ±flushed or marked with rose.　　**3. simia**
　Distal branches of labellum oblong, shorter than and 3–4 times as wide as the basal lobes, red-purple

or rose; helmet ±flushed or marked with pale purple.　　**2. militaris**
6 Bracts 1-veined or the lowest sometimes 3-veined; spur equalling or exceeding the ovary; labellum shallowly 3-lobed, the central lobe longest; throughout British Is.　　**7. mascula**
　Bracts 3-veined; spur shorter than the ovary; labellum with the central lobe shorter or 0; only in Channel Is.　　**6. laxiflora**

Section 1. *Orchis.* Labellum shaped like a man, its central lobe longer than the lateral lobes and forked distally, commonly with a short tooth in the sinus; per.-segs connivent to form a helmet over the column, the outer usually ±coherent below; bracts usually short, membranous, 1-veined; hay-scented when drying.

1. O. purpurea Hudson　　Lady Orchid.

Tubers ovoid. Stem 20–40(–80) cm, stout. Lvs 7–15 × 3–5 cm, 3–5(–7), ovate-oblong, blunt, the uppermost lanceolate; all unspotted, glabrous and very shining above, paler beneath. Spike 5–10(–15) cm, dense, almost black in bud; bracts very small. *Helmet* 10–12 mm, ovate, *dark reddish-purple*, becoming paler. Outer per.-segs ovate, hooded, coherent below, pale greenish with purple mottling within; inner lateral per.-segs linear, acute. *Labellum* 10–15 mm, hanging, whitish flushed with violet or rose, especially at the edges, and *freely dotted with tufts of reddish-purple papillae*; basal lobes linear, curved, dotted with red-purple; *central lobe broad*, widening downwards and *forking into* 2 *broadly rhombic segments* 4–5 times as broad as the lateral lobes, crenulate at their rounded or truncate apices and usually with a small tooth between them. Spur nearly half as long as the ovary, cylindrical, curved downwards and forwards, enlarged and truncate or notched at the tip. Fl. 5. 2n = 40, 42. Grt.

Native. Copses and open woods, rarely in open grassland, on chalk or limestone. Rare and now apparently confined to Kent and Sussex but may persist in Surrey. Europe from Denmark southwards to N. Spain, N. Greece and Crimea; Caucasus; Asia Minor.

2. O. militaris L.　　Soldier Orchid.

Tubers ovoid to subglobose. Stem 20–45 cm. Lvs 3–10 × 0.5–2.5 cm, elliptical-oblong, the uppermost narrowly lanceolate, blunt, concave or folded, unspotted. Spike 3–8 cm, dense, blunt; bracts very small. *Helmet* c. 12 mm, ovate-lanceolate, *ash-grey ±flushed with rose or violet*. Outer per.-segs acuminate, coherent below, veined and spotted with red-violet within; inner linear, acute. Labellum 12–15 mm, whitish, dotted with tufts of violet papillae and flushed with violet towards the edges; basal lobes linear, rose or violet; *central lobe narrowly oblong, forking into* 2 *widely divergent oblong segments* 2–4 times as broad as the basal lobes, entire or denticulate at their apices, with a short narrow acute tooth between them. Spur about half as long as the ovary, cylindrical, curved downwards and forwards, blunt. Fl. 5–6. 2n = 42. Grt.

Native. Open woods or wood-margins, bushy places, dry banks and rough pastures, always over chalk. Very rare; now known only in Bucks and in W. Suffolk but formerly in several localities in S.E. England and the S. Midlands. C. Europe from S. Sweden, Gotland and Estonia southwards to C. Spain, C. Italy, Bulgaria and S. Russia; C. Asia and Asia Minor.

3. O. simia Lam.   Monkey Orchid.

O. tephrosanthos Vill.

Tubers ovoid or subglobose. Stem 15–30 cm, more slender than in *O. militaris*, which it closely resembles. Lvs as in *O. militaris* but somewhat smaller. Helmet whitish faintly streaked with rose, or pale violet, with minute violet dots and 2–3 raised white veins. Labellum 14–16 mm, white, rose, or deep crimson, paler below; *basal lobes* linear, *crimson*; *central lobe* narrow, up to three times the width of the basal lobes, dotted with small tufts of violet papillae, *forking into* 2 *linear* crimson *segments*, as narrow as the basal lobes, rounded or truncate at their apices, with a conspicuous tooth between them. Spur as in *O. militaris*. Fl. 5–6. $2n = 42$. Grt.

Three forms have been described from Britain. That from Kent has a whitish hood and a very narrow deep crimson central lobe of the labellum; one from Oxford differs in its broader, white red-dotted central labellum-lobe; while the other (var. *macra* Lindley) has a violet helmet, pink crimson-dotted central labellum lobe, and purplish (not crimson) basal lobes and distal segments.

Native. Grassy hills, bushy places and field borders on chalk, very rare. Apparently now confined to Oxford (Goring Gap) and Kent, with a single plant in E. Yorks. C. and S. Europe from Belgium to Spain, Italy, N. Balkan peninsula and Crimea; Caucasus; N. Africa; Near East.

4. O. ustulata L.   Dark-winged Orchid, Burnt Orchid.

Tubers ovoid. Stem 8–20(–30) cm, stout. Lvs 2.5–10 × 0.5–1 cm elliptical-oblong to lanceolate, acute. *Spike* 2–5 cm, dense, small-fld, at first conical, *dark maroon* when the fls are unopened, *becoming paler and then whitish*; bracts at least half as long as the ovary. Helmet c. 5 mm, roundish, dark brownish-purple (maroon), ultimately white. Outer per.-segs free to the base, ovate, subacute; inner narrower. Labellum c. 5–6 mm, white dotted with red-purple; basal lobes narrowly oblong, entire or notched at the apex; central lobe oblong, forking distally into 2 oblong segments about equalling the basal lobes, usually crenulate at their apices and commonly with a short tooth between them. Spur about ¼ as long as the ovary, cylindrical, curved downwards and forwards, blunt at the tip. Fl. 5–6. $2n = 42$. Grt.

Native. In short grassland over chalk and limestone; local and decreasing. Widely scattered through England from Somerset to Kent and northwards to Cumbria and Durham; not in Ireland; reaches 250 m in Bucks. C. Europe from Denmark, Gotland and Estonia southwards to N. Spain, S. Italy, N. Balkan peninsula and Ukraine; Caucasus; Siberia.

Section 2. *Morio* Dumort

Labellum 3-lobed, not shaped like a man, the central lobe truncate, emarginate or 2-lobed; outer per.-segs not coherent; bracts about equalling the ovary, membranous, 1-veined; not hay-scented on drying.

5. O. morio L.   Green-winged Orchid.

Tubers ±globose. Stem 10–40 cm. Lvs 3–9 × 0.5–1.5 cm, elliptical-oblong to lanceolate, the lower ones spreading or recurved, the uppermost ±appressed, blunt or acute, unspotted. Spike 2.5–8 cm, rather lax; bracts coloured. Per.-segs connivent in a *helmet, purple with conspicuous green veins*, rarely flesh-coloured or white. Outer per.-segs ovate, blunt; inner narrow, blunt. Labellum hardly exceeding the helmet, broader than long, 3-lobed; the lateral lobes reddish-purple, ±folded back, broadly cuneate, crenulate; central lobe pale reddish-purple dotted with deep red-purple, notched and crenulate, about equalling the lateral lobes. Spur almost equalling the ovary, cylindrical, horizontal or ascending, almost straight, enlarged and blunt at the apex. Fl. 5–6. $2n = 36$. Grt.

Native. In short grassland of old meadows and pastures, banks by roads and railways, etc., usually on calcareous or circumneutral soil; locally abundant through much of lowland England and Wales but becoming rare in the north and with only a single record from Scotland; frequent in a broad band across the central plain of Ireland; Channel Is.; reaches 300 m in Ireland. Europe southwards from S. Norway, Gotland, Estonia and C. Russia; Caucasus; Asia Minor; Siberia.

The above description is of the native subsp. **morio,** which covers most of the range of the species.

6. O. laxiflora Lam. subsp. **laxiflora**   Jersey Orchid.

Tubers ±globose. Stem 20–50(–90) cm, robust. Lvs 7–18 × 0.8–2 cm, lanceolate or linear-lanceolate, acute, keeled and ±folded, the uppermost bract-like. *Spike* 7–22 cm, *lax, cylindrical, dark purple*; bracts somewhat exceeding the ovary, ±membranous, coloured. Outer per.-segs ovate, blunt, dark crimson-purple, the lateral pair held erect and back to back, the upper concave, curving upwards; inner shorter, narrower, blunt, forming a hood. Labellum 10–15 mm, about as broad as long, dark crimson-purple, whitish at the base, with 2 large, rounded, crenate or toothed, ±folded-back lateral lobes; central lobe usually shorter than laterals and truncate, or 0. Spur up to ⅔ as long as the ovary, cylindrical, straight or slightly curving upwards, horizontal or ascending, enlarged and truncate or notched at the tip. Fl. 5–6. $2n = 42$. Grt.

Native. Wet meadows. Frequent in the Channel Is. Europe from Belgium southwards; N. Africa; W. Asia.

7. O. mascula (L.) L. subsp. **mascula**
                              Early Purple Orchid, Blue Butcher.

Tubers ovoid to subglobose. Stem 15–60 cm, ±stout. Lvs 5–20 × 0.5–3 cm, broadly to narrowly oblanceolate-oblong, bluntish, usually with rounded black-purple spots. Spike 4–15 cm, rather lax; bracts about equalling the ovary, coloured, membranous, 1-veined. Outer per.-segs ovate, ±acute, the lateral pair at first spreading, later folded back, the upper connivent with the ±ovate subacute inner per.-segs; all purple to pale pink. Labellum 8–12 mm, about as broad as long, purplish-crimson, paler at the base and dotted with darker purple, 3-lobed; lateral lobes crenate, ±folded back; central lobe slightly larger, crenate with a ±distinct central notch and sometimes with a lateral notch on each side. *Spur usually as long as the ovary, stout*, cylindrical, straight or curved upwards, *horizontal or ascending*, blunt or truncate. Fl. 4–6. 2*n* = 42. Grt.

Native. Woods, copses and open grassland, chiefly on nutrient-rich soils; common. Throughout the British Is., reaching 884 m in Scotland. Mainly W. and W.C. Europe and reaching 70° N. in Norway; N. Africa; N. and W. Asia.

21. ACERAS R.BR.

A small *Orchis*-like plant with entire root-tubers, lfy stems, and long narrow spikes of greenish fls. All per.-segs except the labellum connivent to form a hood over the column. *Labellum shaped like a man*, with long slender lateral lobes (the arms) near its base, and a longer narrow central lobe which forks distally into 2 slender segments (legs) rather shorter than the lateral lobes; *spur 0*, nectar being secreted from 2 tiny depressions at the base of the labellum. Column very short; stigma transversely elongated, forming the roofs and walls of a cavity at the base of the column; rostellum inconspicuous. Anther adnate to the top of the column; pollinia 2, narrowed downwards into caudicles whose bases are attached to the 2 *contiguous ±globose viscidia enclosed in a common pouch*; the viscidia are often coherent, and the pollinia are usually removed together, but occasionally they may be withdrawn separately; pollen in packets of tetrads united by elastic threads.

One sp., in Europe, W. Asia and N. Africa.

Distinguished only by the absence of a spur from *Orchis* section *Orchis* (*O. simia, militaris, purpurea*, etc.), sharing with them the shape of the labellum and the development of a hay-scent (coumarin) on drying, and hybridizing with them.

1. A. anthropophorum (L.) Aiton fil.   Man Orchid.

Root-tubers ovoid or ±globose. Stem 20–40(–60) cm, glabrous, slightly ridged above, with sheathing scales at the base. Lower lvs several, crowded, 6–12 × 1.5–2.5 cm, oblong to oblong-lanceolate, ±acute, keeled, glossy on both sides; upper lvs smaller, erect, clasping the stem, grading into bracts; all unspotted. Spike up to half the length of the stem, narrowly cylindrical, many-fld, becoming lax; bracts membranous, shorter than the ovary. Fls greenish-yellow, often edged or tinged with reddish-brown. Outer per.-segs forming a semi-globose hood c. 6–7 mm, ovate-lanceolate, blunt;

inner per.-segs somewhat shorter and much narrower. Labellum 12–15 mm, hanging almost vertically, greenish-yellow often tinged with maroon, rarely pure yellow; lateral lobes and distal segments of central lobe linear; sometimes with a small tooth in the sinus between the distal segments. Fl. 6–7. Cross-pollinated by small insects. 2*n* = 42. G.

Native. Grassy slopes, field borders, etc., and rarely in scrub and open woodland, always on chalk or limestone; also in old chalk-pits and limestone quarries. Local and decreasing from Hants and Oxford to Kent and northwards in scattered localities to Northampton, Lincoln and Derby. Europe northwards to C. England and the Netherlands; Cyprus; N. Africa.

22. DACTYLORHIZA Necker ex Nevski

Like *Orchis* but, in all native members, with 2–3 *root-tubers usually palmately and deeply 2–5-lobed, the lobes narrowed and prolonged distally*; basal lvs not in a definite rosette at time of flowering; emerging spike not enclosed in spathe-like lvs; *bracts always herbaceous*; *outer lateral per.-segs ±erect or spreading, not contributing to a helmet over the column*; *spur always ±descending*, usually less than 10 mm.

Thirty or more spp. in Europe, temperate Asia, N. Africa and Canary Is.

1 Stem usually hollow for much of its length, with wide or narrow cavity; upper non-sheathing bract-like lvs, transitional to bracts, 0–1(–3); lower bracts usually more than 3 mm wide; outer lateral per.-segs ±erect to near-horizontal; spur usually over 2 mm diam.   *2*

Stem solid for much of its length; upper non-sheathing bract-like lvs, transitional to bracts, (1–)2–6 or more; lower bracts at most 3 mm wide; outer lateral per.-segs patent or drooping; spur slender, up to 2 mm diam.: **maculata** group   *3*

2 Stem-cavity wide, usually more than ½ total diam.; lvs erect, broadest close to base (but at c. ⅓ up from base in subsp. *cruenta*), usually distinctly and narrowly hooded (cucullate) at tip, often yellowish-green, usually unspotted (but in subsp. *cruenta* spotted above and beneath); sides of labellum often reflexed.   **1. incarnata**

Stem-cavity medium or small, often less than ½ total diam.; lvs often spreading, broadest well above base, not or broadly hooded at tip, usually dark green, unspotted or variously marked; labellum usually with sides not or little reflexed.   **4. majalis**

3 Green lvs (excl. bracts) 4–8 or more, all ±narrowly oblong and ±acute, usually with small solid circular spots, sometimes unspotted; labellum with central lobe much smaller and usually shorter than side-lobes; spur 1 mm or less in diam.   **2. maculata**

Green lvs (excl. bracts) 7–12 or more, basal lf widest and obovate-oblong with broad rounded tip, next 2–3 longest, the rest progressively shorter, narrower and more acute, all usually with solid and ±transversely elongated spots, sometimes unspotted; labellum with 3 subequal lobes, the central usually longer than laterals; spur 1–2 mm diam.   **3. fuchsii**

1. D. incarnata (L.) Soó
Early Marsh Orchid, Meadow Orchid.

Orchis incarnata L. sec. Vermeulen; *O. latifolia* L. sec. Pugsley; *O. strictifolia* Opiz

Stem (5–)15–60(–90) cm, *with central cavity* usually wider than half total diam. *Lvs*, excl. basal scales, (3–)4–7(–10), 10–20(–30) × 1–2(–3) cm, *erect to erect-spreading, narrowed gradually from near the wide base, usually distinctly but narrowly hooded* (cucullate) *at the tip, unspotted* or less commonly with spots on upper surface (but note that subsp. *cruenta* has lvs often arcuate-recurved above, narrowed from c. ⅓ up from base and commonly with spots on both surfaces); lvs typically *yellowish-green* and with loose-fitting sheaths. *Lvs transitional between true lvs and bracts commonly 0–1. Fl.-spike* cylindrical, dense, many-fld, *commonly flesh-coloured* but ranging from very pale yellow to brick-red or magenta-purple; lowest bracts more than 3 mm wide and exceeding fls. Outer lateral per.-segs erect at base then somewhat spreading, inner connivent with outer median in a helmet over column; *labellum* 5–7.5(–9) mm, ovate-rhombic to ±orbicular, entire to 3-lobed, *with sides often strongly reflexed* so that it appears quite narrow from the front, variously marked with spots, lines and loops; *spur* 6–9 mm, cylindrical-conical, *descending*. Fl. 5–8. Visited especially by humble-bees. $2n = 40^*$. Grt.

Very variable. J. Heslop-Harrison (*Ber. Geobot. Forsch. Rübel Zurich*, 1954, 55–82; *Proc. Linn. Soc., London*, 1956, **166**, 51–82), largely following Pugsley (*Bot. J. Linn. Soc., London*, 1935, **49**, 553–92), has recognized the following native subspp. which are on the whole distinct, though mixed populations and intergrading forms occur not infrequently:

1 Lvs commonly spotted both above and beneath, spots often elongated parallel to lf-margin; lvs widest at c. ⅓ up from base; bracts, upper stem and ovaries with dark flecks; fls palish lilac to deep magenta.
 subsp. **cruenta** (O. F. Mueller) P. D. Sell
 Lvs usually unspotted, rarely with spots on upper surface only; fls very pale yellow, flesh-pink, lilac, brick-red or purple. 2
2 Fls cream or pale yellowish, lacking red or purple pigment; robust plants up to 50 cm or more, the large fls with deeply 3-lobed labellum c. 7 × 9 mm, the lateral lobes notched.
 subsp. **ochroleuca** (Boll) P. F. Hunt & Summer-
 hayes
 Fls with some red or purple pigment, rarely cream or very pale yellow and then plants not conspicuously robust and labellum not deeply lobed with notched lateral lobes. 3
3 Fls fresh-pink. 4
 Fls purple, magenta or ±brick-red. 5
4 Plants usually not exceeding 40 cm and with only 4–5 lvs; labellum less than 8 mm wide, usually marked with lines. subsp. **incarnata**
 Plants robust, often much exceeding 40 cm and with 6 or more lvs; labellum commonly more than 9 mm

wide and marked with dots.
 subsp. **gemmana** (Pugsley) Heslop-Harrison fil.
5 Fls brick-, ruby- or crimson-red; plants commonly only 5–20 cm and with rather broad lvs; mainly in dune-slacks and -hollows.
 subsp. **coccinea** (Pugsley) Soó
 Fls purple or magenta. 6
6 Plants robust, often much exceeding 40 cm and with 6 or more lvs; labellum commonly more than 9 mm wide and marked with dots.
 subsp. **gemmana** (Pugsley) Heslop-Harrison fil.
 Plants usually not exceeding 40 cm and with 4–5 lvs; labellum 6–8.5 mm wide, variously marked, usually with lines. subsp. **pulchella** (Druce) Soó

Four of the above subspp. are recognized as such in *Flora Europaea*, **5** (1980) and subsp. *ochroleuca* is mentioned but with the statement that 'Variants with yellowish-white flowers . . . occur sporadically throughout the range of the species', and it is omitted from the key to subspp. and the accompanying descriptions. There is no mention of subsp. *gemmana*.

Native. Members of the sp. are found in wet meadows and marshes, on fen-peat ranging from highly calcareous to mildly acid and on soligenous bog-peat of pH down to c. 5.0, also in dune-slacks, the subspp., as far as can be judged from the available information, differing but overlapping in both ecological range and countrywide distribution. Subsp. *incarnata*, with fls flesh- or salmon-pink of paler or deeper shades, is characteristic of ±waterlogged soils of high mineral content (i.e. of marshes rather than fens) and of pH c.6.0–7.5 or higher. It is the most frequent subsp. in S. and S.E. England but falls off in abundance towards the north and west. It certainly reaches Argyll on the Scottish mainland and Rhum and Skye in the Inner Hebrides and may well extend much further north. Populations with fls of broadly similar but much deeper colour, variously described as pinkish brick-red, ruby-red, crimson-red, deep indian-red or maroon, are referred to subsp. *coccinea* and favour moist grassland in coastal sand dunes, especially in hollows or 'slacks', though also found very locally inland, as near the shores of some loughs in C. Ireland. Apart from the difference in fl.-colour the plants also tend to be smaller with relatively broader lvs than in *incarnata*, often somewhat recurved, but there is a wide range of variation and some intergrading with *incarnata* and *pulchella*. It is found locally on sand dunes of the west coast of Great Britain from Devon to Sutherland and also in the Inner and Outer Hebrides and in Ireland. Subsp. *pulchella*, with reddish-purple or magenta fls but otherwise much like *incarnata* and often accompanying it in S. Britain, largely replaces it in the north and west. Throughout its range its main habitats are fen or marsh with *Schoenus nigricans* and pH of substratum 6.2–7.5, poor fen with neither *Schoenus* nor *Sphagnum* spp. and pH 5.2–6.8, and soligenous bog with a *Sphagnum* carpet and pH c. 5.0–6.0, and it has been recorded from suitable localities throughout Great Britain, the Inner Hebrides and Ireland. It is variable in stature, in shape, dimensions and marking of the labellum, etc., but no named local variants seem sufficiently distinct to merit subspecific rank; some populations include intergrades with other subspp. Subsp. *cruenta*, formerly treated as a separate sp., has the erect or slightly recurved, hooded and usually yellowish-green lvs, the ±erect outer lateral per.-segs and strongly reflexed labellum and the diploid chromosome number ($2n = 40$) of *D. incarnata* sens. lat. It is exceptional in having a majority of plants with lvs spotted or blotched both above and, less heavily, beneath and also with dark-flecked bracts. The small to medium-sized fls are usually varying shades of lilac-purple. It is so far known only in a few highly calcareous fens in Counties Mayo, Galway and Clare in W. Ireland and in a small area in W. Ross in N.W. Scotland, where it reaches 450 m. Subsp. *ochroleuca* has robust plants with large creamy or pale yellowish fls lacking any red or purple pigment and with large deeply 3-lobed labellum, c. 7 × 9 mm, with notched lateral lobes: albino forms of other subspp. are less robust and the labellum of their fls lacks the special features of *ochroleuca*, which is known only from reed-swamp on calcareous fen-peat in a few localities in East Anglia and a single locality in Wales. Finally, subsp. *gemmana* is typically even more robust, with plants reaching 80 cm or more and lvs up to 30 cm and having long spikes of large bracts and fls, the fls magenta or flesh-pink with labellum averaging 8 × 10 mm and the spurs 8.5 mm. It is known from a few areas of riverside swamp and sedge-meadow on peat of pH 6.4–7.0 in S.E. England from N. Norfolk to S. Hants and from single localities in C. Wicklow and W. Galway.

D. incarnata sens. lat. is native in most of Europe northwards to c. 70° N. in Norway but is rare in the Mediterranean region, and subsp. *incarnata* occurs throughout its range. The distribution of our remaining subspp. is still unclear and one or two may be confined to the British Is., though closely comparable variants have been described from parts of continental Europe. Subsp. *cruenta* is certainly in parts of N. and C. Europe, including the Alps; subsp. *ochroleuca* seems widely scattered through Scandinavia to 62° N., the Baltic states, Germany and the Alps and perhaps further eastwards also, and subsp. *gemmana* may perhaps occur also in the Netherlands and Germany.

(2–3.) D. maculata group   Spotted Orchids.

Stem usually *solid*, 15–60 cm, ±erect. *Lvs obliquely erect* or somewhat recurved, *usually spotted; transitional lvs 1–6 or more*, erect and stem-appressed. *Infl. conical* at least until all fls have opened; bracts mostly shorter than fls, the lower less than 3 mm wide. Fl.-colour ranging from white to magenta-pink or deep lilac. Fls with *outer lateral per.-segs narrow*, ±spreading or drooping, *outer median and inner lateral convergent into a helmet* over the column; *labellum 7–11 × 9–13 mm*, *3-lobed; spur ±slender*, cylindrical to conical, straight, *shorter than to about equalling labellum*. Fl. 5–8. Visited by a variety

of insects seeking the sugary fluid in the spur-wall. $2n = 40, 80$. Grt.

The two native spp. in the group are distinguishable with little difficulty but each shows much variability, both between and within natural populations, in plant dimensions, number, shape and spotting of lvs, fl.-colour, shape, depth of lobing and marking of labellum, etc. This has led to the recognition of several infraspecific taxa largely based on geographical distribution but overlapping in morphological features because of the high intrapopulation variability. Only those deemed to merit subspecific rank are included in the following account.

2. D. maculata (L.) Soó   Heath Spotted Orchid.

Orchis maculata L.

The common and widespread British plant differs appreciably from what is regarded as the Linnean type and has been placed in a different subsp.:

Subsp. **ericetorum** (E. F. Linton) P. F. Hunt & Summerhayes

Stem 15–50 cm, slender. Lvs, excl. basal scales, 3–8 of which the uppermost 1–3(–4) are transitional to bracts; *true foliage lvs* commonly 2–4, the lowest shortest and, like the succeeding lvs, *narrowly lanceolate to linear and ±acute, usually lightly marked with small ±circular spots, but these faint or 0 in many plants*; heavy marking, and transverse elongation of spots as in *D. fuchsii*, both very rare. *Fl.-spike* rather *few-fld and tending to remain pyramidal. Fls variable in colour* from white to deep lilac-pink, *but predominantly pale*; outer lateral per.-segs long and narrow, often drooping; *labellum broader than long*, averaging c. 7 × 10 mm, usually *shallowly lobed, lateral lobes broadly rounded* and entire or crenulate, *central lobe much smaller, triangular and usually shorter than laterals*; labellum marked with irregular double loops often broken into bars or dots; *spur 3–7 mm, shorter than labellum, filiform* with diam. not exceeding 1 mm, hardly tapering. Fl. 5–7(–8), somewhat earlier than *D. fuchsii*. $2n = 80*$. Grt.

Native. The spotted orchid of peaty substrata over a wide range of moisture content from quite dry heaths to damp moorland, but not found on the deep peat of raised or blanket bogs; often in open oak, birch or pine woodland. Throughout the British Is. northwards to Shetland but commonest in the north and west where suitable habitats are more widespread; reaches 915 m in Scotland. Probably native also in N. France, Belgium, the Netherlands and Germany and at high altitudes in C. Sweden, but subsp. *maculata*, with broader and more heavily spotted lvs and purplish fls with longer and stouter spurs about equalling the labellum, is the common variant in S. Sweden and over much of the wide European range of the sp. The relationship with subsp. *elodes* (Griseb.) Soó, based on plants from the Netherlands, is still uncertain. Subsp. *maculata* is tetraploid ($2n = 80$) like subsp. *ericetorum*, but subsp. *elodes* is reported to have both diploid and tetraploid representatives.

Subsp. **rhoumensis** (H.-Harrison) Soó closely resembles subsp. *ericetorum* in vegetative features, its foliage lvs being narrow, keeled and folded and mostly lightly spotted. The infl. has 10–15 pale purple fls rather more deeply coloured than in typical *ericetorum*; labellum about as broad as long with the small central lobe often shorter than the laterals; spur slender, pointed. Fl. 7–8, 4–6 weeks later than *ericetorum*. $2n = 40^*$, and therefore, unexpectedly, diploid, for which reason it was initially placed as a subsp. under *D. fuchsii*.

Native. Known only from Rhum (Inner Hebrides), where it does not grow on peat but on mineral soil influenced by blown sand and of pH 4.9–5.5, though it is absent from the main areas of machair.

3. D. fuchsii (Druce) Soó   Common Spotted Orchid.

D. maculata subsp. *fuchsii* (Druce) Hyl.

Stem (5–)15–50(–70) cm. Lvs keeled but not or slightly folded; *lowest lf widest, broadly elliptical to obovate-oblong, blunt*, usually shorter than the next 2–3 lvs which, like the *remainder*, are *oblong-lanceolate, sub-acute*, and become *successively narrower up the stem; all marked with ±transversely elongated dark spots or blotches, less commonly unmarked. Fl.-spike* initially conical but *becoming ±cylindrical when all fls have opened; fl.-colour ranging from white to bright reddish-purple. Labellum* distinctly and *often deeply and narrowly divided into 3 subequal lobes, the middle lobe triangular and usually longer than the* laterals; usually clearly marked with a symmetrical pattern of ±continuous reddish lines on a paler pink or whitish background. Spur 5.5–10 mm, narrow (1–2 mm diam.), cylindrical-conical, blunt. Fl. 6–8. $2n = 40^*$. Grt.

Native. British representatives are commonly placed in one or other of the following 3 subspp. which differ in geographical distribution and are most obviously distinct in modal fl.-colour but also in range and means of stature, heaviness of lf-spotting, dimensions of labellum and spur, and time of flowering. There is, however, much overlapping variation in these features both within and between populations, so that individual plants may not be assignable with certainty to a particular subsp.

1 Most plants with small unmarked white fls and unspotted slightly glaucous lvs. On calcareous soils and limestone pavement in W. Ireland, Isle of Man and very locally in W. Scotland.
       subsp. **okellyi** (Druce) Soó
   Fls of most plants palest pink to bright reddish-purple, some occasionally ±white; a majority of plants with spotted lvs.   2
2 Plant commonly 15–30 cm; fls ranging from pale rose-pink to deepish red-purple, often a ±pale magenta; labellum 10–15 mm wide; spur 7.0–8.5 × 1.65–1.85 mm. Mainly in Inner and Outer Hebrides but very locally in Cornwall, W. Ireland, N.W. Scotland and Orkney.
       subsp. **hebridensis** (Wilmott) Soó
   Plant 8–60 cm or more; fls ranging from white to deep lilac-pink; labellum 8–12 mm wide; spur 5.5–6.5 × 1.3–1.6 mm. Widespread through the British Is. but locally replaced by subsp. *hebridensis* in the extreme west.   subsp. **fuchsii**

Subsp. **fuchsii** varies in height from only c. 10 cm to 60 cm or more and in total number of green lvs (excl. bracts) from 5 to 10 or more, the upper small and narrow, transitional to bracts. When growing in favourable habitats it is a taller and more lfy plant than either of the western subspp., but in several respects it is intermediate. Its fls are on average larger than in *okellyi* but smaller than in *hebridensis*; labellum 8–12 mm wide, against 5–9(–11) mm in *okellyi* and 10–15 mm in *hebridensis*. Modal fl.-colour, too, is typically between the white or near-white of *okellyi* and the palish magenta frequent in *hebridensis*, and modal lf-marking between the usual absence of spots in *okellyi* and the quite heavy spotting in *hebridensis*, but all features show a considerable overlap in variation-ranges for the different subspp. The lateral lobes of the labellum are straight-sided and ±rhombic, and the triangular central lobe equals or exceeds them. In S. and C. England plants of woods and scrub are commonly tall and very lfy and the central lobe of the labellum is distinctly longer than the laterals, whereas those of open chalk or limestone grassland are smaller and have a shorter fl.-spike with the central lobe of the labellum about equalling or only slightly longer than the laterals. Subsp. *fuchsii* is frequent in grassland and open woodland or scrub on ±moist calcareous or circumneutral soils through most of Great Britain and Ireland, but where such soils are uncommon, as in Cornwall and Devon, parts of W. Wales and much of C. and N. Scotland, it is much more local. Its countrywide distribution is thus largely complementary to that of *D. maculata* subsp. *ericetorum*, which favours heath and moorland. It reaches Orkney but is absent from most of the Western Isles where it is replaced by subsp. *hebridensis*, and is absent or infrequent in those limestone areas of W. Ireland characterized by subsp. *okellyi*.

Subsp. **hebridensis** (Wilmott) Soó comprises populations growing in the Outer and Inner Hebrides and a few localities in Sutherland and Cornwall and also in scattered coastal areas of W. Ireland in Counties Kerry, Galway and Donegal. Its height, commonly 15–30 cm, is less than in *fuchsii* but it is usually stouter and often has arcuate-recurved lvs. Lf-spotting tends to be heavier and the rather larger fls are more deeply coloured than in *fuchsii*, more markedly so in the Outer Hebrides and in Cornwall (*Orchis maculata* var. *cornubiensis* Pugsley seems to fall within the variation-range of *hebridensis*) than in the Inner Hebrides and W. Ireland. The more extreme populations grow in grasslands of stabilized sand dunes, of uncultivated 'machair' within the influence of blown calcareous sea-sand, and cliff-tops, with the soil pH at tuber depth commonly 5.5–6.0.

Subsp. **okellyi** (Druce) Soó consists similarly of scattered populations in the extreme west of the British Is., being first based on material collected on limestone pavement of the Burren Hills, Co. Clare, and subsequently found in a few other areas in W. Ireland, in the Isle of Man and in Argyll and Sutherland on the western seaboard of Scotland. Commonly but not invariably on shallow calcareous soils of limestone pavement, where it is typically small and slender, its mean height 15–20 cm, but sometimes 7 cm or less. Lvs few, narrow, the lowest usually longer and narrower than in subsp. *fuchsii* but like it in its broadly rounded tip; lvs unmarked or with very faint spots. Fls small, often fragrant, ±white or cream and unmarked or with very faint pinkish markings on the labellum (rarely more heavily marked). Labellum 5–9(–11) mm wide with 3 ±equally long lobes, the laterals rounded with strongly curving outer margins, the central triangular.

Plants of subsp. *fuchsii* which lack anthocyanin pigment are sometimes wrongly assigned to *okellyi* but can be distinguished by the differences in lf-width, fl.-size and shape of lateral lobes of labellum.

4. D. majalis (Reichenb.) P. F. Hunt & Summerhayes, s.l.　　　　　　　　　　　　　　Marsh Orchids.

Orchis latifolia L., p.p.; *O. majalis* Reichenb.

Root-tubers deeply 2–5-fid. Stem 5–60 cm or more, usually with central cavity of medium to small width, often less than $\frac{1}{2}$ total diam. Green *lvs* (excl. bracts) 4–8, *broadest below the middle but well above the closely sheathing base* and *not or slightly hooded at the subacute tip*, erect or somewhat arcuate; non-sheathing transitional lvs (0–)1–3; lvs unmarked or spotted to varying degrees, sometimes ring-spotted, marking often ±restricted to apical half of lvs. *Fl.-spike* 2–8(–12) cm, *ovoid to cylindrical*, with 6–25(–60) fls; *lowest bracts exceeding fls* and 4–6 mm wide. *Fls usually pale to rich reddish-purple*; outer lateral per.-segs ±erect to almost horizontal; *labellum* 5–14 × 6–15 mm, *usually broader than long*, *entire or 3-lobed* with triangular central lobe shorter to longer than the laterals; spur 6.5–10.5 mm, straight or slightly down-curved, conical-cylindrical. Fl. 5–8. $2n = 80^*$. Grt.

The native British representatives of this aggregate fall into four main groups which have often been regarded as separate spp. They are highly variable in vegetative features as well as fls and variation-ranges overlap so that individual plants may not be certainly assignable to a particular group. All are tetraploid with $2n = 80$, and artificial crossing has shown that a member of any group sets viable seed when crossed in either direction with a member of any other group. Presumed natural F_1 hybrids have, moreover, set viable seed when back-crossed with members of the putative parent groups. These facts explain why there is some intergrading when members of two or more groups grow in close association. In many circumstances, nevertheless, they remain distinct, presumably through the operation of geographical, ecological and phenological differences and perhaps also of internal isolating factors. All things considered it has seemed appropriate to follow those who treat all 4 groups as subspp. of *D. majalis*.

1 Green lvs (excl. bracts) 3–5, linear- to oblong-lanceolate, mostly less than 1.5 cm wide, commonly unmarked or with narrow transverse bars in apical half only, rarely more heavily marked; infl. often lax, usually with fewer than 18 fls; labellum deltate or obcordate, 3-lobed with bluntly triangular central lobe commonly longer than laterals.
　　　　　　　　　　　　subsp. **traunsteinerioides**
　Green lvs (excl. bracts) usually 6 or more, the lower often more than 1.5 cm wide, markings various or 0; infl. usually dense and often with more than 18 fls; labellum diamond-shaped or transversely elliptical, subentire to ±distinctly 3-lobed.　　2
2 Labellum subentire, ovate or diamond-shaped, less than 8 mm long and wide, dark reddish-purple with heavy and irregular lines and bars; lvs unspotted or with fine spots mostly towards apex.
　　　　　　　　　　　　subsp. **purpurella**
　Labellum obscurely to distinctly 3-lobed, not diamond-shaped, usually more than 8 × 10 mm, pale pink ot magenta, marked with dots or a pattern

of bars or lines; lvs unmarked or variously marked.
　　　　　　　　　　　　　　　　　　　　　3
3 Labellum transversely elliptical, ±obscurely 3-lobed but sometimes distinctly so, the lateral lobes rounded, central short, blunt; lvs unmarked or locally with ring-spots.　　subsp. **praetermissa**
　Labellum usually distinctly 3-lobed, the rounded lateral lobes notched or incised and usually exceeded by the triangular central lobe; lvs usually marked with variously shaped spots, bars, blotches or, occasionally, rings, sometimes unmarked.
　　　　　　　　　　　　subsp. **occidentalis**

Subsp. **occidentalis** (Pugsley) P. D. Sell
　　　　　　　　　　　Western Marsh Orchid.

Incl. *Orchis kerryensis* Wilmott, *Dactylorchis majalis* subsp. *cambrensis* R. H. Roberts, and *Dactylorhiza majalis* subsp. *scotica* Nelson

Stem usually 10–30 cm. Green sheathing lvs usually 4 or more, often ±crowded towards base of stem, longest up to 12 cm, widest usually more than 1.5 cm across; non-sheathing transitional lvs often 2; *lvs usually marked, often heavily, with solid spots or blotches*, occasionally ring-spotted. *Fl.-spike* rarely exceeding 7 cm, *±dense-fld; fls* usually *deep red-purple*, sometimes paler. Outer lateral per.-segs ±widely spreading. *Labellum* commonly more than 7.5 × 9.5 mm, *distinctly 3-lobed* with *lateral lobes notched or dentate* and *often reflexed* and the smaller triangular *central lobe often equalling or exceeding them*; labellum usually ±heavily marked with dark loops often broken into lines or dots, but sometimes merely with dots on a pale background. Spur 7–9 mm. Fl. 5–6(–7). Marshes, fens, bogs and wet meadows,
mainly on calcareous or near-neutral substrata. Locally frequent in W. Ireland, rare in W. and N.W. Wales and in N. Scotland, also in Inner and Outer Hebrides, Orkney and Shetland. Populations in S.W. and W. Ireland commonly include few to very many plants having unmarked lvs and pale fls, the labellum ±flat and marked only with dots (var. *kerryensis* (Wilmott) Bateman & Denholm). Other variants, within the normal variation-range of the subsp. but predominating locally, have also been named as varieties. There is some intergrading with subspp. *purpurella* and *traunsteinerioides*.

Subsp. **purpurella** (T. & T. A. Stephenson) D. Moresby Moore & Soó
　　　Northern Marsh Orchid, Dwarf Purple Orchid.

Orchis purpurella T. & T. A. Stephenson

Stem (5–)10–25(–40) cm, ±stiffly erect. Green sheathing lvs 4(–5), longest up to 12 cm or more, widest usually 1.5–3 cm across; non-sheathing transitional lvs usually 1–2; *lower internodes often very short; lvs mostly unmarked*, sometimes with few small round spots especially near lf-tips, rarely with abundant small spots. Fl.-spike rarely more than 7 cm, ±dense-fld; fls usually rich red-purple, sometimes magenta or pinkish. Outer lateral per.-segs obliquely spreading to near-vertical. *Labellum* usually not exceeding 7.5–9.5 mm, *±flat*, *broadly diamond-shaped and subentire or ±obscurely 3-lobed* with lateral lobes sometimes notched; *commonly marked with dark loops* interrupted or replaced by dashes or spots. Spur 6–9 mm. Fl. 6–7 (latest-flowering of the subspp.).

Marshes, fens, soligenous bogs and other wet places on calcareous to slightly acid substrata. Locally frequent north of a line from S.W. Wales to the Humber, incl. Anglesey and Isle of Man, Inner and Outer Hebrides, Orkney and Shetland; but with 2 adjacent outliers near Southampton; local in Ireland and absent from the Central Plain and the south-west.

Subsp. **praetermissa** (Druce) D. Moresby Moore & Soó
Southern Marsh Orchid, Common Marsh Orchid, Fen Orchid.

Stem 15–40(–60) cm or more. Green sheathing lvs usually 4–6, longest usually more than 12 cm and widest 1.5–2.5 cm across, all oblong-lanceolate and tapering gradually to the tip; non-sheathing lvs commonly 2; *lvs unmarked* except in var. *junialis*, which has large transversely elongated and predominantly annular markings. Fl.-spike occasionally exceeding 7 cm, usually ±dense-fld; *fls a fairly rich lilac- or pale magenta-pink*, sometimes brighter magenta but still *redder and paler than typical purpurella*. Outer lateral per.-segs obliquely erect. *Labellum transversely elliptical*, usually 9–14 mm wide and *distinctly broader than long*, usually *subentire or ±obscurely 3-lobed, flat or with margins slightly upturned*, but the rounded and rarely notched lateral lobes may reflex at a late stage, *central lobe usually short with blunt or broadly rounded tip*, sometimes longer than laterals; *labellum marked with dots and sometimes fine lines*, these *often only within a central elliptic area* (in var. *junialis* heavy loops and dashes). Spur 6–9 mm, *conical and usually quite stout*, with blunt ±oblique tip. Fl. 6–7.

Found in a variety of wet places on calcareous to slightly acid mineral soils and peat from the margins of reed-swamps to wet meadows, marshes, fens and soligenous bogs, and also in dune-slacks. Locally frequent south of a line from the Ribble estuary (C. Lancs) to Flamborough head (Humberside) with outliers near the East coast northwards to Northumberland, thus overlapping the near-complementary range of subsp. *purpurella* only in small areas of Wales, S. Lancs, Durham and S. Northumberland; Channel Is.; not in Scotland or Ireland.

Subsp. **traunsteinerioides** (Pugsley) Bateman & Denholm
 Narrow-leaved Marsh Orchid, Pugsley's Marsh Orchid.

Orchis majalis Reichenb. subsp. *traunsteinerioides* Pugsley

Stem (9–)15–30(–45) cm, slender. Green sheathing lvs 3–5, *linear-oblong to narrowly lanceolate*, 7–12 × 0.6–1.5 cm; non-sheathing lvs usually 0–1; lvs unmarked or with round to ±transversely elongated

spots especially towards lf-tips, markings sometimes annular in var. *eborensis* (see below). *Fl.-spike* up to 7 cm, *lax*, usually with 8–20 fls; *fls pale to deep red-purple*, but less deep than in *purpurella* or much *occidentalis*. Outer lateral per.-segs obliquely spreading. *Labellum* 6.5–9 × 7–11 mm, varying from *deltate to obcordate or transversely elliptical*, distinctly to obscurely 3-lobed, rarely subentire; *lateral lobes* occasionally notched, *usually reflexed*; *central lobe often longer than laterals*; *labellum commonly marked with irregular dots, dashes or loops covering most of the surface* (in contrast with their restricted area in *praetermissa*). Spur commonly 8–9.5 × 2–3 mm, straight, somewhat longer on average than in the other subspp. but somewhat less stout than in *praetermissa*. Fl. 5–6, much the same as in *occidentalis* and earlier than *purpurella* and *praetermissa*, but overlapping both.

Usually in habitats flushed with calcareous water and with *Schoenus nigricans* dominant or abundant. Local in Ireland from Wicklow westwards to Clare and further north in Antrim and Fermanagh; in Great Britain also widely scattered from N. Hants and Berks (Cothill), a few fens in East Anglia and in N. Wales including Anglesey northwards to N. Yorks and Durham and an outlier at Knapdale (Argyll). Variable, especially in lf-marking and shape of labellum. The Yorks populations consist of small plants rarely exceeding 15 cm, commonly with lvs marked with dots or transverse bars in their apical half and not infrequently with ±transversely elongated ring-markings. The fls are smaller, with variously shaped labellum, sometimes barely 3-lobed, and a spur below average length for the subsp. They have been named var. *eborensis* (Godfery) Bateman & Denholm.

D. majalis is native in much of Europe except the Mediterranean region, but the type subsp. *majalis* is not recognized as extending to the British Is., being replaced here by the endemic subsp. *occidentalis*. Subspp. *purpurella* and *praetermissa* have limited ranges in N.W. Europe. Subsp. *traunsteinerioides* is certainly very closely related to the continental European *D. traunsteineri* (Sauter) Soó of N. and C. Europe and extending eastwards to Poland and Russia. Our plant most closely resembles those of mountains of C. Europe, including the Alps, but it is not yet certain whether they should be placed in the same subsp. of *majalis*.

The 4 subspp. hybridize not only with each other (p. 580) but also with the similarly tetraploid *D. maculata* subsp. *ericetorum*, though intergrading here too is of quite local occurrence. They also cross freely with the diploid *D. fuchsii* but the fertility of the triploid F_1 is very low, so that the appearance of hybrid swarms' is misleading: they are in fact mixtures of parents with 'a highly sterile, very variable first generation of triploids' (Heslop-Harrison, *Jahresber. Naturwiss. Ver. Wuppertal*, 1968, **21–22**, 20–27). Hybrids with *D. incarnata* and with *Gymnadenia conopsea* and *Coeloglossum viride* have also been reported from a number of localities, usually as single plants.

23. ANACAMPTIS L. C. M. Richard

An *Orchis*-like herb with entire root-tubers, lfy stems and conical spikes of small fls. Outer lateral per.-segs spreading. *Labellum* deeply 3-lobed *with 2 obliquely erect 'guide-plates'* decurrent on its base from the lateral lobes of the column; spur long, slender, with no free nectar. Column short; *stigmas 2, on the rounded lateral lobes of the column* which are continued downwards as 'guide plates'; rostellum a centrally-placed pouch between the bases of the stigmas and partially closing the entrance to the spur. Anther adnate to the top of the column; pollinia 2, narrowed downwards into caudicles which are attached by their bases to a single transversely elongated narrowly saddle-shaped viscidium enclosed in the rostellum-pouch; pollen in packets of tetrads united by elastic threads.

One sp. in Europe and N. Africa.

Resembles *Gymnadenia* in the position of the 2 stigmas on lateral lobes of the column, and *Orchis* in the protection of the viscidium in a pouch, but differs from both in the attachment of both pollinia to a single viscidium, and in the 'guide-plates' at the base of the labellum.

1. A. pyramidalis (L.) L. C. M. Richard
　　　　　　　　　　　　　Pyramidal Orchid.

Orchis pyramidalis L.
Root-tubers ovoid or subglobose. Stem 20–50(–75) cm, glabrous, slightly angled above, with 2–3 brown sheathing scales below. Lower lvs 8–15 cm narrowly oblong-lanceolate, acute, keeled; upper lvs smaller, acuminate, grading into bracts; all unspotted. *Spikes 2–5 cm, at first markedly conical*, dense-fld, with a foxy smell; bracts linear-lanceolate, slightly exceeding the ovary. Outer lateral per.-segs broadly lanceolate, curved, spreading; outer median and inner lateral segments connivent into a hood; all deep rosy-purple, becoming paler. Labellum 6–8 mm, pale rose, broadly cuneate, with 3 subequal, oblong, truncate or rounded, ±entire lobes c. 4 mm; lobes sometimes 0; 'guide-plates' convergent towards the base of the column. Spur c. 12 mm, *filiform*, acute, *equalling or exceeding the ovary*. Fl. 6–8. Pollinated by day-and night-flying Lepidoptera, the viscidium coiling tightly round their proboscides after withdrawal. $2n = 36$. Grt.

Native. Grassland and open grassy scrub or woodland on chalk or limestone; calcareous dunes; to 244 m in England. Locally frequent in suitable habitats throughout Great Britain northwards to Fife; Colonsay and Tiree (Inner Hebrides); Barra and Fuday (Outer Hebrides); throughout Ireland; Channel Is. Europe northwards to S. Scandinavia and C. Russia; W. Asia; N. Africa.

145. ARACEAE

Herbs, often with tuberous or elongated rhizomes, rarely woody and climbing. Raphides often present. Fls small, usually crowded, arranged in a compact spike (*spadix*) which is generally enclosed in a lf-like or petaloid bract (*spathe*), either hermaphrodite or unisexual. Unisexual fls with the males on the upper part of the spadix, the females below. Perianth present in hermaphrodite fls; segments 4–6, free or connate into a truncate cup, usually 0 in unisexual fls. Stamens 1–6; anthers opening by pores or slits, free or connate. Staminodes sometimes present. Ovary superior or sunk in the spadix, 1–3-celled. Fr. a fleshy or rather dry berry.

About 115 genera and about 2000 spp., widely distributed throughout the world but by far the greatest number in the tropics.

1 Lvs linear to ensiform; spadix apparently lateral.
　　　　　　　　　　　　　　1. ACORUS
　Lvs ovate or hastate; spadix obviously terminal. 　2
2 Lvs hastate to sagittate; fls confined to lower part
　　of spadix. 　　　　　　　　2. ARUM
　Lvs ovate to orbicular, with cordate to truncate base;
　fls covering the whole spadix. 　　　　3
3 Lvs 5–12 cm; spathe white on inner surface; berries
　red. 　　　　　　　　　　CALLA
　Lvs 40–120 cm; spathe bright yellow; berries green.
　　　　　　　　　　　　LYSICHITON

1. ACORUS L.

Rhizomatous herbs with *linear lvs without petioles*. Scape flattened. *Spadix apparently lateral*, the base of the lf-like spathe continuing the line of the scape, terete, entirely covered with fls. *Fls all hermaphrodite. Per.-segs* 6, free, membranous. Stamens 6.

Two spp. in Europe, temperate Asia and America.

*1. A. calamus L. 　　　　　　　Sweet Flag.

A stout glabrous rhizomatous aromatic perennial up to c. 1 m. Lvs 1–2 cm wide, crowded, distichous, ensiform, acuminate, smelling of tangerines when bruised, midrib thick, often eccentric; margins waved. Scape reddish at base, ending in a long lf-like spathe above the spadix. Spadix 5–9 cm, making an angle of c. 45° with the scape, tapering upwards, obtuse. Fls yellowish, tightly packed and completely covering the spadix. Fr. unknown in Britain. Fl. 5–7. $2n = 24, 36$. Hel. or Hyd.

Introduced. In shallow water at margins of ponds, rivers and canals, local. Native of S. and E. Asia; widely naturalized in North America. Introduced into Europe by 1557 and recorded as naturalized in England by 1660.

*Lysichiton americanus Hultén & St John, a rhizomatous marsh plant with shortly petiolate ovate-oblong lvs 40–120 cm. Spathes up to 25 cm, yellow, breaking off after fl. Fls all hermaphrodite. Fr. a green berry with 2 seeds. Fl. 4. Grown in gardens and naturalized in a few localities in England and Ireland. Western North America.

*Calla palustris L. A glabrous creeping rhizomatous perennial 15–30 cm. Rhizome stout, green, jointed and scaly. Lvs broadly cordate, cuspidate, entire; petiole with a long sheath. Spathe persistent, flat, not enclosing the spadix, oval, cuspidate, white within. Spadix 5–12 cm, stout, greenish entirely covered with fls. Fr a berry, red when ripe. Fl. 6–7. $2n = 36; 72$. Hel.

Introduced. Naturalized in swamps and wet woods near ponds; first planted in 1861. Very local. N., C. and E. Europe; N. Asia; North America.

2. ARUM L.

Perennial herbs with tuberous stocks. Lvs net-veined, petioles sheathing at base. Spadix terminal, terete, *the upper part without fls*; *spathe convolute*, the *margins not connate. Fls all unisexual*, female below, the upper sterile; male above, the upper sterile. Per.-segs 0. Stamens 3–4. Ovary 1-celled, stigmas sessile; ovules orthotropous.

About 15 spp. in Europe and N. Africa.

Lvs appearing in spring; terminal portion of spadix dull purple, rarely yellow; spathe usually twice as long as spadix (common). **1. maculatum**
Lvs well developed by November; terminal portion of spadix always yellow; spathe usually three times as long as spadix (S. coast, very local). **2. italicum**

1. A. maculatum L.   Lords-and-Ladies, Cuckoo-pint.

An erect glabrous perennial 30–50 cm. Tuber c. 2 cm, horizontal, a fresh one produced from the base of the stem each year. *Lvs appearing in spring*, long-petiolate; blade 7–20 cm, often blackish-spotted, *triangular-hastate*, lobes acute or obtuse. *Spathe* 10–25 cm, *erect*, pale yellow-green, edged and sometimes spotted with purple. *Spadix* c. half as long as spathe, *upper part dull purple*, rarely yellow. Fr. scarlet, fleshy; *fr. spike* 3–4 cm, bursting the persistent base of the spathe. Fl. 4–5. Protogynous and cross-pollinated by small flies, particularly midges (for mechanism see Church, *Types*

of Floral Mechanism, Oxford University Press (1908), pp. 70–4). Fr. 7–8. $2n = 56^*$. Gt.

Native. In woods and shady hedgebanks, especially on base-rich substrata, occasionally becoming a persistent weed in gardens; very shade-tolerant. Generally distributed throughout England, Wales and Ireland, less frequent in Scotland and not native in the north. W., C. and S. Europe northwards to Scotland and eastwards to W. Ukraine; N. Africa.

2. A. italicum Miller   Italian Lords-and-Ladies.

Incl. *A. neglectum* (Townsend) Ridley

An erect glabrous perennial 20–30 cm. Tuber c. 5 cm. Lvs well developed by November, long-petiolate, spotted or not; winter lvs 15–30 cm, triangular-hastate; autumn lvs smaller. Spathe up to c. 40 cm, pale green, tip deflexed. Spadix c. $\frac{1}{3}$ length of spathe, upper part yellow or orange-yellow. Fr. scarlet, fleshy; fr. spike 10–15 cm. Fl. 4–5, but rather later than *A. maculatum*. Fr. 8–9. $2n = 84^*$. Gt.

Native. In light shade, often under brambles and in stony ground, usually within 2 km of the sea. S. coast of England from Cornwall to E. Kent, Glamorgan; Channel Is., very local. S. and W. Europe from Brittany southwards; N. Africa; Macaronesia.

Subsp. **italicum**: Lvs with conspicuous whitish veins; lateral lobes divergent. Seeds 3–4 in each berry. Sometimes cultivated for ornament and escaping.

Subsp. **neglectum** (Townsend) Prime: Lvs without conspicuous whitish veins, sometimes with dark spots; lateral lobes convergent and sometimes overlapping. Seeds 1–2 in each berry.

146. LEMNACEAE

Small floating aquatic monoecious herbs. Roots simple or 0. Fls naked, or at first enclosed in a sheath; perianth 0. Male fls consisting of 1–2 stamens with 1–2-celled anthers. Female fls consisting of a solitary, sessile, 1-celled ovary; ovules 1–7.

Three genera with about 40 spp., cosmopolitan in fresh waters.

Thallus ±flattened, with roots.   1. LEMNA
Thallus subglobose, rootless.   2. WOLFFIA

1. LEMNA L.   Duckweed.

Small aquatic herbs frequently forming a green carpet on the surface of stagnant water. Infl. minute, borne in a pocket in the margin of the thallus and consisting of 1 female and 2 male fls enclosed in a sheath. Anthers bilocular. Raphides occur in all spp., particularly abundantly in *L. minor* and *L. trisulca*. The thallus has been variously interpreted as a modified stem, a lf or partly lf and partly stem (see Arber, A., *Water Plants*, Cambridge University Press (1920), p. 73).

About 16 spp. distributed throughout the world.

1 Plants floating on the surface; thallus not stalked.   2
  Plants floating submerged; old thalli distinctly stalked.   **2. trisulca**
2 Several roots to each thallus.   **1. polyrhiza**
  One root to each thallus.   3
3 Thallus nearly flat on both sides, light green above.   **3. minor**
  Thallus convex, typically swollen below, grey-green or reddish-brown.   **4. gibba**

1. L. polyrhiza L.   Greater Duckweed.

Spirodela polyrhiza (L.) Schleiden

Thalli 5–8(–10) mm diam., flat and shiny, often *purplish below*, ovate or almost orbicular, with 5–9 veins, *each with usually 5–15 roots* up to 3 cm; thalli produce towards the end of summer purplish-brown reniform turions 2–4 mm in diam., which become detached and often sink, rising to the surface in spring. Fl. 7 (recorded only from Somerset). $2n = 40^*$. Hyd.

Native. In still waters in ditches and ponds, local. England, except the extreme north and south-west; E. Wales; S. Scotland, rare; Ireland: mainly around Lough Neagh, but extending to Limerick and Clare. Europe, except the extreme north and south; Madeira; Africa;

Asia; America; Australia; rather uncommon through-out most of its range.

2. L. trisulca L.　　　　　　Ivy-leaved Duckweed.

Thalli submerged, translucent, 3-veined, *tapering at the base into a stalk when mature, several thalli attached to each other by their stalks.* Two young thalli arise on opposite sides of, at right angles to, and in the same plane as each old one. Thallus (5–)7–12(–15) mm, ±acute and serrulate at apex, narrowing abruptly to the stalk. Roots 1 to each thallus often ±hooked. Fertile thalli floating, smaller than sterile, pale green and often lacking the characteristic branching, ovate, producing stomata on the upper surface. Seed c. 1 mm, 12–15-ribbed. Fl. 5–7. 2*n* = 40, 44*. Hyd.

Native. In ponds and ditches. Great Britain, rare in the W. and N. but extending to Caithness; Ireland, mainly in the East and Centre. Europe except some of the islands; N. Africa; Asia; North America; Australia.

3. L. minor L.　　Common Duckweed, Duck's-meat.

Thalli floating, opaque, obovate or suborbicular, entire, subapiculate at point of attachment to parent thallus, 1.5–5 mm diam., *slightly convex beneath*, 3-*veined, pale green above with 10–20 meshes beneath* each with a single root up to 15 cm. Fls not uncommon, usually in shallow ditches fully exposed to the sun. Seed c. 0.6 mm. Fl. 6–7. Germ. 2. 2*n* = 40*. Hyd. By far the commonest sp.

Native. In still waters. Generally distributed through-out the British Is. except for much of N. Scotland and Shetland. Cosmopolitan except for the polar regions and tropics.

***L. minuscula** Herter is widespread and increasing rapidly in S. & E. England. It is like *L. minor* but has only 1 obscure vein in the thallus, which is thin and flat on the lower side. It is native in temperate South America and naturalized in a number of European countries.

4. L. gibba L.　　　　　　Fat Duckweed.

Thallus floating, convex above with reticulate markings just visible to the naked eye over most of the upper surface, grey-green or reddish-brown typically strongly swollen, with 40–50 meshes beneath. Summer *thallus* 3–5 mm, ovate, usually asymmetrical and rounded at base. Roots 1 to each thallus, up to 6 cm. Winter thalli dark green, rooted, not gibbous below, formed chiefly after flowering. Fls less frequently produced than in *L. minor.* 2*n* = 64*. Hyd.

Native. In still waters, local. England to N.E. Yorks and S. Lancs, absent from the extreme south-west; Wales: Monmouth, Glamorgan, Flint; Ireland: scat-tered localities, mainly in the east. Europe to 63°N., local in C. Europe, becoming commoner in the Mediter-ranean region; India, ascending to c. 3050 m in Hima-laya; extratropical Africa; North and South America, absent from tropics; Australia and New Zealand; Madeira, Canaries.

2. WOLFFIA Horkel ex Schleiden

Minute rootless floating aquatic herbs. Infl. borne in a hollow in the upper surface of the ovoid thallus, sheath 0; stamen 1; anthers unilocular.

About 12 spp., distributed throughout the world.

1. W. arrhiza (L.) Horkel ex Wimmer

　　　　　　Rootless Duckweed

Lemna arrhiza L.

Thallus 0.5–1 mm, *ovoid to ellipsoid* or occasionally nearly globose, producing daughter thalli by budding from one end. Fl. unknown in Britain. No special resting thalli are produced, but the ordinary ones sink in winter. The smallest British flowering plant. 2*n* = 50*. Hyd.

Native. In still waters, rare. S. England, from Somer-set to Kent, very local. Europe from Lithuania to W.C. Portugal, Sicily and Bulgaria, local; Africa; Asia; America; Australia.

147. SPARGANIACEAE

Rhizomatous perennial aquatic herbs. Stems simple or branched, lfy. Lvs elongate-linear, sheathing at base, erect or floating. *Fls unisexual, crowded in separate glo-bose heads* (capitula), the female towards the base in each infl. *Perianth of 3–6 membranous spathulate scales.* Male fls with 3 or more stamens, the filaments some-times partially united; female with a 1–2-celled *sessile ovary. Fr.* dry, *indehiscent*, narrowed below, the exo-carp spongy. Wind-pollinated.

One genus with about 20 spp., generally distributed, but absent from Africa and South America.

1. SPARGANIUM L.　　　　Bur-reed.

The only genus.

1 Infl. branched, male capitula on lateral branches; per.-segs thick, black-tipped; seed with 6–10 longi-tudinal ridges.　　　　　　　**1. erectum**

Infl. unbranched, all male capitula on the main axis; per.-segs thin, not black-tipped; seed smooth.　　*2*

2 Cauline lvs triangular in section near the base; male capitula more than 3, distant.　　**2. emersum**

Cauline lvs flat in section near the base; male capitula not more than 3, crowded.　　*3*

3 Lf-like bract of lowest female capitulum 10–60 cm, at least twice as long as the whole infl.; male cap-itula usually 2.　　　　　　**3. angustifolium**

Lf-like bract of lowest female capitulum 1–5(–8) cm, barely exceeding the infl.; male capitulum usually solitary　　　　　　　　**4. minimum**

Subgenus 1. SPARGANIUM. Per.-segs rather thick and firm, black-tipped. Seeds longitudinally ridged.

1. S. erectum L.　　　　Branched Bur-reed.

An erect glabrous perennial (30–)50–150(–200) cm. Lvs usually all erect, triangular in section, 10–15 mm wide.

Infl. branched (rarely simple) with the *male capitula* borne above the female *on the branches*. Fl. 6–8. Hyd. or Hel.

Native. On mud or in shallow water in ponds, ditches and slow-flowing rivers and on ungrazed marshland. Common and widely distributed throughout the British Is., except Shetland. Northern temperate regions.

Four subspp., differing in fr. shape and size and, to some extent, in distribution, occur in Britain. They may be distinguished as follows:

1 Fr. with a distinct shoulder, the upper part dark brown
   or black.   2
   Fr. globose to ellipsoid, without a distinct shoulder,
   uniformly light brown and shiny.   3
2 Fr. (5–)6–8(–10) mm (excluding style) and (3–)4–
   6(–7) mm wide at shoulder; upper part flattened.
                                                   subsp. **erectum**
   Fr. 6–7(–8) mm (excluding style) and 2.5–4.5 mm
   wide at shoulder; upper part domed and wrinkled
   below style.   subsp. **microcarpum**
3 Fr. 7–9 × 2–3.5 mm, ellipsoid.   subsp. **neglectum**
   Fr. 5–8 × 4–7 mm, ±globose.   subsp. **oocarpum**

Subsp. **erectum** (*S. ramosum* Hudson; *S. polyedrum* (Ascherson & Graebner) Juzepczuk; *?S. eurycarpum* Engelm.). Ovary usually bilocular. $2n = 30$. England south of the Wash. Much of Europe, except the north, east to C. Siberia; N. Africa.

Subsp. **microcarpum** (Neuman) Domin (*S. neglectum* subsp. *microcarpum* (Neuman) Ascherson & Grabner). Ovary usually unilocular. Throughout the British Is. Almost throughout the range of the sp.

Subsp. **neglectum** (Beeby) Schinz & Thell. (*S. neglectum* Beeby). Ovary unilocular. $2n = 30$. In England common south of the Wash, but extending north to Westmorland. S. Sweden to N. Africa and east to the Caucasus.

Subsp. **oocarpum** (Čelak.) Domin. Ovary unilocular, rarely bilocular; fertility low and perhaps subsp. *erectum* × *neglectum*. England south of the Wash. Distribution imperfectly known, but extending to N. Africa and Anatolia.

Subgenus 2. Xanthosparganium Holmberg. Per.-segs thin, uniformly pale. Seeds smooth.

2. S. emersum Rehmann   Unbranched Bur-reed.

S. simplex Hudson, pro parte

An erect or floating perennial 20–60 cm. Lvs 3–12 mm wide, erect ones triangular in section; floating lvs usually present, keeled, sheathing at base but not inflated. *Infl. simple, with 3–10 remote male capitula on the main axis; anthers 6–8 times as long as broad*; female capitula 3–6, the lower often pedunculate. Fr. 4–5 × 2–2.5 mm, ellipsoid, often constricted in the middle. Fl. 6–7. $2n = 30$. Hyd. or Hel.

Native. In shallow water in rivers, ponds and lakes, absent from very acid waters and from unsilted lakes. Throughout the British Is. north to Shetland, though local in Scotland. Throughout north temperate regions from the Arctic Circle to 40° N.

3. S. angustifolium Michx   Floating Bur-reed.

S. affine Schnizlein

Perennial with long slender floating (rarely erect) stems. Lvs flat, sheathing and inflated at base. Infl. simple, with (1–)2(–3) *male capitula* on the main axis; *anthers 3–4 times as long as broad*; female capitula 2–4, the lower pedunculate. *Fr. c.* 8 mm, ellipsoid. Fl. 8–9. $2n = 30$. Hyd.

Native. In peaty lakes, mainly in mountainous districts. Fairly widely distributed in the north and west of Great Britain, absent from much of the south and east and from most of C. and S. Ireland. Iceland to N. Portugal, the Alps and Macedonia, east to Kamchatka; North America.

4. S. minimum Wallr.   Least Bur-reed.

Stems 8–30 cm, usually floating. Lvs usually 2–6 mm wide, flat, scarcely inflated at base. Infl. simple, *with* 1 (rarely 2 confluent) *male capitulum*; female capitula (1–)2–3(–4), usually sessile. *Fr.* 3.5–4.5 mm, obovoid. Fl. 6–7. $2n = 30$. Hyd.

Native. In lakes, pools and ditches on acid or alkaline substrata with a high proportion of organic matter. Throughout the British Is. in suitable habitats, rare in the south but extending north to Shetland. Throughout Europe south to the Pyrenees and S. Bulgaria; North America.

148. TYPHACEAE

Stout rhizomatous herbs growing in shallow water. Stems erect, simple. Lvs distichous, coriaceous, or thick and spongy, elongate-linear, slightly spirally twisted, sheathing at base and mostly basal. *Fls* unisexual, very numerous, *densely crowded on a terminal spadix*, male above, female below. *Fls surrounded by slender jointed hairs or spathulate scales*, sometimes interpreted as per.-segs. Male fls with 2–5 often monadelphous stamens; pollen often in tetrads; female with a 1-celled, *stipitate ovary. Fr.* dry, *at length splitting*. Wind-pollinated.

One genus with 10 spp., throughout the world from the Arctic Circle to 30° S.

1. TYPHA L.   Reedmace, Bulrush.

The only genus.

Lvs 8–20 mm wide; male and female parts of spadix usually
   contiguous; female fls without scales.   **1. latifolia**
Lvs 3–6 mm wide; male and female parts of spadix usually
   distant; female fls with scales.   **2. angustifolia**

1. T. latifolia L.   Bulrush, Cat's-tail.

A robust perennial 1.5–2.5 m. *Lvs 8–20 mm wide*, linear, over-topping the infl. *Male and female parts of infl. usually contiguous. Female fls without scales.* Fr. cylindrical, tapering at base into a slender stalk. Fl. 6–7. Seeds shed 2–3. $2n = 30$. Hyd.

Native. In reed-swamps, often dominant, especially on inorganic substrata or where there is silting and rapid decay of organic matter, in lakes, ponds, canals and slow-flowing rivers. Generally distributed in suitable habitats throughout the British Is., though less frequent in the north and west. From the Arctic Circle to 30° S., except for C. and S. Africa, S. Asia, Australia and Polynesia.

2. T. angustifolia L.   Lesser Bulrush.

A robust perennial usually not more than 2 m. *Lvs 3–6 mm wide*, convex on the back. *Male and female parts of infl.* about equal in length, *remote* (1–9 cm apart),

very rarely less. Scales *of female fls shorter than the stigmas*. Fl. 6–7. $2n = 30$. Hyd.

Native. In reed-swamps on ±organic soils in lakes, ponds, canals and slow-flowing rivers. Locally common, but much less generally distributed than *T. latifolia* north to the Firth of Forth; local and rare in Ireland. Distribution of the genus. Our plant (subsp. *angustifolia*) has a wide distribution but is absent from America south of Louisiana and California and from Africa.

T. minima Hoppe has been reported from Britain from time to time since 1640, but none of these records has been substantiated and it is unlikely from its known distribution that the plant would occur in this country.

149. CYPERACEAE

Usually perennial, often rhizomatous herbs, most often growing in wet places. Stems usually solid, often trigonous. Lvs usually linear, some or all often reduced to sheaths. Ligule sometimes present. Fls hermaphrodite or unisexual, arising in the axil of a bracteole (glume) and arranged in 1–many-fld spikelets. Spikelets solitary, terminal, or grouped in branched or spike-like infl. often subtended by bracts. Female fl. sometimes enclosed by a modified glume (bracteole, perigynium) fused round it. Perianth of 1–many bristles or scales or, more often, 0. Stamens (1–)2–3(–6); anthers basifixed. Style simple; stigmas 3 or 2, linear, papillose. Ovary unilocular; ovule solitary, erect. Fr. indehiscent, globose or trigonous in plants with 3 stigmas, biconvex in plants with 2 stigmas. Seed erect, embryo small, endosperm abundant. Chromosomes often very small, little longer than broad. Fls wind-pollinated.

About 100 genera and 4000 spp., in all parts of the world.

The genera and spp. of Cyperaceae show a considerable diversity in both their vegetative and reproductive parts. Reduction appears to have occurred independently in many directions and any linear arrangement of genera is consequently bound to be artifical. The arrangement adopted here is based on the assumption that the following features are more likely to be primitive than the other variants which occur: lfy stems; many-fld spikelets; few or no sterile glumes; spiral arrangement of glumes; hermaphrodite fls; presence of perianth; 3 stamens; 3 stigmas; trigonous nut not enclosed by a perigynium or closely enfolding glume.

1 Fls all unisexual; male and female in separate spikes
   or separate parts of the same spike; female fls
   enclosed in perigynia or a closely enfolding inner
   glume.   2
  Fls predominantly hermaphrodite; female fls without
   perigynia or closely enfolding inner glume.   3
2 Female fls enclosed in perigynia, which often end in
   a beak (many spp., several common).   15. CAREX
  Female fls without perigynia but closely enfolded by
   an inner glume (1 sp., very local).   14. KOBRESIA
3 Stem hollow; margins and keel of lf very rough,
   readily cutting the skin.   13. CLADIUM

Stem solid; margins and keel of lf not rough enough
   to cut the skin.   4
4 Bristles long and silky, more than 6   1. ERIOPHORUM
  Bristles shorter than glumes, rarely longer and then
   only 6.   5
5 Infl. of a solitary terminal spikelet; bract not forming
   an apparent prolongation of stem.   6
  Infl. of 2 or more spikelets, or (rarely) spikelet solitary
   but with a small stem-like bract beside it.   8
6 Water plant with elongate slender branched lfy stems.
                             9. ELEOGITON
  Bog or marsh plants; tufted or with creeping rhizome;
   stems nearly or quite lfless.   7
7 Uppermost sheath on flowering stems with a short
   blade.   2. TRICHOPHORUM
  Uppermost sheath on flowering stems without a
   blade, usually conspicuous, rarely very thin and
   delicate.   3. ELEOCHARIS
8 Infl. with several flat or keeled lf-like bracts close
   at base.   9
  Bracts not flat and lf-like or else solitary.   10
9 Spikelets terete, ovoid; glumes not distichous.
                              4. SCIRPUS
  Spikelets flattened, linear or oblong; glumes distichous (rare).   10. CYPERUS
10 Spikelets distichously arranged in a compressed
   oblong head.   6. BLYSMUS
  Spikelets spirally arranged, or solitary; infl. often
   apparently lateral, branched, or if a terminal head,
   then not compressed.   11
11 Infl. a compact blackish head, encircled at base by
   lowest bract; spikelets flattened; glumes distichous.
                         11. SCHOENUS
  Infl. reddish-brown or greenish; spikelets terete;
   glumes spirally arranged.   12
12 Stems lfy only at base or sometimes quite lfless; bracts
   stem-like; infl. apparently lateral.   13
  Stems lfy above base; bracts lf-like; infl. terminal.
                  12. RHYNCHOSPORA
13 Plant slender, seldom exceeding 15 cm; infl. of 1–3
   spikelets.   8. ISOLEPIS
  Plant stout, seldom less than 50 cm; infl. of numerous
   spikelets.   14
14 Spikelets 5 mm or more, few together, reddish-
   brown.   7. SCHOENOPLECTUS
  Spikelets 2–3 mm, crowded into dense heads, greenish (very local).   5. HOLOSCHOENUS

1. ERIOPHORUM L. Cotton-grass.

Perennial rhizomatous herbs, either tufted or with a far-creeping rhizome. Stems lfy. Lvs remaining green through the winter, but usually nearly or quite dead at flowering time. Spikes many-fld, solitary or forming an umbellate infl. Fls hermaphrodite. Glumes spirally arranged, silver-slaty, membranous, lowest sterile. *Perianth of numerous bristles which elongate and become cottony after flowering.* Stamens 3. Stigmas 3; style-base not swollen. Nut compressed-trigonous.

About 20 spp., mainly in arctic and north temperate regions.

1 Spikes several, ±nodding.	*2*
Spike solitary, erect.	**4. vaginatum**
2 Peduncles smooth; stems subterete, except at apex; anthers 2.5–5 mm.	**1. angustifolium**
Peduncles scabrid; stems trigonous throughout; anthers 1.5–2 mm.	*3*
3 Plant rhizomatous; lvs 1–2 mm wide, the cauline with a short ligule; glumes with several veins.	**2. gracile**
Plant tufted; lvs 3–8 mm wide, the uppermost cauline without a ligule; glumes 1-veined.	**3. latifolium**

1. E. angustifolium Honckeny Common Cotton-grass.

E. polystachion L., p.p.

An *extensively creeping* rhizomatous perennial 20–60 cm. *Stem subterete* (when fresh), smooth. *Lvs 3–6 mm wide, channelled, narrowed into a long triquetrous point. Uppermost lf with a ±inflated or funnel-shaped sheath and a short ligule.* Spikes (1–)3–7, ±nodding. *Peduncles smooth,* unequal. Bracts 1–2, shortly sheathing. *Glumes* c. 7 mm, 1-*veined, lanceolate, acuminate,* brownish below, slate-coloured above; *margin broadly hyaline. Bristles* up to c. 4 cm, *smooth at tip* (microscope). Anthers 2.5–5 mm. Nut 1–3 mm, obovoid-trigonous, blackish-brown. Fl. 5–6. Fr. 6–7. $2n = 58^*$. Hel.

Native. In wet bogs, shallow bog pools and acid fens. Recorded from every vice-county in the British Is., but now very rare or even extinct in some, owing to drainage; locally abundant, particularly in the north and west. Europe (except the southern Mediterranean region); arctic regions; Siberia; North America, Greenland.

2. E. gracile Roth Slender Cotton-grass

An extensively creeping rhizomatous perennial similar to *E. angustifolium* but much more slender. Stems triquetrous (when fresh). Lvs 1–2 mm wide, short, obtuse; sheaths not inflated. Spikes 3–6. Peduncles rough with short forward-directed hairs. Glumes c. 5 mm, several-veined, ovate, subacute; margin not hyaline. Anthers 1.5–2 mm. Nut 2.5 mm, yellowish-brown. Fl. 6. Fr. 7–8. $2n = 60, 76$. Hel.

Native. In wet almost neutral fens. Very local in S. and E. England, N. Wales and Ireland. C. and N. Europe, very local in the south and absent from the Mediterranean region and Hungarian plain.

3. E. latifolium Hoppe Broad-leaved Cotton-grass.

E. paniculatum Druce

A *tufted* rhizomatous perennial 20–60 cm, similar in general appearance to *E. angustifolium*. Stems triquetrous (when fresh). Lvs 3–8 mm wide, flat except for the short triquetrous point. *Uppermost lf without a ligule,* its sheath close-fitting and cylindrical. Spikes 2–12. *Peduncles rough* with short, forward-directed hairs. *Glumes* 4–5 mm, 1-veined, lanceolate, acuminate, blackish *with very narrow hyaline margin. Bristles papillose at tip* (microscope). Anthers 1.5–2 mm. Nut 3–3.5 mm, narrowly obovoid-trigonous, reddish-brown. Fl. 5–6. Fr. 6–7. $2n = 54; 72$. Hel.

Native. In wet places on base-rich soils. Scattered throughout the British Is., but local and much less common than *E. angustifolium*. Throughout most of Europe, Asia Minor, Caucasus, Siberia, North America.

4. E. vaginatum L. Cotton-grass, Hare's-tail.

A *tussock-forming* rhizomatous perennial, 30–60 cm. Stems smooth, terete below, trigonous above. Lvs ±setaceous, up to 1 mm wide, triquetrous. Stem-lvs 2–3, *blade* short, that *of uppermost lf almost 0; sheaths strongly inflated but narrowed at mouth. Spike solitary,* terminal, c. 2 cm at flowering, ovoid or almost globose, *broad and rounded at base. Bracts 0. Glumes* c. 7 mm, 1-veined, ovate-lanceolate, acuminate, silvery below, slaty-black above, translucent. *Bristles* c. 2 cm, smooth, *pure white.* Anthers 2.5–3 mm. Nut 2–3 mm, rather broadly obovoid-trigonous, yellowish-brown. Fl. 4–5. Fr. 5–6. $2n = 58$. Hel.

Native. In damp peaty places, especially on blanket bogs, locally abundant. Throughout most of the British Is., but decreasing in the south and extinct in a number of counties. N. and C. Europe, southwards in the mountains to S. Spain and Macedonia.

E. brachyantherum Trautv. & C. A. Meyer, similar to *E. vaginatum* but with the upper stem-lf with a distinct blade, and yellowish bristles, has been reported from W. Ross, apparently in error.

2. TRICHOPHORUM Pers.

Perennial herbs. Stem terete or rarely trigonous. Transverse section of stem with air canals bounded by very thick-walled cells. *Lower sheaths lfless, only the uppermost with a short blade. Spikelet solitary, terminal,* the lowest glume generally fertile, though usually larger than the others. Fls hermaphrodite. Perianth bristles present, not more than 6, shorter or rarely longer than the glumes. Stamens 3. Stigmas 3.

Plant densely tufted; stems terete; bristles shorter than glumes.	**2. cespitosum**
Plant shortly creeping; stems trigonous; bristles longer than glumes (probably extinct).	**1. alpinum**

1. T. alpinum (L.) Pers.

Eriophorum alpinum L.; *Scirpus hudsonianus* (Michx) Fernald

A *shortly creeping* perennial 10–40 cm. *Stems* slender,

trigonous, *rough*. Lvs setaceous, keeled. Spikelet 5–7 mm, 8–12-fld. Glumes obtuse, yellowish-brown, midrib green. *Bristles 4–6, up to 25 mm, white, crumpled*. Nut 1 mm, obovoid-trigonous. Fl. 4–5, Fr. 6. $2n = 58$. Hel.

Closely resembles *T. cespitosum* when in fl., but may always be distinguished by the distinctly creeping rhizome.

Native. Formerly in a bog in Angus, but now apparently extinct through drainage. Europe, southwards in the mountains to C. Spain and S. Ural; Siberia; North America.

2. T. cespitosum (L.) Hartman   Deer-grass.

Scirpus caespitosus L.

A *densely tufted* perennial 5–35 cm. *Stems* slender, *terete*, *smooth*. Lower sheaths lfless, light brown, shiny. Spikelet 3–6 mm. Glumes subacute, the two lower larger than the rest. *Bristles somewhat longer than fr. but shorter than glumes, brownish*. Nut 2 mm, ovoid, trigonous. Fl. 5–6. Fr. 7–8. Hs. or Hel.

Subsp. cespitosum

Basal sheaths shining; uppermost sheath (Fig. 66A) fitting tightly round the stem (at least in fresh material), the opening c. 1 mm, hyaline margin narrow. Spike with 3–10 fls. Glumes brown with a yellowish-brown midrib, the lowest ending in a short, stout green point. $2n = 104$.

Fig. 66. Uppermost sheaths of *Trichophorum cespitosum*. A, subsp. *cespitosum*; B, subsp. *germanicum*. ×2·5.

Subsp. germanicum (Palla) Hegi

T. germanicum Palla; *Scirpus germanicus* (Palla) Lindman

Basal sheaths scarcely shining; uppermost sheath (Fig. 66B) fitting loosely round the stem, the opening 2–3 mm, with broad hyaline margin. Spike with 8–20 fls. Glumes brown with a green midrib, the lowest ending in a stout, green, often almost lf-like, point which usually equals or exceeds the spikelet.

Native. In damp acid peaty places, particularly blanket bogs and heaths, locally dominant. The distribution of the subsp. is not known in detail, but subsp. *germanicum* is much the commoner; subsp. *cespitosum* is rare and its distribution is imperfectly known. The sp. is scat-

tered throughout much of the British Is., but absent from base-rich soils. Europe, southwards to C̆ Spain and S. Bulgaria; Himalaya; North America; Greenland.

3. Eleocharis R.Br.

Perennial herbs. Stems terete or rarely 4-angular. Transverse section of stem with numerous approximately equal air canals without vascular bundles at the intersections of the strips of tissue separating the canals. *At least the upper sheaths entirely lfless. Spikelet solitary, terminal*, the lowest glume usually sterile, or at least different in shape from the others. Fls hermaphrodite. Perianth bristles 0–8, shorter than or not much exceeding nut. Stamens 3 or 2. Stigmas 3 or 2. Style usually with a swollen persistent base. The measurements of nut-length given in the descriptions exclude the style-base. About 200 spp., cosmopolitan.

1 Lowest glume at least $\frac{1}{2}$ as long as spikelet. 2
  Lowest glume much less than $\frac{1}{2}$ as long as spikelet. 4
2 Upper sheath conspicuous, brownish; spikelet brown; style-base enlarged, persistent. 3
  Upper sheath very delicate and inconspicuous; spikelet greenish; style-base not enlarged. **1. parvula**
3 Stems 4-angled; glumes 2 mm, the lowest sterile (wet sandy places, sometimes submerged). **2. acicularis**
  Stems terete; glumes 5 mm, the lowest fertile (damp peaty places). **3. quinqueflora**
4 Plant densely tufted; upper sheath obliquely truncate; stigmas 3; nut triquetrous. **4. multicaulis**
  Plant not densely tufted; upper sheath almost transversely truncate; stigmas 2; nut biconvex. 5
5 The 2 lowest glumes empty, not more than $\frac{1}{2}$ encircling base of spikelet. 6
  The lowest glume empty, ±completely encircling base of spikelet. **7. uniglumis**
6 Bristles 4 (very rarely 0), style-base broader, obviously constricted at junction with nut. **5. palustris**
  Bristles (4–)5(–6), style base long and narrow, hardly constricted at junction with nut. **6. austriaca**

1. E. parvula (Roemer & Schultes) Link ex Bluff, Nees & Schauer

Scirpus parvulus Roemer & Schultes; *S. nanus* Sprengel, non Poiret

A slender glabrous perennial 2–8 cm. *Runners capillary, whitish*, forming small whitish easily detached tubers at their tips. Stems tufted, grooved, setaceous, soft; uppermost sheath very thin, brownish, lfless. Lvs setaceous, channelled, translucent and septate, about equalling stems. *Spikelet 2–3 mm, 3–5-fld. Lowest glume* sterile, c. $\frac{3}{4}$ *as long as spikelet*, encircling it at base. *Glumes* c. 2 mm, ovate, obtuse, *greenish-hyaline*. Bristles longer than fr. Stamens 3. Stigmas 3. Nut c. 1 mm, triquetrous, yellowish. Fl. 8–9. $2n = 10^*$. Hel.

Native. In wet muddy places near the sea, very local. Devon, S. Hants; Merioneth, Caernarvon; N. Kerry, Wicklow, Londonderry. C. Scandinavia southwards to Portugal and S.E. Russia; N. and S. Africa; Japan; America.

2. E. acicularis (L.) Roemer & Schultes

Needle Spike-rush.

A slender rhizomatous perennial 2–10 cm, when submerged and sterile up to 50 cm. *Rhizome creeping, brown. Stems few together, setaceous, 4-angled*, rather stiff; *sheaths brown, lfless. Spikelet 2–4 mm, 4–11-fld. Lowest glume sterile, $\frac{1}{2}$ as long as spikelet. Glumes* c. 2 mm, ovate, *obtuse, reddish-brown.* Bristles 0–1, deciduous. Stamens 3. Stigmas 3. Nut c. 1 mm, finely longitudinally ribbed; swollen style-base separated from top of nut by deep groove. Fl. 8–10. $2n = 20$. Hel.

Native. In wet sandy and muddy places at margins of lakes and pools, sometimes submerged and then persistently sterile. Scattered throughout the British Is. north to W. Inverness, rather local. Most of Europe; N. Asia; Australia; North and South America; Greenland.

3. E. quinqueflora (F. X. Hartmann) O. Schwarz

Few-flowered Spike-rush.

Scirpus pauciflorus Lightf.; *E. pauciflora* (Lightf.) Link

A somewhat tufted glabrous perennial 5–30 cm. *Rhizome short, stout, producing slender runners.* Stems slender (usually less than 1 mm diam.); sheaths lfless, pale reddish-brown, *the upper rather obliquely truncate and obtuse.* Spikelet 4–10 mm, 3–7-fld. *Lowest glume often fertile*, obtuse, *more than $\frac{1}{2}$ as long as spikelet*, encircling it at base. *Glumes* 5 mm, ovate, *acuminate*, reddish-brown with broad hyaline margins. Bristles 4–6 about as long as fr. Stamens 3. Stigmas 3. Nut 1.8–2.2 mm, trigonous, finely longitudinally ribbed, grey when dry, black when fresh; swollen style-base elongate, trigonous, slightly constricted at junction with nut. Fl. 6–7. Fr. 7–8. $2n = 20^*$; 80, c. 100. Hel.

Somewhat resembles *Trichophorum cespitosum*, but is readily distinguished by the completely lfless sheaths.

Native. In damp peaty places on moors and in fens, always(?) with a fairly good supply of bases, local. Scattered throughout the British Is., commoner in the north than the south. Most of Europe, but rare in the south; temperate Asia; North America; Greenland.

4. E. multicaulis (Sm.) Sm.   Many-stalked Spike-rush.

A densely tufted glabrous perennial 15–30 cm. Stems rather slender (c. 1–1.5 mm diam.); sheaths all lfless, pale reddish or brownish, often straw-coloured, ±hyaline, *the upper very obliquely truncate and acute.* Spikelet 5–15 mm, 10–30-fld, often proliferating vegetatively (viviparous). *Lowest glume* sterile, c. $\frac{1}{4}$ *as long as spikelet*, encircling it at base. Glumes c. 5 mm, ovate-oblong, obtuse, reddish-brown, ±hyaline, the midrib green. Bristles 4–6, longer than fr. Stamens 3. *Stigmas* 3. *Nut* 1.2–1.5 mm, *triquetrous*, smooth, yellowish; swollen style-base forming a triquetrous cap to nut. Fl. 7–8. Fr. 8–10. $2n = 20^*$. Hel.

Native. In wet, usually peaty places, particularly in acid bogs and on wet sandy heaths. Scattered throughout the British Is. but mainly in the south and west,

locally common. W. Europe, north to 61° 50′ in Norway and east to S.E. Sweden, S.W. Poland and N.W. Yugoslavia; western N. Africa; Azores.

5. E. palustris (L.) Roemer & Schultes

Common Spike-rush.

E. eupalustris H. Lindb. fil.

A glabrous perennial 10–60 cm. *Rhizome far-creeping, producing single stems in the first season, then many small tufts.* Stems stout to slender (1–4 mm diam.), reddish at base; sheaths all lfless, yellowish-brown, *the upper nearly transversely truncate.* Spikelet 5–20 mm, many-fld. Two *lowest glumes* sterile, much shorter than spikelet, *not more than $\frac{1}{2}$ encircling base of spikelet.* Glumes ovate, margins hyaline. Bristles 0–4, shorter or longer than fr. Stamens 3. *Stigmas* 2. *Nut biconvex*, finely punctate or nearly smooth, yellow to deep brown; swollen style-base constricted at base. Fl. 5–7. Fr. 6–8. Hel.

Native. In marshes, ditches and at margins of ponds. Generally distributed and common throughout the British Is. North temperate regions.

Subsp. **palustris**: Spikelets usually with 40–70 florets. Glumes from middle of spikelet 2.75–3.5 mm. Nut 1.3–1.4(–1.5) mm. Stomatal length 35–56 μm. $2n = 16^*$. Less common than subsp. *vulgaris* in the British Is.

Subsp. **vulgaris** Walters: Spikelets usually with 20–40 florets. Glumes from middle of spikelet 3.5–4.5 mm. Nut (1.3–)1.5–2 mm. Stomatal length 50–77 μm. $2n = (37)$ 38* (39, 40). Common and widely distributed.

6. E. austriaca Hayek   Northern Spike-rush.

Like *E. palustris* but stems weaker with fewer more widely-spaced vascular bundles, and with very little reddish coloration at base; spikelet short, dense, often conical; glumes small, readily caducous; ripe nut 1.2–1.4 mm, yellowish, with very narrow style-base, barely constricted at junction with nut; bristles (4–)5(–6), usually much exceeding nut, but very fragile.

Probably native. Beside upland rivers in Yorks., Northumberland and Cumberland. Scattered localities from N. Norway to N. Spain, C. Yugoslavia and S. Ural.

7. E. uniglumis (Link) Schultes

Incl. *E. watsonii* Bab.

Differs from *E. palustris* as follows: Stems slender, usually shiny; lower sheaths reddish-tinged. *Lowest glume sterile, ±encircling base of spikelet*; glumes usually darker brown with narrower hyaline margins. Bristles frequently 0. Nut 1.4–2.2 mm, usually rather coarsely punctate-striate. $2n = 16, 32; 40^*; 46^*$; c. 68^*, 92^*. Hel.

A much less variable plant than *E. palustris*, often confused with that sp. and with *E. multicaulis*. From the former it may be distinguished by the lowest glume ±encircling the base of the spikelet, and from the latter by the obtuse sheath, 2 stigmas and biconvex nut.

Native. In marshes with rather open vegetation, not in ponds. Frequent near the south and west coasts; scattered and very local inland. Most of Europe; W. Asia; N. Africa.

4. SCIRPUS L.

Stout perennial herbs with lfy stems. *Lvs with flat or keeled, well-developed blades. Infl. terminal, mostly much-branched.* Bracts several, lf-like. Perianth of 1–6 bristles, rough with backward-directed short hairs. Stamens 3. Stigmas 3.

Infl. dense; bracts much exceeding infl.; spikelets
    8–45 mm, red-brown.   **1. maritimus**
Infl. lax; bracts about equalling infl.; spikelets 3–4 mm,
    green or greenish-brown.   **2. sylvaticus**

1. S. maritimus L.   Sea Club-rush.

A stout glabrous perennial, 30–100 cm. Rhizome producing short runners, tuberous at tip. Stems triquetrous, rough towards top, lfy. Lvs up to c. 10 mm wide, margins rough. *Infl.* c. 5 cm, *dense*, corymbose. *Bracts* lf-like to setaceous, the larger *much longer than infl. Spikelets* 8–45 mm, rather few, ovoid, *red-brown*, sessile or in groups of 2–5 at the ends of branches. *Glumes* 5–7.5 mm, *ovate, apex emarginate, awned from the sinus.* Bristles shorter to longer than nut, brown. Stamens 3, filaments flattened. *Nut* 3 mm, *broadly obovate* from a cuneate base, *plano-convex, brown, shiny.* Fl. 7–8, Fr. 8–9. $2n = 80$; c.104. Hel.

Native. In shallow water at the muddy margins of tidal rivers and in ditches and ponds near the sea, locally abundant, rarely inland. Around the coasts of the British Is. north to E. Ross and the Outer Hebrides. Most of Europe and temperate Asia.

2. S. sylvaticus L.   Wood Club-rush.

A stout glabrous perennial 30–120 cm. Rhizome creeping. Stems trigonous, smooth, lfy. Lvs up to 20 mm wide, margins rough. *Infl.* up to 15 cm, *lax, spreading. Bracts* lf-like or the shorter ones setaceous, the larger *shorter than or about equalling infl. Spikelets* 3–4 mm, very numerous, ovoid, *green or greenish-brown*, solitary or in small dense clusters at ends of branches. *Glumes* 1.5–2.9 mm, ovate; *apex entire*, obtuse or subacute. Bristles equalling or longer than nut, brown, rough. Stamens 3, filaments flattened. *Nut* 1 mm, *ovoid or subglobose, compressed-trigonous, yellowish, not shiny.* Fl. 6–7. Fr. 7–8. $2n = 64$; 62. Hel.

Native. In marshes, wet places in woods and beside streams, local. Scattered throughout the British Is. north to Banff and Argyll. Most of Europe; Caucasus; Siberia.

5. HOLOSCHOENUS Link

Tall perennial herbs. Infl. apparently lateral, the stem being continued by a ½-terete bract. *Spikes several, globose, pedunculate, each of numerous spirally arranged spikelets.* Perianth bristles usually 0. Stamens 3. Stigmas 3.

1. H. vulgaris Link   Round-headed Club-rush.
Scirpus holoschoenus L.

A densely tufted perennial 50–150 cm. Stems terete, 2–4 mm diam., smooth. Sheaths mostly without blades, the upper with ½-terete rigid blades shorter than stem, margins rough. Infl. ±umbellate, subtended by (1–)2 long green bracts, one forming an apparent continuation of the stem. Heads 5–10 mm diam., some pedunculate, others sessile. Peduncles stout, flattened, margins rough. Glumes 1.5–3 mm, obovate, mucronate or almost 3-lobed, ciliate. Nut 1 mm, ovoid-trigonous, crowned by the persistent style-base. Fl. 8–9.

Native. On damp sandy flats by the sea, very local. N. Devon, N. Somerset, introduced in Glamorgan. Europe north to S.W. England and White Russia, N.W. Africa; Siberia; Canaries.

6. BLYSMUS Panzer

Perennial herbs. Stems subterete, lfy. *Infl. a terminal spike of several distichous spikelets.* Fls hermaphrodite, spirally arranged. Perianth of 3–6 bristles. Stamens 3. Stigmas 2; style persistent.

Four spp. in temperate Europe and Asia.

Lvs flat, keeled, rough; bracts (except lowest) many-
    ribbed, shorter than spikelet; glumes 3 mm, acute,
    reddish-brown.   **1. compressus**
Lvs involute, ±rush-like, smooth; bracts (except lowest)
    1–3-ribbed, equalling spikelets; glumes 5 mm, obtuse,
    blackish-brown.   **2. rufus**

1. B. compressus (L.) Panzer ex Link   Broad Blysmus.

Scirpus planifolius Grimm; *S. caricis* Retz.; *S. compressus* (L.) Pers., non Moench

A glabrous perennial with a far-creeping rhizome. Stems 10–35 cm, smooth. *Lvs* 1–2 mm wide, rather shorter than stems, tapering from the base, *flat, keeled, margins rough.* Spikelets 5–7 mm, reddish-brown, 10–12 (rarely fewer). Infl. c. 2 cm. Lowest bract green, longer or shorter than infl., *other bracts several-ribbed, shorter than spikelet. Glumes* 3 mm, ovate, *acute, reddish-brown* with a pale midrib and narrow hyaline margin. *Bristles* brown, rough, *longer than nut*, persistent. Nut c. 2 mm, suborbicular, plano-convex, shortly stipitate, blackish. Fl. 6–7. Fr. 8–9. $2n = 44$. Hel.

Native. In marshy places, usually in rather open communities, locally abundant. Scattered throughout England and S. Scotland extending northwards to the Hebrides. Europe; Atlas Mts.; temperate Asia.

2. B. rufus (Hudson) Link   Narrow Blysmus.
Scirpus rufus (Hudson) Schrader

Similar to *B. compressus* but differing as follows: Lvs involute, ±rush-like, scarcely tapering, smooth. Spikelets dark brown, 5–8 in a spike. Bracts (except the lowest) 1–3-ribbed, equalling spikelet. Glumes 5 mm, ovate, obtuse, blackish-brown. Bristles white, short, caducous. Nut 4 mm, ovate, plano-convex, light brown. Fl. 6–7. Fr. 8–9. $2n = 40$. Hel.

Native. Among short grass in salt-marshes, locally abundant. Scattered round the coasts of the British Is. from N. Lincoln and Kerry north to Shetland but very local in the south. Coasts of N. Europe and temperate Asia; inland in N. Germany.

7. SCHOENOPLECTUS (Reichenb.) Palla

Mostly stout perennial herbs. *Stems* triquetrous or terete, *usually nearly or quite lfless.* Transverse section of stem with numerous approximately equal air canals; vascular bundles occur at the intersections of the strips of tissue separating the canals. *Lower bract making what appears to be a continuation of the stem beyond the infl.* Infl. apparently lateral, sessile, capitate or with short branches. Perianth of 6 (rarely fewer or 0) bristles rough with short downward-directed hairs. Stamens 3. Stigmas 3 or 2.

1 Stems triquetrous.	2
Stems terete.	**3. lacustris**
2 Upper sheath usually with a short blade; glumes with obtuse lateral lobes; bristles about equalling nut (very local).	**1. triqueter**
Two or 3 uppermost sheaths with blades up to 30 cm; glumes with acute lateral lobes; bristles much shorter than nut (Jersey).	**2. pungens**

1. S. triqueter (L.) Palla   Triangular Club-rush.

Scirpus triqueter L.

A stout glabrous perennial, 50–150 cm. Rhizome creeping. *Stems triquetrous, the upper sheath usually with a short blade.* Infl. a dense head, usually with some lateral branches up to c. 4 cm. *Lower bract up to twice as long as mature infl.* Spikelets 5–10 mm, ovoid, reddishbrown. *Glumes* 4 mm, broadly obovate, fringed, brownish-hyaline, shallowly emarginate *with rounded lateral lobes*, mucronate; midrib green. *Bristles about equalling nut.* Stigmas 2. Nut c. 2.5 mm, compressed, reddishbrown, shiny. Fl. 8–9. $2n = 40$. Hel.

Native. Muddy banks of tidal rivers, very local. Tamar, Arun, Thames, Medway and Shannon. W., C and S. Europe (local in the north of its range); W. Asia.

2. S. pungens Vahl   Sharp Club-rush.

Scirpus americanus auct. eur., non Pers.

A rather slender glabrous perennial 30–60 cm. *Stems triquetrous. Lvs up to* 30 cm, 2–3, linear. Infl. a dense head of 2–6 sessile spikelets. *Lower bract* up to 15 cm, *much exceeding infl. Glumes* 4 mm, broadly ovate, emarginate *with acute lateral lobes*, mucronate. *Bristles much shorter than nut*, often 0. Stigmas 2. Nut 2.5 mm, compressed, brownish, shiny. Fl. 6–7. Fr. 8–9. $2n = $ c. 80. Hel.

Native. Margins of ponds near the sea, very local. Jersey, introduced in Lancs. W. and C. Europe extending to C. Italy.

3. S. lacustris (L.) Palla   Bulrush.

A stout glabrous rhizomatous perennial 50–300 cm. Stems up to 1.5 cm diam., terete. Submerged lvs linear;

sheaths on aerial stems without blades, except for the uppermost. Lower bract usually shorter than the mature infl. Infl. a dense head or an umbel with unequal rays. Spikelets 3–10 mm, ovoid. Glumes 3–4 mm, broadly obovate, emarginate, with obtuse lateral lobes and an often papillose mucro. Bristles 5–6, scabrid, at least as long as the fr. Fl. 6–7. Fr. 8–9. Hyd.

Native. In rivers, lakes and ponds, usually where there is abundant silt.

Subsp. **lacustris**; Stems up to 300 cm, green. Glumes ±smooth. Stigmas usually 3; nut 2.5–3 mm, compressed-trigonous. $2n = 42$. Scattered throughout the British Is. in fresh water; commonest in S.E. England and C. Ireland. Almost throughout Europe; the same or closely related taxa in Asia, Africa, North and Central America and Australia.

Subsp. **tabernaemontani** (C. C. Gmelin) Á. & D. Löve (*Scirpus tabernaemontani* C. C. Gmelin,): Stems not more than 150 cm, ±glaucous; glumes with numerous reddish papillae, especially near the apex; stigmas usually 2; nut 2–2.5 mm, planoconvex or biconvex. $2n = 42, 44$. Scattered throughout the British Is., often in brackish water and usually near the coast.

S. × **carinatus** (Sm.) Palla (*S. lacustris* × *S. triqueter*) occurs occasionally with the parents. It may be recognized by the stems being trigonous towards the top and terete below.

8. ISOLEPIS R.Br.

Slender herbs. Stems terete. Lvs few, filiform, channelled. *Infl. apparently lateral*, a subterete green bract appearing as a prolongation of the stem. *Spikelets small*, 1–4 *together*. Fls spirally arranged, hermaphrodite. Perianth bristles 0. Stamens 1–2. Stigmas 2–3; style-base persistent.

Bract usually longer than infl.; nut longitudinally ribbed, shiny.	**1. setacea**
Bract usually shorter than infl.; nut smooth, not shiny.	**2. cernua**

1. I. setacea (L.) R.Br.   Bristle Club-rush.

Scirpus setaceus L.

A slender tufted annual or perennial herb with filiform stems 3–15(–30) cm. Lvs 1–2, usually shorter than stems. *Bract usually much exceeding infl.* Spikelets usually 2–4, up to 5 mm, ovoid. Glumes 1.25–2.1 mm, ovate, mucronate, purple-brown, midrib green, margins hyaline. *Nut* 0.5–1.2 mm *trigonous-obovoid, dark brown, ±shiny, longitudinally ribbed* and transversely striate . Fl. 5–7. Fr. 6–9. $2n = 28$. Hel. or Th.

Native. In damp places, sometimes among taller herbage in marshy meadows, more often in bare sandy or gravelly places and beside lakes. Throughout the British Is., in suitable localities. Most of Europe, except the north-east; N. Africa; temperate Asia; Madeira, Azores.

2. I. cernua (Vahl) Roemer & Schultes
Slender Club-rush.

Scirpus cernuus Vahl; *S. filiformis* Savi, non Burm. fil.; *S. pygmaeus* (Vahl) A. Gray, non Lam.; *S. savii* Sebastiani & Mauri

Like *I. setacea* but differs as follows: Annual. Bract shorter or slightly longer than infl. Spikelets usually solitary. Glumes greenish, often with a pale brown or reddish-brown (not purplish) spot on either side of the midrib. Nut broadly trigonous-obovoid or suborbicular, reddish-brown, smooth, not shiny. Fl. 6–8.

Native. In wet places, especially in ±bare sandy or peaty habitats near the sea, local. Great Britain from the Isle of Wight to the Outer Hebrides; inland in Suffolk and Norfolk; in all the maritime counties of Ireland, but rarely far from tidal water. S. and W. Europe; N. Africa; Macaronesia.

9. ELEOGITON Link

Like *Isolepis*, but stem elongated, lfy, usually devoid of lfless sheaths, and spike always solitary, terminal and not over-topped by a stem-like bract.

1. E. fluitans (L.) Link
Floating Club-rush.

Scirpus fluitans L.

A slender, floating lfy perennial 15–40 cm. Stems compressed, branched. Lvs up to 10 cm × 2 mm. Spike 2–5 mm, narrow-ovoid, 3–5-fld, solitary, terminal. Peduncles up to 10 cm, smooth, terete below, triquetrous above. Glumes c. 2 mm, ovate, subacute, pale greenish-hyaline. Bristles 0. Stamens 3. Stigmas 2. Nut 1.5 mm, obovate, compressed-trigonous, pale, crowned by the persistent style-base. Fl. 6–9. Fr. 7–10. $2n = 60$. Hyd.

Native. In ditches, streams and ponds, particularly those with peaty water. Widely distributed but local, mainly in the south and west of the British Is. W. and W.C. Europe extending to S. Sweden and Italy.

10. CYPERUS L.

Annual or perennial rhizomatous herbs. Stems usually lfy, sometimes winged. *Infl. umbellate or capitate. Bracts lf-like.* Spikelets many-fld. *Fls distichous*, hermaphrodite, rarely some male or lowest sterile. Perianth 0. Stamens 3, sometimes 2 or 1; filaments not elongating after flowering. Stigmas 2–3. Nut trigonous or, in spp. with 2 stigmas, lenticular.

About 550 spp. in all the warmer parts of the world, rare in colder regions. *C. papyrus* L. is the source of the papyrus of the ancient Egyptians. The swollen tuberous rhizomes of several spp. are used as food (tiger nuts).

Perennial 20–150 cm; lvs (2–)4–10 mm wide; rhachilla winged; glumes 2–3 mm   **1. longus**
Annual 5–20 cm; lvs 1–3 mm wide; rhachilla not winged; glumes 1–1.3 mm.   **2. fuscus**

1. C. longus L.
Galingale.

An erect glabrous *perennial* 20–150 cm. Rhizome shortly creeping, sympodial, not or scarcely swollen. Stems tri-

gonous, smooth. *Lvs* (2–)4–10 mm *wide, shorter than or equalling stems. Infl. a simple or compound umbel.* The larger bracts much longer than infl. *Primary branches* very unequal, the longer c. 10 cm, each surrounded at base by a cylindrical, loose-fitting membranous sheath obliquely truncate at mouth. Bracteoles small, setaceous. Secondary branches surrounded at base by sheaths similar to those of primary branches. Spikelets 4–25 mm, linear or linear-lanceolate, compressed, distichous, crowded. Lower glumes sterile, c. 1 mm, ovate, ±truncate, hyaline. *Fertile glumes* 2–3 mm, distichous, *closely imbricate*, lanceolate to ovate, *obtuse, keeled*, reddish-brown, keel green, *margins* hyaline towards base and *strongly decurrent forming a delicate wing to the rhachilla.* Stamens 3. Stigmas 3. Nut c. 1 mm, compressed-trigonous, reddish-brown. Fl. 8–9. Hel.

Native. In marshy places beside ponds and in ditches, very local. W. Cornwall, N. Devon, N. Somerset, S. Hants, Isle of Wight, E. Kent, Pembroke, Caernarvon; Channel Is. S., W. and C. Europe, northwards to Wales; W. and C. Asia; N. Africa.

2. C. fuscus L.
Brown Galingale.

A glabrous *annual* 5–20 cm. Stems triquetrous, soft, smooth. *Lvs* 1–5 mm *wide, usually shorter than stems*, soon withering. *Infl. subcapitate or a small dense umbel with few short branches up to c.* 1 cm. Bracts 2–4 mm wide, much exceeding infl. Branches with very loose membranous sheaths split almost to base. Spikelets up to c. 6 mm, oblong, compressed. *Glumes* usually all fertile, *spreading, scarcely imbricate, c.* 1 mm, ovate, *acute, scarcely keeled, dark brown* with broad reddish-brown middle portion, *margins not hyaline and decurrent.* Stamens usually 2. Stigmas 3. Nut 0.75 mm, compressed-trigonous, yellowish. Fl. 7–9. Fr. 8–10. $2n = 72$. Th.

Native. In damp places, especially on bare ground left by the drying up of ponds and ditches, very local. N. Somerset, S. Hants, Surrey, Berks, Middx. Most of Europe; Asia; N. Africa; Madeira, Tenerife.

11. SCHOENUS L.

Perennial herbs. Stems terete. Upper sheaths with longer or shorter blades. *Spikelets 1–4-fld, in compressed, bracteate, terminal heads.* Bract of lowest spikelet encircling base of whole infl. *Glumes distichous, several lower sterile.* Fls hermaphrodite. Perianth of (0–)3–6 bristles. Stamens 3. Stigma 3; style somewhat thickened at base, usually deciduous. Nut trigonous or nearly globose.

About 100 spp., mainly in Australia and New Zealand, a few in Europe, Asia and America.

Lvs at least ½ as long as stems; lf-like point of lowest bract longer than its sheathing base.   **1. nigricans**
Lvs not more than ⅓ as long as stems; lf-like point of lowest bract not longer than its sheathing base.   **2. ferrugineus**

1. S. nigricans L. Bog-rush.

A densely tufted glabrous perennial 15–75 cm. Stems smooth, tough, wiry, lfy only at base. *Lvs at least ½ as long as stem*, wiry, subterete, margins involute. Lower sheaths dark reddish-brown or almost black, tough, shiny. Infl. 1–1.5 cm, dense, ovoid, blackish. *Lowest bract with a lf-like point longer* than its sheathing base. *Spikelets* 5–8 mm, flattened, *5 or more in an infl.* Glumes distichous, keeled, slightly rough on keel. Bristles absent or very short. Nut 1.5 mm, white, shiny, globose. Fl. 5–6. Fr. 7–8. $2n = 44^*$, 54. Hs.

Native. In damp, usually peaty, base-rich places especially near the sea, sometimes in salt-marshes; on blanket bog in W. Ireland; locally abundant. widely distributed throughout the British Is. from Cornwall to Shetland but absent from many inland areas. Most of Europe eastwards to Estonia and the Crimea; N. Africa; W. Asia.

2. S. ferrugineus L. Brown Bog-rush.

Like *S. nigricans* but smaller and more slender. Stems 10–40 cm. Lvs usually less than ⅓ as long as stem. Infl. up to c. 1 cm, rather lax and narrow. Lowest bract with a lf-like point not longer than its sheathing base. Spikelets 1–3 in an infl. Glumes quite smooth on keel. Bristles 6, longer than nut. Nut trigonous. Fl. 7. $2n = 76$. Hs.

Native. Wet peaty places in base-rich flushes, very local; Perthshire. From Fennoscandia and N.W. Russia to S.E. France, and S.E. Russia.

12. RHYNCHOSPORA Vahl

Lfy perennial herbs. Infl. of a compact terminal head or short spike with or without 1 or 2 ±distant, long-pedunculate lateral heads. Bracts lf-like, the lower sheathing. *Spikelets* 1–2-*fld.* Fls hermaphrodite or the upper sometimes unisexual. *Glumes spirally arranged,* imbricate, *several sterile. Perianth of 5–13 bristles.* Stamens 3 or 2. Stigmas usually 2, rarely 3; style-base enlarged, persistent, forming a beak to the nut. Nut biconvex or rarely trigonous.

About 200 spp., cosmopolitan but mainly tropical.

Plant without a creeping rhizome; spikelets whitish or pale brown; bracts not or little longer than terminal head; bristles 9–13. **1. alba**
Plant with a creeping rhizome; spikelets dark reddish-brown; bracts 2–4 times as long as terminal head; bristles 5–6. **2. fusca**

1. R. alba (L.) Vahl White Beak-sedge.

A slender glabrous ±tufted perennial 10–50 cm. Stems terete or trigonous above, lfy. Lvs shorter than or about equalling stems 1–2 mm wide, channelled, margins rough. *Lower sheaths* lfless, *often bearing bulbils in their axils. Bracts not or slightly exceeding terminal head.* *Spikelets* 4–5 mm, *whitish*, becoming pale reddish-brown, usually 2-fld. Terminal cluster of spikelets as broad as or broader than long. Glumes usually 4–5, one or 2 lowest sterile, 2–3 mm, the next fertile, the next sterile and the upper fertile, 4–5 mm. *Bristles* 9–13,

shorter than or equalling fr. (including beak), rough with minute downward-directed hairs. Nut 1.5–2 mm, obovate, biconvex or trigonous, beaked. Fl. 7–8. Fr. 8–9. $2n = 26$; 42. Hel.

Native. In wet, usually peaty places on acid soils, local. scattered throughout the British Is. from Cornwall to Shetland, mainly in the north and west. Europe, except the Mediterranean region and the south-east; Siberia; North America.

2. R. fusca (L.) Aiton fil. Brown Beak-sedge.

A slender glabrous perennial 10–30 cm, like *R. alba* but rhizome far-creeping; lvs usually much shorter than stems; lower sheaths mostly with short blades, not bearing bulbils in their axils; bracts 2–4 times as long as terminal heads; spikelets dark reddish-brown; terminal cluster of spikelets usually longer than broad; bristles 5–6, longer than fr., rough with minute upward-directed hairs; beak of fr. minutely pubescent. Fl. 5–6. Fr. 8–9. $2n = 32$. Hs.

Native. On damp peaty soils on heaths and at margins of bogs, rare and local. S. Devon, Dorset, S. Hants, Surrey; Cardigan; W. Ross, W. Inverness; W. Ireland from W. Cork to Sligo and east to Kildare and Cavan. N., W. and C. Europe extending eastwards to N. Italy.

13. CLADIUM Browne

Perennial rhizomatous herbs. Stems terete or nearly so, usually lfy. *Spikelets terete,* 1–3-*fld.* Fls hermaphrodite, or the upper or lower sometimes male. *Glumes* imbricate, *lower sterile. Perianth* 0. Stamens 2(–3). Stigmas (2–)3; *style-base small, persistent.* Nut trigonous or nearly globose.

About 60 spp. in tropical and temperate regions.

1. C. mariscus (L.) Pohl Great Sedge.

C. jamaicense Crantz; *Mariscus mariscus* (L.) Borbás

A stout harsh perennial 70–300 cm. Rhizome creeping. Stems hollow, terete or bluntly trigonous, 1–4 cm diam. at base (including sheaths). Lvs up to 200 × 1.5 cm, evergreen, growing from base and dying away at top, tough, grey-green, keeled, serrate on margins and keel, ending in a long triquetrous point. Sheaths yellowish-brown, not shiny, very tough. Infl. 30–70 × 5–12 cm, much-branched, each branch terminated by a dense head of 3–10 spikelets, 5–10 mm diam. Spikelets 3–4 mm, 1–3-fld, reddish-brown. Glumes smooth, rounded on back, imbricate, lower 2–4 small and sterile, the remainder fertile. Uppermost fl. sometimes male. Stamens usually 2. Stigmas usually 3; style-base small, persistent. Nut 3 mm, ovoid, acuminate, dark brown, shiny. Fl. 7–8. Fr. 8–9. $2n = 36$, c. 60. Hel.

Native. Forming dense pure stands in reed-swamp and in fens, usually on neutral or alkaline soils, locally abundant. Thinly scattered over the British Is. north to W. Sutherland and the Outer Hebrides; frequent only in Norfolk and in W. and C. Ireland. Europe, N. to C. Finland; N. Africa; Asia.

14. KOBRESIA Willd.

Perennial herbs. Lvs mostly basal. *Spikelets 1-fld, unisexual*, arranged in spikes, the male fls at top, the female below. *Female spikelets with 2 glumes, the inner enfolding the flower*. Perianth 0. Stamens 3. Stigmas 3; style-base not enlarged. Nut trigonous.

About 50 spp. in north temperate regions.

1. K. simpliciuscula (Wahlenb.) Mackenzie

False Sedge.

K. caricina Willd.

A densely tufted glabrous perennial 5–20 cm. Stems trigonous, ±rough, stiff, erect. Lvs plicate, shorter than stems. Infl. up to 2 cm, of 3–10 crowded spikes. Spikes male at top, female below. Stigmas 3. Nuts 2–3 mm, trigonous, pale brown, not shiny. Fl. 6–7. Fr. 7–8. $2n = 72$. Hs.

Native. On moors and damp banks in mountain regions, very local. N.W. Yorks, Durham, Westmorland; Argyll, Perth. N. Europe; Pyrenees, Alps, Carpathians; Caucasus, Altai; Greenland; North America.

15. CAREX L.

Perennial rhizomatous herbs. Stems solid, usually lfy, often triangular in section. Lvs usually linear, ±keeled or involute, less often flat; lf-base usually sheathing; sheaths entire; ligule present at junction of lf and sheath. Infl. of 1 or more spike-like panicles. *Fls unisexual, borne in 1-fld spikelets*, each subtended by a glume. Male fls with 2–3 stamens; perianth 0. *Female fls surrounded by a globose, trigonous or compressed sac (perigynium, utricle)* usually with a longer or shorter beak from which the stigmas project. Ovary trigonous and stigmas 3, or biconvex and stigmas 2. Fr. a trigonous or biconvex nut enclosed within the perigynium. Axis of spikelet occasionally prolonged beyond base of ovary (e.g. *C. microglochin*).

The male and female fls are variously arranged in the infl. Our spp. are, with one exception, monoecious. In the majority of spp. the terminal spike and sometimes some of the upper lateral spikes are male and the rest female. The female spikes frequently have a few male fls at the top and the male spikes less frequently a few female fls at the base. The other spp. have male and female fls in the same spike, the male fls being either at the top or base of the spike.

Probably about 2000 spp. throughout the world often in wet places and specially abundant in the Arctic.

In the following descriptions the fr. includes the nut and the perigynium surrounding it, and measurements of length include the beak. Our Carices fall into 37 sections; it is probable that the majority of these groups are not very closely related and consequently the sequence followed is often arbitrary.

Further details, alternative keys and illustrations are to be found in A. C. Jermy, A. O. Chater & R. W. David, *Sedges of the British Isles* (Bot. Soc. Br. Is. Handbook No. 1, London, 1982).

C. capitata L. has been recorded from South Uist in the Outer Hebrides and *C. bicolor* All. and *C. glacialis* Mackenzie from the island of Rhum; they have not been seen recently and were probably introduced. *C. crawfordii* Fernald and *C. vulpinoidea* Michx have occurred as introductions from North America but have not persisted.

The following hybrids occur fairly frequently where the parents grow in proximity; they may be recognized by their partial or complete sterility and their morphology, which is ±intermediate between that of the parents (for further details and an account of the rarer hybrids see Stace, 1975): *C. acuta* × *nigra*; *C. demissa* × *hostiana*; *C. demissa* × *lepidocarpa*; *C. hostiana* × *lepidocarpa*; *C. otrubae* × *remota* (=*C.* × *pseudaxillaris* K. Richter, *C.* × *axillaris* auct., non Good.); *C. paniculata* × *remota* (=*C.* × *boenninghausiana* Weihe); *C. rostrata* × *vesicaria*.

1 Spikes more than 1, the lateral sessile or stalked.   *2*
   Spike 1, terminal.   *69*
2 Spikes dissimilar in appearance, one or more upper male (sometimes ±concealed among the female), some or all the lateral wholly or mainly female; fruiting spikes sometimes close together but not numerous, squarrose and not forming a subglobose or ovoid ±lobed head.   *3*
   Spikes all similar in appearance, usually all with male and female fls; fruiting spikes often numerous and either small and squarrose or forming a subglobose or ovoid, lobed head.   *50*
3 Fr. hairy or papillose, at least towards the top.   *4*
   Fr. glabrous and smooth.   *15*
4 Fl.-stems lateral, lfless, with a few sheaths surrounding their bases; female spikes 2–3 mm wide, overtopping the slender male spike.   *5*
   Fl.-stems terminal, with lvs in the lower part.   *6*
5 Fr. 3–4 mm, slightly longer than glume; basal sheaths purplish.   **48. digitata**
   Fr. 2.5–3 mm, twice as long as glume; basal sheaths yellowish-brown.   **49. ornithopoda**
6 Lvs longer than stems; female spikes very slender, ±concealed in sheathing bracts.   **50. humilis**
   Lvs usually shorter than stems; female spikes 3–6 mm wide, not concealed by sheathing bracts.   *7*
7 Male spikes 2–3.   *8*
   Male spike 1.   *9*
8 Fr. 5–7 mm; sheaths hairy, at least at top.   **21. hirta**
   Fr. 3.5–5 mm; sheaths glabrous.   **22. lasiocarpa**
9 Lvs distinctly glaucous beneath; fr. shortly hairy or merely papillose at top.   **33. flacca**
   Lvs not or scarcely glaucous; fr. hairy all over.   *10*
10 Female spikes not clustered round base of male, ±distant from each other.   *11*
   Female spikes clustered round base of male, only the lowest sometimes rather remote from the rest.   *12*
11 Fr. 3.5–5 mm; lvs involute; stems bluntly angled.   **22. lasiocarpa**
   Fr. 2–3 mm; lvs flat; stems sharply angled.   **52. tomentosa**
12 Lower bract sheathing.   **51. caryophyllea**
   Lower bract not or scarcely sheathing.   *13*
13 Lowest bract green, lf-like; glumes acuminate.   **55. pilulifera**

Lowest bract brown; glumes not acuminate, sometimes mucronate. 14

14 Stems almost smooth; glumes obtuse, broadly
scarious and finely ciliolate at apex. **53. ericetorum**
Stems scabrid at top; glumes mucronate, not ciliolate.
54. montana

15 Lower sheaths and lvs ±hairy; lowest bract crimped
at base. **47. pallescens**
Lower sheaths and lvs glabrous; lowest bract not
crimped at base. 16

16 Lvs involute; stems scarcely angled, curved (brackish
marshes and dune slacks). **41. extensa**
Lvs not involute; stems sharply angled, straight. 17

17 Stigmas 3; nut trigonous; ripe fr. usually inflated or
trigonous. 18
Stigmas 2; nut biconvex; ripe fr. usually flattened,
often biconvex. 42

18 Female spikes purplish-black, ovoid, ±nodding; male
spikes similar but narrower. **56. atrofusca**
Not as above. 19

19 Male spikes 2 or more. 20
Male spike 1, rarely with a second much smaller close
to its base. 24

20 Fr. broadest above the middle, rounded and minutely
hairy or papillose at the top; beak very short.
33. flacca
Fr. broadest at or below the middle, tapering into
a distinct beak. 21

21 Margins of lvs inrolled. **26. rostrata**
Margins of lvs flat. 22

22 Male glumes obtuse to subacute; male spikes 2–3(–4). 23
Male glumes acuminate; male spikes commonly 5–6.
24. riparia

23 Male spikes 5–7 mm wide; fr. 3.5–5 mm; beak short,
scarcely notched. **23. acutiformis**
Male spikes 2–3 mm wide; fr. 4–6 mm; beak long,
deeply notched. **27. vesicaria**

24 Female spikes with not more than 6 fls; fr. 7–9 mm
36. depauperata
Female spikes usually with more than 6 fls; fr. not
more than 6 mm. 25

25 Plant with creeping rhizomes, not forming dense tussocks. 26
Plant forming dense tussocks. 32

26 Female spikes nodding on slender peduncles. 27
Female spikes erect, or the lowest somewhat nodding;
peduncles stout. 29

27 Female spikes 3–4 mm wide, with 5–8 fls.
59. rariflora
Female spikes 5–7 mm wide, with 7–20 fls. 28

28 Lvs 1–1.5(–2) mm wide; glumes less than 1½ times as
long as and at least as wide as fr.; lateral spikes
entirely female. **57. limosa**
Lvs 2–4 mm wide; glumes more than 1½ times as long
as and c. ⅔ as wide as fr.; lateral spikes male at
base. **58. magellanica**

29 Beak of fr. notched; fr. symmetrical, distinctly
veined. 30
Beak of fr. ±truncate; fr. asymmetrical, not veined. 31

30 Glumes with broad scarious margins; fr. scarcely
trigonous, with several prominent veins.
42. hostiana
Glumes without scarious margins; fr. distinctly trigonous, with 2 conspicuous lateral veins.
40. binervis

31 Plant glaucous; beak of fr. entire and transversely
truncate; lowest bract with a close-fitting narrow
sheath. **34. panicea**
Plant green or yellowish-green; beak of fr. slightly
notched and obliquely truncate; lowest bract with
a loose, wide sheath. **35. vaginata**

32 Plant 50–150 cm; lower lvs mostly 10–20 mm wide;
female spikes usually 5–15 cm × 5–10 mm in fruit. 33
Not as above. 34

33 Female spikes 5–26 cm; fr. c. 3 mm, erect.
29. pendula
Female spikes 2–6 cm; fr. 4–5.5 mm, deflexed.
25. pseudocyperus

34 Female spikes 2(–3) mm wide in fr. 35
Female spikes (3–)4–12 mm wide in fr. 36

35 Female spikes (25–)40–80 mm, ±erect. **32. strigosa**
Female spikes up to 15 mm, pendent. **31. capillaris**

36 Female spikes ovoid, scarcely longer than broad.
(43–46) flava group
Female spikes cylindrical, at least twice as long as
wide. 37

37 Peduncle of lowest female spike at least twice as long
as the spike (in fr.); all spikes ±pendent, 3–5 mm
wide. **30. sylvatica**
Peduncle of lowest female spike shorter to little longer
than the spike (in fr.); at least the upper spikes
erect, 5 mm or more wide. 38

38 Lvs 6–12 mm wide; ligule 7–15 mm; glumes acuminate. **37. laevigata**
Lvs 2–5(–8) mm wide; ligule not more than 3 mm;
glumes usually obtuse and mucronate. 39

39 Fr. 3–4 mm, shining, with slender veins and smooth
beak. **39. punctata**
Fr. 4–6 mm, not shining, prominently veined, with
scabrid beak. 40

40 Glumes reddish-brown with a wide scarious margin.
42. hostiana
Glumes pale brown to dark purplish-brown without
a scarious margin. 41

41 Lower sheaths orange-brown; glumes dark reddish-
or purplish-brown. **40. binervis**
Lower sheaths pale to dark brown; glumes pale brown
to pale reddish-brown. **38. distans**

42 Fr. inflated, spreading ±horizontally. **28. saxatilis**
Fr. not inflated, ±erect. 43

43 Glumes (at least in the lower part of the spikes) with
a long excurrent midrib. **66. recta**
Glumes acute or acuminate, but midrib not excurrent. 44

44 Plant tussock-forming. 45
Plant not tussock-forming. 46

45 Margins of lvs rolling inwards on drying; lower
sheaths brown or reddish-brown. **67. nigra**
Margins of lvs rolling outwards on drying; lower
sheaths yellowish-brown. **63. elata**

46 Fr. without veins. 47
Fr. with veins. 48

47 Stems sharply angled, tough; lowest bract not longer
than infl. **68. bigelowii**
Stems bluntly angled, fragile; lowest bract longer than
infl. **65. aquatilis**

48 Fr. 3.5–5 mm; beak prominent, emarginate.
23. acutiformis
Fr. 2–3.5 mm; beak small, entire. 49

49 Lvs (3–)5–10 mm wide, the margins rolling outwards
on drying; male spikes usually 2–4 **64. acuta**

Lvs 1–3(–5) mm wide, the margins rolling inwards on drying; male spikes usually 1.   **67. nigra**
50 Spikes few, either pedunculate or distant from one another; glumes usually blackish and fr. pale green.   *51*
Spikes usually numerous, either small and squarrose or forming a subglobose to ovoid ±lobed head; glumes brown or greenish.   *53*
51 Spikes ±evenly spaced, not or scarcely overlapping.   **60. buxbaumii**
Spikes clustered, all overlapping, except sometimes the lowest.   *52*
52 Infl. nodding; lowest spike with a peduncle half as long as itself; lowest bract usually longer than infl.   **61. atrata**
Infl. erect; spikes all sessile or subsessile; lowest bract shorter than infl.   **62. norvegica**
53 At least some spikes with male fls (usually easily recognized by the white filaments or narrower glumes) at top.   *54*
At least the upper spikes with male fls at base.   *64*
54 Rhizomes long; stems not tufted.   *55*
Rhizomes absent; stems densely tufted.   *59*
55 Spikes forming a subglobose, scarcely lobed head; bracts glume-like, without a setaceous or lf-like apex.   *56*
Spikes forming an ovoid to oblong, distinctly lobed head; bracts with a setaceous or lf-like apex.   *57*
56 Lvs as long as or longer than the curved stems (sandy shores in the north).   **13. maritima**
Lvs much shorter than the straight stems (very wet bogs).   **12. chordorrhiza**
57 Terminal spike entirely male.   **10. arenaria**
Terminal spike female or mixed.   *58*
58 Stems rather stout, not wiry; infl. 3–7(–10) cm.   **9. disticha**
Stems slender, wiry; infl. usually 1.5–2.5 cm.   **11. divisa**
59 Lvs mostly more than 4 mm wide.   *60*
Lvs 1–3(–4) mm wide.   *62*
60 Plant forming large tussocks; lvs rounded on the back.   **1. paniculata**
Plant not forming conspicuous tussocks; lvs keeled on the back.   *61*
61 Ligule 5–10 mm, longer than broad; at least some bracts setaceous, conspicuous; fr. not dropping readily at maturity.   **4. otrubae**
Ligule 2–5 mm, broader than long; bracts short and inconspicuous; fr. dropping readily at maturity.   **5. vulpina**
62 Fr. (3–)4–6 mm.   **(6–8) muricata group**
Fr. 2–3 mm.   *63*
63 Plant densely tufted; lvs yellowish-green; beak of fr. entire.   **2. appropinquata**
Plant loosely tufted; lvs greyish-green; beak of fr. split down one side, the margins overlapping.   **3. diandra**
64 At least the lower bracts lf-like.   **14. remota**
Bracts not lf-like.   *65*
65 Spikes with up to c. 10 fruits; fr. spreading stellately.   **16. echinata**
Spikes with more than 10 fruits; fr. usually erect.   *66*
66 Spikes more than 4, in an interrupted oblong infl.   *67*
Spikes 3–4, in a dense ±lobed infl.   *68*
67 Spikes pale greenish-white; fr. 2–2.5 mm; beak scabrid.   **19. curta**

Spikes brown; fr. 3–4 mm; beak smooth.   **18. elongata**
68 Stems usually 10–60 cm; fr. 3.5–5 mm, almost winged; beak scabrid.   **15. ovalis**
Stems usually 10–20 cm; fr. 2.5–3 mm; beak smooth.   **20. lachenalii**
69 Spike unisexual; plant usually dioecious.   **17. dioica**
Spike male at top, female below.   *70*
70 Glumes persistent; fr. not deflexed when ripe; lvs usually much curled near the top.   **71. rupestris**
Glumes soon falling; fr. deflexed when ripe; lvs straight.   *71*
71 Stigmas 2; fr. distinctly flattened, lanceolate, dark brown.   **72. pulicaris**
Stigmas 3; fr. not flattened, very narrow, yellowish.   *72*
72 Spike with 3–12 fruits; a bristle arising from the base of the nut and protruding along with stigmas from top of fr.   **69. microglochin**
Spike with 2–5 fruits; persistent stigmas alone protruding from top of fr.   **70. pauciflora**

Subgenus VIGNEA (Beauv. ex Lestib.) Kük.

Monoecious or rarely dioecious. Spikes 1–many, hermaphrodite or unisexual; spp. with a solitary spike dioecious. Stigmas usually 2.

1. C. paniculata L.   Greater Tussock-sedge.

A large *densely tufted* glabrous perennial often *building big tussocks* up to c. 1 m diam. Roots stout, felted. Stems 60–150 cm, triquetrous, rough, dark green, spreading. *Lvs 5–7 mm wide*, shorter than stems, incurled or semi-cylindrical, stiff, *dark green*, the margins serrulate; ligule 2–5 mm, rounded or nearly truncate. Lower sheaths dark brown, not fibrous. Infl. 5–15(–20) cm, branched or rarely nearly simple, spikes numerous, few-fld, male at top or the lower entirely female. Bracts glumaceous with setaceous points. *Female glumes 3–4 mm, triangular-ovate*, sometimes mucronate, brownish-hyaline, embracing the fr. *Fr.* (Fig. 67A) 3–4 mm, ovoid-trigonous, strongly corky at base, lateral angles acute, serrulate above, dorsal rounded, dark brown *with many faint veins at base*; beak 1–1.5 mm, broad, *bifid to base*, serrulate or ciliate on margins. Nut 1.5 mm, ovate, biconvex, almost truncate at top. Stigmas 2. Fl. 5–6. Fr. 7. $2n = 60, 62, 64$. Hs. or Hel.

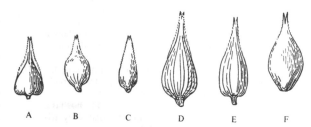

Fig. 67. Fruits of *Carex*. A, *C. paniculata*; B, *C. appropinquata*; C, *C. diandra*; D, *C. otrubae*; E, *C. vulpina*; F, *C. divulsa*. ×5.

Native. In wet, often shady places on peaty base-rich soils. Scattered throughout the British Is. north to Shetland. Europe to 63° N. in Russia, rare in the south; Caucasus; N.W. Africa.

2. C. appropinquata Schumacher

Fibrous Tussock-sedge.

C. paradoxa Willd., non J. F. Gmelin

A *densely tufted* glabrous perennial 30–80 cm, rather similar to *C. paniculata* but smaller. Roots stout or slender, but not felted. Stems triquetrous, rough, *faces flat*. Lvs 1–3 mm wide, shorter than or equalling stems, ±flat, rough, *yellow-green*; ligule c. 2 mm, rounded. Lower sheaths lfless, fibrous, blackish-brown. Infl. 4–8 cm, of 4–6 sessile or nearly sessile spikes. Bracts setaceous, shorter than spikes. *Female glumes* c. 3 mm, *ovate, acuminate, reddish-brown, with wide scarious margin*. Fr. (Fig. 67B) 3–4 mm, broadly ovate or suborbicular, *rather suddenly contracted into the beak*, plano-convex, *distinctly 3–7-veined in lower half, some of the veins reaching base of beak*; beak 1–1.5 mm, *notched*, margins serrulate. Nut c. 1.75 mm, broadly ovate, plano-convex, shortly stipitate. Stigmas 2. Fl. 5–6. Fr. 6–7. $2n = 64$. Hs.

Native. In fens and damp places, on base-rich peaty soils; local. East Anglia; S.E. and Mid-West Yorks; Pembroke; Roxburgh, Selkirk; C. Ireland; formerly in Middx, Herts, Bucks; ascends to 380 m at Malham, Yorks. Europe, except the south; temperate Asia.

3. C. diandra Schrank

Lesser Tussock-sedge

C. teretiuscula Good.

A *shortly creeping* glabrous perennial 20–60 cm. Roots slender, not felted. *Stems slender*, triquetrous, nearly smooth, *faces somewhat convex*. Lvs 1–2(–5) mm wide, shorter than or equalling stems, ±flat, smooth below, serrulate at tips, *grey-green*; ligule very short, truncate. Lower sheaths lfless, ±fibrous, brown. Infl. 1–4 cm, of several sessile spikes. Bracts glumaceous, the lowest sometimes with a setaceous point. *Female glumes* c. 3 mm, *broadly ovate, acute or mucronate*, brownish-hyaline. *Fr.* (Fig. 67C) 3–4 mm, broadly ovate or suborbicular, narrowed into beak, plano-convex, *strongly veined near the base*; beak 1.5–2 mm, broad, notched, *split at back*, the halves overlapping, margins serrulate. Nut c. 2 mm, turbinate, plano-convex, stipitate. Stigmas 2. Fl. 5–6, Fr. 6–7. $2n = 60$. Hs.

Native. In damp meadows and peaty places beside pools, local. Scattered throughout the British Is. north to Orkney. Scattered in Europe (except much of the south); temperate Asia; North America; Japan.

4. C. otrubae Podp.

False Fox-sedge.

C. vulpina auct. occid., non L.

A stout tufted glabrous perennial up to 100 cm. *Stems sharply triquetrous but not winged*, smooth below, rough above, spreading, *faces nearly flat*. Lvs 4–10 mm

wide, shorter than stems, *bright green becoming grey-green when dry and orange-tinged when dead*, channelled, margins rough; *ligule 5–10 mm, longer than broad, not overlapping lf-margins*, ovate, ±acute. *Hyaline front of sheaths neither gland-dotted nor wrinkled*. Lower sheaths brownish, soon decaying. Infl. compound, *branches sessile*, spikes numerous, up to 1 cm. male at top, *yellowish-green or light brown*. Female glumes 4–5 mm, ovate, acuminate, hyaline, with brownish margins and green midrib. Fr. (Fig. 67D) 5–6 mm, ovate, plano-convex, *greenish, becoming dark brown*, distinctly ribbed, *smooth, not readily dropping at maturity*, tapering gradually into the bifid beak which is not slit down the back. Nut 2.5 mm, oblong-ovoid, strongly compressed, shortly stipitate. Fl. 6–7. Fr. 7–9. $2n = 58, 60$. Hs.

Native. On clayey soils, usually in damp grassy places, more rarely in drier places by roads and in hedgebanks. Common and generally distributed throughout most of the British Is., northwards to Outer Hebrides and Sutherland, mainly coastal in the north. Europe, Caucasus; Asia Minor; C. Asia.

5. C. vulpina L.

True Fox-sedge.

Like *C. otrubae*, but stems robust, very sharply angled or almost winged, faces ±concave; lvs bright dark green, even when dry; ligule 2–5 mm, broader than long, overlapping lf-margins, truncate-deltate; hyaline front of lf-sheaths gland-dotted or transversely wrinkled; infl. a warm, slightly reddish, brown; bracts short and inconspicuous with prominent dark auricles; fr. (Fig. 67E) brown, ribs less conspicuous than in *C. otrubae* and usually 0 on the flat face, minutely papillose, readily dropping at maturity. Fl. 5–6. Fr. 6–7. $2n = 68$. Hs.

Native. In damp places near rivers and in ditches, local. S.E. and S.C. England extending northwards to Yorks. Europe, but very local in W. and S. Siberia.

C. muricata group (spp. 6–8).

Glabrous tufted perennials 20–100 cm. Stems triquetrous, rough. Lvs 2–4 mm wide, usually shorter than stems, channelled. Lower sheaths lfless, brown, becoming fibrous in decay. Infl. spike-like, rarely with 1–2 short branches at base, spikes ±crowded, sessile. Bracts setaceous. Glumes 3–4 mm, ovate, acuminate or aristate, pale brown with a green midrib. Fr. lanceolate to ovate, plano-convex, narrowed into a deeply-notched beak. Nut biconvex, shortly stipulate and almost truncate at top. Stigmas 2. Fl. 6–7. Fr. 7–8. Hs.

1 Roots and often basal sheaths tinged with purplish-red; fr. corky and thickened at the base, greenish when ripe. **7. spicata**
Roots and basal sheaths not tinged with purplish-red; fr. not corky and thickened at the base, dark brown when ripe. 2
2 Infl. 3–10(–20) cm; fr. ±equally narrowed at both ends. **6. divulsa**
Infl. 2–3(–4) cm; fr. truncate or rounded at base. **8. muricata**

6. C. divulsa Stokes   Grey Sedge.

Roots and basal sheaths brown. Ligule obtuse, about as wide as or wider than long. Infl. 3–10(–20) cm, sometimes with 1 or 2 short branches at the base. Female glumes whitish or pale greenish-brown. Fr. (Fig. 67F) 3.5–5(–5.5) mm, almost equally narrowed at both ends, not corky and thickened at the base, brown when ripe. $2n = 56, 58*$.

Native. In rough pastures, open woods and hedgebanks. Common in England and Wales, extending northwards to S.E. Scotland; local in Ireland and mainly in the south and east. Much of Europe, north to 59° N. in Sweden; N. Africa; Asia; introduced in North America.

Subsp. **divulsa**; Lvs 2–3 mm wide; ligule about as wide as long. Lowest 3–4 spikes or branches separated from each other by a gap of much more than their own length. Fr. 3.5–4(–4.5) mm, pale, appressed or somewhat spreading. Usually in shady grassy places, commonest in the west.

Subsp. **leersii** (Kneucker) Walo Koch (*C. polyphylla* Kar. & Kir.): Lvs 3–4(–5) mm wide; ligule usually wider than long. Lower spikes separated from each other by a gap of not more than their own length. Fr. 4.5–5(–5.5) mm, dark brown, spreading. In dry calcareous grassland, mainly in the south and east.

7. C. spicata Hudson   Spiked Sedge.

C. contigua Hoppe; *C. muricata* auct. angl., pro parte

Roots and basal sheaths purplish-brown. ligule longer than broad, acute. Infl. 2–4 cm, spikes contiguous or the lowest slightly distant from the next, sessile. Bracts and lf-bases often tinged with purplish-red. Fr. (Fig. 68A) 4–5 mm, greenish when ripe, corky at base, gradually narrowed into a long (c. 2 mm) beak; nut c. 2.2 mm. $2n = $ c. 58.

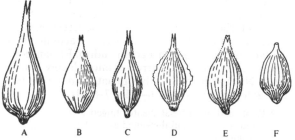

Fig. 68. Fruits of *Carex*. A, *C. spicata*; B, *C. muricata*; C, *C. disticha*; D, *C. arenaria*; E, *C. divisa*; F, *C. chordorrhiza*. ×5.

Native. In marshes or beside ponds or in grassland and on hedgebanks, on basic or acid but base-rich soils. Common in S. and C. England and E. Wales, rarer elsewhere and perhaps not native north of the Ribble and Tees and in Ireland. Europe to 62° N. in Norway, but absent from much of the south; N. Africa, Madeira; W. Asia; introduced in North America.

8. C. muricata L.   Prickly Sedge.

C. pairaei F. W. Schultz

Like *C. spicata* but roots brown; ligule about as broad as long, obtuse, bracts and lf-bases never purplish-red; fr. (Fig. 68B) 3–4 mm, dark brown or almost black when ripe, not corky at base, rather abruptly narrowed into a short (c. 0.7 mm) beak and truncate or rounded at base. Nut 1.7 mm.

Native. Dry grassy places. Scattered throughout England and Wales, local; less frequent in Scotland and Ireland and absent from the north. Most of Europe; W. Asia; N. Africa; Macaronesia; introduced in North America.

Subsp. **muricata**: stems usually strongly scabrid above. Spikes globose. Female glumes much darker and shorter than the greenish or brownish fr. Fr. (3.5–)4–4.5 mm. Calcicole and rare.

Subsp. **lamprocarpa** Čelak. (*C. pairaei* F. W. Schultz): Stems weakly scabrid. Spikes ovoid. Female glumes like or paler than and almost as long as the fr. Fr. 3–3.5 mm. Calcifuge and much commoner than subsp. **muricata**.

9. C. disticha Hudson   Brown Sedge.

An extensively creeping glabrous perennial 20–80 cm. *Stems* triquetrous, rough, *rather stout*. Lvs 2–4 mm wide, about equalling stems, nearly flat; ligule 3–7 mm, tubular, obtuse. Lower sheaths lfless, not fibrous. Spikes numerous, contiguous, sessile, forming a ±dense infl. up to 7(–10) cm. *Terminal spike female or with a few male fls at top, intermediate male, lower female*. Lower bracts glumaceous or sometimes narrowly lf-like and then exceeding infl. Female glumes c.4 mm, ovate, acute, brownish-hyaline. *Fr.* (Fig. 68c) 4–5 mm, ovate, brown plano-convex, distinctly ribbed, *narrowly winged and serrate in upper half*; beak c. 1 mm, bifid. Nut c. 1.75 mm, oval, biconvex, narrowed at top and bottom. Stigmas 2. Fl. 6–7. Fr. 7–8. $2n = 62*$. Hs.

Native. In damp grassy places, fens, marshes and wet meadows. Throughout the British Is. north to Caithness, not uncommon but rather local. C. and N. Europe (except the Arctic), rare in S. Europe; temperate Asia; introduced in North America.

10. C. arenaria L.   Sand Sedge.

An *extensively creeping* glabrous perennial 10–40(–90) cm. Stems triquetrous, rough, often curved. Lvs 1.5–4 mm wide, shorter than or equalling stems, nearly flat; ligule 3–5 mm, triangular, acute. Sheaths on horizontal rhizome soon becoming fibrous; lower sheaths on erect shoots lfless, brown, seldom fibrous. Spikes 5–15, contiguous, sessile, forming a ±dense infl. up to 8 cm. *Upper spikes male, middle ones sometimes male at top, lower female*. Lower bracts glumaceous with setaceous points. Female glumes 5–6 mm, ovate, acute or acuminate, brownish-hyaline. *Fr.* (Fig. 68D) 4–5.5 mm, ovate, plano-convex, distinctly ribbed, *broadly winged and serrate or ciliolate, at least in upper half*; beak 1–1.5 mm, bifid. Nut c. 1.75 mm, compressed-cylindrical, abruptly contracted at top and bottom. Stigmas 2. Fl. 6–7. Fr. 7–8. 2*n* = 58, 64. Hs.

Native. In sandy places usually near the sea, especially on fixed dunes, common. Coasts of the British Is. and locally inland. Coasts of N., N.C. and W. Europe eastwards to near Leningrad; North America (perhaps introduced).

11. C. divisa Hudson   Divided Sedge.

A *creeping* glabrous perennial (15–)30–60(–80) cm. *Stems* triquetrous, ±rough, at least at top, *slender and wiry*. Lvs 1–3 mm wide, shorter or sometimes longer than stems, nearly flat or ±involute; ligule 2–3 mm, ovate, obtuse. Lower sheaths dark brown, lfless, soon decaying. *Spikes 3–8, contiguous or the lower 1–2 somewhat remote, the terminal often male*. Infl. 1–3 cm. Lower bract lf-like or setaceous, shorter than to several times longer than infl. Female glumes 3.5–5 mm, ovate, shortly aristate, brownish-hyaline. *Fr.* (Fig. 68E) 3.5–4 mm, broadly ovate or oval, plano-convex, *faintly veined; beak* 0.5–0.75 mm, *bifid, smooth, not winged*. Nut 2 mm, suborbicular. Stigmas 2. Fl. 5–6. Fr. 7–8. Hs.

Native. In grassy places and beside ditches near the sea or by estuaries, locally abundant but rarely inland. South-east of a line from the Humber to the Bristol Channel; Glamorgan; Holy Island. S., S.C. and W. Europe, mainly coastal in the north; W. Asia, N. Africa.

12. C. chordorrhiza L. fil.   String Sedge.

A glabrous extensively creeping perennial 15–30 cm. *Rhizome obliquely ascending*, giving off lateral non-flowering shoots and terminating in a flowering stem. Stems stout, trigonous, smooth. *Lvs* 1–2 mm wide, those *on the flowering stems few and much shorter than the stems, straight*, flat or ±involute; ligule 1–2 mm, rounded. Lower sheaths several, rather distant, lfless, not fibrous. *Spikes 2–4, male at top, sessile, contiguous, forming a small almost capitate infl.* c. 1 cm. Bracts glumaceous. Female glumes 4 mm, broadly ovate, acute or acuminate, brownish-hyaline. *Fr.* (Fig. 68F) 3.5–4.5 mm, ovoid, *slightly compressed*, shortly stipitate, distinctly veined; *beak* c. 0.5 mm, *±entire*. Nut 2 mm, cylindrical, tapered below, truncate at apex. Stigmas 2. Fl. 6–7. Fr. 7–8. Hel.

Native. In spongy bogs; very rare. W. Sutherland and Easterness. Scattered throughout C. and N. Europe to the Pyrenees and east to C. Ukraine; N. Asia; North America.

13. C. maritima Gunnerus   Curved Sedge.
C. incurva Lightf.

An extensively creeping glabrous perennial. *Stems* 1–18 cm, *curved*, terete; smooth. *Lvs* 1–2 mm wide, *curved, equalling or exceeding stems, involute*, serrulate,

obtuse; ligule 0.5–1 mm, truncate. Lower sheaths brown or blackish, ±fibrous. *Spikes contiguous, forming an ovoid infl.*, 5–15 × 5–10 mm, male at top, male fls almost hidden. Bracts 0. Female glumes 3–4 mm, ovate, acute or obtuse and mucronate, brown with a pale midrib and broad hyaline margins. Fr. (Fig. 69A) 4–4.5 mm, ovoid, brown or almost black; beak 0.5–1 mm, usually smooth. Nut 2 mm, elliptic, biconvex. Stigmas 2, persistent. Fl. 6. Fr. 7. 2*n* = 60. Hs.

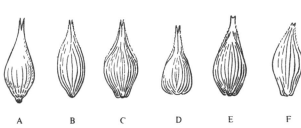

Fig. 69. Fruits of *Carex*. A, *C. maritima*; B, *C. remota*; C, *C. ovalis*; D, *C. echinata*; E, *C. dioica*; F, *C. elongata*. ×5.

Native. In damp hollows on fixed dunes, local. Northumberland; N. Lancs; east and north coasts of Scotland from Angus northwards, Outer Hebrides, Orkney and Shetland. Coasts of arctic and N.W. Europe; Alps; higher mountains of N. Africa, Asia, North and South America.

14. C. remota L.   Remote Sedge

A densely tufted glabrous perennial 30–60(–100) cm. Stems trigonous, smooth below, scabrid above, spreading. Lvs 1.5–2 mm wide, nearly equalling stems, bright green, channelled; ligule 1–2 mm, triangular, acute or obtuse. Lower sheaths straw-coloured, not fibrous. *Spikes* usually 4–7, *remote*, male at base, upper mostly male, lower mostly female; lower 7–10 × 3–4 mm, upper smaller. *Lower bracts lf-like, exceeding infl.*, upper glumaceous. Female glumes c. 2.5 mm, lanceolate to ovate, acute, hyaline, midrib green below, brown above. Fr. (Fig. 69B) 2.5–3.5 mm, ovate, plano-convex, greenish; *beak short, broad, serrulate, notched*. Nut 2 mm, ovate, biconvex. Stigmas 2. Fl. 6. Fr. 7. 2*n* = 62*. Hs.

Native. In damp shady places. Throughout most of the British Is., north to Sutherland. Europe to 63° N. in Norway, rare in the south; Caucasus; N.W. Africa.

15. C. ovalis Good.   Oval Sedge.
C. leporina auct., non L.

A densely tufted glabrous *perennial* 10–90 cm. Stems trigonous, smooth below, rough at top, stiff, often ±curved, lfy only near base. Lvs 1–3 mm wide, shorter than stems, ±flat, bright green, margins rough; ligule c. 1 mm, tubular. *Spikes* (1–)3–9, c. 10 × 5 mm, male at base, *contiguous*, sessile. Lower bracts often setaceous, exceeding spike, upper glumaceous, not more than half as long as spike. *Female glumes* 3–4.5 mm, lanceolate, acute, *brownish-hyaline*. *Fr.* (Fig. 69C)

4–5 mm, erect, elliptical-ovate, plano-convex, *narrowly winged*, distinctly veined, light brown, margins rough near top; *beak* c. 1 mm, bifid, *rough*, green. Nut c. 2 mm, obovate- or elliptical-oblong, plano-convex, shortly stipitate, ±shiny. Stigmas 2. Fl. 7–8. $2n = 64*$; 66, 68. Hs.

Native. In rough grassy places on acid soils, locally common, especially along tracks. Throughout the British Is., but rather local in the south and east. Europe, except the Arctic and on mountains in the south; mountains of N. Africa; temperate Asia; introduced in North America.

16. C. echinata Murray   Star Sedge.
C. stellulata Good.

A tufted glabrous perennial 10–40 cm. Stems trigonous, rather slender, spreading, smooth or scabrid above. Lvs 1–2.5 mm wide, rather shorter than stems, channelled; ligule c. 1 mm, tubular. Lower sheaths soon decaying. *Spikes* 2–5, somewhat remote, 3–6 mm, sessile, *fr. spreading stellately*. Bracts small, glumaceous. Female glumes 2–2.5 mm, broadly ovate, acute, embracing the lower part of the fr., brown with a green midrib and broadly hyaline margins. Fr. (Fig. 69D) 4 mm, yellow-brown, ovate, plano-convex, tapering gradually into the broad *notched scabrid beak*. Nut c. 2 mm, obovoid, somewhat compressed. Stigmas 2. Fl. 5–6. Fr. 6–7. $2n = 56$, 58. Hs. or Hel.

Native. In damp meadows and boggy places, on acid, humus-rich soils. Throughout the British Is., local in the south and east, common in the north and west. Most of Europe, temperate Asia, Siberia; Japan. Mountains of N. Africa; North America.

17. C. dioica L.   Dioecious Sedge.

A shortly creeping *usually dioecious* glabrous perennial 5–15(–40) cm. Stems terete, smooth, rigid, erect. Lvs 0.5–1 mm wide, shorter than stems, channelled, dark green, rather rigid; ligule very short, truncate. Lower sheaths dark brown, not fibrous. Male spike 10–20 × 2–3 mm, subclavate, acute, rarely female at base. Glumes 3–4 mm, ovate to oblong, acute or obtuse, brownish-hyaline. Female spike 5–20 × 5–7 mm, rather dense, 20–30-fld. Bracts 0. *Glumes* 2.5–3.5 mm, *persistent*, ovate, acute, brown with a paler patch surrounding the dark midrib, margins hyaline. *Fr.* (Fig. 69E) 2.5–3.5 mm, spreading horizontally or slightly deflexed when ripe, compressed-ovoid, *tapering to the broad serrulate notched blackish beak*, greenish-brown *with numerous dark brown veins*. Stigmas 2. Nut c. 1.3 mm, suborbicular, compressed. Fl. 5. Fr. 7. $2n = 52*$. Hs.

Native. In fens, base-rich flushes, etc. Scattered throughout the British Is.; local, diminishing through drainage in the south and Midlands. Ascends to 990 m on Ben Lawers. Europe; Siberia.

C. davalliana Sm.   Davall's Sedge.
A densely tufted dioecious perennial 15–25 cm. Lvs setaceous, rough. Female spike c. 15–20-fld. Glumes dark brown, except

for the margins, persistent. Fr. 3.5–4.5 mm, dark brown, somewhat curved and deflexed when ripe. Stigmas 2. $2n = 46$. Hs.

Formerly grew on Lansdown, near Bath, but now long lost through drainage. C. Europe, eastwards to Macedonia.

18. C. elongata L.   Elongated Sedge.

A tufted glabrous perennial 30–70(100) cm. Stems triquetrous, rough. Lvs 2–5 mm wide, usually about equalling stems, nearly flat; ligule 4–8 mm, acute. Lower sheaths lfless, not fibrous. *Spikes* 5–15, c. 1 cm, male at base, or terminal one rarely male at top, sessile, *brown, forming a ±lax infl.* 3–7 cm. Bracts glumaceous. Female glumes 2 mm, obovate, acute or obtuse, brownish-hyaline, midrib green. *Fr.* (Fig. 69F) 3–4 mm, lanceolate, plano-convex, often curved, many ribbed, rounded at base, tapering upwards into the *smooth entire beak*. Nut 2 mm, compressed-cylindrical, narrowed at base, tapering slightly upwards and then truncate. Stigmas 2. Fl. 5. Fr. 6. $2n = 56$. Hs. or Hel.

Native. In marshes and damp woods; local and rather rare. Dorset and E. Kent to Cumberland, Yorks and north to Loch Lomond; Cavan, Roscommon, Lough Neagh and Fermanagh. C. and N. Europe (except the Arctic), rare in S. Europe; Caucasus; N.W. Africa.

19. C. curta Good.   White Sedge.
C. canescens auct., non L.

A tufted glabrous perennial 20–50(–70) cm. Stems triquetrous, rough above. Lvs 2–3 mm wide, usually about equalling stems, nearly flat, pale green; ligule 2–3 mm, lanceolate, acute. Lower sheaths lfless, brown. Infl. 3–5 cm, of 4–8 ±remote spikes 5–8 mm. *Bracts small, glumaceous*, the lower with a short setaceous point. *Female glumes* 2 mm, obovate to suborbicular, cuspidate, *hyaline with a green midrib. Fr.* (Fig. 70A) 2–3 mm, ovate, plano-convex, *pale yellow-green with distinct yellow ribs; beak* c. 0.5 mm, *emarginate, rough*. Nut 1.5–2 mm, elliptic or obovate, biconvex. Stigmas 2. Fl. 7–8. Fr. 8–9. $2n = 56$. Hs. or Hel.

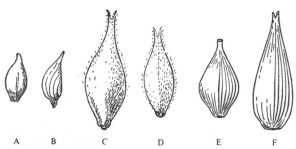

Fig. 70. Fruits of *Carex*. A, *C. curta*; B, *C. lachenalii*; C, *C. hirta*; D, *C. lasiocarpa*; E, *C. acutiformis*; F, *C. riparia*. ×5.

Native. In bogs, acid fens and marshes, but never on limey soils, locally common. Scattered throughout the British Is. and common in the north and west. Europe, south to the Pyrenees and C. Ukraine; temperate Asia; Japan; North and South America.

20. C. lachenalii Schkuhr Hare's-foot Sedge.

C. tripartita auct., non All.; *C. lagopina* Wahlenb.; *C. leporina* L., nom. ambig.

A shortly creeping glabrous *perennial* 10–20(–50) cm. Stems trigonous, often curved. Lvs 1–2 mm wide, shorter than stems, nearly flat; ligule c. 1 mm, ±truncate. Lower sheaths lfless. *Spikes 2–5, contiguous* or the lowest somewhat remote, sessile, forming a dense infl., 1–2 cm. Lower bracts glumaceous. *Female glumes* c. 2.5 mm, broadly ovate or ±rhombic, acute, *reddish-brown* with a hyaline margin and green midrib. *Fr.* (Fig. 70B) 3–4 mm, ovate, narrowed at each end, plano-convex, distinctly ribbed; *beak* c. 0.5 mm, *smooth*, split down the back, the halves overlapping. Nut c. 2 mm, oval, biconvex. Stigmas 2. Fl. 6–7. Fr. 7–8. $2n = 58, 64$. Hs. or Hel.

Native. In flushes and on wet rock-ledges, rare. Cairngorms, Ben Nevis, Glencoe; 750–1140 m. Circumpolar; mountains of C. and S.W. Europe and North America.

Subgenus. CAREX

Monoecious. Spikes 2 or more, differing markedly from one another in appearance and usually unisexual, often pedunculate and subtended by bracts. Stigmas 3 or 2.

21. C. hirta L. Hairy Sedge, Hammer Sedge.

A shortly creeping somewhat hairy perennial 15–70 cm. Stems triquetrous, glabrous, shiny. *Lvs* 2–4 mm wide, shorter than stems, channelled, *hairy on both surfaces*; *sheaths hairy, often densely so*; ligule 1–2 mm, rounded. Lower sheaths mostly with a short blade, reddish, nearly glabrous. Male spikes 2–3, 10–20 × 3–4 mm, fusiform. Male glumes 4–5 mm, obovate, mucronate, ±hairy, reddish-brown; midrib broad, pale; margins and tip hyaline. *Female spikes 2–3(–5)*, 10–30 × 5–7 mm, *erect, distant, the lowest often near base of stem*, peduncles mostly included, glabrous. Bracts similar to lvs, all much longer than spikes, upper about equalling infl., all sheathing. Female glumes 6–8 mm, ovate or oblong, tapering rather abruptly into a long ciliate awn, pale greenish-hyaline. *Fr.* (Fig. 70c) 5–7 mm, ovoid, many-ribbed, *pubescent*, greenish; *beak* 1.5–2 mm, *bifid*, rough inside and outside notch. Nut 2–3 mm, obovoid-trigonous, often stipitate. Stigmas 3. Fl. 5–6. Fr. 6–7. $2n = 112^*$. Hs.

Native. In rough grassy places, woods, damp meadows and damp sandy hollows, common. Throughout the British Is., north to Moray and E. Inverness. Europe, except the north; mountains of N. Africa; Caucasus; introduced in North America.

22. C. lasiocarpa Ehrh. Slender Sedge.

C. filiformis sensu Good., non L.

A rhizomatous nearly *glabrous* perennial 45–120 cm. Stems trigonous, slender, rigid, smooth or slightly rough above. *Lvs* 1–2 mm wide, shorter than stem, *channelled*, stiff, grey-green; ligule 2–3 mm, ovate, obtuse, dark purplish at tip. Lower sheaths many, lfless, dark purplish-

brown, margins sometimes becoming filamentous in decay. Male spikes 1–3, 30–70 × 2–3 mm. Male glumes 4–5 mm, lanceolate, acute, purplish-brown, midrib pale. Female spikes 1–3, 10–30 × 5–8 mm, ±remote, erect. Peduncles short or 0. Bracts lf-like, lower often exceeding infl., very shortly sheathing. Female glumes 3.5–5 mm, lanceolate, acuminate, chestnut-brown; midrib pale. *Fr.* (Fig. 70D) 3.5–4.5 mm, ovoid, subtrigonous, *densely greyish-tomentose*; *beak* 0.5–1 mm, deeply bifid. Nut c. 3 mm, ovoid-trigonous, stipitate. Stigmas 3. Fl. 6–7. Fr. 7–8. Hel.

Native. In reed-swamp in shallow water and in wet acid or alkaline peaty places, locally common; ascends to 595 m. Scattered throughout the British Is. from N. Devon to Orkney. N. and C. Europe, southward to the Pyrenees, C. Italy and S. Ukraine; N. Asia; North America.

23. C. acutiformis Ehrh. Lesser Pond-sedge.

C. paludosa Good.

A shortly creeping glabrous somewhat glaucous perennial 60–150 cm. Stems sharply triquetrous, smooth below, rough above. *Lvs* 7–10 mm wide, those of sterile shoots equalling or exceeding flowering stems, ±keeled; ligule 5–15 mm, lanceolate, acute, margins often purplish. Lower sheaths lfless, brown, not fibrous. *Male spikes 2–3*, contiguous, 10–40 × 3–4 mm. *Male glumes* 5–6 mm, oblong-lanceolate, *obtuse or subacute*, dark brown with a slightly paler midrib. Female spikes 3–4, 20–40 × 7–8 mm, erect, often male at top. Peduncles 0 or short, smooth. Bracts lf-like, lowest broad, exceeding infl., not or shortly sheathing. Female glumes 4–5 mm, oblong-lanceolate with a long, often serrulate acumen, purple-brown with a pale midrib. *Fr.* (Fig. 70E) 3.5–5 mm, elliptical, flattened, shortly stipitate, many-veined, pale green; beak c. 0.5 mm, emarginate or notched. Nut 2 mm, obovoid-trigonous. Stigmas 3 (2). Fl. 6–7. Fr. 7–8. $2n = c. 38, 78$. Hel.

Native. Beside slow-flowing rivers, canals and ponds on clayey or peaty base-rich soils. Scattered throughout the British Is. north to Moray, locally abundant. Europe, south to C. Spain, Sicily and S. Bulgaria; N. Africa; temperate Asia; introduced in North America.

24. C. riparia Curtis Great Pond-sedge.

A large tufted glabrous perennial 100–160 cm. Stems sharply triquetrous, rough. *Lvs up to 15 mm wide*, longer than stem, sharply keeled, *glaucous*; ligule 5–10 mm, ovate, obtuse. *Male spikes* several, *often 5–6. Male glumes* c. 8 mm, oblong, *acuminate* or with an excurrent midrib, dark brown with pale margins and midrib. Female spikes 1–5, distant, upper suberect, lower nodding, (3–)6–10 × 1–1.5 cm, sometimes male at top, acute. Upper peduncles short or almost 0, lower rather long. Bracts lf-like, lowest shortly sheathing, over-topping stem. Female glumes 7–10 mm, oblong-lanceolate to oblong-ovate, brown; midrib strongly excurrent, pale. *Fr.* (Fig. 70F) 5–8 mm, ovoid, strongly convex on the back, weakly so on the front; beak

c. 1.5 mm, bifid, smooth. Nut 2.5–3 mm, ovoid- or ob-ovoid-trigonous, stipitate. Stigmas 3. Fl. 5–6. Fr. 7–8. $2n = 72$. Hel. or Hs.

Native. By slow-flowing rivers, in ditches and ponds, more rarely on drier ground. Common and generally distributed in S. England and the Midlands, local in the west and north to Fife; very local in Ireland. Europe, north to 62° N. in Finland; N. Africa; Caucasus; W. Asia.

25. C. pseudocyperus L.　　　　　Cyperus Sedge.

A glabrous tufted perennial 30–100 cm. Stems stout, sharply triquetrous, angles very rough. Lvs 5–12 mm wide, longer than stem, bright green, margins rough; ligule 10–15 mm, triangular, obtuse. Lower sheaths dark brown, not fibrous. Male spike 1, 2–6 cm, sometimes female at top. Male glumes c. 6 mm, narrowly obovate with a long acumen, light brown with a paler midrib. *Female spikes 3–5, upper clustered, lowest ±remote*, 2–10 × 1 cm, *nodding*. Peduncles slender, rough. Bracts lf-like, lowest shortly sheathing, all but the uppermost exceeding infl. Female glumes 5–8 mm, similar in shape to those of the male fls, broadly hyaline with a narrow green area surrounding the slender whitish midrib; acumen serrulate. *Fr.* (Fig. 71A) 5–6 mm, broader than glumes, *ovoid, tapering into the beak, ±asymmetrical*, green, many-ribbed, *shining, spreading or deflexed*; beak c. 2 mm, deeply notched, usually smooth. Nut 1.5 mm, ovoid-trigonous. Stigmas 3. Fl. 5–6. Fr. 7–8. $2n = 66$. Hel.

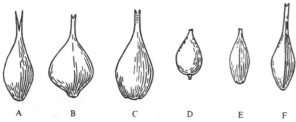

Fig. 71. Fruits of *Carex*. A, *C. pseudocyperus*; B, *C. rostrata*; C, *C. vesicaria*; D, *C. saxatilis*; E, *C. pendula*; F, *C. sylvatica*. ×5.

Native. By slow-flowing rivers, in ditches, ponds and stagnant water in woods; local. England, northwards to N. Lancs, local in Wales and Ireland; very local in Scotland. Europe from S.C. Finland to C. Spain and Macedonia; Algeria; temperate Asia to N. Japan; North America.

26. C. rostrata Stokes　　Beaked Sedge, Bottle Sedge.

C. ampullacea Good.; *C. inflata* auct., non Hudson

A shortly creeping glabrous *rather glaucous* perennial 20–100 cm. *Stems erect, smooth and ½-terete below*, rather rough and trigonous above. Lvs 2–7 mm wide, longer than stems, inrolled or bluntly keeled, very glaucous above when young, rough; *ligule* 2–3 mm, *roundish-truncate*. Lower sheaths soon decaying, not fibrous. Male spikes 2–4, 20–70 × 1.5–2.5 mm, lower with a setaceous bract. Male glumes 5–6 mm, narrowly obovate, acute or subacute, light brown with a pale midrib. Female spikes 2–5, 3–8 × 1 cm, suberect, upper often male at top. Peduncles short, smooth. Bracts lf-like, lower shortly sheathing, equalling or exceeding infl. *Female glumes* 3–5 mm, oblong-lanceolate, *acute*, brown with a pale midrib. Fr. (Fig. 71B) 3.5–6 mm, wider than glumes, ovoid, rather inflated, yellow-green; beak 1–1.5 mm, bifid, smooth. Nut 1.75 mm, obovoid-trigonous. Stigmas 3. Fl. 6–7. Fr. 7–8. $2n = $ c. 60, 76, 82. Hel.

Native. In wet peaty places with a constantly high-water level, local. Generally distributed throughout the British Is., but local in the south. Most of Europe but rare in the south; W. Asia to Altai; Japan; North America.

27. C. vesicaria L.　　　　　Bladder Sedge.

A shortly creeping glabrous *dark green* perennial 30–120 cm. *Stems* erect, *triquetrous*, smooth below, *scabrid on the angles above*. Lvs 4–8 mm wide, longer than stems, keeled, margins serrulate; ligule 5–8 mm, *ovate, acute*. Lower sheaths lfless, acute, developing a fibrous network with age, often reddish or purplish. Male spikes 2–4, 10–40 × 1–2 mm, the lower with a setaceous bract. Male glumes 4–6 mm, oblanceolate, subacute, dark brown with a pale midrib. Female spikes 2–3, remote from the male, upper erect, lower ±nodding, 15–35 × 7–10 mm. Peduncles filiform, smooth, the lower up to 4 cm. *Bracts* lf-like, lower *exceeding infl., sheathing* or not. *Female glumes* 4–6 mm, lanceolate, *acuminate*, dark brown with a pale midrib and hyaline acumen. Fr. (Fig. 71c) 6–8 mm, much wider and longer than the glumes, ovoid, inflated, yellowish- or brownish-green, shiny; beak c. 2 mm, bifid, smooth. Nut c. 2 mm, obovoid-triquetrous. Stigmas 3. Fl. 6, Fr. 7. $2n = 70$, 74, 82, 86, 88. Hel.

Native. In damp places liable to flooding, usually on humus-rich soils. Scattered through the British Is. north to Shetland, local. Europe south to N.E. Spain and S. Bulgaria; temperate Asia; North America.

28. C. saxatilis L.　　　　　Russet Sedge.

C. pulla Good.

A glabrous perennial 10–30 cm. Stems trigonous, smooth below; usually rough above, ±decumbent at base. Lvs 2–4 mm wide, concave; *ligule* 2–4 mm, broadly ovate, *rounded*. Lower sheaths lfless, persistent. Male spike 1(–2), 10–15 × 3 mm, fusiform. Male glumes 3–4 mm, ovate, obtuse, dark purple or almost black. Female spikes 1–2(–3), ±contiguous, 5–20 × 4–5 mm, erect, lower sometimes shortly pedunculate, upper nearly or quite sessile. *Lower bract* lf-like, *about as long as infl.*, not sheathing. Female glumes 2–3 mm, ovate, acute, dark purplish-brown, shorter than fr. *Fr.* (Fig. 71D) 3–3.5 mm, *smooth*, inflated, dark in upper half, pale below; beak c. 0.5 mm, notched. Nut 2 mm, sub-globose. Stigmas 3. Fl. 7. $2n = 80$. Hs.

Native. In moderately base-rich flushes at 750–960 m on the higher Scottish mountains. Local, from Argyll to Sutherland. N. Europe, south to 56°N. in Scotland; circumpolar.

C. × grahamii Boott is probably of hybrid origin. It is largely sterile and of uncertain parentage, but is probably *C. saxatilis* × *binervis* or × *hostiana*. It has been confused with the fully fertile sp., *C. stenolepis* Less., which occurs mainly in the Arctic.

29. C. pendula Hudson — Pendulous Sedge.

A stout tufted glabrous perennial 60–180 cm. Stems triquetrous, smooth. *Lvs 10–20 mm wide*, shorter than stems, ±keeled, yellow-green above, somewhat glaucous beneath; ligule 30–60 mm, lanceolate, acute. Lower sheaths soon decaying. Male spike 1–2, 6–10 cm × 5–7 mm. Male glumes 6–8 mm, lanceolate, acuminate, brownish-hyaline. *Female spikes 4–5, distant, pendent*, 7–16 cm × 5–7 mm, tapering somewhat towards base, dense-fld, often male at top. Lower peduncles included, upper 0. Lower bracts lf-like, about equalling infl., long-sheathing; upper glumaceous with setaceous points, scarcely sheathing. Female glumes 2–2.5 mm, ovate, acute or acuminate, reddish-brown. Fr. (Fig. 71E) c. 3 mm, ellipsoid or ovoid, trigonous, green-brown; *beak c. 0.2 mm, slightly notched*. Nut 2 mm, trigonous, shortly stipitate. Stigmas 3. Fl. 5–6. Fr. 6–7. $2n = 58, 60$. Hs.

One of our most handsome spp., readily recognized by its large size, broad lvs, and long pendent spikes.

Native. In damp woods and on shady stream banks, usually on clayey soils. Scattered throughout the British Is. north to Moray and Mull, local W., C. and S. Europe; Caucasus; N. Africa; Madeira, Azores.

30. C. sylvatica Hudson — Wood Sedge.

A tufted glabrous perennial 10–70 cm. Stems slender, spreading or ±nodding, trigonous, smooth. Lvs 3–6 mm wide, shorter than stems, slightly keeled, smooth, dark green; ligule c. 2 mm, triangular, subacute. Lower sheaths brown, fibrous. Male spike 1, 10–40 × 1–2 mm, tapering downwards. Male glumes 4–5 mm, narrowly obovate, obtuse or subacute, brownish-hyaline. *Female spikes 3–5, distant, ±nodding*, 20–65 × 3–4 mm, *rather lax-fld*. Peduncles filiform, rough, lowest often long-exserted. Lower bracts lf-like, sheathing, shorter than infl., upper setaceous. Female glumes 3–5 mm, ovate, acute, brown with hyaline margins and green midrib. *Fr.* (Fig. 71F) 3–5 mm, ellipsoid- or obovoid-trigonous, *green*; *beak* 1–1.5 mm, *bifid, smooth*. Nut 2–3 mm, ellipsoid- or obovoid-trigonous. Stigmas 3. Fl. 5–7. Fr. 7–9. $2n = 58*$. Hs.

Native. On clayey soils in woods, and in grassland apparently as a woodland relict. Generally distributed and common throughout the British Is., except for the north of Scotland, Orkney, Shetland and Outer

Hebrides. Most of Europe north to 65°N. in Norway; N. Africa; temperate Asia; introduced in North America.

31. C. capillaris L. — Hair Sedge.

A tufted glabrous perennial up to 40 cm. Stems trigonous, smooth, rigid. Lvs 0.5–2.5 mm wide, short, flat, recurved, all ±basal; ligule c. 1 mm, truncate. Lower sheaths mostly with lvs, ±fibrous, dark brown. *Male spike* 1, 5–10 × 1 mm, *usually overtopped by female spikes*, pedunculate. Male glumes 2–3 mm, oblong, obtuse, brownish-hyaline. *Female spikes 2–4*, 5–25 × 1 mm, *few-fld, all apparently arising from the axil of one bract. Peduncles slender, arcuate*, scabrid. Bracts lf-like, long-sheathing, lowest exceeding infl. *Female glumes* 2–3 mm, broadly ovate, mucronate, brownish-hyaline, *caducous*. Fr. (Fig. 72A) c. 3 mm, dark brown, narrowly ovoid, tapering into the short slender smooth nearly entire beak. Nut 2 mm, obovoid-trigonous. Stigmas 3. Fl. 7. $2n = 54$. Hs.

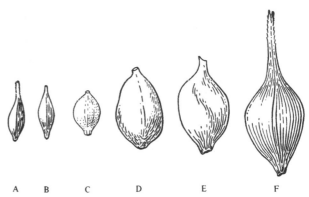

Fig. 72. Fruits of *Carex*. A, *C. capillaris*; B, *C. strigosa*; C, *C. flacca*; D, *C. panicea*; E, *C. vaginata*; F, *C. depauperata*. ×5.

Native. In moist turf, on hummocks in flushes and on rock-ledges of mountains on base-rich soils; very local. Mountain districts of N. England and Scotland from Caernarvon and Mid-west Yorks to Caithness. N. Europe, mountains of C. and S. Europe and N. Africa; N. Asia; North America.

32. C. strigosa Hudson — Thin-spiked Wood-sedge.

A shortly creeping glabrous perennial 30–70 cm. Stems trigonous, smooth, spreading and often nodding. *Lvs 6–10 mm wide*, much shorter than stems, ±keeled, bright green, soft; ligule 5–8 mm, triangular, acute. Lowest sheaths lfless, brown, ±fibrous. Male spike 1, 30–40 × 2 mm, stalked. Male glumes 4.5–5.5 mm, narrowly obovate, brownish with a green midrib. *Female spikes* (3–)5–6, *distant*, the lowest near base of stem, 25–80 × 1.5–2 mm, lax-fld, *erect. Peduncles* slender, *smooth*, nearly or quite included. Lower bracts lf-like, longer than spikes, long-sheathing; upper setaceous, shortly sheathing. Female glumes c. 2.5 mm, ovate or

ovate-lanceolate, acute, green becoming brown. Fr.
(Fig. 72B) 3–4 mm, oblong or narrowly lanceolate, trigonous, rather suddenly narrowed at both ends, often
±curved, green; *beak* c. 0.2 mm, emarginate. Nut
c. 1.5 mm, oblong, trigonous. Stigmas 3. Fl. 5–6. Fr.
8–9. $2n = 66^*$. Hs.

Native. In somewhat open places in damp woods,
especially beside streams on base-rich soils; local. S.
England, north to E. Yorks; a few localities in Wales;
N. and E. Ireland. S., C., and W. Europe from Den-
mark southward and eastward to Bulgaria and the
Caucasus.

33. C. flacca Schreber
Glaucous Sedge, Carnation-grass.
C. glauca Scop.; *C. diversicolor* Crantz, p.p.

A shortly creeping glabrous *glaucous* perennial
10–40(–90) cm. Stems erect, bluntly trigonous or subterete, smooth. Lvs 1.5–4 mm wide, shorter than stems,
slightly keeled, becoming flat, pale green above, glaucous beneath, rough; ligule 2–3 mm, rounded. Lower
sheaths not fibrous, reddish. *Male spikes* 2–3, very
seldom 1, 10–35 × 1–2 mm. Male glumes 3–4 mm, narrowly obovate, rounded or subacute at apex, purple-
brown, midrib broad, whitish, margins narrowly hyaline. *Female spikes* 1–5, ±remote, erect or nodding,
15–55 × 4–6 mm, *dense-fld*, or rather lax towards base,
often male at top. Peduncles slender, nearly smooth,
variable in length. Bracts lf-like, shorter than infl.,
±sheathing. Female glumes 2–3 mm, oblong-ovate,
abruptly contracted to the mucronate apex, purplish-
brown or black, margins hyaline, midrib often pale. *Fr.*
(Fig. 72c) 2–3 mm, ellipsoid to obovoid, rather asymmetrical, *minutely papillose*, yellow-green, reddish or
almost black; beak 0.2 mm, entire. Nut c. 1.5 mm,
obovoid-trigonous, faces ±concave. Stigmas 3. Fl. 5–6.
Fr. 7–8. $2n = 76^*$, 90. Hs.

Native. In dry calcareous grassland, damp clayey
woods, marshes, fens and bogs; common. Generally distributed throughout the British Is. Europe, except the
north-east; N. Africa; introduced in North America.
Represented in the British Is. by the typical subsp.

34. C. panicea L.   Carnation-grass.
A shortly creeping glabrous *glaucous* perennial
10–40(–90) cm. Stems erect, trigonous, smooth. Lvs
1.5–5 mm wide, shorter than stems, nearly flat, glaucous, slightly rough; ligule 1.5–2 mm truncate. Lower
sheaths dull brown, not fibrous. *Male spike* 1,
10–20 × 3–4 mm. Male glumes 3–4.5 mm, ovate, subacute, brown with a pale midrib and hyaline margins.
Female spikes 1–2(–3), distant, erect, 10–15(–25) ×
4–6 mm, *rather few-fld* (up to c. 20), sometimes male
at top. Peduncles rigid, shortly exserted or, more often,
quite included. *Lower bract* lf-like, shorter or somewhat
longer than spike, *closely sheathing*. Female glumes
3–4 mm, broadly ovate, acute, clasping base of fr.,
brown with a pale midrib and hyaline margins. *Fr.* (Fig.
72D) 3–4 mm, *ovoid, inflated, asymmetrical, smooth*,

olive-green, brownish or purplish; *beak short* (less than
0.5 mm), *entire*. Nut 2 mm, broadly obovoid-trigonous.
Stigmas 3. Fl. 5–6, Fr. 6–7, $2n = 32^*$. Hs.

Resembles *C. flacca*, with which it often grows, in
many respects, but may be distinguished by the solitary
male spike, larger swollen fr., fewer-fld spikes, and the
truncate, not rounded, ligule.

Native. In wet grassy places and in fens. Common
and generally distributed throughout the British Is.;
ascends to 1215 m in Scotland. Europe, rare in the south,
temperate Asia; N. Africa; introduced in North
America.

35. C. vaginata Tausch   Sheathed Sedge.
Like *C. panicea* but lvs usually 3–6 mm wide, not glaucous; male spike 7–15 mm; female spikes 5–25 mm; lowest bract often shorter than the spike, with an inflated
sheath; fr. (Fig. 72E) up to 5 mm, less asymmetrical,
gradually narrowed into a distinctly emarginate beak
c. 1 mm. Fl. 7. Fr. 8–9. $2n = 32$. Hs.

Native. In wet grassy places, flushes and on rock-
ledges above 600 m on Scottish mountains. On the
higher mountains from Dumfries to Sutherland, local.
N. Europe, locally in the mountains of C. and S.W.
Europe; Siberia; North America.

36. C. depauperata Curtis ex With.
Starved Wood-sedge.
A shortly creeping glabrous perennial 30–60 cm, forming large loose tufts. Stems trigonous, smooth, rather
weak and slender. Lvs 2–4 mm wide, shorter than stems,
nearly flat, rough above; ligule 2–3 mm, ovate, obtuse.
Lower sheaths lfless, fibrous, purplish. Male spike 1,
18–30 × 2–3 mm. Male glumes 5–6 mm, narrowly obovate, obtuse, brown with broad hyaline margins and
pale midrib. *Female spikes* 2–4, 10–20 × 7–8 mm,
2–4(–6)-fld, erect, rather distant. Peduncles mostly
longer than spikes, partly included, rough. Bracts lf-
like, all longer than spikes and lower often exceeding
infl., long-sheathing. Female glumes 4.5–6 mm, lanceolate, acuminate or aristate, brown with broad hyaline
margins and green midrib. *Fr.* (Fig. 72F) 7–9 mm, *rhom-
boid, very bluntly trigonous*, narrowed at base, rather
abruptly contracted into beak, *pale brown or greenish*,
with many equal ribs; *beak* 3 mm, straight, trigonous-
cylindrical, angles rough, obliquely truncate, split in
front. Nut 4 mm, broadly obovoid-trigonous. Stigmas
3. Fl. 5. Fr. 6. $2n = 44^*$. Hs.

A very distinct sp., readily recognized by the few-fld
spikes and large fr.

Native. In dry calcareous woods and hedgebanks,
very rare. N. Somerset and Mid-Cork, perhaps in Dorset; formerly in Kent, Surrey and Anglesey. W. and
S. Europe; Caucasus.

37. C. laevigata Sm.   Smooth-stalked Sedge.
C. helodes Link

A rather stout tufted glabrous perennial 30–120 cm.
Stems erect, trigonous, slightly rough above. *Lvs*

5–10 *mm wide*, shorter than stems, shallowly keeled, bright green, smooth; ligule 7–15 mm, ovate or triangular, ±obtuse. Lower sheaths lfless, brown, not fibrous. Male spikes 1(–2), 2–6 cm, slender, tapering below. Male glumes 5–6 mm, oblong-lanceolate, subacute or acute, brownish-hyaline. Female spikes 2–4, remote, 20–60 × 5–10 mm, lower pendent, upper suberect. Peduncles filiform, smooth or slightly and remotely toothed. Bracts lf-like, all longer than their spikes but not exceeding infl., long-sheathing. Female *glumes* 3–5 mm, ovate, *acuminate*, brownish with a green midrib. *Fr.* (Fig. 73A) 4–6 *mm*, ovoid or subglobose, rather

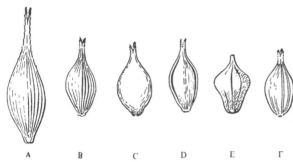

Fig. 73. Fruits of *Carex*. A, *C. laevigata*; B, *C. distans*; C, *C. punctata*; D, *C. binervis*; E, *C. extensa*; F, *C. hostiana*. ×5.

inflated, green; *beak* c. 1.5 *mm*, bifid, smooth or with a few slender teeth. Nut c. 2 mm, trigonous-globose, faces slightly concave, shortly stipitate. Stigmas 3. Fl. 6. Fr. 7–8. $2n = 72^*$. Hs.

Native. In marshes and damp woods, usually on acid but base-rich soils, local. Scattered throughout the British Is. north to W. Sutherland, but absent from areas of low rainfall. W. Europe; N.W. Africa.

38. C. distans L.   Distant Sedge.

A densely tufted glabrous perennial 15–100 cm. Stems trigonous, rigid, smooth. Lvs 2–6 mm wide, usually much shorter than stems, nearly flat, grey-green; ligule 2–3 mm, obtuse. Lower sheaths brown, scarcely fibrous. Male spike 1(–2), 15–30 × 2–3 mm, nearly cylindrical. Male glumes 3–4 mm, ovate, obtuse or subacute, brownish-hyaline. Female spikes 2–4, 10–20 × 4–5 mm, distant, erect. Peduncles slender, ±included. Bracts lf-like, shorter than infl., long-sheathing. Female *glumes* 2.5–3.5 mm, ovate, mucronate, brown or greenish-brown with a pale midrib, *conspicuous in fr.* Fr. (Fig. 73B) 3.5–4.5 mm, ellipsoid, *trigonous, equally many-ribbed*, not glossy, ±*erect*; beak 0.75 mm, bifid, rough. Nut 2.5 mm, elliptical, trigonous. Stigmas 3. Fl. 5–6. Fr. 6–7. $2n = 72, 74^*$. Hs.

Native. In marshes and in cracks of wet rocks. Scattered throughout the British Is. from Cornwall to Caithness, mainly coastal except in the south. Most of Europe except the north-east, W. Asia; mountains of N. Africa.

39. C. punctata Gaudin   Dotted Sedge.

A glabrous rather tufted perennial, somewhat resembling *C. distans*, but usually shorter and more stiffly erect with lvs longer in proportion to the stems. Stems trigonous, smooth. Lvs 2–5 mm wide, nearly flat, pale green, all ±basal; ligule 3 mm, almost truncate. Lower sheaths mostly with lvs, dark brown; sheath of uppermost lf on flowering stems truncate, with two small appendages at sides. Male spike 1, 10–25 × 1–1.5 mm, stalked. Male glumes 3–4 mm, oblong-obovate, brownish-hyaline. Female spikes 2–4, distant, 5–25 × 4–5 mm. Peduncles long, included. Bracts lf-like, longer or shorter than infl., all long-sheathing. Female *glumes* 2.5–3.5 mm, broadly ovate, mucronate, pale reddish-brown, ±hyaline, *inconspicuous in fr. Fr.* (Fig. 73c) 3–4 mm, ovoid, *turgid, glossy*, pellucid-punctulate, *pale greenish, spreading or deflexed*; beak 0.75 mm, bifid, smooth. Nut c. 2 mm, obovoid-trigonous, shining and punctate. Stigmas 3. Fl. 6–7. Fr. 7–8. $2n = 68^*$. Hs.

Native. In marshes and cracks in wet rocks by the sea, very local. South and west coasts from Hants to the Solway Firth; Cork, Kerry, Galway. W. and S. Europe locally north-eastwards to S.W. Sweden and N. Poland; N. Africa.

40. C. binervis Sm.   Green-ribbed Sedge.

A shortly creeping glabrous perennial 15–150 cm. Stems erect, trigonous, nearly smooth. Lvs 2–6 mm wide, shorter than stems, ±keeled, smooth; ligule 1–2 mm, truncate. Lower sheaths brown, not fibrous. Male spike 1, 20–45 × 3–4 mm, narrowly clavate, sometimes with a few female fls at base. Male glumes 4–4.5 mm, oblong-obovate, obtuse, purplish-hyaline with a pale midrib. Female spikes 2–3, very distant, 15–45 × 5–7 mm, cylindrical, ±erect. Lower peduncles long, smooth, often long-exserted; upper short, often quite included. Lower bracts lf-like, long-sheathing, longer than spikes; uppermost glumaceous with a setaceous point, shortly sheathing, shorter than spike. Female glumes 3–4 mm, ovate, mucronate, dark purple-brown with a greenish midrib. *Fr.* (Fig. 73D) 3.5–4.5 mm, ovoid, *subtrigonous with broad sharp lateral angles*, pale green or purplish, *with two prominent dark green submarginal veins*; beak 1–1.5 mm, broad, flattened, bifid, slightly rough at edges. Nut c. 2 mm, ellipsoid or obovoid, trigonous. Stigmas 3. Fl. 6. Fr. 7. $2n = 74^*$. Hs.

Native. On heaths and moors and in rough pastures on acid soils. Throughout the British Is., common in suitable habitats. W. Europe northwards to c. 64° N. in Norway; N.W. Africa.

41. C. extensa Good.   Long-bracted Sedge.

A glabrous rather rigid, densely tufted perennial 5–40 cm. Stems trigonous or subterete. Smooth, slender or rarely stout, sometimes curved. *Lvs* 2–3 mm wide or less, *channelled*, all ±basal; ligule c. 2 mm, ovate or truncate. Lower sheaths lfless, persistent, blackish. Male spike 1(–3). 5–25 × 1–2 mm, sessile. Male glumes 3–3.5 mm, obovate, brownish-hyaline. *Female spikes*

2–4, *contiguous, or the lower somewhat distant*, 5–20 × 4–6 mm. Peduncles of lower spikes short, included. *Bracts lf-like, many times as long as spikes*, narrow, rigid, *spreading or deflexed*. Female glumes 1.5–2 mm, broadly ovate or suborbicular, mucronate, straw-coloured with brownish patches. *Fr.* (Fig. 73E) 3(–4) mm, *ovoid-trigonous*, weakly veined, *greenish* or light brown; beak c. 0.5 mm, notched, smooth. Nut c. 2(–3) mm, broadly ovoid-trigonous. Stigmas 3. Fl. 6–7. Fr. 7–8. $2n = 60^*$. Hs.

Native. In grassy salt-marshes and on damp maritime cliffs and rocks. Around the coasts of the British Is. north of Orkney, locally common. Europe to c. 61° N. in Sweden; W. Asia; N. Africa; introduced in North America.

42. C. hostiana DC.　　　　　　Tawny Sedge.

C. fulva sensu Host; *C. hornschuchiana* Hoppe

A shortly creeping glabrous perennial 15–50 cm. Stems erect, trigonous, nearly smooth. Lvs 2–5 mm wide, much shorter than stems, nearly flat, margins ±rough; ligule 1 mm, truncate. Lower sheaths brown, fibrous in decay. Male spike 1(–2), 10–20 × 1.5–2 mm, cylindrical, tapering at both ends. Male glumes 3.5–4.5 mm, ovate, obtuse or subacute, brown, margins broadly hyaline at top. Female spikes (1–)2–3, very distant, 8–12(–20) × 6–9 mm, ovoid or shortly cylindrical, erect, often male at top. Lower peduncles long, smooth, often long-exserted, upper short, ±included. Lower bracts lf-like, long-sheathing, longer than spikes; uppermost glumaceous with a setaceous point, shortly sheathing, shorter than spike. *Female glumes* 2.5–3.5 mm, broadly ovate, acute, dark brown *with a broad hyaline margin* and pale midrib. *Fr.* (Fig. 73F) 4–5 mm, ovoid, *many-ribbed, rather inflated*; beak c. 1 mm slender, ±flattened, margins serrulate. Nut c. 2 mm, obovoid, trigonous. Stigmas 3. Fl. 6. Fr. 7. $2n = 56^*$. Hs.

Native. In fens and base-rich flushes, not uncommon. Throughout most of the British Is., but very local in the south and east. Europe, except the north-east and the southern Mediterranean region; North America.

C. flava group (43–46)　　　　　Yellow Sedge.

Tufted. Stems smooth; lower sheaths pale greyish or brownish, breaking down into fibres and soon decaying. Male spike 1; female spikes (1–)2–5, ovoid, contiguous or remote, the lowest pedunculate. Lowest bract lf-like, ±sheathing. Female glumes often yellowish and with a narrow scarious margin. Fr. with a nearly or quite smooth emarginate to bifid beak. Stigmas 3.

The spp. are variable and often hybridize when they grow in proximity. The following treatment is based on that in *Flora Europaea*, 5, 309–310; the extra-European distribution is not accurately known.

1 Fr. with curved or deflexed beak at least ½ as long as the rest of the fr. 2
　Fr. with straight beak less than ½ as long as the rest of fr. 3

2 Fr. 5–7 mm; male spike usually subsessile; lvs 3–7 mm wide. **43. flava**
　Fr. (3–)3.5–5 mm; male spike with peduncle 5–30 mm; lvs 2–3(–4) mm wide. **44. lepidocarpa**
3 Fr. 3.5–4 mm; male spike pedunculate; lowest female spike often very remote from the rest. **45. demissa**
　Fr. 1.75–3.5 mm; male spike usually sessile; female spikes usually crowded. **46. serotina**

43. C. flava L.

Stems (10–)20–50(–95) cm. Lvs 3–7 mm wide, shorter than or equalling the stems. Male spike usually subsessile. Female spikes 10–15 × 10–12 mm, densely crowded or the lower somewhat remote; lowest bract much longer than the infl., spreading or deflexed, shortly sheathing. Female glumes 3.5 mm, lanceolate, acute, brownish with green midrib. Fr. (Fig. 74A) 5–7 mm, yellowish-green to yellow, all but the upper arcuate-deflexed at maturity, strongly veined, narrowed into a curved beak 2–2.5 mm. Fl. 6. Fr. 7. $2n = 30, 33, 60^*, 64$. Hs.

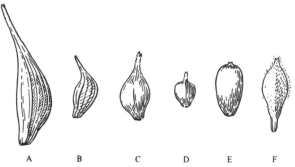

A　　　B　　　C　　　D　　　E　　　F

Fig. 74. Fruits of *Carex*. A, *C. flava*; B, *C. lepidocarpa*; C, *C. demissa*; D, *C. serotina*; E, *C. pallescens*; F, *C. digitata*. ×5.

Native. On damp peaty soils; calcicole. N. Lancs, W. Ireland, rare and very local. Most of Europe, but absent from much of the Mediterranean region; North America.

44. C. lepidocarpa Tausch

Stems (10–)20–60(–75) cm. Lvs 2–3(–4) mm wide, c. ½ as long as the stems. Male spike pedunculate. Female spikes 6–13 × 6–8 mm, the upper densely crowded, the lower somewhat remote; lowest bract equalling or longer than the infl., spreading or deflexed. Female glumes 2.5–4 mm, ovate, acute, brownish becoming yellow. Fr. (Fig. 74B) (3–)3.5–5 mm, greenish- or brownish-yellow, the middle spreading, the lower deflexed, all curved, strongly veined, rather abruptly contracted into a recurved or deflexed beak 1.5–2 mm. Fl. 5–6. Fr. 7–8. $2n = 58, 68^*$. Hs.

Native. In damp, usually peaty places or on calcareous marls; calcicole. Scattered throughout the British Is., locally common. Europe, except for most of the south; North America.

45. C. demissa Hornem.

Stems 5–20(–40) cm, curved. Lvs 2–5 mm wide, at least as long as the stems. Male spike pedunculate. Female spikes 7–13 × 6–8 mm, the upper crowded, the lowest often very remote; lowest bract usually much longer than the infl., erect to spreading. Female glumes c. 3 mm, ovate, ±acute, brown with a green midrib. Fr. (Fig. 74c) (3–)3.5–4 mm, usually green, all spreading or ascending, straight, strongly veined, gradually narrowed into a straight beak c. 1 mm. Fl. 6. Fr. 7–8. 2n = 70*. Hs.

Native. In damp grassy and boggy places, stony margins of lakes, etc., on moderately acid soils. Throughout most of the British Is., but local in the Midlands and East Anglia. Europe, southwards to N.W. Portugal and eastwards to S. Finland; North America.

46. C. serotina Mérat

Incl. *C. scandinavica* E. W. Davies

Stems 2–20(–40) cm. Lvs 1–3(–4) mm wide, usually at least as long as the stems. Male spike usually sessile. Female spikes 3–8 × 3–10 mm, crowded or the lowest sometimes remote; lowest bract much longer than the infl. Female glumes 2–3 mm, ovate, subacute, yellowish-brown with a green midrib. Fr. (Fig. 74D) 1.75–3.5 mm, yellowish-green, mostly spreading or ±erect, straight, with slender veins and a straight beak 0.25–1 mm. Fl. 6–8. Fr. 7–9. 2n = 68*, 70*. Hs.

Native. In damp, usually rather open habitats on base-rich soils, local. Scattered throughout the British Is., particularly near the coasts. Most of Europe, eastwards to c. 40° E. in C. Russia.

47. C. pallescens L.   Pale Sedge.

A tufted perennial 20–60 cm. *Stems* sharply triquetrous, somewhat rough and *shortly hairy*, especially on the angles. *Lvs* up to 5 mm wide, weakly keeled, ±*hairy beneath*, particularly on the ribs and margins; ligule c. 5 mm, ovate, acute or obtuse, margins usually brown and ±ciliate. *Sheaths* ±*hairy*, the lowest brown, lfless, acute. *Male spike* 1, c. 8 mm, ±*concealed by female spikes*. Male glumes 3–4 mm, obovate-long, subacute or mucronate, pale brown with a slender dark midrib. *Female spikes* 2–3, *contiguous* or the lower remote, 5–20 × 5–6 mm, sometimes with a few male fls at top, suberect or lower ±nodding. Peduncles slender, the lowest up to c. 2 cm. Lowest bract lf-like, crimped at the base, often overtopping infl., upper small, shorter than spike, often setaceous. Female glumes 3–4 mm, ovate, acuminate, hyaline with a broad green excurrent midrib. *Fr.* (Fig. 74E) c. 4 mm, *ovoid-oblong*, *convex on both sides*, *bright green*, ±distinctly veined; *beak very short* or almost 0. Nut 2 mm, ellipsoid-trigonous. Stigmas 3. Fl. 5–6. 2n = 62, 64*; 66. Hs.

Native. In damp woods, less frequently in damp grassy places. Scattered throughout the British Is. north to Caithness; rather local but not uncommon. Most of Europe, but absent from parts of the south and north-east; temperate Asia; North America.

48. C. digitata L.   Fingered Sedge.

A nearly glabrous tufted perennial. *Stems* 5–15(–25) cm, slender, flexuous, bluntly trigonous, *lateral*, *with a number of lfless sheaths enfolding the base*. Lvs 1.5–5 mm wide, about as long as the stems, flat, yellowish-green, sometimes shortly and sparsely hairy; ligule 0.5–1.5 mm, ±triangular, obtuse or acute. Lower sheaths blackish, fibrous. *Male spike* 1, 8–15 × 1–1.5 mm, *few-fld*, *overtopped by the upper 1–2 female spikes and therefore appearing lateral*. Male glumes 5 mm, oblong, obtuse or emarginate, whitish above, purplish-brown below, margins incurved. *Female spikes* 1–3, *rather distant*, 10–15 × 2 mm, very lax, 5–10-fld. Peduncles slender, rather short. Bracts glumaceous, sheathing, ±enclosing the peduncles and sometimes the base of the spike. Female *glumes* 3–4.5 mm, *enclosing the fr.*, *broadly obovate*, ±*emarginate*, mucronulate, pale brown and shiny, midrib green. Fr. (Fig. 74F) 3–4.5 mm, obovoid-trigonous, puberulent, brown and rather shiny; beak c. 0.4 mm, nearly entire. Nut c. 2.5 mm, obovoid, stipitate, trigonous, faces nearly flat. Stigmas 3. Fl. 4–5. Fr. 6–7. 2n = 48*, 50*; 52; 54. Hs.

Native. In rather open woods on chalk or limestone, usually in cracks of rock outcrops, or in cracks in ±wooded limestone pavement; very local and rather rare. N. Somerset, N. Wilts, Gloucester, Monmouth, Hereford, Nottingham and Derby to N. Yorks and Westmorland. Most of Europe but absent from parts of the south; temperate Asia.

49. C. ornithopoda Willd.

Like **48** but stems usually 5–15 cm; lvs shorter than the stems, 1–3 mm wide; male spike 4–8 × 1–1.5 mm; female spikes 5–10(–15) × 2–3 mm; female glumes ½–⅔ as long as fr.; fr. (Fig. 75A) 2.5–3 mm. Fl. 5. Fr. 6–7. 2n = c. 46, 52, 54. Hs.

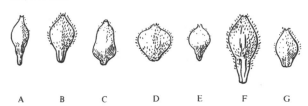

Fig. 75. Fruits of *Carex*. A, *C. ornithopoda*; B, *C. humilis*; C, *C. caryophyllea*; D, *C. tomentosa*; E, *C. ericetorum*; F, *C. montana*; G, *C. pilulifera*. ×5.

Native. In dry grassland on limestone slopes or in cracks in limestone pavement, rarely (?never) in shade; very local. Derby, Yorks, Westmorland. Most of Europe but absent from the Mediterranean region and the south-east; Asia Minor.

Represented in the British Is. by the typical subsp.

50. C. humilis Leysser   Dwarf Sedge.

C. clandestina Good.

A small nearly glabrous tufted perennial. Stems 2–5(–15) cm, slender, flexuous, terminal. Lvs 1–1.5 mm wide, longer than and ±concealing the stems, at first flat, later channelled, rough; ligule 0.5–1 mm, truncate. Lower sheaths dark red-brown, fibrous. Male spike 1, 10–15 × 2–4 mm, tapering at both ends. Male glumes c. 6 mm, oblong, obtuse or subacute, purplish-brown with very broad hyaline margins. *Female spikes* 2–4, distant, 4–10 × 1–2 mm, erect, *very lax*, 2–4-*fld. Peduncles short. Bracts glumaceous, sheathing, ±enclosing the spikes*. Female glumes 2–3 mm, ovate or suborbicular, ±acute, purplish-hyaline, margins incurved. Fr. (Fig. 75B) 2.5 mm, broader than glumes, obovoid-trigonous, puberulent; beak very short, entire. Nut 2 mm, obovoid-trigonous. Stigmas 3. Fl. 3–5. Fr. 5–7. $2n = 36^*$; 38. Hs.

Native. In short turf on chalk and limestone, locally abundant. N. Somerset, W. Gloucester, Hereford, Wilts, Dorset. C. and S. Europe; Caucasus; W. Siberia.

51. C. caryophyllea Latourr.   Spring Sedge.

C. verna Chaix, non Lam.; *C. praecox* Jacq., non Schreber

A nearly glabrous creeping perennial 5–15(–30) cm. Stems triquetrous, slender. Lvs c. 2 mm wide, shorter than stems, nearly flat, rough; ligule 1–2 mm, ovate, obtuse. Lower sheaths dark brown or blackish, fibrous. Male spike 1, 10–15 × 2–3 mm, clavate, occasionally with a few female fls at base. Male glumes 4–5 mm, oblong, subacute or mucronate, thin, brown above, hyaline below. Female spikes 1–3, contiguous or rarely the lowest distant, 5–12 × 3–4 mm, rarely male at top, erect. Peduncles 0 or rarely the lowest up to c. 1 cm. *Bracts* setaceous, *shortly sheathing*. Female glumes 2–2.5 mm, broadly ovate, obtuse and mucronate or ±acute, brown and shining, midrib green. *Fr*. (Fig. 75C) c. 2.5 mm, obovoid or ellipsoid, narrowed below, trigonous, pubescent, *olive-green*; beak c. 0.2 mm, conical, slightly notched. Nut 2 mm, broadly obovoid-trigonous, stipulate, crowned by swollen style-base. Stigmas 3. Fl. 4–5. Fr. 6–7. $2n = 64^*, 66^*, 68^*; 62$. Hs.

Native. In dry grassland, particularly calcareous, locally common. Throughout the British Is. except Orkney, Shetland and Outer Hebrides. Europe, north to c. 63° N. in Sweden; introduced in North America.

52. C. tomentosa L.   Downy Sedge.

C. filiformis auct., non L.

A creeping nearly glabrous perennial 20–50 cm. Stems triquetrous, rough towards top, slender, rigid, erect. *Lvs* 1.5–2 mm wide, shorter than stems, *nearly flat*, rough, rather glaucous; ligule 1–2 mm, ovate. Lower sheaths lfless, purplish, persistent. Male spike usually 1, 12–25 × 1.5–2 mm, shortly stalked. Male glumes 4–5 mm, obovate-oblong, obtuse, apiculate, brownish-hyaline. *Female spikes* 1–2, 8–14 x 4–5 mm, *dense-fld, rather distant*. Peduncles very short. Lower bracts lf-like, equalling or exceeding spike, not or shortly sheathing. Female glumes 2–3 mm, broadly ovate,

mucronate, purplish-brown, midrib green. *Fr*. (Fig. 75D) 2–3 mm, broadly pear-shaped, trigonous, *shortly tomentose*; *beak very short*, slightly emarginate. Nut 1.5 mm, pyriform, trigonous. Stigmas 3. Fl. 5–6. Fr. 6–7. $2n = 48^*$. Hs.

Native. In damp meadows and pastures, very local. N. Wilts, Surrey, Middx, Oxford, E. Gloucester. Europe, from Estonia to E. Spain and C. Greece; Caucasus; Siberia.

53. C. ericetorum Pollich   Rare Spring Sedge.

A nearly glabrous tufted perennial 5–20 cm. *Stems* slender, *smooth*, bluntly trigonous. *Lvs up to* 4 mm *wide*, shorter than the stems, nearly flat; ligule very short, truncate. Lower sheaths brown, fibrous. Male spike 1, 6–12 × 2 mm, clavate. Male glumes 2–3 mm, obovate-oblong, obtuse, purplish-brown with a fringed hyaline border at top. *Female spikes* 1–2(–3), contiguous, 5–12 × 3–4 mm, *dense-fld*. Peduncles 0. *Bracts glumaceous*, small, the lower with a short setaceous point. *Female glumes* 2–2.5 mm, broadly obovate-oblong, dark purplish-brown with a slender pale midrib; *apex rounded*, hyaline, *very finely ciliate. Fr*. (Fig. 75E) c. 2.5 mm, obovoid-trigonous, pubescent, green below, *dark brown above*; beak c. 0.3 mm, rather stout, somewhat notched. Nut 1.5 mm, broadly obovoid-trigonous. Stigmas 3. Fl. 4–5. Fr. 6–7. $2n = 30^*$. Hs.

Native. In dry grassland on chalk and limestone, very local. W. Suffolk, W. Norfolk, Cambridge, Lincoln, Derby, Yorks, Durham, Westmorland. Most of Europe but absent from the Mediterranean region; Caucasus; Siberia.

54. C. montana L.   Soft-leaved Sedge.

A nearly glabrous shortly creeping and tufted perennial 10–30 cm. *Stems* slender, not rigid, *rough at top*. Lvs c. 2 mm wide, shorter or longer than stems, sparsely pubescent beneath becoming glabrous when old; ligule c. 1 mm, ovate, ±acute. Lower sheaths dark reddish-brown, fibrous. Male spike 1, 10–20 mm, subclavate, acute. Male glumes c. 5 mm, oblong, obtuse or subacute, purplish-hyaline with a pale midrib often excurrent in a dark mucro. Female spikes (1–)2(–4), contiguous, 6–9 × 4–6 mm, erect, few fld. Peduncles 0. *Bracts* glumaceous, the lower sometimes with a setaceous point, *not sheathing. Female glumes* 3–5 mm, broadly ovate or obovate, obtuse or retuse, *mucronate*, dark purple-brown with a pale midrib. *Fr*. (Fig. 75F) 3.5–4 mm, exceeding glumes, ellipsoid or obovoid, trigonous, pubescent, *pale brown or blackish* on the exposed face, stipitate; stalk variable, stout; beak c. 0.2 mm, dark, notched. Nut 2.5 mm, ovoid-trigonous, stipitate. Stigmas 3. Fl. 5. $2n = 38^*$. Hs.

Native. In grassy and heathy places and in open woods, chiefly on limestone, locally abundant. England and Wales, north to Derby; ascends to 240 m in the Mendips (Somerset). Europe, from S. Sweden and C. Russia to N.W. Spain and S.W. Bulgaria; temperate Asia.

55. C. pilulifera L. — Pill Sedge.

A nearly glabrous densely tufted rather rigid perennial 10–30 cm. Stems sharply triquetrous, slightly rough above, ±incurved. Lvs 1.5–2 mm wide, shorter than the stems, recurved, yellow-green, nearly flat, rough especially near the top; ligule 0.5–1 mm, truncate or broadly triangular. Lower sheaths fibrous when old. Male spike 1, 8–15 mm, slender. Male glumes 3.5–4 mm, lanceolate, acute, brown with a pale or green midrib. *Female spikes 2–4, contiguous* or the lowest somewhat distant, 5–6(–8) × 4–6 mm, erect, *few fld.* Peduncles 0. *Bracts short or the lowest somewhat exceeding infl., narrowly lf-like, green,* not sheathing. Female glumes 3–3.5 mm, broadly ovate, acuminate, brown with a green midrib. *Fr.* (Fig. 75G) 1–3.5 mm, *almost globose, puberulent,* ribbed, grey-green, stipitate; stalk variable in length, stout; beak 0.3–0.5 mm, obliquely truncate becoming slightly notched, dark. Nut 1.5 mm, broadly obovoid, rounded-trigonous. Stigmas 3. Fl. 5–6. $2n = 18^*$. Hs.

Var. *longebracteata* Lange has slender, few (3–5)-fld female spikes, the lower 1 or 2 distinctly pedunculate; lowest bract lf-like, about as long as infl.

Native. In grassy or heathy places, sometimes in open woods, on sandy or peaty acid soils; locally common. Throughout the British Is. Europe eastwards to Leningrad and the E. Carpathians.

56. C. atrofusca Schkuhr — Scorched Alpine Sedge.
C. ustulata Wahlenb.

A shortly creeping glabrous perennial 5–35 cm. Stems trigonous, smooth, often nodding at top. Lvs 2–5 mm wide, much shorter than stems, nearly flat, rather soft; ligule 2–3 mm, rounded. Lower sheaths soon decaying. Male spike 1, 5–10 × 3–4 mm, sometimes female at base. Male glumes 3–3.5 mm, lanceolate, mucronate, brown, midrib pale. *Female spikes 2–4, 5–12 × 7 mm, contiguous, ovoid,* dense-fld, *nodding. Peduncles very slender, smooth.* Bracts setaceous, sheathing. *Female glumes* c. 3 mm, lanceolate, acuminate, *purple-black,* midrib slender, pale. Fr. (Fig. 76A) 4–4.5 mm, elliptical, compressed-trigonous, purple-black; beak 0.3 mm, notched, ±rough. Nut 1.5 mm, elliptical or obovoid, trigonous, long-stipitate. Stigmas 3. Fl. 7. Fr. 9. $2n = 38$, 40. Hs.

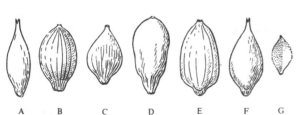

Fig. 76. Fruits of *Carex*. A, *C. atrofusca*; B, *C. limosa*; C, *C. magellanica*; D, *C. rariflora*; E, *C. buxbaumii*; F, *C. atrata*; G, *C. norvegica*. ×5.

Native. In micaceous mountain bogs, 540–1050 m, very rare. Mid-Perth, W. Inverness and Argyll. Alps; N. and W. Scandinavia; N. Ural; Greenland.

57. C. limosa L. — Mud Sedge.

A creeping glabrous perennial up to 40 cm. *Stems* triquetrous, *rough,* slender, rigid. *Lvs 1–2 mm wide,* shorter than stems, *channelled, the greater part of the margins serrulate;* ligule 3–4 mm, acute. Lower sheaths lfless, reddish, not fibrous. Male spike 1, 10–20 × 2 mm, erect. *Male glumes* 3–4 mm, lanceolate, *acute or subacute,* reddish-brown. *Female spikes 1–2(–3),* 7–20 × 5–7 mm, nodding, *up to 20-fld.* Peduncles 0.5–2 cm, very slender, smooth. *Bracts with a lf-like point shorter than infl. or rarely equalling it,* shortly sheathing. *Female glumes* 3.5–4.5 mm, *ovate, acute,* brown or purplish with a green midrib. *Fr.* (Fig. 76B) 3–3.5 mm, elliptical, biconvex, *slightly broader than glumes,* strongly ribbed, pale brownish- or bluish-green; beak very short, entire. Nut 2–2.5 mm, obovoid-trigonous, shiny. Stigmas 3. Fl. 5–6. Fr. 7–8. $2n = 56$, 64. Hel.

Native. In very wet bogs, rooting in peat detritus in shallow water, locally common. Scattered throughout the British Is. from Dorset and Hants northwards, more frequent in the north and west. N. and C. Europe, south to the Pyrenees and S.E. Russia; N. Asia; North America.

58. C. magellanica Lam. — Bog Sedge.
C. paupercula Michx; incl. *C. irrigua* (Wahlenb.) Sm. ex Hoppe.

A shortly creeping glabrous perennial up to 40 cm, like *C. limosa* but stems smooth; lvs 1.5–4 mm wide, the greater part of the margins smooth; ligule c. 5 mm, acute; female spikes 2–4, 5–12 × 4–8 mm, with up to 10 fls; lower bracts exceeding infl.; female glumes 5–6.5 mm, lanceolate, acuminate or aristate, caducous, brown or purplish; fr. (Fig. 76c) 3–3.5 mm, ovate or suborbicular, biconvex, much broader than glumes, inconspicuously ribbed. Fl. 5–6. Fr. 6–7. $2n = 58$. Hel.

Native. In very wet bogs, rare. From the Lake District to N. Uist and W. Sutherland; N. Wales; Antrim. Circumpolar; Mont Cenis; South America.

Subsp. **irrigua** (Wahlenb.) Hiitonen in Europe.

59. C. rariflora (Wahlenb.) Sm.
— Loose-flowered Alpine Sedge.

A shortly creeping glabrous perennial up to c. 20 cm. *Stems trigonous, smooth.* Lvs 1–2 mm wide, shorter than stems, *flat, the greater part of the margins smooth;* ligule c. 3 mm, acute, upper part of sheaths broadly membranous. Lower sheaths soon decaying. Male spike 1, 8–12 × 2 mm, erect. Male glumes 4 mm, ovate, ±truncate and apiculate, dark purple with narrow pale midrib. Female spikes 2, rarely 1 or 3, c. 10 × 4 mm, nodding, lax, up to 8-fld. Peduncles slender. Lower bracts shorter than spike and usually shorter than peduncles. *Female glumes* 3–4 mm, *obovate, almost truncate,* reddish-purple, midrib narrow, pale. *Fr.* (Fig. 76D) 3–4 mm, ellipsoid, *narrower and shorter than glumes*, strongly ribbed; beak very short, entire. Nut 2 mm, ellipsoid, trigonous. Stigmas 3. Fl. 6. Fr. 7. 2*n* = 50; 54. Hel.

Native. In small bogs and wet peaty places at 750–1050 m on the higher Scottish mountains, very local. Perth, Angus, Aberdeen and Inverness. Circumpolar.

60. C. buxbaumii Wahlenb.   Club Sedge.

C. canescens L., nom ambig.; *C. polygama* Schkuhr

A shortly creeping glabrous perennial 30–60 cm. Stems triquetrous, smooth, stiff. Lvs 1.5–2 mm wide, shorter than stems, nearly flat, rather glaucous; ligule c. 3 mm, lanceolate, acute. Lower sheaths lfless, reddish; lf-sheaths reddish, margins often filamentous. *Spikes* 2–5, 7–15 × 5 mm, *erect*, ±*remote*, lower shortly stalked, upper sessile, the terminal female at top, male below. Bracts lf-like, lowest sometimes exceeding infl. Female glumes 3–5 mm, ovate or oblong-ovate, dark purplish-brown with a pale *midrib excurrent in an arista*. Fr. (Fig. 76E) 3–4.5 mm, ovate, much broader than glumes, becoming trigonous when ripe, indistinctly veined, pale greenish; beak very short, notched. Nut 2.5 mm, elliptical, trigonous. Stigmas 3. Fl. 6–7. Fr. 7–8. 2*n* = c. 74. Hel.

Native. In wet spongy fens in two localities in Inverness. N. and C. Europe, local; Siberia; North America; Japan.

61. C. atrata L.   Black Sedge.

A glabrous perennial 30–50 cm. Stems triquetrous, smooth, often ±nodding at top. Lvs 2–6 mm wide, keeled, rather glaucous; ligule 2–3 mm, ovate or ±truncate. Lower sheaths lfless, persistent, dark brown. *Spikes* 3–5, *contiguous*, *lower eventually nodding*, 10–20 × 3–5 mm, terminal one male at base only. Peduncles of lower spikes 1–2 cm. *Lowest bract often exceeding infl.*, others glumaceous. *Female glumes* 3.5–4.5 mm, ovate, *acute*, purple-black with pale midrib. Fr. (Fig. 76F) 3–4 mm, narrowly elliptical or obovate, compressed, ±trigonous, minutely punctate; beak 0.3–0.5 mm, notched. Nut c. 2 mm, ovoid-trigonous. Stigmas 3. Fl. 6–7. 2*n* = 54*; 56. Hs.

Native. On wet rock-ledges on mountains at 730–1120 m, local. Caernarvon, Lake District and the higher Scottish mountains from Dumfries to Ross. N. Europe and the higher mountains in the south; northern Asia and on mountains south to the Caucasus, Turkestan and Baikal; North America: Rocky Mountains south to Utah.

62. C. norvegica Retz.   Close-headed Alpine Sedge.

C. halleri Gunn., p.p; *C. vahlii* Schkuhr; *C. alpina* Liljeblad, non Schrank.

A glabrous perennial 15–30 cm, forming small tufts. Stems triquetrous, rough above, rigid. Lvs 1.5–3 mm wide, shorter than stems, nearly flat; ligule 0.5–1 mm, rounded. Lower sheaths mostly with lvs, reddish, not fibrous. *Spikes* 1–4, *contiguous*, erect, 5–8 × 4–5 mm,

the lowest shortly stalked, the others sessile; terminal spike female at top, male below. *Lower bract short, lf-like.* Female glumes 1.5–2.5 mm, ovate, acute, purplish-black with pale midrib. Fr. (Fig. 76G) 2–2.5 mm, obovoid-trigonous, minutely papillose, especially at the top; beak c. 0.25 mm, emarginate. Nut 1.5 mm, obovoid-trigonous. Stigmas 3. Fl. 6–7. Fr. 7–8. 2*n* = 56. Hs. or Hel.

Native. Among short grass on wet rock-ledges and by mountain streams, c. 610–980 m. Mid Perth, Angus and S. Aberdeen. N. Europe, Alps, Arctic Siberia, North America.

63. C. elata All.   Tufted Sedge.

C. stricta Good., non Lam.; *C. hudsonii* Ar. Benn.

A tufted glabrous perennial up to c. 90 cm. Stems sharply triquetrous, slightly rough. Lvs (3–)4–6 mm wide, sharply keeled, rather glaucous, the margins rolling outwards on drying; ligule 5–10 mm, ovate, acute. *Lowest sheaths lfless*, reddish, acuminate, *edges becoming filamentous* in decay. Male spikes 1–2(–3), 20–50 mm. Male glumes c. 5 mm, narrowly obovate, obtuse, dark purplish-brown with a pale midrib and narrow hyaline border at top. Female spikes usually 2, rather distant, 15–70 × 5–7 mm, often male at top, erect; glumes and fr. arranged in conspicuous longitudinal rows. Peduncles 0 or very short. *Bracts setaceous, lower less than half as long as infl.* Female glumes 3–4 mm, ovate, subacute, *nerve ceasing below or at apex*, purple-brown with paler midrib and *hyaline margins*. Fr. (Fig. 77A) 3–4 mm, broadly ovate or suborbicular, flattened; beak c. 0.2 mm, rather stout, entire, smooth. Nut 2 mm, obovate, biconvex. Stigmas 2. Fl. 5–6. 2*n* = 74*–78; 80. Hs. or Hel.

A   B   C   D   E   F

Fig. 77. Fruits of *Carex*. A, *C. elata*; B, *C. acuta*; C, *C. aquatilis*; D, *C. recta*; E, *C. nigra*; F, *C. bigelowii*. ×5.

Native. By fen ditches and in wet places beside rivers and lakes; locally common. Scattered throughout England and Wales, mainly in the east; local in Ireland and Scotland. Europe north to C. Finland, but absent from much of the south; W. Siberia; N.W. Africa; North America.

64. C. acuta L.   Slender Tufted Sedge.

C. gracilis Curtis

A tufted or shortly creeping glabrous perennial 30–120 cm. *Stems sharply triquetrous*, somewhat rough. Lvs 5–10 mm wide, broadly and bluntly keeled, the margins rolling inwards on drying; *ligule 4–6 mm*, broadly triangular or ±truncate, margins often brownish. *Lowest sheaths bearing short lvs*, not webbed. Male spikes 1–4, 20–60 mm. Male glumes 4.5–5.5 mm, oblong or obovate, obtuse or ±lacerate, purplish with pale midrib, black at tip. Female spikes 2–4, rather distant, 30–100 × 4–5 mm, often male at top, nodding in fl., erect in fr. Peduncles short or upper 0. *Bracts lf-like, the lowest as long as or longer than infl.* Female glumes 2.5–4 mm, ovate, acute, sometimes obtuse, *midrib usually shortly excurrent*, purple-black with pale midrib in lower half. Fr. (Fig. 77B) 2–3.5 mm, obovate or almost elliptical, flattened, rarely (var. *sphaerocarpa* Uecht.) 2 mm, orbicular; beak c. 0.2 mm, slender, entire, smooth. Nut 2 mm, obovate or suborbicular, biconvex. Stigmas 2. Fl. 5–6. $2n = 74$–76, 78, 82–85. Hs. or Hel.

Native. Beside water and in wet grassy places. Scattered throughout the British Is. north to S. Aberdeen; common in many districts. Most of Europe; Siberia.

65. C. aquatilis Wahlenb.   Water Sedge.

A shortly creeping, rather rigid glabrous perennial 20–100 cm. *Stems bluntly trigonous*, smooth, *brittle*. Lvs 3–5 mm wide, concave, dark green and shiny beneath, light green above, the margins rolling inwards on drying; ligule 5–10 mm, ovate, acute. *Lower sheaths lfless*, persistent, often reddish, not fibrous. Male spikes 2–4, 15–40 × 1–2.5 mm. Male glumes 3–4 mm, obovate, obtuse, brown with a pale midrib and hyaline margin at top. *Female spikes* (2–)3–4(–5), rather distant, the lowest sometimes very distant, 20–60 × 3–4 mm, *tapering downwards from about the middle*, sometimes male at top. Peduncles of lower spikes short, of upper 0. *Bracts all lf-like, that of lowest spike longer than infl.* (except when lowest spike is very distant), *of next spike equalling or exceeding infl.* Female glumes c. 2.5 mm, lanceolate to ovate, acute or subacute, dark purple-brown with a pale midrib and hyaline tip. Fr. (Fig. 77c) 2–2.5 mm, broader than glumes, elliptical or obovate-elliptical, somewhat biconvex; beak c. 0.2 mm, entire, smooth. Nut 1.75 mm, obovate, biconvex. Stigmas 2. Fl. 7, $2n = 76$; 77; 84. Hel.

Native. Margins of lakes and streams in mountainous districts; local. Cardigan, Pembroke and Merioneth; Durham and Westmorland, northward to Caithness; scattered throughout Ireland south to Kerry. North Europe southwards to White Russia; Siberia; North America.

66. C. recta Boott   Estuarine Sedge.

C. kattegatensis Fries ex Krecz.; *C. salina* auct., non Wahlenb.

A shortly creeping glabrous perennial 30–100 cm. Stems trigonous, smooth. *Lvs* 3–5 mm wide, shorter than or about equalling stems, weakly keeled, serrulate; ligule 2–3 mm, triangular, acute. Lower sheaths lfless, blackish-brown, not fibrous. Male spikes 1–4, 10–40 × 3–4 mm, lateral often pedunculate. Male glumes 4–5 mm, oblong or lanceolate, obtuse or erose, brown with a pale midrib. Female spikes 2–4,

30–70 × 4–6 mm, erect, or spreading in fr. *Peduncles* 1–3 cm, rough. Bracts lf-like, equalling or exceeding infl., not sheathing. *Female glumes* 4–5 mm, ovate or lanceolate, acute or obtuse *with a long excurrent midrib*, dark brown or purple, midrib pale. Fr. (Fig. 77D) 2.5–3 mm, obovate or suborbicular, flattened, faintly veined; beak c. 0.2 mm, entire. Nut c. 2 mm, orbicular or obovate, shortly stipitate, rather asymmetrical. Stigmas 2. Fl. 7. Fr. 8–9. $2n = 84$. Hel.

Native. Estuaries of rivers in N.E. Scotland, on sandbanks with other sedges, very local. E. Inverness, E. Ross, E. Sutherland, Caithness. Doubtfully recorded from Norway and Faeroes; Atlantic coast of North America from Labrador to Massachusetts; apparently always very local. Probably of hybrid origin (*C. aquatilis* × *C. paleacea*).

67. C. nigra (L.) Reichard   Common Sedge.

C. angustifolia Sm.; *C. goodenovii* Gay; *C. vulgaris* Fr.; *C. cespitosa* auct., non L.; *C. fusca* All.; incl. *C. eboracensis* Nelmes

A creeping or occasionally densely tufted rhizomatous perennial 7–70 cm. Stems slender, not more than 2 mm diam., triquetrous, smooth below, rough above. *Lvs* 1–3(–5) *mm wide*, longer or shorter than stems, margins rolling inwards on drying; ligule 1–3 mm, ovate, rounded. Lower sheaths usually with lvs and soon decaying into a blackish fibrous mass. *Male spikes* 1–2, 5–30 mm, *lower usually much smaller than upper*. Male glumes 3–5 mm, oblong or narrowly obovate, obtuse or subacute, purplish with a pale midrib or rarely brown. Female spikes (1–)2–3(–4), ±contiguous, 7–50 × 4–5 mm, sometimes male at top. *Peduncles usually very short* or 0. Lowest bract lf-like, shorter or somewhat longer than infl., upper much shorter. *Female glumes* 2.5–3.5 mm, lanceolate to obovate-oblong, obtuse to acute or acuminate, black, sometimes with a pale midrib and narrow hyaline border at top, rarely brown. Fr. (Fig. 77E) 2.5–3.5 mm, broader than glumes, ovate to suborbicular, plano-convex, green or purplish; beak very short, entire. Nut 2 mm, elliptical, biconvex. Stigmas 2. Fl. 5–7. Fr. 6–8. $2n = 82$–85. Hs. or Hel.

A very variable and perhaps composite sp. which needs further investigation.

Native. In wet grassy places and beside water, on acid or sometimes basic soils. Widely distributed throughout the British Is.; common in suitable habitats; ascends to 1010 m on Lochnagar. Europe; N. Asia; mountains of N.W. Africa; North America.

68. C. bigelowii Torrey ex Schweinitz   Stiff Sedge.

C. rigida Good., non Schrank; ?*C. concolor* R.Br.; *C. hyperborea* Drejer

A shortly *creeping* rigid glabrous perennial 5–30 cm. Stems sharply triquetrous, usually rough towards the top. *Lvs* 2–4(–7) *mm wide*, shorter than stem, *recurved*, keeled, shortly acuminate, margins rolling outwards on drying; ligule 1–2 mm, ±triangular, acute. Lower sheaths lfless, reddish, persistent. *Male spike* 1(–2),

5–20 mm. Male glumes 3–4 mm, obovate, purplish or almost black with pale base and midrib. Female spikes 2–3, contiguous or the lower distant, 5–15 × 3–5 mm, erect. Peduncles 0 or very short. Lower bract lf-like, shorter than infl., other bracts small, all with large brownish auricles. Female glumes 2.5–3.5 mm, broadly ovate, obtuse, purple or blackish with a pale, sometimes inconspicuous midrib. Fr. (Fig. 77F) 2.5–3 mm, little broader than glumes, ovate, obovate or suborbicular, weakly biconvex; beak very short, entire, smooth. Nut 2 mm, ovate, biconvex. Stigmas 2. Fl. 6–7. $2n = 68$–71. Hs.

Native. In damp stony places on mountains. Mountain regions of N. Wales, N. England and Scotland; Ireland, chiefly in the south and west; from 30 m in Shetland to 1326 m on Ben Nevis. Circumpolar and on mountains southwards to S. Ural and Japan.

Subgenus PRIMOCAREX Kük.

Monoecious. Spike solitary, male above, female below. Stigmas 2 or 3. A very heterogeneous collection of spp.

69. C. microglochin Wahlenb.

A shortly creeping glabrous perennial 5–12 cm. Stems trigonous, smooth, stiff, straight. Lvs 0.5–1 mm wide, usually shorter than stems, nearly flat; ligule c. 0.5 mm, tubular. Lower sheaths soon decaying. *Spike* 5–10 mm, *with 4–12 fr. which are deflexed when ripe.* Male glumes 2.5–3 mm, lanceolate, acute or obtuse, reddish-brown with pale midrib. *Female glumes* c. 2 mm, *caducous,* ovate, obtuse, reddish-brown, midrib pale, tip hyaline. *Fr.* (Fig. 78A) 3.5–4.5(–6) mm, *narrowly conical, abruptly contracted at base, yellowish;* beak c. 1 mm, stout, obliquely truncate. *A stout bristle arises near the base of the nut and protrudes from the top of the beak* together with the stigmas. Nut c. 1.5 mm. Stigmas 3. Fl. 7–8. Fr. 8–9. $2n = 48, 58$. Hel.

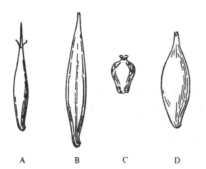

Fig. 78. Fruits of *Carex.* A, *C. microglochin*; B, *C. pauciflora*; C, *C. rupestris*; D, *C. pulicaris.* ×5.

Native. In small micaceous bogs, 600–900 m, very rare. Mid-Perth. Circumpolar, southwards to the Alps in Europe; Caucasus, Altai, Himalaya, Tibet, southernmost South America.

70. C. pauciflora Lightf.   Few-flowered Sedge.

A shortly creeping glabrous perennial 7–25 cm, somewhat similar to *C. microglochin.* Rhizome slender, branched, stems distant. Lvs 1–2 mm wide, those of fertile shoots ±flat, of vegetative shoots often setaceous. *Spike* c. 5 mm, *with 2–4 or rarely more fr.* which are deflexed when ripe. Male glumes 3.5–5 mm, lanceolate, ±acute, reddish-brown with hyaline margins. *Female glumes* c. 4 mm, similar to, but rather broader than male, caducous. *Fr.* (Fig. 78B) 5–7 mm, narrow, *tapered at both ends,* yellowish. Nut 2 mm, oblong. Stigmas 3. The persistent style protrudes from the top of the fr. and should not be confused with the bristle in *C. microglochin.* Fl. 5–6. Fr. 6–7. Hel.

Native. On wet moors and in bogs. Caernarvon; N.E. Yorks and Cumberland to Caithness; Antrim, rather local. Circumpolar.

71. C. rupestris All.   Rock Sedge.

A shortly creeping tufted glabrous perennial 5–20 cm. Stems triquetrous, smooth. *Lvs* 1–1.5 mm wide, about equalling stems, folded or involute, smooth, *usually much curled*; ligule 1–2 mm, ovate, obtuse. Lower sheaths lfless, brown, not fibrous. Spike 10–15 × 2 mm, few-fld, male at top. Bracts glumaceous. Male glumes 2.5–3 mm. *Female glumes* 2.5–3.5 mm, broadly ovate, obtuse, sometimes mucronate, dark brown, *persistent. Fr.* (Fig. 78C) c. 3 mm, obovoid-trigonous, brown, *erect*; beak very short. Nut broadly elliptic, triquetrous. Stigmas 3. Fl. 6–7. Fr. 7–8. $2n = 50, 52$. Hs.

Native. On narrow ledges and in cracks of base-rich rocks in C. and N.W. Scotland, rare and local. Perth, Angus, Aberdeen, Ross, Sutherland and Caithness. North Europe and mountains south to the Alps, Pyrenees, Carpathians, and S. Ural; Siberia; Greenland; North America.

72. C. pulicaris L.   Flea Sedge.

A shortly creeping glabrous perennial 10–30 cm, forming dense patches. Stems terete, slender, rigid, smooth. Lvs 0.5–1 mm wide, nearly equalling stems, channelled, dark green, rather rigid; ligule very short, ovate, rounded. Lower sheaths soon decaying, upper auricled. *Spike* 10–25 mm, *top half male,* 1 mm or less wide, acuminate; *female half lax,* 3–10-fld. Male glumes 4.5–5 mm, oblong, obtuse or subacute, brown. *Female glumes* 3.5–4 mm, *caducous. Fr.* (Fig. 78D) 4–6 mm, lanceolate, flattened, dark brown, *shiny,* shortly and stoutly pedicellate, at length deflexed; *beak almost* 0. Nut c. 2.5 mm, compressed-cylindrical, almost truncate at top, narrowed to a short thick stalk at base. Stigmas 2. Fl. 5–6. Fr. 6–7. $2n = 58, 60$. Hs. or Hel.

Native. In damp calcareous grassland, fens, base-rich flushes, etc.; locally common. Throughout the British Is., but uncommon in the south and east and now lost through drainage in many places. N., C. and W. Europe, southwards to N. Spain and eastwards to Estonia.

150. GRAMINEAE (*Poaceae*)

Annual or perennial herbs, sometimes woody. Lvs with silicified cells (silica cells) on the lower epidermis in 1–2 rows beneath the vascular bundles. Stems usually branched at the base and, in perennials, of two sorts, flowering stems or *culms* and non-flowering or vegetative shoots; in annuals all the stems bear fls. Culms cylindrical or sometimes flattened, solid at the nodes and usually hollow in the internodes. Lvs solitary at the nodes and consisting of a *sheath*, *ligule* and *blade*; sheath

and the upper (*palea*)* generally thin and delicate, sometimes 0, the whole forming the floret. Florets 1 to many, distichous, ±imbricate and sessile on a short slender axis (*rhachilla*) usually with 2 bracts at the base (*glumes*), the whole forming a *spikelet*. One or more of the florets sometimes reduced to an empty lemma and palea or only a lemma, which may be very small. Spikelets pedicellate in panicles or racemes, or sessile in spikes; panicles sometimes spike-like with short branches not readily apparent, bearing crowded spikelets. Fr. usually a caryopsis. A very large family with c. 620 genera and 10 000 spp., generally distributed.

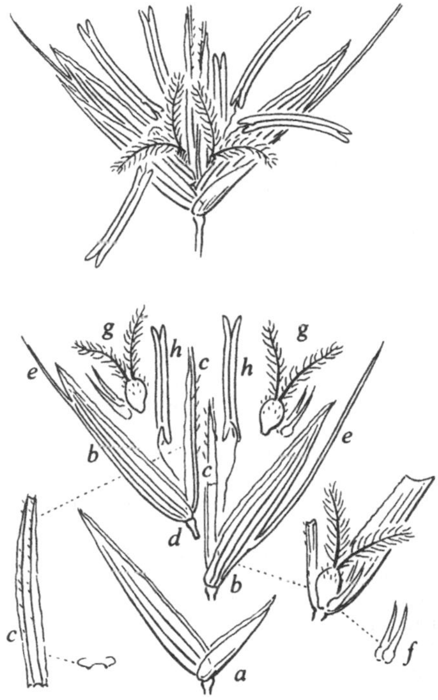

Fig. 79. A grass spikelet. a, glumes; b, lemma; c, palea; d, rhachilla; e, awn; f, lodicules; g, ovary and stigmas; h, stamen. ×5.

encircling the stem, with overlapping or sometimes connate margins; ligule a small flap of tissue at the junction of sheath and blade, sometimes replaced by a ring of hairs, rarely 0; blade usually long and narrow, rarely broad, passing gradually into the sheath, sometimes with thickened projections (*auricles*) at each side of the base. Fls usually hermaphrodite, consisting of 3 stamens (rarely 2 or 1), a 1-celled ovary with a solitary anatropous ovule and, usually, 2 styles with generally plumose stigmas and 2 small delicate scales (*lodicules*), which are occasionally absent. These are enclosed between 2 bracts, the lower (*lemma*)* membranous to coriaceous

1 Ligule a ring of hairs.	2
Ligule membranous or 0.	6
2 Infl. a panicle with distinctly pedicellate spikelets.	3
Infl. of 2–several spikes; spikelets sessile or nearly so.	5
3 Plant large, reed-like; sheaths loose; florets with silky hairs at base.	49. PHRAGMITES
Not as above.	4
4 Stem-base swollen; spikelets numerous; glumes much shorter than spikelet.	51. MOLINIA
Stem-base not swollen; spikelets few; glumes nearly as long as spikelet.	50. DANTHONIA
5 Stout, tufted; spikelets at least 10 mm.	54. SPARTINA
Slender, creeping; spikelets c. 2 mm.	53. CYNODON
6 Spikelets obviously of 2 or more florets.	7
Spikelet of 1 floret, occasionally with vestiges of other florets.	45
7 Infl. a simple spike with the sessile or subsessile spikelets inserted singly or in pairs on the rhachis.	8
Infl. a panicle, or spikelets clustered.	15
8 Glume 1 in lateral spikelets, 2 in terminal.	9
Glumes always 2.	10
9 Glume 1 in all the lateral spikelets.	4. LOLIUM
Glumes 2 in some of the lateral spikelets, though the one next the axis often small.	2. ×FESTULOLIUM
10 Tall perennials; plant tufted or with a far-creeping rhizome.	11
Small annuals; plant neither tufted nor rhizomatous.	14
11 Spikelets edgeways on to rhachis or else terete.	12
Spikelets broadside on to the rhachis, strongly flattened laterally.	13
12 Lemmas awned.	19. BRACHYPODIUM
Lemmas awnless.	2. ×FESTULOLIUM
13 Spikelets 2–3 cm, in pairs.	20. LEYMUS
Spikelets (excluding ±awns) c. 1 cm.	21. ELYMUS
14 Glumes subequal; plant stiff and wiry.	6. DESMAZERIA
Glumes unequal; plant not wiry.	5. VULPIA
15 Spikelets viviparous (mountain districts).	16
Spikelets not viviparous.	18
16 Lvs setaceous.	1. FESTUCA
Lvs flat.	17
17 Lemmas awned.	31. DESCHAMPSIA
Lemmas unawned.	7. POA
18 Glumes largely hyaline, greenish-golden; uppermost lf usually less than 5 mm; margins of lemmas of	

* = flowering glume, lower pale, or valve.

* = pale, upper pale, or valvule.

2 lower florets fringed with hairs; plant smelling strongly of coumarin when crushed (rare).
33. HIEROCHLOË

Not as above. 19

19 Lemma unawned or with a straight awn from the apex, or from a notch at the apex. 20

Lemma of at least some florets with an awn sticking out from the back below the apex; awn usually geniculate. 38

20 Spikelets silvery; upper glume about equalling 1st floret; tufted perennials with narrow dense ±parallel-sided panicles. 28. KOELERIA

Not as above. 21

21 Spikelets crowded in dense one-sided masses towards the ends of the panicle-branches; coarse tufted plant with strongly compressed vegetative shoots.
9. DACTYLIS

Not as above. 22

22 Infl. a ±spreading panicle; at least some branches easily seen. 23

Infl. spike-like; spikelets crowded in groups on short branches. 34

23 Lemmas awned or at least with a bristle-like point. 24
Lemmas unawned. 26

24 Spikelets ovoid, subterete. 18. BROMUS

Spikelets lanceolate, widening upwards or ±parallel-sided, ±strongly compressed. 25

25 Spikelets up to 15 mm, lanceolate, somewhat compressed. 1. FESTUCA

Spikelets rarely less than 20 mm, parallel-sided or widening upwards, strongly compressed.
18. BROMUS

26 Veins of the lemma ±parallel, prominent and extending nearly or quite to the apex (±aquatic). 27

Veins of the lemma converging towards the tip or else inconspicuous or disappearing well below the apex. 28

27 Spikelets 5 mm or more; lemma 7–9-veined.
17. GLYCERIA

Spikelets c. 2 mm; lemma 3-veined. 11. CATABROSA

28 Glumes nearly as long as spikelet; spikelets rather few; margins of sheaths connate. 16. MELICA

Glumes much shorter than spikelet; spikelets usually numerous; margins of sheaths free. 29

29 Spikelets broadly conical or ovate, solitary on long, very slender pedicels; lemma cordate at base.
14. BRIZA

Not as above. 30

30 Lemmas linear-lanceolate, acuminate. 1. FESTUCA
Lemmas broader, acute or obtuse. 31

31 Spikelets 10 mm or more. 32
Spikelets rarely more than 5 mm. 33

32 Spikelets subterete; ovary without a hairy appendage.
1. FESTUCA

Spikelets strongly flattened; ovary with a hairy appendage. 18. BROMUS

33 Spikelets flattened; lemmas keeled; lvs flat when fresh (not salt-marshes). 7. POA

Spikelets subterete; lemmas rounded on back; lvs usually rolled up (salt- and brackish-marshes, rarely elsewhere). 8. PUCCINELLIA

34 Spikelets of 2 kinds; sterile lemmas stiff and distichous, fertile lemmas awned. 10. CYNOSURUS

Spikelets all similar. 35

35 Lemmas ending in 4–5 small points; infl. bluish and

shiny (local). 15. SESLERIA
Not as above. 36

36 Lemmas awned. 5. VULPIA
Lemmas unawned. 37

37 Glumes subequal; wiry annuals 6. DESMAZERIA
Glumes very unequal; soft perennial.
2. ×FESTULOLIUM

38 Spikelets (excluding awns) 9 mm or more. 39
Spikelets (excluding awns) 5 mm or less. 41

39 Upper glume 2 cm or more; spikelets eventually pendent; annual. 24. AVENA

Upper glume rarely more than 1.5 cm; spikelets erect; perennial. 40

40 Lvs long, soft; lower floret usually male, awned; upper floret female or hermaphrodite, usually unawned. 26. ARRHENATHERUM

Lvs short, often rather stiff, 2–4 lower florets all hermaphrodite and awned. 25. AVENULA

41 Spikelets shiny. 42
Spikelets whitish or pinkish, not shiny. 44

42 Annual; 5–20(–30) cm; lvs mostly dead at fl. or soon after; non-flowering shoots 0. 32. AIRA

Perennial; normally taller; lvs green at fl. and fr.; non-flowering shoots present. 43

43 Apex of lemma ±truncate and torn; spikelets silvery or purplish. 31. DESCHAMPSIA

Apex of lemma acute, ending in 2 bristle points; spikelets usually yellowish. 29. TRISETUM

44 Awn tapering to a sharp point; lvs flat, not glaucous; plant ±hairy. 35. HOLCUS

Awn thickened towards the apex; lvs setaceous, very glaucous; plant glabrous (very local).
36. CORYNEPHORUS

45 Spikelets inserted in 2 rows on the rhachis in groups of 2 or 3 at each node. 46

Spikelets solitary or not inserted in 2 rows on the rhachis. 47

46 Central spikelet hermaphrodite, lateral male or sterile. 22. HORDEUM

Central spikelet male or hermaphrodite, lateral hermaphrodite. 23. HORDELYMUS

47 Spikelets with conspicuous persistent bristles on their pedicels. 58. SETARIA

No bristles on pedicels. 48

48 Spikelets sunk in the rhachis of the solitary spike; small annuals; (clayey ground by the sea).
46. PARAPHOLIS

Not as above. 49

49 Infl. ovoid, softly and densely hairy; annual.
30. LAGURUS

Not as above. 50

50 Infl. of 1 or more spikes, the spikelets sessile or nearly so and arranged in 1 or 2 rows. 51

Infl. not as above. 54

51 Spike solitary. 52
Spikes more than 1. 53

52 Lemma awnless; small annual 13. MIBORA
Lemma awned; wiry perennial. 52. NARDUS

53 Spikes arranged in a raceme; spikelets usually awned.
56. ECHINOCHLOA

Spikes digitate or nearly so; spikelets not awned.
57. DIGITARIA

54 Spikelets clustered on short branches and forming a dense cylindrical or ovoid spike-like infl. 55

Spikelets ±spread out along the easily apparent

branches of the panicle.   64

55 Tall sand-binding grasses; spikelets more than 10 mm.   56
   Not as above; spikelets less than 10 mm.   57
56 Infl. whitish, obtuse.   41. Ammophila
   Infl. purplish-brown, acute.
           42. ×Ammocalamagrostis
57 Glumes very unequal, the lower much shorter than
   the spikelet; plant smelling of coumarin when
   crushed.   34. Anthoxanthum
   Glumes subequal, both about as long as or longer
   than the lemma.   58
58 Glumes flattened, whitish with a green line, keel
   winged.   47. Phalaris
   Not as above.   59
59 Glumes awned.   60
   Glumes unawned.   61
60 Awns conspicuous, 5 mm or more; glumes deciduous.
           39. Polypogon
   Awns c. 1 mm; glumes persistent.   40. ×Agropogon
61 Glumes narrow, acuminate, swollen and rounded at
   the base.   38 Gastridium
   Glumes not swollen and rounded at the base.   62
62 Infl. very dense and regular, cylindrical or ovoid;
   glumes ciliate or silky.   63
   Infl. rather lax and irregular; glumes rough on their
   surfaces, falling with the fr.   39. Polypogon
63 Glumes tapering and curving inwards at the apex;
   awn of lemma exserted or spikelets silky; palea 0.
           45. Alopecurus
   Glumes abruptly contracted to the outward-curving
   apex, or, if as above, then lemma unawned and
   spikelets not silky; palea present.   44. Phleum
64 Ligule long; infl. purplish; tall reed-like grass in damp
   places.   47. Phalaris
   Not as above.   65
65 Glumes 0; spikelets strongly flattened (wet places,
   rare).   55. Leersia
   Glumes 2.   66
66 Lower glume small, ±encircling the base of spikelet;
   upper glume large, often awned and resembling the
   lemma.   56. Echinochloa
   Glumes ±equal.   67
67 Glumes rough on their surfaces, falling with the fr.
           39. Polypogon
   Glumes smooth, persistent.   68
68 Glumes ovate, obtuse or subacute; lemma shiny,
   becoming very hard in fr.   48. Milium
   Glumes lanceolate, acuminate; lemma tough but not
   hard in fr.   69
69 Lemma awned; awn c. 5 mm; annuals.   12. Apera
   Lemma awned or not; awn not more than 2 mm; perennials.   70
70 Spikelets with callus hairs at least ⅔ as long as lemma.
           43. Calamagrostis
   Spikelets without or with very short callus hairs.
           37. Agrostis

Tribe. Poeae. Lvs with oblong or rounded silica cells; 2-celled hairs 0; green tissue uniformly distributed between the vascular bundles; 1st foliage lf of seedling narrow, erect. Ligule membranous, glabrous. Spikelets of (1–)2–many florets, laterally compressed or subterete, usually in panicles, rarely in distichous spike-like racemes. Rhachilla usually disarticulating above the glumes. Glumes 2, rarely 1, usually shorter than the 1st floret on the same side. Lemma (3–)5–13-veined, the veins convergent, often herbaceous to membranous, unawned, or awned from the entire or 2-lobed apex; awn not geniculate. Lodicules 2, thin, lobed. Stamens 3, rarely 2 or 1. Ovary glabrous or hairy at apex, without an appendage; styles 2. Fr. with linear to punctiform hilum and compound starch grains. Chromosomes large, basic number 7, rarely 5, 6 or 11.

1. Festuca L.

Perennial herbs. Spikelets subterete, with 3 or more florets. Glumes membranous, the lower 1-, the upper 1–3-veined, shorter than spikelet, unequal or subequal, acute. Lemma membranous or chartaceous, rounded on the back, usually mucronate or awned from the apex. Palea thin, about equalling the lemma, 2-veined and shortly bifid. Lodicules 2, acute. Stamens 3. Ovary glabrous or rarely pubescent at the top; *styles* short, *terminal*.

About 300 spp. in temperate regions and on the higher mountains of the tropics.

1 Wider lvs at least 5 mm wide, flat or canaliculate, convolute in bud.   2
   All lvs less than 5 mm wide, usually plicate.   5
2 No auricles; ligule of uppermost lf 1–5 mm.
           4. altissima
   Auricles conspicuous; ligule a minute rim.   3
3 Lvs flat, glossy beneath; panicle-branches nodding;
   awn at least 10 mm.   **3. gigantea**
   Lvs canaliculate, not glossy beneath; panicle-
   branches not nodding; awn not more than 4 mm.   4
4 Basal sheaths thin, brownish; panicle-branches 2 at
   lowest node, one with 1–3, the other with 4–6 spikelets.   **1. pratensis**
   Basal sheaths coriaceous, whitish, panicle-branches
   2–3 at lowest node, each with 5–15 spikelets.
           2. arundinacea
5 Auricles absent; non-flowering shoots usually extravaginal.   6
   Auricles conspicuous; non-flowering shoots intravaginal.   9
6 Sheaths not decaying into fibres; upper lvs flat when
   fresh; top of ovary hairy.   **5. heterophylla**
   Sheaths decaying into fibres; upper lvs usually setaceous to junciform; top of ovary glabrous.   7
7 Rhizomes absent; plant densely caespitose; awn at
   least half as long as lemma.   **6. nigrescens**
   Rhizomes present; plant ±laxly caespitose; awn less
   than half as long as lemma.   8
8 Rhizomes short; lower glume not more than 4.1 mm.
           7. rubra
   Rhizomes long; lower glume more than 4.1 mm.
           8. juncifolia
9 Lvs laterally compressed; sclerenchyma in a ±inter-
   rupted unequally thickened ring. **17. trachyphylla**
   Lvs not compressed; sclerenchyma in 3 separate
   strands.   10
10 Most lvs acute; pedicels thickened at apex.
           16. indigesta
   Most lvs obtuse; pedicels not thickened at apex.   11

11 Spikelets proliferating.   **15. vivipara**
   Spikelets not proliferating.   *12*
12 Lvs usually 0.2–0.6 mm wide.   *13*
   Lvs usually 0.5–0.85 mm wide.   *15*
13 Lemma unawned, though sometimes mucronate.
                                            9. tenuifolia
   Lemma awned.   *14*
14 Lvs scabrid throughout; sheaths open almost or quite
   to base.   **10. ovina**
   Lvs smooth or weakly scabrid near apex; sheaths
   closed.   **11. armoricana**
15 Lvs smooth.   **14. longifolia**
   Lvs scabrid.   *16*
16 Lvs usually with 1 rib; stems usually scabrid or pubescent above.   **12. guestfalica**
   Lvs usually with 2–3 ribs; stems usually glabrous
   above.   **13. lemanii**

The taxonomy of spp. **6–17** is difficult and debatable; this account is based largely on that by I. Markgraf-Dannenberg in *Flora Europaea*, 5, to which reference should be made for further information.

The length of the spikelets is measured from the base of the lower glume to the apex of the fourth lemma, excluding the awn.

1. F. pratensis Hudson   Meadow Fescue.

F. elatior L., pro parte

An erect glabrous perennial 30–120 cm. Stem-base usually clothed in decaying remains of *dark brown sheaths. Lvs seldom more than* 4 mm *wide*, smooth, margins serrulate. Sheaths smooth; auricles glabrous, inconspicuous. Panicle 8–20 cm, slender, nodding, secund; *lowest node with 2 branches, the longer with* 4–6, *the shorter with* 1–3 *spikelets*; rarely spike-like with solitary subsessile distichous spikelets; *spikelets* 10–15 mm, *linear to lanceolate*, of 5–12 rather distant florets. Glumes unequal, linear-lanceolate, obtuse, lower 1-, upper 3-veined. Lemma lanceolate, hyaline at the apex, smooth, obscurely veined. Palea finely ciliate on the veins. Anthers 3 mm. Ovary glabrous. Fl. 6. 2*n* = 14*. Hs. or Chh.

Native. In meadows and grassy places. Throughout the British Is., rare in northern Scotland. Most of Europe; N. Africa; temperate Asia. Introduced in North America.

2. F. arundinacea Schreber   Tall Fescue.

A stout tufted glabrous perennial 30–200 cm. Stem-base with *tough whitish, sometimes scale-like sheaths. Lvs* flat or ±involute, *up to* 7 mm *wide*, smooth, margins serrulate. Sheaths smooth; auricles ciliate, prominent. Panicle (8–)20–30(–40) cm; *branches* stout, 2–3 *together at the lowest node, with* 5–15 *spikelets. Spikelets* 10–15 mm, *lanceolate to ovate*, of (4–)5–6(–8) usually closely imbricate florets. Glumes unequal, margins hyaline; lower acute, 1-veined; upper lanceolate or oblong, acute or obtuse, 3-veined. Lemma lanceolate, obscurely veined, middle-vein often shortly excurrent. Palea finely ciliate on the veins. Anthers 4 mm. Ovary glabrous. Fl. 6–8, 2*n* = 42*. Hs.

Larger in all its parts than *F. pratensis* with which it hybridizes.

Native. In grassy places. Generally distributed in the British Is. Most of Europe; western Siberia; N. Africa.

3. F. gigantea (L.) Vill.   Giant Fescue, Tall Brome.

Bromus giganteus L.

A stout erect glabrous perennial 50–150 cm. *Lvs* 6–18 mm wide, *scabrid from above downwards, glossy beneath. Sheaths often scabrid from below upwards; auricles prominent*, reddish; ligule up to 2.5 mm. Panicle 10–50 cm, very lax, nodding; branches stout, long, bare below, scabrid, 1–2 together. Spikelets 8–13 mm, of 3–10 florets, lanceolate. Glumes subequal, lanceolate, acuminate, margins broadly hyaline; lower 1-, upper 3-veined. Lemma lanceolate, scaberulous, *awn* 10–18 mm, *slender*. Palea glabrous. Anthers 2 mm. Ovary glabrous. Fl. 6–7. 2*n* = 42. Hs.

Native. In woods and shady hedgebanks. Throughout the British Is. except the extreme north of Scotland. Most of Europe, except the Arctic, rarely south of the Alps; Asia.

4. F. altissima All.   Wood Fescue.

F. sylvatica (Pollich) Vill., non Hudson

An erect glabrous tufted perennial 50–200 cm. *Stem-base clothed with scale-like sheaths*. Lvs 4–14 mm wide, smooth or slightly scabrid, margins finely serrate. Sheaths scabrid; auricles distinct. *Uppermost ligule up to* 5 mm, torn. Panicle 10–16 cm, spreading or ±contracted in fr., subsecund, erect; branches slender, smooth, 1–2 together. *Spikelets* 5–8 mm, of 2–5 florets, *ovate. Glumes subequal*, 1-veined. Lemma lanceolate, acuminate, scabrid, unawned. Palea scabrid. Anthers 2.75–3 mm. *Ovary setose at the top*. Fl. 6–7. 2*n* = 14, 42. Hs.

Native. In rocky woods mainly in the north, usually beside streams. Sussex; Monmouth to Worcester; Cardigan, Derby; N. Lancs and N.E. Yorks to Cumberland and Durham, very local; more generally distributed in Scotland and Ireland though local and often rare. N. and C. Europe southwards to N. Spain and S.C. Russia; W. Asia.

5. F. heterophylla Lam.   Various-leaved Fescue.

An erect *densely tufted* perennial 60–120 cm. Culms smooth. Basal lvs long, setaceous, 0.4–0.6 mm wide, margins serrate; *cauline lvs flat*, 2–3 mm wide, with short spreading hairs on the veins above. Sheaths smooth. Ligule very short. *Panicle* 7–18 cm, slender, *nodding*, ±spreading. *Spikelets* 7–12 mm, of 3–6 florets, *shiny*; florets rather distant. Glumes unequal, subulate, margins hyaline, lower 1-, upper 3-veined. Lemma lanceolate, thin, scaberulous, obscurely veined, tapering into a slender awn up to 6 mm. Palea glabrous. Anthers 2–3 mm. *Ovary hairy at the top*. Fl. 6–8. 2*n* = 28, 42. Hs.

Native. In dry shady places. Scattered and local in southern England; in a few localities elsewhere. Europe from S. England and Poland to N. Spain and Greece.

6. F. nigrescens Lam.

F. rubra subsp. *fallax* (Thuill.) Hayek

A densely caespitose perennial 30–90 cm. Lvs soft, with 5–7 veins, dark green, usually smooth; sheaths closed to the mouth, pinkish, decaying into fibres; auricles 0. Panicle 4–10 cm, the branches scabrid, patent only at flowering. Spikelets (6.5–)7–9.5 mm; upper glume 1.3–1.6 mm wide; lemma 4.6–6.2 × 2.1–2.3 mm, usually glabrous, rather long-acuminate; awn c. $\frac{1}{2}$ as long as lemma. $2n = 28, 42$.

Native. Grassy places. Distribution imperfectly known. S., W. and C. Europe, northwards to S. Sweden.

The above description refers to subsp. **nigrescens**.

7. F. rubra L.   Red Fescue.

A laxly to rather densely caespitose perennial up to 110 cm. Lvs of non-flowering shoots with 3–9 ribs; cauline lvs often flat; sheaths closed to the mouth, pinkish, decaying into fibres; auricles 0. Panicle ±erect, the branches hairy or glabrous, patent at flowering. Lemma 4–9 mm; awn not more than $\frac{1}{2}$ as long as lemma.

Very variable.

1 Plant rather densely tufted; rhizomes short.   *2*
　　Plant loosely tufted; rhizomes long.   *3*
2 Lvs glabrous or subglabrous.   **(d) subsp. pruinosa**
　　Lvs usually with dense, rather long hairs.
　　　　　　　　　　　　　　　　　　　　　(e) subsp. juncea
3 Lvs rigid; spikelets 9–13 mm; lemma usually tomentose.   **(c) subsp. arenaria**
　　Lvs not rigid; spikelets 7–11 mm; lemma not tomentose.   *4*
4 Culms up to 80(–110) cm; plicate lvs 0.65–0.85 mm wide; spikelets 7–10 mm.   **(a) subsp. rubra**
　　Culms not more than 45 cm; plicate lvs 0.5–0.7 mm wide; spikelets 10–11 mm.   **(b) subsp. litoralis**

(a) Subsp. rubra: Lvs smooth, ±hairy, rarely scabrid near the apex; veins (5–)7; sheaths glabrous to sparsely hairy. Panicle 7–14 cm, lax. Spikelets bright green or glaucous, rarely pruinose; upper glume 1.2–1.4(–1.6) mm wide; lemma 5–7 × 1.9–2.3 mm, glabrous or subglabrous. $2n = 14, 28$. Almost throughout the range of the sp.

(b) Subsp. litoralis (G. F. W. Meyer) Auquier: Lvs smooth, glabrous or very shortly hairy. Panicle 6–8 cm, rather dense. Spikelets (8–)10–11 mm, bright green or glaucous, rarely pruinose; upper glume 1.6–2.1 mm wide; lemma 6.2–7.1 × 1–2.3 mm. $2n = 42$. Salt-marshes and damp sandy places by the sea. Coasts of W. Europe and the Baltic.

(c) Subsp. arenaria (Osbeck) Syme; Culms up to 70 cm, stout, rigid. Lvs smooth, densely hairy, rigid, obtuse. Panicle 7–16 cm. Spikelets 6–13 mm, glaucous; upper glume 1.4–1.8 mm wide; lemma 6.9–9.7 × 2.1–2.6 mm. $2n = 56$. Maritime sands. Coasts of N.W. Europe and the Baltic.

(d) Subsp. pruinosa (Hackel) Piper: Culms not more than 45 cm. Lvs glabrous or subglabrous, rather rigid. Panicle 2–6 cm, rather dense. Spikelets 7.6–10 mm; upper glume 1.3–1.7 mm wide; lemma 4.5–6.8 × 2–2.1 mm, glabrous or hairy, usually pruinose. $2n = 42$. Atlantic coasts of Europe and North America.

(e) Subsp. juncea (Hackel) Soó: Culms up to 65 cm, stout. Lvs smooth, usually with dense long hairs, obtuse, rigid. Panicle (5–)6.5–12 cm, erect, rather dense. Spikelets 8.5–12 mm, glaucous; upper glume 1.6–2 mm wide, lemma 5–6.9 × 2.1–2.6 mm, glabrous or shortly hairy. $2n = 42$. Usually on disturbed ground. Throughout most of the range of the sp., except the USSR.

8. F. juncifolia St-Amans   Rush-leaved Fescue.

Like *F. rubra* but with long rhizomes and plant often more robust and often taller; lvs 0.7–1.5 mm wide, smooth, rigid, the ribs very stout and with long hairs; lower glume more than 4.1 mm. $2n = 56$.

Native. Dunes, mainly on the south and east coasts of England, but extending northwards in a few scattered localities to E. Sutherland; 1 locality in S. Wales. W. Europe from N. Spain to the Netherlands.

9. F. tenuifolia Sibth.   Fine-leaved Sheep's-fescue.

A densely caespitose perennial 20–30(–50) cm. Lvs with (5–)7 veins and 1 rib; sheaths open to the base; auricles present. Panicle (2–)4–8 cm, lax, the branches scabrid. Spikelets usually 4–5.2 mm, greenish; upper glume 0.8–1.2 mm wide; lemma 2.3–3.6(–4.4) × 1.4–1.5(–1.7) mm, scabrid or sometimes hairy, unawned though sometimes mucronate. $2n = 14, 28$.

Native. Sandy or peaty soils, usually calcifuge. Scattered throughout Great Britain north to Orkney but rare in the west; apparently rare in Ireland and only in the east. W. and C. Europe.

10. F. ovina L.   Sheep's-fescue.

A densely caespitose perennial up to 70 cm. Lvs with 5–7 veins and 1 rib; sheaths usually open to the base; auricles present. Panicle 2–12 cm, rather dense, the branches scabrid. Spikelets usually 4.8–6.3 mm, green, glaucous or violet-tinged; upper glume 0.8–1.3 mm wide; lemma (2.6–)3.5–5.1 × 1.4–2 mm, glabrous, scabrid or hairy; awn $\frac{1}{4}$–$\frac{1}{3}$ as long as lemma. $2n = 14$.

Native. Grassy places on shallow dry soils. Throughout the British Is. N. and C. Europe.

11. F. armoricana Kerguélen

A densely caespitose perennial up to 36 cm. Lvs with (5–)7(–9) veins; sheaths usually closed in the lower $\frac{1}{3}$–$\frac{1}{2}$, glabrous or sparsely hairy; auricles present. Panicle 1.5–7 cm, rather dense, the branches ±hairy. Spikelets 6–7.4 mm, glaucous; upper glume 1–1.3 mm wide; lemma (3.8–)4.4–4.6(–5) × 1.4–2 mm; awn usually 1.3–1.7 mm. $2n = 28$.

Native. Maritime heaths and rocks. Channel Is. N.W. France.

12. E. guestfalica Boenn. ex Reichenb.   Hard Fescue.

A densely caespitose perennial 30–70 cm. Lvs with 7 veins; sheaths open to the base, glabrous or densely villous; auricles present. Panicle 5–12 cm, lax, the branches scabrid. Spikelets (5.5–)6–7.5 mm; upper glume 1–1.4(–1.6) mm wide; lemma 3.6–5.3 × (1.4–)1.6–2.3 mm, ciliate, weakly scabrid or ±pubescent above; awn up to $\frac{1}{2}$ as long as lemma. $2n = 28$.

Native. Dry grassy places. N.W. and C. Europe.

13. F. lemanii Bast.   Hard Fescue.

F. longifolia auct., non Thuill., *F. duriuscula* auct.,? an L.

Caespitose perennial usually 25–50 cm. Lvs rigid, with (5–)7–9 veins and (1–)2–3 ribs, glaucous, subpruinose, sometimes puberulent near the base, usually scabrid above; sheaths open nearly to the base, glabrous or puberulent; auricles present. Panicle (2–)4–7(–9) cm, usually lax, the branches scabrid. Spikelets (5.6–)6.5–7.2(–8.3) mm; lemma usually 4.2–4.6 × c. 1.8 mm; awn less than $\frac{1}{2}$ as long as lemma. $2n = 42$.

Native. Grassy places. W. Europe.

14. F. longifolia Thuill.   Blue Fescue.

F. caesia Sm.

A densely caespitose perennial 8–35(–45) cm. Lvs with 7 veins and 3–5 ribs, pale green, pruinose, smooth; sheaths open to the base, glabrous; auricles present. Panicle 2.5–7 cm, ±dense, the branches scabridulous. Spikelets 5.5–7.5 mm; upper glume 1–1.1 mm wide; lemma (3.5–)4.2–5 × 1.4–2 mm; awn $\frac{1}{4}$–$\frac{1}{3}$ as long as lemma. $2n = 14$.

Native. Dry grassland in East Anglia. N. and N.C. France.

15. F. vivipara (L.) Sm.   Viviparous Fescue.

A densely caespitose perennial 10–40 cm. Lvs with 5–7(–9) veins and 1 rib, smooth or scabrid, green or glaucous; sheaths usually closed in the lower $\frac{1}{4}$–$\frac{2}{3}$, usually glabrous; auricles present. Panicle usually 3–6 cm, the branches scabrid or pubescent. Spikelets with at least the upper florets proliferating (viviparous), the lower sometimes sexual, with awned lemmas. $2n = 28$ (21, 35, 42, 49, 56).

Native. Common on mountains and down to sea-level in N. and W. Scotland. Mountain districts of the British Is. Arctic and N.E. Europe, southwards to S.W. Ireland.

16. F. indigesta Boiss.

A densely caespitose perennial up to c. 20 cm. Lvs with (5–)7(–9) veins, glaucous, scabrid above and pubescent below, short, recurved, hard; sheaths closed in the lower $\frac{1}{4}$–$\frac{1}{3}$, glabrous or hairy; auricles present. Panicle 2–3 cm, dense, the branches shortly hairy. Spikelets 6–7 mm; upper glume 1.4–1.5 mm wide; lemma 3.9–4.2 × 1.75–2.1 mm, hispid-scabrid and ciliate above; awn 1.5–2 mm.

Subsp. **molinieri** (Litard.) Kerguélen, to which the above description refers, is reported from W. Ireland, but further investigation is needed. Pyrenees to ? W. Ireland.

*17. F. trachyphylla (Hackel) Krajina

Culms 30–75 cm, scabrid above. Lvs with 7 veins and (3–)5–7 ribs, strongly scabrid throughout, sometimes tomentose below; sheaths closed at the base only at least some densely pubescent; auricles present. Panicle 4.5–13 cm, ±interrupted, the branches strongly scabrid. Spikelets (6.2–)7–7.5(–10.8) mm, glaucous; upper glume 1.3–1.4 mm wide; lemma 4.2–4.9(–6.5) × 1.9–2 mm, often pubescent; awn c. $\frac{1}{2}$ as long as lemma. $2n = 42$.

Introduced. Dry places. C. Europe, widely introduced elsewhere.

2. ×FESTULOLIUM Ascherson & Graebner

Hybrids between *Festuca* and *Lolium*, combining some of the characters of both genera.

1. ×F. loliaceum (Hudson) P. Fourn.

Festuca pratensis × Lolium perenne

Spikelets nearly sessile in a simple or more rarely somewhat branched raceme, resembling those of *Festuca pratensis* but more compressed. Glume next to the axis usually present though much reduced in the lateral spikelets and occasionally 0 in some.

Native. In meadows, locally frequent with the parents.

The following hybrids also occur, though more rarely: *Festuca arundinacea × Lolium multiflorum*; *F. pratensis × L. perenne*; *F. gigantea × L. multiflorum*; *F. gigantea × L. perenne*; *F. pratensis × L. multiflorum*; *F. rubra × L. perenne*.

3. ×FESTULPIA Melderis ex Stace & Cotton

Hybrids between *Festuca* and *Vulpia*. Perennials, with characters intermediate between those of the parents.

1. ×F. hubbardii Stace & Cotton

Festuca rubra × Vulpia fasciculata

Perennial, with both extra- and intravaginal branches. Culms 15–40 cm, ±erect. Sheaths closed, soon breaking down into fibres; leaves conduplicate, keeled, smooth; ligule up to 0.3 mm. Panicle 2.5–8.5 cm, simple above, with short branches in the lower half. Spikelets 8.5–12(–15) mm (excl. awns). *Lower glume* 2.4–4.4 mm, *the upper* 5.2–8 mm (incl. awn). *Lemma* 6–9.5 mm (excl. awn), *glabrous or almost so*. Sterile. $2n = 35$.

Native. On ±stable dunes, with the parents; rare.

2. ×F. melderisii Stace & Cotton

Festuca juncifolia × Vulpia fasciculata

Like **1** but more robust; panicle up to c. 12 cm; lower glume 3.5–7.2 mm, the upper 8–11.5 mm (incl. awn); lemma 9.5–11.5 mm (excl. awn), slightly to densely pubescent. Sterile. $2n = 42$.

Native. On mobile dunes, with the parents; rare.

Festuca rubra × *Vulpia bromoides* and *F. rubra* × *V. myuros* also occur, but very rarely.

4. LOLIUM L.

Glabrous herbs. *Infl.* normally *a simple spike* with the *spikelets edgeways on to the axis*. *Spikelets* of many florets, compressed, the *lateral ones with the lower glume suppressed*, the terminal one with 2 glumes. Glumes longer or shorter than the florets, membranous or chartaceous, 5–9-veined, unawned. Lemma membranous, sometimes tumid in fr., unawned or awned, 5-veined. Palea hyaline and tough, about equalling the lemma, 2-veined and usually shortly ciliate on the veins. Lodicules 2, lanceolate to ovate, acute or acuminate. Stamens 3. Ovary sparsely hairy at the top; styles distant, very short. *L. multiflorum* and *L. remotum* are always cultivated or relics of cultivation, at least in Britain.

About 12 spp., in Europe, N. Africa and Asia; introduced elsewhere.

Perennial, with non-flowering shoots at anthesis; lvs flat or folded when young; lemma usually unawned.
          1. perenne
Usually annual or biennial, without non-flowering shoots at anthesis; lvs convolute when young; lemma usually awned.   **2. multiflorum**

1. L. perenne L.   Rye-grass, Ray-grass.

A wiry *perennial* (3–)8–90 cm; culms with 2–4 nodes. *Lvs flat or folded when young*; ligule up to 2.5 cm; truncate. Infl. usually 8–15 cm, with a flexuous slender rhachis. Spikelets 5–23 mm, with 2–10 florets. Glumes linear-lanceolate, ⅓ as long to slight longer than the florets. *Lemma* oblong to lanceolate, usually unawned, acute or subacute. Fl. 5–8. 2*n* = 14. Hs.

Native. In waste places, and sown for fodder. Generally distributed throughout the British Is. Europe, except the Arctic; temperate Asia; N. Africa; introduced in North America and Australia.

Variants with the infl. branched, the rhachis shortened and the spikelets consequently crowded, or with viviparous spikelets occur not uncommonly, particularly late in the flowering season.

1 × 2 (**L. × hybridum** Hausskn.) is common, fertile and often sown in leys.

*2. L. multiflorum Lam.   Italian Rye-grass.

L. perenne subsp. *multiflorum* (Lam.) Husnot

Like **1** but annual, biennial or rarely a short-lived perennial up to 130 cm; culms with 4–5 nodes; lvs convolute when young; ligule up to 4 mm; infl. usually 15–33 cm; lemma usually awned. Fl. 5–9. 2*n* = 14. Th or Hs.

Introduced. Widely cultivated and well naturalized in temperate regions. Perhaps native in S. Europe.

*L. temulentum L. is a casual on rubbish tips and is distinguished from the foregoing by the lemma being turgid and the caryopsis not more than 3 times as long as wide when mature. It is an annual, probably native to S.E. Europe and temperate Asia.

*L. rigidum Gaudin, another casual annual sp., has the florets ±concealed by the glume and sunk in concavities of the rhachis. It is native to S. Europe, N. Africa and S.W. Asia.

5. VULPIA C. C. Gmelin

Glabrous *annuals with short convolute leaves and* slender, *little-branched secund panicles*. Spikelets of 4–6-florets, shortly stalked or almost sessile. *Glumes unequal; lower sometimes minute*. Lemma tough, convolute, veins usually obscure; awn scabrid, longer than the lemma which tapers gradually into it. Palea about equalling the lemma, thin, bifid at the apex, 2-veined, finely ciliate in the upper part. Stamens 1(–3). Fls often cleistogamous.

About 20 spp. in temperate regions, particularly the Mediterranean region and eastwards to Pakistan; N. Africa.; America.

1 Lemma with pointed callus; ovary hairy at apex.
          1. fasciculata
  Lemma with rounded callus; ovary glabrous.   2
2 Anthers 3, 0.7–1.3(–1.9) mm, exserted at anthesis; lemma 3–5 mm.   **5. unilateralis**
  Anthers usually 1, 0.4–0.8 mm, usually included at anthesis; lemma 4–7.8 mm   3
3 Spikelets with 1–3 fertile florets and 3–7 distal sterile florets; lemma of fertile florets 3(–5) veined.
          4. ciliata
  Spikelets with 2–5 fertile florets and 1–2 distal sterile florets; lemma of fertile florets 5-veined.   4
4 Lemma 1.3–1.9 mm wide; the lower glume 2.5–5 mm, ½–¾ as long as the upper.   **2. bromoides**
  Lemma 0.8–1.3 mm wide; the lower glume 0.5–3 mm, usually less than ½ as long as the upper.   **3. myuros**

1. V. fasciculata (Forskål) Samp.

V. uniglumis (Aiton) Dumort; *Festuca uniglumis* Aiton; *V. membranacea* auct., non (L.) Dumort.

Stems 10–50 cm, usually somewhat decumbent at the base. *Sheaths inflated*, upper usually some distance below the panicle at flowering. Panicle 2–12 cm, erect, dense, simple or rarely slightly branched at the base. Spikelets 10–15 mm. *Lower glume* 0.1–2.6 mm; *upper* 10–30 mm, linear-lanceolate, *nearly equalling the first floret, broadly membranous*, *obscurely* 3-*veined*, serrulate on the keel, acuminate or with an awn 3–12 mm. *Lemma* 8–18 mm, subulate, chartaceous, *with 1 prominent vein*, *compressed and keeled in the upper part*, margins convolute; awn up to twice as long as lemma; callus pointed. Stamens 1–3. Fl. 6. 2*n* = 28*. Th.

Native. On dunes, usually in hollows, very local. Coasts of Great Britain, north to Angus; E. coast of Ireland; Channel Is. W. Europe and the Mediterranean region.

2. V. bromoides (L.) S. F. Gray   Squirrel-tail Fescue.

Festuca sciuroides Roth

Stems 6–60 cm, usually somewhat decumbent at the base. Sheaths not inflated, upper usually some distance below the panicle at flowering. Panicle 1–10 cm, erect, rather lax. Spikelets 6–10 mm. Glumes green with a narrow hyaline margin; lower $\frac{1}{2}$–$\frac{3}{4}$ as long as upper, subulate; *upper lanceolate*, acuminate or with an awn up to 2 mm, shorter than the first floret, *with 3 prominent veins*. Lemma chartaceous, obscurely 5-veined, terete, margins convolute; awn usually as long as lemma; callus rounded. Stamens usually 1. Fl. 5–7. $2n = 14^*$. Th.

Native. In dry places on heaths, sandy and rocky places, walls and waste ground. Throughout the British Is., locally common. S., W. and C. Europe.

3. V. myuros (L.) C. C. Gmelin　　Rat's-tail Fescue.

Festuca myuros L.; incl. *V. megalura* (Nutt.) Rydb.

An erect annual 10–60 cm. Sheaths not inflated, upper usually reaching or enclosing the lower part of the panicle at flowering. *Panicle or raceme* 4–20 cm, ±*nodding*, lax, interrupted. Spikelets 7–10 mm. *Glumes* setaceous, or upper subulate, the veins obscure, margins hyaline, *lower* $\frac{1}{10}$–$\frac{1}{25}$ *as long as upper*. Lemma tough, terete, obscurely 5-veined, ±scabrid, sometimes ciliate on the margin towards the top; awn 2–3 times as long as lemma; callus rounded. Stamens usually 1. Fl. 5–7. $2n = 42^*$. Th.

Native. In sandy places and on walls. England, Wales, Ireland, mainly in the south, local; Scotland: probably introduced. W., C. and S. Europe; S.W. Asia; N. Africa; Macaronesia; introduced in S. Africa, North and South America and Australia.

4. V. ciliata Dumort.　　Bearded Fescue.

incl. *V. ambigua* (Le Gall) More

Stems 3–45 cm. Panicle or raceme 3–20 cm, erect or slightly nodding. Spikelets 5–10.5 mm (excluding awns), often purplish. Lower glume less than $\frac{1}{4}$ as long as upper. Fertile lemmas 3(–5)-veined; awn 1–2 times as long as lemma; callus rounded. Stamens 1(–3), included at flowering. Fl. 5–6. $2n = 28^*$. Th.

On sandy heaths and near the sea in S. and E. England. W. and S. Europe; N. Africa; S.W. and C. Asia.

*Subsp. **ciliata**: Lemmas conspicuously ciliate. A casual or locally naturalized in S. England. Throughout the range of the sp., except the north-west.

Subsp. **ambigua** (Le Gall) Stace & Auquier: Lemmas glabrous. Native. Local in S. and E. England, northwards to the Wash.

5. V. unilateralis (L.) Stace　　Mat-grass Fescue.

Nardurus maritimus (L.) Murb.

Stems 4–40 cm. Infl. a rather rigid secund raceme 1–10 cm, sometimes with a few short branches at the lower nodes. Spikelets 4–8 mm (excluding awns). Lower glume $\frac{1}{2}$–$\frac{3}{4}$ as long as upper. Lemma 3–5 mm, obscurely 5-veined, usually with an awn as long as or longer than itself, glabrous or pubescent; callus rounded. Fl. 5. Fr. 6. Germ. autumn. $2n = 14^*$. Th.

Native. In ±open calcareous grassland. S.E. England northwards to S. Lincoln, occasionally introduced elsewhere. S. and W. Europe; N. Africa; S.W. Asia to the Himalaya.

6. DESMAZERIA Dumort.
(*Catapodium* Link)

Glabrous annual herbs. Panicle rigid, simple or branched, secund. Spikelets somewhat compressed. Glumes chartaceous, subequal, 1–3-veined. Lemma coriaceous, 5-veined with 1–5 prominent veins, rounded on the back or keeled distally. Palea shorter than or about equalling lemma, thin; lodicules lobed. Three spp. in the Mediterranean region extending to W. Europe and Afghanistan.

Infl. usually somewhat branched and at least some of the
　　spikelets distinctly stalked; rhachis slender, angled;
　　lower glume 2–3 mm, the upper 2.3–3.3 mm.　**1. rigida**
Infl. usually unbranched and spikelets sessile, sometimes
　　shortly stalked; rhachis stout, flattened; lower glume
　　1.3–2 mm, the upper 1.5–2.3 mm.　　　　　**2. marina**

1. D. rigida (L.) Tutin　　Fern-grass.

Poa rigida L.; *Festuca rigida* (L.) Raspail, non Roth; *Sclerochloa rigida* (L.) Link; *Scleropoa rigida* (L.) Griseb.; *Catapodium rigidum* (L.) C. E. Hubbard

A rigid, glabrous annual 5–35 cm. Lvs narrow, flat or convolute. Sheaths with broad hyaline margins towards the top. Ligule 1–3 mm, ovate. Culms smooth, often somewhat geniculate. Panicle 2–12 cm, strict, extremely rigid; *pedicels* 0.5–3 mm. *Spikelets* 2–4 mm, of 3–6 florets. *Glumes* subequal, the lower 2–3 mm, the upper 2.3–3.3 mm. Lemma 2–2.6(–3) mm ovate, obtuse. Fl. 5–6. Germ. autumn. $2n = 14$. Th.

Native. On dry rocks, walls, banks, and sometimes on arable land; mainly in calcareous districts. Scattered throughout the British Is. but absent from most of Scotland, locally common in the south. S.W. Europe; N. Africa; S.W. Asia; Macaronesia.

2. D. marina (L.) Druce　　Sea Fern-grass.

Festuca marina L.; *F. rottboellioides* Kunth; *Poa loliacea* Hudson; *Catapodium loliaceum* Link; *C. marinum* (L.) C. E. Hubbard; *Desmazeria loliacea* Nyman.

A stout, erect to decumbent or almost prostrate annual up to 25 cm. Lvs narrow, flat or convolute. Ligule rather variable in length, ovate. Culms broadly ribbed, ribs often shiny. Panicle 0.5–4.5 cm; *pedicels* 0.7–2 mm; rhachis flattened. *Spikelets* 4–8 mm, of 3–10 florets, distichous, all directed to one side; florets imbricate. Glumes lanceolate, margins hyaline; *lower* 2–3 mm, *acute*, keeled, narrower than upper; *upper* 2.3–3.3 mm, *obtuse*, rounded on back. Lemma ovate, obtuse or mucronulate, veins distinct, margins hyaline. Fl. 6–7. $2n = 14$. Th.

Native. By the sea on sand, shingle and rocks. Coasts of the British Is., local. S. and W. Europe; N. Africa; Macaronesia.

7. POA L.

Glabrous annual or perennial herbs with compound panicles. Spikelets compressed, of (1–)2–5(–10) florets. Glumes ±unequal, soft, keeled, lower 1–3-veined, upper 3-veined, unawned. *Lemma* keeled, 5-veined, herbaceous, *often with a tuft of long cottony hairs at base, apex usually hyaline, unawned.* Palea 2-veined, the keels often aculeolate or shortly ciliate. Lodicules 2. Stamens 3; anthers usually several times longer than broad. *Ovary glabrous*; styles short, terminal.

About 300 spp. in cold and temperate regions and on mountains in the tropics.

1 Lvs at least 6 mm wide; panicle 10 cm or more; culms up to 150 cm, densely tufted.   **15. chaixii**

   Lvs normally up to c. 3 mm wide; culms usually shorter.   2

2 Plant with distinct creeping rhizomes.   3

   Plant without rhizomes, tufted.   4

3 Stems and sheaths strongly compressed.   **9. compressa**

   Stems and sheaths ±terete.   **10–12. pratensis group**

4 Base of stem bulbous or clothed with persistent ±fibrous and shining sheaths; spikelets often viviparous.   5

   Base of stem slender; spikelets never viviparous.   8

5 Stem bulbous at base; fl. 4–5; plant lfless June–Oct (sandy places or dry shallow soils usually by the sea).   **3. bulbosa**

   Stem not bulbous; fl. 7–8; plant green all summer (damp ledges and screes on higher mountains).   6

6 Lvs (or some of them) usually at least 3 mm wide, narrowed rather abruptly at apex; lower glume ⅔ length of upper.   **4. alpina**

   Lvs usually not exceeding 2 mm wide, tapering gradually at apex; glumes subequal (very rare).   7

7 Spikelets viviparous.   **5. ×jemtlandica**

   Spikelets not viviparous.   **6. flexuosa**

8 Annuals or short-lived perennials; keels of palea with flexuous or crispate hairs along their whole length.   9

   Perennials; keels of palea aculeolate or ciliate only in the lower ½.   10

9 Lower panicle-branches patent or deflexed after anthesis; florets closely imbricate; anthers 0.6–0.8(–1) mm, at least twice as long as wide.   **1. annua**

   Lower panicle-branches erecto-patent after anthesis; florets rather distant; anthers 0.2–0.5 mm, scarcely longer than wide.   **2. infirma**

10 Ligule of uppermost lf 1 mm or more.   11

   Ligule of uppermost lf not more than 0.5 mm.   **7. nemoralis**

11 Ligule of uppermost lf 3.5–10 mm.   **13. trivialis**

   Ligule of uppermost lf not more than 2.5 mm.   12

12 Plant usually intensely glaucous; culms stiffly erect (mountains).   **8. glauca**

   Plant not glaucous; culms not conspicuously stiff (damp grassy places).   **14. palustris**

1. P. annua L.   Annual Meadow-grass

An erect or decumbent tufted annual or short-lived perennial 5–30 cm. Stems sometimes creeping and rooting at the nodes. *Lvs* flat, slightly keeled, smooth, abruptly contracted at the apex, *often transversely wrinkled.* Sheaths smooth, somewhat compressed. Ligule 2–3 mm. Culms smooth, with broad flat shining ridges. *Panicle* 1–8 cm, spreading or compact, ±*triangular*; *branches* smooth, 1–2(–4) together, *the lower spreading or deflexed after flowering.* Spikelets 3–5 mm, lanceo-

late, of 3–5 florets; *florets closely imbricate.* Glumes herbaceous, unequal, boat-shaped, acute, lower 1-, upper 3-veined. Lemma ovate, 5-veined, glabrous or pubescent on the veins towards the base, margins and apex broadly hyaline. Palea nearly equalling lemma, shortly ciliate on the veins. Unopened anthers 0.6–0.8(–1) mm, 2–4 times as long as broad. Fl. 1–12. $2n = 28^*$. Th. or Hs.

Native. In waste places, gardens, cultivated land, grassland, on mountains and by water. Throughout the British Is., common; ascends to 1213 m on Ben Lawers. Throughout nearly the whole world, but in the tropics mainly on mountains.

2. P. infirma Kunth   Early Meadow-grass.

Like *P. annua* L. but panicle small oval or oblong with the lower branches not deflexed after flowering, spikelets long-linear of 5–6 distant or slightly overlapping florets, and unopened anthers 0.2–0.5 mm, about as long as broad. $2n = 14^*$. Th.

Native. In sandy places near the sea. Cornwall, Channel and Scilly Is. S.W. Europe; N. Africa; S.W. Asia.

3. P. bulbosa L.   Bulbous Meadow-grass.

An erect, tufted perennial 10–30 cm. *Stems swollen and ±bulbous at the base*, never creeping. Lvs usually narrow, flat, tapering to a long point. Basal sheaths light brown or purplish, persistent. *Ligule up to 3 mm*, acute. Culms usually ±geniculate below, smooth, obscurely ribbed. *Panicle compact, ovate*, 2–6 cm; branches ±scabrid, 2(–3) together. Spikelets often viviparous; non-viviparous spikelets 3–5 mm, ovate, often purplish or glaucous, usually of 3–4 florets. Glumes subequal, ovate, acute or acuminate, 3-veined ±hyaline. Lemma ±hyaline, ovate, acute, with long cottony hairs on the veins towards the base. Palea about ¾ the length of the lemma, shortly ciliate on the veins. Fl. 4–5. $2n = 28^*$; 45*, 35 (viviparous). Hs. The lvs and stems soon wither after fl. and the 'bulbs' lie loose on or near the surface until autumn.

Native. On sandy shores and on limestone near the coast, inland in Oxford. South and east coasts of England to N. Lincoln; Glamorgan; very local. Europe, except the extreme west and parts of the north; temperate Asia; N. Africa; Macaronesia; introduced in North America.

4. P. alpina L.   Alpine Meadow-grass.

An erect, tufted perennial 10–40 cm. *Base of stem stout, clothed with the persistent fibrous remains of basal sheaths. Lvs usually short, stiff, broad (up to 4 mm), abruptly contracted at the apex*, mucronate, folded at the apex or throughout, often distichous, ±glaucous. Sheaths smooth, uppermost usually far distant from the panicle. *Ligule* 3–5 mm, ovate, ±laciniate. Culms smooth, terete. Panicle 2–5 cm, erect, ±lax, ovoid or oblong; branches 1–2 together, nearly smooth. Spikelets usually viviparous in this country; non-viviparous spikelets 4–7 mm, ovate, of 2–4 florets. Glumes thin, ovate, acute, *lower ⅔ the length of the upper*, scabrid on keel, obscurely 3-veined, broadly hyaline and often purplish. Lemma thin, ovate, ±acute, ±pubescent on keel and 2 lateral veins, margins and apex hyaline. Palea nearly as long as, to slightly longer than, the lemma, shortly ciliate on the veins below. Fl. 7–8. $2n = 28^*, 35^*$ (viviparous); 42; 32–4; 38. Hs.

Native. In gullies and on rocks on mountains from 300 to 1210 m. Caernarvon; mid-W. Yorks, and Lake District; Scottish Highlands; Ireland: Kerry and Sligo only. Arctic Europe; Asia; America; N. Africa; on mountains, except in the north.

5. P. × jemtlandica (Almq.) K. Richter (*P. alpina × flexuosa*)

A perennial resembling *P. alpina* in general appearance, but usually smaller and more slender. Base of stem with persistent sheaths, but not as stout as in **4**. Lvs 1–2 mm wide, flat or folded, tapering gradually to the apex. Panicle slender, purplish, viviparous, spreading in fl., ±nodding; rhachis and branches not flexuous. Glumes subequal. Fl. 7–8. Hs.

Native. On damp stony slopes and rock-ledges, very rare. Ben Nevis and Lochnàgar. N.W. Fennoscandia.

6. P. flexuosa Sm.　　　　Wavy Meadow-grass.

P. laxa subsp. *flexuosa* (Sm.) Hyl.

A tufted perennial 6–20 cm. *Base of stem rather slender, clothed with the fibrous, persistent remains of sheaths.* Stem shortly creeping and rooting at the nodes, then ascending or erect. *Lvs thin*, narrow, *tapering gradually to the ±involute, often obliquely mucronate apex.* Sheaths smooth, compressed, uppermost usually shorter than its lf and often reaching nearly to the base of the panicle. Ligule up to 3.5 mm, acute, ±laciniate. Panicle 2–4 cm, erect, lax, ovoid; branches 1–2 together, flexuous at least at the base. *Spikelets* 3–5 mm, *never viviparous*, ovate, of 2–3 florets. *Glumes* ovate-lanceolate, acuminate or mucronate, *subequal*, smooth; margins and apex hyaline, sometimes purplish. Lemma thin, ovate-lanceolate, obtuse, ±pubescent on keel and 2 lateral veins, margins and apex hyaline. Palea ¾ the length of lemma, aculeolate on the veins. Fl. 7–8. $2n = 42^*$. Hs.

Native. On scree on the higher Scottish mountains, very rare. Ben Nevis, Lochnagar, Cairn Toul, Cairngorm. Mountains of N.W. Europe.

7. P. nemoralis L.　　　　Wood Meadow-grass.

An erect, slender perennial 30–90 cm. Stem shortly creeping at the base. *Lvs* narrow, smooth, *flaccid*. *Ligule very short or 0.* Culms with blackish nodes, terete, smooth. *Panicle* 5–10 cm, *usually nodding*, lax, narrow; branches long, slender, scabrid, 2–5 together at the lower nodes. Spikelets 2–4 mm, ovate, of 1–5 florets. Glumes unequal, lanceolate, acuminate, obscurely veined, broadly hyaline, ±pubescent on the keel. Palea rather shorter than the lemma, veins glabrous. Fl. 6–7. $2n = 28$–33, $42^*, 56^*$. Hs.

Native. In shady places. Throughout most of the British Is., absent from the north of Scotland and local in Ireland, mainly in the east. Europe, only on mountains in the south; temperate Asia; N. Africa; Japan; North America.

8. P. glauca Vahl　　　　Glaucous Meadow-grass.

P. caesia Sm.; *P. balfourii* Parnell

An erect, *usually intensely glaucous*, *stiff*, densely tufted perennial 10–40 cm. Lvs ±involute, rough or smooth, tapering. Uppermost sheath rather longer than or about equalling its lf and far below the panicle. *Ligule* c. 2 mm, truncate. *Panicle* 2–10 cm, lax or dense, *erect*, narrow; branches scabrid, strict or spreading at flowering, 2–3(–4) together at the lower nodes. Spikelets 4–6 mm, ovate, of 2–4 florets. *Glumes* subequal, *broadly lanceolate*, *or upper ovate*, acute or acuminate, scabrid on the keel, broadly hyaline and often purplish. *Lemma ovate*, *±acute*, obscurely veined, margins and apex hyaline, keel and 2 lateral veins pubescent. Palea rather shorter than lemma. Fl. 7–8. $2n = 42^*, 44, 49, 56^*, 63$. Hs.

Native. Damp rock-ledges. On the higher mountains of England, Scotland and Wales, from about 600 to 920 m. N. Europe and south to the Pyrenees and N. Greece; North and South America; N. Asia; Japan.

9. P. compressa L.　　　　Flattened Meadow-grass.

A stiff, ±glaucous, erect or ascending perennial 20–40 cm. *Stem compressed*, arising obliquely from the rhizome, usually geniculate. Lvs flat, smooth. *Sheaths* smooth, *compressed*. Ligule 1 mm or less, truncate. *Culms strongly compressed*, smooth. *Panicle* 2–7 cm, *narrow*, compact or rarely spreading; *branches* short, strict, scabrid, 1–3 together. Spikelets 3–6 mm, lanceolate, of 2–7 florets. Glumes subequal, tough, lanceolate, acute or mucronate, scabrid on keel, margin and apex hyaline. *Lemma oblong*, *obtuse*, keel minutely scabrid with a tuft of cottony hairs at the base, *obscurely veined*; apex hyaline. Palea with aculeolate keels. Fl. 6–8. $2n = 35, 42, 49; 56$. Hs.

Native. On dry banks and walls. Throughout most of the British Is. except the north of Scotland and Ireland, where it is rare and doubtfully native. Most of Europe; Siberia; introduced in North America.

10–12. P. pratensis group. Rhizomatous perennials, usually ±caespitose. Stems terete. Lvs flat or plicate; ligule 1(–3) mm truncate. Panicle ovoid to pyramidal, ±lax. Spikelets not viviparous.

1 Culms usually solitary, with few remains of lf-sheaths at the base; glumes subequal, acuminate, usually both 3-veined.　　　　**12. subcaerulea**
　Culms usually clustered, with numerous remains of lf-sheaths at the base; glumes unequal, acute, the lower usually 1-veined.　　　　*2*

2 Rhizomes long; stems in small tussocks; basal lvs ±flat, shorter than the culms; ligule decurrent on the sheath-margin.　　　　**10. pratensis**
　Rhizomes short; stems in large dense tussocks; basal lvs plicate, nearly as long as the culms; ligule not decurrent.　　　　**11. angustifolia**

10. P. pratensis L.   Smooth Meadow-grass.

Stems (15–)20–50 cm, in small loose tufts. Lvs 2–3 mm wide, flat or canaliculate, the basal distinctly shorter than the culms, ligule c. 1 mm, decurrent on the sheath margin. Panicle 6–15 cm, the branches usually 3–5 at the lower nodes. Glumes unequal, acute, the lower 1-veined. Fl. 5–7. $2n = 42, 50–78, 91, 98$.

Native. In meadows and other grassy places. Throughout the British Is. Europe; S.W. Asia; North America.

11. P. angustifolia L.   Narrow-leaved Meadow-grass.

P. pratensis subsp. *angustifolia* (L.) Gaudin

Densely tufted. Stems (30–)50–70 cm. Lvs 0.8–1.5(–2) mm wide, plicate, the basal nearly as long as the culms, wiry; ligule up to 3 mm, not decurrent. Panicle 6–10 cm, the branches 3–5 at the lower nodes. Glumes unequal, acute, the lower 1-veined. Fl. 5–7. $2n = 46–63$.

Native. In dry grassland, etc. Mainly in S.E. and S.C. England, rare in the west and north; absent from Ireland. Much of Europe and S.W. Asia.

12. P. subcaerulea Sm.   Spreading Meadow-grass.

P. irrigata Lindman, *P. pratensis* subsp. *irrigata* (Lindman) Lindb. fil.

Plant somewhat glaucous; stems (15–)20–30(–45) cm, usually solitary. Lvs 1.5–2.5 mm wide, ±flat, the basal shorter than the culms, fringed with hairs at the mouth of the sheath; ligule c. 1 mm, not decurrent, hairy on the adaxial surface. Panicle 3–6.5 cm, the branches mostly in pairs. Glumes subequal, acuminate, both 3-veined. Fl. 5–7. $2n = 38–117$.

Native. Damp grassy places. Scattered throughout the British Is., common in the north and west. N. and C. Europe.

13. P. trivialis L.   Rough Meadow-grass.

A tufted yellow-green perennial 20–60(–90) cm, with rather weak culms and flaccid lvs. Lvs flat, tapering gradually. Sheaths rough or smooth. *Ligule up to 10 mm, acute.* Culms smooth or scabridulous, terete. Panicle 5–10 cm, lax, broadly ovoid to oblong; branches scabrid, 3–5 together at lower nodes. Spikelets 2–4 mm, ovate, of 2–4 florets. Glumes thin, lanceolate, acuminate, scabrid on keel, lower ¾ length of upper. Lemma lanceolate, acuminate, distinctly 5-veined, keel silky with a mass of cottony hairs at the base, *marginal veins glabrous.* Palea about equalling lemma. Fl. 6. $2n = 14^*$. Hs. or Chh.

Native. In meadows and waste places. Throughout the British Is., common. Europe, temperate Asia; N. Africa; Macaronesia; North America.

14. P. palustris L.   Swamp Meadow-grass.

P. serotina Ehrh. ex Hoffm.

An erect, tufted perennial (15–)40–80 cm. Stems shortly creeping at the base. *Lvs flat, rather narrow, tapering,* ±scabrid. *Sheaths smooth.* Ligule 1–2.5 mm, truncate. Culms smooth, terete. Panicle 10–20 cm, effuse,

narrowly pyramidal; branches scabrid, up to 5 together at the lower nodes. Spikelets 3–5 mm, ovate, of 2–5 florets. Glumes thin, subequal, lanceolate, acuminate, obscurely veined, keel smooth, margins broadly hyaline. Lemma lanceolate, ±acute, obscurely veined, *keel silky with a mass of cottony hairs at the base, marginal* veins hairy, the apex hyaline and often brownish. Palea and lemma subequal. Fl. 6–7. $2n = 28$. Hs.

Native. In a few fens in East Anglia, very local; naturalized by rivers and in waste places in a number of scattered localities. Most of Europe, except much of the west and parts of the south; temperate Asia; North. America.

*15. P. chaixii Vill.   Broad-leaved Meadow-grass.

A tall, stout, *densely tufted* perennial 50–120 cm. Stems shortly creeping at the base. *Lvs 6–10(–15) mm wide, cucullate and apiculate, margin and midrib rough. Sheaths rough.* Ligule 1–2 mm, truncate. *Culms smooth, 2-edged.* Panicle 10–20 cm, narrow; branches scabrid, short, 3–5 together at the lower nodes. Spikelets 4–6 mm, ovate, of 3–5 florets. Glumes unequal, lanceolate, acute, keel scabrid at the top. *Lemma* lanceolate, acute 5-veined, aculeolate on the keel, *not lanate at base.* Palea rather shorter than the lemma. Fl. 6. $2n = 14$. Hs.

Introduced. Naturalized in woods. Scattered through the British Is., local. C. and S. Europe, mainly in mountain districts.

8. PUCCINELLIA Parl.

Tufted or creeping perennial or annual herbs of varied habit. Panicle of many subterete spikelets. Glumes unequal, shorter than the 1st lemma, membranous with hyaline margins, 1–3-veined, bluntly keeled. *Lemma* membranous with hyaline apex, oblong or ovate, rounded on the back, obtuse, 5-veined, veins stopping short of apex or the middle one shortly excurrent. Palea 2-veined, about equalling the lemma. Lodicules 2, membranous, lanceolate, acute. Stamens 3. Ovary glabrous.

About 100 spp. in temperate regions; in muddy places near the sea, rarely inland. Hybrids have been recorded between several of the spp.

1 Lemma 3–4 mm.   2
  Lemma less than 2.8 mm.   3
2 Stoloniferous; panicle with long, not rigid branches; anthers 1.5–2.5 mm.   **1. maritima**
  No stolons; panicle with short, very rigid branches; anthers 0.75–1 mm.   **4. rupestris**
3 Panicle branches 4–6 together, usually all long and bare below; midrib of lemma not reaching tip.
                                                          2. distans
  Panicle branches 2–4 together, at least the very short ones with spikelets to base; midrib of lemma reaching tip.   **3. fasciculata**

1. P. maritima (Hudson) Parl.

                               Common Saltmarsh-grass.

Glyceria maritima (Hudson) Wahlenb.

A ±tufted stoloniferous perennial 10–80 cm. *Lvs* smooth, *narrow*, flat or plicate, somewhat obtuse.

Ligule c. 1 mm, ovate, obtuse. *Panicle* 2–25 cm, narrow; *branches strict or somewhat spreading*, 2–3 *at each node*. Spikelets 7–12 mm, of 4–10 florets. Glumes broadly ovate. Lemma (2.7–)3–4 mm, oblong, obtuse, or apiculate, slightly silky towards the base. Palea equalling or slightly exceeding the lemma, ovate, spinulose on veins. Anthers 2 mm. Fl. 6–7. $2n = 14^*, 49^*, 63^*, 77^*$; 56; 60; 70. Hs.

Native. In salt-marshes and muddy estuaries. Generally distributed around the coasts of the British Is. and common or dominant in suitable habitats. Western coasts of Europe; North America.

2. P. distans (Jacq.) Parl.   Reflexed Saltmarsh-grass.

Glyceria distans (Jacq.) Wahlenb.

A tufted perennial 15–60 cm. *Stolons* 0. Lvs smooth or scabridulous, narrow, flat, acute. Ligule c. 1 mm, ovate, obtuse. *Panicle* 4–18 cm, lanceolate to triangular; *branches* strict before flowering, *spreading horizontally at flowering and ultimately deflexed*, (2–)4–6 *together*, usually all long and devoid of spikelets below. *Spikelets* 4–6 mm, usually of 3–6 florets, ±*uniformly spaced*. Glumes acute or subacute. *Lemma* 1.5–2.8(–4) mm, broadly ovate, obtuse or subacute. Palea equalling lemma, lanceolate to ovate, glabrous or finely and shortly ciliate on veins in upper part. *Anthers* 0.5–0.75 mm. Fl. 6–7. $2n = 28; 14; 42$. Hs.

Native. In salt-marshes, muddy estuaries and occasionally on sandy ground inland. Generally distributed around the S. and E. coasts of the British Is., very local elsewhere. Europe, W. Siberia, N. Africa; introduced in North America.

Subsp. **distans**: Lvs flat. Infl.-branches long, scabrid; spikelets green. Lower glume 1–1.5 mm, the upper 1.5–2.5 mm; lemma 1.5–2.8 mm. $2n = 42$. Throughout the range of the sp.

Subsp. **borealis** (Holmberg) W. E. Hughes (*P. retroflexa* subsp. *borealis* Holmberg; *P. capillaris* (Liljeblad) Jansen): Like subsp. *distans* but lvs often plicate; infl.-branches usually short; lemma 2.2–4 mm. $2n = 42$. Salt-marshes in Scotland. N. and W. Europe, southwards to the Netherlands.

Sterile plants intermediate between *P. distans* and *P. fasciculata* and probably hybrids between the two, occur occasionally where these spp. grow together.

3. P. fasciculata (Torrey) E. P. Bicknell
      Borrer's Saltmarsh-grass.

Glyceria borreri (Bab.) Bab.; *Poa fasciculata* Torrey; *Puccinellia pseudodistans* (Crépin) Jansen & Wachter

A tufted perennial 5–50 cm. Stolons 0. Lvs minutely scabrid, narrow, flat or plicate. Ligule very short. *Panicle* 3–16 cm, lanceolate, secund; *branches spreading*, 2–4 *together*, at least *the* 1 or 2 *very short ones covered with spikelets to the base. Spikelets* 4–7 mm, usually of 4–8 florets, *crowded in groups*. Glumes ovate, obtuse

or subacute. Lemma 1.5–2.5 mm, broadly ovate, obtuse. Palea ovate, finely ciliate on the margins, equalling lemma. *Anthers* 0.75–1 mm. Fl. 6. $2n = 28^*$. Hs.

Native. In muddy places near the sea; very local. Coasts of S. and E. England and Wales from N. Norfolk to Glamorgan. Coasts of W. Europe, extending to the Netherlands, N.E. Italy and N.W. Yugoslavia; North America.

4. P. rupestris (With.) Fernald & Weatherby
      Stiff Saltmarsh-grass.

Glyceria rupestris (With.) E. S. Marshall; *G. procumbens* (Curtis) Dumort.; *Sclerochloa procumbens* (Curtis) Beauv.

A procumbent, rarely ascending or erect annual or biennial 4–40 cm. *Stolons* 0. *Lvs rough*, flat, short and rather broad. Upper sheaths ±inflated. Ligule c. 1 mm, obtuse. *Panicle* 2–8 cm, oblong to ovate, *stiff*; branches spreading, 2–3 together, short, rigid. Spikelets 5–7 mm, of 3–5 florets, crowded on the branches. Glumes ovate, subacute, both strongly veined. Lemma 3–4 mm, ovate, obtuse, strongly 5-veined. Palea equalling lemma, lanceolate, shortly ciliate on the veins. *Anthers* 0.75–1 mm. Fl. 5–7. Th. or Hs.

Native. On clayey and stony seashores and beside brackish ditches. Coasts of England and Wales, northwards to N.E. Yorks, local. W. Europe, from the Netherlands to C. Spain.

9. DACTYLIS L.

Perennials. Vegetative shoots strongly compressed. Panicle compound, lower branches usually long. *Spikelets* with 2–5 florets, compressed, shortly pedicellate and *crowded in dense masses at the ends of the branches*. Glumes subequal, the lower 1-, the upper 3-veined, keeled. *Lemma* 5-veined, keeled, *shortly awned*. Palea equalling lemma, acuminate, the veins finely ciliate. Lodicules 2, linear, acute, about $\frac{1}{2}$ length of palea. Ovary glabrous; styles very short, terminal.

About 3 spp., in Europe, N. Africa and temperate Asia.

1. D. glomerata L.   Cock's-foot.

A coarse, tufted, glabrous, usually glaucous, erect or decumbent perennial up to 2 m. Lvs flat, rough, ±keeled. Sheaths of lvs of vegetative shoots strongly plicate and flattened, of others keeled, rough to nearly smooth. Ligule 2–10 mm, acute, torn. Culms smooth. Panicle 3–15 cm, erect, the lower branches usually long, distant, horizontal or reflexed in fl., erect in fr., the upper very short. Spikelets 5–9 mm, secund, crowded at the ends of the branches, green and violet. Glumes lanceolate, mucronate. Lemma lanceolate; awn short, subapical. Fl. 5–8. $2n = 28^*$. Hs.

Native. In meadows, waste places, by roads and on downs. Generally distributed and common throughout the British Is. Europe; temperate Asia; N. Africa; Macaronesia; introduced in other temperate lands.

Subsp. **glomerata**: Lvs usually stiff and glaucous; lemma ciliate on the keel. $2n = 28$. Native. Widely distributed in meadows, waste places and by roads.

The infl. is sometimes reduced to a dense ovoid head with the spikelets subsessile on the rhachis (var. *collina* Schlecht. (*congesta* Gren. & Godron)).

*Subsp. **aschersoniana** (Graebner) Thell. (*D. polygama* Horvátovszky): Lvs usually soft, not glaucous; lemma not ciliate. $2n = 14$. Introduced. Naturalized in woods in a few places in S.E. England. W. and C. Europe.

10. CYNOSURUS L. Dog's-tail Grass

Erect, glabrous, annual or perennial herbs. *Panicle spike-like. Spikelets* nearly sessile, *dimorphic*; *upper spikelets on each branch fertile*, of few florets; *lower spikelets sterile, with rigid, distichous lemmas.* Fertile spikelets: glumes thin, subequal, acute, keeled. Lemmas terete, coriaceous, awned. Palea 2-veined, shortly bifid at the tip, about as long as the lemma. Lodicules 2. Stamens 3. Ovary glabrous; styles short, terminal.

About 5 spp. in temperate regions of the Old World.

Perennial; uppermost sheath not inflated; panicle narrowly oblong. **1. cristatus**
Annual; uppermost sheath inflated; panicle ovoid, squarrose. **2. echinatus**

1. C. cristatus L.   Crested Dog's-tail.

A wiry, erect, tufted perennial 15–75 cm. Lvs flat, smooth, c. 0.5–2 mm wide. *Uppermost sheath not inflated*, smooth. *Panicle* up to 7 cm, erect, dense and spike-like, *narrowly oblong*; rhachis flattened, zig-zag. Spikelets 3–5 mm, distichous, compressed, dimorphic, the sterile pectinate. Glumes and lemmas of sterile spikelets linear-lanceolate, shortly awned, membranous, with green ciliate keel, the upper lemmas similar to but shorter than the lower. Lemmas of fertile spikelets lanceolate, obscurely 5-veined, shortly awned, ±hairy. Fl. 6–8. $2n = 14$. Hs.

Native. In grassland on acid and basic soils. Common and generally distributed throughout the British Is. Europe, except the extreme north; Caucasus; Madeira; Azores; introduced in North America.

*2. C. echinatus L.   Rough Dog's-tail.

An erect annual 10–50 cm. Lvs flat, smooth, 3–9 mm wide. Uppermost *sheaths inflated*, smooth. *Panicle* up to 4 cm (excl. awns), erect, *dense, ovoid*, squarrose, shining. Spikelets 8–10 mm. Glumes and lemmas of sterile spikelets narrow, the upper lemmas similar to the lower, all with short awns. Lemmas of fertile spikelets ovate-lanceolate, obscurely 5-veined, with awns 6–15 mm. Fl. 6–7. $2n = 14$. Th.

Introduced. A casual; in waste places and on sandy shores. Not infrequent in S. England, becoming rarer northwards and absent from Ireland. S. Europe; N. Africa; S.W. Asia; Macaronesia.

11. CATABROSA Beauv.

A glabrous perennial herb. Sheaths with connate margins. Panicle lax, branches in half whorls of 3–5, successive whorls alternating. Spikelets subterete, usually of 2 florets. Glumes thin, much shorter than the spikelet, unequal; lower apparently without veins, upper prominently 3-veined. Lemma firm with a truncate or erose hyaline apex, prominently 3-veined, veins ±parallel. Palea as long as lemma. Lodicules 2, short, obovate, emarginate, connate below. Stamens 3. Ovary glabrous; styles terminal, very short.

Two spp. in north temperate regions.

1. C. aquatica (L.) Beauv.   Water Whorl-grass.

An ascending rather soft perennial 5–70 cm. Stem creeping and often floating below, rooting at the nodes. Lvs smooth, soft, obtuse, often broad (up to 10 mm), lower often floating. Ligule c. 5 mm, ovate obtuse. Panicle up to 25 cm, lax, oblong, erect; branches smooth. Spikelets 3–5 mm, of (1–)2(–3) rather distant florets. Glumes hyaline or purplish, ovate, lower about $\frac{1}{2}$ length of upper. Lemma 2.5–3.5 mm, ovate-oblong, truncate with erose hyaline apex. Palea elliptical, glabrous or sparsely ciliate on veins. Fl. 6–8. $2n = 20^*$. Hel.

Native. In shallow streams and ditches, at the muddy margins of ponds and in wet sandy places near the sea. Widely scattered throughout the British Is., but local or rare in most districts; apparently less abundant than formerly. Most of Europe; N. and W. Asia; N. Africa; North America (?a different subsp.).

12. APERA Adanson

Annuals. Infl. a panicle with slender scabrid branches, in alternating half-whorls. Spikelets small, shining, laterally compressed, of 1 floret. Glumes membranous, unequal, acute, keeled, lower 1-veined, upper 3-veined, about as long as the floret. *Lemma* chartaceous, rounded on the back, shortly bifid *with a long awn from the sinus.* Veins obscure. Palea equalling lemma, hyaline, entire. Rhachilla shortly produced.

Three spp. in Europe and Asia.

Panicle usually purplish, the branches long, spreading; anthers 1–2 mm. **1. spica-venti**
Panicle usually greenish, the branches short, erect; anthers 0.3–0.4 mm. **2. interrupta**

1. A. spica-venti (L.) Beauv.   Loose Silky-bent.

Agrostis spica-venti L.

A ±erect annual 20–100 cm. *Lvs flat, scabrid, long.* Ligule 3–12 mm. *Panicle* 10–25 cm, *effuse*, lanceolate, often purplish; *branches long*, spreading. Spikelets 2–3 mm. Glumes lanceolate, acute, scabrid on keel, margins hyaline. Lemma scabrid towards the top, shortly bifid; awn 5–12 mm, from the sinus, slender, scabrid, ±flexuous. *Anthers* 1–2 mm. Fl. 6–7. $2n = 14$. Th.

Native. In dry sandy arable fields in East Anglia; elsewhere as a casual. Most of Europe; Siberia; introduced in North America.

2. A. interrupta (L.) Beauv.   Dense Silky-bent.

A ±erect, rather tufted annual 15–60 cm. *Lvs ±convolute, smooth, narrow, short*. Ligule 2–5 mm, truncate, toothed. *Panicle* 3–15(–20) cm, *narrow, ±interrupted*, green; *branches short*, strict. Spikelets 2–2.5 mm, very similar to those of *A. spica-venti* but glumes more broadly hyaline, awn 4–10 mm, and *anthers* 0.3–0.4 mm. Fl. 6–7. 2n = 14*. Th.

?Introduced. In dry sandy arable fields in East Anglia and elsewhere, often only as a casual. W. and C. Europe and in scattered localities eastwards to S.E. Russia and C. Asia; N. Africa; introduced in North America.

13. MIBORA Adanson

A *small annual. Infl. spike-like, linear*, secund. Spikelets of 1 floret, compressed. Glumes membranous, subequal, exceeding the lemma, concave, not keeled, 1-veined, the upper next the rhachis. Lemma 5-veined, hairy, thinner than the glumes. Palea equalling lemma. Rhachilla disarticulating above the glumes.

One sp. in Europe and N. Africa.

1. M. minima (L.) Desv.   Early Sand-grass.

M. verna Beauv.; *Chamagrostis minima* Borkh.

A small, tufted, glabrous annual 2–15 cm. Lvs mostly basal, flat or convolute, obtuse; ligule up to 1 mm, truncate. Culms filiform. Infl. 0.5–2 cm. Glumes truncate, erose and hyaline at the apex. Lemma ovate, obtuse, densely hairy. Palea narrow, densely hairy. Fl. 4–5 (sometimes again 8–9). 2n = 14*. Th.

Native. In wet sandy places near the sea. Channel Is. and Anglesey; well naturalized in a few scattered localities in S. England. S.W., and W.C. Europe; N.W. Africa.

14. BRIZA L. Quaking Grass

Erect, subglabrous, annual or perennial herbs. Panicle ±branched; pedicels slender. *Spikelets ovoid or broadly triangular*, compressed, *often pendent. Glumes ovate-orbicular, cordate, obtuse, awnless*, ±obscurely 3–9-veined and rounded on the back, veins not reaching the apex. *Lemma suborbicular, cordate at the base, usually saccate*, obscurely 7–9-veined, veins not reaching the apex, *unawned*. Palea shorter than the lemma or nearly equalling it, veins shortly ciliate. Lodicules 2, lanceolate. Stamens 3. Ovary glabrous; styles short, terminal.

About 20 spp., in Europe, temperate Africa, Asia and South America.

1 Perennial; ligule 0.5–1.5 mm, truncate.   **1. media**
  Annual; ligule at least 3 mm.   2
2 Spikelets 3–5 mm, numerous.   **2. minor**
  Spikelets 10 mm or more, not more than 12.
                                  3. maxima

1. B. media L.   Quaking Grass, Doddering Dillies.

An erect, somewhat tufted *perennial* 15–60 cm. *Stock shortly creeping*. Lvs flat, 2–4 mm wide, acute. *Ligule 0.5–1.5 mm, truncate. Culms solitary*. Panicle 5–8 cm, effuse, compound; pedicels very slender, smooth, slightly thickened below the spikelets, and mostly longer

than them. *Spikelets* 4–7 mm, ovoid, obtuse, usually purplish. Glumes unequal, making less than a right-angle with the pedicel, shorter than the lowest floret, ovate, obtuse, boat-shaped with a distinct midrib and fainter lateral veins, margins white and shining. Lemma strongly cordate at the base, bluntly keeled in the lower half, the coriaceous shining back surrounded by a thinner purplish area with a broad thin hyaline margin. *Palea almost as long as the lemma*, thin, ovate, obtuse. Fl. 6–7. 2n = 14. Hs.

Native. In meadows and grassy places. Occurring in varied habitats, ranging from wet and acid to dry and calcareous. Generally distributed throughout the British Is. except N. Scotland. Most of Europe, except the Arctic and parts of the south; temperate Asia; introduced in North America.

2. B. minor L.   Lesser Quaking Grass.

A slender, erect annual 10–50 cm. Lvs 3–10 mm wide, flat, acute. *Ligule 3–6 mm, lanceolate, acute. Culms tufted*. Panicle 5–15 cm, diffuse; pedicels very slender, smooth, slightly thickened below the spikelets and longer than them. *Spikelets* 3–5 mm, numerous, broadly triangular, obtuse, green. *Glumes subequal*, spreading at right angles to the pedicels, *equalling or exceeding the first floret*, broadly saccate, obtuse, obscurely veined, margins broadly hyaline. Lemma with an indurated shining area on the back, cordate at the base but otherwise similar to the glumes. *Palea somewhat shorter than the lemma*, thin, ovate. Fl. 7. 2n = 10. Th.

Native. In dry arable fields in the south, rare and very local. Cornwall, Dorset and Hants. S. and W. Europe; N. Africa; S.W. Asia; Macaronesia.

*3. B. maxima L.   Great Quaking Grass.

An erect or ascending annual 10–60 cm. Lvs flat, 3–8 mm wide, tapering to a long fine point. *Ligule* 2–5 mm, *obtuse and ±toothed or torn. Panicle usually of 3–8 large ovoid spikelets*, secund, drooping at apex; pedicels very slender, smooth or slightly aculeolate in the lower part, swollen immediately below the spikelet, at least the lower longer than the spikelets. *Spikelets* 10–25 mm. *Glumes* making less than a right angle with the pedicels, shorter than the first floret, *hyaline*, shining, broadly ovate, obtuse, with 2 rounded keels; lower 5-veined; *upper 9-veined. Lemma hyaline*, shining, broadly ovate; obtuse, 7–9-veined. *Palea up to $\frac{2}{3}$ as long as the lemma*, hyaline, tough, broadly ovate from a narrow base. Fl. 6–7. 2n = 14. Th.

Introduced. Naturalized in waste places; Channel Is., Cornwall. Often cultivated for ornament. Mediterranean region, Macaronesia.

Tribe 2. SESLERIEAE. Like *Poeae* but silica-cells elliptical or saddle-shaped; infl. a dense cylindrical to globose panicle subtended by glume-like bracts; lemma membranous, often hairy or scabrid, 4–5-dentate and often shortly awned at apex; ovary hairy or glabrous at apex; fr. with basal punctiform hilum.

15. SESLERIA Scop.

Perennials. *Panicle ovoid and spike-like. Bracts sheathing base of lower panicle branches.* Spikelets shortly stalked and somewhat compressed, of 2–5 florets. Glumes subequal, keeled, longer than lemmas. Lemma boat-shaped, keeled, 4–5-veined, some of the *veins shortly excurrent.* Palea equalling lemma. Lodicules 2. Stamens 3. Ovary pubescent at top.

About 35 spp. in Europe and W. Asia.

1. S. albicans Kit. ex Schultes   Blue Moor-grass.

S. caerulea (L.) Ard. subsp. *calcarea* (Čelak.) Hegi

An erect, wiry, tufted perennial 15–40 cm. Stems shortly creeping, then ascending. Lvs usually 2.5–3 mm wide, flat, keeled, glaucous, smooth but scabrid on the margin, apex mucronate; uppermost lf usually not more than 1 cm and far below the infl. Sheaths keeled. Ligule very short. Panicle 1–3 cm, ovoid, blue-grey and glistening, with a small scale at its base. Spikelets 5–8 mm. Glumes hyaline, lanceolate, lower acuminate, upper mucronate. Lemma boat-shaped, pubescent and ciliate on the margins, slaty-blue and purplish towards the top, apex with 3–5 small points. Palea slaty-blue or purplish towards the top, ciliate on the veins. Fl. 4–5. $2n - 28$. Hs.

Native. Abundant on calcareous hills and pastures in northern England and western Ireland, and rare on micaceous schists in Scotland; ?extinct in Brecon. C. and W. Europe, Iceland.

S. caerulea (L.) Ard. in fens, etc. from C. Sweden and N.W. Russia to Bulgaria.

Tribe 3. MELICEAE. Like *Poeae* but silica-cells elongate, with rounded ends; lf-sheaths closed to the mouth; infl. a panicle; spikelets slightly compressed laterally, with 1–several fertile florets and (1–)2–3 sterile lemmas; glumes about equalling florets; lemma ±coriaceous, rounded on the back; lodicules truncate, connate laterally; chromosomes small or medium-sized; basic number 9.

16. MELICA L.

Slender perennial herbs with the margins of the sheaths connate. Infl. a panicle or raceme. *Spikelets* of 1–3 fertile florets, *terete, and* 2–3 *sterile lemmas forming a club-shaped structure.* Glumes subequal, thin, unawned, 3–5-veined, nearly as long as the florets. *Lemma* coriaceous, obtuse, *rounded on the back*, 5–9-veined. Palea membranous, 2-veined. Lodicules very short, truncate, connate. Stamens 3. Ovary glabrous; styles short, spreading.

About 70 spp. in temperate regions and on mountains in the tropics.

Spikelets erect, with 1 fertile floret; sheaths with a bristle 1–4 mm at the mouth on the opposite side to the blade. **1. uniflora**
Spikelets nodding, with 2–3 fertile florets; sheaths without a bristle at the mouth. **2. nutans**

1. M. uniflora Retz.   Wood Melick.

A rhizomatous perennial 20–60 cm. *Lvs* flat, with long, scattered hairs above, *rough beneath.* Sheaths pubescent or glabrous with a bristle 1–4 mm projecting at the mouth, opposite the blade. *Panicle* 10–20 cm, lax, *spreading, ovate*, lowest branch usually far distant from others. *Spikelets* 4–7 mm, *erect*, oblong; lower spikelets mostly much shorter than pedicels. *Florets* 2, only the lower fertile. Glumes unequal, purplish-brown, elliptical, subacute, rounded on the back. Lemma oblong, obtuse, obscurely veined, margins narrowly hyaline in upper part. Fl. 56. $2n = 18$. Hp.

Native. In shady hedgebanks and woods. Scattered throughout the British Is. and locally dominant. Most of Europe; Caucasus; N. Africa.

2. M. nutans L.   Mountain Melick.

An erect or ascending perennial 20–60 cm. *Lvs* flat, sparsely hairy above, *smooth beneath. Sheaths glabrous*, somewhat rough, the *lower* purplish, *with reduced lvs. Panicle* 4–20 cm, *nearly simple, linear, secund*, drooping. *Spikelets* 6–7 mm, *drooping*, ovate. Fertile florets 2–3. *Glumes* subequal, purplish-brown and broadly hyaline, oblong, *obtuse.* Lemma ovate with an obtuse, hyaline apex, strongly veined. Fl. 5–6. $2n = 18$. Hp.

Native. In limestone woods and cracks in limestone pavement. In suitable habitats throughout Great Britain, local and rather rare. Most of Europe eastwards to C. Asia and Japan.

Tribe 4. GLYCERIEAE. Perennial ±aquatic herbs. Lf-sheaths with connate margins. Lvs with elongate silica cells with several deep constrictions and stout unicellular papillae; 2-celled hairs 0; green tissue uniformly distributed between the vascular bundles; 1st foliage lf of seedling narrow and erect. Ligule glabrous. Spikelets with numerous florets, laterally compressed or terete. Glumes small. Lemma awnless, prominently 5–11-veined; veins parallel. Lodicules connate, fleshy and swollen towards the top, truncate. Ovary glabrous, without an appendage. Fr. with linear hilum and compound starch grains. Chromosomes small; basic number 5.

17. GLYCERIA R.Br.

Perennial, *aquatic*, glabrous herbs. Leaves flat; sheaths with connate margins. Panicle simple or compound. *Spikelets* subterete, *of many florets.* Glumes hyaline, unequal, shorter than the 1st floret, 1-veined. *Lemma membranous, prominently 7–11-veined*, apex hyaline, veins parallel, not quite reaching apex. Palea tough, 2-veined, bifid, nearly or quite equalling lemma. Lodicules 2, ±connate, truncate, fleshy. Stamens 3. Ovary glabrous; styles terminal, short.

About 40 spp. in temperate regions.

1 Spikelets 5–8 mm, ovate in outline; keels of palea unwinged. **4. maxima**
  Spikelets 8–30 mm, oblong to linear in outline, keels of palea winged in upper part. 2
2 Rhachilla not disarticulating at maturity; plant sterile. ×**pedicellata**

Rhachilla disarticulating at maturity; plant fertile.   3
3 Lemma with 3–5 distinct teeth at apex; palea deeply
     2-fid, distinctly exceeding lemma; anthers
     0.6–1.1 mm.   **3. declinata**
   Lemma without distinct teeth at apex; palea 2-
     denticulate and not or scarcely exceeding lemma;
     anthers 0.7–2.5 mm.   4
4 Lemma 5.5–7.5 mm; longer panicle-branches with
     1–4, the shorter with 1(–2) spikelets.   **1. fluitans**
   Lemma 3–4 mm; longer panicle-branches with 5–16,
     the shorter with 1–6 spikelets.   **2. plicata**

1. G. fluitans (L.) R.Br.

Floating Sweet-grass, Flote-grass.

An ascending or erect, glabrous perennial 25–90 cm. Stems creeping and rooting at the nodes, floating and then ascending or erect. Lvs flat, ±rough, acuminate. Ligule up to 15 mm. Panicle 10–50 cm, contracted in fr.; longer branches with 2–4 spikelets. Spikelets 15–30 mm, linear. Glumes oblong, lower 2–3 mm, acute, upper 3–4.5 mm, obtuse. *Lemma* 5.5–7.5 mm, oblong, scabrid, *apex* hyaline, *entire, acute or subobtuse. Palea equalling lemma*, lanceolate, with 2 short bristle points, shortly ciliate on margins towards top. Anthers 1.5–2.5 mm, usually violet. Fl. 5–8. $2n = 40^*$. Hel. or Hyd.

Native. In stagnant or slow-flowing shallow water. Generally distributed throughout the British Is. Most of Europe; W. Siberia; N. Africa; North America.

2. G. plicata Fr.

Plicate Sweet-grass.

Like *G. fluitans* but panicle-branches patent in fr., the longer with 5–15(–19), the shorter with 1–5 spikelets; glumes very obtuse, the lower 1.5–2.3 mm, the upper 2.5–4 mm; lemma 3.5–4.5 mm, broadly elliptical and hyaline at apex; palea shorter than or equalling lemma; anthers 0.7–1.3 mm, yellow. Fl. 5–6. $2n = 40^*$. Hel. or Hyd.

Native. In streams and ditches. Generally distributed throughout the British Is. but rare in the north. Most of Europe; W. Asia.

G. × pedicellata Towns. (*G. fluitans* (L.) R.Br. × *plicata* Fr.)

Panicle ±branched. Lemma 5–5.5 mm, ±toothed, strongly veined. Palea ovate, minutely ciliate on margins towards top, equalling lemma. Anthers 1.2–1.8 mm, indehiscent. Sterile. Fl. 6–7. Not uncommon beside slow-flowing rivers. Through England, also in Scotland and Ireland. Spreads vegetatively and is frequently abundant in the absence of the parents.

3. G. declinata Bréb.

Small Sweet-grass

Like *G. fluitans* but lvs abruptly contracted at apex; panicle with few spikelets, contracted in fr.; longer branches with 1–6 spikelets; lower glume 1.5–2.5 mm, the upper 2.5–4 mm; lemma 3.5–4.5 mm, narrowly hyaline and with 3–5 distinct teeth at apex; palea distinctly exceeding lemma, with 2 somewhat divergent aristate teeth at apex; anthers 0.6–1.1 mm, usually violet. Fl. 6–9. $2n = 20^*$. Hel.

Native. In swamps and muddy margins of ponds. Fairly widely distributed though less common in most localities than *G. fluitans* and *G. plicata*. W. and C. Europe; Macaronesia.

4. G. maxima (Hartm.) Holmberg   Reed Sweet-grass.

G. aquatica (L.) Wahlenb., non J. & C. Presl

A stout, erect perennial 60–200 cm. Lvs flat or ±folded, up to c. 2 cm wide, smooth, rather abruptly contracted, acute; margins thickened, serrate. *Sheaths* rough, *with a reddish-brown band at the junction with the lf.* Ligule c. 5 mm, obtuse. Panicle 15–30 cm, compound, spreading. *Spikelets* 5–8 mm, *narrowly ovate*. Glumes keeled, subacute. *Lemma* 3–4 mm, ovate, *entire*, obtuse, scabrid, narrowly hyaline at margins and apex, 9-veined. Palea about equalling lemma, ovate, obtuse, very shortly bifid. Anthers 1.2–1.8 mm. Fl. 7–8. $2n = 60^*$. Hel. or Hyd.

Native. In rivers, canals and large ponds, usually in deeper water than the other spp. Scattered throughout the British Is. and abundant beside most lowland rivers; becoming rare in Scotland and absent from the north; Ireland, chiefly in the southern part of the central plain. Most of Europe; temperate Asia; Canada (probably introduced).

Tribe 5. BROMEAE. Annuals or perennials. Lvs with oblong silica-cells and no 2-celled hairs. First foliage lf of seedling narrow and erect. Spikelets of several florets, laterally compressed or terete, in panicles. Rhachilla disarticulating above the glumes and below each floret. Glumes shorter than the spikelet, unequal. Lemma herbaceous to coriaceous, with 5–13 veins, notched at apex and usually awned from below the sinus, rounded or keeled on the back. Lodicules 2, entire, connate below, glabrous. Stamens 3 or 2. Ovary with a terminal, fleshy, hairy appendage; styles 2, inserted laterally. Fr. with a linear hilum as long as the grain; starch-grains simple. Chromosomes large; basic number 7. About 50 spp., mainly temperate regions.

18. BROMUS L.

Annuals, biennials or perennials. Lvs usually flat. Infl. a panicle. Spikelets usually with many florets, subterete to strongly compressed. Glumes 1–9-veined, unequal, the upper usually the larger. Lemma unawned or with a subapical awn.

About 50 spp., mainly temperate regions.

1 Lower glume 1-veined, the upper 3-veined; spikelets
     parallel-sided or becoming wider towards the apex.   2
   Lower glume 3–7-veined, the upper 5–9-veined;
     spikelets ovate to lanceolate in outline, tapering
     towards the apex.   10
2 Annuals; awn usually longer than the lemma; spike-
     lets widest at the apex.   3
   Perennials; awn not longer than the lemma, rarely
     0; spikelets parallel-sided.   7
3 Lemma 20–35 mm.   4
   Lemma 9–20 mm.   5
4 Panicle usually lax, with spreading branches; callus-
     scar almost circular.   **3. diandrus**
   Panicle usually dense, with stiffly erect branches;
     callus-scar ±elliptical.   **4. rigidus**

5 Longer panicle-branches with 4 or more spikelets; lemma 9–13 mm. **5. tectorum**
  Panicle-branches with 1(–3) spikelets; lemma (12–)14–20 mm. 6
6 Panicle drooping, lax, most of the branches as long as or longer than the spikelets. **1. sterilis**
  Panicle erect, most of the branches shorter than the spikelets. **2. madritensis**
7 Lemma unawned or with an awn not more than 1.5 mm; plant with long creeping rhizomes. **6. inermis**
  Lemma with an awn more than 3 mm; plant caespitose. 8
8 Panicle-branches erect; lower lvs 2–3 mm wide, the upper wider than the lower. **7. erectus**
  Panicle-branches spreading; lvs 5–15 mm wide, the upper narrower than the lower. 9
9 Panicle-branches 2 at the lowest node, each with several spikelets; scale at lowest panicle-node ciliate. **8. ramosus**
  Panicle-branches 3–5 at the lowest node, the shorter with 1 spikelet; scale at lowest panicle-node not ciliate. **9. benekenii**
10 Perennial; spikelets strongly compressed; awn shorter than lemma or absent. 11
  Annual; spikelets subterete; awn usually about as long as lemma. 12
11 Glumes cucullate; lemma mucronate or with an awn up to 1 mm; palea c. ½ as long as lemma. **18. willdenowii**
  Glumes acute; lemma with an awn 4–10 mm; palea about as long as lemma. **19. carinatus**
12 Caryopsis and lemma scarcely imbricate and inrolled longitudinally when mature; rhachilla disarticulating tardily. 13
  Caryopsis and lemma imbricate, flat when mature; rhachilla disarticulating readily. 14
13 Lemma 5–6 mm; lower lf-sheaths softly hairy. **16. pseudosecalinus**
  Lemma 6.5–9 mm; lower lf-sheaths usually glabrous. **15. secalinus**
14 Spikelets often subsessile in groups of 3 at the ends of panicle-branches; palea split almost to base. **12. interruptus**
  Spikelets distinctly stalked, not in groups of 3; palea entire or shallow notched. 15
15 Lemma 4.5–6.5 mm, with wide, hyaline strongly angled margin; ripe caryopsis exceeding the palea. **11. lepidus**
  Lemma usually more than 6.5 mm, with narrow, weakly angled or rounded hyaline margin; ripe caryopsis not exceeding the palea. 16
16 Awn flattened at base, arising more than 1.5 mm from apex of lemma, divaricate at maturity. **10. hordeaceus**
  Awn terete at base, arising not more than 1.5 mm from apex of lemma, not divaricate at maturity. 17
17 Panicle usually dense; pedicels mostly shorter than spikelets; lemma with prominent veins. **10. hordeaceus**
  Panicle lax; at least some pedicels longer than spikelets; lemma with obscure veins. 18
18 Panicle up to 30 cm; anthers (3–)4–5 mm; palea about as long as lemma. **17. arvensis**
  Panicle usually not more than 15 cm; anthers not more

than 3 mm; palea shorter than lemma. 19
19 Lemma 6.5–8 mm, with rounded margin; lowest rhachilla-segment 0.5–1 mm. **13. racemosus**
  Lemma 8–11.5 mm, with bluntly angled margin; lowest rhachilla-segment 1.5–1.75 mm. **14. commutatus**

1. B. sterilis L.   Barren Brome.

Anisantha sterilis (L.) Nevski

An untidy annual with erect or ±decumbent stems 5–100 cm. Lvs soft, flat and downy, lower soon withering. Lower sheaths with short, often downward-directed hairs. *Culms glabrous. Panicle* 10–15 cm, *drooping*, simple or slightly branched; *branches usually much longer than the spikelets.* Spikelets 20–35 mm, strongly compressed. Glumes hairy or glabrous, somewhat unequal, margin broadly hyaline; lower almost linear, acuminate, scabrid on keel; upper linear-lanceolate. Lemma 14–20 mm, linear-lanceolate, rounded and scabrid on the back, margin hyaline, lobes 1–2 mm; awn usually longer than lemma, scabrid. Palea hyaline, almost equalling lemma. Anthers 1 mm. Fl. 5–7. Fr. animal dispersed; germ. autumn. $2n = 14$. Th.

Native. In waste places, by roads and as garden weed. Widely distributed and common throughout most of Great Britain, except N. Scotland; Ireland mainly on limestone, rare in the north-west. S., W. and C. Europe north to S. Sweden; N. Africa; S.W. Asia; introduced in North America.

2. B. madritensis L.   Compact Brome.

Anisantha madritensis (L.) Nevski

An erect or ±decumbent annual 10–60 cm. Lvs flat, glabrous or puberulent, lower soon withering. Sheaths glabrous or puberulent. *Culms usually glabrous. Panicle* 4–15 cm, *erect; branches up to* 3 cm. Spikelets 30–50 mm, strongly compressed. Glumes unequal, pubescent or glabrous, narrowly hyaline; lower glume subulate; upper glume linear-lanceolate, shorter than 1st floret. Lemma 12–20 mm, linear-lanceolate, glabrous or pubescent, narrowly hyaline, lobes 1–2 mm; awn as long as or somewhat longer than lemma. Palea shorter than lemma, delicately pectinate-ciliate on the keels. Stamens 2. Anthers 0.5–1 mm. Fl. 6–7. Chasmogamous. $2n = 28, 42$. Th.

?Native. In dry rather open habitats on limestone and sand. Glamorgan, Pembroke, Bristol, Channel Is., very local. S. and W. Europe; N. Africa; S.W. Asia; Macaronesia; introduced in North and South America, S. Africa and Australia.

3. B. diandrus Roth   Great Brome.

Anisantha diandra (Roth) Tutin; *A. gussonii* (Parl.) Nevski; *B. maximus* auct. angl., pro parte; *B. gussonii* Parl.

An erect, usually stout, ±pubescent annual 30–90 cm. Lvs flat, sparsely pubescent, with long hairs on both surfaces. Sheaths with patent hairs. *Culms shortly pubescent at least near the top. Panicle* up to c. 25 cm, erect, *at length nodding, spreading*; *branches* 2–4 at each node, many *longer than the spikelets* and with 1 spikelet. Spikelets 50–70 mm, of 4–8 florets, compressed. Glumes unequal, lanceolate-acuminate, glabrous, *margins hyaline*, upper equalling or nearly equalling the 1st floret. *Lemma* 25–30 mm, lanceolate, scabrid or shortly hairy, veins equally spaced, 3 middle ones usually prominent, margins hyaline, lobes 4–7 mm; awn twice as long as lemma. Palea lanceolate, acuminate, veins shortly pectinate-ciliate. *Stamens* 3 *or* 2. Anthers 1.25–4 mm. Chasmogamous or cleistogamous. Fl. 5–6. 2n = 56. Th.

It is suggested that this sp. arose by hybridization between *B. sterilis* and *B. rigidus*.

Native on sandy shores in the Channel Is.; as a casual in waste places in S. and E. England. S. and W. Europe; N. Africa; S.W. Asia; Macaronesia.

4. B. rigidus Roth

B. maximus Desf.; *Anisantha rigida* (Roth) Hyl.

A stout erect annual rather smaller than but similar in general appearance to *B. diandrus*. Lvs rather densely hairy with short hairs above. *Panicle* 15–20 cm, *stiff, dense*, erect; *branches strict, shorter than the spikelets*. Spikelets 25–35 mm, of 4–5 florets. *Glumes* unequal, narrowly lanceolate, acuminate, *almost entirely hyaline*, except for the veins. Lemma 22–25 mm, broadly hyaline, nearly smooth. *Stamens* 2. Fl. 5–6. Cleistogamous. 2n = 42. Th.

Native. On sandy shores. Channel Is., occasionally as a casual elsewhere. S. and W. Europe; Caucasus; N. Africa; Macaronesia.

*5. B. tectorum L.   Drooping Brome.

A somewhat tufted annual 10–90 cm. Lvs flat, hairy. Panicle 5–15 cm, secund, drooping; *branches* ±pubescent, slender, the longer *bearing* 4 *or more spikelets*. Spikelets 10–20 mm, compressed, narrow. Glumes unequal, pubescent, the margin and apex hyaline; lower glume subulate, the upper lanceolate. Lemma c. 10 mm, with scattered short hairs and a hyaline margin; awn slender, as long as or longer than lemma. Palea c. ¾ the length of lemma. Fl. 5–7. 2n = 14. Th.

Introduced. Naturalized near Thetford (Norfolk); a casual elsewhere in waste places. Most of Europe; Asia; N. Africa; Canary Is.

*6. B. inermis Leysser   Hungarian Brome.

Zerna inermis (Leysser) Lindman

An erect perennial up to c. 150 cm. Lvs up to 5 mm wide, flat. Panicle 15–25 cm, spreading. Spikelets 15–30 mm, lanceolate, compressed. Glumes unequal, *at least the upper with an obtuse hyaline apex*. *Lemma* c. 10 mm, *unawned or with an awn up to* 1.5 mm, *apex obtuse, hyaline*. Palea and lemma subequal. Anthers 4–5 mm. Fl. 6. 2n = 56. H.

Introduced. A rare alien, naturalized in a few places. Most of Europe; temperate Asia to China; introduced in North America.

7. B. erectus Hudson   Upright Brome.

Zerna erecta (Hudson) S. F. Gray

An erect perennial 15–100 cm. *Lower lvs convolute, upper flat, broader than lower*, glabrous or ±hairy. Sheaths glabrous or with spreading hairs. Culms glabrous. *Panicle* 10–15 cm, *erect*, usually reddish or purplish; branches short, slightly scabrid, each with 1–4 spikelets. Spikelets 15–25 mm, narrowly oblong, slightly compressed. Glumes linear-lanceolate, subequal, acuminate, keeled and slightly serrate on the keel. *Lemma* 10–15 mm, *usually glabrous*, 7-veined; margin broadly hyaline in upper ⅓; awn 2–8 mm. Palea slightly shorter than lemma. Anthers 4–7 mm, orange. Fl. 6–7. 2n = 42, 56*. Hs.

Native. On dry grassy banks and downs, preferring chalk or limestone. S., C. and E. England, very local in Wales, S. Scotland and Ireland. S., W. and C. Europe; N. Africa; W. Asia.

8. B. ramosus Hudson   Hairy Brome.

Bromus asper Murray; *Zerna ramosa* (Hudson) Lindman

An erect perennial commonly 100–140(–190) cm. *Lvs flat, lower broader than upper, sparsely hairy. All sheaths clothed with long downward-pointing hairs*. Culms puberulent. Panicle 15–40 cm, often dark purplish or glaucous; *branches in pairs, the lowest pair ±equal in length and each with 5–9 spikelets*, nodding, the lowest pair with a *ciliate scale* at its base. Spikelets 20–40 mm, linear, compressed, pendent. Glumes unequal, often purplish, with broad hyaline margins; lower linear-lanceolate, acuminate; upper lanceolate, mucronate or shortly awned, veins making prominent scabrid ridges towards the base. *Lemma* 10–13 mm, lanceolate, *broadest above the middle and rather abruptly narrowed* into an awn 4–7 mm, 7-veined, hairy on veins and margin. Palea half the length of the lemma, hyaline with green veins. Anthers 2.5–4 mm. Fl. 7–8. 2n = 14, 42*. Hp.

Native. In hedges and woods. Throughout the British Is., except the Channel Is., Outer Hebrides, Orkney and Shetland. W., C. and S. Europe, eastwards to Gotland and E. Romania; N. Africa.

9. B. benekenii (Lange) Trimen   Lesser Hairy Brome.

Zerna benekenii (Lange) Lindman

Like 8 but culms 50–120 cm; panicle 12–20 cm, nodding at the apex; branches (2–)3–5 at a node, with 1–5 spikelets, the lowest branches with a glabrous or pubescent scale which is not ciliate; lemma 11–14 mm. 2n = 28.

Native. In hedges and woods. Scattered localities in Great Britain, very local but probably overlooked. Much of Europe and temperate Asia; N. Africa.

10. B. hordeaceus L.   Lop-grass.

B. mollis L.

Annual to biennial. Stems up to 80 cm, erect, procumbent or ascending. Lvs flat, soft, ±hairy; sheaths hairy; ligule short, truncate, hairy. Panicle (1–)3–10 cm, erect, usually dense, sometimes reduced to a single spikelet. Pedicels mostly shorter than the spikelets. Spikelets 8–25 mm, compressed, glabrous or pubescent, not in groups of 3 at the ends of branches. Glumes usually unequal. Lemma 6.5–11 mm, chartaceous, with prominent veins and rounded margin. Anthers usually less than 1 mm. Caryopsis shorter than palea. $2n = 28$.

Native. In meadows, waste places and on dunes, shingle banks and cliffs, common throughout the British Is., but much less common in the north than the south. Almost all Europe; Caucasus; N. Africa; introduced in North America.

1 Awn flattened at the base, divaricate at maturity; panicles very dense.   (c) subsp. **ferronii**
  Awn terete, straight or slightly curved at maturity; panicles open, sometimes becoming contracted in fr.   2
2 Stems 3–80 cm, usually erect; lemma 8–11 mm; awn straight.   (a) subsp. **hordeaceus**
  Stems 1–8(–12) cm, procumbent or ascending; lemma 6.5–7.5 mm; awn often curving outwards.
                                     (b) subsp. **thominii**

(a) Subsp. **hordeaceus**: Stems 3–80 cm, usually erect. Panicle (3–)5–10 cm; spikelets with 6–12 florets, usually hairy. Lemma 8–11 mm; awn 4–11 mm, straight. Common throughout the range of the sp.

(b) Subsp. **thominii** (Hard.) Maire & Weiller: Stems 1–8(–12) cm, procumbent or ascending. Panicle 1–3 cm; spikelets with 3–5 florets, hairy or glabrous. Lemma 6.5–7.5 mm; awn 3–7 mm, often curving outwards. By the sea, usually on dunes, very local.

(c) Subsp. **ferronii** (Mabille) P. M. Sm.: Stems 2–15(–20) cm, usually erect. Panicle 2–5 cm; spikelets with 4–7 florets, densely hairy. Lemma 6.5–8.5 mm; awn 2–5.5 mm, flattened at the base, divaricate at maturity. On cliffs by the sea, very local.

Hybrids between **10a** and **11**, **B. × pseudothominii** P. M. Sm. (*B. thominii* auct., non Hard.) are common, particularly on disturbed ground, by roads, etc. They are ±intermediate between the parents, but usually have glabrous spikelets.

11. B. lepidus Holmberg   Slender Soft-brome.

B. britannicus I. A. Williams

An erect biennial 15–70 cm. Culms often somewhat decumbent below and rooting at the lower nodes. Lvs flat, hairy; *ligule hairy*; sheaths glabrous or pubescent. Culms puberulent between the ridges. *Panicle 3–10 cm, erect*, linear-lanceolate to broadly lanceolate, lax or dense. Pedicels mostly much shorter than their spikelets. Spikelets 5–15 mm, of (5–)7–9(–11) florets, glabrous (rarely hairy, var. *micromollis* (Krosche) C. E. Hubbard), compressed; *florets at first imbricate, later somewhat spreading. Glumes unequal, ovate, acute or mucronate, margins broadly hyaline. Lemma* 5.5–6.5 mm, 5–7-veined, *sharply angled* about $\frac{2}{3}$ of the way from the base, upper $\frac{1}{2}$ with a broad hyaline shining margin; awn 2–5.5 mm, straight. Palea shorter than

grain; grain about equalling lemma. Anthers 1–2 mm. Fl. 6–8. $2n = 28^*$.

?Native. In waste places, less commonly in grassland. Fairly widely scattered throughout the British Is. N. W. and N. C. Europe.

12. B. interruptus (Hackel) Druce Interrupted Brome.

An erect or somewhat decumbent annual or biennial 20–100 cm. Lvs flat, ±hairy, broad (up to 5 mm); sheaths ±hairy. Panicle 2–8 cm, erect, usually interrupted , the *branches very short. Spikelets* 10–17 mm, hairy, often *subsessile in groups of 3 at ends of branches*, with 5–10 florets; fls closely imbricate. Glumes somewhat unequal, acute or mucronate, margins hyaline; lower glume lanceolate, upper ovate. *Lemma* 7.5–9 mm, broadly ovate, *rounded or very bluntly angled, veins prominent*; awn 3–8 mm. *Palea bifid to base.* Anthers 1–1.5 mm. Fl. 6–7. $2n = 28^*$. Th. or H.

?Native. In arable fields, often associated with *Onobrychis viciifolia.* Scattered throughout S. and E. England; now apparently extinct, except in cultivation.

13. B. racemosus L.   Smooth Brome.

An erect annual 20–100 cm. Culms rather slender, glabrous or rarely puberulent. Lvs flat, soft, ±hairy; sheaths ±hairy. *Panicle 4–15 cm, erect, narrow.* Spikelets 15 mm, of 5–7 florets, at least some branches and pedicels longer than the spikelets, lanceolate to linear-lanceolate, glabrous or nearly so; *florets closely imbricate*, 0.5–1 mm *apart on the rhachilla.* Glumes subequal, lower lanceolate, upper ovate, both acute. *Lemma* 6.5–8 mm, ovate, *veins obscure, margins* hyaline, rounded; awn 5–9 mm. Palea shorter than lemma. Anthers 1.5–3 mm. Fl. 6. $2n = 28^*$. Th.

Native. In meadows and grassy places, sometimes on arable land. Scattered throughout England, Wales and S. Scotland, but rather uncommon; rare in Ireland. Nearly the whole of Europe except for most of Russia.

14. B. commutatus Schrader   Meadow Brome.

B. pratensis Ehrh. ex Hoffm., non Lam.

An erect annual 40–120 cm. Culms usually rather stout, glabrous. Lvs flat, soft. *Ligule glabrous*; sheaths ±hairy. *Panicle 7–20 cm, usually nodding, broader than in B. racemosus*, usually simple; pedicels and panicle branches mostly longer than the spikelets. Spikelets 15–25 mm, of 5–8 florets, ovate-lanceolate, glabrous or nearly so (rarely pubescent); *florets 1.5–2 mm apart on rhachilla.* Glumes unequal, lower lanceolate, upper ovate, both acute. *Lemma* 8–11.5 mm, broadly ovate, *veins obscure*, margins broadly hyaline, bluntly angled; awn 3–10 mm. Palea distinctly shorter than lemma. Anthers 1–1.5 mm. Fl. 6. $2n = 14, 28, 56$. Th.

Native. In meadows, grassy places and on arable land. Scattered throughout the British Is. and commoner than *B. racemosus*. Nearly the whole of Europe except Fennoscandia and N.E. Russia; N. Africa; S.W. Asia; introduced in North America.

*15. B. secalinus L.   Rye-Brome.

An erect annual 20–120 cm. Lvs flat, ±pubescent. Sheaths pubescent or glabrous. Culms glabrous. *Panicle 5–20 cm, secund, nodding*, lax. Spikelets 12–20 mm, compressed, glabrous or (var. *hirtus* (F. Schultz) Ascherson & Graebner) densely pubescent; florets at first imbricate, later spreading. Glumes unequal, margins hyaline; lower lanceolate, upper ovate. *Lemma 6.5–9 mm, ovate, curled round the rolled caryopsis and so strongly convex in fr.*; margins uniformly rounded; veins obscure; awn up to 8 mm. Palea almost equalling lemma. Anthers 1–2 mm. Fl. 6–7. $2n = 28$. Th.

Introduced. A casual in arable land, usually among winter wheat; much less frequent than formerly. Scattered throughout the British Is., rare in the north; E. Ireland. S. and S.C. Europe; W. Asia; introduced in N. Africa and North America.

16. B. pseudosecalinus P. M. Sm.

Stems 30–60 cm, erect, slender; sheaths hairy. Panicle 5–10 cm, lax at flowering, contracted and somewhat nodding later. Spikelets 8–12 mm, glabrous; florets at first imbricate, later spreading. Glumes and lemma obscurely veined. Lemma 5–6 mm, bluntly angled, curled round the rolled caryopsis; awn 2–6 mm. Palea shorter than the lemma. Anthers 1.25–1.75 mm. Fl. 6–7. $2n = 14$.

?Native. Roadsides, clover fields, etc. Scattered localities in England, Wales, S. Scotland and Ireland. Distribution imperfectly known.

*17. B. arvensis L.   Field Brome.

An erect annual 30–110 cm. Lvs flat, hairy. *Panicle 7–30 cm, spreading*, most of *the branches much longer than their spikelets*. Spikelets 10–25 mm, compressed, glabrous and shiny; florets 4–10, imbricate. *Lemma 7–9 mm, bluntly angled*, margins broadly hyaline in the upper ½; apex bifid; awn 6–10 mm. Palea usually equalling lemma. *Fls chasmogamous*, rarely cleistogamous. *Anthers up to 5 mm, never less than ½ as long as lemma.* Fl. 6–7. $2n = 14$. Th.

Introduced. A rare casual in waste places or hayfields, occasionally ±naturalized. S. and S.C. Europe; W. Asia.

B. japonicus Thunb. (*B. patulus* Mert. & Koch) is a rare casual. It has a very lax panicle and pedicels mostly longer than the spikelets.

*18. B. willdenowii Kunth   Rescue Brome.

Bromus unioloides Kunth; *B. catharticus* Vahl; *Ceratochloa unioloides* (Willd.) Beauv.

An erect or ascending perennial 20–100 cm. Upper sheaths glabrous, lower pubescent. Panicle 5–30 cm, ±spreading or lower branches deflexed. *Spikelets 20–40 mm, lanceolate. Glumes hooded at the apex, strongly keeled.* Lemma 14–18 mm, 13-veined. uniformly rounded, mucronate or very shortly awned. Palea c. ½ as long as lemma. Fl. 6–9. Usually cleistogamous. $2n = 42$. H.

Introduced. A casual in waste places. South America; introduced in all temperate regions.

*19. B. carinatus Hooker & Arnott   California Brome.

Ceratochloa carinata (Hooker & Arnott) Tutin

An erect perennial up to c. 80 cm. Culms stout, glabrous. *Lvs up to 20 mm wide*, glabrous, *tough*, tapering to a long point. Sheaths pubescent at the mouth. Ligule c. 2 mm, truncate and torn. Panicle 15–30 cm, ±erect, branches long, spreading. *Spikelets 25–45 mm, of (4–)8–12 florets, linear-lanceolate. Glumes* lanceolate, *acute.* Lemma 15–17 mm, lanceolate, scabrid; margins hyaline; awn 4–10 mm, almost terminal. Palea somewhat shorter than lemma. Fls cleistogamous or chasmogamous, rather distant (2–3 mm apart on the rhachilla). Fl. 6–8. $2n = 56$. Hp.

Introduced. Well naturalized in some localities, particularly along the Thames near Kew and Oxford. Western North America.

Tribe 6. BRACHYPODIEAE. Like *Bromeae* but silica-cells variable; infl. spike-like; spikelets subsessile, terete, with the side of the lemma next to the rhachis; lemma with 5–7 veins and entire apex, rounded on the back; lodicules ciliate; ovary with the outermost layer of nucellus cells becoming thick-walled in fr.; chromosomes small; basic number 5, 7, 9.

19. BRACHYPODIUM Beauv.

Tufted perennial herbs with spike-like infl. *Spikelets subsessile, distichous, terete*, linear-lanceolate, *with the side of the lemma to the axis*, of many florets. Glumes chartaceous, shorter than the 1st floret, unequal, prominently 5–7-veined. Lemma chartaceous, acuminate or awned from the tip, 5–7-veined. Palea nearly as long as the lemma, tough, blunt or emarginate. Ovary pubescent at the top.

About 10 spp. in Europe, temperate Asia, N. and S. Africa.

Lvs soft, yellow-green; awn equalling or exceeding lemma.
                                        1. sylvaticum
Lvs rigid, green or glaucous; awn not more than ½ as long as lemma   **2. pinnatum**

1. B. sylvaticum (Hudson) Beauv.   False-brome.

A tufted ±erect and *pubescent* perennial 30–90 cm. *Lvs* flat, broad (up to 13 mm), *soft, ±drooping, yellow green*, ±scabrid, sparsely pubescent and ciliate. Sheaths pubescent. Ligule 1–5 mm, laciniate. Infl. 6–20 cm. Spikelets 17–25 mm, nearly straight, of 7–12 florets. Glumes lanceolate, acuminate. Lemma 7–12 mm, linear-lanceolate, acute; *awn as long as or longer than lemma.* Palea oblong, emarginate or rounded at apex. Lodicules broad at the base, linear-lanceolate above. Fl. 7. $2n = 18^*$. Hs.

Native. In woods and hedges, sometimes in grassland, then often relict from woodland. Throughout the British Is., except Shetland. Europe; N. Africa; Macaronesia; W. Asia to N.W. Himalaya; Japan.

2. B. pinnatum (L.) Beauv.
                    Tor-grass, Heath False-brome.

A tufted erect, *subglabrous* somewhat rhizomatous perennial 40–120 cm. *Lvs* ±involute, rarely as much as 5 mm wide, often *stiff and erect*, ±scabrid, nearly glabrous. Sheaths glabrous or pubescent. Ligule c. 2 mm. Infl. 6–20 cm. Spikelets 20–40 mm, usually curved away from rhachis, of 8–24 florets. Glumes linear-lanceolate, acute, glabrous, lower 5-, upper 7-veined. *Lemma* linear-lanceolate, *acuminate or usually with an awn* 1–6 mm. Palea oblong, emarginate or rounded at apex. Lodicules narrowly oblong, pubescent. Fl. 7. 2n = 28. Hs. or Chh.

Native. In grassland on chalk and limestone, locally dominant. Mainly in S. and C. England, absent from Wales and most of Scotland; rare and often not native in Ireland. Europe to c. 62° N.; N. Africa; Siberia.

Tribe 7. TRITICEAE. Annual or perennial herbs. Lvs with oblong or elliptical silica cells; 2-celled hairs 0, green tissue uniformly distributed between the vascular bundles; 1st foliage lf of seedling narrow and erect. Ligule membranous. Spikelets all alike or some male or sterile, of 1–many florets, solitary or in groups of 2–6, sessile or subsessile on alternate sides of the solitary spike, broadside on to the axis. Glumes ½ as long to almost as long as spikelet, coriaceous, strongly veined, inserted laterally on the axis. Lemma chartaceous to coriaceous, usually 5–9-veined, unawned or awned from the tip, keeled or rounded on the back. Lodicules 2, ciliate. Stamens 3. Styles free, very short. Ovary hairy at apex. Fr. with long linear hilum; starch grains simple. Chromosomes large, basic number 7.

20. LEYMUS Hochst.
Elymus auct., non L.

Tall stout perennials. *Spikelets* 2 at each node of the rhachis, broadside on to the rhachis, of 3–4 florets. Rhachilla pubescent, disarticulating above the glumes and below each floret. *Glumes* equal, coriaceous, 3–4-veined, *about equalling the spikelet and often placed side by side in front of it.* Lemma coriaceous, 5-veined, rounded on the back, and bluntly keeled towards the apex. Palea tough, equalling lemma. Ovary hirsute.

About 70 spp. in temperate regions of the northern hemisphere.

1. L. arenarius (L.) Hochst.   Lyme-grass.

A stout, erect, glaucous perennial 1–2 m. Stem creeping and rooting freely below. Lvs 8–15 mm wide, shortly hairy on the veins, rigid, pungent. Sheaths glabrous, smooth, ridged. Ligule very short. Spike 15–35 cm × 10–25 mm. Spikelets c. 20 mm, in pairs at the nodes. Glumes lanceolate, acuminate, ±pubescent on the back. Lemma pubescent, lanceolate, acute or obtuse. Palea linear-lanceolate, obtuse or shortly bifid, sparsely ciliate on the keels, near the apex. Fl. 7–8. 2n = 56. Hp.

Native. On dunes, often with *Ammophila arenaria.* Around the coasts of the British Is., local. N. and W. Europe, from the Arctic to N.W. Spain.

21. ELYMUS L.
Agropyron auct., non Gaertner

Tough, usually rhizomatous, perennial herbs. *Infl. a spike of many distichous spikelets. Spikelets of 2–11 florets, solitary at the nodes of the rhachis and broadside on to it.* Glumes somewhat unequal to subequal, membranous to chartaceous, 1–11-veined, *shorter than spikelet.* Lemma chartaceous, 5-veined, awned or unawned. Ovary pubescent.

About 100 spp. in temperate regions.

1 Plant densely tufted; glumes persistent when fr. is shed.   **1. caninus**
   Plant with long rhizomes; glumes falling with the fr.   *2*
2 Veins of lvs slender, numerous; lvs usually flat when dry, with scattered long hairs on upper surface.
                                                      2. repens
   Veins of lvs prominent, broad, ±concealing upper surface of lf; lvs convolute when dry, hairs 0 or very short.   *3*
3 Veins of lvs with numerous short spreading hairs on upper surface.   **4. farctus**
   Veins of lvs glabrous but sometimes scabrid.   *4*
4 Veins of lvs scabrid; most spikelets overlapping by at least half their length.   **3. pycnanthus**
   Veins of lvs smooth; spikelets small, distant.
                                                      2. repens

E. farctus × repens and *F. pycnanthus × repens* are local and have lvs intermediate between those of the parents, and sterile pollen.

1. E. caninus (L.) L.   Bearded Couch.

An erect, *bright green perennial* 30–100 cm. *Rhizomes* 0. Culms smooth, *some of the nodes or culm near the nodes finely pubescent*, sometimes rather sparsely so. Lvs flat, 4–10 mm wide, scabrid and glabrous below, scaberulous, usually with scattered hairs above. Sheaths glabrous or the lower hairy. Spike 10–20 cm, often slender, ±flexuous and nodding, rhachis ciliate. Spikelets 10–17 mm, of 2–5 florets, lanceolate. Glumes membranous, unequal, glabrous, lanceolate, prominently veined; lower acuminate; upper not more than ⅔ length of spikelet. Lemma lanceolate; awn 7–18 mm, very rarely 0, slender, flexuous, scabrid, usually exceeding the lemma. *Anthers* 2–3.5 mm. Fl. 7. 2n = 28*. Hp.

Native. In hedges and woods. Scattered throughout the British Is. to Orkney though absent from most of N. Scotland; very local in Ireland. Europe, rare in the south; temperate Asia.

The rare variant with unawned lemmas has been described as *Agropyron donianum* F. B. White. It is interfertile with the usual awned variant and does not appear to differ from it in any other respect.

2. E. repens (L.) Gould   Couch-grass, Scutch, Twitch.

An erect, *dull-green or ±glaucous perennial* 30–120 cm. *Rhizomes abundant and far-creeping. Culms and nodes glabrous.* Lvs similar to those of *E. caninus.* Sheaths smooth, glabrous, or the lower pubescent. Spike 5–15(–20) cm, usually stiff and erect. Spikelets 8–17 mm, of 3–7 florets. Glumes subequal, ⅗–¾ length of spikelet, lanceolate, acute, chartaceous, margins hyaline. Lemma lanceolate, obtuse and apiculate, acute, or awned. *Anthers* 4–5.5 mm. Very variable. Fl. 6–9. $2n = 42^*$. Hp.

Native. In fields and waste places. A noxious and persistent weed. Throughout the British Is., common. Europe; N. Africa; Macaronesia; Siberia; North America.

Subsp. **repens**: Stems erect. Lvs flat, soft, with slender veins and usually with scattered long hairs above. Glumes 7–12 mm, 5–9-veined; lemma 8–11 mm. Anthers 5–5.5 mm. Waste places and widely distributed as a weed of cultivated land.

Subsp. **arenosus** (Petif) Melderis (*Agropyron maritimum* (Koch & Ziz) Jansen & Wachter): Stems geniculate at the base. Lvs glaucous, narrow, convolute, usually glabrous, with prominent crowded veins. Glumes 5–6 mm, often 3-veined; lemma 6–7 mm. Anthers c. 4 mm. Dunes on the South and East coasts of England. Coasts of N.W. Europe.

3. E. pycnanthus (Godron) Melderis Sea Couch-grass.

Agropyron pungens auct., non (Pers.) Roemer & Schultes

A rather tufted, often glaucous, glabrous perennial 10–120 cm. Rhizomes far-creeping. *Flowering and non-flowering shoots erect. Lvs flat or convolute*, rigid, glabrous; veins *broad and prominent, minutely scabrid*. Sheaths ciliate. *Spike* 4–20 cm, stout, stiff, erect, *resembling an ear of wheat; rhachis* flattened, *tough, not disarticulating at the nodes, margins serrate. Spikelets* 10–20 mm, of 3–10 florets, most of them overlapping each other by at least half their length. Glumes subequal, $\frac{1}{2}$–$\frac{2}{3}$ length of spikelet, lanceolate, acute ±keeled, 4–7-veined, sometimes scabridulous on keel. Lemma lanceolate, obtuse, and apiculate or shortly awned. Palea emarginate, equalling lemma, finely pubescent on veins. Anthers 5–7 mm. Fl. 7–9. $2n = 42^*$. Hp.

Native. Locally dominant on dunes and in saltmarshes. Coasts and estuaries of England and Wales; Dumfriesshire; south and east coasts of Ireland. W. and S. Europe; N. Africa.

4. E. farctus (Viv.) Runemark ex Melderis
　　　　　　　　　　　　　　　Sand Couch-grass.

Agropyron junceiforme (Á. & D. Löve) Á & D. Löve, *A. junceum* auct., non (L.) Beauv.

A glabrous, glaucous perennial 20–60 cm. Rhizomes abundant and far-creeping. Culms stout, smooth, *nodes frequently pruinose. Flowering stems* ±erect, *non-flowering shoots decumbent*. Lvs convolute, flat when damp, stiff, smooth; *veins broad and prominent above, nearly concealing the intervening lamina, densely but shortly pubescent with spreading hairs*. Basal sheaths whitish and persistent, the upper often purplish. Spike 5–15(–20) cm, stout, stiff, erect; *rhachis smooth, flattened, fragile and disarticulating readily at the nodes*

when mature. Glumes somewhat unequal, tough, lanceolate, obtuse. Lemma lanceolate, obtuse and apiculate, emarginate and mucronulate, pubescent above towards the top. Palea spinose-ciliate on the keels throughout its length. Anthers 6–8 mm. Fl. 6–8. $2n = 28^*$. Hp.

Native. On young dunes. Sandy coasts of the British Is. north to Shetland. Western and northern Europe to c. 63°N.

The British plant is subsp. **boreali-atlanticus** (Simonet & Guinochet) Melderis.

E. farctus × pycnanthus (*A. acutum* auct. brit.) is a common hybrid. Sterile shoots decumbent and ascending. Flowering stems erect. Lvs convolute or some flat, ribs broad with short, spreading points. Spike 10–20 cm, erect, slender to stout; rhachis slightly serrate, sometimes fragile. Spikelets up to 20 mm, rather distant and spreading. Anthers 5–6 mm, sterile and rarely exserted. Fl. 6–9. $2n = 35^*$.

Native. With the parents, often forming an intermediate zone between *E. farctus* on the young dunes and *E. pycnanthus* on the older ones.

Spp. of *Triticum* (Wheat), particularly ***T. aestivum** L. (*T. vulgare* Vill.; $2n = 42$), and **T. turgidum** L. ($2n = 28$), are widely cultivated in temperate regions throughout the world. *Triticum* resembles *Elymus* but has 1–2 hermaphrodite florets at the base of the spikelets and a number of male or sterile florets above, and the veins of the lemma are not convergent. The spp. are annual or rarely biennial. The cultivated variants do not become naturalized and seldom persist for more than a year.

*Secale cereale L., Rye, is another closely related plant which is extensively cultivated. It has 2 hermaphrodite florets in the spikelet and the veins of the lemma converge to form a long awn.

22. HORDEUM L.

Annual or perennial herbs. Ligule very short. Infl. a compressed or subterete spike. *Spikelets in alternate distichous groups of 3*, subsessile and broadside on to the rhachis, *of 1 floret; the central one in each group* hermaphrodite, the lateral male or sterile; all hermaphrodite in some cultivated races. Glumes equal, narrow, 1-veined, awned, placed side by side in front of the florets, free to the base. Lemma about equalling the glumes, tough, rounded or dorsally compressed, awned, awn terminal. Palea as long as the lemma, ±2-keeled. Lodicules 2, oblong or lanceolate, narrowed below, delicately fimbriate.

About 20 spp. in temperate regions.

1 Sheath of uppermost lf not inflated; perennial.
　　　　　　　　　　　　　　1. secalinum
　Sheath of uppermost lf inflated; annual.　　　　2
2 Glumes of central spikelet ciliate.　**2. murinum**
　Glumes of central spikelet scabrid.　**3. marinum**

1. H. secalinum Schreber　　　　　Meadow Barley.

H. pratense Hudson; *H. nodosum* auct., non L.

A slender erect perennial 15–70 cm. Lvs 2–5 mm wide, flat, or the lower ±convolute, somewhat rough. Sheaths not inflated. Culms and nodes glabrous. Spike 2–5 cm, compressed. *Glumes setaceous, scabrid. Lemma 3–6 mm, lanceolate* subterete, obscurely veined; *awn short.* Palea with 2 rounded keels. Lemma of sterile spikelets stalked, subulate, ending in a short point. Fl. 6–7. $2n = 14, 28^*$. Hp.

A hybrid with *Elymus repens* occurs rarely.

Native. In meadows. Mainly S. and E. England from the Bristol Channel to the Humber, locally abundant. S. Sweden to C. Spain, Sicily and Bulgaria; N. Africa.

2. H. murinum L.   Wall Barley.

A stout annual 20–60 cm, ±decumbent below. Lvs flat, 2–8 mm wide. Upper sheath glabrous and inflated, the lower sometimes pubescent. Culms and nodes glabrous. Spike 2–7(–12) cm, compressed. Spikelets 8–12 mm. *Glumes of hermaphrodite spikelets subulate, ciliate; outer glumes of lateral spikelets setaceous, scabrid; inner subulate, sparsely ciliate; all awned. Lemma ovate,* flattened and rounded on the back, obscurely veined; awn 2–4 times as long as lemma. Palea lanceolate, obtuse or acute, distinctly veined. Lemma of lateral fls lanceolate. Fl. 6–7. $2n = 14$. Th.

Native. In waste places, especially near the sea. Great Britain north to Caithness, but absent from much of the west and north; Ireland, not native though not uncommon in the east. C. and S. Europe, introduced in Fennoscandia; N. Africa.

3. H. marinum Hudson

Sea Barley, Squirrel-tail Grass.

H. maritimum Stokes

A glaucous annual 5–60 cm, ±decumbent and geniculate below. Lvs flat, 1.5–4 mm wide, stiff, short, scabrid. Upper sheaths inflated, glabrous. Culms and nodes glabrous. *Spike* 1.5–5 cm, *oblong, subterete. Glumes of central spikelet setaceous; inner glumes of lateral spikelets half ovate, outer setaceous; all scabrid,* not ciliate. *Lemma lanceolate,* flattened, obscurely veined. Fl. 6. $2n = 14, 28$. Th.

Native. In grassy places near the sea. Bristol Channel to the Wash, very local; Durham, N. Northumberland. S. Europe and coasts of W. Europe, north to England.

*H. jubatum L., Foxtail Barley, a perennial with inflated sheaths and very long, silky, nearly smooth, spreading awns and small (c. 5 mm) spikelets, occurs occasionally as a casual. N. America.

*H. distichon L., 2-rowed Barley, and ***H. vulgare** L., 6-rowed Barley, occur as relics of cultivation but do not persist or become naturalized.

23. HORDELYMUS (Jessen) Harz

A perennial herb with the characters of *Hordeum* except that the rhachis is always tough, the glumes connate at the base and the *lateral spikelets of each group are hermaphrodite* and the *central one hermaphrodite or sometimes male.* One species in Europe and Asia Minor.

1. H. europaeus (L.) Harz   Wood Barley.

Hordeum europaeum (L.) All.; *H. sylvaticum* Hudson; *Elymus europaeus* L.

A stout, ±tufted, erect perennial 40–120 cm. Lvs up to c. 14 mm wide, flat, usually rough. Sheaths pubescent, upper sometimes glabrous, not inflated. Culms glabrous, nodes pubescent. Spike 4–12 cm, compressed. Glumes stiff, subulate, terminated by a bristle-like point. Lemma 8–10 mm, lanceolate, flattened on the back, margins inrolled, scabrid or shortly pilose; awn 15–20 mm. Palea lanceolate, acute, prominently 2-veined; veins scabrid. Fl. 6–7. $2n = 28$. Hs.

Native. In woods and shady places. England and Wales, local and usually on chalk or limestone. Europe, from C. Spain and the Crimea north to S. Sweden; mountains of N. Africa.

Tribe 8. AVENEAE. Annual or perennial herbs. Lvs with oblong or elliptical silica cells and green tissue uniformly distributed between the vascular bundles; 2-celled hairs 0; 1st foliage lf of seedling narrow and erect. Ligule membranous. Infl a panicle. Spikelets slightly compressed laterally, of 1–5 florets, often shining; florets hermaphrodite or rarely the upper or lower male. Rhachilla usually disarticulating above the glumes, usually produced beyond the uppermost floret. Glumes tough, at least the upper as long as the spikelet and ±enclosing the florets, rarely shorter, but then usually equalling the lowest floret (*Koeleria*). Lemma membranous or coriaceous, usually with a dorsal, often geniculate, awn. Palea 2-keeled, rarely very short or 0. Lodicules 2. Stamens 3. Ovary hairy all over or glabrous, styles free. Fr. with a linear to ovate hilum; starch grains compound. Chromosomes large, basic number 7.

24. AVENA L.

Stout *annual* herbs. Panicle spreading or secund. *Spikelets* large, slightly compressed laterally, of 2–5 florets, *eventually pendent.* Glumes membranous, subequal, both with several veins. Lemma coriaceous, with a stout, scabrid, dorsal, geniculate awn from about the middle. Awn much exceeding the spikelet. Palea tough, rather shorter than the lemma. Rhachilla ±silky, and produced beyond the uppermost floret. Lodicules 2, lanceolate, acute. Stamens 3. Ovary pubescent.

About 20 spp. in temperate regions.

1 Lemma with 2 bristles 5–9 mm at apex.   **4. strigosa**
  Lemma with 2 teeth not more than 1.5 mm at apex.   2
2 Spikelets 17–20 mm; lemma glabrous, usually unawned.   **2. sativa**
  Spikelets 18–30 mm; lemma often hairy, at least the lowest awned.   3
3 Spikelets 18–25 mm; rhachilla disarticulating above the glumes and between the florets at maturity.   **1. fatua**
  Spikelets 25–30 mm; rhachilla disarticulating above the glumes but not between the florets at maturity.   **3. sterilis**

*1. A. fatua L. Wild Oat.

An erect annual 30–100 cm. Lvs flat, slightly scabrid; sheaths smooth; ligule 3–6 mm. Culms smooth, glabrous. Panicle up to 40 cm; branches scabrid, spreading. Spikelets 18–25 mm. Glumes subequal, exceeding the florets, lanceolate, acuminate. Lemma ovate-lanceolate, hyaline; awn 25–40 mm, geniculate. Palea shorter than the lemma, subobtuse. Rhachilla and lower $\frac{1}{3}$ of lemma often clothed with silky usually fulvous hairs, sometimes glabrous. Rhachilla articulated between the florets, the second with a callus-scar at its base. Fl. 7–9. $2n = 42$. Th.

Introduced. Arable fields, very common in S. and E. England, scattered elsewhere. Most of Europe, but native only in the south.

*2. A. sativa L. Oat.

Like **1** but lemma glabrous, the lowest sometimes awned, the others unawned; awn usually straight; rhachilla not articulated but eventually breaking below each floret and the second floret therefore without a callus-scar at the base. $2n = 42$. Th.

Widely cultivated and sometimes persisting for a short time.

*3. A. sterilis L. subsp. **ludoviciana** (Durieu) Nyman Wild Oat.

A. ludoviciana Durieu

Like **1** but spikelets 25–32 mm; lemma narrowly lanceolate, with rigid hairs in the lower $\frac{2}{3}$; awn 30–60 mm; rhachilla disarticulating above the glumes but not between the florets. $2n = 42$. Th.

A weed of arable fields, usually on heavy soils. Locally abundant in S. England. Native of S. Europe, N. Africa and S.W. Asia.

*4. A. strigosa Schreber Black Oat.

Like **1** but the panicle often contracted and ±secund; lemma with 2 straight or flexuous apical bristles 5–9 mm; rhachilla not disarticulating at maturity. $2n = 42$. Th.

Introduced. A rather rare casual. Still sometimes cultivated in upland areas of Wales, Scotland and the Hebrides. Widely distributed in N., W. and C. Europe.

25. AVENULA (Dumort.) Dumort.
Helictotrichon auct., non Besser

Erect, stout perennials. Infl. a panicle. *Spikelets large, erect*, of 2–3(–7) florets, subterete. Glumes somewhat unequal, nearly as long as the spikelet, membranous, bluntly keeled, lower 1–3-veined, the upper 5-veined. Lemma coriaceous below, thin and hyaline above, rounded on the back, 5-veined; awn dorsal, geniculate, stout, scabrid, much exceeding the spikelet. Palea thin, hyaline, shortly bifid at the apex. Rhachilla silky. Lodicules 2. Stamens 3. Ovary pubescent at the top.

About 50 spp. in temperate regions of both hemispheres.

Sheaths glabrous; lvs stiff, glaucous. **1. pratensis**
At least the lower sheaths pubescent; lvs soft, green.
 2. pubescens

1. A. pratensis (L.) Dumort. Meadow Oat-grass.

Avena pratensis L.; *Helictotrichon pratense* (L.) Pilger

An erect, glabrous perennial 30–80 cm. *Lvs ±channelled, glaucous, stiff*, subobtuse, the upper cauline obtuse. *Sheaths glabrous*, basal strict. Ligule 3–5 mm, acute. *Panicle 5–16 cm, strict, narrow*, nearly simple, *lower branches 1–2 together.* Spikelets 12–20(–25) mm. Glumes lanceolate, long-acuminate, both 3-veined, the margin and apex hyaline. *Lemma* lanceolate, *bi-aristate at the tip*, awn from about the middle. Palea shorter than lemma, lanceolate. Rhachilla silky, particularly at the joints. Fl. 6. $2n = 126$*. Hp.

Native. In short turf on chalk and limestone, rarely on other formations. In suitable habitats throughout Great Britain. W. and C. Europe, extending southwards to N.E. Spain and the Apennines; Caucasus.

2. A. pubescens (Hudson) Dumort. Downy Oat-grass.

Avena pubescens Hudson; *Helictotrichon pubescens* (Hudson) Pilger

An erect perennial 30–100 cm. *Lvs flat, soft, ±pubescent*, obtuse. *Sheaths pubescent.* Ligule 5–8 mm, ±acute. *Panicle 6–14 cm, ±spreading*, acute; *lower branches usually 5 together.* Spikelets 10–17 mm. Glumes hyaline, lanceolate, acuminate, the lower 1-veined, the upper 3-veined. Lemma linear-lanceolate, subterete; awn 12–22 mm, from the middle. Palea nearly equalling lemma. Rhachilla villous. Fl. 6–7. $2n = 14$*. Hp.

Native. On basic soils, usually in longer, rougher turf than *H. pratense*; occasionally on other formations. Generally distributed throughout the British Is. and locally abundant. From arctic Norway to N. Portugal and Bulgaria; Siberia; introduced in North America.

26. ARRHENATHERUM Beauv.

Erect perennial herbs. Panicle lax, nodding. Spikelets slightly compressed, usually of 2 florets and a rudiment, the *lower floret male, the upper female or hermaphrodite*, awned, sometimes both hermaphrodite. Glumes hyaline, unequal, keeled, lower 1-veined, $\frac{2}{3}$ the length of the upper; upper as long as the florets, 3-veined. Lemma membranous, 7–9-veined; *lower floret with a long geniculate awn* from near the apex. Palea hyaline, 2-veined. *Rhachilla silky* with a slender prolongation beyond the upper floret. Lodicules 2, lanceolate. Stamens 3. Ovary pubescent.

About 6 spp. in Europe, N. Africa, and W. Asia.

1. A. elatius (L.) Beauv. ex J. & C. Presl
 False Oat-grass.

A. avenaceum Beauv.

An erect perennial 50–150 cm. Stems swollen or not at base; nodes glabrous or pubescent. Lvs flat, ±scabrid. Sheaths smooth. Ligule 1–3 mm. Panicle 10–20 cm, lax, nodding, rather narrow. Spikelets 7–10 mm. Glumes lanceolate, acuminate, unequal. Lemma lanceolate, acute, broadly hyaline; awn of lower lemma 10–20 mm. A variable plant. Fl. 6–7(–11). $2n = 28*$. Hp.

Native. In rough grassy places, on scree and shingle. Common and generally distributed throughout the British Is. Most of Europe, only on mountains in the south; N. Africa; W. Asia; Macaronesia; introduced in North and South America and Australia.

Subsp. **elatius**: Basal internodes not swollen; nodes usually glabrous. Common and widely distributed.

Subsp. **bulbosum** (Willd.) Schübler & Martens: Stems with (1–)2–6(–8) swollen globose basal internodes 6–10 mm diam.; nodes usually hairy. Mainly in the south and west, where it is often commoner than subsp. *elatius*.

27. GAUDINIA Beauv.

Annuals or biennials. Lvs flat. Infl. a distichous spike; rhachis fragile, disarticulating at maturity above the insertion of the spikelet. Spikelets sessile, ±appressed to concavities in the rhachis, with 3–11 florets. Glumes unequal, the lower with 3(–5), the upper with 5–7 prominent veins; lemma obscurely veined, coriaceous, with a geniculate dorsal awn.

Four spp. in the Mediterranean region and the Azores.

*1. G. fragilis (L.) Beauv.

Stems 15–120 cm, erect or ascending, smooth and shiny. Lvs and sheaths ±villous. Infl. up to 35 cm. Spikelets 10–18 mm. Glumes shorter than the spikelets, glabrous and scabrid on the veins or sometimes villous, the lower lanceolate, acute, the upper oblong, obtuse. Lemma c. 7 mm, lanceolate, glabrous or sometimes villous, with an awn c. 10 mm. Fl. 5–7. $2n = 14 + 0–2B$.

Introduced. Well established in damp meadows, particularly in Hants, Isle of Wight, Channel Is. and W. Ireland; also as a casual in many scattered localities. Mediterranean region.

28. KOELERIA Pers.

Perennial herbs. Panicle narrow, ±shining. Spikelets laterally compressed, of 2–3(–5) florets. *Glumes* unequal, *firm*, *upper about equalling the first floret*, keeled, acute or aristate. *Lemma* tough, *keeled*, usually obtuse, often awned. Palea hyaline, about as long as the lemma, bifid at the apex and shortly ciliate on the veins. Rhachilla pubescent, prolonged beyond the uppermost perfect floret, and bearing 1(–2) rudimentary florets. Lodicules somewhat falcate and toothed. Stamens 3. Ovary glabrous.

About 60 spp. in temperate regions of both hemispheres.

1 Base of stem conspicuously thickened; sheaths of basal lvs persistent, reticulately fibrous.
　　　　　　　　　　　　　　　　　　　　　1. vallesiana
　Base of stem not thickened; sheaths of basal lvs not reticulately fibrous. 　　　　　　　　　　2
2 Lvs of non-flowering shoots flat or folded, ±smooth above. 　　　　　　　　　　**2. macrantha**
　Lvs of non-flowering shoots usually convolute, densely silvery-scabrid above. 　　　**3. glauca**

1. K. vallesiana (Honckeny) Gaudin
　　　　　　　　　　　　　　　　Somerset Hair-grass.

A densely tufted, glabrous or slightly pubescent perennial 10–40 cm. *Stems woody and conspicuously bulbous-thickened at the base, clothed with the fibrous-reticulate remains of the sheaths.* Lvs convolute, setaceous, those of the culms often flat, glabrous, short, rigid and glaucous. Sheaths glabrous or slightly pubescent. Panicle 1.5–7 cm, compact, scarcely lobed, usually broader than in the other spp., sometimes ovate. Spikelets 4–6 mm, shining green or pale brown, subsessile. Glumes ±hyaline, lanceolate-acuminate or shortly aristate, glabrous or hirsute, ciliate on the keel. Fl. 6–8. $2n = 42*$. Ch.

Native. On rocky limestone slopes. Brean Down, Uphill, Worle Hill, Purn Hill and Crook Peak, N. Somerset. W. Europe; N. Apennines; N. Africa.

2. K. macrantha (Ledeb.) Schultes
　　　　　　　　　　　　　　　　Crested Hair-grass.

K. cristata (L.) Pers., pro parte, *K. gracilis* Pers.

Stems 10–40 cm. Lvs of non-flowering shoots up to 15 cm, flat or folded, green or glaucous; basal sheaths not reticulately fibrous. Panicle 2–8(–10) cm, narrowly oblong, ±lobed, especially below. Spikelets 2–5(–6) mm. Glumes acuminate, mucronate or shortly awned, green, brown, or purplish with silvery margin. Lemma acuminate, glabrous. Fl. 6–7. $2n = 28$. Hp.

Native. In turf on well-drained base-rich soils, particularly on chalk and limestone. Locally common throughout the British Is. Most of Europe; north temperate Asia.

3. K. glauca (Schrader) DC.

Like **2** but lvs of non-flowering shoots not more than 5 cm, convolute, densely silvery-scabrid, at least above; lemma subobtuse, sometimes mucronulate, puberulent at least towards the base. Fl. 6–7. $2n = 14$. Hp.

Native. On sandy soils. Widely distributed throughout the British Is., extending to N. Scotland. N. Spain, Hungary and the Crimea, northwards to 61° 30′ N. in Russia and eastwards to C. Asia.

29. TRISETUM Pers.

Perennials. Infl. a panicle; spikelets shining, compressed, of 2–several florets. Glumes unequal, hyaline, firm, keeled, 1–3-veined. *Lemma* strongly keeled, 5-veined, apex with 2 *bristle points*; *awn from above the middle*, geniculate. Palea slightly shorter than lemma, hyaline; veins excurrent as 2 bristle points. Rhachilla silky, prolonged beyond the uppermost floret. Lodicules 2, lanceolate. Stamens 3.

About 75 spp. in temperate regions and on mountains in the tropics.

1. T. flavescens (L.) Beauv. 　　Yellow Oat-grass.

Avena flavescens L.

An erect stoloniferous perennial 20–80 cm. Lvs flat, rough or smooth beneath, villous above. Lower sheaths pubescent, the upper often glabrous. Culms smooth. Ligule very short, truncate. Panicle 5–16 cm, oblong or ovate, branches ±rough. Spikelets 5–7.5 mm, shining, usually yellowish; rhachilla and callus hairy. Glumes ±scabrid on the keel; lower linear-lanceolate, acuminate, 1-veined, about $\frac{2}{3}$ length of upper; upper broadly lanceolate, acuminate, 3-veined, almost equalling spikelet. Lemma lanceolate, scabrid on keel, margins broadly hyaline; awn 4.5–6.5 mm scabrid. Anthers 1.5–2.5 mm. The size of the spikelet, glumes and lemma is unusually variable. Fl. 5–6. $2n = 28$. Hp.

Native. In meadows and grassy places, especially on dry calcareous soils. Generally distributed throughout the British Is., but rarer in the north and absent from much of N. Scotland. Most of Europe; Caucasus; N. Africa; introduced in North America.

30. LAGURUS L.

A downy annual. Panicle usually ovoid, exceedingly soft and woolly. Spikelets laterally compressed, with 1 floret. Glumes 1-veined, with a long densely ciliate setaceous apex, subequal. Lemma membranous, 5-veined, nearly as long as the glumes, with 2 long apical setae and a dorsal geniculate awn. Palea somewhat shorter than the lemma.

One sp. S. Europe; N. Africa.

1. L. ovatus L.   Hare's-tail.

A downy, tufted, erect annual 7–50 cm. Culms slender, often branched from the lower nodes. Lvs flat, short, up to 10 mm wide, downy, upper often triangular. Sheaths much inflated, woolly. Ligule up to 3 mm, truncate. Panicle usually 1–2 cm, dense, woolly, greyish. Spikelets 7–9 mm. Glumes very narrow, pectinate-ciliate. Lemma lanceolate, bi-aristate, nearly glabrous; awn from below the tip, about twice as long as lemma. Fl. 6–8. $2n = 14$. Th.

?Native. In sandy places in the Channel Is. Naturalized in a few places in southern England. S. Europe, Mediterranean region, Madeira, Canary Is.

31. DESCHAMPSIA Beauv.

Tufted, glabrous, *perennial* herbs. Infl. a panicle. Spikelets of 2 florets, laterally compressed, shining. Glumes subequal, about as long as the florets, 1–3-veined, keeled, ±hyaline. *Lemma* subterete, *obscurely 5-veined, truncate and jagged at the apex*; awn dorsal, straight, or geniculate; both florets hermaphrodite. Palea as long as the lemma, hyaline. *Rhachilla* silky, *prolonged beyond* the *upper floret*. Lodicules 2; stamens 3. Ovary glabrous.

About 50 spp. in temperate regions.

1 Lvs flat; awns not or scarcely exceeding glumes.
                              1. cespitosa
  Lvs setaceous; awn of lower floret distinctly longer
    than glume.   2
2 Ligule 1–3 mm, truncate; lemma with central tooth
  the longest.   **2. flexuosa**

Ligule 3–8 mm, acute; lemma with lateral teeth the longest   **3. setacea**

1. D. cespitosa (L.) Beauv.   Tufted Hair-grass.

Aira cespitosa L.

A stout densely tufted perennial 10–200 cm. Lvs flat, scabrid above, smooth beneath. Sheaths smooth. Ligule 3–15 mm, acute. Panicle 5–50 cm, lax; branches long, slender. Spikelets 4–6 mm, silvery or purplish. Glumes subequal, firm, hyaline, lanceolate, acute. *Lemma* hyaline, lanceolate, *truncate and jagged at the apex*, the outer teeth not longer than the central tooth; awn straight, about equalling the lemma. Palea truncate and jagged at the apex. Fl. 6–8. Hs.

Native. In damp meadows and woods, usually on badly drained clayey soils. Generally distributed and common throughout the British Is. Most of Europe (only on mountains in the south); circumboreal; mountains of N. Africa.

Var. *parviflora* (Thuill.) C. E. Hubbard has the spikelets 2–3 mm long and is common in woodland, especially in S. England on heavy soils.

Subsp. **cespitosa**: Panicle-branches and keel of lower glume at least sparsely aculeolate. Spikelets silvery and purple, rarely viviparous. Awn arising from near the base of the lemma. Stems usually more than 30 cm. $2n = 24$, $26 + 0$–7B. Common and widely distributed, especially in badly drained grassland.

Subsp. **alpina** (L.) Tzvelev (*D. alpina* (L.) Roemer & Schultes): Panicle-branches and keel of lower glume smooth. Spikelets purple and yellowish, usually viviparous. Awn arising from the middle of the lemma or above. Stems not more than 40 cm. $2n = 52$, 56. Damp stony places and rocks on the higher mountains of N. Wales and Scotland, from 900–1350 m.

2. D. flexuosa (L.) Trin.   Wavy Hair-grass.

Aira flexuosa L.

A tufted, *rather slender perennial* 20–40(–100) cm. Lvs setaceous, scabrid on the margins. *Upper sheaths rough. Ligule 1–3 mm, truncate.* Panicle 5–15 cm, lax; branches long, flexuous, slightly rough. Spikelets 4–7 mm, silvery or purplish. Glumes rather unequal, ovate, acuminate, hyaline, the basal smooth. Lemma lanceolate, tapering to a slightly jagged apex, *the central tooth the longest*, hyaline in the upper half; *awn* geniculate, *arising near the base of the lemma* and exceeding it. Palea lanceolate, acute. Rhachilla short, *florets close together*. Var. *montana* Hudson is smaller (10–30 cm) with shorter lvs and a small (5 cm or less) compact panicle with few rather large spikelets. Fl. 6–7. $2n = 26$, 28, 56. Hs.

Native. On acid heaths and moors and in open acid woods; var. *montana* in alpine pastures. Generally distributed throughout the British Is. in suitable habitats. Most of Europe; temperate Asia; Caucasus; Japan; North America; only on mountains in the south.

3. D. setacea (Hudson) Hackel   Bog Hair-grass.

Aira uliginosa Weihe & Boenn.

A tufted slender perennial 20–70 cm. Like *D. flexuosa*, but with smooth sheaths, linear-lanceolate, acute ligule, 3–8 mm, lemma with the lateral points at the truncate apex longer than the central point, and a long rhachilla separating the florets by at least ⅓ the length of the lower lemma. Fl. 6–7. $2n = 14^*$. Hs.

Native. In wet turfy bogs and at edges of pools. Scattered in suitable habitats throughout Great Britain, but very local and mainly in the south and west. W. Europe from N. Spain to S.W. Norway and W. Poland.

32. AIRA L.

Slender glabrous, *annual* herbs with short, narrow lvs. Infl. a panicle. Spikelets somewhat compressed laterally, of 2 florets. Glumes equal, as long as, or longer than the florets and enclosing them, firm, 1–3-veined, slightly keeled. *Lemma shorter than the glumes, firm, subterete, 5-veined; apex with 2 setaceous points*, base with a tuft of hairs. Awn dorsal, geniculate, arising below the middle of the lemma. Palea thin, shorter than the lemma, bifid. *Rhachilla not produced beyond the second floret.* Stamens 3. Ovary glabrous.

About 10 spp. mainly in Europe and temperate Asia.

Sheaths smooth; pedicels mostly shorter than the spikelets.   **1. praecox**
Sheaths scabrid; pedicels mostly longer than the spikelets.   **2. caryophyllea**

1. A. praecox L.   Early Hair-grass.

An annual 2–12(–20) cm. Lvs obtuse, sheaths smooth, ligule 2–3 mm, acute. *Panicle compact, oblong*, 0.5–5 cm; *branches little longer than spikelets*. Spikelets 2.5–3.5 mm, crowded. *Glumes* as long as the spikelets, lanceolate, acute, *smooth*. Lemma lanceolate, awn arising ⅓ the distance from base to apex. Fl. 4–5. $2n = 14^*$. Th.

Native. On dry rocky slopes, heaths and in dry fields especially on a sandy soil. Generally distributed throughout the British Is. in suitable habitats. W. Europe extending eastwards to S.W. Finland, W. Czechoslovakia and N.W. Italy; introduced in North America.

2. A. caryophyllea L.   Silvery Hair-grass.

Incl. *A. armoricana* Albers

An annual 10–50 cm. Lvs ±oblique at tip; sheaths scabrid; ligule up to 5 mm, acute. *Panicle* 1–12 cm, *effuse, broadly ovate; branches very much longer than spikelets.* Spikelets 2–3.5 mm. Glumes longer than florets, ovate-lanceolate, acute. *Lemma scabrid towards the top;* awn arising about ⅓ the distance from base to apex. Fl. 5. $2n = 14$. Th.

Native. In dry gravelly and sandy places, shallow soils round rocks, and on walls. Generally distributed throughout the British Is. in suitable habitats, but usually less common than *A. praecox*. S., W. and C. Europe; N. Africa; Macaronesia; introduced in North America.

Subsp. **caryophyllea**: Plant usually 5–35 cm; spikelets 2.5–3.5 mm; longer pedicels usually more than 5 mm. $2n = 14$.

Subsp. **multiculmis** (Dumort.) Bonnier & Layens: Plant usually 20–50 cm; spikelets 2–2.5 mm; longer pedicels usually less than 5 mm. $2n = 28$.

33. HIEROCHLOË R.Br.

Perennial, smelling of coumarin when crushed. Panicle ovoid or pyramidal; branches 1–2 at each node. *Spikelets laterally compressed, shining, brownish, of 3 florets, the uppermost hermaphrodite, the 2 lower male.* Glumes membranous, subequal and equalling the florets, keeled, 3-veined, lateral veins short. Lemma of fertile floret tough, 5-veined, keeled, awned or not. Palea of male florets 2-veined; of hermaphrodite floret 1-veined. Lodicules 2, lanceolate, acuminate. Stamens of male florets 3, of hermaphrodite floret 2. Ovary glabrous.

About 30 spp. in temperate regions.

1. H. odorata (L.) Beauv.   Holy-grass.

H. borealis (Schrader) Roemer & Schultes; *Savastana odorata* (L.) Scribner

A tufted, glabrous perennial, 25–50 cm. Lvs flat, acute, upper very short. Sheaths smooth, persistent, upper far distant from panicle. Ligule c. 2 mm, ovate, obtuse. Culms minutely scabrid. Panicle 5–8 cm, pyramidal. Spikelets 4–5 mm, ovate. Glumes ovate, acute. Lemma of male florets ovate, acute, hyaline towards the top, very scabrid or shortly hispid, sparsely ciliate; of hermaphrodite floret similar but narrower, glabrous and shining below, appressed hairy above. Fl. 4–5. $2n = 28^*$, 42. H.

Native. On wet banks in Caithness, Kirkcudbright and Renfrew; Ireland: Lough Neagh. N. and C. Europe, southwards to the Alps; circumboreal.

34. ANTHOXANTHUM L.

Annual or perennial herbs, smelling of coumarin. Panicle compact, ovoid or oblong. *Spikelets lanceolate, acute, compressed, of 2 sterile florets and a terminal hermaphrodite one.* Glumes thin, very unequal, lower 1-veined, upper 3-veined, shortly aristate, longer than the floret and enfolding it. *Sterile lemmas thin, obtuse, bifid or toothed at apex, 3-veined, with a dorsal awn.* Fertile lemma thin but firm, half as long as sterile one, almost orbicular, 5–7-veined, unawned. Palea shorter than lemma, lanceolate, 1-veined. Lodicules 0. Stamens 2. Ovary glabrous.

About 20 spp. in temperate Eurasia and mountains of tropical Africa.

Perennial with non-flowering shoots; awn about as long as upper glume.   **1. odoratum**
Annual; awn distinctly longer than upper glume.   **2. aristatum**

1. A. odoratum L.   Sweet Vernal-grass.

A tufted perennial usually 20–50 cm. Lvs flat, short, sparsely hairy, acuminate. Sheaths smooth, glabrous or pubescent. Ligule up to c. 4 mm. Panicle (2–)4–6(–9) cm, compact, oblong, sometimes lobed below. Spikelets 7–9 mm. *Glumes* hyaline, with a green keel, *pubescent*; lower ovate, acute, 1-veined, ½ length of upper; upper ovate-lanceolate, mucronate, exceeding the floret and enfolding it, 3-veined. *Sterile lemmas oblong, obtuse, bifid*, with brown silky hairs in lower half; *awn of upper about equalling upper glume*, of lower short. Fertile lemma glabrous, almost orbicular, half as long as sterile, unawned, 5–7-veined. Palea shorter than lemma, lanceolate. Anthers 4 mm. Fl. 4–6. Protogynous. $2n = 20^*$. Hp.

Smells strongly of coumarin, which gives the characteristic odour to new-mown hay.

Native. In pastures and meadows and on heaths and moors; equally commonly on acid or basic soils. Generally distributed and common throughout the British Is. Throughout Europe; N. Africa; Macaronesia; circumboreal; only on mountains in the south; introduced in North and South America, Australia and Tasmania.

2. A. aristatum Boiss.   Annual Vernal-grass.

A. puelii Lecoq & Lamotte

A slender, often much-branched annual (5–)10–20 cm. Lvs flat, short, glabrous or hairy, scabrid, acuminate. Sheaths smooth, often somewhat inflated. Ligule up to 4 mm. Panicle 1–3 cm, rather lax, ovoid. Spikelets 6–7 mm. *Glumes glabrous*, broadly hyaline with a green keel, ovate; lower acuminate; upper shortly aristate. *Sterile lemmas tapering to a blunt 3–4-toothed apex*, sparsely silky on back; *awn of upper always long-exserted*. Fertile lemma and palea similar to *A. odoratum*. Anthers 2.5–3 mm. Fl. 6–10. $2n = 10$. Th. *Smell faint.*

Introduced. In sandy fields. Naturalized in a few localities in England, casual elsewhere. S. Europe; N. Africa.

35. HOLCUS L.

More or less pubescent perennial herbs. Infl. a panicle. Spikelets laterally compressed, of 2–3 florets. Glumes membranous, subequal, longer than the florets, strongly keeled, lower 1-, upper 3-veined. Lemma coriaceous, shining, obscurely 5-veined, strongly keeled. *Lower floret hermaphrodite*, unawned; *upper male, with a dorsal awn* from just below the tip. Palea thin, slightly shorter than the lemma. Rhachilla shortly prolonged beyond the second floret. Lodicules 2. Stamens 3. Ovary glabrous.

About 9 spp. in Europe, temperate Asia, N. Africa; Macaronesia.

Nodes not bearded; awn hooked, not exserted. **1. lanatus**
Nodes bearded; awn not hooked, exserted.   **2. mollis**

1. H. lanatus L.   Yorkshire Fog.

A soft, pubescent, tufted, short-lived perennial 20–100 cm. Lvs flat, acute, pubescent on both surfaces. *Sheaths pubescent*, upper inflated. Ligule 1–4 mm, truncate. *Culms and nodes puberulous or subglabrous*. Panicle 3–20 cm, ovate, ±lobed, whitish to dark purple; branches short, pubescent. Spikelets 4–6 mm, crowded. Glumes pubescent and ciliate on the keel; lower lanceolate, upper ovate, both mucronate or with an awn up to 1 mm. *Lemma* coriaceous, smooth *with a few silky hairs at the base*, shining, keeled, lanceolate, ±acute; *awn short, smooth, hooked, included within the glumes.* Fl. 6–9. $2n = 14^*$. Hp.

Native. In waste places, fields and woods. Generally distributed and often abundant throughout the British Is. Europe, except the Arctic; temperate Asia; N. Africa; introduced in North and South America.

2. H. mollis L.   Creeping Soft-grass.

A rather stiff, sparsely pubescent rhizomatous perennial 20–100 cm. Lvs flat, acute, rough. *Sheaths glabrous or lower ±pubescent*, upper somewhat inflated. Ligule 1–5 mm, truncate. *Culms glabrous*, rather slender, *nodes with a tuft of downward-directed hairs*. Panicle 4–12 cm, rather lax, usually brownish or purplish. Spikelets 4–6 mm. Glumes glabrescent, spinose-ciliate on the keels; lower lanceolate; upper ovate, obscurely veined; both acuminate. *Lemma* coriaceous, smooth but *conspicuously silky at the base*, shining, keel scabrid; *awn* geniculate, *exserted*. Plants with short, stiff lvs and small (c. 2 cm) panicles of few spikelets are found. Fl. 6–7. $2n = 28^*, 35^* (42^*, 49^*)$. Hp.

Native. On moderately acid soils; usually in woods. Generally distributed throughout the British Is., though less frequent than *H. lanatus*, and rare in calcareous districts. Most of Europe; N. Africa; introduced in North America.

The pentaploid variant ($2n = 35$) seems to be commoner than the tetraploid in this country; it is sterile but spreads vegetatively. *H. lanatus × mollis* occurs where the parents grow in proximity. It has $2n = 21$, is sterile and closely resembles *H. mollis*.

36. CORYNEPHORUS Beauv.

Spikelets of 2 florets, compressed laterally. Infl. a panicle. Glumes subequal, membranous, exceeding the florets. Lemma firm, hyaline, rounded on the back, 1-veined; *awn from near the base, clavate, bearded and geniculate about the middle*. Palea thin, shorter than lemma. Rhachilla silky, produced beyond the 2nd floret. Lodicules 2. Stamens 3. Ovary glabrous.

Three spp. in Europe, N. Africa and W. Asia.

1. C. canescens (L.) Beauv.   Grey Hair-grass.

Aira canescens L.; *Weingaertneria canescens* (L.) Bernh.

A tufted, very glaucous perennial 10–30 cm. Lvs setaceous, rigid, scabrid, pungent. Sheaths smooth, often purplish, inflated; nodes blackish. Ligule 2–4 mm, acute. Culms slender, often geniculate. Panicle 1.5–10 cm, spreading at flowering. Spikelets 3.5–4 mm, purple and white. Glumes lanceolate, slightly scabrid on keel, broadly hyaline. Lemma lanceolate, subobtuse; lower half of awn stout, chestnut-brown, upper half slender, club-shaped, white or purplish, awn bearded and geniculate at junction of 2 halves, exceeding lemma but shorter than glumes. Palea obtuse. Fl. 6–7. $2n = 14$. Hs.

Native. In sandy places behind the seaward ridge of dunes. Coasts of Norfolk, Suffolk and Channel Is.; perhaps not native in Glamorgan; introduced in S. Lancs, Moray and W. Inverness. S. Norway and Baltic USSR southwards to N. Italy and N.W. Africa; introduced in North America.

37. AGROSTIS L.

Tufted or creeping *perennials*. *Spikelets small*, of 1 floret. Infl. a panicle. Glumes equal or subequal, usually 1-veined, membranous. Lemma shorter than the glumes, ovate, truncate, or obtuse, 3–5-veined, lateral veins usually excurrent; *awn dorsal or* 0. Palea shorter than lemma, sometimes minute or 0, hyaline, 2-veined or veinless. Rhachilla sometimes produced, shortly bearded or glabrous, disarticulating above the glumes. Lodicules 2. Stamens 3. Ovary glabrous.

About 200 spp., mostly in temperate regions.

For a full account of the British spp. of the genus, their varieties and hybrids, as well as the aliens, see Philipson, W. R., *J. Linn. Soc. Bot.* LI, 73–151 (1937–38).

1 Palea less than $\frac{1}{5}$ as long as lemma; ligule of uppermost lf usually acute. *2*
Palea at least $\frac{1}{2}$ as long as lemma; ligule of uppermost lf usually truncate. *4*
2 Most basal lvs less than 0.3 mm wide; glumes rough. **1. curtisii**
Most basal lvs more than 0.5 mm wide; glumes smooth, except on keel. *3*
3 Rhizomes absent; stolons often present. **2. canina**
Rhizomes present; stolons absent. **3. vinealis**
4 Rhizomes with more than 3 scale-lvs present; panicle usually not contracted after flowering. *5*
Rhizomes absent or short and usually with not more than 3 scale-lvs; panicle usually contracted after flowering. *6*
5 Ligules on non-flowering shoots usually wider than long; panicle-branches nearly or quite smooth. **4. capillaris**
Ligules on non-flowering shoots usually longer than wide; panicle-branches very rough. **5. gigantea**
6 Most panicle-branches branched 1–2 times and with spikelets in the lower half; palea $\frac{3}{5}$–$\frac{4}{5}$ as long as lemma. **6. stolonifera**
Most panicle-branches branched 2–3 times and without spikelets in the lower half; palea $\frac{1}{2}$–$\frac{3}{5}$ as long as lemma. **7. castellana**

1. A. curtisii Kerguélen — Bristle Agrostis.

A. setacea Curtis, non Vill.

A *densely tufted* perennial 20–60 cm. *Lvs setaceous*, scabrid, *stiff and glaucous*. Sheaths slightly scabrid. Ligule up to 4 mm, acute, ±torn. Culms scabrid, especially above. Panicle 3–10 cm, narrow, ±spreading at flowering. Spikelets 3–4 mm. *Glumes* lanceolate, acute, *finely scabrid*, keel serrate towards the top. Lemma c. $\frac{2}{3}$ length of glumes, ovate, truncate, 5-nerved; awn from near the base, usually geniculate and exceeding the glumes. Palea minute. Rhachilla shortly bearded. Anthers 1.5–2 mm. Fl. 6–7. $2n = 14$*. Hs.

Native. On dry sandy and peaty heaths and downs in S. and S.W. England and S. Wales. S.W. Europe; N.W. Africa.

2. A. canina L. — Brown Bent-grass.

A loosely tufted perennial 10–70 cm; stems often rooting at the nodes and producing short lfy axillary shoots. Lvs usually flat, 1–2 mm wide; ligule 1.5–4.5 mm, acute to acuminate. Panicle 2–20 cm, lax, usually contracted before and after flowering. Spikelets 1.6–2.5 mm, brownish. Glumes subequal, lanceolate, acute. Lemma c. $\frac{2}{3}$ as long as glumes; awn up to 4.5 mm, arising just below the middle, rarely 0. Palea very short. Fl. 6–7. $2n = 14$. Hp.

Native. In damp acid habitats throughout most of the British Is. Most of Europe and temperate Asia.

3. A. vinealis Schreber — Brown Bent-grass.

A. canina subsp. *montana* (Hartman) Hartman

Like **2** but usually densely tufted; scaly rhizomes present; stolons absent; stems never producing short lfy axillary shoots; ligule acute to obtuse; panicle often dense. $2n = 28$.

Native. In dry acid grassland. Widely distributed in N., W. and C. Europe and probably elsewhere. Distribution imperfectly known through confusion with the preceding.

4. A. capillaris L. — Common Bent-grass.

A. tenuis Sibth., *A. vulgaris* With.

A ±tufted perennial, with rhizomes bearing more than 3 scale-lvs; stems 10–70 cm. Lvs usually flat, up to 4 mm wide; ligule 0.5–2 mm, usually wider than long, truncate. Panicle 1–20 cm, the branches patent during and after flowering, smooth or slightly rough, without spikelets in their lower half; swollen apex of pedicels smooth. Spikelets 2–3.5 mm, greenish to purplish-brown. Glumes ±equal, lanceolate, acute. Lemma $\frac{2}{3}$–$\frac{3}{4}$ as long as glumes, usually unawned. Palea $\frac{1}{2}$–$\frac{2}{3}$ as long as lemma. Fl. 6–8. $2n = 28$. Hp.

Native. Common and widely distributed, particularly in acid grassland. Generally distributed in the British Is. Europe, N. Asia and N. Africa. Introduced in North America, Australia, New Zealand and Tasmania.

The hybrid with *A. stolonifera* is common.

5. A. gigantea Roth — Common Bent-grass.

A. nigra With.

Like **4** but ligules on non-flowering shoots usually longer than wide; panicle-branches strongly aculeolate; swollen apex of pedicels aculeolate in the lower part. Fl. 6–8. $2n = 42 + 0$–$4B$. Hp.

Native. Grassy places, open woodland and a weed of arable land, especially on light sandy or gravelly soils. Throughout the British Is., but commonest in the south and east. Most of Europe; N. Africa; S.W. Asia; North America.

6. A. stolonifera L.   Fiorin.

A. alba auct., non L.: *A. palustris* Hudson

A stoloniferous perennial, sometimes with short rhizomes bearing not more than 3 scale-lvs; stems (5–)15–100(–150) cm, smooth; stolons up to 200 cm. Lvs flat or involute; ligule 1.5–7 mm. Panicle 1–30 cm, the branches mostly branched 1–2 times, patent only at flowering, usually rough, very unequal, usually bearing spikelets in the lower ½. Spikelets 2–3 mm, greenish to purplish. Glumes lanceolate, acute. Lemma ⅗–⅘ as long as glumes, rarely with a short awn from near the apex. Palea ½–⅔ as long as lemma. Fl. 7–8. $2n = 28$ (30, 32, 35, 42, 44, 46). Hp.

Native. In grassy and waste places. Generally distributed in lowland districts in the British Is. Europe; temperate Asia; N. Africa; North America. Introduced in Australia, New Zealand, S. Africa, South America and the Falkland Is.

*7. A. castellana Boiss. & Reuter   Highland Bent.

Like **6** but not usually stoloniferous; ligule 1–3 mm; panicle-branches mostly branched 2–3 times and without spikelets in the lower ½; palea ½–⅔ as long as lemma.

Introduced. Often sown on roadsides, etc. Native of S. Europe and N. Africa.

38. GASTRIDIUM Beauv.

Annuals. *Panicle ±spike-like*, dense and shining, pedicels enlarged and flattened at the top. *Spikelets* of 1 floret, strongly compressed laterally, *swollen and nit-like below*. Glumes coriaceous, subequal, ventricose below, compressed and keeled above, 1-veined. *Lemma tough, hyaline, ⅓–¼ length of lower glume*, truncate, 5-veined, rounded on back; awn geniculate, sometimes 0. Palea thin, hyaline, equalling lemma. Lodicules 2. Stamens 3. Ovary glabrous.

Two spp. in the Mediterranean region, S. and W. Europe, S.W. Asia and N.E. Africa.

1. G. ventricosum (Gouan) Schinz & Thell.   Nitgrass.

G. lendigerum (L.) Desv.; *G. australe* Beauv.

A glabrous tufted annual 10–35 cm. Stems erect or ascending, slender, smooth and shining. Lvs flat, scabrid; ligule 1–3 mm, acute. Sheaths smooth, ±inflated. Panicle 2–10 cm, lanceolate, sometimes lobed; branches scabrid. Spikelets (2–)3–4 mm, acuminate, swollen and nit-like below. Glumes lanceolate, acuminate, ventricose part very polished, keel serrate. Lemma c. 1 mm, ovate, apex 4-toothed, awn from just below the top, exceeding the glumes or sometimes 0. Fl. 6–8. $2n = 14^*$. Th.

Native. In sandy places and on carboniferous limestone, usually near the sea. Sometimes as a casual in arable fields. S. England and S. Wales. S. and W. Europe; N. Africa; Macaronesia.

39. POLYPOGON Desf.

Glabrous annual or perennial herbs. *Panicle ±spike-like, dense. Spikelets falling entire when ripe*, small, of 1 floret, laterally compressed. *Glumes* chartaceous,

equal, *much exceeding the lemma*, obtuse and ±bifid, 1-veined, usually *awned from the sinus*. Lemma firm and silvery, obscurely veined, notched or toothed and usually shortly awned from near the top. Palea hyaline, narrow, 2-veined. *Rhachilla disarticulating below the glumes*. Lodicules 2. Stamens 1–3. Ovary glabrous.

About 15 spp. in warmer temperate regions.

Annual; glumes and lemma awned; glumes hairy.
                                          1. monspeliensis
Perennial; glumes and lemma unawned; glumes scabrid.
                                                2. viridis

1. P. monspeliensis (L.) Desf.   Annual Beardgrass.

A glabrous, ±tufted *annual* 5–80 cm. Lvs ±scabrid, flat, rather broad, acuminate. Sheaths smooth, upper somewhat inflated. Ligule up to 10 mm, obtuse. Panicle 2–16 cm, spike-like, or somewhat lobed, dense, yellowish and silky. Spikelets c. 2 mm. *Glumes* diverging at apex, 2–4 times as long as lemma, *hairy*, narrowly oblong, concave, shortly bifid; *awn* 5–7 mm, slender, scabrid, ±flexuous. Lemma c. 1 mm, oblong, truncate, toothed, shining, with an awn 1–1.5 mm. Palea equalling lemma. Fl. 6–7. $2n = 28$. Th.

Native. In damp pastures near the sea in S. England and the Channel Is. As a casual elsewhere. S. and W. Europe; N. Africa; S.W. Asia; Macaronesia; introduced in North America.

*2. P. viridis (Gouan) Breistr.
                     Water Bent, Beardless Beardgrass.

P. semiverticillatus (Forskål) Hyl.; *Agrostis semiverticillata* (Forskål) C. Christ.; *A. verticillata* Vill.

A tufted or stoloniferous *perennial* 10–100 cm. Lvs flat, short, scabrid. Sheaths smooth. Ligule up to 5 mm, truncate and toothed. Culms smooth, often rooting at the lower nodes. Panicle up to 15 cm, pyramidal, lobed, dense. Spikelets 2–2.5 mm, readily deciduous. *Glumes* lanceolate, acute, *scabrid* at least near the keel; *awn* 0. Lemma c. 1 mm, ovate, truncate, smooth; awn 0. Palea equalling lemma. Fl. 6–7. $2n = 28$. Hp.

Introduced. Naturalized in waste places, Channel Is. As a casual elsewhere in England. S. Europe; S.W. Asia; N. India; N. Africa; Macaronesia. Introduced in W. France, North and South America, S. Africa and S. Australia.

40. ×AGROPOGON P. Fourn.

Hybrid between *Agrostis* and *Polypogon* with characters ±intermediate between these genera.

1. ×A. littoralis (Sm.) C. E. Hubbard (*Agrostis stolonifera* × *Polypogon monspeliensis*).
                                          Perennial Beardgrass.

Polypogon littoralis Sm.; *P. lutosus* auct.

A glabrous tufted *perennial* 8–50 cm. Lvs ±scabrid, flat. Sheaths smooth, upper somewhat inflated. Ligule up to 5 mm, obtuse. Panicle 2–12 cm, narrow, dense, usually somewhat lobed, greenish or purplish, not silky. Spikelets 2–3 mm. Glumes not deciduous, diverging at the tips, slightly hairy, shiny, narrowly oblong, acute and not notched; awns equalling or shorter than the glumes. Lemma c. 1.5 mm, lanceolate notched; *awn shorter than in Polypogon monspeliensis*, sometimes shortly exserted. Palea nearly as long as lemma. Fl. 7–8. $2n = 28^*$. Hp.

Native. In salt marshes. Rare and local from Dorset to Norfolk; Gloucester and Glamorgan. S. and W. Europe, Canary Is.

41. AMMOPHILA Host

Almost glabrous erect, rhizomatous perennials, with convolute lvs. *Panicle dense, ±cylindrical. Spikelets large*, of one floret, strongly compressed laterally. Glumes longer than the lemma, membranous, keeled; lower 1-, upper 3 veined. Lemma coriaceous, 3–5-veined, the tip bifid, with a very short subterminal awn. Palea firm, nearly equalling the lemma. *Rhachilla silky, produced*. Lodicules 2. Stamens 3. Ovary glabrous.

One sp. around the coasts of Europe and N. Africa, and another on the coasts of North America.

1. A. arenaria (L.) Link   Marram Grass.

A. arundinacea Host; *Psamma arenaria* (L.) Roemer & Schultes

A stout, erect perennial 60–120 cm. Rhizome extensively creeping and rooting at the nodes, binding the sand. Lvs convolute, terete, rigid, pungent, polished without and glaucous, strongly ribbed and puberulent within. Sheaths smooth, persistent, upper somewhat inflated. Ligule 10–30 mm, acuminate. *Panicle* 7–20(–30) cm, dense, stout, spike-like, cylindrical, obtuse, *whitish*. Spikelets 12–14 mm. Glumes unequal, whitish, linear-lanceolate, acuminate, keel serrate in upper part, margins hyaline. Lemma 10–14 mm, *the hairs at its base less than ½ as long,* scabrid, compressed, half-lanceolate, tip hyaline with 2 short bristle points; awn stout, not exceeding the bristles. Palea lanceolate, acute, the 2 keels very close together, shortly ciliate. Fl. 7–8. $2n = 28, 56$. Hp.

Native. Abundant and often dominant on dunes. Round the coasts of the British Is. Coasts of W. Europe to c. 62°N. in Norway; introduced in North America.

42. ×AMMOCALAMAGROSTIS P. Fourn.

Hybrids between *Ammophila* and *Calamagrostis*, combining some of the characters of both but more closely resembling *Ammophila* in general appearance.

1. ×A. baltica (Flügge ex Schrader) P. Fourn.
(*Ammophila arenaria × Calamagrostis epigejos*)

Ammophila baltica (Flügge ex Schrader) Dumort.

Like *Ammophila arenaria* in habit and vegetative characters, but taller and stouter. Panicle ±lobed and interrupted, purplish, acute, with longer branches than in *Ammophila*. Lemma 4.5–9 mm, ±rounded on the back; awn slender, just exceeding the 2 bristle points. *Lemma 1–2 times as long as the hairs at its base.* Sterile. Fl. 7–8. $2n = 28, 42$. Hp.

Native. On dunes. Norfolk, Suffolk, Northumberland and W. Sutherland, locally abundant. Coasts of Baltic and North Sea; N. France.

43. CALAMAGROSTIS Adanson

Erect, *perennial* herbs, often found in damp places. Infl. a panicle. Spikelets subterete or ±compressed, of 1 floret. Glumes longer than floret, chartaceous, subequal, lower 1-, upper 3-veined. *Floret with a profusion of silky hairs at its base. Lemma* membranous, 3–5-veined, bifid, *awned from the back or sinus*; awn slender, short. Palea shorter than lemma, thin, hyaline. Rhachilla not produced beyond the floret (*Epigejos* Koch), or with a slender, silky prolongation (*Deyeuxia* Beauv.). Lodicules 2. Stamens 3. Ovary glabrous. Probably a highly heterogeneous genus.

About 280 spp. in temperate regions.

1 Hairs at base of lemma longer than floret.		2
Hairs at base of lemma shorter than floret.		3
2 Leaves scabrid but not hairy above.	**1. epigejos**	
Leaves hairy above.	**2. canescens**	
3 Spikelets 3–4(–4.5) mm; glumes acute.	**3. stricta**	
Spikelet 4.5–6 mm; glumes acuminate.	**4. scotica**	

1. C. epigejos (L.) Roth   Wood Small-reed, Bushgrass.

A stout, erect, glabrous perennial 60–200 cm. Lvs flat, *scabrid*, with a long and slender point. Sheaths smooth. *Ligule up to* 12 mm, acute, torn. Culms scabrid just below the panicle; nodes 2–4. Panicle 15–30 cm, ±spreading, purplish-brown; branches and *pedicels scabrid*. Spikelets 4–10, subterete. Glumes 2–3 times as long as floret, subulate, keeled in the upper part; lower rather broader than upper. *Hairs at base of floret longer than the floret. Lemma* smooth, hyaline, lanceolate, acuminate, bifid, 3-veined; *awn from the middle or above,* slender, *exceeding the lemma*. Palea ⅔ the length of the lemma. Fl. 7–8. $2n = 28^*$; 42, 56. Hp.

Native. In damp woods, ditches and fens. Widely distributed in England; local in Wales and Scotland, north to Sutherland; W. and N. Ireland, very rare. Most of Europe; temperate Asia; Japan; introduced in North America.

2. C. canescens (Weber) Roth   Purple Small-reed.

C. lanceolata Roth

A rather slender, erect perennial 50–150 cm. Lvs ±convolute, long, *with scattered long hairs above*, slightly scabrid beneath. Sheaths and *culms smooth*; nodes 4–6. Ligule 1–4 mm, obtuse and torn. Panicle 5–20 cm, ±spreading, light brown or purplish; branches smooth or scabrid, *pedicels smooth*. Spikelets 4–6 mm, subterete. Glumes 2–3 times as long as floret, lanceolate. *Hairs at base of floret little longer than the lemma. Lemma* smooth, hyaline, lanceolate, *notched, with a very short slender awn from the sinus, 5-veined*. Palea ½ as long as lemma. Fl. 6–7. $2n = 28$. Hp.

Native. In damp shady places and fens. England, local and mainly in the east; Scotland: South Aberdeen. C. and N. Europe southwards to N. Spain, N. Italy and Bulgaria; W. Siberia.

3. C. stricta (Timm) Koeler   Narrow Small-reed.

C. neglecta auct., non (Ehrh.) Beauv.

A slender erect glabrous perennial 10–100 cm. Lvs flat or convolute, usually rather short, glabrous or hairy above, strongly scabrid. Sheaths smooth. Ligule 1–2 mm, obtuse. Culms scabrid just below the panicle; nodes 2–3. Panicle 5–20 cm, narrow. *Spikelets* 3–4(–4.5) mm, pale purplish and green becoming light brown when old. *Glumes lanceolate*, *acute* or sometimes acuminate, exceeding the lemma. Lemma scabrid, hyaline, truncate and bifid at the apex; *awn from about the middle*, *equalling the lemma*. Hairs at the base of the lemma $\frac{2}{3}$ length of the lemma. Rhachilla with a slender, silky prolongation. Var. *hookeri* Syme has a dense reddish-brown panicle usually less than 10 cm and ±lobed, the uppermost ligule rather longer than broad and subacute, and the hairs $\frac{3}{4}$ length of lemma. Fl. 6–7. $2n = 28$. Hp.

Native. In bogs and marshes, rare. W. Norfolk, Cheshire, Yorks, Ayrshire, mid-Perth and Caithness; extinct in Angus. Var. *hookeri*: W. Norfolk and Lough Neagh. N. and C. Europe: N. and C. Asia; Japan; North and South America.

4. C. scotica (Druce) Druce   Scottish Small-reed.

C. strigosa auct. angl., non Hartman

Like *C. stricta* but panicle up to 10 cm, rather dense; spikelets 4.5–6 mm, rather darker than those of *C. stricta*; glumes narrowly lanceolate, long-acuminate, somewhat exceeding the lemma; lemma awned from the base. Fl. 7.

Native. In bogs, very rare. Caithness. Endemic.

44. Phleum L.

Annuals or perennials. *Panicle cylindrical or ovoid, dense and spike-like.* Spikelets of 1 floret, strongly compressed laterally. *Glumes* membranous, longer than lemma, *strongly keeled* and *pectinate-ciliate on keel*, 3-veined, *shortly aristate*, margins overlapping along most of their length. *Lemma thin, truncate or obtuse*, 1–7-veined. Palea equalling lemma, thin, 2-veined. Lodicules 2. Stamens 3. Ovary glabrous.

About 15 spp. in temperate regions.

1 Annual; glumes tapering gradually to apex.
                                                  4. arenarium
  Perennial; glumes truncate or obliquely truncate.   2
2 Infl. ovoid or broadly cylindrical; awn more than
  2 mm.   **2. alpinum**
  Infl. cylindrical, rarely narrowly ovoid; awn 2 mm or
  less.   3
3 Panicle-branches almost completely adnate to the
  rhachis.   **1. pratense**
  Panicle-branches free.   **3. phleoides**

1. P. pratense L.   Cat's-tail.

Perennial up to 150 cm; lower nodes sometimes swollen and tuberous. Lvs up to 10 mm wide; scabrid, at least towards the apex; ligule 1–6 mm; sheaths smooth, the lower blackish, ±fibrous, the upper somewhat inflated. Panicle (1–)2–11(–30) cm, cylindrical; branches adnate to the rhachis. Spikelets 2–5.5 mm. Glumes truncate,

densely pectinate-ciliate on the keel; awn 0.2–2 mm; lower glume softly hairy on the margin, the upper glabrous. Lemma $\frac{2}{3}$–$\frac{3}{4}$ as long as glumes, minutely hairy. Palea as long as the lemma. Fl. 7. Hp.

Native. In grassland; throughout the British Is., except on moorland and the higher mountains. North temperate regions, but introduced in North America.

Subsp. **pratense**: Stems 20–150 cm. Ligule obtuse. Panicle 6–11(–30) cm × 6–8 mm. Awn 1–2 mm. $2n = 42$ (21, 35, 36, 49, 56, 63, 70, 84). Commonly cultivated as a fodder plant and widely naturalized.

Subsp. **bertolonii** (DC.) Bornm. (*P. bertolonii* DC., *P. nodosum* L.): Stems 6–70(–100) cm. Ligule acute. Panicle 1–8(–10) cm × 3–5 mm. Awn 0.2–1.2 mm. $2n = 14$. In pastures and short rough grassland. Almost throughout the range of the sp.

2. P. alpinum L.   Alpine Cat's-tail.

P. commutatum Gaudin

A rather stout perennial 15–50 cm. Lvs smooth. Sheaths smooth, the *basal dark brown or blackish*, the *upper strongly inflated*. Ligule up to 2 mm, truncate. Panicle 1–6 cm × 7–15 mm, ovoid or broadly cylindrical, obtuse, often purplish; branches very short. Spikelets 5–8.5 mm. Glumes oblong, truncate, *awn 1.5–4 mm, longer than in any other British sp.*, scabrid; keel long pectinate-ciliate to base, margins scarcely hyaline, that of the lower glume softly hairy on the margin. *Lemma* c. $\frac{2}{3}$ as long as glumes, minutely hairy, broadly ovate, *truncate*, toothed, 5-veined. Fl. 7–8. $2n = 28$. Hp.

Native. In damp places on the higher mountains. Westmorland, Cumberland, Dunbarton, Perth, Angus, Aberdeen, Banff, Inverness and E. Ross. Arctic regions and higher mountains of Europe and Asia; North America; Andes of South America, South Georgia.

3. P. phleoides (L.) Karsten   Purple-stem Cat's-tail.

P. boehmeri Wibel

An erect perennial 10–40 cm. Lvs scabrid. Sheaths smooth, lower brown or purplish, upper slightly inflated. Ligule 2 mm or less, truncate. *Panicle* 1.5–14 cm × 4–10 mm, cylindrical, ±interrupted, tapering and often subacute; branches free from rhachis. Spikelets c. 3 mm. Glumes narrowly oblong, obliquely truncate, aristate, glabrous, the keel sparsely ciliate-pectinate or scabrid, *hyaline margins very silvery*. *Lemma* $\frac{2}{3}$–$\frac{3}{4}$ length of the glumes, narrow-elliptic, *obtuse, glabrous or with scattered appressed hairs*. Palea lanceolate, acute, silvery. Fl. 6–7. $2n = 14, 28$. Hp.

Native. In dry sandy and chalky pastures in East Anglia and Bedford. Much of Europe; temperate Asia; N. Africa.

4. P. arenarium L.   Sand Cat's tail.

An *annual* 3–15(–30) cm. Lvs short, scabrid on the veins. Sheaths smooth, *lower whitish, upper inflated, fusiform*. Ligule up to 7 mm, acute. Panicle 0.5–5.5 cm × 3–7 mm, narrowly ovoid to cylindrical, blunt. Spikelets

2.2–4.4 mm. *Glumes* lanceolate, shortly aristate, pectinate-ciliate on the keel and shortly ciliate on the margins and *diverging in the upper part*, broadly hyaline. Lemma $\frac{1}{3}$ as long as glumes, glabrous or pubescent, broadly oblong, truncate. Fl. 5–6. $2n = 14$. Th.

Native. On dunes and in sandy fields. Around the coasts of the British Is. except N. Scotland, locally common. S. and W. Europe north-eastward to Gotland; mountains of N. Africa; S.W. Asia.

45. ALOPECURUS L.

Nearly glabrous annual or perennial herbs. *Spikelets* strongly compressed, crowded in narrow spike-like panicles, of 1 floret, *readily deciduous in fr.*; *rhachilla disarticulating below the glumes. Glumes* tough, subequal, equalling or slightly exceeding the floret, *often connate below the middle*, 3-veined. Lemma hyaline, 3-veined, usually awned from the back, margins often connate below. *Palea* 0. Lodicules 0. Stamens 3. Ovary glabrous.

About 50 spp. in temperate regions.

1 Awn scarcely exceeding lemma or 0.   2
  Awn at least twice as long as lemma.   3
2 Panicle narrow-cylindrical, not conspicuously hairy.
                4. aequalis
  Panicle ovoid or broadly cylindrical, conspicuously hairy.   **6. alpinus**
3 Annual; glumes connate for $\frac{1}{3}$–$\frac{1}{2}$ their length.
                1. myosuroides
  Perennial; glumes free nearly to base.   4
4 Stem swollen and bulbous at base (salt-marshes).
                5. bulbosus
  Stem not bulbous at base (meadows and wet places).   5
5 Glumes acute.   6
  Glumes obtuse.   7
6 Lemma acute.   **2. pratensis**
  Lemma obtuse.   ×**brachystylus**
7 Spikelet 2–3 mm; lemma truncate.   **3. geniculatus**
  Spikelet 3.5–4.5 mm; lemma obtuse.   ×**brachystylus**

1. A. myosuroides Hudson   Black grass, Black Twitch.

A. agrestis L.

A tufted *annual* 20–70 cm. Stems decumbent at base. Lvs 2–9 mm wide, scabrid above, smooth beneath. Upper sheaths somewhat inflated. Ligule up to 5 mm, obtuse. *Panicle* 4–12 cm, narrow, *tapering towards the apex*. Spikelets 4.5–7.5 mm. *Glumes connate for* $\frac{1}{3}$–$\frac{1}{2}$ *their length*, *shortly ciliate on veins and keel* and at their base, keel narrowly winged, oblong to lanceolate, acute, narrowly hyaline in upper part. Lemma acute, margins connate for $\frac{1}{3}$–$\frac{1}{2}$ their length; awn from near the base, geniculate, twice as long as lemma. Fl. 6–7. $2n = 14$. Th.

Native. A weed in arable fields and waste places. Scattered throughout Great Britain, abundant in the southeast; a casual in the north; a rare casual in Ireland. S. and W. Europe; N. Africa; W. Asia; introduced in North America and New Zealand.

2. A. pratensis L.   Meadow Foxtail.

Stems 30–110 cm, ±geniculate at the lowest node. Lvs

4–8 mm wide, ±scabrid. Upper sheaths inflated. Ligule 1.5–2.5 mm, truncate. Panicle 3–6(–11) cm, tapering somewhat but obtuse. Spikelets 4–6 mm. *Glumes* ±hyaline, *long-ciliate on keel*, shortly ciliate or subglabrous on lateral veins, lanceolate, acute or acuminate, connate for $\frac{1}{4}$ of their length. *Lemma acute*, margins connate for $\frac{1}{3}$ their length; *awn* from near the base, geniculate, *twice the length of lemma*. Fl. 4–6. Protogynous. $2n = 28$. Hp.

Native. In damp meadows, pastures and grassy places. Generally distributed and common in most of the British Is., except Ireland, where it is local. Most of Europe; W. Siberia; Caucasus.

A. × brachystylus Peterm. (*A.* × *hybridus* Wimmer, *A. geniculatus* × *pratensis*) has the stems ±creeping at base and rooting at nodes; the spikelets 3.5–4.5 mm; the glumes subacute, slaty-grey at tips with hyaline margins and the lemma obtuse. Fl. 5–6. With the parents, in wet places. Usually sterile.

3. A. geniculatus L.   Marsh Foxtail.

A perennial 15–40 cm. Stems ±creeping, rooting at the nodes, geniculate and ascending. Lvs 4–8 mm wide, scabrid above, smooth beneath. Upper sheaths distinctly inflated. Ligule up to 5 mm. Panicle 2–6.5 cm blunt. *Spikelets* 2–3.5 mm. *Glumes* free nearly to base, with silky hairs on the veins, lanceolate, obtuse, *broadly hyaline at the top. Lemma* truncate, *margins free*; *awn from near the base, geniculate, twice as long as lemma. Anthers brown at maturity.* Fl. 6–7. $2n = 28$. Hp.

Native. In wet meadows and at edges of ponds and ditches. Generally distributed throughout the British Is., frequent. Most of Europe, W. Siberia; introduced in North and South America.

4. A. aequalis Sobol.   Orange Foxtail.

A. fulvus Sm.

A ±tufted annual or short-lived perennial, 10–40 cm. Stems decumbent at base. Lvs scabrid above, smooth beneath. Sheaths glaucous, the upper strongly inflated. Ligule up to c. 4 mm, ±truncate and torn. Panicle 1.5–6 cm, narrow, cylindrical, obtuse. *Spikelets* 1.5–3 mm, ovate. Glumes free nearly to base, shortly silky, long-ciliate on keel, lanceolate, obtuse, *margins narrowly hyaline. Lemma* elliptical, obtuse, *margins connate for* $\frac{1}{3}$–$\frac{1}{2}$ *their length*; *awn from just below the middle, straight, not or slightly exceeding lemma. Anthers orange at maturity.* Fl. 5–7. $2n = 14$. ?Th. or Hp.

Native. In similar situations to *A. geniculatus* but far less common. England, from Wilts and Kent to Cheshire and N.W. Yorks, local. Most of Europe, but absent from many islands; Siberia; North America; N. Africa.

5. A. bulbosus Gouan   Bulbous Foxtail.

A slender, tufted perennial 25–50 cm. *Stems swollen and bulbous at base*, erect or ±geniculate. *Lvs smooth*. Upper sheaths somewhat inflated. Ligule 5 mm, obtuse. Panicle 1–3 cm, cylindrical, obtuse. Spikelets 2–3.5 mm. *Glumes* free nearly to base, shortly ciliate on keel and lateral veins, lanceolate, *acute*. *Lemma* oblong, truncate, *margins free*; *awn from near the base, geniculate*, c. 3 *times as long as lemma*, pale below, dark above. Fl. 6. $2n = 14^*$. Hp.

Native. In grassy salt-marshes. From the Bristol Channel to the Humber; very local and apparently diminishing. Coasts of N. and W. Europe eastwards to N.W. Yugoslavia.

6. A. alpinus Sm.   Alpine Foxtail.

An erect perennial 10–60 cm. Stems shortly creeping then erect. Lvs 2.5–7 mm wide, smooth, short and broad. Sheaths inflated. Ligule 2–3 mm, obtuse. *Panicle 1–3 cm, ovoid or broadly cylindrical. Spikelets 3–5 mm*, ovate, obtuse, *silky. Glumes* half-ovate, acute, *clothed with long silky hairs*, free nearly to base. Lemma membranous, ovate, obtuse, *margins fringed in upper part and connate at the base; awn from the middle or above, short, sometimes* 0. Fl. 7–8. $2n = 112$–130. H.

Native. In cold, moderately base-rich springs and rills on the higher Scottish mountains from Dumfries and the Cheviots to E. Ross; N. England; very local. Arctic Russia, Spitsbergen, Novaya Zemlya; Urals; Greenland.

Tribe 9. HAINARDIEAE. Annuals. Lvs with orbicular, oblong or elliptical silica-cells and no 2-celled hairs; green tissue uniformly distributed between the vascular bundles. First foliage lf of seedling narrow and vertical. Ligule membranous. Infl. a slender cylindrical spike. Spikelets with 1(–2) florets, all hermaphrodite, sessile, alternate, sunk in cavities in the rhachis. Glumes 2, inserted laterally or 1 (except in the terminal spikelet), longer than the lemma, strongly veined, appressed to and covering the cavity, except at flowering, flat, rounded or with an asymmetrical keel on the back. Lemma hyaline, 1–3-veined, unawned. Palea 2-veined. Lodicules 2, lanceolate, entire. Stamens 3. Ovary usually with a glabrous apical appendage. Fr. with a linear to elliptical basal hilum; starch-grains compound. Chromosomes large; basic number 7, 13, 19.

46. PARAPHOLIS C. E. Hubbard

Slender glabrous annuals. Lvs short, narrow. Spikelets of 1 floret, placed broadside on to the rhachis and embedded in its concavities. Glumes equal, placed side by side in front of the spikelet. Lemma and palea hyaline, 1-veined. Stamens 3. Ovary glabrous.

Five spp. in W. Europe and the Mediterranean region to C. Asia.

Anthers at least 2 mm; spikes straight or nearly so; uppermost sheath not inflated.   **1. strigosa**
Anthers not more than 1 mm; at least some spikes distinctly curved; uppermost sheath ±inflated. **2. incurva**

1. P. strigosa (Dumort.) C. E. Hubbard   Hard-grass.

Lepturus strigosus Dumort.; *Lepturus filiformis* auct., non (Roth) Trin.

A slender annual 15–40 cm. Stems usually decumbent at base, geniculate and ascending, rarely erect, freely branched. *Uppermost sheaths not inflated*. Culms shining. Ligule less than 0.5 mm. Spikelets 4–6 mm, rounded on the back. Glumes linear-lanceolate, ±asymmetrical, acuminate, 3–5-veined, margins hyaline. Lemma and palea very delicate, lanceolate, acute, equal or subequal. Lodicules lanceolate, acute. Anthers 2–2.5 mm. Fl. 6–8. $2n = 28$. Th.

Native. In salt-marsh turf and waste places by the sea. Widely distributed but local round the coasts of the British Is., north to E. Lothian and Ayr. W. Europe from Denmark southward and east to N. Yugoslavia.

2. P. incurva (L.) C. E. Hubbard   Curved Hard-grass.

Lepturus incurvus (L.) Druce; *Pholiurus incurvus* (L.) Schinz & Thell.

Like *P. strigosa*, but commonly, 5–10 cm. *Uppermost sheaths inflated*. Culms usually not conspicuously shining. Spikes mostly strongly curved. Anthers 0.5–0.9 mm. Fl. 6–7. $2n = 38$. Th.

Native. In bare places, usually on clayey or muddy shingle, among taller vegetation by the sea. S. England from the Wash to the Bristol Channel, extending locally to N. Wales and 1 loc. in Dublin. S. and W. Europe; S.W. Asia; N. Africa; Macaronesia.

Tribe 10. PHALARIDEAE. Lvs linear; silica-bodies oblong; 2-celled micro-hairs absent; ligule membranous. Infl. a panicle. Spikelets strongly compressed laterally, usually with 3 florets, and 2 lower reduced to a small lemma, or 1 or both absent. Rhachilla disarticulating above the glumes. Glumes longer than the florets. Fertile lemma indurate at maturity, unawned. Lodicules 2. Stamens 3. Starch-grains compound. Chromosomes large; basic number 6 or 7.

47. PHALARIS L.

Glabrous annual or perennial herbs. Panicle dense, ovoid or cylindrical, ±lobed and sometimes with branches spreading at anthesis. *Spikelets strongly compressed, with 1–2-rudimentary lower florets and a terminal hermaphrodite one*, or sometimes all florets sterile or male. *Glumes keeled, usually winged from the keel*, chartaceous, 3–5-veined, subequal and exceeding the lemma, often straw-coloured with a green band on the keel. Lemma coriaceous, 5-veined, keeled, ±enclosing the palea. Palea similar to lemma but slightly shorter. Lodicules 2, lanceolate, acuminate. Stamens 3. Ovary glabrous.

About 20 spp., mainly in warm temperate regions.

1 Plant tall and reed-like, growing in damp places or in water; panicle distinctly lobed; branches spreading at flowering.   **1. arundinacea**
  Plant smaller, annual, not reed-like, growing in dry places, usually on waste ground; panicle ovoid or cylindrical.   2
2 Spikelets in groups of 5–7, 1 hermaphrodite, the rest male or sterile.   **4. paradoxa**
  Spikelets all hermaphrodite, except sometimes the basal.   3

3 Glumes with an entire wing.   **2. canariensis**
  Glumes with a toothed wing.   **3. minor**

1. P. arundinacea L   Reed-grass.

Digraphis arundinacea (L.) Trin.

A stout erect reed 60–200 cm. *Rhizomes far-creeping*. Lvs flat, smooth, 6–18 mm wide, acuminate. Sheaths smooth. Ligule 6–10 mm. *Panicle* 5–20 cm, narrowly oblong, *lobed*, purplish; *branches spreading at flowering*. Spikelets c. 5 mm. Glumes lanceolate, acuminate, keeled, but not winged. Lemma broadly lanceolate, acute, hairy towards the top. Fl. 6–7. $2n = 28^*$; 27–31, 35; 42. Hel.

The dead lvs persist throughout the winter (compare *Phragmites*).

Native. In wet places. Generally distributed throughout the British Is. in suitable habitats. Most of Europe; circumboreal; Japan; Macaronesia.

*2. P. canariensis L.   Canary Grass.

An annual 20–60 cm. Lvs flat, scabrid, up to 10 mm wide, acuminate. Sheaths ±smooth, the upper strongly inflated. Ligule up to 8 mm, obtuse. Panicle 1.5–4 cm, dense, ovoid. Spikelets 7–9 mm, ovate, acute. *Glumes* oblanceolate, acute or apiculate, 3-veined, and *strongly winged in upper half*. Fertile lemma coriaceous, shining, silky, ovate, acute, nearly enclosing the palea, 5-veined. Palea similar to lemma but slightly shorter, with hyaline margins. The 2 *lower florets represented by lanceolate, coriaceous, 1-veined lemmas half as long as the fertile one*. Fl. 6–7. $2n = 12^*$. Th.

Introduced. A casual, in waste places. Widely distributed in Great Britain in the lowlands but not common; rare in Ireland. N.W. Africa and the Canary Is., widely naturalized and often cultivated for bird-seed elsewhere.

*3. P. minor Retz.   Lesser Canary Grass.

A rather slender annual 10–30 cm, similar to *P. canariensis*. Lvs slightly scabrid, rather narrow. Upper sheaths somewhat inflated. Panicle 1.5–2.5 cm, ovoid or almost cylindrical. Spikelets 4.5–5.5 mm. *Glumes* lanceolate, acute, *with a rather narrow wing toothed near the top*. Fertile lemma ovate, silky, acute, completely enclosing palea, obscurely veined. *Sterile lemmas* 2, 1 *minute and scale-like, the other subulate, shortly silky*, ⅓ the length of the fertile lemma. Fl. 6–7. Th.

Introduced and naturalized or perhaps native in sandy places in the Channel Is.; a casual elsewhere. Mediterranean region and W. Europe to N.W. France, S.W. Asia.

*4. P. paradoxa L.

An annual 25–85 cm. Lvs up to 7 mm wide; sheaths ±smooth, the upper inflated. Panicle 3–6(–9) cm, widest near the apex. Spikelets in groups of 5–7, the central one hermaphrodite, the others male or sterile, each group falling entire. Fertile spikelets 6–8 mm; glumes with a spinose apex and a wing projecting from the middle of the keel, with 1 prominent tooth near its apex. $2n = 14$.

Introduced. An arable weed, mainly in S.E. England. Mediterranean region; S.W. Europe; Macaronesia.

Tribe 11. MILIEAE. Annuals or perennials. Lvs with few or no silica-cells and no 2-celled hairs; green tissue uniformly distributed between the vascular bundles. First foliage lf of seedling narrow and vertical. Ligule membranous. Infl. a panicle. Spikelets all hermaphrodite, dorsally compressed, with 1 floret. Glumes longer than the floret. Lemma chartaceous or hyaline, becoming strongly indurate in fr., unawned. Palea similar in texture to the lemma, becoming chartaceous. Lodicules 2, entire, acute. Stamens 3. Ovary glabrous. Fr. with linear or oblong hilum; starch-grains compound. Chromosomes large; basic number 4 or 7.

48. MILIUM L.

Annual or perennial glabrous herbs. *Spikelets of 1 floret*, dorsally compressed; rhachilla shortly produced. Glumes subequal, membranous, exceeding the floret, 3-veined. *Lemma* coriaceous, *becoming extremely hard and shining in fr.*, margins folded round the palea, 5-veined, unawned. Palea similar to lemma but smaller and 2-veined. Lodicules 2, acute, toothed on one side. Stamens 3. Ovary glabrous.

About 4 spp. in Europe, Asia and North America.

Perennial 50–120 cm; sheaths smooth; panicle spreading.
                                        1. effusum
Annual 2.5–15 cm; sheaths scabrid; panicle contracted.
                                       2. vernale

1. M. effusum L.   Wood Millet.

An erect, tufted perennial 50–180 cm. Culms smooth. Lvs flat, 5–15 mm wide, thin, scabrid, acute. Sheaths smooth. Ligule 3–8 mm, obtuse. Panicle 10–40 cm, very lax, *ovoid or pyramidal*; *branches* capillary, *spreading or deflexed*, up to 6 together. Spikelets 3–4 mm. Glumes ovate, acute, scaberulous, green. Lemma coriaceous and shining, ovate, subacute. Fl. 6. $2n = 28^*$. Hp.

Native. In damp, shady woods. Throughout the British Is., local and perhaps less frequent than formerly. Most of Europe; temperate Asia; Japan; North America.

2. M. vernale Bieb.   Early Millet.

Incl. *M. scabrum* L. C. M. Richard

An *annual*, usually ±prostrate. Culms 2.5–15 cm, scabrid. Lower lvs 10–20 × 2 mm, uppermost 3–5 mm. *Sheaths scabrid*, somewhat inflated, often purplish. Ligule 2–4 mm, acute. *Panicle* 15–35 mm, *very narrow*, even at flowering; *branches* short, slender, flexuous, *erect*, 2–4 together. Spikelets 2.5–3.5 mm, green, sometimes purplish-tinged. Glumes ovate, acute, scaberulous, usually green with hyaline margins. Lemma coriaceous and shining, ovate, obtuse. Fl. 4. $2n = 8^*$. Th.

Native. In nearly closed turf on fixed dunes, very local. Guernsey. W. and S. Europe; temperate Asia; N. Africa.

*Piptatherum miliaceum (L.) Cosson (*Oryzopsis miliacea* (L.) Bentham & Hooker ex Ascherson & Graebner), differing from *Milium* in the longer more numerous whorled panicle-branches and the narrower acute spikelets with awned lemmas, is sometimes cultivated and ±naturalized in a few places. Mediterranean region.

Tribe 12. ARUNDINEAE. Lvs broadly linear; silicabodies cross-shaped or rounded, with a mixture of somewhat crescentic to saddle-shaped cells; 2-celled microhairs often present; ligule a ring of hairs. Infl. a panicle. Spikelets laterally compressed, with up to 10 florets. Rhachilla disarticulating above the glumes and between the florets, with abundant long hairs. Glumes unequal, shorter than the florets. Lemma membranous, unawned, rounded on the back. Lodicules 2, glabrous. Stamens 3. Starch-grains compound. Chromosomes small; basic number 12.

49. PHRAGMITES Adanson

A stout, extensively creeping reed. Ligule a ring of hairs. Panicle large, lax, nodding. Spikelets, slender, acuminate, subterete. Glumes shorter than 1st floret, 3-veined, upper twice as long as lower. Lemma 3-veined, twice as long as upper glume. Palea not more than ⅓ the length of lemma. All the florets except the lowest with a tuft of long, silky hairs at the base. Stamens 1–3 in lowest floret, 3 in others. Lodicules, 2, oblong, truncate. Ovary glabrous; styles short.

Three spp., cosmopolitan.

1. P. australis (Cav.) Trin. ex Steudel　Common Reed.

P. communis Trin., *Arundo phragmites* L.

A stout, erect, rhizomatous reed 8–350 cm. Rhizome stout, extensively creeping. Lvs flat, 10–50 mm wide, smooth, glaucous beneath, tapering to long slender points and with 3 'tooth-marks' ⅓ of the way from base to apex, deciduous in winter. Sheaths smooth, loose so that all the lvs point one way in the wind; auricles prominent. Ligule a ring of hairs. Panicle 15–30 cm, lax, nodding, soft, dull purple; branches smooth, usually with scattered groups of a few long silky hairs. Spikelets 10–15 mm, of (1–)2–10 florets; rhachilla hairy. Glumes lanceolate, acute. Lemma linear-lanceolate, acute to acuminate. Palea ciliate in upper part. Silky hairs about as long as the lemma. The lvs break off at the junction with the sheaths in autumn; the dead culms and panicles stand throughout the winter. Fl. 8–9. 2*n* = 36, 44, 46, 48, 49, 50, 51, 52, 54, 96. Hel. or Hyd.

Native. In swamps and shallow water, but absent from extremely poor and acid habitats. In suitable habitats throughout the whole British Is. Cosmopolitan, except for a few tropical regions, e.g. the Amazon basin.

*Cortaderia selloana (Schultes & Schultes fil.) Ascherson & Graebner, Pampas Grass, a very large densely tufted perennial with silvery panicles produced in late summer, is often cultivated and sometimes persists as a throw-out from gardens.

Tribe 13. DANTHONIEAE. Leaves linear; silica-bodies rounded or dumbbell-shaped; 2-celled microhairs present; ligule a ring of hairs. Spikelets somewhat compressed laterally, with 2–5 florets. Rhachilla disarticulating above the glumes and between the florets. Glumes subequal, about as long as the florets. Lemma coriaceous, 2-dentate at apex, awned or mucronate from the sinus. Lodicules 2. Stamens 3. Starch-grains compound. Chromosomes small; basic number 12.

50. DANTHONIA DC.
(incl. *Sieglingia* Bernh.)

Perennial. Spikelets subterete, with 2–5 florets. *Glumes about equalling the spikelet*, chartaceous, subequal, bluntly keeled, 5–7-veined, lateral veins about ½ as long as glume. *Lemma* coriaceous, *rounded on back*, 7–9-veined, *apex with 3* short obtuse points. Palea coriaceous, shorter than lemma. Rhachilla produced beyond the uppermost perfect floret and bearing a sterile rudiment. Lodicules 2, ovate, obtuse. Stamens 3. Ovary glabrous, stipitate; *styles* short, spreading, *inserted far apart*.

About 10 spp, cosmopolitan.

1. D. decumbens (L.) DC.　　　　Heath Grass.

Triodia decumbens (L.) Beauv.; *Sieglingia decumbens* (L.) Bernh.

A tufted, ±decumbent perennial 10–60 cm. Lvs flat, ±glabrous. Sheaths glabrous or slightly pubescent, with a ring of hairs at the mouth. Panicle 2–7 cm, of few (up to 15) spikelets. Spikelets 6–15 mm. Glumes lanceolate, obtuse. Lemma ovate, silky on the margin in the lower half, with a tuft of spreading silky hairs at the base. Palea ovate, obtuse, ciliate. Fls cleistogamous or rarely chasmogamous. White compressed cleistogamous subterranean spikelets are borne singly in the axils of the basal sheaths. Fl. 7. 2*n* = 24, 36*, (124*). Hs.

Native. In acid grassland; locally on damp base-rich soils. Generally distributed throughout the British Is. in less peaty habitats than *Nardus*. Most of Europe; Caucasus; N. Africa; North America; Madeira; only on mountains in the south.

Tribe 14. MOLINIEAE. Perennials. Lvs deciduous, with dumbbell-shaped silica-cells and 2-celled hairs; green tissue uniformly distributed between the vascular bundles. First foliage lf of seedling narrow and vertical. Ligule a ring of hairs. Spikelets laterally compressed, with 1–4 florets. Rhachilla disarticulating above the glumes. Glumes shorter than the florets. Lemma membranous, 3-veined, rounded on the back, unawned. Ovary glabrous, without an appendage. Fr. with a linear hilum; starch-grains compound but readily separating. Chromosomes small; basic number 9.

51. MOLINIA Schrank

Glabrous perennial herbs. Ligule a ring of hairs. Panicle strict, branches long, slender. Spikelets subterete, tapering to a long point. Glumes subequal, membranous, about ½ length of 1st floret, 1–3-veined. *Lemma 3-veined*. Palea equalling lemma, tough, obtuse. Lodicules 2, obovate. Stamens 3. Ovary glabrous; styles terminal very short.

Two to three spp., N. temperate.

1. M. caerulea (L.) Moench   Purple moor-grass.

An erect, wiry perennial 15–150 cm, often forming large tussocks. Stock ±creeping, roots very stout; culms with 1 node towards the base, the basal internode up to 5 cm, usually swollen and clavate. Lvs flat, tapering from near the base, sparsely pilose, completely deciduous in winter. Panicle erect, 3–40 cm, green or purplish; pedicels short, finely ciliate. Spikelets 4–9 mm, tapering to a long point; florets 1–4. Glumes lanceolate, acute, 1-veined or upper 3-veined. Lemma tapering to an obtuse apex, bluntly 3-keeled. Anthers large, violet-brown. Fl. 6–8. 2n = 18, 36, 90. Hs.

Native. In damp or wet places in fens and heaths and on mountains. Throughout the British Is., locally abundant. Most of Europe; N. Africa; Caucasus; Siberia; introduced in North America.

Tribe 15. NARDEAE. Lvs setaceous; silica-bodies saddle-shaped or rounded; slender 2-celled microhairs present; ligule membranous. Infl. a unilateral spike. Spikelets triangular in section, with 1 floret. Glumes persistent, the lower very small, the upper usually absent. Lemma chartaceous, with an apical awn. Lodicules absent. Stamens 3. Style 1; stigma papillose. Starch-grains compound. Chromosomes large; basic number 13.

52. NARDUS L.

The only genus. One species.

1. N. stricta L.   Mat-grass.

A wiry, densely tufted perennial usually 10–40 cm. *Lvs setaceous, very hard*, scabrid, *erect when young, later spreading at right angles* to the sheaths, *whitish and persistent when dead*. Sheaths strict, smooth, persistent, whitish, the lower ones lfless. Ligule up to 2 mm. Culms slender, smooth. Spike 2–10 cm. Spikelets 5–9 mm, in 2 rows along 1 side of the rhachis, narrow, acute. Lower glume very small, upper usually 0. Lemma 3-veined, subulate, with a terminal awn 1–3 mm. Fl. 6–8. Apomictic. 2n = 26*. Hs.

Native. Abundant on the poorer siliceous and peaty soils, covering great areas on moors and mountains. Rejected by sheep on account of the harsh foliage and therefore especially abundant in overgrazed areas. Throughout most of the British Is. but absent from some lowland and calcareous districts and only abundant on moors and mountains. Europe; temperate Asia; Caucasus; Greenland; only on mountains in the southern part of its range and always calcifuge; doubtfully native in North America.

Tribe 16. CHLORIDEAE. Lvs linear; silica-bodies saddle-shaped; short 2-celled microhairs present; ligule a ring of multicellular hairs. Infl. of usually 3–5 digitately arranged secund spikes. Spikelets laterally compressed, with 1 fertile and 1 or more sterile florets. Rhachilla disarticulating above the glumes. Glumes subequal, shorter than the floret. Lemma chartaceous, unawned.

Lodicules 2, truncate. Stamens 3. Starch-grains compound. Chromosomes small; basic number 9, 10, 12.

53. CYNODON L. C. M. Richard

Perennials. *Infl. digitate*, of usually 4–5 *unilateral spikes*. Spikelets of 1 floret, laterally compressed. Glumes equal, shorter than the floret, membranous, 1-veined. Lemma coriaceous, strongly compressed, obscurely 3-veined, folded round the palea. Rhachilla glabrous, produced beyond the floret, disarticulating above the glumes. Lodicules 2, short, truncate. Stamens 3. Ovary glabrous; styles long.

About 10 spp. in the warmer regions of the world, mainly in S. Africa and Australia.

1. C. dactylon (L.) Pers.   Bermuda-grass.

Capriola dactylon (L.) O. Kuntze

An extensively creeping perennial 10–30 cm, with stolons and scaly rhizomes. Culms ascending or erect. Lvs short, stiff, scabrid, acuminate, sparsely hairy. Sheaths smooth, persistent. Ligule a ring of hairs. Infl. 1–5 cm, of (3–)4–5(–6) purplish spikes all arising at the same point. Spikelets c. 2 mm. Glumes spreading, subulate. Lemma broadly ovate, acute, sparingly villous on keel and margin. Fl. 8–9. 2n = 36*. Hp.

Native. On sandy shores in scattered localities in S. England and S. Wales; Channel Is. Elsewhere as a casual. W. and S. Europe northwards to England and N. Ukraine; temperate Asia; N. Africa; Macaronesia; introduced in North America.

Tribe 17. SPARTINEAE. Like *Chlorideae* but silica-bodies rounded; 2-celled microhairs globose, sunk in pits in the epidermis; ligule a ring of unicellular hairs; infl. of racemosely arranged spikes; rhachilla disarticulating below the glumes; lodicules absent.

54. SPARTINA Schreber

Stout, erect, glabrous rhizomatous perennials with tough, yellow-green lvs and yellowish or golden infl. *Infl. of a number of erect spikes arranged in a raceme. Spikelets of 1 floret*, large, *distichous* and alternate on the flattened rhachis and pressed close to it, flattened on the side next to the axis. Glumes membranous, keeled, unequal, the upper 3-veined, equalling the lemma. Lemma similar to upper glume, 3–5-veined. Palea thin, hyaline, completely enfolding the stamens and ovary, 2-veined, exceeding the lemma. Lodicules 0. Stamens 3. Ovary glabrous; styles very long, connate below. On tidal mud flats. The spp. are all protogynous.

About 17 spp., mainly in salt-marshes in temperate regions.

1 Glumes with sparse, very short hairs.   **4. alterniflora**
  Glumes rather densely hairy, with hairs up to 0.5 mm.   2
2 Anthers 8–13 mm; hairs of the ligule usually 2–3 mm.   **3. anglica**
  Anthers not more than 8 mm; hairs of the ligule not more than 2 mm.   3
3 Anthers indehiscent, without fertile pollen; lvs 4–12 mm wide.   **2. ×townsendii**

Anthers dehiscent, with fertile pollen; lvs not more
than 6 mm wide.　　　　　　　　　　**1. maritima**

1. S. maritima (Curtis) Fernald　　　　Small Cord-grass.
S. stricta (Aiton) Roth

A stout, erect perennial 15–70 cm, with short creeping
rhizome. Lvs ±erect, rather soft, often purplish, c. 4 mm
wide, falling short of the infl., tapering to a rather stout
point. Sheaths smooth. Hairs of the ligule 0.2–0.5 mm.
Infl. 4–10 cm, *of* (1–)2–3(–5) *erect spikes* arranged in
a raceme. Spikelets 10–14 mm, golden or yellow-brown,
distichous and alternate on the flattened *rhachis* which
is *prolonged up to* 14 mm *beyond the spikelets*. Glumes
keeled, rather densely hairy, with hairs up to 0.5 mm;
lower narrowly lanceolate to almost subulate, shorter
than the lemma, acuminate; upper 3-veined, longer than
the palea, nearly enfolding the floret, subacute, margins
hyaline. Lemma similar to upper glume but shorter.
Palea hyaline, longer than lemma, acute or subacute,
glabrous. Anthers 4–6 mm. Fl. 8–9. 2*n* = 60*. Hel.
　　Native. On tidal mud-flats. S. Devon to the Wash;
Wexford; local. S. and W. Europe from the Netherlands
to Yugoslavia; N. Africa.

2. S. × townsendii H. & J. Groves
　　　　　　　　　　　　Townsend's Cord-grass.

S. alterniflora × maritima

Like **1** but stems 30–130 cm, more robust; lvs 4–12 mm
wide, usually yellowish-green; hairs of the ligule
1–2 mm; infl. 12–35 cm, with usually 2–4 spikes; rhachis
prolonged up to 40 mm beyond the spikelets; spikelets
12–18 mm; upper glume 3–6-veined; anthers 5–8 mm,
indehiscent, without fertile pollen. Fl. 6–8. 2*n* = 62. Hel.
　　Native. Abundant on tidal mud-flats in S. England
and frequently planted elsewhere as a mud-binder.
Introduced in many temperate parts of the world, such
as W. Europe, North and South America, Australia
and New Zealand.

3. S. anglica C. E. Hubbard　　　Common Cord-grass.

Like **1** but stems 30–130 cm, more robust; lvs 6–15 mm
wide, yellowish-green; hairs of the ligule 2–3 mm; infl.
12–35 cm, with 3–6(–12) spikes; rhachis prolonged
up to 50 mm beyond the spikelets; spikelets
(14–)16–21 mm, anthers 8–13 mm, dehiscent, with fertile pollen. Fl. 6–8. 2*n* = 120, 122, 124. Hel.
　　Native. An amphidiploid presumed to have been derived from **2** and growing in similar situations. Distribution probably the same, but this sp. and the hybrid are
often not distinguished.

*4. S. alterniflora Loisel.　　　　Smooth Cord-grass.

Laxly caespitose, with robust stems 40–100 cm. Lvs
5–10 mm wide, green; hairs of the ligule c. 1 mm. Infl.
10–25 cm, with (3–)5–13 rather distant spikes; rhachis
of spikes prolonged up to 27 mm beyond the spikelets.
Spikelets 10–15 mm. Lower glume linear, the upper
ovate-lanceolate, 3–5-veined, with sparse very short

hairs. Lemma ovate-lanceolate, glabrous. Anthers
5–6 mm. Fl. 6–8. 2*n* = 62. Hel.
　　Introduced. Naturalized on mud-flats in Southampton
Water. Native of North America.

Tribe 18. ORYZEAE. Lvs linear; silica-bodies dumbbell-shaped, transversely arranged; slender 2-celled
microhairs present; ligule chartaceous. Infl. a panicle.
Spikelets laterally compressed, with 1 floret. Glumes
absent. Lemma herbaceous, unawned. Lodicules 2,
entire or 2-lobed. Stamens 1–6. Starch-grains compound. Chromosomes small; basic number 12.

55. LEERSIA Swartz

Perennial herbs. Spikelets in a spreading panicle, each
of 1 floret, strongly compressed, apparently falling
entire. Glumes 0. Lemma tough, 5-veined. Palea tough,
3-veined, equalling lemma. Lodicules 2; stamens 1–6.
Ovary glabrous; styles very short. Fr. strongly compressed.
　　About 15 spp. in tropical and warmer temperate
regions.

1. L. oryzoides (L.) Swartz　　　　　　Cut-grass.
Oryza oryzoides (L.) Brand

An erect perennial, 30–100 cm. Stock creeping. Culms
smooth, nodes shortly bearded. Lvs flat, scabrid,
4–6 mm wide, acuminate, abruptly contracted at base.
Sheaths smooth, more or less inflated, the upper usually
enclosing the panicle. Ligule c. 1 mm, truncate. Panicle
5–15 cm, branched; branches slender, smooth and flexuous. Spikelets 5–6 mm. Lemma half-ovate, apiculate,
ciliate on the keel, margins thickened and ±adnate to
those of the palea. Palea subulate, usually ciliate on
the keel. Anthers 0.5 mm. Fl. 8–10. Chasmogamous and
with anthers c. 3 mm or cleistogamous and with panicle
partially or wholly enclosed in uppermost lf-sheath,
depending on climatic conditions. 2*n* = 48. Hel.
　　Native. In wet meadows and beside rivers and ditches.
From Somerset and Dorset to Surrey and W. Sussex.
S. and C. Europe, north to S. Sweden and Finland;
temperate Asia; North America.

Tribe 19. PANICEAE. Lvs linear to lanceolate or ovate;
silica-bodies nodular, dumbbell- or cross-shaped;
slender 2-celled microhairs present; ligule absent or
membranous or a ring of hairs. Infl. a panicle or of
variously arranged racemes or spikes. Spikelets with 2
florets, ±compressed dorsally and falling entire, the
upper floret hermaphrodite, the lower male or sterile.
Glumes and lower lemma similar in texture, the upper
lemma firmer. Lodicules usually 2. Stamens 3. Starch-
grains simple. Chromosomes small; basic number 9, 10,
15, 17, 19, rarely 7.

Two spp. of **Panicum** L., stout annuals whose panicles have
numerous long branches, occur locally as casuals.

*P. miliaceum L., up to 1 m with ascending rigid panicle-
branches, often nodding at the tip, and mostly short-pedicellate
spikelets 4–5.5 mm, is found on rubbish-dumps, etc. Widely
cultivated.

*P. capillare L., usually rather smaller, with widely spreading slender panicle-branches and mostly long-pedicellate spikelets 2–3 mm, sometimes occurs on rubbish dumps and abundantly as a weed in carrot fields in East Anglia. North America.

56. ECHINOCHLOA Beauv.

Ligule 0. Infl. of unilateral racemes arranged along a central axis; rhachis triquetrous. Spikelets plano-convex, ±awned, shortly pedicellate *in clusters* along one side of the branches. Glumes unequal, membranous, 3-veined. *Sterile lemma* membranous, 5-veined, ±*awned.* Fertile lemma coriaceous. Palea similar. Lodicules 2, fleshy. Stamens 3.

About 30 spp., in tropical and warm temperate regions.

*1. E. crus-galli (L.) Beauv.   Cockspur.

Panicum crus-galli L.

An erect glabrous annual 25–100 cm. Lvs smooth, margins thickened and finely serrate. Sheaths smooth. Ligule 0. Infl. 6–25 cm, rhachis flattened, ciliate, pedicels pubescent. Spikelets 3–4 mm, ovate. Lower glume hispid, embracing the base of the spikelet and about half its length; upper as long as the lemmas, ±awned, hispid above. Sterile lemma hispid above. Fertile lemma white or straw-coloured, shiny, ovate, acuminate. Fl. 8–9. $2n = 36, 54$. Th.

Introduced. A frequent casual in cultivated ground and waste places, local. Warmer regions.

57. DIGITARIA Haller

Annual or perennial herbs. *Infl. of shortly racemose or subdigitate spikes. Spikelets* small, dorsally compressed, falling entire, *arranged in 2 rows along one side of the flattened rhachis, usually in pairs*, one on a longer pedicel than the other, each of 2 florets, lower floret sterile and represented by a lemma, upper hermaphrodite. Glumes thin, membranous, very unequal, lower scale-like and minute or suppressed, upper ovate or lanceolate, 3-veined. Sterile lemma membranous, as long as the fertile lemma, distinctly 5-leaved. Fertile lemma coriaceous, very obscurely 3-veined, shiny and minutely ribbed, enfolding the palea. Palea similar to fertile lemma. Lodicules 2, fleshy, truncate. Stamens 3. Ovary glabrous.

About 380 spp., in the warmer regions.

Lvs and sheaths all glabrous; upper glume as long as lemmas.   **1. ischaemum**
At least lower lvs and sheaths hairy; upper glume up to half length of lemmas.   **2. sanguinalis**

1. D. ischaemum (Schreber) Muhl.   Red Millet.

D. humifusa Pers.; *Panicum glabrum* Gaudin

A procumbent tufted annual 10–25 cm. Culms slender, branched. *Lvs glabrous*, short *without a white margin.* Sheaths smooth. Ligule short, truncate. Infl. of usually 3, sometimes more, shortly racemose or subdigitate spreading racemes. *Spikelets* 1.8–2 mm, in groups of 3, plano-convex, ovate, acute. *Lower glume minute or 0,*

upper as long as lemmas, prominently 3-veined, ovate, acute, *downy.* Lower lemma similar to upper glume, 5-veined. Fl. 8. $2n = 36$. Th.

Native? In sandy fields. Possibly native in a few localities in S. England, introduced elsewhere; rare. Warm and warm temperate regions.

*2. D. sanguinalis (L.) Scop.   Crab-grass.

Panicum sanguinale L.

An erect or ascending annual 20–60 cm. Culms ±branched, nodes often bearded. *Lvs without white margin*, flat, at least the lower *hirsute. Lower sheaths hirsute* with many of the hairs arising from tubercles, upper glabrous, or hairy. Ligule short, truncate. Infl. of 2–16, usually 4, spreading racemes 3–20 cm, all arising at the same point or in 2 distinct whorls. *Spikelets* 2.5–3.5 mm, in groups of 2, strongly compressed, lanceolate, acute. *Lower glume scale-like, upper linear-lanceolate, up to ½ as long as the lemma*, 3-veined, *glabrous*, margins fringed. Lower lemma lanceolate, acute, 5-veined, with very short hairs on the veins. Fl. 8–9. $2n = 36$. Th.

Introduced. A casual; in waste places, near ports and in arable fields, rare. Warm and warm temperate regions.

58. SETARIA Beauv.   Bristle-grass.

Annual or perennial herbs. Lvs flat, sheaths smooth, ligule 0. *Panicle spike-like and cylindrical. Spikelets with* 1–*several long bristles from their pedicels*, plano-convex, of 2 florets, lower floret sterile, upper hermaphrodite. Glumes very unequal, membranous, lower 3-, upper 5-veined. Lower lemma membranous, 5-veined. Upper lemma coriaceous, minutely or coarsely rugose, margins folded round palea. Palea similar to upper lemma. Lodicules, stamens and ovary as in *Digitaria*.

About 140 spp. in the warmer regions.

1 Upper glume ½–⅔ as long as the upper lemma; upper lemma conspicuously rugose.   **3. pumila**
  Upper glume about as long as upper lemma; upper lemma minutely rugose.   *2*
2 Teeth of bristles forward-pointing.   **1. viridis**
  Teeth of bristles backward-pointing.   **2. verticillata**

*1. S. viridis (L.) Beauv.   Green Bristle-grass.

Panicum viride L.

An erect, ±tufted annual 15–100 cm. Lvs scabrid, margins thickened. Sheaths smooth, margins ±ciliate. Panicle 1–10 cm. Spikelets 1.8–2.2 mm, ovate, obtuse, shortly pedicellate and subtended by 1–3 flexuous *bristles* 5–10 mm. Bristles with *forward-pointing* teeth. Lower glume ⅓ length of spikelet, triangular, acute; upper equalling lemma, ovate, acute. Lower lemma ovate, obtuse. *Upper lemma* coriaceous, ovate, obtuse, *minutely rugose.* Fl. 7–8. $2n = 18$. Th.

Introduced. A casual, in waste places and near ports, uncommon. Widely distributed in warm temperate and tropical countries.

*S. italica** (L.) Beauv., Millet, differing from *S. viridis* in its large, lobed or interrupted panicle and in the persistent glumes and lower lemma, occurs as a casual. Widely cultivated.

*2. S. verticillata (L.) Beauv.

Panicum verticillatum L.

Similar in general appearance to *S. viridis*. Panicle usually narrower. *Bristles usually solitary, their teeth backward-pointing. Upper lemma minutely rugose.* Fl. 7–8. $2n = 18, 36$. Th.

   Introduced. A casual, in waste places. Warmer regions.

*3. S. pumila (Poiret) Schultes   Yellow Bristle-grass.

S. glauca auct., *S. lutescens* F. T. Hubbard

Similar in general appearance to *S. viridis*. Stems less tufted, often solitary. Sheaths bearded at the mouth. Lvs with occasional, long scattered hairs. Sheaths glabrous. Panicle 1–15 cm. Bristles, 3–8 mm, usually 4–12, yellowish, teeth forward-pointing. Spikelets 2.7–3.3 mm. Upper glume up to ⅔ as long as lemma. *Upper lemma with conspicuous transverse ridges.* Fl. 7–8. $2n = 36, 72$. Th.

   Introduced. A casual, in waste places. Warmer regions.

Bibliography

In addition to the publications referred to in the text, the following books have been freely consulted and many will complement the information provided in this *Flora*. The numerous invaluable county Floras and monographs of particular plant-groups now available are not cited.

Bean, W. J., *Trees and Shrubs Hardy in the British Isles*, Edn 8, 4 vols. London: John Murray. (1970–1980)

Dandy, J. E., *List of British Vascular Plants*. London: British Museum (Nat. Hist.). (1958)

Dony, J. G., Perring, F. H. & Rob, C. M., *English Names of Wild Flowers*. London: Botanical Society of the British Isles. (1980)

Federov, An. A. (ed.), *Chromosome Numbers of Flowering Plants*. Leningrad: Akademija Nauk, SSSR. (1969)

Howes, F. N., *A Dictionary of Useful and Everyday Plants and their Common Names*. Cambridge University Press. (1974)

Hyde, H. A., Wade, A. E. & Morrison, S. G., *Welsh Ferns*, Edn 6. Cardiff: National Museum of Wales. (1978)

Jalas, J. & Suominen, J. (eds), *Atlas Florae Europaeae*, vols 1–6. Helsinki: Committee for Mapping Flora of Europe. (1972–1983)

Jermy, A. C., Arnold, H. R., Farrell, L. & Perring, F. H., *Atlas of Ferns of the British Isles*. London: Botanical Society of the British Isles. (1978)

Kuijt, J., *The Biology of Parasitic Flowering Plants*. Berkeley: University of California Press. (1969)

Mitchell, A. F., *A Field Guide to the Trees of Britain and Northern Europe*. London: Collins. (1974)

Moore, D. M., *Flora Europaea Check-list and Chromosome Index*. Cambridge University Press. (1982)

Page, C. N., *The Ferns of Britain and Ireland*. Cambridge University Press. (1982)

Perring, F. H. & Walters, S. M., *Atlas of the British Flora*, Edn 2. Wakefield: E.P. Publishing Ltd. (1976)

Perring, F. H., *Critical Supplement to the Atlas of the British Flora*. Wakefield: E.P. Publishing Ltd. (1978)

Praeger, R. L., *The Botanist in Ireland*. Dublin: Hodges, Figgis & Co. (1934)

Roles, S. J., *Flora of the British Isles – Illustrations*, 4 vols. Cambridge University Press. (1957–1965).

Scannell, M. J. P. & Synnott, D. M., *Census Catalogue of the Flora of Ireland*. Dublin: Stationery Office. (1972)

Stace, C. A. (ed.), *Hybridization and the Flora of the British Isles*. London: Academic Press. (1975)

Tansley, A. G., *The British Isles and their Vegetation*. Cambridge University Press. (1939)

Tutin, T. G., Heywood, V. H., Burges, N. A., Moore, D. M., Valentine, D. H., Walters, S. M. & Webb, D. A. (eds), *Flora Europaea*, 5 vols. Cambridge University Press. (1964–1980)

Walters, S. M., Brady, A., Brickell, C. D., Cullen, J., Green, P. S., Lewis, J., Matthews, V. A., Webb, D. A., Yeo, P. F. & Alexander, J. C. M. (eds), *The European Garden Flora*, Vol. 2, part 2. Cambridge University Press. (1984)

Watsonia. (Journal of the Botanical Society of the British Isles.) Arbroath & London, 1949–

White, J. (ed.), *Studies on Irish Vegetation*. Dublin: Royal Dublin Society. (1982)

Willis, J. C., *A Dictionary of Flowering Plants and Ferns*, Edn 8 (by H. K. Airy Shaw). Cambridge University Press. (1973)

Authorship of families

The author responsible for the account of each family is given in the following list:

Acanthaceae	D.M.M.	Empetraceae	D.M.M.	Paeoniaceae	A.R.C.
Aceraceae	D.M.M.	Equisetaceae	D.M.M.	Papaveraceae	A.R.C.
Adiantaceae	D.M.M.	Ericaceae	D.M.M.	Parnassiaceae	D.M.M.
Adoxaceae	D.M.M.	Eriocaulaceae	T.G.T.	Phytolaccaceae	D.M.M.
Aizoaceae	T.G.T.	Escalloniaceae	D.M.M.	Pinaceae	D.M.M.
Alismataceae	T.G.T.	Euphorbiaceae	D.M.M.	Pittosporaceae	D.M.M.
Amaranthaceae	T.G.T.	Fagaceae	D.M.M.	Plantaginaceae	T.G.T.
Amaryllidaceae	D.M.M.	Frankeniaceae	T.G.T.	Platanaceae	D.M.M.
Apocynaceae	A.R.C.	Fumariaceae	D.M.M.	Plumbaginaceae	A.R.C.
Aponogetonaceae	T.G.T.	Gentianaceae	D.M.M.	Polemoniaceae	A.R.C.
Aquifoliaceae	D.M.M.	Geraniaceae	D.M.M.	Polygalaceae	T.G.T.
Araceae	T.G.T.	Gramineae	T.G.T.	Polygonaceae	T.G.T.
Araliaceae	D.M.M.	Grossulariaceae	D.M.M.	Polypodiaceae	D.M.M.
Aristolochiaceae	D.M.M.	Haloragidaceae	A.R.C.	Pontederiaceae	D.M.M.
Aspidiaceae	D.M.M.	Hippocastanaceae	D.M.M.	Portulacaceae	A.R.C.
Aspleniaceae	D.M.M.	Hippuridaceae	A.R.C.	Potamogetonaceae	A.R.C.
Athyriaceae	D.M.M.	Hydrangeaceae	D.M.M.	Primulaceae	T.G.T.
Azollaceae	D.M.M.	Hydrocharitaceae	T.G.T.	Pyrolaceae	D.M.M.
Balsaminaceae	D.M.M.	Hymenophyllaceae	D.M.M.	Ranunculaceae	A.R.C.
Berberidaceae	D.M.M.	Hypericaceae	T.G.T.	Resedaceae	A.R.C.
Betulaceae	D.M.M.	Hypolepidaceae	D.M.M.	Rhamnaceae	T.G.T.
Blechnaceae	D.M.M.	Iridaceae	T.G.T.	Rosaceae	D.M.M.
Boraginaceae	T.G.T.	Isoetaceae	T.G.T.	Rubiaceae	A.R.C.
Buddlejaceae	D.M.M.	Juglandaceae	D.M.M.	Ruppiaceae	T.G.T.
Butomaceae	T.G.T.	Juncaceae	P.W.Richards	Salicaceae	D.M.M.
Buxaceae	D.M.M.	Juncaginaceae	T.G.T.	Santalaceae	T.G.T.
Callitrichaceae	A.R.C.	Labiatae	D.M.M.	Sarraceniaceae	T.G.T.
Campanulaceae	T.G.T.	Leguminosae	T.G.T.	Saxifragaceae	D.M.M.
Cannabaceae	T.G.T.	Lemnaceae	T.G.T.	Scheuchzeriaceae	T.G.T.
Caprifoliaceae	D.M.M.	Lentibulariaceae	D.M.M.	Scrophulariaceae	D.M.M.
Caryophyllaceae	A.R.C.	Liliaceae	D.M.M.	Selaginellaceae	D.M.M.
Celastraceae	D.M.M.	Linaceae	D.M.M.	Simaroubaceae	D.M.M.
Ceratophyllaceae	T.G.T.	Loranthaceae	T.G.T.	Solanaceae	T.G.T.
Chenopodiaceae	T.G.T.	Lycopodiaceae	D.M.M.	Sparganiaceae	T.G.T.
Cistaceae	D.M.M.	Lythraceae	T.G.T.	Staphyleaceae	D.M.M.
Compositae	A.R.C.	Malvaceae	A.R.C.	Tamaricaceae	D.M.M.
Convolvulaceae	T.G.T.	Marsiliaceae	D.M.M.	Taxaceae	D.M.M.
Cornaceae	D.M.M.	Menyanthaceae	D.M.M.	Thelypteridaceae	D.M.M.
Corylaceae	D.M.M.	Monotropaceae	D.M.M.	Thymelaeaceae	D.M.M.
Crassulaceae	D.M.M.	Moraceae	D.M.M.	Tiliaceae	A.R.C.
Cruciferae	A.R.C.	Myricaceae	D.M.M.	Typhaceae	T.G.T.
Cucurbitaceae	A.R.C.	Najadaceae	T.G.T.	Ulmaceae	T.G.T.
Cupressaceae	D.M.M.	Nymphaeaceae	A.R.C.	Umbelliferae	T.G.T.
Cyperaceae	T.G.T.	Oleaceae	T.G.T.	Urticaceae	T.G.T.
Diapensiaceae	D.M.M.	Onagraceae	A.R.C.	Valerianaceae	T.G.T.
Dioscoreaceae	A.R.C.	Ophioglossaceae	D.M.M.	Verbenaceae	A.R.C.
Dipsacaceae	A.R.C.	Orchidaceae	A.R.C.	Violaceae	D.M.M.
Droseraceae	T.G.T.	Orobanchaceae	A.R.C.	Vitaceae	T.G.T.
Elaeagnaceae	D.M.M.	Osmundaceae	D.M.M.	Zannichelliaceae	A.R.C.
Elatinaceae	T.G.T.	Oxalidaceae	D.M.M.	Zosteraceae	T.G.T.

Accounts of families for which Professor D. M. Moore is responsible are based on those written by the late Dr E. F. Warburg in earlier editions.

Note on life forms

The Danish botanist Raunkiaer has classified plants, according to the position of the resting buds or persistent stem apices in relation to soil level, into a number of Life Forms, which have in turn been subdivided so as to convey more information about the plants included in the different groups. Life Forms are a convenient method of indicating how a plant passes the unfavourable season, and are also of interest because they show a correlation with climate.

The primary classes are:

1. Phanerophytes – woody plants with buds more than 25 cm above soil level.
2. Chamaephytes – woody or herbaceous plants with buds above the soil surface but below 25 cm.
3. Hemicryptophytes – herbs (very rarely woody plants) with buds at soil level.
4. Geophytes – herbs with buds below the soil surface.
5. Helophytes – marsh plants.
6. Hydrophytes – water plants.
7. Therophytes – plants which pass the unfavourable season as seeds.

The subdivisions and abbreviations used in this book are as follows:

Phanerophytes

MM. Mega- and mesophanerophytes, from 8 m upwards.
M. Microphanerophytes, 2–8 m.
N. Nanophanerophytes, 25 cm–2 m.

Chamaephytes (Ch.)

Chw. Woody chamaephytes.
Chh. Herbaceous chamaephytes.
Chc. Cushion plants.

Hemicryptophytes (H.)

Hp. Protohemicryptophytes, with uniformly lfy stems, but the basal lvs usually smaller than the rest.
Hs. Semi-rosette hemicryptophytes, with lfy stems but the lower lvs larger than the upper ones and the basal internodes shortened.
Hr. Rosette hemicryptophytes, with lfless flowering stems and a basal rosette of lvs.

Geophytes (G.)

Gb. Geophytes with bulbs.
Gr. Geophytes with buds on roots.
Grh. Geophytes with rhizomes.
Grt. Geophytes with root tubers.
Gt. Geophytes with stem tubers or corms.

Helophytes (Hel.)

Not subdivided.

Hydrophytes (Hyd.)

Not subdivided.

Therophytes (Th.)

Not subdivided.

Glossary

---

accrescent   Becoming larger after flowering (usually applied to the calyx).

achene   A small dry indehiscent single-seeded fr.

acicle   See p. 207.

actinomorphic   Radially symmetrical; having more than one plane of symmetry.

acuminate   Fig. 80N.   acumen, the point of such a lf.

acute   Fig. 80D.

adnate   Joined to another organ of a different kind.

alien   Believed on good evidence to have been introduced by man and now ±naturalized.

allopolyploid   A polyploid derived by hybridization between two different spp. with doubling of the chromosome number.

alternate   Strictly, arranged in 2 rows but not opposite; commonly used (as often here) also to include spiral arrangement.

amphimixis   Reproducing by seed resulting from a normal sexual fusion; adj. amphimictic.

amplexicaul   Clasping the stem.

anastomosing   Joining up to form loops.

anatropous (ovule)   Bent over against the stalk.

andromomoecious   Having male and hermaphrodite fls on the same plant.

anemophilous   Wind-pollinated.

angustiseptate   See p. 65.

annual   Completing its life-cycle within 12 months from germination (cf. biennial).

annular   Ring-shaped.

annulus   Special thick-walled cells forming part of the opening mechanism of a fern sporangium, often forming a ring.

antesepalous   Inserted opposite the insertion of the sepals.

anther   The part of the stamen containing the pollen grains (Fig. 81M).

antheridium   The structure containing the male sexual cells.

apiculate   With a small broad point at the apex.

apocarpous (ovary)   Having the carpels free from one another.

apomixis   Reproducing by seed not formed from a sexual fusion; adj. apomictic.

appressed   Pressed close to another organ but not united with it.

arachnoid   Appearing as if covered with cobwebs.

archegonium   The structure containing the female sexual cell in many land plants.

arcuate   Curved so as to form about $\frac{1}{4}$ of a circle or more.

aristate   Awned (Fig. 81c).

ascending   Sloping or curving upwards.

asperous   Rough to the touch.

attenuate   Gradually tapering.

auricles   Small ear-like projections at base of lf (especially in grasses).

autopolyploid   A polyploid derived from one diploid sp. by multiplication of its chromosome sets.

autotrophic   Neither parasitic nor saprophytic.

awn   A stiff bristle-like projection from the tip or back of the lemma in grasses, or from a fr. (usually the indurated style, e.g. Erodium), or, less frequently, the tip of a lf (Fig. 81c).

axile   See placentation.

axillary   Arising in the axil of a lf or bract.

base-rich   Soils containing a relatively large amount of free basic ions, e.g. calcium, magnesium, etc.

basic number   See chromosomes.

basifixed (of anthers)   Joined by the base to the filament and not capable of independent movement.

berry   A fleshy fr., usually several-seeded, without a stony layer surrounding the seeds.

biennial   Completing its life-cycle within two years (but not within one year), not flowering in the first.

bifid   Split deeply in two.

bog ('moss')   A community on wet very acid peat.

bract   Fig. 81P.

bracteole   Fig. 81P.

bulb   An underground organ consisting of a short stem bearing a number of swollen fleshy lf-bases or scale lvs, with or without a tunic, the whole enclosing the next year's bud.

bulbil   A small bulb or tuber arising in the axil of a lf or in an infl. on the aerial part of a plant.

caducous   Falling off at an early stage.

caespitose   Tufted.

calcicole   More frequently found upon or confined to soils containing free calcium carbonate.

calcifuge   Not normally found on soils containing free calcium carbonate.

calyx   The sepals as a whole (Fig. 82A).

campanulate   Bell-shaped.

campylotropous (ovule)   Bent so that the stalk appears to be attached to the side midway between the micropyle and chalaza.

capillary   Hair-like.

capitate   Head-like.

capsule   A dry dehiscent fr. composed of more than one carpel.

carpel   One of the units of which the gynoecium is composed. In a septate ovary the number of divisions usually corresponds to the number of carpels (Fig. 81L).

carpophore   See p. 277.

cartilaginous   Resembling cartilage in consistency.

caryopsis   A fr. (achene) with ovary wall and seed-coat united (Gramineae).

casual   An introduced plant which has not become established though it occurs in places where it is not cultivated.

cauline (of lvs)   Borne on the aerial part of the stem, especially the upper part, but not subtending a fl. or infl.

chartaceous   Of papery texture.

chasmogamous   Of fls which open normally (opposite of cleistogamous).

chlorophyll   The green colouring matter of lvs, etc.

chromosomes   Small deeply staining bodies, found in all nuclei, which determine most or all of the inheritable characters of organisms. Two similar sets of these are normally present in all vegetative cells, the number

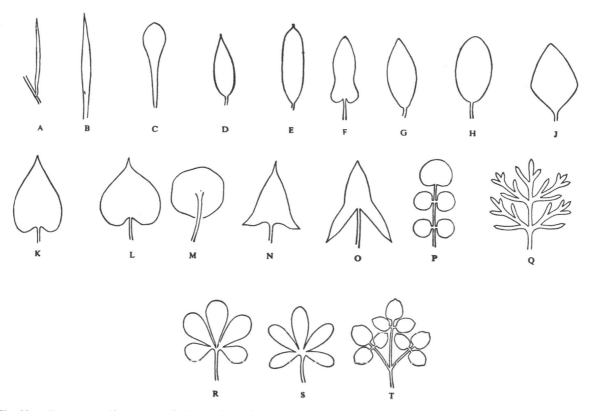

Fig. 80. A, linear; B, ensiform; C, spathulate; D, lanceolate, apex acute; E, oblong, apex mucronate; F, panduriform; G, elliptical; H, oval; J, rhombic, base cuneate; K, ovate, base cordate; L, suborbicular, apex cuspidate; M, peltate; N, triangular, hastate, apex acuminate; O, sagittate; P, simply pinnate, segments orbicular; Q, bipinnatifid; R, pedate, leaflets obovate; S, palmate; T, biternate.

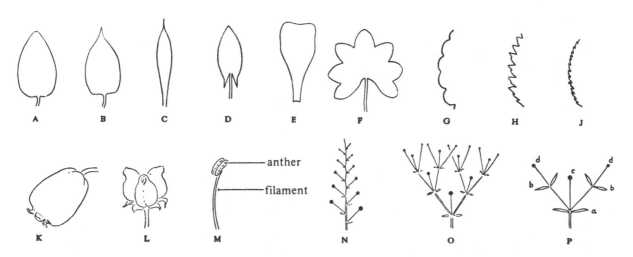

Fig. 81. A, leaf with rounded base and obtuse apex; B, leaf with truncate base and cuspidate apex; C, aristate apex; D, sagittate leaf with subacute apex; E, leaf with retuse apex; F, lobed margin; G, crenate margin; H, dentate margin; J, serrate margin; K, urceolate corolla; L, carpels; M, stamen; N, racemose inflorescence; O, cymose inflorescence; P, bracts and bracteoles: a, bracts of flower d and bracteoles of flower c; b, bracteoles of flower d.

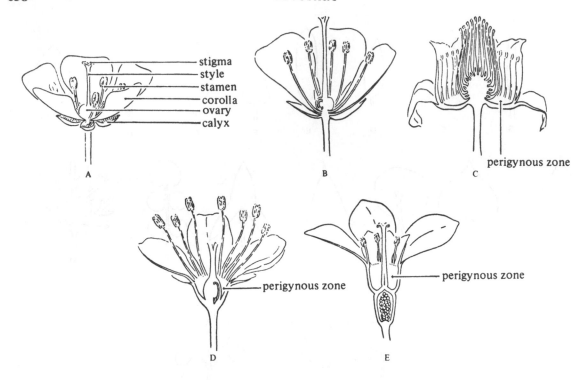

Fig. 82. A, parts of the flower; B, hypogynous flower (ovary superior); C, D, perigynous flowers (ovary superior); E, epigynous flower (ovary inferior).

(*diploid number*, 2*n*) usually being constant for a given sp. The sexual reproductive cells normally contain half this number (*haploid number*, *n*). Closely related spp. often have the same number, or a multiple of a common *basic number*.

ciliate With regularly arranged hairs projecting from the margin.

cilium, cilia A small whip-like structure by means of which some sexual reproductive cells swim.

circumscissile Dehiscing transversely, the top of the capsule coming off as a lid.

cladode A green ±lf-like lateral shoot.

clavate Club-shaped.

claw The narrow lower part of some petals.

cleistogamous Of fls which never open and are self-pollinated (opposite of chasmogamous).

commissure (in Umbelliferae) The faces by which the two carpels are joined together.

compound Of an infl., with the axis branched; of a lf, made up of several distinct lflets.

compressed Flattened.

concolorous Of approximately the same colour throughout.

connate Organs of the same kind growing together and becoming joined, though distinct in origin.

connivent Of two or more organs with their bases wide apart but their apices approaching one another.

contiguous Touching each other at the edges.

contorted (of perianth lobes in bud) With each lobe overlapping the next with the same edge and appearing twisted.

converging, convergent Of two or more organs with their apices closer together than their bases.

convolute Rolled together, coiled.

cordate Fig. 80K.

coriaceous Of a leathery texture.

corm A short, usually erect and tunicated, swollen underground stem of one year's duration, that of the next year arising at the top of the old one and close to it; cf. **tuber**.

corolla The petals as a whole (Fig. 82A).

corymb A raceme with the pedicels becoming shorter towards the top, so that all the fls are at approximately the same level; adj. *corymbose*.

corymbose cyme A flat-topped cyme, thus resembling a corymb in appearance though not in development.

cotyledon The first lf or lvs of a plant, already present in the seed and usually differing in shape from the later lvs. Cotyledons may remain within the testa or may be raised above the ground and become green during germination.

crenate Fig. 81G; dimin. *crenulate*.

crisped Curled.

cultivated ground (or **land**) Includes arable land, gardens, allotments, etc.

cuneate Fig. 80J.

cuneiform Wedge-shaped with the thin end at the base.

cuspidate Fig. 80L, 81B.

cymose Of an infl., usually obconical in outline, whose growing points are each in turn terminated by a fl., so that the continued growth of the infl. depends on the production of new lateral growing points. A consequence of this mode of growth is that the oldest branches or fls are normally at the apex (Fig. 81o).

deciduous Losing its lvs in autumn; dropping off.

decumbent (of stems) Lying on the ground and tending to rise at the end.

decurrent Having the base prolonged down the axis, as in lvs where the blade is continued downwards as a wing on petiole or stem.

decussate (of lvs) Opposite, but successive pairs orientated at right angles to each other.

deflexed Bent sharply downwards.

dehiscent Opening to shed its seeds or spores.

deltate Shaped like the Greek letter Δ.

dentate Fig. 81H; dimin. *denticulate*.

diadelphous See p. 183.

dichasium A cyme in which the branches are opposite and approximately equal; adj. *dichasial*.

didymous Formed of two similar parts attached to each other by a small portion of their surface.

digitate See **palmate**.

dioecious Having the sexes on different plants.

diploid A plant having 2 sets of chromosomes in its nuclei; similarly tetraploid, etc., plants having 4, etc., sets of chromosomes in their nuclei. See also **chromosomes**.

disk The fleshy, sometimes nectar-secreting, portion of the receptacle, surrounding or surmounting the ovary.

distichous Arranged in two diametrically opposite rows.

divaricate Diverging at a wide angle.

diverging, divergent Of two or more organs with their apices wider apart than their bases.

dominant The chief constituents of a particular plant community, e.g. oaks in an oakwood or heather on a moor.

dorsifixed (of anthers) Attached by the back.

drupe A ±fleshy fr. with one or more seeds each surrounded by a stony layer, e.g. sloe, ivy.

ebracteate Devoid of bracts.

ectotrophic See **mycorrhiza**.

effuse Spreading widely.

eglandular Without glands.

ellipsoid Of a solid object elliptic in longitudinal section.

elliptical Fig. 80G.

emarginate Shallowly notched at the apex.

endemic Native only in one country or other small area. If used without qualification in this book means confined to the British Is.

endosperm Nutritive tissue surrounding the embryo in the seed of a flowering plant, formed after fertilization.

endotrophic See **mycorrhiza**.

ensiform Fig. 80B.

entire Not toothed or cut.

entomophilous Insect-pollinated.

epicalyx A calyx-like structure outside but close to the true calyx.

epigeal Above ground; epigeal germination – when cotyledons are raised above the ground.

epigynous Of the whole fl., or the perianth and stamens of a fl., with an inferior ovary (Fig. 82E).

epipetalous Inserted upon the corolla.

equitant Of distichous lvs folded longitudinally and overlapping in their lower parts.

erose Appearing as if gnawed.

escape A plant growing outside a garden, but not well naturalized, derived from cultivated specimens either by vegetative spread or by seed.

exserted Protruding.

exstipulate Without stipules.

extravaginal Of branches which break through the sheaths of the subtending leaves.

extrorse (of anthers) Opening towards the outside of the fl.

falcate Sickle-shaped.

fen A community on alkaline, neutral, or slightly acid wet peat.

fertile Producing seed capable of germination; or (of anthers) containing viable pollen.

filament The stalk of the anther, the two together forming the stamen (Fig. 81M).

filiform Thread-like.

fimbriate With the margin divided into a fringe.

fistular Hollow and cylindrical; tube-like.

flexuous Wavy (of a stem or other axis).

floccose See p. 499.

flush Wet ground, often on hillsides, where water flows but not in a definite channel.

follicle A dry dehiscent fr. formed of one carpel, dehiscing along one side.

fruit (fr.) The ripe seeds and structure surrounding them, whether fleshy or dry; strictly the ovary and seeds, but often used to include other associated parts such as the fleshy receptacle, as in rose and strawberry.

fugacious Withering or falling off very rapidly.

funicle The stalk of the ovule.

fusiform Spindle-shaped.

gamopetalous Having the petals joined into a tube, at least at the base.

geniculate Bent abruptly to make a 'knee'.

gibbous With a rounded swelling on one side, as the base of the calyx in *Acinos*.

glabrescent Becoming glabrous.

glabrous Without hairs.

gland A small globose or oblong vesicle containing oil, resin or other liquid, sunk in, on the surface of, or protruding from any part of a plant. When furnished with a slender stalk usually known as a glandular hair.

glandular Furnished with glands.

glaucous Bluish.

glumaceous Resembling a glume.

glume See p. 613.

gynobasic A style which, because of the infolding of the ovary wall, appears to be inserted at the base of the ovary.

gynodioecious Having female and hermaphrodite fls on separate plants.

gynoecium The female part of the fl., made up of one or more ovaries with their styles and stigmas.

gynomonoecious Having female and hermaphrodite fls on the same plant.

hastate Fig. 80N.

heath A lowland community dominated by heaths or ling, usually on sandy soils with a shallow layer of peat.

hemiparasite See under **parasite**.

herb Any vascular plant which is not woody.

herbaceous (of a plant organ) Soft, green, having the texture of lvs.

hermaphrodite Containing both functional stamens and ovary.

heterochlamydeous Having the per.-segs in two distinct series which differ from one another.

heterosporous Having spores of two distinct sizes.

heterostylous The length of the style in relation to the other parts of the fl. differing in the fls of different plants.

hexaploid See under **diploid**.

hilum   The scar left on the seed by the stalk of the ovule.

hirsute   Clothed with long, not very stiff, hairs.

hispid   Coarsely and stiffly hairy, as many Boraginaceae.

homochlamydeous, homoiochlamydeous   Having all the per.-segs similar.

homogamous   The anthers and stigmas maturing simultaneously.

homosporous   Having all spores of approximately the same size.

hyaline   Thin and translucent.

hybrid   A plant originating by the fertilization of one sp. by another sp.

hybrid swarm   A series of plants originating by hybridization between two (or more) spp. and subsequently recrossing with the parents and between themselves, so that a continuous series of forms arises.

hypogeal   Below ground; hypogeal germination – when the cotyledons remain below the ground.

hypogynous   Of fls in which the stamens are inserted close to and beside or beneath the base of the ovary (Fig. 82B).

imbricate, imbricating   Of organs with their edges overlapping, when in bud, like the tiles on a roof.

imparipinnate   Of a pinnate lf with a terminal unpaired lflet.

impressed   Sunk below the surface.

incurved   Bent gradually inwards.

indehiscent   Not opening to release its seeds or spores.

indumentum   The hairy covering as a whole.

indurated   Hardened and toughened.

indusium   A piece of tissue ±covering or enclosing a sporangium or group of sporangia.

inferior (ovary)   With perianth inserted round the top, the ovary being apparently sunk in and fused with the receptacle (Fig. 82E).

inflexed   Bent inwards.

inflorescence (infl.)   Flowering branch, or portion of the stem above the last stem lvs, including its branches, bracts and fls.

intercalary lf   See p. 383.

internode   The length of stem between two adjacent nodes.

interpetiolar   Between the petioles.

interrupted   Not continuous.

intrapetiolar   Between the petiole and stem.

intravaginal   Within the sheath.

introduced   Not native; known to have been, or strongly suspected of having been, brought into the British Is. accidentally or intentionally by man within historic times.

introrse (of anthers)   Opening towards the middle of the fl.

involucel   See p. 438.

involucral   Forming an involucre.

involucre   Bracts forming a ±calyx-like structure round or just below the base of a usually condensed infl. (e.g. *Anthyllis*, Compositae); adj. *involucrate*.

involute   With the margins rolled upwards.

isomerous   The number of parts in two or more different floral whorls being the same, e.g.: 5 petals, 5 stamens, and 5 carpels.

jaculator   See p. 401.

keel   A sharp edge resembling the keel of a boat; the lower petal or petals when shaped like the keel of a boat (*Fumaria*, Leguminoseae).

lacerate   Deeply and irregularly divided and appearing as if torn.

laciniate   Deeply and irregularly divided into narrow segments.

lamina   The blade of a lf or petal; a thin flat piece of tissue.

lanceolate   Fig. 80D.

lanuginose   With woolly indumentum.

latex   Milky juice.

latiseptate   See p. 65.

lax   Loose; not dense.

lemma   See p. 613.

lenticular   Convex on both faces and ±circular in outline.

ligulate   Strap-shaped.

ligule   A small flap of tissue or a scale borne on the surface of a lf or per.-seg. near its base; see also p. 613 (Compositae).

limb   The flattened expanded part of a calyx or corolla the base of which is tubular.

linear   Fig. 80A.

lingulate   Tongue-shaped.

lip   A group of per.-segs ±united and sharply divided from the remaining per.-segs.

lobed (of lvs)   Divided, but not into separate lflets (Fig. 81F).

loculicidal   Splitting down the middle of each cell of the ovary.

lodicule   See p. 613.

lyrate   Shaped ±like a lyre.

marsh   A community on wet or periodically wet but not peaty soils.

meadow   A grassy field cut for hay.

measurements   See Abbreviations (p. xxix).

megaspore   A large spore giving rise to a prothallus bearing archegonia.

membranous   Thin, dry and flexible, not green.

mericarp   A 1-seeded portion split off from a syncarpous ovary at maturity.

-merous   E.g. in 5-merous (=pentamerous); having the parts in fives.

microspore   A small spore giving rise to a prothallus bearing antheridia.

moder   A top layer of partly decomposed plant remains with associated mineral particles, looser, less acid and less nutrient-poor than mor.

monadelphous (of stamens)   United into a single bundle by the fusion of the filaments.

monochasium   A cyme in which the branches are spirally arranged or alternate or one is more strongly developed than the other; adj. *monochasial*.

monochlamydeous   Having only one series of per.-segs.

monoecious   Having unisexual fls, but both sexes on the same plant.

monopodial   Of a stem in which growth is continued from year to year by the same apical growing point, cf. racemose.

moor   Upland communities, often dominated by heather, on dry or damp but not wet peat.

mor   The compact and tenacious layer of partly decomposed plant remains that overlies the mineral soil in a podsol. Also termed 'raw humus', and normally very acid and nutrient-poor.

mucronate   Provided with a short narrow point (*mucro*), Fig. 80E; dimin. *mucronulate*.

mull soil   A fertile woodland soil with no raw humus layer.

muricate   Rough with short firm projections.

muticous   Without an awn or mucro.

mycorrhiza   An association of roots with a fungus which may form a layer outside the root (ectotrophic) or within the outer tissues (endotrophic).

naked   Devoid of hair or scales.

native   Not known to have been introduced by human agency.

nerve   A strand of strengthening and conducting tissue running through a lf or modified lf.

node A point on the stem where one or more lvs arise.

nodule A small ±globose swelling; adj. *nodular*.

nucellus The tissue between embryo and integument in an ovule.

ob- (in combinations, e.g. obovate) Inverted; an obovate lf is broadest above the middle, an ovate one below the middle (Fig. 80R).

obdiplostemonous The stamens in 2 whorls, the outer opposite the petals, the inner opposite the sepals.

oblong Fig. 80E.

obtuse Blunt (Fig. 81A).

ochrea See p. 300.

octoploid See under **diploid**.

opposite Of two similar organs arising at the same level on opposite sides of the stem.

orbicular Rounded, with length and breadth about the same (Fig. 80P).

orthotropous (ovule) Straight and with the axis of the ovule in the same line as that of the funicle.

oval Fig. 80H.

ovary That part of the gynoecium enclosing the ovules (Fig. 82A).

ovate Fig. 80K.

ovoid Of a solid object which is ovate in longitudinal section; egg-shaped.

ovule A structure containing the egg and developing into the seed after fertilization.

palate See p. 371.

palea See p. 613.

palmate (of a lf) Consisting of more than 3 lflets arising from the same point (Fig. 80s).

panduriform Fig. 80F.

panicle Strictly a branched racemose infl., though often applied to any branched infl.

papillae Small elongated projections; adj. *papillose*.

parasite A plant which derives its food wholly or partially (hemiparasite) from other living plants to which it is attached.

paripinnate Of a pinnate lf with no terminal lflet.

partial infl. Any distinct portion of a branched infl.

pasture Grassy field grazed during summer.

pectinate Lobed, with the lobes resembling and arranged like the teeth of a comb.

pedate Fig. 80R.

pedicel The stalk of a single fl.

peduncle The stalk of an infl. or partial infl.

peltate Of a flat organ with its stalk inserted on the under surface, not at the edge (Fig. 80M).

pentaploid See under **diploid**.

perennating Surviving the winter after flowering.

perennial Living for more than 2 years and usually flowering each year.

perianth The floral lvs as a whole, including sepals and petals if both are present.

perianth-segment (per.-seg.) The separate lvs of which the perianth is made up, especially when petals and sepals cannot be distinguished.

perigynous Of fls in which there is an annular region, flat or concave, between the base of the gynoecium and the insertion of the other floral parts (Fig. 82C,D).

perigynous zone The annular region between the insertion of the gynoecium and of the other floral parts in perigynous or epigynous fls (Fig. 82C-E).

perisperm The nutritive tissue derived from the nucellus in some seeds.

perispore A membrane surrounding a spore.

petal A member of the inner series of per.-segs, if differing from the outer series, and especially if brightly coloured.

petaloid Brightly coloured and resembling petals.

petiole The stalk of a lf.

phyllode A green, flattened petiole resembling a leaf.

pilose Hairy with rather long soft hairs.

pinnate A lf composed of more than 3 lflets arranged in two rows along a common stalk or rhachis (Fig. 80P); bipinnate (2-pinnate), a lf in which the primary divisions are themselves pinnate. Similarly, 3-pinnate, etc.

pinnatifid Pinnately cut, but not into separate portions, the lobes connected by lamina as well as midrib or stalk (Fig. 80Q).

pinnatisect Like pinnatifid but with some of the lower divisions reaching very nearly or quite to the midrib.

placenta The part of the ovary to which the ovules are attached.

placentation The position of the placentae in the ovary. The chief types of placentation are: *apical*, at the apex of the ovary; *axile*, in the angles formed by the meeting of the septa in the middle of the ovary; *basal*, at the base of the ovary; *free-central*, on a column or projection arising from the base in the middle of the ovary, not connected with the wall by septa; *parietal*, on the wall of the ovary or on an intrusion from it; *superficial*, when the ovules are scattered uniformly all over the inner surface of the wall of the ovary.

pollen The microspores of a flowering plant or Conifer.

pollinia Regularly shaped masses of pollen formed by a large number of pollen grains cohering.

polygamous Having male, female and hermaphrodite fls on the same or different plants.

polyploid A plant having a chromosome number which is a multiple, greater than two, of the basic number of its group (see under **chromosomes**).

pome A fr. in which the seeds are surrounded by a tough but not woody or stony layer, derived from the inner part of the fr. wall, and the whole fused with the deeply cup-shaped fleshy receptacle (e.g. apple).

porrect Directed outwards and forwards.

premorse Ending abruptly and appearing as if bitten off at the lower end.

prickle A sharp relatively stout outgrowth from the outer layers. Prickles (unlike thorns) are usually irregularly arranged.

pricklet See p. 207.

procumbent Lying loosely along the surface of the ground.

prostrate Lying rather closely along the surface of the ground.

protandrous Stamens maturing before the ovary.

prothallus A small plant formed by the germination of a spore and bearing antheridia or archegonia or both.

protogynous Ovary maturing before the stamens.

pruinose Having a whitish 'bloom'; appearing as if covered with hoar frost.

pubescent Shortly and softly hairy; dimin. *puberulent*, *puberulous*.

punctate Dotted or shallowly pitted, often with glands.

punctiform Small and ±circular, resembling a dot.

pungent Sharply and stiffly pointed so as to prick.

raceme An unbranched racemose infl. in which the fls are borne on pedicels.

racemose Of an infl., usually conical in outline, whose growing points commonly continue to add to the infl. and in which there is usually no terminal fl. A consequence of this mode of growth is that the youngest and smallest branches or fls are normally nearest the apex (Fig. 81N).

radical (of lvs)   Arising from the base of the stem or from a rhizome.

raphe   The united portions of the funicle and outer integument in an anatropous ovule.

ray   The stalk of a partial umbel.

ray-floret   See p. 441.

receptacle   That flat, concave, or convex upper part of the stem from which the parts of the fl. arise; often used to include the perigynous zone.

recurved   Bent backwards in a curve.

regular   **Actinomorphic** (q.v.).

reniform   Kidney shaped.

replum   See p. 65.

resilient   Springing sharply back when bent out of position.

reticulate   Marked with a network, usually of veins.

retuse   Obtuse or truncate and slightly indented (Fig. 81E).

revolute   Rolled downwards.

rhachilla   See p. 613.

rhachis   The axis of a pinnate lf or an infl.

rhizome   An underground stem lasting more than one growing season; adj. *rhizomatous*.

rhombic   Having ±the shape of a diamond in a pack of playing cards (Fig. 80J).

rotate (of a corolla)   With the petals or lobes spreading out at right angles to the axis, like a wheel.

rounded (of lf-base)   Fig. 81A.

rugose   Wrinkled; dimin. *rugulose*.

ruminate   Looking as though chewed.

runcinate   Pinnately lobed with the lobes directed backwards, towards the base of the lf.

runner   A special form of stolon consisting of an aerial branch rooting at the end and forming a new plant which eventually becomes detached from the parent.

saccate   Pouched.

sagittate   Figs 80O, 81D.

salt-marsh   The series of communities growing on intertidal mud or sandy mud in sheltered places on coasts and in estuaries.

samara   A dry indehiscent fr. part of the wall of which forms a flattened wing.

saprophyte   A plant which derives its food wholly or partially (partial saprophyte) from dead organic matter.

scabrid   Rough to the touch; dimin. *scaberulous*.

scape   The flowering stem of a plant all the foliage lvs of which are radical; adj. *scapigerous*.

scarious   Thin, not green, rather stiff and dry.

schizocarp   A syncarpous ovary which splits up into separate 1-seeded portions (mericarps) when mature; adj. *schizocarpic*.

scrub (incl. thicket)   Any community dominated by shrubs.

secund   All directed towards one side.

seed   A reproductive unit formed from a fertilized ovule.

sepal   A member of the outer series of per.-segs, especially when green and ±lf-like.

sepaloid   Resembling sepals.

septicidal   Dehiscing along the septa of the ovary.

septum   A partition; adj. *septate*.

serrate   Toothed like a saw (Fig. 81J); dimin. *serrulate*.

sessile   Without a stalk.

setaceous   Shaped like a bristle, but not necessarily rigid.

shrub   A woody plant branching abundantly from the base and not reaching a very large size.

silicula   See p. 65.

siliqua   See p. 65.

simple   Not compound.

sinuate   Having a wavy outline.

sinus   The depression between two lobes or teeth.

sorus   A circumscribed group of sporangia.

spathulate   Paddle-shaped (Fig. 80C).

spermatozoid   A male reproductive cell capable of moving by means of cilia.

spike   A simple racemose infl. with sessile fls or spikelets; adj. *spicate*.

spikelet   See p. 613.

spine   A stiff straight sharp-pointed structure.

sporangiophore   A structure, not lf-like, bearing sporangia.

sporangium   A structure containing spores; plur. *sporangia*.

spore   A small asexual reproductive body, usually unicellular and always without tissue differentiation.

sporophyll   A lf-like structure, or one regarded as homologous with a lf, bearing sporangia.

spur   A hollow usually ±conical slender projection from the base of a per.-seg., often a petal, or of a corolla; adj. *spurred*.

stamen   One of the male reproductive organs of the plant (Figs 81M, 82A).

staminode   An infertile, often reduced, stamen.

standard   See p. 183.

stellate   Star-shaped.

sterile   Not producing seed capable of germination; or (of stamens) viable pollen.

stigma   The receptive surface of the gynoecium to which the pollen grains adhere (Fig. 82A).

stipel   A structure similar to a stipule but at the base of the lflets of a compound lf.

stipitate   Having a short stalk or stalk-like base.

stipule   A scale-like or lf-like appendage usually at the base of the petiole, sometimes adnate to it.

stolon   A creeping stem of short duration produced by a plant which has a central rosette or erect stem; when used without qualification is above ground; adj. *stoloniferous*.

stoma   A pore in the epidermis which can be closed by changes in shape of the surrounding cells; plur. *stomata*.

stomium   The part of the sporangium wall (in the ferns) which ruptures during dehiscence.

striate   Marked with long narrow depressions or ridges.

strict   Growing upwards at a small angle to the vertical.

strigose   With stiff appressed hairs.

strophiole   A small hard appendage outside the testa of a seed.

style   The part of the gynoecium connecting the ovary with the stigma (Fig. 82A).

stylopodium   See p. 277.

sub- (in combinations, e.g. subcordate)   Not quite, nearly, e.g. subacute (Fig. 81D), suborbicular (Fig. 80L).

subulate   Awl-shaped, narrow, pointed and ±flattened.

sucker   A shoot arising adventitiously from a root of a tree or shrub often at some distance from the main stem.

suffruticose   Adjective from *suffrutex*, a dwarf shrub or undershrub.

superior (ovary)   With perianth inserted round the base, the ovary being free (Fig. 82B–D).

suture   The line of junction of two carpels.

sympodial   Of a stem in which the growing point either terminates in an infl. or dies each year, growth being continued by a new lateral growing point, cf. **cymose**.

syncarpous (ovary)   Having the carpels united to one another.

taxon   A taxonomic entity of any rank.

tendril   A climbing organ formed from the whole or a part of a stem, lf or petiole. Most frequently the terminal portion of a pinnate lf, as in many Leguminosae.

terete   Not ridged, grooved or angled.

terminal   Borne at the end of a stem and limiting its growth.

ternate lf   A compound lf divided into 3 ±equal parts, which may themselves be similarly divided (2- or 3-ternate); Fig. 80F).

testa   The skin or outer coat of a seed.

tetrad   A group of 4 spores cohering in a tetrahedral shape or as a flat plate and originating from a single spore mother-cell.

tetraploid   See under **diploid**.

thallus   The plant body when not differentiated into stem, lf, etc.

thorn   A woody sharp-pointed structure formed from a modified branch.

tomentum   A dense covering of short cottony hairs; adj. *tomentose*.

tree   A woody plant with normally a single main stem (trunk) bearing lateral branches and often attaining a considerable size.

triangular   Having ±the shape of a triangle. Fig. 80N.

trifid   Split into three but not to the base.

trigonous   Of a solid body triangular in section but obtusely angled.

triploid   See under **diploid**.

triquetrous   Of a solid body triangular in section and acutely angled.

trisect   Cut into 3 almost separate parts.

truncate   Fig. 81B.

tube   The fused part of a corolla or calyx, or a hollow, cylindrical, empty prolongation of an anther.

tuber   A swollen portion of a stem or root of one year's duration, those of successive years not arising directly from the old ones nor bearing any constant relation to them; cf. corm.

tuberculate   With small blunt projections, warty.

tubercle   A ±spherical or ovoid swelling; see p. 305.

tunic   A dry, usually brown and ±papery covering round a bulb or corm; adj. *tunicated*.

turbinate   Top-shaped.

turion   A detachable winter-bud by means of which many water plants perennate.

umbel   An infl. in which the pedicels all arise from the top of the main stem. Also used of compound umbels in which the peduncles also arise from the same point. An umbrella-shaped infl.

unarmed   Devoid of thorns, spines or prickles.

undulate   Wavy in a plane at right angles to the surface.

unilocular   Having a single cavity; similarly bilocular, etc., having 2, etc., cavities.

urceolate (corolla)   ±globular to subcylindrical but strongly contracted at the mouth (Fig. 81K).

valvate   Of per.-segs with their edges in contact but not overlapping in bud.

vein   See **nerve**.

versatile   With the filament attached near the middle of the anther so as to allow of movement.

villous   Shaggy.

viscid   Sticky.

vitta   See p. 277.

viviparous   With the fls proliferating vegetatively and not forming seed.

waste place   Uncultivated ±open habitat much influenced by man. [NOTE: not used in Hooker's sense of almost any uncultivated place.]

whorl   More than two organs of the same kind arising at the same level; see also p. 402; adj. *whorled*.

wing   The lateral petals in the fls of many Leguminosae and Fumariaceae.

zygomorphic   Having only one plane of symmetry.

Index